Tortora's

Principles of ANATOMY & PHYSIOLOGY

Global Edition

GERARD J. TORTORA
Bergen Community College

BRYAN DERRICKSON
Valencia College

WILEY

Cover image: ©Mopic/Shutterstock.

Founded in 1807, John Wiley & Sons, Inc. has been a valued source of knowledge and understanding for more than 200 years, helping people around the world meet their needs and fulfill their aspirations. Our company is built on a foundation of principles that include responsibility to the communities we serve and where we live and work. In 2008, we launched a Corporate Citizenship Initiative, a global effort to address the environmental, social, economic, and ethical challenges we face in our business. Among the issues we are addressing are carbon impact, paper specifications and procurement, ethical conduct within our business and among our vendors, and community and charitable support. For more information, please visit our website: www.wiley.com/go/citizenship.

ISBN:

Tortora PAP SET 978-1-119-40006-6

PAP Textbook 978-1-119-38292-8

PAP Study Guide 978-1-119-39993-3

Printed in Markono Print Media Pte Ltd

10 9 8 7 6 5 4 3 2 1

About the Authors

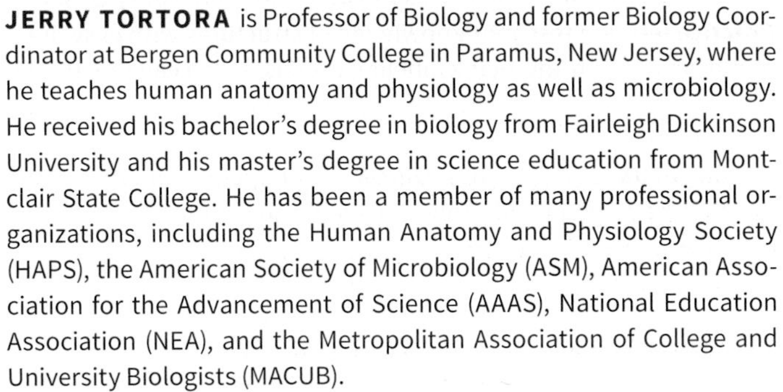

JERRY TORTORA is Professor of Biology and former Biology Coordinator at Bergen Community College in Paramus, New Jersey, where he teaches human anatomy and physiology as well as microbiology. He received his bachelor's degree in biology from Fairleigh Dickinson University and his master's degree in science education from Montclair State College. He has been a member of many professional organizations, including the Human Anatomy and Physiology Society (HAPS), the American Society of Microbiology (ASM), American Association for the Advancement of Science (AAAS), National Education Association (NEA), and the Metropolitan Association of College and University Biologists (MACUB).

Above all, Jerry is devoted to his students and their aspirations. In recognition of this commitment, Jerry was the recipient of MACUB's 1992 President's Memorial Award. In 1996, he received a National Institute for Staff and Organizational Development (NISOD) excellence award from the University of Texas and was selected to represent Bergen Community College in a campaign to increase awareness of the contributions of community colleges to higher education.

Jerry is the author of several best-selling science textbooks and laboratory manuals, a calling that often requires an additional 40 hours per week beyond his teaching responsibilities. Nevertheless, he still makes time for four or five weekly aerobic workouts that include biking and running. He also enjoys attending college basketball and professional hockey games and performances at the Metropolitan Opera House.

BRYAN DERRICKSON is Professor of Biology at Valencia College in Orlando, Florida, where he teaches human anatomy and physiology as well as general biology and human sexuality. He received his bachelor's degree in biology from Morehouse College and his Ph.D. in cell biology from Duke University. Bryan's study at Duke was in the Physiology Division within the Department of Cell Biology, so while his degree is in cell biology, his training focused on physiology. At Valencia, he frequently serves on faculty hiring committees. He has served as a member of the Faculty Senate, which is the governing body of the college, and as a member of the Faculty Academy Committee (now called the Teaching and Learning Academy), which sets the standards for the acquisition of tenure by faculty members. Nationally, he is a member of the Human Anatomy and Physiology Society (HAPS) and the National Association of Biology Teachers (NABT). Bryan has always wanted to teach. Inspired by several biology professors while in college, he decided to pursue physiology with an eye to teaching at the college level. He is completely dedicated to the success of his students. He particularly enjoys the challenges of his diverse student population, in terms of their age, ethnicity, and academic ability, and finds being able to reach all of them, despite their differences, a rewarding experience. His students continually recognize Bryan's efforts and care by nominating him for a campus award known as the "Valencia Professor Who Makes Valencia a Better Place to Start." Bryan has received this award three times.

To all my children: Lynne, Gerard Jr., Kenneth, Anthony, and Drew, whose love and support have been the wind beneath my wings. GJT

To my family: Rosalind, Hurley, Cherie, and Robb. Your support and motivation have been invaluable to me. B.H.D.

Preface

Welcome to your course in anatomy and physiology! Many of you are taking this course because you hope to pursue a career in one of the allied health fields or nursing. Or perhaps you are simply interested in learning more about your own body. Whatever your motivation, ***Principles of Anatomy and Physiology, Global edition*** and ***WileyPLUS*** have all the content and tools that you need to successfully navigate what can be a very challenging course.

In this global edition, we have made every effort to provide you with an accurate, clearly written, and expertly illustrated presentation of the structure and function of the human body and to explore the practical and relevant applications of your knowledge to everyday life and career development. This edition remains true to these goals. It distinguishes itself from original editions with updated and new illustrations and enhanced digital online learning resources.

Engaging Digitally

The content in ***Principles of Anatomy and Physiology*** is completely integrated into **WileyPLUS**. This allows you to create a personalized study plan, assess your progress along the way, and access the content and resources you need to master the material. WileyPLUS provides immediate insight into your strengths and problem areas with visual reports that highlight what's most important for you to act on.

Many dynamic programs integrated into the course help build your knowledge and understanding, and keep you motivated. Fifteen **3-D Physiology** animations were developed around the most difficult physiological concepts to help students like you understand them more effectively. **Muscles in Motion** are animations of the seven major joints of the body, helping you learn origin, insertion, and movements of muscles surrounding those joints. **Real Anatomy** is 3-D imaging software that allows you to dissect through multiple layers of a real human body to study and learn the anatomical structures of all body systems. And **Anatomy Drill and Practice** lets you test your knowledge of structures with easy drag-and-drop or fill-in-the-blank labeling exercises. You can practice labeling illustrations, cadaver photographs, histology micrographs, or anatomical models.

WileyPLUS also includes ***ORION*** – integrated adaptive practice that helps you build proficiency and use your study time most effectively.

Acknowledgments

Principles of Human Anatomy & Physiology Global edition and ***WileyPLUS with ORION*** would not be possible without the help of many, particularly the academic colleagues who collaborated with us along the way. We are very grateful that Wiley has commissioned a board of advisors in anatomy and physiology to act as a sounding board on course issues, challenges, and solutions. In particular we thank those members of the board with expertise in the 2-semester A&P course: DJ Hennager, Kirkwood Community College; Heather Labbe, University of Montana-Missoula; Tom Lancraft, St. Petersburg College; Russel Nolan, Baton Rouge Community College; and Terry Thompson, Wor-Wic Community College.

We wish to especially thank several academic colleagues for their helpful contributions to this edition, particularly to WileyPLUS with ORION. The improvements and enhancements for this edition are possible in large part because of the expertise and input of the following group of people:

Matthew Abbott, Des Moines Area Community College
Ayanna Alexander-Street, Lehman College of New York
Donna Balding, Macon State College
Celina Bellanceau, Florida Southern College
Dena Berg, Tarrant County College
Betsy Brantley, Valencia Community College
Susan Burgoon, Armadillo College
Steven Burnett, Clayton State University
Heidi Bustamante, University of Colorado, Boulder
Anthony Contento, Colorado State University
Liz Csikar, Mesa Community College
Kent Davis, Brigham Young University, Idaho
Kathryn Durham, Lorain County Community College
Kaushik Dutta, University of New England
Karen Eastman, Chattanooga State Community College
John Erickson, Ivy Tech Community College of Indiana
Tara Fay, University of Scranton
John Fishback, Ozark Tech Community College
Linda Flora, Delaware County Community College
Aaron Fried, Mohawk Valley Community College
Sophia Garcia, Tarrant County Community College
Lynn Gargan, Tarrant County Community College
Caroline Garrison, Carroll Community College
Harold Grau, Christopher Newport University
Mark Hubley, Prince George's Community College
Jason Hunt, Brigham Young University, Idaho
Lena Garrison, Carroll Community College
Geoffrey Goellner, Minnesota State University, Mankato

DJ Hennager, Kirkwood Community College
Lisa Hight, Baptist College of Health Sciences
Alexander Imholtz, Prince George's Community College
Michelle Kettler, University of Wisconsin
Cynthia Kincer, Wytheville Community College
Tom Lancraft, St. Petersburg College
Claire Leonard, William Paterson University
Jerri Lindsey, Tarrant County Community College
Alice McAfee, University of Toledo
Shannon Meadows, Roane State Community College
Shawn Miller, University of Utah
Erin Morrey, Georgia Perimeter College
Qian Moss, Des Moines Area Community College
Mark Nielsen, University of Utah
Margaret Ott, Tyler Junior College
Eileen Preseton, Tarrant County College
Saeed Rahmanian, Roane State Community College
Sandra Reznik, St. John's University
Laura Ritt, Burlington Community College
Amanda Rosenzweig, Delgado Community College
Jeffrey Spencer, University of Akron
Sandy Stewart, Vincennes University
Jane Torrie, Tarrant County College
Maureen Tubbiola, St. Cloud State
Jamie Weiss, William Paterson University

Finally, our hats are off to everyone at Wiley. We enjoy collaborating with this enthusiastic, dedicated, and talented team of publishing professionals. Our thanks to the entire team: Maria Guarascio, Senior Editor; Linda Muriello, Senior Product Designer; Lindsey Myers, Assistant Development Editor; MaryAlice Skidmore, Editorial Assistant; Trish McFadden, Content Management Editor; Mary Ann Price, Photo Manager; Tom Nery, Designer; and Alan Halfen, Senior Marketing Manager.

GERARD J. TORTORA
Department of Science and Health, S229
Bergen Community College
400 Paramus Road
Paramus, NJ 07652
gjtauthor01@optonline.com

BRYAN DERRICKSON
Department of Science, PO Box 3028
Valencia College
Orlando, FL 32802
bderrickson@valenciacollege.edu

Contents

An Introduction to the Human Body

CHAPTER **1**

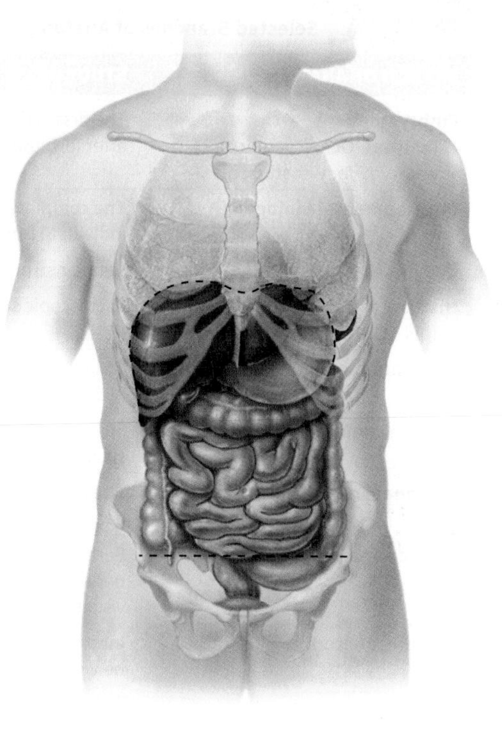

The Human Body and Homeostasis

Humans have many ways to maintain homeostasis, the state of relative stability of the body's internal environment. Disruptions to homeostasis often set in motion corrective cycles, called feedback systems, that help restore the conditions needed for health and life.

Our fascinating journey through the human body begins with an overview of the meanings of anatomy and physiology, followed by a discussion of the organization of the human body and the properties that it shares with all living things. Next, you will discover how the body regulates its own internal environment; this unceasing process, called homeostasis, is a major theme in every chapter of this book. Finally, we introduce the basic vocabulary that will help you speak about the body in a way that is understood by scientists and health-care professionals alike.

Q Did you ever wonder why an autopsy is performed?

1.1 Anatomy and Physiology Defined

OBJECTIVE

- **Define** anatomy and physiology, and name several branches of these sciences.

Two branches of science—anatomy and physiology—provide the foundation for understanding the body's parts and functions. **Anatomy** (a-NAT-ō-mē; *ana-* = up; *-tomy* = process of cutting) is the science of body *structures* and the relationships among them. It was first studied by **dissection** (dis-SEK-shun; *dis-* = apart; *-section* = act of cutting), the careful cutting apart of body structures to study their relationships. Today, a variety of imaging techniques (see **Table 1.3**) also contribute to the advancement of anatomical knowledge. Whereas anatomy deals with structures of the body, **physiology** (fiz′-ē-OL-ō-jē; *physio-* = nature; *-logy* = study of) is the science of body *functions*—how the body parts work. **Table 1.1** describes several branches of anatomy and physiology.

Because structure and function are so closely related, you will learn about the human body by studying its anatomy and physiology together. The structure of a part of the body often reflects its functions. For example, the bones of the skull join tightly to form a rigid case that protects the brain. The bones of the fingers are more loosely joined to allow a variety of movements. The walls of the air sacs in the lungs are very thin, permitting rapid movement of inhaled oxygen into the blood.

1.2 Levels of Structural Organization and Body Systems

OBJECTIVES

- **Describe** the body's six levels of structural organization.
- **List** the 11 systems of the human body, representative organs present in each, and their general functions.

The levels of organization of a language—letters, words, sentences, paragraphs, and so on—can be compared to the levels of organization of the human body. Your exploration of the human body will extend from atoms and molecules to the whole person. From the smallest to the largest, six levels of organization will help you to understand anatomy and physiology: the chemical, cellular, tissue, organ, system, and organismal levels of organization (**Figure 1.1**).

TABLE 1.1 Selected Branches of Anatomy and Physiology

BRANCH OF ANATOMY	STUDY OF
Embryology (em′-brē-OL-ō-jē; *embry-* = embryo; *-logy* = study of)	The first eight weeks of development after fertilization of a human egg.
Developmental biology	The complete development of an individual from fertilization to death.
Cell biology	Cellular structure and functions.
Histology (his-TOL-ō -jē; *hist-* = tissue)	Microscopic structure of tissues.
Gross anatomy	Structures that can be examined without a microscope.
Systemic anatomy	Structure of specific systems of the body such as the nervous or respiratory systems.
Regional anatomy	Specific regions of the body such as the head or chest.
Surface anatomy	Surface markings of the body to understand internal anatomy through visualization and palpation (gentle touch).
Imaging anatomy	Internal body structures that can be visualized with techniques such as x-rays, MRI, CT scans, and other technologies for clinical analysis and medical intervention.
Pathological anatomy (path′-ō-LOJ-i-kal; *path-* = disease)	Structural changes (gross to microscopic) associated with disease.

BRANCH OF PHYSIOLOGY	STUDY OF
Molecular physiology	Functions of individual molecules such as proteins and DNA.
Neurophysiology (NOOR-ō-fiz-ē-ol′-ō-jē; *neuro-* = nerve)	Functional properties of nerve cells.
Endocrinology (en′-dō-kri-NOL-ō-jē; *endo-* = within; *-crin* = secretion)	Hormones (chemical regulators in the blood) and how they control body functions.
Cardiovascular physiology (kar-dē-ō-VAS-kū-lar; *cardi-* = heart; *vascular* = blood vessels)	Functions of the heart and blood vessels.
Immunology (im′-ū-NOL-ō-jē; *immun-* = not susceptible)	The body's defenses against disease-causing agents.
Respiratory physiology (RES-pi-ra-tōr-ē; *respira-* = to breathe)	Functions of the air passageways and lungs.
Renal physiology (RĒ-nal; *ren-* = kidney)	Functions of the kidneys.
Exercise physiology	Changes in cell and organ functions due to muscular activity.
Pathophysiology (Path-ō-fiz-ē-ol′-ō-jē)	Functional changes associated with disease and aging.

1. **Chemical level.** This very basic level can be compared to the *letters of the alphabet* and includes **atoms**, the smallest units of matter that participate in chemical reactions, and **molecules**, two or more atoms joined together. Certain atoms, such as carbon (C), hydrogen (H), oxygen (O), nitrogen (N), phosphorus (P), calcium (Ca), and sulfur (S), are essential for maintaining life. Two familiar molecules found in the body are deoxyribonucleic acid (DNA), the genetic material passed from one generation to the next, and glucose, commonly known as blood sugar. Chapters 2 and 25 focus on the chemical level of organization.

2. **Cellular level.** Molecules combine to form **cells**, the basic structural and functional units of an organism that are composed of chemicals. Just as *words* are the smallest elements of language that make sense, cells are the smallest living units in the human body. Among the many kinds of cells in your body are muscle cells, nerve cells, and epithelial cells. **Figure 1.1** shows a smooth muscle cell, one of the three types of muscle cells in the body. The cellular level of organization is the focus of Chapter 3.

3. **Tissue level. Tissues** are groups of cells and the materials surrounding them that work together to perform a particular function, similar to the way words are put together to form *sentences*. There are just four basic types of tissues in your body: epithelial tissue, connective tissue, muscular tissue, and nervous tissue. *Epithelial tissue* covers body surfaces, lines hollow organs and cavities, and forms glands. *Connective tissue* connects, supports, and protects body organs while distributing blood vessels to other tissues. *Muscular tissue* contracts to make body parts move and generates heat. *Nervous tissue* carries information from one part of the body to another through nerve impulses. Chapter 4 describes the tissue level of organization in greater detail. Shown in **Figure 1.1** is smooth muscle tissue, which consists of tightly packed smooth muscle cells.

4. **Organ level.** At the organ level, different types of tissues are joined together. Similar to the relationship between sentences and *paragraphs*, **organs** are structures that are composed of two or more different types of tissues; they have specific functions and usually have recognizable shapes. Examples of organs are the stomach, skin, bones, heart, liver, lungs, and brain. **Figure 1.1** shows how several tissues make up the stomach. The stomach's outer covering is a layer of epithelial tissue and connective tissue that reduces friction when the stomach moves and rubs against other organs. Underneath are three layers of a type of muscular tissue called *smooth muscle tissue*, which contracts to churn and mix food and then push it into the next digestive organ, the small intestine. The innermost lining is an *epithelial tissue layer* that produces fluid and chemicals responsible for digestion in the stomach.

5. **System** (*organ-system*) **level.** A **system** (or *chapter*, in our language analogy) consists of related organs (*paragraphs*) with a common function. An example of the system level, also called the *organ-system level*, is the digestive system, which breaks down and absorbs food. Its organs include the mouth, salivary glands, pharynx (throat), esophagus (food tube), stomach, small intestine, large intestine,

FIGURE 1.1 Levels of structural organization in the human body.

The levels of structural organization are chemical, cellular, tissue, organ, system, and organismal.

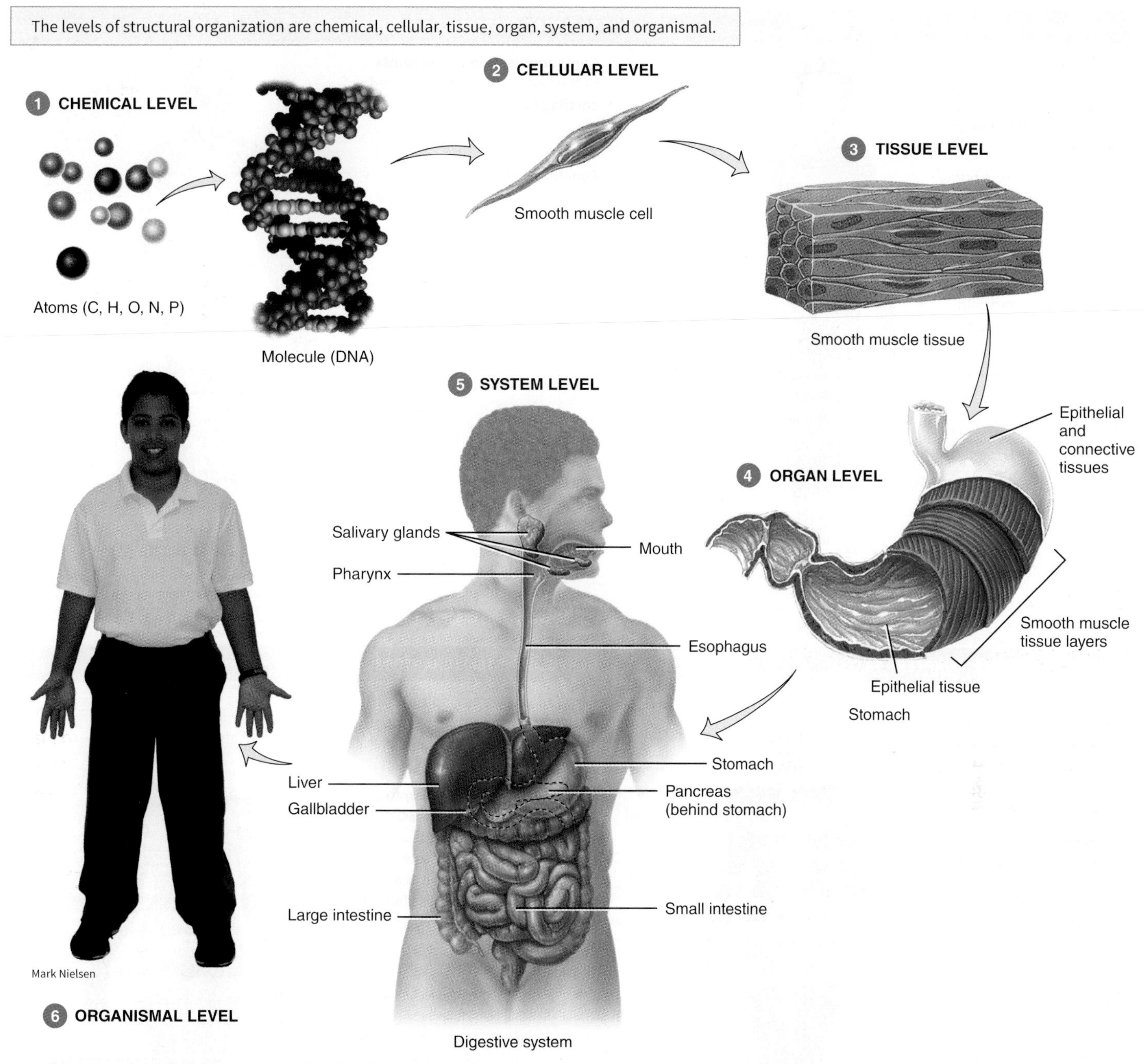

Q Which level of structural organization is composed of two or more different types of tissues that work together to perform a specific function?

liver, gallbladder, and pancreas. Sometimes an organ is part of more than one system. The pancreas, for example, is part of both the digestive system and the hormone-producing endocrine system.

6 **Organismal level.** An **organism** (OR-ga-nizm), any living individual, can be compared to a *book* in our analogy. All the parts of the human body functioning together constitute the total organism.

In the chapters that follow, you will study the anatomy and physiology of the body systems. **Table 1.2** lists the components and introduces the functions of these systems. You will also discover that all body systems influence one another. As you study each of the body systems in more detail, you will discover how they work together to maintain health, provide protection from disease, and allow for reproduction of the human species.

See Clinical Connection: Noninvasive Diagnostic Techniques

TABLE 1.2 The Eleven Systems of the Human Body

INTEGUMENTARY SYSTEM (CHAPTER 5)

Components: **Skin** and associated structures, such as **hair, fingernails** and **toenails, sweat glands**, and **oil glands**.

Functions: Protects body; helps regulate body temperature; eliminates some wastes; helps make vitamin D; detects sensations such as touch, pain, warmth, and cold; stores fat and provides insulation.

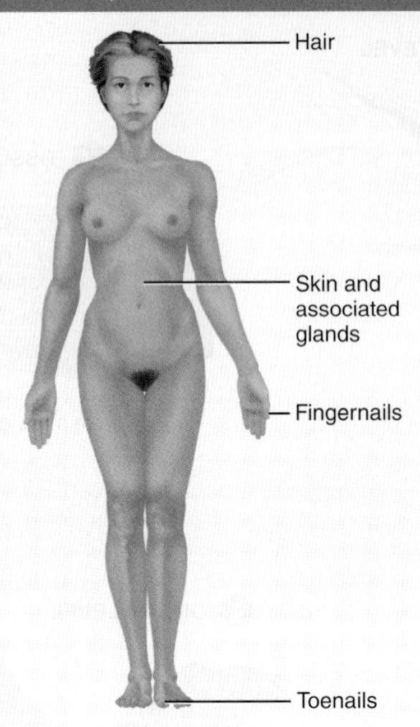

SKELETAL SYSTEM (CHAPTERS 6–9)

Components: **Bones** and **joints** of the body and their associated **cartilages**.

Functions: Supports and protects body; provides surface area for muscle attachments; aids body movements; houses cells that produce blood cells; stores minerals and lipids (fats).

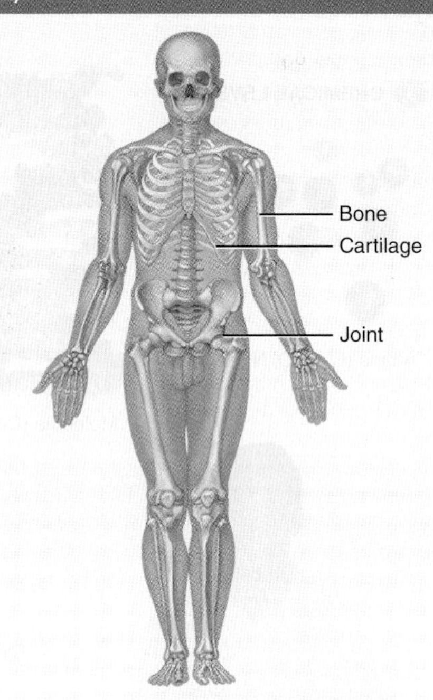

MUSCULAR SYSTEM (CHAPTERS 10, 11)

Components: Specifically, **skeletal muscle tissue**—muscle usually attached to bones (other muscle tissues include smooth and cardiac).

Functions: Participates in body movements, such as walking; maintains posture; produces heat.

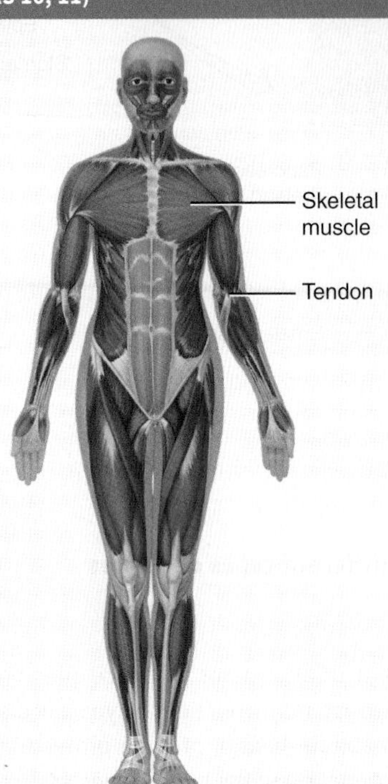

NERVOUS SYSTEM (CHAPTERS 12–17)

Components: **Brain, spinal cord, nerves**, and special sense organs, such as **eyes** and **ears**.

Functions: Generates action potentials (nerve impulses) to regulate body activities; detects changes in body's internal and external environments, interprets changes, and responds by causing muscular contractions or glandular secretions.

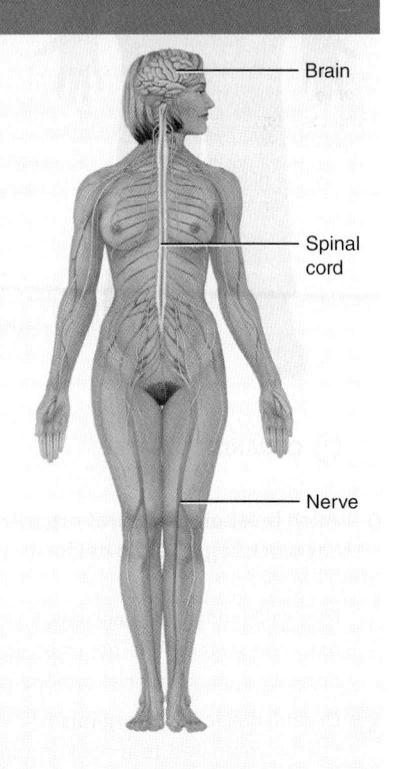

ENDOCRINE SYSTEM (CHAPTER 18)

Components: Hormone-producing glands (**pineal gland, hypothalamus, pituitary gland, thymus, thyroid gland, parathyroid glands, adrenal glands, pancreas, ovaries**, and **testes**) and hormone-producing cells in several other organs.

Functions: Regulates body activities by releasing hormones (chemical messengers transported in blood from endocrine gland or tissue to target organ).

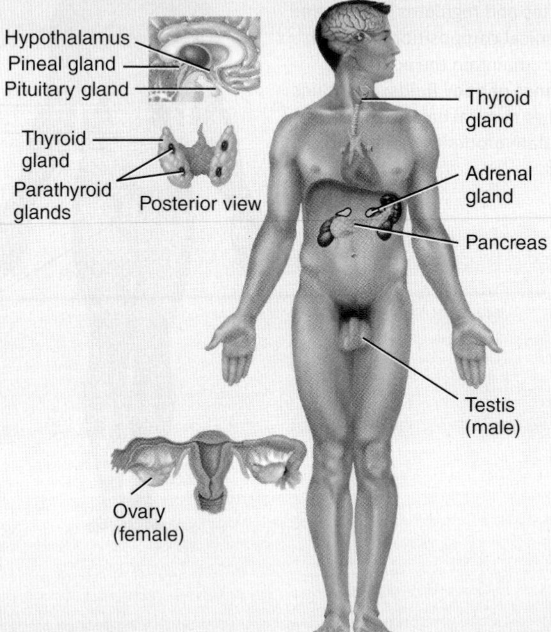

CARDIOVASCULAR SYSTEM (CHAPTERS 19–21)

Components: **Blood, heart**, and **blood vessels**.

Functions: Heart pumps blood through blood vessels; blood carries oxygen and nutrients to cells and carbon dioxide and wastes away from cells and helps regulate acid–base balance, temperature, and water content of body fluids; blood components help defend against disease and repair damaged blood vessels.

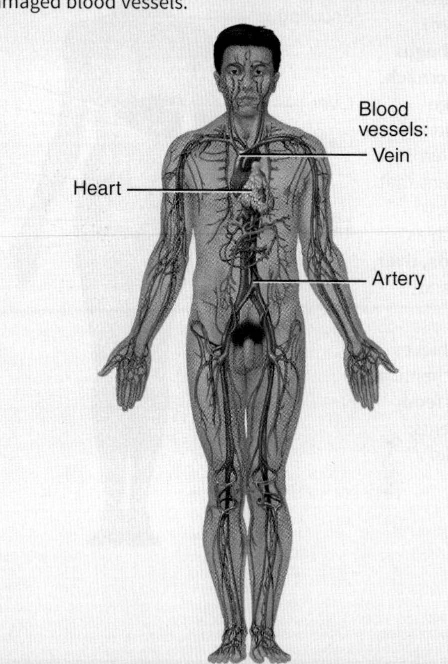

LYMPHATIC SYSTEM AND IMMUNITY (CHAPTER 22)

Components: **Lymphatic fluid** and **vessels; spleen, thymus, lymph nodes,** and **tonsils**; cells that carry out immune responses (**B cells, T cells**, and others).

Functions: Returns proteins and fluid to blood; carries lipids from gastrointestinal tract to blood; contains sites of maturation and proliferation of B cells and T cells that protect against disease-causing microbes.

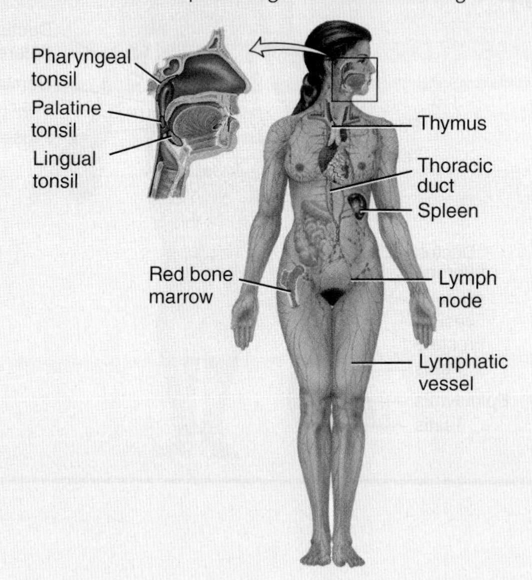

RESPIRATORY SYSTEM (CHAPTER 23)

Components: **Lungs** and air passageways such as the **pharynx** (*throat*), **larynx** (*voice box*), **trachea** (*windpipe*), and **bronchial tubes** leading into and out of lungs.

Functions: Transfers oxygen from inhaled air to blood and carbon dioxide from blood to exhaled air; helps regulate acid–base balance of body fluids; air flowing out of lungs through vocal cords produces sounds.

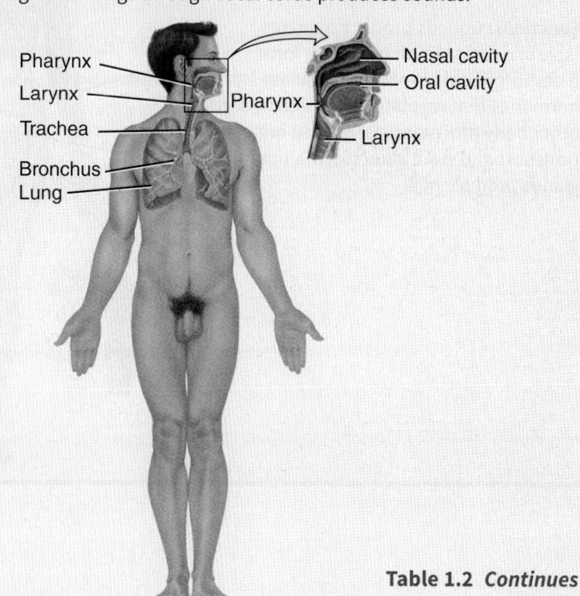

Table 1.2 ***Continues***

TABLE 1.2 The Eleven Systems of the Human Body (*Continued*)

DIGESTIVE SYSTEM (CHAPTER 24)

Components: Organs of gastrointestinal tract, a long tube that includes the **mouth, pharynx** (*throat*), **esophagus** (*food tube*), **stomach, small** and **large intestines**, and **anus**; also includes accessory organs that assist in digestive processes, such as **salivary glands, liver, gallbladder**, and **pancreas**.

Functions: Achieves physical and chemical breakdown of food; absorbs nutrients; eliminates solid wastes.

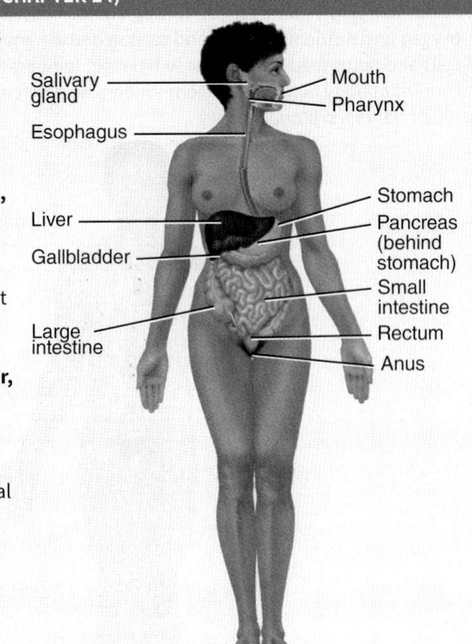

URINARY SYSTEM (CHAPTER 26)

Components: **Kidneys, ureters, urinary bladder**, and **urethra**.

Functions: Produces, stores, and eliminates urine; eliminates wastes and regulates volume and chemical composition of blood; helps maintain the acid–base balance of body fluids; maintains body's mineral balance; helps regulate production of red blood cells.

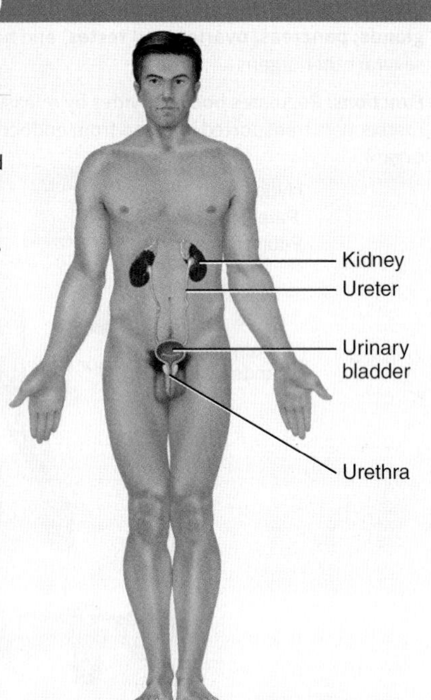

REPRODUCTIVE SYSTEMS (CHAPTER 28)

Components: **Gonads** (**testes** in males and **ovaries** in females) and associated organs (**uterine tubes** or *fallopian tubes*, **uterus, vagina**, and **mammary glands** in females and **epididymis, ductus** or (*vas*) **deferens, seminal vesicles, prostate**, and **penis** in males).

Functions: Gonads produce gametes (sperm or oocytes) that unite to form a new organism; gonads also release hormones that regulate reproduction and other body processes; associated organs transport and store gametes; mammary glands produce milk

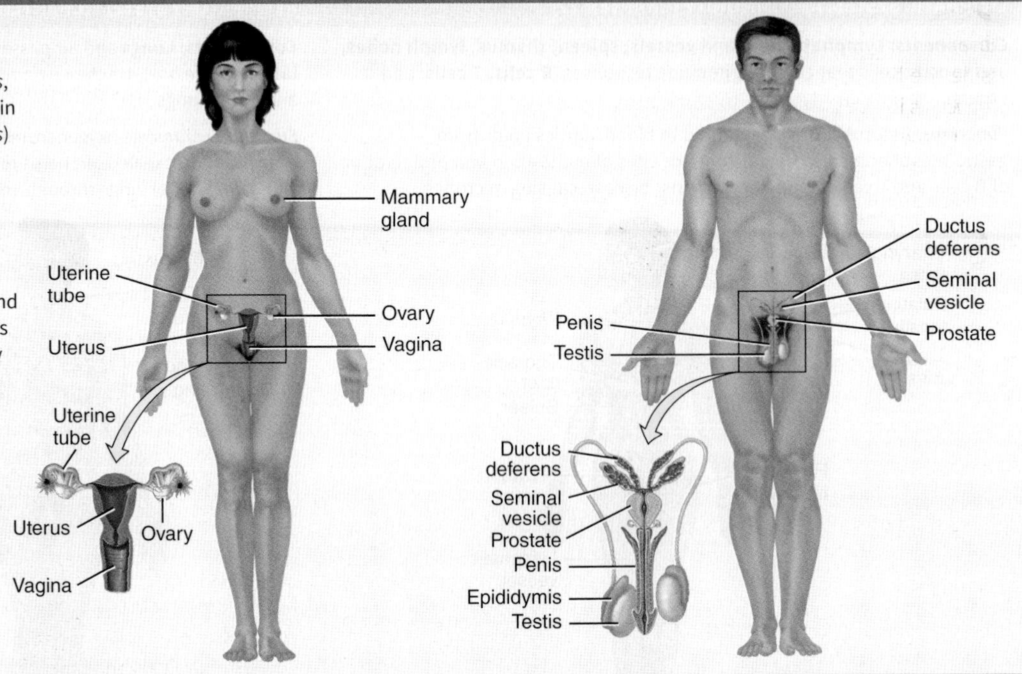

1.3 Characteristics of the Living Human Organism

OBJECTIVE

- **Define** the important life processes of the human body.

Basic Life Processes

Certain processes distinguish organisms, or living things, from nonliving things. Following are the six most important life processes of the human body:

1. **Metabolism** (me-TAB-ō-lizm) is the sum of all chemical processes that occur in the body. One phase of metabolism is **catabolism** (ka-TAB-ō-lizm; *catabol-* = throwing down; *-ism* = a condition), the breakdown of complex chemical substances into simpler components. The other phase of metabolism is **anabolism** (a-NAB-ō-lizm; *anabol-* = a raising up), the building up of complex chemical substances from smaller, simpler components. For example, digestive processes catabolize (split) proteins in food into amino acids. These amino acids are then used to anabolize (build) new proteins that make up body structures such as muscles and bones.
2. **Responsiveness** is the body's ability to detect and respond to changes. For example, an increase in body temperature during a fever represents a change in the internal environment (within the body), and turning your head toward the sound of squealing brakes is a response to a change in the external environment (outside the body) to prepare the body for a potential threat. Different cells in the body respond to environmental changes in characteristic ways. Nerve cells respond by generating electrical signals known as nerve impulses (action potentials). Muscle cells respond by contracting, which generates force to move body parts.
3. **Movement** includes motion of the whole body, individual organs, single cells, and even tiny structures inside cells. For example, the coordinated action of leg muscles moves your whole body from one place to another when you walk or run. After you eat a meal that contains fats, your gallbladder contracts and releases bile into the gastrointestinal tract to help digest them. When a body tissue is damaged or infected, certain white blood cells move from the bloodstream into the affected tissue to help clean up and repair the area. Inside the cell, various parts, such as secretory vesicles (see **Figure 3.20**), move from one position to another to carry out their functions.
4. **Growth** is an increase in body size that results from an increase in the size of existing cells, an increase in the number of cells, or both. In addition, a tissue sometimes increases in size because the amount of material between cells increases. In a growing bone, for example, mineral deposits accumulate between bone cells, causing the bone to grow in length and width.
5. **Differentiation** (dif′-er-en-shē-Ā-shun) is the development of a cell from an unspecialized to a specialized state. Such precursor cells, which can divide and give rise to cells that undergo differentiation, are known as **stem cells**. As you will see later in the text, each type of cell in the body has a specialized structure or function that differs from that of its precursor (ancestor) cells. For example, red blood cells and several types of white blood cells all arise from the same unspecialized precursor cells in red bone marrow. Also through differentiation, a single fertilized human egg (ovum) develops into an embryo, and then into a fetus, an infant, a child, and finally an adult.
6. **Reproduction** (rē-prō-DUK-shun) refers either to (1) the formation of new cells for tissue growth, repair, or replacement, or (2) the production of a new individual. The formation of new cells occurs through cell division. The production of a new individual occurs through the fertilization of an ovum by a sperm cell to form a zygote, followed by repeated cell divisions and the differentiation of these cells.

When any one of the life processes ceases to occur properly, the result is death of cells and tissues, which may lead to death of the organism. Clinically, loss of the heartbeat, absence of spontaneous breathing, and loss of brain functions indicate death in the human body.

See Clinical Connection: Autopsy

1.4 Homeostasis

OBJECTIVES

- **Define** homeostasis.
- **Describe** the components of a feedback system.
- **Contrast** the operation of negative and positive feedback systems.
- **Explain** how homeostatic imbalances are related to disorders.

Homeostasis (hō′-mē-ō-STĀ-sis; *homeo-* = sameness; *-stasis* = standing still) is the maintenance of relatively stable conditions in the body's internal environment. It occurs because of the ceaseless interplay of the body's many regulatory systems. Homeostasis is a dynamic condition. In response to changing conditions, the body's parameters can shift among points in a narrow range that is compatible with maintaining life. For example, the level of glucose in blood normally stays between 70 and 110 milligrams of glucose per 100 milliliters of blood.* Each structure, from the cellular level to the system level, contributes in some way to keeping the internal environment of the body within normal limits.

Homeostasis and Body Fluids

An important aspect of homeostasis is maintaining the volume and composition of **body fluids**, dilute, watery solutions containing dissolved chemicals that are found inside cells as well as surrounding them (See **Figure 27.1**). The fluid within cells is **intracellular fluid** (*intra-* = inside), abbreviated *ICF*. The fluid outside body cells is **extracellular fluid** (*ECF*) (*extra-* = outside). The ECF that fills the narrow

*Appendix A describes metric measurements.

spaces between cells of tissues is known as **interstitial fluid** (in′-ter-STISH-al; *inter-* = between). As you progress with your studies, you will learn that the ECF differs depending on where it occurs in the body: ECF within blood vessels is termed **blood plasma**, within lymphatic vessels it is called **lymph**, in and around the brain and spinal cord it is known as **cerebrospinal fluid**, in joints it is referred to as **synovial fluid**, and the ECF of the eyes is called **aqueous humor** and **vitreous body**.

The proper functioning of body cells depends on precise regulation of the composition of their surrounding fluid. Because extracellular fluid surrounds the cells of the body, it serves as the body's *internal environment.* By contrast, the *external environment* of the body is the space that surrounds the entire body.

Figure 1.2 is a simplified view of the body that shows how a number of organ systems allow substances to be exchanged between the

FIGURE 1.2 A simplified view of exchanges between the external and internal environments. Note that the linings of the respiratory, digestive, and urinary systems are continuous with the external environment.

The internal environment of the body refers to the extracellular fluid (interstitial fluid and plasma) that surrounds body cells.

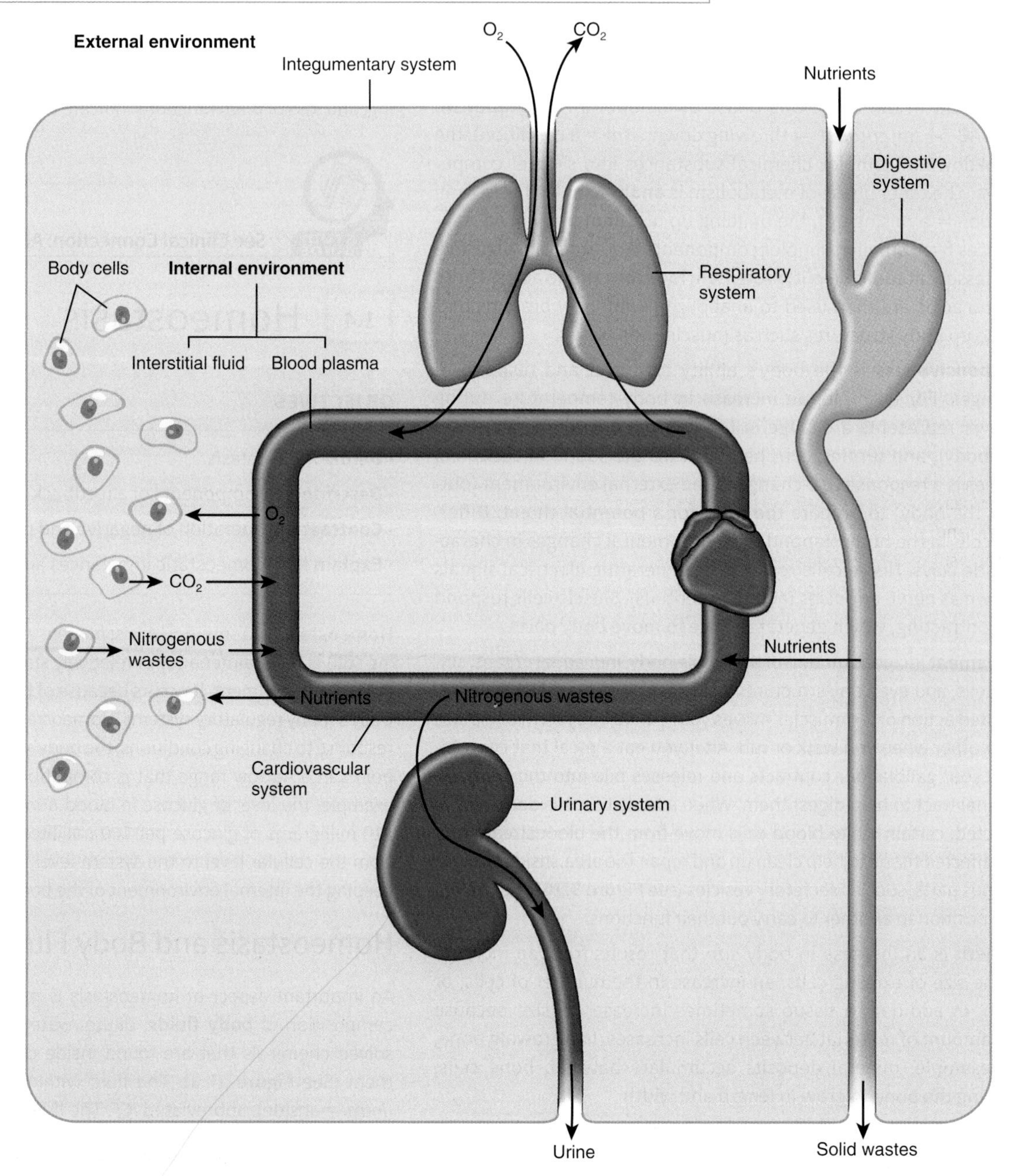

Q How does a nutrient in the external environment reach a body cell?

external environment, internal environment, and body cells in order to maintain homeostasis. Note that the integumentary system covers the outer surface of the body. Although this system does not play a major role in the exchange of materials, it protects the internal environment from damaging agents in the external environment. From the external environment, oxygen enters plasma through the respiratory system and nutrients enter plasma through the digestive system. After entering plasma, these substances are transported throughout the body by the cardiovascular system. Oxygen and nutrients eventually leave plasma and enter interstitial fluid by crossing the walls of blood capillaries, the smallest blood vessels of the body. Blood capillaries are specialized to allow the transfer of material between plasma and interstitial fluid. From interstitial fluid, oxygen and nutrients are taken up by cells and metabolized for energy. During this process, the cells produce waste products, which enter interstitial fluid and then move across blood capillary walls into plasma. The cardiovascular system transports these wastes to the appropriate organs for elimination from the body into the external environment. The waste product CO_2 is removed from the body by the respiratory system; nitrogen-containing wastes, such as urea and ammonia, are eliminated from the body by the urinary system.

Control of Homeostasis

Homeostasis in the human body is continually being disturbed. Some disruptions come from the external environment in the form of physical insults such as the intense heat of a hot summer day or a lack of enough oxygen for that two-mile run. Other disruptions originate in the internal environment, such as a blood glucose level that falls too low when you skip breakfast. Homeostatic imbalances may also occur due to psychological stresses in our social environment—the demands of work and school, for example. In most cases the disruption of homeostasis is mild and temporary, and the responses of body cells quickly restore balance in the internal environment. However, in some cases the disruption of homeostasis may be intense and prolonged, as in poisoning, overexposure to temperature extremes, severe infection, or major surgery.

Fortunately, the body has many regulating systems that can usually bring the internal environment back into balance. Most often, the nervous system and the endocrine system, working together or independently, provide the needed corrective measures. The nervous system regulates homeostasis by sending electrical signals known as *nerve impulses (action potentials)* to organs that can counteract changes from the balanced state. The endocrine system includes many glands that secrete messenger molecules called *hormones* into the blood. Nerve impulses typically cause rapid changes, but hormones usually work more slowly. Both means of regulation, however, work toward the same end, usually through negative feedback systems.

Feedback Systems The body can regulate its internal environment through many feedback systems. A **feedback system** or, *feedback loop*, is a cycle of events in which the status of a body condition is monitored, evaluated, changed, remonitored, reevaluated, and so on. Each monitored variable, such as body temperature, blood pressure, or blood glucose level, is termed a *controlled condition* (*controlled variable*). Any disruption that changes a controlled condition is called a *stimulus*. A feedback system includes three basic components: a receptor, a control center, and an effector (**Figure 1.3**).

1. A **receptor** is a body structure that monitors changes in a controlled condition and sends input to a control center. This pathway is called an *afferent pathway* (AF-er-ent; *af-* = toward; *-ferrent* = carried), since the information flows *toward* the control center. Typically, the *input* is in the form of nerve impulses or chemical signals. For example, certain nerve endings in the skin sense temperature and can detect changes, such as a dramatic drop in temperature.
2. A **control center** in the body, for example, the brain, sets the narrow range or *set point* within which a controlled condition should be maintained, evaluates the input it receives from receptors, and

FIGURE 1.3 **Operation of a feedback system.**

The three basic components of a feedback system are the receptor, control center, and effector.

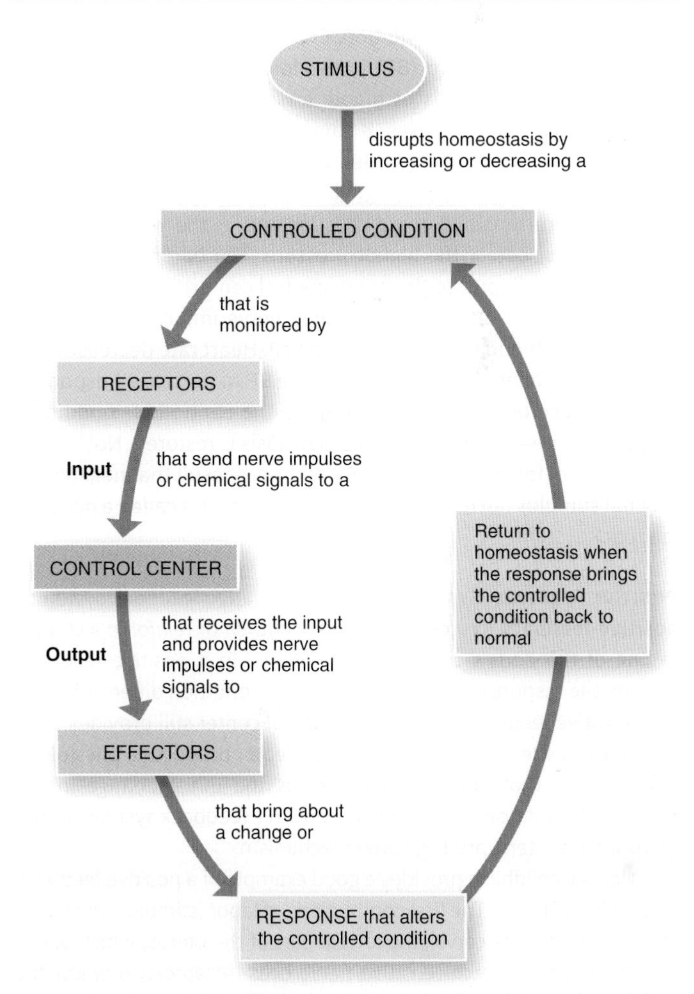

Q What is the main difference between negative and positive feedback systems?

generates output commands when they are needed. *Output* from the control center typically occurs as nerve impulses, or hormones or other chemical signals. This pathway is called an *efferent pathway* (EF-er-ent; *ef-* = away from), since the information flows *away from* the control center. In our skin temperature example, the brain acts as the control center, receiving nerve impulses from the skin receptors and generating nerve impulses as output.

3. An **effector** (e-FEK-tor) is a body structure that receives output from the control center and produces a **response** or effect that changes the controlled condition. Nearly every organ or tissue in the body can behave as an effector. When your body temperature drops sharply, your brain (control center) sends nerve impulses (output) to your skeletal muscles (effectors). The result is shivering, which generates heat and raises your body temperature.

A group of receptors and effectors communicating with their control center forms a feedback system that can regulate a controlled condition in the body's internal environment. In a feedback system, the response of the system "feeds back" information to change the controlled condition in some way, either negating it (negative feedback) or enhancing it (positive feedback).

Negative Feedback Systems A **negative feedback system** *reverses* a change in a controlled condition. Consider the regulation of blood pressure. Blood pressure (BP) is the force exerted by blood as it presses against the walls of blood vessels. When the heart beats faster or harder, BP increases. If some internal or external stimulus causes blood pressure (controlled condition) to rise, the following sequence of events occurs (**Figure 1.4**). *Baroreceptors* (the receptors), pressure-sensitive nerve cells located in the walls of certain blood vessels, detect the higher pressure. The baroreceptors send nerve impulses (input) to the brain (control center), which interprets the impulses and responds by sending nerve impulses (output) to the heart and blood vessels (the effectors). Heart rate decreases and blood vessels dilate (widen), which cause BP to decrease (response). This sequence of events quickly returns the controlled condition—blood pressure—to normal, and homeostasis is restored. Notice that the activity of the effector causes BP to drop, a result that negates the original stimulus (an increase in BP). This is why it is called a negative feedback system.

Positive Feedback Systems Unlike a negative feedback system, a **positive feedback system** tends to *strengthen* or *reinforce* a change in one of the body's controlled conditions. In a positive feedback system, the response affects the controlled condition differently than in a negative feedback system. The control center still provides commands to an effector, but this time the effector produces a physiological response that adds to or *reinforces* the initial change in the controlled condition. The action of a positive feedback system continues until it is interrupted by some mechanism.

Normal childbirth provides a good example of a positive feedback system (**Figure 1.5**). The first contractions of labor (stimulus) push part of the fetus into the cervix, the lowest part of the uterus, which opens into the vagina. Stretch-sensitive nerve cells (receptors) monitor the amount of stretching of the cervix (controlled condition). As stretching increases, they send more nerve impulses (input) to the brain (control center), which in turn causes the pituitary gland to release the hormone

FIGURE 1.4 Homeostatic regulation of blood pressure by a negative feedback system. The broken return arrow with a negative sign surrounded by a circle symbolizes negative feedback.

If the response reverses the stimulus, a system is operating by negative feedback.

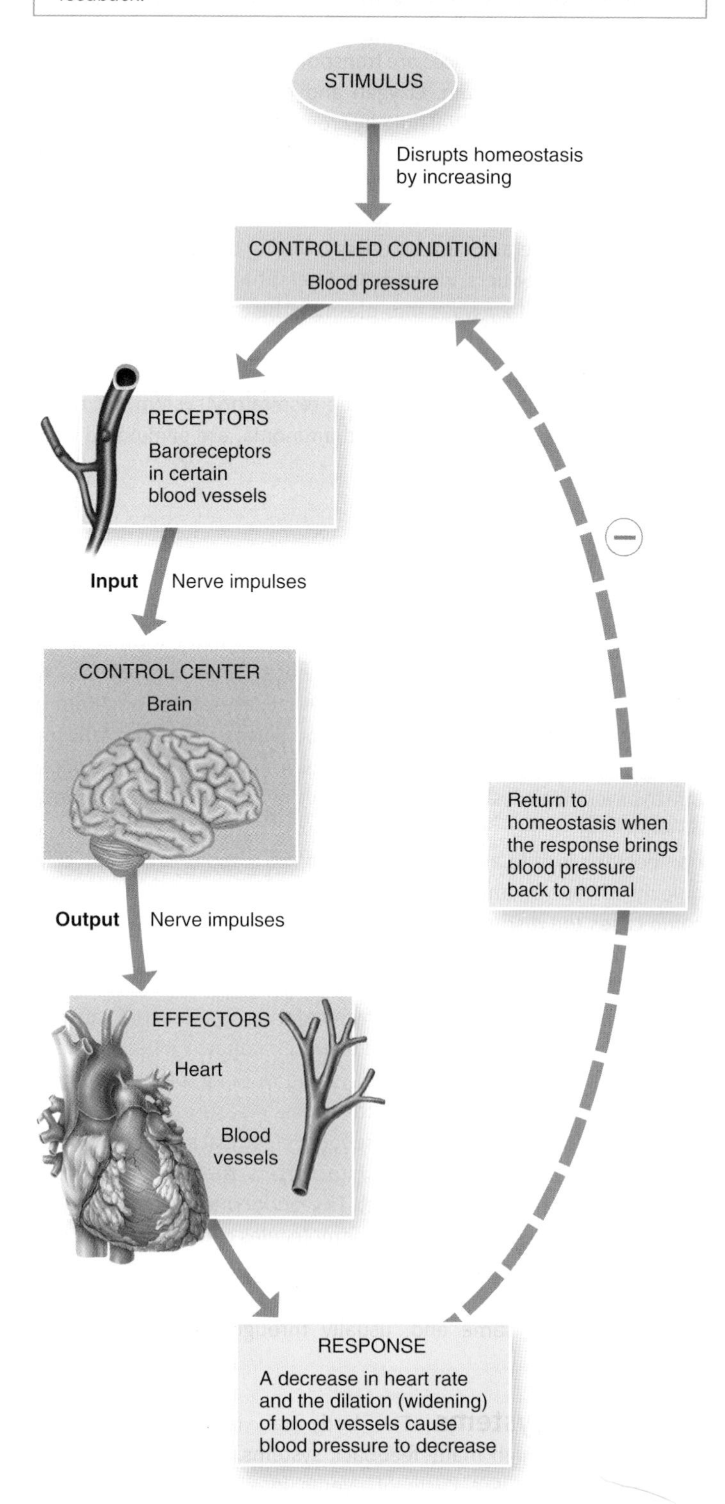

Q What would happen to heart rate if some stimulus caused blood pressure to decrease? Would this occur by way of positive or negative feedback?

FIGURE 1.5 Positive feedback control of labor contractions during birth of a baby. The broken return arrow with a positive sign surrounded by a circle symbolizes positive feedback.

If the response enhances or intensifies the stimulus, a system is operating by positive feedback.

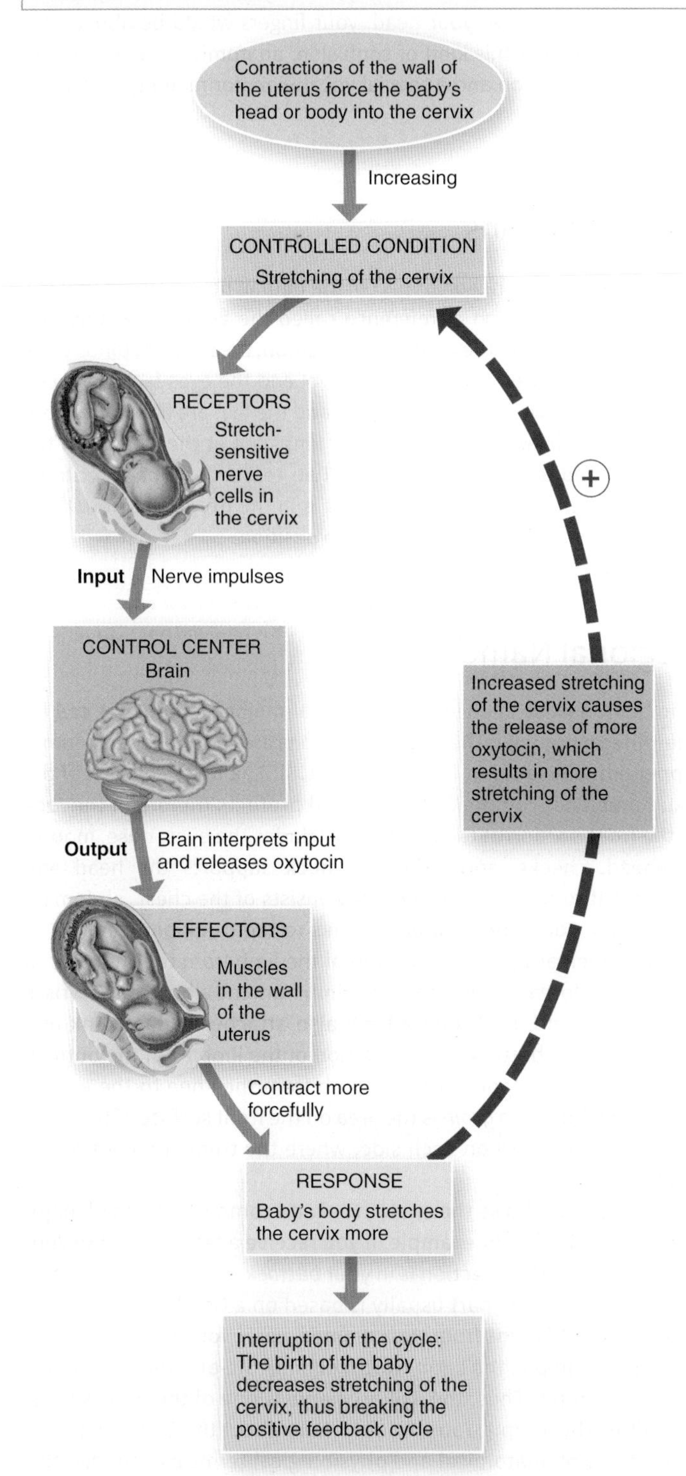

Q Why do positive feedback systems that are part of a normal physiological response include some mechanism that terminates the system?

oxytocin (output) into the blood. Oxytocin causes muscles in the wall of the uterus (effector) to contract even more forcefully. The contractions push the fetus farther down the uterus, which stretches the cervix even more. The cycle of stretching, hormone release, and ever-stronger contractions is interrupted only by the birth of the baby. Then, stretching of the cervix ceases and oxytocin is no longer released.

Another example of positive feedback is what happens to your body when you lose a great deal of blood. Under normal conditions, the heart pumps blood under sufficient pressure to body cells to provide them with oxygen and nutrients to maintain homeostasis. Upon severe blood loss, blood pressure drops and blood cells (including heart cells) receive less oxygen and function less efficiently. If the blood loss continues, heart cells become weaker, the pumping action of the heart decreases further, and blood pressure continues to fall. This is an example of a positive feedback cycle that has serious consequences and may even lead to death if there is no medical intervention. As you will see in Chapter 19, blood clotting is also an example of a positive feedback system.

These examples suggest some important differences between positive and negative feedback systems. Because a positive feedback system continually reinforces a change in a controlled condition, some event outside the system must shut it off. If the action of a positive feedback system is not stopped, it can "run away" and may even produce life-threatening conditions in the body. The action of a negative feedback system, by contrast, slows and then stops as the controlled condition returns to its normal state. Usually, positive feedback systems reinforce conditions that do not happen very often, and negative feedback systems regulate conditions in the body that remain fairly stable over long periods.

Homeostatic Imbalances

You've seen homeostasis defined as a condition in which the body's internal environment remains relatively stable. The body's ability to maintain homeostasis gives it tremendous healing power and a remarkable resistance to abuse. The physiological processes responsible for maintaining homeostasis are in large part also responsible for your good health.

For most people, lifelong good health is not something that happens effortlessly. The many factors in this balance called health include the following:

- The environment and your own behavior.
- Your genetic makeup.
- The air you breathe, the food you eat, and even the thoughts you think.

The way you live your life can either support or interfere with your body's ability to maintain homeostasis and recover from the inevitable stresses life throws your way.

Many diseases are the result of years of poor health behavior that interferes with the body's natural drive to maintain homeostasis. An obvious example is smoking-related illness. Smoking tobacco exposes sensitive lung tissue to a multitude of chemicals that cause cancer and damage the lung's ability to repair itself. Because diseases such as emphysema and lung cancer are difficult to treat and are very rarely cured, it is much wiser to quit smoking—or never start—than to hope a doctor can "fix" you once you are diagnosed with a lung disease. Developing a lifestyle that works with, rather than against, your body's homeostatic processes helps you maximize your personal potential for optimal health and well-being.

As long as all of the body's controlled conditions remain within certain narrow limits, body cells function efficiently, homeostasis is maintained, and the body stays healthy. Should one or more components of the body lose their ability to contribute to homeostasis, however, the normal balance among all of the body's processes may be disturbed. If the homeostatic imbalance is moderate, a disorder or disease may occur; if it is severe, death may result.

A **disorder** is any abnormality of structure or function. **Disease** is a more specific term for an illness characterized by a recognizable set of signs and symptoms. A *local disease* affects one part or a limited region of the body (for example, a sinus infection); a *systemic disease* affects either the entire body or several parts of it (for example, influenza). Diseases alter body structures and functions in characteristic ways. A person with a disease may experience **symptoms**, *subjective* changes in body functions that are not apparent to an observer. Examples of symptoms are headache, nausea, and anxiety. *Objective* changes that a clinician can observe and measure are called **signs**. Signs of disease can be either anatomical, such as swelling or a rash, or physiological, such as fever, high blood pressure, or paralysis.

The science that deals with why, when, and where diseases occur and how they are transmitted among individuals in a community is known as **epidemiology** (ep′-i-dē-mē-OL-ō-jē; *epi-* = upon; *-demi* = people). **Pharmacology** (far′-ma-KOL-ō-jē; *pharmac-* = drug) is the science that deals with the effects and uses of drugs in the treatment of disease.

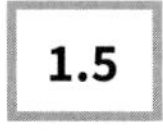

See Clinical Connection: Diagnosis of Disease

1.5 Basic Anatomical Terminology

OBJECTIVES

- **Describe** the anatomical position.
- **Relate** the anatomical names and the corresponding common names for various regions of the human body.
- **Define** the anatomical planes, anatomical sections, and directional terms used to describe the human body.
- **Outline** the major body cavities, the organs they contain, and their associated linings.

Scientists and health-care professionals use a common language of special terms when referring to body structures and their functions. The language of anatomy they use has precisely defined meanings that allow us to communicate clearly and precisely. For example, is it correct to say, "The wrist is above the fingers"? This might be true if your upper limbs (described shortly) are at your sides. But if you hold your hands up above your head, your fingers would be above your wrists. To prevent this kind of confusion, anatomists use a standard anatomical position and a special vocabulary for relating body parts to one another.

Body Positions

Descriptions of any region or part of the human body assume that it is in a standard position of reference called the **anatomical position** (an′-a-TOM-i-kal). In the anatomical position, the subject stands erect facing the observer, with the head level and the eyes facing directly forward. The lower limbs are parallel and the feet are flat on the floor and directed forward, and the upper limbs are at the sides with the palms turned forward (**Figure 1.6**). Two terms describe a reclining body. If the body is lying facedown, it is in the **prone** position. If the body is lying faceup, it is in the **supine** position.

Regional Names

The human body is divided into several major regions that can be identified externally. The principal regions are the head, neck, trunk, upper limbs, and lower limbs (**Figure 1.6**). The **head** consists of the skull and face. The *skull* encloses and protects the brain; the *face* is the front portion of the head that includes the eyes, nose, mouth, forehead, cheeks, and chin. The **neck** supports the head and attaches it to the trunk. The **trunk** consists of the chest, abdomen, and pelvis. Each **upper limb** attaches to the trunk and consists of the shoulder, armpit, arm (portion of the limb from the shoulder to the elbow), forearm (portion of the limb from the elbow to the wrist), wrist, and hand. Each **lower limb** also attaches to the trunk and consists of the buttock, thigh (portion of the limb from the buttock to the knee), leg (portion of the limb from the knee to the ankle), ankle, and foot. The *groin* is the area on the front surface of the body marked by a crease on each side, where the trunk attaches to the thighs.

Figure 1.6 shows the anatomical and common names of major parts of the body. For example, if you receive a tetanus shot in your *gluteal region*, the injection is in your *buttock*. Because the anatomical term for a body part usually is based on a Greek or Latin word, it may look different from the common name for the same part or area. For example, the Latin word *axilla* (ak-SIL-a) is the anatomical term for armpit. Thus, the axillary nerve is one of the nerves passing within the armpit. You will learn more about the Greek and Latin word roots of anatomical and physiological terms as you read this book.

FIGURE 1.6 The anatomical position. The anatomical names and corresponding common names (in parentheses) are indicated for specific body regions. For example, the cephalic region is the head.

In the anatomical position, the subject stands erect facing the observer with the head level and the eyes facing forward. The lower limbs are parallel and the feet are flat on the floor and directed forward, and the upper limbs are at the sides with the palms facing forward.

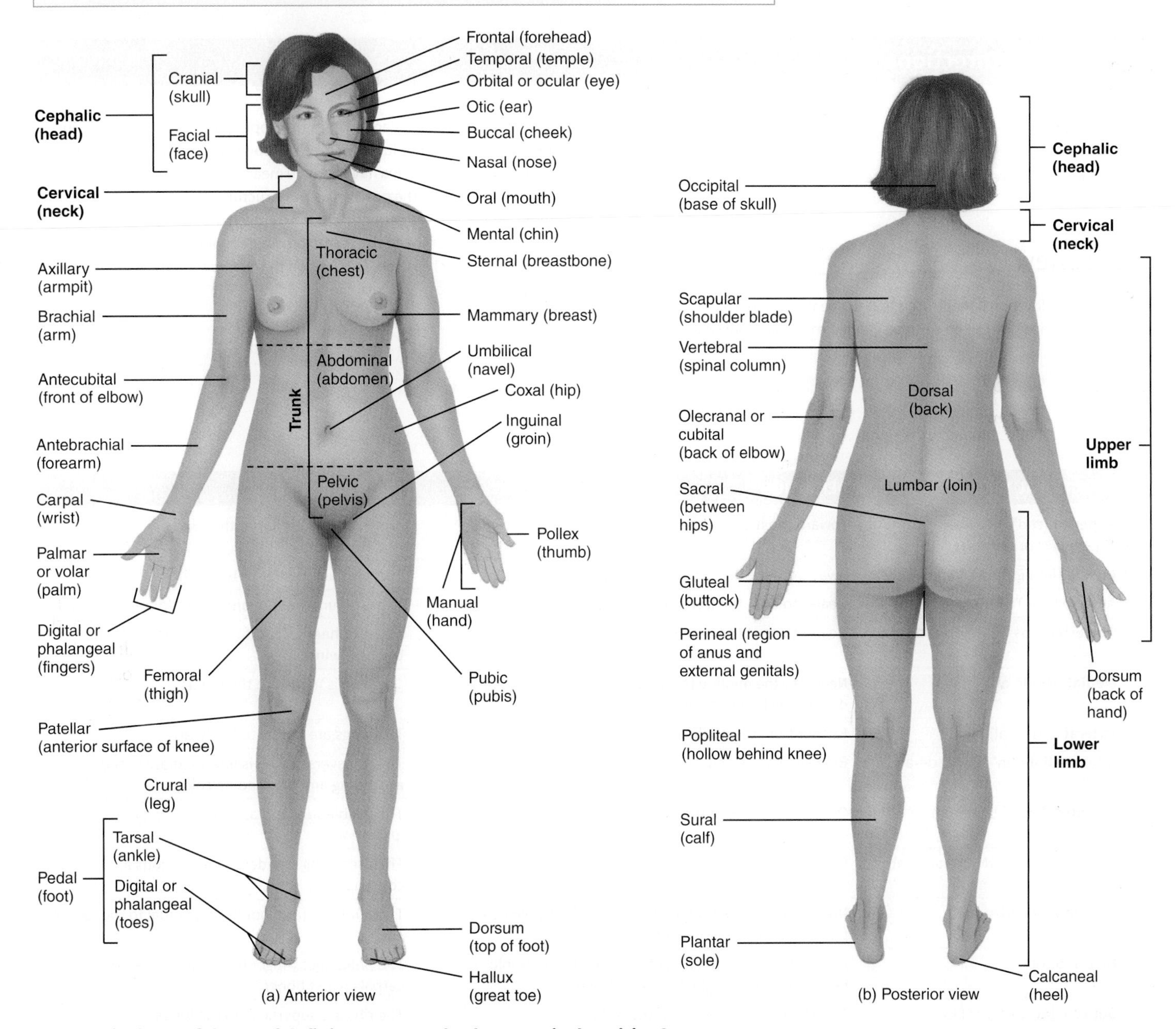

Q What is the usefulness of defining one standard anatomical position?

Directional Terms

To locate various body structures, anatomists use specific **directional terms**, words that describe the position of one body part relative to another. Several directional terms are grouped in pairs that have opposite meanings, such as anterior (front) and posterior (back). **Exhibit 1** and **Figure 1.7** present the main directional terms.

EXHIBIT 1 Directional Terms *(Figure 1.7)*

OBJECTIVE

- **Define** each directional term used to describe the human body.

Overview

Most of the directional terms used to describe the relationship of one part of the body to another can be grouped into pairs that have opposite meanings. For example, **superior** means toward the upper part of the body, and **inferior** means toward the lower part of the body. It is important to understand that directional terms have relative meanings; they make sense only when used to describe the position of one structure relative to another. For example, your knee is superior to your ankle, even though both are located in the inferior half of the body. Study the directional terms below and the example of how each is used. As you read the examples, look at **Figure 1.7** to see the location of each structure.

DIRECTIONAL TERM	DEFINITION	EXAMPLE OF USE
Superior (soo'-PĒR-ē-or) (*cephalic* or *cranial*)	Toward the head, or the upper part of a structure.	The heart is superior to the liver.
Inferior (in-FĒ-rē-or) (*caudal*)	Away from the head, or the lower part of a structure.	The stomach is inferior to the lungs.
Anterior (an-TĒR-ē-or) (*ventral*)*	Nearer to or at the front of the body.	The sternum (breastbone) is anterior to the heart.
Posterior (pos-TĒR-ē-or) (*dorsal*)	Nearer to or at the back of the body.	The esophagus (food tube) is posterior to the trachea (windpipe).
Medial (MĒ-dē-al)	Nearer to the midline (an imaginary vertical line that divides the body into equal right and left sides).	The ulna is medial to the radius.
Lateral (LAT-er-al)	Farther from the midline.	The lungs are lateral to the heart.
Intermediate (in'-ter-MĒ-dē-at)	Between two structures.	The transverse colon is intermediate to the ascending and descending colons.
Ipsilateral (ip-si-LAT-er-al)	On the same side of the body as another structure.	The gallbladder and ascending colon are ipsilateral.
Contralateral (KON-tra-lat-er-al)	On the opposite side of the body from another structure.	The ascending and descending colons are contralateral.
Proximal (PROK-si-mal)	Nearer to the attachment of a limb to the trunk; nearer to the origination of a structure.	The humerus (arm bone) is proximal to the radius.
Distal (DIS-tal)	Farther from the attachment of a limb to the trunk; farther from the origination of a structure.	The phalanges (finger bones) are distal to the carpals (wrist bones).
Superficial (soo'-per-FISH-al) (*external*)	Toward or on the surface of the body.	The ribs are superficial to the lungs.
Deep (*Internal*)	Away from the surface of the body.	The ribs are deep to the skin of the chest and back.

*Note that the terms *anterior* and *ventral* mean the same thing in humans. However, in four-legged animals *ventral* refers to the belly side and is therefore *inferior*. Similarly, the terms *posterior* and *dorsal* mean the same thing in humans, but in four-legged animals *dorsal* refers to the back side and is therefore *superior*.

FIGURE 1.7 **Directional terms.**

Directional terms precisely locate various parts of the body relative to one another.

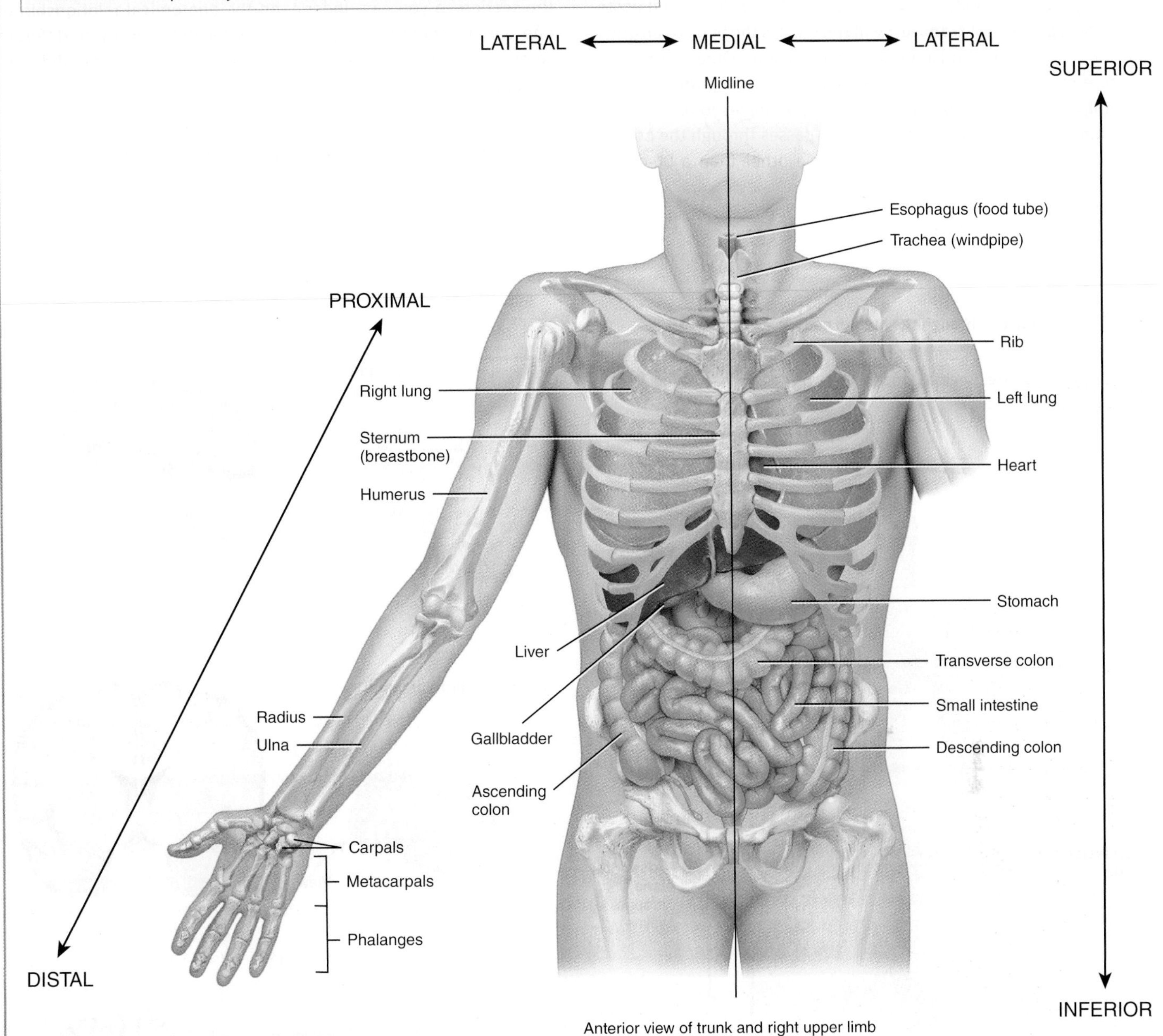

Anterior view of trunk and right upper limb

Q Is the radius proximal to the humerus? Is the esophagus anterior to the trachea? Are the ribs superficial to the lungs? Is the urinary bladder medial to the ascending colon? Is the sternum lateral to the descending colon?

Planes and Sections

You will also study parts of the body relative to **planes**, imaginary flat surfaces that pass through the body parts (**Figure 1.8**). A **sagittal plane** (SAJ-i-tal; *sagitt-* = arrow) is a vertical plane that divides the body or an organ into right and left sides. More specifically, when such a plane passes through the midline of the body or an organ and divides it into *equal* right and left sides, it is called a **midsagittal plane** or a *median plane*. The **midline** is an imaginary vertical line that divides the body into equal left and right sides. If the sagittal plane does not pass through the midline but instead divides the body or an organ into *unequal* right and left sides, it is called a **parasagittal plane** (*para-* = near). A **frontal** or *coronal plane* (kō-RŌ-nal; *corona* = crown) divides the body or an organ into anterior (front) and posterior (back) portions. A **transverse plane** divides the body or an organ into superior (upper) and inferior (lower) portions. Other names for a transverse plane are a *cross-sectional* or *horizontal plane*. Sagittal, frontal, and transverse planes are all at right angles to one another. An **oblique plane** (ō-BLĒK), by contrast, passes through the body or an organ at an oblique angle (any angle other than a 90-degree angle).

When you study a body region, you often view it in section. A **section** is a cut of the body or one of its organs made along one of the planes just described. It is important to know the plane of the section so you can understand the anatomical relationship of one part to another. Figure 1.9a–c indicates how three different sections—*midsagittal*, *frontal*, and *transverse*—provide different views of the brain.

FIGURE 1.8 Planes through the human body.

Frontal, transverse, sagittal, and oblique planes divide the body in specific ways.

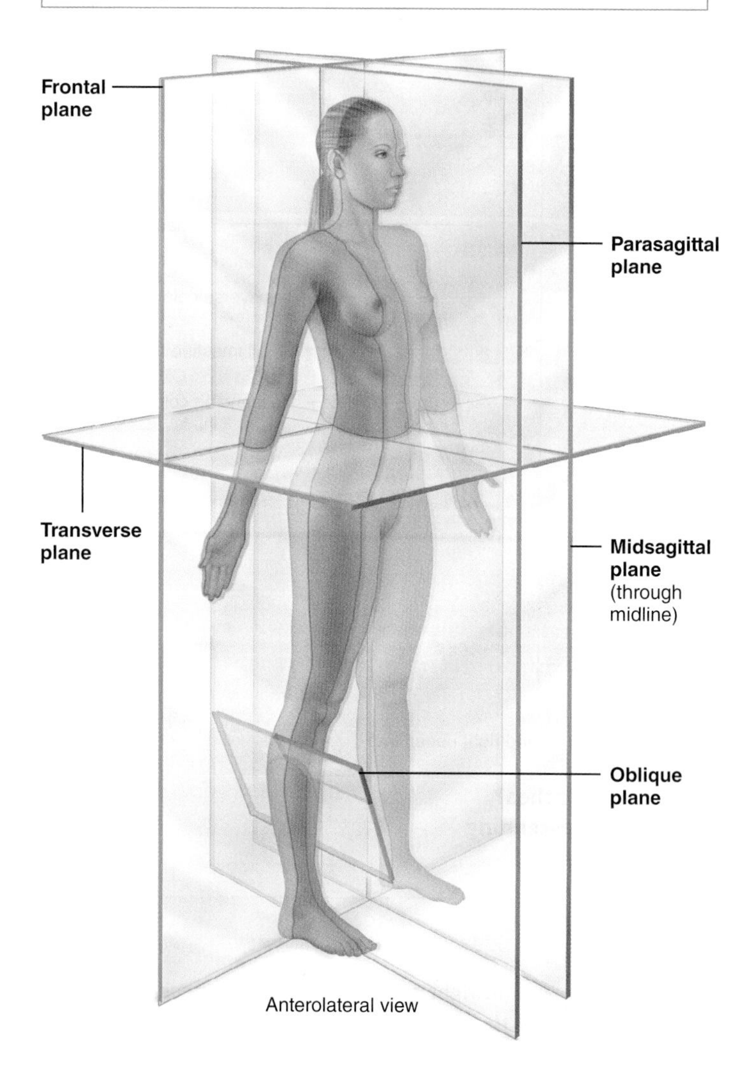

Q Which plane divides the heart into anterior and posterior portions?

FIGURE 1.9 Planes and sections through different parts of the brain. The diagrams (left) show the planes, and the photographs (right) show the resulting sections. Note: The "view" arrows in the diagrams indicate the direction from which each section is viewed. This aid is used throughout the book to indicate viewing perspectives.

Planes divide the body in various ways to produce sections.

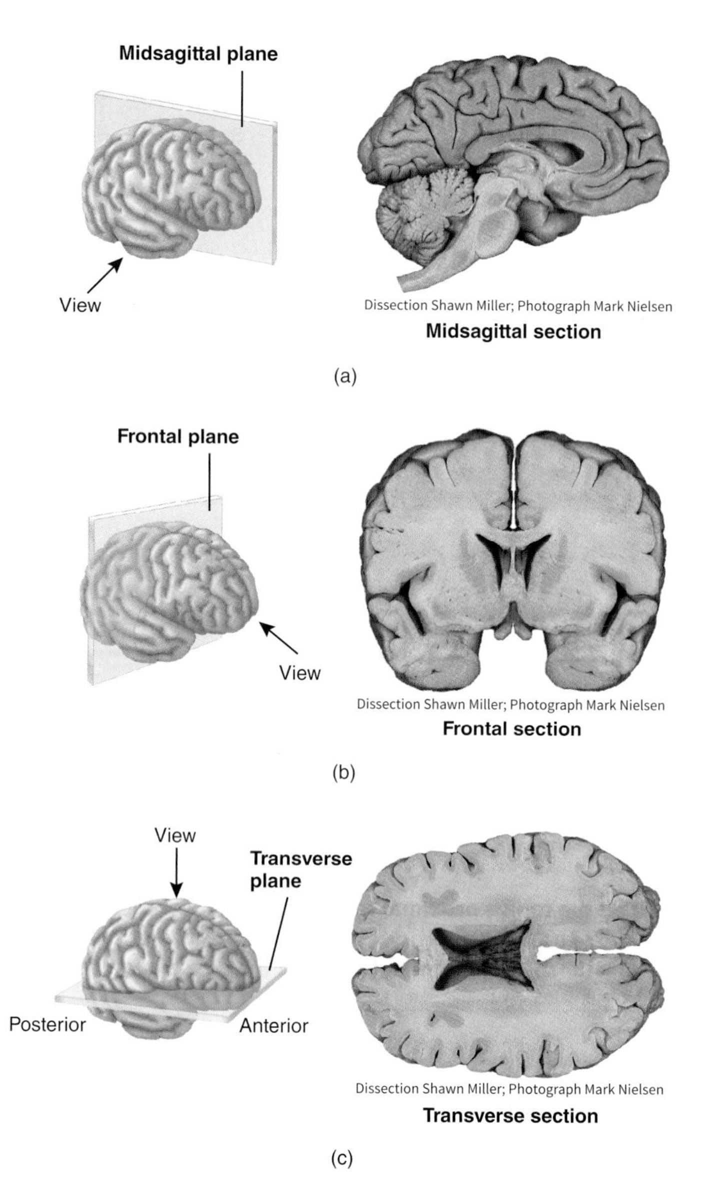

Q Which plane divides the brain into unequal right and left portions?

Body Cavities

Body cavities are spaces that enclose internal organs. Bones, muscles, ligaments, and other structures separate the various body cavities from one another. Here we discuss several body cavities (Figure 1.10).

The cranial bones form a hollow space of the head called the **cranial cavity** (KRĀ-nē-al), which contains the brain. The bones of the vertebral column (backbone) form the **vertebral** (*spinal*) **canal** (VER-te-bral), which contains the spinal cord. The cranial cavity and vertebral canal are continuous with one another. Three layers of protective tissue, the **meninges** (me-NIN-jēz), and a shock-absorbing fluid surround the brain and spinal cord.

The major body cavities of the trunk are the thoracic and abdominopelvic cavities. The **thoracic cavity** (thor-AS-ik; *thorac-* = chest) or chest cavity (Figure 1.11) is formed by the ribs, the muscles of the chest, the sternum (breastbone), and the thoracic portion of the vertebral column. Within the thoracic cavity are the **pericardial cavity** (per′-i-KAR-dē-al; *peri-* = around; *-cardial* = heart), a fluid-filled space that surrounds the heart, and two fluid-filled spaces called **pleural cavities** (PLOOR-al; *pleur-* = rib or side), one around each lung. The central part of the thoracic cavity is an anatomical region called the **mediastinum** (mē′-dē-as-TĪ-num; *media-* = middle; *-stinum* = partition). It is between the lungs, extending from the sternum to the vertebral column and from the first rib to the diaphragm (Figure 1.11a, b). The mediastinum contains all thoracic organs except the lungs themselves. Among the structures in the mediastinum are the heart, esophagus, trachea, thymus, and several large blood vessels that enter and exit the heart. The **diaphragm** (DĪ-a-fram = partition or wall) is a dome-shaped muscle that separates the thoracic cavity from the abdominopelvic cavity.

FIGURE 1.10 Body cavities. The black dashed line in (a) indicates the border between the abdominal and pelvic cavities.

The major cavities of the trunk are the thoracic and abdominopelvic cavities.

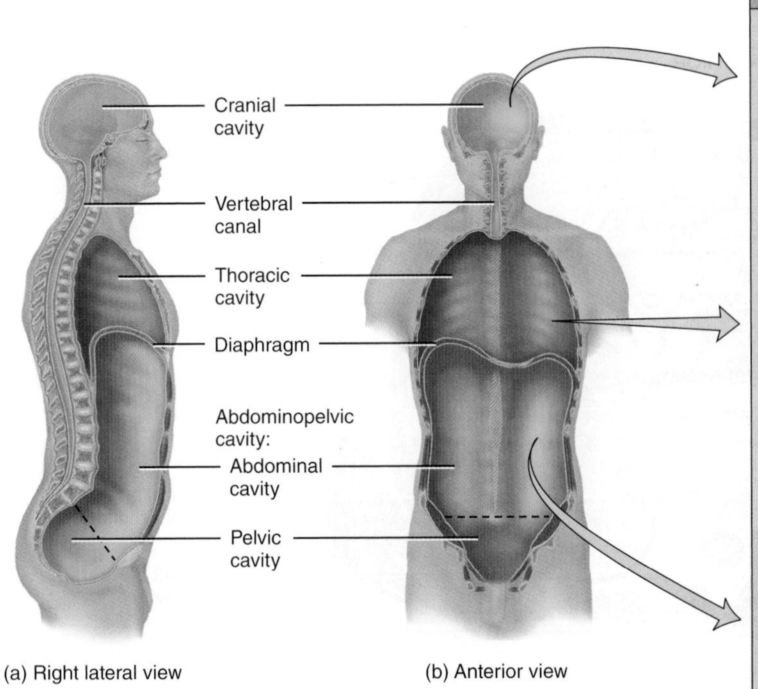

CAVITY	COMMENTS
Cranial cavity	Formed by cranial bones and contains brain.
Vertebral canal	Formed by vertebral column and contains spinal cord and the beginnings of spinal nerves.
Thoracic cavity*	Chest cavity; contains pleural and pericardial cavities and the mediastinum.
Pleural cavity	A potential space between the layers of the pleura that surrounds a lung.
Pericardial cavity	A potential space between the layers of the pericardium that surrounds the heart.
Mediastinum	Central portion of thoracic cavity between the lungs; extends from sternum to vertebral column and from first rib to diaphragm; contains heart, thymus, esophagus, trachea, and several large blood vessels.
Abdominopelvic cavity	Subdivided into abdominal and pelvic cavities.
Abdominal cavity	Contains stomach, spleen, liver, gallbladder, small intestine, and most of large intestine; the serous membrane of the abdominal cavity is the peritoneum.
Pelvic cavity	Contains urinary bladder, portions of large intestine, and internal organs of reproduction.

*See Figure 1.11 for details of the thoracic cavity.

Q In which cavities are the following organs located: urinary bladder, stomach, heart, small intestine, lungs, internal female reproductive organs, thymus, spleen, liver? Use the following symbols for your responses: T = thoracic cavity, A = abdominal cavity, or P = pelvic cavity.

FIGURE 1.11 The thoracic cavity. The black dashed lines indicate the borders of the mediastinum. Note: When transverse sections are viewed inferiorly (from below), the anterior aspect of the body appears on top and the left side of the body appears on the right side of the illustration.

The thoracic cavity contains three smaller cavities and the mediastinum.

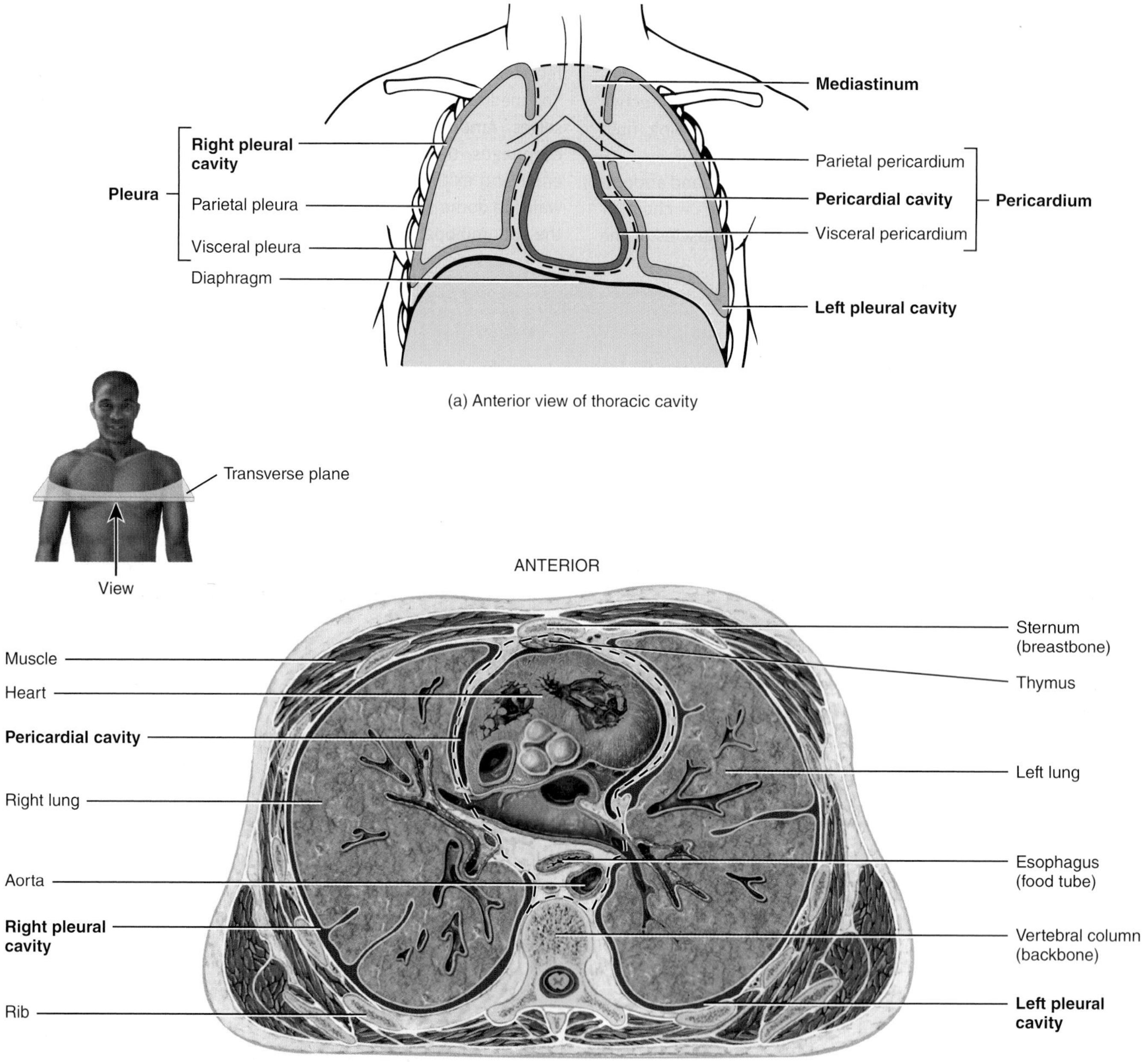

(a) Anterior view of thoracic cavity

(b) Inferior view of transverse section of thoracic cavity

Q What is the name of the cavity that surrounds the heart? Which cavities surround the lungs?

The **abdominopelvic cavity** (ab-dom′-i-nō-PEL-vik; see Figure 1.10) extends from the diaphragm to the groin and is encircled by the abdominal muscular wall and the bones and muscles of the pelvis. As the name suggests, the abdominopelvic cavity is divided into two portions, even though no wall separates them (Figure 1.12). The superior portion, the **abdominal cavity** (ab-DOM-i-nal; *abdomin-* = belly), contains the stomach, spleen, liver, gallbladder, small intestine, and most of the large intestine. The inferior portion, the **pelvic cavity** (PEL-vik; *pelv-* = basin), contains the urinary bladder, portions of the large intestine, and internal organs of the reproductive

FIGURE 1.12 **The abdominopelvic cavity.** The black dashed lower line shows the approximate boundary between the abdominal and pelvic cavities.

The abdominopelvic cavity extends from the diaphragm to the groin.

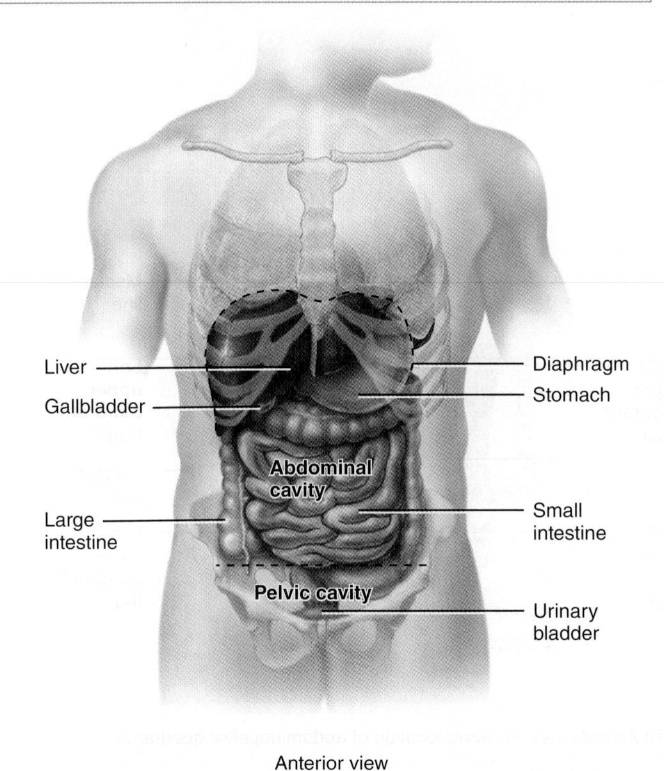

Q To which body systems do the organs shown here within the abdominal and pelvic cavities belong? (*Hint:* Refer to Table 1.2.)

system. Organs inside the thoracic and abdominopelvic cavities are called **viscera** (VIS-er-a).

Thoracic and Abdominal Cavity Membranes A **membrane** is a thin, pliable tissue that covers, lines, partitions, or connects structures. One example is a slippery, double-layered membrane associated with body cavities that does not open directly to the exterior called a **serous membrane** (SĒR-us). It covers the viscera within the thoracic and abdominal cavities and also lines the walls of the thorax and abdomen. The parts of a serous membrane are (1) the *parietal layer* (pa-RĪ-e-tal), a thin epithelium that lines the walls of the cavities, and (2) the *visceral layer* (VIS-er-al), a thin epithelium that covers and adheres to the viscera within the cavities. Between the two layers is a potential space that contains a small amount of lubricating fluid (*serous fluid*). The fluid allows the viscera to slide somewhat during movements, such as when the lungs inflate and deflate during breathing.

The serous membrane of the pleural cavities is called the **pleura** (PLOO-ra). The *visceral pleura* clings to the surface of the lungs, and the *parietal pleura* lines the chest wall, covering the superior surface of the diaphragm (see Figure 1.11a). In between is the *pleural cavity*, filled with a small amount of lubricating serous fluid (see Figure 1.11). The serous membrane of the pericardial cavity is the **pericardium** (per′-i-KAR-dē-um). The *visceral pericardium* covers the surface of the heart; the *parietal pericardium* lines the chest wall. Between them is the *pericardial cavity*, filled with a small amount of lubricating serous fluid (see Figure 1.11). The **peritoneum** (per′-i-tō-NĒ-um) is the serous membrane of the abdominal cavity. The *visceral peritoneum* covers the abdominal viscera, and the *parietal peritoneum* lines the abdominal wall, covering the inferior surface of the diaphragm. Between them is the *peritoneal cavity*, which contains a small amount of lubricating serous fluid. Most abdominal organs are surrounded by the peritoneum. Some are not surrounded by the peritoneum; instead they are posterior to it. Such organs are said to be *retroperitoneal* (re′-trō-per-i-tō-NĒ-al; *retro-* = behind). The kidneys, adrenal glands, pancreas, duodenum of the small intestine, ascending and descending colons of the large intestine, and portions of the abdominal aorta and inferior vena cava are retroperitoneal.

In addition to the major body cavities just described, you will also learn about other body cavities in later chapters. These include the *oral (mouth) cavity*, which contains the tongue and teeth (see Figure 24.5); the *nasal cavity* in the nose (see Figure 23.1); the *orbital cavities (orbits)*, which contain the eyeballs (see Figure 7.3); the *middle ear cavities (middle ears)*, which contain small bones (see Figure 17.19); and the *synovial cavities*, which are found in freely movable joints and contain synovial fluid (see Figure 9.3).

A summary of the major body cavities and their membranes is presented in the table included in Figure 1.10.

Abdominopelvic Regions and Quadrants

To describe the location of the many abdominal and pelvic organs more easily, anatomists and clinicians use two methods of dividing the abdominopelvic cavity into smaller areas. In the first method, two horizontal and two vertical lines, aligned like a tic-tac-toe grid, partition this cavity into nine **abdominopelvic regions** (Figure 1.13a). The superior horizontal line, the *subcostal line* (*sub* = below; *costal* = rib), passes across the lowest level of the 10th costal cartilages (see also Figure 7.22b); the inferior horizontal line, the *transtubercular line* (trans-too-BER-kū-lar), passes across the superior margins of the iliac crests of the right and left hip bone (see Figure 8.9). Two vertical lines, the left and right *midclavicular lines* (mid-kla-VIK-ū-lar), are drawn through the midpoints of the clavicles (collar bones), just medial to the nipples. The four lines divide the abdominopelvic cavity into a larger middle section and smaller left and right sections. The names of the nine abdominopelvic regions are **right hypochondriac** (hī′-pō-KON-drē-ak), **epigastric** (ep-i-GAS-trik), **left hypochondriac, right lumbar, umbilical** (um-BIL-i-kal), **left lumbar, right inguinal** (*iliac*) (IN-gwi-nal), **hypogastric** (*pubic*), and **left inguinal** (*iliac*).

The second method is simpler and divides the abdominopelvic cavity into **quadrants** (KWOD-rantz; *quad-* = one-fourth), as shown in Figure 1.13b. In this method, a midsagittal line (the *median line*) and a transverse line (the *transumbilical line*) are passed through the **umbilicus** (um-BI-li-kus; *umbilic-* = navel) or *belly button.* The names of the abdominopelvic quadrants are **right upper quadrant (RUQ), left upper quadrant (LUQ), right lower quadrant (RLQ)**, and **left lower quadrant (LLQ)**. The nine-region division is more widely used for anatomical

FIGURE 1.13 **Regions and quadrants of the abdominopelvic cavity.**

The nine-region designation is used for anatomical studies; the quadrant designation is used to locate the site of pain, tumors, or some other abnormality.

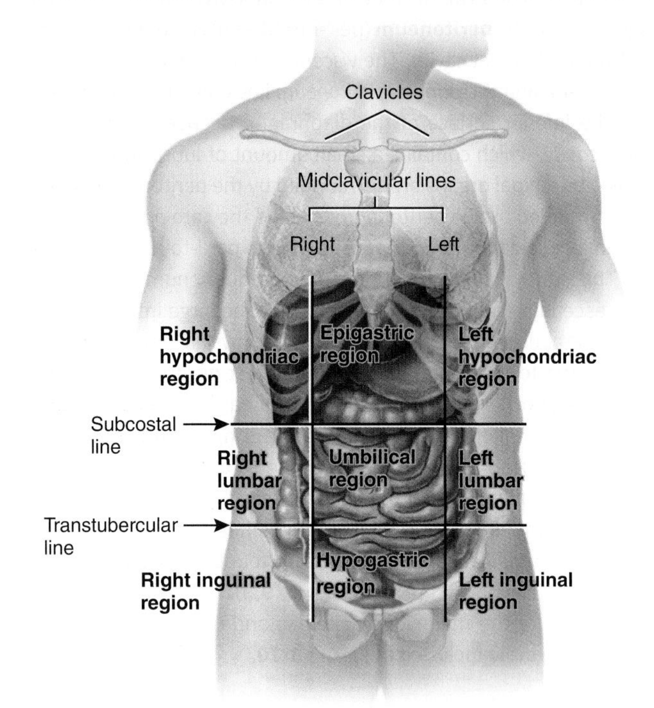

(a) Anterior view showing location of abdominopelvic regions

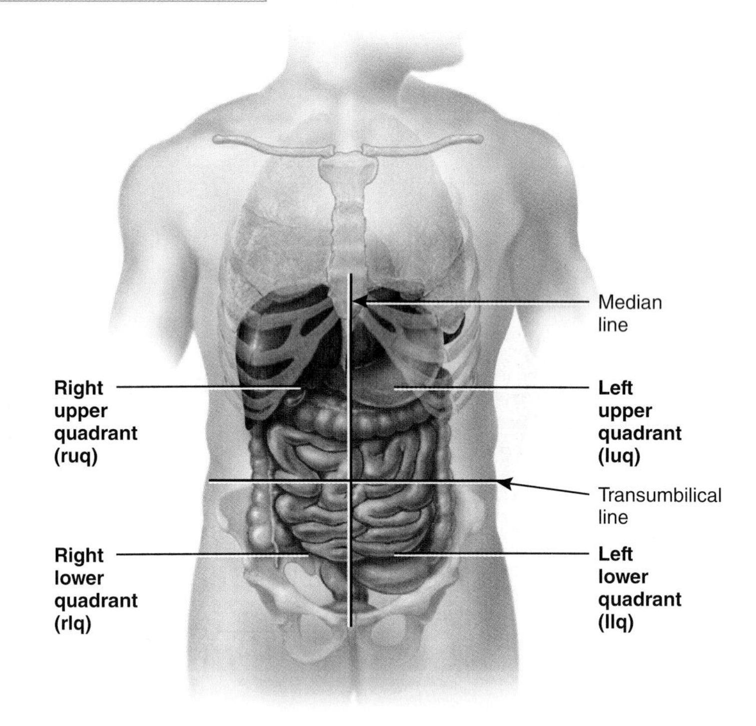

(b) Anterior view showing location of abdominopelvic quadrants

Q In which abdominopelvic region is each of the following found: most of the liver, ascending colon, urinary bladder, and most of the small intestine? In which abdominopelvic quadrant would pain from appendicitis (inflammation of the appendix) be felt?

studies, and quadrants are more commonly used by clinicians for describing the site of abdominopelvic pain, a tumor, or another abnormality.

1.6 Aging and Homeostasis

OBJECTIVE

- **Describe** some of the general anatomical and physiological changes that occur with aging.

As you will see later, **aging** is a normal process characterized by a progressive decline in the body's ability to restore homeostasis. Aging produces observable changes in structure and function and increases vulnerability to stress and disease. The changes associated with aging are apparent in all body systems. Examples include wrinkled skin, gray hair, loss of bone mass, decreased muscle mass and strength, diminished reflexes, decreased production of some hormones, increased incidence of heart disease, Increased susceptibility to infections and cancer, decreased lung capacity, less efficient functioning of the digestive system, decreased kidney function, menopause, and enlarged prostate, These and other effects of aging will be discussed in details in later chapters.

1.7 Medical Imaging

OBJECTIVE

- **Describe** the principles and importance of medical imaging procedures in the evaluation of organ functions and the diagnosis of disease.

Medical imaging refers to techniques and procedures used to create images of the human body. Various types of medical imaging allow visualization of structures inside our bodies and are increasingly helpful for precise diagnosis of a wide range of anatomical and physiological disorders. The grandparent of all medical imaging techniques is conventional radiography (x-rays), in medical use since the late 1940s. The newer imaging technologies not only contribute to diagnosis of disease, but they also are advancing our understanding of normal anatomy and physiology. **Table 1.3** describes some commonly used medical imaging techniques. Other imaging methods, such as cardiac catheterization, will be discussed in later chapters.

TABLE 1.3 Common Medical Imaging Procedures

RADIOGRAPHY

Procedure: A single barrage of x-rays passes through the body, producing an image of interior structures on x-ray–sensitive film. The resulting two-dimensional image is a *radiograph* (RĀ-dē-ō-graf′), commonly called an x-ray.

Comments: Relatively inexpensive, quick, and simple to perform; usually provides sufficient information for diagnosis. X-rays do not easily pass through dense structures, so bones appear white. Hollow structures, such as the lungs, appear black. Structures of intermediate density, such as skin, fat, and muscle, appear as varying shades of gray. At low doses, x-rays are useful for examining soft tissues such as the breast (**mammography**) and for determining bone density (**bone densitometry** or **DEXA scan**).

It is necessary to use a substance called a contrast medium to make hollow or fluid-filled structures visible (appear white) in radiographs. X-rays make structures that contain contrast media appear white. The medium may be introduced by injection, orally, or rectally, depending on the structure to be imaged. Contrast x-rays are used to image blood vessels (**angiography**), the urinary system (**intravenous urography**), and the gastrointestinal tract (**barium contrast x-ray**).

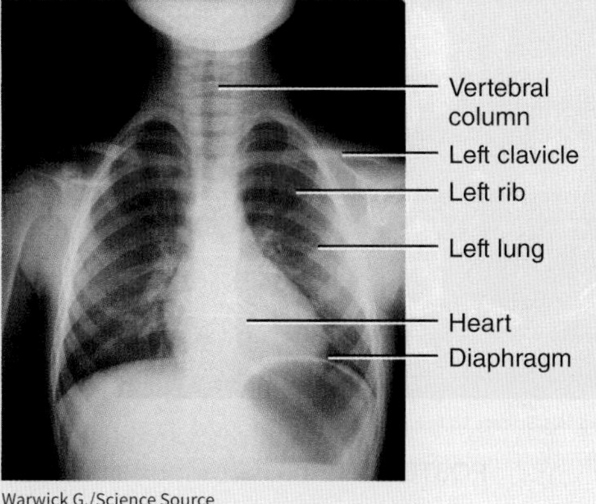

Warwick G./Science Source

Radiograph of thorax in anterior view

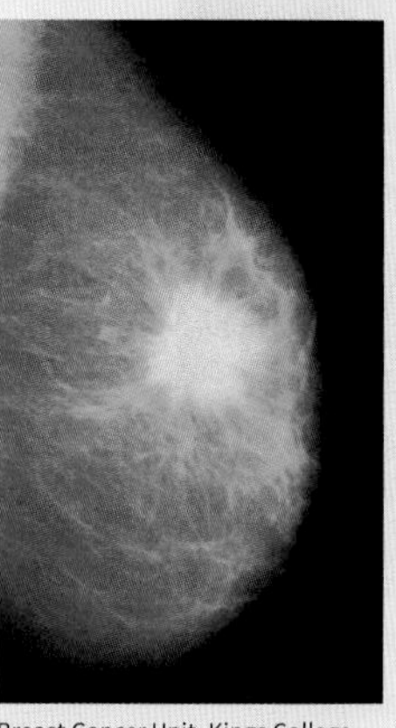

Breast Cancer Unit, Kings College Hospital, London/Science Source

Mammogram of female breast showing cancerous tumor (white mass with uneven border)

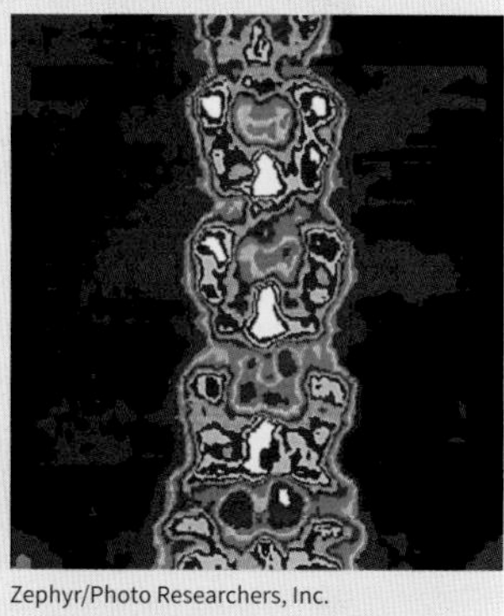

Zephyr/Photo Researchers, Inc.

Bone densitometry scan of lumbar spine in anterior view

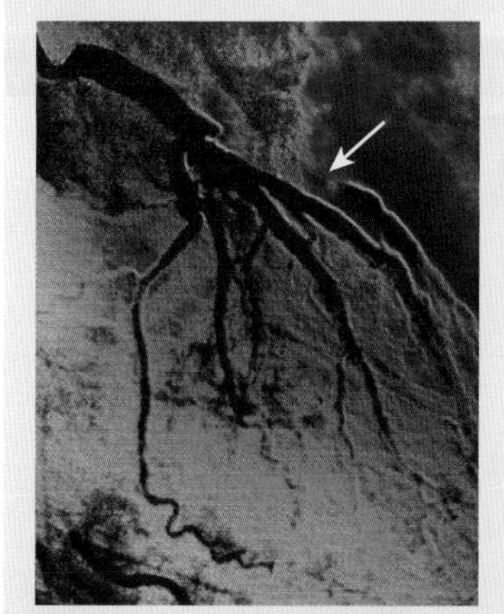

Cardio-Thoracic Centre, Freeman Hospital, Newcastle-Upon-Tyne/Science Source

Angiogram of adult human heart showing blockage in coronary artery (arrow)

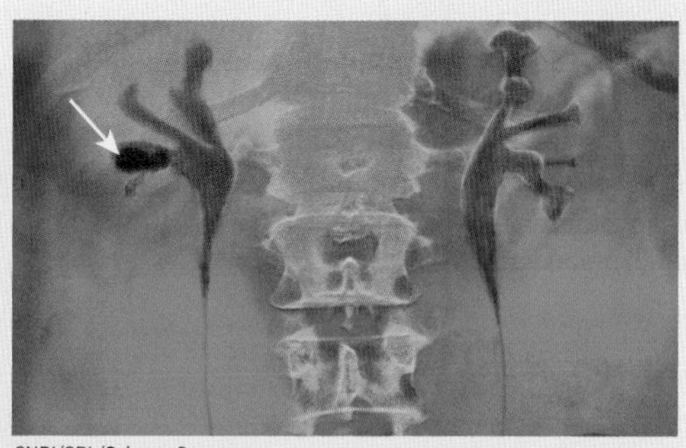

CNRI/SPL/Science Source

Intravenous urogram showing kidney stone (arrow) in right kidney

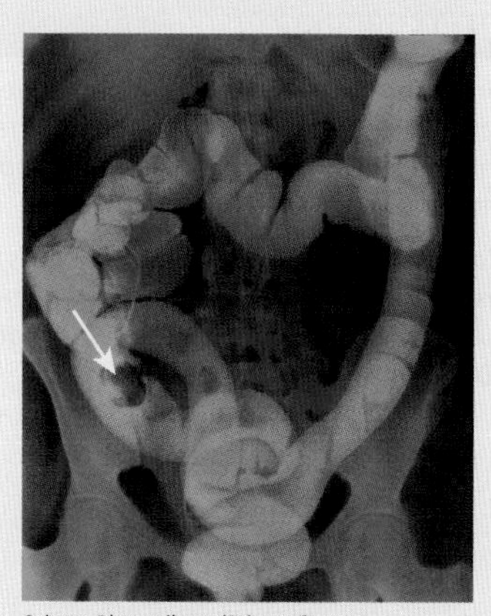

Science Photo Library/Science Source

Barium contrast x-ray showing cancer of the ascending colon (arrow)

Table 1.3 ***Continues***

TABLE 1.3 Common Medical Imaging Procedures (*Continued*)

MAGNETIC RESONANCE IMAGING (MRI)

Procedure: The body is exposed to a high-energy magnetic field, which causes protons (small positive particles within atoms, such as hydrogen) in body fluids and tissues to arrange themselves in relation to the field. Then a pulse of radio waves "reads" these ion patterns, and a color-coded image is assembled on a video monitor. The result is a two- or three-dimensional blueprint of cellular chemistry.

Comments: Relatively safe but cannot be used on patients with metal in their bodies. Shows fine details for soft tissues but not for bones. Most useful for differentiating between normal and abnormal tissues. Used to detect tumors and artery-clogging fatty plaques; reveal brain abnormalities; measure blood flow; and detect a variety of musculoskeletal, liver, and kidney disorders.

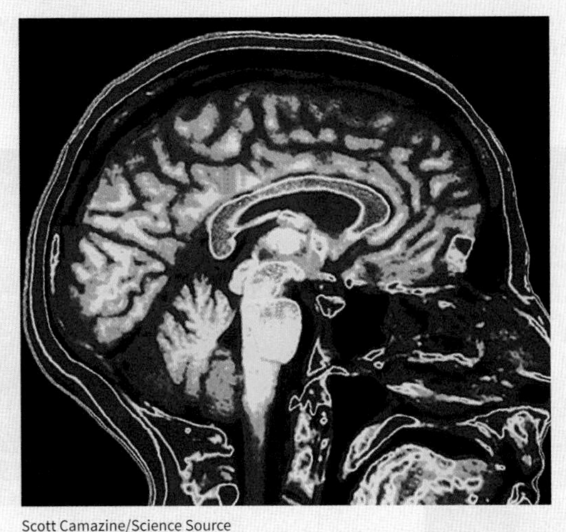

Scott Camazine/Science Source

Magnetic resonance image of brain in sagittal section

COMPUTED TOMOGRAPHY (CT)

[formerly called computerized axial tomography (CAT) scanning]

Procedure: In this form of computer-assisted radiography, an x-ray beam traces an arc at multiple angles around a section of the body. The resulting transverse section of the body, called a *CT scan*, is shown on a video monitor.

Comments: Visualizes soft tissues and organs with much more detail than conventional radiographs. Differing tissue densities show up as various shades of gray. Multiple scans can be assembled to build three-dimensional views of structures (described next). Whole-body CT scanning typically targets the torso and appears to provide the most benefit in screening for lung cancers, coronary artery disease, and kidney cancers.

ANTERIOR

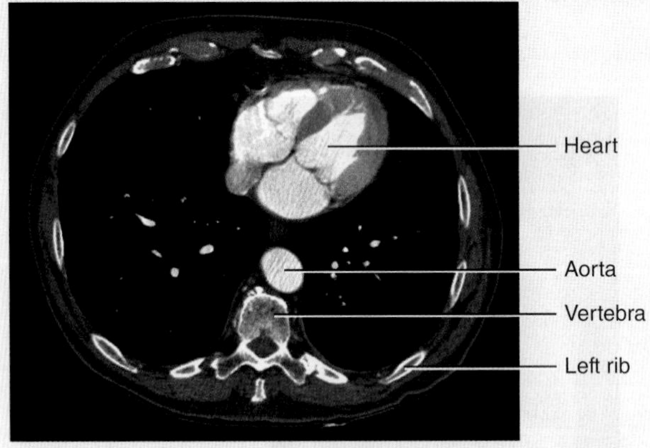

Living Art Enterprises/Science Source

POSTERIOR

Computed tomography scan of thorax in inferior view

ULTRASOUND SCANNING

Procedure: High-frequency sound waves produced by a handheld wand reflect off body tissues and are detected by the same instrument. The image, which may be still or moving, is called a *sonogram* (SON-ō-gram) and is shown on a video monitor.

Comments: Safe, noninvasive, painless, and uses no dyes. Most commonly used to visualize the fetus during pregnancy. Also used to observe the size, location, and actions of organs and blood flow through blood vessels (**Doppler ultrasound**).

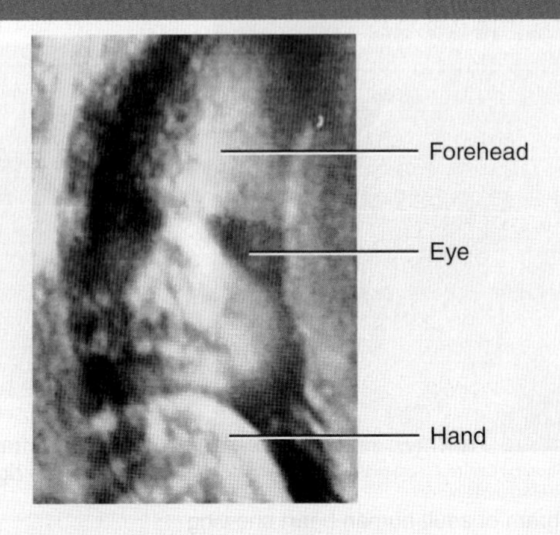

Sonogram of fetus (Courtesy of Andrew Joseph Tortora and Damaris Soler)

CORONARY (CARDIAC) COMPUTED TOMOGRAPHY ANGIOGRAPHY (CCTA) SCAN

Procedure: In this form of computer-assisted radiography, an iodine-containing contrast medium is injected into a vein and a beta blocker is given to decrease heart rate. Then, numerous x-ray beams trace an arc around the heart and a scanner detects the x-ray beams and transmits them to a computer, which transforms the information into a three-dimensional image of the coronary blood vessels on a monitor. The image produced is called a *CCTA scan* and can be generated in less than 20 seconds.

Comments: Used primarily to determine if there are any coronary artery blockages (for example, atherosclerotic plaque or calcium) that may require an intervention such as angioplasty or stent. The CCTA scan can be rotated, enlarged, and moved at any angle. The procedure can take thousands of images of the heart within the time of a single heartbeat, so it provides a great amount of detail about the heart's structure and function.

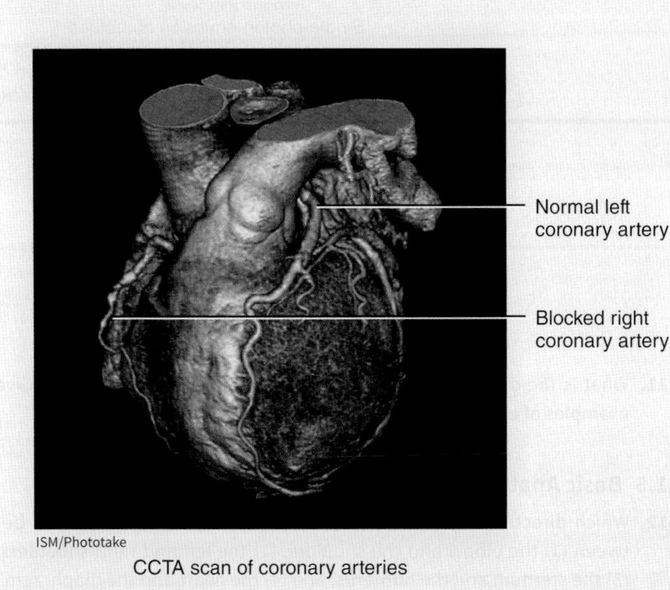

ISM/Phototake

CCTA scan of coronary arteries

POSITRON EMISSION TOMOGRAPHY (PET)

Procedure: A substance that emits positrons (positively charged particles) is injected into the body, where it is taken up by tissues. The collision of positrons with negatively charged electrons in body tissues produces gamma rays (similar to x-rays) that are detected by gamma cameras positioned around the subject. A computer receives signals from the gamma cameras and constructs a *PET scan* image, displayed in color on a video monitor. The PET scan shows where the injected substance is being used in the body. In the PET scan image shown here, the black and blue colors indicate minimal activity; the red, orange, yellow, and white colors indicate areas of increasingly greater activity.

Comments: Used to study the physiology of body structures, such as metabolism in the brain or heart.

ANTERIOR

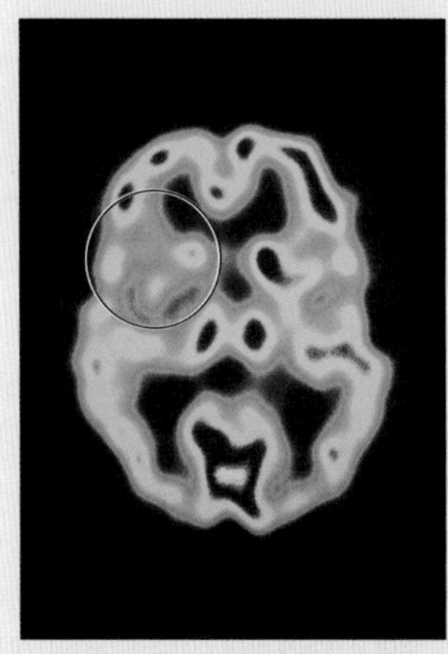

Department of Nuclear Medicine, Charing Cross Hospital /Photo Researchers, Inc.

POSTERIOR

Positron emission tomography scan of transverse section of brain (circled area at upper left indicates where a stroke has occurred)

ENDOSCOPY

Procedure: Endoscopy involves the visual examination of the inside of body organs or cavities using a lighted instrument with lenses called an *endoscope*. The image is viewed through an eyepiece on the endoscope or projected onto a monitor.

Comments: Examples include *colonoscopy* (used to examine the interior of the colon, which is part of the large intestine), *laparoscopy* (used to examine the organs within the abdominopelvic cavity), and *arthroscopy* (used to examine the interior of a joint, usually the knee).

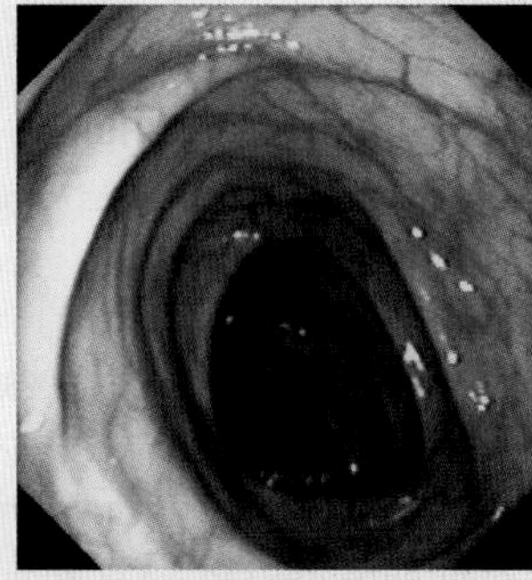

© Camal/Phototake

Interior view of colon as shown by colonoscopy

Table 1.3 ***Continues***

TABLE 1.3 **Common Medical Imaging Procedures (*Continued*)**

RADIONUCLIDE SCANNING

Procedure: A *radionuclide* (radioactive substance) is introduced intravenously into the body and carried by the blood to the tissue to be imaged. Gamma rays emitted by the radionuclide are detected by a gamma camera outside the subject, and the data are fed into a computer. The computer constructs a *radionuclide image* and displays it in color on a video monitor. Areas of intense color take up a lot of the radionuclide and represent high tissue activity; areas of less intense color take up smaller amounts of the radionuclide and represent low tissue activity. **Single-photon-emission computed tomography (SPECT) scanning** is a specialized type of radionuclide scanning that is especially useful for studying the brain, heart, lungs, and liver.

Comments: Used to study activity of a tissue or organ, such as searching for malignant tumors in body tissue or scars that may interfere with heart muscle activity.

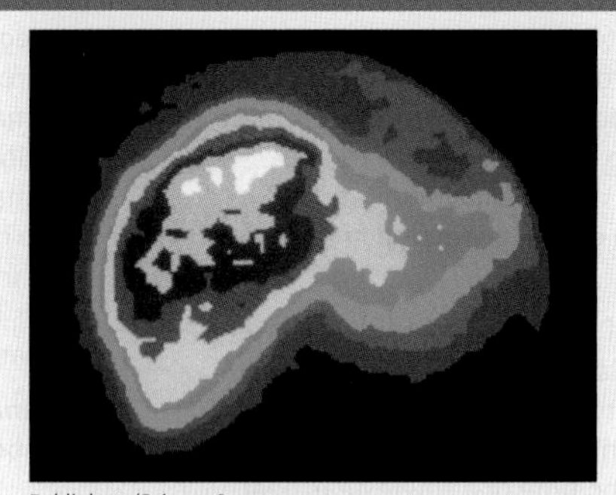

Publiphoto/Science Source

Radionuclide (nuclear) scan of normal human liver

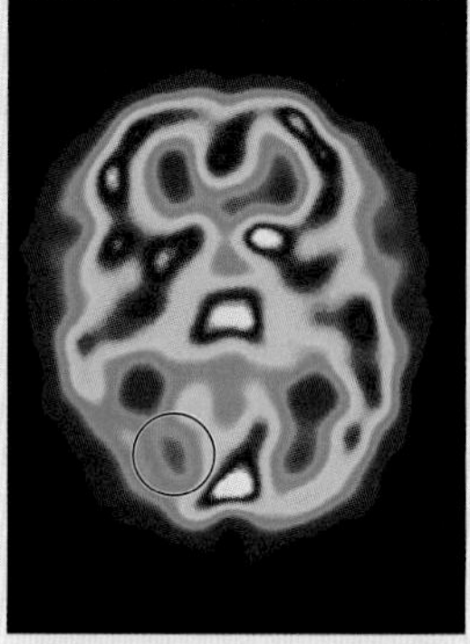

Dept. of Nuclear Medicine, Charing Cross Hospital/Science Source

Single-photon-emission computed tomography (SPECT) scan of transverse section of the brain (the almost all green area at lower left indicates migraine attack)

Review Questions

1.1 Anatomy and Physiology Defined

1. What body function might a respiratory therapist strive to improve? What structures are involved?

2. Give your own example of how the structure of a part of the body is related to its function.

1.2 Levels of Structural Organization and Body Systems

3. Define the following terms: atom, molecule, cell, tissue, organ, system, and organism.

4. At what levels of organization would an exercise physiologist study the human body? (*Hint: Refer to* ***Table 1.1***.)

5. Referring to **Table 1.2**, which body systems help eliminate wastes?

1.3 Characteristics of the Living Human Organism

6. List the six most important life processes in the human body.

1.4 Homeostasis

7. Describe the locations of intracellular fluid, extracellular fluid, interstitial fluid, and blood plasma.

8. Why is extracellular fluid called the internal environment of the body?

9. What types of disturbances can act as stimuli that initiate a feedback system?

10. Define receptor, control center, and effector.

11. What is the difference between symptoms and signs of a disease? Give examples of each.

1.5 Basic Anatomical Terminology

12. Which directional terms can be used to specify the relationships between (1) the elbow and the shoulder, (2) the left and right shoulders, (3) the sternum and the humerus, and (4) the heart and the diaphragm?

13. Locate each region shown in **Figure 1.6** on your own body, and then identify it by its anatomical name and the corresponding common name.

14. What structures separate the various body cavities from one another?

15. Locate the nine abdominopelvic regions and the four abdominopelvic quadrants on yourself, and list some of the organs found in each.

1.6 Aging and Homeostasis

16. What are some of the signs of aging?

1.7 Medical Imaging

17. Which forms of medical imaging would be used to show a blockage in an artery of the heart?

18. Of the medical imaging techniques outlined in **Table 1.3**, which one best reveals the physiology of a structure?

19. Which medical imaging technique would you use to determine whether a bone was broken?

Critical Thinking Questions

1. You are studying for your first anatomy and physiology exam and want to know which areas of your brain are working hardest as you study. Your classmate suggests that you could have a computed tomography (CT) scan done to assess your brain activity. Would this be the best way to determine brain activity levels? Why or why not?
2. There is much interest in using stem cells to help in the treatment of diseases such as type 1 diabetes, which is due to a malfunction of some of the normal cells in the pancreas. What would make stem cells useful in disease treatment?
3. On her first anatomy and physiology exam, Heather defined homeostasis as "the condition in which the body approaches room temperature and stays there." Do you agree with Heather's definition?

The Chemical Level of Organization

CHAPTER 2

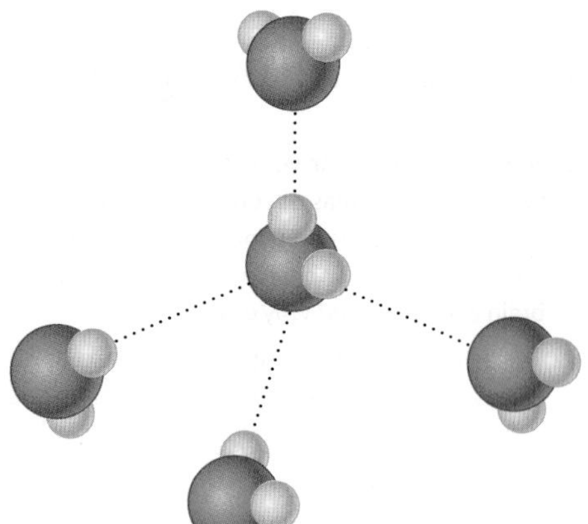

Chemistry and Homeostasis

Maintaining the proper assortment and quantity of thousands of different chemicals in your body, and monitoring the interactions of these chemicals with one another, are two important aspects of homeostasis.

You learned in Chapter 1 that the chemical level of organization, the lowest level of structural organization, consists of atoms and molecules. These letters of the anatomical alphabet ultimately combine to form body organs and systems of astonishing size and complexity. In this chapter, we consider how atoms bond together to form molecules, and how atoms and molecules release or store energy in processes known as chemical reactions. You will also learn about the vital importance of water—which accounts for nearly two-thirds of your body weight—in chemical reactions and the maintenance of homeostasis. Finally, we present several groups of molecules whose unique properties contribute to the assembly of your body's structures and help power the processes that enable you to live.

Q Did you ever wonder how fatty acids relate to health and disease?

2.1 How Matter Is Organized

OBJECTIVES

- **Identify** the main chemical elements of the human body.
- **Describe** the structures of atoms, ions, molecules, free radicals, and compounds.

Chemistry (KEM-is-trē) is the science of the structure and interactions of matter. All living and nonliving things consist of **matter**, which is anything that occupies space and has **mass**. Mass is the amount of matter in any object, which does not change. **Weight**, the force of gravity acting on matter, does change. When objects are farther from Earth, the pull of gravity is weaker; this is why the weight of an astronaut is close to zero in outer space.

Chemical Elements

Matter exists in three states: solid, liquid, and gas. *Solids*, such as bones and teeth, are compact and have a definite shape and volume. *Liquids*, such as blood plasma, have a definite volume and assume the shape of their container. *Gases*, like oxygen and carbon dioxide, have neither a definite shape nor volume. *All* forms of matter—both living and nonliving—are made up of a limited number of building blocks called **chemical elements**. Each element is a substance that cannot be split into a simpler substance by ordinary chemical means. Scientists now recognize 118 elements. Of these, 92 occur naturally on Earth. The rest have been produced from the natural elements using particle accelerators or nuclear reactors. Each named element is designated by a **chemical symbol**, one or two letters of the element's name in English, Latin, or another language. Examples of chemical symbols are H for hydrogen, C for carbon, O for oxygen, N for nitrogen, Ca for calcium, and Na for sodium (*natrium* = sodium).*

Twenty-six different chemical elements normally are present in your body. Just four elements, called the **major elements**, constitute about 96% of the body's mass: oxygen, carbon, hydrogen, and nitrogen. Eight others, the **lesser elements**, contribute about 3.6% to the body's mass: calcium, phosphorus (P), potassium (K), sulfur (S), sodium, chlorine (Cl), magnesium (Mg), and iron (Fe; *ferrum* = iron). An additional 14 elements—the **trace elements**—are present in tiny amounts. Together, they account for the remaining body mass, about 0.4%. Several trace elements have important functions in the body. For example, iodine is needed to make thyroid hormones. The functions of some trace elements are unknown. Table 2.1 lists the main chemical elements of the human body.

Structure of Atoms

Each element is made up of **atoms**, the smallest units of matter that retain the properties and characteristics of the element. Atoms are extremely small. Two hundred thousand of the largest atoms would fit on the period at the end of this sentence. Hydrogen atoms, the smallest atoms, have a diameter less than 0.1 nanometer (0.1×10^{-9} m = 0.0000000001 m), and the largest atoms are only five times larger.

Dozens of different **subatomic particles** compose individual atoms. However, only three types of subatomic particles are important for understanding the chemical reactions in the human body: protons,

*The periodic table of elements, which lists all of the known chemical elements, can be found in Appendix B.

TABLE 2.1 Main Chemical Elements in the Body

CHEMICAL ELEMENT (SYMBOL)	% OF TOTAL BODY MASS	SIGNIFICANCE
MAJOR ELEMENTS	(about 96)	
Oxygen (O)	65.0	Part of water and many organic (carbon-containing) molecules; used to generate ATP, a molecule used by cells to temporarily store chemical energy.
Carbon (C)	18.5	Forms backbone chains and rings of all organic molecules: carbohydrates, lipids (fats), proteins, and nucleic acids (DNA and RNA).
Hydrogen (H)	9.5	Constituent of water and most organic molecules; ionized form (H^+) makes body fluids more acidic.
Nitrogen (N)	3.2	Component of all proteins and nucleic acids.
LESSER ELEMENTS	(about 3.6)	
Calcium (Ca)	1.5	Contributes to hardness of bones and teeth; ionized form (Ca^{2+}) needed for blood clotting, release of some hormones, contraction of muscle, and many other processes.
Phosphorus (P)	1.0	Component of nucleic acids and ATP; required for normal bone and tooth structure.
Potassium (K)	0.35	Ionized form (K^+) is the most plentiful cation (positively charged particle) in intracellular fluid; needed to generate action potentials.
Sulfur (S)	0.25	Component of some vitamins and many proteins.
Sodium (Na)	0.2	Ionized form (Na^+) is the most plentiful cation in extracellular fluid; essential for maintaining water balance; needed to generate action potentials.
Chlorine (Cl)	0.2	Ionized form (Cl^-) is the most plentiful anion (negatively charged particle) in extracellular fluid; essential for maintaining water balance.
Magnesium (Mg)	0.1	Ionized form (Mg^{2+}) needed for action of many enzymes (molecules that increase the rate of chemical reactions in organisms).
Iron (Fe)	0.005	Ionized forms (Fe^{2+} and Fe^{3+}) are part of hemoglobin (oxygen-carrying protein in red blood cells) and some enzymes.
TRACE ELEMENTS	(about 0.4)	Aluminum (Al), boron (B), chromium (Cr), cobalt (Co), copper (Cu), fluorine (F), iodine (I), manganese (Mn), molybdenum (Mo), selenium (Se), silicon (Si), tin (Sn), vanadium (V), and zinc (Zn).

neutrons, and electrons (**Figure 2.1**). The dense central core of an atom is its **nucleus**. Within the nucleus are positively charged **protons** (p^+) and uncharged (neutral) **neutrons** (n^0). The tiny, negatively charged **electrons** (e^-) move about in a large space surrounding the nucleus. They do not follow a fixed path or orbit but instead form a negatively charged "cloud" that envelops the nucleus (**Figure 2.1a**).

Even though their exact positions cannot be predicted, specific groups of electrons are most likely to move about within certain regions around the nucleus. These regions, called **electron shells**, may be depicted as simple circles around the nucleus. Because each electron shell can hold a specific number of electrons, the electron shell model best conveys this aspect of atomic structure (**Figure 2.1b**). The first electron shell (nearest the nucleus) never holds more than 2 electrons. The second shell holds a maximum of 8 electrons, and the third can hold up to 18 electrons. The electron shells fill with electrons in a specific order, beginning with the first shell. For example, notice in **Figure 2.2** that sodium (Na), which has 11 electrons total, contains 2 electrons in the first shell, 8 electrons in the second shell, and 1 electron in the third shell. The most massive element present in the human body is iodine, which has a total of 53 electrons: 2 in the first shell, 8 in the second shell, 18 in the third shell, 18 in the fourth shell, and 7 in the fifth shell.

FIGURE 2.1 Two representations of the structure of an atom. Electrons move about the nucleus, which contains neutrons and protons. (a) In the electron cloud model of an atom, the shading represents the chance of finding an electron in regions outside the nucleus. (b) In the electron shell model, filled circles represent individual electrons, which are grouped into concentric circles according to the shells they occupy. Both models depict a carbon atom, with six protons, six neutrons, and six electrons.

An atom is the smallest unit of matter that retains the properties and characteristics of its element.

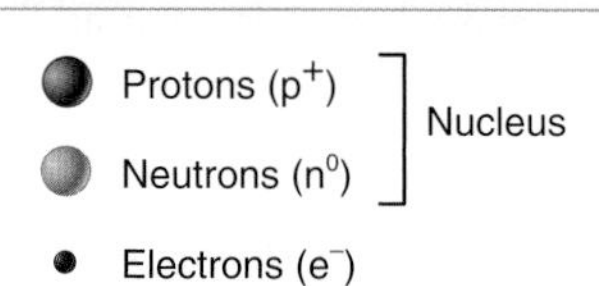

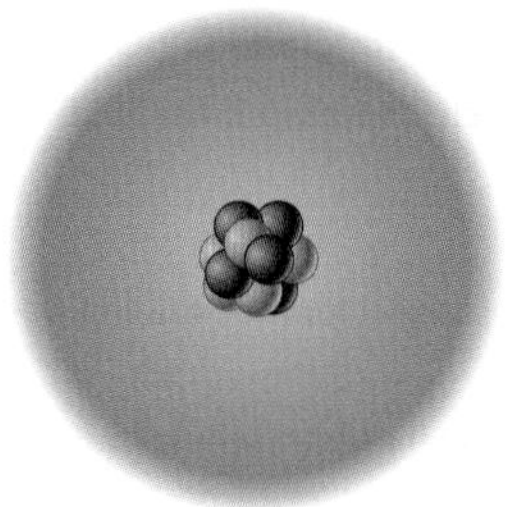

(a) Electron cloud model

(b) Electron shell model

Q How are the electrons of carbon distributed between the first and second electron shells?

The number of electrons in an atom of an element always equals the number of protons. Because each electron and proton carries one charge, the negatively charged electrons and the positively charged protons balance each other. Thus, each atom is electrically neutral; its total charge is zero.

Atomic Number and Mass Number

The *number of protons* in the nucleus of an atom is an atom's **atomic number**. Atoms of different elements have different atomic numbers because they have different numbers of protons. For example, oxygen has an atomic number of 8 because its nucleus has 8 protons, and sodium has an atomic number of 11 because its nucleus has 11 protons.

The **mass number** of an atom is the sum of its protons and neutrons. Because sodium has 11 protons and 12 neutrons, its mass number is 23 (**Figure 2.2**). Although all atoms of one element have the same number of protons, they may have different numbers of neutrons and thus different mass numbers. **Isotopes** are atoms of an element that have different numbers of neutrons and therefore different mass numbers. In a sample of oxygen, for example, most atoms have 8 neutrons, and a few have 9 or 10, but all have 8 protons and 8 electrons. Most isotopes are stable, which means that their nuclear structure does not change over time. The stable isotopes of oxygen are designated ^{16}O, ^{17}O, and ^{18}O (or O-16, O-17, and O-18). As you already may have determined, the numbers indicate the mass number of each isotope. As you will discover shortly, the number of electrons of an atom determines its chemical properties. Although the isotopes of an element have different numbers of neutrons, they have identical chemical properties because they have the same number of electrons.

Certain isotopes called **radioactive isotopes** *(radioisotopes)* are unstable; their nuclei decay (spontaneously change) into a stable configuration. Examples are H-3, C-14, O-15, and O-19. As they decay, these atoms emit radiation—either subatomic particles or packets of energy—and in the process often transform into a different element. For example, the radioactive isotope of carbon, C-14, decays to N-14. The decay of a radioisotope may be as fast as a fraction of a second or as slow as millions of years. The **half-life** of an isotope is the time required for half of the radioactive atoms in a sample of that isotope to decay into a more stable form. The half-life of C-14, which is used to determine the age of organic samples, is about 5730 years; the half-life of I-131, an important clinical tool, is 8 days.

See Clinical Connection: Harmful and Beneficial Effects of Radiation

Atomic Mass

The standard unit for measuring the mass of atoms and their subatomic particles is a **dalton**, also known as an *atomic mass unit (amu).* A neutron has a mass of 1.008 daltons, and a proton has a mass of 1.007 daltons. The mass of an electron, at 0.0005 dalton, is almost 2000 times smaller than the mass of a neutron or proton. The **atomic mass** (also called the *atomic weight*) of an element is the average mass of all its naturally occurring isotopes. Typically, the atomic mass of an element is close to the mass number of its most abundant isotope.

Ions, Molecules, and Compounds

As we discussed, atoms of the same element have the same number of protons. The atoms of each element have a characteristic way of losing, gaining, or sharing their electrons when interacting with other atoms to achieve stability. The way that electrons behave enables atoms in the body to exist in electrically charged forms called ions, or to join with each other into complex combinations called molecules. If an atom either *gives up* or *gains* electrons, it becomes an ion. An **ion** is an atom that has a positive or negative charge because it has unequal numbers of protons and electrons. *Ionization* is the process of giving up or gaining electrons. An ion of an atom is symbolized by writing its chemical symbol followed by the number of its positive (+) or negative (−) charges. Thus, Ca^{2+} stands for a calcium ion that has two positive charges because it has lost two electrons.

When two or more atoms *share* electrons, the resulting combination is called a **molecule** (MOL-e-kūl). A *molecular formula* indicates the elements and the number of atoms of each element that make up a molecule. A molecule may consist of two atoms of the same kind, such as an oxygen molecule (**Figure 2.3a**). The molecular formula for a molecule of oxygen is O_2. The subscript 2 indicates that the molecule contains two atoms of oxygen. Two or more different kinds of atoms may also form a molecule, as in a water molecule (H_2O). In H_2O one atom of oxygen shares electrons with two atoms of hydrogen.

A **compound** is a substance that contains atoms of two or more different elements. Most of the atoms in the body are joined into

FIGURE 2.2 **Atomic structures of several stable atoms.**

The atoms of different elements have different atomic numbers because they have different numbers of protons.

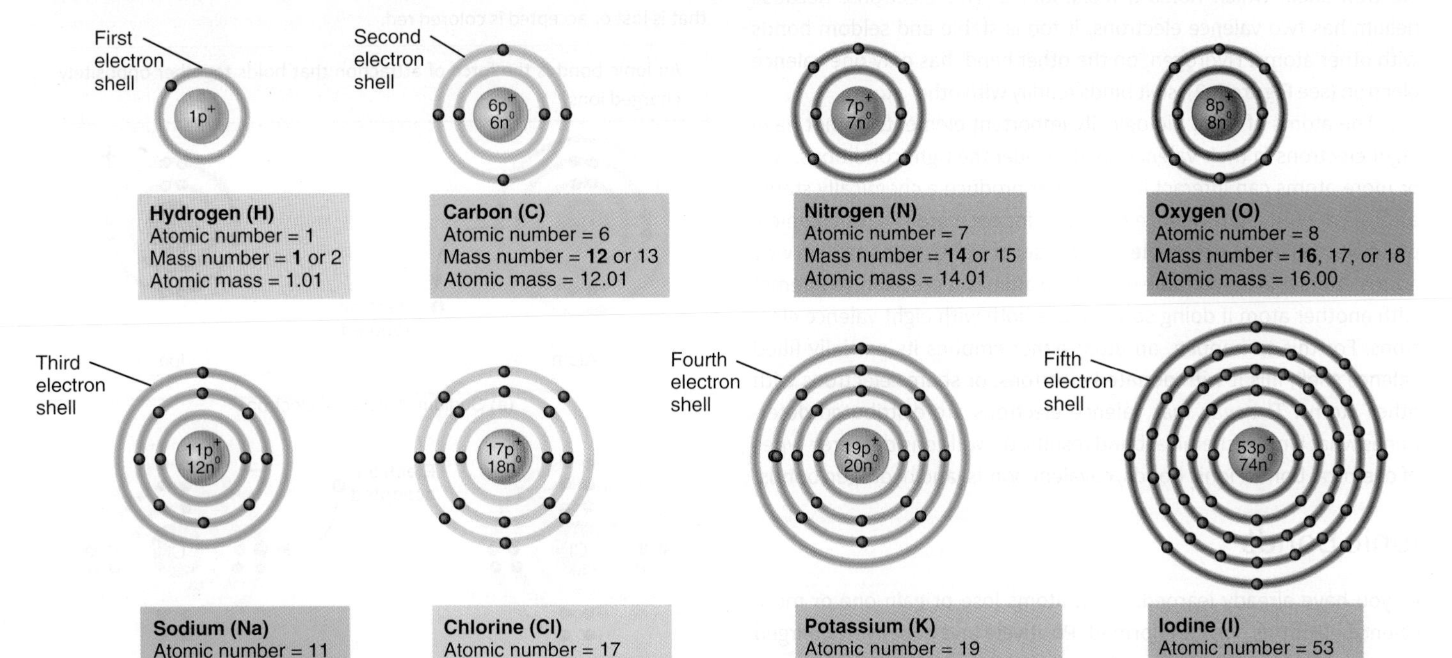

Atomic number = number of protons in an atom
Mass number = number of protons and neutrons in an atom (boldface indicates most common isotope)
Atomic mass = average mass of all stable atoms of a given element in daltons

Q Which four of these elements are present most abundantly in living organisms?

compounds. Water (H_2O) and sodium chloride (NaCl), common table salt, are compounds. However, a molecule of oxygen (O_2) is not a compound because it consists of atoms of only one element.

A **free radical** is an atom or group of atoms with an unpaired electron in the outermost shell. A common example is superoxide, which is formed by the addition of an electron to an oxygen molecule (**Figure 2.3b**). Having an unpaired electron makes a free radical unstable, highly reactive, and destructive to nearby molecules. Free radicals become stable by either giving up their unpaired electron to, or taking on an electron from, another molecule. In so doing, free radicals may break apart important body molecules.

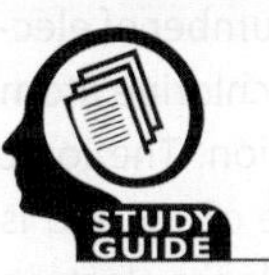

See Clinical Connection: Free Radicals and Antioxidants

FIGURE 2.3 **Atomic structures of an oxygen molecule and a superoxide free radical.**

A free radical has an unpaired electron in its outermost electron shell.

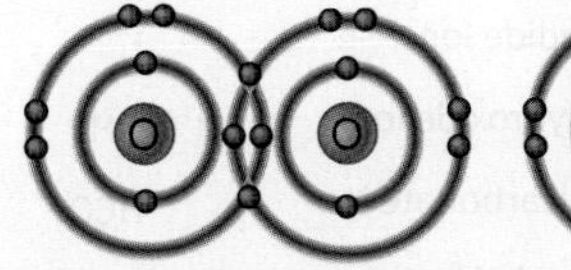

(a) Oxygen molecule (O_2)

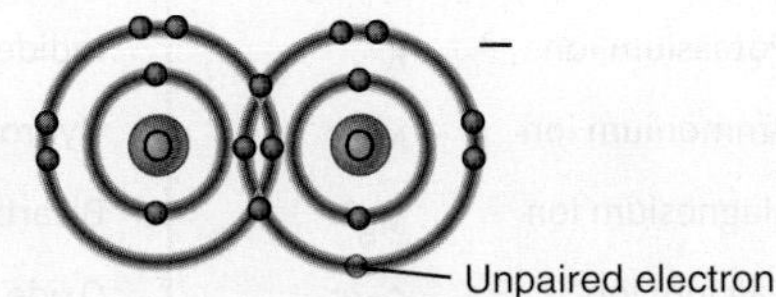

(b) Superoxide free radical (O_2^-)

Q What substances in the body can inactivate oxygen-derived free radicals?

2.2 Chemical Bonds

OBJECTIVES

- **Describe** how valence electrons form chemical bonds.
- **Distinguish** among ionic, covalent, and hydrogen bonds.

The forces that hold together the atoms of a molecule or a compound are **chemical bonds**. The likelihood that an atom will form a chemical bond with another atom depends on the number of electrons in its outermost shell, also called the **valence shell**. An atom with a valence shell holding eight electrons is *chemically stable*, which means it is unlikely to form chemical bonds with other atoms. Neon, for example, has eight electrons in its valence shell, and for this reason it does not bond easily with other atoms. The valence shell of hydrogen and helium is the first electron shell, which holds a maximum of two electrons. Because helium has two valence electrons, it too is stable and seldom bonds with other atoms. Hydrogen, on the other hand, has only one valence electron (see Figure 2.2), so it binds readily with other atoms.

The atoms of most biologically important elements do not have eight electrons in their valence shells. Under the right conditions, two or more atoms can interact in ways that produce a chemically stable arrangement of eight valence electrons for each atom. This chemical principle, called the **octet rule** (*octet* = set of eight), helps explain why atoms interact in predictable ways. One atom is more likely to interact with another atom if doing so will leave both with eight valence electrons. For this to happen, an atom either empties its partially filled valence shell, fills it with donated electrons, or shares electrons with other atoms. The way that valence electrons are distributed determines what kind of chemical bond results. We will consider three types of chemical bonds: ionic bonds, covalent bonds, and hydrogen bonds.

Ionic Bonds

As you have already learned, when atoms lose or gain one or more valence electrons, ions are formed. Positively and negatively charged ions are attracted to one another—opposites attract. The force of attraction that holds together ions with opposite charges is an **ionic bond**. Consider sodium and chlorine atoms, the components of common table salt. Sodium has one valence electron (**Figure 2.4a**). If sodium *loses* this electron, it is left with the eight electrons in its second shell, which becomes the valence shell. As a result, however, the total number of protons (11) exceeds the number of electrons (10). Thus, the sodium atom has become a **cation** (KAT-ī-on), or positively charged ion. A sodium ion has a charge of 1+ and is written Na^+. By contrast, chlorine has seven valence electrons (**Figure 2.4b**). If chlorine *gains* an electron from a neighboring atom, it will have a complete octet in its third electron shell. After gaining an electron, the total number of electrons (18) exceeds the number of protons (17), and the chlorine atom has become an **anion** (AN-ī-on), a negatively charged ion. The ionic form of chlorine is called a *chloride* ion. It has a charge of 1− and is written Cl^-. When an atom of sodium donates its sole valence electron to an atom of chlorine, the resulting positive and negative charges pull both ions tightly together, forming an ionic bond (**Figure 2.4c**). The resulting compound is sodium chloride, written NaCl.

In general, ionic compounds exist as solids, with an orderly, repeating arrangement of the ions, as in a crystal of NaCl (**Figure 2.4d**). A crystal of NaCl may be large or small—the total number of ions can vary—but the ratio of Na^+ to Cl^- is always 1:1. In the body, ionic bonds are found mainly in teeth and bones, where they give great strength to these important structural tissues. An ionic compound that breaks apart into positive and negative ions in solution is called an **electrolyte** (e-LEK-trō-līt). Most ions in the body are dissolved in body fluids as electrolytes, so named because their solutions can conduct an electric current. (In Chapter 27 we will discuss the chemistry and importance of electrolytes.) **Table 2.2** lists the names and symbols of common ions in the body.

FIGURE 2.4 Ions and ionic bond formation. (a) A sodium atom can have a complete octet of electrons in its outermost shell by losing one electron. (b) A chlorine atom can have a complete octet by gaining one electron. (c) An ionic bond may form between oppositely charged ions. (d) In a crystal of NaCl, each Na^+ is surrounded by six Cl^-. In (a), (b), and (c), the electron that is lost or accepted is colored red.

An ionic bond is the force of attraction that holds together oppositely charged ions.

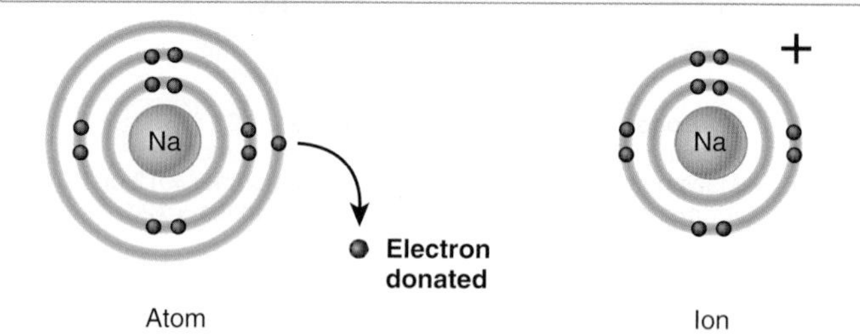

(a) Sodium: 1 valence electron

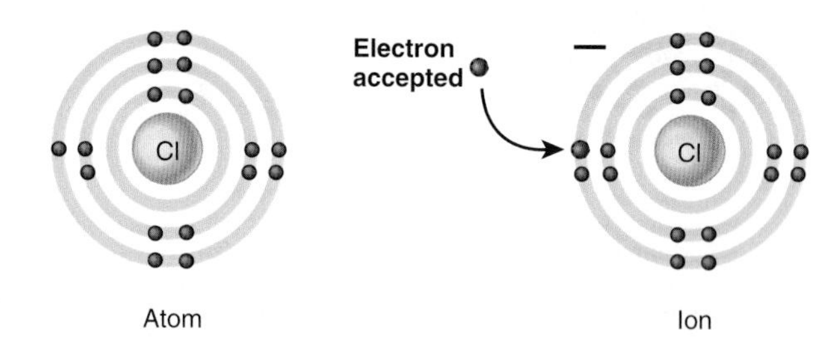

(b) Chlorine: 7 valence electrons

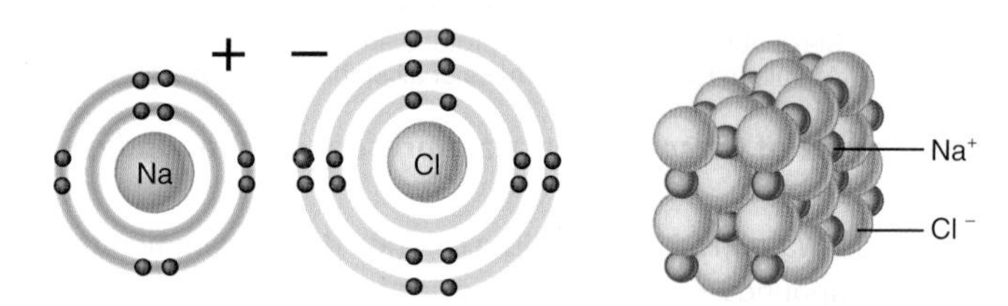

(c) Ionic bond in sodium chloride (NaCl)

(d) Packing of ions in a crystal of sodium chloride

Q What are cations and anions?

TABLE 2.2 Common Ions in the Body

CATIONS		ANIONS	
NAME	**SYMBOL**	**NAME**	**SYMBOL**
Hydrogen ion	H^+	Fluoride ion	F^-
Sodium ion	Na^+	Chloride ion	Cl^-
Potassium ion	K^+	Iodide ion	I^-
Ammonium ion	NH_4^+	Hydroxide ion	OH^-
Magnesium ion	Mg^{2+}	Bicarbonate ion	HCO_3^-
Calcium ion	Ca^{2+}	Oxide ion	O^{2-}
Iron(II) ion	Fe^{2+}	Sulfate ion	SO_4^{2-}
Iron(III) ion	Fe^{3+}	Phosphate ion	PO_4^{3-}

Covalent Bonds

When a **covalent bond** forms, two or more atoms *share* electrons rather than gaining or losing them. Atoms form a covalently bonded molecule by sharing one, two, or three pairs of valence electrons. The larger the number of electron pairs shared between two atoms, the stronger the covalent bond. Covalent bonds may form between atoms of the same element or between atoms of different elements. They are the most common chemical bonds in the body, and the compounds that result from them form most of the body's structures.

A **single covalent bond** results when two atoms share one electron pair. For example, a molecule of hydrogen forms when two hydrogen atoms share their single valence electrons (**Figure 2.5a**), which allows both atoms to have a full valence shell at least part of the time.

FIGURE 2.5 **Covalent bond formation.** The red electrons are shared equally in (a)–(d) and unequally in (e). To the right are simpler ways to represent these molecules. In a structural formula, each covalent bond is denoted by a straight line between the chemical symbols for two atoms. In molecular formulas, the number of atoms in each molecule is noted by subscripts.

In a covalent bond, two atoms share one, two, or three pairs of electrons in the outer shell.

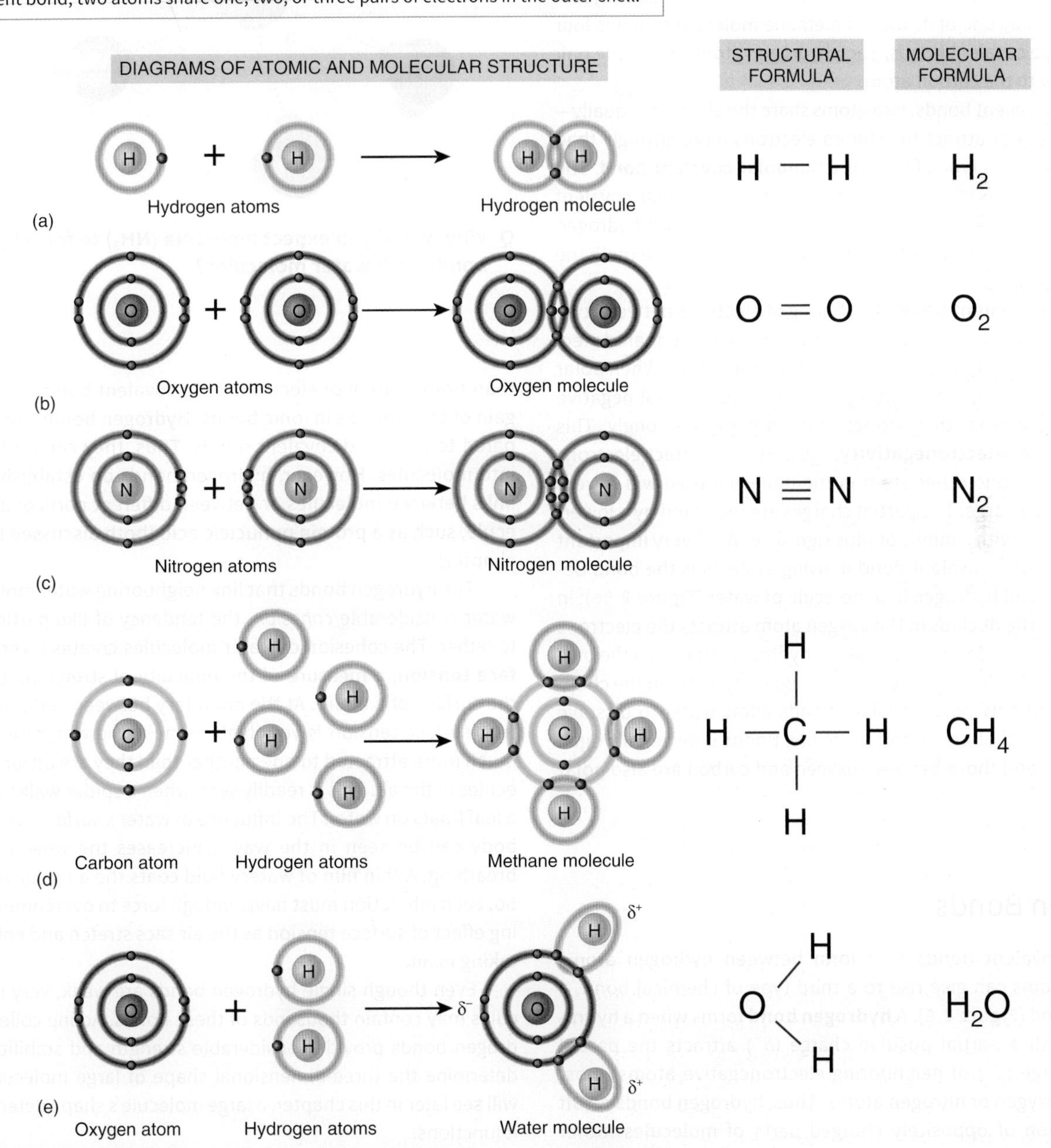

Q What is the main difference between an ionic bond and a covalent bond?

A **double covalent bond** results when two atoms share two pairs of electrons, as happens in an oxygen molecule (Figure 2.5b). A **triple covalent bond** occurs when two atoms share three pairs of electrons, as in a molecule of nitrogen (Figure 2.5c). Notice in the *structural formulas* for covalently bonded molecules in Figure 2.5 that the number of lines between the chemical symbols for two atoms indicates whether the bond is a single (—), double (=), or triple (≡) covalent bond.

The same principles of covalent bonding that apply to atoms of the same element also apply to covalent bonds between atoms of different elements. The gas methane (CH_4) contains covalent bonds formed between the atoms of two different elements, one carbon and four hydrogens (Figure 2.5d). The valence shell of the carbon atom can hold eight electrons but has only four of its own. The single electron shell of a hydrogen atom can hold two electrons, but each hydrogen atom has only one of its own. A methane molecule contains four separate single covalent bonds. Each hydrogen atom shares one pair of electrons with the carbon atom.

In some covalent bonds, two atoms share the electrons equally—one atom does not attract the shared electrons more strongly than the other atom. This type of bond is a **nonpolar covalent bond**. The bonds between two identical atoms are always nonpolar covalent bonds (Figure 2.5a–c). The bonds between carbon and hydrogen atoms are also nonpolar, such as the four C—H bonds in a methane molecule (Figure 2.5d).

In a **polar covalent bond**, the sharing of electrons between two atoms is unequal—the nucleus of one atom attracts the shared electrons more strongly than the nucleus of the other atom. When polar covalent bonds form, the resulting molecule has a partial negative charge near the atom that attracts electrons more strongly. This atom has greater **electronegativity**, the power to attract electrons to itself. At least one other atom in the molecule then will have a partial positive charge. The partial charges are indicated by a lowercase Greek delta with a minus or plus sign: δ^- or δ^+. A very important example of a polar covalent bond in living systems is the bond between oxygen and hydrogen in a molecule of water (Figure 2.5e); in this molecule, the nucleus of the oxygen atom attracts the electrons more strongly than do the nuclei of the hydrogen atoms, so the oxygen atom is said to have greater electronegativity. Later in the chapter, we will see how polar covalent bonds allow water to dissolve many molecules that are important to life. Bonds between nitrogen and hydrogen and those between oxygen and carbon are also polar bonds.

Hydrogen Bonds

The polar covalent bonds that form between hydrogen atoms and other atoms can give rise to a third type of chemical bond, a hydrogen bond (Figure 2.6). A **hydrogen bond** forms when a hydrogen atom with a partial positive charge (δ^+) attracts the partial negative charge (δ^-) of neighboring electronegative atoms, most often larger oxygen or nitrogen atoms. Thus, hydrogen bonds result from attraction of oppositely charged parts of molecules rather

FIGURE 2.6 Hydrogen bonding among water molecules. Each water molecule forms hydrogen bonds (indicated by dotted lines) with three to four neighboring water molecules.

Hydrogen bonds occur because hydrogen atoms in one water molecule are attracted to the partial negative charge of the oxygen atom in another water molecule.

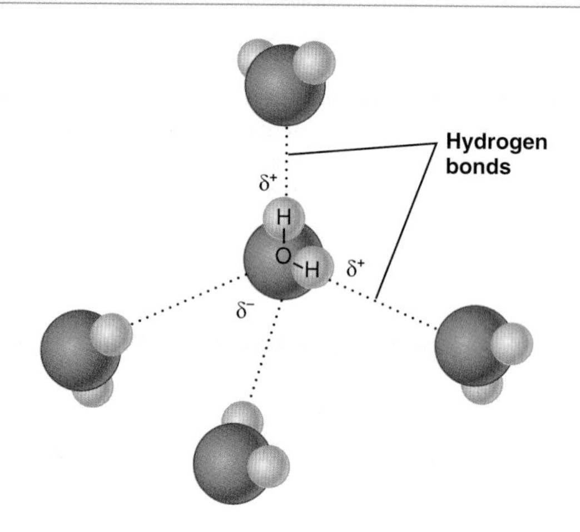

Q Why would you expect ammonia (NH_3) to form hydrogen bonds with water molecules?

than from sharing of electrons as in covalent bonds, or the loss or gain of electrons as in ionic bonds. Hydrogen bonds are weak compared to ionic and covalent bonds. Thus, they cannot bind atoms into molecules. However, hydrogen bonds do establish important links between molecules or between different parts of a large molecule, such as a protein or nucleic acid (both discussed later in this chapter).

The hydrogen bonds that link neighboring water molecules give water considerable *cohesion*, the tendency of like particles to stay together. The cohesion of water molecules creates a very high **surface tension**, a measure of the difficulty of stretching or breaking the surface of a liquid. At the boundary between water and air, water's surface tension is very high because the water molecules are much more attracted to one another than they are attracted to molecules in the air. This is readily seen when a spider walks on water or a leaf floats on water. The influence of water's surface tension on the body can be seen in the way it increases the work required for breathing. A thin film of watery fluid coats the air sacs of the lungs. So, each inhalation must have enough force to overcome the opposing effect of surface tension as the air sacs stretch and enlarge when taking in air.

Even though single hydrogen bonds are weak, very large molecules may contain thousands of these bonds. Acting collectively, hydrogen bonds provide considerable strength and stability and help determine the three-dimensional shape of large molecules. As you will see later in this chapter, a large molecule's shape determines how it functions.

2.3 Chemical Reactions

OBJECTIVES

- **Define** a chemical reaction.
- **Describe** the various forms of energy.
- **Compare** exergonic and endergonic chemical reactions.
- **Explain** the role of activation energy and catalysts in chemical reactions.
- **Describe** synthesis, decomposition, exchange, and reversible reactions.

A **chemical reaction** occurs when new bonds form or old bonds break between atoms. Chemical reactions are the foundation of all life processes, and as we have seen, the interactions of valence electrons are the basis of all chemical reactions. Consider how hydrogen and oxygen molecules react to form water molecules (**Figure 2.7**). The starting substances—two H_2 and one O_2—are known as the **reactants**. The ending substances—two molecules of H_2O—are the **products**. The arrow in the figure indicates the direction in which the reaction proceeds. In a chemical reaction, the total mass of the reactants equals the total mass of the products. Thus, the number of atoms of each element is the same before and after the reaction. However, because the atoms are rearranged, the reactants and products have different chemical properties. Through thousands of different chemical reactions, body structures are built and body functions are carried out. The term **metabolism** refers to all the chemical reactions occurring in the body.

FIGURE 2.7 The chemical reaction between two hydrogen molecules (H_2) and one oxygen molecule (O_2) to form two molecules of water (H_2O). Note that the reaction occurs by breaking old bonds and making new bonds.

The number of atoms of each element is the same before and after a chemical reaction.

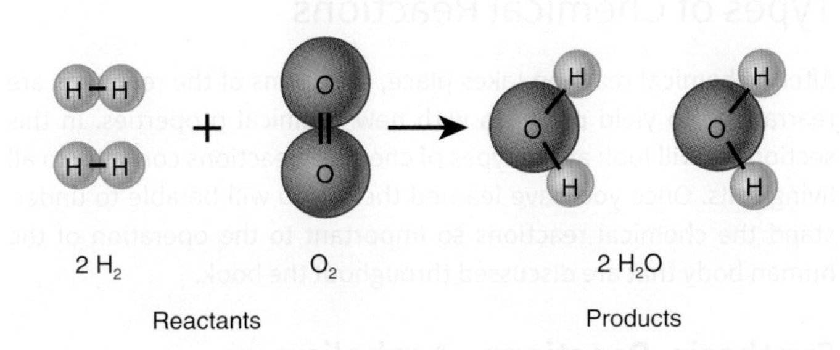

Q Why does this reaction require two molecules of H_2?

Forms of Energy and Chemical Reactions

Each chemical reaction involves energy changes. **Energy** (*en-* = in; *-ergy* = work) is the capacity to do work. Two principal forms of energy are **potential energy**, energy stored by matter due to its position, and **kinetic energy**, the energy associated with matter in motion. For example, the energy stored in water behind a dam or in a person poised to jump down some steps is potential energy. When the gates of the dam are opened or the person jumps, potential energy is converted into kinetic energy. **Chemical energy** is a form of potential energy that is stored in the bonds of compounds and molecules. The total amount of energy present at the beginning and end of a chemical reaction is the same. Although energy can be neither created nor destroyed, it may be converted from one form to another. This principle is known as the **law of conservation of energy**. For example, some of the chemical energy in the foods we eat is eventually converted into various forms of kinetic energy, such as mechanical energy used to walk and talk. Conversion of energy from one form to another generally releases heat, some of which is used to maintain normal body temperature.

Energy Transfer in Chemical Reactions

Chemical bonds represent stored chemical energy, and chemical reactions occur when new bonds are formed or old bonds are broken between atoms. The *overall reaction* may either release energy or absorb energy. **Exergonic reactions** (*ex-* = out) release more energy than they absorb. By contrast, **endergonic reactions** (*end-* = within) absorb more energy than they release.

A key feature of the body's metabolism is the coupling of exergonic reactions and endergonic reactions. Energy released from an exergonic reaction often is used to drive an endergonic one. In general, exergonic reactions occur as nutrients, such as glucose, are broken down. Some of the energy released may be trapped in the covalent bonds of adenosine triphosphate (ATP), which we describe more fully later in this chapter. If a molecule of glucose is completely broken down, the chemical energy in its bonds can be used to produce as many as 32 molecules of ATP. The energy transferred to the ATP molecules is then used to drive endergonic reactions needed to build body structures, such as muscles and bones. The energy in ATP is also used to do the mechanical work involved in the contraction of muscle or the movement of substances into or out of cells.

Activation Energy Because particles of matter such as atoms, ions, and molecules have kinetic energy, they are continuously moving and colliding with one another. A sufficiently forceful collision can disrupt the movement of valence electrons, causing an existing chemical bond to break or a new one to form. The collision energy needed to break the chemical bonds of the reactants is called the **activation energy** of the reaction (**Figure 2.8**). This initial energy "investment" is needed to start a reaction. The reactants must absorb enough energy for their chemical bonds to become unstable and their valence electrons to form new combinations. Then, as new bonds form, energy is released to the surroundings.

FIGURE 2.8 Activation energy.

Activation energy is the energy needed to break chemical bonds in the reactant molecules so a reaction can start.

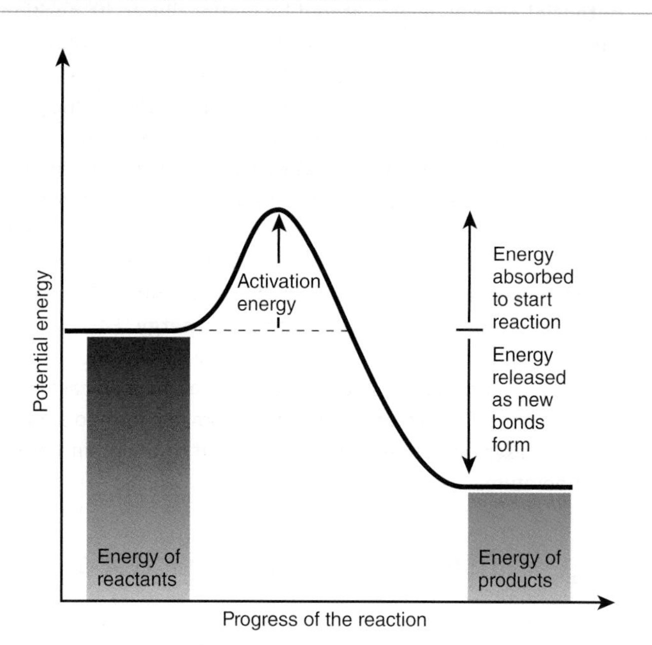

Q Why is the reaction illustrated here exergonic?

FIGURE 2.9 Comparison of energy needed for a chemical reaction to proceed with a catalyst (blue curve) and without a catalyst (red curve).

Catalysts speed up chemical reactions by lowering the activation energy.

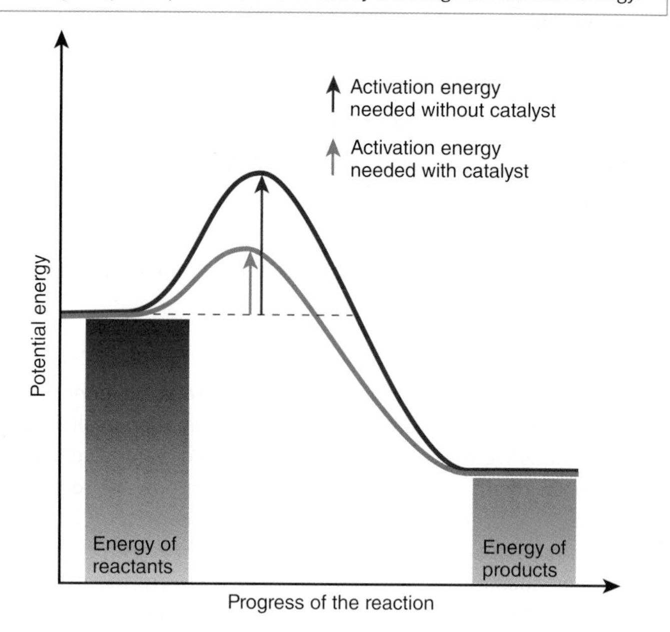

Q Does a catalyst change the potential energies of the products and reactants?

Both the concentration of particles and the temperature influence the chance that a collision will occur and cause a chemical reaction.

- ***Concentration.*** The more particles of matter present in a confined space, the greater the chance that they will collide (think of people crowding into a subway car at rush hour). The concentration of particles increases when more are added to a given space or when the pressure on the space increases, which forces the particles closer together so that they collide more often.
- ***Temperature.*** As temperature rises, particles of matter move about more rapidly. Thus, the higher the temperature of matter, the more forcefully particles will collide, and the greater the chance that a collision will produce a reaction.

Catalysts As we have seen, chemical reactions occur when chemical bonds break or form after atoms, ions, or molecules collide with one another. Body temperature and the concentrations of molecules in body fluids, however, are far too low for most chemical reactions to occur rapidly enough to maintain life. Raising the temperature and the number of reacting particles of matter in the body could increase the frequency of collisions and thus increase the rate of chemical reactions, but doing so could also damage or kill the body's cells.

Substances called catalysts solve this problem. **Catalysts** are chemical compounds that speed up chemical reactions by lowering the activation energy needed for a reaction to occur (Figure 2.9). The most important catalysts in the body are enzymes, which we will discuss later in this chapter.

A catalyst does not alter the difference in potential energy between the reactants and the products. Rather, it lowers the amount of energy needed to start the reaction.

For chemical reactions to occur, some particles of matter—especially large molecules—not only must collide with sufficient force, but they must hit one another at precise spots. A catalyst helps to properly orient the colliding particles. Thus, they interact at the spots that make the reaction happen. Although the action of a catalyst helps to speed up a chemical reaction, the catalyst itself is unchanged at the end of the reaction. A single catalyst molecule can assist one chemical reaction after another.

Types of Chemical Reactions

After a chemical reaction takes place, the atoms of the reactants are rearranged to yield products with new chemical properties. In this section we will look at the types of chemical reactions common to all living cells. Once you have learned them, you will be able to understand the chemical reactions so important to the operation of the human body that are discussed throughout the book.

Synthesis Reactions—Anabolism When two or more atoms, ions, or molecules combine to form new and larger molecules, the processes are called **synthesis reactions**. The word *synthesis* means "to put together." A synthesis reaction can be expressed as follows:

A	+	B	Combine to form →	AB
Atom, ion, or molecule A		Atom, ion, or molecule B		New molecule AB

One example of a synthesis reaction is the reaction between two hydrogen molecules and one oxygen molecule to form two molecules of water (see **Figure 2.7**).

$2\ H_2$	+	O_2	Combine to form →	$2\ H_2O$
Two hydrogen molecules		One oxygen molecule		Two water molecules

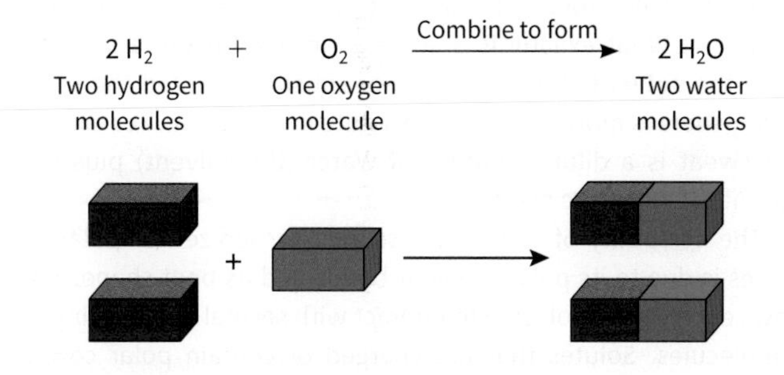

All of the synthesis reactions that occur in your body are collectively referred to as **anabolism** (a-NAB-ō-lizm). Overall, anabolic reactions are usually endergonic because they absorb more energy than they release. Combining simple molecules like amino acids (discussed shortly) to form large molecules such as proteins is an example of anabolism.

Decomposition Reactions—Catabolism

Decomposition reactions split up large molecules into smaller atoms, ions, or molecules. A decomposition reaction is expressed as follows:

AB	Breaks down into →	A	+	B
Molecule AB		Atom, ion, or molecule A		Atom, ion, or molecule B

For example, under proper conditions, a methane molecule can decompose into one carbon atom and two hydrogen molecules:

CH_4	Breaks down into →	C	+	$2\ H_2$
One methane molecule		One carbon atom		Two hydrogen molecules

The decomposition reactions that occur in your body are collectively referred to as **catabolism** (ka-TAB-ō-lizm). Overall, catabolic reactions are usually exergonic because they release more energy than they absorb. For instance, the series of reactions that break down glucose to pyruvic acid, with the net production of two molecules of ATP, are important catabolic reactions in the body. These reactions will be discussed in Chapter 25.

Exchange Reactions

Many reactions in the body are **exchange reactions**; they consist of both synthesis and decomposition reactions. One type of exchange reaction works like this:

$$AB + CD \longrightarrow AD + BC$$

The bonds between A and B and between C and D break (decomposition), and new bonds then form (synthesis) between A and D and between B and C. An example of an exchange reaction is

HCl	+	$NaHCO_3$	→	H_2CO_3	+	NaCl
Hydrochloric acid		Sodium bicarbonate		Carbonic acid		Sodium chloride

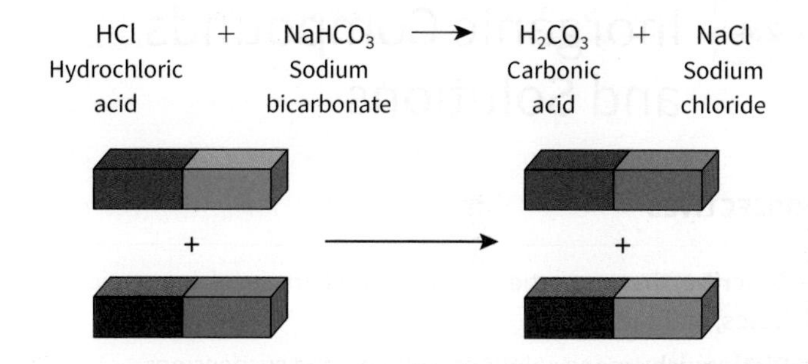

Notice that the ions in both compounds have "switched partners": The hydrogen ion (H^+) from HCl has combined with the bicarbonate ion (HCO_3^-) from $NaHCO_3$, and the sodium ion (Na^+) from $NaHCO_3$ has combined with the chloride ion (Cl^-) from HCl.

Reversible Reactions

Some chemical reactions proceed in only one direction, from reactants to products, as previously indicated by the single arrows. Other chemical reactions may be reversible. In a **reversible reaction**, the products can revert to the original reactants. A reversible reaction is indicated by two half-arrows pointing in opposite directions:

$$AB \underset{\text{Combines to form}}{\overset{\text{Breaks down into}}{\rightleftharpoons}} A + B$$

Some reactions are reversible only under special conditions:

$$AB \underset{\text{Heat}}{\overset{\text{Water}}{\rightleftharpoons}} A + B$$

In that case, whatever is written above or below the arrows indicates the condition needed for the reaction to occur. In these reactions, AB breaks down into A and B only when water is added, and A and B react to produce AB only when heat is applied. Many reversible reactions in the body require catalysts called enzymes. Often, different enzymes guide the reactions in opposite directions.

Oxidation–Reduction Reactions You will learn in Chapter 25 that chemical reactions called oxidation–reduction reactions are essential to life, since they are the reactions that break down food molecules to produce energy. These reactions are concerned with the transfer of electrons between atoms and molecules. **Oxidation** refers to the loss of electrons; in the process the oxidized substance releases energy. **Reduction** refers to the gain of electrons; in the process the reduced substance gains energy. **Oxidation–reduction reactions** are always parallel; when one substance is oxidized, another is reduced at the same time. When a food molecule, such as glucose, is oxidized, the energy produced is used by a cell to carry out its various functions.

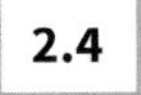

2.4 Inorganic Compounds and Solutions

OBJECTIVES

- **Describe** the properties of water and those of inorganic acids, bases, and salts.
- **Distinguish** among solutions, colloids, and suspensions.
- **Define** pH and explain the role of buffer systems in homeostasis.

Most of the chemicals in your body exist in the form of compounds. Biologists and chemists divide these compounds into two principal classes: inorganic compounds and organic compounds. **Inorganic compounds** usually lack carbon and are structurally simple. Their molecules also have only a few atoms and cannot be used by cells to perform complicated biological functions. They include water and many salts, acids, and bases. Inorganic compounds may have either ionic or covalent bonds. Water makes up 55–60% of a lean adult's total body mass; all other inorganic compounds combined add 1–2%. Inorganic compounds that contain carbon include carbon dioxide (CO_2), bicarbonate ion (HCO_3^-), and carbonic acid (H_2CO_3). **Organic compounds** always contain carbon, usually contain hydrogen, and always have covalent bonds. Most are large molecules, many made up of long carbon atom chains. Organic compounds make up the remaining 38–43% of the human body.

Water

Water is the most important and abundant inorganic compound in all living systems. Although you might be able to survive for weeks without food, without water you would die in a matter of days. Nearly all the body's chemical reactions occur in a watery medium. Water has many properties that make it such an indispensable compound for life. We have already mentioned the most important property of water, its polarity—the uneven sharing of valence electrons that confers a partial negative charge near the one oxygen atom and two partial positive charges near the two hydrogen atoms in a water molecule (see Figure 2.5e). This property makes water an excellent solvent for other ionic or polar substances, gives water molecules cohesion (the tendency to stick together), and allows water to resist temperature changes.

Water as a Solvent In medieval times people searched in vain for a "universal solvent," a substance that would dissolve all other materials. They found nothing that worked as well as water. Although it is the most versatile solvent known, water is not the universal solvent sought by medieval alchemists. If it were, no container could hold it because it would dissolve all potential containers! What exactly is a solvent? In a **solution**, a substance called the **solvent** dissolves another substance called the **solute**. Usually there is more solvent than solute in a solution. For example, your sweat is a dilute solution of water (the solvent) plus small amounts of salts (the solutes).

The versatility of water as a solvent for ionized or polar substances is due to its polar covalent bonds and its bent shape, which allows each water molecule to interact with several neighboring ions or molecules. Solutes that are charged or contain polar covalent bonds are **hydrophilic** (*hydro-* = water; *-philic* = loving), which means they dissolve easily in water. Common examples of hydrophilic solutes are sugar and salt. Molecules that contain mainly nonpolar covalent bonds, by contrast, are **hydrophobic** (*-phobic* = fearing). They are not very water-soluble. Examples of hydrophobic compounds include animal fats and vegetable oils.

To understand the dissolving power of water, consider what happens when a crystal of a salt such as sodium chloride (NaCl) is placed in water (Figure 2.10). The electronegative oxygen atom in water molecules attracts the sodium ions (Na^+), and the electropositive hydrogen atoms in water molecules attract the chloride ions (Cl^-). Soon, water molecules surround and separate Na^+ and Cl^- ions from each other at the surface of the crystal, breaking the ionic bonds that held NaCl together. The water molecules surrounding the ions also lessen the chance that Na^+ and Cl^- will come together and re-form an ionic bond.

The ability of water to form solutions is essential to health and survival. Because water can dissolve so many different substances, it is an ideal medium for metabolic reactions. Water enables dissolved reactants to collide and form products. Water also dissolves waste products, which allows them to be flushed out of the body in the urine.

Water in Chemical Reactions Water serves as the medium for most chemical reactions in the body and participates as a reactant or product in certain reactions. During digestion, for example, decomposition reactions break down large nutrient molecules into smaller molecules by the addition of water molecules. This type of reaction is called **hydrolysis** (hī-DROL-i-sis; *-lysis* = to loosen or break apart). Hydrolysis reactions enable dietary nutrients to be absorbed into the body. By contrast, when two smaller molecules join to form a larger molecule in a **dehydration synthesis reaction** (*de-* = from, down, or out; *hydra-* = water), a water molecule is one of the products

FIGURE 2.10 How polar water molecules dissolve salts and polar substances. When a crystal of sodium chloride is placed in water, the slightly negative oxygen end (red) of water molecules is attracted to the positive sodium ions (Na^+), and the slightly positive hydrogen portions (gray) of water molecules are attracted to the negative chloride ions (Cl^-). In addition to dissolving sodium chloride, water also causes it to dissociate, or separate into charged particles, which is discussed shortly.

Water is a versatile solvent because its polar covalent bonds, in which electrons are shared unequally, create positive and negative regions.

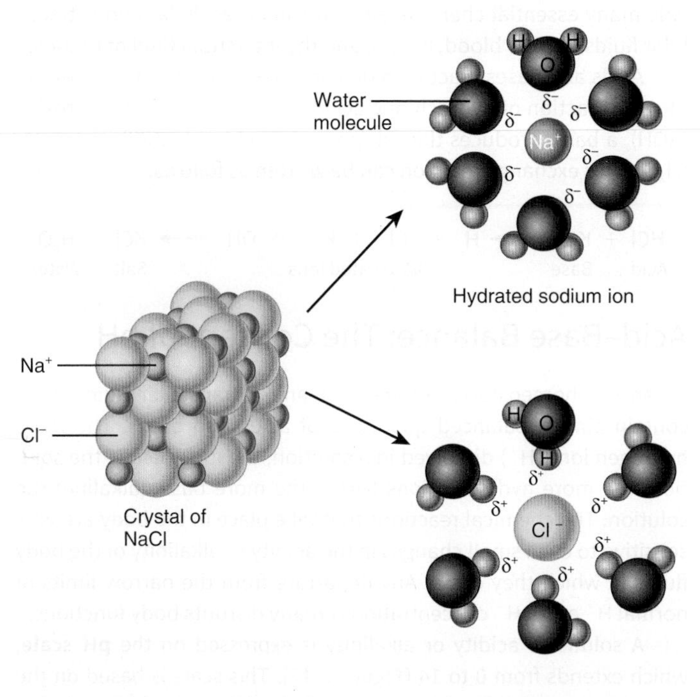

Q Table sugar (sucrose) easily dissolves in water but is not an electrolyte. Is it likely that all the covalent bonds between atoms in table sugar are nonpolar bonds? Why or why not?

formed. As you will see later in the chapter, such reactions occur during synthesis of proteins and other large molecules (for example, see **Figure 2.21**).

Thermal Properties of Water In comparison to most substances, water can absorb or release a relatively large amount of heat with only a modest change in its own temperature. For this reason, water is said to have a high *heat capacity.* The reason for this property is the large number of hydrogen bonds in water. As water absorbs heat energy, some of the energy is used to break hydrogen bonds. Less energy is then left over to increase the motion of water molecules, which would increase the water's temperature. The high heat capacity of water is the reason it is used in automobile radiators; it cools the engine by absorbing heat without its own temperature rising to an unacceptably high level. The large amount of body water has a similar effect: It lessens the impact of environmental temperature changes, helping to maintain body temperature homeostasis.

Water also requires a large amount of heat to change from a liquid to a gas. Its *heat of vaporization* is high. As water evaporates from the surface of the skin, it removes a large quantity of heat, providing an important cooling mechanism.

Water as a Lubricant Water is a major component of mucus and other lubricating fluids throughout the body. Lubrication is especially necessary in the chest (pleural and pericardial cavities) and abdomen (peritoneal cavity), where internal organs touch and slide over one another. It is also needed at joints, where bones, ligaments, and tendons rub against one another. Inside the gastrointestinal tract, mucus and other watery secretions moisten foods, which aids their smooth passage through the digestive system.

Solutions, Colloids, and Suspensions

A **mixture** is a combination of elements or compounds that are physically blended together but not bound by chemical bonds. For example, the air you are breathing is a mixture of gases that includes nitrogen, oxygen, argon, and carbon dioxide. Three common liquid mixtures are solutions, colloids, and suspensions.

Once mixed together, solutes in a solution remain evenly dispersed among the solvent molecules. Because solute particles in a solution are very small, a solution looks transparent.

A **colloid** differs from a solution mainly because of the size of its particles. The solute particles in a colloid are large enough to scatter light, just as water droplets in fog scatter light from a car's headlight beams. For this reason, colloids usually appear translucent or opaque. Milk is an example of a liquid that is both a colloid and a solution: The large milk proteins make it a colloid, whereas calcium salts, milk sugar (lactose), ions, and other small particles are in solution.

The solutes in both solutions and colloids do not settle out and accumulate on the bottom of the container. In a **suspension**, by contrast, the suspended material may mix with the liquid or suspending medium for some time, but eventually it will settle out. Blood is an example of a suspension. When freshly drawn from the body, blood has an even, reddish color. After blood sits for a while in a test tube, red blood cells settle out of the suspension and drift to the bottom of the tube (see **Figure 19.1a**). The upper layer, the liquid portion of blood, appears pale yellow and is called blood plasma. Blood plasma is both a solution of ions and other small solutes and a colloid due to the presence of larger plasma proteins.

The **concentration** of a solution may be expressed in several ways. One common way is by a mass per volume **percentage**, which gives the relative mass of a solute found in a given volume of solution. For example, you may have seen the following on the label of a bottle of wine: "Alcohol 14.1% by volume." Another way expresses concentration in units of **moles per liter (mol/L)**, also called *molarity*, which relate to the total number of molecules in a given volume of solution. A **mole** is

TABLE 2.3 Percentage and Molarity

DEFINITION	EXAMPLE
Percentage (mass per volume) Number of grams of a substance per 100 milliliters (mL) of solution	To make a 10% NaCl solution, take 10 g of NaCl and add enough water to make a total of 100 mL of solution.
Molarity - moles (mol) per liter A 1 molar (1 M) solution = 1 mole of a solute in 1 liter of solution	To make a 1 molar (1 M) solution of NaCl, dissolve 1 mole of NaCl (58.44 g) in enough water to make a total of 1 liter of solution.

the amount of any substance that has a mass in grams equal to the sum of the atomic masses of all its atoms. For example, 1 mole of the element chlorine (atomic mass = 35.45) is 35.45 grams and 1 mole of the salt sodium chloride (NaCl) is 58.44 grams (22.99 for Na + 35.45 for Cl). Just as a dozen always means 12 of something, a mole of anything has the same number of particles: 6.023×10^{23}. This huge number is called *Avogadro's number*. Thus, measurements of substances that are stated in moles tell us about the numbers of atoms, ions, or molecules present. This is important when chemical reactions are occurring because each reaction requires a set number of atoms of specific elements. Table 2.3 describes these ways of expressing concentration.

Inorganic Acids, Bases, and Salts

When inorganic acids, bases, or salts dissolve in water, they **dissociate** (dis′-sō-sē-ĀT); that is, they separate into ions and become surrounded by water molecules. An **acid** (Figure 2.11a) is a substance that dissociates into one or more **hydrogen ions (H^+)** and one or more anions. Because H^+ is a single proton with one positive charge, an acid is also referred to as a **proton donor**. A **base**, by contrast (Figure 2.11b), removes H^+ from a solution and is therefore a **proton acceptor**. Many bases dissociate into one or more **hydroxide ions (OH^-)** and one or more cations.

A **salt**, when dissolved in water, dissociates into cations and anions, neither of which is H^+ or OH^- (Figure 2.11c). In the body, salts such as potassium chloride are electrolytes that are important for carrying electrical currents (ions flowing from one place to another), especially in nerve and muscular tissues. The ions of salts also provide many essential chemical elements in intracellular and extracellular fluids such as blood, lymph, and the interstitial fluid of tissues.

Acids and bases react with one another to form salts. For example, the reaction of hydrochloric acid (HCl) and potassium hydroxide (KOH), a base, produces the salt potassium chloride (KCl) and water (H_2O). This exchange reaction can be written as follows:

$$\underset{\text{Acid}}{HCl} + \underset{\text{Base}}{KOH} \longrightarrow \underbrace{H^+ + Cl^- + K^+ + OH^-}_{\text{Dissociated ions}} \longrightarrow \underset{\text{Salt}}{KCl} + \underset{\text{Water}}{H_2O}$$

FIGURE 2.11 Dissociation of inorganic acids, bases, and salts.

Dissociation is the separation of inorganic acids, bases, and salts into ions in a solution.

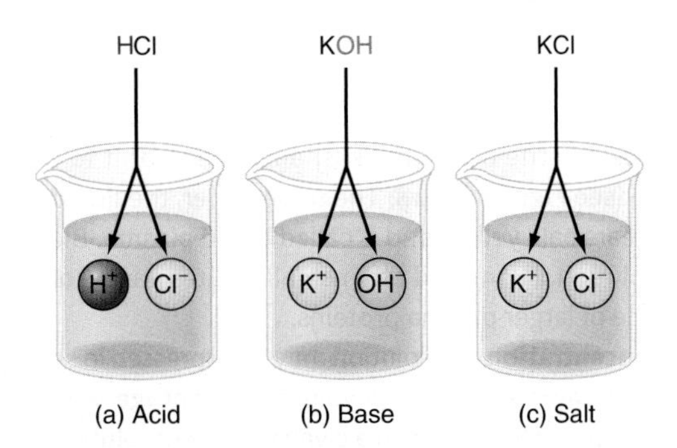

Q The compound $CaCO_3$ (calcium carbonate) dissociates into a calcium ion (Ca^{2+}) and a carbonate ion (CO_3^{2-}). Is it an acid, a base, or a salt? What about H_2SO_4, which dissociates into two H_1 and one SO_4^{2-}?

Acid–Base Balance: The Concept of pH

To ensure homeostasis, intracellular and extracellular fluids must contain almost balanced quantities of acids and bases. The more hydrogen ions (H^+) dissolved in a solution, the more acidic the solution; the more hydroxide ions (OH^-), the more basic (alkaline) the solution. The chemical reactions that take place in the body are very sensitive to even small changes in the acidity or alkalinity of the body fluids in which they occur. Any departure from the narrow limits of normal H^+ and OH^- concentrations greatly disrupts body functions.

A solution's acidity or alkalinity is expressed on the **pH scale**, which extends from 0 to 14 (Figure 2.12). This scale is based on the concentration of H^+ in moles per liter. A pH of 7 means that a solution contains one ten-millionth (0.0000001) of a mole of hydrogen ions per liter. The number 0.0000001 is written as 1×10^{-7} in scientific notation, which indicates that the number is 1 with the decimal point moved seven places to the left. To convert this value to pH, the negative exponent (−7) is changed to a positive number (7). A solution with a H^+ concentration of 0.0001 (10^{-4}) mol/L has a pH of 4; a solution with a H^+ concentration of 0.000000001 (10^{-9}) mol/L has a pH of 9; and so on. It is important to realize that a change of one whole number on the pH scale represents a *tenfold* change in the number of H^+. A pH of 6 denotes 10 times more H^+ than a pH of 7, and a pH of 8 indicates 10 times fewer H^+ than a pH of 7 and 100 times fewer H^+ than a pH of 6.

The midpoint of the pH scale is 7, where the concentrations of H^+ and OH^- are equal. A substance with a pH of 7, such as pure water, is neutral. A solution that has more H^+ than OH^- is an **acidic solution** and has a pH below 7. A solution that has more OH^- than H^+ is a **basic** (*alkaline*) **solution** and has a pH above 7.

Maintaining pH: Buffer Systems

Although the pH of body fluids may differ, as we have discussed, the normal limits for each fluid are quite narrow. Table 2.4 shows the pH values for certain body fluids along with those of some common

FIGURE 2.12 **The pH scale.** A pH below 7 indicates an acidic solution—more H^+ than OH^-. A pH above 7 indicates a basic (alkaline) solution; that is, there are more OH^- than H^+.

The lower the numerical value of the pH, the more acidic is the solution because the H^+ concentration becomes progressively greater. The higher the pH, the more basic the solution.

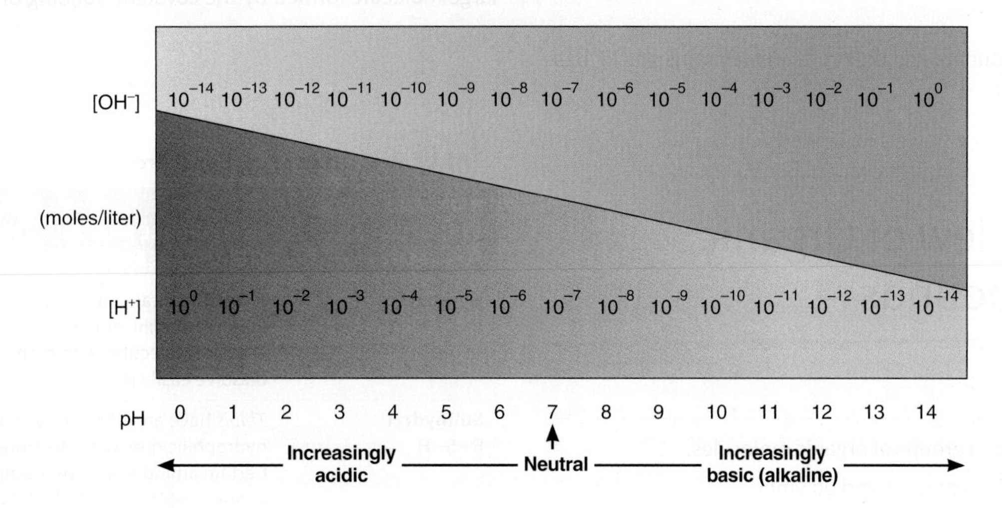

Q At pH 7 (neutrality), the concentrations of H^+ and OH^- are equal (10^{-7} mol/liter). What are the concentrations of H^+ and OH^- at pH 6? Which pH is more acidic, 6.82 or 6.91? Which pH is closer to neutral, 8.41 or 5.59?

TABLE 2.4 pH Values of Selected Substances

SUBSTANCE*	pH VALUE
• Gastric juice (found in the stomach)	1.2–3.0
Lemon juice	2.3
Vinegar	3.0
Carbonated soft drink	3.0–3.5
Orange juice	3.5
• Vaginal fluid	3.5–4.5
Tomato juice	4.2
Coffee	5.0
• Urine	4.6–8.0
• Saliva	6.35–6.85
Milk	6.8
Distilled (pure) water	7.0
• Blood	7.35–7.45
• Semen (fluid containing sperm)	7.20–7.60
• Cerebrospinal fluid (fluid associated with nervous system)	7.4
• Pancreatic juice (digestive juice of the pancreas)	7.1–8.2
• Bile (liver secretion that aids fat digestion)	7.6–8.6
Milk of magnesia	10.5
Lye (sodium hydroxide)	14.0

*Bullets (•) denote substances in the human body.

substances outside the body. Homeostatic mechanisms maintain the pH of blood between 7.35 and 7.45, which is slightly more basic than pure water. You will learn in Chapter 27 that if the pH of blood falls below 7.35, a condition called *acidosis* occurs, and if the pH rises above 7.45, it results in a condition called *alkalosis;* both conditions can seriously compromise homeostasis. Saliva is slightly acidic, and semen is slightly basic. Because the kidneys help remove excess acid from the body, urine can be quite acidic.

Even though strong acids and bases are continually taken into and formed by the body, the pH of fluids inside and outside cells remains almost constant. One important reason is the presence of **buffer systems**, which function to convert strong acids or bases into weak acids or bases. Strong acids (or bases) ionize easily and contribute many H^+ (or OH^-) to a solution. Therefore, they can change pH drastically, which can disrupt the body's metabolism. Weak acids (or bases) do not ionize as much and contribute fewer H^+ (or OH^-). Hence, they have less effect on the pH. The chemical compounds that can convert strong acids or bases into weak ones are called **buffers**. They do so by removing or adding protons (H^+).

One important buffer system in the body is the **carbonic acid–bicarbonate buffer system**. Carbonic acid (H_2CO_3) can act as a weak acid, and the bicarbonate ion (HCO_3^-) can act as a weak base. Hence, this buffer system can compensate for either an excess or a shortage of H^+. For example, if there is an excess of H^+ (an acidic condition), HCO_3^- can function as a weak base and remove the excess H^+, as follows:

$$\underset{\text{Hydrogen ion}}{H^+} + \underset{\text{Bicarbonate ion (weak base)}}{HCO_3^-} \longrightarrow \underset{\text{Carbonic acid}}{H_2CO_3}$$

If there is a shortage of H^+ (an alkaline condition), by contrast, H_2CO_3 can function as a weak acid and provide needed H^+ as follows:

H_2CO_3	$\longrightarrow$	H^+	+	HCO_3^-
Carbonic acid (weak acid)		Hydrogen ion		Bicarbonate ion

Chapter 27 describes buffers and their roles in maintaining acid–base balance in more detail.

2.5 Overview of Organic Compounds

OBJECTIVES

- **Describe** the functional groups of organic molecules.
- **Distinguish** between monomers and polymers.

Many organic molecules are relatively large and have unique characteristics that allow them to carry out complex functions. Important categories of organic compounds include carbohydrates, lipids, proteins, nucleic acids, and adenosine triphosphate (ATP).

Carbon has several properties that make it particularly useful to living organisms. For one thing, it can form bonds with one to thousands of other carbon atoms to produce large molecules that can have many different shapes. Due to this property of carbon, the body can build many different organic compounds, each of which has a unique structure and function. Moreover, the large size of most carbon-containing molecules and the fact that some do not dissolve easily in water make them useful materials for building body structures.

Organic compounds are usually held together by covalent bonds. Carbon has four electrons in its outermost (valence) shell. It can bond covalently with a variety of atoms, including other carbon atoms, to form rings and straight or branched chains. Other elements that most often bond with carbon in organic compounds are hydrogen, oxygen, and nitrogen. Sulfur and phosphorus are also present in organic compounds. The other elements listed in **Table 2.1** are present in a smaller number of organic compounds.

The chain of carbon atoms in an organic molecule is called the **carbon skeleton**. Many of the carbons are bonded to hydrogen atoms, yielding a **hydrocarbon**. Also attached to the carbon skeleton are distinctive **functional groups**, other atoms or molecules bound to the hydrocarbon skeleton. Each type of functional group has a specific arrangement of atoms that confers characteristic chemical properties on the organic molecule attached to it. **Table 2.5** lists the most common functional groups of organic molecules and describes some of their properties. Because organic molecules often are big, there are shorthand methods for representing their structural formulas. **Figure 2.13** shows two ways to indicate the structure of the sugar glucose, a molecule with a ring-shaped carbon skeleton that has several hydroxyl groups attached.

Small organic molecules can combine into very large molecules that are called **macromolecules** (*macro-* = large). Macromolecules are usually **polymers** (*poly-* = many; *-mers* = parts). A polymer is a large molecule formed by the covalent bonding of many identical or

TABLE 2.5 Major Functional Groups of Organic Molecules

NAME AND STRUCTURAL FORMULA*	OCCURRENCE AND SIGNIFICANCE
Hydroxyl R—O—H	*Alcohols* contain an —OH group, which is polar and hydrophilic due to its electronegative O atom. Molecules with many —OH groups dissolve easily in water.
Sulfhydryl R—S—H	*Thiols* have an —SH group, which is polar and hydrophilic due to its electronegative S atom. Certain amino acids (for example, cysteine) contain —SH groups, which help stabilize the shape of proteins.
Carbonyl R—C(=O)—R	*Ketones* contain a carbonyl group within the carbon skeleton. The carbonyl group is polar and hydrophilic due to its electronegative O atom.
or R—C(=O)—H	*Aldehydes* have a carbonyl group at the end of the carbon skeleton.
Carboxyl R—C(=O)—OH or R—C(=O)—O^-	*Carboxylic acids* contain a carboxyl group at the end of the carbon skeleton. All amino acids have a —COOH group at one end. The negatively charged form predominates at the pH of body cells and is hydrophilic.
Ester R—C(=O)—O—R	*Esters* predominate in dietary fats and oils and also occur in our body as triglycerides. Aspirin is an ester of salicylic acid, a pain-relieving molecule found in the bark of the willow tree.
Phosphate R—O—P(=O)(O^-)—O^-	*Phosphates* contain a phosphate group (—PO_4^{2-}), which is very hydrophilic due to the dual negative charges. An important example is adenosine triphosphate (ATP), which transfers chemical energy between organic molecules during chemical reactions.
Amino R—N(H)—H or R—N^+(H)(H)—H	*Amines* have an —NH_2 group, which can act as a base and pick up a hydrogen ion, giving the amino group a positive charge. At the pH of body fluids, most amino groups have a charge of 1^+. All amino acids have an amino group at one end.

*R = variable group.

FIGURE 2.13 **Alternative ways to write the structural formula for glucose.**

In standard shorthand, carbon atoms are understood to be at locations where two bond lines intersect, and single hydrogen atoms are not indicated.

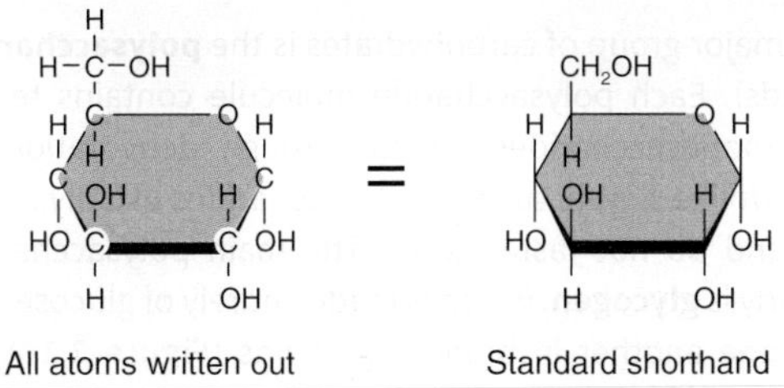

All atoms written out = Standard shorthand

Q How many hydroxyl groups does a molecule of glucose have? How many carbon atoms are part of glucose's carbon skeleton?

TABLE 2.6 **Major Carbohydrate Groups**

TYPE OF CARBOHYDRATE	EXAMPLES
Monosaccharides (simple sugars that contain from 3 to 7 carbon atoms)	Glucose (the main blood sugar). Fructose (found in fruits). Galactose (in milk sugar). Deoxyribose (in DNA). Ribose (in RNA).
Disaccharides (simple sugars formed from the combination of two monosaccharides by dehydration synthesis)	Sucrose (table sugar) = glucose + fructose. Lactose (milk sugar) = glucose + galactose. Maltose = glucose + glucose.
Polysaccharides (from tens to hundreds of monosaccharides joined by dehydration synthesis)	Glycogen (stored form of carbohydrates in animals). Starch (stored form of carbohydrates in plants and main carbohydrates in food). Cellulose (part of cell walls in plants that cannot be digested by humans but aids movement of food through intestines).

similar small building-block molecules called **monomers** (*mono-* = one). Usually, the reaction that joins two monomers is a dehydration synthesis. In this type of reaction, a hydrogen atom is removed from one monomer and a hydroxyl group is removed from the other to form a molecule of water (see **Figure 2.15a**). Macromolecules such as carbohydrates, lipids, proteins, and nucleic acids are assembled in cells via dehydration synthesis reactions.

Molecules that have the same molecular formula but different structures are called **isomers** (Ī-so-merz; *iso-* = equal or the same). For example, the molecular formulas for the sugars glucose and fructose are both $C_6H_{12}O_6$. The individual atoms, however, are positioned differently along the carbon skeleton (see **Figure 2.15a**), giving the sugars different chemical properties.

2.6 Carbohydrates

OBJECTIVES

- **Identify** the building blocks of carbohydrates.
- **Describe** the functions of carbohydrates.

Carbohydrates include sugars, glycogen, starches, and cellulose. Even though they are a large and diverse group of organic compounds and have several functions, carbohydrates represent only 2–3% of your total body mass. In humans and animals, carbohydrates function mainly as a source of chemical energy for generating ATP needed to drive metabolic reactions. Only a few carbohydrates are used for building structural units. One example is deoxyribose, a type of sugar that is a building block of deoxyribonucleic acid (DNA), the molecule that carries inherited genetic information.

Carbon, hydrogen, and oxygen are the elements found in carbohydrates. The ratio of hydrogen to oxygen atoms is usually 2:1, the same as in water. Although there are exceptions, carbohydrates generally contain one water molecule for each carbon atom. This is the reason they are called carbohydrates, which means "watered carbon." The three major groups of carbohydrates, based on their sizes, are monosaccharides, disaccharides, and polysaccharides (**Table 2.6**).

Monosaccharides and Disaccharides: The Simple Sugars

Monosaccharides and disaccharides are known as **simple sugars**. The monomers of carbohydrates, **monosaccharides** (mon′-ō-SAK-a-rīds; *sacchar-* = sugar), contain from three to seven carbon atoms. They are designated by names ending in "-ose" with a prefix that indicates the number of carbon atoms. For example, monosaccharides with three carbons are called *trioses* (*tri-* = three). There are also *tetroses* (four-carbon sugars), *pentoses* (five-carbon sugars), *hexoses* (six-carbon sugars), and *heptoses* (seven-carbon sugars). Examples of pentoses and hexoses are illustrated in **Figure 2.14**. Cells throughout the body break down the hexose glucose to produce ATP.

A **disaccharide** (dī-SAK-a-rīd; *di-* = two) is a molecule formed from the combination of two monosaccharides by dehydration synthesis (**Figure 2.15**). For example, molecules of the monosaccharides glucose and fructose combine to form a molecule of the disaccharide sucrose (table sugar), as shown in **Figure 2.15a**. Glucose and fructose are isomers. As you learned earlier in the chapter, isomers have the same molecular formula, but the relative positions of the oxygen and carbon atoms are different, causing the sugars to have different chemical properties. Notice that the formula for sucrose is $C_{12}H_{22}O_{11}$, not $C_{12}H_{24}O_{12}$, because a molecule of water is removed as the two monosaccharides are joined.

FIGURE 2.14 Monosaccharides. The structural formulas of selected monosaccharides are shown.

Monosaccharides are the monomers used to build carbohydrates.

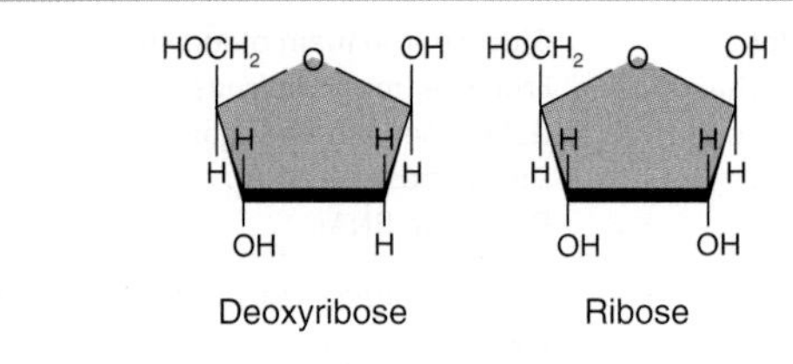

(a) Pentoses

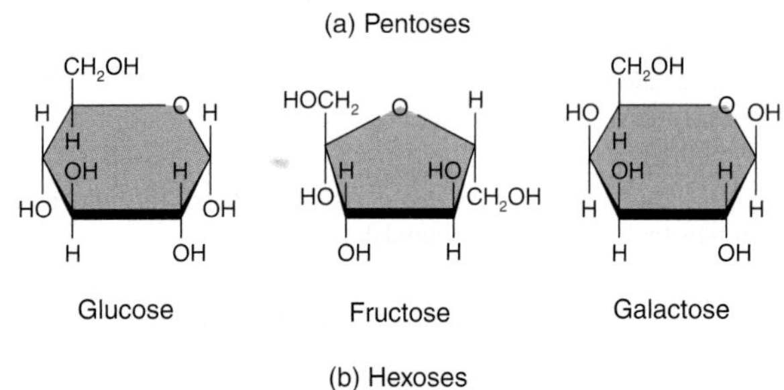

(b) Hexoses

Q Which of these monosaccharides are hexoses?

Disaccharides can also be split into smaller, simpler molecules by hydrolysis. A molecule of sucrose, for example, may be hydrolyzed into its components, glucose and fructose, by the addition of water. **Figure 2.15a** also illustrates this reaction.

Polysaccharides

The third major group of carbohydrates is the **polysaccharides** (pol′-ē-SAK-a-rīds). Each polysaccharide molecule contains tens or hundreds of monosaccharides joined through dehydration synthesis reactions. Unlike simple sugars, polysaccharides usually are insoluble in water and do not taste sweet. The main polysaccharide in the human body is **glycogen**, which is made entirely of glucose monomers linked to one another in branching chains (**Figure 2.16**). A limited amount of carbohydrates is stored as glycogen in the liver and skeletal muscles. **Starches** are polysaccharides formed from glucose by plants. They are found in foods such as pasta and potatoes and are the major carbohydrates in the diet. Like disaccharides, polysaccharides such as glycogen and starches can be broken down into monosaccharides through hydrolysis reactions. For example, when the blood glucose level falls, liver cells break down glycogen into glucose and release it

FIGURE 2.15 Disaccharides. (a) The structural and molecular formulas for the monosaccharides glucose and fructose and the disaccharide sucrose. In dehydration synthesis (read from left to right), two smaller molecules, glucose and fructose, are joined to form a larger molecule of sucrose. Note the loss of a water molecule. In hydrolysis (read from right to left), the addition of a water molecule to the larger sucrose molecule breaks the disaccharide into two smaller molecules, glucose and fructose. Shown in (b) and (c) are the structural formulas of the disaccharides lactose and maltose, respectively.

A disaccharide consists of two monosaccharides that have combined by dehydration synthesis.

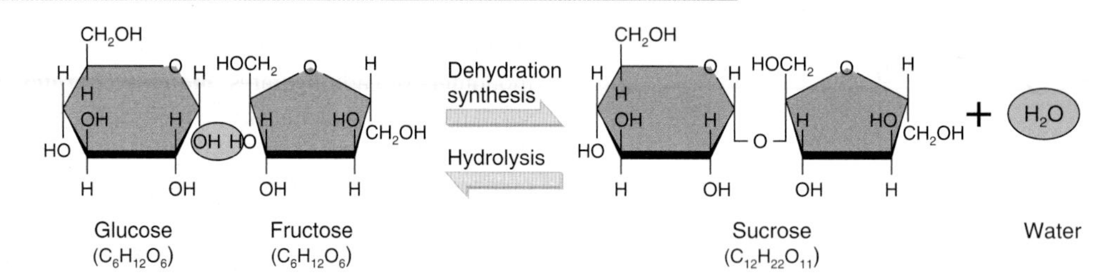

(a) Dehydration synthesis and hydrolysis of sucrose

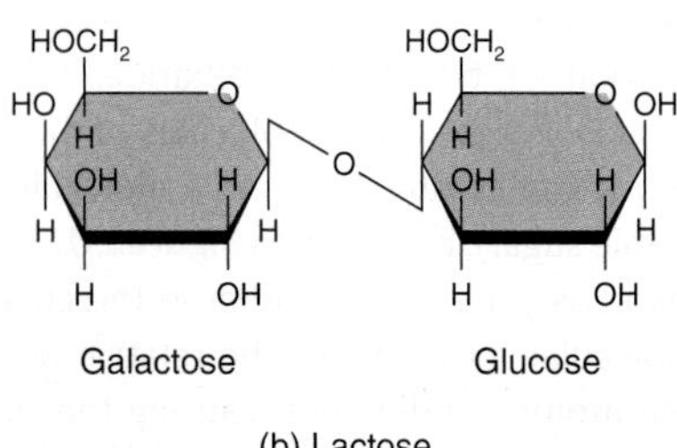

(b) Lactose

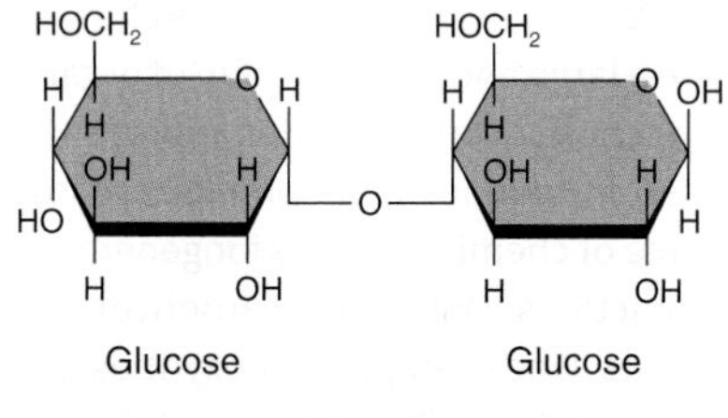

(c) Maltose

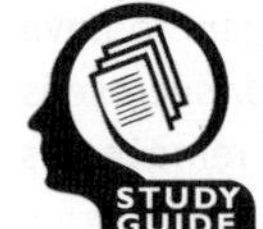

See Clinical Connection: Artificial Sweeteners

Q How many carbon atoms are there in fructose? In sucrose?

FIGURE 2.16 Part of a glycogen molecule, the main polysaccharide in the human body.

Glycogen is made up of glucose monomers and is the stored form of carbohydrate in the human body.

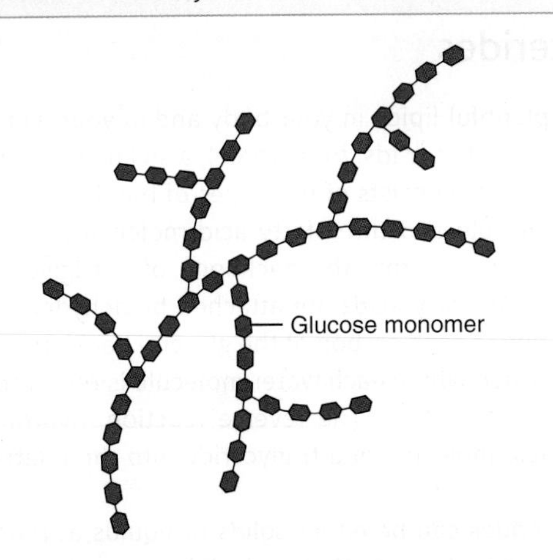

Q Which body cells store glycogen?

into the blood, making it available to body cells, which break it down to synthesize ATP. **Cellulose** is a polysaccharide formed from glucose by plants that cannot be digested by humans but does provide bulk to help eliminate feces.

2.7 Lipids

OBJECTIVES

- **Identify** the different types of lipids.
- **Discuss** the functions of lipids.

A second important group of organic compounds is **lipids** (*lip-* = fat). Lipids make up 18–25% of body mass in lean adults. Like carbohydrates, lipids contain carbon, hydrogen, and oxygen. Unlike carbohydrates, they do not have a 2:1 ratio of hydrogen to oxygen. The proportion of electronegative oxygen atoms in lipids is usually smaller than in carbohydrates, so there are fewer polar covalent bonds. As a result, most lipids are insoluble in polar solvents such as water; they are *hydrophobic.* Because they are hydrophobic, only the smallest lipids (some fatty acids) can dissolve in watery blood plasma. To become more soluble in blood plasma, other lipid molecules join with hydrophilic protein molecules. The resulting lipid–protein complexes are termed **lipoproteins**. Lipoproteins are soluble because the proteins are on the outside and the lipids are on the inside.

The diverse lipid family includes fatty acids, triglycerides (fats and oils), phospholipids (lipids that contain phosphorus), steroids (lipids that contain rings of carbon atoms), eicosanoids (20-carbon lipids), and a variety of other substances, including fat-soluble vitamins (vitamins A, D, E, and K) and lipoproteins. Table 2.7 introduces the various types of lipids and highlights their roles in the human body.

Fatty Acids

Among the simplest lipids are the **fatty acids**, which are used to synthesize triglycerides and phospholipids. Fatty acids can also be catabolized to generate adenosine triphosphate (ATP). A fatty acid consists of a carboxyl group and a hydrocarbon chain (Figure 2.17a). Fatty acids can be either saturated or unsaturated. A **saturated fatty acid** contains only *single covalent bonds* between the carbon atoms of the hydrocarbon chain. Because they lack double bonds, each carbon atom of the hydrocarbon chain is *saturated with hydrogen atoms* (see, for example, palmitic acid in Figure 2.17a). An **unsaturated fatty**

TABLE 2.7 Types of Lipids in the Body

TYPE OF LIPID	FUNCTIONS
Fatty acids	Used to synthesize triglycerides and phospholipids or catabolized to generate adenosine triphosphate (ATP).
Triglycerides (*fats and oils*)	Protection, insulation, energy storage.
Phospholipids	Major lipid component of cell membranes.
Steroids	
Cholesterol	Minor component of all animal cell membranes; precursor of bile salts, vitamin D, and steroid hormones.
Bile salts	Needed for digestion and absorption of dietary lipids.
Vitamin D	Helps regulate calcium level in body; needed for bone growth and repair.
Adrenocortical hormones	Help regulate metabolism, resistance to stress, and salt and water balance.
Sex hormones	Stimulate reproductive functions and sexual characteristics.
Eicosanoids (*prostaglandins and leukotrienes*)	Have diverse effects on modifying responses to hormones, blood clotting, inflammation, immunity, stomach acid secretion, airway diameter, lipid breakdown, and smooth muscle contraction.
Other lipids	
Carotenes	Needed for synthesis of vitamin A (used to make visual pigments in eye); function as antioxidants.
Vitamin E	Promotes wound healing, prevents tissue scarring, contributes to normal structure and function of nervous system, and functions as antioxidant.
Vitamin K	Required for synthesis of blood-clotting proteins.
Lipoproteins	Transport lipids in blood, carry triglycerides and cholesterol to tissues, and remove excess cholesterol from blood.

FIGURE 2.17 **Fatty acid structure and triglyceride synthesis.** Each time a glycerol and a fatty acid are joined in dehydration synthesis (b), a molecule of water is removed. Shown in (c) is a triglyceride molecule that contains two saturated fatty acids and a monounsaturated fatty acid. The kink (bend) in the oleic acid occurs at the double bond.

One glycerol and three fatty acids are the building blocks of triglycerides.

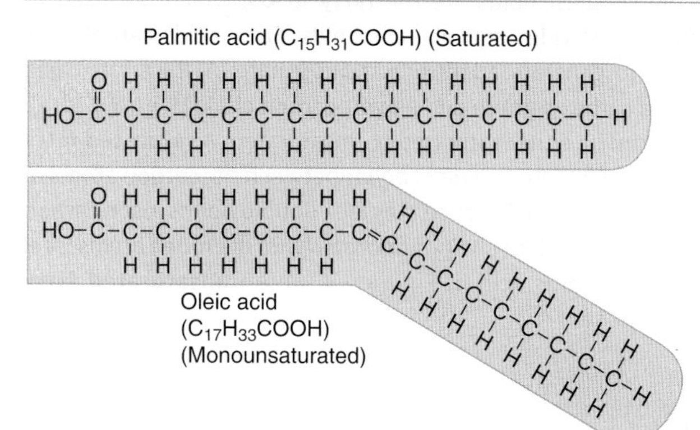

(a) Structures of saturated and unsaturated fatty acids

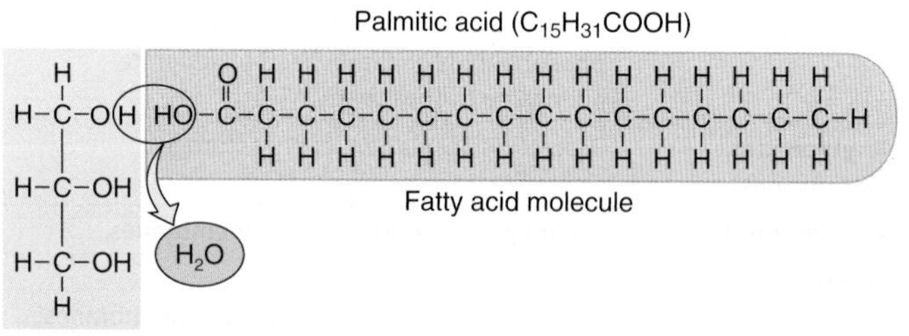

(b) Dehydration synthesis involving glycerol and a fatty acid

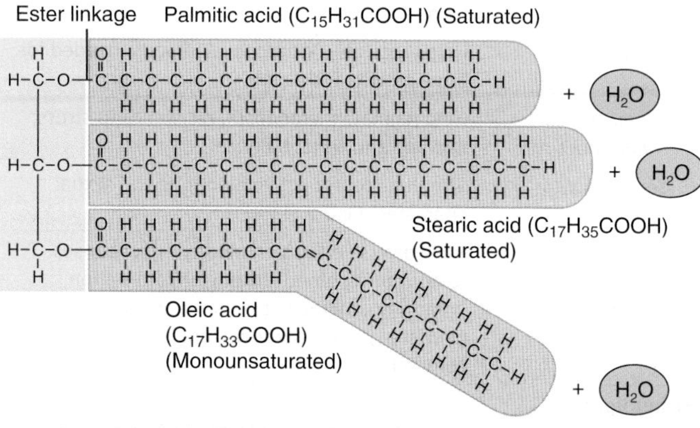

(c) Triglyceride (fat) molecule

Q Does the oxygen in the water molecule removed during dehydration synthesis come from the glycerol or from a fatty acid?

acid contains one or more *double covalent bonds* between the carbon atoms of the hydrocarbon chain. Thus, the fatty acid is not completely saturated with hydrogen atoms (see, for example, oleic acid in Figure 2.17a). The unsaturated fatty acid has a kink (bend) at the site of the double bond. If the fatty acid has just one double bond in the hydrocarbon chain, it is *monounsaturated* and it has just one kink. If a fatty acid has more than one double bond in the hydrocarbon chain, it is *polyunsaturated* and it contains more than one kink.

Triglycerides

The most plentiful lipids in your body and in your diet are the **triglycerides** (trī-GLI-ser-īds; *tri-* = three), also known as *triacylglycerols*. A triglyceride consists of two types of building blocks: a single glycerol molecule and three fatty acid molecules. A three-carbon **glycerol** molecule forms the backbone of a triglyceride (Figure 2.17b, c). Three fatty acids are attached by dehydration synthesis reactions, one to each carbon of the glycerol backbone. The chemical bond formed where each water molecule is removed is an *ester linkage* (see Table 2.5). The reverse reaction, hydrolysis, breaks down a single molecule of a triglyceride into three fatty acids and glycerol.

Triglycerides can be either solids or liquids at room temperature. A **fat** is a triglyceride that is a solid at room temperature. The fatty acids of a fat are mostly saturated. Because these saturated fatty acids lack double bonds in their hydrocarbon chains, they can closely pack together and solidify at room temperature. A fat that mainly consists of saturated fatty acids is called a **saturated fat**. Although saturated fats occur mostly in meats (especially red meats) and nonskim dairy products (whole milk, cheese, and butter), they are also found in a few plant products, such as cocoa butter, palm oil, and coconut oil. Diets that contain large amounts of saturated fats are associated with disorders such as heart disease and colorectal cancer.

An **oil** is a triglyceride that is a liquid at room temperature. The fatty acids of an oil are mostly unsaturated. Recall that unsaturated fatty acids contain one or more double bonds in their hydrocarbon chains. The kinks at the sites of the double bonds prevent the unsaturated fatty acids of an oil from closely packing together and solidifying. The fatty acids of an oil can be either monounsaturated or polyunsaturated. **Monounsaturated fats** contain triglycerides that mostly consist of monounsaturated fatty acids. Olive oil, peanut oil, canola oil, most nuts, and avocados are rich in triglycerides with monounsaturated fatty acids. **Polyunsaturated fats** contain triglycerides that mostly consist of polyunsaturated fatty acids. Corn oil, safflower oil, sunflower oil, soybean oil, and fatty fish (salmon, tuna, and mackerel) contain a high percentage of polyunsaturated fatty acids. Both monounsaturated and polyunsaturated fats are believed to decrease the risk of heart disease.

Triglycerides are the body's most highly concentrated form of chemical energy. Triglycerides provide more than twice as much energy per gram as do carbohydrates and proteins. Our capacity to store triglycerides in adipose (fat) tissue is unlimited for all practical purposes. Excess dietary carbohydrates, proteins, fats, and oils all have the same fate: They are deposited in adipose tissue as triglycerides.

See Clinical Connection: Fatty Acids in Health and Disease

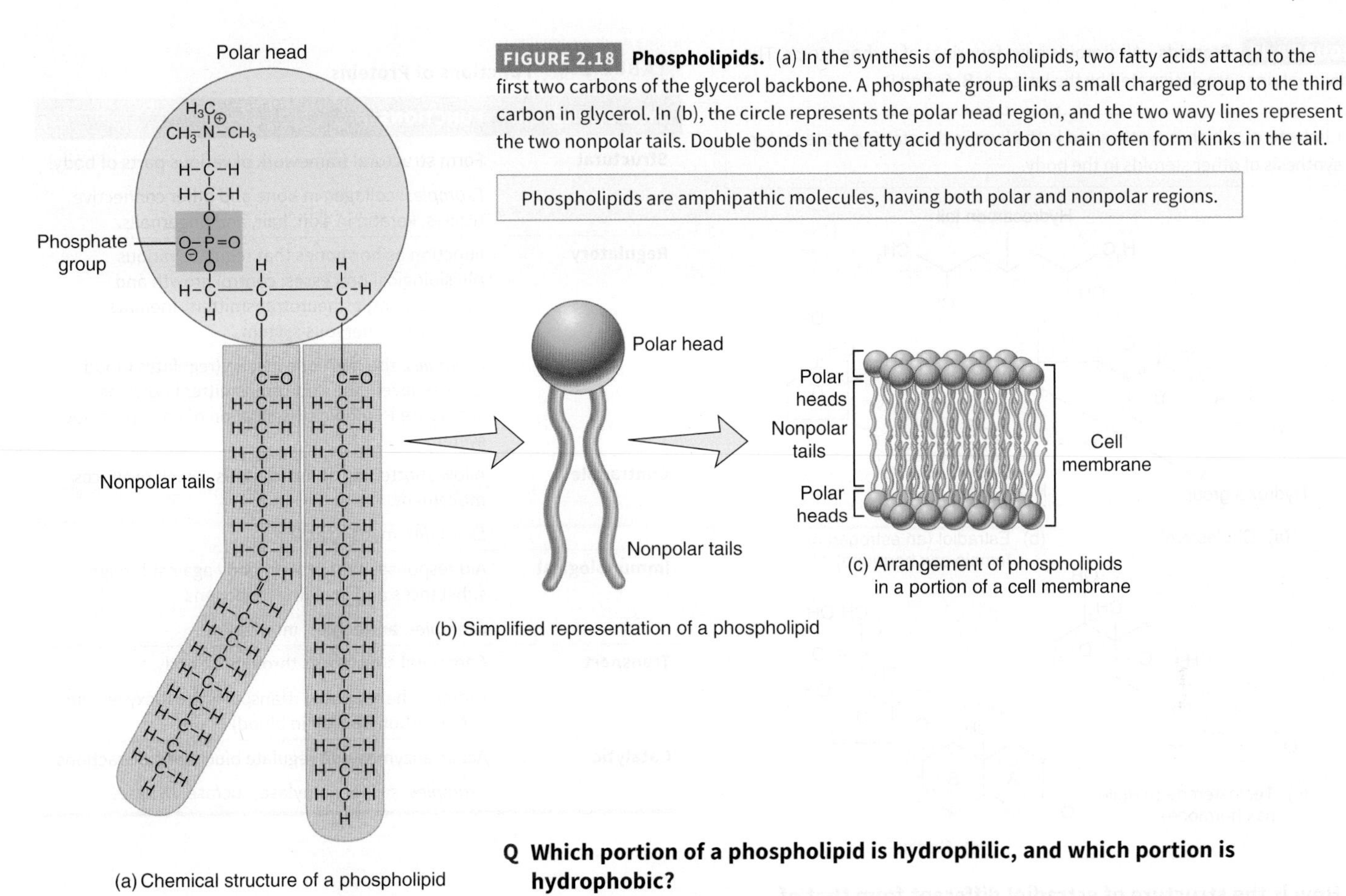

FIGURE 2.18 Phospholipids. (a) In the synthesis of phospholipids, two fatty acids attach to the first two carbons of the glycerol backbone. A phosphate group links a small charged group to the third carbon in glycerol. In (b), the circle represents the polar head region, and the two wavy lines represent the two nonpolar tails. Double bonds in the fatty acid hydrocarbon chain often form kinks in the tail.

Phospholipids are amphipathic molecules, having both polar and nonpolar regions.

Q Which portion of a phospholipid is hydrophilic, and which portion is hydrophobic?

Phospholipids

Like triglycerides, **phospholipids** (**Figure 2.18**) have a glycerol backbone and two fatty acid chains attached to the first two carbons. In the third position, however, a phosphate group (PO_4^{3-}) links a small charged group that usually contains nitrogen (N) to the backbone. This portion of the molecule (the "head") is polar and can form hydrogen bonds with water molecules. The two fatty acids (the "tails"), by contrast, are nonpolar and can interact only with other lipids. Molecules that have both polar and nonpolar parts are said to be **amphipathic** (am-fē-PATH-ik; *amphi-* = on both sides; *-pathic* = feeling). Amphipathic phospholipids line up tail-to-tail in a double row to make up much of the membrane that surrounds each cell (**Figure 2.18c**).

Steroids

The structure of **steroids** differs considerably from that of the triglycerides. Steroids have four rings of carbon atoms (colored gold in **Figure 2.19**). Body cells synthesize other steroids from cholesterol (**Figure 2.19a**), which has a large nonpolar region consisting of the four rings and a hydrocarbon tail. In the body, the commonly encountered steroids, such as cholesterol, estrogens, testosterone, cortisol, bile salts, and vitamin D, are known as **sterols** because they also have at least one hydroxyl (alcohol) group (—OH). The polar hydroxyl groups make sterols weakly amphipathic. Cholesterol is needed for cell membrane structure; estrogens and testosterone are required for regulating sexual functions; cortisol is necessary for maintaining normal blood sugar levels; bile salts are needed for lipid digestion and absorption; and vitamin D is related to bone growth. In Chapter 10, we will discuss the use of anabolic steroids by athletes to increase muscle size, strength, and endurance.

Other Lipids

Eicosanoids (ī-KŌ-sa-noyds; *eicosan-* = twenty) are lipids derived from a 20-carbon fatty acid called arachidonic acid. The two principal subclasses of eicosanoids are the **prostaglandins** (pros′-ta-GLAN-dins) and the **leukotrienes** (loo′-kō-TRĪ-ēnz). Prostaglandins have a wide variety of functions. They modify responses to hormones, contribute to the inflammatory response (Chapter 22), prevent stomach ulcers, dilate (enlarge) airways to the lungs, regulate body temperature, and influence formation of blood clots, to name just a few. Leukotrienes participate in allergic and inflammatory responses.

FIGURE 2.19 Steroids. All steroids have four rings of carbon atoms. The individual rings are designated by the letters A, B, C, and D.

Cholesterol, which is synthesized in the liver, is the starting material for synthesis of other steroids in the body.

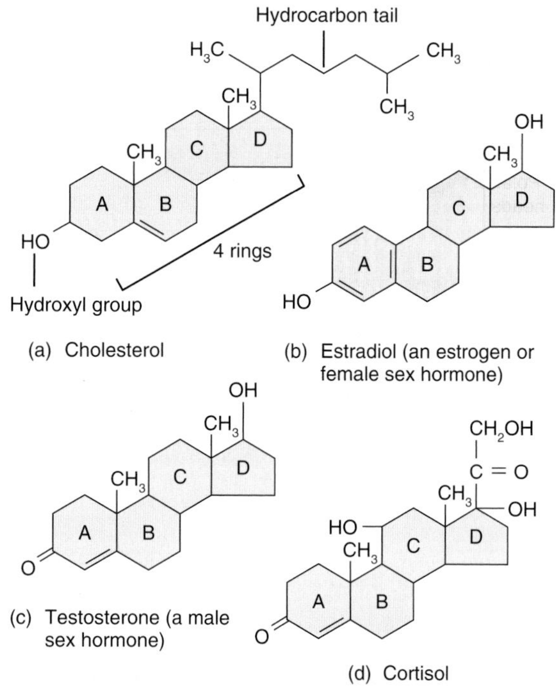

(a) Cholesterol

(b) Estradiol (an estrogen or female sex hormone)

(c) Testosterone (a male sex hormone)

(d) Cortisol

Q How is the structure of estradiol different from that of testosterone?

Other lipids include fat-soluble vitamins such as beta-carotenes (the yellow-orange pigments in egg yolk, carrots, and tomatoes that are converted to vitamin A); vitamins D, E, and K; and lipoproteins.

2.8 Proteins

OBJECTIVES

- **Identify** the building blocks of proteins.
- **Describe** the functional roles of proteins.

Proteins are large molecules that contain carbon, hydrogen, oxygen, and nitrogen. Some proteins also contain sulfur. A normal, lean adult body is 12–18% protein. Much more complex in structure than carbohydrates or lipids, proteins have many roles in the body and are largely responsible for the structure of body tissues. Enzymes are proteins that speed up most biochemical reactions. Other proteins work as "motors" to drive muscle contraction. Antibodies are proteins that defend against invading microbes. Some hormones that regulate homeostasis also are proteins. Table 2.8 describes several important functions of proteins.

TABLE 2.8 Functions of Proteins

TYPE OF PROTEIN	FUNCTIONS
Structural	Form structural framework of various parts of body. *Examples:* collagen in bone and other connective tissues; keratin in skin, hair, and fingernails.
Regulatory	Function as hormones that regulate various physiological processes; control growth and development; as neurotransmitters, mediate responses of nervous system. *Examples:* the hormone insulin (regulates blood glucose level); the neurotransmitter known as substance P (mediates sensation of pain in nervous system).
Contractile	Allow shortening of muscle cells, which produces movement. *Examples:* myosin; actin.
Immunological	Aid responses that protect body against foreign substances and invading pathogens. *Examples:* antibodies; interleukins.
Transport	Carry vital substances throughout body. *Example:* hemoglobin (transports most oxygen and some carbon dioxide in blood).
Catalytic	Act as enzymes that regulate biochemical reactions. *Examples:* salivary amylase; sucrase; ATPase.

Amino Acids and Polypeptides

The monomers of proteins are **amino acids** (a-MĒ-nō). Each of the 20 different amino acids has a hydrogen (H) atom and three important functional groups attached to a central carbon atom (Figure 2.20a): (1) an amino group (—NH_2), (2) an acidic carboxyl group (—COOH), and (3) a side chain (R group). At the normal pH of body fluids, both the amino group and the carboxyl group are ionized (Figure 2.20b). The different side chains give each amino acid its distinctive chemical identity (Figure 2.20c).

A protein is synthesized in stepwise fashion—one amino acid is joined to a second, a third is then added to the first two, and so on. The covalent bond joining each pair of amino acids is a **peptide bond**. It always forms between the carbon of the carboxyl group (—COOH) of one amino acid and the nitrogen of the amino group (—NH_2) of another. As the peptide bond is formed, a molecule of water is removed (Figure 2.21), making this a dehydration synthesis reaction. Breaking a peptide bond, as occurs during digestion of dietary proteins, is a hydrolysis reaction (Figure 2.21).

When two amino acids combine, a **dipeptide** results. Adding another amino acid to a dipeptide produces a **tripeptide**. Further additions of amino acids result in the formation of a chainlike **peptide** (4–9 amino acids) or **polypeptide** (10–2000 or more amino acids). Small

FIGURE 2.20 Amino acids. (a) In keeping with their name, amino acids have an amino group (shaded blue) and a carboxyl (acid) group (shaded red). The side chain (R group) is different in each amino acid. (b) At pH close to 7, both the amino group and the carboxyl group are ionized. (c) Glycine is the simplest amino acid; the side chain is a single H atom. Cysteine is one of two amino acids that contain sulfur (S). The side chain in tyrosine contains a six-carbon ring. Lysine has a second amino group at the end of its side chain.

Body proteins contain 20 different amino acids, each of which has a unique side chain.

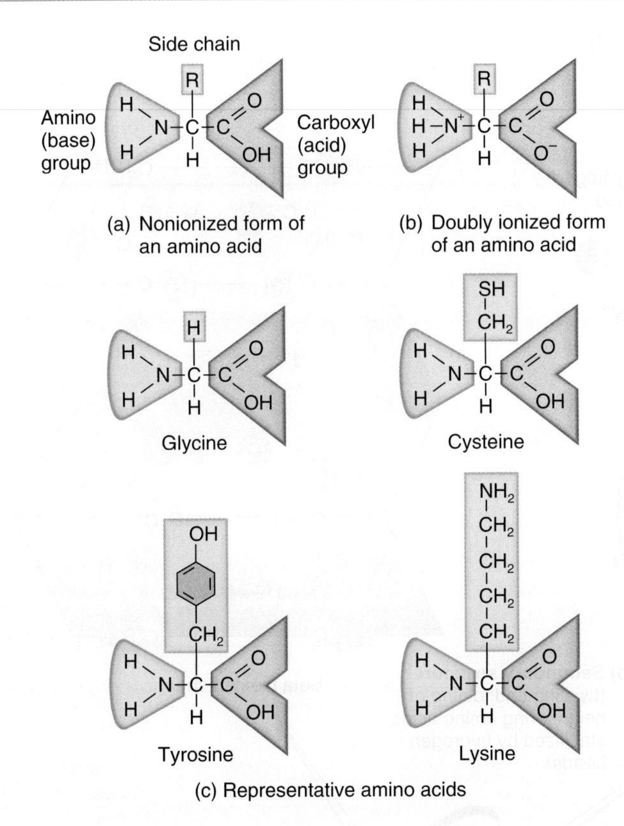

Q In an amino acid, what is the minimum number of carbon atoms? Of nitrogen atoms?

proteins may consist of a single polypeptide chain with as few as 50 amino acids. Larger proteins have hundreds or thousands of amino acids and may consist of two or more polypeptide chains folded together.

Because each variation in the number or sequence of amino acids can produce a different protein, a great variety of proteins is possible. The situation is similar to using an alphabet of 20 letters to form words. Each different amino acid is like a letter, and their various combinations give rise to a seemingly endless diversity of words (peptides, polypeptides, and proteins).

Levels of Structural Organization in Proteins

Proteins exhibit four levels of structural organization. The **primary structure** is the unique sequence of amino acids that are linked by covalent peptide bonds to form a polypeptide chain (**Figure 2.22a**). A protein's primary structure is genetically determined, and any changes in a protein's amino acid sequence can have serious consequences for body cells. In **sickle cell disease**, for example, a nonpolar amino acid (valine) replaces a polar amino acid (glutamate) through two mutations in the oxygen-carrying protein hemoglobin. This change of amino acids diminishes hemoglobin's water solubility. As a result, the altered hemoglobin tends to form crystals inside red blood cells, producing deformed, sickle-shaped cells that cannot properly squeeze through narrow blood vessels. The symptoms and treatment of sickle cell disease are discussed in Disorders: Homeostatic Imbalances in Chapter 19 in Study Guide.

The **secondary structure** of a protein is the repeated twisting or folding of neighboring amino acids in the polypeptide chain (**Figure 2.22b**). Two common secondary structures are *alpha helixes* (clockwise spirals) and *beta pleated sheets.* The secondary structure of a protein is stabilized by hydrogen bonds, which form at regular intervals along the polypeptide backbone.

The **tertiary structure** (TUR-shē-er′-ē) refers to the three-dimensional shape of a polypeptide chain. Each protein has a unique tertiary structure that determines how it will function. The tertiary folding pattern may allow amino acids at opposite ends of the chain to be close neighbors (**Figure 2.22c**). Several types of bonds can

FIGURE 2.21 Formation of a peptide bond between two amino acids during dehydration synthesis. In this example, glycine is joined to alanine, forming a dipeptide (read from left to right). Breaking a peptide bond occurs via hydrolysis (read from right to left).

Amino acids are the monomers used to build proteins.

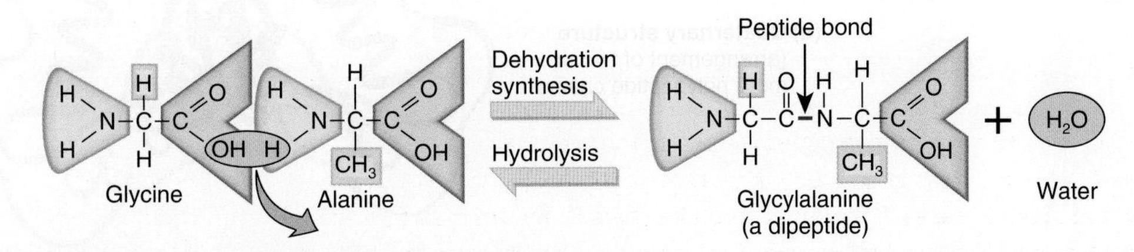

Q What type of reaction takes place during catabolism of proteins?

FIGURE 2.22 Levels of structural organization in proteins. (a) The primary structure is the sequence of amino acids in the polypeptide. (b) Common secondary structures include alpha helixes and beta pleated sheets. For simplicity, the amino acid side groups are not shown here. (c) The tertiary structure is the overall folding pattern that produces a distinctive, three-dimensional shape. (d) The quaternary structure in a protein is the arrangement of two or more polypeptide chains relative to one another.

The unique shape of each protein permits it to carry out specific functions.

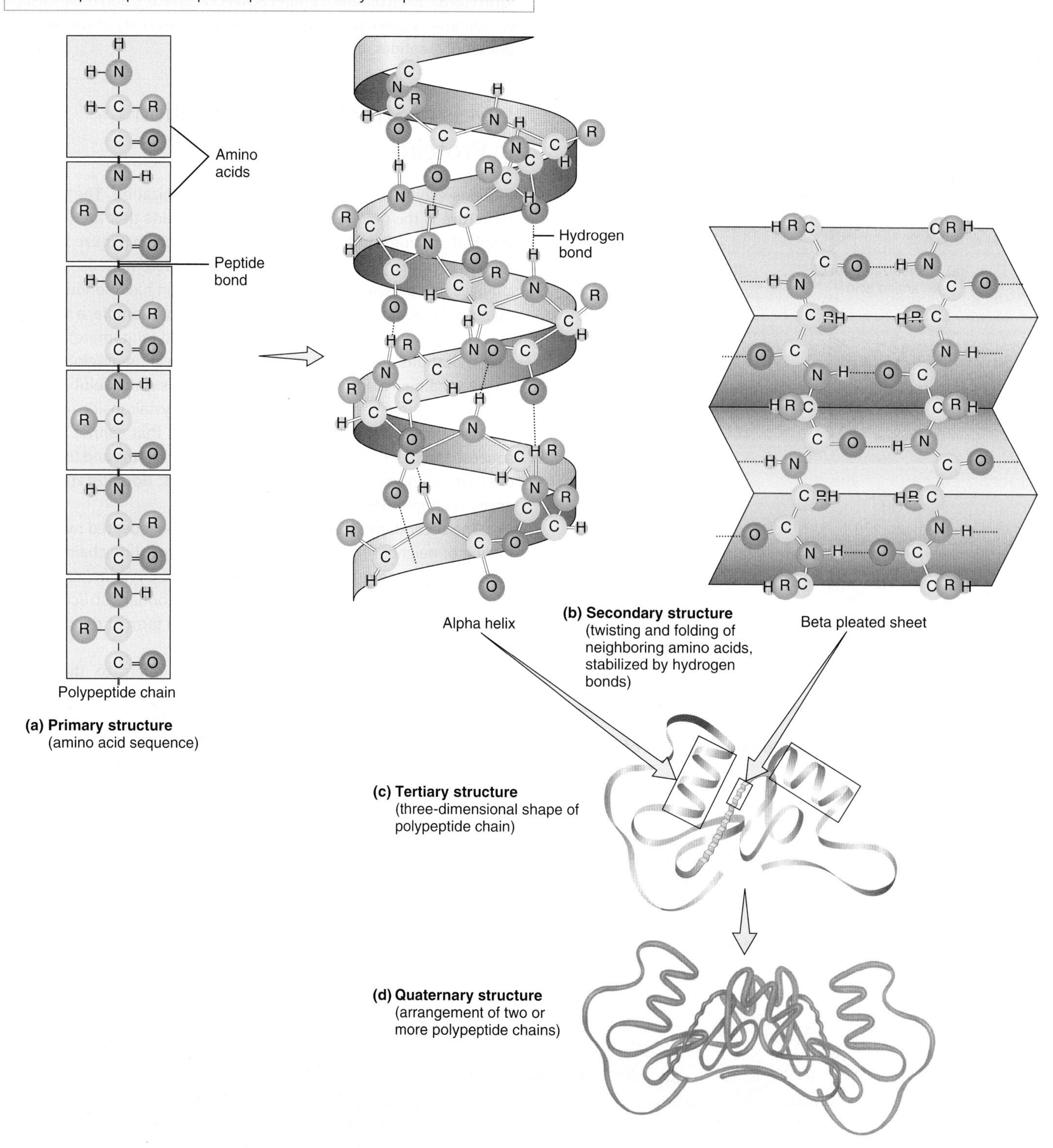

Q Do all proteins have a quaternary structure?

contribute to a protein's tertiary structure. The strongest but least common bonds, S—S covalent bonds called *disulfide bridges*, form between the sulfhydryl groups of two monomers of the amino acid cysteine. Many weak bonds—hydrogen bonds, ionic bonds, and hydrophobic interactions—also help determine the folding pattern. Some parts of a polypeptide are attracted to water (hydrophilic), and other parts are repelled by it (hydrophobic). Because most proteins in our body exist in watery surroundings, the folding process places most amino acids with hydrophobic side chains in the central core, away from the protein's surface. Often, helper molecules known as *chaperones* aid the folding process.

In those proteins that contain more than one polypeptide chain (not all of them do), the arrangement of the individual polypeptide chains relative to one another is the **quaternary structure** (KWA-ter-ner′-ē; Figure 2.22d). The bonds that hold polypeptide chains together are similar to those that maintain the tertiary structure.

Proteins vary tremendously in structure. Different proteins have different architectures and different three-dimensional shapes. This variation in structure and shape is directly related to their diverse functions. In practically every case, the function of a protein depends on its ability to recognize and bind to some other molecule. Thus, a hormone binds to a specific protein on a cell in order to alter its function, and an antibody protein binds to a foreign substance (antigen) that has invaded the body. A protein's unique shape permits it to interact with other molecules to carry out a specific function.

On the basis of overall shape, proteins are classified as fibrous or globular. **Fibrous proteins** are insoluble in water and their polypeptide chains form long strands that are parallel to each other. Fibrous proteins have many structural functions. Examples include *collagen* (strengthens bones, ligaments, and tendons), *elastin* (provides stretch in skin, blood vessels, and lung tissue), *keratin* (forms structure of hair and nails and waterproofs the skin), *dystrophin* (reinforces parts of muscle cells), *fibrin* (forms blood clots), and *actin* and *myosin* (are involved in contraction of muscle cells, division in all cells, and transport of substances within cells). **Globular proteins** are more or less soluble in water and their polypeptide chains are spherical (globular) in shape. Globular proteins have metabolic functions. Examples include *enzymes*, which function as catalysts; *antibodies* and *complement proteins*, which help protect us against disease; *hemoglobin*, which transports oxygen; *lipoproteins*, which transport lipids and cholesterol; *albumins*, which help regulate blood pH; *membrane proteins*, which transport substances into and out of cells; and some *hormones* such as *insulin*, which helps regulate blood sugar level.

Homeostatic mechanisms maintain the temperature and chemical composition of body fluids, which allow body proteins to keep their proper three-dimensional shapes. If a protein encounters an altered environment, it may unravel and lose its characteristic shape (secondary, tertiary, and quaternary structure). This process is called **denaturation**. Denatured proteins are no longer functional. Although in some cases denaturation can be reversed, a frying egg is a common example of permanent denaturation. In a raw egg the soluble egg-white protein (albumin) is a clear, viscous fluid. When heat is applied to the egg, the protein denatures, becomes insoluble, and turns white.

Enzymes

In living cells, most catalysts are protein molecules called **enzymes** (EN-zīms). Some enzymes consist of two parts—a protein portion, called the **apoenzyme** (ā′-pō-EN-zīm), and a nonprotein portion, called a **cofactor**. The cofactor may be a metal ion (such as iron, magnesium, zinc, or calcium) or an organic molecule called a *coenzyme*. Coenzymes often are derived from vitamins. The names of enzymes usually end in the suffix-*ase*. All enzymes can be grouped according to the types of chemical reactions they catalyze. For example, *oxidases* add oxygen, *kinases* add phosphate, *dehydrogenases* remove hydrogen, *ATPases* split ATP, *anhydrases* remove water, *proteases* break down proteins, and *lipases* break down triglycerides.

Enzymes catalyze specific reactions. They do so with great efficiency and with many built-in controls. Three important properties of enzymes are as follows:

1. ***Enzymes are highly specific.*** Each particular enzyme binds only to specific **substrates**—the reactant molecules on which the enzyme acts. Of the more than 1000 known enzymes in your body, each has a characteristic three-dimensional shape with a specific surface configuration, which allows it to recognize and bind to certain substrates. In some cases, the part of the enzyme that catalyzes the reaction, called the **active site**, is thought to fit the substrate like a key fits in a lock. In other cases the active site changes its shape to fit snugly around the substrate once the substrate enters the active site. This change in shape is known as an *induced fit.*

   Not only is an enzyme matched to a particular substrate; it also catalyzes a specific reaction. From among the large number of diverse molecules in a cell, an enzyme must recognize the correct substrate and then take it apart or merge it with another substrate to form one or more specific products.
2. ***Enzymes are very efficient.*** Under optimal conditions, enzymes can catalyze reactions at rates that are from 100 million to 10 billion times more rapid than those of similar reactions occurring without enzymes. The number of substrate molecules that a single enzyme molecule can convert to product molecules in one second is generally between 1 and 10,000 and can be as high as 600,000.
3. ***Enzymes are subject to a variety of cellular controls.*** Their rate of synthesis and their concentration at any given time are under the control of a cell's genes. Substances within the cell may either enhance or inhibit the activity of a given enzyme. Many enzymes have both active and inactive forms in cells. The rate at which the inactive form becomes active or vice versa is determined by the chemical environment inside the cell.

Enzymes lower the activation energy of a chemical reaction by decreasing the "randomness" of the collisions between molecules. They also help bring the substrates together in the proper orientation so that the reaction can occur. Figure 2.23a depicts how an enzyme works:

1. The substrates make contact with the active site on the surface of the enzyme molecule, forming a temporary intermediate compound called the **enzyme–substrate complex**. In this reaction the two substrate molecules are sucrose (a disaccharide) and water.

FIGURE 2.23 **How an enzyme works.**

An enzyme speeds up a chemical reaction without being altered or consumed.

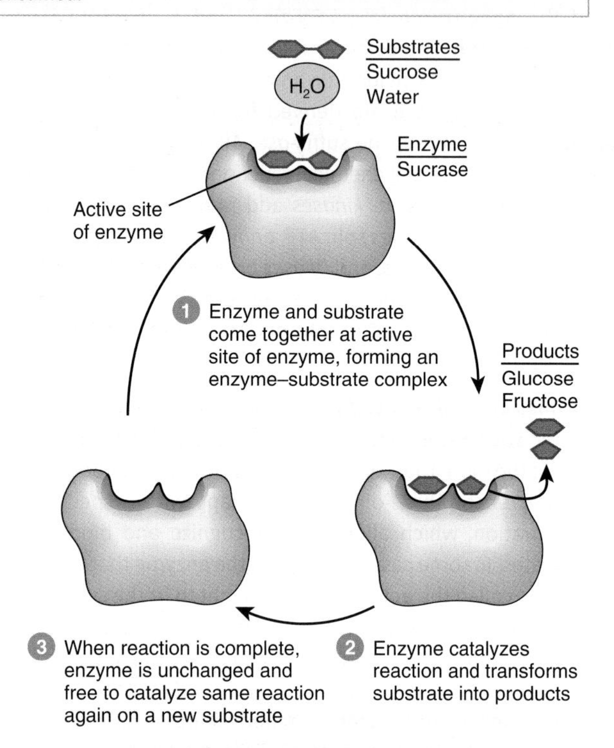

(a) Mechanism of enzyme action

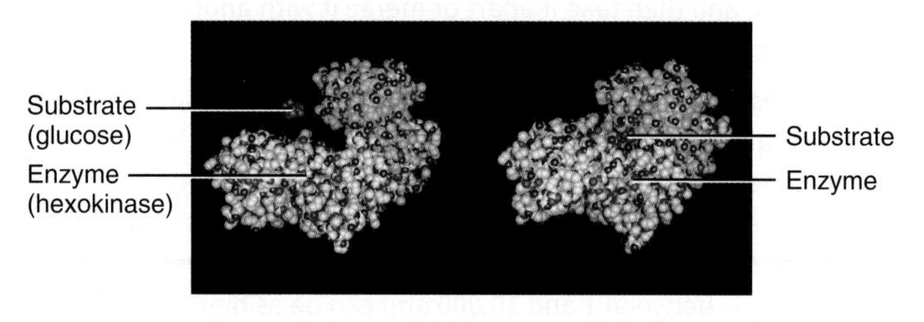

(b) Molecular model of enzyme and substrate uncombined (left) and enzyme–substrate complex (right)

Q Why is it that sucrase cannot catalyze the formation of sucrose from glucose and fructose?

2 The substrate molecules are transformed by the rearrangement of existing atoms, the breakdown of the substrate molecule, or the combination of several substrate molecules into the products of the reaction. Here the products are two monosaccharides: glucose and fructose.

3 After the reaction is completed and the reaction products move away from the enzyme, the unchanged enzyme is free to attach to other substrate molecules.

Sometimes a single enzyme may catalyze a reversible reaction in either direction, depending on the relative amounts of the substrates and products. For example, the enzyme *carbonic anhydrase* catalyzes the following reversible reaction:

$$\underset{\text{Carbon dioxide}}{CO_2} + \underset{\text{Water}}{H_2O} \xrightleftharpoons{\text{Carbonic anhydrase}} \underset{\text{Carbonic acid}}{H_2CO_3}$$

During exercise, when more CO_2 is produced and released into the blood, the reaction flows to the right, increasing the amount of carbonic acid in the blood. Then, as you exhale CO_2, its level in the blood falls and the reaction flows to the left, converting carbonic acid to CO_2 and H_2O.

2.9 Nucleic Acids

OBJECTIVES

- **Distinguish** between DNA and RNA.
- **Describe** the components of a nucleotide.

Nucleic acids (noo-KLĒ-ik), so named because they were first discovered in the nuclei of cells, are huge organic molecules that contain carbon, hydrogen, oxygen, nitrogen, and phosphorus. Nucleic acids are of two varieties. The first, **deoxyribonucleic acid (DNA)** (dē-ok′-sē-rī-bō-nū-KLĒ-ik), forms the inherited genetic material inside each human cell. In humans, each **gene** (JĒN) is a segment of a DNA molecule. Our genes determine the traits we inherit, and by controlling protein synthesis they regulate most of the activities that take place in body cells throughout our lives. When a cell divides, its hereditary information passes on to the next generation of cells. **Ribonucleic acid (RNA)**, the second type of nucleic acid, relays instructions from the genes to guide each cell's synthesis of proteins from amino acids.

A nucleic acid is a chain of repeating monomers called **nucleotides**. Each nucleotide of DNA consists of three parts (**Figure 2.24**):

1. ***Nitrogenous base.*** DNA contains four different nitrogenous bases, which contain atoms of C, H, O, and N. In DNA the four **nitrogenous bases** are adenine (A), thymine (T), cytosine (C), and guanine (G). Adenine and guanine are larger, double-ring bases called **purines** (PŪR-ēnz); thymine and cytosine are smaller, single-ring bases called **pyrimidines** (pī-RIM-i-dēnz). The nucleotides are named according to the base that is present. For instance, a nucleotide containing thymine is called a thymine nucleotide, one containing adenine is called an adenine nucleotide, and so on.

FIGURE 2.24 Components of a nucleotide. A DNA nucleotide is shown.

Nucleotides are the repeating units of nucleic acids. Each nucleotide consists of a nitrogenous base, a pentose sugar, and a phosphate group.

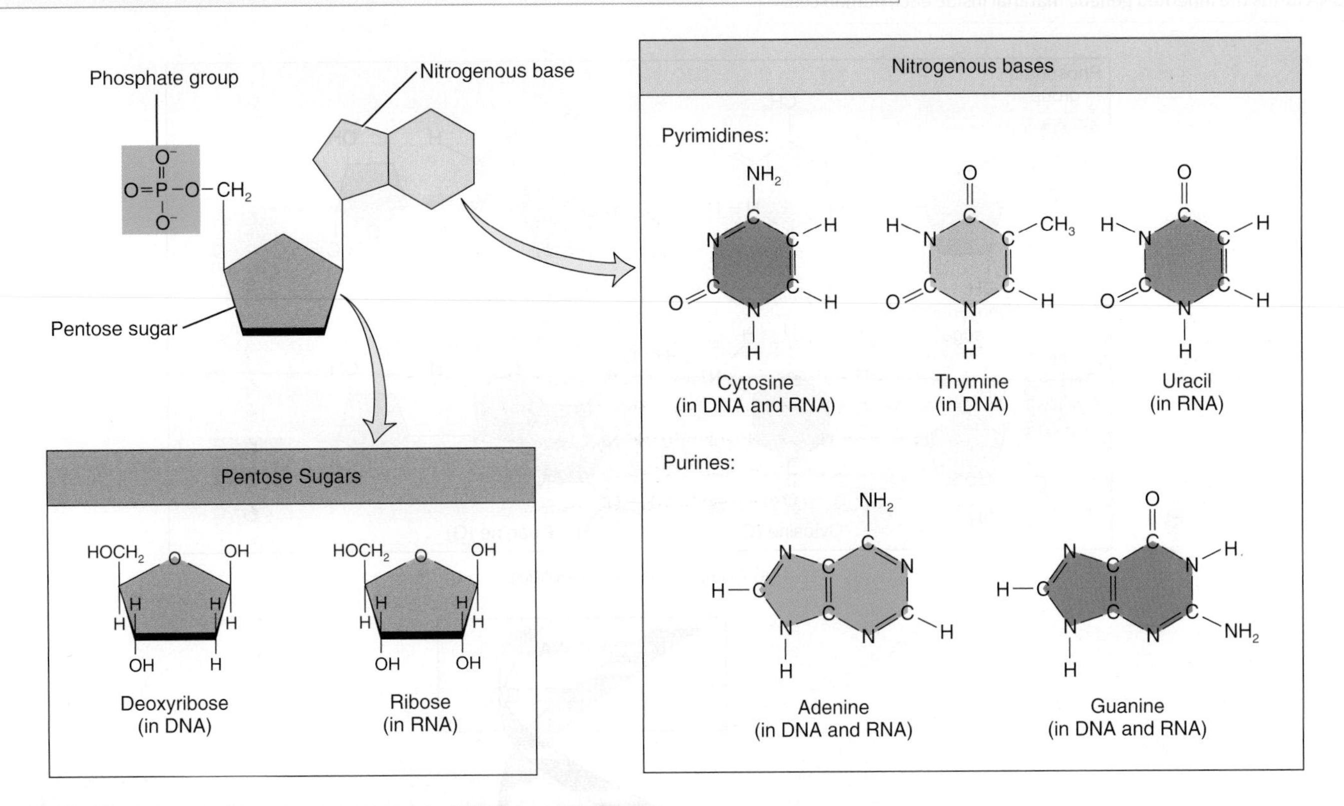

Q Which nitrogenous bases are present in DNA? In RNA?

2. ***Pentose sugar.*** A five-carbon sugar called **deoxyribose** attaches to each base in DNA.
3. ***Phosphate group.*** Phosphate groups (PO_4^{3-}) alternate with pentose sugars to form the "backbone" of a DNA strand; the bases project inward from the backbone chain (see **Figure 2.25**).

In 1953, F.H.C. Crick of Great Britain and J.D. Watson, a young American scientist, published a brief paper describing how these three components might be arranged in DNA. Their insights into data gathered by others led them to construct a model so elegant and simple that the scientific world immediately knew it was correct! In the Watson–Crick **double helix** model, DNA resembles a spiral ladder (**Figure 2.25**). Two strands of alternating phosphate groups and deoxyribose sugars form the uprights of the ladder. Paired bases, held together by hydrogen bonds, form the rungs. Because adenine always pairs with thymine, and cytosine always pairs with guanine, if you know the sequence of bases in one strand of DNA, you can predict the sequence on the complementary (second) strand. Each time DNA is copied, as when living cells divide to increase their number, the two strands unwind. Each strand serves as the template or mold on which to construct a new second strand. Any change that occurs in the base sequence of a DNA strand is called a *mutation.* Some mutations can result in the death of a cell, cause cancer, or produce genetic defects in future generations.

RNA, the second variety of nucleic acid, differs from DNA in several respects. In humans, RNA is single-stranded. The sugar in the RNA nucleotide is the pentose **ribose**, and RNA contains the pyrimidine base uracil (U) instead of thymine. Cells contain three different kinds of RNA: messenger RNA, ribosomal RNA, and transfer RNA. Each has a specific role to perform in carrying out the instructions coded in DNA (see **Figure 3.29**).

A summary of the major differences between DNA and RNA is presented in **Table 2.9**.

See Clinical Connection: DNA Fingerprinting

FIGURE 2.25 DNA molecule. DNA is arranged in a double helix. The paired bases project toward the center of the double helix. The structure of the DNA helix is stabilized by hydrogen bonds (dotted lines) between each base pair.

DNA forms the inherited genetic material inside each human cell.

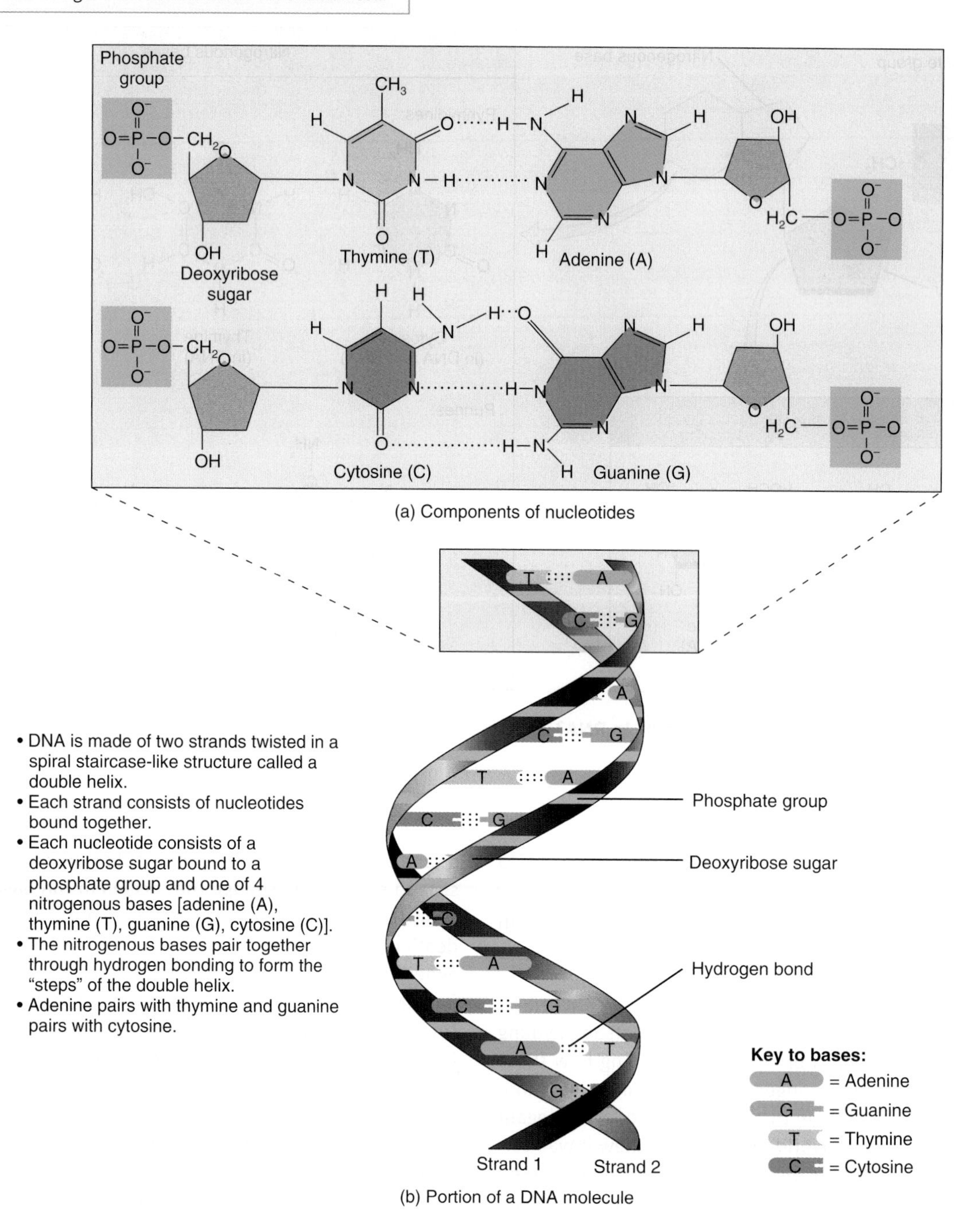

Q Which bases always pair with each other?

TABLE 2.9 Comparison between DNA and RNA

FEATURE	DNA	RNA
Nitrogenous bases	Adenine (A), cytosine (C), guanine (G), thymine (T).*	Adenine (A), cytosine (C), guanine (G), uracil (U).
Sugar in nucleotides	Deoxyribose.	Ribose.
Number of strands	Two (double-helix, like a twisted ladder).	One.
Nitrogenous base pairing (number of hydrogen bonds)	A with T (2), G with C (3).	A with U (2), G with C (3).
How is it copied?	Self-replicating.	Made by using DNA as a blueprint.
Function	Encodes information for making proteins.	Carries the genetic code and assists in making proteins.
Types	Nuclear, mitochondrial.†	Messenger RNA (mRNA), transfer RNA (tRNA), ribosomal RNA (rRNA).‡

*Letters and words in blue emphasize the differences between DNA and RNA.

†The nucleus and mitochondria are cellular organelles, which will be discussed in Chapter 3.

‡These RNAs participate in the process of protein synthesis, which will also be discussed in Chapter 3.

2.10 Adenosine Triphosphate

OBJECTIVE

- **Describe** the functional role of adenosine triphosphate.

Adenosine triphosphate (ATP) (a-DEN-ō-sēn trī-FOS-fāt) is the "energy currency" of living systems (**Figure 2.26**). ATP transfers the energy liberated in exergonic catabolic reactions to power cellular activities that require energy (endergonic reactions). Among these cellular activities are muscular contractions, movement of chromosomes during cell division, movement of structures within cells, transport of substances across cell membranes, and synthesis of larger molecules from smaller ones. As its name implies, ATP consists of three phosphate groups attached to adenosine, a unit composed of adenine and the five-carbon sugar ribose.

When a water molecule is added to ATP, the third phosphate group (PO_4^{3-}), symbolized by Ⓟ in the following discussion, is removed, and the overall reaction liberates energy. The enzyme that catalyzes the hydrolysis of ATP is called *ATPase*. Removal of the third phosphate group produces a molecule called **adenosine diphosphate (ADP)** in the following reaction:

$$\underset{\text{Adenosine triphosphate}}{\text{ATP}} + \underset{\text{Water}}{H_2O} \xrightarrow{\textit{ATPase}} \underset{\text{Adenosine diphosphate}}{\text{ADP}} + \underset{\text{Phosphate group}}{Ⓟ} + \underset{\text{Energy}}{E}$$

As noted previously, the energy supplied by the catabolism of ATP into ADP is constantly being used by the cell. As the supply of ATP at any given time is limited, a mechanism exists to replenish it: The enzyme *ATP synthase* catalyzes the addition of a phosphate group to ADP in the following reaction:

$$\underset{\text{Adenosine diphosphate}}{\text{ADP}} + \underset{\text{Phosphate group}}{Ⓟ} + \underset{\text{Energy}}{E} \xrightarrow{\textit{ATP synthase}} \underset{\text{Adenosine triphosphate}}{\text{ATP}} + \underset{\text{Water}}{H_2O}$$

FIGURE 2.26 Structures of ATP and ADP. "Squiggles" (~) indicate the two phosphate bonds that can be used to transfer energy. Energy transfer typically involves hydrolysis of the last phosphate bond of ATP.

ATP transfers chemical energy to power cellular activities.

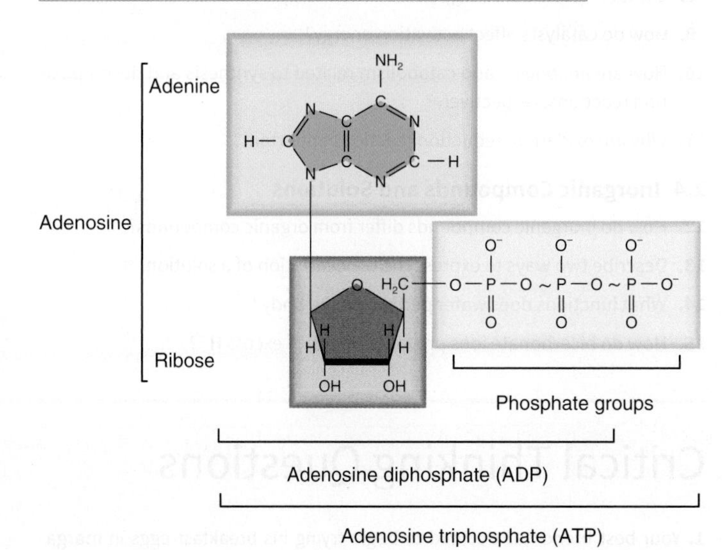

Q What are some cellular activities that depend on energy supplied by ATP?

Where does the cell get the energy required to produce ATP? The energy needed to attach a phosphate group to ADP is supplied mainly by the catabolism of glucose in a process called cellular respiration. Cellular respiration has two phases, anaerobic and aerobic:

1. ***Anaerobic phase.*** In a series of reactions that do not require oxygen, glucose is partially broken down by a series of catabolic reactions into pyruvic acid. Each glucose molecule that is converted into a pyruvic acid molecule yields two molecules of ATP.
2. ***Aerobic phase.*** In the presence of oxygen, glucose is completely broken down into carbon dioxide and water. These reactions generate heat and 30 or 32 ATP molecules.

Chapters 10 and 25 cover the details of cellular respiration.

In Chapter 1 you learned that the human body comprises various levels of organization; this chapter has just showed you the alphabet of atoms and molecules that is the basis for the language of the body. Now that you have an understanding of the chemistry of the human body, you are ready to form words; in Chapter 3 you will see how atoms and molecules are organized to form structures of cells and perform the activities of cells that contribute to homeostasis.

Review Questions

2.1 How Matter Is Organized

1. List the names and chemical symbols of the 12 most abundant chemical elements in the human body.
2. What are the atomic number, mass number, and atomic mass of carbon? How are they related?
3. Define isotopes and free radicals.

2.2 Chemical Bonds

4. Which electron shell is the valence shell of an atom, and what is its significance?
5. Compare the properties of ionic, covalent, and hydrogen bonds.
6. What information is conveyed when you write the molecular or structural formula for a molecule?

2.3 Chemical Reactions

7. What is the relationship between reactants and products in a chemical reaction?
8. Compare potential energy and kinetic energy.
9. How do catalysts affect activation energy?
10. How are anabolism and catabolism related to synthesis and decomposition reactions, respectively?
11. Why are oxidation–reduction reactions important?

2.4 Inorganic Compounds and Solutions

12. How do inorganic compounds differ from organic compounds?
13. Describe two ways to express the concentration of a solution.
14. What functions does water perform in the body?
15. How do bicarbonate ions prevent buildup of excess H^+?

2.5 Overview of Organic Compounds

16. Which functional group helps stabilize the shape of proteins?
17. What is an isomer?

2.6 Carbohydrates

18. How are carbohydrates classified?
19. How are dehydration synthesis and hydrolysis reactions related?

2.7 Lipids

20. What is the importance to the body of triglycerides, phospholipids, steroids, lipoproteins, and eicosanoids?
21. Distinguish among saturated, monounsaturated, and polyunsaturated fats.

2.8 Proteins

22. Define a protein. What is a peptide bond?
23. What are the different levels of structural organization in proteins?
24. Why are enzymes important?

2.9 Nucleic Acids

25. How do DNA and RNA differ?
26. What is a nitrogenous base?

2.10 Adenosine Triphosphate

27. In the reaction catalyzed by ATP synthase, what are the substrates and products? Is this an exergonic or endergonic reaction?

Critical Thinking Questions

1. Your best friend has decided to begin frying his breakfast eggs in margarine instead of butter because he has heard that eating butter is bad for his heart. Has he made a wise choice? Are there other alternatives?
2. A 4-month-old baby is admitted to the hospital with a temperature of 102°F (38.9°C). Why is it critical to treat the fever as quickly as possible?
3. During chemistry lab, Maria places sucrose (table sugar) in a glass beaker, adds water, and stirs. As the table sugar disappears, she loudly proclaims that she has chemically broken down the sucrose into fructose and glucose. Is Maria's chemical analysis correct?

The Cellular Level of Organization

CHAPTER 3

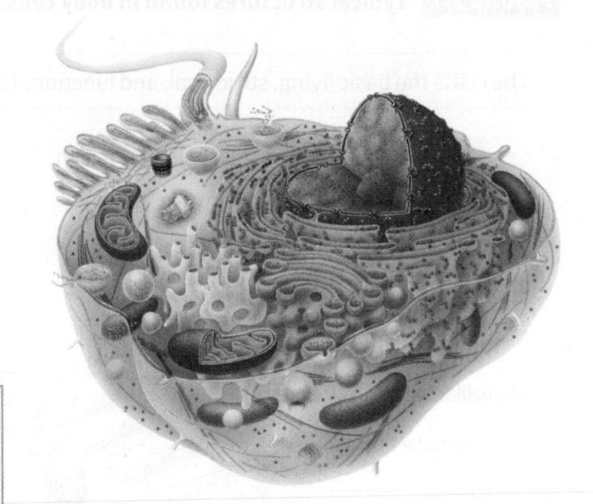

Cells and Homeostasis

Cells carry out a multitude of functions that help each system contribute to the homeostasis of the entire body. At the same time, all cells share key structures and functions that support their intense activity.

In the previous chapter you learned about the atoms and molecules that compose the alphabet of the language of the human body. These are combined into about 200 different types of words called cells. All cells arise from existing cells in which one cell divides into two identical cells. Different types of cells fulfill unique roles that support homeostasis and contribute to the many functional capabilities of the human organism. As you study the various parts of a cell and their relationships to one another, you will learn that cell structure and function are intimately related. In this chapter, you will learn that cells carry out a dazzling array of chemical reactions to create and maintain life processes—in part, by isolating specific types of chemical reactions within specialized cellular structures. Although isolated, the chemical reactions are coordinated to maintain life in a cell, tissue, organ, system, and organism.

Q Did you ever wonder why cancer is so difficult to treat?

3.1 Parts of a Cell

OBJECTIVE

- **Name** and **describe** the three main parts of a cell.

The average adult human body consists of more than 100 trillion cells. **Cells** are the basic, living, structural, and functional units of the body. The scientific study of cells is called **cell biology** or *cytology*.

Figure 3.1 provides an overview of the typical structures found in body cells. Most cells have many of the structures shown in this diagram. For ease of study, we divide the cell into three main parts: plasma membrane, cytoplasm, and nucleus.

1. The **plasma membrane** forms the cell's flexible outer surface, separating the cell's *internal environment* (everything inside the cell) from the *external environment* (everything outside the cell). It is a selective barrier that regulates the flow of materials into and out of a cell. This selectivity helps establish and maintain the appropriate environment for normal cellular activities. The plasma membrane also plays a key role in communication among cells and between cells and their external environment.
2. The **cytoplasm** (SĪ-tō-plasm; *-plasm* = formed or molded) consists of all the cellular contents between the plasma membrane and the nucleus. This compartment has two components: cytosol and organelles. **Cytosol** (SĪ-tō-sol), the fluid portion of cytoplasm, also called *intracellular fluid*, contains water, dissolved solutes, and suspended particles. Within the cytosol are several different types of **organelles** (or-gan-ELZ = little organs). Each type of organelle has a characteristic shape and specific functions. Examples include the cytoskeleton, ribosomes, endoplasmic reticulum, Golgi complex, lysosomes, peroxisomes, and mitochondria.
3. The **nucleus** (NOO-klē-us = nut kernel) is a large organelle that houses most of a cell's DNA. Within the nucleus, each **chromosome** (KRŌ-mō-sōm; *chromo-* = colored), a single molecule of DNA associated with several proteins, contains thousands of hereditary units called **genes** that control most aspects of cellular structure and function.

3.2 The Plasma Membrane

OBJECTIVES

- **Distinguish** between cytoplasm and cytosol.
- **Explain** the concept of selective permeability.
- **Define** the electrochemical gradient and **describe** its components.

The **plasma membrane**, a flexible yet sturdy barrier that surrounds and contains the cytoplasm of a cell, is best described by using a structural model called the **fluid mosaic model**. According to this model, the molecular arrangement of the plasma membrane resembles a continually moving sea of fluid lipids that contains a mosaic of many different proteins (**Figure 3.2**). Some proteins float freely like icebergs

FIGURE 3.1 Typical structures found in body cells.

The cell is the basic living, structural, and functional unit of the body.

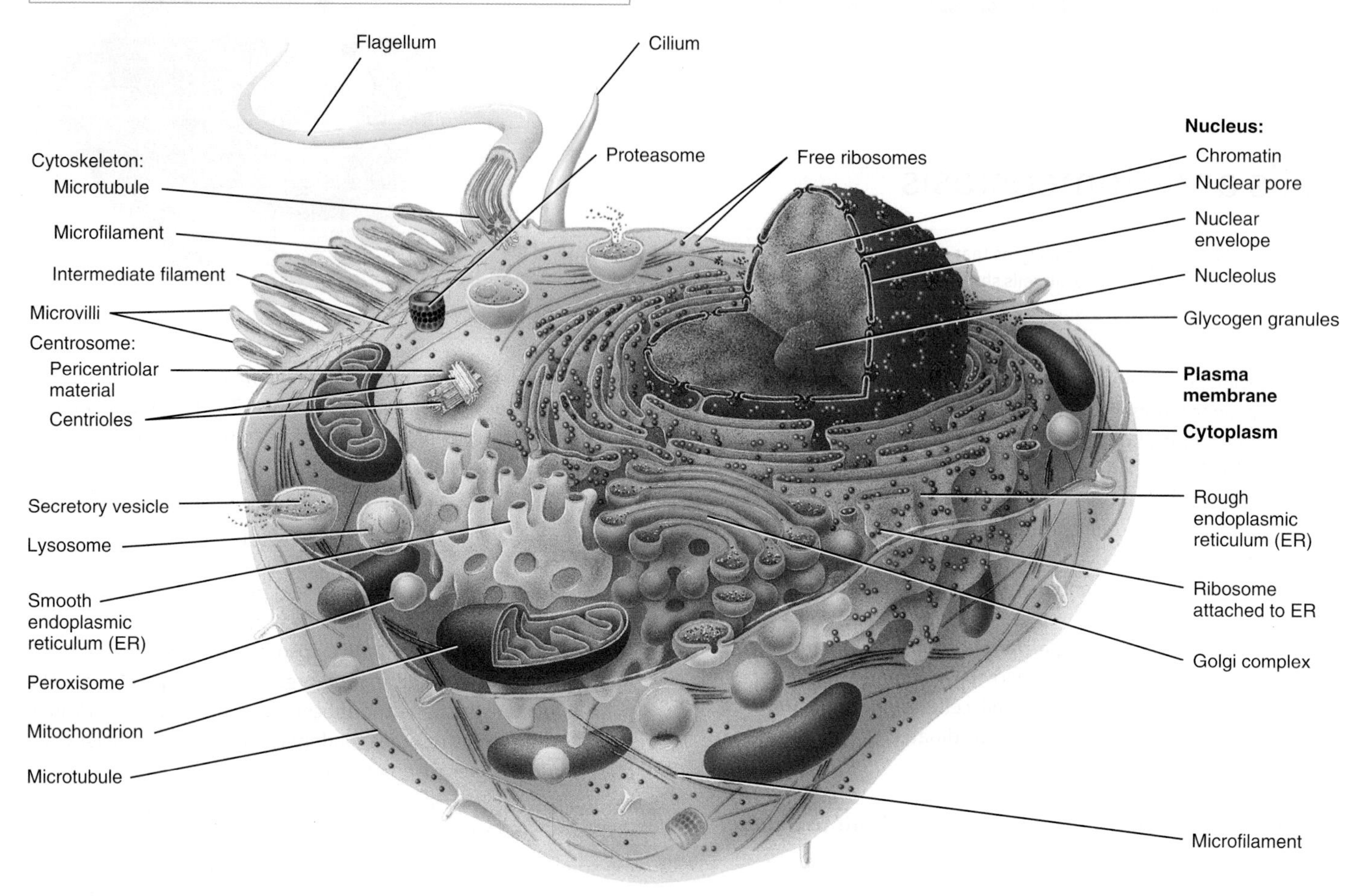

Q What are the three principal parts of a cell?

in the lipid sea, whereas others are anchored at specific locations like islands. The membrane lipids allow passage of several types of lipid-soluble molecules but act as a barrier to the entry or exit of charged or polar substances. Some of the proteins in the plasma membrane allow movement of polar molecules and ions into and out of the cell. Other proteins can act as signal receptors or as molecules that link the plasma membrane to intracellular or extracellular proteins.

Structure of the Plasma Membrane

The Lipid Bilayer The basic structural framework of the plasma membrane is the **lipid bilayer**, two back-to-back layers made up of three types of lipid molecules—phospholipids, cholesterol, and glycolipids (Figure 3.2). About 75% of the membrane lipids are **phospholipids**, lipids that contain phosphorus. Present in smaller amounts are **cholesterol** (about 20%), a steroid with an attached —OH (hydroxyl) group, and various **glycolipids** (about 5%), lipids with attached carbohydrate groups.

The bilayer arrangement occurs because the lipids are **amphipathic** (am-fē-PATH-ik) molecules, which means that they have both polar and nonpolar parts. In phospholipids (see Figure 2.18), the polar part is the phosphate-containing "head," which is *hydrophilic* (*hydro-* = water; *-philic* = loving). The nonpolar parts are the two long fatty acid "tails," which are *hydrophobic* (*-phobic* = fearing) hydrocarbon chains. Because "like seeks like," the phospholipid molecules orient themselves in the bilayer with their hydrophilic heads facing outward. In this way, the heads face a watery fluid on either side—cytosol on the inside and extracellular fluid on the outside. The hydrophobic fatty acid tails in each half of the bilayer point toward one another, forming a nonpolar, hydrophobic region in the membrane's interior.

Cholesterol molecules are weakly amphipathic (see Figure 2.19a) and are interspersed among the other lipids in both layers of the membrane. The tiny —OH group is the only polar region of cholesterol, and it forms hydrogen bonds with the polar heads of phospholipids and glycolipids. The stiff steroid rings and hydrocarbon tail of cholesterol are nonpolar; they fit among the fatty acid tails of the phospholipids and glycolipids. The carbohydrate groups of

FIGURE 3.2 The fluid mosaic arrangement of lipids and proteins in the plasma membrane.

Membranes are fluid structures because the lipids and many of the proteins are free to rotate and move sideways in their own half of the bilayer.

Functions of the Plasma Membrane

1. Acts as a barrier separating inside and outside of the cell.
2. Controls the flow of substances into and out of the cell.
3. Helps identify the cell to other cells (e.g., immune cells).
4. Participates in intercellular signaling.

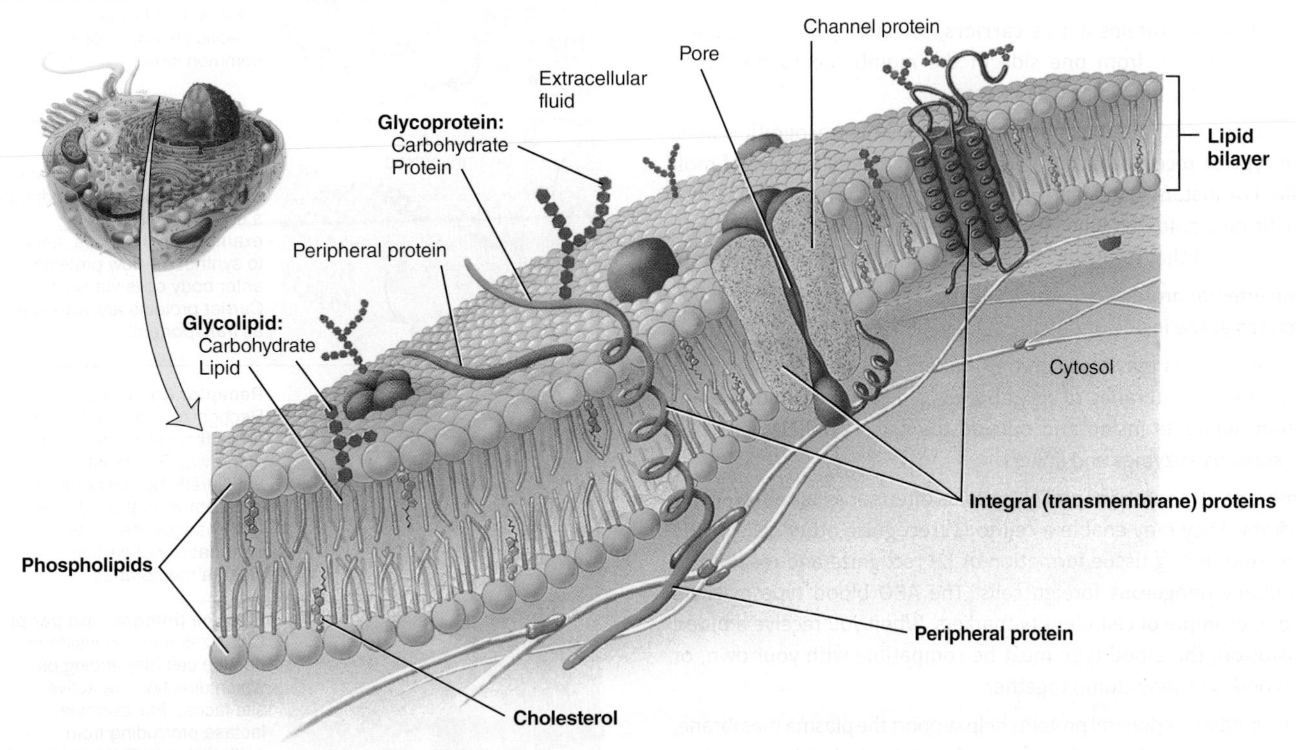

Q What is the glycocalyx?

glycolipids form a polar "head"; their fatty acid "tails" are nonpolar. Glycolipids appear only in the membrane layer that faces the extracellular fluid, which is one reason the two sides of the bilayer are asymmetric, or different.

Arrangement of Membrane Proteins

Membrane proteins are classified as integral or peripheral according to whether they are firmly embedded in the membrane (Figure 3.2). **Integral proteins** extend into or through the lipid bilayer and are firmly embedded in it. Most integral proteins are **transmembrane proteins**, which means that they span the entire lipid bilayer and protrude into both the cytosol and extracellular fluid. A few integral proteins are tightly attached to one side of the bilayer by covalent bonding to fatty acids. Like membrane lipids, integral membrane proteins are amphipathic. Their hydrophilic regions protrude into either the watery extracellular fluid or the cytosol, and their hydrophobic regions extend among the fatty acid tails.

As their name implies, **peripheral proteins** (pe-RIF-er-al) are not as firmly embedded in the membrane. They are attached to the polar heads of membrane lipids or to integral proteins at the inner or outer surface of the membrane.

Many integral proteins are **glycoproteins**, proteins with carbohydrate groups attached to the ends that protrude into the extracellular fluid. The carbohydrates are *oligosaccharides* (*oligo-* = few; *-saccharides* = sugars), chains of 2 to 60 monosaccharides that may be straight or branched. The carbohydrate portions of glycolipids and glycoproteins form an extensive sugary coat called the **glycocalyx** (glī-kō-KĀL-iks). The pattern of carbohydrates in the glycocalyx varies from one cell to another. Therefore, the glycocalyx acts like a molecular "signature" that enables cells to recognize one another. For example, a white blood cell's ability to detect a "foreign" glycocalyx is one basis of the immune response that helps us destroy invading organisms. In addition, the glycocalyx enables cells to adhere to one another in some tissues and protects cells from being digested by enzymes in the extracellular fluid. The hydrophilic properties of the glycocalyx attract a film of fluid to the surface of many cells. This action makes red blood cells slippery as they flow through narrow blood vessels and protects cells that line the airways and the gastrointestinal tract from drying out.

Functions of Membrane Proteins

Generally, the types of lipids in cellular membranes vary only slightly. In contrast, the membranes of different cells and various intracellular organelles have remarkably different assortments of proteins that determine many of the membrane's functions (**Figure 3.3**).

- Some integral proteins form **ion channels**, *pores* or holes that specific ions, such as potassium ions (K^+), can flow through to get into or out of the cell. Most ion channels are *selective*; they allow only a single type of ion to pass through.
- Other integral proteins act as **carriers**, selectively moving a polar substance or ion from one side of the membrane to the other. Carriers are also known as *transporters*.
- Integral proteins called **receptors** serve as cellular recognition sites. Each type of receptor recognizes and binds a specific type of molecule. For instance, insulin receptors bind the hormone insulin. A specific molecule that binds to a receptor is called a **ligand** (LĪ-gand; *liga* = tied) of that receptor.
- Some integral proteins are **enzymes** that catalyze specific chemical reactions at the inside or outside surface of the cell.
- Integral proteins may also serve as **linkers** that anchor proteins in the plasma membranes of neighboring cells to one another or to protein filaments inside and outside the cell. Peripheral proteins also serve as enzymes and linkers.
- Membrane glycoproteins and glycolipids often serve as **cell-identity markers**. They may enable a cell to (1) recognize other cells of the same kind during tissue formation or (2) recognize and respond to potentially dangerous foreign cells. The ABO blood type markers are one example of cell-identity markers. When you receive a blood transfusion, the blood type must be compatible with your own, or red blood cells may clump together.

In addition, peripheral proteins help support the plasma membrane, anchor integral proteins, and participate in mechanical activities such as moving materials and organelles within cells, changing cell shape during cell division and in muscle cells, and attaching cells to one another.

Membrane Fluidity

Membranes are fluid structures; that is, most of the membrane lipids and many of the membrane proteins easily rotate and move sideways in their own half of the bilayer. Neighboring lipid molecules exchange places about 10 million times per second and may wander completely around a cell in only a few minutes! Membrane fluidity depends both on the number of double bonds in the fatty acid tails of the lipids that make up the bilayer, and on the amount of cholesterol present. Each double bond puts a "kink" in the fatty acid tail (see **Figure 2.18**), which increases membrane fluidity by preventing lipid molecules from packing tightly in the membrane. Membrane fluidity is an excellent compromise for the cell; a rigid membrane would lack mobility, and a completely fluid membrane would lack the structural organization and mechanical support required by the cell. Membrane fluidity allows interactions to occur within the plasma membrane, such as the assembly of membrane proteins. It also enables the movement of the membrane components

FIGURE 3.3 Functions of membrane proteins.

Membrane proteins largely reflect the functions a cell can perform.

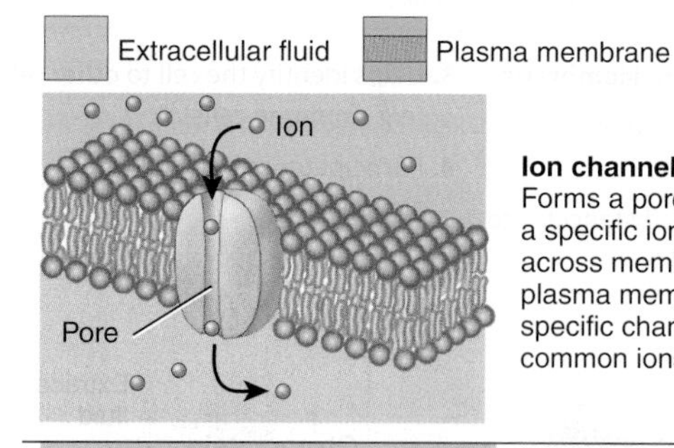

Ion channel (integral)
Forms a pore through which a specific ion can flow to get across membrane. Most plasma membranes include specific channels for several common ions.

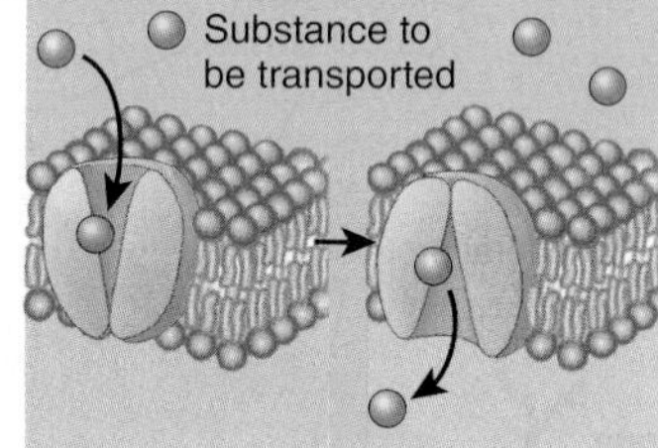

Carrier (integral)
Transports a specific substance across membrane by undergoing a change in shape. For example, amino acids, needed to synthesize new proteins, enter body cells via carriers. Carrier proteins are also known as *transporters*.

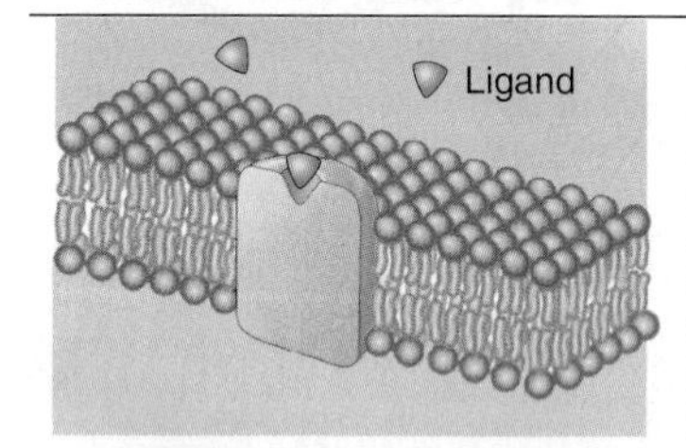

Receptor (integral)
Recognizes specific ligand and alters cell's function in some way. For example, antidiuretic hormone binds to receptors in the kidneys and changes the water permeability of certain plasma membranes.

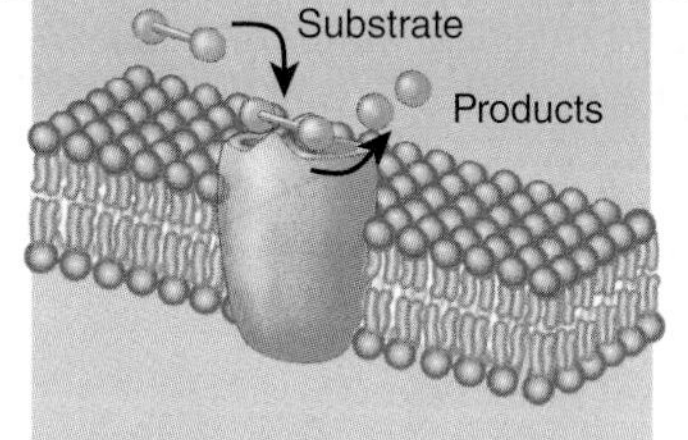

Enzyme (integral and peripheral)
Catalyzes reaction inside or outside cell (depending on which direction the active site faces). For example, lactase protruding from epithelial cells lining your small intestine splits the disaccharide lactose in the milk you drink.

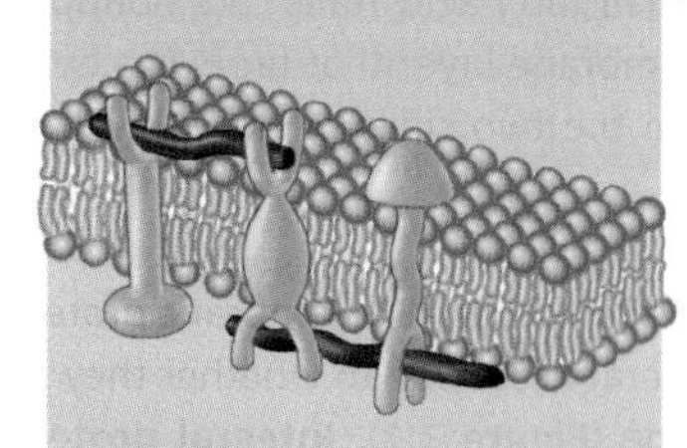

Linker (integral and peripheral)
Anchors filaments inside and outside the plasma membrane, providing structural stability and shape for the cell. May also participate in movement of the cell or link two cells together.

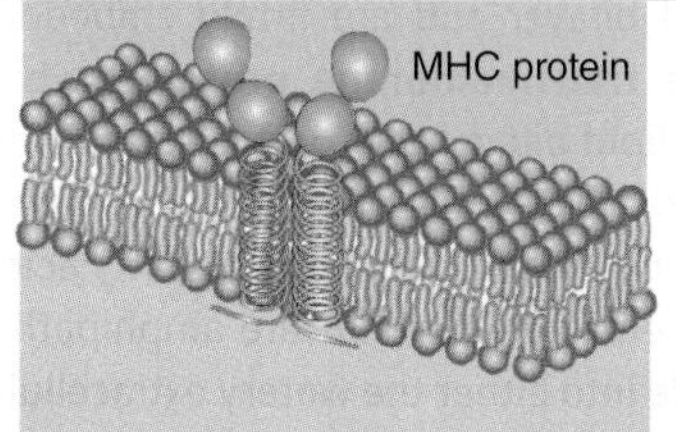

Cell identity marker (glycoprotein)
Distinguishes your cells from anyone else's (unless you are an identical twin). An important class of such markers are the major histocompatibility (MHC) proteins.

Q When stimulating a cell, the hormone insulin first binds to a protein in the plasma membrane. This action best represents which membrane protein function?

responsible for cellular processes such as cell movement, growth, division, and secretion, and the formation of cellular junctions. Fluidity allows the lipid bilayer to self-seal if torn or punctured. When a needle is pushed through a plasma membrane and pulled out, the puncture site seals spontaneously, and the cell does not burst. This property of the lipid bilayer allows a procedure called intracytoplasmic sperm injection to help infertile couples conceive a child; scientists can fertilize an oocyte by injecting a sperm cell through a tiny syringe. It also permits removal and replacement of a cell's nucleus in cloning experiments, such as the one that created Dolly, the famous cloned sheep.

Despite the great mobility of membrane lipids and proteins in their own half of the bilayer, they seldom flip-flop from one half of the bilayer to the other, because it is difficult for hydrophilic parts of membrane molecules to pass through the hydrophobic core of the membrane. This difficulty contributes to the asymmetry of the membrane bilayer.

Because of the way it forms hydrogen bonds with neighboring phospholipid and glycolipid heads and fills the space between bent fatty acid tails, cholesterol makes the lipid bilayer stronger but less fluid at normal body temperature. At low temperatures, cholesterol has the opposite effect—it increases membrane fluidity.

Membrane Permeability

The term *permeable* means that a structure permits the passage of substances through it, while *impermeable* means that a structure does not permit the passage of substances through it. The permeability of the plasma membrane to different substances varies. Plasma membranes permit some substances to pass more readily than others. This property of membranes is termed **selective permeability** (per′-mē-a-BIL-i-tē).

The lipid bilayer portion of the plasma membrane is highly permeable to nonpolar molecules such as oxygen (O_2), carbon dioxide (CO_2), and steroids; moderately permeable to small, uncharged polar molecules, such as water and urea (a waste product from the breakdown of amino acids); and impermeable to ions and large, uncharged polar molecules, such as glucose. The permeability characteristics of the plasma membrane are due to the fact that the lipid bilayer has a nonpolar, hydrophobic interior (see **Figure 2.18c**). So, the more hydrophobic or lipid-soluble a substance, the greater the membrane's permeability to that substance. Thus, the hydrophobic interior of the plasma membrane allows nonpolar molecules to rapidly pass through, but prevents passage of ions and large, uncharged polar molecules. The permeability of the lipid bilayer to water and urea is an unexpected property given that they are polar molecules. These two molecules are thought to pass through the lipid bilayer in the following way: As the fatty acid tails of membrane phospholipids and glycolipids randomly move about, small gaps briefly appear in the hydrophobic environment of the membrane's interior. Because water and urea are small polar molecules that have no overall charge, they can move from one gap to another until they have crossed the membrane.

Transmembrane proteins that act as channels and carriers increase the plasma membrane's permeability to a variety of ions and uncharged polar molecules that, unlike water and urea molecules, cannot cross the lipid bilayer unassisted. Channels and carriers are very selective. Each one helps a specific molecule or ion to cross the membrane. Macromolecules, such as proteins, are so large that they are unable to pass across the plasma membrane except by endocytosis and exocytosis (discussed later in this chapter).

Gradients across the Plasma Membrane

The selective permeability of the plasma membrane allows a living cell to maintain different concentrations of certain substances on either side of the plasma membrane. A **concentration gradient** is a difference in the concentration of a chemical from one place to another, such as from the inside to the outside of the plasma membrane. Many ions and molecules are more concentrated in either the cytosol or the extracellular fluid. For instance, oxygen molecules and sodium ions (Na^+) are more concentrated in the extracellular fluid than in the cytosol; the opposite is true of carbon dioxide molecules and potassium ions (K^+).

The plasma membrane also creates a difference in the distribution of positively and negatively charged ions between the two sides of the plasma membrane. Typically, the inner surface of the plasma membrane is more negatively charged and the outer surface is more positively charged. A difference in electrical charges between two regions constitutes an **electrical gradient**. Because it occurs across the plasma membrane, this charge difference is termed the **membrane potential**.

As you will see shortly, the concentration gradient and electrical gradient are important because they help move substances across the plasma membrane. In many cases a substance will move across a plasma membrane *down its concentration gradient.* That is to say, a substance will move "downhill," from where it is more concentrated to where it is less concentrated, to reach equilibrium. Similarly, a positively charged substance will tend to move toward a negatively charged area, and a negatively charged substance will tend to move toward a positively charged area. The combined influence of the concentration gradient and the electrical gradient on movement of a particular ion is referred to as its **electrochemical gradient**.

3.3 Transport across the Plasma Membrane

OBJECTIVE

- **Describe** the processes that transport substances across the plasma membrane.

Transport of materials across the plasma membrane is essential to the life of a cell. Certain substances must move into the cell to support metabolic reactions. Other substances that have been produced by the cell for export or as cellular waste products must move out of the cell.

Substances generally move across cellular membranes via transport processes that can be classified as passive or active, depending on whether they require cellular energy. In **passive processes**, a substance moves down its concentration or electrical gradient to cross the membrane using only its own kinetic energy (energy of motion). Kinetic energy is intrinsic to the particles that are moving. There is no input of energy from the cell. An example is simple diffusion. In **active processes**, cellular energy is used to drive the substance "uphill" against its concentration or electrical gradient. The cellular energy used is usually in the form of adenosine triphosphate (ATP). An example is active transport. Another way that some substances may enter and leave cells is an active process in which tiny, spherical membrane sacs referred to as **vesicles** are used. Examples include endocytosis, in which vesicles detach from the plasma membrane while bringing materials into a cell, and exocytosis, the merging of vesicles with the plasma membrane to release materials from the cell.

Passive Processes

The Principle of Diffusion

Learning why materials diffuse across membranes requires an understanding of how diffusion occurs in a solution. **Diffusion** (di-FŪ-zhun; *diffus-* = spreading) is a passive process in which the random mixing of particles in a solution occurs because of the particles' kinetic energy. Both the *solutes,* the dissolved substances, and the *solvent,* the liquid that does the dissolving, undergo diffusion. If a particular solute is present in high concentration in one area of a solution and in low concentration in another area, solute molecules will diffuse toward the area of lower concentration—they move *down their concentration gradient.* After some time, the particles become evenly distributed throughout the solution and the solution is said to be at equilibrium. The particles continue to move about randomly due to their kinetic energy, but their concentrations do not change.

For example, when you place a crystal of dye in a water-filled container (**Figure 3.4**), the color is most intense in the area closest to the dye because its concentration is higher there. At increasing distances, the color is lighter and lighter because the dye concentration is lower. Some time later, the solution of water and dye will have a uniform color, because the dye molecules and water molecules have diffused down their concentration gradients until they are evenly mixed in solution—they are at equilibrium (ē-kwi-LIB-rē-um).

In this simple example, no membrane was involved. Substances may also diffuse through a membrane, if the membrane is permeable to them. Several factors influence the diffusion rate of substances across plasma membranes:

- ***Steepness of the concentration gradient.*** The greater the difference in concentration between the two sides of the membrane, the higher the rate of diffusion. When charged particles are diffusing, the steepness of the electrochemical gradient determines the diffusion rate across the membrane.
- ***Temperature.*** The higher the temperature, the faster the rate of diffusion. All of the body's diffusion processes occur more rapidly in a person with a fever.
- ***Mass of the diffusing substance.*** The larger the mass of the diffusing particle, the slower its diffusion rate. Smaller molecules diffuse more rapidly than larger ones.

FIGURE 3.4 Principle of diffusion. At the beginning of our experiment, a crystal of dye placed in a cylinder of water dissolves (a) and then diffuses from the region of higher dye concentration to regions of lower dye concentration (b). At equilibrium (c), the dye concentration is uniform throughout, although random movement continues.

In diffusion, a substance moves down its concentration gradient.

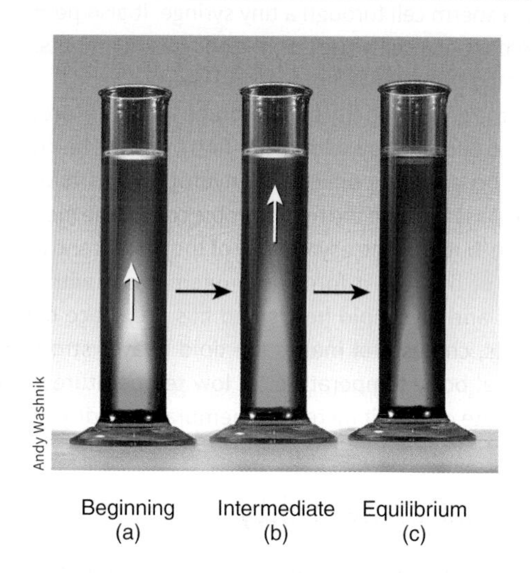

Q How would having a fever affect body processes that involve diffusion?

- ***Surface area.*** The larger the membrane surface area available for diffusion, the faster the diffusion rate. For example, the air sacs of the lungs have a large surface area available for diffusion of oxygen from the air into the blood. Some lung diseases, such as emphysema, reduce the surface area. This slows the rate of oxygen diffusion and makes breathing more difficult.
- ***Diffusion distance.*** The greater the distance over which diffusion must occur, the longer it takes. Diffusion across a plasma membrane takes only a fraction of a second because the membrane is so thin. In pneumonia, fluid collects in the lungs; the additional fluid increases the diffusion distance because oxygen must move through both the built-up fluid and the membrane to reach the bloodstream.

Now that you have a basic understanding of the nature of diffusion, we will consider three types of diffusion: simple diffusion, facilitated diffusion, and osmosis.

Simple Diffusion

Simple diffusion is a passive process in which substances move freely through the lipid bilayer of the plasma membranes of cells without the help of membrane transport proteins (**Figure 3.5**). Nonpolar, hydrophobic molecules move across the lipid bilayer through the process of simple diffusion. Such molecules include oxygen, carbon dioxide, and nitrogen gases; fatty acids; steroids; and fat-soluble vitamins (A, D, E, and K). Small, uncharged polar molecules such as water, urea, and small alcohols also pass through the lipid bilayer by simple diffusion. Simple diffusion through the lipid bilayer is important in the movement of oxygen and carbon dioxide between blood and body cells, and between blood and air

FIGURE 3.5 Simple diffusion, channel-mediated facilitated diffusion, and carrier-mediated facilitated diffusion.

In simple diffusion, a substance moves across the lipid bilayer of the plasma membrane without the help of membrane transport proteins. In facilitated diffusion, a substance moves across the lipid bilayer aided by a channel protein or a carrier protein.

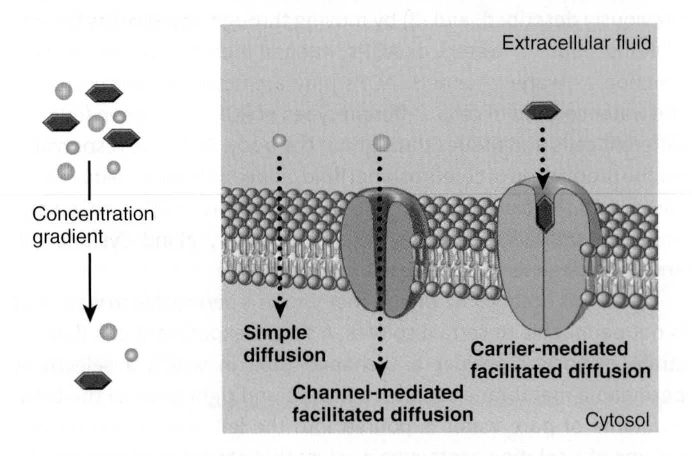

Q What types of molecules move across the lipid bilayer of the plasma membrane via simple diffusion?

FIGURE 3.6 Channel-mediated facilitated diffusion of potassium ions (K^+) through a gated K^+ channel. A gated channel is one in which a portion of the channel protein acts as a gate to open or close the channel's pore to the passage of ions.

Channels are integral membrane proteins that allow specific, small, inorganic ions to pass across the membrane by facilitated diffusion.

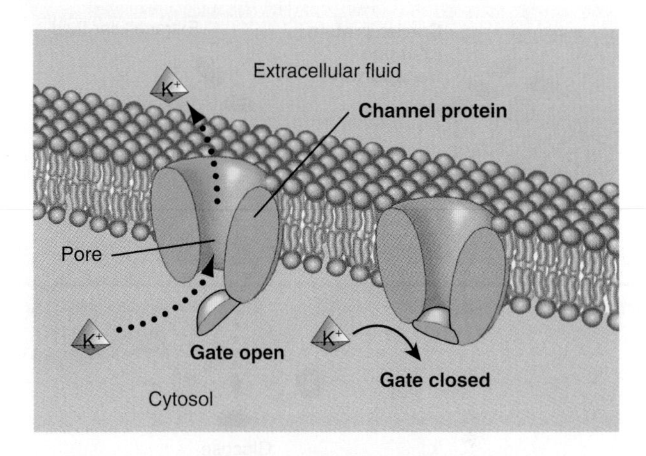

Details of the K^+ channel

Q Is the concentration of K^+ in body cells higher in the cytosol or in the extracellular fluid?

within the lungs during breathing. It also is the route for absorption of some nutrients and excretion of some wastes by body cells.

Facilitated Diffusion

Solutes that are too polar or highly charged to move through the lipid bilayer by simple diffusion can cross the plasma membrane by a passive process called **facilitated diffusion**. In this process, an integral membrane protein assists a specific substance across the membrane. The integral membrane protein can be either a membrane channel or a carrier.

Channel-Mediated Facilitated Diffusion In **channel-mediated facilitated diffusion**, a solute moves down its concentration gradient across the lipid bilayer through a membrane channel (**Figure 3.5**). Most membrane channels are *ion channels,* integral transmembrane proteins that allow passage of small, inorganic ions that are too hydrophilic to penetrate the nonpolar interior of the lipid bilayer. Each ion can diffuse across the membrane only at certain sites. In typical plasma membranes, the most numerous ion channels are selective for K^+ (potassium ions) or Cl^- (chloride ions); fewer channels are available for Na^+ (sodium ions) or Ca^{2+} (calcium ions). Diffusion of ions through channels is generally slower than free diffusion through the lipid bilayer because channels occupy a smaller fraction of the membrane's total surface area than lipids. Still, facilitated diffusion through channels is a very fast process: More than a million potassium ions can flow through a K^+ channel in one second!

A channel is said to be *gated* when part of the channel protein acts as a "plug" or "gate," changing shape in one way to open the pore and in another way to close it (**Figure 3.6**). Some gated channels randomly alternate between the open and closed positions; others are regulated by chemical or electrical changes inside and outside the cell. When the gates of a channel are open, ions diffuse into or out of cells, down their electrochemical gradients. The plasma membranes of different types of cells may have different numbers of ion channels and thus display different permeabilities to various ions.

Carrier-Mediated Facilitated Diffusion In **carrier-mediated facilitated diffusion**, a *carrier* (also called a *transporter*) moves a solute down its concentration gradient across the plasma membrane (see **Figure 3.5**). Since this is a passive process, no cellular energy is required. The solute binds to a specific carrier on one side of the membrane and is released on the other side after the carrier undergoes a change in shape. The solute binds more often to the carrier on the side of the membrane with a higher concentration of solute. Once the concentration is the same on both sides of the membrane, solute molecules bind to the carrier on the cytosolic side and move out to the extracellular fluid as rapidly as they bind to the carrier on the extracellular side and move into the cytosol. The rate of carrier-mediated facilitated diffusion (how quickly it occurs) is determined by the steepness of the concentration gradient across the membrane.

The number of carriers available in a plasma membrane places an upper limit, called the *transport maximum,* on the rate at which facilitated diffusion can occur. Once all of the carriers are occupied, the transport maximum is reached, and a further increase in the concentration gradient does not increase the rate of facilitated diffusion. Thus, much like a completely saturated sponge can absorb no more water, the process of carrier-mediated facilitated diffusion exhibits *saturation*.

FIGURE 3.7 Carrier-mediated facilitated diffusion of glucose across a plasma membrane. The carrier protein binds to glucose in the extracellular fluid and releases it into the cytosol.

Carriers are integral membrane proteins that undergo changes in shape in order to move substances across the membrane by facilitated diffusion.

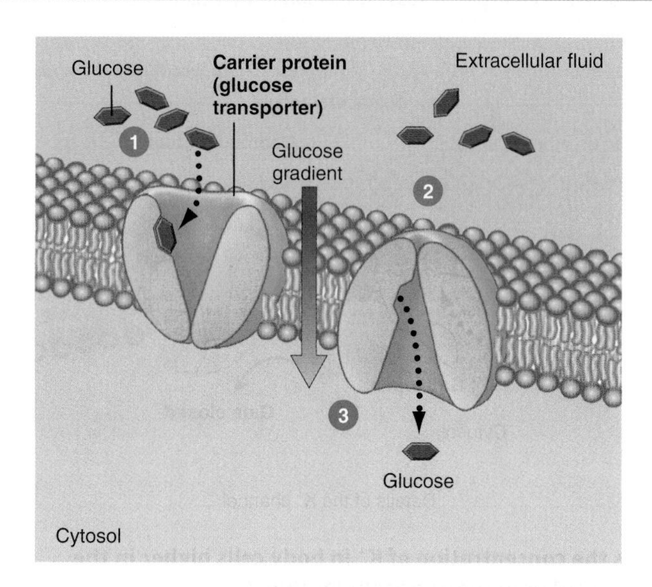

Q Does insulin alter glucose transport by facilitated diffusion?

Substances that move across the plasma membrane by carrier-mediated facilitated diffusion include glucose, fructose, galactose, and some vitamins. Glucose, the body's preferred energy source for making ATP, enters many body cells by carrier-mediated facilitated diffusion as follows (**Figure 3.7**):

1. Glucose binds to a specific type of carrier protein called the *glucose transporter* (GluT) on the outside surface of the membrane.
2. As the transporter undergoes a change in shape, glucose passes through the membrane.
3. The transporter releases glucose on the other side of the membrane.

The selective permeability of the plasma membrane is often regulated to achieve homeostasis. For instance, the hormone insulin, via the action of the insulin receptor, promotes the insertion of many copies of glucose transporters into the plasma membranes of certain cells. Thus, the effect of insulin is to elevate the transport maximum for facilitated diffusion of glucose into cells. With more glucose transporters available, body cells can pick up glucose from the blood more rapidly. An inability to produce or utilize insulin is called diabetes mellitus (Chapter 18).

Osmosis

Osmosis (oz-MŌ-sis) is a type of diffusion in which there is net movement of a solvent through a selectively permeable membrane. Like the other types of diffusion, osmosis is a passive process. In living systems, the solvent is water, which moves by osmosis across plasma membranes from an area of *higher water concentration* to an area of *lower water concentration*. Another way to understand this idea is to consider the solute concentration: In osmosis, water moves through a selectively permeable membrane from an area of *lower solute concentration* to an area of *higher solute concentration*. During osmosis, water molecules pass through a plasma membrane in two ways: (1) by moving between neighboring phospholipid molecules in the lipid bilayer via simple diffusion, as previously described, and (2) by moving through **aquaporins** (ak-wa-POR-ins; *aqua-* = water), or **AQPs**, integral membrane proteins that function as water channels. AQPs play a critical role in controlling the water content of cells. Different types of AQPs have been found in different cells and tissues throughout the body. AQPs are responsible for the production of cerebrospinal fluid, aqueous humor, tears, sweat, saliva, and the concentration of urine. Mutations of AQPs have been linked to cataracts, diabetes insipidus, salivary gland dysfunction, and neurodegenerative diseases.

Osmosis occurs only when a membrane is permeable to water but is not permeable to certain solutes. A simple experiment can demonstrate osmosis. Consider a U-shaped tube in which a selectively permeable membrane separates the left and right arms of the tube. A volume of pure water is poured into the left arm, and the same volume of a solution containing a solute that cannot pass through the membrane is poured into the right arm (**Figure 3.8a**). Because the *water* concentration is higher on the left and lower on the right, net movement of water molecules—osmosis—occurs from left to right, so that the water is moving down its concentration gradient. At the same time, the membrane prevents diffusion of the solute from the right arm into the left arm. As a result, the volume of water in the left arm decreases, and the volume of solution in the right arm increases (**Figure 3.8b**).

You might think that osmosis would continue until no water remained on the left side, but this is *not* what happens. In this experiment, the higher the column of solution in the right arm becomes, the more pressure it exerts on its side of the membrane. Pressure exerted in this way by a liquid, known as **hydrostatic pressure**, forces water molecules to move back into the left arm. Equilibrium is reached when just as many water molecules move from right to left due to the hydrostatic pressure as move from left to right due to osmosis (**Figure 3.8b**).

To further complicate matters, the solution with the impermeable solute also exerts a force, called the **osmotic pressure**. The osmotic pressure of a solution is proportional to the concentration of the solute particles that cannot cross the membrane—the higher the solute concentration, the higher the solution's osmotic pressure. Consider what would happen if a piston were used to apply more pressure to the fluid in the right arm of the tube in **Figure 3.8**. With enough pressure, the volume of fluid in each arm could be restored to the starting volume, and the concentration of solute in the right arm would be the same as it was at the beginning of the experiment (**Figure 3.8c**). The amount of pressure needed to restore the starting condition equals the osmotic pressure. So, in our experiment, osmotic pressure is the pressure needed to stop the movement of water from the left tube into the right tube. Notice that the osmotic pressure of a solution does not produce the movement of water during osmosis. Rather it is the pressure that would *prevent* such water movement.

FIGURE 3.8 **Principle of osmosis.** Water molecules move through the selectively permeable membrane; solute molecules cannot. (a) Water molecules move from the left arm into the right arm, down the water concentration gradient. (b) The volume of water in the left arm has decreased and the volume of solution in the right arm has increased. (c) Pressure applied to the solution in the right arm restores the starting conditions.

Osmosis is the movement of water molecules through a selectively permeable membrane.

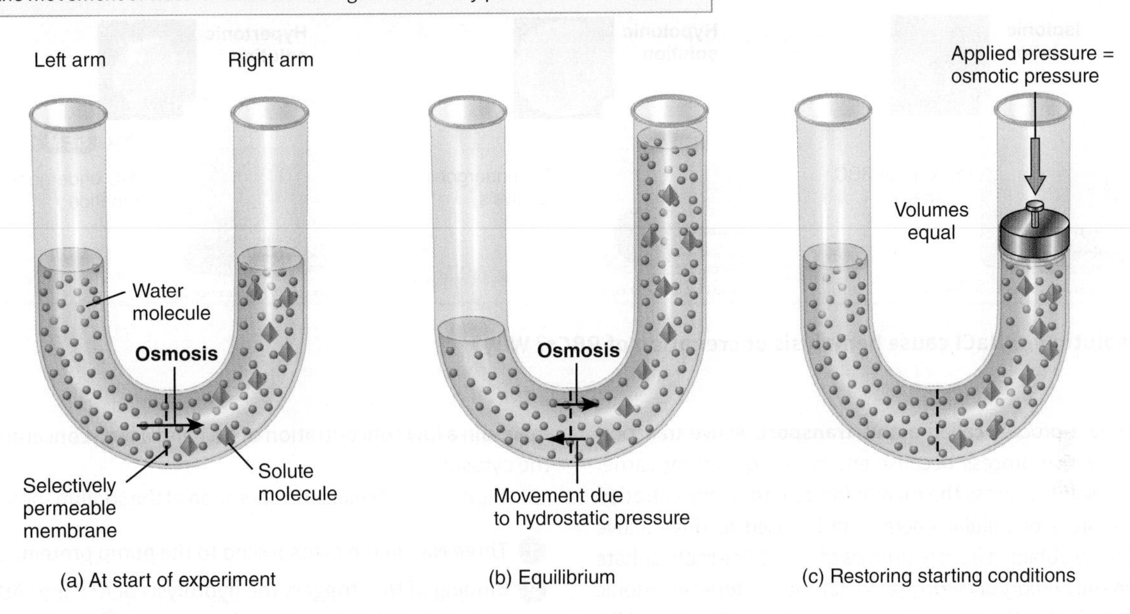

Q Will the fluid level in the right arm rise until the water concentrations are the same in both arms?

Normally, the osmotic pressure of the cytosol is the same as the osmotic pressure of the interstitial fluid outside cells. Because the osmotic pressure on both sides of the plasma membrane (which is selectively permeable) is the same, cell volume remains relatively constant. When body cells are placed in a solution having a different osmotic pressure than cytosol, however, the shape and volume of the cells change. As water moves by osmosis into or out of the cells, their volume increases or decreases. A solution's **tonicity** (tō-NIS-i-tē; *tonic* = tension) is a measure of the solution's ability to change the volume of cells by altering their water content.

Any solution in which a cell—for example, a red blood cell (RBC)—maintains its normal shape and volume is an **isotonic solution** (ī′-sō-TON-ik; *iso-* = same) (**Figure 3.9**). The concentrations of solutes that cannot cross the plasma membrane are the same on both sides of the membrane in this solution. For instance, a 0.9% NaCl solution (0.9 gram of sodium chloride in 100 mL of solution), called a *normal (physiological) saline solution,* is isotonic for RBCs. The RBC plasma membrane permits the water to move back and forth, but it behaves as though it is impermeable to Na^+ and Cl^-, the solutes. (Any Na^+ or Cl^- ions that enter the cell through channels or transporters are immediately moved back out by active transport or other means.) When RBCs are bathed in 0.9% NaCl, water molecules enter and exit at the same rate, allowing the RBCs to keep their normal shape and volume.

A different situation results if RBCs are placed in a **hypotonic solution** (hī′-pō-TON-ik; *hypo-* = less than), a solution that has a *lower* concentration of solutes than the cytosol inside the RBCs (**Figure 3.9**). In this case, water molecules enter the cells faster than they leave, causing the RBCs to swell and eventually to burst. The rupture of RBCs in this manner is called **hemolysis** (hē-MOL-i-sis; *hemo-* = blood; *-lysis* = to loosen or split apart); the rupture of other types of cells due to placement in a hypotonic solution is referred to simply as **lysis**. Pure water is very hypotonic and causes rapid hemolysis.

A **hypertonic solution** (hī′-per-TON-ik; *hyper-* = greater than) has a *higher* concentration of solutes than does the cytosol inside RBCs (**Figure 3.9**). One example of a hypertonic solution is a 2% NaCl solution. In such a solution, water molecules move out of the cells faster than they enter, causing the cells to shrink. Such shrinkage of cells is called **crenation** (kre-NĀ-shun).

See Clinical Connection: Medical Uses of Isotonic, Hypertonic, and Hypotonic Solutions

Active Processes

Active Transport Some polar or charged solutes that must enter or leave body cells cannot cross the plasma membrane through any form of passive transport because they would need to move "uphill," *against* their concentration gradients. Such solutes may be able to cross

FIGURE 3.9 Tonicity and its effects on red blood cells (RBCs). The arrows indicate the direction and degree of water movement into and out of the cells.

Cells placed in an isotonic solution maintain their shape because there is no net water movement into or out of the cells.

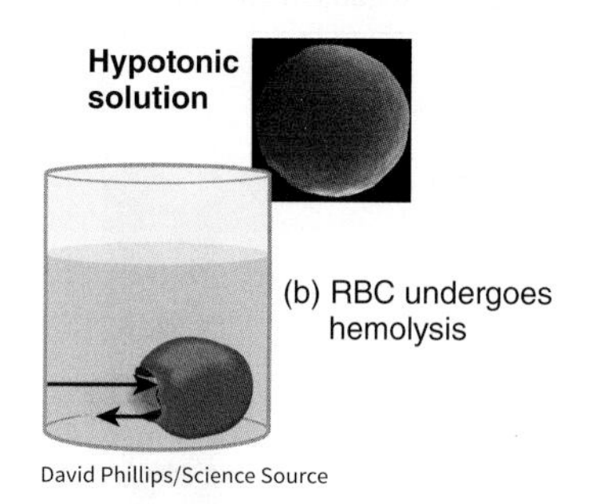

Q Will a 2% solution of NaCl cause hemolysis or crenation of RBCs? Why?

the membrane by a process called **active transport**. Active transport is considered an active process because energy is required for carrier proteins to move solutes across the membrane against a concentration gradient. Two sources of cellular energy can be used to drive active transport: (1) Energy obtained from hydrolysis of adenosine triphosphate (ATP) is the source in *primary active transport;* (2) energy stored in an ionic concentration gradient is the source in *secondary active transport.* Like carrier-mediated facilitated diffusion, active transport processes exhibit a transport maximum and saturation. Solutes actively transported across the plasma membrane include several ions, such as Na^+, K^+, H^+, Ca^{2+}, I^- (iodide ions), and Cl^-; amino acids; and monosaccharides. (Note that some of these substances also cross the membrane via facilitated diffusion when the proper channel proteins or carriers are present.)

Primary Active Transport In **primary active transport**, energy derived from hydrolysis of ATP changes the shape of a carrier protein, which "pumps" a substance across a plasma membrane against its concentration gradient. Indeed, carrier proteins that mediate primary active transport are often called **pumps**. A typical body cell expends about 40% of the ATP it generates on primary active transport. Chemicals that turn off ATP production—for example, the poison cyanide—are lethal because they shut down active transport in cells throughout the body.

The most prevalent primary active transport mechanism expels sodium ions (Na^+) from cells and brings potassium ions (K^+) in. Because of the specific ions it moves, this carrier is called the **sodium–potassium pump**. Because a part of the sodium–potassium pump acts as an *ATPase,* an enzyme that hydrolyzes ATP, another name for this pump is **Na^+–K^+ ATPase**. All cells have thousands of sodium–potassium pumps in their plasma membranes. These sodium–potassium pumps maintain a low concentration of Na^+ in the cytosol by pumping these ions into the extracellular fluid against the Na^+ concentration gradient. At the same time, the pumps move K^+ into cells against the K^+ concentration gradient. Because K^+ and Na^+ slowly leak back across the plasma membrane down their electrochemical gradients—through passive transport or secondary active transport—the sodium–potassium pumps must work nonstop to maintain a low concentration of Na^+ and a high concentration of K^+ in the cytosol.

Figure 3.10 depicts the operation of the sodium–potassium pump:

1. Three Na^+ in the cytosol bind to the pump protein.
2. Binding of Na^+ triggers the hydrolysis of ATP into ADP, a reaction that also attaches a phosphate group Ⓟ to the pump protein. This chemical reaction changes the shape of the pump protein, expelling the three Na^+ into the extracellular fluid. Now the shape of the pump protein favors binding of two K^+ in the extracellular fluid to the pump protein.
3. The binding of K^+ triggers release of the phosphate group from the pump protein. This reaction again causes the shape of the pump protein to change.
4. As the pump protein reverts to its original shape, it releases K^+ into the cytosol. At this point, the pump is again ready to bind three Na^+, and the cycle repeats.

The different concentrations of Na^+ and K^+ in cytosol and extracellular fluid are crucial for maintaining normal cell volume and for the ability of some cells to generate electrical signals such as action potentials. Recall that the tonicity of a solution is proportional to the concentration of its solute particles that cannot penetrate the membrane. Because sodium ions that diffuse into a cell or enter through secondary active transport are immediately pumped out, it is as if they never entered. In effect, sodium ions behave as if they cannot penetrate the membrane. Thus, sodium ions are an important contributor to the tonicity of the extracellular fluid. A similar condition holds for K^+ in the cytosol. By helping to maintain normal tonicity on each side of the plasma membrane, the sodium–potassium pumps ensure that cells neither shrink nor swell due to the movement of water by osmosis out of or into cells.

Secondary Active Transport In **secondary active transport**, the energy stored in a Na^+ or H^+ concentration gradient is used to drive

FIGURE 3.10 The sodium–potassium pump (Na^+–K^+ATPase) expels sodium ions (Na^+) and brings potassium ions (K^+) into the cell.

Sodium–potassium pumps maintain a low intracellular concentration of sodium ions.

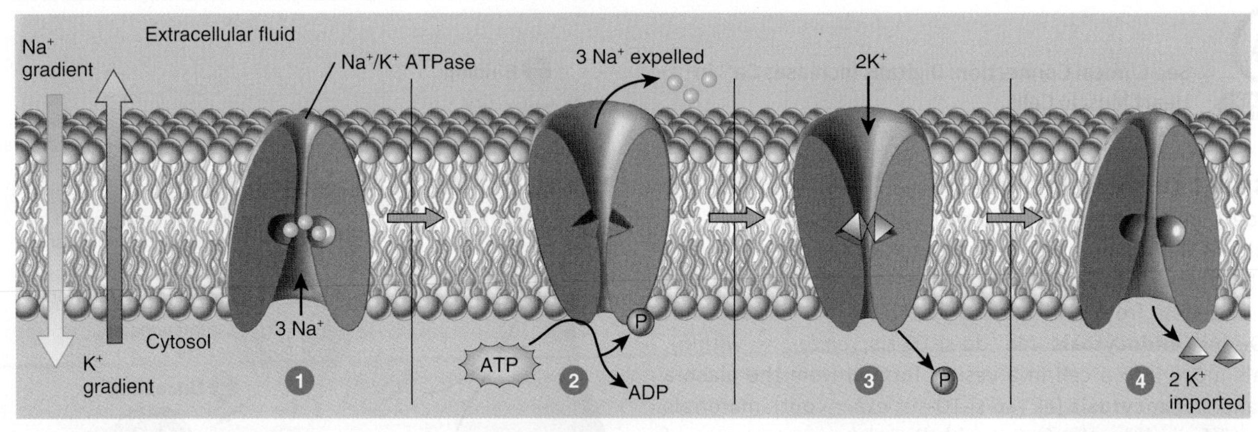

Q What is the role of ATP in the operation of this pump?

other substances across the membrane against their own concentration gradients. Because a Na^+ or H^+ gradient is established by primary active transport, secondary active transport *indirectly* uses energy obtained from the hydrolysis of ATP.

The sodium–potassium pump maintains a steep concentration gradient of Na^+ across the plasma membrane. As a result, the sodium ions have stored or potential energy, just like water behind a dam. Accordingly, if there is a route for Na^+ to leak back in, some of the stored energy can be converted to kinetic energy (energy of motion) and used to transport other substances *against their concentration gradients.* In essence, secondary active transport proteins harness the energy in the Na^+ concentration gradient by providing routes for Na^+ to leak into cells. In secondary active transport, a carrier protein simultaneously binds to Na^+ and another substance and then changes its shape so that both substances cross the membrane at the same time. If these transporters move two substances in the same direction they are called **symporters** (sim-PORT-ers; *sym-* = same); **antiporters** (an′-tē-PORT-ers), by contrast, move two substances in opposite directions across the membrane (*anti-* = against).

Plasma membranes contain several antiporters and symporters that are powered by the Na^+ gradient (**Figure 3.11**). For instance, the concentration of calcium ions (Ca^{2+}) is low in the cytosol because Na^+–Ca^{2+} antiporters eject calcium ions. Likewise, Na^+–H^+ antiporters help regulate the cytosol's pH (H^+ concentration) by expelling excess H^+. By contrast, dietary glucose and amino acids are absorbed into cells that line the small intestine by Na^+–glucose and Na^+–amino acid symporters (**Figure 3.11b**). In each case, sodium ions are moving down their concentration gradient while the other

FIGURE 3.11 Secondary active transport mechanisms. (a) Antiporters carry two substances across the membrane in opposite directions. (b) Symporters carry two substances across the membrane in the same direction.

Secondary active transport mechanisms use the energy stored in an ionic concentration gradient (here, for Na^+). Because primary active transport pumps that hydrolyze ATP maintain the gradient, secondary active transport mechanisms consume ATP indirectly.

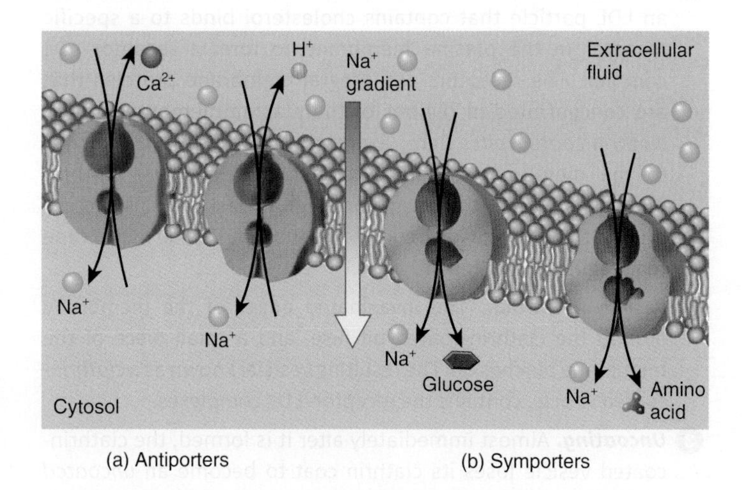

Q What is the main difference between primary and secondary active transport mechanisms?

solutes move "uphill," against their concentration gradients. Keep in mind that all these symporters and antiporters can do their job because the sodium–potassium pumps maintain a low concentration of Na^+ in the cytosol.

See Clinical Connection: Digitalis Increases Ca^{2+} in Heart Muscle Cells

Transport in Vesicles A **vesicle** (VES-i-kul = little blister or bladder), as noted earlier, is a small, spherical sac. As you will learn later in this chapter, a variety of substances are transported in vesicles from one structure to another within cells. Vesicles also import materials from and release materials into extracellular fluid. During **endocytosis** (en′-dō-sī-TŌ-sis; *endo-* = within), materials move into a cell in a vesicle formed from the plasma membrane. In **exocytosis** (ek′-sō-sī-TŌ-sis; *exo-* = out), materials move out of a cell by the fusion with the plasma membrane of vesicles formed inside the cell. Both endocytosis and exocytosis require energy supplied by ATP. Thus, transport in vesicles is an active process.

Endocytosis Here we consider three types of endocytosis: receptor-mediated endocytosis, phagocytosis, and bulk-phase endocytosis. **Receptor-mediated endocytosis** is a highly selective type of endocytosis by which cells take up specific ligands. (Recall that ligands are molecules that bind to specific receptors.) A vesicle forms after a receptor protein in the plasma membrane recognizes and binds to a particular particle in the extracellular fluid. For instance, cells take up cholesterol-containing low-density lipoproteins (LDLs), transferrin (an iron-transporting protein in the blood), some vitamins, antibodies, and certain hormones by receptor-mediated endocytosis. Receptor-mediated endocytosis of LDLs (and other ligands) occurs as follows (**Figure 3.12**):

1. ***Binding.*** On the extracellular side of the plasma membrane, an LDL particle that contains cholesterol binds to a specific receptor in the plasma membrane to form a receptor–LDL complex. The receptors are integral membrane proteins that are concentrated in regions of the plasma membrane called *clathrin-coated pits.* Here, a protein called *clathrin* attaches to the membrane on its cytoplasmic side. Many clathrin molecules come together, forming a basketlike structure around the receptor–LDL complexes that causes the membrane to invaginate (fold inward).
2. ***Vesicle formation.*** The invaginated edges of the membrane around the clathrin-coated pit fuse, and a small piece of the membrane pinches off. The resulting vesicle, known as a *clathrin-coated vesicle,* contains the receptor–LDL complexes.
3. ***Uncoating.*** Almost immediately after it is formed, the clathrin-coated vesicle loses its clathrin coat to become an *uncoated vesicle.* Clathrin molecules either return to the inner surface of the plasma membrane or help form coats on other vesicles inside the cell.

FIGURE 3.12 Receptor-mediated endocytosis of a low-density lipoprotein (LDL) particle.

Receptor-mediated endocytosis imports materials that are needed by cells.

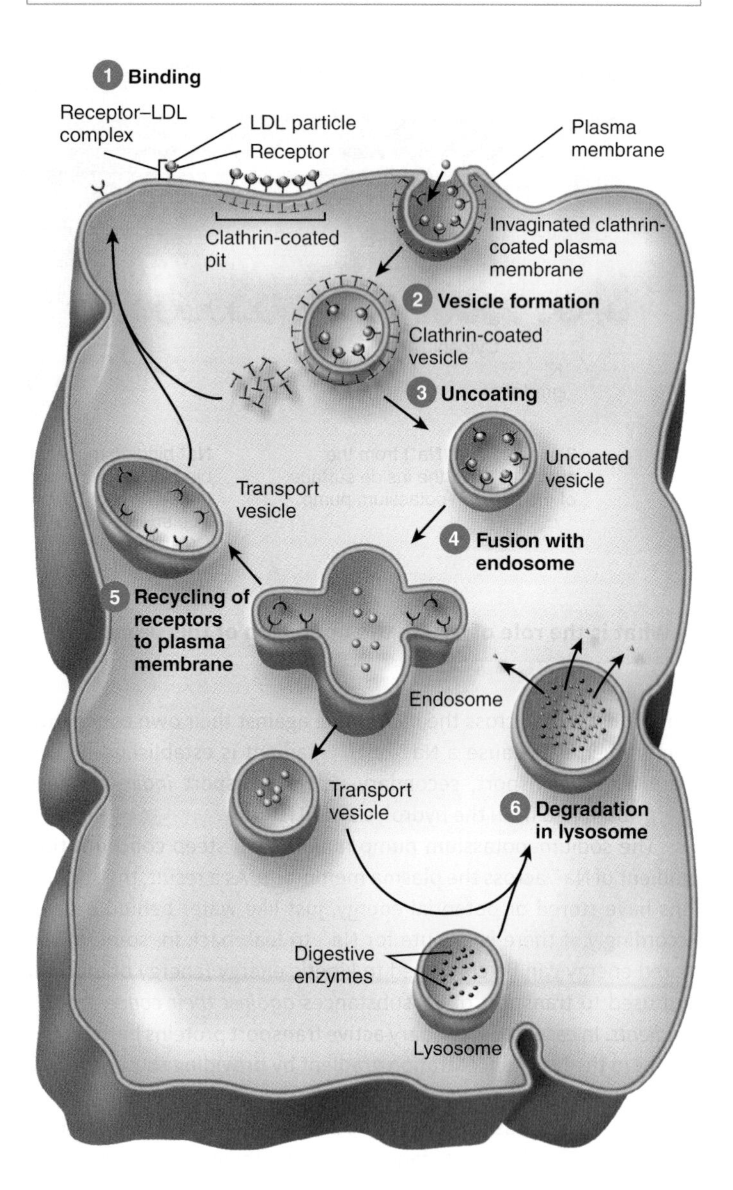

Q What are several other examples of ligands that can undergo receptor-mediated endocytosis?

4. ***Fusion with endosome.*** The uncoated vesicle quickly fuses with a vesicle known as an *endosome.* Within an endosome, the LDL particles separate from their receptors.
5. ***Recycling of receptors to plasma membrane.*** Most of the receptors accumulate in elongated protrusions of the endosome (the arms of the cross-shaped vesicle at the center of the figure). These pinch off, forming transport vesicles that return the receptors to the plasma membrane. An LDL receptor

is returned to the plasma membrane about 10 minutes after it enters a cell.

6 ***Degradation in lysosomes.*** Other transport vesicles, which contain the LDL particles, bud off the endosome and soon fuse with a *lysosome*. Lysosomes contain many digestive enzymes. Certain enzymes break down the large protein and lipid molecules of the LDL particle into amino acids, fatty acids, and cholesterol. These smaller molecules then leave the lysosome. The cell uses cholesterol for rebuilding its membranes and for synthesis of steroids, such as estrogen. Fatty acids and amino acids can be used for ATP production or to build other molecules needed by the cell.

See Clinical Connection: Viruses and Receptor-Mediated Endocytosis

Phagocytosis (fag′-ō-sī-TŌ-sis; *phago-* = to eat) or "cell eating" is a form of endocytosis in which the cell engulfs large solid particles, such as worn-out cells, whole bacteria, or viruses (**Figure 3.13**). Only a few body cells, termed **phagocytes** (FAG-ō-sīts), are able to carry out phagocytosis. Two main types of phagocytes are *macrophages,* located in many body tissues, and *neutrophils,* a type of white blood cell. Phagocytosis begins when the particle binds to a plasma membrane receptor on the phagocyte, causing it to extend **pseudopods** (SOO-dō-pods; *pseudo-* = false; *-pods* = feet), projections of its plasma membrane and cytoplasm. Pseudopods surround the particle outside the cell, and the membranes fuse to form a vesicle called a *phagosome,* which enters the cytoplasm. The phagosome fuses with one or more lysosomes, and lysosomal enzymes break down the ingested material. In most cases, any undigested materials in the phagosome remain indefinitely in a vesicle called a *residual body.* The residual bodies are then either secreted by the cell via exocytosis or they remain stored in the cell as lipofuscin granules.

Most body cells carry out **bulk-phase endocytosis**, also called *pinocytosis* (pi-nō-sī-TŌ-sis; *pino-* = to drink) or "cell drinking," a form of endocytosis in which tiny droplets of extracellular fluid are taken up (**Figure 3.14**). No receptor proteins are involved; all solutes dissolved in the extracellular fluid are brought into the cell. During bulk-phase endocytosis, the plasma membrane folds inward and forms a vesicle containing a droplet of extracellular fluid. The vesicle detaches or "pinches off" from the plasma membrane and enters the cytosol. Within the cell, the vesicle fuses with a lysosome, where enzymes degrade the engulfed solutes. The resulting smaller molecules, such as amino acids and fatty acids, leave the lysosome to be used elsewhere in the cell. Bulk-phase endocytosis occurs in most cells, especially absorptive cells in the intestines and kidneys.

EXOCYTOSIS In contrast with endocytosis, which brings materials into a cell, exocytosis releases materials from a cell. All cells carry out exocytosis, but it is especially important in two types of cells: (1) secretory cells that liberate digestive enzymes, hormones, mucus, or other secretions and (2) nerve cells that release substances called

FIGURE 3.13 Phagocytosis. Pseudopods surround a particle, and the membranes fuse to form a phagosome.

Phagocytosis is a vital defense mechanism that helps protect the body from disease.

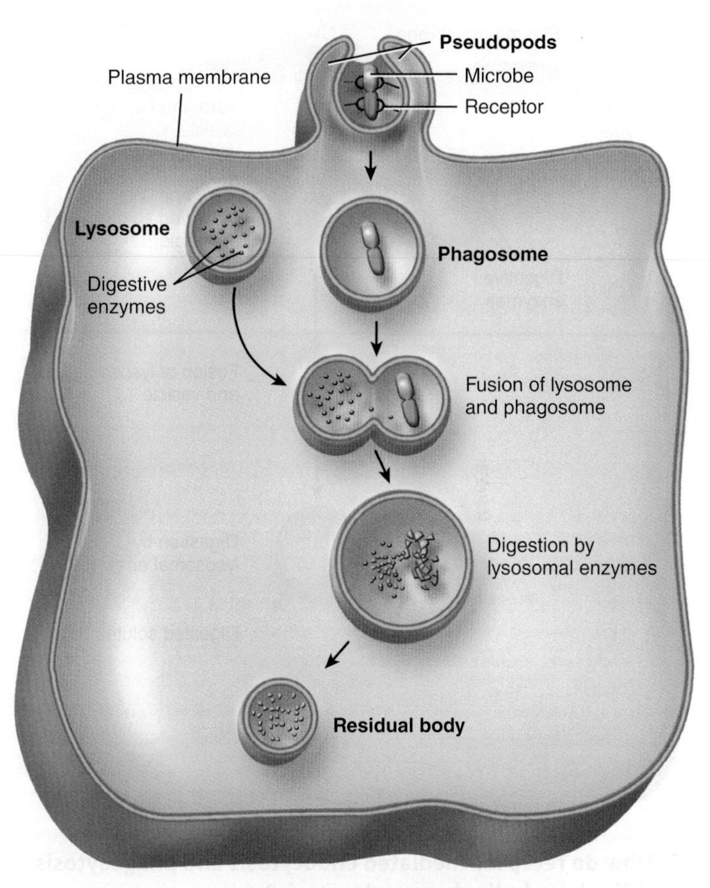

(a) Diagram of the process

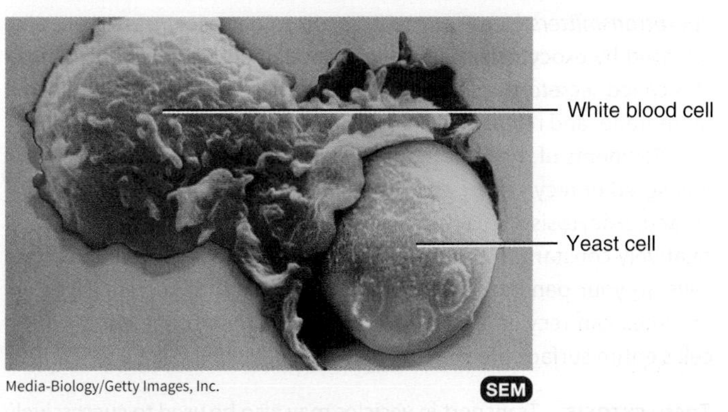

Media-Biology/Getty Images, Inc.

(b) White blood cell engulfing a yeast cell

See Clinical Connection: Phagocytosis and Microbes

Q What triggers pseudopod formation?

FIGURE 3.14 **Bulk-phase endocytosis.** The plasma membrane folds inward, forming a vesicle.

Most body cells carry out bulk-phase endocytosis, the nonselective uptake of tiny droplets of extracellular fluid.

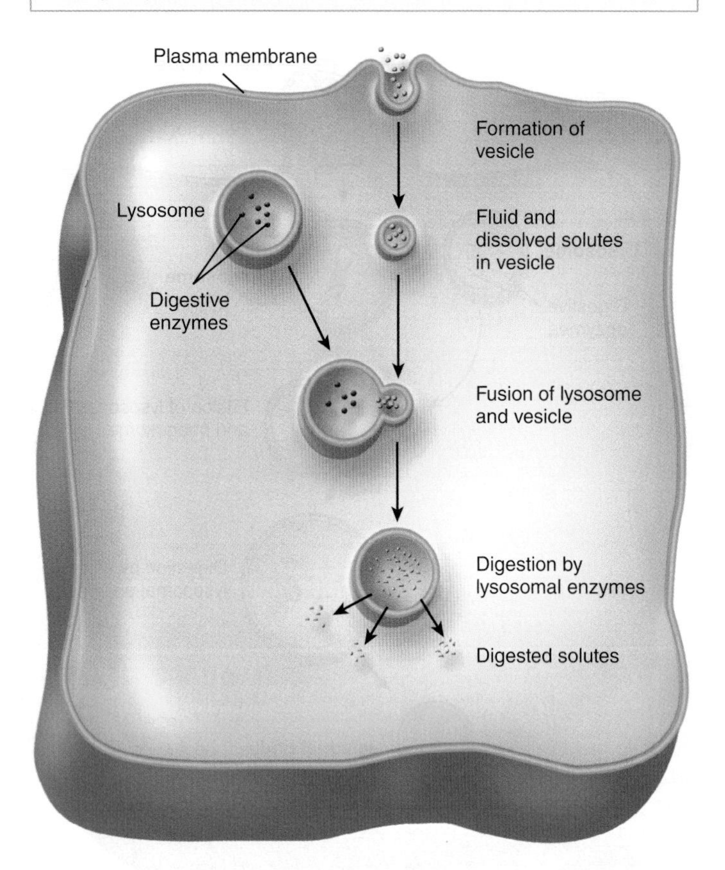

Q How do receptor-mediated endocytosis and phagocytosis differ from bulk-phase endocytosis?

neurotransmitters (see **Figure 12.23**). In some cases, wastes are also released by exocytosis. During exocytosis, membrane-enclosed vesicles called *secretory vesicles* form inside the cell, fuse with the plasma membrane, and release their contents into the extracellular fluid.

Segments of the plasma membrane lost through endocytosis are recovered or recycled by exocytosis. The balance between endocytosis and exocytosis keeps the surface area of a cell's plasma membrane relatively constant. Membrane exchange is quite extensive in certain cells. In your pancreas, for example, the cells that secrete digestive enzymes can recycle an amount of plasma membrane equal to the cell's entire surface area in 90 minutes.

TRANSCYTOSIS Transport in vesicles may also be used to successively move a substance into, across, and out of a cell. In this active process, called transcytosis (tranz'-sī-TŌ-sis), vesicles undergo endocytosis on one side of a cell, move across the cell, and then undergo exocytosis on the opposite side. As the vesicles fuse with the plasma membrane, the vesicular contents are released into the extracellular fluid. Transcytosis occurs most often across the endothelial cells that line blood vessels and is a means for materials to move between blood plasma and interstitial fluid. For instance, when a woman is pregnant, some of her antibodies cross the placenta into the fetal circulation via transcytosis.

Table 3.1 summarizes the processes by which materials move into and out of cells.

3.4 Cytoplasm

OBJECTIVE

- **Describe** the structure and function of cytoplasm, cytosol, and organelles.

Cytoplasm consists of all the cellular contents between the plasma membrane and the nucleus, and has two components: (1) the cytosol and (2) organelles, tiny structures that perform different functions in the cell.

Cytosol

The **cytosol** (*intracellular fluid*) is the fluid portion of the cytoplasm that surrounds organelles (see **Figure 3.1**) and constitutes about 55% of total cell volume. Although it varies in composition and consistency from one part of a cell to another, cytosol is 75–90% water plus various dissolved and suspended components. Among these are different types of ions, glucose, amino acids, fatty acids, proteins, lipids, ATP, and waste products, some of which we have already discussed. Also present in some cells are various organic molecules that aggregate into masses for storage. These aggregations may appear and disappear at different times in the life of a cell. Examples include *lipid droplets* that contain triglycerides, and clusters of glycogen molecules called *glycogen granules* (see **Figure 3.1**).

The cytosol is the site of many chemical reactions required for a cell's existence. For example, enzymes in cytosol catalyze *glycolysis*, a series of 10 chemical reactions that produce two molecules of ATP from one molecule of glucose (see **Figure 25.4**). Other types of cytosolic reactions provide the building blocks for maintenance of cell structures and for cell growth.

The **cytoskeleton** is a network of protein filaments that extends throughout the cytosol (see **Figure 3.1**). Three types of filaments contribute to the cytoskeleton's structure, as well as the structure of other organelles. In the order of their increasing diameter, these structures are microfilaments, intermediate filaments, and microtubules.

MICROFILAMENTS **Microfilaments** (mī-krō-FIL-a-ments) are the thinnest elements of the cytoskeleton. They are composed of the proteins *actin* and *myosin* and are most prevalent at the edge of a cell (**Figure 3.15a**). Microfilaments have two general functions: They help generate movement and provide mechanical support. With respect to movement, microfilaments are involved in muscle contraction, cell division, and cell locomotion, such as occurs during the migration of embryonic cells during development, the invasion of tissues by white blood cells to fight infection, or the migration of skin cells during wound healing.

TABLE 3.1 Transport of Materials into and out of Cells

TRANSPORT PROCESS	DESCRIPTION	SUBSTANCES TRANSPORTED
PASSIVE PROCESSES	Movement of substances down a concentration gradient until equilibrium is reached; do not require cellular energy in the form of ATP.	
Diffusion	Movement of molecules or ions down a concentration gradient due to their kinetic energy until they reach equilibrium.	
Simple diffusion	Passive movement of a substance down its concentration gradient through the lipid bilayer of the plasma membrane without the help of membrane transport proteins.	Nonpolar, hydrophobic solutes: oxygen, carbon dioxide, and nitrogen gases; fatty acids; steroids; and fat-soluble vitamins. Polar molecules such as water, urea, and small alcohols.
Facilitated diffusion	Passive movement of a substance down its concentration gradient through the lipid bilayer by transmembrane proteins that function as channels or carriers.	Polar or charged solutes: glucose; fructose; galactose; some vitamins; and ions such as K^+, Cl^-, Na^+, and Ca^{2+}.
Osmosis	Passive movement of water molecules across a selectively permeable membrane from an area of higher to lower water concentration until equilibrium is reached.	Solvent: water in living systems.
ACTIVE PROCESSES	Movement of substances against a concentration gradient; requires cellular energy in the form of ATP.	
Active Transport	Active process in which a cell expends energy to move a substance across the membrane against its concentration gradient by transmembrane proteins that function as carriers.	Polar or charged solutes.
Primary active transport	Active process in which a substance moves across the membrane against its concentration gradient by pumps (carriers) that use energy supplied by hydrolysis of ATP.	Na^+, K^+, Ca^{2+}, H^+, I^-, Cl^-, and other ions.
Secondary active transport	Coupled active transport of two substances across the membrane using energy supplied by a Na^+ or H^+ concentration gradient maintained by primary active transport pumps. Antiporters move Na^+ (or H^+) and another substance in opposite directions across the membrane; symporters move Na^+ (or H^+) and another substance in the same direction across the membrane.	Antiport: Ca^{2+}, H^+ out of cells. Symport: glucose, amino acids into cells.
Transport in Vesicles	Active process in which substances move into or out of cells in vesicles that bud from plasma membrane; requires energy supplied by ATP.	
Endocytosis	Movement of substances into a cell in vesicles.	
Receptor-mediated endocytosis	Ligand–receptor complexes trigger infolding of a clathrin-coated pit that forms a vesicle containing ligands.	Ligands: transferrin, low-density lipoproteins. (LDLs), some vitamins, certain hormones, and antibodies.
Phagocytosis	"Cell eating"; movement of a solid particle into a cell after pseudopods engulf it to form a phagosome.	Bacteria, viruses, and aged or dead cells.
Bulk-phase endocytosis	"Cell drinking"; movement of extracellular fluid into a cell by infolding of plasma membrane to form a vesicle.	Solutes in extracellular fluid.
Exocytosis	Movement of substances out of a cell in secretory vesicles that fuse with the plasma membrane and release their contents into the extracellular fluid.	Neurotransmitters, hormones, and digestive enzymes.
Transcytosis	Movement of a substance through a cell as a result of endocytosis on one side and exocytosis on the opposite side.	Substances, such as antibodies, across endothelial cells. This is a common route for substances to pass between blood plasma and interstitial fluid.

Microfilaments provide much of the mechanical support that is responsible for the basic strength and shapes of cells. They anchor the cytoskeleton to integral proteins in the plasma membrane. Microfilaments also provide mechanical support for cell extensions called **microvilli** (mī-krō-VIL-ī; *micro-* = small; *-villi* = tufts of hair; singular is *microvillus*), nonmotile, microscopic fingerlike projections of the plasma membrane. Within each microvillus is a core of parallel microfilaments that supports it. Because they greatly increase the surface area of the cell, microvilli are abundant on cells involved in absorption, such as the epithelial cells that line the small intestine.

INTERMEDIATE FILAMENTS As their name suggests, **intermediate filaments** are thicker than microfilaments but thinner than microtubules (**Figure 3.15b**). Several different proteins can compose intermediate filaments, which are exceptionally strong. They are found in parts of cells subject to mechanical stress; they help stabilize the position of organelles such as the nucleus and help attach cells to one another.

MICROTUBULES **Microtubules** (mī-krō-TOO-būls′), the largest of the cytoskeletal components, are long, unbranched hollow tubes composed mainly of the protein *tubulin*. The assembly of microtubules begins in an organelle called the centrosome (discussed shortly). The microtubules

FIGURE 3.15 **Cytoskeleton.**

The cytoskeleton is a network of three types of protein filaments—microfilaments, intermediate filaments, and microtubules—that extend throughout the cytoplasm.

Functions of the Cytoskeleton

1. Serves as a scaffold that helps determine a cell's shape and organize the cellular contents.
2. Aids movement of organelles within the cell, of chromosomes during cell division, and of whole cells such as phagocytes.

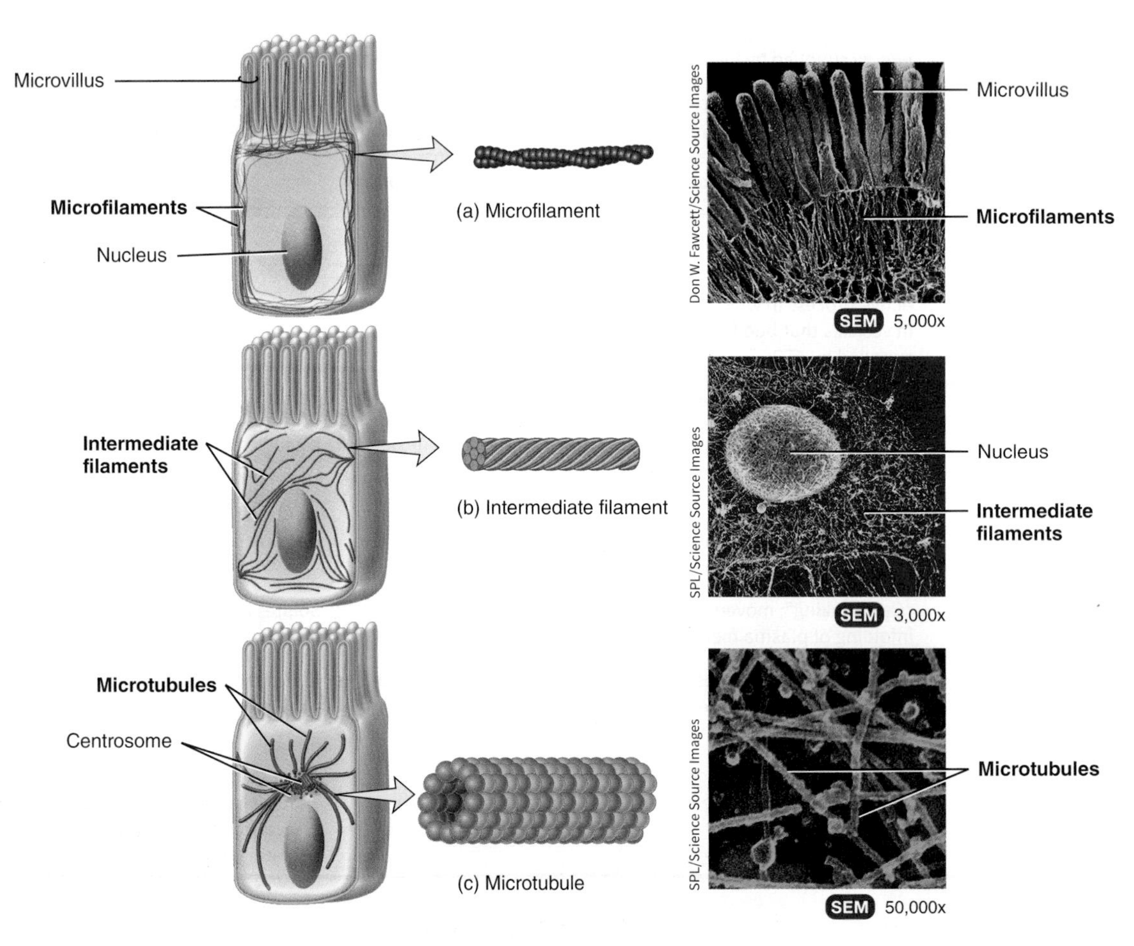

Q Which cytoskeletal component helps form the structure of centrioles, cilia, and flagella?

grow outward from the centrosome toward the periphery of the cell (**Figure 3.15c**). Microtubules help determine cell shape. They also function in the movement of organelles such as secretory vesicles, of chromosomes during cell division, and of specialized cell projections, such as cilia and flagella.

Organelles

As noted earlier, **organelles** are specialized structures within the cell that have characteristic shapes, and they perform specific functions in cellular growth, maintenance, and reproduction. Despite the many chemical reactions going on in a cell at any given time, there is little interference among reactions because they are confined to different organelles. Each type of organelle has its own set of enzymes that carry out specific reactions, and serves as a functional compartment for specific biochemical processes. The numbers and types of organelles vary in different cells, depending on the cell's function. Although they have different functions, organelles often cooperate to maintain homeostasis. Even though the nucleus is a large organelle, it is discussed in a separate section because of its special importance in directing the life of a cell.

FIGURE 3.16 Centrosome.

Located near the nucleus, the centrosome consists of a pair of centrioles and the pericentriolar matrix.

Functions of the Centrosomes

1. The pericentriolar matrix of the centrosome contains tubulins that build microtubules in nondividing cells.
2. The pericentriolar matrix of the centrosome forms the mitotic spindle during cell division.

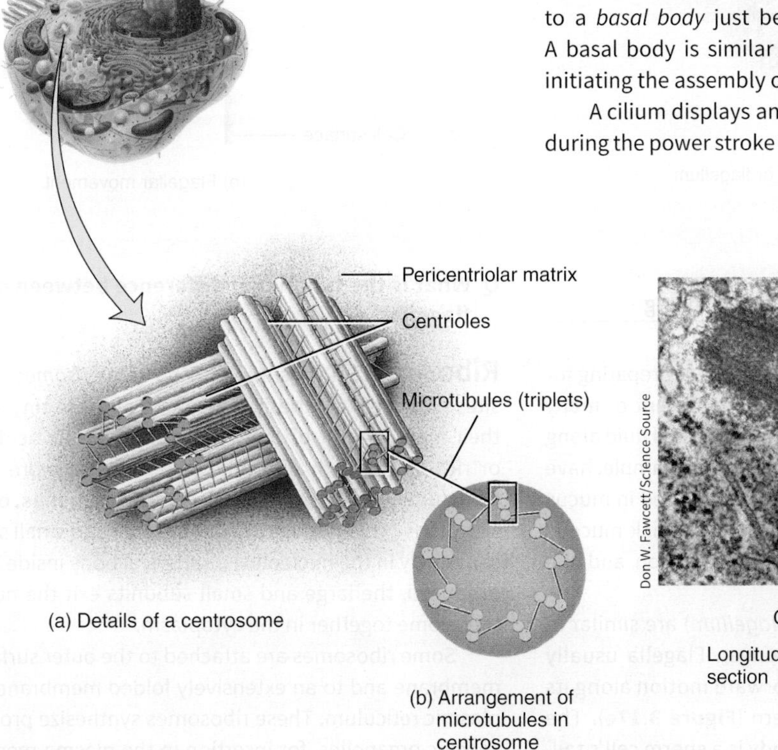

(a) Details of a centrosome

(b) Arrangement of microtubules in centrosome

(c) Centrioles

Q If you observed that a cell did not have a centrosome, what could you predict about its capacity for cell division?

Centrosome The **centrosome** (SEN-trō-sōm), or *microtubule organizing center*, located near the nucleus, consists of two components: a pair of centrioles and the pericentriolar matrix (**Figure 3.16a**). The two **centrioles** (SEN-trē-ōls) are cylindrical structures, each composed of nine clusters of three microtubules (triplets) arranged in a circular pattern (**Figure 3.16b**). The long axis of one centriole is at a right angle to the long axis of the other (**Figure 3.16c**). Surrounding the centrioles is the **pericentriolar matrix** (per′-ē-sen′-trē-Ō-lar), which contains hundreds of ring-shaped complexes composed of the protein *tubulin*. These tubulin complexes are the organizing centers for growth of the mitotic spindle, which plays a critical role in cell division, and for microtubule formation in nondividing cells. During cell division, centrosomes replicate so that succeeding generations of cells have the capacity for cell division.

Cilia and Flagella Microtubules are the dominant components of cilia and flagella, which are motile projections of the cell surface. **Cilia** (SIL-ē-a = eyelashes; singular is *cilium*) are numerous, short, hairlike projections that extend from the surface of the cell (see **Figures 3.1** and **3.17b**). Each cilium contains a core of 20 microtubules surrounded by plasma membrane (**Figure 3.17a**). The microtubules are arranged such that one pair in the center is surrounded by nine clusters of two fused microtubules (doublets). Each cilium is anchored to a *basal body* just below the surface of the plasma membrane. A basal body is similar in structure to a centriole and functions in initiating the assembly of cilia and flagella.

A cilium displays an oarlike pattern of beating; it is relatively stiff during the power stroke (oar digging into the water), but more flexible

FIGURE 3.17 **Cilia and flagella.**

A cilium contains a core of microtubules with one pair in the center surrounded by nine clusters of doublet microtubules.

Functions of the Cilia and Flagella

1. Cilia move fluids along a cell's surface.

2. A flagellum moves an entire cell.

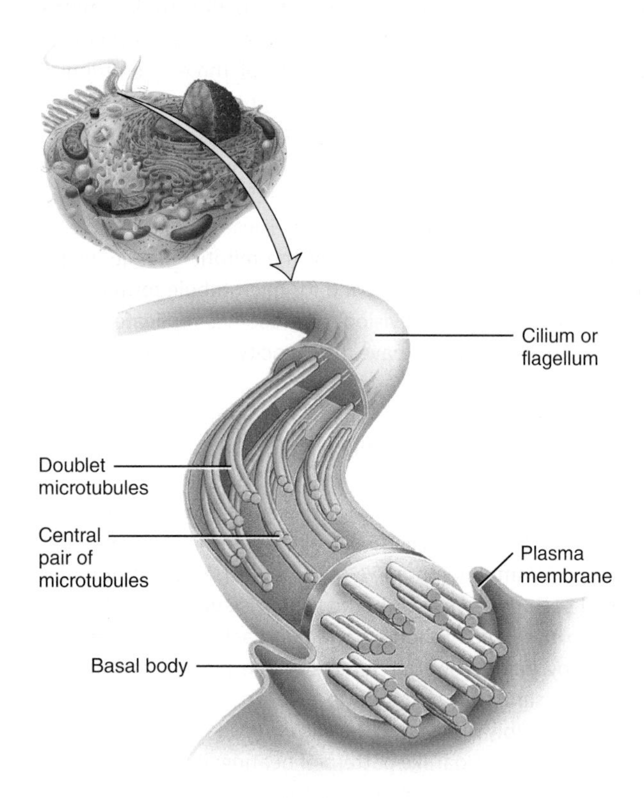

(a) Arrangement of microtubules in a cilium or flagellum

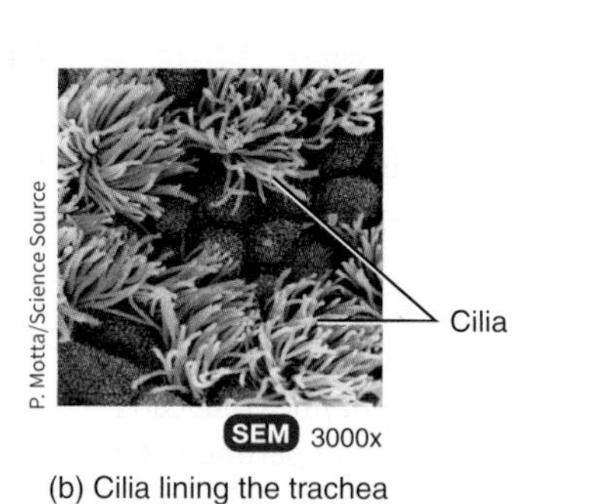

(b) Cilia lining the trachea

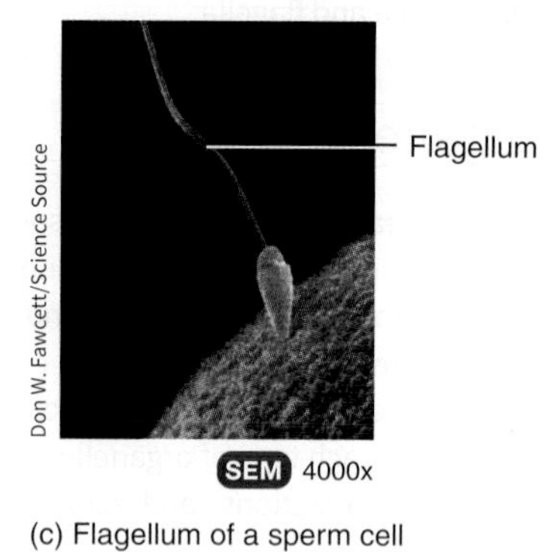

(c) Flagellum of a sperm cell

Movement of liquid

Cilium

1 2 3 4 5 6 7 8

Cell surface

Power stroke

Recovery stroke

(d) Ciliary movement

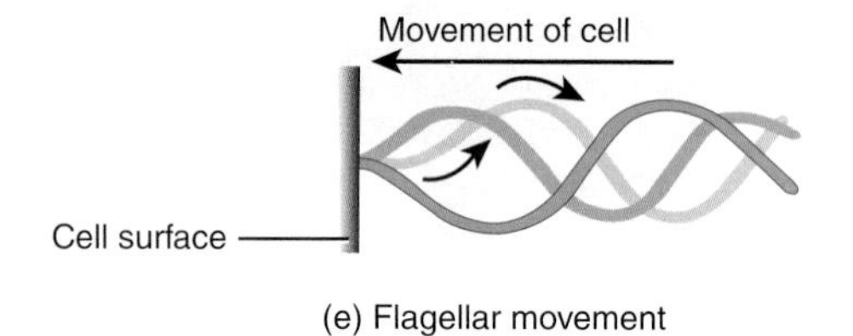

(e) Flagellar movement

See Clinical Connection: Cilia and Smoking

Q What is the functional difference between cilia and flagella?

during the recovery stroke (oar moving above the water preparing for a new stroke) (**Figure 3.17d**). The coordinated movement of many cilia on the surface of a cell causes the steady movement of fluid along the cell's surface. Many cells of the respiratory tract, for example, have hundreds of cilia that help sweep foreign particles trapped in mucus away from the lungs. In cystic fibrosis, the extremely thick mucous secretions that are produced interfere with ciliary action and the normal functions of the respiratory tract.

Flagella (fla-JEL-a = whip; singular is *flagellum*) are similar in structure to cilia but are typically much longer. Flagella usually move an entire cell. A flagellum generates forward motion along its axis by rapidly wiggling in a wavelike pattern (**Figure 3.17e**). The only example of a flagellum in the human body is a sperm cell's tail, which propels the sperm toward the oocyte in the uterine tube (**Figure 3.17c**).

Ribosomes

Ribosomes (RĪ-bō-sōms; *-somes* = bodies) are the sites of protein synthesis. The name of these tiny structures reflects their high content of one type of ribonucleic acid (ribosomal RNA, or rRNA), but each ribosome also includes more than 50 proteins. Structurally, a ribosome consists of two subunits, one about half the size of the other (**Figure 3.18**). The large and small subunits are made separately in the nucleolus, a spherical body inside the nucleus. Once produced, the large and small subunits exit the nucleus separately, then come together in the cytoplasm.

Some ribosomes are attached to the outer surface of the nuclear membrane and to an extensively folded membrane called the endoplasmic reticulum. These ribosomes synthesize proteins destined for specific organelles, for insertion in the plasma membrane, or for export from the cell. Other ribosomes are "free" or unattached to other cytoplasmic structures. Free ribosomes synthesize proteins used in

FIGURE 3.18 **Ribosomes.**

Ribosomes are the sites of protein synthesis.

Functions of Ribosomes

1. Ribosomes associated with endoplasmic reticulum synthesize proteins destined for insertion in the plasma membrane or secretion from the cell.
2. Free ribosomes synthesize proteins used in the cytosol.

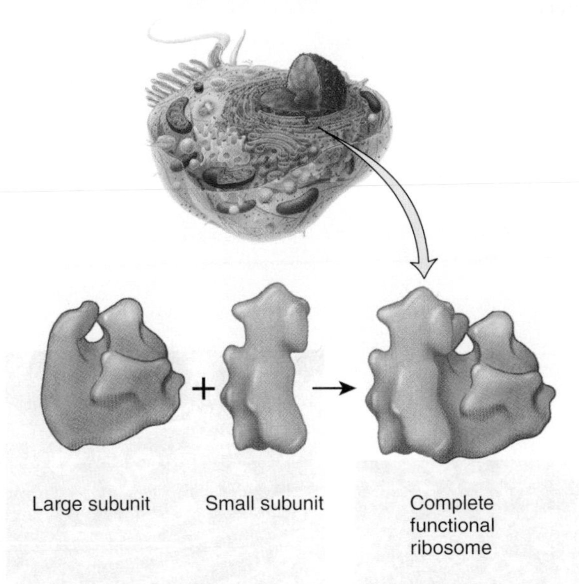

(a) Details of ribosomal subunits

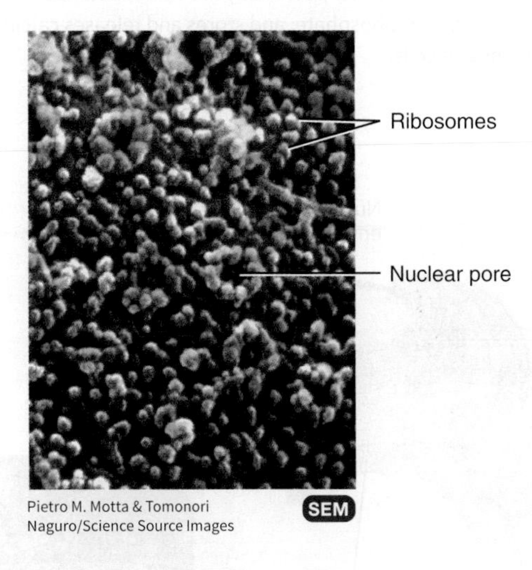

Pietro M. Motta & Tomonori Naguro/Science Source Images

(b) SEM of ribosomes & pores on nuclear membrane

Q Where are subunits of ribosomes synthesized and assembled?

the cytosol. Ribosomes are also located within mitochondria, where they synthesize mitochondrial proteins.

Endoplasmic Reticulum

The **endoplasmic reticulum (ER)** (en′-dō-PLAS-mik re-TIK-ū-lum; *-plasmic* = cytoplasm; *reticulum* = network) is a network of membranes in the form of flattened sacs or tubules (**Figure 3.19**). The ER extends from the nuclear envelope (membrane around the nucleus), to which it is connected and projects throughout the cytoplasm. The ER is so extensive that it constitutes more than half of the membranous surfaces within the cytoplasm of most cells.

Cells contain two distinct forms of ER, which differ in structure and function. **Rough ER** is continuous with the nuclear membrane and usually is folded into a series of flattened sacs. The outer surface of rough ER is studded with ribosomes, the sites of protein synthesis. Proteins synthesized by ribosomes attached to rough ER enter spaces within the ER for processing and sorting. In some cases, enzymes attach the proteins to carbohydrates to form glycoproteins. In other cases, enzymes attach the proteins to phospholipids, also synthesized by rough ER. These molecules (glycoproteins and phospholipids) may be incorporated into the membranes of organelles, inserted into the plasma membrane, or secreted via exocytosis. Thus rough ER produces secretory proteins, membrane proteins, and many organellar proteins.

Smooth ER extends from the rough ER to form a network of membrane tubules (**Figure 3.19**). Unlike rough ER, smooth ER does not have ribosomes on the outer surfaces of its membrane. However, smooth ER contains unique enzymes that make it functionally more diverse than rough ER. Because it lacks ribosomes, smooth ER does not synthesize proteins, but it does synthesize fatty acids and steroids, such as estrogens and testosterone. In liver cells, enzymes of the smooth ER help release glucose into the bloodstream and inactivate or detoxify lipid-soluble drugs or potentially harmful substances, such as alcohol, pesticides, and *carcinogens* (cancer-causing agents). In liver, kidney, and intestinal cells, a smooth ER enzyme removes the phosphate group from glucose-6-phosphate, which allows the "free" glucose to enter the bloodstream. In muscle cells, the calcium ions (Ca^{2+}) that trigger contraction are released from the sarcoplasmic reticulum, a form of smooth ER.

See Clinical Connection: Smooth ER and Drug Tolerance

Golgi Complex

Most of the proteins synthesized by ribosomes attached to rough ER are ultimately transported to other regions of the cell. The first step in the transport pathway is through an organelle called the **Golgi complex** (GOL-jē). It consists of 3 to 20 **cisterns** (sis-TER-nē = cavities; singular is *cistern*), small, flattened membranous sacs with bulging edges that resemble a stack of pita bread

FIGURE 3.19 **Endoplasmic reticulum.**

The endoplasmic reticulum is a network of membrane-enclosed sacs or tubules that extend throughout the cytoplasm and connect to the nuclear envelope.

Functions of Endoplasmic Reticulum

1. Rough ER synthesizes glycoproteins and phospholipids that are transferred into cellular organelles, inserted into the plasma membrane, or secreted during exocytosis.
2. Smooth ER synthesizes fatty acids and steroids, such as estrogens and testosterone; inactivates or detoxifies drugs and other potentially harmful substances; removes the phosphate group from glucose-6-phosphate; and stores and releases calcium ions that trigger contraction in muscle cells.

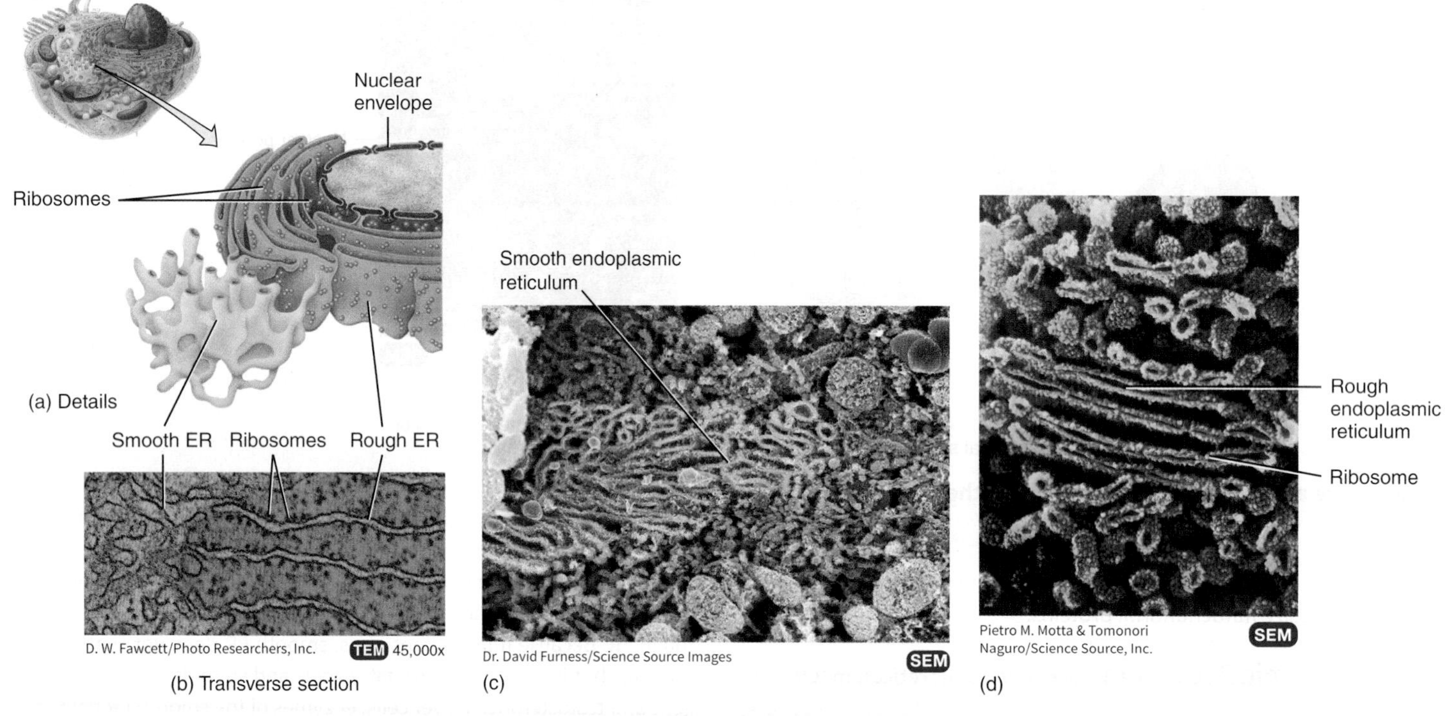

Q What are the structural and functional differences between rough and smooth ER?

(Figure 3.20). The cisterns are often curved, giving the Golgi complex a cuplike shape. Most cells have several Golgi complexes, and Golgi complexes are more extensive in cells that secrete proteins, a clue to the organelle's role in the cell.

The cisterns at the opposite ends of a Golgi complex differ from each other in size, shape, and enzymatic activity. The convex **entry *(cis)* face** is a cistern that faces the rough ER. The concave **exit *(trans)* face** is a cistern that faces the plasma membrane. Sacs between the entry and exit faces are called **medial cisterns**. Transport vesicles (described shortly) from the ER merge to form the entry face. From the entry face, the cisterns are thought to mature, in turn becoming medial and then exit cisterns.

Different enzymes in the entry, medial, and exit cisterns of the Golgi complex permit each of these areas to modify, sort, and package proteins into vesicles for transport to different destinations. The entry face receives and modifies proteins produced by the rough ER. The medial cisterns add carbohydrates to proteins to form glycoproteins and lipids to proteins to form lipoproteins. The exit face modifies the molecules further and then sorts and packages them for transport to their destinations.

Proteins arriving at, passing through, and exiting the Golgi complex do so through maturation of the cisternae and exchanges that occur via transfer vesicles (Figure 3.21):

1. Proteins synthesized by ribosomes on the rough ER are surrounded by a piece of the ER membrane, which eventually buds from the membrane surface to form transport vesicles.
2. Transport vesicles move toward the entry face of the Golgi complex.
3. Fusion of several transport vesicles creates the entry face of the Golgi complex and releases proteins into its lumen (space).
4. The proteins move from the entry face into one or more medial cisterns. Enzymes in the medial cisterns modify the proteins to form glycoproteins, glycolipids, and lipoproteins. **Transfer vesicles** that bud from the edges of the cisterns move specific

FIGURE 3.20 **Golgi complex.**

The opposite faces of a Golgi complex differ in size, shape, content, and enzymatic activities.

Functions of the Golgi Complex

1. Modifies, sorts, packages, and transports proteins received from the rough ER.
2. Forms secretory vesicles that discharge processed proteins via exocytosis into extracellular fluid; forms membrane vesicles that ferry new molecules to the plasma membrane; forms transport vesicles that carry molecules to other organelles, such as lysosomes.

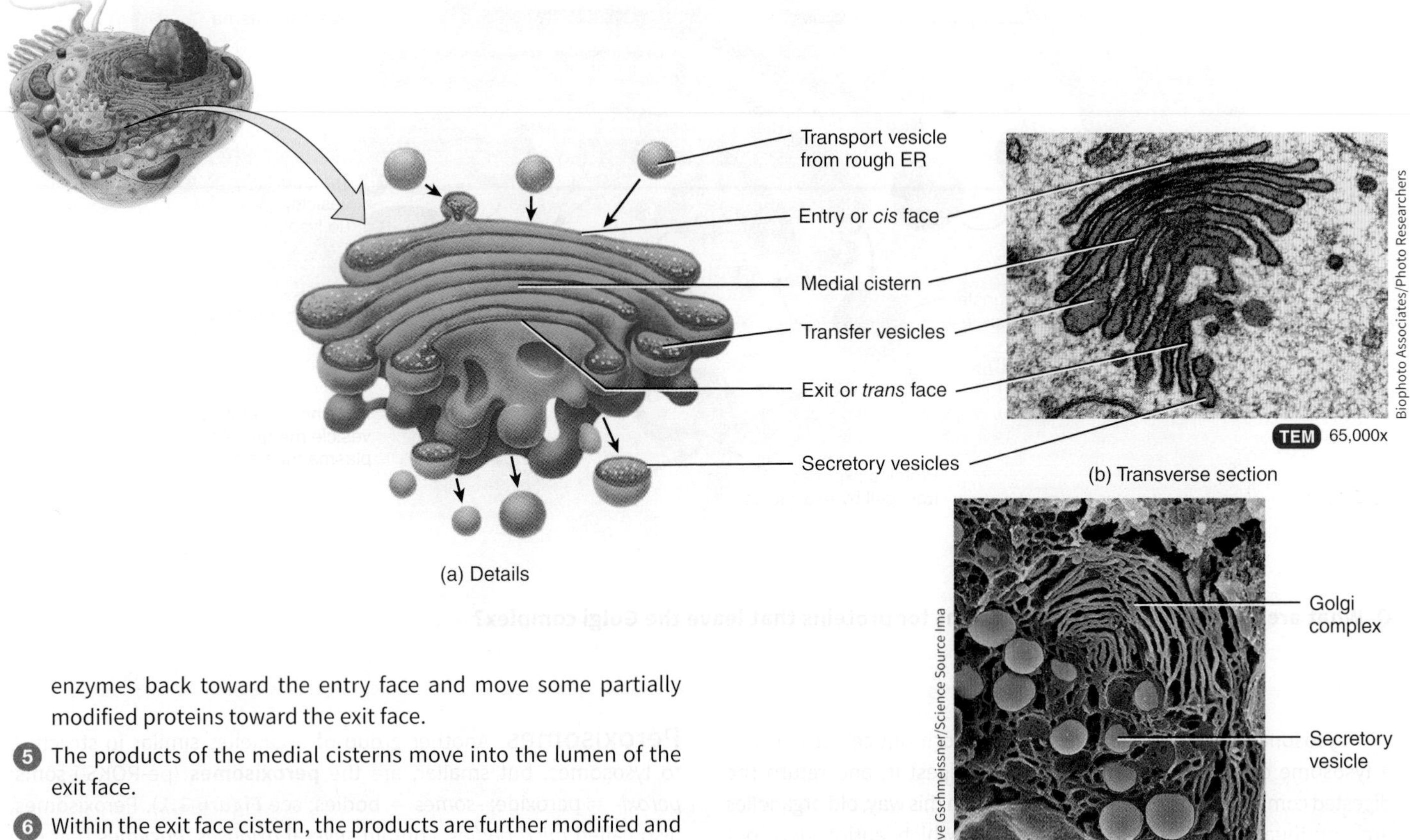

Q How do the entry and exit faces differ in function?

enzymes back toward the entry face and move some partially modified proteins toward the exit face.

5. The products of the medial cisterns move into the lumen of the exit face.
6. Within the exit face cistern, the products are further modified and are sorted and packaged.
7. Some of the processed proteins leave the exit face and are stored in **secretory vesicles**. These vesicles deliver the proteins to the plasma membrane, where they are discharged by exocytosis into the extracellular fluid. For example, certain pancreatic cells release the hormone insulin in this way.
8. Other processed proteins leave the exit face in **membrane vesicles** that deliver their contents to the plasma membrane for incorporation into the membrane. In doing so, the Golgi complex adds new segments of plasma membrane as existing segments are lost and modifies the number and distribution of membrane molecules.
9. Finally, some processed proteins leave the exit face in transport vesicles that will carry the proteins to another cellular destination. For instance, transport vesicles carry digestive enzymes to lysosomes; the structure and functions of these important organelles are discussed next.

Lysosomes

Lysosomes (LĪ-sō-sōms; *lyso-* = dissolving; *-somes* = bodies) are membrane-enclosed vesicles that form from the Golgi complex (**Figure 3.22**). They can contain as many as 60 kinds of powerful digestive and hydrolytic enzymes that can break down a wide variety of molecules once lysosomes fuse with vesicles formed during endocytosis. Because lysosomal enzymes work best at an acidic pH, the lysosomal membrane includes active transport pumps that import hydrogen ions (H^+). Thus, the lysosomal interior has a pH of 5, which is 100 times more acidic than the pH of the cytosol (pH 7). The lysosomal membrane also includes transporters that move the final products of digestion, such as glucose, fatty acids, and amino acids, into the cytosol.

FIGURE 3.21 Processing and packaging of proteins by the Golgi complex.

All proteins exported from the cell are processed in the Golgi complex.

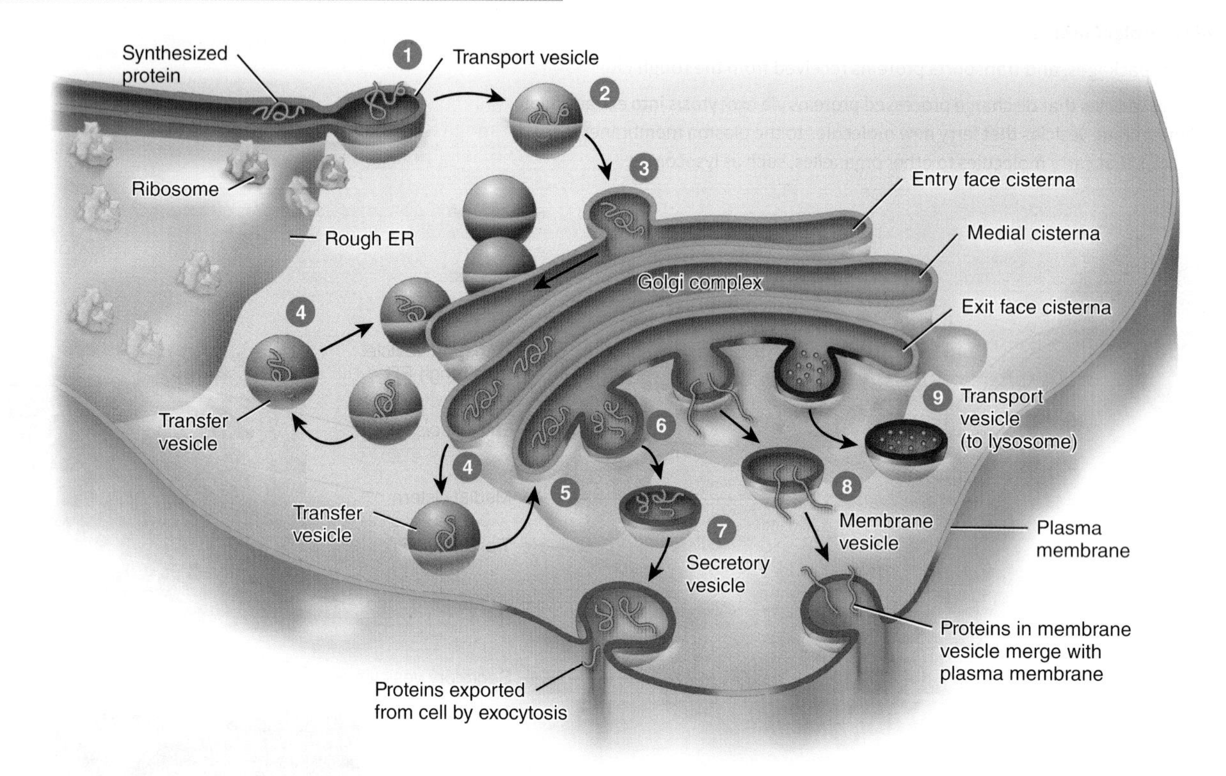

Q What are the three general destinations for proteins that leave the Golgi complex?

Lysosomal enzymes also help recycle worn-out cell structures. A lysosome can engulf another organelle, digest it, and return the digested components to the cytosol for reuse. In this way, old organelles are continually replaced. The process by which entire worn-out organelles are digested is called **autophagy** (aw-TOF-a-jē; *auto-* = self; *-phagy* = eating). In autophagy, the organelle to be digested is enclosed by a membrane derived from the ER to create a vesicle called an **autophagosome** (aw-tō-FĀ-gō-sōm); the vesicle then fuses with a lysosome. In this way, a human liver cell, for example, recycles about half of its cytoplasmic contents every week. Autophagy is also involved in cellular differentiation, control of growth, tissue remodeling, adaptation to adverse environments, and cell defense. Lysosomal enzymes may also destroy the entire cell that contains them, a process known as **autolysis** (aw-TOL-i-sis). Autolysis occurs in some pathological conditions and also is responsible for the tissue deterioration that occurs immediately after death.

As we just discussed, most lysosomal enzymes act within a cell. However, some operate in extracellular digestion. One example occurs during fertilization. The head of a sperm cell releases lysosomal enzymes that aid its penetration of the oocyte by dissolving its protective coating in a process called the acrosomal reaction (see Section 29.1).

Peroxisomes Another group of organelles similar in structure to lysosomes, but smaller, are the **peroxisomes** (pe-ROKS-i-sōms; *peroxi-* = peroxide; *-somes* = bodies; see **Figure 3.1**). Peroxisomes, also called *microbodies*, contain several *oxidases*, enzymes that can oxidize (remove hydrogen atoms from) various organic substances. For instance, amino acids and fatty acids are oxidized in peroxisomes as part of normal metabolism. In addition, enzymes in peroxisomes oxidize toxic substances, such as alcohol. Thus, peroxisomes are very abundant in the liver, where detoxification of alcohol and other damaging substances occurs. A by-product of the oxidation reactions is hydrogen peroxide (H_2O_2), a potentially toxic compound, and associated free radicals such as superoxide. However, peroxisomes also contain the enzyme *catalase*, which decomposes H_2O_2. Because production and degradation of H_2O_2 occur within the same organelle, peroxisomes protect other parts of the cell from the toxic effects of H_2O_2. Peroxisomes also contain enzymes that destroy superoxide. Without peroxisomes, by-products of metabolism could accumulate inside a cell and result in cellular death. Peroxisomes can self-replicate. New peroxisomes may form from preexisting ones by enlarging and dividing. They may also form by a process in which components accumulate at a given site in the cell and then assemble into a peroxisome.

FIGURE 3.22 Lysosomes.

Lysosomes contain several types of powerful digestive enzymes.

Functions of Lysosomes

1. Digest substances that enter a cell via endocytosis and transport final products of digestion into cytosol.
2. Carry out autophagy, the digestion of worn-out organelles.
3. Implement autolysis, the digestion of an entire cell.
4. Accomplish extracellular digestion.

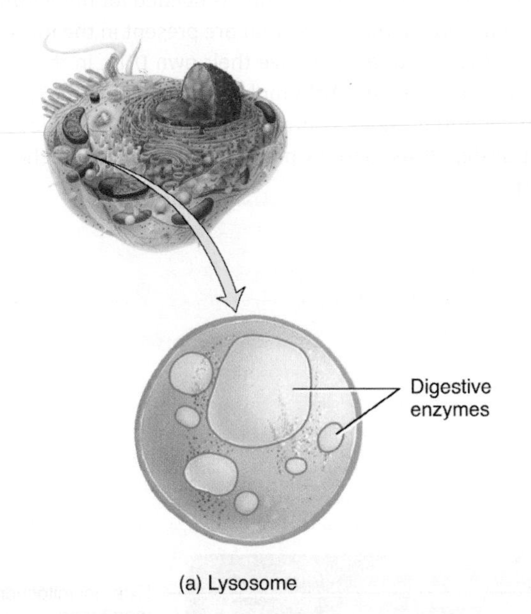

(a) Lysosome

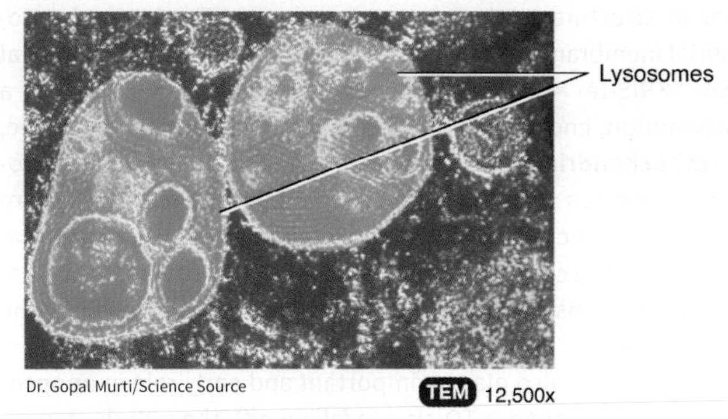

Dr. Gopal Murti/Science Source

(b) Several lysosomes

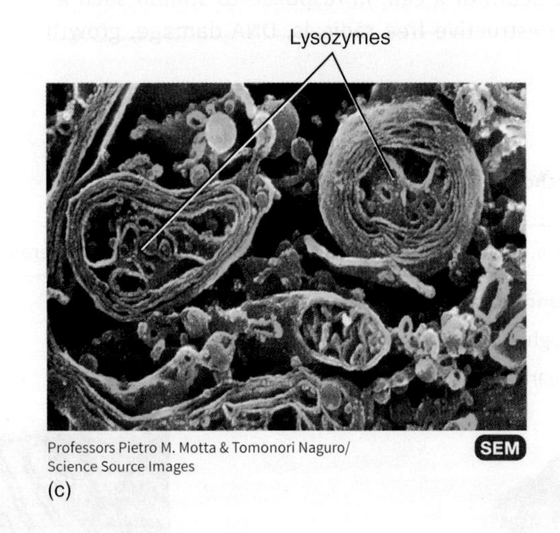

Professors Pietro M. Motta & Tomonori Naguro/ Science Source Images

(c)

See Clinical Connection: Tay-Sachs Disease

Q What is the name of the process by which worn-out organelles are digested by lysosomes?

Proteasomes As you have just learned, lysosomes degrade proteins delivered to them in vesicles. Cytosolic proteins also require disposal at certain times in the life of a cell. Continuous destruction of unneeded, damaged, or faulty proteins is the function of tiny barrel-shaped structures consisting of four stacked rings of proteins around a central core called **proteasomes** (PRŌ-tē-a-sōms = protein bodies). For example, proteins that are part of metabolic pathways need to be degraded after they have accomplished their function. Such protein destruction plays a part in negative feedback by halting a pathway once the appropriate response has been achieved. A typical body cell contains many thousands of proteasomes, in both the cytosol and the nucleus. Discovered only recently because they are far too small to discern under the light microscope and do not show up well in electron micrographs, proteasomes were so named because they contain myriad *proteases*, enzymes that cut proteins into small peptides. Once the enzymes of a proteasome have chopped up a protein into smaller chunks, other enzymes then break down the peptides into amino acids, which can be recycled into new proteins.

See Clinical Connection: Proteasomes and Disease

Mitochondria Because they generate most of the ATP through aerobic (oxygen-requiring) respiration, **mitochondria** (mī-tō-KON-drē-a; *mito-* = thread; *-chondria* = granules; singular is *mitochondrion*) are referred to as the "powerhouses" of the cell. A cell may have as few as a hundred or as many as several thousand mitochondria, depending on its activity. Active cells that use ATP at a high rate—such as those found in the muscles, liver, and kidneys—have a large number of mitochondria. For example, regular exercise can lead to an increase in the number of mitochondria in muscle cells, which allows muscle cells to function more efficiently. Mitochondria are usually located within the cell where oxygen enters the cell or where the ATP is used, for example, among the contractile proteins in muscle cells.

A mitochondrion consists of an **external mitochondrial membrane** and an **internal mitochondrial membrane** with a small fluid-filled space between them (**Figure 3.23**). Both membranes are similar in structure to the plasma membrane. The internal mitochondrial membrane contains a series of folds called **mitochondrial cristae** (KRIS-tē = ridges). The central fluid-filled cavity of a mitochondrion, enclosed by the internal mitochondrial membrane, is the **mitochondrial matrix**. The elaborate folds of the cristae provide an enormous surface area for the chemical reactions that are part of the aerobic phase of *cellular respiration,* the reactions that produce most of a cell's ATP (see Chapter 25). The enzymes that catalyze these reactions are located on the cristae and in the matrix of the mitochondria.

Mitochondria also play an important and early role in **apoptosis** (ap′-ōp-TŌ-sis or ap-ō-TŌ-sis = a falling off), the orderly, genetically programmed death of a cell. In response to stimuli such as large numbers of destructive free radicals, DNA damage, growth factor deprivation, or lack of oxygen and nutrients, certain chemicals are released from mitochondria following the formation of a pore in the outer mitochondrial membrane. One of the chemicals released into the cytosol of the cell is cytochrome *c*, which while inside the mitochondria is involved in aerobic cellular respiration. In the cytosol, however, cytochrome *c* and other substances initiate a cascade of activation of protein-digesting enzymes that bring about apoptosis.

Like peroxisomes, mitochondria self-replicate, a process that occurs during times of increased cellular energy demand or before cell division. Synthesis of some of the proteins needed for mitochondrial functions occurs on the ribosomes that are present in the mitochondrial matrix. Mitochondria even have their own DNA, in the form of multiple copies of a circular DNA molecule that contains 37 genes. These mitochondrial genes control the synthesis of 2 ribosomal RNAs, 22 transfer RNAs, and 13 proteins that build mitochondrial components.

FIGURE 3.23 Mitochondria.

Within mitochondria, chemical reactions of aerobic cellular respiration generate ATP.

Functions of Mitochondria

1. Generate ATP through reactions of aerobic cellular respiration.
2. Play an important early role in apoptosis.

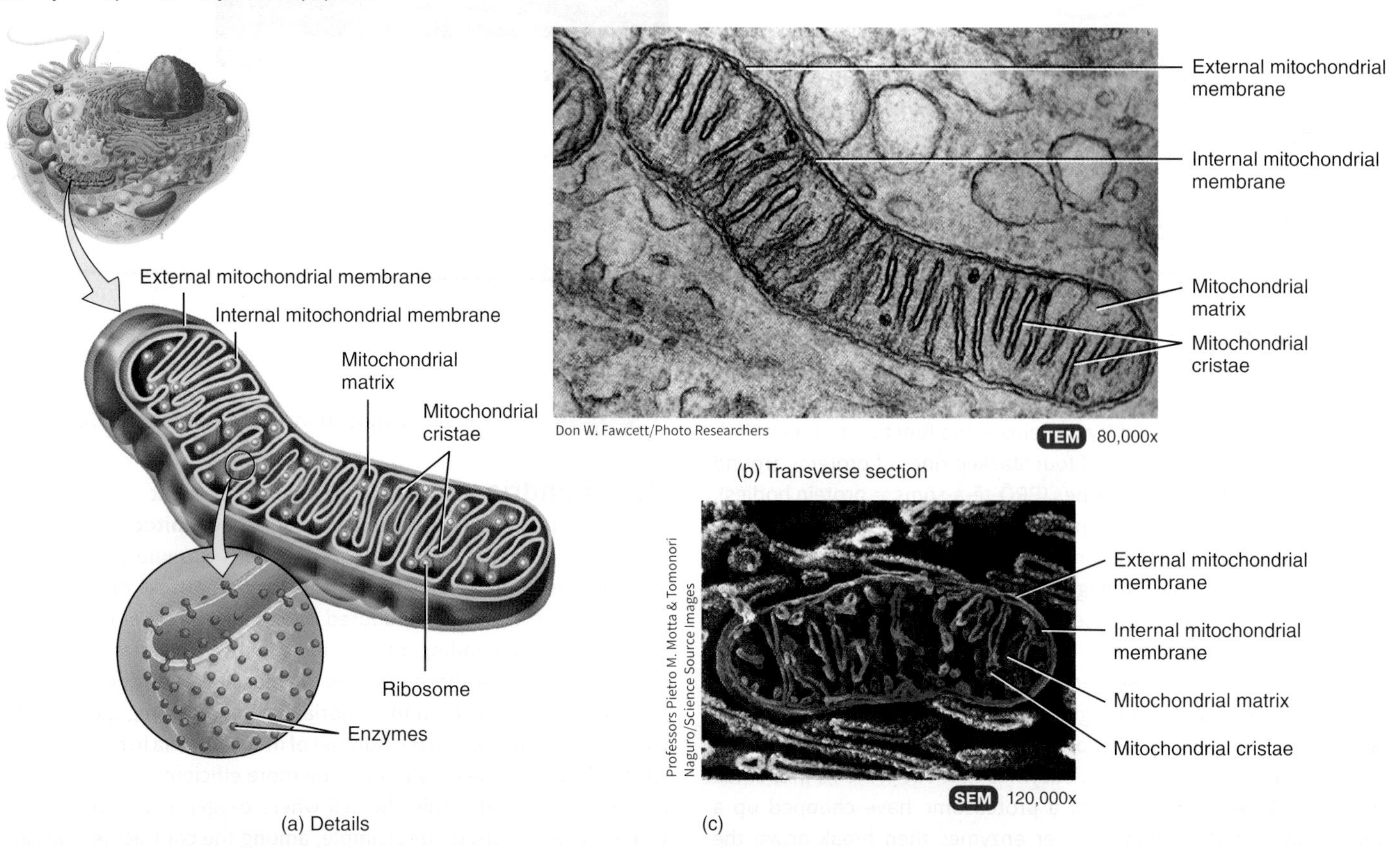

Q How do the mitochondrial cristae contribute to its ATP-producing function?

Although the nucleus of each somatic cell contains genes from both your mother and your father, mitochondrial genes are inherited only from your mother. This is due to the fact that all mitochondria in a cell are descendants of those that were present in the oocyte (egg) during the fertilization process. The head of a sperm (the part that penetrates and fertilizes an oocyte) normally lacks most organelles, such as mitochondria, ribosomes, endoplasmic reticulum, and the Golgi complex, and any sperm mitochondria that do enter the oocyte are soon destroyed. Since all mitochondrial genes are inherited from the maternal parent, mitochondrial DNA can be used to trace maternal lineage (in other words, to determine whether two or more individuals are related through their mother's side of the family).

3.5 Nucleus

OBJECTIVE

- **Describe** the structure and function of the nucleus.

The **nucleus** is a spherical or oval-shaped structure that usually is the most prominent feature of a cell (**Figure 3.24**). Most cells have a single nucleus, although some, such as mature red blood cells, have none.

FIGURE 3.24 Nucleus.

The nucleus contains most of the cell's genes, which are located on chromosomes.

Functions of the Nucleus

1. Controls cellular structure.
2. Directs cellular activities.
3. Produces ribosomes in nucleoli.

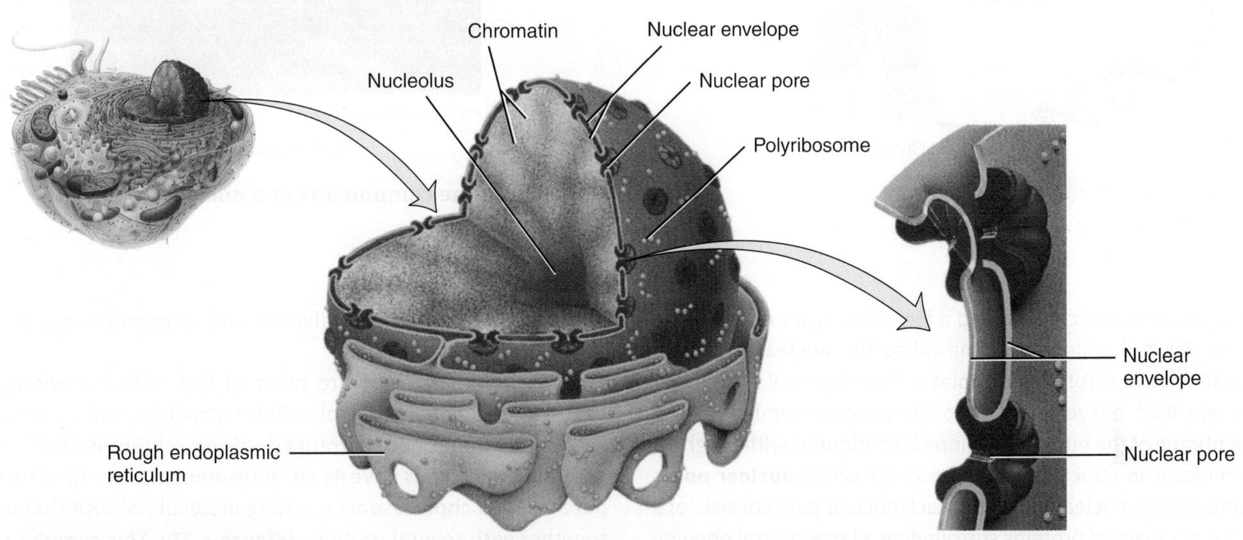

(a) Details of the nucleus

(b) Details of the nuclear envelope

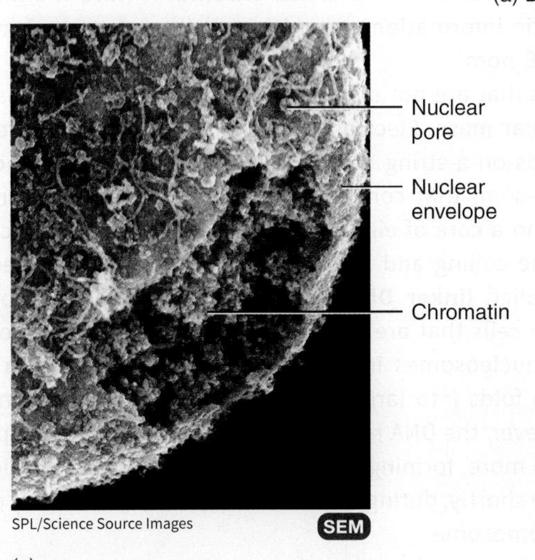

SPL/Science Source Images

(c)

Q What is chromatin?

See Clinical Connection: Genomics

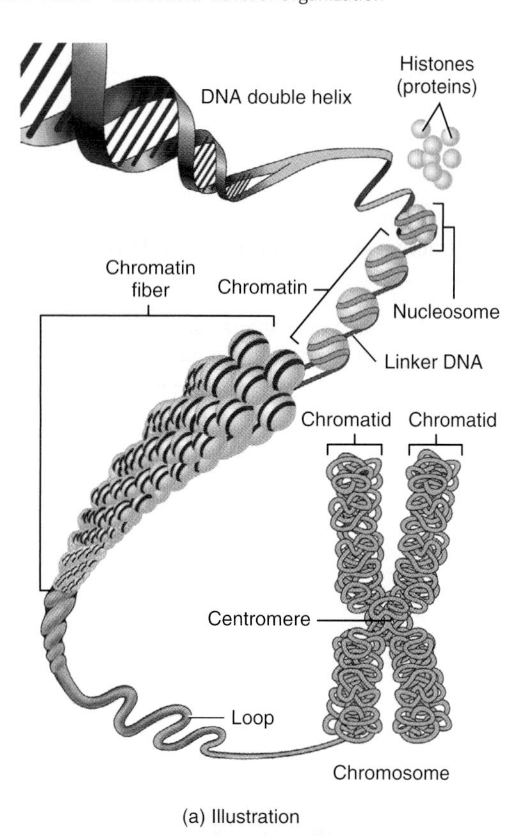

(a) Illustration

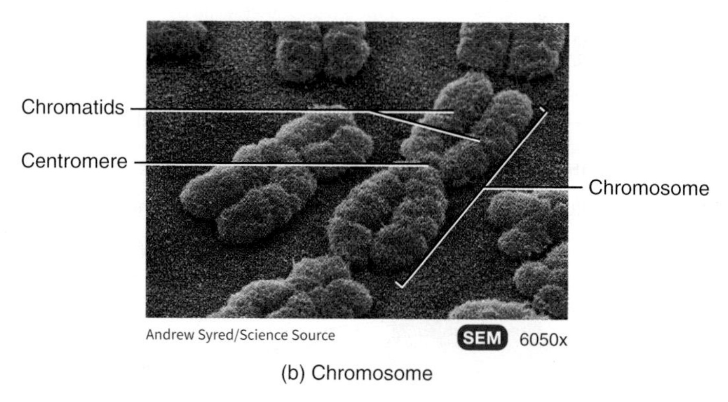

(b) Chromosome

FIGURE 3.25 Packing of DNA into a chromosome in a dividing cell. When packing is complete, two identical DNA molecules and their histones form a pair of chromatids, which are held together by a centromere.

A chromosome is a highly coiled and folded DNA molecule that is combined with protein molecules.

Q What are the components of a nucleosome?

In contrast, skeletal muscle cells and a few other types of cells have multiple nuclei. A double membrane called the **nuclear envelope** separates the nucleus from the cytoplasm. Both layers of the nuclear envelope are lipid bilayers similar to the plasma membrane. The outer membrane of the nuclear envelope is continuous with rough ER and resembles it in structure. Many openings called **nuclear pores** extend through the nuclear envelope. Each nuclear pore consists of a circular arrangement of proteins surrounding a large central opening that is about 10 times wider than the pore of a channel protein in the plasma membrane.

Nuclear pores control the movement of substances between the nucleus and the cytoplasm. Small molecules and ions move through the pores passively by diffusion. Most large molecules, such as RNAs and proteins, cannot pass through the nuclear pores by diffusion. Instead, their passage involves an active transport process in which the molecules are recognized and selectively transported through the nuclear pore into or out of the nucleus. For example, proteins needed for nuclear functions move from the cytosol into the nucleus; newly formed RNA molecules move from the nucleus into the cytosol in this manner.

Inside the nucleus are one or more spherical bodies called **nucleoli** (noo′-KLĒ-ō-li; singular is *nucleolus*) that function in producing ribosomes. Each nucleolus is simply a cluster of protein, DNA, and RNA; it is not enclosed by a membrane. Nucleoli are the sites of synthesis of rRNA and assembly of rRNA and proteins into ribosomal subunits. Nucleoli are quite prominent in cells that synthesize large amounts of protein, such as muscle and liver cells. Nucleoli disperse and disappear during cell division and reorganize once new cells are formed.

Within the nucleus are most of the cell's hereditary units, called *genes*, which control cellular structure and direct cellular activities. Genes are arranged along chromosomes. Human somatic (body) cells have 46 chromosomes, 23 inherited from each parent. Each chromosome is a long molecule of DNA that is coiled together with several proteins (Figure 3.25). This complex of DNA, proteins, and some RNA is called **chromatin** (KRŌ-ma-tin). The total genetic information carried in a cell or an organism is its **genome** (JĒ-nōm).

In cells that are not dividing, the chromatin appears as a diffuse, granular mass. Electron micrographs reveal that chromatin has a beads-on-a-string structure. Each bead is a **nucleosome** (NOO-klē-ō-sōm) that consists of double-stranded DNA wrapped twice around a core of eight proteins called **histones**, which help organize the coiling and folding of DNA. The string between the beads is called **linker DNA**, which holds adjacent nucleosomes together. In cells that are not dividing, another histone promotes coiling of nucleosomes into a larger-diameter **chromatin fiber**, which then folds into large loops. Just before cell division takes place, however, the DNA replicates (duplicates) and the loops condense even more, forming a pair of **chromatids** (KRŌ-ma-tids). As you will see shortly, during cell division a pair of chromatids constitutes a chromosome.

The main parts of a cell, their structure, and their functions are summarized in Table 3.2.

TABLE 3.2 Cell Parts and Their Functions

PART	DESCRIPTION	FUNCTIONS
PLASMA MEMBRANE	Fluid mosaic lipid bilayer (phospholipids, cholesterol, and glycolipids) studded with proteins; surrounds cytoplasm.	Protects cellular contents; makes contact with other cells; contains channels, transporters, receptors, enzymes, cell-identity markers, and linker proteins; mediates entry and exit of substances.
CYTOPLASM	Cellular contents between plasma membrane and nucleus—cytosol and organelles.	Site of all intracellular activities except those occurring in the nucleus.
Cytosol	Composed of water, solutes, suspended particles, lipid droplets, and glycogen granules.	Fluid in which many of cell's metabolic reactions occur.
	The cytoskeleton is a network in the cytoplasm composed of three protein filaments: microfilaments, intermediate filaments, and microtubules.	The cytoskeleton maintains shape and general organization of cellular contents; responsible for cell movements.
Organelles	Specialized structures with characteristic shapes.	Each organelle has specific functions.
Centrosome	Pair of centrioles plus pericentriolar matrix.	The pericentriolar matrix contains tubulins, which are used for growth of the mitotic spindle and microtubule formation.
Cilia and flagella	Motile cell surface projections that contain 20 microtubules and a basal body.	Cilia: move fluids over cell's surface; flagella: move entire cell.
Ribosome	Composed of two subunits containing ribosomal RNA and proteins; may be free in cytosol or attached to rough ER.	Protein synthesis.
Endoplasmic reticulum (ER)	Membranous network of flattened sacs or tubules. Rough ER is covered by ribosomes and is attached to the nuclear envelope; smooth ER lacks ribosomes.	Rough ER: synthesizes glycoproteins and phospholipids that are transferred to cellular organelles, inserted into plasma membrane, or secreted during exocytosis; smooth ER: synthesizes fatty acids and steroids, inactivates or detoxifies drugs, removes phosphate group from glucose-6-phosphate, and stores and releases calcium ions in muscle cells.
Golgi complex	Consists of 3–20 flattened membranous sacs called cisternae; structurally and functionally divided into entry (*cis*) face, medial cisternae, and exit (*trans*) face.	Entry (*cis*) face accepts proteins from rough ER; medial cisternae form glycoproteins, glycolipids, and lipoproteins; exit (*trans*) face modifies molecules further, then sorts and packages them for transport to their destinations.
Lysosome	Vesicle formed from Golgi complex; contains digestive enzymes.	Fuses with and digests contents of endosomes, phagosomes, and vesicles formed during bulk-phase endocytosis and transports final products of digestion into cytosol; digests worn-out organelles (autophagy), entire cells (autolysis), and extracellular materials.
Peroxisome	Vesicle containing oxidases (oxidative enzymes) and catalase (decomposes hydrogen peroxide); new peroxisomes bud from preexisting ones.	Oxidizes amino acids and fatty acids; detoxifies harmful substances, such as hydrogen peroxide and associated free radicals.
Proteasome	Tiny barrel-shaped structure that contains proteases (proteolytic enzymes).	Degrades unneeded, damaged, or faulty proteins by cutting them into small peptides.
Mitochondrion	Consists of an external and an internal mitochondrial membrane, cristae, and matrix; new mitochondria form from preexisting ones.	Site of aerobic cellular respiration reactions that produce most of a cell's ATP. Plays an important early role in apoptosis.
NUCLEUS	Consists of a nuclear envelope with pores, nucleoli, and chromosomes, which exist as a tangled mass of chromatin in interphase cells.	Nuclear pores control the movement of substances between the nucleus and cytoplasm, nucleoli produce ribosomes, and chromosomes consist of genes that control cellular structure and direct cellular functions.

3.6 Protein Synthesis

OBJECTIVE

- **Describe** the sequence of events in protein synthesis.

Although cells synthesize many chemicals to maintain homeostasis, much of the cellular machinery is devoted to synthesizing large numbers of diverse proteins. The proteins in turn determine the physical and chemical characteristics of cells and, therefore, of the organisms formed from them. Some proteins help assemble cellular structures such as the plasma membrane, the cytoskeleton, and other organelles. Others serve as hormones, antibodies, and contractile elements in muscular tissue. Still others act as enzymes, regulating the rates of the numerous chemical reactions that occur in cells, or as transporters, carrying various materials in the blood. Just as genome means all of the genes in an organism, **proteome** (PRŌ-tē-ōm) refers to all of an organism's proteins.

In the process called **gene expression**, a gene's DNA is used as a template for synthesis of a specific protein. First, in a process aptly named *transcription,* the information encoded in a specific region of DNA is *transcribed* (copied) to produce a specific molecule of RNA (ribonucleic acid). In a second process, referred to as translation, the RNA attaches to a ribosome, where the information contained in RNA is *translated* into a corresponding sequence of amino acids to form a new protein molecule (**Figure 3.26**).

FIGURE 3.26 **Overview of gene expression.** Synthesis of a specific protein requires transcription of a gene's DNA into RNA and translation of RNA into a corresponding sequence of amino acids.

Transcription occurs in the nucleus; translation occurs in the cytoplasm.

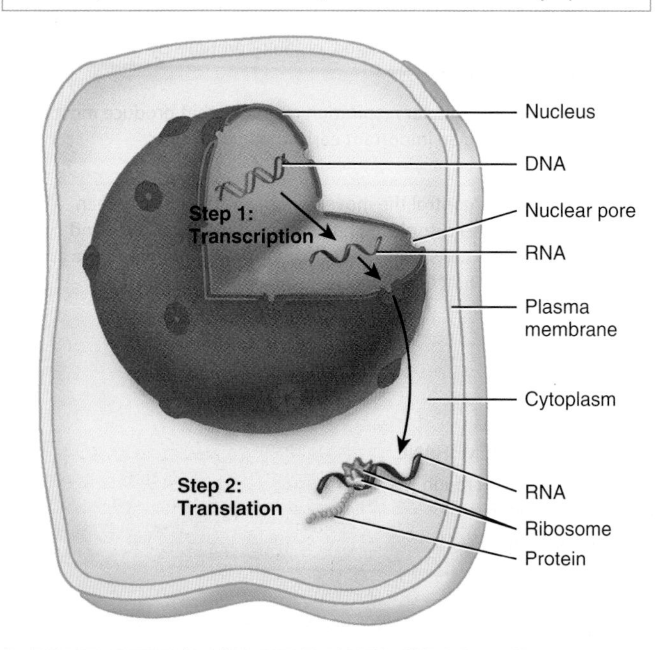

Q Why are proteins important in the life of a cell?

DNA and RNA store genetic information as sets of three nucleotides. A sequence of three such nucleotides in DNA is called a **base triplet**. Each DNA base triplet is transcribed as a complementary sequence of three nucleotides, called a **codon**. A given codon specifies a particular amino acid. The **genetic code** is the set of rules that relate the base triplet sequence of DNA to the corresponding codons of RNA and the amino acids they specify.

Transcription

During **transcription**, which occurs in the nucleus, the genetic information represented by the sequence of base triplets in DNA serves as a template for copying the information into a complementary sequence of codons. Three types of RNA are made from the DNA template:

1. **Messenger RNA (mRNA)** directs the synthesis of a protein.
2. **Ribosomal RNA (rRNA)** joins with ribosomal proteins to make ribosomes.
3. **Transfer RNA (tRNA)** binds to an amino acid and holds it in place on a ribosome until it is incorporated into a protein during translation. One end of the tRNA carries a specific amino acid, and the opposite end consists of a triplet of nucleotides called an **anticodon**. By pairing between complementary bases, the tRNA anticodon attaches to the mRNA codon. Each of the more than 20 different types of tRNA binds to only one of the 20 different amino acids.

The enzyme **RNA polymerase** (po-LIM-er-ās) catalyzes transcription of DNA. However, the enzyme must be instructed where to start the transcription process and where to end it. Only one of the two DNA strands serves as a template for RNA synthesis. The segment of DNA where transcription begins, a special nucleotide sequence called a **promoter**, is located near the beginning of a gene (**Figure 3.27a**). This is where RNA polymerase attaches to the DNA. During transcription, bases pair in a complementary manner: The bases cytosine (C), guanine (G), and thymine (T) in the DNA template pair with guanine, cytosine, and adenine (A), respectively, in the RNA strand (**Figure 3.27b**). However, adenine in the DNA template pairs with uracil (U), not thymine, in RNA:

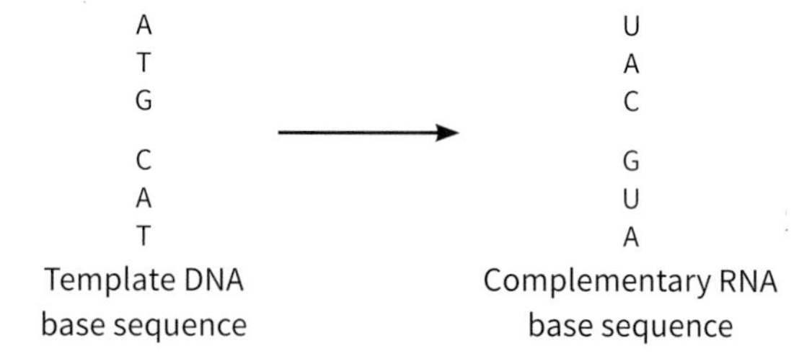

Transcription of the DNA strand ends at another special nucleotide sequence called a **terminator**, which specifies the end of the gene (**Figure 3.27a**). When RNA polymerase reaches the terminator, the enzyme detaches from the transcribed RNA molecule and the DNA strand.

Not all parts of a gene actually code for parts of a protein. Regions within a gene called **introns** *do not* code for parts of proteins. They are located between regions called **exons** that *do* code for segments of a protein. Immediately after transcription, the transcript includes information from both introns and exons and is called **pre-mRNA**. The introns are removed from pre-mRNA by **small nuclear ribonucleoproteins** (snRNPs,

FIGURE 3.27 **Transcription.** DNA transcription begins at a promoter and ends at a terminator.

During transcription, the genetic information in DNA is copied to RNA.

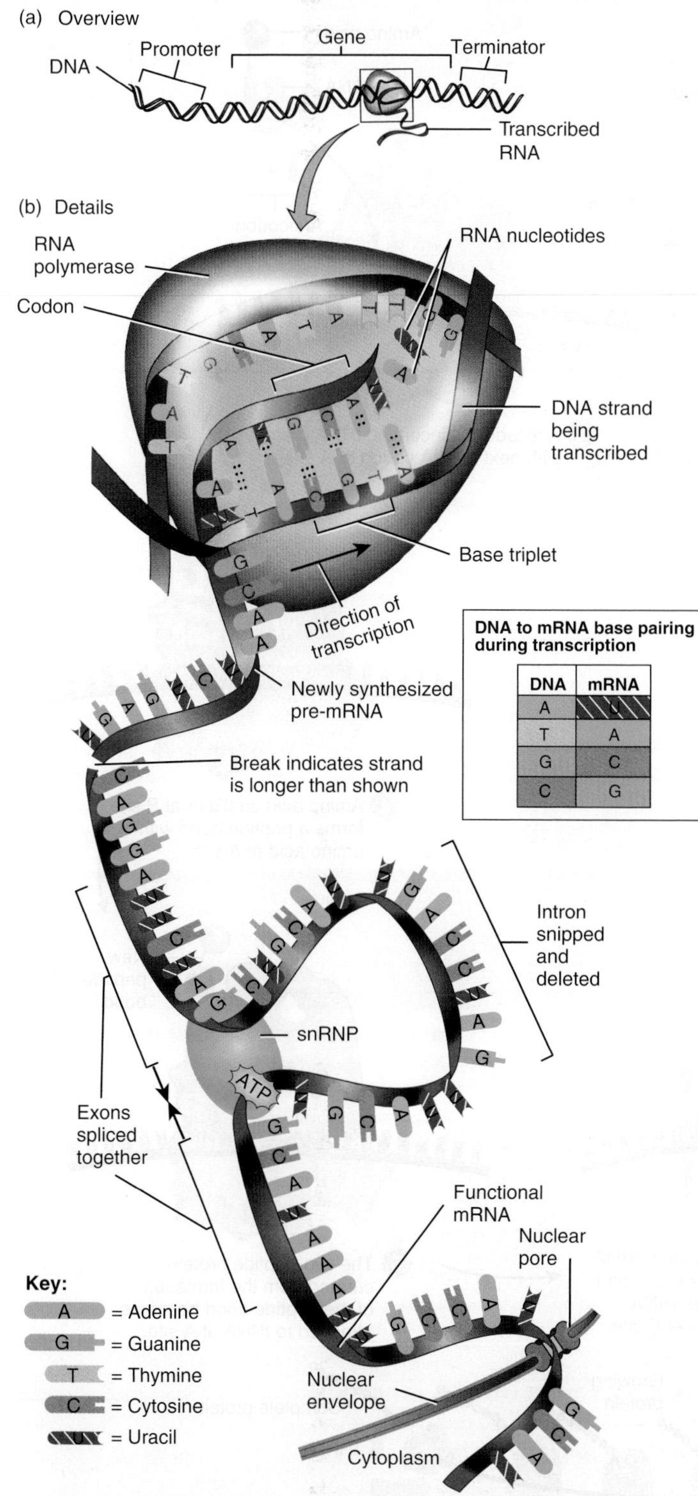

Q If the DNA template had the base sequence AGCT, what would be the mRNA base sequence, and what enzyme would catalyze DNA transcription?

pronounced "snurps"; **Figure 3.27b**). The snRNPs are enzymes that cut out the introns and splice together the exons. The resulting product is a functional mRNA molecule that passes through a pore in the nuclear envelope to reach the cytoplasm, where translation takes place.

Although the human genome contains around 30,000 genes, there are probably 500,000 to 1 million human proteins. How can so many proteins be coded for by so few genes? Part of the answer lies in **alternative splicing** of mRNA, a process in which the pre-mRNA transcribed from a gene is spliced in different ways to produce several different mRNAs. The different mRNAs are then translated into different proteins. In this way, one gene may code for 10 or more different proteins. In addition, chemical modifications are made to proteins after translation, for example, as proteins pass through the Golgi complex. Such chemical alterations can produce two or more different proteins from a single translation.

Translation

In the process of **translation**, the nucleotide sequence in an mRNA molecule specifies the amino acid sequence of a protein. Ribosomes in the cytoplasm carry out translation. The small subunit of a ribosome has a binding site for mRNA; the larger subunit has three binding sites for tRNA molecules: a P site, A site, and E site (**Figure 3.28**). The **P (peptidyl) site** binds the tRNA carrying the growing polypeptide chain. The **A (aminoacyl) site** binds the tRNA carrying the next amino acid to be added to the growing polypeptide. The **E (exit) site** binds tRNA just before it is released from the ribosome. Translation occurs in the following way (**Figure 3.29**):

1. An mRNA molecule binds to the small ribosomal subunit at the mRNA binding site. A special tRNA, called *initiator tRNA*, binds to the start codon (AUG) on mRNA, where translation begins. The tRNA anticodon (UAC) attaches to the mRNA codon (AUG) by pairing between the complementary bases. Besides being the start

FIGURE 3.28 **Translation.** During translation, an mRNA molecule binds to a ribosome. Then, the mRNA nucleotide sequence specifies the amino acid sequence of a protein.

Ribosomes have a binding site for mRNA and a P site, A site, and E site for attachment of tRNAs.

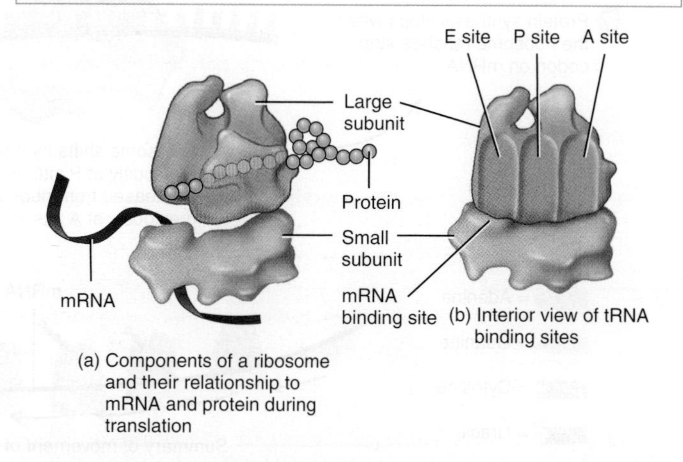

Q What roles do the P and A sites serve?

FIGURE 3.29 Protein elongation and termination of protein synthesis during translation.

During protein synthesis, the small and large ribosomal subunits join to form a functional ribosome. When the process is complete, they separate.

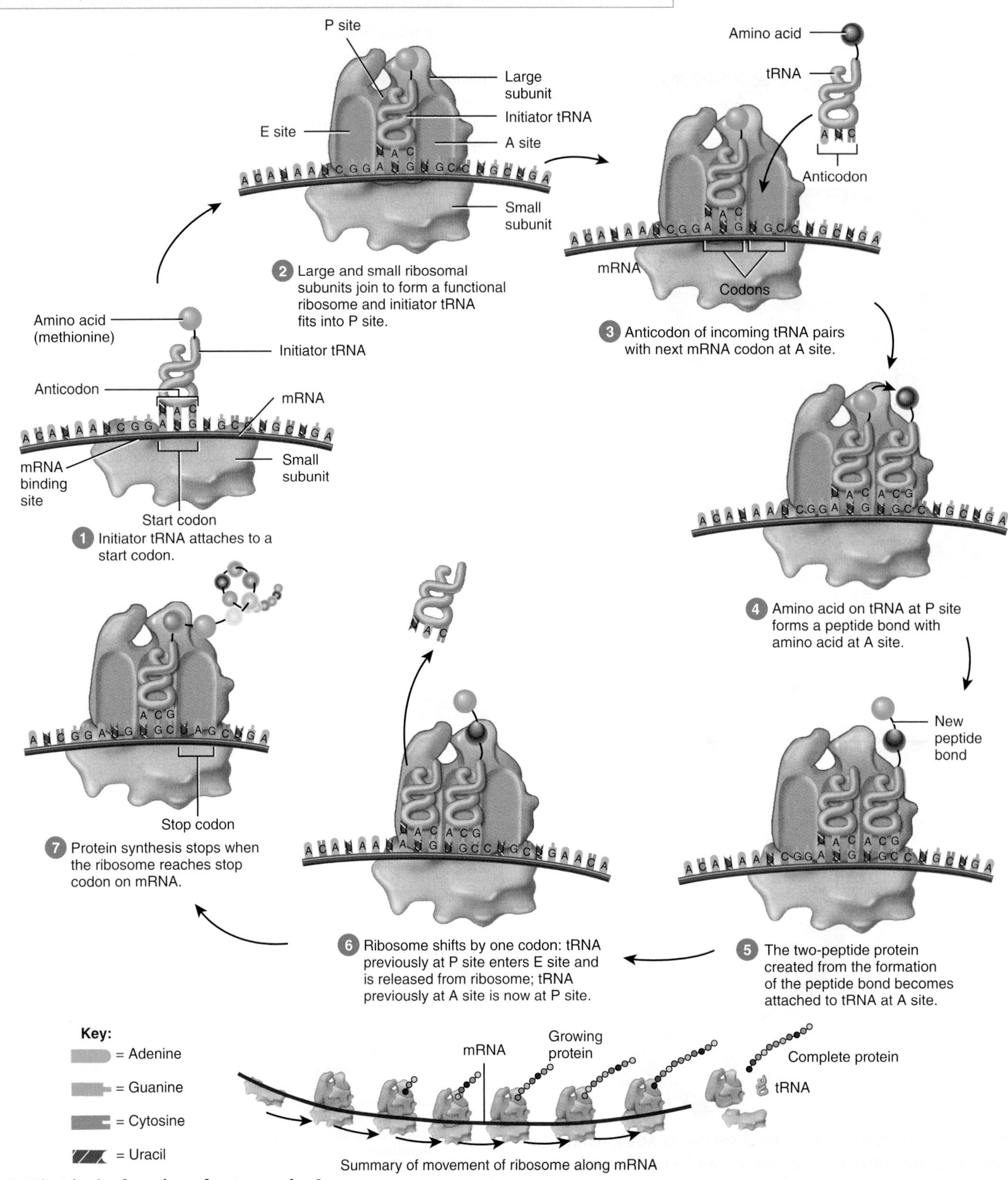

Q What is the function of a stop codon?

codon, AUG is also the codon for the amino acid methionine. Thus, methionine is always the first amino acid in a growing polypeptide.

2 Next, the large ribosomal subunit attaches to the small ribosomal subunit–mRNA complex, creating a functional ribosome. The initiator tRNA, with its amino acid (methionine), fits into the P site of the ribosome.

3 The anticodon of another tRNA with its attached amino acid pairs with the second mRNA codon at the A site of the ribosome.

4 A component of the large ribosomal subunit catalyzes the formation of a peptide bond between methionine and the amino acid carried by the tRNA at the A site.

5 Following the formation of the peptide bond, the resulting two-peptide protein becomes attached to the tRNA at the A site.

6 After peptide bond formation, the ribosome shifts the mRNA strand by one codon. The tRNA in the P site enters the E site and is subsequently released from the ribosome. The tRNA in the A site bearing the two-peptide protein shifts into the P site, allowing another tRNA with its amino acid to bind to a newly exposed codon at the A site. Steps 3 through 6 occur repeatedly, and the protein lengthens progressively.

7 Protein synthesis ends when the ribosome reaches a stop codon at the A site, which causes the completed protein to detach from the final tRNA. In addition, tRNA vacates the P site and the ribosome splits into its large and small subunits.

Protein synthesis progresses at a rate of about 15 peptide bonds per second. As the ribosome moves along the mRNA and before it completes synthesis of the whole protein, another ribosome may attach behind it and begin translation of the same mRNA strand. Several ribosomes attached to the same mRNA constitute a **polyribosome**. The simultaneous movement of several ribosomes along the same mRNA molecule permits the translation of one mRNA into several identical proteins at the same time.

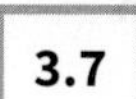

See Clinical Connection: Recombinant DNA

3.7 Cell Division

OBJECTIVES

- **Discuss** the stages, events, and significance of somatic and reproductive cell division.
- **Describe** the signals that induce somatic cell division.

Most cells of the human body undergo **cell division**, the process by which cells reproduce themselves. The two types of cell division—somatic cell division and reproductive cell division—accomplish different goals for the organism.

A **somatic cell** (sō-MAT-ik; *soma* = body) is any cell of the body other than a germ cell. A **germ cell** is a gamete (sperm or oocyte) or any precursor cell destined to become a gamete. In **somatic cell division**, a cell undergoes a nuclear division called **mitosis** (mī-TŌ-sis; *mitos* = thread) and a cytoplasmic division called **cytokinesis** (sī-tō-ki-NĒ-sis; *cyto-* = cell; *-kinesis* = movement) to produce two genetically identical cells, each with the same number and kind of chromosomes as the original cell. Somatic cell division replaces dead or injured cells and adds new ones during tissue growth.

Reproductive cell division is the mechanism that produces gametes, the cells needed to form the next generation of sexually reproducing organisms. This process consists of a special two-step division called *meiosis*, in which the number of chromosomes in the nucleus is reduced by half.

Somatic Cell Division

The **cell cycle** is an orderly sequence of events in which a somatic cell duplicates its contents and divides in two. Some cells divide more than others. Human cells, such as those in the brain, stomach, and kidneys, contain 23 pairs of chromosomes, for a total of 46. One member of each pair is inherited from each parent. The two chromosomes that make up each pair are called **homologous chromosomes** (hō-MOL-ō-gus; *homo-* = same) or *homologs*; they contain similar genes arranged in the same (or almost the same) order. When examined under a light microscope, homologous chromosomes generally look very similar. The exception to this rule is one pair of chromosomes called the **sex chromosomes**, designated X and Y. In females the homologous pair of sex chromosomes consists of two large X chromosomes; in males the pair consists of an X and a much smaller Y chromosome. Because somatic cells contain two sets of chromosomes, they are called **diploid (2*n*) cells** (DIP-loyd; *dipl-* = double; *-oid* = form).

When a cell reproduces, it must replicate (duplicate) all its chromosomes to pass its genes to the next generation of cells. The cell cycle consists of two major periods: interphase, when a cell is not dividing, and the mitotic (M) phase, when a cell is dividing (**Figure 3.30**).

Interphase During **interphase** (IN-ter-fāz) the cell replicates its DNA through a process that will be described shortly. It also produces additional organelles and cytosolic components in anticipation of cell division. Interphase is a state of high metabolic activity; it is during this time that the cell does most of its growing. Interphase consists of three phases: G_1, S, and G_2 (**Figure 3.30**). The S stands for *synthesis* of DNA. Because the G phases are periods when there is no activity related to DNA duplication, they are thought of as *gaps* or interruptions in DNA duplication.

The **G_1 phase** is the interval between the mitotic phase and the S phase. During G_1, the cell is metabolically active; it replicates most of its organelles and cytosolic components but not its DNA. Replication of centrosomes also begins in the G_1 phase. Virtually all of the cellular activities described in this chapter happen during G_1. For a cell with a total cell cycle time of 24 hours, G_1 lasts 8 to 10 hours. However, the duration of this phase is quite variable. It is very short in many embryonic cells or cancer cells. Cells that remain in G_1 for a very long time, perhaps destined never to divide again, are said to be in the

FIGURE 3.30 The cell cycle. Not illustrated is cytokinesis (division of the cytoplasm), which usually occurs during late anaphase of the mitotic phase.

In a complete cell cycle, a starting cell duplicates its contents and divides into two identical cells.

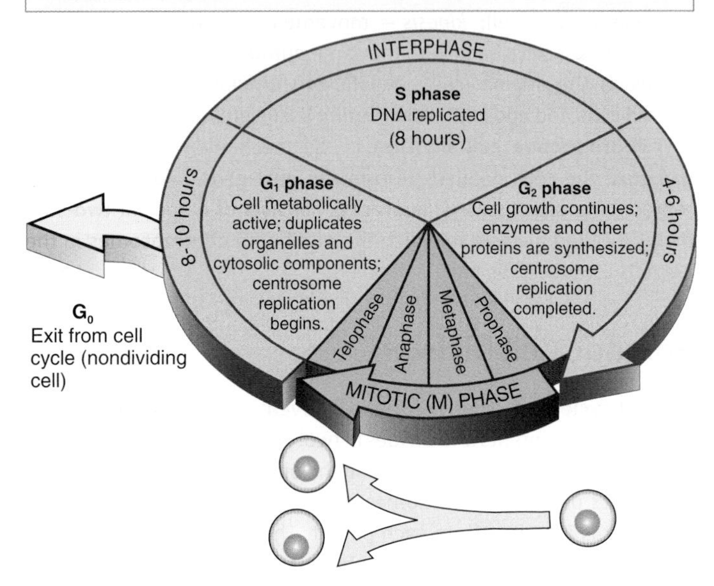

Q During which phase of the cell cycle does DNA replication occur?

G_0 phase. Most nerve cells are in the G_0 phase. Once a cell enters the S phase, however, it is committed to go through the rest of the cell cycle.

The **S phase**, the interval between G_1 and G_2, lasts about 8 hours. During the S phase, DNA replication occurs. As a result of DNA replication, the two identical cells formed during cell division later in the cell cycle will have the same genetic material. The **G_2 phase** is the interval between the S phase and the mitotic phase. It lasts 4 to 6 hours. During G_2, cell growth continues, enzymes and other proteins are synthesized in preparation for cell division, and replication of centrosomes is completed. When DNA replicates during the S phase, its helical structure partially uncoils, and the two strands separate at the points where hydrogen bonds connect base pairs (**Figure 3.31**). Each exposed base of the old DNA strand then pairs with the complementary base of a newly synthesized nucleotide. A new DNA strand takes shape as chemical bonds form between neighboring nucleotides. The uncoiling and complementary base pairing continues until each of the two original DNA strands is joined with a newly formed complementary DNA strand. The original DNA molecule has become two identical DNA molecules.

A microscopic view of a cell during interphase shows a clearly defined nuclear envelope, a nucleolus, and a tangled mass of chromatin (**Figure 3.32a**). Once a cell completes its activities during the G_1, S, and G_2 phases of interphase, the mitotic phase begins.

Mitotic Phase

The **mitotic (M) phase** of the cell cycle, which results in the formation of two identical cells, consists of a nuclear division (mitosis) and a cytoplasmic division (cytokinesis) to form two identical cells. The events that occur during mitosis and cytokinesis are plainly visible under a microscope because chromatin condenses into discrete chromosomes.

FIGURE 3.31 Replication of DNA. The two strands of the double helix separate by breaking the hydrogen bonds (shown as dotted lines) between nucleotides. New, complementary nucleotides attach at the proper sites, and a new strand of DNA is synthesized alongside each of the original strands. Arrows indicate hydrogen bonds forming again between pairs of bases.

Replication doubles the amount of DNA.

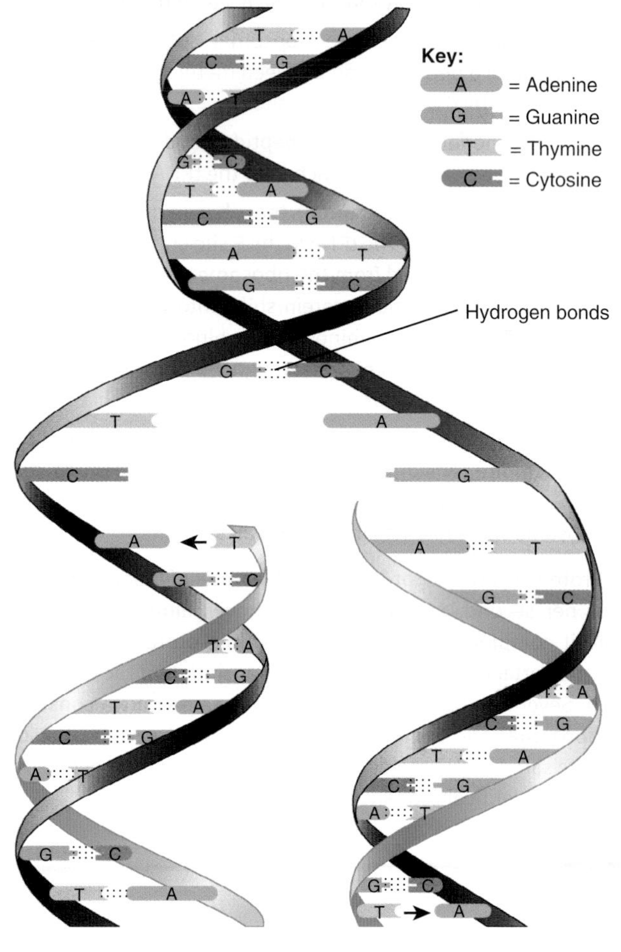

Q Why must DNA replication occur before cytokinesis in somatic cell division?

NUCLEAR DIVISION: MITOSIS Mitosis, as noted earlier, is the distribution of two sets of chromosomes into two separate nuclei. The process results in the *exact* partitioning of genetic information. For convenience, biologists divide the process into four stages: prophase, metaphase, anaphase, and telophase. However, mitosis is a continuous process; one stage merges seamlessly into the next.

1. **Prophase** (PRŌ-fāz). During early prophase, the chromatin fibers condense and shorten into chromosomes that are visible under the light microscope (**Figure 3.32b**). The condensation process may prevent entangling of the long DNA strands as they move during mitosis. Because longitudinal DNA replication took place during the S phase of interphase, each prophase chromosome consists of

FIGURE 3.32 Cell division: mitosis and cytokinesis. Begin the sequence at 1 at the top of the figure and read clockwise to complete the process.

In somatic cell division, a single starting cell divides to produce two identical diploid cells.

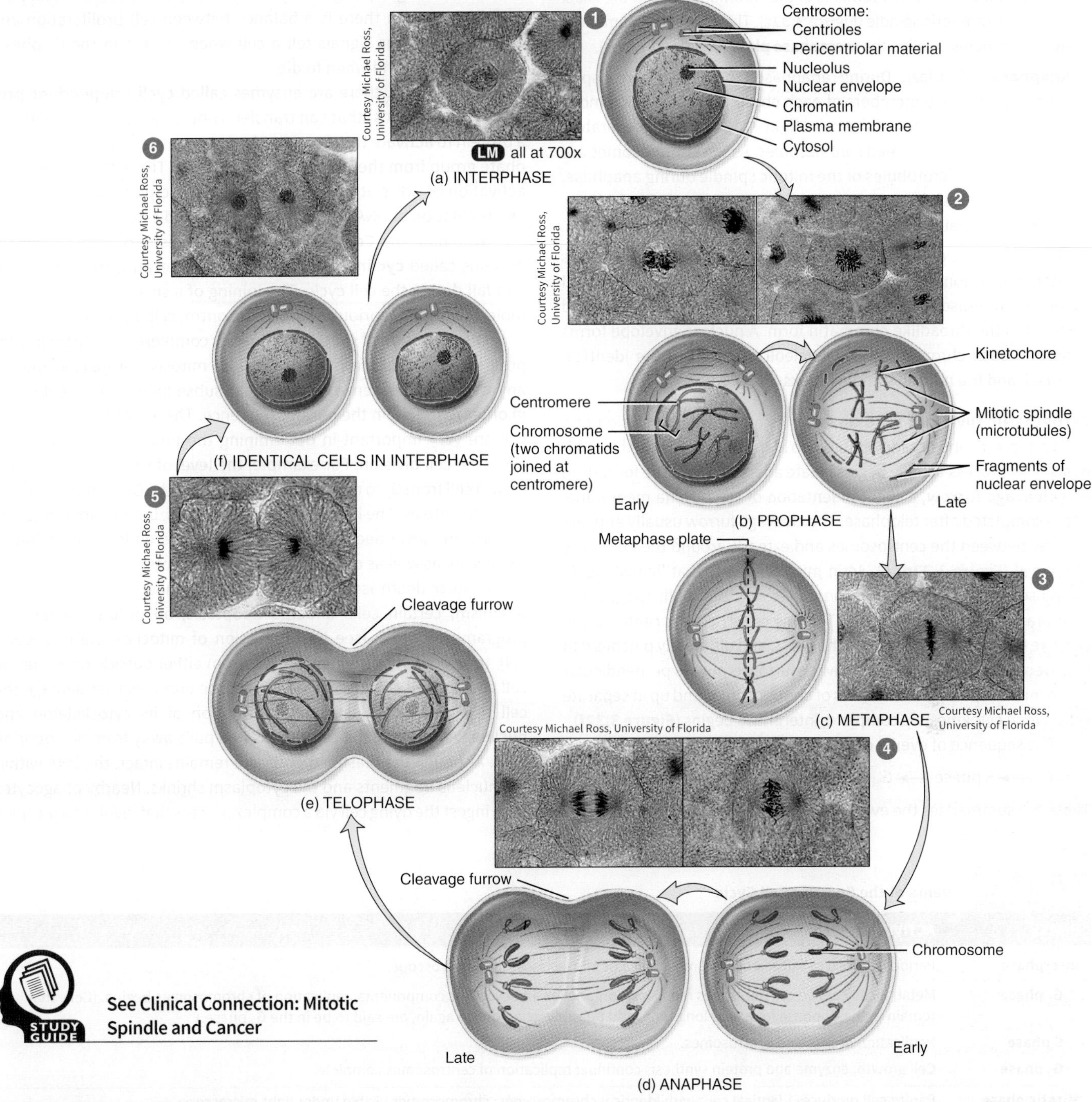

See Clinical Connection: Mitotic Spindle and Cancer

Q When does cytokinesis begin?

a pair of identical strands called *chromatids*. A constricted region called a **centromere** (SEN-trō-mēr) holds the chromatid pair together. At the outside of each centromere is a protein complex known as the **kinetochore** (ki-NET-ō-kor). Later in prophase, tubulins in the pericentriolar material of the centrosomes start to form the **mitotic spindle**, a football-shaped assembly of microtubules that attach to the kinetochore (**Figure 3.32b**). As the microtubules lengthen, they push the centrosomes to the poles (ends) of the cell so that the spindle extends from pole to pole. The mitotic spindle is responsible for the separation of chromatids to opposite poles of the cell. Then, the nucleolus disappears and the nuclear envelope breaks down.

2. **Metaphase** (MET-a-fāz). During metaphase, the microtubules of the mitotic spindle align the centromeres of the chromatid pairs at the exact center of the mitotic spindle (Figure 3.32c). This plane of alignment of the centromeres is called the **metaphase plate** (*equatorial plane*).
3. **Anaphase** (AN-a-fāz). During anaphase, the centromeres split, separating the two members of each chromatid pair, which move toward opposite poles of the cell (Figure 3.32d). Once separated, the chromatids are termed *chromosomes.* As the chromosomes are pulled by the microtubules of the mitotic spindle during anaphase, they appear V-shaped because the centromeres lead the way, dragging the trailing arms of the chromosomes toward the pole.
4. **Telophase** (TEL-ō-fāz). The final stage of mitosis, telophase, begins after chromosomal movement stops (Figure 3.32e). The identical sets of chromosomes, now at opposite poles of the cell, uncoil and revert to the threadlike chromatin form. A nuclear envelope forms around each chromatin mass, nucleoli reappear in the identical nuclei, and the mitotic spindle breaks up.

CYTOPLASMIC DIVISION: CYTOKINESIS As noted earlier, division of a cell's cytoplasm and organelles into two identical cells is called **cytokinesis**. This process usually begins in late anaphase with the formation of a **cleavage furrow**, a slight indentation of the plasma membrane, and is completed after telophase. The cleavage furrow usually appears midway between the centrosomes and extends around the periphery of the cell (Figure 3.32d, e). Actin microfilaments that lie just inside the plasma membrane form a *contractile ring* that pulls the plasma membrane progressively inward. The ring constricts the center of the cell, like tightening a belt around the waist, and ultimately pinches it in two. Because the plane of the cleavage furrow is always perpendicular to the mitotic spindle, the two sets of chromosomes end up in separate cells. When cytokinesis is complete, interphase begins (Figure 3.32f).

The sequence of events can be summarized as

$$G_1 \longrightarrow \text{S phase} \longrightarrow G_2 \text{ phase} \longrightarrow \text{mitosis} \longrightarrow \text{cytokinesis}$$

Table 3.3 summarizes the events of the cell cycle in somatic cells.

Control of Cell Destiny

A cell has three possible destinies: (1) to remain alive and functioning without dividing, (2) to grow and divide, or (3) to die. Homeostasis is maintained when there is a balance between cell proliferation and cell death. Various signals tell a cell when to exist in the G_0 phase, when to divide, and when to die.

Within a cell, there are enzymes called **cyclin-dependent protein kinases (Cdk's)** that can transfer a phosphate group from ATP to a protein to activate the protein; other enzymes can remove the phosphate group from the protein to deactivate it. The activation and deactivation of Cdk's at the appropriate time is crucial in the initiation and regulation of DNA replication, mitosis, and cytokinesis.

Switching the Cdk's on and off is the responsibility of cellular proteins called **cyclins** (SĪK-lins), so named because their levels rise and fall during the cell cycle. The joining of a specific cyclin and Cdk molecule triggers various events that control cell division.

The activation of specific cyclin–Cdk complexes is responsible for progression of a cell from G_1 to S to G_2 to mitosis in a specific order. If any step in the sequence is delayed, all subsequent steps are delayed in order to maintain the normal sequence. The levels of cyclins in the cell are very important in determining the timing and sequence of events in cell division. For example, the level of the cyclin that helps drive a cell from G_2 to mitosis rises throughout the G_1, S, and G_2 phases and into mitosis. The high level triggers mitosis, but toward the end of mitosis, the level declines rapidly and mitosis ends. Destruction of this cyclin, as well as others in the cell, is by proteasomes.

Cellular death is also regulated. Throughout the lifetime of an organism, certain cells undergo apoptosis, an orderly, genetically programmed death (see the discussion of mitochondria in Section 3.4). In apoptosis, a triggering agent from either outside or inside the cell causes "cell-suicide" genes to produce enzymes that damage the cell in several ways, including disruption of its cytoskeleton and nucleus. As a result, the cell shrinks and pulls away from neighboring cells. Although the plasma membrane remains intact, the DNA within the nucleus fragments and the cytoplasm shrinks. Nearby phagocytes then ingest the dying cell via a complex process that involves a receptor

TABLE 3.3 Events of the Somatic Cell Cycle

PHASE	ACTIVITY
Interphase	Period between cell divisions; chromosomes not visible under light microscope.
G_1 phase	Metabolically active cell duplicates most of its organelles and cytosolic components; replication of chromosomes begins. (Cells that remain in the G_1 phase for a very long time, and possibly never divide again, are said to be in the G_0 phase.)
S phase	Replication of DNA and centrosomes.
G_2 phase	Cell growth, enzyme and protein synthesis continue; replication of centrosomes complete.
Mitotic phase	Parent cell produces identical cells with identical chromosomes; chromosomes visible under light microscope.
Mitosis	Nuclear division; distribution of two sets of chromosomes into separate nuclei.
Prophase	Chromatin fibers condense into paired chromatids; nucleolus and nuclear envelope disappear; each centrosome moves to an opposite pole of the cell.
Metaphase	Centromeres of chromatid pairs line up at metaphase plate.
Anaphase	Centromeres split; identical sets of chromosomes move to opposite poles of cell.
Telophase	Nuclear envelopes and nucleoli reappear; chromosomes resume chromatin form; mitotic spindle disappears.
Cytokinesis	Cytoplasmic division; contractile ring forms cleavage furrow around center of cell, dividing cytoplasm into separate and equal portions.

protein in the plasma membrane of the phagocyte that binds to a lipid in the plasma membrane of the dying cell. Apoptosis removes unneeded cells during fetal development, such as the webbing between digits. It continues to occur after birth to regulate the number of cells in a tissue and eliminate potentially dangerous cells such as cancer cells.

Apoptosis is a normal type of cell death; in contrast, **necrosis** (ne-KRŌ-sis = death) is a pathological type of cell death that results from tissue injury. In necrosis, many adjacent cells swell, burst, and spill their cytoplasm into the interstitial fluid. The cellular debris usually stimulates an inflammatory response by the immune system, a process that does not occur in apoptosis.

Reproductive Cell Division

In the process called sexual reproduction, each new organism is the result of the union of two different gametes (fertilization), one produced by each parent. If gametes had the same number of chromosomes as somatic cells, the number of chromosomes would double at fertilization. **Meiosis** (mī-Ō-sis; *mei-* = lessening; *-osis* = condition of), the reproductive cell division that occurs in the gonads (ovaries and testes), produces gametes in which the number of chromosomes is reduced by half. As a result, gametes contain a single set of 23 chromosomes and thus are **haploid (*n*) cells** (HAP-loyd; *hapl-* = single). Fertilization restores the diploid number of chromosomes.

Meiosis

Unlike mitosis, which is complete after a single round, meiosis occurs in two successive stages: **meiosis I** and **meiosis II**. During the interphase that precedes meiosis I, the chromosomes of the diploid cell start to replicate. As a result of replication, each chromosome consists of two sister (genetically identical) chromatids, which are attached at their centromeres. This replication of chromosomes is similar to the one that precedes mitosis in somatic cell division.

MEIOSIS I Meiosis I, which begins once chromosomal replication is complete, consists of four phases: prophase I, metaphase I, anaphase I, and telophase I (**Figure 3.33a**). Prophase I is an extended phase in which the chromosomes shorten and thicken, the nuclear envelope and nucleoli disappear, and the mitotic spindle forms. Two events that are not seen in mitotic prophase occur during prophase I of meiosis (**Figure 3.33b**). First, the two sister chromatids of each pair of homologous chromosomes pair off, an event called **synapsis** (sin-AP-sis). The resulting four chromatids form a structure called a **tetrad** (TE-trad; *tetra* = four). Second, parts of the chromatids of two homologous chromosomes may be exchanged with one another. Such an exchange between parts of nonsister (genetically different) chromatids is called **crossing-over**. This process, among others, permits an exchange of genes between chromatids of homologous chromosomes. Due to crossing-over, the resulting cells are genetically unlike each other and genetically unlike the starting cell that produced them. Crossing-over results in **genetic recombination**—that is, the formation of new combinations of genes—and accounts for part of the great genetic variation among humans and other organisms that form gametes via meiosis.

In metaphase I, the tetrads formed by the homologous pairs of chromosomes line up along the metaphase plate of the cell, with homologous chromosomes side by side (**Figure 3.33a**). During anaphase I, the members of each homologous pair of chromosomes separate as they are pulled to opposite poles of the cell by the microtubules attached to the centromeres. The paired chromatids, held by a centromere, remain together. (Recall that during mitotic anaphase, the centromeres split and the sister chromatids separate.) Telophase I and cytokinesis of meiosis are similar to telophase and cytokinesis of mitosis. The net effect of meiosis I is that each resulting cell contains the haploid number of chromosomes because it contains only one member of each pair of the homologous chromosomes present in the starting cell.

MEIOSIS II The second stage of meiosis, meiosis II, also consists of four phases: prophase II, metaphase II, anaphase II, and telophase II (**Figure 3.33a**). These phases are similar to those that occur during mitosis; the centromeres split, and the sister chromatids separate and move toward opposite poles of the cell.

In summary, meiosis I begins with a diploid starting cell and ends with two cells, each with the haploid number of chromosomes. During meiosis II, each of the two haploid cells formed during meiosis I divides; the net result is four haploid gametes that are genetically different from the original diploid starting cell.

Figure 3.34 and **Table 3.4** compare the events of mitosis and meiosis.

TABLE 3.4 Comparison between Mitosis and Meiosis

POINT OF COMPARISON	MITOSIS	MEIOSIS
Cell type	Somatic.	Gamete.
Number of divisions	1	2
Stages	Interphase.	Interphase I only.
	Prophase.	Prophase I and II.
	Metaphase.	Metaphase I and II.
	Anaphase.	Anaphase I and II.
	Telophase.	Telophase I and II.
Copy DNA?	Yes, interphase.	Yes, interphase I; No, interphase II.
Tetrads?	No.	Yes.
Number of cells	2.	4.
Number of chromosomes per cell.	46, or two sets of 23; this makeup, called diploid (2*n*), is identical to the chromosomes in the starting cell.	One set of 23; this makeup, called haploid (*n*), represents half of the chromosomes in the starting cell.

FIGURE 3.33 Meiosis, reproductive cell division. Details of each of the stages are discussed in the text.

In reproductive cell division, a single diploid starting cell undergoes meiosis I and meiosis II to produce four haploid gametes that are genetically different from the starting cell that produced them.

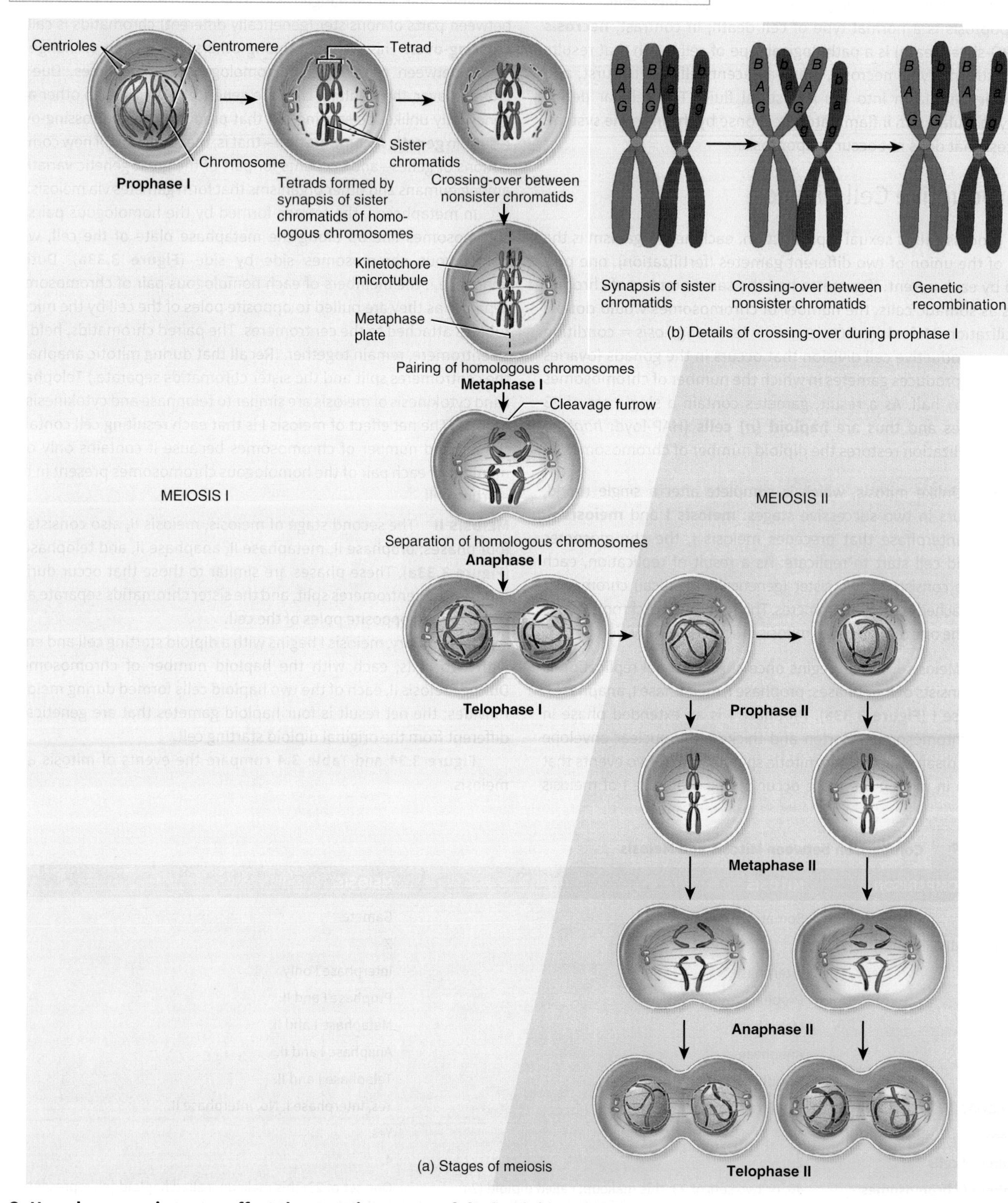

Q How does crossing-over affect the genetic content of the haploid gametes?

FIGURE 3.34 Comparison between mitosis (left) and meiosis (right) in which the starting cell has two pairs of homologous chromosomes.

The phases of meiosis II and mitosis are similar.

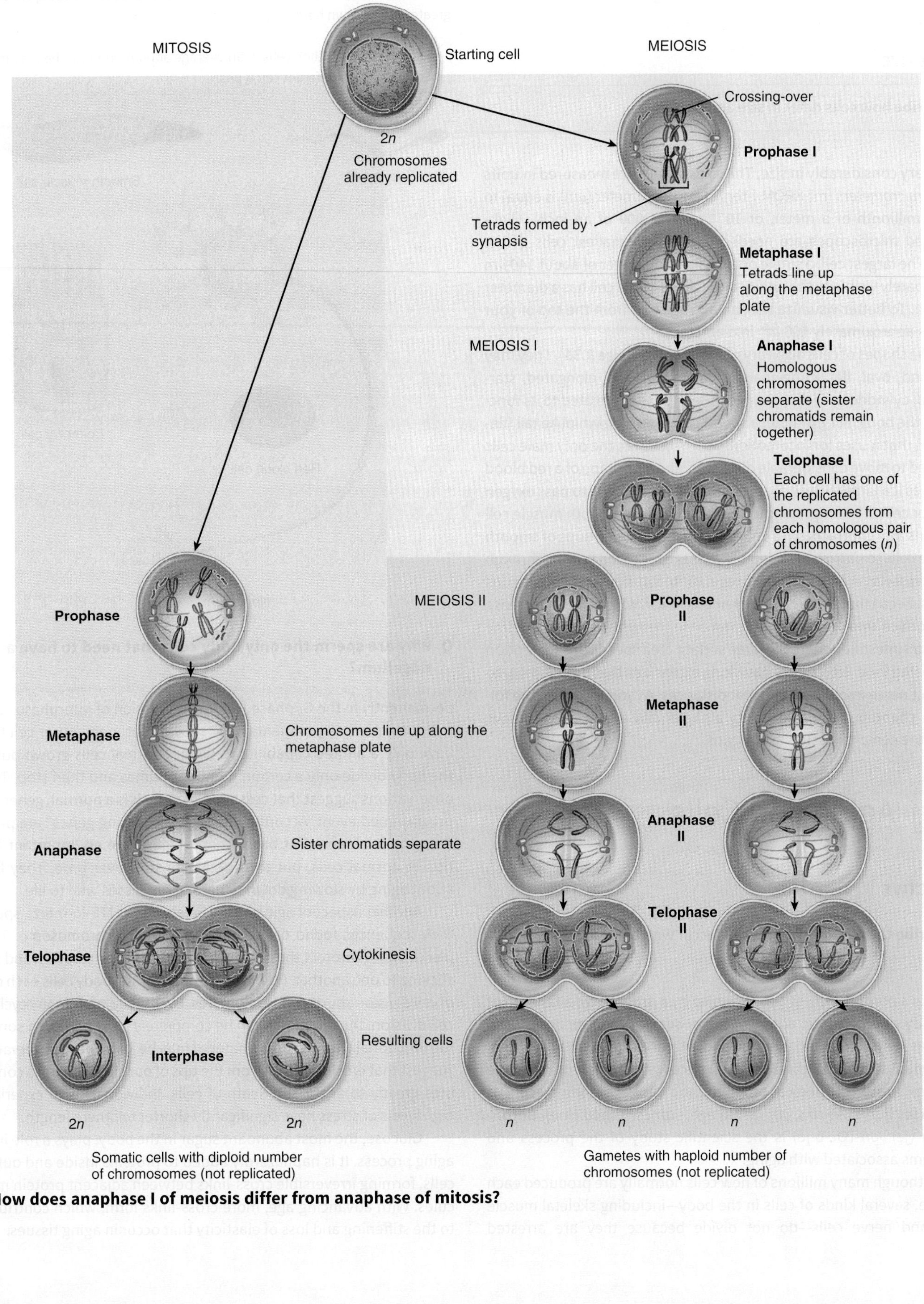

MITOSIS

Starting cell

2n

Chromosomes already replicated

MEIOSIS

Crossing-over

Prophase I

Tetrads formed by synapsis

Metaphase I
Tetrads line up along the metaphase plate

MEIOSIS I

Anaphase I
Homologous chromosomes separate (sister chromatids remain together)

Telophase I
Each cell has one of the replicated chromosomes from each homologous pair of chromosomes (*n*)

Prophase

MEIOSIS II

Prophase II

Metaphase

Chromosomes line up along the metaphase plate

Metaphase II

Anaphase

Sister chromatids separate

Anaphase II

Telophase

Cytokinesis

Telophase II

Resulting cells

Interphase

2n 2n

n *n* *n* *n*

Somatic cells with diploid number of chromosomes (not replicated)

Gametes with haploid number of chromosomes (not replicated)

Q How does anaphase I of meiosis differ from anaphase of mitosis?

3.8 Cellular Diversity

OBJECTIVE

- **Describe** how cells differ in size and shape.

Cells vary considerably in size. The sizes of cells are measured in units called *micrometers* (mī-KROM-i-ters). One micrometer (μm) is equal to 1 one-millionth of a meter, or 10^{-6} m (1/25,000 of an inch). High-powered microscopes are needed to see the smallest cells of the body. The largest cell, a single oocyte, has a diameter of about 140 μm and is barely visible to the unaided eye. A red blood cell has a diameter of 8 μm. To better visualize this, an average hair from the top of your head is approximately 100 μm in diameter.

The shapes of cells also vary considerably (**Figure 3.35**). They may be round, oval, flat, cube-shaped, column-shaped, elongated, star-shaped, cylindrical, or disc-shaped. A cell's shape is related to its function in the body. For example, a sperm cell has a long whiplike tail (flagellum) that it uses for locomotion. Sperm cells are the only male cells required to move considerable distances. The disc shape of a red blood cell gives it a large surface area that enhances its ability to pass oxygen to other cells. The long, spindle shape of a relaxed smooth muscle cell shortens as it contracts. This change in shape allows groups of smooth muscle cells to narrow or widen the passage for blood flowing through blood vessels. In this way, they regulate blood flow through various tissues. Recall that some cells contain microvilli, which greatly increase their surface area. Microvilli are common in the epithelial cells that line the small intestine, where the large surface area speeds the absorption of digested food. Nerve cells have long extensions that permit them to conduct nerve impulses over great distances. As you will see in the following chapters, cellular diversity also permits organization of cells into more complex tissues and organs.

FIGURE 3.35 Diverse shapes and sizes of human cells. The relative difference in size between the smallest and largest cells is actually much greater than shown here.

The nearly 100 trillion cells in an average adult human can be classified into about 200 different cell types.

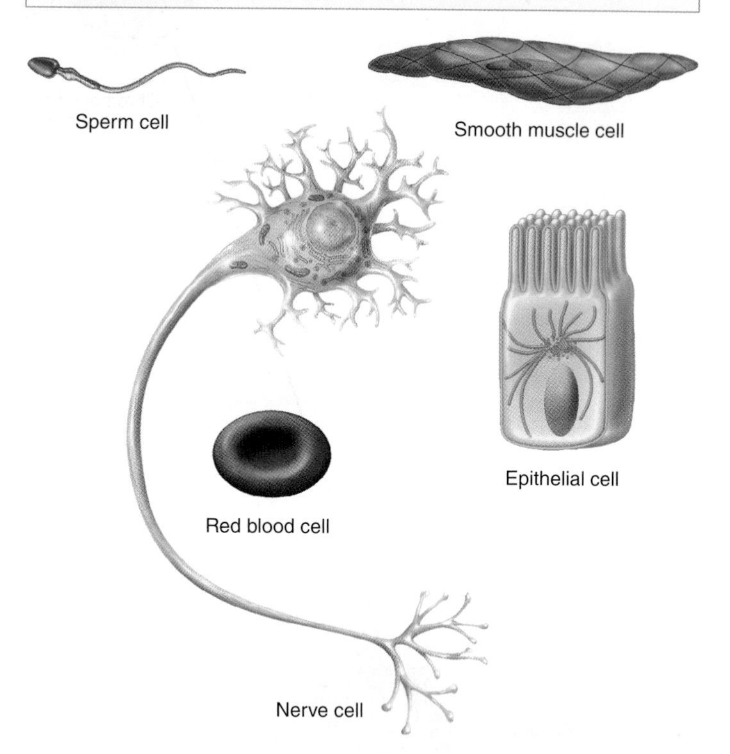

Q Why are sperm the only body cells that need to have a flagellum?

3.9 Aging and Cells

OBJECTIVE

- **Describe** the cellular changes that occur with aging.

Aging is a normal process accompanied by a progressive alteration of the body's homeostatic adaptive responses. It produces observable changes in structure and function and increases vulnerability to environmental stress and disease. The specialized branch of medicine that deals with the medical problems and care of elderly persons is **geriatrics** (jer′-ē-AT-riks; *ger-* = old age; *-iatrics* = medicine). **Gerontology** (jer′-on-TOL-ō-jē) is the scientific study of the process and problems associated with aging.

Although many millions of new cells normally are produced each minute, several kinds of cells in the body—including skeletal muscle cells and nerve cells—do not divide because they are arrested permanently in the G_0 phase (see the discussion of interphase earlier in the chapter). Experiments have shown that many other cell types have only a limited capability to divide. Normal cells grown outside the body divide only a certain number of times and then stop. These observations suggest that cessation of mitosis is a normal, genetically programmed event. According to this view, "aging genes" are part of the genetic blueprint at birth. These genes have an important function in normal cells, but their activities slow over time. They bring about aging by slowing down or halting processes vital to life.

Another aspect of aging involves **telomeres** (TĒ-lō-mērz), specific DNA sequences found only at the tips of each chromosome. These pieces of DNA protect the tips of chromosomes from erosion and from sticking to one another. However, in most normal body cells each cycle of cell division shortens the telomeres. Eventually, after many cycles of cell division, the telomeres can be completely gone and even some of the functional chromosomal material may be lost. These observations suggest that erosion of DNA from the tips of our chromosomes contributes greatly to aging and death of cells. Individuals who experience high levels of stress have significantly shorter telomere length.

Glucose, the most abundant sugar in the body, plays a role in the aging process. It is haphazardly added to proteins inside and outside cells, forming irreversible cross-links between adjacent protein molecules. With advancing age, more cross-links form, which contributes to the stiffening and loss of elasticity that occur in aging tissues.

Some theories of aging explain the process at the cellular level, while others concentrate on regulatory mechanisms operating within the entire organism. For example, the immune system may start to attack the body's own cells. This *autoimmune response* might be caused by changes in cell-identity markers at the surface of cells that cause antibodies to attach to and mark the cell for destruction. As changes in the proteins on the plasma membrane of cells increase, the autoimmune response intensifies, producing the well-known signs of aging. In the chapters that follow, we will discuss the effects of aging on each body system in sections similar to this one.

See Clinical Connection: Free Radicals and Antioxidants

Review Questions

3.1 Parts of a Cell

1. List the three main parts of a cell and explain their functions.

3.2 The Plasma Membrane

2. How do hydrophobic and hydrophilic regions govern the arrangement of membrane lipids in a bilayer?
3. What substances can and cannot diffuse through the lipid bilayer?
4. "The proteins present in a plasma membrane determine the functions that a membrane can perform." Is this statement true or false? Explain your answer.
5. How does cholesterol affect membrane fluidity?
6. Why are membranes said to have selective permeability?
7. What factors contribute to an electrochemical gradient?

3.3 Transport across the Plasma Membrane

8. What factors can increase the rate of diffusion?
9. How does simple diffusion compare with facilitated diffusion?
10. What is osmotic pressure?
11. Distinguish among isotonic, hypotonic, and hypertonic solutions.
12. What is the key difference between passive and active processes?
13. How do symporters and antiporters carry out their functions?
14. What are the sources of cellular energy for active transport?
15. In what ways are endocytosis and exocytosis similar and different?

3.4 Cytoplasm

16. What are some of the chemicals present in cytosol?
17. What is the function of cytosol?
18. Define an organelle.
19. Which organelles are surrounded by a membrane and which are not?
20. Which organelles contribute to synthesizing protein hormones and packaging them into secretory vesicles?
21. What happens on the cristae and in the matrix of mitochondria?

3.5 Nucleus

22. How do large particles enter and exit the nucleus?
23. Where are ribosomes produced?
24. How is DNA packed in the nucleus?

3.6 Protein Synthesis

25. What is meant by the term gene expression?
26. What is the difference between transcription and translation?

3.7 Cell Division

27. Distinguish between somatic and reproductive cell division and explain the importance of each.
28. What is the significance of interphase?
29. Outline the major events of each stage of the mitotic phase of the cell cycle.
30. How are apoptosis and necrosis similar? How do they differ?
31. How are haploid and diploid cells different?
32. What are homologous chromosomes?

3.8 Cellular Diversity

33. How is cell shape related to function? Give several of your own examples.

3.9 Aging and Cells

34. What is one reason that some tissues become stiffer as they age?

Critical Thinking Questions

1. Mucin is a protein present in saliva and other secretions. When mixed with water, it becomes the slippery substance known as mucus. Trace the route taken by mucin through the cell, from its synthesis to its secretion, listing all organelles and processes involved.
2. Sam does not consume alcohol, whereas his brother Sebastian regularly drinks large quantities of alcohol. If we could examine the liver cells of each of these brothers, would we see a difference in smooth ER and peroxisomes? Explain.
3. Marathon runners can become dehydrated due to the extreme physical activity. What types of fluids should they consume in order to rehydrate their cells?

The Tissue Level of Organization

CHAPTER **4**

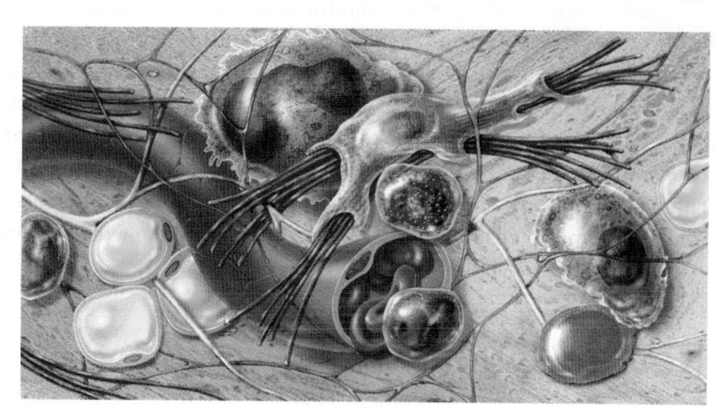

Tissues and Homeostasis

The four basic types of tissues in the human body contribute to homeostasis by providing diverse functions including protection, support, communication among cells, and resistance to disease, to name just a few.

As you learned in Chapter 3, a cell is a complex collection of compartments, each of which carries out a host of biochemical reactions that make life possible. However, a cell seldom functions as an isolated unit in the body. Instead, cells usually work together in groups called tissues. The structure and properties of a specific tissue are influenced by factors such as the nature of the extracellular material that surrounds the tissue cells and the connections between the cells that compose the tissue. Tissues may be hard, semisolid, or even liquid in their consistency, a range exemplified by bone, fat, and blood. In addition, tissues vary tremendously with respect to the kinds of cells present, how the cells are arranged, and the type of extracellular material.

Q Did you ever wonder whether the complications of liposuction outweigh the benefits?

4.1 Types of Tissues

OBJECTIVE

- **Name** the four basic types of tissues that make up the human body, and **state** the characteristics of each.

A **tissue** is a group of cells that usually have a common origin in an embryo and function together to carry out specialized activities. **Histology** (his′-TOL-ō-jē; *histo-* = tissue; *-logy* = study of) is the science that deals with the study of tissues. A **pathologist** (pa-THOL-ō-jist; *patho-* = disease) is a physician who examines cells and tissues to help other physicians make accurate diagnoses. One of the principal functions of a pathologist is to examine tissues for any changes that might indicate disease.

Body tissues can be classified into four basic types according to their structure and function (**Figure 4.1**):

1. **Epithelial tissue** covers body surfaces and lines hollow organs, body cavities, and ducts; it also forms glands. This tissue allows the body to interact with both its internal and external environments.
2. **Connective tissue** protects and supports the body and its organs. Various types of connective tissues bind organs together, store energy reserves as fat, and help provide the body with immunity to disease-causing organisms.
3. **Muscular tissue** is composed of cells specialized for contraction and generation of force. In the process, muscular tissue generates heat that warms the body.
4. **Nervous tissue** detects changes in a variety of conditions inside and outside the body and responds by generating electrical signals called nerve action potentials (nerve impulses) that activate muscular contractions and glandular secretions.

Epithelial tissue and most types of connective tissue, except cartilage, bone, and blood, are more general in nature and have a wide distribution in the body. These tissues are components of most body organs and have a wide range of structures and functions. We will look at epithelial tissue and connective tissue in some detail in this chapter. The general features of bone tissue and blood will be introduced here, but their detailed discussion is presented in Chapters 6 and 19, respectively. Similarly, the structure and function of muscular tissue and nervous tissue are introduced here and examined in detail in Chapters 10 and 12, respectively.

Normally, most cells within a tissue remain anchored to other cells or structures. Only a few cells, such as phagocytes, move freely through the body, searching for invaders to destroy. However, many cells migrate extensively during the growth and development process before birth.

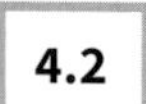

See Clinical Connection: Biopsy

4.2 Cell Junctions

OBJECTIVE

- **Describe** the structure and functions of the five main types of cell junctions.

Before looking more specifically at the types of tissues, we will first examine how cells are held together to form tissues. Most

FIGURE 4.1 **Types of tissues.**

Each of the four types of tissues has different cells that vary in shape, structure, function, and distribution.

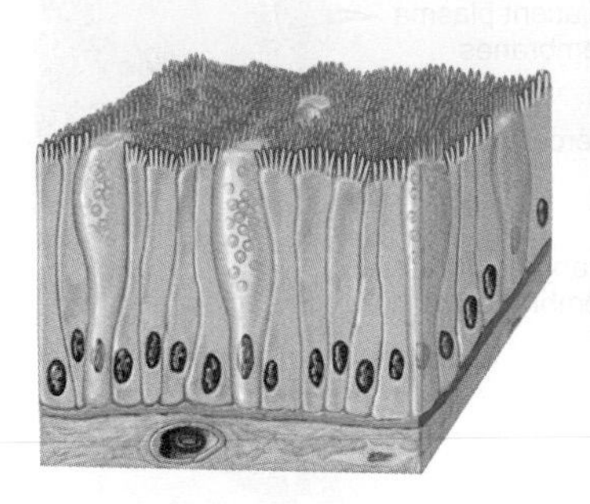

(a) Epithelial tissue

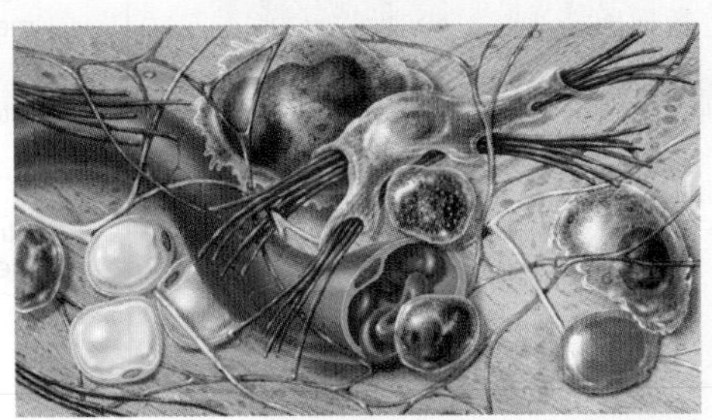

(b) Connective tissue

(c) Muscular tissue

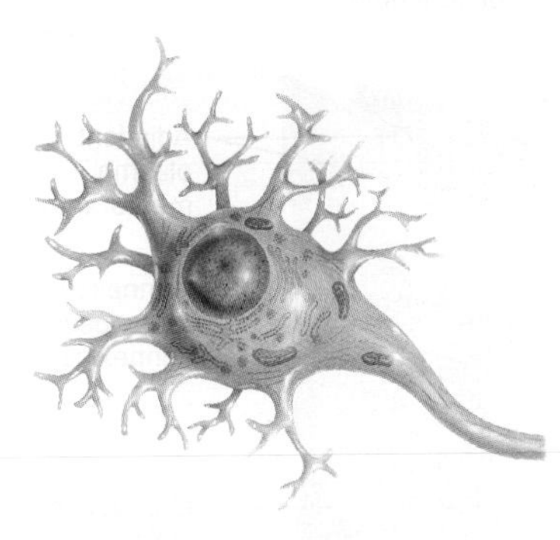

(d) Nervous tissue

Q What are some key differences in function among the four tissue types?

epithelial cells and some muscle and nerve cells are tightly joined into functional units. **Cell junctions** are contact points between the plasma membranes of tissue cells. Here we consider the five most important types of cell junctions: tight junctions, adherens junctions, desmosomes, hemidesmosomes, and gap junctions (**Figure 4.2**).

Tight Junctions

Tight junctions consist of weblike strands of transmembrane proteins that fuse together the outer surfaces of adjacent plasma membranes to seal off passageways between adjacent cells (**Figure 4.2a**). Cells of epithelial tissue that lines the stomach, intestines, and urinary bladder have many tight junctions. They inhibit the passage of substances between cells and prevent the contents of these organs from leaking into the blood or surrounding tissues.

Adherens Junctions

Adherens junctions (ad-HĒR-ens) contain *plaque* (PLAK), a dense layer of proteins on the inside of the plasma membrane that attaches both to membrane proteins and to microfilaments of the cytoskeleton (**Figure 4.2b**). Transmembrane glycoproteins called **cadherins** join the cells. Each cadherin inserts into the plaque from the opposite side of the plasma membrane, partially crosses the intercellular space (the space between the cells), and connects to cadherins of an adjacent cell. In epithelial cells, adherens junctions often form extensive zones called **adhesion belts** because they encircle the cell similar to the way a belt encircles your waist. Adherens junctions help epithelial surfaces resist separation during various contractile activities, as when food moves through the intestines.

Desmosomes

Like adherens junctions, **desmosomes** (DEZ-mō-sōms; *desmo-* = band) contain plaque and have transmembrane glycoproteins (cadherins) that extend into the intercellular space between adjacent cell membranes and attach cells to one another (**Figure 4.2c**). However, unlike adherens junctions, the plaque of desmosomes does not attach to microfilaments. Instead, a desmosome plaque attaches to elements of the cytoskeleton known as intermediate filaments, which consist of the protein keratin. The intermediate filaments extend from desmosomes on one side of the cell across the cytosol to desmosomes on the opposite side of the cell. This structural arrangement contributes to the stability of the cells and tissue. These spot weld–like junctions are common among the cells that make up the epidermis (the outermost layer of the skin) and among cardiac muscle cells in the heart. Desmosomes prevent epidermal cells from separating under tension and cardiac muscle cells from pulling apart during contraction.

Hemidesmosomes

Hemidesmosomes (*hemi-* = half) resemble desmosomes, but they do not link adjacent cells. The name arises from the fact that they look like half of a desmosome (**Figure 4.2d**). However, the transmembrane glycoproteins in hemidesmosomes are **integrins** rather than cadherins. On the inside of the plasma membrane, integrins attach to intermediate filaments made of the protein keratin. On the outside of the plasma membrane, the integrins attach to the protein *laminin*, which is present in the basement membrane (discussed shortly). Thus, hemidesmosomes anchor cells not to each other but to the basement membrane.

Gap Junctions

At **gap junctions**, membrane proteins called **connexins** form tiny fluid-filled tunnels called *connexons* that connect neighboring cells

FIGURE 4.2 **Cell junctions.**

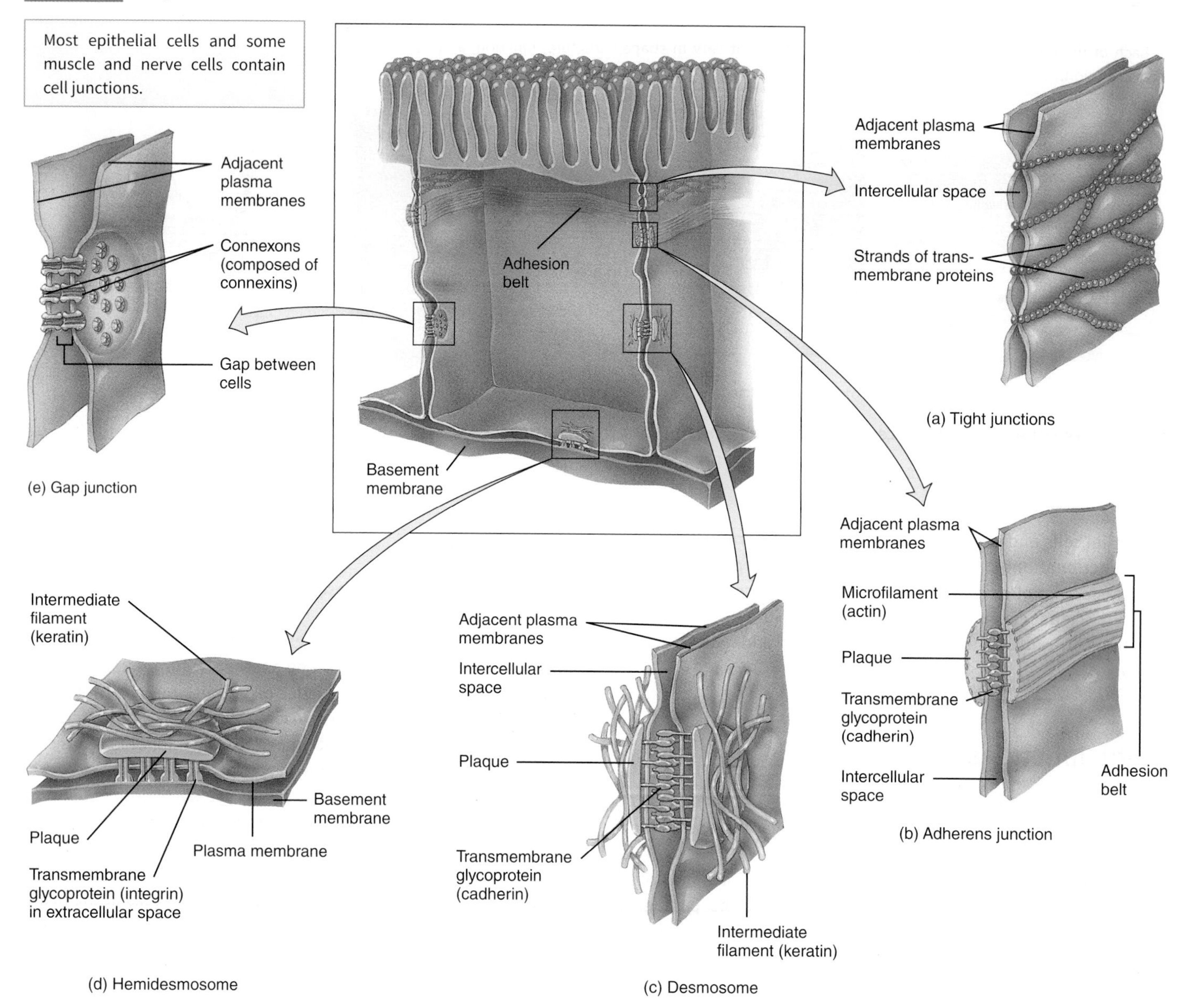

Q Which type of cell junction functions in communication between adjacent cells?

(Figure 4.2e). The plasma membranes of gap junctions are not fused together as in tight junctions but are separated by a very narrow intercellular gap (space). Through the connexons, ions and small molecules can diffuse from the cytosol of one cell to another, but the passage of large molecules such as vital intracellular proteins is prevented. The transfer of nutrients, and perhaps wastes, takes place through gap junctions in avascular tissues such as the lens and cornea of the eye. Gap junctions allow the cells in a tissue to communicate with one another. In a developing embryo, some of the chemical and electrical signals that regulate growth and cell differentiation travel via gap junctions. Gap junctions also enable nerve or muscle impulses to spread rapidly among cells, a process that is crucial for the normal operation of some parts of the nervous system and for the contraction of muscle in the heart, gastrointestinal tract, and uterus.

4.3 Comparison between Epithelial and Connective Tissues

OBJECTIVE

- **State** the main differences between epithelial and connective tissues.

Before examining epithelial tissue and connective tissue in more detail, let's compare these two widely distributed tissues (Figure 4.3).

FIGURE 4.3 Comparison between epithelial tissue and connective tissue.

The ratio of cells to extracellular matrix is a major difference between epithelial tissue and connective tissue.

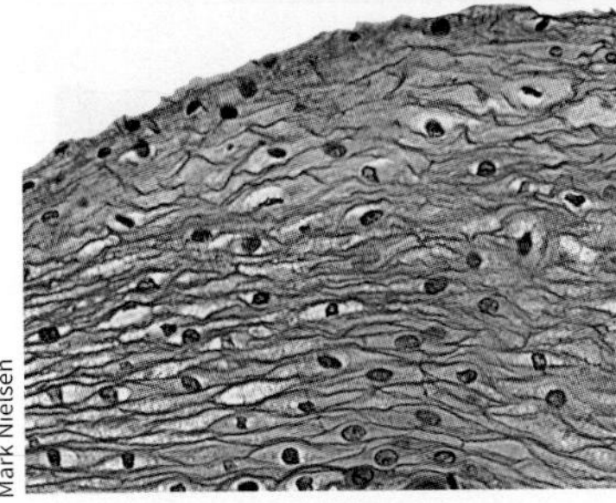

(a) Epithelial tissue with many cells tightly packed together and little to no extracellular matrix

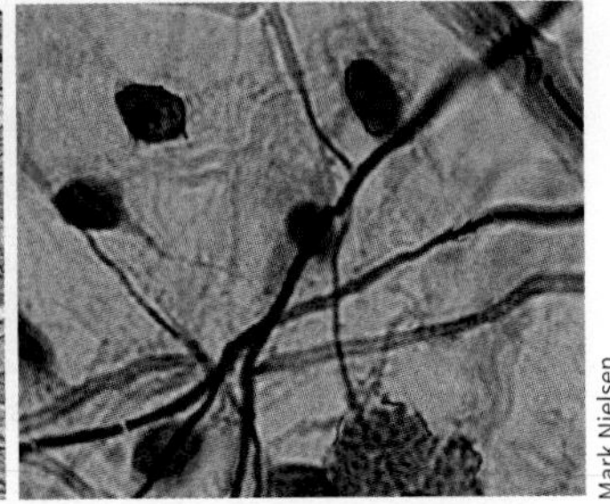

(b) Connective tissue with a few scattered cells surrounded by large amounts of extracellular matrix

Q What relationship between epithelial tissue and connective tissue is important for the survival and function of epithelial tissues?

Major structural differences between an epithelial tissue and a connective tissue are immediately obvious under a light microscope. The first obvious difference is the number of cells in relation to the extracellular matrix (the substance between cells). In an epithelial tissue many cells are tightly packed together with little or no extracellular matrix, whereas in a connective tissue a large amount of extracellular material separates cells that are usually widely scattered. The second obvious difference is that an epithelial tissue has no blood vessels, whereas most connective tissues have significant networks of blood vessels. Another key difference is that epithelial tissue almost always forms surface layers and is not covered by another tissue. An exception is the epithelial lining of blood vessels where blood constantly passes over the epithelium. While these key structural distinctions account for some of the major functional differences between these tissue types, they also lead to a common bond. Because epithelial tissue lacks blood vessels and forms surfaces, it is always found immediately adjacent to blood vessel–rich connective tissue, which enables it to make the exchanges with blood necessary for the delivery of oxygen and nutrients and the removal of wastes that are critical processes for its survival and function.

4.4 Epithelial Tissue

OBJECTIVES

- **Describe** the general features of epithelial tissue.
- **List** the location, structure, and function of each different type of epithelial tissue.

An **epithelial tissue** (ep-i-THĒ-lē-al) or *epithelium* (plural is *epithelia*) consists of cells arranged in continuous sheets, in either single or multiple layers. Because the cells are closely packed and are held tightly together by many cell junctions, there is little intercellular space between adjacent plasma membranes. Epithelial tissue is arranged in two general patterns in the body: (1) covering and lining various surfaces and (2) forming the secreting portions of glands. Functionally, epithelial tissue protects, secretes (mucus, hormones, and enzymes), absorbs (nutrients in the gastrointestinal tract), and excretes (various substances in the urinary tract).

The various surfaces of covering and lining epithelial cells often differ in structure and have specialized functions. The **apical** (*free*) **surface** of an epithelial cell faces the body surface, a body cavity, the lumen (interior space) of an internal organ, or a tubular duct that receives cell secretions (**Figure 4.4**). Apical surfaces may contain cilia or microvilli. The **lateral surfaces** of an epithelial cell, which face the adjacent cells on either side, may contain tight junctions, adherens junctions, desmosomes, and/or gap junctions. The **basal surface** of an epithelial cell is opposite the apical surface. The basal surfaces of the deepest layer of epithelial cells adhere to extracellular materials such as the basement membrane. Hemidesmosomes in the basal surfaces of the deepest layer of epithelial cells anchor the epithelium to the basement membrane (described next). In discussing epithelia with multiple layers, the term *apical layer* refers to the most superficial layer of cells, and the *basal layer* is the deepest layer of cells.

The **basement membrane** is a thin extracellular layer that commonly consists of two layers, the basal lamina and reticular lamina. The *basal lamina* (*lamina* = thin layer) is closer to—and secreted by—

FIGURE 4.4 Surfaces of epithelial cells and the structure and location of the basement membrane.

The basement membrane is found between an epithelial tissue and a connective tissue.

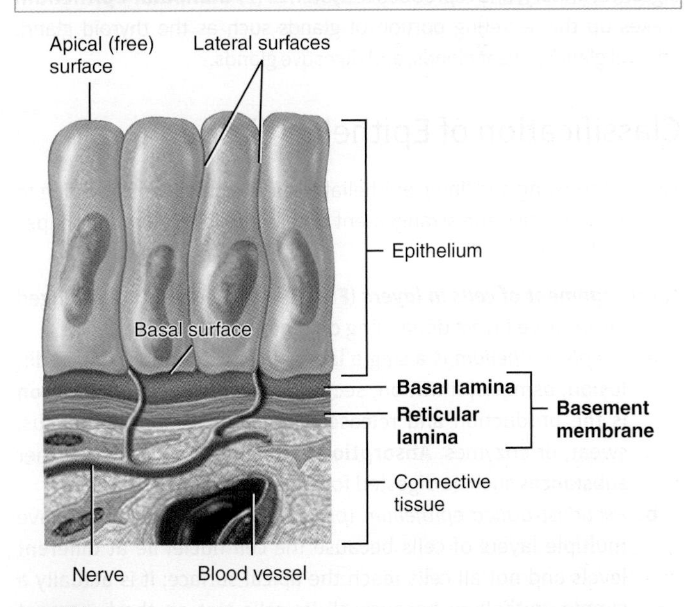

See Clinical Connection: Basement Membranes and Disease

Q What are the functions of the basement membrane?

the epithelial cells. It contains proteins such as laminin and collagen (described shortly), as well as glycoproteins and proteoglycans (also described shortly). As you have already learned, the laminin molecules in the basal lamina adhere to integrins in hemidesmosomes and thus attach epithelial cells to the basement membrane (see **Figure 4.2d**). The *reticular lamina* is closer to the underlying connective tissue and contains proteins such as collagen produced by connective tissue cells called *fibroblasts* (see **Figure 4.8**). In addition to attaching to and anchoring the epithelium to its underlying connective tissue, basement membranes have other functions. They form a surface along which epithelial cells migrate during growth or wound healing, restrict passage of larger molecules between epithelium and connective tissue, and participate in filtration of blood in the kidneys.

Epithelial tissue has its own nerve supply but, as mentioned previously, is **avascular** (*a-* = without; *-vascular* = vessel), relying on the blood vessels of the adjacent connective tissue to bring nutrients and remove wastes. Exchange of substances between an epithelial tissue and connective tissue occurs by diffusion.

Because epithelial tissue forms boundaries between the body's organs, or between the body and the external environment, it is repeatedly subjected to physical stress and injury. A high rate of cell division allows epithelial tissue to constantly renew and repair itself by sloughing off dead or injured cells and replacing them with new ones. Epithelial tissue has many different roles in the body; the most important are protection, filtration, secretion, absorption, and excretion. In addition, epithelial tissue combines with nervous tissue to form special organs for smell, hearing, vision, and touch.

Epithelial tissue may be divided into two types. (1) **Covering and lining epithelium**, also called **surface epithelium**, forms the *outer* covering of the skin and some internal organs. It also forms the *inner* lining of blood vessels, ducts, body cavities, and the interior of the respiratory, digestive, urinary, and reproductive systems. (2) **Glandular epithelium** makes up the secreting portion of glands such as the thyroid gland, adrenal glands, sweat glands, and digestive glands.

Classification of Epithelial Tissue

Types of covering and lining epithelial tissue are classified according to two characteristics: the arrangement of cells into layers and the shapes of the cells (**Figure 4.5**).

1. ***Arrangement of cells in layers*** (**Figure 4.5**). The cells are arranged in one or more layers depending on function:
   a. *Simple epithelium* is a single layer of cells that functions in diffusion, osmosis, filtration, secretion, or absorption. **Secretion** is the production and release of substances such as mucus, sweat, or enzymes. **Absorption** is the intake of fluids or other substances such as digested food from the intestinal tract.
   b. *Pseudostratified epithelium* (*pseudo-* = false) appears to have multiple layers of cells because the cell nuclei lie at different levels and not all cells reach the apical surface; it is actually a simple epithelium because all its cells rest on the basement membrane. Cells that do extend to the apical surface may contain cilia; others (goblet cells) secrete mucus.
   c. *Stratified epithelium* (*stratum* = layer) consists of two or more layers of cells that protect underlying tissues in locations where there is considerable wear and tear.

FIGURE 4.5 Cell shapes and arrangement of layers for covering and lining epithelium.

Cell shapes and arrangement of layers are the bases for classifying covering and lining epithelium.

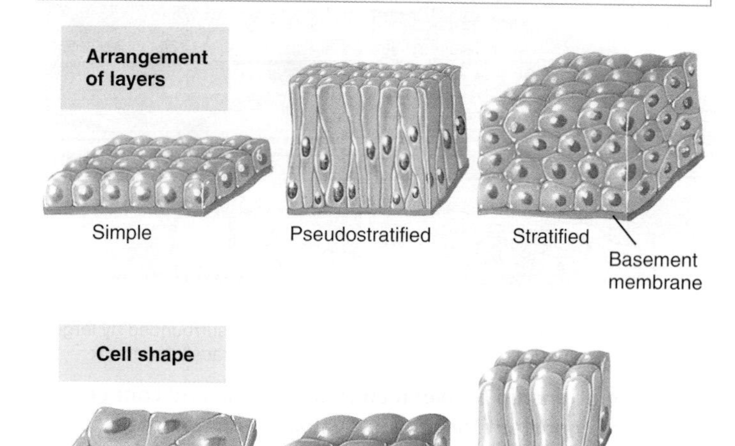

Q Which cell shape is best adapted for the rapid movement of substances from one cell to another?

2. ***Cell shapes*** (**Figure 4.5**). Epithelial cells vary in shape depending on their function:
   a. *Squamous* cells (SKWĀ-mus = flat) are thin, which allows for the rapid passage of substances through them.
   b. *Cuboidal* cells are as tall as they are wide and are shaped like cubes or hexagons. They may have microvilli at their apical surface and function in either secretion or absorption.
   c. *Columnar* cells are much taller than they are wide, like columns, and protect underlying tissues. Their apical surfaces may have cilia or microvilli, and they often are specialized for secretion and absorption.
   d. *Transitional* cells change shape, from squamous to cuboidal and back, as organs such as the urinary bladder stretch (distend) to a larger size and then collapse to a smaller size.

When we combine the two characteristics (arrangements of layers and cell shapes), we come up with the following types of epithelial tissues:

I. Simple epithelium
- A. Simple squamous epithelium
  1. Endothelium (lines heart, blood vessels, lymphatic vessels)
  2. Mesothelium (forms epithelial layer of serous membranes)
- B. Simple cuboidal epithelium
- C. Simple columnar epithelium
  1. Nonciliated (lacks cilia)
  2. Ciliated (contains cilia)
- D. Pseudostratified columnar epithelium
  1. Nonciliated (lacks cilia)
  2. Ciliated (contains cilia)

II. Stratified epithelium

A. Stratified squamous epithelium*
   1. Nonkeratinized (lacks keratin)
   2. Keratinized (contains keratin)

B. Stratified cuboidal epithelium*

C. Stratified columnar epithelium*

D. Transitional epithelium or urothelium (lines most of urinary tract)

*This classification is based on the shape of the cells at the apical surface.

Covering and Lining Epithelium

As noted earlier, covering and lining epithelium forms the outer covering of the skin and some internal organs. It also forms the inner lining of blood vessels, ducts, and body cavities, and the interior of the respiratory, digestive, urinary, and reproductive systems. **Table 4.1** describes covering and lining epithelium in more detail. The discussion of each type

TABLE 4.1 Epithelial Tissue: Covering and Lining Epithelium

A. SIMPLE SQUAMOUS EPITHELIUM

Description	**Simple squamous epithelium** is a single layer of flat cells that resembles a tiled floor when viewed from apical surface; centrally located nucleus that is flattened and oval or spherical in shape.
Location	Most commonly (1) lines the cardiovascular and lymphatic system (heart, blood vessels, lymphatic vessels), where it is known as **endothelium** (en′-dō-THĒ-lē-um; *endo-* = within; *-thelium* = covering), and (2) forms the epithelial layer of serous membranes (peritoneum, pleura, pericardium), where it is called **mesothelium** (mez′-ō-THĒ-lē-um; *meso-* = middle). Also found in air sacs of lungs, glomerular (Bowman's) capsule of kidneys, inner surface of tympanic membrane (eardrum).
Function	Present at sites of filtration (such as blood filtration in kidneys) or diffusion (such as diffusion of oxygen into blood vessels of lungs) and at site of secretion in serous membranes. Not found in body areas subject to mechanical stress (wear and tear).

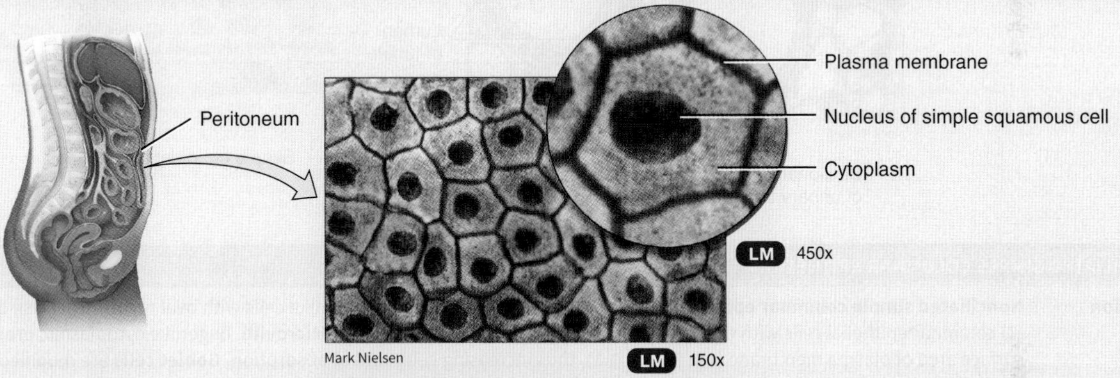

Surface view of simple squamous epithelium of mesothelial lining of peritoneum

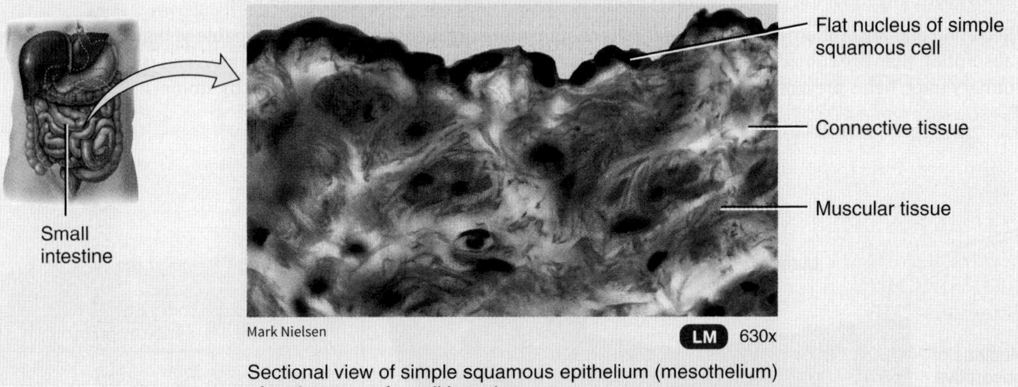

Sectional view of simple squamous epithelium (mesothelium) of peritoneum of small intestine

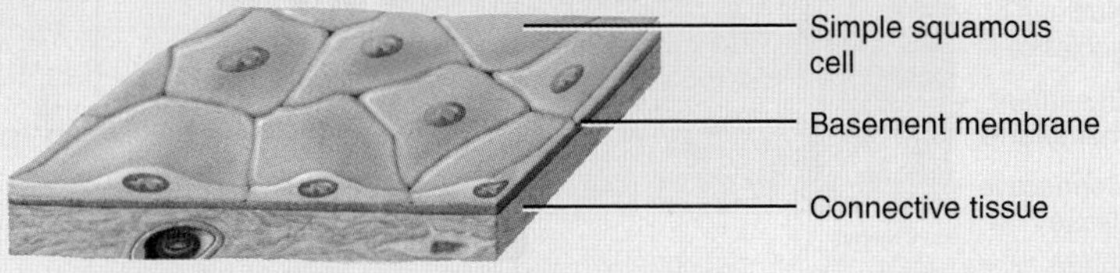

Simple squamous epithelium

Table 4.1 ***Continues***

TABLE 4.1 **Epithelial Tissue: Covering and Lining Epithelium (*Continued*)**

B. SIMPLE CUBOIDAL EPITHELIUM

Description	**Simple cuboidal epithelium** is a single layer of cube-shaped cells; round, centrally located nucleus. Cuboidal cell shape is obvious when tissue is sectioned and viewed from the side. (Note: Strictly cuboidal cells could not form small tubes; these cuboidal cells are more pie-shaped but still nearly as high as they are wide at the base.)
Location	Covers surface of ovary; lines anterior surface of capsule of lens of the eye; forms pigmented epithelium at posterior surface of retina of the eye; lines kidney tubules and smaller ducts of many glands; makes up secreting portion of some glands such as thyroid gland and ducts of some glands such as pancreas.
Function	Secretion and absorption.

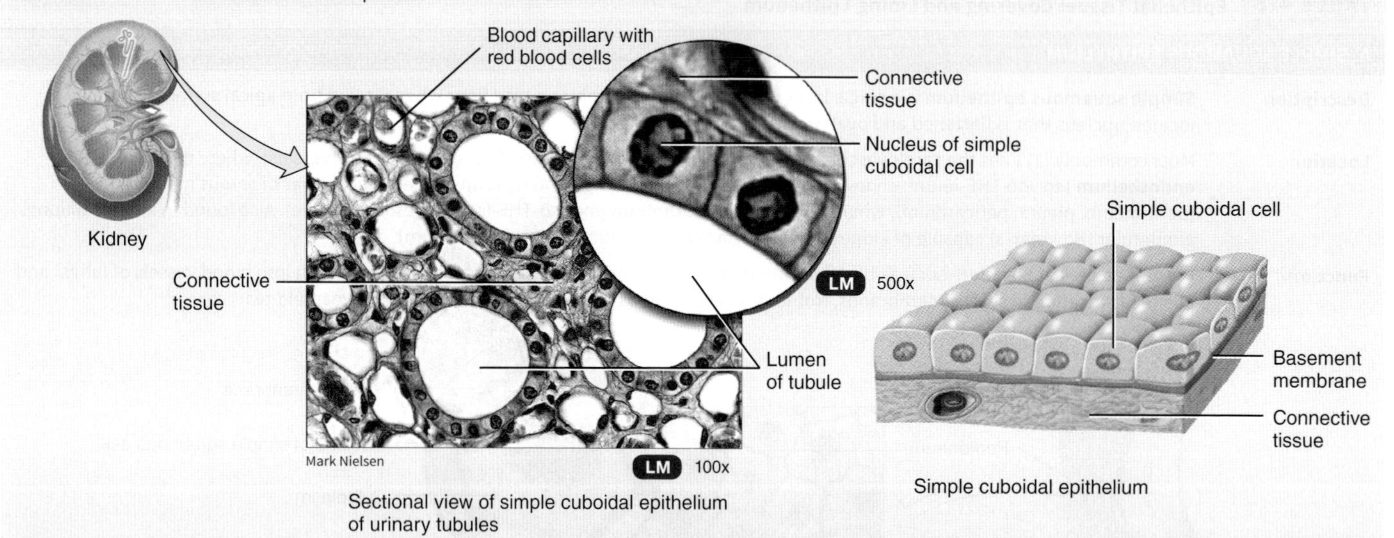

Sectional view of simple cuboidal epithelium of urinary tubules

Simple cuboidal epithelium

C. NONCILIATED SIMPLE COLUMNAR EPITHELIUM

Description	**Nonciliated simple columnar epithelium** is a single layer of nonciliated columnlike cells with oval nuclei near base of cells; contains (1) columnar epithelial cells with microvilli at apical surface and (2) goblet cells. **Microvilli**, fingerlike cytoplasmic projections, increase surface area of plasma membrane (see Figure 3.1), thus increasing cell's rate of absorption. **Goblet cells** are modified columnar epithelial cells that secrete mucus, a slightly sticky fluid, at their apical surfaces. Before release, mucus accumulates in upper portion of cell, causing it to bulge and making the whole cell resemble a goblet or wine glass.
Location	Lines gastrointestinal tract (from stomach to anus), ducts of many glands, and gallbladder.
Function	Secretion and absorption; larger columnar cells contain more organelles and thus are capable of higher level of secretion and absorption than are cuboidal cells. Secreted mucus lubricates linings of digestive, respiratory, and reproductive tracts, and most of urinary tract; helps prevent destruction of stomach lining by acidic gastric juice secreted by stomach.

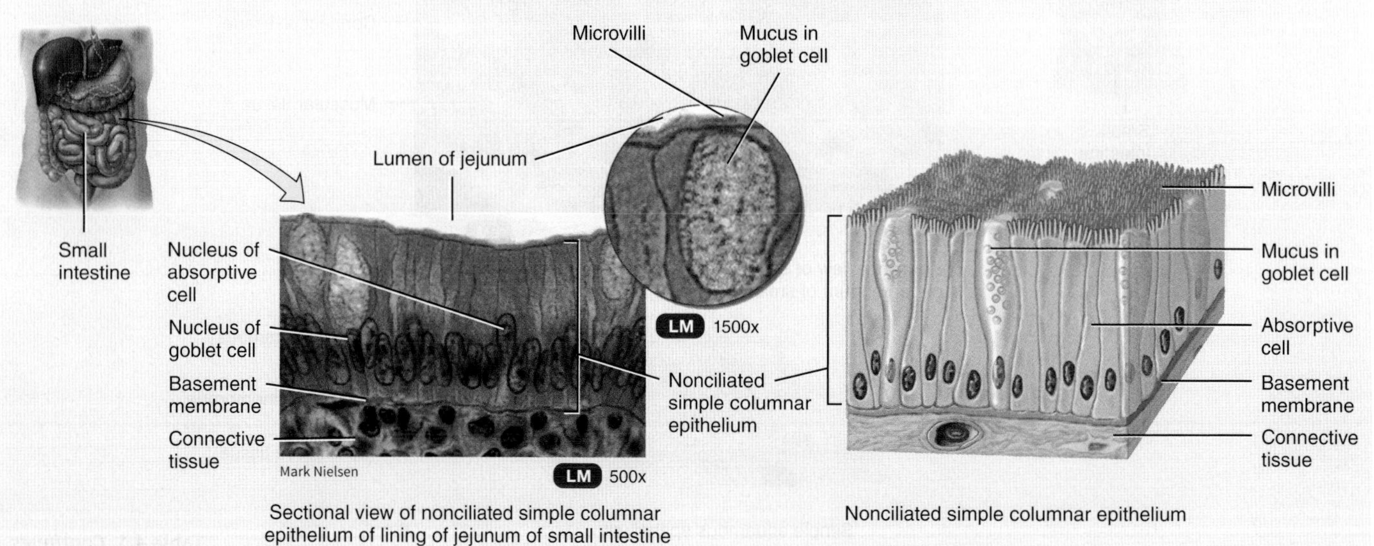

Sectional view of nonciliated simple columnar epithelium of lining of jejunum of small intestine

Nonciliated simple columnar epithelium

D. CILIATED SIMPLE COLUMNAR EPITHELIUM

Description **Ciliated simple columnar epithelium** is a single layer of ciliated columnlike cells with oval nuclei near base of cells. Goblet cells are usually interspersed.

Location Lines some bronchioles (small tubes) of respiratory tract, uterine (fallopian) tubes, uterus, some paranasal sinuses, central canal of spinal cord, and ventricles of brain.

Function Cilia beat in unison, moving mucus and foreign particles toward throat, where they can be coughed up and swallowed or spit out. Coughing and sneezing speed up movement of cilia and mucus. Cilia also help move oocytes expelled from ovaries through uterine (fallopian) tubes into uterus.

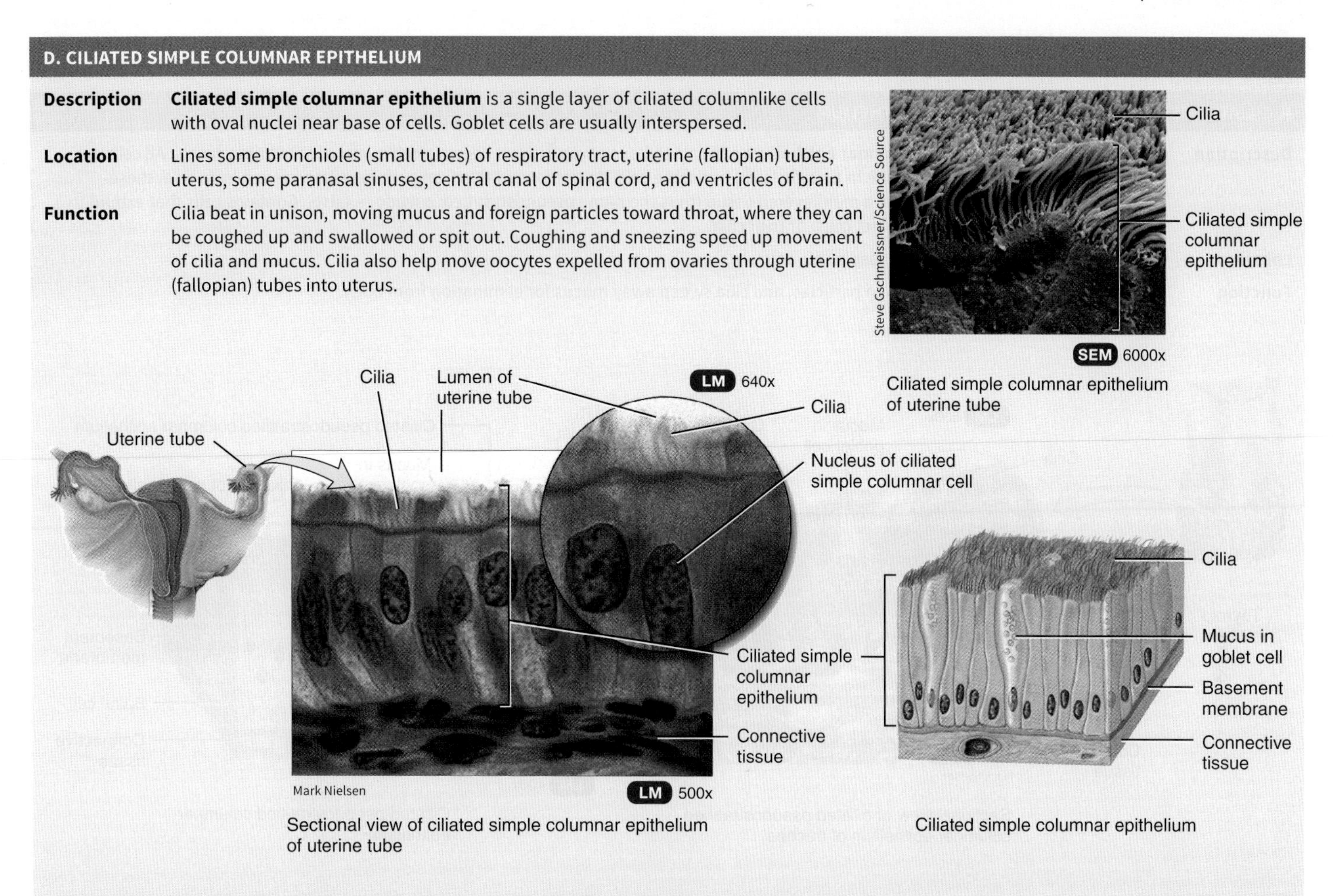

Ciliated simple columnar epithelium of uterine tube

Sectional view of ciliated simple columnar epithelium of uterine tube

Ciliated simple columnar epithelium

E. NONCILIATED PSEUDOSTRATIFIED COLUMNAR EPITHELIUM

Description **Nonciliated pseudostratified columnar epithelium** appears to have several layers because the nuclei of the cells are at various levels. Even though all the cells are attached to the basement membrane in a single layer, some cells do not extend to the apical surface. When viewed from the side, these features give the false impression of a multilayered tissue—thus the name pseudostratified epithelium (*pseudo-* = false). Contains cells without cilia and also lacks goblet cells.

Location Lines epididymis, larger ducts of many glands, and parts of male urethra.

Function Absorption and secretion.

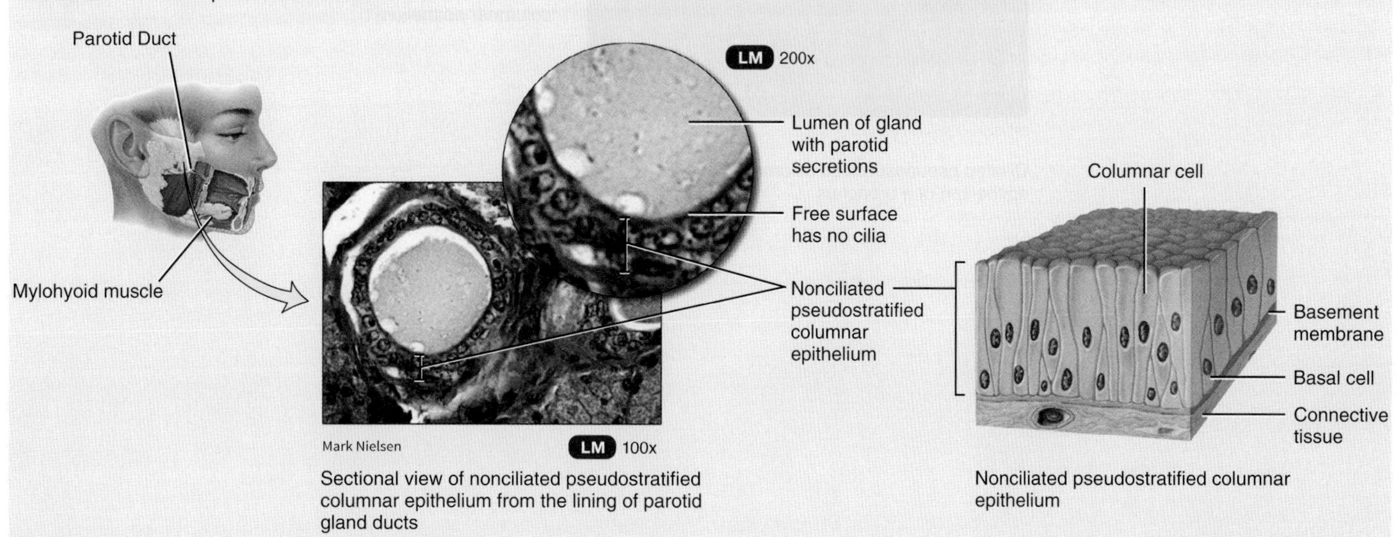

Sectional view of nonciliated pseudostratified columnar epithelium from the lining of parotid gland ducts

Nonciliated pseudostratified columnar epithelium

Table 4.1 ***Continues***

TABLE 4.1 Epithelial Tissue: Covering and Lining Epithelium (*Continued*)

F. CILIATED PSEUDOSTRATIFIED COLUMNAR EPITHELIUM

Description	**Ciliated pseudostratified columnar epithelium** appears to have several layers because cell nuclei are at various levels. All cells are attached to basement membrane in a single layer, but some cells do not extend to apical surface. When viewed from side, these features give false impression of a multilayered tissue (thus the name pseudostratified; *pseudo* = false). Contains cells that extend to surface and secrete mucus (globlet cells) or bear cilia.
Location	Lines airways of most of upper respiratory tract.
Function	Secretes mucus that traps foreign particles, and cilia sweep away mucus for elimination from body.

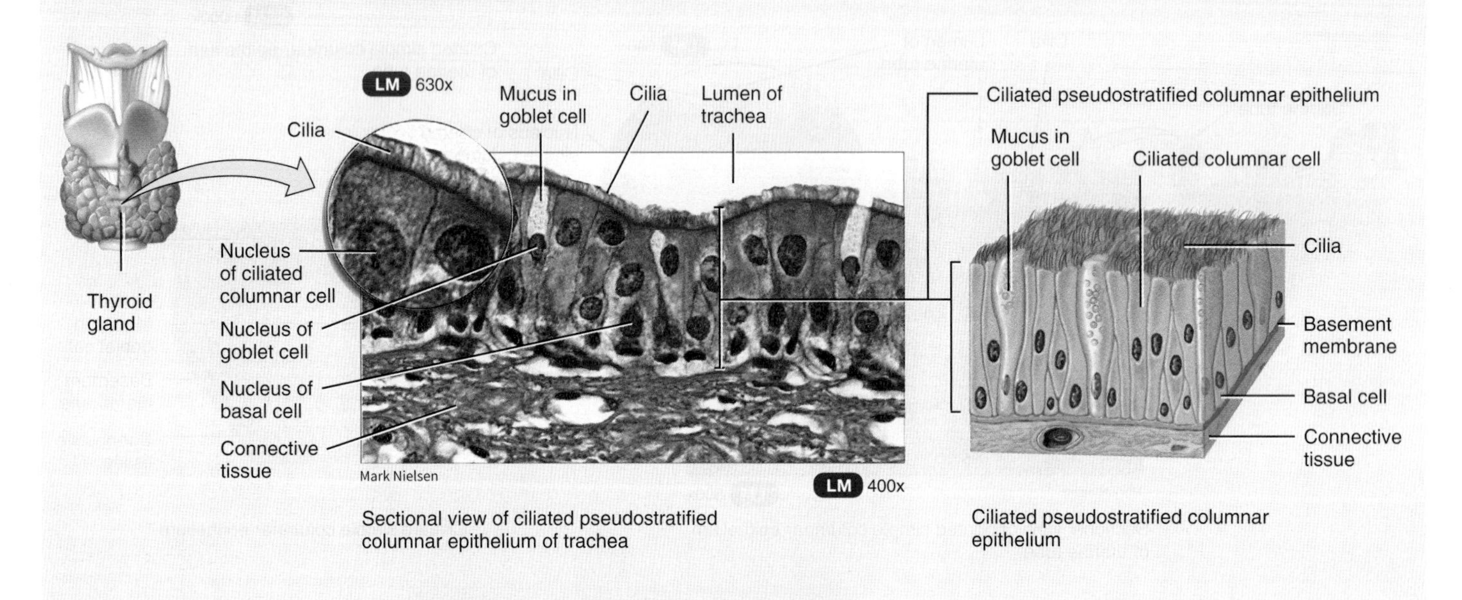

Sectional view of ciliated pseudostratified columnar epithelium of trachea

Ciliated pseudostratified columnar epithelium

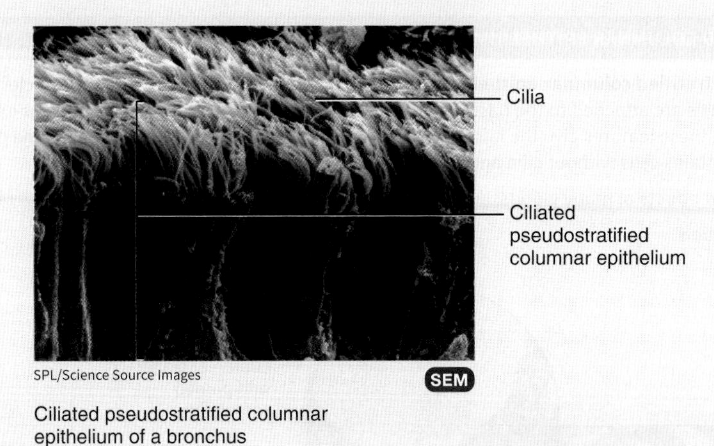

Ciliated pseudostratified columnar epithelium of a bronchus

G. STRATIFIED SQUAMOUS EPITHELIUM

Description

Stratified squamous epithelium has two or more layers of cells; cells in apical layer and several layers deep to it are squamous; cells in deeper layers vary from cuboidal to columnar. As basal cells divide, daughter cells arising from cell divisions push upward toward apical layer. As they move toward surface and away from blood supply in underlying connective tissue, they become dehydrated and less metabolically active. Tough proteins predominate as cytoplasm is reduced, and cells become tough, hard structures that eventually die. At apical layer, after dead cells lose cell junctions they are sloughed off, but they are replaced continuously as new cells emerge from basal cells.

Keratinized stratified squamous epithelium develops tough layer of keratin in apical layer of cells and several layers deep to it (see Figure 5.3). (**Keratin** is a tough, fibrous intracellular protein that helps protect skin and underlying tissues from heat, microbes, and chemicals.) Relative amount of keratin increases in cells as they move away from nutritive blood supply and organelles die.

Nonkeratinized stratified squamous epithelium does not contain large amounts of keratin in apical layer and several layers deep and is constantly moistened by mucus from salivary and mucous glands; organelles are not replaced.

Location

Keratinized variety forms superficial layer of skin; nonkeratinized variety lines wet surfaces (lining of mouth, esophagus, part of epiglottis, part of pharynx, and vagina) and covers tongue.

Function

Protection against abrasion, water loss, ultraviolet radiation, and foreign invasion. Both types form first line of defense against microbes.

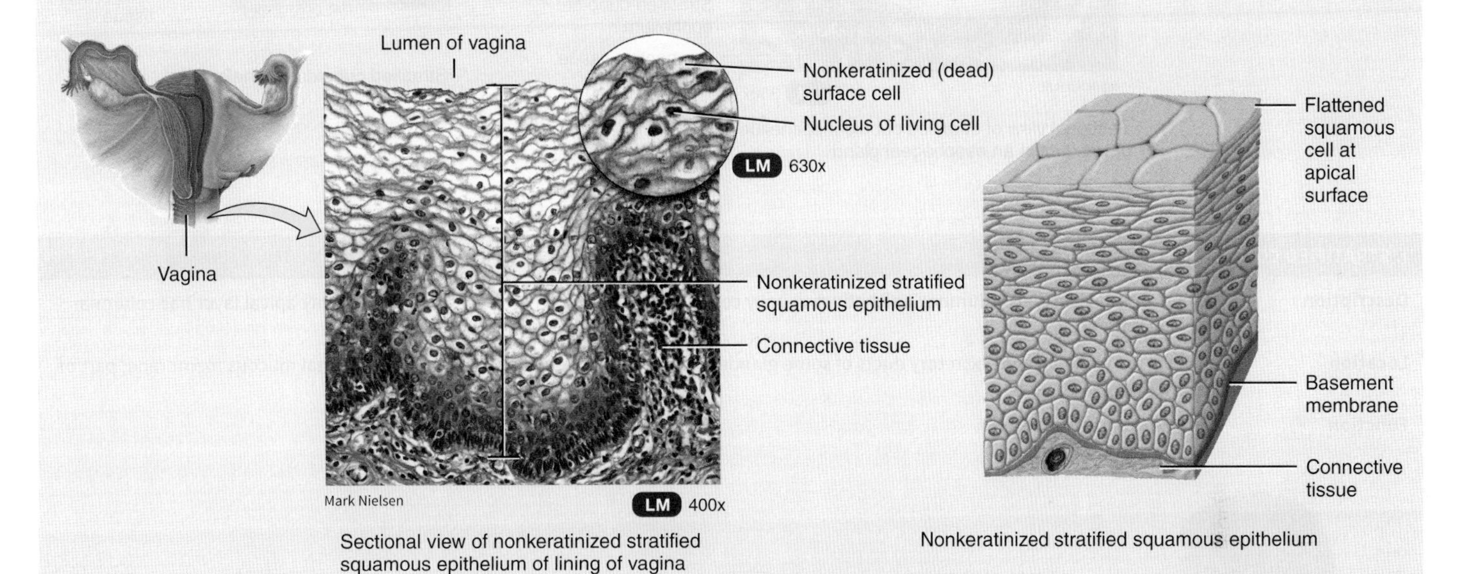

Sectional view of nonkeratinized stratified squamous epithelium of lining of vagina

Nonkeratinized stratified squamous epithelium

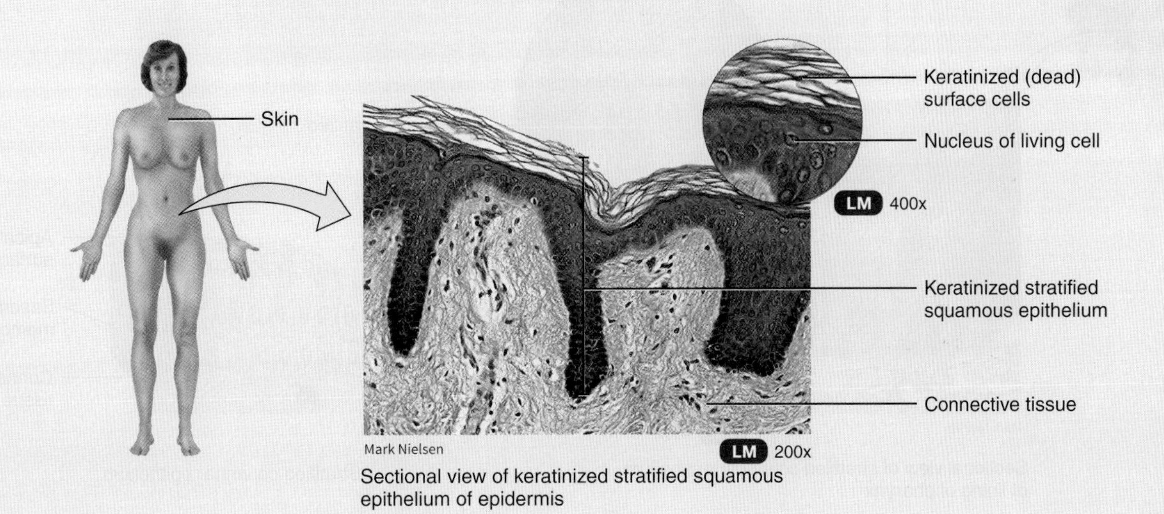

Sectional view of keratinized stratified squamous epithelium of epidermis

Table 4.1 ***Continues***

TABLE 4.1 Epithelial Tissue: Covering and Lining Epithelium (*Continued*)

H. STRATIFIED CUBOIDAL EPITHELIUM

Description **Stratified cuboidal epithelium** has two or more layers of cells; cells in apical layer are cube-shaped; fairly rare type.

Location Ducts of adult sweat glands and esophageal glands, part of male urethra.

Function Protection; limited secretion and absorption.

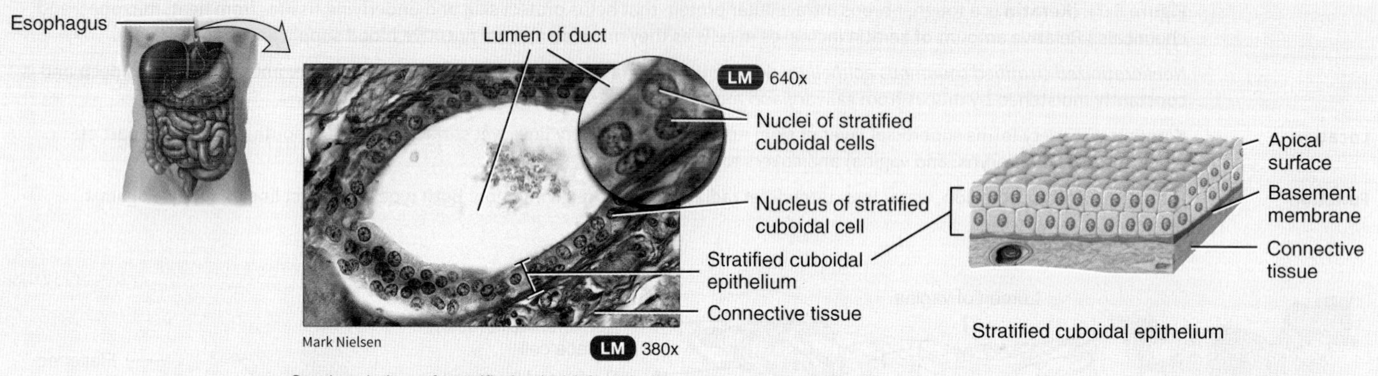

Sectional view of stratified cuboidal epithelium of the duct of an esophageal gland

Stratified cuboidal epithelium

I. STRATIFIED COLUMNAR EPITHELIUM

Description Basal layers in **stratified columnar epithelium** usually consist of shortened, irregularly shaped cells; only apical layer has columnar cells; uncommon.

Location Lines part of urethra; large excretory ducts of some glands, such as esophageal glands; small areas in anal mucous membrane; part of conjunctiva of eye.

Function Protection and secretion.

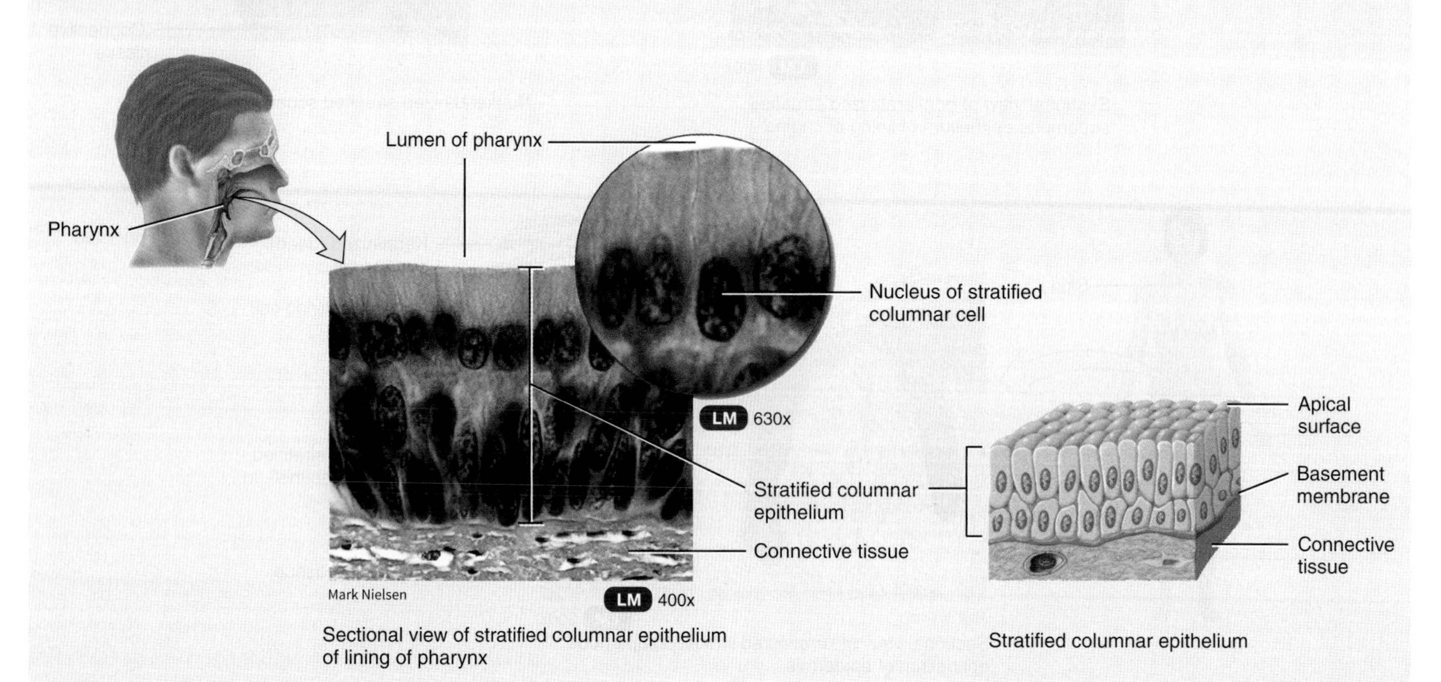

Sectional view of stratified columnar epithelium of lining of pharynx

Stratified columnar epithelium

J. TRANSITIONAL EPITHELIUM (UROTHELIUM)

Description **Transitional epithelium (urothelium)** has a variable appearance (transitional). In relaxed or unstretched state, looks like stratified cuboidal epithelium, except apical layer cells tend to be large and rounded. As tissue is stretched, cells become flatter, giving the appearance of stratified squamous epithelium. Multiple layers and elasticity make it ideal for lining hollow structures (urinary bladder) subject to expansion from within.

Location Lines urinary bladder and portions of ureters and urethra.

Function Allows urinary organs to stretch and maintain protective lining while holding variable amounts of fluid without rupturing.

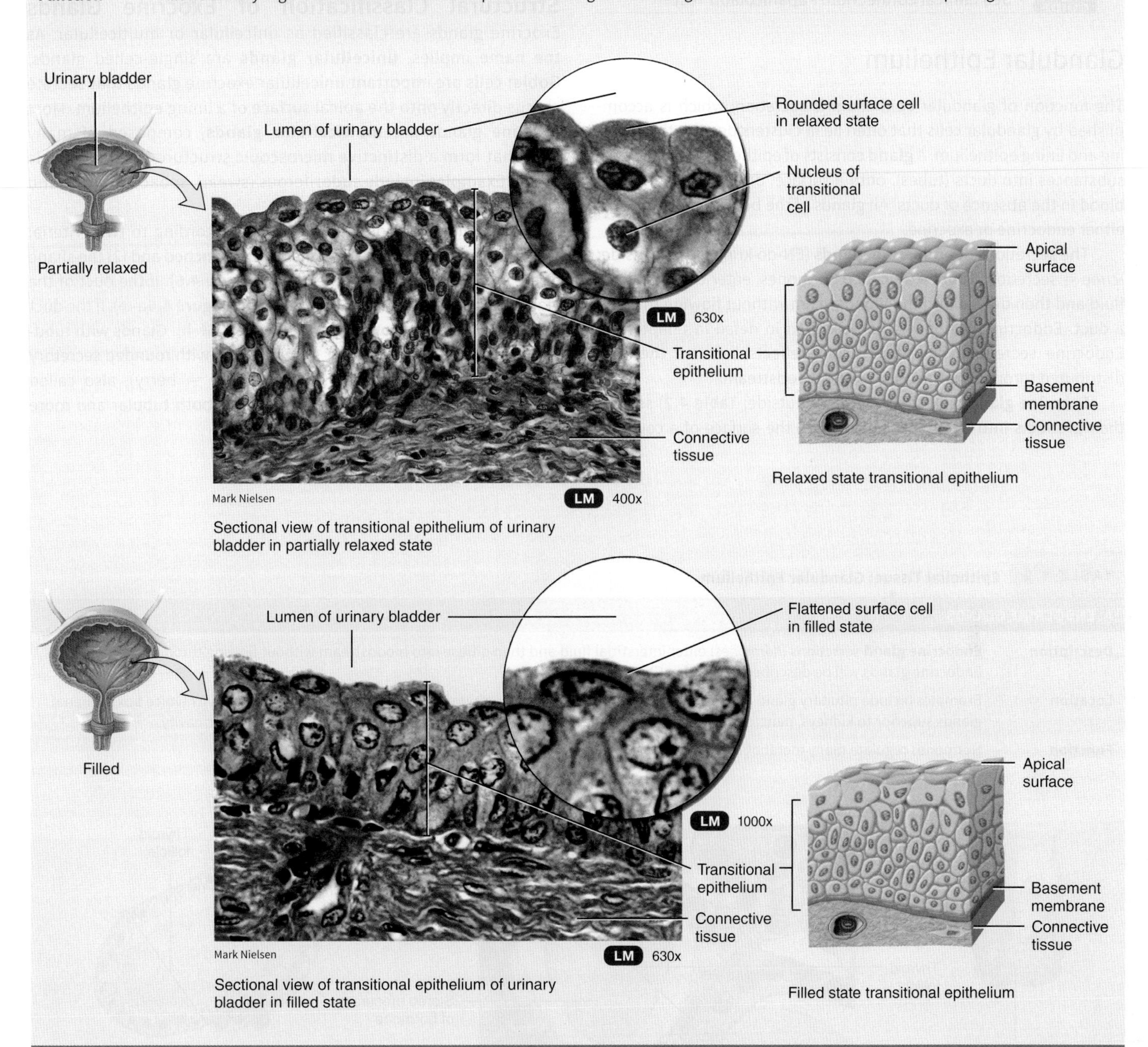

Sectional view of transitional epithelium of urinary bladder in partially relaxed state

Relaxed state transitional epithelium

Sectional view of transitional epithelium of urinary bladder in filled state

Filled state transitional epithelium

consists of a photomicrograph, a corresponding diagram, and an inset that identifies a major location of the tissue in the body. Descriptions, locations, and functions of the tissues accompany each illustration.

See Clinical Connection: Papanicolaou Test

Glandular Epithelium

The function of glandular epithelium is secretion, which is accomplished by glandular cells that often lie in clusters deep to the covering and lining epithelium. A **gland** consists of epithelium that secretes substances into ducts (tubes), onto a surface, or eventually into the blood in the absence of ducts. All glands of the body are classified as either endocrine or exocrine.

The secretions of **endocrine glands** (EN-dō-krin; *endo-* = inside; *-crine* = secretion; Table 4.2), called hormones, enter the interstitial fluid and then diffuse into the bloodstream without flowing through a duct. Endocrine glands will be described in detail in Chapter 18. Endocrine secretions have far-reaching effects because they are distributed throughout the body by the bloodstream.

Exocrine glands (EK-sō-krin; *exo-* = outside; Table 4.2) secrete their products into ducts that empty onto the surface of a covering and lining epithelium such as the skin surface or the lumen of a hollow organ. The secretions of exocrine glands have limited effects and some of them would be harmful if they entered the bloodstream. As you will learn later in the text, some glands of the body, such as the pancreas, ovaries, and testes, are mixed glands that contain both endocrine and exocrine tissue.

Structural Classification of Exocrine Glands

Exocrine glands are classified as unicellular or multicellular. As the name implies, **unicellular glands** are single-celled glands. Goblet cells are important unicellular exocrine glands that secrete mucus directly onto the apical surface of a lining epithelium. Most exocrine glands are **multicellular glands**, composed of many cells that form a distinctive microscopic structure or macroscopic organ. Examples include sudoriferous (sweat), sebaceous (oil), and salivary glands.

Multicellular glands are categorized according to two criteria: (1) whether their ducts are branched or unbranched and (2) the shape of the secretory portions of the gland (Figure 4.6). If the duct of the gland does not branch, it is a **simple gland** (Figure 4.6a–e). If the duct branches, it is a **compound gland** (Figure 4.6f–h). Glands with tubular secretory parts are **tubular glands**; those with rounded secretory portions are **acinar glands** (AS-i-nar; *acin-* = berry), also called *alveolar glands*. **Tubuloacinar glands** have both tubular and more rounded secretory parts.

TABLE 4.2 Epithelial Tissue: Glandular Epithelium

A. ENDOCRINE GLANDS	
Description	**Endocrine gland** secretions (*hormones*) enter interstitial fluid and then diffuse into bloodstream without flowing through a duct. Endocrine glands will be described in detail in Chapter 18.
Location	Examples include pituitary gland at base of brain, pineal gland in brain, thyroid and parathyroid glands near larynx (voice box), adrenal glands superior to kidneys, pancreas near stomach, ovaries in pelvic cavity, testes in scrotum, thymus in thoracic cavity.
Function	Hormones regulate many metabolic and physiological activities to maintain homeostasis.

Sectional view of endocrine gland (thyroid gland)

Endocrine gland (thyroid gland)

B. EXOCRINE GLANDS

Description **Exocrine gland** secretory products are released into ducts that empty onto surface of a covering and lining epithelium, such as skin surface or lumen of hollow organ.

Location Sweat, oil, and earwax glands of skin; digestive glands such as salivary glands (secrete into mouth cavity) and pancreas (secretes into small intestine).

Function Produce substances such as sweat to help lower body temperature, oil, earwax, saliva, or digestive enzymes.

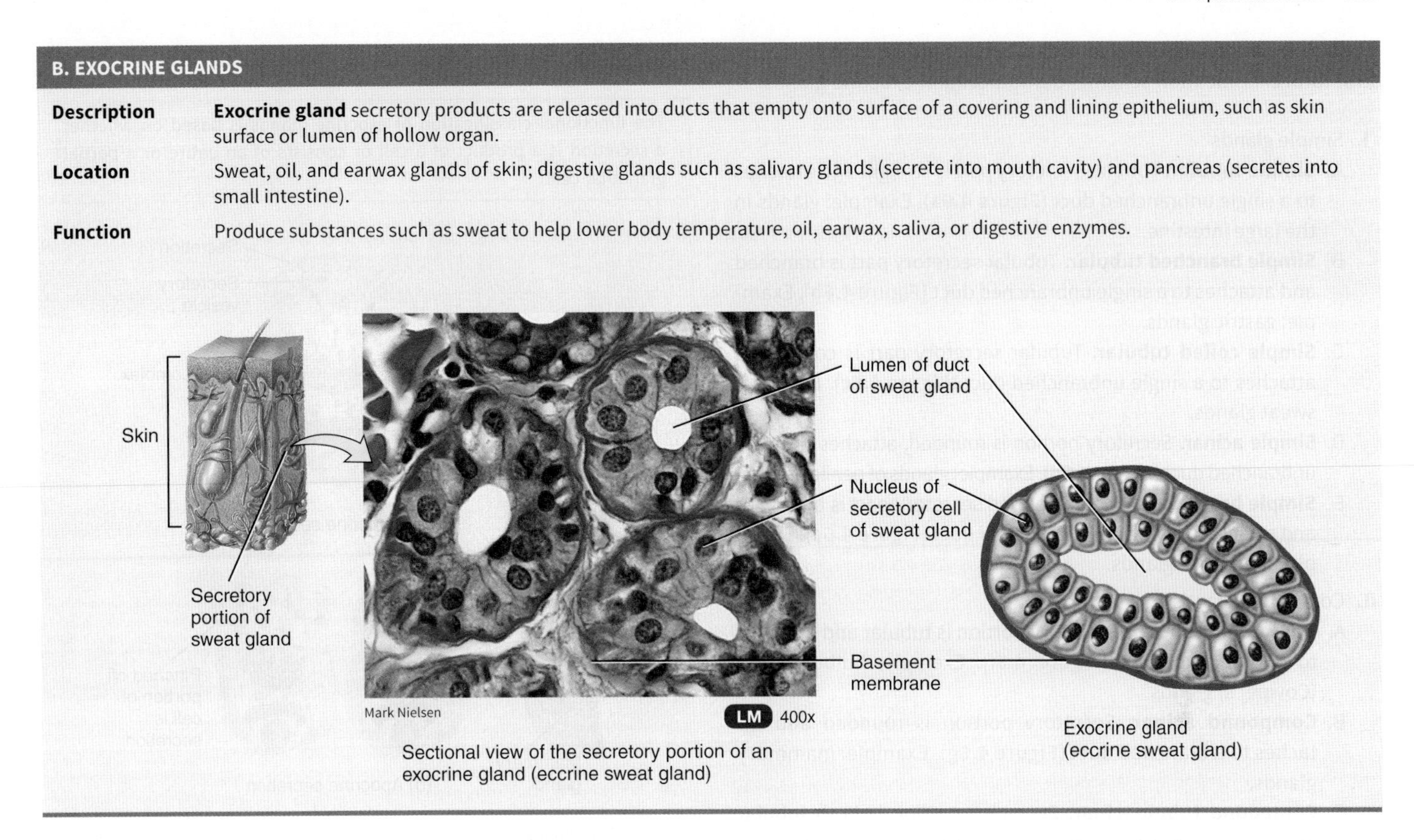

Sectional view of the secretory portion of an exocrine gland (eccrine sweat gland)

FIGURE 4.6 **Multicellular exocrine glands.** Pink represents the secretory portion; lavender represents the duct.

Structural classification of multicellular exocrine glands is based on the branching pattern of the duct and the shape of the secreting portion.

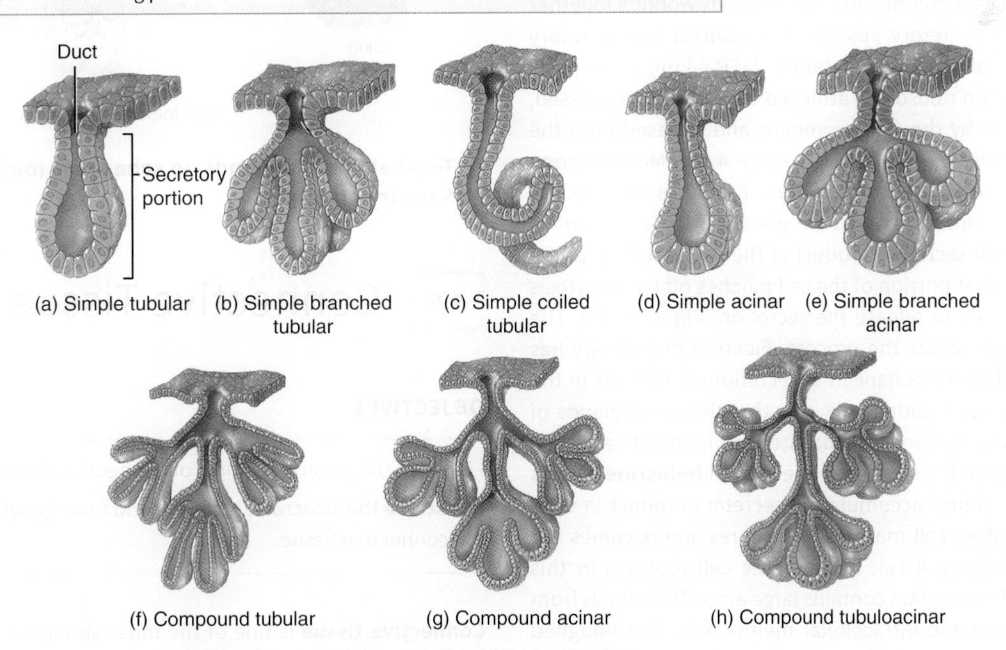

Q How do simple multicellular exocrine glands differ from compound ones?

Combinations of these features are the criteria for the following structural classification scheme for multicellular exocrine glands:

I. Simple glands
   A. **Simple tubular.** Tubular secretory part is straight and attaches to a single unbranched duct (**Figure 4.6a**). Example: glands in the large intestine.
   B. **Simple branched tubular.** Tubular secretory part is branched and attaches to a single unbranched duct (**Figure 4.6b**). Example: gastric glands.
   C. **Simple coiled tubular.** Tubular secretory part is coiled and attaches to a single unbranched duct (**Figure 4.6c**). Example: sweat glands.
   D. **Simple acinar.** Secretory portion is rounded, attaches to single unbranched duct (**Figure 4.6d**). Example: glands of penile urethra.
   E. **Simple branched acinar.** Rounded secretory part is branched and attaches to a single unbranched duct (**Figure 4.6e**). Example: sebaceous glands.

II. Compound glands
   A. **Compound tubular.** Secretory portion is tubular and attaches to a branched duct (**Figure 4.6f**). Example: bulbourethral (Cowper's) glands.
   B. **Compound acinar.** Secretory portion is rounded and attaches to a branched duct (**Figure 4.6g**). Example: mammary glands.
   C. **Compound tubuloacinar.** Secretory portion is both tubular and rounded and attaches to a branched duct (**Figure 4.6h**). Example: acinar glands of the pancreas.

Functional Classification of Exocrine Glands The functional classification of exocrine glands is based on how their secretions are released. Each of these secretory processes begins with the endoplasmic reticulum and Golgi complex working together to form intracellular secretory vesicles that contain the secretory product. Secretions of **merocrine glands** (MER-ō-krin; *mero-* = a part) are synthesized on ribosomes attached to rough ER; processed, sorted, and packaged by the Golgi complex; and released from the cell in secretory vesicles via exocytosis (**Figure 4.7a**). Most exocrine glands of the body are merocrine glands. Examples include the salivary glands and pancreas. **Apocrine glands** (AP-ō-krin; *apo-* = from) accumulate their secretory product at the apical surface of the secreting cell. Then, that portion of the cell pinches off by exocytosis from the rest of the cell to release the secretion (**Figure 4.7b**). The cell repairs itself and repeats the process. Electron microscopy has confirmed that this is the mechanism of secretion of milk fats in the mammary glands. Recent evidence reveals that the sweat glands of the skin, named apocrine sweat glands after this mode of secretion, actually undergo merocrine secretion. The cells of **holocrine glands** (HŌ-lō-krin; *holo-* = entire) accumulate a secretory product in their cytosol. As the secretory cell matures, it ruptures and becomes the secretory product (**Figure 4.7c**). Because the cell ruptures in this mode of secretion, the secretion contains large amounts of lipids from the plasma membrane and intracellular membranes. The sloughed off cell is replaced by a new cell. One example of a holocrine gland is a sebaceous gland of the skin.

FIGURE 4.7 **Functional classification of multicellular exocrine glands.**

The functional classification of exocrine glands is based on whether a secretion is a product of a cell or consists of an entire or a partial glandular cell.

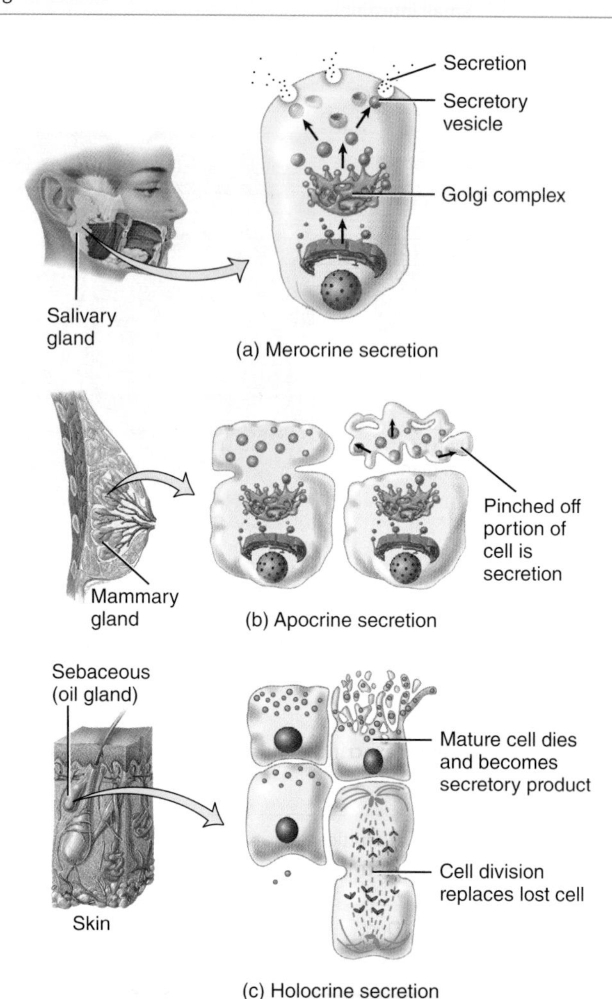

Q To what class of glands do sebaceous (oil) glands belong? Salivary glands?

4.5 Connective Tissue

OBJECTIVES

- **Explain** the general features of connective tissue.
- **Describe** the structure, location, and function of the various types of connective tissue.

Connective tissue is one of the most abundant and widely distributed tissues in the body. In its various forms, connective tissue has a variety of functions. It binds together, supports, and strengthens other body tissues; protects and insulates internal organs; compartmentalizes structures such as skeletal muscles; serves as the major transport system within the body (blood, a fluid connective tissue); is the primary location of stored energy reserves (adipose, or fat, tissue); and is the main source of immune responses.

General Features of Connective Tissue

Connective tissue consists of two basic elements: extracellular matrix and cells. A connective tissue's **extracellular matrix** (MĀ-triks) is the material located between its widely spaced cells. The extracellular matrix consists of *protein fibers* and *ground substance*, the material between the cells and the fibers. The extracellular fibers are secreted by the connective tissue cells and account for many of the functional properties of the tissue in addition to controlling the surrounding watery environment via specific proteoglycan molecules (described shortly). The structure of the extracellular matrix determines much of the tissue's qualities. For instance, in cartilage, the extracellular matrix is firm but pliable. The extracellular matrix of bone, by contrast, is hard and inflexible.

Recall that, in contrast to epithelial tissue, connective tissue does not usually occur on body surfaces. Also, unlike epithelial tissue, connective tissue usually is highly vascular; that is, it has a rich blood supply. Exceptions include cartilage, which is avascular, and tendons, with a scanty blood supply. Except for cartilage, connective tissue, like epithelial tissue, is supplied with nerves.

Connective Tissue Cells

Embryonic cells called mesenchymal cells give rise to the cells of connective tissue. Each major type of connective tissue contains an immature class of cells with a name ending in *-blast*, which means "to bud or sprout." These immature cells are called *fibroblasts* in loose and dense connective tissue (described shortly), *chondroblasts* in cartilage, and *osteoblasts* in bone. Blast cells retain the capacity for cell division and secrete the extracellular matrix that is characteristic of the tissue. In some connective tissues, once the extracellular matrix is produced, the immature cells differentiate into mature cells with names ending in *-cyte*, namely, *fibrocytes*, *chondrocytes*, and *osteocytes*. Mature cells have reduced capacities for cell division and extracellular matrix formation and are mostly involved in monitoring and maintaining the extracellular matrix.

Connective tissue cells vary according to the type of tissue and include the following (**Figure 4.8**):

1. **Fibroblasts** (FĪ-brō-blasts; *fibro-* = fibers) are large, flat cells with branching processes. They are present in all the general connective tissues, and usually are the most numerous.
2. **Macrophages** (MAK-rō-fā-jez; *macro-* = large; *-phages* = eaters) are phagocytes that develop from *monocytes,* a type of white blood cell. *Fixed macrophages* reside in a particular tissue; examples include alveolar macrophages in the lungs or splenic macrophages in the spleen. *Wandering macrophages* have the ability to move throughout the tissue and gather at sites of infection or inflammation to carry on phagocytosis.
3. **Plasma cells** (*plasmocytes*) are found in many places in the body, but most plasma cells reside in connective tissue, especially in the gastrointestinal and respiratory tracts.
4. **Mast cells** (*mastocytes*) are involved in the inflammatory response, the body's reaction to injury or infection and can also bind to, ingest, and kill bacteria.
5. **Adipocytes** (AD-i-pō-sīts) are fat cells or *adipose* cells, connective tissue cells that store triglycerides (fats). They are found deep to the skin and around organs such as the heart and kidneys.
6. **Leukocytes** (white blood cells) are not found in significant numbers in normal connective tissue. However, in response to certain conditions they migrate from blood into connective tissue. For example, *neutrophils* gather at sites of infection, and *eosinophils* migrate to sites of parasitic invasions and allergic responses.

Connective Tissue Extracellular Matrix

Each type of connective tissue has unique properties, based on the specific extracellular materials between the cells. The extracellular matrix consists of two major components: (1) the ground substance and (2) the fibers.

Ground Substance

As noted earlier, the **ground substance** is the component of a connective tissue between the cells and fibers. The ground substance may be fluid, semifluid, gelatinous, or calcified. It supports cells, binds them together, stores water, and provides a medium for exchange of substances between the blood and cells. It plays an active role in how tissues develop, migrate, proliferate, and change shape, and in how they carry out their metabolic functions.

Ground substance contains water and an assortment of large organic molecules, many of which are complex combinations of polysaccharides and proteins. The polysaccharides include hyaluronic acid, chondroitin sulfate, dermatan sulfate, and keratan sulfate. Collectively, they are referred to as **glycosaminoglycans (GAGs)** (glī-kōs-a-mē′-nō-GLĪ-kans). Except for hyaluronic acid, the GAGs are associated with proteins called **proteoglycans** (prō-tē-ō-GLĪ-kans). The proteoglycans form a core protein and the GAGs project from the protein like the bristles of a brush. One of the most important properties of GAGs is that they trap water, making the ground substance more jellylike.

Hyaluronic acid (hī′-a-loo-RON-ik) is a viscous, slippery substance that binds cells together, lubricates joints, and helps maintain the shape of the eyeballs. White blood cells, sperm cells, and some bacteria produce *hyaluronidase*, an enzyme that breaks apart hyaluronic acid, thus causing the ground substance of connective tissue to become more liquid. The ability to produce hyaluronidase helps white blood cells move more easily through connective tissue to reach sites of infection and aids penetration of an oocyte by a sperm cell during fertilization. It also accounts for the rapid spread of bacteria through connective tissue. **Chondroitin sulfate** (kon-DROY-tin) provides support and adhesiveness in cartilage, bone, skin, and blood vessels. The skin, tendons, blood vessels, and heart valves contain **dermatan sulfate**; bone, cartilage, and the cornea of the eye

FIGURE 4.8 **Representative cells and fibers present in connective tissues.**

Fibroblasts are usually the most numerous connective tissue cells.

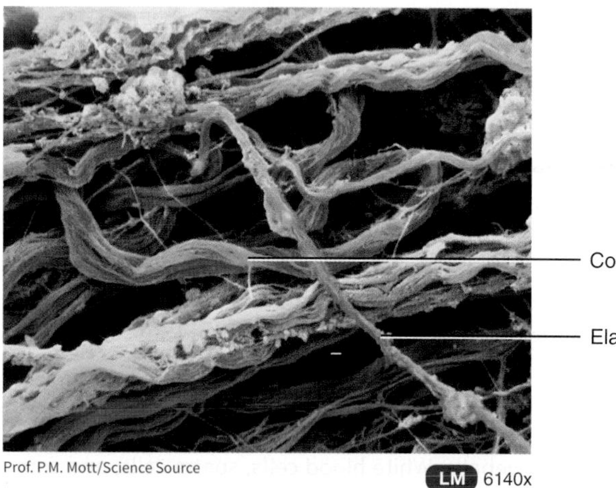

Prof. P.M. Mott/Science Source

Q What is the function of fibroblasts?

contain **keratan sulfate**. Also present in the ground substance are **adhesion proteins**, which are responsible for linking components of the ground substance to one another and to the surfaces of cells. The main adhesion protein of connective tissues is **fibronectin**, which binds to both collagen fibers (discussed shortly) and ground substance, linking them together. Fibronectin also attaches cells to the ground substance.

See Clinical Connection: Chondroitin Sulfate, Glucosamine, and Joint Disease

Fibers Three types of **fibers** are embedded in the extracellular matrix between the cells: collagen fibers, elastic fibers, and reticular fibers (**Figure 4.8**). They function to strengthen and support connective tissues.

Collagen fibers (KOL-a-jen; *colla* = glue) are very strong and resist pulling or stretching, but they are not stiff, which allows tissue flexibility. The properties of different types of collagen fibers vary from tissue to tissue. For example, the collagen fibers found in cartilage and bone form different associations with surrounding molecules. As a result of these associations, the collagen fibers in cartilage are surrounded by more water molecules than those in bone, which gives cartilage a more cushioning effect. Collagen fibers often occur in parallel bundles (see **Table 4.5A**, dense regular connective tissue). The bundle arrangement adds great tensile strength to the tissue. Chemically, collagen fibers consist of the protein *collagen,* which is the most abundant protein in your body, representing about 25% of the total. Collagen fibers are found in most types of connective tissues, especially bone, cartilage, tendons (which attach muscle to bone), and ligaments (which attach bone to bone).

See Clinical Connection: Sprain

Elastic fibers, which are smaller in diameter than collagen fibers, branch and join together to form a fibrous network within a connective tissue. An elastic fiber consists of molecules of the protein *elastin* surrounded by a glycoprotein named *fibrillin*, which adds strength and stability. Because of their unique molecular structure, elastic fibers are strong but can be stretched up to 150% of their relaxed length without breaking. Equally important, elastic fibers have the ability to return to their original shape after being stretched, a property called *elasticity.* Elastic fibers are plentiful in skin, blood vessel walls, and lung tissue.

Reticular fibers (*reticul-* = net), consisting of *collagen* arranged in fine bundles with a coating of glycoprotein, provide support in the walls of blood vessels and form a network around the cells in some tissues, such as areolar connective tissue (a-RĒ-ō-lar; *areol* = small space), adipose tissue, nerve fibers, and smooth muscle tissue. Produced by fibroblasts, reticular fibers are much thinner than collagen fibers and form branching networks. Like collagen fibers, reticular fibers provide support and strength. Reticular fibers are plentiful in reticular connective tissue, which forms the **stroma** (supporting framework) of many soft organs, such as the spleen and lymph nodes. These fibers also help form the basement membrane.

Classification of Connective Tissue

Because of the diversity of cells and extracellular matrix and the differences in their relative proportions, the classification of connective tissue is not always clear-cut and several classifications exist. We offer the following classification scheme:

I. Embryonic connective tissue
- A. Mesenchyme
- B. Mucous (mucoid) connective tissue

II. Mature connective tissue
- A. Connective tissue proper
  - 1. Loose connective tissue
    - a. Areolar connective tissue
    - b. Adipose tissue
    - c. Reticular connective tissue
  - 2. Dense connective tissue
    - a. Dense regular connective tissue
    - b. Dense irregular connective tissue
    - c. Elastic connective tissue
- B. Supporting connective tissue
  - 1. Cartilage
    - a. Hyaline cartilage
    - b. Fibrocartilage
    - c. Elastic cartilage
  - 2. Bone tissue
    - a. Compact bone
    - b. Spongy bone
- C. Liquid connective tissue
  - 1. Blood
  - 2. Lymph

Before examining each of the connective tissues in detail, it will be helpful to first describe the overall basis for the classification scheme that we are using. **Embryonic connective tissue** refers to connective tissue present in an embryo or a fetus. **Mature connective tissue** refers to connective tissue that is present at birth and persists throughout life. One category of mature connective tissue is **connective tissue proper**, which is flexible and contains a viscous ground substance with abundant fibers. A second category of mature connective tissue is **supporting connective tissue**, which protects and supports soft tissues of the body. The third category of mature connective tissue is **liquid connective tissue**, which means that the extracellular matrix is liquid.

Embryonic Connective Tissue

Note that our classification scheme has two major subclasses of connective tissue: embryonic and mature. **Embryonic connective tissue** is of two types: **mesenchyme** and **mucous connective tissue**. Mesenchyme is present primarily in the *embryo,* the developing human from fertilization through the first two months of pregnancy, and in the *fetus,* the developing human from the third month of pregnancy (**Table 4.3**).

Mature Connective Tissue

The first type of mature connective tissue we will consider is connective tissue proper.

Connective Tissue Proper

This type of connective tissue is flexible and has a viscous ground substance with abundant fibers.

Loose Connective Tissue The fibers of **loose connective tissue** are *loosely* arranged between cells. The types of loose connective tissue are areolar connective tissue, adipose tissue, and reticular connective tissue (**Table 4.4**).

See Clinical Connection: Liposuction and Cryolipolysis

Dense Connective Tissue **Dense connective tissue** is a second type of connective tissue proper that contains more fibers, which are thicker and more *densely* packed, but have considerably fewer cells than loose connective tissue. There are three types: dense regular

TABLE 4.3 Embryonic Connective Tissues

A. MESENCHYME

Description	**Mesenchyme** has irregularly shaped mesenchymal cells embedded in semifluid ground substance that contains delicate reticular fibers.
Location	Almost exclusively under skin and along developing bones of embryo; some in adult connective tissue, especially along blood vessels.
Function	Forms almost all other types of connective tissue.

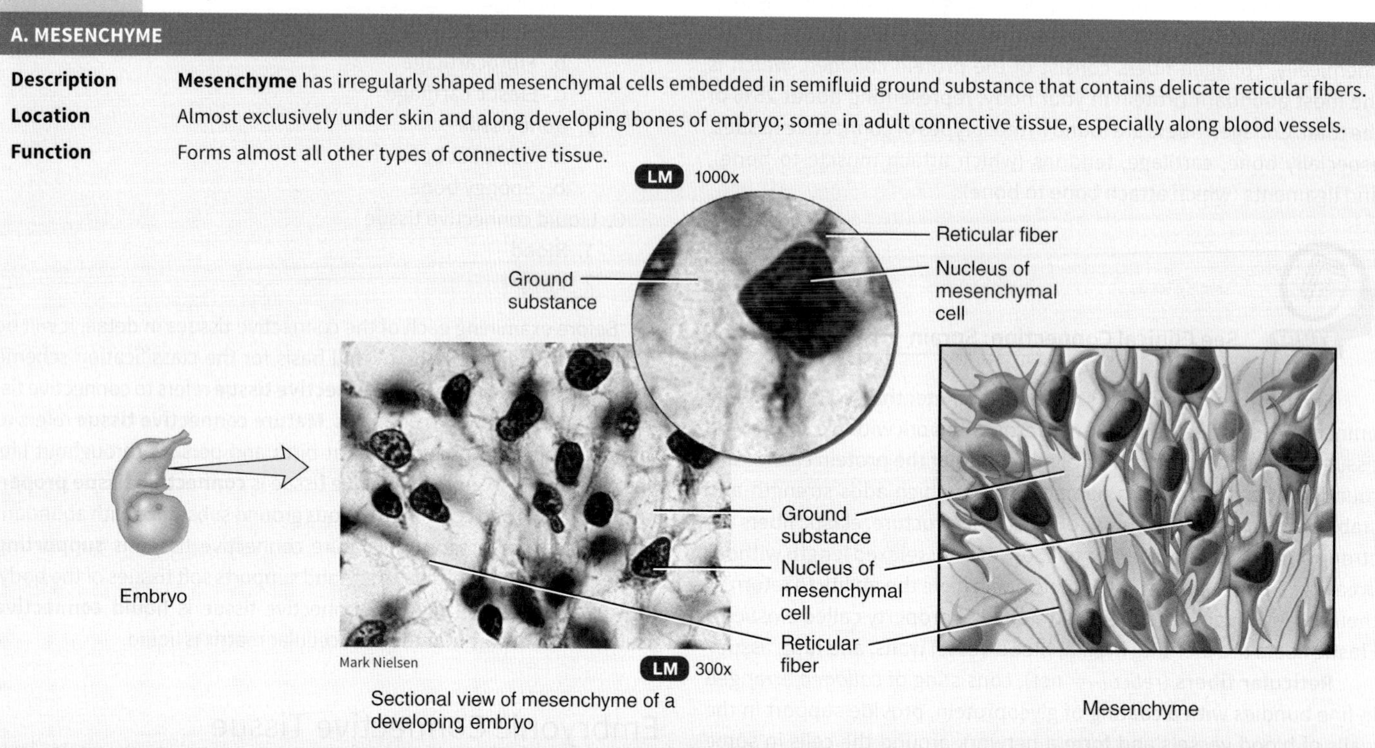

Sectional view of mesenchyme of a developing embryo

Mesenchyme

B. MUCOUS (MUCOID) CONNECTIVE TISSUE

Description	**Mucous (mucoid) connective tissue** has widely scattered fibroblasts embedded in viscous, jellylike ground substance that contains fine collagen fibers.
Location	Umbilical cord of fetus.
Function	Support.

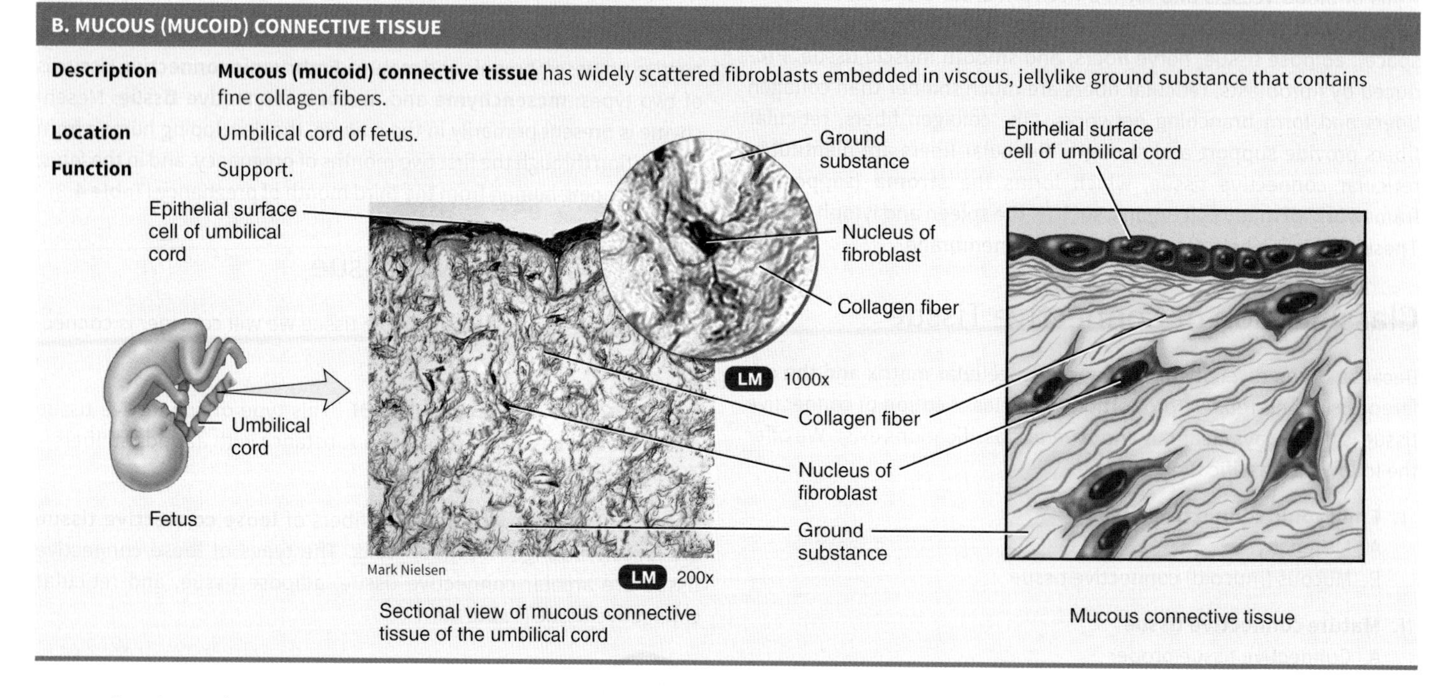

Sectional view of mucous connective tissue of the umbilical cord

Mucous connective tissue

connective tissue, dense irregular connective tissue, and elastic connective tissue (**Table 4.5**).

Supporting Connective Tissue This type of mature connective tissue includes cartilage and bone.

CARTILAGE **Cartilage** (KAR-ti-lij) consists of a dense network of collagen fibers and elastic fibers firmly embedded in chondroitin sulfate, a gel-like component of the ground substance. Cartilage can endure considerably more stress than loose and dense connective tissues. The strength of cartilage is due to its collagen fibers, and its *resilience* (ability to assume its original shape after deformation) is due to chondroitin sulfate.

Like other connective tissue, cartilage has few cells and large quantities of extracellular matrix. It differs from other connective tissue, however, in not having nerves or blood vessels in its extracellular

TABLE 4.4 Mature Connective Tissue: Connective Tissue Proper—Loose Connective Tissue

A. AREOLAR CONNECTIVE TISSUE

Description **Areolar connective tissue** is one of the most widely distributed connective tissues; consists of fibers (collagen, elastic, reticular) arranged randomly and several kinds of cells (fibroblasts, macrophages, plasma cells, adipocytes, mast cells, and a few white blood cells) embedded in semifluid ground substance (hyaluronic acid, chondroitin sulfate, dermatan sulfate, and keratan sulfate).

Location In and around nearly every body structure (thus, called "packing material" of the body): in subcutaneous layer deep to skin; papillary (superficial) region of dermis of skin; lamina propria of mucous membranes; around blood vessels, nerves, and body organs.

Function Strength, elasticity, support.

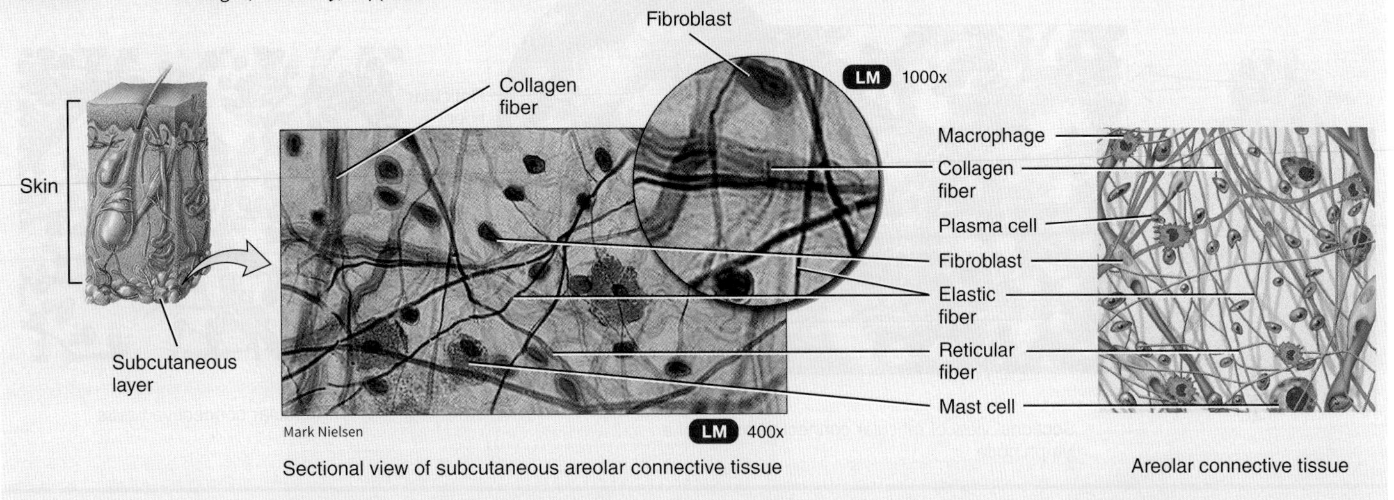

Sectional view of subcutaneous areolar connective tissue

Areolar connective tissue

B. ADIPOSE TISSUE

Description **Adipose tissue** has cells derived from fibroblasts (called *adipocytes*) that are specialized for storage of triglycerides (fats) as a large, centrally located droplet. Cell fills up with a single, large triglyceride droplet, and cytoplasm and nucleus are pushed to periphery of cell. With weight gain, amount of adipose tissue increases and new blood vessels form. Thus, an obese person has many more blood vessels than does a lean person, a situation that can cause high blood pressure, since the heart has to work harder. Most adipose tissue in adults is *white adipose tissue* (just described). *Brown adipose tissue* (BAT) is darker due to very rich blood supply and numerous pigmented mitochondria that participate in aerobic cellular respiration. BAT is widespread in the fetus and infant; adults have only small amounts.

Location Wherever areolar connective tissue is located: subcutaneous layer deep to skin, around heart and kidneys, yellow bone marrow, padding around joints and behind eyeball in eye socket.

Function Reduces heat loss through skin; serves as an energy reserve; supports and protects organs. In newborns, BAT generates heat to maintain proper body temperature. Adipose tissue is also an excellent source of stem cells, which are used in rejuvenation medicine to repair or replace damaged tissue.

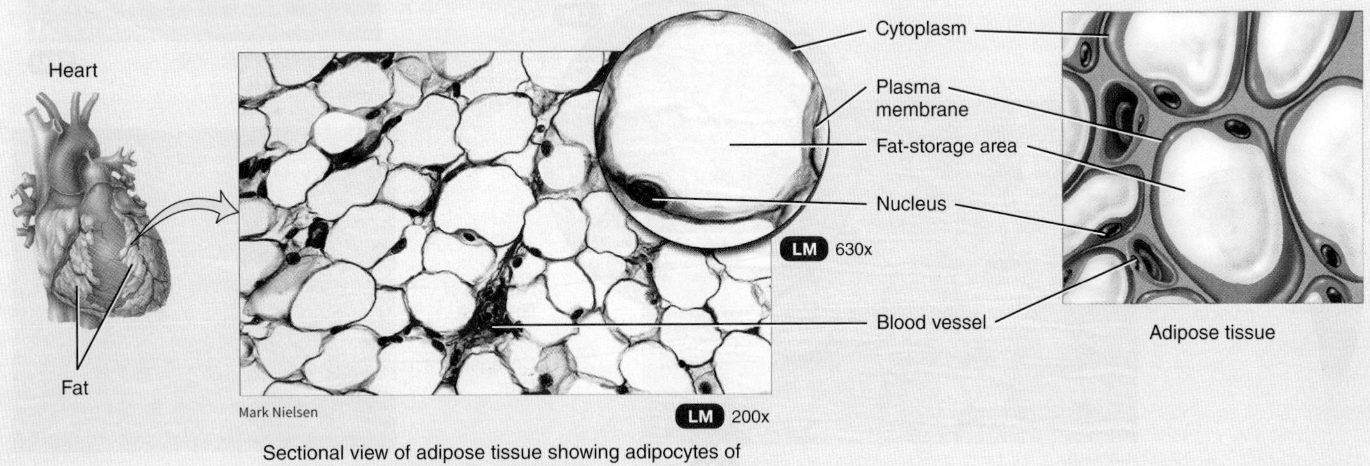

Sectional view of adipose tissue showing adipocytes of white fat and details of an adipocyte

Adipose tissue

Table 4.1 *Continues*

TABLE 4.4 Mature Connective Tissue: Connective Tissue Proper—Loose Connective Tissue (*Continued*)

C. RETICULAR CONNECTIVE TISSUE

Description	**Reticular connective tissue** is a fine interlacing network of reticular fibers (thin form of collagen fiber) and reticular cells.
Location	Stroma (supporting framework) of liver, spleen, lymph nodes; red bone marrow; reticular lamina of basement membrane; around blood vessels and muscles.
Function	Forms stroma of organs; binds smooth muscle tissue cells; filters and removes worn-out blood cells in spleen and microbes in lymph nodes.

Lymph node

LM 640x

Reticular fiber

Nucleus of reticular cell

Reticular fiber

Mark Nielsen

LM 400x

Sectional view of reticular connective tissue of a lymph node

Reticular connective tissue

TABLE 4.5 Mature Connective Tissue: Connective Tissue Proper—Dense Connective Tissue

A. DENSE REGULAR CONNECTIVE TISSUE

Description	**Dense regular connective tissue** forms shiny white extracellular matrix; mainly collagen fibers *regularly* arranged in bundles with fibroblasts in rows between them. Collagen fibers (protein structures secreted by fibroblasts) are not living, so damaged tendons and ligaments heal slowly.
Location	Forms tendons (attach muscle to bone), most ligaments (attach bone to bone), and aponeuroses (sheetlike tendons that attach muscle to muscle or muscle to bone).
Function	Provides strong attachment between various structures. Tissue structure withstands pulling (tension) along long axis of fibers.

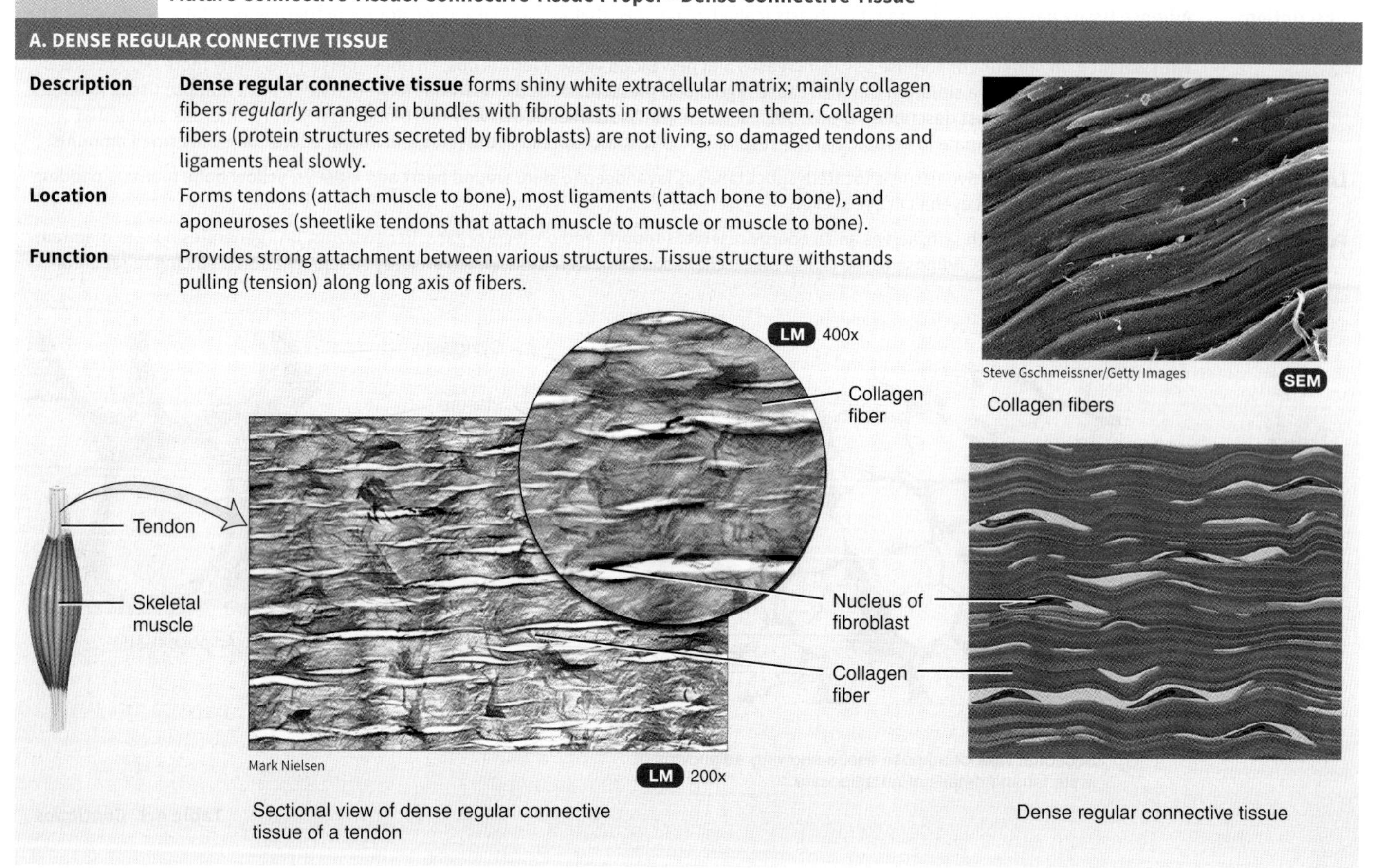

Collagen fibers

Sectional view of dense regular connective tissue of a tendon

Dense regular connective tissue

B. DENSE IRREGULAR CONNECTIVE TISSUE

Description **Dense irregular connective tissue** is made up of collagen fibers; usually *irregularly* arranged with a few fibroblasts.

Location Often occurs in sheets, such as fasciae (tissue beneath skin and around muscles and other organs), reticular (deeper) region of dermis of skin, fibrous pericardium of heart, periosteum of bone, perichondrium of cartilage, joint capsules, membrane capsules around various organs (kidneys, liver, testes, lymph nodes); also in heart valves.

Function Provides tensile (pulling) strength in many directions.

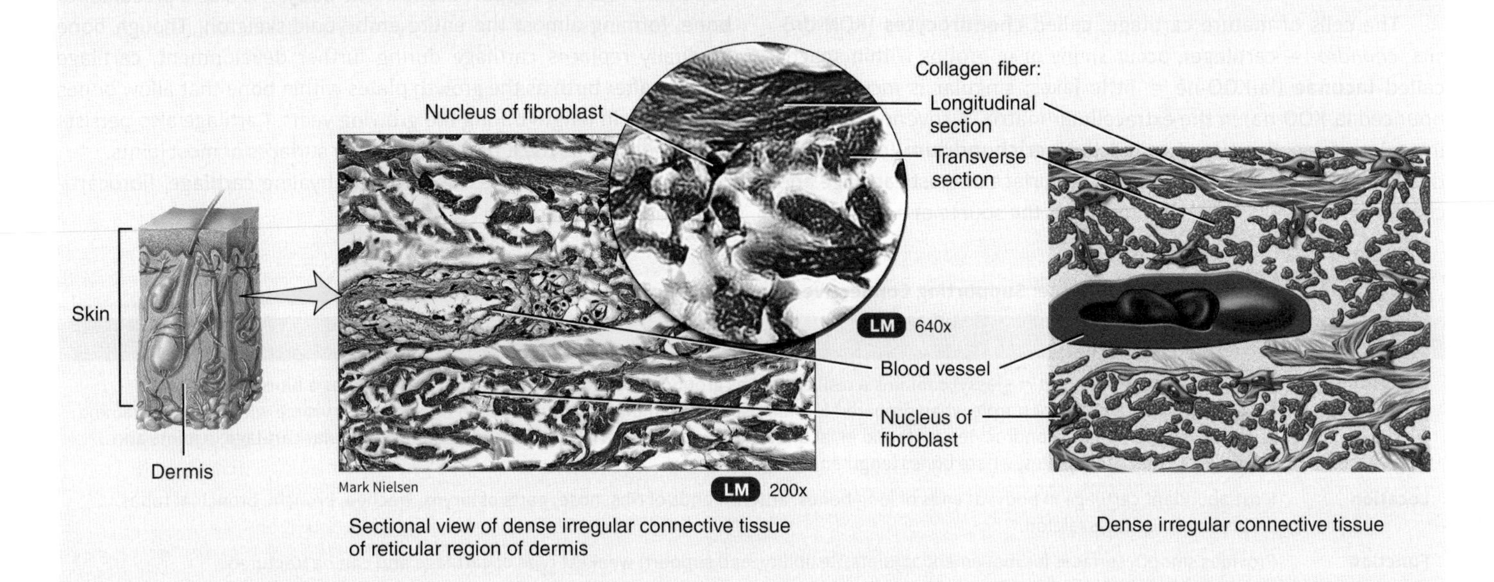

Sectional view of dense irregular connective tissue of reticular region of dermis

Dense irregular connective tissue

C. ELASTIC CONNECTIVE TISSUE

Description **Elastic connective tissue** contains predominantly elastic fibers with fibroblasts between them; unstained tissue is yellowish.

Location Lung tissue, walls of elastic arteries, trachea, bronchial tubes, true vocal cords, suspensory ligaments of penis, some ligaments between vertebrae.

Function Allows stretching of various organs; is strong and can recoil to original shape after being stretched. Elasticity is important to normal functioning of lung tissue (recoils in exhaling) and elastic arteries (recoil between heartbeats to help maintain blood flow).

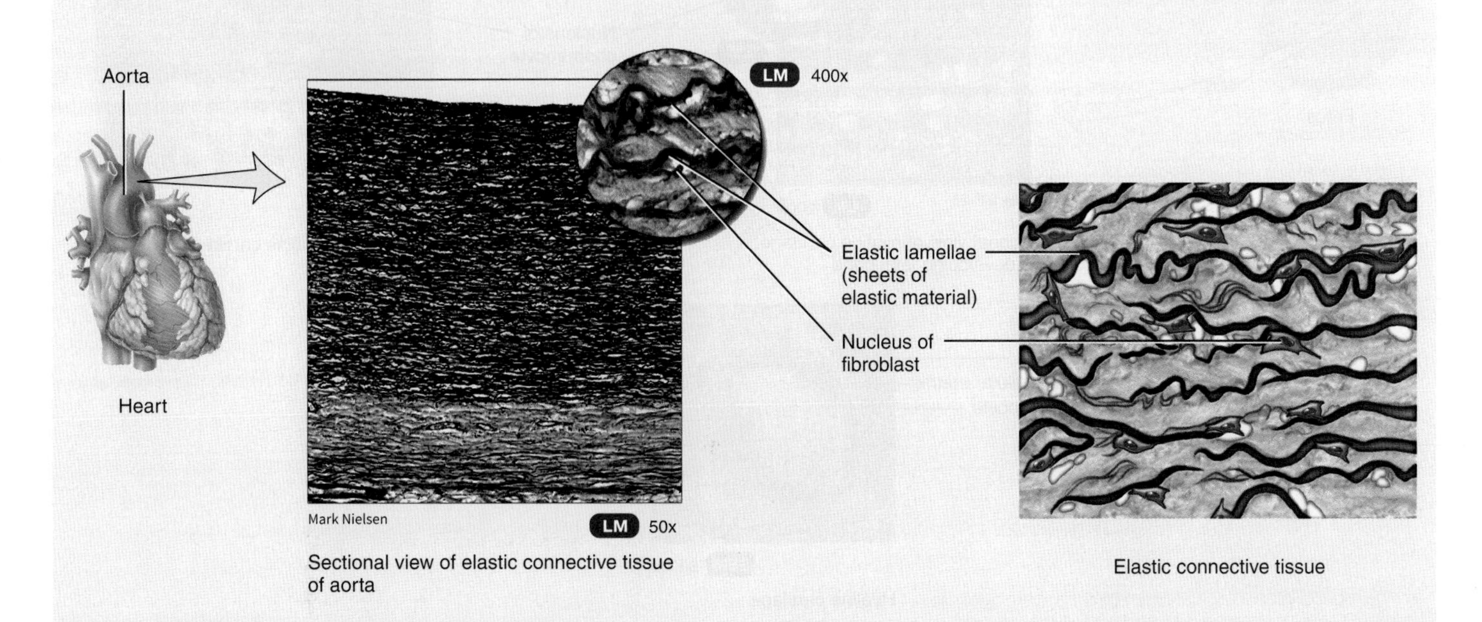

Sectional view of elastic connective tissue of aorta

Elastic connective tissue

matrix. Interestingly, cartilage does not have a blood supply because it secretes an *antiangiogenesis factor* (an′-tī-an′-jē-ō-JEN-e-sis; *anti-* = against; *angio-* = vessel; *-genesis* = production), a substance that prevents blood vessel growth. Because of this property, antiangiogenesis factor is being studied as a possible cancer treatment. If cancer cells can be stopped from promoting new blood vessel growth, their rapid rate of cell division and expansion can be slowed or even halted.

The cells of mature cartilage, called **chondrocytes** (KON-drō-sīts; *chondro-* = cartilage), occur singly or in groups within spaces called **lacunae** (la-KOO-nē = little lakes; singular is *lacuna,* pronounced la-KOO-na) in the extracellular matrix. A covering of dense irregular connective tissue called the **perichondrium** (per′-i-KON-drē-um; *peri-* = around) surrounds the surface of most cartilage and contains blood vessels and nerves and is the source of new cartilage cells. Since cartilage has no blood supply, it heals poorly following an injury.

The cells and collagen-embedded extracellular matrix of cartilage form a strong, firm material that resists tension (stretching), compression (squeezing), and shear (pushing in opposite directions). The chondroitin sulfate in the extracellular matrix is largely responsible for cartilage's resilience. Because of these properties, cartilage plays an important role as a support tissue in the body. It is also a precursor to bone, forming almost the entire embryonic skeleton. Though bone gradually replaces cartilage during further development, cartilage persists after birth as the growth plates within bone that allow bones to increase in length during the growing years. Cartilage also persists throughout life as the lubricated articular surfaces of most joints.

There are three types of cartilage: hyaline cartilage, fibrocartilage, and elastic cartilage (**Table 4.6**).

TABLE 4.6 Mature Connective Tissue: Supporting Connective Tissue—Cartilage

A. HYALINE CARTILAGE

Description	**Hyaline cartilage** (*hyalinos* = glassy) contains a resilient gel as ground substance and appears in the body as a bluish-white, shiny substance (can stain pink or purple when prepared for microscopic examination; fine collagen fibers are not visible with ordinary staining techniques); prominent chondrocytes are found in lacunae surrounded by perichondrium (exceptions: articular cartilage in joints and cartilage of epiphyseal plates, where bones lengthen during growth).
Location	Most abundant cartilage in body; at ends of long bones, anterior ends of ribs, nose, parts of larynx, trachea, bronchi, bronchial tubes, embryonic and fetal skeleton.
Function	Provides smooth surfaces for movement at joints, flexibility, and support; weakest type of cartilage and can be fractured.

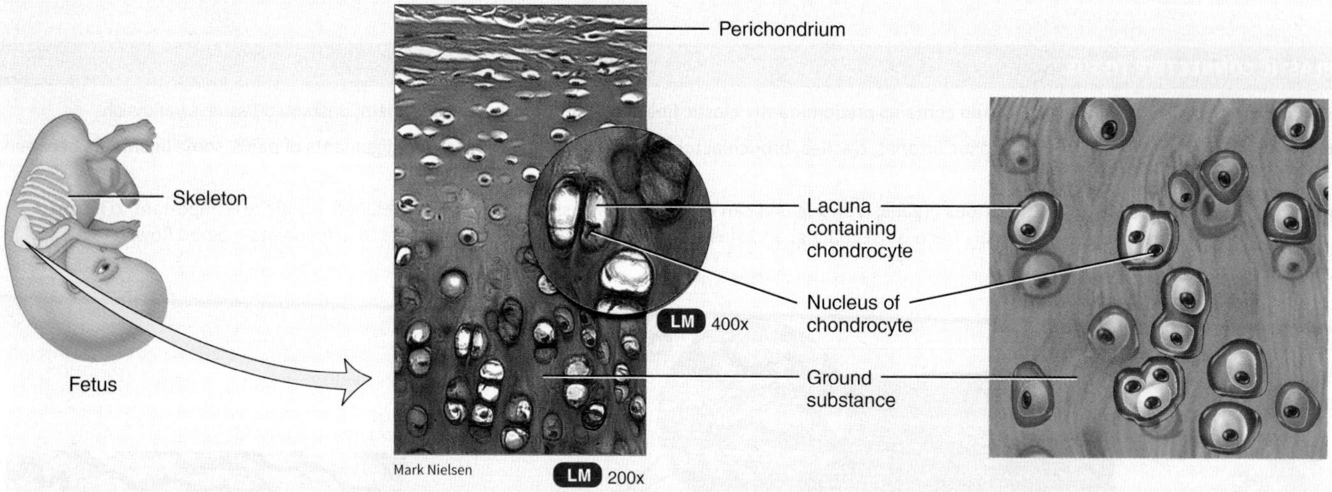

Mark Nielsen

Sectional view of hyaline cartilage of a developing fetal bone

Hyaline cartilage

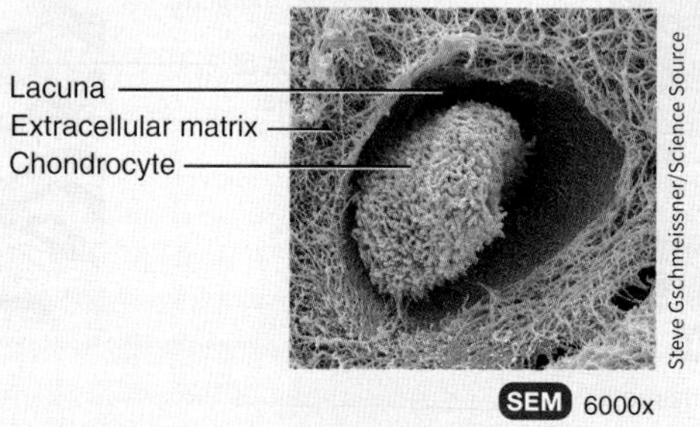

Steve Gschmeissner/Science Source

Hyaline cartilage

B. FIBROCARTILAGE

Description **Fibrocartilage** has chondrocytes among clearly visible thick bundles of collagen fibers within extracellular matrix; lacks perichondrium.

Location Pubic symphysis (where hip bones join anteriorly), intervertebral discs, menisci (cartilage pads) of knee, portions of tendons that insert into cartilage.

Function Support and joining structures together. Strength and rigidity make it the strongest type of cartilage.

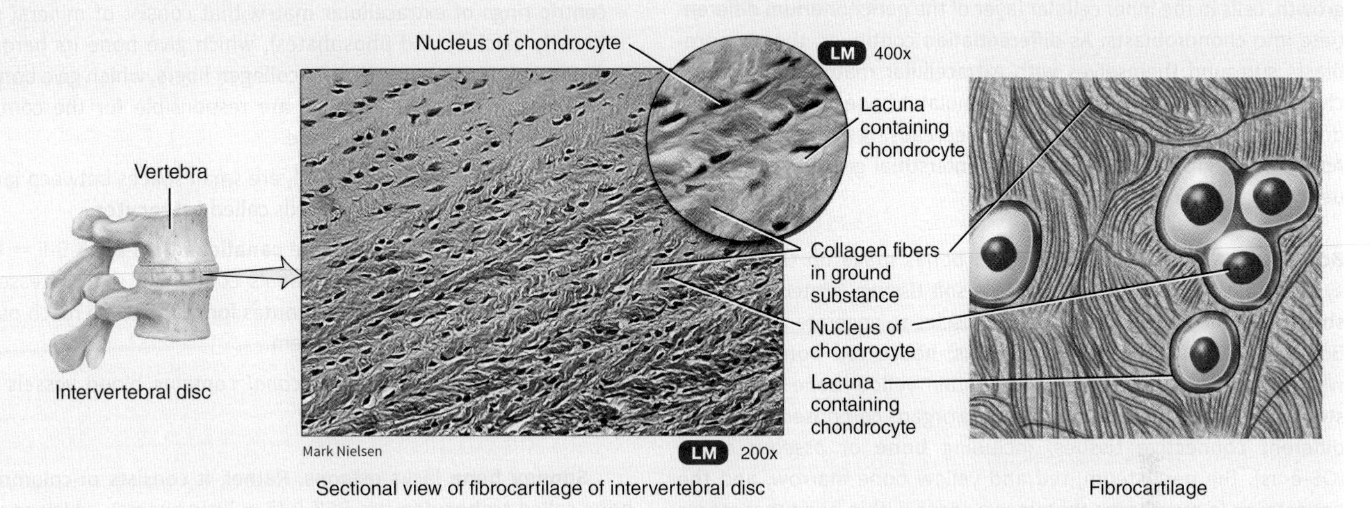

Sectional view of fibrocartilage of intervertebral disc

Fibrocartilage

C. ELASTIC CARTILAGE

Description **Elastic cartilage** has chondrocytes in threadlike network of elastic fibers within extracellular matrix; perichondrium present.

Location Lid on top of larynx (epiglottis), part of external ear (auricle), auditory (eustachian) tubes.

Function Provides strength and elasticity; maintains shape of certain structures.

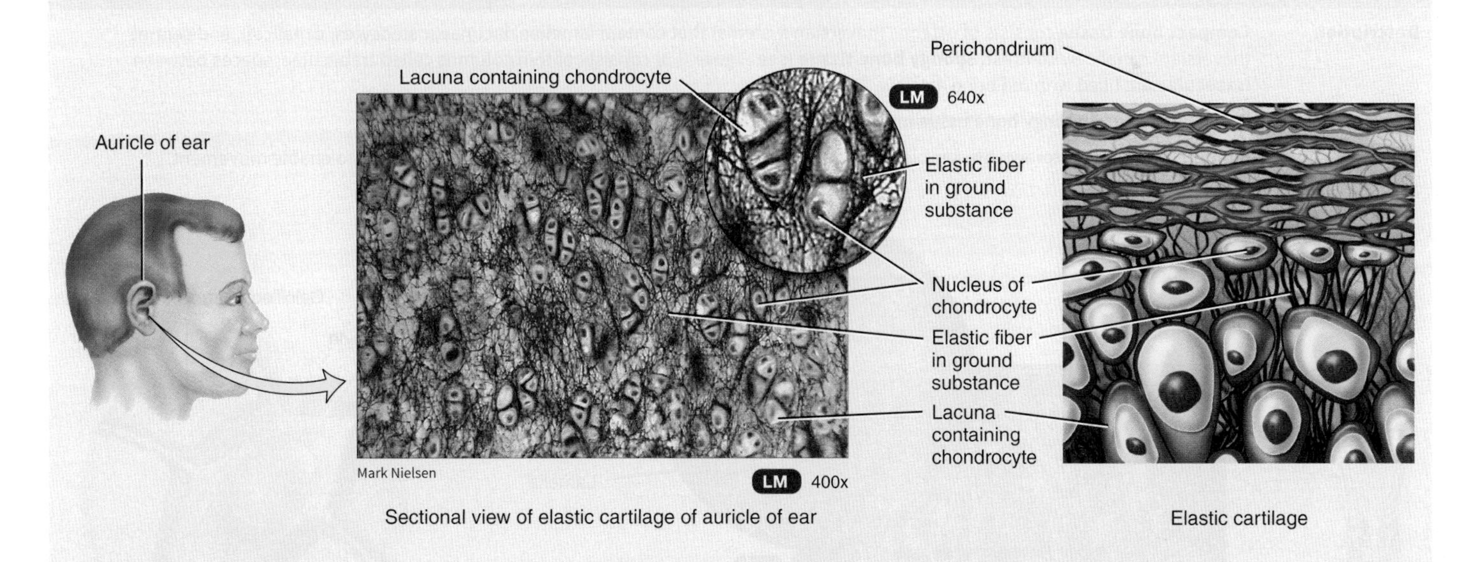

Sectional view of elastic cartilage of auricle of ear

Elastic cartilage

Metabolically, cartilage is a relatively inactive tissue that grows slowly. When injured or inflamed, cartilage repair proceeds slowly, in large part because cartilage is avascular. Substances needed for repair and blood cells that participate in tissue repair must diffuse or migrate into the cartilage. The growth of cartilage follows two basic patterns: interstitial growth and appositional growth.

In **interstitial growth** (in′-ter-STISH-al), there is growth from within the tissue. When cartilage grows by interstitial growth, the cartilage increases rapidly in size due to the division of existing chondrocytes and the continuous deposition of increasing amounts of extracellular matrix by the chondrocytes. As the chondrocytes synthesize new matrix, they are pushed away from each other. These events cause the cartilage to expand from within like bread rising, which is the reason for the term *inter*stitial. This growth pattern occurs while the cartilage is young and pliable, during childhood and adolescence.

In **appositional growth** (a-pō-ZISH-un-al), there is growth at the outer surface of the tissue. When cartilage grows by appositional growth, cells in the inner cellular layer of the perichondrium differentiate into chondroblasts. As differentiation continues, the chondroblasts surround themselves with extracellular matrix and become chondrocytes. As a result, matrix accumulates beneath the perichondrium on the outer surface of the cartilage, causing it to grow in width. Appositional growth starts later than interstitial growth and continues through adolescence.

Bone Tissue Cartilage, joints, and bones make up the skeletal system. The skeletal system supports soft tissues, protects delicate structures, and works with skeletal muscles to generate movement. Bones store calcium and phosphorus; house red bone marrow, which produces blood cells; and contain yellow bone marrow, a storage site for triglycerides. Bones are organs composed of several different connective tissues, including **bone** or *osseous tissue* (OS-ē-us), the periosteum, red and yellow bone marrow, and the endosteum (a membrane that lines a space within bone that stores yellow bone marrow). Bone tissue is classified as either compact or spongy, depending on how its extracellular matrix and cells are organized.

The basic unit of **compact bone** is an **osteon** or *haversian system* (Table 4.7). Each osteon has four parts:

1. The **lamellae** (la-MEL-lē = little plates; singular is *lamella*) are concentric rings of extracellular matrix that consist of mineral salts (mostly calcium and phosphates), which give bone its hardness and compressive strength, and collagen fibers, which give bone its tensile strength. The lamellae are responsible for the compact nature of this type of bone tissue.
2. **Lacunae**, as already mentioned, are small spaces between lamellae that contain mature bone cells called **osteocytes**.
3. Projecting from the lacunae are **canaliculi** (kan-a-LIK-ū-lī = little canals), networks of minute canals containing the processes of osteocytes. Canaliculi provide routes for nutrients to reach osteocytes and for wastes to leave them.
4. A **central canal** or *haversian canal* contains blood vessels and nerves.

Spongy bone lacks osteons. Rather, it consists of columns of bone called **trabeculae** (tra-BEK-ū-lē = little beams), which contain lamellae, osteocytes, lacunae, and canaliculi. Spaces between

TABLE 4.7 Mature Connective Tissue: Supporting Connective Tissue—Bone Tissue

Description	**Compact bone tissue** consists of osteons (haversian systems) that contain lamellae, lacunae, osteocytes, canaliculi, and central (haversian) canals. By contrast, **spongy bone tissue** (see Figure 6.3) consists of thin columns called trabeculae; spaces between trabeculae are filled with red bone marrow.
Location	Both compact and spongy bone tissue make up the various parts of bones of the body.
Function	Support, protection, storage; houses blood-forming tissue; serves as levers that act with muscle tissue to enable movement.

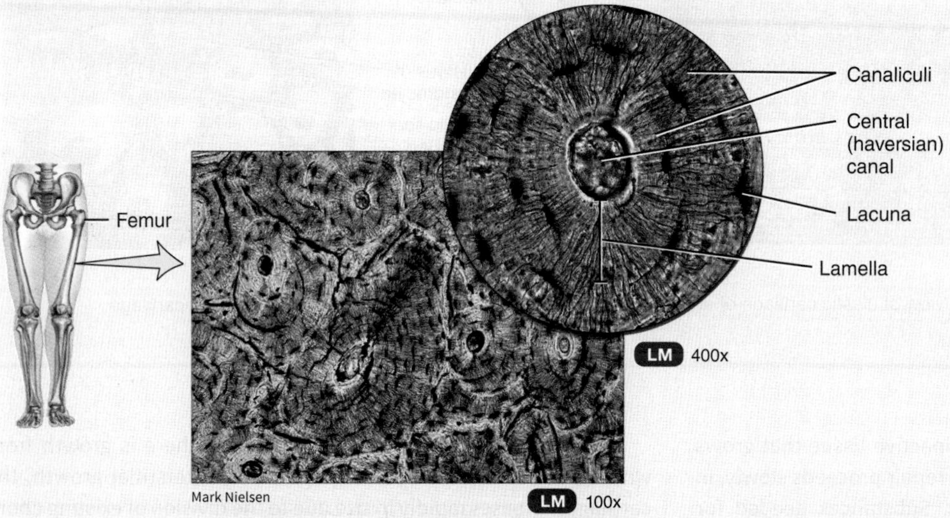

Sectional view of several osteons (haversian systems) of femur (thigh bone)

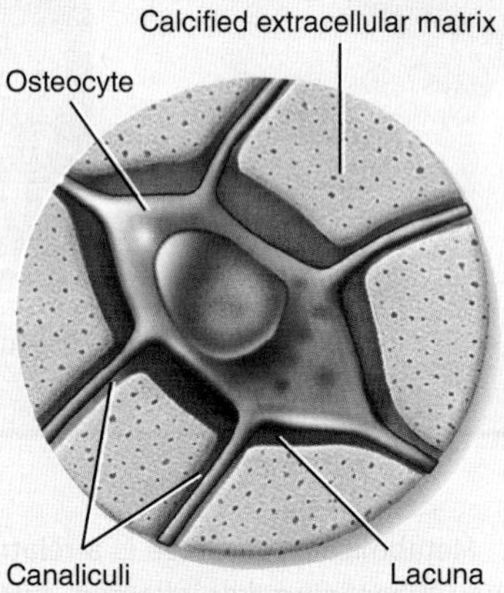

Details of an osteocyte

TABLE 4.8 Mature Connective Tissue: Liquid Connective Tissue—Blood

Description	**Blood** consists of blood plasma and formed elements: red blood cells (erythrocytes), white blood cells (leukocytes), platelets (thrombocytes).
Location	Within blood vessels (arteries, arterioles, capillaries, venules, veins), within chambers of heart.
Function	Red blood cells: transport oxygen and some carbon dioxide; white blood cells: carry on phagocytosis and mediate allergic reactions and immune system responses; platelets: essential for blood clotting.

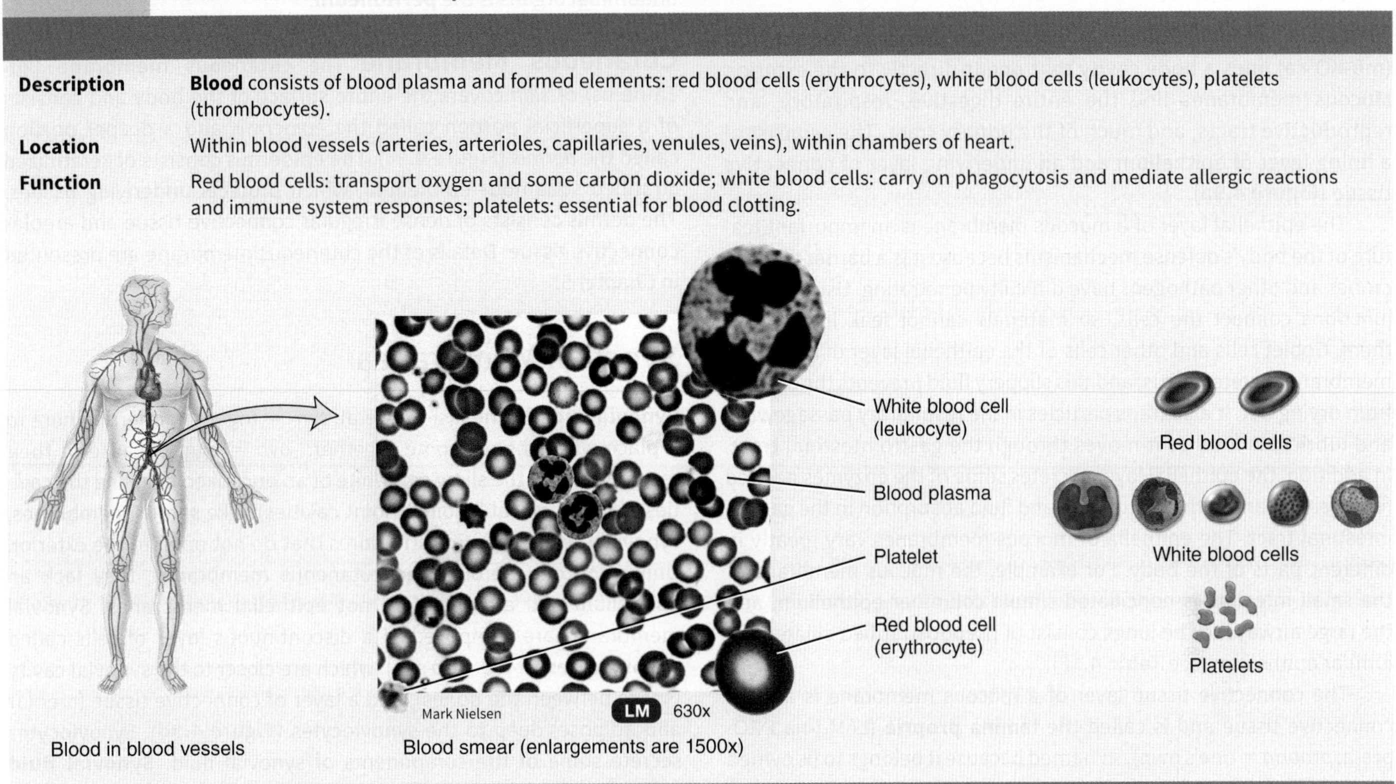

Blood in blood vessels

Blood smear (enlargements are 1500x)

trabeculae are filled with red bone marrow. Chapter 6 presents bone tissue histology in more detail.

Liquid Connective Tissue This is the final type of mature connective tissue. **A liquid connective tissue** has a liquid as its extracellular matrix.

Blood Tissue **Blood**, one of the liquid connective tissues has a liquid extracellular matrix called blood plasma and formed elements. **Blood plasma** is a pale yellow fluid that consists mostly of water with a wide variety of dissolved substances— nutrients, wastes, enzymes, plasma proteins, hormones, respiratory gases, and ions. Suspended in the blood plasma are **formed elements**—red blood cells (erythrocytes), white blood cells (leukocytes), and platelets (thrombocytes) (Table 4.8). **Red blood cells** transport oxygen to body cells and remove some carbon dioxide from them. **White blood cells** are involved in phagocytosis, immunity, and allergic reactions. **Platelets** (PLĀT-lets) participate in blood clotting. The details of blood are considered in Chapter 19.

Lymph **Lymph** is the extracellular fluid that flows in lymphatic vessels. It is a liquid connective tissue that consists of several types of cells in a clear liquid extracellular matrix that is similar to blood plasma but with much less protein. The composition of lymph varies from one part of the body to another. For example, lymph leaving lymph nodes includes many lymphocytes, a type of white blood cell, in contrast to lymph from the small intestine, which has a high content of newly absorbed dietary lipids. The details of lymph are considered in Chapter 22.

4.6 Membranes

OBJECTIVES

- **Define** a membrane.
- **Describe** the classification of membranes.

Membranes are flat sheets of pliable tissue that cover or line a part of the body. The majority of membranes consist of an epithelial layer and an underlying connective tissue layer and are called **epithelial membranes**. The principal epithelial membranes of the body are mucous membranes, serous membranes, and the cutaneous membrane, or skin. Another type of membrane, a synovial membrane, lines joints and contains connective tissue but no epithelium.

Epithelial Membranes

Mucous Membranes A **mucous membrane** or *mucosa* (mū-KŌ-sa) lines a body cavity that opens directly to the exterior. Mucous membranes line the entire digestive, respiratory, and reproductive tracts, and much of the urinary tract. They consist of a lining layer of epithelium and an underlying layer of connective tissue (**Figure 4.9a**).

The epithelial layer of a mucous membrane is an important feature of the body's defense mechanisms because it is a barrier that microbes and other pathogens have difficulty penetrating. Usually, tight junctions connect the cells, so materials cannot leak in between them. Goblet cells and other cells of the epithelial layer of a mucous membrane secrete mucus, and this slippery fluid prevents the cavities from drying out. It also traps particles in the respiratory passageways and lubricates food as it moves through the gastrointestinal tract. In addition, the epithelial layer secretes some of the enzymes needed for digestion and is the site of food and fluid absorption in the gastrointestinal tract. The epithelia of mucous membranes vary greatly in different parts of the body. For example, the mucous membrane of the small intestine is nonciliated simple columnar epithelium, and the large airways to the lungs consist of pseudostratified ciliated columnar epithelium (see **Table 4.1F**).

The connective tissue layer of a mucous membrane is areolar connective tissue and is called the **lamina propria** (LAM-i-na PRŌ-prē-a; *propria* = one's own), so named because it belongs to (is owned by) the mucous membrane. The lamina propria supports the epithelium, binds it to the underlying structures, allows some flexibility of the membrane, and affords some protection for underlying structures. It also holds blood vessels in place and is the vascular source for the overlying epithelium. Oxygen and nutrients diffuse from the lamina propria to the covering epithelium; carbon dioxide and wastes diffuse in the opposite direction.

Serous Membranes A **serous membrane** (SĒR-us = watery) or *serosa* lines a body cavity that does not open directly to the exterior (thoracic or abdominal cavities), and it covers the organs that are within the cavity. Serous membranes consist of areolar connective tissue covered by mesothelium (simple squamous epithelium) (**Figure 4.9b**). You will recall from Chapter 1 that serous membranes have two layers: The layer attached to and lining the cavity wall is called the **parietal layer** (pa-RĪ-e-tal; *pariet-* = wall); the layer that covers and adheres to the organs within the cavity is the **visceral layer** (*viscer-* = body organ) (see **Figure 1.10a**). The mesothelium of a serous membrane secretes **serous fluid**, a watery lubricant that allows organs to glide easily over one another or to slide against the walls of cavities.

Recall from Chapter 1 that the serous membrane lining the thoracic cavity and covering the lungs is the **pleura**. The serous membrane lining the heart cavity and covering the heart is the **pericardium**. The serous membrane lining the abdominal cavity and covering the abdominal organs is the **peritoneum**.

Cutaneous Membrane The **cutaneous membrane** (kū-TĀ-nē-us) or *skin* covers the entire surface of the body and consists of a superficial portion called the *epidermis* and a deeper portion called the *dermis* (**Figure 4.9c**). The epidermis consists of keratinized stratified squamous epithelium, which protects underlying tissues. The dermis consists of dense irregular connective tissue and areolar connective tissue. Details of the cutaneous membrane are presented in Chapter 5.

Synovial Membranes

Synovial membranes (si-NŌ-vē-al; *syn-* = together, referring here to a place where bones come together; *-ova* = egg, because of their resemblance to the slimy egg white of an uncooked egg) line the cavities of freely movable joints (joint cavities). Like serous membranes, synovial membranes line structures that do not open to the exterior. Unlike mucous, serous, and cutaneous membranes, they lack an epithelium and are therefore not epithelial membranes. Synovial membranes are composed of a discontinuous layer of cells called **synoviocytes** (si-NŌ-vē-ō-sīts), which are closer to the synovial cavity (space between the bones), and a layer of connective tissue (areolar and adipose) deep to the synoviocytes (**Figure 4.9d**). Synoviocytes secrete some of the components of synovial fluid. **Synovial fluid** lubricates and nourishes the cartilage covering the bones at movable joints and contains macrophages that remove microbes and debris from the joint cavity.

4.7 Muscular Tissue

OBJECTIVES

- **Describe** the general features of muscular tissue.
- **Contrast** the structure, location, and mode of control of skeletal, cardiac, and smooth muscle tissue.

Muscular tissue consists of elongated cells called *muscle fibers* or *myocytes* that can use ATP to generate force. As a result, muscular tissue produces body movements, maintains posture, and generates heat. It also provides protection. Based on location and certain structural and functional features, muscular tissue is classified into three types: **skeletal, cardiac,** and **smooth** (**Table 4.9**).

Chapter 10 provides a more detailed discussion of muscular tissue.

FIGURE 4.9 **Membranes.**

A membrane is a flat sheet of pliable tissues that covers or lines a part of the body.

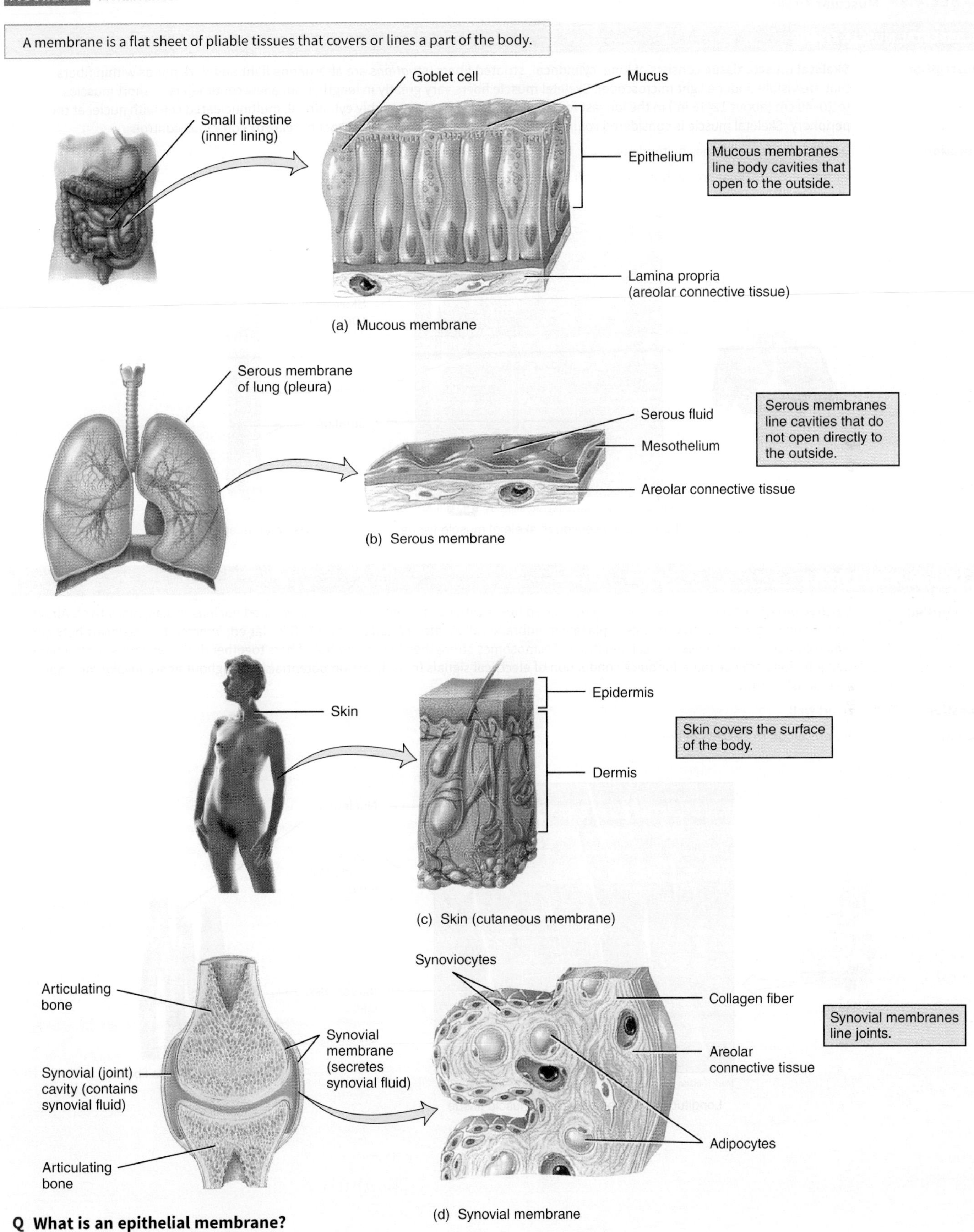

Q What is an epithelial membrane?

TABLE 4.9 Muscular Tissue

A. SKELETAL MUSCLE TISSUE

Description **Skeletal muscle tissue** consists of long, cylindrical, striated fibers (*striations* are alternating light and dark bands within fibers that are visible under a light microscope). Skeletal muscle fibers vary greatly in length, from a few centimeters in short muscles to 30–40 cm (about 12–16 in.) in the longest muscles. A muscle fiber is a roughly cylindrical, multinucleated cell with nuclei at the periphery. Skeletal muscle is considered *voluntary* because it can be made to contract or relax by conscious control.

Location Usually attached to bones by tendons.

Function Motion, posture, heat production, protection.

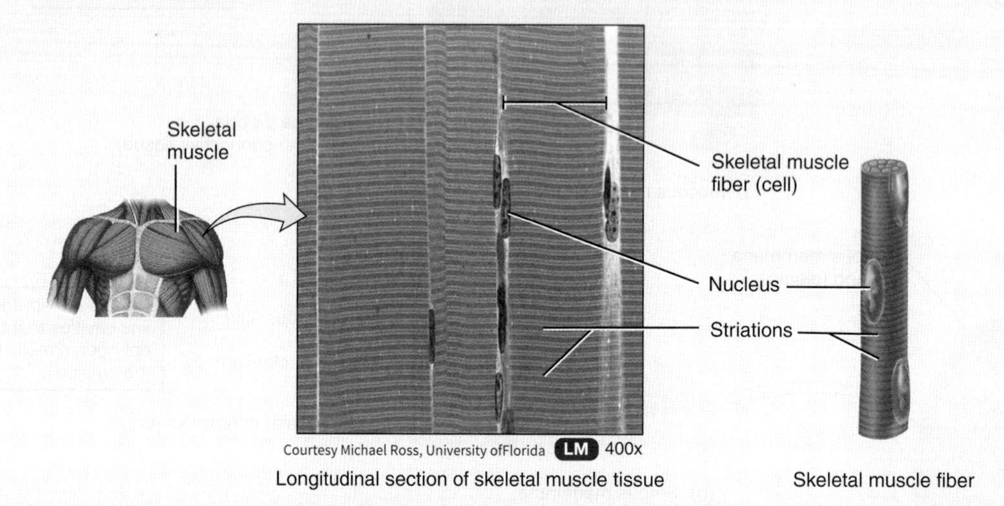

Longitudinal section of skeletal muscle tissue

Skeletal muscle fiber

B. CARDIAC MUSCLE TISSUE

Description **Cardiac muscle tissue** consists of branched, striated fibers with usually only one centrally located nucleus (occasionally two). Attach end to end by transverse thickenings of plasma membrane called *intercalated discs* (in-TER-ka-lāt-ed; *intercalate* = to insert between), which contain desmosomes and gap junctions. Desmosomes strengthen tissue and hold fibers together during vigorous contractions. Gap junctions provide route for quick conduction of electrical signals (muscle action potentials) throughout heart. *Involuntary* (not conscious) control.

Location Heart wall.

Function Pumps blood to all parts of body.

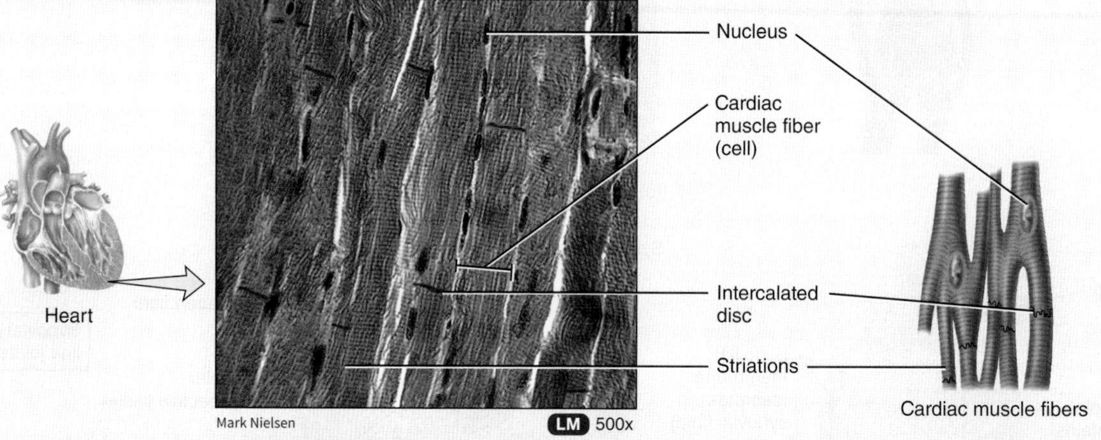

Longitudinal section of cardiac muscle tissue

Cardiac muscle fibers

C. SMOOTH MUSCLE TISSUE

Description	**Smooth muscle tissue** consists of nonstriated fibers (lack striations, hence the term *smooth*). Smooth muscle fiber is a small spindle-shaped cell thickest in middle, tapering at each end, and containing a single, centrally located nucleus. Gap junctions connect many individual fibers in some smooth muscle tissue (for example, in wall of intestines). Usually involuntary; can produce powerful contractions as many muscle fibers contract in unison. Where gap junctions are absent, such as iris of eye, smooth muscle fibers contract individually, like skeletal muscle fibers.
Location	Iris of eyes; walls of hollow internal structures such as blood vessels, airways to lungs, stomach, intestines, gallbladder, urinary bladder, and uterus.
Function	Motion (constriction of blood vessels and airways, propulsion of foods through gastrointestinal tract, contraction of urinary bladder and gallbladder).

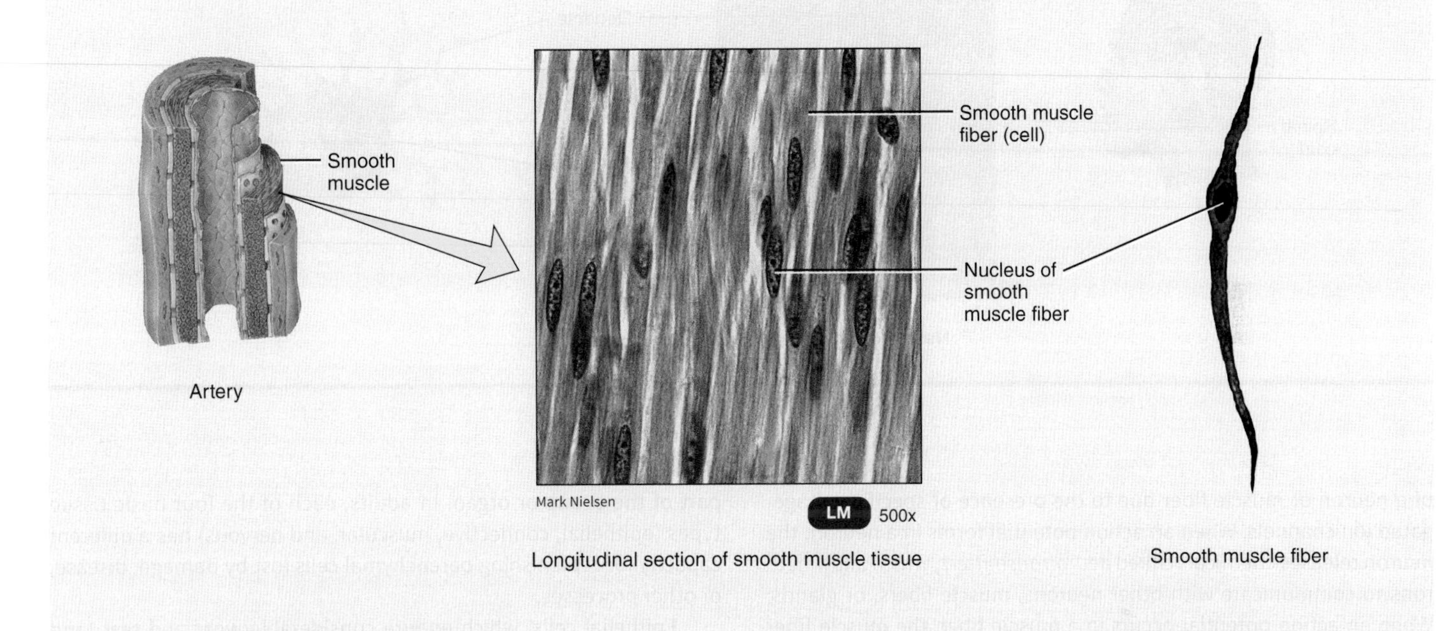

Longitudinal section of smooth muscle tissue

Smooth muscle fiber

4.8 Nervous Tissue

OBJECTIVE

- **Describe** the structural features and functions of nervous tissue.

Despite the awesome complexity of the nervous system, **nervous tissue** consists of only two principal types of cells: neurons and neuroglia. **Neurons** (NOO-rons; *neuro-* = nerve), or *nerve cells*, are sensitive to various stimuli. They convert stimuli into electrical signals called **nerve action potentials** (*nerve impulses*) and conduct these action potentials to other neurons, to muscle tissue, or to glands. Most neurons consist of three basic parts: a cell body and two kinds of cell processes—dendrites and axons (**Table 4.10**). The **cell body** contains the nucleus and other organelles. **Dendrites** (*dendr-* = tree) are tapering, highly branched, and usually short cell processes (extensions). They are the major receiving or input portion of a neuron. The **axon** (*axo-* = axis) of a neuron is a single, thin, cylindrical process that may be very long. It is the output portion of a neuron, conducting nerve impulses toward another neuron or to some other tissue.

Even though **neuroglia** (noo-RŌG-lē-a; *-glia* = glue) do not generate or conduct nerve impulses, these cells do have many important supportive functions. The detailed structure and function of neurons and neuroglia are considered in Chapter 12.

4.9 Excitable Cells

OBJECTIVE

- **Explain** the concept of electrical excitability.

Neurons and muscle fibers are considered **excitable cells** because they exhibit **electrical excitability**, the ability to respond to certain stimuli by producing electrical signals such as *action potentials*. Action potentials can propagate (travel) along the plasma membrane

TABLE 4.10 Nervous Tissue

Description	**Nervous tissue** consists of (1) neurons (nerve cells), which consist of cell body and processes extending from cell body (one to multiple dendrites and a single axon); and (2) neuroglia, which do not generate or conduct nerve impulses but have other important supporting functions.
Location	Nervous system.
Function	Exhibits sensitivity to various types of stimuli; converts stimuli into nerve impulses (action potentials); conducts nerve impulses to other neurons, muscle fibers, or glands.

Spinal cord

Dendrite

Nucleus of neuroglial cell

Nucleus in cell body

Axon

Mark Nielsen

LM 400x

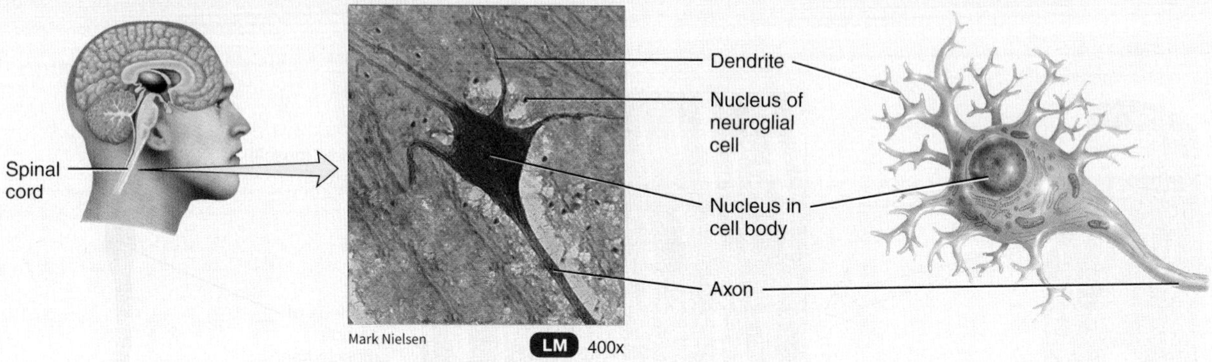

Neuron of spinal cord

of a neuron or muscle fiber due to the presence of specific voltage-gated ion channels. When an action potential forms in a neuron, the neuron releases chemicals called *neurotransmitters*, which allow neurons to communicate with other neurons, muscle fibers, or glands. When an action potential occurs in a muscle fiber, the muscle fiber contracts, resulting in activities such as movement of the limbs, propulsion of food through the small intestine, and movement of blood out of the heart and into the blood vessels of the body. The muscle action potential and the nerve action potential are discussed in detail in Chapters 10 and 12, respectively.

4.10 Tissue Repair: Restoring Homeostasis

OBJECTIVE

- **Describe** the role of tissue repair in restoring homeostasis.

Tissue repair is the replacement of worn-out, damaged, or dead cells. New cells originate by cell division from the **stroma**, (STRŌ-ma = bed or covering), the supporting connective tissue, or from the **parenchyma** (pa-RENG-ki-ma), cells that constitute the functioning part of the tissue or organ. In adults, each of the four basic tissue types (epithelial, connective, muscular, and nervous) has a different capacity for replenishing parenchymal cells lost by damage, disease, or other processes.

Epithelial cells, which endure considerable wear and tear (and even injury) in some locations, have a continuous capacity for renewal. In some cases, immature, undifferentiated cells called **stem cells** divide to replace lost or damaged cells. For example, stem cells reside in protected locations in the epithelia of the skin and gastrointestinal tract to replenish cells sloughed from the apical layer, and stem cells in red bone marrow continually provide new red and white blood cells and platelets. In other cases, mature, differentiated cells can undergo cell division; examples include hepatocytes (liver cells) and endothelial cells in blood vessels.

Some connective tissues also have a continuous capacity for renewal. One example is bone, which has an ample blood supply. Connective tissues such as cartilage can replenish cells much less readily, in part because of a smaller blood supply.

Muscular tissue has a relatively poor capacity for renewal of lost cells. Even though skeletal muscle tissue contains stem cells called *satellite cells,* they do not divide rapidly enough to replace extensively damaged muscle fibers. Cardiac muscle tissue lacks satellite cells, and existing cardiac muscle fibers do not undergo mitosis to form new cells. Recent evidence suggests that stem cells do migrate into the heart from the blood. There, they can differentiate and replace a limited number of cardiac muscle fibers and endothelial cells in heart blood vessels. Smooth muscle fibers can proliferate to some extent, but they do so much more slowly than the cells of epithelial or connective tissues.

Nervous tissue has the poorest capacity for renewal. Although experiments have revealed the presence of some stem cells in the brain, they normally do not undergo mitosis to replace damaged neurons. Discovering why this is so is a major goal of researchers who seek ways to repair nervous tissue damaged by injury or disease.

The restoration of an injured tissue or organ to normal structure and function depends entirely on whether parenchymal cells are active in the repair process. If parenchymal cells accomplish the repair, **tissue regeneration** is possible, and a near-perfect reconstruction of the injured tissue may occur. However, if fibroblasts of the stroma are active in the repair, the replacement tissue will be a new connective tissue. The fibroblasts synthesize collagen and other extracellular matrix materials that aggregate to form scar tissue, a process known as **fibrosis**. Because scar tissue is not specialized to perform the functions of the parenchymal tissue, the original function of the tissue or organ is impaired.

When tissue damage is extensive, as in large, open wounds, both the connective tissue stroma and the parenchymal cells are active in repair; fibroblasts divide rapidly, and new collagen fibers are manufactured to provide structural strength. Blood capillaries also sprout new buds to supply the healing tissue with the materials it needs. All these processes create an actively growing connective tissue called **granulation tissue** (gran-ū-LĀ-shun). This new tissue forms across a wound or surgical incision to provide a framework (stroma) that supports the epithelial cells that migrate into the open area and fill it. The newly formed granulation tissue also secretes a fluid that kills bacteria.

At times, a small but significant number of patients develop a complication of surgery called **wound dehiscence** (dē-HISS-ens), the partial or complete separation of the outer layers of a sutured incision. A common cause is surgical error in which sutures or staples are placed too far apart, too close to the incision edges, or under too much pressure. It can also occur if sutures are removed too early or if there is a deep wound infection. Other contributing factors are age, chemotherapy, coughing, straining, vomiting, obesity, smoking, and use of anticoagulants such as aspirin. A major complication of wound dehiscence is the protrusion of an organ through the open wound, especially the intestines. This can lead to peritonitis (inflammation of the peritoneum) and septic shock (shock that results from bacterial toxins due to vasodilation).

Three factors affect tissue repair: nutrition, blood circulation, and age. Nutrition is vital because the healing process places a great demand on the body's store of nutrients. Adequate protein in the diet is important because most of the structural components of a tissue are proteins. Several vitamins also play a direct role in wound healing and tissue repair. For example, vitamin C directly affects the normal production and maintenance of matrix materials, especially collagen, and strengthens and promotes the formation of new blood vessels. In a person with vitamin C deficiency, even superficial wounds fail to heal, and the walls of the blood vessels become fragile and are easily ruptured.

Proper blood circulation is essential to transport oxygen, nutrients, antibodies, and many defensive cells to the injured site. The blood also plays an important role in the removal of tissue fluid, bacteria, foreign bodies, and debris, elements that would otherwise interfere with healing. The third factor in tissue repair, age, is the topic of the next section.

See Clinical Connection: Adhesions

4.11 Aging and Tissues

OBJECTIVE

- **Describe** the effects of aging on tissues.

In later chapters, the effects of aging on specific body systems will be considered. With respect to tissues, epithelial tissues get progressively thinner and connective tissues become more fragile with aging. This is evidenced by an increased incidence of skin and mucous membrane disorders, wrinkles, more susceptibility to bruises, increased loss of bone density, higher rates of bone fractures, and increased episodes of joint pain and disorders. There is also an effect of aging on muscle tissue as evidenced by loss of skeletal muscle mass and strength, decline in the efficiency of pumping action of the heart, and decreased activity of smooth muscle–containing organs, for example, organs of the gastrointestinal tract.

Generally, tissues heal faster and leave less obvious scars in the young than in the aged. In fact, surgery performed on fetuses leaves no scars at all. The younger body is generally in a better nutritional state, its tissues have a better blood supply, and its cells have a higher metabolic rate. Thus, its cells can synthesize needed materials and divide more quickly. The extracellular components of tissues also change with age. Glucose, the most abundant sugar in the body, plays a role in the aging process. As the body ages, glucose is haphazardly added to proteins inside and outside cells, forming irreversible cross-links between adjacent protein molecules. With advancing age, more cross-links form, which contributes to the stiffening and loss of elasticity that occur in aging tissues. Collagen fibers, responsible for the strength of tendons, increase in number and change in quality with aging. Changes in the collagen of arterial walls affect the flexibility of arteries as much as the fatty deposits associated with atherosclerosis (see Coronary Artery Disease in the Disorders: Homeostatic Imbalances section of Chapter 20). Elastin, another extracellular component, is responsible for the elasticity of blood vessels and skin. It thickens, fragments, and acquires a greater affinity for calcium with age—changes that may also be associated with the development of atherosclerosis.

Review Questions

4.1 Types of Tissues

1. Define a tissue.
2. What are the four basic types of human tissues?

4.2 Cell Junctions

3. Which type of cell junction prevents the contents of organs from leaking into surrounding tissues?
4. Which types of cell junctions are found in epithelial tissue?

4.3 Comparison between Epithelial and Connective Tissues

5. Why are epithelial and connective tissues found adjacent to each other?

4.4 Epithelial Tissue

6. Describe the various layering arrangements and cell shapes of epithelial tissue.
7. What characteristics are common to all epithelial tissues?
8. How is the structure of the following epithelial tissues related to their functions: simple squamous, simple cuboidal, simple columnar (ciliated and nonciliated), pseudostratified columnar (ciliated and nonciliated), stratified squamous (keratinized and nonkeratinized), stratified cuboidal, stratified columnar, and transitional?
9. Where are endothelium and mesothelium located?
10. What is the difference between endocrine glands and exocrine glands? Name and give examples of the three functional classes of exocrine glands based on how their secretions are released.

4.5 Connective Tissue

11. In what ways does connective tissue differ from epithelial tissue?
12. What are the features of the cells, ground substance, and fibers that make up connective tissue?
13. How are connective tissues classified? List the various types.
14. Describe how the structure of the following connective tissue is related to its function: areolar connective tissue, adipose tissue, reticular connective tissue, dense regular connective tissue, dense irregular connective tissue, elastic connective tissue, hyaline cartilage, fibrocartilage, elastic cartilage, bone tissue, blood tissue, and lymph.
15. What is the difference between interstitial and appositional growth of cartilage?

4.6 Membranes

16. Define the following kinds of membranes: mucous, serous, cutaneous, and synovial. How do they differ from one another?
17. Where is each type of membrane located in the body? What are their functions?

4.7 Muscular Tissue

18. Which types of muscular tissue are striated? Which is smooth?
19. Which types of muscular tissue have gap junctions?

4.8 Nervous Tissue

20. What are the functions of the dendrites, cell body, and axon of a neuron?

4.9 Excitable Cells

21. Why is electrical excitability important to neurons and muscle fibers?

4.10 Tissue Repair: Restoring Homeostasis

22. How are stromal and parenchymal repair of a tissue different?
23. What is the importance of granulation tissue?

4.11 Aging and Tissues

24. What common changes occur in epithelial and connective tissues with aging?

Critical Thinking Questions

1. Imagine that you live 50 years in the future, and that you can custom-design a human to suit the environment. Your assignment is to customize the human's tissues so that the individual can survive on a large planet with gravity, a cold, dry climate, and a thin atmosphere. What adaptations would you incorporate into the structure and/or amount of tissues, and why?
2. You are entering a "Cutest Baby Contest" and have asked your colleagues to help you choose the most adorable picture of yourself as a baby. One of your colleagues rudely points out that you were quite chubby as an infant. You, however, are not offended. Explain to your colleague the benefit of that "baby fat."
3. You've been on a "bread-and-water" diet for 3 weeks and have noticed that a cut on your shin won't heal and bleeds easily. Why?

The Integumentary System

CHAPTER **5**

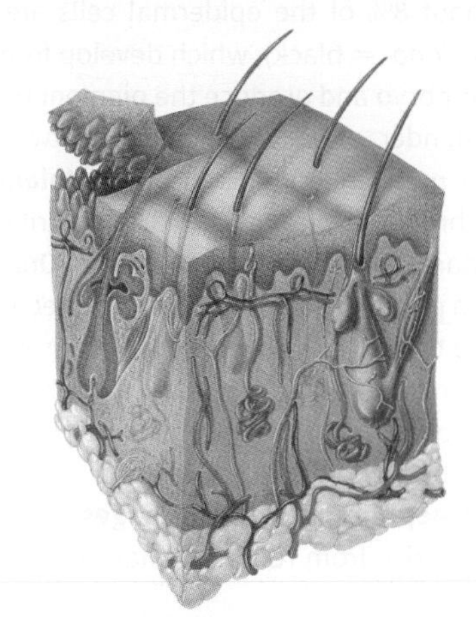

The Integumentary System and Homeostasis

The integumentary system contributes to homeostasis by protecting the body and helping regulate body temperature. It also allows you to sense pleasurable, painful, and other stimuli in your external environment.

The integumentary system helps maintain a constant body temperature, protects the body, and provides sensory information about the surrounding environment. Of all of the body's organs, none is more easily inspected or more exposed to infection, disease, and injury than the skin. Although its location makes it vulnerable to damage from trauma, sunlight, microbes, and pollutants in the environment, the skin's protective features ward off such damage. Because of its visibility, skin reflects our emotions (frowning, blushing) and some aspects of normal physiology (such as sweating). Changes in skin color may also indicate homeostatic imbalances in the body. For example, the bluish skin color associated with hypoxia (oxygen deficiency at the tissue level) is one sign of heart failure as well as other disorders. Abnormal skin eruptions or rashes such as chickenpox, cold sores, or measles may reveal systemic infections or diseases of internal organs, whereas other conditions, such as warts, age spots, or pimples, may involve the skin alone. So important is the skin to self-image that many people spend a great deal of time and money to restore it to a more normal or youthful appearance.

Q Did you ever wonder why it is so difficult to save the life of someone with extensive third-degree burns?

5.1 Structure of the Skin

OBJECTIVES

- **Describe** the layers of the epidermis and the cells that compose them.
- **Compare** the composition of the papillary and reticular regions of the dermis.
- **Explain** the basis for different skin colors.

Recall from Chapter 1 that a system consists of a group of organs working together to perform specific activities. The **integumentary system** (in-teg-ū-MEN-tar-ē; *in-* = inward; *-tegere* = to cover) is composed of the skin, hair, oil and sweat glands, nails, and sensory receptors.

Dermatology (der′-ma-TOL-ō-jē; *dermato-* = skin; *-logy* = study of) is the medical specialty that deals with the structure, function, and disorders of the integumentary system.

The **skin**, also known as the *cutaneous membrane* (kū-TĀ-nē-us), covers the external surface of the body and is the largest organ of the body in weight. In adults, the skin covers an area of about 2 square meters (22 square feet) and weighs 4.5–5 kg (10–11 lb), about 7% of total body weight. It ranges in thickness from 0.5 mm (0.02 in.) on the eyelids to 4.0 mm (0.16 in.) on the heels. Over most of the body it is 1–2 mm (0.04–0.08 in.) thick. The skin consists of two main parts (**Figure 5.1**). The superficial, thinner portion, which is composed of *epithelial tissue*, is the **epidermis** (ep′-i-DERM-is; *epi-* = above). The deeper, thicker *connective tissue* portion is the **dermis**. While the epidermis is avascular, the dermis is vascular. For this reason, if you cut the epidermis there is no bleeding, but if the cut penetrates to the dermis there is bleeding.

Deep to the dermis, but not part of the skin, is the **subcutaneous** (*subQ*) **layer**. Also called the *hypodermis* (*hypo-* = below), this layer consists of areolar and adipose tissues. Fibers that extend from the dermis anchor the skin to the subcutaneous layer, which in turn attaches to underlying fascia, the connective tissue around muscles and bones. The subcutaneous layer serves as a storage depot for fat and contains large blood vessels that supply the skin. This region (and sometimes the dermis) also contains nerve endings called **lamellated corpuscles** or *pacinian corpuscles* (pa-SIN-ē-an) that are sensitive to pressure (**Figure 5.1**).

Epidermis

The epidermis is composed of keratinized stratified squamous epithelium. It contains four principal types of cells: keratinocytes, melanocytes, intraepidermal macrophages, and tactile epithelial cells (**Figure 5.2**). About 90% of epidermal cells are **keratinocytes** (ker-a-TIN-ō-sīts′; *keratino-* = hornlike; *-cytes* = cells), which are arranged in four or five layers and produce the protein **keratin** (KER-a-tin) (**Figure 5.2a**). Recall from Chapter 4 that keratin is a tough, fibrous protein that helps protect the skin and underlying tissues from abrasions, heat, microbes, and chemicals. Keratinocytes also produce lamellar granules, which release a water-repellent sealant that decreases water entry and loss and inhibits the entry of foreign materials.

About 8% of the epidermal cells are **melanocytes** (MEL-a-nō-sīts′; *melano-* = black), which develop from the ectoderm of a developing embryo and produce the pigment melanin (Figure 5.2b). Their long, slender projections extend between the keratinocytes and transfer melanin granules to them. **Melanin** (MEL-a-nin) is a yellow-red or brown-black pigment that contributes to skin color and absorbs damaging ultraviolet (UV) light. Once inside keratinocytes, the melanin granules cluster to form a protective veil over the nucleus, on the side toward the skin surface. In this way, they shield the nuclear DNA from damage by UV light. Although their melanin granules effectively protect keratinocytes, melanocytes themselves are particularly susceptible to damage by UV light.

Intraepidermal macrophages or *Langerhans cells* (LANG-er-hans) arise from red bone marrow and migrate to the epidermis (Figure 5.2c), where they constitute a small fraction of the epidermal cells. They participate in immune responses mounted against microbes that invade the skin, and are easily damaged by UV light. Their role in the immune response is to help other cells of the immune system recognize an invading microbe and destroy it.

Tactile epithelial cells, or *Merkel cells* (MER-kel), are the least numerous of the epidermal cells. They are located in the deepest layer of the epidermis, where they contact the flattened process of a sensory neuron (nerve cell), a structure called a **tactile disc** or *Merkel disc* (Figure 5.2d). Tactile epithelial cells and their associated tactile discs detect touch sensations.

Several distinct layers of keratinocytes in various stages of development form the epidermis (Figure 5.3). In most regions of the body the epidermis has four strata (STRĀ-ta) or layers—stratum basale,

FIGURE 5.1 Components of the integumentary system. The skin consists of a superficial, thin epidermis and a deep, thicker dermis. Deep to the skin is the subcutaneous layer, which attaches the dermis to underlying fascia.

The integumentary system includes the skin, hair, oil and sweat glands, nails, and sensory receptors.

Functions of the Integumentary System

1. Regulates body temperature.
2. Stores blood.
3. Protects body from external environment.
4. Detects cutaneous sensations.
5. Excretes and absorbs substances.
6. Synthesizes vitamin D.

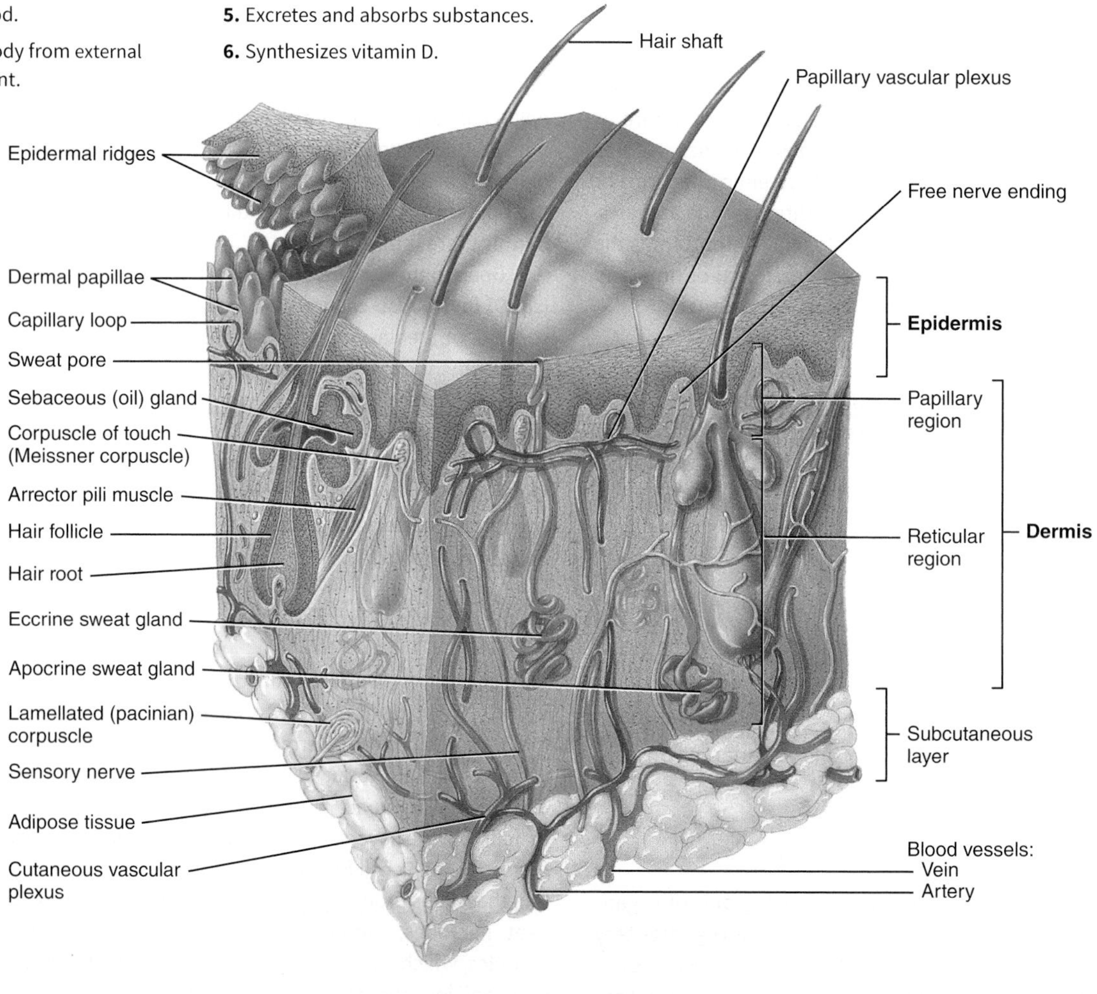

(a) Sectional view of skin and subcutaneous layer

FIGURE 5.1 Continued

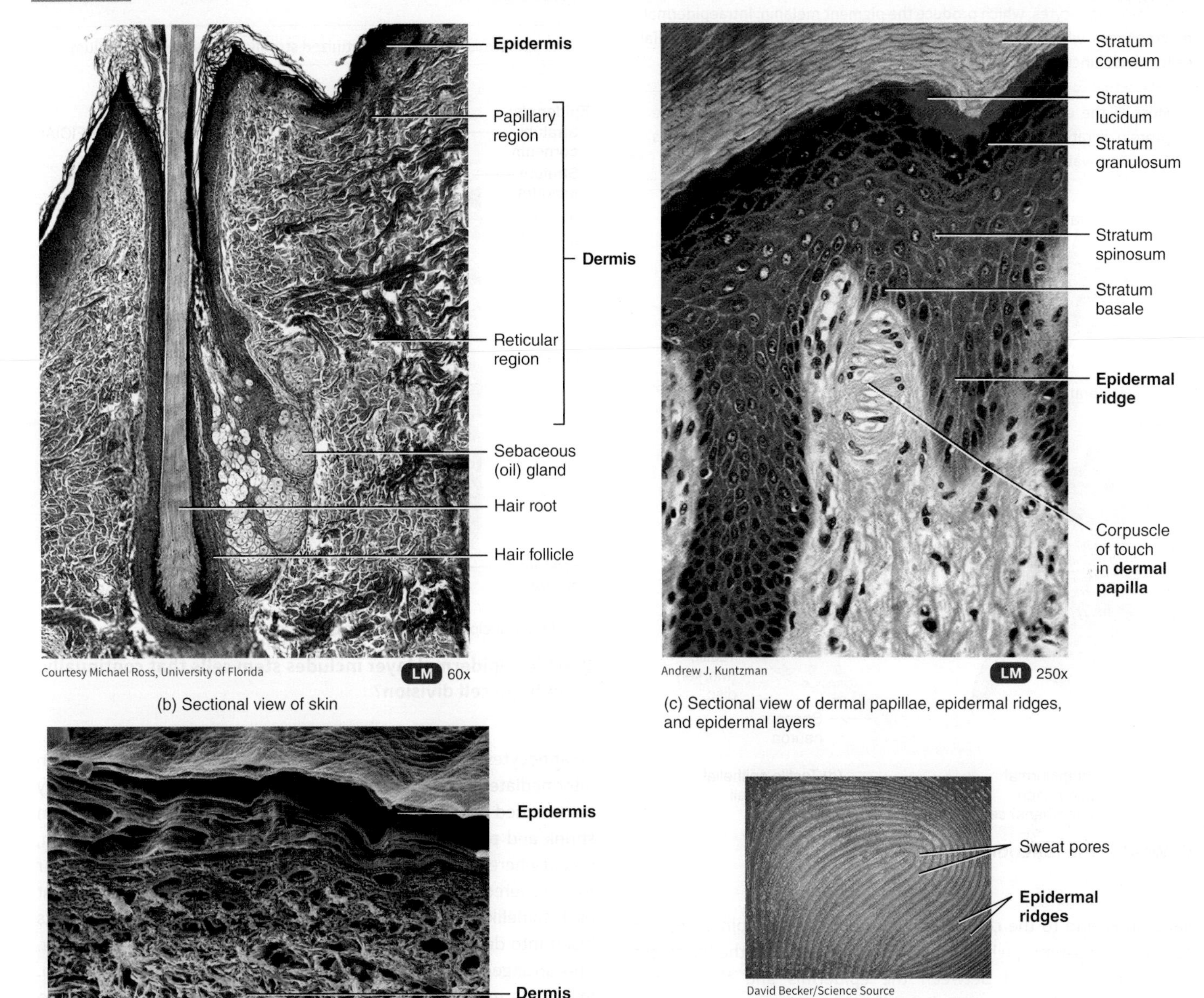

(b) Sectional view of skin

(c) Sectional view of dermal papillae, epidermal ridges, and epidermal layers

(d) Epidermal ridges and sweat pores

Q What types of tissues make up the epidermis and the dermis?

stratum spinosum, stratum granulosum, and a thin stratum corneum. This is called **thin skin**. Where exposure to friction is greatest, such as in the fingertips, palms, and soles, the epidermis has five layers—stratum basale, stratum spinosum, stratum granulosum, stratum lucidum, and a thick stratum corneum. This is called **thick skin**. The details of thin and thick skin are discussed later in the chapter (see Section 5.3).

Stratum Basale The deepest layer of the epidermis is the **stratum basale** (ba-SA-lē; *basal* = base), composed of a single row of cuboidal or columnar keratinocytes. Some cells in this layer are *stem cells* that undergo cell division to continually produce new keratinocytes. The nuclei of keratinocytes in the stratum basale are large, and their cytoplasm contains many ribosomes, a small Golgi complex, a few mitochondria, and some rough endoplasmic reticulum. The cytoskeleton within keratinocytes of the stratum basale includes scattered intermediate filaments, called *keratin intermediate filaments (tonofilaments)*. The keratin intermediate filaments form the tough protein keratin in its more superficial epidermal layers. Keratin protects the deeper layers from injury. Keratin intermediate filaments attach to desmosomes, which bind cells of the stratum basale to

FIGURE 5.2 **Cells in the epidermis.** Besides keratinocytes, the epidermis contains melanocytes, which produce the pigment melanin; intraepidermal macrophages, which participate in immune responses; and tactile epithelial cells, which function in the sensation of touch.

Most of the epidermis consists of keratinocytes, which produce the protein keratin (protects underlying tissues), and lamellar granules (contain a waterproof sealant).

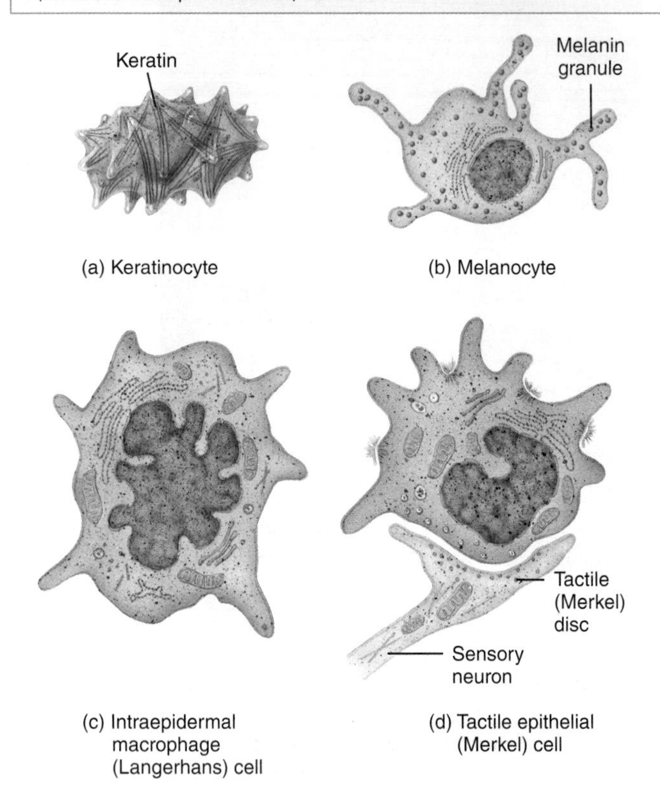

Q What is the function of melanin?

FIGURE 5.3 **Layers of the epidermis.** See also Fig. 5.1d.

The epidermis consists of keratinized stratified squamous epithelium.

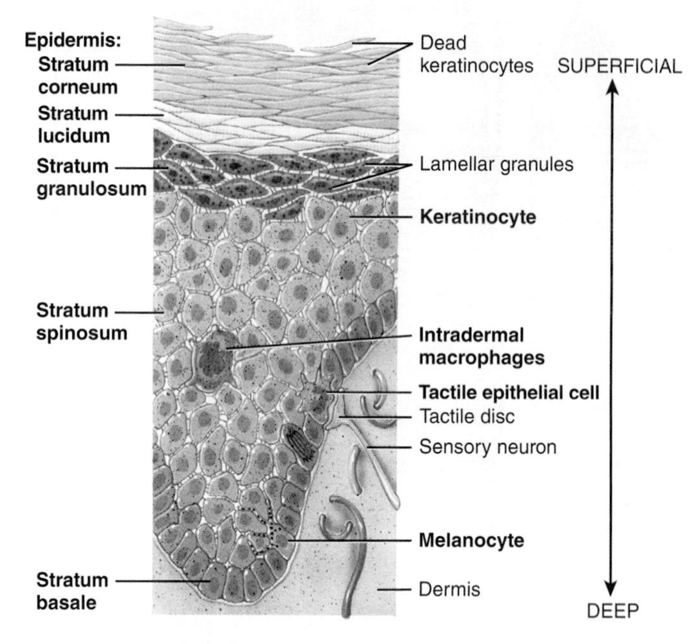

Four principal cell types in epidermis of thick skin

Q Which epidermal layer includes stem cells that continually undergo cell division?

each other and to the cells of the adjacent stratum spinosum, and to hemidesmosomes, which bind the keratinocytes to the basement membrane positioned between the epidermis and the dermis. Melanocytes and tactile epithelial cells with their associated tactile discs are scattered among the keratinocytes of the basal layer. The stratum basale is also known as the *stratum germinativum* (jer′-mi-na-TĒ-vum; *germ-* = sprout) to indicate its role in forming new cells.

See Clinical Connection: Skin Grafts

Stratum Spinosum Superficial to the stratum basale is the **stratum spinosum** (spi-NŌ-sum; *spinos-* = thornlike). This stratum mainly consists of numerous keratinocytes arranged in 8–10 layers. Cells in the more superficial layers become somewhat flattened. The keratinocytes in the stratum spinosum, which are produced by the stem cells in the basal layer, have the same organelles as cells of the stratum basale, and some retain their ability to divide. The keratinocytes of this layer produce coarser bundles of keratin in intermediate filaments than those of the basal layer. Although they are rounded and larger in living tissue, cells of the stratum spinosum shrink and pull apart when prepared for microscopic examination, except where the membranes join at desmosomes, so that they appear to be covered with thornlike spines (thus, the name) (Figure 5.3). At each spinelike projection, bundles of keratin intermediate filaments insert into desmosomes, which tightly join the cells to one another. This arrangement provides both strength and flexibility to the skin. Intraepidermal macrophages and projections of melanocytes are also present in the stratum spinosum.

Stratum Granulosum At about the middle of the epidermis, the **stratum granulosum** (gran-ū-LŌ-sum; *granulos-* = little grains) consists of three to five layers of flattened keratinocytes that are undergoing apoptosis. (Recall from Chapter 3 that *apoptosis* is an orderly, genetically programmed cell death in which the nucleus fragments before the cells die.) The nuclei and other organelles of these cells begin to degenerate as they move farther from their source of nutrition (the dermal blood vessels). Even though keratin intermediate filaments are no longer being produced by these cells, they become more apparent because the organelles in the cells are regressing. A distinctive feature of cells in this layer is the presence of darkly staining granules of a protein called **keratohyalin** (ker′-a-tō-HĪ-a-lin), which assembles keratin intermediate filaments into keratin. Also present in the keratinocytes are membrane-enclosed

lamellar granules (la-MEL-ar), which fuse with the plasma membrane and release a lipid-rich secretion. This secretion is deposited in the spaces between cells of the stratum granulosum, stratum lucidum, and stratum corneum. The lipid-rich secretion acts as a water-repellent sealant, retarding loss and entry of water and entry of foreign materials. As their nuclei break down during apoptosis, the keratinocytes of the stratum granulosum can no longer carry on vital metabolic reactions, and they die. Thus, the stratum granulosum marks the transition between the deeper, metabolically active strata and the dead cells of the more superficial strata.

Stratum Lucidum The **stratum lucidum** (LOO-si-dum; *lucid-* = clear) is present only in the thick skin of areas such as the fingertips, palms, and soles. It consists of four to six layers of flattened clear, dead keratinocytes that contain large amounts of keratin and thickened plasma membranes. This probably provides an additional level of toughness in this region of thick skin.

Stratum Corneum The **stratum corneum** (KOR-nē-um; *corne-* = horn or horny) consists on average of 25 to 30 layers of flattened dead keratinocytes, but can range in thickness from a few cells in thin skin to 50 or more cell layers in thick skin. The cells are extremely thin, flat, plasma membrane–enclosed packages of keratin that no longer contain a nucleus or any internal organelles. They are the final product of the differentiation process of the keratinocytes. The cells within each layer overlap one another like the scales on the skin of a snake. Neighboring layers of cells also form strong connections with one another. The plasma membranes of adjacent cells are arranged in complex, wavy folds that fit together like pieces of a jigsaw puzzle to hold the layers together. In this outer stratum of the epidermis, cells are continuously shed and replaced by cells from the deeper strata. Its multiple layers of dead cells help the stratum corneum to protect deeper layers from injury and microbial invasion. Constant exposure of skin to friction stimulates increased cell production and keratin production that results in the formation of a **callus**, an abnormal thickening of the stratum corneum.

Keratinization and Growth of the Epidermis

Newly formed cells in the stratum basale are slowly pushed to the surface. As the cells move from one epidermal layer to the next, they accumulate more and more keratin, a process called **keratinization** (ker′-a-tin-i-ZĀ-shun). Then they undergo apoptosis. Eventually the keratinized cells slough off and are replaced by underlying cells that in turn become keratinized. The whole process by which cells form in the stratum basale, rise to the surface, become keratinized, and slough off takes about four to six weeks in an average epidermis of 0.1 mm (0.004 in.) thickness. Nutrients and oxygen diffuse to the avascular epidermis from blood vessels in the dermis. The epidermal cells of the stratum basale are closest to these blood vessels and receive most of the nutrients and oxygen. These cells are the most active metabolically and continuously undergo cell division to produce new keratinocytes. As the new keratinocytes are pushed farther from the blood supply by continuing cell division, the epidermal strata above the basale receive fewer nutrients, and the cells become less active and eventually die. The rate of cell division in the stratum basale increases when the outer layers of the epidermis are stripped away, as occurs in abrasions and burns. The mechanisms that regulate this remarkable growth are not well understood, but hormonelike proteins such as **epidermal growth factor (EGF)** play a role. An excessive amount of keratinized cells shed from the skin of the scalp is called **dandruff**.

Table 5.1 summarizes the distinctive features of the epidermal strata.

TABLE 5.1 Summary of Epidermal Strata (see Figure 5.3)

STRATUM	DESCRIPTION
Basale	Deepest layer, composed of single row of cuboidal or columnar keratinocytes that contain scattered keratin intermediate filaments (tonofilaments); stem cells undergo cell division to produce new keratinocytes; melanocytes and tactile epithelial cells associated with tactile discs are scattered among keratinocytes.
Spinosum	Eight to ten rows of many-sided keratinocytes with bundles of keratin intermediate filaments; contains projections of melanocytes and intraepidermal macrophages.
Granulosum	Three to five rows of flattened keratinocytes, in which organelles are beginning to degenerate; cells contain the protein keratohyalin (converts keratin intermediate filaments into keratin) and lamellar granules (release lipid-rich, water-repellent secretion).
Lucidum	Present only in skin of fingertips, palms, and soles; consists of four to six rows of clear, flat, dead keratinocytes with large amounts of keratin.
Corneum	Few to 50 or more rows of dead, flat keratinocytes that contain mostly keratin.

See Clinical Connection: Psoriasis

Dermis

The second, deeper part of the skin, the *dermis*, is composed of dense irregular connective tissue containing collagen and elastic fibers. This woven network of fibers has great tensile strength (resists pulling or stretching forces). The dermis also has the ability to stretch and recoil easily. It is much thicker than the epidermis, and this thickness varies from region to region in the body, reaching its greatest thickness on the palms and soles. Leather, which we use for belts, shoes, baseball gloves, and basketballs, is the dried and treated dermis of other animals. The few cells present in the dermis include predominantly fibroblasts, with some macrophages, and a few adipocytes near its boundary with the subcutaneous layer. Blood vessels, nerves, glands, and hair follicles (epithelial invaginations of the epidermis) are embedded in the dermal layer. The dermis is essential to the survival of the epidermis, and these adjacent layers form many important structural and functional relations. Based on its tissue structure, the dermis can be divided into a thin superficial papillary region and a thick deeper reticular region.

The **papillary region** makes up about one-fifth of the thickness of the total layer (see Figure 5.1). It contains thin collagen and fine elastic fibers. Its surface area is greatly increased by **dermal papillae** (pa-PIL-ē = nipples), small, nipple-shaped structures that project into the undersurface of the epidermis. All dermal papillae contain **capillary loops** (blood vessels). Some also contain tactile receptors called **corpuscles of touch** or *Meissner corpuscles* (MĪS-ner), nerve endings that are sensitive to touch. Still other dermal papillae also contain **free nerve endings**, dendrites that lack any apparent structural specialization. Different free nerve endings initiate signals that give rise to sensations of warmth, coolness, pain, tickling, and itching.

The **reticular region** (*reticul-* = netlike), which is attached to the subcutaneous layer, contains bundles of thick collagen fibers, scattered fibroblasts, and various wandering cells (such as macrophages). Some adipose cells can be present in the deepest part of the layer, along with some coarse elastic fibers (see Figure 5.1). The collagen fibers in the reticular region are arranged in a netlike manner and have a more regular arrangement than those in the papillary region. The more regular orientation of the large collagen fibers helps the skin resist stretching. Blood vessels, nerves, hair follicles, sebaceous (oil) glands, and sudoriferous (sweat) glands occupy the spaces between fibers.

The combination of collagen and elastic fibers in the reticular region provides the skin with strength, **extensibility** (ek-sten′-si-BIL-i-tē), the ability to stretch, and **elasticity** (e-las-TIS-i-tē), the ability to return to original shape after stretching. The extensibility of skin can be readily seen around joints and in pregnancy and obesity.

See Clinical Connection: Stretch Marks

The surfaces of the palms, fingers, soles, and toes have a series of ridges and grooves. They appear either as straight lines or as a pattern of loops and whorls, as on the tips of the digits. These **epidermal ridges** are produced during the third month of fetal development as downward projections of the epidermis into the dermis between the dermal papillae of the papillary region (see Figure 5.1). The epidermal ridges create a strong bond between the epidermis and dermis in a region of high mechanical stress. The epidermal ridges also increase the surface area of the epidermis and thus increase the grip of the hand or foot by increasing friction. Finally, the epidermal ridges greatly increase surface area, which increases the number of corpuscles of touch and thus increases tactile sensitivity. Because the ducts of sweat glands open on the tops of the epidermal ridges as sweat pores, the sweat and ridges form **fingerprints** (or **footprints**) on touching a smooth object. The epidermal ridge pattern is in part genetically determined and is unique for each individual. Even identical twins have different patterns. Normally, the ridge pattern does not change during life, except to enlarge, and thus can serve as the basis for identification. The study of the pattern of epidermal ridges is called **dermatoglyphics** (der′-ma-tō-GLIF-iks; *glyphe* = carved work).

In addition to forming epidermal ridges, the complex papillary surface of the dermis has other functional properties. The dermal papillae greatly increase the surface contact between the dermis and epidermis. This increased dermal contact surface, with its extensive network of small blood vessels, serves as an important source of nutrition for the overlying epidermis. Molecules diffuse from the small blood capillaries in the dermal papillae to the cells of the stratum basale, allowing the basal epithelial stem cells to divide and the keratinocytes to grow and develop. As keratinocytes push toward the surface and away from the dermal blood source, they are no longer able to obtain the nutrition they require, leading to the eventual breakdown of their organelles.

The dermal papillae fit together with the complementary epidermal ridge to form an extremely strong junction between the two layers. This jigsaw puzzle–like connection strengthens the skin against shearing forces (forces that laterally shift in relation to each other) that attempt to separate the epidermis from the dermis.

Table 5.2 summarizes the structural features of the papillary and reticular regions of the dermis.

TABLE 5.2 Summary of Papillary and Reticular Regions of the Dermis (see Figure 5.1b)

REGION	DESCRIPTION
Papillary	Superficial portion of dermis (about one-fifth); consists of areolar connective tissue with thin collagen and fine elastic fibers; contains dermal ridges that house blood capillaries, corpuscles of touch, and free nerve endings.
Reticular	Deeper portion of dermis (about four-fifths); consists of dense irregular connective tissue with bundles of thick collagen and some coarse elastic fibers. Spaces between fibers contain some adipose cells, hair follicles, nerves, sebaceous glands, and sudoriferous glands.

See Clinical Connection: Tension Lines and Surgery

The Structural Basis of Skin Color

Melanin, hemoglobin, and carotene are three pigments that impart a wide variety of colors to skin. The amount of melanin causes the skin's color to vary from pale yellow to reddish-brown to black. The difference between the two forms of melanin, *pheomelanin* (fē-ō-MEL-a-nin) (yellow to red) and *eumelanin* (ū-MEL-a-nin) (brown to black), is most apparent in the hair. Melanocytes, the melanin-producing cells, are most plentiful in the epidermis of the penis, nipples of the breasts, area just around the nipples (areolae), face, and limbs. They are also present in mucous membranes. Because the *number* of melanocytes is about the same in all people, differences in skin color are due mainly to the *amount of pigment* the melanocytes produce and transfer to keratinocytes. In some people who are genetically predisposed, melanin accumulates in patches called **freckles**. Freckles typically are reddish or brown and tend to be more visible in the summer than the winter. As a person ages, **age** (*liver*) **spots** may develop. These flat blemishes have nothing to do with the liver. They look like freckles and range in color from light brown to black. Like freckles, age spots are accumulations of melanin. Age spots are darker than freckles and build up over time due to exposure to sunlight. Age spots do not fade away during the winter months and are more common in adults over 40. A round, flat, or raised area that represents a benign localized overgrowth of melanocytes and usually develops in childhood or adolescence is called a **nevus** (NĒ-vus), or a *mole*.

Melanocytes synthesize melanin from the amino acid *tyrosine* in the presence of an enzyme called *tyrosinase.* Synthesis occurs in an organelle called a **melanosome** (MEL-an-ō-sōm). Exposure to ultraviolet (UV) light increases the enzymatic activity within melanosomes and thus increases melanin production. Both the amount and darkness of melanin increase on UV exposure, which gives the skin a tanned appearance and helps protect the body against further UV radiation. Melanin absorbs UV radiation, prevents damage to DNA in epidermal cells, and neutralizes free radicals that form in the skin following damage by UV radiation. Thus, within limits, melanin serves a protective function. In response to DNA damage, melanin production increases. As you will see later, exposing the skin to a *small* amount of UV light is actually necessary for the skin to begin the process of vitamin D synthesis. However, repeatedly exposing the skin to a *large* amount of UV light may cause skin cancer. A tan is lost when the melanin-containing keratinocytes are shed from the stratum corneum.

See Clinical Connection: Albinism and Vitiligo

Dark-skinned individuals have large amounts of melanin in the epidermis, so their skin color ranges from yellow to reddish-brown to black. Light-skinned individuals have little melanin in the epidermis. Thus, the epidermis appears translucent, and skin color ranges from pink to red depending on the oxygen content of the blood moving through capillaries in the dermis. The red color is due to **hemoglobin** (hē-mō-GLŌ-bin), the oxygen-carrying pigment in red blood cells.

Carotene (KAR-ō-tēn; *carot* = carrot) is a yellow-orange pigment that gives egg yolks and carrots their color. This precursor of vitamin A, which is used to synthesize pigments needed for vision, is stored in the stratum corneum and fatty areas of the dermis and subcutaneous layer in response to excessive dietary intake. In fact, so much carotene may be deposited in the skin after eating large amounts of carotene-rich foods that the skin actually turns orange, which is especially apparent in light-skinned individuals. Decreasing carotene intake eliminates the problem.

See Clinical Connection: Skin Color as a Diagnostic Clue

Tattooing and Body Piercing

Tattooing is a permanent coloration of the skin in which a foreign pigment is deposited with a needle into macrophages in the dermis. It is believed that the practice originated in ancient Egypt between 4000 and 2000 B.C. Today, tattooing is performed in one form or another by nearly all peoples of the world, and it is estimated that about one in three U.S. college students has one or more tattoos. They are created by injecting ink with a needle that punctures the epidermis, moves between 50 and 3000 times per minute, and deposits the ink in the dermis. Since the dermis is stable (unlike the epidermis, which is shed about every four to six weeks), tattoos are permanent. However, they can fade over time due to exposure to sunlight, improper healing, picking scabs, and flushing away of ink particles by the lymphatic system. Sometimes tattoos are used as landmarks for radiation and as permanent makeup (eyeliner, lip liner, lipstick, blush, and eyebrows). Among the risks of tattoos are infections (staph, impetigo, and cellulitis.) Tattoos can be removed by lasers, which use concentrated beams of light. In the procedure, which requires a series of treatments, the tattoo inks and pigments selectively absorb the high-intensity laser light without destroying normal surrounding skin tissue. The laser causes the tattoo to dissolve into small ink particles that are eventually removed by the immune system. Laser removal of tattoos involves a considerable investment in time and money, can be quite painful, and may result in scarring and discoloration.

Body piercing, the insertion of jewelry through an artificial opening, is also an ancient practice employed by Egyptian pharaohs and Roman soldiers and is a current tradition among many Americans. Today it is estimated that about one in two U.S. college students has had a body piercing. For most piercing locations, the piercer cleans the skin with an antiseptic, retracts the skin with forceps, and pushes a needle through the skin. Then the jewelry is connected to the needle and pushed through the skin. Total healing can take up to a year. Among the sites that are pierced are the ears, nose, eyebrows, lips, tongue, nipples, navel, and genitals. Potential complications of body piercing are infections, allergic reactions, and anatomical damage (such as nerve damage or cartilage deformation). In addition, body piercing jewelry may interfere with certain medical procedures such as masks used for resuscitation, airway management procedures, urinary catheterization, radiographs, and delivery of a baby. For this reason, body piercing jewelries must be removed prior to certain medical procedures.

5.2 Accessory Structures of the Skin

OBJECTIVE

- **Contrast** the structure, distribution, and functions of hair, skin glands, and nails.

Accessory structures of the skin—hair, skin glands, and nails—develop from the embryonic epidermis. They have a host of important functions. For example, hair and nails protect the body, and sweat glands help regulate body temperature.

Hair

Hairs, or *pili* (PĪ-lī), are present on most skin surfaces except the palms, palmar surfaces of the fingers, the soles, and plantar surfaces of the feet. In adults, hair usually is most heavily distributed across the scalp, in the eyebrows, in the axillae (armpits), and around the external genitalia. Genetic and hormonal influences largely determine the thickness and the pattern of hair distribution.

Although the protection it offers is limited, hair on the head guards the scalp from injury and the sun's rays. It also decreases heat loss from the scalp. Eyebrows and eyelashes protect the eyes from foreign particles, similar to the way hair in the nostrils and in the external ear canal defends those structures. Touch receptors (hair root plexuses) associated with hair follicles are activated whenever a hair is moved even slightly. Thus, hairs also function in sensing light touch.

Anatomy of a Hair

Each hair is composed of columns of dead, keratinized epidermal cells bonded together by extracellular proteins. The **hair shaft** is the superficial portion of the hair, which projects above the surface of the skin (Figure 5.4a). The **hair root** is the portion of the hair deep to the shaft that penetrates into the dermis, and sometimes into the subcutaneous layer. The shaft and root of the hair both consist of three concentric layers of cells: medulla, cortex, and cuticle of the hair (Figure 5.4c, d). The inner *medulla*, which may be lacking in thinner hair, is composed of two or three rows of irregularly shaped cells that contain large amounts of pigment granules in dark hair, small amounts of pigment granules in gray hair, and a lack of pigment granules and the presence of air bubbles in white hair. The middle *cortex* forms the major part of the shaft and consists of elongated cells. The *cuticle of the hair*, the outermost layer, consists of a single layer of thin, flat cells that are the most heavily keratinized. Cuticle cells on the shaft are arranged like shingles on the side of a house, with their free edges pointing toward the end of the hair (Figure 5.4b).

Surrounding the root of the hair is the **hair follicle** (FOL-i-kul), which is made up of an external root sheath and an internal root sheath (Figure 5.4c, d). The *external root sheath* is a downward continuation of the epidermis. The *internal root sheath* is produced by the matrix (described shortly) and forms a cellular tubular sheath of epithelium between the external root sheath and the hair. Together, the external and internal root sheath are referred to as the **epithelial root sheath**. The dense dermis surrounding the hair follicle is called the **dermal root sheath**.

The base of each hair follicle and its surrounding dermal root sheath is an onion-shaped structure, the **hair bulb** (Figure 5.4c). This structure houses a nipple-shaped indentation, the **papilla of the hair**, which contains areolar connective tissue and many blood vessels that nourish the growing hair follicle. The bulb also contains a germinal layer of cells called the **hair matrix**. The hair matrix cells arise from the stratum basale, the site of cell division. Hence, hair matrix cells are responsible for the growth of existing hairs, and they produce new hairs when old hairs are shed. This replacement process occurs within the same follicle. Hair matrix cells also give rise to the cells of the internal root sheath.

See Clinical Connection: Hair Removal

Sebaceous (oil) glands (discussed shortly) and a bundle of smooth muscle cells are also associated with hairs (Figure 5.4a). The smooth muscle is the **arrector pili** (a-REK-tor PĪ-lī; *arrect-* = to raise). It extends from the superficial dermis of the skin to the dermal root sheath around the side of the hair follicle. In its normal position, hair emerges at a less than 90-degree angle to the surface of the skin. Under physiological or emotional stress, such as cold or fright, autonomic nerve endings stimulate the arrector pili muscles to contract, which pulls the hair shafts perpendicular to the skin surface. This action causes "goose bumps" or "gooseflesh" because the skin around the shaft forms slight elevations.

Surrounding each hair follicle are dendrites of neurons that form a **hair root plexus** (PLEK-sus), which is sensitive to touch (Figure 5.4a). The hair root plexuses generate nerve impulses if their hair shafts are moved.

Hair Growth

Each hair follicle goes through a growth cycle, which consists of a growth stage, a regression stage, and a resting stage. During the **growth stage**, cells of the hair matrix divide. As new cells from the hair matrix are added to the base of the hair root, existing cells of the hair root are pushed upward and the hair grows longer. While the cells of the hair are being pushed upward, they become keratinized and die. Following the growth stage is the **regression stage**, when the cells of the hair matrix stop dividing, the hair follicle atrophies (shrinks), and the hair stops growing. After the regression stage, the hair follicle enters a **resting stage**. Following the resting stage, a new growth cycle begins. The old hair root falls out or is pushed out of the hair follicle, and a new hair begins to grow in its place. Scalp hair is in the growth stage for 2 to 6 years, the regression stage for 2 to 3 weeks, and the resting stage for about 3 months. At any time, about 85% of scalp hairs are in the growth stage. Visible hair is dead, but until the hair is pushed out of its follicle by a new hair, portions of its root within the scalp are alive.

Normal hair loss in the adult scalp is about 70–100 hairs per day. Both the rate of growth and the replacement cycle may be altered by illness, radiation therapy, chemotherapy (described next), age, genetics, gender, and severe emotional stress. Rapid weight-loss diets that severely restrict calories or protein increase hair loss. The rate of shedding also increases for three to four months after childbirth. **Alopecia** (al′-ō-PĒ-shē-a), the partial or complete lack of hair, may result from genetic factors, aging, endocrine disorders, chemotherapy, or skin disease.

See Clinical Connection: Chemotherapy and Hair Loss

See Clinical Connection: Hair and Hormones

Types of Hairs

Hair follicles develop at about 12 weeks after fertilization. Usually by the fifth month of development, the follicles produce very fine, nonpigmented, downy hairs called **lanugo** (la-NOO-gō = wool or down) that cover the body of the fetus. Prior to birth, the lanugo of the eyebrows, eyelashes, and scalp are shed and

FIGURE 5.4 Hair.

Hairs are growths of epidermis composed of dead, keratinized epidermal cells.

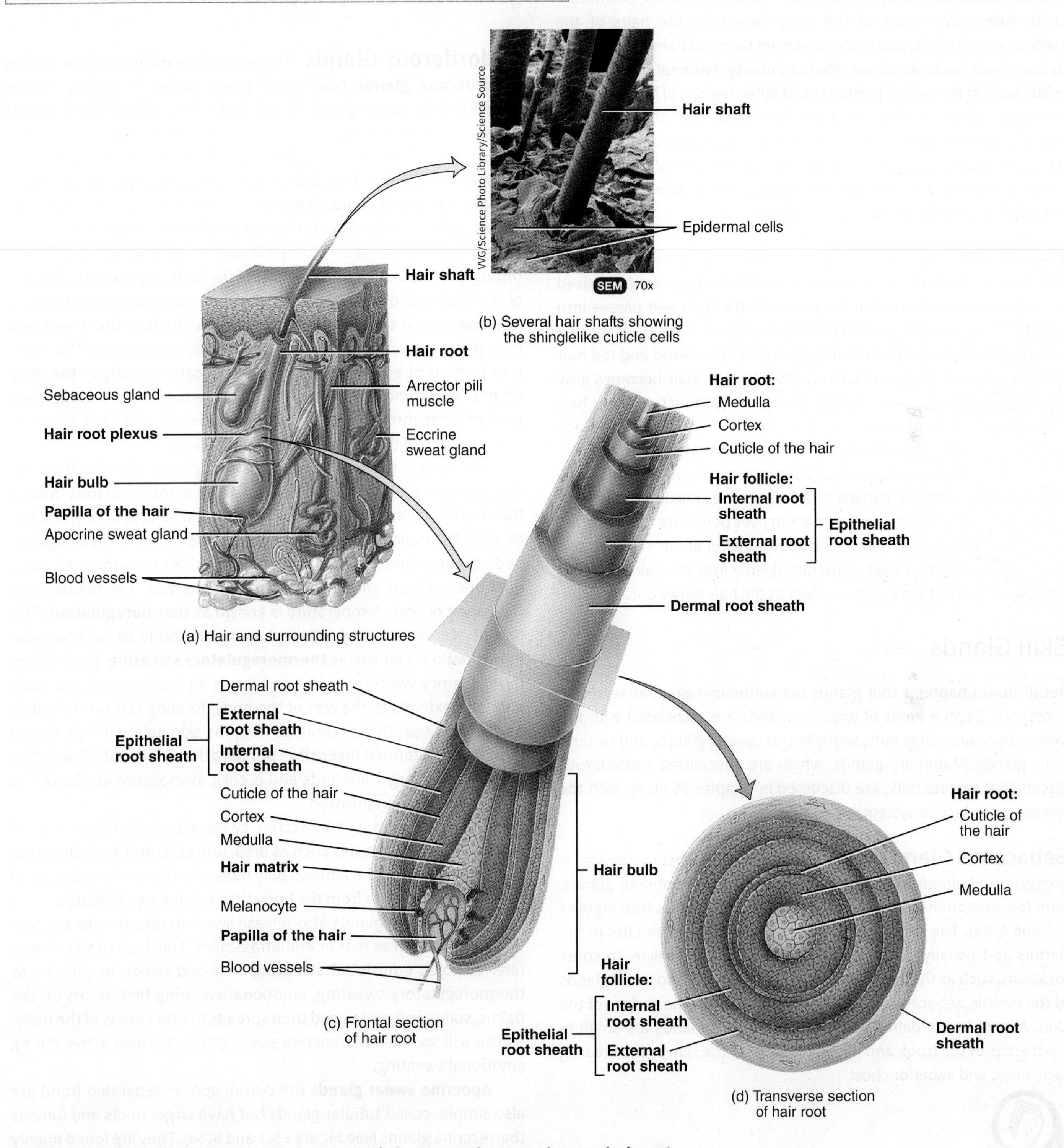

(a) Hair and surrounding structures

(b) Several hair shafts showing the shinglelike cuticle cells

(c) Frontal section of hair root

(d) Transverse section of hair root

Q Why does it hurt when you pluck out a hair but not when you have a haircut?

replaced by long, coarse, heavily pigmented hairs called **terminal hairs**. The lanugo of the rest of the body are replaced by **vellus hairs** (VEL-us = fleece), commonly called "peach fuzz," which are short, fine, pale hairs that are barely visible to the naked eye. During childhood, vellus hairs cover most of the body except for the hairs of the eyebrows, eyelashes, and scalp, which are terminal hairs. In response to hormones (androgens) secreted at puberty, terminal hairs replace vellus hairs in the axillae (armpits) and pubic regions of boys and girls and they replace vellus hairs on the face, limbs, and chests of boys, which leads to the formation of a mustache, a beard, hairy arms and legs, and a hairy chest. During adulthood, about 95% of body hair on males is terminal hair and 5% is vellus hair; on females, about 35% of body hair is terminal hair and 65% is vellus hair.

Hair Color The color of hair is due primarily to the amount and type of melanin in its keratinized cells. Melanin is synthesized by melanocytes scattered in the matrix of the bulb and passes into cells of the cortex and medulla of the hair (Figure 5.4c). Dark-colored hair contains mostly eumelanin (brown to black); blond and red hair contain variants of pheomelanin (yellow to red). Hair becomes gray because of a progressive decline in melanin production; gray hair contains only a few melanin granules. White hair results from the lack of melanin and the accumulation of air bubbles in the shaft.

Hair coloring is a process that adds or removes pigment. Temporary hair dyes coat the surface of a hair shaft and usually wash out within 2 or 3 shampoos. Semipermanent dyes penetrate the hair shaft moderately and do fade and wash out of hair after about 5 to 10 shampoos. Permanent hair dyes penetrate deeply into the hair shaft and don't wash out but are eventually lost as the hair grows out.

Skin Glands

Recall from Chapter 4 that glands are epithelial cells that secrete a substance. Several kinds of exocrine glands are associated with the skin: sebaceous (oil) glands, sudoriferous (sweat) glands, and ceruminous glands. Mammary glands, which are specialized sudoriferous glands that secrete milk, are discussed in Chapter 28 along with the female reproductive system.

Sebaceous Glands **Sebaceous glands** (se-BĀ-shus; *sebace-* = greasy) or *oil glands* are simple, branched acinar (rounded) glands. With few exceptions, they are connected to hair follicles (see Figures 5.1 and 5.4a). The secreting portion of a sebaceous gland lies in the dermis and usually opens into the neck of a hair follicle. In some locations, such as the lips, glans penis, labia minora, and tarsal glands of the eyelids, sebaceous glands open directly onto the surface of the skin. Absent in the palms and soles, sebaceous glands are small in most areas of the trunk and limbs, but large in the skin of the breasts, face, neck, and superior chest.

See Clinical Connection: Acne

Sebaceous glands secrete an oily substance called **sebum** (SĒ-bum), a mixture of triglycerides, cholesterol, proteins, and inorganic salts. Sebum coats the surface of hairs and helps keep them from drying and becoming brittle. Sebum also prevents excessive evaporation of water from the skin, keeps the skin soft and pliable, and inhibits the growth of some (but not all) bacteria.

Sudoriferous Glands There are three million to four million **sudoriferous glands** (soo′-dor-IF-er-us; *sudor-* = sweat; *-ferous* = bearing) or *sweat glands* in the body. The cells of these glands release sweat, or perspiration, into hair follicles or onto the skin surface through pores. Sweat glands are divided into two main types, eccrine and apocrine, based on their structure and type of secretion.

Eccrine sweat glands (EK-rin; *eccrine* = secreting outwardly) are simple, coiled tubular glands that are much more common than apocrine sweat glands (see Figures 5.1 and 5.4a). They are distributed throughout the skin of most regions of the body, especially in the skin of the forehead, palms, and soles. Eccrine sweat glands are not present, however, in the margins of the lips, nail beds of the fingers and toes, glans penis, glans clitoris, labia minora, or eardrums. The secretory portion of eccrine sweat glands is located mostly in the deep dermis (sometimes in the upper subcutaneous layer). The excretory duct projects through the dermis and epidermis and ends as a pore at the surface of the epidermis (see Figure 5.1).

The sweat produced by eccrine sweat glands (about 600 mL per day) consists primarily of water, with small amounts of ions (mostly Na^+ and Cl^-), urea, uric acid, ammonia, amino acids, glucose, and lactic acid. The main function of eccrine sweat glands is to help regulate body temperature through evaporation. As sweat evaporates, large quantities of heat energy leave the body surface. The homeostatic regulation of body temperature is known as **thermoregulation**. This role of eccrine sweat glands in helping the body to achieve thermoregulation is known as **thermoregulatory sweating**. During thermoregulatory sweating, sweat first forms on the forehead and scalp and then extends to the rest of the body, forming last on the palms and soles. Sweat that evaporates from the skin before it is perceived as moisture is termed **insensible perspiration** (*in-* = not). Sweat that is excreted in larger amounts and is seen as moisture on the skin is called **sensible perspiration**.

The sweat produced by eccrine sweat glands also plays a small role in eliminating wastes such as urea, uric acid, and ammonia from the body. However, the kidneys play more of a role in the excretion of these waste products from the body than eccrine sweat glands.

Eccrine sweat glands also release sweat in response to an emotional stress such as fear or embarrassment. This type of sweating is referred to as **emotional sweating** or a *cold sweat.* In contrast to thermoregulatory sweating, emotional sweating first occurs on the palms, soles, and axillae and then spreads to other areas of the body. As you will soon learn, apocrine sweat glands are also active during emotional sweating.

Apocrine sweat glands (AP-ō-krin; *apo-* = separated from) are also simple, coiled tubular glands but have larger ducts and lumens than eccrine glands (see Figures 5.1 and 5.4a). They are found mainly in the skin of the axilla (armpit), groin, areolae (pigmented areas around the nipples) of the breasts, and bearded regions of the face in adult males. These glands were once thought to release their secretions in an apocrine manner (see text coverage in Chapter 4 and

Figure 4.7b)—by pinching off a portion of the cell. We now know, however, that their secretion is via exocytosis, which is characteristic of eccrine glands (see Figure 5.4a). Nevertheless, the term *apocrine* is still used. The secretory portion of these sweat glands is located in the lower dermis or upper subcutaneous layer, and the excretory duct opens into hair follicles (see Figure 5.1).

Compared to eccrine sweat, apocrine sweat appears milky or yellowish in color. Apocrine sweat contains the same components as eccrine sweat plus lipids and proteins. Sweat secreted from apocrine sweat glands is odorless. However, when apocrine sweat interacts with bacteria on the surface of the skin, the bacteria metabolize its components, causing apocrine sweat to have a musky odor that is often referred to as body odor. Eccrine sweat glands start to function soon after birth, but apocrine sweat glands do not begin to function until puberty.

Apocrine sweat glands, along with eccrine sweat glands, are active during emotional sweating. In addition, apocrine sweat glands secrete sweat during sexual activities. In contrast to eccrine sweat glands, apocrine sweat glands are not active during thermoregulatory sweating and, therefore, do not play a role in thermoregulation.

Ceruminous Glands Modified sweat glands in the external ear, called **ceruminous glands** (se-RŪ-mi-nus; *cer-* = wax), produce a waxy lubricating secretion. The secretory portions of ceruminous glands lie in the subcutaneous layer, deep to sebaceous glands. Their excretory ducts open either directly onto the surface of the external auditory canal (ear canal) or into ducts of sebaceous glands. The combined secretion of the ceruminous and sebaceous glands is a yellowish material called **cerumen** (se-ROO-men), or earwax. Cerumen, together with hairs in the external auditory canal, provides a sticky barrier that impedes the entrance of foreign bodies and insects. Cerumen also waterproofs the canal and prevents bacteria and fungi from entering cells.

Table 5.3 presents a summary of skin glands.

See Clinical Connection: Impacted Cerumen

Nails

Nails are plates of tightly packed, hard, dead, keratinized epidermal cells that form a clear, solid covering over the dorsal surfaces of the distal portions of the digits. Each nail consists of a nail body, a free edge, and a nail root (Figure 5.5). The **nail body** (*plate*) is the visible portion of the nail. It is comparable to the stratum corneum of the epidermis of the skin, with the exception that its flattened, keratinized cells fill with a harder type of keratin and the cells are not shed. Below the nail body is a region of epithelium and a deeper layer of dermis. Most of the nail body appears pink because of blood flowing through the capillaries in the underlying dermis. The **free edge** is the part of

TABLE 5.3 Summary of Skin Glands (see Figures 5.1 and 5.4a)

FEATURE	SEBACEOUS (OIL) GLANDS	ECCRINE SWEAT GLANDS	APOCRINE SWEAT GLANDS	CERUMINOUS GLANDS
Distribution	Largely in lips, glans penis, labia minora, and tarsal glands; small in trunk and limbs; absent in palms and soles.	Throughout skin of most regions of body, especially skin of forehead, palms, and soles.	Skin of axillae, groin, areolae, bearded regions of face, clitoris, and labia minora.	External auditory canal.
Location of secretory portion	Dermis.	Mostly in deep dermis (sometimes in upper subcutaneous layer).	Mostly in deep dermis and upper subcutaneous layer.	Subcutaneous layer.
Termination of excretory duct	Mostly connected to hair follicle.	Surface of epidermis.	Hair follicles.	Surface of external auditory canal or into ducts of sebaceous glands.
Secretion	Sebum (mixture of triglycerides, cholesterol, proteins, and inorganic salts).	Perspiration, which consists of water, ions (Na^+, Cl^-), urea, uric acid, ammonia, amino acids, glucose, and lactic acid.	Perspiration, which consists of same components as eccrine sweat glands plus lipids and proteins.	Cerumen, a waxy material.
Functions	Prevent hairs from drying out, prevent water loss from skin, keep skin soft, inhibit growth of some bacteria.	Regulation of body temperature, waste removal, stimulated during emotional stress.	Stimulated during emotional stress and sexual excitement.	Impede entrance of foreign bodies and insects into external ear canal, waterproof canal, prevent microbes from entering cells.
Onset of function	Relatively inactive during childhood; activated during puberty.	Soon after birth.	Puberty.	Soon after birth.

FIGURE 5.5 **Nails.** Shown is a fingernail.

Nail cells arise by transformation of superficial cells of the nail matrix.

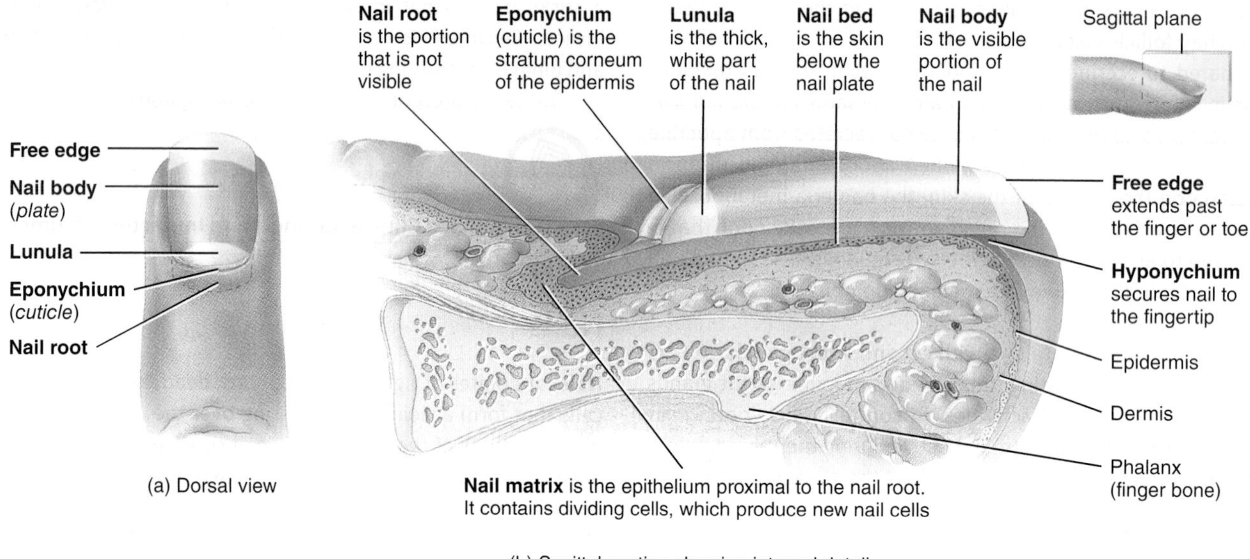

(a) Dorsal view

(b) Sagittal section showing internal detail

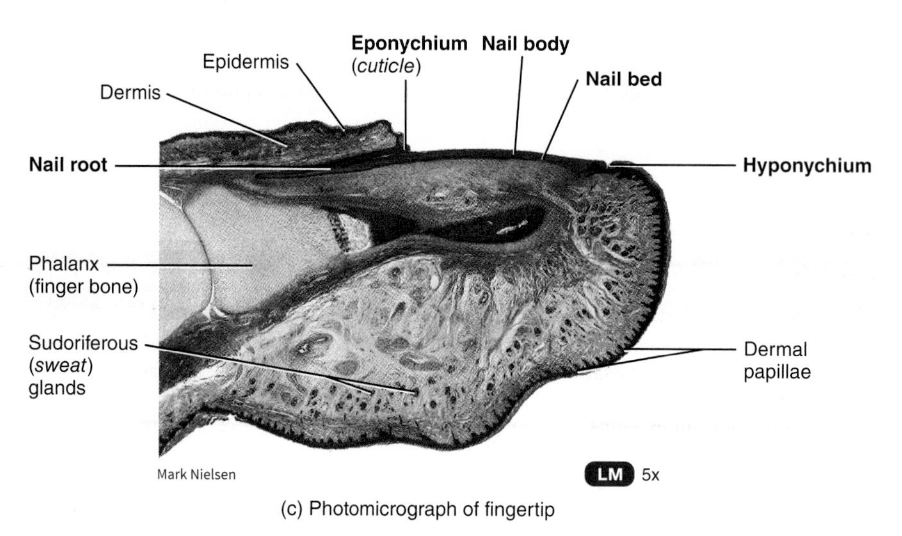

(c) Photomicrograph of fingertip

Q Why are nails so hard?

the nail body that may extend past the distal end of the digit. The free edge is white because there are no underlying capillaries. The **nail root** is the portion of the nail that is buried in a fold of skin. The whitish, crescent-shaped area of the proximal end of the nail body is called the **lunula** (LOO-noo-la = little moon). It appears whitish because the vascular tissue underneath does not show through due to a thickened region of epithelium in the area. Beneath the free edge is a thickened region of stratum corneum called the **hyponychium** (hī′-pō-NIK-ē-um; *hypo-* = below; *-onych* = nail). It is the junction between the free edge and skin of the fingertip and secures the nail to the fingertip. The **nail bed** is the skin below the nail plate that extends from the lunula to the hyponychium. The epidermis of the nail bed lacks a stratum granulosum. The **eponychium** (ep′-ō-NIK-ē-um; *ep-* = above) or *cuticle* is a narrow band of epidermis that extends from and adheres to the margin (lateral border) of the nail wall. It occupies the proximal border of the nail and consists of stratum corneum. You might be surprised to know that a **hangnail** has nothing to do with the nail itself. It is a small torn piece of skin at the side or base of a fingernail or toenail, usually caused by dryness of the eponychium.

The portion of the epithelium proximal to the nail root is the **nail matrix**. The superficial nail matrix cells divide mitotically to produce new nail cells. The growth rate of nails is determined by the rate of mitosis in matrix cells, which is influenced by factors such as a person's age, health, and nutritional status. Nail growth also varies according to the season, the time of day, and environmental temperature. The average growth in the length of fingernails is about 1 mm (0.04 in.) per week. The growth rate is somewhat slower in toenails. The longer the digit the faster the nail grows.

Nails have a variety of functions:

1. They protect the distal end of the digits.
2. They provide support and counterpressure to the palmar surface of the fingers to enhance touch perception and manipulation.
3. They allow us to grasp and manipulate small objects, and they can be used to scratch and groom the body in various ways.

5.3 Types of Skin

OBJECTIVE

- **Compare** structural and functional differences in thin and thick skin.

Although the skin over the entire body is similar in structure, there are quite a few local variations related to thickness of the epidermis, strength, flexibility, degree of keratinization, distribution and type of hair, density and types of glands, pigmentation, vascularity (blood supply), and innervation (nerve supply). Two major types of skin are recognized on the basis of certain structural and functional properties: thin (hairy) skin and thick (hairless) skin (see also Section 5.1). The greatest contributor to epidermal thickness is the increased number of layers in the stratum corneum. This arises in response to the greater mechanical stress in regions of thick skin.

Table 5.4 presents a comparison of the features of thin and thick skin.

5.4 Functions of the Skin

OBJECTIVE

- **Describe** how the skin contributes to the regulation of body temperature, storage of blood, protection, sensation, excretion and absorption, and synthesis of vitamin D.

Now that you have a basic understanding of the structure of the skin, you can better appreciate its many functions, which were introduced at the beginning of this chapter. The numerous functions of the integumentary system (mainly the skin) include thermoregulation, storage of blood, protection, cutaneous sensations, excretion and absorption, and synthesis of vitamin D.

Thermoregulation

Recall that **thermoregulation** is the homeostatic regulation of body temperature. The skin contributes to thermoregulation in two ways: by liberating sweat at its surface and by adjusting the flow of blood in the dermis. In response to high environmental temperature or heat produced by exercise, sweat production from eccrine sweat glands increases; the evaporation of sweat from the skin surface helps lower body temperature. In addition, blood vessels in the dermis of the skin dilate (become wider); consequently, more blood flows through the dermis, which increases the amount of heat loss from the body (see **Figure 25.19**). In response to low environmental temperature, production of sweat from eccrine sweat glands is decreased, which helps conserve heat. Also, the blood vessels in the dermis of the skin constrict (become narrow), which decreases blood flow through the skin and reduces heat loss from the body. And, skeletal muscle contractions generate body heat.

TABLE 5.4 Comparison of Thin and Thick Skin

FEATURE	THIN SKIN	THICK SKIN
Distribution	All parts of body except areas such as palms, palmar surface of digits, and soles.	Areas such as palms, palmar surface of digits, and soles.
Epidermal thickness	0.10–0.15 mm (0.004–0.006 in.).	0.6–4.5 mm (0.024–0.18 in.), due mostly to a thicker stratum corneum.
Epidermal strata	Stratum lucidum essentially lacking; thinner strata spinosum and corneum.	Stratum lucidum present; thicker strata spinosum and corneum.
Epidermal ridges	Lacking due to poorly developed, fewer, and less-well-organized dermal papillae.	Present due to well-developed and more numerous dermal papillae organized in parallel rows.
Hair follicles and arrector pili muscles	Present.	Absent.
Sebaceous glands	Present.	Absent.
Sudoriferous glands	Fewer.	More numerous.
Sensory receptors	Sparser.	Denser.

Blood Reservoir

The dermis houses an extensive network of blood vessels that carry 8–10% of the total blood flow in a resting adult. For this reason, the skin acts as a **blood reservoir**.

Protection

The skin provides **protection** to the body in various ways. Keratin protects underlying tissues from microbes, abrasion, heat, and chemicals, and the tightly interlocked keratinocytes resist invasion by microbes. Lipids released by lamellar granules inhibit evaporation of water from the skin surface, thus guarding against dehydration; they also retard entry of water across the skin surface during showers and swims. The oily sebum from the sebaceous glands keeps skin and hairs from drying out and contains *bactericidal chemicals* (substances that kill bacteria). The acidic pH of perspiration retards the growth of some microbes. The pigment melanin helps shield against the damaging effects of ultraviolet light. Two types of cells carry out protective functions that are immunological in nature. Intraepidermal macrophages alert the immune system to the presence of potentially harmful microbial invaders by recognizing and processing them, and macrophages in the dermis phagocytize bacteria and viruses that manage to bypass the intraepidermal macrophages of the epidermis.

Cutaneous Sensations

Cutaneous sensations are sensations that arise in the skin, including tactile sensations—touch, pressure, vibration, and tickling—as well as thermal sensations such as warmth and coolness. Another cutaneous sensation, pain, usually is an indication of impending or actual tissue damage. There is a wide variety of nerve endings and receptors distributed throughout the skin, including the tactile discs of the epidermis, the corpuscles of touch in the dermis, and hair root plexuses around each hair follicle. Chapter 16 provides more details on the topic of cutaneous sensations.

Excretion and Absorption

The skin normally has a small role in **excretion**, the elimination of substances from the body, and **absorption**, the passage of materials from the external environment into body cells. Despite the almost waterproof nature of the stratum corneum, about 400 mL of water evaporates through it daily. A sedentary person loses an additional 200 mL per day as sweat; a physically active person loses much more. Besides removing water and heat from the body, sweat also is the vehicle for excretion of small amounts of salts, carbon dioxide, and two organic molecules that result from the breakdown of proteins—ammonia and urea.

The absorption of water-soluble substances through the skin is negligible, but certain lipid-soluble materials do penetrate the skin. These include fat-soluble vitamins (A, D, E, and K), certain drugs, and the gases oxygen and carbon dioxide. Toxic materials that can be absorbed through the skin include organic solvents such as acetone (in some nail polish removers) and carbon tetrachloride (dry-cleaning fluid); salts of heavy metals such as lead, mercury, and arsenic; and the substances in poison ivy and poison oak. Since topical (applied to the skin) steroids, such as cortisone, are lipid-soluble, they move easily into the papillary region of the dermis. Here, they exert their anti-inflammatory properties by inhibiting histamine production by mast cells (recall that histamine contributes to inflammation). Certain drugs that are absorbed by the skin may be administered by applying adhesive patches to the skin.

See Clinical Connection: Transdermal Drug Administration

Synthesis of Vitamin D

Synthesis of vitamin D requires activation of a precursor molecule in the skin by ultraviolet (UV) rays in sunlight. Enzymes in the liver and kidneys then modify the activated molecule, finally producing *calcitriol*, the most active form of vitamin D. Calcitriol is a hormone that aids in the absorption of calcium from foods in the gastrointestinal tract into the blood. Only a small amount of exposure to UV light (about 10 to 15 minutes at least twice a week) is required for vitamin D synthesis. People who avoid sun exposure and individuals who live in colder, northern climates may require vitamin D supplements to avoid vitamin D deficiency. Most cells of the immune system have vitamin D receptors, and the cells activate vitamin D in response to an infection, especially a respiratory infection, such as influenza. Vitamin D is believed to enhance phagocytic activity, increase the production of antimicrobial substances in phagocytes, regulate immune functions, and help reduce inflammation.

5.5 Maintaining Homeostasis: Skin Wound Healing

OBJECTIVE

- **Explain** how epidermal wounds and deep wounds heal.

Skin damage sets in motion a sequence of events that repairs the skin to its normal (or near-normal) structure and function. Two kinds of wound-healing processes can occur, depending on the depth of the injury. Epidermal wound healing occurs following wounds that affect only the epidermis; deep wound healing occurs following wounds that penetrate the dermis.

Epidermal Wound Healing

Even though the central portion of an epidermal wound may extend to the dermis, the edges of the wound usually involve only slight damage to superficial epidermal cells. Common types of epidermal wounds include abrasions, in which a portion of skin has been scraped away, and minor burns.

In response to an epidermal injury, basal cells of the epidermis surrounding the wound break contact with the basement membrane. The cells then enlarge and migrate across the wound (**Figure 5.6a**). The cells appear to migrate as a sheet until advancing cells from opposite sides of the wound meet. When epidermal cells encounter one another, they stop migrating due to a cellular response called **contact inhibition**. Migration of the epidermal cells stops completely when each is finally in contact with other epidermal cells on all sides.

As the basal epidermal cells migrate, a hormone called *epidermal growth factor* stimulates basal stem cells to divide and replace the ones that have moved into the wound. The relocated basal epidermal cells divide to build new strata, thus thickening the new epidermis (**Figure 5.6b**).

Deep Wound Healing

Deep wound healing occurs when an injury extends to the dermis and subcutaneous layer. Because multiple tissue layers must be repaired, the healing process is more complex than in epidermal wound healing. In addition, because scar tissue is formed, the healed tissue loses

FIGURE 5.6 **Skin wound healing.**

In an epidermal wound, the injury is restricted to the epidermis; in a deep wound, the injury extends deep into the dermis.

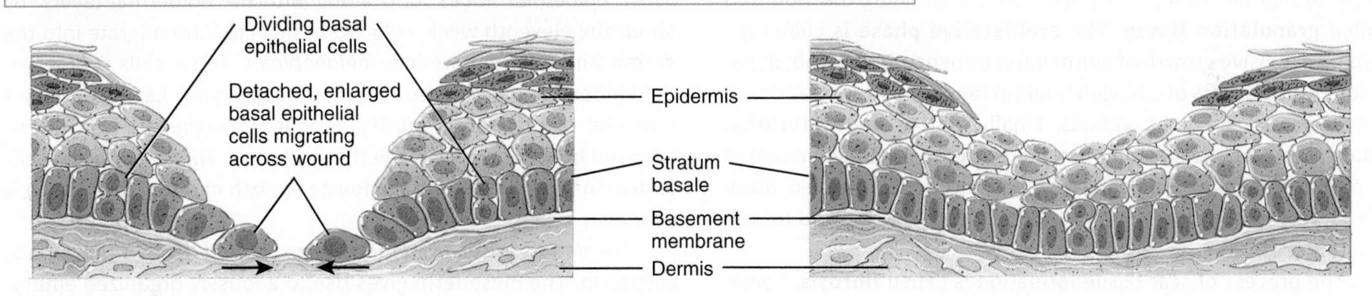

(a) Division of stratum basale cells and migration across wound

(b) Thickening of epidermis

Epidermal wound healing

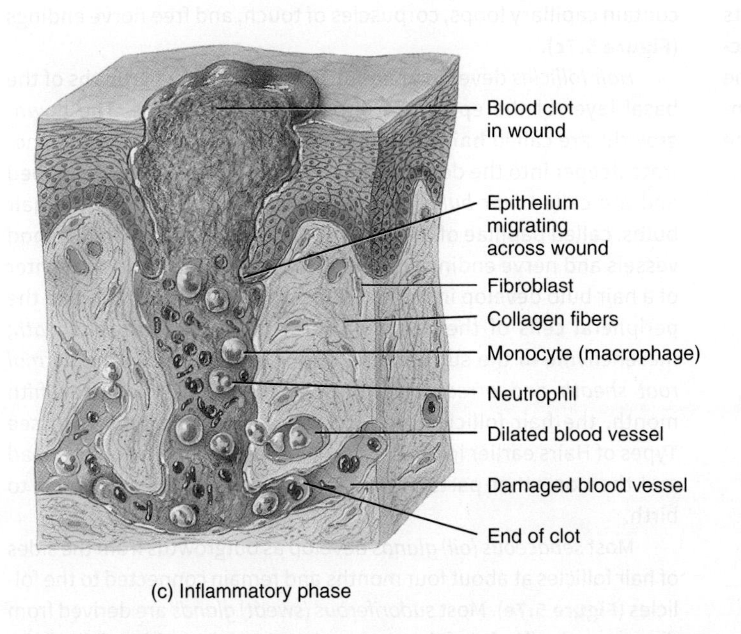

(c) Inflammatory phase

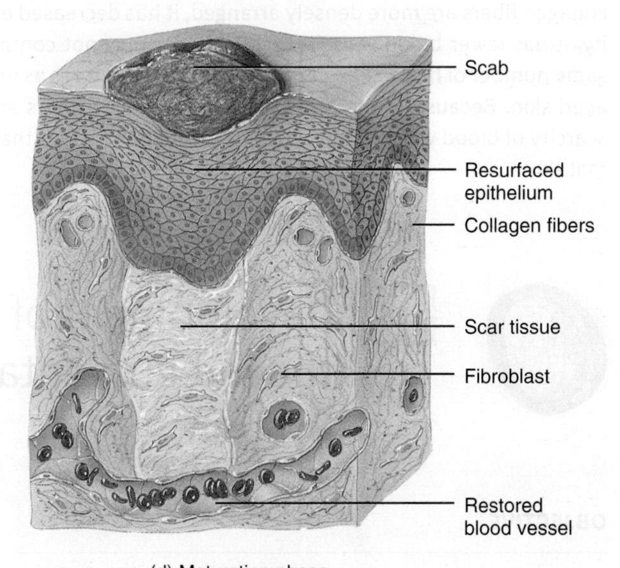

(d) Maturation phase

Deep wound healing

Q Would you expect an epidermal wound to bleed? Why or why not?

some of its normal function. Deep wound healing occurs in four phases: an inflammatory phase, a migratory phase, a proliferative phase, and a maturation phase.

During the **inflammatory phase**, a blood clot forms in the wound and loosely unites the wound edges (Figure 5.6c). As its name implies, this phase of deep wound healing involves **inflammation**, a vascular and cellular response that helps eliminate microbes, foreign material, and dying tissue in preparation for repair. The vasodilation and increased permeability of blood vessels associated with inflammation enhance delivery of helpful cells. These include phagocytic white blood cells called neutrophils; monocytes, which develop into macrophages that phagocytize microbes; and mesenchymal cells, which develop into fibroblasts.

The three phases that follow do the work of repairing the wound. In the **migratory phase**, the clot becomes a scab, and epithelial cells migrate beneath the scab to bridge the wound. Fibroblasts migrate along fibrin threads and begin synthesizing scar tissue (collagen fibers and glycoproteins), and damaged blood vessels begin to regrow. During this phase, the tissue filling the wound is called **granulation tissue**. The **proliferative phase** is characterized by extensive growth of epithelial cells beneath the scab, deposition by fibroblasts of collagen fibers in random patterns, and continued growth of blood vessels. Finally, during the **maturation phase**, the scab sloughs off once the epidermis has been restored to normal thickness. Collagen fibers become more organized, fibroblasts decrease in number, and blood vessels are restored to normal (Figure 5.6d).

The process of scar tissue formation is called **fibrosis**. Sometimes, so much scar tissue is formed during deep wound healing that a raised scar—one that is elevated above the normal epidermal surface—results. If such a scar remains within the boundaries of the original wound, it is a **hypertrophic scar**. If it extends beyond the boundaries into normal surrounding tissues, it is a **keloid scar**, also called a *cheloid scar*. Scar tissue differs from normal skin in that its collagen fibers are more densely arranged, it has decreased elasticity, it has fewer blood vessels, and it may or may not contain the same number of hairs, skin glands, or sensory structures as undamaged skin. Because of the arrangement of collagen fibers and the scarcity of blood vessels, scars usually are lighter in color than normal skin.

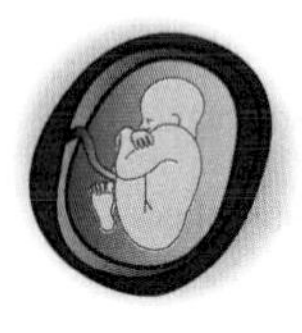

5.6 Development of the Integumentary System

OBJECTIVE

- **Describe** the development of the epidermis, its accessory structures, and the dermis.

The epidermis is derived from the **ectoderm**, which covers the surface of the embryo. Initially, at about the fourth week after fertilization, the epidermis consists of only a single layer of ectodermal cells (Figure 5.7a). At the beginning of the seventh week the single layer, called the **basal layer**, divides and forms a superficial protected layer of flattened cells called the **periderm** (Figure 5.7b). The peridermal cells are continuously sloughed off, and by the fifth month of development secretions from sebaceous glands mix with them and hairs to form a fatty substance called **vernix caseosa** (VER-niks KĀ-sē-ō-sa; *vernix* = varnish; *caseosa* = cheese). This substance covers and protects the skin of the fetus from the constant exposure to the amniotic fluid in which it is bathed. In addition, the vernix caseosa facilitates the birth of the fetus because of its slippery nature and protects the skin from being damaged by the nails.

By about 11 weeks (Figure 5.7c), the basal layer forms an **intermediate layer** of cells. Proliferation of the basal cells eventually forms all layers of the epidermis, which are present at birth (Figure 5.7h). *Epidermal ridges* form along with the epidermal layers. By about the eleventh week, cells from the ectoderm migrate into the dermis and differentiate into *melanoblasts*. These cells soon enter the epidermis and differentiate into *melanocytes.* Later in the first trimester of pregnancy, *intraepidermal macrophages*, which arise from red bone marrow, invade the epidermis. *Tactile epithelial cells* appear in the epidermis in the fourth to sixth months; their origin is unknown.

The *dermis* arises from **mesoderm** located deep to the surface ectoderm. The mesoderm gives rise to a loosely organized embryonic connective tissue called **mesenchyme** (MEZ-en-kīm; see Figure 5.7a). By 11 weeks, the mesenchymal cells differentiate into fibroblasts and begin to form collagen and elastic fibers. As the epidermal ridges form, parts of the superficial dermis project into the epidermis and develop into the *dermal papillae*, which contain capillary loops, corpuscles of touch, and free nerve endings (Figure 5.7c).

Hair follicles develop at about 12 weeks as downgrowths of the basal layer of the epidermis into the deeper dermis. The downgrowths are called **hair buds** (Figure 5.7d). As the hair buds penetrate deeper into the dermis, their distal ends become club-shaped and are called *hair bulbs* (Figure 5.7e). Invaginations of the hair bulbs, called papillae of the hair, fill with mesoderm in which blood vessels and nerve endings develop (Figure 5.7f). Cells in the center of a hair bulb develop into the *matrix*, which forms the *hair*, and the peripheral cells of the hair bulb form the *epithelial root sheath*; mesenchyme in the surrounding dermis develops into the *dermal root sheath* and *arrector pili muscle* (Figure 5.7g). By the fifth month, the hair follicles produce lanugo (delicate fetal hair; see Types of Hairs earlier in the chapter). It is produced first on the head and then on other parts of the body, and is usually shed prior to birth.

Most *sebaceous (oil) glands* develop as outgrowths from the sides of hair follicles at about four months and remain connected to the follicles (Figure 5.7e). Most *sudoriferous (sweat) glands* are derived from downgrowths (**buds**) of the stratum basale of the epidermis into the dermis (Figure 5.7d). As the buds penetrate into the dermis, the

FIGURE 5.7 **Development of the integumentary system.**

The epidermis develops from ectoderm, and the dermis develops from mesoderm.

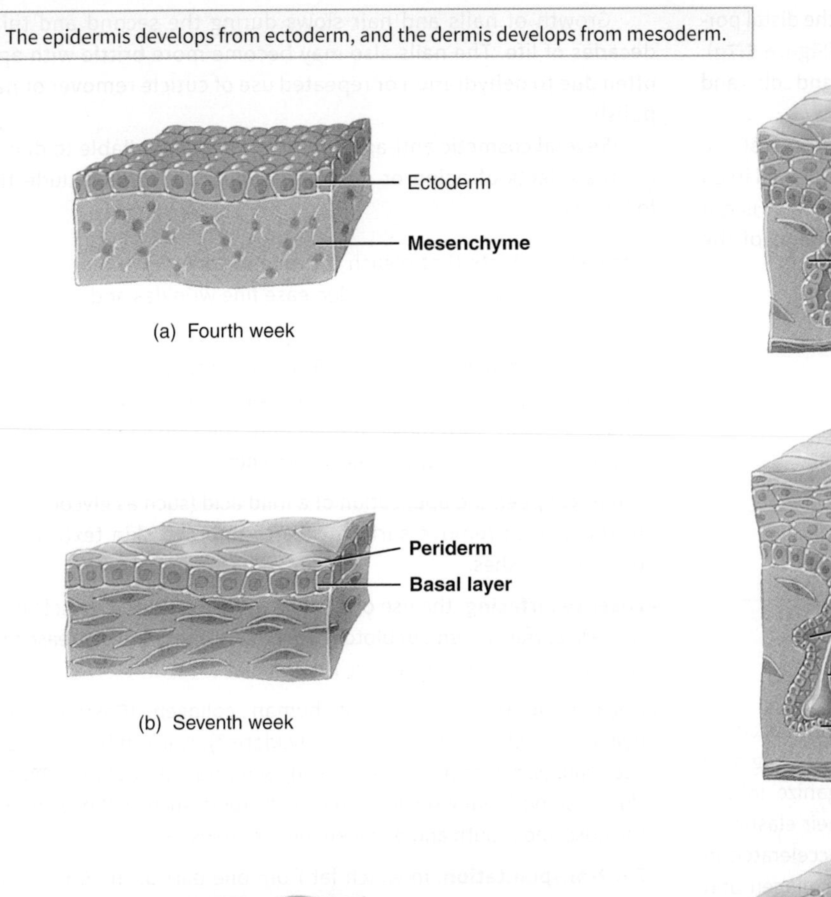

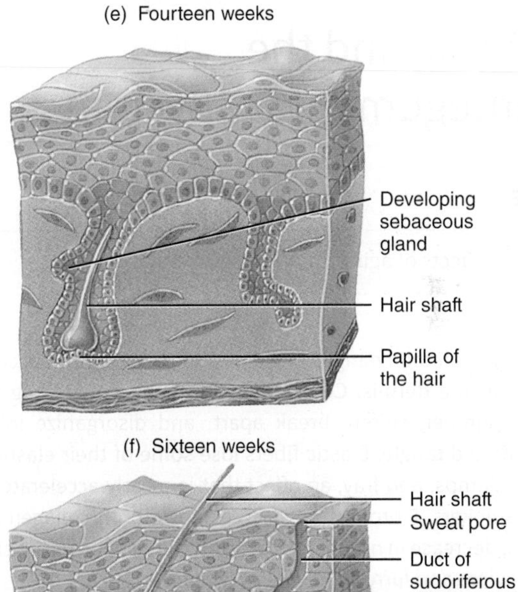

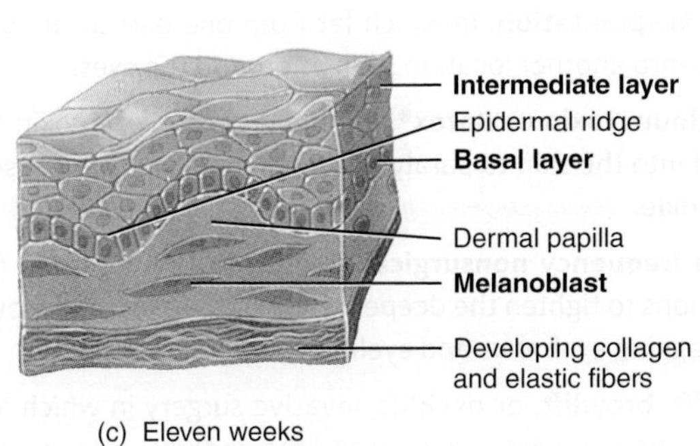

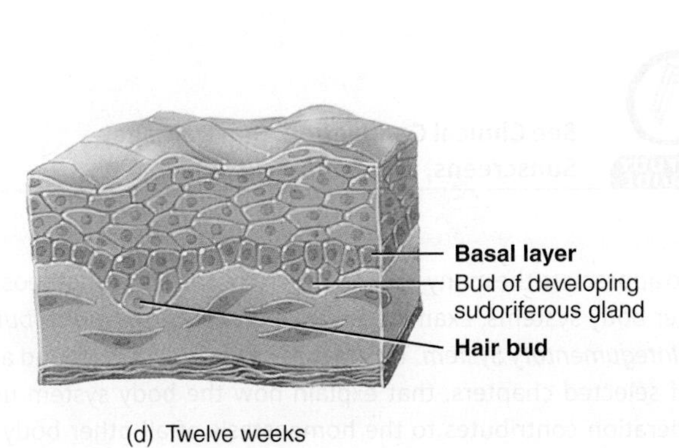

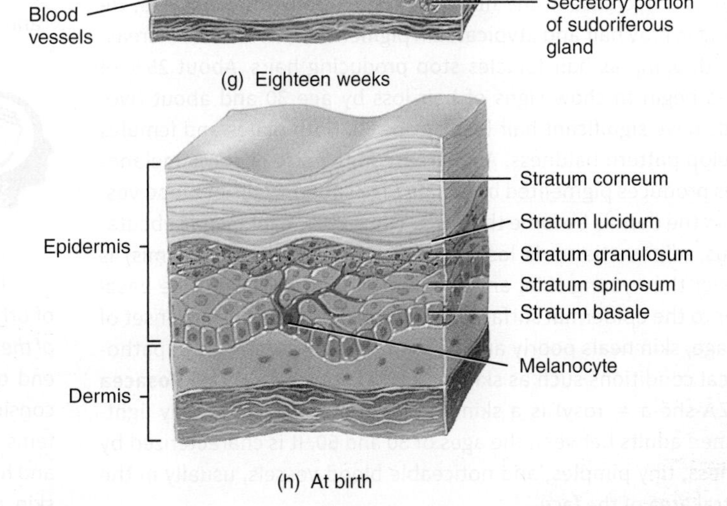

Q What is the composition of vernix caseosa?

proximal portion forms the duct of the sweat gland and the distal portion coils and forms the secretory portion of the gland (Figure 5.7g). Sweat glands appear at about five months on the palms and soles and a little later in other regions.

Nails are developed at about 10 weeks. Initially they consist of a thick layer of epithelium called the **primary nail field**. The nail itself is keratinized epithelium and grows distally from its base. It is not until the ninth month that the nails actually reach the tips of the digits.

5.7 Aging and the Integumentary System

OBJECTIVE

- **Describe** the effects of aging on the integumentary system.

Most of the age-related changes begin at about age 40 and occur in the proteins in the dermis. Collagen fibers in the dermis begin to decrease in number, stiffen, break apart, and disorganize into a shapeless, matted tangle. Elastic fibers lose some of their elasticity, thicken into clumps, and fray, an effect that is greatly accelerated in the skin of smokers. Fibroblasts, which produce both collagen and elastic fibers, decrease in number. As a result, the skin forms the characteristic crevices and furrows known as *wrinkles.*

The pronounced effects of skin aging do not become noticeable until people reach their late 40s. Intraepidermal macrophages dwindle in number and become less efficient phagocytes, thus decreasing the skin's immune responsiveness. Moreover, decreased size of sebaceous glands leads to dry and broken skin that is more susceptible to infection. Production of sweat diminishes, which probably contributes to the increased incidence of heat stroke in the elderly. There is a decrease in the number of functioning melanocytes, resulting in gray hair and atypical skin pigmentation. Hair loss increases with aging as hair follicles stop producing hairs. About 25% of males begin to show signs of hair loss by age 30 and about two-thirds have significant hair loss by age 60. Both males and females develop pattern baldness. An increase in the size of some melanocytes produces pigmented blotching (age spots). Walls of blood vessels in the dermis become thicker and less permeable, and subcutaneous adipose tissue is lost. Aged skin (especially the dermis) is thinner than young skin, and the migration of cells from the basal layer to the epidermal surface slows considerably. With the onset of old age, skin heals poorly and becomes more susceptible to pathological conditions such as skin cancer and pressure sores. **Rosacea** (ro-ZĀ-shē-a = rosy) is a skin condition that affects mostly light-skinned adults between the ages of 30 and 60. It is characterized by redness, tiny pimples, and noticeable blood vessels, usually in the central area of the face.

Growth of nails and hair slows during the second and third decades of life. The nails also may become more brittle with age, often due to dehydration or repeated use of cuticle remover or nail polish.

Several cosmetic anti-aging treatments are available to diminish the effects of aging or sun-damaged skin. These include the following:

- **Topical products** that bleach the skin to tone down blotches and blemishes (hydroquinone) or decrease fine wrinkles and roughness (retinoic acid).
- **Microdermabrasion** (mī-krō-DER-ma-brā′-zhun; *mikros-* = small; *-derm-* = skin; *-abrasio* = to wear away), the use of tiny crystals under pressure to remove and vacuum the skin's surface cells to improve skin texture and reduce blemishes.
- **Chemical peel**, the application of a mild acid (such as glycolic acid) to the skin to remove surface cells to improve skin texture and reduce blemishes.
- **Laser resurfacing**, the use of a laser to clear up blood vessels near the skin surface, even out blotches and blemishes, and decrease fine wrinkles. An example is the IPL Photofacial®.
- **Dermal fillers**, injections of human collagen (Cosmoderm®), hyaluronic acid (Restylane® and Juvaderm®), calcium hydroxylapatite (Radiesse®), or poly-L-lactic acid (Sculptra®) that plumps up the skin to smooth out wrinkles and fill in furrows, such as those around the nose and mouth and between the eyebrows.
- **Fat transplantation**, in which fat from one part of the body is injected into another location such as around the eyes.
- **Botulinum toxin** or **Botox®**, a diluted version of a toxin that is injected into the skin to paralyze skeletal muscles that cause the skin to wrinkle.
- **Radio frequency nonsurgical facelift**, the use of radio frequency emissions to tighten the deeper layers of the skin of the jowls, neck, and sagging eyebrows and eyelids.
- **Facelift**, **browlift**, or **necklift**, invasive surgery in which loose skin and fat are removed surgically and the underlying connective tissue and muscle are tightened.

See Clinical Connection: Sun Damage, Sunscreens, and Sunblocks

To appreciate the many ways that skin contributes to homeostasis of other body systems, examine *Focus on Homeostasis: Contributions of the Integumentary System.* This feature is the first of 11, found at the end of selected chapters, that explain how the body system under consideration contributes to the homeostasis of all other body systems. Next, in Chapter 6, we will explore how bone tissue is formed and how bones are assembled into the skeletal system, which, like the skin, protects many of our internal organs.

FOCUS on HOMEOSTASIS

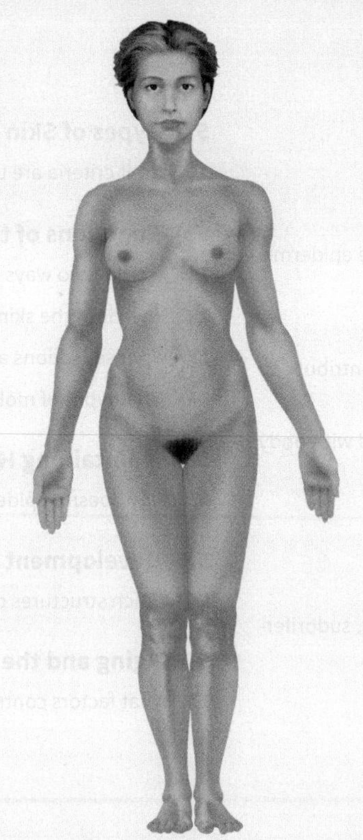

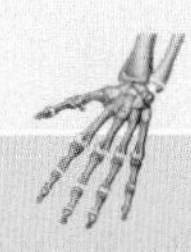

SKELETAL SYSTEM

- Skin helps activate vitamin D, needed for proper absorption of dietary calcium and phosphorus to build and maintain bones

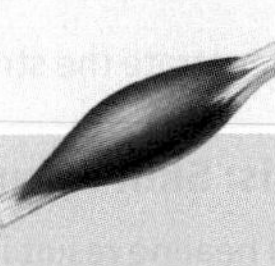

MUSCULAR SYSTEM

- Skin helps provide calcium ions, needed for muscle contraction

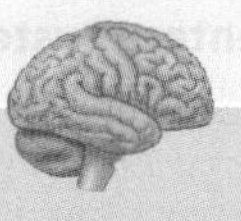

NERVOUS SYSTEM

- Nerve endings in skin and subcutaneous tissue provide input to brain for touch, pressure, thermal, and pain sensations

ENDOCRINE SYSTEM

- Keratinocytes in skin help activate vitamin D to calcitriol, a hormone that aids absorption of dietary calcium and phosphorus

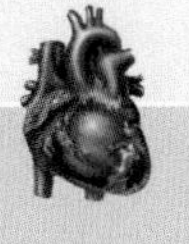

CARDIOVASCULAR SYSTEM

- Local chemical changes in dermis cause widening and narrowing of skin blood vessels, which help adjust blood flow to skin

CONTRIBUTIONS OF THE INTEGUMENTARY SYSTEM FOR ALL BODY SYSTEMS

- Skin and hair provide barriers that protect all internal organs from damaging agents in external environment
- Sweat glands and skin blood vessels regulate body temperature, needed for proper functioning of other body systems

LYMPHATIC SYSTEM and IMMUNITY

- Skin is "first line of defense" in immunity, providing mechanical barriers and chemical secretions that discourage penetration and growth of microbes
- Intraepidermal macrophages in epidermis participate in immune responses by recognizing and processing foreign antigens
- Macrophages in dermis phagocytize microbes that penetrate skin surface

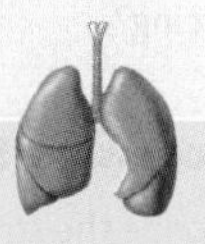

RESPIRATORY SYSTEM

- Hairs in nose filter dust particles from inhaled air
- Stimulation of pain nerve endings in skin may alter breathing rate

DIGESTIVE SYSTEM

- Skin helps activate vitamin D to the hormone calcitriol, which promotes absorption of dietary calcium and phosphorus in small intestine

URINARY SYSTEM

- Kidney cells receive partially activated vitamin D hormone from skin and convert it to calcitriol
- Some waste products are excreted from body in sweat, contributing to excretion by urinary system

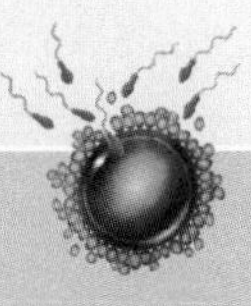

REPRODUCTIVE SYSTEMS

- Nerve endings in skin and subcutaneous tissue respond to erotic stimuli, thereby contributing to sexual pleasure
- Suckling of a baby stimulates nerve endings in skin, leading to milk ejection
- Mammary glands (modified sweat glands) produce milk
- Skin stretches during pregnancy as fetus enlarges

Review Questions

5.1 Structure of the Skin

1. What structures are included in the integumentary system?
2. How does the process of keratinization occur?
3. What are the structural and functional differences between the epidermis and dermis?
4. How are epidermal ridges formed?
5. What are the three pigments in the skin, and how do they contribute to skin color?
6. What is a tattoo? What are some potential problems associated with body piercing?

5.2 Accessory Structures of the Skin

7. Describe the structure of a hair. What causes "goose bumps"?
8. Contrast the locations and functions of sebaceous (oil) glands, sudoriferous (sweat) glands, and ceruminous glands.
9. Describe the parts of a nail.

5.3 Types of Skin

10. What criteria are used to distinguish thin and thick skin?

5.4 Functions of the Skin

11. In what two ways does the skin help regulate body temperature?
12. How does the skin serve as a protective barrier?
13. What sensations arise from stimulation of neurons in the skin?
14. What types of molecules can penetrate the stratum corneum?

5.5 Maintaining Homeostasis: Skin Wound Healing

15. Why doesn't epidermal wound healing result in scar formation?

5.6 Development of the Integumentary System

16. Which structures develop as downgrowths of the stratum basale?

5.7 Aging and the Integumentary System

17. What factors contribute to the susceptibility of aging skin to infection?

Critical Thinking Questions

1. The amount of dust that collects in a house with an assortment of dogs, cats, and people is truly amazing. A lot of these dust particles had a previous "life" as part of the home's living occupants. Where did the dust originate on the human body?
2. Josie reassures her mother that the tattoo she received at the tattoo parlor will eventually disappear. She knows this because she has learned in biology class that skin cells are shed every four weeks. Is Josie correct?
3. Six months ago, Chef Eduardo sliced through the end of his right thumbnail. Although the surrounding nail grows normally, this part of his nail remains split and doesn't seem to want to "heal." What has happened to cause this?

The Skeletal System: Bone Tissue

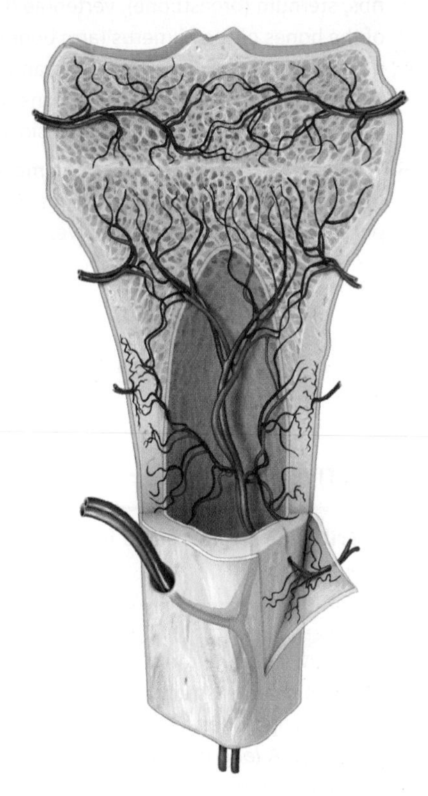

Bone Tissue and Homeostasis

Bone tissue is continuously growing, remodeling, and repairing itself. It contributes to homeostasis of the body by providing support and protection, producing blood cells, and storing minerals and triglycerides.

Bone tissue is a complex and dynamic living tissue. It continually engages in a process called *bone remodeling*—the building of new bone tissue and breaking down of old bone tissue. In the early days of space exploration, young, healthy men in prime physical shape returned from their space flights only to alarm their physicians. Physical examinations of the astronauts revealed that they had lost up to 20% of their total bone density during their extended stay in space. The zero-gravity (weightless) environment of space, coupled with the fact that the astronauts traveled in small capsules that greatly limited their movement for extended periods of time, placed minimal strain on their bones. In contrast, athletes subject their bones to great forces, which place significant strain on the bone tissue. Accomplished athletes show an increase in overall bone density. How is bone capable of changing in response to the different mechanical demands placed on it? Why do high activity levels that strain bone tissue greatly improve bone health? This chapter surveys the various components of bones to help you understand how bones form, how they age, and how exercise affects their density and strength.

Q Did you ever wonder why more females than males are affected by osteoporosis?

6.1 Functions of Bone and the Skeletal System

OBJECTIVE

- **Describe** the six main functions of the skeletal system.

A **bone** is an organ made up of several different tissues working together: bone (osseous) tissue, cartilage, dense connective tissue, epithelium, adipose tissue, and nervous tissue. The entire framework of bones and their cartilages constitute the **skeletal system**. The study of bone structure and the treatment of bone disorders is referred to as **osteology** (os-tē-OL-o-jē; *osteo-* = bone; *-logy* = study of). The skeletal system performs several basic functions:

1. ***Support.*** The skeleton serves as the structural framework for the body by supporting soft tissues and providing attachment points for the tendons of most skeletal muscles.
2. ***Protection.*** The skeleton protects the most important internal organs from injury. For example, cranial bones protect the brain, and the rib cage protects the heart and lungs.
3. ***Assistance in movement.*** Most skeletal muscles attach to bones; when they contract, they pull on bones to produce movement. This function is discussed in detail in Chapter 10.
4. ***Mineral homeostasis (storage and release).*** Bone tissue makes up about 18% of the weight of the human body. It stores several minerals, especially calcium and phosphorus, which contribute to the strength of bone. Bone tissue stores about 99% of the body's calcium. On demand, bone releases minerals into the blood to maintain critical mineral balances (homeostasis) and to distribute the minerals to other parts of the body.
5. ***Blood cell production.*** Within certain bones, a connective tissue called **red bone marrow** produces red blood cells, white blood cells, and platelets, a process called **hemopoiesis** (hēm-ō-poy-ē-sis; *hemo-* = blood; *-poiesis* = making). Red bone marrow consists of developing blood cells, adipocytes, fibroblasts, and macrophages within a network of reticular fibers. It is present in developing bones of the fetus and in some adult bones, such as the hip (pelvic) bones,

ribs, sternum (breastbone), vertebrae (backbones), skull, and ends of the bones of the humerus (arm bone) and femur (thigh bone). In a newborn, all bone marrow is red and is involved in hemopoiesis. With increasing age, much of the bone marrow changes from red to yellow. Blood cell production is considered in detail in Section 19.2.

6. ***Triglyceride storage.*** **Yellow bone marrow** consists mainly of adipose cells, which store triglycerides. The stored triglycerides are a potential chemical energy reserve.

6.2 Structure of Bone

OBJECTIVE

- **Describe** the structure and functions of each part of a long bone.

We will now examine the structure of bone at the macroscopic level. Macroscopic bone structure may be analyzed by considering the parts of a long bone, such as the humerus (the arm bone) shown in Figure 6.1a. A *long bone* is one that has greater length than width. A typical long bone consists of the following parts:

1. The **diaphysis** (dī-AF-i-sis = growing between) is the bone's shaft or body—the long, cylindrical, main portion of the bone.
2. The **epiphyses** (e-PIF-i-sēz = growing over; singular is *epiphysis*) are the proximal and distal ends of the bone.
3. The **metaphyses** (me-TAF-i-sēz; *meta-* = between; singular is *metaphysis*) are the regions between the diaphysis and the epiphyses. In a growing bone, each metaphysis contains an *epiphyseal (growth) plate* (ep′-i-FIZ-ē-al), a layer of hyaline cartilage that allows the diaphysis of the bone to grow in length (described later in the chapter). When a bone ceases to grow in length at about ages 14–24, the cartilage in the epiphyseal plate is replaced by bone; the resulting bony structure is known as the *epiphyseal line.*
4. The **articular cartilage** is a thin layer of hyaline cartilage covering the part of the epiphysis where the bone forms an articulation (joint) with another bone. Articular cartilage reduces friction and absorbs shock at freely movable joints. Because articular cartilage lacks a perichondrium and lacks blood vessels, repair of damage is limited.
5. The **periosteum** (per-ē-OS-tē-um; *peri-* = around) is a tough connective tissue sheath and its associated blood supply that surrounds the bone surface wherever it is not covered by articular cartilage. It is composed of an *outer fibrous layer* of dense irregular connective tissue and an *inner osteogenic layer* that consists of cells. Some of the cells enable bone to grow in thickness, but not in length. The periosteum also protects the bone, assists in fracture repair, helps nourish bone tissue, and serves as an attachment point for ligaments and tendons. The periosteum is attached to the underlying bone by **perforating fibers** or *Sharpey's fibers*, thick bundles of collagen that extend from the periosteum into the bone extracellular matrix.
6. The **medullary cavity** (MED-ul-er-ē; *medulla-* = marrow, pith), or *marrow cavity*, is a hollow, cylindrical space within the diaphysis that contains fatty yellow bone marrow and numerous blood vessels in adults. This cavity minimizes the weight of the bone by reducing the dense bony material where it is least needed. The long bones' tubular design provides maximum strength with minimum weight.
7. The **endosteum** (end-OS-tē-um; *endo-* = within) is a thin membrane that lines the medullary cavity. It contains a single layer of bone-forming cells and a small amount of connective tissue.

6.3 Histology of Bone Tissue

OBJECTIVES

- **Explain** why bone tissue is classified as a connective tissue.
- **Describe** the cellular composition of bone tissue and the functions of each type of cell.
- **Compare** the structural and functional differences between compact and spongy bone tissue.

We will now examine the structure of bone at the microscopic level. Like other connective tissues, **bone**, or *osseous tissue* (OS-ē-us), contains an abundant extracellular matrix that surrounds widely separated cells. The extracellular matrix is about 15% water, 30% collagen fibers, and 55% crystallized mineral salts. The most abundant mineral salt is calcium phosphate [$Ca_3(PO_4)_2$]. It combines with another mineral salt, calcium hydroxide [$Ca(OH)_2$], to form crystals of **hydroxyapatite** [$Ca_{10}(PO_4)_6(OH)_2$] (hī-drok-sē-AP-a-tīt). As the crystals form, they combine with still other mineral salts, such as calcium carbonate ($CaCO_3$), and ions such as magnesium, fluoride, potassium, and sulfate. As these mineral salts are deposited in the framework formed by the collagen fibers of the extracellular matrix, they crystallize and the tissue hardens. This process, called **calcification** (kal′-si-fi-KĀ-shun), is initiated by bone-building cells called osteoblasts (described shortly).

It was once thought that calcification simply occurred when enough mineral salts were present to form crystals. We now know that the process requires the presence of collagen fibers. Mineral salts first begin to crystallize in the microscopic spaces between collagen fibers. After the spaces are filled, mineral crystals accumulate around the collagen fibers. The combination of crystallized salts and collagen fibers is responsible for the characteristics of bone.

Although a bone's *hardness* depends on the crystallized inorganic mineral salts, a bone's *flexibility* depends on its collagen fibers. Like reinforcing metal rods in concrete, collagen fibers and other organic molecules provide *tensile strength*, resistance to being

FIGURE 6.1 **Parts of a long bone.** The spongy bone tissue of the epiphyses and metaphyses contains red bone marrow, and the medullary cavity of the diaphysis contains yellow bone marrow (in adults).

A long bone is covered by articular cartilage at the articular surfaces of its proximal and distal epiphyses and by periosteum around all other parts of the bone.

Functions of Bone Tissue

1. Supports soft tissue and provides attachment for skeletal muscles.
2. Protects internal organs.
3. Assists in movement, along with skeletal muscles.
4. Stores and releases minerals.
5. Contains red bone marrow, which produces blood cells.
6. Contains yellow bone marrow, which stores triglycerides (fats).

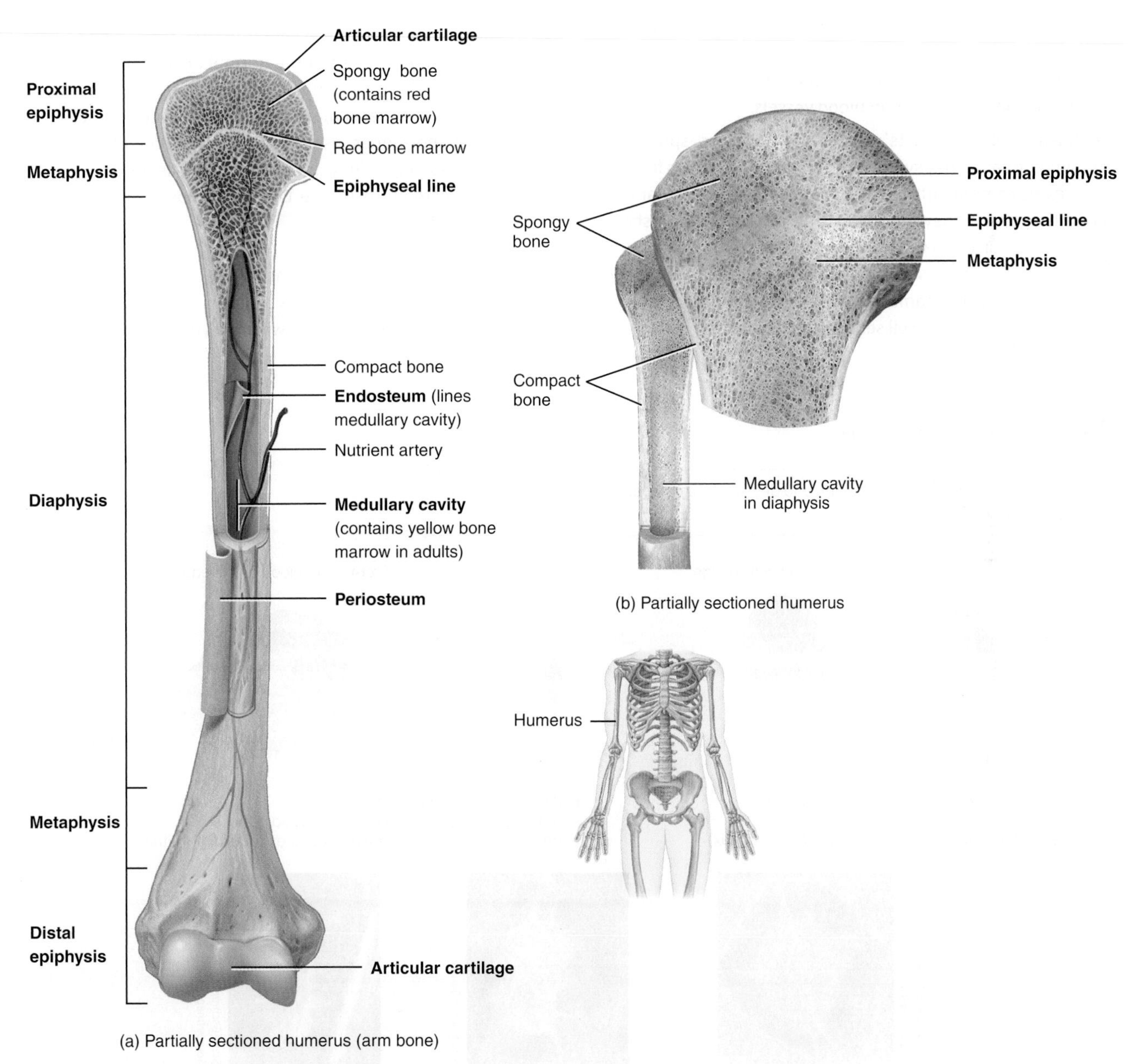

(a) Partially sectioned humerus (arm bone)

(b) Partially sectioned humerus

Q What is the functional significance of the periosteum?

stretched or torn apart. Soaking a bone in an acidic solution, such as vinegar, dissolves its mineral salts, causing the bone to become rubbery and flexible. As you will see shortly, when the need for particular minerals arises or as part of bone formation or breakdown, bone cells called osteoclasts secrete enzymes and acids that break down both the mineral salts and the collagen fibers of the extracellular matrix of bone.

Four types of cells are present in bone tissue: osteoprogenitor cells, osteoblasts, osteocytes, and osteoclasts (**Figure 6.2**).

1. **Osteoprogenitor cells** (os′-tē-ō-prō-JEN-i-tor; *-genic* = producing) are unspecialized bone stem cells derived from mesenchyme, the tissue from which almost all connective tissues are formed. They are the only bone cells to undergo cell division; the resulting cells develop into osteoblasts. Osteoprogenitor cells are found along the inner portion of the periosteum, in the endosteum, and in the canals within bone that contain blood vessels.
2. **Osteoblasts** (OS-tē-ō-blasts′; *-blasts* = buds or sprouts) are bone-building cells. They synthesize and secrete collagen fibers and other organic components needed to build the extracellular matrix of bone tissue, and they initiate calcification (described shortly). As osteoblasts surround themselves with extracellular matrix, they become trapped in their secretions and become osteocytes. (Note: The ending *-blast* in the name of a bone cell or any other connective tissue cell means that the cell secretes extracellular matrix.)
3. **Osteocytes** (OS-tē-ō-sīts′; *-cytes* = cells), mature bone cells, are the main cells in bone tissue and maintain its daily metabolism, such as the exchange of nutrients and wastes with the blood. Like osteoblasts, osteocytes do not undergo cell division. (Note: The ending *-cyte* in the name of a bone cell or any other tissue cell means that the cell maintains and monitors the tissue.)
4. **Osteoclasts** (OS-tē-ō-klasts′; *-clast* = break) are huge cells derived from the fusion of as many as 50 monocytes (a type of white blood cell) and are concentrated in the endosteum. On the side of the cell that faces the bone surface, the osteoclast's plasma membrane is deeply folded into a *ruffled border*. Here the cell releases powerful lysosomal enzymes and acids that digest the protein and mineral components of the underlying extracellular bone matrix. This breakdown of bone extracellular matrix, termed **bone resorption** (rē-SORP-shun), is part of the normal development, maintenance, and repair of bone. (Note: The ending *-clast* means that the cell breaks down extracellular matrix.) As you will see later, in response to certain hormones, osteoclasts help regulate blood calcium level (see Section 6.7). They are also target cells for drug therapy used to treat osteoporosis (see Disorders: Homeostatic Imbalances in Study Guide).

You may find it convenient to use an aid called a mnemonic device (ne-MON-ik = memory) to learn new or unfamiliar information. One such mnemonic that will help you remember the difference

FIGURE 6.2 **Types of cells in bone tissue.**

Osteoprogenitor cells undergo cell division and develop into osteoblasts, which secrete bone extracellular matrix.

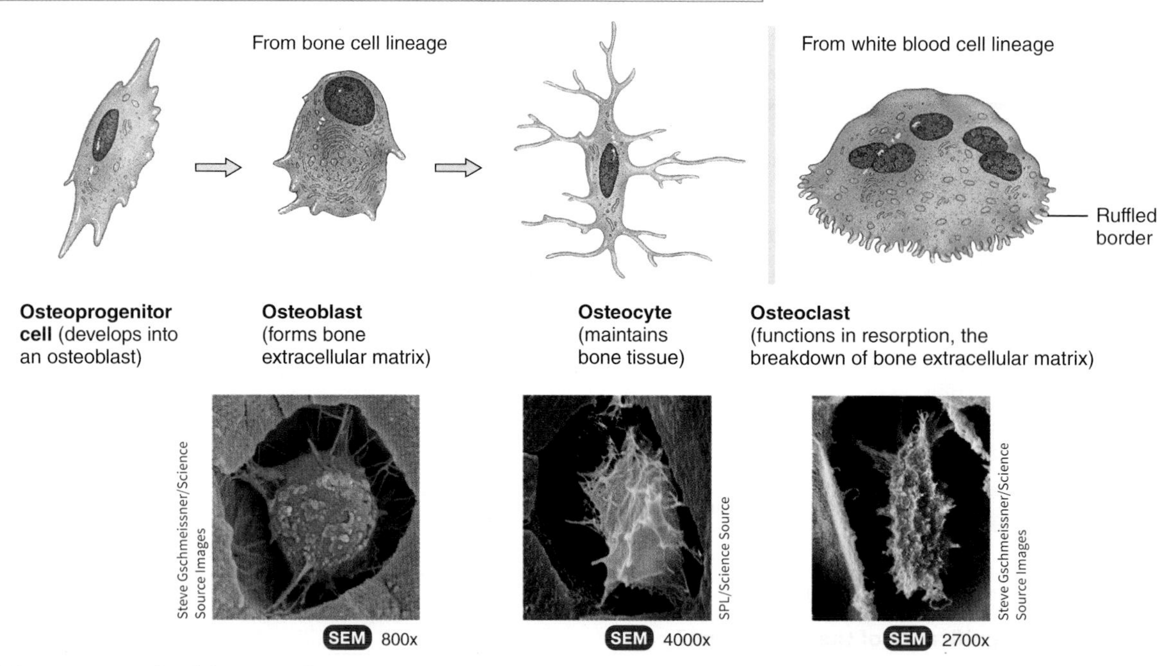

Q Why is bone resorption important?

between the function of osteoblasts and osteoclasts is as follows: osteo**B**lasts **B**uild bone, while osteo**C**lasts **C**arve out bone.

Bone is not completely solid but has many small spaces between its cells and extracellular matrix components. Some spaces serve as channels for blood vessels that supply bone cells with nutrients. Other spaces act as storage areas for red bone marrow. Depending on the size and distribution of the spaces, the regions of a bone may be categorized as compact or spongy (see **Figure 6.1**). Overall, about 80% of the skeleton is compact bone and 20% is spongy bone.

Compact Bone Tissue

Compact bone tissue contains few spaces (**Figure 6.3a**) and is the strongest form of bone tissue. It is found beneath the periosteum of all bones and makes up the bulk of the diaphyses of long bones. Compact bone tissue provides protection and support and resists the stresses produced by weight and movement.

Compact bone tissue is composed of repeating structural units called **osteons**, or *haversian systems* (ha-VER-shan). Each osteon consists of concentric lamellae arranged around an **osteonic** (haversian or central) **canal**. Resembling the growth rings of a tree, the **concentric lamellae** (la-MEL-ē) are circular plates of mineralized extracellular matrix of increasing diameter, surrounding a small network of blood vessels and nerves located in the central canal (**Figure 6.3a**). These tubelike units of bone generally form a series of parallel cylinders that, in long bones, tend to run parallel to the long axis of the bone. Between the concentric lamellae are small spaces called **lacunae** (la-KOO-nē = little lakes; singular is *lacuna*), which contain osteocytes. Radiating in all directions from the lacunae are tiny **canaliculi** (kan-a-LIK-ū-lī = small channels), which are filled with extracellular fluid. Inside the canaliculi are slender fingerlike processes of osteocytes (see inset at right of **Figure 6.3a**). Neighboring osteocytes communicate via gap junctions (see Section 4.2). The canaliculi connect lacunae with one another and with the central canals, forming an intricate, miniature system of interconnected canals throughout the bone. This system provides many routes for nutrients and oxygen to reach the osteocytes and for the removal of wastes.

Osteons in compact bone tissue are aligned in the same direction and are parallel to the length of the diaphysis. As a result, the shaft of a long bone resists bending or fracturing even when considerable force is applied from either end. Compact bone tissue tends to be thickest in those parts of a bone where stresses are applied in relatively few directions. The lines of stress in a bone are not static. They change as a person learns to walk and in response to repeated strenuous physical activity, such as weight training. The lines of stress in a bone also can change because of fractures or physical deformity. Thus, the organization of osteons is not static but changes over time in response to the physical demands placed on the skeleton.

The areas between neighboring osteons contain lamellae called **interstitial lamellae** (in′-ter-STISH-al), which also have lacunae with osteocytes and canaliculi. Interstitial lamellae are fragments of older osteons that have been partially destroyed during bone rebuilding or growth.

Blood vessels and nerves from the periosteum penetrate the compact bone through transverse **interosteonic** (Volkmann's or *perforating*) **canals**. The vessels and nerves of the interosteonic canals connect with those of the medullary cavity, periosteum, and central canals.

Arranged around the entire outer and inner circumference of the shaft of a long bone are lamellae called **circumferential lamellae** (ser′-kum-fer-EN-shē-al). They develop during initial bone formation. The circumferential lamellae directly deep to the periosteum are called *external circumferential lamellae*. They are connected to the periosteum by **perforating** (Sharpey's) **fibers**. The circumferential lamellae that line the medullary cavity are called *internal circumferential lamellae* (**Figure 6.3a**).

Spongy Bone Tissue

In contrast to compact bone tissue, **spongy bone tissue**, also referred to as *trabecular* or *cancellous bone tissue*, does not contain osteons (**Figure 6.3b, c**). Spongy bone tissue is always located in the *interior* of a bone, protected by a covering of compact bone. It consists of lamellae that are arranged in an irregular pattern of thin columns called **trabeculae** (tra-BEK-ū-lē = little beams; singular is *trabecula*). Between the trabeculae are spaces that are visible to the unaided eye. These macroscopic spaces are filled with red bone marrow in bones that produce blood cells, and yellow bone marrow (adipose tissue) in other bones. Both types of bone marrow contain numerous small blood vessels that provide nourishment to the osteocytes. Each trabecula consists of concentric lamellae, osteocytes that lie in lacunae, and canaliculi that radiate outward from the lacunae.

Spongy bone tissue makes up most of the interior bone tissue of short, flat, sesamoid, and irregularly shaped bones. In long bones it forms the core of the epiphyses beneath the paper-thin layer of compact bone, and forms a variable narrow rim bordering the medullary cavity of the diaphysis. Spongy bone is always covered by a layer of compact bone for protection.

At first glance, the trabeculae of spongy bone tissue may appear to be less organized than the osteons of compact bone tissue. However, they are precisely oriented along lines of stress, a characteristic that helps bones resist stresses and transfer force without breaking. Spongy bone tissue tends to be located where bones are not heavily stressed or where stresses are applied from many directions. The trabeculae do not achieve their final arrangement until locomotion is completely learned. In fact, the arrangement can even be altered as lines of stress change due to a poorly healed fracture or a deformity.

Spongy bone tissue is different from compact bone tissue in two respects. First, spongy bone tissue is light, which reduces the overall weight of a bone. This reduction in weight allows the bone to move more readily when pulled by a skeletal muscle. Second, the trabeculae of spongy bone tissue support and protect the red bone marrow. Spongy bone in the hip bones, ribs, sternum (breastbone), vertebrae, and the proximal ends of the humerus and femur is the only site where red bone marrow is stored and, thus, the site where hemopoiesis (blood cell production) occurs in adults.

FIGURE 6.3 Histology of compact and spongy bone. (a) Sections through the diaphysis of a long bone, from the surrounding periosteum on the right, to compact bone in the middle, to spongy bone and the medullary cavity on the left. The inset at the upper right shows an osteocyte in a lacuna. (b, c) Details of spongy bone. See **Table 4.7** for a photo-micrograph of compact bone tissue and **Figure 6.11a** for a scanning electron micrograph of spongy bone tissue.

Bone tissue is organized in concentric lamellae around an osteonic canal in compact bone and in irregularly arranged lamellae in the trabeculae in spongy bone.

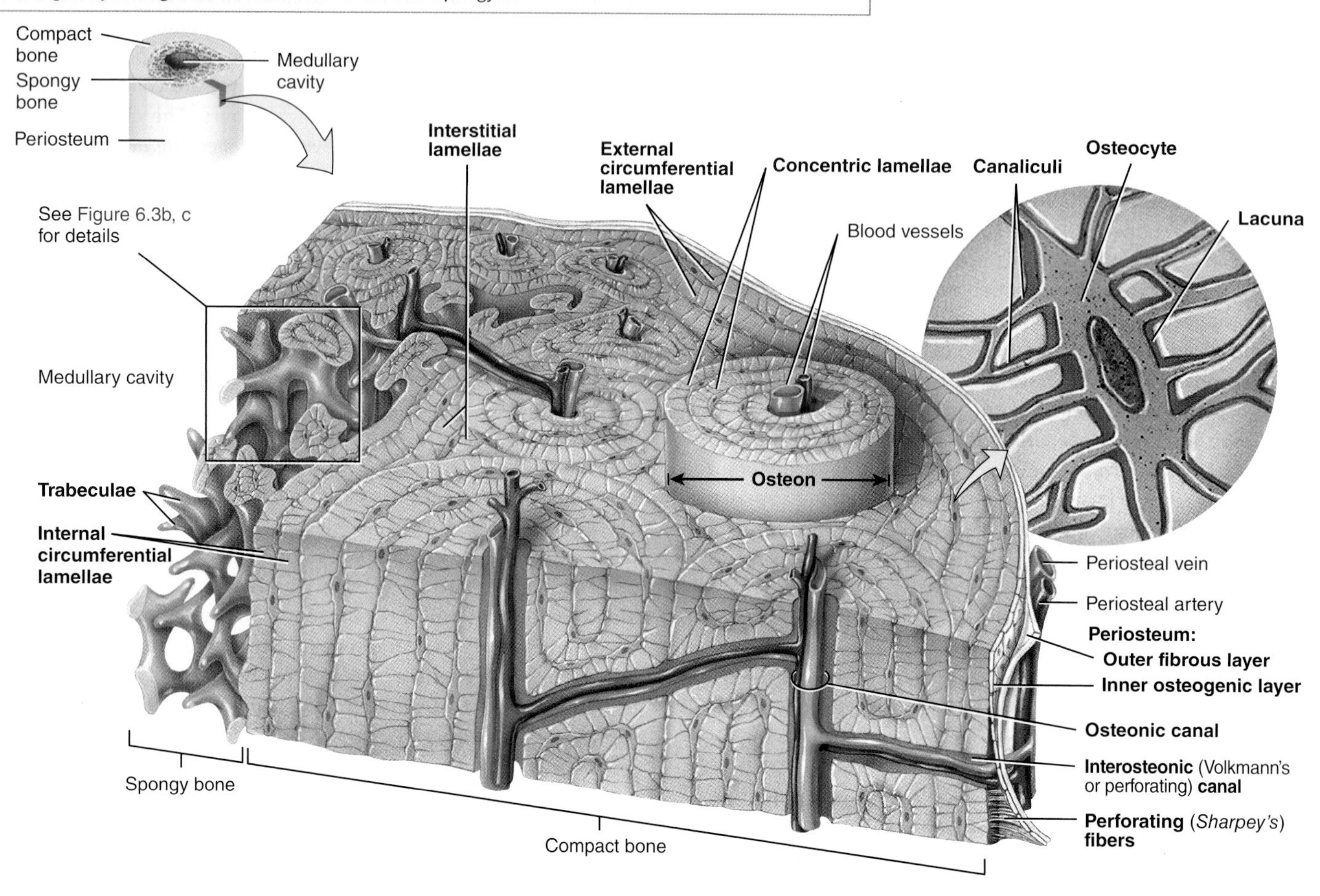

(a) Osteons (haversian systems) in compact bone and trabeculae in spongy bone

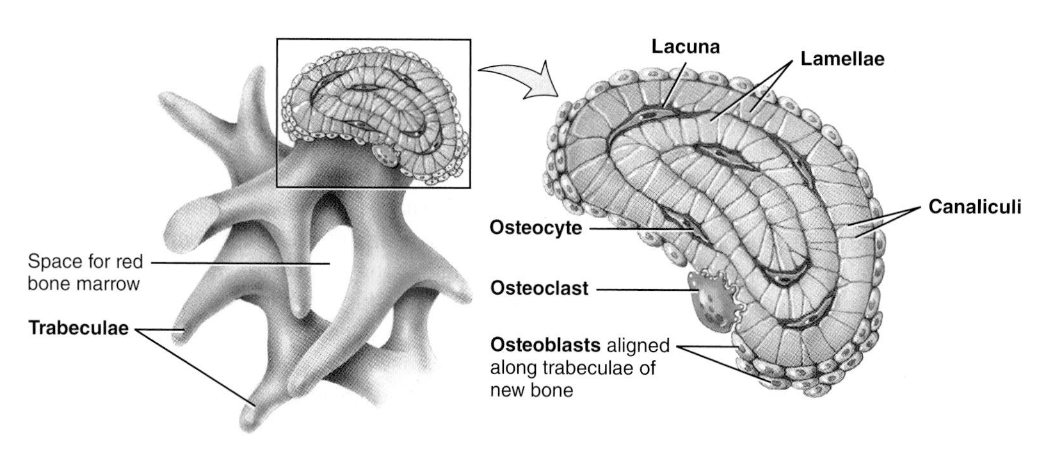

(b) Enlarged aspect of spongy bone trabeculae

(c) Details of a section of a trabecula

Q As people age, some osteonic (haversian) canals may become blocked. What effect would this have on the surrounding osteocytes?

6.4 Blood and Nerve Supply of Bone

OBJECTIVE

- **Describe** the blood and nerve supply of bone.

Bone is richly supplied with blood. Blood vessels, which are especially abundant in portions of bone containing red bone marrow, pass into bones from the periosteum. We will consider the blood supply of a long bone such as the mature tibia (shin bone) shown in Figure 6.4.

Periosteal arteries (per-ē-OS-tē-al), small arteries accompanied by nerves, enter the diaphysis through many interosteonic (Volkmann's or perforating) canals and supply the periosteum and outer part of the compact bone (see Figure 6.3a). Near the center of the diaphysis, a large **nutrient artery** passes through a hole in compact bone called the **nutrient foramen** (*foramina* is plural). On entering the medullary cavity, the nutrient artery divides into proximal and distal branches that course toward each end of the bone. These branches supply both the inner part of compact bone tissue of the diaphysis and the spongy bone tissue and red bone marrow as far as the epiphyseal plates (or lines). Some bones, like the tibia, have only one nutrient artery; others, like the femur (thigh bone), have several. The ends of long bones are supplied by the metaphyseal and epiphyseal arteries, which arise from arteries that supply the associated joint. The **metaphyseal arteries** (met-a-FIZ-ē-al) enter the metaphyses of a long bone and, together with the nutrient artery, supply the red bone marrow and bone tissue of the metaphyses. The **epiphyseal arteries** (ep′-i-FIZ-ē-al) enter the epiphyses of a long bone and supply the red bone marrow and bone tissue of the epiphyses.

FIGURE 6.4 **Blood supply of a mature long bone.**

Bone is richly supplied with blood vessels.

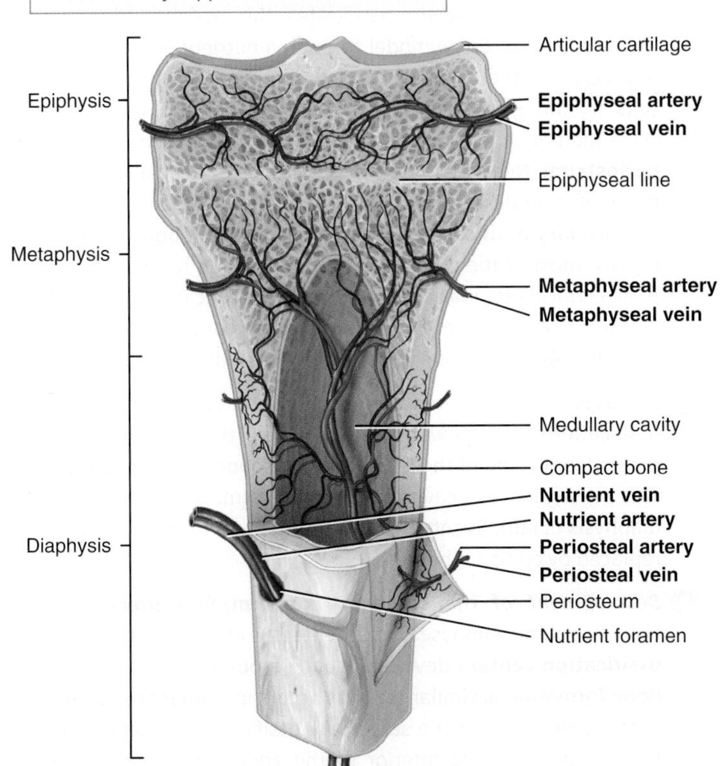

Partially sectioned tibia (shin bone)

Q Where do periosteal arteries enter bone tissue?

Veins that carry blood away from long bones are evident in three places: (1) One or two **nutrient veins** accompany the nutrient artery and exit through the diaphysis; (2) numerous **epiphyseal veins** and **metaphyseal veins** accompany their respective arteries and exit through the epiphyses and metaphyses, respectively; and (3) many small **periosteal veins** accompany their respective arteries and exit through the periosteum.

Nerves accompany the blood vessels that supply bones. The periosteum is rich in sensory nerves, some of which carry pain sensations. These nerves are especially sensitive to tearing or tension, which explains the severe pain resulting from a fracture or a bone tumor. For the same reason, there is some pain associated with a bone marrow needle biopsy. In this procedure, a needle is inserted into the middle of the bone to withdraw a sample of red bone marrow to examine it for conditions such as leukemias, metastatic neoplasms, lymphoma, Hodgkin's disease, and aplastic anemia. As the needle penetrates the periosteum, pain is felt. Once it passes through, there is little pain.

6.5 Bone Formation

OBJECTIVES

- **Describe** the steps of intramembranous and endochondral ossification.
- **Explain** how bone grows in length and thickness.
- **Describe** the process involved in bone remodeling.

The process by which bone forms is called **ossification** (os′-i-fi-KĀ-shun; *ossi-* = bone; *-fication* = making) or *osteogenesis* (os′-tē-ō-JEN-e-sis). Bone formation occurs in four principal situations: (1) the initial formation of bones in an embryo and fetus, (2) the growth of bones during infancy, childhood, and adolescence until their adult sizes are reached, (3) the remodeling of bone (replacement of old bone by new bone tissue throughout life), and (4) the repair of fractures (breaks in bones) throughout life.

Initial Bone Formation in an Embryo and Fetus

We will first consider the initial formation of bone in an embryo and fetus. The embryonic "skeleton," initially composed of mesenchyme in the general shape of bones, is the site where cartilage formation and ossification occur during the sixth week of embryonic development. Bone formation follows one of two patterns.

The two patterns of bone formation, which both involve the replacement of a preexisting connective tissue with bone, do not lead to differences in the structure of mature bones, but are simply different methods of bone development. In the first type of ossification, called **intramembranous ossification** (in′-tra-MEM-bra-nus; *intra-* = within; *-membran-* = membrane), bone forms directly within mesenchyme, which is arranged in sheetlike layers that resemble membranes. In the second type, **endochondral ossification** (en′-dō-KON-dral; *endo-* = within; *-chondral* = cartilage), bone forms within hyaline cartilage that develops from mesenchyme.

Intramembranous Ossification

Intramembranous ossification is the simpler of the two methods of bone formation. The flat bones of the skull, most of the facial bones, mandible (lower jawbone), and the medial part of the clavicle (collar bone) are formed in this way. Also, the "soft spots" that help the fetal skull pass through the birth canal later harden as they undergo intramembranous ossification, which occurs as follows (**Figure 6.5**):

1. ***Development of the ossification center.*** At the site where the bone will develop, specific chemical messages cause the cells of the mesenchyme to cluster together and differentiate, first into osteoprogenitor cells and then into osteoblasts. The site of such a cluster is called an **ossification center**. Osteoblasts secrete the organic extracellular matrix of bone until they are surrounded by it.
2. ***Calcification.*** Next, the secretion of extracellular matrix stops, and the cells, now called osteocytes, lie in lacunae and extend their narrow cytoplasmic processes into canaliculi that radiate in all directions. Within a few days, calcium and other mineral salts are deposited and the extracellular matrix hardens or calcifies (calcification).
3. ***Formation of trabeculae.*** As the bone extracellular matrix forms, it develops into trabeculae that fuse with one another to form spongy bone around the network of blood vessels in the tissue. Connective tissue associated with the blood vessels in the trabeculae differentiates into red bone marrow.
4. ***Development of the periosteum.*** In conjunction with the formation of trabeculae, the mesenchyme condenses at the periphery of the bone and develops into the periosteum. Eventually, a thin layer of compact bone replaces the surface layers of the spongy bone, but spongy bone remains in the center. Much of the newly formed bone is remodeled (destroyed and reformed) as the bone is transformed into its adult size and shape.

Endochondral Ossification

The replacement of cartilage by bone is called endochondral ossification. Although most bones of the body are formed in this way, the process is best observed in a long bone. It proceeds as follows (**Figure 6.6**):

1. ***Development of the cartilage model.*** At the site where the bone is going to form, specific chemical messages cause the cells in mesenchyme to crowd together in the general shape of the future bone, and then develop into chondroblasts. The chondroblasts secrete cartilage extracellular matrix, producing a **cartilage model** (future diaphysis) consisting of hyaline cartilage. A covering called the **perichondrium** (per′-i-KON-drē-um) develops around the cartilage model.
2. ***Growth of the cartilage model.*** Once chondroblasts become deeply buried in the cartilage extracellular matrix, they are called chondrocytes. The cartilage model grows in length by continual cell division of chondrocytes, accompanied by further secretion of the cartilage extracellular matrix. This type of cartilaginous growth, called **interstitial** (*endogenous*) **growth** (growth from within), results in an increase in length. In contrast, growth of the cartilage in thickness is due mainly to the deposition of extracellular matrix material on the cartilage surface of the model by new chondroblasts that develop from the perichondrium. This process is called **appositional** (*exogenous*) **growth** (a-pō-ZISH-o-nal), meaning growth at the outer surface. Interstitial growth and appositional growth of cartilage are described in more detail in Section 4.5.

   As the cartilage model continues to grow, chondrocytes in its midregion hypertrophy (increase in size) and the surrounding cartilage extracellular matrix begins to calcify. Other chondrocytes within the calcifying cartilage die because nutrients can no longer diffuse quickly enough through the extracellular matrix. As these chondrocytes die, the spaces left behind by dead chondrocytes merge into small cavities called lacunae.
3. ***Development of the primary ossification center.*** Primary ossification proceeds *inward* from the external surface of the bone. A nutrient artery penetrates the perichondrium and the calcifying cartilage model through a nutrient foramen in the midregion of the cartilage model, stimulating osteoprogenitor cells in the perichondrium to differentiate into osteoblasts. Once the perichondrium starts to form bone, it is known as the **periosteum**. Near the middle of the model, periosteal capillaries grow into the disintegrating calcified cartilage, inducing growth of a **primary ossification center**, a region where bone tissue will replace most of the cartilage. Osteoblasts then begin to deposit bone extracellular matrix over the remnants of calcified cartilage, forming spongy bone trabeculae. Primary ossification spreads from this central location toward both ends of the cartilage model.
4. ***Development of the medullary (marrow) cavity.*** As the primary ossification center grows toward the ends of the bone, osteoclasts break down some of the newly formed spongy bone trabeculae. This activity leaves a cavity, the medullary (marrow) cavity, in the diaphysis (shaft). Eventually, most of the wall of the diaphysis is replaced by compact bone.
5. ***Development of the secondary ossification centers.*** When branches of the epiphyseal artery enter the epiphyses, **secondary ossification centers** develop, usually around the time of birth. Bone formation is similar to what occurs in primary ossification centers. However, in the secondary ossification centers spongy bone remains in the interior of the epiphyses (no medullary cavities are formed here). In contrast to primary ossification, secondary ossification proceeds *outward* from the center of the epiphysis toward the outer surface of the bone.

FIGURE 6.5 Intramembranous ossification. Refer to this figure as you read the corresponding numbered paragraphs in the text. Illustrations ① and ② show a smaller field of vision at higher magnification than illustrations ③ and ④.

Intramembranous ossification involves the formation of bone within mesenchyme arranged in sheetlike layers that resemble membranes.

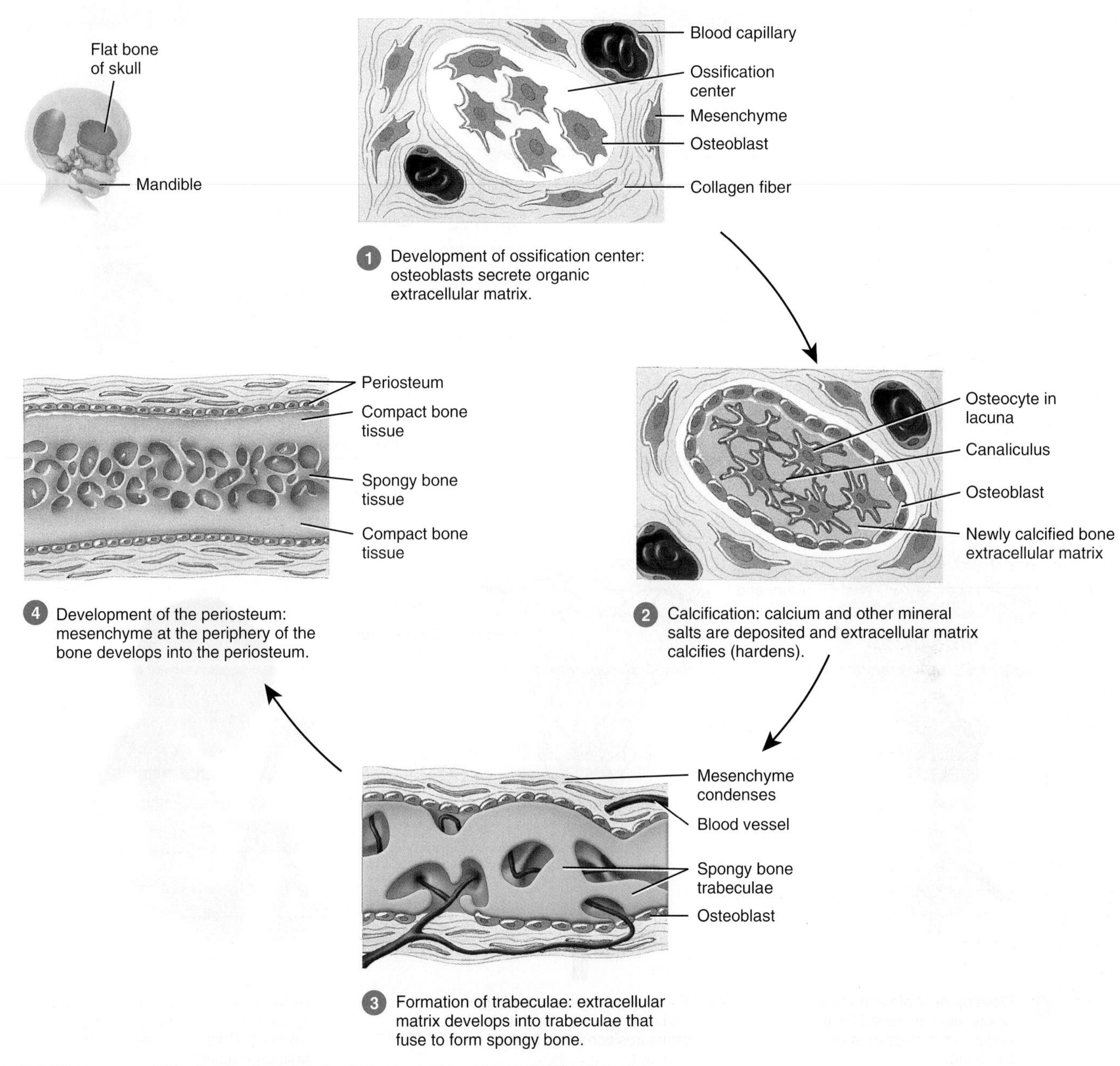

Q Which bones of the body develop by intramembranous ossification?

FIGURE 6.6 Endochondral ossification.

During endochondral ossification, bone gradually replaces a cartilage model.

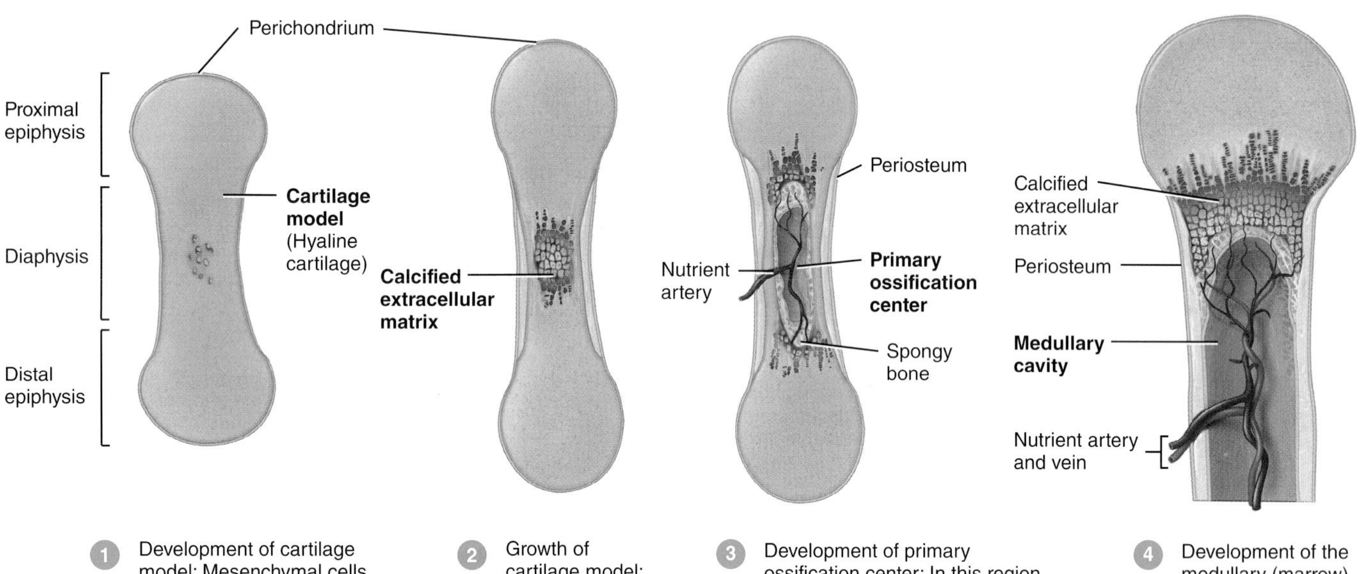

1. Development of cartilage model: Mesenchymal cells develop into chondroblasts, which form the cartilage model.
2. Growth of cartilage model: Growth occurs by cell division of chondrocytes.
3. Development of primary ossification center: In this region of the diaphysis, bone tissue has replaced most of the cartilage.
4. Development of the medullary (marrow) cavity: Bone breakdown by osteoclasts forms the medullary cavity.

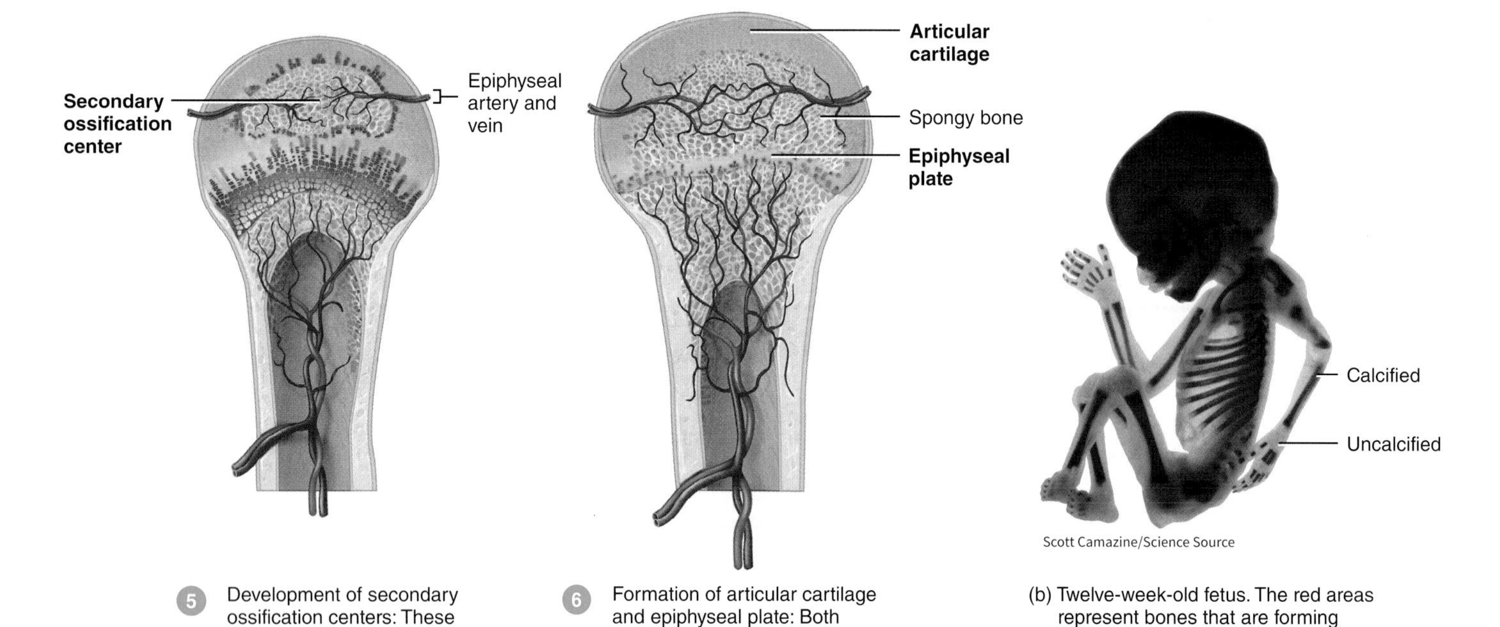

5. Development of secondary ossification centers: These occur in the epiphyses of the bone.
6. Formation of articular cartilage and epiphyseal plate: Both structures consist of hyaline cartilage.

(a) Sequence of events

(b) Twelve-week-old fetus. The red areas represent bones that are forming (calcified). Clear areas represent cartilage (uncalcified).

Q Where in the cartilage model do secondary ossification centers develop during endochondral ossification?

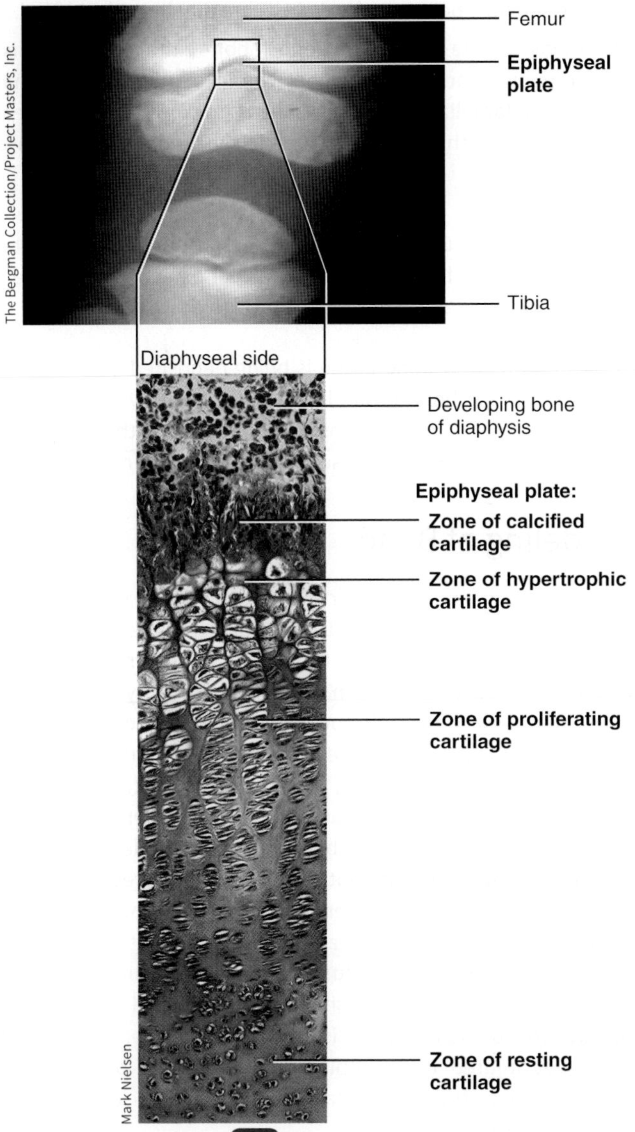

(b) Histology of the epiphyseal plate

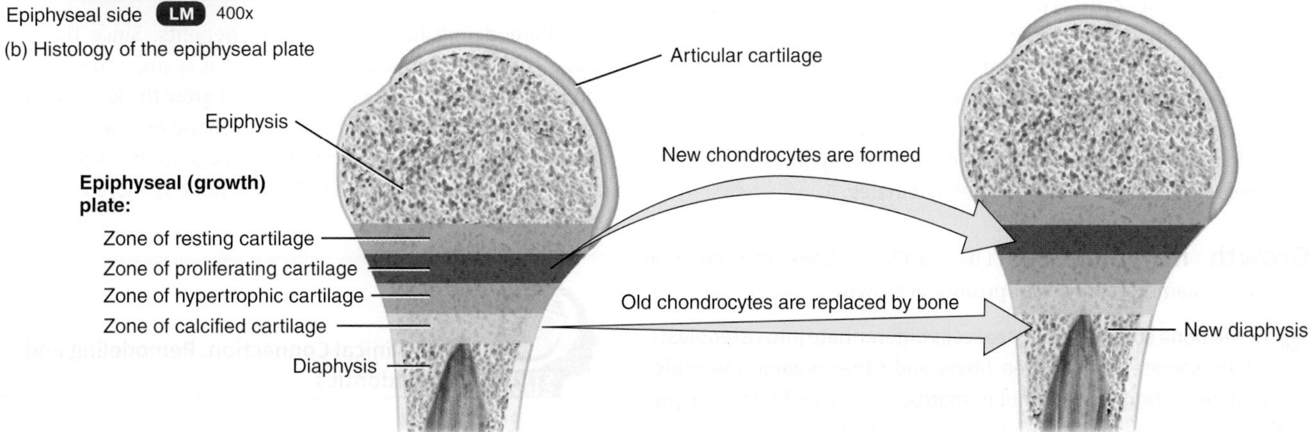

(c) Lengthwise growth of bone at epiphyseal plate

Q How does the epiphyseal (growth) plate account for the lengthwise growth of the diaphysis?

FIGURE 6.7 Epiphyseal (growth) plate. The epiphyseal (growth) plate appears as a dark band between whiter calcified areas in the radiograph (x-ray) shown in part (a).

> The epiphyseal (growth) plate allows the diaphysis of a bone to increase in length.

6. ***Formation of articular cartilage and the epiphyseal (growth) plate.*** The hyaline cartilage that covers the epiphyses becomes the articular cartilage. Prior to adulthood, hyaline cartilage remains between the diaphysis and epiphysis as the epiphyseal (growth) plate, the region responsible for the lengthwise growth of long bones that you will learn about next.

Bone Growth during Infancy, Childhood, and Adolescence

During infancy, childhood, and adolescence, bones throughout the body grow in thickness by appositional growth, and long bones lengthen by the addition of bone material on the diaphyseal side of the epiphyseal plate by interstitial growth.

Growth in Length The growth in length of long bones involves the following two major events: (1) interstitial growth of cartilage on the epiphyseal side of the epiphyseal plate and (2) replacement of cartilage on the diaphyseal side of the epiphyseal plate with bone by endochondral ossification.

To understand how a bone grows in length, you need to know some of the details of the structure of the epiphyseal plate. The **epiphyseal** (*growth*) **plate** (ep-i-FIZ-ē-al) is a layer of hyaline cartilage in the metaphysis of a growing bone that consists of four zones (**Figure 6.7b**):

1. ***Zone of resting cartilage.*** This layer is nearest the epiphysis and consists of small, scattered chondrocytes. The term "resting" is used because the cells do not function in bone growth. Rather, they anchor the epiphyseal plate to the epiphysis of the bone.

2. ***Zone of proliferating cartilage.*** Slightly larger chondrocytes in this zone are arranged like stacks of coins. These chondrocytes undergo interstitial growth as they divide and secrete extracellular matrix. The chondrocytes in this zone divide to replace those that die at the diaphyseal side of the epiphyseal plate.
3. ***Zone of hypertrophic cartilage*** (hī-per-TRŌ-fik). This layer consists of large, maturing chondrocytes arranged in columns.
4. ***Zone of calcified cartilage.*** The final zone of the epiphyseal plate is only a few cells thick and consists mostly of chondrocytes that are dead because the extracellular matrix around them has calcified. Osteoclasts dissolve the calcified cartilage, and osteoblasts and capillaries from the diaphysis invade the area. The osteoblasts lay down bone extracellular matrix, replacing the calcified cartilage by the process of endochondral ossification. Recall that endochondral ossification is the replacement of cartilage with bone. As a result, the zone of calcified cartilage becomes the "new diaphysis" that is firmly cemented to the rest of the diaphysis of the bone.

The activity of the epiphyseal plate is the only way that the diaphysis can increase in length. As a bone grows, chondrocytes proliferate on the epiphyseal side of the plate. New chondrocytes replace older ones, which are destroyed by calcification. Thus, the cartilage is replaced by bone on the diaphyseal side of the plate. In this way the thickness of the epiphyseal plate remains relatively constant, but the bone on the diaphyseal side increases in length (**Figure 6.7c**). If a bone fracture damages the epiphyseal plate, the fractured bone may be shorter than normal once adult stature is reached. This is because damage to cartilage, which is avascular, accelerates closure of the epiphyseal plate due to the cessation of cartilage cell division, thus inhibiting lengthwise growth of the bone.

When adolescence comes to an end (at about age 18 in females and age 21 in males), the epiphyseal plates close; that is, the epiphyseal cartilage cells stop dividing and bone replaces all remaining cartilage. The epiphyseal plate fades, leaving a bony structure called the **epiphyseal line**. With the appearance of the epiphyseal line, bone growth in length stops completely.

Closure of the epiphyseal plate is a gradual process and the degree to which it occurs is useful in determining bone age, predicting adult height, and establishing age at death from skeletal remains, especially in infants, children, and adolescents. For example, an open epiphyseal plate indicates a younger person, while a partially closed epiphyseal plate or a completely closed one indicates an older person. It should also be kept in mind that closure of the epiphyseal plate, on average, takes place 1–2 years earlier in females.

Growth in Thickness Like cartilage, bone can grow in thickness (diameter) only by appositional growth (**Figure 6.8a**):

1. At the bone surface, periosteal cells differentiate into osteoblasts, which secrete the collagen fibers and other organic molecules that form bone extracellular matrix. The osteoblasts become surrounded by extracellular matrix and develop into osteocytes. This process forms bone ridges on either side of a periosteal blood vessel. The ridges slowly enlarge and create a groove for the periosteal blood vessel.
2. Eventually, the ridges fold together and fuse, and the groove becomes a tunnel that encloses the blood vessel. The former periosteum now becomes the endosteum that lines the tunnel.
3. Osteoblasts in the endosteum deposit bone extracellular matrix, forming new concentric lamellae. The formation of additional concentric lamellae proceeds inward toward the periosteal blood vessel. In this way, the tunnel fills in, and a new osteon is created.
4. As an osteon is forming, osteoblasts under the periosteum deposit new circumferential lamellae, further increasing the thickness of the bone. As additional periosteal blood vessels become enclosed as in step 1, the growth process continues.

Recall that as new bone tissue is being deposited on the outer surface of bone, the bone tissue lining the medullary cavity is destroyed by osteoclasts in the endosteum. In this way, the medullary cavity enlarges as the bone increases in thickness (**Figure 6.8b**).

Remodeling of Bone

Like skin, bone forms before birth but continually renews itself thereafter. **Bone remodeling** is the ongoing replacement of old bone tissue by new bone tissue. It involves **bone resorption**, the removal of minerals and collagen fibers from bone by osteoclasts, and **bone deposition**, the addition of minerals and collagen fibers to bone by osteoblasts. Thus, bone resorption results in the destruction of bone extracellular matrix, while bone deposition results in the formation of bone extracellular matrix. At any given time, about 5% of the total bone mass in the body is being remodeled. Remodeling also takes place at different rates in different regions of the body. The distal portion of the femur is replaced about every four months. By contrast, bone in certain areas of the shaft of the femur will not be replaced completely during an individual's life. Even after bones have reached their adult shapes and sizes, old bone is continually destroyed and new bone is formed in its place. Remodeling also removes injured bone, replacing it with new bone tissue. Remodeling may be triggered by factors such as exercise, sedentary lifestyle, and changes in diet.

Remodeling has several other benefits. Since the strength of bone is related to the degree to which it is stressed, if newly formed bone is subjected to heavy loads, it will grow thicker and therefore be stronger than the old bone. Also, the shape of a bone can be altered for proper support based on the stress patterns experienced during the remodeling process. Finally, new bone is more resistant to fracture than old bone.

See Clinical Connection: Remodeling and Orthodontics

During the process of bone resorption, an osteoclast attaches tightly to the bone surface at the endosteum or periosteum and forms a leakproof seal at the edges of its ruffled border (see **Figure 6.2**).

FIGURE 6.8 **Bone growth in thickness.**

As new bone is deposited on the outer surface of bone by osteoblasts, the bone tissue lining the medullary cavity is destroyed by osteoclasts in the endosteum.

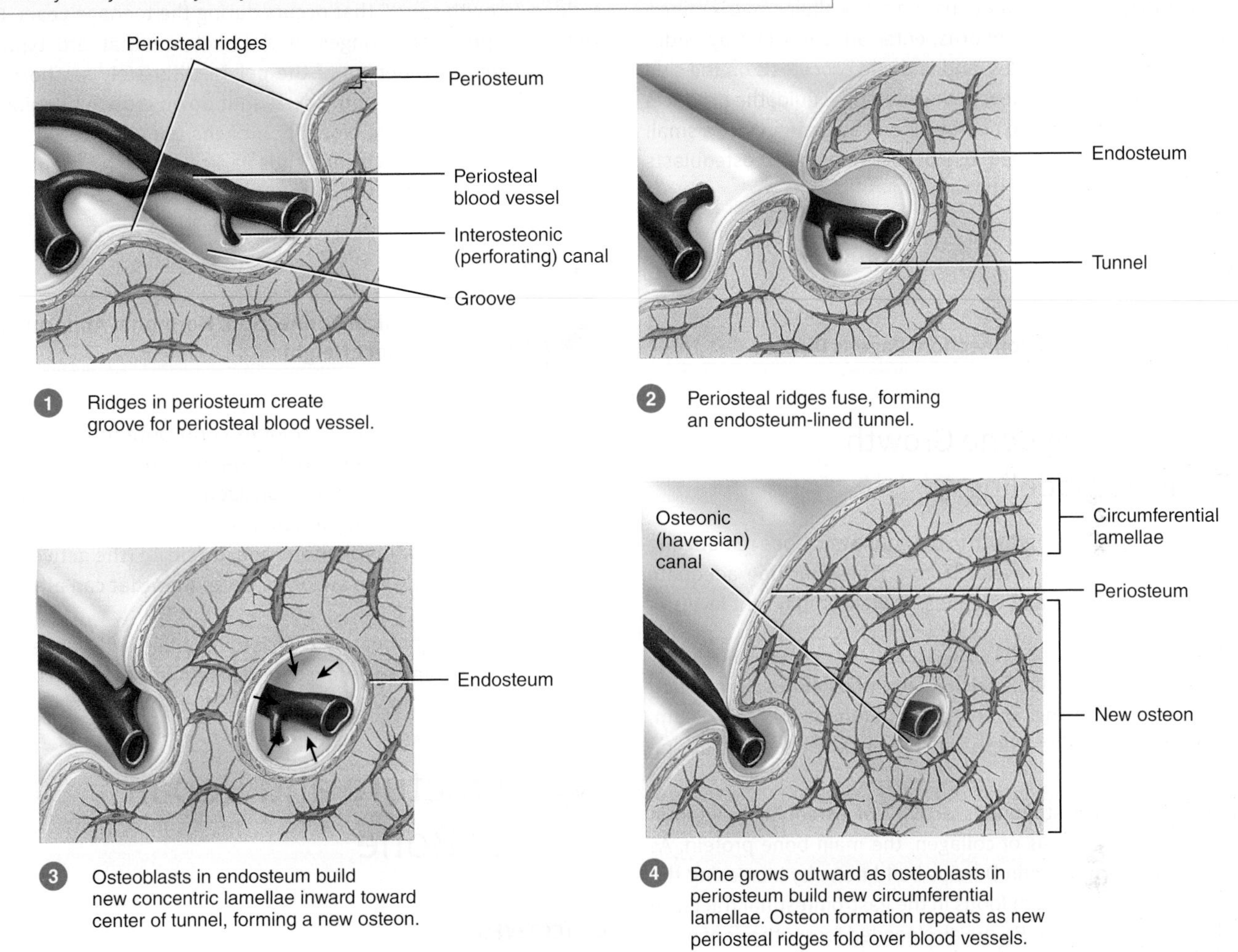

(a) Microscopic details

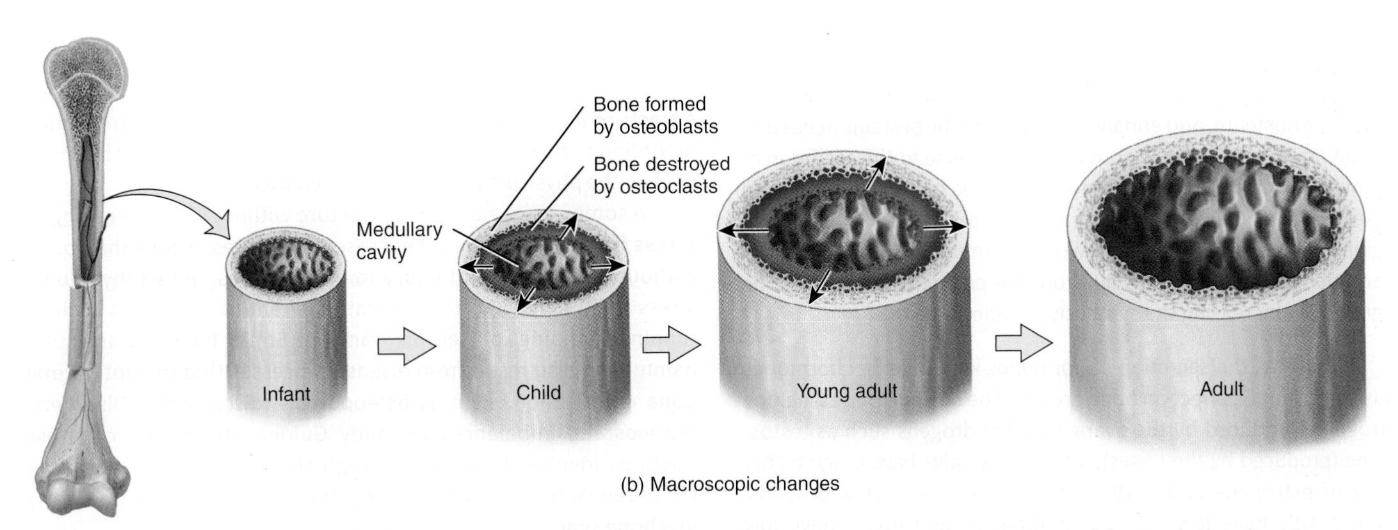

(b) Macroscopic changes

Q How does the medullary cavity enlarge during growth in thickness?

Then it releases protein-digesting lysosomal enzymes and several acids into the sealed pocket. The enzymes digest collagen fibers and other organic substances while the acids dissolve the bone minerals. Working together, several osteoclasts carve out a small tunnel in the old bone. The degraded bone proteins and extracellular matrix minerals, mainly calcium and phosphorus, enter an osteoclast by endocytosis, cross the cell in vesicles, and undergo exocytosis on the side opposite the ruffled border. Now in the interstitial fluid, the products of bone resorption diffuse into nearby blood capillaries. Once a small area of bone has been resorbed, osteoclasts depart and osteoblasts move in to rebuild the bone in that area.

See Clinical Connection: Paget's Disease

Factors Affecting Bone Growth and Bone Remodeling

Normal bone metabolism—growth in the young and bone remodeling in the adult—depends on several factors. These include adequate dietary intake of minerals and vitamins, as well as sufficient levels of several hormones.

1. ***Minerals.*** Large amounts of calcium and phosphorus are needed while bones are growing, as are smaller amounts of magnesium, fluoride, and manganese. These minerals are also necessary during bone remodeling.
2. ***Vitamins.*** Vitamin A stimulates activity of osteoblasts. Vitamin C is needed for synthesis of collagen, the main bone protein. As you will soon learn, vitamin D helps build bone by increasing the absorption of calcium from foods in the gastrointestinal tract into the blood. Vitamins K and B_{12} are also needed for synthesis of bone proteins.
3. ***Hormones.*** During childhood, the hormones most important to bone growth are the insulin-like growth factors (IGFs), which are produced by the liver and bone tissue (see Section 18.6). IGFs stimulate osteoblasts, promote cell division at the epiphyseal plate and in the periosteum, and enhance synthesis of the proteins needed to build new bone. IGFs are produced in response to the secretion of growth hormone (GH) from the anterior lobe of the pituitary gland (see Section 18.6). Thyroid hormones (T_3 and T_4) from the thyroid gland also promote bone growth by stimulating osteoblasts. In addition, the hormone insulin from the pancreas promotes bone growth by increasing the synthesis of bone proteins.

At puberty, the secretion of hormones known as sex hormones causes a dramatic effect on bone growth. The **sex hormones** include estrogens (produced by the ovaries) and androgens such as testosterone (produced by the testes). Although females have much higher levels of estrogens and males have higher levels of androgens, females also have low levels of androgens, and males have low levels of estrogens. The adrenal glands of both sexes produce androgens, and other tissues, such as adipose tissue, can convert androgens to estrogens. These hormones are responsible for increased osteoblast activity, synthesis of bone extracellular matrix, and the sudden "growth spurt" that occurs during the teenage years. Estrogens also promote changes in the skeleton that are typical of females, such as widening of the pelvis. Ultimately sex hormones, especially estrogens in both sexes, shut down growth at epiphyseal (growth) plates, causing elongation of the bones to cease. Lengthwise growth of bones typically ends earlier in females than in males due to their higher levels of estrogens.

See Clinical Connection: Hormonal Abnormalities That Affect Height

During adulthood, sex hormones contribute to bone remodeling by slowing resorption of old bone and promoting deposition of new bone. One way that estrogens slow resorption is by promoting apoptosis (programmed death) of osteoclasts. As you will see shortly, parathyroid hormone, calcitriol (the active form of vitamin D), and calcitonin are other hormones that can affect bone remodeling.

Moderate weight-bearing exercises maintain sufficient strain on bones to increase and maintain their density.

Fracture and Repair of Bone

OBJECTIVES

- **Describe** several common types of fractures.
- **Explain** the sequence of events involved in fracture repair.

A **fracture** (FRAK-choor) is any break in a bone. Fractures are named according to their severity, the shape or position of the fracture line, or even the physician who first described them.

In some cases, a bone may fracture without visibly breaking. A **stress fracture** is a series of microscopic fissures in bone that forms without any evidence of injury to other tissues. In healthy adults, stress fractures result from repeated, strenuous activities such as running, jumping, or aerobic dancing. Stress fractures are quite painful and also result from disease processes that disrupt normal bone calcification, such as osteoporosis (discussed in Disorders: Homeostatic Imbalances in Study Guide). About 25% of stress fractures involve the tibia. Although standard x-ray images often fail to reveal the presence of stress fractures, they show up clearly in a bone scan.

FIGURE 6.9 **Steps in repair of a bone fracture.**

Bone heals more rapidly than cartilage because its blood supply is more plentiful.

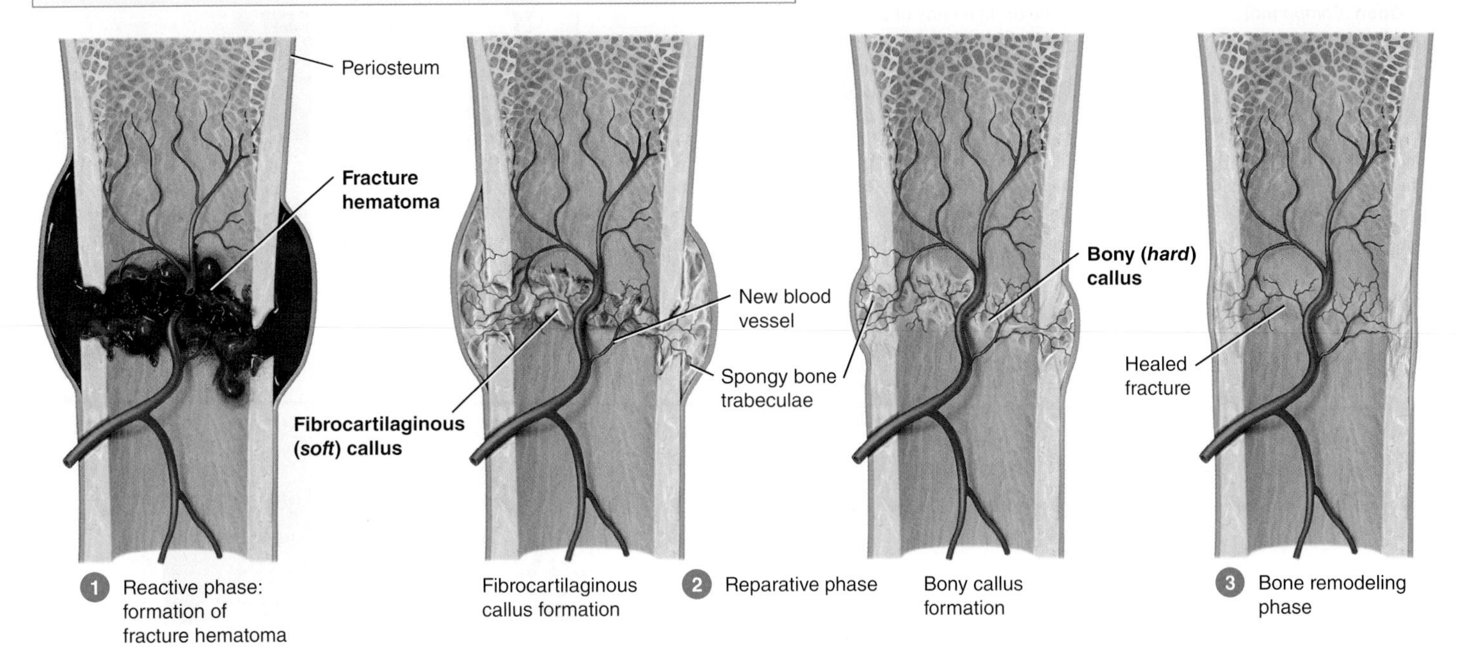

Q Why does it sometimes take months for a fracture to heal?

The repair of a bone fracture involves the following phases (**Figure 6.9**):

1 ***Reactive phase.*** This phase is an early inflammatory phase. Blood vessels crossing the fracture line are broken. As blood leaks from the torn ends of the vessels, a mass of blood (usually clotted) forms around the site of the fracture. This mass of blood, called a **fracture hematoma** (hē′-ma-TŌ-ma; *hemat-* = blood; *-oma* = tumor), usually forms 6 to 8 hours after the injury. Because the circulation of blood stops at the site where the fracture hematoma forms, nearby bone cells die. Swelling and inflammation occur in response to dead bone cells, producing additional cellular debris. Phagocytes (neutrophils and macrophages) and osteoclasts begin to remove the dead or damaged tissue in and around the fracture hematoma. This stage may last up to several weeks.

2a ***Reparative phase: Fibrocartilaginous callus formation.*** The reparative phase is characterized by two events: the formation of a fibrocartilaginous callus, and a bony callus to bridge the gap between the broken ends of the bones.

Blood vessels grow into the fracture hematoma and phagocytes begin to clean up dead bone cells. Fibroblasts from the periosteum invade the fracture site and produce collagen fibers. In addition, cells from the periosteum develop into chondroblasts and begin to produce fibrocartilage in this region. These events lead to the development of a **fibrocartilaginous** (*soft*) **callus** (fi-brō-kar-ti-LAJ-i-nus), a mass of repair tissue consisting of collagen fibers and cartilage that bridges the broken ends of the bone. Formation of the fibrocartilaginous callus takes about 3 weeks.

2b ***Reparative phase: Bony callus formation.*** In areas closer to well-vascularized healthy bone tissue, osteoprogenitor cells develop into osteoblasts, which begin to produce spongy bone trabeculae. The trabeculae join living and dead portions of the original bone fragments. In time, the fibrocartilage is converted to spongy bone, and the callus is then referred to as a **bony** (*hard*) **callus**. The bony callus lasts about 3 to 4 months.

3 ***Bone remodeling phase.*** The final phase of fracture repair is bone remodeling of the callus. Dead portions of the original fragments of broken bone are gradually resorbed by osteoclasts. Compact bone replaces spongy bone around the periphery of the fracture. Sometimes, the repair process is so thorough that the fracture line is undetectable, even in a radiograph (x-ray). However, a thickened area on the surface of the bone remains as evidence of a healed fracture.

See Clinical Connection: Treatments for Fractures

Although bone has a generous blood supply, healing sometimes takes months. The calcium and phosphorus needed to strengthen and harden new bone are deposited only gradually, and bone cells generally grow and reproduce slowly. The temporary disruption in their blood supply also helps explain the slowness of healing of severely fractured bones. Some of the common types of fractures are shown in **Table 6.1.**

TABLE 6.1 Some Common Fractures

FRACTURE	DESCRIPTION	ILLUSTRATION	RADIOGRAPH
Open (*Compound*)	The broken ends of the bone protrude through the skin. Conversely, a *closed (simple) fracture* does not break the skin.	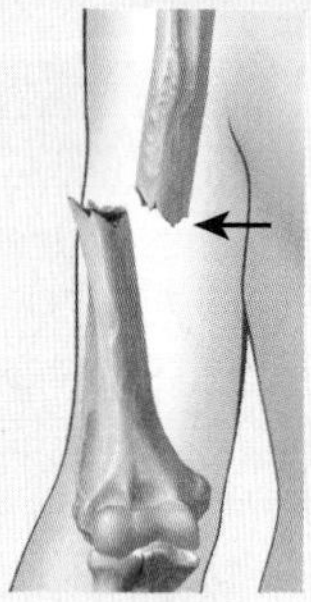 Humerus	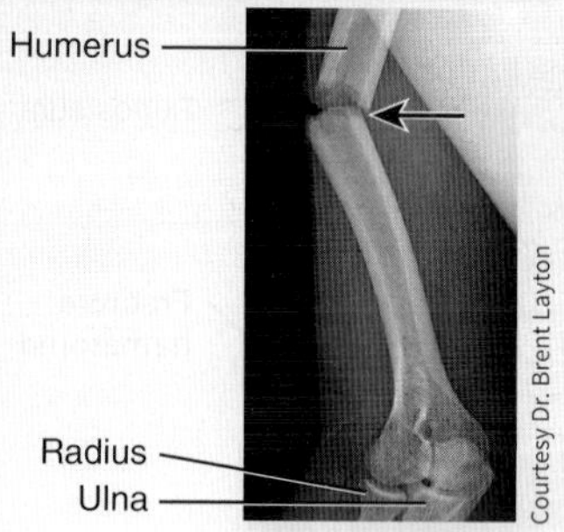 Radius; Ulna; Courtesy Dr. Brent Layton
Comminuted (KOM-i-noo-ted; *com-* = together; *-minuted* = crumbled)	The bone is splintered, crushed, or broken into pieces at the site of impact, and smaller bone fragments lie between the two main fragments.	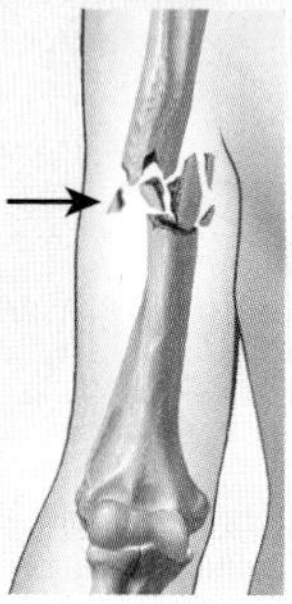 Humerus	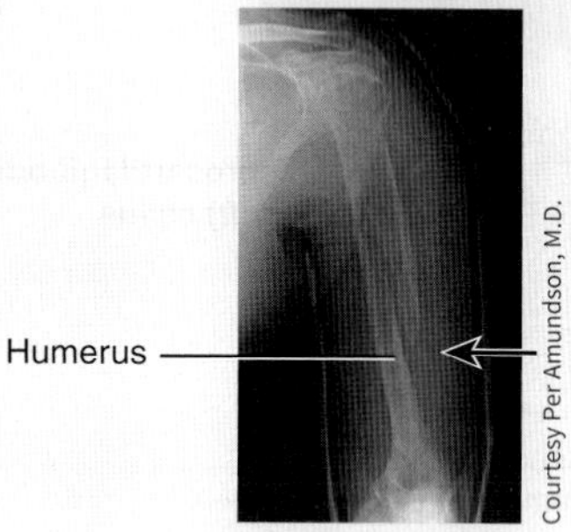 Courtesy Per Amundson, M.D.
Greenstick	A partial fracture in which one side of the bone is broken and the other side bends; similar to the way a green twig breaks on one side while the other side stays whole, but bends; occurs only in children, whose bones are not fully ossified and contain more organic material than inorganic material.	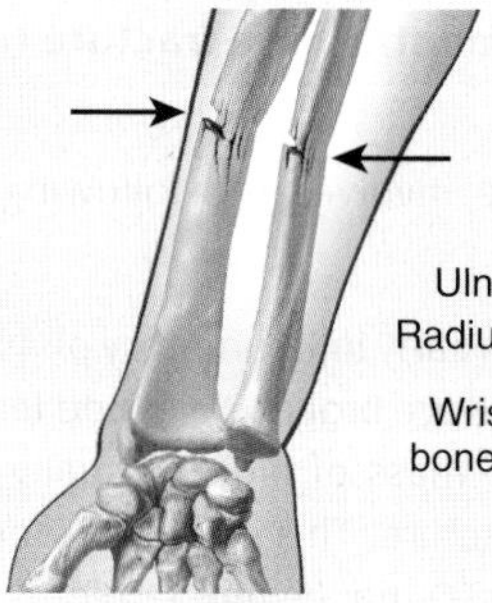 Ulna; Radius; Wrist bones	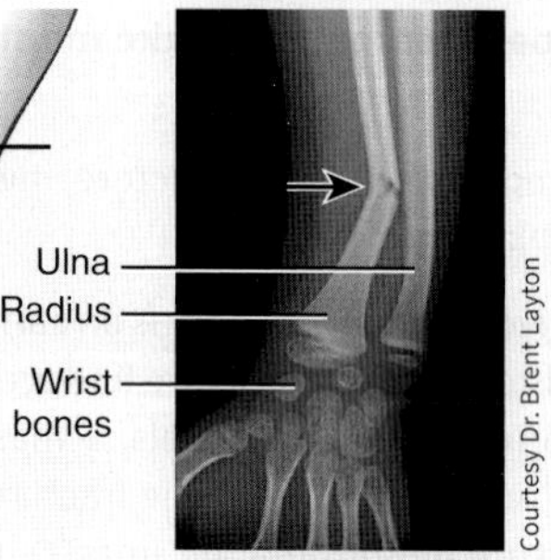 Courtesy Dr. Brent Layton
Impacted	One end of the fractured bone is forcefully driven into the interior of the other.	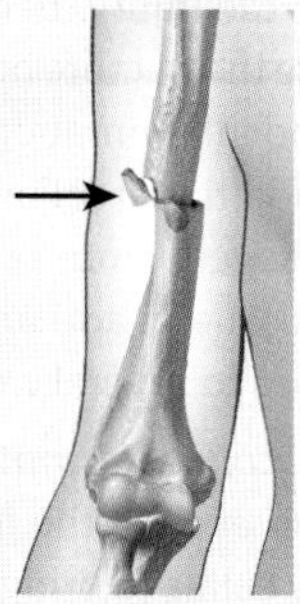 Humerus	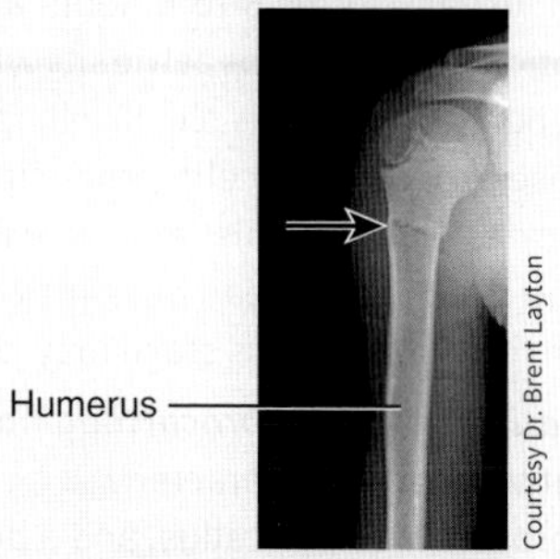 Courtesy Dr. Brent Layton
Pott	Fracture of the distal end of the lateral leg bone (fibula), with serious injury of the distal tibial articulation.	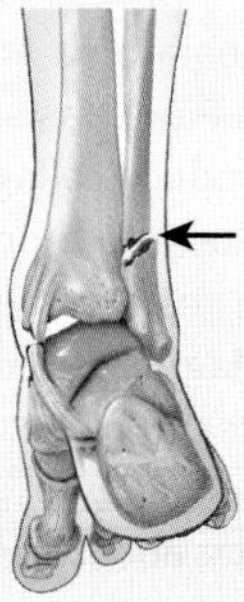 Tibia; Fibula; Ankle bones	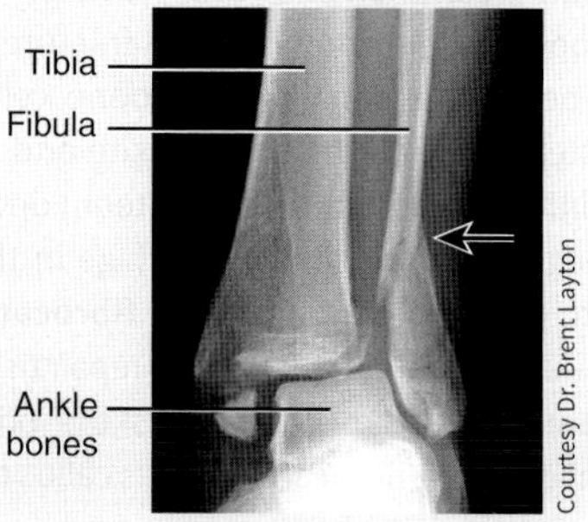 Courtesy Dr. Brent Layton

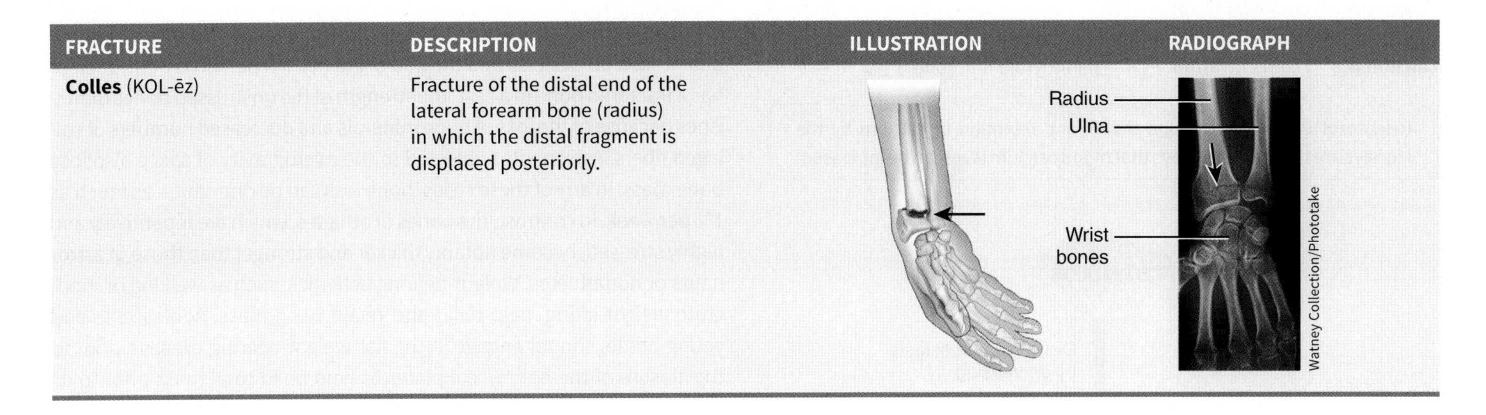

FRACTURE	DESCRIPTION	ILLUSTRATION	RADIOGRAPH
Colles (KOL-ēz)	Fracture of the distal end of the lateral forearm bone (radius) in which the distal fragment is displaced posteriorly.		

6.7 Bone's Role in Calcium Homeostasis

OBJECTIVES

- **Describe** the importance of calcium in the body.
- **Explain** how blood calcium level is regulated.

Bone is the body's major calcium reservoir, storing 99% of total body calcium. One way to maintain the level of calcium in the blood is to control the rates of calcium resorption from bone into blood and of calcium deposition from blood into bone. Both nerve and muscle cells depend on a stable level of calcium ions (Ca^{2+}) in extracellular fluid to function properly. Blood clotting also requires Ca^{2+}. Also, many enzymes require Ca^{2+} as a cofactor (an additional substance needed for an enzymatic reaction to occur). For this reason, the blood plasma level of Ca^{2+} is very closely regulated between 9 and 11 mg/100 mL. Even small changes in Ca^{2+} concentration outside this range may prove fatal—the heart may stop (cardiac arrest) if the concentration goes too high, or breathing may cease (respiratory arrest) if the level falls too low. The role of bone in calcium homeostasis is to help "buffer" the blood Ca^{2+} level, releasing Ca^{2+} into blood plasma (using osteoclasts) when the level decreases, and absorbing Ca^{2+} (using osteoblasts) when the level rises.

Ca^{2+} exchange is regulated by hormones, the most important of which is **parathyroid hormone (PTH)** secreted by the parathyroid glands (see Figure 18.13). This hormone increases blood Ca^{2+} level. PTH secretion operates via a negative feedback system (Figure 6.10). If some stimulus causes the blood Ca^{2+} level to decrease, parathyroid gland cells (receptors) detect this change and increase their production of a molecule known as cyclic adenosine monophosphate (cyclic AMP). The gene for PTH within the nucleus of a parathyroid gland cell (the control center) detects the intracellular increase in cyclic AMP (the input). As a result, PTH synthesis speeds up, and more PTH (the output) is released into the blood. The presence of higher levels of PTH increases the number and activity of osteoclasts (effectors), which step up the pace of bone resorption. The resulting release of Ca^{2+} from bone into blood returns the blood Ca^{2+} level to normal.

PTH also acts on the kidneys (effectors) to decrease loss of Ca^{2+} in the urine, so more is retained in the blood. And PTH stimulates formation of **calcitriol** (the active form of vitamin D), a hormone that promotes absorption of calcium from foods in the gastrointestinal tract into the blood. Both of these actions also help elevate blood Ca^{2+} level.

Another hormone works to decrease blood Ca^{2+} level. When blood Ca^{2+} rises above normal, *parafollicular cells* in the thyroid gland secrete **calcitonin (CT)** (kal-si-TŌ-nin). CT inhibits activity of osteoclasts, speeds blood Ca^{2+} uptake by bone, and accelerates Ca^{2+} deposition into bones. The net result is that CT promotes bone formation and decreases blood Ca^{2+} level. Despite these effects, the role of CT in normal calcium homeostasis is uncertain because it can be completely absent without causing symptoms. Nevertheless, calcitonin harvested from salmon (Miacalcin®) is an effective drug for treating osteoporosis because it slows bone resorption.

Figure 18.14 summarizes the roles of parathyroid hormone, calcitriol, and calcitonin in regulation of blood Ca^{2+} level.

6.8 Exercise and Bone Tissue

OBJECTIVE

- **Describe** how exercise and mechanical stress affect bone tissue.

Within limits, bone tissue has the ability to alter its strength in response to changes in mechanical stress. When placed under stress, bone tissue becomes stronger through increased deposition of mineral salts and production of collagen fibers by osteoblasts. Without mechanical stress, bone does not remodel normally because bone resorption occurs more quickly than bone formation. Research has shown that high-impact intermittent strains more strongly influence bone deposition as compared with lower-impact constant strains. Therefore, running and jumping stimulate bone remodeling more dramatically than walking.

FIGURE 6.10 Negative feedback system for the regulation of blood calcium (Ca^{2+}) concentration. PTH = parathyroid hormone.

Release of calcium from bone matrix and retention of calcium by the kidneys are the two main ways that blood calcium level can be increased.

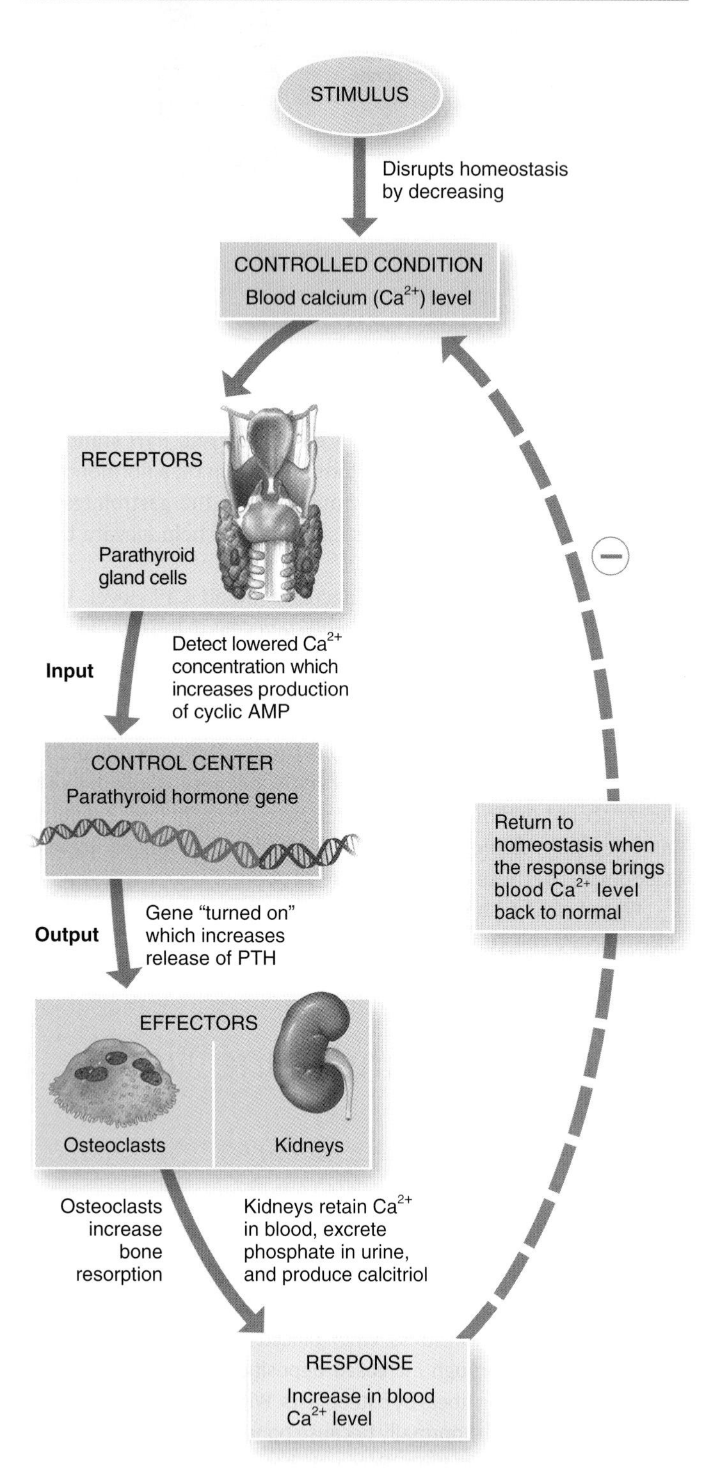

Q What body functions depend on proper levels of Ca^{2+}?

The main mechanical stresses on bone are those that result from the pull of skeletal muscles and the pull of gravity. If a person is bedridden or has a fractured bone in a cast, the strength of the unstressed bones diminishes because of the loss of bone minerals and decreased numbers of collagen fibers. Astronauts subjected to the microgravity of space also lose bone mass. In any of these cases, bone loss can be dramatic—as much as 1% per week. In contrast, the bones of athletes, which are repetitively and highly stressed, become notably thicker and stronger than those of astronauts or nonathletes. Weight-bearing activities, such as walking or moderate weight lifting, help build and retain bone mass. Adolescents and young adults should engage in regular weight-bearing exercise prior to the closure of the epiphyseal plates to help build total mass prior to its inevitable reduction with aging. However, people of all ages can and should strengthen their bones by engaging in weight-bearing exercise.

6.9 Aging and Bone Tissue

OBJECTIVE

- **Describe** the effects of aging on bone tissue.

From birth through adolescence, more bone tissue is produced than is lost during bone remodeling. In young adults, the rates of bone deposition and resorption are about the same. As the level of sex hormones diminishes during middle age, especially in women after menopause, a decrease in bone mass occurs because bone resorption by osteoclasts outpaces bone deposition by osteoblasts. In old age, loss of bone through resorption occurs more rapidly than bone gain. Because women's bones generally are smaller and less massive than men's bones to begin with, loss of bone mass in old age typically has a greater adverse effect in females. These factors contribute to the higher incidence of osteoporosis in females.

There are two principal effects of aging on bone tissue: loss of bone mass and brittleness. Loss of bone mass results from **demineralization** (dē-min′-er-al-i-ZĀ-shun), the loss of calcium and other minerals from bone extracellular matrix. This loss usually begins after age 30 in females, accelerates greatly around age 45 as levels of estrogens decrease, and continues until as much as 30% of the calcium in bones is lost by age 70. Once bone loss begins in females, about 8% of bone mass is lost every 10 years. In males, calcium loss typically does not begin until after age 60, and about 3% of bone mass is lost every 10 years. The loss of calcium from bones is one of the problems in osteoporosis (see Disorders section in Study Guide).

The second principal effect of aging on the skeletal system, brittleness, results from a decreased rate of protein synthesis. Recall that the organic part of bone extracellular matrix, mainly collagen fibers, gives bone its tensile strength. The loss of tensile strength causes the bones to become very brittle and susceptible to fracture. In some elderly people, collagen fiber synthesis slows, in part due to diminished production of growth hormone. In addition to increasing the susceptibility to fractures, loss of bone mass also leads to deformity, pain, loss of height, and loss of teeth.

Table 6.2 summarizes the factors that influence bone metabolism.

TABLE 6.2 Summary of Factors That Affect Bone Growth

FACTOR	COMMENT
MINERALS	
Calcium and phosphorus	Make bone extracellular matrix hard.
Magnesium	Helps form bone extracellular matrix.
Fluoride	Helps strengthen bone extracellular matrix.
Manganese	Activates enzymes involved in synthesis of bone extracellular matrix.
VITAMINS	
Vitamin A	Needed for the activity of osteoblasts during remodeling of bone; deficiency stunts bone growth; toxic in high doses.
Vitamin C	Needed for synthesis of collagen, the main bone protein; deficiency leads to decreased collagen production, which slows down bone growth and delays repair of broken bones.
Vitamin D	Active form (calcitriol) is produced by the kidneys; helps build bone by increasing absorption of calcium from gastrointestinal tract into blood; deficiency causes faulty calcification and slows down bone growth; may reduce the risk of osteoporosis but is toxic if taken in high doses. People who have minimal exposure to ultraviolet rays or do not take vitamin D supplements may not have sufficient vitamin D to absorb calcium. This interferes with calcium metabolism.
Vitamins K and B_{12}	Needed for synthesis of bone proteins; deficiency leads to abnormal protein production in bone extracellular matrix and decreased bone density.
HORMONES	
Growth hormone (GH)	Secreted by the anterior lobe of the pituitary gland; promotes general growth of all body tissues, including bone, mainly by stimulating production of insulin-like growth factors.
Insulin-like growth factors (IGFs)	Secreted by the liver, bones, and other tissues on stimulation by growth hormone; promotes normal bone growth by stimulating osteoblasts and by increasing the synthesis of proteins needed to build new bone.
Thyroid hormones (T_3 and T_4)	Secreted by thyroid gland; promote normal bone growth by stimulating osteoblasts.
Insulin	Secreted by the pancreas; promotes normal bone growth by increasing the synthesis of bone proteins.
Sex hormones (estrogens and testosterone)	Secreted by the ovaries in women (estrogens) and by the testes in men (testosterone); stimulate osteoblasts and promote the sudden "growth spurt" that occurs during the teenage years; shut down growth at the epiphyseal plates around age 18–21, causing lengthwise growth of bone to end; contribute to bone remodeling during adulthood by slowing bone resorption by osteoclasts and promoting bone deposition by osteoblasts.
Parathyroid hormone (PTH)	Secreted by the parathyroid glands; promotes bone resorption by osteoclasts; enhances recovery of calcium ions from urine; promotes formation of the active form of vitamin D (calcitriol).
Calcitonin (CT)	Secreted by the thyroid gland; inhibits bone resorption by osteoclasts.
EXERCISE	Weight-bearing activities stimulate osteoblasts and, consequently, help build thicker, stronger bones and retard loss of bone mass that occurs as people age.
AGING	As the level of sex hormones diminishes during middle age to older adulthood, especially in women after menopause, bone resorption by osteoclasts outpaces bone deposition by osteoblasts, which leads to a decrease in bone mass and an increased risk of osteoporosis.

Review Questions

6.1 Functions of Bone and the Skeletal System

1. How does the skeletal system function in support, protection, movement, and storage of minerals?
2. Describe the role of bones in blood cell production.
3. Which bones contain red bone marrow?
4. How do red bone marrow and yellow bone marrow differ in composition and function?

6.2 Structure of Bone

5. Diagram the parts of a long bone, and list the functions of each part.

6.3 Histology of Bone Tissue

6. Why is bone considered a connective tissue?
7. What factors contribute to the hardness and tensile strength of bone?
8. List the four types of cells in bone tissue and their functions.
9. What is the composition of the extracellular matrix of bone tissue?
10. How are compact and spongy bone tissues different in microscopic appearance, location, and function?
11. What is a bone scan and how is it used clinically?

6.4 Blood and Nerve Supply of Bone

12. Explain the location and roles of the nutrient arteries, nutrient foramina, epiphyseal arteries, and periosteal arteries.
13. Which part of a bone contains sensory nerves associated with pain?
14. Describe one situation in which these sensory neurons are important.
15. How is a bone marrow needle biopsy performed? What conditions are diagnosed through this procedure?

6.5 Bone Formation

16. What are the major events of intramembranous ossification and endochondral ossification, and how are they different?
17. Describe the zones of the epiphyseal (growth) plate and their functions, and the significance of the epiphyseal line.
18. Explain how bone growth in length differs from bone growth in thickness.
19. How could the metaphyseal area of a bone help determine the age of a skeleton?
20. Define remodeling, and describe the roles of osteoblasts and osteoclasts in the process.
21. What factors affect bone growth and bone remodeling?

6.6 Fracture and Repair of Bone

22. List the types of fractures and outline the four steps involved in fracture repair.
23. Define each of the common fractures.

6.7 Bone's Role in Calcium Homeostasis

24. How do hormones act on bone to regulate calcium homeostasis?

6.8 Exercise and Bone Tissue

25. How do mechanical stresses strengthen bone tissue?
26. Would children raised in space ever be able to return to Earth?
27. Why is it important to engage in weight-bearing exercises before the epiphyseal plates close?

6.9 Aging and Bone Tissue

28. What is demineralization, and how does it affect the functioning of bone?
29. What changes occur in the organic part of bone extracellular matrix with aging?

Critical Thinking Questions

1. Taryn is a high school senior who is undergoing a strenuous running regimen for several hours a day in order to qualify for her state high school track meet. Lately she has experienced intense pain in her right leg that is hindering her workouts. Her physician performs an examination of her right leg. The doctor doesn't notice any outward evidence of injury; he then orders a bone scan. What does her doctor suspect the problem is?
2. While playing basketball, nine-year-old Marcus fell and broke his left arm. The arm was placed in a cast and appeared to heal normally. As an adult, Marcus was puzzled because it seemed that his right arm was longer than his left arm. He measured both arms and he was correct—his right arm is longer! How would you explain to Marcus what happened?
3. Astronauts in space exercise as part of their daily routine, yet they still have problems with bone weakness after prolonged stays in space. Why does this happen?

The Skeletal System: The Axial Skeleton

CHAPTER 7

The Axial Skeleton and Homeostasis

The bones of the axial skeleton contribute to homeostasis by protecting many of the body's organs such as the brain, spinal cord, heart, and lungs. They are also important in support and calcium storage and release.

Without bones, you could not survive. You would be unable to perform movements such as walking or grasping, and the slightest blow to your head or chest could damage your brain or heart. Because the skeletal system forms the framework of the body, a familiarity with the names, shapes, and positions of individual bones will help you locate and name many other anatomical features. For example, the radial artery, the site where the pulse is usually taken, is named for its closeness to the radius, the lateral bone of the forearm. The ulnar nerve is named for its proximity to the ulna, the medial bone of the forearm. The frontal lobe of the brain lies deep to the frontal (forehead) bone. The tibialis anterior muscle lies along the anterior surface of the tibia (shin bone). Parts of certain bones also serve to locate structures within the skull and to outline the lungs, heart, and abdominal and pelvic organs.

Q Did you ever wonder what causes people to become measurably shorter as they age?

7.1 Divisions of the Skeletal System

OBJECTIVE

- **Describe** how the skeleton is organized into axial and appendicular divisions.

Movements such as throwing a ball, biking, and walking require interactions between bones and muscles. To understand how muscles produce different movements, from high fives to three-point shots, you will need to learn where the muscles attach on individual bones and what types of joints are involved. Together, the bones, muscles, and joints form an integrated system called the **musculoskeletal system**. The branch of medical science concerned with the prevention or correction of disorders of the musculoskeletal system is called **orthopedics** (or′-thō-PĒ-diks; *ortho-* = correct; *-pedi* = child).

The adult human skeleton consists of 206 named bones, most of which are paired, with one member of each pair on the right and left sides of the body. The skeletons of infants and children have more than 206 bones because some of their bones fuse later in life. Examples are the hip bones and some bones (sacrum and coccyx) of the vertebral column (backbone).

Bones of the adult skeleton are grouped into two principal divisions: the **axial skeleton** and the **appendicular skeleton** (*appendic-* = to hang onto). Table 7.1 presents the 80 bones of the axial skeleton and the 126 bones of the appendicular skeleton. Figure 7.1 shows how both divisions join to form the complete skeleton (the bones of the axial skeleton are shown in blue). You can remember the names of the divisions if you think of the axial skeleton as consisting of the bones that lie around the longitudinal *axis* of the human body, an imaginary vertical line that runs through the body's center of gravity from the head to the space between the feet: skull bones, auditory ossicles (ear bones), hyoid bone (see Figure 7.5), ribs, sternum (breastbone), and bones of the vertebral column. The appendicular skeleton consists of the bones of the **upper** and **lower limbs** (*extremities* or *appendages*), plus the bones forming the **girdles** that connect the limbs to the axial skeleton. Functionally, the auditory ossicles in the middle ear, which vibrate in response to sound waves that strike the eardrum, are not part of either the axial or appendicular skeleton, but they are grouped with the axial skeleton for convenience (see Chapter 17).

We will organize our study of the skeletal system around the two divisions of the skeleton, with emphasis on how the many bones of the body are interrelated. In this chapter we focus on the axial skeleton, looking first at the skull and then at the bones of the vertebral column and the chest. In Chapter 8 we explore the appendicular skeleton, examining in turn the bones of the pectoral (shoulder) girdle and upper limbs, and then the pelvic (hip) girdle and the lower limbs. Before we examine the axial skeleton, we direct your attention to some general characteristics of bones.

TABLE 7.1 The Bones of the Adult Skeletal System

DIVISION OF THE SKELETON	STRUCTURE	NUMBER OF BONES
Axial skeleton	**Skull**	
	Cranium	8
	Face	14
	Hyoid bone	1
	Auditory ossicles (see Figure 7.5)	6
	Vertebral column	26
	Thorax	
	Sternum	1
	Ribs	24
	Number of bones =	**80**
Appendicular skeleton	**Pectoral** (*shoulder*) **girdles**	
	Clavicle	2
	Scapula	2
	Upper limbs	
	Humerus	2
	Ulna	2
	Radius	2
	Carpals	16
	Metacarpals	10
	Phalanges	28
	Pelvic (*hip*) **girdle**	
	Hip, pelvic, or coxal bone	2
	Lower limbs	
	Femur	2
	Patella	2
	Fibula	2
	Tibia	2
	Tarsals	14
	Metatarsals	10
	Phalanges	28
	Number of bones =	**126**
	Total bones in an adult skeleton =	**206**

7.2 Types of Bones

OBJECTIVE

- **Classify** bones based on their shape or location.

Almost all bones of the body can be classified into five main types based on shape: long, short, flat, irregular, and sesamoid (Figure 7.2). As you learned in Chapter 6, **long bones** have greater length than width, consist of a shaft and a variable number of extremities or epiphyses (ends), and are slightly curved for strength. A curved bone absorbs the stress of the body's weight at several different points, so that it is evenly distributed. If bones were straight, the weight of the body would be unevenly distributed, and the bone would fracture more easily. Long bones consist mostly of *compact bone tissue* in their diaphyses but have considerable amounts of *spongy bone tissue* in their epiphyses. Long bones vary tremendously in size and include those in the femur (thigh bone), tibia and fibula (leg bones), humerus (arm bone), ulna and radius (forearm bones), and phalanges (finger and toe bones).

Short bones are somewhat cube-shaped and are nearly equal in length and width. They consist of spongy bone tissue except at the surface, which has a thin layer of compact bone tissue. Examples of short bones are most carpal (wrist) bones and most tarsal (ankle) bones.

Flat bones are generally thin and composed of two nearly parallel plates of compact bone tissue enclosing a layer of spongy bone tissue. Flat bones afford considerable protection and provide extensive areas for muscle attachment. Flat bones include the cranial bones, which protect the brain; the sternum (breastbone) and ribs, which protect organs in the thorax; and the scapulae (shoulder blades).

Irregular bones have complex shapes and cannot be grouped into any of the previous categories. They vary in the amount of spongy and compact bone present. Such bones include the vertebrae (backbones), hip bones, certain facial bones, and the calcaneus.

Sesamoid bones (SES-a-moyd = shaped like a sesame seed) develop in certain tendons where there is considerable friction, tension, and physical stress, such as the palms and soles. They may vary in number from person to person, are not always completely ossified, and typically measure only a few millimeters in diameter. Notable exceptions are the two patellae (kneecaps), large sesamoid bones located in the quadriceps femoris tendon (see Figure 11.20a) that are normally present in everyone. Functionally, sesamoid bones protect tendons from excessive wear and tear, and they often change the direction of pull of a tendon, which improves the mechanical advantage at a joint.

An additional type of bone is classified by location rather than shape. **Sutural bones** (SOO-chur-al; *sutur-* = seam) are small bones located in sutures (joints) between certain cranial bones (see Figure 7.6). Their number varies greatly from person to person.

Recall from Chapter 6 that in adults, red bone marrow is restricted to flat bones such as the ribs, sternum (breastbone), and skull; irregular bones such as vertebrae (backbones) and hip bones; long bones such as the proximal epiphyses of the femur (thigh bone) and humerus (arm bone); and some short bones.

FIGURE 7.1 Divisions of the skeletal system. The axial skeleton is indicated in blue. (Note the position of the hyoid bone in **Figure 7.5**.)

The adult human skeleton consists of 206 bones grouped into two divisions: the axial skeleton and the appendicular skeleton.

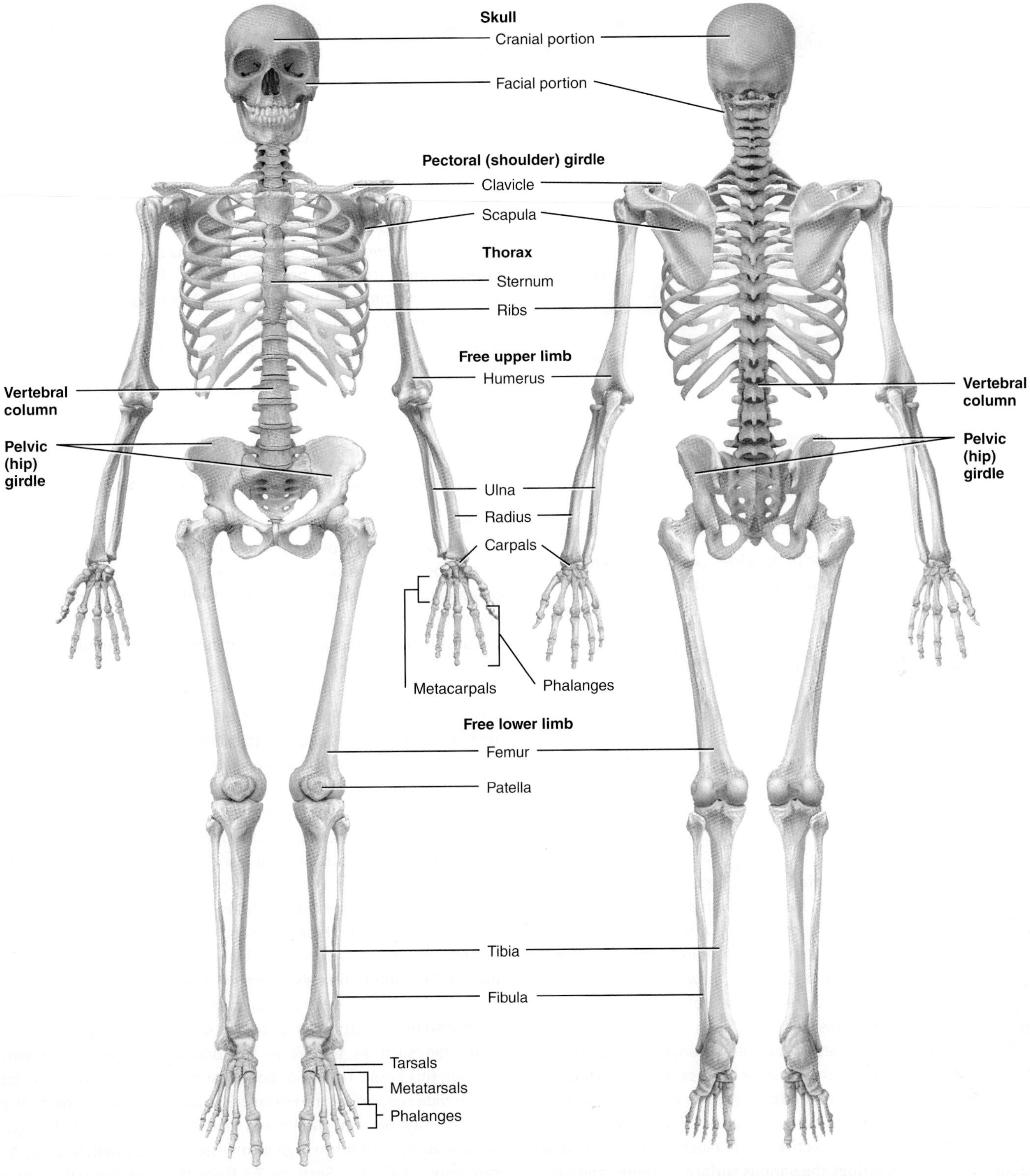

(a) Anterior view

(b) Posterior view

Q Which of the following structures are part of the axial skeleton, and which are part of the appendicular skeleton? Skull, clavicle, vertebral column, shoulder girdle, humerus, pelvic girdle, and femur.

FIGURE 7.2 **Types of bones based on shape.** The bones are not drawn to scale.

The shapes of bones largely determine their functions.

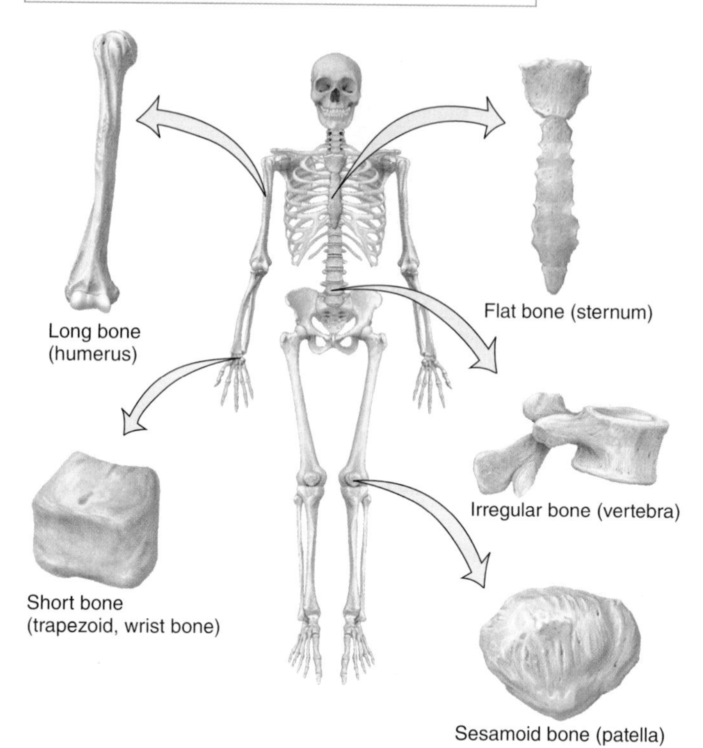

Q Which type of bone primarily provides protection and a large surface area for muscle attachment?

7.3 Bone Surface Markings

OBJECTIVE

- **Describe** the principal surface markings on bones and the functions of each.

Bones have characteristic **surface markings**, structural features adapted for specific functions. Most are not present at birth but develop in response to certain forces and are most prominent in the adult skeleton. In response to tension on a bone surface from tendons, ligaments, aponeuroses, and fasciae, new bone is deposited, resulting in raised or roughened areas. Conversely, compression on a bone surface results in a depression.

There are two major types of surface markings: (1) *depressions and openings*, which allow the passage of soft tissues (such as blood vessels, nerves, ligaments, and tendons) or form joints, and (2) *processes*, projections or outgrowths that either help form joints or serve as attachment points for connective tissue (such as ligaments and tendons). Table 7.2 describes the various surface markings and provides examples of each.

7.4 Skull: An Overview

OBJECTIVE

- **Name** the cranial and facial bones and indicate whether they are paired or single.

Components of the Skull

The **skull** is the bony framework of the head. It contains 22 bones (not counting the bones of the middle ears) and rests on the superior end of the vertebral column (backbone). The bones of the skull are grouped into two categories: cranial bones and facial bones. The **cranial bones** (*crani-* = brain case) form the cranial cavity, which encloses and protects the brain. The eight cranial bones are the frontal bone, two parietal bones, two temporal bones, the occipital bone, the sphenoid bone, and the ethmoid bone. Fourteen **facial bones** form the face: two nasal bones, two maxillae (or maxillas), two zygomatic bones, the mandible, two lacrimal bones, two palatine bones, two inferior nasal conchae, and the vomer.

General Features and Functions of the Skull

Besides forming the large cranial cavity, the skull also forms several smaller cavities, including the nasal cavity and orbits (eye sockets), which open to the exterior. Certain skull bones also contain cavities called paranasal sinuses that are lined with mucous membranes and open into the nasal cavity. Also within the skull are small middle ear cavities in the temporal bones that house the structures that are involved in hearing and equilibrium (balance).

Other than the auditory ossicles (tiny bones involved in hearing), which are located within the temporal bones, the mandible is the only movable bone of the skull. Joints called sutures attach most of the skull bones together and are especially noticeable on the outer surface of the skull.

The skull has many surface markings, such as foramina (rounded passageways) and fissures (slitlike openings) through which blood vessels and nerves pass. You will learn the names of important skull bone surface markings as we describe each bone.

In addition to protecting the brain, the cranial bones stabilize the positions of the brain, blood vessels, lymphatic vessels, and nerves through the attachment of their inner surfaces to meninges (membranes). The outer surfaces of cranial bones provide large areas of attachment for muscles that move various parts of the head. The bones also provide attachment for some muscles that produce facial expression such as the frown of concentration you wear when studying this book. The facial bones form the framework of the face and provide support for the entrances to the digestive and respiratory systems. Together, the cranial and facial bones protect and support the delicate special sense organs for vision, taste, smell, hearing, and equilibrium (balance). Sections 7.5 through 7.7 describe the various bones that comprise the skull.

TABLE 7.2 Bone Surface Markings

MARKING	DESCRIPTION	EXAMPLE
DEPRESSIONS AND OPENINGS: SITES ALLOWING THE PASSAGE OF SOFT TISSUE (NERVES, BLOOD VESSELS, LIGAMENTS, TENDONS) OR FORMATION OF JOINTS		
Fissure (FISH-ur)	Narrow slit between adjacent parts of bones through which blood vessels or nerves pass.	Superior orbital fissure of sphenoid bone (**Figure 7.12**).
Foramen (fō-RĀ-men = hole; plural is *foramina*, fō-RĀM-i-na)	Opening through which blood vessels, nerves, or ligaments pass.	Optic foramen of sphenoid bone (**Figure 7.12**).
Fossa (FOS-a = trench; plural is *fossae*, FOS-ē)	Shallow depression.	Coronoid fossa of humerus (**Figure 8.4a**).
Sulcus (SUL-kus = groove; plural is sulci, SUL-sī)	Furrow along bone surface that accommodates blood vessel, nerve, or tendon.	Intertubercular sulcus of humerus (**Figure 8.4a**).
Meatus (mē-Ā-tus = passageway; plural is *meati*, mē-Ā-tī)	Tubelike opening.	External auditory meatus of temporal bone (**Figure 7.4a**).
PROCESSES: PROJECTIONS OR OUTGROWTHS ON BONE THAT FORM JOINTS OR ATTACHMENT POINTS FOR CONNECTIVE TISSUE, SUCH AS LIGAMENTS AND TENDONS		
Processes that form joints		
Condyle (KON-dīl; *condylus* = knuckle)	Large, round protuberance with a smooth articular surface at end of bone.	Lateral condyle of femur (**Figure 8.11a**).
Facet (FAS-et or fa-SET)	Smooth, flat, slightly concave or convex articular surface.	Superior articular facet of vertebra (**Figure 7.18d**).
Head	Usually rounded articular projection supported on neck (constricted portion) of bone.	Head of femur (**Figure 8.11a**).
Processes that form attachment points for connective tissue		
Crest	Prominent ridge or elongated projection.	Iliac crest of hip bone (**Figure 8.9b**).
Epicondyle (*epi-* = above)	Typically roughened projection above condyle.	Medial epicondyle of femur (**Figure 8.11a**).
Line (*linea*)	Long, narrow ridge or border (less prominent than crest).	Linea aspera of femur (**Figure 8.11b**).
Spinous process	Sharp, slender projection.	Spinous process of vertebra (**Figure 7.17**).
Trochanter (trō-KAN-ter)	Very large projection.	Greater trochanter of femur (**Figure 8.11b**).
Tubercle (TOO-ber-kul; *tuber-* = knob)	Variably sized rounded projection.	Greater tubercle of humerus (**Figure 8.4a**).
Tuberosity	Variably sized projection that has a rough, bumpy surface.	Ischial tuberosity of hip bone (**Figure 8.9b**).

7.5 Cranial Bones

OBJECTIVE

- **Describe** the following cranial bones and their main features: frontal, parietal, temporal, occipital, sphenoid, and ethmoid.

Frontal Bone

The **frontal bone** forms the forehead (the anterior part of the cranium), the roofs of the *orbits* (eye sockets), and most of the anterior part of the cranial floor (**Figure 7.3**). Soon after birth, the left and right sides of the frontal bone are united by the *metopic suture*, which usually disappears between the ages of six and eight.

Note the *frontal squama*, a scalelike plate of bone that forms the forehead of the skull (**Figure 7.3**). It gradually slopes inferiorly from the coronal suture, on the top of the skull, then angles abruptly and becomes almost vertical above the orbits. At the superior border of the orbits, the frontal bone thickens, forming the *supraorbital margin* (*supra-* = above; *-orbi*= circle). From this margin, the frontal bone extends posteriorly to form the roof of the orbit, which is part of the floor of the cranial cavity. Within the supraorbital margin, slightly medial to its midpoint, is a hole called the *supraorbital foramen.* Sometimes the foramen is incomplete and is called the *supraorbital notch.* As you read about each foramen associated with a cranial bone, refer to **Table 7.3** to note which structures pass through it. The *frontal sinuses* lie deep to the frontal squama. Sinuses, or, more technically, paranasal sinuses, are mucous membrane–lined cavities within certain skull bones that will be discussed later.

FIGURE 7.3 Anterior view of the skull.

The skull consists of cranial bones and facial bones.

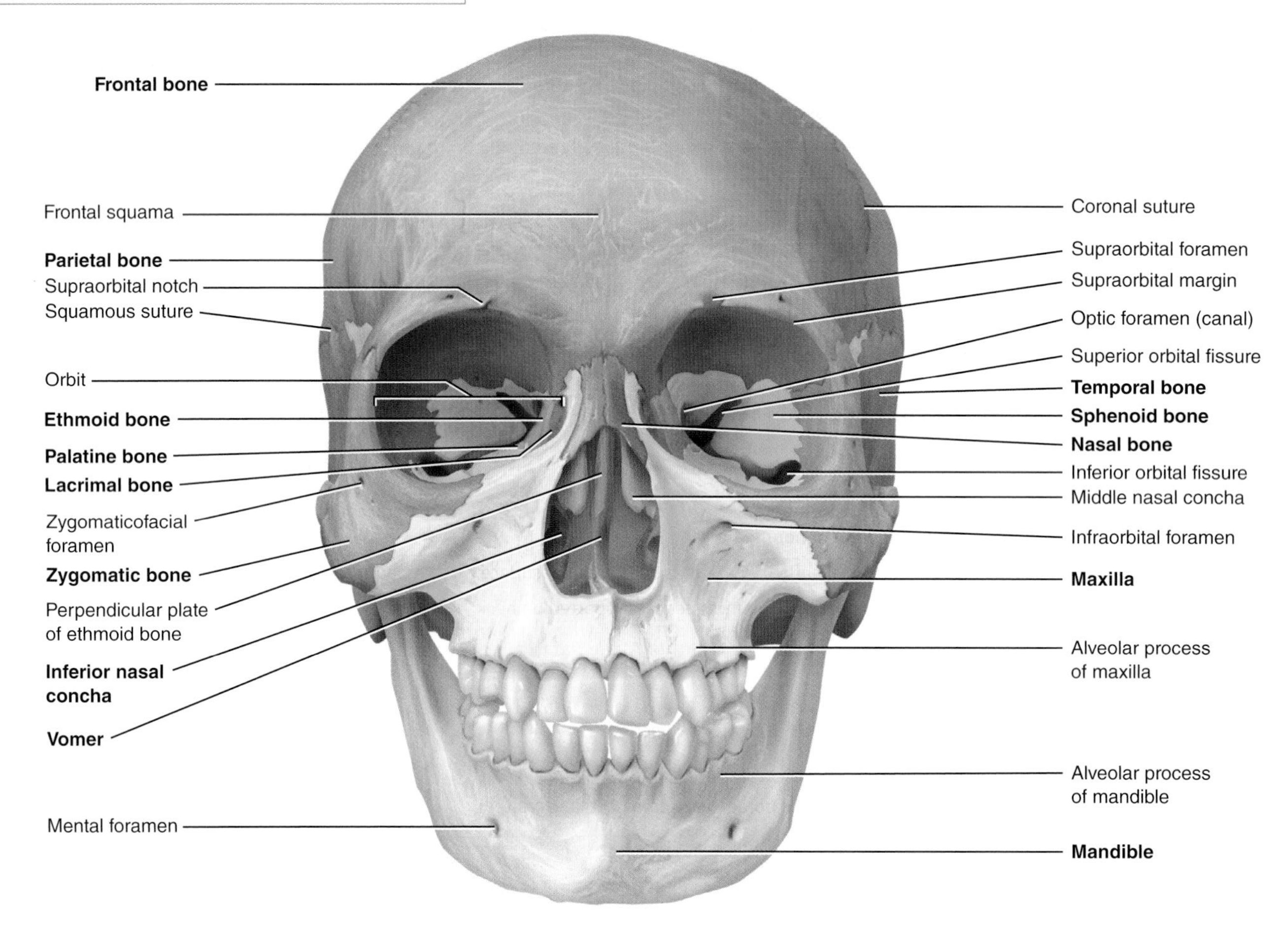

Q Which of the bones shown here are cranial bones?

See Clinical Connection: Black Eye

Parietal Bones

The two **parietal bones** (pa-RĪ-e-tal; *pariet-* = wall) form the greater portion of the sides and roof of the cranial cavity (Figure 7.4). The internal surfaces of the parietal bones contain many protrusions and depressions that accommodate the blood vessels supplying the dura mater, the superficial connective tissue (meninx) covering of the brain.

Temporal Bones

The paired **temporal bones** (*tempor-* = temple) form the inferior lateral aspects of the cranium and part of the cranial floor. In Figure 7.4a, note the *temporal squama* (= scale), the thin, flat part of the temporal bone that forms the anterior and superior part of the *temple* (the region of the cranium around the ear). Projecting from the inferior portion of the temporal squama is the *zygomatic process*, which articulates (forms a joint) with the temporal process of the zygomatic (cheek) bone. Together, the zygomatic process of the temporal bone and the temporal process of the zygomatic bone form the *zygomatic arch*.

A socket called the *mandibular fossa* is located on the inferior posterior surface of the zygomatic process of each temporal bone. Anterior to the mandibular fossa is a rounded elevation, the *articular tubercle* (Figure 7.4a). The mandibular fossa and articular tubercle articulate with the mandible (lower jawbone) to form the *temporomandibular joint (TMJ)*.

The *mastoid portion* (*mastoid* = breast-shaped; Figure 7.4a) of the temporal bone is located posterior and inferior to the *external auditory meatus* (*meatus* = passageway), or ear canal, which directs

FIGURE 7.4 **Superior and right lateral views of the skull.**

The zygomatic arch is formed by the zygomatic process of the temporal bone and the temporal process of the zygomatic bone.

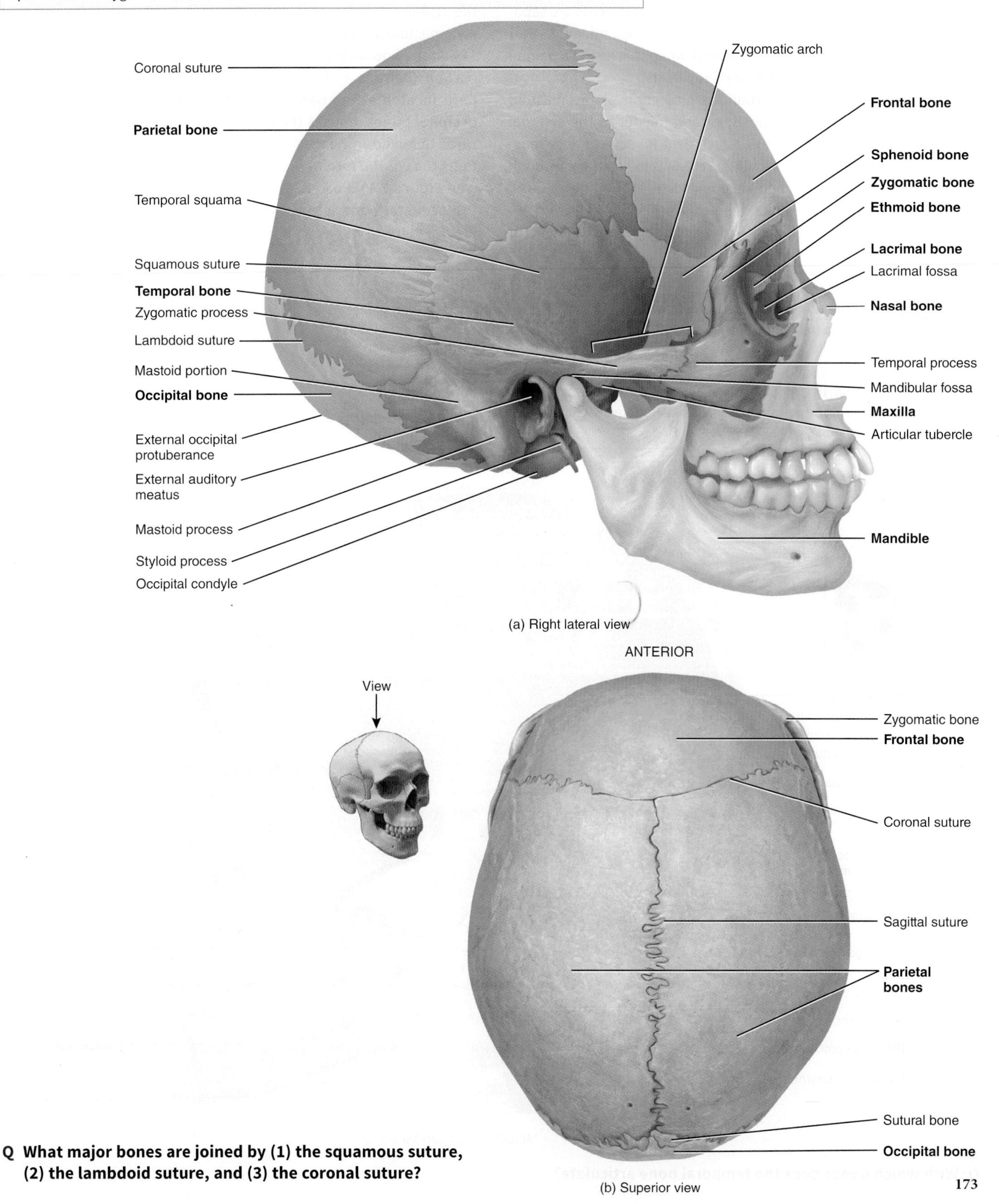

Q What major bones are joined by (1) the squamous suture, (2) the lambdoid suture, and (3) the coronal suture?

sound waves into the ear. In an adult, this portion of the bone contains several *mastoid air cells* that communicate with the hollow space of the middle ear. These tiny air-filled compartments are separated from the brain by thin bony partitions. Middle ear infections that go untreated can spread into the mastoid air cells, causing a painful inflammation called **mastoiditis** (mas′-toy-DĪ-tis).

The *mastoid process* is a rounded projection of the mastoid portion of the temporal bone posterior and inferior to the external auditory meatus. It is the point of attachment for several neck muscles. The *internal auditory meatus* (**Figure 7.5**) is the opening through which the facial (VII) nerve and vestibulocochlear (VIII) nerve pass. The *styloid process* (*styl-* = stake or pole) projects inferiorly from the inferior surface of the temporal bone and serves as a point of attachment for muscles and ligaments of the tongue and neck (see **Figure 7.4a**). Between the styloid process and the mastoid process is the *stylomastoid foramen*, through which the facial (VII) nerve and stylomastoid artery pass (see **Figure 7.7**).

At the floor of the cranial cavity (see **Figure 7.8a**) is the *petrous portion* (*petrous* = rock) of the temporal bone. This triangular part, located at the base of the skull between the sphenoid and occipital bones, houses the internal ear and the middle ear, structures involved in hearing and equilibrium (balance). It also contains the *carotid foramen*, through which the carotid artery passes (see **Figure 7.7**). Posterior to the carotid foramen and anterior to the occipital bone is the *jugular foramen*, a passageway for the jugular vein.

FIGURE 7.5 Medial view of sagittal section of the skull. Although the hyoid bone is not part of the skull, it is included here for reference. The location of the auditory ossicles (incus, malleus, and stapes) is also shown.

The cranial bones are the frontal, parietal, temporal, occipital, sphenoid, and ethmoid bones. The facial bones are the nasal bones, maxillae, zygomatic bones, lacrimal bones, palatine bones, inferior nasal conchae, mandible, and vomer.

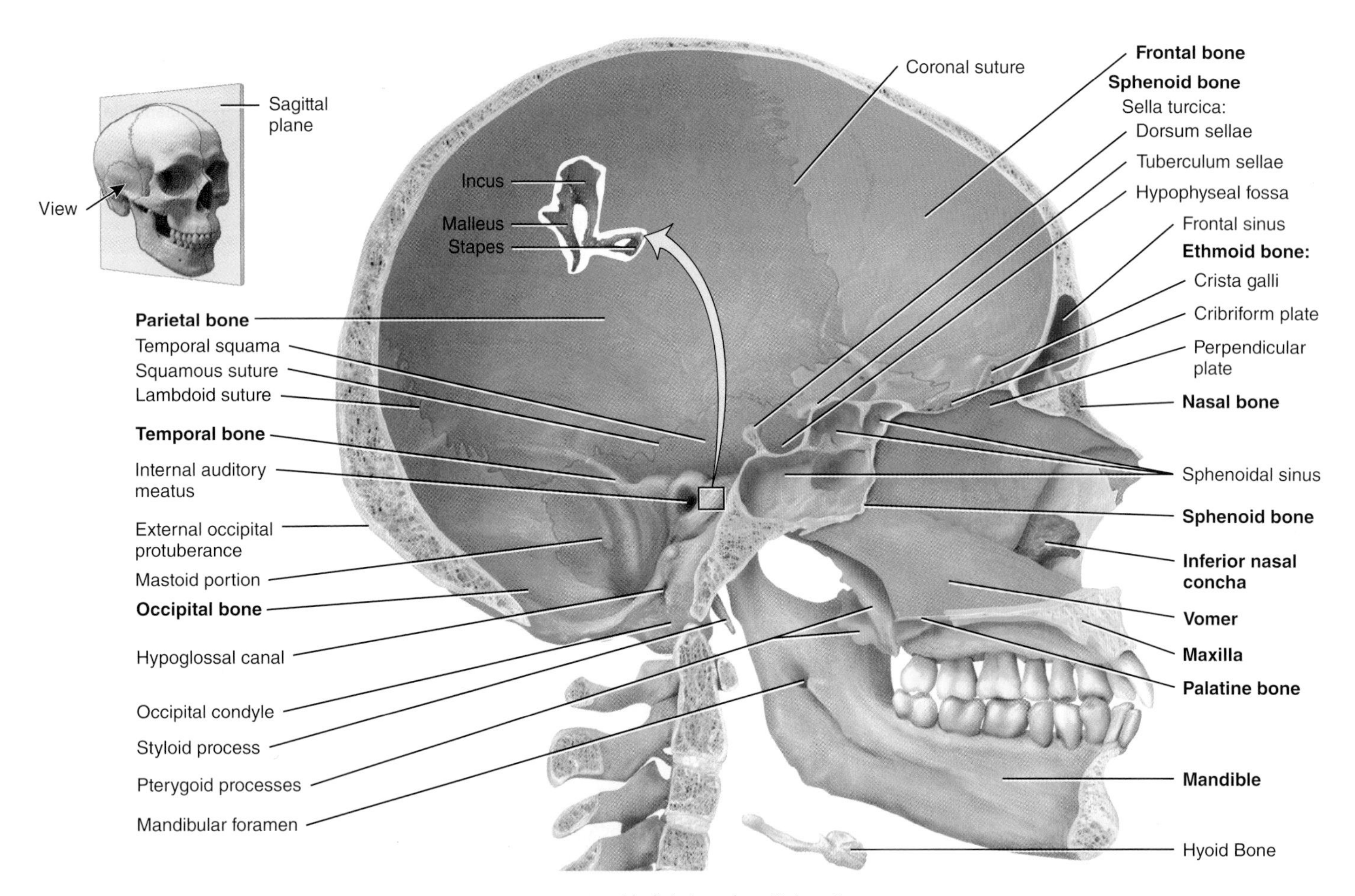

Q With which bones does the temporal bone articulate?

Occipital Bone

The **occipital bone** (ok-SIP-i-tal; *occipit-* = back of head) forms the posterior part and most of the base of the cranium (**Figure 7.6**; also see **Figure 7.4**). Also view the occipital bone and surrounding structures in the inferior view of the skull in **Figure 7.7**. The *foramen magnum* (= large hole) is in the inferior part of the bone. The medulla oblongata (inferior part of the brain) connects with the spinal cord within this foramen, and the vertebral and spinal arteries also pass through it along with the accessory (XI) nerve. The *occipital condyles*, oval processes with convex surfaces on either side of the foramen magnum (**Figure 7.7**), articulate with depressions on the first cervical vertebra (atlas) to form the *atlanto-occipital joint*, which allows you to nod your head "yes." Superior to each occipital condyle on the inferior surface of the skull is the *hypoglossal canal* (*hypo-* = under; *-glossal* = tongue). (See **Figure 7.5**.)

The *external occipital protuberance* is the most prominent midline projection on the posterior surface of the bone just above the foramen magnum. You may be able to feel this structure as a bump on the back of your head, just above your neck. (See **Figure 7.4a**.) A large fibrous, elastic ligament, the *ligamentum nuchae* (*nucha-* = nape of neck), extends from the external occipital protuberance to the seventh cervical vertebra to help support the head. Extending laterally from the protuberance are two curved ridges, the *superior nuchal lines*, and below these are two *inferior nuchal lines*, which are areas of muscle attachment (**Figure 7.7**).

FIGURE 7.6 Posterior view of the skull. The sutures are exaggerated for emphasis.

The occipital bone forms most of the posterior and inferior portions of the cranium.

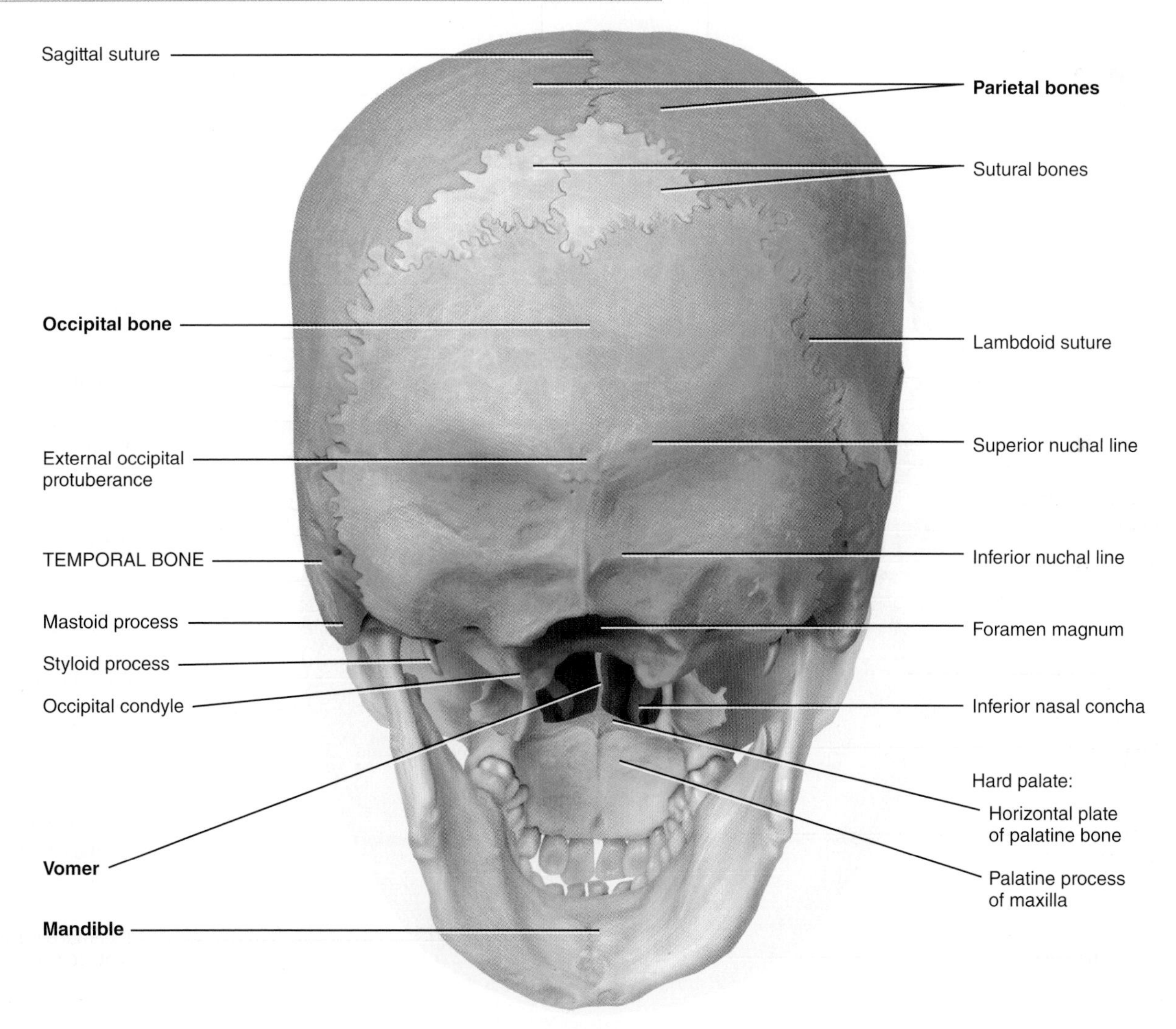

Q Which bones form the posterior, lateral portion of the cranium?

Sphenoid Bone

The **sphenoid bone** (SFĒ-noyd = wedge-shaped) lies at the middle part of the base of the skull (**Figures 7.7** and **7.8**). This bone is called the keystone of the cranial floor because it articulates with all the other cranial bones, holding them together. View the floor of the cranium superiorly (**Figure 7.8a**) and note the sphenoid articulations. The sphenoid bone joins anteriorly with the frontal and ethmoid bones, laterally with the temporal bones, and posteriorly with the occipital bone. The sphenoid lies posterior and slightly superior to the nasal cavity and forms part of the floor, side walls, and rear wall of the orbit (see **Figure 7.12**).

The shape of the sphenoid resembles a butterfly with outstretched wings (**Figure 7.8b**). The *body* of the sphenoid is the hollowed cubelike medial portion between the ethmoid and occipital bones. The space inside the body is the *sphenoidal sinus*, which drains into the nasal cavity (see **Figure 7.13**). The *sella turcica* (SEL-a TUR-si-ka; *sella* = saddle; *turcica* = Turkish) is a bony saddle-shaped

FIGURE 7.7 **Inferior view of the skull.** The mandible (lower jawbone) has been removed.

The occipital condyles of the occipital bone articulate with the first cervical vertebra to form the atlanto-occipital joint.

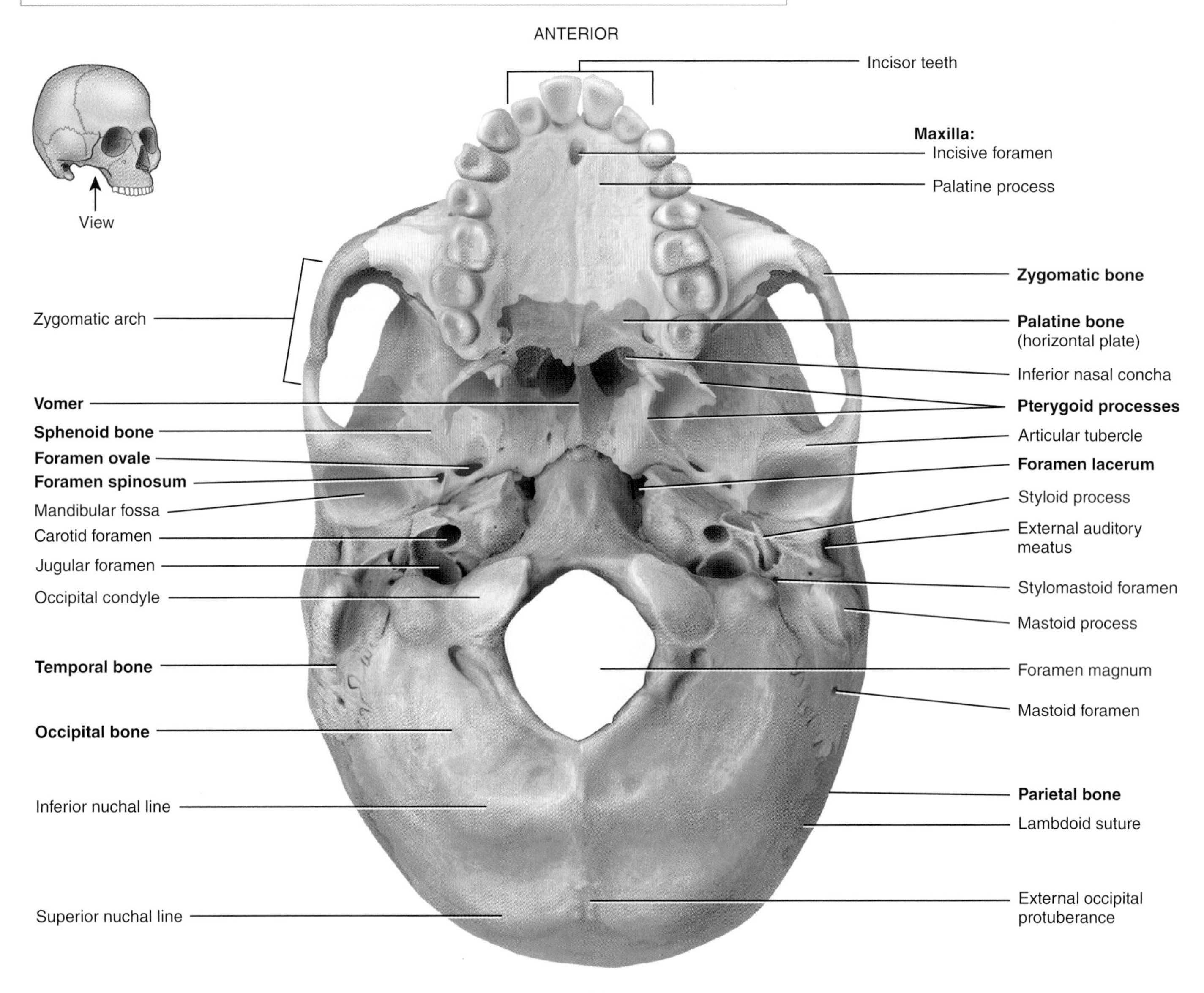

Q What organs of the nervous system join together within the foramen magnum?

FIGURE 7.8 Sphenoid bone.

The sphenoid bone is called the keystone of the cranial floor because it articulates with all other cranial bones, holding them together.

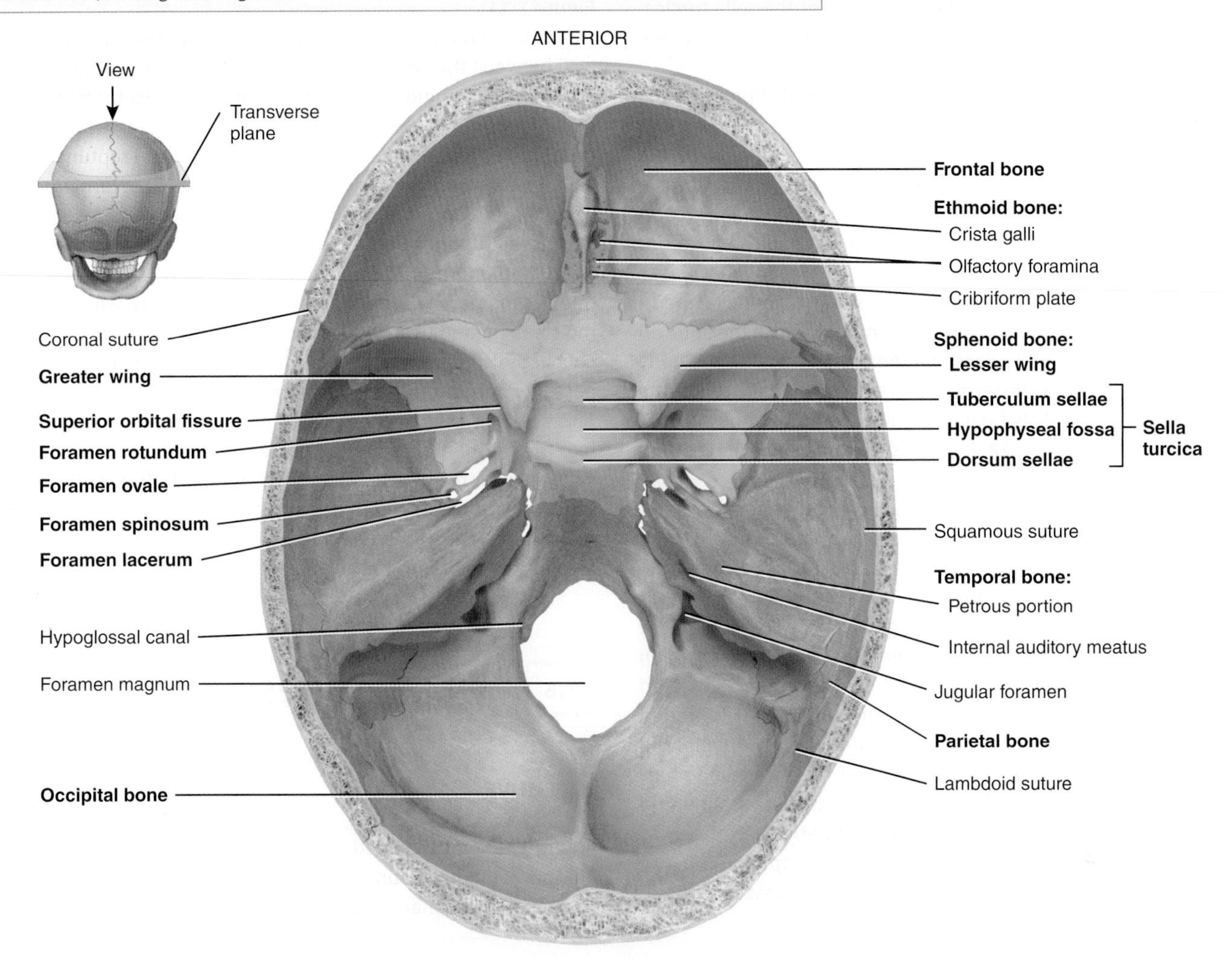

(a) Superior view of sphenoid bone in floor of cranium

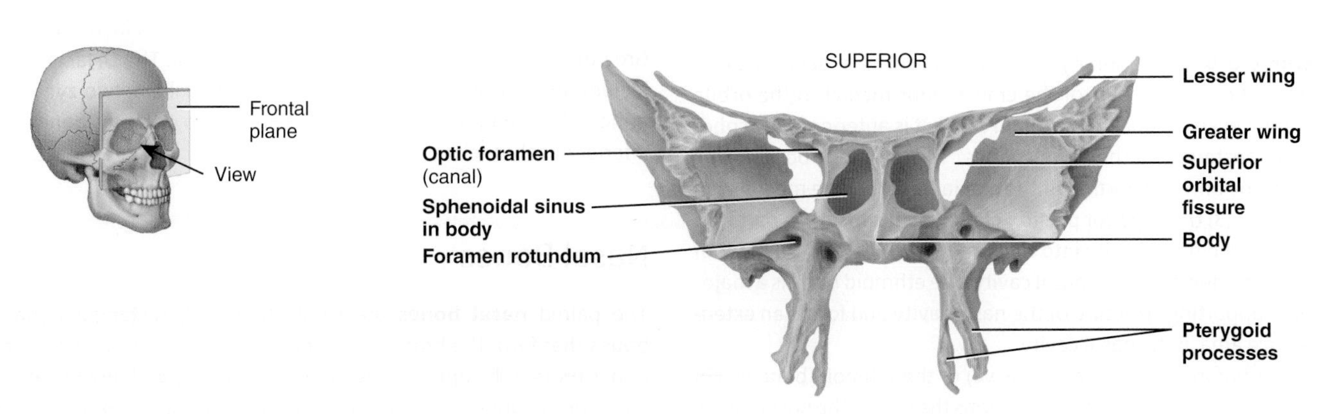

(b) Anterior view of sphenoid bone

Q Name the bones that articulate with the sphenoid bone, starting at the crista galli of the ethmoid bone and going in a clockwise direction.

structure on the superior surface of the body of the sphenoid (**Figure 7.8a**). The anterior part of the sella turcica, which forms the horn of the saddle, is a ridge called the *tuberculum sellae.* The seat of the saddle is a depression, the *hypophyseal fossa* (hī-pō-FIZ-ē-al), which contains the pituitary gland. The posterior part of the sella turcica, which forms the back of the saddle, is another ridge called the *dorsum sellae.*

The *greater wings* of the sphenoid project laterally from the body and form the anterolateral floor of the cranium. The greater wings also form part of the lateral wall of the skull just anterior to the temporal bone and can be viewed externally. The *lesser wings*, which are smaller, form a ridge of bone anterior and superior to the greater wings. They form part of the floor of the cranium and the posterior part of the orbit of the eye.

Between the body and lesser wing just anterior to the sella turcica is the *optic foramen* or *canal* (*optic* = eye), through which the optic (II) nerve and ophthalmic artery pass into the orbit. Lateral to the body between the greater and lesser wings is a triangular slit called the *superior orbital fissure.* This fissure may also be seen in the anterior view of the orbit in **Figure 7.12**. Blood vessels and cranial nerves pass through this fissure.

The *pterygoid processes* (TER-i-goyd = winglike) project inferiorly from the points where the body and greater wings of the sphenoid bone unite; they form the lateral posterior region of the nasal cavity (see **Figures 7.7** and **7.8b**). Some of the muscles that move the mandible attach to the pterygoid processes. At the base of the lateral pterygoid process in the greater wing is the *foramen ovale* (= oval hole). The *foramen lacerum* (= lacerated), covered in part by a layer of fibrocartilage in living subjects, is bounded anteriorly by the sphenoid bone and medially by the sphenoid and occipital bones. It transmits a branch of the ascending pharyngeal artery. Another foramen associated with the sphenoid bone is the *foramen rotundum* (= round hole) located at the junction of the anterior and medial parts of the sphenoid bone. The maxillary branch of the trigeminal (V) nerve passes through the foramen rotundum.

Ethmoid Bone

The **ethmoid bone** (ETH-moyd = like a sieve) is a delicate bone located in the anterior part of the cranial floor medial to the orbits and is spongelike in appearance (**Figure 7.9**). It is anterior to the sphenoid and posterior to the nasal bones. The ethmoid bone forms (1) part of the anterior portion of the cranial floor; (2) the medial wall of the orbits; (3) the superior portion of the **nasal septum**, a partition that divides the nasal cavity into right and left sides; and (4) most of the superior sidewalls of the nasal cavity. The ethmoid bone is a major superior supporting structure of the nasal cavity and forms an extensive surface area in the nasal cavity.

The *cribriform plate* (*cribri-* = sieve) of the ethmoid bone lies in the anterior floor of the cranium and forms the roof of the nasal cavity. The cribriform plate contains the *olfactory foramina* (*olfact-* = to smell) through which the olfactory nerves pass. Projecting superiorly from the cribriform plate is a triangular process called the *crista galli* (*crista* = crest; *galli* = cock), which serves as a point of attachment for the falx cerebri, the membrane that separates the two sides of the brain. Projecting inferiorly from the cribriform plate is the *perpendicular plate*, which forms the superior portion of the nasal septum (see **Figure 7.11**).

The *lateral masses* of the ethmoid bone compose most of the wall between the nasal cavity and the orbits. They contain 3 to 18 air spaces called ethmoidal cells. The ethmoidal cells together form the *ethmoidal sinuses* (see **Figure 7.13**). The lateral masses contain two thin, scroll-shaped projections lateral to the nasal septum. These are called the *superior nasal concha* (KON-ka = shell) or *turbinate* and the *middle nasal concha (turbinate).* The plural form is *conchae* (KON-kē). A third pair of conchae, the inferior nasal conchae, are separate bones (discussed shortly). The conchae greatly increase the vascular and mucous membrane surface area in the nasal cavity, which warms and moistens (humidifies) inhaled air before it passes into the lungs. The conchae also cause inhaled air to swirl; as a result, many inhaled particles become trapped in the mucus that lines the nasal cavity. This action of the conchae helps cleanse inhaled air before it passes into the rest of the respiratory passageways. The superior nasal conchae are near the olfactory foramina of the cribriform plate where the sensory receptors for olfaction (smell) terminate in the mucous membrane of the superior nasal conchae. Thus, they increase the surface area for the sense of smell.

7.6 Facial Bones

OBJECTIVE

- **Identify** the location and surface features of the following bones: nasal, lacrimal, palatine, inferior nasal conchae, vomer, maxillae, zygomatic, and mandible.

The shape of the face changes dramatically during the first two years after birth. The brain and cranial bones expand, the first set of teeth form and erupt (emerge), and the paranasal sinuses increase in size. Growth of the face ceases at about 16 years of age. The 14 facial bones include two nasal bones, two maxillae (or maxillas), two zygomatic bones, the mandible, two lacrimal bones, two palatine bones, two inferior nasal conchae, and the vomer.

Nasal Bones

The paired **nasal bones** are small, flattened, rectangular-shaped bones that form the bridge of the nose (see **Figure 7.3**). These small bones protect the upper entry to the nasal cavity and provide attachment for a couple of thin muscles of facial expression. For those of you who wear glasses, they are the bones that form the resting place for the bridge of the glasses. The major structural portion of the nose consists of cartilage.

FIGURE 7.9 **Ethmoid bone.**

The ethmoid bone forms part of the anterior portion of the cranial floor, the medial wall of the orbits, the superior portions of the nasal septum, and most of the side walls of the nasal cavity.

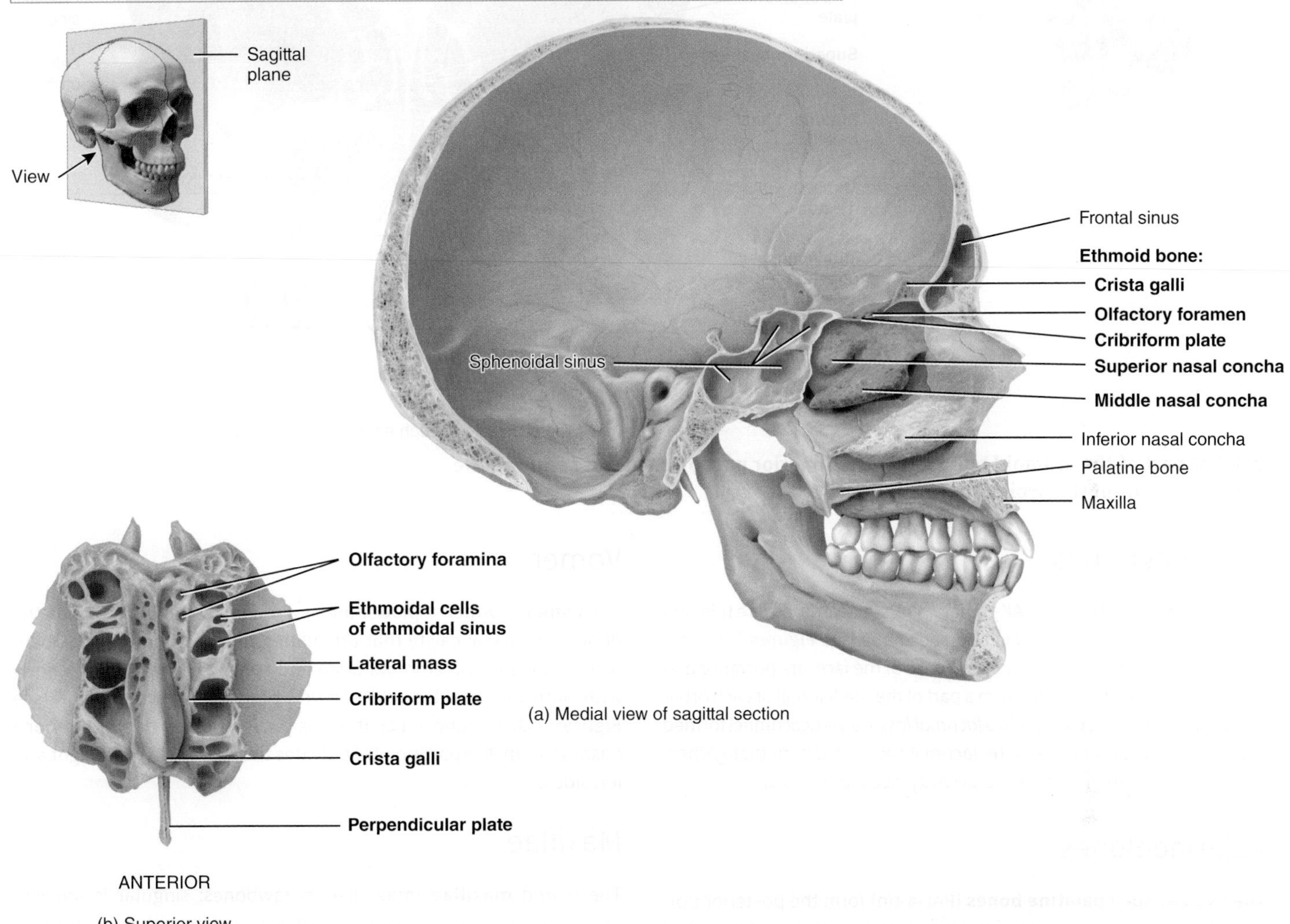

(a) Medial view of sagittal section

(b) Superior view

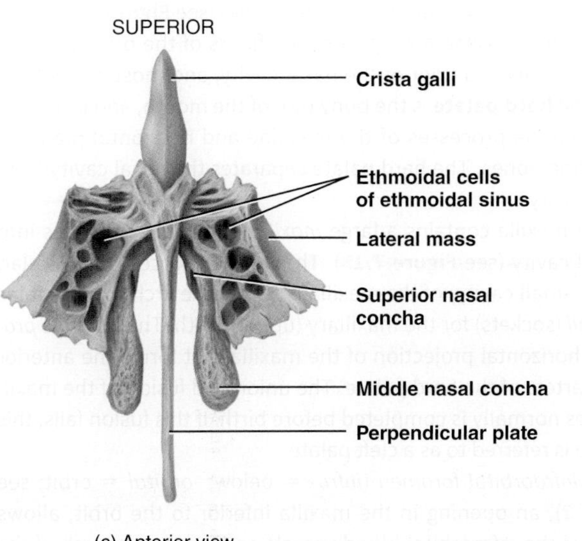

(c) Anterior view

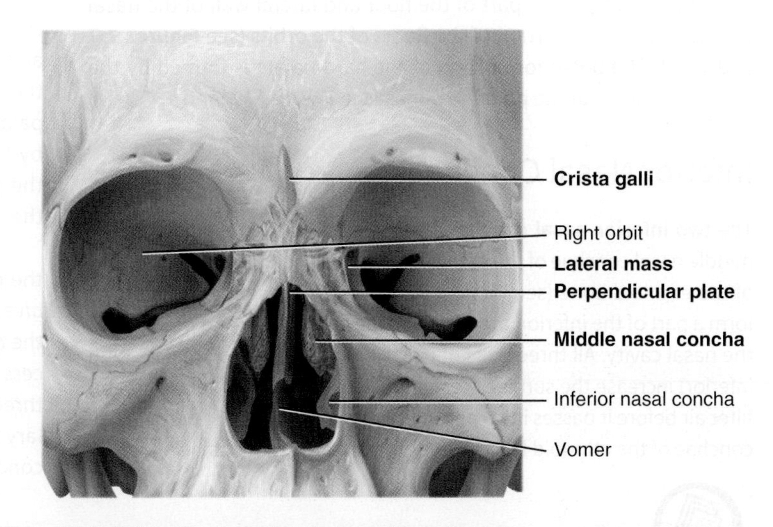

(d) Anterior view of position of ethmoid bone in skull (projected to the surface)

Figure 7.9 *Continues*

FIGURE 7.9 Continued

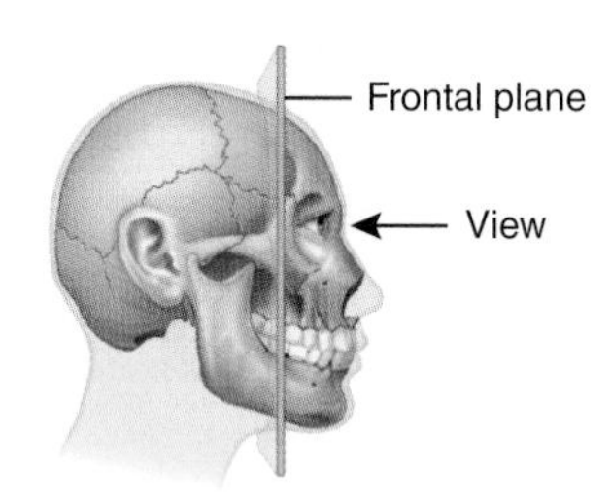

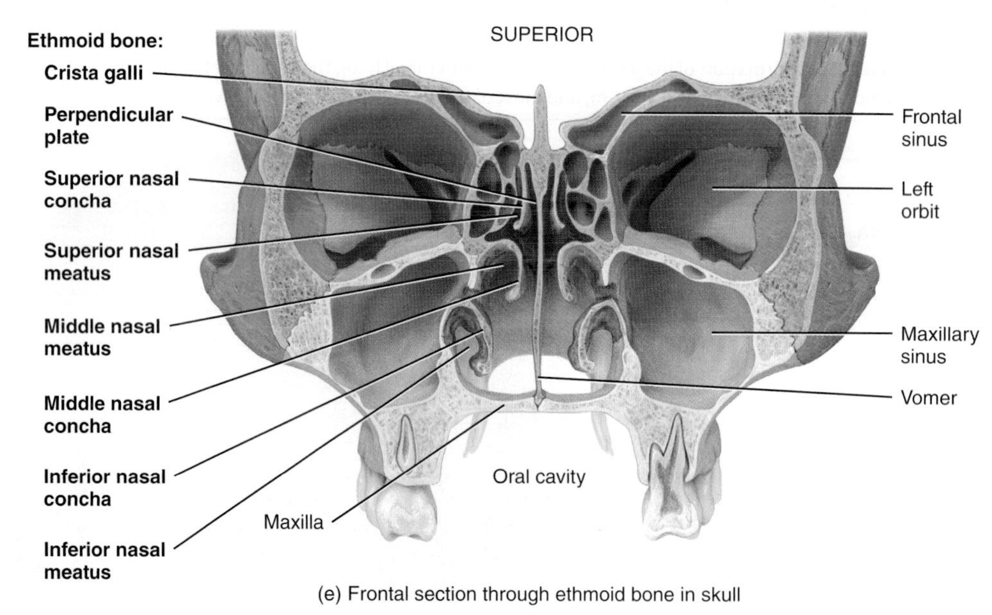

(e) Frontal section through ethmoid bone in skull

Q What part of the ethmoid bone forms the superior part of the nasal septum? The medial walls of the orbits?

Lacrimal Bones

The paired **lacrimal bones** (LAK-ri-mal; *lacrim-* = teardrops) are thin and roughly resemble a fingernail in size and shape (see Figures 7.3, 7.4a, and 7.12). These bones, the smallest bones of the face, are posterior and lateral to the nasal bones and form a part of the medial wall of each orbit. The lacrimal bones each contain a *lacrimal fossa*, a vertical tunnel formed with the maxilla, that houses the lacrimal sac, a structure that gathers tears and passes them into the nasal cavity (see Figure 7.12).

Palatine Bones

The two L-shaped **palatine bones** (PAL-a-tīn) form the posterior portion of the hard palate, part of the floor and lateral wall of the nasal cavity, and a small portion of the floors of the orbits (see Figures 7.7 and 7.12). The posterior portion of the hard palate is formed by the *horizontal plates* of the palatine bones (see Figures 7.6 and 7.7).

Inferior Nasal Conchae

The two **inferior nasal conchae** (*turbinates*), which are inferior to the middle nasal conchae of the ethmoid bone, are separate bones, not part of the ethmoid bone (see Figures 7.3 and 7.9). These scroll-like bones form a part of the inferior lateral wall of the nasal cavity and project into the nasal cavity. All three pairs of nasal conchae (superior, middle, and inferior) increase the surface area of the nasal cavity and help swirl and filter air before it passes into the lungs. However, only the superior nasal conchae of the ethmoid bone are involved in the sense of smell.

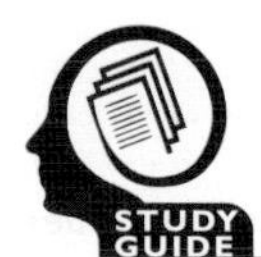

See Clinical Connection: Cleft Palate and Cleft Lip

Vomer

The **vomer** (VŌ-mer = plowshare) is a roughly triangular bone on the floor of the nasal cavity that articulates superiorly with the perpendicular plate of the ethmoid bone and sphenoid bone and inferiorly with both the maxillae and palatine bones along the midline (see Figures 7.3, 7.7, and 7.11). It forms the inferior portion of the bony nasal septum, the partition that divides the nasal cavity into right and left sides.

Maxillae

The paired **maxillae** (mak-SIL-ē = jawbones; singular is *maxilla*) unite to form the upper jawbone. They articulate with every bone of the face except the mandible (lower jawbone) (see Figures 7.3, 7.4a, and 7.7). The maxillae form part of the floors of the orbits, part of the lateral walls and floor of the nasal cavity, and most of the hard palate. The **hard palate** is the bony roof of the mouth, and is formed by the palatine processes of the maxillae and horizontal plates of the palatine bones. The hard palate separates the nasal cavity from the oral cavity.

Each maxilla contains a large *maxillary sinus* that empties into the nasal cavity (see Figure 7.13). The *alveolar process* (al-VĒ-ō-lar; *alveol-* = small cavity) of the maxilla is a ridgelike arch that contains the *alveoli* (sockets) for the maxillary (upper) teeth. The *palatine process* is a horizontal projection of the maxilla that forms the anterior three-quarters of the hard palate. The union and fusion of the maxillary bones normally is completed before birth. If this fusion fails, this condition is referred to as a cleft palate.

The *infraorbital foramen* (*infra-* = below; *-orbital* = orbit; see Figure 7.3), an opening in the maxilla inferior to the orbit, allows passage of the infraorbital blood vessels and nerve, a branch of the maxillary division of the trigeminal (V) nerve. Another prominent foramen in the maxilla is the *incisive foramen* (= incisor teeth) just posterior to the incisor teeth (see Figure 7.7). It transmits branches of the greater palatine blood vessels and nasopalatine nerve. A final structure associated with the maxilla and sphenoid bone is the *inferior orbital fissure*, located between the greater wing of the sphenoid and the maxilla (see Figure 7.12).

Zygomatic Bones

The two **zygomatic bones** (*zygo-* = yokelike), commonly called cheekbones, form the prominences of the cheeks and part of the lateral wall and floor of each orbit (see Figure 7.12). They articulate with the frontal, maxilla, sphenoid, and temporal bones.

The *temporal process* of the zygomatic bone projects posteriorly and articulates with the zygomatic process of the temporal bone to form the *zygomatic arch* (see Figure 7.4a).

Mandible

The **mandible** (*mand-* = to chew), or lower jawbone, is the largest, strongest facial bone (Figure 7.10). It is the only movable skull bone (other than the auditory ossicles, the small bones of the ear). In the lateral view, you can see that the mandible consists of a curved, horizontal portion, the *body*, and two perpendicular portions, the *rami* (RĀ-mī = branches; singular is *ramus*). The *angle* of the mandible is the area where each *ramus* meets the body. Each ramus has a posterior *condylar process* (KON-di-lar) that articulates with the mandibular fossa and articular tubercle of the temporal bone (see Figure 7.4a) to form the **temporomandibular joint (TMJ)**, and an anterior *coronoid process* (KOR-ō-noyd) to which the temporalis muscle attaches. The depression between the coronoid and condylar processes is called the *mandibular notch*. The *alveolar process* is the ridgelike arch containing the *alveoli* (sockets) for the mandibular (lower) teeth.

The *mental foramen* (*ment-* = chin) is approximately inferior to the second premolar tooth. It is near this foramen that dentists reach the mental nerve when injecting anesthetics. Another foramen associated with the mandible is the *mandibular foramen* on the medial surface of each ramus, another site often used by dentists to inject anesthetics. The mandibular foramen is the beginning of the *mandibular canal*, which runs obliquely in the ramus and anteriorly to the body. Through the canal pass the inferior alveolar nerves and blood vessels, which are distributed to the mandibular teeth.

FIGURE 7.10 Mandible.

The mandible is the largest and strongest facial bone.

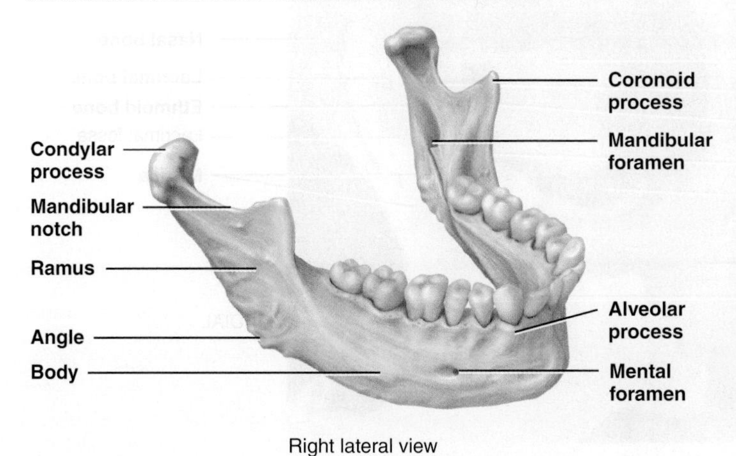

Right lateral view

Q What is the distinctive functional feature of the mandible among almost all the other skull bones?

See Clinical Connection: Temporomandibular Joint Syndrome

7.7 Special Features of the Skull

OBJECTIVE

- **Describe** the following special features of the skull: sutures, paranasal sinuses, and fontanels.

In addition to cranial bones and facial bones, the skull contains other components: the nasal septum, orbits, foramina, sutures, paranasal sinuses, and fontanels.

Nasal Septum

The nasal cavity is a space inside the skull that is divided into right and left sides by a vertical partition called the **nasal septum**, which consists of bone and cartilage. The three components of the nasal septum are the vomer, septal cartilage, and the perpendicular plate of the ethmoid bone (Figure 7.11). The anterior border of the vomer articulates with the septal cartilage, which is hyaline cartilage, to form the anterior portion of the septum. The superior border of the vomer articulates with the perpendicular plate of the ethmoid bone to form the remainder of the nasal septum. The term "broken nose," in most cases, refers to damage to the septal cartilage rather than the nasal bones themselves.

See Clinical Connection: Deviated Nasal Septum

Orbits

Seven bones of the skull join to form each **orbit** (eye socket) or *orbital cavity*, which contains the eyeball and associated structures (Figure 7.12). The three cranial bones of the orbit are the frontal,

FIGURE 7.11 **Nasal septum.**

The structures that form the nasal septum are the perpendicular plate of the ethmoid bone, the vomer, and septal cartilage.

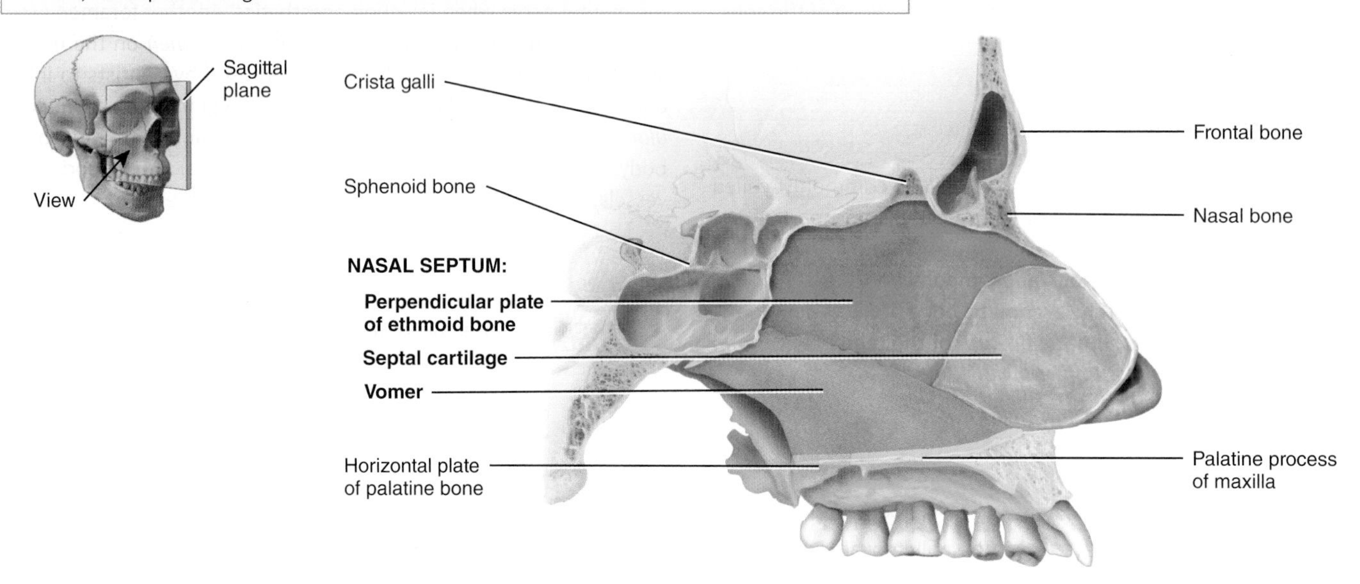

Sagittal section

Q What is the function of the nasal septum?

FIGURE 7.12 **Details of the orbit (eye socket).**

The orbit is a pyramid-shaped structure that contains the eyeball and associated structures.

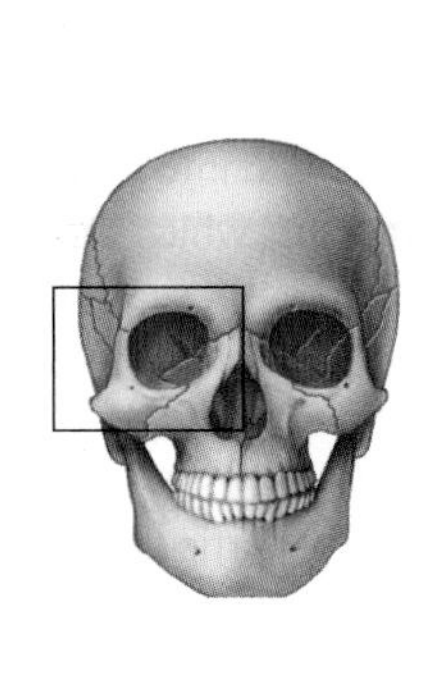

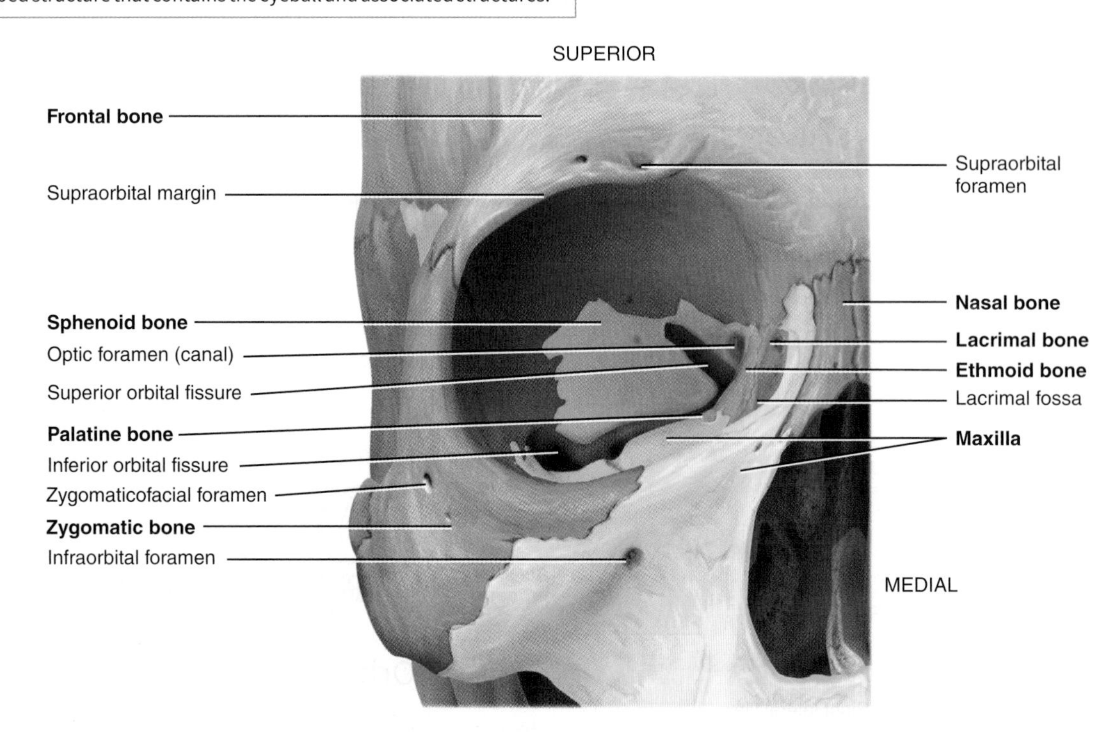

Anterior view showing the bones of the right orbit

Q Which seven bones form the orbit?

sphenoid, and ethmoid; the four facial bones are the palatine, zygomatic, lacrimal, and maxilla. Each pyramid-shaped orbit has four regions that converge posteriorly:

1. Parts of the frontal and sphenoid bones comprise the *roof* of the orbit.
2. Parts of the zygomatic and sphenoid bones form the *lateral wall* of the orbit.
3. Parts of the maxilla, zygomatic, and palatine bones make up the *floor* of the orbit.
4. Parts of the maxilla, lacrimal, ethmoid, and sphenoid bones form the *medial wall* of the orbit.

Associated with each orbit are five openings:

1. The *optic foramen* (*canal*) is at the junction of the roof and medial wall.
2. The *superior orbital fissure* is at the superior lateral angle of the apex.
3. The *inferior orbital fissure* is at the junction of the lateral wall and floor.
4. The *supraorbital foramen* is on the medial side of the supraorbital margin of the frontal bone.
5. The *lacrimal fossa* is in the lacrimal bone.

Foramina

We mentioned most of the **foramina** (openings for blood vessels, nerves, or ligaments; singular is *foramen*) of the skull in the descriptions of the cranial and facial bones that they penetrate. As preparation for studying other systems of the body, especially the nervous and cardiovascular systems, these foramina and the structures passing through them are listed in **Table 7.3**. For your convenience and for future reference, the foramina are listed alphabetically.

TABLE 7.3 Principal Foramina of the Skull

FORAMEN	LOCATION	STRUCTURES PASSING THROUGH*
Carotid (relating to carotid artery in neck)	Petrous portion of temporal bone (**Figure 7.8a**).	Internal carotid artery, sympathetic nerves for eyes.
Hypoglossal canal (*hypo-* = under; *-glossus* = tongue)	Superior to base of occipital condyles (**Figure 7.5a**).	Hypoglossal (XII) nerve, branch of ascending pharyngeal artery.
Infraorbital (*infra-* = below)	Inferior to orbit in maxilla (**Figure 7.12**).	Infraorbital nerve and blood vessels, branch of maxillary division of trigeminal (V) nerve.
Jugular (*jugul-* = throat)	Posterior to carotid canal between petrous portion of temporal bone and occipital bone (**Figure 7.7a**).	Internal jugular vein; glossopharyngeal (IX), vagus (X), and accessory (XI) nerves.
Lacerum (*lacerum* = lacerated)	Bounded anteriorly by sphenoid bone, posteriorly by petrous portion of temporal bone, medially by sphenoid and occipital bones (**Figure 7.8a**).	Branch of ascending pharyngeal artery.
Magnum (= large)	Occipital bone (**Figure 7.7**).	Medulla oblongata and its membranes (meninges), accessory (XI) nerve, vertebral and spinal arteries.
Mandibular (*mand-* = to chew)	Medial surface of ramus of mandible (**Figure 7.10**).	Inferior alveolar nerve and blood vessels.
Mastoid (= breast-shaped)	Posterior border of mastoid process of temporal bone (**Figure 7.7**).	Emissary vein to transverse sinus, branch of occipital artery to dura mater.
Mental (*ment-* = chin)	Inferior to second premolar tooth in mandible (**Figure 7.10**).	Mental nerve and vessels.
Olfactory (*olfact-* = to smell)	Cribriform plate of ethmoid bone (**Figure 7.8a**).	Olfactory (I) nerve.
Optic (= eye)	Between superior and inferior portions of small wing of sphenoid bone (**Figure 7.12**).	Optic (II) nerve, ophthalmic artery.
Ovale (= oval)	Greater wing of sphenoid bone (**Figure 7.8a**).	Mandibular branch of trigeminal (V) nerve.
Rotundum (= round)	Junction of anterior and medial parts of sphenoid bone (**Figure 7.8a, b**).	Maxillary branch of trigeminal (V) nerve.
Stylomastoid (*stylo-* = stake or pole)	Between styloid and mastoid processes of temporal bone (**Figure 7.7**).	Facial (VII) nerve, stylomastoid artery.
Supraorbital (*supra-* = above)	Supraorbital margin of orbit in frontal bone (**Figure 7.12**).	Supraorbital nerve and artery.

*The cranial nerves listed here (roman numerals I–XII) are described in **Table 14.4**.

Unique Features of the Skull

The skull exhibits several unique features not seen in other bones of the body. These include sutures, paranasal sinuses, and fontanels.

Sutures A **suture** (SOO-chur = seam) is an immovable joint (in most cases in an adult skull) that holds most skull bones together. Sutures in the skulls of infants and children, however, often are movable and function as important growth centers in the developing skull. The names of many sutures reflect the bones they unite. For example, the frontozygomatic suture is between the frontal bone and the zygomatic bone. Similarly, the sphenoparietal suture is between the sphenoid bone and the parietal bone. In other cases, however, the names of sutures are not so obvious. Of the many sutures found in the skull, we will identify only four prominent ones:

1. The **coronal suture** (KO-rō-nal; *coron-* = relating to the frontal or coronal plane) unites the frontal bone and both parietal bones (see Figure 7.4).
2. The **sagittal suture** (SAJ-i-tal; *sagitt-* = arrow) unites the two parietal bones on the superior midline of the skull (see Figure 7.4b). The sagittal suture is so named because in the infant, before the bones of the skull are firmly united, the suture and the fontanels (soft spots) associated with it resemble an arrow.
3. The **lambdoid suture** (LAM-doyd) unites the two parietal bones to the occipital bone. This suture is so named because of its resemblance to the capital Greek letter lambda (Λ), as can be seen in Figure 7.6 (with the help of a little imagination). Sutural bones may occur within the sagittal and lambdoid sutures.
4. The two **squamous sutures** (SKWĀ-mus; *squam-* = flat, like the flat overlapping scales of a snake) unite the parietal and temporal bones on the lateral aspects of the skull (see Figure 7.4a).

Paranasal Sinuses The **paranasal sinuses** (par′-a-NĀ-zal SĪ-nus-ez; *para-* = beside) are cavities within certain cranial and facial bones near the nasal cavity. They are most evident in a sagittal section of the skull (Figure 7.13c). The paranasal sinuses are lined with mucous membranes that are continuous with the lining of the nasal cavity. Secretions produced by the mucous membranes of the paranasal sinuses drain into the lateral wall of the nasal cavity. Paranasal sinuses are quite small or absent at birth, but increase in size during two periods of facial enlargement—during the eruption of the teeth and at the onset of puberty. They arise as outgrowths of the nasal mucosa that project into the surrounding bones. Skull bones containing the paranasal sinuses are the frontal, sphenoid, ethmoid, and maxillae. The paranasal sinuses allow the skull to increase in size without a change in the mass (weight) of the bone. The paranasal sinuses increase the surface area of the nasal mucosa, thus increasing the production of mucus to help moisten and cleanse inhaled air. In addition, the paranasal sinuses serve as resonating (echo) chambers within the skull that intensify and prolong sounds, thereby enhancing the quality of the voice. The influence of the paranasal sinuses on your voice becomes obvious when you have a cold; the passageways

FIGURE 7.13 **Paranasal sinuses projected to the surface.**

Paranasal sinuses are mucous membrane–lined spaces in the frontal, sphenoid, ethmoid, and maxillary bones that connect to the nasal cavity.

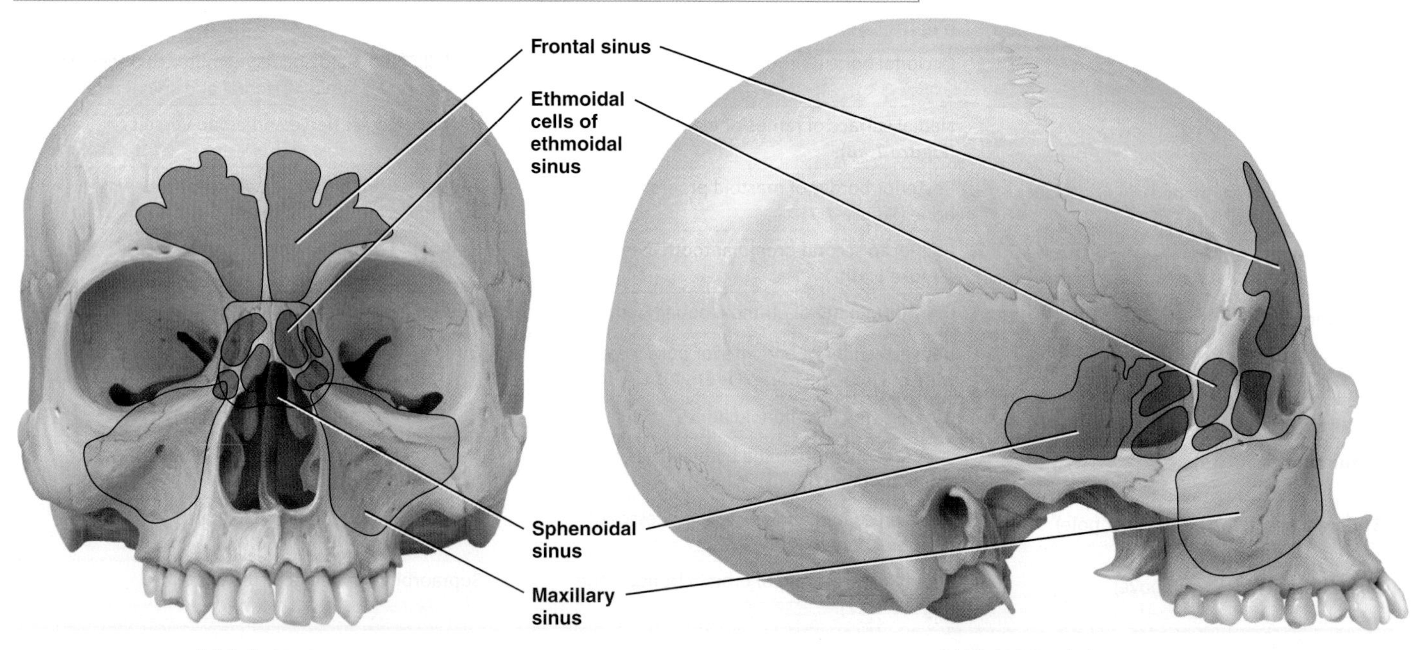

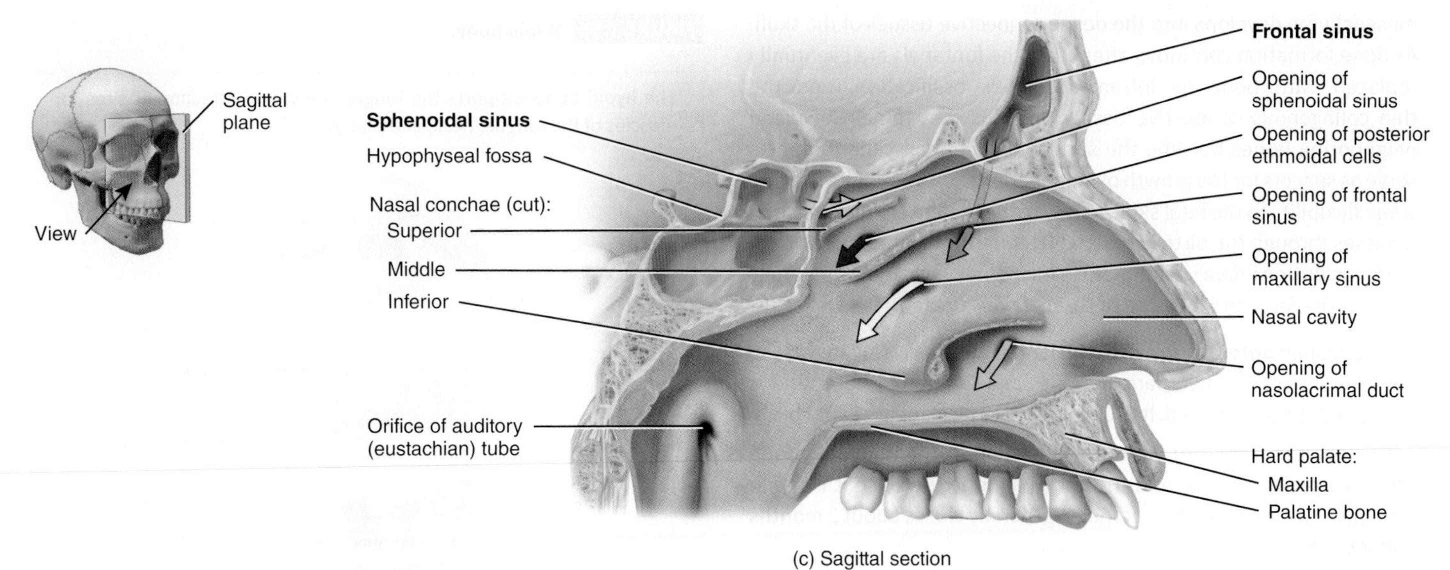

(c) Sagittal section

Q What are the functions of the paranasal sinuses?

through which sound travels into and out of the paranasal sinuses become blocked by excess mucus production, changing the quality of your voice.

See Clinical Connection: Sinusitis

Fontanels The skull of a developing embryo consists of cartilage and mesenchyme arranged in thin plates around the developing brain. Gradually, ossification occurs, and bone slowly replaces most of the cartilage and mesenchyme. At birth, bone ossification is incomplete, and the mesenchyme-filled spaces become dense connective tissue regions between incompletely developed cranial bones called **fontanels** (fon-ta-NELZ = little fountains), commonly called "soft spots" (**Figure 7.14**). Fontanels are areas where unossified

FIGURE 7.14 Fontanels at birth.

Fontanels are mesenchyme-filled spaces between cranial bones that are present at birth.

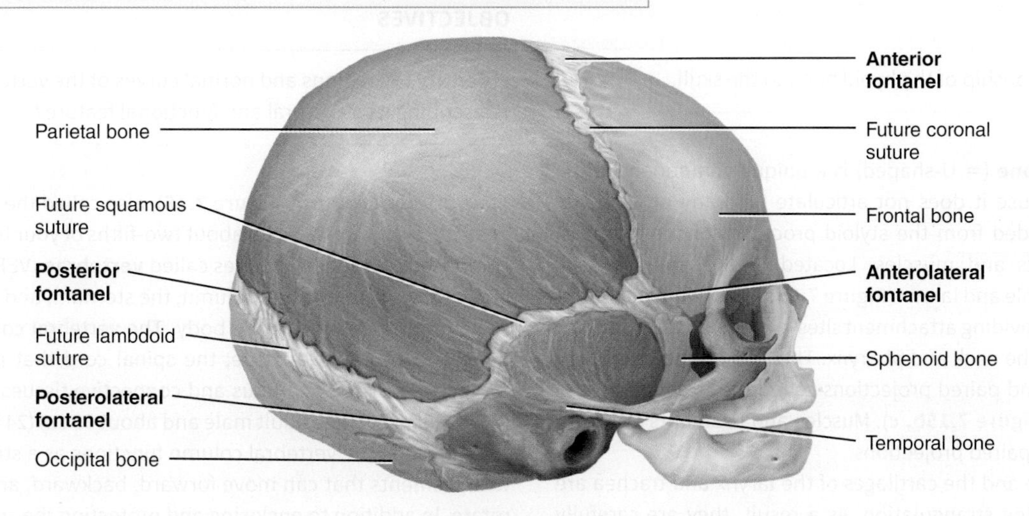

Right lateral view

Q Which fontanel is bordered by four different skull bones?

mesenchyme develops into the dense connective tissues of the skull. As bone formation continues after birth, the fontanels are eventually replaced with bone by intramembranous ossification, and the thin collagenous connective tissue junctions that remain between neighboring bones become the sutures. Functionally, the fontanels serve as spacers for the growth of neighboring skull bones and provide some flexibility to the fetal skull, allowing the skull to change shape as it passes through the birth canal and later permitting rapid growth of the brain during infancy. Although an infant may have many fontanels at birth, the form and location of six are fairly constant:

- The unpaired **anterior fontanel**, the largest fontanel, is located at the midline among the two parietal bones and the frontal bone, and is roughly diamond-shaped. It usually closes 18 to 24 months after birth.
- The unpaired **posterior fontanel** is located at the midline among the two parietal bones and the occipital bone. Because it is much smaller than the anterior fontanel, it generally closes about 2 months after birth.
- The paired **anterolateral fontanels**, located laterally among the frontal, parietal, temporal, and sphenoid bones, are small and irregular in shape. Normally, they close about 3 months after birth.
- The paired **posterolateral fontanels**, located laterally among the parietal, occipital, and temporal bones, are irregularly shaped. They begin to close 1 to 2 months after birth, but closure is generally not complete until 12 months.

The amount of closure in fontanels helps a physician gauge the degree of brain development. In addition, the anterior fontanel serves as a landmark for withdrawal of blood for analysis from the superior sagittal sinus (a large midline vein within the covering tissues that surround the brain). (See **Figure 21.24**.)

FIGURE 7.15 **Hyoid bone.**

The hyoid bone supports the tongue, providing attachment sites for muscles of the tongue, neck, and pharynx.

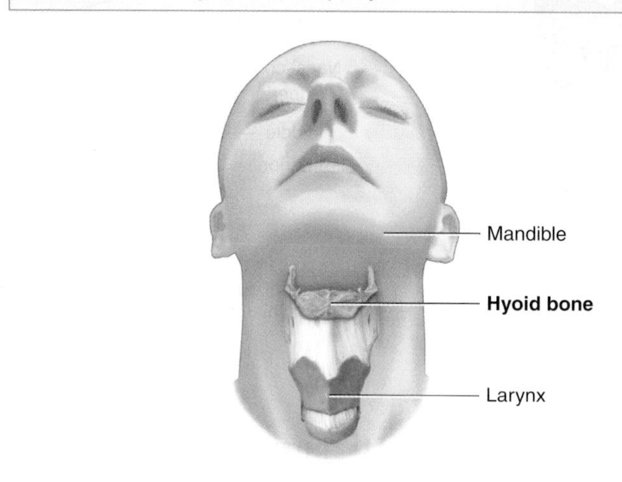

(a) Position of hyoid bone

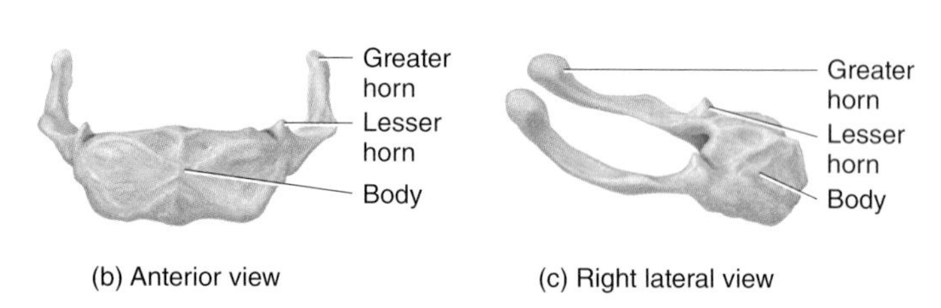

(b) Anterior view

(c) Right lateral view

Q In what way is the hyoid bone different from all other bones of the axial skeleton?

7.8 Hyoid Bone

OBJECTIVE

- **Describe** the relationship of the hyoid bone to the skull.

The single **hyoid bone** (= U-shaped) is a unique component of the axial skeleton because it does not articulate with any other bone. Rather, it is suspended from the styloid processes of the temporal bones by ligaments and muscles. Located in the anterior neck between the mandible and larynx (**Figure 7.15a**), the hyoid bone supports the tongue, providing attachment sites for some tongue muscles and for muscles of the neck and pharynx. The hyoid bone consists of a horizontal *body* and paired projections called the *lesser horns* and the *greater horns* (**Figure 7.15b, c**). Muscles and ligaments attach to the body and these paired projections.

The hyoid bone and the cartilages of the larynx and trachea are often fractured during strangulation. As a result, they are carefully examined at autopsy when manual strangulation is a suspected cause of death.

7.9 Vertebral Column

OBJECTIVES

- **Identify** the regions and normal curves of the vertebral column, describing its structural and functional features.

The **vertebral column** (**Figure 7.16**), also called the *spine*, *backbone*, or *spinal column*, makes up about two-fifths of your total height and is composed of a series of bones called **vertebrae** (VER-te-brē; singular is *vertebra*). The vertebral column, the sternum, and the ribs form the skeleton of the trunk of the body. The vertebral column consists of bone and connective tissue; the spinal cord that it surrounds and protects consists of nervous and connective tissues. At about 71 cm (28 in.) in an average adult male and about 61 cm (24 in.) in an average adult female, the vertebral column functions as a strong, flexible rod with elements that can move forward, backward, and sideways, and rotate. In addition to enclosing and protecting the spinal cord, it supports the head and serves as a point of attachment for the ribs, pelvic girdle, and muscles of the back and upper limbs.

FIGURE 7.16 Vertebral column. The numbers in parentheses in part (a) indicate the number of vertebrae in each region. In part (d), the relative size of the disc has been enlarged for emphasis.

The adult vertebral column typically contains 26 vertebrae.

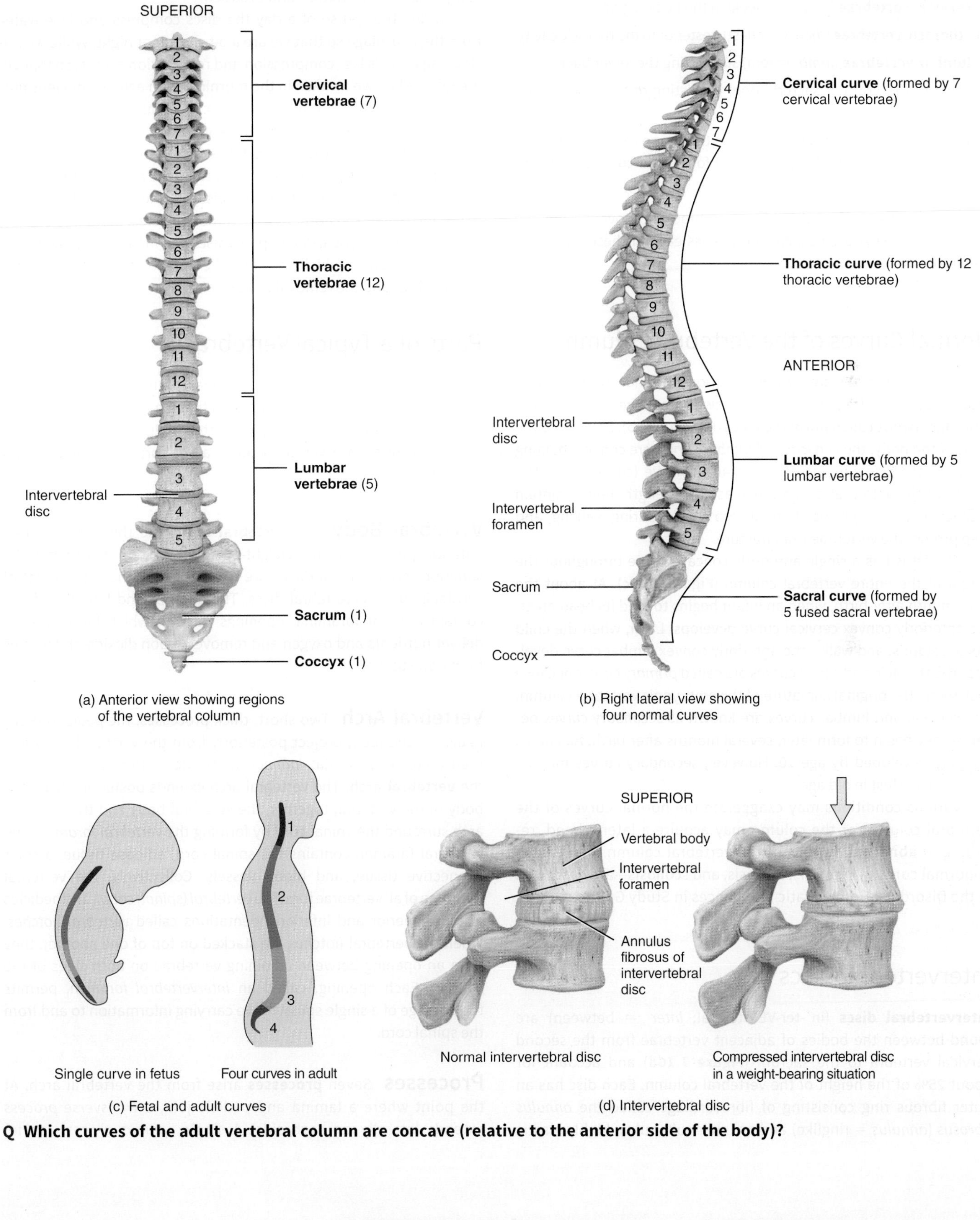

(a) Anterior view showing regions of the vertebral column

(b) Right lateral view showing four normal curves

(c) Fetal and adult curves

(d) Intervertebral disc

Q Which curves of the adult vertebral column are concave (relative to the anterior side of the body)?

The total number of vertebrae during early development is 33. As a child grows, several vertebrae in the sacral and coccygeal regions fuse. As a result, the adult vertebral column typically contains 26 vertebrae (Figure 7.16a). These are distributed as follows:

- 7 **cervical vertebrae** (*cervic-* = neck) in the neck region.
- 12 **thoracic vertebrae** (*thorax* = chest) posterior to the thoracic cavity.
- 5 **lumbar vertebrae** (*lumb-* = loin) supporting the lower back.
- 1 **sacrum** (SĀ-krum = sacred bone) consisting of five fused sacral vertebrae.
- 1 **coccyx** (KOK-siks = cuckoo, because the shape resembles the bill of a cuckoo bird) usually consisting of four fused **coccygeal vertebrae** (kok-SIJ-ē-al).

The cervical, thoracic, and lumbar vertebrae are movable, but the sacrum and coccyx are not. We will discuss each of these regions in detail shortly.

Normal Curves of the Vertebral Column

When viewed from the anterior or posterior, a normal adult vertebral column appears straight. But when viewed from the side, it shows four slight bends called **normal curves** (Figure 7.16b). Relative to the front of the body, the *cervical* and *lumbar curves* are convex (bulging out); the *thoracic* and *sacral curves* are concave (cupping in). The curves of the vertebral column increase its strength, help maintain balance in the upright position, absorb shocks during walking, and help protect the vertebrae from fracture.

The fetus has a single anteriorly concave curve throughout the length of the entire vertebral column (Figure 7.16c). At about the third month after birth, when an infant begins to hold its head erect, the anteriorly convex cervical curve develops. Later, when the child sits up, stands, and walks, the anteriorly convex lumbar curve develops. The thoracic and sacral curves are called *primary curves* because they retain the original curvature of the embryonic vertebral column. The cervical and lumbar curves are known as *secondary curves* because they begin to form later, several months after birth. All curves are fully developed by age 10. However, secondary curves may be progressively lost in old age.

Various conditions may exaggerate the normal curves of the vertebral column, or the column may acquire a lateral bend, resulting in **abnormal curves** of the vertebral column. Three such abnormal curves—kyphosis, lordosis, and scoliosis—are described in the Disorders: Homeostatic Imbalances in Study Guide.

Intervertebral Discs

Intervertebral discs (in′-ter-VER-te-bral; *inter-* = between) are found between the bodies of adjacent vertebrae from the second cervical vertebra to the sacrum (Figure 7.16d) and account for about 25% of the height of the vertebral column. Each disc has an outer fibrous ring consisting of fibrocartilage called the *annulus fibrosus* (*annulus* = ringlike) and an inner soft, pulpy, highly elastic substance called the *nucleus pulposus* (*pulposus* = pulplike). The superior and inferior surfaces of the disc consist of a thin plate of hyaline cartilage. The discs form strong joints, permit various movements of the vertebral column, and absorb vertical shock. Under compression, they flatten and broaden.

During the course of a day the discs compress and lose water from their cartilage so that we are a bit shorter at night. While we are sleeping there is less compression and rehydration occurs, so that we are taller when we awaken in the morning. With age, the nucleus pulposus hardens and becomes less elastic. Decrease in vertebral height with age results from bone loss in the vertebral bodies and not a decrease in thickness of the intervertebral discs.

Since intervertebral discs are avascular, the annulus fibrosus and nucleus pulposus rely on blood vessels from the bodies of vertebrae to obtain oxygen and nutrients and remove wastes. Certain stretching exercises, such as yoga, decompress discs and increase general blood circulation, both of which speed up the uptake of oxygen and nutrients by discs and the removal of wastes.

Parts of a Typical Vertebra

Vertebrae in different regions of the spinal column vary in size, shape, and detail, but they are similar enough that we can discuss the structures (and the functions) of a typical vertebra (Figure 7.17). Vertebrae typically consist of a vertebral body, a vertebral arch, and several processes.

Vertebral Body

The **vertebral body**, the thick, disc-shaped anterior portion, is the weight-bearing part of a vertebra. Its superior and inferior surfaces are roughened for the attachment of cartilaginous intervertebral discs. The anterior and lateral surfaces contain nutrient foramina, openings through which blood vessels deliver nutrients and oxygen and remove carbon dioxide and wastes from bone tissue.

Vertebral Arch

Two short, thick processes, the *pedicles* (PED-i-kuls = little feet), project posteriorly from the vertebral body and then unite with the flat *laminae* (LAM-i-nē = thin layers) to form the **vertebral arch**. The vertebral arch extends posteriorly from the body of the vertebra; together, the vertebral body and the vertebral arch surround the spinal cord by forming the *vertebral foramen.* The vertebral foramen contains the spinal cord, adipose tissue, areolar connective tissue, and blood vessels. Collectively, the vertebral foramina of all vertebrae form the *vertebral (spinal) canal*. The pedicles exhibit superior and inferior indentations called *vertebral notches.* When the vertebral notches are stacked on top of one another, they form an opening between adjoining vertebrae on both sides of the column. Each opening, called an *intervertebral foramen*, permits the passage of a single spinal nerve carrying information to and from the spinal cord.

Processes

Seven **processes** arise from the vertebral arch. At the point where a lamina and pedicle join, a *transverse process* extends laterally on each side. A single *spinous process* (*spine*)

FIGURE 7.17 Structure of a typical vertebra, as illustrated by a thoracic vertebra. In part (b), only one spinal nerve has been included, and it has been extended beyond the intervertebral foramen for clarity.

A vertebra consists of a vertebral body, a vertebral arch, and several processes.

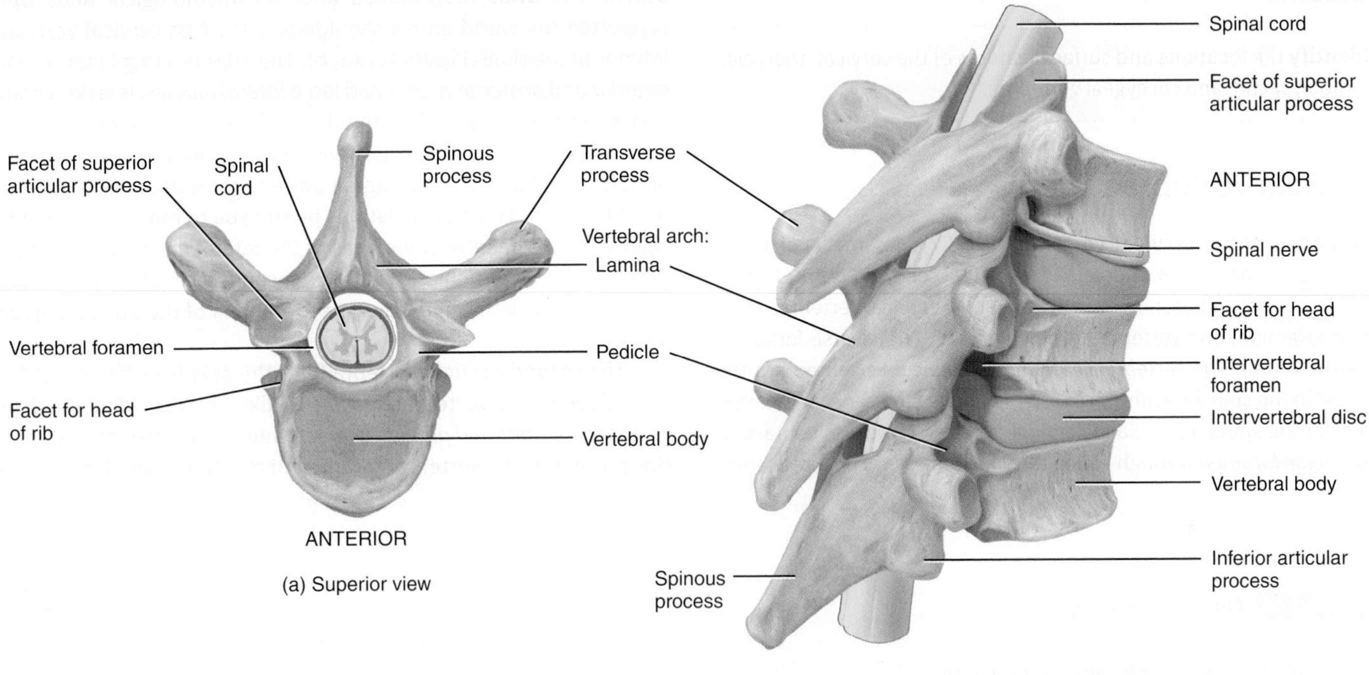

(a) Superior view

(b) Right posterolateral view of articulated vertebrae

Q What are the functions of the vertebral and intervertebral foramina?

projects posteriorly from the junction of the laminae. These three processes serve as points of attachment for muscles. The remaining four processes form joints with other vertebrae above or below. The two *superior articular processes* of a vertebra articulate (form joints) with the two inferior articular processes of the vertebra immediately above them. In turn, the two *inferior articular processes* of that vertebra articulate with the two superior articular processes of the vertebra immediately below them, and so on. The articulating surfaces of the articular processes, which are referred to as *facets* (FAS-ets or fa-SETS = little faces), are covered with hyaline cartilage. The articulations formed between the vertebral bodies and articular facets of successive vertebrae are termed *intervertebral joints.*

Regions of the Vertebral Column

Section 7.10 present the five regions of the vertebral column, beginning superiorly and moving inferiorly. The regions are the cervical, thoracic, lumbar, sacral, and coccygeal. Note that vertebrae in each region are numbered in sequence, from superior to inferior. When you actually view the bones of the vertebral column, you will notice that the transition from one region to the next is not abrupt but gradual, a feature that helps the vertebrae fit together.

Age-Related Changes in the Vertebral Column

With advancing age the vertebral column undergoes changes that are characteristic of the skeletal system in general. These changes include reduction in the mass and density of the bone along with a reduction in the collagen-to-mineral content within the bone, changes that make the bones more brittle and susceptible to damage. The articular surfaces, those surfaces where neighboring bones move against one another, lose their covering cartilage as they age; in their place rough bony growths form that lead to arthritic conditions. In the vertebral column, bony growths around the intervertebral discs, called *osteophytes*, can lead to a narrowing (stenosis) of the vertebral canal. This narrowing can lead to compression of spinal nerves and the spinal cord, which can manifest as pain and decreased muscle function in the back and lower limbs.

7.10 Vertebral Regions

OBJECTIVE

- **Identify** the locations and surface features of the cervical, thoracic, lumbar, sacral, and coccygeal vertebrae.

Cervical Vertebrae

The bodies of the **cervical vertebrae** (C1–C7) are smaller than all other vertebrae except those that form the coccyx (**Figure 7.18a**). Their vertebral arches, however, are larger. All cervical vertebrae have three foramina: one vertebral foramen and two transverse foramina (**Figure 7.18c**). The *vertebral foramina* of cervical vertebrae are the largest in the spinal column because they house the cervical enlargement of the spinal cord. Each cervical transverse process contains a *transverse foramen* through which the vertebral artery and its accompanying vein and nerve fibers pass. The spinous processes of C2 through C6 are often *bifid*—that is, they branch into two small projections at the tips (**Figure 7.18a, c**).

The first two cervical vertebrae differ considerably from the others. The **atlas** (C1), named after the mythological Atlas who supported the world on his shoulders, is the first cervical vertebra inferior to the skull (**Figure 7.18a, b**). The atlas is a ring of bone with *anterior* and *posterior arches* and large *lateral masses.* It lacks a body and a spinous process. The superior surfaces of the lateral masses, called *superior articular facets*, are concave. They articulate with the occipital condyles of the occipital bone to form the paired *atlanto-occipital joints.* These articulations permit you to move your head to signify "yes." The inferior surfaces of the lateral masses, the *inferior articular facets*, articulate with the second cervical vertebra. The transverse processes and transverse foramina of the atlas are quite large.

The second cervical vertebra (C2), the **axis** (see **Figure 7.18a, d, e**), does have a vertebral body. A peglike process called the *dens* (= tooth) or *odontoid process* projects superiorly through the anterior portion of the vertebral foramen of the atlas. The dens makes

FIGURE 7.18 **Cervical vertebrae.**

The cervical vertebrae are found in the neck region.

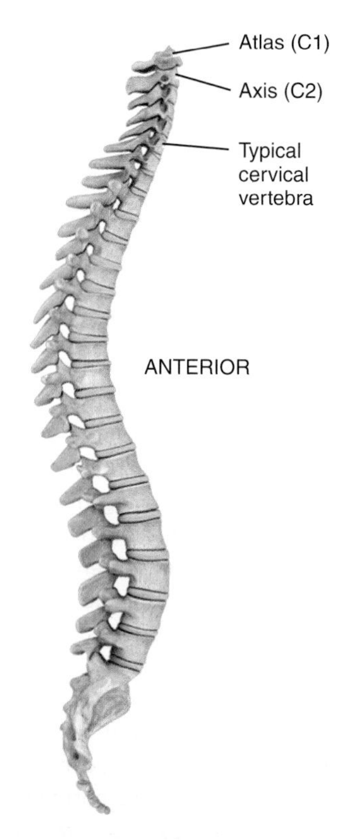

Location of cervical vertebrae

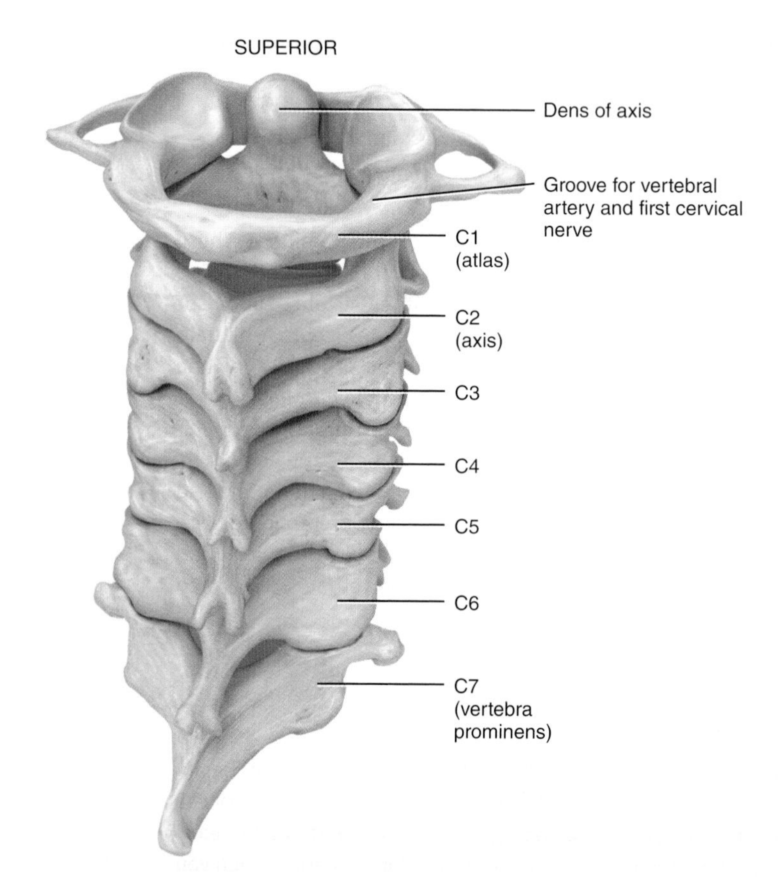

(a) Posterior view of articulated cervical vertebrae

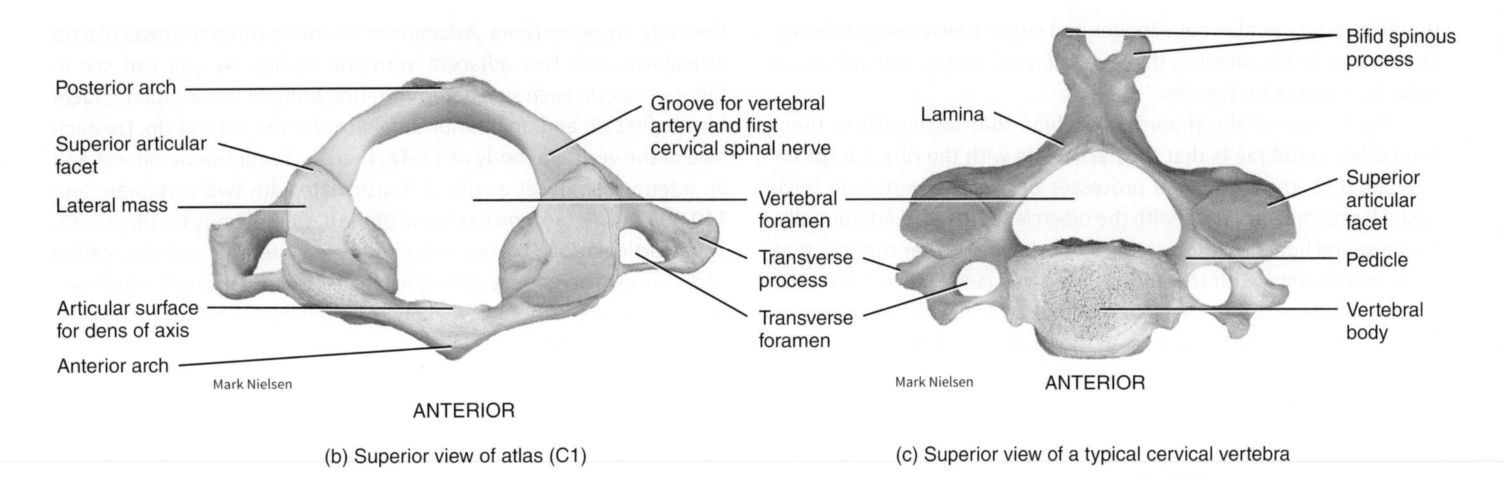

(b) Superior view of atlas (C1)

(c) Superior view of a typical cervical vertebra

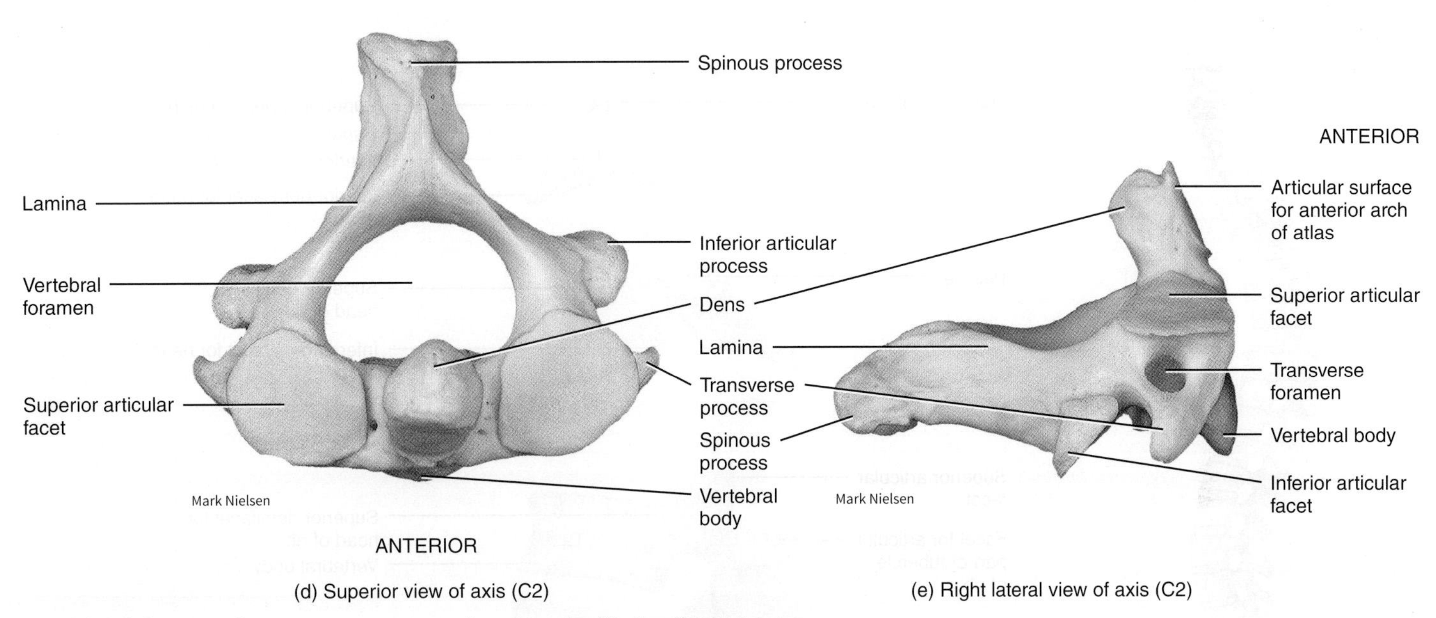

(d) Superior view of axis (C2)

(e) Right lateral view of axis (C2)

Q Which joint permits you to move your head to signify "no"? Which bones are involved?

a pivot on which the atlas and head rotate. This arrangement permits side-to-side movement of the head, as when you move your head to signify "no." The articulation formed between the anterior arch of the atlas and dens of the axis, and between their articular facets, is called the *atlanto-axial joint.* In some instances of trauma, the dens of the axis may be driven into the medulla oblongata of the brain. This type of injury is the usual cause of death from whiplash injuries.

The third through sixth cervical vertebrae (C3–C6), represented by the vertebra in **Figure 7.18c**, correspond to the structural pattern of the typical cervical vertebra previously described. The seventh cervical vertebra (C7), called the *vertebra prominens*, is somewhat different (see **Figure 7.18a**). It has a large, nonbifid spinous process that may be seen and felt at the base of the neck, but otherwise is typical.

Thoracic Vertebrae

Thoracic vertebrae (T1–T12; **Figure 7.19**) are considerably larger and stronger than cervical vertebrae. In addition, the spinous processes on T1 through T10 are long, laterally flattened, and directed inferiorly. In contrast, the spinous processes on T11 and T12 are shorter, broader, and directed more posteriorly. Compared to cervical vertebrae, thoracic vertebrae also have longer and larger transverse processes. They are easily identified by their *costal facets* (*cost-* = rib), which are articular surfaces for the ribs.

The feature of the thoracic vertebrae that distinguishes them from other vertebrae is that they articulate with the ribs. Except for T11 and T12, the transverse processes of thoracic vertebrae have costal facets that articulate with the *tubercles* of the ribs. Additionally, the vertebral bodies of thoracic vertebrae have articular surfaces that form articulations with the *heads* of the ribs (see **Figure 7.23**). The articular surfaces on the vertebral bodies are called either facets or demifacets. A *facet* is formed when the head of a rib articulates with the body of one vertebra. A *demifacet* is formed when the head of a rib articulates with two adjacent vertebral bodies. As you can see in **Figure 7.19**, on each side of the vertebral body T1 has a superior facet for the first rib and an inferior demifacet for the second rib. On each side of the vertebral body of T2–T8, there is a superior demifacet and an inferior demifacet as ribs 2–9 articulate with two vertebrae, and T10–T12 have a facet on each side of the vertebral body for ribs 10–12. These articulations between the thoracic vertebrae and ribs, called *vertebrocostal joints*, are distinguishing features of thoracic vertebrae. Movements of the thoracic region are limited by the attachment of the ribs to the sternum.

FIGURE 7.19 Thoracic vertebrae.

The thoracic vertebrae are found in the chest region and articulate with the ribs.

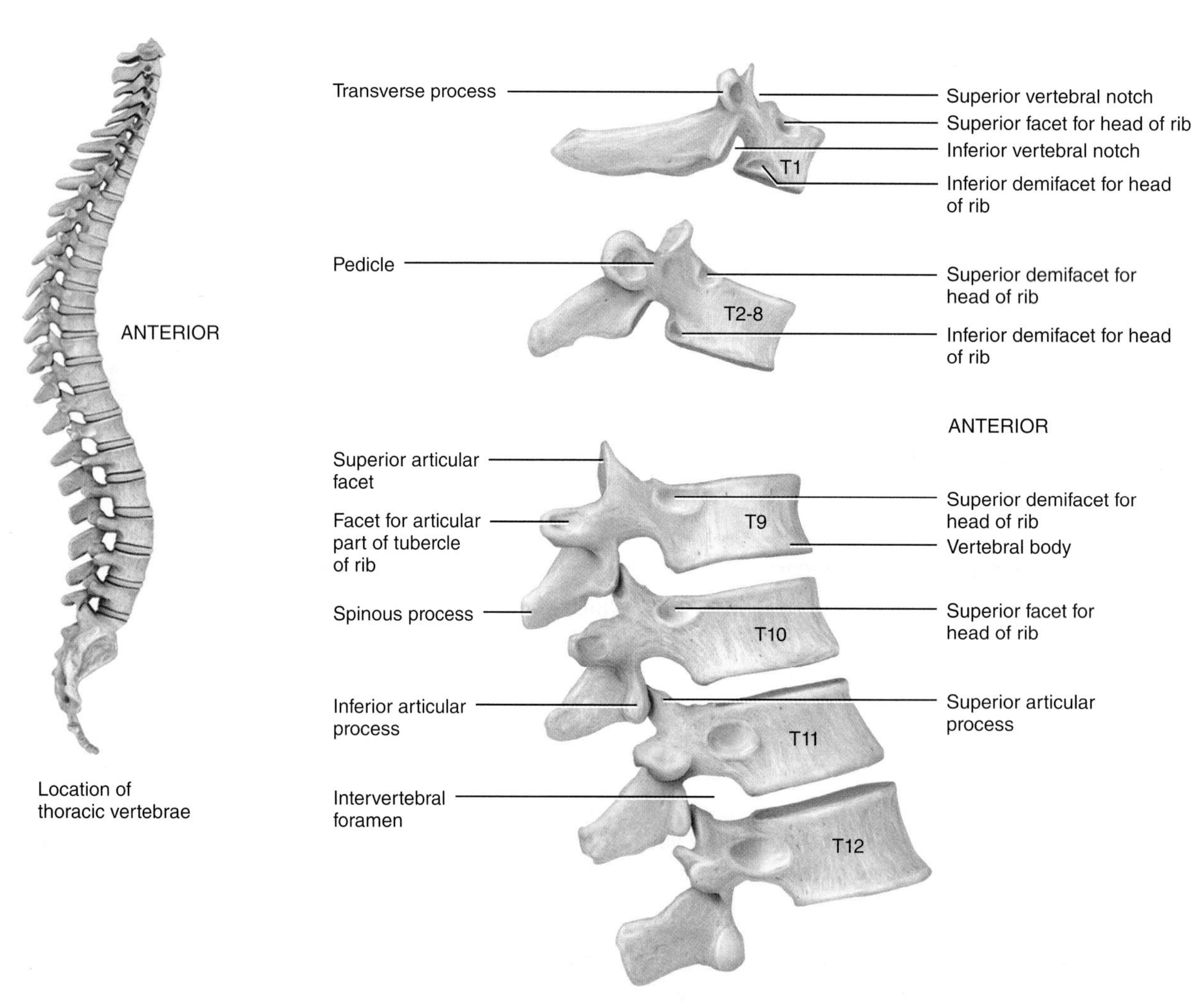

(a) Right lateral view of several articulated thoracic vertebrae

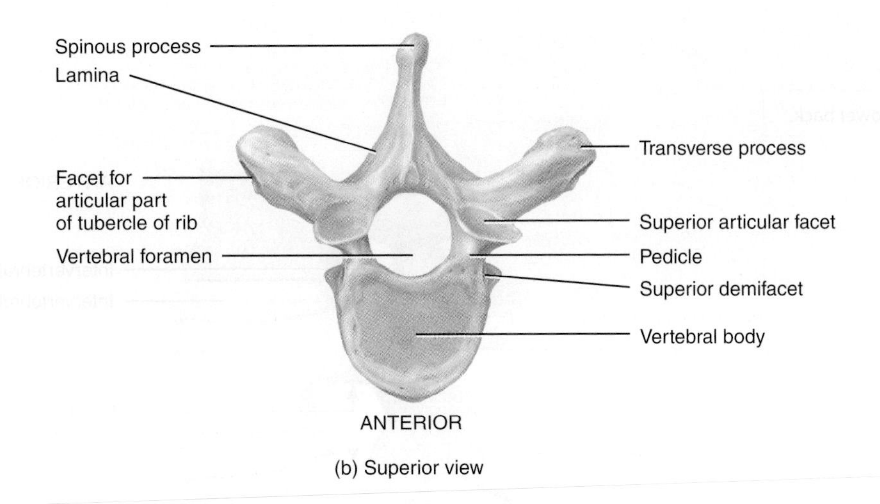

(b) Superior view

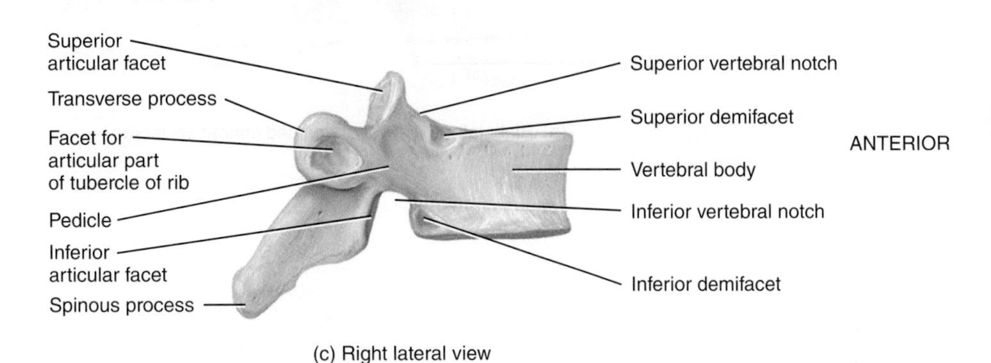

(c) Right lateral view

Q Which parts of the thoracic vertebrae articulate with the ribs?

Lumbar Vertebrae

The **lumbar vertebrae** (L1–L5) are the largest and strongest of the unfused bones in the vertebral column (**Figure 7.20**) because the amount of body weight supported by the vertebrae increases toward the inferior end of the backbone. Their various projections are short and thick. The superior articular processes are directed medially instead of superiorly, and the inferior articular processes are directed laterally instead of inferiorly. The spinous processes are quadrilateral in shape, are thick and broad, and project nearly straight posteriorly. The spinous processes are well adapted for the attachment of the large back muscles.

A summary of the major structural differences among cervical, thoracic, and lumbar vertebrae is presented in **Table 7.4**.

Sacral and Coccygeal Vertebrae

Sacrum The **sacrum** (SĀ-krum) is a triangular bone formed by the union of five sacral vertebrae (S1–S5) (**Figure 7.21a**). The sacral vertebrae begin to fuse in individuals between 16 and 18 years of age, a process usually completed by age 30. Positioned at the posterior portion of the pelvic cavity medial to the two hip bones, the sacrum serves as a strong foundation for the pelvic girdle. The female sacrum is shorter, wider, and more curved between S2 and S3 than the male sacrum (see **Table 8.1**).

The concave anterior side of the sacrum faces the pelvic cavity. It is smooth and contains four *transverse lines* (*ridges*) that mark the joining of the sacral vertebral bodies (**Figure 7.21a**). At the ends of these lines are four pairs of *anterior sacral foramina.* The lateral portion of the superior surface of the sacrum contains a smooth surface called the *sacral ala* (ĀL-a = wing; plural is *alae*, ĀL-ē), which is formed by the fused transverse processes of the first sacral vertebra (S1).

The convex, posterior surface of the sacrum contains a *median sacral crest*, the fused spinous processes of the upper sacral vertebrae; a *lateral sacral crest*, the fused transverse processes of the sacral vertebrae; and four pairs of *posterior sacral foramina* (**Figure 7.21b**). These foramina connect with anterior sacral foramina to allow passage of nerves and blood vessels. The *sacral canal* is a continuation of the vertebral cavity. The laminae of the fifth sacral vertebra, and

See Clinical Connection: Caudal Anesthesia

FIGURE 7.20 **Lumbar vertebrae.**

Lumbar vertebrae are found in the lower back.

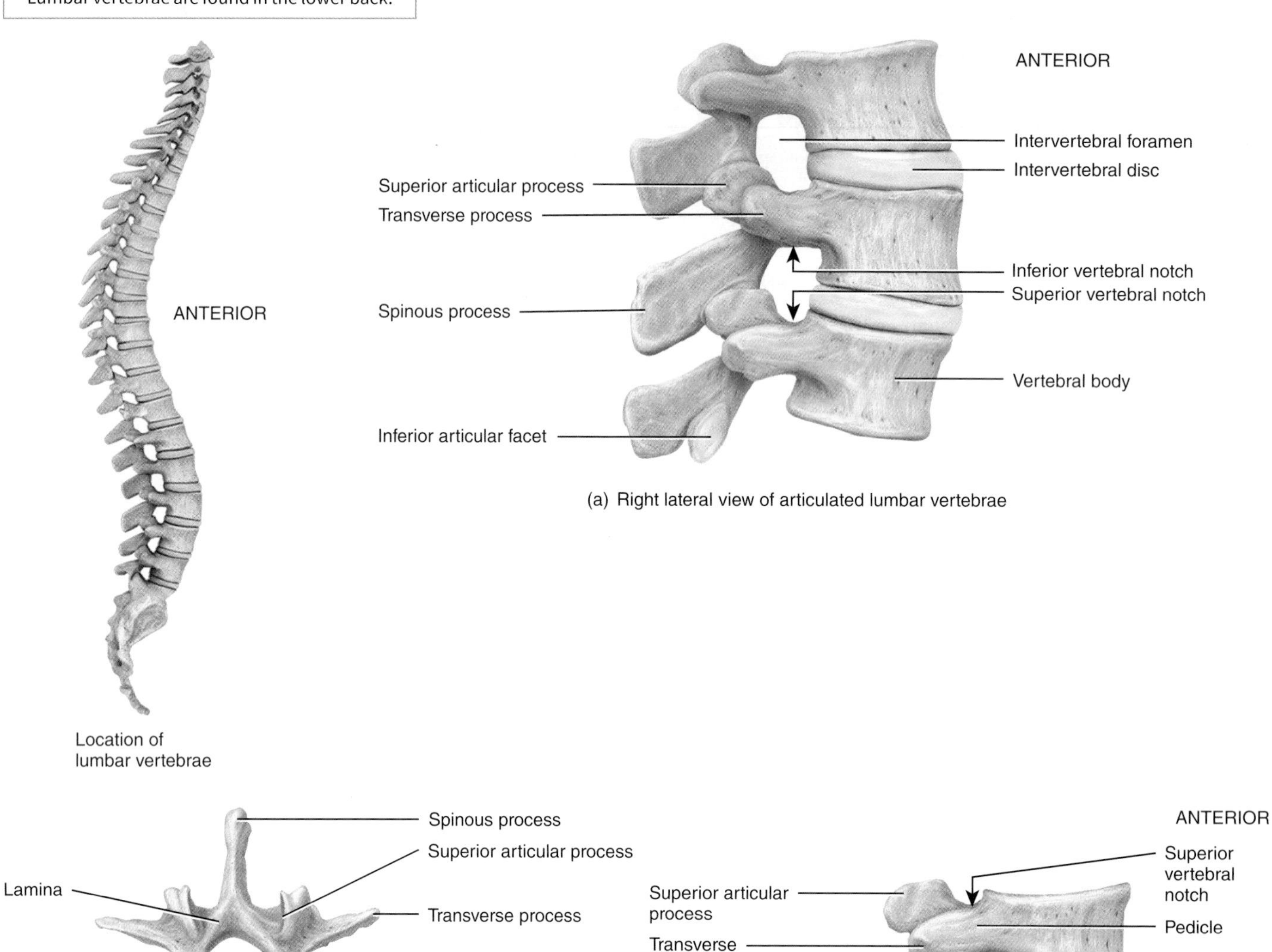

Q Why are the lumbar vertebrae the largest and strongest in the vertebral column?

sometimes the fourth, fail to meet. This leaves an inferior entrance to the vertebral canal called the *sacral hiatus* (hī-Ā-tus = opening). On either side of the sacral hiatus is a *sacral cornu* (KOR-noo; *cornu* = horn; plural is *cornua*, KOR-noo-a), an inferior articular process of the fifth sacral vertebra. They are connected by ligaments to the coccyx.

The narrow inferior portion of the sacrum is known as the *apex.* The broad superior portion of the sacrum is called the *base.* The anteriorly projecting border of the base, called the *sacral promontory* (PROM-on-tō-rē), is one of the points used for measurements of the pelvis. On both lateral surfaces the sacrum has a large ear-shaped *auricular surface* that articulates with the ilium of each hip bone to form the *sacroiliac joint* (see **Figure 8.8**). Posterior to the auricular surface is a roughened surface, the *sacral tuberosity*, which contains depressions for the attachment of ligaments. The sacral tuberosity unites with the hip bones to form the sacroiliac joints. The *superior articular processes* of the sacrum articulate with the inferior articular processes of the fifth lumbar vertebra, and the base of the sacrum articulates with the body of the fifth lumbar vertebra to form the *lumbosacral joint.*

TABLE 7.4 Comparison of Major Structural Features of Cervical, Thoracic, and Lumbar Vertebrae

CHARACTERISTIC	CERVICAL	THORACIC	LUMBAR
Overall structure			
Size	Small.	Larger.	Largest.
Foramina	One vertebral and two transverse.	One vertebral.	One vertebral.
Spinous processes	Slender, often bifid (C2–C6).	Long, fairly thick (most project inferiorly).	Short, blunt (project posteriorly rather than inferiorly).
Transverse processes	Small.	Fairly large.	Large and blunt.
Articular facets for ribs	Absent.	Present.	Absent.
Direction of articular facets			
Superior	Posterosuperior.	Posterolateral.	Medial.
Inferior	Anteroinferior.	Anteromedial.	Lateral.
Size of intervertebral discs	Thick relative to size of vertebral bodies.	Thin relative to size of vertebral bodies.	Thickest.

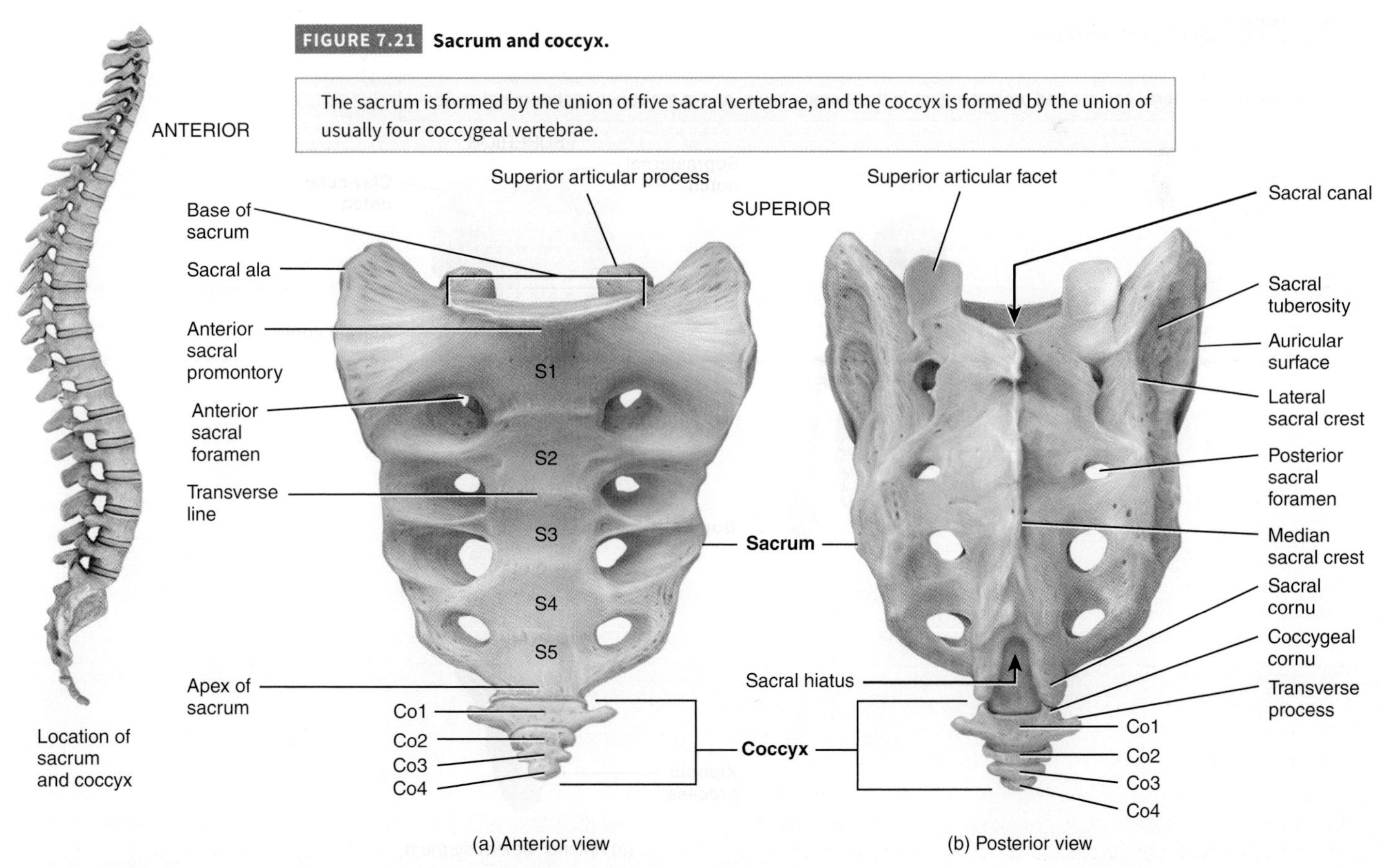

FIGURE 7.21 Sacrum and coccyx.

The sacrum is formed by the union of five sacral vertebrae, and the coccyx is formed by the union of usually four coccygeal vertebrae.

Q How many foramina pierce the sacrum, and what is their function?

Coccyx The **coccyx**, like the sacrum, is triangular in shape. It is formed by the fusion of usually four coccygeal vertebrae, indicated in Figure 7.21a as Co1–Co4. The coccygeal vertebrae fuse somewhat later than the sacral vertebrae, between the ages of 20 and 30. The dorsal surface of the body of the coccyx contains two long *coccygeal cornua* that are connected by ligaments to the sacral cornua. The coccygeal cornua are the pedicles and superior articular processes of the first coccygeal vertebra. They are on the lateral surfaces of the coccyx, formed by a series of *transverse processes*; the first pair are the largest. The coccyx articulates superiorly with the apex of the sacrum. In females, the coccyx points inferiorly to allow the passage of a baby during birth; in males, it points anteriorly (see Table 8.1).

7.11 Thorax

OBJECTIVE

- **Identify** the bones of the thorax, including sternum and ribs, and their functions.

The term **thorax** refers to the entire chest region. The skeletal part of the thorax, the **thoracic cage**, is a bony enclosure formed by the sternum, ribs and their costal cartilages, and the bodies of the thoracic vertebrae. (The costal cartilages attach the ribs to the sternum. The thoracic cage is narrower at its superior end and broader at its inferior end and is flattened from front to back. It encloses and protects the organs in the thoracic and superior abdominal cavities, provides support for the bones of the upper limbs, and, as you will see in Chapter 23, plays a role in breathing.

Sternum

The **sternum**, or breastbone, is a flat, narrow bone located in the center of the anterior thoracic wall that measures about 15 cm (6 in.) in length and consists of three parts (Figure 7.22). The superior part is the **manubrium** (ma-NOO-brē-um = handlelike); the middle and largest part is the **body**; and the inferior, smallest part is the **xiphoid process** (ZĪ-foyd = sword-shaped). The segments of the sternum typically fuse by age 25, and the points of fusion are marked by transverse ridges.

The junction of the manubrium and body forms the *sternal angle.* The manubrium has a depression on its superior surface, the *suprasternal notch.* Lateral to the suprasternal notch are *clavicular notches* that articulate with the medial ends of the clavicles to form

FIGURE 7.22 Skeleton of the thorax.

The bones of the thorax enclose and protect organs in the thoracic cavity and in the superior abdominal cavity.

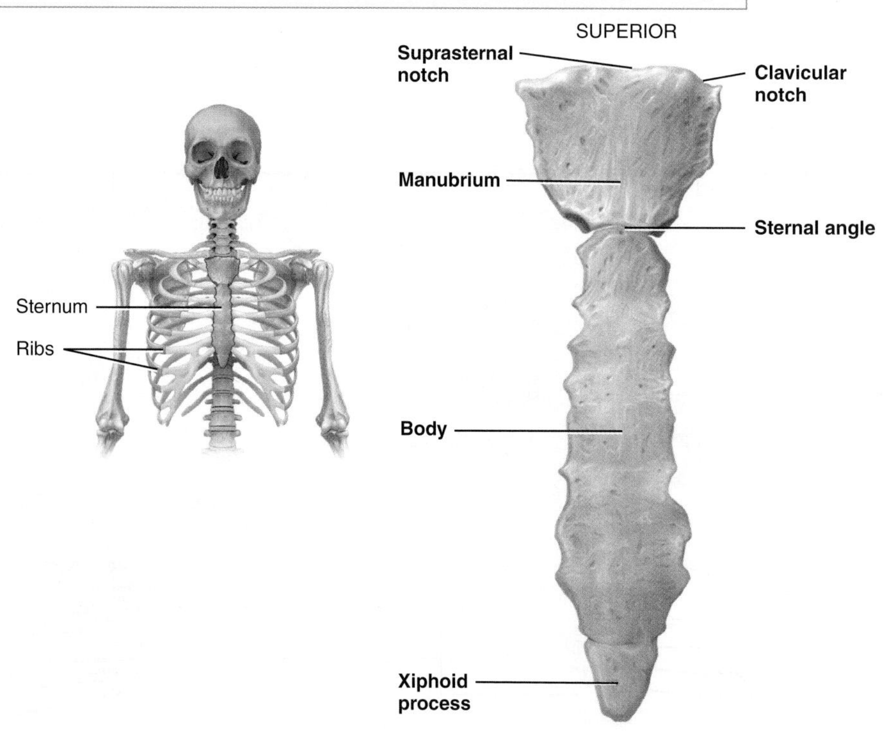

(a) Anterior view of sternum

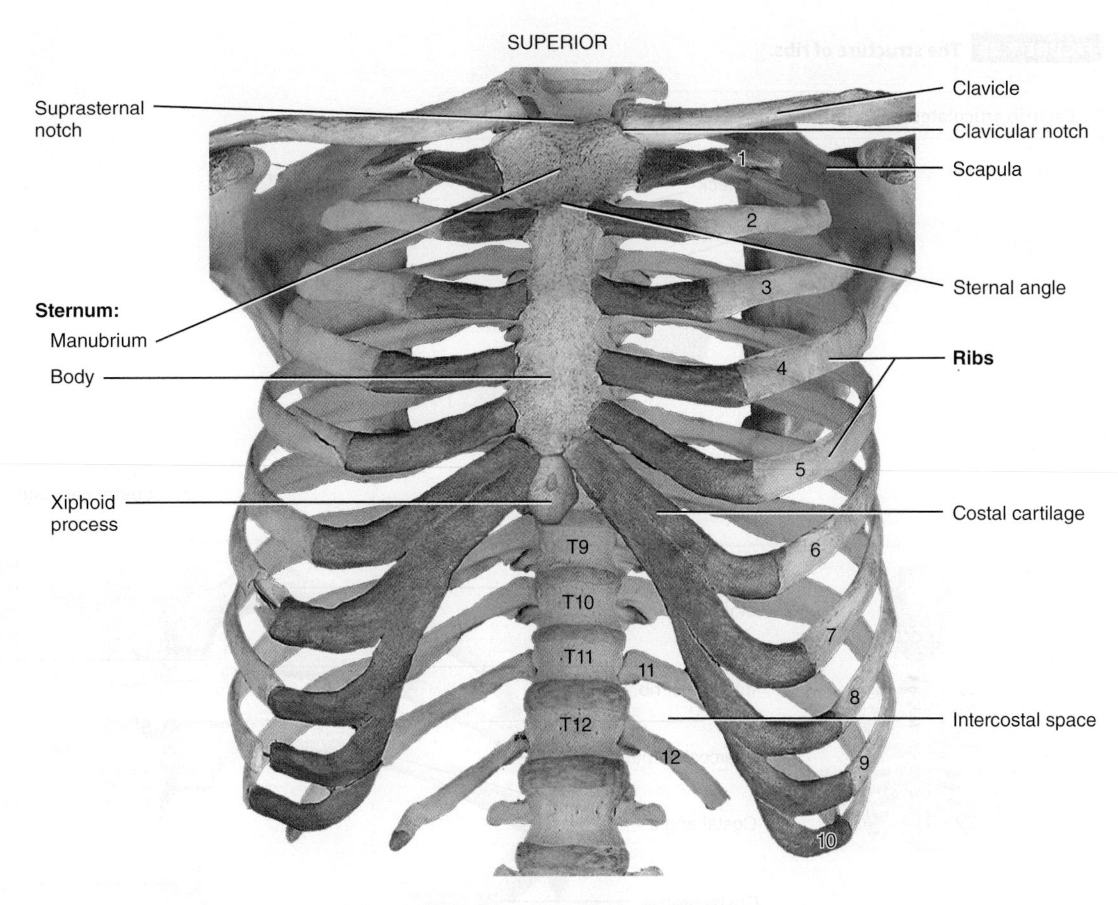

(b) Anterior view of skeleton of thorax

Q With which ribs does the body of the sternum articulate?

the *sternoclavicular joints.* The manubrium also articulates with the costal cartilages of the first and second ribs. The body of the sternum articulates directly or indirectly with the costal cartilages of the second through tenth ribs. The xiphoid process consists of hyaline cartilage during infancy and childhood and does not completely ossify until about age 40. No ribs are attached to it, but the xiphoid process provides attachment for some abdominal muscles. Incorrect positioning of the hands of a rescuer during cardiopulmonary resuscitation (CPR) may fracture the xiphoid process, driving it into internal organs. During thoracic surgery, the sternum may be split along the midline and the halves spread apart to allow surgeons access to structures in the thoracic cavity such as the thymus, heart, and great vessels of the heart. After surgery, the halves of the sternum are held together with wire sutures.

Ribs

Twelve pairs of **ribs**, numbered 1–12 from superior to inferior, give structural support to the sides of the thoracic cavity (**Figure 7.22b**). The ribs increase in length from the first through seventh, and then decrease in length to rib 12. Each rib articulates posteriorly with its corresponding thoracic vertebra.

The first through seventh pairs of ribs have a direct anterior attachment to the sternum by a strip of hyaline cartilage called *costal cartilage* (*cost-* = rib). The costal cartilages contribute to the elasticity of the thoracic cage and prevent various blows to the chest from fracturing the sternum and/or ribs. The ribs that have costal cartilages and attach directly to the sternum are called *true* (*vertebrosternal*) *ribs.* The articulations formed between the true ribs and the sternum are called *sternocostal joints.* The remaining five pairs of ribs are termed *false ribs* because their costal cartilages either attach indirectly to the sternum or do not attach to the sternum at all. The cartilages of the eighth, ninth, and tenth pairs of ribs attach to one another and then to the cartilages of the seventh pair of ribs. These false ribs are called *vertebrochondral ribs.* The eleventh and twelfth pairs of ribs are false ribs designated as *floating* (*vertebral*) *ribs* because the costal cartilages at their anterior ends do not attach to the sternum at all. These ribs attach only posteriorly to the thoracic vertebrae. Inflammation of one or more costal cartilages, called *costochondritis*, is characterized by local tenderness and pain in the anterior chest wall that may radiate. The symptoms mimic the chest pain (angina pectoris) associated with a heart attack.

Figure 7.23a shows the parts of a typical (third through ninth) rib. The *head* is a projection at the posterior end of the rib that

FIGURE 7.23 The structure of ribs.

Each rib articulates posteriorly with its corresponding thoracic vertebra.

Head
Neck
Tubercle
Articular facet
Costal angle
Costal groove
Superior facet
Inferior facet
Body
Facet for costal cartilage

(a) Posterior view of left rib

Transverse process of vertebra
Facet for tubercle of rib
Body
Intercostal space
Costal angle
Costal groove
Head of rib (seen through transverse process):
Superior facet
Inferior facet
Spinous process of vertebra (shortened)
Sternum
Costal cartilage

(b) Posterior view of left ribs articulated with thoracic vertebrae and sternum

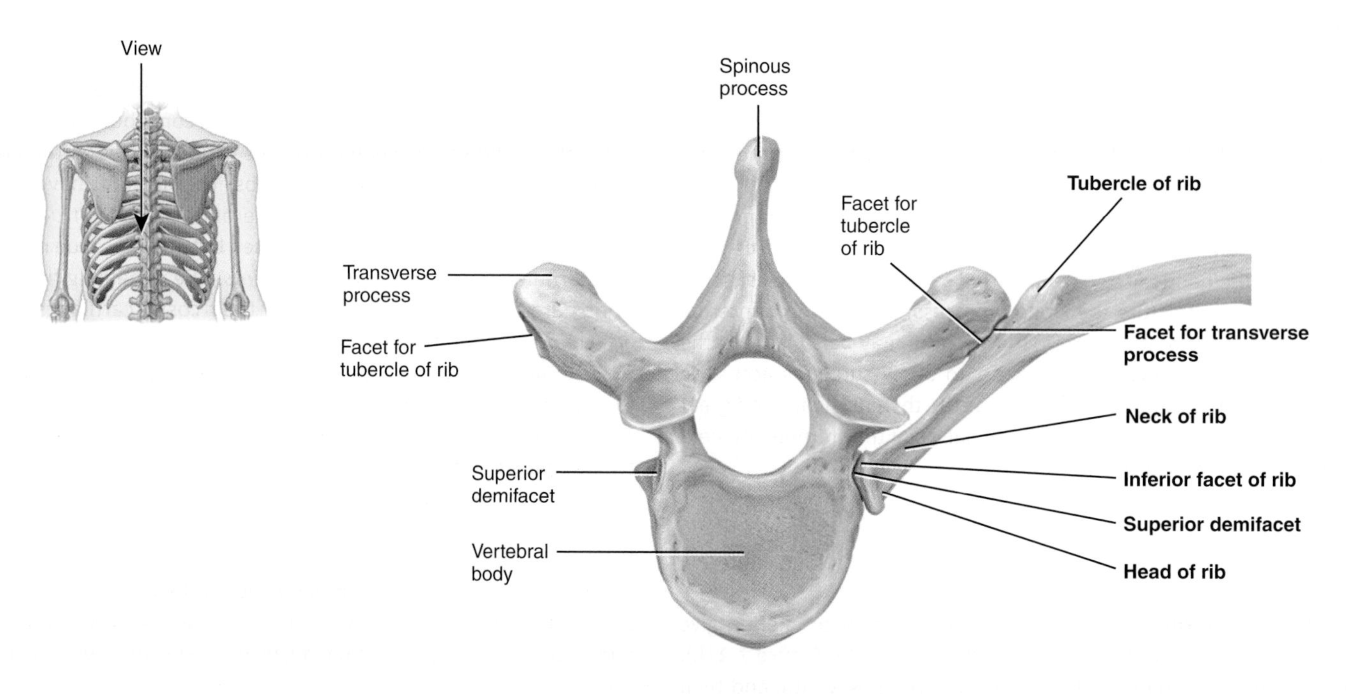

(c) Superior view of left rib articulated with thoracic vertebra

Q How does a rib articulate with a thoracic vertebra?

contains a pair of articular *facets* (superior and inferior). The facet of the head may fit either into a facet on the body of a single vertebra or into the demifacets of two adjoining vertebrae to form a *vertebrocostal joint.* The *neck* is a constricted portion of a rib just lateral to the head. A knoblike structure on the posterior surface, where the neck joins the body, is called a *tubercle* (TOO-ber-kul). The *nonarticular part* of the tubercle attaches to the transverse process of a vertebra by a ligament (lateral costotransverse ligament). The *articular part* of the tubercle articulates with the facet of a transverse process of a vertebra (**Figure 7.23c**) to form vertebrocostal joints. The *body* (*shaft*) is the main part of the rib. A short distance beyond the tubercle, an abrupt change in the curvature of the shaft occurs. This point is called the *costal angle.* The inner surface of the rib has a *costal groove* that protects the intercostal blood vessels and a small nerve.

Spaces between ribs, called *intercostal spaces*, are occupied by intercostal muscles, blood vessels, and nerves. Surgical access to the lungs or other structures in the thoracic cavity is commonly obtained through an intercostal space. Special rib retractors are used to create a wide separation between ribs. The costal cartilages are sufficiently elastic in younger individuals to permit considerable bending without breaking.

In summary, the posterior portion of the rib connects to a thoracic vertebra by its head and the articular part of a tubercle. The facet of the head fits into either a facet on the body of one vertebra (T1 only) or into the demifacets of two adjoining vertebrae. The articular part of the tubercle articulates with the facet of the transverse process of the vertebra.

See Clinical Connection: Rib Fractures, Dislocations, and Separations

Review Questions

7.1 Divisions of the Skeletal System

1. On what basis is the skeleton grouped into the axial and appendicular divisions?

7.2 Types of Bones

2. Give examples of long, short, flat, and irregular bones.

7.3 Bone Surface Markings

3. What are surface markings? What are their general functions?

7.4 Skull: An Overview

4. What is the purpose of the skull?

7.5 Cranial Bones

5. What structures pass through the supraorbital foramen?

6. How do the parietal bones relate to the cranial cavity?

7. What structures form the zygomatic arch?

8. What structures pass through the hypoglossal canal?

9. Why is the sphenoid bone called the keystone of the cranial floor?

10. The ethmoid bone forms which other cranial structures?

7.6 Facial Bones

11. Which bones form the hard palate? Which bones form the nasal septum?

7.7 Special Features of the Skull

12. What structures make up the nasal septum?

13. Which foramina and fissures are associated with the orbit?

14. Define the following: foramen, suture, paranasal sinus, and fontanel.

7.8 Hyoid Bone

15. What are the functions of the hyoid bone?

7.9 Vertebral Column

16. What are the functions of the vertebral column?

17. Describe the four curves of the vertebral column.

18. What are the three main parts of a typical vertebra?

19. What are the principal distinguishing characteristics of the bones of the various regions of the vertebral column?

7.10 Vertebral Regions

20. How do the atlas and axis differ from the other cervical vertebrae?

21. Describe several distinguishing features of thoracic vertebrae.

22. What are the distinguishing features of the lumbar vertebrae?

23. How many vertebrae fuse to form the sacrum and coccyx?

7.11 Thorax

24. What bones form the skeleton of the thorax?

25. What are the functions of the bones of the thorax?

26. What is the clinical significance of the xiphoid process?

27. How are ribs classified?

Critical Thinking Questions

1. Jimmy is in a car accident. He can't open his mouth and has been told that he suffers from the following: black eye, broken nose, broken cheek, broken upper jaw, damaged eye socket, and punctured lung. Describe *exactly* what structures have been affected by his car accident.
2. Bubba is a tug-of-war expert. He practices day and night by pulling on a rope attached to an 800-lb anchor. What kinds of changes in his bone structure would you expect him to develop?
3. A new mother brings her newborn infant home and has been told by her well-meaning friend not to wash the baby's hair for several months because the water and soap could "get through that soft area in the top of the head and cause brain damage." Explain to her why this is not true.

The Skeletal System: The Appendicular Skeleton

CHAPTER 8

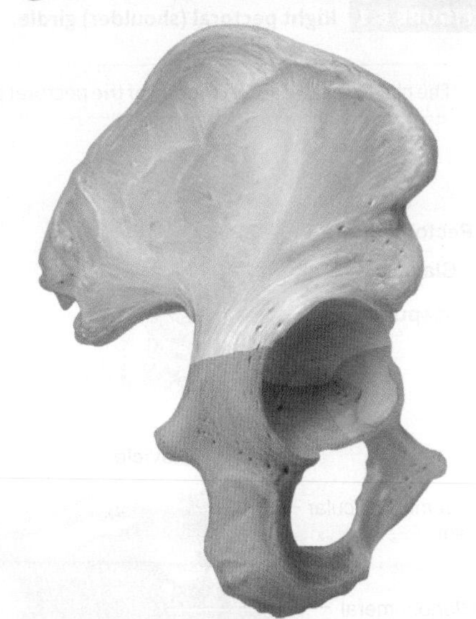

The Appendicular Skeleton and Homeostasis

The bones of the appendicular skeleton contribute to homeostasis by providing attachment points and leverage for muscles, which aids body movements; by providing support and protection of internal organs, such as the reproductive organs; and by storing and releasing calcium.

As noted in Chapter 7, the two main divisions of the skeletal system are the axial skeleton and the appendicular skeleton. As you learned in that chapter, the general function of the axial skeleton is the protection of internal organs; the primary function of the appendicular skeleton, the focus of this chapter, is movement. The appendicular skeleton includes the bones that make up the upper and lower limbs as well as the bones of the two girdles that attach the limbs to the axial skeleton. The bones of the appendicular skeleton are connected with one another and with skeletal muscles, permitting you to do things such as walk, write, use a computer, dance, swim, and play a musical instrument.

Q Did you ever wonder what causes runner's knee?

8.1 Pectoral (Shoulder) Girdle

OBJECTIVE

- **Identify** the bones of the pectoral (shoulder) girdle, their functions, and their principal markings.

The human body has two **pectoral** (*shoulder*) **girdles** (PEK-tō-ral) that attach the bones of the upper limbs to the axial skeleton (**Figure 8.1**). Each of the two pectoral girdles consists of a clavicle and a scapula. The *clavicle* is the anterior bone and articulates with the manubrium of the sternum at the *sternoclavicular joint.* The scapula articulates with the clavicle at the *acromioclavicular joint* and with the humerus at the *glenohumeral (shoulder) joint.* The pectoral girdles do not articulate with the vertebral column and are held in position and stabilized by a group of large muscles that extend from the vertebral column and ribs to the scapula.

Clavicle

Each slender, **S**-shaped **clavicle** (KLAV-i-kul = key), or *collarbone*, lies horizontally across the anterior part of the thorax superior to the first rib (**Figure 8.2**). It is subcutaneous (under the skin) and easily palpable along its length. The bone is **S**-shaped because the medial half is convex anteriorly (curves toward you when viewed in the anatomical position), and the lateral half is concave anteriorly (curves away from you). It is rougher and more curved in males.

The medial end, called the *sternal end*, is rounded and articulates with the manubrium of the sternum to form the *sternoclavicular joint.* The broad, flat, lateral end, the *acromial end* (a-KRŌ-mē-al), articulates with the acromion of the scapula to form the *acromioclavicular joint* (see **Figure 8.1**). The *conoid tubercle* (KŌ-noyd = conelike) on the inferior surface of the lateral end of the bone is a point of attachment for the conoid ligament, which attaches the clavicle and scapula. As its name implies, the *impression for the costoclavicular ligament* on the inferior surface of the sternal end is a point of attachment for the costoclavicular ligament (**Figure 8.2b**), which attaches the clavicle and first rib.

See Clinical Connection: Fractured Clavicle

Scapula

Each **scapula** (SCAP-ū-la; plural is *scapulae*), or *shoulder blade*, is a large, triangular, flat bone situated in the superior part of the posterior thorax between the levels of the second and seventh ribs (**Figure 8.3**).

FIGURE 8.1 **Right pectoral (shoulder) girdle.**

The clavicle is the anterior bone of the pectoral girdle, and the scapula is the posterior bone.

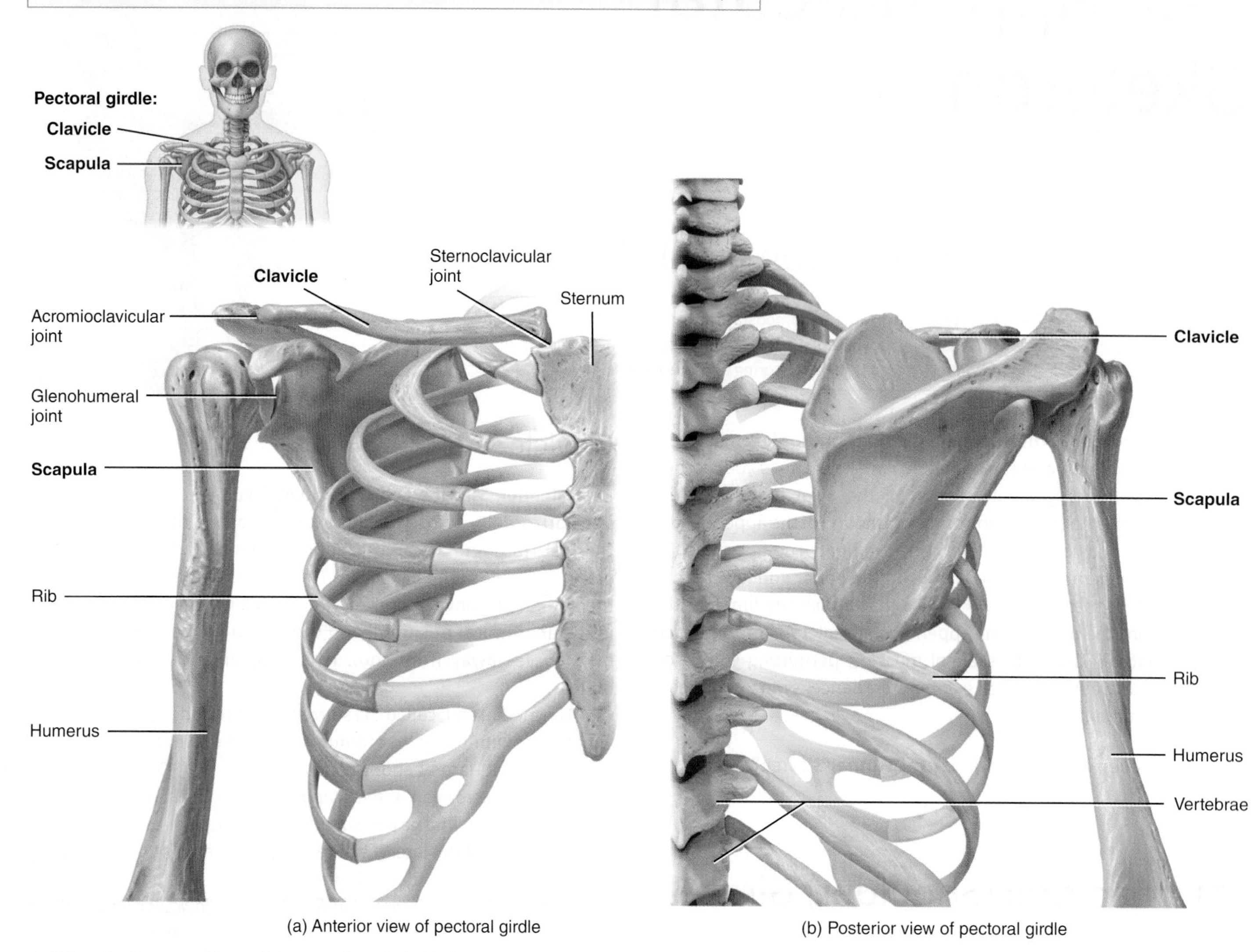

(a) Anterior view of pectoral girdle

(b) Posterior view of pectoral girdle

Q What is the function of the pectoral girdle?

A prominent ridge called the *spine* runs diagonally across the posterior surface of the scapula. The lateral end of the spine projects as a flattened, expanded process called the *acromion* (a-KRŌ-mē-on; *acrom-* = topmost; *-omos* = shoulder), easily felt as the high point of the shoulder. Tailors measure the length of the upper limb from the acromion. As noted earlier, the acromion articulates with the acromial end of the clavicle to form the *acromioclavicular joint.* Inferior to the acromion is a shallow depression, the *glenoid cavity*, that accepts the head of the humerus (arm bone) to form the *glenohumeral (shoulder) joint* (see **Figure 8.1**).

The thin edge of the scapula closer to the vertebral column is called the *medial (vertebral) border.* The thick edge of the scapula closer to the arm is called the *lateral (axillary) border.* The medial and lateral borders join at the *inferior angle.* The superior edge of the scapula, called the *superior border*, joins the medial border at the *superior angle.* The *scapular notch* is a prominent indentation along the superior border through which the suprascapular nerve passes.

At the lateral end of the superior border of the scapula is a projection of the anterior surface called the *coracoid process* (KOR-a-koyd = like a crow's beak), to which the tendons of muscles (pectoralis minor, coracobrachialis, and biceps brachii) and ligaments (coracoacromial, conoid, and trapezoid) attach. Superior and inferior to the spine on the posterior surface of the scapula are two fossae: The *supraspinous*

FIGURE 8.2 **Right clavicle.**

The clavicle articulates medially with the manubrium of the sternum and laterally with the acromion of the scapula.

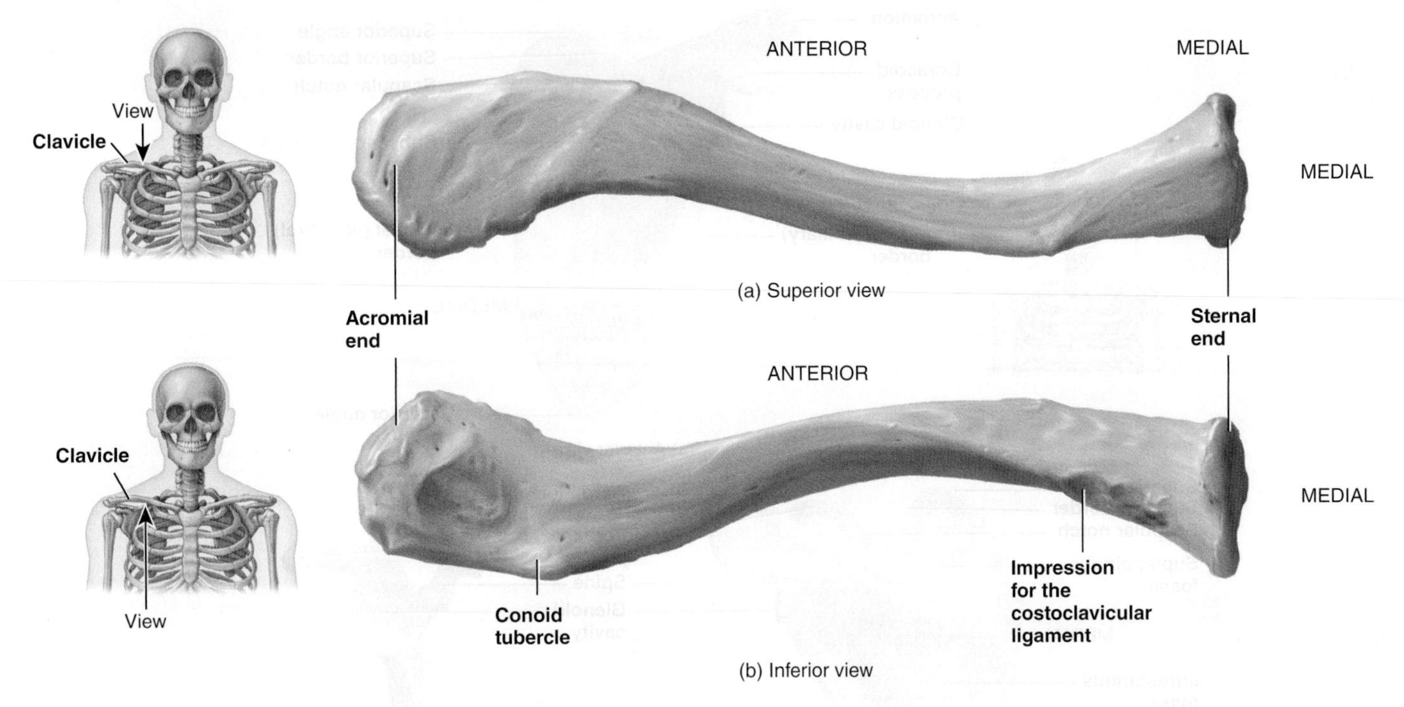

Q Which part of the clavicle is its weakest point?

fossa (sū-pra-SPĪ-nus) is a surface of attachment for the supraspinatus muscle of the shoulder, and the *infraspinous fossa* (in-fra-SPĪ-nus) serves as a surface of attachment for the infraspinatus muscle of the shoulder. On the anterior surface of the scapula is a slightly hollowed-out area called the *subscapular fossa*, a surface of attachment for the subscapularis muscle.

8.2 Upper Limb (Extremity)

OBJECTIVE

- **Identify** the bones of the upper limb and their principal markings.

Each **upper limb** (*upper extremity*) has 30 bones in three locations—(1) the humerus in the arm; (2) the ulna and radius in the forearm; and (3) the 8 carpals in the carpus (wrist), the 5 metacarpals in the metacarpus (palm), and the 14 phalanges (bones of the digits) in the hand (see **Figures 8.4** and **8.5**).

Skeleton of the Arm—Humerus

The **humerus** (HŪ-mer-us), or arm bone, is the longest and largest bone of the upper limb (**Figure 8.4**). It articulates proximally with the scapula and distally with two bones, the ulna and the radius, to form the elbow joint.

The proximal end of the humerus features a rounded *head* that articulates with the glenoid cavity of the scapula to form the *glenohumeral (shoulder) joint.* Distal to the head is the *anatomical neck*, which is visible as an oblique groove. It is the former site of the epiphyseal (growth) plate in an adult humerus. The *greater tubercle* is a lateral projection distal to the anatomical neck. It is the most laterally palpable bony landmark of the shoulder region and is immediately inferior to the palpable acromion of the scapula mentioned earlier. The *lesser tubercle* projects anteriorly. Between the two tubercles there is a groove named the *intertubercular sulcus*. The *surgical neck* is a constriction in the humerus just distal to the tubercles, where the head tapers to the shaft; it is so named because fractures often occur here.

The *body (shaft)* of the humerus is roughly cylindrical at its proximal end, but it gradually becomes triangular until it is flattened and broad at its distal end. Laterally, at the middle portion of the shaft, there is a roughened, V-shaped area called the *deltoid tuberosity*.

FIGURE 8.3 **Right scapula (shoulder blade).**

The glenoid cavity of the scapula articulates with the head of the humerus to form the glenohumeral (shoulder) joint.

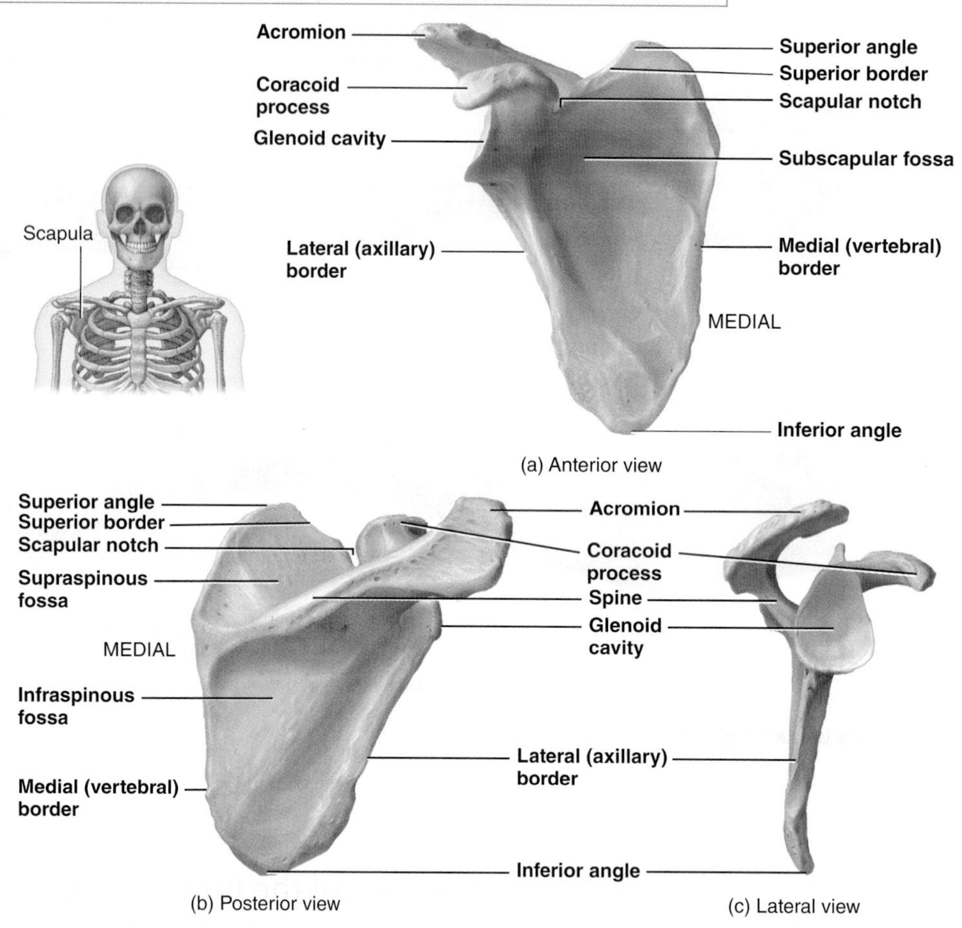

Q Which part of the scapula forms the high point of the shoulder?

This area serves as a point of attachment for the tendons of the deltoid muscle. On the posterior surface of the humerus is the *radial groove*, which runs along the deltoid tuberosity and contains the radial nerve.

Several prominent features are evident at the distal end of the humerus. The *capitulum* (ka-PIT-ū-lum; *capit-* = head) is a rounded knob on the lateral aspect of the bone that articulates with the head of the radius. The *radial fossa* is an anterior depression above the capitulum that articulates with the head of the radius when the forearm is flexed (bent). The *trochlea* (TROK-lē-a = pulley), located medial to the capitulum, is a spool-shaped surface that articulates with the trochlear notch of the ulna. The *coronoid fossa* (KOR-ō-noyd = crown-shaped) is an anterior depression that receives the coronoid process of the ulna when the forearm is flexed. The *olecranon fossa* (ō-LEK-ra-non = elbow) is a large posterior depression that receives the olecranon of the ulna when the forearm is extended (straightened). The *medial epicondyle* and *lateral epicondyle* are rough projections on either side of the distal end of the humerus to which the tendons of most muscles of the forearm are attached. The ulnar nerve may be palpated by rolling a finger over the skin surface above the posterior surface of the medial epicondyle. This nerve is the one that makes you feel a very severe pain when you hit your elbow, which for some reason is commonly referred to as the funnybone, even though this event is anything but funny.

Skeleton of the Forearm—Ulna and Radius

The **ulna** is located on the medial aspect (the little-finger side) of the forearm and is longer than the radius (**Figure 8.5**). A convenient mnemonic to help you remember the location of the ulna in relation to the hand is "p.u." (the **p**inky is on the **u**lna side).

At the proximal end of the ulna (**Figure 8.5b**) is the *olecranon*, which forms the prominence of the elbow. With the olecranon, an

FIGURE 8.4 Right humerus in relation to the scapula, ulna, and radius.

The humerus is the longest and largest bone of the upper limb.

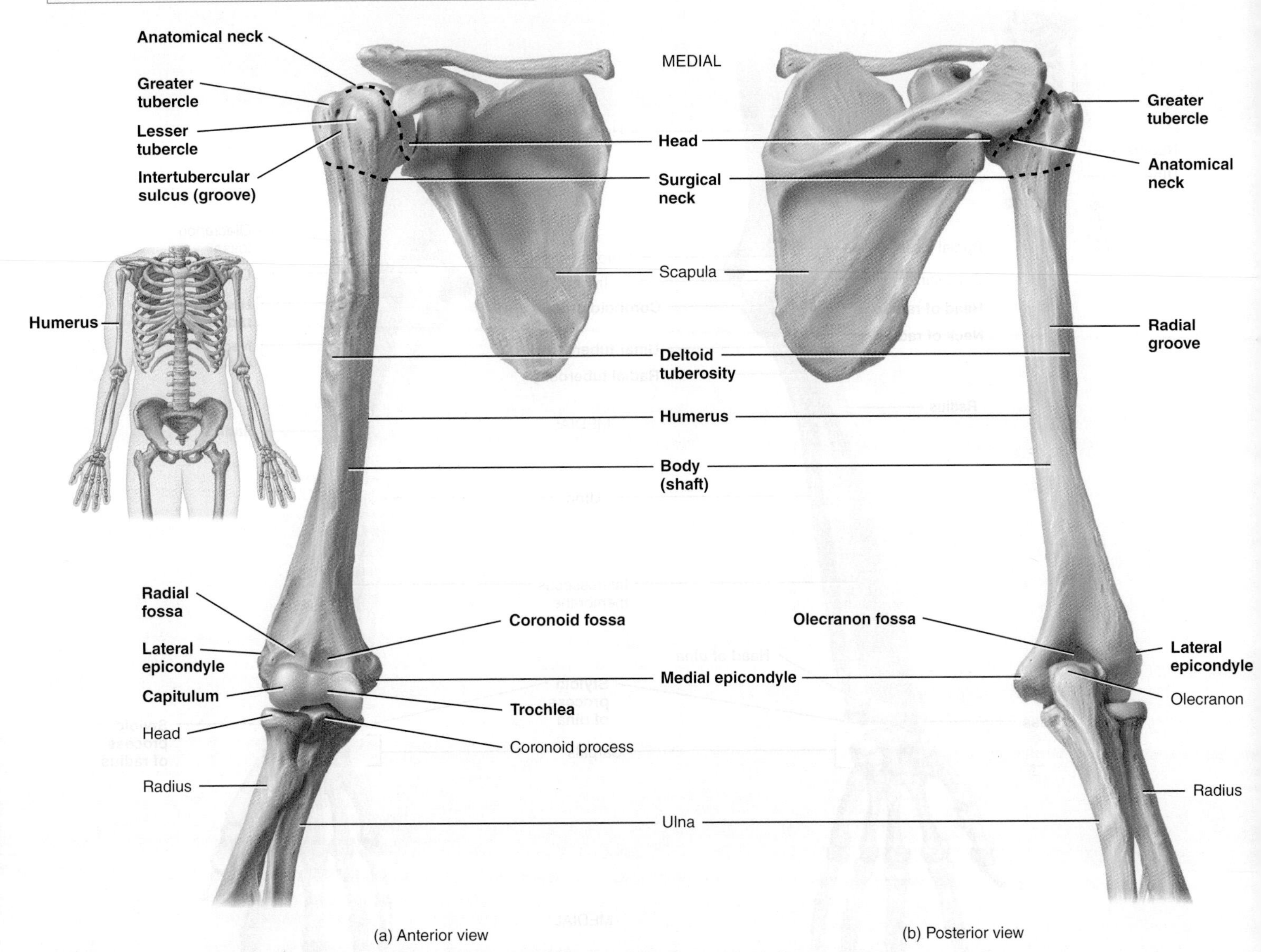

Q Which parts of the humerus articulate with the radius at the elbow? With the ulna at the elbow?

anterior projection called the *coronoid process* (**Figure 8.5a**) articulates with the trochlea of the humerus. The *trochlear notch* is a large curved area between the olecranon and coronoid process that forms part of the elbow joint (see **Figure 8.6b**). Lateral and inferior to the trochlear notch is a depression, the *radial notch*, which articulates with the head of the radius. Just inferior to the coronoid process is the *ulnar tuberosity*, to which the biceps brachii muscle attaches. The distal end of the ulna consists of a *head* that is separated from the wrist by a disc of fibrocartilage. A *styloid process* is located on the posterior side of the ulna's distal end. It provides attachment for the ulnar collateral ligament to the wrist.

The **radius** is the smaller bone of the forearm and is located on the lateral aspect (thumb side) of the forearm (**Figure 8.5a**). In contrast to the ulna, the radius is narrow at its proximal end and widens at its distal end.

The proximal end of the radius has a disc-shaped *head* that articulates with the capitulum of the humerus and the radial notch of the ulna. Inferior to the head is the constricted *neck*. A roughened area inferior to the neck on the anteromedial side, called the *radial tuberosity*, is a point of attachment for the tendons of the biceps brachii muscle. The shaft of the radius widens distally to form a *styloid process* on the lateral side, which can be felt proximal to the thumb. The

FIGURE 8.5 Right ulna and radius in relation to the humerus and carpals.

In the forearm, the longer ulna is on the medial side, and the shorter radius is on the lateral side.

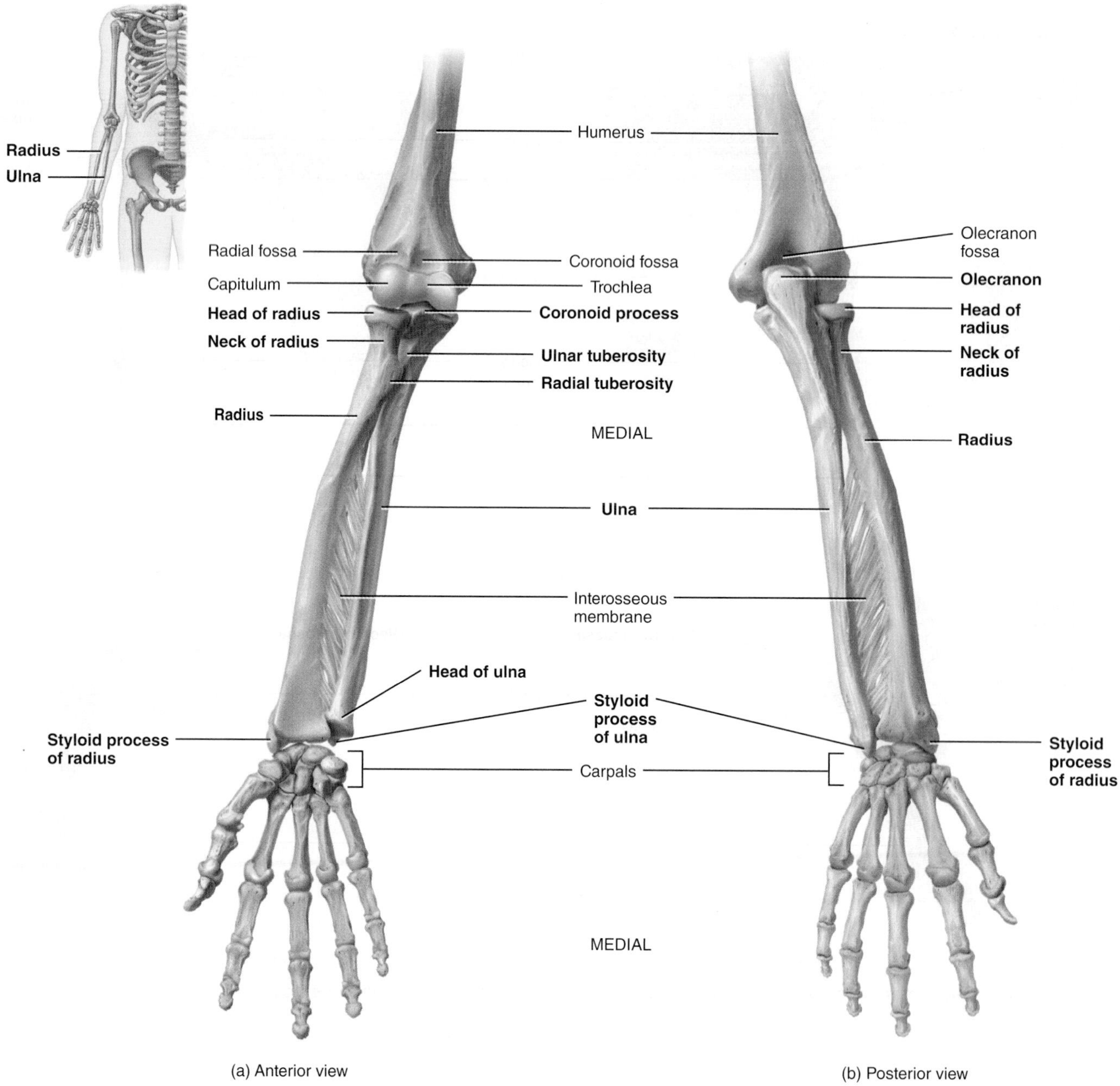

(a) Anterior view

(b) Posterior view

Q What part of the ulna is called the "elbow"?

distal end of the radius contains a narrow concavity, the *ulnar notch*, which articulates with the head of the ulna. The styloid process provides attachment for the brachioradialis muscle and for attachment of the radial collateral ligament to the wrist. Fracture of the distal end of the radius is the most common fracture in adults older than 50, typically occurring during a fall.

The ulna and radius articulate with the humerus at the *elbow joint.* The articulation occurs in two places (**Figure 8.6a, b**): where the head of the radius articulates with the capitulum of the humerus, and where the trochlear notch of the ulna articulates with the trochlea of the humerus.

The ulna and the radius connect with one another at three sites. First, a broad, flat, fibrous connective tissue called the *interosseous membrane* (in-ter-OS-ē-us; *inter-* = between, *-osse* = bone) joins the shafts of the two bones (see **Figure 8.5**). This membrane also provides a site of attachment for some of the deep

FIGURE 8.6 Articulations formed by the ulna and radius. (a) Elbow joint. (b) Joint surfaces at proximal end of the ulna. (c) Joint surfaces at distal ends of radius and ulna.

The elbow joint is formed by two articulations: (1) the trochlear notch of the ulna with the trochlea of the humerus and (2) the head of the radius with the capitulum of the humerus.

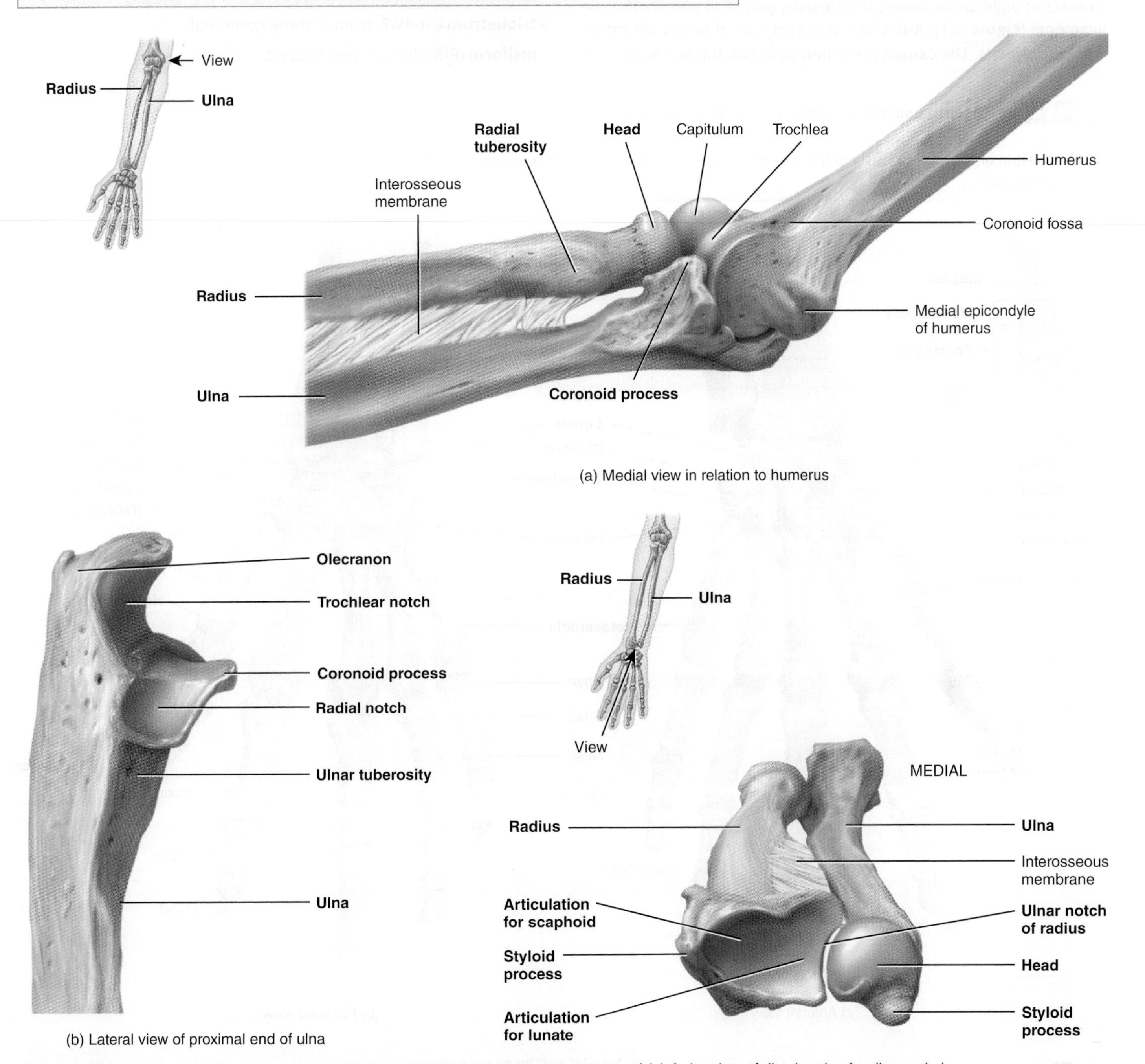

(a) Medial view in relation to humerus

(b) Lateral view of proximal end of ulna

(c) Inferior view of distal ends of radius and ulna

Q How many points of attachment are there between the radius and ulna?

skeletal muscles of the forearm. The ulna and radius articulate directly at their proximal and distal ends (**Figure 8.6b, c**). Proximally, the head of the radius articulates with the ulna's radial notch. This articulation is the *proximal radioulnar joint.* Distally, the head of the ulna articulates with the *ulnar notch* of the radius. This articulation is the *distal radioulnar joint.* Finally, the distal end of the radius articulates with three bones of the wrist—the lunate, the scaphoid, and the triquetrum—to form the *radiocarpal (wrist) joint.*

Skeleton of the Hand—Carpals, Metacarpals, and Phalanges

Carpals The **carpus** (*wrist*) is the proximal region of the hand and consists of eight small bones, the **carpals**, joined to one another by ligaments (**Figure 8.7**). Articulations among carpal bones are called *intercarpal joints*. The carpals are arranged in two transverse rows of four bones each. Their names reflect their shapes. The carpals in the proximal row, from lateral to medial, are the

- **scaphoid** (SKAF-oyd = boatlike)
- **lunate** (LOO-nāt = moon-shaped)
- **triquetrum** (trī-KWĒ-trum = three-cornered)
- **pisiform** (PĪS-i-form = pea-shaped).

FIGURE 8.7 **Right wrist and hand in relation to the ulna and radius.**

The skeleton of the hand consists of the proximal carpals, the intermediate metacarpals, and the distal phalanges.

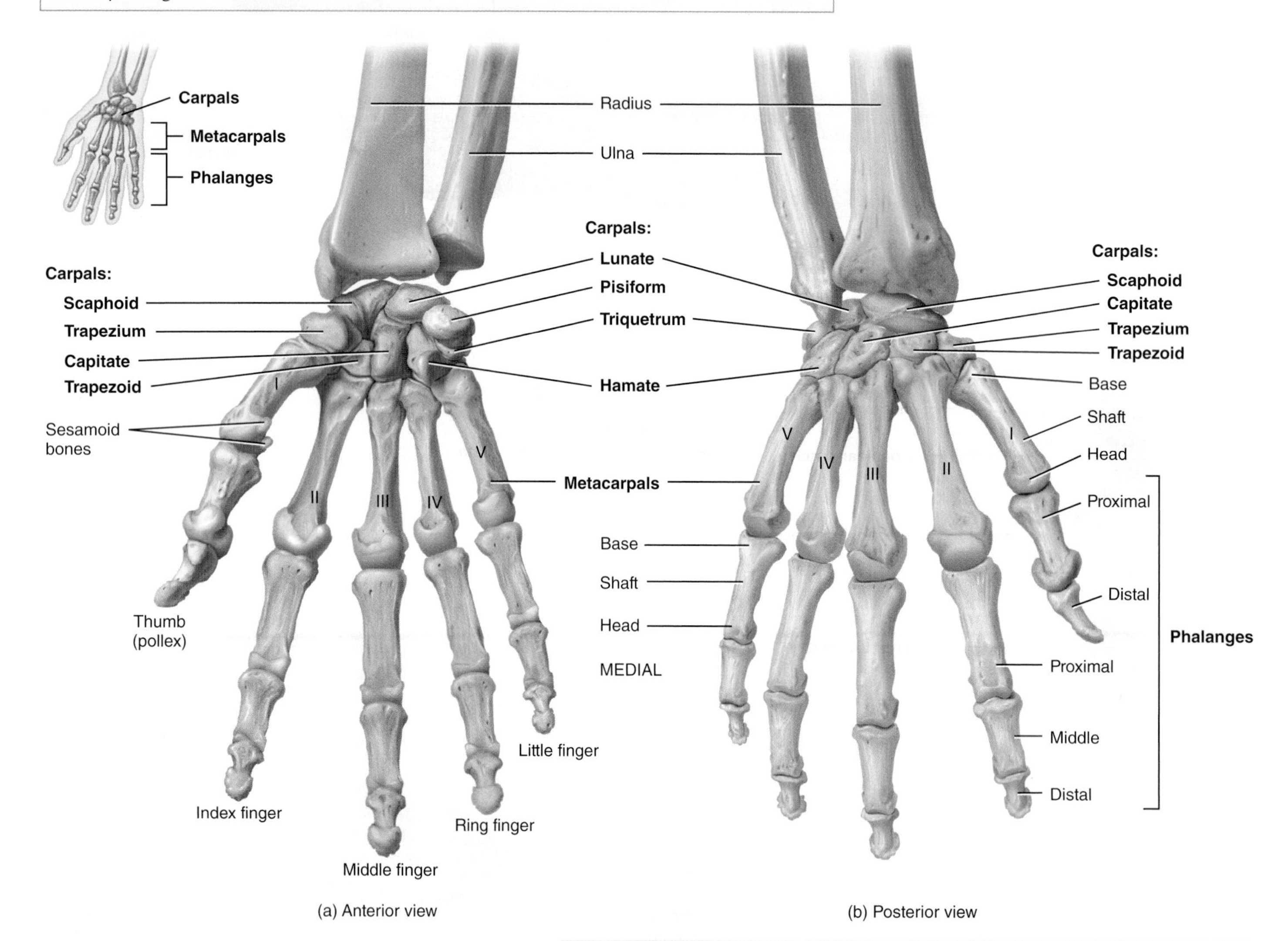

(a) Anterior view

(b) Posterior view

See Clinical Connection: Boxer's Fracture

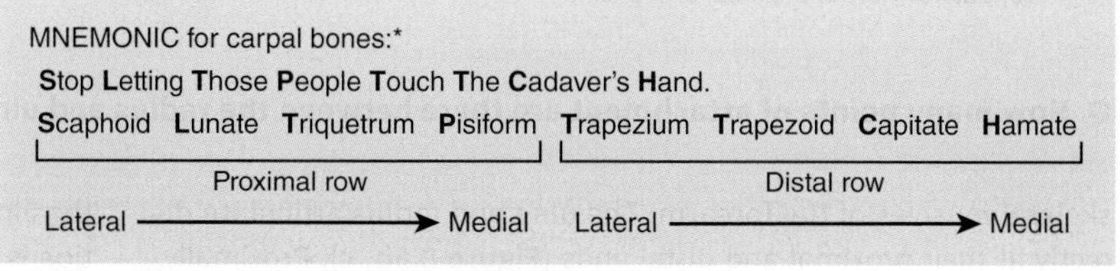

* *Edward Tanner, University of Alabama, SOM*

Q Which is the most frequently fractured wrist bone?

The proximal row of carpals articulates with the distal ends of the ulna and radius to form the *wrist joint*. The carpals in the distal row, from lateral to medial, are the

- **trapezium** (tra-PĒ-zē-um = four-sided figure with no two sides parallel)
- **trapezoid** (TRAP-e-zoyd = four-sided figure with two sides parallel)
- **capitate** (KAP-i-tāt = head-shaped)
- **hamate** (HAM-āt = hooked).

The capitate is the largest carpal bone; its rounded projection, the head, articulates with the lunate. The hamate is named for a large hook-shaped projection on its anterior surface. In about 70% of carpal fractures, only the scaphoid is broken. This is because the force of a fall on an outstretched hand is transmitted from the capitate through the scaphoid to the radius.

The anterior concave space formed by the pisiform and hamate (on the ulnar side), and the scaphoid and trapezium (on the radial side), with the rooflike covering of the *flexor retinaculum* (strong fibrous bands of connective tissue) is the **carpal tunnel**. The long flexor tendons of the digits and thumb and the median nerve pass through the carpal tunnel. Narrowing of the carpal tunnel, due to such factors as inflammation, may give rise to a condition called *carpal tunnel syndrome* (described in Clinical Connection: Carpal Tunnel Syndrome in Section 11.18).

There is a useful mnemonic for learning the names of the carpal bones provided in **Figure 8.7**. The first letter of the carpal bones from lateral to medial (proximal row, then distal row) corresponds to the first letter of each word in the mnemonic.

Metacarpals

The **metacarpus** (*meta-* = beyond), or *palm*, is the intermediate region of the hand and consists of five bones called **metacarpals**.

Each metacarpal bone consists of a proximal *base*, an intermediate *shaft*, and a distal *head* (**Figure 8.7b**). The metacarpal bones are numbered I to V (or 1–5), starting with the thumb, from lateral to medial. The bases articulate with the distal row of carpal bones to form the *carpometacarpal joints*. The heads articulate with the proximal phalanges to form the *metacarpophalangeal joints*. The heads of the metacarpals, commonly called "knuckles," are readily visible in a clenched fist.

Phalanges

The **phalanges** (fa-LAN-jēz; *phalan-* = a battle line), or bones of the *digits*, make up the distal part of the hand. There are 14 phalanges in the five digits of each hand and, like the metacarpals, the digits are numbered I to V (or 1–5), beginning with the thumb, from lateral to medial. A single bone of a digit is referred to as a *phalanx* (FĀ- lanks).

Each phalanx consists of a proximal *base*, an intermediate *shaft*, and a distal *head*. The *thumb* (*pollex*) has two phalanges called *proximal* and *distal phalanges*. The other four digits have three phalanges called *proximal*, *middle*, and *distal phalanges*. In order from the thumb, these other four digits are commonly referred to as the *index finger*, *middle finger*, *ring finger*, and *little finger*. The proximal phalanges of all digits articulate with the metacarpal bones. The middle phalanges of the fingers (II–V) articulate with their distal phalanges. (The proximal phalanx of the thumb [I] articulates with its distal phalanx.) Joints between phalanges are called *interphalangeal joints*.

8.3 Pelvic (Hip) Girdle

OBJECTIVE

- **Identify** the bones of the pelvic girdle and their principal markings.

The **pelvic** (*hip*) **girdle** consists of the two **hip bones**, also called **coxal** (KOK-sal; *cox-* = hip) or **pelvic bones** or **os coxa** (**Figure 8.8**).

FIGURE 8.8 **Bony pelvis.** Shown here is the female bony pelvis.

The hip bones unite anteriorly at the pubic symphysis and posteriorly at the sacrum to form the bony pelvis.

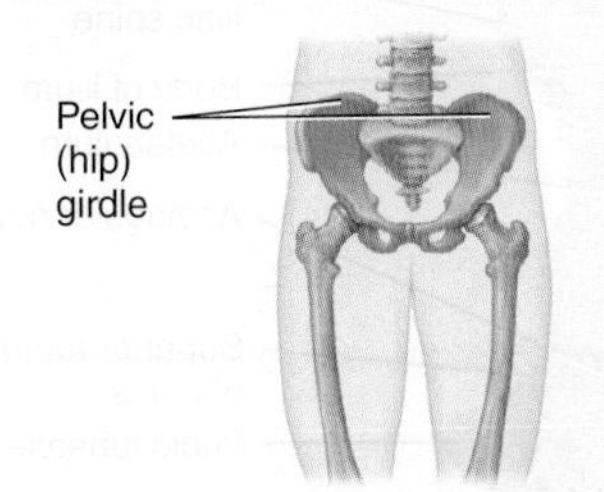

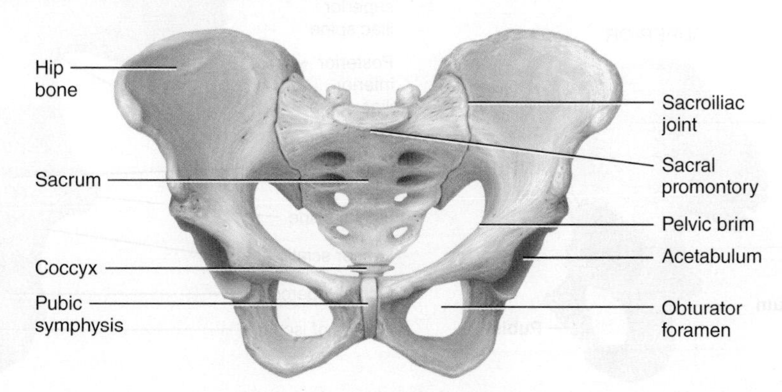

Anterosuperior view of pelvic girdle

Q What are the functions of the bony pelvis?

The hip bones unite anteriorly at a joint called the **pubic symphysis** (PŪ-bik SIM-fi-sis). They unite posteriorly with the sacrum at the *sacroiliac joints.* The complete ring composed of the hip bones, pubic symphysis, sacrum, and coccyx forms a deep, basinlike structure called the **bony pelvis** (*pelv-* = basin). The plural is *pelves* (PEL-vēz) or *pelvises.* Functionally, the bony pelvis provides a strong and stable support for the vertebral column and pelvic and lower abdominal organs. The pelvic girdle of the bony pelvis also connects the bones of the lower limbs to the axial skeleton.

Each of the two hip bones of a newborn consists of three bones separated by cartilage: a superior *ilium*, an inferior and anterior *pubis*, and an inferior and posterior *ischium*. By age 23, the three separate bones fuse together (**Figure 8.9a**). Although the hip bones function as single bones, anatomists commonly discuss each hip bone as three separate bones.

Ilium

The **ilium** (IL-ē-um = flank), the largest of the three components of the hip bone (**Figure 8.9b**), is composed of a superior *ala* (= wing) and an inferior *body*. The body is one of the components of the *acetabulum*, the socket for the head of the femur. The superior border of the ilium, the *iliac crest*, ends anteriorly in a blunt *anterior superior iliac spine.* Bruising of the anterior superior iliac spine and associated soft tissues, such as occurs in body contact sports, is called a **hip pointer**. Below this spine is the *anterior inferior iliac spine*. Posteriorly, the iliac crest ends in a sharp *posterior superior iliac spine.* Below this spine is the *posterior inferior iliac spine*. The spines serve as points of attachment for the tendons of the muscles of the trunk, hip, and thighs. Below the posterior inferior iliac spine is the *greater sciatic notch* (sī-AT-ik), through which the sciatic nerve (the longest nerve in the body) passes, along with other nerve and muscles.

The medial surface of the ilium contains the *iliac fossa*, a concavity where the tendon of the iliacus muscle attaches. Posterior to this fossa are the *iliac tuberosity*, a point of attachment for the sacroiliac ligament, and the *auricular surface* (*auric-* = ear-shaped), which articulates with the sacrum to form the *sacroiliac joint* (see **Figure 8.8**). Projecting anteriorly and inferiorly from the auricular surface is a ridge called the *arcuate line* (AR-kū-āt; *arc-* = bow).

The other conspicuous markings of the ilium are three arched lines on its lateral surface called the *posterior gluteal line* (*glut-* = buttock), the *anterior gluteal line*, and the *inferior gluteal line.* The gluteal muscles attach to the ilium between these lines.

FIGURE 8.9 Right hip bone. The lines of fusion of the ilium, ischium, and pubis depicted in part (a) and (b) are not always visible in an adult.

The acetabulum is the socket for the head of the femur, where the three parts of the hip bone converge and ossify.

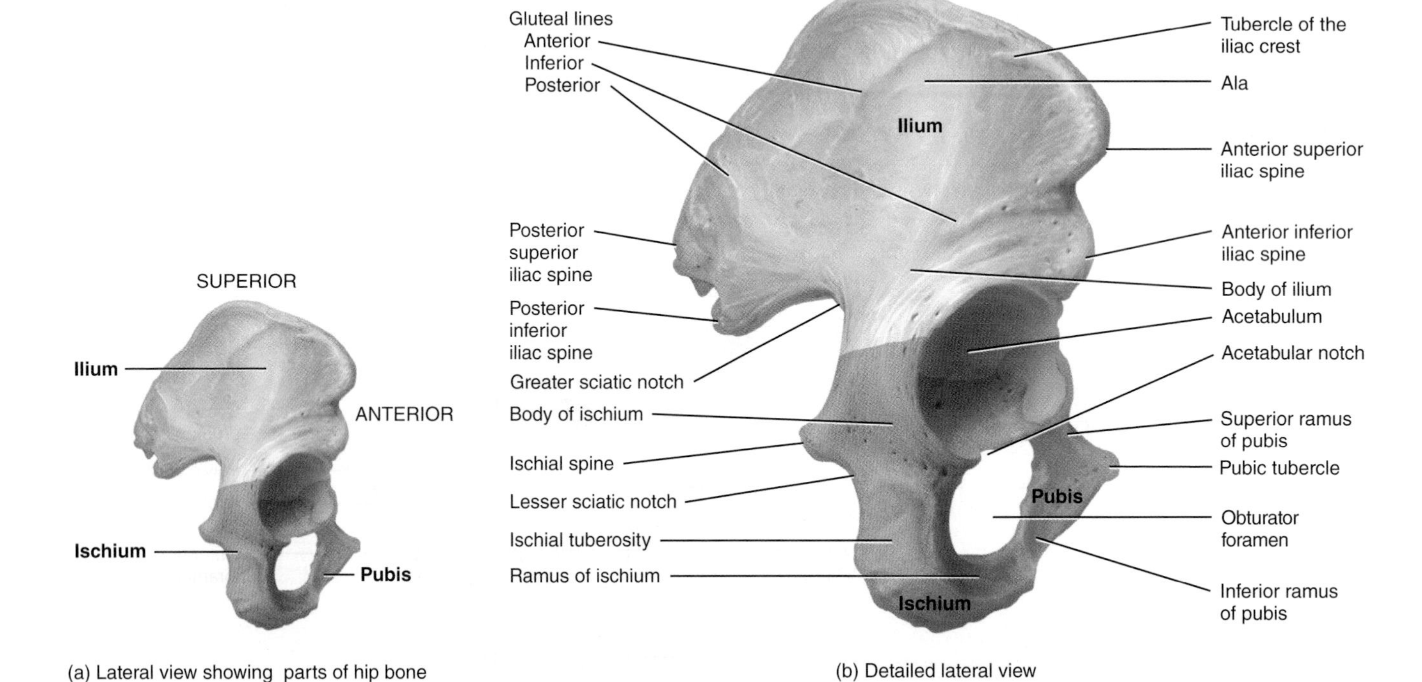

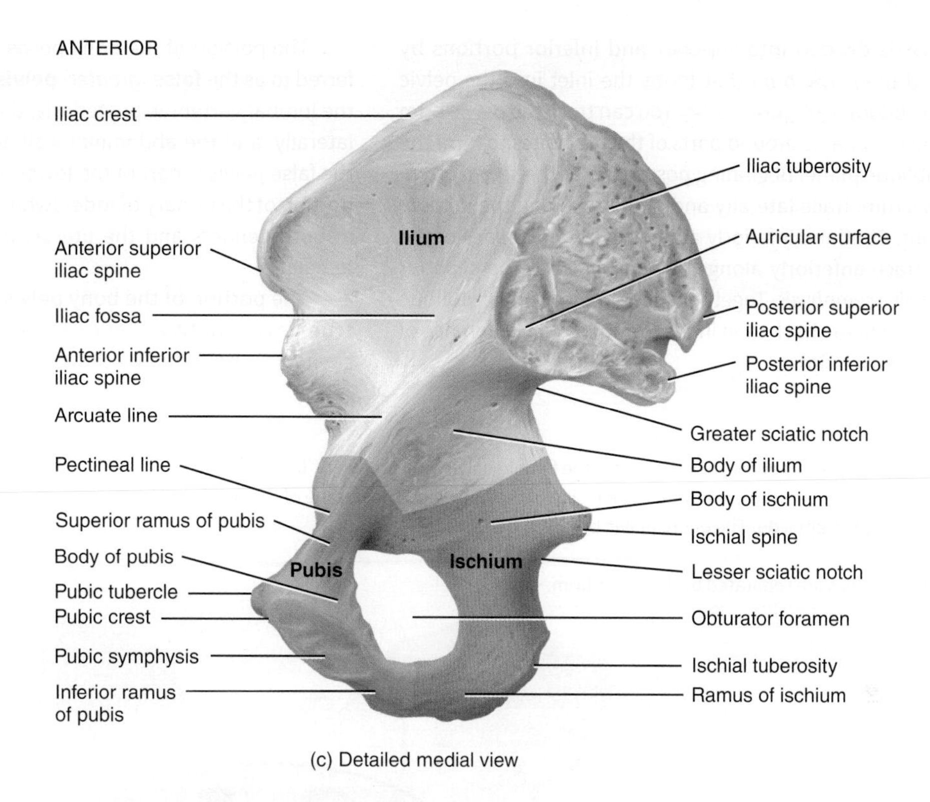

(c) Detailed medial view

Q Which part of the hip bone articulates with the femur? With the sacrum?

Ischium

The **ischium** (IS-kē-um = hip), the inferior, posterior portion of the hip bone (**Figure 8.9b, c**), comprises a superior *body* and an inferior *ramus* (*ram-* = branch; plural is *rami*). The ramus is the portion of the ischium that fuses with the pubis. Features of the ischium include the prominent *ischial spine*, a *lesser sciatic notch* below the spine, and a rough and thickened *ischial tuberosity*. Because this prominent tuberosity is just deep to the skin, it commonly begins hurting after a relatively short time when you sit on a hard surface. Together, the ramus and the pubis surround the *obturator foramen* (OB-too-rā-tōr; *obtur-* = closed up), the largest foramen in the skeleton. The foramen is so named because, even though blood vessels and nerves pass through it, it is nearly completely closed by the fibrous *obturator membrane*.

Pubis

The **pubis** (PŪ-bis; plural is *pubes*), meaning pubic bone, is the anterior and inferior part of the hip bone (**Figure 8.9b, c**). A *superior ramus*, an *inferior ramus*, and a *body* between the rami make up the pubis. The anterior, superior border of the body is the *pubic crest*, and at its lateral end is a projection called the *pubic tubercle.* This tubercle is the beginning of a raised line, the *pectineal line* (pek-TIN-ē-al), which extends superiorly and laterally along the superior ramus to merge with the arcuate line of the ilium. These lines, as you will see shortly, are important landmarks for distinguishing the superior (false) and inferior (true) portions of the bony pelvis.

The *pubic symphysis* is the joint between the two pubes of the hip bones (see **Figure 8.8**). It consists of a disc of fibrocartilage. Inferior to this joint, the inferior rami of the two pubic bones converge to form the *pubic arch.* In the later stages of pregnancy, the hormone relaxin (produced by the ovaries and placenta) increases the flexibility of the pubic symphysis to ease delivery of the baby. Weakening of the joint, together with an already altered center of gravity due to an enlarged uterus, also changes the mother's gait during pregnancy.

The *acetabulum* (as-e-TAB-ū-lum = vinegar cup) is a deep fossa formed by the ilium, ischium, and pubis. It functions as the socket that accepts the rounded head of the femur. Together, the acetabulum and the femoral head form the *hip (coxal) joint.* On the inferior side of the acetabulum is a deep indentation, the *acetabular notch*, that forms a foramen through which blood vessels and nerves pass and serves as a point of attachment for ligaments of the femur (for example, the ligament of the head of the femur).

8.4 False and True Pelves

OBJECTIVES

- **Distinguish** between the false and true pelves.
- **Explain** why the false and true pelves are important clinically.

The bony pelvis is divided into superior and inferior portions by boundary called the *pelvic brim* that forms the inlet into the pelvic cavity from the abdomen (**Figure 8.10a**). You can trace the pelvic brim by following the landmarks around parts of the hip bones to form the outline of an oblique plane. Beginning posteriorly at the *sacral promontory* of the sacrum, trace laterally and inferiorly along the *arcuate lines* of the ilium. Continue inferiorly along the *pectineal lines* of the pubis. Finally, trace anteriorly along the *pubic crest* to the superior portion of the *pubic symphysis.* Together, these points form an oblique plane that is higher in the back than in the front. The circumference of this plane is the pelvic brim.

The portion of the bony pelvis superior to the pelvic brim is referred to as the **false** (*greater*) **pelvis** (**Figure 8.10b**). It is bordered by the lumbar vertebrae posteriorly, the upper portions of the hip bones laterally, and the abdominal wall anteriorly. The space enclosed by the false pelvis is part of the lower abdomen; it contains the superior portion of the urinary bladder (when it is full) and the lower intestines in both genders and the uterus, ovaries, and uterine tubes of the female.

The portion of the bony pelvis inferior to the pelvic brim is the **true** (*lesser*) **pelvis** (**Figure 8.10b**). It has an inlet, an outlet, and a cavity. It is bounded by the sacrum and coccyx posteriorly, inferior

FIGURE 8.10 True and false pelves. Shown here is the female pelvis. For simplicity, in part (a) the landmarks of the pelvic brim are shown only on the left side of the body, and the outline of the pelvic brim is shown only on the right side. The entire pelvic brim is shown in **Table 8.1**.

The true and false pelves are separated by the pelvic brim.

(a) Anterosuperior view of pelvic girdle

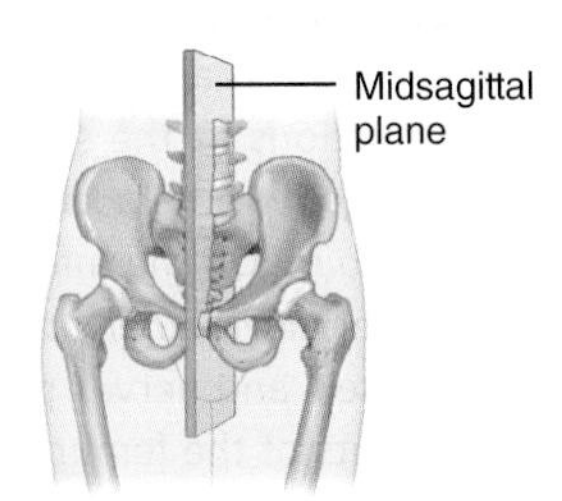

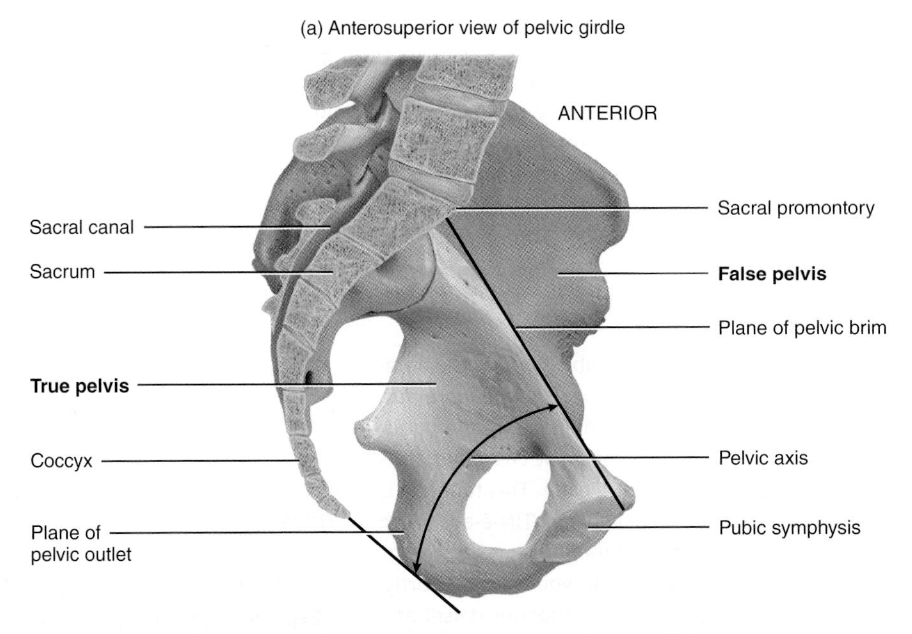

(b) Midsagittal section indicating locations of true (blue) and false (pink) pelves

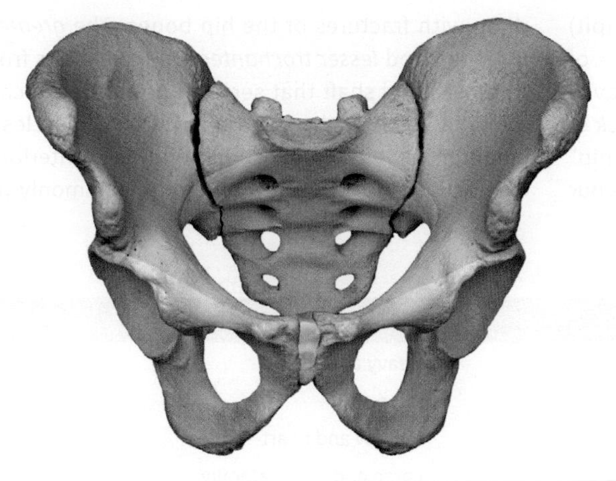
(c) Anterosuperior view of false pelvis (pink)

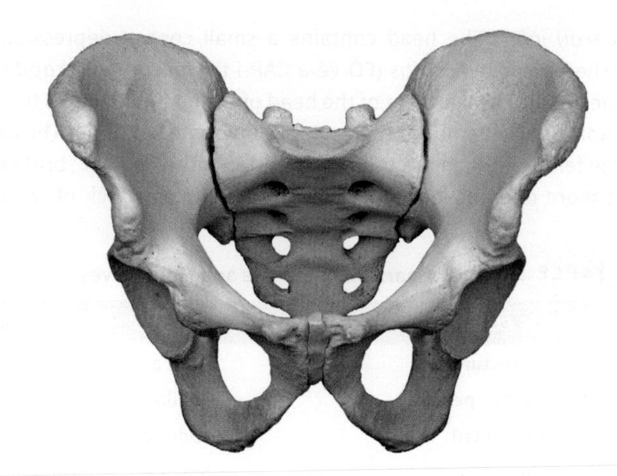
(d) Anterosuperior view of true pelvis (blue)

Q What is the significance of the pelvic axis?

portions of the ilium and ischium laterally, and the pubic bones anteriorly. The true pelvis surrounds the pelvic cavity, which was described in Chapter 1 (see **Figure 1.9**). The true pelvis contains the rectum and urinary bladder in both genders, the vagina and cervix of the uterus in females, and the prostate in males. The superior opening of the true pelvis, bordered by the pelvic brim, is called the *pelvic inlet*; the inferior opening of the true pelvis is the *pelvic outlet*, which is covered by the muscle at the floor of the pelvis. The *pelvic axis* is an imaginary line that curves through the true pelvis from the central point of the plane of the pelvic inlet to the central point of the plane of the pelvic outlet. During childbirth the pelvic axis is the route taken by the baby's head as it descends through the pelvis.

STUDY GUIDE See Clinical Connection: Pelvimetry

8.5 Comparison of Female and Male Pelves

OBJECTIVE

- **Compare** the principal differences between female and male pelves.

Generally, the bones of males are larger and heavier and possess larger surface markings than those of females of comparable age and physical stature. Sex-related differences in the features of bones are readily apparent when comparing the adult female and male pelves. Most of the structural differences in the pelves are adaptations to the requirements of pregnancy and childbirth. The female's pelvis is wider and shallower than the male's. Consequently, there is more space in the true pelvis of the female, especially in the pelvic inlet and pelvic outlet, to accommodate the passage of the infant's head at birth. Other significant structural differences between the pelves of females and males are listed and illustrated in **Table 8.1**.

8.6 Lower Limb (Extremity)

OBJECTIVE

- **Identify** the bones of the lower limb and their principal markings.

Each **lower limb** (*lower extremity*) has 30 bones in four locations—(1) the femur in the thigh; (2) the patella (kneecap); (3) the tibia and fibula in the leg; and (4) the 7 tarsals in the tarsus (ankle), the 5 metatarsals in the metatarsus, and the 14 phalanges (bones of the digits) in the foot (see **Figures 8.11** and **8.13**).

Skeleton of the Thigh—Femur and Patella

Femur The **femur**, or thigh bone, is the longest, heaviest, and strongest bone in the body (**Figure 8.11**). Its proximal end articulates with the acetabulum of the hip bone. Its distal end articulates with the tibia and patella. The *body* (*shaft*) of the femur angles medially and, as a result, the knee joints are closer to the midline than the hip joints. This angle of the femoral shaft (*angle of convergence*) is greater in females because the female pelvis is broader.

The proximal end of the femur consists of a rounded *head* that articulates with the acetabulum of the hip bone to form the *hip (coxal) joint.* The head contains a small central depression (pit) called the *fovea capitis* (FŌ-vē-a CAP-i-tis; *fovea* = pit, *capitis* = of the head). The ligament of the head of the femur connects the fovea capitis of the femur to the acetabulum of the hip bone. The *neck* of the femur is a constricted region distal to the head. A "broken hip" is more often associated with a fracture in the neck of the femur than with fractures of the hip bones. The *greater trochanter* (trō-KAN-ter) and *lesser trochanter* are projections from the junction of the neck and shaft that serve as points of attachment for the tendons of some of the thigh and buttock muscles. The greater trochanter is the prominence felt and seen anterior to the hollow on the side of the hip. It is a landmark commonly used to locate the

TABLE 8.1 Comparison of Female and Male Pelves

POINT OF COMPARISON	FEMALE	MALE
General structure	Light and thin.	Heavy and thick.
False (greater) pelvis	Shallow.	Deep.
Pelvic brim (inlet)	Wide and more oval.	Narrow and heart-shaped.
Acetabulum	Small and faces anteriorly.	Large and faces laterally.
Obturator foramen	Oval.	Round.
Pubic arch	Greater than 90° angle.	Less than 90° angle.

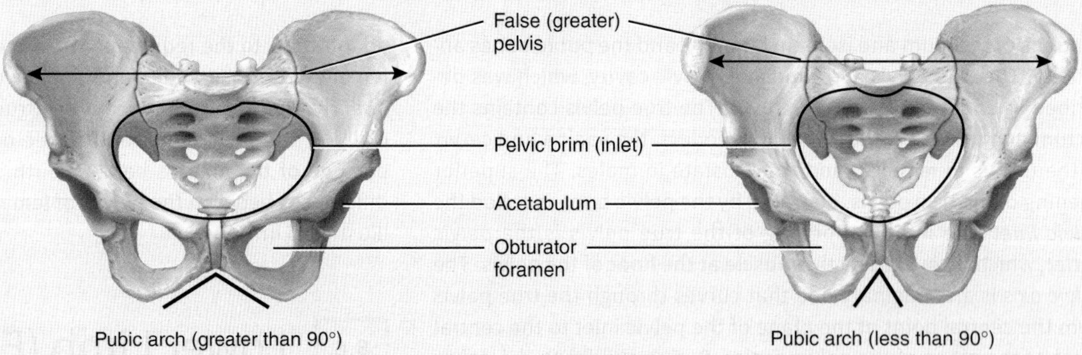

Anterior views

Iliac crest	Less curved.	More curved.
Ilium	Less vertical.	More vertical.
Greater sciatic notch	Wide (almost 90°).	Narrow (about 70°; inverted V).
Sacrum	Shorter, wider (see anterior views), and less curved anteriorly.	Longer, narrower (see anterior views), and more curved anteriorly.

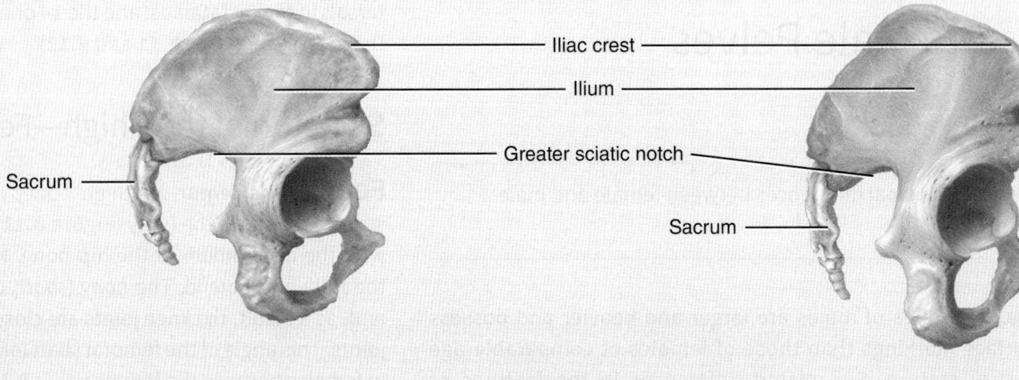

Right lateral views

POINT OF COMPARISON	FEMALE	MALE
Pelvic outlet	Wider.	Narrower.
Ischial tuberosity	Shorter, farther apart, and more medially projecting.	Longer, closer together, and more laterally projecting.

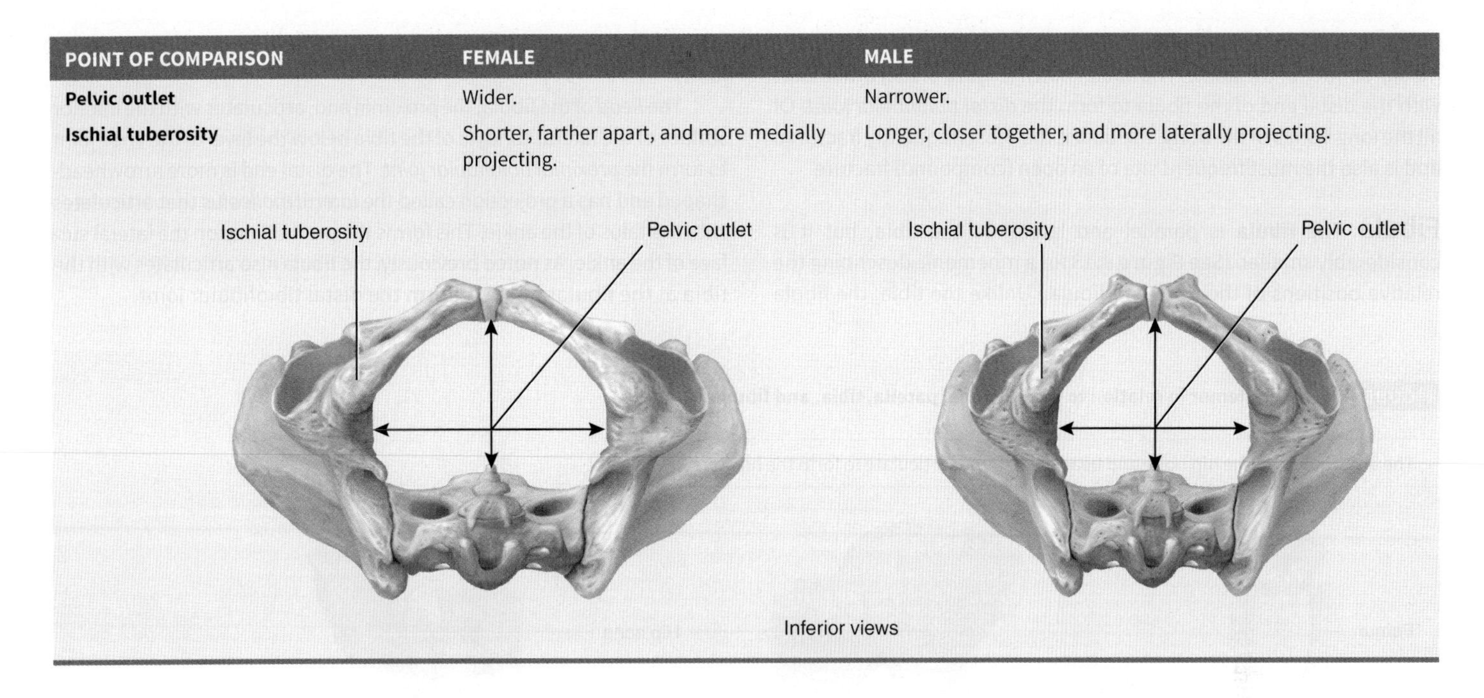

site for intramuscular injections into the lateral surface of the thigh. The lesser trochanter is inferior and medial to the greater trochanter. Between the anterior surfaces of the trochanters is a narrow *intertrochanteric line* (**Figure 8.11a**). A ridge called the *intertrochanteric crest* appears between the posterior surfaces of the trochanters (**Figure 8.11b**).

Inferior to the intertrochanteric crest on the posterior surface of the body of the femur is a vertical ridge called the *gluteal tuberosity.* It blends into another vertical ridge called the *linea aspera* (LIN-ē-a AS-per-a; *asper* = rough). Both ridges serve as attachment points for the tendons of several thigh muscles.

The expanded distal end of the femur includes the *medial condyle* (= knuckle) and the *lateral condyle.* These articulate with the medial and lateral condyles of the tibia. Superior to the condyles are the *medial epicondyle* and the *lateral epicondyle,* to which ligaments of the knee joint attach. A depressed area between the condyles on the posterior surface is called the *intercondylar fossa* (in-ter-KON-di-lar). The *patellar surface* is located between the condyles on the anterior surface. Just superior to the medial epicondyle is the *adductor tubercle,* a roughened projection that is a site of attachment for the adductor magnus muscle.

Patella The **patella** (= little dish), or kneecap, is a small, triangular bone located anterior to the knee joint (**Figure 8.12**). The broad proximal end of this sesamoid bone, which develops in the tendon of the quadriceps femoris muscle, is called the *base;* the pointed distal end is referred to as the *apex.* The posterior surface contains two *articular facets,* one for the medial condyle of the femur and another for the lateral condyle of the femur. The patellar ligament attaches the patella to the tibial tuberosity. The *patellofemoral joint,* between the posterior surface of the patella and the patellar surface of the femur, is the intermediate component of the *tibiofemoral (knee) joint.* The patella increases the leverage of the tendon of the quadriceps femoris muscle, maintains the position of the tendon when the knee is bent (flexed), and protects the knee joint.

See Clinical Connection: Patellofemoral Stress Syndrome

Skeleton of the Leg—Tibia and Fibula

Tibia The **tibia**, or shin bone, is the larger, medial, weight-bearing bone of the leg (**Figure 8.13**). The term *tibia* means flute, because the tibial bones of birds were used in ancient times to make musical instruments. The tibia articulates at its proximal end with the femur and fibula, and at its distal end with the fibula and the talus bone of the ankle. The tibia and fibula, like the ulna and radius, are connected by an interosseous membrane.

The proximal end of the tibia is expanded into a *lateral condyle* and a *medial condyle.* These articulate with the condyles of the femur to form the lateral and medial *tibiofemoral (knee) joints.* The inferior surface of the lateral condyle articulates with the head of the fibula. The slightly concave condyles are separated by an upward projection called the *intercondylar eminence* (**Figure 8.13b**). The *tibial tuberosity* on the anterior surface is a point of attachment for the patellar ligament. Inferior to and continuous with the tibial tuberosity is a sharp ridge that can be felt below the skin, known as the *anterior border (crest)* or *shin.*

The medial surface of the distal end of the tibia forms the *medial malleolus* (mal-LĒ-ō-lus = hammer). This structure articulates with the talus of the ankle and forms the prominence that can be felt on the medial surface of the ankle. The *fibular notch* (**Figure 8.13c**) articulates with the distal end of the fibula to form the *distal tibiofibular joint.* Of all the long bones of the body, the tibia is the most frequently fractured and is also the most frequent site of an open (compound) fracture.

Fibula The **fibula** is parallel and lateral to the tibia, but it is considerably smaller. (See **Figure 8.13** for a mnemonic describing the relative positions of the tibia and fibula.) Unlike the tibia, the fibula does not articulate with the femur, but it does help stabilize the ankle joint.

The *head* of the fibula, the proximal end, articulates with the inferior surface of the lateral condyle of the tibia below the level of the knee joint to form the *proximal tibiofibular joint.* The distal end is more arrowhead-shaped and has a projection called the *lateral malleolus* that articulates with the talus of the ankle. This forms the prominence on the lateral surface of the ankle. As noted previously, the fibula also articulates with the tibia at the fibular notch to form the distal tibiofibular joint.

FIGURE 8.11 Right femur in relation to the hip bone, patella, tibia, and fibula.

The acetabulum of the hip bone and head of the femur articulate to form the hip joint.

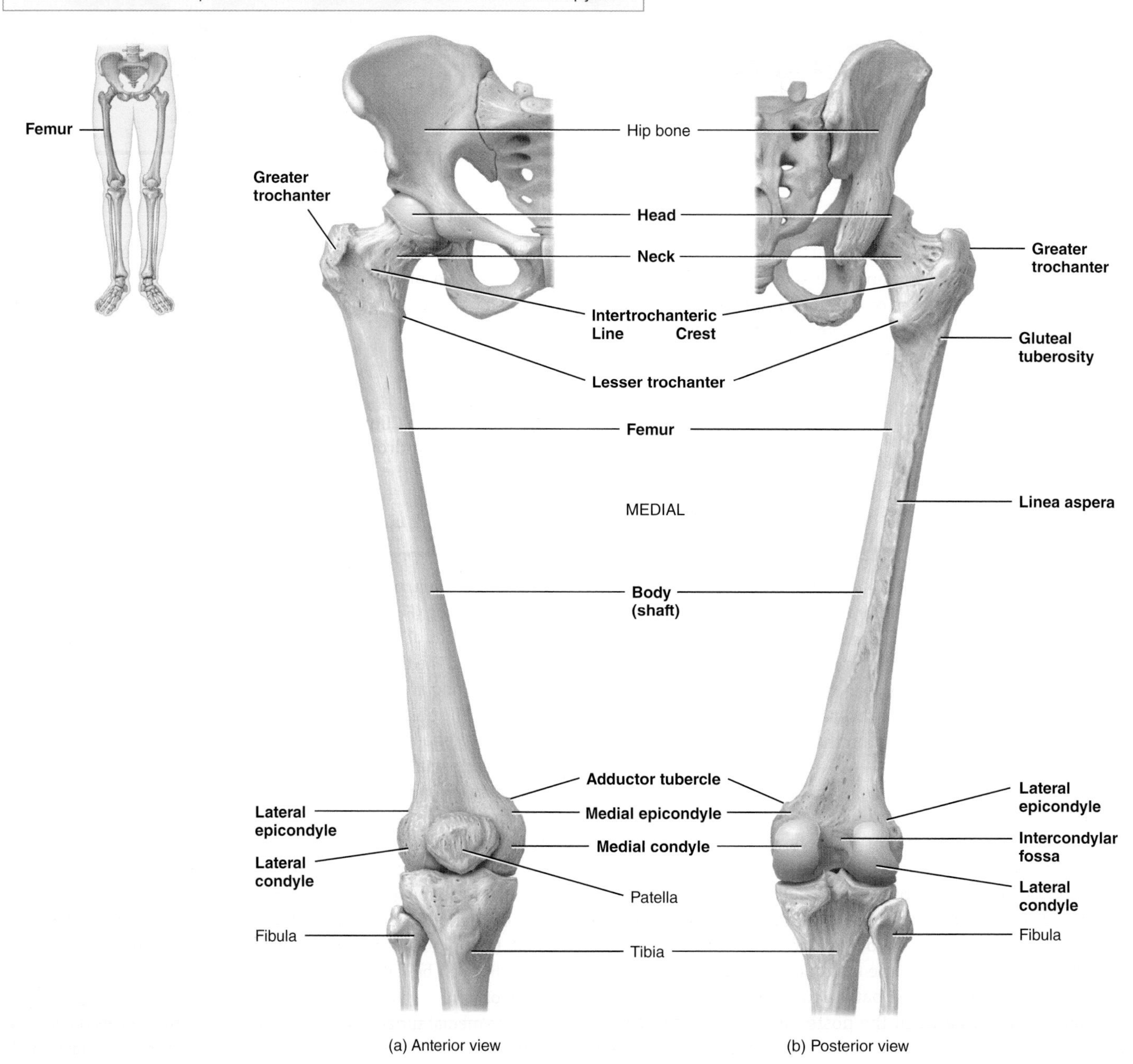

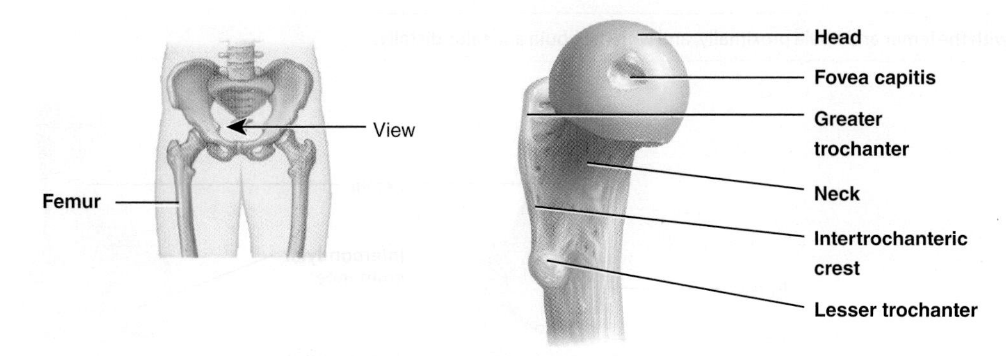

(c) Medial view of proximal end of femur

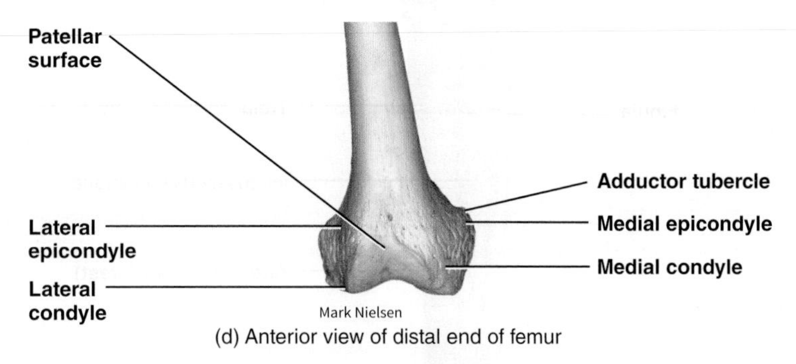

(d) Anterior view of distal end of femur

Q Why is the angle of convergence of the femurs greater in females than males?

Skeleton of the Foot-Tarsals, Metatarsals, and Phalanges

The **tarsus** (ankle) is the proximal region of the foot and consists of seven **tarsal bones** (Figure 8.14). They include the **talus** (TĀ-lus = ankle bone) and **calcaneus** (kal-KĀ-nē-us = heel), located in the posterior part of the foot. The calcaneus is the largest and strongest tarsal bone. The anterior tarsal bones are the **navicular** (na-VIK-ū-lar = like a little boat), three **cuneiform bones** (KŪ-nē-i-form = wedge-shaped) called the **third** (*lateral*), **second** (*intermediate*), and **first** (*medial*) **cuneiforms**, and the **cuboid** (KŪ-boyd = cube-shaped). (A mnemonic to help you remember the names of the tarsal bones is included in Figure 8.14.) Joints between tarsal bones are called *intertarsal joints.* The talus, the most superior tarsal bone, is the only bone of the foot that articulates with the fibula and tibia. It articulates on one side with the medial malleolus of the tibia and on the other side with the lateral malleolus of the fibula. These articulations form the *talocrural*

FIGURE 8.12 Right patella.

The patella articulates with the lateral and medial condyles of the femur.

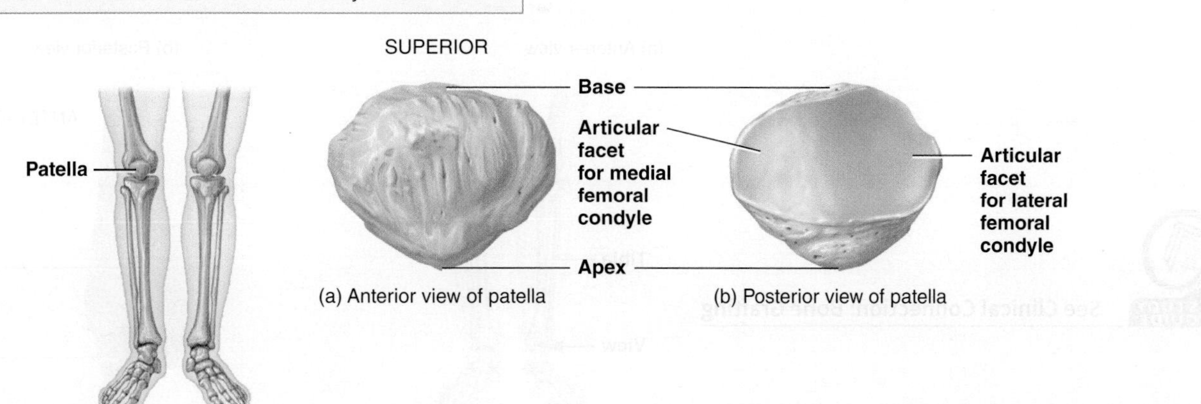

(a) Anterior view of patella

(b) Posterior view of patella

Q The patella is classified as which type of bone? Why?

FIGURE 8.13 **Right tibia and fibula in relation to the femur, patella, and talus.**

The tibia articulates with the femur and fibula proximally, and with the fibula and talus distally.

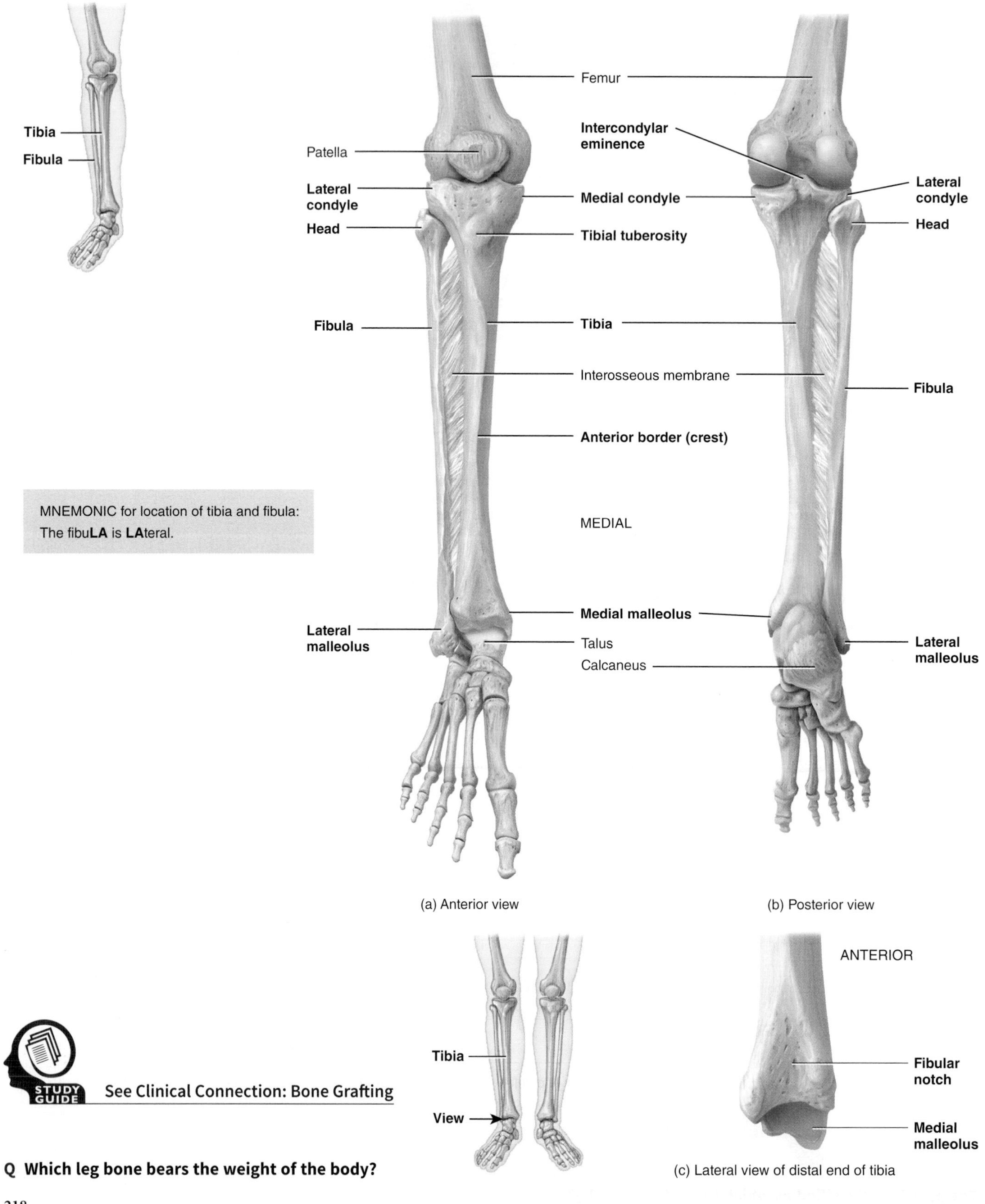

(a) Anterior view

(b) Posterior view

(c) Lateral view of distal end of tibia

Q Which leg bone bears the weight of the body?

FIGURE 8.14 **Right foot.**

The skeleton of the foot consists of the proximal tarsals, the intermediate metatarsals, and the distal phalanges.

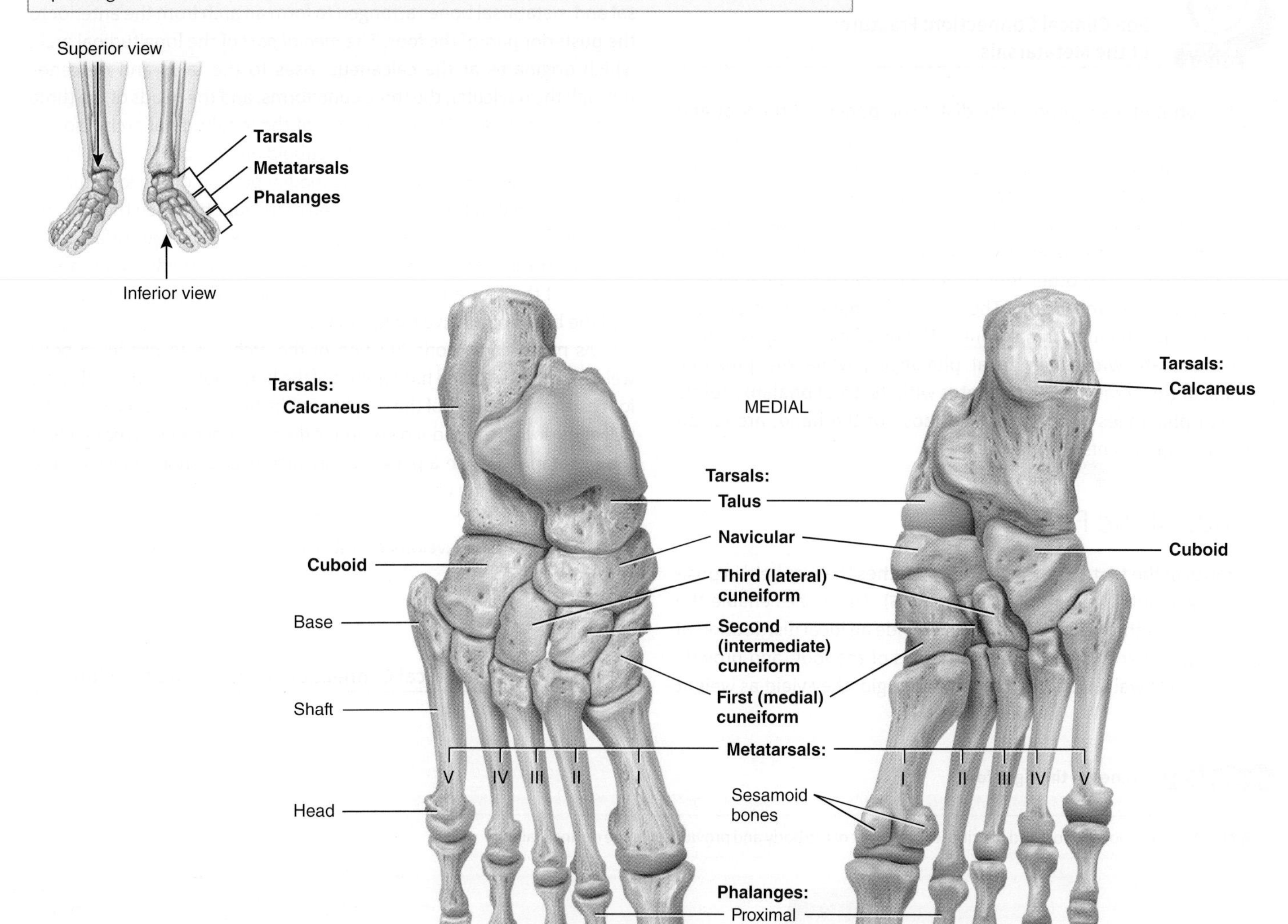

(a) Superior view

(b) Inferior view

MNEMONIC for tarsals:

Tall	Centers	Never	Take	Shots	From	Corners.
Talus	Calcaneus	Navicular	Third cuneiform	Second cuneiform	First cuneiform	Cuboid

Q Which tarsal bone articulates with the tibia and fibula?

(ankle) joint. During walking, the talus transmits about half the weight of the body to the calcaneus. The remainder is transmitted to the other tarsal bones.

The **metatarsus**, the intermediate region of the foot, consists of five **metatarsal bones** numbered I to V (or 1–5) from the medial to lateral position (**Figure 8.14**). Like the metacarpals of the palm of the hand, each metatarsal consists of a proximal *base*, an intermediate *shaft*, and a distal *head.* The metatarsals articulate proximally with the first, second, and third cuneiform bones and with the cuboid to form the *tarsometatarsal joints.* Distally, they articulate with the proximal row of phalanges to form the *metatarsophalangeal joints*. The first metatarsal is thicker than the others because it bears more weight.

See Clinical Connection: Fractures of the Metatarsals

The **phalanges** comprise the distal component of the foot and resemble those of the hand both in number and arrangement. The toes are numbered I to V (or 1–5) beginning with the great toe, from medial to lateral. Each *phalanx* (singular) consists of a proximal *base*, an intermediate *shaft*, and a distal *head*. The great or big toe (*hallux*; HAL-eks) has two large, heavy phalanges called *proximal* and *distal phalanges*. The other four toes each have three phalanges—*proximal*, *middle*, and *distal*. The proximal phalanges of all toes articulate with the metatarsal bones. The middle phalanges of toes (II–V) articulate with their distal phalanges, while the proximal phalanx of the great toe (I) articulates with its distal phalanx. Joints between phalanges of the foot, like those of the hand, are called *interphalangeal joints*.

Arches of the Foot

The bones of the foot are arranged in two **arches** that are held in position by ligaments and tendons (**Figure 8.15**). The arches enable the foot to support the weight of the body, provide an ideal distribution of body weight over the soft and hard tissues of the foot, and provide leverage while walking. The arches are not rigid; they yield as weight is applied and spring back when the weight is lifted, thus storing energy for the next step and helping to absorb shocks. Usually, the arches are fully developed by age 12 or 13.

The **longitudinal arch** has two parts, both of which consist of tarsal and metatarsal bones arranged to form an arch from the anterior to the posterior part of the foot. The *medial part* of the longitudinal arch, which originates at the calcaneus, rises to the talus and descends through the navicular, the three cuneiforms, and the heads of the three medial metatarsals. The *lateral part* of the longitudinal arch also begins at the calcaneus. It rises at the cuboid and descends to the heads of the two lateral metatarsals. The medial portion of the longitudinal arch is so high that the medial portion of the foot between the ball and heel does not touch the ground when you walk on a hard surface.

The **transverse arch** is found between the medial and lateral aspects of the foot and is formed by the navicular, three cuneiforms, and the bases of the five metatarsals.

As noted earlier, one function of the arches is to distribute body weight over the soft and hard tissues of the body. Normally, the ball of the foot carries about 40% of the weight and the heel carries about 60%. The ball of the foot is the padded portion of the sole superficial to the heads of the metatarsals. When a person wears high-heeled shoes, however, the distribution of weight changes so that the ball of the foot may carry up to 80% and the heel 20%. As a result, the fat pads at the ball of the foot are damaged, joint pain develops, and structural changes in bones may occur.

See Clinical Connection: Flatfoot and Clawfoot

FIGURE 8.15 **Arches of the right foot.**

Arches help the foot support and distribute the weight of the body and provide leverage during walking.

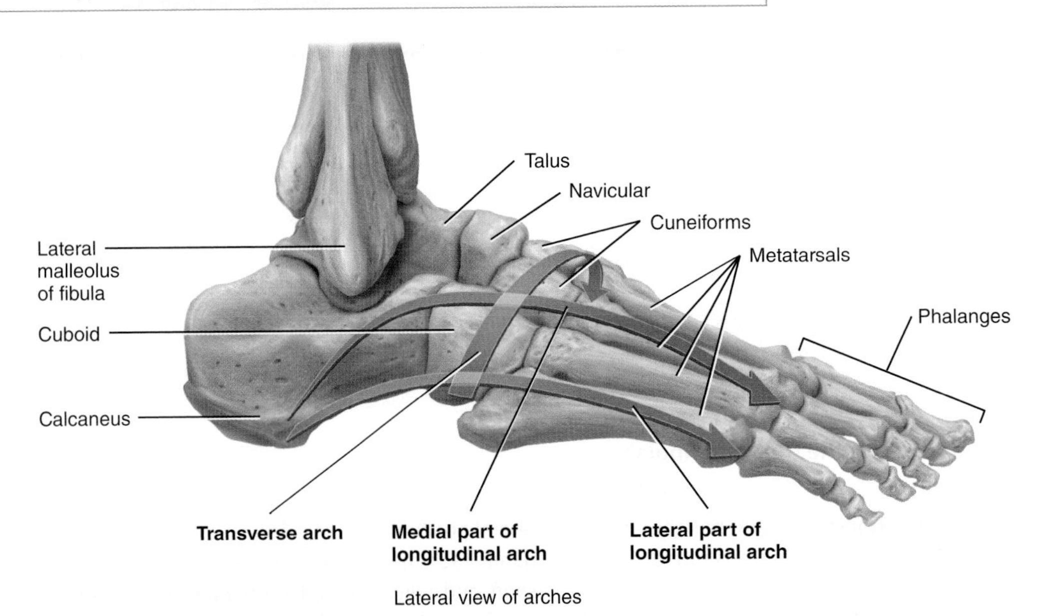

Q What structural feature of the arches allows them to absorb shocks?

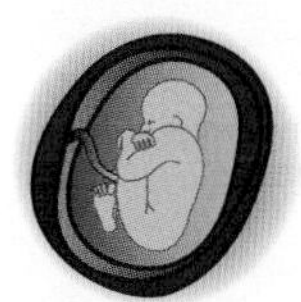

8.7 Development of the Skeletal System

OBJECTIVE

- **Describe** the development of the skeletal system.

Most skeletal tissue arises from *mesenchymal cells*, connective tissue cells derived from **mesoderm**. However, much of the skeleton of the skull arises from **ectoderm**. The mesenchymal cells condense and form models of bones in areas where the bones themselves will ultimately form. In some cases, the bones form directly within the mesenchyme (intramembranous ossification; see **Figure 6.5**). In other cases, the bones form within hyaline cartilage that develops from mesenchyme (endochondral ossification; see **Figure 6.6**).

The *skull* begins development during the fourth week after fertilization. It develops from mesenchyme around the developing brain and consists of two major portions: **neurocranium** (mesodermal in origin), which forms the bones of the skull, and **viscerocranium** (ectodermal in origin), which forms the bones of the face (**Figure 8.16a**). The neurocranium is divided into two parts:

1. The **cartilaginous neurocranium** consists of hyaline cartilage developed from mesenchyme at the base of the developing skull. It later undergoes endochondral ossification to form the *bones at the base of the skull.*
2. The **membranous neurocranium** consists of mesenchyme and later undergoes intramembranous ossification to form the *flat bones that make up the roof and sides of the skull.* During fetal life and infancy the flat bones are separated by membrane-filled spaces called fontanels (see **Figure 7.14**).

The viscerocranium, like the neurocranium, is divided into two parts:

1. The **cartilaginous viscerocranium** is derived from the cartilage of the first two pharyngeal (branchial) arches (see **Figure 29.13**). Endochondral ossification of these cartilages forms the *ear bones* and *hyoid bone.*
2. The **membranous viscerocranium** is derived from mesenchyme in the first pharyngeal arch and, following intramembranous ossification, forms the facial bones.

Vertebrae and *ribs* are derived from portions of cube-shaped masses of mesoderm called somites (see **Figure 10.17**). Mesenchymal cells from these regions surround the notochord (see **Figure 10.17**) at about 4 weeks after fertilization. The **notochord** is a solid cylinder of mesodermal cells that induces (stimulates) the mesenchymal cells to form the *vertebral bodies, costal (rib) centers,* and *vertebral arch centers.* Between the vertebral bodies, the notochord induces mesenchymal cells to form the *nucleus pulposus* of an intervertebral disc and surrounding mesenchymal cells form the *annulus fibrosus* of an intervertebral disc. As development continues,

FIGURE 8.16 Development of the skeletal system. Bones that develop from the cartilaginous neurocranium are indicated in light blue; from the cartilaginous viscerocranium in dark blue; from the membranous neurocranium in dark red; and from the membranous viscerocranium in light red.

After the limb buds develop, endochondral ossification of the limb bones begins by the end of the eighth embryonic week.

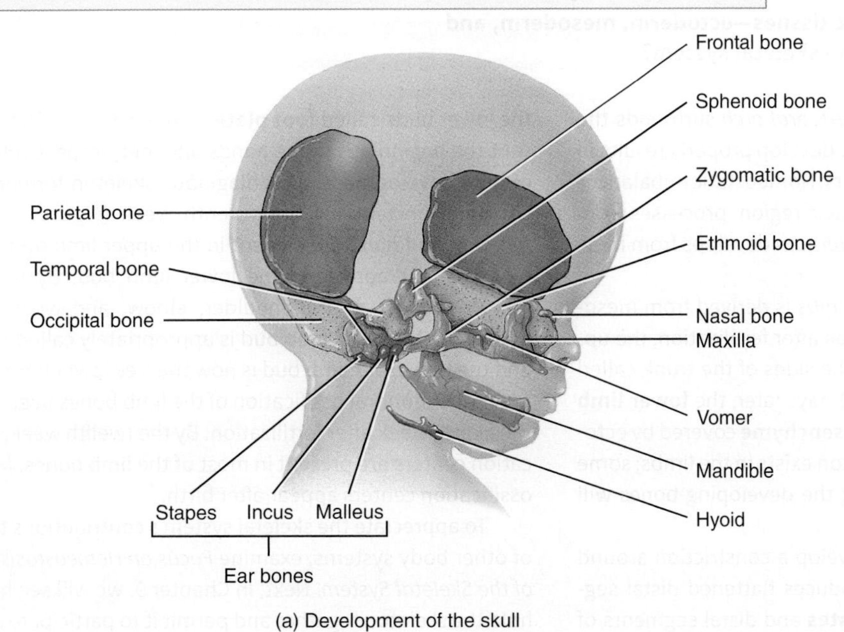

(a) Development of the skull

Figure 8.16 ***Continues***

FIGURE 8.16 Continued

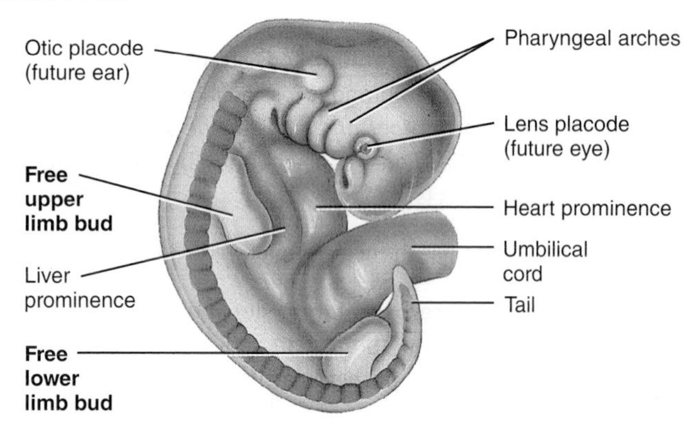

(b) Four-week embryo showing development of free limb buds

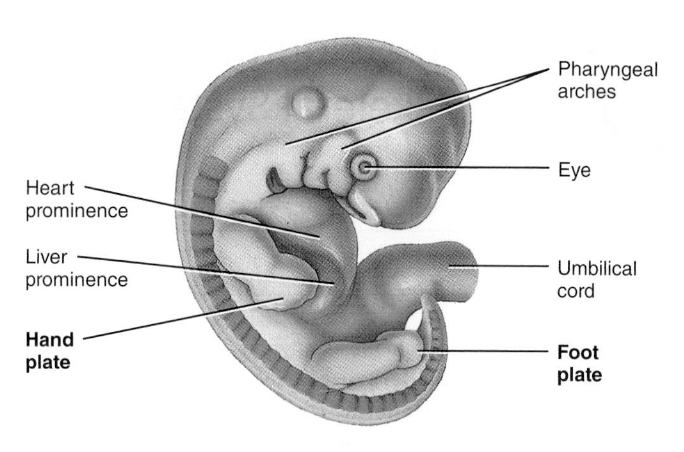

(c) Six-week embryo showing development of hand and foot plates

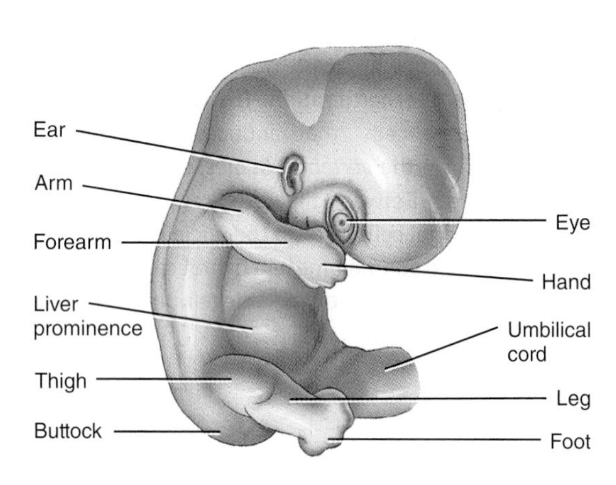

(d) Seven-week embryo showing development of arm, forearm, and hand in free upper limb bud and thigh, leg, and foot in free lower limb bud

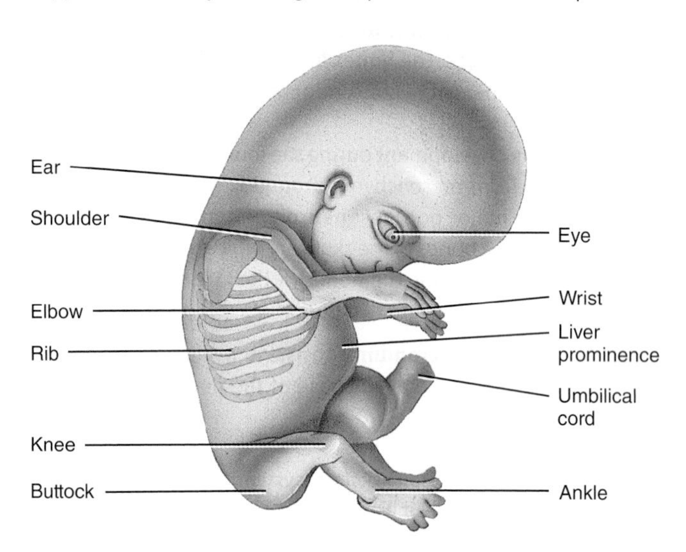

(e) Eight-week embryo in which free limb buds have developed into free upper and lower limbs

Q Which of the three basic embryonic tissues—ectoderm, mesoderm, and endoderm—gives rise to most of the skeletal system?

other parts of a vertebra form and the *vertebral arch* surrounds the spinal cord (failure of the vertebral arch to develop properly results in a condition called spina bifida; see Disorders: Homeostatic Imbalances in Chapter 7 in Study Guide). In the thoracic region, processes from the vertebrae develop into the *ribs.* The *sternum* develops from mesoderm in the anterior body wall.

The *skeleton of the limb girdles and limbs* is derived from mesoderm. During the middle of the fourth week after fertilization, the upper limbs appear as small elevations at the sides of the trunk called **upper limb buds** (Figure 8.16b). About 2 days later, the **lower limb buds** appear. The limb buds consist of **mesenchyme** covered by ectoderm. At this point, a mesenchymal skeleton exists in the limbs; some of the masses of mesoderm surrounding the developing bones will become the skeletal muscles of the limbs.

By the sixth week, the limb buds develop a constriction around the middle portion. The constriction produces flattened distal segments of the upper buds called **hand plates** and distal segments of the lower buds called **foot plates** (Figure 8.16c). These plates represent the beginnings of the hands and feet, respectively. At this stage of limb development, a cartilaginous skeleton formed from mesenchyme is present. By the seventh week (Figure 8.16d), the *arm*, *forearm*, and *hand* are evident in the upper limb bud, and the *thigh*, *leg*, and *foot* appear in the lower limb bud. By the eighth week (Figure 8.16e), as the shoulder, elbow, and wrist areas become apparent, the upper limb bud is appropriately called the upper limb, and the free lower limb bud is now the free lower limb.

Endochondral ossification of the limb bones begins by the end of the eighth week after fertilization. By the twelfth week, primary ossification centers are present in most of the limb bones. Most secondary ossification centers appear after birth.

To appreciate the skeletal system's contributions to homeostasis of other body systems, examine *Focus on Homeostasis: Contributions of the Skeletal System.* Next, in Chapter 9, we will see how joints both hold the skeleton together and permit it to participate in movements.

FOCUS on HOMEOSTASIS

CONTRIBUTIONS OF THE SKELETAL SYSTEM FOR ALL BODY SYSTEMS

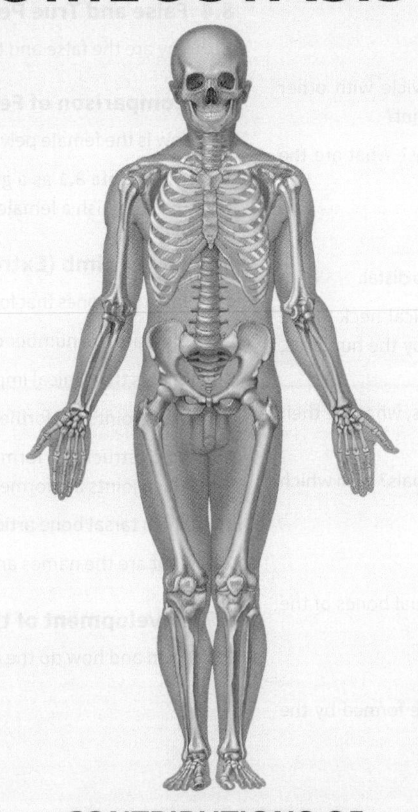

- Bones provide support and protection for internal organs
- Bones store and release calcium, which is needed for proper functioning of most body tissues

INTEGUMENTARY SYSTEM

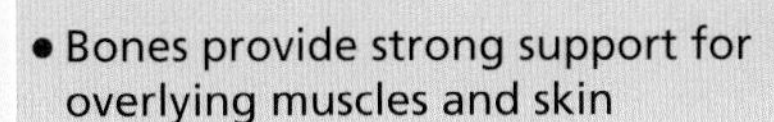

- Bones provide strong support for overlying muscles and skin

MUSCULAR SYSTEM

- Bones provide attachment points for muscles and leverage for muscles to bring about body movements
- Contraction of skeletal muscle requires calcium ions

NERVOUS SYSTEM

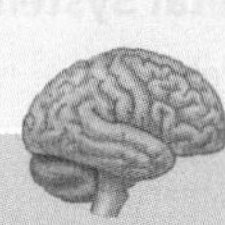

- Skull and vertebrae protect brain and spinal cord
- Normal blood level of calcium is needed for normal functioning of neurons and neuroglia

ENDOCRINE SYSTEM

- Bones store and release calcium, needed during exocytosis of hormone-filled vesicles and for normal actions of many hormones

CARDIOVASCULAR SYSTEM

- Red bone marrow carries out hemopoiesis (blood cell formation)
- Rhythmic beating of the heart requires calcium ions

LYMPHATIC SYSTEM and IMMUNITY

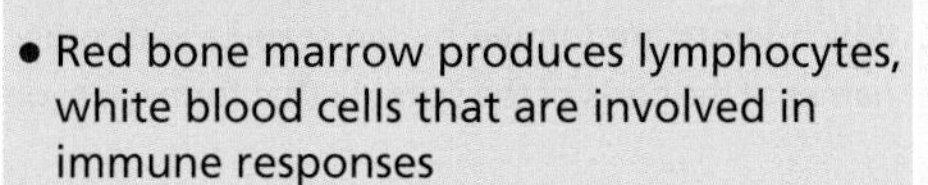

- Red bone marrow produces lymphocytes, white blood cells that are involved in immune responses

RESPIRATORY SYSTEM

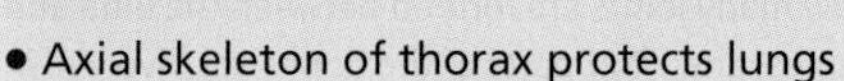

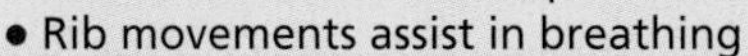

- Axial skeleton of thorax protects lungs
- Rib movements assist in breathing
- Some muscles used for breathing attach to bones via tendons

DIGESTIVE SYSTEM

- Teeth masticate (chew) food
- Rib cage protects esophagus, stomach, and liver
- Pelvis protects portions of the intestines

URINARY SYSTEM

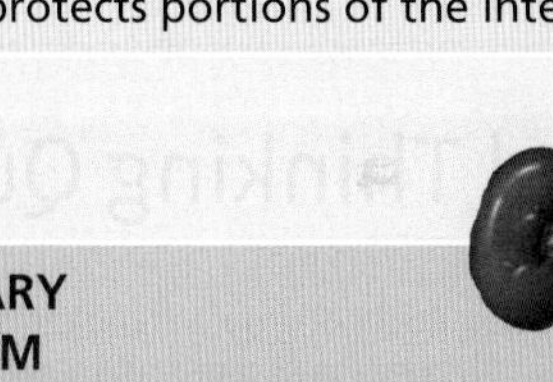

- Ribs partially protect kidneys
- Pelvis protects urinary bladder and urethra

REPRODUCTIVE SYSTEMS

- Pelvis protects ovaries, uterine (fallopian) tubes, and uterus in females
- Pelvis protects part of ductus (vas) deferens and accessory glands in males
- Bones are an important source of calcium needed for milk synthesis during lactation

Review Questions

8.1 Pectoral (Shoulder) Girdle

1. What is the function of the pectoral girdle?
2. Which joints are formed by the articulation of the clavicle with other bones? Which areas of the clavicle are involved in each joint?
3. Which joints are formed by the scapula with other bones? What are the names of the parts of the scapula that form each joint?

8.2 Upper Limb (Extremity)

4. Name the bones that form the upper limb, from proximal to distal.
5. Distinguish between the anatomical neck and the surgical neck of the humerus. Name the proximal and distal points formed by the humerus, and indicate which parts of the bones are involved.
6. How many joints are formed between the ulna and radius, what are their names, and what parts of the bones are involved?
7. Which is more distal, the base or the head of the meta carpals? With which bones do the proximal phalanges articulate?

8.3 Pelvic (Hip) Girdle

8. Describe the distinguishing characteristics of the individual bones of the pelvic girdle.
9. Which bones form the acetabulum? What is its function?
10. Why is the obturator foramen so named? Which joints are formed by the union of the hip bones with other bones?

8.4 False and True Pelves

11. Why are the false and true pelves important clinically?

8.5 Comparison of Female and Male Pelves

12. How is the female pelvis adapted for pregnancy and childbirth?
13. Using Table 8.1 as a guide, select the three ways that are easiest for you to distinguish a female from a male pelvis.

8.6 Lower Limb (Extremity)

14. Name the bones that form the lower limb, from proximal to distal.
15. Compare the number of bones in the carpus and the tarsus.
16. What is the clinical importance of the greater trochanter?
17. Which joints are formed by the femur?
18. Which structures form the medial and lateral prominences of the ankle? Which joints are formed by the tibia and fibula with other bones?
19. Which tarsal bone articulates with both the tibia and fibula?
20. What are the names and functions of the arches of the foot?

8.7 Development of the Skeletal System

21. When and how do the limbs develop?

Critical Thinking Questions

1. Mr. Smith's dog Rover dug up a complete set of human bones in the woods near his house. After examining the scene, the local police collected the bones and transported them to the coroner's office for identification. Later, Mr. Smith read in the newspaper that the bones belonged to an elderly female. How was this determined?
2. A proud dad holds his 5-month-old baby girl upright on her feet while supporting her under her arms. He states that she can never be a dancer because her feet are too flat. Is this true? Why or why not?
3. The local newspaper reported that Farmer White caught his hand in a piece of machinery last Tuesday. He lost the lateral two fingers of his left hand. His daughter, who is taking high school science, reports that Farmer White has three remaining phalanges. Is she correct, or does she need a refresher course in anatomy? Support your answer.

Joints

CHAPTER 9

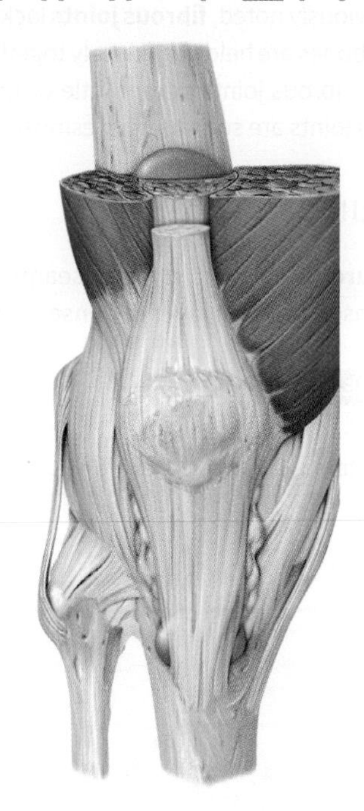

Joints and Homeostasis

The joints of the skeletal system contribute to homeostasis by holding bones together in ways that allow for movement and flexibility.

The human skeleton needs to move, but bones are too rigid to bend without being damaged. Fortunately, flexible connective tissues hold bones together at points of contact called joints while still permitting, in most cases, some degree of movement. Think for a moment about the amazing range of motion and the complexity of the coordinated movements that occur as the bones of the body move against one another; movements such as hitting a golf ball or playing a piano are far more complex than those of almost any machine. Many joint actions are repeated daily and produce continuous work from childhood, into adolescence, and throughout our adult lives. How does the structure of a joint make this incredible staying power possible? Why do joints sometimes fail and cause our movements to become painful? How can we prolong the efficient function of our joints? Read on to answer these questions as you learn about the structure and function of the machinery that allows you to go about your everyday activities.

Q Did you ever wonder why pitchers so often require rotator cuff surgery?

9.1 Joint Classifications

OBJECTIVE

- **Describe** the structural and functional classifications of joints.

A **joint**, also called an **articulation** (ar-tik′-ū-LĀ-shun) or **arthrosis** (ar-THRŌ-sis), is a point of contact between two bones, between bone and cartilage, or between bone and teeth. The scientific study of joints is termed **arthrology** (ar-THROL-ō-jē, *arthr-*=joint; *ology* = study of). The study of motion of the human body is called **kinesiology** (ki-nē-sē-OL-o-jē; *kinesi-* = movement).

Joints are classified structurally, based on their anatomical characteristics, and functionally, based on the type of movement they permit.

The structural classification of joints is based on two criteria: (1) the presence or absence of a space between the articulating bones, called a synovial cavity, and (2) the type of connective tissue that binds the bones together. Structurally, joints are classified as one of the following types:

- **Fibrous joints** (FĪ-brus): There is no synovial cavity, and the bones are held together by dense irregular connective tissue that is rich in collagen fibers.
- **Cartilaginous joints** (kar′-ti-LAJ-i-nus): There is no synovial cavity, and the bones are held together by cartilage.
- **Synovial joints** (si-NŌ-vē-al): The bones forming the joint have a synovial cavity and are united by the dense irregular connective tissue of an articular capsule, and often by accessory ligaments.

The functional classification of joints relates to the degree of movement they permit. Functionally, joints are classified as one of the following types:

- **Synarthrosis** (sin′-ar-THRŌ-sis; *syn-* = together): An immovable joint. The plural is *synarthroses.*
- **Amphiarthrosis** (am′-fē-ar-THRŌ-sis; *amphi-* = on both sides): A slightly movable joint. The plural is *amphiarthroses.*
- **Diarthrosis** (dī-ar-THRŌ-sis = movable joint): A freely movable joint. The plural is *diarthroses.* All diarthroses are synovial joints. They have a variety of shapes and permit several different types of movements.

The following sections present the joints of the body according to their structural classifications. As we examine the structure of each type of joint, we will also outline its functions.

9.2 Fibrous Joints

OBJECTIVE

- **Describe** the structure and functions of the three types of fibrous joints.

As previously noted, **fibrous joints** lack a synovial cavity, and the articulating bones are held very closely together by dense irregular connective tissue. Fibrous joints permit little or no movement. The three types of fibrous joints are sutures, syndesmoses, and interosseous membranes.

Sutures

A **suture** (SOO-chur; *sutur* = seam) (**Figure 9.1a**) is a fibrous joint composed of a thin layer of dense irregular connective tissue; sutures occur only between bones of the skull. An example is the coronal suture between the parietal and frontal bones. The irregular, interlocking edges of sutures give them added strength and decrease their chance of fracturing. Sutures are joints that form as the numerous bones of the skull come in contact during development. They are immovable or slightly movable. In older individuals, sutures are immovable (*synarthroses*), but in infants and children they are slightly movable (*amphiarthroses*) (**Figure 9.1b**). Sutures play important roles in shock absorption in the skull.

FIGURE 9.1 Fibrous joints.

At a fibrous joint the bones are held together by dense irregular connective tissue.

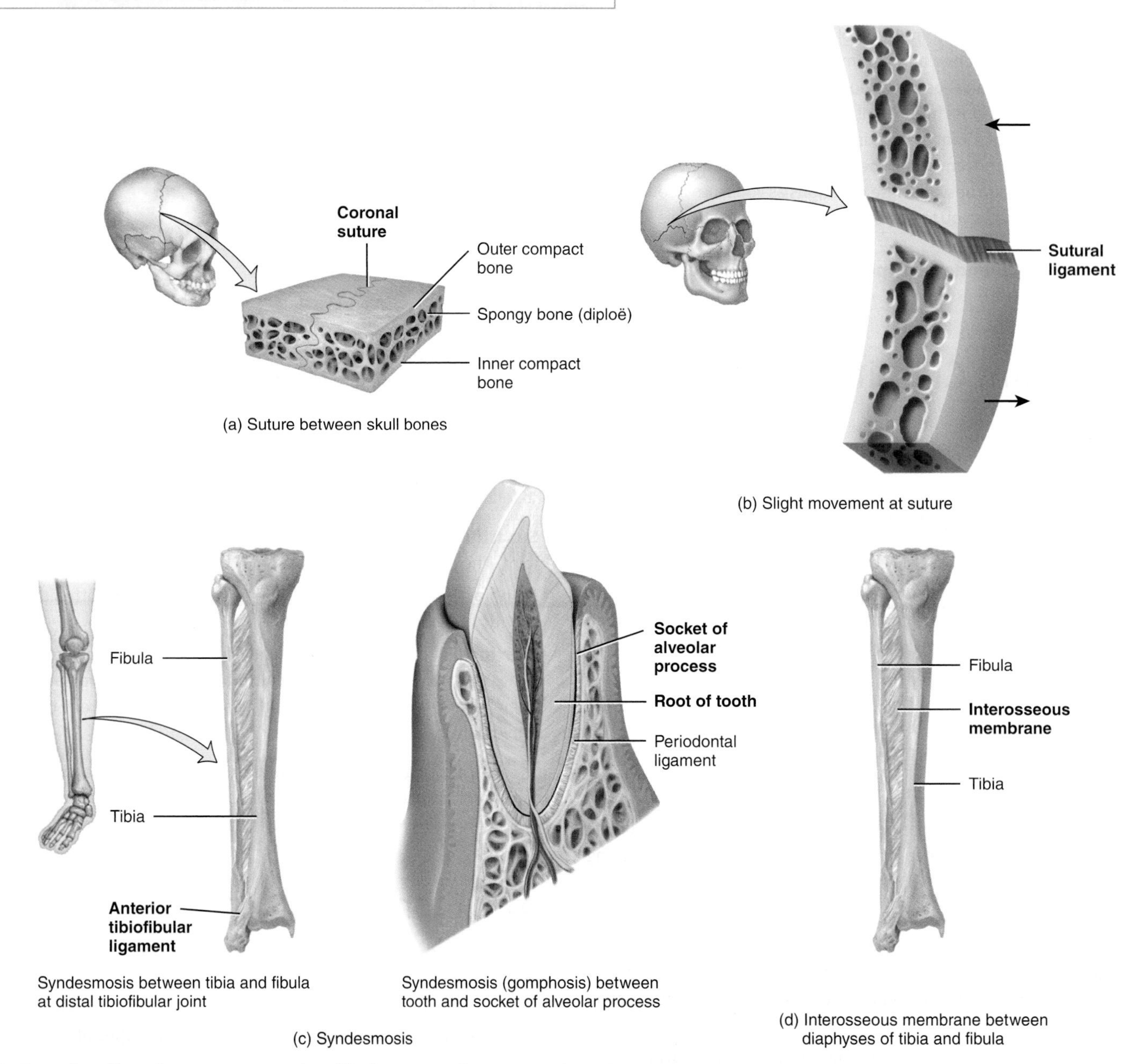

Q Functionally, why are sutures classified as synarthroses, and syndesmoses as amphiarthroses?

Some sutures, although present during growth of the skull, are replaced by bone in the adult. Such a suture is called a **synostosis** (sin′-os-TŌ-sis; *os-* = bone), or bony joint—a joint in which there is a complete fusion of two separate bones into one. For example, the frontal bone grows in halves that join together across a suture line. Usually they are completely fused by age 6 and the suture becomes obscure. If the suture persists beyond age 6, it is called a **frontal** (*metopic*) **suture** (me-TŌ-pik; *metopon* = forehead). A synostosis is classified as a synarthrosis because it is immovable.

Syndesmoses

A **syndesmosis** (sin′-dez-MŌ-sis; *syndesmo-* = band or ligament; plural is *syndesmoses*) is a fibrous joint in which there is a greater distance between the articulating surfaces and more dense irregular connective tissue than in a suture. The dense irregular connective tissue is typically arranged as a bundle (ligament), allowing the joint to permit limited movement. One example of a syndesmosis is the distal tibiofibular joint, where the anterior tibiofibular ligament connects the tibia and fibula (**Figure 9.1c**, left). It permits slight movement (*amphiarthrosis*). Another example of a syndesmosis is called a **gomphosis** (gom-FŌ-sis; *gompbo-* = bolt or nail) or *dentoalveolar joint*, in which a cone-shaped peg fits into a socket. The only examples of gomphoses in the human body are the articulations between the roots of the teeth (cone-shaped pegs) and their sockets (dental alveoli) in the alveolar processes in the maxillae and mandible (**Figure 9.1c**, right). The dense irregular connective tissue between a tooth and its socket is the thin periodontal ligament (membrane). A healthy gomphosis permits minute shock-absorbing movements (*amphiarthrosis*). Inflammation and degeneration of the gums, periodontal ligament, and bone is called *periodontal disease*.

Interosseous Membranes

The final category of fibrous joint is the **interosseous membrane** (in′-ter-OS-ē-us), which is a substantial sheet of dense irregular connective tissue that binds neighboring long bones and permits slight movement (*amphiarthrosis*). There are two principal interosseous membrane joints in the human body. One occurs between the radius and ulna in the forearm (see **Figure 8.5**) and the other occurs between the tibia and fibula in the leg (**Figure 9.1d**). These strong connective tissue sheets not only help hold these adjacent long bones together, they also play an important role in defining the range of motion between the neighboring bones and provide an increased attachment surface for muscles that produce movements of the digits of the hand and foot.

9.3 Cartilaginous Joints

OBJECTIVE

- **Describe** the structure and functions of the three types of cartilaginous joints.

Like a fibrous joint, a **cartilaginous joint** lacks a synovial cavity and allows little or no movement. Here the articulating bones are tightly connected by either hyaline cartilage or fibrocartilage (see **Table 4.6**). The three types of cartilaginous joints are synchondroses, symphyses, and epiphyseal cartilages.

Synchondroses

A **synchondrosis** (sin′-kon-DRŌ-sis; *chondro-* = cartilage; plural is *synchondroses*) is a cartilaginous joint in which the connecting material is hyaline cartilage and is slightly movable (*amphiarthrosis*) to immovable (*synarthrosis*). One example of a synchondrosis is the joint between the first rib and the manubrium of the sternum (**Figure 9.2a**).

FIGURE 9.2 Cartilaginous joints.

At a cartilaginous joint the bones are held together by cartilage.

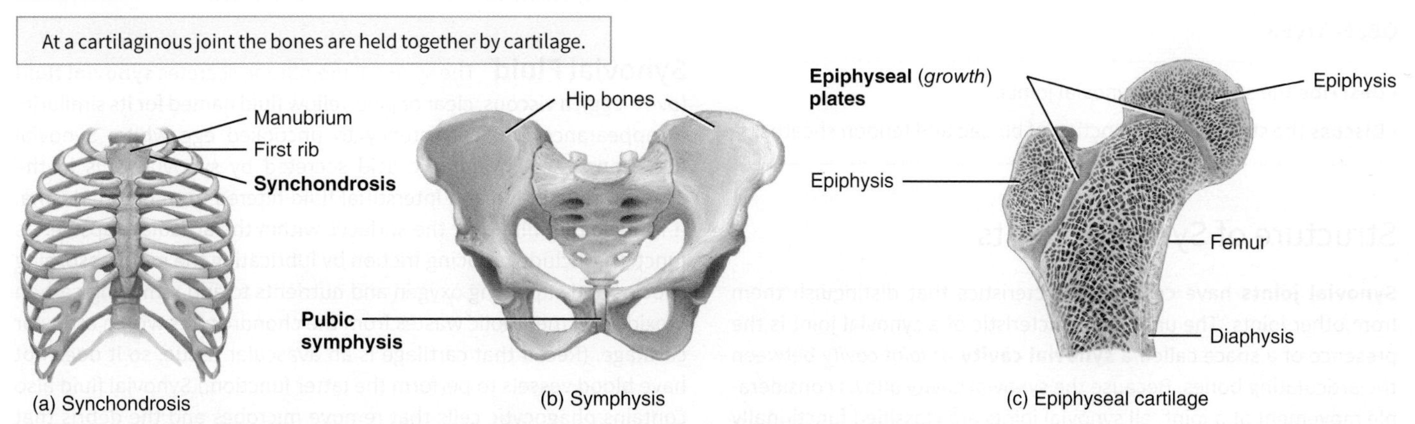

Q What is the structural difference between a synchondrosis, symphysis, and epiphyseal cartilage?

In an x-ray of a young person's skeleton, the synchondroses are easily seen as thin dark areas between the white-appearing bone tissue (see Figure 6.7a). This is how a medical professional can view an x-ray and determine that a person still has some degree of growth remaining. Breaks in a bone that extend into the epiphyseal plate and damage the cartilage of the synchondrosis can affect further growth of the bone, leading to abbreviated development and a bone of shortened length.

Symphyses

A **symphysis** (SIM-fi-sis = growing together; plural is *symphyses*) is a cartilaginous joint in which the ends of the articulating bones are covered with hyaline cartilage, but a broad, flat disc of fibrocartilage connects the bones. All symphyses occur in the midline of the body. The pubic symphysis between the anterior surfaces of the hip bones is one example of a symphysis (Figure 9.2b). This type of joint is also found at the junction of the manubrium and body of the sternum (see Figure 7.22) and at the intervertebral joints between the bodies of vertebrae (see Figure 7.20a). A portion of the intervertebral disc is composed of fibrocartilage. A symphysis is a slightly movable joint (*amphiarthrosis*).

Epiphyseal Cartilages

Epiphyseal cartilages are actually hyaline cartilage growth centers during endochondral bone formation, not joints associated with movements. An example of epiphyseal cartilage is the epiphyseal (growth) plate that connects the epiphysis and diaphysis of a growing bone (Figure 9.2c). A photomicrograph of the epiphyseal plate is shown in Figure 6.7b. Functionally, epiphyseal cartilage is an immovable joint (*synarthrosis*). When bone elongation ceases, bone replaces the hyaline cartilage, and becomes a synostosis, a bony joint.

9.4 Synovial Joints

OBJECTIVES

- **Describe** the structure of synovial joints.
- **Discuss** the structure and function of bursae and tendon sheaths.

Structure of Synovial Joints

Synovial joints have certain characteristics that distinguish them from other joints. The unique characteristic of a synovial joint is the presence of a space called a **synovial cavity** or *joint cavity* between the articulating bones. Because the synovial cavity allows considerable movement at a joint, all synovial joints are classified functionally as freely movable (*diarthroses*). The bones at a synovial joint are covered by a layer of hyaline cartilage called **articular cartilage**. The cartilage covers the articulating surfaces of the bones with a smooth, slippery surface but does not bind them together. Articular cartilage reduces friction between bones in the joint during movement and helps to absorb shock.

See Clinical Connection: Autologous Chondrocyte Implantation

Articular Capsule A sleevelike **articular capsule** or *joint capsule* surrounds a synovial joint, encloses the synovial cavity, and unites the articulating bones. The articular capsule is composed of two layers, an outer fibrous membrane and an inner synovial membrane (Figure 9.3a). The **fibrous membrane** usually consists of dense irregular connective tissue (mostly collagen fibers) that attaches to the periosteum of the articulating bones. In fact, the fibrous membrane is literally a thickened continuation of the periosteum between the bones. The flexibility of the fibrous membrane permits considerable movement at a joint, while its great tensile strength (resistance to stretching) helps prevent the bones from dislocating, the displacement of a bone from a joint. The fibers of some fibrous membranes are arranged as parallel bundles of dense regular connective tissue that are highly adapted for resisting strains. The strength of these fiber bundles, called **ligaments** (*liga-* = bound or tied), is one of the principal mechanical factors that hold bones close together in a synovial joint. Ligaments are often designated by individual names. The inner layer of the articular capsule, the **synovial membrane**, is composed of areolar connective tissue with elastic fibers. At many synovial joints the synovial membrane includes accumulations of adipose tissue, called **articular fat pads**. An example is the infrapatellar fat pad in the knee (see Figure 9.15c).

A **"double-jointed"** person does not really have extra joints. Individuals who are double-jointed have greater flexibility in their articular capsules and ligaments; the resulting increase in range of motion allows them to entertain fellow partygoers with activities such as touching their thumbs to their wrists and putting their ankles or elbows behind their necks. Unfortunately, such flexible joints are less structurally stable and are more easily dislocated.

Synovial Fluid The synovial membrane secretes **synovial fluid** (*ov-* = egg), a viscous, clear or pale yellow fluid named for its similarity in appearance and consistency to uncooked egg white. Synovial fluid consists of hyaluronic acid secreted by synovial cells in the synovial membrane and interstitial fluid filtered from blood plasma. It forms a thin film over the surfaces within the articular capsule. Its functions include reducing friction by lubricating the joint, absorbing shocks, and supplying oxygen and nutrients to and removing carbon dioxide and metabolic wastes from the chondrocytes within articular cartilage. (Recall that cartilage is an avascular tissue, so it does not have blood vessels to perform the latter function.) Synovial fluid also contains phagocytic cells that remove microbes and the debris that results from normal wear and tear in the joint. When a synovial joint is immobile for a time, the fluid becomes quite viscous (gel-like), but

FIGURE 9.3 Structure of a typical synovial joint. Note the two layers of the articular capsule—the fibrous membrane and the synovial membrane. Synovial fluid lubricates the synovial cavity, which is located between the synovial membrane and the articular cartilage.

The distinguishing feature of a synovial joint is the synovial cavity between the articulating bones.

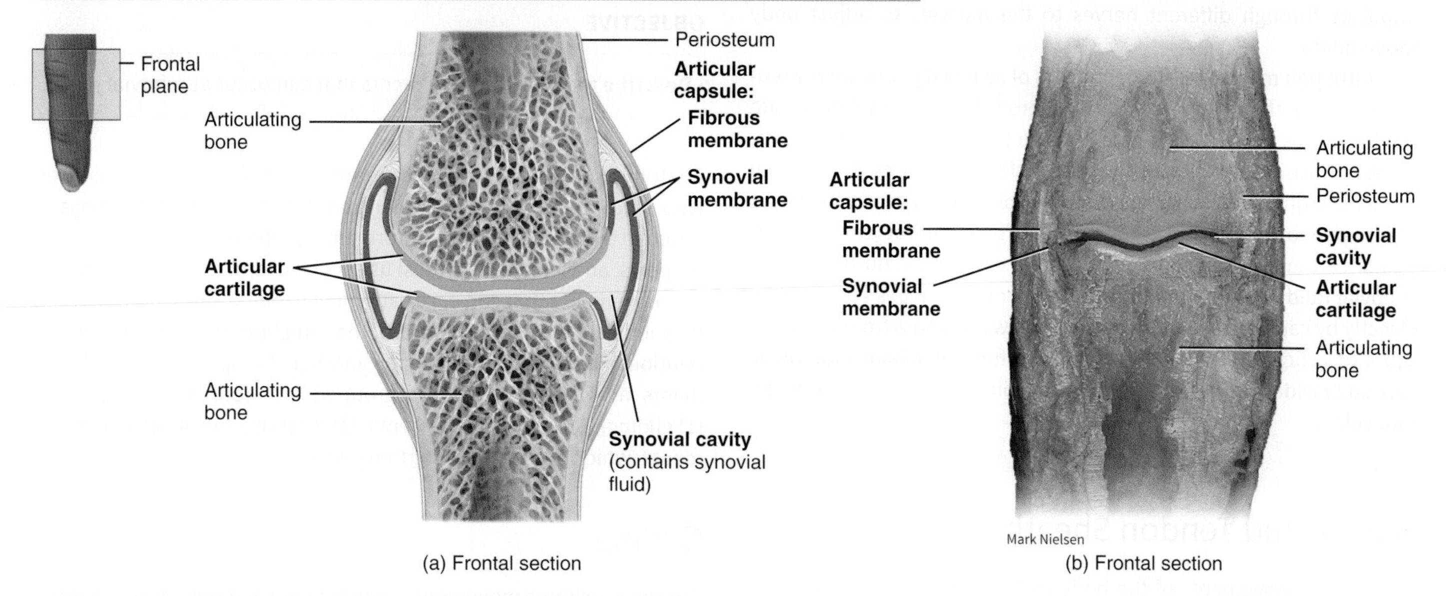

Q What is the functional classification of synovial joints?

as joint movement increases, the fluid becomes less viscous. One of the benefits of warming up before exercise is that it stimulates the production and secretion of synovial fluid; within limits, more fluid means less stress on the joints during exercise.

We are all familiar with the cracking sounds heard as certain joints move, or the popping sounds that arise when a person pulls on his fingers to crack his knuckles. According to one theory, when the synovial cavity expands, the pressure inside the synovial cavity decreases, creating a partial vacuum. The suction draws carbon dioxide and oxygen out of blood vessels in the synovial membrane, forming bubbles in the fluid. When the fingers are flexed (bent) the volume of the cavity decreases and the pressure increases; this bursts the bubbles and creates cracking or popping sounds as the gases are driven back into solution.

Accessory Ligaments, Articular Discs, and Labra

Many synovial joints also contain **accessory ligaments** called extracapsular ligaments and intracapsular ligaments (see **Figure 9.15d**). *Extracapsular ligaments* lie outside the articular capsule. Examples are the fibular and tibial collateral ligaments of the knee joint. *Intracapsular ligaments* occur within the articular capsule but are excluded from the synovial cavity by folds of the synovial membrane. Examples are the anterior and posterior cruciate ligaments of the knee joint.

Inside some synovial joints, such as the knee, crescent-shaped pads of fibrocartilage lie between the articular surfaces of the bones and are attached to the fibrous capsule. These pads are called **articular discs** or *menisci* (me-NIS-sī or me-NIS-kī; singular is *meniscus*). **Figures 9.15c** and **d** depict the lateral and medial menisci in the knee joint. The discs bind strongly to the inside of the fibrous membrane and usually subdivide the synovial cavity into two spaces, allowing separate movements to occur in each space. As you will see later, separate movements also occur in the respective compartments of the temporomandibular joint (TMJ) (see Section 9.9). The functions of the menisci are not completely understood but are known to include the following: (1) shock absorption; (2) a better fit between articulating bony surfaces; (3) providing adaptable surfaces for combined movements; (4) weight distribution over a greater contact surface; and (5) distribution of synovial lubricant across the articular surfaces of the joint.

See Clinical Connection: Torn Cartilage and Arthroscopy

A **labrum** (LĀ-brum; plural is *labra*), prominent in the ball-and-socket joints of the shoulder and hip (see **Figures 9.12c, d; 9.14c**), is the fibrocartilaginous lip that extends from the edge of the joint socket. The labrum helps deepen the joint socket and increases the area of contact between the socket and the ball-like surface of the head of the humerus or femur.

Nerve and Blood Supply

The nerves that supply a joint are the same as those that supply the skeletal muscles that move the joint. Synovial joints contain many nerve endings that are distributed to the articular capsule and associated ligaments. Some of the nerve endings convey information about pain from the joint to the spinal cord and brain for processing. Other nerve endings respond to the degree of movement and stretch at a joint, such as when a physician strikes the tendon below your kneecap to test your reflexes. The spinal cord and brain respond by sending impulses through different nerves to the muscles to adjust body movements.

Although many of the components of synovial joints are avascular, arteries in the vicinity send out numerous branches that penetrate the ligaments and articular capsule to deliver oxygen and nutrients. Veins remove carbon dioxide and wastes from the joints. The arterial branches from several different arteries typically merge around a joint before penetrating the articular capsule. The chondrocytes in the articular cartilage of a synovial joint receive oxygen and nutrients from synovial fluid derived from blood; all other joint tissues are supplied directly by capillaries. Carbon dioxide and wastes pass from chondrocytes of articular cartilage into synovial fluid and then into veins; carbon dioxide and wastes from all other joint structures pass directly into veins.

Bursae and Tendon Sheaths

The various movements of the body create friction between moving parts. Saclike structures called **bursae** (BER-sē = purses; singular is *bursa*) are strategically situated to alleviate friction in some joints, such as the shoulder and knee joints (see **Figures 9.12** and **9.15c**). Bursae are not strictly part of synovial joints, but they do resemble joint capsules because their walls consist of an outer fibrous membrane of thin, dense connective tissue lined by a synovial membrane. They are filled with a small amount of fluid that is similar to synovial fluid. Bursae can be located between the skin and bones, tendons and bones, muscles and bones, or ligaments and bones. The fluid-filled bursal sacs cushion the movement of these body parts against one another.

See Clinical Connection: Bursitis

Structures called tendon sheaths also reduce friction at joints. **Tendon sheaths** or *synovial sheaths* are tubelike bursae; they wrap around certain tendons that experience considerable friction as they pass through tunnels formed by connective tissue and bone. The inner layer of a tendon sheath, the *visceral layer*, is attached to the surface of the tendon. The outer layer, known as the *parietal layer*, is attached to bone (see **Figure 11.18a**). Between the layers is a cavity that contains a film of synovial fluid. A tendon sheath protects all sides of a tendon from friction as the tendon slides back and forth. Tendon sheaths are found where tendons pass through synovial cavities, such as the tendon of the biceps brachii muscle at the shoulder joint (see **Figure 9.12c**). Tendon sheaths are also found at the wrist and ankle, where many tendons come together in a confined space (see **Figure 11.18a**), and in the fingers and toes, where there is a great deal of movement (see **Figure 11.18**).

9.5 Types of Movements at Synovial Joints

OBJECTIVE

- **Describe** the types of movements that can occur at synovial joints.

Anatomists, physical therapists, and kinesiologists (professionals who study the science of human movement and look for ways to improve the efficiency and performance of the human body at work, in sports, and in daily activities) use specific terminology to designate the movements that can occur at synovial joints. These precise terms may indicate the form of motion, the direction of movement, or the relationship of one body part to another during movement. Movements at synovial joints are grouped into four main categories: (1) gliding, (2) angular movements, (3) rotation, and (4) special movements, which occur only at certain joints.

Gliding

Gliding is a simple movement in which nearly flat bone surfaces move back-and-forth and from side-to-side with respect to one another (**Figure 9.4**). There is no significant alteration of the angle between the bones. Gliding movements are limited in range due to the structure of the articular capsule and associated ligaments and bones; however, these sliding movements can also be combined with rotation. The intercarpal and intertarsal joints are examples of articulations where gliding movements occur.

FIGURE 9.4 Gliding movements at synovial joints.

Gliding movements consist of side-to-side and back-and-forth motions.

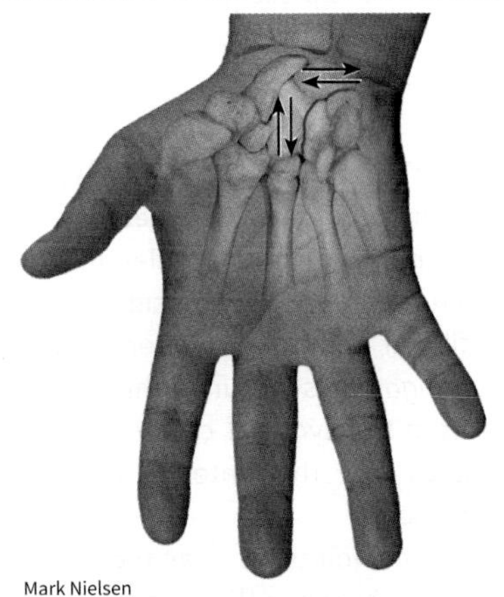

Mark Nielsen

Gliding between carpals (arrows)

Q What are two examples of joints that permit gliding movements?

Angular Movements

In **angular movements**, there is an increase or a decrease in the angle between articulating bones. The major angular movements are flexion, extension, lateral flexion, hyperextension, abduction, adduction, and circumduction. These movements are discussed with respect to the body in the anatomical position (see **Figure 1.5**).

Flexion, Extension, Lateral Flexion, and Hyperextension Flexion and extension are opposite movements. In **flexion** (FLEK-shun; *flex-* = to bend) there is a decrease in the angle between articulating bones; in **extension** (eks-TEN-shun; *exten-* = to stretch out) there is an increase in the angle between articulating bones, often to restore a part of the body to the anatomical position after it has been flexed (**Figure 9.5**). Both movements usually occur along the sagittal plane. All of the following are examples of flexion (as you have probably already guessed, extension is simply the reverse of these movements):

- Bending the head toward the chest at the atlanto-occipital joints between the atlas (the first vertebra) and the occipital bone of the skull, and at the cervical intervertebral joints between the cervical vertebrae (**Figure 9.5a**)
- Bending the trunk forward at the intervertebral joints as in doing a crunch with your abdominal muscles
- Moving the humerus forward at the shoulder joint, as in swinging the arms forward while walking (**Figure 9.5b**)
- Moving the forearm toward the arm at the elbow joint between the humerus, ulna, and radius as in bending your elbow (**Figure 9.5c**)
- Moving the palm toward the forearm at the wrist or radiocarpal joint between the radius and carpals, as in the upward movement when doing wrist curls (**Figure 9.5d**)
- Bending the digits of the hand at the interphalangeal joints between phalanges as in clenching your fingers to make a fist
- Moving the femur forward at the hip joint between the femur and hip bone, as in walking (**Figure 9.5e**)
- Moving the heel toward the buttock at the tibiofemoral joint between the tibia, femur, and patella, as occurs when bending the knee (**Figure 9.5f**)

FIGURE 9.5 Angular movements at synovial joints—flexion, extension, hyperextension, and lateral flexion.

In angular movements, there is an increase or decrease in the angle between articulating bones.

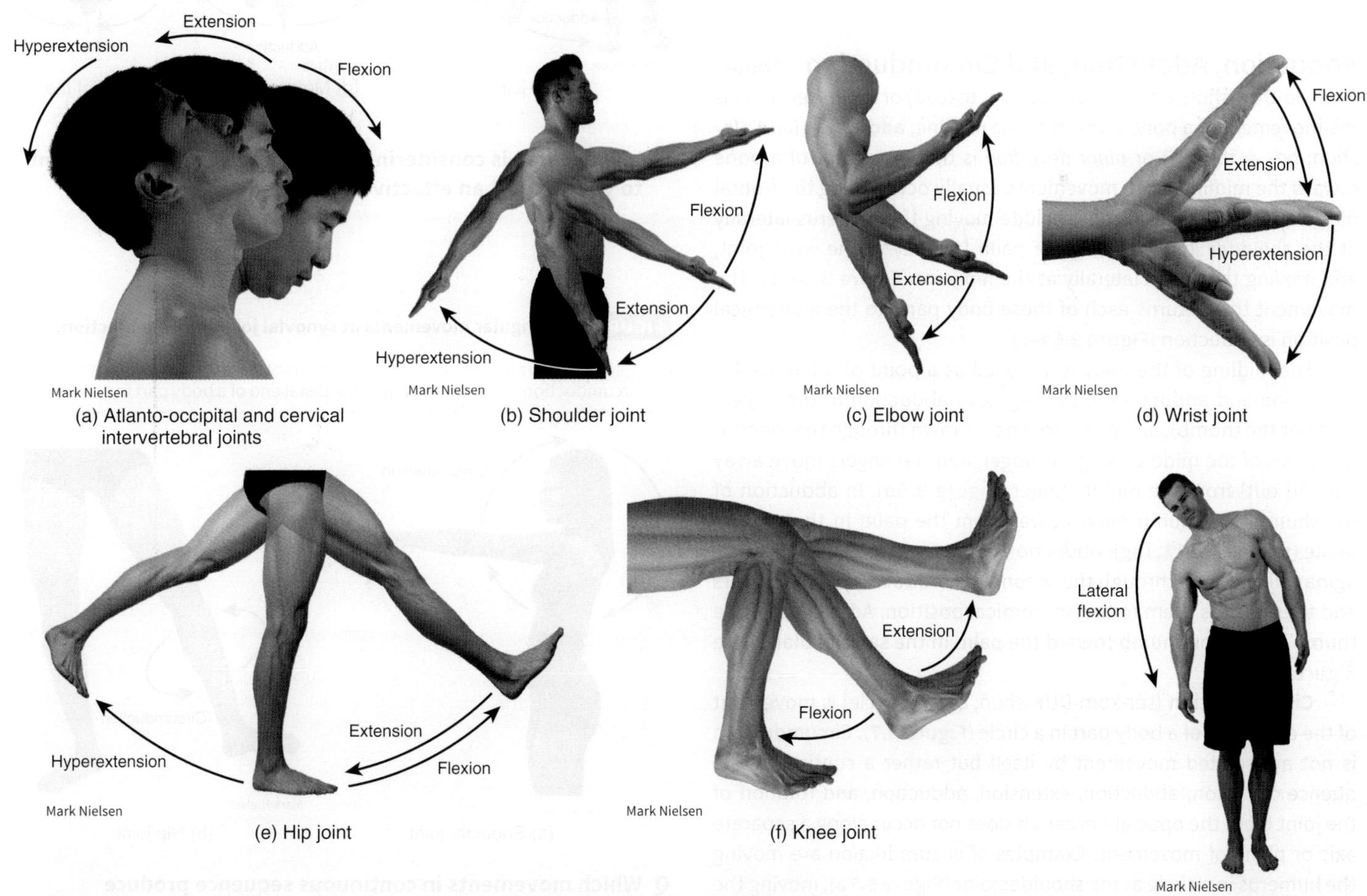

Q What are two examples of flexion that do not occur along the sagittal plane?

Although flexion and extension usually occur along the sagittal plane, there are a few exceptions. For example, flexion of the thumb involves movement of the thumb medially across the palm at the carpometacarpal joint between the trapezium and metacarpal of the thumb, as when you touch your thumb to the opposite side of your palm (see **Figure 11.18g**). Another example is movement of the trunk sideways to the right or left at the waist. This movement, which occurs along the frontal plane and involves the intervertebral joints, is called **lateral flexion** (**Figure 9.5g**).

Continuation of extension beyond the anatomical position is called **hyperextension** (hī-per-ek-STEN-shun; *hyper-* = beyond or excessive). Examples of hyperextension include:

- Bending the head backward at the atlanto-occipital and cervical intervertebral joints as in looking up at stars (**Figure 9.5a**)
- Bending the trunk backward at the intervertebral joints as in a backbend
- Moving the humerus backward at the shoulder joint, as in swinging the arms backward while walking (**Figure 9.5b**)
- Moving the palm backward at the wrist joint as in preparing to shoot a basketball (**Figure 9.5d**)
- Moving the femur backward at the hip joint, as in walking (**Figure 9.5e**)

Hyperextension of hinge joints, such as the elbow, interphalangeal, and knee joints, is usually prevented by the arrangement of ligaments and the anatomical alignment of the bones.

Abduction, Adduction, and Circumduction

Abduction (ab-DUK-shun; *ab-* = away; *-duct-* = to lead) or *radial deviation* is the movement of a bone away from the midline; **adduction** (ad-DUK-shun; *ad-* = toward) or *ulnar deviation* is the movement of a bone toward the midline. Both movements usually occur along the frontal plane. Examples of abduction include moving the humerus laterally at the shoulder joint, moving the palm laterally at the wrist joint, and moving the femur laterally at the hip joint (**Figure 9.6a–c**). The movement that returns each of these body parts to the anatomical position is adduction (**Figure 9.6a–c**).

The midline of the body is *not* used as a point of reference for abduction and adduction of the digits. In abduction of the fingers (but not the thumb), an imaginary line is drawn through the longitudinal axis of the middle (longest) finger, and the fingers move away (spread out) from the middle finger (**Figure 9.6d**). In abduction of the thumb, the thumb moves away from the palm in the sagittal plane (see **Figure 11.18g**). Abduction of the toes is relative to an imaginary line drawn through the second toe. Adduction of the fingers and toes returns them to the anatomical position. Adduction of the thumb moves the thumb toward the palm in the sagittal plane (see **Figure 11.18g**).

Circumduction (ser-kum-DUK-shun; *circ-* = circle) is movement of the distal end of a body part in a circle (**Figure 9.7**). Circumduction is not an isolated movement by itself but rather a continuous sequence of flexion, abduction, extension, adduction, and rotation of the joint (or in the opposite order). It does not occur along a separate axis or plane of movement. Examples of circumduction are moving the humerus in a circle at the shoulder joint (**Figure 9.7a**), moving the hand in a circle at the wrist joint, moving the thumb in a circle at the

FIGURE 9.6 Angular movements at synovial joints—abduction and adduction.

Abduction and adduction usually occur along the frontal plane.

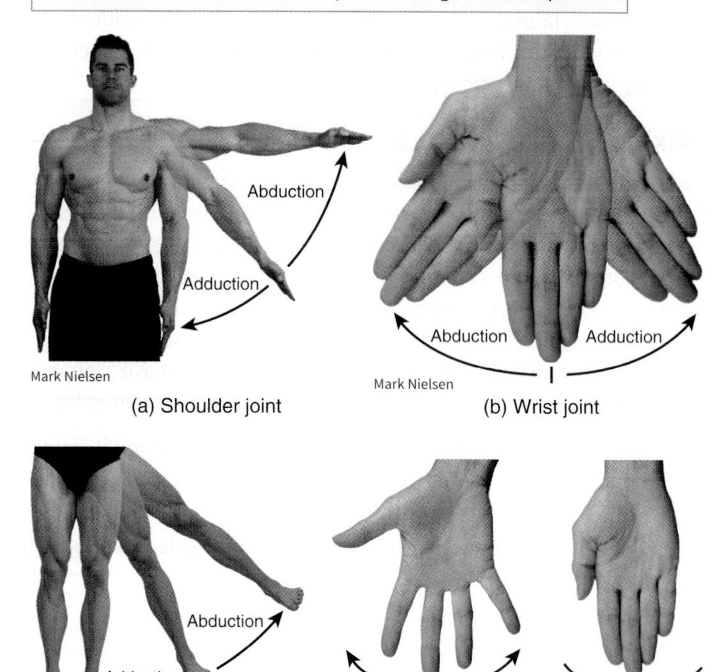

Q In what way is considering adduction as "adding your limb to your trunk" an effective learning device?

FIGURE 9.7 Angular movements at synovial joints—circumduction.

Circumduction is the movement of the distal end of a body part in a circle.

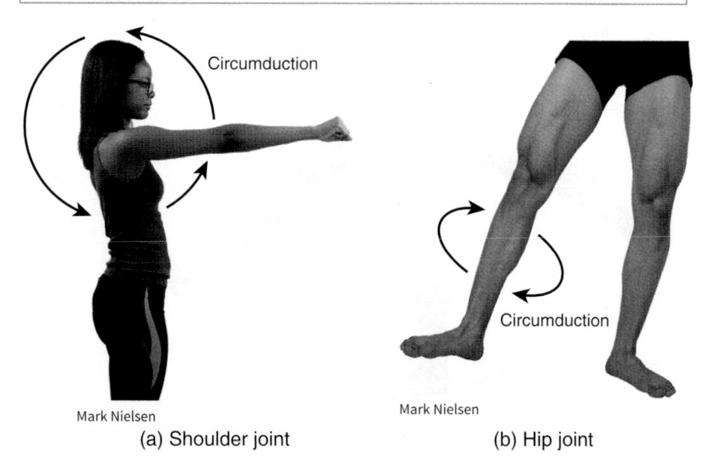

Q Which movements in continuous sequence produce circumduction?

FIGURE 9.8 **Rotation at synovial joints.**

In rotation, a bone revolves around its own longitudinal axis.

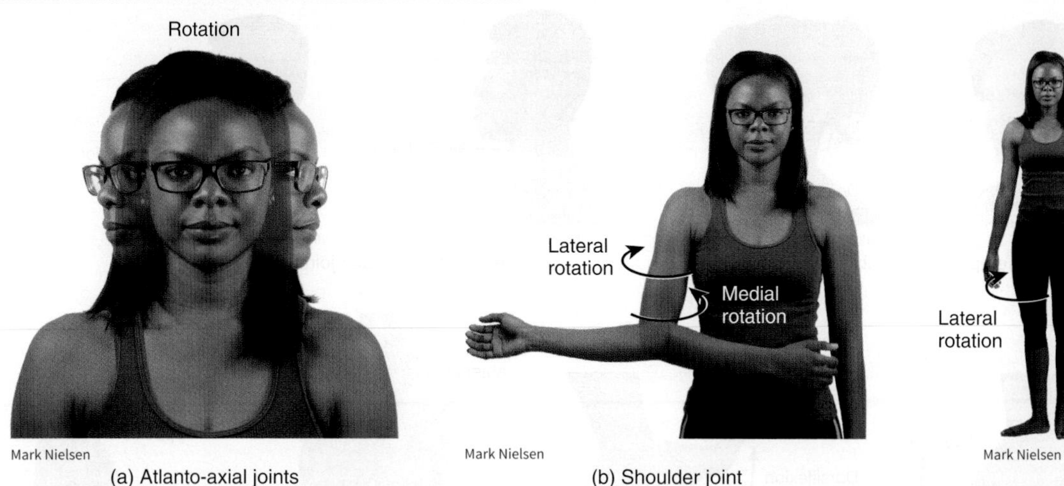

(a) Atlanto-axial joints (b) Shoulder joint (c) Hip joint

Q How do medial and lateral rotation differ?

carpometacarpal joint, moving the fingers in a circle at the metacarpophalangeal joints (between the metacarpals and phalanges), and moving the femur in a circle at the hip joint (**Figure 9.7b**). Both the shoulder and hip joints permit circumduction. Flexion, abduction, extension, and adduction are more limited in the hip joints than in the shoulder joints due to the tension on certain ligaments and muscles and the depth of the acetabulum in the hip joint (see Sections 9.10 and 9.12).

Rotation

In **rotation** (rō-TĀ-shun; *rota-* = revolve), a bone revolves around its own longitudinal axis. One example is turning the head from side to side at the atlanto-axial joint (between the atlas and axis), as when you shake your head "no" (**Figure 9.8a**). Another is turning the trunk from side-to-side at the intervertebral joints while keeping the hips and lower limbs in the anatomical position. In the limbs, rotation is defined relative to the midline, and specific qualifying terms are used. If the anterior surface of a bone of the limb is turned toward the midline, the movement is called *medial (internal) rotation.* You can medially rotate the humerus at the shoulder joint as follows: Starting in the anatomical position, flex your elbow and then move your palm across the chest (**Figure 9.8b**). You can medially rotate the femur at the hip joint as follows: Lie on your back, bend your knee, and then move your leg and foot laterally from the midline. Although you are moving your leg and foot laterally, the femur is rotating medially (**Figure 9.8c**). Medial rotation of the leg at the knee joint can be produced by sitting on a chair, bending your knee, raising your lower limb off the floor, and turning your toes medially. If the anterior surface of the bone of a limb is turned away from the midline, the mov ement is called *lateral (external) rotation* (see **Figure 9.8b, c**).

Special Movements

Special movements occur only at certain joints. They include elevation, depression, protraction, retraction, inversion, eversion, dorsiflexion, plantar flexion, supination, pronation, and opposition (**Figure 9.9**):

- **Elevation** (el-e-VĀ-shun = to lift up) is a superior movement of a part of the body, such as closing the mouth at the temporomandibular joint (between the mandible and temporal bone) to elevate the mandible (**Figure 9.9a**) or shrugging the shoulders at the acromioclavicular joint to elevate the scapula and clavicle. Its opposing movement is depression. Other bones that may be elevated (or depressed) include the hyoid and ribs.
- **Depression** (de-PRESH-un = to press down) is an inferior movement of a part of the body, such as opening the mouth to depress the mandible (**Figure 9.9b**) or returning shrugged shoulders to the anatomical position to depress the scapula and clavicle.
- **Protraction** (prō-TRAK-shun = to draw forth) is a movement of a part of the body anteriorly in the transverse plane. Its opposing movement is retraction. You can protract your mandible at the temporomandibular joint by thrusting it outward (**Figure 9.9c**) or protract your clavicles at the acromioclavicular and sternoclavicular joints by crossing your arms.
- **Retraction** (rē-TRAK-shun = to draw back) is a movement of a protracted part of the body back to the anatomical position (**Figure 9.9d**).
- **Inversion** (in-VER-zhun = to turn inward) is movement of the sole medially at the intertarsal joints (between the tarsals) (**Figure 9.9e**). Its opposing movement is eversion. Physical therapists also refer to inversion combined with plantar flexion of the feet as *supination.*

FIGURE 9.9 **Special movements at synovial joints.**

Special movements occur only at certain synovial joints.

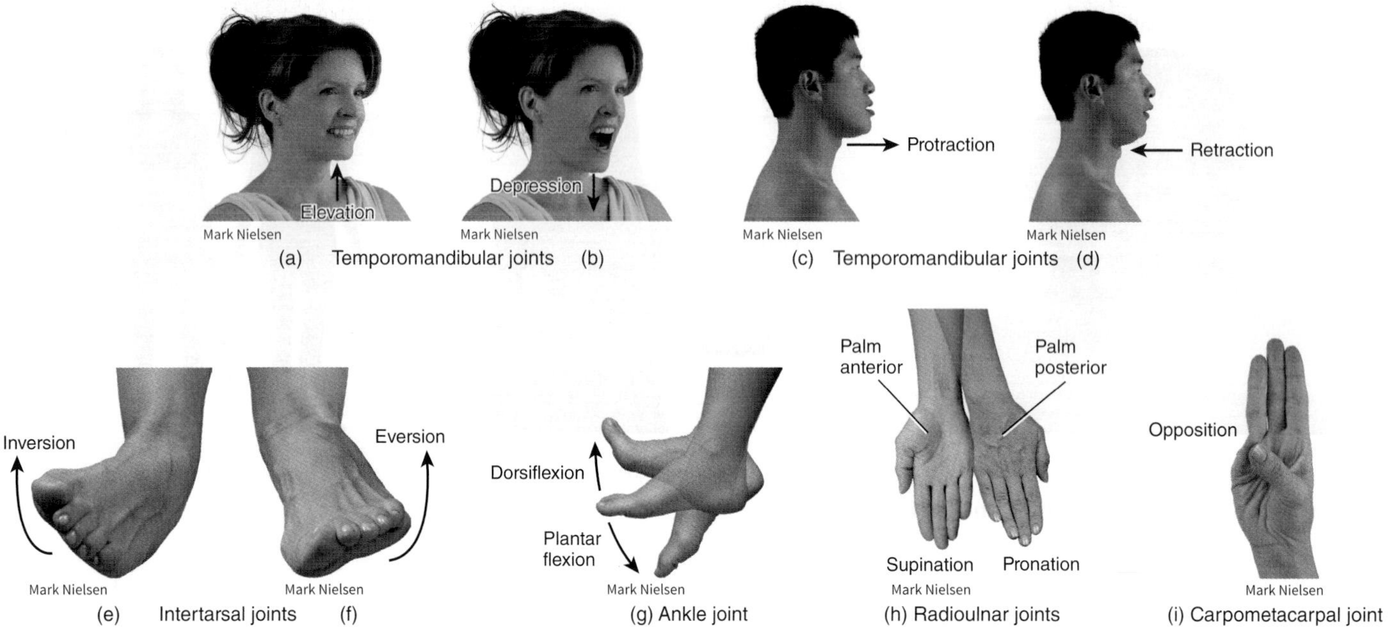

Q What movement of the shoulder girdles occur when you bring your arms forward until the elbows touch?

- **Eversion** (ē-VER-zhun = to turn outward) is a movement of the sole laterally at the intertarsal joints (**Figure 9.9f**). Physical therapists also refer to eversion combined with dorsiflexion of the feet as *pronation.*
- **Dorsiflexion** (dor-si-FLEK-shun) refers to bending of the foot at the ankle or talocrural joint (between the tibia, fibula, and talus) in the direction of the dorsum (superior surface) (**Figure 9.9g**). Dorsiflexion occurs when you stand on your heels. Its opposing movement is plantar flexion.
- **Plantar flexion** (PLAN-tar) involves bending of the foot at the ankle joint in the direction of the plantar or inferior surface (see **Figure 9.9g**), as when you elevate your body by standing on your toes.
- **Supination** (soo-pi-NĀ-shun) is a movement of the forearm at the proximal and distal radioulnar joints in which the palm is turned anteriorly (**Figure 9.9h**). This position of the palms is one of the defining features of the anatomical position. Its opposing movement is pronation.
- **Pronation** (prō-NĀ-shun) is a movement of the forearm at the proximal and distal radioulnar joints in which the distal end of the radius crosses over the distal end of the ulna and the palm is turned posteriorly (**Figure 9.9h**).
- **Opposition** (op-ō-ZISH-un) is the movement of the thumb at the carpometacarpal joint (between the trapezium and metacarpal of the thumb) in which the thumb moves across the palm to touch the tips of the fingers on the same hand (**Figure 9.9i**). These "opposable thumbs" allow the distinctive digital movement that gives humans and other primates the ability to grasp and manipulate objects very precisely.

A summary of the movements that occur at synovial joints is presented in **Table 9.1**.

9.6 Types of Synovial Joints

OBJECTIVE

- **Describe** the six subtypes of synovial joints.

Although all synovial joints have many characteristics in common, the shapes of the articulating surfaces vary; thus, many types of movements are possible. Synovial joints are divided into six categories based on type of movement: plane, hinge, pivot, condyloid, saddle, and ball-and-socket.

Plane Joints

The articulating surfaces of bones in a **plane joint** (PLĀN), also called a *planar joint* (PLĀ-nar), are flat or slightly curved (**Figure 9.10a**). Plane joints primarily permit back-and-forth and side-to-side movements between the flat surfaces of bones, but they may also rotate against one another. Many plane joints are *biaxial*, meaning that they permit movement in two axes. An *axis* is a straight line around which a bone rotates (revolves) or slides. If plane joints rotate in addition to

TABLE 9.1 Summary of Movements at Synovial Joints

MOVEMENT	DESCRIPTION	MOVEMENT	DESCRIPTION
Gliding	Movement of relatively flat bone surfaces back-and-forth and side-to-side over one another; little change in angle between bones.	**Rotation**	Movement of bone around longitudinal axis; in limbs, may be medial (toward midline) or lateral (away from midline).
Angular	Increase or decrease in angle between bones.	**Special**	Occurs at specific joints.
Flexion	Decrease in angle between articulating bones, usually in sagittal plane.	Elevation	Superior movement of body part.
		Depression	Inferior movement of body part.
Lateral flexion	Movement of trunk in frontal plane.	Protraction	Anterior movement of body part in transverse plane.
Extension	Increase in angle between articulating bones, usually in sagittal plane.	Retraction	Posterior movement of body part in transverse plane.
		Inversion	Medial movement of sole.
Hyperextension	Extension beyond anatomical position.	Eversion	Lateral movement of sole.
Abduction	Movement of bone away from midline, usually in frontal plane.	Dorsiflexion	Bending foot in direction of dorsum (superior surface).
		Plantar flexion	Bending foot in direction of plantar surface (sole).
Adduction	Movement of bone toward midline, usually in frontal plane.	Supination	Movement of forearm that turns palm anteriorly.
		Pronation	Movement of forearm that turns palm posteriorly.
Circumduction	Flexion, abduction, extension, adduction, and rotation in succession (or in the opposite order); distal end of body part moves in circle.	Opposition	Movement of thumb across palm to touch fingertips on same hand.

sliding, then they are *triaxial (multiaxial)*, permitting movement in three axes. Some examples of plane joints are the intercarpal joints (between carpal bones at the wrist), intertarsal joints (between tarsal bones at the ankle), sternoclavicular joints (between the manubrium of the sternum and the clavicle), acromioclavicular joints (between the acromion of the scapula and the clavicle), sternocostal joints (between the sternum and ends of the costal cartilages at the tips of the second through seventh pairs of ribs), and vertebrocostal joints (between the heads and tubercles of ribs and bodies and transverse processes of thoracic vertebrae).

Hinge Joints

In a **hinge joint**, or *ginglymus joint* (JIN-gli-mus), the convex surface of one bone fits into the concave surface of another bone (**Figure 9.10b**). As the name implies, hinge joints produce an angular, opening-and-closing motion like that of a hinged door. In most joint movements, one bone remains in a fixed position while the other moves around an axis. Hinge joints are *uniaxial (monaxial)* because they typically allow motion around a single axis. Hinge joints permit only flexion and extension. Examples of hinge joints are the knee (actually a modified hinge joint, which will be described later), elbow, ankle, and interphalangeal joints (between the phalanges of the fingers and toes).

Pivot Joints

In a **pivot joint**, or *trochoid joint* (TRŌ-koyd), the rounded or pointed surface of one bone articulates with a ring formed partly by another bone and partly by a ligament (**Figure 9.10c**). A pivot joint is *uniaxial* because it allows rotation only around its own longitudinal axis. Examples of pivot joints are the atlanto-axial joint, in which the atlas rotates around the axis and permits the head to turn from side-to-side as when you shake your head "no" (see **Figure 9.8a**), and the radioulnar joints that enable the palms to turn anteriorly and posteriorly as the head of the radius pivots around its long axis in the radial notch of the ulna (see **Figure 9.9h**).

Condyloid Joints

In a **condyloid joint** (KON-di-loyd; *condyl-* = knuckle) or *ellipsoidal joint*, the convex oval-shaped projection of one bone fits into the oval-shaped depression of another bone (**Figure 9.10d**). A condyloid joint is *biaxial* because the movement it permits is around two axes (flexion–extension and abduction–adduction), plus limited circumduction (remember that circumduction is not an isolated movement). Examples of condyloid joints are the radiocarpal (wrist) and metacarpophalangeal joints (between the metacarpals and proximal phalanges) of the second through fifth digits.

Saddle Joints

In a **saddle joint** or *sellar joint* (SEL-ar), the articular surface of one bone is saddle-shaped, and the articular surface of the other bone fits into the "saddle" as a sitting rider would sit (**Figure 9.10e**). The movements at a saddle joint are the same as those at a condyloid joint: *biaxial* (flexion–extension and abduction–adduction) plus limited circumduction. An example of a saddle joint is the carpometacarpal joint between the trapezium of the carpus and metacarpal of the thumb.

FIGURE 9.10 **Types of synovial joints.** For each type, a drawing of the actual joint and a simplified diagram are shown.

Synovial joints are classified into six principal types based on the shapes of the articulating bone surfaces.

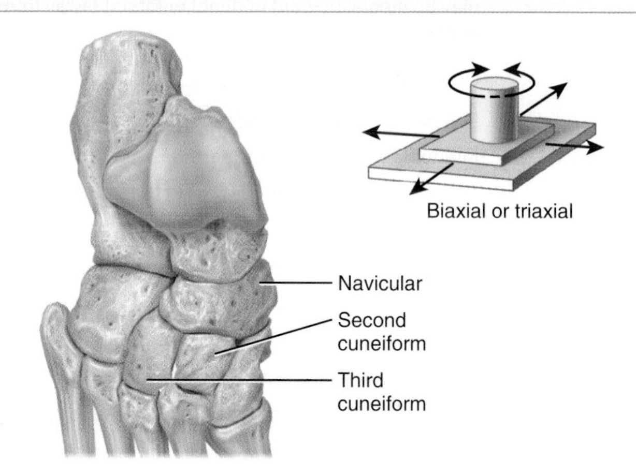

(a) **Plane joint** between navicular and second and third cuneiforms of tarsus in foot

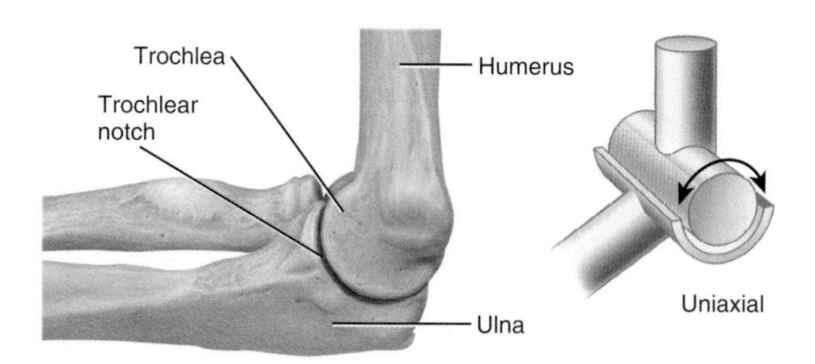

(b) **Hinge joint** between trochlea of humerus and trochlear notch of ulna at the elbow

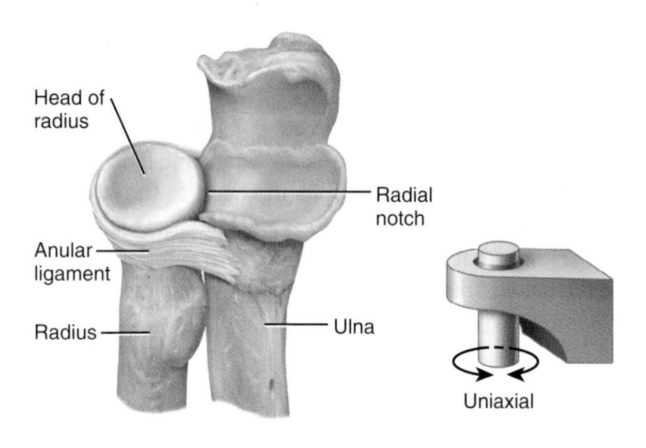

(c) **Pivot joint** between head of radius and radial notch of ulna

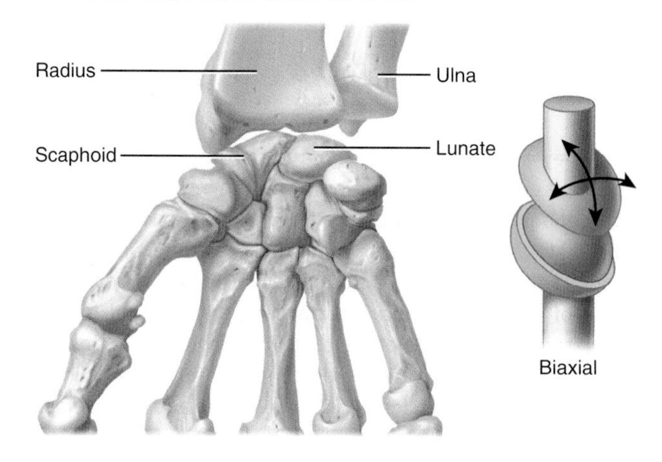

(d) **Condyloid joint** between radius and scaphoid and lunate bones of carpus (wrist)

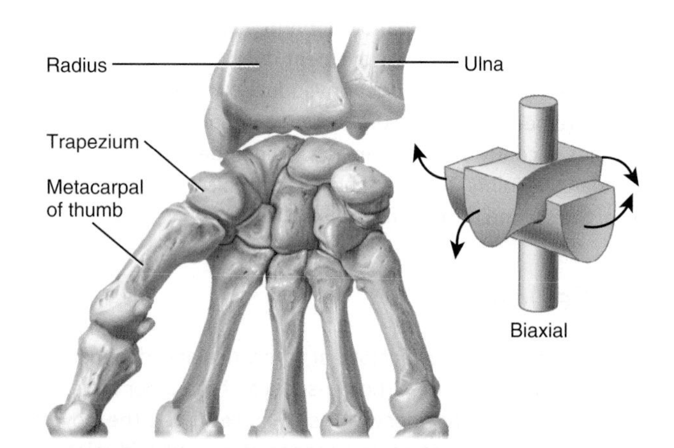

(e) **Saddle joint** between trapezium of carpus (wrist) and metacarpal of thumb

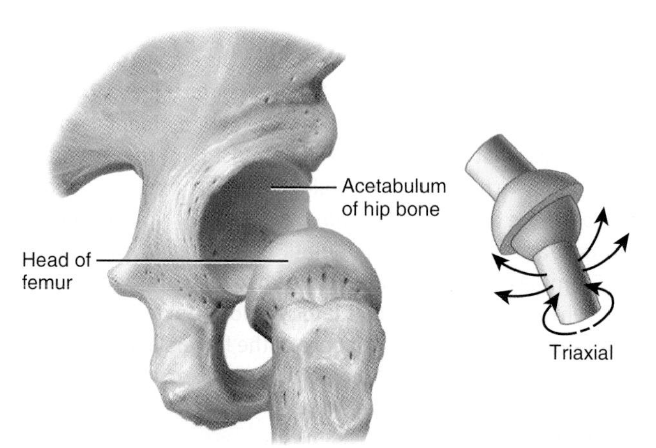

(f) **Ball-and-socket joint** between head of femur and acetabulum of hip bone

Q What are other examples of pivot joints (besides the one shown in this figure)?

TABLE 9.2 Summary of Structural and Functional Classifications of Joints

STRUCTURAL CLASSIFICATION	DESCRIPTION	FUNCTIONAL CLASSIFICATION	EXAMPLE
FIBROUS	**No Synovial Cavity; Articulating Bones Held Together by Fibrous Connective Tissue.**		
Suture	Articulating bones united by thin layer of dense irregular connective tissue, found between skull bones; with age, some sutures replaced by synostosis (separate cranial bones fuse into single bone).	Synarthrosis (immovable) and amphiarthrosis (slightly movable).	Coronal suture.
Syndesmosis	Articulating bones united by more dense irregular connective tissue, usually a ligament.	Amphiarthrosis (slightly movable).	Distal tibiofibular joint.
Interosseous membrane	Articulating bones united by substantial sheet of dense irregular connective tissue.	Amphiarthrosis (slightly movable).	Between tibia and fibula.
CARTILAGINOUS	**No Synovial Cavity; Articulating Bones United by Hyaline Cartilage or Fibrocartilage.**		
Synchondrosis	Connecting material: hyaline cartilage.	Amphiarthrosis (slightly movable) to synarthrosis (immovable).	Between first rib and manubrium of sternum.
Symphysis	Connecting material: broad, flat disc of fibrocartilage.	Amphiarthrosis (slightly movable).	Pubic symphysis and intervertebral joints.
Epiphyseal cartilage	A hyaline cartilage growth center, not actually a joint.	Synarthrosis (immovable).	Epiphyseal plate between diaphysis and epiphysis of long bone.
SYNOVIAL	**Characterized by Synovial Cavity, Articular Cartilage, and Articular (Joint) Capsule; May Contain Accessory Ligaments, Articular Discs, and Bursae.**		
Plane	Articulated surfaces flat or slightly curved.	Many biaxial diarthroses (freely movable): back-and-forth and side-to-side movements. Some triaxial diarthroses: back-and-forth, side-to-side, rotation.	Intercarpal, intertarsal, sternocostal (between sternum and second to seventh pairs of ribs), and vertebrocostal joints.
Hinge	Convex surface fits into concave surface.	Uniaxial diarthrosis: flexion–extension.	Knee (modified hinge), elbow, ankle, and interphalangeal joints.
Pivot	Rounded or pointed surface fits into ring formed partly by bone and partly by ligament.	Uniaxial diarthrosis: rotation.	Atlanto-axial and radioulnar joints.
Condyloid	Oval-shaped projection fits into oval-shaped depression.	Biaxial diarthrosis: flexion–extension, abduction–adduction.	Radiocarpal and metacarpophalangeal joints.
Saddle	Articular surface of one bone is saddle-shaped; articular surface of other bone "sits" in saddle.	Biaxial diarthrosis: flexion–extension, abduction–adduction.	Carpometacarpal joint between trapezium and metacarpal of thumb.
Ball-and-socket	Ball-like surface fits into cuplike depression.	Triaxial diarthrosis: flexion–extension, abduction–adduction, rotation.	Shoulder and hip joints.

Ball-and-Socket Joints

A **ball-and-socket joint** or *spheroid joint* (SFĒ-royd) consists of the ball-like surface of one bone fitting into a cuplike depression of another bone (**Figure 9.10f**). Such joints are *triaxial (multiaxial)*, permitting movements around three axes (flexion–extension, abduction–adduction, and rotation). Examples of ball-and-socket joints are the shoulder and hip joints. At the shoulder joint, the head of the humerus fits into the glenoid cavity of the scapula. At the hip joint, the head of the femur fits into the acetabulum of the hip (coxal) bone.

Table 9.2 summarizes the structural and functional categories of joints.

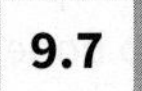

9.7 Factors Affecting Contact and Range of Motion at Synovial Joints

OBJECTIVE

- **Describe** six factors that influence the type of movement and range of motion possible at a synovial joint.

The articular surfaces of synovial joints contact one another and determine the type and possible range of motion. **Range of motion (ROM)** refers to the range, measured in degrees of a circle, through which the bones of a joint can be moved. The following factors contribute to keeping the articular surfaces in contact and affect range of motion:

1. ***Structure or shape of the articulating bones.*** The structure or shape of the articulating bones determines how closely they can fit together. The articular surfaces of some bones have a complementary relationship. This spatial relationship is very obvious at the hip joint, where the head of the femur articulates with the acetabulum of the hip bone. An interlocking fit allows rotational movement.
2. ***Strength and tension (tautness) of the joint ligaments.*** The different components of a fibrous capsule are tense or taut only when the joint is in certain positions. Tense ligaments not only restrict the range of motion but also direct the movement of the articulating bones with respect to each other. In the knee joint, for example, the anterior cruciate ligament is taut and the posterior cruciate ligament is loose when the knee is straightened, and the reverse occurs when the knee is bent. In the hip joint, certain ligaments become taut when standing and more firmly attach the head of the femur to the acetabulum of the hip bone.
3. ***Arrangement and tension of the muscles.*** Muscle tension reinforces the restraint placed on a joint by its ligaments, and thus restricts movement. A good example of the effect of muscle tension on a joint is seen at the hip joint. When the thigh is flexed with the knee extended, the flexion of the hip joint is restricted by the tension of the hamstring muscles on the posterior surface of the thigh, so most of us can't raise a straightened leg more than a 90-degree angle from the floor. But if the knee is also flexed, the tension on the hamstring muscles is lessened, and the thigh can be raised farther, allowing you to raise your thigh to touch your chest.
4. ***Contact of soft parts.*** The point at which one body surface contacts another may limit mobility. For example, if you bend your arm at the elbow, it can move no farther after the anterior surface of the forearm meets with and presses against the biceps brachii muscle of the arm. Joint movement may also be restricted by the presence of adipose tissue.
5. ***Hormones.*** Joint flexibility may also be affected by hormones. For example, relaxin, a hormone produced by the placenta and ovaries, increases the flexibility of the fibrocartilage of the pubic symphysis and loosens the ligaments between the sacrum, hip bone, and coccyx toward the end of pregnancy. These changes permit expansion of the pelvic outlet, which assists in delivery of the baby.
6. ***Disuse.*** Movement at a joint may be restricted if a joint has not been used for an extended period. For example, if an elbow joint is immobilized by a cast, range of motion at the joint may be limited for a time after the cast is removed. Disuse may also result in decreased amounts of synovial fluid, diminished flexibility of ligaments and tendons, and *muscular atrophy*, a reduction in size or wasting of a muscle.

9.8 Selected Joints of the Body

OBJECTIVE

- **Identify** the major joints of the body by location, classification, and movements.

In Chapters 7 and 8, we discussed the major bones and their markings. In this chapter we have examined how joints are classified according to their structure and function, and we have introduced the movements that occur at joints. **Table 9.3** (selected joints of the axial skeleton) and **Table 9.4** (selected joints of the appendicular skeleton) will help you integrate the information you have learned in all three chapters. These tables list some of the major joints of the body according to their articular components (the bones that enter into their formation), their structural and functional classification, and the type(s) of movement that occur(s) at each joint.

Over the next few sections in this chapter, we examine in detail several selected joints of the body. Each section considers a specific synovial joint and contains (1) a definition—a description of the type of joint and the bones that form the joint; (2) the anatomical components—a description of the major connecting ligaments, articular disc (if present), articular capsule, and other distinguishing features of the joint; and (3) the joint's possible movements. Each section also refers you to a figure that illustrates the joint. The joints described are the temporomandibular joint (TMJ), shoulder (humeroscapular or glenohumeral) joint, elbow joint, hip (coxal) joint, and knee (tibiofemoral) joint. Because these joints are described in Sections 9.9–9.13, they are not included in **Tables 9.3** and **9.4**.

9.9 Temporomandibular Joint

OBJECTIVE

- **Describe** the anatomical components of the temporomandibular joint and explain the movements that can occur at this joint.

Definition

The **temporomandibular joint (TMJ)** (tem′-po-rō-man-DIB-ū-lar) is a combined hinge and plane joint formed by the condylar process of the mandible and the mandibular fossa and articular tubercle of the temporal bone. The temporomandibular joint is the only freely movable joint between skull bones (with the exception of the ear ossicles); all other skull joints are sutures and therefore immovable or slightly movable.

TABLE 9.3 Selected Joints of the Axial Skeleton

JOINT	ARTICULAR COMPONENTS	CLASSIFICATION	MOVEMENTS
Suture	Between skull bones.	*Structural:* fibrous. *Functional:* amphiarthrosis and synarthrosis.	None.
Atlanto-occipital	Between superior articular facets of atlas and occipital condyles of occipital bone.	*Structural:* synovial (condyloid). *Functional:* diarthrosis.	Flexion and extension of head; slight lateral flexion of head to either side.
Atlanto-axial	(1) Between dens of axis and anterior arch of atlas; (2) between lateral masses of atlas and axis.	*Structural:* synovial (pivot) between dens and anterior arch; synovial (planar) between lateral masses. *Functional:* diarthrosis.	Rotation of head.
Intervertebral	(1) Between vertebral bodies; (2) between vertebral arches.	*Structural:* cartilaginous (symphysis) between vertebral bodies; synovial (planar) between vertebral arches. *Functional:* amphiarthrosis between vertebral bodies; diarthrosis between vertebral arches.	Flexion, extension, lateral flexion, and rotation of vertebral column.
Vertebrocostal	(1) Between facets of heads of ribs and facets of bodies of adjacent thoracic vertebrae and intervertebral discs between them; (2) between articular part of tubercles of ribs and facets of transverse processes of thoracic vertebrae.	*Structural:* synovial (planar). *Functional:* diarthrosis.	Slight gliding.
Sternocostal	Between sternum and first seven pairs of ribs.	*Structural:* cartilaginous (synchondrosis) between sternum and first pair of ribs; synovial (plane) between sternum and second through seventh pairs of ribs. *Functional:* synarthrosis between sternum and first pair of ribs; diarthrosis between sternum and second through seventh pairs of ribs.	None between sternum and first pair of ribs; slight gliding between sternum and second through seventh pairs of ribs.
Lumbosacral	(1) Between body of fifth lumbar vertebra and base of sacrum; (2) between inferior articular facets of fifth lumbar vertebra and superior articular facets of first vertebra of sacrum.	*Structural:* cartilaginous (symphysis) between body and base; synovial (planar) between articular facets. *Functional:* amphiarthrosis between body and base; diarthrosis between articular facets.	Flexion, extension, lateral flexion, and rotation of vertebral column.

Anatomical Components

1. ***Articular disc (meniscus).*** Fibrocartilage disc that separates the synovial cavity into superior and inferior compartments, each with a synovial membrane (**Figure 9.11c**).
2. ***Articular capsule.*** Thin, fairly loose envelope around the circumference of the joint (**Figure 9.11a, b**).
3. ***Lateral ligament.*** Two short bands on the lateral surface of the articular capsule that extend inferiorly and posteriorly from the inferior border and tubercle of the zygomatic process of the temporal bone to the lateral and posterior aspect of the neck of the mandible. The lateral ligament is covered by the parotid gland and helps strengthen the joint laterally and prevent displacement of the mandible (**Figure 9.11a**).
4. ***Sphenomandibular ligament*** (sfē-nō-man-DIB-ū-lar). Thin band that extends inferiorly and anteriorly from the spine of the sphenoid bone to the ramus of the mandible (**Figure 9.11b**). It does not contribute significantly to the strength of the joint.
5. ***Stylomandibular ligament*** (stī-lō-man-DIB-ū-lar). Thickened band of deep cervical fascia that extends from the styloid process of the temporal bone to the inferior and posterior border of the ramus of the mandible. This ligament separates the parotid gland from the submandibular gland and limits movement of the mandible at the TMJ (**Figure 9.11a, b**).

TABLE 9.4 Selected Joints of the Appendicular Skeleton

JOINT	ARTICULAR COMPONENTS	CLASSIFICATION	MOVEMENTS
Sternoclavicular	Between sternal end of clavicle, manubrium of sternum, and first costal cartilage.	*Structural:* synovial (plane, pivot). *Functional:* diarthrosis.	Gliding, with limited movements in nearly every direction.
Acromioclavicular	Between acromion of scapula and acromial end of clavicle.	*Structural:* synovial (plane). *Functional:* diarthrosis.	Gliding and rotation of scapula on clavicle.
Radioulnar	Proximal radioulnar joint between head of radius and radial notch of ulna; distal radioulnar joint between ulnar notch of radius and head of ulna.	*Structural:* synovial (pivot). *Functional:* diarthrosis.	Rotation of forearm.
Wrist (radiocarpal)	Between distal end of radius and scaphoid, lunate, and triquetrum of carpus.	*Structural:* synovial (condyloid). *Functional:* diarthrosis.	Flexion, extension, abduction, adduction, circumduction, and slight hyperextension of wrist.
Intercarpal	Between proximal row of carpal bones, distal row of carpal bones, and between both rows of carpal bones (midcarpal joints).	*Structural:* synovial (plane), except for hamate, scaphoid, and lunate (midcarpal) joint, which is synovial (saddle). *Functional:* diarthrosis.	Gliding plus flexion, extension, abduction, adduction, and slight rotation at midcarpal joints.
Carpometacarpal	Carpometacarpal joint of thumb between trapezium of carpus and first metacarpal; carpometacarpal joints of remaining digits formed between carpus and second through fifth metacarpals.	*Structural:* synovial (saddle) at thumb; synovial (plane) at remaining digits. *Functional:* diarthrosis.	Flexion, extension, abduction, adduction, and circumduction at thumb; gliding at remaining digits.
Metacarpophalangeal and metatarsophalangeal	Between heads of metacarpals (or metatarsals) and bases of proximal phalanges.	*Structural:* synovial (condyloid). *Functional:* diarthrosis.	Flexion, extension, abduction, adduction, and circumduction of phalanges.
Interphalangeal	Between heads of phalanges and bases of more distal phalanges.	*Structural:* synovial (hinge). *Functional:* diarthrosis.	Flexion and extension of phalanges.
Sacroiliac	Between auricular surfaces of sacrum and ilia of hip bones.	*Structural:* synovial (plane). *Functional:* diarthrosis.	Slight gliding (even more so during pregnancy).
Pubic symphysis	Between anterior surfaces of hip bones.	*Structural:* cartilaginous (symphysis). *Functional:* amphiarthrosis.	Slight movements (even more so during pregnancy).
Tibiofibular	Proximal tibiofibular joint between lateral condyle of tibia and head of fibula; distal tibiofibular joint between distal end of fibula and fibular notch of tibia.	*Structural:* synovial (plane) at proximal joint; fibrous (syndesmosis) at distal joint. *Functional:* diarthrosis at proximal joint; amphiarthrosis at distal joint.	Slight gliding at proximal joint; slight rotation of fibula during dorsiflexion of foot.
Ankle (talocrural)	(1) Between distal end of tibia and its medial malleolus and talus; (2) between lateral malleolus of fibula and talus.	*Structural:* synovial (hinge). *Functional:* diarthrosis.	Dorsiflexion and plantar flexion of foot.
Intertarsal	Subtalar joint between talus and calcaneus of tarsus; talocalcaneonavicular joint between talus and calcaneus and navicular of tarsus; calcaneocuboid joint between calcaneus and cuboid of tarsus.	*Structural:* synovial (plane) at subtalar and calcaneocuboid joints; synovial (saddle) at talocalcaneonavicular joint. *Functional:* diarthrosis.	Inversion and eversion of foot.
Tarsometatarsal	Between three cuneiforms of tarsus and bases of five metatarsal bones.	*Structural:* synovial (plane). *Functional:* diarthrosis.	Slight gliding.

FIGURE 9.11 **Right temporomandibular joint (TMJ).**

The TMJ is the only movable joint between skull bones.

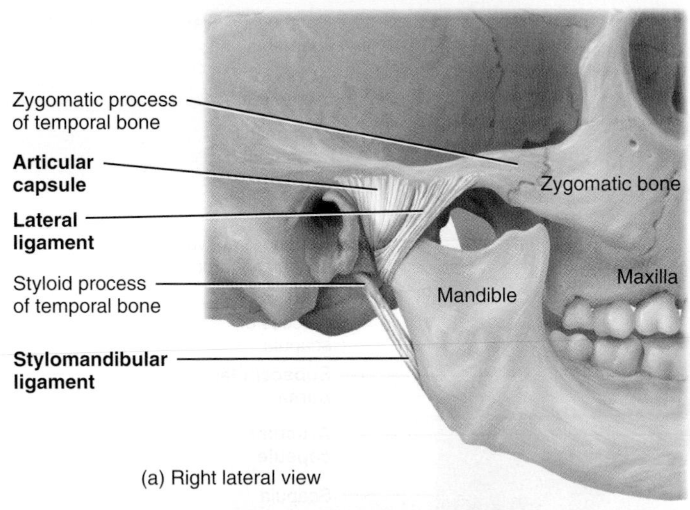

(a) Right lateral view

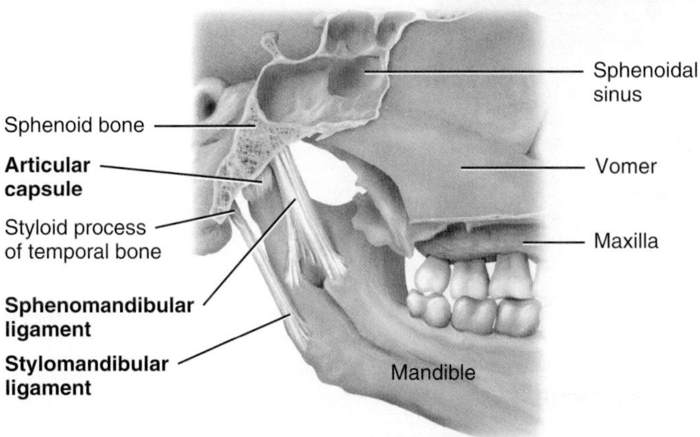

(b) Left medial view

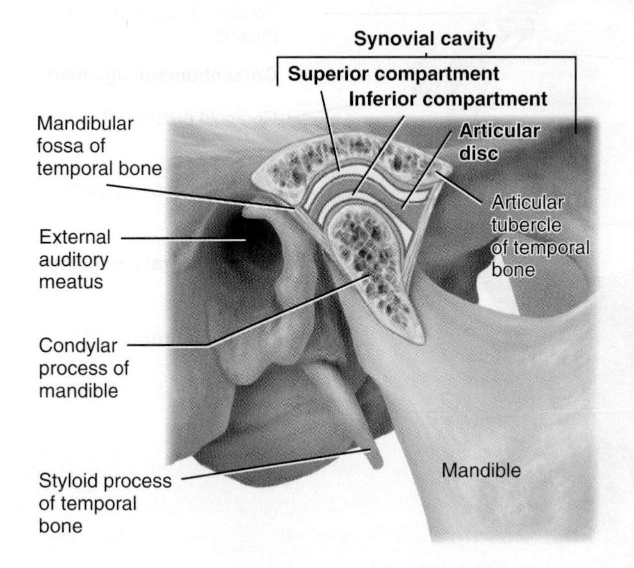

(c) Sagittal section viewed from right

Q Which ligament prevents displacement of the mandible?

Movements

In the temporomandibular joint, only the mandible moves because the temporal bone is firmly anchored to other bones of the skull by sutures. Accordingly, the mandible may function in depression (jaw opening) and elevation (jaw closing), which occurs in the inferior compartment, and protraction, retraction, lateral displacement, and slight rotation, which occur in the superior compartment (see Figure 9.9a–d).

9.10 Shoulder Joint

OBJECTIVE

- **Describe** the anatomical components of the shoulder joint and the movements that can occur at this joint.

Definition

The **shoulder joint** is a ball-and-socket joint formed by the head of the humerus and the glenoid cavity of the scapula. It is also referred to as the *humeroscapular* or *glenohumeral joint* (glē-no-HŪ-mer-al).

Anatomical Components

1. ***Articular capsule.*** Thin, loose sac that completely envelops the joint and extends from the glenoid cavity to the anatomical neck of the humerus. The inferior part of the capsule is its weakest area (Figure 9.12).
2. ***Coracohumeral ligament*** (kor′-a-kō-HŪ-mer-al). Strong, broad ligament that strengthens the superior part of the articular capsule and extends from the coracoid process of the scapula to the greater tubercle of the humerus (Figure 9.12a, b). The ligament strengthens the superior part of the articular capsule and reinforces the anterior aspect of the articular capsule.
3. ***Glenohumeral ligaments.*** Three thickenings of the articular capsule over the anterior surface of the joint that extend from the glenoid cavity to the lesser tubercle and anatomical neck of the humerus. These ligaments are often indistinct or absent and provide only minimal strength (Figure 9.12a, b). They play a role in joint stabilization when the humerus approaches or exceeds its limits of motion.
4. ***Transverse humeral ligament.*** Narrow sheet extending from the greater tubercle to the lesser tubercle of the humerus (Figure 9.12a). The ligament functions as a retinaculum (retaining band of connective tissue) to hold the long head of the biceps brachii muscle.
5. ***Glenoid labrum.*** Narrow rim of fibrocartilage around the edge of the glenoid cavity that slightly deepens and enlarges the glenoid cavity (Figure 9.12b, c).
6. ***Bursae.*** Four *bursae* (see Section 9.4) are associated with the shoulder joint. They are the *subscapular bursa* (Figure 9.12a), *subdeltoid bursa*, not labeled in *subacromial bursa* (Figure 9.12a–c), and *subcoracoid bursa*.

FIGURE 9.12 **Right shoulder (humeroscapular or glenohumeral) joint.**

Most of the stability of the shoulder joint results from the arrangement of the rotator cuff muscles.

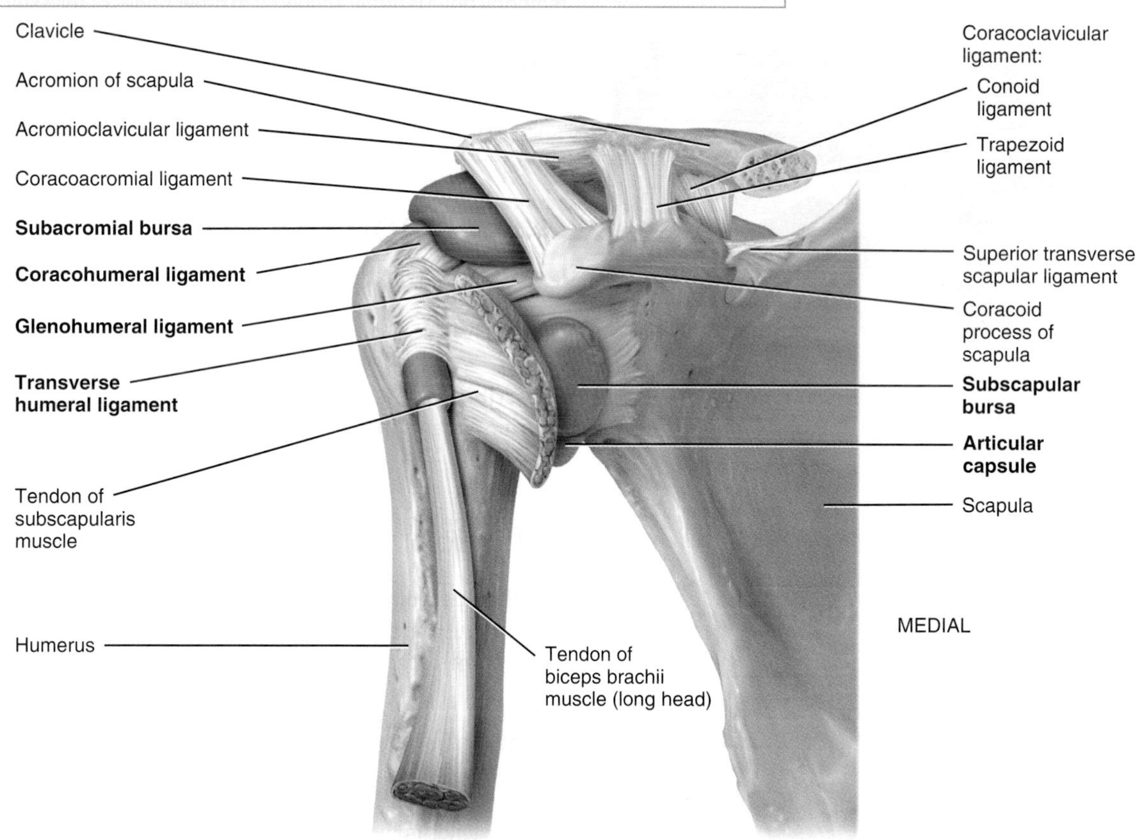

(a) Anterior view

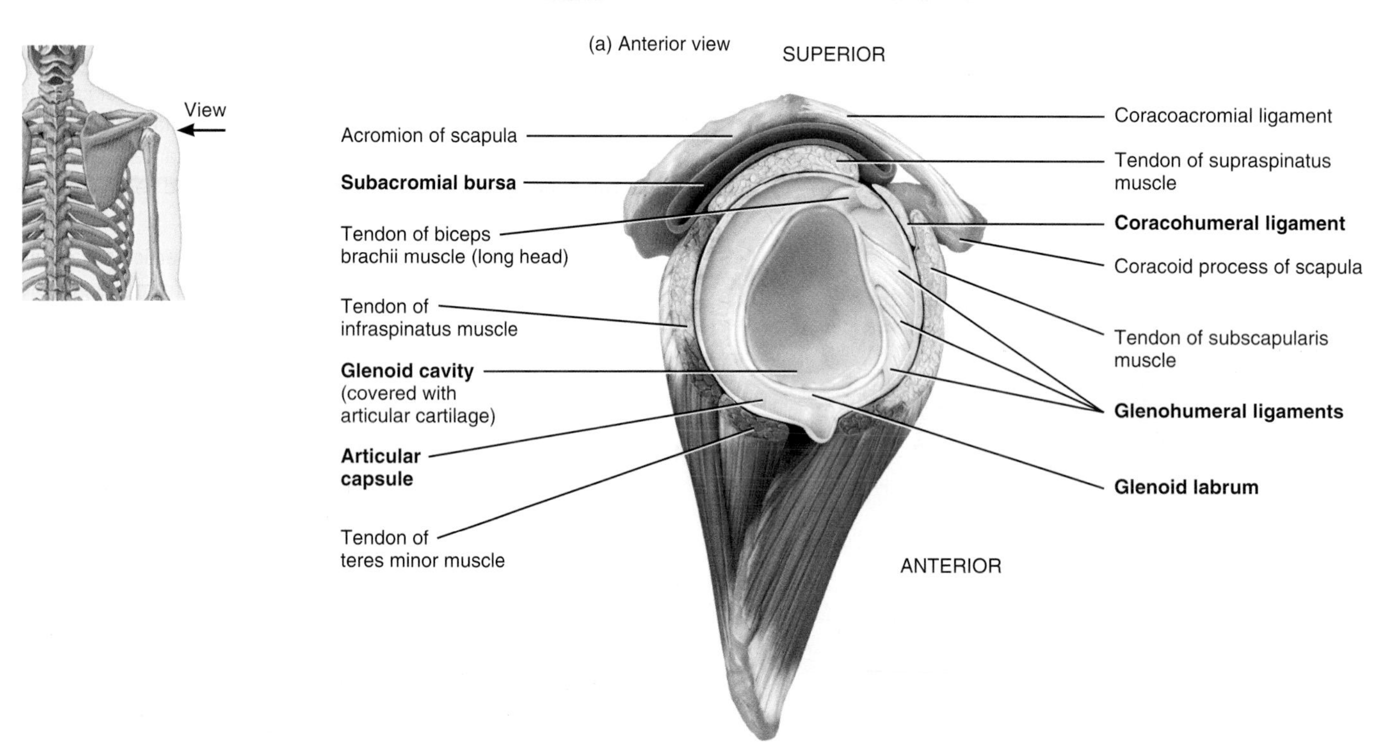

(b) Lateral view (opened)

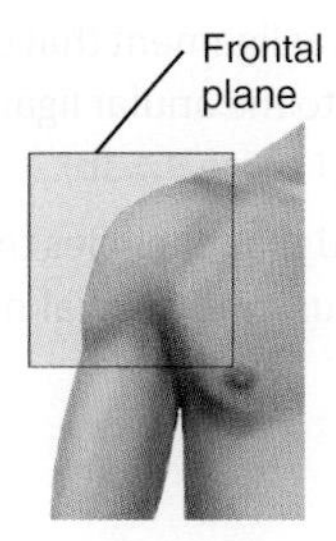

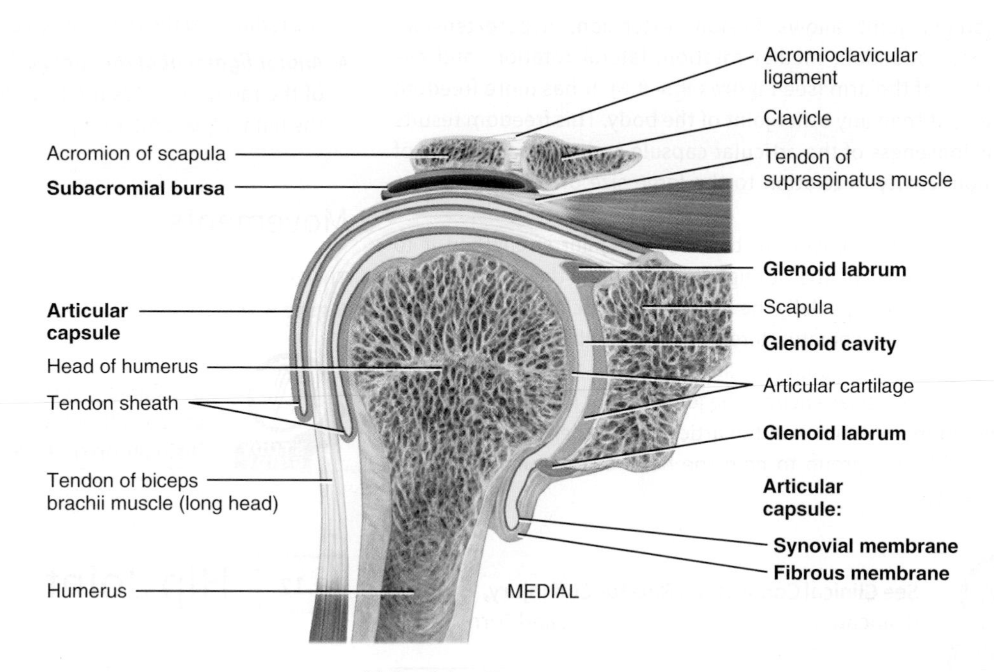

(c) Frontal section

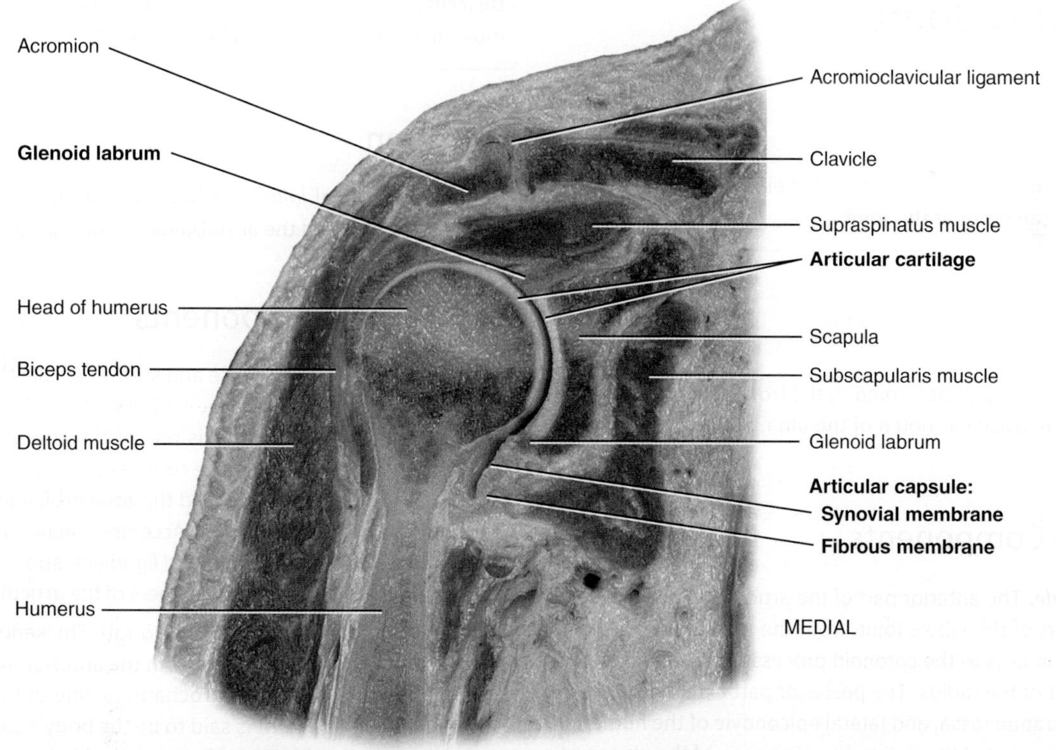

Dissection Shawn Miller, Photograph Mark Nielsen

(d) Frontal section

Q Why does the shoulder joint have more freedom of movement than any other joint of the body?

Movements

The shoulder joint allows flexion, extension, hyperextension, abduction, adduction, medial rotation, lateral rotation, and circumduction of the arm (see **Figures 9.5–9.8**). It has more freedom of movement than any other joint of the body. This freedom results from the looseness of the articular capsule and the shallowness of the glenoid cavity in relation to the large size of the head of the humerus.

Although the ligaments of the shoulder joint strengthen it to some extent, most of the strength results from the muscles that surround the joint, especially the *rotator cuff muscles.* These muscles (supraspinatus, infraspinatus, teres minor, and subscapularis) anchor the humerus to the scapula (see also **Figure 11.15**). The tendons of the rotator cuff muscles encircle the joint (except for the inferior portion) and intimately surround the articular capsule. The rotator cuff muscles work as a group to hold the head of the humerus in the glenoid cavity.

See Clinical Connection: Rotator Cuff Injury, Dislocated and Separated Shoulder, and Torn Glenoid Labrum

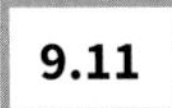

9.11 Elbow Joint

OBJECTIVE

- **Describe** the anatomical components of the elbow joint and the movements that can occur at this joint.

Definition

The **elbow joint** is a hinge joint formed by the trochlea and capitulum of the humerus, the trochlear notch of the ulna, and the head of the radius.

Anatomical Components

1. ***Articular capsule.*** The anterior part of the articular capsule covers the anterior part of the elbow joint, from the radial and coronoid fossae of the humerus to the coronoid process of the ulna and the anular ligament of the radius. The posterior part extends from the capitulum, olecranon fossa, and lateral epicondyle of the humerus to the anular ligament of the radius, the olecranon of the ulna, and the ulna posterior to the radial notch (**Figure 9.13a, b**).
2. ***Ulnar collateral ligament.*** Thick, triangular ligament that extends from the medial epicondyle of the humerus to the coronoid process and olecranon of the ulna (**Figure 9.13a**). Part of this ligament deepens the socket for the trochlea of the humerus.
3. ***Radial collateral ligament.*** Strong, triangular ligament that extends from the lateral epicondyle of the humerus to the anular ligament of the radius and the radial notch of the ulna (**Figure 9.13b**).
4. ***Anular ligament of the radius.*** Strong band that encircles the head of the radius. It holds the head of the radius in the radial notch of the ulna (**Figure 9.13a, b**).

Movements

The elbow joint allows flexion and extension of the forearm (see **Figure 9.5c**).

See Clinical Connection: Tennis Elbow, Little League Elbow, and Tommy John Surgery Dislocation of the Radial Head

9.12 Hip Joint

OBJECTIVE

- **Describe** the anatomical components of the hip joint and the movements that can occur at this joint.

Definition

The **hip joint** (*coxal joint*) is a ball-and-socket joint formed by the head of the femur and the acetabulum of the hip bone.

Anatomical Components

1. ***Articular capsule.*** Very dense and strong capsule that extends from the rim of the acetabulum to the neck of the femur (**Figure 9.14c**). With its accessory ligaments, this is one of the strongest structures of the body. The articular capsule consists of circular and longitudinal fibers. The circular fibers, called the zona orbicularis, form a collar around the neck of the femur. Accessory ligaments known as the *iliofemoral ligament, pubofemoral ligament*, and *ischiofemoral ligament* reinforce the longitudinal fibers of the articular capsule.
2. ***Iliofemoral ligament*** (il′-ē-ō-FEM-ō-ral). Thickened portion of the articular capsule that extends from the anterior inferior iliac spine of the hip bone to the intertrochanteric line of the femur (**Figure 9.14a, b**). This ligament is said to be the body's strongest and prevents hyperextension of the femur at the hip joint during standing.
3. ***Pubofemoral ligament*** (pū′-bō-FEM-ō-ral). Thickened portion of the articular capsule that extends from the pubic part of the rim of the acetabulum to the neck of the femur (**Figure 9.14a**). This ligament prevents overabduction of the femur at the hip joint and strengthens the articular capsule.

FIGURE 9.13 **Right elbow joint.**

The elbow joint is formed by parts of three bones: humerus, ulna, and radius.

(a) Medial aspect

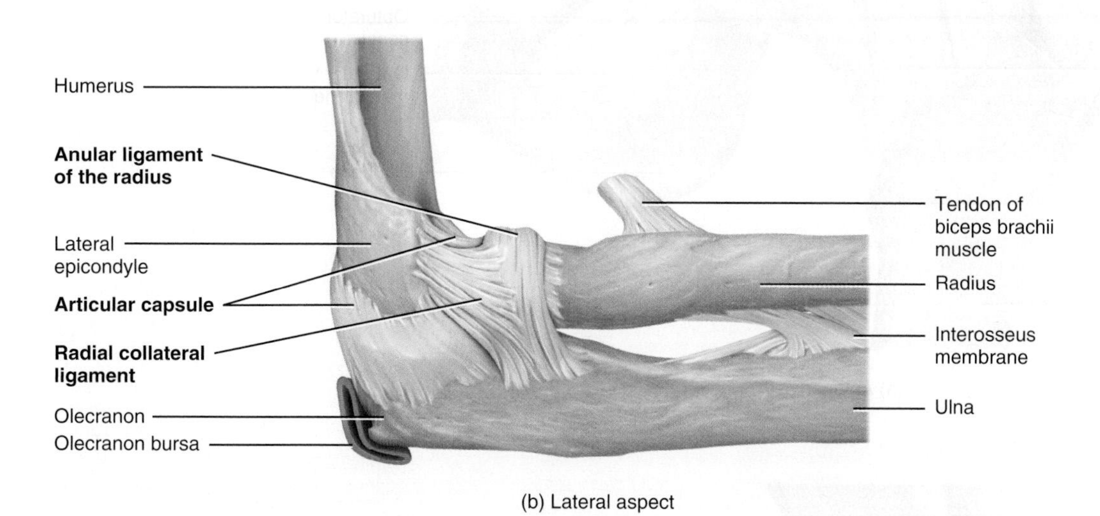

(b) Lateral aspect

Q Which movements are possible at a hinge joint?

4. ***Ischiofemoral ligament*** (is′-kē-ō-FEM-ō-ral). Thickened portion of the articular capsule that extends from the ischial wall bordering the acetabulum to the neck of the femur (**Figure 9.14b**). This ligament slackens during adduction, tenses during abduction, and strengthens the articular capsule.
5. ***Ligament of the head of the femur.*** Flat, triangular band (primarily a synovial fold) that extends from the fossa of the acetabulum to the fovea capitis of the head of the femur (**Figure 9.14c**). The ligament usually contains a small artery that supplies the head of the femur.
6. ***Acetabular labrum*** (as-e-TAB-ū-lar LĀ-brum). Fibrocartilage rim attached to the margin of the acetabulum that enhances the depth of the acetabulum (**Figure 9.14c**). As a result, dislocation of the femur is rare.
7. ***Transverse ligament of the acetabulum.*** Strong ligament that crosses over the acetabular notch. It supports part of the acetabular labrum and is connected with the ligament of the head of the femur and the articular capsule (**Figure 9.14c**).

Movements

The hip joint allows flexion, extension, abduction, adduction, lateral rotation medial rotation, and circumduction of the thigh (see **Figures 9.5–9.8**). The extreme stability of the hip joint is related to the very strong articular capsule and its accessory ligaments, the manner in which the femur fits into the acetabulum, and the muscles surrounding the joint. Although the shoulder and hip joints are both ball-and-socket joints, the hip joints do not have as wide a range of motion. Flexion is limited by the anterior surface of the thigh coming into contact with the anterior abdominal wall when the knee is flexed and by tension of the hamstring muscles when the knee is extended. Extension is limited by tension of the iliofemoral, pubofemoral, and ischiofemoral ligaments. Abduction is limited by the tension of the pubofemoral ligament, and adduction is limited by contact with the opposite limb and tension in the ligament of the head of the femur. Medial rotation is limited by the tension in the ischiofemoral ligament, and lateral rotation is limited by tension in the iliofemoral and pubofemoral ligaments.

FIGURE 9.14 **Right hip (coxal) joint.**

The articular capsule of the hip joint is one of the strongest structures in the body.

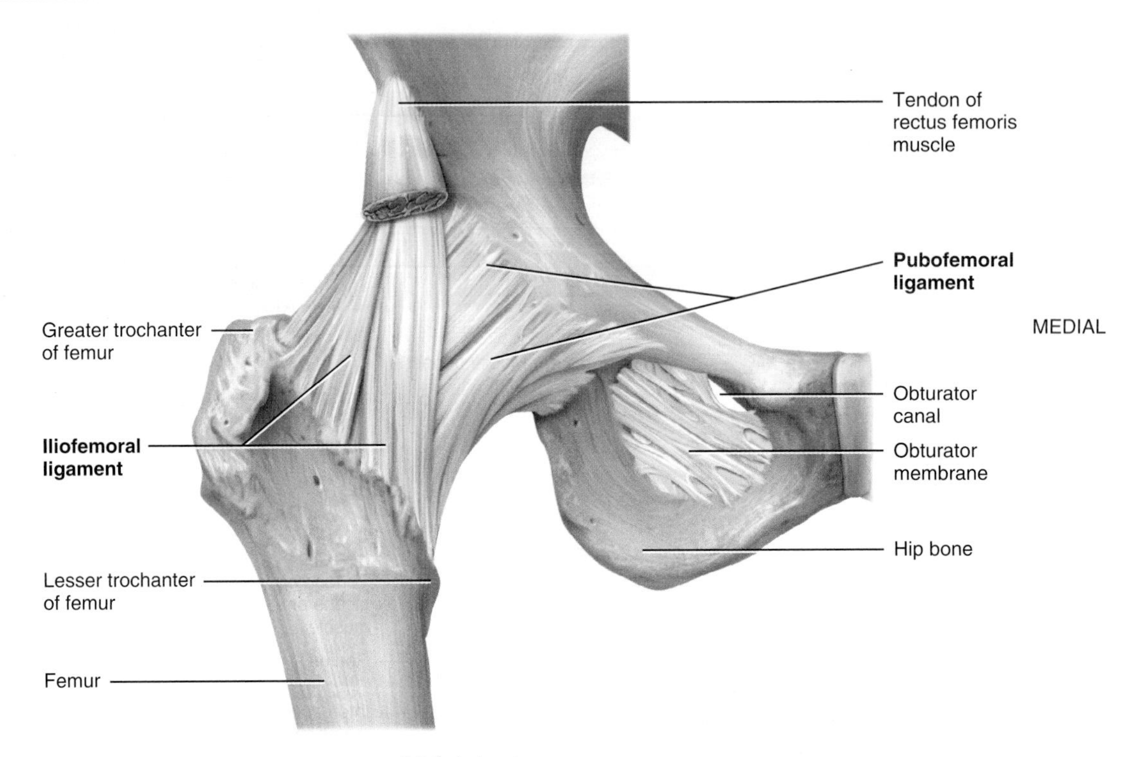

(a) Anterior view

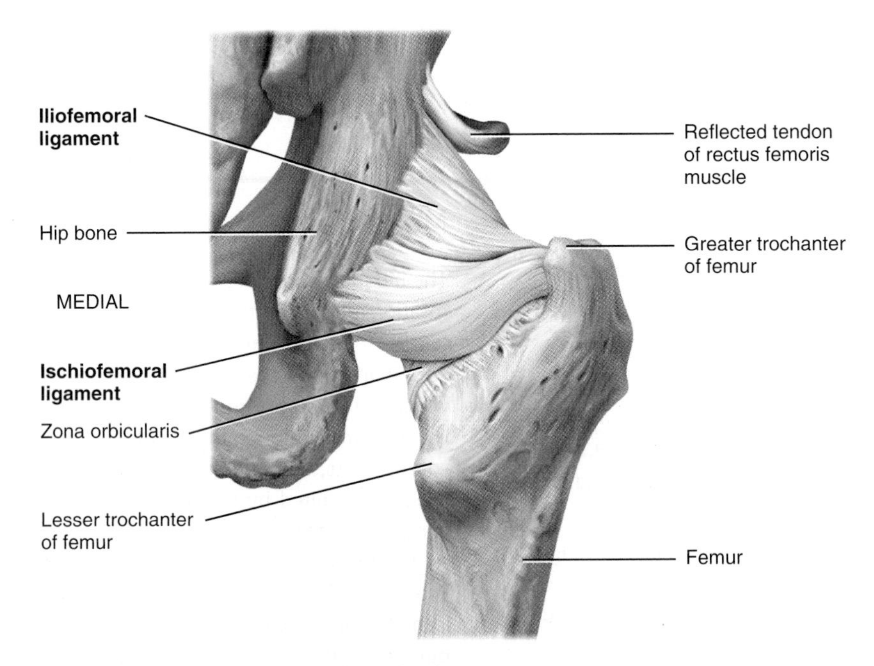

(b) Posterior view

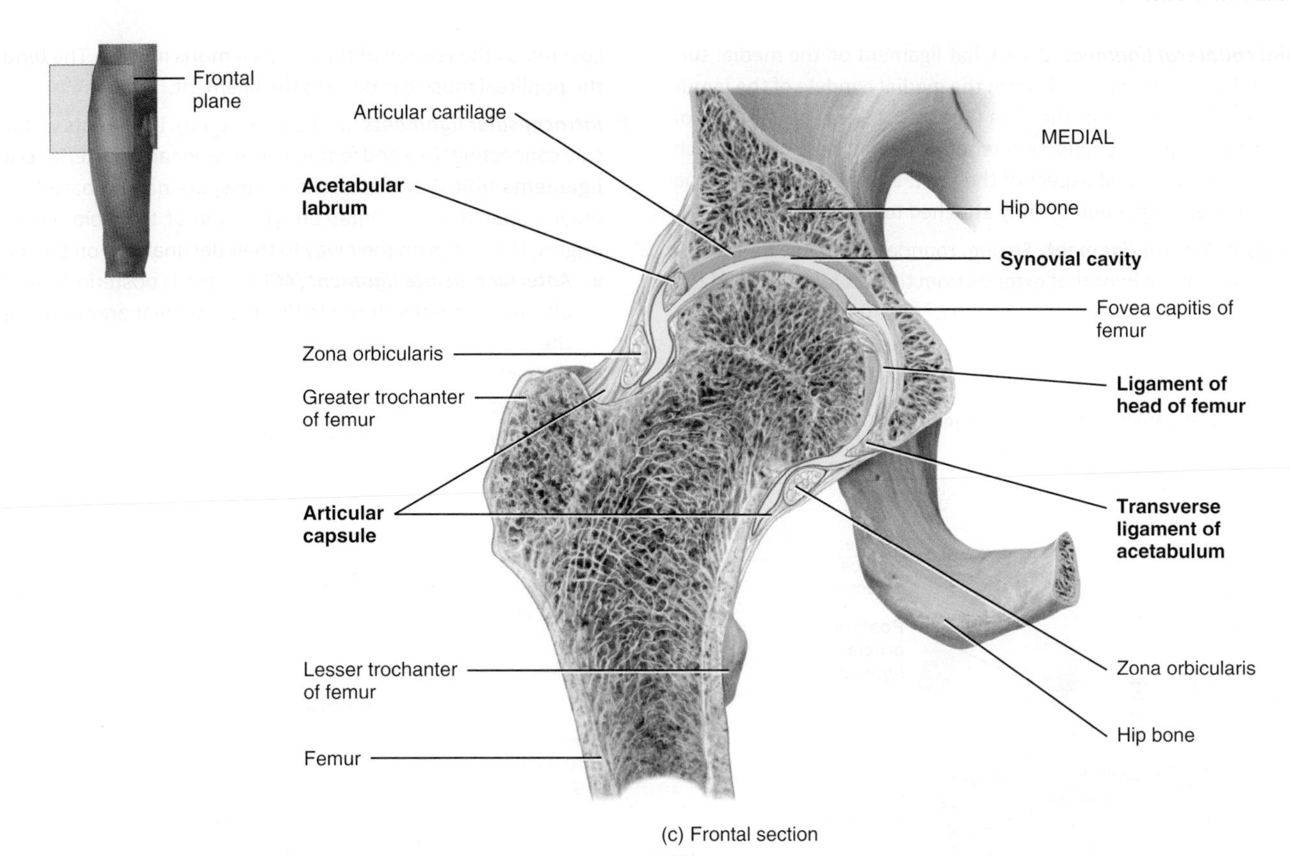

(c) Frontal section

Q Which ligaments limit the degree of extension that is possible at the hip joint?

9.13 Knee Joint

OBJECTIVE

- **Describe** the main anatomical components of the knee joint and explain the movements that can occur at this joint.

Definition

The **knee joint** *(tibiofemoral joint)* is the largest and most complex joint of the body (**Figure 9.15**). It is a modified hinge joint (because its primary movement is a uniaxial hinge movement) that consists of three joints within a single synovial cavity:

1. Laterally is a *tibiofemoral joint*, between the lateral condyle of the femur, lateral meniscus, and lateral condyle of the tibia, which is the weight-bearing bone of the leg.
2. Medially is another *tibiofemoral joint*, between the medial condyle of the femur, medial meniscus, and medial condyle of the tibia.
3. An intermediate *patellofemoral joint* is between the patella and the patellar surface of the femur.

Anatomical Components

1. ***Articular capsule.*** No complete, independent capsule unites the bones of the knee joint. The ligamentous sheath surrounding the joint consists mostly of muscle tendons or their expansions (**Figure 9.15e–g**). There are, however, some capsular fibers connecting the articulating bones.
2. ***Medial and lateral patellar retinacula*** (ret′-i-NAK-ū-la). Fused tendons of insertion of the quadriceps femoris muscle and the fascia lata (fascia of thigh) that strengthen the anterior surface of the joint (**Figure 9.15e**).
3. ***Patellar ligament.*** Continuation of common tendon of insertion of quadriceps femoris muscle that extends from the patella to the tibial tuberosity. Also strengthens the anterior surface of the joint. Posterior surface of the ligament is separated from the synovial membrane of the joint by an infrapatellar fat pad (**Figure 9.15c–e**).
4. ***Oblique popliteal ligament*** (pop-LIT-ē-al). Broad, flat ligament that extends from the intercondylar fossa and lateral condyle of the femur to the head and medial condyle of the tibia (**Figure 9.15f, h**). The ligament strengthens the posterior surface of the joint.
5. ***Arcuate popliteal ligament.*** Extends from lateral condyle of femur to styloid process of the head of the fibula. Strengthens the lower lateral part of the posterior surface of the joint (**Figure 9.15f**).

6. ***Tibial collateral ligament.*** Broad, flat ligament on the medial surface of the joint that extends from the medial condyle of the femur to the medial condyle of the tibia (**Figure 9.15a, e–h**). Tendons of the sartorius, gracilis, and semitendinosus muscles, all of which strengthen the medial aspect of the joint, cross the ligament. The tibial collateral ligament is firmly attached to the medial meniscus.
7. ***Fibular collateral ligament.*** Strong, rounded ligament on the lateral surface of the joint that extends from the lateral condyle of the femur to the lateral side of the head of the fibula (**Figure 9.15a, e–h**). It strengthens the lateral aspect of the joint. The ligament is covered by the tendon of the biceps femoris muscle. The tendon of the popliteal muscle is deep to the ligament.
8. ***Intracapsular ligaments*** (in′-tra-KAP-sū-lar). Ligaments within capsule connecting tibia and femur. The anterior and posterior **cruciate ligaments** (KROO-shē-āt = like a cross) are named based on their origins relative to the intercondylar area of the tibia. From their origins, they cross on their way to their destinations on the femur.
   a. ***Anterior cruciate ligament (ACL).*** Extends posteriorly and laterally from a point anterior to the intercondylar area of the tibia to the posterior part of the medial surface of the lateral condyle of the femur (**Figure 9.15a, b, h**). The ACL limits hyperextension of the knee (which normally does not occur at this joint) and prevents the anterior sliding of the tibia on the femur. This ligament is stretched or torn in about 70% of all serious knee injuries.

FIGURE 9.15 **Right knee (tibiofemoral) joint.**

The knee joint is the largest and most complex joint in the body.

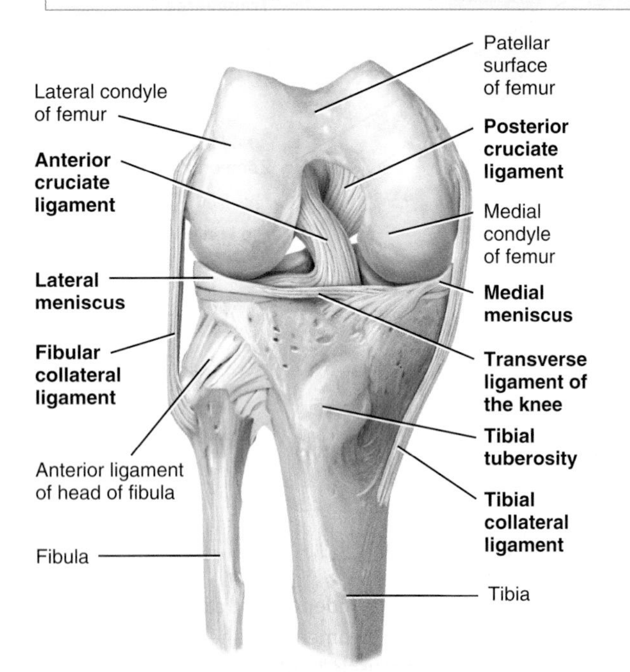

(a) Anterior deep view

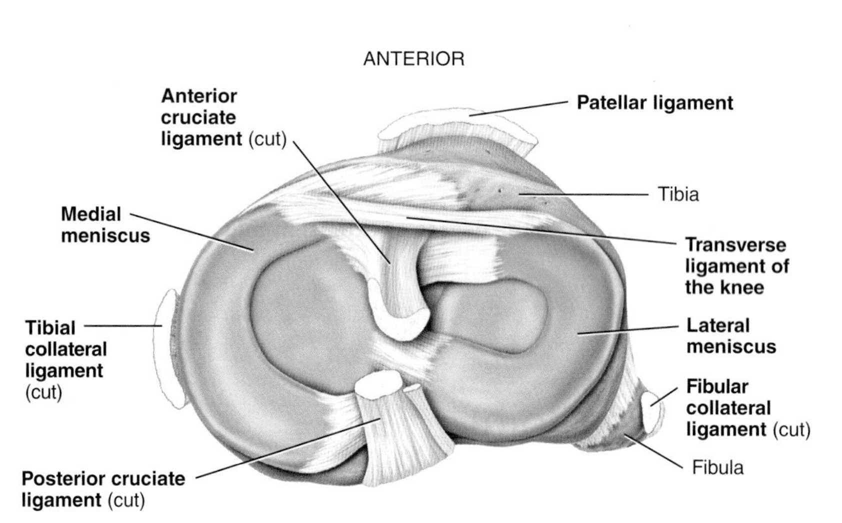

(b) Superior view of menisci

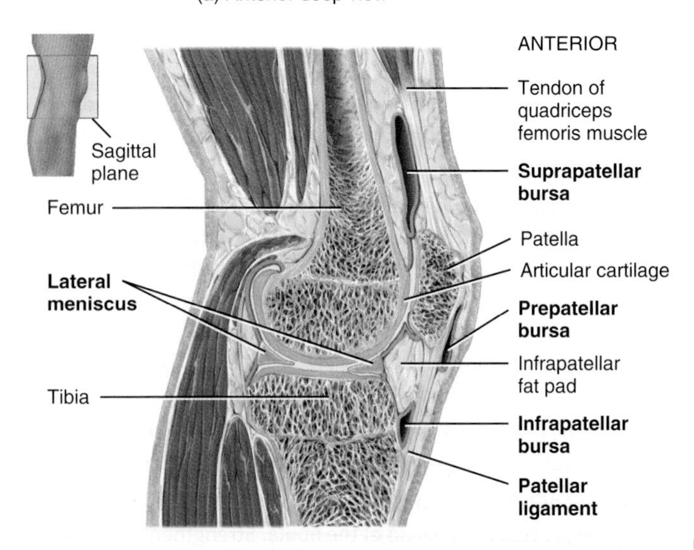

(c) Sagittal section

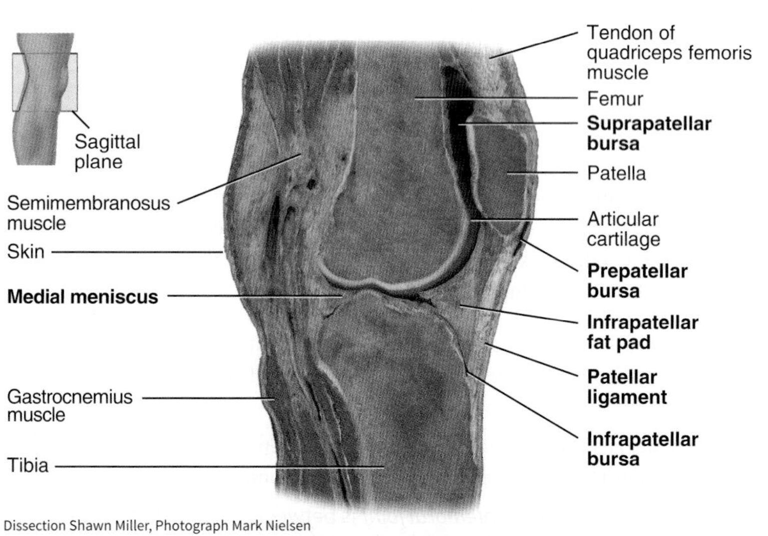

Dissection Shawn Miller, Photograph Mark Nielsen

(d) Sagittal section

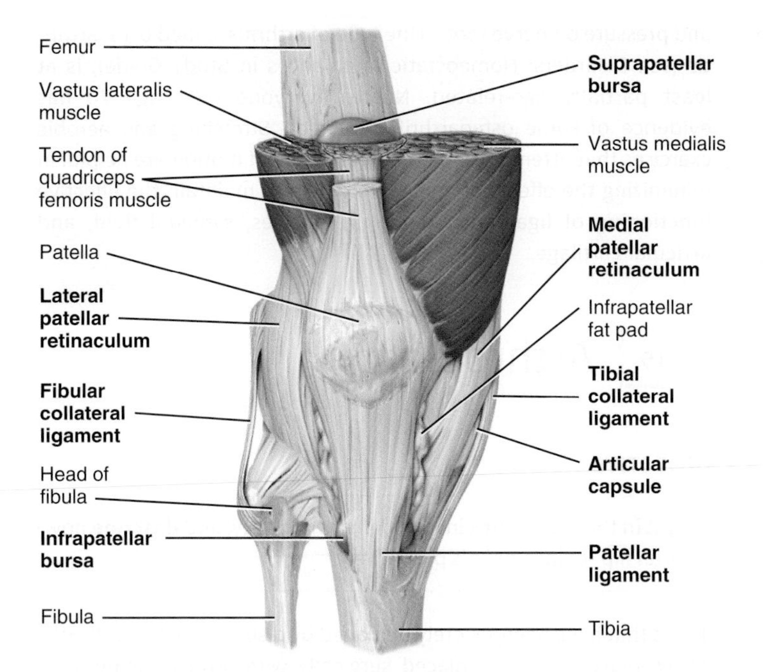

(e) Anterior superficial view

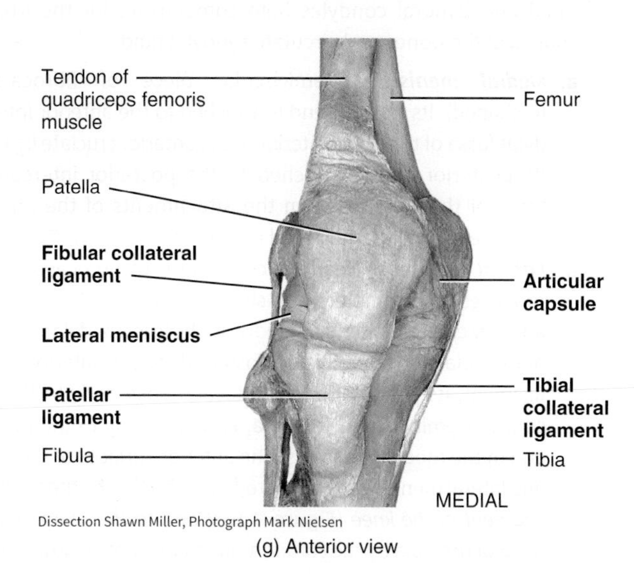

Dissection Shawn Miller, Photograph Mark Nielsen

(g) Anterior view

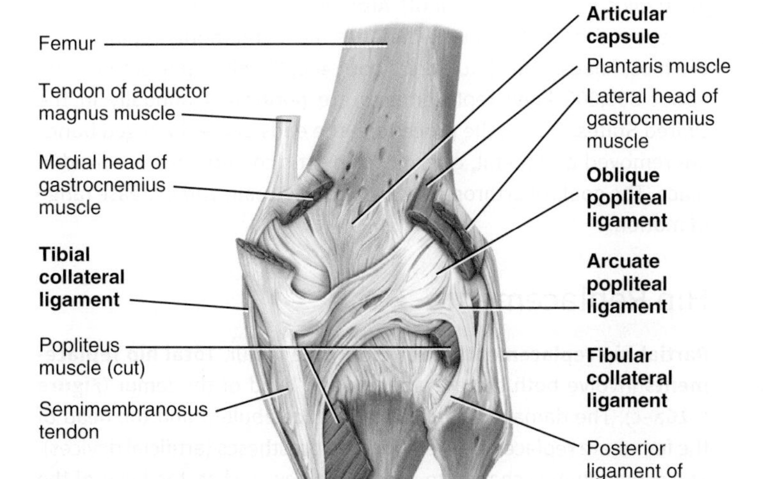

(f) Posterior deep view

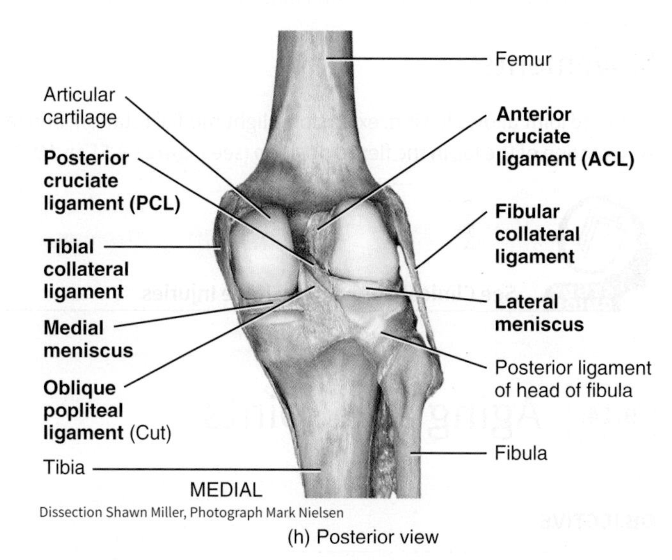

Dissection Shawn Miller, Photograph Mark Nielsen

(h) Posterior view

Q What movement occurs at the knee joint when the quadriceps femoris (anterior thigh) muscles contract?

ACL injuries are much more common in females than males, perhaps as much as 3 to 6 times. The reasons are unclear but may be related to less space between the femoral condyle in females so that the space for ACL movement is limited; the wider pelvis of females that creates a greater angle between the femur and tibia and increases the risk for an ACL tear; female hormones that allow for greater flexibility of ligaments, muscles, and tendons but which do not permit them to absorb the stresses put on them, thus transferring the stresses to the ACL; and females' lesser muscle strength, causing them to rely more on the ACL to hold the knee in place.

b. *Posterior cruciate ligament (PCL).* Extends anteriorly and medially from a depression on the posterior intercondylar area of the tibia and lateral meniscus to the anterior part of the lateral surface of the medial condyle of the femur (**Figure 9.15a, b, h**). The PCL prevents the posterior sliding of the tibia (and anterior sliding of the femur) when the knee is flexed. This is very important when walking down stairs or a steep incline.

9. ***Articular discs (menisci).*** Two fibrocartilage discs between the tibial and femoral condyles help compensate for the irregular shapes of the bones and circulate synovial fluid.
   a. ***Medial meniscus.*** Semicircular piece of fibrocartilage (C-shaped). Its anterior end is attached to the anterior intercondylar fossa of the tibia, anterior to the anterior cruciate ligament. Its posterior end is attached to the posterior intercondylar fossa of the tibia between the attachments of the posterior cruciate ligament and lateral meniscus (**Figure 9.15a, b, d, h**).
   b. ***Lateral meniscus.*** Nearly circular piece of fibrocartilage (approaches an incomplete O in shape) (**Figure 9.15a, b, d, h**). Its anterior end is attached anteriorly to the intercondylar eminence of the tibia, and laterally and posteriorly to the anterior cruciate ligament. Its posterior end is attached posteriorly to the intercondylar eminence of the tibia, and anteriorly to the posterior end of the medial meniscus. The anterior surfaces of the medial and lateral menisci are connected to each other by the *transverse ligament of the knee* (**Figure 9.15a**) and to the margins of the head of the tibia by the *coronary ligaments* (not illustrated).
10. The more important bursae of the knee include the following:
   a. *Prepatellar bursa* between the patella and skin (**Figure 9.15c, d**).
   b. *Infrapatellar bursa* between superior part of tibia and patellar ligament (**Figure 9.15c–e**).
   c. *Suprapatellar bursa* between inferior part of femur and deep surface of quadriceps femoris muscle (**Figure 9.15c–e**).

Movements

The knee joint allows flexion, extension, slight medial rotation, and lateral rotation of the leg in the flexed position (see **Figures 9.5f and 9.8c**).

See Clinical Connection: Knee Injuries

9.14 Aging and Joints

OBJECTIVE

- **Explain** the effects of aging on joints.

Aging usually results in decreased production of synovial fluid in joints. In addition, the articular cartilage becomes thinner with age, and ligaments shorten and lose some of their flexibility. The effects of aging on joints are influenced by genetic factors and by wear and tear, and vary considerably from one person to another. Although degenerative changes in joints may begin as early as age 20, most changes do not occur until much later. By age 80, almost everyone develops some type of degeneration in the knees, elbows, hips, and shoulders. It is also common for elderly individuals to develop degenerative changes in the vertebral column, resulting in a hunched-over posture and pressure on nerve roots. One type of arthritis, called osteoarthritis (see Disorders: Homeostatic Imbalances in Study Guide), is at least partially age-related. Nearly everyone over age 70 has evidence of some osteoarthritic changes. Stretching and aerobic exercises that attempt to maintain full range of motion are helpful in minimizing the effects of aging. They help to maintain the effective functioning of ligaments, tendons, muscles, synovial fluid, and articular cartilage.

9.15 Arthroplasty

OBJECTIVE

- **Explain** the procedures involved in arthroplasty, and **describe** how a total hip replacement is performed.

Joints that have been severely damaged by diseases such as arthritis, or by injury, may be replaced surgically with artificial joints in a procedure referred to as **arthroplasty** (AR-thrō-plas′-tē; *arthr-* = joint; *-plasty* = plastic repair of). Although most joints in the body can be repaired by arthroplasty, the ones most commonly replaced are the hips, knees, and shoulders. About 400,000 hip replacements and about 300,000 knee replacements are performed annually in the United States. During the procedure, the ends of the damaged bones are removed and metal, ceramic, or plastic components are fixed in place. The goals of arthroplasty are to relieve pain and increase range of motion.

Hip Replacements

Partial hip replacements involve only the femur. **Total hip replacements** involve both the acetabulum and head of the femur (**Figure 9.16a–c**). The damaged portions of the acetabulum and the head of the femur are replaced by prefabricated prostheses (artificial devices). The acetabulum is shaped to accept the new socket, the head of the femur is removed, and the center of the femur is shaped to fit the femoral component. The acetabular component consists of a plastic such as polyethylene, and the femoral component is composed of a metal such as cobalt-chrome, titanium alloys, or stainless steel. These materials are designed to withstand a high degree of stress and to prevent a response by the immune system. Once the appropriate acetabular and femoral components are selected, they are attached to the healthy portion of bone with acrylic cement, which forms an interlocking mechanical bond.

Knee Replacements

Knee replacements are actually a resurfacing of cartilage and, like hip replacements, may be partial or total. In a **total knee replacement**, the damaged cartilage is removed from the distal end of the femur, the proximal end of the tibia, and the back surface of the patella (if the back surface of the patella is not badly damaged, it may

FIGURE 9.16 **Total hip and knee replacement.**

In a total hip replacement, damaged portions of the acetabulum and the head of the femur are replaced by prostheses.

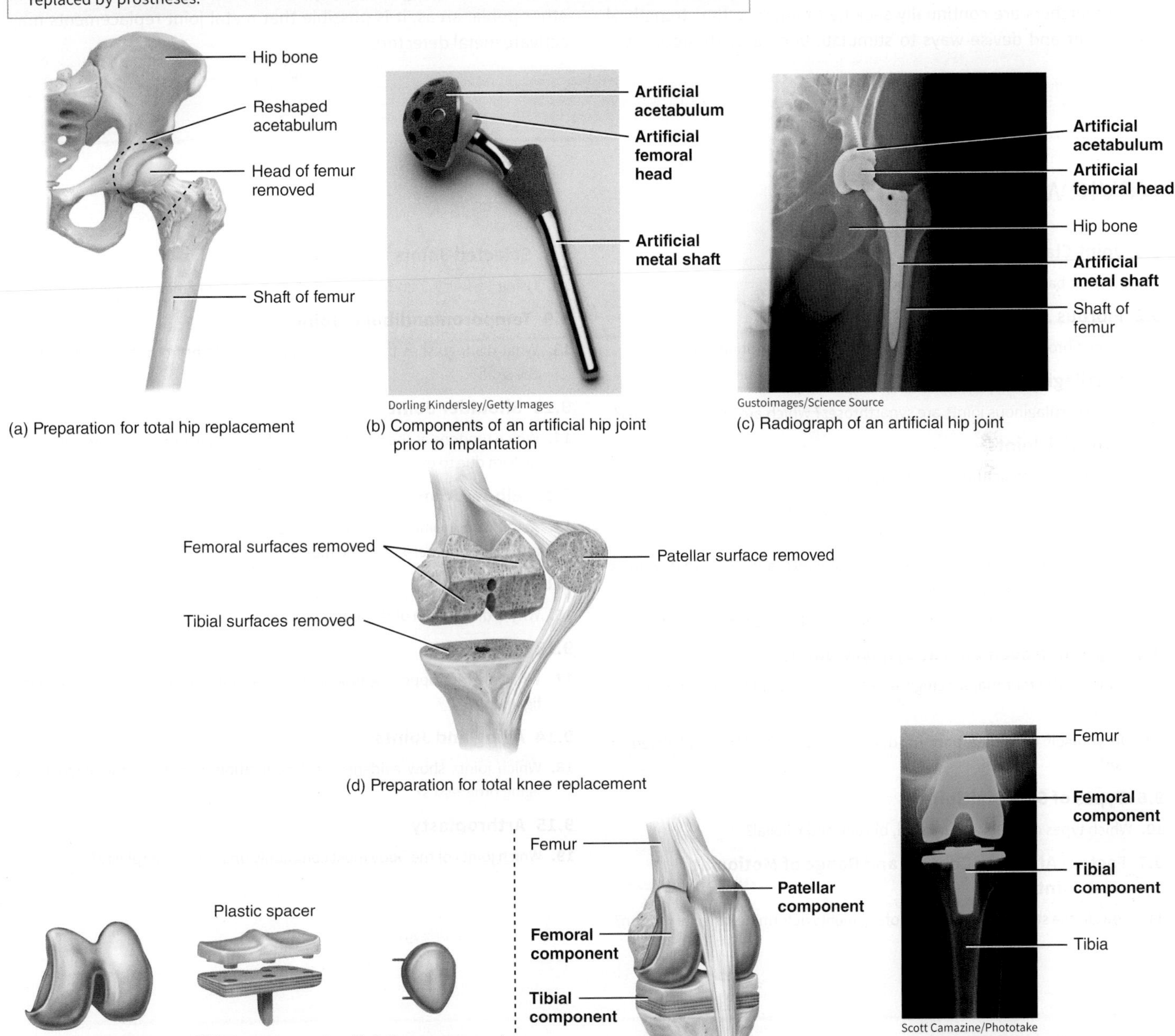

Dorling Kindersley/Getty Images

Gustoimages/Science Source

Scott Camazine/Phototake

(a) Preparation for total hip replacement

(b) Components of an artificial hip joint prior to implantation

(c) Radiograph of an artificial hip joint

(d) Preparation for total knee replacement

(e) Components of artificial knee joint prior to implantation (left) and implanted (right)

(f) Radiograph of total knee replacement

Q What is the purpose of arthroplasty?

be left intact) (**Figure 9.16d–f**). The femur is reshaped and fitted with a metal femoral component and cemented in place. The tibia is reshaped and fitted with a plastic tibial component that is cemented in place. If the back surface of the patella is badly damaged, it is replaced with a plastic patellar implant.

In a **partial knee replacement**, also called a *unicompartmental knee replacement*, only one side of the knee joint is replaced. Once the damaged cartilage is removed from the distal end of the femur, the femur is reshaped and a metal femoral component is cemented in place. Then the damaged cartilage from the proximal end of the tibia is removed, along with the meniscus. The tibia is reshaped and fitted with a plastic tibial component that is cemented in place. If the back surface of the patella is badly damaged, it is replaced with a plastic patellar component.

Researchers are continually seeking to improve the strength of the cement and devise ways to stimulate bone growth around the implanted area. Potential complications of arthroplasty include infection, blood clots, loosening or dislocation of the replacement components, and nerve injury.

With increasing sensitivity of metal detectors at airports and other public areas, it is possible that metal joint replacements may activate metal detectors.

Review Questions

9.1 Joint Classifications

1. On what basis are joints classified?

9.2 Fibrous Joints

2. Which fibrous joints are synarthroses? Which are amphiarthroses?

9.3 Cartilaginous Joints

3. Which cartilaginous joints are synarthroses? Which are amphiarthroses?

9.4 Synovial Joints

4. How does the structure of synovial joints classify them as diarthroses?
5. What are the functions of articular cartilage, synovial fluid, and articular discs?
6. What types of sensations are perceived at joints, and from what sources do joints receive nourishment?
7. In what ways are bursae similar to joint capsules? How do they differ?

9.5 Types of Movements at Synovial Joints

8. What are the four major categories of movements that occur at synovial joints?
9. On yourself or with a partner, demonstrate each movement listed in Table 9.1.

9.6 Types of Synovial Joints

10. Which types of joints are uniaxial, biaxial, and triaxial?

9.7 Factors Affecting Contact and Range of Motion at Synovial Joints

11. How do the strength and tension of ligaments determine range of motion?

9.8 Selected Joints of the Body

12. Using Tables 9.3 and 9.4 as a guide, identify only the cartilaginous joints.

9.9 Temporomandibular Joint

13. What distinguishes the temporomandibular joint from the other joints of the skull?

9.10 Shoulder Joint

14. Which tendons at the shoulder joint of a baseball pitcher are most likely to be torn due to excessive circumduction?

9.11 Elbow Joint

15. At the elbow joint, which ligaments connect (a) the humerus and the ulna, and (b) the humerus and the radius?

9.12 Hip Joint

16. Why is dislocation of the femur so rare?

9.13 Knee Joint

17. What are the opposing functions of the anterior and posterior cruciate ligaments?

9.14 Aging and Joints

18. Which joints show evidence of degeneration in nearly all individuals as aging progresses?

9.15 Arthroplasty

19. Which joints of the body most commonly undergo arthroplasty?

Critical Thinking Questions

1. Katie loves pretending that she's a human cannonball. As she jumps off the diving board, she assumes the proper position before she pounds into the water: head and thighs tucked against her chest; back rounded; arms pressed against her sides while her forearms, crossed in front of her shins, hold her legs tightly folded against her chest. Use the proper anatomical terms to describe the position of Katie's back, head, and free limbs.
2. During football practice, Jeremiah was tackled and twisted his leg. There was a sharp pain, followed immediately by swelling of the knee joint. The pain and swelling worsened throughout the remainder of the afternoon until Jeremiah could barely walk. The coach told Jeremiah to see a doctor who might want to "drain the water off his knee." What was the coach referring to and what specifically do you think happened to Jeremiah's knee joint to cause these symptoms?
3. After lunch, during a particularly long and dull class video, Antonio became sleepy and yawned. To his dismay, he was then unable to close his mouth. Explain what happened and what should be done to correct this problem.

Muscular Tissue

CHAPTER **10**

Muscular Tissue and Homeostasis

Muscular tissue contributes to homeostasis by producing body movements, moving substances through the body, and producing heat to maintain normal body temperature.

Although bones provide leverage and form the framework of the body, they cannot move body parts by themselves. Motion results from the alternating contraction and relaxation of muscles, which make up 40–50% of total adult body weight (depending on the percentage of body fat, gender, and exercise regimen). Your muscular strength reflects the primary function of muscle—the transformation of chemical energy into mechanical energy to generate force, perform work, and produce movement. In addition, muscle tissues stabilize body position, regulate organ volume, generate heat, and propel fluids and food matter through various body systems.

Q Did you ever wonder what causes rigor mortis?

10.1 Overview of Muscular Tissue

OBJECTIVES

- **Explain** the structural differences among the three types of muscular tissue.
- **Compare** the functions and special properties of the three types of muscular tissue.

Types of Muscular Tissue

The three types of muscular tissue—skeletal, cardiac, and smooth—were introduced in Chapter 4 (see **Table 4.9**). The scientific study of muscles is known as **myology** (mī-OL-ō-jē; *myo-* = muscle; *-logy* = study of). Although the different types of muscular tissue share some properties, they differ from one another in their microscopic anatomy and location, and in how they are controlled by the nervous and endocrine systems.

Skeletal muscle tissue is so named because most skeletal muscles move the bones of the skeleton. (A few skeletal muscles attach to and move the skin or other skeletal muscles.) Skeletal muscle tissue is *striated:* Alternating light and dark protein bands (*striations*) are seen when the tissue is examined with a microscope (see **Table 4.9**). Skeletal muscle tissue works mainly in a *voluntary* manner. Its activity can be consciously controlled by neurons (nerve cells) that are part of the somatic (voluntary) division of the nervous system. (**Figure 12.10** depicts the divisions of the nervous system.) Most skeletal muscles also are controlled subconsciously to some extent. For example, your diaphragm continues to alternately contract and relax without conscious control so that you don't stop breathing. Also, you do not need to consciously think about contracting the skeletal muscles that maintain your posture or stabilize body positions.

Only the heart contains **cardiac muscle tissue**, which forms most of the heart wall. Cardiac muscle is also *striated*, but its action is *involuntary.* The alternating contraction and relaxation of the heart is not consciously controlled. Rather, the heart beats because it has a natural pacemaker that initiates each contraction. This built-in rhythm is termed **autorhythmicity** (aw′-tō-rith-MISS-i-tē). Several hormones and neurotransmitters can adjust heart rate by speeding or slowing the pacemaker.

Smooth muscle tissue is located in the walls of hollow internal structures, such as blood vessels, airways, and most organs in the abdominopelvic cavity. It is also found in the skin, attached to hair follicles. Under a microscope, this tissue lacks the striations of skeletal and cardiac muscle tissue. For this reason, it looks *nonstriated*, which is why it is referred to as *smooth.* The action of smooth muscle is usually *involuntary*, and some smooth muscle tissue, such as the muscles that propel food through your gastrointestinal tract, has autorhythmicity. Both cardiac muscle and smooth muscle are regulated by neurons that are part of the autonomic (involuntary) division of the nervous system and by hormones released by endocrine glands.

Functions of Muscular Tissue

Through sustained contraction or alternating contraction and relaxation, muscular tissue has four key functions: producing body movements, stabilizing body positions, storing and moving substances within the body, and generating heat.

1. ***Producing body movements.*** Movements of the whole body such as walking and running, and localized movements such as grasping

a pencil, keyboarding, or nodding the head rely on the integrated functioning of skeletal muscles, bones, and joints.

2. ***Stabilizing body positions.*** Skeletal muscle contractions stabilize joints and help maintain body positions, such as standing or sitting. Postural muscles contract continuously when you are awake; for example, sustained contractions of your neck muscles hold your head upright when you are listening intently to your anatomy and physiology lecture.
3. ***Storing and moving substances within the body.*** Storage is accomplished by sustained contractions of ringlike bands of smooth muscle called *sphincters*, which prevent outflow of the contents of a hollow organ. Temporary storage of food in the stomach or urine in the urinary bladder is possible because smooth muscle sphincters close off the outlets of these organs. Cardiac muscle contractions of the heart pump blood through the blood vessels of the body. Contraction and relaxation of smooth muscle in the walls of blood vessels help adjust blood vessel diameter and thus regulate the rate of blood flow. Smooth muscle contractions also move food and substances such as bile and enzymes through the gastrointestinal tract, push gametes (sperm and oocytes) through the passageways of the reproductive systems, and propel urine through the urinary system. Skeletal muscle contractions promote the flow of lymph and aid the return of blood in veins to the heart.
4. ***Generating heat.*** As muscular tissue contracts, it produces heat, a process known as **thermogenesis** (ther′-mō-JEN-e-sis). Much of the heat generated by muscle is used to maintain normal body temperature. Involuntary contractions of skeletal muscles, known as *shivering*, can increase the rate of heat production.

Properties of Muscular Tissue

Muscular tissue has four special properties that enable it to function and contribute to homeostasis:

1. **Electrical excitability** (ek-sīt′-a-BIL-i-tē), a property of both muscle and nerve cells that was introduced in Chapter 4, is the ability to respond to certain stimuli by producing electrical signals called **action potentials** (*impulses*). Action potentials in muscles are referred to as *muscle action potentials*; those in nerve cells are called *nerve action potentials*. Chapter 12 provides more detail about how action potentials arise (see Section 12.3). For muscle cells, two main types of stimuli trigger action potentials. One is autorhythmic *electrical signals* arising in the muscular tissue itself, as in the heart's pacemaker. The other is *chemical stimuli*, such as neurotransmitters released by neurons, hormones distributed by the blood, or even local changes in pH.
2. **Contractility** (kon′-trak-TIL-i-tē) is the ability of muscular tissue to contract forcefully when stimulated by an action potential. When a skeletal muscle contracts, it generates tension (force of contraction) while pulling on its attachment points. If the tension generated is great enough to overcome the resistance of the object to be moved, the muscle shortens and movement occurs.
3. **Extensibility** (ek-sten′-si-BIL-i-tē) is the ability of muscular tissue to stretch, within limits, without being damaged. The connective tissue within the muscle limits the range of extensibility and keeps it within the contractile range of the muscle cells. Normally, smooth muscle is subject to the greatest amount of stretching. For example, each time your stomach fills with food, the smooth muscle in the wall is stretched. Cardiac muscle also is stretched each time the heart fills with blood.
4. **Elasticity** (e-las-TIS-i-tē) is the ability of muscular tissue to return to its original length and shape after contraction or extension.

Skeletal muscle is the focus of much of this chapter. Cardiac muscle and smooth muscle are described briefly here. Cardiac muscle is discussed in more detail in Chapter 20 (the heart), and smooth muscle is included in Chapter 15 (the autonomic nervous system), as well as in discussions of the various organs containing smooth muscle.

10.2 Structure of Skeletal Muscle Tissue

OBJECTIVES

- **Explain** the importance of connective tissue components, blood vessels, and nerves to skeletal muscles.
- **Describe** the microscopic anatomy of a skeletal muscle fiber.
- **Distinguish** thick filaments from thin filaments.
- **Describe** the functions of skeletal muscle proteins.

Each of your skeletal muscles is a separate organ composed of hundreds to thousands of cells, which are called **muscle fibers** (*myocytes*) because of their elongated shapes. Thus, *muscle cell* and *muscle fiber* are two terms for the same structure. Skeletal muscle also contains connective tissues surrounding muscle fibers, and blood vessels and nerves (**Figure 10.1**). To understand how contraction of skeletal muscle can generate tension, you must first understand its gross and microscopic anatomy.

Connective Tissue Components

Connective tissue surrounds and protects muscular tissue. The **subcutaneous layer** or *hypodermis*, which separates muscle from skin (see **Figure 11.21**), is composed of areolar connective tissue and adipose tissue. It provides a pathway for nerves, blood vessels, and lymphatic vessels to enter and exit muscles. The adipose tissue of the subcutaneous layer stores most of the body's triglycerides, serves as an insulating layer that reduces heat loss, and protects muscles from physical trauma. **Fascia** (FASH-ē-a = bandage) is a dense sheet or broad band of irregular connective tissue that lines the body wall and limbs and supports and surrounds muscles and other organs of the body. As you will see, fascia holds muscles with similar functions together (see **Figure 11.21**). Fascia allows free

FIGURE 10.1 **Organization of skeletal muscle and its connective tissue coverings.**

A skeletal muscle consists of individual muscle fibers (cells) bundled into fascicles and surrounded by three connective tissue layers that are extensions of the fascia.

Functions of Muscular Tissues

1. Producing motions.
2. Stabilizing body positions.
3. Storing and moving substances within the body.
4. Generating heat (thermogenesis).

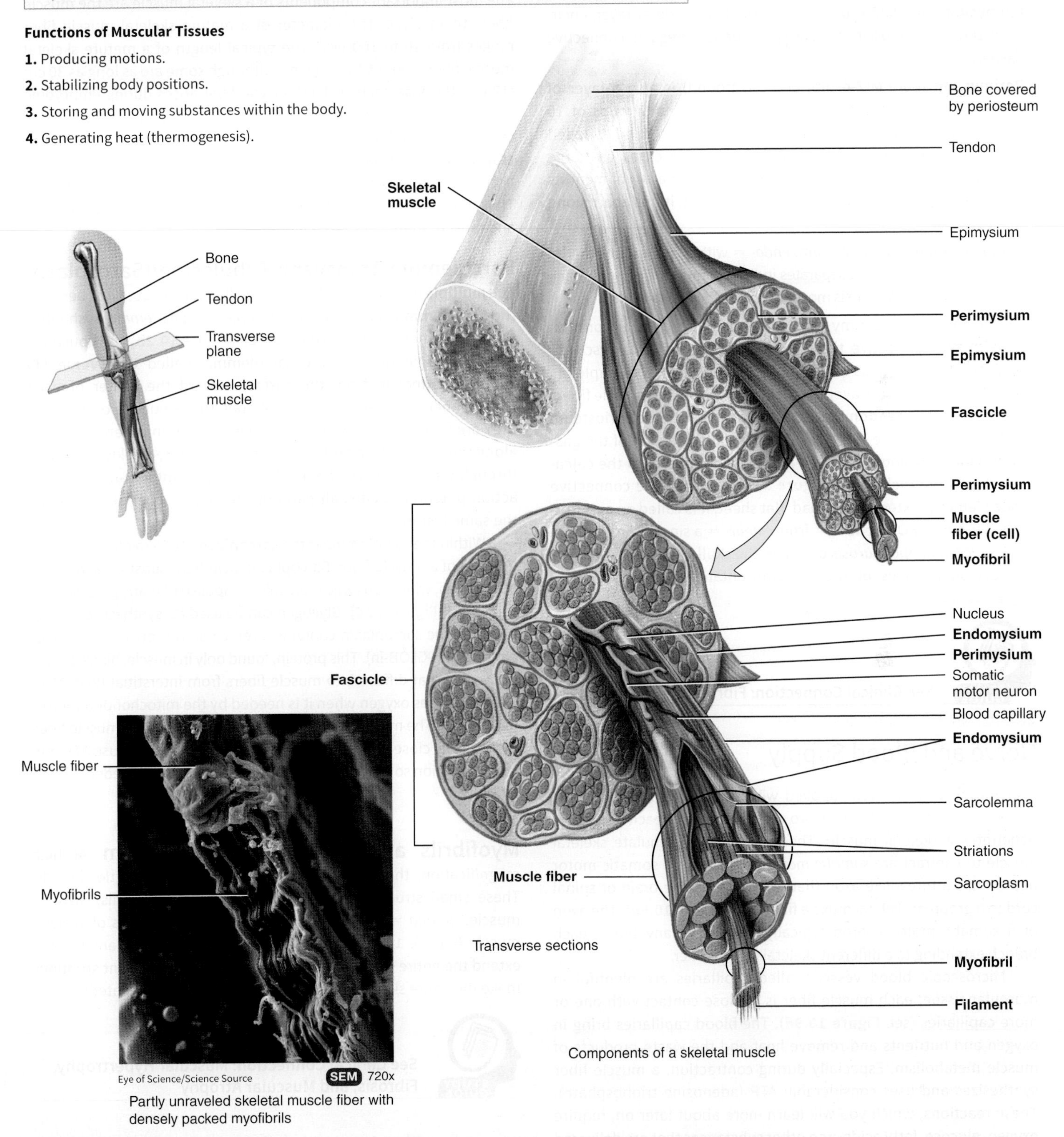

Components of a skeletal muscle

Partly unraveled skeletal muscle fiber with densely packed myofibrils

Q Which connective tissue coat surrounds groups of muscle fibers, separating them into fascicles?

movement of muscles; carries nerves, blood vessels, and lymphatic vessels; and fills spaces between muscles.

Three layers of connective tissue extend from the fascia to protect and strengthen skeletal muscle (**Figure 10.1**):

- **Epimysium** (ep-i-MĪZ-ē-um; *epi-* = upon) is the outer layer, encircling the entire muscle. It consists of dense irregular connective tissue.
- **Perimysium** (per-i-MĪZ-ē-um; *peri-* = around) is also a layer of dense irregular connective tissue, but it surrounds groups of 10 to 100 or more muscle fibers, separating them into bundles called **fascicles** (FAS-i-kuls = little bundles). Many fascicles are large enough to be seen with the naked eye. They give a cut of meat its characteristic "grain"; if you tear a piece of meat, it rips apart along the fascicles.
- **Endomysium** (en′-dō-MĪZ-ē-um; *endo-* = within) penetrates the interior of each fascicle and separates individual muscle fibers from one another. The endomysium is mostly reticular fibers.

The epimysium, perimysium, and endomysium are all continuous with the connective tissue that attaches skeletal muscle to other structures, such as bone or another muscle. For example, all three connective tissue layers may extend beyond the muscle fibers to form a ropelike **tendon** that attaches a muscle to the periosteum of a bone. An example is the *calcaneal (Achilles) tendon* of the gastrocnemius (calf) muscle, which attaches the muscle to the calcaneus (heel bone) (shown in **Figure 11.22c**). When the connective tissue elements extend as a broad, flat sheet, it is called an **aponeurosis** (ap-ō-noo-RŌ-sis; *apo-* = from; *-neur-* = a sinew). An example is the *epicranial aponeurosis* on top of the skull between the frontal and occipital bellies of the occipitofrontalis muscle (shown in **Figure 11.4a, c**).

See Clinical Connection: Fibromyalgia

Nerve and Blood Supply

Skeletal muscles are well supplied with nerves and blood vessels. Generally, an artery and one or two veins accompany each nerve that penetrates a skeletal muscle. The neurons that stimulate skeletal muscle to contract are *somatic motor neurons.* Each somatic motor neuron has a threadlike axon that extends from the brain or spinal cord to a group of skeletal muscle fibers (see **Figure 10.9d**). The axon of a somatic motor neuron typically branches many times, each branch extending to a different skeletal muscle fiber.

Microscopic blood vessels called capillaries are plentiful in muscular tissue; each muscle fiber is in close contact with one or more capillaries (see **Figure 10.9d**). The blood capillaries bring in oxygen and nutrients and remove heat and the waste products of muscle metabolism. Especially during contraction, a muscle fiber synthesizes and uses considerable ATP (adenosine triphosphate). These reactions, which you will learn more about later on, require oxygen, glucose, fatty acids, and other substances that are delivered to the muscle fiber in the blood.

Microscopic Anatomy of a Skeletal Muscle Fiber

The most important components of a skeletal muscle are the muscle fibers themselves. The diameter of a mature skeletal muscle fiber ranges from 10 to 100 μm.* The typical length of a mature skeletal muscle fiber is about 10 cm (4 in.), although some are as long as 30 cm (12 in.). Because each skeletal muscle fiber arises during embryonic development from the fusion of a hundred or more small mesodermal cells called *myoblasts* (MĪ-ō-blasts) (**Figure 10.2a**), each mature skeletal muscle fiber has a hundred or more nuclei. Once fusion has occurred, the muscle fiber loses its ability to undergo cell division. Thus, the number of skeletal muscle fibers is set before you are born, and most of these cells last a lifetime.

Sarcolemma, Transverse Tubules, and Sarcoplasm

The multiple nuclei of a skeletal muscle fiber are located just beneath the **sarcolemma** (sar′-kō-LEM-ma; *sarc-* = flesh; *-lemma* = sheath), the plasma membrane of a muscle cell (**Figure 10.2b, c**). Thousands of tiny invaginations of the sarcolemma, called **transverse (T) tubules**, tunnel in from the surface toward the center of each muscle fiber. Because T tubules are open to the outside of the fiber, they are filled with interstitial fluid. Muscle action potentials travel along the sarcolemma and through the T tubules, quickly spreading throughout the muscle fiber. This arrangement ensures that an action potential excites all parts of the muscle fiber at essentially the same instant.

Within the sarcolemma is the **sarcoplasm** (SAR-kō-plazm), the cytoplasm of a muscle fiber. Sarcoplasm includes a substantial amount of glycogen, which is a large molecule composed of many glucose molecules (see **Figure 2.16**). Glycogen can be used for synthesis of ATP. In addition, the sarcoplasm contains a red-colored protein called **myoglobin** (mī-ō-GLŌB-in). This protein, found only in muscle, binds oxygen molecules that diffuse into muscle fibers from interstitial fluid. Myoglobin releases oxygen when it is needed by the mitochondria for ATP production. The mitochondria lie in rows throughout the muscle fiber, strategically close to the contractile muscle proteins that use ATP during contraction so that ATP can be produced quickly as needed (**Figure 10.2c**).

Myofibrils and Sarcoplasmic Reticulum At high magnification, the sarcoplasm appears stuffed with little threads. These small structures are the **myofibrils** (mī-ō-FĪ-brils; *myo-* = muscle; *-fibrilla* = little fiber), the contractile organelles of skeletal muscle (**Figure 10.2c**). Myofibrils are about 2 μm in diameter and extend the entire length of a muscle fiber. Their prominent striations make the entire skeletal muscle fiber appear striped (striated).

See Clinical Connection: Muscular Hypertrophy, Fibrosis, and Muscular Atrophy

*One micrometer (μm) = 10^{-6} meter (1/25,000 in.).

FIGURE 10.2 Microscopic organization of skeletal muscle. (a) During embryonic development, many myoblasts fuse to form one skeletal muscle fiber. Once fusion has occurred, a skeletal muscle fiber loses the ability to undergo cell division, but satellite cells retain this ability. (b–d) The sarcolemma of the fiber encloses sarcoplasm and myofibrils, which are striated. Sarcoplasmic reticulum wraps around each myofibril. Thousands of transverse tubules, filled with interstitial fluid, invaginate from the sarcolemma toward the center of the muscle fiber. A photomicrograph of skeletal muscle tissue is shown in **Table 4.9**.

The contractile elements of muscle fibers, the myofibrils, contain overlapping thick and thin filaments.

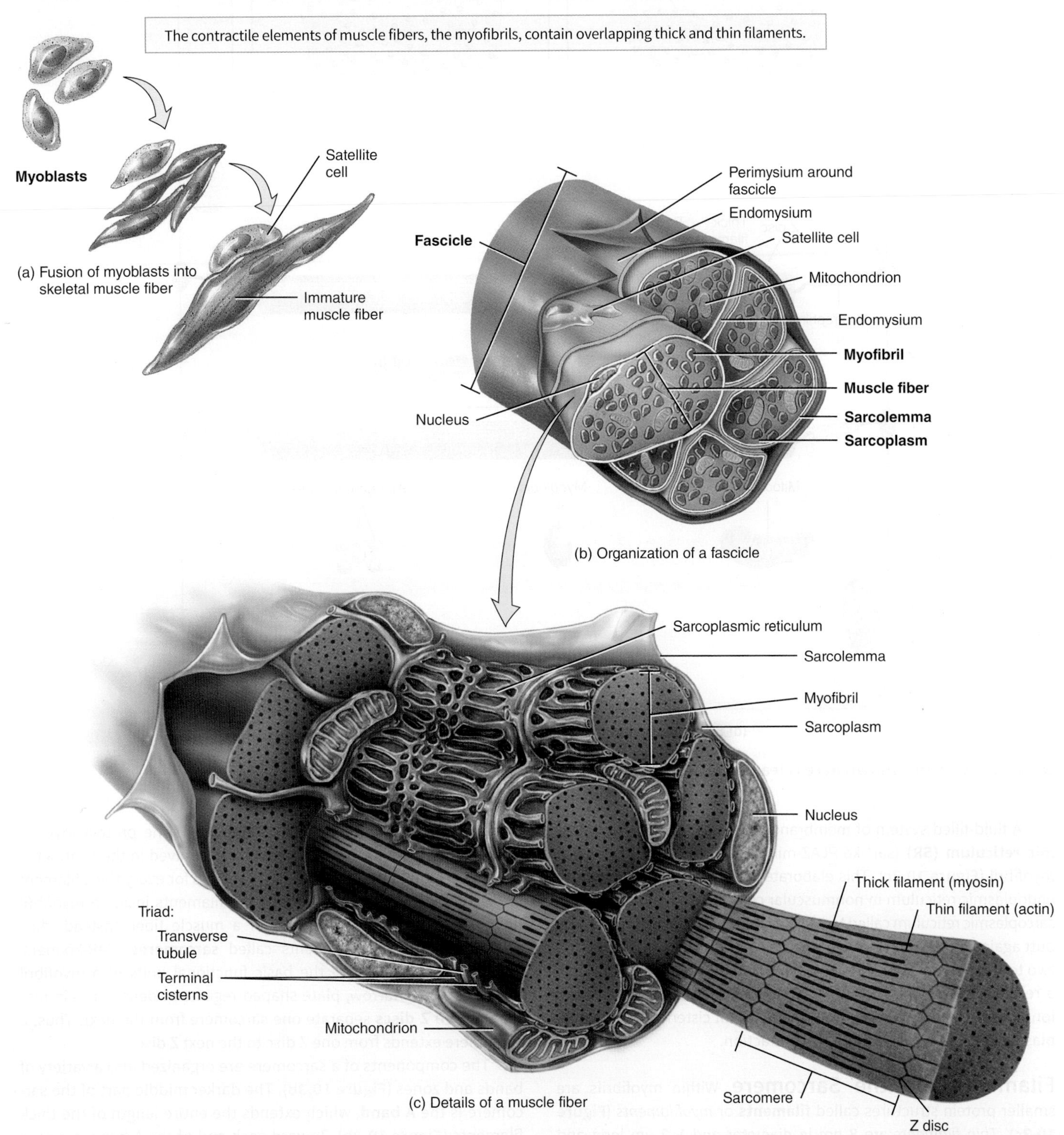

Figure 10.2 ***Continues***

FIGURE 10.2 **Continued**

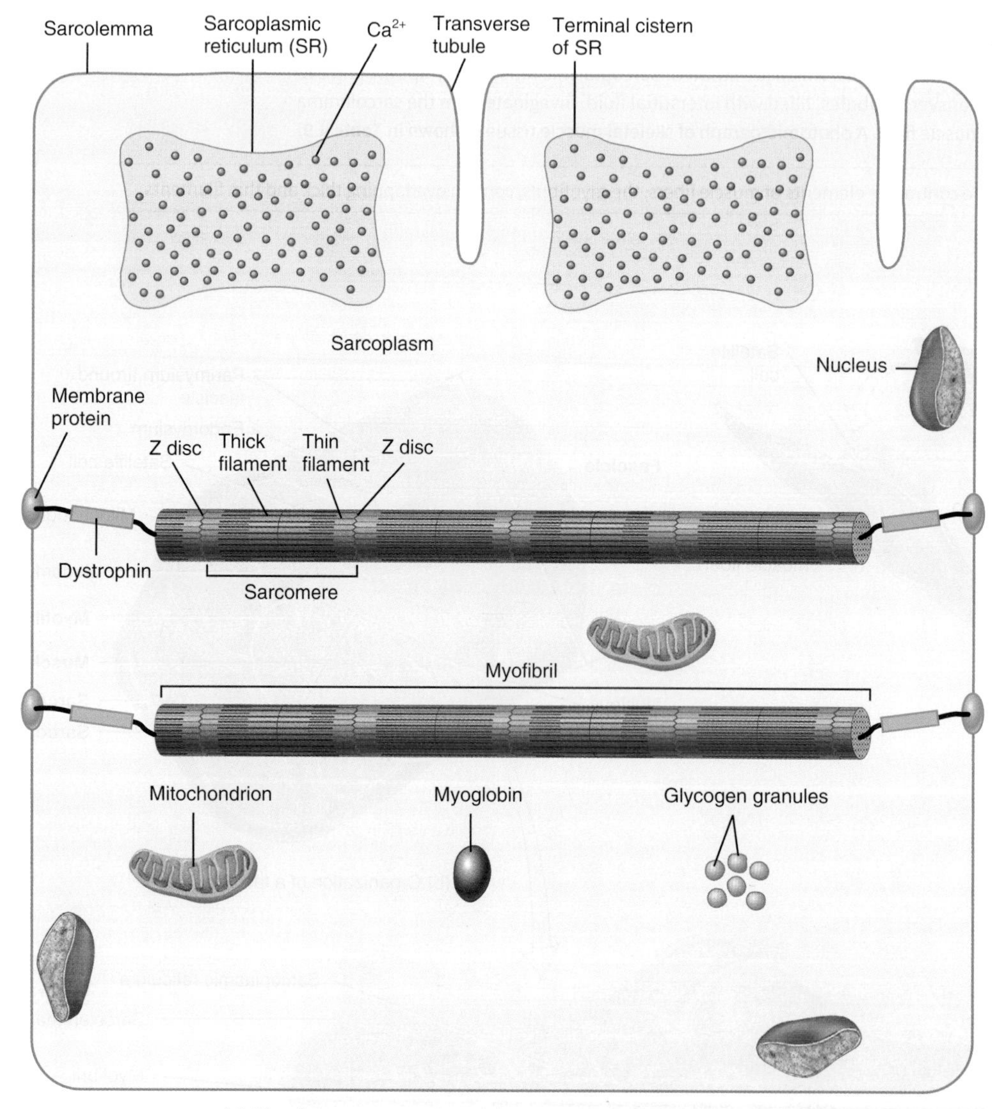

(d) Simplistic representation of the components of a muscle fiber

Q Which structure shown here releases calcium ions to trigger muscle contraction?

A fluid-filled system of membranous sacs called the **sarcoplasmic reticulum (SR)** (sar′-kō-PLAZ-mik re-TIK-ū-lum) encircles each myofibril (**Figure 10.2c**). This elaborate system is similar to smooth endoplasmic reticulum in nonmuscular cells. Dilated end sacs of the sarcoplasmic reticulum called **terminal cisterns** (SIS-terns = reservoirs) butt against the T tubule from both sides. A transverse tubule and the two terminal cisterns on either side of it form a **triad** (*tri-* = three). In a relaxed muscle fiber, the sarcoplasmic reticulum stores calcium ions (Ca^{2+}). Release of Ca^{2+} from the terminal cisterns of the sarcoplasmic reticulum triggers muscle contraction.

Filaments and the Sarcomere

Within myofibrils are smaller protein structures called **filaments** or *myofilaments* (**Figure 10.2c**). *Thin filaments* are 8 nm in diameter and 1–2 μm long and composed of the protein actin, while *thick filaments* are 16 nm in diameter and 1–2 μm long and composed of the protein myosin. Both thin and thick filaments are directly involved in the contractile process. Overall, there are two thin filaments for every thick filament in the regions of filament overlap. The filaments inside a myofibril do not extend the entire length of a muscle fiber. Instead, they are arranged in compartments called **sarcomeres** (SAR-kō-mērs; *-mere* = part), which are the basic functional units of a myofibril (**Figure 10.3a**). Narrow, plate-shaped regions of dense protein material called **Z discs** separate one sarcomere from the next. Thus, a sarcomere extends from one Z disc to the next Z disc.

The components of a sarcomere are organized into a variety of bands and zones (**Figure 10.3b**). The darker middle part of the sarcomere is the **A band**, which extends the entire length of the thick filaments (**Figure 10.3b**). Toward each end of the A band is a *zone of overlap*, where the thick and thin filaments lie side by side. The

FIGURE 10.3 The arrangement of filaments within a sarcomere. A sarcomere extends from one Z disc to the next.

Myofibrils contain two types of filaments: thick filaments and thin filaments.

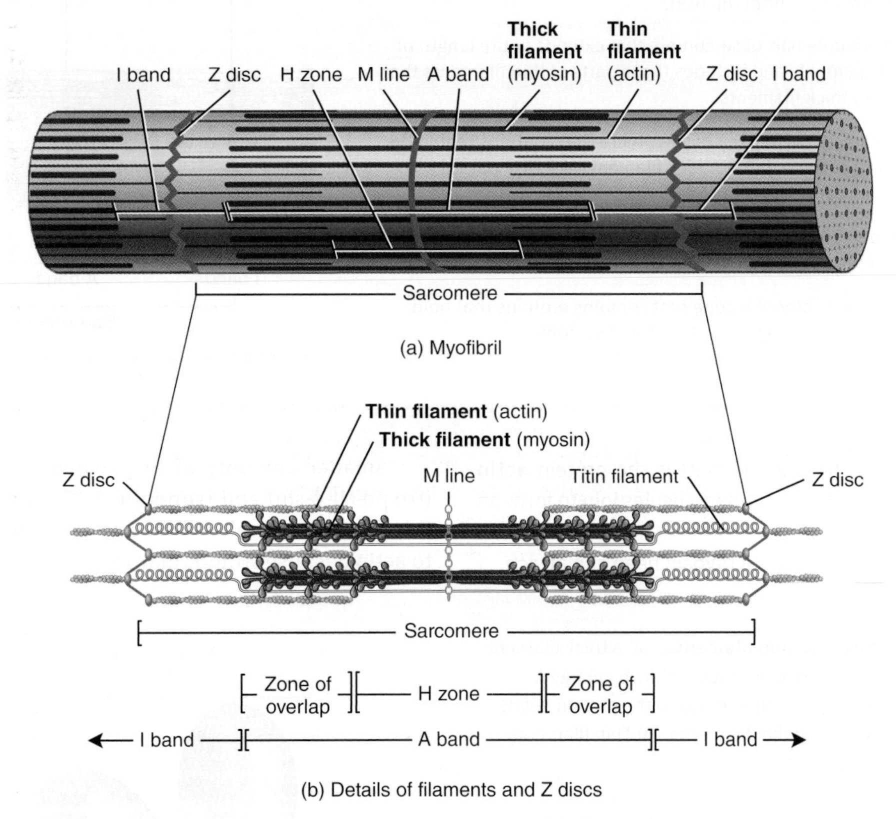

Q Which of the following is the smallest: muscle fiber, thick filament, or myofibril? Which is largest?

I band is a lighter, less dense area that contains the rest of the thin filaments but no thick filaments (**Figure 10.3b**), and a Z disc passes through the center of each I band. The alternating dark A bands and light I bands create the striations that can be seen in both myofibrils and in whole skeletal and cardiac muscle fibers. A narrow **H zone** in the center of each A band contains thick but not thin filaments. A mnemonic that will help you to remember the composition of the I and H bands is as follows: the letter I is thin (contains thin filaments), while the letter H is thick (contains thick filaments). Supporting proteins that hold the thick filaments together at the center of the H zone form the **M line**, so named because it is at the *middle* of the sarcomere. **Table 10.1** summarizes the components of the sarcomere.

Muscle Proteins

Myofibrils are built from three kinds of proteins: (1) contractile proteins, which generate force during contraction; (2) regulatory proteins, which help switch the contraction process on and off; and (3) structural proteins, which keep the thick and thin filaments in the proper alignment, give the myofibril elasticity and extensibility, and link the myofibrils to the sarcolemma and extracellular matrix.

The two *contractile proteins* in muscle are myosin and actin, components of thick and thin filaments, respectively. **Myosin** (MĪ-ō-sin) is the main component of thick filaments and functions as a motor protein in all three types of muscle tissue. *Motor proteins* pull various cellular structures to achieve movement by converting the chemical energy in ATP to the mechanical energy of motion, that is, the production of force. In skeletal muscle, about 300 molecules of myosin form a single thick filament. Each myosin molecule is shaped like two golf clubs twisted together (**Figure 10.4a**). The *myosin tail* (twisted golf club handles) points toward the M line in the center of the sarcomere. Tails of neighboring myosin molecules lie parallel to one another, forming the shaft of the thick filament. The two projections of each myosin molecule (golf club heads) are called *myosin heads.* Each myosin head has two binding sites (**Figure 10.4a**): (1) an *actin-binding site* and (2) an *ATP-binding site*. The ATP-binding site also functions as an *ATPase*—an enzyme that hydrolyzes ATP to generate energy for muscle contraction. The heads project outward from the shaft in a spiraling fashion, each extending toward one of the six thin filaments that surround each thick filament.

TABLE 10.1 Components of a Sarcomere

COMPONENT	DESCRIPTION
Z discs	Narrow, plate-shaped regions of dense material that separate one sarcomere from the next.
A band	Dark, middle part of sarcomere that extends entire length of thick filaments and includes those parts of thin filaments that overlap thick filaments.
I band	Lighter, less dense area of sarcomere that contains remainder of thin filaments but no thick filaments. A Z disc passes through center of each I band.
H zone	Narrow region in center of each A band that contains thick filaments but no thin filaments.
M line	Region in center of H zone that contains proteins that hold thick filaments together at center of sarcomere.

Z disc M line Z disc H zone I band A band I band Sarcomere

Courtesy Hiroyouki Sasaki, Yale E.Goldman and Clara Franzini-Armstrong TEM 21,600x

The main component of the thin filament is the protein **actin** (AK-tin) (see **Figure 10.3b**). Individual actin molecules join to form an actin filament that is twisted into a helix (**Figure 10.4b**). On each actin molecule is a *myosin-binding site*, where a myosin head can attach.

Smaller amounts of two *regulatory proteins*—**tropomyosin** (trō-pō-MĪ-ō-sin) and **troponin** (TRŌ-pō-nin)—are also part of the thin filament. In relaxed muscle, myosin is blocked from binding to actin because strands of tropomyosin cover the myosin-binding

FIGURE 10.4 Structure of thick and thin filaments. (a) A thick filament contains about 300 myosin molecules, one of which is shown enlarged. The myosin tails form the shaft of the thick filament, and the myosin heads project outward toward the surrounding thin filaments. (b) Thin filaments contain actin, troponin, and tropomyosin.

Contractile proteins (myosin and actin) generate force during contraction; regulatory proteins (troponin and tropomyosin) help switch contraction on and off.

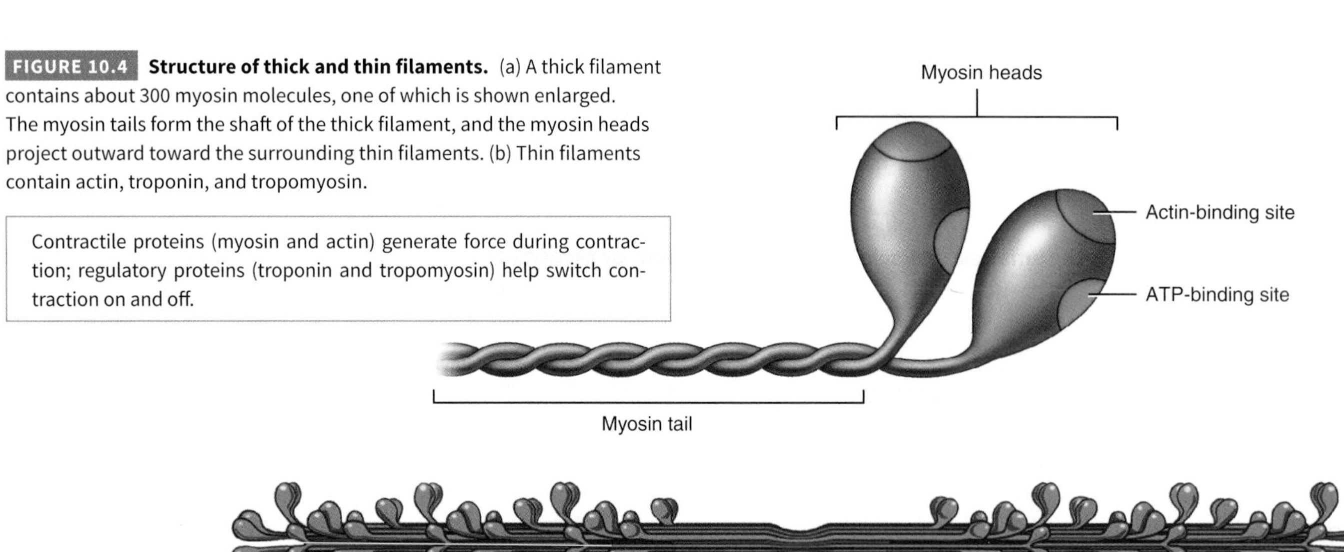

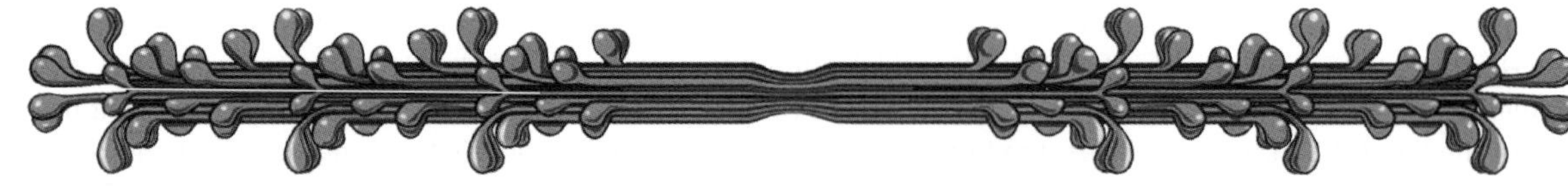

(a) Thick filament (below) and myosin molecule (above)

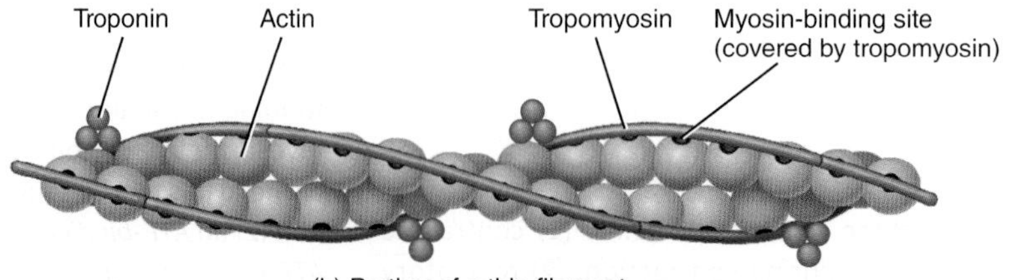

(b) Portion of a thin filament

Q Which proteins connect into the Z disc? Which proteins are present in the A band? In the I band?

sites on actin. The tropomyosin strands in turn are held in place by troponin molecules. You will soon learn that when calcium ions (Ca^{2+}) bind to troponin, troponin undergoes a conformational change (change in shape); this change moves tropomyosin away from myosin-binding sites on actin, and muscle contraction subsequently begins as myosin binds to actin.

Besides contractile and regulatory proteins, muscle contains about a dozen *structural proteins*, which contribute to the alignment, stability, elasticity, and extensibility of myofibrils. Several key structural proteins are titin, α-actinin, myomesin, nebulin, and dystrophin. **Titin** (*titan* = gigantic) is the third most plentiful protein in skeletal muscle (after actin and myosin). This molecule's name reflects its huge size. With a molecular mass of about 3 million daltons, titin is 50 times larger than an average-sized protein. Each titin molecule spans half a sarcomere, from a Z disc to an M line (see **Figure 10.3b**), a distance of 1 to 1.2 μm in relaxed muscle. Each titin molecule connects a Z disc to the M line of the sarcomere, thereby helping stabilize the position of the thick filament. The part of the titin molecule that extends from the Z disc is very elastic. Because it can stretch to at least four times its resting length and then spring back unharmed, titin accounts for much of the elasticity and extensibility of myofibrils. Titin probably helps the sarcomere return to its resting length after a muscle has contracted or been stretched, may help prevent overextension of sarcomeres, and maintains the central location of the A bands.

The dense material of the Z discs contains molecules of **α-actinin**, which bind to actin molecules of the thin filament and to titin. Molecules of the protein **myomesin** (mī-ō-MĒ-sin) form the M line. The M line proteins bind to titin and connect adjacent thick filaments to one another. Myomesin holds the thick filaments in alignment at the M line. **Nebulin** (NEB-ū-lin) is a long, nonelastic protein wrapped around the entire length of each thin filament. It helps anchor the thin filaments to the Z discs and regulates the length of thin filaments during development. **Dystrophin** (dis-TRŌ-fin) links thin filaments of the sarcomere to integral membrane proteins of the sarcolemma, which are attached in turn to proteins in the connective tissue extracellular matrix that surrounds muscle fibers (see **Figure 10.2d**). Dystrophin and its associated proteins are thought to reinforce the sarcolemma and help transmit the tension generated by the sarcomeres to the tendons. The relationship of dystrophin to muscular dystrophy is discussed in Disorders: Homeostatic Imbalances in Study Guide.

Table 10.2 summarizes the different types of skeletal muscle fiber proteins, and **Table 10.3** summarizes the levels of organization within a skeletal muscle.

TABLE 10.2 Summary of Skeletal Muscle Fiber Proteins

TYPE OF PROTEIN	DESCRIPTION
Contractile proteins	Proteins that generate force during muscle contractions.
Myosin	Contractile protein that makes up thick filament; molecule consists of a tail and two myosin heads, which bind to myosin-binding sites on actin molecules of thin filament during muscle contraction.
Actin	Contractile protein that is the main component of thin filament; each actin molecule has a myosin-binding site where myosin head of thick filament binds during muscle contraction.
Regulatory proteins	Proteins that help switch muscle contraction process on and off.
Tropomyosin	Regulatory protein that is a component of thin filament; when skeletal muscle fiber is relaxed, tropomyosin covers myosin-binding sites on actin molecules, thereby preventing myosin from binding to actin.
Troponin	Regulatory protein that is a component of thin filament; when calcium ions (Ca^{2+}) bind to troponin, it changes shape; this conformational change moves tropomyosin away from myosin-binding sites on actin molecules, and muscle contraction subsequently begins as myosin binds to actin.
Structural proteins	Proteins that keep thick and thin filaments of myofibrils in proper alignment, give myofibrils elasticity and extensibility, and link myofibrils to sarcolemma and extracellular matrix.
Titin	Structural protein that connects Z disc to M line of sarcomere, thereby helping to stabilize thick filament position; can stretch and then spring back unharmed, and thus accounts for much of the elasticity and extensibility of myofibrils.
α-Actinin	Structural protein of Z discs that attaches to actin molecules of thin filaments and to titin molecules.
Myomesin	Structural protein that forms M line of sarcomere; binds to titin molecules and connects adjacent thick filaments to one another.
Nebulin	Structural protein that wraps around entire length of each thin filament; helps anchor thin filaments to Z discs and regulates length of thin filaments during development.
Dystrophin	Structural protein that links thin filaments of sarcomere to integral membrane proteins in sarcolemma, which are attached in turn to proteins in connective tissue matrix that surrounds muscle fibers; thought to help reinforce sarcolemma and help transmit tension generated by sarcomeres to tendons.

TABLE 10.3 Levels of Organization within a Skeletal Muscle

LEVEL	DESCRIPTION
Skeletal muscle 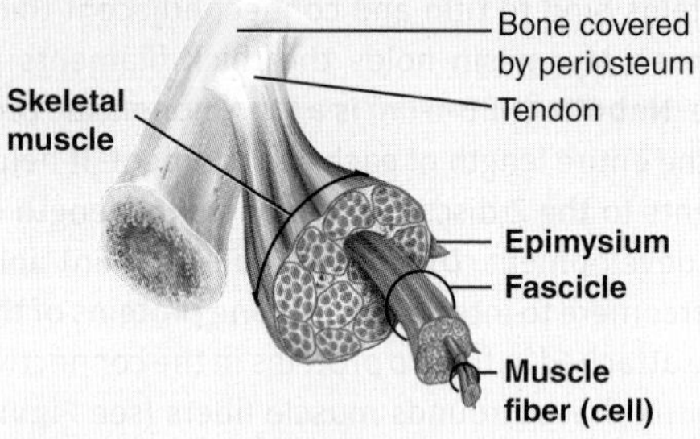	Organ made up of fascicles that contain muscle fibers (cells), blood vessels, and nerves; wrapped in epimysium.
Fascicle 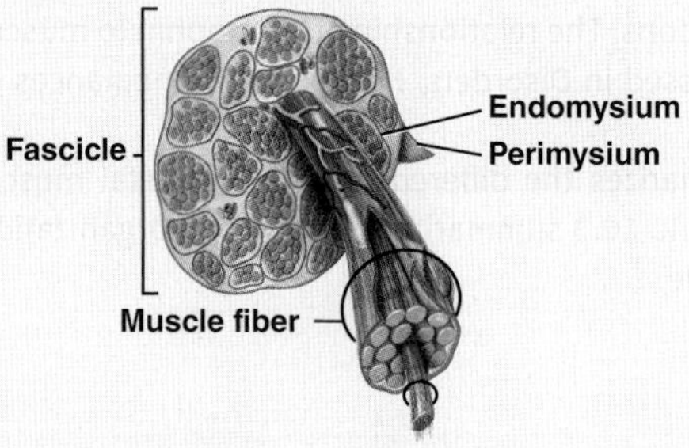	Bundle of muscle fibers wrapped in perimysium.
Muscle fiber (cell) 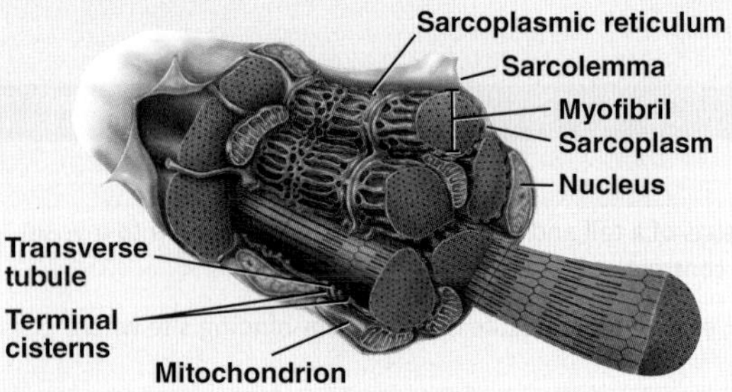	Long cylindrical cell covered by endomysium and sarcolemma; contains sarcoplasm, myofibrils, many peripherally located nuclei, mitochondria, transverse tubules, sarcoplasmic reticulum, and terminal cisterns. The fiber has a striated appearance.
Myofibril 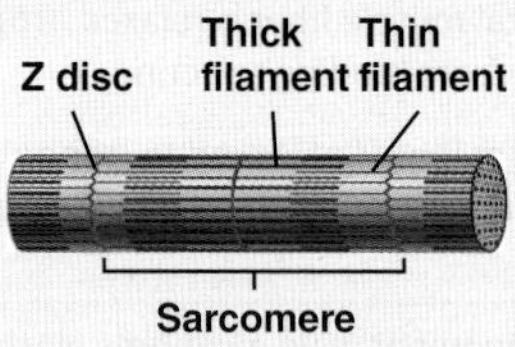	Threadlike contractile elements within sarcoplasm of muscle fiber that extend entire length of fiber; composed of filaments.
Filaments (myofilaments) 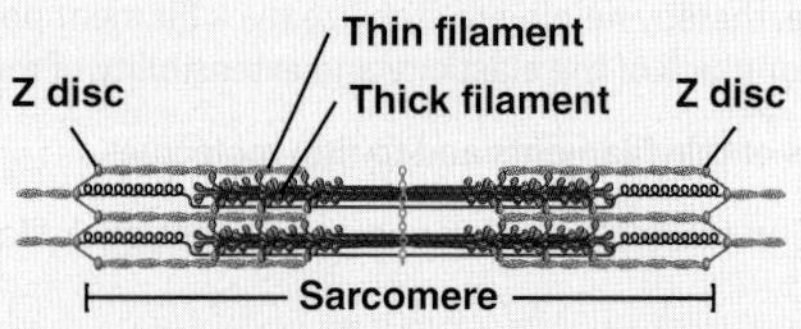	Contractile proteins within myofibrils that are of two types: thick filaments composed of myosin and thin filaments composed of actin, tropomyosin, and troponin; sliding of thin filaments past thick filaments produces muscle shortening.

10.3 Contraction and Relaxation of Skeletal Muscle Fibers

OBJECTIVES

- **Outline** the steps involved in the sliding filament mechanism of muscle contraction.
- **Describe** how muscle action potentials arise at the neuromuscular junction.

When scientists examined the first electron micrographs of skeletal muscle in the mid-1950s, they were surprised to see that the lengths of the thick and thin filaments were the same in both relaxed and contracted muscle. It had been thought that muscle contraction must be a folding process, somewhat like closing an accordion. Instead, researchers discovered that skeletal muscle shortens during contraction because the thick and thin filaments slide past one another. The model describing this process is known as the **sliding filament mechanism**.

The Sliding Filament Mechanism

Muscle contraction occurs because myosin heads attach to and "walk" along the thin filaments at both ends of a sarcomere, progressively pulling the thin filaments toward the M line (**Figure 10.5**). As a result, the thin filaments slide inward and meet at the center of a sarcomere. They may even move so far inward that their ends overlap (**Figure 10.5c**). As the thin filaments slide inward, the I band and H zone narrow and eventually disappear altogether when the muscle is maximally contracted. However, the width of the A band and the individual lengths of the thick and thin filaments remain unchanged. Since the thin filaments on each side of the sarcomere are attached to Z discs, when the thin filaments slide inward, the Z discs come closer together, and the sarcomere shortens. Shortening of the sarcomeres causes shortening of the whole muscle fiber, which in turn leads to shortening of the entire muscle.

FIGURE 10.5 Sliding filament mechanism of muscle contraction, as it occurs in two adjacent sarcomeres.

During muscle contractions, thin filaments move toward the M line of each sarcomere.

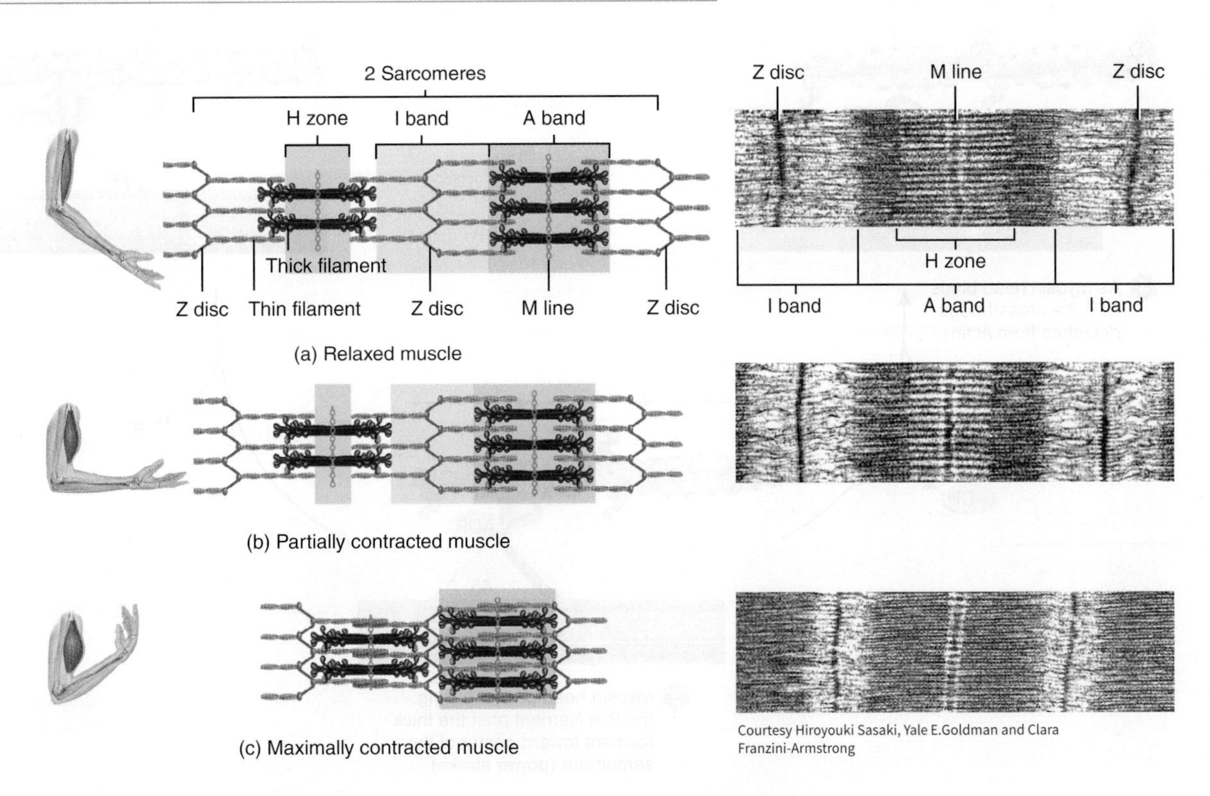

Courtesy Hiroyouki Sasaki, Yale E.Goldman and Clara Franzini-Armstrong

Q What happens to the I band and H zone as muscle contracts? Do the lengths of the thick and thin filaments change?

The Contraction Cycle At the onset of contraction, the sarcoplasmic reticulum releases calcium ions (Ca^{2+}) into the sarcoplasm. There, they bind to troponin. Troponin then moves tropomyosin away from the myosin-binding sites on actin. Once the binding sites are "free," the **contraction cycle**—the repeating sequence of events that causes the filaments to slide—begins. The contraction cycle consists of four steps (**Figure 10.6**):

1. ***ATP hydrolysis.*** As mentioned earlier, a myosin head includes an ATP-binding site that functions as an ATPase—an enzyme that hydrolyzes ATP into ADP (adenosine diphosphate) and a phosphate group. The energy generated from this hydrolysis reaction is stored in the myosin head for later use during the contraction cycle. The myosin head is said to be *energized* when it contains stored energy. The energized myosin head assumes

FIGURE 10.6 The contraction cycle. Sarcomeres exert force and shorten through repeated cycles during which the myosin heads attach to actin (forming cross-bridges), rotate, and detach.

During the power stroke of contraction, cross-bridges rotate and move the thin filaments past the thick filaments toward the center of the sarcomere.

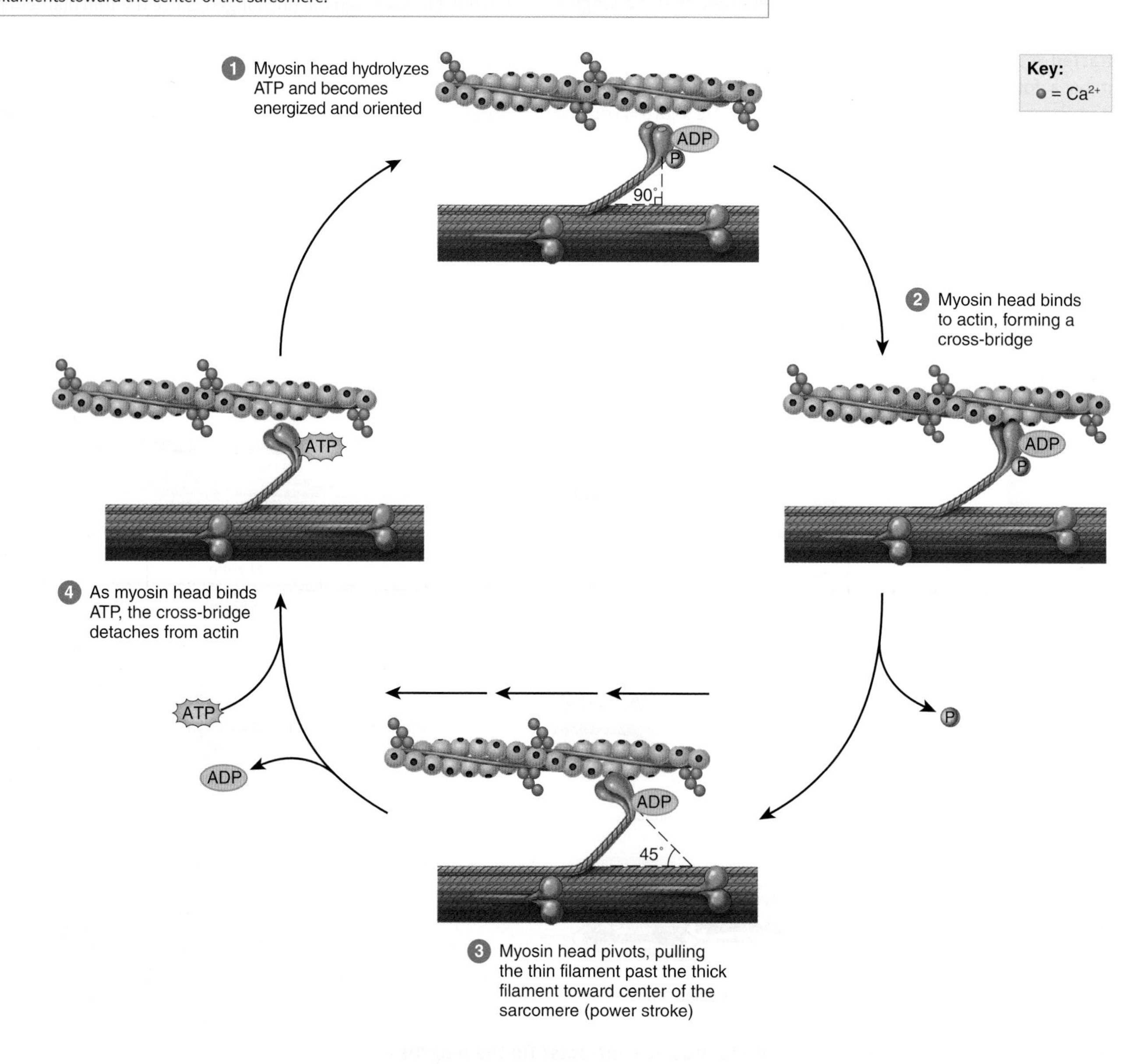

Q What would happen if ATP suddenly were not available after the sarcomere had started to shorten?

a "cocked" position, like a stretched spring. In this position, the myosin head is perpendicular (at a 90° angle) relative to the thick and thin filaments and has the proper orientation to bind to an actin molecule. Notice that the products of ATP hydrolysis—ADP and a phosphate group—are still attached to the myosin head.

2 ***Attachment of myosin to actin.*** The energized myosin head attaches to the myosin-binding site on actin and releases the previously hydrolyzed phosphate group. When a myosin head attaches to actin during the contraction cycle, the myosin head is referred to as a **cross-bridge**. Although a single myosin molecule has a double head, only one head binds to actin at a time.

3 ***Power stroke.*** After a cross-bridge forms, the myosin head pivots, changing its position from a 90° angle to a 45° angle relative to the thick and thin filaments. As the myosin head changes to its new position, it pulls the thin filament past the thick filament toward the center of the sarcomere, generating tension (force) in the process. This event is known as the **power stroke**. The energy required for the power stroke is derived from the energy stored in the myosin head from the hydrolysis of ATP (see step 1). Once the power stroke occurs, ADP is released from the myosin head.

4 ***Detachment of myosin from actin.*** At the end of the power stroke, the cross-bridge remains firmly attached to actin until it binds another molecule of ATP. As ATP binds to the ATP-binding site on the myosin head, the myosin head detaches from actin.

The contraction cycle repeats as the myosin ATPase hydrolyzes the newly bound molecule of ATP, and continues as long as ATP is available and the Ca^{2+} level near the thin filament is sufficiently high. The cross-bridges keep rotating back and forth with each power stroke, pulling the thin filaments toward the M line. Each of the 600 cross-bridges in one thick filament attaches and detaches about five times per second. At any one instant, some of the myosin heads are attached to actin, forming cross-bridges and generating force, and other myosin heads are detached from actin, getting ready to bind again.

As the contraction cycle continues, movement of cross-bridges applies the force that draws the Z discs toward each other, and the sarcomere shortens. During a maximal muscle contraction, the distance between two Z discs can decrease to half the resting length. The Z discs in turn pull on neighboring sarcomeres, and the whole muscle fiber shortens. Some of the components of a muscle are elastic: They stretch slightly before they transfer the tension generated by the sliding filaments. The elastic components include titin molecules, connective tissue around the muscle fibers (endomysium, perimysium, and epimysium), and tendons that attach muscle to bone. As the cells of a skeletal muscle start to shorten, they first pull on their connective tissue coverings and tendons. The coverings and tendons stretch and then become taut, and the tension passed through the tendons pulls on the bones to which they are attached. The result is movement of a part of the body. You will soon learn, however, that the contraction cycle does not always result in shortening of the muscle fibers and the whole muscle. In some contractions, the cross-bridges rotate and generate tension, but the thin filaments cannot slide inward because the tension they generate is not large enough to move the load on the muscle (such as trying to lift a whole box of books with one hand).

Excitation–Contraction Coupling

An increase in Ca^{2+} concentration in the sarcoplasm starts muscle contraction, and a decrease stops it. When a muscle fiber is relaxed, the concentration of Ca^{2+} in its sarcoplasm is very low, only about 0.1 micromole per liter (0.1 μmol/L). However, a huge amount of Ca^{2+} is stored inside the sarcoplasmic reticulum (**Figure 10.7a**). As a muscle action potential propagates along the sarcolemma and into the T tubules, it causes the release of Ca^{2+} from the SR into the sarcoplasm and this triggers muscle contraction. The sequence of events that links excitation (a muscle action potential) to contraction (sliding of the filaments) is referred to as **excitation–contraction coupling**.

Excitation-contraction coupling occurs at the triads of the skeletal muscle fiber. Recall that a *triad* consists of a transverse (T) tubule and two opposing terminal cisterns of the sarcoplasmic reticulum (SR). At a given triad, the T tubule and terminal cisterns are mechanically linked together by two groups of integral membrane proteins: voltage-gated Ca^{2+} channels and Ca^{2+} release channels (**Figure 10.7a**). **Voltage-gated Ca^{2+} channels** are located in the T tubule membrane; they arranged in clusters of four known as *tetrads*. The main role of these voltage-gated Ca^{2+} channels in excitation-contraction coupling is to serve as voltage sensors that trigger the opening of the Ca^{2+} release channels. **Ca^{2+} release channels** are present in the terminal cisternal membrane of the SR. When a skeletal muscle fiber is at rest, the part of the Ca^{2+} release channel that extends into the sarcoplasm is blocked by a given cluster of voltage-gated Ca^{2+} channels, preventing Ca^{2+} from leaving the SR (**Figure 10.7a**). When a skeletal muscle fiber is excited and an action potential travels along the T tubule, the voltage-gated Ca^{2+} channels detect the change in voltage and undergo a conformational change that ultimately causes the Ca^{2+} release channels to open (**Figure 10.7b**). Once these channels open, large amounts of Ca^{2+} flow out of the SR into the sarcoplasm around the thick and thin filaments. As a result, the Ca^{2+} concentration in the sarcoplasm rises tenfold or more. The released calcium ions combine with troponin, which in turn undergoes a conformational change that causes tropomyosin to move away from the myosin-binding sites on actin. Once these binding sites are free, myosin heads bind to them to form cross-bridges, and the muscle fiber contracts.

The terminal cisternal membrane of the sarcoplasmic reticulum also contains **Ca^{2+}-ATPase pumps** that use ATP to constantly transport Ca^{2+} from the sarcoplasm into the SR (**Figure 10.7a,b**). As long as muscle action potentials continue to propagate along the T tubules, the Ca^{2+} release channels remain open and Ca^{2+} flows into the sarcoplasm faster than it is transported back into the SR by the Ca^{2+}-ATPase pumps. After the last action potential has propagated throughout the T tubules, the Ca^{2+} release channels close. As the Ca^{2+}-ATPase pumps move Ca^{2+} back into the SR, the Ca^{2+} level in the sarcoplasm rapidly decreases.

Inside the SR, molecules of a protein known as **calsequestrin** bind to Ca^{2+}, allowing even more Ca^{2+} to be sequestered (stored) within the SR. In a relaxed muscle fiber, the concentration of Ca^{2+} is 10,000 times higher in the SR than in the sarcoplasm. As the Ca^{2+} level in the sarcoplasm decreases, Ca^{2+} is released from troponin, tropomyosin covers the myosin-binding sites on actin, and the muscle fiber relaxes.

See Clinical Connection: Rigor Mortis

FIGURE 10.7 Mechanism of excitation-contraction coupling in a skeletal muscle fiber. (a) During relaxation, the level of Ca^{2+} in the sarcoplasm is low, only 0.1 μM (0.0001 mM), because calcium ions are pumped into the sarcoplasmic reticulum by Ca^{2+}-ATPase pumps. (b) A muscle action potential propagating along a transverse tubule causes voltage-gated Ca^{2+} channels to undergo a conformational change that opens Ca^{2+} release channels in the sarcoplasmic reticulum, calcium ions flow into the sarcoplasm, and contraction begins.

An increase in the Ca^{2+} level in the sarcoplasm starts the sliding of thin filaments. When the level of Ca^{2+} in the sarcoplasm declines, sliding stops.

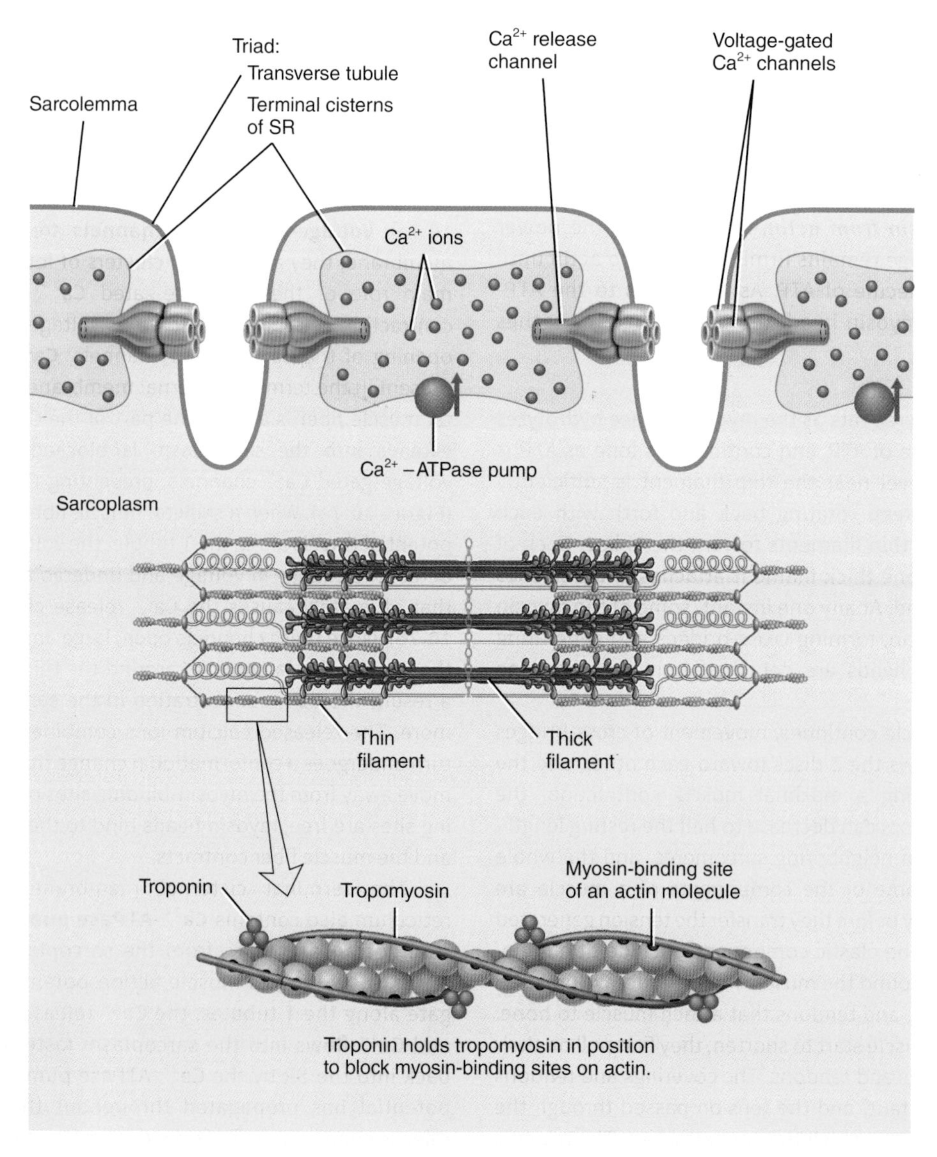

(a) Relaxation

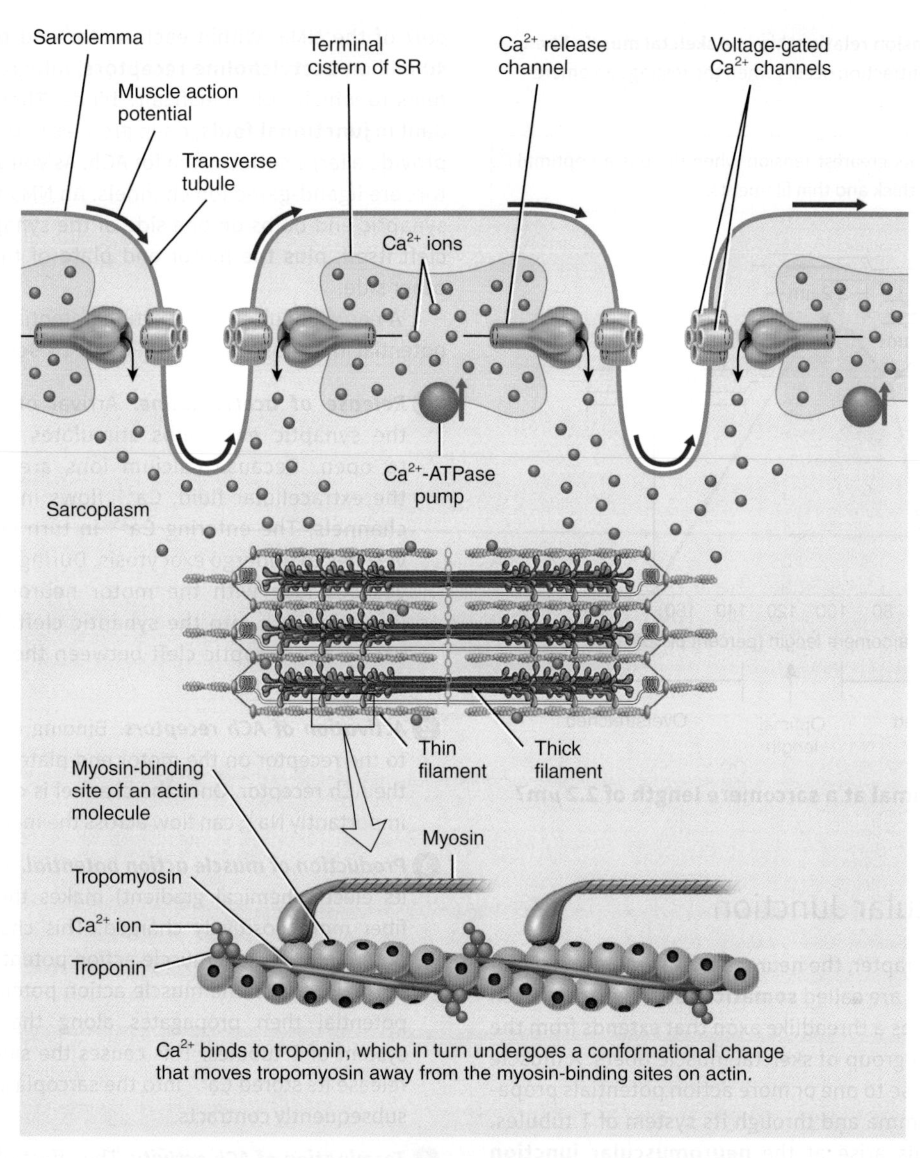

(b) Contraction

Q What are the three functions of ATP in muscle contraction?

Length–Tension Relationship Figure 10.8 shows the **length–tension relationship** for skeletal muscle, which indicates how the forcefulness of muscle contraction depends on the length of the sarcomeres within a muscle *before contraction begins.* At a sarcomere length of about 2.0–2.4 μm (which is very close to the resting length in most muscles), the zone of overlap in each sarcomere is optimal, and the muscle fiber can develop maximum tension. Notice in Figure 10.8 that maximum tension (100%) occurs when the zone of overlap between a thick and thin filament extends from the edge of the H zone to one end of a thick filament.

As the sarcomeres of a muscle fiber are stretched to a longer length, the zone of overlap shortens, and fewer myosin heads can make contact with thin filaments. Therefore, the tension the fiber can produce decreases. When a skeletal muscle fiber is stretched to 170% of its optimal length, there is no overlap between the thick and thin filaments. Because none of the myosin heads can bind to thin filaments, the muscle fiber cannot contract, and tension is zero. As sarcomere lengths become increasingly shorter than the optimum, the tension that can develop again decreases. This is because thick filaments crumple as they are compressed by the Z discs, resulting in fewer myosin heads making contact with thin filaments. Normally, resting muscle fiber length is held very close to the optimum by firm attachments of skeletal muscle to bones (via their tendons) and to other inelastic tissues.

FIGURE 10.8 Length–tension relationship in a skeletal muscle fiber. Maximum tension during contraction occurs when the resting sarcomere length is 2.0–2.4 μm.

A muscle fiber develops its greatest tension when there is an optimal zone of overlap between thick and thin filaments.

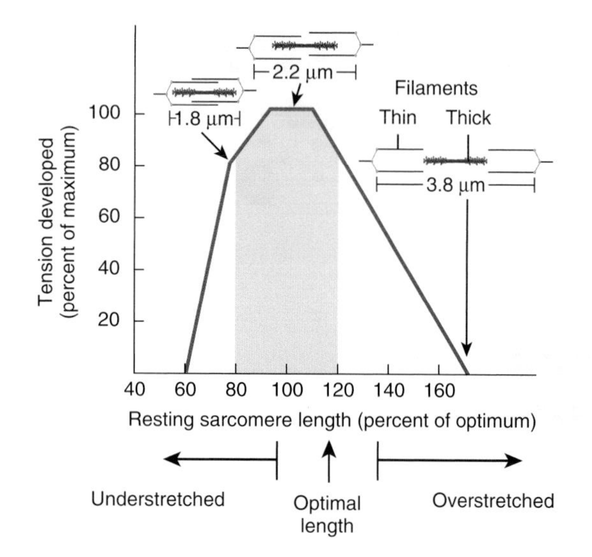

Q Why is tension maximal at a sarcomere length of 2.2 μm?

The Neuromuscular Junction

As noted earlier in the chapter, the neurons that stimulate skeletal muscle fibers to contract are called **somatic motor neurons**. Each somatic motor neuron has a threadlike axon that extends from the brain or spinal cord to a group of skeletal muscle fibers. A muscle fiber contracts in response to one or more action potentials propagating along its sarcolemma and through its system of T tubules. Muscle action potentials arise at the **neuromuscular junction (NMJ)** (noo-rō-MUS-kū-lar), the synapse between a somatic motor neuron and a skeletal muscle fiber (Figure 10.9a). A **synapse** is a region where communication occurs between two neurons, or between a neuron and a target cell—in this case, between a somatic motor neuron and a muscle fiber. At most synapses a small gap, called the **synaptic cleft**, separates the two cells. Because the cells do not physically touch, the action potential cannot "jump the gap" from one cell to another. Instead, the first cell communicates with the second by releasing a chemical messenger called a **neurotransmitter**.

At the NMJ, the end of the motor neuron, called the **axon terminal**, divides into a cluster of **synaptic end bulbs** (Figure 10.9a, b), the *neural part* of the NMJ. Suspended in the cytosol within each synaptic end bulb are hundreds of membrane-enclosed sacs called **synaptic vesicles**. Inside each synaptic vesicle are thousands of molecules of **acetylcholine (ACh)** (as′-ē-til-KŌ-lēn), the neurotransmitter released at the NMJ.

The region of the sarcolemma opposite the synaptic end bulbs, called the **motor end plate** (Figure 10.9b, c), is the *muscular part* of the NMJ. Within each motor end plate are 30 million to 40 million **acetylcholine receptors**, integral transmembrane proteins to which ACh specifically binds. These receptors are abundant in **junctional folds**, deep grooves in the motor end plate that provide a large surface area for ACh. As you will see, the ACh receptors are ligand-gated ion channels. An NMJ thus includes all of the synaptic end bulbs on one side of the synaptic cleft, the synaptic cleft itself, plus the motor end plate of the muscle fiber on the other side.

A nerve impulse (nerve action potential) elicits a muscle action potential in the following way (Figure 10.9c):

1. ***Release of acetylcholine.*** Arrival of the nerve impulse at the synaptic end bulbs stimulates voltage-gated channels to open. Because calcium ions are more concentrated in the extracellular fluid, Ca^{2+} flows inward through the open channels. The entering Ca^{2+} in turn stimulates the synaptic vesicles to undergo exocytosis. During exocytosis, the synaptic vesicles fuse with the motor neuron's plasma membrane, liberating ACh into the synaptic cleft. The ACh then diffuses across the synaptic cleft between the motor neuron and the motor end plate.
2. ***Activation of ACh receptors.*** Binding of two molecules of ACh to the receptor on the motor end plate opens an ion channel in the ACh receptor. Once the channel is open, small cations, most importantly Na^+, can flow across the membrane.
3. ***Production of muscle action potential.*** The inflow of Na^+ (down its electrochemical gradient) makes the inside of the muscle fiber more positively charged. This change in the membrane potential triggers a muscle action potential. Each nerve impulse normally elicits one muscle action potential. The muscle action potential then propagates along the sarcolemma into the system of T tubules. This causes the sarcoplasmic reticulum to release its stored Ca^{2+} into the sarcoplasm, and the muscle fiber subsequently contracts.
4. ***Termination of ACh activity.*** The effect of ACh binding lasts only briefly because ACh is rapidly broken down by an enzyme called **acetylcholinesterase (AChE)** (as′-ē-til-kō′-lin-ES-ter-ās). This enzyme is located on the extracellular side of the motor end plate membrane. AChE breaks down ACh into acetyl and choline, products that cannot activate the ACh receptor.

If another nerve impulse releases more acetylcholine, steps 2 and 3 repeat. When action potentials in the motor neuron cease, ACh is no longer released, and AChE rapidly breaks down the ACh already present in the synaptic cleft. This ends the production of muscle action potentials, the Ca^{2+} moves from the sarcoplasm of the muscle fiber back into the sarcoplasmic reticulum, and the Ca^{2+} release channels in the sarcoplasmic reticulum membrane close.

A skeletal muscle fiber has only one NMJ and it is usually located near the midpoint of the fiber. Muscle action potentials that arise at the NMJ propagate toward both ends of the fiber. This arrangement permits nearly simultaneous activation (and thus contraction) of all parts of the muscle fiber.

FIGURE 10.9 **Structure of the neuromuscular junction (NMJ).**

Synaptic end bulbs at the tips of axon terminals contain synaptic vesicles filled with acetylcholine (ACh).

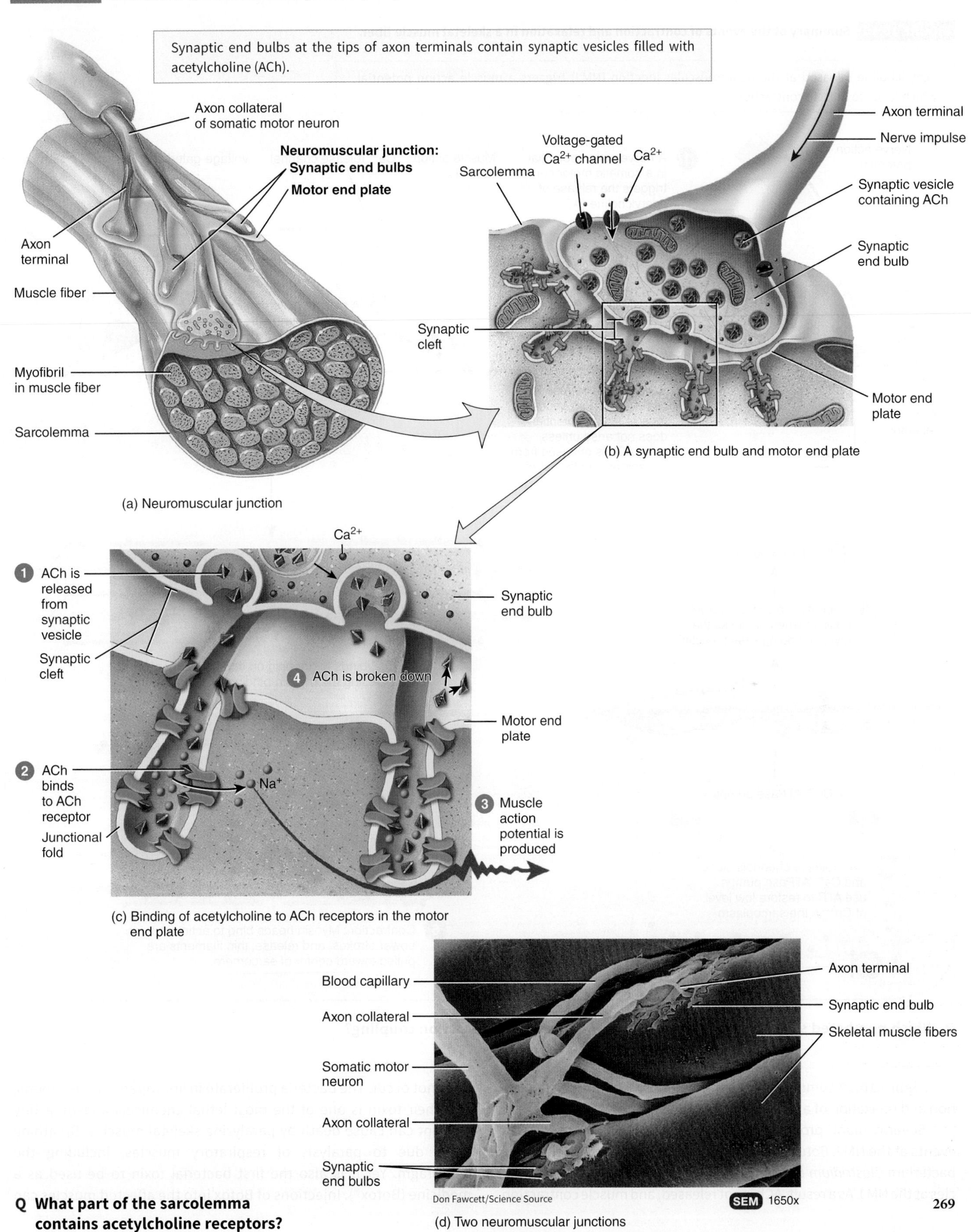

(a) Neuromuscular junction

(b) A synaptic end bulb and motor end plate

(c) Binding of acetylcholine to ACh receptors in the motor end plate

(d) Two neuromuscular junctions

Q What part of the sarcolemma contains acetylcholine receptors?

FIGURE 10.10 Summary of the events of contraction and relaxation in a skeletal muscle fiber.

Acetylcholine released at the neuromuscular junction (NMJ) triggers a muscle action potential, which leads to muscle contraction.

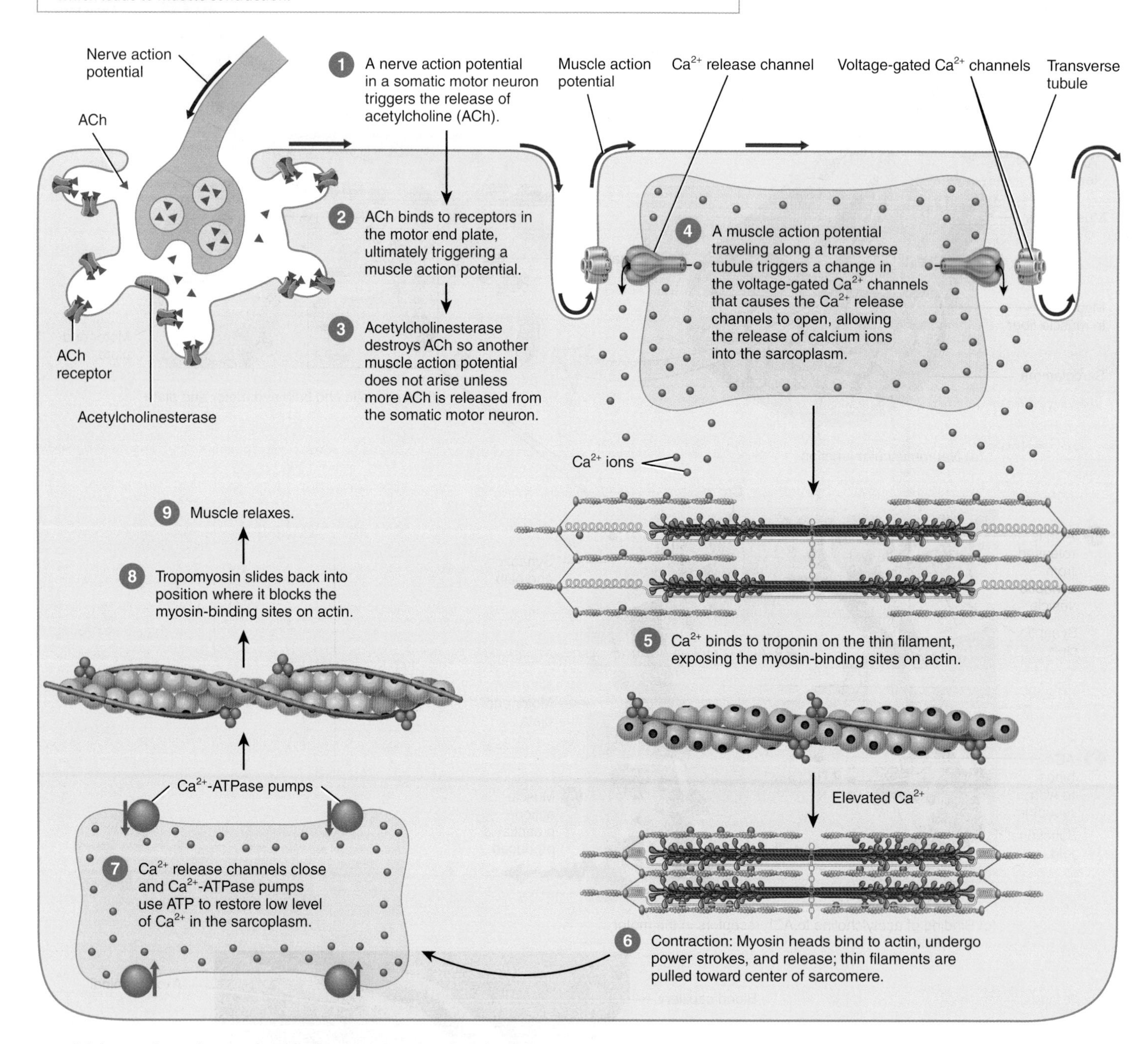

Q Which numbered steps in this figure are part of excitation–contraction coupling?

Figure 10.10 summarizes the events that occur during contraction and relaxation of a skeletal muscle fiber.

Several plant products and drugs selectively block certain events at the NMJ. *Botulinum toxin* (bot-u-LĪN-um), produced by the bacterium *Clostridium botulinum*, blocks exocytosis of synaptic vesicles at the NMJ. As a result, ACh is not released, and muscle contraction does not occur. The bacteria proliferate in improperly canned foods, and their toxin is one of the most lethal chemicals known. A tiny amount can cause death by paralyzing skeletal muscles. Breathing stops due to paralysis of respiratory muscles, including the diaphragm. Yet it is also the first bacterial toxin to be used as a medicine (Botox®). Injections of Botox into the affected muscles can help patients who have strabismus (crossed eyes), blepharospasm (uncontrollable blinking), or spasms of the vocal cords that interfere with speech. It is also used to alleviate chronic back pain due to muscle spasms in the lumbar region and as a cosmetic treatment to relax muscles that cause facial wrinkles.

The plant derivative *curare*, a poison used by South American Indians on arrows and blowgun darts, causes muscle paralysis by binding to and blocking ACh receptors. In the presence of curare, the ion channels do not open. Curare-like drugs are often used during surgery to relax skeletal muscles.

A family of chemicals called *anticholinesterase agents* has the property of slowing the enzymatic activity of acetylcholinesterase, thus slowing removal of ACh from the synaptic cleft. At low doses, these agents can strengthen weak muscle contractions. One example is neostigmine, which is used to treat patients with myasthenia gravis (see the Disorders: Homeostatic Imbalances in Study Guide). Neostigmine is also used as an antidote for curare poisoning and to terminate the effects of curare-like drugs after surgery.

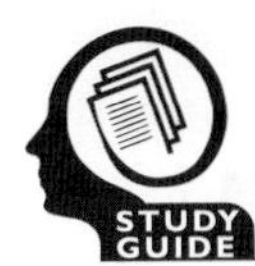

See Clinical Connection: Electromyography

10.4 Muscle Metabolism

OBJECTIVES

- **Describe** the reactions by which muscle fibers produce ATP.
- **Distinguish** between anaerobic glycolysis and aerobic respiration.
- **Describe** the factors that contribute to muscle fatigue.

Production of ATP in Muscle Fibers

Unlike most cells of the body, skeletal muscle fibers often switch between a low level of activity, when they are relaxed and using only a modest amount of ATP, and a high level of activity, when they are contracting and using ATP at a rapid pace. A huge amount of ATP is needed to power the contraction cycle, to pump Ca^{2+} into the sarcoplasmic reticulum, and for other metabolic reactions involved in muscle contraction. However, the ATP present inside muscle fibers is enough to power contraction for only a few seconds. If muscle contractions continue past that time, the muscle fibers must make more ATP. Muscle fibers have three ways to produce ATP: (1) from creatine phosphate, (2) by anaerobic glycolysis, and (3) by aerobic respiration (**Figure 10.11**). The use of creatine phosphate for ATP production is unique to muscle fibers, but all body cells can make ATP by the reactions of anaerobic glycolysis and aerobic respiration. We consider the events of glycolysis and aerobic respiration briefly in this chapter and in more detail in Chapter 25.

Creatine Phosphate While muscle fibers are relaxed, they produce more ATP than they need for resting metabolism. Most of the excess ATP is used to synthesize **creatine phosphate** (KRĒ-a-tēn), an energy-rich molecule that is found in muscle fibers (**Figure 10.11a**). The enzyme creatine kinase (CK) catalyzes the transfer of one of the high-energy phosphate groups from ATP to creatine, forming creatine phosphate and ADP. **Creatine** is a small, amino acid–like molecule that is synthesized in the liver, kidneys, and pancreas and then transported to muscle fibers. Creatine phosphate is three to six times more plentiful than ATP in the sarcoplasm of a relaxed muscle fiber. When contraction begins and the ADP level starts to rise, CK catalyzes the transfer of a high-energy phosphate group from creatine phosphate back to ADP. This direct phosphorylation reaction quickly generates new ATP molecules. Since the formation of ATP from creatine phosphate occurs very rapidly, creatine phosphate is the first source of energy when muscle contraction begins. The other energy-generating mechanisms in a muscle fiber (the pathways of anaerobic glycolysis and aerobic respiration) take a relatively longer period of time to produce ATP compared to creatine phosphate. Together, stores of creatine phosphate and ATP provide enough energy for muscles to contract maximally for about 15 seconds.

See Clinical Connection: Creatine Supplementation

Anaerobic Glycolysis When muscle activity continues and the supply of creatine phosphate within the muscle fiber is depleted, glucose is catabolized to generate ATP. Glucose passes easily from the blood into contracting muscle fibers via facilitated diffusion, and it is also produced by the breakdown of glycogen within muscle fibers (**Figure 10.11b**). Then, a series of reactions known as *glycolysis* quickly breaks down each glucose molecule into two molecules of pyruvic acid. Glycolysis occurs in the cytosol and produces a net gain of two molecules of ATP. Because glycolysis does not require oxygen, it can occur whether oxygen is present (aerobic conditions) or absent (anaerobic conditions).

Ordinarily, the pyruvic acid formed by glycolysis in the cytosol enters mitochondria, where it undergoes a series of oxygen-requiring reactions called aerobic respiration (described next) that produce a large amount of ATP. During heavy exercise, however, not enough oxygen is available to skeletal muscle fibers. Under these anaerobic conditions, the pyruvic acid generated from glycolysis is converted to lactic acid. The entire process by which the breakdown of glucose gives rise to lactic acid when oxygen is absent or at a low concentration is referred to as **anaerobic glycolysis** (**Figure 10.11b**). Each molecule of glucose catabolized via anaerobic glycolysis yields 2 molecules of lactic acid and 2 molecules of ATP. Most of the lactic acid produced by this process diffuses out of the skeletal muscle fiber into the blood. Liver cells can take up some of the lactic acid molecules from the bloodstream and convert them back to glucose. In addition to providing new glucose molecules, this conversion reduces the acidity of the blood. When produced at a rapid rate, lactic acid can accumulate in active skeletal muscle fibers and in the bloodstream. This buildup is thought

FIGURE 10.11 Production of ATP for muscle contraction. (a) Creatine phosphate, formed from ATP while the muscle is relaxed, transfers a high-energy phosphate group to ADP, forming ATP during muscle contraction. (b) Breakdown of muscle glycogen into glucose and production of pyruvic acid from glucose via glycolysis produce both ATP and lactic acid. Because no oxygen is needed, this is an anaerobic pathway. (c) Within mitochondria, pyruvic acid, fatty acids, and amino acids are used to produce ATP via aerobic respiration, an oxygen-requiring set of reactions.

During a long-term event such as a marathon race, most ATP is produced aerobically.

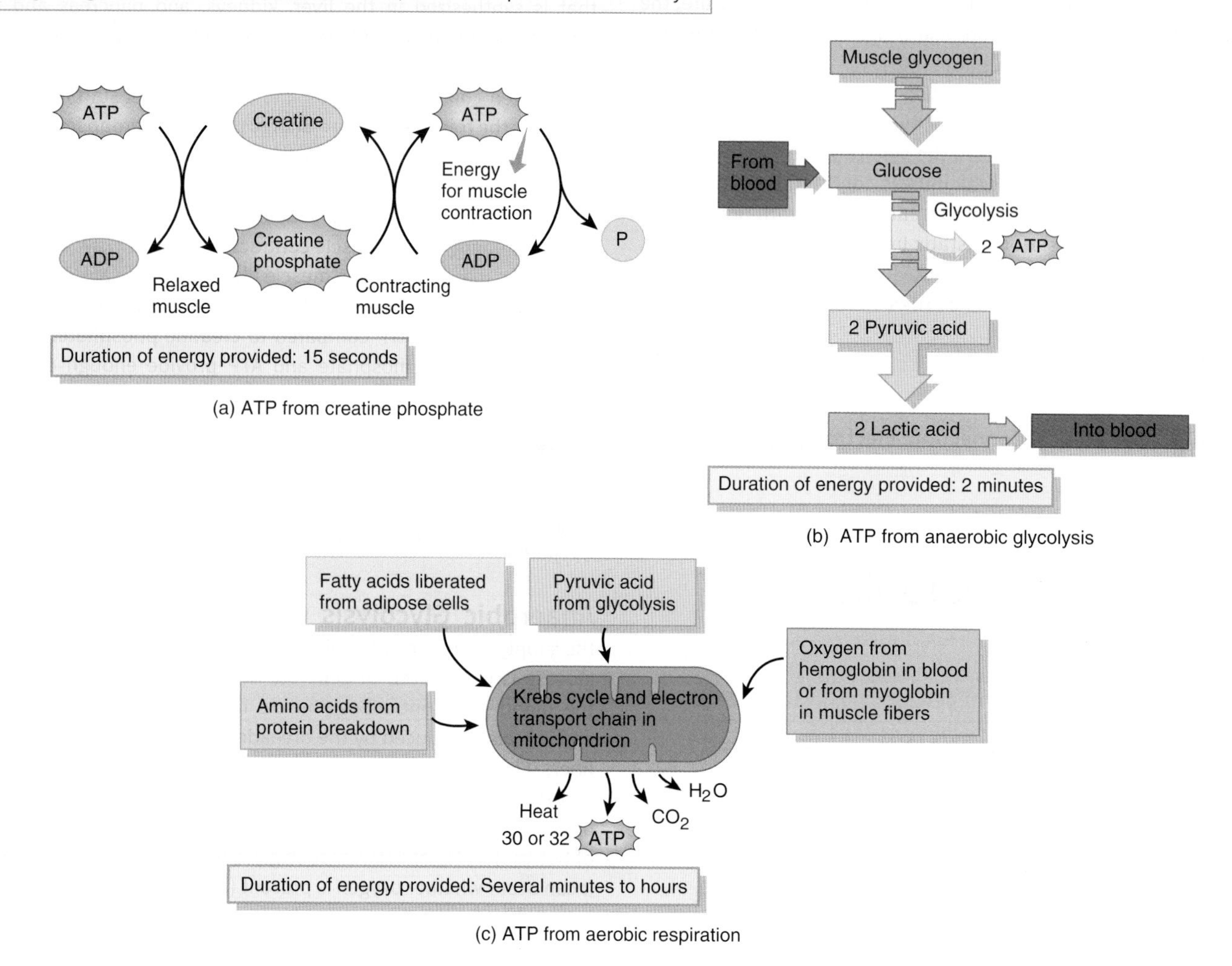

Q Where inside a skeletal muscle fiber are the events shown here occurring?

to be responsible for the muscle soreness that is felt during strenuous exercise. Compared to aerobic respiration, anaerobic glycolysis produces fewer ATPs, but it is faster and can occur when oxygen levels are low. Anaerobic glycolysis provides enough energy for about 2 minutes of maximal muscle activity.

Aerobic Respiration If sufficient oxygen is present, the pyruvic acid formed by glycolysis enters the mitochondria, where it undergoes **aerobic respiration**, a series of oxygen-requiring reactions (the *Krebs cycle* and the *electron transport chain*) that produce ATP, carbon dioxide, water, and heat (**Figure 10.11c**). Thus, when oxygen is present, glycolysis, the Krebs cycle, and the electron transport chain occur. Although aerobic respiration is slower than anaerobic glycolysis, it yields much more ATP. Each molecule of glucose catabolized under aerobic conditions yields about 30 or 32 molecules of ATP.

Muscular tissue has two sources of oxygen: (1) oxygen that diffuses into muscle fibers from the blood and (2) oxygen released by myoglobin within muscle fibers. Both myoglobin (found only in muscle cells) and hemoglobin (found only in red blood cells) are oxygen-binding proteins. They bind oxygen when it is plentiful and release oxygen when it is scarce.

Aerobic respiration supplies enough ATP for muscles during periods of rest or light to moderate exercise provided sufficient oxygen and nutrients are available. These nutrients include the pyruvic acid obtained from the glycolysis of glucose, fatty acids from the breakdown of triglycerides, and amino acids from the breakdown of proteins. In activities that last from several minutes to an hour or more, aerobic respiration provides nearly all of the needed ATP.

Muscle Fatigue

The inability of a muscle to maintain force of contraction after prolonged activity is called **muscle fatigue** (fa-TĒG). Fatigue results mainly from changes within muscle fibers. Even before actual muscle fatigue occurs, a person may have feelings of tiredness and the desire to cease activity; this response, called *central fatigue*, is caused by changes in the central nervous system (brain and spinal cord). Although its exact mechanism is unknown, it may be a protective mechanism to stop a person from exercising before muscles become damaged. As you will see, certain types of skeletal muscle fibers fatigue more quickly than others.

Although the precise mechanisms that cause muscle fatigue are still not clear, several factors are thought to contribute. One is inadequate release of calcium ions from the SR, resulting in a decline of Ca^{2+} concentration in the sarcoplasm. Depletion of creatine phosphate also is associated with fatigue, but surprisingly, the ATP levels in fatigued muscle often are not much lower than those in resting muscle. Other factors that contribute to muscle fatigue include insufficient oxygen, depletion of glycogen and other nutrients, buildup of lactic acid and ADP, and failure of action potentials in the motor neuron to release enough acetylcholine.

Oxygen Consumption after Exercise

During prolonged periods of muscle contraction, increases in breathing rate and blood flow enhance oxygen delivery to muscle tissue. After muscle contraction has stopped, heavy breathing continues for a while, and oxygen consumption remains above the resting level. Depending on the intensity of the exercise, the recovery period may be just a few minutes, or it may last as long as several hours. The term **oxygen debt** has been used to refer to the added oxygen, over and above the resting oxygen consumption, that is taken into the body after exercise. This extra oxygen is used to "pay back" or restore metabolic conditions to the resting level in three ways: (1) to convert lactic acid back into glycogen stores in the liver, (2) to resynthesize creatine phosphate and ATP in muscle fibers, and (3) to replace the oxygen removed from myoglobin.

The metabolic changes that occur *during exercise* can account for only some of the extra oxygen used *after exercise*. Only a small amount of glycogen resynthesis occurs from lactic acid. Instead, most glycogen is made much later from dietary carbohydrates. Much of the lactic acid that remains after exercise is converted back to pyruvic acid and used for ATP production via aerobic respiration in the heart, liver, kidneys, and skeletal muscle. Oxygen use after exercise also is boosted by ongoing changes. First, the elevated body temperature after strenuous exercise increases the rate of chemical reactions throughout the body. Faster reactions use ATP more rapidly, and more oxygen is needed to produce the ATP. Second, the heart and the muscles used in breathing are still working harder than they were at rest, and thus they consume more ATP. Third, tissue repair processes are occurring at an increased pace. For these reasons, **recovery oxygen uptake** is a better term than oxygen debt for the elevated use of oxygen after exercise.

10.5 Control of Muscle Tension

OBJECTIVES

- **Describe** the structure and function of a motor unit, and define motor unit recruitment.
- **Explain** the phases of a twitch contraction.
- **Describe** how frequency of stimulation affects muscle tension, and how muscle tone is produced.
- **Distinguish** between isotonic and isometric contractions.

A single nerve impulse in a somatic motor neuron elicits a single muscle action potential in all skeletal muscle fibers with which it forms synapses. Action potentials always have the same size in a given neuron or muscle fiber. In contrast, the force of muscle fiber contraction does vary; a muscle fiber is capable of producing a much greater force than the one that results from a single action potential. The total force or tension that a single muscle fiber can produce depends mainly on the rate at which nerve impulses arrive at the neuromuscular junction. The number of impulses per second is the *frequency of stimulation.* Maximum tension is also affected by the amount of stretch before contraction (see **Figure 10.8**) and by nutrient and oxygen availability. The total tension a whole muscle can produce depends on the number of muscle fibers that are contracting in unison.

Motor Units

Even though each skeletal muscle fiber has only a single neuromuscular junction, the axon of a somatic motor neuron branches out and forms neuromuscular junctions with many different muscle fibers. A **motor unit** consists of a somatic motor neuron plus all of the skeletal muscle fibers it stimulates (**Figure 10.12**). A single somatic

FIGURE 10.12 **Motor units.** Two somatic motor neurons (one purple and one green) are shown, each supplying the muscle fibers of its motor unit.

A motor unit consists of a somatic motor neuron plus all of the muscle fibers it stimulates.

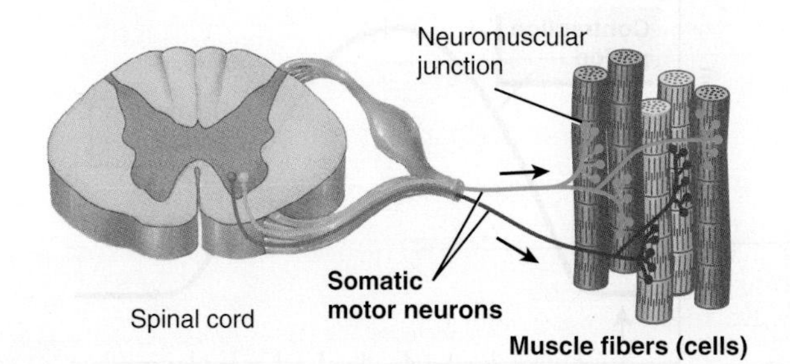

Q What is the effect of the size of a motor unit on its strength of contraction? (Assume that each muscle fiber can generate about the same amount of tension.)

motor neuron makes contact with an average of 150 skeletal muscle fibers, and all of the muscle fibers in one motor unit contract in unison. Typically, the muscle fibers of a motor unit are dispersed throughout a muscle rather than clustered together.

Whole muscles that control precise movements consist of many small motor units. For instance, muscles of the larynx (voice box) that control voice production have as few as two or three muscle fibers per motor unit, and muscles controlling eye movements may have 10 to 20 muscle fibers per motor unit. In contrast, skeletal muscles responsible for large-scale and powerful movements, such as the biceps brachii muscle in the arm and the gastrocnemius muscle in the calf of the leg, have as many as 2000 to 3000 muscle fibers in some motor units. Because all of the muscle fibers of a motor unit contract and relax together, the total strength of a contraction depends, in part, on the size of the motor units and the number that are activated at a given time.

Twitch Contraction

A **twitch contraction** is the brief contraction of all muscle fibers in a motor unit in response to a single action potential in its motor neuron. In the laboratory, a twitch can be produced by direct electrical stimulation of a motor neuron or its muscle fibers. The record of a muscle contraction, called a **myogram** (MĪ-ō-gram), is shown in Figure 10.13. Twitches of skeletal muscle fibers last anywhere from 20 to 200 msec. This is very long compared to the brief 1–2 msec* that a muscle action potential lasts.

Note that a brief delay occurs between application of the stimulus (time zero on the graph) and the beginning of contraction. The delay, which lasts about 2 msec, is termed the **latent period**. During the latent period, the muscle action potential sweeps over the sarcolemma and calcium ions are released from the sarcoplasmic reticulum. The second phase, the **contraction period**, lasts 10–100 msec. During this time, Ca^{2+} binds to troponin, myosin-binding sites on actin are exposed, and cross-bridges form. Peak tension develops in the muscle fiber. During the third phase, the **relaxation period**, also lasting 10–100 msec, Ca^{2+} is actively transported back into the sarcoplasmic reticulum, myosin-binding sites are covered by tropomyosin, myosin heads detach from actin, and tension in the muscle fiber decreases. The actual duration of these periods depends on the type of skeletal muscle fiber. Some fibers, such as the fast-twitch fibers that move the eyes (described shortly), have contraction periods as brief as 10 msec and equally brief relaxation periods. Others, such as the slow-twitch fibers that move the legs, have contraction and relaxation periods of about 100 msec each.

If two stimuli are applied, one immediately after the other, the muscle will respond to the first stimulus but not to the second. When a muscle fiber receives enough stimulation to contract, it temporarily loses its excitability and cannot respond for a time. The period of lost excitability, called the **refractory period** (rē-FRAK-tō-rē), is a characteristic of all muscle and nerve cells. The duration of the refractory period varies with the muscle involved. Skeletal muscle has a short refractory period of about 1 msec; cardiac muscle has a longer refractory period of about 250 msec.

*One millisecond (msec) = 10^{-3} second (0.001 sec).

FIGURE 10.13 Myogram of a twitch contraction. The arrow indicates the time at which the stimulus occurred.

A myogram is a record of a muscle contraction.

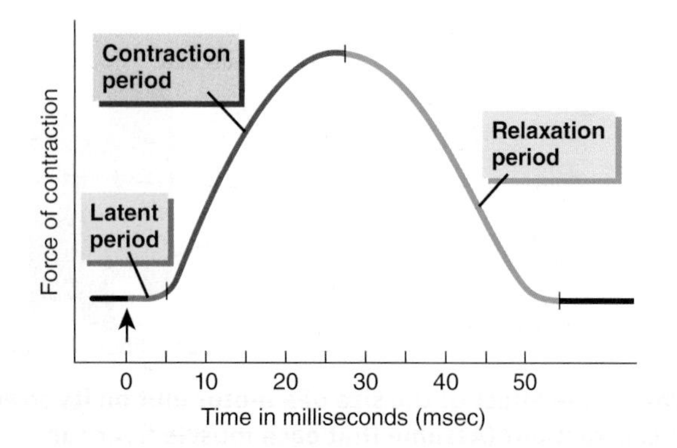

Q What events occur during the latent period?

Frequency of Stimulation

When a second stimulus occurs after the refractory period of the first stimulus is over, but before the skeletal muscle fiber has relaxed, the second contraction will actually be stronger than the first (Figure 10.14b). This phenomenon, in which stimuli arriving at different times cause larger contractions, is called **wave summation**. When a skeletal muscle fiber is stimulated at a rate of 20 to 30 times per second, it can only partially relax between stimuli. The result is a sustained but wavering contraction called **unfused** (*incomplete*) **tetanus** (*tetan-* = rigid, tense; Figure 10.14c). When a skeletal muscle fiber is stimulated at a higher rate of 80 to 100 times per second, it does not relax at all. The result is **fused** (*complete*) **tetanus**, a sustained contraction in which individual twitches cannot be detected (Figure 10.14d).

Wave summation and both kinds of tetanus occur when additional Ca^{2+} is released from the sarcoplasmic reticulum by subsequent stimuli while the levels of Ca^{2+} in the sarcoplasm are still elevated from the first stimulus. Because of the buildup in the Ca^{2+} level, the peak tension generated during fused tetanus is 5 to 10 times larger than the peak tension produced during a single twitch. Even so, smooth, sustained voluntary muscle contractions are achieved mainly by out-of-synchrony unfused tetanus in different motor units.

The stretch of elastic components, such as tendons and connective tissues around muscle fibers, also affects wave summation. During wave summation, elastic components are not given much time to spring back between contractions, and thus remain taut. While in this state, the elastic components do not require very much stretching before the beginning of the next muscular contraction. The combination of the tautness of the elastic components and the partially contracted state of the filaments enables the force of another contraction to be greater than the one before.

Motor Unit Recruitment

The process in which the number of active motor units increases is called **motor unit recruitment**. Typically, the different motor units

FIGURE 10.14 Myograms showing the effects of different frequencies of stimulation. (a) Single twitch. (b) When a second stimulus occurs before the muscle fiber has relaxed, the second contraction is stronger than the first, a phenomenon called wave summation. (The solid black line indicates the force of contraction expected in a single twitch.) (c) Unfused tetanus produces a jagged curve due to partial relaxation of the muscle fiber between stimuli. (d) In fused tetanus, which occurs when there are 80–100 stimuli per second, the myogram line, like the contraction force, is steady and sustained.

Due to wave summation, the tension produced during a sustained contraction is greater than that produced by a single twitch.

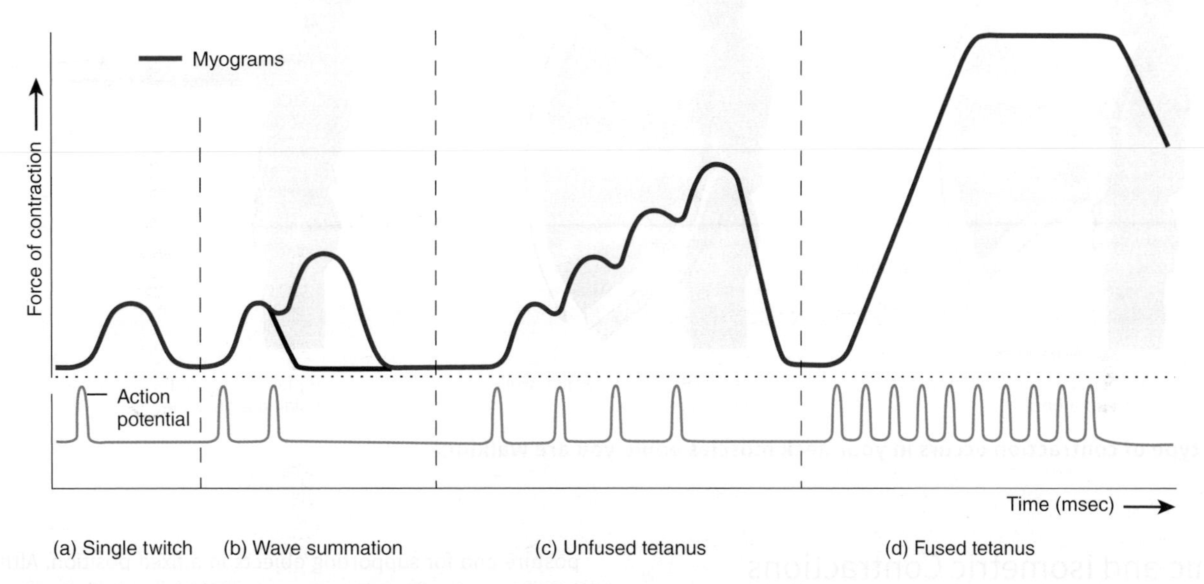

Q Would the peak force of the second contraction in part (b) be larger or smaller if the second stimulus were applied a few milliseconds later?

of an entire muscle are not stimulated to contract in unison. While some motor units are contracting, others are relaxed. This pattern of motor unit activity delays muscle fatigue and allows contraction of a whole muscle to be sustained for long periods. The weakest motor units are recruited first, with progressively stronger motor units added if the task requires more force.

Recruitment is one factor responsible for producing smooth movements rather than a series of jerks. As mentioned, the number of muscle fibers innervated by one motor neuron varies greatly. Precise movements are brought about by small changes in muscle contraction. Therefore, the small muscles that produce precise movements are made up of small motor units. For this reason, when a motor unit is recruited or turned off, only slight changes occur in muscle tension. By contrast, large motor units are active when a large amount of tension is needed and precision is less important.

See Clinical Connection: Anaerobic Training versus Aerobic Training

Muscle Tone

Even at rest, a skeletal muscle exhibits **muscle tone** (*tonos* = tension), a small amount of tautness or tension in the muscle due to weak, involuntary contractions of its motor units. Recall that skeletal muscle contracts only after it is activated by acetylcholine released by nerve impulses in its motor neurons. Hence, muscle tone is established by neurons in the brain and spinal cord that excite the muscle's motor neurons. When the motor neurons serving a skeletal muscle are damaged or cut, the muscle becomes **flaccid** (FLAK-sid or FLAS-sid = flabby), a state of limpness in which muscle tone is lost. To sustain muscle tone, small groups of motor units are alternatively active and inactive in a constantly shifting pattern. Muscle tone keeps skeletal muscles firm, but it does not result in a force strong enough to produce movement. For example, when you are awake, the muscles in the back of the neck are in normal tonic contraction; they keep the head upright and prevent it from slumping forward on the chest. Muscle tone also is important in smooth muscle tissues, such as those found in the gastrointestinal tract, where the walls of the digestive organs maintain a steady pressure on their contents. The tone of smooth muscle fibers in the walls of blood vessels plays a crucial role in maintaining blood pressure.

See Clinical Connection: Hypotonia and Hypertonia

FIGURE 10.15 Comparison between isotonic (concentric and eccentric) and isometric contractions. (a and b) Isotonic contractions of the biceps brachii muscle in the arm. (c) Isometric contraction of shoulder and arm muscles.

In an isotonic contraction, tension remains constant as muscle length decreases or increases; in an isometric contraction, tension increases greatly without a change in muscle length.

(a) Concentric contraction while picking up a book

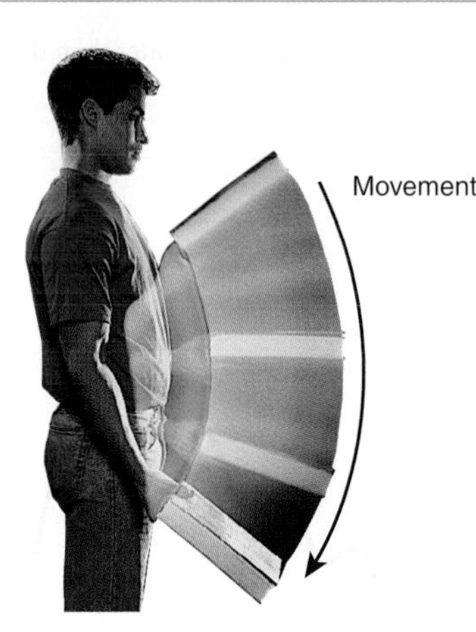

(b) Eccentric contraction while lowering a book

(c) Isometric contraction while holding a book steady

Q What type of contraction occurs in your neck muscles while you are walking?

Isotonic and Isometric Contractions

Muscle contractions may be either isotonic or isometric. In an **isotonic contraction** (ī′-sō-TON-ik; *iso-* = equal; *-tonic* = tension), the *tension* (force of contraction) developed in the muscle remains almost constant while the muscle changes its length. Isotonic contractions are used for body movements and for moving objects. The two types of isotonic contractions are concentric and eccentric. If the tension generated in a **concentric isotonic contraction** (kon-SEN-trik) is great enough to overcome the resistance of the object to be moved, the muscle shortens and pulls on another structure, such as a tendon, to produce movement and to reduce the angle at a joint. Picking up a book from a table involves concentric isotonic contractions of the biceps brachii muscle in the arm (**Figure 10.15a**). By contrast, as you lower the book to place it back on the table, the previously shortened biceps lengthens in a controlled manner while it continues to contract. When the length of a muscle increases during a contraction, the contraction is an **eccentric isotonic contraction** (ek-SEN-trik) (**Figure 10.15b**). During an eccentric contraction, the tension exerted by the myosin cross-bridges resists movement of a load (the book, in this case) and slows the lengthening process. For reasons that are not well understood, repeated eccentric isotonic contractions (for example, walking downhill) produce more muscle damage and more delayed-onset muscle soreness than do concentric isotonic contractions.

In an **isometric contraction** (ī′-sō-MET-rik; *metro* = measure or length), the tension generated is not enough to exceed the resistance of the object to be moved, and the muscle does not change its length. An example would be holding a book steady using an outstretched arm (**Figure 10.15c**). These contractions are important for maintaining posture and for supporting objects in a fixed position. Although isometric contractions do not result in body movement, energy is still expended. The book pulls the arm downward, stretching the shoulder and arm muscles. The isometric contraction of the shoulder and arm muscles counteracts the stretch. Isometric contractions are important because they stabilize some joints as others are moved. Most activities include both isotonic and isometric contractions.

10.6 Types of Skeletal Muscle Fibers

OBJECTIVE

- **Compare** the structure and function of the three types of skeletal muscle fibers.

Skeletal muscle fibers are not all alike in composition and function. For example, muscle fibers vary in their content of myoglobin, the red-colored protein that binds oxygen in muscle fibers. Skeletal muscle fibers that have a high myoglobin content are termed *red muscle fibers* and appear darker (the dark meat in chicken legs and thighs); those that have a low content of myoglobin are called *white muscle fibers* and appear lighter (the white meat in chicken breasts). Red muscle fibers also contain more mitochondria and are supplied by more blood capillaries.

Skeletal muscle fibers also contract and relax at different speeds, and vary in which metabolic reactions they use to generate ATP and in how quickly they fatigue. For example, a fiber is categorized as either slow or fast depending on how rapidly the ATPase in its myosin heads hydrolyzes ATP. Based on all these structural and functional characteristics, skeletal muscle fibers are classified into three main types: (1) slow oxidative fibers, (2) fast oxidative–glycolytic fibers, and (3) fast glycolytic fibers.

Slow Oxidative Fibers

Slow oxidative (SO) fibers appear dark red because they contain large amounts of myoglobin and many blood capillaries. Because they have many large mitochondria, SO fibers generate ATP mainly by aerobic respiration, which is why they are called oxidative fibers. These fibers are said to be "slow" because the ATPase in the myosin heads hydrolyzes ATP relatively slowly, and the contraction cycle proceeds at a slower pace than in "fast" fibers. As a result, SO fibers have a slow speed of contraction. Their twitch contractions last from 100 to 200 msec, and they take longer to reach peak tension. However, slow fibers are very resistant to fatigue and are capable of prolonged, sustained contractions for many hours. These slow-twitch, fatigue-resistant fibers are adapted for maintaining posture and for aerobic, endurance-type activities such as running a marathon.

Fast Oxidative–Glycolytic Fibers

Fast oxidative–glycolytic (FOG) fibers are typically the largest fibers. Like slow oxidative fibers, they contain large amounts of myoglobin and many blood capillaries. Thus, they also have a dark red appearance. FOG fibers can generate considerable ATP by aerobic respiration, which gives them a moderately high resistance to fatigue. Because their intracellular glycogen level is high, they also generate ATP by anaerobic glycolysis. FOG fibers are "fast" because the ATPase in their myosin heads hydrolyzes ATP three to five times faster than the myosin ATPase in SO fibers, which makes their speed of contraction faster. Thus, twitches of FOG fibers reach peak tension more quickly than those of SO fibers but are briefer in duration—less than 100 msec. FOG fibers contribute to activities such as walking and sprinting.

Fast Glycolytic Fibers

Fast glycolytic (FG) fibers have low myoglobin content, relatively few blood capillaries, and few mitochondria, and appear white in color. They contain large amounts of glycogen and generate ATP mainly by glycolysis. Due to their ability to hydrolyze ATP rapidly, FG fibers contract strongly and quickly. These fast-twitch fibers are adapted for intense anaerobic movements of short duration, such as weight lifting or throwing a ball, but they fatigue quickly. Strength training programs that engage a person in activities requiring great strength for short times increase the size, strength, and glycogen content of fast glycolytic fibers. The FG fibers of a weight lifter may be 50% larger than those of a sedentary person or an endurance athlete because of increased synthesis of muscle proteins. The overall result is muscle enlargement due to hypertrophy of the FG fibers.

Distribution and Recruitment of Different Types of Fibers

Most skeletal muscles are a mixture of all three types of skeletal muscle fibers; about half of the fibers in a typical skeletal muscle are SO fibers. However, the proportions vary somewhat, depending on the action of the muscle, the person's training regimen, and genetic factors. For example, the continually active postural muscles of the neck, back, and legs have a high proportion of SO fibers. Muscles of the shoulders and arms, in contrast, are not constantly active but are used briefly now and then to produce large amounts of tension, such as in lifting and throwing. These muscles have a high proportion of FG fibers. Leg muscles, which not only support the body but are also used for walking and running, have large numbers of both SO and FOG fibers.

Within a particular motor unit, all of the skeletal muscle fibers are of the same type. The different motor units in a muscle are recruited in a specific order, depending on need. For example, if weak contractions suffice to perform a task, only SO motor units are activated. If more force is needed, the motor units of FOG fibers are also recruited. Finally, if maximal force is required, motor units of FG fibers are also called into action with the other two types. Activation of various motor units is controlled by the brain and spinal cord.

Table 10.4 summarizes the characteristics of the three types of skeletal muscle fibers.

10.7 Exercise and Skeletal Muscle Tissue

OBJECTIVE

- **Describe** the effects of exercise on different types of skeletal muscle fibers.

The relative ratio of fast glycolytic (FG) and slow oxidative (SO) fibers in each muscle is genetically determined and helps account for individual differences in physical performance. For example, people with a higher proportion of FG fibers (see **Table 10.4**) often excel in activities that require periods of intense activity, such as weight lifting or sprinting. People with higher percentages of SO fibers are better at activities that require endurance, such as long-distance running.

Although the total number of skeletal muscle fibers usually does not increase with exercise, the characteristics of those present can change to some extent. Various types of exercises can induce changes in the fibers in a skeletal muscle. Endurance-type (aerobic) exercises, such as running or swimming, cause a gradual transformation of some FG fibers into fast oxidative–glycolytic (FOG) fibers. The

TABLE 10.4 Characteristics of the Three Types of Skeletal Muscle Fibers

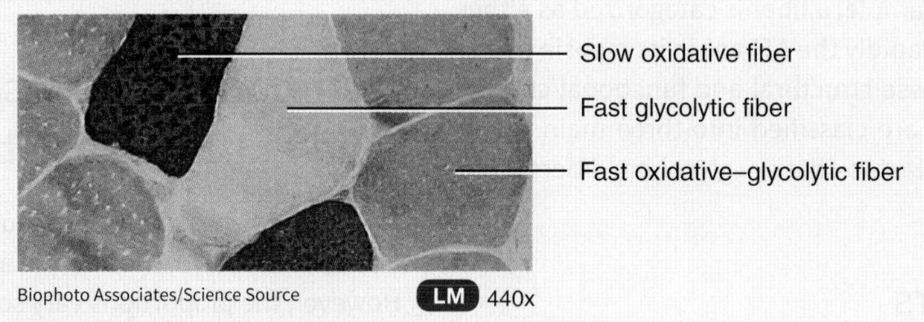

Biophoto Associates/Science Source

Transverse section of three types of skeletal muscle fibers

	SLOW OXIDATIVE (SO) FIBERS	FAST OXIDATIVE–GLYCOLYTIC (FOG) FIBERS	FAST GLYCOLYTIC (FG) FIBERS
STRUCTURAL CHARACTERISTIC			
Myoglobin content	Large amount.	Large amount.	Small amount.
Mitochondria	Many.	Many.	Few.
Capillaries	Many.	Many.	Few.
Color	Red.	Red-pink.	White (pale).
FUNCTIONAL CHARACTERISTIC			
Capacity for generating ATP and method used	High, by aerobic respiration.	Intermediate, by both aerobic respiration and anaerobic glycolysis.	Low, by anaerobic glycolysis.
Rate of ATP hydrolysis by myosin ATPase	Slow.	Fast.	Fast.
Contraction velocity	Slow.	Fast.	Fast.
Fatigue resistance	High.	Intermediate.	Low.
Creatine kinase	Lowest amount.	Intermediate amount.	Highest amount.
Glycogen stores	Low.	Intermediate.	High.
Order of recruitment	First.	Second.	Third.
Location where fibers are abundant	Postural muscles such as those of neck.	Lower limb muscles.	Extraocular muscles.
Primary functions of fibers	Maintaining posture and aerobic endurance activities.	Walking, sprinting.	Rapid, intense movements of short duration.

transformed muscle fibers show slight increases in diameter, number of mitochondria, blood supply, and strength. Endurance exercises also result in cardiovascular and respiratory changes that cause skeletal muscles to receive better supplies of oxygen and nutrients but do not increase muscle mass. By contrast, exercises that require great strength for short periods produce an increase in the size and strength of FG fibers. The increase in size is due to increased synthesis of thick and thin filaments. The overall result is muscle enlargement (hypertrophy), as evidenced by the bulging muscles of body builders.

A certain degree of elasticity is an important attribute of skeletal muscles and their connective tissue attachments. Greater elasticity contributes to a greater degree of flexibility, increasing the range of motion of a joint. When a relaxed muscle is physically stretched, its ability to lengthen is limited by connective tissue structures, such as fasciae. Regular stretching gradually lengthens these structures, but the process occurs very slowly. To see an improvement in flexibility, stretching exercises must be performed regularly—daily, if possible—for many weeks.

Effective Stretching

Stretching cold muscles does not increase flexibility and may cause injury. Tissues stretch best when slow, gentle force is applied at elevated tissue temperatures. An external source of heat, such as hot packs or ultrasound, may be used, but 10 or more minutes of muscular contraction is also a good way to raise muscle temperature. Exercise heats muscle more deeply and thoroughly than external measures. That's where the term "warm-up" comes from. Many people stretch before they engage in exercise, but it's important to warm up (for example, walking, jogging, easy swimming, or easy aerobics) *before* stretching to avoid injury.

Strength Training

Strength training refers to the process of exercising with progressively heavier resistance for the purpose of strengthening the musculoskeletal system. This activity results not only in stronger muscles, but in many other health benefits as well. Strength training also helps to increase bone strength by increasing the deposition of bone minerals in young adults and helping to prevent, or at least slow, their loss in later life. By increasing muscle mass, strength training raises resting metabolic rate, the amount of energy expended at rest, so a person can eat more food without gaining weight. Strength training helps to prevent back injury and other injuries from participation in sports and other physical activities. Psychological benefits include reductions in feelings of stress and fatigue. As repeated training builds exercise tolerance, it takes increasingly longer before lactic acid is produced in the muscle, resulting in a reduced probability of muscle spasms.

See Clinical Connection: Anabolic Steroids

10.8 Cardiac Muscle Tissue

OBJECTIVE

- **Describe** the main structural and functional characteristics of cardiac muscle tissue.

The principal tissue in the heart wall is **cardiac muscle tissue** (described in more detail in Chapter 20 and illustrated in Figure 20.9). Between the layers of cardiac muscle fibers (the contractile cells of the heart) are sheets of connective tissue that contain blood vessels, nerves, and the conduction system of the heart. Cardiac muscle fibers have the same arrangement of actin and myosin and the same bands, zones, and Z discs as skeletal muscle fibers. However, *intercalated discs* (in-TER-ka-lāt-ed; *intercal-* = to insert between) are unique to cardiac muscle fibers. These microscopic structures are irregular transverse thickenings of the sarcolemma that connect the ends of cardiac muscle fibers to one another. The discs contain *desmosomes*, which hold the fibers together, and *gap junctions*, which allow muscle action potentials to spread from one cardiac muscle fiber to another (see Figure 4.2e). Cardiac muscle tissue has an endomysium and perimysium, but lacks an epimysium.

In response to a single action potential, cardiac muscle tissue remains contracted 10 to 15 times longer than skeletal muscle tissue (see Figure 20.11). The long contraction is due to prolonged delivery of Ca^{2+} into the sarcoplasm. In cardiac muscle fibers, Ca^{2+} enters the sarcoplasm both from the sarcoplasmic reticulum (as in skeletal muscle fibers) and from the interstitial fluid that bathes the fibers. Because the channels that allow inflow of Ca^{2+} from interstitial fluid stay open for a relatively long time, a cardiac muscle contraction lasts much longer than a skeletal muscle twitch.

We have seen that skeletal muscle tissue contracts only when stimulated by acetylcholine released by a nerve impulse in a motor neuron. In contrast, cardiac muscle tissue contracts when stimulated by its own autorhythmic muscle fibers. Under normal resting conditions, cardiac muscle tissue contracts and relaxes about 75 times a minute. This continuous, rhythmic activity is a major physiological difference between cardiac and skeletal muscle tissue. The mitochondria in cardiac muscle fibers are larger and more numerous than in skeletal muscle fibers. This structural feature correctly suggests that cardiac muscle depends largely on aerobic respiration to generate ATP, and thus requires a constant supply of oxygen. Cardiac muscle fibers can also use lactic acid produced by skeletal muscle fibers to make ATP, a benefit during exercise. Like skeletal muscle, cardiac muscle fibers can undergo hypertrophy in response to an increased workload. This is called a *physiological enlarged heart* and it is why many athletes have enlarged hearts. By contrast, a *pathological enlarged heart* is related to significant heart disease.

10.9 Smooth Muscle Tissue

OBJECTIVE

- **Describe** the main structural and functional characteristics of smooth muscle tissue.

Like cardiac muscle tissue, **smooth muscle tissue** is usually activated involuntarily. Of the two types of smooth muscle tissue, the more common type is **visceral (*single-unit*) smooth muscle tissue** (Figure 10.16a). It is found in the skin and in tubular arrangements that form part of the walls of small arteries and veins and of hollow organs such as the stomach, intestines, uterus, and urinary bladder. Like cardiac muscle, visceral smooth muscle is autorhythmic. The fibers connect to one another by gap junctions, forming a network through which muscle action potentials can spread. When a neurotransmitter, hormone, or autorhythmic signal stimulates one fiber, the muscle action potential is transmitted to neighboring fibers, which then contract in unison, as a single unit.

The second type of smooth muscle tissue, **multi-unit smooth muscle tissue** (Figure 10.16b), consists of individual fibers, each with its own motor neuron terminals and with few gap junctions between neighboring fibers. Stimulation of one visceral muscle fiber causes contraction of many adjacent fibers, but stimulation of one multi-unit fiber causes contraction of that fiber only. Multi-unit smooth muscle tissue is found in the walls of large arteries, in airways to the lungs, in the arrector pili muscles that attach to hair follicles, in the muscles of the iris that adjust pupil diameter, and in the ciliary body that adjusts focus of the lens in the eye.

Microscopic Anatomy of Smooth Muscle

A single relaxed smooth muscle fiber is 30–200 μm long. It is thickest in the middle (3–8 μm) and tapers at each end (Figure 10.16c). Within

FIGURE 10.16 Smooth muscle tissue. (a) One autonomic motor neuron synapses with several visceral smooth muscle fibers, and action potentials spread to neighboring fibers through gap junctions. (b) Three autonomic motor neurons synapse with individual multi-unit smooth muscle fibers; stimulation of one multi-unit fiber causes contraction of that fiber only. (c) Relaxed and contracted smooth muscle fiber. A photomicrograph of smooth muscle tissue is shown in Table 4.9.

Visceral smooth muscle fibers connect to one another by gap junctions and contract as a single unit. Multi-unit smooth muscle fibers lack gap junctions and contract independently.

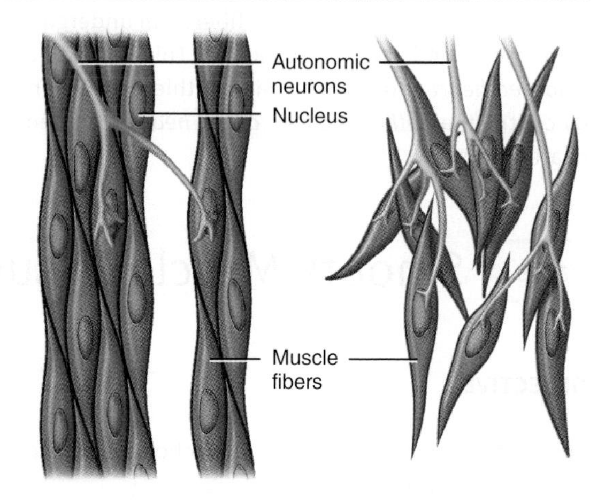

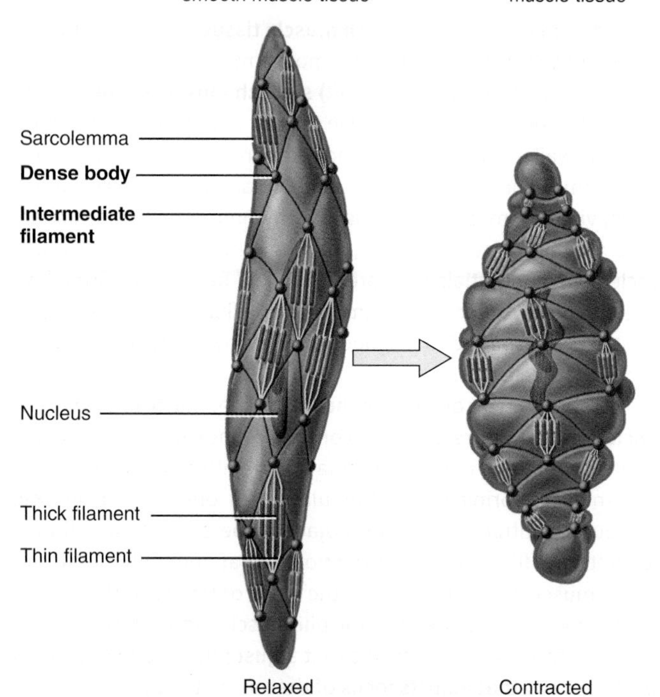

(c) Microscopic anatomy of a relaxed and contracted smooth muscle fiber

Q Which type of smooth muscle is more like cardiac muscle than skeletal muscle, with respect to both its structure and function?

each fiber is a single, oval, centrally located nucleus. The sarcoplasm of smooth muscle fibers contains both *thick filaments* and *thin filaments*, in ratios between 1:10 and 1:15, but they are not arranged in orderly sarcomeres as in striated muscle. Smooth muscle fibers also contain **intermediate filaments**. Because the various filaments have no regular pattern of overlap, smooth muscle fibers do not exhibit striations (see Table 4.9), causing a smooth appearance. Smooth muscle fibers also lack transverse tubules and have only a small amount of sarcoplasmic reticulum for storage of Ca^{2+}. Although there are no transverse tubules in smooth muscle tissue, there are small pouchlike invaginations of the plasma membrane called **caveolae** (kav′-ē-Ō-lē; *cavus* = space) that contain extracellular Ca^{2+} that can be used for muscular contraction.

In smooth muscle fibers, the thin filaments attach to structures called **dense bodies**, which are functionally similar to Z discs in striated muscle fibers. Some dense bodies are dispersed throughout the sarcoplasm; others are attached to the sarcolemma. Bundles of intermediate filaments also attach to dense bodies and stretch from one dense body to another (Figure 10.16c). During contraction, the sliding filament mechanism involving thick and thin filaments generates tension that is transmitted to intermediate filaments. These in turn pull on the dense bodies attached to the sarcolemma, causing a lengthwise shortening of the muscle fiber. As a smooth muscle fiber contracts, it rotates as a corkscrew turns. The fiber twists in a helix as it contracts, and rotates in the opposite direction as it relaxes.

Physiology of Smooth Muscle

Although the principles of contraction are similar, smooth muscle tissue exhibits some important physiological differences from cardiac and skeletal muscle tissue. Contraction in a smooth muscle fiber starts more slowly and lasts much longer than skeletal muscle fiber contraction. Another difference is that smooth muscle can both shorten and stretch to a greater extent than the other muscle types.

An increase in the concentration of Ca^{2+} in the sarcoplasm of a smooth muscle fiber initiates contraction, just as in striated muscle. Sarcoplasmic reticulum (the reservoir for Ca^{2+} in striated muscle) is found in small amounts in smooth muscle. Calcium ions flow into smooth muscle sarcoplasm from both the interstitial fluid and sarcoplasmic reticulum. Because there are no transverse tubules in smooth muscle fibers (there are caveolae instead), it takes longer for Ca^{2+} to reach the filaments in the center of the fiber and trigger the contractile process. This accounts, in part, for the slow onset of contraction of smooth muscle.

Several mechanisms regulate contraction and relaxation of smooth muscle cells. In one such mechanism, a regulatory protein called **calmodulin** (cal-MOD-ū-lin) binds to Ca^{2+} in the sarcoplasm. (Recall that troponin takes this role in striated muscle fibers.) After binding to Ca^{2+}, calmodulin activates an enzyme called *myosin light chain kinase*. This enzyme uses ATP to add a phosphate group to a portion of the myosin head. Once the phosphate group is attached, the myosin head can bind to actin, and contraction can occur. Because myosin light chain kinase works rather slowly, it contributes to the slowness of smooth muscle contraction.

Not only do calcium ions enter smooth muscle fibers slowly, they also move slowly out of the muscle fiber, which delays relaxation. The prolonged presence of Ca^{2+} in the cytosol provides for **smooth muscle tone**, a state of continued partial contraction. Smooth muscle tissue can thus sustain long-term tone, which is important in the gastrointestinal tract, where the walls maintain a steady pressure on the contents of the tract, and in the walls of blood vessels called arterioles, which maintain a steady pressure on blood.

Most smooth muscle fibers contract or relax in response to action potentials from the autonomic nervous system. In addition, many smooth muscle fibers contract or relax in response to stretching, hormones, or local factors such as changes in pH, oxygen and carbon dioxide levels, temperature, and ion concentrations. For example, the hormone epinephrine, released by the adrenal medulla, causes relaxation of smooth muscle in the airways and in some blood vessel walls (those that have so-called β_2 receptors; see **Table 15.2**).

Unlike striated muscle fibers, smooth muscle fibers can stretch considerably and still maintain their contractile function. When smooth muscle fibers are stretched, they initially contract, developing increased tension. Within a minute or so, the tension decreases. This phenomenon, which is called the **stress–relaxation response**, allows smooth muscle to undergo great changes in length while retaining the ability to contract effectively. Thus, even though smooth muscle in the walls of blood vessels and hollow organs such as the stomach, intestines, and urinary bladder can stretch, the pressure on the contents within them changes very little. After the organ empties, the smooth muscle in the wall rebounds, and the wall retains its firmness.

10.10 Regeneration of Muscular Tissue

OBJECTIVE

- **Explain** how muscle fibers regenerate.

Because mature skeletal muscle fibers have lost the ability to undergo cell division, growth of skeletal muscle after birth is due mainly to **hypertrophy** (hī-PER-trō-fē), the enlargement of existing cells, rather than to **hyperplasia** (hī-per-PLĀ-zē-a), an increase in the number of fibers. Satellite cells divide slowly and fuse with existing fibers to assist both in muscle growth and in repair of damaged fibers. Thus, skeletal muscle tissue can regenerate only to a limited extent.

Until recently it was believed that damaged cardiac muscle fibers could not be replaced and that healing took place exclusively by fibrosis, the formation of scar tissue. New research described in Chapter 20 indicates that, under certain circumstances, cardiac muscle tissue can regenerate. In addition, cardiac muscle fibers can undergo hypertrophy in response to increased workload. Hence, many athletes have enlarged hearts.

Smooth muscle tissue, like skeletal and cardiac muscle tissue, can undergo hypertrophy. In addition, certain smooth muscle fibers, such as those in the uterus, retain their capacity for division and thus can grow by hyperplasia. Also, new smooth muscle fibers can arise from cells called *pericytes*, stem cells found in association with blood capillaries and small veins. Smooth muscle fibers can also proliferate in certain pathological conditions, such as occur in the development of atherosclerosis (see Disorders: Homeostatic Imbalances in Chapter 20 in Study Guide). Compared with the other two types of muscle tissue, smooth muscle tissue has considerably greater powers of regeneration. Such powers are still limited when compared with other tissues, such as epithelium.

Table 10.5 summarizes the major characteristics of the three types of muscular tissue.

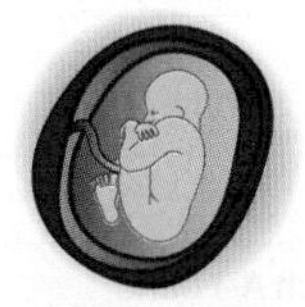

10.11 Development of Muscle

OBJECTIVE

- **Describe** the development of muscles.

Except for muscles such as those of the iris of the eyes and the arrector pili muscles attached to hairs, all muscles of the body are derived from **mesoderm**. As the mesoderm develops, part of it becomes arranged in dense columns on either side of the developing nervous system. These columns of mesoderm undergo segmentation into a series of cube-shaped structures called **somites** (SŌ-mīts) (**Figure 10.17a**). The first pair of somites appears on the 20th day of embryonic development. Eventually, 42 to 44 pairs of somites are formed by the end of the fifth week. The number of somites can be correlated to the approximate age of the embryo.

The cells of a somite differentiate into three regions: (1) a **myotome** (MĪ-ō-tōm), which, as the name suggests, forms the skeletal muscles of the head, neck, and limbs; (2) a **dermatome** (DER-ma-tōm), which forms the connective tissues, including the dermis of the skin; and (3) a **sclerotome** (SKLE-rō-tōm), which gives rise to the vertebrae (**Figure 10.17b**).

Cardiac muscle develops from mesodermal cells that migrate to and envelop the developing heart while it is still in the form of endocardial heart tubes (see **Figure 20.19**).

Smooth muscle develops from **mesodermal cells** that migrate to and envelop the developing gastrointestinal tract and viscera.

TABLE 10.5 **Summary of the Major Features of the Three Types of Muscular Tissue**

CHARACTERISTIC	SKELETAL MUSCLE	CARDIAC MUSCLE	SMOOTH MUSCLE
Microscopic appearance and features	Long cylindrical fiber with many peripherally located nuclei; unbranched; striated.	Branched cylindrical fiber with one centrally located nucleus; intercalated discs join neighboring fibers; striated.	Fiber thickest in middle, tapered at each end, and with one centrally positioned nucleus; not striated.
Location	Most commonly attached by tendons to bones.	Heart.	Walls of hollow viscera, airways, blood vessels, iris and ciliary body of eye, arrector pili muscles of hair follicles.
Fiber diameter	Very large (10–100 μm).	Large (10–20 μm).	Small (3–8 μm).
Connective tissue components	Endomysium, perimysium, and epimysium.	Endomysium and perimysium.	Endomysium.
Fiber length	Very large (100 μm–30 cm = 12 in.).	Large (50–100 μm).	Intermediate (30–200 μm).
Contractile proteins organized into sarcomeres	Yes.	Yes.	No.
Sarcoplasmic reticulum	Abundant.	Some.	Very little.
Transverse tubules present	Yes, aligned with each A–I band junction.	Yes, aligned with each Z disc.	No.
Junctions between fibers	None.	Intercalated discs contain gap junctions and desmosomes.	Gap junctions in visceral smooth muscle; none in multi-unit smooth muscle.
Autorhythmicity	No.	Yes.	Yes, in visceral smooth muscle.
Source of Ca^{2+} for contraction	Sarcoplasmic reticulum.	Sarcoplasmic reticulum and interstitial fluid.	Sarcoplasmic reticulum and interstitial fluid.
Regulator proteins for contraction	Troponin and tropomyosin.	Troponin and tropomyosin.	Calmodulin and myosin light chain kinase.
Speed of contraction	Fast.	Moderate.	Slow.
Nervous control	Voluntary (somatic nervous system).	Involuntary (autonomic nervous system).	Involuntary (autonomic nervous system).
Contraction regulation	Acetylcholine released by somatic motor neurons.	Acetylcholine and norepinephrine released by autonomic motor neurons; several hormones.	Acetylcholine and norepinephrine released by autonomic motor neurons; several hormones; local chemical changes; stretching.
Capacity for regeneration	Limited, via satellite cells.	Limited, under certain conditions.	Considerable (compared with other muscle tissues, but limited compared with epithelium), via pericytes.

FIGURE 10.17 Location and structure of somites, key structures in the development of the muscular system.

Most muscles are derived from mesoderm.

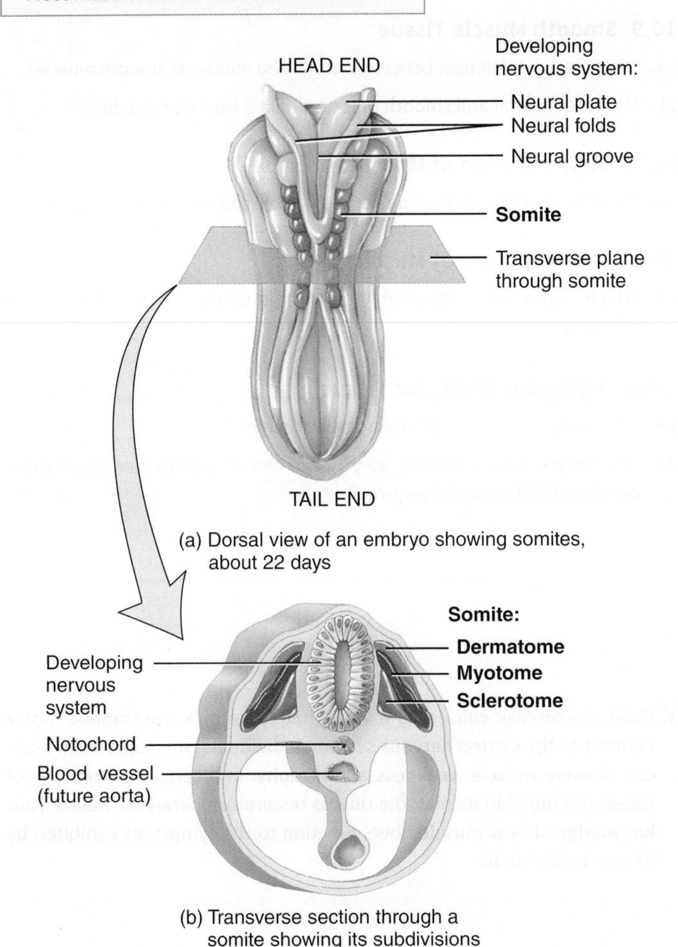

Q Which part of a somite differentiates into skeletal muscle?

10.12 Aging and Muscular Tissue

OBJECTIVE

- **Explain** the effects of aging on skeletal muscle.

Between the ages of 30 and 50, humans undergo a slow, progressive loss of skeletal muscle mass that is replaced largely by fibrous connective tissue and adipose tissue. An estimated 10% of muscle mass is lost during these years. In part, this decline may be due to decreased levels of physical activity. Accompanying the loss of muscle mass is a decrease in maximal strength, a slowing of muscle reflexes, and a loss of flexibility. With aging, the relative number of slow oxidative (SO) fibers appears to increase. This could be due to either the atrophy of the other fiber types or their conversion into slow oxidative fibers. Another 40% of muscle is typically lost between the ages of 50 and 80. Loss of muscle strength is usually not perceived by persons until they reach the age of 60 to 65. At that point it is most common for muscles of the lower limbs to weaken before those of the upper limbs. Thus the independence of the elderly may be affected when it becomes difficult to climb stairs or get up from a seated position.

Assuming that there is not a chronic medical condition for which exercise is contraindicated, exercise has been shown to be effective at any age. Aerobic activities and strength training programs are effective in older people and can slow or even reverse the age-associated decline in muscular performance.

Review Questions

10.1 Overview of Muscular Tissue

1. What features distinguish the three types of muscular tissue?
2. List the general functions of muscular tissue.
3. Describe the four properties of muscular tissue.

10.2 Structure of Skeletal Muscle Tissue

4. What types of fascia cover skeletal muscles?
5. Why is a rich blood supply important for muscle contraction?
6. How are the structures of thin and thick filaments different?

10.3 Contraction and Relaxation of Skeletal Muscle Fibers

7. What roles do contractile, regulatory, and structural proteins play in muscle contraction and relaxation?
8. How do calcium ions and ATP contribute to muscle contraction and relaxation?
9. How does sarcomere length influence the maximum tension that is possible during muscle contraction?
10. How is the motor end plate different from other parts of the sarcolemma?

10.4 Muscle Metabolism

11. Which ATP-producing reactions are aerobic and which are anaerobic?
12. Which sources provide ATP during a marathon race?
13. What factors contribute to muscle fatigue?
14. Why is the term recovery oxygen uptake more accurate than oxygen debt?

10.5 Control of Muscle Tension

15. How are the sizes of motor units related to the degree of muscular control they allow?

16. What is motor unit recruitment?

17. Why is muscle tone important?

18. Define each of the following terms: concentric isotonic contraction, eccentric isotonic contraction, and isometric contraction.

19. Demonstrate an isotonic contraction. How does it feel? What do you think causes the physical discomfort you are experiencing?

10.6 Types of Skeletal

20. Why are some skeletal muscle fibers classified as "fast" and others are said to be "slow"?

21. In what order are the various types of skeletal muscle fibers recruited when you sprint to make it to the bus stop?

10.7 Exercise and Skeletal Muscle Tissue

22. On a cellular level, what causes muscle hypertrophy?

10.8 Cardiac Muscle Tissue

23. What are the similarities among and differences between skeletal and cardiac muscle?

10.9 Smooth Muscle Tissue

24. What are the differences between visceral and multi-unit smooth muscle?

25. How are skeletal and smooth muscle similar? How do they differ?

10.10 Regeneration of Muscular Tissue

26. Which type of muscular tissue has the highest capacity for regeneration?

10.11 Development of Muscle

27. Which structures develop from myotomes, dermatomes, and sclerotomes?

10.12 Aging and Muscular Tissue

28. Why does muscle strength decrease with aging?

29. Why do you think a healthy 30-year-old can lift a 25-lb load much more comfortably than an 80-year-old?

Critical Thinking Questions

1. Weightlifter Jamal has been practicing many hours a day, and his muscles have gotten noticeably bigger. He tells you that his muscle cells are "multiplying like crazy and making him get stronger and stronger." Do you believe his explanation? Why or why not?
2. Chicken breasts are composed of "white meat," whereas chicken legs are composed of "dark meat." The breasts and legs of migrating ducks are dark meat. The breasts of both chickens and ducks are used in flying. How can you explain the differences in the color of the meat (muscles)? How are they adapted for their particular functions?
3. Polio is a disease caused by a virus that can attack the somatic motor neurons in the central nervous system. Individuals who suffer from polio can develop muscle weakness and atrophy. In a certain percentage of cases, the individuals may die due to respiratory paralysis. Relate your knowledge of how muscle fibers function to the symptoms exhibited by infected individuals.

The Muscular System

CHAPTER 11

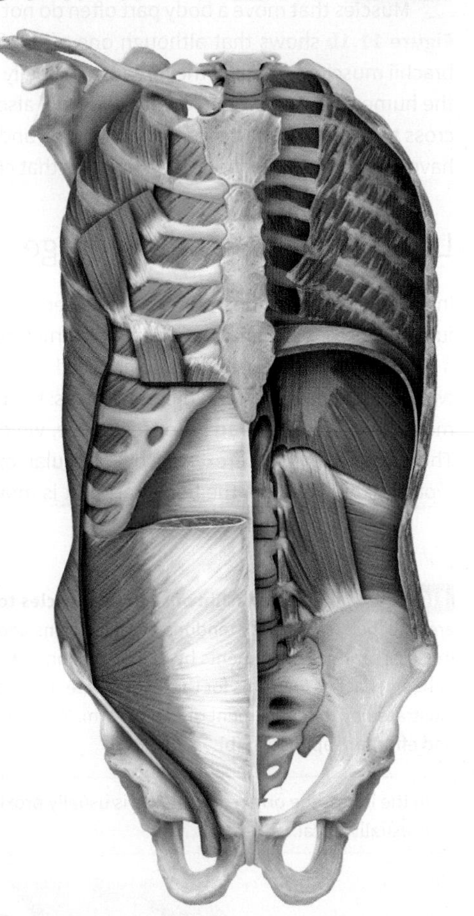

The Muscular System and Homeostasis

The muscular system and muscular tissue of your body contribute to homeostasis by stabilizing body position, producing movements, regulating organ volume, moving substances within the body, and producing heat.

Almost all of the 700 individual muscles that make up the muscular system, such as the biceps brachii muscle, include both skeletal muscle tissue and connective tissue. The function of most muscles is to produce movements of body parts. A few muscles function mainly to stabilize bones so that other skeletal muscles can execute a movement more effectively. This chapter presents many of the major skeletal muscles in the body, most of which are found on both the right and left sides. We will identify the attachment sites and innervation (the nerve or nerves that stimulate contraction) of each muscle described. Developing a working knowledge of these key aspects of skeletal muscle anatomy will enable you to understand how normal movements occur. This knowledge is especially crucial for professionals, such as those in the allied health and physical rehabilitation fields, who work with patients whose normal patterns of movement and physical mobility have been disrupted by physical trauma, surgery, or muscular paralysis.

Q Did you ever wonder how carpal tunnel syndrome occurs?

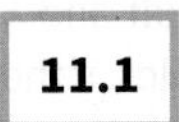

11.1 How Skeletal Muscles Produce Movements

OBJECTIVES

- **Describe** the relationship between bones and skeletal muscles in producing body movements.
- **Define** lever and fulcrum, and **compare** the three types of levers based on location of the fulcrum, effort, and load.
- **Identify** the types of fascicle arrangements in a skeletal muscle, and **relate** the arrangements to strength of contraction and range of motion.
- **Explain** how the prime mover, antagonist, synergist, and fixator in a muscle group work together to produce movement.

Muscle Attachment Sites: Origin and Insertion

Together, the voluntarily controlled muscles of your body compose the **muscular system**. Those skeletal muscles that produce movements do so by exerting force on tendons, which in turn pull on bones or other structures (such as skin). Most muscles cross at least one joint and are usually attached to articulating bones that form the joint (**Figure 11.1a**).

When a skeletal muscle contracts, it moves one of the articulating bones. The two articulating bones usually do not move equally in response to contraction. One bone remains stationary or near its original position, either because other muscles stabilize that bone by contracting and pulling it in the opposite direction or because its structure makes it less movable. Ordinarily, the attachment of a muscle's tendon to the stationary bone is called the **origin** (OR-i-jin); the attachment of the muscle's other tendon to the movable bone is called the **insertion** (in-SER-shun). A good analogy is a spring on a door. In this example, the part of the spring attached to the frame is the origin; the part attached to the door represents the insertion. A useful rule of thumb is that the origin is usually proximal and the insertion distal; the insertion is usually pulled toward the origin. The fleshy portion of the muscle between the tendons is called the **belly** (*body*), the coiled middle portion of the spring in our example. The **actions** of a muscle are the main movements that occur when the muscle contracts. In our spring example, this would be the closing of the door. Certain muscles are also capable of **reverse muscle action (RMA)**. This means that during specific movements of the body the actions are reversed; therefore, the positions of the origin and insertion of a specific muscle are switched.

Muscles that move a body part often do not cover the moving part. Figure 11.1b shows that although one of the functions of the biceps brachii muscle is to move the forearm, the belly of the muscle lies over the humerus, not over the forearm. You will also see that muscles that cross two joints, such as the rectus femoris and sartorius of the thigh, have more complex actions than muscles that cross only one joint.

Lever Systems and Leverage

In producing movement, bones act as levers, and joints function as the fulcrums of these levers. A **lever** is a rigid structure that can move around a fixed point called a **fulcrum**, symbolized by F. A lever is acted on at two different points by two different forces: the **effort** (E), which causes movement, and the **load** L or *resistance*, which opposes movement. The effort is the force exerted by muscular contraction; the load is typically the weight of the body part that is moved or some resistance that the moving body part is trying to overcome (such as the weight of a book you might be picking up). Motion occurs when the effort applied to the bone at the insertion exceeds the load. Consider the biceps brachii flexing the forearm at the elbow as an object is lifted (Figure 11.1b). When the forearm is raised, the elbow is the fulcrum. The weight of the forearm plus the weight of the object in the hand is the load. The force of contraction of the biceps brachii pulling the forearm up is the effort.

The relative distance between the fulcrum and load and the point at which the effort is applied determine whether a given lever operates at a mechanical advantage or a mechanical disadvantage. For example, if the load is closer to the fulcrum and the effort farther from the fulcrum, then only a relatively small effort is required to move a large load over a small distance. This is called a **mechanical advantage**. If, instead, the load is farther from the fulcrum and the effort is applied closer to the fulcrum, then a relatively large effort is required to move a small load (but at greater speed). This is called a **mechanical disadvantage**. Compare chewing something hard (the load) with your front teeth to chewing it with the teeth in the back of your mouth. It is much easier to crush the hard food item with the back teeth because they are closer to the fulcrum (the jaw or temporomandibular joint) than are the front teeth. Here is one more example you can try. Straighten out a paper clip. Now get a pair of scissors and try to cut the paper clip with the tip of the scissors (mechanical disadvantage) versus near the pivot point of the scissors (mechanical advantage).

Levers are categorized into three types according to the positions of the fulcrum, the effort, and the load:

1. The fulcrum is between the effort and the load in **first-class levers** (Figure 11.2a). (Think EFL.) Scissors and seesaws are examples of first-class levers. A first-class lever can produce either a mechanical advantage or a mechanical disadvantage depending on whether the effort or the load is closer to the fulcrum. (Think of an adult and a child on a seesaw.) As we have seen in the preceding examples, if the effort (child) is farther from the fulcrum than the load (adult), a heavy load can be moved, but not very far or fast. If the effort is closer to the

FIGURE 11.1 Relationship of skeletal muscles to bones. Muscles are attached to bones by tendons at their origins and insertions. Skeletal muscles produce movements by pulling on bones. Bones serve as levers, and joints act as fulcrums for the levers. Here the lever–fulcrum principle is illustrated by the movement of the forearm. Note where the load (resistance) and effort are applied in (b).

In the limbs, the origin of a muscle is usually proximal and the insertion is usually distal.

Shoulder joint
Scapula
Origins from scapula and humerus
Origins from scapula
Tendons
Belly of triceps brachii muscle
Belly of biceps brachii muscle
Humerus
Tendon
Insertion on ulna
Tendon
Elbow joint
Insertion on radius
Ulna
Radius

(a) Origin and insertion of a skeletal muscle

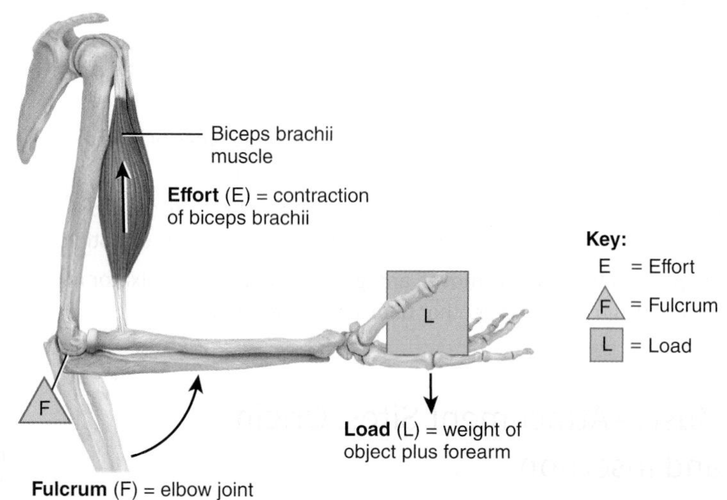

(b) Movement of the forearm lifting a weight

Q Where is the belly of the muscle that extends the forearm located?

FIGURE 11.2 **Lever structure and types of levers.**

Levers are divided into three types based on the placement of the fulcrum, effort, and load (resistance).

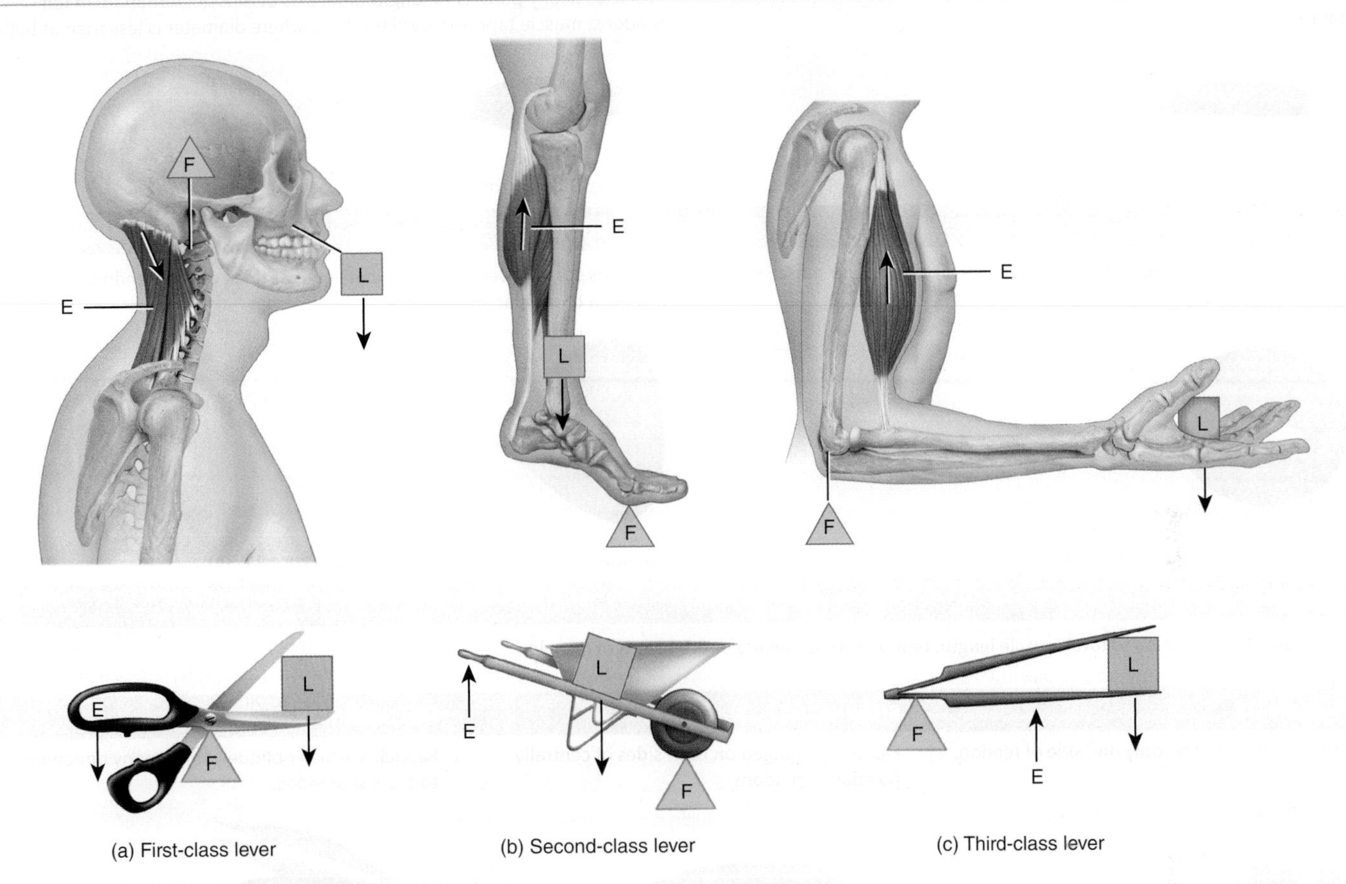

(a) First-class lever

(b) Second-class lever

(c) Third-class lever

Q Which type of lever produces the most force?

fulcrum than the load, only a lighter load can be moved, but it moves far and fast. There are few first-class levers in the body. One example is the lever formed by the head resting on the vertebral column (**Figure 11.2a**). When the head is raised, the contraction of the posterior neck muscles provides the effort (E), the joint between the atlas and the occipital bone (atlanto-occipital joint) forms the fulcrum F, and the weight of the anterior portion of the skull is the load L.

2. The load is between the fulcrum and the effort in **second-class levers** (**Figure 11.2b**). (Think E*L*F.) Second-class levers operate like a wheelbarrow. They always produce a mechanical advantage because the load is always closer to the fulcrum than the effort. This arrangement sacrifices speed and range of motion for force; this type of lever produces the most force. This class of lever is uncommon in the human body. An example is standing up on your toes. The fulcrum F is the ball of the foot. The load L is the weight of the body. The effort (E) is the contraction of the muscles of the calf, which raise the heel off the ground.
3. The effort is between the fulcrum and the load in **third-class levers** (**Figure 11.2c**). (Think F*E*L.) These levers operate like a pair of forceps and are the most common levers in the body. Third-class levers always produce a mechanical disadvantage because the effort is always closer to the fulcrum than the load. In the body, this arrangement favors speed and range of motion over force. The elbow joint, the biceps brachii muscle, and the bones of the arm and forearm are one example of a third-class lever (**Figure 11.2c**). As we have seen, in flexing the forearm at the elbow, the elbow joint is the fulcrum F, the contraction of the biceps brachii muscle provides the effort (E) and the weight of the hand and forearm is the load L.

Effects of Fascicle Arrangement

Recall from Chapter 10 that the skeletal muscle fibers (cells) within a muscle are arranged in bundles known as **fascicles** (FAS-i-kuls). Within a fascicle, all muscle fibers are parallel to one another. The fascicles, however, may form one of five patterns with respect to the tendons: parallel, fusiform (spindle-shaped, narrow toward the ends and wide in the middle), circular, triangular, or pennate (shaped like a feather) (**Table 11.1**).

Fascicular arrangement affects a muscle's power and range of motion. As a muscle fiber contracts, it shortens to about 70% of its resting length. The longer the fibers in a muscle, the greater the range

TABLE 11.1 Arrangement of Fascicles

PARALLEL	FUSIFORM
Fascicles parallel to longitudinal axis of muscle; terminate at either end in flat tendons.	Fascicles nearly parallel to longitudinal axis of muscle; terminate in flat tendons; muscle tapers toward tendons, where diameter is less than at belly.

Example: Sternohyoid muscle (see **Figure 11.8a**)	*Example:* Digastric muscle (see **Figure 11.8a**)
CIRCULAR	**TRIANGULAR**
Fascicles in concentric circular arrangements form sphincter muscles that enclose an orifice (opening).	Fascicles spread over broad area converge at thick central tendon; gives muscle a triangular appearance.
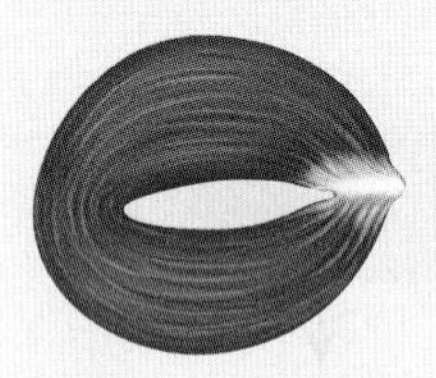	
Example: Orbicularis oculi muscle (see **Figure 11.4a**)	*Example:* Pectoralis major muscle (see **Figure 11.3a**)

PENNATE		
Short fascicles in relation to total muscle length; tendon extends nearly entire length of muscle.		
UNIPENNATE	**BIPENNATE**	**MULTIPENNATE**
Fascicles arranged on only one side of tendon.	Fascicles arranged on both sides of centrally positioned tendons.	Fascicles attach obliquely from many directions to several tendons.
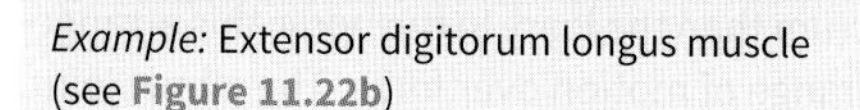	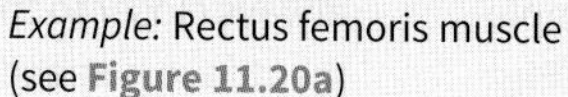 	
Example: Extensor digitorum longus muscle (see **Figure 11.22b**)	*Example:* Rectus femoris muscle (see **Figure 11.20a**)	*Example:* Deltoid muscle (see **Figure 11.10a**)

of motion it can produce. However, the power of a muscle depends not on length but on its total cross-sectional area, because a short fiber can contract as forcefully as a long one. So the more fibers per unit of cross-sectional area a muscle has, the more power it can produce. Fascicular arrangement often represents a compromise between power and range of motion. Pennate muscles, for instance, have a large number of short-fibered fascicles distributed over their tendons, giving them greater power but a smaller range of motion. In contrast, parallel muscles have comparatively fewer fascicles, but they have long fibers that extend the length of the muscle, giving them a greater range of motion but less power.

See Clinical Connection: Intramuscular Injections

Coordination among Muscles

Movements often are the result of several skeletal muscles acting as a group. Most skeletal muscles are arranged in opposing (antagonistic) pairs at joints—that is, flexors–extensors, abductors–adductors, and so on. Within opposing pairs, one muscle, called the **prime mover** or *agonist* (= leader), contracts to cause an action while the other muscle, the **antagonist** (*anti-* = against), stretches and yields to the effects of the prime mover. In the process of flexing the forearm at the elbow, for instance, the biceps brachii is the prime mover, and the triceps brachii is the antagonist (see **Figure 11.1a**). The antagonist and prime mover are usually located on opposite sides of the bone or joint, as is the case in this example.

With an opposing pair of muscles, the roles of the prime mover and antagonist can switch for different movements. For example, while extending the forearm at the elbow against resistance (i.e., lowering the load shown in **Figure 11.2c**), the triceps brachii becomes the prime mover, and the biceps brachii is the antagonist. If a prime mover and its antagonist contract at the same time with equal force, there will be no movement.

Sometimes a prime mover crosses other joints before it reaches the joint at which its primary action occurs. The biceps brachii, for example, spans both the shoulder and elbow joints, with primary action on the forearm. To prevent unwanted movements at intermediate joints or to otherwise aid the movement of the prime mover, muscles called **synergists** (SIN-er-jists; *syn-* = together; *-ergon* = work) contract and stabilize the intermediate joints. As an example, muscles that flex the fingers (prime movers) cross the intercarpal and radiocarpal joints (intermediate joints). If movement at these intermediate joints were unrestrained, you would not be able to flex your fingers without flexing the wrist at the same time. Synergistic contraction of the wrist extensor muscles stabilizes the wrist joint and prevents unwanted movement, while the flexor muscles of the fingers contract to bring about the primary action, efficient flexion of the fingers. Synergists are usually located close to the prime mover.

Some muscles in a group also act as **fixators**, stabilizing the origin of the prime mover so that the prime mover can act more efficiently. Fixators steady the proximal end of a limb while movements occur at the distal end. For example, the scapula is a freely movable bone that serves as the origin for several muscles that move the arm. When the arm muscles contract, the scapula must be held steady. In abduction of the arm, the deltoid muscle serves as the prime mover, and fixators (pectoralis minor, trapezius, subclavius, serratus anterior muscles, and others) hold the scapula firmly against the back of the chest (see **Figure 11.14a, b**). The insertion of the deltoid muscle pulls on the humerus to abduct the arm. Under different conditions—that is, for different movements—and at different times, many muscles may act as prime movers, antagonists, synergists, or fixators.

In the limbs, a **compartment** is a group of skeletal muscles, their associated blood vessels, and associated nerves, all of which have a common function. In the upper limbs, for example, flexor compartment muscles are anterior, and extensor compartment muscles are posterior.

See Clinical Connection: Benefits of Stretching

11.2 How Skeletal Muscles Are Named

OBJECTIVE

- **Explain** seven features used in naming skeletal muscles.

The names of most of the skeletal muscles contain combinations of the word roots of their distinctive features. This works two ways. You can learn the names of muscles by remembering the terms that refer to muscle features, such as the pattern of the muscle's fascicles; the size, shape, action, number of origins, and location of the muscle; and the sites of origin and insertion of the muscle. Knowing the names of a muscle will then give you clues about its features. Study **Table 11.2** to become familiar with the terms used in muscle names.

11.3 Overview of the Principal Skeletal Muscles

OBJECTIVE

- **Describe** why organizing muscles into groups is beneficial.

The various muscles of the body are often organized into groups that perform certain functions. Most muscle groups share many features in common. Grouping muscles is a powerful tool to help you simplify the learning process. For example, the muscles within a group can share common attachments to bones, have common actions at joints, and be innervated by the same nerve. Grouping muscles by shared features reduces the amount of detailed information that you have to consume as you realize an attachment or action can be applied to a group of muscles. Sections 11.4–11.23 will assist you in learning about the principal skeletal muscles of the body. Each of these sections contains the following elements:

- ***Objective.*** This statement describes what you should learn in that section.
- ***Overview.*** These paragraphs provide a general introduction to the muscles under consideration and emphasize how the muscles are organized within various regions. The discussion also highlights any distinguishing features of the muscles.
- ***Muscle names.*** The word roots indicate how the muscles are named. As noted previously, once you have mastered the naming of the muscles, you can more easily understand their actions.
- ***Origins, insertions, and actions.*** You are given the origin, insertion, and actions of each muscle.
- ***Innervation.*** You are also given the nerve or nerves that cause contraction of each muscle. In general, cranial nerves, which arise from the lower parts of the brain, serve muscles in the head region. Spinal nerves, which arise from the spinal cord within the vertebral column, innervate muscles in the rest of the body. Cranial nerves are designated by both a name and a roman numeral: the facial (VII) nerve, for example. Spinal nerves are numbered in groups according to the part of the spinal cord from which they arise: C = cervical (neck region), T = thoracic (chest region), L = lumbar (lower-back region), and S = sacral (buttocks region). An example is T1, the first thoracic spinal nerve.
- ***Relating muscles to movements.*** These exercises will help you organize the muscles in the body region under consideration according to the actions they produce.
- ***Questions.*** These knowledge checkpoints relate specifically to information in each section, and take the form of review, critical thinking, and/or application questions.
- ***Clinical Connections.*** Selected sections include clinical applications, which explore the clinical, professional, or everyday relevance

TABLE 11.2 Characteristics Used to Name Muscles

NAME	MEANING	EXAMPLE	FIGURE
DIRECTION: Orientation of muscle fascicles relative to the body's midline			
Rectus	Parallel to midline	Rectus abdominis	11.10b
Transverse	Perpendicular to midline	Transversus abdominis	11.10b
Oblique	Diagonal to midline	External oblique	11.10a
SIZE: Relative size of the muscle			
Maximus	Largest	Gluteus maximus	11.20c
Minimus	Smallest	Gluteus minimus	11.20d
Longus	Long	Adductor longus	11.20a
Brevis	Short	Adductor brevis	11.20b
Latissimus	Widest	Latissimus dorsi	11.15b
Longissimus	Longest	Longissimus capitis	11.19a
Magnus	Large	Adductor magnus	11.20b
Major	Larger	Pectoralis major	11.10a
Minor	Smaller	Pectoralis minor	11.14a
Vastus	Huge	Vastus lateralis	11.20a
SHAPE: Relative shape of the muscle			
Deltoid	Triangular	Deltoid	11.15b
Trapezius	Trapezoid	Trapezius	11.3b
Serratus	Saw-toothed	Serratus anterior	11.14b
Rhomboid	Diamond-shaped	Rhomboid major	11.15c
Orbicularis	Circular	Orbicularis oculi	11.4a
Pectinate	Comblike	Pectineus	11.20a
Piriformis	Pear-shaped	Piriformis	11.20d
Platys	Flat	Platysma	11.4c
Quadratus	Square, four-sided	Quadratus femoris	11.20d
Gracilis	Slender	Gracilis	11.20a
ACTION: Principal action of the muscle			
Flexor	Decreases joint angle	Flexor carpi radialis	11.17a
Extensor	Increases joint angle	Extensor carpi ulnaris	11.17d
Abductor	Moves bone away from midline	Abductor pollicis longus	11.17e
Adductor	Moves bone closer to midline	Adductor longus	11.20a
Levator	Raises or elevates body part	Levator scapulae	11.14a
Depressor	Lowers or depresses body part	Depressor labii inferioris	11.4a
Supinator	Turns palm anteriorly	Supinator	11.17c
Pronator	Turns palm posteriorly	Pronator teres	11.17a
Sphincter	Decreases size of an opening	External anal sphincter	11.12
Tensor	Makes body part rigid	Tensor fasciae latae	11.20a
Rotator	Rotates bone around longitudinal axis	Rotatore	11.19b
NUMBER OF ORIGINS: Number of tendons of origin			
Biceps	Two origins	Biceps brachii	11.16a
Triceps	Three origins	Triceps brachii	11.16b
Quadriceps	Four origins	Quadriceps femoris	11.20a
LOCATION: Structure near which a muscle is found			
Example: Temporalis, muscle near temporal bone.			11.4c
ORIGIN AND INSERTION: Sites where muscle originates and inserts			
Example: Sternocleidomastoid, originating on sternum and clavicle and inserting on mastoid process of temporal bone.			11.3a

of a particular muscle or its function through descriptions of disorders or clinical procedures.

- ***Figures.*** The figures may present superficial and deep, anterior and posterior, or medial and lateral views to show each muscle's position as clearly as possible. The muscle names in all capital letters are specifically referred to in the tabular part of the section.

As you study groups of muscles in sections 11.4–11.23, refer to **Figure 11.3** to see how each group is related to the others.

FIGURE 11.3 Principal superficial skeletal muscles.

Most movements require several skeletal muscles acting in groups rather than individually.

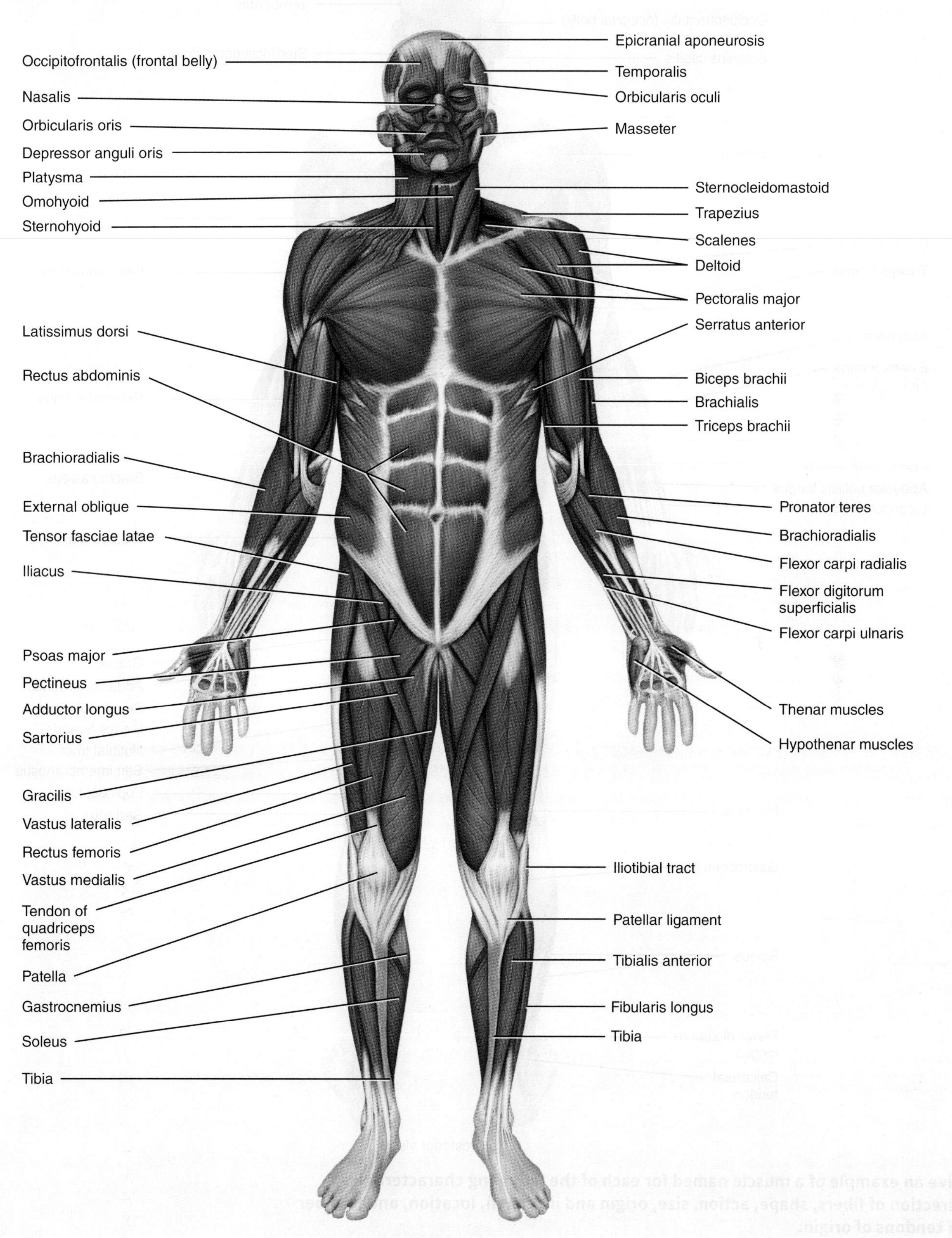

(a) Anterior view

Figure 11.3 *Continues*

FIGURE 11.3 Continued

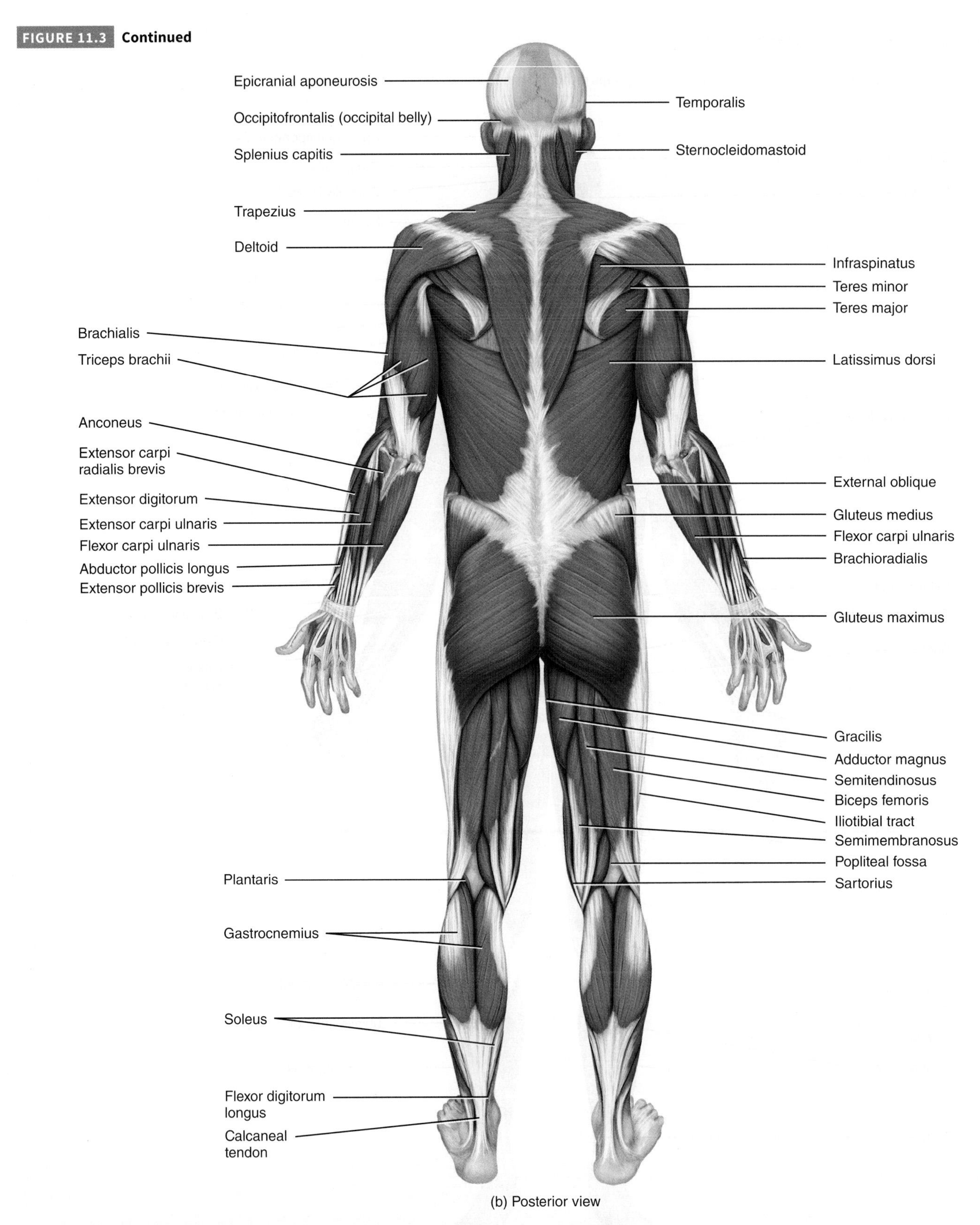

(b) Posterior view

Q Give an example of a muscle named for each of the following characteristics: direction of fibers, shape, action, size, origin and insertion, location, and number of tendons of origin.

11.4 Muscles of the Head That Produce Facial Expressions

OBJECTIVE

- **Describe** the origin, insertion, action, and innervation of the muscles of facial expression.

The muscles of facial expression, which provide us with the ability to express a wide variety of emotions, lie within the subcutaneous layer (**Figure 11.4**). They usually originate in the fascia or bones of the skull and insert into the skin. Because of their insertions, the muscles of facial expression move the skin rather than a joint when they contract.

Among the noteworthy muscles in this group are those surrounding the orifices (openings) of the head such as the eyes, nose, and mouth. These muscles function as *sphincters* (SFINGK-ters), which close the orifices, and *dilators* (DĪ-lā-tors), which dilate or

FIGURE 11.4 **Muscles of the head that produce facial expressions.**

When they contract, muscles of facial expression move the skin rather than a joint.

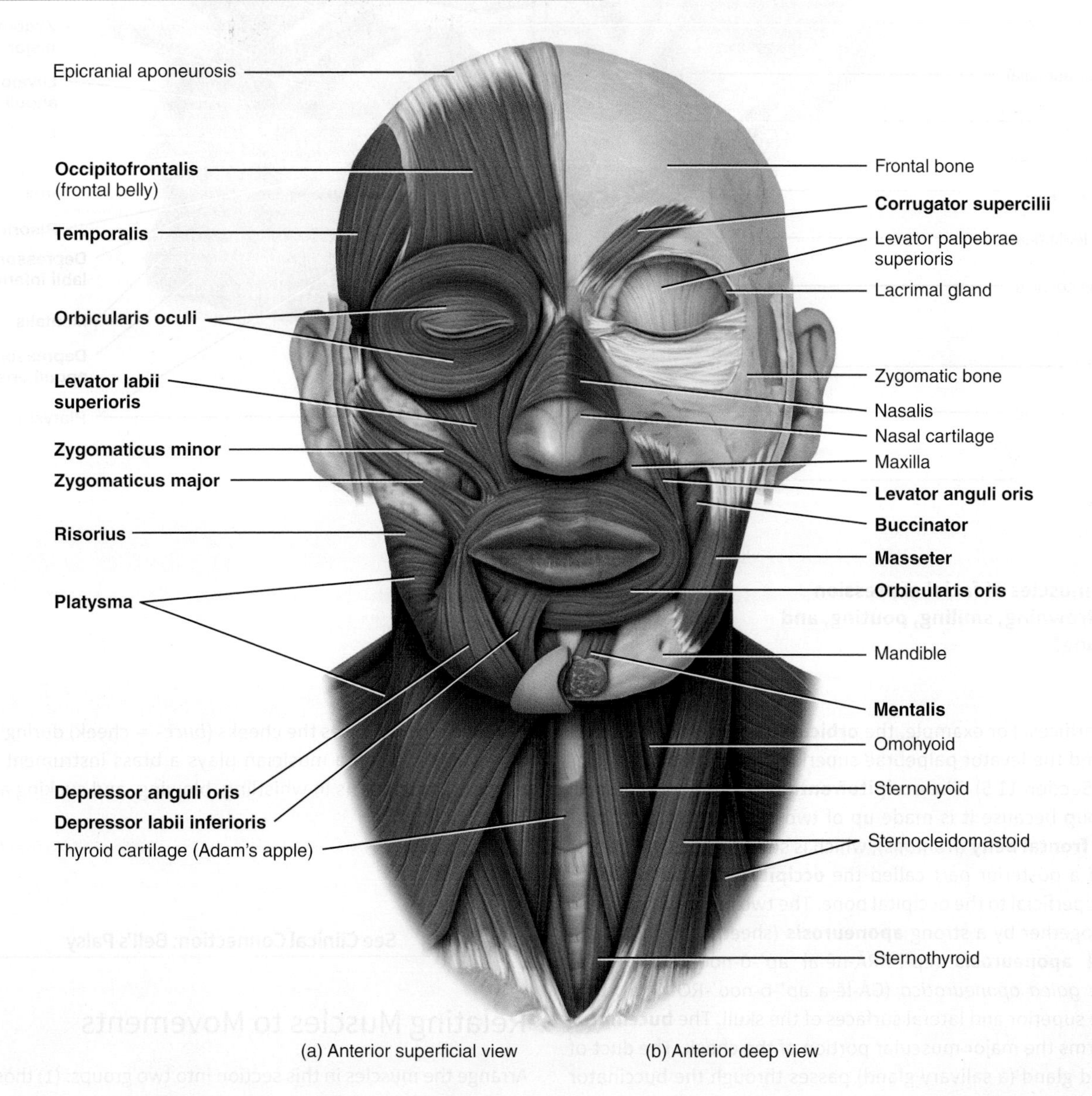

(a) Anterior superficial view (b) Anterior deep view

Figure 11.4 ***Continues***

FIGURE 11.4 **Continued**

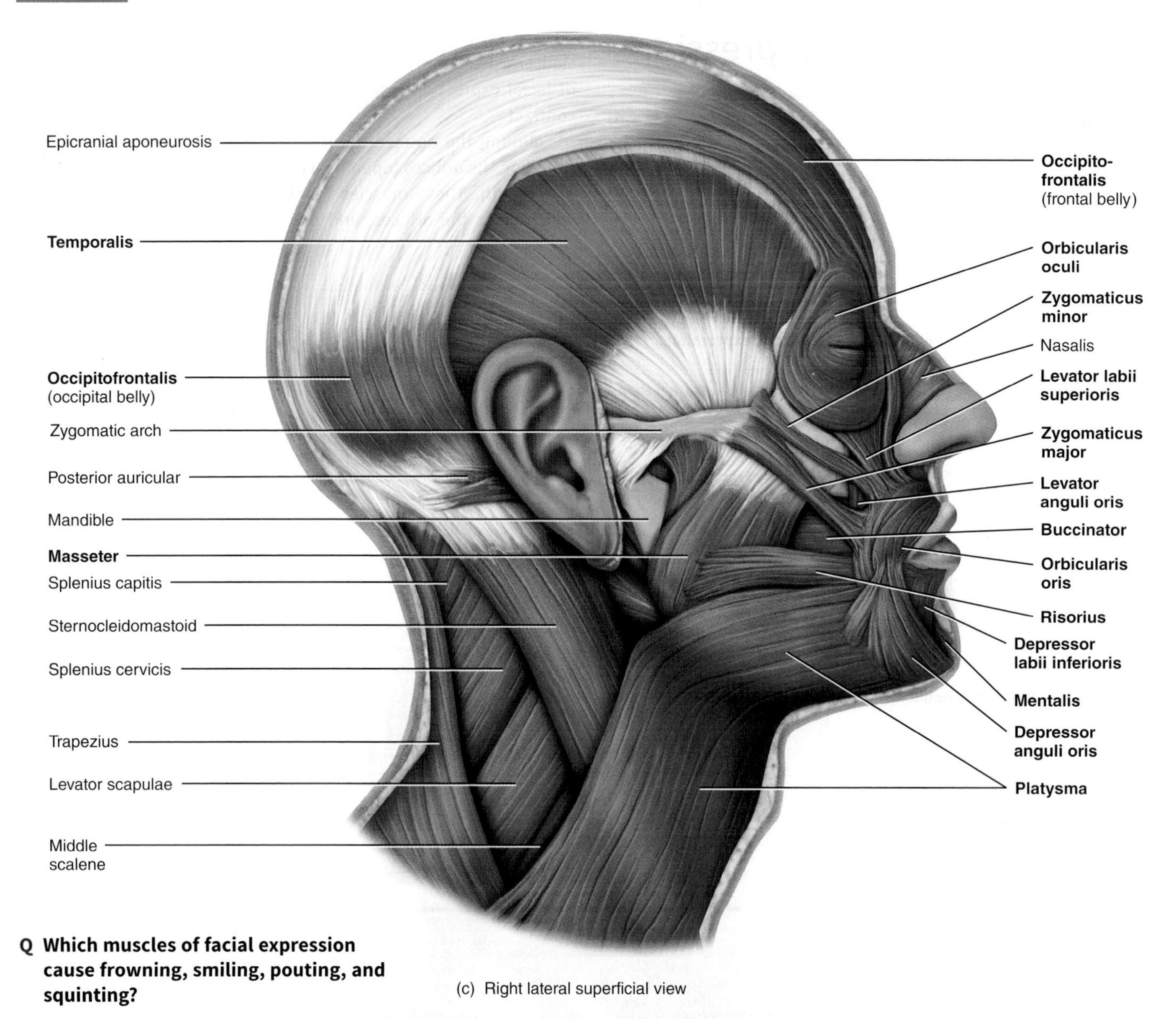

(c) Right lateral superficial view

Q Which muscles of facial expression cause frowning, smiling, pouting, and squinting?

open the orifices. For example, the **orbicularis oculi** muscle closes the eye, and the levator palpebrae superioris muscle opens it (discussed in Section 11.5). The **occipitofrontalis** is an unusual muscle in this group because it is made up of two parts: an anterior part called the **frontal belly** (*frontalis*), which is superficial to the frontal bone, and a posterior part called the **occipital belly** (*occipitalis*), which is superficial to the occipital bone. The two muscular portions are held together by a strong **aponeurosis** (sheetlike tendon), the **epicranial aponeurosis** (ep-i-KRĀ-nē-al ap′-ō-noo-RŌ-sis), also called the *galea aponeurotica* (GĀ-lē-a ap′-ō-noo′-RŌ-ti-ka), that covers the superior and lateral surfaces of the skull. The **buccinator** muscle forms the major muscular portion of the cheek. The duct of the parotid gland (a salivary gland) passes through the buccinator muscle to reach the oral cavity. The buccinator muscle is so named because it compresses the cheeks (*bucc-* = cheek) during blowing—for example, when a musician plays a brass instrument such as a trumpet. It functions in whistling, blowing, and sucking and assists in chewing.

See Clinical Connection: Bell's Palsy

Relating Muscles to Movements

Arrange the muscles in this section into two groups: (1) those that act on the mouth and (2) those that act on the eyes.

MUSCLE	ORIGIN	INSERTION	ACTION	INNERVATION
SCALP MUSCLES				
Occipitofrontalis (ok-sip′-i-tō-frun-TĀ-lis)				
Frontal belly	Epicranial aponeurosis.	Skin superior to supraorbital margin.	Draws scalp anteriorly, raises eyebrows, and wrinkles skin of forehead horizontally as in look of surprise.	Facial (VII) nerve.
Occipital belly (*occipit-* = back of the head)	Occipital bone and mastoid process of temporal bone.	Epicranial aponeurosis.	Draws scalp posteriorly.	Facial (VII) nerve.
MOUTH MUSCLES				
Orbicularis oris (or-bi′-kū-LAR-is OR-is; *orb-* = circular; *oris* = of the mouth)	Muscle fibers surrounding opening of mouth.	Skin at corner of mouth.	Closes and protrudes lips, as in kissing; compresses lips against teeth; and shapes lips during speech.	Facial (VII) nerve.
Zygomaticus major (zī-gō-MA-ti-kus; *zygomatic* = cheek bone; *major* = greater)	Zygomatic bone.	Skin at angle of mouth and orbicularis oris.	Draws angle of mouth superiorly and laterally, as in smiling.	Facial (VII) nerve.
Zygomaticus minor (*minor* = lesser)	Zygomatic bone.	Upper lip.	Raises (elevates) upper lip, exposing maxillary (upper) teeth.	Facial (VII) nerve.
Levator labii superioris (le-VĀ-tor LĀ-bē-ī soo-per′-ē-OR-is; *levator* = raises or elevates; *labii* = lip; *superioris* = upper)	Superior to infraorbital foramen of maxilla.	Skin at angle of mouth and orbicularis oris.	Raises upper lip.	Facial (VII) nerve.
Depressor labii inferioris (de-PRE-sor LĀ-bē-ī; *depressor* = depresses or lowers; *inferioris* = lower)	Mandible.	Skin of lower lip.	Depresses (lowers) lower lip.	Facial (VII) nerve.
Depressor anguli oris (ANG-ū-lī; *angul* = angle or corner; *oris* = mouth)	Mandible.	Angle of mouth.	Draws angle of mouth laterally and inferiorly, as in opening mouth.	Facial (VII) nerve.
Levator anguli oris	Inferior to infraorbital foramen.	Skin of lower lip and orbicularis oris.	Draws angle of mouth laterally and superiorly.	Facial (VII) nerve.
Buccinator (BUK-si-nā′-tor; *bucc-* = cheek)	Alveolar processes of maxilla and mandible and pterygomandibular raphe (fibrous band extending from pterygoid process of sphenoid bone to mandible).	Orbicularis oris.	Presses cheeks against teeth and lips, as in whistling, blowing, and sucking; draws corner of mouth laterally; and assists in mastication (chewing) by keeping food between the teeth (and not between teeth and cheeks).	Facial (VII) nerve.
Risorius (ri-ZOR-ē-us; *risor* = laughter)	Fascia over parotid (salivary) gland.	Skin at angle of mouth.	Draws angle of mouth laterally, as in grimacing.	Facial (VII) nerve.
Mentalis (men-TĀ-lis; *mental* = chin)	Mandible.	Skin of chin.	Elevates and protrudes lower lip and pulls skin of chin up, as in pouting.	Facial (VII) nerve.
NECK MUSCLES				
Platysma (pla-TIZ-ma; *platys* = flat, broad)	Fascia over deltoid and pectoralis major muscles.	Mandible, blends with muscles around angle of mouth, and skin of lower face.	Draws outer part of lower lip inferiorly and posteriorly as in pouting; depresses mandible.	Facial (VII) nerve.
ORBIT AND EYEBROW MUSCLES				
Orbicularis oculi (OK-ū-lī = eye)	Medial wall of orbit.	Circular path around orbit.	Closes eye.	Facial (VII) nerve.
Corrugator supercilii (KOR-u-gā′-tor soo-per-SIL-ē-ī; *corrugat* = wrinkle; *supercilii* = eyebrow)	Medial end of superciliary arch of frontal bone.	Skin of eyebrow.	Draws eyebrow inferiorly and wrinkles skin of forehead vertically as in frowning.	Facial (VII) nerve.

11.5 Muscles of the Head That Move the Eyeballs (Extrinsic Eye Muscles) and Upper Eyelids

OBJECTIVE

- **Describe** the origin, insertion, action, and innervation of the extrinsic eye muscles that move the eyeballs and upper eyelids.

Muscles that move the eyeballs are called **extrinsic eye muscles** because they originate outside the eyeballs (in the orbit) and insert on the outer surface of the sclera ("white of the eye") (**Figure 11.5**). The extrinsic eye muscles are some of the fastest contracting and most precisely controlled skeletal muscles in the body.

Three pairs of extrinsic eye muscles control movements of the eyeballs: (1) superior and inferior recti, (2) lateral and medial recti, and (3) superior and inferior obliques. The four recti muscles (superior, inferior, lateral, and medial) arise from a tendinous ring in the orbit and insert into the sclera of the eye. As their names imply, the **superior** and **inferior recti** move the eyeballs superiorly and inferiorly; the **lateral** and **medial recti** move the eyeballs laterally and medially, respectively.

FIGURE 11.5 Muscles of the head that move the eyeballs (extrinsic eye muscles) and upper eyelid.

The extrinsic muscles of the eyeball are among the fastest contracting and most precisely controlled skeletal muscles in the body.

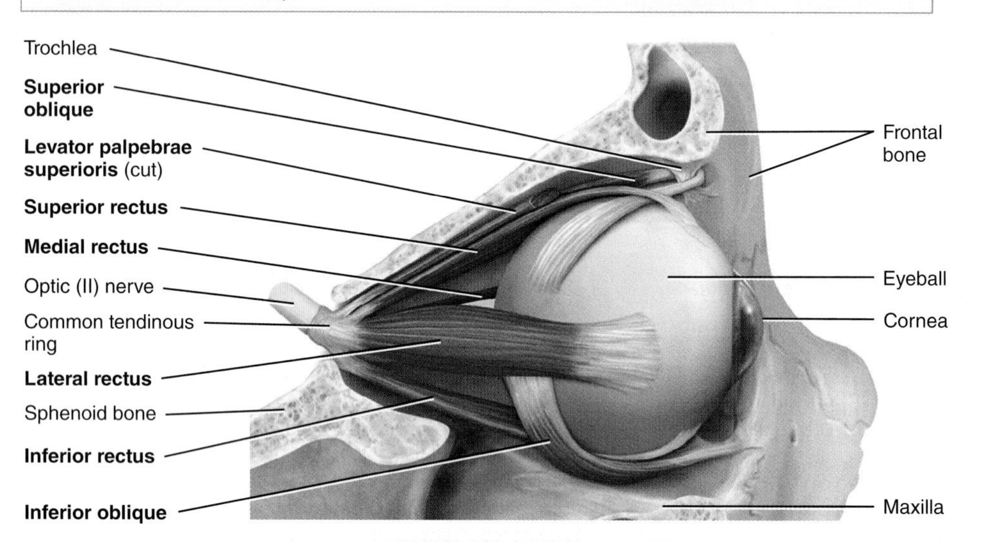

(a) Lateral view of right eyeball

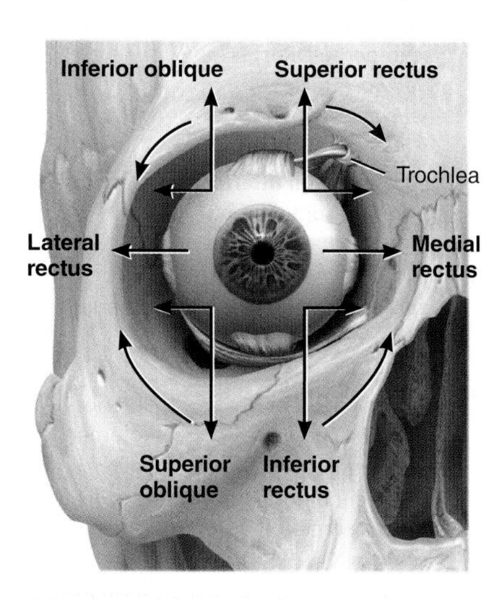

(b) Movements of right eyeball in response to contraction of extrinsic muscles

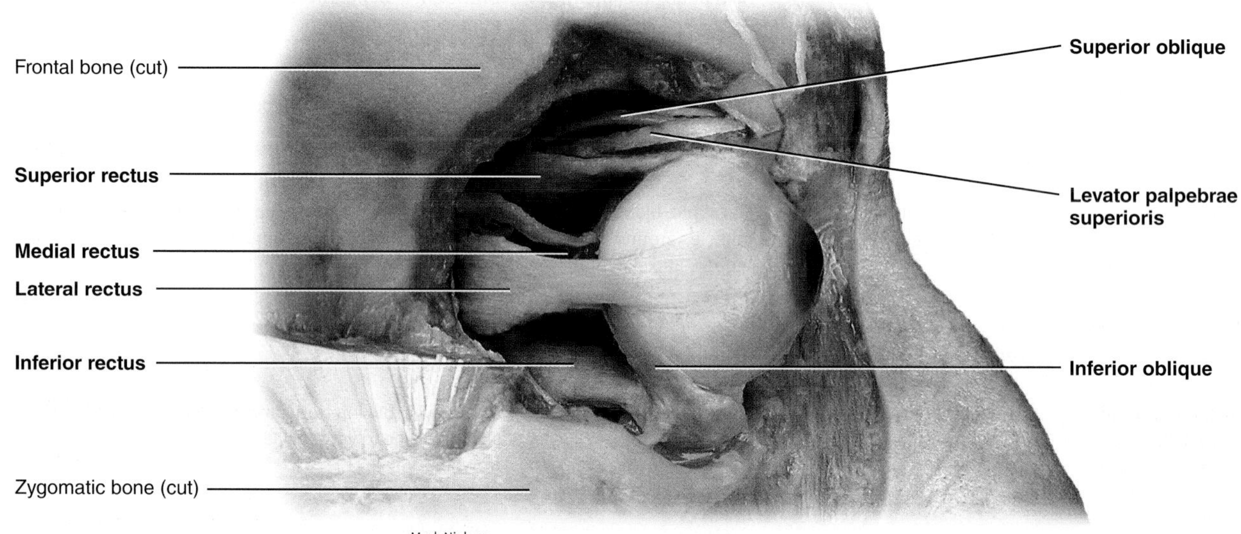

(c) Right lateral view

Q How does the inferior oblique muscle move the eyeball superiorly and laterally?

MUSCLE	ORIGIN	INSERTION	ACTION	INNERVATION
Superior rectus (*rectus* = fascicles parallel to midline)	Common tendinous ring (attached to orbit around optic foramen).	Superior and central part of eyeballs.	Moves eyeballs superiorly (elevation) and medially (adduction), and rotates them medially.	Oculomotor (III) nerve.
Inferior rectus	Same as above.	Inferior and central part of eyeballs.	Moves eyeballs inferiorly (depression) and medially (adduction), and rotates them laterally.	Oculomotor (III) nerve.
Lateral rectus	Same as above.	Lateral side of eyeballs.	Moves eyeballs laterally (abduction).	Abducens (VI) nerve.
Medial rectus	Same as above.	Medial side of eyeballs.	Moves eyeballs medially (adduction).	Oculomotor (III) nerve.
Superior oblique (*oblique* = fascicles diagonal to midline)	Sphenoid bone, superior and medial to common tendinous ring in orbit.	Eyeball between superior and lateral recti. Muscle inserts into superior and lateral surfaces of eyeball via tendon that passes through trochlea.	Moves eyeballs inferiorly (depression) and laterally (abduction), and rotates them medially.	Trochlear (IV) nerve.
Inferior oblique	Maxilla in floor of orbit.	Eyeballs between inferior and lateral recti.	Moves eyeballs superiorly (elevation) and laterally (abduction), and rotates them laterally.	Oculomotor (III) nerve.
Levator palpebrae superioris (le-VĀ-tor PAL-pe-brē soo-per′-ē-OR-is; *palpebrae* = eyelids)	Roof of orbit (lesser wing of sphenoid bone).	Skin and tarsal plate of upper eyelids.	Elevates upper eyelids (opens eyes).	Oculomotor (III) nerve.

The actions of the oblique muscles cannot be deduced from their names. The **superior oblique** muscle originates posteriorly near the tendinous ring, then passes anteriorly superior to the medial rectus muscle, and ends in a round tendon. The tendon extends through a pulleylike loop of fibrocartilaginous tissue called the *trochlea* (= pulley) on the anterior and medial part of the roof of the orbit. Finally, the tendon turns and inserts on the posterolateral aspect of the eyeball. Accordingly, the superior oblique muscle moves the eyeballs inferiorly and laterally. The **inferior oblique** muscle originates on the maxilla at the anteromedial aspect of the floor of the orbit. It then passes posteriorly and laterally and inserts on the posterolateral aspect of the eyeball. Because of this arrangement, the inferior oblique muscle moves the eyeballs superiorly and laterally.

Unlike the recti and oblique muscles, the **levator palpebrae superioris** does not move the eyeballs, since its tendon passes the eyeball and inserts into the upper eyelid. Rather, it raises the upper eyelids, that is, opens the eyes. It is therefore an antagonist to the orbicularis oculi, which closes the eyes.

See Clinical Connection: Strabismus

Relating Muscles to Movements

Arrange the muscles in this section according to their actions on the eyeballs: (1) elevation, (2) depression, (3) abduction, (4) adduction, (5) medial rotation, and (6) lateral rotation. The same muscle may be mentioned more than once.

11.6 Muscles That Move the Mandible and Assist in Mastication and Speech

OBJECTIVE

- **Describe** the origin, insertion, action, and innervation of the muscles that move the mandible and assist in mastication and speech.

The muscles that move the mandible (lower jawbone) at the temporomandibular joint (TMJ) are known as the **muscles of mastication** (*chewing*) (**Figure 11.6**). Of the four pairs of muscles involved in mastication, three are powerful closers of the jaw and account for the strength of the bite: **masseter, temporalis**, and **medial pterygoid**. Of these, the masseter is the strongest muscle of mastication. The medial and **lateral pterygoid** muscles assist in mastication by moving the mandible from side to side to help grind food. Additionally, the lateral pterygoid muscles protract (protrude) the mandible. The masseter muscle has been removed in **Figure 11.6** to illustrate the deeper pterygoid muscles; the masseter can be seen in **Figure 11.4c**. Note the enormous bulk of the temporalis and masseter muscles compared to the smaller mass of the two pterygoid muscles.

Note: A mnemonic for muscles of mastication is **T**eeny **M**ice **M**ake **P**etite **L**ittle **P**rints = **T**emporalis, **M**asseter, **M**edial **P**terygoid, and **L**ateral **P**terygoid.

MUSCLE	ORIGIN	INSERTION	ACTION	INNERVATION
Masseter (MA-se-ter = chewer) (see **Figure 11.4c**)	Maxilla and zygomatic arch.	Angle and ramus of mandible.	Elevates mandible, as in closing mouth.	Mandibular division of trigeminal (V) nerve.
Temporalis (tem′-pō-RĀ-lis; *tempor-* = time or temples)	Temporal bone.	Coronoid process and ramus of mandible.	Elevates and retracts mandible.	Mandibular division of trigeminal (V) nerve.
Medial pterygoid (TER-i-goyd; *medial* = closer to midline; *pterygoid* = winglike)	Medial surface of lateral portion of pterygoid process of sphenoid bone; maxilla.	Angle and ramus of mandible.	Elevates and protracts (protrudes) mandible and moves mandible from side to side.	Mandibular division of trigeminal (V) nerve.
Lateral pterygoid (*lateral* = farther from midline)	Greater wing and lateral surface of lateral portion of pterygoid process of sphenoid bone.	Condyle of mandible; temporomandibular joint (TMJ).	Protracts mandible, depresses mandible as in opening mouth, and moves mandible from side to side.	Mandibular division of trigeminal (V) nerve.

FIGURE 11.6 Muscles that move the mandible (lower jawbone) and assist in mastication (chewing) and speech.

The muscles that move the mandible are also known as muscles of mastication.

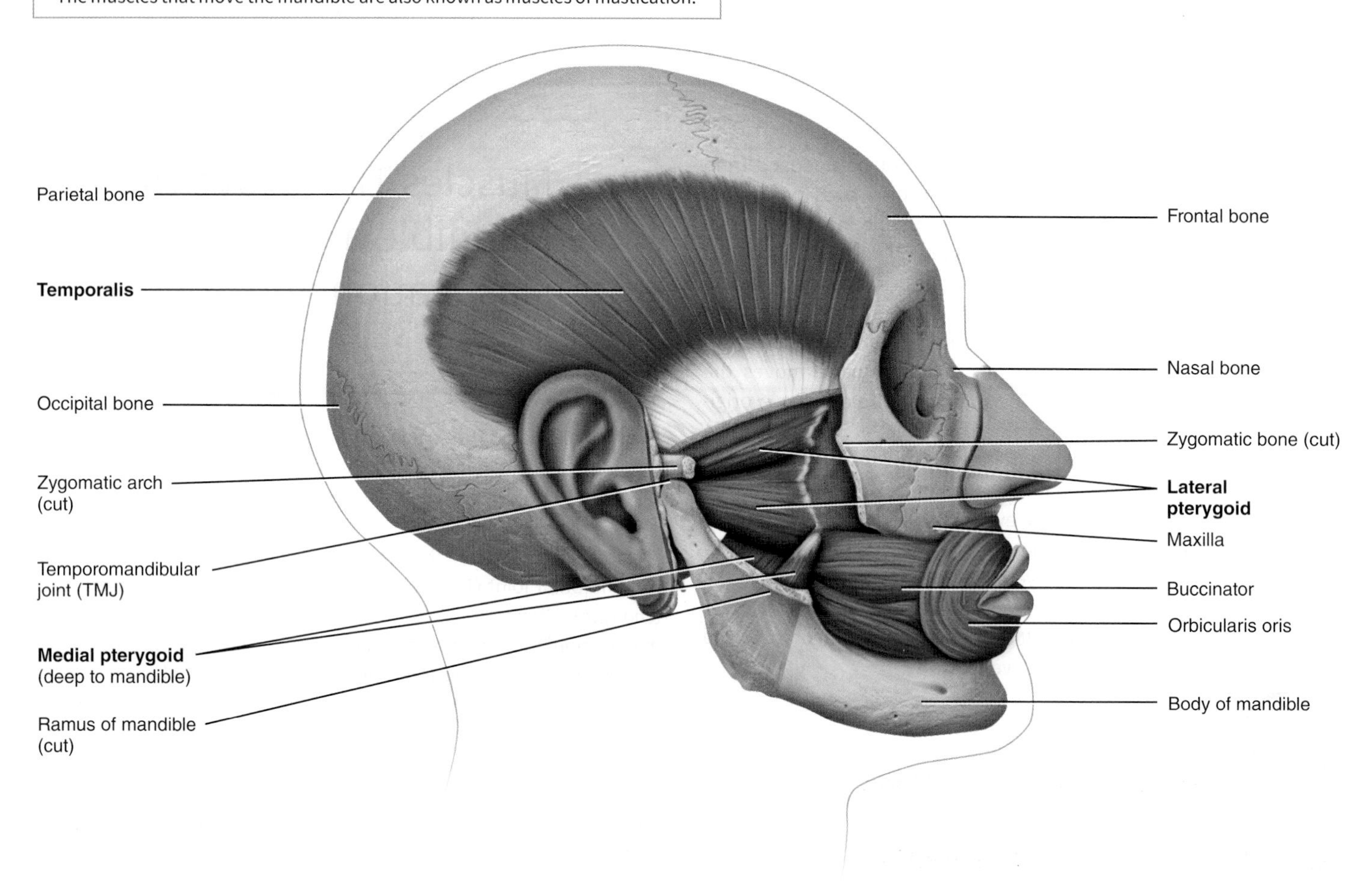

Q Which is the strongest muscle of mastication?

See Clinical Connection: Gravity and the Mandible

Relating Muscles to Movements

Arrange the muscles in this section according to their actions on the mandible: (1) elevation, (2) depression, (3) retraction, (4) protraction, and (5) side-to-side movement. The same muscle may be mentioned more than once.

11.7 Muscles of the Head That Move the Tongue and Assist in Mastication and Speech

OBJECTIVE

- **Describe** the origin, insertion, action, and innervation of the muscles that move the tongue and assist in mastication and speech.

The tongue is a highly mobile structure that is vital to digestive functions such as *mastication* (chewing), detection of taste, and *deglutition* (swallowing). It is also important in speech. The tongue's mobility is greatly aided by its attachment to the mandible, styloid process of the temporal bone, and hyoid bone.

The tongue is divided into lateral halves by a median fibrous septum. The septum extends throughout the length of the tongue. Inferiorly, the septum attaches to the hyoid bone. Muscles of the tongue are of two principal types: extrinsic and intrinsic. **Extrinsic tongue muscles** originate outside the tongue and insert into it (**Figure 11.7**). They move the entire tongue in various directions, such as anteriorly, posteriorly, and laterally. **Intrinsic tongue muscles** originate and insert within the tongue. These muscles alter the shape of the tongue rather than moving the entire tongue. The extrinsic and intrinsic muscles of the tongue insert into both lateral halves of the tongue.

When you study the extrinsic tongue muscles, you will notice that all of their names end in *glossus*, meaning tongue. You will also notice that the actions of the muscles are obvious, considering the positions of the mandible, styloid process, hyoid bone, and soft palate, which serve as origins for these muscles. For example, the **genioglossus** (origin: the mandible) pulls the tongue downward and forward, the **styloglossus** (origin: the styloid process) pulls the tongue upward and backward, the **hyoglossus** (origin: the hyoid bone) pulls the tongue downward and flattens it, and the **palatoglossus** (origin: the soft palate) raises the back portion of the tongue.

See Clinical Connection: Intubation during Anesthesia

Relating Muscles to Movements

Arrange the muscles in this section according to the following actions on the tongue: (1) depression, (2) elevation, (3) protraction, and (4) retraction. The same muscle may be mentioned more than once.

FIGURE 11.7 Muscles of the head that move the tongue and assist in mastication (chewing) and speech—extrinsic tongue muscles.

The extrinsic and intrinsic muscles of the tongue are arranged in both lateral halves of the tongue.

Q What are the functions of the tongue?

MUSCLE	ORIGIN	INSERTION	ACTION	INNERVATION
Genioglossus (jē′-nē-ō-GLOS-us; *genio-* = chin; *-glossus* = tongue)	Mandible.	Undersurface of tongue and hyoid bone.	Depresses tongue and thrusts it anteriorly (protraction).	Hypoglossal (XII) nerve.
Styloglossus (stī′-lō-GLOS-us; *stylo-* = stake or pole; styloid process of temporal bone)	Styloid process of temporal bone.	Side and undersurface of tongue.	Elevates tongue and draws it posteriorly (retraction).	Hypoglossal (XII) nerve.
Hyoglossus (hī′-ō-GLOS-us; *hyo-* = U-shaped)	Greater horn and body of hyoid bone.	Side of tongue.	Depresses tongue and draws down its sides.	Hypoglossal (XII) nerve.
Palatoglossus (pal′-a-tō-GLOS-us; palato- = roof of mouth or palate)	Anterior surface of soft palate.	Side of tongue.	Elevates posterior portion of tongue and draws soft palate down on tongue.	Pharyngeal plexus, which contains axons from the vagus (X) nerve.

11.8 Muscles of the Anterior Neck That Assist in Deglutition and Speech

OBJECTIVE

- **Describe** the origin, insertion, action, and innervation of the muscles of the anterior neck that assist in deglutition and speech.

Two groups of muscles are associated with the anterior aspect of the neck: (1) the **suprahyoid muscles**, so called because they are located superior to the hyoid bone, and (2) the **infrahyoid muscles**, named for their position inferior to the hyoid bone (Figure 11.8). Both groups of muscles stabilize the hyoid bone, allowing it to serve as a firm base on which the tongue can move.

As a group, the suprahyoid muscles elevate the hyoid bone, floor of the oral cavity, and tongue during deglutition (swallowing). As its name suggests, the **digastric** muscle has two bellies, anterior and posterior, united by an intermediate tendon that is held in position by a fibrous loop. This muscle elevates the hyoid bone and larynx (voice box) during swallowing and speech. In a *reverse muscle action (RMA)*,

FIGURE 11.8 **Muscles of the anterior neck that assist in deglutition (swallowing) and speech.**

The suprahyoid muscles elevate the hyoid bone, the floor of the oral cavity, and the tongue during swallowing.

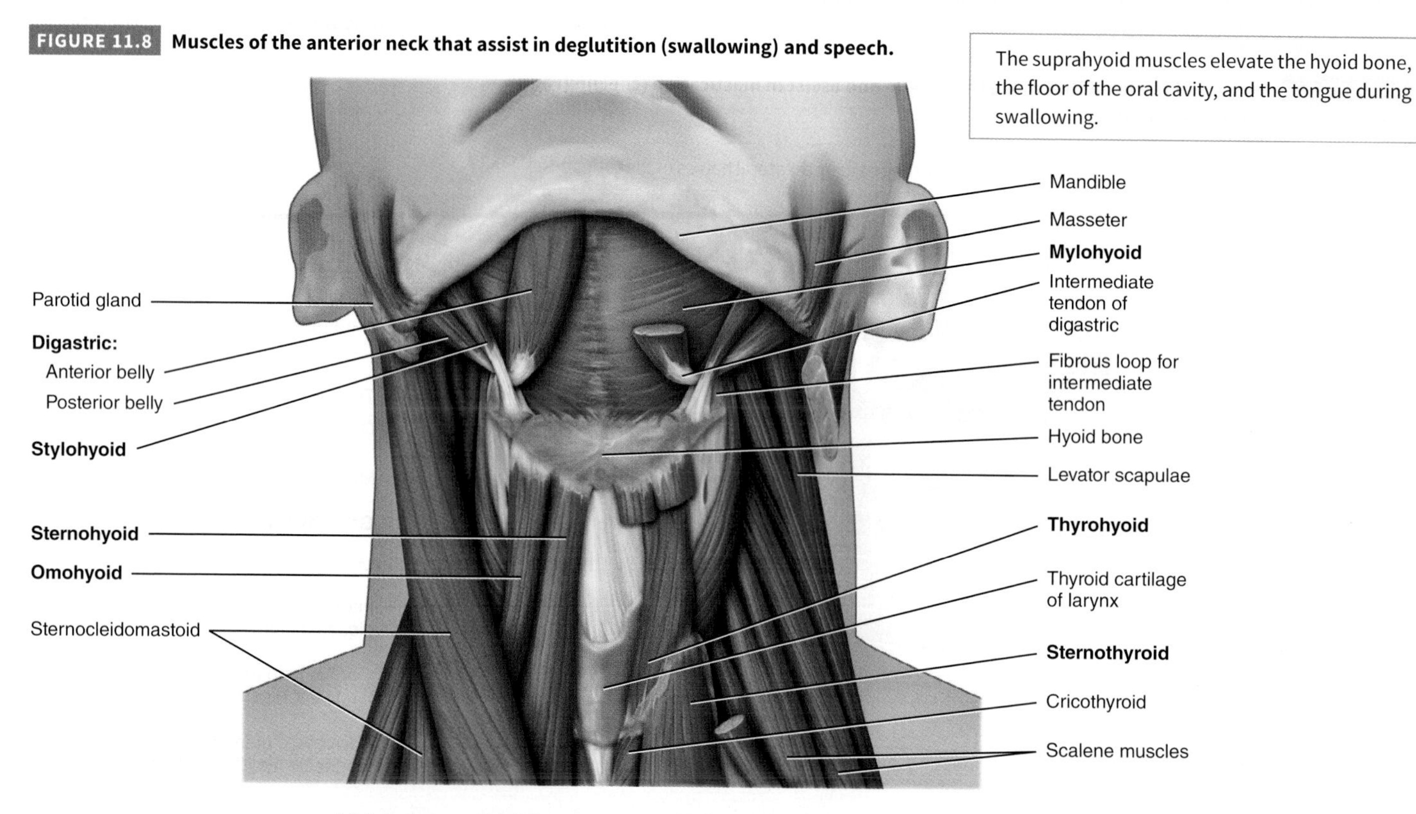

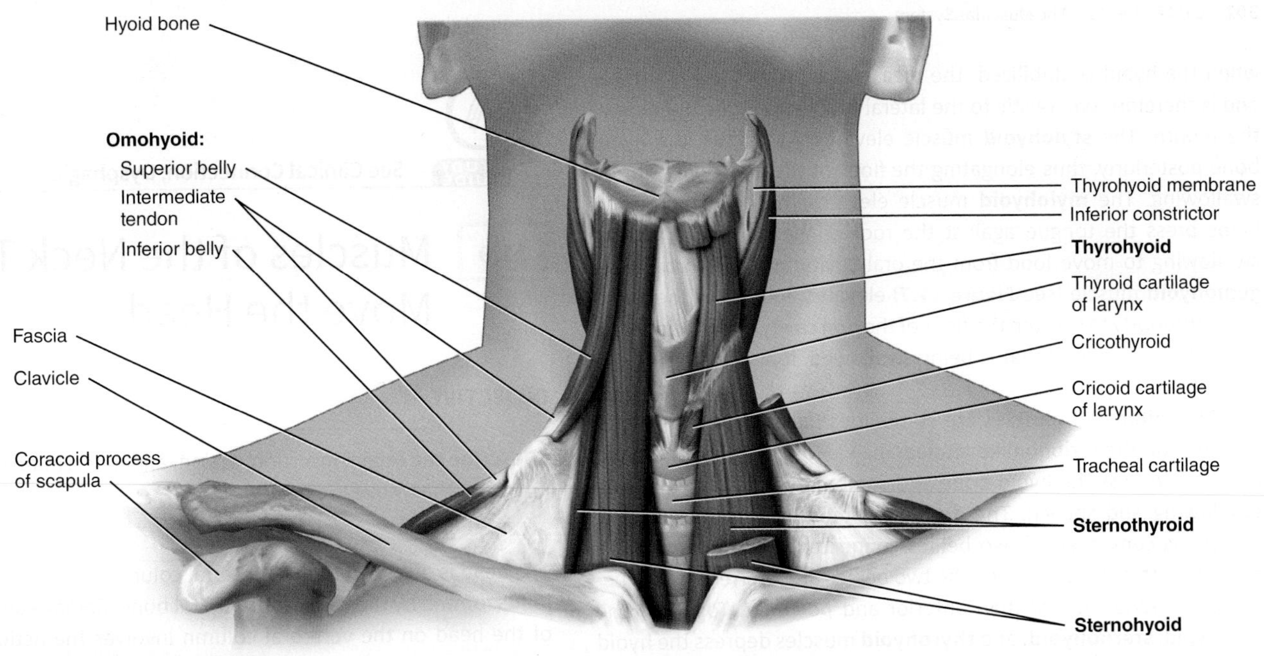

Anterior superficial view (c) Anterior deep view

Q What is the combined action of the suprahyoid and infrahyoid muscles?

MUSCLE	ORIGIN	INSERTION	ACTION	INNERVATION
SUPRAHYOID MUSCLES				
Digastric (dī′-GAS-trik; *di-* = two; *-gastr-* = belly)	Anterior belly from inner side of inferior border of mandible; posterior belly from temporal bone.	Body of hyoid bone via an intermediate tendon.	Elevates hyoid bone. RMA: Depresses mandible, as in opening mouth.	Anterior belly: mandibular division of trigeminal (V) nerve. Posterior belly: facial (VII) nerve.
Stylohyoid (stī′-lō-HĪ-oyd; *stylo-* = stake or pole, styloid process of temporal bone; *-hyo-* = U-shaped, pertaining to hyoid bone)	Styloid process of temporal bone.	Body of hyoid bone.	Elevates hyoid bone and draws it posteriorly.	Facial (VII) nerve.
Mylohyoid (mī′-lō-HĪ-oyd; *mylo-* = mill)	Inner surface of mandible.	Body of hyoid bone.	Elevates hyoid bone and floor of mouth and depresses mandible.	Mandibular division of trigeminal (V) nerve.
Geniohyoid (jē′-nē-ō-HĪ-oyd; *genio-* = chin) (see **Figure 11.7**)	Inner surface of mandible.	Body of hyoid bone.	Elevates hyoid bone, draws hyoid bone and tongue anteriorly. Depresses mandible.	First cervical spinal nerve (C1).
INFRAHYOID MUSCLES				
Omohyoid (ō-mō-HĪ-oyd; *omo-* = relationship to shoulder)	Superior border of scapula and superior transverse ligament.	Body of hyoid bone.	Depresses hyoid bone.	Branches of spinal nerves C1–C3.
Sternohyoid (ster′-nō-HĪ-oyd; *sterno-* = sternum)	Medial end of clavicle and manubrium of sternum.	Body of hyoid bone.	Depresses hyoid bone.	Branches of spinal nerves C1–C3.
Sternothyroid (ster′-nō-THĪ-royd; *thyro-* = thyroid gland)	Manubrium of sternum.	Thyroid cartilage of larynx.	Depresses thyroid cartilage of larynx.	Branches of spinal nerves C1–C3.
Thyrohyoid (thī′-rō-HĪ-oyd)	Thyroid cartilage of larynx.	Greater horn of hyoid bone.	Depresses hyoid bone. RMA: Elevates thyroid cartilage.	Branches of spinal nerves C1–C2 and descending hypoglossal (XII) nerve.

when the hyoid is stabilized, the digastric depresses the mandible and is therefore synergistic to the lateral pterygoid in the opening of the mouth. The **stylohyoid** muscle elevates and draws the hyoid bone posteriorly, thus elongating the floor of the oral cavity during swallowing. The **mylohyoid** muscle elevates the hyoid bone and helps press the tongue against the roof of the oral cavity during swallowing to move food from the oral cavity into the throat. The **geniohyoid** muscle (see Figure 11.7) elevates and draws the hyoid bone anteriorly to shorten the floor of the oral cavity and to widen the throat to receive food that is being swallowed. It also depresses the mandible.

The infrahyoid muscles are sometimes called "strap" muscles because of their ribbonlike appearance. Most of the infrahyoid muscles depress the hyoid bone and some move the larynx during swallowing and speech. The **omohyoid** muscle, like the digastric muscle, is composed of two bellies connected by an intermediate tendon. In this case, however, the two bellies are referred to as *superior* and *inferior*, rather than anterior and posterior. Together, the omohyoid, **sternohyoid**, and **thyrohyoid** muscles depress the hyoid bone. In addition, the **sternothyroid** muscle depresses the thyroid cartilage (Adam's apple) of the larynx to produce low sounds; the RMA of the thyrohyoid muscle elevates the thyroid cartilage to produce high sounds.

Relating Muscles to Movements

Arrange the muscles in this section according to the following actions on the hyoid bone: (1) elevating it, (2) drawing it anteriorly, (3) drawing it posteriorly, and (4) depressing it; and on the thyroid cartilage: (1) elevating it and (2) depressing it. The same muscle may be mentioned more than once.

See Clinical Connection: Dysphagia

11.9 Muscles of the Neck That Move the Head

OBJECTIVE

- **Describe** the origin, insertion, action, and innervation of the muscles that move the head.

The head is attached to the vertebral column at the atlanto-occipital joints formed by the atlas and occipital bone. Balance and movement of the head on the vertebral column involves the action of several neck muscles. For example, acting together (bilaterally), contraction of the two **sternocleidomastoid (SCM)** muscles flexes the cervical portion of the vertebral column and flexes the head. Acting singly (unilaterally), each sternocleidomastoid muscle laterally flexes and rotates the head. Each SCM consists of two bellies (Figure 11.9c); they are more evident near the anterior attachments. The separation of the two bellies is variable and thus more evident in some persons than in others. The two bellies insert as the **sternal head** and the **clavicular head** of the SCM. The bellies also function differently; muscular spasm in the two bellies causes somewhat different symptoms. Bilateral contraction of the **spenalis capitis**, **semispinalis capitis**, **splenius capitis**, and **longissimus capitis** muscles extends the head (Figure 11.9a, b). However, when these same muscles contract

MUSCLE	ORIGIN	INSERTION	ACTION	INNERVATION
Sternocleidomastoid (ster′-nō-klī′-dō-MAS-toyd; *sterno-* = breastbone; *-cleido-* = clavicle; *-mastoid* = mastoid process of temporal bone)	Sternal head: manubrium of sternum; clavicular head: medial third of clavicle.	Mastoid process of temporal bone and lateral half of superior nuchal line of occipital bone.	Acting together (bilaterally), flex cervical portion of vertebral column, extend head at atlanto-occipital joints; acting singly (unilaterally), laterally flex neck and head to same side and rotate head to side opposite contracting muscle. Laterally rotate and flex head to opposite side of contracting muscle. Posterior fibers of muscle can assist in extension of head. RMA: Elevate sternum during forced inhalation.	Accessory (XI) nerve, C2, and C3.
Semispinalis capitis (se′-mē-spi-NĀ-lis KAP-i-tis; *semi-* = half; *spine* = spinous process; *capit* = head)	Articular processes of C4–C6 and transverse processes of C7–T7.	Occipital bone between superior and inferior nuchal lines.	Acting together, extend head and vertebral column; acting singly, rotate head to side opposite contracting muscle.	Cervical spinal nerves.
Splenius capitis (SPLĒ-nē-us KAP-i-tis; *splenium* = bandage)	Ligamentum nuchae and spinous processes of C7–T4.	Occipital bone and mastoid process of temporal bone.	Extend head; acting together, muscle of each region (cervical and thoracic) extend vertebral column of their respective regions.	Cervical spinal nerves.
Longissimus capitis (lon-JIS-i-mus KAP-i-tis; *longissimus* = longest)	Articular processes of T1–T4.	Mastoid process of temporal bone.	Acting together, extend head and vertebral column; acting singly, laterally flex and rotate head to same side as contracting muscle.	Cervical spinal nerves.
Spinalis capitis (spi-NĀ-lis KAP-i-tis; *spinal* = vertebral column)	Often absent or very small; arises with semispinalis capitis.	Occipital bone.	Extends head and vertebral column.	Cervical spinal nerves.

FIGURE 11.9 **Muscles of the neck that move the head.**

The sternocleidomastoid muscle divides the neck into two principal triangles: anterior and posterior.

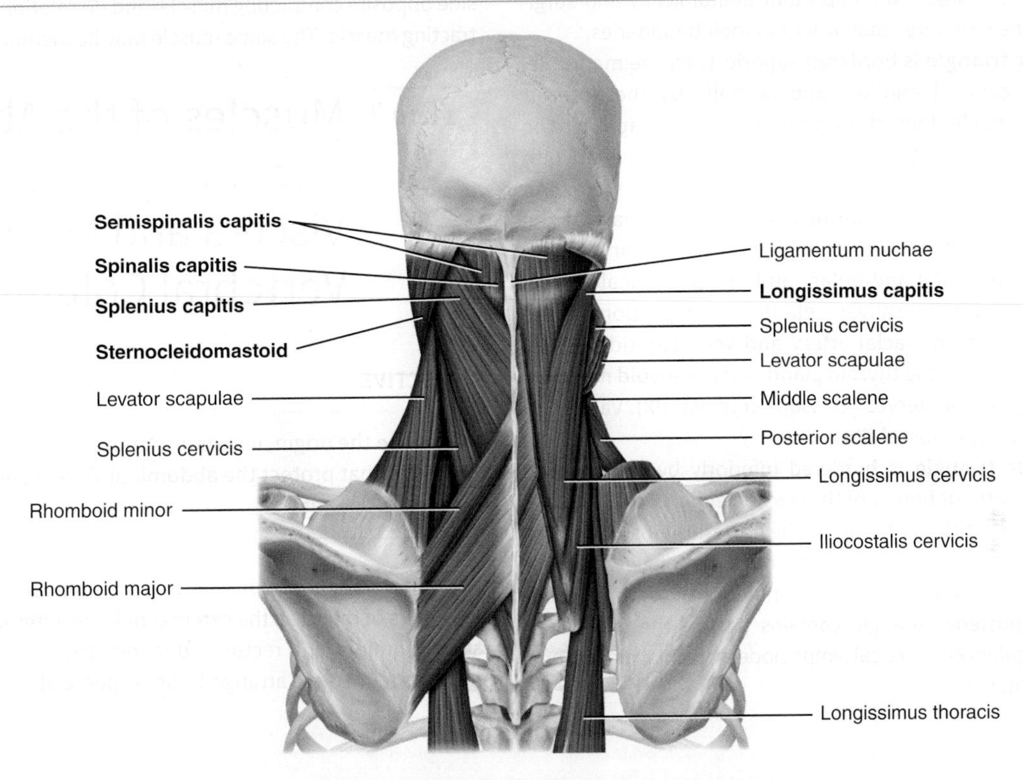

(a) Posterior superficial view (b) Posterior deep view

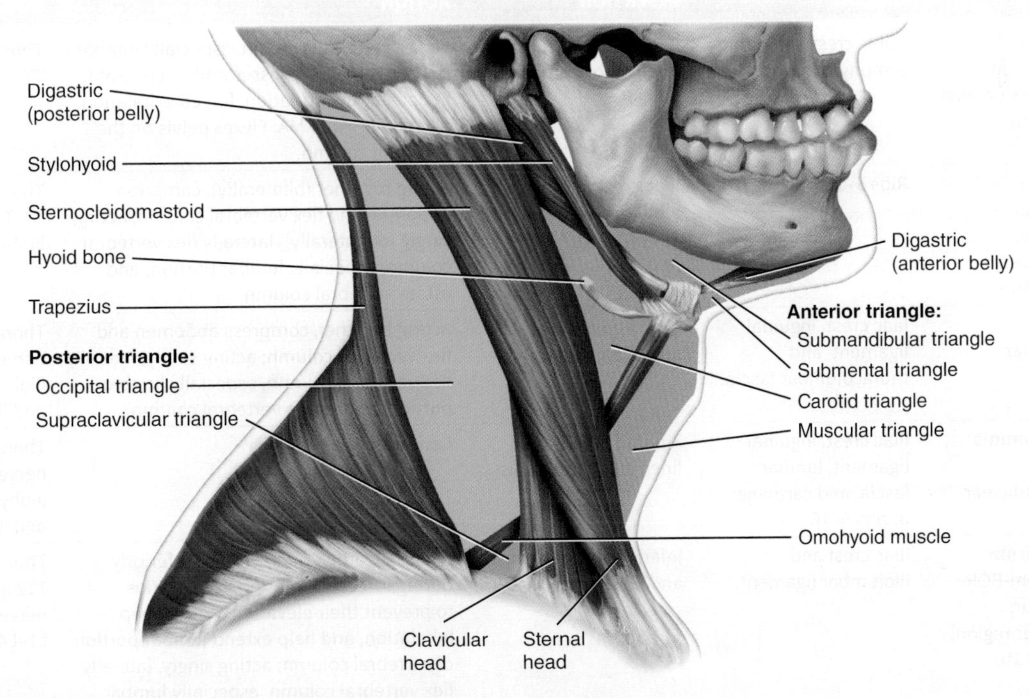

(c) Right lateral view of triangles of neck

Q Why are triangles of the neck important?

unilaterally, their actions are quite different, involving primarily rotation of the head.

The sternocleidomastoid muscle is an important landmark that divides the neck into two major triangles: anterior and posterior (**Figure 11.9c**). The triangles are important anatomically and surgically because of the structures that lie within their boundaries.

The **anterior triangle** is bordered superiorly by the mandible, medially by the cervical midline, and laterally by the anterior border of the sternocleidomastoid muscle. It has its apex at the sternum (**Figure 11.9c**). The anterior triangle is subdivided into three paired triangles: *submandibular*, *carotid*, and *muscular*. An unpaired *submental triangle* is formed by the upper part of the combined right and left anterior triangles. The anterior triangle contains submental, submandibular, and deep cervical lymph nodes; the submandibular salivary gland and a portion of the parotid salivary gland; the facial artery and vein; carotid arteries and internal jugular vein; the thyroid gland and infrahyoid muscles; and the following cranial nerves: glossopharyngeal (IX), vagus (X), accessory (XI), and hypoglossal (XII).

The **posterior triangle** is bordered inferiorly by the clavicle, anteriorly by the posterior border of the sternocleidomastoid muscle, and posteriorly by the anterior border of the trapezius muscle (**Figure 11.9c**). The posterior triangle is subdivided into two triangles, *occipital* and *supraclavicular (omoclavicular)*, by the inferior belly of the omohyoid muscle. The posterior triangle contains part of the subclavian artery, external jugular vein, cervical lymph nodes, brachial plexus, and the accessory (XI) nerve.

Relating Muscles to Movements

Arrange the muscles in this section according to the following actions on the head: (1) flexion, (2) lateral flexion, (3) extension, (4) rotation to side opposite contracting muscle, and (5) rotation to same side as contracting muscle. The same muscle may be mentioned more than once.

11.10 Muscles of the Abdomen That Protect Abdominal Viscera and Move the Vertebral Column

OBJECTIVE

- **Describe** the origin, insertion, action, and innervation of the muscles that protect the abdominal viscera and move the vertebral column.

The anterolateral abdominal wall is composed of skin, fascia, and four pairs of muscles: the external oblique, internal oblique, transversus abdominis, and rectus abdominis (**Figure 11.10**). The first three muscles named are arranged from superficial to deep.

MUSCLE	ORIGIN	INSERTION	ACTION	INNERVATION
Rectus abdominis (REK-tus ab-DOM-in-is; *rectus* = fascicles parallel to midline; *abdominis* = abdomen)	Pubic crest and pubic symphysis.	Cartilage of ribs 5–7 and xiphoid process.	Flexes vertebral column, especially lumbar portion, and compresses abdomen to aid in defecation, urination, forced exhalation, and childbirth. RMA: Flexes pelvis on the vertebral column.	Thoracic spinal nerves T7–T12.
External oblique (ō-BLĒK; *external* = closer to surface; *oblique* = fascicles diagonal to midline)	Ribs 5–12.	Iliac crest and linea alba.	Acting together (bilaterally), compress abdomen and flex vertebral column; acting singly (unilaterally), laterally flex vertebral column, especially lumbar portion, and rotate vertebral column.	Thoracic spinal nerves T7–T12 and the iliohypogastric nerve.
Internal oblique (*internal* = farther from surface)	Iliac crest, inguinal ligament, and thoracolumbar fascia.	Cartilage of ribs 7–10 and linea alba.	Acting together, compress abdomen and flex vertebral column; acting singly, laterally flex vertebral column, especially lumbar portion, and rotate vertebral column.	Thoracic spinal nerves T8–T12, the iliohypogastric nerve, and ilioinguinal nerve.
Transversus abdominis (tranz-VER-sus = fascicles perpendicular to midline)	Iliac crest, inguinal ligament, lumbar fascia, and cartilages of ribs 5–10.	Xiphoid process, linea alba, and pubis.	Compresses abdomen.	Thoracic spinal nerves T8–T12, iliohypogastric nerve, and ilioinguinal nerve.
Quadratus lumborum (kwod-RĀ-tus lum-BŌR-um; *quad-* = four; *lumbo-* = lumbar region) (see **Figure 11.11b**)	Iliac crest and iliolumbar ligament.	Inferior border of rib 12 and L1–L4.	Acting together, pull 12th ribs inferiorly during forced exhalation, fix 12th ribs to prevent their elevation during deep inhalation, and help extend lumbar portion of vertebral column; acting singly, laterally flex vertebral column, especially lumbar portion. RMA: Elevates hip bone, commonly on one side.	Thoracic spinal nerves T12 and lumbar spinal nerves L1–L3 or L1–L4.

FIGURE 11.10 Muscles of the abdomen that protect abdominal viscera and move the vertebral column (backbone).

The anterolateral abdominal muscles protect the abdominal viscera, move the vertebral column, and assist in forced exhalation, defecation, urination, and childbirth.

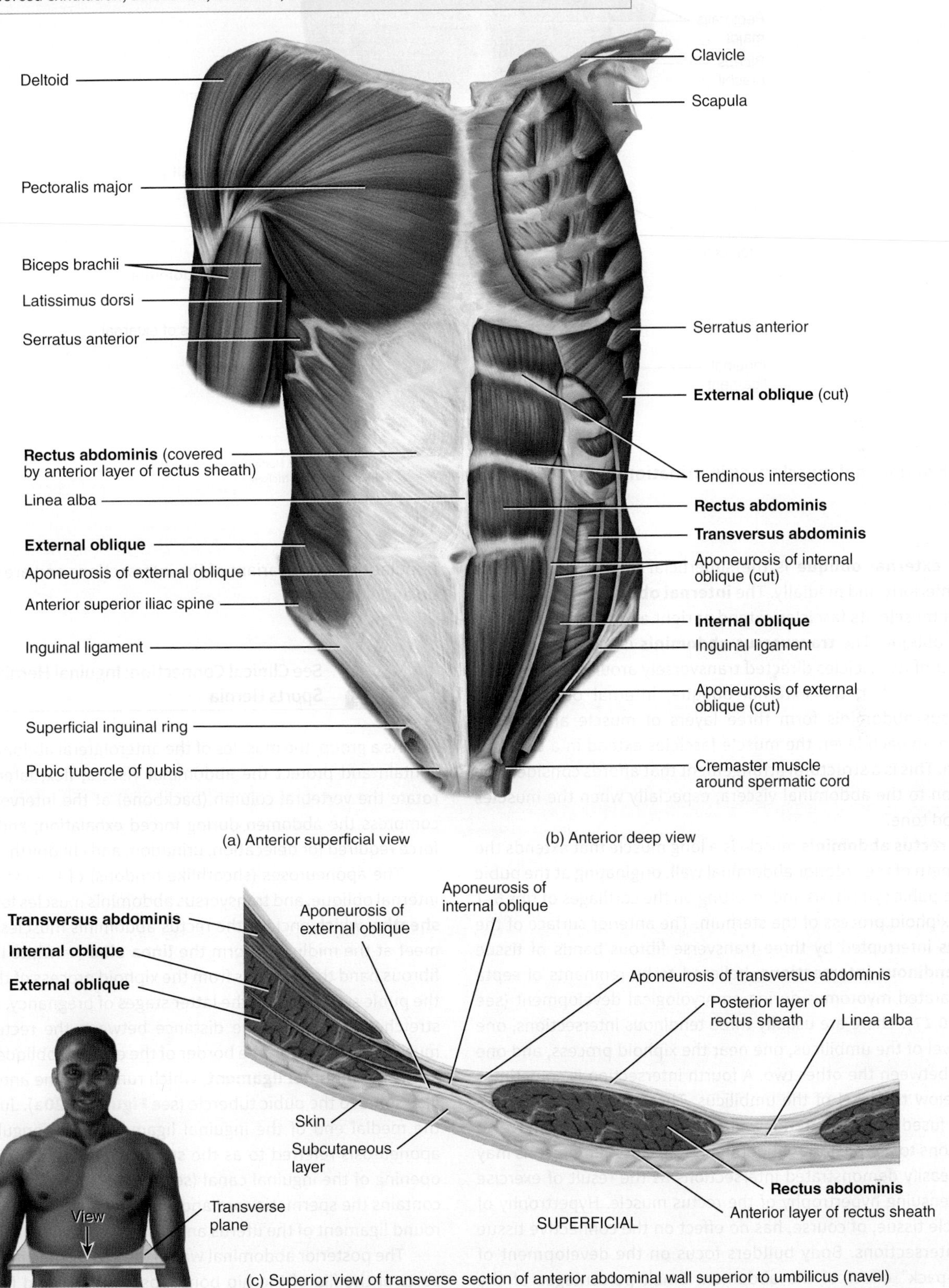

(a) Anterior superficial view

(b) Anterior deep view

(c) Superior view of transverse section of anterior abdominal wall superior to umbilicus (navel)

Figure 11.10 *Continues*

FIGURE 11.10 Continued

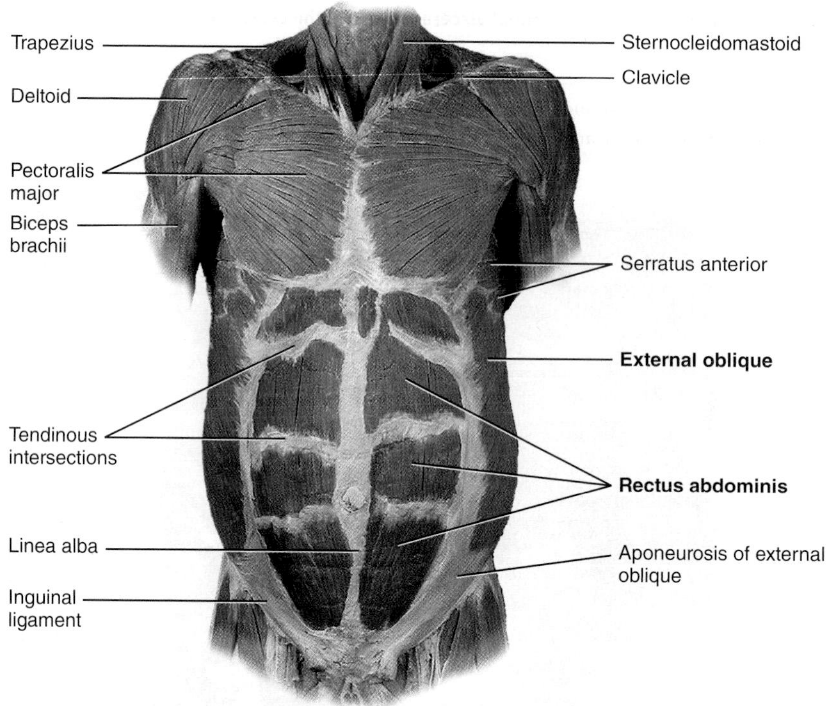

(d) Anterior view

Dissection Nathan Mortensen and Shawn Miller; Photograph Mark Nielsen

Q Which abdominal muscle aids in urination?

The **external oblique** is the superficial muscle. Its fascicles extend inferiorly and medially. The **internal oblique** is the intermediate flat muscle. Its fascicles extend at right angles to those of the external oblique. The **transversus abdominis** is the deep muscle, with most of its fascicles directed transversely around the abdominal wall. Together, the external oblique, internal oblique, and transversus abdominis form three layers of muscle around the abdomen. In each layer, the muscle fascicles extend in a different direction. This is a structural arrangement that affords considerable protection to the abdominal viscera, especially when the muscles have good tone.

The **rectus abdominis** muscle is a long muscle that extends the entire length of the anterior abdominal wall, originating at the pubic crest and pubic symphysis and inserting on the cartilages of ribs 5–7 and the xiphoid process of the sternum. The anterior surface of the muscle is interrupted by three transverse fibrous bands of tissue called **tendinous intersections**, believed to be remnants of septa that separated myotomes during embryological development (see Figure 10.17). There are usually three tendinous intersections, one at the level of the umbilicus, one near the xiphoid process, and one midway between the other two. A fourth intersection is sometimes found below the level of the umbilicus. These tendinous intersections are fused with the anterior wall of the rectus sheath but have no connections to the posterior abdominal wall. Muscular persons may possess easily demonstrated intersections as the result of exercise and the ensuing hypertrophy of the rectus muscle. Hypertrophy of the muscle tissue, of course, has no effect on the connective tissue of the intersections. Body builders focus on the development of the "six-pack" effect of the abdomen. Small percentages of the population have a variant of the intersections and are able to develop an "eight-pack."

See Clinical Connection: Inguinal Hernia and Sports Hernia

As a group, the muscles of the anterolateral abdominal wall help contain and protect the abdominal viscera; flex, laterally flex, and rotate the vertebral column (backbone) at the intervertebral joints; compress the abdomen during forced exhalation; and produce the force required for defecation, urination, and childbirth.

The aponeuroses (sheathlike tendons) of the external oblique, internal oblique, and transversus abdominis muscles form the **rectus sheaths**, which enclose the rectus abdominis muscles. The sheaths meet at the midline to form the **linea alba** (= white line), a tough, fibrous band that extends from the xiphoid process of the sternum to the pubic symphysis. In the latter stages of pregnancy, the linea alba stretches to increase the distance between the rectus abdominis muscles. The inferior free border of the external oblique aponeurosis forms the **inguinal ligament**, which runs from the anterior superior iliac spine to the pubic tubercle (see Figure 11.20a). Just superior to the medial end of the inguinal ligament is a triangular slit in the aponeurosis referred to as the **superficial inguinal ring**, the outer opening of the inguinal canal (see Figure 28.2). The **inguinal canal** contains the spermatic cord and ilioinguinal nerve in males, and the round ligament of the uterus and ilioinguinal nerve in females.

The posterior abdominal wall is formed by the lumbar vertebrae, parts of the ilia of the hip bones, psoas major and iliacus muscles (described in Section 11.20), and quadratus lumborum muscle. The anterolateral abdominal wall can contract and distend; the posterior abdominal wall is bulky and stable by comparison.

Relating Muscles to Movements

Arrange the muscles in this section according to the following actions on the vertebral column: (1) flexion, (2) lateral flexion, (3) extension, and (4) rotation. The same muscle may be mentioned more than once.

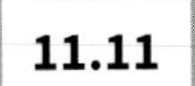

11.11 Muscles of the Thorax That Assist in Breathing

OBJECTIVE

- **Describe** the origin, insertion, action, and innervation of the muscles of the thorax that assist in breathing.

The muscles of the thorax (chest) alter the size of the thoracic cavity so that breathing can occur. Inhalation (breathing in) occurs when the thoracic cavity increases in size, and exhalation (breathing out) occurs when the thoracic cavity decreases in size.

The dome-shaped **diaphragm** is the most important muscle that powers breathing. It also separates the thoracic and abdominal cavities. The diaphragm has a convex superior surface that forms the floor of the thoracic cavity (Figure 11.11b) and a concave, inferior surface that forms the roof of the abdominal cavity (Figure 11.11b). The **peripheral muscular portion** of the diaphragm originates on the xiphoid process of the sternum, the inferior six ribs and their costal cartilages, and the lumbar vertebrae and their intervertebral discs and the twelfth rib (Figure 11.11d). From their various origins, the fibers of the muscular portion converge and insert into the **central tendon**, a strong aponeurosis located near the center of the muscle (Figure 11.11b–d). The central tendon fuses with the inferior surface of the pericardium (covering of the heart) and the pleurae (coverings of the lungs).

The diaphragm has three major openings through which various structures pass between the thorax and abdomen. These structures include the aorta, along with the thoracic duct and azygous vein, which pass through the **aortic hiatus**; the esophagus with accompanying vagus (X) nerves, which pass through the **esophageal hiatus**; and the inferior vena cava, which passes through the **caval opening** (*foramen for the vena cava*). In a condition called a *hiatus hernia*, the stomach protrudes superiorly through the esophageal hiatus.

Movements of the diaphragm also help return venous blood passing through abdominal veins to the heart. Together with the anterolateral abdominal muscles, the diaphragm helps to increase intra-abdominal pressure to evacuate the pelvic contents during defecation, urination, and childbirth. This mechanism is further assisted when you take a deep breath and close the rima glottidis (the space between the vocal folds). The trapped air in the respiratory system prevents the diaphragm from elevating. The increase in intra-abdominal pressure also helps support the vertebral column and helps prevent flexion during weight lifting. This greatly assists the back muscles in lifting a heavy weight.

Other muscles involved in breathing, called **intercostals**, span the *intercostal spaces*, the spaces between ribs. These muscles are arranged in three layers, only two of which are discussed here. The 11 pairs of **external intercostals** occupy the superficial layer, and their fibers run in an oblique direction interiorly and anteriorly from the rib above to the rib below. They elevate the ribs during inhalation to help expand the thoracic cavity. The 11 pairs of **internal intercostals** occupy the intermediate layer of the intercostal spaces. The fibers of these muscles run at right angles to the external intercostals, in an oblique direction interiorly and posteriorly from the inferior border of the rib above to the superior border of the rib below. They draw adjacent ribs together during forced exhalation to help decrease the size of the thoracic cavity.

Note: A mnemonic for the action of the intercostal muscles is singing "Old MacDonald had a farm, **E, I, E, I, O**" = **E**xternal **I**ntercostals **E**levate during **I**nhalation, **O**h!"

MUSCLE	ORIGIN	INSERTION	ACTION	INNERVATION
Diaphragm (DĪ-a-fram; *dia-* = *across*; *-phragm* = wall)	Xiphoid process of sternum, costal cartilages and adjacent portions of ribs 7–12, lumbar vertebrae and their intervertebral discs.	Central tendon.	Contraction of diaphragm causes it to flatten and increases vertical dimension of thoracic cavity, resulting in inhalation; relaxation of diaphragm causes it to move superiorly and decreases vertical dimension of thoracic cavity, resulting in exhalation.	Phrenic nerve, which contains axons from cervical spinal nerves (C3–C5).
External intercostals (in′-ter-KOS-tals; *external* = closer to surface; *inter-* = between; *-costa* = rib)	Inferior border of rib above.	Superior border of rib below.	Contraction elevates ribs and increases anteroposterior and lateral dimensions of thoracic cavity, resulting in inhalation; relaxation depresses ribs and decreases anteroposterior and lateral dimensions of thoracic cavity, resulting in exhalation.	Thoracic spinal nerves T2–T12.
Internal intercostals (*internal* = farther from surface)	Superior border of rib below.	Inferior border of rib above.	Contraction draws adjacent ribs together to further decrease anteroposterior and lateral dimensions of thoracic cavity during forced exhalation.	Thoracic spinal nerves T2–T12.

As you will see in Chapter 23, the diaphragm and external intercostal muscles are used during quiet inhalation and exhalation. However, during deep, forceful inhalation (during exercise or playing a wind instrument), the sternocleidomastoid, scalene, and pectoralis minor muscles are also used; during deep, forceful exhalation, the external oblique, internal oblique, transversus abdominis, rectus abdominis, and internal intercostals are also used.

Relating Muscles to Movements

Arrange the muscles in this section according to the following actions: (1) increase in vertical length, (2) increase in lateral and anteroposterior dimensions, and (3) decrease in lateral and anteroposterior dimensions of the thorax.

FIGURE 11.11 **Muscles of the thorax (chest) that assist in breathing.**

Openings in the diaphragm permit the passage of the aorta, esophagus, and inferior vena cava.

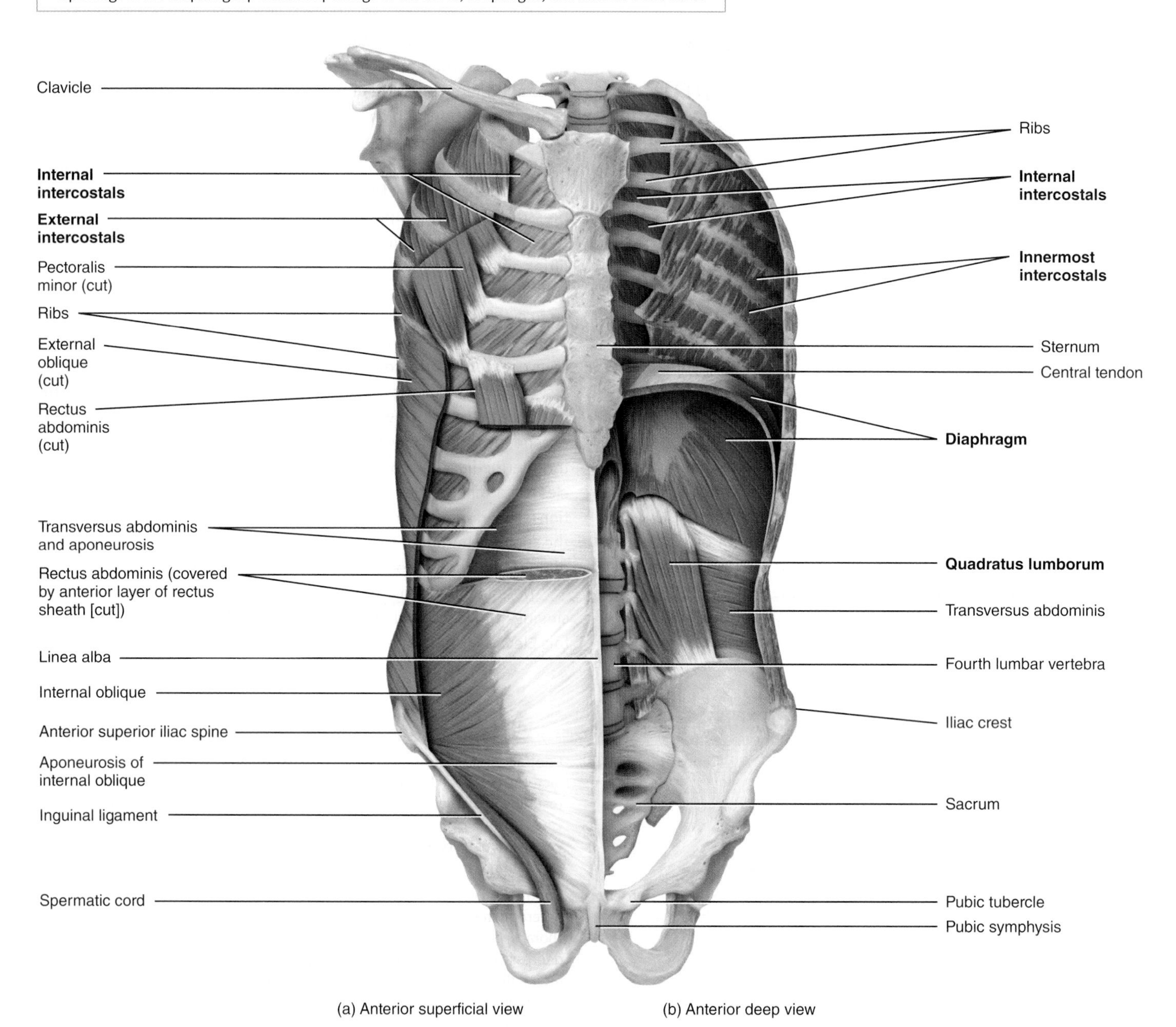

(a) Anterior superficial view (b) Anterior deep view

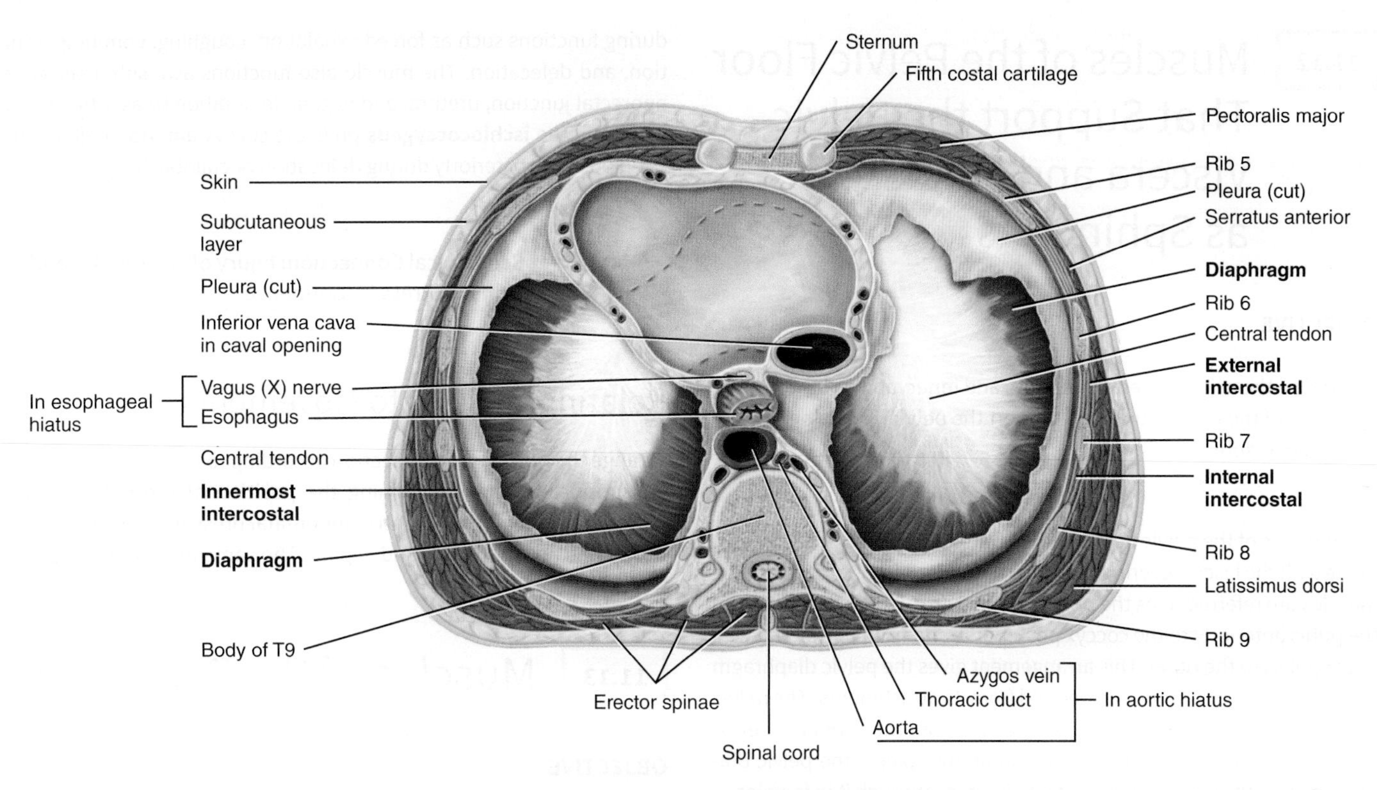

(c) Superior view of diaphragm

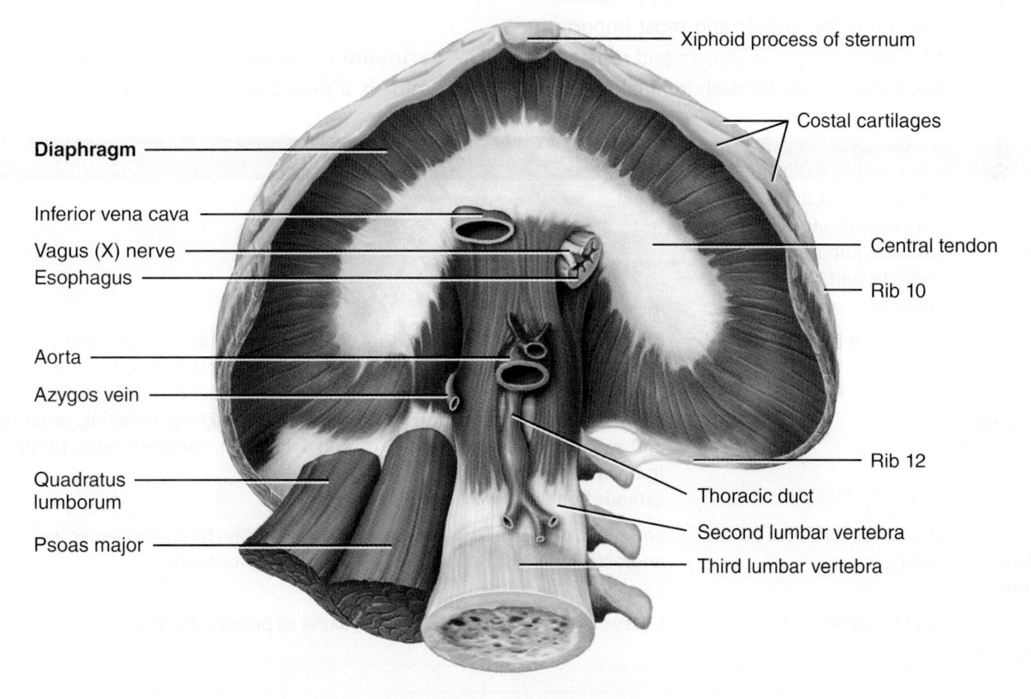

(d) Inferior view of diaphragm

Q Which muscle associated with breathing is innervated by the phrenic nerve?

11.12 Muscles of the Pelvic Floor That Support the Pelvic Viscera and Function as Sphincters

OBJECTIVE

- **Describe** the origin, insertion, action, and innervation of the muscles of the pelvic floor that support the pelvic viscera and function as sphincters.

The muscles of the pelvic floor are the levator ani and ischiococcygeus. Along with the fascia covering their internal and external surfaces, these muscles are referred to as the **pelvic diaphragm**, which stretches from the pubis anteriorly to the coccyx posteriorly, and from one lateral wall of the pelvis to the other. This arrangement gives the pelvic diaphragm the appearance of a funnel suspended from its attachments. The pelvic diaphragm separates the pelvic cavity above from the perineum below (see Section 11.13). The anal canal and urethra pierce the pelvic diaphragm in both sexes, and the vagina also goes through it in females.

The three components of the **levator ani** muscle are the **pubococcygeus, puborectalis**, and **iliococcygeus**. Figure 11.12 shows these muscles in the female and Figure 11.13 in Section 11.13 illustrates them in the male. The levator ani is the largest and most important muscle of the pelvic floor. It supports the pelvic viscera and resists the inferior thrust that accompanies increases in intra-abdominal pressure during functions such as forced exhalation, coughing, vomiting, urination, and defecation. The muscle also functions as a sphincter at the anorectal junction, urethra, and vagina. In addition to assisting the levator ani, the **ischiococcygeus** pulls the coccyx anteriorly after it has been pushed posteriorly during defecation or childbirth.

See Clinical Connection: Injury of Levator Ani and Urinary Stress Incontinence

Relating Muscles to Movements

Arrange the muscles in this section according to the following actions: (1) supporting and maintaining the position of the pelvic viscera; (2) resisting an increase in intra-abdominal pressure; and (3) constriction of the anus, urethra, and vagina. The same muscle may be mentioned more than once.

11.13 Muscles of the Perineum

OBJECTIVE

- **Describe** the origin, insertion, action, and innervation of the muscles of the perineum.

The **perineum** is the region of the trunk inferior to the pelvic diaphragm. It is a diamond-shaped area that extends from the pubic

MUSCLE	ORIGIN	INSERTION	ACTION	INNERVATION
Levator ani (le-VĀ-tor Ā-nē; *levator* = raises; *ani* = anus)	Muscle is divisible into three parts: pubococcygeus muscle, puborectalis muscle, and iliococcygeus muscle.			
Pubococcygeus (pū′-bō-kok-SIJ-ē-us; *pubo-* = pubis; *-coccygeus* = coccyx)	Pubis and ischial spine.	Coccyx, urethra, anal canal, perineal body of perineum (wedge-shaped mass of fibrous tissue in center of perineum), and anococcygeal ligament (narrow fibrous band that extends from anus to coccyx).	Supports and maintains position of pelvic viscera; resists increase in intra-abdominal pressure during forced exhalation, coughing, vomiting, urination, and defecation; constricts anus, urethra, and vagina.	Sacral spinal nerves S2–S4.
Puborectalis (pū-bō-rek-TĀ-lis; *rectal* = rectum)	Posterior surface of pubic body.	Forms a sling posterior to the anorectal junction.	Helps maintain fecal continence and assists in defecation.	Sacral spinal nerves S2–S4.
Iliococcygeus (il′-ē-ō-kok-SIJ-ē-us; *ilio-* = ilium)	Ischial spine.	Coccyx.	Same as pubococcygeus	Sacral spinal nerves S2–S4.
Ischiococcygeus (is′-kē-ō-kok-SIJ-ē-us; *ischio-*= hip)	Ischial spine.	Lower sacrum and upper coccyx.	Supports and maintains position of pelvic viscera; resists increase in intra-abdominal pressure during forced exhalation, coughing, vomiting, urination, and defecation; pulls coccyx anteriorly following defecation or childbirth.	Sacral spinal nerves S4–S5.

FIGURE 11.12 **Muscles of the pelvic floor that support the pelvic viscera, assist in resisting intra-abdominal pressure, and function as sphincters.**

The pelvic diaphragm supports the pelvic viscera.

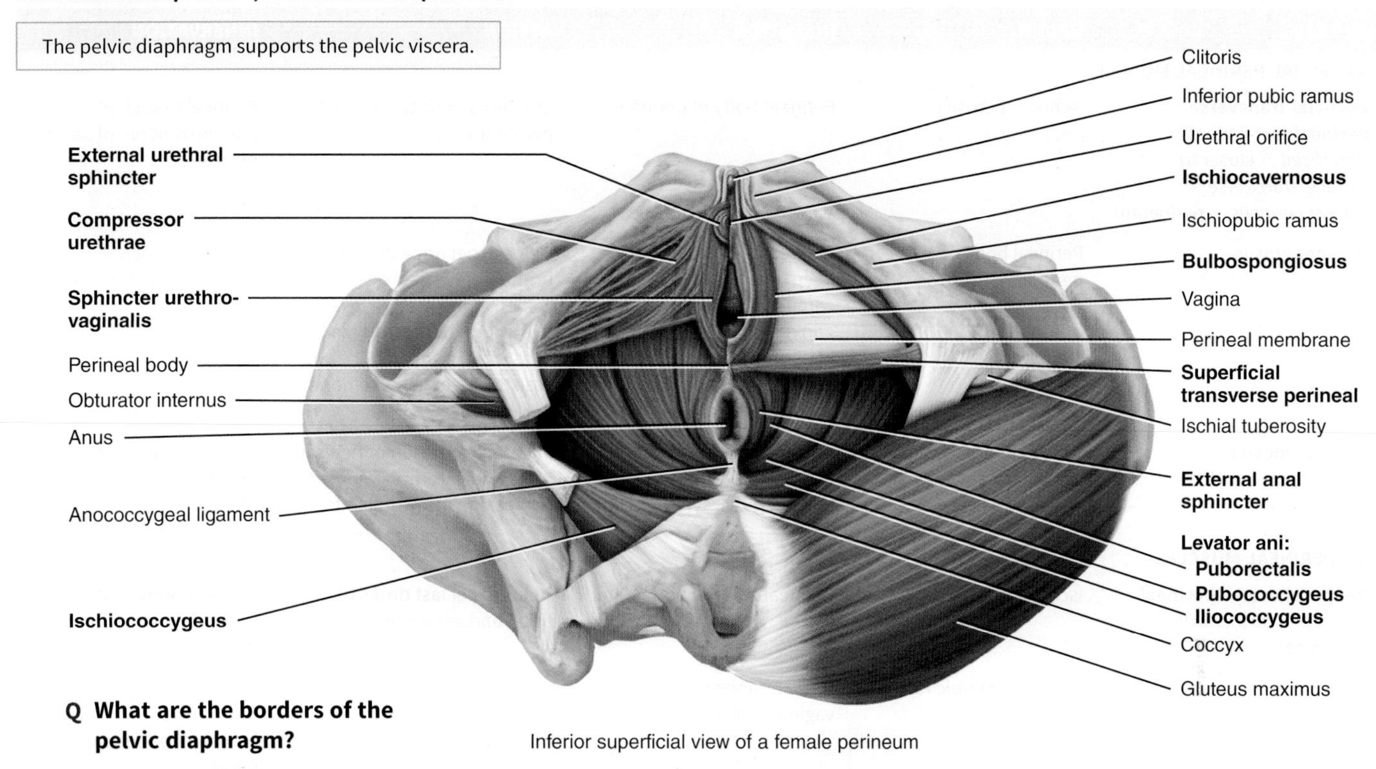

Inferior superficial view of a female perineum

Q What are the borders of the pelvic diaphragm?

FIGURE 11.13 **Muscles of the perineum.**

The urogenital diaphragm assists in urination in females and males, plays a part in ejaculation in males, and helps strengthen the pelvic floor.

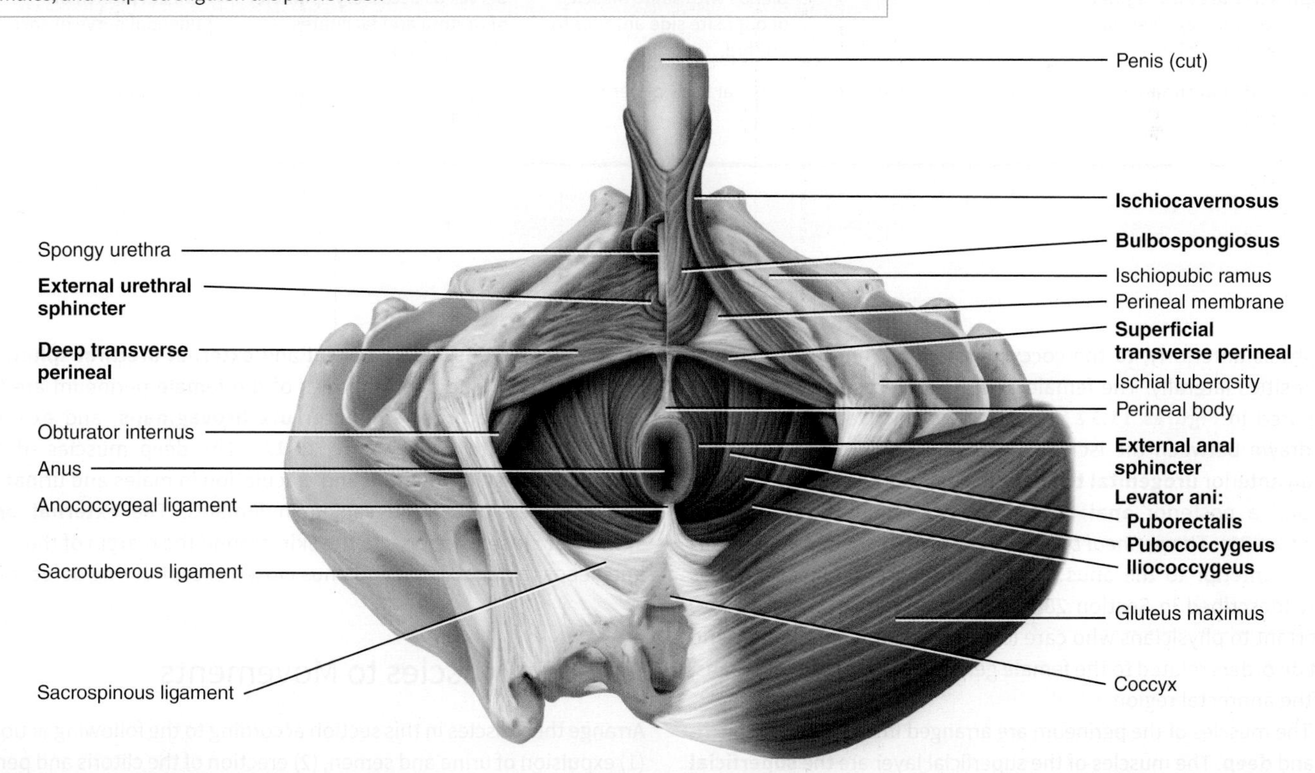

Inferior superficial view of male perineum

Q What are the borders of the perineum?

MUSCLE	ORIGIN	INSERTION	ACTION	INNERVATION
SUPERFICIAL PERINEAL MUSCLES				
Superficial transverse perineal (per-i-NĒ-al; *superficial* = closer to surface; *transverse* = across; *perineus* = perineum)	Ischial tuberosity.	Perineal body of perineum.	Stabilizes perineal body of perineum.	Perineal branch of pudendal nerve of sacral plexus.
Bulbospongiosus (bul′-bō-spon′-jē-Ō-sus; *bulb-* = bulb; *-spongio-* = sponge)	Perineal body of perineum.	Perineal membrane of deep muscles of perineum, corpus spongiosum of penis, and deep fascia on dorsum of penis in male; pubic arch and root and dorsum of clitoris in female.	Helps expel urine during urination, helps propel semen along urethra, assists in erection of penis in male; constricts vaginal orifice and assists in erection of clitoris in female.	Perineal branch of pudendal nerve of sacral plexus.
Ischiocavernosus (is′-kē-ō-ka′-ver-NŌ-sus; *ischio-* = hip)	Ischial tuberosity and ischial and pubic rami.	Corpora cavernosa of penis in male and clitoris in female; pubic symphysis.	Maintains erection of penis in male and clitoris in female by decreasing urine drainage.	Perineal branch of pudendal nerve of sacral plexus.
DEEP PERINEAL MUSCLES				
Deep transverse perineal (*deep* = farther from surface)	Ischial ramus.	Perineal body of perineum.	Helps expel last drops of urine and semen in male.	Perineal branch of pudendal nerve of sacral plexus.
External urethral sphincter (ū-RĒ-thral SFINGK-ter)	Ischial and pubic rami.	Median raphe in male and vaginal wall in female.	Helps expel last drops of urine and semen in male and urine in female.	Sacral spinal nerve S4 and inferior rectal branch of pudendal nerve.
Compressor urethrae (ū-RĒ-thrē) (see Figure 11.12)	Ischiopubic ramus.	Blends with same muscle of opposite side anterior to urethra.	Serves as accessory sphincter of urethra.	Perineal branch of pudendal nerve of sacral plexus.
Sphincter urethrovaginalis (ū-RĒ-thrō-vaj-i-NAL-is) (see Figure 11.12)	Perineal body.	Blends with same muscle of opposite side anterior to urethra.	Serves as accessory sphincter of urethra and facilitates closing of vagina.	Perineal branch of pudendal nerve of sacral plexus.
External anal sphincter (Ā-nal)	Anococcygeal ligament.	Perineal body of perineum.	Keeps anal canal and anus closed.	Sacral spinal nerve S4 and inferior rectal branch of pudendal nerve.

symphysis anteriorly, to the coccyx posteriorly, and to the ischial tuberosities laterally. The female and the male perineums may be compared in Figures 11.12 and 11.13, respectively. A transverse line drawn between the ischial tuberosities divides the perineum into an anterior **urogenital triangle** that contains the external genitals and a posterior **anal triangle** that contains the anus (see Figure 28.21). The *perineal body* of the perineum, a muscular intersection anterior to the anus into which several perineal muscles insert (described in Section 28.1). Clinically, the perineum is very important to physicians who care for women during pregnancy and treat disorders related to the female genital tract, urogenital organs, and the anorectal region.

The muscles of the perineum are arranged in two layers: **superficial** and **deep**. The muscles of the superficial layer are the **superficial transverse perineal**, the **bulbospongiosus**, and the **ischiocavernosus** (Figures 11.12 and 11.13). The deep muscles of the male perineum are the **deep transverse perineal** and **external urethral sphincter** (Figure 11.13). The deep muscles of the female perineum are the **compressor urethrae, sphincter urethrovaginalis**, and external urethral sphincter (see Figure 11.12). The deep muscles of the perineum assist in urination and ejaculation in males and urination and compression of the vagina in females. The **external anal sphincter** closely adheres to the skin around the margin of the anus and keeps the anal canal and anus closed except during defecation.

Relating Muscles to Movements

Arrange the muscles in this section according to the following actions; (1) expulsion of urine and semen, (2) erection of the clitoris and penis, (3) closure of the anal orifice, and (4) constriction of the vaginal orifice. The same muscle may be mentioned more than once.

11.14 Muscles of the Thorax That Move the Pectoral Girdle

OBJECTIVE

- **Describe** the origin, insertion, action, and innervation of the muscles of the thorax that move the pectoral girdle.

The main action of the muscles that move the pectoral (shoulder) girdle (clavicle and scapula) is to stabilize the scapula so it can function as a steady origin for most of the muscles that move the humerus. Because scapular movements usually accompany humeral movements in the same direction, the muscles also move the scapula to increase the range of motion of the humerus. For example, it would not be possible to raise the arm above the head if the scapula did not move with the humerus. During abduction, the scapula follows the humerus by rotating upward.

Muscles that move the pectoral girdle can be classified into two groups based on their location in the thorax: **anterior** and **posterior thoracic muscles** (Figure 11.14). The anterior thoracic muscles are the subclavius, pectoralis minor, and serratus anterior. The **subclavius** is a small, cylindrical muscle under the clavicle that extends from the clavicle to the first rib. It steadies the clavicle during movements of the pectoral girdle. The **pectoralis minor** is a thin, flat, triangular muscle that is deep to the pectoralis major. Besides its role in movements of the scapula, the pectoralis minor muscle assists in forced inhalation. The **serratus anterior** is a large, flat, fan-shaped muscle between the ribs and scapula. It is so named because of the saw-toothed appearance of its origins on the ribs.

The posterior thoracic muscles are the trapezius, levator scapulae, rhomboid major, and rhomboid minor. The **trapezius** is a large, flat, triangular sheet of muscle extending from the skull and vertebral column medially to the pectoral girdle laterally. It is the most superficial back muscle and covers the posterior neck region and superior portion of the trunk. The two trapezius muscles form a trapezoid (diamond-shaped quadrangle)—hence its name. The **levator scapulae** is a narrow, elongated muscle in the posterior portion of the neck. It is deep to the sternocleidomastoid and trapezius muscles. As its name

MUSCLE	ORIGIN	INSERTION	ACTION	INNERVATION
ANTERIOR THORACIC MUSCLES				
Subclavius (sub-KLĀ-vē-us; *sub-* = under; *-clavius* = clavicle)	Rib 1.	Clavicle.	Depresses and moves clavicle anteriorly and helps stabilize pectoral girdle.	Subclavian nerve.
Pectoralis minor (pek′-tō-RĀ-lis; *pector* = breast, chest, thorax; *minor* = lesser)	Ribs 2–5, 3–5, or 2–4.	Coracoid process of scapula.	Abducts scapula and rotates it downward. RMA: Elevates ribs 3–5 during forced inhalation when scapula is fixed.	Medial pectoral nerve.
Serratus anterior (ser-Ā-tus; *serratus* = saw-toothed; *anterior* = front)	Ribs 1–8 or 1–9.	Vertebral border and inferior angle of scapula.	Abducts scapula and rotates it upward. RMA: Elevates ribs when scapula is stabilized. Known as "boxer's muscle" because it is important in horizontal arm movements such as punching and pushing.	Long thoracic nerve.
POSTERIOR THORACIC MUSCLES				
Trapezius (tra-PĒ-zē-us; *trapezi* = trapezoid-shaped)	Superior nuchal line of occipital bone, ligamentum nuchae, and spines of C7–T12.	Clavicle and acromion and spine of scapula.	Superior fibers upward rotate scapula; middle fibers adduct scapula; inferior fibers depress and upward rotate scapula; superior and inferior fibers together rotate scapula upward; stabilizes scapula. RMA: Superior fibers can help extend head.	Accessory (XI) nerve and cervical spinal nerves C3–C5.
Levator scapulae (le-VĀ-tor SKA-pū-lē; *levator* = raises; *scapulae* = scapula)	Transverse processes of C1–C4.	Superior vertebral border of scapula.	Elevates scapula and rotates it downward.	Dorsal scapular nerve and cervical spinal nerves C3–C5.
Rhomboid major (rom-BOYD; *rhomboid* = rhomboid or diamond-shaped) (see Figure 11.15c)	Spines of T2–T5.	Vertebral border of scapula inferior to spine.	Elevates and adducts scapula and rotates it downward; stabilizes scapula.	Dorsal scapular nerve.
Rhomboid minor (see Figure 11.15c)	Spines of C7–T1.	Vertebral border of scapula superior to spine.	Elevates and adducts scapula and rotates it downward; stabilizes scapula.	Dorsal scapular nerve.

FIGURE 11.14 Muscles of the thorax (chest) that move the pectoral (shoulder) girdle (clavicle and scapula).

Muscles that move the pectoral girdle originate on the axial skeleton and insert on the clavicle or scapula.

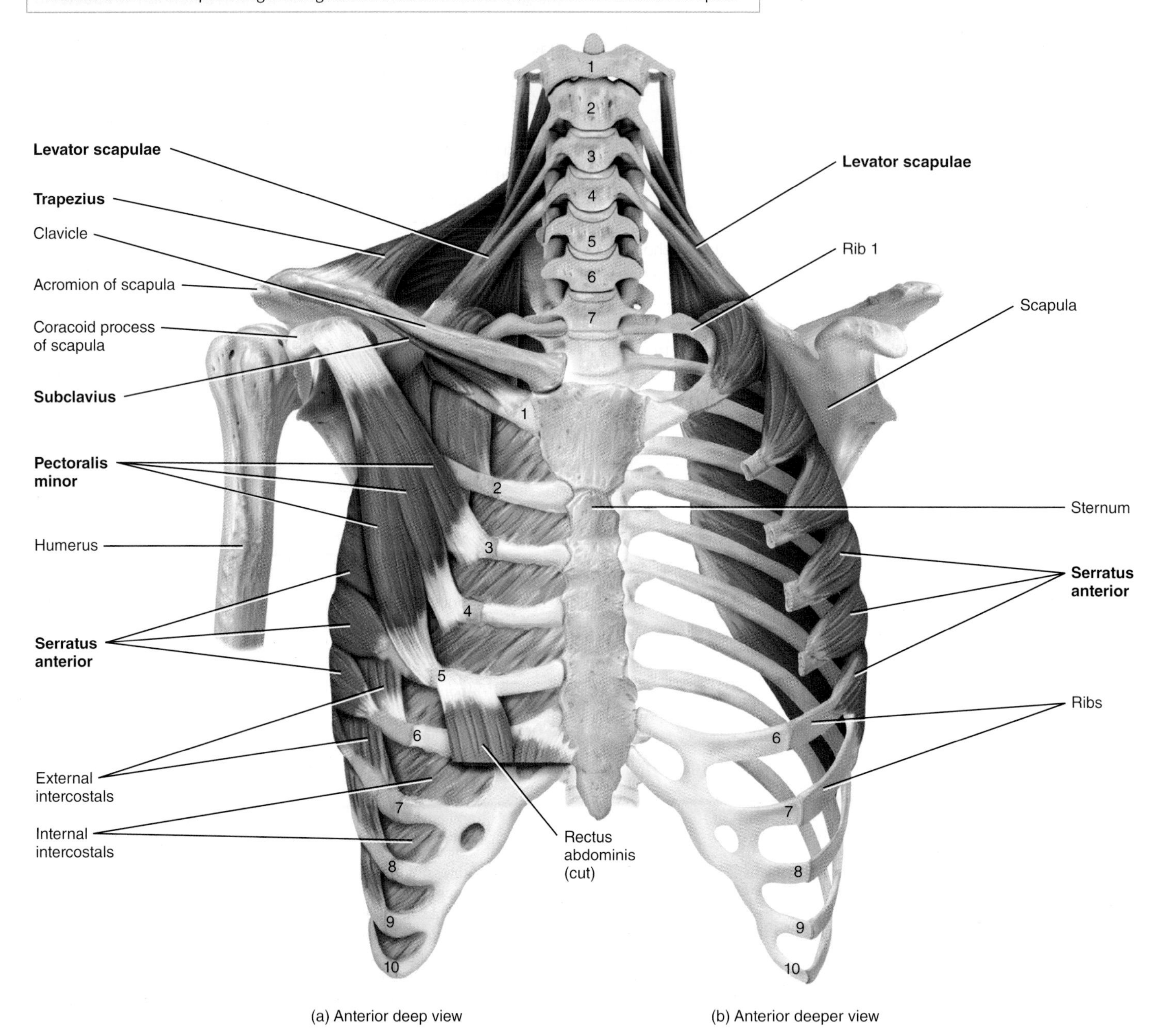

Q What is the main action of the muscles that move the pectoral girdle?

suggests, one of its actions is to elevate the scapula (see Figure 11.15c). The **rhomboid major** and **rhomboid minor** lie deep to the trapezius and are not always distinct from each other. They appear as parallel bands that pass inferiorly and laterally from the vertebrae to the scapula (see Figure 11.15c). Their names are based on their shape—that is, a rhomboid (an oblique parallelogram). The rhomboid major is about two times wider than the rhomboid minor. Both muscles are used when forcibly lowering the raised upper limbs, as in driving a stake with a sledgehammer.

To understand the actions of muscles that move the scapula, it is first helpful to review the various movements of the scapula:

- **Elevation:** superior movement of the scapula, such as shrugging the shoulders or lifting a weight over the head.
- **Depression:** inferior movement of the scapula, as in pulling down on a rope attached to a pulley.
- **Abduction** (*protraction*): movement of the scapula laterally and anteriorly, as in doing a "push-up" or punching.

- **Adduction** (*retraction*): movement of the scapula medially and posteriorly, as in pulling the oars in a rowboat.
- **Upward rotation:** movement of the inferior angle of the scapula laterally so that the glenoid cavity is moved upward. This movement is required to move the humerus past the horizontal, as in raising the arms in a "jumping jack."
- **Downward rotation:** movement of the inferior angle of the scapula medially so that the glenoid cavity is moved downward. This movement is seen when a gymnast on parallel bars supports the weight of the body on the hands.

Relating Muscles to Movements

Arrange the muscles in this section according to the following actions on the scapula: (1) depression, (2) elevation, (3) abduction, (4) adduction, (5) upward rotation, and (6) downward rotation. The same muscle may be mentioned more than once.

11.15 Muscles of the Thorax and Shoulder That Move the Humerus

OBJECTIVE

- **Describe** the origin, insertion, action, and innervation of the muscles of the thorax that move the humerus.

Of the nine muscles that cross the shoulder joint, all except the pectoralis major and latissimus dorsi originate on the scapula (shoulder blade). The pectoralis major and latissimus dorsi thus are called **axial muscles**, because they originate on the axial skeleton. The remaining seven muscles, the **scapular muscles**, arise from the scapula (Figure 11.15).

MUSCLE	ORIGIN	INSERTION	ACTION	INNERVATION
AXIAL MUSCLES THAT MOVE THE HUMERUS				
Pectoralis major (pek′-tō-RĀ-lis; *pector* = chest; *major* = larger) (see also Figure 11.10a)	Clavicle (clavicular head), sternum, and costal cartilages of ribs 2–6 and sometimes ribs 1–7 (sternocostal head).	Greater tubercle and lateral lip of intertubercular sulcus of humerus.	As a whole, adducts and medially rotates arm at shoulder joint; clavicular head flexes arm, and sternocostal head extends flexed arm to side of trunk.	Medial and lateral pectoral nerves
Latissimus dorsi (la-TIS-i-mus DOR-sī; *latissimus* = widest; *dorsi* = of the back)	Spines of T7–L5, lumbar vertebrae, crests of sacrum and ilium, ribs 9–12 via thoracolumbar fascia.	Intertubercular sulcus of humerus.	Extends, adducts, and medially rotates arm at shoulder joint; draws arm inferiorly and posteriorly. RMA: Elevates vertebral column and torso.	Thoracodorsal nerve.
SCAPULAR MUSCLES THAT MOVE THE HUMERUS				
Deltoid (DEL-toyd = triangularly shaped)	Acromial extremity of clavicle (anterior fibers), acromion of scapula (lateral fibers), and spine of scapula (posterior fibers).	Deltoid tuberosity of humerus.	Lateral fibers abduct arm at shoulder joint; anterior fibers flex and medially rotate arm at shoulder joint; posterior fibers extend and laterally rotate arm at shoulder joint.	Axillary nerve.
Subscapularis (sub-scap′-ū-LĀ-ris; *sub-* = below; *-scapularis* = scapula)	Subscapular fossa of scapula.	Lesser tubercle of humerus.	Medially rotates arm at shoulder joint.	Upper and lower subscapular nerve.
Supraspinatus (soo-pra-spī-NĀ-tus; *supra-* = above; *-spina* = spine [of the scapula])	Supraspinous fossa of scapula.	Greater tubercle of humerus.	Assists deltoid muscle in abducting arm at shoulder joint.	Suprascapular nerve.
Infraspinatus (in′-fra-spī-NĀ-tus; *infra-* = below)	Infraspinous fossa of scapula.	Greater tubercle of humerus.	Laterally rotates arm at shoulder joint.	Suprascapular nerve.
Teres major (TE-rēz; *teres* = long and round)	Inferior angle of scapula.	Medial lip of intertubercular sulcus of humerus.	Extends arm at shoulder joint and assists in adduction and medial rotation of arm at shoulder joint.	Lower subscapular nerve.
Teres minor	Inferior lateral border of scapula.	Greater tubercle of humerus.	Laterally rotates and extends arm at shoulder joint.	Axillary nerve.
Coracobrachialis (kor′-a-kō-brā-kē-Ā-lis; *coraco-* = coracoid process [of the scapula]; *-brachi-* = arm)	Coracoid process of scapula.	Middle of medial surface of shaft of humerus.	Flexes and adducts arm at shoulder joint.	Musculocutaneous nerve.

FIGURE 11.15 **Muscles of the thorax (chest) and shoulder that move the humerus (arm bone).**

The strength and stability of the shoulder joint are provided by the tendons that form the rotator cuff.

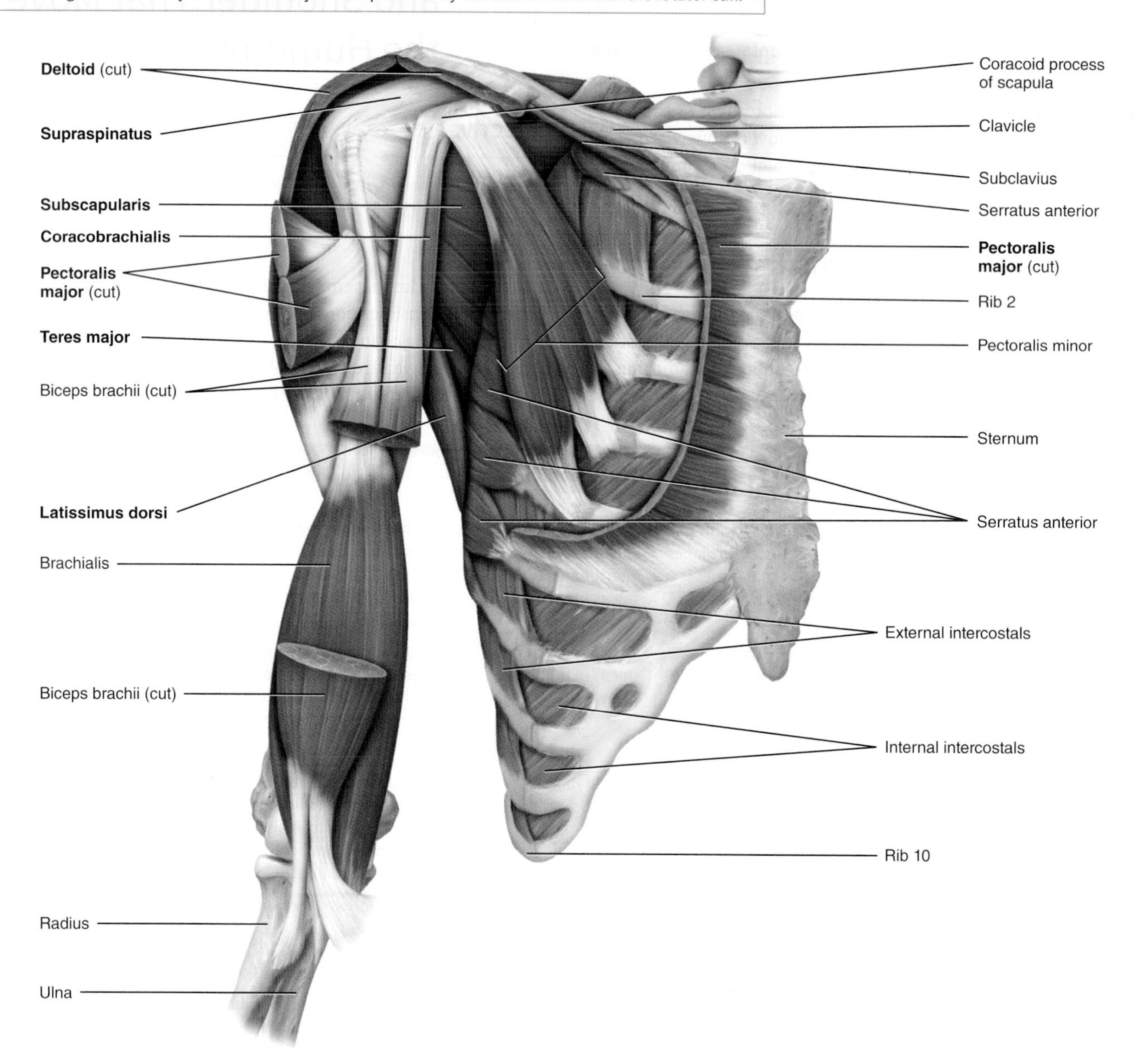

(a) Anterior deep view (the intact pectoralis major muscle is shown in Figure 11.3a)

See Clinical Connection: Impingement Syndrome

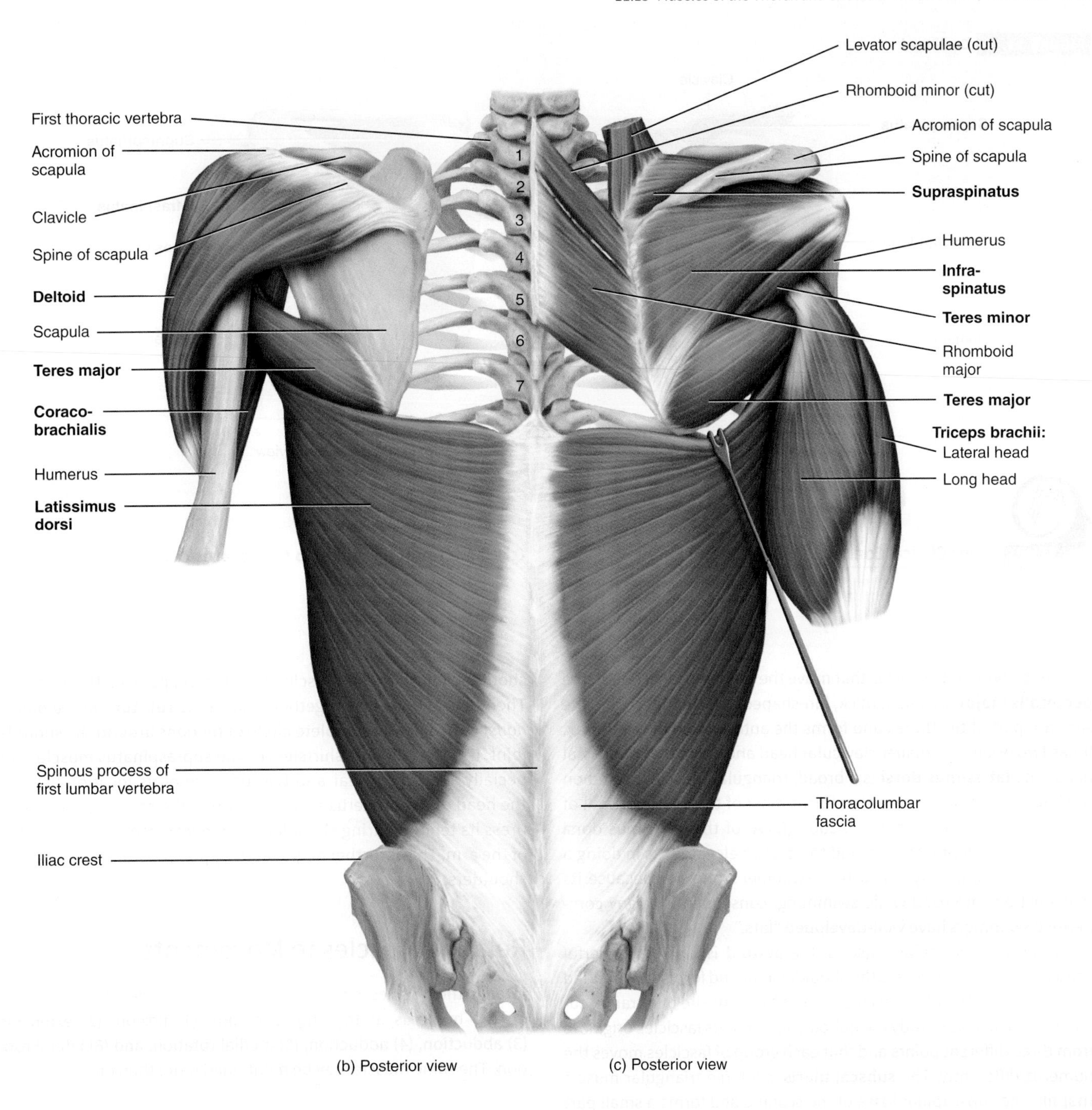

(b) Posterior view

(c) Posterior view

Figure 11.15 ***Continues***

FIGURE 11.15 **Continued**

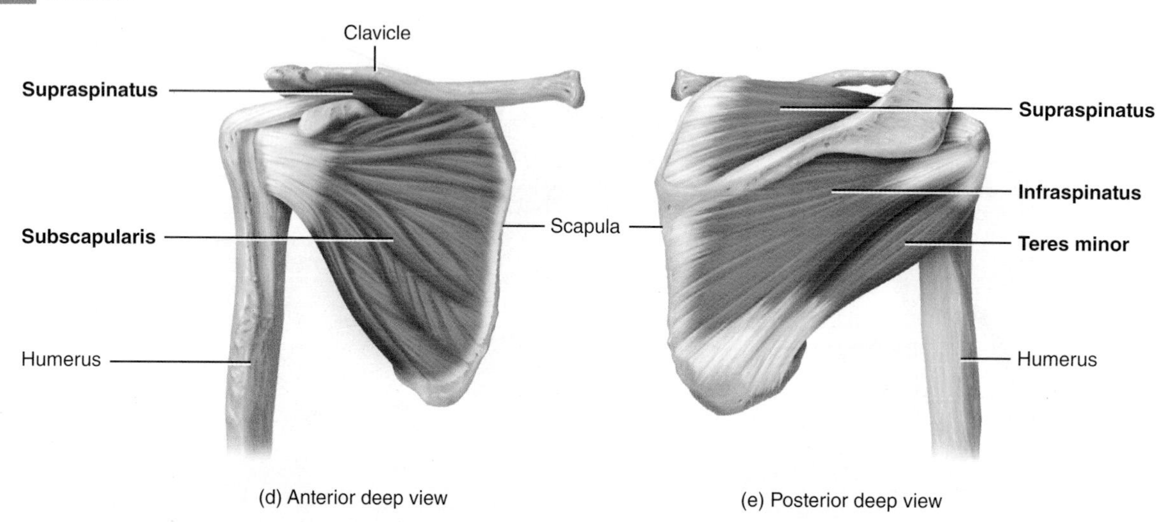

See Clinical Connection: Rotator Cuff Injury

Q Which tendons make up the rotator cuff?

Of the two axial muscles that move the humerus (arm bone), the **pectoralis major** is a large, thick, fan-shaped muscle that covers the superior part of the thorax and forms the anterior fold of the thorax. It has two origins: a smaller clavicular head and a larger sternocostal head. The **latissimus dorsi** is a broad, triangular muscle located on the inferior part of the back that forms most of the posterior wall of the axilla. The reverse muscle action (RMA) of the latissimus dorsi enables the vertebral column and torso to be elevated, as in doing a pullup. It is commonly called the "swimmer's muscle" because its many actions are used while swimming; consequently, many competitive swimmers have well-developed "lats."

Among the scapular muscles, the **deltoid** is a thick, powerful shoulder muscle that covers the shoulder joint and forms the rounded contour of the shoulder. This muscle is a frequent site of intramuscular injections. As you study the deltoid, note that its fascicles originate from three different points and that each group of fascicles moves the humerus differently. The **subscapularis** is a large triangular muscle that fills the subscapular fossa of the scapula and forms a small part in the apex of the posterior wall of the axilla. The **supraspinatus**, a rounded muscle named for its location in the supraspinous fossa of the scapula, lies deep to the trapezius. The **infraspinatus** is a triangular muscle, also named for its location in the infraspinous fossa of the scapula. The **teres major** is a thick, flattened muscle inferior to the teres minor that also helps form part of the posterior wall of the axilla. The **teres minor** is a cylindrical, elongated muscle, often inseparable from the infraspinatus, which lies along its superior border. The **coracobrachialis** is an elongated, narrow muscle in the arm.

Four deep muscles of the shoulder—subscapularis, supraspinatus, infraspinatus, and teres minor—strengthen and stabilize the shoulder joint. These muscles join the scapula to the humerus. Their flat tendons fuse together to form the **rotator** *(musculotendinous)* **cuff**, a nearly complete circle of tendons around the shoulder joint, like the cuff on a shirtsleeve. The supraspinatus muscle is especially subject to wear and tear because of its location between the head of the humerus and acromion of the scapula, which compress its tendon during shoulder movements, especially abduction of the arm. This is further aggravated by poor posture with slouched shoulders.

Relating Muscles to Movements

Arrange the muscles in this section according to the following actions on the humerus at the shoulder joint: (1) flexion, (2) extension, (3) abduction, (4) adduction, (5) medial rotation, and (6) lateral rotation. The same muscle may be mentioned more than once.

11.16 Muscles of the Arm That Move the Radius and Ulna

OBJECTIVE

- **Describe** the origin, insertion, action, and innervation of the muscles of the arm that move the radius and ulna.

Most of the muscles that move the radius and ulna (forearm bones) cause flexion and extension at the elbow, which is a hinge joint. The biceps brachii, brachialis, and brachioradialis muscles are the flexor muscles. The extensor muscles are the triceps brachii and the anconeus (Figure 11.16).

The **biceps brachii** is the large muscle located on the anterior surface of the arm. As indicated by its name, it has two heads of origin (long and short), both from the scapula. The muscle spans both the shoulder and elbow joints. In addition to its role in flexing the forearm at the elbow joint, it also supinates the forearm at the radioulnar

MUSCLE	ORIGIN	INSERTION	ACTION	INNERVATION
FOREARM FLEXORS				
Biceps brachii (BĪ-seps BRĀ-kē-ī; *biceps* = two heads of origin; *brachii* = arm)	Long head originates from tubercle above glenoid cavity of scapula (supraglenoid tubercle). Short head originates from coracoid process of scapula.	Radial tuberosity of radius and bicipital aponeurosis.*	Flexes forearm at elbow joint, supinates forearm at radioulnar joints, and flexes arm at shoulder joint.	Musculocutaneous nerve.
Brachialis (brā-kē-Ā-lis)	Distal, anterior surface of humerus.	Ulnar tuberosity and coronoid process of ulna.	Flexes forearm at elbow joint.	Musculocutaneous and radial nerves.
Brachioradialis (brā′-kē-ō-rā-dē-Ā-lis; *radi* = radius)	Lateral border of distal end of humerus.	Superior to styloid process of radius.	Flexes forearm at elbow joint; supinates and pronates forearm at radioulnar joints to neutral position.	Radial nerve.
FOREARM EXTENSORS				
Triceps brachii (TRĪ-seps = three heads of origin)	Long head originates from infraglenoid tubercle, a projection inferior to glenoid cavity of scapula. Lateral head originates from lateral and posterior surface of humerus. Medial head originates from entire posterior surface of humerus inferior to a groove for the radial nerve.	Olecranon of ulna.	Extends forearm at elbow joint and extends arm at shoulder joint.	Radial nerve.
Anconeus (an-KŌ-nē-us; *ancon* = elbow)	Lateral epicondyle of humerus.	Olecranon and superior portion of shaft of ulna.	Extends forearm at elbow joint.	Radial nerve.
FOREARM PRONATORS				
Pronator teres (PRŌ-nā-tor TE-rēz; *pronator* = turns palm posteriorly; *tero* = round and long) (see also Figure 11.17a)	Medial epicondyle of humerus and coronoid process of ulna.	Midlateral surface of radius.	Pronates forearm at radioulnar joints and weakly flexes forearm at elbow joint.	Median nerve.
Pronator quadratus (PRŌ-nā-tor kwod-RĀ-tus; *quadratus* = square, four-sided) (see also Figure 11.17a–c)	Distal portion of shaft of ulna.	Distal portion of shaft of radius.	Pronates forearm at radioulnar joints.	Median nerve.
FOREARM SUPINATOR				
Supinator (SOO-pi-nā-tor = turns palm anteriorly) (see also Figure 11.17b,c)	Lateral epicondyle of humerus and ridge near radial notch of ulna (supinator crest).	Lateral surface of proximal one-third of radius.	Supinates forearm at radioulnar joints.	Deep radial nerve.

*The bicipital aponeurosis is a broad aponeurosis from the tendon of insertion of the biceps brachii muscle that descends medially across the brachial artery and fuses with deep fascia over the forearm flexor muscles (see Figure 11.17a). It also helps to protect the median nerve and brachial artery.

joints and flexes the arm at the shoulder joint. The **brachialis** is deep to the biceps brachii muscle. It is the most powerful flexor of the forearm at the elbow joint. For this reason, it is the "workhorse" of the elbow flexors. The **brachioradialis** flexes the forearm at the elbow joint, especially when a quick movement is required or when a weight is lifted slowly during flexion of the forearm.

The **triceps brachii** is the large muscle located on the posterior surface of the arm. It is the more powerful of the extensors of the forearm at the elbow joint. As its name implies, it has three heads of origin, one from the scapula (long head) and two from the humerus (lateral and medial heads). The long head crosses the shoulder joint; the other heads do not. The **anconeus** is a small muscle located on the lateral part of the posterior aspect of the elbow that assists the triceps brachii in extending the forearm at the elbow joint.

Some muscles that move the radius and ulna are involved in pronation and supination at the radioulnar joints. The pronators, as suggested by their names, are the **pronator teres** and **pronator quadratus** muscles. The supinator of the forearm is aptly named the **supinator** muscle. You use the powerful action of the supinator when you twist a corkscrew or turn a screw with a screwdriver.

In the limbs, functionally related skeletal muscles and their associated blood vessels and nerves are grouped together by fascia into regions called **compartments**. In the arm, the biceps brachii, brachialis, and coracobrachialis muscles compose the **anterior** (*flexor*)

FIGURE 11.16 **Muscles of the arm that move the radius and ulna (forearm bones).**

The anterior arm muscles flex the forearm, and the posterior arm muscles extend it.

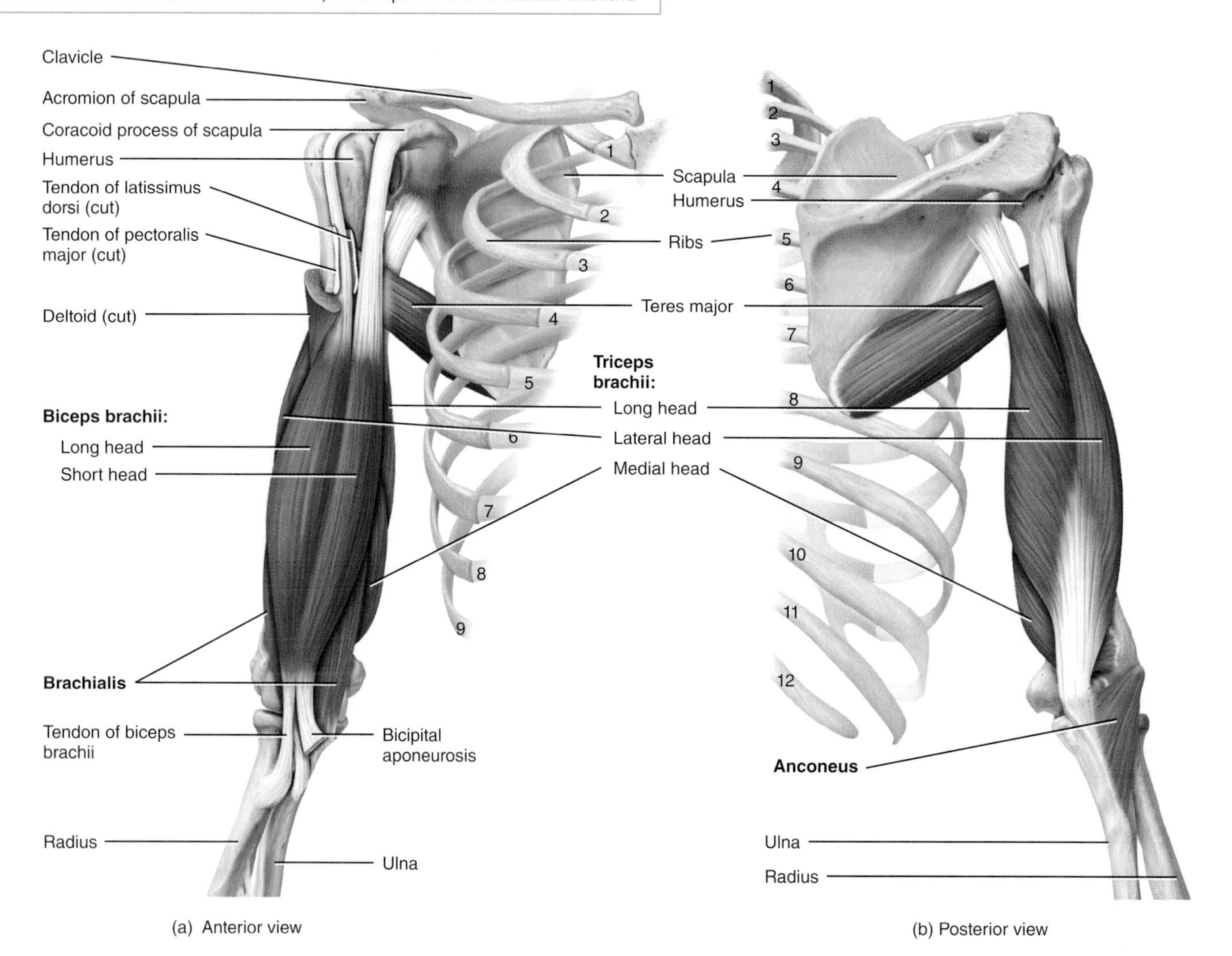

(a) Anterior view

(b) Posterior view

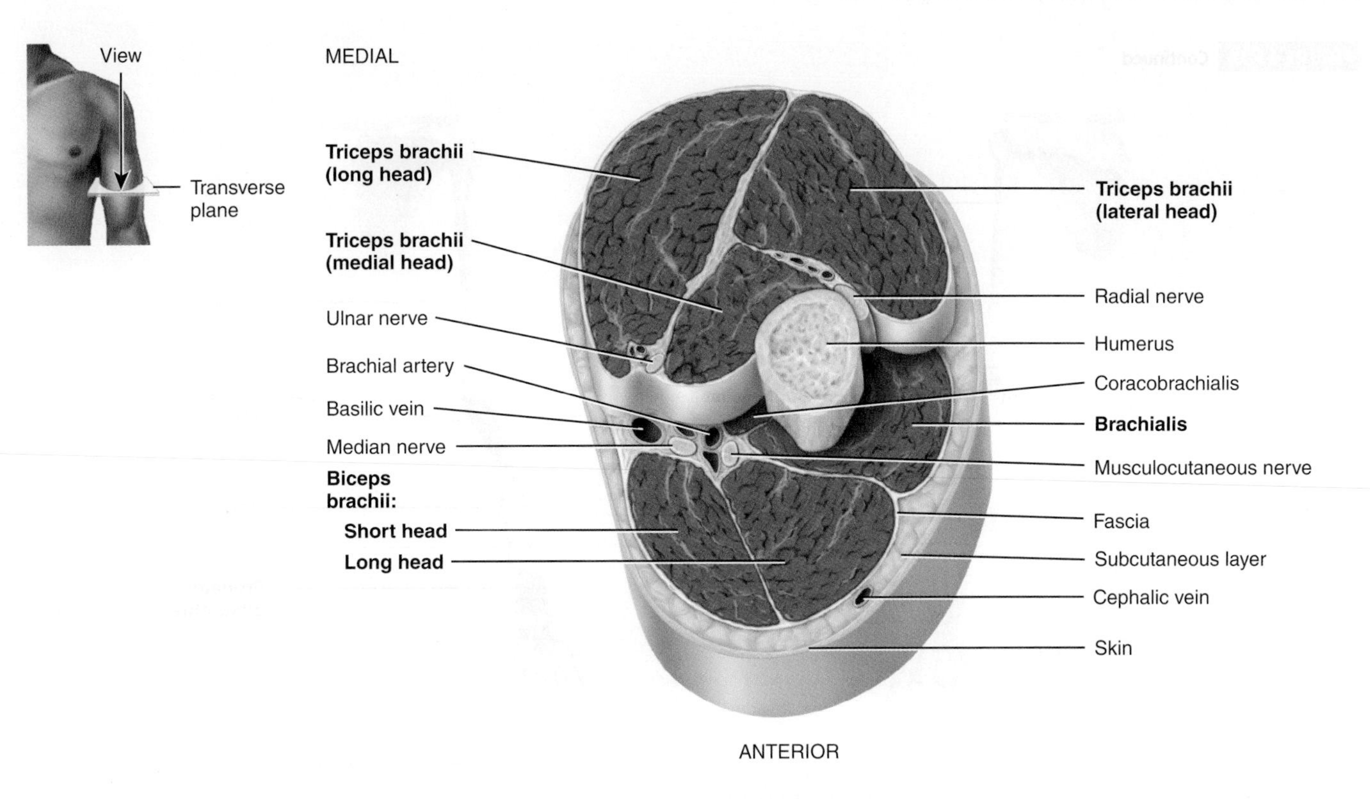

(c) Superior view of transverse section of arm

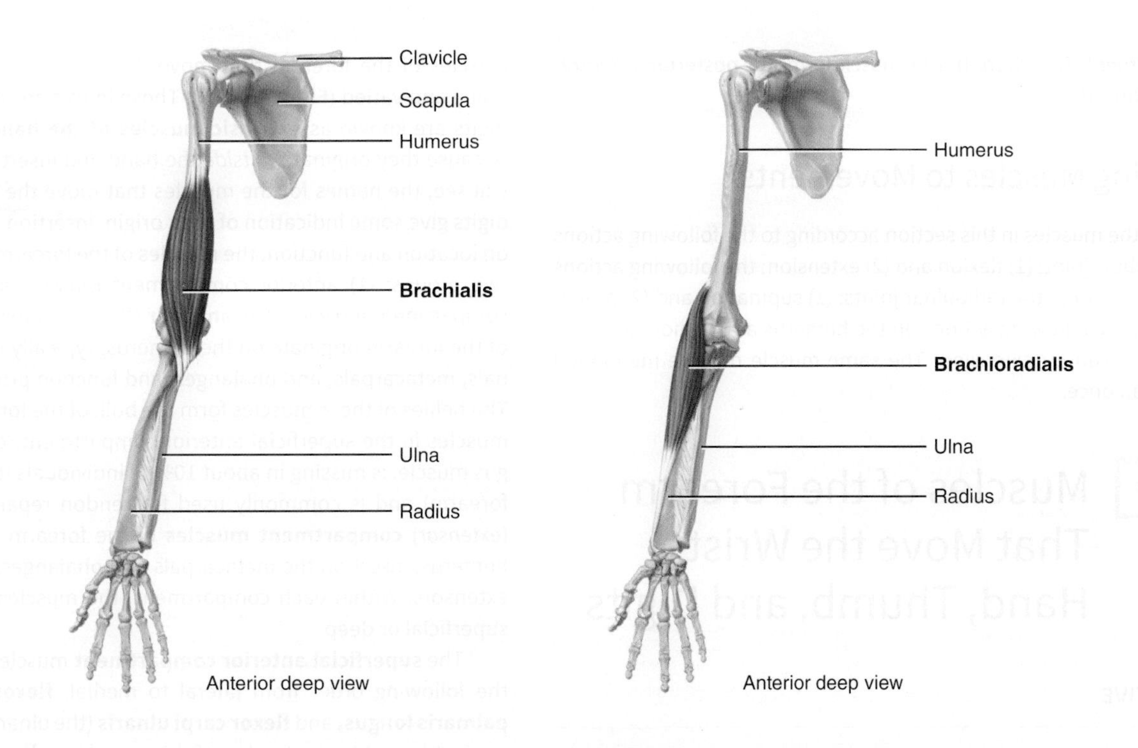

Figure 11.16 ***Continues***

FIGURE 11.16 Continued

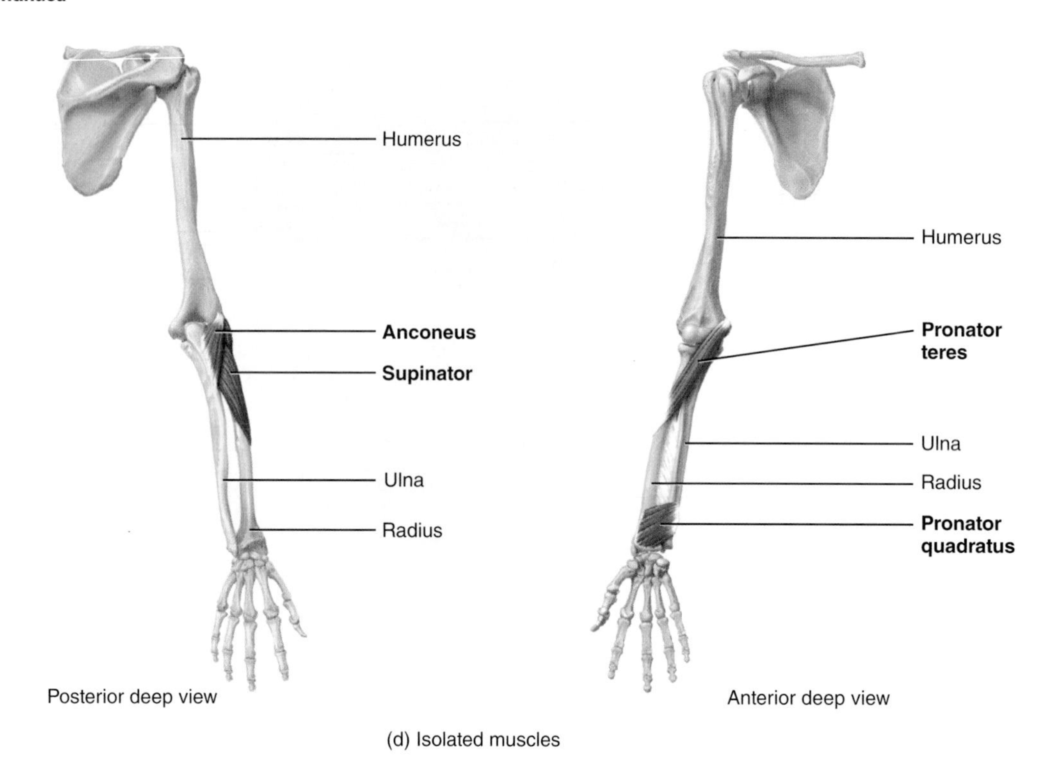

(d) Isolated muscles

Q Which muscles are the most powerful flexor and the most powerful extensor of the forearm?

compartment. The triceps brachii muscle forms the **posterior** (*extensor*) **compartment**.

Relating Muscles to Movements

Arrange the muscles in this section according to the following actions on the elbow joint: (1) flexion and (2) extension; the following actions on the forearm at the radioulnar joints: (1) supination and (2) pronation; and the following actions on the humerus at the shoulder joint: (1) flexion and (2) extension. The same muscle may be mentioned more than once.

11.17 Muscles of the Forearm That Move the Wrist, Hand, Thumb, and Digits

OBJECTIVE

- **Describe** the origin, insertion, action, and innervation of the muscles of the forearm that move the wrist, hand, and digits.

Muscles of the forearm that move the wrist, hand, and digits are many and varied (**Figure 11.17**). Those in this group that act on the digits are known as **extrinsic muscles of the hand** (*ex-*= outside) because they originate *outside* the hand and insert within it. As you will see, the names for the muscles that move the wrist, hand, and digits give some indication of their origin, insertion, or action. Based on location and function, the muscles of the forearm are divided into two groups: (1) anterior compartment muscles and (2) posterior compartment muscles. The **anterior** (*flexor*) **compartment muscles** of the forearm originate on the humerus; typically insert on the carpals, metacarpals, and phalanges; and function primarily as flexors. The bellies of these muscles form the bulk of the forearm. One of the muscles in the superficial anterior compartment, the palmaris longus muscle, is missing in about 10% of individuals (usually in the left forearm) and is commonly used for tendon repair. The **posterior** (*extensor*) **compartment muscles** of the forearm originate on the humerus, insert on the metacarpals and phalanges, and function as extensors. Within each compartment, the muscles are grouped as superficial or deep.

The **superficial anterior compartment** muscles are arranged in the following order from lateral to medial: **flexor carpi radialis, palmaris longus,** and **flexor carpi ulnaris** (the ulnar nerve and artery are just lateral to the tendon of this muscle at the wrist). The **flexor digitorum superficialis** muscle is deep to the other three muscles and is the largest superficial muscle in the forearm.

The **deep anterior compartment** muscles are arranged in the following order from lateral to medial: **flexor pollicis longus** (the only flexor of the distal phalanx of the thumb) and **flexor digitorum profundus** (ends in four tendons that insert into the distal phalanges of the fingers).

The **superficial posterior compartment** muscles are arranged in the following order from lateral to medial: **extensor carpi radialis longus, extensor carpi radialis brevis, extensor digitorum** (occupies most of the posterior surface of the forearm and divides into four tendons that insert into the middle and distal phalanges of the fingers), **extensor digiti minimi** (a slender muscle usually connected to the extensor digitorum), and **extensor carpi ulnaris**.

STUDY GUIDE See Clinical Connection: Golfer's Elbow

The **deep posterior compartment** muscles are arranged in the following order from lateral to medial: **abductor pollicis longus, extensor pollicis brevis, extensor pollicis longus,** and **extensor indicis.**

The tendons of the muscles of the forearm that attach to the wrist or continue into the hand, along with blood vessels and nerves, are held close to bones by strong fasciae. The tendons are also surrounded by tendon sheaths. At the wrist, the deep fascia is thickened into fibrous bands called **retinacula** (*retinacul* = holdfast). The **flexor retinaculum** is located over the palmar surface of the carpal bones. The long flexor tendons of the digits and wrist and the median nerve pass deep to the flexor retinaculum. The flexor retinaculum and carpal bones form a narrow space called the **carpal tunnel**. Through this tunnel pass the median nerve and tendons of the flexor digitorum superficialis, flexor digitorum profundus, and flexor pollicis longus muscles (**Figure 11.17f**). The **extensor retinaculum** is located over the dorsal surface of the carpal bones. The extensor tendons of the wrist and digits pass deep to it.

MUSCLE	ORIGIN	INSERTION	ACTION	INNERVATION
SUPERFICIAL ANTERIOR (FLEXOR) COMPARTMENT OF THE FOREARM				
Flexor carpi radialis (FLEK-sor KAR-pē-rā′-dē-Ā-lis; *flexor* = decreases angle at joint; *carpi* = wrist; *radi* = radius)	Medial epicondyle of humerus.	Metacarpals II and III.	Flexes and abducts hand (*radial deviation*) at wrist joint.	Median nerve.
Palmaris longus (pal-MA-ris LON-gus; *palma* = palm; *longus* = long)	Medial epicondyle of humerus.	Flexor retinaculum and *palmar aponeurosis* (fascia in center of palm).	Weakly flexes hand at wrist joint.	Median nerve.
Flexor carpi ulnaris (ŭl-NAR-is = ulna)	Medial epicondyle of humerus and superior posterior border of ulna.	Pisiform, hamate, and base of metacarpal V.	Flexes and adducts hand (*ulnar deviation*) at wrist joint.	Ulnar nerve.
Flexor digitorum superficialis (di-ji-TOR-um soo′-per-fish′-ē-Ā-lis; *digit* = finger or toe; *superficialis* = closer to surface)	Medial epicondyle of humerus, coronoid process of ulna, and ridge along lateral margin or anterior surface (anterior oblique line) of radius.	Middle phalanx of each finger.*	Flexes middle phalanx of each finger at proximal interphalangeal joint, proximal phalanx of each finger at metacarpophalangeal joint, and hand at wrist joint.	Median nerve.
DEEP ANTERIOR (FLEXOR) COMPARTMENT OF THE FOREARM				
Flexor pollicis longus (POL-li-sis = thumb)	Anterior surface of radius and *interosseous membrane* (sheet of fibrous tissue that holds shafts of ulna and radius together).	Base of distal phalanx of thumb.	Flexes distal phalanx of thumb at interphalangeal joint.	Median nerve.
Flexor digitorum profundus (prō-FUN-dus = deep)	Anterior medial surface of body of ulna.	Base of distal phalanx of each finger.	Flexes distal and middle phalanges of each finger at interphalangeal joints, proximal phalanx of each finger at metacarpophalangeal joint, and hand at wrist joint.	Median and ulnar nerves.

Continues

MUSCLE	ORIGIN	INSERTION	ACTION	INNERVATION
SUPERFICIAL POSTERIOR (EXTENSOR) COMPARTMENT OF THE FOREARM				
Extensor carpi radialis longus (eks-TEN-sor = increases angle at joint)	Lateral supracondylar ridge of humerus.	Metacarpal II.	Extends and abducts hand at wrist joint (ulnar deviation).	Radial nerve.
Extensor carpi radialis brevis (*brevis* = short)	Lateral epicondyle of humerus.	Metacarpal III.	Extends and abducts hand at wrist joint.	Radial nerve.
Extensor digitorum	Lateral epicondyle of humerus.	Distal and middle phalanges of each finger.	Extends distal and middle phalanges of each finger at interphalangeal joints, proximal phalanx of each finger at metacarpophalangeal joint, and hand at wrist joint.	Radial nerve.
Extensor digiti minimi (DIJ-i-tē MIN-i-mē; *minimi* = smallest)	Lateral epicondyle of humerus.	Tendon of extensor digitorum on phalanx V.	Extends proximal phalanx of little finger at metacarpophalangeal joint and hand at wrist joint.	Deep radial nerve.
Extensor carpi ulnaris	Lateral epicondyle of humerus and posterior border of ulna.	Metacarpal V.	Extends and adducts hand at wrist joint (ulnar deviation).	Deep radial nerve.
DEEP POSTERIOR (EXTENSOR) COMPARTMENT OF THE FOREARM				
Abductor pollicis longus (ab-DUK-tor = moves part away from midline)	Posterior surface of middle of radius and ulna and interosseous membrane.	Metacarpal I.	Abducts and extends thumb at carpometacarpal joint and abducts hand at wrist joint.	Deep radial nerve.
Extensor pollicis brevis	Posterior surface of middle of radius and interosseous membrane.	Base of proximal phalanx of thumb.	Extends proximal phalanx of thumb at metacarpophalangeal joint, first metacarpal of thumb at carpometacarpal joint, and hand at wrist joint.	Deep radial nerve.
Extensor pollicis longus	Posterior surface of middle of ulna and interosseous membrane.	Base of distal phalanx of thumb.	Extends distal phalanx of thumb at interphalangeal joint, extends first metacarpal of thumb at carpometacarpal joint, and abducts hand at wrist joint.	Deep radial nerve.
Extensor indicis (IN-di-kis = index)	Posterior surface of ulna and interosseous membrane.	Tendon of extensor digitorum of index finger.	Extends distal and middle phalanges of index finger at interphalangeal joints, proximal phalanx of index finger at metacarpophalangeal joint, and hand at wrist joint.	Deep radial nerve.

*Reminder: The thumb or pollex; numbered I or 1, is the first digit and has two phalanges: proximal and distal. The remaining digits, the fingers, are numbered II–V (2–5), and each has three phalanges: proximal, middle, and distal.

Relating Muscles to Movements

Arrange the muscles in this section according to the following actions on the wrist joint: (1) flexion, (2) extension, (3) abduction (radial deviation), and (4) adduction (ulnar deviation); the following actions on the fingers at the metacarpophalangeal joints: (1) flexion and (2) extension; the following actions on the fingers at the interphalangeal joints: (1) flexion and (2) extension; the following actions on the thumb at the carpometacarpal, metacarpophalangeal, and interphalangeal joints: (1) extension and (2) abduction; and the following action on the thumb at the interphalangeal joint: flexion. The same muscle may be mentioned more than once.

FIGURE 11.17 Muscles of the forearm that move the wrist, hand, thumb, and digits.

The anterior compartment muscles function as flexors, and the posterior compartment muscles function as extensors.

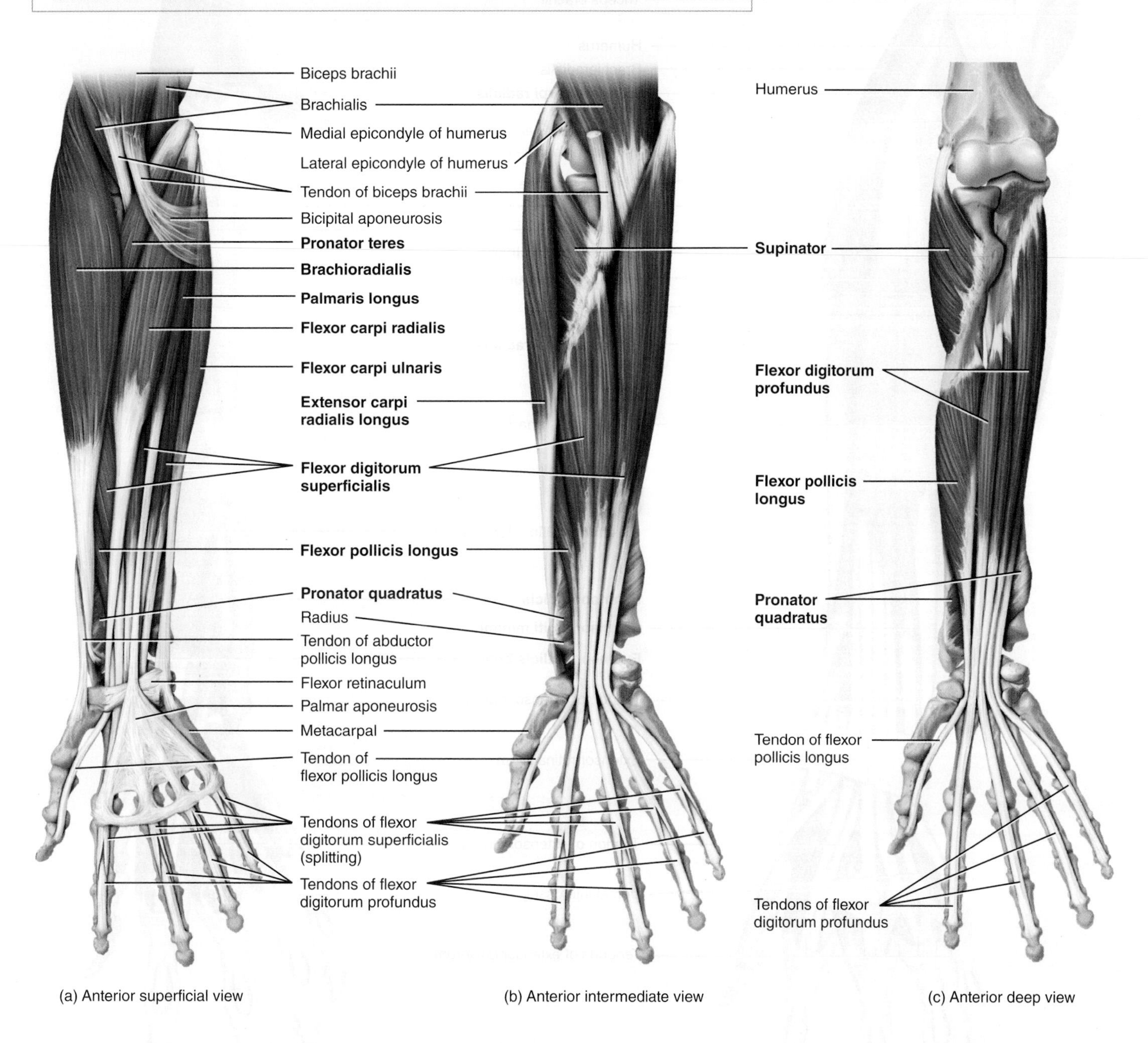

(a) Anterior superficial view

(b) Anterior intermediate view

(c) Anterior deep view

Figure 11.17 *Continues*

FIGURE 11.17 Continued

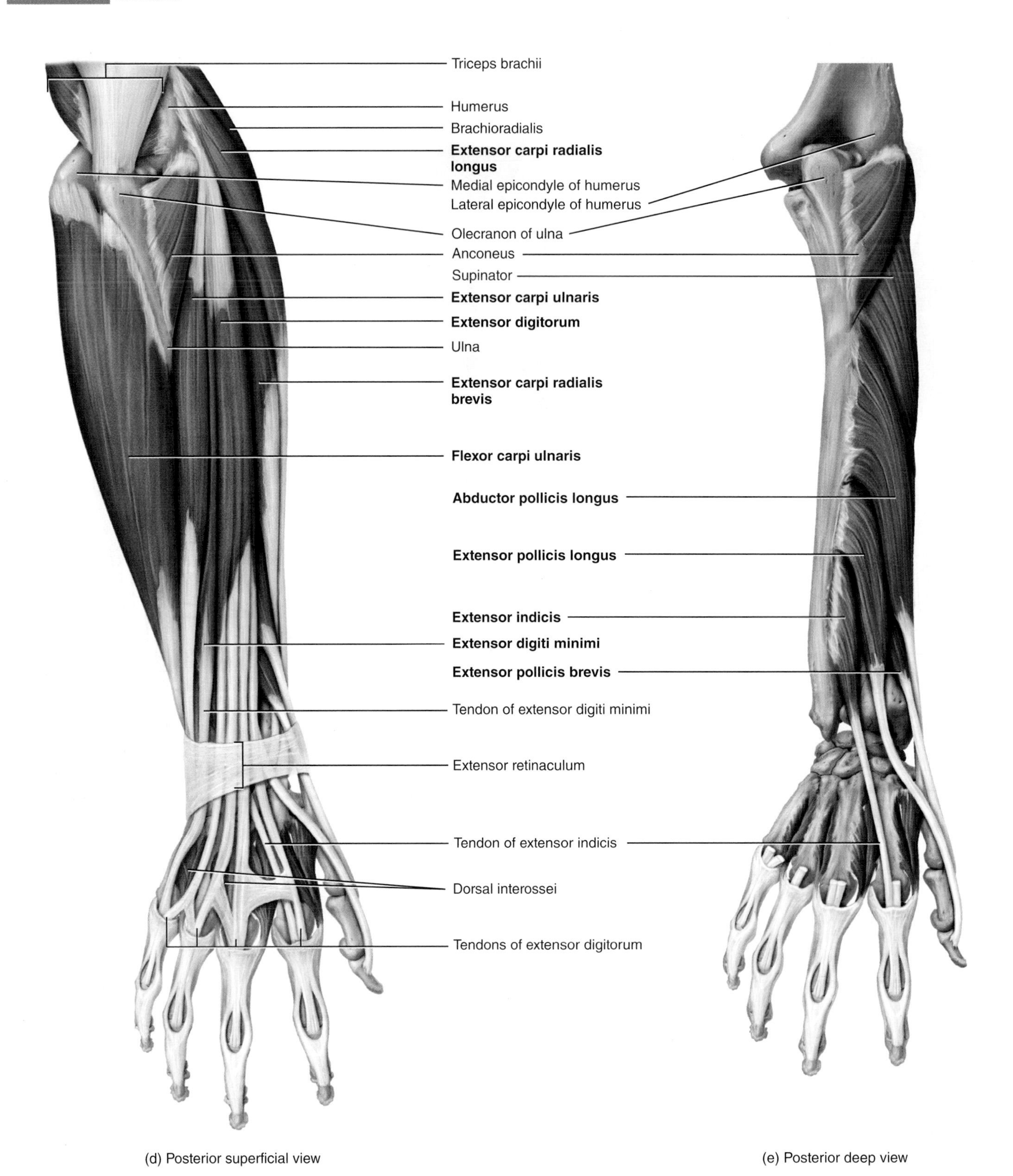

(d) Posterior superficial view

(e) Posterior deep view

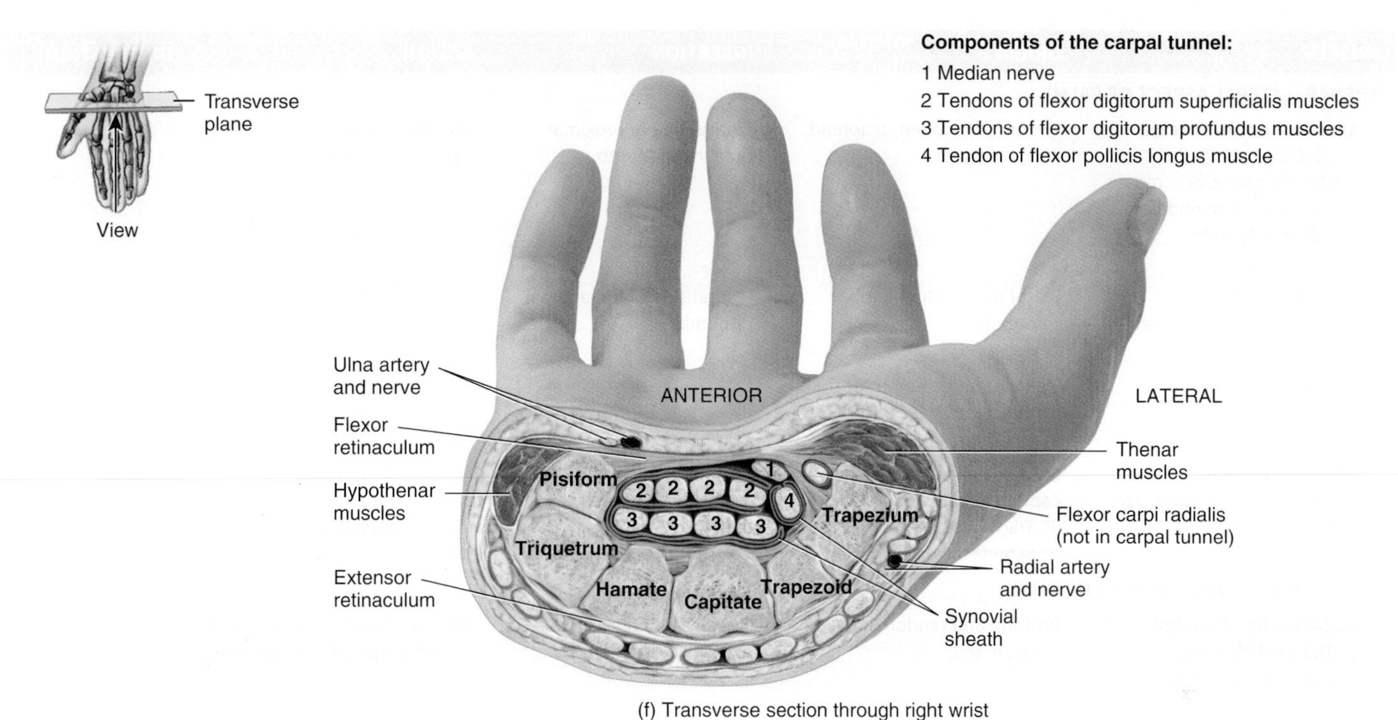

(f) Transverse section through right wrist

Q What structures pass deep to the flexor retinaculum?

11.18 Muscles of the Palm That Move the Digits—Intrinsic Muscles of the Hand

OBJECTIVE

- **Describe** the origin, insertion, action, and innervation of the muscles of the palm that move the digits (the intrinsic muscles of the hand).

Several of the muscles discussed in Section 11.17 move the digits in various ways and are known as extrinsic muscles of the hand. They produce the powerful but crude movements of the digits. The **intrinsic muscles of the hand** in the palm produce the weak but intricate and precise movements of the digits that characterize the human hand (**Figure 11.18**). The muscles in this group are so named because their origins and insertions are *within* the hand.

The intrinsic muscles of the hand are divided into three groups: (1) **thenar**, (2) **hypothenar**, and (3) **intermediate**. The thenar muscles include the abductor pollicis brevis, opponens pollicis, flexor pollicis brevis, and adductor pollicis (acts on the thumb but is not in the thenar eminence). The **abductor pollicis brevis** is a thin, short, relatively broad superficial muscle on the lateral side of the thenar eminence. The **flexor pollicis brevis** is a short, wide muscle that is medial to the abductor pollicis brevis muscle. The **opponens pollicis** is a small, triangular muscle that is deep to the flexor pollicis brevis and abductor pollicis brevis muscles. The **adductor pollicis** is fan-shaped and has two heads (oblique and transverse) separated by a gap through which the radial artery passes. The thenar muscles plus the adductor pollicis form the **thenar eminence**, the lateral rounded contour on the palm that is also called the *ball of the thumb.*

The three hypothenar muscles act on the little finger and form the **hypothenar eminence**, the medial rounded contour on the palm that is also called the ball of the little finger. The hypothenar muscles are the abductor digiti minimi, flexor digiti minimi brevis, and opponens digiti minimi. The **abductor digiti minimi** is a short, wide muscle and is the most superficial of the hypothenar muscles. It is a powerful muscle that plays an important role in grasping an object with outspread fingers. The **flexor digiti minimi brevis** muscle is also short and wide and is lateral to the abductor digiti minimi muscle. The **opponens digiti minimi** muscle is triangular and deep to the other two hypothenar muscles.

The 11 or 12 intermediate (midpalmar) muscles include the lumbricals, palmar interossei, and dorsal interossei. The **lumbricals**, as their name indicates, are worm-shaped. They originate from and insert into the tendons of other muscles (flexor digitorum profundus and extensor digitorum). The **palmar interossei** are the smallest and more anterior of the interossei muscles. The **dorsal interossei** are

MUSCLE	ORIGIN	INSERTION	ACTION	INNERVATION
THENAR (LATERAL ASPECT OF PALM)				
Abductor pollicis brevis (ab-DUK-tor POL-li-sis BREV-is; *abductor* = moves part away from middle; *pollicis* = thumb; *brevis* = short)	Flexor retinaculum, scaphoid, and trapezium.	Lateral side of proximal phalanx of thumb.	Abducts thumb at carpometacarpal joint.	Median nerve.
Opponens pollicis (op-PŌ-nenz = opposes)	Flexor retinaculum and trapezium.	Lateral side of metacarpal I (thumb).	Moves thumb across palm to meet any finger (opposition) at carpometacarpal joint.	Median nerve.
Flexor pollicis brevis (FLEK-sor = decreases angle at joint)	Flexor retinaculum, trapezium, capitate, and trapezoid.	Lateral side of proximal phalanx of thumb.	Flexes thumb at carpometacarpal and metacarpophalangeal joints.	Median and ulnar nerves.
Adductor pollicis (ad-DUK-tor = moves part toward midline)	Oblique head originates from capitate and metacarpal II and III. Transverse head originates from metacarpal III.	Medial side of proximal phalanx of thumb by tendon containing sesamoid bone.	Adducts thumb at carpometacarpal and metacarpophalangeal joints.	Ulnar nerve.
HYPOTHENAR (MEDIAL ASPECT OF PALM)				
Abductor digiti minimi (DIJ-i-tē MIN-i-mē; *digit* = finger or toe; *minimi* = smallest)	Pisiform and tendon of flexor carpi ulnaris.	Medial side of proximal phalanx of little finger.	Abducts and flexes little finger at metacarpophalangeal joint.	Ulnar nerve.
Flexor digiti minimi brevis	Flexor retinaculum and hamate.	Medial side of proximal phalanx of little finger.	Flexes little finger at carpometacarpal and metacarpophalangeal joints.	Ulnar nerve.
Opponens digiti minimi	Flexor retinaculum and hamate.	Medial side of metacarpal V (little finger).	Moves little finger across palm to meet thumb (opposition) at carpometacarpal joint.	Ulnar nerve.
INTERMEDIATE (MIDPALMAR)				
Lumbricals (LUM-bri-kals; *lumbric* = earthworm) (four muscles)	Lateral sides of tendons and flexor digitorum profundus of each finger.	Lateral sides of tendons of extensor digitorum on proximal phalanges of each finger.	Flex each finger at metacarpophalangeal joints and extend each finger at interphalangeal joints.	Median and ulnar nerves.
Palmar interossei (PAL-mar in′-ter-OS-ē-i; *palma* = palm; *inter-* = between; *-ossei* = bones) (three distinct muscles but some describe four)	Sides of shafts of metacarpals of all digits (except III).	Sides of bases of proximal phalanges of all fingers (except III).	Adduct and flex each finger (except III) at metacarpophalangeal joints and extend these digits at interphalangeal joints.	Ulnar nerve.
Dorsal interossei (DOR-sal = back surface) (four muscles)	Adjacent sides of metacarpals.	Proximal phalanx of fingers II–IV.	Abduct fingers II–IV at metacarpophalangeal joints, flex fingers II–IV at metacarpophalangeal joints, and extend fingers II–IV at interphalangeal joints.	Ulnar nerve.

the most posterior of this series of muscles. Both sets of interossei muscles are located between the metacarpals and are important in abduction, adduction, flexion, and extension of the fingers, and in movements in skilled activities such as writing, typing, and playing a piano.

The functional importance of the hand is readily apparent when you consider that certain hand injuries can result in permanent disability. Most of the dexterity of the hand depends on movements of the thumb. The general activities of the hand are free motion, power grip (forcible movement of the fingers and thumb against the palm, as in squeezing), precision handling (a change in position of a handled object that requires exact control of finger and thumb positions, as in winding a watch or threading a needle), and pinch (compression between the thumb and index finger or between the thumb and first two fingers).

FIGURE 11.18 Muscles of the palm that move the digits—intrinsic muscles of the hand.

The intrinsic muscles of the hand produce the intricate and precise movements of the digits that characterize the human hand.

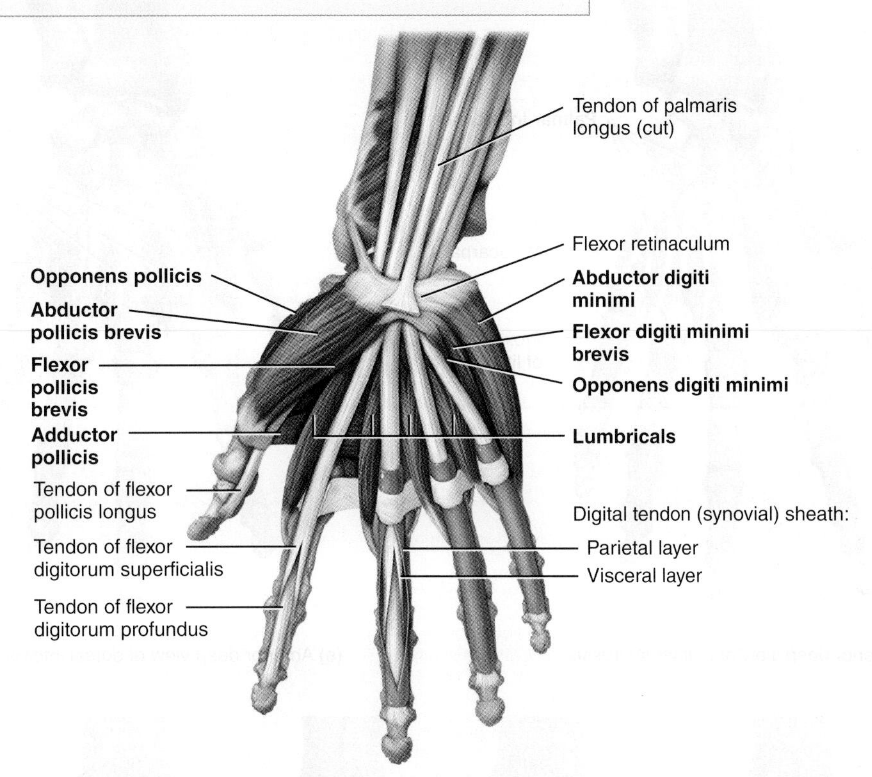

(a) Anterior superficial view

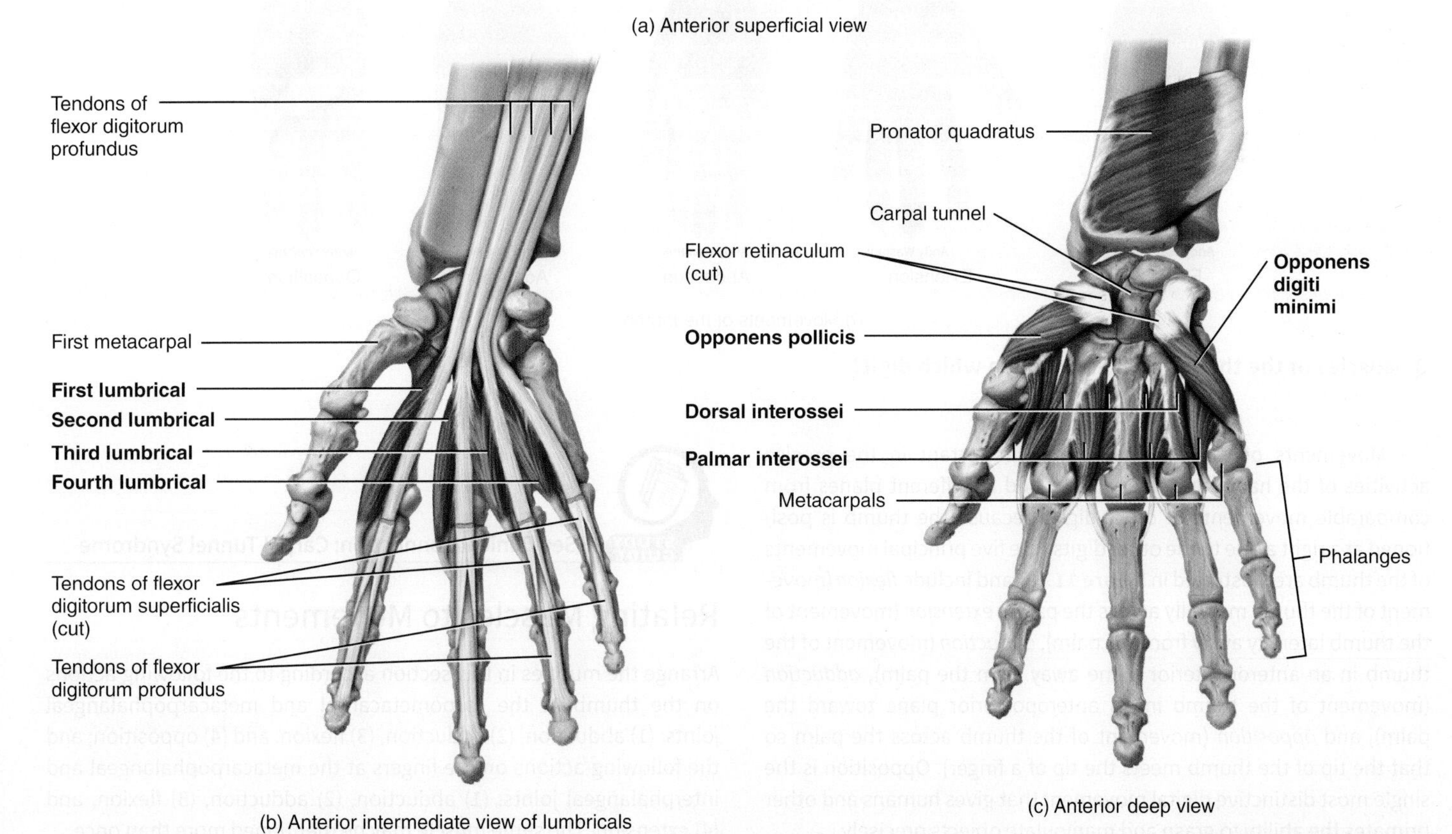

(b) Anterior intermediate view of lumbricals

(c) Anterior deep view

Figure 11.18 *Continues*

FIGURE 11.18 Continued

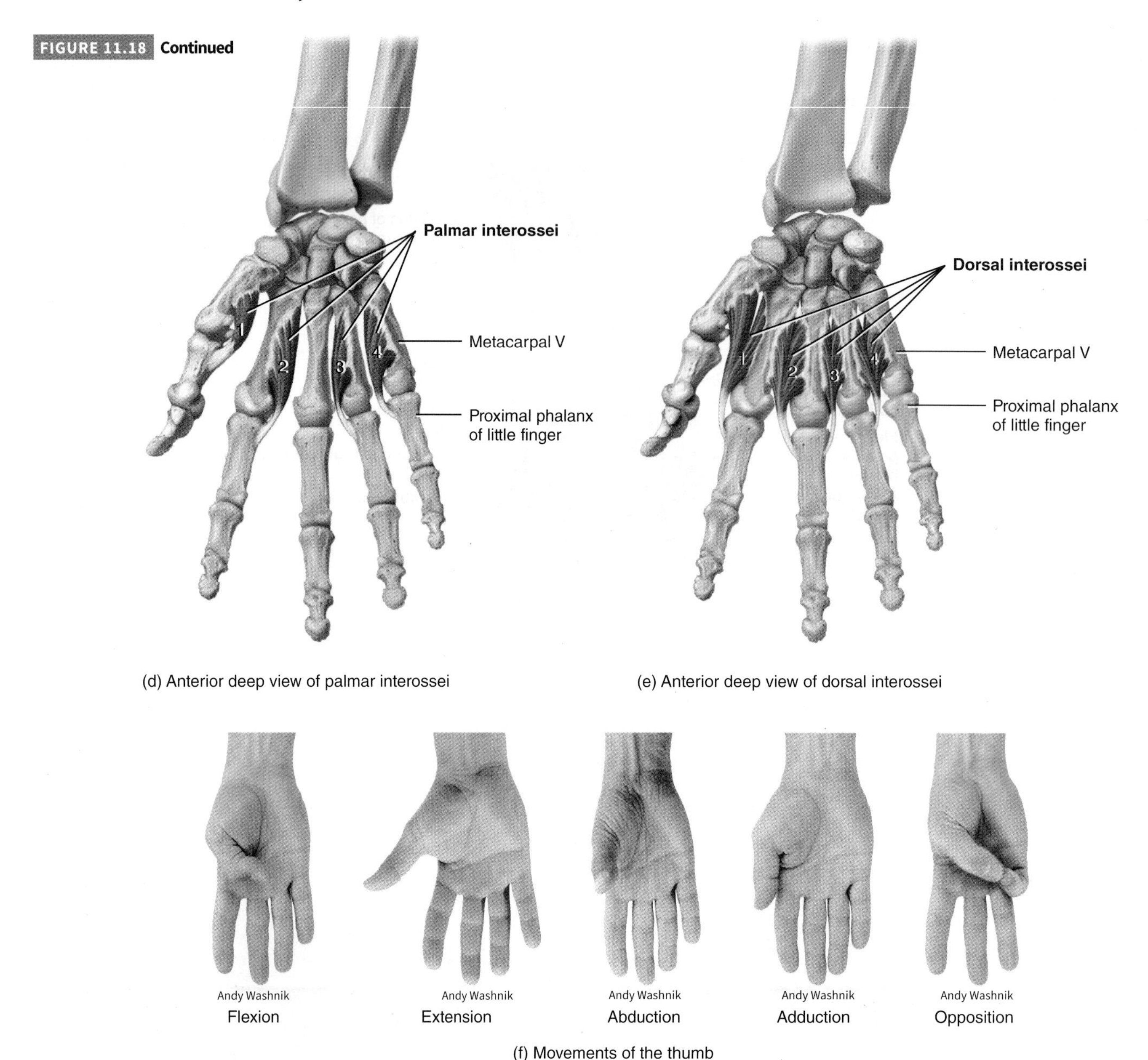

(d) Anterior deep view of palmar interossei

(e) Anterior deep view of dorsal interossei

(f) Movements of the thumb

Q Muscles of the thenar eminence act on which digit?

Movements of the thumb are very important in the precise activities of the hand, and they are defined in different planes from comparable movements of other digits because the thumb is positioned at a right angle to the other digits. The five principal movements of the thumb are illustrated in Figure 11.18f and include *flexion* (movement of the thumb medially across the palm), *extension* (movement of the thumb laterally away from the palm), *abduction* (movement of the thumb in an anteroposterior plane away from the palm), *adduction* (movement of the thumb in an anteroposterior plane toward the palm), and *opposition* (movement of the thumb across the palm so that the tip of the thumb meets the tip of a finger). Opposition is the single most distinctive digital movement that gives humans and other primates the ability to grasp and manipulate objects precisely.

See Clinical Connection: Carpal Tunnel Syndrome

Relating Muscles to Movements

Arrange the muscles in this section according to the following actions on the thumb at the carpometacarpal and metacarpophalangeal joints: (1) abduction, (2) adduction, (3) flexion, and (4) opposition; and the following actions on the fingers at the metacarpophalangeal and interphalangeal joints: (1) abduction, (2) adduction, (3) flexion, and (4) extension. The same muscle may be mentioned more than once.

11.19 Muscles of the Neck and Back That Move the Vertebral Column

OBJECTIVE

- **Describe** the origin, insertion, action, and innervation of the muscles that move the vertebral column.

The muscles that move the vertebral column (backbone) are quite complex because they have multiple origins and insertions and there is considerable overlap among them. One way to group the muscles is on the basis of the general direction of the muscle bundles and their approximate lengths. For example, the splenius muscles arise from the midline and extend laterally and superiorly to their insertions (**Figure 11.19a**). The erector spinae muscle group (consisting of the iliocostalis, longissimus, and spinalis muscles) arises either from the midline or more laterally but usually runs almost longitudinally, with neither a significant lateral nor medial direction as it is traced superiorly. The muscles of the transversospinalis group (semispinalis, multifidus, rotatores) arise laterally but extend toward the midline as they are traced superiorly. Deep to these three muscle groups are small segmental muscles that extend between spinous processes or transverse processes of vertebrae. Note in Section 11.10 that the rectus abdominis, external oblique, internal oblique, and quadratus lumborum muscles also play a role in moving the vertebral column.

The bandagelike **splenius** muscles are attached to the sides and back of the neck. The two muscles in this group are named on the basis of their superior attachments (insertions): **splenius capitis**

MUSCLE	ORIGIN	INSERTION	ACTION	INNERVATION
SPLENIUS				
Splenius capitis (SPLĒ-nē-us KAP-i-tis; *splenium* = bandage; *capit-* = head)	Ligamentum nuchae and spinous processes of C7–T4.	Occipital bone and mastoid process of temporal bone.	Acting together (bilaterally), extend head and extend vertebral column; acting singly (unilaterally), laterally flex and/or rotate head to same side as contracting muscle.	Middle cervical spinal nerves.
Splenius cervicis (SER-vi-cis; *cervic-* = neck)	Spinous processes of T3–T6.	Transverse processes of C1–C2 or C1–C4.	Acting together, extend head; acting singly, laterally flex and/or rotate head to same side as contracting muscle.	Inferior cervical spinal nerves.
ERECTOR SPINAE (e-REK-tor SPĪ-nē) Consists of iliocostalis muscles (lateral), longissimus muscles (intermediate), and spinalis muscles (medial).				
ILIOCOSTALIS GROUP (LATERAL)				
Iliocostalis cervicis (il′-ē-ō-kos-TĀL-is; *ilio-* = flank; *-costa-* = rib)	Ribs 1–6.	Transverse processes of C4–C6.	Acting together, muscles of each region (cervical, thoracic, and lumbar) extend and maintain erect posture of vertebral column of their respective regions; acting singly, laterally flex vertebral column of their respective regions to same side as contracting muscle.	Cervical and thoracic spinal nerves.
Iliocostalis thoracis (thō-RĀ-sis = chest)	Ribs 7–12.	Ribs 1–6.		Thoracic spinal nerves.
Iliocostalis lumborum (lum-BOR-um = loin)	Iliac crest..	Ribs 7–12.		Lumbar spinal nerves.
LONGISSIMUS GROUP (Intermediate)				
Longissimus capitis (lon-JIS-i-mus = longest)	Articular processes of C4–C7 and transverse processes of T1–T4.	Mastoid process of temporal bone.	Acting together, both longissimus capitis muscles extend head and extend vertebral column; acting singly, rotate head to same side as contracting muscle.	Middle and inferior cervical spinal nerves.
Longissimus cervicis	Transverse processes of T4–T5.	Transverse processes of C2–C6.	Acting together, longissimus cervicis and both longissimus thoracis muscles extend vertebral column of their respective regions; acting singly, laterally flex vertebral column of their respective regions.	Cervical and superior thoracic spinal nerves. Thoracic and lumbar spinal nerves.
Longissimus thoracis	Transverse processes of lumbar vertebrae.	Transverse processes of all thoracic and superior lumbar vertebrae and ribs 9 and 10.		

Continues

MUSCLE	ORIGIN	INSERTION	ACTION	INNERVATION
SPINALIS GROUP (MEDIAL)				
Spinalis capitis (spi-NĀ-lis = vertebral column)	Often absent or very small. Arises with semispinalis capitis.	Occipital bone.	Acting together, muscles of each region (cervical, thoracic, and lumbar) extend vertebral column of their respective regions and extend head.	Cervical spinal nerves.
Spinalis cervicis	Ligamentum nuchae and spinous process of C7.	Spinous process of axis.		Inferior cervical and thoracic spinal nerves.
Spinalis thoracis	Spinous processes of T10–L2.	Spinous processes of superior thoracic vertebrae.		Thoracic spinal nerves.
TRANSVERSOSPINALES (trans-ver-sō-spi-NĀ-lēz)				
Semispinalis capitis (sem′-ē-spi-NĀ-lis; *semi-* = partially or one half)	Articular processes of C4–C6 and transverse processes of C7–T7.	Occipital bone between superior and inferior nuchal lines.	Acting together, extend head and vertebral column; acting singly, rotate head to side opposite contracting muscle.	Cervical and thoracic spinal nerves.
Semispinalis cervicis	Transverse processes of T1–T5.	Spinous processes of C1–C5.	Acting together, both semispinalis cervicis and both semispinalis thoracis muscles extend vertebral column of their respective regions; acting singly, rotate head to side opposite contracting muscle.	Cervical and thoracic spinal nerves.
Semispinalis thoracis	Transverse processes of T6–T10.	Spinous processes of C6–T4.		Thoracic spinal nerves.
Multifidus (mul-TIF-i-dus; *multi-* = many; *-fid-* = segmented)	Sacrum; ilium; transverse processes of L1–L5, T1–T12, and C4–C7.	Spinous process of a more superior vertebra.	Acting together, extend vertebral column; acting singly, weakly laterally flex vertebral column and weakly rotate vertebral column to side opposite contracting muscle.	Cervical, thoracic, and lumbar spinal nerves.
Rotatores (rō′-ta-TŌ-rēz; singular is **rotatore;** *rotatore* = to rotate)	Transverse processes of all vertebrae.	Spinous process of vertebra superior to the one of origin.	Acting together, weakly extend vertebral column; acting singly, weakly rotate vertebral column to side opposite contracting muscle.	Cervical, thoracic, and lumbar spinal nerves.
SEGMENTAL (seg-MEN-tal)				
Interspinales (in-ter-spi-NĀ-lēz; *inter-* = between)	Superior surface of all spinous processes.	Inferior surface of spinous process of vertebra superior to the one of origin.	Acting together, weakly extend vertebral column; acting singly, stabilize vertebral column during movement.	Cervical, thoracic, and lumbar spinal nerves.
Intertransversarii (in′-ter-trans-vers-AR-ē-i; singular is **intertransversarius**)	Transverse process of all vertebrae.	Transverse processes of vertebra superior to the one of origin.	Acting together, weakly extend vertebral column; acting singly, weakly laterally flex vertebral column and stabilize it during movements.	Cervical, thoracic, and lumbar spinal nerves.
SCALENES (SKĀ-lēnz)				
Anterior scalene (SKĀ-lēn; *anterior* = front; *scalene* = uneven)	Transverse processes of C3–C6.	Rib 1.	Acting together, right and left anterior scalene and middle scalene muscles elevate first ribs during deep inhalation.	Cervical spinal nerves.
Middle scalene	Transverse processes of C2–C7.	Rib 1.	RMA: Flex cervical vertebrae; acting singly, laterally flex and slightly rotate cervical vertebrae.	Cervical spinal nerves.
Posterior scalene	Transverse processes of C4–C6.	Rib 2.	Acting together, right and left posterior scalene elevate second ribs during deep inhalation. RMA: Flex cervical vertebrae; acting singly, laterally flex and slightly rotate cervical vertebrae.	Cervical spinal nerves. Cervical spenial nerves

FIGURE 11.19 Muscles of the neck and back that move the vertebral column (backbone).
The trapezius and occipitofrontalis muscles have been removed.

The erector spinae group (iliocostalis, longissimus, and spinalis muscles) is the largest muscular mass of the back and is the chief extensor of the vertebral column.

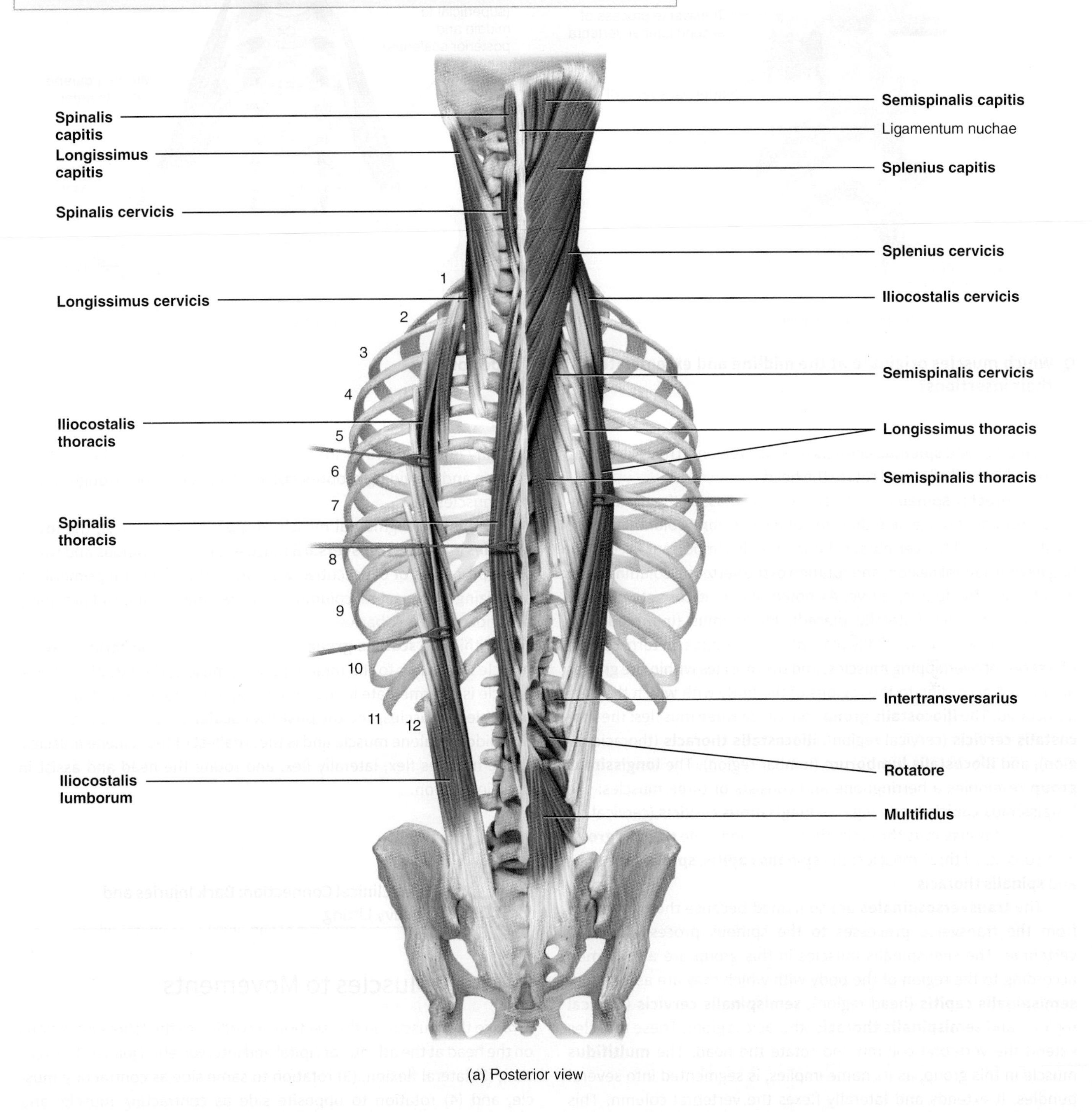

(a) Posterior view

Figure 11.19 ***Continues***

FIGURE 11.19 **Continued**

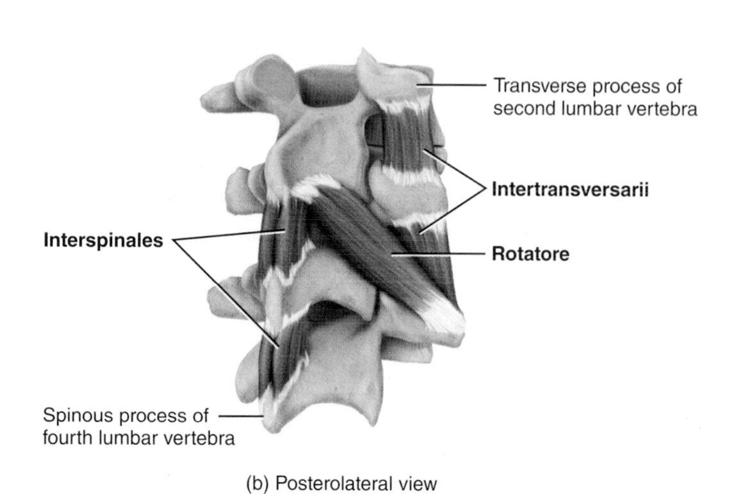

(b) Posterolateral view

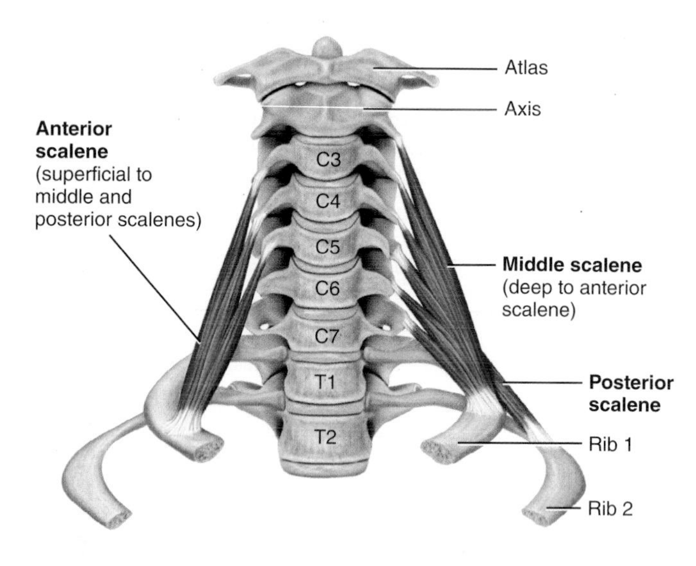

(c) Anterior view

Q Which muscles originate at the midline and extend laterally and superiorly to their insertions?

(head region) and **splenius cervicis** (cervical region). They extend the head and laterally flex and rotate the head.

The **erector spinae** is the largest muscle mass of the back, forming a prominent bulge on either side of the vertebral column. It is the chief extensor of the vertebral column. It is also important in controlling flexion, lateral flexion, and rotation of the vertebral column and in maintaining the lumbar curve. As noted above, it consists of three groups: iliocostalis (laterally placed), longissimus (intermediately placed), and spinalis (medially placed). These groups, in turn, consist of a series of overlapping muscles, and the muscles within the groups are named according to the regions of the body with which they are associated. The **iliocostalis group** consists of three muscles: the **iliocostalis cervicis** (cervical region), **iliocostalis thoracis** (thoracic region), and **iliocostalis lumborum** (lumbar region). The **longissimus group** resembles a herringbone and consists of three muscles: the **longissimus capitis** (head region), **longissimus cervicis** (cervical region), and **longissimus thoracis** (thoracic region). The **spinalis group** also consists of three muscles: the **spinalis capitis**, **spinalis cervicis**, and **spinalis thoracis**.

The **transversospinales** are so named because their fibers run from the transverse processes to the spinous processes of the vertebrae. The semispinalis muscles in this group are also named according to the region of the body with which they are associated: **semispinalis capitis** (head region), **semispinalis cervicis** (cervical region), and **semispinalis thoracis** (thoracic region). These muscles extend the vertebral column and rotate the head. The **multifidus** muscle in this group, as its name implies, is segmented into several bundles. It extends and laterally flexes the vertebral column. This muscle is large and thick in the lumbar region and is important in maintaining the lumbar curve. The **rotatores** muscles of this group are short and are found along the entire length of the vertebral column. These small muscles contribute little to vertebral movement but play important roles in monitoring the position of the vertebral column and providing proprioceptive feedback to the stronger vertebral muscles.

Within the segmental muscle group (Figure 11.19b), the **interspinales** and **intertransversarii** muscles unite the spinous and transverse processes of consecutive vertebrae. They function primarily in stabilizing the vertebral column during its movements, and providing proprioceptive feedback.

Within the **scalene group** (Figure 11.19c), the **anterior scalene** muscle is anterior to the middle scalene muscle, the **middle scalene** muscle is intermediate in placement and is the longest and largest of the scalene muscles, and the **posterior scalene** muscle is posterior to the middle scalene muscle and is the smallest of the scalene muscles. These muscles flex, laterally flex, and rotate the head and assist in deep inhalation.

See Clinical Connection: Back Injuries and Heavy Lifting

Relating Muscles to Movements

Arrange the muscles in this section according to the following actions on the head at the atlanto-occipital and intervertebral joints: (1) extension, (2) lateral flexion, (3) rotation to same side as contracting muscle, and (4) rotation to opposite side as contracting muscle; and arrange the muscles according to the following actions on the vertebral column at the intervertebral joints: (1) flexion, (2) extension, (3) lateral flexion, (4) rotation, and (5) stabilization. The same muscle may be mentioned more than once.

11.20 Muscles of the Gluteal Region That Move the Femur

OBJECTIVE

- **Describe** the origin, insertion, action, and innervation of the muscles of the gluteal region that move the femur.

As you will see, muscles of the lower limbs are larger and more powerful than those of the upper limbs because of differences in function. While upper limb muscles are characterized by versatility of movement, lower limb muscles function in stability, locomotion, and maintenance of posture. In addition, muscles of the lower limbs often cross two joints and act equally on both.

The majority of muscles that move the femur (thigh bone) originate on the pelvic girdle and insert on the femur (**Figure 11.20**). The **psoas major** and **iliacus** muscles share a common insertion (lesser trochanter of femur) and are collectively known as the **iliopsoas** muscle. There are three gluteal muscles: gluteus maximus, gluteus medius, and gluteus minimus. The **gluteus maximus** is the largest and heaviest of the three muscles and is one of the largest muscles in the body. It is the chief extensor of the femur. In its reverse muscle action (RMA), it is a powerful extensor of the torso at the hip joint. The **gluteus medius** is mostly deep to the gluteus maximus and is a powerful abductor of the femur at the hip joint. It is a common site for intramuscular injection. The **gluteus minimus** is the smallest of the gluteal muscles and lies deep to the gluteus medius.

The **tensor fasciae latae** muscle is located on the lateral surface of the thigh. The *fascia lata* is a layer of deep fascia, composed of dense connective tissue, that encircles the entire thigh. It is well developed laterally where, together with the tendons of the tensor fasciae latae and gluteus maximus muscles, it forms a structure called the **iliotibial tract**. The tract inserts into the lateral condyle of the tibia.

MUSCLE	ORIGIN	INSERTION	ACTION	INNERVATION
Iliopsoas (il-ē-ō-SŌ-as) **Psoas major** (SŌ-as MĀ-jor; *psoa* = a muscle of the loin; *major* = larger)	Transverse processes and bodies of lumbar vertebrae.	With iliacus into lesser trochanter of femur.	Psoas major and iliacus muscles acting together flex thigh at hip joint, rotate thigh laterally, and flex trunk on hip as in sitting up from supine position.	Lumbar spinal nerves L2–L3.
Iliacus (il′-ē-A-cus; *iliac* = ilium)	Iliac fossa and sacrum.	With psoas major into lesser trochanter of femur.		Femoral nerve.
Gluteus maximus (GLOO-tē-us MAK-si-mus; *glute* = rump or buttock; *maximus* = largest)	Iliac crest, sacrum, coccyx, and aponeurosis of sacrospinalis.	Iliotibial tract of fascia lata and superior lateral part of linea aspera (gluteal tuberosity) under greater trochanter of femur.	Extends thigh at hip joint and laterally rotates thigh; helps lock knee in extension. RMA: Extends torso.	Inferior gluteal nerve.
Gluteus medius (MĒ-dē-us = middle)	Ilium.	Greater trochanter of femur.	Abducts thigh at hip joint and medially rotates thigh.	Superior gluteal nerve.
Gluteus minimus (MIN-i-mus = smallest)	Ilium.	Greater trochanter of femur.	Abducts thigh at hip joint and medially rotates thigh.	Superior gluteal nerve.
Tensor fasciae latae (TEN-sor FA-shē-ē LĀ-tē; *tensor* = makes tense; *fasciae* = band; *lat* = wide)	Iliac crest.	Tibia by way of iliotibial tract.	Flexes and abducts thigh at hip joint.	Superior gluteal nerve.
Piriformis (pir-i-FOR-mis; *piri-* = pear; *-form-* = shape)	Anterior sacrum.	Superior border of greater trochanter of femur.	Laterally rotates and abducts thigh at hip joint.	Sacral spinal nerves S1 or S2, mainly S1.
Obturator internus (OB-too-rā′-tor in-TER-nus; *obturator* = obturator foramen; *intern-* = inside)	Inner surface of obturator foramen, pubis, and ischium.	Medial surface of greater trochanter of femur.	Laterally rotates and abducts thigh at hip joint.	Nerve to obturator internus.

MUSCLE	ORIGIN	INSERTION	ACTION	INNERVATION
Obturator externus (ex-TER-nus; *extern-* = outside)	Outer surface of obturator membrane.	Deep depression inferior to greater trochanter (trochanteric fossa) of femur.	Laterally rotates and abducts thigh at hip joint.	Obturator nerve.
Superior gemellus (jem-EL-lus; *superior* = above; *gemell* = twins)	Ischial spine.	Medial surface of greater trochanter of femur.	Laterally rotates and abducts thigh at hip joint.	Nerve to obturator internus.
Inferior gemellus (*inferior* = below)	Ischial tuberosity.	Medial surface of greater trochanter of femur.	Laterally rotates and abducts thigh at hip joint.	Nerve to quadratus femoris.
Quadratus femoris (kwod-RĀ-tus FEM-or-is; *quad* = square, four-sided; *femoris* = femur)	Ischial tuberosity.	Elevation superior to mid-portion of intertrochanteric crest (quadrate tubercle) on posterior femur.	Laterally rotates and stabilizes hip joint.	Nerve to quadratus femoris.
Adductor longus (ad-DUK-tor LONG-us; *adductor* = moves part closer to midline; *longus* = long)	Pubic crest and pubic symphysis.	Linea aspera of femur.	Adducts and flexes thigh at hip joint and rotates thigh.* RMA: Extends thigh.	Obturator nerve.
Adductor brevis (BREV-is = short)	Inferior ramus of pubis.	Superior half of linea aspera of femur.	Adducts and flexes thigh at hip joint and rotates thigh.* RMA: Extends thigh.	Obturator nerve.
Adductor magnus (MAG-nus = large)	Inferior ramus of pubis and ischium to ischial tuberosity.	Linea aspera of femur.	Adducts thigh at hip joint and rotates thigh; anterior part flexes thigh at hip joint, and posterior part extends thigh at hip joint.*	Obturator and sciatic nerves.
Pectineus (pek-TIN-ē-us; *pectin* = comb)	Superior ramus of pubis.	Pectineal line of femur, between lesser trochanter and linea aspera.	Flexes and adducts thigh at hip joint.	Femoral nerve.

*All adductors are unique muscles that cross the thigh joint obliquely from an anterior origin to a posterior insertion. As a result, they laterally rotate the hip joint when the foot is off the ground but medially rotate the hip joint when the foot is on the ground.

The **piriformis, obturator internus, obturator externus, superior gemellus, inferior gemellus,** and **quadratus femoris** muscles are all deep to the gluteus maximus muscle and function as lateral rotators of the femur at the hip joint.

Three muscles on the medial aspect of the thigh are the **adductor longus, adductor brevis,** and **adductor magnus**. They originate on the pubic bone and insert on the femur. These three muscles adduct the thigh and are unique in their ability to both medially and laterally rotate the thigh. When the foot is on the ground, these muscles medially rotate the thigh, but when the foot is off the ground, they are lateral rotators of the thigh. This results from their oblique orientation, from an anterior origin to a posterior insertion. In addition, the adductor longus flexes the thigh and the adductor magnus extends the thigh. The pectineus muscle also adducts and flexes the femur at the hip joint.

Technically, the adductor muscles and pectineus muscles are components of the medial compartment of the thigh and could be included in Section 11.21. However, they are included here because they act on the femur.

At the junction between the trunk and lower limb is a space called the **femoral triangle**. The base is formed superiorly by the inguinal ligament, medially by the lateral border of the adductor longus muscle, and laterally by the medial border of the sartorius muscle. The apex is formed by the crossing of the adductor longus by the sartorius muscle (**Figure 11.20a**). The contents of the femoral triangle, from lateral to medial, are the femoral nerve and its branches, the femoral artery and several of its branches, the femoral vein and its proximal tributaries, and the deep inguinal lymph nodes. The femoral artery is easily accessible within the triangle and is the site for insertion of catheters that may extend into the aorta and ultimately into the coronary vessels of the heart. Such catheters are utilized during cardiac catheterization, coronary angiography, and other procedures involving the heart. Inguinal hernias frequently appear in this area.

See Clinical Connection: Groin Pull

FIGURE 11.20 **Muscles of the gluteal region that move the femur (thigh bone).**

Most muscles that move the femur originate on the pelvic (hip) girdle and insert on the femur.

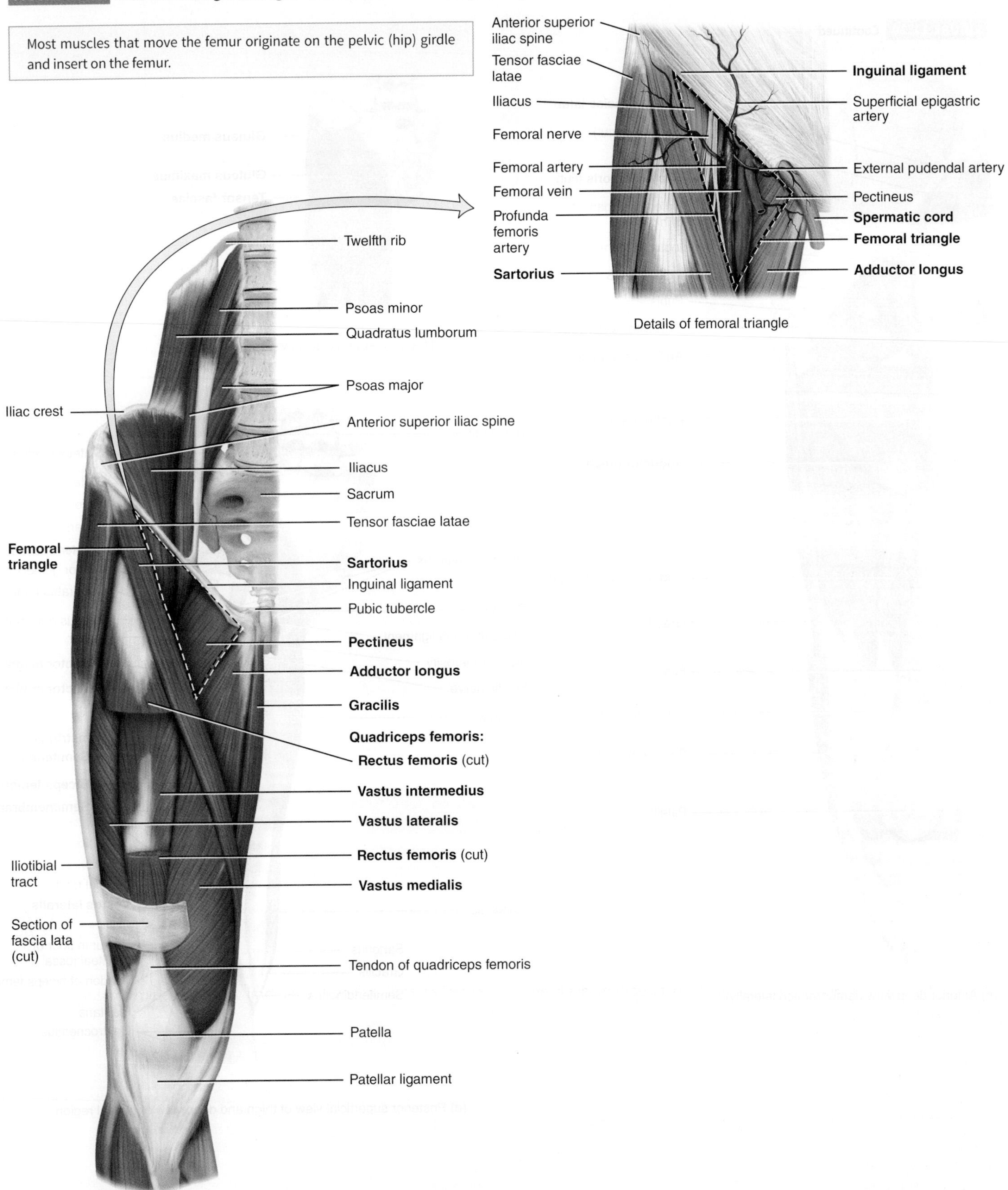

(a) Anterior superficial view (the femoral triangle is indicated by a dashed line)

Figure 11.20 ***Continues***

FIGURE 11.20 Continued

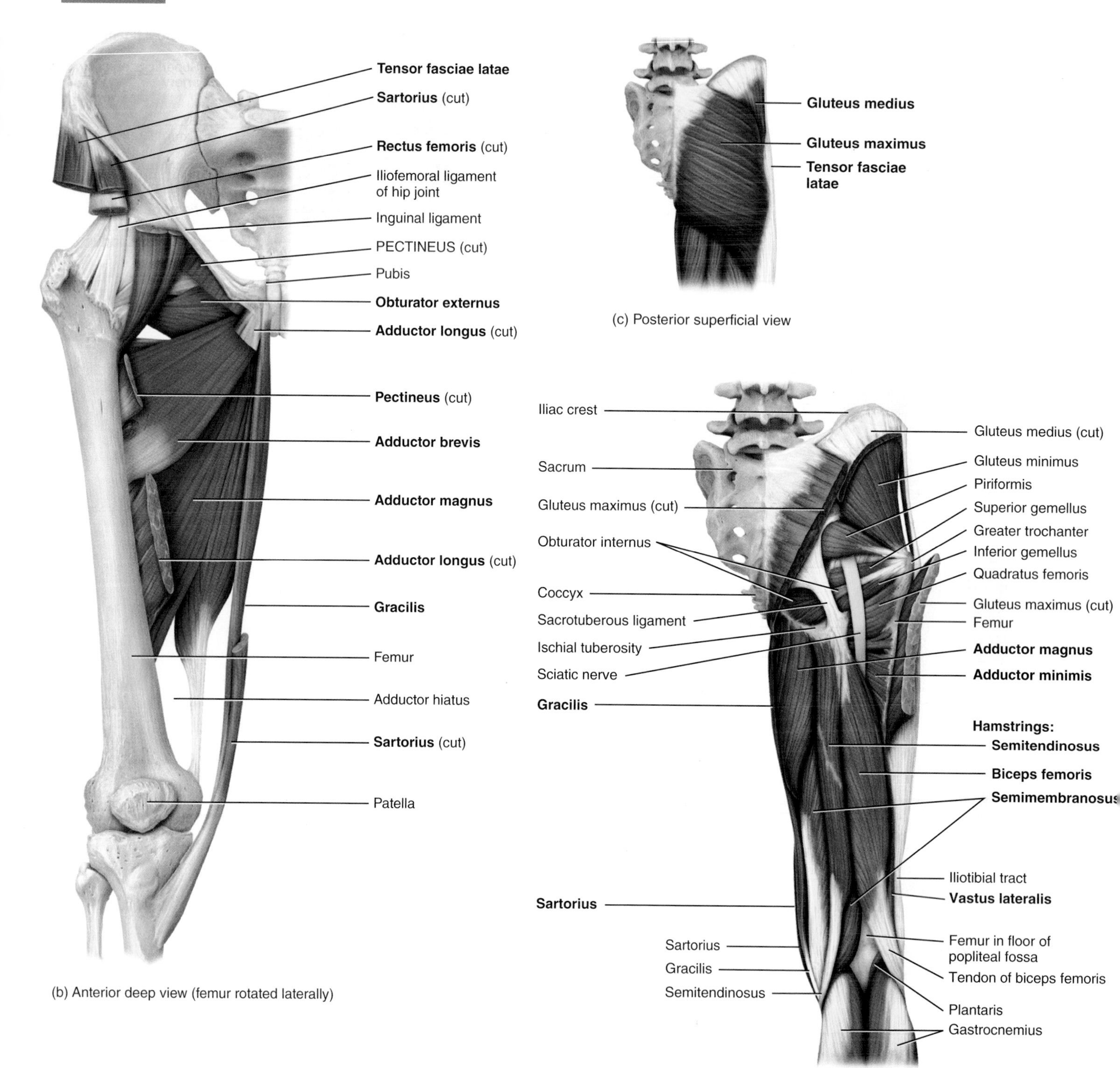

(b) Anterior deep view (femur rotated laterally)

(c) Posterior superficial view

(d) Posterior superficial view of thigh and deep view of gluteal region

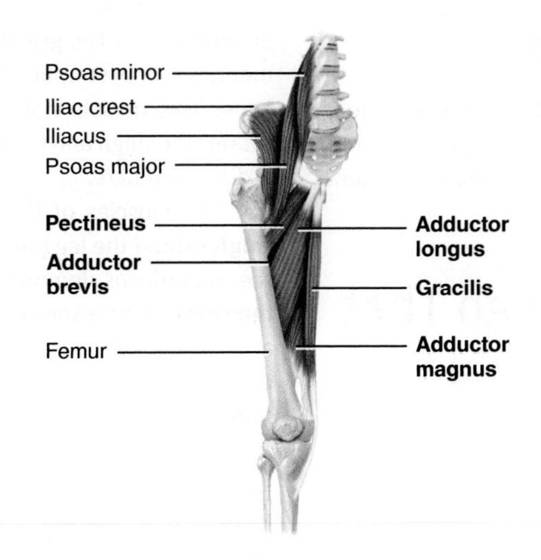

Anterior deep view

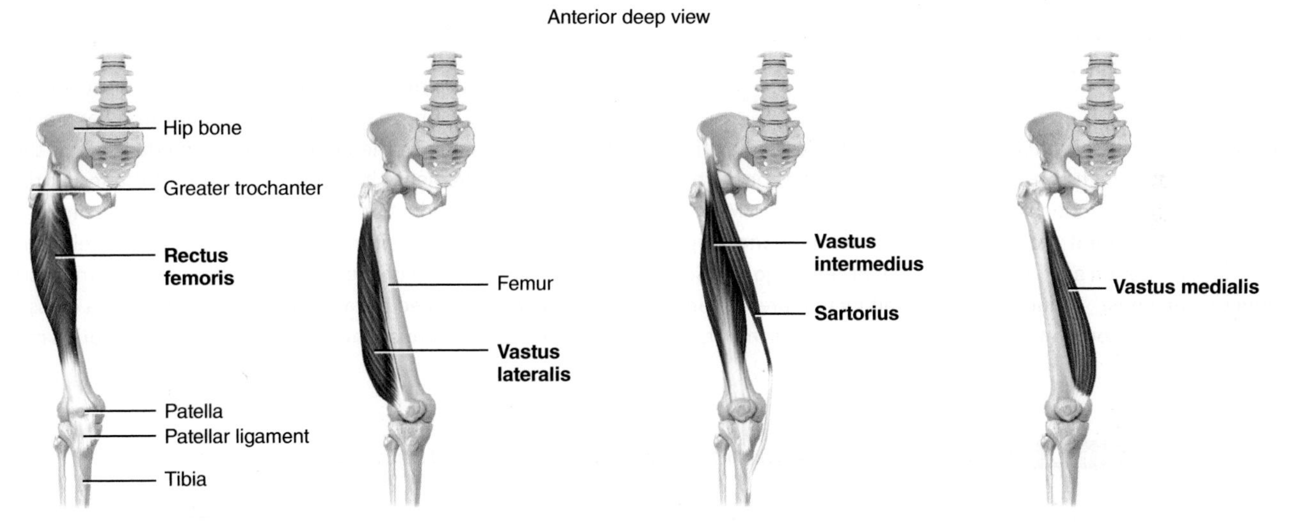

Anterior views

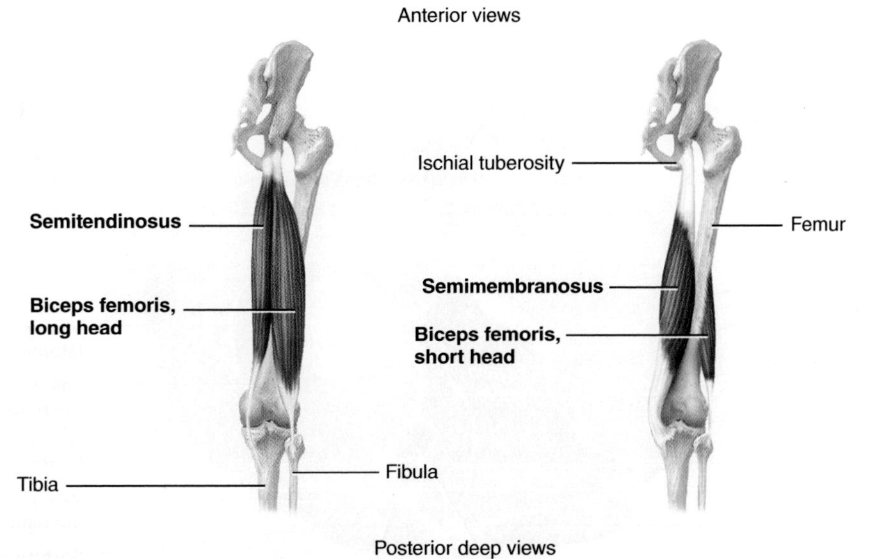

Posterior deep views

(e) Isolated muscles

Q What are the principal differences between the muscles of the free upper and lower limbs?

Relating Muscles to Movements

Arrange the muscles in this section according to the following actions on the thigh at the hip joint: (1) flexion, (2) extension, (3) abduction, (4) adduction, (5) medial rotation, and (6) lateral rotation. The same muscle may be mentioned more than once.

11.21 Muscles of the Thigh That Move the Femur, Tibia, and Fibula

OBJECTIVE

- **Describe** the origin, insertion, action, and innervation of the muscles that move the femur, tibia, and fibula.

Deep fascia (*intermuscular septum*) separates the muscles of the thigh that act on the femur (thigh bone) and tibia and fibula (leg bones) into medial, anterior, and posterior compartments (Figure 11.21). Most of the muscles of the **medial** (*adductor*) **compartment of the thigh** have a similar orientation and adduct the femur at the hip joint. (See the adductor magnus, adductor longus, adductor brevis, and pectineus, which are components of the medial compartment, in Section 11.20.) The **gracilis**, the other muscle in the medial compartment, is a long, straplike muscle on the medial aspect of the thigh and knee. This muscle not only adducts the thigh, but also medially rotates the thigh and flexes the leg at the knee joint. For this reason, it is discussed here.

The muscles of the **anterior** (*extensor*) **compartment of the thigh** extend the leg (and flex the thigh). This compartment contains the quadriceps femoris and sartorius muscles. The **quadriceps femoris** muscle is the largest muscle in the body, covering most of the anterior surface and sides of the thigh. The muscle is actually a composite muscle, usually described as four separate muscles: (1) **rectus femoris**, on the anterior aspect of the thigh; (2) **vastus lateralis**, on the lateral aspect of the thigh; (3) **vastus medialis**, on the medial aspect of the thigh; and (4) **vastus intermedius**, located deep to the rectus femoris between the vastus lateralis and vastus medialis. The common tendon for the four muscles, known as the **quadriceps tendon**, inserts into the patella. The tendon continues below the patella as the **patellar ligament**, which attaches to the tibial tuberosity. The quadriceps femoris muscle is the great extensor muscle of the leg. The **sartorius** is a long, narrow muscle that forms a band across the thigh from the ilium of the hip bone to the medial side of the tibia. The various movements it produces (flexion of the leg at the knee joint and flexion, abduction, and lateral rotation at the hip joint) help effect the cross-legged sitting position in which the heel of one limb is placed on the knee of the opposite limb. Its name means *tailor's muscle*; it was so called because tailors often assume a cross-legged sitting position. (Because the major action of the sartorius muscle is

FIGURE 11.21 Muscles of the thigh that move the femur (thigh bone) and tibia and fibula (leg bones).

Muscles that act on the leg originate in the hip and thigh and are separated into compartments by deep fascia.

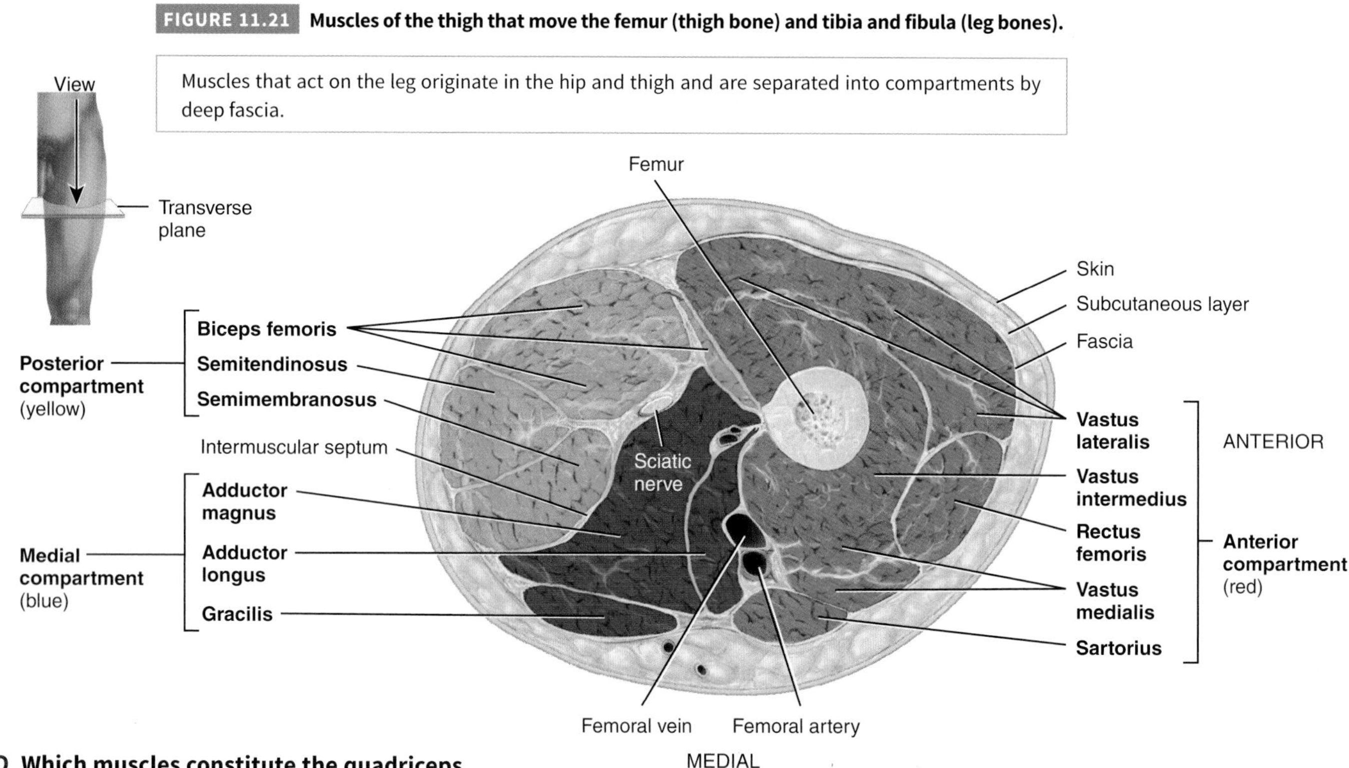

Superior view of transverse section of thigh

Q Which muscles constitute the quadriceps femoris and hamstring muscles?

MUSCLE	ORIGIN	INSERTION	ACTION	INNERVATION
MEDIAL (ADDUCTOR) COMPARTMENT OF THE THIGH				
Adductor magnus (ad-DUK-tor MAG-nus)	See Section 11.20.			
Adductor longus (LONG-us)	See Section 11.20.			
Adductor brevis (BREV-is)	See Section 11.20.			
Pectineus (pek-TIN-ē-us)	See Section 11.20.			
Gracilis (GRAS-i-lis = slender) (see also Figure 11.20a)	Body and inferior ramus of pubis.	Medial surface of body of tibia.	Adducts thigh at hip joint, medially rotates thigh, and flexes leg at knee joint.	Obturator nerve.
ANTERIOR (EXTENSOR) COMPARTMENT OF THE THIGH (see also Figure 11.20a)				
Quadriceps femoris (KWOD-ri-seps FEM-or-is; *quadriceps* = four heads [of origin]; *femoris* = femur)				
Rectus femoris (REK-tus = fascicles parallel to midline)	Anterior inferior iliac spine.	Patella via quadriceps tendon and then tibial tuberosity via patellar ligament.	All four heads extend leg at knee joint; rectus femoris muscle acting alone also flexes thigh at hip joint.	Femoral nerve.
Vastus lateralis (VAS-tus lat′-e-RĀ-lis; *vast* = huge; *lateralis* = lateral)	Greater trochanter and linea aspera of femur.	Patella via quadriceps tendon and then tibial tuberosity via patellar ligament.	All four heads extend leg at knee joint; rectus femoris muscle acting alone also flexes thigh at hip joint.	Femoral nerve.
Vastus medialis (mē-dē-Ā-lis = medial)	Linea aspera of femur.	Patella via quadriceps tendon and then tibial tuberosity via patellar ligament.	All four heads extend leg at knee joint; rectus femoris muscle acting alone also flexes thigh at hip joint.	Femoral nerve.
Vastus intermedius (in′-ter-MĒ-dē-us = middle)	Anterior and lateral surfaces of body of femur.	Patella via quadriceps tendon and then tibial tuberosity via patellar ligament.	All four heads extend leg at knee joint; rectus femoris muscle acting alone also flexes thigh at hip joint.	Femoral nerve.
Sartorius (sar-TOR-ē-us; *sartor* = tailor; longest muscle in body)	Anterior superior iliac spine.	Medial surface of body of tibia.	Weakly flexes leg at knee joint; weakly flexes, abducts, and laterally rotates thigh at hip joint.	Femoral nerve.
POSTERIOR (FLEXOR) COMPARTMENT OF THE THIGH (see also Figure 11.20d)				
Hamstrings A collective designation for three separate muscles.				
Biceps femoris (BĪ-seps = two heads of origin)	Long head arises from ischial tuberosity; short head arises from linea aspera of femur.	Head of fibula and lateral condyle of tibia.	Flexes leg at knee joint and extends thigh at hip joint.	Tibial and fibular nerves from sciatic nerve.
Semitendinosus (sem′-ē-ten-di-NŌ-sus; *semi-* = half; *-tendo-* = tendon)	Ischial tuberosity.	Proximal part of medial surface of shaft of tibia.	Flexes leg at knee joint and extends thigh at hip joint.	Tibial nerve from sciatic nerve.
Semimembranosus (sem′-ē-mem-bra-NŌ-sus; *membran-* = membrane)	Ischial tuberosity.	Medial condyle of tibia.	Flexes leg at knee joint and extends thigh at hip joint.	Tibial nerve from sciatic nerve.

to move the thigh rather than the leg, it could have been included in Section 11.20.)

The muscles of the **posterior** (*flexor*) **compartment of the thigh** flex the leg (and extend the thigh). This compartment is composed of three muscles collectively called the **hamstrings:** (1) **biceps femoris**, (2) **semitendinosus**, and (3) **semimembranosus**. The hamstrings are so named because their tendons are long and stringlike in the popliteal area. Because the hamstrings span two joints (hip and knee), they are both extensors of the thigh and flexors of the leg. The **popliteal fossa** is a diamond-shaped space on the posterior aspect of the knee bordered laterally by the tendons of the biceps femoris muscle and medially by the tendons of the semitendinosus and semimembranosus muscles.

See Clinical Connection: Pulled Hamstrings

Relating Muscles to Movements

Arrange the muscles in this section according to the following actions on the thigh at the hip joint: (1) abduction, (2) adduction, (3) lateral rotation, (4) flexion, and (5) extension; and according to the following actions on the leg at the knee joint: (1) flexion and (2) extension. The same muscle may be mentioned more than once.

11.22 Muscles of the Leg That Move the Foot and Toes

OBJECTIVE

- **Describe** the origin, insertion, action, and innervation of the muscles of the leg that move the foot and toes.

Muscles that move the foot and toes are located in the leg (Figure 11.22). The muscles of the leg, like those of the thigh, are divided by deep fascia into three compartments: anterior, lateral, and posterior. The **anterior compartment of the leg** consists of muscles that dorsiflex the foot. In a situation analogous to the wrist, the tendons of the muscles of the anterior compartment are held firmly to the ankle by thickenings of deep fascia called the **superior extensor retinaculum** (*transverse ligament of the ankle*) and **inferior extensor retinaculum** (*cruciate ligament of the ankle*).

Within the anterior compartment, the **tibialis anterior** is a long, thick muscle against the lateral surface of the tibia, where it is easy to palpate (feel). The **extensor hallucis longus** is a thin muscle between and partly deep to the tibialis anterior and **extensor digitorum longus** muscles. This featherlike muscle is lateral to the tibialis anterior muscle, where it can also be palpated easily. The **fibularis** (*peroneus*) **tertius** muscle is part of the extensor digitorum longus, with which it shares a common origin.

The **lateral** (*fibular*) **compartment of the leg** contains two muscles that plantar flex and evert the foot: the **fibularis** (*peroneus*) **longus** and **fibularis** (*peroneus*) **brevis.**

The **posterior compartment of the leg** consists of muscles in superficial and deep groups. The superficial muscles share a common tendon of insertion, the **calcaneal** (*Achilles*) **tendon**, the strongest tendon of the body. It inserts into the calcaneal bone of the ankle. The superficial and most of the deep muscles plantar flex the foot at the ankle joint. The superficial muscles of the posterior compartment are the gastrocnemius, soleus, and plantaris—the so-called calf muscles. The large size of these muscles is directly related to the characteristic upright stance of humans. The **gastrocnemius** is the most superficial

MUSCLE	ORIGIN	INSERTION	ACTION	INNERVATION
ANTERIOR COMPARTMENT OF THE LEG				
Tibialis anterior (tib′-ē-Ā-lis an-TĒR-ē-or; *tibialis* = tibia; *anterior* = front)	Lateral condyle and body of tibia and interosseous membrane (sheet of fibrous tissue that holds shafts of tibia and fibula together).	Metatarsal I and first (medial) cuneiform.	Dorsiflexes foot at ankle joint and inverts (supinates) foot at intertarsal joints.	Deep fibular (peroneal) nerve.
Extensor hallucis longus (eks-TEN-sor HAL-ū-sis LON-gus; *extensor* = increases angle at joint; *hallucis* = hallux or great toe; *longus* = long)	Anterior surface of middle third of fibula and interosseous membrane.	Distal phalanx of great toe.	Dorsiflexes foot at ankle joint and extends proximal phalanx of great toe at metatarsophalangeal joint.	Deep fibular (peroneal) nerve.
Extensor digitorum longus (di′-ji-TOR-um; *digit-* = finger or toe)	Lateral condyle of tibia, anterior surface of fibula, and interosseous membrane.	Middle and distal phalanges of toes II–V.*	Dorsiflexes foot at ankle joint and extends distal and middle phalanges of each toe at interphalangeal joints and proximal phalanx of each toe at metatarsophalangeal joint.	Deep fibular (peroneal) nerve.
Fibularis (peroneus) tertius (fib-ū-LĀ-ris; per′-Ō-nē-us TER-shus; *peron* = fibula; *tertius* = third)	Distal third of fibula and interosseous membrane.	Base of metatarsal V.	Dorsiflexes foot at ankle joint and everts (pronates) foot at intertarsal joints.	Deep fibular (peroneal) nerve.
LATERAL (FIBULAR) COMPARTMENT OF THE LEG				
Fibularis (peroneus) longus	Head and body of fibula.	Metatarsal I and first cuneiform.	Plantar flexes foot at ankle joint and everts (pronates) foot at intertarsal joints.	Superficial fibular (peroneal) nerve.
Fibularis (peroneus) brevis (BREV-is = short)	Distal half of body of fibula.	Base of metatarsal V.	Plantar flexes foot at ankle joint and everts (pronates) foot at intertarsal joints.	Superficial fibular (peroneal) nerve.

MUSCLE	ORIGIN	INSERTION	ACTION	INNERVATION
SUPERFICIAL POSTERIOR COMPARTMENT OF THE LEG				
Gastrocnemius (gas′-trok-NĒ-mē-us; *gastro-* = belly; *cnem-* = leg)	Lateral and medial condyles of femur and capsule of knee.	Calcaneus by way of calcaneal (Achilles) tendon.	Plantar flexes foot at ankle joint and flexes leg at knee joint.	Tibial nerve.
Soleus (SŌ-lē-us; *sole* = type of flat fish)	Head of fibula and medial border of tibia.	Calcaneus by way of calcaneal (Achilles) tendon.	Plantar flexes foot at ankle joint.	Tibial nerve.
Plantaris (plan-TĀR-is = sole)	Lateral epicondyle of femur.	Calcaneus medial to calcaneal (Achilles) tendon (occasionally fused with calcaneal tendon).	Plantar flexes foot at ankle joint and flexes leg at knee joint.	Tibial nerve.
DEEP POSTERIOR COMPARTMENT OF THE LEG				
Popliteus (pop-LIT-ē-us; *poplit* = back of knee)	Lateral condyle of femur.	Proximal tibia.	Flexes leg at knee joint and medially rotates tibia to unlock the extended knee.	Tibial nerve.
Tibialis posterior (tib′-ē-Ā-lis; *posterior* = back)	Proximal tibia, fibula, and interosseous membrane.	Metatarsals II–IV; navicular; and all three cuneiforms.	Plantar flexes foot at ankle joint and inverts (supinates) foot at intertarsal joints.	Tibial nerve.
Flexor digitorum longus (FLEK-sor = decreases angle at point)	Middle third of posterior surface of tibia.	Distal phalanges of toes II–V.	Plantar flexes foot at ankle joint; flexes distal and middle phalanges of toes II–V at interphalangeal joints and proximal phalanx of toes II–V at metatarsophalangeal joint.	Tibial nerve.
Flexor hallucis longus	Inferior two-thirds of posterior portion of fibula.	Distal phalanx of great toe.	Plantar flexes foot at ankle joint; flexes distal phalanx of great toe at interphalangeal joint and proximal phalanx of great toe at metatarsophalangeal joint.	Tibial nerve.

*Reminder: The great toe or hallux is the first toe and has two phalanges: proximal and distal. The remaining toes are numbered II–V (2–5), and each has three phalanges: proximal, middle, and distal.

muscle and forms the prominence of the calf. The **soleus**, which lies deep to the gastrocnemius, is broad and flat. It derives its name from its resemblance to a flat fish (sole). The **plantaris** is a small muscle that may be absent; conversely, sometimes there are two of them in each leg. It runs obliquely between the gastrocnemius and soleus muscles.

The deep muscles of the posterior compartment are the popliteus, tibialis posterior, flexor digitorum longus, and flexor hallucis longus. The **popliteus** is a triangular muscle that forms the floor of the popliteal fossa. The **tibialis posterior** is the deepest muscle in the posterior compartment. It lies between the flexor digitorum longus and flexor hallucis longus muscles. The **flexor digitorum longus** is smaller than the **flexor hallucis longus**, even though the former flexes four toes and the latter flexes only the great toe at the interphalangeal joint.

See Clinical Connection: Shin Splint Syndrome

Relating Muscles to Movements

Arrange the muscles in this section according to the following actions on the foot at the ankle joint: (1) dorsiflexion and (2) plantar flexion; according to the following actions on the foot at the intertarsal joints: (1) inversion and (2) eversion; and according to the following actions on the toes at the metatarsophalangeal and interphalangeal joints: (1) flexion and (2) extension. The same muscle may be mentioned more than once.

FIGURE 11.22 **Muscles of the leg that move the foot and toes.**

The superficial muscles of the posterior compartment share a common tendon of insertion, the calcaneal (Achilles) tendon, which inserts into the calcaneal bone of the ankle.

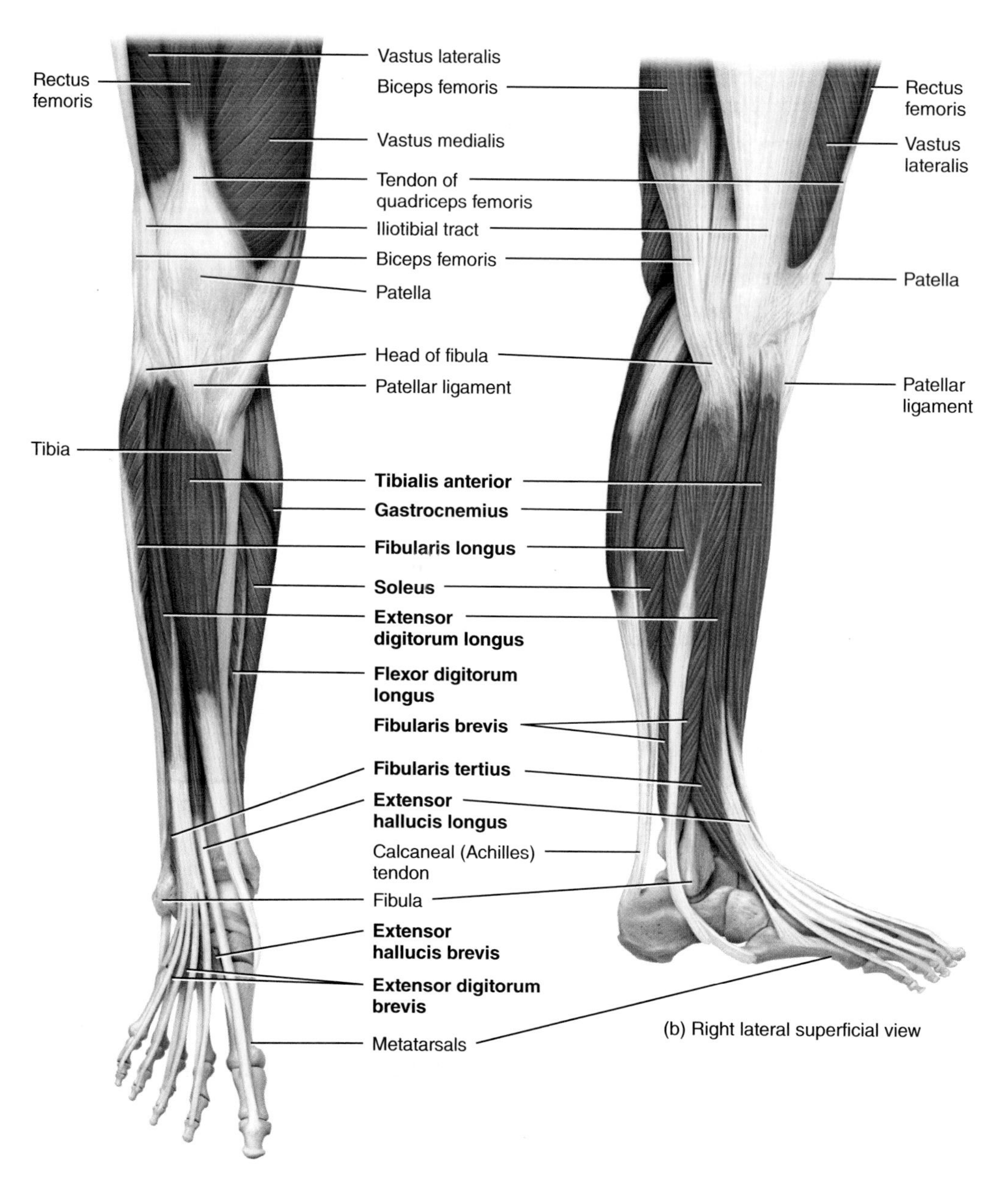

(a) Anterior superficial view

(b) Right lateral superficial view

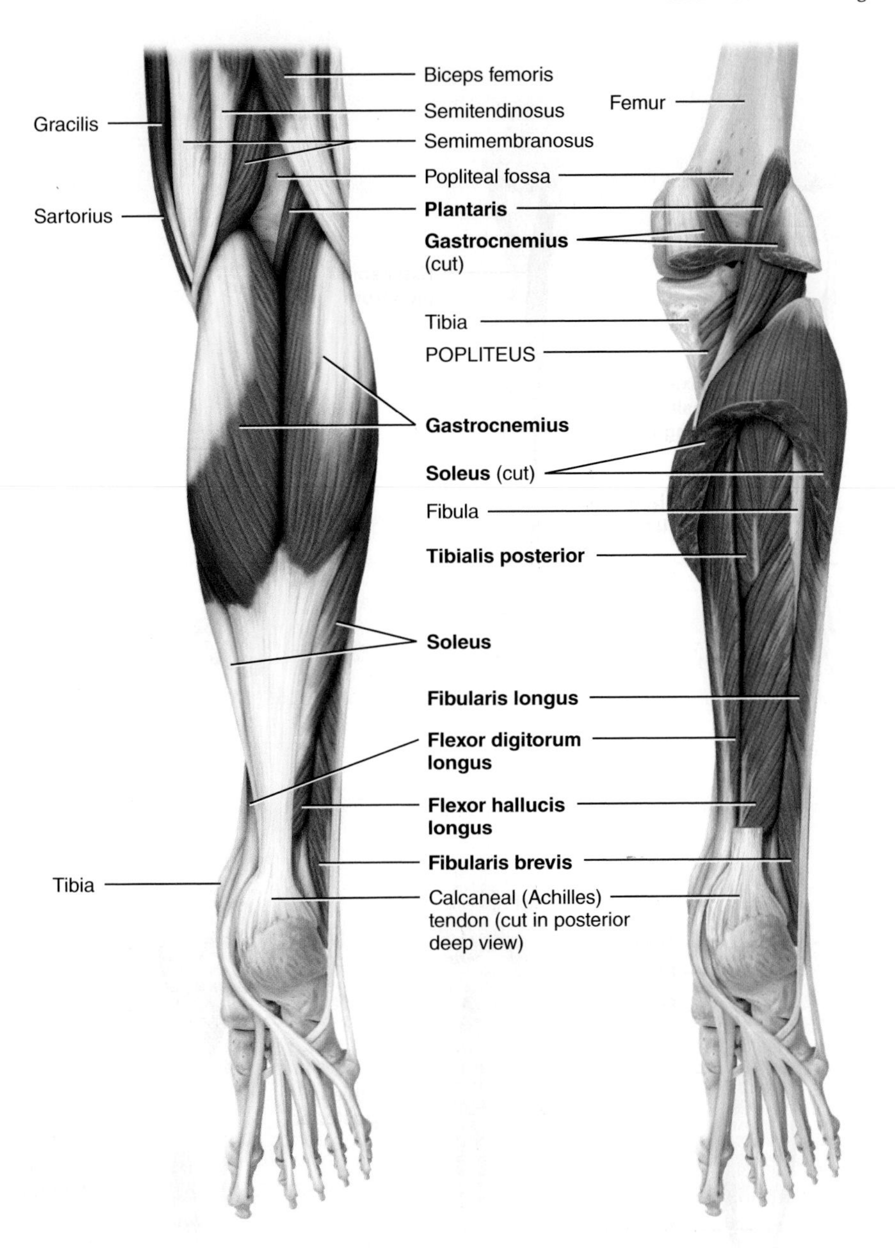

(c) Posterior superficial view

(d) Posterior deep view

Figure 11.22 ***Continues***

FIGURE 11.22 Continued

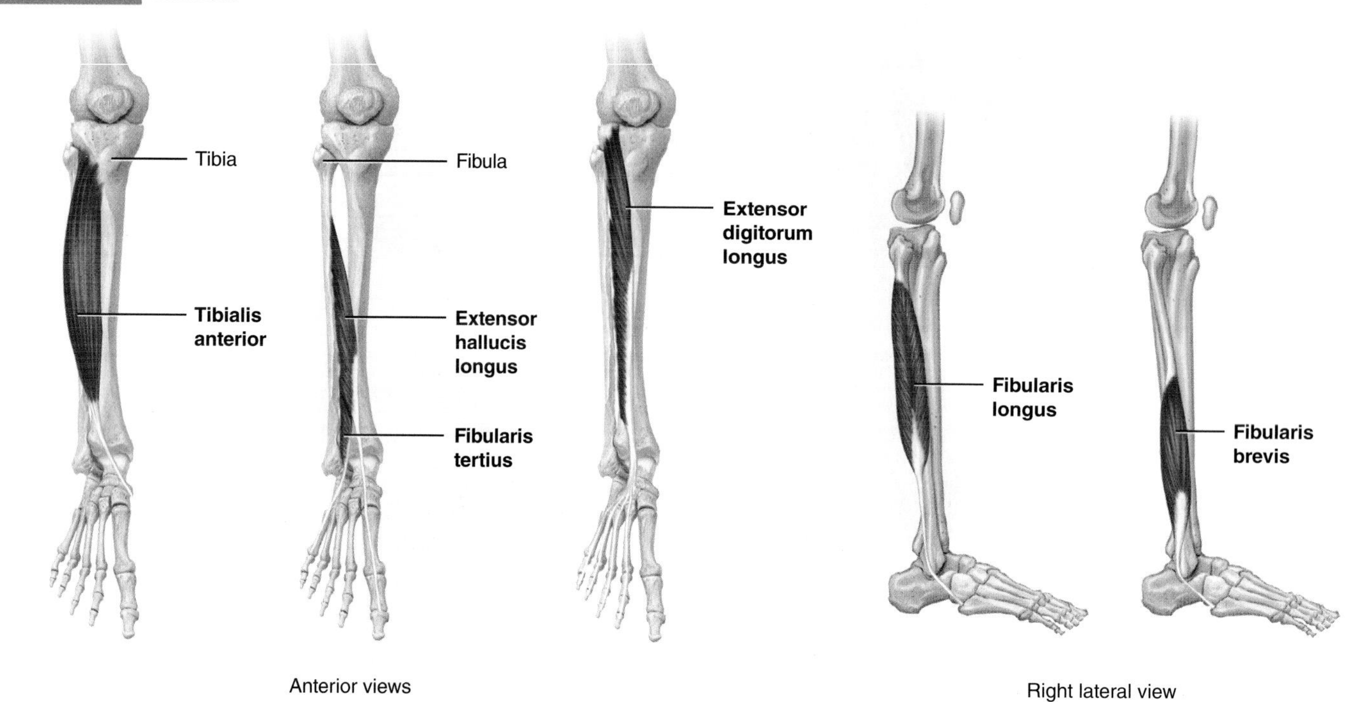

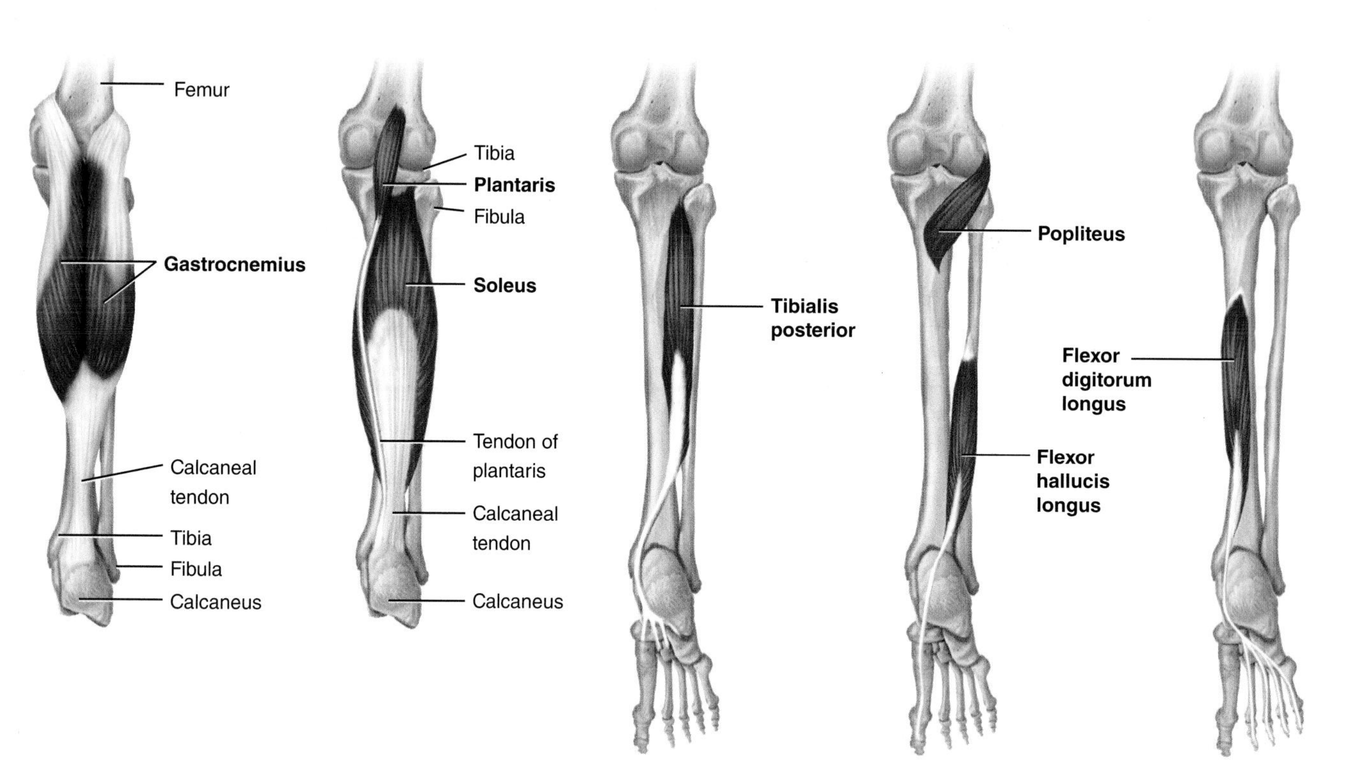

(e) Isolated muscles

Q What structures firmly hold the tendons of the anterior compartment muscles to the ankle?

11.23 Intrinsic Muscles of the Foot That Move the Toes

OBJECTIVE

- **Describe** the origin, insertion, action, and innervation of the intrinsic muscles of the foot that move the toes.

The muscles in this exhibit are termed **intrinsic muscles of the foot** because they originate and insert *within* the foot (Figure 11.23). The muscles of the hand are specialized for precise and intricate movements, but those of the foot are limited to support and locomotion. The deep fascia of the foot forms the **plantar aponeurosis** (*fascia*) that extends from the calcaneus bone to the phalanges of the toes. The aponeurosis supports the longitudinal arch of the foot and encloses the flexor tendons of the foot.

The intrinsic muscles of the foot are divided into two groups: **dorsal muscles of the foot** and **plantar muscles of the foot**. There are two dorsal muscles, the **extensor hallucis brevis** and the **extensor digitorum brevis**. The latter is a four-part muscle deep to the tendons of the extensor digitorum longus muscle, which extends toes II–V at the metatarsophalangeal joints.

The plantar muscles are arranged in four layers. The most superficial layer, called the first layer, consists of three muscles. The **abductor hallucis**, which lies along the medial border of the sole and is comparable to the abductor pollicis brevis in the hand, abducts the great toe at the metatarsophalangeal joint. The **flexor digitorum brevis**, which lies in the middle of the sole, flexes toes II–V at the interphalangeal and metatarsophalangeal joints. The **abductor digiti minimi**, which lies along the lateral border of the sole and is comparable to the same muscle in the hand, abducts the little toe.

The second layer consists of the **quadratus plantae**, a rectangular muscle that arises by two heads and flexes toes II–V at the metatarsophalangeal joints, and the **lumbricals**, four small muscles that are similar to the lumbricals in the hands. They flex the proximal phalanges and extend the distal phalanges of toes II–V.

Three muscles compose the third layer. The **flexor hallucis brevis**, which lies adjacent to the plantar surface of the metatarsal of the great toe and is comparable to the same muscle in the hand, flexes the great toe. The **adductor hallucis**, which has an oblique and transverse head like the adductor pollicis in the hand, adducts the great toe. The **flexor digiti minimi brevis**, which lies superficial to the metatarsal of the little toe and is comparable to the same muscle in the hand, flexes the little toe.

The fourth layer is the deepest and consists of two muscle groups. The **dorsal interossei** are four muscles that abduct toes II–IV, flex the proximal phalanges, and extend the distal phalanges. The three **plantar interossei** abduct toes III–V, flex the proximal phalanges, and extend the distal phalanges. The interossei of the feet are similar to those of the hand. However, their actions are relative to the midline of the second digit rather than the third digit as in the hand.

Relating Muscles to Movements

Arrange the muscles in this section according to the following actions on the great toe at the metatarsophalangeal joint: (1) flexion, (2) extension, (3) abduction, and (4) adduction; and according to the following actions on toes II–V at the metatarsophalangeal and interphalangeal joints: (1) flexion, (2) extension, (3) abduction, and (4) adduction. The same muscle may be mentioned more than once.

MUSCLE	ORIGIN	INSERTION	ACTION	INNERVATION
DORSAL				
Extensor hallucis brevis (eks-TEN-sor HAL-ū-sis BREV-is; *extensor* = increases angle at joint; *hallucis* = hallux or great toe; *brevis* = short) (see **Figure 11.22a**)	Calcaneus and inferior extensor retinaculum.	Proximal phalanx of great toe.	Extends great toe at metatarsophalangeal joint.	Deep fibular (peroneal) nerve.
Extensor digitorum brevis (di′-ji-TOR-um; *digit* = finger or toe) (see **Figure 11.22a**)	Calcaneus and inferior extensor retinaculum.	Middle phalanges of toes II–IV.	Extends toes II–IV at interphalangeal joints.	Deep fibular (peroneal) nerve.
PLANTAR				
First layer (most superficial)				
Abductor hallucis (*abductor* = moves part away from midline)	Calcaneus, plantar aponeurosis, and flexor retinaculum.	Medial side of proximal phalanx of great toe with the tendon of flexor hallucis brevis.	Abducts and flexes great toe at metatarsophalangeal joint.	Medial plantar nerve.
Flexor digitorum brevis (*flexor* = decreases angle at joint)	Calcaneus, plantar aponeurosis, and flexor retinaculum.	Sides of middle phalanx of toes II–V.	Flexes toes II–V at proximal interphalangeal and metatarsophalangeal joints.	Medial plantar nerve.
Abductor digiti minimi (DIJ-i-tē MIN-i-mē; *minimi* = smallest)	Calcaneus, plantar aponeurosis, and flexor retinaculum.	Lateral side of proximal phalanx of little toe with tendon of flexor digiti minimi brevis.	Abducts and flexes little toe at metatarsophalangeal joint.	Lateral plantar nerve.
Second layer				
Quadratus plantae (kwod-RĀ-tus PLAN-tē; *quad* = square, four-sided; *planta* = sole)	Calcaneus.	Tendon of flexor digitorum longus.	Assists flexor digitorum longus to only flex toes II–V at interphalangeal and metatarsophalangeal joints.	Lateral plantar nerve.
Lumbricals (LUM-bri-kals; *lumbric* = earthworm)	Tendons of flexor digitorum longus.	Tendons of extensor digitorum longus on proximal phalanges of toes II–V.	Extend toes II–V at interphalangeal joints and flex toes II–V at metatarsophalangeal joints.	Medial and lateral plantar nerves.
Third layer				
Flexor hallucis brevis	Cuboid and third (lateral) cuneiform.	Medial and lateral sides of proximal phalanx of great toe via tendon containing sesamoid bone.	Flexes great toe at metatarsophalangeal joint.	Medial plantar nerve.
Adductor hallucis (ad-DUK-tor = moves part closer to midline)	Metatarsals II–IV, ligaments of metatarsals III–V at metatarsophalangeal joints, and tendon of fibularis (peroneus) longus.	Lateral side of proximal phalanx of great toe.	Adducts and flexes great toe at metatarsophalangeal joint.	Lateral plantar nerve.
Flexor digiti minimi brevis	Metatarsal V and tendon of fibularis (peroneus) longus.	Lateral side of proximal phalanx of little toe.	Flexes little toe at metatarsophalangeal joint.	Lateral plantar nerve.
Fourth layer (deepest)				
Dorsal interossei (in-ter-OS-ē-ī)	Adjacent side of all metatarsals.	Proximal phalanges: both sides of toe II and lateral side of toes III and IV.	Abduct and flex toes II–IV at metatarsophalangeal joints and extend toes at interphalangeal joints.	Lateral plantar nerve.
Plantar interossei (PLAN-tar)	Metatarsals III–V.	Medial side of proximal phalanges of toes III–V.	Adduct and flex proximal metatarsophalangeal joints and extend toes at interphalangeal joints.	Lateral plantar nerve.

FIGURE 11.23 **Intrinsic muscles of the foot that move the toes.**

The muscles of the hand are specialized for precise and intricate movements; those of the foot are limited to support and movement.

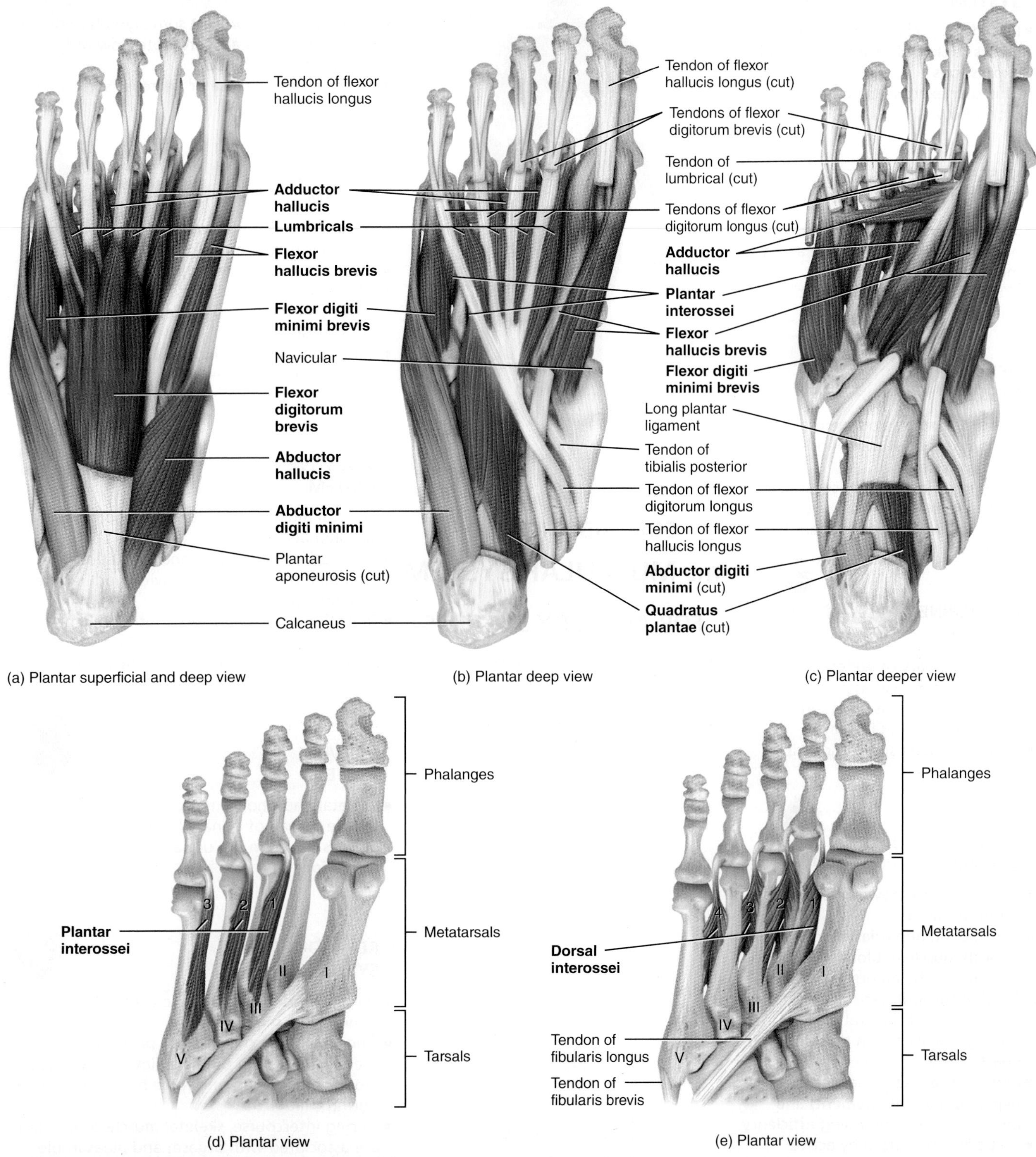

(a) Plantar superficial and deep view

(b) Plantar deep view

(c) Plantar deeper view

(d) Plantar view

(e) Plantar view

Q What structure supports the longitudinal arch and encloses the flexor tendons of the foot?

FOCUS on HOMEOSTASIS

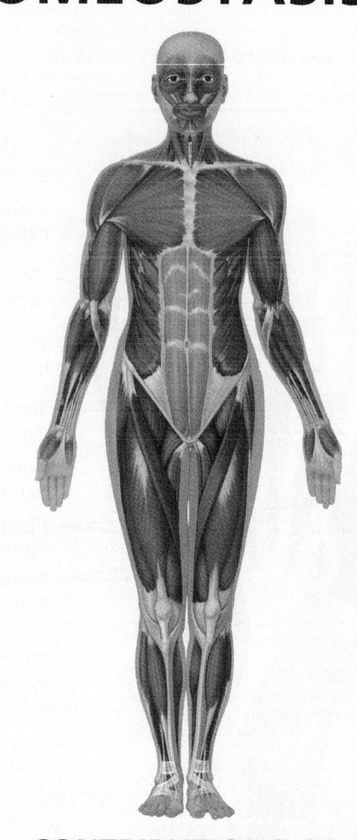

CONTRIBUTIONS OF THE MUSCULAR SYSTEM FOR ALL BODY SYSTEMS

- Produces body movements
- Stabilizes body positions
- Moves substances within the body
- Produces heat that helps maintain normal body temperature

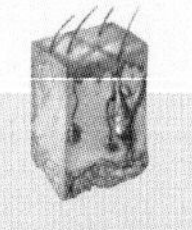

INTEGUMENTARY SYSTEM

- Pull of skeletal muscles on attachments to skin of face causes facial expressions
- Muscular exercise increases skin blood flow

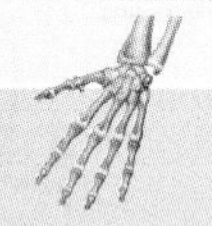

SKELETAL SYSTEM

- Skeletal muscle causes movement of body parts by pulling on attachments to bones
- Skeletal muscle provides stability for bones and joints

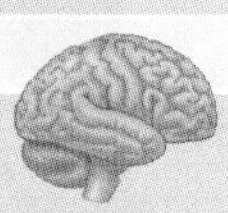

NERVOUS SYSTEM

- Smooth, cardiac, and skeletal muscles carry out commands for the nervous system
- Shivering—involuntary contraction of skeletal muscles that is regulated by the brain—generates heat to raise body temperature

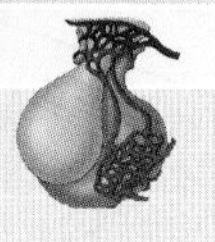

ENDOCRINE SYSTEM

- Regular activity of skeletal muscles (exercise) improves action and signaling mechanisms of some hormones, such as insulin
- Muscles protect some endocrine glands

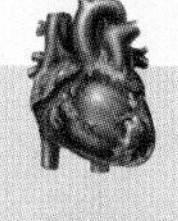

CARDIOVASCULAR SYSTEM

- Cardiac muscle powers pumping action of heart
- Contraction and relaxation of smooth muscle in blood vessel walls help adjust the amount of blood flowing through various body tissues
- Contraction of skeletal muscles in the legs assists return of blood to the heart
- Regular exercise causes cardiac hypertrophy (enlargement) and increases heart's pumping efficiency
- Lactic acid produced by active skeletal muscles may be used for ATP production by the heart

LYMPHATIC SYSTEM and IMMUNITY

- Skeletal muscles protect some lymph nodes and lymphatic vessels and promote flow of lymph inside lymphatic vessels
- Exercise may increase or decrease some immune responses

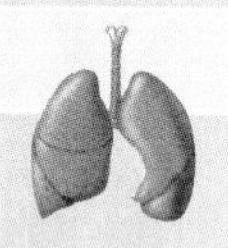

RESPIRATORY SYSTEM

- Skeletal muscles involved with breathing cause air to flow into and out of the lungs
- Smooth muscle fibers adjust size of airways
- Vibrations in skeletal muscles of larynx control air flowing past vocal cords, regulating voice production
- Coughing and sneezing, due to skeletal muscle contractions, help clear airways
- Regular exercise improves efficiency of breathing

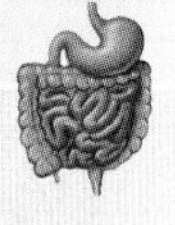

DIGESTIVE SYSTEM

- Skeletal muscles protect and support organs in the abdominal cavity
- Alternating contraction and relaxation of skeletal muscles power chewing and initiate swallowing
- Smooth muscle sphincters control volume of organs of the gastrointestinal (GI) tract
- Smooth muscles in walls of GI tract mix and move its contents through the tract

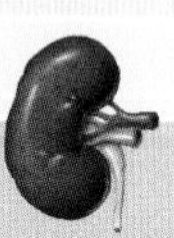

URINARY SYSTEM

- Skeletal and smooth muscle sphincters and smooth muscle in wall of urinary bladder control whether urine is stored in the urinary bladder or voided (urination)

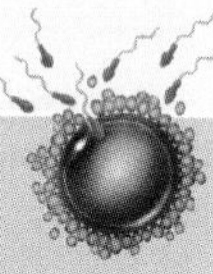

REPRODUCTIVE SYSTEMS

- Skeletal and smooth muscle contractions eject semen from male
- Smooth muscle contractions propel oocyte along uterine tube, help regulate flow of menstrual blood from uterus, and force baby from uterus during childbirth
- During intercourse, skeletal muscle contractions are associated with orgasm and pleasurable sensations in both sexes

Review Questions

11.1 How Skeletal Muscles Produce Movements

1. Using the terms origin, insertion, and belly, describe how skeletal muscles produce body movements by pulling on bones.
2. List the three types of levers, and give an example of a first-, second-, and third-class lever found in the body.
3. Define the roles of the prime mover (agonist), antagonist, synergist, and fixator in producing various movements of the free upper limb.
4. What is a muscle compartment?

11.2 How Skeletal Muscles Are Named

5. Select 10 muscles in Figure 11.3 and identify the features on which their names are based. (*Hint*: Use the prefix, suffix, and root of each muscle's name as a guide.)

11.3 Overview of the Principal Skeletal Muscles

6. List the different features most muscle groups share.

11.4 Muscles of the Head That Produce Facial Expressions

7. Why do the muscles of facial expression move the skin rather than a joint?

11.5 Muscles of the Head That Move the Eyeballs (Extrinsic Eye Muscles) and Upper Eyelids

8. Which muscles that move the eyeballs contract and relax as you look to your left without moving your head?

11.6 Muscles That Move the Mandible and Assist in Mastication and Speech

9. What would happen if you lost tone in the masseter and temporalis muscles?

11.7 Muscles of the Head That Move the Tongue and Assist in Mastication and Speech

10. When your physician says, "Open your mouth, stick out your tongue, and say ahh," to examine the inside of your mouth for possible signs of infection, which muscles do you contract?

11.8 Muscles of the Anterior Neck That Assist in Deglutition and Speech

11. Which tongue, facial, and mandibular muscles do you use for chewing?

11.9 Muscles of the Neck That Move the Head

12. What muscles do you contract to signify "yes" and "no"?

11.10 Muscles of the Abdomen That Protect Abdominal Viscera and Move the Vertebral Column

13. Which muscles do you contract when you "suck in your gut," thereby compressing the anterior abdominal wall?

11.11 Muscles of the Thorax That Assist in Breathing

14. What are the names of the three openings in the diaphragm, and which structures pass through each?

11.12 Muscles of the Pelvic Floor That Support the Pelvic Viscera and Function as Sphincters

15. Which muscles are strengthened by Kegel exercises?

11.13 Muscles of the Perineum

16. What are the borders and contents of the urogenital triangle and the anal triangle?

11.14 Muscles of the Thorax That Move the Pectoral Girdle

17. What muscles in this exhibit are used to raise your shoulders, lower your shoulders, join your hands behind your back, and join your hands in front of your chest?

11.15 Muscles of the Thorax and Shoulder That Move the Humerus

18. Why are the two muscles that cross the shoulder joint called axial muscles, and the seven others called scapular muscles?

11.16 Muscles of the Arm That Move the Radius and Ulna

19. Flex your arm. Which group of muscles is contracting? Which group of muscles must relax so that you can flex your arm?

11.17 Muscles of the Forearm That Move the Wrist, Hand, Thumb, and Digits

20. Which muscles and actions of the wrist, hand, thumb, and fingers are used when writing?

11.18 Muscles of the Palm That Move the Digits—Intrinsic Muscles of the Hand

21. How do the actions of the extrinsic and intrinsic muscles of the hand differ?

11.19 Muscles of the Neck and Back That Move the Vertebral Column

22. What is the largest muscle group of the back?

11.20 Muscles of the Gluteal Region That Move the Femur

23. What is the origin of most muscles that move the femur?

11.21 Muscles of the Thigh That Move the Femur, Tibia, and Fibula

24. Which muscles are part of the medial, anterior, and posterior compartments of the thigh?

11.22 Muscles of the Leg That Move the Foot and Toes

25. What are the superior extensor retinaculum and inferior extensor retinaculum?

11.23 Intrinsic Muscles of the Foot That Move the Toes

26. How do the intrinsic muscles of the hand and foot differ in function?

Critical Thinking Questions

1. During a facelift, the cosmetic surgeon accidentally severs the facial nerve on the right side of the face. What are some of the effects this would have on the patient, and what muscles are involved?
2. While taking the bus to the supermarket, 11-year-old Desmond informs his mother that he has to "go to the bathroom" (urinate). His mother tells him he must "hold it" until they arrive at the store. What muscles must remain contracted in order for him to prevent urination?
3. Minor-league pitcher José has been throwing a hundred pitches a day in order to perfect his curve ball. Lately he has experienced pain in his pitching arm. The doctor diagnosed a torn rotator cuff. José was confused because he thought cuffs were only found on shirt sleeves, not inside his shoulder. Explain to José what the doctor means and how this injury could affect his arm movement.

Nervous Tissue

CHAPTER **12**

Nervous Tissue and Homeostasis

The excitable characteristic of nervous tissue allows for the generation of nerve impulses (action potentials) that provide communication with and regulation of most body organs.

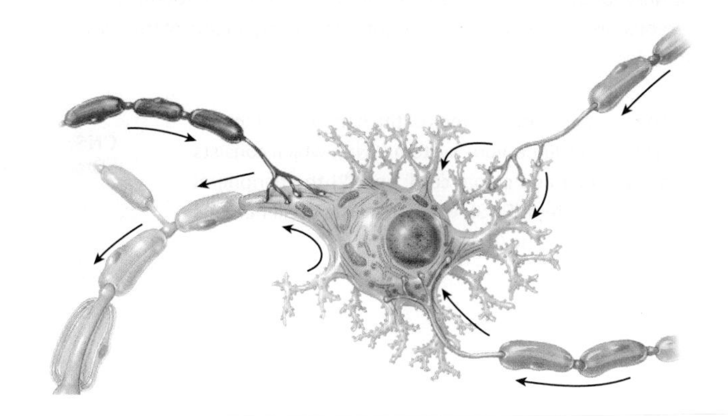

Both the nervous and endocrine systems have the same objective: to keep controlled conditions within limits that maintain life. The nervous system regulates body activities by responding rapidly using nerve impulses; the endocrine system responds by releasing hormones. Chapter 18 compares the roles of both systems in maintaining homeostasis.

The nervous system is also responsible for our perceptions, behaviors, and memories, and it initiates all voluntary movements. Because this system is quite complex, we discuss its structure and function in several chapters. This chapter focuses on the organization of the nervous system and the properties of neurons (nerve cells) and neuroglia (cells that support the activities of neurons). We then examine the structure and functions of the spinal cord and spinal nerves (Chapter 13), and of the brain and cranial nerves (Chapter 14). The autonomic nervous system, which operates without voluntary control, will be covered in Chapter 15. Chapter 16 will discuss the somatic senses—touch, pressure, warmth, cold, pain, and others—and their sensory and motor pathways to show how nerve impulses pass into the spinal cord and brain or from the spinal cord and brain to muscles and glands. Exploration of the nervous system concludes with a discussion of the special senses: smell, taste, vision, hearing, and equilibrium (Chapter 17).

Q Did you ever wonder how the human nervous system coordinates and integrates all body systems so rapidly and efficiently?

12.1 Overview of the Nervous System

OBJECTIVES

- **Describe** the organization of the nervous system.
- **Describe** the three basic functions of the nervous system.

Organization of the Nervous System

With a mass of only 2 kg (4.5 lb), about 3% of total body weight, the **nervous system** is one of the smallest and yet the most complex of the 11 body systems. This intricate network of billions of neurons and even more neuroglia is organized into two main subdivisions: the central nervous system and the peripheral nervous system. **Neurology** deals with normal functioning and disorders of the nervous system. A **neurologist** (noo-ROL-ō-jist) is a physician who diagnoses and treats disorders of the nervous system.

Central Nervous System The **central nervous system (CNS)** consists of the brain and spinal cord (**Figure 12.1a**). The brain is the part of the CNS that is located in the skull and contains about 85 billion neurons. The spinal cord is connected to the brain through the foramen magnum of the occipital bone and is encircled by the bones of the vertebral column. The spinal cord contains about 100 million neurons. The CNS processes many different kinds of incoming sensory information. It is also the source of thoughts, emotions, and memories. Most signals that stimulate muscles to contract and glands to secrete originate in the CNS.

Peripheral Nervous System The **peripheral nervous system (PNS)** (pe-RIF-e-ral) consists of all nervous tissue outside the CNS (**Figure 12.1a**). Components of the PNS include nerves and sensory receptors. A **nerve** is a bundle of hundreds to thousands of axons plus associated connective tissue and blood vessels that lies outside the brain and spinal cord. Twelve pairs of **cranial nerves** emerge from the brain and thirty-one pairs of **spinal nerves** emerge from the spinal cord. Each nerve follows a defined path and serves a specific region of the body. The term **sensory receptor** refers to a structure of the nervous system that monitors changes in the external or internal environment. Examples of sensory receptors include touch receptors in the skin, photoreceptors in the eye, and olfactory (smell) receptors in the nose.

The PNS is divided into sensory and motor divisions (**Figure 12.1b**). The **sensory** or *afferent* **division** of the PNS conveys input into

FIGURE 12.1 Organization of the nervous system. (a) Subdivisions of the nervous system. (b) Nervous system organizational chart; blue boxes represent sensory components of the peripheral nervous system, red boxes represent motor components of the PNS, and green boxes represent effectors (muscles and glands).

The two main subdivisions of the nervous system are (1) the central nervous system (CNS), which consists of the brain and spinal cord, and (2) the peripheral nervous system (PNS), which consists of all nervous tissue outside the CNS.

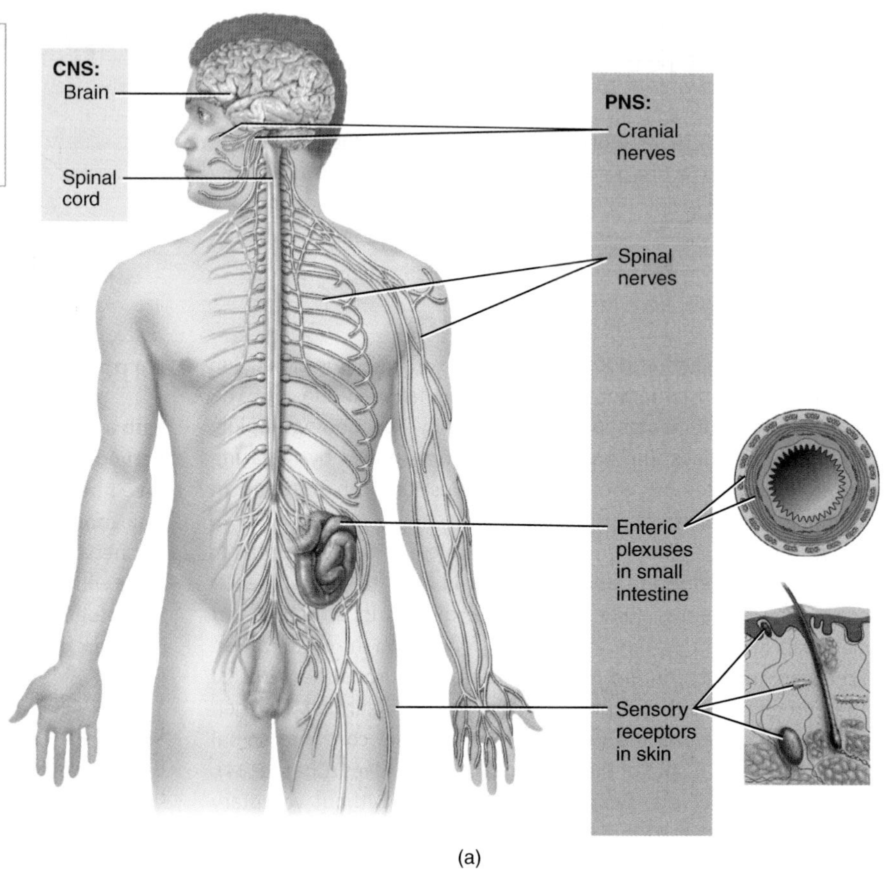

(a)

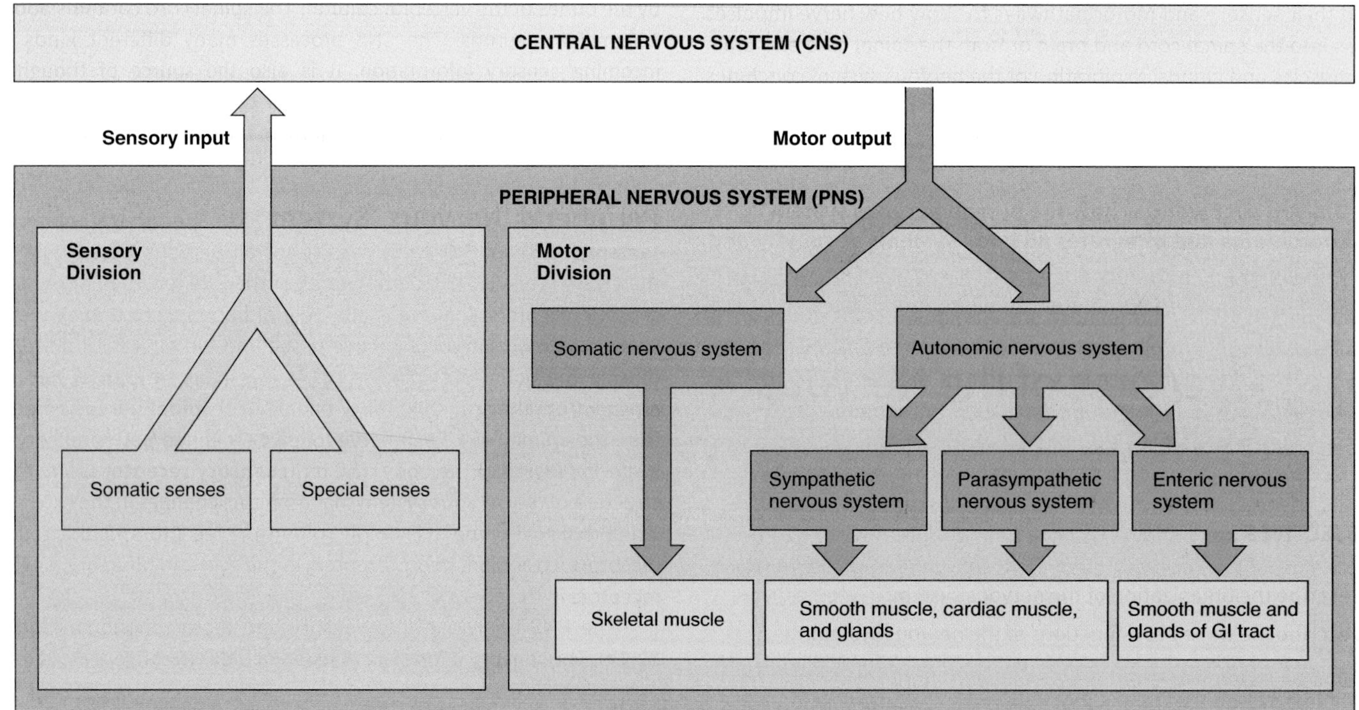

(b)

Q What are some of the functions of the CNS?

the CNS from sensory receptors in the body. This division provides the CNS with sensory information about the *somatic senses* (tactile, thermal, pain, and proprioceptive sensations) and *special senses* (smell, taste, vision, hearing, and equilibrium).

The **motor** or *efferent* **division** of the PNS conveys output from the CNS to effectors (muscles and glands). This division is further subdivided into a somatic nervous system and an autonomic nervous system (Figure 12.1b). The **somatic nervous system (SNS)** (sō-MAT-ik; *soma* = body) conveys output from the CNS to *skeletal muscles* only. Because its motor responses can be consciously controlled, the action of this part of the PNS is *voluntary*. The **autonomic nervous system (ANS)** (aw′-tō-NOM-ik; *auto-* = self; *-nomic* = law) conveys output from the CNS to *smooth muscle*, *cardiac muscle*, and *glands*. Because its motor responses are not normally under conscious control, the action of the ANS is *involuntary*. The ANS is comprised of two main branches, the **sympathetic nervous system** and the **parasympathetic nervous system**. With a few exceptions, effectors receive innervation from both of these branches, and usually the two branches have opposing actions. For example, neurons of the sympathetic nervous system increase heart rate, and neurons of the parasympathetic nervous system slow it down. In general, the parasympathetic nervous system takes care of "rest-and-digest" activities, and the sympathetic nervous system helps support exercise or emergency actions—the so-called "fight-or-flight" responses. A third branch of the autonomic nervous system is the **enteric nervous system (ENS)** (en-TER-ik; *enteron* = intestines), an extensive network of over 100 million neurons confined to the wall of the gastrointestinal (GI) tract. The ENS helps regulate the activity of the smooth muscle and glands of the GI tract. Although the ENS can function independently, it communicates with and is regulated by the other branches of the ANS.

Functions of the Nervous System

The nervous system carries out a complex array of tasks. It allows us to sense various smells, produce speech, and remember past events; in addition, it provides signals that control body movements and regulates the operation of internal organs. These diverse activities can be grouped into three basic functions: sensory (input), integrative (process), and motor (output).

- **Sensory function.** Sensory receptors *detect* internal stimuli, such as an increase in blood pressure, or external stimuli (for example, a raindrop landing on your arm). This sensory information is then carried into the brain and spinal cord through cranial and spinal nerves.
- **Integrative function.** The nervous system *processes* sensory information by analyzing it and making decisions for appropriate responses—an activity known as **integration**.
- **Motor function.** Once sensory information is integrated, the nervous system *may elicit an appropriate motor response* by activating **effectors** (muscles and glands) through cranial and spinal nerves. Stimulation of the effectors causes muscles to contract and glands to secrete.

The three basic functions of the nervous system occur, for example, when you answer your cell phone after hearing it ring. The sound of the ringing cell phone stimulates sensory receptors in your ears (sensory function). This auditory information is subsequently relayed into your brain where it is processed and the decision to answer the phone is made (integrative function). The brain then stimulates the contraction of specific muscles that will allow you to grab the phone and press the appropriate button to answer it (motor function).

12.2 Histology of Nervous Tissue

OBJECTIVES

- **Contrast** the histological characteristics and the functions of neurons and neuroglia.
- **Distinguish** between gray matter and white matter.

Nervous tissue comprises two types of cells—*neurons* and *neuroglia*. These cells combine in a variety of ways in different regions of the nervous system. In addition to forming the complex processing networks within the brain and spinal cord, neurons also connect all regions of the body to the brain and spinal cord. As highly specialized cells capable of reaching great lengths and making extremely intricate connections with other cells, neurons provide most of the unique functions of the nervous system, such as sensing, thinking, remembering, controlling muscle activity, and regulating glandular secretions. As a result of their specialization, most neurons have lost the ability to undergo mitotic divisions. Neuroglia are smaller cells, but they greatly outnumber neurons—perhaps by as much as 25 times. Neuroglia support, nourish, and protect neurons, and maintain the interstitial fluid that bathes them. Unlike neurons, neuroglia continue to divide throughout an individual's lifetime. Both neurons and neuroglia differ structurally depending on whether they are located in the central nervous system or the peripheral nervous system. These differences in structure correlate with the differences in function of the central nervous system and the peripheral nervous system.

Neurons

Like muscle cells, **neurons** (*nerve cells*) (NOO-rons) possess **electrical excitability** (ek-sīt′-a-BIL-i-tē), the ability to respond to a stimulus and convert it into an action potential. A **stimulus** is any change in the environment that is strong enough to initiate an action potential. An **action potential** (*nerve impulse*) is an electrical signal that propagates (travels) along the surface of the membrane of a neuron. It begins and travels due to the movement of ions (such as sodium and potassium) between interstitial fluid and the inside of a neuron through specific ion channels in its plasma membrane.

Once begun, a nerve impulse travels rapidly and at a constant strength.

Some neurons are tiny and propagate impulses over a short distance (less than 1 mm) within the CNS. Others are the longest cells in the body. The neurons that enable you to wiggle your toes, for example, extend from the lumbar region of your spinal cord (just above waist level) to the muscles in your foot. Some neurons are even longer. Those that allow you to feel a feather tickling your toes stretch all the way from your foot to the lower portion of your brain. Nerve impulses travel these great distances at speeds ranging from 0.5 to 130 meters per second (1 to 290 mi/hr).

Parts of a Neuron Most neurons have three parts: (1) a cell body, (2) dendrites, and (3) an axon (Figure 12.2). The **cell body**, also known as the *perikaryon* (per′-i-KAR-ē-on) or *soma*, contains a nucleus surrounded by cytoplasm that includes typical cellular organelles such as lysosomes, mitochondria, and a Golgi complex. Neuronal cell bodies also contain free ribosomes and prominent clusters of rough endoplasmic reticulum, termed **Nissl bodies** (NIS-el). The ribosomes are the sites of protein synthesis. Newly synthesized proteins produced by Nissl bodies are used to replace cellular components, as material for growth of neurons, and to regenerate damaged axons in the PNS. The cytoskeleton includes both **neurofibrils** (noo-rō-FĪ-brils), composed of bundles of intermediate filaments that provide the cell shape and support, and **microtubules** (mī-krō-TOO-būls′), which assist in moving materials between the cell body and axon. Aging neurons also contain **lipofuscin** (līp′-o-FYŪS-īn), a pigment that occurs as clumps of yellowish brown granules in the cytoplasm. Lipofuscin is a product of neuronal lysosomes that accumulates as the neuron ages, but does not seem to harm the neuron. A collection of neuron cell bodies outside the CNS is called a **ganglion** (GANG-lē-on = sculling or knot; *ganglia* is plural).

A **nerve fiber** is a general term for any neuronal process (extension) that emerges from the cell body of a neuron. Most neurons have two kinds of processes: multiple dendrites and a single axon. **Dendrites** (DEN-drīts = little trees) are the receiving or input portions of a neuron. The plasma membranes of dendrites (and cell bodies) contain numerous receptor sites for binding chemical messengers from other cells. Dendrites usually are short, tapering, and highly branched. In many neurons the dendrites form a tree-shaped array of processes extending from the cell body. Their cytoplasm contains Nissl bodies, mitochondria, and other organelles.

The single **axon** (= axis) of a neuron propagates nerve impulses toward another neuron, a muscle fiber, or a gland cell. An axon is a long, thin, cylindrical projection that often joins to the cell body at a cone-shaped elevation called the **axon hillock** (HIL-lok = small hill). The part of the axon closest to the axon hillock is the **initial segment**. In most neurons, nerve impulses arise at the junction of the axon hillock and the initial segment, an area called the **trigger zone**, from which they travel along the axon to their destination. An axon contains mitochondria, microtubules, and neurofibrils. Because rough endoplasmic reticulum is not present, protein synthesis does not occur in the axon. The cytoplasm of an axon, called **axoplasm**, is surrounded by a plasma membrane known as the **axolemma** (*lemma* = sheath or husk). Along the length of an axon, side branches called **axon collaterals** may branch off, typically at a right angle to the axon. The axon and its collaterals end by dividing into many fine processes called **axon terminals** or *axon telodendria* (tēl′-ō-DEN-drē-a).

The site of communication between two neurons or between a neuron and an effector cell is called a **synapse** (SIN-aps). The tips of some axon terminals swell into bulb-shaped structures called **synaptic end bulbs**; others exhibit a string of swollen bumps called **varicosities** (var′-i-KOS-i-tēz). Both synaptic end bulbs and varicosities contain many tiny membrane-enclosed sacs called **synaptic vesicles** that store a chemical called a **neurotransmitter** (noo′-rō-trans′-MIT-ter). A neurotransmitter is a molecule released from a synaptic vesicle that excites or inhibits another neuron, muscle fiber, or gland cell. Many neurons contain two or even three types of neurotransmitters, each with different effects on the postsynaptic cell.

Because some substances synthesized or recycled in the neuron cell body are needed in the axon or at the axon terminals, two types of transport systems carry materials from the cell body to the axon terminals and back. The slower system, which moves materials about 1–5 mm per day, is called **slow axonal transport**. It conveys axoplasm in one direction only—from the cell body toward the axon terminals. Slow axonal transport supplies new axoplasm to developing or regenerating axons and replenishes axoplasm in growing and mature axons.

Fast axonal transport, which is capable of moving materials a distance of 200–400 mm per day, uses proteins that function as "motors" to move materials along the surfaces of microtubules of the neuron's cytoskeleton. Fast axonal transport moves materials in both directions—away from and toward the cell body. Fast axonal transport that occurs in an **anterograde** (forward) direction moves organelles and synaptic vesicles from the cell body to the axon terminals. Fast axonal transport that occurs in a **retrograde** (backward) direction moves membrane vesicles and other cellular materials from the axon terminals to the cell body to be degraded or recycled. Substances that enter the neuron at the axon terminals are also moved to the cell body by fast retrograde transport. These substances include trophic chemicals such as nerve growth factor and harmful agents such as tetanus toxin and the viruses that cause rabies, herpes simplex, and polio.

Structural Diversity in Neurons Neurons display great diversity in size and shape. For example, their cell bodies range in diameter from 5 micrometers (μm) (slightly smaller than a red blood cell) up to 135 μm (barely large enough to see with the unaided eye). The pattern of dendritic branching is varied and distinctive for neurons in different parts of the nervous system. A few small neurons lack an axon, and many others have very short axons. As we have already discussed, the longest axons are almost as long as a person is tall, extending from the toes to the lowest part of the brain.

Classification of Neurons Both structural and functional features are used to classify the various neurons in the body.

FIGURE 12.2 Structure of a multipolar neuron. A multipolar neuron has a cell body, several short dendrites, and a single long axon. Arrows indicate the direction of information flow: dendrites → cell body → axon → axon terminals.

The basic parts of a neuron are dendrites, a cell body, and an axon.

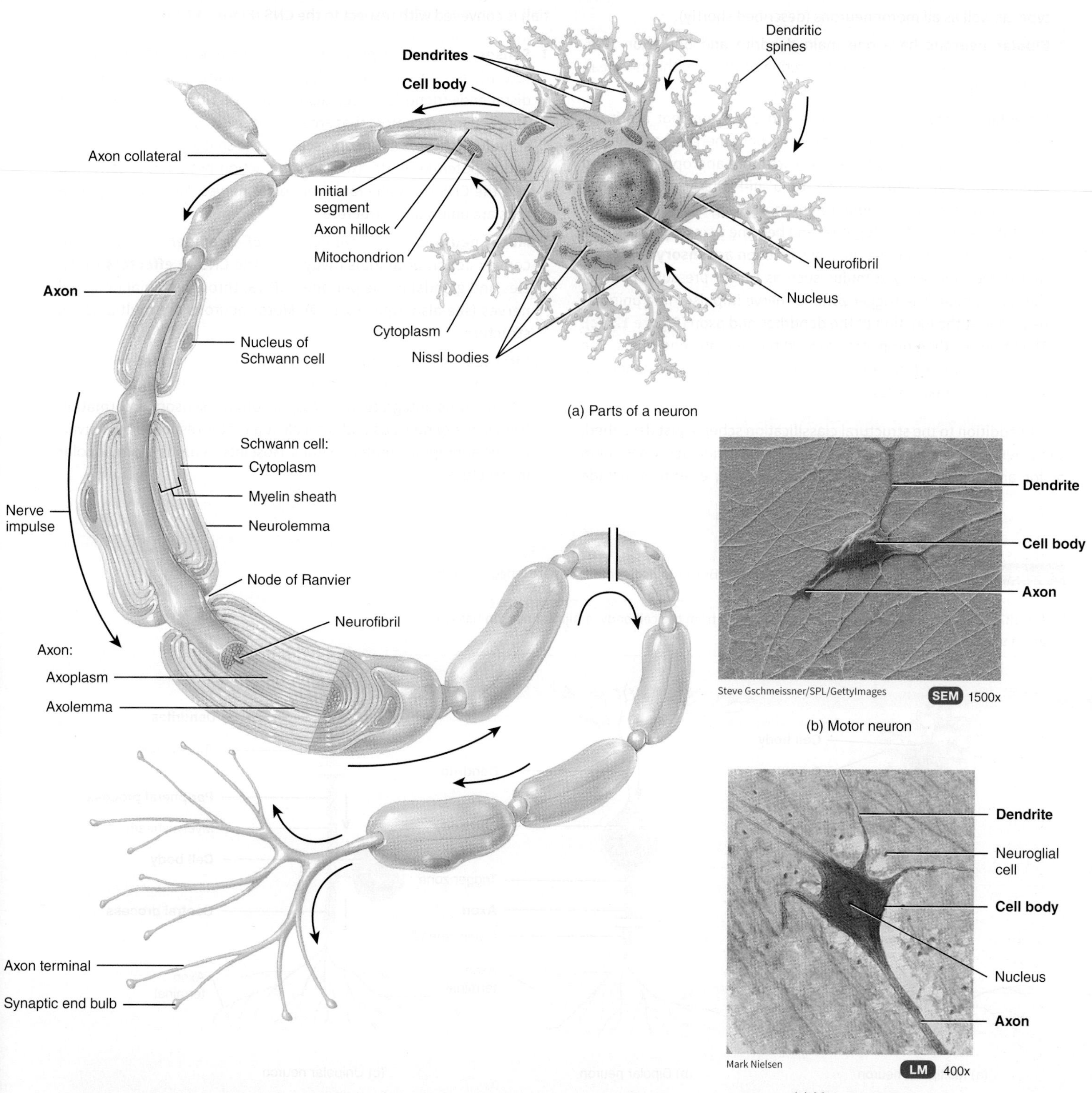

Q What roles do the dendrites, cell body, and axon play in communication of signals?

STRUCTURAL CLASSIFICATION Structurally, neurons are classified according to the number of processes extending from the cell body (Figure 12.3):

1. **Multipolar neurons** usually have several dendrites and one axon (Figure 12.3a). Most neurons in the brain and spinal cord are of this type, as well as all motor neurons (described shortly).
2. **Bipolar neurons** have one main dendrite and one axon (Figure 12.3b). They are found in the retina of the eye, the inner ear, and the olfactory area (*olfact* = to smell) of the brain.
3. **Unipolar neurons** have dendrites and one axon that are fused together to form a continuous process that emerges from the cell body (Figure 12.3c). These neurons are more appropriately called **pseudounipolar neurons** (soo′-dō-ū′-ni-PŌ-lar) because they begin in the embryo as bipolar neurons. During development, the dendrites and axon fuse together and become a single process. The dendrites of most unipolar neurons function as **sensory receptors** that detect a sensory stimulus such as touch, pressure, pain, or thermal stimuli. The trigger zone for nerve impulses in a unipolar neuron is at the junction of the dendrites and axon (Figure 12.3c). The impulses then propagate toward the synaptic end bulbs. The cell bodies of most unipolar neurons are located in the ganglia of spinal and cranial nerves.

In addition to the structural classification scheme just described, some neurons are named for the histologist who first described them or for an aspect of their shape or appearance; examples include **Purkinje cells** (pur-KIN-jē) in the cerebellum and **pyramidal cells** (pi-RAM-i-dal), found in the cerebral cortex of the brain, which have pyramid-shaped cell bodies (Figure 12.4).

FUNCTIONAL CLASSIFICATION Functionally, neurons are classified according to the direction in which the nerve impulse (action potential) is conveyed with respect to the CNS (Figure 12.5).

1. **Sensory neurons** or *afferent neurons* (AF-er-ent NOO-ronz; *af-* = toward; *-ferrent* = carried) either contain sensory receptors at their distal ends (dendrites) (see also Figure 12.10) or are located just after sensory receptors that are separate cells. Once an appropriate stimulus activates a sensory receptor, the sensory neuron forms an action potential in its axon and the action potential is conveyed *into* the CNS through cranial or spinal nerves. Most sensory neurons are unipolar in structure.
2. **Motor neurons** or *efferent neurons* (EF-e-rent; *ef-* = away from) convey action potentials *away* from the CNS to **effectors** (muscles and glands) in the periphery (PNS) through cranial or spinal nerves (see also Figure 12.10). Motor neurons are multipolar in structure.
3. **Interneurons** or *association neurons* are mainly located within the CNS between sensory and motor neurons (see also Figure 12.10). Interneurons integrate (process) incoming sensory information from sensory neurons and then elicit a motor response by activating the appropriate motor neurons. Most interneurons are multipolar in structure.

FIGURE 12.3 **Structural classification of neurons.** Breaks indicate that axons are longer than shown.

A multipolar neuron has many processes extending from the cell body, a bipolar neuron has two, and a unipolar neuron has one.

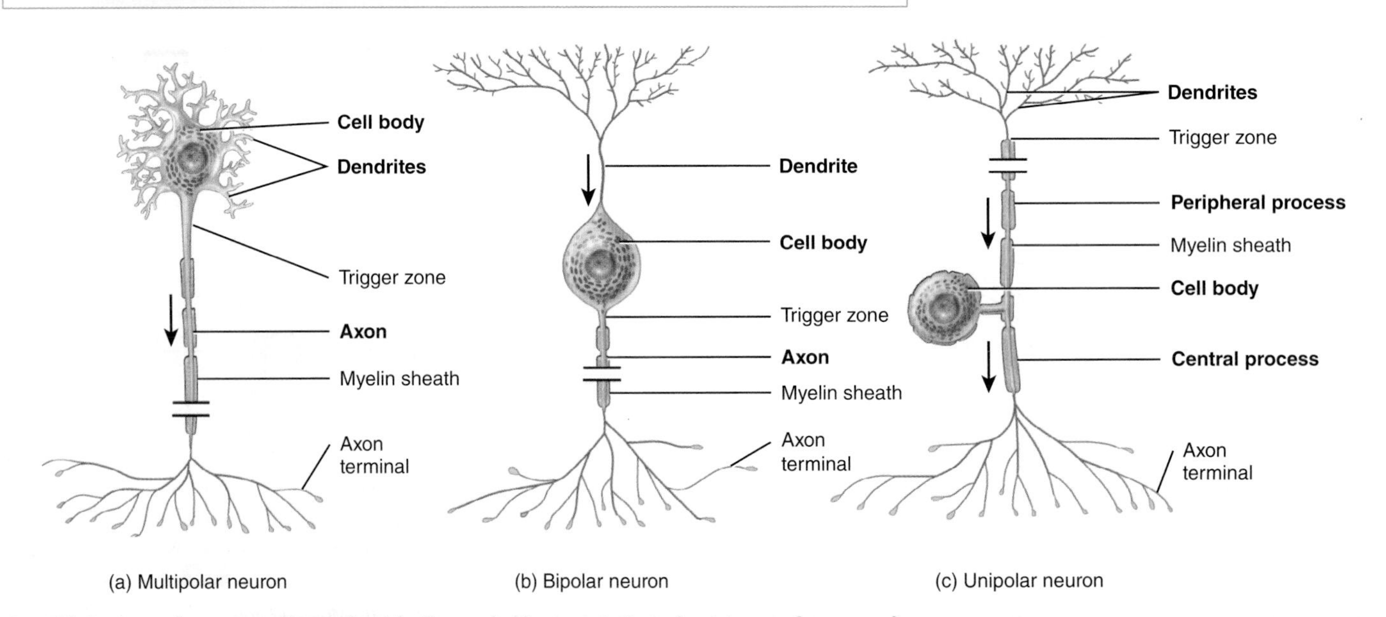

Q Which type of neuron shown in this figure is the most abundant type of neuron in the CNS?

FIGURE 12.4 Two examples of CNS neurons. Arrows indicate the direction of information flow.

The dendritic branching pattern often is distinctive for a particular type of neuron.

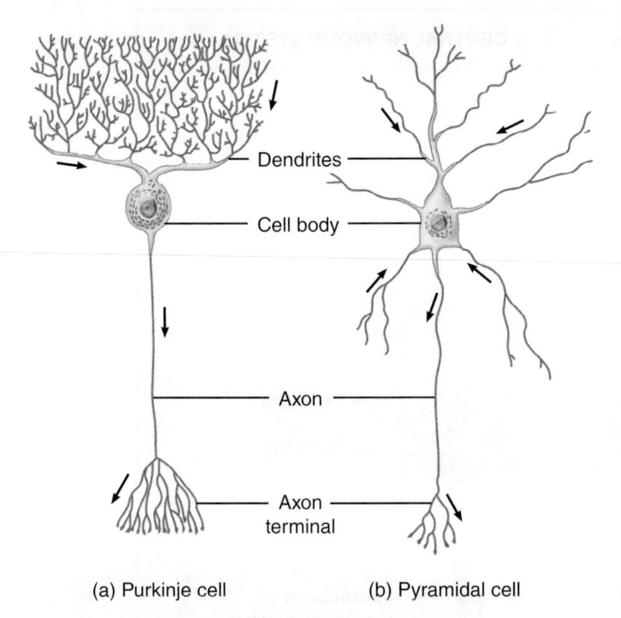

Q How did the pyramidal cells get their name?

Neuroglia

Neuroglia (noo-RŌG-lē-a; *-glia* = glue) or *glia* (GLĒ-a) make up about half the volume of the CNS. Their name derives from the idea of early histologists that they were the "glue" that held nervous tissue together. We now know that neuroglia are not merely passive bystanders but rather actively participate in the activities of nervous tissue. Generally, neuroglia are smaller than neurons, and they are 5 to 25 times more numerous. In contrast to neurons, glia do not generate or propagate action potentials, and they can multiply and divide in the mature nervous system. In cases of injury or disease, neuroglia multiply to fill in the spaces formerly occupied by neurons. Brain tumors derived from glia, called **gliomas** (glē-Ō-mas), tend to be highly malignant and to grow rapidly. Of the six types of neuroglia, four—astrocytes, oligodendrocytes, microglia, and ependymal cells—are found only in the CNS. The remaining two types—Schwann cells and satellite cells—are present in the PNS.

Neuroglia of the CNS

Neuroglia of the CNS can be classified on the basis of size, cytoplasmic processes, and intracellular organization into four types: astrocytes, oligodendrocytes, microglial cells, and ependymal cells (**Figure 12.6**).

ASTROCYTES These star-shaped cells have many processes and are the largest and most numerous of the neuroglia. There are two types of **astrocytes** (AS-trō-sīts; *astro-* = star; *-cyte* = cell). *Protoplasmic astrocytes* have many short branching processes and are found in gray matter (described shortly). *Fibrous astrocytes* have many long unbranched processes and are located mainly in white matter (also described shortly). The processes of astrocytes make contact with blood capillaries, neurons, and the pia mater (a thin membrane around the brain and spinal cord).

The functions of astrocytes include the following:

1. Astrocytes contain microfilaments that give them considerable strength, which enables them to support neurons.
2. Processes of astrocytes wrapped around blood capillaries isolate neurons of the CNS from various potentially harmful substances in blood by secreting chemicals that maintain the unique selective permeability characteristics of the endothelial cells of the capillaries. In effect, the endothelial cells create a *blood–brain barrier*, which restricts the movement of substances between the blood and interstitial fluid of the CNS. Details of the blood–brain barrier are discussed in Chapter 14.
3. In the embryo, astrocytes secrete chemicals that appear to regulate the growth, migration, and interconnection among neurons in the brain.
4. Astrocytes help to maintain the appropriate chemical environment for the generation of nerve impulses. For example, they regulate the concentration of important ions such as K^+; take up excess neurotransmitters; and serve as a conduit for the passage of nutrients and other substances between blood capillaries and neurons.
5. Astrocytes may also play a role in learning and memory by influencing the formation of neural synapses (see Section 16.5).

OLIGODENDROCYTES These resemble astrocytes but are smaller and contain fewer processes. Processes of **oligodendrocytes** (OL-i-gō-den′-drō-sīts; *oligo-* = few; *-dendro-* = tree) are responsible for forming and maintaining the myelin sheath around CNS axons. As you will see shortly, the **myelin sheath** is a multilayered lipid and protein covering around some axons that insulates them and increases the speed of nerve impulse conduction. Such axons are said to be *myelinated* (MĪ-e-li-nā-ted).

MICROGLIAL CELLS OR MICROGLIA These neuroglia are small cells with slender processes that give off numerous spinelike projections. **Microglial cells** or *microglia* (mī-KROG-lē-a; *micro-* = small) function as phagocytes. Like tissue macrophages, they remove cellular debris formed during normal development of the nervous system and phagocytize microbes and damaged nervous tissue.

EPENDYMAL CELLS **Ependymal cells** (ep-EN-de-mal; *epen-* = above; *-dym-* = garment) are cuboidal to columnar cells arranged in a single layer that possess microvilli and cilia. These cells line the ventricles of the brain and central canal of the spinal cord (spaces filled with cerebrospinal fluid, which protects and nourishes the brain and spinal cord). Functionally, ependymal cells produce, possibly monitor, and assist in the circulation of cerebrospinal fluid. They also form the blood–cerebrospinal fluid barrier, which is discussed in Chapter 14.

FIGURE 12.5 Functional classification of neurons.

Neurons are divided into three functional classes: sensory neurons, interneurons, and motor neurons.

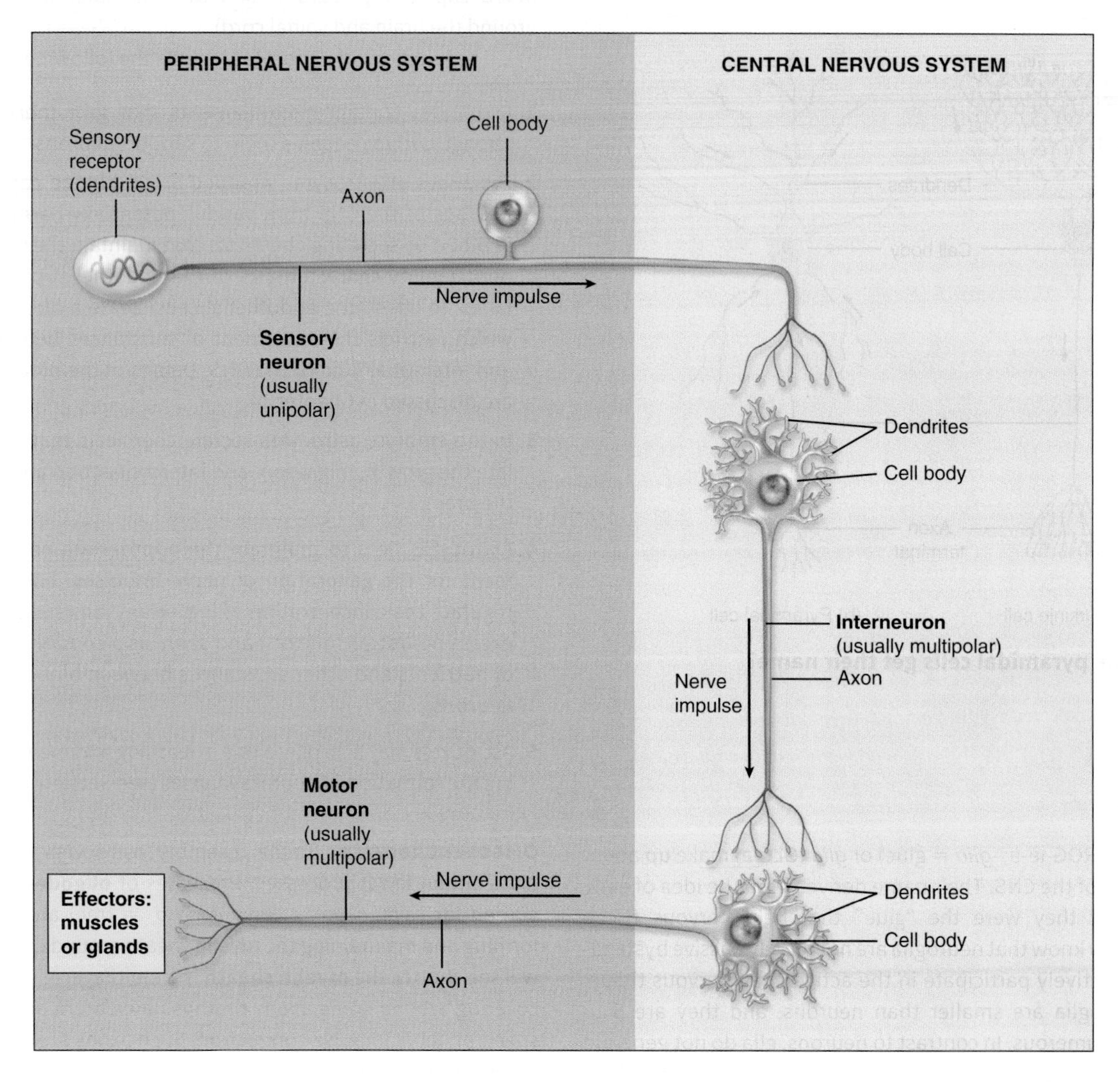

Q Which functional class of neurons is responsible for integration?

Neuroglia of the PNS

Neuroglia of the PNS completely surround axons and cell bodies. The two types of glial cells in the PNS are Schwann cells and satellite cells (Figure 12.7).

SCHWANN CELLS These cells encircle PNS axons. Like oligodendrocytes, they form the myelin sheath around axons. A single oligodendrocyte myelinates several axons, but each **Schwann cell** (SCHVON or SCHWON) myelinates a single axon (Figure 12.7a; see also Figure 12.8a, c). A single Schwann cell can also enclose as many as 20 or more unmyelinated axons (axons that lack a myelin sheath) (Figure 12.7b). Schwann cells participate in axon regeneration, which is more easily accomplished in the PNS than in the CNS.

SATELLITE CELLS These flat cells surround the cell bodies of neurons of PNS ganglia (Figure 12.7c). Besides providing structural support, **satellite cells** (SAT-i-līt) regulate the exchanges of materials between neuronal cell bodies and interstitial fluid.

Myelination

As you have already learned, axons surrounded by a multilayered lipid and protein covering, called the **myelin sheath**, are said to be **myelinated** (Figure 12.8a). The sheath electrically insulates the axon of a neuron and increases the speed of nerve impulse conduction. Axons without such a covering are said to be **unmyelinated** (Figure 12.8b).

Two types of neuroglia produce myelin sheaths: Schwann cells (in the PNS) and oligodendrocytes (in the CNS). Schwann cells begin to form myelin sheaths around axons during fetal development. Each

FIGURE 12.6 Neuroglia of the central nervous system.

Neuroglia of the CNS are distinguished on the basis of size, cytoplasmic processes, and intracellular organization.

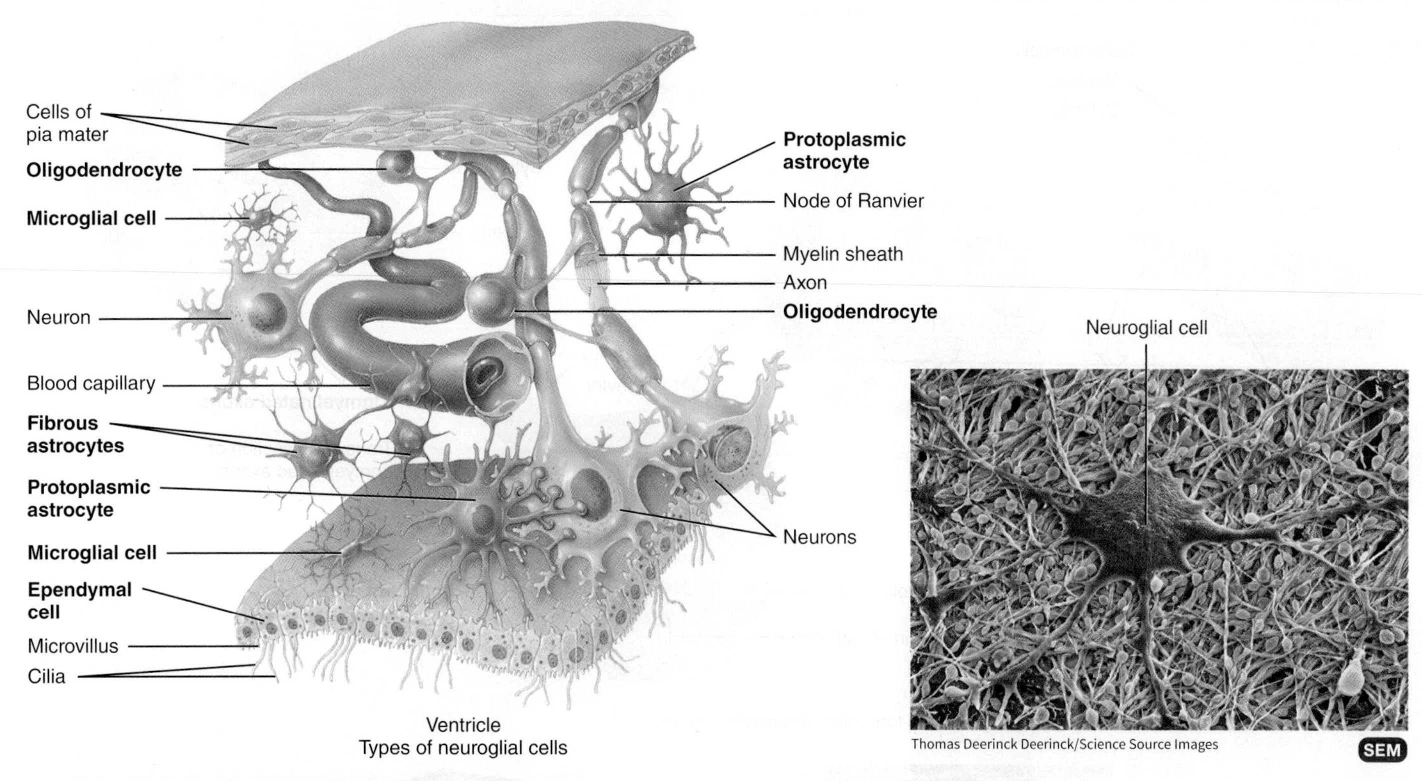

Q Which CNS neuroglia function as phagocytes?

FIGURE 12.7 Neuroglia of the peripheral nervous system.

Neuroglia of the PNS completely surround axons and cell bodies of neurons.

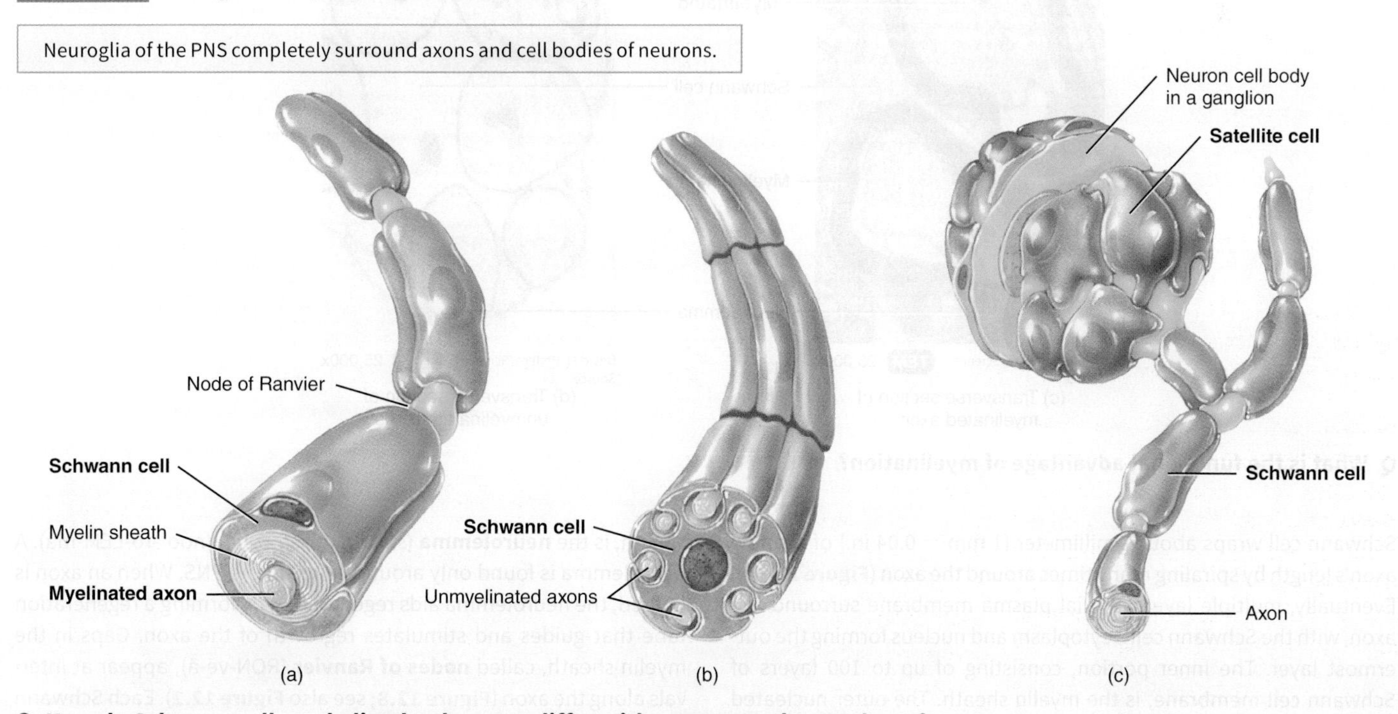

Q How do Schwann cells and oligodendrocytes differ with respect to the number of axons they myelinate?

FIGURE 12.8 Myelinated and unmyelinated axons. Notice that one layer of Schwann cell plasma membrane surrounds unmyelinated axons.

Axons surrounded by a myelin sheath produced either by Schwann cells in the PNS or by oligodendrocytes in the CNS are said to be myelinated.

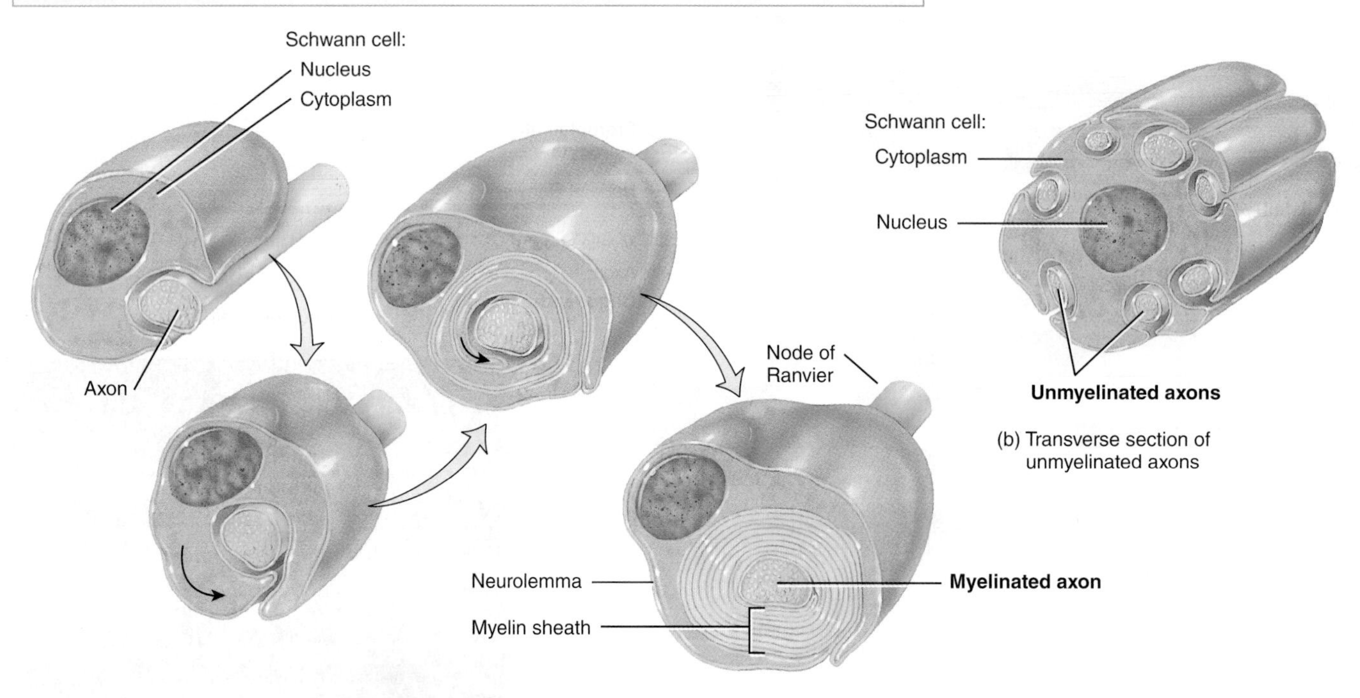

(a) Transverse sections of stages in the formation of a myelin sheath

(b) Transverse section of unmyelinated axons

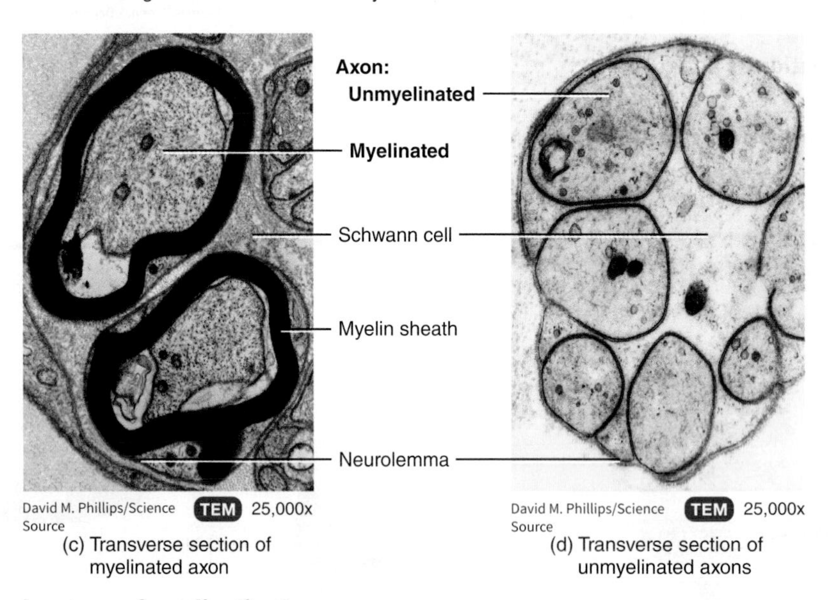

David M. Phillips/Science Source

(c) Transverse section of myelinated axon

David M. Phillips/Science Source

(d) Transverse section of unmyelinated axons

Q What is the functional advantage of myelination?

Schwann cell wraps about 1 millimeter (1 mm = 0.04 in.) of a single axon's length by spiraling many times around the axon (**Figure 12.8a**). Eventually, multiple layers of glial plasma membrane surround the axon, with the Schwann cell's cytoplasm and nucleus forming the outermost layer. The inner portion, consisting of up to 100 layers of Schwann cell membrane, is the myelin sheath. The outer nucleated cytoplasmic layer of the Schwann cell, which encloses the myelin sheath, is the **neurolemma** (*sheath of Schwann*) (noo′-rō-LEM-ma). A neurolemma is found only around axons in the PNS. When an axon is injured, the neurolemma aids regeneration by forming a regeneration tube that guides and stimulates regrowth of the axon. Gaps in the myelin sheath, called **nodes of Ranvier** (RON-vē-ā), appear at intervals along the axon (**Figure 12.8; see also Figure 12.2**). Each Schwann cell wraps one axon segment between two nodes.

In the CNS, an oligodendrocyte myelinates parts of several axons. Each oligodendrocyte puts forth about 15 broad, flat processes that spiral around CNS axons, forming a myelin sheath. A neurolemma is not present, however, because the oligodendrocyte cell body and nucleus do not envelop the axon. Nodes of Ranvier are present, but they are fewer in number. Axons in the CNS display little regrowth after injury. This is thought to be due, in part, to the absence of a neurolemma, and in part to an inhibitory influence exerted by the oligodendrocytes on axon regrowth.

The amount of myelin increases from birth to maturity, and its presence greatly increases the speed of nerve impulse conduction. An infant's responses to stimuli are neither as rapid nor as coordinated as those of an older child or an adult, in part because myelination is still in progress during infancy.

Collections of Nervous Tissue

The components of nervous tissue are grouped together in a variety of ways. Neuronal cell bodies are often grouped together in clusters. The axons of neurons are usually grouped together in bundles. In addition, widespread regions of nervous tissue are grouped together as either gray matter or white matter.

Clusters of Neuronal Cell Bodies Recall that a **ganglion** (plural is *ganglia*) refers to a cluster of neuronal cell bodies located in the PNS. As mentioned earlier, ganglia are closely associated with cranial and spinal nerves. By contrast, a **nucleus** is a cluster of neuronal cell bodies located in the CNS.

Bundles of Axons Recall that a **nerve** is a bundle of axons that is located in the PNS. Cranial nerves connect the brain to the periphery, whereas spinal nerves connect the spinal cord to the periphery. A **tract** is a bundle of axons that is located in the CNS. Tracts interconnect neurons in the spinal cord and brain.

Gray and White Matter In a freshly dissected section of the brain or spinal cord, some regions look white and glistening, and others appear gray (Figure 12.9). **White matter** is composed primarily of myelinated axons. The whitish color of myelin gives white matter its name. The **gray matter** of the nervous system contains neuronal cell

FIGURE 12.9 Distribution of gray matter and white matter in the spinal cord and brain.

White matter consists primarily of myelinated axons of many neurons. Gray matter consists of neuron cell bodies, dendrites, unmyelinated axons, axon terminals, and neuroglia.

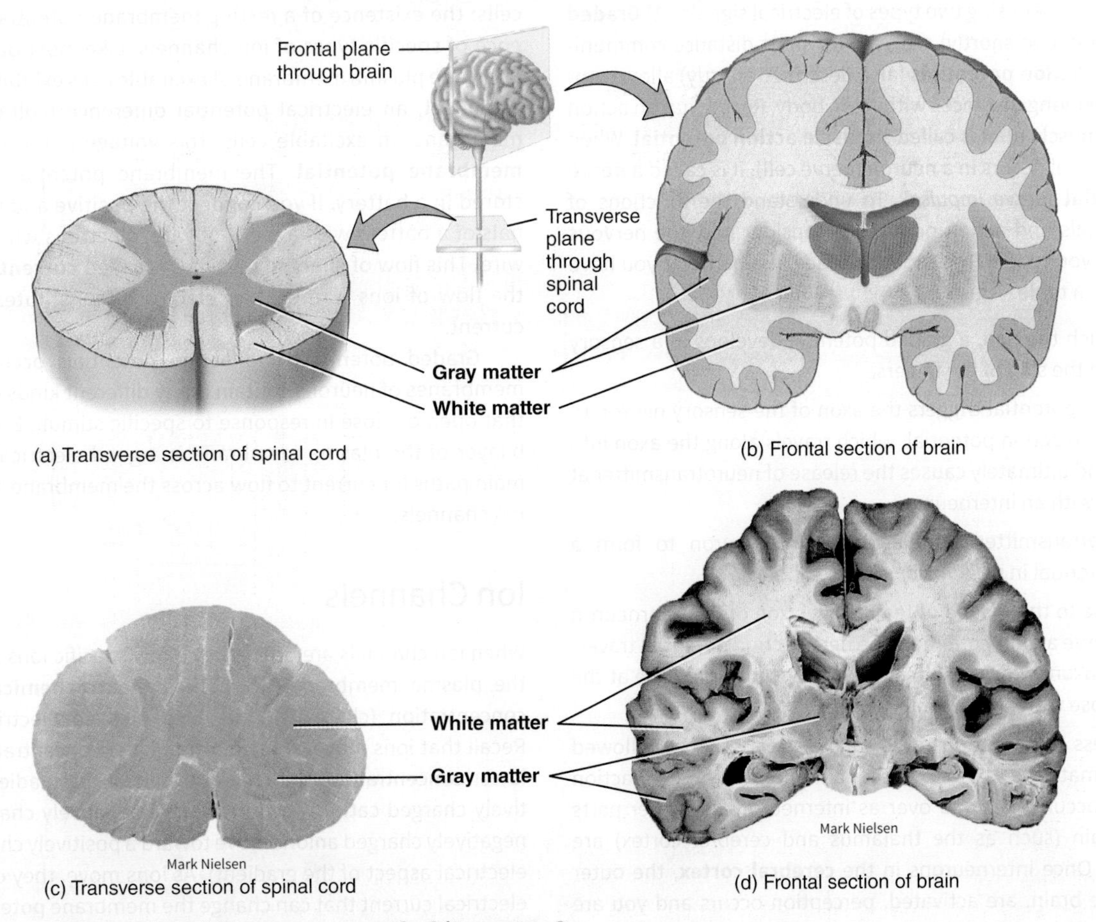

Q What is responsible for the white appearance of white matter?

bodies, dendrites, unmyelinated axons, axon terminals, and neuroglia. It appears grayish, rather than white, because the Nissl bodies impart a gray color and there is little or no myelin in these areas. Blood vessels are present in both white and gray matter. In the spinal cord, the white matter surrounds an inner core of gray matter that, depending on how imaginative you are, is shaped like a butterfly or the letter H in transverse section; in the brain, a thin shell of gray matter covers the surface of the largest portions of the brain, the cerebrum and cerebellum (**Figure 12.9**). The arrangement of gray matter and white matter in the spinal cord and brain is discussed more extensively in Chapters 13 and 14, respectively.

12.3 Electrical Signals in Neurons: An Overview

OBJECTIVES

- **Describe** the cellular properties that permit communication among neurons and effectors.
- **Compare** the basic types of ion channels, and **explain** how they relate to graded potentials and action potentials.

Like muscle fibers, neurons are electrically excitable. They communicate with one another using two types of electrical signals: (1) **Graded potentials** (described shortly) are used for short-distance communication only. (2) **Action potentials** (also described shortly) allow communication over long distances within the body. Recall that an action potential in a muscle fiber is called a **muscle action potential**. When an action potential occurs in a neuron (nerve cell), it is called a **nerve action potential** (*nerve impulse*). To understand the functions of graded potentials and action potentials, consider how the nervous system allows you to feel the smooth surface of a pen that you have picked up from a table (**Figure 12.10**):

1. As you touch the pen, a graded potential develops in a sensory receptor in the skin of the fingers.
2. The graded potential triggers the axon of the sensory neuron to form a nerve action potential, which travels along the axon into the CNS and ultimately causes the release of neurotransmitter at a synapse with an interneuron.
3. The neurotransmitter stimulates the interneuron to form a graded potential in its dendrites and cell body.
4. In response to the graded potential, the axon of the interneuron forms a nerve action potential. The nerve action potential travels along the axon, which results in neurotransmitter release at the next synapse with another interneuron.
5. This process of neurotransmitter release at a synapse followed by the formation of a graded potential and then a nerve action potential occurs over and over as interneurons in higher parts of the brain (such as the thalamus and cerebral cortex) are activated. Once interneurons in the **cerebral cortex**, the outer part of the brain, are activated, perception occurs and you are able to feel the smooth surface of the pen touch your fingers. As you will learn in Chapter 14, perception, the conscious awareness of a sensation, is primarily a function of the cerebral cortex.

Suppose that you want to use the pen to write a letter. The nervous system would respond in the following way (**Figure 12.10**):

6. A stimulus in the brain causes a graded potential to form in the dendrites and cell body of an **upper motor neuron**, a type of motor neuron that synapses with a lower motor neuron farther down in the CNS in order to contract a skeletal muscle. The graded potential subsequently causes a nerve action potential to occur in the axon of the upper motor neuron, followed by neurotransmitter release.
7. The neurotransmitter generates a graded potential in a **lower motor neuron**, a type of motor neuron that directly supplies skeletal muscle fibers. The graded potential triggers the formation of a nerve action potential and then release of the neurotransmitter at neuromuscular junctions formed with skeletal muscle fibers that control movements of the fingers.
8. The neurotransmitter stimulates the muscle fibers that control finger movements to form muscle action potentials. The muscle action potentials cause these muscle fibers to contract, which allows you to write with the pen.

The production of graded potentials and action potentials depends on two basic features of the plasma membrane of excitable cells: the existence of a resting membrane potential and the presence of specific types of ion channels. Like most other cells in the body, the plasma membrane of excitable cells exhibits a **membrane potential**, an electrical potential difference (voltage) across the membrane. In excitable cells, this voltage is termed the **resting membrane potential**. The membrane potential is like voltage stored in a battery. If you connect the positive and negative terminals of a battery with a piece of wire, electrons will flow along the wire. This flow of charged particles is called **current**. In living cells, the flow of ions (rather than electrons) constitutes the electrical current.

Graded potentials and action potentials occur because the membranes of neurons contain many different kinds of ion channels that open or close in response to specific stimuli. Because the lipid bilayer of the plasma membrane is a good electrical insulator, the main paths for current to flow across the membrane are through the ion channels.

Ion Channels

When ion channels are open, they allow specific ions to move across the plasma membrane, down their **electrochemical gradient**—a concentration (chemical) difference plus an electrical difference. Recall that ions move from areas of higher concentration to areas of lower concentration (the chemical part of the gradient). Also, positively charged cations move toward a negatively charged area, and negatively charged anions move toward a positively charged area (the electrical aspect of the gradient). As ions move, they create a flow of electrical current that can change the membrane potential.

Ion channels open and close due to the presence of "gates." The gate is a part of the channel protein that can seal the channel pore shut or move aside to open the pore (see **Figure 3.6**). The electrical signals produced by neurons and muscle fibers rely on four types of ion channels: leak channels, ligand-gated channels, mechanically-gated channels, and voltage-gated channels:

1. The gates of **leak channels** randomly alternate between open and closed positions (**Figure 12.11a**). Typically, plasma membranes have many more potassium ion (K^+) leak channels than sodium ion (Na^+) leak channels, and the potassium ion leak channels are leakier than the sodium ion leak channels. Thus, the membrane's permeability to K^+ is much higher than its permeability to Na^+. Leak

FIGURE 12.10 **Overview of nervous system functions.**

Graded potentials and nerve and muscle action potentials are involved in the relay of sensory stimuli, integrative functions such as perception, and motor activities.

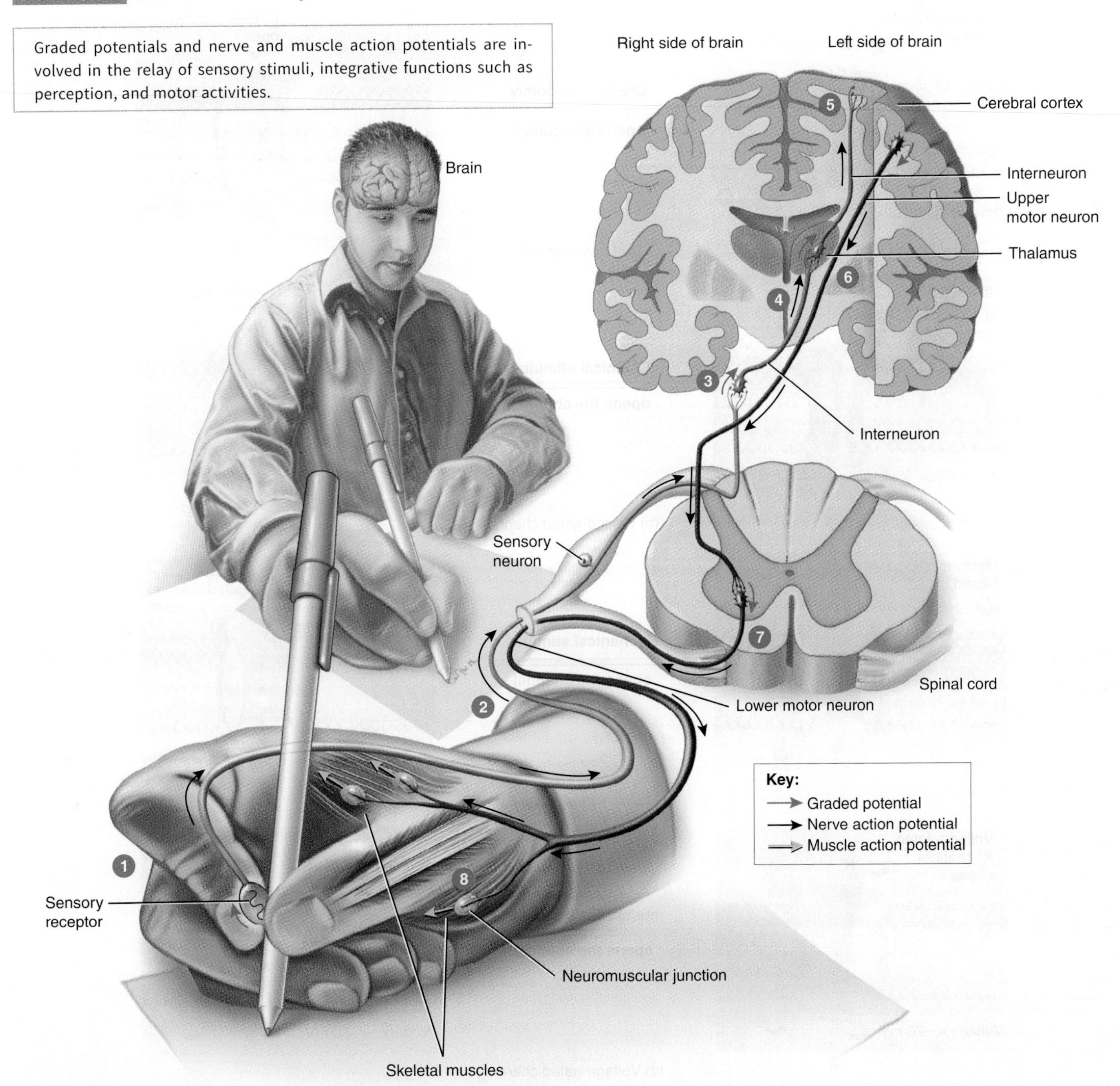

Q In which region of the brain does perception primarily occur?

FIGURE 12.11 Ion channels in the plasma membrane. (a) Leak channels randomly open and close. (b) A chemical stimulus—here, the neurotransmitter acetylcholine—opens a ligand-gated channel. (c) A mechanical stimulus opens a mechanically-gated channel. (d) A change in membrane potential opens voltage-gated K^+ channels during an action potential.

The electrical signals produced by neurons and muscle fibers rely on four types of ion channels: leak channels, ligand-gated channels, mechanically-gated channels, and voltage-gated channels.

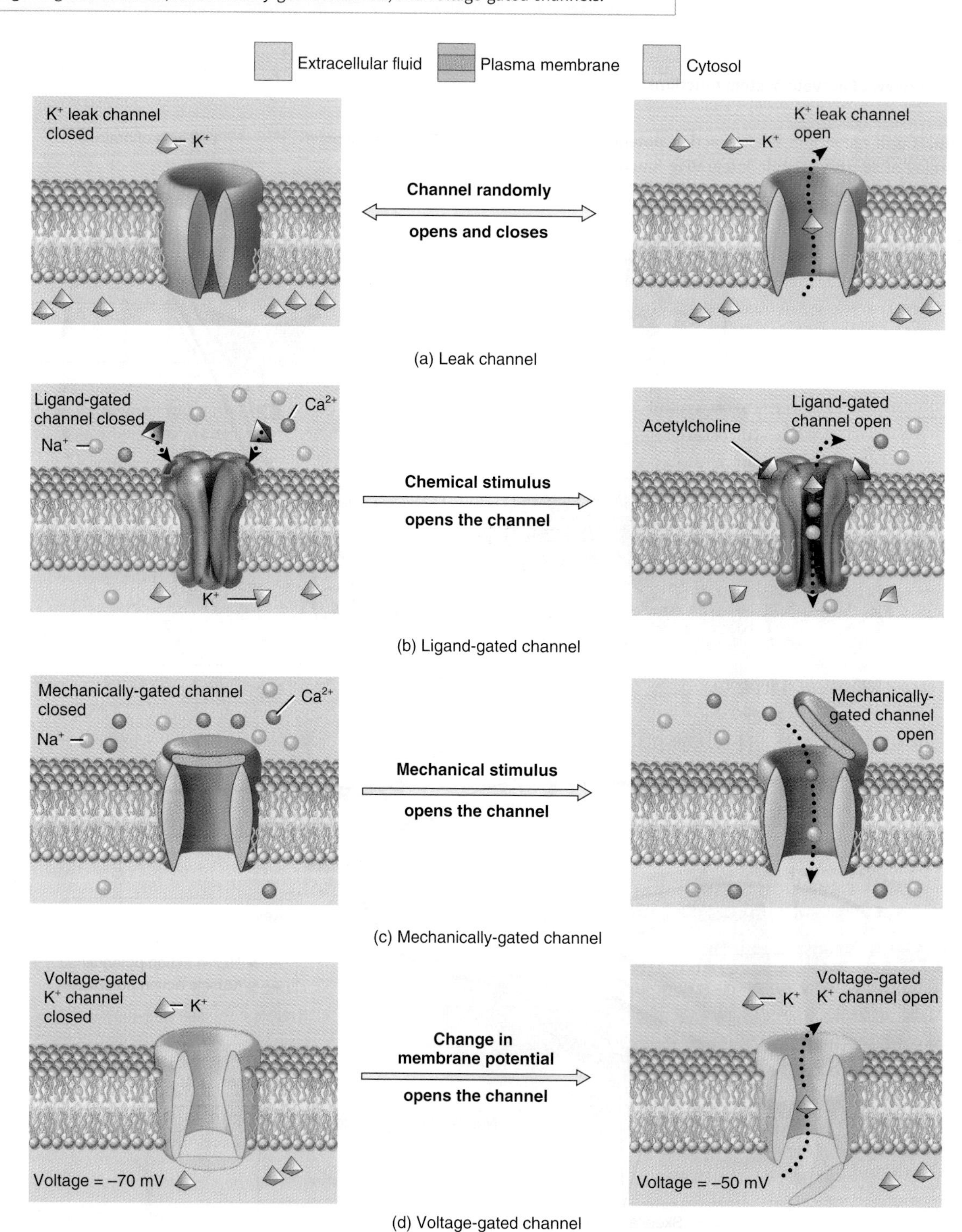

Q What type of gated channel is activated by a touch on the arm?

TABLE 12.1 Ion Channels in Neurons

TYPE OF ION CHANNEL	DESCRIPTION	LOCATION
Leak channels	Gated channels that randomly open and close.	Found in nearly all cells, including dendrites, cell bodies, and axons of all types of neurons.
Ligand-gated channels	Gated channels that open in response to binding of ligand (chemical) stimulus.	Dendrites of some sensory neurons such as pain receptors and dendrites and cell bodies of interneurons and motor neurons.
Mechanically-gated channels	Gated channels that open in response to mechanical stimulus (such as touch, pressure, vibration, or tissue stretching).	Dendrites of some sensory neurons such as touch receptors, pressure receptors, and some pain receptors.
Voltage-gated channels	Gated channels that open in response to voltage stimulus (change in membrane potential).	Axons of all types of neurons.

channels are found in nearly all cells, including the dendrites, cell bodies, and axons of all types of neurons.

2. A **ligand-gated channel** opens and closes in response to the binding of a ligand (chemical) stimulus. A wide variety of chemical ligands—including neurotransmitters, hormones, and particular ions—can open or close ligand-gated channels. The neurotransmitter acetylcholine, for example, opens cation channels that allow Na^+ and Ca^{2+} to diffuse inward and K^+ to diffuse outward (**Figure 12.11b**). Ligand-gated channels are located in the dendrites of some sensory neurons, such as pain receptors, and in dendrites and cell bodies of interneurons and motor neurons.
3. A **mechanically-gated channel** opens or closes in response to mechanical stimulation in the form of vibration (such as sound waves), touch, pressure, or tissue stretching (**Figure 12.11c**). The force distorts the channel from its resting position, opening the gate. Examples of mechanically-gated channels are those found in auditory receptors in the ears, in receptors that monitor stretching of internal organs, and in touch receptors and pressure receptors in the skin.
4. A **voltage-gated channel** opens in response to a change in membrane potential (voltage) (**Figure 12.11d**). Voltage-gated channels participate in the generation and conduction of action potentials in the axons of all types of neurons.

Table 12.1 presents a summary of the four major types of ion channels in neurons.

12.4 Resting Membrane Potential

OBJECTIVE

- **Describe** the factors that maintain a resting membrane potential.

The resting membrane potential exists because of a small buildup of negative ions in the cytosol along the inside of the membrane, and an equal buildup of positive ions in the extracellular fluid (ECF) along the outside surface of the membrane (**Figure 12.12a**). Such a separation of positive and negative electrical charges is a form of potential energy, which is measured in volts or millivolts (1 mV = 0.001 V). The greater the difference in charge across the membrane, the larger the membrane potential (voltage). Notice in **Figure 12.12a** that the buildup of charge occurs only very close to the membrane. The cytosol or extracellular fluid elsewhere in the cell contains equal numbers of positive and negative charges and is electrically neutral.

The resting membrane potential of a cell can be measured in the following way: The tip of a recording microelectrode is inserted inside the cell, and a reference electrode is placed outside the cell in the extracellular fluid. *Electrodes* are devices that conduct electrical charges. The recording microelectrode and the reference electrode are connected to an instrument known as a *voltmeter*, which detects the electrical difference (voltage) across the plasma membrane (**Figure 12.12b**). In neurons, the resting membrane potential ranges from −40 to −90 mV. A typical value is −70 mV. The minus sign indicates that the inside of the cell is negative relative to the outside. A cell that exhibits a membrane potential is said to be **polarized**. Most body cells are polarized; the membrane potential varies from +5 mV to −100 mV in different types of cells.

The resting membrane potential arises from three major factors:

1. ***Unequal distribution of ions in the ECF and cytosol.*** A major factor that contributes to the resting membrane potential is the unequal distribution of various ions in extracellular fluid and cytosol (**Figure 12.13**). Extracellular fluid is rich in Na^+ and chloride ions (Cl^-). In cytosol, however, the main cation is K^+, and the two dominant anions are phosphates attached to molecules, such as the three phosphates in ATP, and amino acids in proteins. Because the plasma membrane typically has more K^+ leak channels than Na^+ leak channels, the number of potassium ions that diffuse down their concentration gradient out of the cell into the ECF is greater than the number of sodium ions that diffuse down their concentration gradient from the ECF into the cell. As more and more positive potassium ions exit, the inside of the membrane becomes increasingly negative, and the outside of the membrane becomes increasingly positive.

FIGURE 12.12 Resting membrane potential. To measure resting membrane potential, the tip of the recording microelectrode is inserted inside the neuron, and the reference electrode is placed in the extracellular fluid. The electrodes are connected to a voltmeter that measures the difference in charge across the plasma membrane (in this case −70 mV, indicating that the inside of the cell is negative relative to the outside).

The resting membrane potential is an electrical potential difference (voltage) that exists across the plasma membrane of an excitable cell under resting conditions.

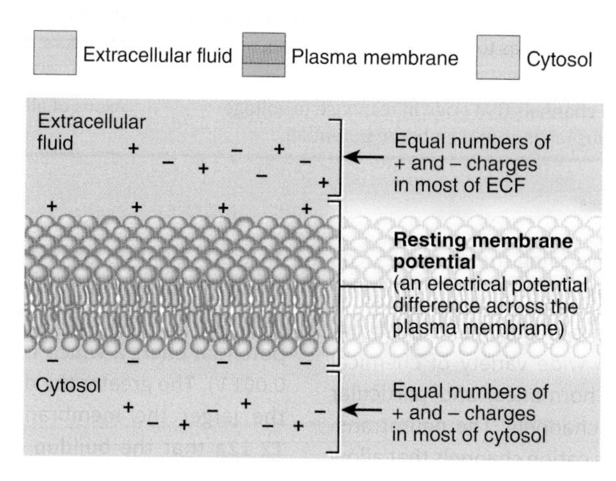

(a) Distribution of charges that produce the resting membrane potential of a neuron

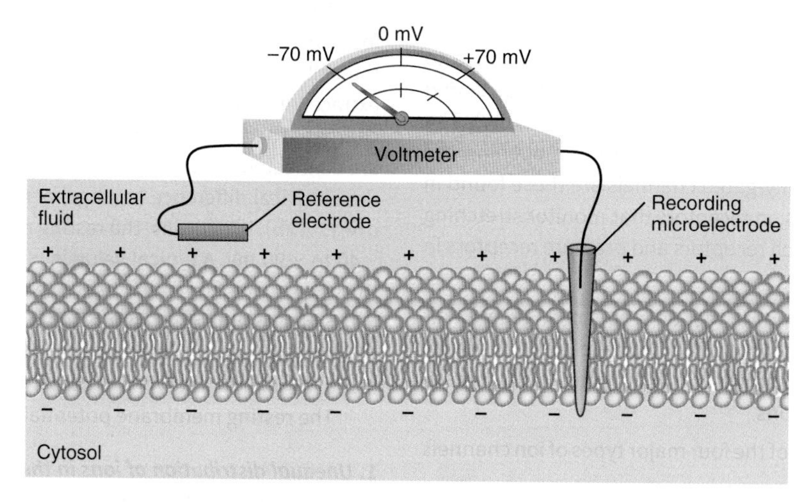

(b) Measurement of the resting membrane potential of a neuron

Q The resting membrane potential of a neuron typically is −70 mV. What does this mean?

2. ***Inability of most anions to leave the cell.*** Another factor contributes to the negative resting membrane potential: Most anions inside the cell are not free to leave (**Figure 12.13**). They cannot follow the K^+ out of the cell because they are attached to nondiffusible molecules such as ATP and large proteins.

3. ***Electrogenic nature of the Na^+–K^+ ATPases.*** Membrane permeability to Na^+ is very low because there are only a few sodium leak channels. Nevertheless, sodium ions do slowly diffuse inward, down their concentration gradient. Left unchecked, such inward leakage of Na^+ would eventually destroy the resting membrane potential. The small inward Na^+ leak and outward K^+ leak are offset by the Na^+–K^+ ATPases (sodium–potassium pumps) (**Figure 12.13**). These pumps help maintain the resting membrane potential by pumping out Na^+ as fast as it leaks in. At the same time, the Na^+–K^+ ATPases bring in K^+. However, the potassium ions eventually leak back out of the cell as they move down their concentration gradient.

FIGURE 12.13 Three factors that contribute to the resting membrane potential. (1) Because the plasma membrane has more K^+ leak channels (blue) than Na^+ leak channels (rust), the number of K^+ ions that leave the cell is greater than the number of Na^+ ions that enter the cell. As more and more K^+ ions leave the cell, the inside of the membrane becomes increasingly negative and the outside of the membrane becomes increasingly positive. (2) Trapped anions (turquoise and red) cannot follow K^+ out of the cell because they are attached to nondiffusible molecules such as ATP and large proteins. (3) The electrogenic Na^+–K^+ ATPase (purple) expels 3 Na^+ ions for every 2 K^+ ions imported.

The resting membrane potential is determined by three major factors: (1) unequal distribution of ions in the ECF and cytosol, (2) inability of most anions to leave the cell, and (3) the electrogenic nature of the Na^+–K^+ ATPases.

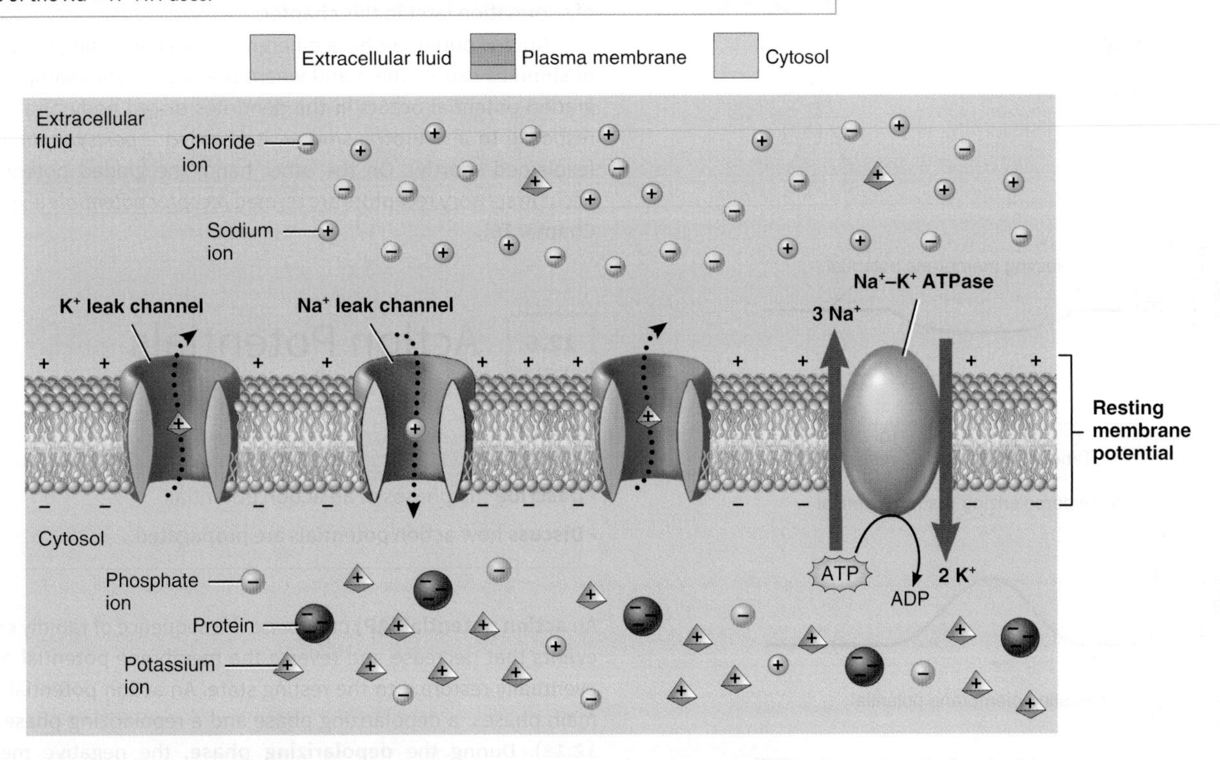

Q Suppose that the plasma membrane of a neuron has more Na^+ leak channels than K^+ leak channels. What effect would this have on the resting membrane potential?

Recall that the Na^+–K^+ ATPases expel three Na^+ for each two K^+ imported (see **Figure 3.10**). Since these pumps remove more positive charges from the cell than they bring into the cell, they are *electrogenic*, which means they contribute to the negativity of the resting membrane potential. Their total contribution, however, is very small: only −3 mV of the total −70 mV resting membrane potential in a typical neuron.

12.5 Graded Potentials

OBJECTIVE

- **Describe** how a graded potential is generated.

A **graded potential** is a small deviation from the resting membrane potential that makes the membrane either more polarized (inside more negative) or less polarized (inside less negative). When the response makes the membrane more polarized (inside more negative), it is termed a **hyperpolarizing graded potential** (hī-per-PŌ-lar-ī′-zing) (**Figure 12.14a**). When the response makes the membrane less polarized (inside less negative), it is termed a **depolarizing graded potential** (**Figure 12.14b**).

A graded potential occurs when a stimulus causes mechanically-gated or ligand-gated channels to open or close in an excitable cell's plasma membrane (**Figure 12.15**). Typically, mechanically-gated channels and ligand-gated channels can be present in the dendrites of sensory neurons, and ligand-gated channels are numerous in the dendrites and cell bodies of interneurons and motor neurons. Hence, graded potentials occur mainly in the dendrites and cell body of a neuron.

To say that these electrical signals are *graded* means that they vary in amplitude (size), depending on the strength of the stimulus

FIGURE 12.14 **Graded potentials.** Most graded potentials occur in the dendrites and cell body (areas colored blue).

During a hyperpolarizing graded potential, the membrane potential is inside more negative than the resting level; during a depolarizing graded potential, the membrane potential is inside less negative than the resting level.

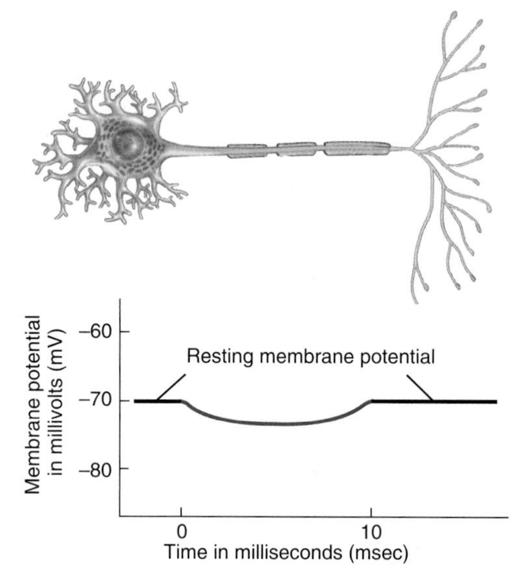

(a) Hyperpolarizing graded potential

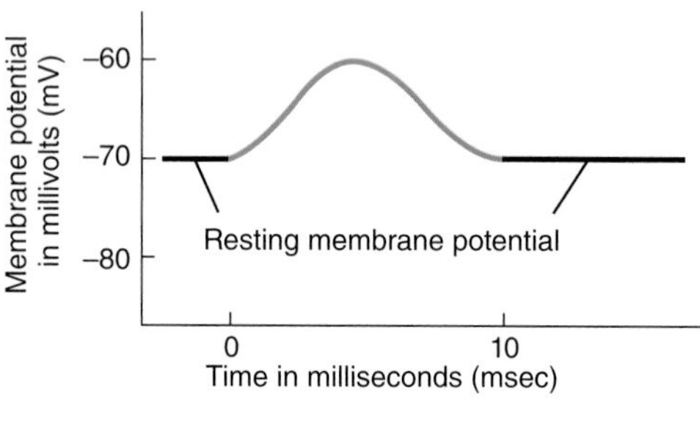

(b) Depolarizing graded potential

Q What kind of graded potential describes a change in membrane potential from −70 to −60 mV? From −70 to −80 mV?

(Figure 12.16). They are larger or smaller depending on how many ligand-gated or mechanically-gated channels have opened (or closed) and how long each remains open. The opening or closing of these ion channels alters the flow of specific ions across the membrane, producing a flow of current that is *localized*, which means that it spreads to adjacent regions along the plasma membrane in either direction from the stimulus source for a short distance and then gradually dies out as the charges are lost across the membrane through leak channels. This mode of travel by which graded potentials die out as they spread along the membrane is known as **decremental conduction** (dek-re-MENT-al). Because they die out within a few millimeters of their point of origin, graded potentials are useful for short-distance communication only.

Although an individual graded potential undergoes decremental conduction, it can become stronger and last longer by summating with other graded potentials. **Summation** is the process by which graded potentials add together. If two depolarizing graded potentials summate, the net result is a larger depolarizing graded potential (Figure 12.17). If two hyperpolarizing graded potentials summate, the net result is a larger hyperpolarizing graded potential. If two equal but opposite graded potentials summate (one depolarizing and the other hyperpolarizing), then they cancel each other out and the overall graded potential disappears. You will learn more about the process of summation later in this chapter.

Graded potentials have different names depending on which type of stimulus causes them and where they occur. For example, when a graded potential occurs in the dendrites or cell body of a neuron in response to a neurotransmitter, it is called a *postsynaptic potential* (explained shortly). On the other hand, the graded potentials that occur in sensory receptors are termed *receptor potentials* (explained in Chapter 16).

12.6 Action Potentials

OBJECTIVES

- **Describe** the phases of an action potential.
- **Discuss** how action potentials are propagated.

An **action potential (AP)** or *impulse* is a sequence of rapidly occurring events that decrease and reverse the membrane potential and then eventually restore it to the resting state. An action potential has two main phases: a depolarizing phase and a repolarizing phase (Figure 12.18). During the **depolarizing phase**, the negative membrane potential becomes less negative, reaches zero, and then becomes positive. During the **repolarizing phase**, the membrane potential is restored to the resting state of −70 mV. Following the repolarizing phase there may be an **after-hyperpolarizing phase**, during which the membrane potential temporarily becomes more negative than the resting level. Two types of voltage-gated channels open and then close during an action potential. These channels are present mainly in the axon plasma membrane and axon terminals. The first channels that open, the voltage-gated Na^+ channels, allow Na^+ to rush into the cell, which causes the depolarizing phase. Then voltage-gated K^+ channels open, allowing K^+ to flow out, which produces the repolarizing phase. The after-hyperpolarizing phase occurs when the voltage-gated K^+ channels remain open after the repolarizing phase ends.

An action potential occurs in the membrane of the axon of a neuron when depolarization reaches a certain level termed the **threshold** (about −55 mV in many neurons). Different neurons may have different thresholds for generation of an action potential, but the threshold in a particular neuron usually is constant. The generation of an action potential depends on whether a particular stimulus is able to bring the membrane potential to threshold (Figure 12.19). An action potential will not occur in response to a **subthreshold stimulus**, a weak

FIGURE 12.15 Generation of graded potentials in response to the opening of mechanically-gated channels or ligand-gated channels. (a) A mechanical stimulus (pressure) opens a mechanically-gated channel that allows passage of cations (mainly Na^+ and Ca^{2+}) into the cell, causing a depolarizing graded potential. (b) The neurotransmitter acetylcholine (a ligand stimulus) opens a cation channel that allows passage of Na^+, K^+, and Ca^{2+}; Na^+ inflow is greater than either Ca^{2+} inflow or K^+ outflow, causing a depolarizing graded potential. (c) The neurotransmitter glycine (a ligand stimulus) opens a Cl^- channel that allows passage of Cl^- ions into the cell, causing a hyperpolarizing graded potential.

A graded potential forms in response to the opening of mechanically-gated channels or ligand-gated channels.

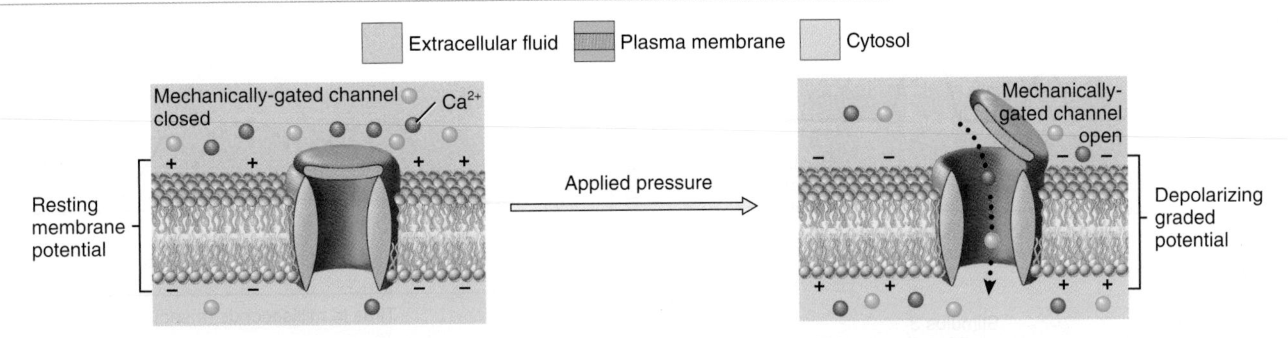

(a) Depolarizing graded potential caused by pressure, a mechanical stimulus

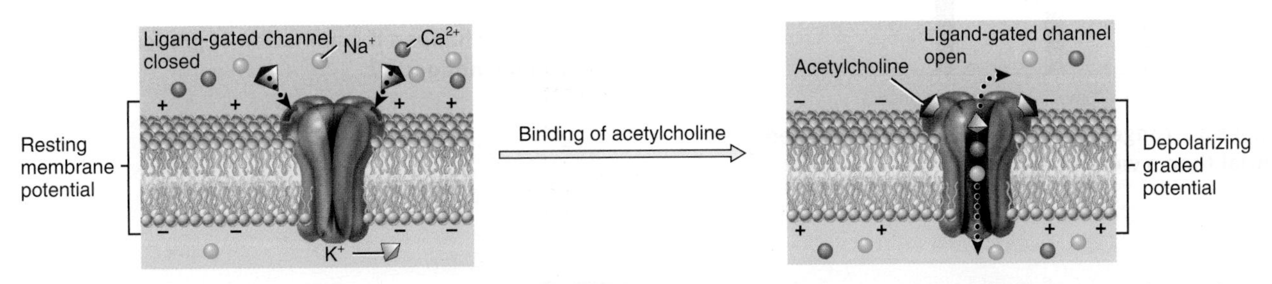

(b) Depolarizing graded potential caused by the neurotransmitter acetylcholine, a ligand stimulus

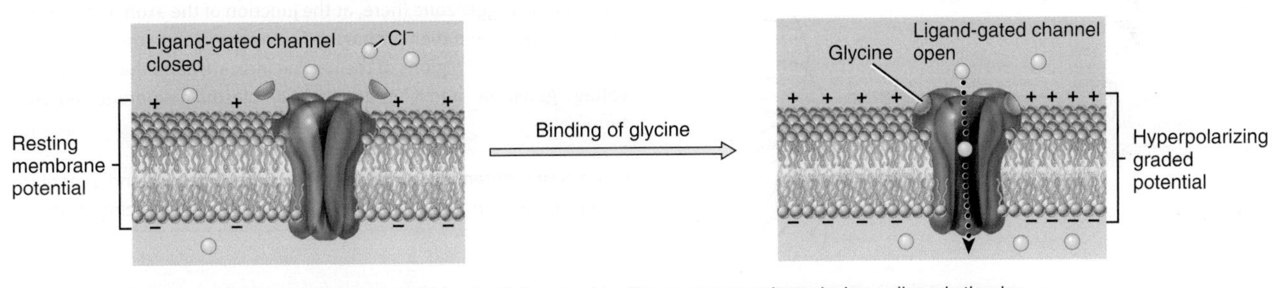

(c) Hyperpolarizing graded potential caused by the neurotransmitter glycine, a ligand stimulus

Q Which parts of a neuron contain mechanically-gated channels? Ligand-gated channels?

depolarization that cannot bring the membrane potential to threshold. However, an action potential will occur in response to a **threshold stimulus**, a stimulus that is just strong enough to depolarize the membrane to threshold. Several action potentials will form in response to a **suprathreshold stimulus**, a stimulus that is strong enough to depolarize the membrane *above* threshold. Each of the action potentials caused by a suprathreshold stimulus has the same amplitude (size) as an action potential caused by a threshold stimulus. Therefore, once an action potential is generated, the amplitude of an action potential is always the same and does not depend on stimulus intensity. Instead, the greater the stimulus strength above threshold, the greater the frequency of the action potentials until a maximum frequency is reached as determined by the absolute refractory period (described shortly).

As you have just learned, an action potential is generated in response to a threshold stimulus but does not form when there is a subthreshold stimulus. In other words, an action potential either occurs completely or it does not occur at all. This characteristic of an action potential is known as the **all-or-none principle**. The all-or-none principle of the action potential is similar to pushing the first

FIGURE 12.16 The graded nature of graded potentials. As stimulus strength increases (stimuli 1, 2, and 3), the amplitude (size) of each resulting depolarizing graded potential increases. Although not shown, a similar relationship exists between stimulus strength and the amplitude of a hyperpolarizing graded potential.

The amplitude of a graded potential depends on the stimulus strength. The greater the stimulus strength, the larger the amplitude of the graded potential.

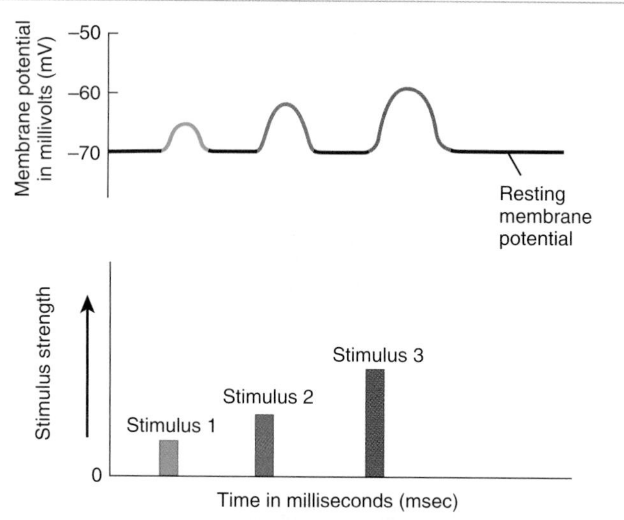

Q Why does a stronger stimulus cause a larger graded potential than a weaker stimulus?

FIGURE 12.17 Summation of graded potentials. Summation of two depolarizing graded potentials happens in response to two stimuli of the same strength that occur very close together in time. The dotted lines represent the individual depolarizing graded potentials that would form if summation did not occur.

Summation occurs when two or more graded potentials add together to become larger in amplitude.

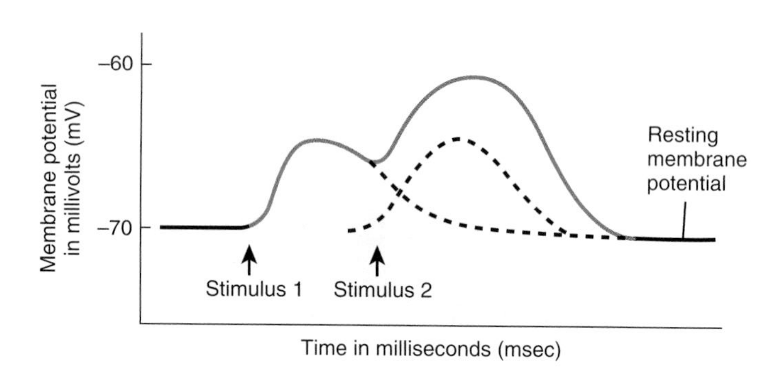

Q What would happen if summation of graded potentials in a neuron did not occur?

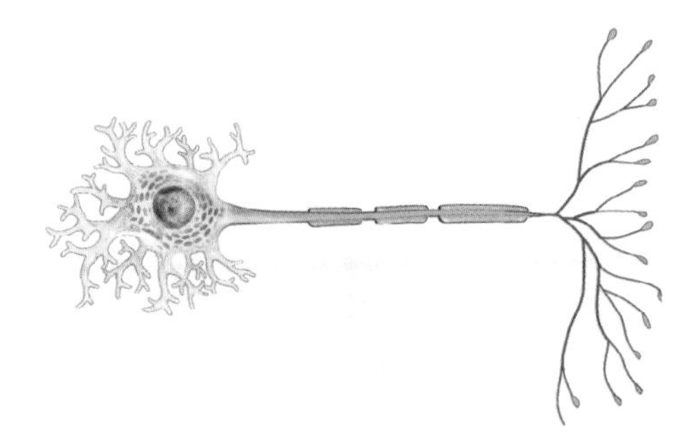

FIGURE 12.18 Action potential (AP) or impulse. The action potential arises at the trigger zone (here, at the junction of the axon hillock and the initial segment) and then propagates along the axon to the axon terminals. The green-colored regions of the neuron indicate parts that typically have voltage-gated Na^+ and K^+ channels (axon plasma membrane and axon terminals).

An action potential consists of a depolarizing phase and a repolarizing phase, which may be followed by an after-hyperpolarizing phase.

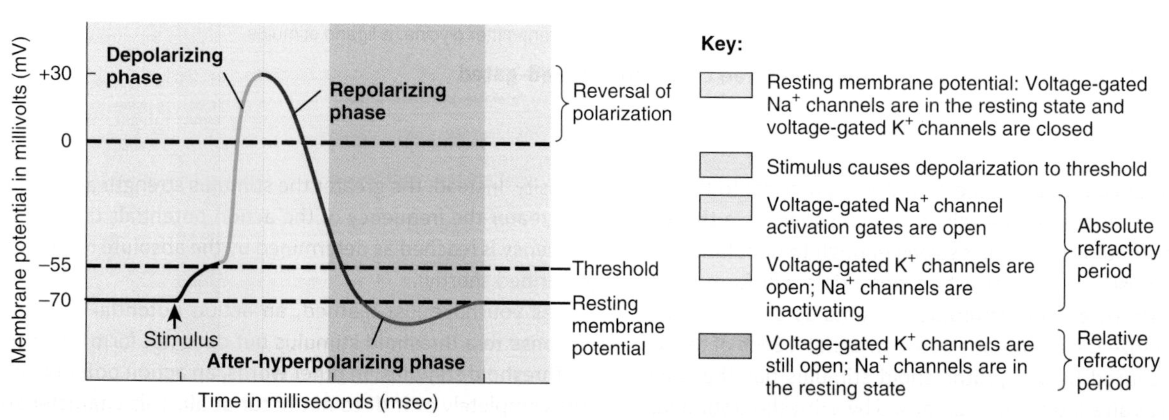

Q Which channels are open during the depolarizing phase? During the repolarizing phase?

FIGURE 12.19 Stimulus strength and action potential generation. A subthreshold stimulus does not cause an action potential. An action potential does occur in response to a threshold stimulus, which is just strong enough to depolarize the membrane to threshold. Several action potentials form in response to a suprathreshold stimulus. Each of the action potentials caused by the suprathreshold stimulus has the same amplitude (size) as the action potential caused by the threshold stimulus. For simplicity, the after-hyperpolarizing phase of the action potential is not shown.

An action potential will occur only once the membrane potential reaches threshold.

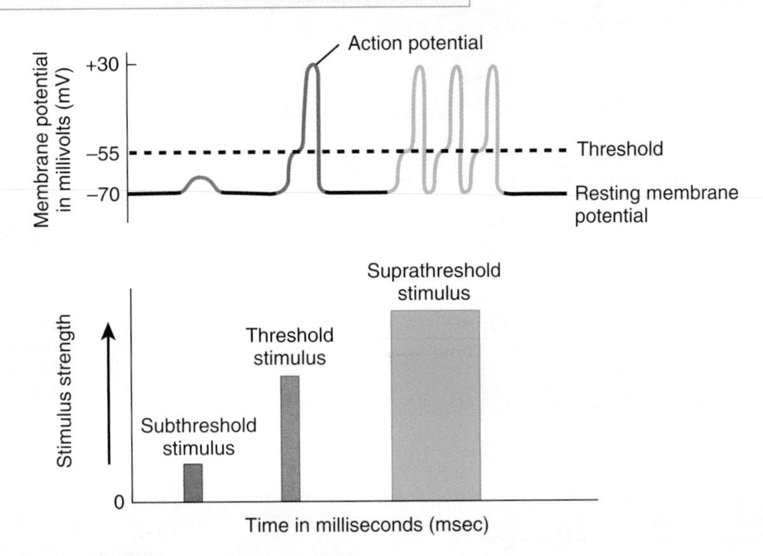

Q Will an action potential occur in response to a hyperpolarizing graded potential that spreads from the dendrites or cell body to the trigger zone of the axon of a neuron? Why or why not?

domino in a long row of standing dominoes. When the push on the first domino is strong enough (when depolarization reaches threshold), that domino falls against the second domino, and the *entire* row topples (an action potential occurs). Stronger pushes on the first domino produce the identical effect—toppling of the entire row. Thus, pushing on the first domino produces an all-or-none event: The dominoes all fall or none fall.

Depolarizing Phase

When a depolarizing graded potential or some other stimulus causes the membrane of the axon to depolarize to threshold, voltage-gated Na^+ channels open rapidly. Both the electrical and the chemical gradients favor inward movement of Na^+, and the resulting inrush of Na^+ causes the depolarizing phase of the action potential (see Figure 12.18). The inflow of Na^+ changes the membrane potential from −55 mV to +30 mV. At the peak of the action potential, the inside of the membrane is 30 mV more positive than the outside.

Each voltage-gated Na^+ channel has two separate gates, an *activation gate* and an *inactivation gate.* In the *resting state* of a voltage-gated Na^+ channel, the inactivation gate is open, but the activation gate is closed (step 1 in Figure 12.20). As a result, Na^+ cannot move into the cell through these channels. At threshold, voltage-gated Na^+ channels are activated. In the *activated state* of a voltage-gated Na^+ channel, both the activation and inactivation gates in the channel are open and Na^+ inflow begins (step 2 in Figure 12.20). As more channels open, Na^+ inflow increases, the membrane depolarizes further, and more Na^+ channels open. This is an example of a positive feedback mechanism. During the few ten-thousandths of a second that the voltage-gated Na^+ channel is open, about 20,000 Na^+ flow across the membrane and change the membrane potential considerably. However, the concentration of Na^+ hardly changes because of the millions of Na^+ present in the extracellular fluid. The sodium–potassium pumps easily bail out the 20,000 or so Na^+ that enter the cell during a single action potential and maintain the low concentration of Na^+ inside the cell.

Repolarizing Phase

Shortly after the activation gates of the voltage-gated Na^+ channels open, the inactivation gates close (step 3 in Figure 12.20). Now the voltage-gated Na^+ channel is in an *inactivated state.* In addition to opening voltage-gated Na^+ channels, a threshold-level depolarization also opens voltage-gated K^+ channels (steps 3 and 4 in Figure 12.20). Because the voltage-gated K^+ channels open more slowly, their opening occurs at about the same time the voltage-gated Na^+ channels are closing. The slower opening of voltage-gated K^+ channels and the closing of previously open voltage-gated Na^+ channels produce the repolarizing phase of the action potential. As the Na^+ channels are inactivated, Na^+ inflow slows. At the same time, the K^+ channels are opening, accelerating K^+ outflow. Slowing of Na^+ inflow

FIGURE 12.20 Changes in ion flow through voltage-gated channels during the depolarizing and repolarizing phases of an action potential. Leak channels and sodium–potassium pumps are not shown.

Inflow of sodium ions (Na^+) causes the depolarizing phase, and outflow of potassium ions (K^+) causes the repolarizing phase of an action potential.

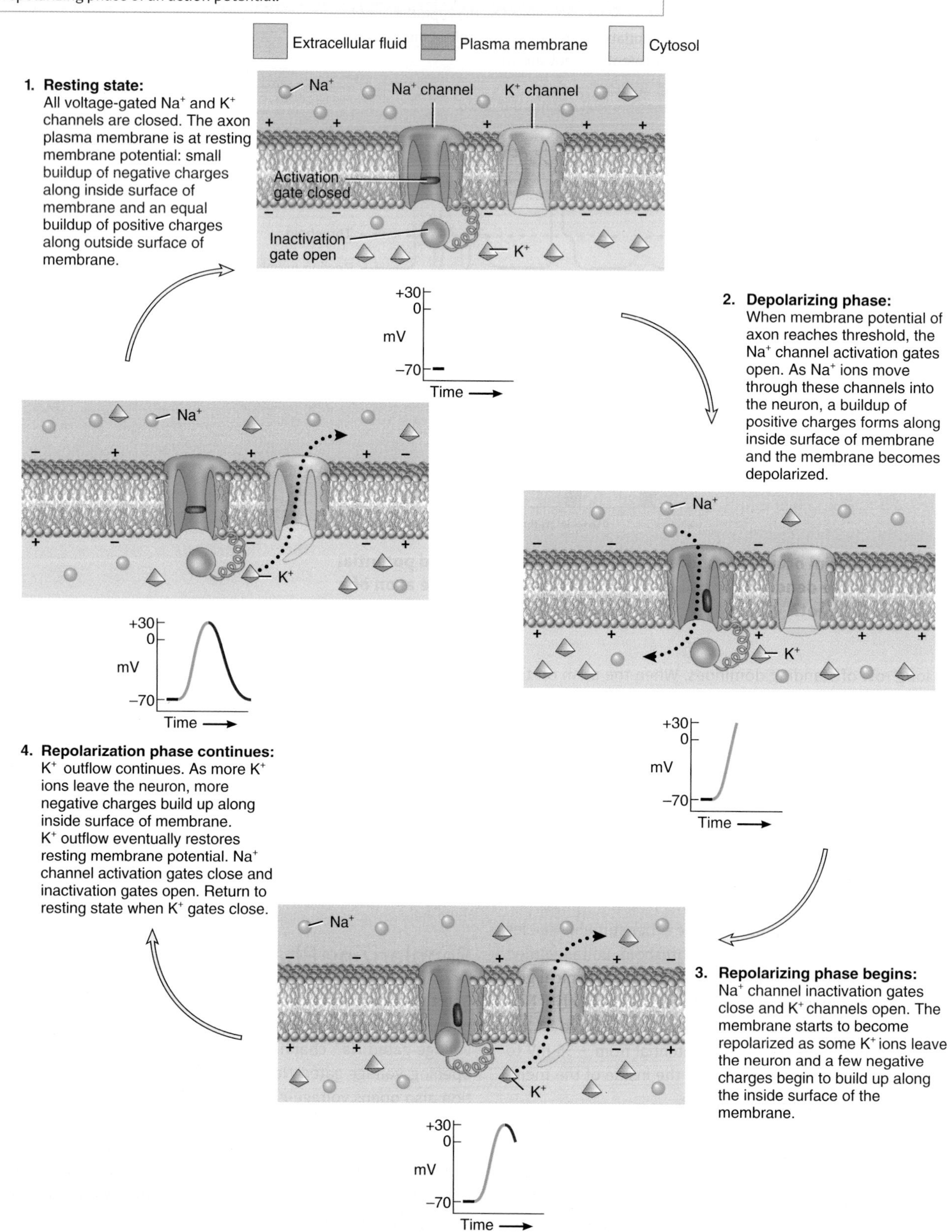

Q Given the existence of leak channels for both K^+ and Na^+, could the membrane repolarize if the voltage-gated K^+ channels did not exist?

and acceleration of K^+ outflow cause the membrane potential to change from +30 mV to −70 mV. Repolarization also allows inactivated Na^+ channels to revert to the resting state.

After-Hyperpolarizing Phase

While the voltage-gated K^+ channels are open, outflow of K^+ may be large enough to cause an after-hyperpolarizing phase of the action potential (see Figure 12.18). During this phase, the voltage-gated K^+ channels remain open and the membrane potential becomes even more negative (about −90 mV). As the voltage-gated K^+ channels close, the membrane potential returns to the resting level of −70 mV. Unlike voltage-gated Na^+ channels, most voltage-gated K^+ channels do not exhibit an inactivated state. Instead, they alternate between closed (resting) and open (activated) states.

Refractory Period

The period of time after an action potential begins during which an excitable cell cannot generate another action potential in response to a *normal* threshold stimulus is called the **refractory period** (rē-FRAK-tor-ē) (see key in Figure 12.18). During the **absolute refractory period**, even a very strong stimulus cannot initiate a second action potential. This period coincides with the period of Na^+ channel activation and inactivation (steps 2–4 in Figure 12.20). Inactivated Na^+ channels cannot reopen; they first must return to the resting state (step 1 in Figure 12.20). In contrast to action potentials, graded potentials do not exhibit a refractory period.

Large-diameter axons have a larger surface area and have a brief absolute refractory period of about 0.4 msec. Because a second nerve impulse can arise very quickly, up to 1000 impulses per second are possible. Small-diameter axons have absolute refractory periods as long as 4 msec, enabling them to transmit a maximum of 250 impulses per second. Under normal body conditions, the maximum frequency of nerve impulses in different axons ranges between 10 and 1000 per second.

The **relative refractory period** is the period of time during which a second action potential can be initiated, but only by a larger-than-normal stimulus. It coincides with the period when the voltage-gated K^+ channels are still open after inactivated Na^+ channels have returned to their resting state (see Figure 12.18).

Propagation of Action Potentials

To communicate information from one part of the body to another, action potentials in a neuron must travel from where they arise at the trigger zone of the axon to the axon terminals. In contrast to the graded potential, an action potential is not decremental (it does not die out). Instead, an action potential keeps its strength as it spreads along the membrane. This mode of conduction is called **propagation** (prop′-a-GĀ-shun), and it depends on positive feedback. As you have already learned, when sodium ions flow in, they cause voltage-gated Na^+ channels in adjacent segments of the membrane to open. Thus, the action potential travels along the membrane rather like the activity of that long row of dominoes. In actuality, it is not the same action potential that propagates along the entire axon. Instead, the action potential regenerates over and over at adjacent regions of membrane from the trigger zone to the axon terminals. In a neuron, an action potential can propagate in this direction only—it cannot propagate back toward the cell body because any region of membrane that has just undergone an action potential is temporarily in the absolute refractory period and cannot generate another action potential. Because they can travel along a membrane without dying out, action potentials function in communication over long distances.

See Clinical Connection: Neurotoxins and Local Anesthetics

Continuous and Saltatory Conduction There are two types of propagation: continuous conduction and saltatory conduction. The type of action potential propagation described so far is **continuous conduction**, which involves step-by-step depolarization and repolarization of each adjacent segment of the plasma membrane (Figure 12.21a). In continuous conduction, ions flow through their voltage-gated channels in each adjacent segment of the membrane. Note that the action potential propagates only a relatively short distance in a few milliseconds. Continuous conduction occurs in unmyelinated axons and in muscle fibers.

Action potentials propagate more rapidly along myelinated axons than along unmyelinated axons. If you compare parts (a) and (b) in Figure 12.21 you will see that the action potential propagates much farther along the myelinated axon in the same period of time. **Saltatory conduction** (SAL-ta-tō-rē; *saltat-* = leaping), the special mode of action potential propagation that occurs along myelinated axons, occurs because of the uneven distribution of voltage-gated channels. Few voltage-gated channels are present in regions where a myelin sheath covers the axolemma. By contrast, at the nodes of Ranvier (where there is no myelin sheath), the axolemma has many voltage-gated channels. Hence, current carried by Na^+ and K^+ flows across the membrane mainly at the nodes.

When an action potential propagates along a myelinated axon, an electric current (carried by ions) flows through the extracellular fluid surrounding the myelin sheath and through the cytosol from one node to the next. The action potential at the first node generates ionic currents in the cytosol and extracellular fluid that depolarize the membrane to threshold, opening voltage-gated Na^+ channels at the second node. The resulting ionic flow through the opened channels constitutes an action potential at the second node. Then, the action potential at the second node generates an ionic current that opens voltage-gated Na^+ channels at the third node, and so on. Each node repolarizes after it depolarizes.

The flow of current across the membrane only at the nodes of Ranvier has two consequences:

1. The action potential appears to "leap" from node to node as each nodal area depolarizes to threshold, thus the name "saltatory." Because an action potential leaps across long segments of the myelinated axolemma as current flows from one node to the next, it travels much faster than it would in an unmyelinated axon of the same diameter.

FIGURE 12.21 Propagation of an action potential in a neuron after it arises at the trigger zone. Dotted lines indicate ionic current flow. The insets show the path of current flow. (a) In continuous conduction along an unmyelinated axon, ionic currents flow across each adjacent segment of the membrane. (b) In saltatory conduction along a myelinated axon, the action potential (nerve impulse) at the first node generates ionic currents in the cytosol and interstitial fluid that open voltage-gated Na^+ channels at the second node, and so on at each subsequent node.

Unmyelinated axons exhibit continuous conduction; myelinated axons exhibit saltatory conduction.

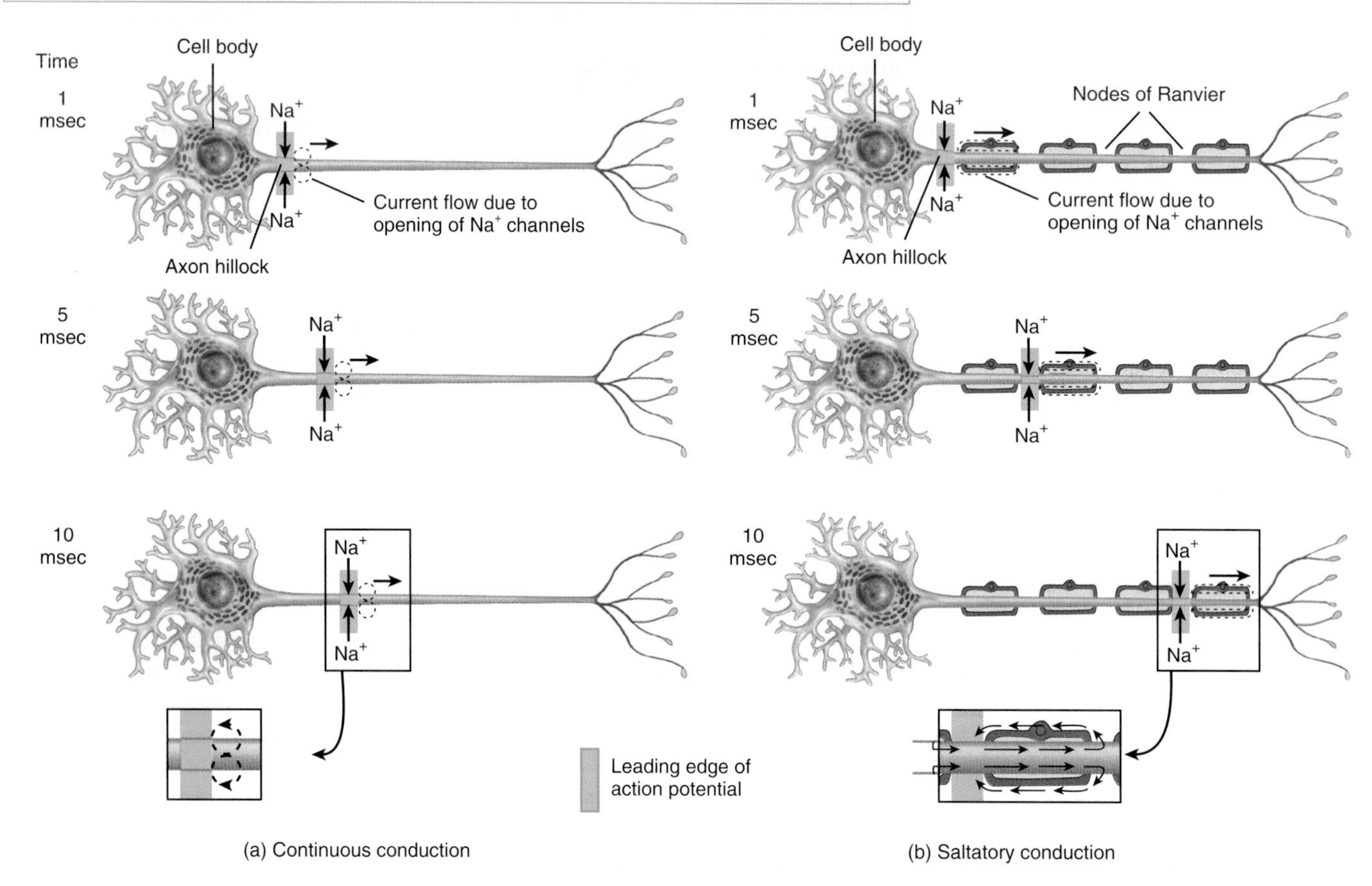

Q What factors determine the speed of propagation of an action potential?

2. Opening a smaller number of channels only at the nodes, rather than many channels in each adjacent segment of membrane, represents a more energy-efficient mode of conduction. Because only small regions of the membrane depolarize and repolarize, minimal inflow of Na^+ and outflow of K^+ occurs each time an action potential passes by. Thus, less ATP is used by sodium–potassium pumps to maintain the low intracellular concentration of Na^+ and the low extracellular concentration of K^+.

Factors That Affect the Speed of Propagation

The speed of propagation of an action potential is affected by three major factors: amount of myelination, axon diameter, and temperature.

1. ***Amount of myelination.*** As you have just learned, action potentials propagate more rapidly along myelinated axons than along unmyelinated axons.
2. ***Axon diameter.*** Larger diameter axons propagate action potentials faster than smaller ones due to their larger surface areas.
3. ***Temperature.*** Axons propagate action potentials at lower speeds when cooled.

Classification of Nerve Fibers

Axons can be classified into three major groups based on the amount of myelination, their diameters, and their propagation speeds:

- **A fibers** are the largest diameter axons (5–20 μm) and are myelinated. A fibers have a brief absolute refractory period and conduct nerve impulses (action potentials) at speeds of 12 to 130 m/sec (27–290 mi/hr). The axons of sensory neurons that propagate impulses associated with touch, pressure, position of joints, and some thermal and pain sensations are A fibers, as are the axons of motor neurons that conduct impulses to skeletal muscles.
- **B fibers** are axons with diameters of 2–3 μm. Like A fibers, B fibers are myelinated and exhibit saltatory conduction at speeds up to 15 m/sec (34 mi/hr). B fibers have a somewhat longer absolute refractory period than A fibers. B fibers conduct sensory nerve impulses from

TABLE 12.2 Comparison of Graded Potentials and Action Potentials in Neurons

CHARACTERISTIC	GRADED POTENTIALS	ACTION POTENTIALS
Origin	Arise mainly in dendrites and cell body.	Arise at trigger zones and propagate along axon.
Types of channels	Ligand-gated or mechanically-gated ion channels.	Voltage-gated channels for Na^+ and K^+.
Conduction	Decremental (not propagated); permit communication over short distances.	Propagate and thus permit communication over longer distances.
Amplitude (size)	Depending on strength of stimulus, varies from less than 1 mV to more than 50 mV.	All or none; typically about 100 mV.
Duration	Typically longer, ranging from several milliseconds to several minutes.	Shorter, ranging from 0.5 to 2 msec.
Polarity	May be hyperpolarizing (inhibitory to generation of action potential) or depolarizing (excitatory to generation of action potential).	Always consist of depolarizing phase followed by repolarizing phase and return to resting membrane potential.
Refractory period	Not present; summation can occur.	Present; summation cannot occur.

the viscera to the brain and spinal cord. They also constitute all of the axons of the autonomic motor neurons that extend from the brain and spinal cord to the ANS relay stations called autonomic ganglia.

- **C fibers** are the smallest diameter axons (0.5–1.5 μm) and all are unmyelinated. Nerve impulse propagation along a C fiber ranges from 0.5 to 2 m/sec (1–4 mi/hr). C fibers exhibit the longest absolute refractory periods. These unmyelinated axons conduct some sensory impulses for pain, touch, pressure, heat, and cold from the skin, and pain impulses from the viscera. Autonomic motor fibers that extend from autonomic ganglia to stimulate the heart, smooth muscle, and glands are C fibers. Examples of motor functions of B and C fibers are constricting and dilating the pupils, increasing and decreasing the heart rate, and contracting and relaxing the urinary bladder.

Encoding of Stimulus Intensity

How can your sensory systems detect stimuli of differing intensities if all nerve impulses are the same size? Why does a light touch feel different from firmer pressure? The main answer to this question is the *frequency of action potentials*—how often they are generated at the trigger zone. A light touch generates a low frequency of action potentials. A firmer pressure elicits action potentials that pass down the axon at a higher frequency. In addition to this "frequency code," a second factor is the number of sensory neurons recruited (activated) by the stimulus. A firm pressure stimulates a larger number of pressure-sensitive neurons than does a light touch.

Comparison of Electrical Signals Produced by Excitable Cells

We have seen that excitable cells—neurons and muscle fibers—produce two types of electrical signals: graded potentials and action potentials (impulses). One obvious difference between them is that the propagation of action potentials permits communication over long distances, but graded potentials can function only in short-distance communication because they are not propagated. Table 12.2 presents a summary of the differences between graded potentials and action potentials.

As we discussed in Chapter 10, propagation of a muscle action potential along the sarcolemma and into the T tubule system initiates the events of muscle contraction. Although action potentials in muscle fibers and in neurons are similar, there are some notable differences. The typical resting membrane potential of a neuron is -70 mV, but it is closer to -90 mV in skeletal and cardiac muscle fibers. The duration of a nerve impulse is 0.5–2 msec, but a muscle action potential is considerably longer—about 1.0–5.0 msec for skeletal muscle fibers and 10–300 msec for cardiac and smooth muscle fibers. Finally, the propagation speed of action potentials along the largest diameter myelinated axons is about 18 times faster than the propagation speed along the sarcolemma of a skeletal muscle fiber.

12.7 Signal Transmission at Synapses

OBJECTIVES

- **Explain** the events of signal transmission at electrical and chemical synapses.
- **Distinguish** between spatial and temporal summation.
- **Give** examples of excitatory and inhibitory neurotransmitters, and describe how they act.

Recall from Chapter 10 that a **synapse** (SIN-aps) is a region where communication occurs between two neurons or between a neuron and an effector cell (muscle cell or glandular cell). The term **presynaptic neuron** (*pre-* = before) refers to a nerve cell that carries a nerve impulse toward a synapse. It is the cell that sends a signal. A **postsynaptic cell** is the cell that receives a signal. It may be a nerve cell called a **postsynaptic neuron** (*post-* = after) that carries a nerve impulse away from a synapse or an **effector cell** that responds to the impulse at the synapse.

Most synapses between neurons are **axodendritic** (ak′-so-den-DRIT-ik = from axon to dendrite), while others are **axosomatic** (ak′-sō-sō-MAT-ik = from axon to cell body) or **axoaxonic** (ak′-so-ak-SON-ik = from axon to axon) (**Figure 12.22**). In addition, synapses may be electrical or chemical, and they differ both structurally and functionally.

In Chapter 10 we described the events occurring at one type of synapse, the neuromuscular junction. Our focus in this chapter is on synaptic communication among the billions of neurons in the nervous system. Synapses are essential for homeostasis because they allow information to be filtered and integrated. During learning, the structure and function of particular synapses change. The changes may allow some signals to be transmitted while others are blocked. For example, the changes in your synapses from studying will determine how well you do on your anatomy and physiology tests! Synapses are also important because some diseases and neurological disorders result from disruptions of synaptic communication, and many therapeutic and addictive chemicals affect the body at these junctions.

Electrical Synapses

At an **electrical synapse**, action potentials (impulses) conduct directly between the plasma membranes of adjacent neurons through structures called **gap junctions**. Each gap junction contains a hundred or so tubular *connexons*, which act like tunnels to connect the cytosol of the two cells directly (see **Figure 4.2e**). As ions flow from one cell to the next through the connexons, the action potential spreads from cell to cell. Gap junctions are common in visceral smooth muscle, cardiac muscle, and the developing embryo. They also occur in the brain.

Electrical synapses have two main advantages:

1. ***Faster communication.*** Because action potentials conduct directly through gap junctions, electrical synapses are faster than chemical synapses. At an electrical synapse, the action potential passes directly from the presynaptic cell to the postsynaptic cell. The events that occur at a chemical synapse take some time and delay communication slightly.
2. ***Synchronization.*** Electrical synapses can synchronize (coordinate) the activity of a group of neurons or muscle fibers. In other words, a large number of neurons or muscle fibers can produce action potentials in unison if they are connected by gap junctions. The value of synchronized action potentials in the heart or in visceral smooth muscle is coordinated contraction of these fibers to produce a heartbeat or move food through the gastrointestinal tract.

Chemical Synapses

Although the plasma membranes of presynaptic and postsynaptic neurons in a **chemical synapse** are close, they do not touch. They are separated by the **synaptic cleft**, a space of 20–50 nm* that is filled with interstitial fluid. Nerve impulses cannot conduct across the synaptic cleft, so an alternative, indirect form of communication occurs. In response to a nerve impulse, the presynaptic neuron releases a neurotransmitter that diffuses through the fluid in the synaptic cleft and binds to

*1 nanometer (nm) = 10^{-9} (0.000000001) meter.

FIGURE 12.22 Examples of synapses between neurons. Arrows indicate the direction of information flow: presynaptic neuron → postsynaptic neuron. Presynaptic neurons can form a synapse with the axon (axoaxonic: red), a dendrite (axodendritic; blue), or the cell body (axosomatic; green) of a postsynaptic neuron.

Neurons communicate with other neurons at synapses, which are junctions between one neuron and a second neuron or an effector cell.

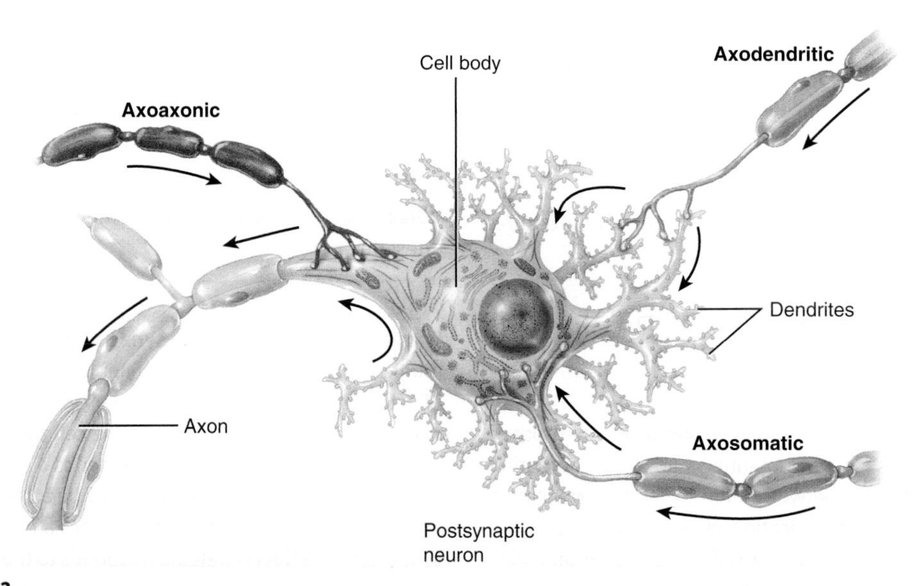

Q What is a synapse?

receptors in the plasma membrane of the postsynaptic neuron. The postsynaptic neuron receives the chemical signal and in turn produces a **postsynaptic potential**, a type of graded potential. Thus, the presynaptic neuron converts an electrical signal (nerve impulse) into a chemical signal (released neurotransmitter). The postsynaptic neuron receives the chemical signal and in turn generates an electrical signal (postsynaptic potential). The time required for these processes at a chemical synapse, a **synaptic delay** of about 0.5 msec, is the reason that chemical synapses relay signals more slowly than electrical synapses.

A typical chemical synapse transmits a signal as follows (**Figure 12.23**):

1. A nerve impulse arrives at a synaptic end bulb (or at a varicosity) of a presynaptic axon.
2. The depolarizing phase of the nerve impulse opens **voltage-gated Ca^{2+} channels**, which are present in the membrane of synaptic end bulbs. Because calcium ions are more concentrated in the extracellular fluid, Ca^{2+} flows inward through the opened channels.
3. An increase in the concentration of Ca^{2+} inside the presynaptic neuron serves as a signal that triggers exocytosis of the synaptic vesicles. As vesicle membranes merge with the plasma membrane, neurotransmitter molecules within the vesicles are released into the synaptic cleft. Each synaptic vesicle contains several thousand molecules of neurotransmitter.
4. The neurotransmitter molecules diffuse across the synaptic cleft and bind to **neurotransmitter receptors** in the postsynaptic neuron's plasma membrane. The receptor shown in **Figure 12.23**

FIGURE 12.23 Signal transmission at a chemical synapse. Through exocytosis of synaptic vesicles, a presynaptic neuron releases neurotransmitter molecules. After diffusing across the synaptic cleft, the neurotransmitter binds to receptors in the plasma membrane of the postsynaptic neuron and produces a postsynaptic potential.

At a chemical synapse, a presynaptic neuron converts an electrical signal (nerve impulse) into a chemical signal (neurotransmitter release). The postsynaptic neuron then converts the chemical signal back into an electrical signal (postsynaptic potential).

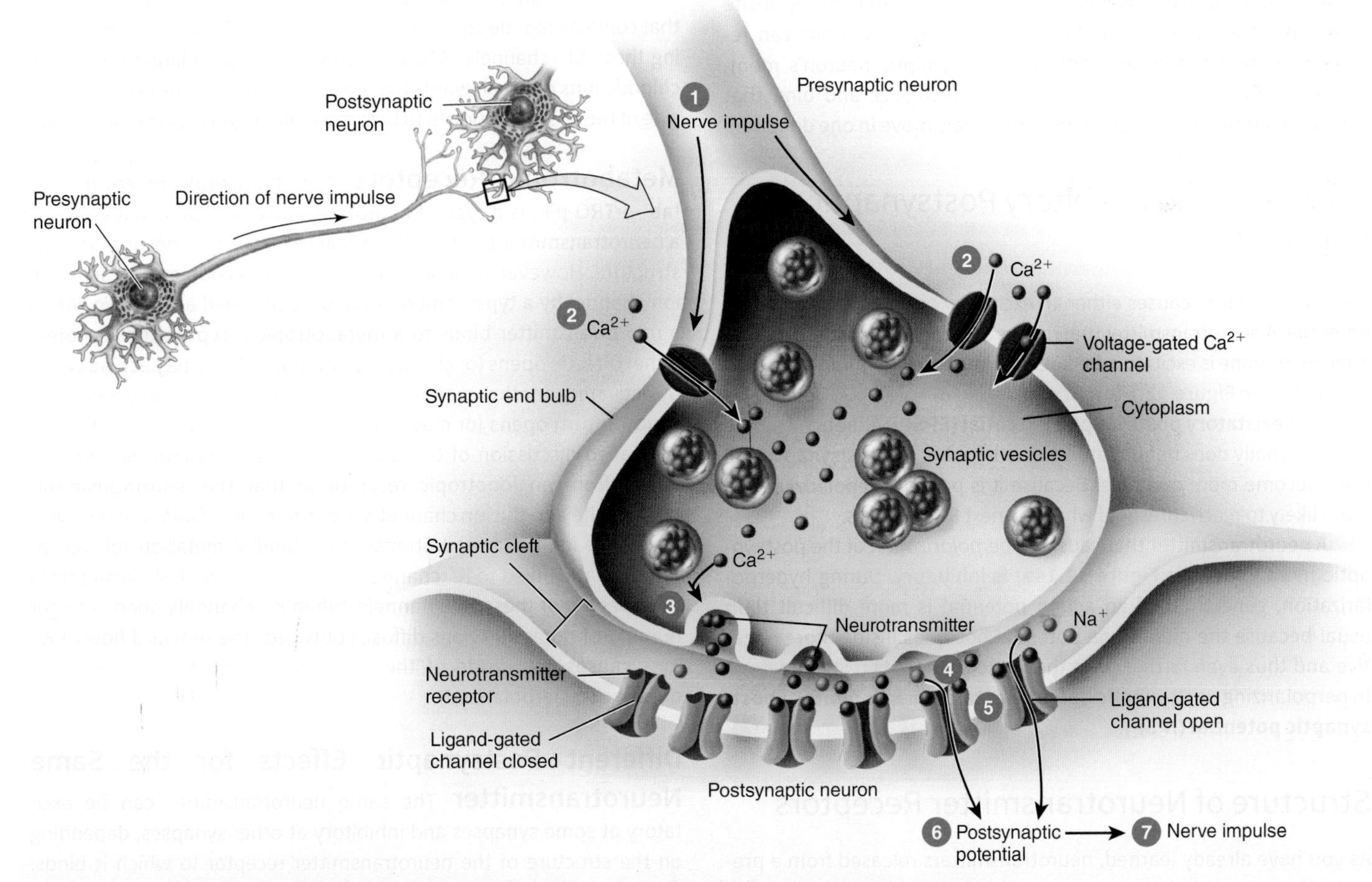

Q Why may electrical synapses work in two directions, but chemical synapses can transmit a signal in only one direction?

is part of a ligand-gated channel (see Figure 12.11b); you will soon learn that this type of neurotransmitter receptor is called an *ionotropic receptor*. Not all neurotransmitters bind to ionotropic receptors; some bind to *metabotropic receptors* (described shortly).

5. Binding of neurotransmitter molecules to their receptors on ligand-gated channels opens the channels and allows particular ions to flow across the membrane.
6. As ions flow through the opened channels, the voltage across the membrane changes. This change in membrane voltage is a **postsynaptic potential**. Depending on which ions the channels admit, the postsynaptic potential may be a depolarization (excitation) or a hyperpolarization (inhibition). For example, opening of Na^+ channels allows inflow of Na^+, which causes depolarization. However, opening of Cl^- or K^+ channels causes hyperpolarization. Opening Cl^- channels permits Cl^- to move into the cell, while opening the K^+ channels allows K^+ to move out—in either event, the inside of the cell becomes more negative.
7. When a depolarizing postsynaptic potential reaches threshold, it triggers an action potential in the axon of the postsynaptic neuron.

At most chemical synapses, only *one-way information transfer* can occur—from a presynaptic neuron to a postsynaptic neuron or an effector, such as a muscle fiber or a gland cell. For example, synaptic transmission at a neuromuscular junction (NMJ) proceeds from a somatic motor neuron to a skeletal muscle fiber (but not in the opposite direction). Only synaptic end bulbs of presynaptic neurons can release neurotransmitter, and only the postsynaptic neuron's membrane has the receptor proteins that can recognize and bind that neurotransmitter. As a result, action potentials move in one direction.

Excitatory and Inhibitory Postsynaptic Potentials

A neurotransmitter causes either an excitatory or an inhibitory graded potential. A neurotransmitter that causes *depolarization* of the postsynaptic membrane is excitatory because it brings the membrane closer to threshold (see Figure 12.14b). A depolarizing postsynaptic potential is called an **excitatory postsynaptic potential (EPSP)**. Although a single EPSP normally does not initiate a nerve impulse, the postsynaptic cell does become more excitable. Because it is partially depolarized, it is more likely to reach threshold when the next EPSP occurs.

A neurotransmitter that causes *hyperpolarization* of the postsynaptic membrane (see Figure 12.14a) is inhibitory. During hyperpolarization, generation of an action potential is more difficult than usual because the membrane potential becomes inside more negative and thus even farther from threshold than in its resting state. A hyperpolarizing postsynaptic potential is termed an **inhibitory postsynaptic potential (IPSP)**.

Structure of Neurotransmitter Receptors

As you have already learned, neurotransmitters released from a presynaptic neuron bind to **neurotransmitter receptors** in the plasma membrane of a postsynaptic cell. Each type of neurotransmitter receptor has one or more neurotransmitter binding sites where its specific neurotransmitter binds. When a neurotransmitter binds to the correct neurotransmitter receptor, an ion channel opens and a postsynaptic potential (either an EPSP or IPSP) forms in the membrane of the postsynaptic cell. Neurotransmitter receptors are classified as either ionotropic receptors or metabotropic receptors based on whether the neurotransmitter binding site and the ion channel are components of the same protein or are components of different proteins.

Ionotropic Receptors An **ionotropic receptor** (ī-on-ō-TROP-ik) is a type of neurotransmitter receptor that contains a neurotransmitter binding site and an ion channel. In other words, the neurotransmitter binding site and the ion channel are components of the *same* protein. An ionotropic receptor is a type of ligand-gated channel (see Figure 12.11b). In the absence of neurotransmitter (the ligand), the ion channel component of the ionotropic receptor is closed. When the correct neurotransmitter binds to the ionotropic receptor, the ion channel opens, and an EPSP or IPSP occurs in the postsynaptic cell.

Many excitatory neurotransmitters bind to ionotropic receptors that contain cation channels (Figure 12.24a). EPSPs result from opening these cation channels. When cation channels open, they allow passage of the three most plentiful cations (Na^+, K^+, and Ca^{2+}) through the postsynaptic cell membrane, but Na^+ inflow is greater than either Ca^{2+} inflow or K^+ outflow, and the inside of the postsynaptic cell becomes less negative (depolarized).

Many inhibitory neurotransmitters bind to ionotropic receptors that contain chloride channels (Figure 12.24b). IPSPs result from opening these Cl^- channels. When Cl^- channels open, a larger number of chloride ions diffuse inward. The inward flow of Cl^- ions causes the inside of the postsynaptic cell to become more negative (hyperpolarized).

Metabotropic Receptors A **metabotropic receptor** (me-tab′-ō-TRO-pik) is a type of neurotransmitter receptor that contains a neurotransmitter binding site but lacks an ion channel as part of its structure. However, a metabotropic receptor is coupled to a separate ion channel by a type of membrane protein called a *G protein*. When a neurotransmitter binds to a metabotropic receptor, the G protein either directly opens (or closes) the ion channel or it may act indirectly by activating another molecule, a "second messenger," in the cytosol, which in turn opens (or closes) the ion channel (see Section 18.4 for a detailed discussion of G proteins). Thus, a metabotropic receptor differs from an ionotropic receptor in that the neurotransmitter binding site and the ion channel are components of *different* proteins.

Some inhibitory neurotransmitters bind to metabotropic receptors that are linked to K^+ channels (Figure 12.24c). IPSPs result from the opening of these K^+ channels. When K^+ channels open, a larger number of potassium ions diffuses outward. The outward flow of K^+ ions causes the inside of the postsynaptic cell to become more negative (hyperpolarized).

Different Postsynaptic Effects for the Same Neurotransmitter The same neurotransmitter can be excitatory at some synapses and inhibitory at other synapses, depending on the structure of the neurotransmitter receptor to which it binds. For example, at some excitatory synapses acetylcholine (ACh) binds to ionotropic receptors that contain cation channels that open and subsequently generate EPSPs in the postsynaptic cell (Figure 12.24a).

FIGURE 12.24 Ionotropic and metabotropic neurotransmitter receptors. (a) The ionotropic acetylcholine (ACh) receptor contains two binding sites for the neurotransmitter ACh and a cation channel. Binding of ACh to this receptor causes the cation channel to open, allowing passage of the three most plentiful cations and an excitatory postsynaptic potential (EPSP) to be generated. (b) The ionotropic gamma-aminobutyric acid (GABA) receptor contains two binding sites for the neurotransmitter GABA and a Cl^- channel. Binding of GABA to this receptor causes the Cl^- channel to open, allowing a larger number of chloride ions to diffuse inward and an inhibitory postsynaptic potential (IPSP) to be generated. (c) The metabotropic acetylcholine (ACh) receptor contains a binding site for the neurotransmitter ACh. Binding of ACh to this receptor activates a G protein, which in turn opens a K^+ channel, allowing a larger number of potassium ions to diffuse out of the cell and an IPSP to form.

An ionotropic receptor is a type of neurotransmitter receptor that contains a neurotransmitter binding site and an ion channel; a metabotropic receptor is a type of neurotransmitter receptor that contains a neurotransmitter binding site and is coupled to a separate ion channel by a G protein.

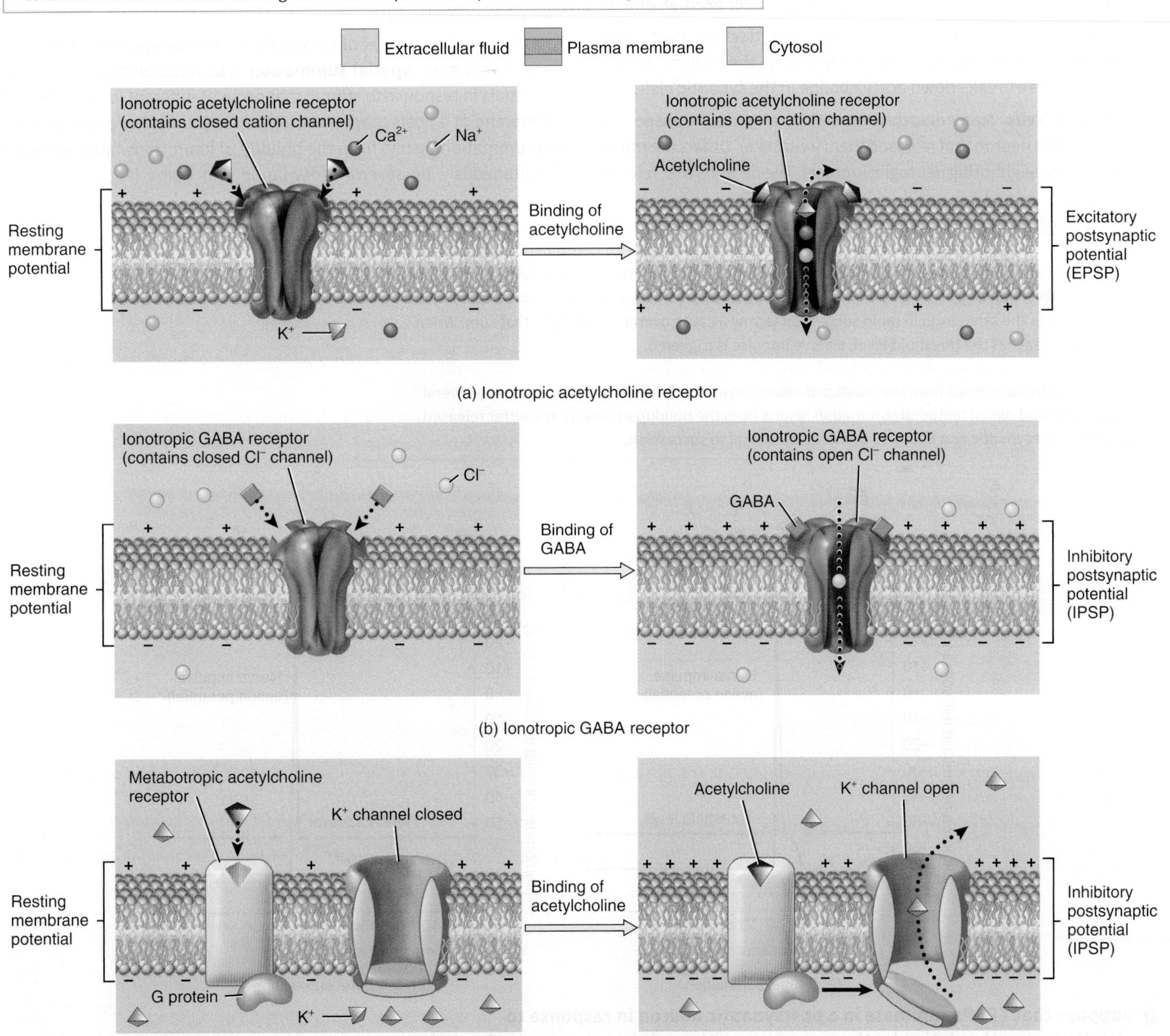

Q How can the neurotransmitter acetylcholine (ACh) be excitatory at some synapses and inhibitory at others?

By contrast, at some inhibitory synapses ACh binds to metabotropic receptors coupled to G proteins that open K^+ channels, resulting in the formation of IPSPs in the postsynaptic cell (Figure 12.24c).

Removal of Neurotransmitter

Removal of the neurotransmitter from the synaptic cleft is essential for normal synaptic function. If a neurotransmitter could linger in the synaptic cleft, it would influence the postsynaptic neuron, muscle fiber, or gland cell indefinitely. Neurotransmitter is removed in three ways:

1. ***Diffusion.*** Some of the released neurotransmitter molecules diffuse away from the synaptic cleft. Once a neurotransmitter molecule is out of reach of its receptors, it can no longer exert an effect.
2. ***Enzymatic degradation.*** Certain neurotransmitters are inactivated through enzymatic degradation. For example, the enzyme acetylcholinesterase breaks down acetylcholine in the synaptic cleft.
3. ***Uptake by cells.*** Many neurotransmitters are actively transported back into the neuron that released them (reuptake). Others are transported into neighboring neuroglia (uptake). The neurons that release norepinephrine, for example, rapidly take up the norepinephrine and recycle it into new synaptic vesicles. The membrane proteins that accomplish such uptake are called *neurotransmitter transporters.*

Spatial and Temporal Summation of Postsynaptic Potentials

A typical neuron in the CNS receives input from 1000 to 10,000 synapses. Integration of these inputs involves summation of the postsynaptic potentials that form in the postsynaptic neuron. Recall that summation is the process by which graded potentials add together. The greater the summation of EPSPs, the greater the chance that threshold will be reached. At threshold, one or more nerve impulses (action potentials) arise.

There are two types of summation: spatial summation and temporal summation. **Spatial summation** is summation of postsynaptic potentials in response to stimuli that occur at different *locations* in the membrane of a postsynaptic cell at the same time. For example, spatial summation results from the buildup of neurotransmitter released simultaneously by *several* presynaptic end bulbs (Figure 12.25a).

FIGURE 12.25 Spatial and temporal summation. (a) When presynaptic neurons 1 and 2 separately cause EPSPs (arrows) in postsynaptic neuron 3, the threshold level is not reached in neuron 3. Spatial summation occurs only when neurons 1 and 2 act simultaneously on neuron 3; their EPSPs sum to reach the threshold level and trigger a nerve impulse (action potential). (b) Temporal summation occurs when stimuli applied to the same axon in rapid succession (arrows) cause overlapping EPSPs that sum. When depolarization reaches the threshold level, a nerve impulse is triggered.

Spatial summation results from the buildup of neurotransmitter released simultaneously by several presynaptic end bulbs; temporal summation results from the buildup of neurotransmitter released by a single presynaptic end bulb two or more times in rapid succession.

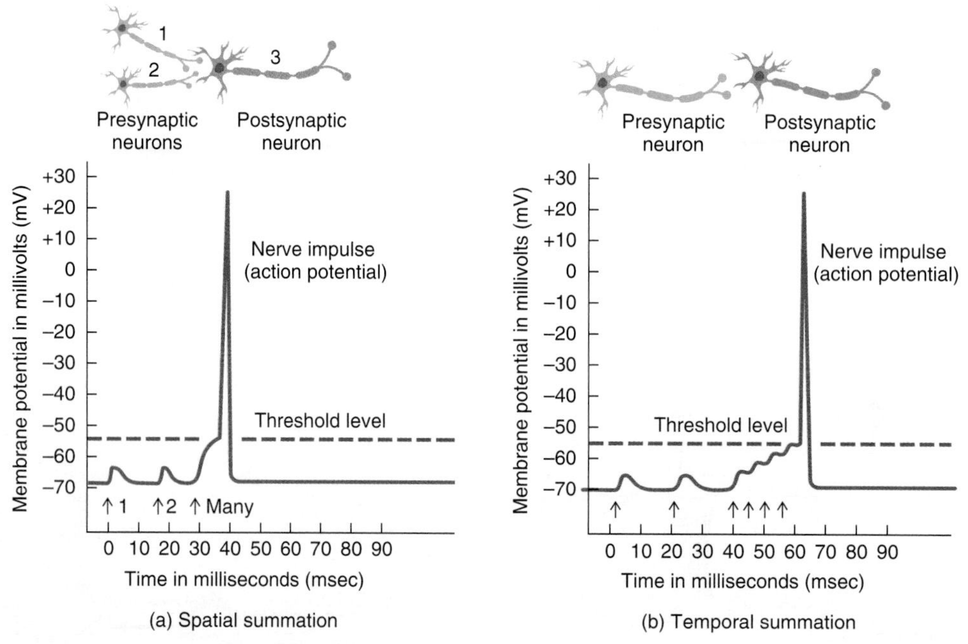

Q Suppose that EPSPs summate in a postsynaptic neuron in response to simultaneous stimulation by the neurotransmitters glutamate, serotonin, and acetylcholine released by three separate presynaptic neurons. Is this an example of spatial or temporal summation?

Temporal summation is summation of postsynaptic potentials in response to stimuli that occur at the same location in the membrane of the postsynaptic cell but at different *times*. For example, temporal summation results from buildup of neurotransmitter released by a *single* presynaptic end bulb two or more times in rapid succession (**Figure 12.25b**). Because a typical EPSP lasts about 15 msec, the second (and subsequent) release of neurotransmitter must occur soon after the first one if temporal summation is to occur. Summation is rather like a vote on the Internet. Many people voting "yes" or "no" on an issue at the same time can be compared to spatial summation. One person voting repeatedly and rapidly is like temporal summation. Most of the time, spatial and temporal summations are acting together to influence the chance that a neuron fires an action potential.

A single postsynaptic neuron receives input from many presynaptic neurons, some of which release excitatory neurotransmitters and some of which release inhibitory neurotransmitters (**Figure 12.26**). The sum of all the excitatory and inhibitory effects at any given time determines the effect on the postsynaptic neuron, which may respond in the following ways:

1. ***EPSP.*** If the total excitatory effects are greater than the total inhibitory effects but less than the threshold level of stimulation, the result is an EPSP that does not reach threshold. Following an EPSP, subsequent stimuli can more easily generate a nerve impulse through summation because the neuron is partially depolarized.
2. ***Nerve impulse(s).*** If the total excitatory effects are greater than the total inhibitory effects and threshold is reached, one or more nerve impulses (action potentials) will be triggered. Impulses continue to be generated as long as the EPSP is at or above the threshold level.
3. ***IPSP.*** If the total inhibitory effects are greater than the excitatory effects, the membrane hyperpolarizes (IPSP). The result is inhibition of the postsynaptic neuron and an inability to generate a nerve impulse.

See Clinical Connection: Strychnine Poisoning

Table 12.3 summarizes the structural and functional elements of a neuron.

FIGURE 12.26 Summation of postsynaptic potentials at the trigger zone of a postsynaptic neuron. Presynaptic neurons 1, 3, and 5 release excitatory neurotransmitters (purple dots) that generate excitatory postsynaptic potentials (EPSPs) (purple arrows) in the membrane of a postsynaptic neuron. Presynaptic neurons 2 and 4 release inhibitory neurotransmitters (red dots) that generate inhibitory postsynaptic potentials (IPSPs) (red arrows) in the membrane of the postsynaptic neuron. The net summation of these EPSPs and IPSPs determines whether an action potential will be generated at the trigger zone of the postsynaptic neuron.

If the net summation of EPSPs and IPSPs is a depolarization that reaches threshold, then an action potential will occur at the trigger zone of a postsynaptic neuron.

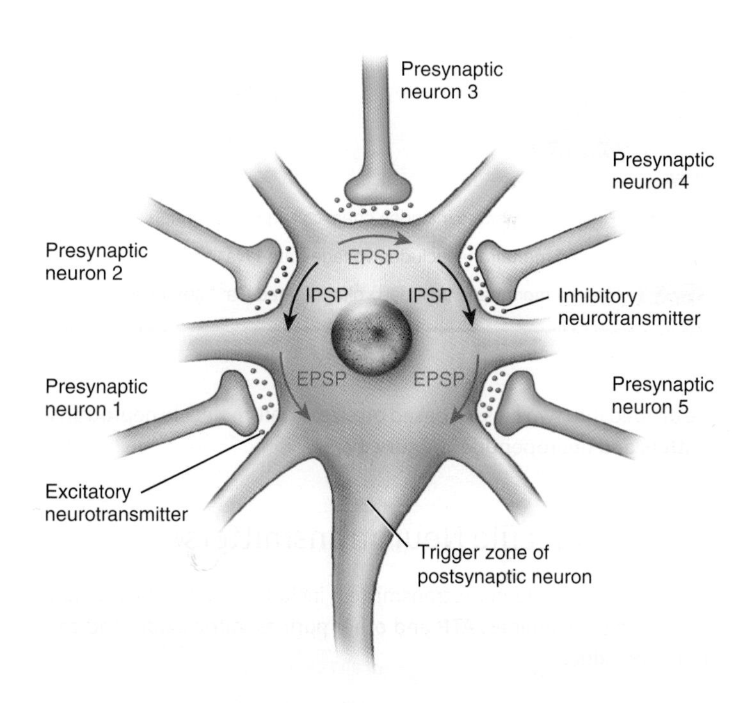

Q Suppose that the net summation of the EPSPs and IPSPs shown in this figure is a depolarization that brings the membrane potential of the trigger zone of the postsynaptic neuron to −60 mV. Will an action potential occur in the postsynaptic neuron?

12.8 Neurotransmitters

OBJECTIVE

- **Describe** the classes and functions of neurotransmitters.

About 100 substances are either known or suspected to be neurotransmitters. Some neurotransmitters bind to their receptors and act quickly to open or close ion channels in the membrane. Others act more slowly via second-messenger systems to influence chemical reactions inside cells. The result of either process can be excitation or inhibition of postsynaptic neurons. Many neurotransmitters are also hormones released into the bloodstream by endocrine cells in organs throughout the body. Within the brain, certain neurons, called **neurosecretory cells,** also secrete hormones. Neurotransmitters can

TABLE 12.3 Summary of Neuronal Structure and Function

STRUCTURE	FUNCTIONS
Dendrites	Receive stimuli through activation of ligand-gated or mechanically-gated ion channels; in sensory neurons, produce generator or receptor potentials; in motor neurons and interneurons, produce excitatory and inhibitory postsynaptic potentials (EPSPs and IPSPs).
Cell body	Receives stimuli and produces EPSPs and IPSPs through activation of ligand-gated ion channels.
Junction of **axon hillock** and **initial segment** of axon	Trigger zone in many neurons; integrates EPSPs and IPSPs and, if sum is depolarization that reaches threshold, initiates action potential (nerve impulse).
Axon	Propagates nerve impulses from initial segment (or from dendrites of sensory neurons) to axon terminals in self-regenerating manner; impulse amplitude does not change as it propagates along axon.
Axon terminals and synaptic end bulbs (or varicosities)	Inflow of Ca^{2+} caused by depolarizing phase of nerve impulse triggers exocytosis of neurotransmitter from synaptic vesicles.

Key:

Plasma membrane includes chemically gated channels

Plasma membrane includes voltage-gated Na^+ and K^+ channels

Plasma membrane includes voltage-gated Ca^{2+} channels

be divided into two classes based on size: small-molecule neurotransmitters and neuropeptides (Figure 12.27).

Small-Molecule Neurotransmitters

The small-molecule neurotransmitters include acetylcholine, amino acids, biogenic amines, ATP and other purines, nitric oxide, and carbon monoxide.

Acetylcholine The best-studied neurotransmitter is **acetylcholine (ACh)** (a-sē′-til-KŌ-lēn), which is released by many PNS neurons and by some CNS neurons. ACh is an excitatory neurotransmitter at some synapses, such as the neuromuscular junction, where the binding of ACh to ionotropic receptors opens cation channels (see Figure 12.24a). It is also an inhibitory neurotransmitter at other synapses, where it binds to metabotropic receptors coupled to G proteins that open K^+ channels (see Figure 12.24c). For example, ACh slows heart rate at inhibitory synapses made by parasympathetic neurons of the vagus (X) nerve. The enzyme *acetylcholinesterase (AChE)* (a-sē′-til-kō′-lin-ES-ter-ās) inactivates ACh by splitting it into acetate and choline fragments.

Amino Acids Several amino acids are neurotransmitters in the CNS. **Glutamate** (gloo-TA-māt) (glutamic acid) and **aspartate** (as-PAR-tāt) (aspartic acid) have powerful excitatory effects. Most excitatory neurons in the CNS and perhaps half of the synapses in the brain communicate via glutamate. At some glutamate synapses, binding of the neurotransmitter to ionotropic receptors opens cation channels. The consequent inflow of cations (mainly Na^+ ions) produces an EPSP. Inactivation of glutamate occurs via reuptake. Glutamate transporters actively transport glutamate back into the synaptic end bulbs and neighboring neuroglia.

Gamma-aminobutyric acid (GABA) (GAM-ma am-i-nō-bu-TIR-ik) and **glycine** are important inhibitory neurotransmitters. At many synapses, the binding of GABA to ionotropic receptors opens Cl^- channels (see Figure 12.24b). GABA is found only in the CNS, where it is the most common inhibitory neurotransmitter. As many as one-third of all brain synapses use GABA. Antianxiety drugs such as diazepam (Valium®) enhance the action of GABA. Like GABA, the binding of glycine to ionotropic receptors opens Cl^- channels. About half of the inhibitory synapses in the spinal cord use the amino acid glycine; the rest use GABA.

Biogenic Amines Certain amino acids are modified and decarboxylated (carboxyl group removed) to produce biogenic amines. Those that are prevalent in the nervous system include norepinephrine, epinephrine, dopamine, and serotonin. Most biogenic amines bind to metabotropic receptors; there are many different types of metabotropic receptors for each biogenic amine. Biogenic amines may cause either excitation or inhibition, depending on the type of metabotropic receptor at the synapse.

Norepinephrine (NE) (nor′-ep-i-NEF-rin) plays roles in arousal (awakening from deep sleep), dreaming, and regulating mood. A smaller number of neurons in the brain use **epinephrine** as a neurotransmitter. Both epinephrine and norepinephrine also serve as hormones. Cells of the adrenal medulla, the inner portion of the adrenal gland, release them into the blood.

SMALL-MOLECULE NEUROTRANSMITTERS

Acetylcholine

$H_3C-N^+(CH_3)_2-CH_2-CH_2-O-C(=O)-CH_3$

Nitric oxide

$N{=}O$

Carbon monoxide

$C{\equiv}O$

Amino Acids

Glutamate

$H_3N^+-CH(COO^-)-CH_2-CH_2-COO^-$

Aspartate

$H_3N^+-CH(COO^-)-CH_2-COO^-$

Gamma aminobutyric acid (GABA)

$H_3N^+-CH_2-CH_2-CH_2-COO^-$

Glycine

$H_3N^+-CH_2-COO^-$

Biogenic Amines

Norepinephrine

HO, HO (catechol ring) $-CH(OH)-CH_2-NH_3^+$

Epinephrine

HO, HO (catechol ring) $-CH(OH)-CH_2-NH_2^+-CH_3$

Dopamine

HO, HO (catechol ring) $-CH_2-CH_2-NH_3^+$

Serotonin

HO (indole ring, N–H) $-CH_2-CH_2-NH_3^+$

Purines

Example: ATP

Adenosine: Adenine (NH_2, purine ring with N, C, H) + Ribose (H, H, H, H, OH, OH)

$H_2C-O-P(O^-)(=O)-O\sim P(O^-)(=O)-O\sim P(O^-)(=O)-O^-$

Phosphate groups

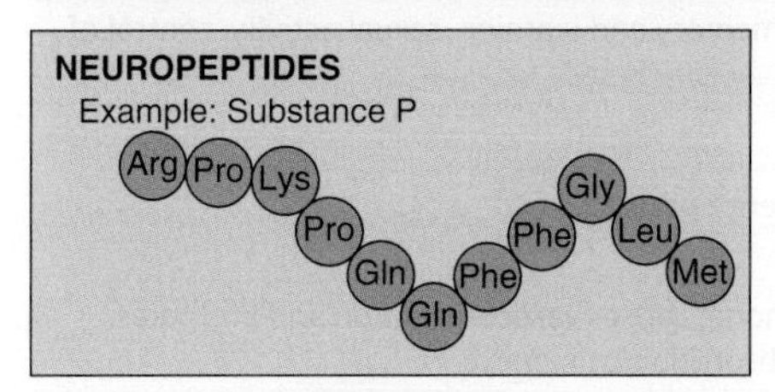

Q Why are norepinephrine, epinephrine, dopamine, and serotonin classified as biogenic amines?

FIGURE 12.27 Neurotransmitters. Neurotransmitters are divided into two major classes based on size: small-molecule neurotransmitters and neuropeptides. The neuropeptide shown is substance P, which consists of 11 amino acids linked by peptide bonds in the following order: arginine (Arg), proline (Pro), lysine (Lys), proline, glutamine (Gln), glutamine, phenylalanine (Phe), phenylalanine, glycine (Gly), leucine (Leu), and methionine (Met).

Neurotransmitters are chemical substances that neurons use to communicate with other neurons, muscle fibers, and glands.

Brain neurons containing the neurotransmitter **dopamine (DA)** (DŌ-pa-mēn) are active during emotional responses, addictive behaviors, and pleasurable experiences. In addition, dopamine-releasing neurons help regulate skeletal muscle tone and some aspects of movement due to contraction of skeletal muscles. The muscular stiffness that occurs in Parkinson's disease is due to degeneration of neurons that release dopamine (see Disorders: Homeostatic Imbalances in Chapter 16 in Study Guide). One form of schizophrenia is due to accumulation of excess dopamine.

Norepinephrine, dopamine, and epinephrine are classified chemically as **catecholamines** (kat-e-KŌL-a-mēns). They all have an amino group (—NH_2) and a catechol ring composed of six carbons and two adjacent hydroxyl (—OH) groups. Catecholamines are synthesized from the amino acid tyrosine. Inactivation of catecholamines occurs via reuptake into synaptic end bulbs. Then they are either recycled back into the synaptic vesicles or destroyed by enzymes. The two enzymes that break down catecholamines are **catechol-*O*-methyltransferase** (**COMT**) (kat′-e-kōl-ō-meth-il-TRANS-fer-ās), and **monoamine oxidase** (**MAO**) (mon-ō-AM-īn OK-si-dās).

Serotonin (ser′-ō-TŌ-nin), which is also known as *5-hydroxytryptamine (5-HT)*, is concentrated in the neurons in a part of the brain called the raphe nucleus. It is thought to be involved in sensory perception, temperature regulation, control of mood, appetite, and the induction of sleep.

ATP and Other Purines

The characteristic ring structure of the adenosine portion of ATP (Figure 12.27) is called a purine ring. Adenosine itself, as well as its triphosphate, diphosphate, and monophosphate derivatives (ATP, ADP, and AMP), is an excitatory neurotransmitter in both the CNS and the PNS. Most of the synaptic vesicles that contain ATP also contain another neurotransmitter. In the PNS, ATP and norepinephrine are released together from some sympathetic neurons; some parasympathetic neurons release ATP and acetylcholine in the same vesicles.

Nitric Oxide

The simple gas **nitric oxide (NO)** is an important excitatory neurotransmitter secreted in the brain, spinal cord, adrenal glands, and nerves to the penis and has widespread effects throughout the body. NO contains a single nitrogen atom, in contrast to nitrous oxide (N_2O), or laughing gas, which has two nitrogen atoms. N_2O is sometimes used as an anesthetic during dental procedures.

The enzyme **nitric oxide synthase (NOS)** catalyzes formation of NO from the amino acid arginine. Based on the presence of NOS, it is estimated that more than 2% of the neurons in the brain produce NO. Unlike all previously known neurotransmitters, NO is not synthesized in advance and packaged into synaptic vesicles. Rather, it is formed on demand and acts immediately. Its action is brief because NO is a highly reactive free radical. It exists for less than 10 seconds before it combines with oxygen and water to form inactive nitrates and nitrites. Because NO is lipid-soluble, it diffuses from cells that produce it into neighboring cells, where it activates an enzyme for production of a second messenger called cyclic GMP. Some research suggests that NO plays a role in memory and learning.

The first recognition of NO as a regulatory molecule was the discovery in 1987 that a chemical called EDRF (endothelium-derived relaxing factor) was actually NO. Endothelial cells in blood vessel walls release NO, which diffuses into neighboring smooth muscle cells and causes relaxation. The result is vasodilation, an increase in blood vessel diameter. The effects of such vasodilation range from a lowering of blood pressure to erection of the penis in males. Sildenafil (Viagra®) alleviates erectile dysfunction (impotence) by enhancing the effect of NO. In larger quantities, NO is highly toxic. Phagocytic cells, such as macrophages and certain white blood cells, produce NO to kill microbes and tumor cells.

Carbon Monoxide Carbon monoxide (CO), like NO, is not produced in advance and packaged into synaptic vesicles. It too is formed as needed and diffuses out of cells that produce it into adjacent cells. CO is an excitatory neurotransmitter produced in the brain and in response to some neuromuscular and neuroglandular functions. CO might protect against excess neuronal activity and might be related to dilation of blood vessels, memory, olfaction (sense of smell), vision, thermoregulation, insulin release, and anti-inflammatory activity.

Neuropeptides

Neurotransmitters consisting of 3 to 40 amino acids linked by peptide bonds called **neuropeptides** (noor-ō-PEP-tīds) are numerous and widespread in both the CNS and PNS. Neuropeptides bind to metabotropic receptors and have excitatory or inhibitory actions, depending on the type of metabotropic receptor at the synapse. Neuropeptides are formed in the neuron cell body, packaged into vesicles, and transported to axon terminals. Besides their role as neurotransmitters, many neuropeptides serve as hormones that regulate physiological responses elsewhere in the body.

Scientists discovered that certain brain neurons have plasma membrane receptors for opiate drugs such as morphine and heroin. The quest to find the naturally occurring substances that use these receptors brought to light the first neuropeptides: two molecules, each a chain of five amino acids, named **enkephalins** (en-KEF-a-lins). Their potent analgesic (pain-relieving) effect is 200 times stronger than morphine. Other so-called *opioid peptides* include the **endorphins** (en-DOR-fins) and **dynorphins** (dī-NOR-fins). It is thought that opioid peptides are the body's natural painkillers. Acupuncture may produce analgesia (loss of pain sensation) by increasing the release of opioids. These neuropeptides have also been linked to improved memory and learning; feelings of pleasure or euphoria; control of body temperature; regulation of hormones that affect the onset of puberty, sexual drive, and reproduction; and mental illnesses such as depression and schizophrenia.

Another neuropeptide, **substance P**, is released by neurons that transmit pain-related input from peripheral pain receptors into the central nervous system, enhancing the perception of pain. Enkephalin and endorphin suppress the release of substance P, thus decreasing the number of nerve impulses being relayed to the brain for pain sensations. Substance P has also been shown to counter the effects of certain nerve-damaging chemicals, prompting speculation that it might prove useful as a treatment for nerve degeneration.

See Clinical Connection: Modifying the Effects of Neurotransmitters

Table 12.4 provides brief descriptions of these neuropeptides, as well as others that will be discussed in later chapters.

TABLE 12.4 Neuropeptides

SUBSTANCE	DESCRIPTION
Substance P	Found in sensory neurons, spinal cord pathways, and parts of brain associated with pain; enhances perception of pain.
Enkephalins	Inhibit pain impulses by suppressing release of substance P; may have role in memory and learning, control of body temperature, sexual activity, and mental illness.
Endorphins	Inhibit pain by blocking release of substance P; may have role in memory and learning, sexual activity, control of body temperature, and mental illness.
Dynorphins	May be related to controlling pain and registering emotions.
Hypothalamic releasing and inhibiting hormones	Produced by hypothalamus; regulate release of hormones by anterior pituitary.
Angiotensin II	Stimulates thirst; may regulate blood pressure in brain. As a hormone, causes vasoconstriction and promotes release of aldosterone, which increases rate of salt and water reabsorption by kidneys.
Cholecystokinin (CCK)	Found in brain and small intestine; may regulate feeding as a "stop eating" signal. As a hormone, regulates pancreatic enzyme secretion during digestion, and contraction of smooth muscle in gastrointestinal tract.
Neuropeptide Y	Stimulates food intake; may play a role in the stress response.

12.9 Neural Circuits

OBJECTIVE

- **Identify** the various types of neural circuits in the nervous system.

The CNS contains billions of neurons organized into complicated networks called **neural circuits**, functional groups of neurons that process specific types of information. In a **simple series circuit**, a presynaptic neuron stimulates a single postsynaptic neuron. The second neuron then stimulates another, and so on. However, most neural circuits are more complex.

A single presynaptic neuron may synapse with several postsynaptic neurons. Such an arrangement, called **divergence,** permits one presynaptic neuron to influence several postsynaptic neurons (or several muscle fibers or gland cells) at the same time. In a **diverging circuit**, the nerve impulse from a single presynaptic neuron causes the stimulation of increasing numbers of cells along the circuit (**Figure 12.28a**). For example, a small number of neurons in the brain that govern a particular body movement stimulate a much larger number of neurons in the spinal cord. Sensory signals are also arranged in diverging circuits, allowing a sensory impulse to be relayed to several regions of the brain. This arrangement amplifies the signal.

In another arrangement, called **convergence**, several presynaptic neurons synapse with a single postsynaptic neuron. This arrangement permits more effective stimulation or inhibition of the postsynaptic neuron. In a **converging circuit** (**Figure 12.28b**), the postsynaptic neuron receives nerve impulses from several different sources. For example, a single motor neuron that synapses with skeletal muscle fibers at neuromuscular junctions receives input from several pathways that originate in different brain regions.

Some circuits are organized so that stimulation of the presynaptic cell causes the postsynaptic cell to transmit a series of nerve impulses. One such circuit is called a **reverberating circuit** (**Figure 12.28c**). In this pattern, the incoming impulse stimulates the first neuron, which stimulates the second, which stimulates the third, and so on. Branches from later neurons synapse with earlier ones. This arrangement sends impulses back through the circuit again and again. The output signal

FIGURE 12.28 **Examples of neural circuits.**

A neural circuit is a functional group of neurons that processes a specific kind of information.

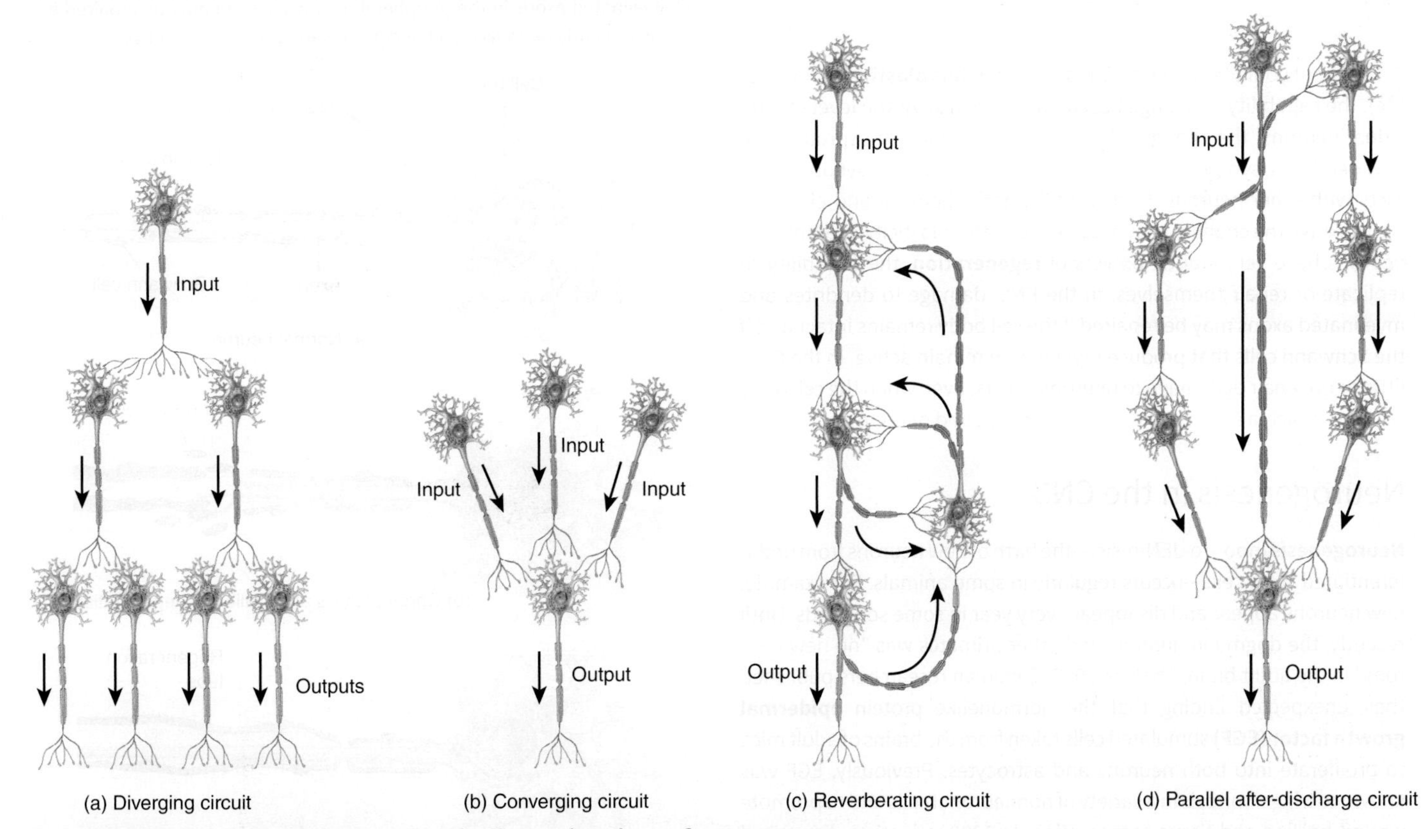

Q A motor neuron in the spinal cord typically receives input from neurons that originate in several different regions of the brain. Is this an example of convergence or divergence?

may last from a few seconds to many hours, depending on the number of synapses and the arrangement of neurons in the circuit. Inhibitory neurons may turn off a reverberating circuit after a period of time. Among the body responses thought to be the result of output signals from reverberating circuits are breathing, coordinated muscular activities, waking up, and short-term memory.

A fourth type of circuit is the **parallel after-discharge circuit** (Figure 12.28d). In this circuit, a single presynaptic cell stimulates a group of neurons, each of which synapses with a common postsynaptic cell. A differing number of synapses between the first and last neurons imposes varying synaptic delays, so that the last neuron exhibits multiple EPSPs or IPSPs. If the input is excitatory, the postsynaptic neuron then can send out a stream of impulses in quick succession. Parallel after-discharge circuits may be involved in precise activities such as mathematical calculations.

12.10 Regeneration and Repair of Nervous Tissue

OBJECTIVES

- **Define** plasticity and neurogenesis.
- **Describe** the events involved in damage and repair of peripheral nerves.

Throughout your life, your nervous system exhibits **plasticity** (plas-TIS-i-tē), the capability to change based on experience. At the level of individual neurons, the changes that can occur include the sprouting of new dendrites, synthesis of new proteins, and changes in synaptic contacts with other neurons. Undoubtedly, both chemical and electrical signals drive the changes that occur. Despite this plasticity, mammalian neurons have very limited powers of **regeneration**, the capability to replicate or repair themselves. In the PNS, damage to dendrites and myelinated axons may be repaired if the cell body remains intact and if the Schwann cells that produce myelination remain active. In the CNS, little or no repair of damage to neurons occurs. Even when the cell body remains intact, a severed axon cannot be repaired or regrown.

Neurogenesis in the CNS

Neurogenesis (noo′-rō-JEN-e-sis)—the birth of new neurons from undifferentiated stem cells—occurs regularly in some animals. For example, new neurons appear and disappear every year in some songbirds. Until recently, the dogma in humans and other primates was “no new neurons” in the adult brain. Then, in 1992, Canadian researchers published their unexpected finding that the hormonelike protein **epidermal growth factor (EGF)** stimulated cells taken from the brains of adult mice to proliferate into both neurons and astrocytes. Previously, EGF was known to trigger mitosis in a variety of nonneuronal cells and to promote wound healing and tissue regeneration. In 1998, scientists discovered that significant numbers of new neurons do arise in the adult human hippocampus, an area of the brain that is crucial for learning.

The nearly complete lack of neurogenesis in other regions of the brain and spinal cord seems to result from two factors: (1) inhibitory influences from neuroglia, particularly oligodendrocytes, and (2) absence of growth-stimulating cues that were present during fetal development. Axons in the CNS are myelinated by oligodendrocytes rather than Schwann cells, and this CNS myelin is one of the factors inhibiting regeneration of neurons. Perhaps this same mechanism stops axonal growth once a target region has been reached during development. Also, after axonal damage, nearby astrocytes proliferate rapidly, forming a type of scar tissue that acts as a physical barrier to regeneration. Thus, injury of the brain or spinal cord usually is permanent. Ongoing research seeks ways to improve the environment for existing spinal cord axons to bridge the injury gap. Scientists also are trying to find ways to stimulate dormant stem cells to replace neurons lost through damage or disease and to develop tissue-cultured neurons that can be used for transplantation purposes.

Damage and Repair in the PNS

Axons and dendrites that are associated with a neurolemma may undergo repair if the cell body is intact, if the Schwann cells are functional, and if scar tissue formation does not occur too rapidly (Figure 12.29). Most nerves in the PNS consist of processes that are covered

FIGURE 12.29 **Damage and repair of a neuron in the PNS.**

Myelinated axons in the peripheral nervous system may be repaired if the cell body remains intact and if Schwann cells remain active.

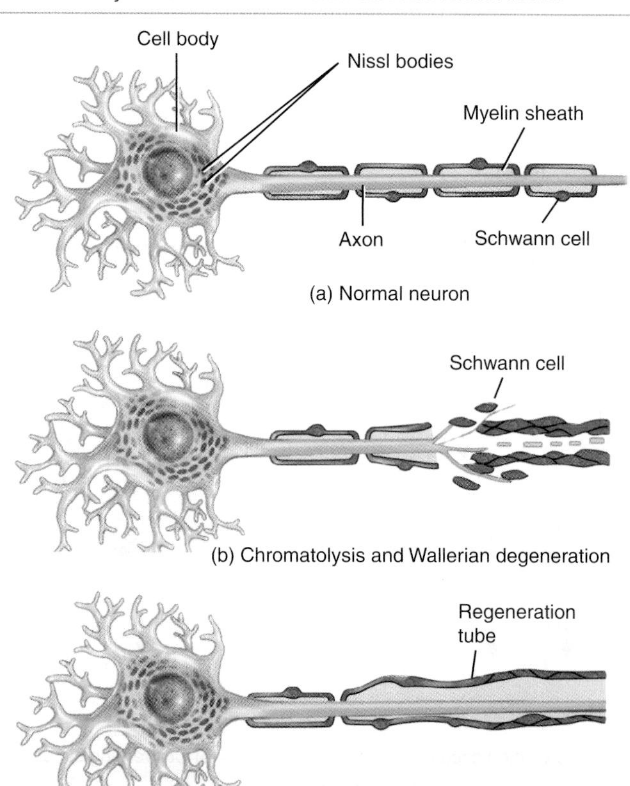

Q What is the role of the neurolemma in regeneration?

with a neurolemma. A person who injures axons of a nerve in an upper limb, for example, has a good chance of regaining nerve function.

When there is damage to an axon, changes usually occur both in the cell body of the affected neuron and in the portion of the axon distal to the site of injury. Changes also may occur in the portion of the axon proximal to the site of injury.

About 24 to 48 hours after injury to a process of a normal peripheral neuron (Figure 12.29a), the Nissl bodies break up into fine granular masses. This alteration is called **chromatolysis** (krō′-ma-TOL-i-sis; *chromato-* = color; *-lysis* = destruction). By the third to fifth day, the part of the axon distal to the damaged region becomes slightly swollen and then breaks up into fragments; the myelin sheath also deteriorates (Figure 12.29b). Even though the axon and myelin sheath degenerate, the neurolemma remains. Degeneration of the distal portion of the axon and myelin sheath is called **Wallerian degeneration** (waw-LE-rē′-an).

Following chromatolysis, signs of recovery in the cell body become evident. Macrophages phagocytize the debris. Synthesis of RNA and protein accelerates, which favors rebuilding or regeneration of the axon. The Schwann cells on either side of the injured site multiply by mitosis, grow toward each other, and may form a **regeneration tube** across the injured area (Figure 12.29c). The tube guides growth of a new axon from the proximal area across the injured area into the distal area previously occupied by the original axon. However, new axons cannot grow if the gap at the site of injury is too large or if the gap becomes filled with collagen fibers.

During the first few days following damage, buds of regenerating axons begin to invade the tube formed by the Schwann cells (Figure 12.29b). Axons from the proximal area grow at a rate of about 1.5 mm (0.06 in.) per day across the area of damage, find their way into the distal regeneration tubes, and grow toward the distally located receptors and effectors. Thus, some sensory and motor connections are reestablished and some functions are restored. In time, the Schwann cells form a new myelin sheath.

Review Questions

12.1 Overview of the Nervous system

1. What is the purpose of a sensory receptor?

2. What are the components and functions of the SNS and ANS?

3. Which subdivisions of the PNS control voluntary actions? Involuntary actions?

4. Explain the concept of integration and provide an example.

12.2 Histology of Nervous Tissue

5. Describe the parts of a neuron and the functions of each.

6. Give several examples of the structural and functional classifications of neurons.

7. What is a neurolemma, and why is it important?

8. With reference to the nervous system, what is a nucleus?

12.3 Electrical Signals in Neurons: An Overview

9. What types of electrical signals occur in neurons?

10. Why are voltage-gated channels important?

12.4 Resting Membrane Potential

11. What is the typical resting membrane potential of a neuron?

12. How do leak channels contribute to resting membrane potential?

12.5 Graded Potentials

13. What is a hyperpolarizing graded potential?

14. What is a depolarizing graded potential?

12.6 Action Potentials

15. What happens during the depolarizing phase of an action potential?

16. How is saltatory conduction different from continuous conduction?

17. What effect does myelination have on the speed of propagation of an action potential?

12.7 Signal Transmission at Synapses

18. How is neurotransmitter removed from the synaptic cleft?

19. How are excitatory and inhibitory postsynaptic potentials similar and different?

20. Why are action potentials said to be "all-or-none," and EPSPs and IPSPs are described as "graded"?

12.8 Neurotransmitters

21. Which neurotransmitters are excitatory, and which are inhibitory? How do they exert their effects?

22. In what ways is nitric oxide different from all previously known neurotransmitters?

12.9 Neural Circuits

23. What is a neural circuit?

24. What are the functions of diverging, converging, reverberating, and parallel after-discharge circuits?

12.10 Regeneration and Repair of Nervous Tissue

25. What factors contribute to a lack of neurogenesis in most parts of the brain?

26. What is the function of the regeneration tube in repair of neurons?

Critical Thinking Questions

1. The buzzing of the alarm clock woke Carrie. She stretched, yawned, and started to salivate as she smelled the brewing coffee. She could feel her stomach rumble. List the divisions of the nervous system that are involved in each of these actions.
2. Baby Ming is learning to crawl. He also likes to pull himself onto window sills, gnawing on the painted wood of his century-old home as he looks out the windows. Lately his mother, an anatomy and physiology student, has noticed some odd behavior and took Ming to the pediatrician. Blood work determined that Ming had a high level of lead in his blood, ingested from the old leaded paint on the window sill. The doctor indicated that lead poisoning is a type of demyelination disorder. Why should Ming's mother be concerned?
3. As a torture procedure for his enemies, mad scientist Dr. Moro is trying to develop a drug that will enhance the effects of substance P. What cellular mechanisms could he enlist to design such a drug?

The Spinal Cord and Spinal Nerves

CHAPTER 13

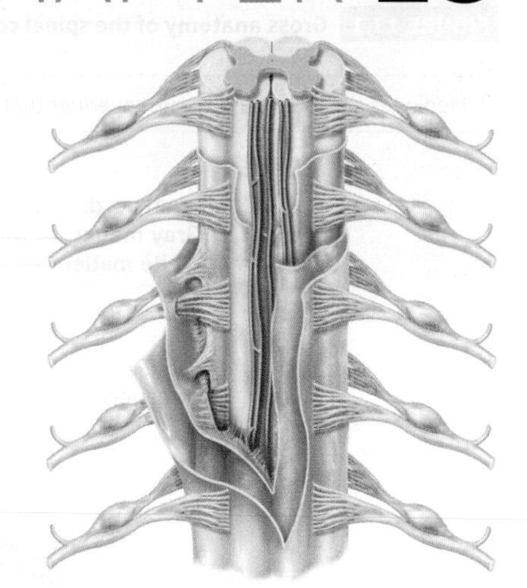

The Spinal Cord and Spinal Nerves and Homeostasis

The spinal cord and spinal nerves contribute to homeostasis by providing quick, reflexive responses to many stimuli. The spinal cord is the pathway for sensory input to the brain and motor output from the brain.

About 100 million neurons and even more neuroglia compose the spinal cord, the part of the central nervous system that extends from the brain. The spinal cord and its associated spinal nerves contain neural circuits that control some of your most rapid reactions to environmental changes. If you pick up something hot, the grasping muscles may relax and you may drop the hot object even before you are consciously aware of the extreme heat or pain. This is an example of a spinal cord reflex—a quick, automatic response to certain kinds of stimuli that involves neurons only in the spinal nerves and spinal cord. Besides processing reflexes, the gray matter of the spinal cord also is a site for integration (summing) of excitatory postsynaptic potentials (EPSPs) and inhibitory postsynaptic potentials (IPSPs), which you learned about in Chapter 12. These graded potentials arise as neurotransmitter molecules interact with their receptors at synapses in the spinal cord. The white matter of the spinal cord contains a dozen major sensory and motor tracts, which function as the "highways" along which sensory input travels to the brain and motor output travels from the brain to skeletal muscles and other effectors. Recall that the spinal cord is continuous with the brain and that together they make up the central nervous system (CNS).

Q Did you ever wonder why spinal cord injuries can have such widespread effects on the body?

13.1 Spinal Cord Anatomy

OBJECTIVES

- **Describe** the protective structures and the gross anatomical features of the spinal cord.
- **Explain** how spinal nerves are connected to the spinal cord.

Protective Structures

Recall from the previous chapter that the nervous tissue of the central nervous system is very delicate and does not respond well to injury or damage. Accordingly, nervous tissue requires considerable protection. The first layer of protection for the central nervous system is the hard bony skull and vertebral column. The skull encases the brain and the vertebral column surrounds the spinal cord, providing strong protective defenses against damaging blows or bumps. The second protective layer is the meninges, three membranes that lie between the bony encasement and the nervous tissue in both the brain and spinal cord. Finally, a space between two of the meningeal membranes contains cerebrospinal fluid, a buoyant liquid that suspends the central nervous tissue in a weightless environment while surrounding it with a shock-absorbing, hydraulic cushion.

Vertebral Column The spinal cord is located within the vertebral canal of the vertebral column. As you learned in Chapter 7, the vertebral foramina of all of the vertebrae, stacked one on top of the other, form the vertebral canal. The surrounding vertebrae provide a sturdy shelter for the enclosed spinal cord (see **Figure 13.1b**). The vertebral ligaments, meninges, and cerebrospinal fluid provide additional protection.

Meninges The **meninges** (me-NIN-jēz; singular is meninx [MĒ-ninks]) are three protective, connective tissue coverings that encircle the spinal cord and brain. From superficial to deep they are the (1) dura mater, (2) arachnoid mater, and (3) pia mater. The **spinal meninges** surround the spinal cord (**Figure 13.1a**) and are continuous with the **cranial meninges**, which encircle the brain (shown in **Figure 14.2a**). All three spinal meninges cover the spinal nerves up to the point where they exit the spinal column through the intervertebral foramina. The spinal cord is also protected by a cushion of fat and connective tissue located in the **epidural space** (ep′-i-DOO-ral), a space between the dura mater and the wall of the vertebral canal (**Figure 13.1b**). Following is a description of each meningeal layer.

1. **Dura mater** (DOO-ra MĀ-ter = tough mother). The most superficial of the three spinal meninges is a thick strong layer composed of

FIGURE 13.1 **Gross anatomy of the spinal cord.**

Meninges are connective tissue coverings that surround the spinal cord and brain.

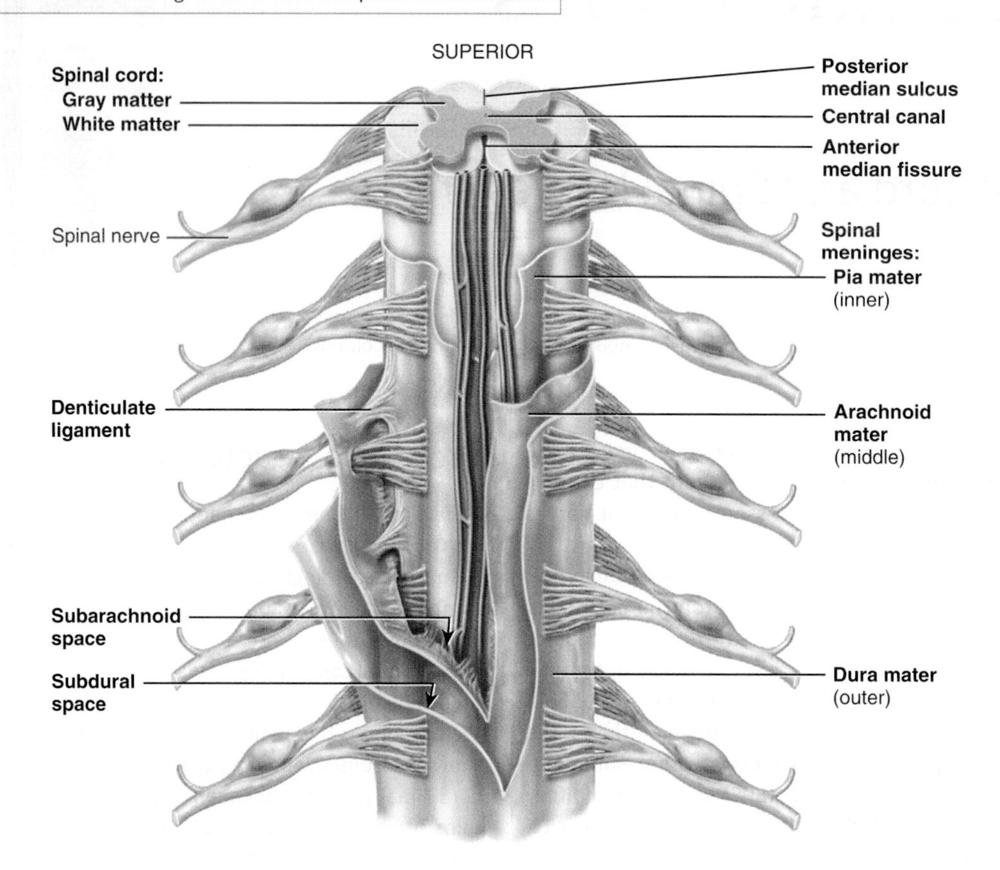

(a) Anterior view and transverse section through spinal cord

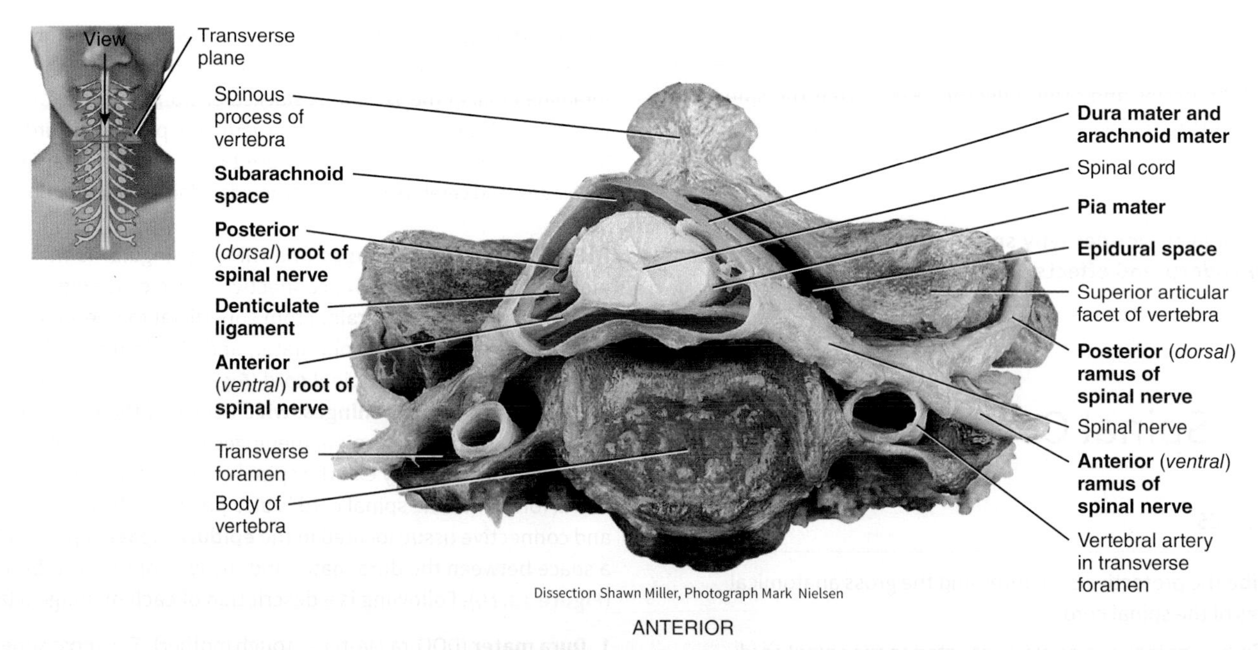

(b) Transverse section of the spinal cord within a cervical vertebra

Q What are the superior and inferior boundaries of the spinal dura mater?

dense irregular connective tissue. The dura mater forms a sac from the level of the foramen magnum in the occipital bone, where it is continuous with the meningeal dura mater of the brain, to the second sacral vertebra. The dura mater is also continuous with the epineurium, the outer covering of spinal and cranial nerves.

2. **Arachnoid mater** (a-RAK-noyd MĀ-ter; *arachn-* = spider; *-oid* = similar to). This layer, the middle of the meningeal membranes, is a thin, avascular covering comprised of cells and thin, loosely arranged collagen and elastic fibers. It is called the arachnoid mater because of its spider's web arrangement of delicate collagen fibers and some elastic fibers. It is deep to the dura mater and is continuous through the foramen magnum with the arachnoid mater of the brain. Between the dura mater and the arachnoid mater is a thin **subdural space**, which contains interstitial fluid.
3. **Pia mater** (PĒ-a MĀ-ter; *pia* = delicate). This innermost meninx is a thin transparent connective tissue layer that adheres to the surface of the spinal cord and brain. It consists of thin squamous to cuboidal cells within interlacing bundles of collagen fibers and some fine elastic fibers. Within the pia mater are many blood vessels that supply oxygen and nutrients to the spinal cord. Triangular-shaped membranous extensions of the pia mater suspend the spinal cord in the middle of its dural sheath. These extensions, called **denticulate ligaments** (den-TIK-ū-lāt = small tooth), are thickenings of the pia mater. They project laterally and fuse with the arachnoid mater and inner surface of the dura mater between the anterior and posterior nerve roots of spinal nerves on either side (Figure 13.1a, b). Extending along the entire length of the spinal cord, the denticulate ligaments protect the spinal cord against sudden displacement that could result in shock. Between the arachnoid mater and pia mater is a space, the **subarachnoid space**, which contains shock-absorbing cerebrospinal fluid.

See Clinical Connection: Spinal Tap

External Anatomy of the Spinal Cord

The **spinal cord** is roughly oval in shape, being flattened slightly anteriorly and posteriorly. In adults, it extends from the medulla oblongata, the inferior part of the brain, to the superior border of the second lumbar vertebra (Figure 13.2). In newborn infants, it extends to the third or fourth lumbar vertebra. During early childhood, both the spinal cord and the vertebral column grow longer as part of overall body growth. Elongation of the spinal cord stops around age 4 or 5, but growth of the vertebral column continues. Thus, the spinal cord does not extend the entire length of the adult vertebral column. The length of the adult spinal cord ranges from 42 to 45 cm (16–18 in.). Its maximum diameter is approximately 1.5 cm (0.6 in.) in the lower cervical region and is smaller in the thoracic region and at its inferior tip.

When the spinal cord is viewed externally, two conspicuous enlargements can be seen. The superior enlargement, the **cervical enlargement**, extends from the fourth cervical vertebra (C4) to the first thoracic vertebra (T1). Nerves to and from the upper limbs arise from the cervical enlargement. The inferior enlargement, called the **lumbar enlargement**, extends from the ninth to the twelfth thoracic vertebra. Nerves to and from the lower limbs arise from the lumbar enlargement.

Inferior to the lumbar enlargement, the spinal cord terminates as a tapering, conical structure called the **conus medullaris** (KŌ-nus med-ū-LAR-is; *conus* = cone), which ends at the level of the intervertebral disc between the first and second lumbar vertebrae (L1–L2) in adults. Arising from the conus medullaris is the **filum terminale** (FĪ-lum ter-mi-NAL-ē = terminal filament), an extension of the pia mater that extends inferiorly, fuses with the arachnoid mater, and dura mater, and anchors the spinal cord to the coccyx.

Spinal nerves are the paths of communication between the spinal cord and specific regions of the body. The spinal cord appears to be segmented because the 31 pairs of spinal nerves emerge at regular intervals from intervertebral foramina (Figure 13.2). Indeed, each pair of spinal nerves is said to arise from a *spinal segment*. Within the spinal cord there is no obvious segmentation but, for convenience, the naming of spinal nerves is based on the segment in which they are located. There are 8 pairs of *cervical nerves* (represented in Figure 13.2 as C1–C8), 12 pairs of *thoracic nerves* (T1–T12), 5 pairs of *lumbar nerves* (L1–L5), 5 pairs of *sacral nerves* (S1–S5), and 1 pair of *coccygeal nerves* (Co1).

Two bundles of axons, called **roots**, connect each spinal nerve to a segment of the cord by even smaller bundles of axons called **rootlets** (see Figure 13.3a). The **posterior** (*dorsal*) **root** and rootlets contain only sensory axons, which conduct nerve impulses from sensory receptors in the skin, muscles, and internal organs into the central nervous system. Each posterior root has a swelling, the **posterior** (*dorsal*) **root ganglion**, which contains the cell bodies of sensory neurons. The **anterior** (*ventral*) **root** and rootlets contain axons of motor neurons, which conduct nerve impulses from the CNS to effectors (muscles and glands).

As spinal nerves branch from the spinal cord, they pass laterally to exit the vertebral canal through the intervertebral foramina between adjacent vertebrae. However, because the spinal cord is shorter than the vertebral column, nerves that arise from the lumbar, sacral, and coccygeal regions of the spinal cord do not leave the vertebral column at the same level they exit the cord. The roots of these lower spinal nerves angle inferiorly alongside the filum terminale in the vertebral canal like wisps of hair. Accordingly, the roots of these nerves are collectively named the **cauda equina** (KAW-da ē-KWĪ-na), meaning "horse's tail" (Figure 13.2).

Internal Anatomy of the Spinal Cord

A transverse section of the spinal cord reveals regions of white matter that surround an inner core of gray matter (Figure 13.3). The white matter of the spinal cord consists primarily of bundles of myelinated axons of neurons. Two grooves penetrate the white matter of the spinal cord and divide it into right and left sides. The **anterior median fissure** is a wide groove on the anterior (ventral) side. The **posterior median sulcus** is a narrow furrow on the posterior (dorsal) side. The gray matter of the spinal cord is shaped like the letter H or a butterfly; it consists of dendrites and cell bodies of neurons, unmyelinated axons, and

FIGURE 13.2 **External anatomy of the spinal cord and spinal nerves.**

The spinal cord extends from the medulla oblongata of the brain to the superior border of the second lumbar vertebra.

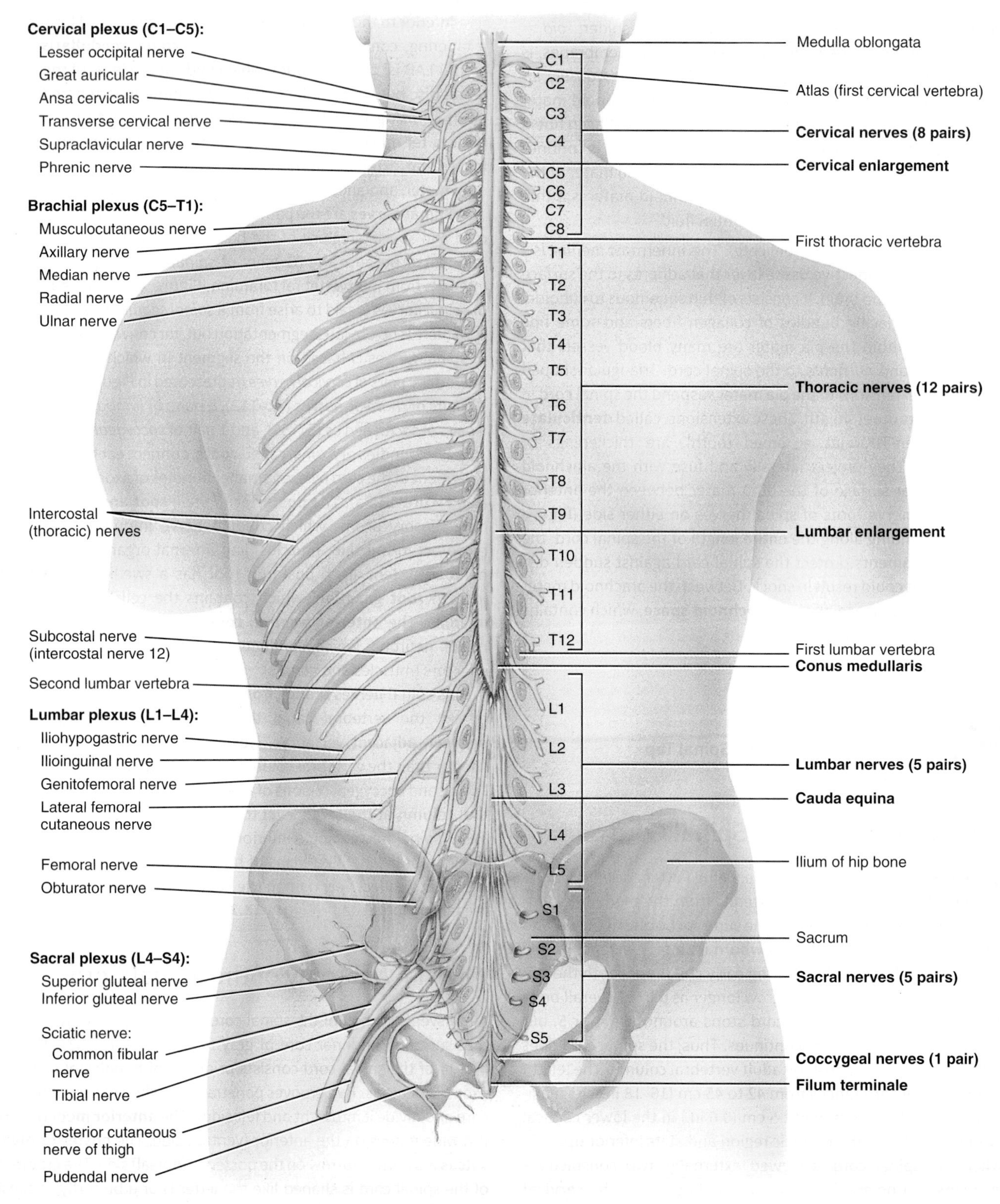

(a) Posterior view of entire spinal cord and portions of spinal nerves

Q What portion of the spinal cord connects with nerves of the upper limbs?

FIGURE 13.3 Internal anatomy of the spinal cord: the organization of gray matter and white matter. For simplicity, dendrites are not shown in this and several other illustrations of transverse sections of the spinal cord. Blue, red, and green arrows indicate the direction of nerve impulse propagation.

The posterior gray horn contains axons of sensory neurons and cell bodies of interneurons; the lateral gray horn contains cell bodies of autonomic motor neurons; and the anterior gray horn contains cell bodies of somatic motor neurons.

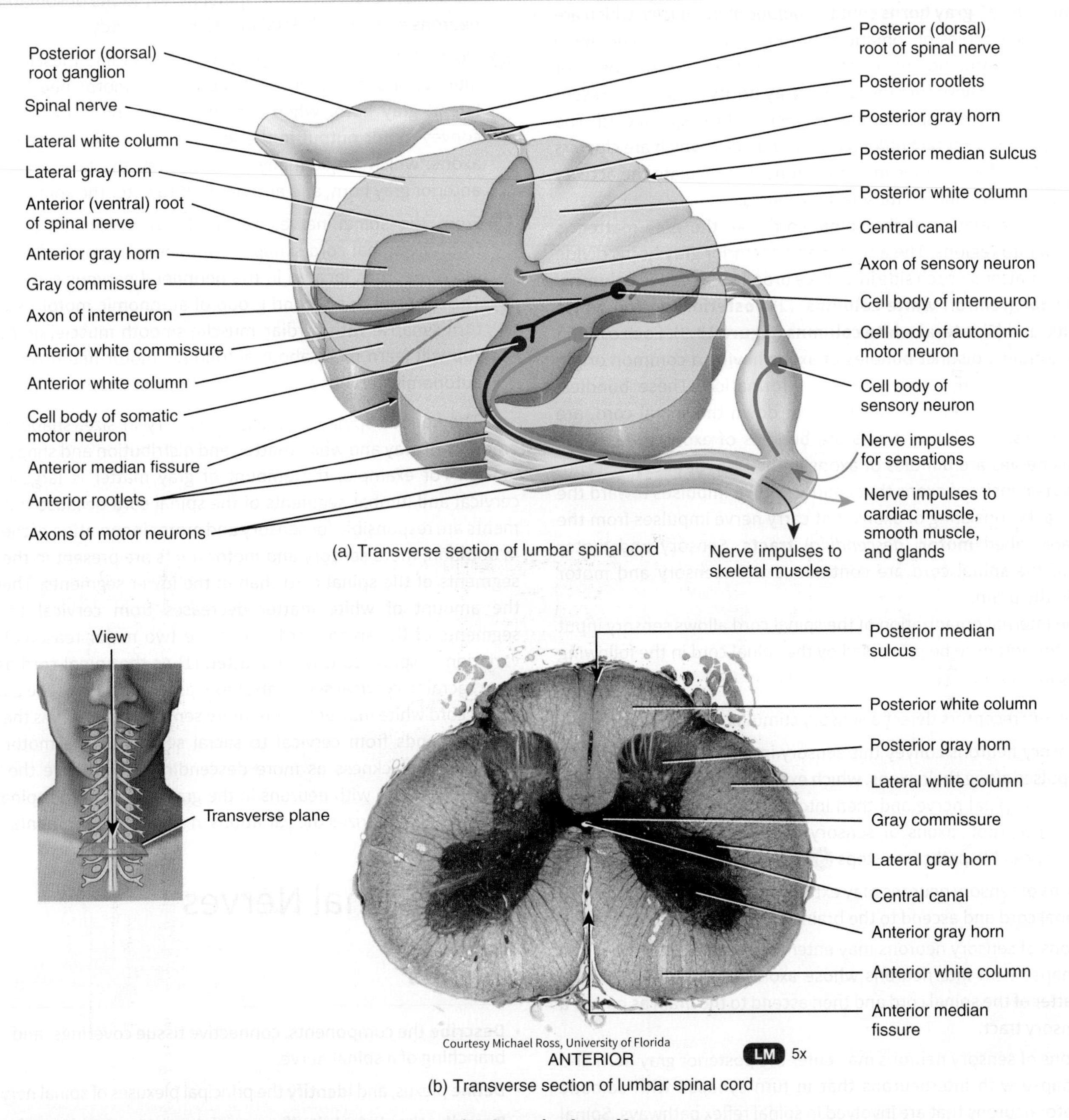

(a) Transverse section of lumbar spinal cord

Courtesy Michael Ross, University of Florida

(b) Transverse section of lumbar spinal cord

Q What is the difference between a horn and a column in the spinal cord?

neuroglia. The **gray commissure** (KOM-mi-shur) forms the crossbar of the H. In the center of the gray commissure is a small space called the **central canal**; it extends the entire length of the spinal cord and is filled with cerebrospinal fluid. At its superior end, the central canal is continuous with the fourth ventricle (a space that contains cerebrospinal fluid) in the medulla oblongata of the brain. Anterior to the gray commissure is the **anterior** (*ventral*) **white commissure**, which connects the white matter of the right and left sides of the spinal cord.

In the gray matter of the spinal cord and brain, clusters of neuronal cell bodies form functional groups called **nuclei**. *Sensory* *nuclei* receive input from receptors via sensory neurons, and *motor nuclei* provide output to effector tissues via motor neurons. The gray matter on each side of the spinal cord is subdivided into regions called **horns** (Figure 13.3). The **posterior** (*dorsal*) **gray horns** contain axons of incoming sensory neurons as well as cell bodies and axons of interneurons. Recall that cell bodies of sensory neurons are located in the posterior (dorsal) root ganglion of a spinal nerve. The **anterior** (*ventral*) **gray horns** contain *somatic motor nuclei*, which are clusters of cell bodies of somatic motor neurons that provide nerve impulses for contraction of skeletal muscles. Between the posterior and anterior gray horns are the **lateral gray horns**, which are present only in thoracic and upper lumbar segments of the spinal cord. The lateral gray horns contain *autonomic motor nuclei,* which are clusters of cell bodies of autonomic motor neurons that regulate the activity of cardiac muscle, smooth muscle, and glands.

The white matter of the spinal cord, like the gray matter, is organized into regions. The anterior and posterior gray horns divide the white matter on each side into three broad areas called **columns**: (1) **anterior** (*ventral*) **white columns**, (2) **posterior** (*dorsal*) **white columns**, and (3) **lateral white columns** (Figure 13.3). Each column in turn contains distinct bundles of axons having a common origin or destination and carrying similar information. These bundles, which may extend long distances up or down the spinal cord, are called **tracts**. Recall that tracts are bundles of axons in the CNS, whereas nerves are bundles of axons in the PNS. **Sensory** (*ascending*) **tracts** consist of axons that conduct nerve impulses toward the brain. Tracts consisting of axons that carry nerve impulses from the brain are called **motor** (*descending*) **tracts**. Sensory and motor tracts of the spinal cord are continuous with sensory and motor tracts in the brain.

The internal organization of the spinal cord allows sensory input and motor output to be processed by the spinal cord in the following way (Figure 13.4):

1. Sensory receptors detect a sensory stimulus.
2. Sensory neurons convey this sensory input in the form of nerve impulses along their axons, which extend from sensory receptors into the spinal nerve and then into the posterior root. From the posterior root, axons of sensory neurons may proceed along three possible paths (see steps 3, 4, and 5).
3. Axons of sensory neurons may extend into the white matter of the spinal cord and ascend to the brain as part of a sensory tract.
4. Axons of sensory neurons may enter the posterior gray horn and synapse with interneurons whose axons extend into the white matter of the spinal cord and then ascend to the brain as part of a sensory tract.
5. Axons of sensory neurons may enter the posterior gray horn and synapse with interneurons that in turn synapse with somatic motor neurons that are involved in spinal reflex pathways. Spinal cord reflexes are described in more detail later in this chapter.
6. Motor output from the spinal cord to skeletal muscles involves somatic motor neurons of the anterior gray horn. Many somatic motor neurons are regulated by the brain. Axons from higher brain centers form motor tracts that descend from the brain into the white matter of the spinal cord. There they synapse with somatic motor neurons either directly or indirectly by first synapsing with interneurons that in turn synapse with somatic motor neurons.
7. When activated, somatic motor neurons convey motor output in the form of nerve impulses along their axons, which sequentially pass through the anterior gray horn and anterior root to enter the spinal nerve. From the spinal nerve, axons of somatic motor neurons extend to skeletal muscles of the body.
8. Motor output from the spinal cord to cardiac muscle, smooth muscle, and glands involves autonomic motor neurons of the lateral gray horn. When activated, autonomic motor neurons convey motor output in the form of nerve impulses along their axons, which sequentially pass through the lateral gray horn, anterior gray horn, and anterior root to enter the spinal nerve.
9. From the spinal nerve, axons of autonomic motor neurons from the spinal cord synapse with another group of autonomic motor neurons located in the peripheral nervous system (PNS). The axons of this second group of autonomic motor neurons in turn synapse with cardiac muscle, smooth muscle, and glands. You will learn more about autonomic motor neurons when the autonomic nervous system is described in Chapter 15.

The various spinal cord segments vary in size, shape, relative amounts of gray and white matter, and distribution and shape of gray matter. For example, the amount of gray matter is largest in the cervical and lumbar segments of the spinal cord because these segments are responsible for sensory and motor innervation of the limbs. In addition, more sensory and motor tracts are present in the upper segments of the spinal cord than in the lower segments. Therefore, the amount of white matter decreases from cervical to sacral segments of the spinal cord. There are two major reasons for this variation in spinal cord white matter: (1) As the spinal cord ascends from sacral to cervical segments, more ascending axons are added to spinal cord white matter to form more sensory tracts. (2) As the spinal cord descends from cervical to sacral segments, the motor tracts decrease in thickness as more descending axons leave the motor tracts to synapse with neurons in the gray matter of the spinal cord. Table 13.1 summarizes the variations in spinal cord segments.

13.2 Spinal Nerves

OBJECTIVES

- **Describe** the components, connective tissue coverings, and branching of a spinal nerve.
- **Define** plexus, and **identify** the principal plexuses of spinal nerves.
- **Describe** the clinical significance of dermatomes.

Spinal nerves are associated with the spinal cord and, like all nerves of the peripheral nervous system (PNS), are parallel bundles of axons and their associated neuroglial cells wrapped in several layers of connective tissue. Spinal nerves connect the CNS to sensory receptors,

FIGURE 13.4 Processing of sensory input and motor output by the spinal cord.

Sensory input is conveyed from sensory receptors to the posterior gray horns of the spinal cord, and motor output is conveyed from the anterior and lateral gray horns of the spinal cord to effectors (muscles and glands).

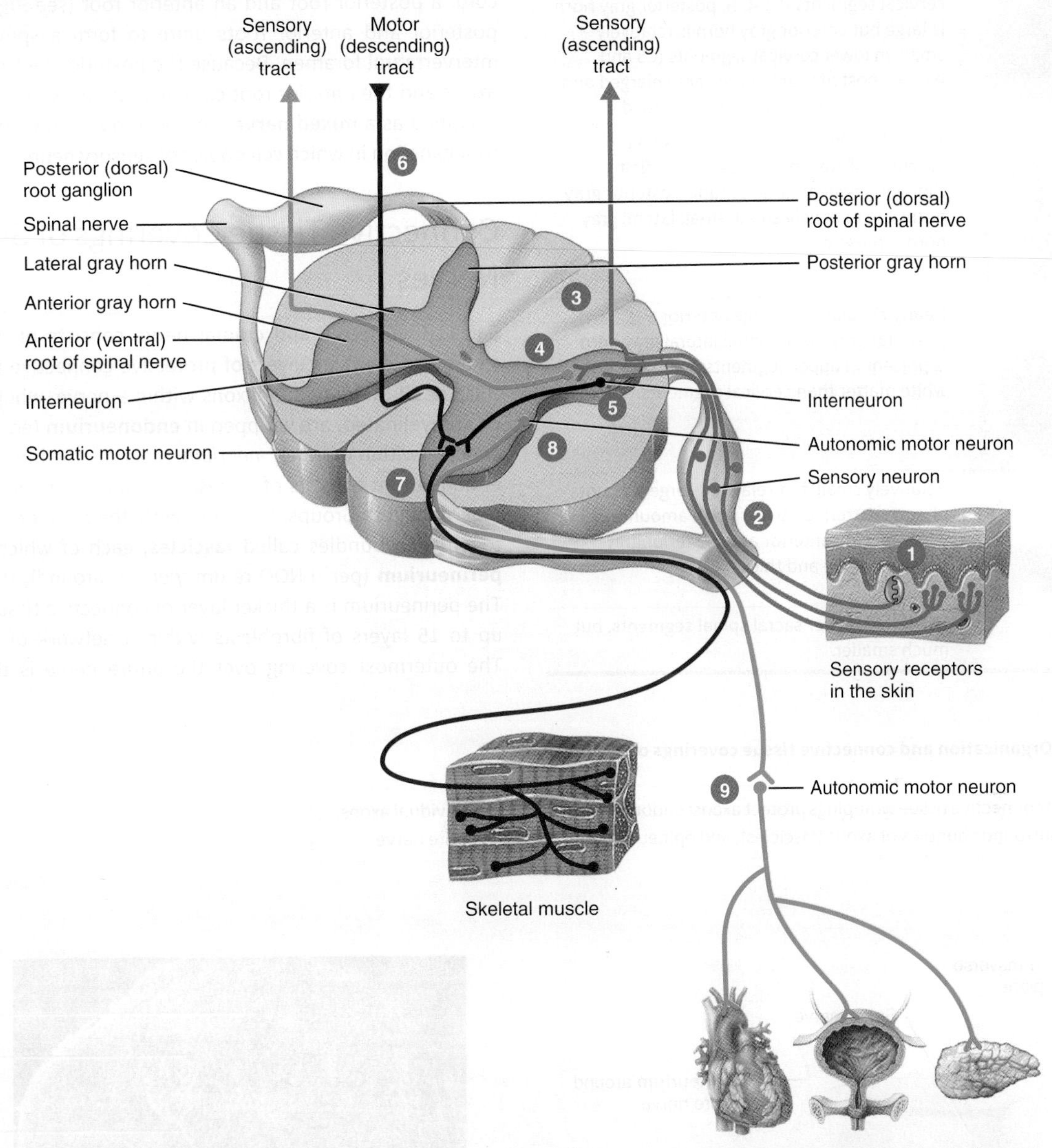

Q Lateral gray horns are found in which segments of the spinal cord?

muscles, and glands in all parts of the body. The 31 pairs of spinal nerves are named and numbered according to the region and level of the vertebral column from which they emerge (see **Figure 13.2**). Not all spinal cord segments are aligned with their corresponding vertebrae. Recall that the spinal cord ends near the level of the superior border of the second lumbar vertebra (L2), and that the roots of the lumbar, sacral, and coccygeal nerves descend at an angle to reach their respective foramina before emerging from the vertebral column. This arrangement constitutes the cauda equina.

The first cervical pair of spinal nerves emerges from the spinal cord between the occipital bone and the atlas (first cervical vertebra, or C1). Most of the remaining spinal nerves emerge from the spinal cord through the intervertebral foramina between adjoining vertebrae. Spinal nerves C1–C7 exit the vertebral canal *above* their corresponding vertebrae. Spinal nerve C8 exits the vertebral canal between vertebrae C7 and T1. Spinal nerves T1–L5 exit the vertebral canal *below* their corresponding vertebrae. From the spinal cord, the roots of the sacral spinal nerves (S1–S5) and the coccygeal spinal nerves (Co1) enter the

TABLE 13.1 Comparison of Various Spinal Cord Segments

SEGMENT	DISTINGUISHING CHARACTERISTICS
Cervical (Mark Nielsen)	Relatively large diameter, relatively large amounts of white matter, oval; in upper cervical segments (C1–C4), posterior gray horn is large but anterior gray horn is relatively small; in lower cervical segments (C5 and below), posterior gray horns are enlarged and anterior gray horns are well developed.
Thoracic (Mark Nielsen)	Small diameter due to relatively small amounts of gray matter; except for first thoracic segment, anterior and posterior gray horns are relatively small; small lateral gray horn is present.
Lumbar (Mark Nielsen)	Nearly circular; very large anterior and posterior gray horns; small lateral gray horn is present in upper segments; relatively less white matter than cervical segments.
Sacral (Mark Nielsen)	Relatively small, but relatively large amounts of gray matter; relatively small amounts of white matter; anterior and posterior gray horns are large and thick.
Coccygeal	Resembles lower sacral spinal segments, but much smaller.

sacral canal, the part of the vertebral canal in the sacrum (see **Figure 7.21**). Subsequently, spinal nerves S1–S4 exit the sacral canal via the four pairs of anterior and posterior sacral foramina, and spinal nerves S5 and Co1 exit the sacral canal via the sacral hiatus.

As noted earlier, a typical spinal nerve has two connections to the cord: a posterior root and an anterior root (see **Figure 13.3a**). The posterior and anterior roots unite to form a spinal nerve at the intervertebral foramen. Because the posterior root contains sensory axons and the anterior root contains motor axons, a spinal nerve is classified as a **mixed nerve**. The posterior root contains a posterior root ganglion in which cell bodies of sensory neurons are located.

Connective Tissue Coverings of Spinal Nerves

Each spinal nerve and cranial nerve consists of many individual axons and contains layers of protective connective tissue coverings (**Figure 13.5**). Individual axons within a nerve, whether myelinated or unmyelinated, are wrapped in **endoneurium** (en′-dō-NOO-rē-um; *endo-* = within or inner; *-neurium* = nerve), the innermost layer. The endoneurium consists of a mesh of collagen fibers, fibroblasts, and macrophages. Groups of axons with their endoneurium are held together in bundles called **fascicles**, each of which is wrapped in **perineurium** (per′-i-NOO-rē-um; *peri-* = around), the middle layer. The perineurium is a thicker layer of connective tissue. It consists of up to 15 layers of fibroblasts within a network of collagen fibers. The outermost covering over the entire nerve is the **epineurium**

FIGURE 13.5 Organization and connective tissue coverings of a spinal nerve.

Three layers of connective tissue wrappings protect axons: Endoneurium surrounds individual axons, perineurium surrounds bundles of axons (fascicles), and epineurium surrounds an entire nerve.

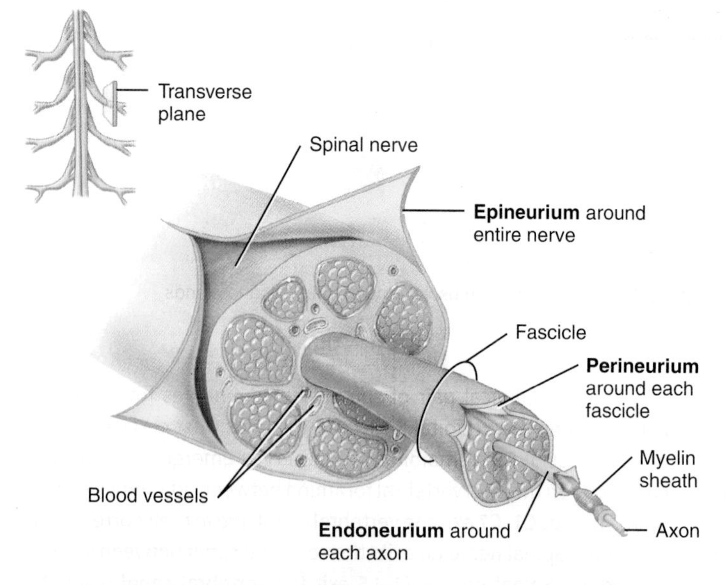

(a) Transverse section showing the coverings of a spinal nerve

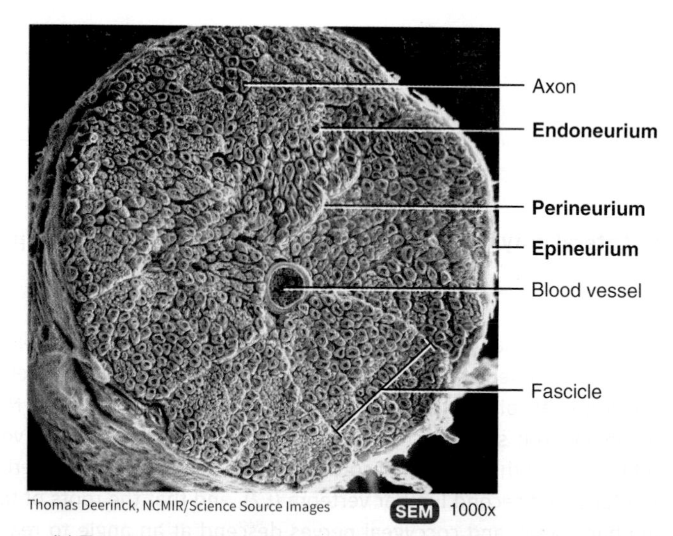

Thomas Deerinck, NCMIR/Science Source Images

(b) Transverse section of several nerve fascicles

Q Why are all spinal nerves classified as mixed nerves?

(ep′-i-NOO-rē-um; *epi-* = over). It consists of fibroblasts and thick collagen fibers. Extensions of the epineurium also fill the spaces between fascicles. The dura mater of the spinal meninges fuses with the epineurium as the nerve passes through the intervertebral foramen. Note the presence of blood vessels, which nourish the spinal meninges (Figure 13.5b). You may recall from Chapter 10 that the connective tissue coverings of skeletal muscles—endomysium, perimysium, and epimysium—are similar in organization to those of nerves.

Distribution of Spinal Nerves

Branches A short distance after passing through its intervertebral foramen, a spinal nerve divides into several branches (Figure 13.6). These branches are known as **rami** (RĀ-mī = branches). The **posterior** (*dorsal*) **ramus** (RĀ-mus; singular form) serves the deep muscles and skin of the posterior surface of the trunk. The **anterior** (*ventral*) **ramus** serves the muscles and structures of the upper and lower limbs and the skin of the lateral and anterior surfaces of the trunk. In addition to posterior and anterior rami, spinal nerves also give off a **meningeal branch** (me-NIN-jē′-al). This branch reenters the vertebral cavity through the intervertebral foramen and supplies the vertebrae, vertebral ligaments, blood vessels of the spinal cord, and meninges. Other branches of a spinal nerve are the **rami communicantes** (kō-mū-ni-KAN-tēz), components of the autonomic nervous system that will be discussed in Chapter 15.

Plexuses Axons from the anterior rami of spinal nerves, except for thoracic nerves T2–T12, do not go directly to the body structures they supply. Instead, they form networks on both the left and right sides of the body by joining with various numbers of axons from anterior rami of adjacent nerves. Such a network of axons is called a **plexus** (PLEK-sus = braid or network). The principal plexuses are the **cervical plexus, brachial plexus, lumbar plexus,** and **sacral plexus.** A smaller **coccygeal plexus** is also present. Refer to Figure 13.2 to see

FIGURE 13.6 Branches of a typical spinal nerve, shown in transverse section through the thoracic portion of the spinal cord. (See also Figure 13.1b.)

The branches of a spinal nerve are the posterior ramus, the anterior ramus, the meningeal branch, and the rami communicantes.

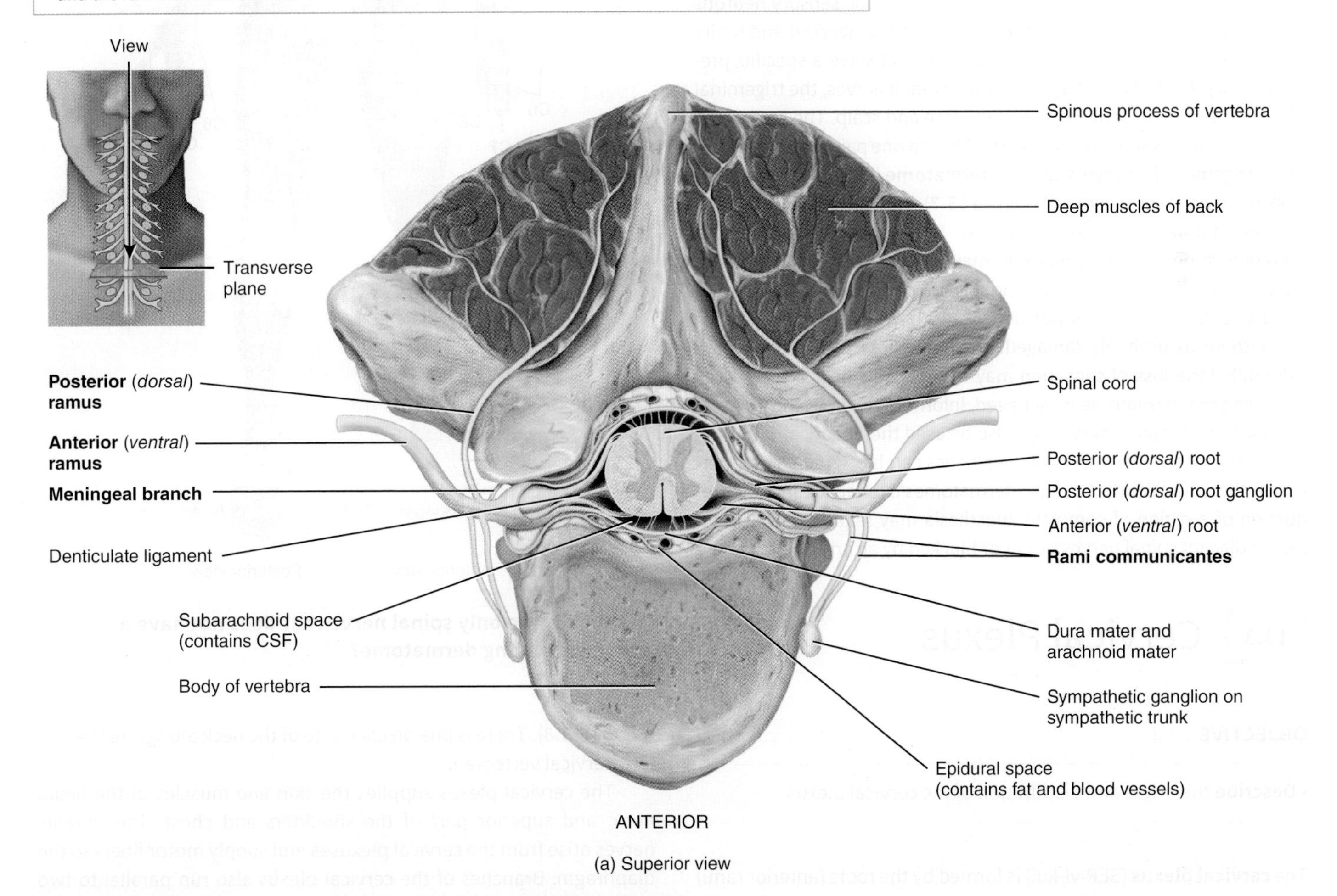

(a) Superior view

Q Which spinal nerve branches serve the upper and lower limbs?

their relationships to one another. Emerging from the plexuses are nerves bearing names that are often descriptive of the general regions they serve or the course they take. Each of the nerves in turn may have several branches named for the specific structures they innervate.

Sections 13.3–13.6 summarize the principal plexuses. The anterior rami of spinal nerves T2–T12 are called intercostal nerves and will be discussed next.

Intercostal Nerves The anterior rami of spinal nerves T2–T12 do not enter into the formation of plexuses and are known as **intercostal nerves** or *thoracic nerves*. These nerves directly connect to the structures they supply in the intercostal spaces. After leaving its intervertebral foramen, the anterior ramus of nerve T2 innervates the intercostal muscles of the second intercostal space and supplies the skin of the axilla and posteromedial aspect of the arm. Nerves T3–T6 extend along the costal grooves of the ribs and then to the intercostal muscles and skin of the anterior and lateral chest wall. Nerves T7–T12 supply the intercostal muscles and abdominal muscles, along with the overlying skin. The posterior rami of the intercostal nerves supply the deep back muscles and skin of the posterior aspect of the thorax.

Dermatomes

The skin over the entire body is supplied by somatic sensory neurons that carry nerve impulses from the skin into the spinal cord and brain. Each spinal nerve contains sensory neurons that serve a specific, predictable segment of the body. One of the cranial nerves, the trigeminal (V) nerve, serves most of the skin of the face and scalp. The area of the skin that provides sensory input to the CNS via one pair of spinal nerves or the trigeminal (V) nerve is called a **dermatome** (DER-ma-tōm; *derma-* = skin; *-tome* = thin segment) (Figure 13.7). The nerve supply in adjacent dermatomes overlaps somewhat. Knowing which spinal cord segments supply each dermatome makes it possible to locate damaged regions of the spinal cord. If the skin in a particular region is stimulated but the sensation is not perceived, the nerves supplying that dermatome are probably damaged. In regions where the overlap is considerable, little loss of sensation may result if only one of the nerves supplying the dermatome is damaged. Information about the innervation patterns of spinal nerves can also be used therapeutically. Cutting posterior roots or infusing local anesthetics can block pain either permanently or transiently. Because dermatomes overlap, deliberate production of a region of complete anesthesia may require that at least three adjacent spinal nerves be cut or blocked by an anesthetic drug.

FIGURE 13.7 **Distribution of dermatomes.**

A dermatome is an area of skin that provides sensory input to the CNS via the posterior roots of one pair of spinal nerves or via the trigeminal (V) nerve.

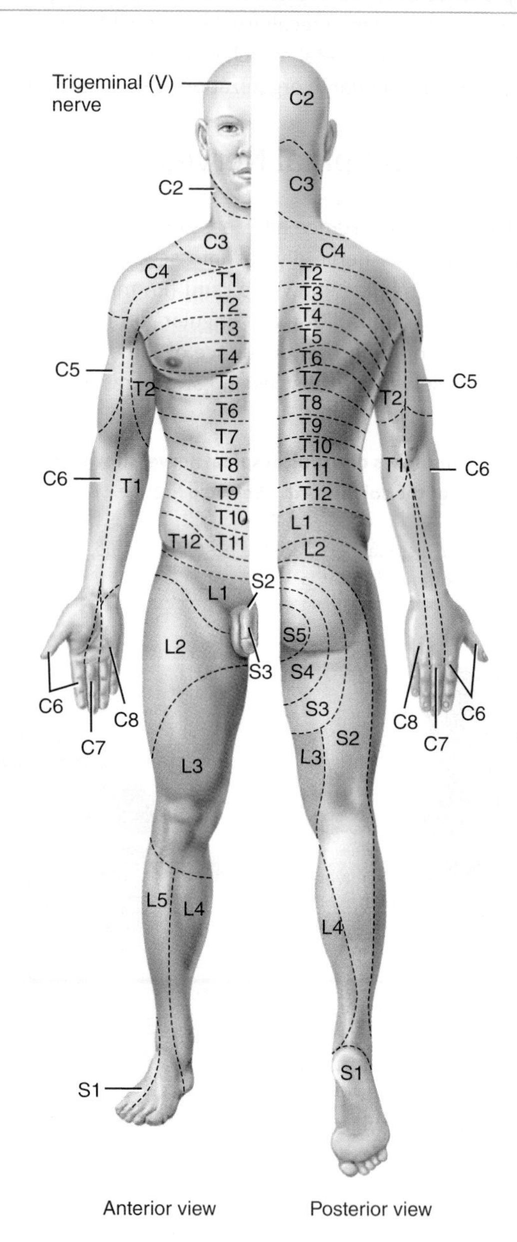

Q Which is the only spinal nerve that does not have a corresponding dermatome?

13.3 Cervical Plexus

OBJECTIVE

- **Describe** the origin and distribution of the cervical plexus.

The **cervical plexus** (SER-vi-kul) is formed by the roots (anterior rami) of the first four cervical nerves (C1–C4), with contributions from C5 (Figure 13.8). There is one on each side of the neck alongside the first four cervical vertebrae.

The cervical plexus supplies the skin and muscles of the head, neck, and superior part of the shoulders and chest. The phrenic nerves arise from the cervical plexuses and supply motor fibers to the diaphragm. Branches of the cervical plexus also run parallel to two cranial nerves, the accessory (XI) nerve and hypoglossal (XII) nerve.

NERVE	ORIGIN	DISTRIBUTION
SUPERFICIAL (SENSORY) BRANCHES		
Lesser occipital	C2	Skin of scalp posterior and superior to ear.
Great auricular (aw-RIK-ū-lar)	C2–C3	Skin anterior, inferior, and over ear, and over parotid glands.
Transverse cervical	C2–C3	Skin over anterior and lateral aspect of neck.
Supraclavicular	C3–C4	Skin over superior portion of chest and shoulder.
DEEP (LARGELY MOTOR) BRANCHES		
Ansa cervicalis (AN-sa ser-vi-KAL-is)		Divides into superior and inferior roots.
Superior root	C1	Infrahyoid and geniohyoid muscles of neck.
Inferior root	C2–C3	Infrahyoid muscles of neck.
Phrenic (FREN-ik)	C3–C5	Diaphragm.
Segmental branches	C1–C5	Prevertebral (deep) muscles of neck, levator scapulae, and middle scalene muscles.

FIGURE 13.8 Cervical plexus in anterior view.

The cervical plexus supplies the skin and muscles of the head, neck, superior portion of the shoulders and chest, and diaphragm.

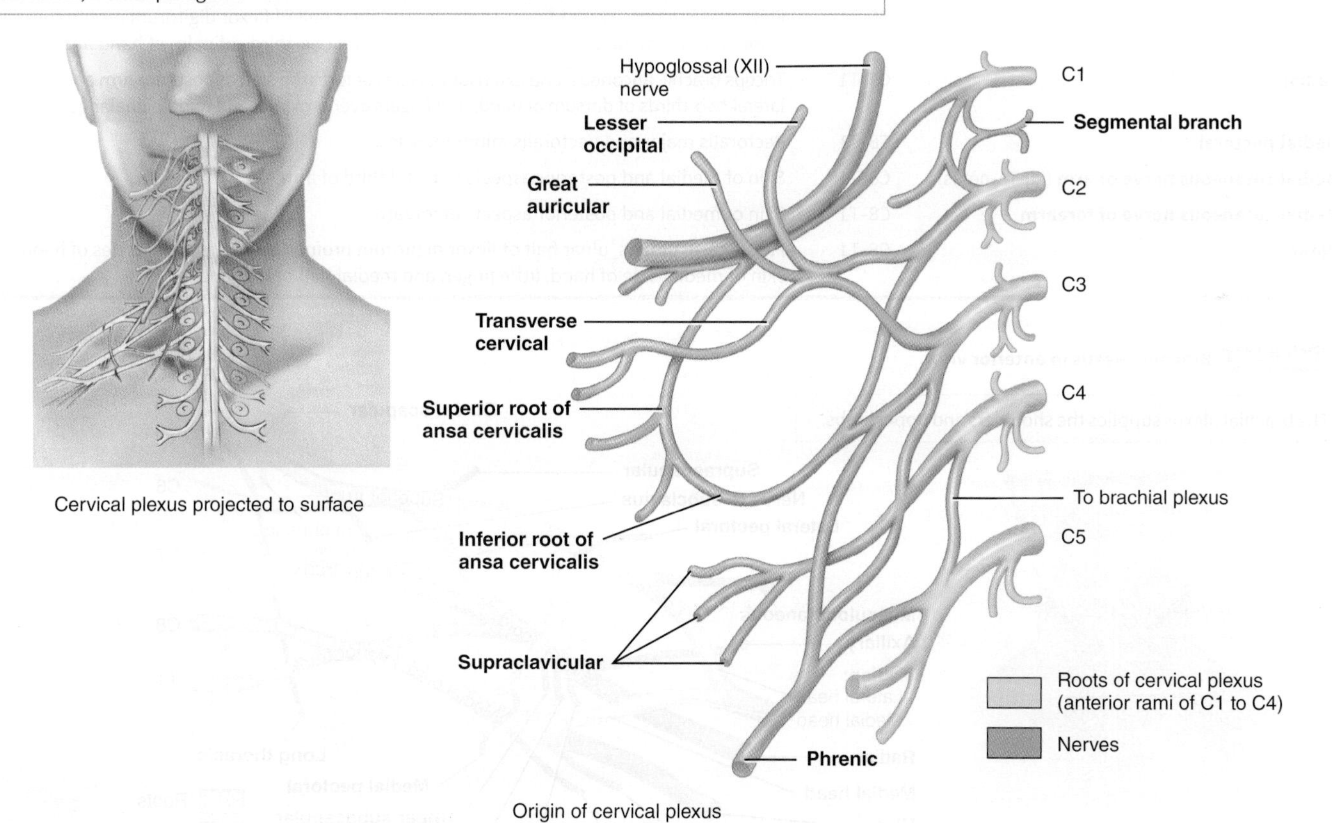

Q Why does complete severing of the spinal cord at level C2 cause respiratory arrest?

See Clinical Connection: Injuries to the Phrenic Nerves

13.4 Brachial Plexus

OBJECTIVE

- **Describe** the origin, distribution, and effects of damage to the brachial plexus.

The roots (anterior rami) of spinal nerves C5–C8 and T1 form the **brachial plexus** (BRĀ-kē-al), which extends inferiorly and laterally on either side of the last four cervical and first thoracic vertebrae (Figure 13.9a). It passes above the first rib posterior to the clavicle and then enters the axilla.

Since the brachial plexus is so complex, an explanation of its various parts is helpful. As with the cervical and other plexuses, the **roots** are the anterior rami of the spinal nerves. The roots of several spinal

NERVE	ORIGIN	DISTRIBUTION
Dorsal scapular (SKAP-ū-lar)	C5	Levator scapulae, rhomboid major, and rhomboid minor muscles.
Long thoracic (thō-RAS-ik)	C5–C7	Serratus anterior muscle.
Nerve to subclavius (sub-KLĀ-vē-us)	C5–C6	Subclavius muscle.
Suprascapular	C5–C6	Supraspinatus and infraspinatus muscles.
Musculocutaneous (mus′-kū-lō-kū-TĀN-ē-us)	C5–C7	Coracobrachialis, biceps brachii, and brachialis muscles.
Lateral pectoral (PEK-tō-ral)	C5–C7	Pectoralis major muscle.
Upper subscapular	C5–C6	Subscapularis muscle.
Thoracodorsal (thō-RĀ-kō-dor-sal)	C6–C8	Latissimus dorsi muscle.
Lower subscapular	C5–C6	Subscapularis and teres major muscles.
Axillary (AK-si-lar-ē)	C5–C6	Deltoid and teres minor muscles; skin over deltoid and superior posterior aspect of arm.
Median	C5–T1	Flexors of forearm, except flexor carpi ulnaris; ulnar half of flexor digitorum profundus, and some muscles of hand (lateral palm); skin of lateral two-thirds of palm of hand and fingers.
Radial	C5–T1	Triceps brachii, anconeus, and extensor muscles of forearm; skin of posterior arm and forearm, lateral two-thirds of dorsum of hand, and fingers over proximal and middle phalanges.
Medial pectoral	C8–T1	Pectoralis major and pectoralis minor muscles.
Medial cutaneous nerve of arm (kū-TĀ-nē-us)	C8–T1	Skin of medial and posterior aspects of distal third of arm.
Medial cutaneous nerve of forearm	C8–T1	Skin of medial and posterior aspects of forearm.
Ulnar	C8–T1	Flexor carpi ulnaris, ulnar half of flexor digitorum profundus, and most muscles of hand; skin of medial side of hand, little finger, and medial half of ring finger.

FIGURE 13.9 Brachial plexus in anterior view.

The brachial plexus supplies the shoulders and upper limbs.

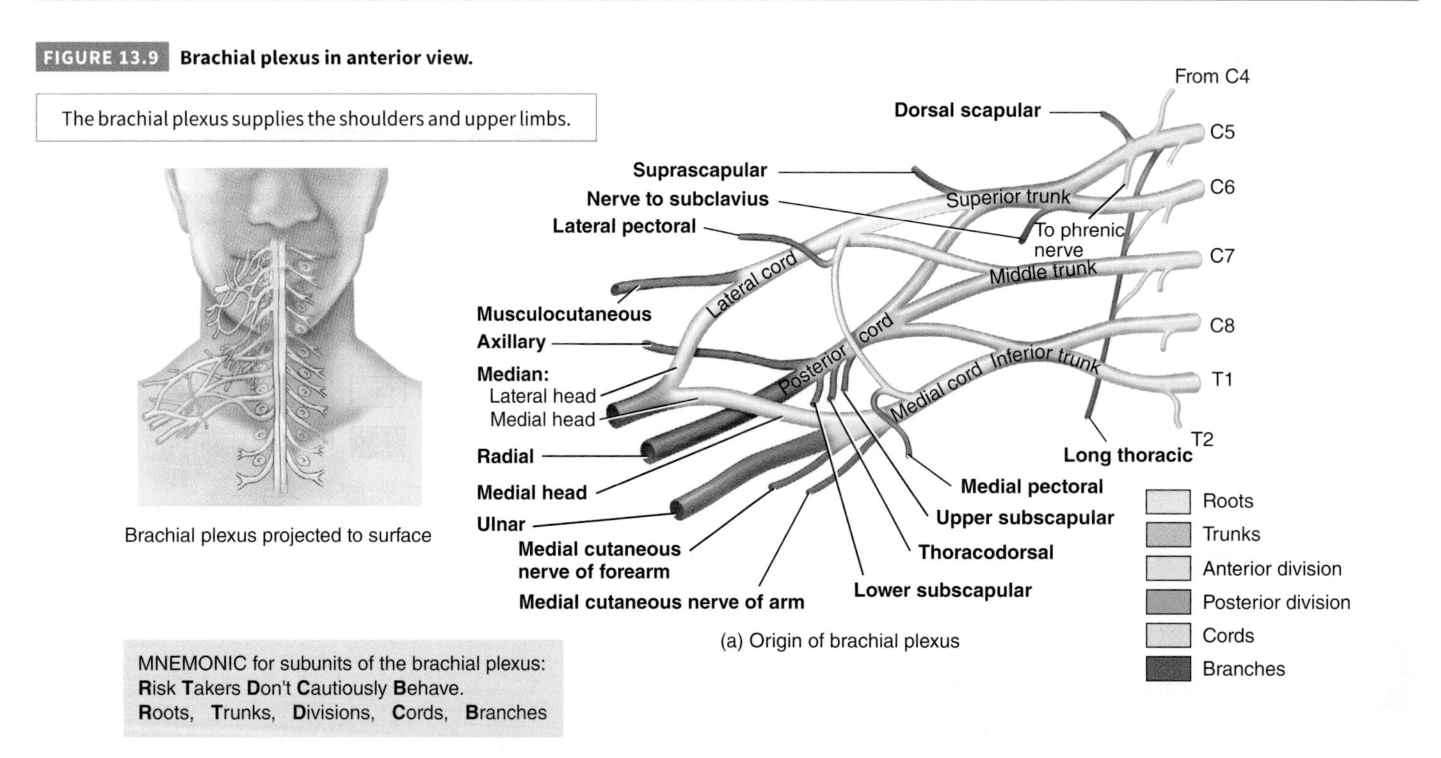

(a) Origin of brachial plexus

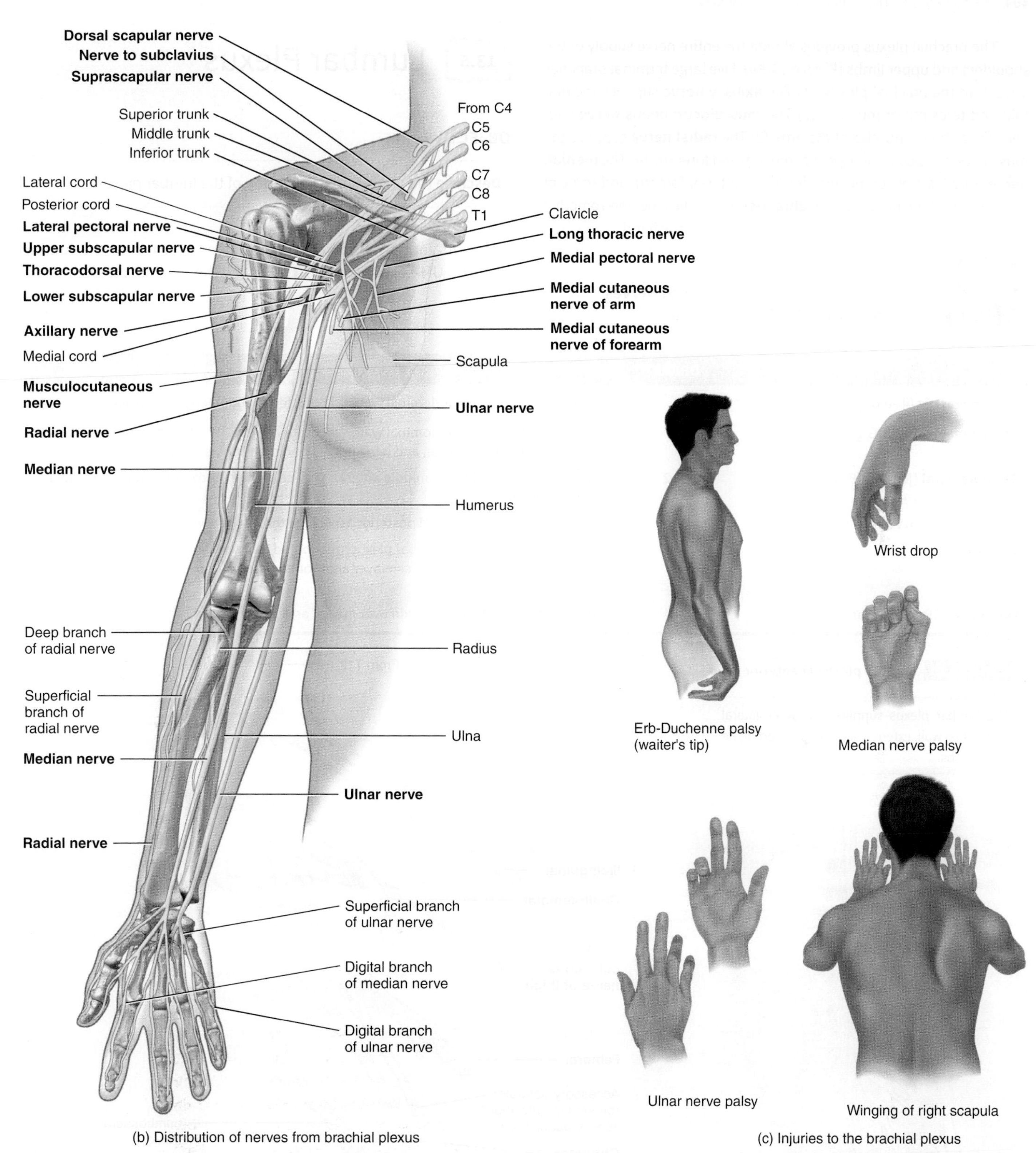

(b) Distribution of nerves from brachial plexus

(c) Injuries to the brachial plexus

Q What five important nerves arise from the brachial plexus?

nerves unite to form **trunks** in the inferior part of the neck. These are the *superior*, *middle*, and *inferior trunks*. Posterior to the clavicles, the trunks diverge into **divisions**, called the *anterior* and *posterior divisions.* In the axillae, the divisions unite to form **cords** called the *lateral*, *medial*, and *posterior cords.* The cords are named for their relationship to the axillary artery, a large artery that supplies blood to the upper limb. The **branches** of the brachial plexus form the principal nerves of the brachial plexus.

The brachial plexus provides almost the entire nerve supply of the shoulders and upper limbs (Figure 13.9b). Five large terminal branches arise from the brachial plexus: (1) The **axillary nerve** supplies the deltoid and teres minor muscles. (2) The **musculocutaneous nerve** supplies the anterior muscles of the arm. (3) The **radial nerve** supplies the muscles on the posterior aspect of the arm and forearm. (4) The **median nerve** supplies most of the muscles of the anterior forearm and some of the muscles of the hand. (5) The **ulnar nerve** supplies the anteromedial muscles of the forearm and most of the muscles of the hand.

See Clinical Connection: Injuries to Nerves Emerging from the Brachial Plexus

13.5 Lumbar Plexus

OBJECTIVE

- **Describe** the origin and distribution of the lumbar plexus.

The roots (anterior rami) of spinal nerves L1–L4 form the **lumbar plexus** (LUM-bar) (Figure 13.10). Unlike the brachial plexus, there is

NERVE	ORIGIN	DISTRIBUTION
Iliohypogastric (il′-ē-ō-hī-pō-GAS-trik)	L1	Muscles of anterolateral abdominal wall; skin of inferior abdomen and buttock.
Ilioinguinal (il′-ē-ō-ING-gwi-nal)	L1	Muscles of anterolateral abdominal wall; skin of superior and medial aspect of thigh, root of penis and scrotum in male, and labia majora and mons pubis in female.
Genitofemoral (jen′-i-tō-FEM-or-al)	L1–L2	Cremaster muscle; skin over middle anterior surface of thigh, scrotum in male, and labia majora in female.
Lateral cutaneous nerve of thigh	L2–L3	Skin over lateral, anterior, and posterior aspects of thigh.
Femoral	L2–L4	Largest nerve arising from lumbar plexus; distributed to flexor muscles of hip joint and extensor muscles of knee joint, skin over anterior and medial aspect of thigh and medial side of leg and foot.
Obturator (OB-too-rā′-tor)	L2–L4	Adductor muscles of hip joint; skin over medial aspect of thigh.

FIGURE 13.10 Lumbar plexus in anterior view.

The lumbar plexus supplies the anterolateral abdominal wall, external genitals, and part of the lower limbs.

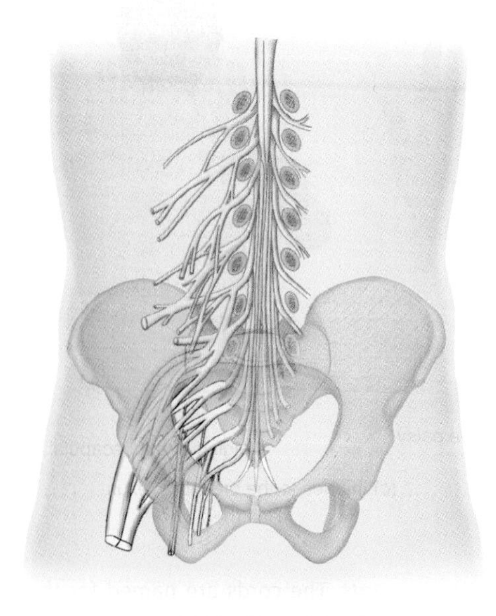

Lumbar plexus projected to surface

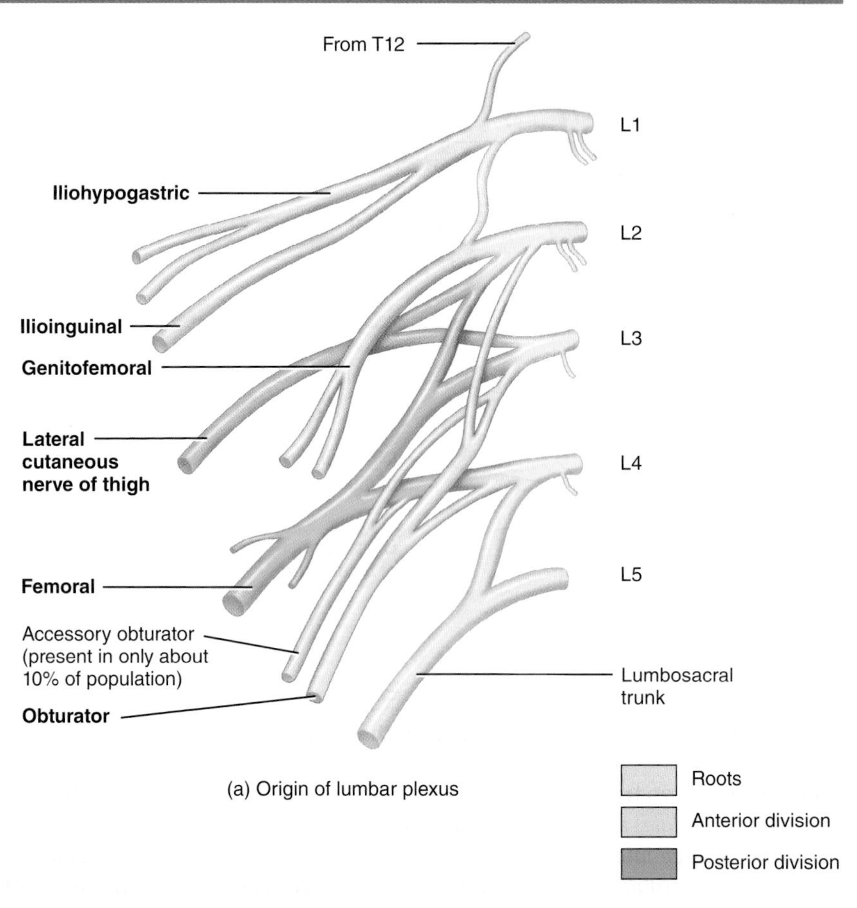

(a) Origin of lumbar plexus

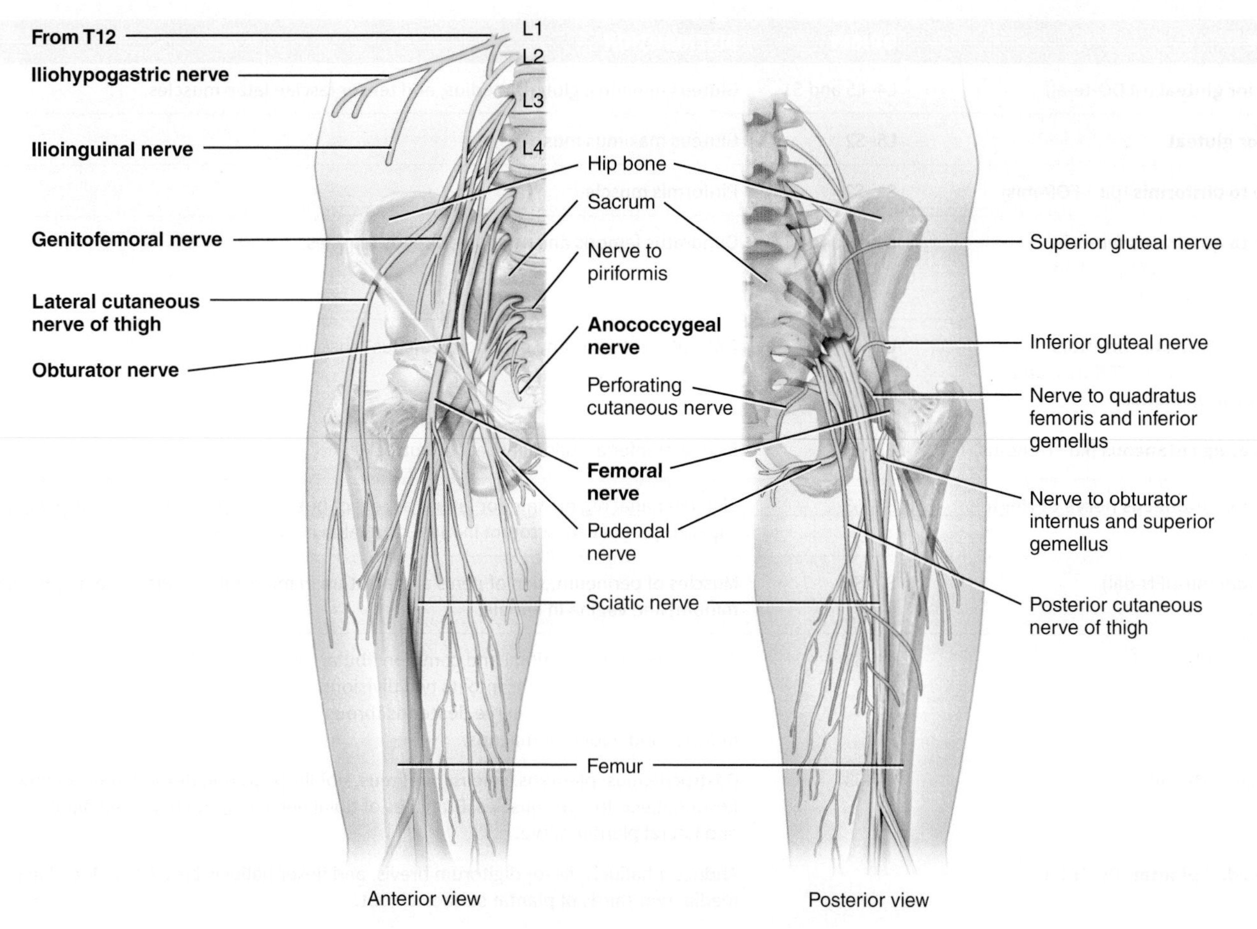

(b) Distribution of nerves from lumbar plexus

Q What are the signs of femoral nerve injury?

minimal intermingling of fibers in the lumbar plexus. On either side of the first four lumbar vertebrae, the lumbar plexus passes obliquely outward, between the superficial and deep heads of the psoas major muscle and anterior to the quadratus lumborum muscle. Between the heads of the psoa major, the roots of the lumbar plexuses split into anterior and posterior divisions, which then give rise to the peripheral branches of the plexas.

The lumbar plexus supplies the anterolateral abdominal wall, external genitals, and part of the lower limbs.

See Clinical Connection: Injuries to the Lumbar Plexus

13.6 Sacral and Coccygeal Plexuses

OBJECTIVE

- **Describe** the origin and distribution of the sacral and coccygeal plexuses.

The roots (anterior rami) of spinal nerves L4–L5 and S1–S4 form the **sacral plexus** (SĀ-kral) (**Figure 13.11**). This plexus is situated largely anterior to the sacrum. The sacral plexus supplies the buttocks,

NERVE	ORIGIN	DISTRIBUTION
Superior gluteal (GLOO-tē-al)	L4–L5 and S1	Gluteus minimus, gluteus medius, and tensor fasciae latae muscles.
Inferior gluteal	L5–S2	Gluteus maximus muscle.
Nerve to piriformis (pir-i-FOR-mis)	S1–S2	Piriformis muscle.
Nerve to quadratus femoris (quod-RĀ-tus FEM-or-is) and **inferior gemellus** (jem-EL-us)	L4–L5 and S1	Quadratus femoris and inferior gemellus muscles.
Nerve to obturator internus (OB-too-rā′-tor in-TER-nus) and **superior gemellus**	L5–S2	Obturator internus and superior gemellus muscles.
Perforating cutaneous (kū′-TĀ-nē-us)	S2–S3	Skin over inferior medial aspect of buttock.
Posterior cutaneous nerve of thigh	S1–S3	Skin over anal region, inferior lateral aspect of buttock, superior posterior aspect of thigh, superior part of calf, scrotum in male, and labia majora in female.
Pudendal (pū-DEN-dal)	S2–S4	Muscles of perineum; skin of penis and scrotum in male and clitoris, labia majora, labia minora, and vagina in female.
Sciatic (sī-AT-ik)	L4–S3	Actually two nerves—tibial and common fibular—bound together by common sheath of connective tissue; splits into its two divisions, usually at the knee. (See below for distributions.) As sciatic nerve descends through thigh, it sends branches to hamstring muscles and adductor magnus.
Tibial (TIB-ē-al)	L4–S3	Gastrocnemius, plantaris, soleus, popliteus, tibialis posterior, flexor digitorum longus, and flexor hallucis longus muscles. Branches of tibial nerve in foot are medial plantar nerve and lateral plantar nerve.
Medial plantar (PLAN-tar)		Abductor hallucis, flexor digitorum brevis, and flexor hallucis brevis muscles; skin over medial two-thirds of plantar surface of foot.
Lateral plantar		Remaining muscles of foot not supplied by medial plantar nerve; skin over lateral third of plantar surface of foot.
Common fibular (FIB-ū-lar)	L4–S2	Divides into superficial fibular and deep fibular branch.
Superficial fibular		Fibularis longus and fibularis brevis muscles; skin over distal third of anterior aspect of leg and dorsum of foot.
Deep fibular		Tibialis anterior, extensor hallucis longus, fibularis tertius, and extensor digitorum longus and extensor digitorum brevis muscles; skin on adjacent sides of great and second toes.

perineum, and lower limbs. The largest nerve in the body—the sciatic nerve—arises from the sacral plexus.

The roots (anterior rami) of spinal nerves S4–S5 and the coccygeal nerves form a small **coccygeal plexus** (kok-SIG-ē-al). From this plexus arises the **anococcygeal nerves** (Figure 13.11a), which supply a small area of skin in the coccygeal region.

See Clinical Connection: Injury to the Sciatic Nerve

13.7 Spinal Cord Physiology

OBJECTIVES

- **Describe** the functions of the major sensory and motor tracts of the spinal cord.
- **Describe** the functional components of a reflex arc and the ways reflexes maintain homeostasis.

FIGURE 13.11 Sacral and coccygeal plexuses in anterior view.

The sacral plexus supplies the buttocks, perineum, and lower limbs.

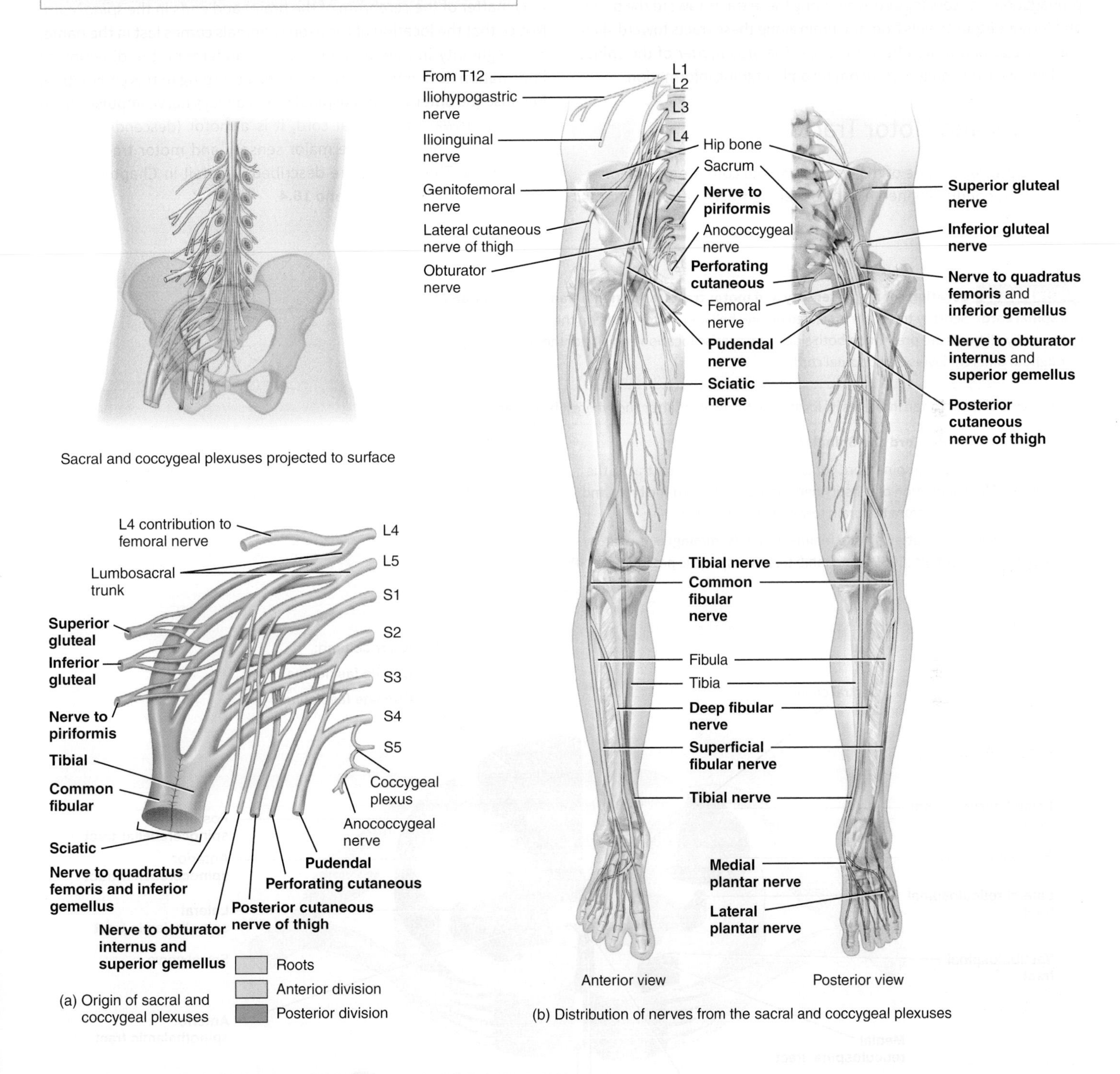

(a) Origin of sacral and coccygeal plexuses

(b) Distribution of nerves from the sacral and coccygeal plexuses

Q What is the origin of the sacral plexus?

The spinal cord has two principal functions in maintaining homeostasis: nerve impulse propagation and integration of information. The *white matter tracts* in the spinal cord are highways for nerve impulse propagation. Sensory input travels along these tracts toward the brain, and motor output travels from the brain along these tracts toward skeletal muscles and other effector tissues. The *gray matter* of the spinal cord receives and integrates incoming and outgoing information.

Sensory and Motor Tracts

As noted previously, one of the ways the spinal cord promotes homeostasis is by conducting nerve impulses along tracts. Often, the name of a tract indicates its position in the white matter and where it begins and ends. For example, the anterior corticospinal tract is located in the *anterior* white column; it begins in the *cerebral cortex* (superficial gray matter of the cerebrum of the brain) and ends in the *spinal cord*. Notice that the location of the axon terminals comes last in the name. This regularity in naming allows you to determine the direction of information flow along any tract named according to this convention. Because the anterior corticospinal tract conveys nerve impulses from the brain toward the spinal cord, it is a motor (descending) tract. **Figure 13.12** highlights the major sensory and motor tracts in the spinal cord. These tracts are described in detail in Chapter 16 and summarized in **Tables 16.3** and **16.4**.

FIGURE 13.12 Locations of major sensory and motor tracts, shown in a transverse section of the spinal cord. Sensory tracts are indicated on one half and motor tracts on the other half of the cord, but actually all tracts are present on both sides. The precise location and size of the tracts changes throughout different levels of the spinal cord.

The name of a tract often indicates its location in the white matter and where it begins and ends.

Functions of the Spinal Cord and Spinal Nerves

1. The white matter of the spinal cord contains sensory and motor tracts, the "highways" for conduction of sensory nerve impulses toward the brain and motor nerve impulses from the brain toward effector tissues.
2. The spinal cord gray matter is a site for integration (summing) of excitatory postsynaptic potentials (EPSPs) and inhibitory postsynaptic potentials (IPSPs).
3. Spinal nerves and the nerves that branch from them connect the CNS to the sensory receptors, muscles, and glands in all parts of the body.

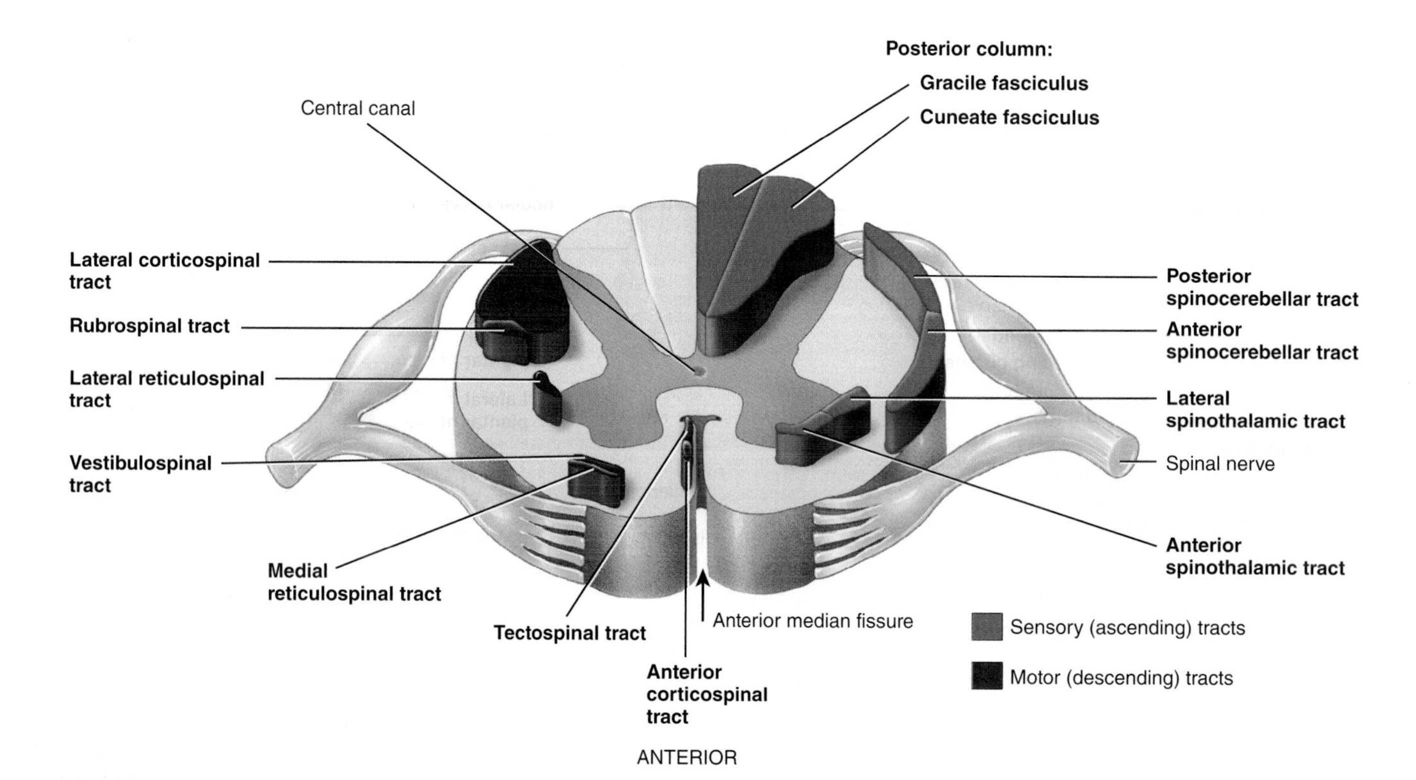

Q Based on its name, list the origin and destination of the spinothalamic tract. Is this a sensory or a motor tract?

Nerve impulses from sensory receptors propagate up the spinal cord to the brain along two main routes on each side: the spinothalamic tract and the posterior column. The **spinothalamic tract** (spī′-nō-tha-LAM-ik) conveys nerve impulses for sensing pain, temperature, itch, and tickle. The **posterior column** consists of two tracts: the **gracile fasciculus** (GRAS-īl fa-SIK-ū-lus) and the **cuneate fasciculus** (KŪ-nē-āt). The posterior column tracts convey nerve impulses for touch, pressure, vibration, and conscious proprioception (the awareness of the positions and movements of muscles, tendons, and joints).

The sensory systems keep the CNS informed of changes in the external and internal environments. The sensory information is integrated (processed) by interneurons in the spinal cord and brain. Responses to the integrative decisions are brought about by motor activities (muscular contractions and glandular secretions). The cerebral cortex, the outer part of the brain, plays a major role in controlling precise voluntary muscular movements. Other brain regions provide important integration for regulation of automatic movements. Motor output to skeletal muscles travels down the spinal cord in two types of descending pathways: direct and indirect. The **direct motor pathways**, also called *pyramidal pathways*, include the **lateral corticospinal** (kor′-ti-kō-SPĪ-nal), **anterior corticospinal**, and **corticobulbar tracts** (kor′-ti-kō-BUL-bar). They convey nerve impulses that originate in the cerebral cortex and are destined to cause *voluntary* movements of skeletal muscles. **Indirect motor pathways**, also called *extrapyramidal pathways*, include the **rubrospinal** (ROO-brō-spī-nal), **tectospinal** (TEK-tō-spī-nal), **vestibulospinal** (ves-TIB-ū-lō-spī-nal), **lateral reticulospinal** (re-TIK-ū-lō-spī-nal), and **medial reticulospinal tracts**. These tracts convey nerve impulses from the brainstem to cause *automatic movements* and help coordinate body movements with visual stimuli. Indirect pathways also maintain skeletal muscle tone, sustain contraction of postural muscles, and play a major role in equilibrium by regulating muscle tone in response to movements of the head.

Reflexes and Reflex Arcs

The second way the spinal cord promotes homeostasis is by serving as an integrating center for some reflexes. A **reflex** is a fast, involuntary, unplanned sequence of actions that occurs in response to a particular stimulus. Some reflexes are inborn, such as pulling your hand away from a hot surface before you even feel that it is hot. Other reflexes are learned or acquired. For instance, you learn many reflexes while acquiring driving expertise. Slamming on the brakes in an emergency is one example. When integration takes place in the spinal cord gray matter, the reflex is a **spinal reflex**. An example is the familiar patellar reflex (knee jerk). If integration occurs in the brainstem rather than the spinal cord, the reflex is called a **cranial reflex**. An example is the tracking movements of your eyes as you read this sentence. You are probably most aware of **somatic reflexes**, which involve contraction of skeletal muscles. Equally important, however, are the **autonomic** (*visceral*) **reflexes**, which generally are not consciously perceived. They involve responses of smooth muscle, cardiac muscle, and glands. As you will see in Chapter 15, body functions such as heart rate, digestion, urination, and defecation are controlled by the autonomic nervous system through autonomic reflexes.

Nerve impulses propagating into, through, and out of the CNS follow specific pathways, depending on the kind of information, its origin, and its destination. The pathway followed by nerve impulses that produce a reflex is a **reflex arc** (*reflex circuit*). A reflex arc includes the following five functional components (**Figure 13.13**):

1. **Sensory receptor.** The distal end of a sensory neuron (dendrite) or an associated sensory structure serves as a sensory receptor. It responds to a specific **stimulus**—a change in the internal or external environment—by producing a graded potential called a generator (or receptor) potential (described in Section 16.1). If a generator potential reaches the threshold level of depolarization, it will trigger one or more nerve impulses in the sensory neuron.

2. **Sensory neuron.** The nerve impulses propagate from the sensory receptor along the axon of the sensory neuron to the axon terminals, which are located in the gray matter of the spinal cord or brainstem. From here, relay neurons send nerve impulses to the area of the brain that allows conscious awareness that the reflex has occurred.

3. **Integrating center.** One or more regions of gray matter within the CNS acts as an integrating center. In the simplest type of reflex, the integrating center is a single synapse between a sensory neuron and a motor neuron. A reflex pathway having only one synapse in the CNS is termed a **monosynaptic reflex arc** (mon′-ō-si-NAP-tik; *mono-* = one). More often, the integrating center consists of one or more interneurons, which may relay impulses to other interneurons as well as to a motor neuron. A **polysynaptic reflex arc** (*poly-* = many) involves more than two types of neurons and more than one CNS synapse.

4. **Motor neuron.** Impulses triggered by the integrating center propagate out of the CNS along a motor neuron to the part of the body that will respond.

5. **Effector.** The part of the body that responds to the motor nerve impulse, such as a muscle or gland, is the effector. Its action is called a reflex. If the effector is skeletal muscle, the reflex is a **somatic reflex**. If the effector is smooth muscle, cardiac muscle, or a gland, the reflex is an **autonomic** (*visceral*) **reflex**.

Because reflexes are normally so predictable, they provide useful information about the health of the nervous system and can greatly aid diagnosis of disease. Damage or disease anywhere along its reflex arc can cause a reflex to be absent or abnormal. For example, tapping the patellar ligament normally causes reflex extension of the knee joint. Absence of the patellar reflex could indicate damage of the sensory or motor neurons, or a spinal cord injury in the lumbar region. Somatic reflexes generally can be tested simply by tapping or stroking the body surface.

Next, we examine four important somatic spinal reflexes: the stretch reflex, the tendon reflex, the flexor (withdrawal) reflex, and the crossed extensor reflex.

FIGURE 13.13 General components of a reflex arc. Arrows show the direction of nerve impulse propagation.

A reflex is a fast, predictable sequence of involuntary actions that occur in response to certain changes in the environment.

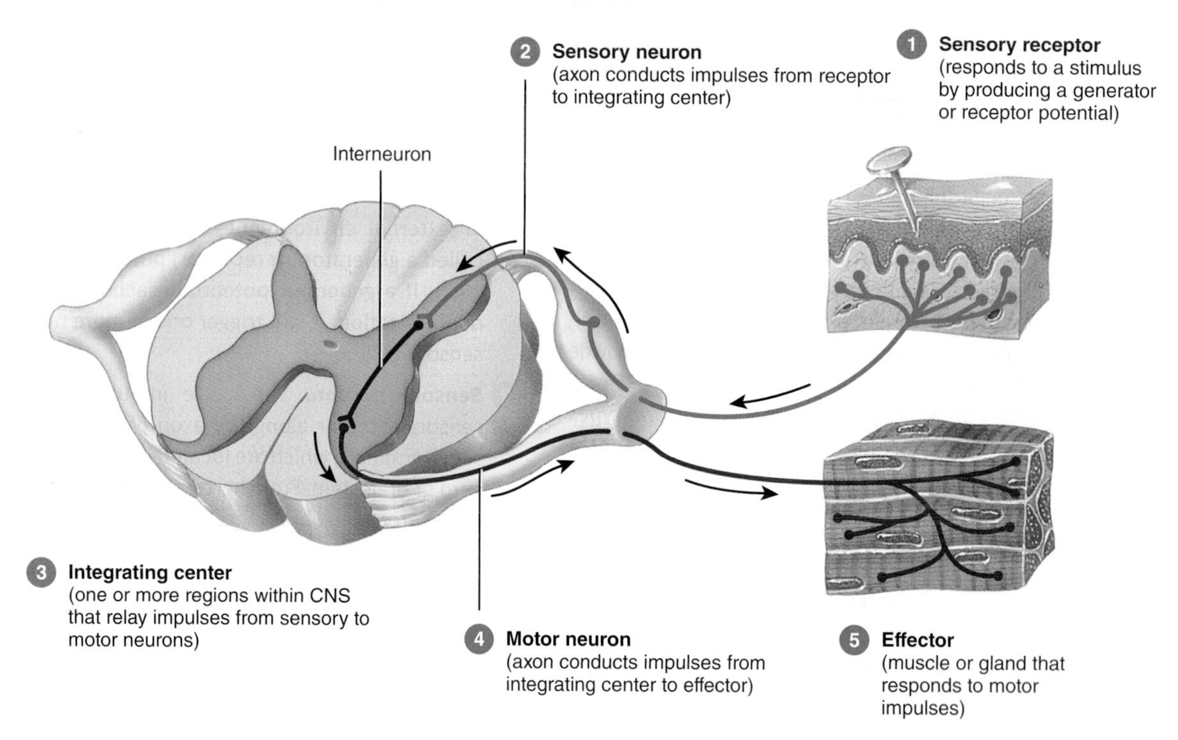

Q What initiates a nerve impulse in a sensory neuron? Which branch of the nervous system includes all integrating centers for reflexes?

The Stretch Reflex

A **stretch reflex** causes contraction of a skeletal muscle (the effector) in response to stretching of the muscle. This type of reflex occurs via a monosynaptic reflex arc. The reflex can occur by activation of a single sensory neuron that forms one synapse in the CNS with a single motor neuron. Stretch reflexes can be elicited by tapping on tendons attached to muscles at the elbow, wrist, knee, and ankle joints. An example of a stretch reflex is the patellar reflex (knee jerk), which is described in Clinical Connection: Reflexes and Diagnosis in Study Guide.

A stretch reflex operates as follows (**Figure 13.14**):

1. Slight stretching of a muscle stimulates sensory receptors in the muscle called **muscle spindles** (shown in more detail in **Figure 16.4**). The spindles monitor changes in the length of the muscle.
2. In response to being stretched, a muscle spindle generates one or more nerve impulses that propagate along a somatic sensory neuron through the posterior root of the spinal nerve and into the spinal cord.
3. In the spinal cord (integrating center), the sensory neuron makes an excitatory synapse with, and thereby activates, a motor neuron in the anterior gray horn.
4. If the excitation is strong enough, one or more nerve impulses arises in the motor neuron and propagates, along its axon, which extends from the spinal cord into the anterior root and through peripheral nerves to the stimulated muscle. The axon terminals of the motor neuron form neuromuscular junctions (NMJs) with skeletal muscle fibers of the stretched muscle.
5. Acetylcholine released by nerve impulses at the NMJs triggers one or more muscle action potentials in the stretched muscle (effector), and the muscle contracts. Thus, muscle stretch is followed by muscle contraction, which relieves the stretching.

In the reflex arc just described, sensory nerve impulses enter the spinal cord on the same side from which motor nerve impulses leave it. This arrangement is called an **ipsilateral reflex** (ip-si-LAT-er-al = same side). All monosynaptic reflexes are ipsilateral.

In addition to the large-diameter motor neurons that innervate typical skeletal muscle fibers, smaller-diameter motor neurons innervate smaller, specialized muscle fibers within the muscle spindles themselves. The brain regulates muscle spindle sensitivity through pathways to these smaller motor neurons. This regulation ensures proper muscle spindle signaling over a wide range of muscle lengths during voluntary and reflex contractions. By adjusting how vigorously a muscle spindle

FIGURE 13.14 **Stretch reflex.** This monosynaptic reflex arc has only one synapse in the CNS—between a single sensory neuron and a single motor neuron. A polysynaptic reflex arc to antagonistic muscles that includes two synapses in the CNS and one interneuron is also illustrated. Plus signs (+) indicate excitatory synapses; the minus sign (−) indicates an inhibitory synapse.

The stretch reflex causes contraction of a muscle that has been stretched.

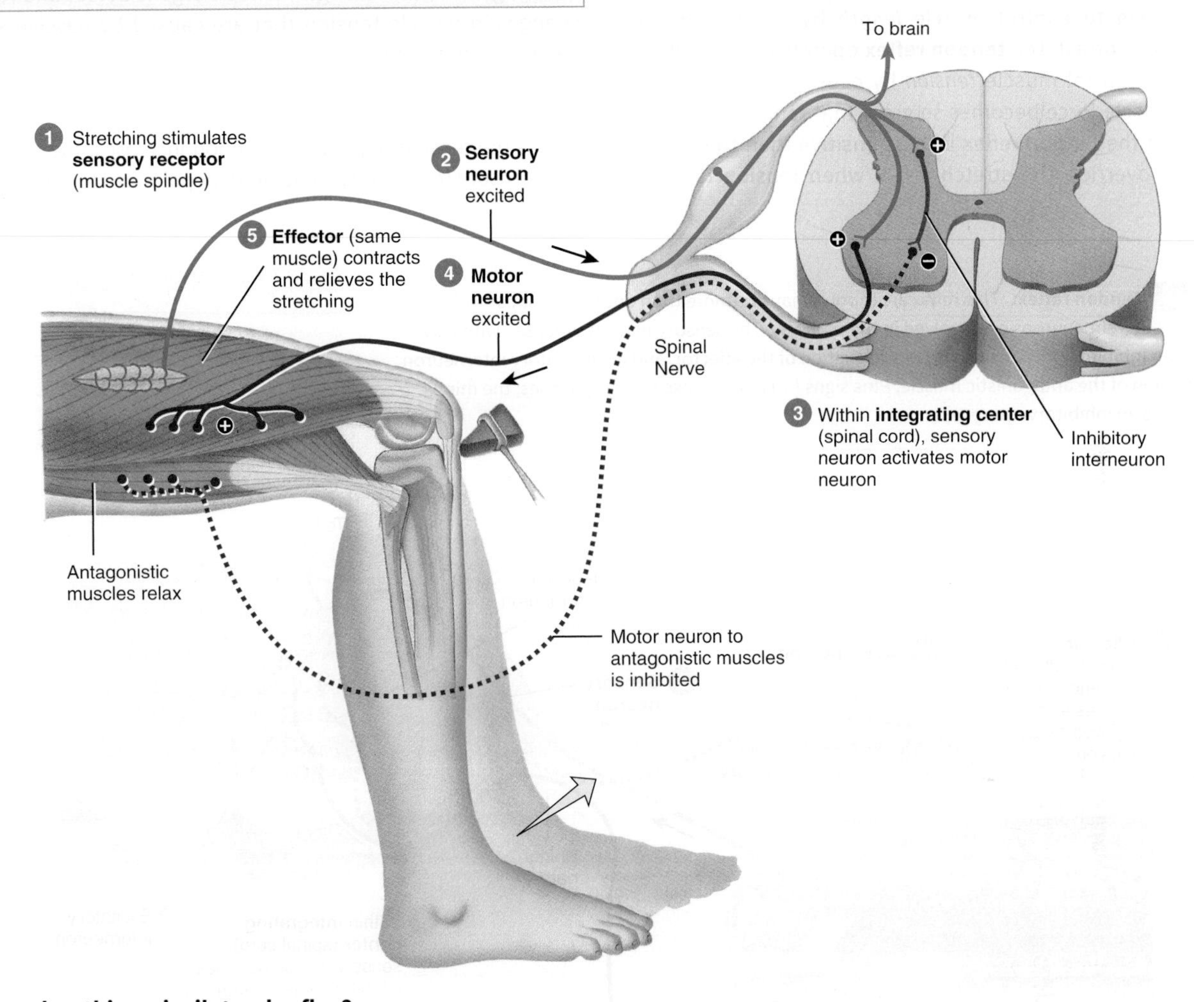

Q What makes this an ipsilateral reflex?

responds to stretching, the brain sets an overall level of **muscle tone**, which is the small degree of contraction present while the muscle is at rest. Because the stimulus for the stretch reflex is stretching of muscle, this reflex helps avert injury by preventing overstretching of muscles.

Although the stretch reflex pathway itself is monosynaptic (just two neurons and one synapse), a polysynaptic reflex arc to the antagonistic muscles operates at the same time. This arc involves three neurons and two synapses. An axon collateral (branch) from the muscle spindle sensory neuron also synapses with an inhibitory interneuron in the integrating center. In turn, the interneuron synapses with and inhibits a motor neuron that normally excites the antagonistic muscles (**Figure 13.14**). Thus, when the stretched muscle contracts during a stretch reflex, antagonistic muscles that oppose the contraction relax. This type of arrangement, in which the components of a neural circuit simultaneously cause contraction of one muscle and relaxation of its antagonists, is termed **reciprocal innervation** (rē-SIP-ro′-kal in′-er-VĀ-shun). Reciprocal innervation prevents conflict between opposing muscles and is vital in coordinating body movements.

Axon collaterals of the muscle spindle sensory neuron also relay nerve impulses to the brain over specific ascending pathways. In this way, the brain receives input about the state of stretch or contraction of skeletal muscles, enabling it to coordinate muscular movements. The nerve impulses that pass to the brain also allow conscious awareness that the reflex has occurred.

The stretch reflex can also help maintain posture. For example, if a standing person begins to lean forward, the gastrocnemius and other calf muscles are stretched. Consequently, stretch reflexes are initiated in these muscles, which cause them to contract and reestablish the body's upright posture. Similar types of stretch reflexes occur in the muscles of the shin when a standing person begins to lean backward.

The Tendon Reflex The stretch reflex operates as a feedback mechanism to control muscle *length* by causing muscle contraction. In contrast, the **tendon reflex** operates as a feedback mechanism to control muscle *tension* by causing muscle relaxation before muscle force becomes so great that tendons might be torn. Although the tendon reflex is less sensitive than the stretch reflex, it can override the stretch reflex when tension is great, making you drop a very heavy weight, for example. Like the stretch reflex, the tendon reflex is ipsilateral. The sensory receptors for this reflex are called **tendon** (*Golgi tendon*) **organs** (shown in more detail in Figure 16.4), which lie within a tendon near its junction with a muscle. In contrast to muscle spindles, which are sensitive to changes in muscle length, tendon organs detect and respond to changes in muscle tension that are caused by passive stretch or muscular contraction.

A tendon reflex operates as follows (Figure 13.15):

1. As the tension applied to a tendon increases, the tendon organ (sensory receptor) is stimulated (depolarized to threshold).

FIGURE 13.15 **Tendon reflex.** This reflex arc is polysynaptic—more than one CNS synapse and more than two different neurons are involved in the pathway. The sensory neuron synapses with two interneurons. An inhibitory interneuron causes relaxation of the effector, and a stimulatory interneuron causes contraction of the antagonistic muscle. Plus signs (+) indicate excitatory synapses; the minus sign (−) indicates an inhibitory synapse.

The tendon reflex causes relaxation of the muscle attached to the stimulated tendon organ.

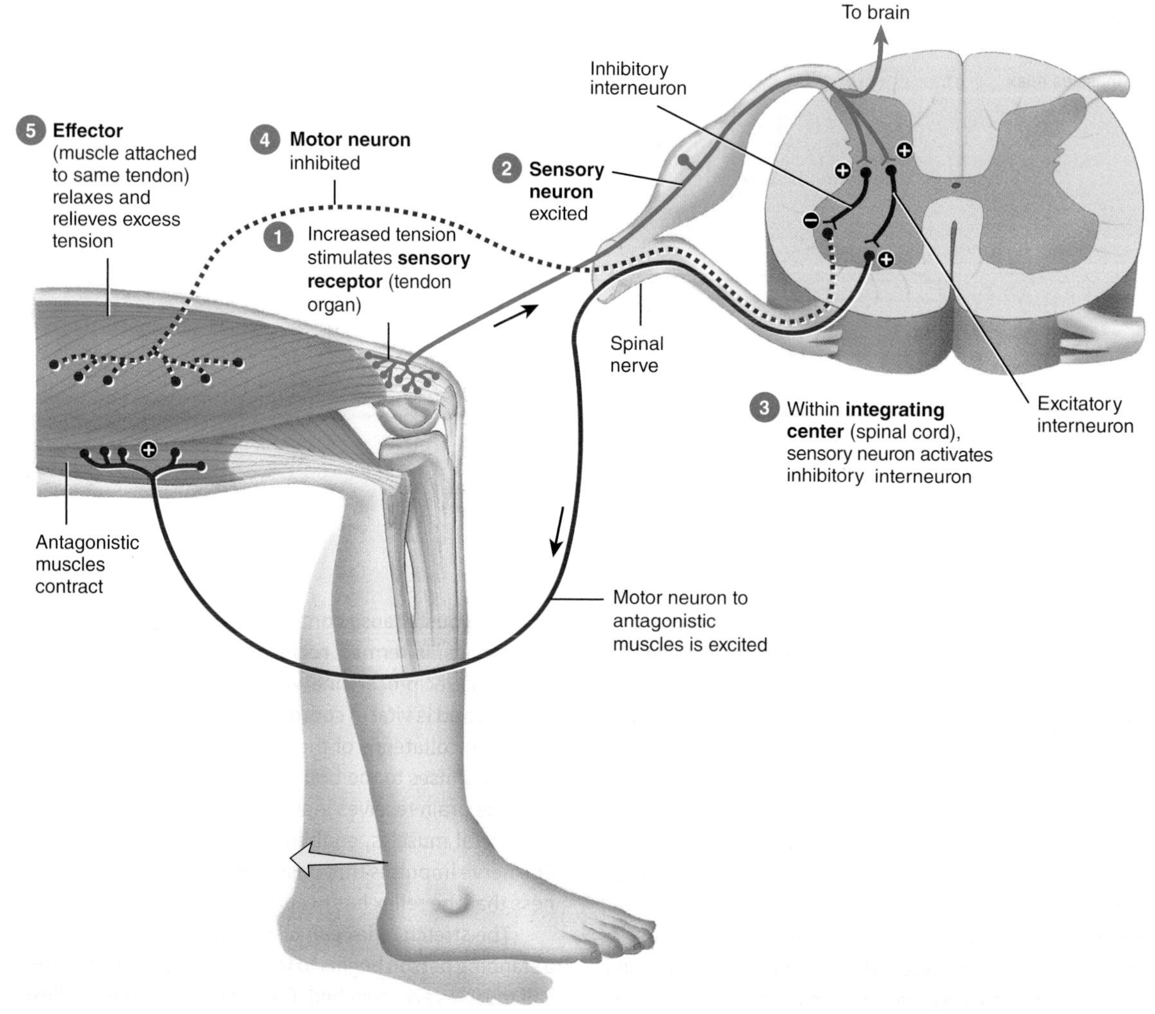

Q What is reciprocal innervation?

2. Nerve impulses arise and propagate into the spinal cord along a sensory neuron.
3. Within the spinal cord (integrating center), the sensory neuron activates an inhibitory interneuron that synapses with a motor neuron.
4. The inhibitory neurotransmitter inhibits (hyperpolarizes) the motor neuron, which then generates fewer nerve impulses.
5. The muscle relaxes and relieves excess tension.

Thus, as tension on the tendon organ increases, the frequency of inhibitory impulses increases; inhibition of the motor neurons to the muscle developing excess tension (effector) causes relaxation of the muscle. In this way, the tendon reflex protects the tendon and muscle from damage due to excessive tension.

Note in Figure 13.15 that the sensory neuron from the tendon organ also synapses with an excitatory interneuron in the spinal cord. The excitatory interneuron in turn synapses with motor neurons controlling antagonistic muscles. Thus, while the tendon reflex brings about relaxation of the muscle attached to the tendon organ, it also triggers contraction of antagonists. Here we have another example of reciprocal innervation. The sensory neuron also relays nerve impulses to the brain by way of sensory tracts, thus informing the brain about the state of muscle tension throughout the body.

The Flexor and Crossed Extensor Reflexes Another reflex involving a polysynaptic reflex arc results when, for instance, you step on a tack. In response to such a painful stimulus, you immediately withdraw your leg. This reflex, called the **flexor reflex** or *withdrawal reflex*, operates as follows (Figure 13.16):

1. Stepping on a tack stimulates the dendrites (sensory receptor) of a pain-sensitive neuron.
2. This sensory neuron then generates nerve impulses, which propagate into the spinal cord.
3. Within the spinal cord (integrating center), the sensory neuron activates interneurons that extend to several spinal cord segments.
4. The interneurons activate motor neurons in several spinal cord segments. As a result, the motor neurons generate nerve impulses, which propagate toward the axon terminals.
5. Acetylcholine released by the motor neurons causes the flexor muscles in the thigh (effectors) to contract, producing withdrawal of the leg. This reflex is protective because contraction of flexor muscles moves a limb away from the source of a possibly damaging stimulus.

The flexor reflex, like the stretch reflex, is ipsilateral—the incoming and outgoing impulses propagate into and out of the same side of the spinal cord. The flexor reflex also illustrates another feature of polysynaptic reflex arcs. Moving your entire lower or upper limb away from a painful stimulus involves contraction of more than one muscle group. Hence, several motor neurons must simultaneously convey impulses to several limb muscles. Because nerve impulses from one sensory neuron ascend and descend in the spinal cord and activate interneurons in several segments of the spinal cord, this type of reflex is called an **intersegmental reflex arc** (in′-ter-seg-MEN-tal; *inter-* = between). Through intersegmental reflex arcs, a single sensory neuron can activate several motor neurons, thereby stimulating more than one effector. The monosynaptic stretch reflex, in contrast, involves muscles receiving nerve impulses from one spinal cord segment only.

Something else may happen when you step on a tack: You may start to lose your balance as your body weight shifts to the other foot. Besides initiating the flexor reflex that causes you to withdraw the limb, the pain impulses from stepping on the tack also initiate a **crossed extensor reflex** to help you maintain your balance; it operates as follows (Figure 13.17):

1. Stepping on a tack stimulates the sensory receptor of a pain-sensitive neuron in the right foot.
2. This sensory neuron then generates nerve impulses, which propagate into the spinal cord.
3. Within the spinal cord (integrating center), the sensory neuron activates several interneurons that synapse with motor neurons on the left side of the spinal cord in several spinal cord segments. Thus, incoming pain signals cross to the opposite side through interneurons at that level, and at several levels above and below the point of entry into the spinal cord.
4. The interneurons excite motor neurons in several spinal cord segments that innervate extensor muscles. The motor neurons in turn generate more nerve impulses, which propagate toward the axon terminals.
5. Acetylcholine released by the motor neurons causes extensor muscles in the thigh (effectors) of the unstimulated left limb to contract, producing extension of the left leg. In this way, weight can be placed on the foot that must now support the entire body. A comparable reflex occurs with painful stimulation of the left lower limb or either upper limb.

See Clinical Connection: Reflexes and Diagnosis

Unlike the flexor reflex, which is an ipsilateral reflex, the crossed extensor reflex involves a **contralateral reflex arc** (kon-tra-LAT-er-al = opposite side): Sensory impulses enter one side of the spinal cord and motor impulses exit on the opposite side. Thus, a crossed extensor reflex synchronizes the extension of the contralateral limb with the withdrawal (flexion) of the stimulated limb. Reciprocal innervation also occurs in both the flexor reflex and the crossed extensor reflex. In the flexor reflex, when the flexor muscles of a painfully stimulated lower limb are contracting, the extensor muscles of the same limb are relaxing to some degree. If both sets of muscles contracted at the same time, the two sets of muscles would pull on the bones in opposite directions, which might immobilize the limb. Because of reciprocal innervation, one set of muscles contracts while the other relaxes.

FIGURE 13.16 **Flexor (withdrawal) reflex.** Plus signs (+) indicate excitatory synapses.

The flexor reflex causes withdrawal of a part of the body in response to a painful stimulus.

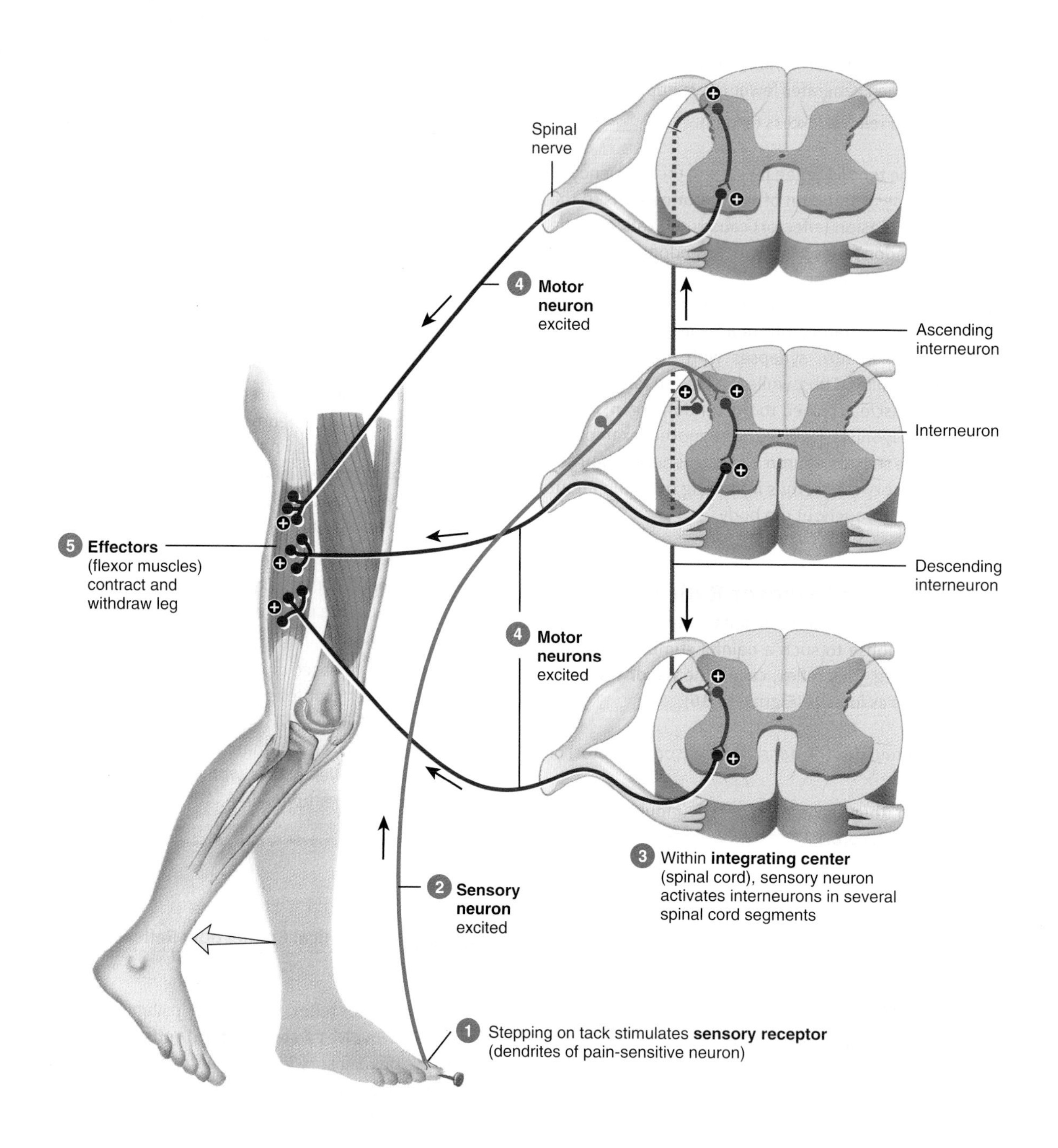

Q Why is the flexor reflex classified as an intersegmental reflex arc?

FIGURE 13.17 **Crossed extensor reflex.** The flexor reflex arc is shown (at left) for comparison with the crossed extensor reflex arc. Plus signs (+) indicate excitatory synapses.

A crossed extensor reflex causes contraction of muscles that extend joints in the limb opposite a painful stimulus.

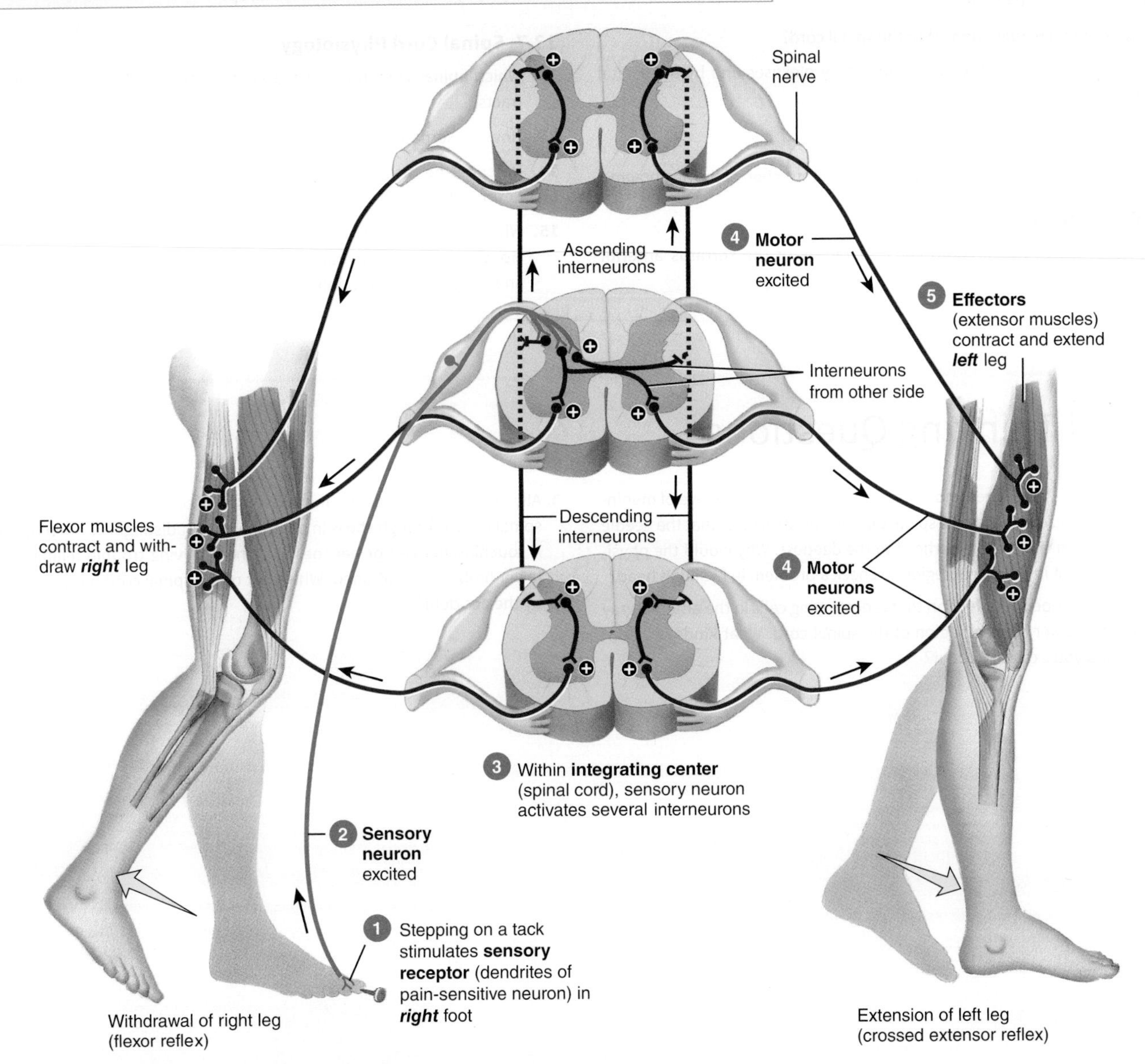

Q Why is the crossed extensor reflex classified as a contralateral reflex arc?

Review Questions

13.1 Spinal Cord Anatomy

1. Where are the spinal meninges located? Where are the epidural, subdural, and subarachnoid spaces located?
2. What are the cervical and lumbar enlargements?
3. Define conus medullaris, filum terminale, and cauda equina. What is a spinal segment? How is the spinal cord partially divided into right and left sides?
4. What does each of the following terms mean? Gray commissure, central canal, anterior gray horn, lateral gray horn, posterior gray horn, anterior

white column, lateral white column, posterior white column, ascending tract, and descending tract.

13.2 Spinal Nerves

5. How are spinal nerves named and numbered? Why are all spinal nerves classified as mixed nerves?
6. How do spinal nerves connect to the spinal cord?
7. Which regions of the body are supplied by plexuses and by intercostal nerves?

13.3 Cervical Plexus

8. Which nerve from the cervical plexus causes contraction of the diaphragm?

13.4 Brachial Plexus

9. Injury of which nerve could cause paralysis of the serratus anterior muscle?

13.5 Lumbar Plexus

10. What is the largest nerve arising from the lumbar plexus?

13.6 Sacral and Coccygeal Plexuses

11. Injury of which nerve causes footdrop?

13.7 Spinal Cord Physiology

12. Which spinal cord tracts are ascending tracts? Which are descending tracts?
13. How are somatic and autonomic reflexes similar and different?
14. Describe the mechanism and function of a stretch reflex, tendon reflex, flexor (withdrawal) reflex, and crossed extensor reflex.
15. What does each of the following terms mean in relation to reflex arcs? Monosynaptic, ipsilateral, polysynaptic, intersegmental, contralateral, and reciprocal innervation.

Critical Thinking Questions

1. Evalina's severe headaches and other symptoms were suggestive of meningitis, so her physician ordered a spinal tap. List the structures that the needle will pierce from the most superficial to the deepest. Why would the physician order a test in the spinal region to check a problem in Evalina's head?
2. Sunil has developed an infection that is destroying cells in the anterior gray horns in the lower cervical region of the spinal cord. What kinds of symptoms would you expect to occur?
3. Allyson is in a car accident and suffers spinal cord compression in the lower spinal cord. Although she is in pain, she cannot distinguish when the doctor is touching her calf or her toes and she is having trouble telling how her lower limbs are positioned. What part of the spinal cord has been affected by the accident?

The Brain and Cranial Nerves

CHAPTER **14**

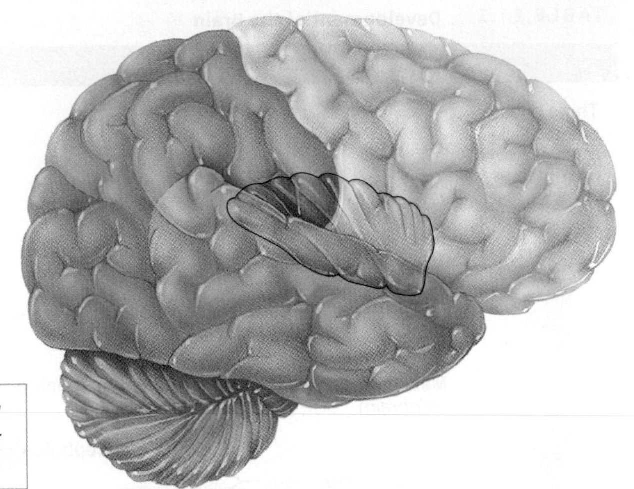

The Brain, Cranial Nerves, and Homeostasis

The brain contributes to homeostasis by receiving sensory input, integrating new and stored information, making decisions, and executing responses through motor activities.

Solving an equation, feeling hungry, laughing—the neural processes needed for each of these activities occur in different regions of the brain, that portion of the central nervous system contained within the cranium. About 85 billion neurons and 10 trillion to 50 trillion neuroglia make up the brain, which has a mass of about 1300 g (almost 3 lb) in adults. On average, each neuron forms 1000 synapses with other neurons. Thus, the total number of synapses, about a thousand trillion or 10^{15}, is larger than the number of stars in our galaxy.

The brain is the control center for registering sensations, correlating them with one another and with stored information, making decisions, and taking actions. It also is the center for the intellect, emotions, behavior, and memory. But the brain encompasses yet a larger domain: It directs our behavior toward others. With ideas that excite, artistry that dazzles, or rhetoric that mesmerizes, one person's thoughts and actions may influence and shape the lives of many others. As you will see shortly, different regions of the brain are specialized for different functions. Different parts of the brain also work together to accomplish certain shared functions. This chapter explores how the brain is protected and nourished, what functions occur in the major regions of the brain, and how the spinal cord and the 12 pairs of cranial nerves connect with the brain to form the control center of the human body.

Q Did you ever wonder how cerebrovascular accidents (strokes) occur and how they are treated?

14.1 Brain Organization, Protection, and Blood Supply

OBJECTIVES

- **Identify** the major parts of the brain.
- **Describe** how the brain is protected.
- **Describe** the blood supply of the brain.

In order to understand the terminology used for the principal parts of the adult brain, it will be helpful to know how the brain develops. The brain and spinal cord develop from the ectodermal **neural tube** (see **Figure 14.27**). The anterior part of the neural tube expands, along with the associated neural crest tissue. Constrictions in this expanded tube soon appear, creating three regions called **primary brain vesicles**: *prosencephalon, mesencephalon*, and *rhombencephalon* (see **Figure 14.28**). Both the prosencephalon and rhombencephalon subdivide further, forming **secondary brain vesicles**. The *prosencephalon* (PROS-en-sef′-a-lon), or forebrain, gives rise to the telencephalon and diencephalon, and the *rhombencephalon* (ROM-ben-sef′-a-lon), or hindbrain, develops into the metencephalon and myelencephalon. The various brain vesicles give rise to the following adult structures:

- The **telencephalon** (tel′-en-SEF-a-lon; *tel-* = distant; *-encephalon* = brain) develops into the *cerebrum* and *lateral ventricles.*
- The **diencephalon** (dī′-en-SEF-a-lon) forms the *thalamus, hypothalamus, epithalamus,* and *third ventricle.*
- The **mesencephalon** (mes′-en-SEF-a-lon; *mes-* = middle), or midbrain, gives rise to the *midbrain* and *aqueduct of the midbrain (cerebral aqueduct).*
- The **metencephalon** (met′-en-SEF-a-lon; *met-* = after) becomes the *pons, cerebellum,* and *upper part of the fourth ventricle.*
- The **myelencephalon** (mī-el-en-SEF-a-lon; *myel-* = marrow) forms the *medulla oblongata* and *lower part of the fourth ventricle.*

The walls of these brain regions develop into nervous tissue, while the hollow interior of the tube is transformed into its various ventricles (fluid-filled spaces). The expanded neural crest tissue becomes prominent in head development. Most of the protective structures of the brain—that is, most of the bones of the skull, associated connective tissues, and meningeal membranes—arise from this expanded neural crest tissue.

These relationships are summarized in **Table 14.1**.

TABLE 14.1 Development of the Brain

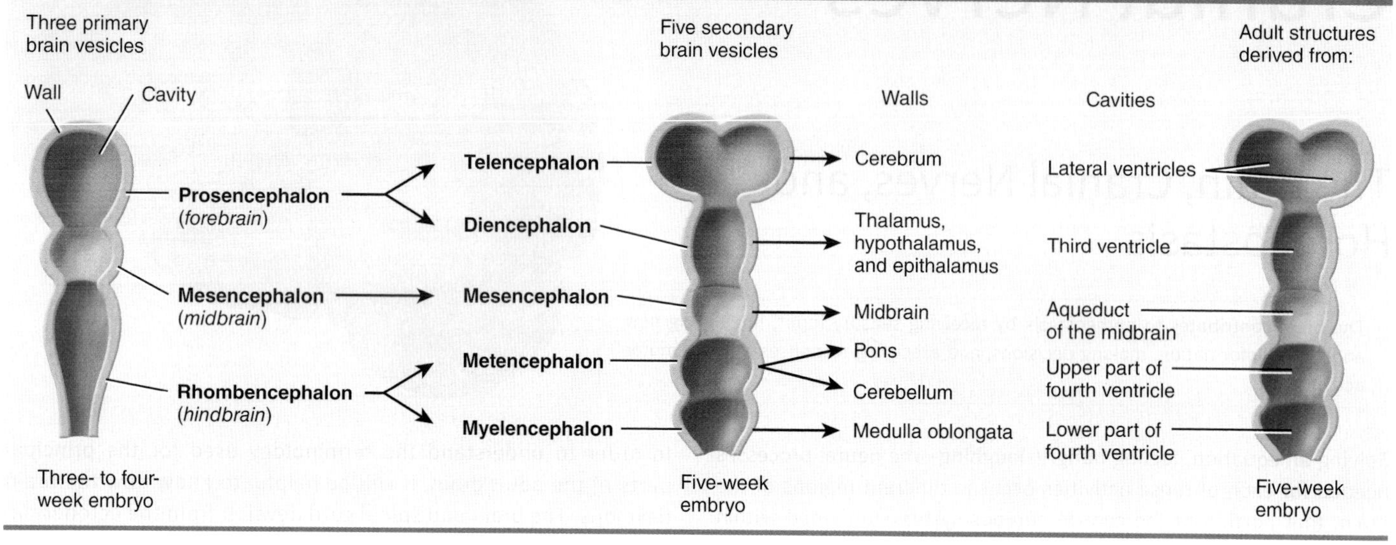

Three primary brain vesicles	Five secondary brain vesicles	Adult structures derived from: Walls	Cavities
Prosencephalon (*forebrain*)	**Telencephalon**	Cerebrum	Lateral ventricles
	Diencephalon	Thalamus, hypothalamus, and epithalamus	Third ventricle
Mesencephalon (*midbrain*)	**Mesencephalon**	Midbrain	Aqueduct of the midbrain
Rhombencephalon (*hindbrain*)	**Metencephalon**	Pons Cerebellum	Upper part of fourth ventricle
	Myelencephalon	Medulla oblongata	Lower part of fourth ventricle
Three- to four-week embryo (Wall, Cavity)	Five-week embryo		Five-week embryo

Major Parts of the Brain

The adult **brain** consists of four major parts: brainstem, cerebellum, diencephalon, and cerebrum (**Figure 14.1**). The **brainstem** is continuous with the spinal cord and consists of the medulla oblongata, pons, and midbrain. Posterior to the brainstem is the **cerebellum** (ser′-e-BEL-um = little brain). Superior to the brainstem is the **diencephalon** (*di-* = through), which consists of the thalamus, hypothalamus, and epithalamus. Supported on the diencephalon and brainstem is the **cerebrum** (se-RĒ-brum = brain), the largest part of the brain.

Protective Coverings of the Brain

The cranium (see **Figure 7.4**) and the cranial meninges surround and protect the brain. The **cranial meninges** (me-NIN-jēz) are continuous with the spinal meninges, have the same basic structure, and bear the same names: the outer **dura mater** (DOO-ra MĀ-ter), the middle **arachnoid mater** (a-RAK-noyd), and the inner **pia mater** (PĒ-a or PĪ-a) (**Figure 14.2**). However, the cranial dura mater has two layers; the spinal dura mater has only one. The two dural layers are called the *periosteal layer* (which is external) and the *meningeal layer* (which is internal). The dural layers around the brain are fused together except where they separate to enclose the dural venous sinuses (endothelial-lined venous channels) that drain venous blood from the brain and deliver it into the internal jugular veins. Also, there is no epidural space around the brain. Blood vessels that enter brain tissue pass along the surface of the brain, and as they penetrate inward they are sheathed by a loose-fitting sleeve of pia mater. Three extensions of the dura mater separate parts of the brain: (1) The **falx cerebri** (FALKS ser-i-BRĒ; *falx* = sickle-shaped) separates the two hemispheres (sides) of the cerebrum. (2) The **falx cerebelli** (ser′-e-BEL-ī) separates the two hemispheres of the cerebellum. (3) The **tentorium cerebelli** (ten-TŌ-rē-um = tent) separates the cerebrum from the cerebellum.

Brain Blood Flow and the Blood–Brain Barrier

Blood flows to the brain mainly via the internal carotid and vertebral arteries (see **Figure 21.19**); the dural venous sinuses drain into the internal jugular veins to return blood from the head to the heart (see **Figure 21.24**).

In an adult, the brain represents only 2% of total body weight, but it consumes about 20% of the oxygen and glucose used by the body, even when you are resting. Neurons synthesize ATP almost exclusively from glucose via reactions that use oxygen. When the activity of neurons and neuroglia increases in a particular region of the brain, blood flow to that area also increases. Even a brief slowing of brain blood flow may cause disorientation or a lack of consciousness, such as when you stand up too quickly after sitting for a long period of time. Typically, an interruption in blood flow for 1 or 2 minutes impairs neuronal function, and total deprivation of oxygen for about 4 minutes causes permanent injury. Because virtually no glucose is stored in the brain, the supply of glucose also must be continuous. If blood entering the brain has a low level of glucose, mental confusion, dizziness, convulsions, and loss of consciousness may occur. People with diabetes must be vigilant about their blood sugar levels because these levels can drop quickly, leading to diabetic shock, which is characterized by seizure, coma, and possibly death.

The **blood–brain barrier (BBB)** consists mainly of tight junctions that seal together the endothelial cells of brain blood capillaries and a thick basement membrane that surrounds the capillaries. As you learned in Chapter 12, astrocytes are one type of neuroglia; the processes of many astrocytes press up against the capillaries and secrete chemicals that maintain the "tightness" of the tight junctions. The BBB allows certain substances in blood to enter brain tissue and prevents passage to others. Lipid-soluble substances (including O_2, CO_2), steroid hormones, alcohol, barbiturates, nicotine,

FIGURE 14.1 **The brain.** The pituitary gland is discussed with the endocrine system in Chapter 18.

The four principal parts of the brain are the brainstem, cerebellum, diencephalon, and cerebrum.

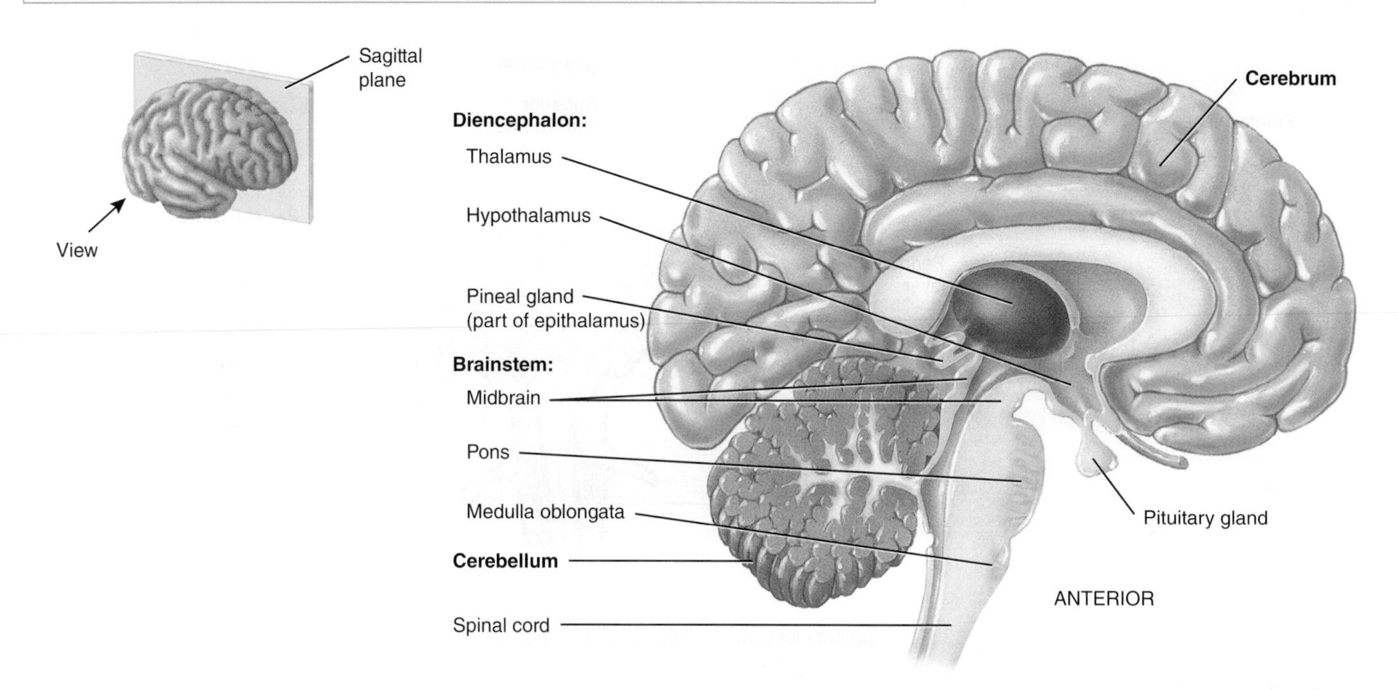

(a) Sagittal section, medial view

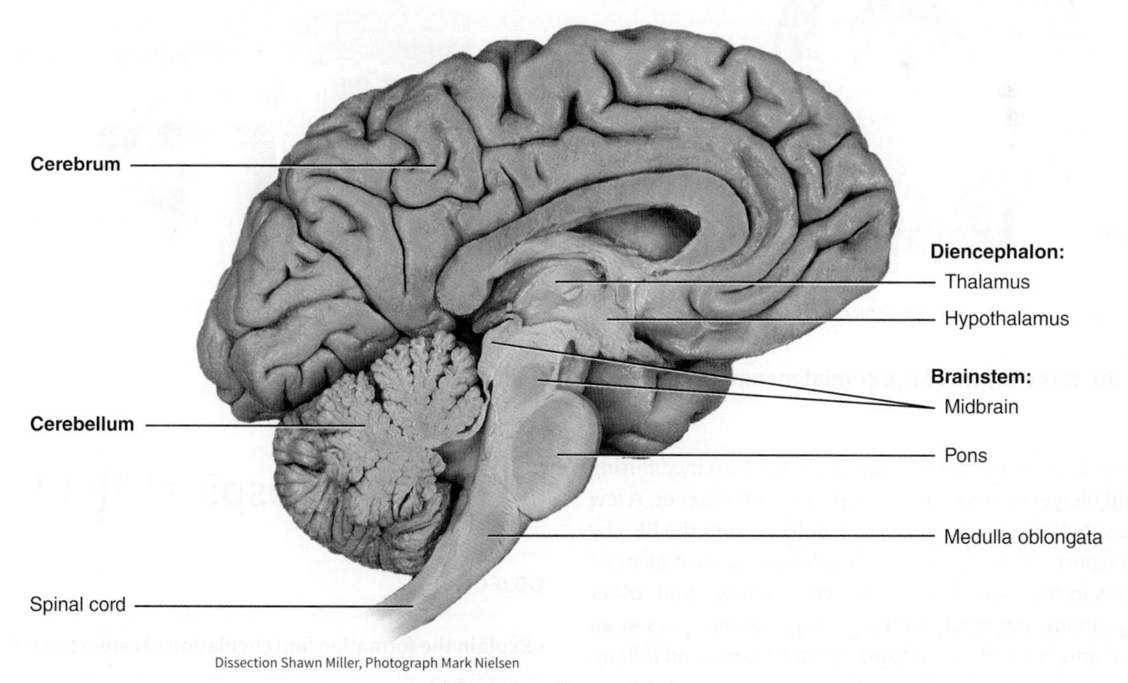

Dissection Shawn Miller, Photograph Mark Nielsen

(b) Sagittal section, medial view

Q Which part of the brain is the largest?

FIGURE 14.2 The protective coverings of the brain.

Cranial bones and cranial meninges protect the brain.

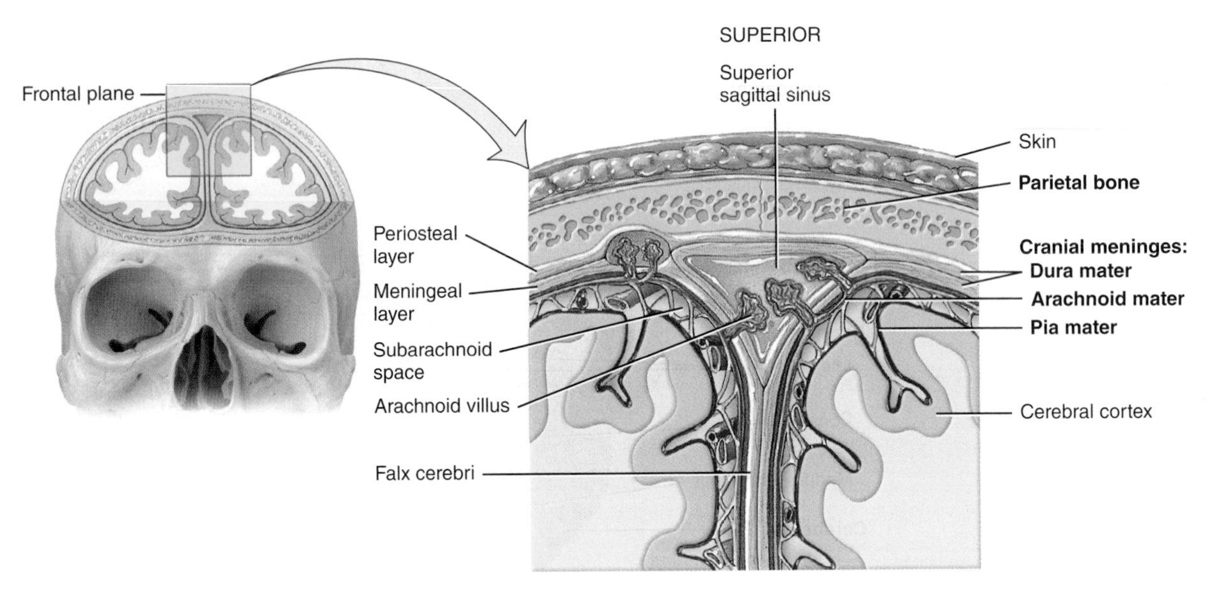

(a) Anterior view of frontal section through skull showing the cranial meninges

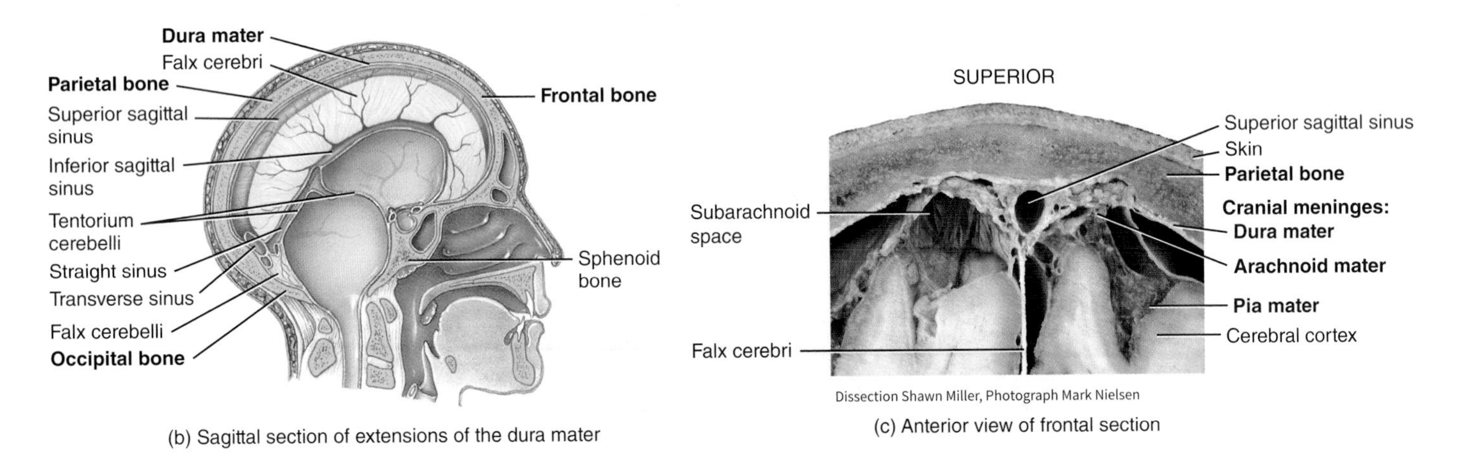

(b) Sagittal section of extensions of the dura mater

(c) Anterior view of frontal section

Q What are the three layers of the cranial meninges, from superficial to deep?

and caffeine) and water molecules easily cross the BBB by diffusing across the lipid bilayer of endothelial cell plasma membranes. A few water-soluble substances, such as glucose, quickly cross the BBB by facilitated transport. Other water-soluble substances, such as most ions, are transported across the BBB very slowly. Still other substances—proteins and most antibiotic drugs—do not pass at all from the blood into brain tissue. Trauma, certain toxins, and inflammation can cause a breakdown of the BBB.

See Clinical Connection: Breaching the Blood–Brain Barrier

14.2 Cerebrospinal Fluid

OBJECTIVE

- **Explain** the formation and circulation of cerebrospinal fluid.

Cerebrospinal fluid (CSF) is a clear, colorless liquid composed primarily of water that protects the brain and spinal cord from chemical and physical injuries. It also carries small amounts of oxygen, glucose, and other needed chemicals from the blood to neurons and neuroglia. CSF continuously circulates through cavities in the brain and spinal cord and around the brain and spinal cord in the subarachnoid space (the space between the arachnoid mater and pia mater). The total volume of CSF is 80 to 150 mL (3 to 5 oz) in an adult. CSF contains small amounts of glucose, proteins, lactic acid, urea, cations (Na^+, K^+, Ca^{2+}, Mg^{2+}), and anions (Cl^- and HCO_3^-); it also contains some white blood cells.

Figure 14.3 shows the four CSF-filled cavities within the brain, which are called **ventricles** (VEN-tri-kuls = little cavities). There is one **lateral ventricle** in each hemisphere of the cerebrum. (Think of them as ventricles 1 and 2.) Anteriorly, the lateral ventricles are separated by a thin membrane, the **septum pellucidum** (SEP-tum pe-LOO-si-dum; *pellucid* = transparent). The **third ventricle** is a narrow, slitlike cavity along the midline superior to the hypothalamus and between the right and left halves of the thalamus. The **fourth ventricle** lies between the brainstem and the cerebellum.

Functions of CSF

The CSF has three basic functions in helping to maintain homeostasis.

1. ***Mechanical protection.*** CSF serves as a shock-absorbing medium that protects the delicate tissues of the brain and spinal cord from jolts that would otherwise cause them to hit the bony walls of the cranial cavity and vertebral canal. The fluid also buoys the brain so that it "floats" in the cranial cavity.
2. ***Chemical protection.*** CSF provides an optimal chemical environment for accurate neuronal signaling. Even slight changes in the ionic composition of CSF within the brain can seriously disrupt production of action potentials and postsynaptic potentials.
3. ***Circulation.*** CSF is a medium for minor exchange of nutrients and waste products between the blood and adjacent nervous tissue.

Formation of CSF in the Ventricles

The majority of CSF production is from the **choroid plexuses** (KŌ-royd = membranelike), networks of blood capillaries in the walls of the ventricles (**Figure 14.4a**). Ependymal cells joined by tight junctions cover the capillaries of the choroid plexuses. Selected substances (mostly water) from the blood plasma, which are filtered from the capillaries, are secreted by the ependymal cells to produce the cerebrospinal fluid. This secretory capacity is bidirectional and accounts for continuous production of CSF and transport of metabolites from the nervous tissue back to the blood. Because of the tight junctions between ependymal cells, materials entering CSF from choroid capillaries cannot leak between these cells; instead, they

FIGURE 14.3 Locations of ventricles within a "transparent" brain. One interventricular foramen on each side connects a lateral ventricle to the third ventricle, and the aqueduct of the midbrain connects the third ventricle to the fourth ventricle.

Ventricles are cavities within the brain that are filled with cerebrospinal fluid.

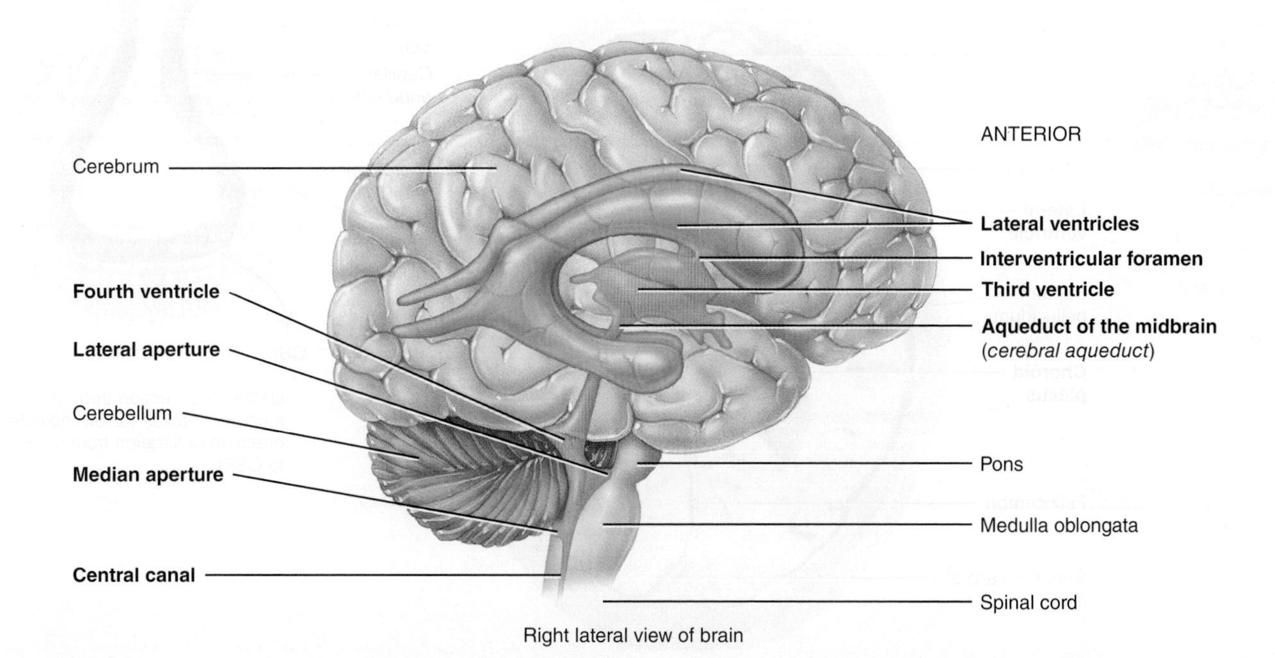

Q Which brain region is anterior to the fourth ventricle? Which is posterior to it?

must pass through the ependymal cells. This **blood–cerebrospinal fluid barrier** permits certain substances to enter the CSF but excludes others, protecting the brain and spinal cord from potentially harmful blood-borne substances. In contrast to the blood–brain barrier, which is formed mainly by tight junctions of brain capillary endothelial cells, the blood–cerebrospinal fluid barrier is formed by tight junctions of ependymal cells.

Circulation of CSF

The CSF formed in the choroid plexuses of each lateral ventricle flows into the third ventricle through two narrow, oval openings, the **interventricular foramina** (in′-ter-ven-TRIK-ū-lar; singular is *foramen*; Figure 14.4b). More CSF is added by the choroid plexus in the roof of the third ventricle. The fluid then flows through the **aqueduct of the midbrain** (*cerebral aqueduct*) (AK-we-dukt), which passes through the midbrain, into the fourth ventricle. The choroid plexus of the fourth ventricle contributes more fluid. CSF enters the subarachnoid space through three openings in the roof of the fourth ventricle: a single **median aperture** (AP-er-chur) and paired **lateral apertures**, one on each side. CSF then circulates in the central canal of the spinal cord and in the subarachnoid space around the surface of the brain and spinal cord.

CSF is gradually reabsorbed into the blood through **arachnoid villi**, fingerlike extensions of the arachnoid mater that project into the dural venous sinuses, especially the **superior sagittal sinus** (see Figure 14.2). (A cluster of arachnoid villi is called an **arachnoid granulation**.) Normally, CSF is reabsorbed as rapidly as it is formed by the choroid plexuses, at a rate of about 20 mL/hr (480 mL/day). Because the rates of formation and reabsorption are the same, the pressure of CSF normally is constant. For the same reason, the volume of CSF remains constant. Figure 14.4d summarizes the production and flow of CSF.

See Clinical Connection: Hydrocephalus

FIGURE 14.4 **Pathways of circulating cerebrospinal fluid.**

CSF is formed from blood plasma by ependymal cells that cover the choroid plexuses of the ventricles.

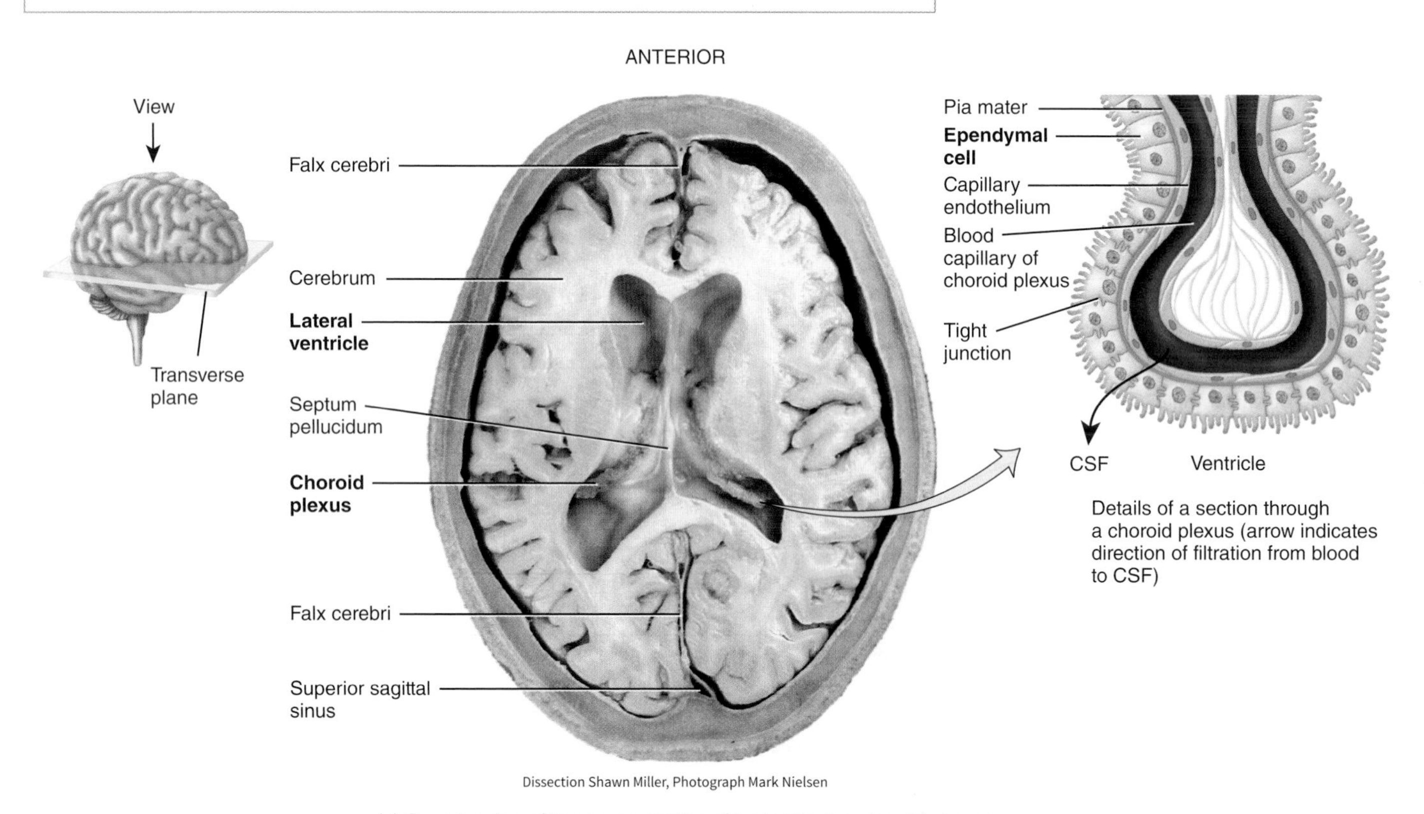

(a) Superior view of transverse section of brain showing choroid plexuses

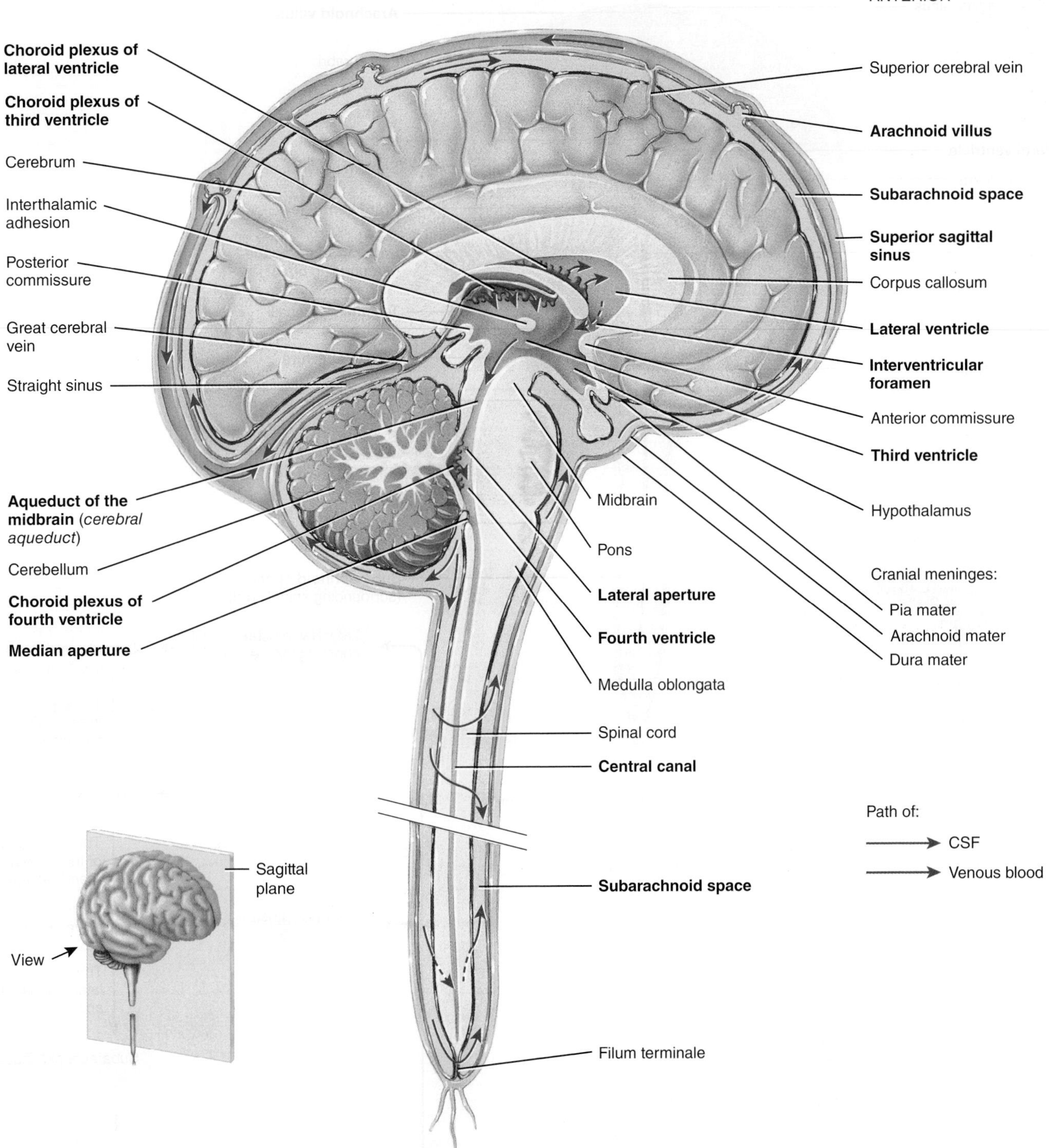

(b) Sagittal section of brain and spinal cord

Figure 14.4 *Continues*

FIGURE 14.4 **Continued**

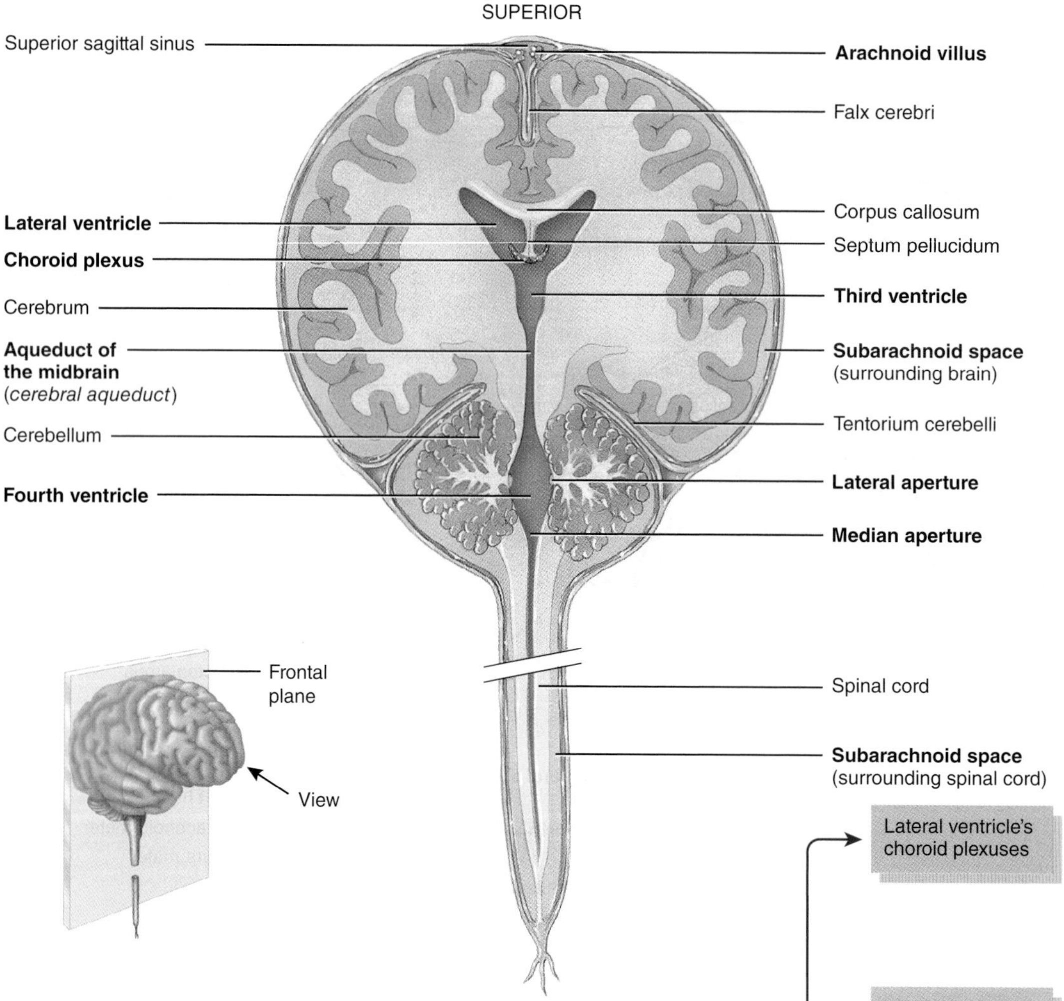

SUPERIOR

Superior sagittal sinus

Arachnoid villus

Falx cerebri

Lateral ventricle

Choroid plexus

Cerebrum

Aqueduct of the midbrain (*cerebral aqueduct*)

Cerebellum

Fourth ventricle

Corpus callosum

Septum pellucidum

Third ventricle

Subarachnoid space (surrounding brain)

Tentorium cerebelli

Lateral aperture

Median aperture

Frontal plane

View

Spinal cord

Subarachnoid space (surrounding spinal cord)

(c) Frontal section of brain and spinal cord

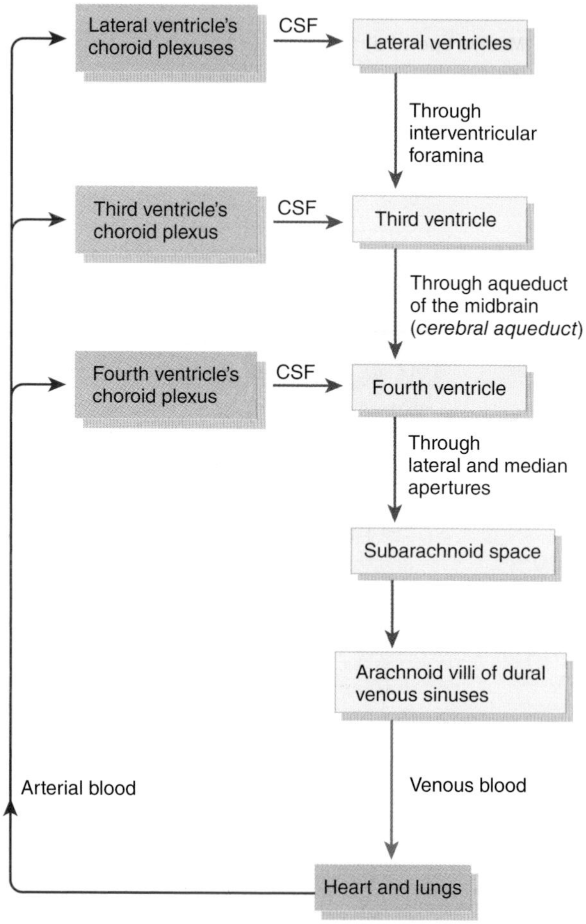

Lateral ventricle's choroid plexuses → CSF → Lateral ventricles

Through interventricular foramina

Third ventricle's choroid plexus → CSF → Third ventricle

Through aqueduct of the midbrain (*cerebral aqueduct*)

Fourth ventricle's choroid plexus → CSF → Fourth ventricle

Through lateral and median apertures

Subarachnoid space

Arachnoid villi of dural venous sinuses

Venous blood

Heart and lungs

Arterial blood

(d) Summary of the formation, circulation, and absorption of cerebrospinal fluid (CSF)

Q Where is CSF reabsorbed?

14.3 The Brainstem and Reticular Formation

OBJECTIVE

- **Describe** the structures and functions of the brainstem and reticular formation.

The brainstem is the part of the brain between the spinal cord and the diencephalon. It consists of three structures: (1) medulla oblongata, (2) pons, and (3) midbrain. Extending through the brainstem is the reticular formation, a netlike region of interspersed gray and white matter.

Medulla Oblongata

The **medulla oblongata** (me-DOOL-la ob′-long-GA-ta), or more simply the *medulla*, is continuous with the superior part of the spinal cord; it forms the inferior part of the brainstem (Figure 14.5; see also Figure 14.1). The medulla begins at the foramen magnum and extends to the inferior border of the pons, a distance of about 3 cm (1.2 in.).

The medulla's white matter contains all sensory (ascending) tracts and motor (descending) tracts that extend between the spinal cord and other parts of the brain. Some of the white matter forms bulges on the anterior aspect of the medulla. These protrusions, called the **pyramids** (Figure 14.6; see also Figure 14.5), are formed by the large corticospinal tracts that pass from the cerebrum to the spinal cord. The corticospinal tracts control voluntary movements of

FIGURE 14.5 **Medulla oblongata in relation to the rest of the brainstem.**

The brainstem consists of the medulla oblongata, pons, and midbrain.

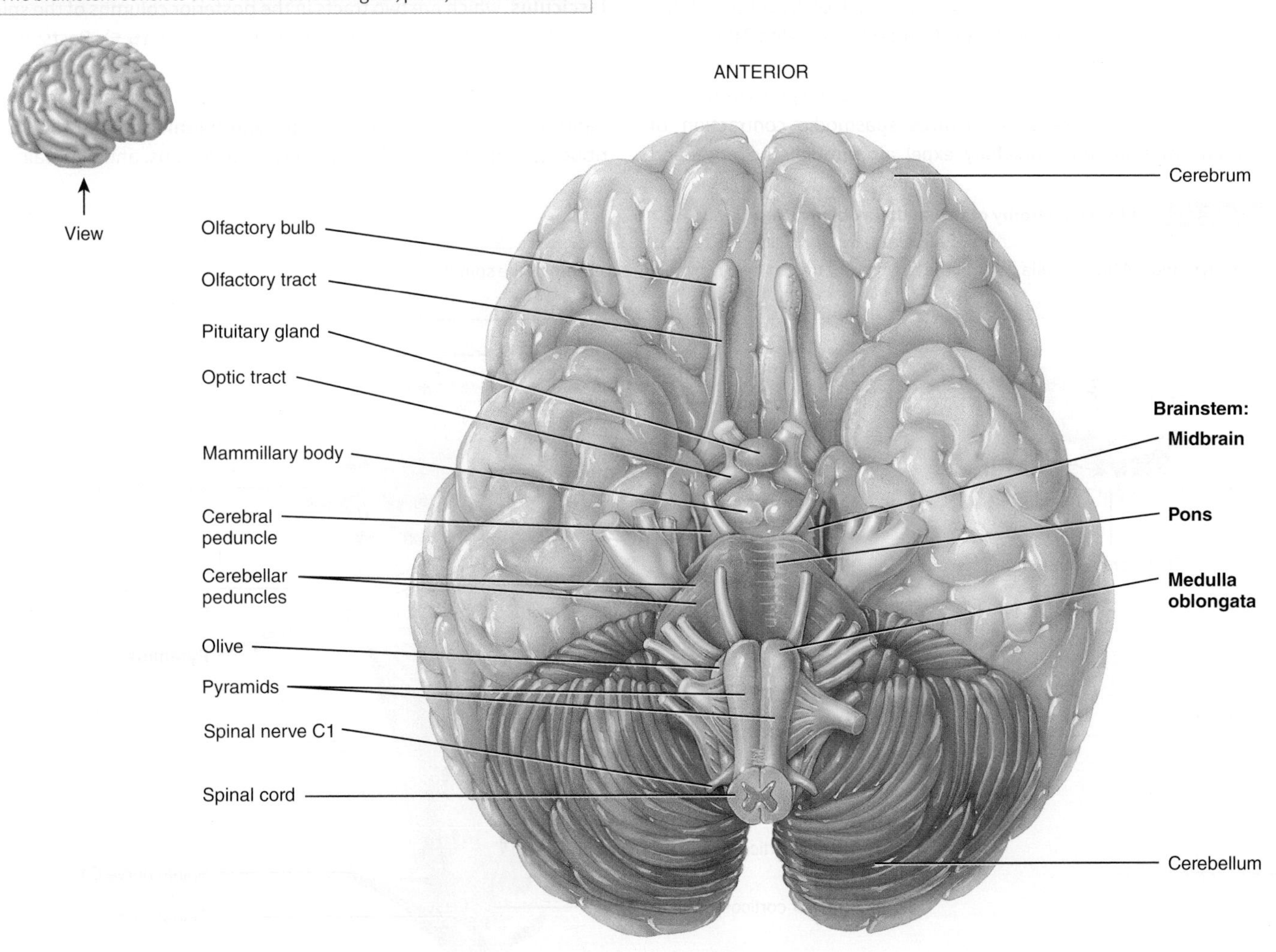

Q What part of the brainstem contains the pyramids? The cerebral peduncles? Literally means "bridge"?

the limbs and trunk (see Figure 16.10). Just superior to the junction of the medulla with the spinal cord, 90% of the axons in the left pyramid cross to the right side, and 90% of the axons in the right pyramid cross to the left side. This crossing is called the **decussation of pyramids** (dē′-ku-SĀ-shun; *decuss* = crossing) and explains why each side of the brain controls voluntary movements on the opposite side of the body.

The medulla also contains several **nuclei**. (Recall that a nucleus is a collection of neuronal cell bodies within the CNS.) Some of these nuclei control vital body functions. Examples of nuclei in the medulla that regulate vital activities include the cardiovascular center and the medullary rhythmicity center. The **cardiovascular (CV) center** regulates the rate and force of the heartbeat and the diameter of blood vessels (see Figure 21.13). The **medullary respiratory center** adjusts the basic rhythm of breathing (see Figure 23.23).

Besides regulating heartbeat, blood vessel diameter, and the normal breathing rhythm, nuclei in the medulla also control reflexes for vomiting, swallowing, sneezing, coughing, and hiccupping. The **vomiting center** of the medulla causes **vomiting**, the forcible expulsion of the contents of the upper gastrointestinal (GI) tract through the mouth (see Section 24.9). The **deglutition center** (dē-gloo-TISH-un) of the medulla promotes **deglutition** (swallowing) of a mass of food that has moved from the oral cavity of the mouth into the pharynx (throat) (see Section 24.8). **Sneezing** involves spasmodic contraction of breathing muscles that forcefully expel air through the nose and mouth. **Coughing** involves a long-drawn and deep inhalation and then a strong exhalation that suddenly sends a blast of air through the upper respiratory passages. **Hiccupping** is caused by spasmodic contractions of the diaphragm (a muscle of breathing) that ultimately result in the production of a sharp sound on inhalation. Sneezing, coughing, and hiccupping are described in more detail in Table 23.2.

Just lateral to each pyramid is an oval-shaped swelling called an **olive** (see Figures 14.5, 14.6). Within the olive is the **inferior olivary nucleus**, which receives input from the cerebral cortex, red nucleus of the midbrain, and spinal cord. Neurons of the inferior olivary nucleus extend their axons into the cerebellum, where they regulate the activity of cerebellar neurons. By influencing cerebellar neuron activity, the inferior olivary nucleus provides instructions that the cerebellum uses to make adjustments to muscle activity as you learn new motor skills.

Nuclei associated with sensations of touch, pressure, vibration, and conscious proprioception are located in the posterior part of the medulla. These nuclei are the right and left **gracile nucleus** (GRAS-il = slender) and **cuneate nucleus** (KŪ-nē-āt = wedge). Ascending sensory axons of the **gracile fasciculus** (fa-SIK-ū-lus) and the **cuneate fasciculus**, which are two tracts in the posterior columns of the spinal cord, form synapses in these nuclei (see Figure 16.5). Postsynaptic neurons then relay the sensory information to the thalamus on the opposite side of the brain. The axons ascend to the thalamus in a band of white matter called the **medial lemniscus** (lem-NIS-kus = ribbon), which extends through the medulla, pons, and midbrain (see

FIGURE 14.6 Internal anatomy of the medulla oblongata.

The pyramids of the medulla contain the large motor tracts that run from the cerebrum to the spinal cord.

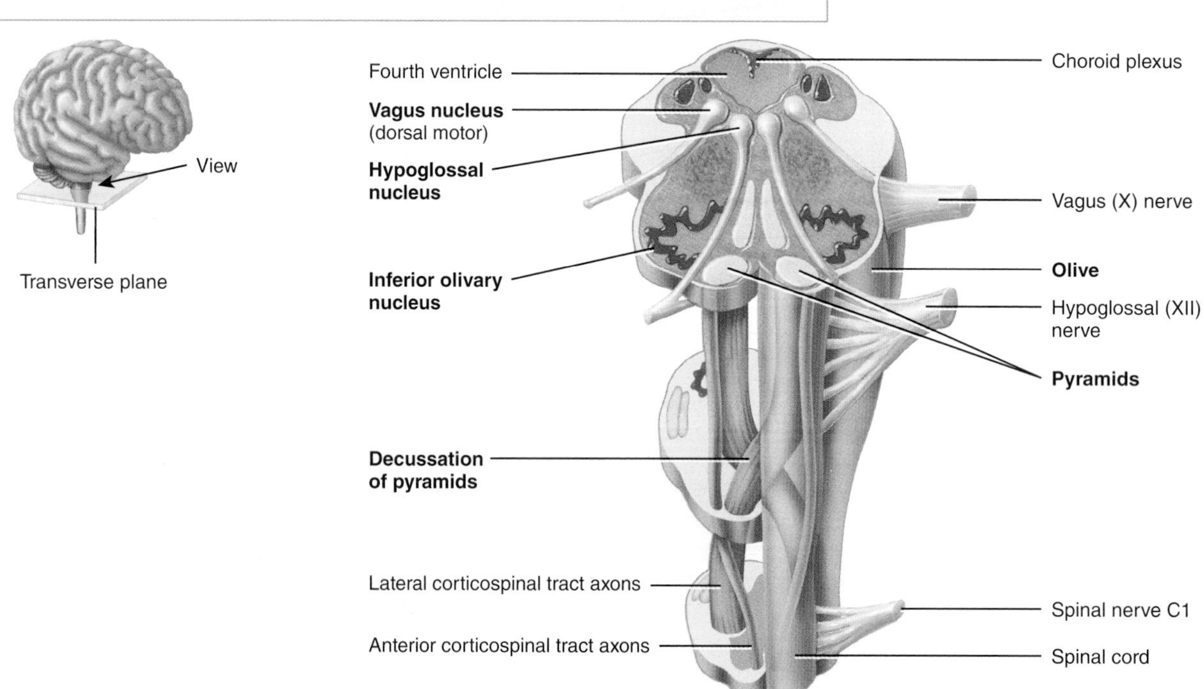

Q What does decussation mean? What is the functional consequence of decussation of the pyramids?

Figure 14.7b). The tracts of the posterior columns and the axons of the medial lemniscus are collectively known as the **posterior column–medial lemniscus pathway**.

The medulla also contains nuclei that are components of sensory pathways for gustation (taste), audition (hearing), and equilibrium (balance). The **gustatory nucleus** (GUS-ta-tō′-rē) of the medulla is part of the gustatory pathway from the tongue to the brain; it receives gustatory input from the taste buds of the tongue (see Figure 17.3e). The **cochlear nuclei** (KOK-lē-ar) of the medulla are part of the auditory pathway from the inner ear to the brain; they receive auditory input from the cochlea of the inner ear (see Figure 17.23). The **vestibular nuclei** (ves-TIB-ū-lar) of the medulla and pons are components of the equilibrium pathway from the inner ear to the brain; they receive sensory information associated with equilibrium from *proprioceptors* (receptors that provide information regarding body position and movements) in the vestibular apparatus of the inner ear (see Figure 17.26).

Finally, the medulla contains nuclei associated with the following five pairs of cranial nerves.

1. **Vestibulocochlear (VIII) nerves.** Several nuclei in the medulla receive sensory input from and provide motor output to the cochlea of the internal ear via the vestibulocochlear nerves. These nerves convey impulses related to hearing.
2. **Glossopharyngeal (IX) nerves.** Nuclei in the medulla relay sensory and motor impulses related to taste, swallowing, and salivation via the glossopharyngeal nerves.
3. **Vagus (X) nerves.** Nuclei in the medulla receive sensory impulses from and provide motor impulses to the pharynx and larynx and many thoracic and abdominal viscera via the vagus nerves.
4. **Accessory (XI) nerves (cranial portion).** These fibers are actually part of the vagus (X) nerves. Nuclei in the medulla are the origin for nerve impulses that control swallowing via the vagus nerves (cranial portion of the accessory nerves).
5. **Hypoglossal (XII) nerves.** Nuclei in the medulla are the origin for nerve impulses that control tongue movements during speech and swallowing via the hypoglossal nerves.

Pons

The **pons** (= bridge) lies directly superior to the medulla and anterior to the cerebellum and is about 2.5 cm (1 in.) long (see Figures 14.1, 14.5). Like the medulla, the pons consists of both nuclei and tracts. As its name implies, the pons is a bridge that connects parts of the brain with one another. These connections are provided by bundles of axons. Some axons of the pons connect the right and left sides of the cerebellum. Others are part of ascending sensory tracts and descending motor tracts.

See Clinical Connection: Injury to the Medulla

The pons has two major structural components: a ventral region and a dorsal region. The ventral region of the pons forms a large synaptic relay station consisting of scattered gray centers called the **pontine nuclei** (PON-tīn). Entering and exiting these nuclei are numerous white matter tracts, each of which provides a connection between the cortex (outer layer) of a cerebral hemisphere and that of the opposite hemisphere of the cerebellum. This complex circuitry plays an essential role in coordinating and maximizing the efficiency of voluntary motor output throughout the body. The dorsal region of the pons is more like the other regions of the brainstem, the medulla and midbrain. It contains ascending and descending tracts along with the nuclei of cranial nerves.

Also within the pons is the **pontine respiratory group**, shown in Figure 23.23. Together with the medullary respiratory center, the pontine respiratory group helps control breathing.

The pons also contains nuclei associated with the following four pairs of cranial nerves.

1. **Trigeminal (V) nerves.** Nuclei in the pons receive sensory impulses for somatic sensations from the head and face and provide motor impulses that govern chewing via the trigeminal nerves.
2. **Abducens (VI) nerves.** Nuclei in the pons provide motor impulses that control eyeball movement via the abducens nerves.
3. **Facial (VII) nerves.** Nuclei in the pons receive sensory impulses for taste and provide motor impulses to regulate secretion of saliva and tears and contraction of muscles of facial expression via the facial nerves.
4. **Vestibulocochlear (VIII) nerves.** Nuclei in the pons receive sensory impulses from and provide motor impulses to the vestibular apparatus via the vestibulocochlear nerves. These nerves convey impulses related to balance and equilibrium.

Midbrain

The **midbrain** or *mesencephalon* extends from the pons to the diencephalon (see Figures 14.1, 14.5) and is about 2.5 cm (1 in.) long. The aqueduct of the midbrain (cerebral aqueduct) passes through the midbrain, connecting the third ventricle above with the fourth ventricle below. Like the medulla and the pons, the midbrain contains both nuclei and tracts (Figure 14.7).

The anterior part of the midbrain contains paired bundles of axons known as the **cerebral peduncles** (pe-DUNK-kuls = little feet; see Figures 14.5, 14.7b). The cerebral peduncles consist of axons of the corticospinal, corticobulbar, and corticopontine tracts, which conduct nerve impulses from motor areas in the cerebral cortex to the spinal cord, medulla, and pons, respectively.

The posterior part of the midbrain, called the **tectum** (TEK-tum = roof), contains four rounded elevations (Figure 14.7a). The two superior elevations, nuclei known as the **superior colliculi** (ko-LIK-ū-lī = little hills; singular is *colliculus*), serve as reflex centers for certain visual activities. Through neural circuits from the retina of the eye to the superior colliculi to the extrinsic eye muscles, visual stimuli elicit eye movements for tracking moving images (such as a moving car) and scanning stationary images (as you are doing to read this sentence). The superior colliculi are also responsible for reflexes that govern movements of the head, eyes, and trunk in response to visual stimuli. The two inferior elevations, the **inferior colliculi**, are part of the auditory pathway, relaying impulses from the receptors for hearing in the inner ear to the brain. These two nuclei are also reflex centers for the *startle reflex*, sudden movements of the head, eyes, and trunk that occur when you are surprised by a loud noise such as a gunshot.

The midbrain contains several other nuclei, including the left and right **substantia nigra** (sub-STAN-shē-a = substance; NĪ-gra = black), which are large and darkly pigmented (**Figure 14.7b**). Neurons that release dopamine, extending from the substantia nigra to the basal nuclei, help control subconscious muscle activities. Loss of these neurons is associated with Parkinson's disease (see Disorders:

FIGURE 14.7 Midbrain.

The midbrain connects the pons to the diencephalon.

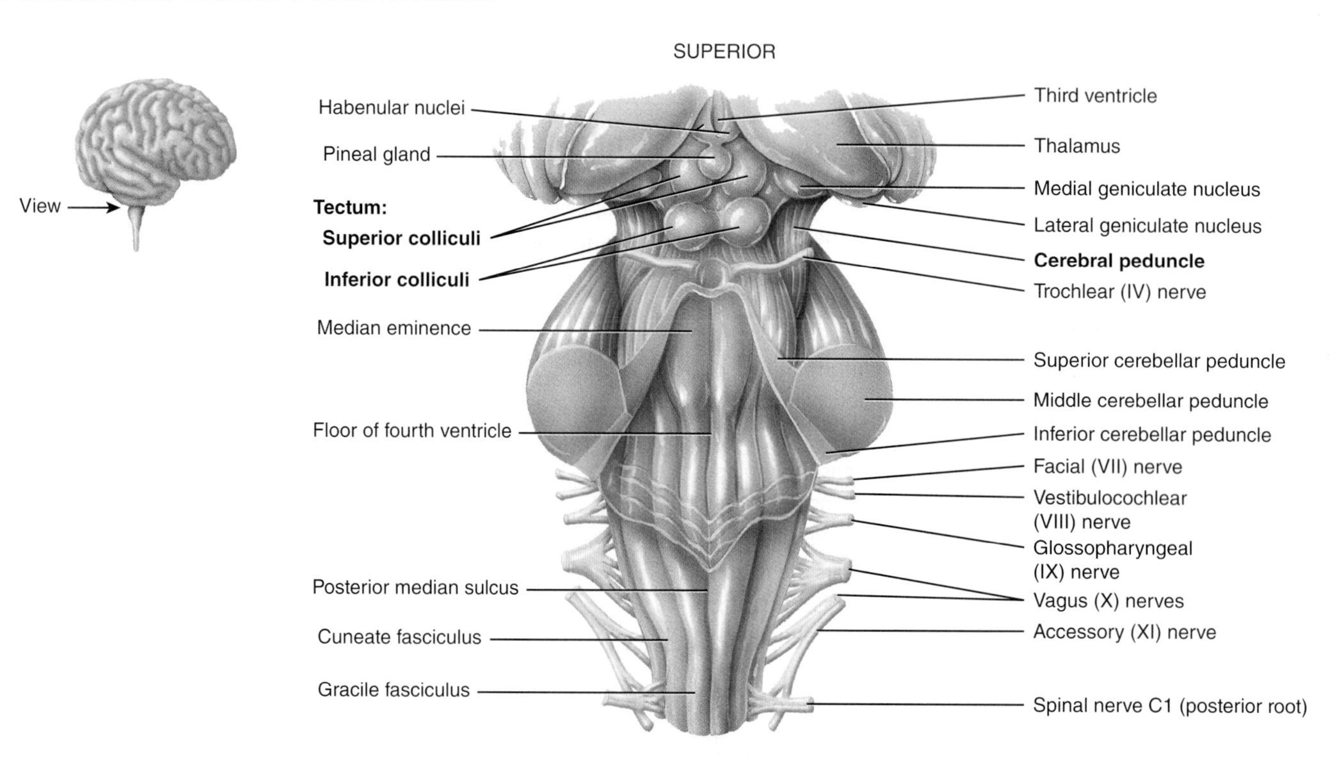

(a) Posterior view of midbrain in relation to brainstem

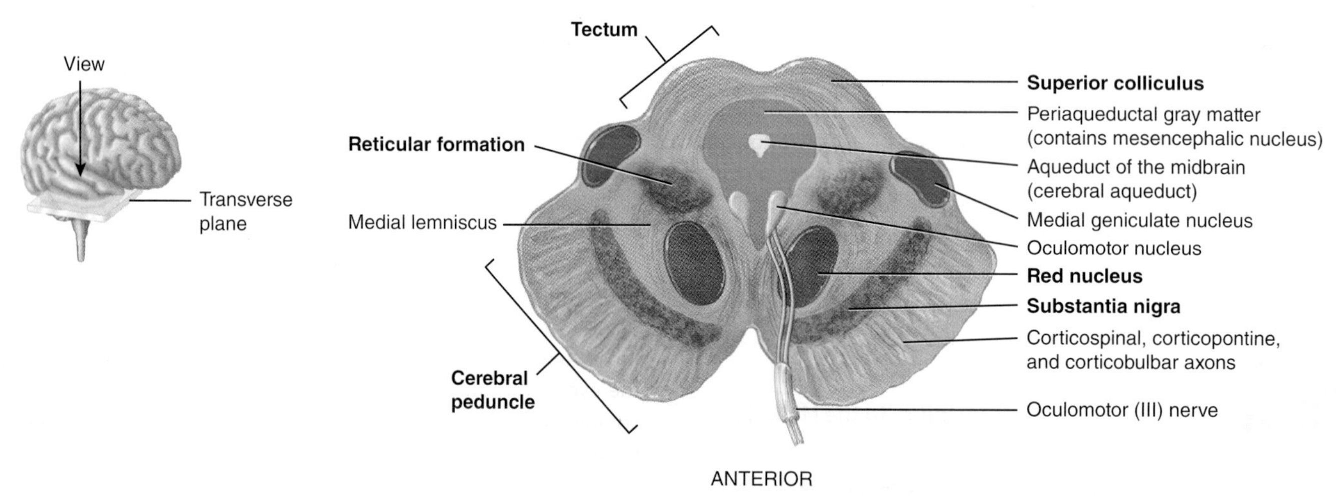

(b) Transverse section of midbrain

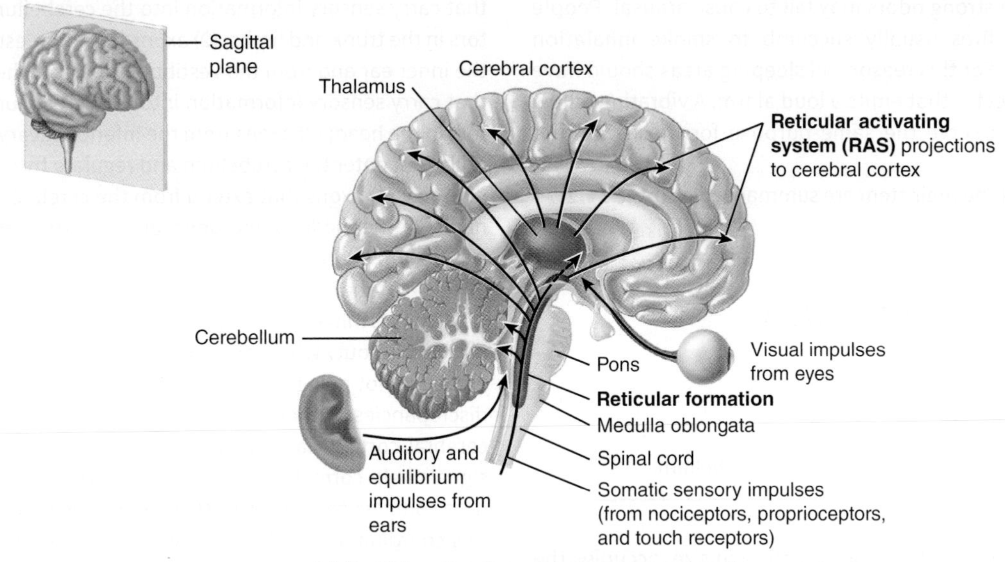

(c) Sagittal section through brain and spinal cord showing the reticular formation

Q What is the importance of the cerebral peduncles?

Homeostatic Imbalances in Chapter 16 of Study Guide). Also present are the left and right **red nuclei**, which look reddish due to their rich blood supply and an iron-containing pigment in their neuronal cell bodies. Axons from the cerebellum and cerebral cortex form synapses in the red nuclei, which help control muscular movements.

Still other nuclei in the midbrain are associated with two pairs of cranial nerves.

1. **Oculomotor (III) nerves.** Nuclei in the midbrain provide motor impulses that control movements of the eyeball, while accessory oculomotor nuclei provide motor control to the smooth muscles that regulate constriction of the pupil and changes in shape of the lens via the oculomotor nerves.
2. **Trochlear (IV) nerves.** Nuclei in the midbrain provide motor impulses that control movements of the eyeball via the trochlear nerves.

Reticular Formation

In addition to the well-defined nuclei already described, much of the brainstem consists of small clusters of neuronal cell bodies (gray matter) interspersed among small bundles of myelinated axons (white matter). The broad region where white matter and gray matter exhibit a netlike arrangement is known as the **reticular formation** (re-TIK-ū-lar; *ret-* = net; **Figure 14.7c**). It extends from the superior part of the spinal cord, throughout the brainstem, and into the inferior part of the diencephalon. Neurons within the reticular formation have both ascending (sensory) and descending (motor) functions.

The ascending portion of the reticular formation is called the **reticular activating system (RAS)**, which consists of sensory axons that project to the cerebral cortex, both directly and through the thalamus. Many sensory stimuli can activate the ascending portion of the RAS. Among these are visual and auditory stimuli; mental activities; stimuli from pain, touch, and pressure receptors; and receptors in our limbs and head that keep us aware of the position of our body parts. Perhaps the most important function of the RAS is **consciousness**, a state of wakefulness in which an individual is fully alert, aware, and oriented. Visual and auditory stimuli and mental activities can stimulate the RAS to help maintain consciousness. The RAS is also active during **arousal**, or awakening from sleep. Another function of the RAS is to help maintain **attention** (concentrating on a single object or thought) and *alertness.* The RAS also prevents **sensory overload** (excessive visual and/or auditory stimulation) by filtering out insignificant information so that it does not reach consciousness. For example, while waiting in the hallway for your anatomy class to begin, you may be unaware of all the noise around you while reviewing your notes for class. Inactivation of the RAS produces **sleep**, a state of partial consciousness from which an individual can be aroused. Damage to the RAS, on the other hand, results in **coma**, a state of unconsciousness from which an individual cannot be aroused. In the lightest stages of coma, brainstem and spinal cord reflexes persist, but in the deepest states even those reflexes are lost, and if respiratory and cardiovascular controls are lost, the patient dies. Drugs such as melatonin affect the RAS by helping to induce sleep, and general anesthetics turn off consciousness via the RAS. The descending portion of the RAS has connections to the cerebellum and spinal cord and helps regulate **muscle tone**, the slight degree of involuntary contraction in normal resting skeletal muscles. This portion of the RAS also assists in the regulation of heart rate, blood pressure, and respiratory rate.

Even though the RAS receives input from the eyes, ears, and other sensory receptors, there is no input from receptors for the sense of smell; even strong odors may fail to cause arousal. People who die in house fires usually succumb to smoke inhalation without awakening. For this reason, all sleeping areas should have a nearby smoke detector that emits a loud alarm. A vibrating pillow or flashing light can serve the same purpose for those who are hearing impaired.

The functions of the brainstem are summarized in Table 14.2.

14.4 The Cerebellum

OBJECTIVE

- **Describe** the structure and functions of the cerebellum.

The **cerebellum**, second only to the cerebrum in size, occupies the inferior and posterior aspects of the cranial cavity. Like the cerebrum, the cerebellum has a highly folded surface that greatly increases the surface area of its outer gray matter cortex, allowing for a greater number of neurons. The cerebellum accounts for about a tenth of the brain mass yet contains nearly half of the neurons in the brain. The cerebellum is posterior to the medulla and pons and inferior to the posterior portion of the cerebrum (see Figure 14.1). A deep groove known as the **transverse fissure**, along with the **tentorium cerebelli**, which supports the posterior part of the cerebrum, separates the cerebellum from the cerebrum (see Figures 14.2b, 14.11b).

In superior or inferior views, the shape of the cerebellum resembles a butterfly. The central constricted area is the **vermis** (= worm), and the lateral "wings" or lobes are the **cerebellar hemispheres** (Figure 14.8a, b). Each hemisphere consists of lobes separated by deep and distinct fissures. The **anterior lobe** and **posterior lobe** govern subconscious aspects of skeletal muscle movements. The **flocculonodular lobe** (flok-ū-lō-NOD-ū-lar; *flocculo-* = wool-like tuft) on the inferior surface contributes to equilibrium and balance.

The superficial layer of the cerebellum, called the **cerebellar cortex**, consists of gray matter in a series of slender, parallel folds called **folia** (= leaves). Deep to the gray matter are tracts of white matter called **arbor vitae** (AR-bor VĪ-tē = tree of life) that resemble branches of a tree. Even deeper, within the white matter, are the **cerebellar nuclei**, regions of gray matter that give rise to axons carrying impulses from the cerebellum to other brain centers.

Three paired **cerebellar peduncles** attach the cerebellum to the brainstem (see Figure 14.8b). These bundles of white matter consist of axons that conduct impulses between the cerebellum and other parts of the brain. The **superior cerebellar peduncles** contain axons that extend from the cerebellum to the red nuclei of the midbrain and to several nuclei of the thalamus. The **middle cerebellar peduncles** are the largest peduncles; their axons carry impulses for voluntary movements from the pontine nuclei (which receive input from motor areas of the cerebral cortex) into the cerebellum. The **inferior cerebellar peduncles** consist of (1) axons of the spinocerebellar tracts that carry sensory information into the cerebellum from proprioceptors in the trunk and limbs; (2) axons from the vestibular apparatus of the inner ear and from the vestibular nuclei of the medulla and pons that carry sensory information into the cerebellum from proprioceptors in the head; (3) axons from the inferior olivary nucleus of the medulla that enter the cerebellum and regulate the activity of cerebellar neurons; (4) axons that extend from the cerebellum to the vestibular nuclei of the medulla and pons; and (5) axons that extend from the cerebellum to the reticular formation.

The primary function of the cerebellum is to evaluate how well movements initiated by motor areas in the cerebrum are actually being carried out. When movements initiated by the cerebral motor areas are not being carried out correctly, the cerebellum detects the discrepancies. It then sends feedback signals to motor areas of the cerebral cortex, via its connections to the thalamus. The feedback signals help correct the errors, smooth the movements, and coordinate complex sequences of skeletal muscle contractions. Aside from this coordination of skilled movements, the cerebellum is the main brain region that regulates posture and balance. These aspects of cerebellar function make possible all skilled muscular activities, from catching a baseball to dancing to speaking. The presence of reciprocal connections between the cerebellum and association areas of the cerebral cortex suggests that the cerebellum may also have nonmotor functions such as cognition (acquisition of knowledge) and language processing. This view is supported by imaging studies using MRI and PET. Studies also suggest that the cerebellum may play a role in processing sensory information.

The functions of the cerebellum are summarized in Table 14.2.

14.5 The Diencephalon

OBJECTIVE

- **Describe** the components and functions of the diencephalon (thalamus, hypothalamus, and epithalamus).

The diencephalon forms a central core of brain tissue just superior to the midbrain. It is almost completely surrounded by the cerebral hemispheres and contains numerous nuclei involved in a wide variety of sensory and motor processing between higher and lower brain centers. The diencephalon extends from the brainstem to the cerebrum and surrounds the third ventricle; it includes the thalamus, hypothalamus, and epithalamus. Projecting from the hypothalamus is the hypophysis, or pituitary gland. Portions of the diencephalon in the wall of the third ventricle are called circumventricular organs and will be discussed shortly. The optic tracts carrying neurons from the retina enter the diencephalon.

Thalamus

The **thalamus** (THAL-a-mus = inner chamber), which measures about 3 cm (1.2 in.) in length and makes up 80% of the diencephalon, consists of paired oval masses of gray matter organized into

FIGURE 14.8 **Cerebellum.**

The cerebellum coordinates skilled movements and regulates posture and balance.

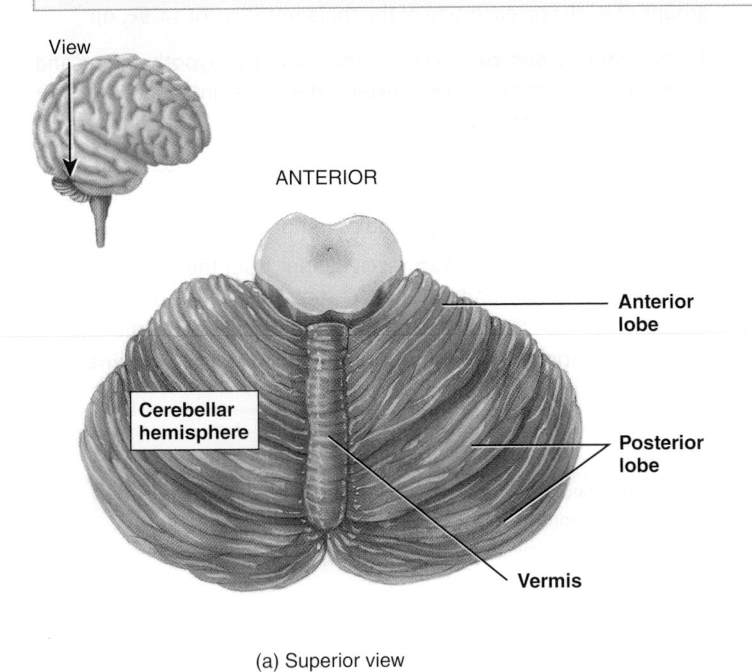

(a) Superior view

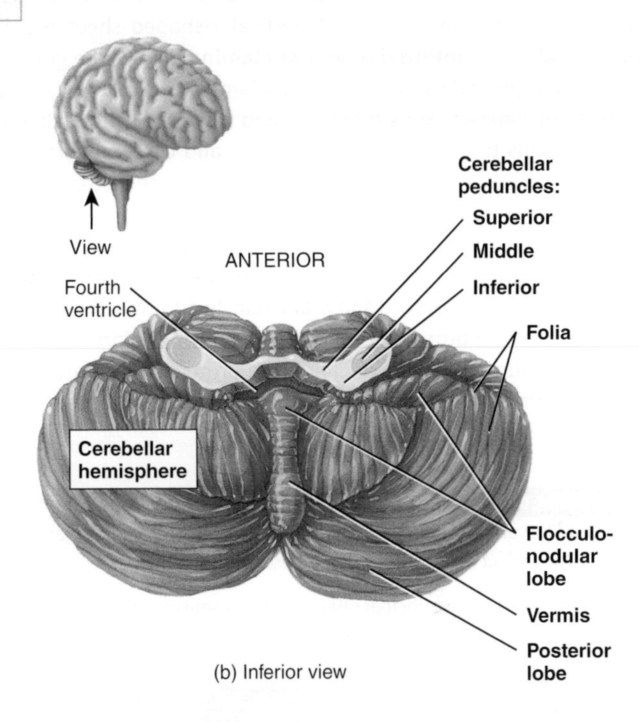

(b) Inferior view

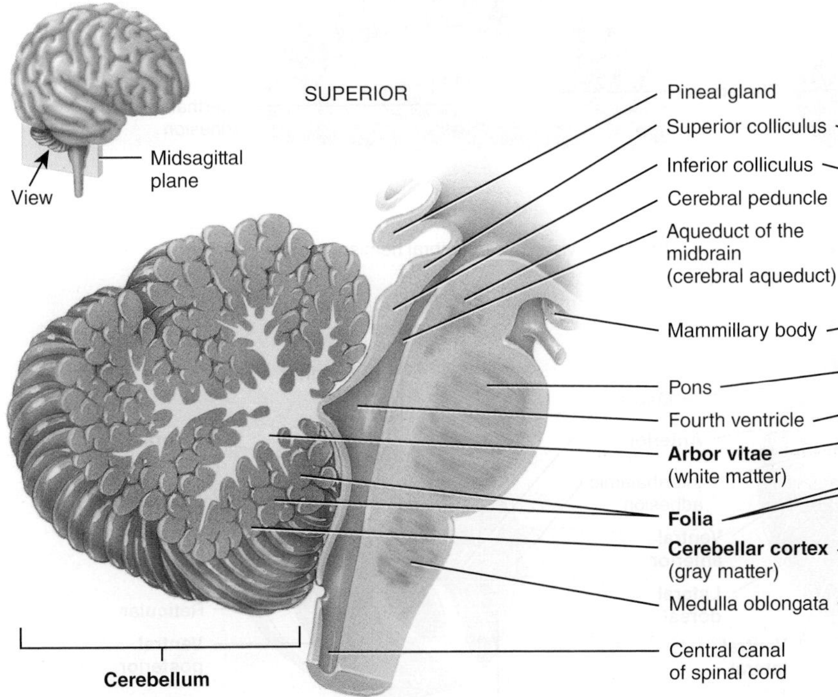

(c) Midsagittal section of cerebellum and brainstem

Dissection Shawn Miller, Photograph Mark Nielsen

(d) Midsagittal section

Q Which structures contain the axons that carry information into and out of the cerebellum?

See Clinical Connection: Ataxia

nuclei with interspersed tracts of white matter (**Figure 14.9**). A bridge of gray matter called the **interthalamic adhesion** (*intermediate mass*) joins the right and left halves of the thalamus in about 70% of human brains. A vertical Y-shaped sheet of white matter called the **internal medullary lamina** divides the gray matter of the right and left sides of the thalamus (**Figure 14.9c**). It consists of myelinated axons that enter and leave the various thalamic nuclei. Axons that connect the thalamus and cerebral cortex pass through the **internal capsule**, a thick band of white matter lateral to the thalamus (see **Figure 14.13b**).

The thalamus is the major relay station for most sensory impulses that reach the primary sensory areas of the cerebral cortex from the spinal cord and brainstem. In addition, the thalamus contributes to motor functions by transmitting information from the cerebellum and basal nuclei to the primary motor area of the cerebral cortex. The thalamus also relays nerve impulses between different areas of the cerebrum and plays a role in the maintenance of consciousness.

Based on their positions and functions, there are seven major groups of nuclei on each side of the thalamus (**Figure 14.9c, d**):

1. The **anterior nucleus** receives input from the hypothalamus and sends output to the limbic system (described in Section 14.6). It functions in emotions and memory.
2. The **medial nuclei** receive input from the limbic system and basal nuclei and send output to the cerebral cortex. They function in emotions, learning, memory, and cognition (thinking and knowing).
3. Nuclei in the **lateral group** receive input from the limbic system, superior colliculi, and cerebral cortex and send output to the cerebral cortex. The **lateral dorsal nucleus** functions in the expression of emotions. The **lateral posterior nucleus** and **pulvinar nucleus** help integrate sensory information.

FIGURE 14.9 Thalamus. Note the position of the thalamus in the lateral view (a) and in the medial view (b). The various thalamic nuclei shown in (c) and (d) are correlated by color to the cortical regions to which they project in (a) and (b).

The thalamus is the principal relay station for sensory impulses that reach the cerebral cortex from other parts of the brain and the spinal cord.

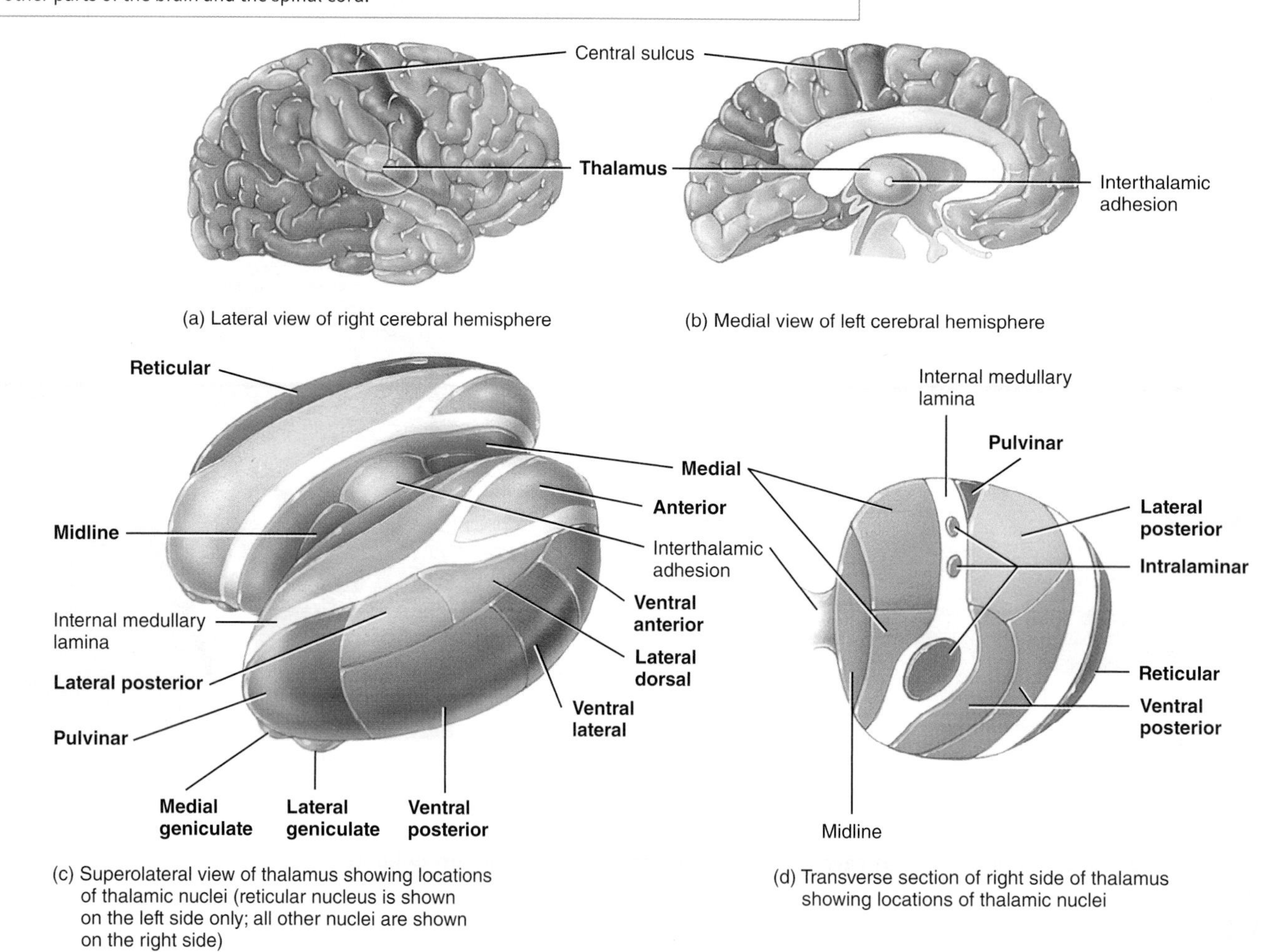

(a) Lateral view of right cerebral hemisphere

(b) Medial view of left cerebral hemisphere

(c) Superolateral view of thalamus showing locations of thalamic nuclei (reticular nucleus is shown on the left side only; all other nuclei are shown on the right side)

(d) Transverse section of right side of thalamus showing locations of thalamic nuclei

Q What structure usually connects the right and left halves of the thalamus?

4. Five nuclei are part of the **ventral group**. The **ventral anterior nucleus** receives input from the basal nuclei and sends output to motor areas of the cerebral cortex; it plays a role in movement control. The **ventral lateral nucleus** receives input from the cerebellum and basal nuclei and sends output to motor areas of the cerebral cortex; it also plays a role in movement control. The **ventral posterior nucleus** relays impulses for somatic sensations such as touch, pressure, vibration, itch, tickle, temperature, pain, and proprioception from the face and body to the cerebral cortex. The **lateral geniculate nucleus** (je-NIK-ū-lat = bent like a knee) relays visual impulses for sight from the retina to the primary visual area of the cerebral cortex. The **medial geniculate nucleus** relays auditory impulses for hearing from the ear to the primary auditory area of the cerebral cortex.
5. **Intralaminar nuclei** (in′-tra-LA-mi′-nar) lie within the internal medullary lamina and make connections with the reticular formation, cerebellum, basal nuclei, and wide areas of the cerebral cortex. They function in arousal (activation of the cerebral cortex from the brainstem reticular formation) and integration of sensory and motor information.
6. The **midline nucleus** forms a thin band adjacent to the third ventricle and has a presumed function in memory and olfaction.
7. The **reticular nucleus** surrounds the lateral aspect of the thalamus, next to the internal capsule. This nucleus monitors, filters, and integrates activities of other thalamic nuclei.

Hypothalamus

The **hypothalamus** (hī′-pō-THAL-a-mus; *hypo-* = under) is a small part of the diencephalon located inferior to the thalamus. It is composed of a dozen or so nuclei in four major regions:

1. The **mammillary region** (MAM-i-ler-ē; *mammill-* = nipple-shaped), adjacent to the midbrain, is the most posterior part of the hypothalamus. It includes the *mammillary bodies* and *posterior hypothalamic nuclei* (Figure 14.10). The **mammillary bodies** are two small, rounded projections that serve as relay stations for reflexes related to the sense of smell.
2. The **tuberal region** (TOO-ber-al), the widest part of the hypothalamus, includes the *dorsomedial nucleus*, *ventromedial nucleus*, and *arcuate nucleus* (AR-kū-āt), plus the stalklike **infundibulum** (in-fun-DIB-ū-lum = funnel), which connects the pituitary gland to the hypothalamus (Figure 14.10). The **median eminence** is a slightly raised region that encircles the infundibulum (see Figure 14.7a).
3. The **supraoptic region** (*supra-* = above; *-optic* = eye) lies superior to the optic chiasm (point of crossing of optic nerves) and contains the *paraventricular nucleus*, *supraoptic nucleus*, *anterior hypothalamic nucleus*, and *suprachiasmatic nucleus* (soo′-pra-kī′-az-MA-tik) (Figure 14.10). Axons from the paraventricular and supraoptic nuclei form the hypothalamohypophyseal tract (hī′-pō-thal′-a-mō-hī-pō-FIZ-ē-al), which extends through the infundibulum to the posterior lobe of the pituitary (see Figure 18.8).

FIGURE 14.10 Hypothalamus. Selected portions of the hypothalamus and a three-dimensional representation of hypothalamic nuclei are shown (after Netter).

The hypothalamus controls many body activities and is an important regulator of homeostasis.

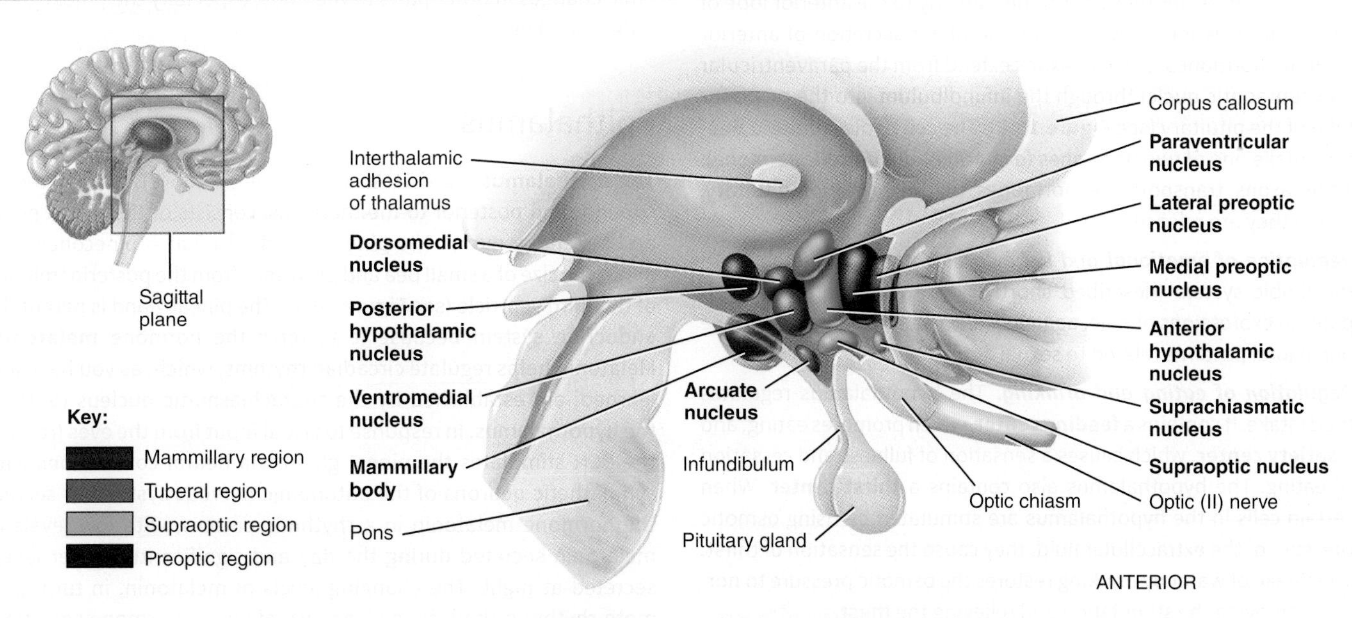

Sagittal section of brain showing hypothalamic nuclei

Q What are the four major regions of the hypothalamus, from posterior to anterior?

4. The **preoptic region** anterior to the supraoptic region is usually considered part of the hypothalamus because it participates with the hypothalamus in regulating certain autonomic activities. The preoptic region contains the *medial* and *lateral preoptic nuclei* (Figure 14.10).

The hypothalamus controls many body activities and is one of the major regulators of homeostasis. Sensory impulses related to both somatic and visceral senses arrive at the hypothalamus, as do impulses from receptors for vision, taste, and smell. Other receptors within the hypothalamus itself continually monitor osmotic pressure, blood glucose level, certain hormone concentrations, and the temperature of blood. The hypothalamus has several very important connections with the pituitary gland and produces a variety of hormones, which are described in more detail in Chapter 18. Some functions can be attributed to specific hypothalamic nuclei, but others are not so precisely localized. Important functions of the hypothalamus include the following:

- ***Control of the ANS.*** The hypothalamus controls and integrates activities of the autonomic nervous system, which regulates contraction of smooth muscle and cardiac muscle and the secretions of many glands. Axons extend from the hypothalamus to parasympathetic and sympathetic nuclei in the brainstem and spinal cord. Through the ANS, the hypothalamus is a major regulator of visceral activities, including regulation of heart rate, movement of food through the gastrointestinal tract, and contraction of the urinary bladder.
- ***Production of hormones.*** The hypothalamus produces several hormones and has two types of important connections with the pituitary gland, an endocrine gland located inferior to the hypothalamus (see Figure 14.1). First, hypothalamic hormones known as *releasing hormones* and *inhibiting hormones* are released into capillary networks in the median eminence (see Figure 18.5). The bloodstream carries these hormones directly to the anterior lobe of the pituitary, where they stimulate or inhibit secretion of anterior pituitary hormones. Second, axons extend from the paraventricular and supraoptic nuclei through the infundibulum into the posterior lobe of the pituitary (see Figure 18.8). The cell bodies of these neurons make one of two hormones (*oxytocin* or *antidiuretic hormone*). Their axons transport the hormones to the posterior pituitary, where they are released.
- ***Regulation of emotional and behavioral patterns.*** Together with the limbic system (described shortly), the hypothalamus participates in expressions of rage, aggression, pain, and pleasure, and the behavioral patterns related to sexual arousal.
- ***Regulation of eating and drinking.*** The hypothalamus regulates food intake. It contains a **feeding center**, which promotes eating, and a **satiety center**, which causes a sensation of fullness and cessation of eating. The hypothalamus also contains a **thirst center**. When certain cells in the hypothalamus are stimulated by rising osmotic pressure of the extracellular fluid, they cause the sensation of thirst. The intake of water by drinking restores the osmotic pressure to normal, removing the stimulation and relieving the thirst.
- ***Control of body temperature.*** The hypothalamus also functions as the body's **thermostat**, which senses body temperature so that it is maintained at a desired setpoint. If the temperature of blood flowing through the hypothalamus is above normal, the hypothalamus directs the autonomic nervous system to stimulate activities that promote heat loss. When blood temperature is below normal, by contrast, the hypothalamus generates impulses that promote heat production and retention.
- ***Regulation of circadian rhythms.*** The suprachiasmatic nucleus (SCN) of the hypothalamus serves as the body's internal biological clock because it establishes **circadian** (*daily*) **rhythms** (ser-KĀ-dē-an), patterns of biological activity (such as the sleep–wake cycle) that occur on a circadian schedule (cycle of about 24 hours). This nucleus receives input from the eyes (retina) and sends output to other hypothalamic nuclei, the reticular formation, and the pineal gland. The visual input to the SCN synchronizes the neurons of the SCN to the light–dark cycle associated with day and night. Without this input, the SCN still promotes biological rhythms, but the rhythms become progressively out of sync with the normal light–dark cycle because the inherent activity of the SCN creates cycles that last about 25 hours instead of 24. Therefore, the SCN must receive light–dark cues from the external environment in order to create rhythms that occur on a 24-hour cycle. The mechanism responsible for the internal clock in an SCN neuron is due to the rhythmic turning on and off of **clock genes** in the cell's nucleus, resulting in alternating levels of **clock proteins** in the cell's cytosol. The clock genes are self-starting. They are turned on automatically and then are transcribed and translated. The resulting clock proteins accumulate in the cytosol and then enter the nucleus to turn off the clock genes. Gradually, the clock proteins degrade, and without these proteins present, the clock genes are activated again and the cycle repeats, with each cycle corresponding to a 24-hours period. The alternating levels of clock proteins causes rhythmic changes in the output of SCN neurons, which in turn causes rhythmic changes in other parts of the body, especially the pineal gland (described next).

Epithalamus

The **epithalamus** (ep′-i-THAL-a-mus; *epi-* = above), a small region superior and posterior to the thalamus, consists of the pineal gland and habenular nuclei. The **pineal gland** (PĪN-ē-al = pineconelike) is about the size of a small pea and protrudes from the posterior midline of the third ventricle (see Figure 14.1). The pineal gland is part of the endocrine system because it secretes the hormone **melatonin**. Melatonin helps regulate circadian rhythms, which, as you have just learned, are established by the suprachiasmatic nucleus (SCN) of the hypothalamus. In response to visual input from the eyes (retina), the SCN stimulates the pineal gland (via neural connections with sympathetic neurons of the autonomic nervous system) to secrete the hormone melatonin in a rhythmic pattern, with low levels of melatonin secreted during the day and significantly higher levels secreted at night. The changing levels of melatonin, in turn, promote rhythmic changes in sleep, wakefulness, hormone secretion, and body temperature. In addition to its role in regulating circadian rhythms, melatonin is involved in other functions. It induces sleep, serves as an antioxidant, and inhibits reproductive functions in certain animals. As more melatonin is liberated during darkness than in light, this hormone is thought to promote sleepiness. When taken orally, melatonin also appears to contribute to the setting of the body's biological clock by inducing sleep and helping the body to adjust to jet lag. The **habenular nuclei** (ha-BEN-ū-lar), shown in Figure 14.7a, are involved in olfaction, especially emotional responses to odors such as a loved one's cologne or Mom's chocolate chip cookies baking in the oven.

The functions of the three parts of the diencephalon are summarized in Table 14.2.

Circumventricular Organs

Parts of the diencephalon, called **circumventricular organs (CVOs)** (ser′-kum-ven-TRIK-ū-lar) because they lie in the wall of the third ventricle, can monitor chemical changes in the blood because they lack a blood–brain barrier. CVOs include part of the hypothalamus, the pineal gland, the pituitary gland, and a few other nearby structures. Functionally, these regions coordinate homeostatic activities of the endocrine and nervous systems, such as the regulation of blood pressure, fluid balance, hunger, and thirst. CVOs are also thought to be the sites of entry into the brain of HIV, the virus that causes AIDS. Once in the brain, HIV may cause dementia (irreversible deterioration of mental state) and other neurological disorders.

14.6 The Cerebrum

OBJECTIVES

- **Describe** the cortex, gyri, fissures, and sulci of the cerebrum.
- **Locate** each of the lobes of the cerebrum.
- **Describe** the tracts that compose the cerebral white matter.
- **Describe** the nuclei that compose the basal nuclei.
- **Describe** the structures and functions of the limbic system.

The **cerebrum** is the "seat of intelligence." It provides us with the ability to read, write, and speak; to make calculations and compose music; and to remember the past, plan for the future, and imagine things that have never existed before. The cerebrum consists of an outer cerebral cortex, an internal region of cerebral white matter, and gray matter nuclei deep within the white matter.

Cerebral Cortex

The **cerebral cortex** (*cortex* = rind or bark) is a region of gray matter that forms the outer rim of the cerebrum (Figure 14.11a). Although only 2–4 mm (0.08–0.16 in.) thick, the cerebral cortex contains billions of neurons arranged in distinct layers. During embryonic development, when brain size increases rapidly, the gray matter of the cortex enlarges much faster than the deeper white matter. As a result, the cortical region rolls and folds on itself. The folds are called **gyri** (JĪ-rī = circles; singular is *gyrus*) or *convolutions* (kon′-vō-LOO-shuns) (Figure 14.11). The deepest grooves between folds are known as **fissures**; the shallower grooves between folds are termed **sulci** (SUL-sī = grooves; singular is *sulcus*). The most prominent fissure, the **longitudinal fissure**, separates the cerebrum into right and left halves called **cerebral hemispheres**. Within the longitudinal fissure between the cerebral hemispheres is the falx cerebri. The cerebral hemispheres are connected internally by the **corpus callosum** (kal-LŌ-sum; *corpus* = body; *callosum* = hard), a broad band of white matter containing axons that extend between the hemispheres (see Figure 14.12).

Lobes of the Cerebrum

Each cerebral hemisphere can be further subdivided into several lobes. The lobes are named after the bones that cover them: frontal, parietal, temporal, and occipital lobes (see Figure 14.11). The **central sulcus** (SUL-kus) separates the **frontal lobe** from the **parietal lobe**. A major gyrus, the **precentral gyrus**—located immediately anterior to the central sulcus—contains the primary motor area of the cerebral cortex. Another major gyrus, the **postcentral gyrus**, which is located immediately posterior to the central sulcus, contains the primary somatosensory area of the cerebral cortex. The **lateral cerebral sulcus** (*fissure*) separates the frontal lobe from the **temporal lobe**. The **parieto-occipital sulcus** separates the parietal lobe from the **occipital lobe**. A fifth part of the cerebrum, the **insula**, cannot be seen at the surface of the brain because it lies within the lateral cerebral sulcus, deep to the parietal, frontal, and temporal lobes (Figure 14.11b).

Cerebral White Matter

The **cerebral white matter** consists primarily of myelinated axons in three types of tracts (Figure 14.12):

1. **Association tracts** contain axons that conduct nerve impulses between gyri in the same hemisphere.
2. **Commissural tracts** (kom′-i-SYUR-al) contain axons that conduct nerve impulses from gyri in one cerebral hemisphere to corresponding gyri in the other cerebral hemisphere. Three important groups of commissural tracts are the **corpus callosum** (the largest fiber bundle in the brain, containing about 300 million fibers), **anterior commissure**, and **posterior commissure**.
3. **Projection tracts** contain axons that conduct nerve impulses from the cerebrum to lower parts of the CNS (thalamus, brainstem, or spinal cord) or from lower parts of the CNS to the cerebrum. An example is the **internal capsule**, a thick band of white matter that contains both ascending and descending axons (see Figure 14.13b).

Basal Nuclei

Deep within each cerebral hemisphere are three nuclei (masses of gray matter) that are collectively termed the **basal nuclei**

FIGURE 14.11 **Cerebrum.** Because the insula cannot be seen externally, it has been projected to the surface in (b).

The cerebrum is the "seat of intelligence"; it provides us with the ability to read, write, and speak; to make calculations and compose music; to remember the past and plan for the future; and to create.

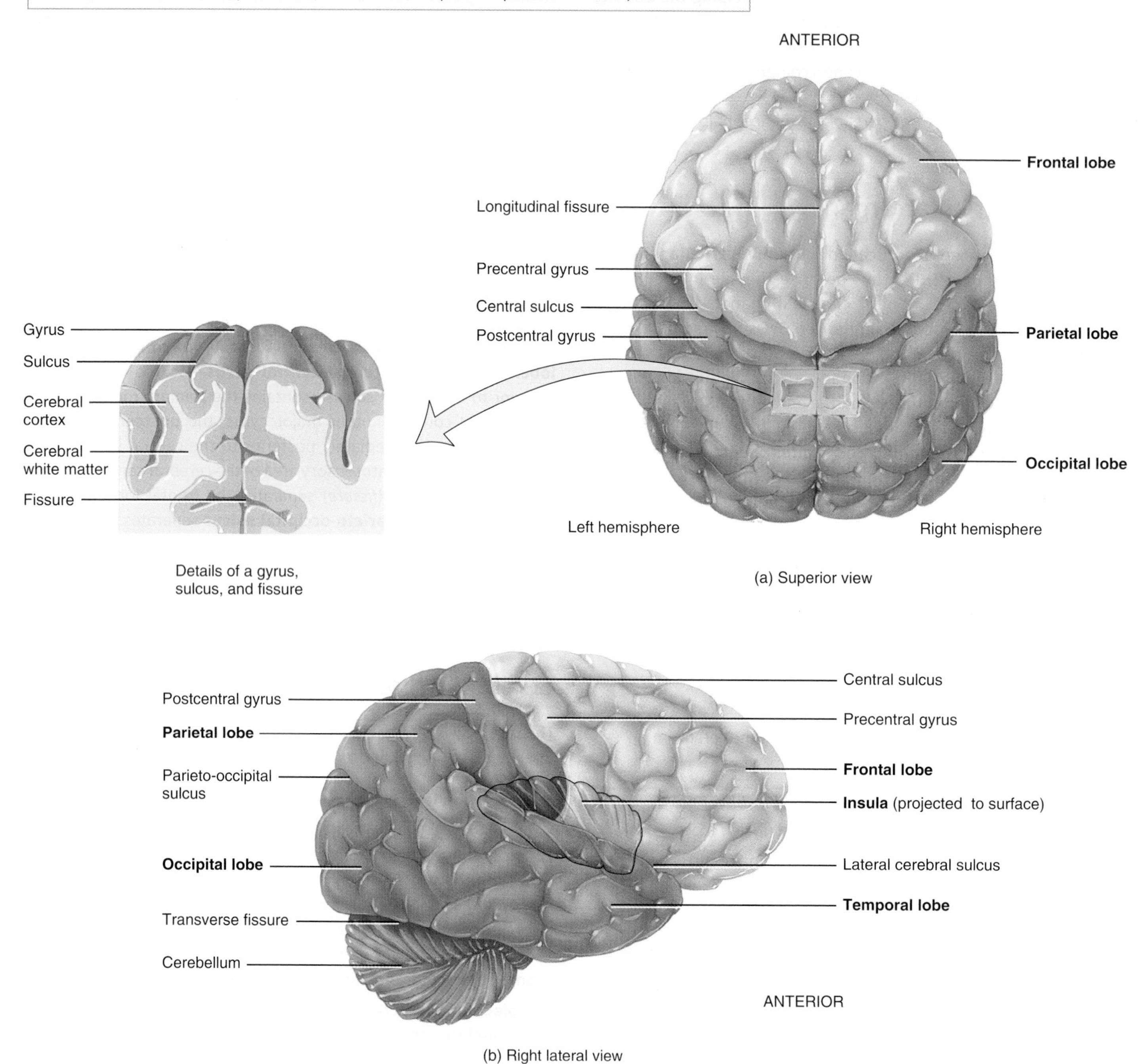

Q During development, does the gray matter or the white matter enlarge more rapidly? What are the brain folds, shallow grooves, and deep grooves called?

FIGURE 14.12 **Organization of white matter tracts of the left cerebral hemisphere.**

Association tracts, commissural tracts, and projection tracts form white matter tracts in the cerebral hemispheres.

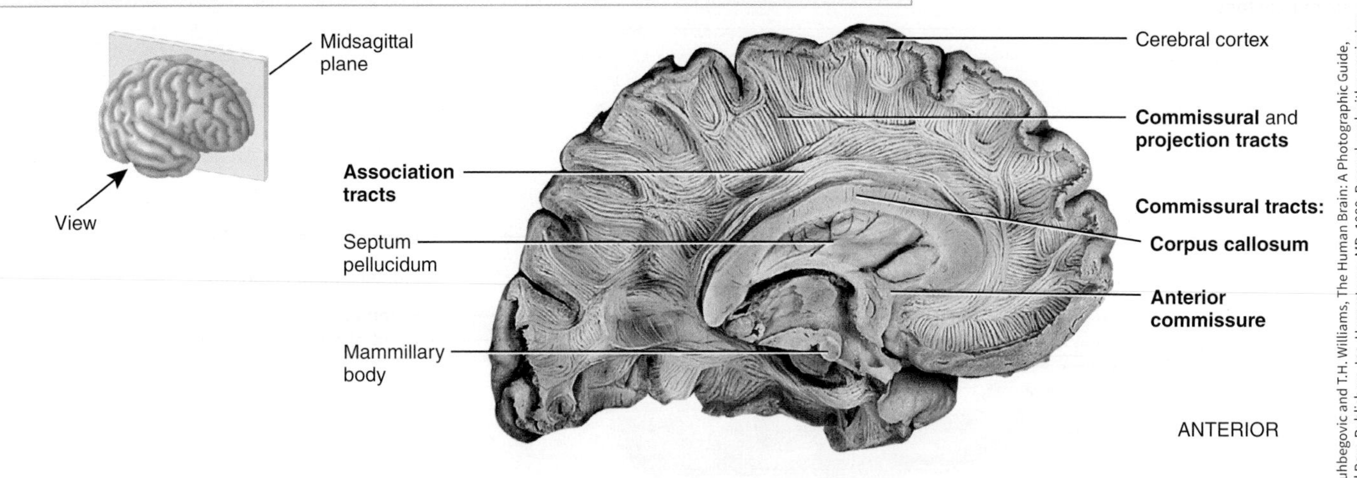

Q Which tracts carry impulses between gyri of the same hemisphere? Between gyri in opposite hemispheres? Between the cerebrum and thalamus, brainstem, and spinal cord?

(**Figure 14.13**). (Historically, these nuclei have been called the *basal ganglia*. However, this is a misnomer because a *ganglion* is an aggregate of neuronal cell bodies in the peripheral nervous system. While both terms still appear in the literature, we use *nuclei*, as this is the correct term as determined by the *Terminologia Anatomica*, the final say on correct anatomical terminology.)

Two of the basal nuclei lie side by side, just lateral to the thalamus. They are the **globus pallidus** (GLŌ-bus PAL-i-dus; *globus* = ball; *pallidus* = pale), which is closer to the thalamus, and the **putamen** (pū-TĀ-men = shell), which is closer to the cerebral cortex. Together, the globus pallidus and putamen are referred to as the **lentiform nucleus** (LEN-ti-form = shaped like a lens). The third of the basal nuclei is the **caudate nucleus** (KAW-dāt; *caud-* = tail), which has a large "head" connected to a smaller "tail" by a long, comma-shaped "body." Together, the lentiform and caudate nuclei are known as the **corpus striatum** (strī-Ā-tum; *corpus* = body; *striatum* = striated). The term corpus striatum refers to the striated (striped) appearance of the internal capsule as it passes among the basal nuclei. Nearby structures that are functionally linked to the basal nuclei are the *substantia nigra* of the midbrain and the *subthalamic nuclei* of the diencephalon (see **Figures 14.7b, 14.13b**). Axons from the substantia nigra terminate in the caudate nucleus and putamen. The subthalamic nuclei interconnect with the globus pallidus.

The **claustrum** (KLAWS-trum) is a thin sheet of gray matter situated lateral to the putamen. It is considered by some to be a subdivision of the basal nuclei. The function of the claustrum in humans has not been clearly defined, but it may be involved in visual attention.

The basal nuclei receive input from the cerebral cortex and provide output to motor parts of the cortex via the medial and ventral group nuclei of the thalamus. In addition, the basal nuclei have extensive connections with one another. A major function of the basal nuclei is to help regulate initiation and termination of movements. Activity of neurons in the putamen precedes or anticipates body movements; activity of neurons in the caudate nucleus occurs prior to eye movements. The globus pallidus helps regulate the muscle tone required for specific body movements. The basal nuclei also control subconscious contractions of skeletal muscles. Examples include automatic arm swings while walking and true laughter in response to a joke (not the kind you consciously initiate to humor your A&P instructor).

In addition to influencing motor functions, the basal nuclei have other roles. They help initiate and terminate some cognitive processes, such as attention, memory, and planning, and may act with the limbic system to regulate emotional behaviors. Disorders such as Parkinson's disease, obsessive–compulsive disorder, schizophrenia, and chronic anxiety are thought to involve dysfunction of circuits between the basal nuclei and the limbic system and are described in more detail in Chapter 16 of Study Guide.

The Limbic System

Encircling the upper part of the brainstem and the corpus callosum is a ring of structures on the inner border of the cerebrum and floor of the diencephalon that constitutes the **limbic system**

FIGURE 14.13 Basal nuclei. In (a) the basal nuclei have been projected to the surface; in both (a) and (b) they are shown in purple.

The basal nuclei help initiate and terminate movements, suppress unwanted movements, and regulate muscle tone.

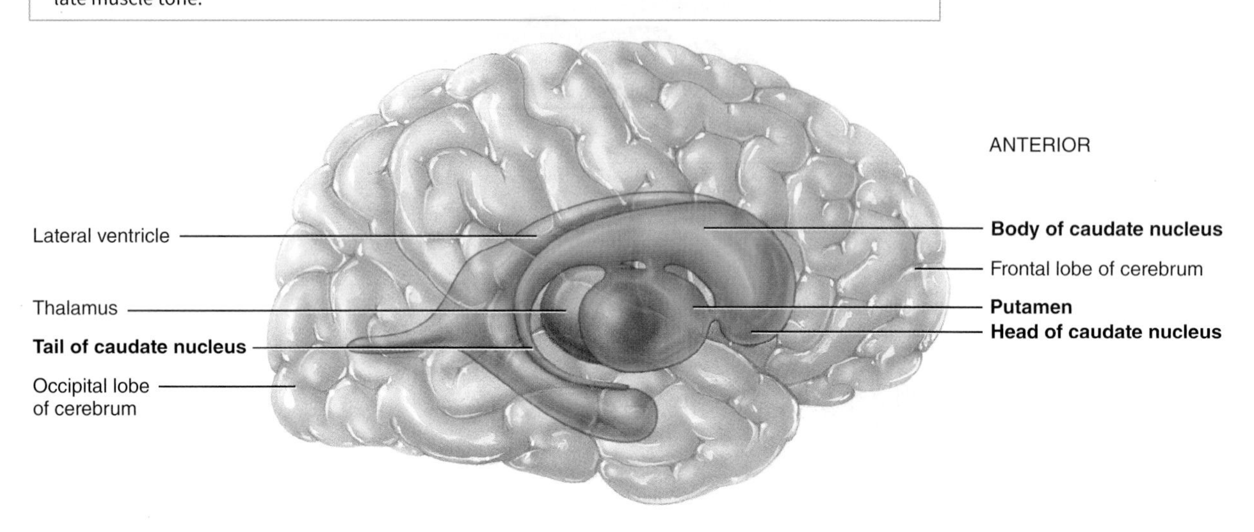

(a) Lateral view of right side of brain

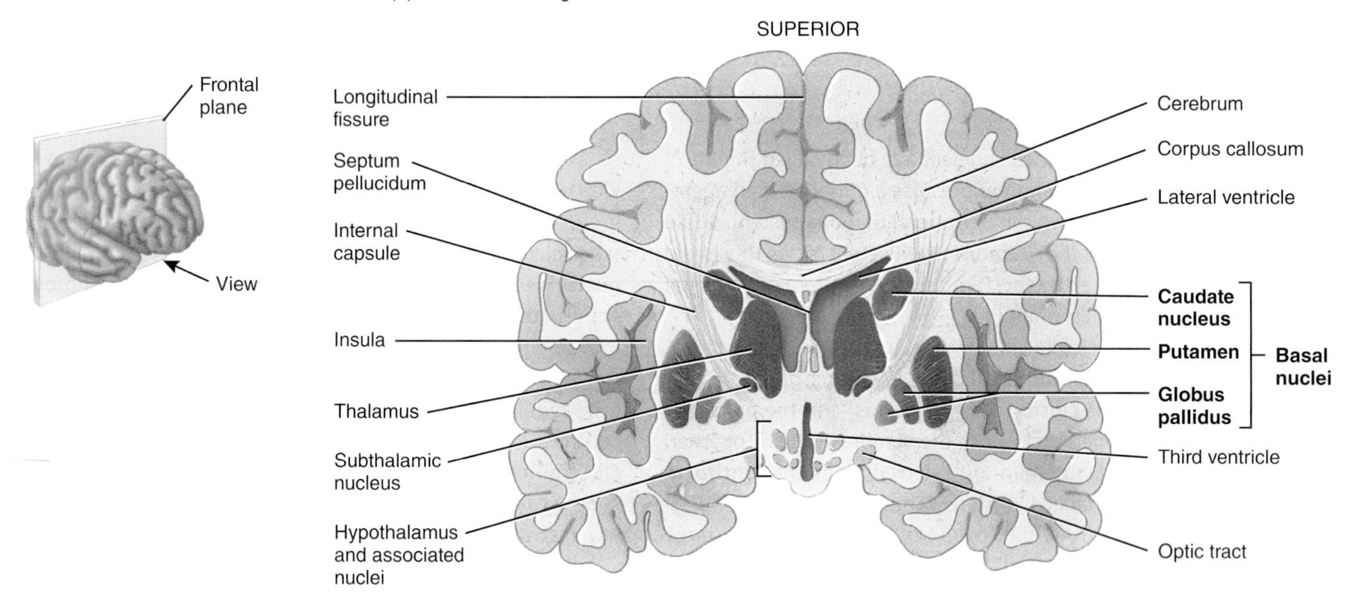

(b) Anterior view of frontal section

Q Where are the basal nuclei located relative to the thalamus?

(*limbic* = border). The main components of the limbic system are as follows (**Figure 14.14**):

- The so-called **limbic lobe** is a rim of cerebral cortex on the medial surface of each hemisphere. It includes the **cingulate gyrus** (SIN-gyu-lat; *cingul-* = belt), which lies above the corpus callosum, and the **parahippocampal gyrus** (par′-a-hip-ō-KAM-pal), which is in the temporal lobe below. The **hippocampus** (hip′-ō-KAM -pus = seahorse) is a portion of the parahippocampal gyrus that extends into the floor of the lateral ventricle.
- The **dentate gyrus** (*dentate* = toothed) lies between the hippocampus and parahippocampal gyrus.
- The **amygdala** (a-MIG-da-la; *amygda-* = almond-shaped) is composed of several groups of neurons located close to the tail of the caudate nucleus.
- The **septal nuclei** are located within the septal area formed by the regions under the corpus callosum and the paraterminal gyrus (a cerebral gyrus).

FIGURE 14.14 **Components of the limbic system (shaded green) and surrounding structures.**

The limbic system governs emotional aspects of behavior.

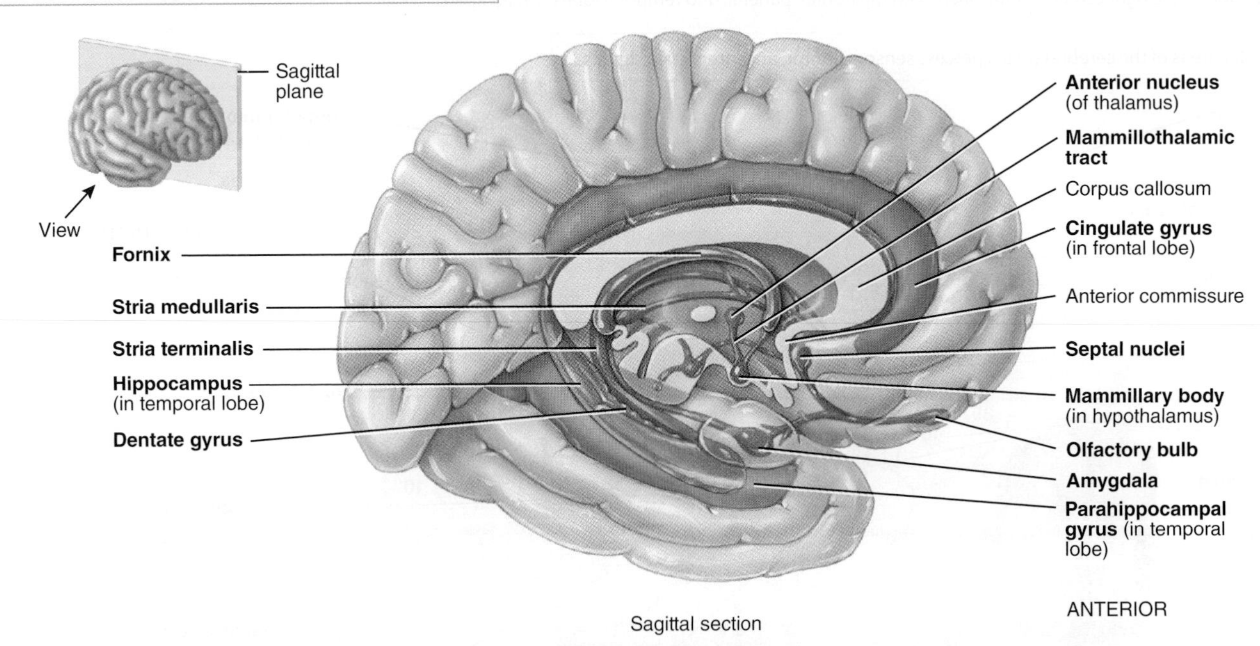

Q Which part of the limbic system functions with the cerebrum in memory?

- The **mammillary bodies** of the hypothalamus are two round masses close to the midline near the cerebral peduncles.
- Two nuclei of the thalamus, the anterior nucleus and the medial nucleus, participate in limbic circuits (see Figure 14.9c, d).
- The **olfactory bulbs** are flattened bodies of the olfactory pathway that rest on the cribriform plate.
- The **fornix**, **stria terminalis**, **stria medullaris**, **medial forebrain bundle**, and **mammillothalamic tract** (mam-i-lō-tha-LAM-ik) are linked by bundles of interconnecting myelinated axons.

The limbic system is sometimes called the "emotional brain" because it plays a primary role in a range of emotions, including pain, pleasure, docility, affection, and anger. It also is involved in olfaction (smell) and memory. Experiments have shown that when different areas of animals' limbic systems are stimulated, the animals' reactions indicate that they are experiencing intense pain or extreme pleasure. Stimulation of other limbic system areas in animals produces tameness and signs of affection. Stimulation of a cat's amygdala or certain nuclei of the hypothalamus produces a behavioral pattern called rage—the cat extends its claws, raises its tail, opens its eyes wide, hisses, and spits. By contrast, removal of the amygdala produces an animal that lacks fear and aggression. Likewise, a person whose amygdala is damaged fails to recognize fearful expressions in others or to express fear in situations where this emotion would normally be appropriate, for example, while being attacked by an animal.

Together with parts of the cerebrum, the limbic system also functions in memory; damage to the limbic system causes memory impairment. One portion of the limbic system, the hippocampus, is seemingly unique among structures of the central nervous system—it has cells reported to be capable of mitosis. Thus, the portion of the brain that is responsible for some aspects of memory may develop new neurons, even in the elderly.

The functions of the cerebrum are summarized in Table 14.2.

14.7 Functional Organization of the Cerebral Cortex

OBJECTIVES

- **Describe** the locations and functions of the sensory, association, and motor areas of the cerebral cortex.
- **Explain** the significance of hemispheric lateralization.
- **Indicate** the significance of brain waves.

Specific types of sensory, motor, and integrative signals are processed in certain regions of the cerebral cortex (Figure 14.15). Generally, **sensory areas** receive sensory information and are involved in **perception**, the conscious awareness of a sensation; **motor areas** control the execution of voluntary movements; and **association areas** deal with more complex integrative functions such as memory, emotions, reasoning, will, judgment, personality traits, and intelligence. In this section we will also discuss hemispheric lateralization and brain waves.

FIGURE 14.15 Functional areas of the cerebrum. Broca's speech area and Wernicke's area are in the left cerebral hemisphere of most people; they are shown here to indicate their relative locations. The numbers, still used today, are from K. Brodmann's map of the cerebral cortex, first published in 1909. The smaller figure shows the lateral view of the right cerebral hemisphere with the frontal, parietal, and temporal lobes spread apart.

Particular areas of the cerebral cortex process sensory, motor, and integrative signals.

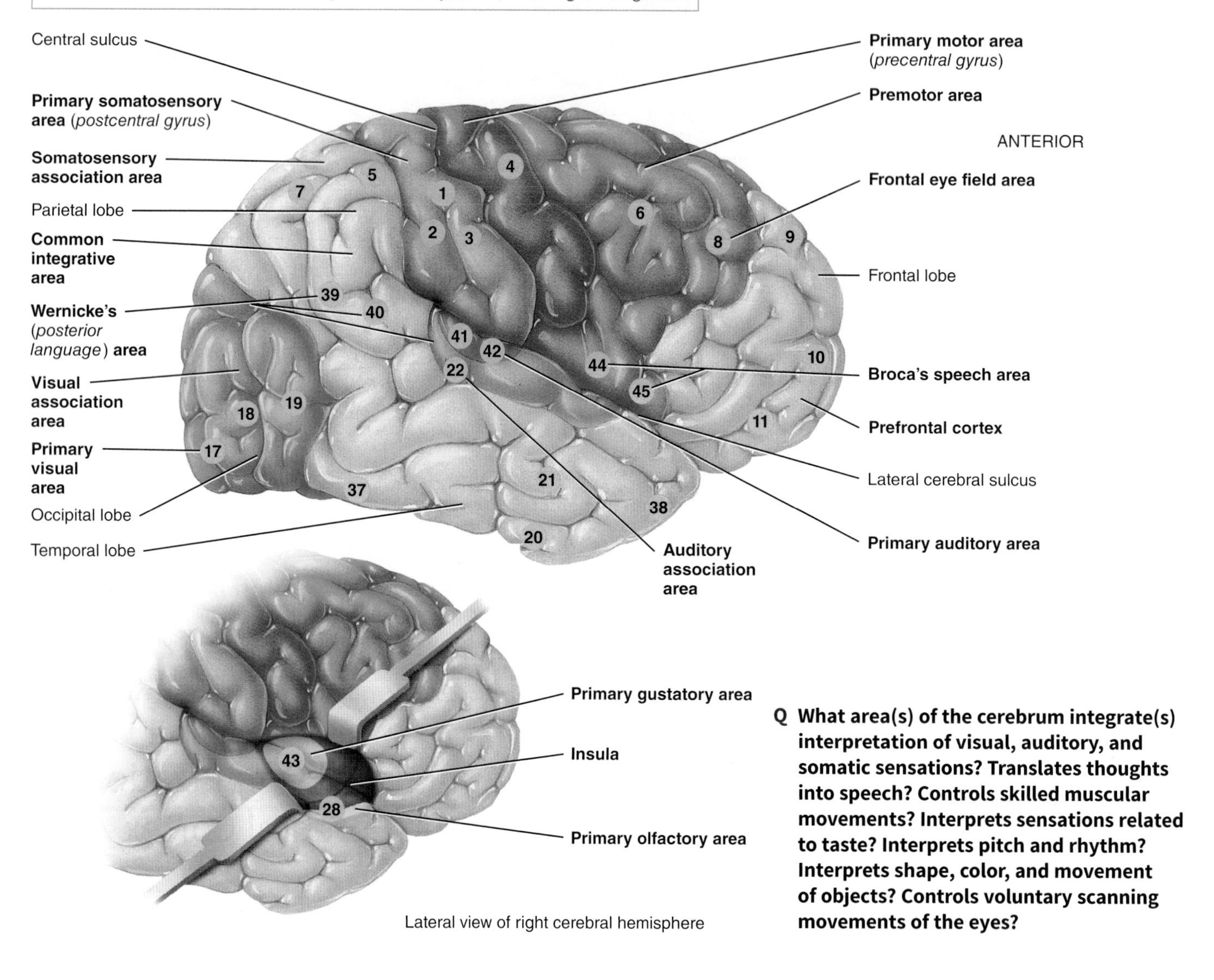

Q What area(s) of the cerebrum integrate(s) interpretation of visual, auditory, and somatic sensations? Translates thoughts into speech? Controls skilled muscular movements? Interprets sensations related to taste? Interprets pitch and rhythm? Interprets shape, color, and movement of objects? Controls voluntary scanning movements of the eyes?

Sensory Areas

Sensory impulses arrive mainly in the posterior half of both cerebral hemispheres, in regions behind the central sulci. In the cerebral cortex, primary sensory areas receive sensory information that has been relayed from peripheral sensory receptors through lower regions of the brain. Sensory association areas often are adjacent to the primary areas. They usually receive input both from the primary areas and from other brain regions. Sensory association areas integrate sensory experiences to generate meaningful patterns of recognition and awareness. For example, a person with damage in the *primary* visual area would be blind in at least part of his visual field, but a person with damage to a visual *association* area might see normally yet be unable to recognize ordinary objects such as a lamp or a toothbrush just by looking at them.

The following are some important sensory areas (**Figure 14.15**; the significance of the numbers in parentheses is explained in the figure caption):

- The **primary somatosensory area** (areas 1, 2, and 3) is located directly posterior to the central sulcus of each cerebral hemisphere in the postcentral gyrus of each parietal lobe. It extends from the lateral cerebral sulcus, along the lateral surface of the parietal lobe to the longitudinal fissure, and then along the medial surface of the parietal lobe within the longitudinal fissure. The primary somatosensory area receives nerve impulses for touch, pressure, vibration, itch, tickle, temperature (coldness and warmth), pain, and proprioception (joint and muscle position) and is involved in the perception of these somatic sensations. A "map" of the entire body is present in the primary somatosensory area: Each point within the area receives impulses from a specific part of the body (see **Figure 16.8a**). The size of the cortical area receiving impulses from a particular part of the body depends on the number of receptors present there rather than on the size of the body part. For example, a larger region of the somatosensory area receives impulses from the lips and fingertips than from the thorax or hip. This distorted somatic sensory map of the body is known as the **sensory homunculus** (*homunculus* = little man). The primary somatosensory area allows you to pinpoint where somatic sensations originate, so that you know exactly where on your body to swat that mosquito.

- The **primary visual area** (area 17), located at the posterior tip of the occipital lobe mainly on the medial surface (next to the longitudinal fissure), receives visual information and is involved in visual perception.
- The **primary auditory area** (areas 41 and 42), located in the superior part of the temporal lobe near the lateral cerebral sulcus, receives information for sound and is involved in auditory perception.
- The **primary gustatory area** (area 43), located in the insula, receives impulses for taste and is involved in gustatory perception and taste discrimination.
- The **primary olfactory area** (area 28), located in the temporal lobe on the medial aspect, receives impulses for smell and is involved in olfactory perception.

Motor Areas

Motor output from the cerebral cortex flows mainly from the anterior part of each hemisphere. Among the most important motor areas are the following (**Figure 14.15**):

- The **primary motor area** (area 4) is located in the precentral gyrus of the frontal lobe. As is true for the primary somatosensory area, a "map" of the entire body is present in the primary motor area: Each region within the area controls voluntary contractions of specific muscles or groups of muscles (see **Figure 16.8b**). Electrical stimulation of any point in the primary motor area causes contraction of specific skeletal muscle fibers on the opposite side of the body. Different muscles are represented unequally in the primary motor area. More cortical area is devoted to those muscles involved in skilled, complex, or delicate movement. For instance, the cortical region devoted to muscles that move the fingers is much larger than the region for muscles that move the toes. This distorted muscle map of the body is called the **motor homunculus**.
- **Broca's speech area** (BRŌ-kaz) (areas 44 and 45) is located in the frontal lobe close to the lateral cerebral sulcus. Speaking and understanding language are complex activities that involve several sensory, association, and motor areas of the cortex. In about 97% of the population, these language areas are localized in the *left* hemisphere. The planning and production of speech occur in the *left* frontal lobe in most people. From Broca's speech area, nerve impulses pass to the premotor regions that control the muscles of the larynx, pharynx, and mouth. The impulses from the premotor area result in specific, coordinated muscle contractions. Simultaneously, impulses propagate from Broca's speech area to the primary motor area. From here, impulses also control the breathing muscles to regulate the proper flow of air past the vocal cords. The coordinated contractions of your speech and breathing muscles enable you to speak your thoughts. People who suffer a cerebrovascular accident (CVA) or stroke in this area can still have clear thoughts but are unable to form words, a phenomenon referred to as *nonfluent aphasia*; see Chapter 16, *Clinical Connection: Aphasia* (in the Study Guide).

Association Areas

The association areas of the cerebrum consist of large areas of the occipital, parietal, and temporal lobes and of the frontal lobes anterior to the motor areas. Association areas are connected with one another by association tracts and include the following (**Figure 14.15**):

- The **somatosensory association area** (areas 5 and 7) is just posterior to and receives input from the primary somatosensory area, as well as from the thalamus and other parts of the brain. This area permits you to determine the exact shape and texture of an object by feeling it, to determine the orientation of one object with respect to another as they are felt, and to sense the relationship of one body part to another. Another role of the somatosensory association area is the storage of memories of past somatic sensory experiences, enabling you to compare current sensations with previous experiences. For example, the somatosensory association area allows you to recognize objects such as a pencil and a paperclip simply by touching them.
- The **visual association area** (areas 18 and 19), located in the occipital lobe, receives sensory impulses from the primary visual area and the thalamus. It relates present and past visual experiences and is essential for recognizing and evaluating what is seen. For example, the visual association area allows you to recognize an object such as a spoon simply by looking at it.
- The **facial recognition area**, corresponding roughly to areas 20, 21, and 37 in the inferior temporal lobe, receives nerve impulses from the visual association area. This area stores information about faces, and it allows you to recognize people by their faces. The facial recognition area in the *right* hemisphere is usually more dominant than the corresponding region in the left hemisphere.
- The **auditory association area** (area 22), located inferior and posterior to the primary auditory area in the temporal cortex, allows you to recognize a particular sound as speech, music, or noise.

- The **orbitofrontal cortex**, corresponding roughly to area 11 along the lateral part of the frontal lobe, receives sensory impulses from the primary olfactory area. This area allows you to identify odors and to discriminate among different odors. During olfactory processing, the orbitofrontal cortex of the *right* hemisphere exhibits greater activity than the corresponding region in the left hemisphere.
- **Wernicke's area** (VER-ni-kēz) (*posterior language area*; area 22, and possibly areas 39 and 40), a broad region in the *left* temporal and parietal lobes, interprets the meaning of speech by recognizing spoken words. It is active as you translate words into thoughts. The regions in the *right* hemisphere that correspond to Broca's and Wernicke's areas in the left hemisphere also contribute to verbal communication by adding emotional content, such as anger or joy, to spoken words. Unlike those who have CVAs in Broca's area, people who suffer strokes in Wernicke's area can still speak, but cannot arrange words in a coherent fashion (fluent aphasia, or "word salad").
- The **common integrative area** (areas 5, 7, 39, and 40) is bordered by somatosensory, visual, and auditory association areas. It receives nerve impulses from these areas and from the primary gustatory area, the primary olfactory area, the thalamus, and parts of the brainstem. This area integrates sensory interpretations from the association areas and impulses from other areas, allowing the formation of thoughts based on a variety of sensory inputs. It then transmits signals to other parts of the brain for the appropriate response to the sensory signals it has interpreted.
- The **prefrontal cortex** (*frontal association area*) is an extensive area in the anterior portion of the frontal lobe that is well developed in primates, especially humans (areas 9, 10, 11, and 12; area 12 is not illustrated since it can be seen only in a medial view). This area has numerous connections with other areas of the cerebral cortex, thalamus, hypothalamus, limbic system, and cerebellum. The prefrontal cortex is concerned with the makeup of a person's personality, intellect, complex learning abilities, recall of information, initiative, judgment, foresight, reasoning, conscience, intuition, mood, planning for the future, and development of abstract ideas. A person with bilateral damage to the prefrontal cortices typically becomes rude, inconsiderate, incapable of accepting advice, moody, inattentive, less creative, unable to plan for the future, and incapable of anticipating the consequences of rash or reckless words or behavior.
- The **premotor area** (area 6) is a motor association area that is immediately anterior to the primary motor area. Neurons in this area communicate with the primary motor cortex, the sensory association areas in the parietal lobe, the basal nuclei, and the thalamus. The premotor area deals with learned motor activities of a complex and sequential nature. It generates nerve impulses that cause specific groups of muscles to contract in a specific sequence, as when you write your name. The premotor area also serves as a memory bank for such movements.
- The **frontal eye field area** (area 8) in the frontal cortex is sometimes included in the premotor area. It controls voluntary scanning movements of the eyes—like those you just used in reading this sentence.

The functions of the various parts of the brain are summarized in Table 14.2.

Hemispheric Lateralization

Although the brain is almost symmetrical on its right and left sides, subtle anatomical differences between the two hemispheres exist. For example, in about two-thirds of the population, the planum temporale, a region of the temporal lobe that includes Wernicke's area, is 50% larger on the left side than on the right side. This asymmetry appears in the human fetus at about 30 weeks gestation. Physiological differences also exist; although the two hemispheres share performance of many functions, each hemisphere also specializes in performing certain unique functions. This functional asymmetry is termed **hemispheric lateralization**.

Despite some dramatic differences in functions of the two hemispheres, there is considerable variation from one person to another. Also, lateralization seems less pronounced in females than in males, both for language (left hemisphere) and for visual and spatial skills (right hemisphere). For instance, females are less likely than males to suffer aphasia after damage to the left hemisphere. A possibly related observation is that the anterior commissure is 12% larger and the corpus callosum has a broader posterior portion in females. Recall that both the anterior commissure and the corpus callosum are commissural tracts that provide communication between the two hemispheres.

Table 14.3 summarizes some of the functional differences between the two cerebral hemispheres.

Brain Waves

At any instant, brain neurons are generating millions of nerve impulses (action potentials). Taken together, these electrical signals are called **brain waves**. Brain waves generated by neurons close to the brain surface, mainly neurons in the cerebral cortex, can be detected by sensors called electrodes placed on the forehead and scalp. A record of such waves is called an **electroencephalogram EEG** (e-lek′-trō-en-SEF-a-lō-gram; *electro-* = electricity; *-gram* = recording).

Patterns of activation of brain neurons produce four types of brain waves (Figure 14.16):

1. **Alpha waves.** These rhythmic waves occur at a frequency of about 8–13 cycles per second. (The unit commonly used to express frequency is the hertz [Hz]. One hertz is one cycle per second.) Alpha waves are present in the EEGs of nearly all normal individuals when they are awake and resting with their eyes closed. These waves disappear entirely during sleep.
2. **Beta waves.** The frequency of these waves is between 14 and 30 Hz. Beta waves generally appear when the nervous system is active—that is, during periods of sensory input and mental activity.
3. **Theta waves.** Theta waves (THĀ-ta) have frequencies of 4–7 Hz. These waves normally occur in children and adults experiencing emotional stress.

TABLE 14.2 Summary of Functions of Principal Parts of the Brain

PART	FUNCTION	PART	FUNCTION
BRAINSTEM		**DIENCEPHALON**	
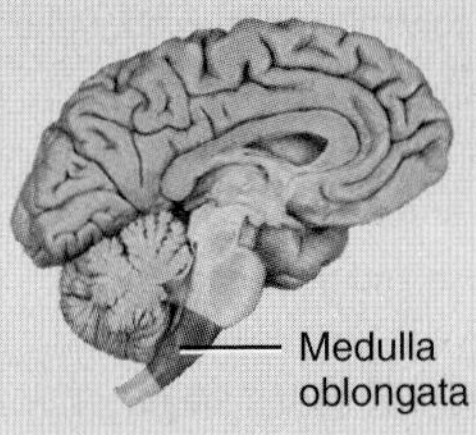	***Medulla oblongata:*** Contains sensory (ascending) and motor (descending) tracts. Cardiovascular center regulates heartbeat and blood vessel diameter. Medullary respiratory center (together with pons) regulates breathing. Contains gracile nucleus, cuneate nucleus, gustatory nucleus, cochlear nuclei, and vestibular nuclei (components of sensory pathways to brain). Inferior olivary nucleus provides instructions that cerebellum uses to adjust muscle activity when learning new motor skills. Other nuclei coordinate vomiting, swallowing, sneezing, coughing, and hiccupping. Contains nuclei of origin for vestibulocochlear (VIII), glossopharyngeal (IX), vagus (X), accessory (XI), and hypoglossal (XII) nerves. Reticular formation (also in pons, midbrain, and diencephalon) functions in consciousness and arousal.	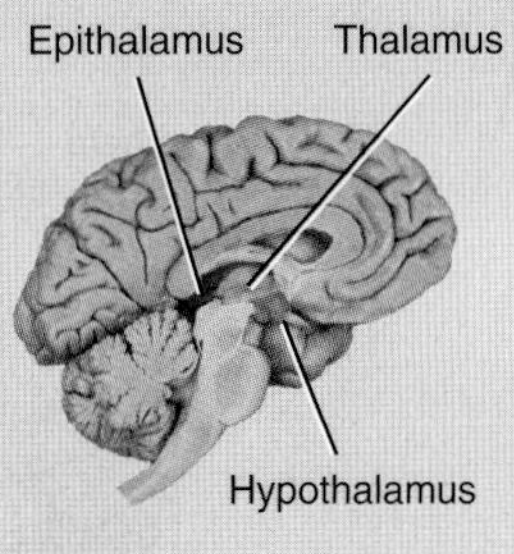	***Thalamus:*** Relays almost all sensory input to cerebral cortex. Contributes to motor functions by transmitting information from cerebellum and basal nuclei to primary motor area of cerebral cortex. Plays role in maintenance of consciousness. ***Hypothalamus:*** Controls and integrates activities of autonomic nervous system. Produces hormones, including releasing hormones, inhibiting hormones, oxytocin, and antidiuretic hormone (ADH). Regulates emotional and behavioral patterns (together with limbic system). Contains feeding and satiety centers (regulate eating), thirst center (regulates drinking), and suprachiasmatic nucleus (regulates circadian rhythms). Controls body temperature by serving as body's thermostat. ***Epithalamus:*** Consists of pineal gland (secretes melatonin) and habenular nuclei (involved in olfaction).
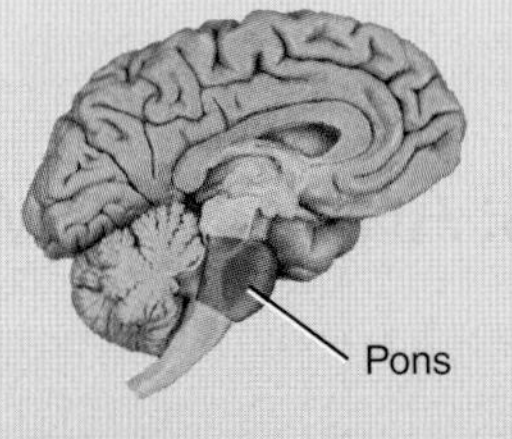	***Pons:*** Contains sensory and motor tracts. Pontine nuclei relay nerve impulses from motor areas of cerebral cortex to cerebellum. Contains vestibular nuclei (along with medulla) that are part of equilibrium pathway to brain. Pontine respiratory group (together with the medulla) helps control breathing. Contains nuclei of origin for trigeminal (V), abducens (VI), facial (VII), and vestibulocochlear (VIII) nerves.	**CEREBRUM** 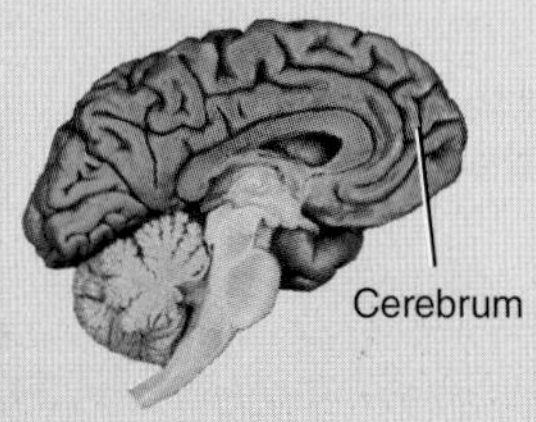	Sensory areas of cerebral cortex are involved in perception of sensory information; motor areas control execution of voluntary movements; association areas deal with more complex integrative functions such as memory, personality traits, and intelligence. Basal nuclei help initiate and terminate movements, suppress unwanted movements, and regulate muscle tone. Limbic system promotes range of emotions, including pleasure, pain, docility, affection, fear, and anger.
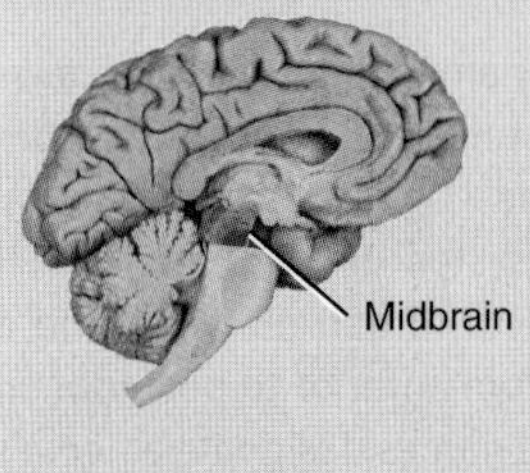	***Midbrain:*** Contains sensory and motor tracts. Superior colliculi coordinate movements of head, eyes, and trunk in response to visual stimuli. Inferior colliculi coordinate movements of head, eyes, and trunk in response to auditory stimuli. Substantia nigra and red nucleus contribute to control of movement. Contains nuclei of origin for oculomotor (III) and trochlear (IV) nerves.		
CEREBELLUM			
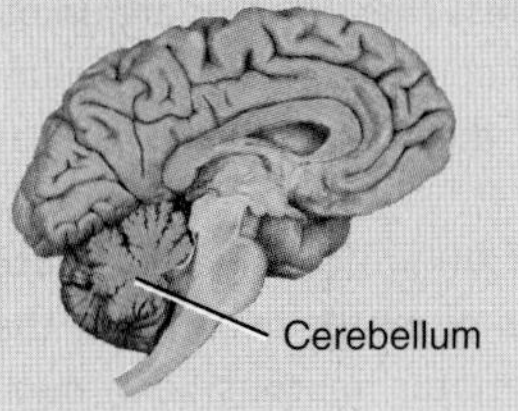	Smooths and coordinates contractions of skeletal muscles. Regulates posture and balance. May have role in cognition and language processing.		

4. **Delta waves.** The frequency of these waves is 1–5 Hz. Delta waves occur during deep sleep in adults, but they are normal in awake infants. When produced by an awake adult, they indicate brain damage.

Electroencephalograms are useful both in studying normal brain functions, such as changes that occur during sleep, and in diagnosing a variety of brain disorders, such as epilepsy, tumors, trauma, hematomas, metabolic abnormalities, sites of trauma, and

TABLE 14.3 Functional Differences between Right and Left Hemispheres

RIGHT HEMISPHERE FUNCTIONS	LEFT HEMISPHERE FUNCTIONS
Receives somatic sensory signals from, and controls muscles on, left side of body.	Receives somatic sensory signals from, and controls muscles on, right side of body.
Musical and artistic awareness.	Reasoning
Space and pattern perception.	Numerical and scientific skills.
Recognition of faces and emotional content of facial expressions.	Ability to use and understand sign language.
Generating emotional content of language.	Spoken and written language.
Generating mental images to compare spatial relationships.	Persons with damage in the left hemisphere often exhibit aphasia.
Identifying and discriminating among odors.	
Patients with damage in right hemisphere regions that correspond to Broca's and Wernicke's areas in the left hemisphere speak in a monotonous voice, having lost the ability to impart emotional inflection to what they say.	

Right hemisphere
Left hemisphere
Anterior view

degenerative diseases. The EEG is also utilized to determine if "life" is present, that is, to establish or confirm that brain death has occurred.

FIGURE 14.16 Types of brain waves recorded in an electroencephalogram (EEG).

Brain waves indicate electrical activity of the cerebral cortex.

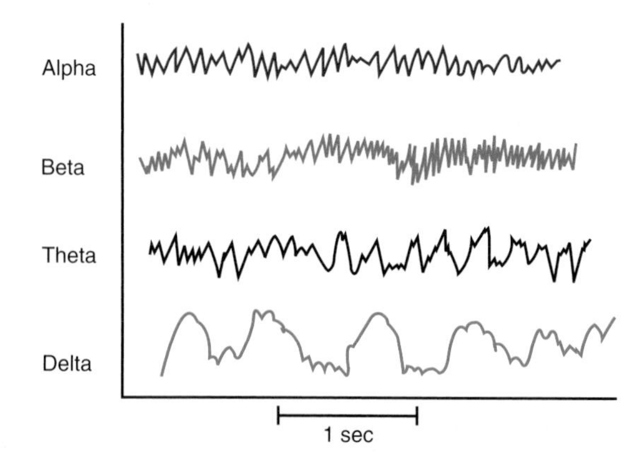

See Clinical Connection: Brain Injuries

Q Which type of brain wave indicates emotional stress?

14.8 Cranial Nerves: An Overview

OBJECTIVE

- **Identify** the cranial nerves by name, number, and type.

The 12 pairs of **cranial nerves** are so named because they pass through various foramina in the bones of the cranium and arise from the brain inside the cranial cavity. Like the 31 pairs of spinal nerves, they are part of the peripheral nervous system (PNS). Each cranial nerve has both a number, designated by a roman numeral, and a name. The numbers indicate the order, from anterior to posterior, in which the nerves arise from the brain. The names designate a nerve's distribution or function.

Three cranial nerves (I, II, and VIII) carry axons of sensory neurons and thus are called **special sensory nerves**. These nerves are unique to the head and are associated with the special senses of smelling, seeing, and hearing. The cell bodies of most sensory neurons are located in ganglia outside the brain.

Five cranial nerves (III, IV, VI, XI, and XII) are classified as **motor nerves** because they contain only axons of motor neurons as they leave the brainstem. The cell bodies of motor neurons lie in nuclei within the brain. Motor axons that innervate skeletal muscles are of two types:

1. *Branchial motor axons* innervate skeletal muscles that develop from the pharyngeal (branchial) arches (see **Figure 14.28**). These neurons leave the brain through the mixed cranial nerves and the accessory nerve.

2. *Somatic motor axons* innervate skeletal muscles that develop from head somites (eye muscles and tongue muscles). These neurons exit the brain through five motor cranial nerves (III, IV, VI, XI, and XII). Motor axons that innervate smooth muscle, cardiac muscle, and glands are called *autonomic motor axons* and are part of the parasympathetic division.

The remaining four cranial nerves (V, VII, IX, and X) are **mixed nerves**—they contain axons both of sensory neurons entering the brainstem and motor neurons leaving the brainstem.

Each cranial nerve is covered in detail in Sections 14.9 through 14.18. Although the cranial nerves are mentioned singly in the exhibits with regard to their type, location, and function, remember that they are paired structures.

Table 14.4 presents a summary of the components and principal functions of the cranial nerves, including a mnemonic to help you remember their names.

See Clinical Connection: Dental Anesthesia

14.9 Olfactory (I) Nerve

OBJECTIVE

- **Identify** the termination of the olfactory (I) nerve in the brain, the foramen through which it passes, and its function.

The **olfactory (I) nerve** (ōl-FAK-tō-rē; *olfact-* = to smell) is entirely sensory; it contains axons that conduct nerve impulses for olfaction, the sense of smell (**Figure 14.17**). The olfactory epithelium occupies the superior part of the nasal cavity, covering the inferior surface of the cribriform plate and extending down along the superior nasal concha. The olfactory receptors within the olfactory epithelium are bipolar neurons. Each has a single odor-sensitive, knob-shaped dendrite projecting from one side of the cell body and an unmyelinated axon extending from the other side. Bundles of axons of olfactory receptors extend through about 20 olfactory foramina in the cribriform plate of the ethmoid bone on each side of the nose. These 40 or so bundles of axons collectively form the right and left olfactory nerves.

FIGURE 14.17 Olfactory (I) nerve.

The olfactory epithelium is located on the inferior surface of the cribriform plate and superior nasal conchae.

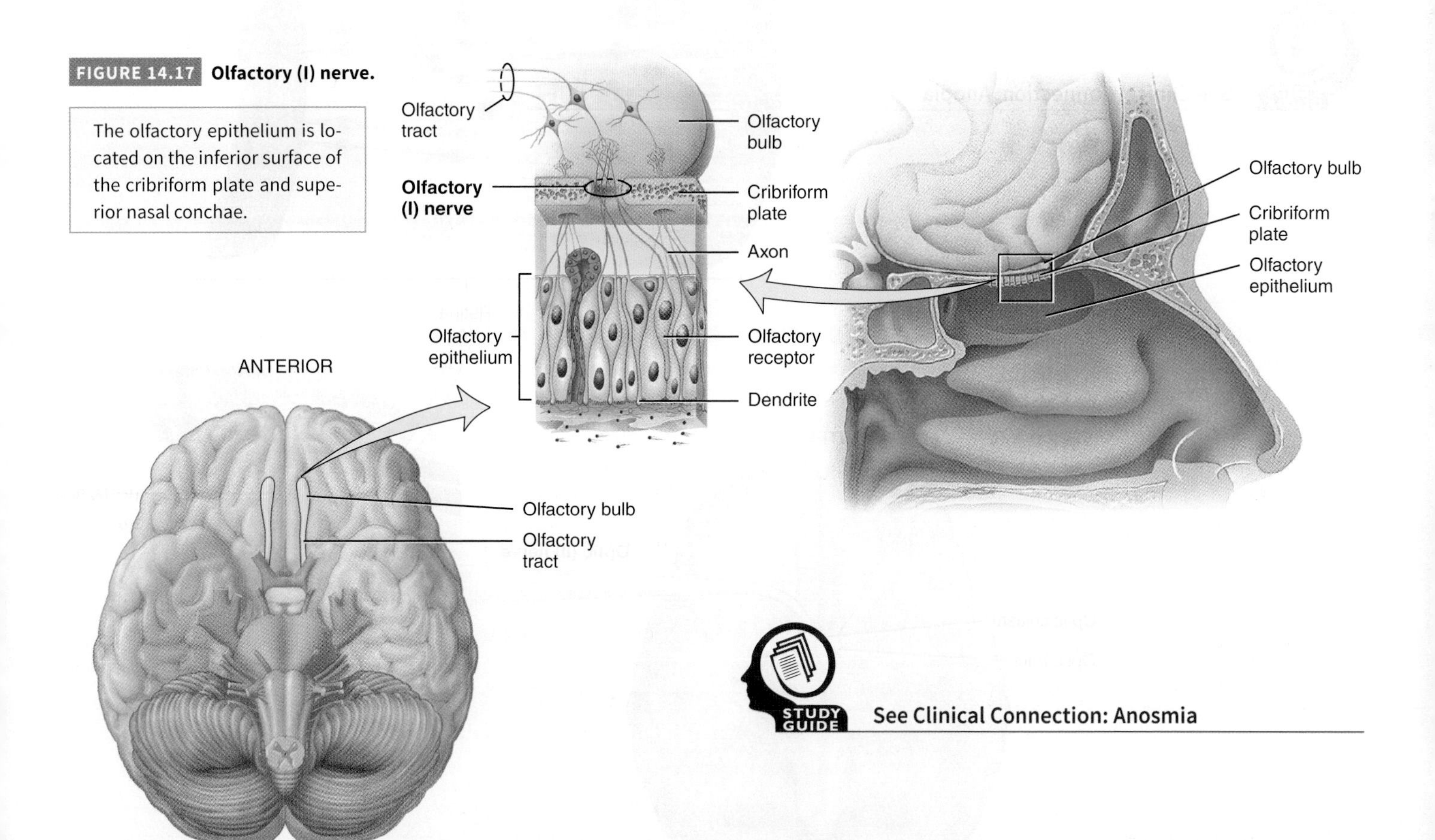

See Clinical Connection: Anosmia

Q Where do axons in the olfactory tracts terminate?

Olfactory nerves end in the brain in paired masses of gray matter called the **olfactory bulbs**, two extensions of the brain that rest on the cribriform plate. Within the olfactory bulbs, the axon terminals of olfactory receptors form synapses with the dendrites and cell bodies of the next neurons in the olfactory pathway. The axons of these neurons make up the **olfactory tracts**, which extend posteriorly from the olfactory bulbs (Figure 14.17). Axons in the olfactory tracts end in the primary olfactory area in the temporal lobe of the cerebral cortex.

14.10 Optic (II) Nerve

OBJECTIVE

- **Identify** the termination of the optic (II) nerve in the brain, the foramen through which it exits the skull, and its function.

The **optic (II) nerve** (OP-tik; *opti-* = the eye, vision) is entirely sensory and is technically a tract of the brain and not a nerve; it contains axons that conduct nerve impulses for vision (Figure 14.18). In the retina, rods and cones initiate visual signals and relay them to bipolar cells, which transmit the signals to ganglion cells. Axons of all ganglion cells in the retina of each eye join to form an optic nerve, which passes through the optic foramen. About 10 mm (0.4 in.) posterior to the eyeball, the two optic nerves merge to form the **optic chiasm** (kī-AZM = a crossover, as in the letter X). Within the chiasm, axons from the medial half of each eye cross to the opposite side; axons from the lateral half remain on the same side. Posterior to the chiasm, the regrouped axons, some from each eye, form the **optic tracts**. Most axons in the optic tracts end in the lateral geniculate nucleus of the thalamus. There they synapse with neurons whose axons extend to the primary visual area in the occipital lobe of the cerebral cortex (area 17 in Figure 14.15). A few axons pass through the lateral geniculate nucleus and then extend to the superior colliculi of the midbrain and to motor nuclei of the brainstem where they synapse with motor neurons that control the extrinsic and intrinsic eye muscles.

FIGURE 14.18 Optic (II) nerve.

In sequence, visual signals are relayed from rods and cones to bipolar cells to ganglion cells.

See Clinical Connection: Anopia

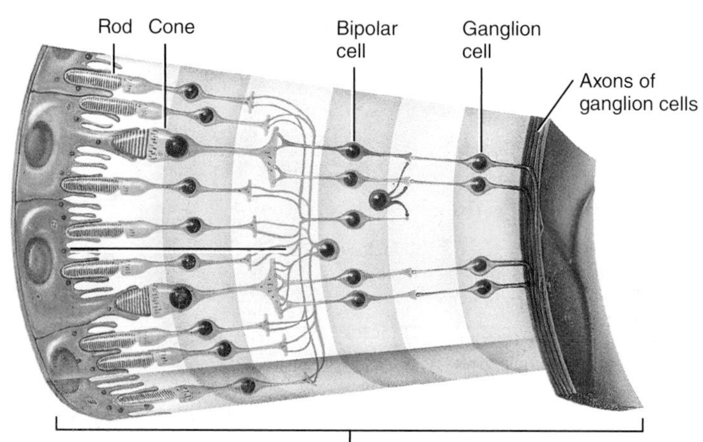

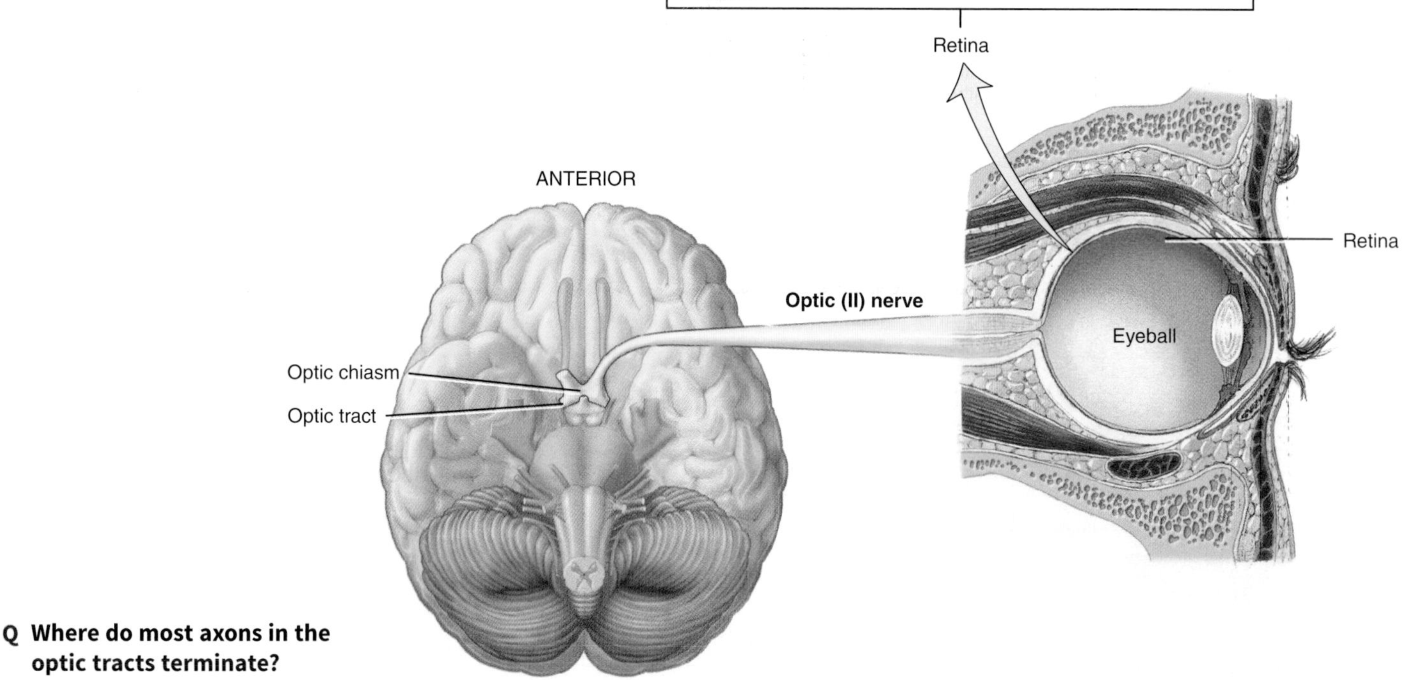

Q Where do most axons in the optic tracts terminate?

14.11 Oculomotor (III), Trochlear (IV), and Abducens (VI) Nerves

OBJECTIVE

- **Identify** the origins of the oculomotor (III), trochlear (IV), and abducens (VI) nerves in the brain, the foramen through which each exits the skull, and their functions.

The oculomotor, trochlear, and abducens nerves are the cranial nerves that control the muscles that move the eyeballs. They are all motor nerves that contain only motor axons as they exit the brainstem. Sensory axons from the extrinsic eyeball muscles begin their course toward the brain in each of these nerves, but eventually these sensory axons leave the nerves to join the ophthalmic branch of the trigeminal nerve. The sensory axons *do not* return to the brain in the oculomotor, trochlear, or abducens nerves. The cell bodies of the unipolar sensory neurons reside in the mesencephalic nucleus and they enter the midbrain via the trigeminal (V) nerve. These axons convey nerve impulses from the extrinsic eyeball muscles for *proprioception*, the perception of the movements and position of the body independent of vision.

The **oculomotor (III) nerve** (ok′-ū-lō-MŌ-tor; *oculo-* = eye; *-motor* = a mover) has its motor nucleus in the anterior part of the midbrain. The oculomotor nerve extends anteriorly and divides into superior and inferior branches, both of which pass through the superior orbital fissure into the orbit (**Figure 14.19a**). Axons in the superior branch innervate the superior rectus (an extrinsic eyeball muscle) and the levator palpebrae superioris (the muscle of the upper eyelid). Axons in the inferior branch supply the medial rectus, inferior rectus, and inferior oblique muscles—all extrinsic eyeball muscles. These somatic motor neurons control movements of the eyeball and upper eyelid.

FIGURE 14.19 Oculomotor (III), trochlear (IV), and abducens (VI) nerves.

The oculomotor (III) nerve has the widest distribution among extrinsic eye muscles.

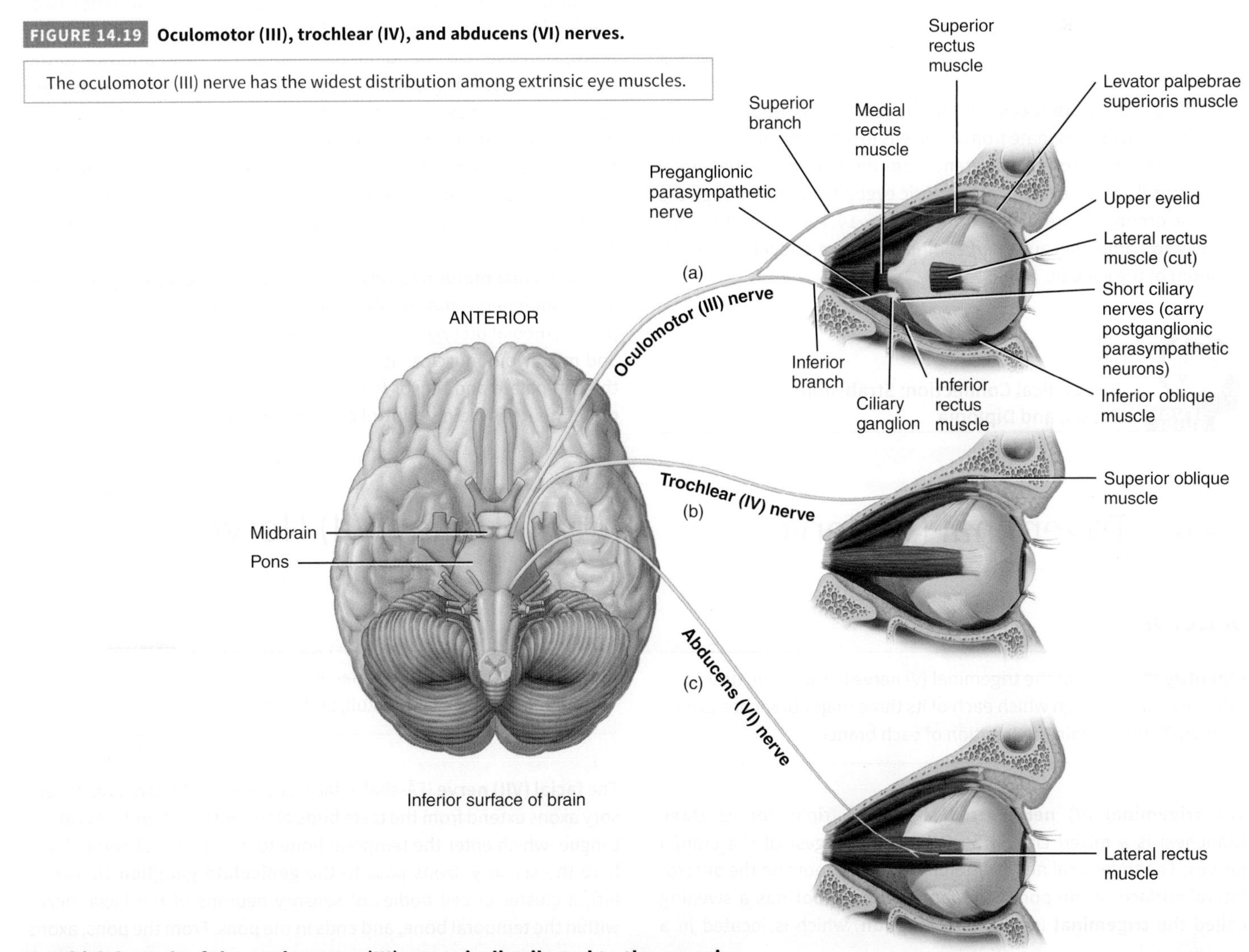

Q Which branch of the oculomotor (III) nerve is distributed to the superior rectus muscle? Which is the smallest cranial nerve?

The inferior branch of the oculomotor nerve also supplies parasympathetic motor axons to intrinsic eyeball muscles, which consist of smooth muscle. They include the ciliary muscle of the eyeball and the circular muscles (sphincter pupillae) of the iris. Parasympathetic impulses propagate from a nucleus in the midbrain (*accessory oculomotor nucleus*) to the **ciliary ganglion**, a synaptic relay center for the two motor neurons of the parasympathetic nervous system. From the ciliary ganglion, parasympathetic motor axons extend to the ciliary muscle, which adjusts the lens for near vision (*accommodation*). Other parasympathetic motor axons stimulate the circular muscles of the iris to contract when bright light stimulates the eye, causing a decrease in the size of the pupil (*constriction*).

The **trochlear (IV) nerve** (TRŌK-lē-ar; *trochle-* = a pulley) is the smallest of the 12 cranial nerves and is the only one that arises from the posterior aspect of the brainstem. The somatic motor neurons originate in a nucleus in the midbrain (trochlear nucleus), and axons from the nucleus cross to the opposite side as they exit the brain on its posterior aspect. The nerve then wraps around the pons and exits through the superior orbital fissure into the orbit. These somatic motor axons innervate the superior oblique muscle of the eyeball, another extrinsic eyeball muscle that controls movement of the eyeball (**Figure 14.19b**).

Neurons of the **abducens (VI) nerve** (ab-DOO-senz; *ab-* = away; *-ducens* = to lead) originate from a nucleus in the pons (abducens nucleus). Somatic motor axons extend from the nucleus to the lateral rectus muscle of the eyeball, an extrinsic eyeball muscle, through the superior orbital fissure of the orbit (**Figure 14.19c**). The abducens nerve is so named because nerve impulses cause abduction (lateral rotation) of the eyeball.

See Clinical Connection: Strabismus, Ptosis, and Diplopia

14.12 Trigeminal (V) Nerve

OBJECTIVE

- **Identify** the origin of the trigeminal (V) nerve in the brain, describe the foramina through which each of its three major branches exits the skull, and explain the function of each branch.

The **trigeminal (V) nerve** (trī-JEM-i-nal = triple, for its three branches) is a mixed cranial nerve and the largest of the cranial nerves. The trigeminal nerve emerges from two roots on the anterolateral surface of the pons. The large sensory root has a swelling called the **trigeminal** (*semilunar*) **ganglion**, which is located in a fossa on the inner surface of the petrous portion of the temporal bone. The ganglion contains cell bodies of most of the primary sensory neurons. Neurons of the smaller motor root originate in a nucleus in the pons.

As indicated by its name, the trigeminal nerve has three branches: ophthalmic, maxillary, and mandibular (**Figure 14.20**). The **ophthalmic nerve** (of-THAL-mik; *ophthalm-* = the eye), the smallest branch, passes into the orbit via the superior orbital fissure. The **maxillary nerve** (*maxilla* = upper jawbone) is intermediate in size between the ophthalmic and mandibular nerves and passes through the foramen rotundum. The **mandibular nerve** (*mandibula* = lower jawbone), the largest branch, passes through the foramen ovale.

Sensory axons in the trigeminal nerve carry nerve impulses for touch, pain, and thermal sensations (heat and cold). The ophthalmic nerve contains sensory axons from the skin over the upper eyelid, cornea, lacrimal glands, upper part of the nasal cavity, side of the nose, forehead, and anterior half of the scalp. The maxillary nerve includes sensory axons from the mucosa of the nose, palate, part of the pharynx, upper teeth, upper lip, and lower eyelid. The mandibular nerve contains sensory axons from the anterior two-thirds of the tongue (not taste), cheek and mucosa deep to it, lower teeth, skin over the mandible and side of the head anterior to the ear, and mucosa of the floor of the mouth. The sensory axons from the three branches enter the trigeminal ganglion, where their cell bodies are located, and terminate in nuclei in the pons. The trigeminal nerve also contains sensory axons from proprioceptors located in the muscles of mastication and extrinsic muscles of the eyeball, but the cell bodies of these neurons are located in the mesencephalic nucleus.

Branchial motor neurons of the trigeminal nerve are part of the mandibular nerve and supply muscles of mastication (masseter, temporalis, medial pterygoid, lateral pterygoid, anterior belly of digastric, and mylohyoid muscles, as well as the tensor veli palatini muscle in the soft palate and tensor tympani muscle in the middle ear). These motor neurons mainly control chewing movements.

14.13 Facial (VII) Nerve

OBJECTIVE

- **Identify** the origins of the facial (VII) nerve in the brain, the foramen through which it exits the skull, and its function.

The **facial (VII) nerve** (FĀ-shal = face) is a mixed cranial nerve. Its sensory axons extend from the taste buds of the anterior two-thirds of the tongue, which enter the temporal bone to join the facial nerve. From here the sensory axons pass to the **geniculate ganglion** (je-NIK-ū-lat), a cluster of cell bodies of sensory neurons of the facial nerve within the temporal bone, and ends in the pons. From the pons, axons

FIGURE 14.20 **Trigeminal (V) nerve.**

The three branches of the trigeminal (V) nerve leave the cranium through the superior orbital fissure, foramen rotundum, and foramen ovale.

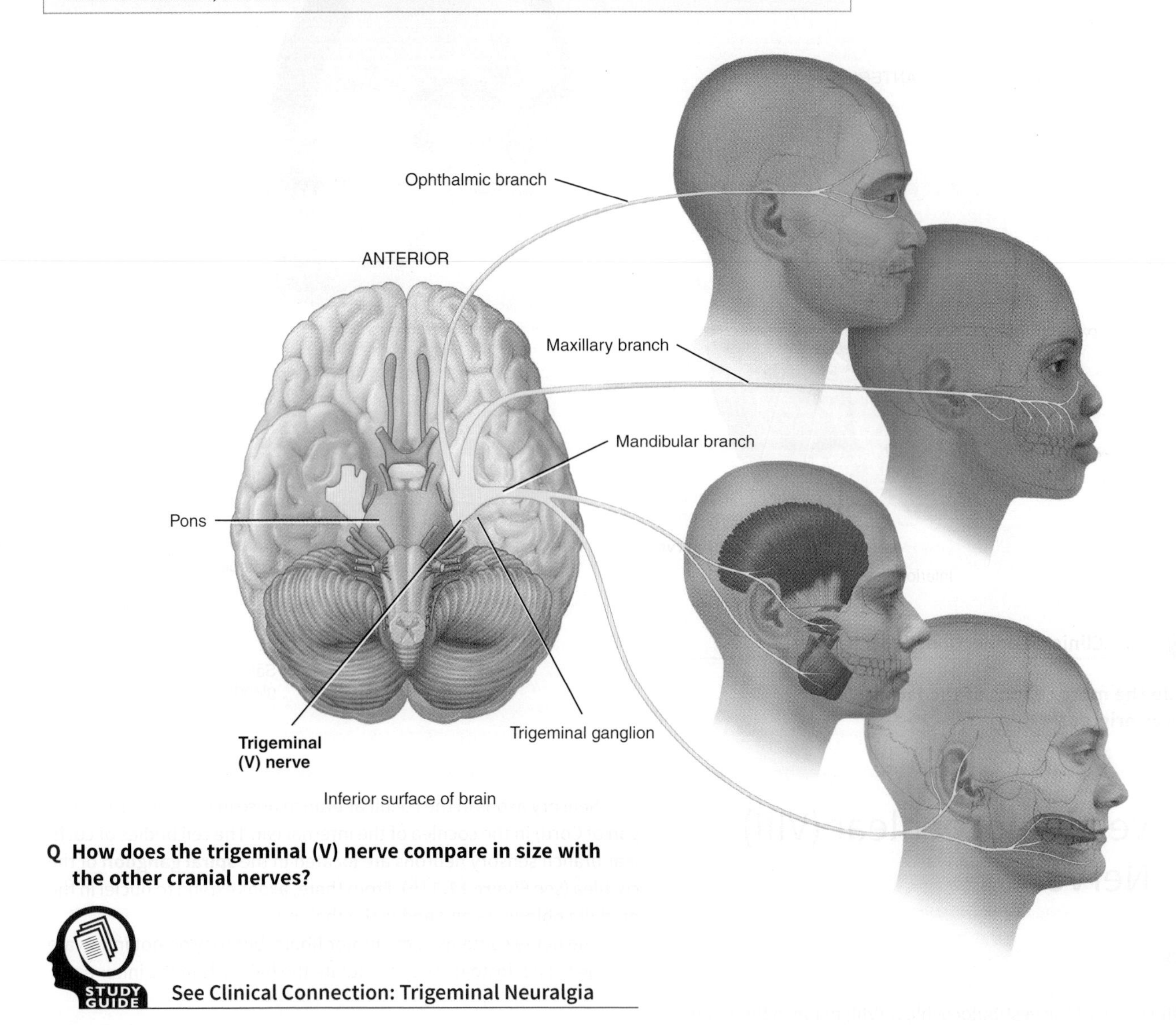

Q How does the trigeminal (V) nerve compare in size with the other cranial nerves?

STUDY GUIDE **See Clinical Connection: Trigeminal Neuralgia**

extend to the thalamus, and then to the gustatory areas of the cerebral cortex (**Figure 14.21**). The sensory portion of the facial nerve also contains axons from skin in the ear canal that relay touch, pain, and thermal sensations. Additionally, proprioceptors from muscles of the face and scalp relay information through their cell bodies in a nucleus in the midbrain (mesencephalic nucleus).

Axons of branchial motor neurons arise from a nucleus in the pons and exit the stylomastoid foramen to innervate middle ear, facial, scalp, and neck muscles. Nerve impulses propagating along these axons cause contraction of the muscles of facial expression plus the stylohyoid muscle, the posterior belly of the digastric muscle, and the stapedius muscle. The facial nerve innervates more named muscles than any other nerve in the body.

Axons of the parasympathetic motor neurons run in branches of the facial nerve and end in two ganglia: the **pterygopalatine ganglion** (ter′-i-gō-PAL-a-tīn) and the **submandibular ganglion**. From synaptic relays in the two ganglia, postganglionic parasympathetic motor axons extend to lacrimal glands (which secrete tears), nasal glands, palatine glands, and saliva-producing sublingual and submandibular glands.

FIGURE 14.21 **Facial (VII) nerve.**

The facial (VII) nerve causes contraction of the muscles of facial expression.

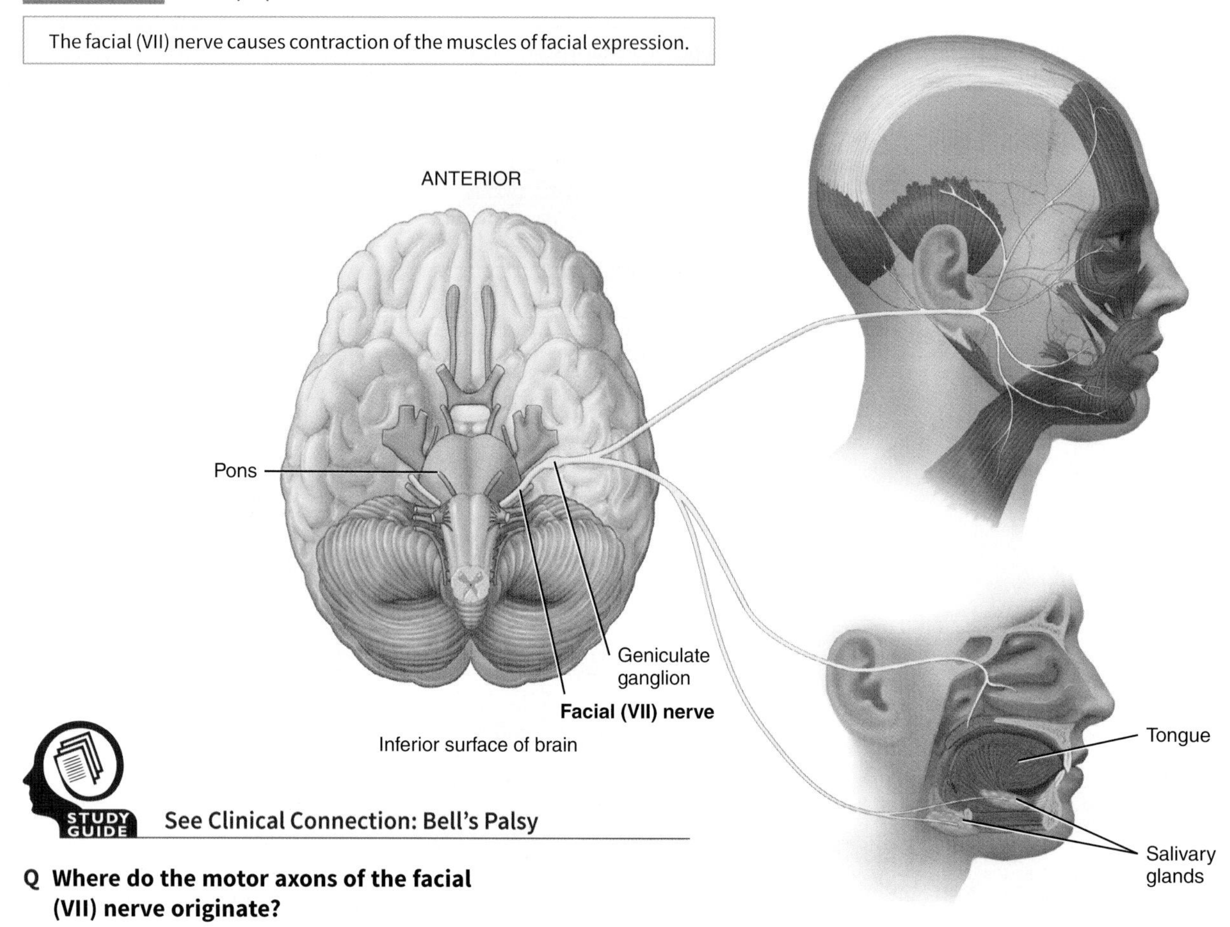

See Clinical Connection: Bell's Palsy

Q Where do the motor axons of the facial (VII) nerve originate?

14.14 Vestibulocochlear (VIII) Nerve

OBJECTIVE

- **Identify** the origin of the vestibulocochlear (VIII) nerve in the brain, the foramen through which it exits the skull, and the functions of each of its branches.

The **vestibulocochlear (VIII) nerve** (ves-tib-ū-lō-KOK-lē-ar; *vestibulo-* = small cavity; *-cochlear* = spiral, snail-like) was formerly known as the *acoustic* or *auditory nerve*. It is a sensory cranial nerve and has two branches, the vestibular branch and the cochlear branch (Figure 14.22). The **vestibular branch** carries impulses for equilibrium and the **cochlear branch** carries impulses for hearing.

Sensory axons in the vestibular branch extend from the semicircular canals, the saccule, and the utricle of the inner ear to the **vestibular ganglia**, where the cell bodies of the neurons are located (see Figure 17.21b), and end in vestibular nuclei in the pons and cerebellum. Some sensory axons also enter the cerebellum via the inferior cerebellar peduncle.

Sensory axons in the cochlear branch arise in the spiral organ (organ of Corti) in the cochlea of the internal ear. The cell bodies of cochlear branch sensory neurons are located in the **spiral ganglion** of the cochlea (see Figure 17.21b). From there, axons extend to nuclei in the medulla oblongata and end in the thalamus.

The nerve contains some motor fibers, but they do not innervate muscle tissue. Instead, they modulate the hair cells in the inner ear.

14.15 Glossopharyngeal (IX) Nerve

OBJECTIVE

- **Identify** the origin of the glossopharyngeal (IX) nerve in the brain, the foramen through which it exits the skull, and its function.

The **glossopharyngeal (IX) nerve** (glos′-ō-fa-RIN-jē-al; *glosso-* = tongue; *-pharyngeal* = throat) is a mixed cranial nerve (Figure 14.23). Sensory axons of the glossopharyngeal nerve arise from (1) taste buds

FIGURE 14.22 **Vestibulocochlear (VIII) nerve.**

The vestibular branch of the vestibulocochlear (VIII) nerve carries impulses for equilibrium, while the cochlear branch carries impulses for hearing.

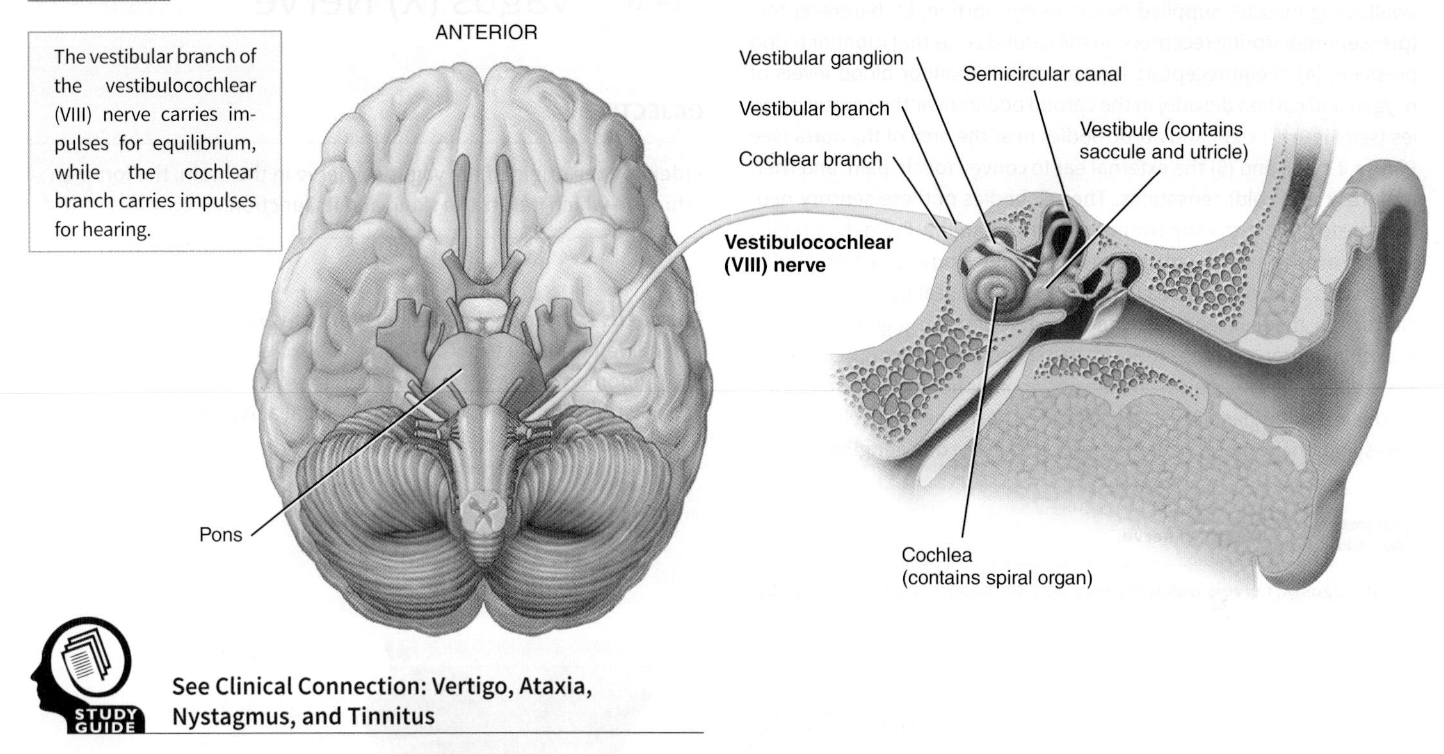

See Clinical Connection: Vertigo, Ataxia, Nystagmus, and Tinnitus

Q What structures are found in the vestibular and spiral ganglia?

FIGURE 14.23 **Glossopharyngeal (IX) nerve.**

Sensory axons in the glossopharyngeal (IX) nerve supply the taste buds.

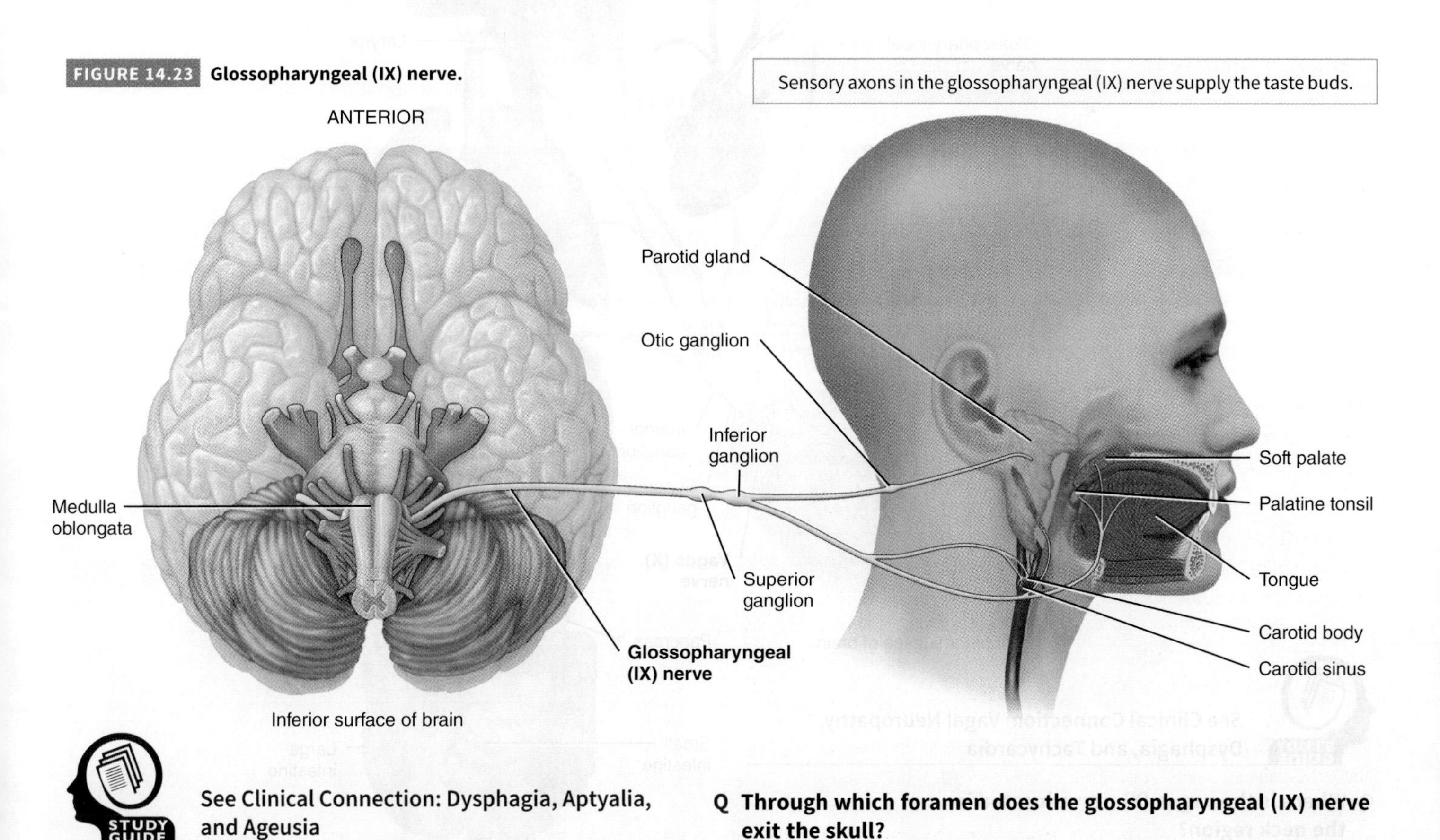

See Clinical Connection: Dysphagia, Aptyalia, and Ageusia

Q Through which foramen does the glossopharyngeal (IX) nerve exit the skull?

on the posterior one-third of the tongue, (2) proprioceptors from some swallowing muscles supplied by the motor portion, (3) baroreceptors (pressure-monitoring receptors) in the carotid sinus that monitor blood pressure, (4) chemoreceptors (receptors that monitor blood levels of oxygen and carbon dioxide) in the carotid bodies near the carotid arteries (see Figure 23.26) and aortic bodies near the arch of the aorta (see Figure 23.26), and (5) the external ear to convey touch, pain, and thermal (heat and cold) sensations. The cell bodies of these sensory neurons are located in the **superior** and **inferior ganglia**. From the ganglia, sensory axons pass through the jugular foramen and end in the medulla.

Axons of motor neurons in the glossopharyngeal nerve arise in nuclei of the medulla and exit the skull through the jugular foramen. Branchial motor neurons innervate the stylopharyngeus muscle, which assists in swallowing, and axons of parasympathetic motor neurons stimulate the parotid gland to secrete saliva. The postganglionic cell bodies of parasympathetic motor neurons are located in the **otic ganglion**.

14.16 Vagus (X) Nerve

OBJECTIVE

- **Identify** the origin of the vagus (X) nerve in the brain, the foramen through which it exits the skull, and its function.

The **vagus (X) nerve** (VĀ-gus = vagrant or wandering) is a mixed cranial nerve that is distributed from the head and neck into the thorax and abdomen (Figure 14.24). The nerve derives its name from its wide distribution. In the neck, it lies medial and posterior to the internal jugular vein and common carotid artery.

FIGURE 14.24 Vagus (X) nerve.

The vagus (X) nerve is widely distributed in the head, neck, thorax, and abdomen.

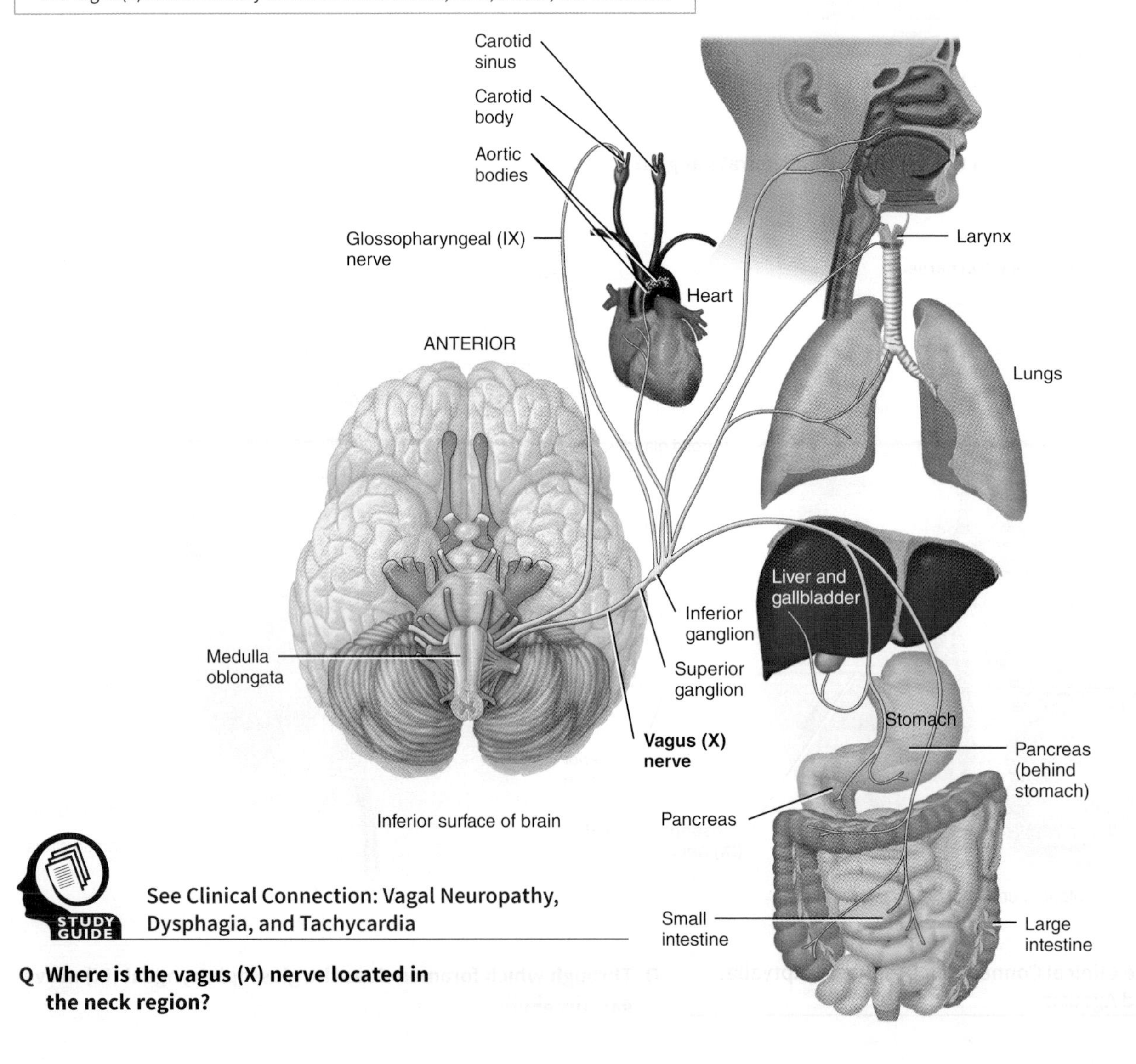

STUDY GUIDE See Clinical Connection: Vagal Neuropathy, Dysphagia, and Tachycardia

Q Where is the vagus (X) nerve located in the neck region?

Sensory axons in the vagus nerve arise from the skin of the external ear for touch, pain, and thermal sensations; a few taste buds in the epiglottis and pharynx; and proprioceptors in muscles of the neck and throat. Also, sensory axons come from baroreceptors in the carotid sinus and chemoreceptors in the carotid and aortic bodies. The majority of sensory neurons come from visceral sensory receptors in most organs of the thoracic and abdominal cavities that convey sensations (such as hunger, fullness, and discomfort) from these organs. The sensory neurons have cell bodies in the **superior** and **inferior ganglia** and then pass through the jugular foramen to end in the medulla and pons.

The branchial motor neurons, which run briefly with the accessory nerve, arise from nuclei in the medulla oblongata and supply muscles of the pharynx, larynx, and soft palate that are used in swallowing, vocalization, and coughing. Historically these motor neurons have been called the cranial accessory nerve, but these fibers actually belong to the vagus (X) nerve.

Axons of parasympathetic motor neurons in the vagus nerve originate in nuclei of the medulla and supply the lungs, heart, glands of the gastrointestinal (GI) tract, and smooth muscle of the respiratory passageways, esophagus, stomach, gallbladder, small intestine, and most of the large intestine (see **Figure 15.3**). Parasympathetic motor axons initiate smooth muscle contractions in the gastrointestinal tract to aid motility and stimulate secretion by digestive glands; activate smooth muscle to constrict respiratory passageways; and decrease heart rate.

14.17 Accessory (XI) Nerve

OBJECTIVE

- **Identify** the origin of the accessory (XI) nerve in the spinal cord, the foramina through which it first enters and then exits the skull, and its function.

The **accessory (XI) nerve** (ak-SES-ō-rē = assisting) is a branchial motor cranial nerve (**Figure 14.25**). Historically it has been divided into two parts, a cranial accessory nerve and a spinal accessory nerve. The cranial accessory nerve actually is part of the vagus (X) nerve (see Section 14.16). The "old" spinal accessory nerve is the accessory nerve we discuss in this exhibit. Its motor axons arise in the anterior gray horn of the first five segments of the cervical portion of the spinal cord. The axons from the segments exit the spinal cord laterally and come together, ascend through the foramen magnum, and then exit through the jugular foramen along with the vagus and glossopharyngeal nerves. The accessory nerve conveys motor impulses to the sternocleidomastoid and trapezius muscles to coordinate head movements. Some sensory axons in the accessory nerve, which originate from proprioceptors in the sternocleidomastoid and trapezius muscles, begin their course toward the brain in the accessory nerve, but eventually leave the nerve

FIGURE 14.25 Accessory (XI) nerve.

The accessory (XI) nerve exits the cranium through the jugular foramen.

Q How does the accessory (XI) nerve differ from the other cranial nerves?

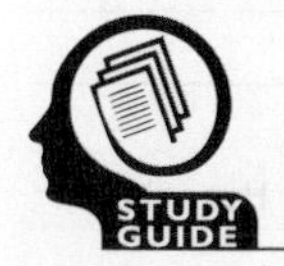

See Clinical Connection: Paralysis of the Sternocleidomastoid and Trapezius Muscles

TABLE 14.4 Summary of Cranial Nerves*

CRANIAL NERVE	COMPONENTS	PRINCIPAL FUNCTIONS
Olfactory (I)	*Special sensory*	Olfaction (smell).
Optic (II)	*Special sensory*	Vision (sight).
Oculomotor (III)	*Motor*	
	Somatic	Movement of eyeballs and upper eyelid.
	Motor (autonomic)	Adjusts lens for near vision (accommodation).
		Constriction of pupil.
Trochlear (IV)	*Motor*	
	Somatic	Movement of eyeballs.
Trigeminal (V)	*Mixed*	
	Sensory	Touch, pain, and thermal sensations from scalp, face, and oral cavity (including teeth and anterior two-thirds of tongue).
	Motor (branchial)	Chewing and controls middle ear muscle.
Abducens (VI)	*Motor*	
	Somatic	Movement of eyeballs.
Facial (VII)	*Mixed*	
	Sensory	Taste from anterior two-thirds of tongue.
		Touch, pain, and thermal sensations from skin in external ear canal.
	Motor (branchial)	Control of muscles of facial expression and middle ear muscle.
	Motor (autonomic)	Secretion of tears and saliva.
Vestibulocochlear (VIII)	*Special sensory*	Hearing and equilibrium.
Glossopharyngeal (IX)	*Mixed*	
	Sensory	Taste from posterior one-third of tongue.
		Proprioception in some swallowing muscles.
		Monitors blood pressure and oxygen and carbon dioxide levels in blood.
		Touch, pain, and thermal sensations from skin of external ear and upper pharynx.
	Motor (branchial)	Assists in swallowing.
	Motor (autonomic)	Secretion of saliva.
Vagus (X)	*Mixed*	
	Sensory	Taste from epiglottis.
		Proprioception from throat and voice box muscles.
		Monitors blood pressure and oxygen and carbon dioxide levels in blood.
		Touch, pain, and thermal sensations from skin of external ear.
		Sensations from thoracic and abdominal organs.
	Motor (branchial)	Swallowing, vocalization, and coughing.
	Motor (autonomic)	Motility and secretion of gastrointestinal organs.
		Constriction of respiratory passageways.
		Decreases heart rate.
Accessory (XI)	*Motor*	
	Branchial	Movement of head and pectoral girdle.
Hypoglossal (XII)	*Motor*	
	Somatic	Speech, manipulation of food, and swallowing.

***MNEMONIC FOR CRANIAL NERVES:**

Oh	Oh	Oh	To	Touch	And	Feel	Very	Green	Vegetables	AH!	
Olfactory	Optic	Oculomotor	Trochlear	Trigeminal	Abducens	Facial	Vestibulocochlear	Glossopharyngeal	Vagus	Accessory	Hypoglossal

to join nerves of the cervical plexus, while others remain in the accessory nerve. From the cervical plexus they enter the spinal cord via the posterior roots of cervical spinal nerves; their cell bodies are located in the posterior root ganglia of those nerves. In the spinal cord the axons ascend to nuclei in the medulla oblongata.

14.18 Hypoglossal (XII) Nerve

OBJECTIVE

- **Identify** the origin of the hypoglossal (XII) nerve in the brain, the foramen through which it exits the skull, and its function.

The **hypoglossal (XII) nerve** (hī′-pō-GLOS-al; *hypo-* = below; *-glossal* = tongue) is a motor cranial nerve. The somatic motor axons originate in a nucleus in the medulla oblongata (hypoglossal nucleus), exit the medulla on its anterior surface, and pass through the hypoglossal canal to supply the muscles of the tongue (**Figure 14.26**). These axons conduct nerve impulses for speech and swallowing. The sensory axons do not return to the brain in the hypoglossal nerve. Instead, sensory axons that originate from proprioceptors in the tongue muscles begin their course toward the brain in the hypoglossal nerve but leave the nerve to join cervical spinal nerves and end in the medulla oblongata, again entering the central nervous system via posterior roots of cervical spinal nerves.

14.19 Development of the Nervous System

OBJECTIVE

- **Describe** how the parts of the brain develop.

Development of the nervous system begins in the third week of gestation with a thickening of the **ectoderm** called the **neural**

FIGURE 14.26 **Hypoglossal (XII) nerve.**

The hypoglossal (XII) nerve exits the cranium through the hypoglossal canal.

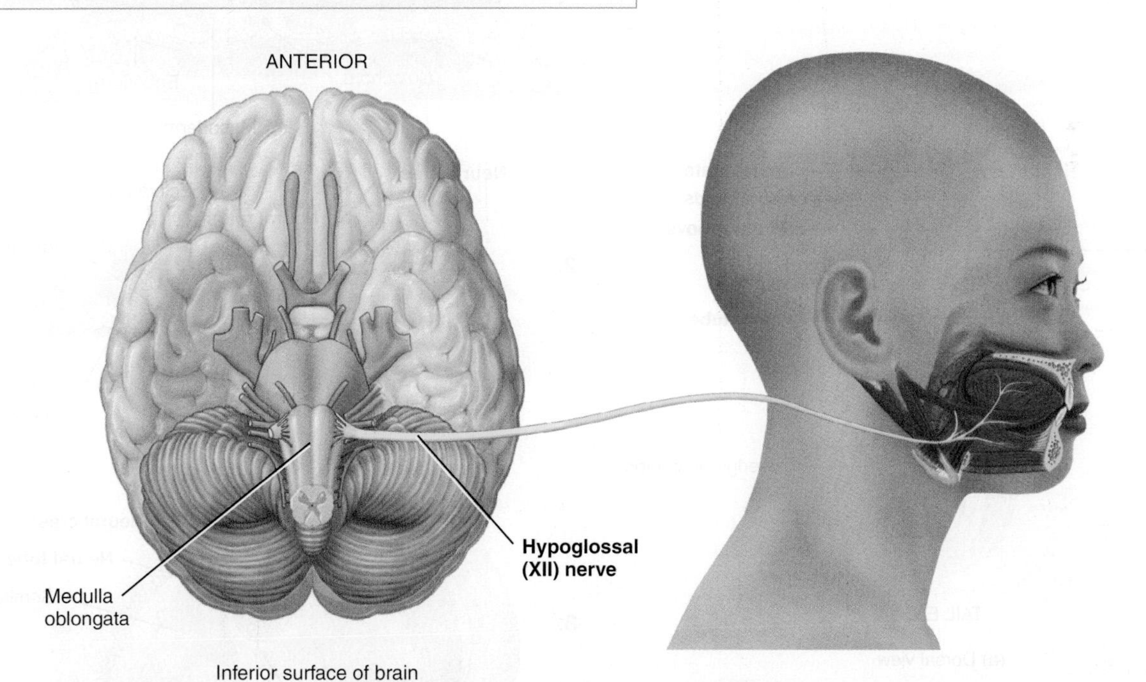

Q What important motor functions are related to the hypoglossal (XII) nerve?

See Clinical Connection: Dysarthria and Dysphagia

plate (Figure 14.27). The plate folds inward and forms a longitudinal groove, the **neural groove**. The raised edges of the neural plate are called **neural folds**. As development continues, the neural folds increase in height and meet to form a tube called the **neural tube**.

Three layers of cells differentiate from the wall that encloses the neural tube. The outer or **marginal layer** cells develop into the *white matter* of the nervous system. The middle or **mantle layer** cells develop into the *gray matter*. The cells of the inner or **ependymal layer** (ep-EN-di-mal) eventually form the *lining of the central canal of the spinal cord* and *ventricles of the brain.*

The **neural crest** is a mass of tissue between the neural tube and the skin ectoderm (Figure 14.27b). It differentiates and eventually forms the *posterior (dorsal) root ganglia of spinal nerves*, *spinal nerves*, *ganglia of cranial nerves*, *cranial nerves*, *ganglia of the autonomic nervous system*, *adrenal medulla*, and *meninges.*

As discussed at the beginning of this chapter, during the third to fourth week of embryonic development, the anterior part of the neural tube develops into three enlarged areas called **primary brain vesicles** that are named for their relative positions. These are the **prosencephalon** (*pros-* = before) or forebrain, **mesencephalon** or midbrain, and **rhombencephalon** (*rhomb-* = behind) or hindbrain (Figure 14.28a; see also Table 14.1). During the fifth week of development **secondary brain vesicles** begin to develop. The prosencephalon develops into two secondary brain vesicles called the **telencephalon** and the **diencephalon** (Figure 14.28b). The rhombencephalon also develops into two secondary brain vesicles called the **metencephalon** and the **myelencephalon**. The area of the neural tube inferior to the myelencephalon gives rise to the *spinal cord.*

The brain vesicles continue to develop as follows (Figure 14.28c, d; see also Table 14.1):

FIGURE 14.27 Origin of the nervous system. (a) Dorsal view of an embryo in which the neural folds have partially united, forming the early neural tube. (b) Transverse sections through the embryo showing the formation of the neural tube.

The nervous system begins developing in the third week from a thickening of ectoderm called the neural plate.

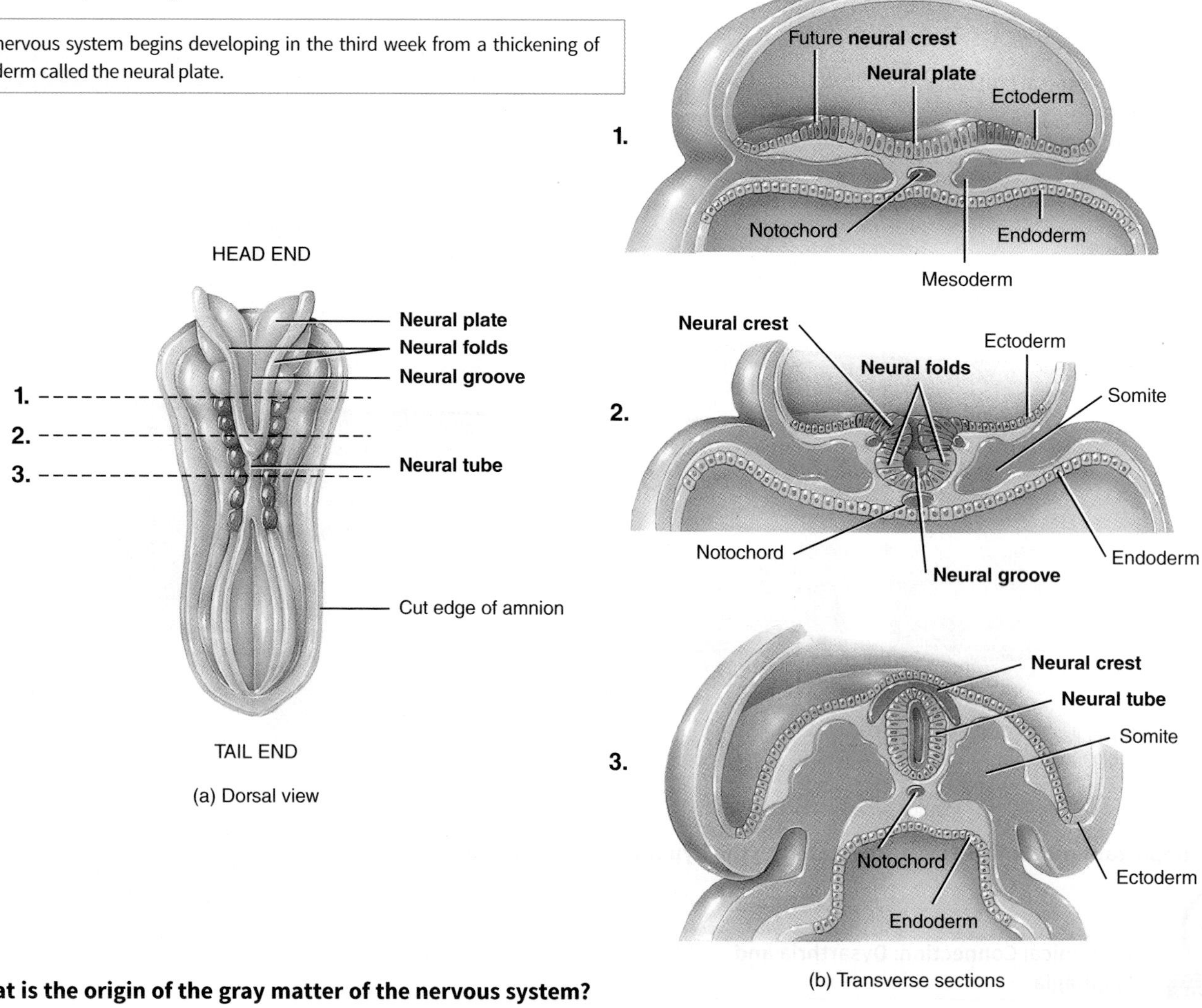

(a) Dorsal view

(b) Transverse sections

Q What is the origin of the gray matter of the nervous system?

FIGURE 14.28 **Development of the brain and spinal cord.**

The various parts of the brain develop from the primary brain vesicles.

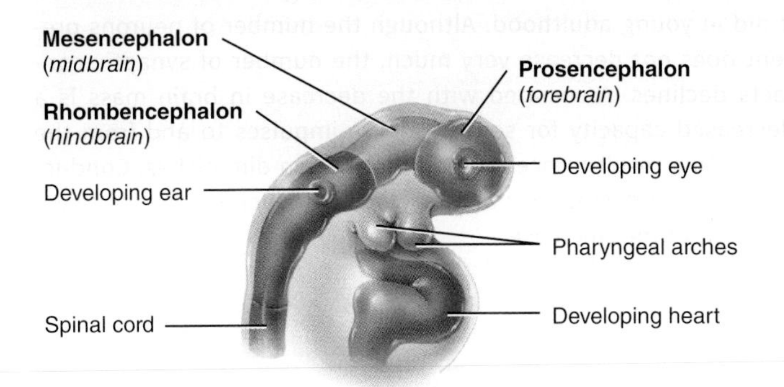

(a) Three–four week embryo showing primary brain vesicles

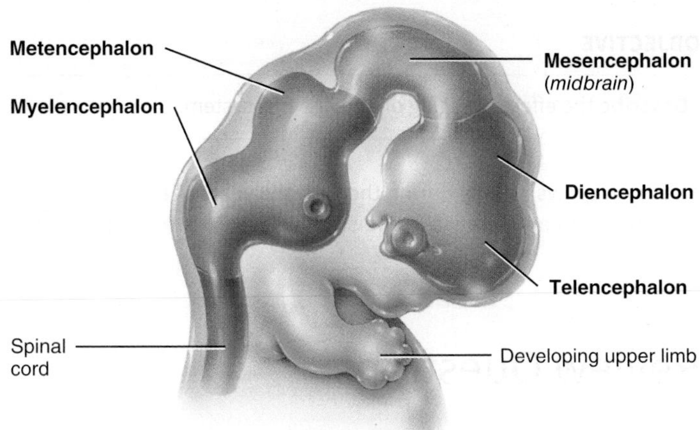

(b) Seven-week embryo showing secondary brain vesicles

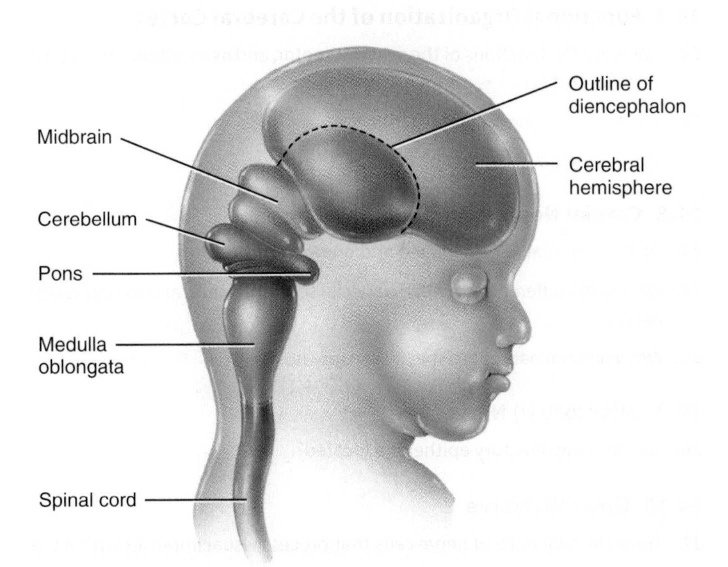

(c) Eleven-week fetus showing expanding cerebral hemispheres overgrowing the diencephalon

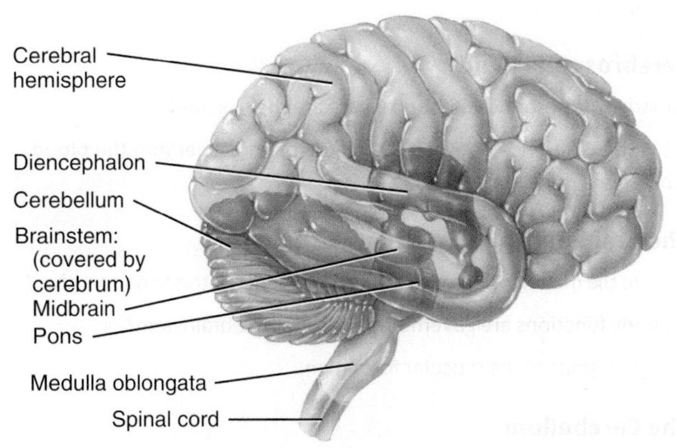

(d) Brain at birth (the diencephalon and superior portion of the brainstem have been projected to the surface)

Q Which primary brain vesicle does not develop into a secondary brain vesicle?

- The telencephalon develops into the *cerebral hemispheres*, including the *basal nuclei*, and houses the paired *lateral ventricles*.
- The diencephalon develops into the *thalamus*, *hypothalamus*, and *epithalamus*.
- The mesencephalon develops into the *midbrain*, which surrounds the *aqueduct of the midbrain* (*cerebral aqueduct*).
- The metencephalon becomes the *pons* and *cerebellum* and houses part of the *fourth ventricle*.
- The myelencephalon develops into the *medulla oblongata* and houses the remainder of the *fourth ventricle*.

Two **neural tube defects**—spina bifida (see Disorders: Homeostatic Imbalances in Chapter 7 of the Study Guide) and anencephaly (absence of the skull and cerebral hemispheres, discussed in Section 29.1)—are associated with low levels of folic acid (folate), one of the B vitamins, in the first few weeks of development. These and other defects occur when the neural tube does not close properly. Many foods, especially grain products such as cereals and bread, are now fortified with folic acid; however, the incidence of both disorders is greatly reduced when women who are or may become pregnant take folic acid supplements.

14.20 Aging and the Nervous System

OBJECTIVE

- **Describe** the effects of aging on the nervous system.

The brain grows rapidly during the first few years of life. Growth is due mainly to an increase in the size of neurons already present, the proliferation and growth of neuroglia, the development of dendritic branches and synaptic contacts, and continuing myelination of axons. From early adulthood onward, brain mass declines. By the time a person reaches 80, the brain weighs about 7% less than it did in young adulthood. Although the number of neurons present does not decrease very much, the number of synaptic contacts declines. Associated with the decrease in brain mass is a decreased capacity for sending nerve impulses to and from the brain. As a result, processing of information diminishes. Conduction velocity decreases, voluntary motor movements slow down, and reflex times increase.

Review Questions

14.1 Brain Organization, Protection, and Blood Supply

1. Compare the sizes and locations of the cerebrum and cerebellum.
2. Describe the locations of the cranial meninges.
3. Explain the blood supply to the brain and the importance of the blood–brain barrier.

14.2 Cerebrospinal Fluid

4. What structures produce CSF, and where are they located?
5. What is the difference between the blood–brain barrier and the blood–cerebrospinal fluid barrier?

14.3 The Brainstem and Reticular Formation

6. Where are the medulla, pons, and midbrain located relative to one another?
7. What body functions are governed by nuclei in the brainstem?
8. List the functions of the reticular formation.

14.4 The Cerebellum

9. Describe the location and principal parts of the cerebellum.
10. Where do the axons of each of the three pairs of cerebellar peduncles begin and end? What are their functions?

14.5 The Diencephalon

11. Why is the thalamus considered a "relay station" in the brain?
12. Why is the hypothalamus considered part of both the nervous system and the endocrine system?
13. What are the functions of the epithalamus?
14. Define a circumventricular organ.

14.6 The Cerebrum

15. List and locate the lobes of the cerebrum. How are they separated from one another? What is the insula?
16. Distinguish between the precentral gyrus and the postcentral gyrus.
17. Describe the organization of cerebral white matter and indicate the function of each major group of fibers.
18. List the basal nuclei. What are the functions of the basal nuclei?
19. Define the limbic system and list several of its functions.

14.7 Functional Organization of the Cerebral Cortex

20. Compare the functions of the sensory, motor, and association areas of the cerebral cortex.
21. What is hemispheric lateralization?
22. What is the diagnostic value of an EEG?

14.8 Cranial Nerves: An Overview

23. How are cranial nerves named and numbered?
24. What is the difference between a special sensory, motor, and mixed cranial nerve?
25. Which cranial nerves are special sensory nerves?

14.9 Olfactory (I) Nerve

26. Where is the olfactory epithelium located?

14.10 Optic (II) Nerve

27. Trace the sequence of nerve cells that process visual impulses within the retina.

14.11 Oculomotor (III), Trochlear (IV), and Abducens (VI) Nerves

28. How are the oculomotor (III), trochlear (IV), and abducens (VI) nerves related functionally?

14.12 Trigeminal (V) Nerve

29. What are the three branches of the trigeminal (V) nerve, and which branch is the largest?

14.13 Facial (VII) Nerve

30. Why is the facial (VII) nerve considered the major motor nerve of the head?

14.14 Vestibulocochlear (VIII) Nerve

31. What are the functions of each of the two branches of the vestibulocochlear (VIII) nerve?

14.15 Glossopharyngeal (IX) Nerve

32. Which other cranial nerves are also distributed to the tongue?

14.16 Vagus (X) Nerve

33. On what basis is the vagus (X) nerve named?

14.17 Accessory (XI) Nerve

34. Where do the motor axons of the accessory (XI) nerve originate?

14.18 Hypoglossal (XII) Nerve

35. In what portion of the brain does the hypoglossal nucleus originate?

14.19 Development of the Nervous System

36. What parts of the brain develop from each primary brain vesicle?

14.20 Aging and the Nervous System

37. How is brain mass related to age?

Critical Thinking Questions

1. An elderly relative suffered a CVA (stroke) and now has difficulty moving her right arm, and she also has speech problems. What areas of the brain were damaged by the stroke?
2. Nicky has recently had a viral infection and now she cannot move the muscles on the right side of her face. In addition, she is experiencing a loss of taste and a dry mouth, and she cannot close her right eye. What cranial nerve has been affected by the viral infection?
3. You have been hired by a pharmaceutical company to develop a drug to regulate a specific brain disorder. What is a major physiological roadblock to developing such a drug, and how can you design a drug to bypass that roadblock so that the drug is delivered to the brain where it is needed?

The Autonomic Nervous System

CHAPTER **15**

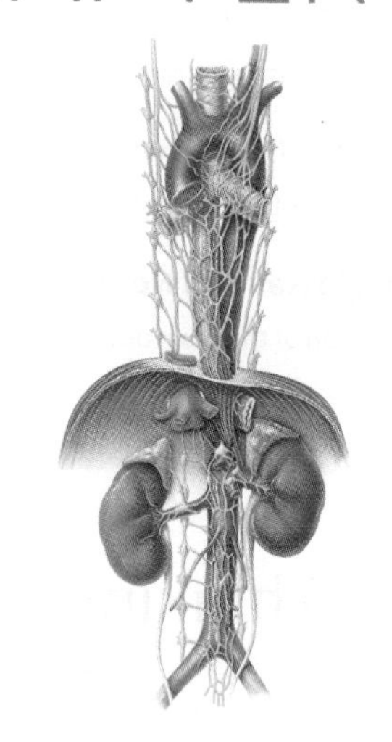

The Autonomic Nervous System and Homeostasis

The autonomic nervous system contributes to homeostasis by conveying motor output from the central nervous system to smooth muscle, cardiac muscle, and glands for appropriate responses to integrated sensory information.

As you learned in Chapter 12, the motor (efferent) division of the peripheral nervous system (PNS) is divided into a somatic nervous system (SNS) and autonomic nervous system (ANS). The ANS usually operates without conscious control. However, centers in the hypothalamus and brainstem do regulate ANS reflexes. In this chapter, we compare structural and functional features of the somatic and autonomic nervous systems. Then we discuss the anatomy of the motor portion of the ANS and compare the organization and actions of its two major parts, the sympathetic and parasympathetic divisions.

Q Did you ever wonder how some blood pressure medications exert their effects through the autonomic nervous system?

15.1 Comparison of Somatic and Autonomic Nervous Systems

OBJECTIVE

- **Compare** the structural and functional differences between the somatic and autonomic parts of the nervous system.

Somatic Nervous System

The **somatic nervous system** consists of somatic motor neurons that innervate the skeletal muscles of the body. When a somatic motor neuron stimulates a skeletal muscle, it contracts; the effect is always excitation. If somatic motor neurons cease to stimulate a skeletal muscle, the result is a paralyzed, limp muscle that has no muscle tone.

The somatic nervous system usually operates under voluntary (conscious) control. Voluntary control of movement involves motor areas of the cerebral cortex that activate somatic motor neurons whenever you have a desire to move. For example, if you want to perform a particular movement (kick a ball, turn a screwdriver, smile for a picture, etc.), neural pathways from the primary motor area of the cerebral cortex activate somatic motor neurons that cause the appropriate skeletal muscles to contract. The somatic nervous system is not always under voluntary control, however. The somatic motor neurons that innervate skeletal muscles involved in posture, balance, breathing, and somatic reflexes (such as the flexor reflex) are involuntarily controlled by integrating centers in the brainstem and spinal cord.

The somatic nervous system can also receive sensory input from sensory neurons that convey information for somatic senses (tactile, thermal, pain, and proprioceptive sensations; see Chapter 16) or the special senses (sight, hearing, taste, smell, and equilibrium; see Chapter 17). All of these sensations normally are consciously perceived. In response to this sensory information, somatic motor neurons cause the appropriate skeletal muscles of the body to contract.

Autonomic Nervous System

The **autonomic nervous system (ANS)** (aw′-tō-NOM-ik) is the part of the nervous system that regulates cardiac muscle, smooth muscle, and glands. These tissues are often referred to as **visceral effectors** because they are usually associated with the viscera (internal organs) of the body. The term *autonomic* is derived from the Latin words *auto-* = self and *-nomic* = law because the ANS was once thought to be self-governing.

The autonomic nervous system consists of autonomic motor neurons that regulate visceral activities by either increasing (exciting) or decreasing (inhibiting) ongoing activities in their effector tissues (cardiac muscle, smooth muscle, and glands). Changes in the diameter of the pupils, dilation and constriction of blood vessels, and adjustment of the rate and force of the heartbeat are examples of autonomic motor responses. Unlike skeletal muscle, tissues innervated by the ANS often function to some extent even if their nerve supply is damaged. The heart continues to beat when it is removed for transplantation into another person, smooth muscle in the lining of the gastrointestinal tract contracts rhythmically on its own, and glands produce some secretions in the absence of ANS control.

The ANS usually operates without conscious control. For example, you probably cannot voluntarily slow down your heart rate; instead, your heart rate is subconsciously regulated. For this reason, some autonomic responses are the basis for *polygraph* ("lie detector") tests. However, practitioners of yoga or other techniques of meditation may learn how to regulate at least some of their autonomic activities through long practice. **Biofeedback**, in which monitoring devices display information about a body function such as heart rate or blood pressure, enhances the ability to learn such conscious control. (For more on biofeedback, see the Medical Terminology section at the end of the chapter).

The ANS can also receive sensory input from sensory neurons associated with **interoceptors** (IN-ter-ō-sep′-tors), sensory receptors located in blood vessels, visceral organs, muscles, and the nervous system that monitor conditions in the *internal* environment. Examples of interoceptors are chemoreceptors that monitor blood CO_2 level and mechanoreceptors that detect the degree of stretch in the walls of organs or blood vessels. Unlike those triggered by a flower's perfume, a beautiful painting, or a delicious meal, these sensory signals are not consciously perceived most of the time, although intense activation of interoceptors may produce conscious sensations. Two examples of perceived visceral sensations are pain sensations from damaged viscera and angina pectoris (chest pain) from inadequate blood flow to the heart. Signals from the somatic senses and special senses, acting via the limbic system, also influence responses of autonomic motor neurons. Seeing a bike about to hit you, hearing squealing brakes of a nearby car, or being grabbed by an attacker would all increase the rate and force of your heartbeat.

The ANS consists of two main division (branches): the **sympathetic nervous system** and the **parasympathetic nervous system**. Most organs receive nerves from both of these divisions, an arrangement known as **dual innervation**. In general, one division stimulates the organ to increase its activity (excitation), and the other division decreases the organ's activity (inhibition). For example, neurons of the sympathetic nervous system increase heart rate, and neurons of the parasympathetic nervous system slow it down. The sympathetic nervous system promotes the *fight-or-flight* response, which prepares the body for emergency situations. By contrast, the parasympathetic nervous system enhances *rest-and-digest* activities, which conserve and restore body energy during times of rest or digesting a meal. Although both the sympathetic and parasympathetic divisions are concerned with maintaining health, they do so in dramatically different ways.

The ANS is also comprised of a third division known as the **enteric nervous system (ENS)**. The ENS consists of millions of neurons in plexuses that extend most of the length of the gastrointestinal tract. Its operation is involuntary. Although the neurons of the ENS can function autonomously, they can also be regulated by the other divisions of the ANS. The ENS contains sensory neurons, interneurons, and motor neurons. Enteric sensory neurons monitor chemical changes within the GI tract as well as the stretching of its walls. Enteric interneurons integrate information from the sensory neurons and provide input to motor neurons. Enteric motor neurons govern contraction of GI tract smooth muscle and secretion of GI tract glands. The ENS is described in greater detail in the discussion of the digestive system in Chapter 24. The rest of this chapter is devoted to the sympathetic and parasympathetic divisions of the ANS.

Recall from Chapter 10 that the axon of a single, myelinated somatic motor neuron extends from the CNS all the way to the skeletal muscle fibers in its motor unit (**Figure 15.1a**). By contrast, most autonomic motor pathways consist of two motor neurons in series; that is, one following the other (**Figure 15.1b**). The first neuron (preganglionic neuron) has its cell body in the CNS; its myelinated axon extends from the CNS to an **autonomic ganglion**. (Recall that a *ganglion* is a collection of neuronal cell bodies In the PNS.) The cell body of the second neuron (postganglionic neuron) is also in that same autonomic ganglion; its unmyelinated axon extends directly from the ganglion to the effector (smooth muscle, cardiac muscle, or a gland). Alternatively, in some autonomic pathways, the first motor neuron extends to specialized cells called *chromaffin cells* in the adrenal medullae (inner portion of the adrenal glands) rather than an autonomic ganglion. Chromaffin cells secrete the neurotransmitters epinephrine and norepinephrine (NE). All somatic motor neurons release only acetylcholine (ACh) as their neurotransmitter, but autonomic motor neurons release either ACh or norepinephrine (NE).

Table 15.1 compares the somatic and autonomic nervous systems.

15.2 Anatomy of Autonomic Motor Pathways

OBJECTIVES

- **Describe** preganglionic and postganglionic neurons of the autonomic nervous system.
- **Compare** the anatomical components of the sympathetic and parasympathetic divisions of the autonomic nervous system.

Anatomical Components

Each division of the ANS has two motor neurons. The first of the two motor neurons in any autonomic motor pathway is called a **preganglionic neuron** (**Figure 15.1b**). Its cell body is in the brain or spinal cord; its axon exits the CNS as part of a cranial or spinal nerve. The axon of a preganglionic neuron is a small-diameter, myelinated type B fiber that usually extends to an autonomic ganglion, where it synapses with a **postganglionic neuron**, the second neuron in the autonomic motor pathway. Note that the postganglionic neuron lies entirely outside the CNS in the PNS. Its cell body and dendrites are located in an **autonomic ganglion**, where it forms synapses with one or more preganglionic axons. The axon of a postganglionic neuron is a small-diameter, unmyelinated type C fiber that terminates in a

FIGURE 15.1 Motor neuron pathways in the (a) somatic nervous system and (b) autonomic nervous system (ANS). Note that autonomic motor neurons release either acetylcholine (ACh) or norepinephrine (NE); somatic motor neurons release only ACh.

Somatic nervous system stimulation always excites its effectors (skeletal muscle fibers); stimulation by the autonomic nervous system either excites or inhibits visceral effectors.

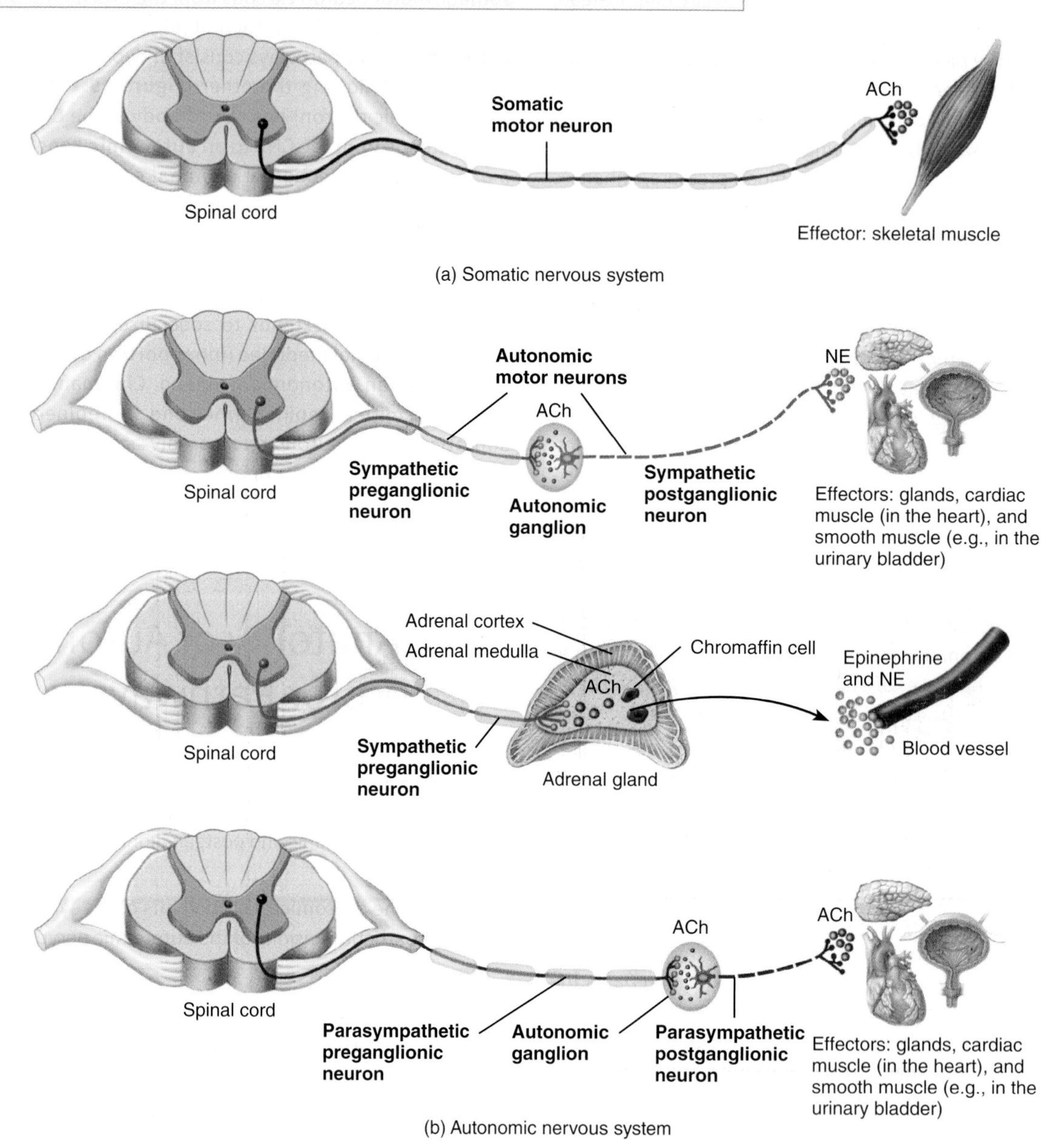

Q What does dual innervation mean?

visceral effector. Thus, preganglionic neurons convey nerve impulses from the CNS to autonomic ganglia, and postganglionic neurons relay the impulses from autonomic ganglia to visceral effectors.

Preganglionic Neurons

In the sympathetic division, the preganglionic neurons have their cell bodies in the lateral horns of the gray matter in the 12 thoracic segments and the first two (and sometimes three) lumbar segments of the spinal cord (Figure 15.2). For this reason, the sympathetic division is also called the **thoracolumbar division** (thōr′-a-kō-LUM-bar), and the axons of the sympathetic preganglionic neurons are known as the **thoracolumbar outflow**.

Cell bodies of preganglionic neurons of the parasympathetic division are located in the nuclei of four cranial nerves in the brainstem (III, VII, IX, and X) and in the lateral gray matter of the second through fourth sacral segments of the spinal cord (Figure 15.3). Hence, the parasympathetic division is also known as the **craniosacral division** (krā′-nē-ō-SĀK-ral), and the axons of the parasympathetic preganglionic neurons are referred to as the **craniosacral outflow**.

TABLE 15.1 Comparison of the Somatic and Autonomic Nervous Systems

	SOMATIC NERVOUS SYSTEM	AUTONOMIC NERVOUS SYSTEM
Sensory input	From somatic senses and special senses.	Mainly from interoceptors; some from somatic senses and special senses.
Control of motor output	Voluntary control from cerebral cortex, with contributions from basal ganglia, cerebellum, brainstem, and spinal cord.	Involuntary control from hypothalamus, limbic system, brainstem, and spinal cord; limited control from cerebral cortex.
Motor neuron pathway	One-neuron pathway: Somatic motor neurons extending from CNS synapse directly with effector.	Usually two-neuron pathway: Preganglionic neurons extending from CNS synapse with postganglionic neurons in autonomic ganglion, and postganglionic neurons extending from ganglion synapse with visceral effector. Alternatively, preganglionic neurons may extend from CNS to synapse with chromaffin cells of adrenal medullae.
Neurotransmitters and hormones	All somatic motor neurons release only acetylcholine (ACh).	All sympathetic and parasympathetic preganglionic neurons release ACh. Most sympathetic postganglionic neurons release NE; those to most sweat glands release ACh. All parasympathetic postganglionic neurons release ACh. Chromaffin cells of adrenal medullae release epinephrine and norepinephrine (NE).
Effectors	Skeletal muscle.	Smooth muscle, cardiac muscle, and glands.
Responses	Contraction of skeletal muscle.	Contraction or relaxation of smooth muscle; increased or decreased rate and force of contraction of cardiac muscle; increased or decreased secretions of glands.

Autonomic Ganglia

There are two major groups of autonomic ganglia: (1) sympathetic ganglia, which are components of the sympathetic division of the ANS, and (2) parasympathetic ganglia, which are components of the parasympathetic division of the ANS.

Sympathetic Ganglia The sympathetic ganglia are the sites of synapses between sympathetic preganglionic and postganglionic neurons. There are two major types of sympathetic ganglia: sympathetic trunk ganglia and prevertebral ganglia. **Sympathetic trunk ganglia** (also called *vertebral chain ganglia* or *paravertebral ganglia*) lie in a vertical row on either side of the vertebral column. These ganglia extend from the base of the skull to the coccyx (Figure 15.2). Postganglionic axons from sympathetic trunk ganglia primarily innervate organs above the diaphragm, such as the head, neck, shoulders, and heart. Sympathetic trunk ganglia in the neck have specific names. They are the **superior**, **middle**, and **inferior cervical ganglia**. The remaining sympathetic trunk ganglia do not have individual names. Because the sympathetic trunk ganglia are near the spinal cord, most sympathetic preganglionic axons are short and most sympathetic postganglionic axons are long.

The second group of sympathetic ganglia, the **prevertebral** (*collateral*) **ganglia**, lies anterior to the vertebral column and close to the large abdominal arteries. In general, postganglionic axons from prevertebral ganglia innervate organs below the diaphragm. There are five major prevertebral ganglia (Figure 15.2; see also Figure 15.5): (1) The **celiac ganglion** (SĒ-lē-ak) is on either side of the celiac trunk, an artery that is just inferior to the diaphragm. (2) The **superior mesenteric ganglion** (MEZ-en-ter′-ik) is near the beginning of the superior mesenteric artery in the upper abdomen. (3) The **inferior mesenteric ganglion** is near the beginning of the inferior mesenteric artery in the middle of the abdomen. (4) The **aorticorenal ganglion** (ā-or′-ti-kō-RĒ-nal) and (5) the **renal ganglion** are near the renal artery of each kidney.

Parasympathetic Ganglia Preganglionic axons of the parasympathetic division synapse with postganglionic neurons in **terminal** (*intramural*) **ganglia**. Most of these ganglia are located close to or actually within the wall of a visceral organ. Terminal ganglia in the head have specific names. They are the **ciliary ganglion, pterygopalatine ganglion** (ter′-i-gō-PAL-a-tīn), **submandibular ganglion,** and **otic ganglion** (Figure 15.3). The remaining terminal ganglia do not have specific names. Because terminal ganglia are located either close to or in the wall of the visceral organ, parasympathetic preganglionic axons are long, in contrast to parasympathetic postganglionic axons, which are short.

Postganglionic Neurons

Once axons of sympathetic preganglionic neurons pass to sympathetic trunk ganglia, they may connect with postganglionic neurons in one of the following ways (Figure 15.4):

1. An axon may synapse with postganglionic neurons in the ganglion it first reaches.
2. An axon may ascend or descend to a higher or lower ganglion before synapsing with postganglionic neurons. The axons of incoming sympathetic preganglionic neurons pass up or down the sympathetic trunk from ganglion to ganglion.
3. An axon may continue, without synapsing, through the sympathetic trunk ganglion to end at a prevertebral ganglion and synapse with postganglionic neurons there.
4. An axon may also pass, without synapsing, through the sympathetic trunk ganglion and a prevertebral ganglion and then extend to chromaffin cells of the adrenal medullae that are functionally similar to sympathetic postganglionic neurons.

FIGURE 15.2 Structure of the sympathetic division of the autonomic nervous system. Solid lines represent preganglionic axons; dashed lines represent postganglionic axons. Although the innervated structures are shown for only one side of the body for diagrammatic purposes, the sympathetic division actually innervates tissues and organs on both sides.

Cell bodies of sympathetic preganglionic neurons are located in the lateral horns of gray matter in the 12 thoracic and first two lumbar segments of the spinal cord.

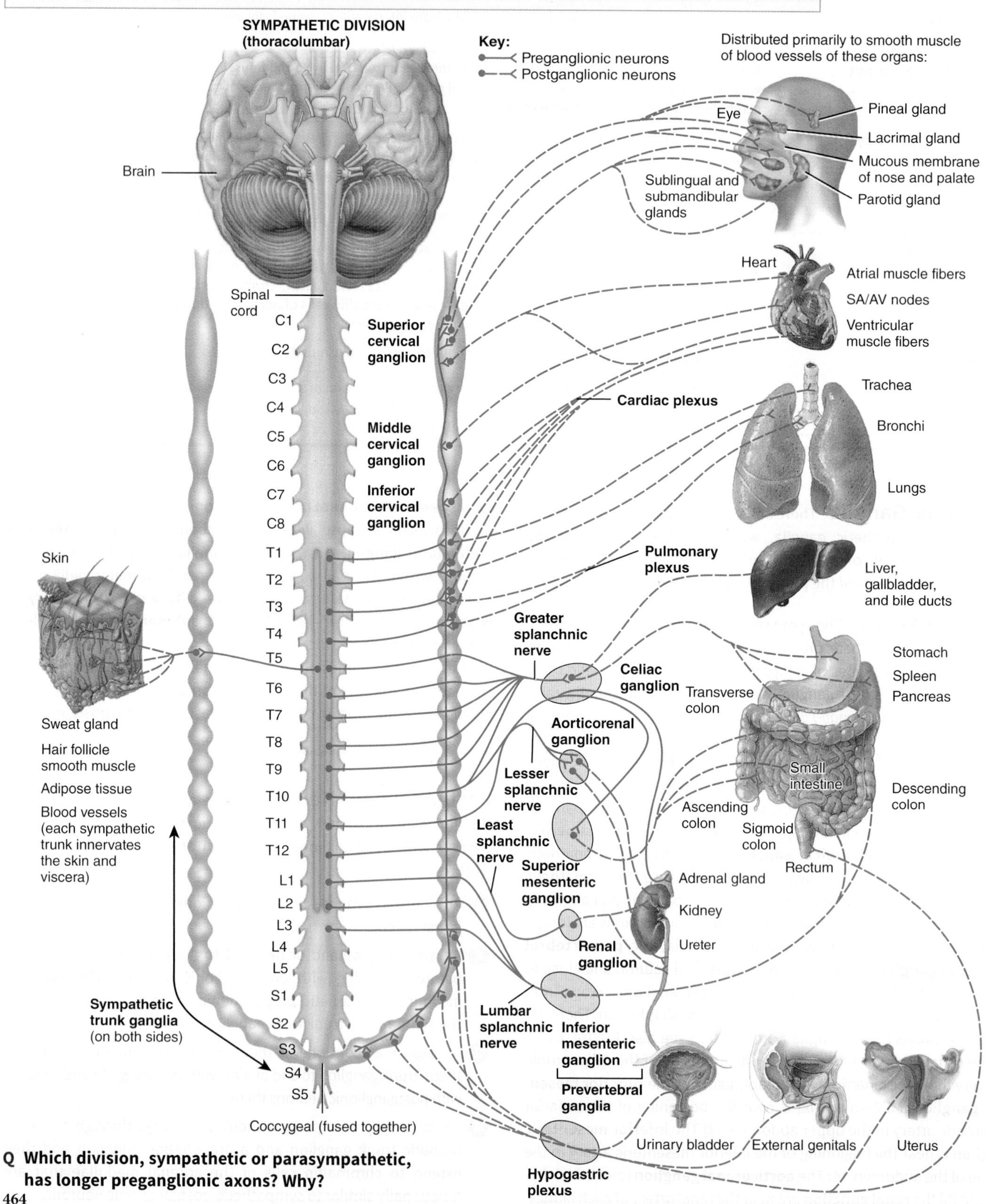

Q Which division, sympathetic or parasympathetic, has longer preganglionic axons? Why?

FIGURE 15.3 Structure of the parasympathetic division of the autonomic nervous system. Solid lines represent preganglionic axons; dashed lines represent postganglionic axons. Although the innervated structures are shown only for one side of the body for diagrammatic purposes, the parasympathetic division actually innervates tissues and organs on both sides.

Cell bodies of parasympathetic preganglionic neurons are located in brainstem nuclei and in the lateral gray matter in the second through fourth sacral segments of the spinal cord.

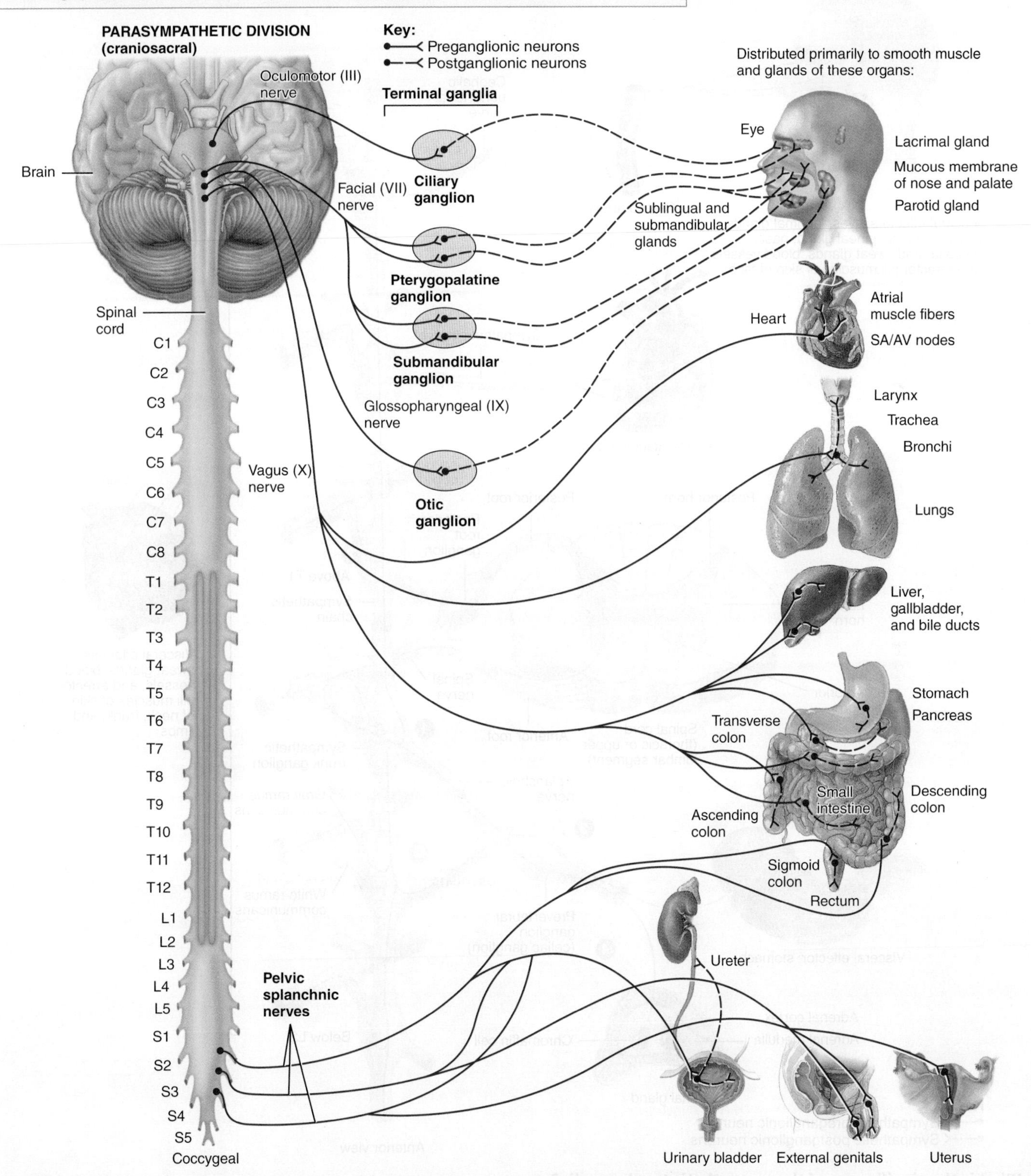

Q Which ganglia are associated with the parasympathetic division? Sympathetic division?

FIGURE 15.4 Types of connections between ganglia and postganglionic neurons in the sympathetic division of the ANS. Also illustrated are the gray and white rami communicantes.

Sympathetic ganglia lie in two chains on either side of the vertebral column (sympathetic trunk ganglia) and near large abdominal arteries anterior to the vertebral column (prevertebral ganglia).

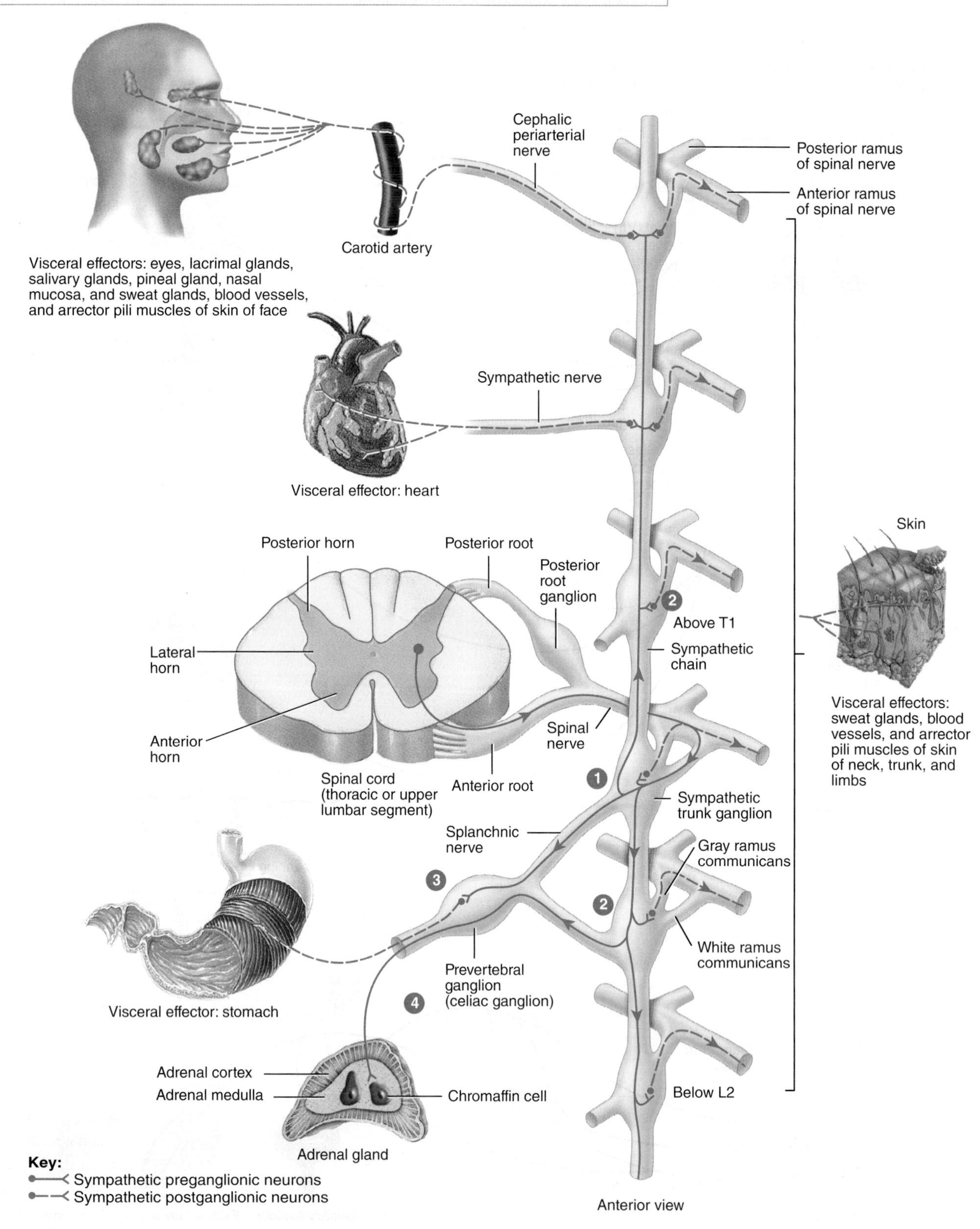

Q What is the significance of the sympathetic trunk ganglia?

A single sympathetic preganglionic fiber has many axon collaterals (branches) and may synapse with 20 or more postganglionic neurons. This pattern of projection is an example of divergence and helps explain why many sympathetic responses affect almost the entire body simultaneously. After exiting their ganglia, the postganglionic axons typically terminate in several visceral effectors (see **Figure 15.2**).

Axons of preganglionic neurons of the parasympathetic division pass to terminal ganglia near or within a visceral effector (see **Figure 15.3**). In the ganglion, the presynaptic neuron usually synapses with only four or five postsynaptic neurons, all of which supply a single visceral effector, allowing parasympathetic responses to be localized to a single effector.

Autonomic Plexuses In the thorax, abdomen, and pelvis, axons of both sympathetic and parasympathetic neurons form tangled networks called **autonomic plexuses**, many of which lie along major arteries. The autonomic plexuses also may contain sympathetic ganglia and axons of autonomic neurons. The major plexuses in the thorax are the **cardiac plexus**, which supplies the heart, and the **pulmonary plexus**, which supplies the bronchial tree (**Figure 15.5**).

The abdomen and pelvis also contain major autonomic plexuses (**Figure 15.5**), and often the plexuses are named after the artery along which they are distributed. The **celiac** (*solar*) **plexus** is the largest autonomic plexus and surrounds the celiac trunk. It contains two large celiac ganglia, two aorticorenal ganglia, and a dense network of autonomic axons and is distributed to the stomach, spleen, pancreas, liver, gallbladder, kidneys, adrenal medullae, testes, and ovaries. The **superior mesenteric plexus** contains the superior mesenteric ganglion and supplies the small and large intestines. The **inferior mesenteric plexus** contains the inferior mesenteric ganglion, which innervates the large intestine. Axons of some sympathetic postganglionic neurons from the inferior mesenteric ganglion also extend through the **hypogastric plexus**, which is anterior to the fifth lumbar vertebra, to supply the pelvic viscera. The **renal plexus** contains the renal ganglion and supplies the renal arteries within the kidneys and ureters.

With this background in mind, we can now examine some of the specific structural features of the sympathetic and parasympathetic divisions of the ANS in more detail.

Structure of the Sympathetic Division

Pathway from Spinal Cord to Sympathetic Trunk Ganglia Cell bodies of sympathetic preganglionic neurons are part of the lateral gray horns of all thoracic segments and of the first two lumbar segments of the spinal cord (see **Figure 15.2**). The preganglionic axons leave the spinal cord along with the somatic motor neurons at the same segmental level. After exiting through the intervertebral foramina, the myelinated preganglionic sympathetic axons pass into the anterior root of a spinal nerve and enter a short pathway called a **white ramus** (RĀ-mus) before passing to the nearest sympathetic trunk ganglion on the same side (see **Figure 15.4**). Collectively, the white rami are called the **white rami communicantes** (kō-mū-ni-KAN-tēz; singular is **ramus communicans**). Thus, white rami communicantes are structures containing sympathetic preganglionic axons that connect the anterior ramus of the spinal nerve with the ganglia of the sympathetic trunk. The "white" in their name indicates that they contain myelinated axons. Only the thoracic and first two or three lumbar nerves have white rami communicantes.

Organization of Sympathetic Trunk Ganglia The paired **sympathetic trunk ganglia** are arranged anterior and lateral to the vertebral column, one on either side. Typically, there are 3 cervical, 11 or 12 thoracic, 4 or 5 lumbar, 4 or 5 sacral sympathetic trunk ganglia, and 1 coccygeal ganglion. The right and left coccygeal ganglia are fused together and usually lie at the midline. Although the sympathetic trunk ganglia extend inferiorly from the neck, chest, and abdomen to the coccyx, they receive preganglionic axons only from the thoracic and lumbar segments of the spinal cord (see **Figure 15.2**).

The cervical portion of each sympathetic trunk is located in the neck and is subdivided into superior, middle, and inferior ganglia (see **Figure 15.2**). Postganglionic neurons leaving the **superior cervical ganglion** serve the head and heart. They are distributed to the sweat glands, smooth muscle of the eye, blood vessels of the face, lacrimal glands, pineal gland, nasal mucosa, salivary glands (which include the submandibular, sublingual, and parotid glands), and heart. Postganglionic neurons leaving the **middle cervical ganglion** and the **inferior cervical ganglion** innervate the heart and blood vessels of the neck, shoulder, and upper limb.

The thoracic portion of each sympathetic trunk lies anterior to the necks of the corresponding ribs. This region of the sympathetic trunk receives most of the sympathetic preganglionic axons. Postganglionic neurons from the thoracic sympathetic trunk innervate the heart, lungs, bronchi, and other thoracic viscera. In the skin, these neurons also innervate sweat glands, blood vessels, and arrector pili muscles of hair follicles. The lumbar portion of each sympathetic trunk lies lateral to the corresponding lumbar vertebrae. The sacral region of the sympathetic trunk lies in the pelvic cavity on the medial side of the anterior sacral foramina.

Pathways from Sympathetic Trunk Ganglia to Visceral Effectors Axons leave the sympathetic trunk in four possible ways: (1) They can enter spinal nerves; (2) they can form cephalic periarterial nerves; (3) they can form sympathetic nerves; and (4) they can form splanchnic nerves.

SPINAL NERVES Recall that some of the incoming sympathetic preganglionic neurons synapse with postganglionic neurons in the sympathetic trunk, either in the ganglion at the level of entry or in a ganglion farther up or down the sympathetic trunk. The axons of some of these postganglionic neurons leave the sympathetic trunk by entering a short pathway called a **gray ramus** and then merge with the anterior ramus of a spinal nerve. Therefore, **gray rami communicantes** are structures containing sympathetic postganglionic axons that connect the ganglia of the sympathetic trunk to spinal nerves (see **Figure 15.4**). The "gray" in their name indicates that they contain unmyelinated axons. Gray rami communicantes outnumber the white rami because there is a gray ramus leading to each of the 31 pairs of spinal nerves. The axons of the postganglionic neurons that leave the

FIGURE 15.5 **Autonomic plexuses in the thorax, abdomen, and pelvis.**

An autonomic plexus is a network of sympathetic and parasympathetic axons that sometimes also includes sympathetic ganglia.

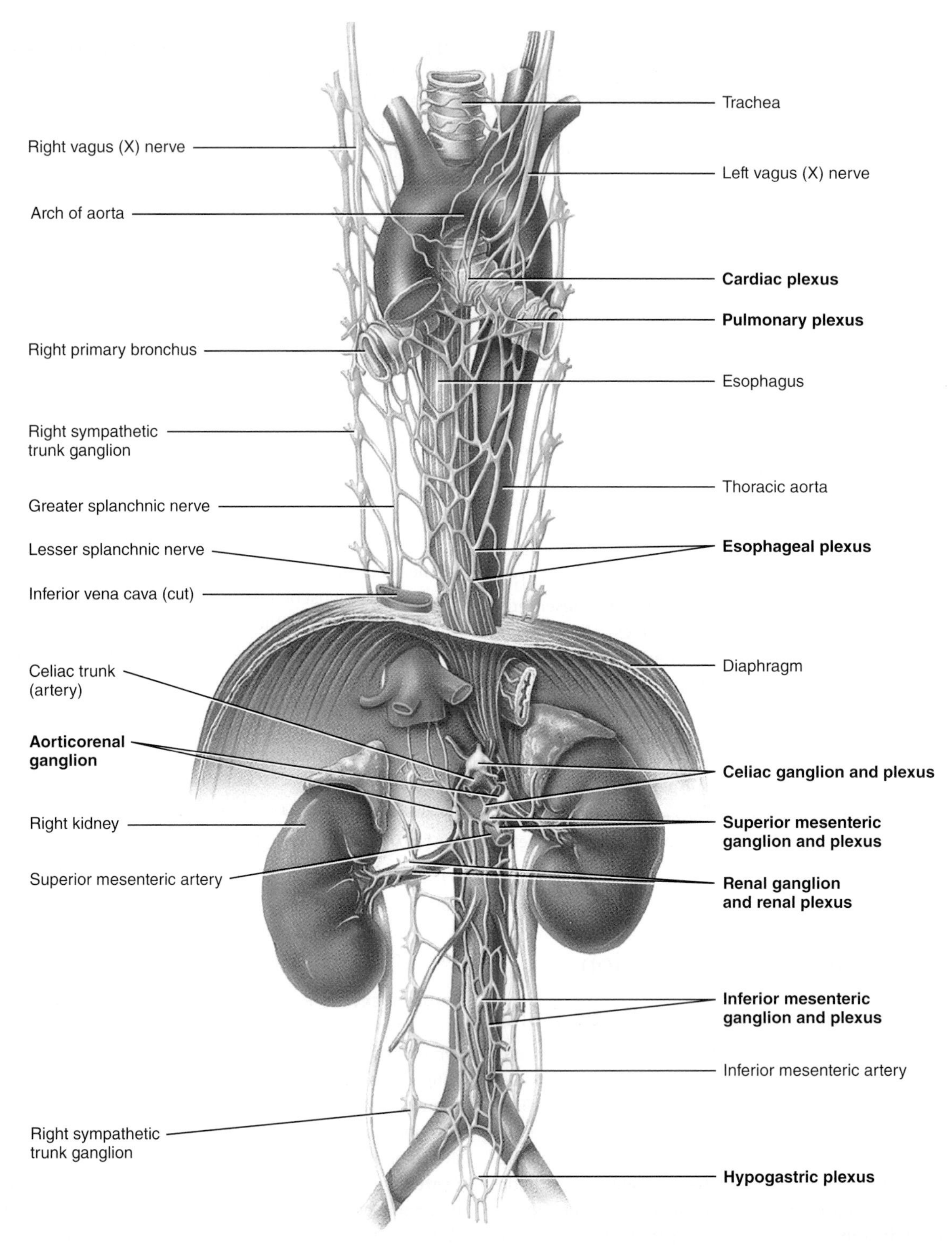

(a) Anterior view

Q Which is the largest autonomic plexus?

sympathetic trunk to enter spinal nerves provide sympathetic innervation to the visceral effectors in the skin of the neck, trunk, and limbs, including sweat glands, smooth muscle in blood vessels, and arrector pili muscles of hair follicles.

Cephalic Periarterial Nerves Some sympathetic preganglionic neurons that enter the sympathetic trunk ascend to the superior cervical ganglion, where they synapse with postganglionic neurons. The axons of some of these postganglionic neurons leave the sympathetic trunk by forming **cephalic periarterial nerves** (per′-ē-ar-TĒ-rē-al), nerves that extend to the head by wrapping around and following the course of various arteries (such as the carotid arteries) that pass from the neck to the head (see Figure 15.4). Cephalic periarterial nerves provide sympathetic innervation to visceral effectors in the skin of the face (sweat glands, smooth muscle of blood vessels, and arrector pili muscles of hair follicles), as well as other visceral effectors of the head (smooth muscle of the eye, lacrimal glands, pineal gland, nasal mucosa, and salivary glands).

Sympathetic Nerves Some of the incoming sympathetic preganglionic neurons synapse with postganglionic neurons in one or more ganglia of the sympathetic trunk. Then the axons of the postganglionic neurons leave the trunk by forming **sympathetic nerves** that extend to visceral effectors in the thoracic cavity (Figure 15.4). Sympathetic nerves provide sympathetic innervation to the heart and lungs.

- ***Sympathetic nerves to the heart.*** Sympathetic innervation of the heart consists of axons of preganglionic neurons that enter the sympathetic trunk and then form synapses with postganglionic neurons in the superior, middle, and inferior cervical ganglia and first through fourth thoracic ganglia (T1–T4). From these ganglia, axons of postganglionic neurons exit the sympathetic trunk by forming sympathetic nerves that enter the cardiac plexus to supply the heart (see Figure 15.2).
- ***Sympathetic nerves to the lungs.*** Sympathetic innervation of the lungs consists of axons of preganglionic neurons that enter the sympathetic trunk and then form synapses with postganglionic neurons in the second through fourth thoracic ganglia (T2–T4). From these ganglia, axons of sympathetic postganglionic neurons exit the trunk by forming sympathetic nerves that enter the pulmonary plexus to supply the smooth muscle of the bronchi and bronchioles of the lungs (see Figure 15.2).

Splanchnic Nerves Recall that some sympathetic preganglionic axons pass through the sympathetic trunk without terminating in it. Beyond the trunk, they form nerves known as **splanchnic nerves** (SPLANGK-nik; see Figures 15.2 and 15.4), which extend to outlying prevertebral ganglia.

- ***Splanchnic nerves to abdominopelvic organs.*** Most sympathetic preganglionic axons that enter splanchnic nerves are destined to synapse with sympathetic postganglionic neurons in the prevertebral ganglia that supply the organs of the abdominopelvic cavity. Preganglionic axons from the fifth through ninth or tenth thoracic ganglia (T5–T9 or T10) form the **greater splanchnic nerve**. It pierces the diaphragm and enters the **celiac ganglion** of the celiac plexus. From there, postganglionic neurons follow and innervate blood vessels to the stomach, spleen, liver, kidneys, and small intestine. Preganglionic axons from the tenth and eleventh thoracic ganglia (T10–T11) form the **lesser splanchnic nerve**. It pierces the diaphragm and passes through the celiac plexus to enter the aorticorenal ganglion and superior mesenteric ganglion of the superior mesenteric plexus. Postganglionic neurons from the superior mesenteric ganglion follow and innervate blood vessels of the small intestine and proximal colon. The **least** (*lowest*) **splanchnic nerve**, which is not always present, is formed by preganglionic axons from the twelfth thoracic ganglia (T12) or a branch of the lesser splanchnic nerve. It pierces the diaphragm and enters the renal plexus near the kidney. Postganglionic neurons from the renal plexus supply kidney arterioles and the ureters. Preganglionic axons that form the **lumbar splanchnic nerve** from the first through fourth lumbar ganglia (L1–L4) enter the inferior mesenteric plexus and terminate in the **inferior mesenteric ganglion**, where they synapse with postganglionic neurons. Axons of postganglionic neurons extend through the inferior mesenteric plexus to supply the distal colon and rectum; they also extend through the hypogastric plexus to supply blood vessels of the distal colon, rectum, urinary bladder, and genital organs. Postganglionic axons leaving the prevertebral ganglia follow the course of various arteries to abdominal and pelvic visceral effectors.
- ***Splanchnic nerves to the adrenal medulla.*** Some sympathetic preganglionic axons pass, without synapsing, through the sympathetic trunk, greater splanchnic nerves, and celiac ganglion, and then extend to **chromaffin cells** in the adrenal medullae of the adrenal glands (see Figures 15.1 and 15.4). Developmentally, the adrenal medullae and sympathetic ganglia are derived from the same tissue, the neural crest (see Figure 14.27b). The adrenal medullae are modified sympathetic ganglia, and the chromaffin cells are similar to sympathetic postganglionic neurons, except they lack dendrites and axons. Rather than extending to another organ, however, these cells release hormones into the blood. On stimulation by sympathetic preganglionic neurons, the chromaffin cells of the adrenal medullae release a mixture of catecholamine hormones—about 80% **epinephrine**, 20% **norepinephrine**, and a trace amount of **dopamine**. These hormones circulate throughout the body and intensify responses elicited by sympathetic postganglionic neurons.

See Clinical Connection: Horner's Syndrome

Structure of the Parasympathetic Division

Cell bodies of parasympathetic preganglionic neurons are found in nuclei in the brainstem and in the lateral gray matter of the second through fourth sacral segments of the spinal cord (see Figure 15.3). Their axons emerge as part of a cranial nerve or as part of the anterior root of a spinal nerve. The **cranial parasympathetic outflow** consists of preganglionic axons that extend from the brainstem in four cranial nerves. The **sacral parasympathetic outflow** consists of preganglionic axons in anterior roots of the second through fourth sacral spinal

FIGURE 15.6 **Pelvic splanchnic nerves.**

Through pelvic splanchnic nerves, axons of parasympathetic preganglionic neurons extend to parasympathetic postganglionic neurons in terminal ganglia in the walls of the colon, ureters, urinary bladder, and reproductive organs.

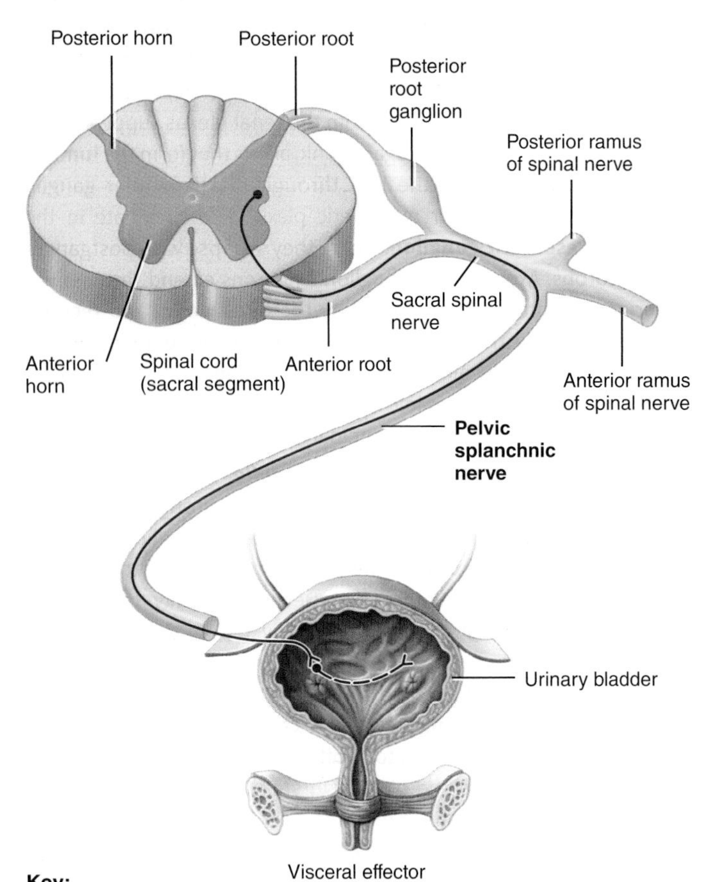

Q Pelvic splanchnic nerves branch from which spinal nerves?

nerves. The preganglionic axons of both the cranial and sacral outflows end in terminal ganglia, where they synapse with postganglionic neurons.

The cranial outflow has four pairs of ganglia and the ganglia associated with the vagus (X) nerve. The four pairs of cranial parasympathetic ganglia innervate structures in the head and are located close to the organs they innervate (see **Figure 15.3**).

1. The **ciliary ganglia** lie lateral to each optic (II) nerve near the posterior aspect of the orbit. Preganglionic axons pass with the oculomotor (III) nerves to the ciliary ganglia. Postganglionic axons from the ganglia innervate smooth muscle fibers in the eyeball.
2. The **pterygopalatine ganglia** are located lateral to the sphenopalatine foramen, between the sphenoid and palatine bones. They receive preganglionic axons from the facial (VII) nerve and send postganglionic axons to the nasal mucosa, palate, pharynx, and lacrimal glands.
3. The **submandibular ganglia** are found near the ducts of the submandibular salivary glands. They receive preganglionic axons from the facial nerves and send postganglionic axons to the submandibular and sublingual salivary glands.
4. The **otic ganglia** are situated just inferior to each foramen ovale. They receive preganglionic axons from the glossopharyngeal (IX) nerves and send postganglionic axons to the parotid salivary glands.

Preganglionic axons that leave the brain as part of the vagus (X) nerves carry nearly 80% of the total craniosacral outflow. Vagal axons extend to many terminal ganglia in the thorax and abdomen. As the vagus nerve passes through the thorax, it sends axons to the heart and the airways of the lungs. In the abdomen, it supplies the liver, gallbladder, stomach, pancreas, small intestine, and part of the large intestine.

The sacral parasympathetic outflow consists of preganglionic axons from the anterior roots of the second through fourth sacral spinal nerves (S2–S4). As the preganglionic axons course through the sacral spinal nerves, they branch off these nerves to form **pelvic splanchnic nerves** (**Figure 15.6**). Pelvic splanchnic nerves synapse with parasympathetic postganglionic neurons located in terminal ganglia in the walls of the innervated viscera. From the terminal ganglia, parasympathetic postganglionic axons innervate smooth muscle and glands in the walls of the colon, ureters, urinary bladder, and reproductive organs.

15.3 ANS Neurotransmitters and Receptors

OBJECTIVE

- **Describe** the neurotransmitters and receptors involved in autonomic responses.

Based on the neurotransmitter they produce and release autonomic neurons are classified as either cholinergic or adrenergic. The receptors for the neurotransmitters are integral membrane proteins located in the plasma membrane of the postsynaptic neuron or effector cell.

Cholinergic Neurons and Receptors

Cholinergic neurons (kō′-lin-ER-jik) release the neurotransmitter **acetylcholine (ACh)**. In the ANS, the cholinergic neurons include (1) all sympathetic and parasympathetic preganglionic neurons, (2) sympathetic postganglionic neurons that innervate most sweat glands, and (3) all parasympathetic postganglionic neurons (**Figure 15.7**).

ACh is stored in synaptic vesicles and released by exocytosis. It then diffuses across the synaptic cleft and binds with specific

FIGURE 15.7 Cholinergic neurons and adrenergic neurons in the sympathetic and parasympathetic divisions.

Cholinergic neurons release acetylcholine; adrenergic neurons release norepinephrine. Cholinergic receptors (nicotinic or muscarinic) and adrenergic receptors are integral membrane proteins located in the plasma membrane of a postsynaptic neuron or an effector cell.

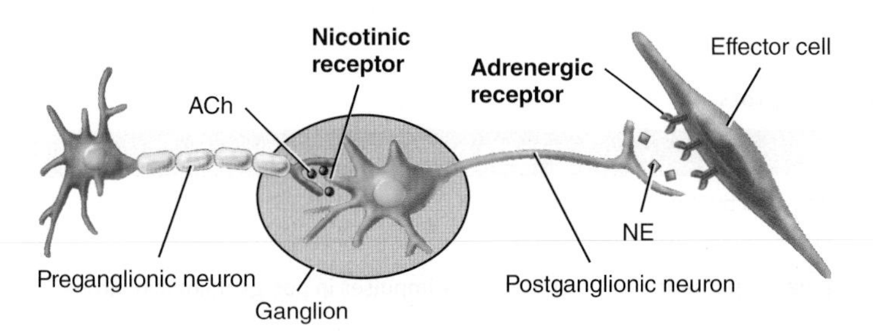

(a) Sympathetic division–innervation to most effector tissues

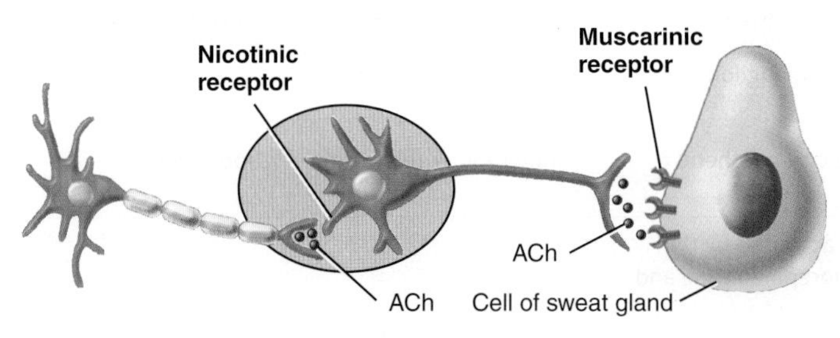

(b) Sympathetic division–innervation to most sweat glands

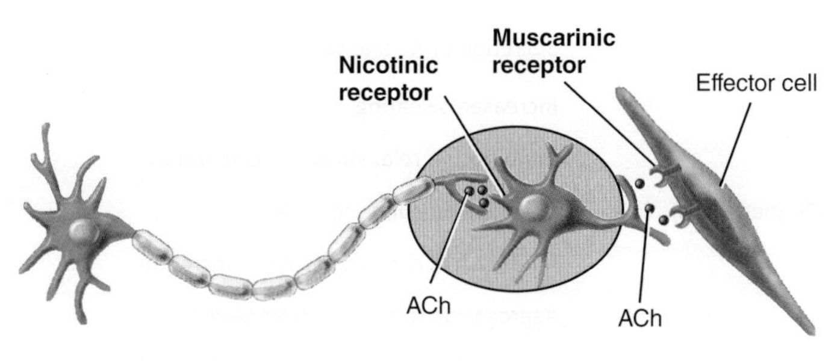

(c) Parasympathetic division

Q Which ANS neurons are adrenergic? What types of effector tissues contain muscarinic receptors?

cholinergic receptors, integral membrane proteins in the *postsynaptic* plasma membrane. The two types of cholinergic receptors, both of which bind ACh, are nicotinic receptors and muscarinic receptors. **Nicotinic receptors** (nik′-ō-TIN-ik) are present in the plasma membrane of dendrites and cell bodies of both sympathetic and parasympathetic postganglionic neurons (Figure 15.7), the plasma membranes of chromaffin cells of the adrenal medullae, and in the motor end plate at the neuromuscular junction. They are so named because nicotine mimics the action of ACh by binding to these receptors. (Nicotine, a natural substance in tobacco leaves, is not a naturally occurring substance in humans and is not normally present in nonsmokers.) **Muscarinic receptors** (mus′-ka-RIN-ik) are present in the plasma membranes of all effectors (smooth muscle, cardiac muscle, and glands) innervated by parasympathetic postganglionic axons. In addition, most sweat glands receive their innervation from *cholinergic* sympathetic postganglionic neurons and possess muscarinic receptors (see Figure 15.7b, c). These receptors are so named because a mushroom poison called muscarine mimics the actions of ACh by binding to them. Nicotine does not activate muscarinic receptors, and muscarine does not activate nicotinic receptors, but ACh does activate both types of cholinergic receptors.

Activation of nicotinic receptors by ACh causes depolarization and thus excitation of the postsynaptic cell, which can be a postganglionic neuron, an autonomic effector, or a skeletal muscle fiber. Activation of muscarinic receptors by ACh sometimes causes depolarization (excitation) and sometimes causes hyperpolarization (inhibition), depending on which particular cell bears the muscarinic receptors. For example, binding of ACh to muscarinic receptors inhibits (relaxes) smooth muscle sphincters in the gastrointestinal tract. By contrast, ACh excites muscarinic receptors in smooth muscle fibers in the circular muscles of the iris of the eye, causing them to contract. Because acetylcholine is quickly inactivated by the enzyme **acetylcholinesterase (AChE)**, effects triggered by cholinergic neurons are brief.

Adrenergic Neurons and Receptors

In the ANS, **adrenergic neurons** (ad′-ren-ER-jik) release **norepinephrine (NE)**, also known as *noradrenalin* (Figure 15.7a). Most sympathetic postganglionic neurons are adrenergic. Like ACh, NE is stored in synaptic vesicles and released by exocytosis. Molecules of NE diffuse across the synaptic cleft and bind to specific adrenergic receptors on the postsynaptic membrane, causing either excitation or inhibition of the effector cell.

Adrenergic receptors bind both norepinephrine and epinephrine. The norepinephrine can either be released as a neurotransmitter by sympathetic postganglionic neurons or released as a hormone into the blood by chromaffin cells of the adrenal medullae; epinephrine is released as a hormone. The two main types of adrenergic receptors are **alpha (α) receptors** and **beta (β) receptors**, which are found on visceral effectors innervated by most sympathetic postganglionic axons. These receptors are further classified into subtypes—α_1, α_2, β_1, β_2, and β_3—based on the specific responses they elicit and by their selective binding of drugs that activate or block them. Although there are some exceptions, activation of α_1 and β_1 receptors generally produces excitation, and activation of α_2 and β_2 receptors causes inhibition of effector tissues. β_3 receptors are present only on cells of brown adipose tissue, where their activation causes *thermogenesis* (heat production). Cells of most effectors contain either alpha or beta receptors; some visceral effector cells contain both. Norepinephrine stimulates alpha receptors more strongly than beta receptors; epinephrine is a potent stimulator of both alpha and beta receptors.

The activity of norepinephrine at a synapse is terminated either when the NE is taken up by the axon that released it or when the NE is enzymatically inactivated by either **catechol-*O*-methyltransferase (COMT)** (kat′-e-kōl-ō-meth-il-TRANS-fer-ās) or **monoamine oxidase**

(MAO) (mon-ō-AM-ēn OK-si-dās). Compared to ACh, norepinephrine lingers in the synaptic cleft for a longer time. Thus, effects triggered by adrenergic neurons typically are longer lasting than those triggered by cholinergic neurons.

Table 15.2 describes the locations of cholinergic and adrenergic receptors and summarizes the responses that occur when each type of receptor is activated.

Receptor Agonists and Antagonists

A large variety of drugs and natural products can selectively activate or block specific cholinergic or adrenergic receptors. An **agonist** (*agon* = a contest) is a substance that binds to and activates a receptor, in the process mimicking the effect of a natural neurotransmitter or hormone. Phenylephrine, an adrenergic agonist at α_1 receptors, is

TABLE 15.2 Location and Responses of Adrenergic and Cholinergic Receptors

TYPE OF RECEPTOR	MAJOR LOCATIONS	EFFECTS OF RECEPTOR ACTIVATION
CHOLINERGIC	Integral proteins in postsynaptic plasma membranes; activated by the neurotransmitter acetylcholine.	
Nicotinic	Plasma membrane of postganglionic sympathetic and parasympathetic neurons.	Excitation → impulses in postganglionic neurons.
	Chromaffin cells of adrenal medullae.	Epinephrine and norepinephrine secretion.
	Sarcolemma of skeletal muscle fibers (motor end plate).	Excitation → contraction.
Muscarinic	Effectors innervated by parasympathetic postganglionic neurons.	In some receptors, excitation; in others, inhibition.
	Sweat glands innervated by cholinergic sympathetic postganglionic neurons.	Increased sweating.
	Skeletal muscle blood vessels innervated by cholinergic sympathetic postganglionic neurons.	Inhibition → relaxation → vasodilation.
ADRENERGIC	Integral proteins in postsynaptic plasma membranes; activated by the neurotransmitter norepinephrine and the hormones norepinephrine and epinephrine.	
α_1	Smooth muscle fibers in blood vessels that serve salivary glands, skin, mucosal membranes, kidneys, and abdominal viscera; radial muscle in iris of eye; sphincter muscles of stomach and urinary bladder.	Excitation → contraction, which causes vasoconstriction, dilation of pupil, and closing of sphincters.
	Salivary gland cells.	Secretion of K^+ and water.
	Sweat glands on palms and soles.	Increased sweating.
α_2	Smooth muscle fibers in some blood vessels.	Inhibition → relaxation → vasodilation.
	Cells of pancreatic islets that secrete the hormone insulin (beta cells).	Decreased insulin secretion.
	Pancreatic acinar cells.	Inhibition of digestive enzyme secretion.
	Platelets in blood.	Aggregation to form platelet plug.
β_1	Cardiac muscle fibers.	Excitation → increased force and rate of contraction.
	Juxtaglomerular cells of kidneys.	Renin secretion.
	Posterior pituitary.	Antidiuretic hormone (ADH) secretion.
	Adipose cells.	Breakdown of triglycerides → release of fatty acids into blood.
β_2	Smooth muscle in walls of airways; in blood vessels that serve heart, skeletal muscle, adipose tissue, and liver; and in walls of visceral organs, such as urinary bladder.	Inhibition → relaxation, which causes dilation of airways, vasodilation, and relaxation of organ walls.
	Ciliary muscle in eye.	Inhibition → relaxation.
	Hepatocytes in liver.	Glycogenolysis (breakdown of glycogen into glucose).
β_3	Brown adipose tissue.	Thermogenesis (heat production).

a common ingredient in cold and sinus medications. Because it constricts blood vessels in the nasal mucosa, phenylephrine reduces production of mucus, thus relieving nasal congestion. An **antagonist** (*anti* = against) is a substance that binds to and blocks a receptor, thereby preventing a natural neurotransmitter or hormone from exerting its effect. For example, atropine blocks muscarinic ACh receptors, dilates the pupils, reduces glandular secretions, and relaxes smooth muscle in the gastrointestinal tract. As a result, it is used to dilate the pupils during eye examinations, in the treatment of smooth muscle disorders such as iritis and intestinal hypermotility, and as an antidote for chemical warfare agents that inactivate acetylcholinesterase.

Propranolol (Inderal®) often is prescribed for patients with hypertension (high blood pressure). It is a nonselective beta blocker, meaning it binds to all types of beta receptors and prevents their activation by epinephrine and norepinephrine. The desired effects of propranolol are due to its *blockade* of β_1 receptors—namely, decreased heart rate and force of contraction and a consequent decrease in blood pressure. Undesired effects due to blockade of β_2 receptors may include hypoglycemia (low blood glucose), resulting from decreased glycogen breakdown and decreased gluconeogenesis (the conversion of a noncarbohydrate into glucose in the liver), and mild bronchoconstriction (narrowing of the airways). If these side effects pose a threat to the patient, a selective β_1 blocker (which binds only to specific beta receptors) such as metoprolol (Lopressor®) can be prescribed instead of propranolol.

15.4 Physiology of the ANS

OBJECTIVE

- **Describe** the major responses of the body to stimulation by the sympathetic and parasympathetic divisions of the ANS.

Autonomic Tone

As noted earlier, most body organs receive innervation from both divisions of the ANS, which typically work in opposition to one another. The balance between sympathetic and parasympathetic activity, called **autonomic tone**, is regulated by the hypothalamus. Typically, the hypothalamus turns up sympathetic tone at the same time it turns down parasympathetic tone, and vice versa. The two divisions can affect body organs differently because their postganglionic neurons release different neurotransmitters and because the effector organs possess different adrenergic and cholinergic receptors. A few structures receive only sympathetic innervation—sweat glands, arrector pili muscles attached to hair follicles in the skin, the kidneys, the spleen, most blood vessels, and the adrenal medullae (see **Figure 15.2**). In these structures there is no opposition from the parasympathetic division. Still, an increase in sympathetic tone has one effect, and a decrease in sympathetic tone produces the opposite effect.

Sympathetic Responses

During physical or emotional stress, the sympathetic division dominates the parasympathetic division. High sympathetic tone favors body functions that can support vigorous physical activity and rapid production of ATP. At the same time, the sympathetic division reduces body functions that favor the storage of energy. Besides physical exertion, various emotions—such as fear, embarrassment, or rage—stimulate the sympathetic division. Visualizing body changes that occur during "E situations" such as exercise, emergency, excitement, and embarrassment will help you remember most of the sympathetic responses. Activation of the sympathetic division and release of hormones by the adrenal medullae set in motion a series of physiological responses collectively called the **fight-or-flight response**, which includes the following effects:

- The pupils of the eyes dilate.
- Heart rate, force of heart contraction, and blood pressure increase.
- The airways dilate, allowing faster movement of air into and out of the lungs.
- The blood vessels that supply the kidneys and gastrointestinal tract constrict, which decreases blood flow through these tissues. The result is a slowing of urine formation and digestive activities, which are not essential during exercise.
- Blood vessels that supply organs involved in exercise or fighting off danger—skeletal muscles, cardiac muscle, liver, and adipose tissue—dilate, allowing greater blood flow through these tissues.
- Liver cells perform glycogenolysis (breakdown of glycogen to glucose), and adipose tissue cells perform lipolysis (breakdown of triglycerides to fatty acids and glycerol).
- Release of glucose by the liver increases blood glucose level.
- Processes that are not essential for meeting the stressful situation are inhibited. For example, muscular movements of the gastrointestinal tract and digestive secretions slow down or even stop.

The effects of sympathetic stimulation are longer lasting and more widespread than the effects of parasympathetic stimulation for three reasons: (1) Sympathetic postganglionic axons diverge more extensively; as a result, many tissues are activated simultaneously. (2) Acetylcholinesterase quickly inactivates acetylcholine, but norepinephrine lingers in the synaptic cleft for a longer period. (3) Epinephrine and norepinephrine secreted into the blood from the adrenal medullae intensify and prolong the responses caused by NE liberated from sympathetic postganglionic axons. These blood-borne hormones circulate throughout the body, affecting all tissues that have alpha and beta receptors. In time, blood-borne NE and epinephrine are destroyed by enzymes in the liver.

Parasympathetic Responses

In contrast to the fight-or-flight activities of the sympathetic division, the parasympathetic division enhances **rest-and-digest** activities. Parasympathetic responses support body functions that conserve and restore body energy during times of rest and recovery. In the quiet intervals between periods of exercise, parasympathetic impulses to the

TABLE 15.3 Comparison of Sympathetic and Parasympathetic Divisions of the ANS

	SYMPATHETIC (THORACOLUMBAR)	PARASYMPATHETIC (CRANIOSACRAL)
Distribution	Wide regions of body: skin, sweat glands, arrector pili muscles of hair follicles, adipose tissue, smooth muscle of blood vessels.	Limited mainly to head and to viscera of thorax, abdomen, and pelvis; some blood vessels.
Location of preganglionic neuron cell bodies and site of outflow	Lateral gray horns of spinal cord segments T1–L2. Axons of preganglionic neurons constitute thoracolumbar outflow.	Nuclei of cranial nerves III, VII, IX, and X and lateral gray matter of spinal cord segments S2–S4. Axons of preganglionic neurons constitute craniosacral outflow.
Associated ganglia	Sympathetic trunk ganglia and prevertebral ganglia.	Terminal ganglia.
Ganglia locations	Close to CNS and distant from visceral effectors.	Typically near or within wall of visceral effectors.
Axon length and divergence	Preganglionic neurons with short axons synapse with many postganglionic neurons with long axons that pass to many visceral effectors.	Preganglionic neurons with long axons usually synapse with four to five postganglionic neurons with short axons that pass to single visceral effector.
White and gray rami communicantes	Both present; white rami communicantes contain myelinated preganglionic axons; gray rami communicantes contain unmyelinated postganglionic axons.	Neither present.
Neurotransmitters	Preganglionic neurons release acetylcholine (ACh), which is excitatory and stimulates postganglionic neurons; most postganglionic neurons release norepinephrine (NE); postganglionic neurons that innervate most sweat glands and some blood vessels in skeletal muscle release ACh.	Preganglionic neurons release ACh, which is excitatory and stimulates postganglionic neurons; postganglionic neurons release ACh.
Physiological effects	Fight-or-flight responses.	Rest-and-digest activities.

digestive glands and the smooth muscle of the gastrointestinal tract predominate over sympathetic impulses. This allows energy-supplying food to be digested and absorbed. At the same time, parasympathetic responses reduce body functions that support physical activity.

The acronym *SLUDD* can be helpful in remembering five parasympathetic responses. It stands for salivation (S), lacrimation (L), urination (U), digestion (D), and defecation (D). All of these activities are stimulated mainly by the parasympathetic division. In addition to the increasing SLUDD responses, other important parasympathetic responses are "three decreases": decreased heart rate, decreased diameter of airways (bronchoconstriction), and decreased diameter (constriction) of the pupils.

Table 15.3 compares the structural and functional features of the sympathetic and parasympathetic divisions of the ANS. Table 15.4 lists the responses of glands, cardiac muscle, and smooth muscle to stimulation by the sympathetic and parasympathetic divisions of the ANS.

15.5 Integration and Control of Autonomic Functions

OBJECTIVES

- **Describe** the components of an autonomic (visceral) reflex.
- **Explain** the relationship of the hypothalamus to the ANS.

Autonomic (Visceral) Reflexes

Autonomic (visceral) reflexes are responses that occur when nerve impulses pass through an autonomic reflex arc. These reflexes play a key role in regulating controlled conditions in the body, such as *blood pressure*, by adjusting heart rate, force of ventricular contraction, and blood vessel diameter; *digestion*, by adjusting the motility (movement) and muscle tone of the gastrointestinal tract; and *defecation* and *urination*, by regulating the opening and closing of sphincters.

The components of an autonomic reflex arc are as follows:

- **Sensory receptor.** Like the receptor in a somatic reflex arc (see Figure 13.14), the sensory receptor in an autonomic reflex arc is the distal end of a sensory neuron, which responds to a stimulus and produces a change that will ultimately trigger nerve impulses. These sensory receptors are mostly associated with interoceptors (which respond to internal stimuli such as stretching of a visceral wall or chemical composition of a body fluid).
- **Sensory neuron.** Conducts nerve impulses from receptors to the CNS.
- **Integrating center.** Interneurons within the CNS relay signals from sensory neurons to motor neurons. The main integrating centers for most autonomic reflexes are located in the hypothalamus and brainstem. Some autonomic reflexes, such as those for urination and defecation, have integrating centers in the spinal cord.
- **Motor neurons.** Nerve impulses triggered by the integrating center propagate out of the CNS along motor neurons to an effector. In an autonomic reflex arc, two motor neurons connect the CNS to an

TABLE 15.4 Effects of Sympathetic and Parasympathetic Divisions of the ANS

VISCERAL EFFECTOR	EFFECT OF SYMPATHETIC STIMULATION (α OR β ADRENERGIC RECEPTORS, EXCEPT AS NOTED)*	EFFECT OF PARASYMPATHETIC STIMULATION (MUSCARINIC ACh RECEPTORS)
GLANDS		
Adrenal medullae	Secretion of epinephrine and norepinephrine (nicotinic ACh receptors).	No known effect.
Lacrimal (tear)	Slight secretion of tears (α).	Secretion of tears.
Pancreas	Inhibits secretion of digestive enzymes and the hormone insulin (α_2); promotes secretion of the hormone glucagon (β_2).	Secretion of digestive enzymes and the hormone insulin.
Posterior pituitary	Secretion of antidiuretic hormone (ADH) (β_1).	No known effect.
Pineal	Increases synthesis and release of melatonin (β).	No known effect.
Sweat	Increases sweating in most body regions (muscarinic ACh receptors); sweating on palms and soles (α_1).	No known effect.
Adipose tissue†	Lipolysis (breakdown of triglycerides into fatty acids and glycerol) (β_1); release of fatty acids into blood (β_1 and β_3).	No known effect.
Liver†	Glycogenolysis (conversion of glycogen into glucose); gluconeogenesis (conversion of noncarbohydrates into glucose); decreased bile secretion (α and β_2).	Glycogen synthesis; increased bile secretion.
Kidney, juxtaglomerular cells†	Secretion of renin (β_1).	No known effect.
CARDIAC (HEART) MUSCLE		
	Increased heart rate and force of atrial and ventricular contractions (β_1).	Decreased heart rate; decreased force of atrial contraction.
SMOOTH MUSCLE		
Iris, radial muscle	Contraction → dilation of pupil (α_1).	No known effect.
Iris, circular muscle	No known effect.	Contraction → constriction of pupil.
Ciliary muscle of eye	Relaxation to adjust shape of lens for distant vision (β_2).	Contraction for close vision.
Lungs, bronchial muscle	Relaxation → airway dilation (β_2).	Contraction → airway constriction.
Gallbladder and ducts	Relaxation to facilitate storage of bile in the gallbladder (β_2).	Contraction → release of bile into small intestine.
Stomach and intestines	Decreased motility and tone (α_1, α_2, β_2); contraction of sphincters (α_1).	Increased motility and tone; relaxation of sphincters.
Spleen	Contraction and discharge of stored blood into general circulation (α_1).	No known effect.
Ureter	Increases motility (α_1).	Increases motility (?).
Urinary bladder	Relaxation of muscular wall (β_2); contraction of internal urethral sphincter (α_1).	Contraction of muscular wall; relaxation of internal urethral sphincter.
Uterus	Inhibits contraction in nonpregnant women (β_2); promotes contraction in pregnant women (α_1).	Minimal effect.
Sex organs	In males: contraction of smooth muscle of ductus (vas) deferens, prostate, and seminal vesicle resulting in ejaculation (α_1).	Vasodilation; erection of clitoris (females) and penis (males).
Hair follicles, arrector pili muscle	Contraction → erection of hairs resulting in goose bumps (α_1).	No known effect.

Table 15.4 *Continues*

TABLE 15.4 Effects of Sympathetic and Parasympathetic Divisions of the ANS (*Continued*)

VISCERAL EFFECTOR	EFFECT OF SYMPATHETIC STIMULATION (α OR β ADRENERGIC RECEPTORS, EXCEPT AS NOTED)*	EFFECT OF PARASYMPATHETIC STIMULATION (MUSCARINIC ACH RECEPTORS)
VASCULAR SMOOTH MUSCLE		
Salivary gland arterioles	Vasoconstriction, which decreases secretion of saliva (α_1).	Vasodilation, which increases secretion of saliva.
Gastric gland arterioles	Vasoconstriction, which inhibits secretion (α_1).	Secretion of gastric juice.
Intestinal gland arterioles	Vasoconstriction, which inhibits secretion (α_1).	Secretion of intestinal juice.
Coronary (heart) arterioles	Relaxation → vasodilation (β_2); contraction → vasoconstriction (α_1, α_2); contraction → vasoconstriction (muscarinic ACh receptors).	Contraction → vasoconstriction.
Skin and mucosal arterioles	Contraction → vasoconstriction (α_1).	Vasodilation, which may not be physiologically significant.
Skeletal muscle arterioles	Contraction → vasoconstriction (α_1); relaxation → vasodilation (β_2); relaxation → vasodilation (muscarinic ACh receptors).	No known effect.
Abdominal viscera arterioles	Contraction → vasoconstriction (α_1, β_2).	No known effect.
Brain arterioles	Slight contraction → vasoconstriction (α_1).	No known effect.
Kidney arterioles	Constriction of blood vessels → decreased urine volume (α_1).	No known effect.
Systemic veins	Contraction → constriction (α_1); relaxation → dilation (β_2).	No known effect.

*Subcategories of α and β receptors are listed if known.

†Grouped with glands because they release substances into the blood.

effector: The preganglionic neuron conducts motor impulses from the CNS to an autonomic ganglion, and the postganglionic neuron conducts motor impulses from an autonomic ganglion to an effector (see **Figure 15.1**).

- **Effector.** In an autonomic reflex arc, the effectors are smooth muscle, cardiac muscle, and glands, and the reflex is called an autonomic reflex.

Autonomic Control by Higher Centers

Normally, we are not aware of muscular contractions of our digestive organs, our heartbeat, changes in the diameter of our blood vessels, and pupil dilation and constriction because the integrating centers for these autonomic responses are in the spinal cord or the lower regions of the brain. Sensory neurons deliver input to these centers, and autonomic motor neurons provide output that adjusts activity in the visceral effector, usually without our conscious perception.

The hypothalamus is the major control and integration center of the ANS. The hypothalamus receives sensory input related to visceral functions, olfaction (smell), and gustation (taste), as well as changes in temperature, osmolarity, and levels of various substances in blood. It also receives input relating to emotions from the limbic system. Output from the hypothalamus influences autonomic centers in both the brainstem (such as the cardiovascular, salivation, swallowing, and vomiting centers) and the spinal cord (such as the defecation and urination reflex centers in the sacral spinal cord).

Anatomically, the hypothalamus is connected to both the sympathetic and parasympathetic divisions of the ANS by axons of neurons with dendrites and cell bodies in various hypothalamic nuclei. The axons form tracts from the hypothalamus to parasympathetic and sympathetic nuclei in the brainstem and spinal cord through relays in the reticular formation. The posterior and lateral parts of the hypothalamus control the sympathetic division. Stimulation of these areas produces an increase in heart rate and force of contraction, a rise in blood pressure due to constriction of blood vessels, an increase in body temperature, dilation of the pupils, and inhibition of the gastrointestinal tract. In contrast, the anterior and medial parts of the hypothalamus control the parasympathetic division. Stimulation of these areas results in a decrease in heart rate, lowering of blood pressure, constriction of the pupils, and increased secretion and motility of the gastrointestinal tract.

• • •

Now that we have discussed the structure and function of the nervous system, you can appreciate the many ways that this system contributes to homeostasis of other body systems by examining *Focus on Homeostasis: Contributions of the Nervous System*.

FOCUS on HOMEOSTASIS

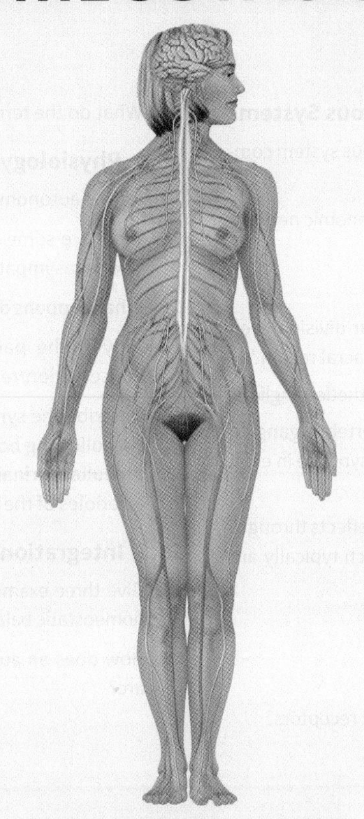

CONTRIBUTIONS OF THE NERVOUS SYSTEM FOR ALL BODY SYSTEMS

- Together with hormones from the endocrine system, nerve impulses provide communication and regulation of most body tissues

INTEGUMENTARY SYSTEM

- Sympathetic nerves of the autonomic nervous system (ANS) control contraction of smooth muscles attached to hair follicles and secretion of perspiration from sweat glands

SKELETAL SYSTEM

- Pain receptors in bone tissue warn of bone trauma or damage

MUSCULAR SYSTEM

- Somatic motor neurons receive instructions from motor areas of the brain and stimulate contraction of skeletal muscles to bring about body movements
- Basal nuclei and reticular formation set level of muscle tone
- Cerebellum coordinates skilled movements

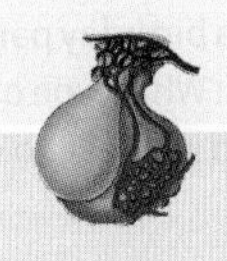

ENDOCRINE SYSTEM

- Hypothalamus regulates secretion of hormones from anterior and posterior pituitary
- ANS regulates secretion of hormones from adrenal medulla and pancreas

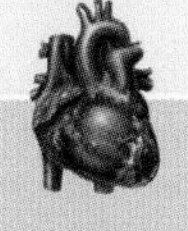

CARDIOVASCULAR SYSTEM

- Cardiovascular center in the medulla oblongata provides nerve impulses to ANS that govern heart rate and the forcefulness of the heartbeat
- Nerve impulses from ANS also regulate blood pressure and blood flow through blood vessels

LYMPHATIC SYSTEM and IMMUNITY

- Certain neurotransmitters help regulate immune responses
- Activity in nervous system may increase or decrease immune responses

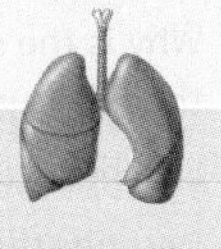

RESPIRATORY SYSTEM

- Respiratory areas in brainstem control breathing rate and depth
- ANS helps regulate diameter of airways

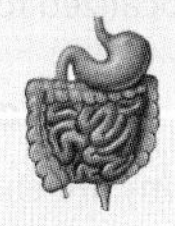

DIGESTIVE SYSTEM

- Enteric division of the ANS helps regulate digestion
- Parasympathetic division of ANS stimulates many digestive processes

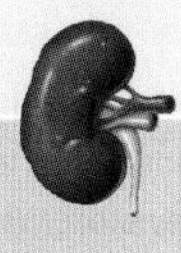

URINARY SYSTEM

- ANS helps regulate blood flow to kidneys, thereby influencing the rate of urine formation
- Brain and spinal cord centers govern emptying of the urinary bladder

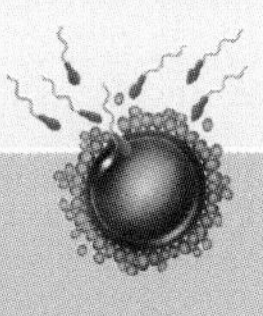

REPRODUCTIVE SYSTEMS

- Hypothalamus and limbic system govern a variety of sexual behaviors
- ANS brings about erection of penis in males and clitoris in females and ejaculation of semen in males
- Hypothalamus regulates release of anterior pituitary hormones that control gonads (ovaries and testes)
- Nerve impulses elicited by touch stimuli from suckling infant cause release of oxytocin and milk ejection in nursing mothers

Review Questions

15.1 Comparison of Somatic and Autonomic Nervous Systems

1. How do the autonomic nervous system and somatic nervous system compare in structure and function?
2. What are the main input and output components of the autonomic nervous system?

15.2 Anatomy of Autonomic Motor Pathways

3. Why is the sympathetic division called the thoracolumbar division even though its ganglia extend from the cervical region to the sacral region?
4. List the organs served by each sympathetic and parasympathetic ganglion.
5. Describe the locations of sympathetic trunk ganglia, prevertebral ganglia, and terminal ganglia. Which types of autonomic neurons synapse in each type of ganglion?
6. Why does the sympathetic division produce simultaneous effects throughout the body, in contrast to parasympathetic effects, which typically are localized to specific organs?

15.3 ANS Neurotransmitters and Receptors

7. Why are cholinergic and adrenergic neurons so named?
8. What neurotransmitters and hormones bind to adrenergic receptors?
9. What do the terms agonist and antagonist mean?

15.4 Physiology of the ANS

10. Define autonomic tone.
11. What are some examples of the antagonistic effects of the sympathetic and parasympathetic divisions of the autonomic nervous system?
12. What happens during the fight-or-flight response?
13. Why is the parasympathetic division of the ANS called an energy conservation/restoration system?
14. Describe the sympathetic response in a frightening situation for each of the following body parts: hair follicles, iris of eye, lungs, spleen, adrenal medullae, urinary bladder, stomach, intestines, gallbladder, liver, heart, arterioles of the abdominal viscera, and arterioles of skeletal muscles.

15.5 Integration and Control of Autonomic Functions

15. Give three examples of controlled conditions in the body that are kept in homeostatic balance by autonomic (visceral) reflexes.
16. How does an autonomic (visceral) reflex arc differ from a somatic reflex arc?

Critical Thinking Questions

1. You've been to the "all-you-can-eat" buffet and have consumed large amounts of food. After returning home, you recline on the couch to watch television. Which division of the nervous system will be handling your body's after-dinner activities? List several organs involved, the major nerve supply to each organ, and the effects of the nervous system on their functions.
2. Ciara is driving home from school, listening to her favorite music, when a dog darts into the street in front of her car. She manages to swerve to avoid hitting the dog. As she continues on her way, she notices her heart is racing, she has goose bumps, and her hands are sweaty. Why is she experiencing these effects?
3. Mrs. Young is experiencing a bout of diarrhea that is keeping her housebound. She would like to go to a birthday party for her brother but is afraid to attend because of her diarrhea. What type of drug, related to the autonomic nervous system function, could she take to help relieve her diarrhea?

Sensory, Motor, and Integrative Systems

CHAPTER 16

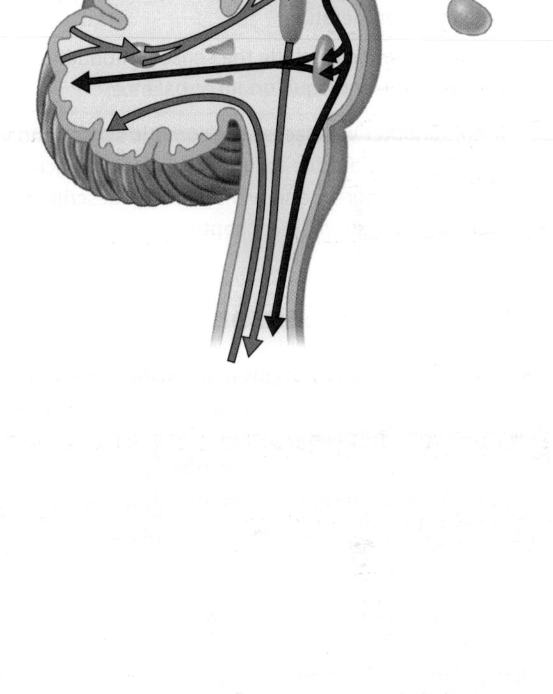

Sensory, Motor, and Integrative Systems and Homeostasis

The sensory and motor pathways of the body provide routes for input into the brain and spinal cord and for output to targeted organs for responses such as muscle contraction.

In the previous four chapters we described the organization of the nervous system. In this chapter, we explore the levels and components of sensation. We also examine the pathways that convey somatic sensory nerve impulses from the body to the brain and the pathways that carry impulses from the brain to skeletal muscles to produce movements. As sensory impulses reach the CNS, they become part of a large pool of sensory input. However, not every nerve impulse transmitted to the CNS elicits a response. Rather, each piece of incoming information is combined with other arriving and previously stored information in a process called *integration*. Integration occurs at many places along pathways in the CNS, such as the spinal cord, brainstem, cerebellum, basal nuclei, and cerebral cortex. You will also learn how the motor responses that govern muscle contraction are modified at several of these levels. To conclude this chapter, we introduce three complex integrative functions of the brain: (1) wakefulness and sleep, (2) learning and memory, and (3) language.

Q Did you ever wonder how drugs such as aspirin and ibuprofen relieve pain?

16.1 Sensation

OBJECTIVES

- **Define** sensation, and discuss the components of sensation.
- **Describe** the different ways to classify sensory receptors.

In its broadest definition, **sensation** is the conscious or subconscious awareness of changes in the external or internal environment. The nature of the sensation and the type of reaction generated vary according to the ultimate destination of nerve impulses (action potentials) that convey sensory information to the CNS. Sensory impulses that reach the spinal cord may serve as input for spinal reflexes, such as the stretch reflex you learned about in Chapter 13. Sensory impulses that reach the lower brainstem elicit more complex reflexes, such as changes in heart rate or breathing rate. When sensory impulses reach the cerebral cortex, we become consciously aware of the sensory stimuli and can precisely locate and identify specific sensations such as touch, pain, hearing, or taste. As you learned in Chapter 14, **perception** is the conscious interpretation of sensations and is primarily a function of the cerebral cortex. We have no perception of some sensory information because it never reaches the cerebral cortex. For example, certain sensory receptors constantly monitor the pressure of blood in blood vessels. Because the nerve impulses conveying blood pressure information propagate to the cardiovascular center in the medulla oblongata rather than to the cerebral cortex, blood pressure is not consciously perceived.

Sensory Modalities

Each unique type of sensation—such as touch, pain, vision, or hearing—is called a **sensory modality** (mō-DAL-i-tē). A given sensory neuron carries information for only one sensory modality. Neurons relaying impulses for touch to the somatosensory area of the cerebral cortex do not transmit impulses for pain. Likewise, nerve impulses from the eyes are perceived as sight, and those from the ears are perceived as sounds.

The different sensory modalities can be grouped into two classes: general senses and special senses.

1. The **general senses** refer to both somatic senses and visceral senses. **Somatic senses** (*somat-* = of the body) include tactile sensations (touch, pressure, vibration, itch, and tickle), thermal sensations (warm and cold), pain sensations, and proprioceptive sensations. Proprioceptive sensations allow perception of both the static (nonmoving) positions of limbs and body parts (joint and muscle position sense) and movements of the limbs and head. **Visceral senses** provide information about conditions within internal organs, for example, pressure, stretch, chemicals, nausea, hunger, and temperature.
2. The **special senses** include the sensory modalities of smell, taste, vision, hearing, and equilibrium or balance.

In this chapter we discuss the somatic senses and visceral pain. The special senses are the focus of Chapter 17. Visceral senses were discussed in Chapter 15 and will be further described in association with individual organs in later chapters.

The Process of Sensation

The process of sensation begins in a **sensory receptor**, which can be either a specialized cell or the dendrites of a sensory neuron. A given sensory receptor responds vigorously to one particular kind of **stimulus**, a change in the environment that can activate certain sensory receptors. A sensory receptor responds only weakly or not at all to other stimuli. This characteristic of sensory receptors is known as **selectivity**.

For a sensation to arise, the following four events typically occur:

1. ***Stimulation of the sensory receptor***. An appropriate stimulus must occur within the sensory receptor's *receptive field*, that is, the body region where stimulation activates the receptor and produces a response.
2. ***Transduction of the stimulus***. A sensory receptor converts the energy in the stimulus into a graded potential, a process known as **transduction**. Recall that graded potentials vary in amplitude (size), depending on the strength of the stimulus that causes them, and are not propagated. (See Section 12.3 to review the differences between action potentials and graded potentials.) Each type of sensory receptor exhibits selectivity: It can transduce (convert) only one kind of stimulus. For example, odorant molecules in the air stimulate olfactory (smell) receptors in the nose, which transduce the molecules' chemical energy into electrical energy in the form of a graded potential.
3. ***Generation of nerve impulses.*** When a graded potential in a sensory neuron reaches threshold, it triggers one or more nerve impulses, which then propagate toward the CNS. Sensory neurons that conduct impulses from the PNS into the CNS are called first-order neurons (see Section 16.3).
4. ***Integration of sensory input.*** A particular region of the CNS receives and integrates (processes) the sensory nerve impulses. Conscious sensations or perceptions are integrated in the cerebral cortex. You seem to see with your eyes, hear with your ears, and feel pain in an injured part of your body because sensory impulses from each part of the body arrive in a specific region of the cerebral cortex, which interprets the sensation as coming from the stimulated sensory receptors.

Sensory Receptors

Types of Sensory Receptors

Several structural and functional characteristics of sensory receptors can be used to group them into different classes. These include (1) microscopic structure, (2) location of the receptors and the origin of stimuli that activate them, and (3) type of stimulus detected.

Microscopic Structure On a microscopic level, sensory receptors may be one of the following: (1) free nerve endings of first-order sensory neurons, (2) encapsulated nerve endings of first-order sensory neurons, or (3) separate cells that synapse with first-order sensory neurons. **Free nerve endings** are bare (not encapsulated) dendrites; they lack any structural specializations that can be seen under a light microscope (**Figure 16.1a**). Receptors for pain, temperature, tickle, itch, and some touch sensations are free nerve endings. Receptors for other somatic and visceral sensations, such as pressure, vibration, and some touch sensations, are **encapsulated nerve endings**. Their dendrites are enclosed in a connective tissue capsule that has a distinctive microscopic structure—for example, lamellated corpuscles (**Figure 16.1b**). The different types of capsules enhance the sensitivity or specificity of the receptor. Sensory receptors for some special senses are specialized, **separate cells** that synapse with sensory neurons. These include *hair cells* for hearing and equilibrium in the inner ear, *gustatory receptors* in taste buds (**Figure 16.1c**) and *photoreceptors* in the retina of the eye for vision. The olfactory receptors for the sense of smell are not separate cells; instead, they are located in olfactory cilia, which are hair like structures that project from the dendrite of an olfactory receptor cell (a type of neuron). You will learn more about the receptors for the special senses in Chapter 17.

A sensory receptor responds to a stimulus by generating a graded potential known as a **receptor potential** (**Figure 16.1a–c**). In sensory receptors that are free nerve endings or encapsulated nerve endings, if the receptor potential is large enough to reach threshold, it triggers one or more nerve impulses in the axon of the sensory neuron (**Figure 16.1a, b**). The nerve impulses then propagate along the axon into the CNS. In sensory receptors that are separate cells, the receptor potential triggers release of neurotransmitter through exocytosis of synaptic vesicles (**Figure 16.1c**). The neurotransmitter molecules liberated from the synaptic vesicles diffuse across the synaptic cleft and produce a postsynaptic potential (PSP), a type of graded potential, in the sensory neuron. If threshold is reached, the PSP will trigger one or more nerve impulses, which propagate along the axon into the CNS.

The amplitude of a receptor potential varies with the intensity of the stimulus, with an intense stimulus producing a large potential and a weak stimulus eliciting a small one. Similarly, large receptor potentials trigger nerve impulses at high frequencies in the first-order neuron, in contrast to small receptor potentials, which trigger nerve impulses at lower frequencies.

FIGURE 16.1 Types of sensory receptors and their relationship to first-order sensory neurons.
(a) Free nerve endings: in this case, a cold-sensitive receptor. These endings are bare dendrites of first-order neurons with no apparent structural specialization. (b) An encapsulated nerve ending: in this case, a vibration-sensitive receptor. Encapsulated nerve endings are dendrites of first-order neurons. (c) A separate receptor cell—here, a gustatory (taste) receptor—and its synapse with a first-order neuron.

Sensory receptors respond to stimuli by generating receptor potentials.

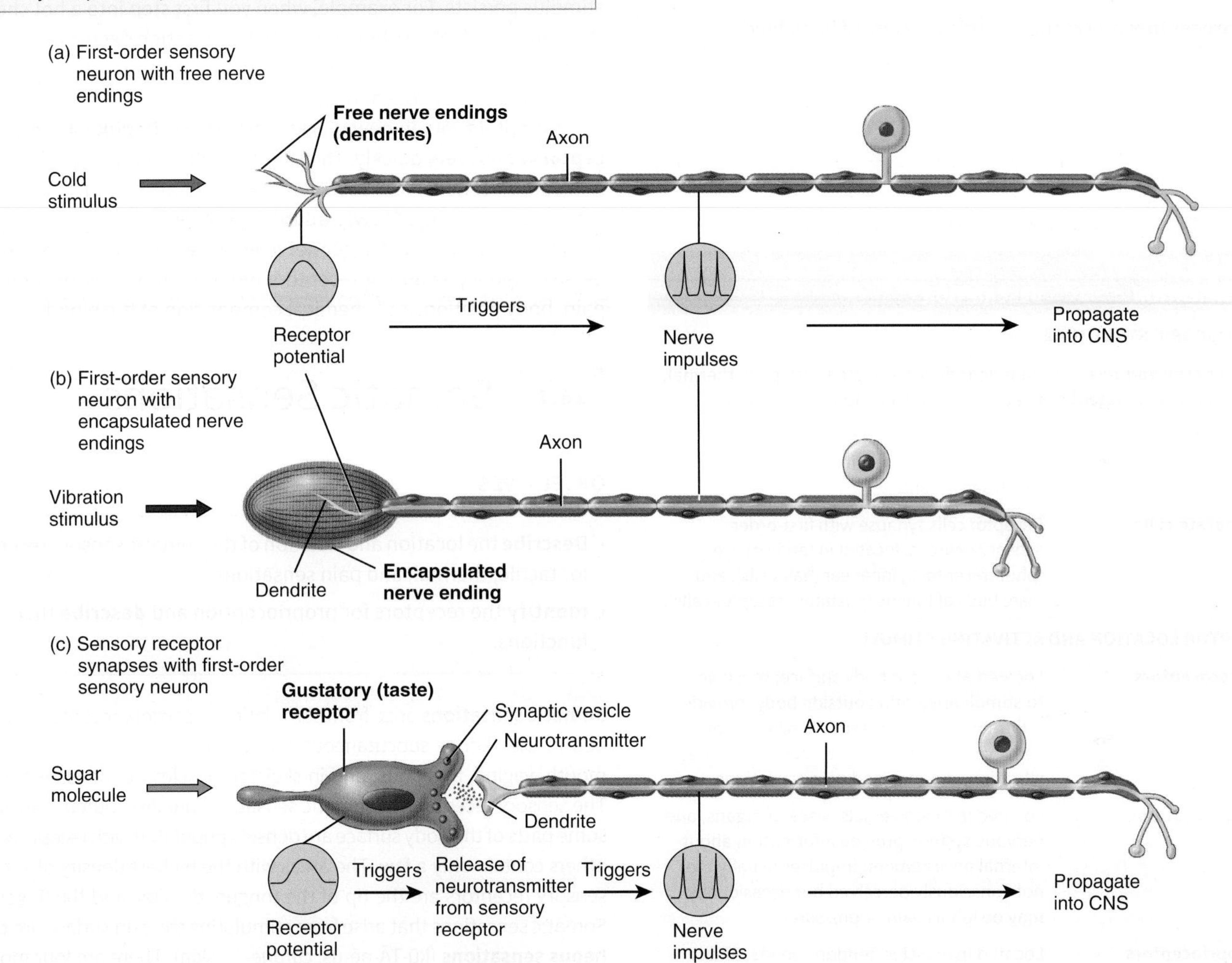

Q Which senses are served by receptors that are separate cells?

Location of Receptors and Origin of Activating Stimuli Another way to group sensory receptors is based on the location of the receptors and the origin of the stimuli that activate them.

- **Exteroceptors** (EKS-ter-ō-sep′-tors) are located at or near the external surface of the body; they are sensitive to stimuli originating outside the body and provide information about the *external* environment. The sensations of hearing, vision, smell, taste, touch, pressure, vibration, temperature, and pain are conveyed by exteroceptors.
- **Interoceptors** (IN-ter-ō-sep′-tors) or *visceroceptors* are located in blood vessels, visceral organs, muscles, and the nervous system and monitor conditions in the *internal* environment. The nerve impulses produced by interoceptors usually are not consciously perceived; occasionally, however, activation of interoceptors by strong stimuli may be felt as pain or pressure.
- **Proprioceptors** (PRŌ-prē-ō-sep′-tors) are located in muscles, tendons, joints, and the inner ear. They provide information about body position, muscle length and tension, and the position and movement of your joints.

Type of Stimulus Detected A third way to group sensory receptors is according to the type of stimulus they detect. Most stimuli are in the form of mechanical energy, such as sound waves or pressure changes; electromagnetic energy, such as light or heat; or chemical energy, such as in a molecule of glucose.

- **Mechanoreceptors** are sensitive to mechanical stimuli such as the deformation, stretching, or bending of cells. Mechanoreceptors provide sensations of touch, pressure, vibration, proprioception, and hearing and equilibrium. They also monitor the stretching of blood vessels and internal organs.

- **Thermoreceptors** detect changes in temperature.
- **Nociceptors** (nō′-sē-SEP-tors; *noci-* = harmful) respond to painful stimuli resulting from physical or chemical damage to tissue.
- **Photoreceptors** detect light that strikes the retina of the eye.
- **Chemoreceptors** detect chemicals in the mouth (taste), nose (smell), and body fluids.
- **Osmoreceptors** detect the osmotic pressure of body fluids.

Table 16.1 summarizes the classification of sensory receptors.

TABLE 16.1 Classification of Sensory Receptors

BASIS OF CLASSIFICATION	DESCRIPTION
MICROSCOPIC STRUCTURE	
Free nerve endings (nonencapsulated)	Bare dendrites associated with pain, thermal, tickle, itch, and some touch sensations.
Encapsulated nerve endings	Dendrites enclosed in connective tissue capsule for pressure, vibration, and some touch sensations.
Separate cells	Receptor cells synapse with first-order sensory neurons; located in retina of eye (photoreceptors), inner ear (hair cells), and taste buds of tongue (gustatory receptor cells).
RECEPTOR LOCATION AND ACTIVATING STIMULI	
Exteroceptors	Located at or near body surface; sensitive to stimuli originating outside body; provide information about external environment; convey visual, smell, taste, touch, pressure, vibration, thermal, and pain sensations.
Interoceptors	Located in blood vessels, visceral organs, and nervous system; provide information about internal environment; impulses usually are not consciously perceived but occasionally may be felt as pain or pressure.
Proprioceptors	Located in muscles, tendons, joints, and inner ear; provide information about body position, muscle length and tension, position and motion of joints, and equilibrium (balance).
TYPE OF STIMULUS DETECTED	
Mechanoreceptors	Detect mechanical stimuli; provide sensations of touch, pressure, vibration, proprioception, and hearing and equilibrium; also monitor stretching of blood vessels and internal organs.
Thermoreceptors	Detect changes in temperature.
Nociceptors	Respond to painful stimuli resulting from physical or chemical damage to tissue.
Photoreceptors	Detect light that strikes the retina of the eye.
Chemoreceptors	Detect chemicals in mouth (taste), nose (smell), and body fluids.
Osmoreceptors	Sense osmotic pressure of body fluids.

Adaptation in Sensory Receptors A characteristic of most sensory receptors is **adaptation,** in which the receptor potential decreases in amplitude during a maintained, constant stimulus. As you may already have guessed, this causes the frequency of nerve impulses in the sensory neuron to decrease. Because of adaptation, the perception of a sensation may fade or disappear even though the stimulus persists. For example, when you first step into a hot shower, the water may feel very hot, but soon the sensation decreases to one of comfortable warmth even though the stimulus (the high temperature of the water) does not change.

Receptors vary in how quickly they adapt. **Rapidly adapting receptors** adapt very quickly. They are specialized for signaling *changes* in a stimulus. Receptors associated with vibration, touch, and smell are rapidly adapting. **Slowly adapting receptors**, by contrast, adapt slowly and continue to trigger nerve impulses as long as the stimulus persists. Slowly adapting receptors monitor stimuli associated with pain, body position, and chemical composition of the blood.

16.2 Somatic Sensations

OBJECTIVES

- **Describe** the location and function of the somatic sensory receptors for tactile, thermal, and pain sensations.
- **Identify** the receptors for proprioception and **describe** their functions.

Somatic sensations arise from stimulation of sensory receptors embedded in the skin or subcutaneous layer; in mucous membranes of the mouth, vagina, and anus; and in skeletal muscles, tendons, and joints. The sensory receptors for somatic sensations are distributed unevenly—some parts of the body surface are densely populated with receptors, and others contain only a few. The areas with the highest density of somatic sensory receptors are the tip of the tongue, the lips, and the fingertips. Somatic sensations that arise from stimulating the skin surface are **cutaneous sensations** (kū-TĀ-nē-us; *cutane-* = skin). There are four modalities of somatic sensation: tactile, thermal, pain, and proprioceptive.

Tactile Sensations

The **tactile sensations** (TAK-tīl; *tact-* = touch) include touch, pressure, vibration, itch, and tickle. Although we perceive differences among these sensations, they arise by activation of some of the same types of receptors. Several types of encapsulated mechanoreceptors attached to large-diameter myelinated A fibers mediate sensations of touch, pressure, and vibration. Other tactile sensations, such as itch and tickle sensations, are detected by free nerve endings attached to small-diameter, unmyelinated C fibers. Recall that larger-diameter, myelinated axons propagate nerve impulses more rapidly than do smaller-diameter, unmyelinated axons. Tactile receptors in the skin or subcutaneous layer include corpuscles of touch, hair root plexuses, type I cutaneous mechanoreceptors, type II cutaneous mechanoreceptors, lamellated corpuscles, and free nerve endings (Figure 16.2).

FIGURE 16.2 **Structure and location of sensory receptors in the skin and subcutaneous layer.**

The somatic sensations of touch, pressure, vibration, warmth, cold, and pain arise from sensory receptors in the skin, subcutaneous layer, and mucous membranes.

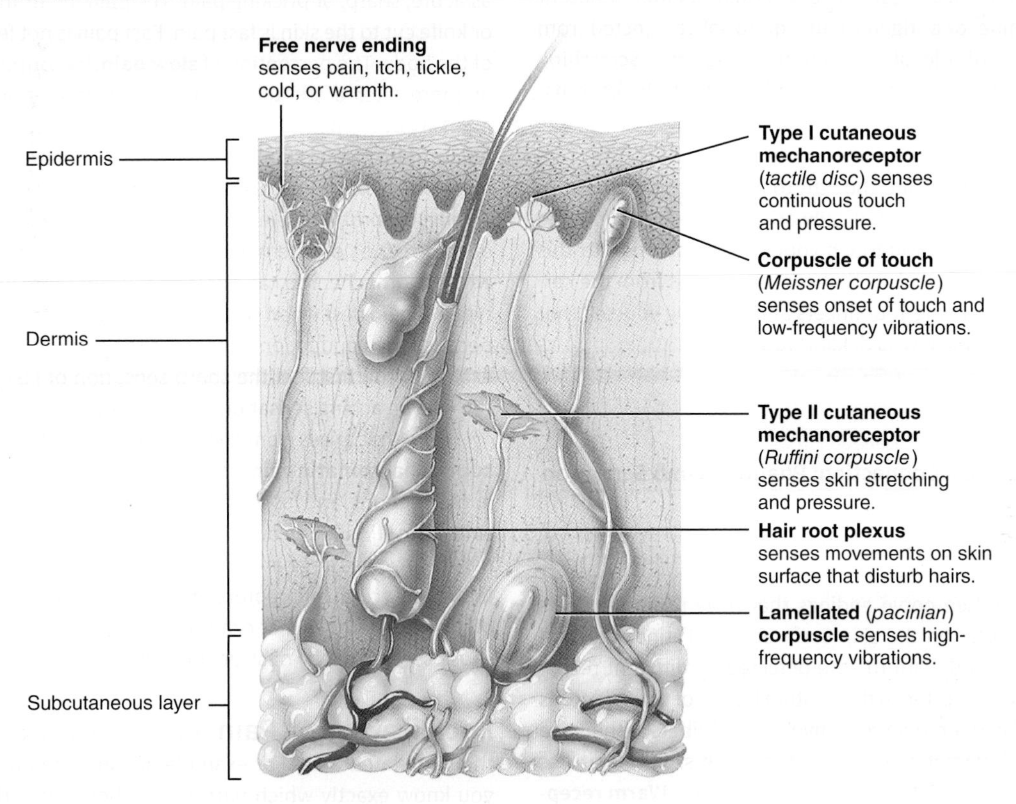

Q Which sensations can arise when free nerve endings are stimulated?

Touch Sensations of **touch** generally result from stimulation of tactile receptors in the skin or subcutaneous layer. There are two types of rapidly adapting touch receptors. **Corpuscles of touch**, or Meissner corpuscles (MĪS-ner), are touch receptors that are located in the dermal papillae of hairless skin. Each corpuscle is an egg-shaped mass of dendrites enclosed by a capsule of connective tissue. Because corpuscles of touch are rapidly adapting receptors, they generate nerve impulses mainly at the onset of a touch. They are abundant in the fingertips, hands, eyelids, tip of the tongue, lips, nipples, soles, clitoris, and tip of the penis. **Hair root plexuses** are rapidly adapting touch receptors found in hairy skin; they consist of free nerve endings wrapped around hair follicles. Hair root plexuses detect movements on the skin surface that disturb hairs. For example, an insect landing on a hair causes movement of the hair shaft that stimulates the free nerve endings.

There are also two types of slowly adapting touch receptors. **Type I cutaneous mechanoreceptors**, also known as *tactile (Merkel) discs*, are saucer-shaped, flattened free nerve endings that make contact with tactile epithelial cells (Merkel cells) of the stratum basale (see **Figure 5.2d**). They are plentiful in the fingertips, hands, lips, and external genitalia. These receptors respond to continuous touch, such as holding an object in your hand for an extended period of time. **Type II cutaneous mechanoreceptors**, or *Ruffini corpuscles*, are elongated, encapsulated receptors located in the dermis, subcutaneous layer, and other tissues of the body. They are highly sensitive to skin stretching, such as when a masseuse stretches your skin during a massage.

Pressure **Pressure**, a sustained sensation that is felt over a larger area than touch, occurs with deeper deformation of the skin and subcutaneous layer. The receptors that contribute to sensations of pressure are type I and type II mechanoreceptors. These receptors are able to respond to a steady pressure stimulus because they are slowly adapting.

Vibration Sensations of **vibration** result from rapidly repetitive sensory signals from tactile receptors. The receptors for vibration sensations are lamellated corpuscles and corpuscles of touch. A **lamellated corpuscle**, or *pacinian corpuscle* (pa-SIN-ē-an), consists of a nerve ending surrounded by a multilayered connective tissue capsule that resembles a sliced onion. Like corpuscles of touch, lamellated corpuscles adapt rapidly. They are found in the dermis, subcutaneous layer, and other body tissues. Lamellated corpuscles respond to high-frequency vibrations, such as the vibrations you feel when you use a power drill or other electric tools. Corpuscles of touch also detect vibrations, but they respond to low-frequency vibrations.

An example is the vibrations you feel when your hand moves across a textured object such as a basket or paneled door.

Itch

The **itch** sensation results from stimulation of free nerve endings by certain chemicals, such as bradykinin (a kinin, a potent vasodilator), histamine, or antigens in mosquito saliva injected from a bite, often because of a local inflammatory response. Scratching usually alleviates itching by activating a pathway that blocks transmission of the itch signal through the spinal cord.

Tickle

Free nerve endings are thought to mediate the **tickle** sensation. This intriguing sensation typically arises only when someone else touches you, not when you touch yourself. The solution to this puzzle seems to lie in the impulses that conduct to and from the cerebellum when you are moving your fingers and touching yourself that don't occur when someone else is tickling you.

See Clinical Connection: Phantom Limb Sensation

Thermal Sensations

Thermoreceptors are free nerve endings that have receptive fields about 1 mm in diameter on the skin surface. Two distinct **thermal sensations**—coldness and warmth—are detected by different receptors. **Cold receptors** are located in the stratum basale of the epidermis and are attached to medium-diameter, myelinated A fibers, although a few connect to small-diameter, unmyelinated C fibers. Temperatures between 10° and 35°C (50–95°F) activate cold receptors. **Warm receptors**, which are not as abundant as cold receptors, are located in the dermis and are attached to small-diameter, unmyelinated C fibers; they are activated by temperatures between 30° and 45°C (86–113°F). Cold and warm receptors both adapt rapidly at the onset of a stimulus, but they continue to generate impulses at a lower frequency throughout a prolonged stimulus. Temperatures below 10°C and above 45°C primarily stimulate pain receptors, rather than thermoreceptors, producing painful sensations, which we discuss next.

Pain Sensations

Pain is indispensable for survival. It serves a protective function by signaling the presence of noxious, tissue-damaging conditions. From a medical standpoint, the subjective description and indication of the location of pain may help pinpoint the underlying cause of disease.

Nociceptors, the receptors for pain, are free nerve endings found in every tissue of the body except the brain (Figure 16.2). Intense thermal, mechanical, or chemical stimuli can activate nociceptors. Tissue irritation or injury releases chemicals such as prostaglandins, kinins, and potassium ions (K^+) that stimulate nociceptors. Pain may persist even after a pain-producing stimulus is removed because pain-mediating chemicals linger, and because nociceptors exhibit very little adaptation. Conditions that elicit pain include excessive distension (stretching) of a structure, prolonged muscular contractions, muscle spasms, or ischemia (inadequate blood flow to an organ).

Types of Pain

There are two types of pain: fast and slow. The perception of **fast pain** occurs very rapidly, usually within 0.1 second after a stimulus is applied, because the nerve impulses propagate along medium-diameter, myelinated A fibers. This type of pain is also known as acute, sharp, or pricking pain. The pain felt from a needle puncture or knife cut to the skin is fast pain. Fast pain is not felt in deeper tissues of the body. The perception of **slow pain**, by contrast, begins a second or more after a stimulus is applied. It then gradually increases in intensity over a period of several seconds or minutes. Impulses for slow pain conduct along small-diameter, unmyelinated C fibers. This type of pain, which may be excruciating, is also referred to as chronic, burning, aching, or throbbing pain. Slow pain can occur both in the skin and in deeper tissues or internal organs. An example is the pain associated with a toothache. You can perceive the difference in onset of these two types of pain best when you injure a body part that is far from the brain because the conduction distance is long. When you stub your toe, for example, you first feel the sharp sensation of fast pain and then feel the slower, aching sensation of slow pain.

Pain that arises from stimulation of receptors in the skin is called **superficial somatic pain**; stimulation of receptors in skeletal muscles, joints, tendons, and fascia causes **deep somatic pain**. **Visceral pain** results from stimulation of nociceptors in visceral organs. If stimulation is *diffuse* (involves large areas), visceral pain can be severe. Diffuse stimulation of visceral nociceptors might result from distension or ischemia of an internal organ. For example, a kidney stone or a gallstone might cause severe pain by obstructing and distending a ureter or bile duct.

Localization of Pain

Fast pain is very precisely localized to the stimulated area. For example, if someone pricks you with a pin, you know exactly which part of your body was stimulated. Somatic slow pain also is well localized but more diffuse (involves large areas); it usually appears to come from a larger area of the skin. In some instances of visceral slow pain, the affected area is where the pain is felt. If the pleural membranes around the lungs are inflamed, for example, you experience chest pain.

However, in many instances of visceral pain, the pain is felt in or just deep to the skin that overlies the stimulated organ, or in a surface area far from the stimulated organ. This phenomenon is called **referred pain**. Figure 16.3 shows skin regions to which visceral pain may be referred. In general, the visceral organ involved and the area to which the pain is referred are served by the same segment of the spinal cord. For example, sensory fibers from the heart, the skin superficial to the heart, and the skin along the medial aspect of the left arm enter spinal cord segments T1 to T5. Thus, the pain of a heart attack typically is felt in the skin over the heart and along the medial aspect of the left arm.

Pain sensations sometimes occur out of proportion to minor damage, persist chronically due to an injury, or even appear for no obvious reason. In such cases, **analgesia** (an-al-JĒ-zē-a; *an-* = without; *-algesia* = pain) or pain relief is needed. Analgesic drugs such as aspirin and ibuprofen (for example, Advil® or Motrin®) block formation of prostaglandins, which stimulate nociceptors. Local anesthetics, such as Novocaine®, provide short-term pain relief by blocking conduction of nerve impulses along the axons of first-order pain neurons. Morphine and other opiate drugs (drugs derived from or containing opium) alter the quality of pain perception in the brain; pain is still

FIGURE 16.3 Distribution of referred pain. Colored parts indicate skin areas to which visceral pain is referred.

Nociceptors are present in almost every tissue of the body.

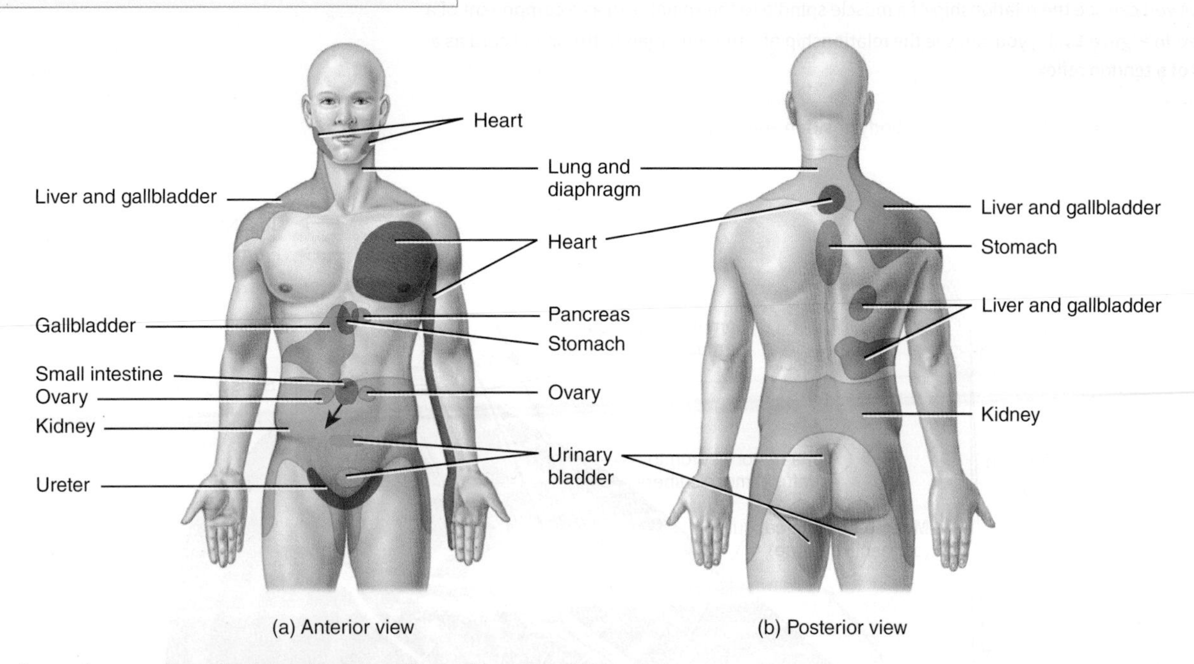

Q Which visceral organ has the broadest area for referred pain?

sensed but it is no longer perceived as being so noxious. Many pain clinics use anticonvulsant and antidepressant medications to treat those suffering from chronic pain.

See Clinical Connection: Acupuncture

Proprioceptive Sensations

Proprioceptive sensations (*proprius* = self or one's own) are also called *proprioception* (prō-prē-ō-SEP-shun). Proprioceptive sensations allow us to recognize that parts of our body belong to us (self). They also allow us to know where our head and limbs are located and how they are moving even if we are not looking at them, so that we can walk, type, or dress without using our eyes. **Kinesthesia** (kin′-es-THĒ-zē-a; *kin-* = motion; *-esthesia* = perception) is the perception of body movements. Proprioceptive sensations arise in receptors termed **proprioceptors**. Those proprioceptors embedded in muscles (especially postural muscles) and tendons inform us of the degree to which muscles are contracted, the amount of tension on tendons, and the positions of joints. Hair cells of the inner ear monitor the orientation of the head relative to the ground and head position during movements. The way they provide information for maintaining balance and equilibrium will be described in Chapter 17. Because most proprioceptors adapt slowly and only slightly, the brain continually receives nerve impulses related to the position of different body parts and makes adjustments to ensure coordination.

Proprioceptors also allow **weight discrimination**, the ability to assess the weight of an object. This type of information helps you to determine the muscular effort necessary to perform a task. For example, as you pick up a shopping bag, you quickly realize whether it contains books or feathers, and you then exert the correct amount of effort needed to lift it.

Here we discuss three types of proprioceptors: muscle spindles, tendon organs, and joint kinesthetic receptors.

Muscle Spindles

Muscle spindles are the proprioceptors that monitor changes in the length of skeletal muscles and participate in stretch reflexes (shown in Figure 13.14). By adjusting how vigorously a muscle spindle responds to stretching of a skeletal muscle, the brain sets an overall level of **muscle tone**, the small degree of contraction that is present while the muscle is at rest.

Each **muscle spindle** consists of several slowly adapting sensory nerve endings that wrap around 3 to 10 specialized muscle fibers, called **intrafusal fibers** (in′-tra-FŪ-sal = within a spindle). A connective tissue capsule encloses the sensory nerve endings and intrafusal fibers and anchors the spindle to the endomysium and perimysium (Figure 16.4). Muscle spindles are interspersed among most skeletal muscle fibers and aligned parallel to them. In muscles that produce finely controlled movements, such as those of the fingers or eyes as you read music and play a musical instrument, muscle spindles are plentiful. Muscles involved in coarser but more forceful movements, like the quadriceps femoris and hamstring muscles of the thigh, have fewer muscle spindles. The only skeletal muscles that lack spindles are the tiny muscles of the middle ear.

FIGURE 16.4 Two types of proprioceptors: a muscle spindle and a tendon organ. In muscle spindles, which monitor changes in skeletal muscle length, sensory nerve endings wrap around the central portion of intrafusal muscle fibers. In tendon organs, which monitor the force of muscle contraction, sensory nerve endings are activated by increasing tension on a tendon. If you examine **Figure 13.14** you can see the relationship of a muscle spindle to the spinal cord as a component of a stretch reflex. In **Figure 13.15**, you can see the relationship of a tendon organ to the spinal cord as a component of a tendon reflex.

Proprioceptors provide information about body position and movement.

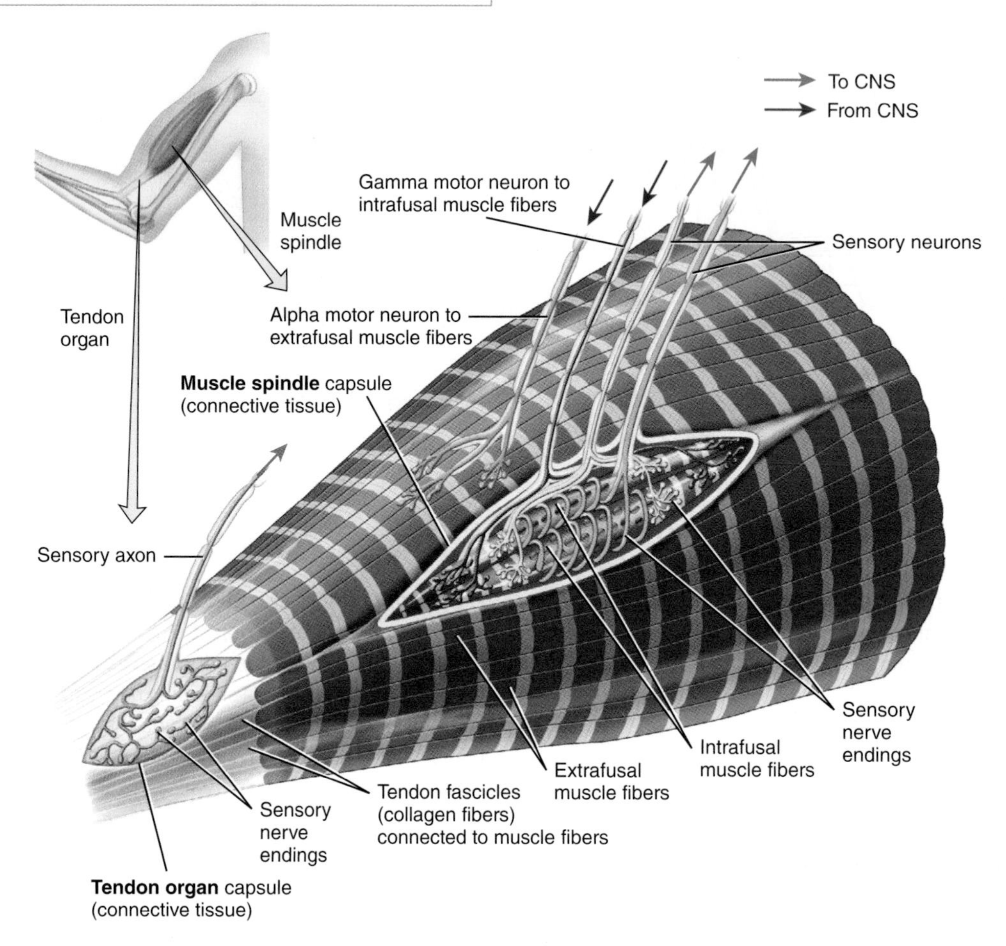

Q How is a muscle spindle activated?

The main function of muscle spindles is to measure *muscle length*—how much a muscle is being stretched. Either sudden or prolonged stretching of the central areas of the intrafusal muscle fibers stimulates the sensory nerve endings. The resulting nerve impulses propagate into the CNS. Information from muscle spindles arrives quickly at the somatic sensory areas of the cerebral cortex, which allows conscious perception of limb positions and movements. At the same time, impulses from muscle spindles pass to the cerebellum, where the input is used to coordinate muscle contractions.

In addition to their sensory nerve endings near the middle of intrafusal fibers, muscle spindles contain motor neurons called **gamma motor neurons**. These motor neurons terminate near both ends of the intrafusal fibers and adjust the tension in a muscle spindle to variations in the length of the muscle. For example, when your biceps muscle shortens in response to lifting a weight, gamma motor neurons stimulate the ends of the intrafusal fibers to contract slightly. This keeps the intrafusal fibers taut even though the contractile muscle fibers surrounding the spindle are reducing spindle tension. This maintains the sensitivity of the muscle spindle to stretching of the muscle. As the frequency of impulses in its gamma motor neuron increases, a muscle spindle becomes more sensitive to stretching of its midregion.

Surrounding muscle spindles are ordinary skeletal muscle fibers, called **extrafusal muscle fibers** (*extrafusal* = outside a spindle), which are supplied by large-diameter A fibers called **alpha motor neurons**. The cell bodies of both gamma and alpha motor neurons are located in the anterior gray horn of the spinal cord (or in the brainstem for muscles in the head). During the stretch reflex, impulses in muscle spindle sensory axons propagate into the spinal cord and brainstem and activate alpha motor neurons that connect to extrafusal muscle fibers in the same muscle. In this way, activation of

TABLE 16.2 Summary of Receptors for Somatic Sensations

RECEPTOR TYPE	RECEPTOR STRUCTURE AND LOCATION	SENSATIONS	ADAPTATION RATE
TACTILE RECEPTORS			
Corpuscles of touch (Meissner corpuscles)	Capsule surrounds mass of dendrites in dermal papillae of hairless skin.	Onset of touch and low-frequency vibrations.	Rapid.
Hair root plexuses	Free nerve endings wrapped around hair follicles in skin.	Movements on skin surface that disturb hairs.	Rapid.
Type I cutaneous mechanoreceptors (tactile discs)	Saucer-shaped free nerve endings make contact with tactile epithelial cells in epidermis.	Continuous touch and pressure.	Slow.
Type II cutaneous mechanoreceptors (Ruffini corpuscles)	Elongated capsule surrounds dendrites deep in dermis and in ligaments and tendons.	Skin stretching and pressure.	Slow.
Lamellated (pacinian) corpuscles	Oval, layered capsule surrounds dendrites; present in dermis and subcutaneous layer, submucosal tissues, joints, periosteum, and some viscera.	High-frequency vibrations.	Rapid.
Itch and tickle receptors	Free nerve endings in skin and mucous membranes.	Itching and tickling.	Both slow and rapid.
THERMORECEPTORS			
Warm receptors and cold receptors	Free nerve endings in skin and mucous membranes of mouth, vagina, and anus.	Warmth or cold.	Initially rapid, then slow.
PAIN RECEPTORS			
Nociceptors	Free nerve endings in every body tissue except brain.	Pain.	Slow.
PROPRIOCEPTORS			
Muscle spindles	Sensory nerve endings wrap around central area of encapsulated intrafusal muscle fibers within most skeletal muscles.	Muscle length.	Slow.
Tendon organs	Capsule encloses collagen fibers and sensory nerve endings at junction of tendon and muscle.	Muscle tension.	Slow.
Joint kinesthetic receptors	Lamellated corpuscles, type II cutaneous mechanoreceptors, tendon organs, and free nerve endings.	Joint position and movement.	Rapid.

its muscle spindles causes contraction of a skeletal muscle, which relieves the stretching.

Tendon Organs **Tendon organs** are slowly adapting receptors located at the junction of a tendon and a muscle. By initiating tendon reflexes (see **Figure 13.15**), tendon organs protect tendons and their associated muscles from damage due to excessive tension. (When a muscle contracts, it exerts a force that pulls the points of attachment of the muscle at either end toward each other. This force is the *muscle tension.*) Each tendon organ consists of a thin capsule of connective tissue that encloses a few *tendon fascicles* (bundles of collagen fibers) (**Figure 16.4**). Penetrating the capsule are one or more sensory nerve endings that entwine among and around the collagen fibers of the tendon. When tension is applied to a muscle, the tendon organs generate nerve impulses that propagate into the CNS, providing information about changes in muscle tension. The resulting tendon reflexes decrease muscle tension by causing muscle relaxation.

Joint Kinesthetic Receptors Several types of **joint kinesthetic receptors** (kin′-es-THET-ik) are present within and around the articular capsules of synovial joints. Free nerve endings and type II cutaneous mechanoreceptors in the capsules of joints respond to pressure. Small lamellated corpuscles in the connective tissue outside articular capsules respond to acceleration and deceleration of joints during movement. Joint ligaments contain receptors similar to tendon organs that adjust reflex inhibition of the adjacent muscles when excessive strain is placed on the joint.

Table 16.2 summarizes the types of somatic sensory receptors and the sensations they convey.

16.3 Somatic Sensory Pathways

OBJECTIVES

- **Describe** the general components of a sensory pathway.
- **Describe** the neuronal components and functions of the posterior column–medial lemniscus, anterolateral, trigeminothalamic, and spinocerebellar pathways.
- **Explain** the basis for mapping the primary somatosensory area.

Somatic sensory (somatosensory) pathways relay information from somatic sensory receptors to the primary somatosensory area (postcentral gyrus) in the parietal lobe of the cerebral cortex and to the cerebellum. A somatic sensory pathway to the cerebral cortex consist of thousands of sets of three neurons: a first-order neuron, a second-order neuron, and a third-order neuron. Integration (processing) of information occurs at each synapse along the pathway.

1. **First-order (primary) neurons** are sensory neurons that conduct impulses from somatic sensory receptors into the brainstem or spinal cord. All other neurons in a somatic sensory pathway are interneurons, which are located completely within the central nervous system (CNS). From the face, nasal cavity, oral cavity, teeth, and eyes, somatic sensory impulses propagate along the *cranial nerves* into the brainstem. From the neck, trunk, limbs, and posterior aspect of the head, somatic sensory impulses propagate along *spinal nerves* into the spinal cord.
2. **Second-order (secondary) neurons** conduct impulses from the brainstem or spinal cord to the thalamus. Axons of second-order neurons **decussate** (cross over to the opposite side) as they course through the brainstem or spinal cord before ascending to the thalamus.
3. **Third-order (tertiary) neurons** conduct impulses from the thalamus to the primary somatosensory area on the same side. As the impulses reach the primary somatosensory area, perception of the sensation occurs. Because the axons of second-order neurons decussate as they pass through the brainstem or spinal cord, somatic sensory information on one side of the body is perceived by the primary somatosensory area on the *opposite* side of the brain.

Regions within the CNS where neurons synapse with other neurons that are a part of a particular sensory or motor pathway are known as **relay stations** because neural signals are being relayed from one region of the CNS to another. For example, the neurons of many sensory pathways synapse with neurons in the thalamus; therefore the thalamus functions as a major relay station. In addition to the thalamus, many other regions of the CNS, including the spinal cord and brainstem, can function as relay stations.

Somatic sensory impulses ascend to the cerebral cortex via three general pathways: (1) the posterior column–medial lemniscus pathway, (2) the anterolateral (spinothalamic) pathway, and (3) the trigeminothalamic pathway. Somatic sensory impulses reach the cerebellum via the spinocerebellar tracts.

Posterior Column–Medial Lemniscus Pathway to the Cerebral Cortex

Nerve impulses for touch, pressure, vibration, and proprioception from the limbs, trunk, neck, and posterior head ascend to the cerebral cortex along the **posterior column–medial lemniscus pathway** (lem-NIS-kus = ribbon) (**Figure 16.5**). The name of the pathway comes from the names of two white-matter tracts that convey the impulses: the posterior column of the spinal cord and the medial lemniscus of the brainstem.

First-order neurons in the posterior column–medial lemniscus pathway extend from sensory receptors in the limbs, trunk, neck, and posterior head into the spinal cord and ascend to the medulla oblongata on the same side of the body. The cell bodies of these first-order neurons are in the posterior (dorsal) root ganglia of spinal nerves. In the spinal cord, their axons form the **posterior** (*dorsal*) **columns**, which consist of two tracts, the **gracile fasciculus** (GRAS-īl fa-SIK-ū-lus) and **cuneate fasciculus** (KŪ-nē-āt). The axons of the first-order neurons synapse with the dendrites of second-order neurons, whose cell bodies are located in the gracile nucleus or cuneate nucleus of the

FIGURE 16.5 The posterior column–medial lemniscus pathway.

The posterior column–medial lemniscus pathway conveys nerve impulses for touch, pressure, vibration, and conscious proprioception from the limbs, trunk, neck, and posterior head to the cerebral cortex.

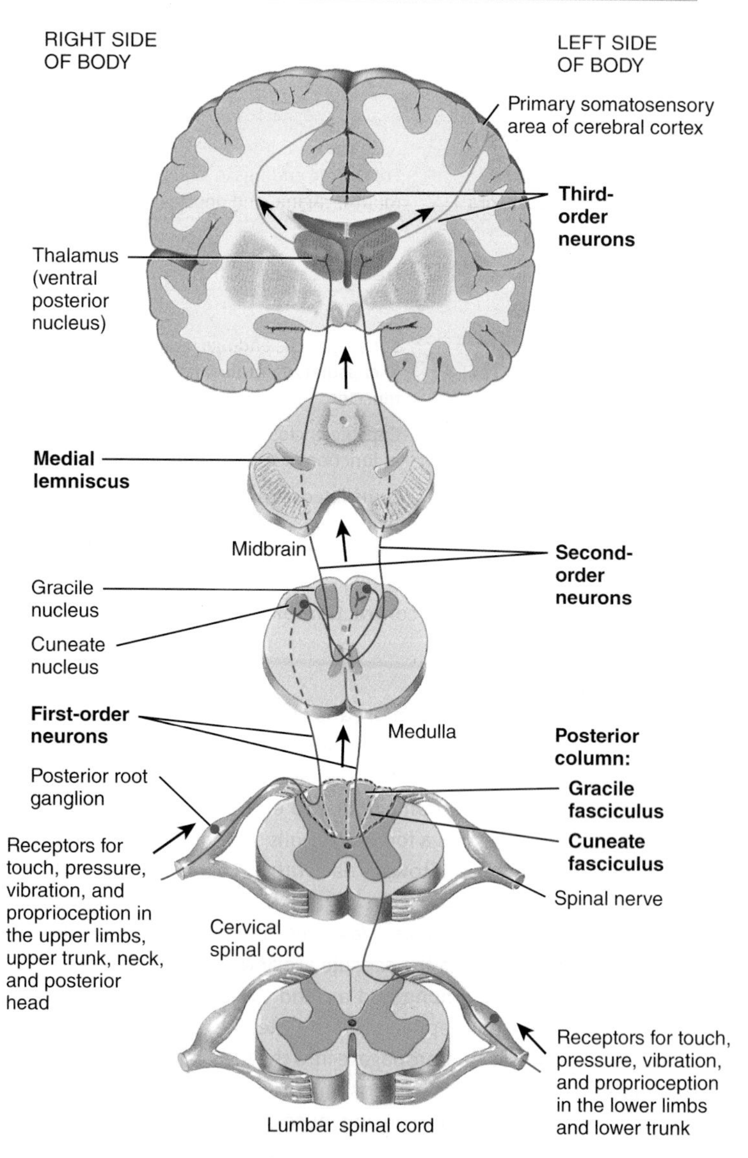

Q What are the two major tracts that form the posterior columns?

medulla. Nerve impulses for touch, pressure, vibration, and conscious proprioception from the upper limbs, upper trunk, neck, and posterior head propagate along axons in the cuneate fasciculus and arrive at the cuneate nucleus. Nerve impulses for touch, pressure, vibration, and conscious proprioception from the lower limbs and lower trunk propagate along axons in the gracile fasciculus and arrive at the gracile nucleus.

The axons of the second-order neurons cross to the opposite side of the medulla and enter the **medial lemniscus**, a thin ribbon-like projection tract that extends from the medulla to the ventral posterior nucleus of the thalamus. In the thalamus, the axon terminals of second-order neurons synapse with third-order neurons, which project their axons to the primary somatosensory area of the cerebral cortex.

Anterolateral (*Spinothalamic*) Pathway to the Cerebral Cortex

Nerve impulses for pain, temperature, itch, and tickle from the limbs, trunk, neck, and posterior head ascend to the cerebral cortex along the **anterolateral** (*spinothalamic*) **pathway** (spī-nō-tha-LAM-ik). First-order neurons of the anterolateral pathway connect a receptor of the limbs, trunk, neck, or posterior head with the spinal cord (**Figure 16.6**). The cell bodies of the first-order neurons are in the posterior root ganglion. The axon terminals of the first-order neurons synapse with second-order neurons, whose cell bodies are located in the posterior gray horn of the spinal cord. The axons of the second-order neurons cross to the opposite side of the spinal cord. Then they pass upward to the brainstem as the **spinothalamic tract**. The axons of the second-order neurons end in the ventral posterior nucleus of the thalamus, where they synapse with the third-order neurons. The axons of the third-order neurons project to the primary somatosensory area on the same side of the cerebral cortex as the thalamus.

Trigeminothalamic Pathway to the Cerebral Cortex

Nerve impulses for most somatic sensations (tactile, thermal, and pain) from the face, nasal cavity, oral cavity, and teeth ascend to the cerebral cortex along the **trigeminothalamic pathway** (trī-jem′-i-nō-tha-LAM-ik). First-order neurons of the trigeminothalamic pathway extend from somatic sensory receptors in the face, nasal cavity, oral cavity, and teeth into the pons through the trigeminal (V) nerves (**Figure 16.7**). The cell bodies of these first-order neurons are in the trigeminal ganglion. The axon terminals of some first-order neurons synapse with second-order neurons in the pons. The axons of other first-order neurons descend into the medulla to synapse with second-order neurons. The axons of the second-order neurons cross to the opposite side of the pons and medulla and then ascend as the **trigeminothalamic tract** to the ventral posterior nucleus of the thalamus. In the thalamus, the axon terminals of the second-order neurons synapse with third-order neurons, which project their axons to the primary somatosensory area on the same side of the cerebral cortex as the thalamus.

Mapping the Primary Somatosensory Area

Specific areas of the cerebral cortex receive somatic sensory input from particular parts of the body. Other areas of the cerebral cortex provide output in the form of instructions for movement of particular parts of the body. The *somatic sensory map* and the *somatic motor map* relate body parts to these cortical areas.

FIGURE 16.6 The anterolateral (spinothalamic) pathway.

The anterolateral pathway conveys nerve impulses for pain, cold, warmth, itch, and tickle from the limbs, trunk, neck, and posterior head to the cerebral cortex.

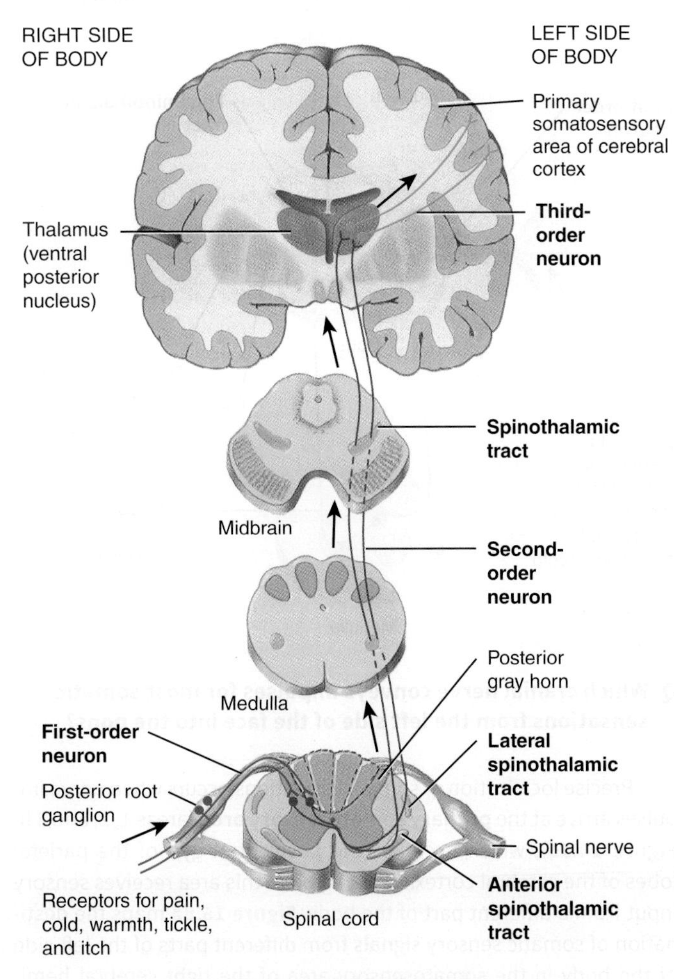

Q What types of sensory deficits could be produced by damage to the right spinothalamic tract?

FIGURE 16.7 The trigeminothalamic pathway.

The trigeminothalamic pathway conveys nerve impulses for most somatic sensations (tactile, thermal, and pain) from the face, nasal cavity, oral cavity, and teeth to the cerebral cortex.

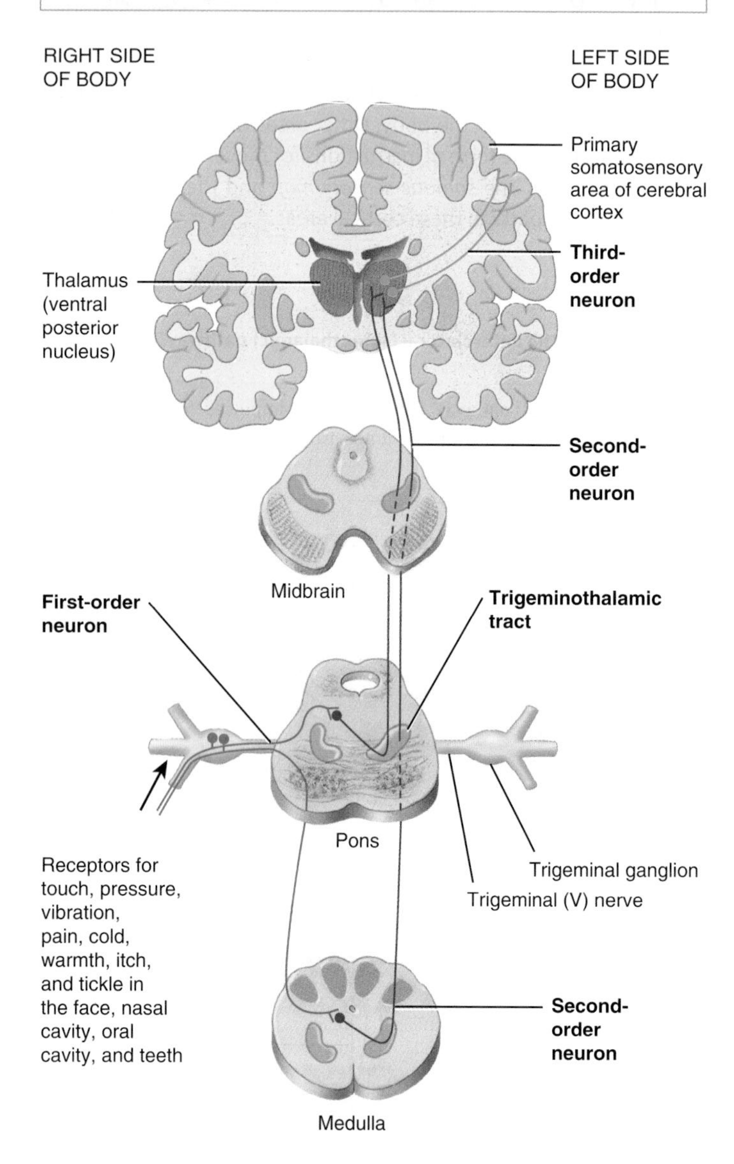

Q Which cranial nerve conveys impulses for most somatic sensations from the left side of the face into the pons?

Precise localization of somatic sensations occurs when nerve impulses arrive at the **primary somatosensory area** (areas 1, 2, and 3 in Figure 14.15), which occupies the postcentral gyri of the parietal lobes of the cerebral cortex. Each region in this area receives sensory input from a different part of the body. Figure 16.8a maps the destination of somatic sensory signals from different parts of the left side of the body in the somatosensory area of the right cerebral hemisphere. The left cerebral hemisphere has a similar primary somatosensory area that receives sensory input from the right side of the body.

Note that some parts of the body—chiefly the lips, face, tongue, and hand—provide input to large regions in the somatosensory area. Other parts of the body, such as the trunk and lower limbs, project to much smaller cortical regions. The relative sizes of these regions in the somatosensory area are proportional to the number of specialized sensory receptors within the corresponding part of the body. For example, there are many sensory receptors in the skin of the lips but few in the skin of the trunk. This distorted somatic sensory map of the body is known as the **sensory homunculus** (hō-MONK-ū-lus = little man). The size of the cortical region that represents a body part may expand or shrink somewhat, depending on the quantity of sensory impulses received from that body part. For example, people who learn to read Braille eventually have a larger cortical region in the somatosensory area to represent the fingertips.

Somatic Sensory Pathways to the Cerebellum

Two tracts in the spinal cord—the **anterior spinocerebellar tract** (spī-nō-ser-e-BEL-ar) and the **posterior spinocerebellar tract**—are the major routes proprioceptive impulses take to reach the cerebellum. Although they are not consciously perceived, sensory impulses conveyed to the cerebellum along these two pathways are critical for posture, balance, and coordination of skilled movements.

Table 16.3 summarizes the major somatic sensory tracts and pathways.

See Clinical Connection: Syphilis

16.4 Control of Body Movement

OBJECTIVES

- **Identify** the locations and functions of the different types of neurons that regulate lower motor neurons.
- **Explain** how the cerebral contex, brainstem, basal nuclei, and cerebellum contribute to body movement.
- **Compare** the locations and functions of the direct and indirect motor pathways.

Neural circuits in the brain and spinal cord orchestrate all voluntary movements. Ultimately, all excitatory and inhibitory signals that control movement converge on the motor neurons that extend out of the brainstem and spinal cord to innervate skeletal muscles in the body. These neurons are known as **lower motor neurons (LMNs)** because they have their cell bodies in the *lower* parts of the CNS (brainstem

FIGURE 16.8 Somatic sensory and somatic motor maps in the cerebral cortex, right hemisphere.
(a) Primary somatosensory area (postcentral gyrus) and (b) primary motor area (precentral gyrus) of the right cerebral hemisphere. The left hemisphere has similar representation. (After Penfield and Rasmussen.)

Each point on the body surface maps to a specific region in both the primary somatosensory area and the primary motor area.

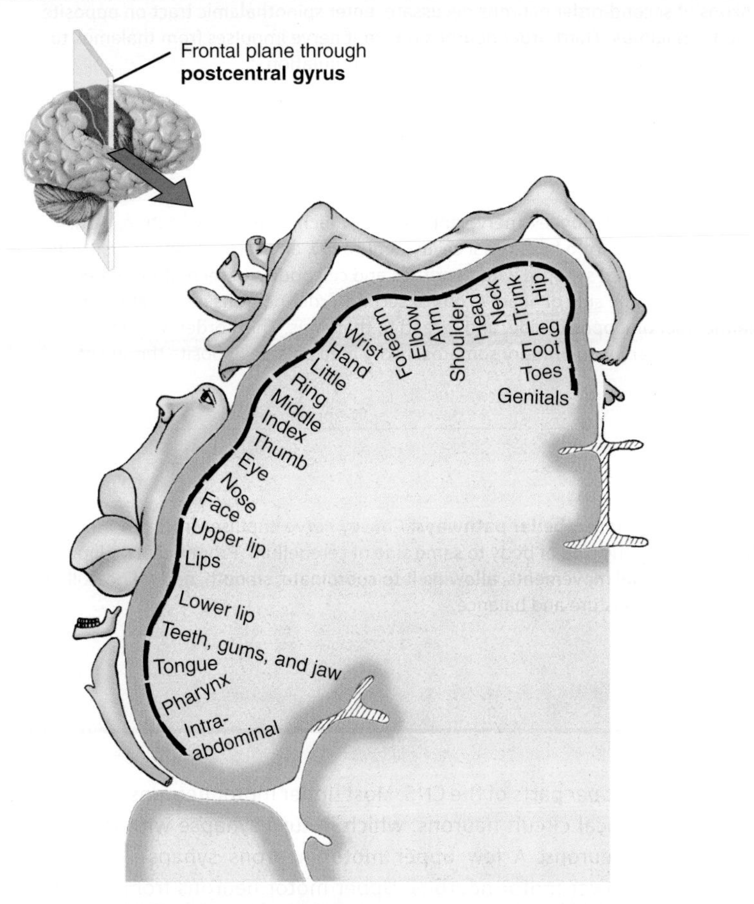

(a) Frontal section of primary somatosensory area in right cerebral hemisphere

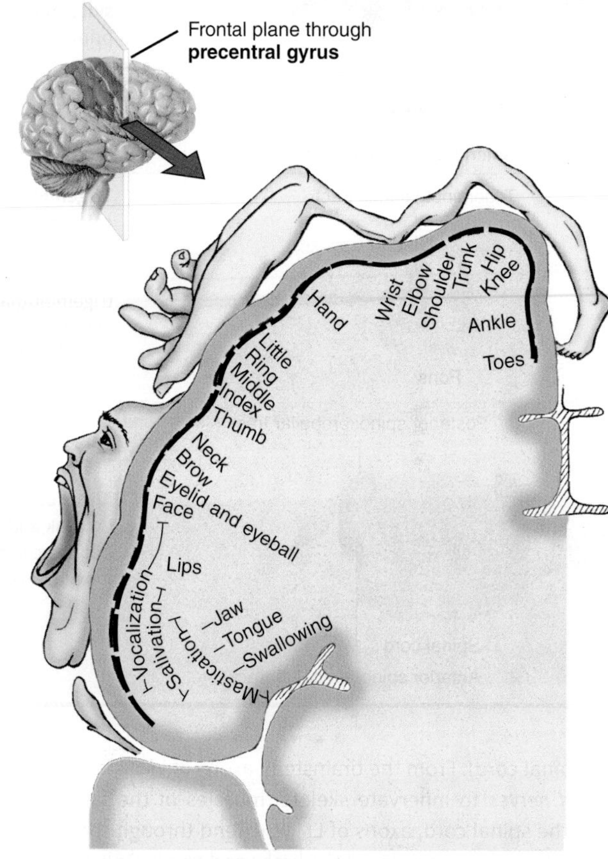

(b) Frontal section of primary motor area in right cerebral hemisphere

Q How do the somatosensory and motor representations compare for the hand, and what does this difference imply?

TABLE 16.3 Major Somatic Sensory Tracts and Pathways

TRACTS AND LOCATIONS	PATHWAY FUNCTIONS
Posterior column: Gracile fasciculus; Cuneate fasciculus; Spinal cord	**Posterior column–medial lemniscus pathway: Cuneate fasciculus** conveys nerve impulses for touch, pressure, vibration, and conscious proprioception from upper limbs, upper trunk, neck, and posterior head, and **gracile fasciculus** conveys nerve impulses for touch, pressure, vibration, and conscious proprioception from lower limbs and lower trunk. Axons of first-order neurons from one side of body form posterior column on same side and end in medulla, where they synapse with dendrites and cell bodies of second-order neurons. Axons of second-order neurons decussate, enter **medial lemniscus** on opposite side, and extend to thalamus. Third-order neurons transmit nerve impulses from thalamus to primary somatosensory area on side opposite the site of stimulation.

Continues

TABLE 16.3 Major Somatic Sensory Tracts and Pathways (*Continued*)

TRACTS AND LOCATIONS	PATHWAY FUNCTIONS
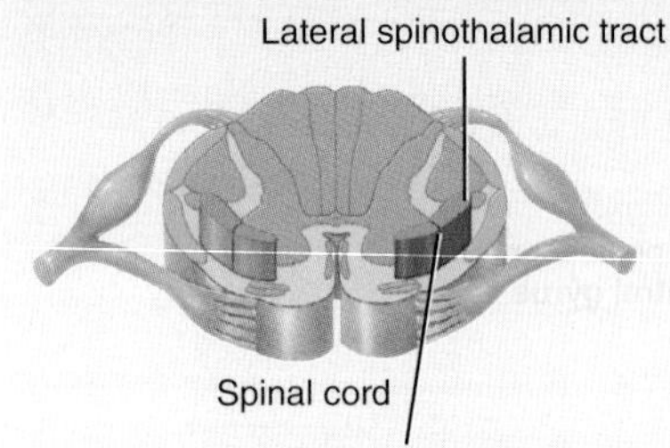	**Anterolateral pathway:** Conveys nerve impulses for pain, cold, warmth, itch, and tickle from limbs, trunk, neck, and posterior head. Axons of first-order neurons from one side of body synapse with dendrites and cell bodies of second-order neurons in posterior gray horn on same side of body. Axons of second-order neurons decussate, enter spinothalamic tract on opposite side, and extend to thalamus. Third-order neurons transmit nerve impulses from thalamus to primary somatosensory area on side opposite the site of stimulation.
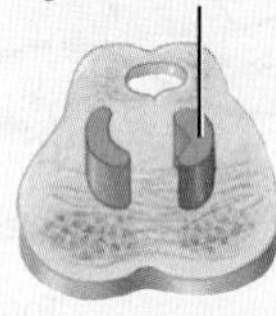	**Trigeminothalamic pathway:** Conveys nerve impulses for touch, pressure, vibration, pain, cold, warmth, itch, and tickle from face, nasal cavity, oral cavity, and teeth. Axons of first-order neurons from one side of head synapse with dendrites and cell bodies of second-order neurons in pons and medulla on same side of head. Axons of second-order neurons decussate, enter trigeminothalamic tract on opposite side, and extend to thalamus. Third-order neurons transmit nerve impulses from thalamus to primary somatosensory area on side opposite the site of stimulation.
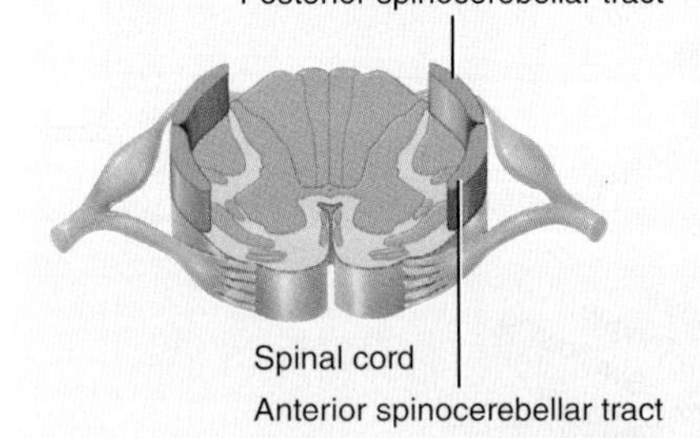	**Anterior and posterior spinocerebellar pathways:** Convey nerve impulses from proprioceptors in trunk and lower limb of one side of body to same side of cerebellum. Proprioceptive input informs cerebellum of actual movements, allowing it to coordinate, smooth, and refine skilled movements and maintain posture and balance.

and spinal cord). From the brainstem, axons of LMNs extend through *cranial nerves* to innervate skeletal muscles of the face and head. From the spinal cord, axons of LMNs extend through *spinal nerves* to innervate skeletal muscles of the limbs and trunk. Only LMNs provide output from the CNS to skeletal muscle fibers. For this reason, they are also called the *final common pathway.*

Neurons in four distinct but highly interactive neural circuits participate in control of movement by providing input to lower motor neurons (Figure 16.9):

1. ***Local circuit neurons.*** Input arrives at lower motor neurons from nearby interneurons called **local circuit neurons**. These neurons are located close to the lower motor neuron cell bodies in the brainstem and spinal cord. Local circuit neurons receive input from somatic sensory receptors, such as nociceptors and muscle spindles, as well as from higher centers in the brain. They help coordinate rhythmic activity in specific muscle groups, such as alternating flexion and extension of the lower limbs during walking.
2. ***Upper motor neurons.*** Both local circuit neurons and lower motor neurons receive input from **upper motor neurons (UMNs)**,* neurons that have cell bodies in motor processing centers in the *upper* parts of the CNS. Most upper motor neurons synapse with local circuit neurons, which in turn synapse with lower motor neurons. A few upper motor neurons synapse directly with lower motor neurons. Upper motor neurons from the cerebral cortex are essential for the planning and execution of voluntary movements of the body. Other upper motor neurons originate in motor centers of the brainstem: the vestibular nuclei, reticular formation, superior colliculus, and red nucleus. Upper motor neurons from the brainstem help regulate posture, balance, muscle tone, and reflexive movements of the head and trunk.
3. ***Basal nuclei neurons.*** **Basal nuclei neurons** assist movement by providing input to upper motor neurons. Neural circuits interconnect the basal nuclei with motor areas of the cerebral cortex (via the thalamus) and the brainstem. These circuits help initiate and terminate movements, suppress unwanted movements, and establish a normal level of muscle tone.
4. ***Cerebellar neurons.*** **Cerebellar neurons** also aid movement by controlling the activity of upper motor neurons. Neural circuits interconnect the cerebellum with motor areas of the cerebral cortex (via the thalamus) and the brainstem. A prime function of the cerebellum is to monitor differences between intended movements and movements actually performed. Then, it issues commands to upper motor neurons to reduce errors in movement. The cerebellum thus coordinates body movements and helps maintain normal posture and balance.

*An upper motor neuron is actually an interneuron and not a true motor neuron: it is so named because the cell originates in the upper part of the CNS and regulates the activity of lower motor neurons. Only a lower motor neuron is a true motor neuron because it conveys action potentials from the CNS to skeletal muscles in the periphery.

FIGURE 16.9 Neural circuits that regulate lower motor neurons. Lower motor neurons receive input directly from (1) local circuit neurons (purple arrow) and (2) upper motor neurons in the cerebral cortex and brainstem (green arrows). Neural circuits involving (3) basal nuclei neurons and (4) cerebellar neurons regulate activity of upper motor neurons (red arrows).

Because lower motor neurons provide all output to skeletal muscles, they are called the final common pathway.

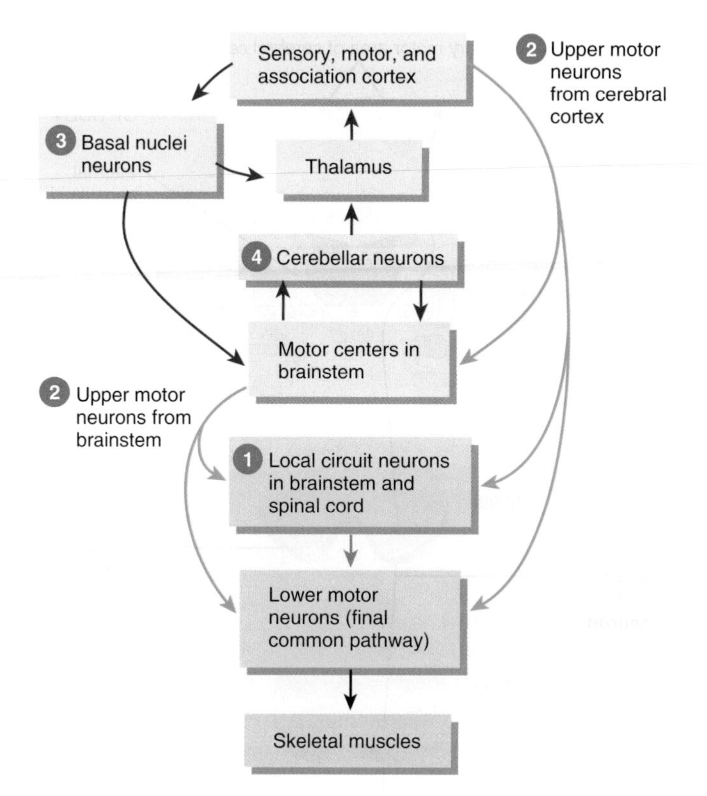

Q How do the functions of upper motor neurons from the cerebral cortex and from the brainstem differ?

See Clinical Connection: Paralysis

Control of Movement by the Cerebral Cortex

Control of body movements involves motor pathways that begin in motor areas of the cerebral cortex. Two such areas are the **premotor area** (area 6 in the frontal lobe shown in **Figure 14.15**) and the **primary motor area** (area 4 in the precentral gyrus of the frontal lobe also shown in **Figure 14.15**).

Premotor Area The role of the premotor area in body movements is as follows. The idea or desire to move a part of the body is generated in one or more cortical association areas, such as the prefrontal cortex, somatosensory association area, auditory association area, or visual association area (see **Figure 14.15**). This information is sent to the basal nuclei, which process the information and sends it to the thalamus and then to the premotor area, where a motor plan is developed. This plan identifies which muscles should contract, how much they need to contract, and in what order. From the premotor area, the plan is transmitted to the primary motor area for execution. The premotor area also stores information about learned motor activities. By activating the appropriate neurons of the primary motor area, the premotor area causes specific groups of muscles to contract in a specific sequence.

Primary Motor Area The primary motor area is the major control region for the execution of voluntary movements. Electrical stimulation of any point in the primary motor area causes contraction of specific muscles on the opposite side of the body. The primary motor area controls muscles by forming descending pathways that extend to the spinal cord and brainstem (described shortly). As is true for somatic sensory representation in the primary somatosensory area, a "map" of the body is present in the primary motor area: Each point within the area controls muscle fibers in a different part of the body. Different muscles are represented unequally in the primary motor area (see **Figure 16.8b**). More cortical area is devoted to those muscles involved in skilled, complex, or delicate movement. Muscles in the thumb, fingers, lips, tongue, and vocal cords have large representations; the trunk has a much smaller representation. This distorted muscle map of the body is called the **motor homunculus**.

Direct Motor Pathways The axons of upper motor neurons extend from the brain to lower motor neurons via two types of pathways—direct and indirect. *Direct motor pathways* provide input to lower motor neurons via axons that extend directly from the cerebral cortex. *Indirect motor pathways* provide input to lower motor neurons from motor centers in the brainstem. Direct and indirect pathways both govern generation of action potentials in the lower motor neurons, the neurons that stimulate contraction of skeletal muscles.

Action potentials for voluntary movements propagate from the cerebral cortex to lower motor neurons via the **direct motor pathways**. Also known as the *pyramidal pathways*, the direct motor pathways consist of axons that descend from pyramidal cells of the primary motor area and premotor area. *Pyramidal cells* are upper motor neurons that have pyramid-shaped cell bodies (see **Figure 12.5b**). They are the main output cells of the cerebral cortex. The direct motor pathways consist of corticospinal pathways and the corticobulbar pathway.

CORTICOSPINAL PATHWAYS The **corticospinal pathways** (kor′-ti-kō-SPĪ-nal) conduct impulses for the control of muscles of the limbs and trunk. Axons of upper motor neurons in the cerebral cortex form the **corticospinal tracts**, which descend through the *internal capsule* of the cerebrum and the cerebral peduncle of the midbrain. In the medulla oblongata, the axon bundles of the corticospinal tracts form the ventral bulges known as the *pyramids.* About 90% of the corticospinal axons *decussate* (cross over) to the *contralateral* (opposite) side in the medulla oblongata and then descend into the spinal cord where they synapse with a local circuit neuron or a lower motor neuron. The 10% that remain on the *ipsilateral* (same) side eventually decussate at the spinal cord levels where they synapse with a local circuit neuron or lower motor neuron. Thus, the right cerebral cortex controls most of the muscles on the left side of the body, and the left cerebral cortex controls most of the muscles on the right side of the body. There are two types of corticospinal tracts: the lateral corticospinal tract and the anterior corticospinal tract.

1. ***Lateral corticospinal tract.*** Corticospinal axons that decussate in the medulla form the **lateral corticospinal tract** in the lateral white column of the spinal cord (**Figure 16.10a**). These axons synapse with local circuit neurons or lower motor neurons in the anterior gray horn of the spinal cord. Axons of these lower motor neurons exit the cord in the anterior roots of spinal nerves and terminate in skeletal muscles that control movements of the distal parts of the limbs. The distal muscles are responsible for precise, agile, and

FIGURE 16.10 Direct motor pathways: the corticospinal pathways.

The corticospinal pathways conduct nerve impulses for the control of muscles of the limbs and trunk.

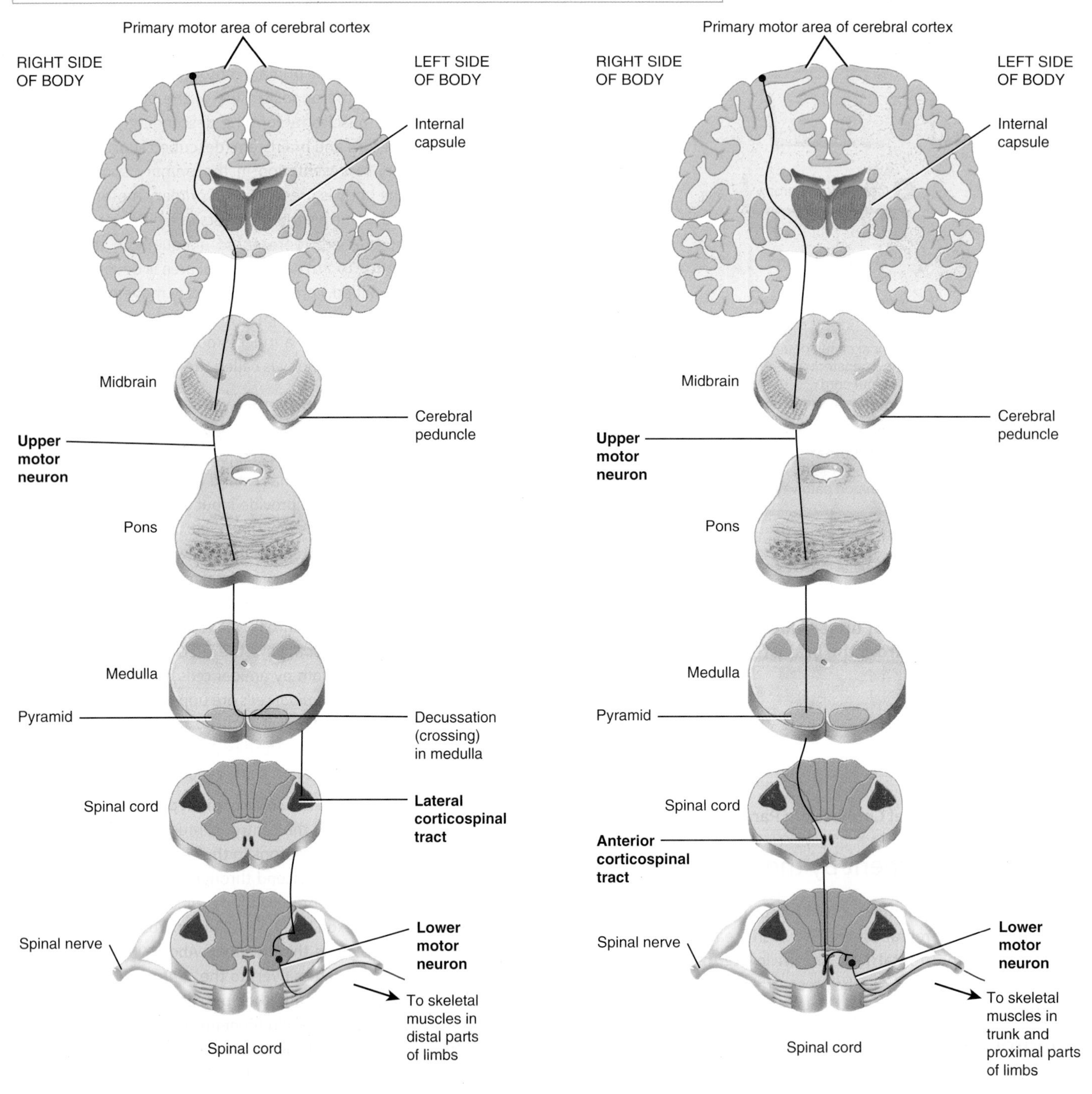

Q Which tract conveys nerve impulses that result in contractions of muscles in the distal parts of the limbs?

highly skilled movements of the hands and feet. Examples include the movements needed to button a shirt or play the piano.

2. ***Anterior corticospinal tract.*** Corticospinal axons that do not decussate in the medulla form the **anterior corticospinal tract** in the anterior white column of the spinal cord (Figure 16.10b). At each spinal cord level, some of these axons decussate via the anterior white commissure. Then, they synapse with local circuit neurons or lower motor neurons in the anterior gray horn. Axons of these lower motor neurons exit the cord in the anterior roots of spinal nerves. They terminate in skeletal muscles that control movements of the trunk and proximal parts of the limbs.

CORTICOBULBAR PATHWAY The **corticobulbar pathway** (kor′-ti-kō-BUL-bar) conducts impulses for the control of skeletal muscles in the head. Axons of upper motor neurons from the cerebral cortex form the **corticobulbar tract**, which descends along with the corticospinal tracts through the internal capsule of the cerebrum and cerebral peduncle of the midbrain (Figure 16.11). Some of the axons of the corticobulbar tract decussate; others do not. The axons terminate in the motor nuclei of nine pairs of cranial nerves in the brain-stem: the oculomotor (III), trochlear (IV), trigeminal (V), abducens (VI), facial (VII), glossopharyngeal (IX), vagus (X), accessory (XI), and hypoglossal (XII). The lower motor neurons of the cranial nerves convey impulses that control precise, voluntary movements of the eyes, tongue, and neck, plus chewing, facial expression, speech, and swallowing.

FIGURE 16.11 Direct motor pathway: the corticobulbar pathway. For simplicity, only two cranial nerves are illustrated.

The corticobulbar pathway conducts nerve impulses for the control of skeletal muscles in the head.

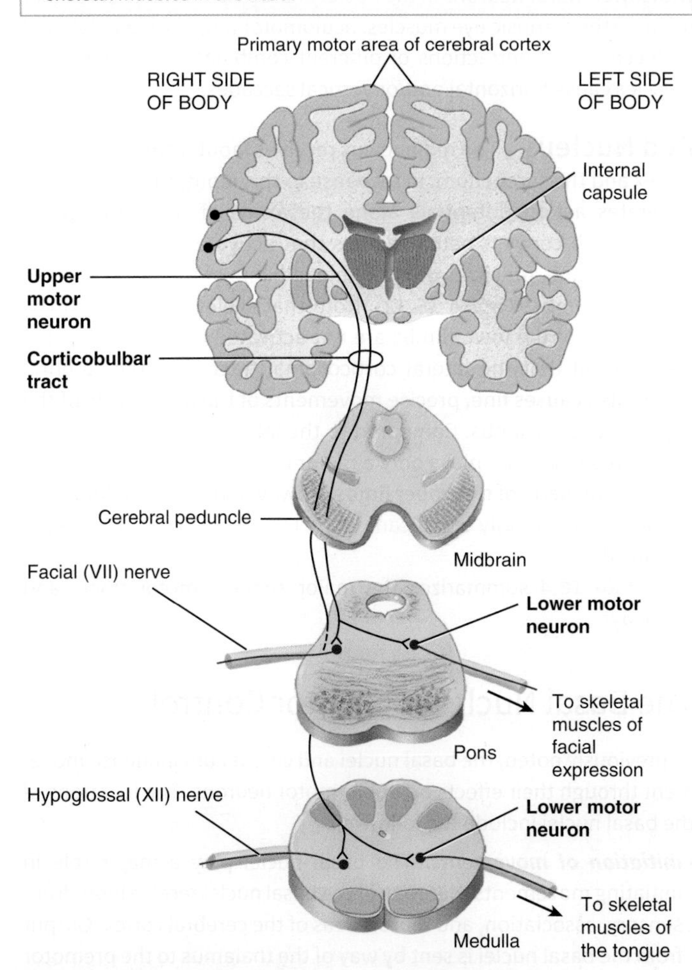

Q The axons of the corticobulbar tract terminate in the motor nuclei of which cranial nerves?

Control of Movement by the Brainstem

The brainstem is another region important to motor control. It contains four major motor centers that help regulate body movements—(1) the **vestibular nuclei** in the medulla and pons; (2) the **reticular formation** located throughout the brainstem; (3) the **superior colliculus** in the midbrain, and (4) the **red nucleus**, also present in the midbrain (Figure 16.12).

See Clinical Connection: Amyotrophic Lateral Sclerosis

FIGURE 16.12 The indirect motor pathways. For simplicity, the vestibular nucleus is shown only in the pons, the reticular formation is shown only in the medulla, and only one reticulospinal tract is shown in the spinal cord.

In general, the indirect motor pathways conduct action potentials to cause involuntary movements that regulate posture, balance, muscle tone, and reflexive movements of the head and trunk.

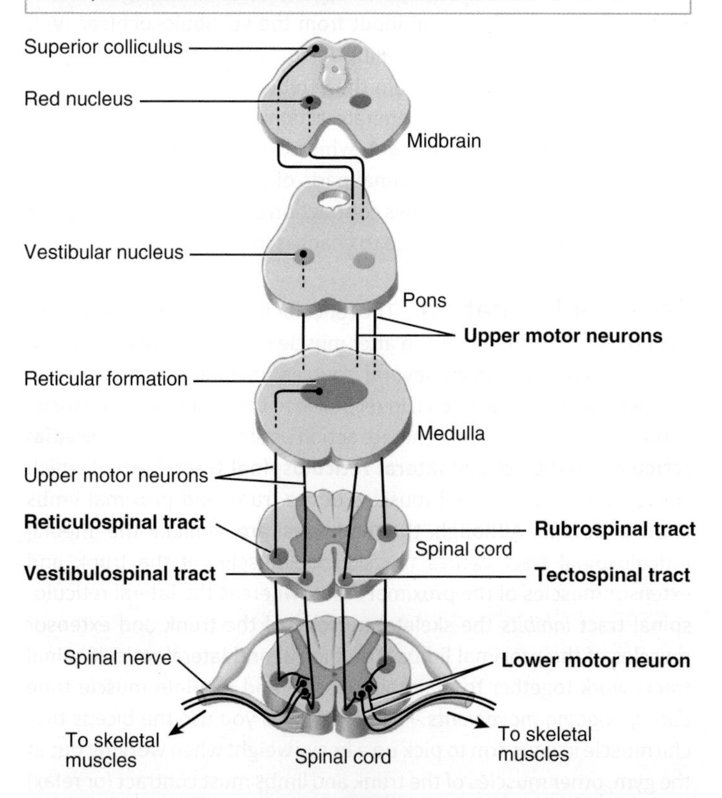

Q How is the rubrospinal tract different from the other tracts of the indirect motor pathways?

Indirect Motor Pathways The brainstem motor centers give rise to the **indirect motor pathways**, also known as *extrapyramidal pathways*, which include all somatic motor tracts other than the corticospinal and corticobulbar tracts. Axons of upper motor neurons descend from the brainstem motor centers into five major tracts of the spinal cord and terminate on local circuit neurons or lower motor neurons. These tracts are the *rubrospinal* (ROO-brō-spī-nal), *tectospinal* (TEK-tō-spī-nal), *vestibulospinal* (ves-TIB-ū-lō-spī-nal), *lateral reticulospinal* (re-TIK-ū-lō-spī-nal), and *medial reticulospinal tracts* (Figure 16.12). In general, the indirect motor pathways convey action potentials from the brainstem to cause involuntary movements that regulate posture, balance, muscle tone, and reflexive movements of the head and trunk. An exception is the rubrospinal tract, which plays an ancillary role to the lateral corticospinal tract in the regulation of voluntary movements of the upper limbs.

Vestibular Nuclei Many postural muscles of the trunk and limbs are reflexively controlled by upper motor neurons in the brainstem. **Postural reflexes** keep the body in an upright and balanced position. Input for postural reflexes comes from three sources: (1) the eyes, which provide visual information about the position of the body in space; (2) the vestibular apparatus of the inner ear, which provides information about the position of the head, and (3) proprioceptors in muscles and joints, which provide information about the position of the limbs. In response to this sensory input, upper motor neurons in the brainstem activate lower motor neurons, which in turn cause the appropriate postural muscles to contract in order keep the body properly oriented in space.

The vestibular nuclei play an important role in the regulation of posture. They receive neural input from the vestibulocochlear (VIII) nerve regarding the state of equilibrium (balance) of the body (mainly the head) and neural input from the cerebellum. In response to this input, the vestibular nuclei generate action potentials along the axons of the **vestibulospinal tract**, which conveys signals to skeletal muscles of the trunk and proximal parts of the limbs (Figure 16.12). The vestibulospinal tract causes contraction of these muscles in order to maintain posture in response to changes in equilibrium.

Reticular Formation The reticular formation also helps control posture. In addition, it can alter muscle tone. The reticular formation receives input from several sources, including the eyes, ear, cerebellum, and basal nuclei. In response to this input, discrete nuclei in the reticular formation generate action potentials along the **medial reticulospinal tract** and **lateral reticulospinal tract**, both of which convey signals to skeletal muscles of the trunk and proximal limbs (Figure 16.12). Although the pathways are similar, the medial reticulospinal tract *excites* the skeletal muscles of the trunk and extensor muscles of the proximal limbs, whereas the lateral reticulospinal tract *inhibits* the skeletal muscles of the trunk and extensor muscles of the proximal limbs. The medial and lateral reticulospinal tracts work together to maintain posture and regulate muscle tone *during ongoing movements*. For example, as you use the biceps brachii muscle in your arm to pick up a heavy weight when working out at the gym, other muscles of the trunk and limbs must contract (or relax) to maintain your posture. Those muscles that need to contract will be activated by the medial reticulospinal tract, whereas those muscles that need to relax will be inhibited by the lateral reticulospinal tract.

Superior Colliculus The superior colliculus receives visual input from the eyes and auditory input from the ears (via connections with the inferior colliculus). When this input occurs in a sudden, unexpected manner, the superior colliculus produces action potentials along the **tectospinal tract**, which conveys neural signals that activate skeletal muscles in the head and trunk (Figure 16.12). This allows the body to turn in the direction of the sudden visual stimulus (such as a bug darting across the floor) or the sudden auditory stimulus (such as a bolt of thunder). These responses serve to protect you from potentially dangerous stimuli.

The superior colliculus is also an integrating center for **saccades** (sa-kādz′), small, rapid jerking movements of the eyes that occur as a person looks at different points in the visual field. Although you typically do not realize it, your eyes are constantly making saccades as you read the sentences on the pages of this book or as you look at different parts of a picture or statue. In addition to upper motor neurons that give rise to the tectospinal tract, the superior colliculus also contains upper motor neurons that synapse with local circuit neurons in the **gaze centers** in the reticular formation of the midbrain and pons. The local circuit neurons in the gaze centers in turn synapse with lower motor neurons in the nuclei of the three cranial nerves that regulate the extrinsic eye muscles: oculomotor (III), trochlear (IV), and abducens (VI). Contractions of different combinations of these eye muscles cause horizontal and/or vertical saccades.

Red Nucleus The red nucleus receives input from the cerebral cortex and the cerebellum. In response to this input, the red nucleus generates action potentials along the axons of the **rubrospinal tract**, which conveys neural signals that activate skeletal muscles that cause fine, precise, voluntary movements of the distal parts of the upper limbs (Figure 16.12). Note that skeletal muscles in the distal parts of the lower limbs are not activated by the rubrospinal tract. Recall that the lateral corticospinal tract from the cerebral cortex also causes fine, precise movements of the distal parts of the *upper* and *lower* limbs. Compared to the lateral corticospinal tract, the rubrospinal tract plays only a minor role in contracting muscles of the distal parts of the upper limbs. However, the rubrospinal tract becomes functionally significant if the lateral corticospinal tract is damaged.

Table 16.4 summarizes the major somatic motor tracts and pathways.

The Basal Nuclei and Motor Control

As previously noted, the basal nuclei and cerebellum influence movement through their effects on upper motor neurons. The functions of the basal nuclei include the following:

- ***Initiation of movements.*** The basal nuclei play a major role in initiating movements. Neurons of the basal nuclei receive input from sensory, association, and motor areas of the cerebral cortex. Output from the basal nuclei is sent by way of the thalamus to the premotor area, which in turn communicates with upper motor neurons in

TABLE 16.4 Major Somatic Motor Tracts and Pathways

TRACTS AND LOCATIONS	PATHWAY FUNCTIONS
DIRECT (PYRAMIDAL) PATHWAYS 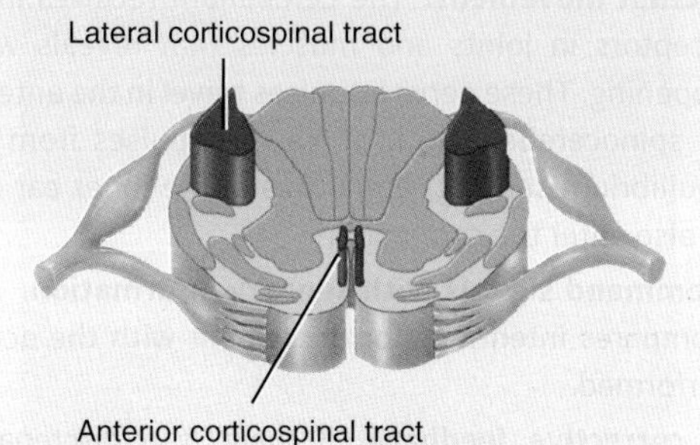	**Lateral corticospinal pathway:** Conveys nerve impulses from motor cortex to skeletal muscles on opposite side of body for precise, voluntary movements of distal parts of limbs. Axons of upper motor neurons (UMNs) descend from precentral gyrus of cortex into medulla. Here 90% decussate (cross over to opposite side) and then enter contralateral side of spinal cord to form this tract. At their level of termination, UMNs end in anterior gray horn on same side. They provide input to lower motor neurons, which innervate skeletal muscles. **Anterior corticospinal pathway:** Conveys nerve impulses from motor cortex to skeletal muscles on opposite side of body for movements of trunk and proximal parts of limbs. Axons of UMNs descend from cortex into medulla. Here the 10% that do not decussate enter the spinal cord and form this tract. At their level of termination, UMNs decussate and end in anterior gray horn on opposite side of body. They provide input to lower motor neurons, which innervate skeletal muscles.
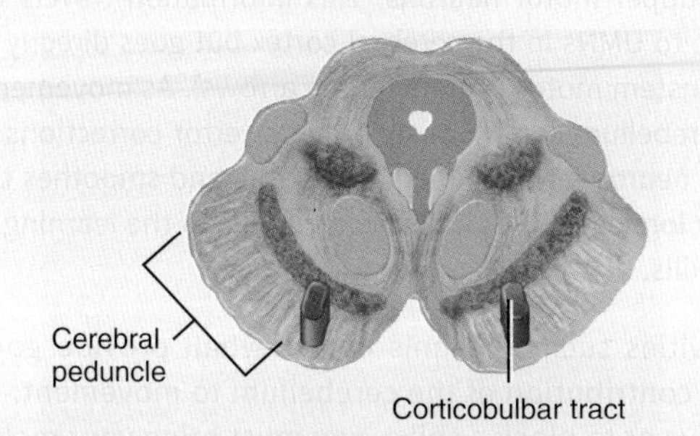	**Corticobulbar pathway:** Conveys nerve impulses from motor cortex to skeletal muscles of head and neck to coordinate precise, voluntary movements. Axons of UMNs descend from cortex into brainstem, where some decussate and others do not. They provide input to lower motor neurons in nuclei of the oculomotor (III), trochlear (IV), trigeminal (V), abducens (VI), facial (VII), glossopharyngeal (IX), vagus (X), accessory (XI), and hypoglossal (XII) nerves, which control voluntary movements of the eyes, tongue, and neck; chewing; facial expression; and speech.
INDIRECT (EXTRAPYRAMIDAL) PATHWAYS 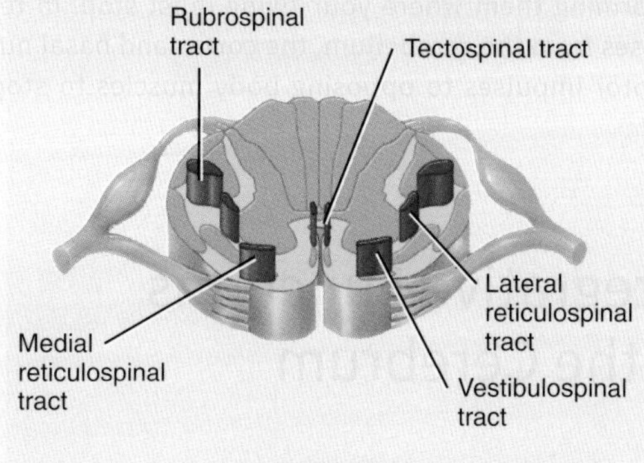	**Rubrospinal pathway:** Conveys nerve impulses from red nucleus (which receives input from cerebral cortex and cerebellum) to contralateral skeletal muscles that govern precise, voluntary movements of distal parts of upper limbs. **Tectospinal pathway:** Conveys nerve impulses from superior colliculus to contralateral skeletal muscles that reflexively move head, eyes, and trunk in response to visual or auditory stimuli. **Vestibulospinal pathway:** Conveys nerve impulses from vestibular nucleus (which receives input about head movements from inner ear) to ipsilateral skeletal muscles of trunk and proximal parts of limbs for maintaining posture and balance in response to head movements. **Lateral and medial reticulospinal pathways:** Conveys nerve impulses from reticular formation to ipsilateral skeletal muscles of trunk and proximal parts of limbs for maintaining posture and regulating muscle tone in response to ongoing body movements.

the primary motor area. The upper motor neurons then activate the corticospinal and corticobulbar tracts to promote movement. Therefore, this circuit—from cortex to basal nuclei to thalamus to cortex—is responsible for the initiation of movements.

- ***Suppression of unwanted movements.*** The basal nuclei suppress unwanted movements by tonically inhibiting the neurons of the thalamus that affect the activity of the upper motor neurons in the motor cortex. When a particular movement is desired, the inhibition of thalamic neurons by the basal nuclei is removed, which allows the thalamic neurons to activate the appropriate upper motor neurons in the motor cortex.
- ***Regulation of muscle tone.*** The basal nuclei influence muscle tone. Neurons of the basal nuclei send action potentials into the reticular formation that reduce muscle tone via the medial and lateral reticulospinal tracts. Damage or destruction of some basal nuclei connections causes a generalized increase in muscle tone.
- ***Regulation of nonmotor processes.*** The basal nuclei influence several nonmotor aspects of cortical function, including sensory, limbic,

cognitive, and linguistic functions. For example, the basal nuclei help initiate and terminate some cognitive processes, such as attention, memory, and planning. In addition, the basal nuclei may act with the limbic system to regulate emotional behaviors.

See Clinical Connection: Disorders of the Basal Nuclei

Modulation of Movement by the Cerebellum

In addition to maintaining proper posture and balance, the cerebellum is active in both learning and performing rapid, coordinated, highly skilled movements such as hitting a golf ball, speaking, and swimming. Cerebellar function involves four activities (**Figure 16.13**):

FIGURE 16.13 Input to and output from the cerebellum.

The cerebellum coordinates and smooths contractions of skeletal muscles during skilled movements and helps maintain posture and balance.

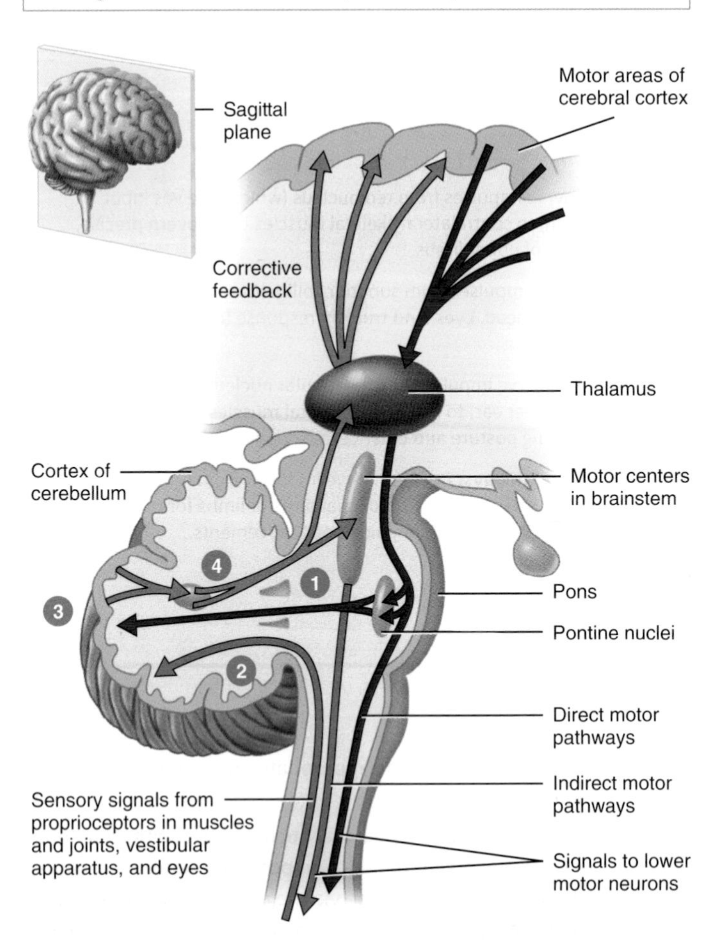

Sagittal section through brain and spinal cord

Q Which tracts carry information from proprioceptors in joints and muscles to the cerebellum?

1. **Monitoring intentions for movement.** The cerebellum receives impulses from the motor cortex and basal nuclei via the pontine nuclei in the pons regarding what movements are planned (red arrows).
2. **Monitoring actual movement.** The cerebellum receives input from proprioceptors in joints and muscles that reveals what actually is happening. These nerve impulses travel in the anterior and posterior spinocerebellar tracts. Nerve impulses from the vestibular (equilibrium-sensing) apparatus in the inner ear and from the eyes also enter the cerebellum.
3. **Comparing command signals with sensory information.** The cerebellum compares intentions for movement with the actual movement performed.
4. **Sending out corrective feedback.** If there is a discrepancy between intended and actual movement, the cerebellum sends feedback to upper motor neurons. This information travels via the thalamus to UMNs in the cerebral cortex but goes directly to UMNs in brainstem motor centers (green arrows). As movements occur, the cerebellum continuously provides error corrections to upper motor neurons, which decreases errors and smoothes the motion. Over longer periods it also contributes to the learning of new motor skills.

Skilled activities such as tennis or volleyball provide good examples of the contribution of the cerebellum to movement. To make a good serve or to block a spike, you must bring your racket or arms forward just far enough to make solid contact. How do you stop at exactly the right point? Before you even hit the ball, the cerebellum has sent nerve impulses to the cerebral cortex and basal nuclei informing them where your swing must stop. In response to impulses from the cerebellum, the cortex and basal nuclei transmit motor impulses to opposing body muscles to stop the swing.

16.5 Integrative Functions of the Cerebrum

OBJECTIVES

- **Compare** the integrative cerebral functions of wakefulness and sleep, coma, learning and memory, and language.
- **Describe** the four stages come of sleep.
- **Explain** the factors that contribute to memory.

We turn now to a fascinating, though incompletely understood, function of the cerebrum: **integration**, the processing of sensory information by analyzing and storing it and making decisions for various responses. The **integrative functions** include cerebral activities such as sleep and wakefulness, learning and memory, and language.

Wakefulness and Sleep

Humans sleep and awaken in a 24-hour cycle called a **circadian rhythm** (ser-KĀ-dē-an; *circa-* = about; *-dia* = a day) that is established by the suprachiasmatic nucleus of the hypothalamus (see **Figure 14.10**). A person who is awake is in a state of readiness and is able to react consciously to various stimuli. EEG recordings show that the cerebral cortex is very active during wakefulness; fewer impulses arise during most stages of sleep.

The Role of the Reticular Activating System in Awakening

How does your nervous system make the transition between these two states? Because stimulation of some of its parts increases activity of the cerebral cortex, a portion of the reticular formation is known as the **reticular activating system (RAS)** (see **Figure 14.7c**). When this area is active, many nerve impulses are transmitted to widespread areas of the cerebral cortex, both directly and via the thalamus. The effect is a generalized increase in cortical activity.

Arousal, or awakening from sleep, also involves increased activity in the RAS. For arousal to occur, the RAS must be stimulated. Many sensory stimuli can activate the RAS: painful stimuli detected by nociceptors, touch and pressure on the skin, movement of the limbs, bright light, or the buzz of an alarm clock. Once the RAS is activated, the cerebral cortex is also activated, and arousal occurs. The result is a state of wakefulness called **consciousness** (KON-shus-nes). Notice in **Figure 14.7c** that even though the RAS receives input from somatic sensory receptors, the eyes, and the ears, there is no input from olfactory receptors; even strong odors may fail to cause arousal. People who die in house fires usually succumb to smoke inhalation without awakening. For this reason, all sleeping areas should have a nearby smoke detector that emits a loud alarm. A vibrating pillow or flashing light can serve the same purpose for those who are hearing impaired.

Sleep

Sleep is a state of altered consciousness or partial unconsciousness from which an individual can be aroused. Although it is essential, the exact functions of sleep are still unclear. Sleep deprivation impairs attention, learning, and performance. Normal sleep consists of two components: non-rapid eye movement (NREM) sleep and rapid eye movement (REM) sleep.

NREM sleep consists of four gradually merging stages:

1. *Stage 1* is a transition stage between wakefulness and sleep that normally lasts 1–7 minutes. The person is relaxed with eyes closed and has fleeting thoughts. People awakened during this stage often say they have not been sleeping.
2. *Stage 2* or *light sleep* is the first stage of true sleep. In it, a person is easy to awaken. Fragments of dreams may be experienced, and the eyes may slowly roll from side to side.
3. *Stage 3* is a period of moderately deep sleep. Body temperature and blood pressure decrease, and it is a little more difficult to awaken the person. This stage occurs about 20 minutes after falling asleep.
4. *Stage 4* is the deepest level of sleep. Although brain metabolism decreases significantly and body temperature drops slightly at this time, most reflexes are intact, and muscle tone is decreased only slightly. During this stage, it is very difficult to awaken a person.

Several physiological changes occur during NREM sleep. There are decreases in heart rate, respiratory rate, and blood pressure. Muscle tone also decreases, out only slightly. As a result, there is a moderate amount of muscle tone during NREM sleep, which allows the sleeping person to shift body positions while in bed. Dreaming sometimes takes place during NREM sleep but only occasionally. You will soon learn that most dreaming occurs during REM sleep. When dreaming does occur during NREM sleep, the dreams are usually less vivid, less emotional, and more logical than REM dreams.

During **REM sleep**, the eyes move rapidly back and forth under closed eyelids. REM sleep is also known as *paradoxical sleep* because EEG readings taken during this time show high-frequency, small-amplitude waves, which are similar to those of a person who is awake. Surprisingly, neuronal activity is high during REM sleep—brain blood flow and oxygen use are actually higher during REM sleep than during intense mental or physical activity while awake! In spite of this high amount of neuronal activity, it is even more difficult to awaken a person during REM sleep than during any of the stages of NREM sleep.

REM sleep is associated with several physiological changes. For example, heart rate, respiratory rate, and blood pressure increase during REM sleep. In addition, most somatic motor neurons are inhibited during REM sleep, which causes a significant decrease in muscle tone and even paralyzes the skeletal muscles. The main exceptions to this inhibition are those somatic motor neurons that govern breathing and eye movements. REM sleep is also the period when most dreaming occurs. Brain imaging studies on people going through REM sleep reveal that there is increased activity in both the visual association area (which is involved in recognition of visual images) and limbic system (which plays a major role in generation of emotions) and decreased activity in the prefrontal cortex (which is concerned with reasoning). These studies help to explain why dreams during REM sleep are often full of vivid imagery, emotional responses, and situations that may be illogical or even bizarre. Erection of the penis and enlargement of the clitoris may also occur during REM sleep, even when dream content is not sexual. The presence of penile erections during REM sleep in a man with erectile dysfunction (inability to attain an erection while awake) indicates that his problem has a psychological, rather than a physical cause.

Intervals of NREM and REM sleep alternate throughout the night. Initially, a person falls asleep by sequentially going through the stages of NREM sleep (from stage 1 to stage 4) in about 45 minutes. Then the person goes through the stages of NREM sleep in reverse order (from stage 4 to stage 1) in about the same amount of time before entering a period of REM sleep. Afterward, the person again descends through the stages of NREM sleep, and then ascends back through the stages of NREM sleep to enter another period of REM sleep. During a typical 8-hour sleep period, there are four or five of these NREM-to-REM cycles. The first episode of REM sleep lasts 10–20 minutes. REM periods, which occur approximately every 90 minutes, gradually lengthen, with the final one lasting about 50 minutes. In adults, REM sleep totals 90–120 minutes during a typical 8-hour sleep period. As a person ages, the average total time spent sleeping decreases, and the percentage of REM sleep declines. As much as 50% of an infant's sleep is REM sleep, as opposed to 35% for 2-year-olds and 25% for adults. Although we do not yet understand the function of REM sleep, the high percentage of REM sleep in infants and children is thought to be important for the maturation of the brain.

Different parts of the brain mediate NREM and REM sleep. NREM sleep is induced by **NREM sleep centers** in the hypothalamus and basal forebrain, whereas REM sleep is promoted by a **REM sleep center** in the pons and midbrain. Several lines of evidence suggest the existence of sleep-inducing chemicals in the brain. One apparent sleep-inducer is adenosine, which accumulates during periods of high usage of ATP (adenosine triphosphate) by the nervous system. Adenosine inhibits neurons of the RAS that participate in arousal. Adenosine binds to specific receptors, called A1 receptors, and inhibits certain cholinergic (acetylcholine-releasing) neurons of the RAS that participate in arousal. Thus, activity in the RAS during sleep is low due to the inhibitory effect of adenosine. Caffeine (in coffee) and theophylline (in tea)—substances known for their ability to maintain wakefulness—bind to and block the A1 receptors, preventing adenosine from binding and inducing sleep.

Sleep is essential to the normal functioning of the body. Studies have shown that sleep deprivation impairs attention, memory, performance, and immunity; if the lack of sleep lasts long enough, it can lead to mood swings, hallucinations, and even death. Although it is essential, the exact functions of sleep are still unclear. There has been considerable debate in the scientific community about the importance of sleep, but some proposed functions of sleep are widely accepted: (1) restoration, providing time for the body to repair itself; (2) consolidation of memories; (3) enhancement of immune system function; and (4) maturation of the brain.

Coma Recall that sleep is state of unconsciousness from which an individual can be aroused by stimuli. By contrast, a **coma** is a state of unconsciousness in which an individual has little or no response to stimuli. Causes of coma include head injuries, damages to the reticular activating system (RAS), brain infections, alcohol intoxication, and drug overdoses. If brain damage is minor or reversible, a person may come out of a coma and recover fully; if brain damage is severe and irreversible, recovery is unlikely.

After a few weeks of being in a coma, some patients enter into a **persistent vegetative state** in which the patient has normal sleep–wake cycles but does not have an awareness of the surroundings. Individuals in this state are unable to speak or to respond to commands. They may smile, laugh, or cry, but do not understand the meaning of these actions.

It is important to point out that people who are in a coma or a persistent vegetative state are not brain dead because their EEGs still exhibit waveform activity. One of the criteria used to confirm that brain death has occurred is the absence of brain waves (flat EEG).

See Clinical Connection: Sleep Disorders

Learning and Memory

Without memory, we would repeat mistakes and be unable to learn. Similarly, we would not be able to repeat our successes or accomplishments, except by chance. Although both learning and memory have been studied extensively, we still have no completely satisfactory explanation for how we recall information or how we remember events. However, we do know something about how information is acquired and stored, and it is clear that there are different categories of memory.

Learning is the ability to acquire new information or skills through instruction or experience. There are two main categories of learning associative learning and nonassociative learning. **Associative learning** occurs when a connection is made between two stimuli. The Russian physiologist Ivan Pavlov provided a classic example of associative learning when he observed that ringing a bell stimulated the salivation reflex in dogs. When he first began this experiment, Pavlov rang the bell and then provided food for the dogs. The presence of the food caused the dogs to salivate. After repeating this activity several times, Pavlov observed that the dogs would still salivate even if he did not provide them with any food, which indicated that the dogs learned to associate food with the bell ringing. **Nonassociative learning** occurs when repeated exposure to a single stimulus causes a change in behavior. There are two types of nonassociative learning; habituation and sensitization. In **habituation**, repeated exposure to an irrelevant stimulus causes a *decreased* behavioral response. For example, when you first hear a loud sound, it may make you jump. However, if this loud sound occurs over and over again, you may eventually stop paying attention to it. Habituation demonstrates that an animal has learned to ignore an unimportant stimulus. In **sensitization**, repeated exposure to a noxious stimulus causes an *increased* behavioral response. For example, if a limb is damaged repeatedly by a painful stimulus, the flexor (withdrawal) reflex for the affected limb becomes more vigorous. Sensitization demonstrates that an animal has learned to respond more quickly to a harmful stimulus.

Memory is the process by which information acquired through learning is stored and retrieved. There are two main types of memory: declarative memory and procedural memory. **Declarative (explicit) memory** is the memory of experiences that can be verbalized (declared) such as facts, events, objects, names, and places. This type of memory requires conscious recall and is stored in the association areas of the cerebral cortex. For example, visual memories are stored in the visual association area, and auditory memories are stored in the auditory association area. **Procedural (implicit) memory** is the memory of motor skills, procedures, and rules. Examples include riding a bike, serving a tennis ball, and performing the steps of your favorite dance. This type of memory does not require conscious recall, and it is stored in the basal nuclei, cerebellum, and premotor area.

Memory, whether declarative or procedural, occurs in stages over a period of time. **Short-term memory** is the temporary ability to recall a few pieces of information for seconds to minutes. One example is when you look up an unfamiliar telephone number, cross the room to the phone, and then dial the new number. If the number has no special significance, it is usually forgotten within a few seconds. Information in short-term memory may later be transformed into a more permanent type of memory, called **long-term memory**, which lasts from days to years. For example, if you use that new telephone number often enough, it becomes part of long-term memory. Although the brain receives many stimuli, you normally pay attention to only a few of them at a time. It has been estimated that only 1% of all of the information that comes to your consciousness is stored as long-term memory. Note that memory does not record every detail as if it were a DVR recorder. Even when details are lost, you can often explain the idea or concept using your own words and ways of viewing things.

Some evidence supports the notion that short-term memory depends more on electrical and chemical events in the brain than on structural changes at synapses Several conditions that inhibit the electrical activity of the brain, such as anesthesia, coma, and electroconvulsive therapy (ECT), disrupt short-term memories without altering previously established long-term memories. Studies also suggest that short-term memory may involve a temporary increase in the activity of preexisting synapses, especially those that are components of reverberating circuits. Recall that, in a reverberating circuit, one neuron stimulates a second neuron, which stimulates a third neuron, and so on. Branches from later neurons synapse with earlier ones. This arrangement sends action potentials back through the circuit again and again (see **Figure 12.28c**).

The process by which a short-term memory is transformed into a long-term memory is called **memory consolidation**. The hippocampus plays a major role in the consolidation of declarative memories. It serves as a temporary storage facility for new long-term declarative memories and then transfers these memories to the appropriate areas of the cerebral cortex for permanent storage. A key factor that contributes to memory consolidation is repetition. Therefore, you remember more information if you review every day for an upcoming physiology exam instead of cramming for the exam the night before!

For an experience to become part of long-term memory, it must produce persistent structural and functional changes that represent the experience in the brain. This capability for change associated with learning is termed **plasticity**. It involves changes in individual neurons as well as changes in the strengths of synaptic connections among neurons. For example, electron micrographs of neurons subjected to prolonged, intense activity reveal an increase in the number of presynaptic terminals and enlargement of synaptic end bulbs in presynaptic neurons, as well as an increase in the number of dendritic branches in postsynaptic neurons. Moreover, neurons grow new synaptic end bulbs with increasing gage, presumably because of increased use. Opposite changes occur when neurons are inactive. For example, the visual area of the cerebral cortex of animals that have lost their eyesight becomes thinner.

A phenomenon called **long-term potentiation (LTP)** (pō-ten′-shē-Ā-shun) is believed to underlie some aspects of memory; transmission at some synapses within the hippocampus is enhanced (potentiated) for hours or weeks after a brief period of high-frequency stimulation. The neurotransmitter released is glutamate, which acts on NMDA* glutamate receptors on the postsynaptic neurons. In some cases, induction of LTP depends on the release of nitric oxide (NO) from the postsynaptic neurons after they have been activated by glutamate. The NO in turn diffuses into the presynaptic neurons and causes LTP.

*Named after the chemical *N*-methyl-D-aspartate, which is used to detect this type of glutamate receptor.

See Clinical Connection: Amnesia

Language

Animals as diverse as ants, birds, whales, and humans have developed ways to communicate with members of their own species. Humans use language to communicate with one another. **Language** is a system of vocal sounds and symbols that conveys information. Most commonly it is spoken and/or written.

The cerebral cortex contains two **language areas**—Wernicke's area and Broca's area, which are usually present only in the *left* cerebral hemisphere (see **Figure 14.15**). *Wernicke's area*, an association area found in the temporal lobe, interprets the meaning of written or spoken words. It essentially translates words into thoughts. Wernicke's area receives input from the primary visual area (for written words) and from the primary auditory area (for spoken words). *Broca's area*, a motor area located in the frontal lobe, is active as you translate thoughts into speech. To accomplish this function, Broca's area receives input from Wernicke's area and then generates a motor pattern for activation of muscles needed for the words that you want to say. The motor pattern is transmitted from Broca's area to the primary motor area, which in turn activates the appropriate speech muscles. The contractions of your speech muscles enable you to speak your thoughts.

To further understand how the language areas function, consider the neural pathways that are used when you see or hear a particular word and then say that word:

1. Information about the word is conveyed to Wernicke's area. If the word is written, Wernicke's area receives input about the word from the primary visual area. If the word is spoken, Wernicke's area receives input about the word from the primary auditory area.
2. Once Wernicke's area receives this information, it translates the written or spoken word into the appropriate thought.
3 For a person to say this word, Wernicke's area transmits information about the word to Broca's area.
4. Broca's area receives this input and then develops a motor pattern for activation of the muscles needed to say the word.
5. The motor pattern is conveyed from Broca's area to the primary motor area, which subsequently activates the appropriate muscles of speech. Contraction of the speech muscles allows the word to be spoken.

See Clinical Connection: Aphasia

Review Questions

16.1 Sensation

1. How is sensation different from perception?
2. What is a sensory modality?
3. What is a receptor potential?
4. What is the difference between rapidly adapting and slowly adapting receptors?

16.2 Somatic Sensations

5. Which somatic sensory receptors are encapsulated?
6. Why do some receptors adapt slowly and others adapt rapidly?
7. Which somatic sensory receptors mediate touch sensations?
8. How does fast pain differ from slow pain?
9. What is referred pain, and how is it useful in diagnosing internal disorders?
10. What aspects of muscle function are monitored by muscle spindles and tendon organs?

16.3 Somatic Sensory Pathways

11. What are the functional differences between the posterior column–medial lemniscus pathway, the anterolateral pathway, and the trigeminothalamic pathway?
12. Which body parts have the largest representation in the primary somatosensory area?
13. What type of sensory information is carried in the spinocerebellar tracts?

16.4 Control of Body Movement

14. Trace the path of a motor impulse from the upper motor neurons through the final common pathway.
15. Which parts of the body have the largest representation in the motor cortex? Which have the smallest?
16. Explain why the two main somatic motor pathways are called "direct" and "indirect."
17. Explain the role of the cerebral cortex, basal nuclei, brainstem, and cerebellum in body movement.

16.5 Integrative Functions of the Cerebrum

18. Describe how sleep and wakefulness are related to the reticular activating system (RAS).
19. What are the four stages of non-rapid eye movement (NREM) sleep? How is NREM sleep distinguished from rapid eye movement (REM) sleep?
20. Define memory. What are the three kinds of memory? What is memory consolidation?
21. What is long-term potentiation?
22. What is language?

Critical Thinking Questions

1. When Joni first stepped onto the sailboat, she smelled the tangy sea air and felt the motion of water beneath her feet. After a few minutes, she no longer noticed the smell, but unfortunately she was aware of the rolling motion for hours. What types of receptors are involved in smell and detection of motion? Why did her sensation of smell fade but the rolling sensation remain?
2. Monique sticks her left hand into a hot tub heated to about 43°C (110°F) in order to decide if she wants to enter. Trace the pathway involved in transmitting the sensation of heat from her left hand to the somatosensory area in the cerebral cortex.
3. Marvin has had trouble sleeping. Last night his mother found him sleepwalking and gently led him back to his bed. When Marvin was awakened by his alarm clock the next day, he had no recollection of sleepwalking and, in fact, told his mother about the vivid dreams he had. What specific stages of sleep did Marvin undergo during the night? What neurological mechanism awakened Marvin in the morning?

The Special Senses

CHAPTER **17**

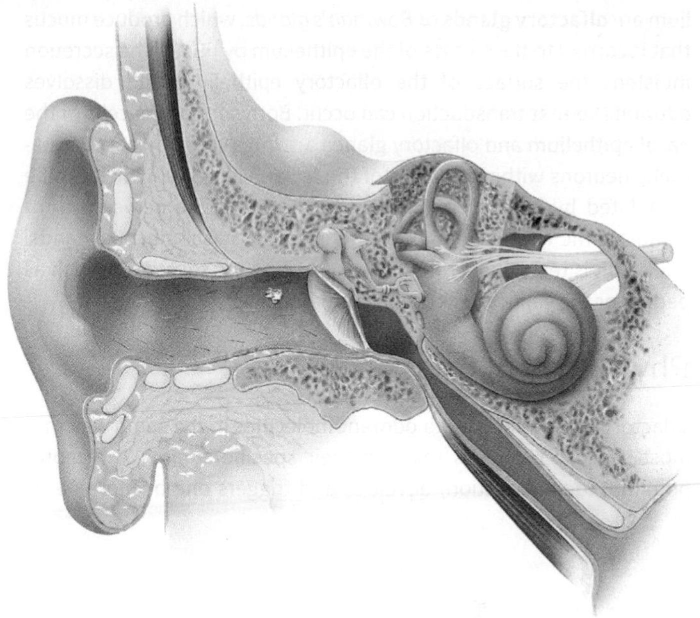

The Special Senses and Homeostasis

Sensory organs have special receptors that allow us to smell, taste, see, hear, and maintain equilibrium or balance. Information conveyed from these receptors to the central nervous system is used to help maintain homeostasis.

Recall from Chapter 16 that the general senses include somatic senses (tactile, thermal, pain, and proprioceptive) and visceral sensations. As you learned in that chapter, receptors for the general senses are scattered throughout the body and are relatively simple in structure. They range from modified dendrites of sensory neurons to specialized structures associated with the ends of dendrites. Receptors for the special senses—smell, taste, vision, hearing, and equilibrium—are anatomically distinct from one another and are concentrated in specific locations in the head. They are usually embedded in the epithelial tissue within complex sensory organs such as the eyes and ears. Neural pathways for the special senses are also more complex than those for the general senses. In this chapter we examine the structure and function of the special sense organs, and the pathways involved in conveying their information to the central nervous system.

Q Did you ever wonder how LASIK is performed?

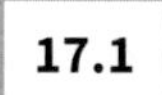

17.1 Olfaction: Sense of Smell

OBJECTIVES

- **Describe** the structure of the olfactory receptors and other cells involved in olfaction.
- **Outline** the neural pathway for olfaction.

Last night as you were studying anatomy and physiology in the lounge, all of a sudden you were surrounded by the smell of freshly baked brownies. When you followed your nose and begged for one, biting into the moist, flavorful treat transported you back 10 years into your mother's kitchen. Both smell and taste are chemical senses; the sensations arise from the interaction of molecules with smell or taste receptors. To be detected by either sense, the stimulating molecules must be dissolved. Because impulses for smell and taste propagate to the limbic system (and to higher cortical areas as well), certain odors and tastes can evoke strong emotional responses or a flood of memories.

Anatomy of Olfactory Receptors

The receptors for the sense of smell or **olfaction** (ōl-FAK-shun; *olfact-* = smell) are located in the olfactory epithelium of the nose. With a total area of 5 cm^2 (a little less than 1 $in.^2$), the **olfactory epithelium** (ōl-FAK-tō-rē) occupies the superior part of the nasal cavity, covering the inferior surface of the cribriform plate and extending along the superior nasal concha (**Figure 17.1a**). The olfactory epithelium consists of three kinds of cells: olfactory receptor cells, supporting cells, and basal cells (**Figure 17.1b**).

Olfactory receptor cells are the first-order neurons of the olfactory pathway. Each olfactory receptor cell is a bipolar neuron with an exposed, knob-shaped dendrite and an axon projecting through the cribriform plate that ends in the olfactory bulb. Extending from the dendrite of an olfactory receptor cell are several nonmotile **olfactory cilia**, which are the sites of olfactory transduction. (Recall that *transduction* is the conversion of stimulus energy into a graded potential in a sensory receptor.) Within the plasma membranes of the olfactory cilia are **olfactory receptor** proteins that detect inhaled chemicals. Chemicals that bind to and stimulate the olfactory receptors in the olfactory cilia are called **odorants**. Olfactory receptor cells respond to the chemical stimulation of an odorant molecule by producing a receptor potential, thus initiating the olfactory response.

Supporting cells are columnar epithelial cells of the mucous membrane lining the nose. They provide physical support, nourishment, and electrical insulation for the olfactory receptor cells and help detoxify chemicals that come in contact with the olfactory epithelium. **Basal cells** are stem cells located between the bases of the supporting cells. They continually undergo cell division to produce new olfactory receptor cells, which live for only about two months before being replaced. This process is remarkable considering that olfactory receptor cells are neurons, and as you have already learned, mature neurons are generally not replaced.

Within the connective tissue that supports the olfactory epithelium are **olfactory glands** or *Bowman's glands*, which produce mucus that is carried to the surface of the epithelium by ducts. The secretion moistens the surface of the olfactory epithelium and dissolves odorants so that transduction can occur. Both supporting cells of the nasal epithelium and olfactory glands are innervated by parasympathetic neurons within branches of the facial (VII) nerve, which can be stimulated by certain chemicals. Impulses in these nerves in turn stimulate the lacrimal glands in the eyes and nasal mucous glands. The result is tears and a runny nose after inhaling substances such as pepper or the vapors of household ammonia.

Physiology of Olfaction

Olfactory receptors react to odorant molecules in the same way that most sensory receptors react to their specific stimuli: A receptor potential (depolarization) develops and triggers one or more nerve impulses. This process, called *olfactory transduction*, occurs in the following way (Figure 17.2): Binding of an odorant to an olfactory receptor protein in an olfactory cilium stimulates a membrane protein called a *G protein*. The G protein, in turn, activates the enzyme *adenylyl cyclase* to produce a substance called *cyclic adenosine monophosphate (cAMP)*, a type of second messenger (see Section 18.4). The cAMP opens a cation channel that allows Na^+ and Ca^{2+} to enter the cytosol, which causes a depolarizing receptor potential to form in the membrane of the olfactory receptor cell. If the depolarization reaches threshold, an action potential is generated along the axon of the olfactory receptor cell.

The human nose contains about 10 million olfactory receptors, of which there are about 400 different functional types. Each type of olfactory receptor can react to only a select group of odorants. Only one type of receptor is found in any given olfactory receptor cell. Therefore, 400 different types of olfactory receptor cells are present in the olfactory epithelium.

FIGURE 17.1 Olfactory epithelium and olfactory pathway. (a) Location of olfactory epithelium in nasal cavity. (b) Details of olfactory epithelium. (c) Histology of the olfactory epithelium. (d) Olfactory pathway.

The olfactory epithelium consists of olfactory receptor cells, supporting cells, and basal cells.

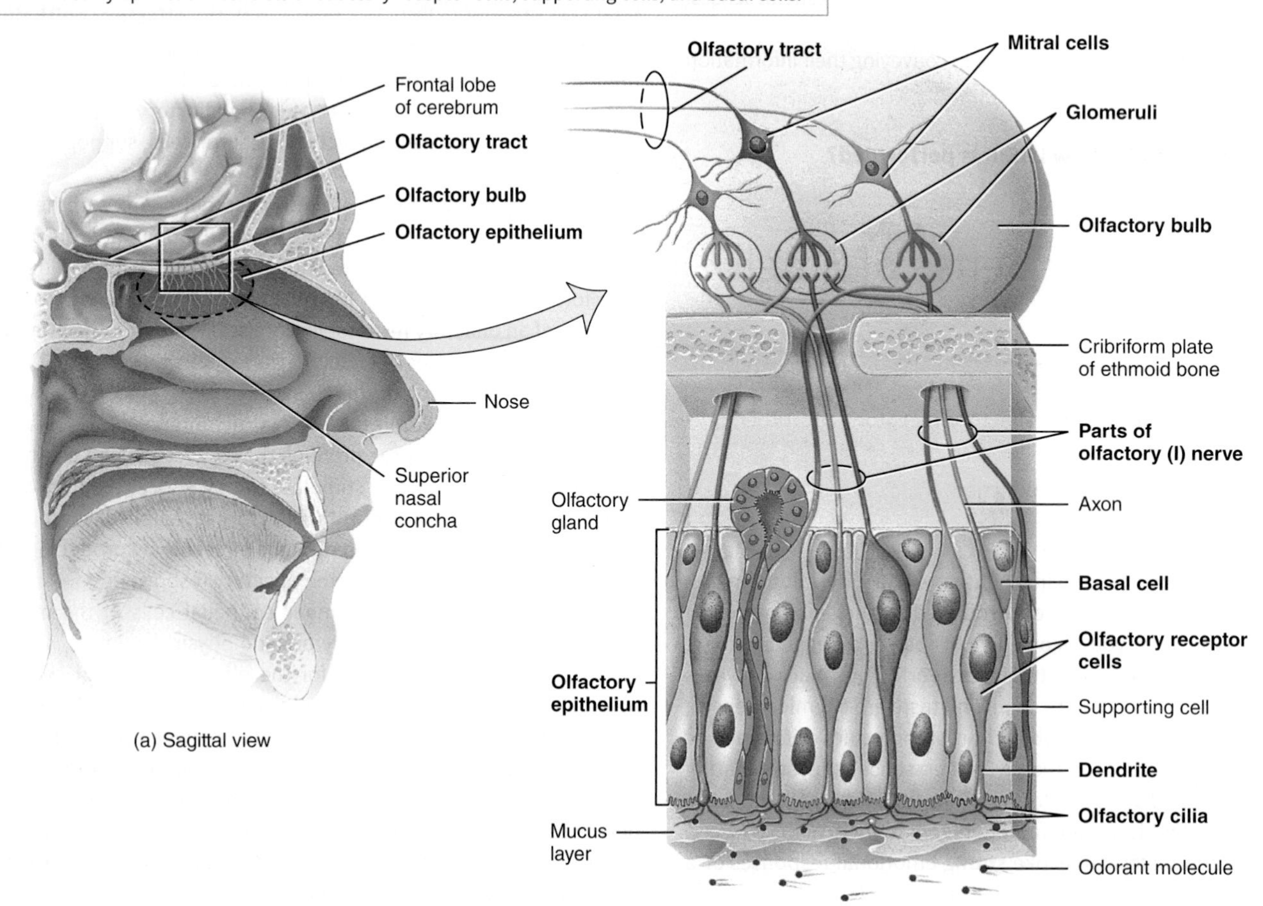

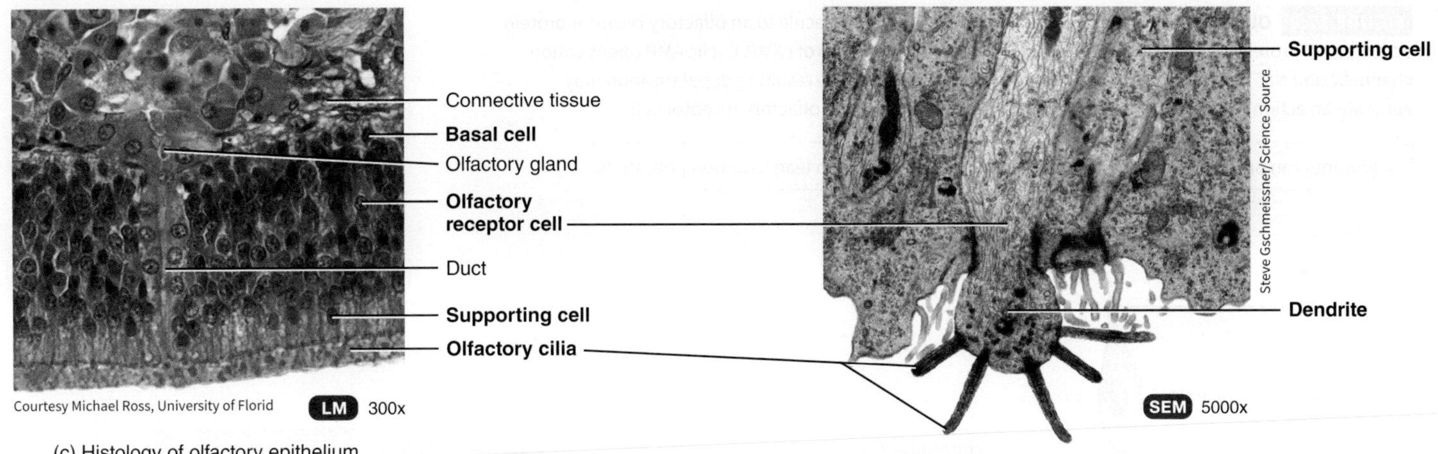

(c) Histology of olfactory epithelium

Olfactory receptor cell (blue)

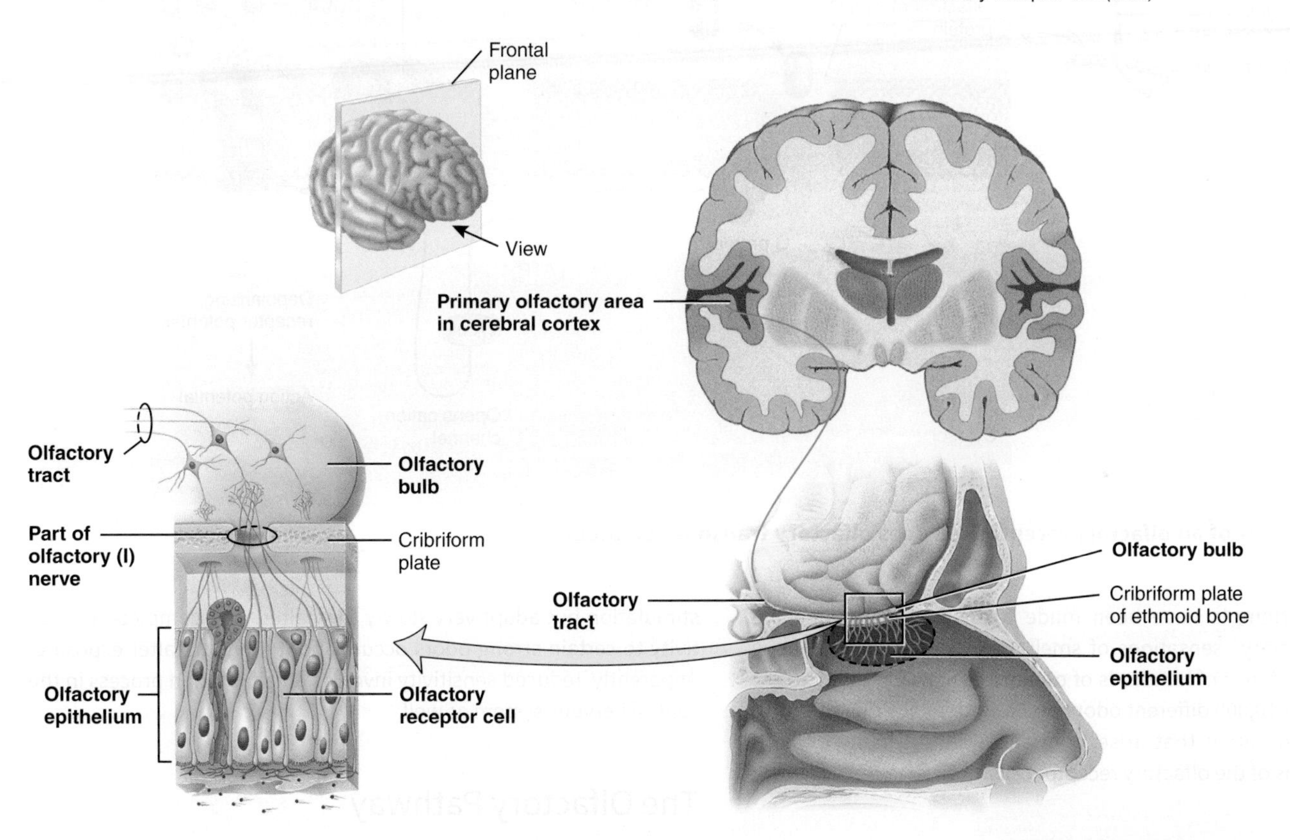

(d) Olfactory pathway

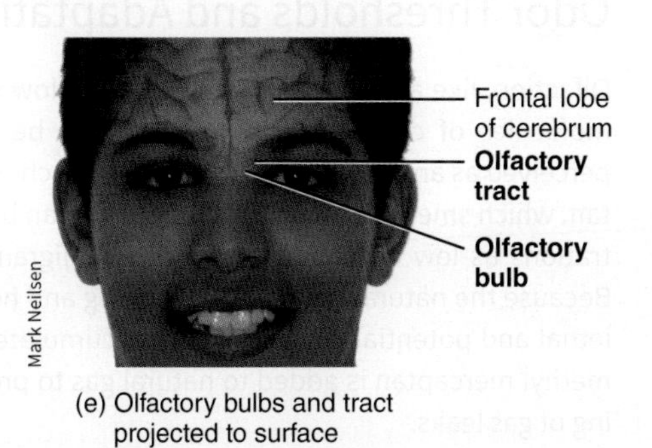

(e) Olfactory bulbs and tract projected to surface

Q What is the life span of an olfactory receptor cell?

FIGURE 17.2 Olfactory transduction. Binding of an odorant molecule to an olfactory receptor protein activates a G protein and adenylyl cyclase, resulting in the production of cAMP. Cyclic AMP opens cation channels, and Na^+ and Ca^{2+} ions enter the olfactory receptor cell. The resulting depolarization may generate an action potential, which propagates along the axon of the olfactory receptor cell.

Odorants can produce depolarizing receptor potentials, which can lead to action potentials.

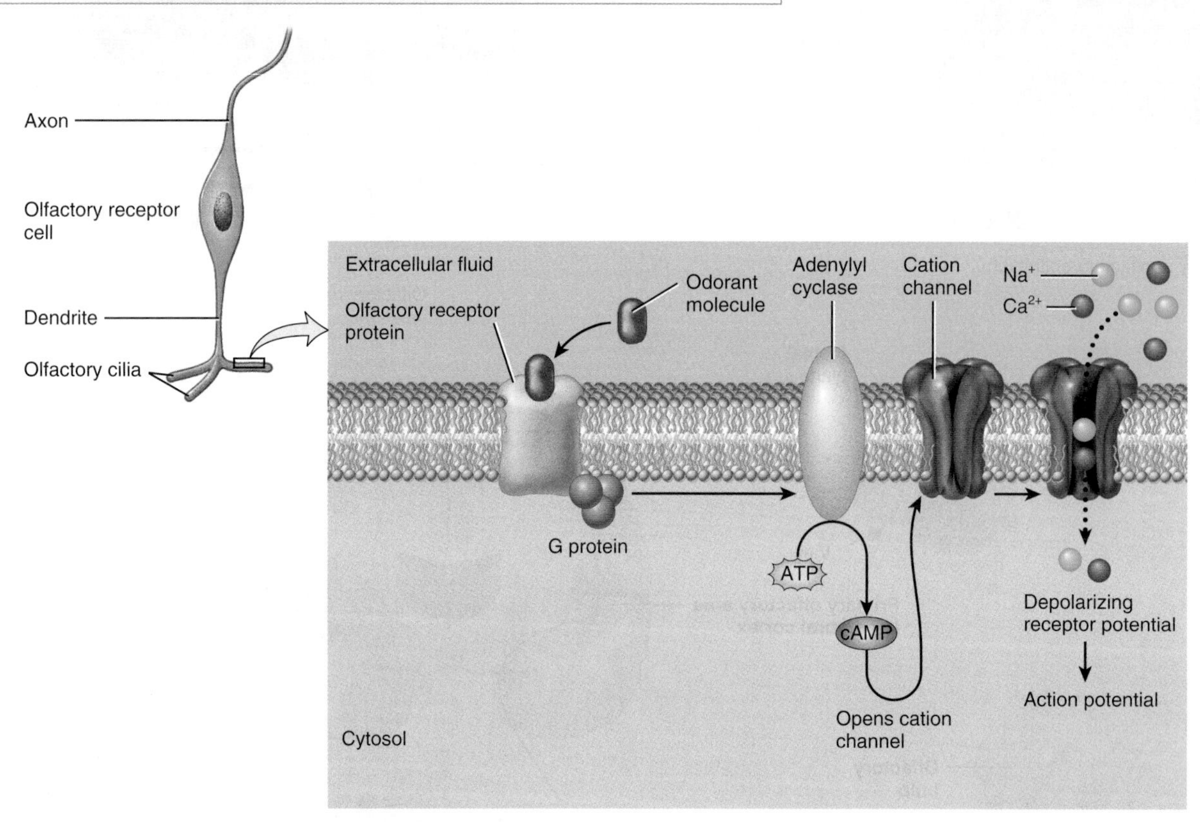

Q In which part of an olfactory receptor cell does olfactory transduction occur?

Many attempts have been made to distinguish among and classify "primary" sensations of smell. Genetic evidence now suggests the existence of hundreds of primary odors. Our ability to recognize about 10,000 different odors probably depends on patterns of activity in the brain that arise from activation of many different combinations of the olfactory receptor cells.

Odor Thresholds and Adaptation

Olfaction, like all the special senses, has a low threshold. Only a few molecules of certain substances need to be present in air to be perceived as an odor. A good example is the chemical methyl mercaptan, which smells like rotten cabbage and can be detected in concentrations as low as 1/25 billionth of a milligram per milliliter of air. Because the natural gas used for cooking and heating is odorless but lethal and potentially explosive if it accumulates, a small amount of methyl mercaptan is added to natural gas to provide olfactory warning of gas leaks.

Adaptation (decreasing sensitivity) to odors occurs rapidly. Olfactory receptors adapt by about 50% in the first second or so after stimulation but adapt very slowly thereafter. Still, complete insensitivity to certain strong odors occurs about a minute after exposure. Apparently, reduced sensitivity involves an adaptation process in the central nervous system as well.

The Olfactory Pathway

On each side of the nose, some 40 or so bundles of axons of olfactory receptor cells form the right and left **olfactory (I) nerves** (see **Figure 17.1a**). The olfactory nerves pass through the olfactory foramina of the cribriform plate of the ethmoid bone and extend to parts of the brain known as the **olfactory bulbs**, which contain ball-like arrangements called **glomeruli** (glō-MER-ū-lī = little balls; singular is *glomerulus*). Within each glomerulus, axons of olfactory receptor cells converge onto **mitral cells**—the second order neurons of the olfactory pathway. Each glomerulus receives input from only one type of olfactory receptor. This allows the mitral cells of a particular glomerulus to convey information about a select group of odorants to the remaining parts of the olfactory pathway. The axons of the mitral cells form the **olfactory tract**. Some of the axons of the olfactory tract project to the

primary olfactory area in the temporal lobe of the cerebral cortex, where conscious awareness of smell occurs (**Figure 17.1d**). Olfactory sensations are the only sensations that reach the cerebral cortex without first synapsing in the thalamus. Other axons of the olfactory tract project to the limbic system; these neural connections account for our emotional responses to odors. From the olfactory cortex, a pathway extends via the thalamus to the **orbitofrontal cortex** in the frontal lobe, where odor identification and discrimination occur (see area II in **Figure 14.15**). People who suffer damage in this area have difficulty identifying different odors. Positron emission tomography (PET) studies suggest some degree of hemispheric lateralization: The orbitofrontal cortex of the *right* hemisphere exhibits greater activity during olfactory processing than the corresponding area in the *left* hemisphere.

See Clinical Connection: Hyposmia

17.2 Gustation: Sense of Taste

OBJECTIVES

- **Identify** the five primary tastes.
- **Explain** the process of taste transduction.
- **Describe** the gustatory pathway to the brain.

Like olfaction, **gustation** (gus-TĀ-shun), or taste, is a chemical sense. However, gustation is much simpler than olfaction in that only five primary tastes can be distinguished: *salty*, *sour*, *sweet*, *bitter*, and *umami* (oo-MAH-mē). Salty taste is caused by the presence of sodium ions (Na^+) in food. A common dietary source of Na^+ is NaCl (table salt). Sour taste is produced by hydrogen ions (H^+) released from acids. Lemons have a sour taste because they contain citric acid. Sweet taste is elicited by sugars such as glucose, fructose, and sucrose and by artificial sweeteners such as saccharin, aspartame, and sucralose. Bitter taste is caused by a wide variety of substances, including caffeine, morphine, and quinine. In addition, many poisonous substances like strychnine have a bitter taste. When something tastes bitter, a natural response is to spit it out, a reaction that serves to protect you from ingesting potentially harmful substances. The umami taste, first reported by Japanese scientists, is described as "meaty" or "savory." It is elicited by amino acids (especially glutamate) that are present in food. This is the reason why the additive monosodium glutamate (MSG) is used as a flavor enhancer in many foods. All other flavors, such as chocolate, pepper, and coffee, are combinations of the five primary tastes, plus any accompanying olfactory, tactile, and thermal sensations. Odors from food can pass upward from the mouth into the nasal cavity, where they stimulate olfactory receptors. Because olfaction is much more sensitive than taste, a given concentration of a food substance may stimulate the olfactory system thousands of times more strongly than it stimulates the gustatory system. When you have a cold or are suffering from allergies and cannot taste your food, it is actually olfaction that is blocked, not taste.

Anatomy of Taste Buds and Papillae

The receptors for sensations of taste are located in the taste buds (**Figure 17.3**). Most of the nearly 10,000 taste buds of a young adult are on the tongue, but some are found on the soft palate (posterior portion of the roof of the mouth), pharynx (throat), and epiglottis (cartilage lid over voice box). The number of taste buds declines with age. Each **taste bud** is an oval body consisting of three kinds of epithelial cells: supporting cells, gustatory receptor cells, and basal cells (see **Figure 17.3c**). The **supporting cells** surround about 50 **gustatory receptor cells** (GUS-ta-tōr-ē) in each taste bud. **Gustatory microvilli** (*gustatory hairs*) project from each gustatory receptor cell to the external surface through the **taste pore**, an opening in the taste bud. **Basal cells**, stem cells found at the periphery of the taste bud near the connective tissue layer, produce supporting cells, which then develop into gustatory receptor cells. Each gustatory receptor cell has a life span of about 10 days. This is why it does not take taste receptors on the tongue too long to recover from being burned by that too hot cup of coffee or cocoa. At their base, the gustatory receptor cells synapse with dendrites of the first-order neurons that form the first part of the gustatory pathway. The dendrites of each first-order neuron branch profusely and contact many gustatory receptor cells in several taste buds.

Taste buds are found in elevations on the tongue called **papillae** (pa-PIL-ē; singular is *papilla*), which increase the surface area and provide a rough texture to the upper surface of the tongue (**Figure 17.3a, b**). Three types of papillae contain taste buds:

1. About 12 very large, circular **vallate papillae** (VAL-āt = wall-like) or *circumvallate papillae* form an inverted V-shaped row at the back of the tongue. Each of these papillae houses 100–300 taste buds.
2. **Fungiform papillae** (FUN-ji-form = mushroomlike) are mushroom-shaped elevations scattered over the entire surface of the tongue that contain about five taste buds each.
3. **Foliate papillae** (FO-lē-āt = leaflike) are located in small trenches on the lateral margins of the tongue, but most of their taste buds degenerate in early childhood.

In addition, the entire surface of the tongue has **filiform papillae** (FIL-i-form = threadlike). These pointed, threadlike structures contain tactile receptors but no taste buds. They increase friction between the tongue and food, making it easier for the tongue to move food in the oral cavity.

Physiology of Gustation

Chemicals that stimulate gustatory receptor cells are known as **tastants**. Once a tastant is dissolved in saliva, it can make contact

FIGURE 17.3 **The relationship of gustatory receptor cells in taste buds to tongue papillae.**

Gustatory receptor cells are located in taste buds.

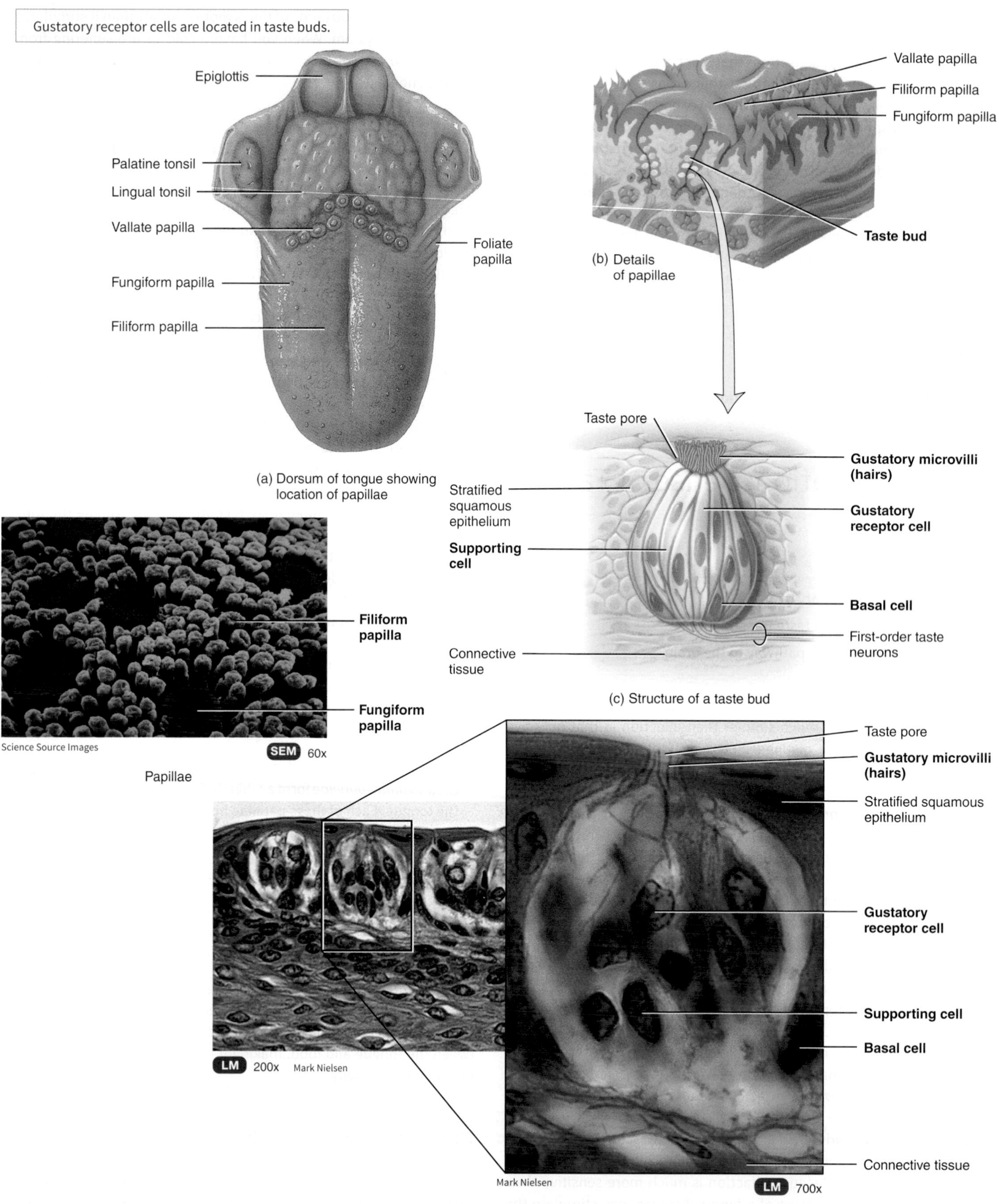

(d) Histology of a taste bud from a vallate papilla

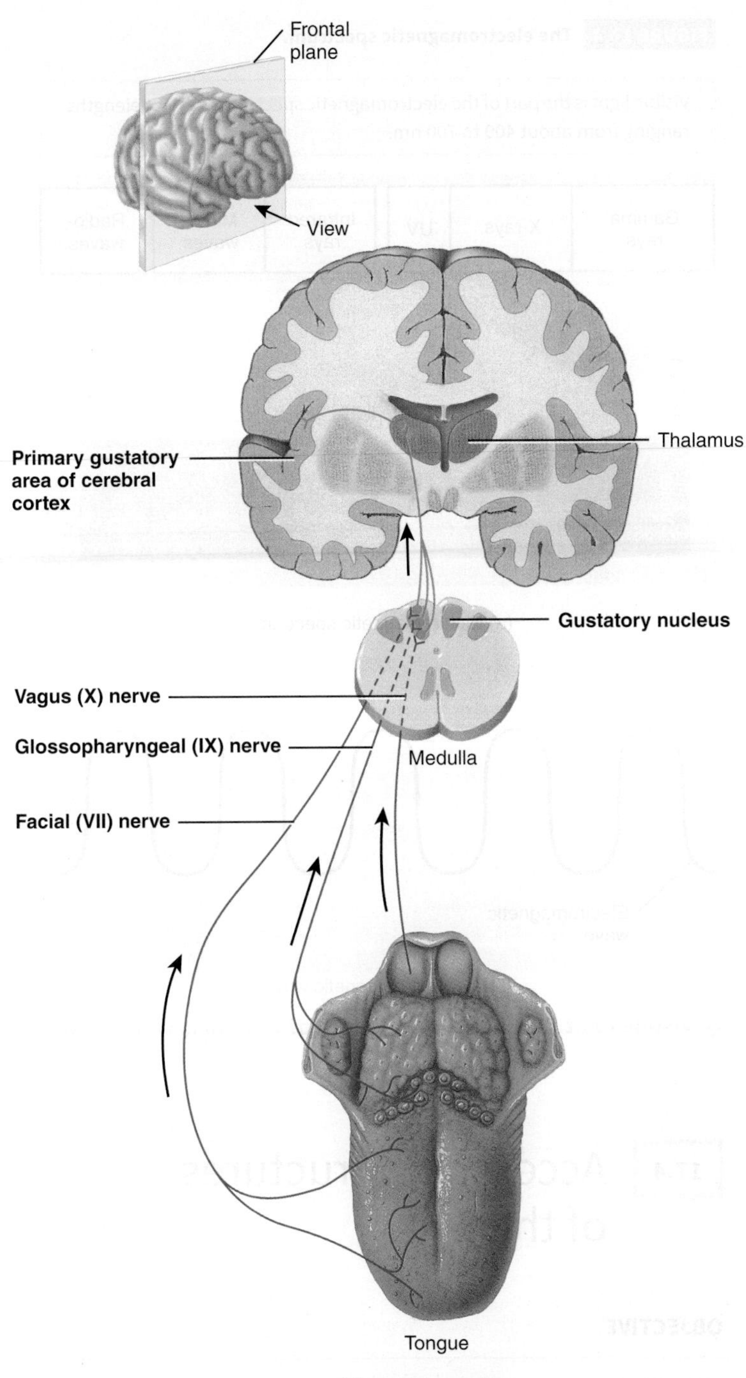

(e) Gustatory pathway

Q What role do basal cells play in taste buds?

with the plasma membranes of the gustatory microvilli, which are the sites of taste transduction. The result is a depolarizing receptor potential that stimulates exocytosis of synaptic vesicles from the gustatory receptor cell. In turn, the liberated neurotransmitter molecules trigger graded potentials that produce nerve impulses in the first-order sensory neurons that synapse with gustatory receptor cells.

The receptor potential arises differently for different tastants. The sodium ions (Na^+) in a salty food enter gustatory receptor cells via Na^+ channels in the plasma membrane. The accumulation of Na^+ inside the cell causes depolarization, which leads to release of neurotransmitter. The hydrogen ions (H^+) in sour tastants flow into gustatory receptor cells via H^+ channels. Again, the result is depolarization and the liberation of neurotransmitter.

Other tastants, responsible for stimulating sweet, bitter, and umami tastes, do not themselves enter gustatory receptor cells. Rather, they bind to receptors on the plasma membrane that are linked to G proteins. The G proteins then activate enzymes that produce the second messenger *inositol trisphosphate (IP_3)* (in-Ō-si-tōl tris-FOS-fāt). IP_3 in turn ultimately causes depolarization of the gustatory receptor cell and release of neurotransmitter.

An individual gustatory receptor cell responds to only one type of tastant. This is due to the fact that the membrane of a gustatory receptor cell has either ion channels or receptors for only one of the primary tastes. For example, a gustatory receptor cell that detects bitter tastants only has receptors for these tastants and cannot respond to salty, sour, sweet, or umami tastants. Thus, each gustatory receptor cell is "tuned" to detect a specific primary taste, and this segregation is maintained as the specific taste information is relayed into the brain. It is also important to mention that a given taste bud contains gustatory receptor cells for each type of tastant, allowing all of the primary tastes to be detected in all parts of the tongue.

If all tastants cause release of neurotransmitter from gustatory receptor cells, why do foods taste different? The answer to this question is thought to lie in the patterns of activity in the brain that arise when gustatory receptor cells are activated. Different tastes arise from activation of different combinations of gustatory receptor cells. For example, the tastants in chocolate activate a certain combination of gustatory receptor cells, and the resultant pattern of activity in the brain is interpreted as the flavor chocolate. By contrast, the tastants in vanilla activate a different combination of gustatory receptor cells, and the resultant pattern of activity in the brain is interpreted as the flavor vanilla.

Taste Thresholds and Adaptation

The threshold for taste varies for each of the primary tastes. The threshold for bitter substances, such as quinine, is lowest. Because poisonous substances often are bitter, the low threshold (or high sensitivity) may have a protective function. The threshold for sour substances (such as lemon), as measured by using hydrochloric acid, is somewhat higher. The thresholds for salty substances (represented by sodium chloride), and for sweet substances (as measured by using sucrose) are similar, and are higher than those for bitter or sour substances.

Complete adaptation to a specific taste can occur in 1–5 minutes of continuous stimulation. Taste adaptation is due to changes that occur in the taste receptors, in olfactory receptors, and in neurons of the gustatory pathway in the CNS.

The Gustatory Pathway

Three cranial nerves contain axons of the first-order gustatory neurons that innervate the taste buds. The **facial (VII) nerve** serves taste buds in the anterior two-thirds of the tongue; the **glossopharyngeal**

(IX) nerve serves taste buds in the posterior one-third of the tongue; and the **vagus (X) nerve** serves taste buds in the throat and epiglottis (**Figure 17.3e**). From the taste buds, nerve impulses propagate along these cranial nerves to the **gustatory nucleus** in the medulla oblongata. From the medulla, some axons carrying taste signals project to the **limbic system** and the **hypothalamus**; others project to the **thalamus**. Taste signals that project from the thalamus to the **primary gustatory area** in the insula of the cerebral cortex (see area 43 in **Figure 14.15**) give rise to the conscious perception of taste and discrimination of taste sensations.

See Clinical Connection: Taste Aversion

17.3 Vision: An Overview

OBJECTIVES

- **Discuss** why vision is important.
- **Define** visible light.

Vision, the act of seeing, is extremely important to human survival because it allows us to view potentially dangerous objects in our surroundings. More than half the sensory receptors in the human body are located in the eyes, and a large part of the cerebral cortex is devoted to processing visual information. In this section of the chapter you will learn about electromagnetic radiation and visible light. In sections 17.4 through 17.6, you will learn about the accessory structures of the eye, the anatomy of the eyeball itself, and the physiology of vision. **Ophthalmology** (of-thal-MOL-ō-jē; *ophthalmo-* = eye; *-logy* = study of) is the science that deals with the eyes and their disorders.

Electromagnetic radiation (e-lek′-trō-mag′-NET-ik) is energy in the form of waves that radiates from the sun. There are many types of electromagnetic radiation, including gamma rays, x-rays, UV rays, visible light, infrared radiation, microwaves, and radio waves. This range of electromagnetic radiation is known as the **electromagnetic spectrum** (**Figure 17.4**). The distance between two consecutive peaks of an electromagnetic wave is the *wavelength*. Wavelengths range from short to long; for example, gamma rays have wavelengths smaller than a nanometer, and most radio waves have wavelengths greater than a meter.

The eyes are responsible for the detection of **visible light**, the part of the electromagnetic spectrum with wavelengths ranging from about 400 to 700 nm. Visible light exhibits colors: The color of visible light depends on its wavelength. For example, light that has a wavelength of 400 nm is violet, and light that has a wavelength of 700 nm is red. An object can absorb certain wavelengths of visible light and reflect others; the object will appear the color of the wavelength that is reflected. For example, a green apple appears green because it reflects mostly green light and absorbs most other wavelengths of visible light. An object appears white because it reflects all wavelengths of visible light. An object appears black because it absorbs all wavelengths of visible light.

FIGURE 17.4 The electromagnetic spectrum.

Visible light is the part of the electromagnetic spectrum with wavelengths ranging from about 400 to 700 nm.

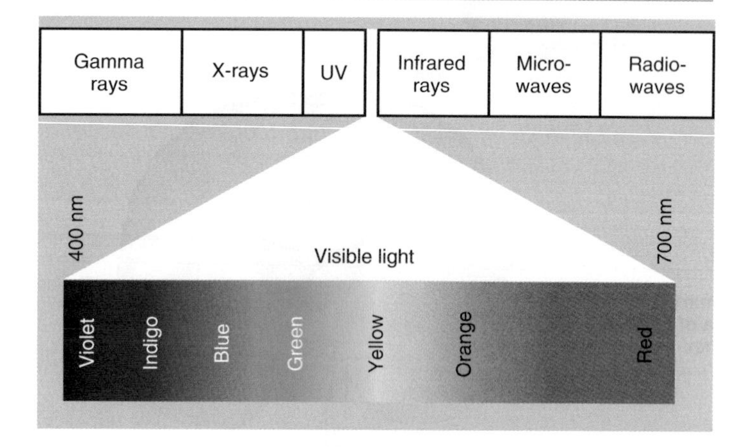

(a) Electromagnetic spectrum

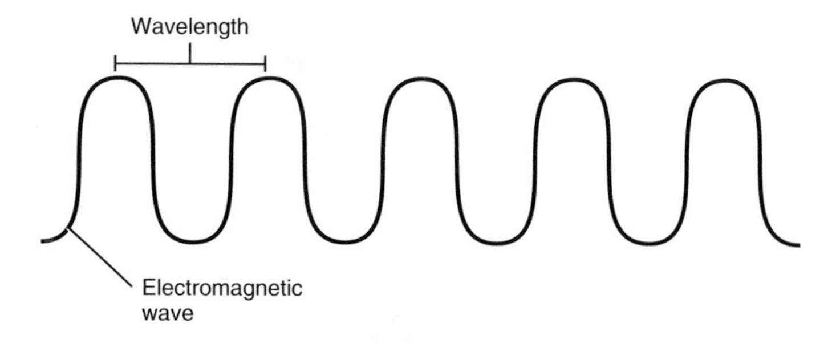

(b) An electromagnetic wave

Q Visible light that has a wavelength of 700 nm is what color?

17.4 Accessory Structures of the Eye

OBJECTIVE

- **Identify** the accessory structures of the eye.

The **accessory structures of the eye** include the eyelids, eyelashes, eyebrows, the lacrimal (tear-producing) apparatus, and extrinsic eye muscles.

Eyelids

The upper and lower **eyelids**, or *palpebrae* (PAL-pe-brē; singular is *palpebra*), shade the eyes during sleep, protect the eyes from excessive light and foreign objects, and spread lubricating secretions over

FIGURE 17.5 **Surface anatomy of the right eye.**

The palpebral fissure is the space between the upper and lower eyelids that exposes the eyeball.

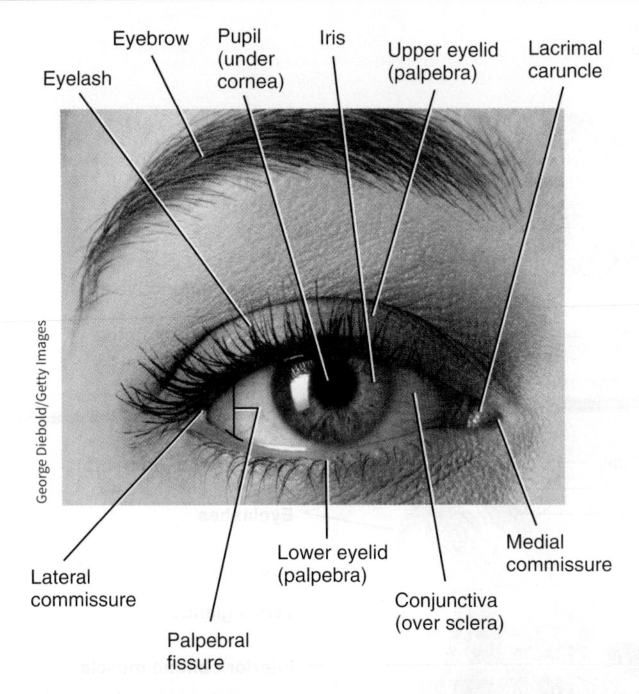

Q Which structure shown here is continuous with the inner lining of the eyelids?

the eyeballs (**Figure 17.5**). The upper eyelid is more movable than the lower and contains in its superior region the **levator palpebrae superioris** muscle (see **Figure 17.6a**). Sometimes a person may experience an annoying *twitch* in an eyelid, an involuntary quivering similar to muscle twitches in the hand, forearm, leg, or foot. Twitches are almost always harmless and usually last for only a few seconds. They are often associated with stress and fatigue. The space between the upper and lower eyelids that exposes the eyeball is the **palpebral fissure** (PAL-pe-bral). Its angles are known as the **lateral commissure** (KOM-i-shur), which is narrower and closer to the temporal bone, and the **medial commissure**, which is broader and nearer the nasal bone. In the medial commissure is a small, reddish elevation, the **lacrimal caruncle** (KAR-ung-kul), which contains sebaceous (oil) glands and sudoriferous (sweat) glands. The whitish material that sometimes collects in the medial commissure comes from these glands.

From superficial to deep, each eyelid consists of epidermis, dermis, subcutaneous tissue, fibers of the orbicularis oculi muscle, a tarsal plate, tarsal glands, and conjunctiva. The **tarsal plate** is a thick fold of connective tissue that gives form and support to the eyelids. Embedded in each tarsal plate is a row of elongated modified sebaceous glands, known as **tarsal glands** or *Meibomian glands* (mī-BŌ-mē-an), that secrete a fluid that helps keep the eyelids from adhering to each other (**Figure 17.6a**). Infection of the tarsal glands produces a tumor or cyst on the eyelid called a **chalazion** (ka-LĀ-zē-on = small bump). The **conjunctiva** (kon′-junk-TĪ-va) is a thin, protective mucous membrane composed of nonkeratinized stratified squamous epithelium with numerous goblet cells that is supported by areolar connective tissue. The **palpebral conjunctiva** lines the inner aspect of the eyelids, and the **bulbar conjunctiva** passes from the eyelids onto the surface of the eyeball, where it covers the sclera (the "white" of the eye) but not the cornea, which is a transparent region that forms the outer anterior surface of the eyeball. Over the sclera, the conjunctiva is vascular. Both the sclera and the cornea will be discussed in more detail shortly. Dilation and congestion of the blood vessels of the bulbar conjunctiva due to local irritation or infection are the cause of **bloodshot eyes**.

Eyelashes and Eyebrows

The **eyelashes**, which project from the border of each eyelid, and the **eyebrows**, which arch transversely above the upper eyelids, help protect the eyeballs from foreign objects, perspiration, and the direct rays of the sun. Sebaceous glands at the base of the hair follicles of the eyelashes, called **sebaceous ciliary glands**, release a lubricating fluid into the follicles. Infection of these glands, usually by bacteria, causes a painful, pus-filled swelling called a **sty**.

The Lacrimal Apparatus

The **lacrimal apparatus** (LAK-ri-mal; *lacrim-* = *tears*) is a group of structures that produces and drains **lacrimal fluid** or *tears* in a process called *lacrimation*. The **lacrimal glands**, each about the size and shape of an almond, secrete lacrimal fluid, which drains into 6–12 **excretory lacrimal ducts** that empty tears onto the surface of the conjunctiva of the upper lid (**Figure 17.6b**). From here the tears pass medially over the anterior surface of the eyeball to enter two small openings called **lacrimal puncta** (singular is *punctum*). Tears then pass into two ducts, the superior and inferior **lacrimal canaliculi**, which lead into the **lacrimal sac** (within the lacrimal fossa) and then into the **nasolacrimal duct**. This duct carries the lacrimal fluid into the nasal cavity just inferior to the inferior nasal concha where it mixes with mucus. An infection of the lacrimal sacs is called **dacryocystitis** (dak′-rē-ō-sis-TĪ-tis; *dacryo-* = lacrimal sac; *-itis* = inflammation of). It is usually caused by a bacterial infection and results in blockage of the nasolacrimal ducts.

The lacrimal glands are supplied by parasympathetic fibers of the facial (VII) nerves. The lacrimal fluid produced by these glands is a watery solution containing salts, some mucus, and **lysozyme** (LĪ-sō-zīm), a protective bactericidal enzyme. The fluid protects, cleans, lubricates, and moistens the eyeball. After being secreted from the lacrimal gland, lacrimal fluid is spread medially over the surface of the eyeball by the blinking of the eyelids. Each gland produces about 1 mL of lacrimal fluid per day.

Normally, tears are cleared away as fast as they are produced, either by evaporation or by passing into the lacrimal canals and then into the nasal cavity. If an irritating substance makes contact with the conjunctiva, however, the lacrimal glands are stimulated to oversecrete, and tears accumulate (watery eyes). Lacrimation is a protective mechanism, as the tears dilute and wash away the irritating substance. Watery eyes also occur when an inflammation of the nasal mucosa, such as occurs with a cold, obstructs the nasolacrimal ducts and blocks drainage of tears. Only humans express emotions, both

FIGURE 17.6 **Accessory structures of the eye.**

Accessory structures of the eye include the eyelids, eyelashes, eyebrows, lacrimal apparatus, and extrinsic eye muscles.

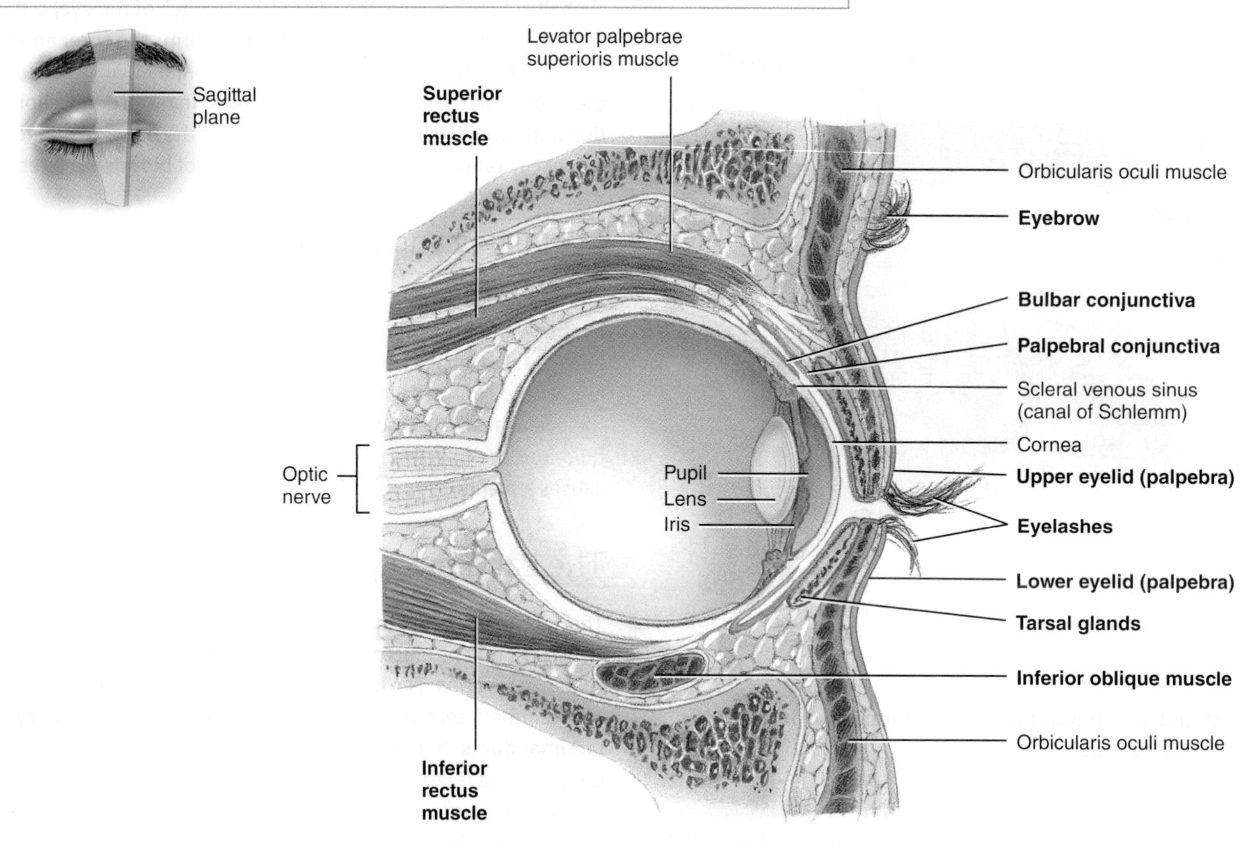

(a) Sagittal section of eye and its accessory structures

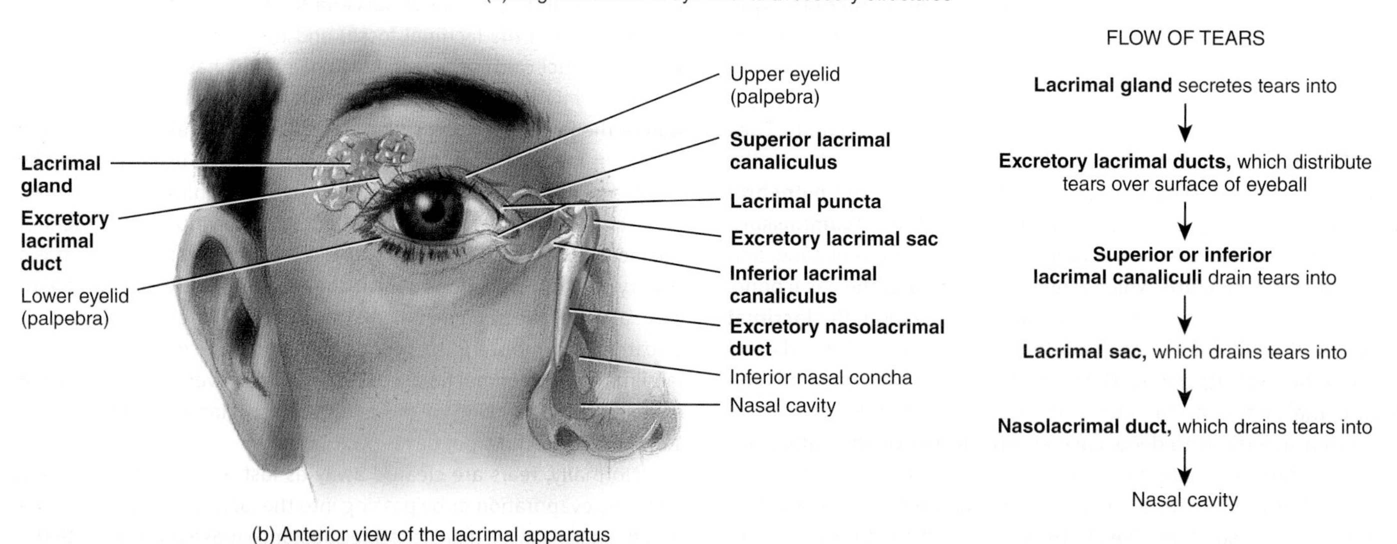

(b) Anterior view of the lacrimal apparatus

Q What is lacrimal fluid, and what are its functions?

happiness and sadness, by **crying**. In response to parasympathetic stimulation, the lacrimal glands produce excessive lacrimal fluid that may spill over the edges of the eyelids and even fill the nasal cavity with fluid. This is how crying produces a runny nose.

Extrinsic Eye Muscles

The eyes sit in the bony depressions of the skull called the *orbits*. The orbits help protect the eyes, stabilize them in three-dimensional space, and anchor them to the muscles that produce their essential movements. The extrinsic eye muscles extend from the walls of the bony orbit to the sclera (white) of the eye and are surrounded in the orbit by a significant quantity of **periorbital fat** (per′-ē-OR-bi-tal). These muscles are capable of moving the eye in almost any direction. Six extrinsic eye muscles move each eye: the **superior rectus, inferior rectus, lateral rectus, medial rectus, superior oblique**, and **inferior oblique** (Figure 17.6a; see also Figure 17.7). They are supplied by the oculomotor (III), trochlear (IV), or abducens (VI) nerves. In general, the motor units in these muscles are small. Some motor neurons serve only two or three muscle fibers—fewer than in any other part of the body except the larynx (voice box). Such small motor units permit smooth, precise, and rapid movement of the eyes. As indicated in Section 11.5, the extrinsic eye muscles move the eyeball laterally, medially, superiorly, and inferiorly. For example, looking to the right requires simultaneous *contraction* of the right lateral rectus and left medial rectus muscles of the eyeball and *relaxation* of the left lateral rectus and right medial rectus of the eyeball. The oblique muscles preserve rotational stability of the eyeball. Neural circuits in the brain stem and cerebellum coordinate and synchronize the movements of the eyes.

17.5 Anatomy of the Eyeball

OBJECTIVES

- **Identify** the components of the eye.
- **Discuss** the functions of these components.

The adult **eyeball** measures about 2.5 cm (1 in.) in diameter. Of its total surface area, only the anterior one-sixth is exposed; the remainder is recessed and protected by the orbit, into which it fits. Anatomically, the wall of the eyeball consists of three layers: (1) fibrous tunic, (2) vascular tunic, and (3) retina (inner tunic).

Fibrous Tunic

The **fibrous tunic** (TOO-nik) is the superficial layer of the eyeball and consists of the anterior cornea and posterior sclera (Figure 17.7). The **cornea** (KOR-nē-a) is a transparent coat that covers the colored iris. Because it is curved, the cornea helps focus light onto the retina. Its outer surface consists of nonkeratinized stratified squamous epithelium. The middle coat of the cornea consists of collagen fibers and fibroblasts, and the inner surface is simple squamous epithelium. Since the central part of the cornea receives oxygen from the outside air, contact lenses that are worn for long periods of time must be permeable to permit oxygen to pass through them. The **sclera** (SKLE-ra; *scler-* = hard), the "white" of the eye, is a layer of dense connective tissue made up mostly of collagen fibers and fibroblasts. The sclera covers the entire eyeball except the cornea; it gives shape to the eyeball, makes it more rigid, protects its inner parts, and serves as a site of attachment for the extrinsic eye muscles. At the junction of the sclera and cornea is an opening known as the **scleral venous sinus** or (*canal of Schlemm*). A fluid called aqueous humor, which will be described later, drains into this sinus (Figure 17.7).

Vascular Tunic

The **vascular tunic** or *uvea* (Ū-ve-a) is the middle layer of the eyeball. It is composed of three parts: choroid, ciliary body, and iris (Figure 17.7). The highly vascularized **choroid** (KŌ-royd), which is the posterior portion of the vascular tunic, lines most of the internal surface of the sclera. Its numerous blood vessels provide nutrients to the posterior surface of the retina. The choroid also contains melanocytes that produce the pigment melanin, which causes this layer to appear dark brown in color. Melanin in the choroid absorbs stray light rays, which prevents reflection and scattering of light within the eyeball. As a result, the image cast on the retina by the cornea and lens remains sharp and clear. Albinos lack melanin in all parts of the body, including the eye. They often need to wear sunglasses, even indoors, because even moderately bright light is perceived as bright glare due to light scattering.

In the anterior portion of the vascular tunic, the choroid becomes the **ciliary body** (SIL-ē-ar′-ē). It extends from the **ora serrata** (Ō-ra ser-RĀ-ta), the jagged anterior margin of the retina, to a point just posterior to the junction of the sclera and cornea. Like the choroid, the ciliary body appears dark brown in color because it contains melanin-producing melanocytes. In addition, the ciliary body consists of ciliary processes and ciliary muscle. The **ciliary processes** are protrusions or folds on the internal surface of the ciliary body. They contain blood capillaries that secrete aqueous humor. Extending from the ciliary process are **zonular fibers** or *suspensory ligaments* that attach to the lens. The fibers consist of thin, hollow fibrils that resemble elastic connective tissue fibers. The **ciliary muscle** is a circular band of smooth muscle. Contraction or relaxation of the ciliary muscle changes the tightness of the zonular fibers, which alters the shape of the lens, adapting it for near or far vision.

The **iris** (= rainbow), the colored portion of the eyeball, is shaped like a flattened donut. It is suspended between the cornea and the lens and is attached at its outer margin to the ciliary processes. It consists of melanocytes and circular and radial smooth muscle fibers. The amount of melanin in the iris determines the eye color. The eyes appear brown to black when the iris contains a large amount of melanin, blue when its melanin concentration is very low, and green when its melanin concentration is moderate.

A principal function of the iris is to regulate the amount of light entering the eyeball through the **pupil** (*pupil* = little person; because this is where you see a reflection of yourself when looking into someone's eyes), the hole in the center of the iris. The pupil appears black because, as you look through the lens, you see the heavily pigmented back of the eye (choroid and retina). However, if bright light is directed into the pupil, the reflected light is red because of the blood vessels on the surface of the retina. It is for this reason that a person's eyes appear red in a photograph ("red eye") when the flash is directed into the pupil. Autonomic reflexes regulate pupil diameter in response to light levels (Figure 17.8). When bright light stimulates the eye, parasympathetic fibers of the oculomotor (III) nerve

FIGURE 17.7 Anatomy of the eyeball.

The wall of the eyeball consists of three layers: the fibrous tunic, the vascular tunic, and the retina.

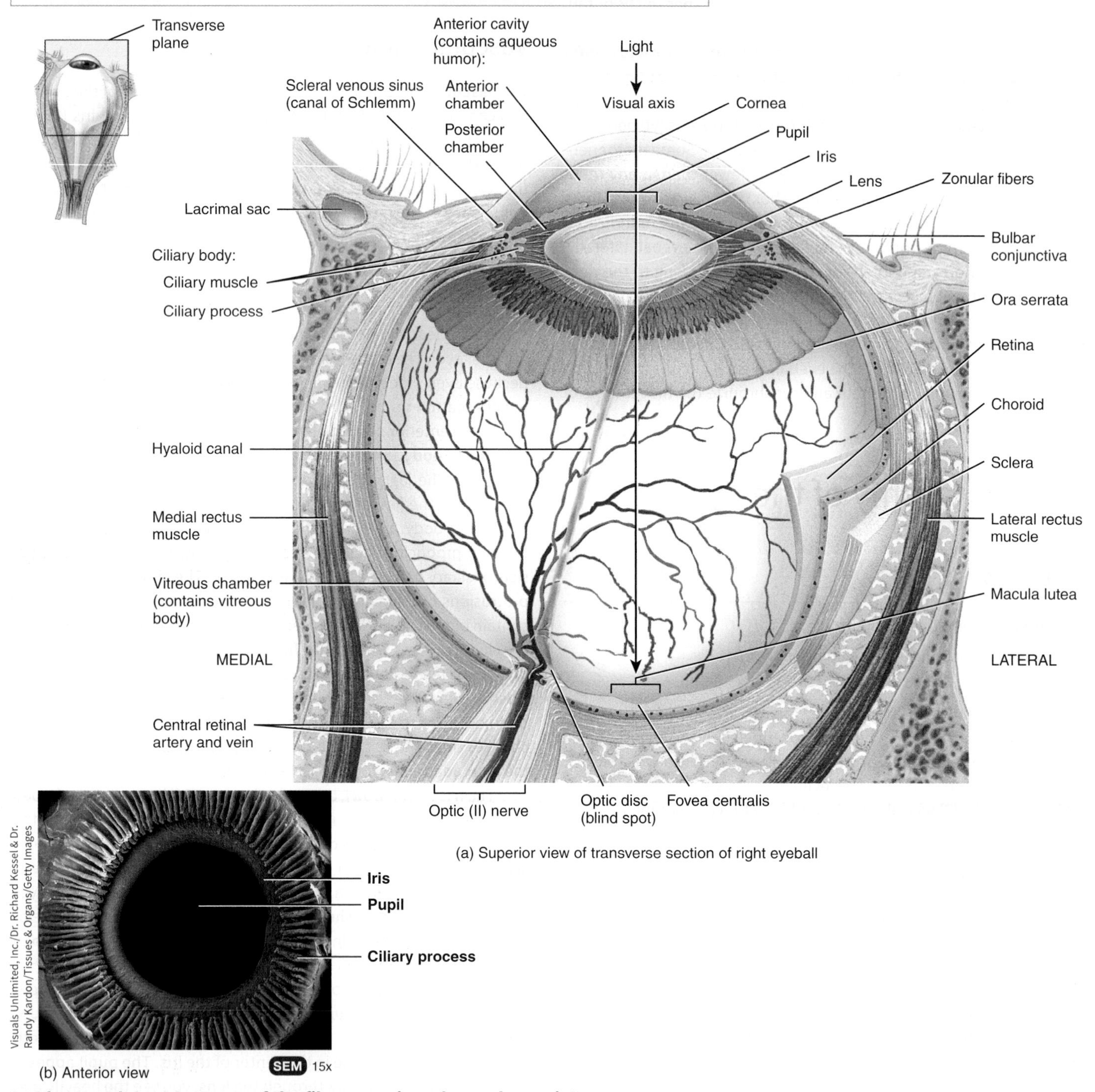

(a) Superior view of transverse section of right eyeball

(b) Anterior view

Q What are the components of the fibrous tunic and vascular tunic?

stimulate the **circular muscles** or *sphincter pupillae* (pu-PIL-ē) of the iris to contract, causing a decrease in the size of the pupil (constriction). In dim light, sympathetic neurons stimulate the **radial muscles** or *dilator pupillae* of the iris to contract, causing an increase in the pupil's size (dilation).

Retina

The third and inner layer of the eyeball, the **retina**, lines the posterior three-quarters of the eyeball and is the beginning of the visual pathway (see **Figure 17.7**). This layer's anatomy can be viewed with an

FIGURE 17.8 Responses of the pupil to light of varying brightness.

Contraction of the circular muscles causes constriction of the pupil; contraction of the radial muscles causes dilation of the pupil.

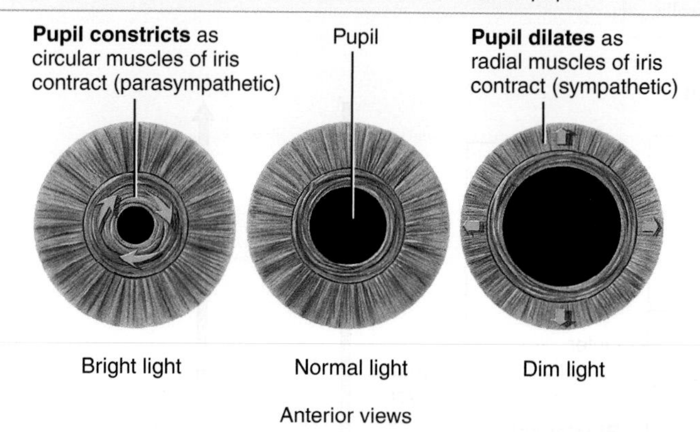

Q Which division of the autonomic nervous system causes pupillary constriction? Which causes pupillary dilation?

ophthalmoscope (of-THAL-mō-skōp; *ophthalmos-* = eye; *-skopeo* = to examine), an instrument that shines light into the eye and allows an observer to peer through the pupil, providing a magnified image of the retina and its blood vessels as well as the optic (II) nerve (**Figure 17.9**). The surface of the retina is the only place in the body where blood vessels can be viewed directly and examined for pathological changes, such as those that occur with hypertension, diabetes mellitus, cataracts, and age-related macular disease. Several landmarks are visible through an ophthalmoscope. The **optic disc** is the site where the optic (II) nerve exits the eyeball. Bundled together with the optic nerve are the **central retinal artery**, a branch of the ophthalmic artery, and the **central retinal vein** (see **Figure 17.7**). Branches of the central retinal artery fan out to nourish the anterior surface of the retina; the central retinal vein drains blood from the retina through the optic disc. Also visible are the macula lutea and fovea centralis, which are described shortly.

FIGURE 17.9 A normal retina, as seen through an ophthalmoscope. Blood vessels in the retina can be viewed directly and examined for pathological changes.

The optic disc is the site where the optic nerve exits the eyeball. The fovea centralis is the area of highest visual acuity.

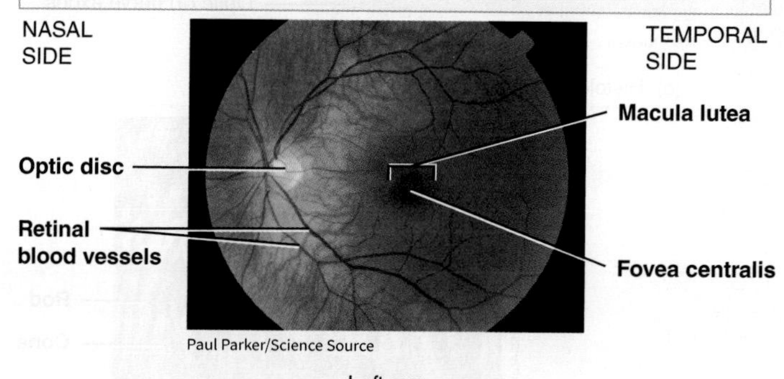

Q Evidence of what diseases may be seen through an ophthalmoscope?

The retina consists of a pigmented layer and a neural layer. The **pigmented layer** is a sheet of melanin-containing epithelial cells located between the choroid and the neural part of the retina. The melanin in the pigmented layer of the retina, as in the choroid, also helps to absorb stray light rays. The **neural** (*sensory*) **layer** of the retina is a multilayered outgrowth of the brain that processes visual data extensively before sending nerve impulses into axons that form the optic nerve. Three distinct layers of retinal neurons—the **photoreceptor cell layer**, the **bipolar cell layer**, and the **ganglion cell layer**—are separated by two zones, the *outer* and *inner synaptic layers*, where synaptic contacts are made (**Figure 17.10**). Note that light passes through the ganglion and bipolar cell layers and both synaptic layers before it reaches the photoreceptor layer. Two other types of cells present in the bipolar cell layer of the retina are called **horizontal cells** and **amacrine cells** (AM-a-krin). These cells form laterally directed neural circuits that modify the signals being transmitted along the pathway from photoreceptors to bipolar cells to ganglion cells.

Photoreceptors are specialized cells in the photoreceptor layer that begin the process by which light rays are ultimately converted to nerve impulses. There are two types of photoreceptors: rods and cones. Each retina has about 6 million cones and 120 million rods. **Rods** allow us to see in dim light, such as moonlight. Because rods do not provide color vision, in dim light we can see only black, white, and all shades of gray in between. Brighter lights stimulate **cones**, which produce color vision. Three types of cones are present in the retina: (1) *blue cones*, which are sensitive to blue light, (2) *green cones*, which are sensitive to green light, and (3) *red cones*, which are sensitive to red light. Color vision results from the stimulation of various combinations of these three types of cones. Most of our experiences are mediated by the cone system, the loss of which produces legal blindness. A person who loses rod vision mainly has difficulty seeing in dim light and thus should not drive at night.

See Clinical Connection: Age-Related Macular Disease

From photoreceptors, information flows through the outer synaptic layer to bipolar cells and then from bipolar cells through the inner synaptic layer to ganglion cells. The axons of ganglion cells extend posteriorly to the optic disc and exit the eyeball as the optic (II) nerve. The optic disc is also called the **blind spot**. Because it contains no rods or cones, we cannot see images that strike the blind spot. Normally, you are not aware of having a blind spot, but you can easily demonstrate its presence. Hold this book about 20 in. from your face with the cross shown at the end of this paragraph directly in front of your right eye. You should be able to see the cross and the square when you close your left eye. Now, keeping the left eye closed, slowly bring the page closer to your face while keeping the right eye on the cross. At a certain distance the square will disappear from your field of vision because its image falls on the blind spot.

+ ■

FIGURE 17.10 Microscopic structure of the retina. The downward blue arrow at right indicates the direction of the signals passing through the neural layer of the retina. Eventually, nerve impulses arise in ganglion cells and propagate along their axons, which make up the optic (II) nerve.

In the retina, visual signals pass from photoreceptors to bipolar cells to ganglion cells.

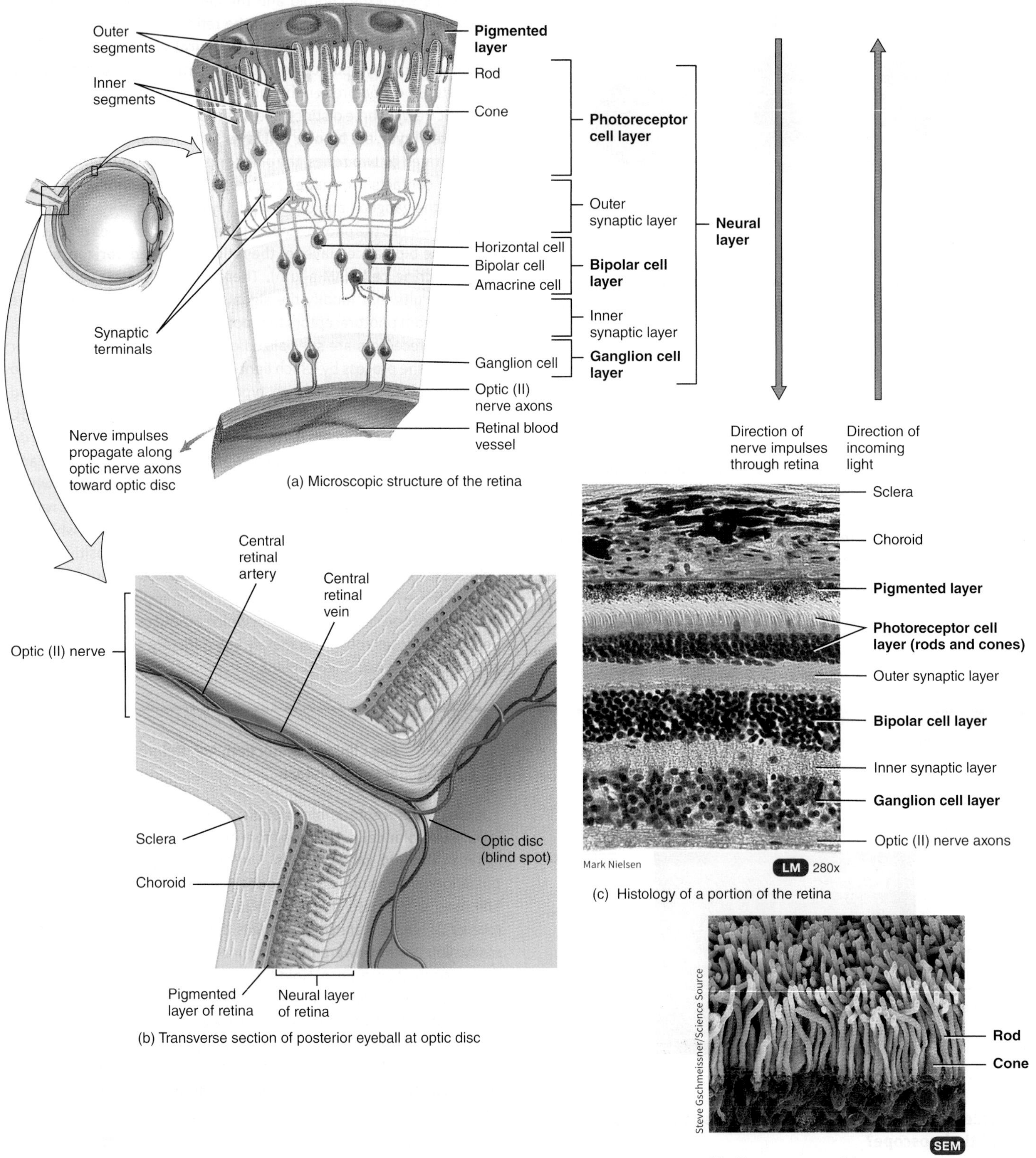

(a) Microscopic structure of the retina

(b) Transverse section of posterior eyeball at optic disc

Mark Nielsen

LM 280x

(c) Histology of a portion of the retina

Steve Gschmeissner/Science Source

SEM

(d) Photoreceptor cell layer

Q What are the two types of photoreceptors, and how do their functions differ?

The **macula lutea** (MAK-ū-la LOO-tē-a; *macula* = a small, flat spot; *lute-* = yellowish) or *yellow spot* is in the exact center of the posterior portion of the retina, at the visual axis of the eye (see Figure 17.9). The **fovea centralis** (FŌ-vē-a) (see Figures 17.7 and 17.9), a small depression in the center of the macula lutea, contains only cones. In addition, the layers of bipolar and ganglion cells, which scatter light to some extent, do not cover the cones here; these layers are displaced to the periphery of the fovea centralis. As a result, the fovea centralis is the area of highest **visual acuity** (a-KU-i-tē) or *resolution* (sharpness of vision). A main reason that you move your head and eyes while looking at something is to place images of interest on your fovea centralis—as you do to read the words in this sentence! Rods are absent from the fovea centralis and are more plentiful toward the periphery of the retina. Because rod vision is more sensitive than cone vision, you can see a faint object (such as a dim star) better if you gaze slightly to one side rather than looking directly at it.

See Clinical Connection: Detached Retina

Lens

Behind the pupil and iris, within the cavity of the eyeball, is the **lens** (see Figure 17.7). Within the cells of the lens, proteins called **crystallins** (KRIS-ta-lins), arranged like the layers of an onion, make up the refractive media of the lens, which normally is perfectly transparent and lacks blood vessels. It is enclosed by a clear connective tissue capsule and held in position by encircling zonular fibers, which attach to the ciliary processes. The lens helps focus images on the retina to facilitate clear vision.

Interior of the Eyeball

The lens divides the interior of the eyeball into two cavities: the anterior cavity and vitreous chamber. The **anterior cavity**—the space anterior to the lens—consists of two chambers. The **anterior chamber** lies between the cornea and the iris. The **posterior chamber** lies behind the iris and in front of the zonular fibers and lens (**Figure 17.11**). Both chambers of the anterior cavity are filled with **aqueous humor** (ĀK-wē-us HŪ-mer; *aqua* = water), a transparent

FIGURE 17.11 The iris separating the anterior and posterior chambers of the anterior cavity of the eye. The section is through the anterior portion of the eyeball at the junction of the cornea and sclera. Arrows indicate the flow of aqueous humor.

The anterior cavity of the eye contains aqueous humor.

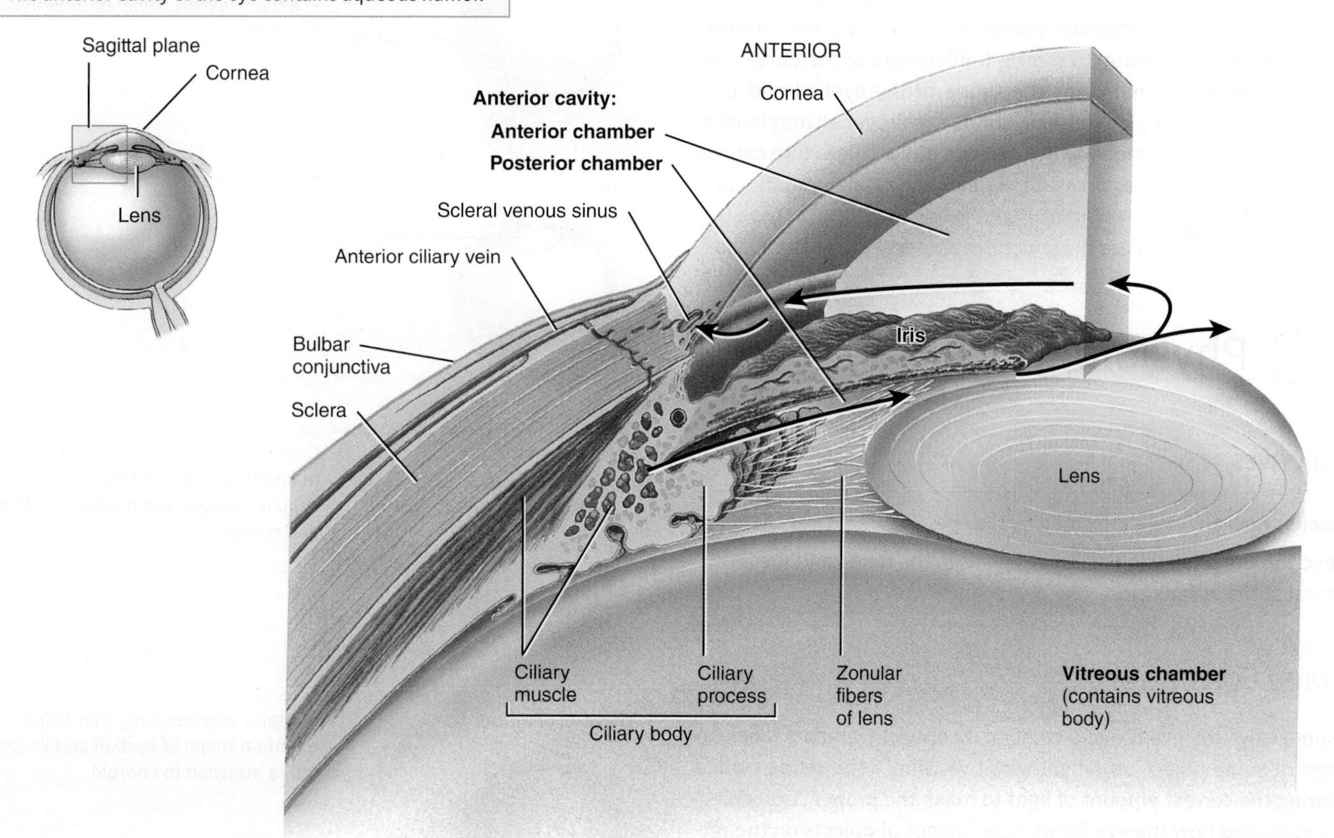

Q Where is aqueous humor produced, what is its circulation path, and where does it drain from the eyeball?

watery fluid that nourishes the lens and cornea. Aqueous humor continually filters out of blood capillaries in the ciliary processes of the ciliary body and enters the posterior chamber. It then flows forward between the iris and the lens, through the pupil, and into the anterior chamber. From the anterior chamber, aqueous humor drains into the scleral venous sinus (canal of Schlemm) and then into the blood. Normally, aqueous humor is completely replaced about every 90 minutes.

The larger posterior cavity of the eyeball is the **vitreous chamber** (VIT-rē-us), which lies between the lens and the retina. Within the vitreous chamber is the **vitreous body**, a transparent jellylike substance that holds the retina flush against the choroid, giving the retina an even surface for the reception of clear images. It occupies about four-fifths of the eyeball. Unlike the aqueous humor, the vitreous body does not undergo constant replacement. It is formed during embryonic life and consists of mostly water plus collagen fibers and hyaluronic acid. The vitreous body also contains phagocytic cells that remove debris, keeping this part of the eye clear for unobstructed vision. Occasionally, collections of debris may cast a shadow on the retina and create the appearance of specks that dart in and out of the field of vision. These *vitreal floaters*, which are more common in older individuals, are usually harmless and do not require treatment. The **hyaloid canal** (HĪ-a-loyd) is a narrow channel that is inconspicuous in adults and runs through the vitreous body from the optic disc to the posterior aspect of the lens. In the fetus, it is occupied by the hyaloid artery (see Figure 17.27d).

The pressure in the eye, called **intraocular pressure**, is produced mainly by the aqueous humor and partly by the vitreous body; normally it is about 16 mmHg (millimeters of mercury). The intraocular pressure maintains the shape of the eyeball and prevents it from collapsing. Puncture wounds to the eyeball may lead to the loss of aqueous humor and the vitreous body. This in turn causes a decrease in intraocular pressure, a detached retina, and in some cases blindness.

Table 17.1 summarizes the structures associated with the eyeball.

TABLE 17.1 Summary of the Structures of the Eyeball

STRUCTURE	FUNCTION
Fibrous tunic	***Cornea:*** Admits and refracts (bends) light. ***Sclera:*** Provides shape and protects inner parts.
Vascular tunic	***Iris:*** Regulates amount of light that enters eyeball. ***Ciliary body:*** Secretes aqueous humor and alters shape of lens for near or far vision (accommodation). ***Choroid:*** Provides blood supply and absorbs scattered light.
Retina 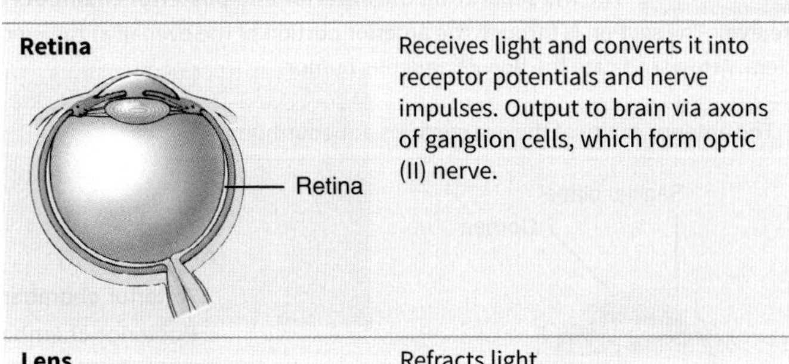	Receives light and converts it into receptor potentials and nerve impulses. Output to brain via axons of ganglion cells, which form optic (II) nerve.
Lens 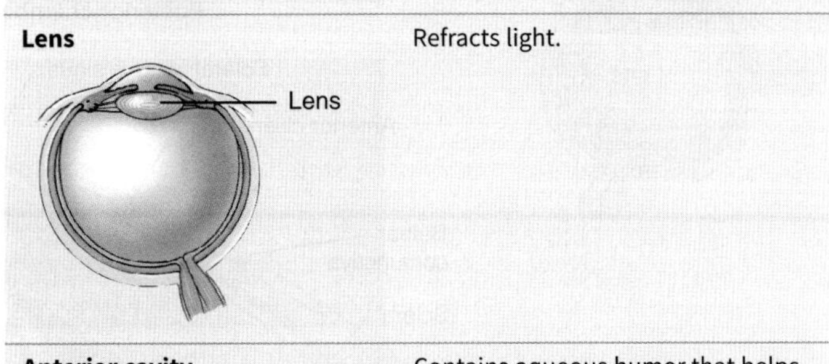	Refracts light.
Anterior cavity 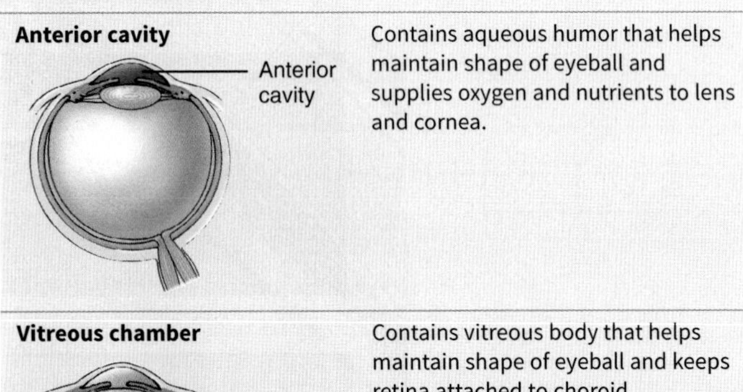	Contains aqueous humor that helps maintain shape of eyeball and supplies oxygen and nutrients to lens and cornea.
Vitreous chamber 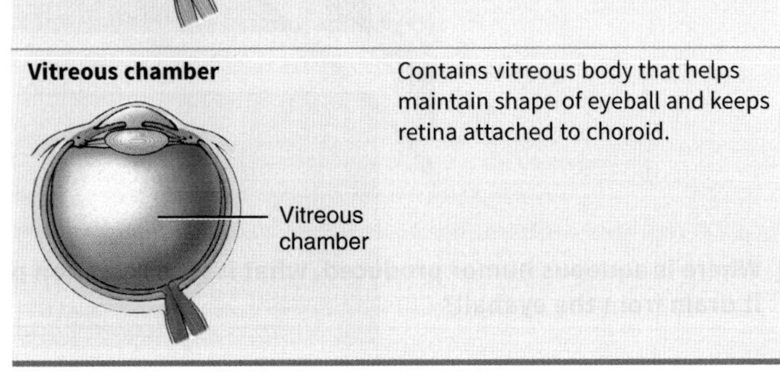	Contains vitreous body that helps maintain shape of eyeball and keeps retina attached to choroid.

17.6 Physiology of Vision

OBJECTIVES

- **Discuss** how an image is formed by the eye.
- **Describe** the processing of visual signals in the retina and the neural pathway for vision.

Image Formation

In some ways the eye is like a camera: Its optical elements focus an image of some object on a light-sensitive "film"—the retina—while ensuring the correct amount of light to make the proper "exposure." To understand how the eye forms clear images of objects on the retina, we must examine three processes: (1) the refraction or bending of light by the lens and cornea; (2) accommodation, the change in shape of the lens; and (3) constriction or narrowing of the pupil.

Refraction of Light Rays When light rays traveling through a transparent substance (such as air) pass into a second transparent substance with a different density (such as water), they bend at the junction between the two substances. This bending is called **refraction** (re-FRAK-shun) (Figure 17.12a). As light rays enter the eye, they are refracted at the anterior and posterior surfaces of the cornea. Both surfaces of the lens of the eye further refract the light rays so they come into exact focus on the retina.

Images focused on the retina are inverted (upside down) (Figure 17.12b, c). They also undergo right-to-left reversal; that is, light from the right side of an object strikes the left side of the retina, and vice versa. The reason the world does not look inverted and reversed is that the brain "learns" early in life to coordinate visual images with the orientations of objects. The brain stores the inverted and reversed images we acquired when we first reached for and touched objects and interprets those visual images as being correctly oriented in space.

About 75% of the total refraction of light occurs at the cornea. The lens provides the remaining 25% of focusing power and also changes the focus to view near or distant objects. When an object is 6 m (20 ft) or more away from the viewer, the light rays reflected from the object are nearly parallel to one another (Figure 17.12b). The lens must bend these parallel rays just enough so that they fall exactly focused on the central fovea, where vision is sharpest. Because light rays that are reflected from objects closer than 6 m (20 ft) are divergent rather than parallel (Figure 17.12c), the rays must be refracted more if they are to be focused on the retina. This additional refraction is accomplished through a process called accommodation.

Accommodation and the Near Point of Vision A surface that curves outward, like the surface of a ball, is said to be *convex.* When the surface of a lens is convex, that lens will refract incoming light rays toward each other, so that they eventually intersect. If the surface of a lens curves inward, like the inside of a hollow ball, the lens is said to be *concave* and causes light rays to refract away from each other. The lens of the eye is convex on both its anterior and posterior surfaces, and its focusing power increases as its curvature becomes greater. When the eye is focusing on a close object, the lens becomes more curved, causing greater refraction of the light rays. This increase in the curvature of the lens for near vision is called **accommodation** (a-kom-a-DĀ-shun) (Figure 17.12c). The **near point of vision** is the minimum distance from the eye that an object can be clearly focused with maximum accommodation. This distance is about 10 cm (4 in.) in a young adult.

How does accommodation occur? When you are viewing distant objects, the ciliary muscle of the ciliary body is relaxed and the lens is flatter because it is stretched in all directions by taut zonular fibers (see Figure 17.12b). When you view a close object, the ciliary muscle contracts, which pulls the ciliary process and choroid forward toward the lens. This action releases tension on the lens and zonular fibers. Because it is elastic, the lens becomes more spherical (more convex), which increases its focusing power and causes greater convergence of the light rays (see Figure 17.12c). Parasympathetic fibers of the oculomotor (III) nerve innervate the ciliary muscle of the ciliary body and, therefore, mediate the process of accommodation.

FIGURE 17.12 Refraction of light rays. (a) Refraction is the bending of light rays at the junction of two transparent substances with different densities. (b) The cornea and lens refract light rays from distant objects so the image is focused on the retina. (c) In accommodation, the lens becomes more spherical, which increases the refraction of light.

Images focused on the retina are inverted and left-to-right reversed.

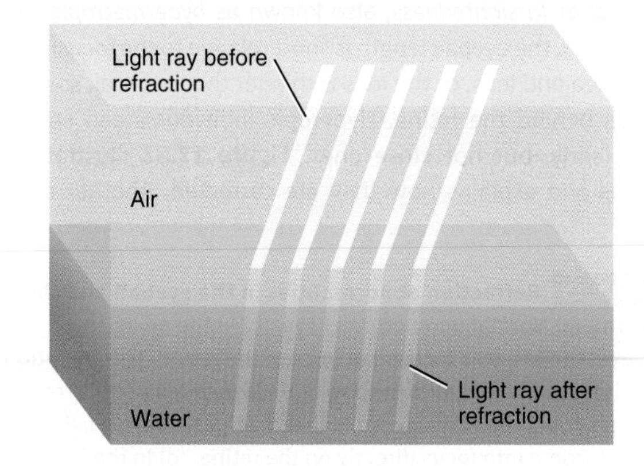

(a) Refraction of light rays

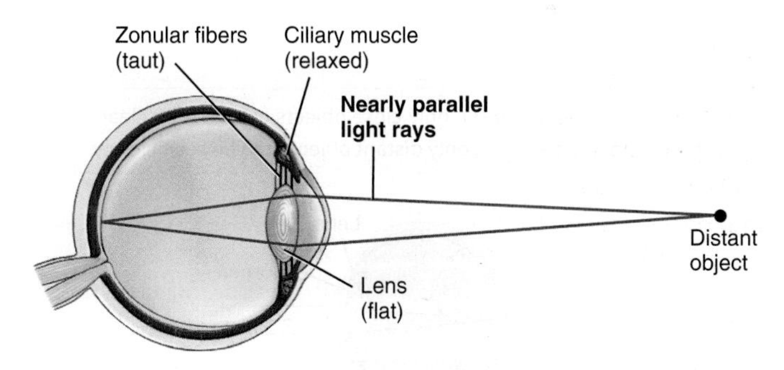

(b) Viewing a distant object

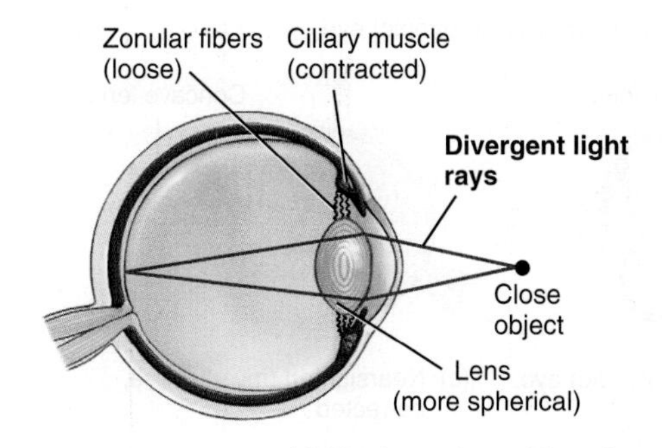

(c) Viewing a close object via accommodation

Q What sequence of events occurs during accommodation?

See Clinical Connection: Presbyopia

Refraction Abnormalities The normal eye, known as an **emmetropic eye** (em′-e-TROP-ik), can sufficiently refract light rays from an object 6 m (20 ft) away so that a clear image is focused on the retina. However, many people lack this ability because of refraction abnormalities. Among these abnormalities are **myopia** (mī-Ō-pē-a), or *nearsightedness,* which occurs when the eyeball is too long relative to the focusing power of the cornea and lens, or when the lens is thicker than normal, so an image converges in front of the retina. Myopic individuals can see close objects clearly, but not distant objects. In **hyperopia** (hī-per-Ō-pē-a) or *farsightedness,* also known as *hypermetropia* (hī′-per-me-TRŌ-pē-a), the eyeball length is short relative to the focusing power of the cornea and lens, or the lens is thinner than normal, so an image converges behind the retina. Hyperopic individuals can see distant objects clearly, but not close ones. Figure 17.13 illustrates these conditions and explains how they are corrected. Another refraction abnormality is **astigmatism** (a-STIG-ma-tizm), in which either the cornea or the lens has an irregular curvature. As a result, parts of the image are out of focus, and thus vision is blurred or distorted.

Most errors of vision can be corrected by eyeglasses, contact lenses, or surgical procedures. A contact lens floats on a film of tears over the cornea. The anterior outer surface of the contact lens corrects the visual defect, and its posterior surface matches the curvature of the cornea. LASIK involves reshaping the cornea to correct refraction abnormalities permanently.

See Clinical Connection: LASIK

FIGURE 17.13 Refraction abnormalities in the eyeball and their correction. (a) Normal (emmetropic) eye. (b) In the nearsighted or myopic eye, the image is focused in front of the retina. The condition may result from an elongated eyeball or thickened lens. (c) Correction of myopia is by use of a concave lens that diverges entering light rays so that they come into focus directly on the retina. (d) In the farsighted or hyperopic eye, the image is focused behind the retina. The condition results from a shortened eyeball or a thin lens. (e) Correction of hyperopia is by a convex lens that converges entering light rays so that they focus directly on the retina.

In myopia (nearsightedness), only close objects can be seen clearly; in hyperopia (farsightedness), only distant objects can be seen clearly.

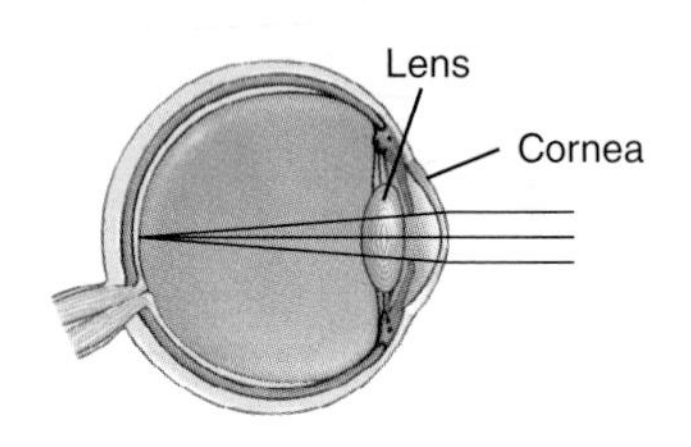

(a) Normal (emmetropic) eye

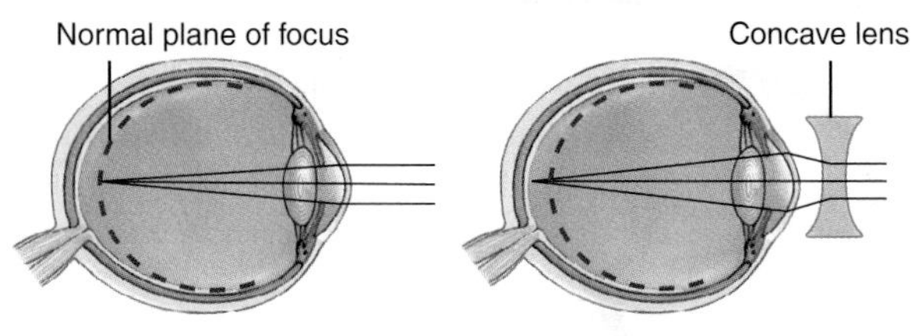

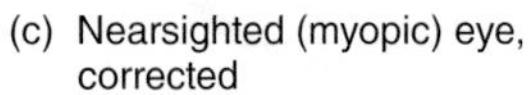

(b) Nearsighted (myopic) eye, uncorrected

(c) Nearsighted (myopic) eye, corrected

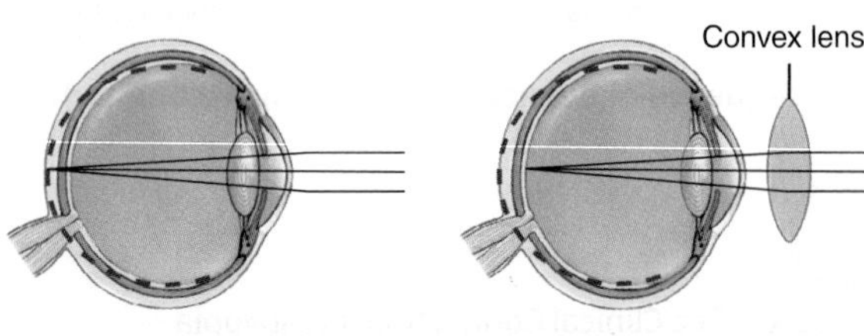

(d) Farsighted (hyperopic) eye, uncorrected

(e) Farsighted (hyperopic) eye, corrected

Q What is presbyopia?

Constriction of the Pupil

The circular muscle fibers of the iris also have a role in the formation of clear retinal images. As you have already learned, **constriction of the pupil** is a narrowing of the diameter of the hole through which light enters the eye due to the contraction of the circular muscles of the iris. This autonomic reflex occurs simultaneously with accommodation and prevents light rays from entering the eye through the periphery of the lens. Light rays entering at the periphery would not be brought to focus on the retina and would result in blurred vision. The pupil, as noted earlier, also constricts in bright light.

Convergence

Because of the position of their eyes in their heads, many animals, such as horses and goats, see one set of objects off to the left through one eye, and an entirely different set of objects off to the right through the other. In humans, both eyes focus on only one set of objects—a characteristic called **binocular vision**. This feature of our visual system allows the perception of depth and an appreciation of the three-dimensional nature of objects.

Binocular vision occurs when light rays from an object strike corresponding points on the two retinas. When we stare straight ahead at a distant object, the incoming light rays are aimed directly at both pupils and are refracted to comparable spots on the retinas of both eyes. As we move closer to an object, however, the eyes must rotate medially if the light rays from the object are to strike the same points on both retinas. The term **convergence** refers to this medial movement of the two eyeballs so that both are directed toward the object being viewed, for example, tracking a pencil moving toward your eyes. The nearer the object, the greater the degree of convergence needed to maintain binocular vision. The coordinated action of the extrinsic eye muscles brings about convergence.

Photoreceptor Function

Photoreceptors and Photopigments

Rods and cones were named for the different appearance of the *outer segment*—the distal end next to the pigmented layer—of each of these types of photoreceptors. The outer segments of rods are cylindrical or rod-shaped; those of cones are tapered or cone-shaped (**Figure 17.14**). Transduction of light energy into a receptor potential occurs in the outer segment of both rods and cones. The photopigments are integral proteins in the plasma membrane of the outer segment. In cones the plasma membrane is folded back and forth in a pleated fashion; in rods the pleats pinch off from the plasma membrane to form discs. The outer segment of each rod contains a stack of about 1000 discs, piled up like coins inside a wrapper.

Photoreceptor outer segments are renewed at an astonishingly rapid pace. In rods, one to three new discs are added to the base of the outer segment every hour while old discs slough off at the tip and are phagocytized by pigment epithelial cells. The *inner segment* contains the cell nucleus, Golgi complex, and many mitochondria. At its proximal end, the photoreceptor expands into bulblike synaptic terminals filled with synaptic vesicles.

FIGURE 17.14 Structure of rod and cone photoreceptors. The inner segments contain the metabolic machinery for synthesis of photopigments and production of ATP. The photopigments are embedded in the membrane discs or folds of the outer segments. New discs in rods and new folds in cones form at the base of the outer segment. Pigmented epithelial cells phagocytize the old discs and folds that slough off the distal tip of the outer segments.

Transduction of light energy into a receptor potential occurs in the outer segments of rods and cones.

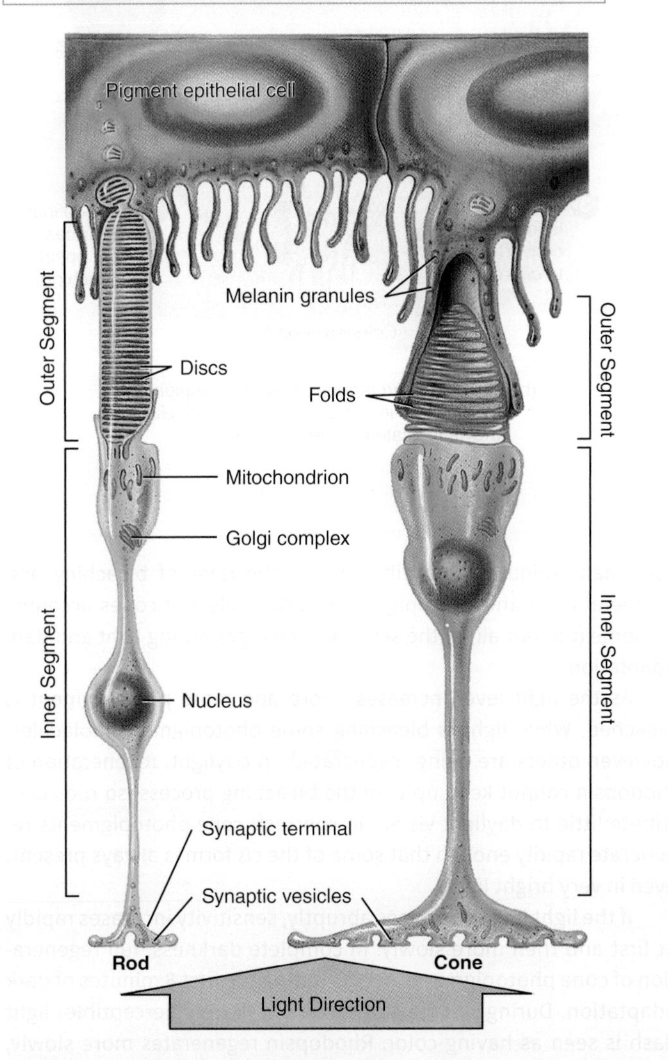

Q What are the functional similarities between rods and cones?

The first step in visual transduction is absorption of light by a **photopigment** (*visual pigment*)**,** a colored protein that undergoes structural changes when it absorbs light, in the outer segment of a photoreceptor. Light absorption initiates the events that lead to the production of a receptor potential. The single type of photopigment in rods is **rhodopsin** (rō-DOP-sin; *rhod-* = rose; *-opsin* = related to vision). Three different **cone photopigments** are present in the retina, one in each of the three types of cones (blue cones, green cones, and red cones). Color vision results from different colors of light selectively activating the different cone photopigments.

See Clinical Connection: Color Blindness and Night Blindness

All photopigments associated with vision contain two parts: a glycoprotein known as **opsin** and a derivative of vitamin A called **retinal** (**Figure 17.15a**). Vitamin A derivatives are formed from carotene, the plant pigment that gives carrots their orange color. Good vision depends on adequate dietary intake of carotene-rich vegetables such as carrots, spinach, broccoli, and yellow squash, or foods that contain vitamin A, such as liver.

Retinal is the light-absorbing part of all visual photopigments. In the human retina, there are four different opsins, three in the cones and one in the rods. Small variations in the amino acid sequences of the different opsins permit the rods and cones to absorb different colors (wavelengths) of incoming light.

Photopigments respond to light in the following cyclical process (**Figure 17.15b**):

1. ***Isomerization.*** In darkness, retinal has a bent shape, called *cis*-retinal, which fits snugly into the opsin portion of the photopigment. When *cis*-retinal absorbs a photon of light, it straightens out to a shape called *trans*-retinal. This *cis*-to-*trans* conversion is called **isomerization** and is the first step in visual transduction. After retinal isomerizes, chemical changes occur in the outer segment of the photoreceptor. These chemical changes lead to the production of a receptor potential (see **Figure 17.16**).
2. ***Bleaching.*** In about a minute, *trans*-retinal completely separates from opsin. Retinal is responsible for the color of the photopigment, so the separation of trans-retinal from opsin causes opsin to look colorless. Because of the color change, this part of the cycle is termed **bleaching** of photopigment.
3. ***Conversion.*** An enzyme called **retinal isomerase** converts *trans*-retinal back to *cis*-retinal.
4. ***Regeneration.*** The *cis*-retinal then can bind to opsin, reforming a functional photopigment. This part of the cycle—resynthesis of a photopigment—is called **regeneration**.

FIGURE 17.15 Photopigments and vision.

Retinal, a derivative of vitamin A, is the light-absorbing part of all visual photopigments.

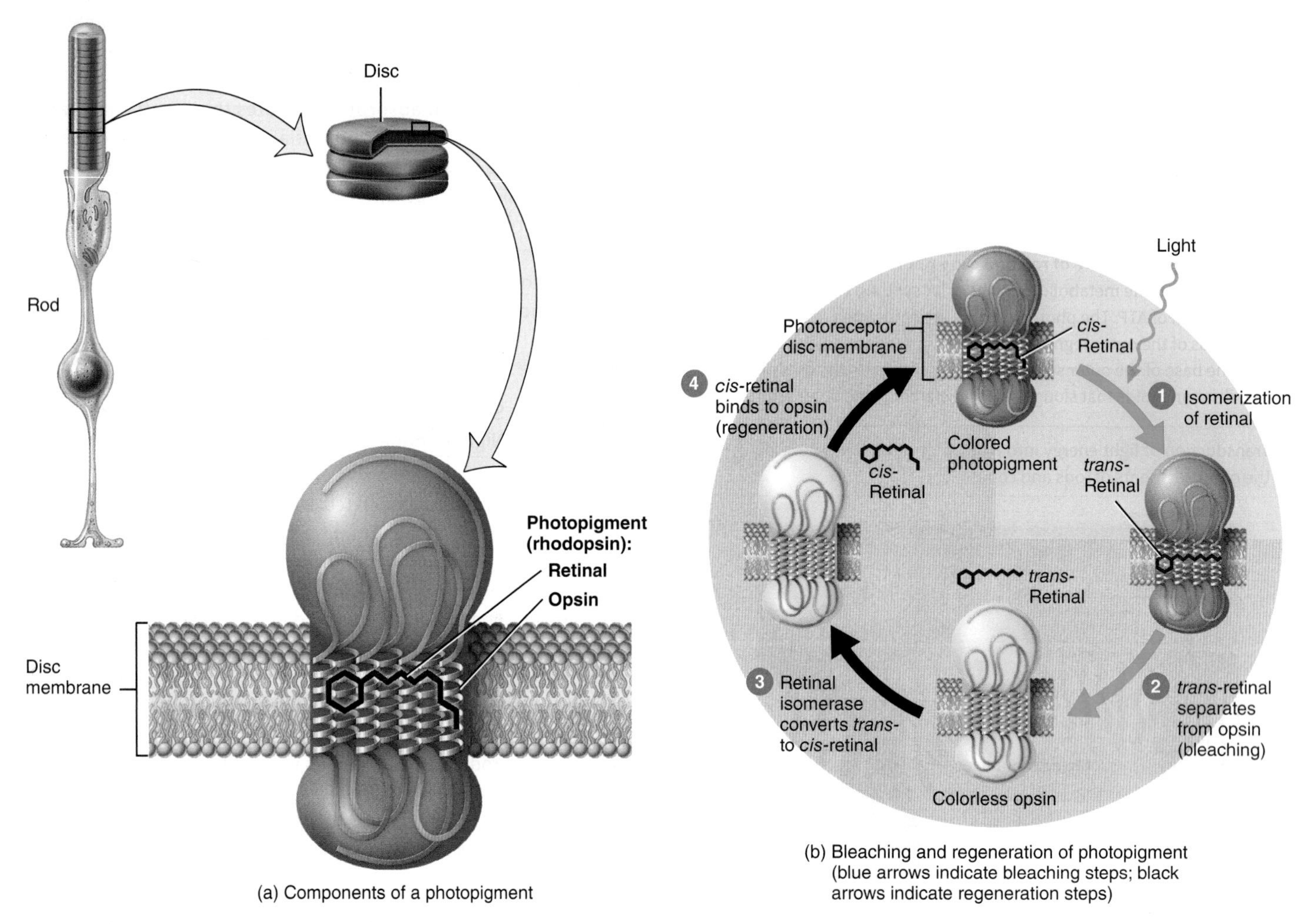

(a) Components of a photopigment

(b) Bleaching and regeneration of photopigment (blue arrows indicate bleaching steps; black arrows indicate regeneration steps)

Q What is the conversion of *cis*-retinal to *trans*-retinal called?

The pigmented layer of the retina adjacent to the photoreceptors stores a large quantity of vitamin A and contributes to the regeneration process in rods. The extent of rhodopsin regeneration decreases drastically if the retina detaches from the pigmented layer. Cone photopigments regenerate much more quickly than the rhodopsin in rods and are less dependent on the pigmented layer. After complete bleaching, regeneration of half of the rhodopsin takes 5 minutes; half of the cone photopigments regenerate in only 90 seconds. Full regeneration of bleached rhodopsin takes 30 to 40 minutes.

Light and Dark Adaptation

When you emerge from dark surroundings (say, a tunnel) into the sunshine, **light adaptation** occurs—your visual system adjusts in seconds to the brighter environment by decreasing its sensitivity. On the other hand, when you enter a darkened room such as a theater, your visual system undergoes **dark adaptation**—its sensitivity increases slowly over many minutes. The difference in the rates of bleaching and regeneration of the photopigments in the rods and cones accounts for some (but not all) of the sensitivity changes during light and dark adaptation.

As the light level increases, more and more photopigment is bleached. While light is bleaching some photopigment molecules, however, others are being regenerated. In daylight, regeneration of rhodopsin cannot keep up with the bleaching process, so rods contribute little to daylight vision. In contrast, cone photopigments regenerate rapidly enough that some of the *cis* form is always present, even in very bright light.

If the light level decreases abruptly, sensitivity increases rapidly at first and then more slowly. In complete darkness, full regeneration of cone photopigments occurs during the first 8 minutes of dark adaptation. During this time, a threshold (barely perceptible) light flash is seen as having color. Rhodopsin regenerates more slowly, and our visual sensitivity increases until even a single photon (the smallest unit of light) can be detected. In that situation, although much dimmer light can be detected, threshold flashes appear gray-white, regardless of their color. At very low light levels, such as starlight, objects appear as shades of gray because only the rods are functioning.

Phototransduction

Phototransduction is the process by which light energy is converted into a receptor potential in the outer segment of a photoreceptor. In most sensory systems, activation of a sensory receptor by its adequate stimulus triggers a depolarizing receptor potential. In the visual system, however, activation of a photoreceptor by its adequate stimulus (light) causes a hyperpolarizing receptor potential. Just as surprising is that, when the photoreceptor is at rest—that is, in the dark—the cell is relatively depolarized. To understand how phototransduction occurs, you must first examine the operation of a photoreceptor in the absence of light (**Figure 17.16a**):

1. In darkness, *cis*-retinal is the form of retinal associated with the photopigment of the photoreceptor. Photopigment molecules are present in the disc membranes of the photoreceptor outer segment.
2. Another important occurrence during darkness is that there is a high concentration of **cyclic GMP (cGMP)** in the cytosol of the photoreceptor outer segment. This is due to the continuous production of cGMP by the enzyme **guanylyl cyclase** in the disc membrane.
3. After it is produced, cGMP binds to and opens nonselective cation channels in the outer segment membrane. These **cGMP-gated channels** mainly allow Na^+ ions to enter the cell.
4. The inflow of Na^+, called the **dark current**, depolarizes the photoreceptor. As a result, in darkness, the membrane potential of a photoreceptor is about −40 mV. This is much closer to zero than a typical neuron's resting membrane potential of −70 mV.
5. The depolarization during darkness spreads from the outer segment to the synaptic terminal, which contains **voltage-gated Ca^{2+} channels** in its membrane. The depolarization keeps these channels open, allowing Ca^{2+} to enter the cell. The entry of Ca^{2+} in turn triggers exocytosis of synaptic vesicles, resulting in tonic release of large amounts of neurotransmitter from the synaptic terminal. The neurotransmitter in rods and cones is the amino acid glutamate (glutamic acid). At synapses between rods and some bipolar cells, glutamate is an inhibitory neurotransmitter: It triggers inhibitory postsynaptic potentials (IPSPs) that hyperpolarize the bipolar cells and prevent them from sending signals on to the ganglion cells.

The absorption of light and isomerization of retinal initiates chemical changes in the photoreceptor outer segment that allow photo transduction to occur (**Figure 17.16b**):

1. When light strikes the retina, *cis*-retinal undergoes isomerization to *trans*-retinal.
2. Isomerization of retinal causes activation of a G protein known as **transducin** that is located in the disc membrane.
3. Transducin in turn activates an enzyme called **cGMP phosphodiesterase**, which is also present in the disc membrane.
4. Once activated, cGMP phosphodiesterase breaks down cGMP. The breakdown of cGMP lowers the concentration of cGMP in the cytosol of the outer segment.
5. As a result, the number of open cGMP-gated channels in the outer segment membrane is reduced and Na^+ inflow decreases.
6. The decreased Na^+ inflow causes the membrane potential to drop to about −65 mV, thereby producing a hyperpolarizing receptor potential.
7. The hyperpolarization spreads from the outer segment to the synaptic terminal, causing a decrease in the number of open voltage-gated Ca^{2+} channels. Ca^{2+} entry into the cell is reduced, which decreases the release of neurotransmitter from the synaptic terminal. Dim lights cause small and brief receptor potentials that partially turn off neurotransmitter release; brighter lights elicit larger and longer receptor potentials that shut down neurotransmitter release more completely. Thus, light excites the bipolar cells that synapse with rods by turning off the release of an inhibitory neurotransmitter! The excited bipolar cells subsequently stimulate the ganglion cells to form action potentials in their axons.

Recall that the discs of rods form by pinching off from the plasma membrane of the outer segment; in cones, the discs are continuous with the outer segment membrane. Therefore, in rods, molecules of photopigment, transducin, cGMP phosphodiesterase, and guanylyl cyclase are located in a different membrane than the cGMP-gated channels; in cones, all of these proteins are located in the same membrane.

Processing of Visual Input in the Retina

Within the neural layer of the retina, certain features of visual input are enhanced while other features may be discarded. Input from several cells may either converge upon a smaller number of postsynaptic neurons (*convergence*) or diverge to a large number (*divergence*). Overall, convergence predominates: There are only 1 million ganglion cells, but 126 million photoreceptors in the human eye.

Once receptor potentials arise in the outer segments of rods and cones, they spread through the inner segments to the synaptic terminals. Neurotransmitter molecules released by rods and cones induce local graded potentials in both bipolar cells and horizontal cells. Between 6 and 600 rods synapse with a single bipolar cell in the outer synaptic layer of the retina; a cone more often synapses with a single bipolar cell. The convergence of many rods onto a single bipolar cell increases the light sensitivity of rod vision but slightly blurs the image that is perceived. Cone vision, although less sensitive, is sharper because of the one-to-one synapses between cones and their bipolar cells.

Synaptic activity between photoreceptors and bipolar cells is influenced by horizontal cells (see **Figure 17.10a**). Horizontal cells form synapses with photoreceptors and have only indirect effects on bipolar cells. In adjacent areas of the retina, one photoreceptor usually forms an excitatory synapse with a horizontal cell, and the horizontal cell in turn forms an inhibitory synapse with the presynaptic terminals of another photoreceptor. In this way, one photoreceptor can excite the horizontal cell, which can then inhibit the other photoreceptor,

FIGURE 17.16 **Phototransduction.**

Light causes a hyperpolarizing receptor potential in photoreceptors, which decreases release of an inhibitory neurotransmitter (glutamate).

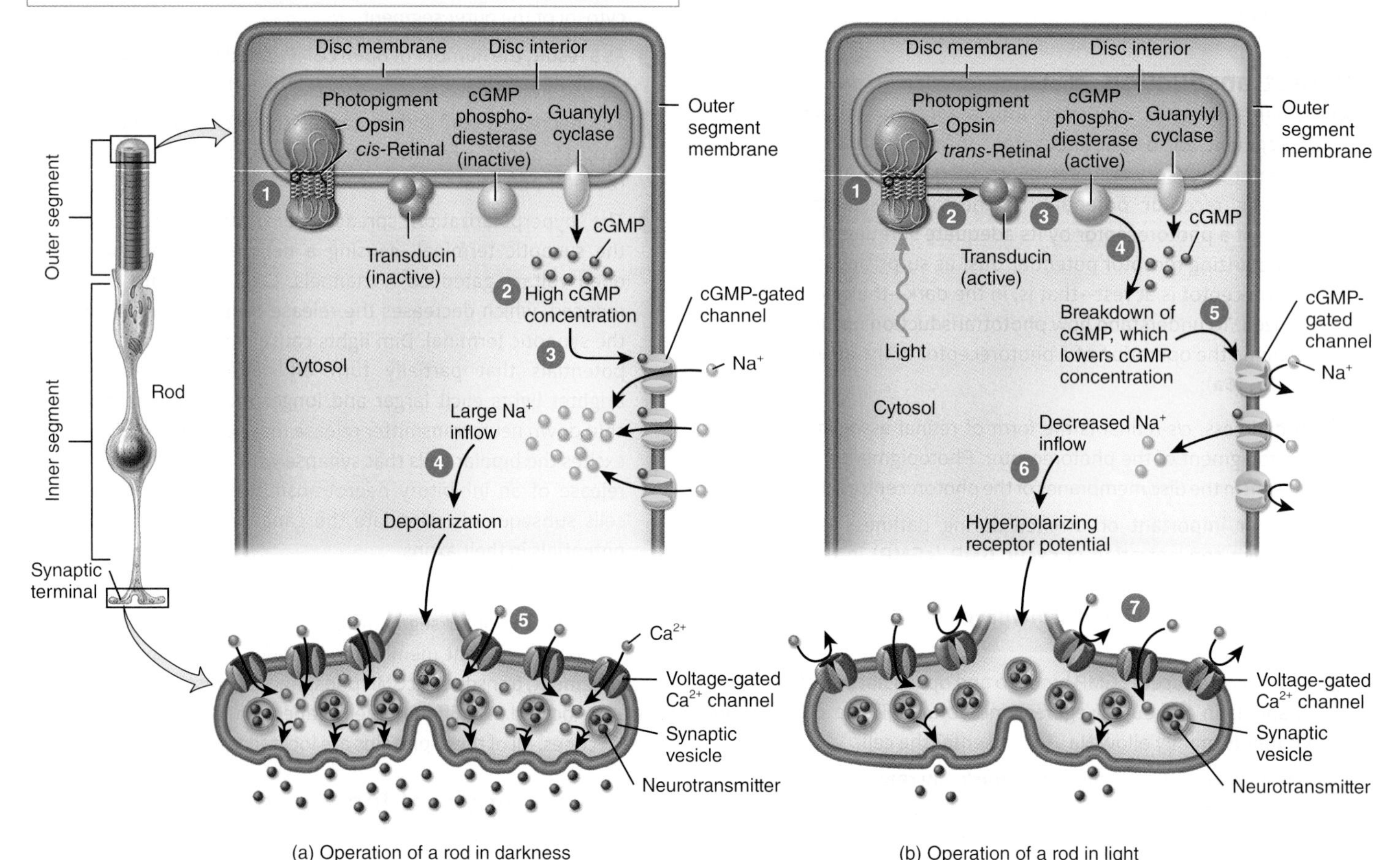

(a) Operation of a rod in darkness

(b) Operation of a rod in light

Q What is the function of cyclic GMP in photoreceptors?

decreasing the amount of neurotransmitter that is released onto a bipolar cell. Hence, horizontal cells can transmit laterally directed inhibitory signals to photoreceptors. This lateral inhibition helps to improve visual contrast between adjacent areas of the retina.

Synaptic activity between bipolar cells and ganglion cells is influenced by amacrine cells (see Figure 17.10a). Amacrine cells transmit laterally directed inhibitory signals (lateral inhibition) at synapses formed with bipolar cells and ganglion cells. There are many different types of amacrine cells and they have a variety of functions. Depending on which amacrine cells are involved, they can respond to a change in the level of illumination in the retina, the onset or offset of a visual signal, or movement of a visual signal in a particular direction.

The Visual Pathway

The axons of the retinal ganglion cells form the **optic (II) nerve** which provide output from the retina to the brain. The optic (II) nerves pass through the **optic chiasm** (KĪ-azm = a crossover, as in the letter X), a crossing point of the optic nerves (Figure 17.17a, b). Some axons cross to the opposite side, but others remain uncrossed. After passing through the optic chiasm, the axons, now part of the **optic tract**, enter the brain and most of them terminate in the **lateral geniculate nucleus** of the thalamus. Here they synapse with neurons whose axons form the **optic radiations**, which project to the **primary visual areas** in the occipital lobes of the cerebral cortex (area 17 in Figure 14.15), and visual perception begins. Some of the fibers in the optic tracts terminate in the **superior colliculi**, which control the extrinsic eye muscles, and the **pretectal nuclei**, which control pupillary and accommodation reflexes.

Everything that can be seen by one eye is that eye's **visual field**. As noted earlier, because our eyes are located anteriorly in our heads, the visual fields overlap considerably (Figure 17.17b). We have binocular vision due to the large region where the visual fields of the two eyes overlap—the **binocular visual field**. The visual field of each eye is divided into two regions: the **nasal** (*central*) **half** and the **temporal** (*peripheral*) **half**. For each eye, light rays from an object in the nasal half of the visual field fall on the temporal half of the retina, and light rays from an object in the temporal half of the visual field fall on the nasal half of the retina. Visual information from the *right* half of each visual field is conveyed to the *left* side of the brain, and visual information from the *left* half of each visual field is conveyed to the *right* side of the brain, as follows (**Figure 17.17c, d**):

1. The axons of all retinal ganglion cells in one eye exit the eyeball at the optic disc and form the optic nerve on that side.
2. At the optic chiasm, axons from the temporal half of each retina do not cross but continue directly to the lateral geniculate nucleus of the thalamus on the same side.
3. In contrast, axons from the nasal half of each retina cross the optic chiasm and continue to the opposite thalamus.
4. Each optic tract consists of crossed and uncrossed axons that project from the optic chiasm to the thalamus on one side.
5. Axon collaterals (branches) of the retinal ganglion cells project to the midbrain, where they participate in neural circuits that govern constriction of the pupils in response to light and coordination of head and eye movements. Collaterals also extend to the suprachiasmatic nucleus of the hypothalamus, which establishes patterns of sleep and other activities that occur on a circadian or daily schedule in response to intervals of light and darkness.
6. The axons of thalamic neurons form the optic radiations as they project from the thalamus to the **primary visual area** in the occipital lobe of the cortex on the same side.

The arrival of action potentials in the primary visual area allows you to perceive light. The primary visual area has a map of visual space: Each region within the cortex receives input from a different part of the retina, which in turn receives input from a particular part of the visual field. A large amount of cortical area is devoted to input from the portion of the visual field that strikes the macula. Recall that the macula contains the fovea, the part of the retina with the highest visual acuity. Relatively smaller amounts of cortical areas are devoted to those portions of the visual field that strike the peripheral parts of the retina.

Input from the primary visual area is conveyed to the **visual association area** in the occipital lobe. There are also areas in the parietal and temporal lobes that receive and process visual input; for simplicity, these areas will be considered as an extension of the visual association area. The visual association area further processes visual input to provide more complex visual patterns, such as three-dimensional position, overall form, motion, and color. In addition, the visual association area stores visual memories and relates past

FIGURE 17.17 The visual pathway. (a) Partial dissection of the brain reveals the optic radiations (axons extending from the thalamus to the occipital lobe). (b) An object in the binocular visual field can be seen with both eyes. In (c) and (d), note that information from the right side of the visual field of each eye projects to the left side of the brain, and information from the left side of the visual field of each eye projects to the right side of the brain.

The axons of ganglion cells in the temporal half of each retina extend to the thalamus on the same side; the axons of ganglion cells in the nasal half of each retina extend to the thalamus on the opposite side.

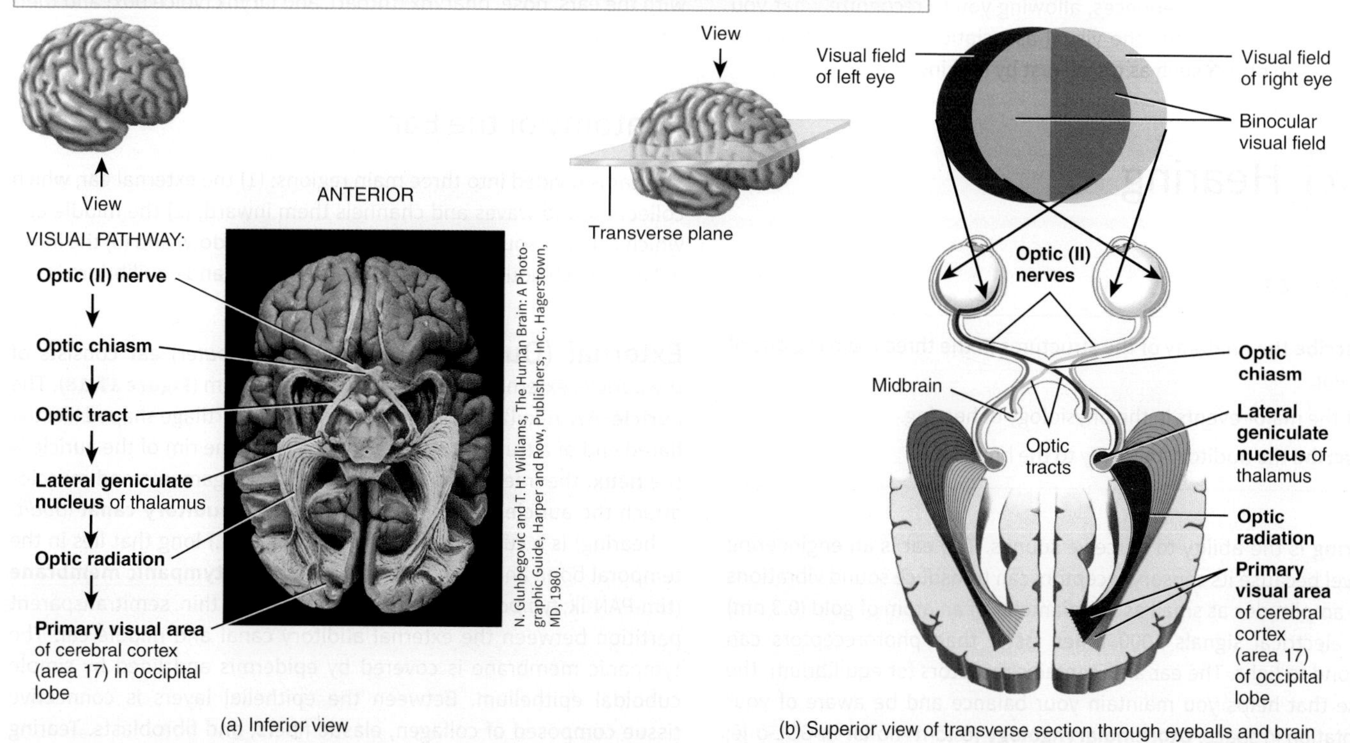

(a) Inferior view

(b) Superior view of transverse section through eyeballs and brain

Figure 17.17 *Continues*

FIGURE 17.17 Continued

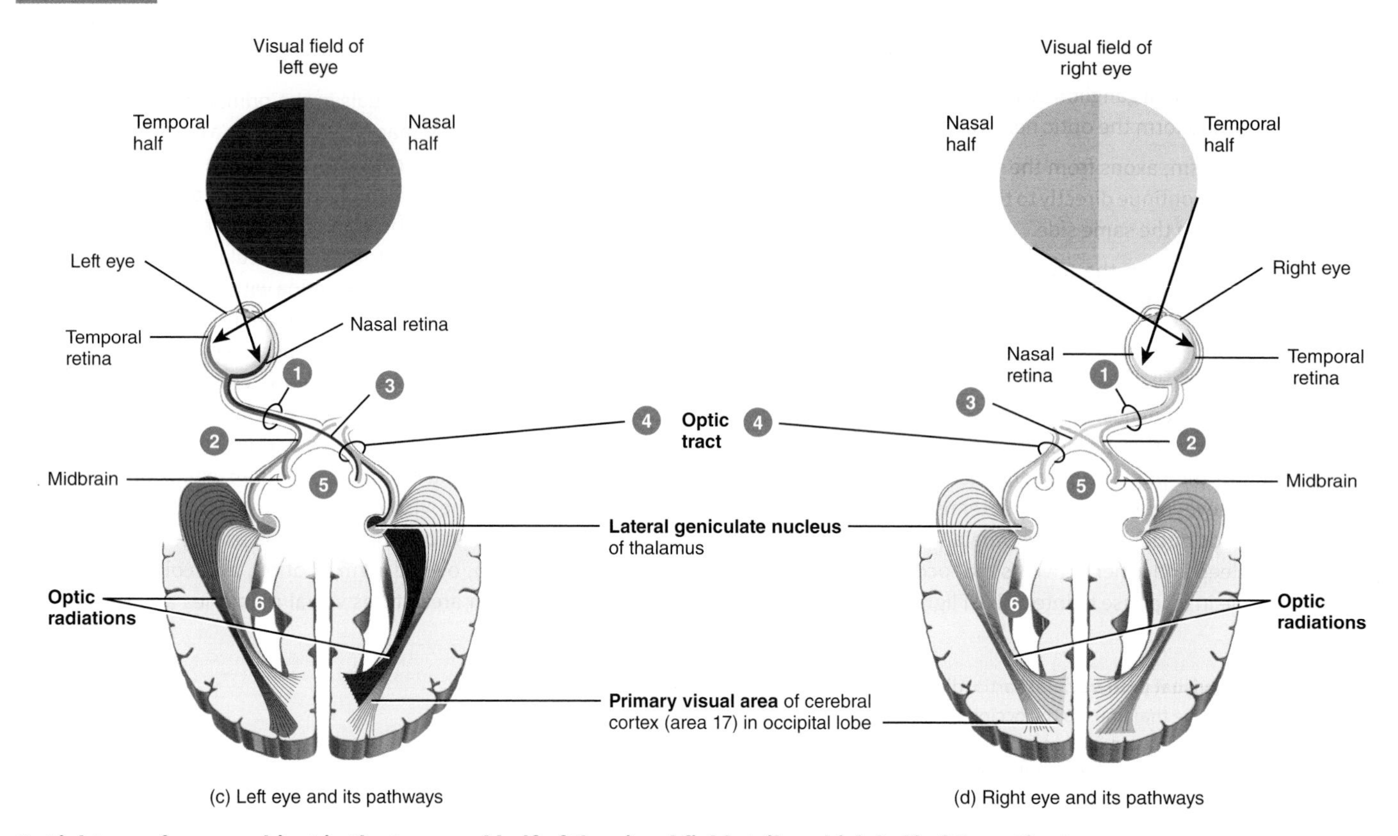

(c) Left eye and its pathways

(d) Right eye and its pathways

Q Light rays from an object in the temporal half of the visual field strike which half of the retina?

and present visual experiences, allowing you to recognize what you are seeing. For example, the visual association area allows you to recognize an object such as pencil just by looking at it.

17.7 Hearing

OBJECTIVES

- **Describe** the anatomy of the structures in the three main regions of the ear.
- **List** the major events in the physiology of hearing.
- **Describe** the auditory pathway to the brain.

Hearing is the ability to perceive sounds. The ear is an engineering marvel because its sensory receptors can transduce sound vibrations with amplitudes as small as the diameter of an atom of gold (0.3 nm) into electrical signals 1000 times faster than photoreceptors can respond to light. The ear also contains receptors for equilibrium, the sense that helps you maintain your balance and be aware of your orientation in space. **Otorhinolaryngology** (ō-tō-rī′-nō-lar-in-GOL-ō-jē; *oto-* = ear; *-rhino-* = nose; *-laryngo-* = larynx) is the science that deals with the ears, nose, pharynx (throat), and larynx (voice box) and their disorders.

Anatomy of the Ear

The ear is divided into three main regions: (1) the external ear, which collects sound waves and channels them inward; (2) the middle ear, which conveys sound vibrations to the oval window; and (3) the internal ear, which houses the receptors for hearing and equilibrium.

External (Outer) Ear The **external** (*outer*) **ear** consists of the auricle, external auditory canal, and eardrum (**Figure 17.18**). The **auricle** (AW-ri-kul) or *pinna* is a flap of elastic cartilage shaped like the flared end of a trumpet and covered by skin. The rim of the auricle is the **helix**; the inferior portion is the **lobule**. Ligaments and muscles attach the auricle to the head. The **external auditory canal** (*audit-* = hearing) is a curved tube about 2.5 cm (1 in.) long that lies in the temporal bone and leads to the eardrum. The **tympanic membrane** (tim-PAN-ik; *tympan-* = a drum) or *eardrum* is a thin, semitransparent partition between the external auditory canal and middle ear. The tympanic membrane is covered by epidermis and lined by simple cuboidal epithelium. Between the epithelial layers is connective tissue composed of collagen, elastic fibers, and fibroblasts. Tearing of the tympanic membrane is called a **perforated eardrum**. It may be

FIGURE 17.18 Anatomy of the ear.

The ear has three principal regions: the external (outer) ear, the middle ear, and the internal (inner) ear. (See key below.)

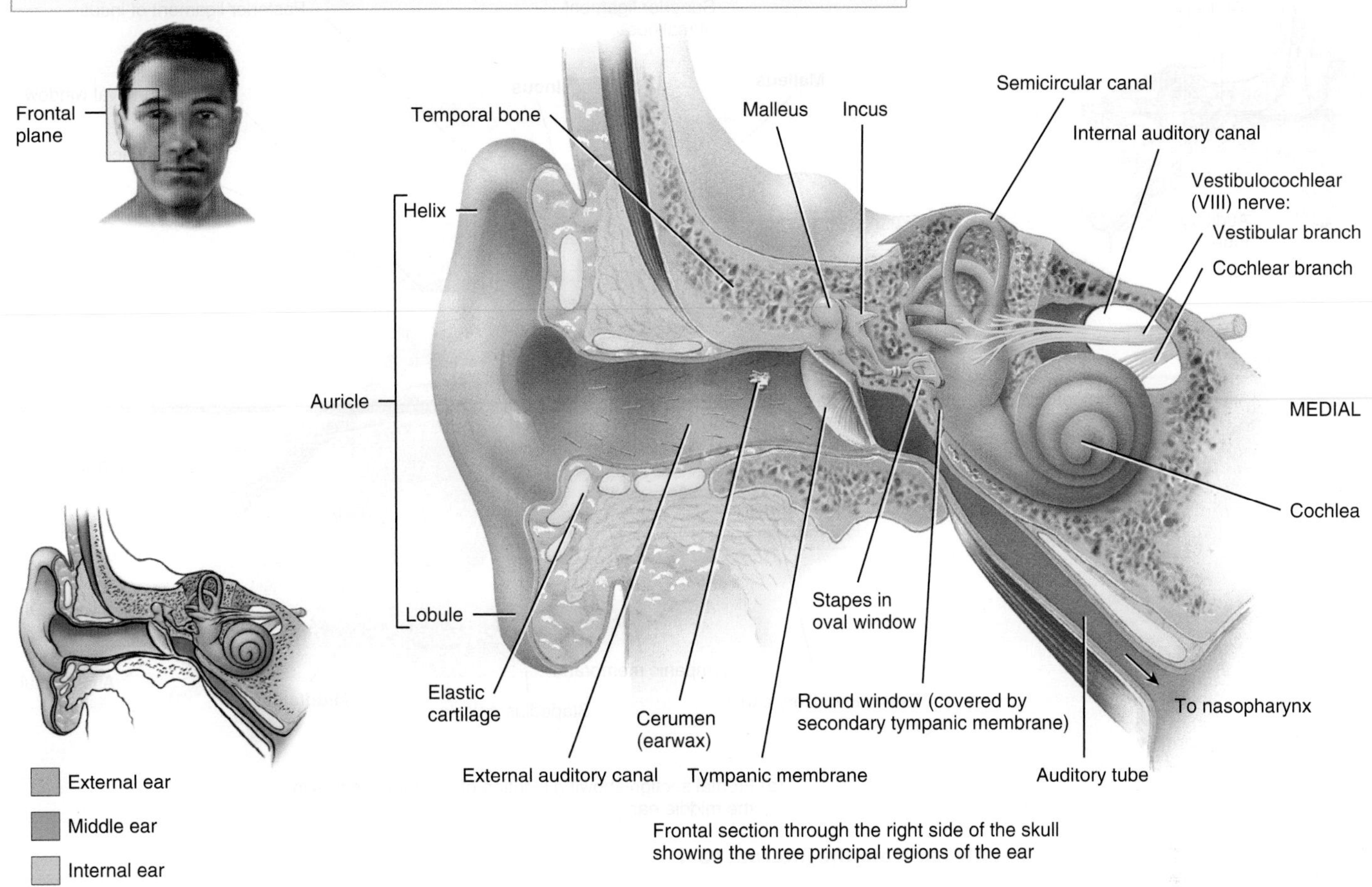

Frontal section through the right side of the skull showing the three principal regions of the ear

Q To which structure of the external ear does the malleus of the middle ear attach?

due to pressure from a cotton swab, trauma, or a middle ear infection, and usually heals within a month. The tympanic membrane may be examined directly by an **otoscope** (Ō-tō-skōp; *oto-* = ear; *-skopeo* = to view), a viewing instrument that illuminates and magnifies the external auditory canal and tympanic membrane.

Near the exterior opening, the external auditory canal contains a few hairs and specialized sweat glands called **ceruminous glands** (se-ROO-mi-nus) that secrete **earwax** or *cerumen* (se-ROO-men). The combination of hairs and cerumen helps prevent dust and foreign objects from entering the ear. Cerumen also prevents damage to the delicate skin of the external ear canal by water and insects. Cerumen usually dries up and falls out of the ear canal. However, some people produce a large amount of cerumen, which can become impacted and can muffle incoming sounds. The treatment for **impacted cerumen** is usually periodic ear irrigation or removal of wax with a blunt instrument by trained medical personnel.

Middle Ear The **middle ear** is a small, air-filled cavity in the petrous portion of the temporal bone that is lined by epithelium (Figure 17.19). It is separated from the external ear by the tympanic membrane and from the internal ear by a thin bony partition that contains two small openings: the oval window and the round window. Extending across the middle ear and attached to it by ligaments are the three smallest bones in the body, the **auditory ossicles** (OS-si-kuls), which are connected by synovial joints. The bones, named for their shapes, are the malleus, incus, and stapes—commonly called the hammer, anvil, and stirrup, respectively. The "handle" of the **malleus** (MAL-ē-us) attaches to the internal surface of the tympanic membrane. The head of the malleus articulates with the body of the incus. The **incus** (ING-kus), the middle bone in the series, articulates with the head of the stapes. The base or footplate of the **stapes** (STĀ-pēz) fits into the **oval window**. Directly below the oval window is another opening, the **round window**, which is enclosed by a membrane called the **secondary tympanic membrane**.

Besides the ligaments, two tiny skeletal muscles also attach to the ossicles (Figure 17.19). The **tensor tympani** (TIM-pan-ē) muscle, which is supplied by the mandibular branch of the trigeminal (V) nerve, limits movement and increases tension on the eardrum to prevent damage to the inner ear from loud noises. The **stapedius** (sta-PĒ-de-us) muscle, which is supplied by the facial (VII) nerve, is

FIGURE 17.19 **The right middle ear and the auditory ossicles.**

Common names for the malleus, incus, and stapes are the hammer, anvil, and stirrup, respectively.

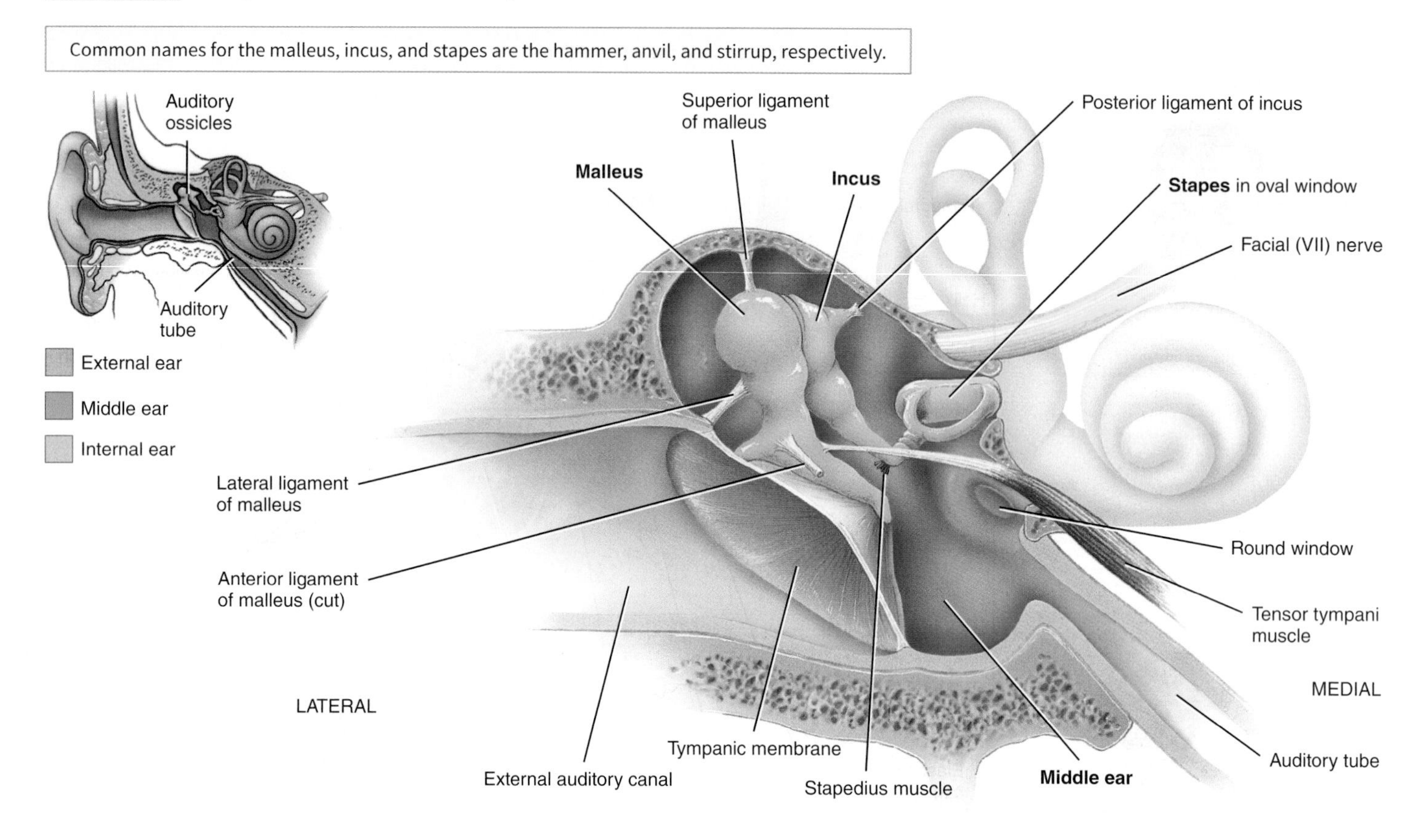

(a) Frontal section showing location of auditory ossicles in the middle ear

Q What structures separate the middle ear from the internal ear?

the smallest skeletal muscle in the human body. By dampening large vibrations of the stapes due to loud noises, it protects the oval window, but it also decreases the sensitivity of hearing. For this reason, paralysis of the stapedius muscle is associated with **hyperacusia** (hī′-per-a-KŪ-sē-a), which is abnormally sensitive hearing. Because it takes a fraction of a second for the tensor tympani and stapedius muscles to contract, they can protect the inner ear from prolonged loud noises but not from brief ones such as a gunshot.

The anterior wall of the middle ear contains an opening that leads directly into the **auditory tube** or *pharyngotympanic tube*, commonly known as the *eustachian tube* (ū′-STĀ-kē-an, ū-STĀ-shun). The auditory tube, which consists of both bone and elastic cartilage, connects the middle ear with the nasopharynx (superior portion of the throat). It is normally closed at its medial (pharyngeal) end. During swallowing and yawning, it opens, allowing air to enter or leave the middle ear until the pressure in the middle ear equals the atmospheric pressure. Most of us have experienced our ears popping as the pressures equalize. When the pressures are balanced, the tympanic membrane vibrates freely as sound waves strike it. If the pressure is not equalized, intense pain, hearing impairment, ringing in the ears, and vertigo could develop. The auditory tube also is a route for pathogens to travel from the nose and throat to the middle ear, causing the most common type of ear infection (see *otitis media* in Disorders: Homeostatic Imbalances in the Study Guide).

Internal (Inner) Ear

The **internal** (*inner*) **ear** is also called the *labyrinth* (LAB-i-rinth) because of its complicated series of canals (**Figure 17.20**). Structurally, it consists of two main divisions: an outer bony labyrinth that encloses an inner membranous labyrinth. It is like long balloons put inside a rigid tube. The **bony labyrinth** is a series of cavities in the petrous portion of the temporal bone divided into three areas: (1) the semicircular canals, (2) the vestibule, and (3) the cochlea. The bony labyrinth is lined with periosteum and contains **perilymph**. This fluid, which is chemically similar to cerebrospinal fluid, surrounds the **membranous labyrinth**, a series of epithelial sacs and tubes inside the bony labyrinth that have the same general form as the bony labyrinth and house the receptors for hearing and equilibrium. The epithelial membranous labyrinth contains **endolymph**. The level of potassium ions (K^+) in endolymph is unusually high for an extracellular fluid, and potassium ions play a role in the generation of auditory signals (described shortly).

FIGURE 17.20 **The right internal ear.** The outer, cream-colored area is part of the bony labyrinth; the inner, pink-colored area is the membranous labyrinth.

The bony labyrinth contains perilymph, and the membranous labyrinth contains endolymph.

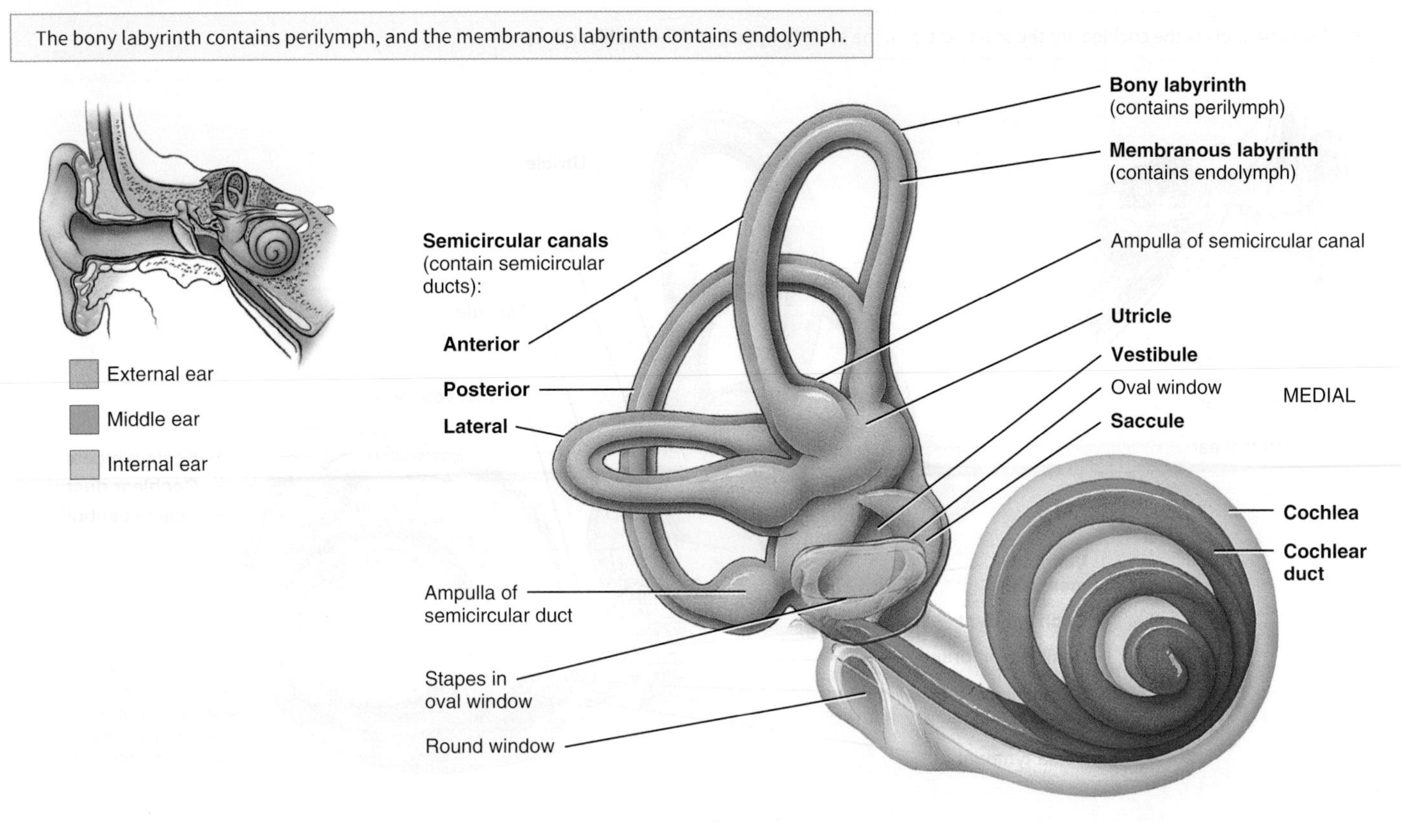

Components of the right internal ear

Q What are the names of the two sacs that lie in the membranous labyrinth of the vestibule?

The **vestibule** (VES-ti-būl) is the oval central portion of the bony labyrinth. The membranous labyrinth in the vestibule consists of two sacs called the **utricle** (Ū-tri-kul = little bag) and the **saccule** (SAK-ūl = little sac), which are connected by a small duct. Projecting superiorly and posteriorly from the vestibule are the three bony **semicircular canals**, each of which lies at approximately right angles to the other two. Based on their positions, they are named the anterior, posterior, and lateral semicircular canals. The anterior and posterior semicircular canals are vertically oriented; the lateral one is horizontally oriented. At one end of each canal is a swollen enlargement called the **ampulla** (am-PUL-la = saclike duct). The portions of the membranous labyrinth that lie inside the bony semicircular canals are called the **semicircular ducts**. These structures connect with the utricle of the vestibule.

The vestibular (ves-TIB-ū-lar) branch of the vestibulocochlear (VIII) nerve consists of *ampullary, utricular*, and *saccular nerves.* These nerves contain both first-order sensory neurons and efferent neurons that synapse with receptors for equilibrium. The first-order sensory neurons carry sensory information from the receptors, and the efferent neurons carry feedback signals to the receptors, apparently to modify their sensitivity. Cell bodies of the sensory neurons are located in the **vestibular ganglia** (see Figure 17.21b).

Anterior to the vestibule is the **cochlea** (KOK-lē-a = snail-shaped), a bony spiral canal (Figure 17.21a) that resembles a snail's shell and makes almost three turns around a central bony core called the **modiolus** (mō-DĪ-ō′-lus; Figure 17.21b). Sections through the cochlea reveal that it is divided into three channels: cochlear duct, scala vestibuli, and scala tympani (Figure 17.21a–c). The **cochlear duct** (KOK-lē-ar) or *scala media* is a continuation of the membranous labyrinth into the cochlea; it is filled with endolymph. The channel above the cochlear duct is the **scala vestibuli**, which ends at the oval window. The channel below is the **scala tympani**, which ends at the round window. Both the scala vestibuli and scala tympani are part of the bony labyrinth of the cochlea; therefore, these chambers are filled with perilymph. The scala vestibuli and scala tympani are completely separated by the cochlear duct, except for an opening at the apex of the cochlea, the **helicotrema** (hel-i-kō-TRĒ-ma; see Figure 17.22). The cochlea adjoins the wall of the vestibule, into which the scala vestibuli opens. The perilymph in the vestibule is continuous with that of the scala vestibuli.

The **vestibular membrane** separates the cochlear duct from the scala vestibuli, and the **basilar membrane** (BĀS-i-lar) separates the cochlear duct from the scala tympani. Resting on the basilar membrane is the **spiral organ** or *organ of Corti* (KOR-tē) (Figure 17.21c, d). The spiral organ is a coiled sheet of epithelial cells, including supporting cells and about 16,000 **hair cells**, which are the receptors for hearing. There are two groups of hair cells: The *inner hair cells* are arranged in a

FIGURE 17.21 Semicircular canals, vestibule, and cochlea of the right ear. Note that the cochlea makes nearly three complete turns.

The three channels in the cochlea are the scala vestibuli, the scala tympani, and the cochlear duct.

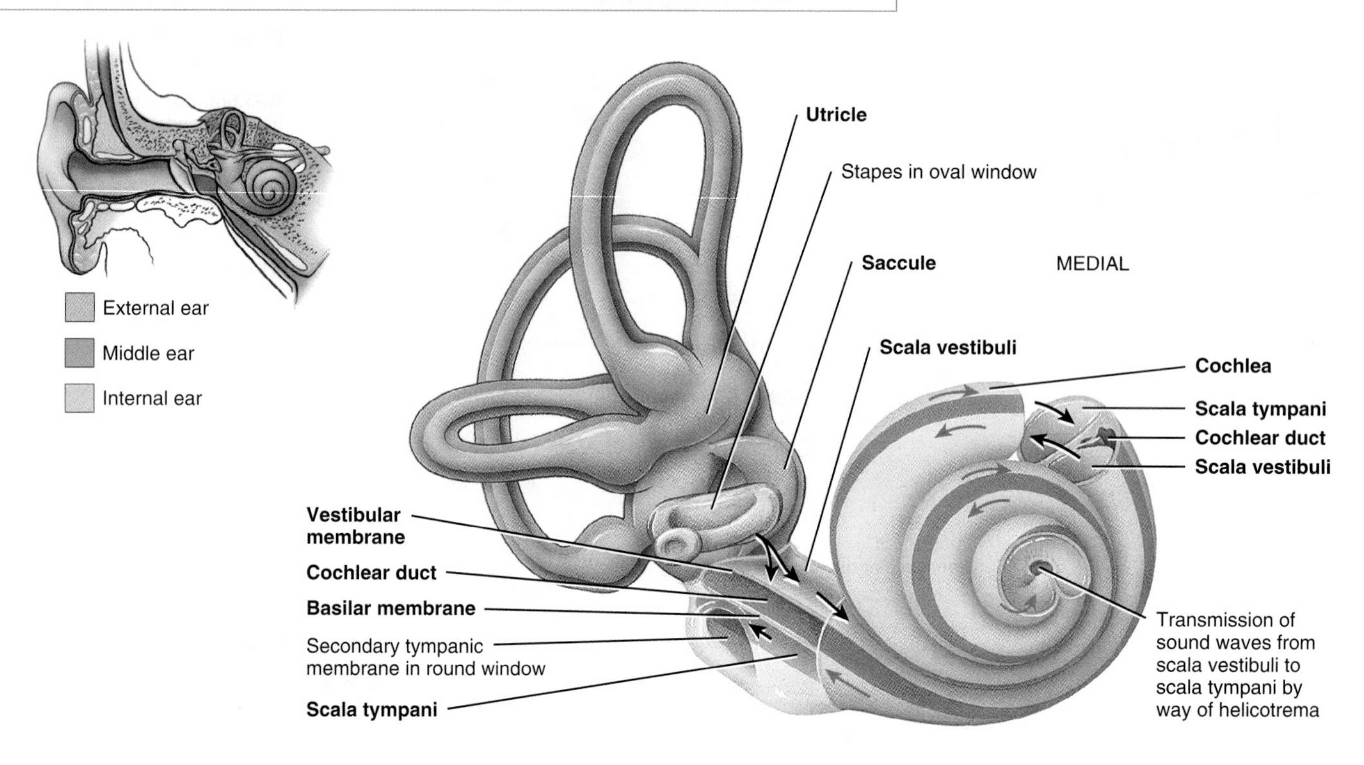

(a) Sections through the cochlea

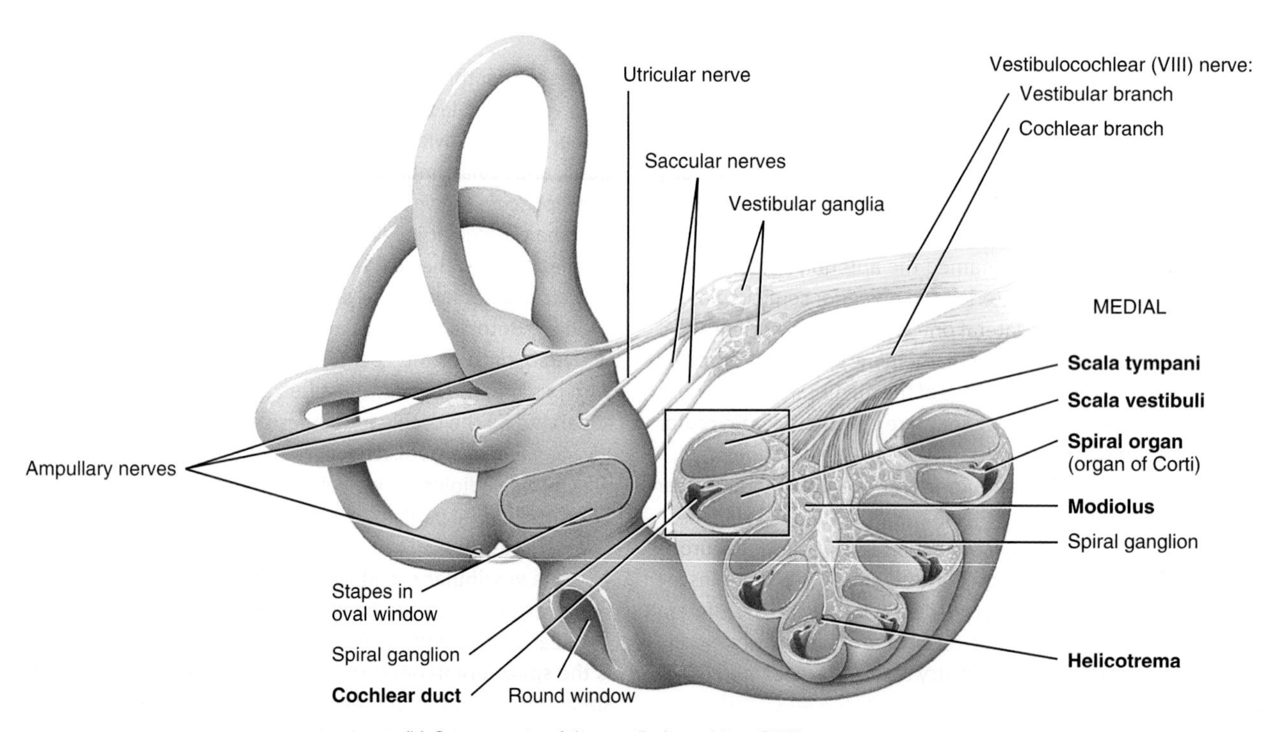

(b) Components of the vestibulocochlear (VIII) nerve

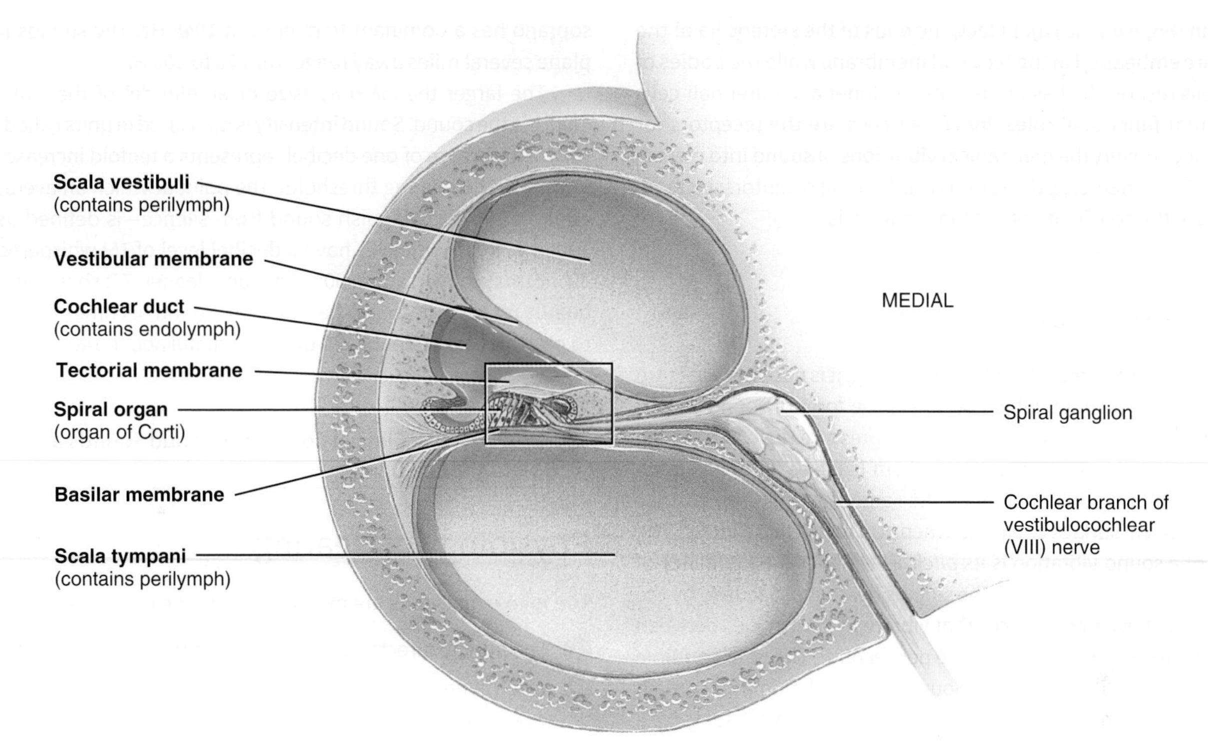

(c) Section through one turn of the cochlea

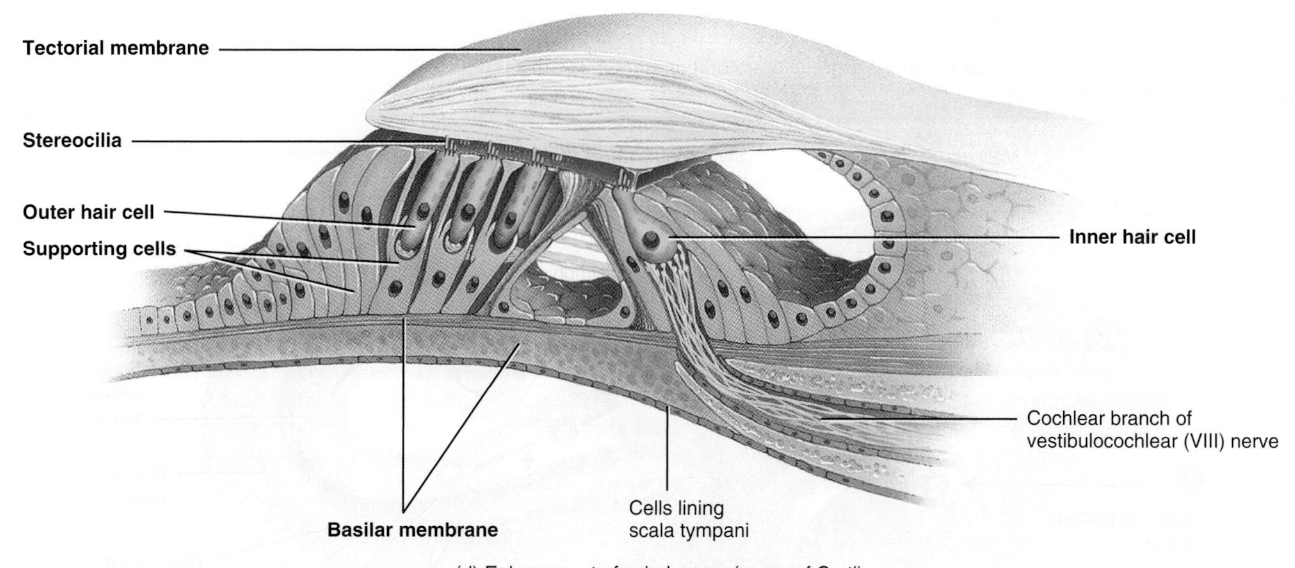

(d) Enlargement of spiral organ (organ of Corti)

Q What are the three subdivisions of the bony labyrinth?

single row, whereas the *outer hair cells* are arranged in three rows. At the apical tip of each hair cell are **stereocilia** that extend into the endolymph of the cochlear duct. Despite their name, stereocilia are actually long, hairlike microvilli arranged in several rows of graded height.

At their basal ends, inner and outer hair cells synapse both with first-order sensory neurons and with motor neurons from the cochlear branch of the vestibulocochlear (VIII) nerve. Cell bodies of the sensory neurons are located in the **spiral ganglion** (**Figure 17.21b, c**). Although outer hair cells outnumber them by 3 to 1, the inner hair cells synapse with 90–95% of the first-order sensory neurons in the cochlear nerve that relay auditory information to the brain. By contrast, 90% of the motor neurons in the cochlear nerve synapse with outer hair cells. The **tectorial membrane** (tek-TŌ-rē-al; *tector-* = covering), a flexible gelatinous membrane, covers the hair cells of the spiral organ (Figure 17.21d). In fact, the ends of the stereocilia of the hair cells are embedded in the tectorial membrane while the bodies of the hair cells rest on the basilar membrane. Inner and outer hair cells have different functional roles. Inner hair cells are the receptors for hearing: They convert the mechanical vibrations of sound into electrical signals. Outer hair cells do not serve as hearing receptors; instead, they increase the sensitivity of the inner hair cells.

The Nature of Sound Waves

In order to understand the physiology of hearing, it is necessary to learn something about its input, which occurs in the form of sound waves. **Sound waves** are alternating high- and low-pressure regions traveling in the same direction through some medium (such as air). They originate from a vibrating object in much the same way that ripples arise and travel over the surface of a pond when you toss a stone into it. The *frequency* of a sound vibration is its *pitch*. The higher the frequency of vibration, the higher is the pitch. The sounds heard most acutely by the human ear are those from sources that vibrate at frequencies between 500 and 5000 **hertz (Hz**; 1 Hz = 1 cycle per second). The entire audible range extends from 20 to 20,000 Hz. Sounds of speech primarily contain frequencies between 100 and 3000 Hz, and the "high C" sung by a soprano has a dominant frequency at 1048 Hz. The sounds from a jet plane several miles away range from 20 to 100 Hz.

The larger the *intensity* (size or amplitude) of the vibration, the *louder* is the sound. Sound intensity is measured in units called **decibels (dB)**. An increase of one decibel represents a tenfold increase in sound intensity. The hearing threshold—the point at which an average young adult can just distinguish sound from silence—is defined as 0 dB at 1000 Hz. Rustling leaves have a decibel level of 15; whispered speech, 30; normal conversation, 60; a vacuum cleaner, 75; shouting, 80; and a nearby motorcycle or jackhammer, 90. Sound becomes uncomfortable to a normal ear at about 120 dB, and painful above 140 dB.

See Clinical Connection: Loud Sounds and Hair Cell Damage

Physiology of Hearing

The following events are involved in hearing (Figure 17.22):

1. The auricle directs sound waves into the external auditory canal.
2. When sound waves strike the tympanic membrane, the alternating waves of high and low pressure in the air cause the tympanic

FIGURE 17.22 Events in the stimulation of auditory receptors in the right ear. The cochlea has been uncoiled to more easily visualize the transmission of sound waves and their distortion of the vestibular and basilar membranes of the cochlear duct.

Hair cells of the spiral organ (organ of Corti) convert a mechanical vibration (stimulus) into an electrical signal (receptor potential).

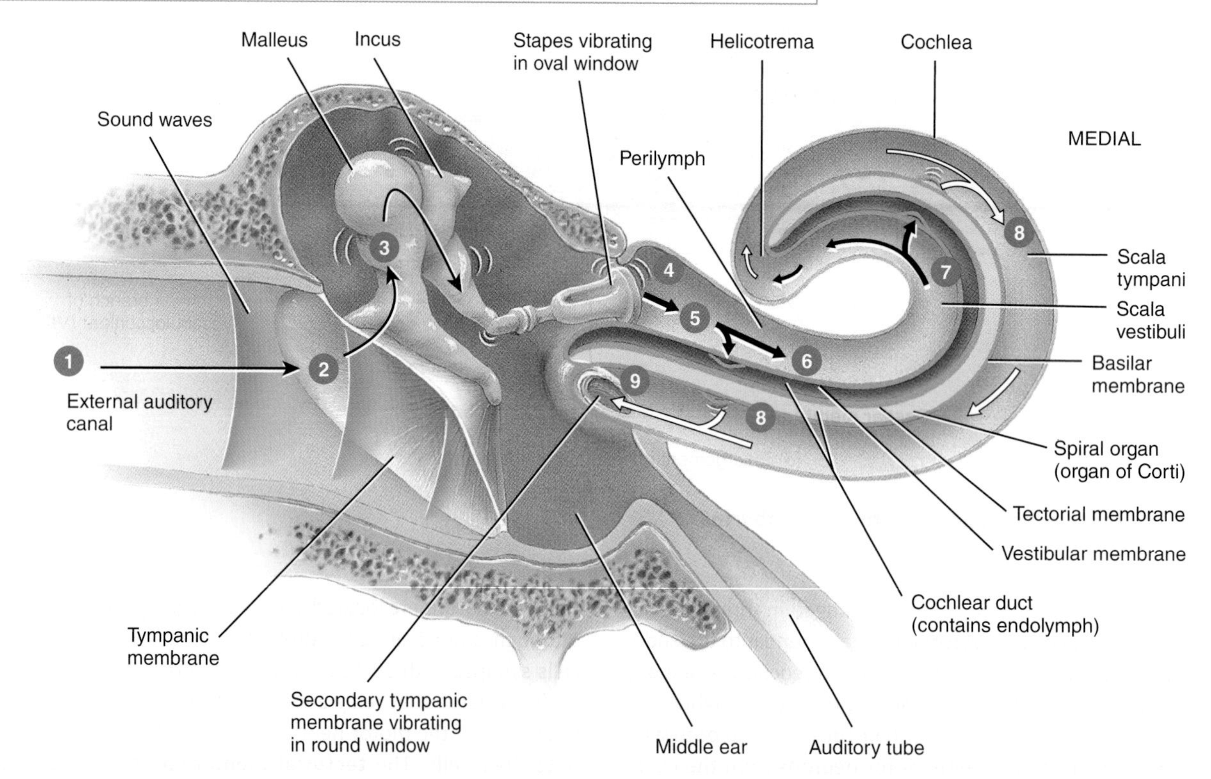

Q Which part of the basilar membrane vibrates most vigorously in response to high-frequency (high-pitched) sounds?

membrane to vibrate back and forth. The tympanic membrane vibrates slowly in response to low-frequency (low-pitched) sounds and rapidly in response to high-frequency (high-pitched) sounds.

3. The central area of the tympanic membrane connects to the malleus, which vibrates along with the tympanic membrane. This vibration is transmitted from the malleus to the incus and then to the stapes.
4. As the stapes moves back and forth, its oval-shaped footplate, which is attached via a ligament to the circumference of the oval window, vibrates in the oval window. The vibrations at the oval window are about 20 times more vigorous than those of the tympanic membrane because the auditory ossicles efficiently transmit small vibrations spread over a large surface area (the tympanic membrane) into larger vibrations at a smaller surface (the oval window).
5. The movement of the stapes at the oval window sets up fluid pressure waves in the perilymph of the cochlea. As the oval window bulges inward, it pushes on the perilymph of the scala vestibuli.
6. Pressure waves are transmitted from the scala vestibuli to the scala tympani and eventually to the round window, causing it to bulge outward into the middle ear. (See 9 in the figure.)
7. As the pressure waves deform the walls of the scalea vestibuli and scala tympani, they also push the vestibular membrane back and forth, creating pressure waves in the endolymph inside the cochlear duct.
8. The pressure waves in the endolymph cause the basilar membrane to vibrate, which moves the hair cells of the spiral organ against the tectorial membrane. This leads to bending of the stereocilia and ultimately to the generation of nerve impulses in first-order neurons in cochlear nerve fibers.

Sound waves of various frequencies cause certain regions of the basilar membrane to vibrate more intensely than other regions. Each segment of the basilar membrane is "tuned" for a particular pitch. Because the membrane is narrower and stiffer at the base of the cochlea (closer to the oval window), high-frequency (high-pitched) sounds induce maximal vibrations in this region. Toward the apex of the cochlea, the basilar membrane is wider and more flexible; low-frequency (low-pitched) sounds cause maximal vibration of the basilar membrane there. Loudness is determined by the intensity of sound waves. High-intensity sound waves cause larger vibrations of the basilar membrane, which leads to a higher frequency of nerve impulses reaching the brain. Louder sounds also may stimulate a larger number of hair cells.

Sound Transduction

Inner hair cells transduce mechanical vibrations into electrical signals (**Figure 17.23**). As the basilar membrane vibrates, the stereocilia at the apex of the hair cell bend back and forth and slide against one another. Mechanically gated cation channels are located in the membrane of the stereocilia. Opening these channels allows cations in the endolymph, primarily K^+, to enter the hair cell cytosol. (Recall that K^+ levels in endolymph are very high, which is not normally the case in other extracellular fluids of the body.) As cations enter, they produce a depolarizing receptor potential. A *tip link* protein connects a mechanically gated cation channel in a stereocilium to the tip of its taller stereocilium neighbor. When the hair cell is at rest, the stereocilia point straight up and the cation channels are in a partially open state (**Figure 17.23a**). This allows a few K^+ to enter the cell, causing a weak depolarizing receptor potential. The weak depolarization

FIGURE 17.23 Sound transduction.

Hair cells of the spiral organ convert a mechanical vibration into a receptor potential.

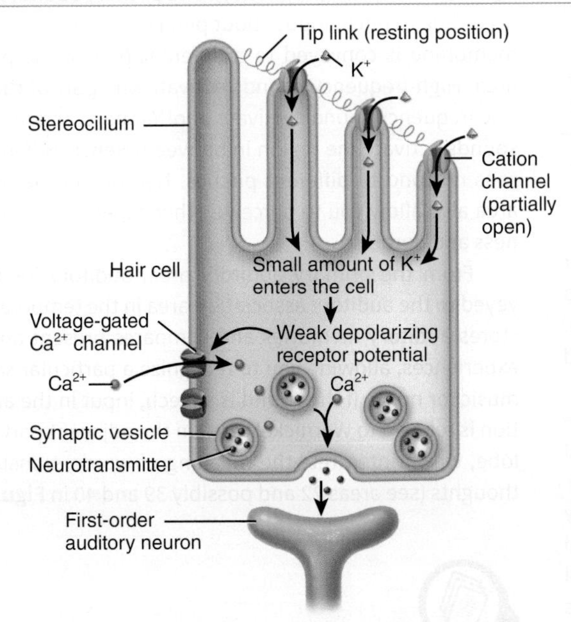

(a) Resting hair cell (weakly depolarized)

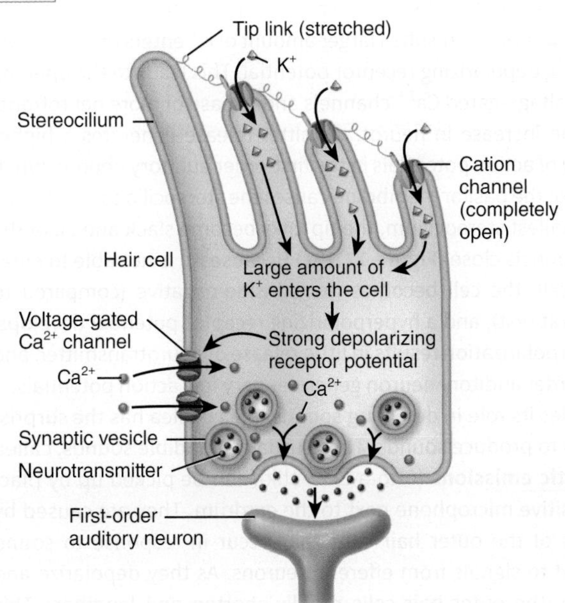

(b) Strongly depolarized hair cell

Figure 17.23 *Continues*

FIGURE 17.23 Continued

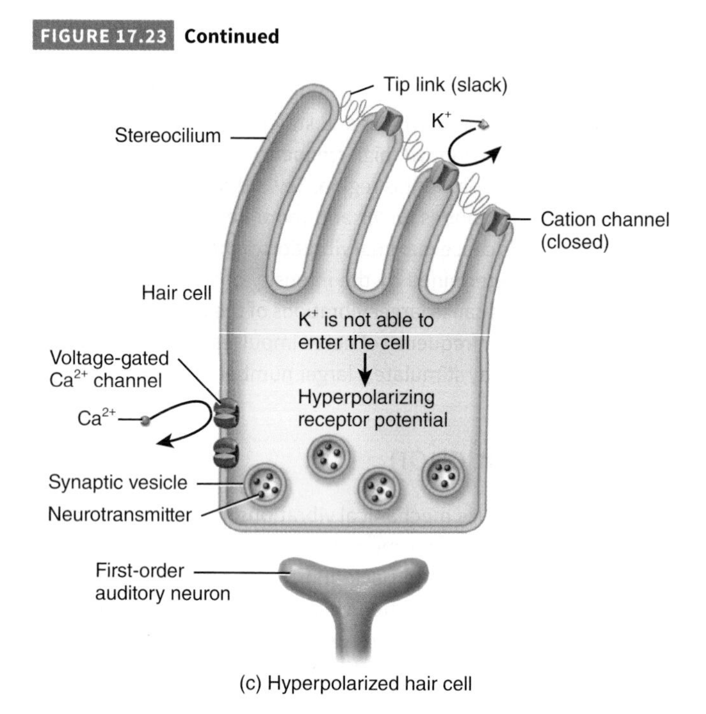

(c) Hyperpolarized hair cell

Q What is the purpose of the tip link proteins associated with the hair cells of the spiral organ.

spreads along the plasma membrane and opens a few voltage-gated Ca^{2+} channels in the base of the cell. As a result, a small amount of Ca^{2+} enters the cell and triggers exocytosis of a small number of synaptic vesicles containing neurotransmitter. The low level of neurotransmitter release generates a low frequency of action potentials in the first-order auditory neuron that synapses with the hair cell. When vibration of the basilar membrane causes the stereocilia to bend toward the tallest stereocilium, the tip links are stretched and tug on the cation channels, causing the cation channels to completely open (Figure 17.23b). As a result, a larger amount of K^+ enters the cell, causing a strong depolarizing receptor potential. This leads to the opening of more voltage-gated Ca^{2+} channels and release of more neurotransmitter. The increase in neurotransmitter release generates a higher frequency of action potentials in the first-order auditory neuron. When vibration of the basilar membrane causes the stereocilia to bend away from the tallest stereocilium, the tip links become slack and all of the cation channels close (Figure 17.23c) Because K^+ is not able to enter the hair cell, the cell becomes more inside-negative (compared to when it is at rest), and a hyperpolarizing receptor potential develops. This hyperpolarization results in little release of neurotransmitter, and the first-order auditory neuron generates very few action potentials.

Besides its role in detecting sounds, the cochlea has the surprising ability to produce sounds. These usually inaudible sounds, called **otoacoustic emissions** (ō-tō-a-KOO-stik), can be picked up by placing a sensitive microphone next to the eardrum. They are caused by vibrations of the outer hair cells that occur in response to sound waves and to signals from efferent neurons. As they depolarize and repolarize, the outer hair cells rapidly shorten and lengthen. This vibratory behavior appears to change the stiffness of the tectorial membrane and is thought to enhance the movement of the basilar membrane, which amplifies the responses of the inner hair cells. At the same time, the outer hair cell vibrations set up a traveling wave that goes back toward the stapes and leaves the ear as an otoacoustic emission. Detection of these inner ear–produced sounds is a fast, inexpensive, and noninvasive way to screen newborns for hearing defects. In deaf babies, otoacoustic emissions are not produced or are greatly reduced in size.

The Auditory Pathway

The release of neurotransmitter from hair cells of the spiral organ ultimately generates action potentials in the first-order auditory neurons that innervate the hair cells. The axons of these neurons form the cochlear branch of the vestibulocochlear (VIII) nerve (Figure 17.24). These axons synapse with neurons in the **cochlear nuclei** in the medulla oblongata. Some of the axons from the cochlear nuclei decussate (cross over) in the medulla, ascend in a tract called the **lateral lemniscus** on the opposite side, and terminate in the **inferior colliculus** of the midbrain. Other axons from the cochlear nuclei end in the **superior olivary nucleus** of the pons. Slight differences in the timing of action potentials arriving from the two ears at the superior olivary nuclei allow us to locate the source of a sound. Axons from the superior olivary nuclei ascend to the midbrain, where they terminate in the inferior colliculi. From each inferior colliculus, axons extend to the **medical geniculate nucleus** of the thalamus. Neurons in the thalamus, in turn, project axons to the **primary auditory area** of the cerebral cortex in the temporal lobe of the cerebrum (see areas 41 and 42 in Figure 14.15) in the temporal lobe, where conscious awareness of sound occurs. From the primary auditory cortex, axons extend to the **auditory association area** of the cerebral cortex in the temporal lobe of the cerebrum (see area 22 in Figure 14.15) for more complex integration of sound input.

The arrival of action potentials in the primary auditory area allows you to perceive sound. One aspect of sound that is perceived by this area is pitch (frequency). The primary auditory area is mapped according to pitch: Input about pitch from each portion of the basilar membrane is conveyed to a different part of the primary auditory area. High-frequency sounds activate one part of the auditory area, low-frequency sounds activate another part, and medium-frequency sounds activate the region in between. Hence, different cortical neurons respond to different pitches. Neurons in the primary auditory area also allow you to perceive other aspects of sound such as loudness and duration.

From the primary auditory area, auditory information is conveyed to the auditory association area in the temporal lobe. This area stores auditory memories and compares present and past auditory experiences, allowing you to recognize a particular sound as speech, music, or noise. If the sound is speech, input in the auditory association is relayed to Wernicke's area in the adjacent part of the temporal lobe, which interprets the meaning of words, translating them into thoughts (see areas 22 and possibly 39 and 40 in Figure 14.15).

See Clinical Connection: Cochlear Implants

FIGURE 17.24 **The auditory pathway.**

From hair cells of the cochlea, auditory information is conveyed along the cochlear branch of the vestibulocochlear (VIII) nerve and then to the brainstem, thalamus, and cerebral cortex.

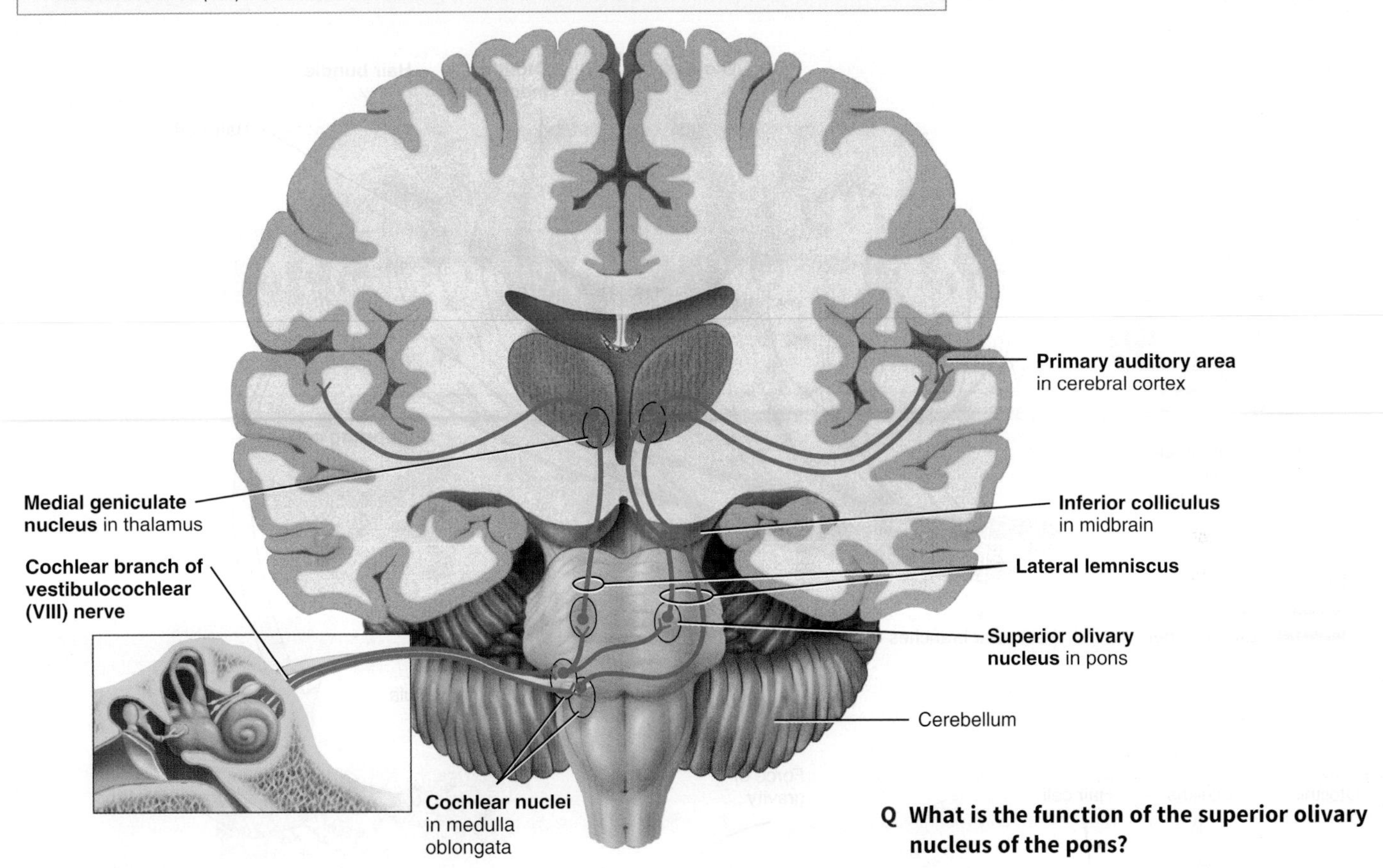

Q What is the function of the superior olivary nucleus of the pons?

17.8 Equilibrium

OBJECTIVES

- **Explain** the function of each of the receptor organs for equilibrium.
- **Describe** the equilibrium pathway to the brain.

The ear not only detects sound, but also changes in **equilibrium** (ē-kwi-LIB-rē-um) or balance. Body movements that stimulate the receptors for equilibrium include linear acceleration or deceleration, such as when a car suddenly takes off or stops; tilting the head forward or backward, as if to say "yes"; and rotational (angular) acceleration or deceleration, such as when a rollercoaster takes a quick curve. Collectively, the receptor organs for equilibrium are called the **vestibular apparatus** (ves-TIB-ū-lar); these include the *utricle* and *saccule* of the vestibule and the *semicircular ducts* of the semicircular canals.

Otolithic Organs: Utricle and Saccule

The two **otolithic organs** are the utricle and saccule. Attached to the inner walls of both the utricle and the saccule is a small, thickened region called the **macula** (MAK-ū-la; **Figure 17.25**). The two *maculae* (plural) (MAK-ū-lē) contain the receptors for linear acceleration or deceleration and the position of the head (head tilt). The maculae consist of two types of cells: **hair cells**, which are the sensory receptors, and **supporting cells**. Hair cells have on their surface stereocilia (which are actually microvilli) of graduated height, plus one *kinocilium*, a conventional cilium that extends beyond the longest stereocilium. As in the cochlea, the stereocilia are connected by tip links. Collectively, the stereocilia and kinocilium are called a **hair bundle**. Scattered among the hair cells are columnar supporting cells that probably secrete the thick, gelatinous, glycoprotein layer, called the **otolithic membrane** (ō-tō-LITH-ik), that rests on the hair cells. A layer of dense calcium carbonate crystals, called **otoliths** (Ō-tō-liths; *oto-* = ear; *-liths* = stones) extends over the entire surface of the otolithic membrane.

The maculae of the utricle and saccule are perpendicular to one another. When the head is in an upright position, the macula of the

FIGURE 17.25 Location and structure of receptors in the maculae of the right ear. Both first-order sensory neurons (blue) and efferent neurons (red) synapse with the hair cells.

The movement of stereocilia initiates depolarizing receptor potentials.

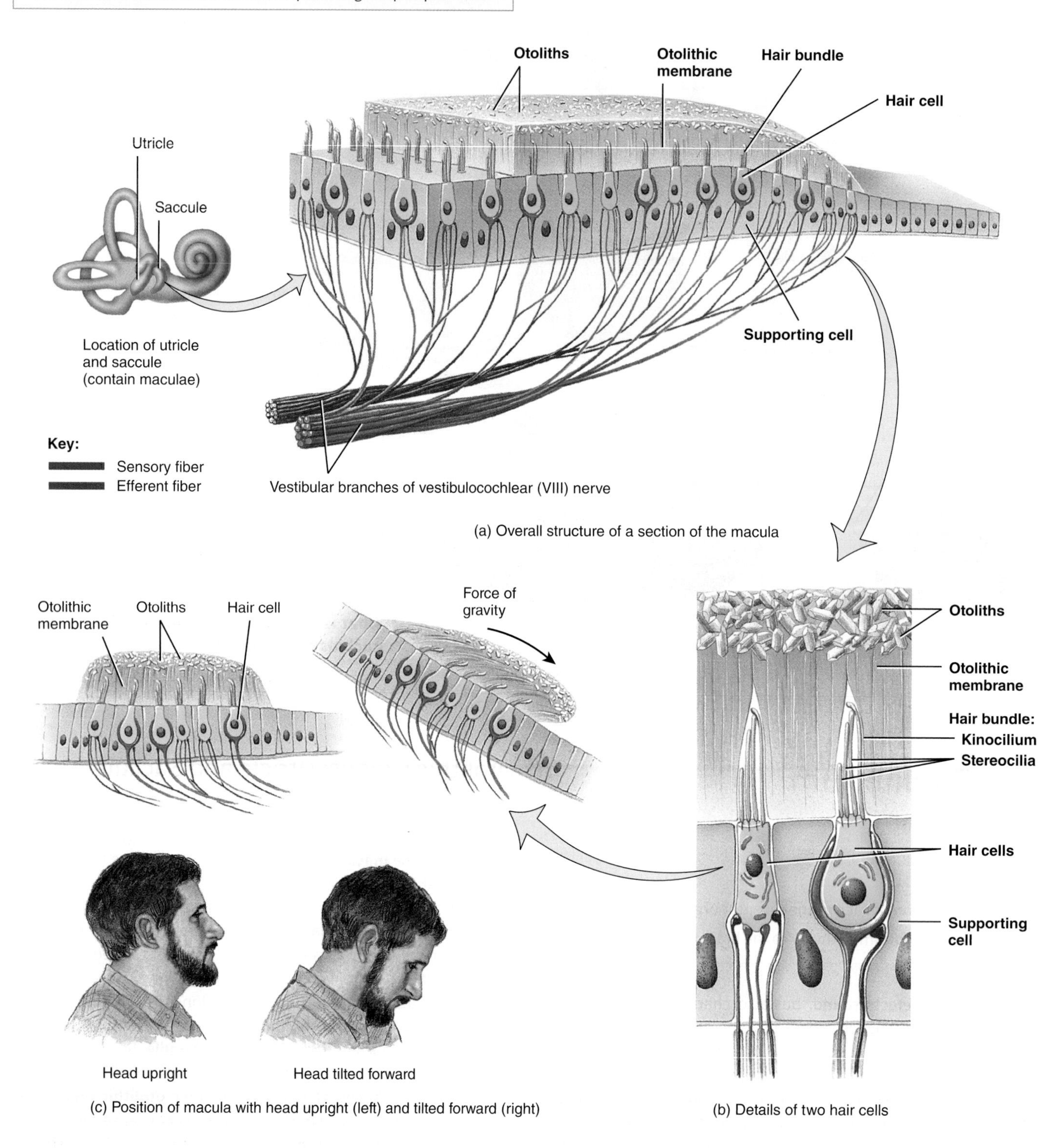

Q What are the functions of the utricle and the saccule?

utricle is oriented horizontally and the macula of the saccule is oriented vertically. Because of these orientations, the utricle and saccule have different functional roles. The utricle responds to linear acceleration or deceleration that occurs in a horizontal direction, such as when the body is being moved in a car that is speeding up or slowing down. The utricle also responds when the head tilts forward or backward. The saccule responds to linear acceleration or deceleration that occurs in a vertical direction, such as when the body is being moved up or down in an elevator.

Because the otolithic membrane sits on top of the macula, if you tilt your head forward, the otolithic membrane (along with the otoliths) is pulled by gravity. It slides "downhill" over the hair cells in the direction of the tilt, bending the hair bundles. However, if you are sitting upright in a car that suddenly jerks forward, the otolithic membrane lags behind the head movement due to inertia, pulls on the hair bundles, and makes them bend in the other direction. Bending of the hair bundles in one direction stretches the tip links, which pulls open cation channels, producing depolarizing receptor potentials; bending in the opposite direction closes the cation channels and produces hyperpolarization.

As the hair cells depolarize and hyperpolarize, they release neurotransmitter at a faster or slower rate. The hair cells synapse with first-order sensory neurons of the vestibular branch of the vestibulocochlear (VIII) nerve (see **Figure 17.21b**). These neurons fire nerve impulses at a slow or rapid pace depending on the amount of neurotransmitter present. Efferent neurons also synapse with the hair cells and sensory neurons. Evidently, the efferent neurons regulate the sensitivity of the hair cells and sensory neurons.

Semicircular Ducts

The three semicircular ducts lie at right angles to one another in three planes (**Figure 17.26**). The two vertical ducts are the anterior and

FIGURE 17.26 Location and structure of the semicircular ducts of the right ear. Both first-order sensory neurons (blue) and efferent neurons (red) synapse with the hair cells. The ampullary nerves are branches of the vestibular division of the vestibulocochlear (VIII) nerve.

The ampulla of each semicircular duct contains a crista that is covered by a cupula.

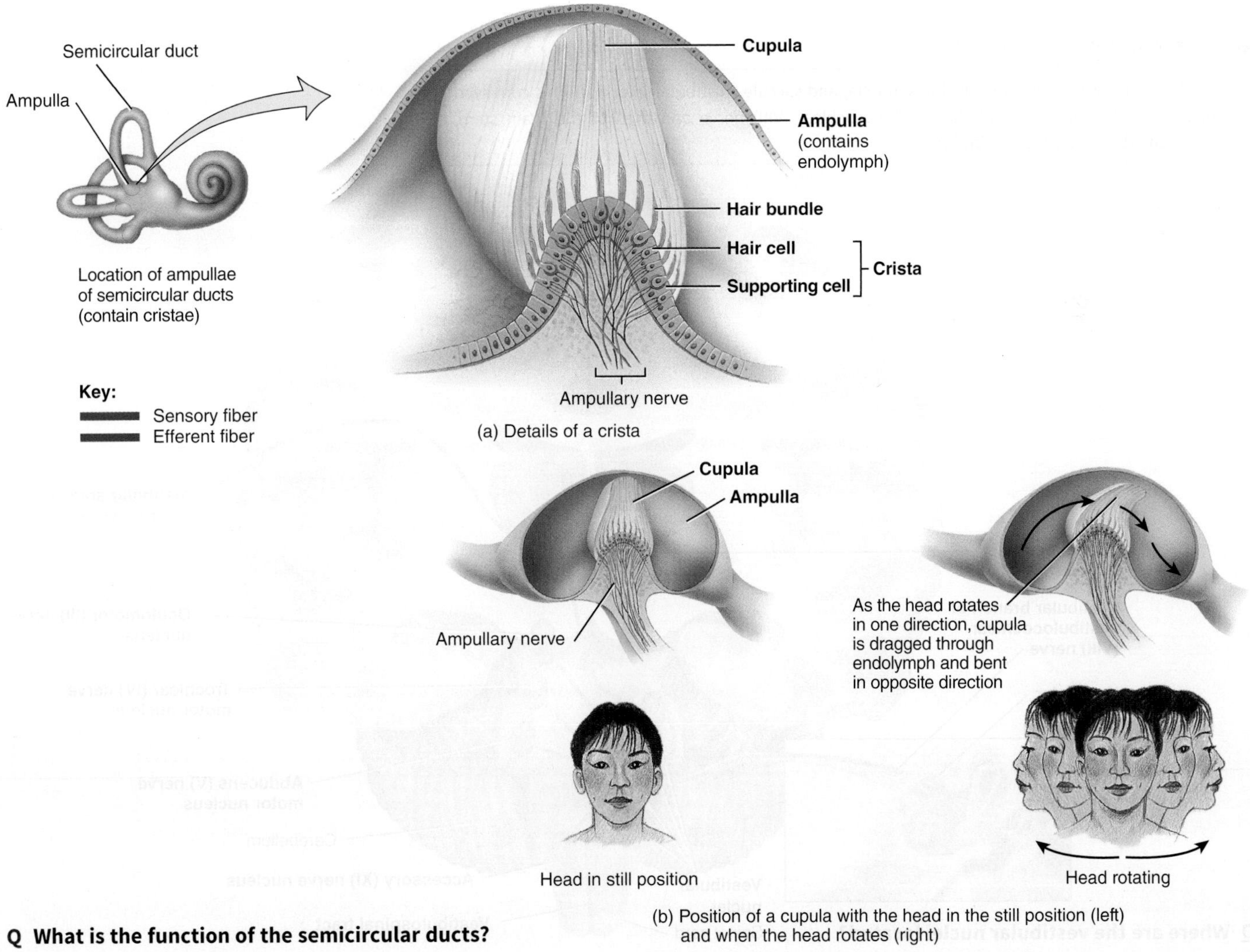

Q What is the function of the semicircular ducts?

posterior semicircular ducts, and the horizontal one is the lateral semicircular duct (see also Figure 17.20). This positioning permits detection of rotational acceleration or deceleration. The dilated portion of each duct, the **ampulla**, contains a small elevation called the **crista** (KRIS-*ta* = *crest*; plural is *cristae*). Each crista consists of a group of **hair cells** and **supporting cells**. The hair cells contain a kinocilium and stereocilia (collectively known as a **hair bundle**), and the stereocilia are interconnected via tip links. Covering the crista is a mass of gelatinous material called the **cupula** (KŪ-pū-la).

When the head rotates, the attached semicircular ducts and hair cells move with it (Figure 17.26). However, the endolymph within the ampulla is not attached and lags behind due to inertia. The drag of the endolymph causes the cupula and the hair bundles that project into it to bend in the direction opposite to that of the head movement. If the head continues to move at a steady pace, the endolymph begins to move at the same rate as the rest of the head. This causes the cupula and its embedded hair bundles to stop bending and to return to their resting positions. Once the head stops moving, the endolymph temporarily keeps moving due to inertia, which causes the cupula and its hair bundles to bend in the same direction as the preceding head movement. At some point the endolymph stops moving and the cupula and its hair bundles return to their resting, unbent positions. Note that bending the hair bundles in one direction depolarizes the hair cells; bending in the opposite direction hyperpolarizes the cells. The hair cells synapse with first-order sensory neurons of the vestibular branch of the vestibulocochlear (VIII) nerve. When hair cells are depolarized, there is a greater frequency of action potentials generated in the vestibulocochlear (VIII) nerve than when hair cells are hyperpolarized.

See Clinical Connection: Motion Sickness

Equilibrium Pathways

Bending of **hair bundles of the hair cells** in the semicircular ducts, utricle, or saccule causes the release of a neurotransmitter (probably glutamate), which generates nerve impulses in the sensory neurons that innervate the hair cells. The cell bodies of sensory neurons are located in the **vestibular ganglia**. Nerve impulses pass along the axons of these neurons, which form the **vestibular branch of the vestibulocochlear (VIII) nerve** (Figure 17.27). Most of these axons

FIGURE 17.27 The equilibrium pathway.

From hair cells of the semicircular ducts, utricle, and saccule, vestibular information is conveyed along the vestibular branch of the vestibulocochlear (VIII) nerve and then to the brain stem, cerebellum, thalamus, and cerebral cortex.

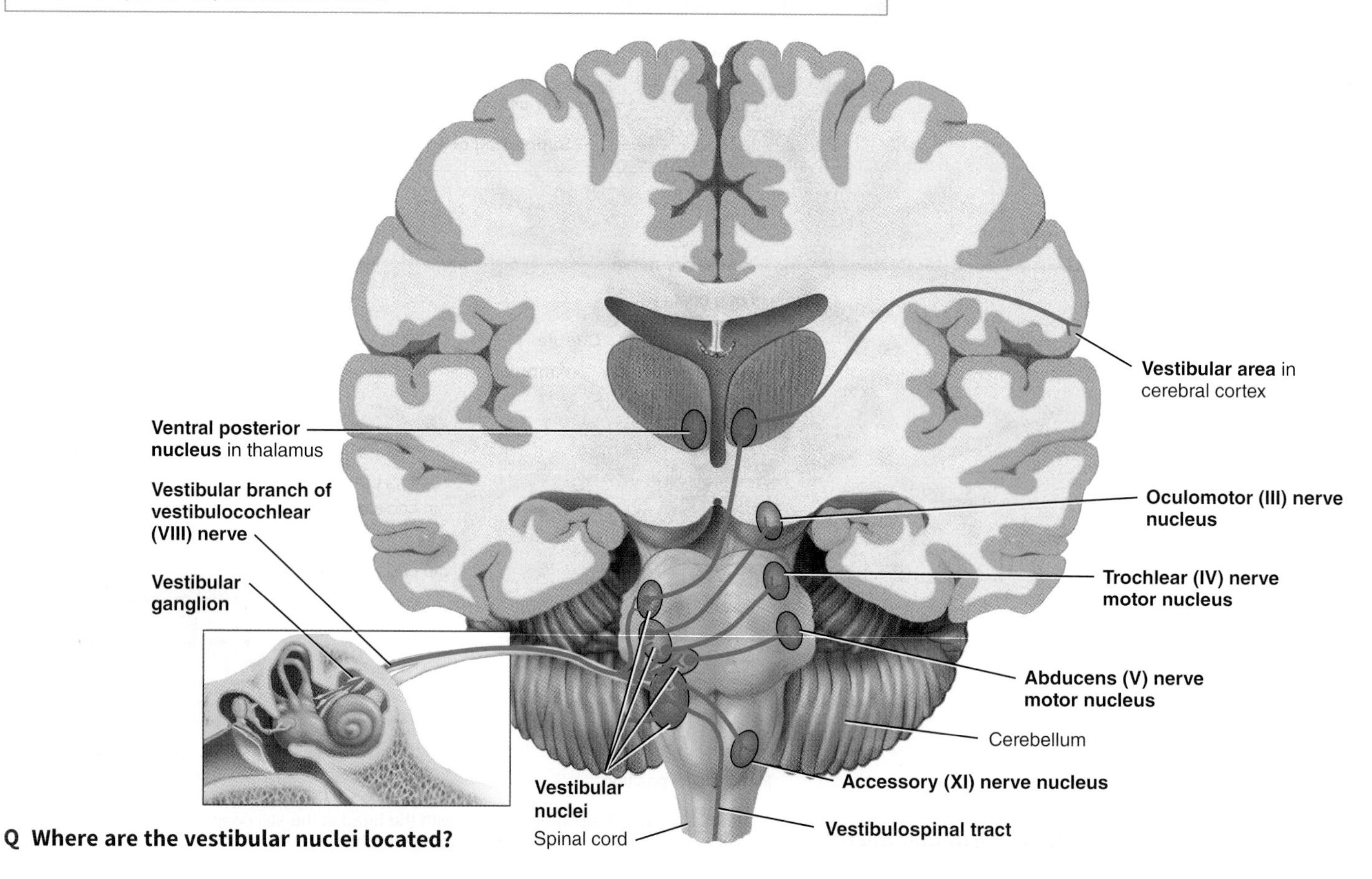

Q Where are the vestibular nuclei located?

synapse with sensory neurons in **vestibular nuclei**, the major integrating centers for equilibrium, in the medulla oblongata and pons. The vestibular nuclei also receive input from the eyes and proprioceptors, especially proprioceptors in the neck and limb muscles that indicate the position of the head and limbs. The remaining axons enter the cerebellum through the **inferior cerebellar peduncles** (see Figure 14.8b). Bidirectional pathways connect the cerebellum and vestibular nuclei.

The vestibular nuclei integrate information from vestibular, visual, and somatic receptors and then send commands to (1) the **nuclei of cranial nerves**—oculomotor (III), trochlear (IV), and abducens (VI)—that control coupled movements of the eyes with those of the head to help maintain focus on the visual field; (2) **nuclei of the accessory (XI) nerves** to help control head and neck movements to assist in maintaining equilibrium; (3) the **vestibulospinal tract**, which conveys impulses down the spinal cord to maintain muscle tone in skeletal muscles to help maintain equilibrium; and (4) the **ventral posterior nucleus** in the thalamus and then to the **vestibular area** in the parietal lobe of the cerebral cortex (which is part of the primary somatosensory area; see areas 1, 2, and 3 in Figure 14.15) to provide us with the conscious awareness of the position and movements of the head and limbs.

Table 17.2 summarizes the structures of the ear related to hearing and equilibrium.

TABLE 17.2 Summary of Structures of the Ear

REGIONS OF THE EAR AND KEY STRUCTURES	FUNCTION
External (outer) ear External auditory canal Auricle Tympanic membrane	***Auricle (pinna):*** Collects sound waves. ***External auditory canal (external auditory meatus):*** Directs sound waves to eardrum. ***Tympanic membrane (eardrum):*** Sound waves cause it to vibrate, which in turn causes malleus to vibrate.
Middle ear 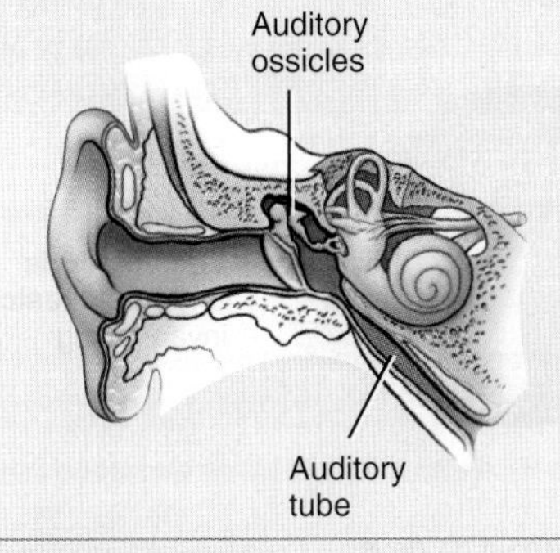 	***Auditory ossicles:*** Transmit and amplify vibrations from tympanic membrane to oval window. ***Auditory tube (eustachian tube):*** Equalizes air pressure on both sides of tympanic membrane.
Internal (inner) ear 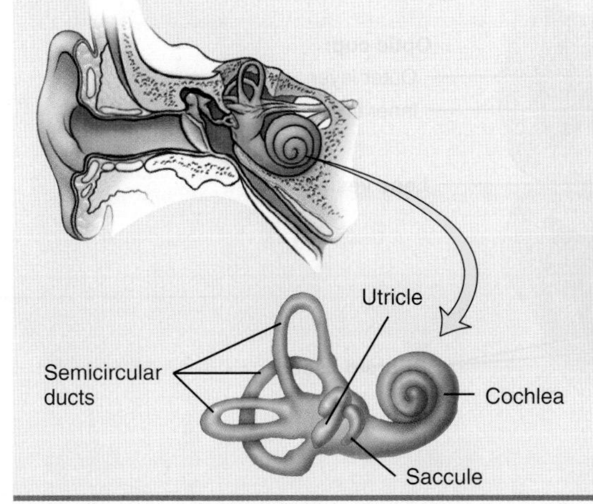 	***Cochlea:*** Contains a series of fluids, channels, and membranes that transmit vibrations to spiral organ (organ of Corti), the organ of hearing; hair cells in spiral organ produce receptor potentials, which elicit nerve impulses in cochlear branch of vestibulocochlear (VIII) nerve. ***Vestibular apparatus:*** Includes semicircular ducts, utricle, and saccule, which generate nerve impulses that propagate along vestibular branch of vestibulocochlear (VIII) nerve. ***Semicircular ducts:*** Detect rotational acceleration or deceleration. ***Utricle:*** Detects linear acceleration or deceleration that occurs in a horizontal direction and also head tilt. ***Saccule:*** Detects linear acceleration or deceleration that occurs in a vertical direction.

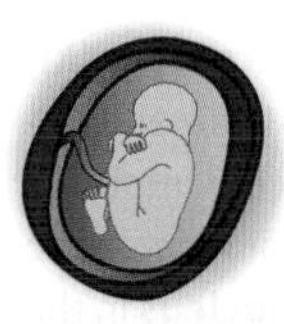

17.9 Development of the Eyes and Ears

OBJECTIVE

- **Describe** the development of the eyes and the ears.

Eyes

The *eyes* begin to develop about 22 days after fertilization when the **ectoderm** of the lateral walls of the prosencephalon (forebrain) bulges out to form a pair of shallow grooves called the **optic grooves**. Within a few days, as the neural tube is closing, the optic grooves enlarge and grow toward the surface ectoderm and become known as the **optic vesicles**. When the optic vesicles reach the surface ectoderm, the surface ectoderm thickens to form the **lens placodes** (PLAK-ods). In addition, the distal portions of the optic vesicles invaginate, forming the **optic cups**; they remain attached to the prosencephalon by narrow, hollow proximal structures called **optic stalks**. Figure 17.28 shows the stages in the development of the eyes.

The lens placodes also invaginate and develop into **lens vesicles** that sit in the optic cups. The lens vesicles eventually develop into the *lenses*. Blood is supplied to the developing lenses (and retina) by the hyaloid arteries. These arteries gain access to the developing eyes through a groove on the inferior surface of the optic cup and optic stalk called the **choroid fissure**. As the lenses mature, part of the hyaloid arteries that pass through the vitreous chamber degenerate; the remaining portions of the hyaloid arteries become the *central retinal arteries*.

FIGURE 17.28 Development of the eyes.

The eyes begin to develop about 22 days after fertilization from ectoderm of the prosencephalon.

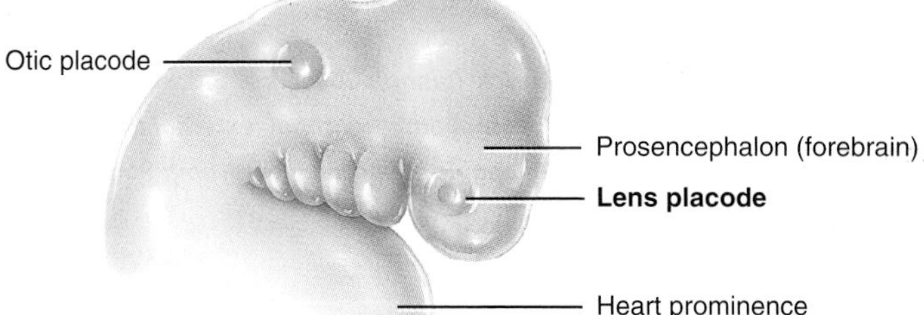

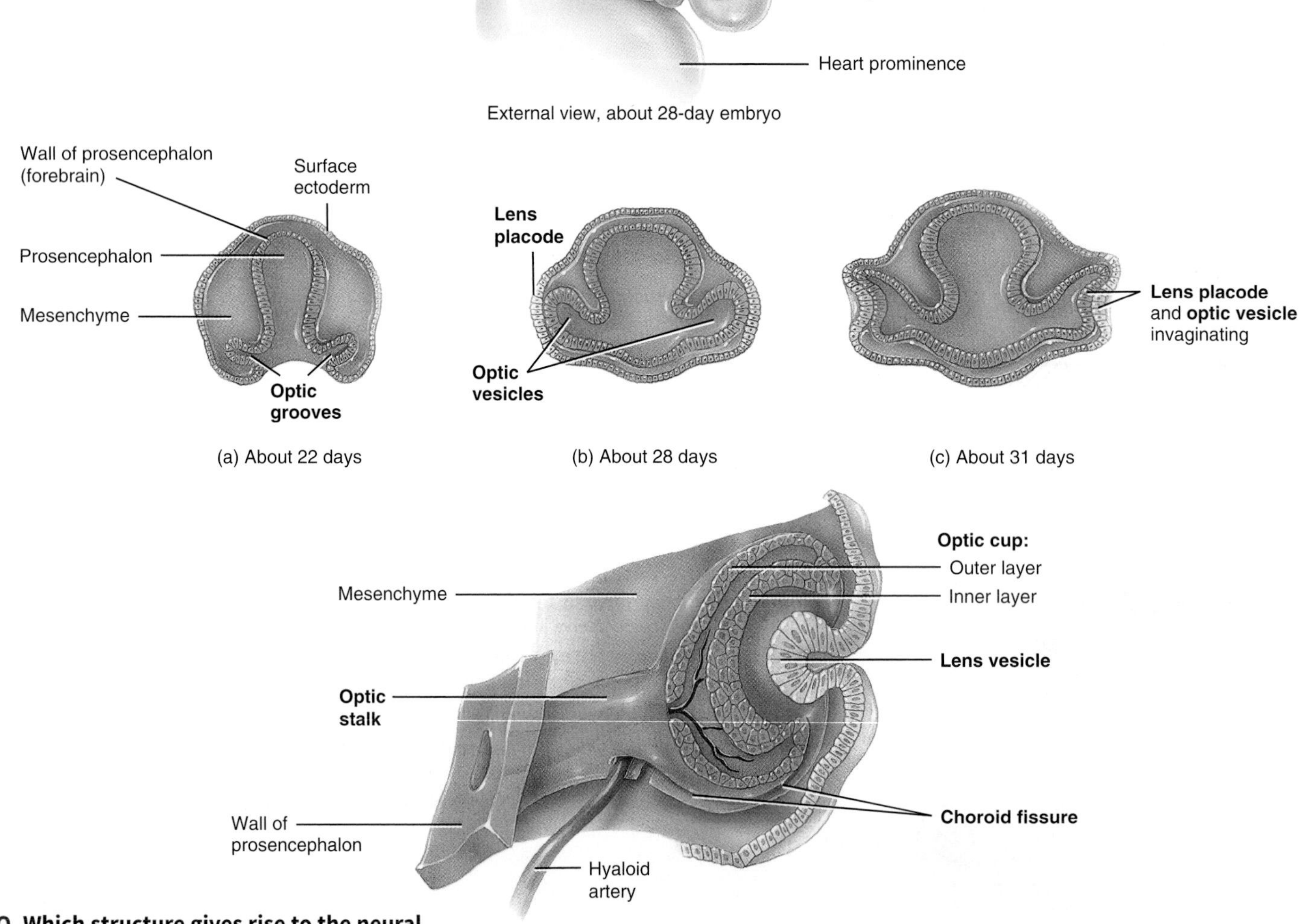

Q Which structure gives rise to the neural and pigmented layers of the retina?

The inner wall of the optic cup forms the *neural layer* of the retina, while the outer layer forms the *pigmented layer* of the retina. Axons from the neural layer grow through the optic stalk to the brain, converting the optic stalk to the *optic (II) nerve.* Although myelination of the optic nerves begins late in fetal life, it is not completed until the 10th week after birth.

The anterior portion of the optic cup forms the epithelium of the *ciliary body, iris,* and *circular and radial muscles* of the iris. The connective tissue of the ciliary body, *ciliary muscle,* and *zonular fibers* of the lens develop from **mesenchyme** around the anterior portion of the optic cup.

Mesenchyme surrounding the optic cup and optic stalk differentiates into an inner layer that gives rise to the *choroid* and an outer layer that develops into the *sclera* and part of the *cornea.* The remainder of the cornea is derived from surface ectoderm.

The *anterior chamber* develops from a cavity that forms in the mesenchyme between the iris and cornea; the *posterior chamber* develops from a cavity that forms in the mesenchyme between the iris and lens.

Some mesenchyme around the developing eye enters the optic cup through the choroid fissure. This mesenchyme occupies the space between the lens and retina and differentiates into a delicate network of fibers. Later the spaces between the fibers fill with a jellylike substance, thus forming the *vitreous body* in the vitreous chamber.

The *eyelids* form from surface ectoderm and mesenchyme. The upper and lower eyelids meet and fuse at about eight weeks of development and remain closed until about 26 weeks of development.

Ears

The first portion of the ear to develop is the *internal ear.* It begins to form about 22 days after fertilization as a thickening of the surface ectoderm, called **otic placodes** (Figure 17.29a), that appears on either side of the rhombencephalon (hindbrain). The otic placodes invaginate quickly (Figure 17.29b) to form the **otic pits** (Figure 17.29c).

FIGURE 17.29 Development of the ears.

The first parts of the ears to develop are the internal ears, which begin to form about 22 days after fertilization as thickenings of surface ectoderm

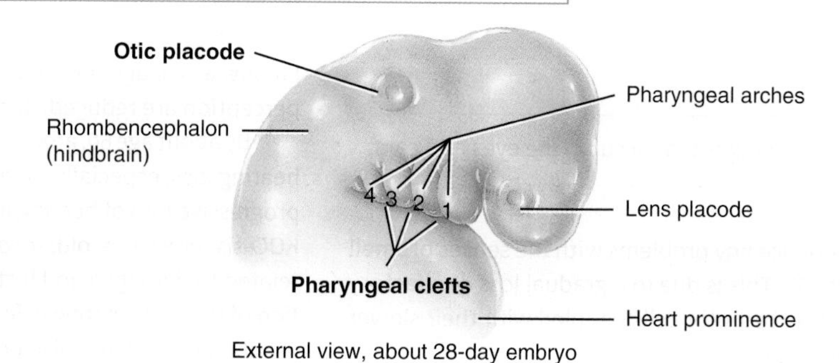

External view, about 28-day embryo

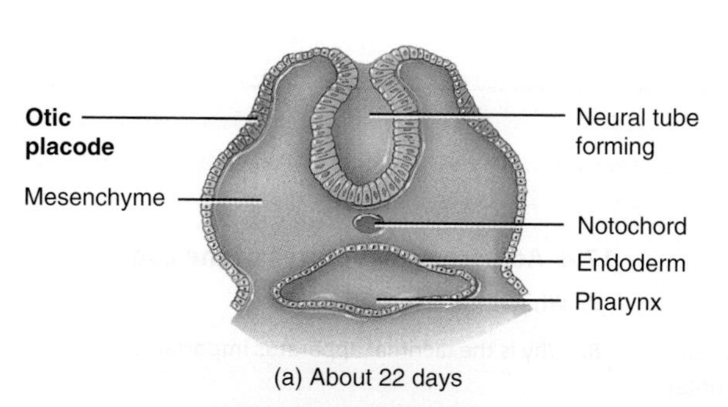

(a) About 22 days

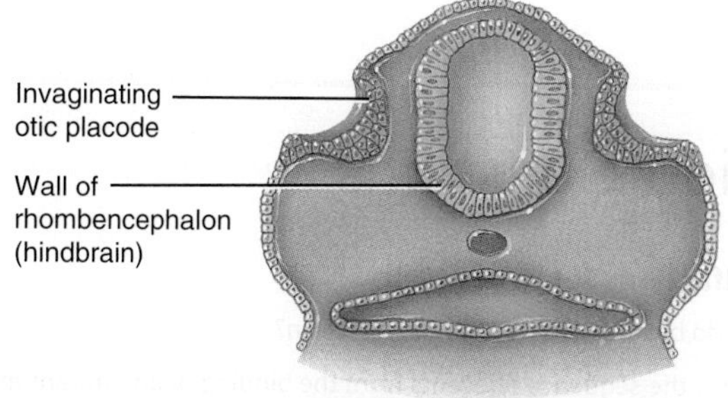

(b) About 24 days

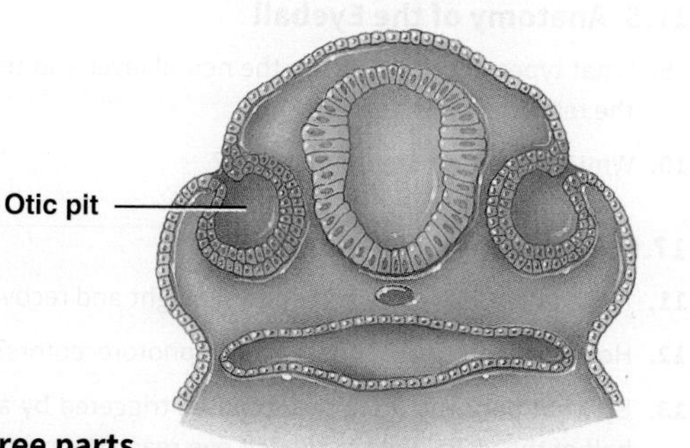

(c) About 27 days

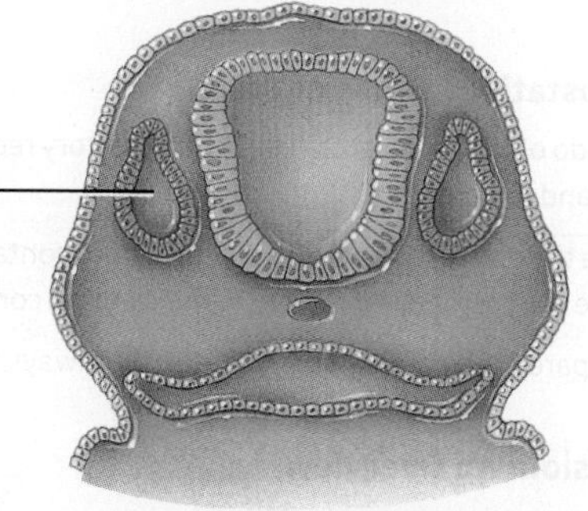

(d) About 32 days

Q How do the three parts of the ear differ in origin?

Next, the otic pits pinch off from the surface ectoderm to form the **otic vesicles** within the mesenchyme of the head (**Figure 17.29d**). During later development, the otic vesicles will form the structures associated with the *membranous labyrinth* of the internal ear. Mesenchyme around the otic vesicles produces cartilage that later ossifies to form the bone associated with the *bony labyrinth* of the internal ear.

The *middle ear* develops from a structure called the first **pharyngeal** (*branchial*) **pouch**, an **endoderm**-lined outgrowth of the primitive pharynx (see **Figure SG18.1a**). The pharyngeal pouches are discussed in detail in Section 29.1. The *auditory ossicles* develop from the first and second pharyngeal arches.

The *external ear* develops from the first **pharyngeal cleft**, an endoderm-lined groove between the first and second pharyngeal arches (see **Figure 17.29**). The pharyngeal clefts are discussed in detail in Section 29.1.

17.10 Aging and the Special Senses

OBJECTIVE

- **Describe** the age-related changes that occur in the eyes and ears.

Most people do not experience any problems with the senses of smell and taste until about age 50. This is due to a gradual loss of olfactory receptor cells and gustatory receptor cells coupled with their slower rate of replacement as we age.

Several age-related changes occur in the eyes. As noted earlier, the lens loses some of its elasticity and thus cannot change shape as easily, resulting in presbyopia (see Section 17.6). Cataracts (loss of transparency of the lenses) also occur with aging (see Disorders: Homeostatic Imbalances in the Study Guide). In old age, the sclera ("white" of the eye) becomes thick and rigid and develops a yellowish or brownish coloration due to many years of exposure to ultraviolet light, wind, and dust. The sclera may also develop random splotches of pigment, especially in people with dark complexions. The iris fades or develops irregular pigment. The muscles that regulate the size of the pupil weaken with age and the pupils become smaller, react more slowly to light, and dilate more slowly in the dark. For these reasons, elderly people find that objects are not as bright, their eyes may adjust more slowly when going outdoors, and they have problems going from brightly lit to darkly lit places. Some diseases of the retina are more likely to occur in old age, including age-related macular disease and detached retina (see reference to Clinical Connections in Sections 17.5 and 17.6). A disorder called glaucoma (see the Study Guide) develops in the eyes of aging people as a result of the buildup of aqueous humor. Tear production and the number of mucous cells in the conjunctiva may decrease with age, resulting in dry eyes. The eyelids lose their elasticity, becoming baggy and wrinkled. The amount of fat around the orbits may decrease, causing the eyeballs to sink into the orbits. Finally, as we age the sharpness of vision decreases, color and depth perception are reduced, and "vitreal floaters" increase.

By about age 60, around 25% of individuals experience a noticeable hearing loss, especially for higher-pitched sounds. The age-associated progressive loss of hearing in both ears is called **presbycusis** (pres-bē-KOO-sis; *presby-* = old; *-acou-* = hearing; *-sis* = condition). It may be related to damaged and lost hair cells in the spiral organ or degeneration of the nerve pathway for hearing. Tinnitus (ringing in the ears) and vestibular imbalance also occur more frequently in the elderly.

Review Questions

17.1 Olfaction: Sense of Smell

1. How do basal cells contribute to olfaction?
2. What is the sequence of events from the binding of an odorant molecule to an olfactory cilium to the arrival of a nerve impulse in the orbitofrontal area?

17.2 Gustation: Sense of Taste

3. How do olfactory receptor cells and gustatory receptor cells differ in structure and function?
4. Trace the path of a gustatory stimulus from contact of a tastant with saliva to the primary gustatory area in the cerebral cortex.
5. Compare the olfactory and gustatory pathways.

17.3 Vision: An Overview

6. What is visible light?

17.4 Accessory Structures of the Eye

7. What is the conjunctiva?
8. Why is the lacrimal apparatus important?

17.5 Anatomy of the Eyeball

9. What types of cells make up the neural layer and the pigmented layer of the retina?
10. Why is aqueous humor important?

17.6 Physiology of Vision

11. How do photopigments respond to light and recover in darkness?
12. How do receptor potentials arise in photoreceptors?
13. By what pathway do nerve impulses triggered by an object in the nasal half of the visual field of the left eye reach the primary visual area of the cortex?

17.7 Hearing

14. How are sound waves transmitted from the auricle to the spiral organ?
15. How do hair cells in the cochlea and vestibular apparatus transduce mechanical vibrations into electrical signals?
16. What is the pathway for auditory impulses from the cochlea to the cerebral cortex?

17.8 Equilibrium

17. Compare the functions of the utricle, saccule, and semicircular ducts.
18. What is the role of vestibular input to the cerebellum?
19. Describe the equilibrium pathways.

17.9 Development of the Eyes and Ears

20. How do the origins of the eyes and ears differ?

17.10 Aging and the Special Senses

21. What changes in the eyes and ears are related to the aging process, and how do they take place?

Critical Thinking Questions

1. Mario has experienced damage to his facial nerve. How would this affect his special senses?
2. The shift nurse brings ailing 80-year-old Gertrude her dinner. As Gertrude eats a small amount of her food, she comments that she isn't hungry and that "hospital food just doesn't taste good!" The nurse gives Gertrude a menu so she can choose her morning breakfast. Gertrude complains that she is having trouble reading the menu and asks the nurse to read it to her. As the nurse begins to read, Gertrude loudly asks her to "speak up and turn off the buzzing." What does the nurse know about aging and the special senses that help to explain Gertrude's comments?
3. As you help your neighbor put drops in her 6-year-old daughter's eyes, the daughter states, "That medicine tastes bad." How do you explain to your neighbor how her daughter can "taste" the eyedrops?

The Endocrine System

CHAPTER **18**

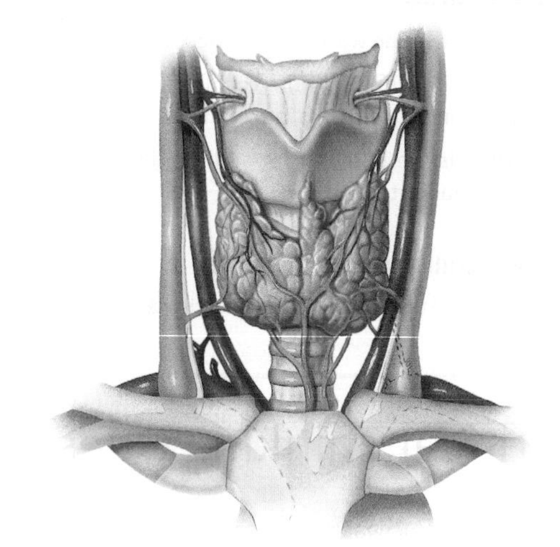

The Endocrine System and Homeostasis

The hormones of the endocrine system contribute to homeostasis by regulating the activity and growth of target cells in your body. Hormones also regulate your metabolism.

As girls and boys enter puberty, they start to develop striking differences in physical appearance and behavior. Perhaps no other period in life so dramatically shows the impact of the endocrine system in directing development and regulating body functions. In girls, estrogens promote accumulation of adipose tissue in the breasts and hips, sculpting a feminine shape. At the same time or a little later, increasing levels of testosterone in boys begin to help build muscle mass and enlarge the vocal cords, producing a lower-pitched voice. These changes are just a few examples of the powerful influence of endocrine secretions. Less dramatically, perhaps, multitudes of hormones help maintain homeostasis on a daily basis. They regulate the activity of smooth muscle, cardiac muscle, and some glands; alter metabolism; spur growth and development; influence reproductive processes; and participate in circadian (daily) rhythms established by the suprachiasmatic nucleus of the hypothalamus.

Q Did you ever wonder why thyroid gland disorders affect all major body systems?

18.1 Comparison of Control by the Nervous and Endocrine Systems

OBJECTIVE

- **Compare** control of body functions by the nervous system and endocrine system.

The nervous and endocrine systems act together to coordinate functions of all body systems. Recall that the nervous system acts through nerve impulses (action potentials) conducted along axons of neurons. At synapses, nerve impulses trigger the release of mediator (messenger) molecules called *neurotransmitters* (shown in **Figure 12.23**). The endocrine system also controls body activities by releasing mediators, called *hormones*, but the means of control of the two systems are very different.

A **hormone** (*hormon* = to excite or get moving) is a molecule that is released in one part of the body but regulates the activity of cells in other parts of the body. Most hormones enter interstitial fluid and then the bloodstream. The circulating blood delivers hormones to cells throughout the body. Both neurotransmitters and hormones exert their effects by binding to receptors on or in their "target" cells. Several chemicals act as both neurotransmitters and hormones. One familiar example is norepinephrine, which is released as a neurotransmitter by sympathetic postganglionic neurons and as a hormone by chromaffin cells of the adrenal medullae.

Responses of the endocrine system often are slower than responses of the nervous system; although some hormones act within seconds, most take several minutes or more to cause a response. The effects of nervous system activation are generally briefer than those of the endocrine system. The nervous system acts on specific muscles and glands. The influence of the endocrine system is much broader; it helps regulate virtually all types of body cells.

We will also have several opportunities to see how the nervous and endocrine systems function together as an interlocking "supersystem." For example, certain parts of the nervous system stimulate or inhibit the release of hormones by the endocrine system.

Table 18.1 compares the characteristics of the nervous and endocrine systems. In this chapter, we focus on the major endocrine glands and hormone-producing tissues and examine how their hormones govern body activities.

18.2 Endocrine Glands

OBJECTIVE

- **Distinguish** between exocrine and endocrine glands.

Recall from Chapter 4 that the body contains two kinds of glands: exocrine glands and endocrine glands. **Exocrine glands** (EKS-ō-krin; *exo-* = outside) secrete their products into ducts that carry the

TABLE 18.1 Comparison of Control by the Nervous and Endocrine Systems

CHARACTERISTIC	NERVOUS SYSTEM	ENDOCRINE SYSTEM
Molecules	Neurotransmitters released locally in response to nerve impulses.	Hormones delivered to tissues throughout body by blood.
Site of action	Close to site of release, at synapse; binds to receptors in postsynaptic membrane.	Far from site of release (usually); binds to receptors on or in target cells.
Types of target cells	Muscle (smooth, cardiac, and skeletal) cells, gland cells, other neurons.	Cells throughout body.
Time to onset of action	Typically within milliseconds (thousandths of a second).	Seconds to hours or days.
Duration of action	Generally briefer (milliseconds).	Generally longer (seconds to days).

secretions into body cavities, into the lumen of an organ, or to the outer surface of the body. Exocrine glands include sudoriferous (sweat), sebaceous (oil), mucous, and digestive glands. **Endocrine glands** (EN-dō-krin; *endo-* = within) secrete their products (hormones) into the interstitial fluid surrounding the secretory cells rather than into ducts. From the interstitial fluid, hormones diffuse into blood capillaries and blood carries them to target cells throughout the body. Because of their dependence on the cardiovascular system to distribute their products, endocrine glands are some of the most vascular tissues of the body. Considering that most hormones are required in very small amounts, circulating levels typically are low.

The endocrine glands include the pituitary, thyroid, parathyroid, adrenal, and pineal glands (**Figure 18.1**). In addition, several organs and tissues are not exclusively classified as endocrine glands but contain cells that secrete hormones. These include the hypothalamus, thymus, pancreas, ovaries, testes, kidneys, stomach, liver, small intestine, skin, heart, adipose tissue, and placenta. Taken together, all endocrine glands and hormone-secreting cells constitute the **endocrine system** (EN′-dō-krin; *endo-* = within; *-crino* = to secrete). The science of the structure and function of the endocrine glands and the diagnosis and treatment of disorders of the endocrine system is known as **endocrinology** (*-logy* = study of).

18.3 Hormone Activity

OBJECTIVES

- **Describe** how hormones interact with target-cell receptors.
- **Compare** the two chemical classes of hormones based on their solubility.

The Role of Hormone Receptors

Although a given hormone travels throughout the body in the blood, it affects only specific target cells. Hormones, like neurotransmitters, influence their target cells by chemically binding to specific protein **receptors**. Only the target cells for a given hormone have receptors that bind and recognize that hormone. For example, thyroid-stimulating hormone (TSH) binds to receptors on cells of the thyroid gland, but it does not bind to cells of the ovaries because ovarian cells do not have TSH receptors.

Receptors, like other cellular proteins, are constantly being synthesized and broken down. Generally, a target cell has 2000 to 100,000 receptors for a particular hormone. If a hormone is present in excess, the number of target-cell receptors may decrease—an effect called **down-regulation**. For example, when certain cells of the testes are exposed to a high concentration of luteinizing hormone (LH), the number of LH receptors decreases. Down-regulation makes a target cell *less sensitive* to a hormone. In contrast, when a hormone is deficient, the number of receptors may increase. This phenomenon, known as **up-regulation**, makes a target cell *more sensitive* to a hormone.

See Clinical Connection: Blocking Hormone Receptors

Circulating and Local Hormones

Most endocrine hormones are **circulating hormones**—they pass from the secretory cells that make them into interstitial fluid and then into the blood (**Figure 18.2a**). Other hormones, termed **local hormones**, act locally on neighboring cells or on the same cell that secreted them without entering the bloodstream (**Figure 18.2b**). Local hormones that act on neighboring cells are called **paracrines** (PAR-a-krins; *para-* = beside or near), and those that act on the same cell that secreted them are called **autocrines** (AW-tō-krins; *auto-* = self). One example of a local hormone is interleukin-2 (IL-2), which is released by helper T cells (a type of white blood cell) during immune responses (see Chapter 22). IL-2 helps activate other nearby immune cells, a paracrine effect. But it also acts as an autocrine by stimulating the same cell that released it to proliferate. This action generates more helper T cells that can secrete even more IL-2 and thus strengthen the immune response. Another example of a local hormone is the gas nitric oxide (NO), which is released by endothelial cells lining blood vessels. NO causes relaxation of nearby smooth muscle fibers in blood vessels, which in turn causes vasodilation (increase in blood vessel diameter). The effects of such vasodilation range from a lowering of blood pressure

FIGURE 18.1 Location of many endocrine glands. Also shown are other organs that contain endocrine cells and associated structures.

Endocrine glands secrete hormones, which circulating blood delivers to target tissues.

Functions of Hormones

1. Help regulate:
   - Chemical composition and volume of internal environment (extracellular fluid).
   - Metabolism and energy balance.
   - Contraction of smooth and cardiac muscle fibers.
   - Glandular secretions.
   - Some immune system activities.
2. Control growth and development.
3. Regulate operation of reproductive systems.
4. Help establish circadian rhythms.

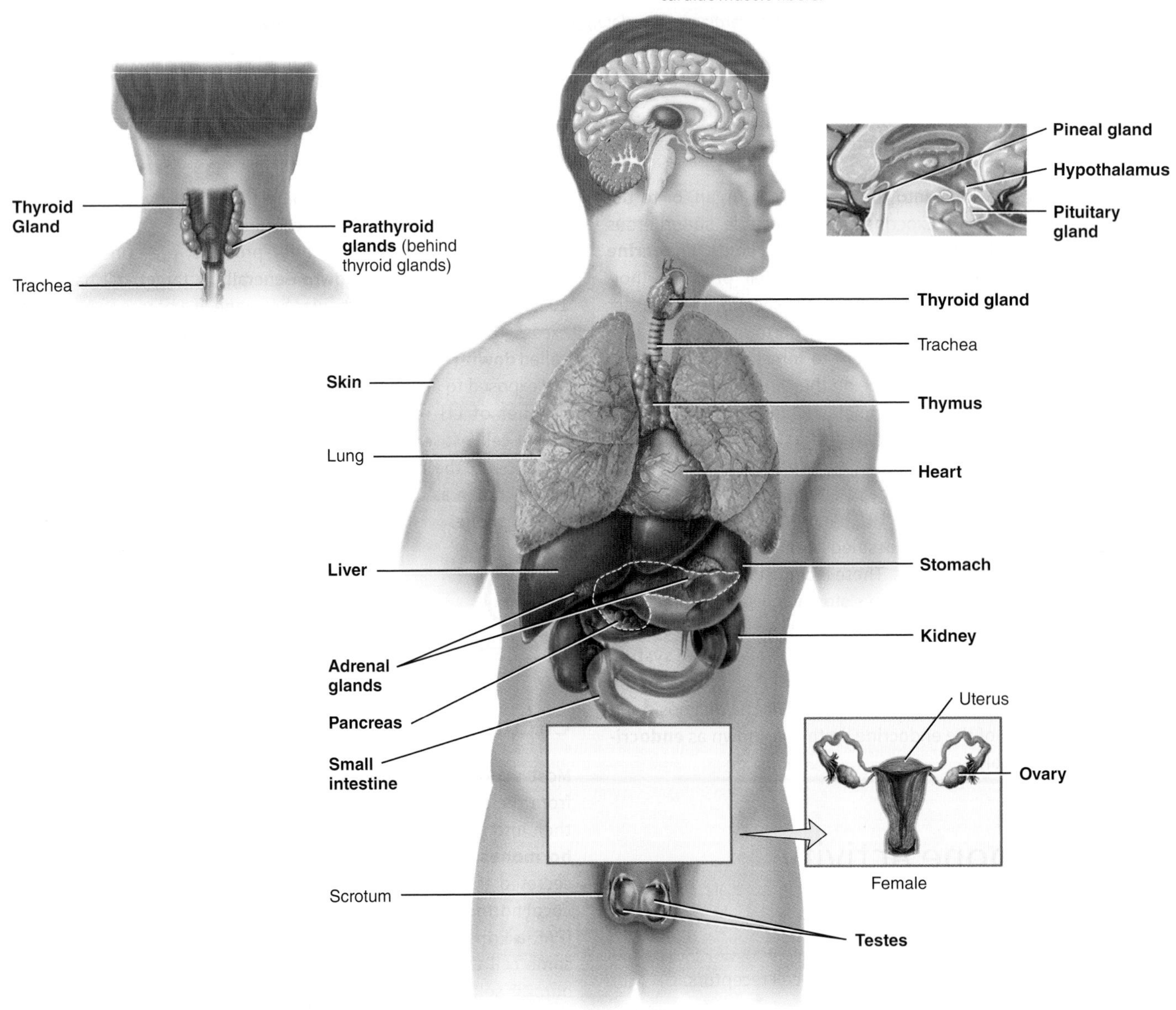

Q What is the basic difference between endocrine glands and exocrine glands?

to erection of the penis in males. The drug Viagra® (sildenafil) enhances the effects stimulated by nitric oxide in the penis.

Local hormones usually are inactivated quickly; circulating hormones may linger in the blood and exert their effects for a few minutes or occasionally for a few hours. In time, circulating hormones are inactivated by the liver and excreted by the kidneys. In cases of kidney or liver failure, excessive levels of hormones may build up in the blood.

FIGURE 18.2 Comparison between circulating hormones and local hormones (autocrines and paracrines).

Circulating hormones are carried through the bloodstream to act on distant target cells. Paracrines act on neighboring cells, and autocrines act on the same cells that produced them.

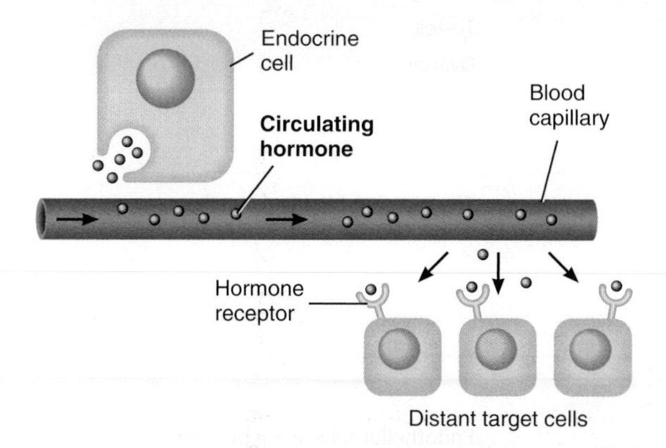

(a) Circulating hormones

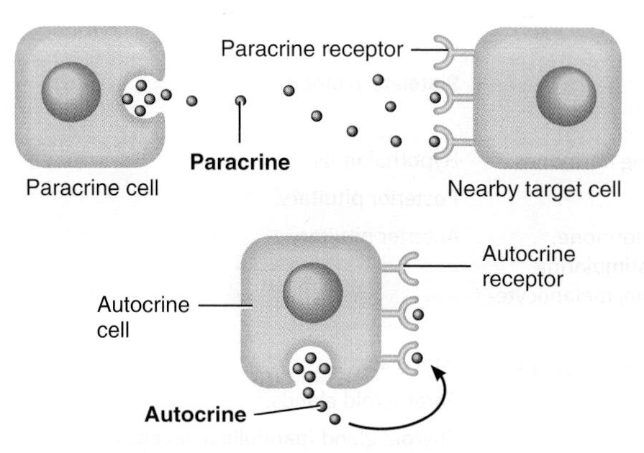

(b) Local hormones (paracrines and autocrines)

Q In the stomach, one stimulus for secretion of hydrochloric acid by parietal cells is the release of histamine by neighboring mast cells. Is histamine an autocrine or a paracrine in this situation?

Chemical Classes of Hormones

Chemically, hormones can be divided into two broad classes: those that are soluble in lipids, and those that are soluble in water. This chemical classification is also useful functionally because the two classes exert their effects differently.

Lipid-Soluble Hormones The lipid-soluble hormones include steroid hormones, thyroid hormones, and nitric oxide.

1. **Steroid hormones** are derived from cholesterol. Each steroid hormone is unique due to the presence of different chemical groups attached at various sites on the four rings at the core of its structure (see **Table 18.2**). These small differences allow for a large diversity of functions.
2. Two **thyroid hormones** (T_3 and T_4) are synthesized by attaching iodine to the amino acid tyrosine. The presence of two benzene rings within a T_3 or T_4 molecule makes these molecules very lipid-soluble (see **Table 18.2**).
3. The gas **nitric oxide (NO)** is both a hormone and a neurotransmitter. Its synthesis is catalyzed by the enzyme nitric oxide synthase.

Water-Soluble Hormones The water-soluble hormones include amine hormones, peptide and protein hormones, and eicosanoid hormones.

1. **Amine hormones** (a-MĒN) are synthesized by decarboxylating (removing a molecule of CO_2) and otherwise modifying certain amino acids. They are called amines because they retain an amino group (—NH_3^+). The catecholamines—epinephrine, norepinephrine, and dopamine—are synthesized by modifying the amino acid tyrosine. Histamine is synthesized from the amino acid histidine by mast cells and platelets. Serotonin and melatonin are derived from tryptophan.
2. **Peptide hormones** and **protein hormones** are amino acid polymers. The smaller peptide hormones consist of chains of 3 to 49 amino acids; the larger protein hormones include 50 to 200 amino acids. Examples of peptide hormones are antidiuretic hormone and oxytocin; protein hormones include growth hormone and insulin. Several of the protein hormones, such as thyroid-stimulating hormone, have attached carbohydrate groups and thus are **glycoprotein hormones**.
3. The **eicosanoid hormones** (ī-KŌ-sa-noyd; *eicos-* = twenty forms; *-oid* = resembling) are derived from arachidonic acid, a 20-carbon fatty acid. The two major types of eicosanoids are **prostaglandins (PGs)** and **leukotrienes (LTs)**. The eicosanoids are important local hormones, and they may act as circulating hormones as well.

Table 18.2 summarizes the classes of lipid-soluble and water-soluble hormones and provides an overview of the major hormones and their sites of secretion.

Hormone Transport in the Blood

Most water-soluble hormone molecules circulate in the watery blood plasma in a "free" form (not attached to other molecules), but most lipid-soluble hormone molecules are bound to **transport proteins**. The transport proteins, which are synthesized by cells in the liver, have three functions:

1. They make lipid-soluble hormones temporarily water-soluble, thus increasing their solubility in blood.
2. They retard passage of small hormone molecules through the filtering mechanism in the kidneys, thus slowing the rate of hormone loss in the urine.
3. They provide a ready reserve of hormone, already present in the bloodstream.

TABLE 18.2 Summary of Hormones by Chemical Class

CHEMICAL CLASS	HORMONES	SITE OF SECRETION
LIPID-SOLUBLE		
Steroid hormones	Aldosterone, cortisol, androgens.	Adrenal cortex.
	Calcitriol (active form of vitamin D).	Kidneys.
	Testosterone.	Testes.
Aldosterone	Estrogens, progesterone.	Ovaries.
Thyroid hormones Triiodothyronine (T_3)	T_3 (triiodothyronine), T_4 (thyroxine).	Thyroid gland (follicular cells).
Gas	Nitric oxide (NO).	Endothelial cells lining blood vessels.
WATER-SOLUBLE		
Amines	Epinephrine, norepinephrine (catecholamines).	Adrenal medulla.
	Melatonin.	Pineal gland.
	Histamine.	Mast cells in connective tissues.
Norepinephrine	Serotonin.	Platelets in blood.
Peptides and proteins	All hypothalamic releasing and inhibiting hormones.	Hypothalamus.
	Oxytocin, antidiuretic hormone.	Posterior pituitary.
	Growth hormone, thyroid-stimulating hormone, adrenocorticotropic hormone, follicle-stimulating hormone, luteinizing hormone, prolactin, melanocyte-stimulating hormone.	Anterior pituitary.
	Insulin, glucagon, somatostatin, pancreatic polypeptide.	Pancreas.
	Parathyroid hormone.	Parathyroid glands.
Oxytocin	Calcitonin.	Thyroid gland (parafollicular cells).
	Gastrin, secretin, cholecystokinin, GIP (glucose-dependent insulinotropic peptide).	Stomach and small intestine (enteroendocrine cells).
	Erythropoietin.	Kidneys.
	Leptin.	Adipose tissue.
Eicosanoids A leukotriene (LTB_4)	Prostaglandins, leukotrienes.	All cells except red blood cells.

In general, 0.1–10% of the molecules of a lipid-soluble hormone are not bound to a transport protein. This **free fraction** diffuses out of capillaries, binds to receptors, and triggers responses. As free hormone molecules leave the blood and bind to their receptors, transport proteins release new ones to replenish the free fraction.

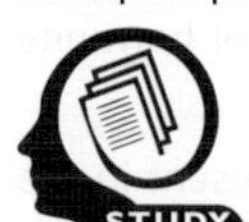

See Clinical Connection: Administering Hormones

18.4 Mechanisms of Hormone Action

OBJECTIVE

- **Describe** the two general mechanisms of hormone action.

The response to a hormone depends on both the hormone itself and the target cell. Various target cells respond differently to the same hormone. Insulin, for example, stimulates synthesis of glycogen in liver cells and synthesis of triglycerides in adipose cells.

The response to a hormone is not always the synthesis of new molecules, as is the case for insulin. Other hormonal effects include changing the permeability of the plasma membrane, stimulating transport of a substance into or out of the target cells, altering the rate of specific metabolic reactions, or causing contraction of smooth muscle or cardiac muscle. In part, these varied effects of hormones are possible because a single hormone can set in motion several different cellular responses. However, a hormone must first "announce its arrival" to a target cell by binding to its receptors. The receptors for lipid-soluble hormones are located inside target cells. The receptors for water-soluble hormones are part of the plasma membrane of target cells.

Action of Lipid-Soluble Hormones

As you just learned, lipid-soluble hormones, including steroid hormones and thyroid hormones, bind to receptors within target cells. Their mechanism of action is as follows (**Figure 18.3**):

1. A free lipid-soluble hormone molecule diffuses from the blood, through interstitial fluid, and through the lipid bilayer of the plasma membrane into a cell.
2. If the cell is a target cell, the hormone binds to and activates receptors located within the cytosol or nucleus. The activated receptor–hormone complex then alters gene expression: It turns specific genes of the nuclear DNA on or off.
3. As the DNA is transcribed, new messenger RNA (mRNA) forms, leaves the nucleus, and enters the cytosol. There, it directs synthesis of a new protein, often an enzyme, on the ribosomes.
4. The new proteins alter the cell's activity and cause the responses typical of that hormone.

FIGURE 18.3 Mechanism of action of the lipid-soluble steroid hormones and thyroid hormones.

Lipid-soluble hormones bind to receptors inside target cells.

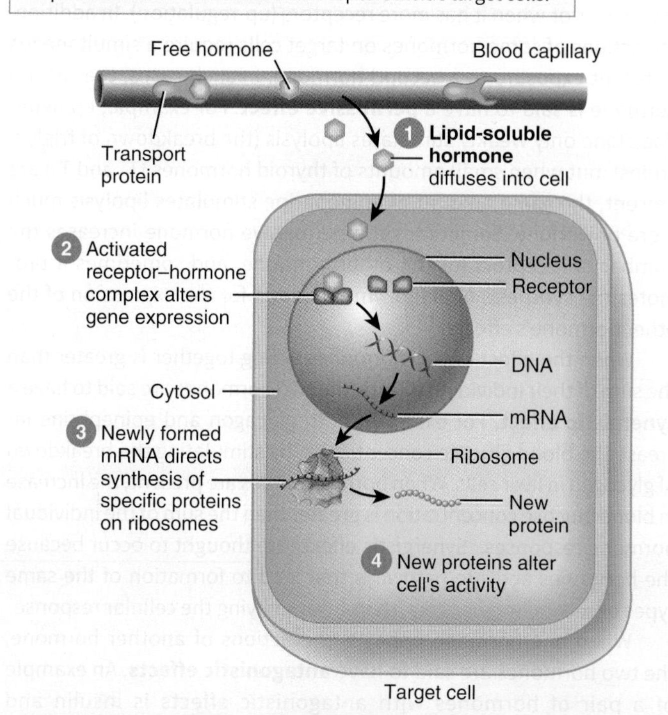

Q What is the action of the receptor–hormone complex?

Action of Water-Soluble Hormones

Because amine, peptide, protein, and eicosanoid hormones are not lipid-soluble, they cannot diffuse through the lipid bilayer of the plasma membrane and bind to receptors inside target cells. Instead, water-soluble hormones bind to receptors that protrude from the target-cell surface. The receptors are integral transmembrane proteins in the plasma membrane. When a water-soluble hormone binds to its receptor at the outer surface of the plasma membrane, it acts as the **first messenger**. The first messenger (the hormone) then causes production of a **second messenger** inside the cell, where specific hormone-stimulated responses take place. One common second messenger is **cyclic adenosine monophosphate**, also known as **cyclic AMP (cAMP)**. Neurotransmitters, neuropeptides, and several sensory transduction mechanisms (for example, vision; see **Figure 17.16**) also act via second-messenger systems.

The action of a typical water-soluble hormone occurs as follows (**Figure 18.4**):

1. A water-soluble hormone (the first messenger) diffuses from the blood through interstitial fluid and then binds to its receptor at the exterior surface of a target cell's plasma membrane. The hormone–receptor complex activates a membrane protein called a **G protein**. The activated G protein in turn activates **adenylyl cyclase** (a-DEN-i-lil SĪ-klās).
2. Adenylyl cyclase converts ATP into cyclic AMP (cAMP). Because the enzyme's active site is on the inner surface of the plasma membrane, this reaction occurs in the cytosol of the cell.
3. Cyclic AMP (the second messenger) activates one or more protein kinases, which may be free in the cytosol or bound to the plasma membrane. A **protein kinase** is an enzyme that phosphorylates (adds a phosphate group to) other cellular proteins (such as enzymes). The donor of the phosphate group is ATP, which is converted to ADP.
4. Activated protein kinases phosphorylate one or more cellular proteins. Phosphorylation activates some of these proteins and inactivates others, rather like turning a switch on or off.
5. Phosphorylated proteins in turn cause reactions that produce physiological responses. Different protein kinases exist within different target cells and within different organelles of the same target cell. Thus, one protein kinase might trigger glycogen synthesis, a second might cause the breakdown of triglyceride, a third may promote protein synthesis, and so forth. As noted in step 4, phosphorylation by a protein kinase can also inhibit certain proteins. For example, some of the kinases unleashed when epinephrine binds to liver cells inactivate an enzyme needed for glycogen synthesis.

FIGURE 18.4 Mechanism of action of the water-soluble hormones (amines, peptides, proteins, and eicosanoids).

Water-soluble hormones bind to receptors embedded in the plasma membranes of target cells.

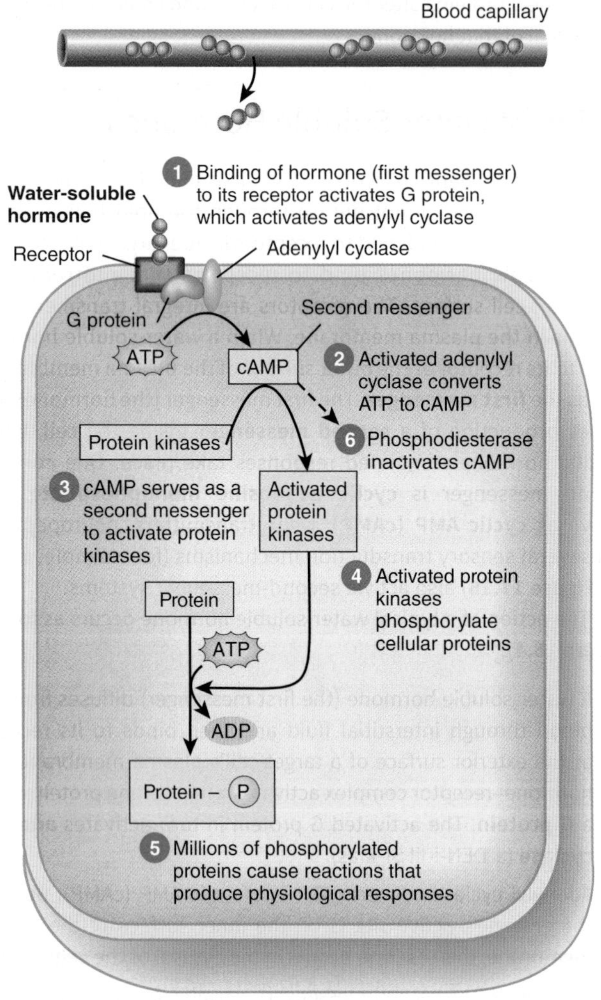

Q Why is cAMP a "second messenger"?

6 After a brief period, an enzyme called **phosphodiesterase** (fos′-fō-dī-ES-ter′-ās) inactivates cAMP. Thus, the cell's response is turned off unless new hormone molecules continue to bind to their receptors in the plasma membrane.

The binding of a hormone to its receptor activates many G-protein molecules, which in turn activate molecules of adenylyl cyclase (as noted in step 1). Unless they are further stimulated by the binding of more hormone molecules to receptors, G proteins slowly inactivate, thus decreasing the activity of adenylyl cyclase and helping to stop the hormone response. G proteins are a common feature of most second-messenger systems.

Many hormones exert at least some of their physiological effects through the *increased* synthesis of cAMP. Examples include antidiuretic hormone (ADH), thyroid-stimulating hormone (TSH), adrenocorticotropic hormone (ACTH), glucagon, epinephrine, and hypothalamic releasing hormones. In other cases, such as growth hormone–inhibiting hormone (GHIH), the level of cyclic AMP *decreases* in response to the binding of a hormone to its receptor. Besides cAMP, other second messengers include calcium ions (Ca^{2+}); **cyclic guanosine monophosphate**, also referred to as **cyclic GMP (cGMP)**, a cyclic nucleotide similar to cAMP; **inositol trisphosphate (IP_3)** (in-Ō-si-tōl tris-FOS-fāt); and **diacylglycerol (DAG)** (dī′-as-il-GLIS-er-ol). A given hormone may use different second messengers in different target cells.

Hormones that bind to plasma membrane receptors can induce their effects at very low concentrations because they initiate a cascade or chain reaction, each step of which multiplies or amplifies the initial effect. For example, the binding of a single molecule of epinephrine to its receptor on a liver cell may activate a hundred or so G proteins, each of which activates an adenylyl cyclase molecule. If each adenylyl cyclase produces even 1000 cAMP, then 100,000 of these second messengers will be liberated inside the cell. Each cAMP may activate a protein kinase, which in turn can act on hundreds or thousands of substrate molecules. Some of the kinases phosphorylate and activate a key enzyme needed for glycogen breakdown. The end result of the binding of a single molecule of epinephrine to its receptor is the breakdown of millions of glycogen molecules into glucose monomers.

Hormone Interactions

The responsiveness of a target cell to a hormone depends on (1) the hormone's concentration in the blood, (2) the abundance of the target cell's hormone receptors, and (3) influences exerted by other hormones. A target cell responds more vigorously when the level of a hormone rises or when it has more receptors (up-regulation). In addition, the actions of some hormones on target cells require a simultaneous or recent exposure to a second hormone. In such cases, the second hormone is said to have a **permissive effect**. For example, epinephrine alone only weakly stimulates lipolysis (the breakdown of triglycerides), but when small amounts of thyroid hormones (T_3 and T_4) are present, the same amount of epinephrine stimulates lipolysis much more powerfully. Sometimes the permissive hormone increases the number of receptors for the other hormone, and sometimes it promotes the synthesis of an enzyme required for the expression of the other hormone's effects.

When the effect of two hormones acting together is greater than the sum of their individual effects, the two hormones are said to have a **synergistic effect**. For example, both glucagon and epinephrine increase the blood glucose concentration by stimulating the breakdown of glycogen in liver cells. When both hormones are present, the increase in blood glucose concentration is greater than the sum of the individual hormone responses. Synergistic effects are thought to occur because the hormones activate pathways that lead to formation of the same types of second messengers, thereby amplifying the cellular response.

When one hormone opposes the actions of another hormone, the two hormones are said to have **antagonistic effects**. An example of a pair of hormones with antagonistic effects is insulin and glucagon: Insulin promotes synthesis of glycogen by liver cells, and glucagon stimulates breakdown of glycogen in the liver. Antagonistic effects occur because the hormones activate pathways that cause opposite cellular responses or one hormone decreases the number of receptors (down-regulation) for the other hormone.

18.5 Control of Hormone Secretion

OBJECTIVE

- **Describe** the mechanisms of control of hormone secretion.

The release of most hormones occurs in short bursts, with little or no secretion between bursts. When stimulated, an endocrine gland will release its hormone in more frequent bursts, increasing the concentration of the hormone in the blood. In the absence of stimulation, the blood level of the hormone decreases. Regulation of secretion normally prevents overproduction or underproduction of any given hormone to help maintain homeostasis.

Hormone secretion is regulated by (1) signals from the nervous system, (2) chemical changes in the blood, and (3) other hormones. For example, nerve impulses to the adrenal medullae regulate the release of epinephrine; blood Ca^{2+} level regulates the secretion of parathyroid hormone; and a hormone from the anterior pituitary (adrenocorticotropic hormone) stimulates the release of cortisol by the adrenal cortex. Most hormonal regulatory systems work via negative feedback (see **Figure 1.4**), but a few operate via positive feedback (see **Figure 1.5**). For example, during childbirth, the hormone oxytocin stimulates contractions of the uterus, and uterine contractions in turn stimulate more oxytocin release, a positive feedback effect.

Now that you have a general understanding of the roles of hormones in the endocrine system, we turn to discussions of the various endocrine glands and the hormones they secrete.

18.6 Hypothalamus and Pituitary Gland

OBJECTIVES

- **Describe** the locations of and relationships between the hypothalamus and pituitary gland.
- **Describe** the location, histology, hormones, and functions of the anterior and posterior pituitary.

For many years, the **pituitary gland** (pi-TOO-i-tār-ē) or *hypophysis* (hī-POF-i-sis) was called the "master" endocrine gland because it secretes several hormones that control other endocrine glands. We now know that the pituitary gland itself has a master—the **hypothalamus**. This small region of the brain below the thalamus is the major link between the nervous and endocrine systems. Cells in the hypothalamus synthesize at least nine different hormones, and the pituitary gland secretes seven. Together, these hormones play important roles in the regulation of virtually all aspects of growth, development, metabolism, and homeostasis.

The pituitary gland is a pea-shaped structure that measures 1–1.5 cm (0.5 in.) in diameter and lies in the hypophyseal fossa of the sella turcica of the sphenoid bone. It attaches to the hypothalamus by a stalk, the **infundibulum** (in-fun-DIB-ū-lum = a funnel; **Figure 18.5a**), and has two anatomically and functionally separate portions: the anterior pituitary and the posterior pituitary. The **anterior pituitary** (*anterior lobe*), also called the *adenohypophysis* (ad′-e-nō-hī-POF-i-sis; *adeno-* = gland; *-hypophysis* = undergrowth), accounts for about 75% of the total weight of the gland and is composed of epithelial tissue. The anterior pituitary consists of two parts in an adult: The **pars distalis** is the larger portion, and the **pars tuberalis** (PARS too-be′-RAL-is) forms a sheath around the infundibulum. The **posterior pituitary** (*posterior lobe*), also called the *neurohypophysis* (noo′-rō-hī-POF-i-sis; *neuro-* = nerve), is composed of neural tissue. It also consists of two parts: the **pars nervosa** (ner-VŌ-sa), the larger bulbar portion, and the infundibulum. A third region of the pituitary gland called the **pars intermedia** atrophies during human fetal development and ceases to exist as a separate lobe in adults (see **Figure 18.20b**). However, some of its cells migrate into adjacent parts of the anterior pituitary, where they persist.

Anterior Pituitary

The **anterior pituitary** secretes hormones that regulate a wide range of bodily activities, from growth to reproduction.

Types of Anterior Pituitary Cells and Their Hormones Five types of anterior pituitary cells—somatotrophs, thyrotrophs, gonadotrophs, lactotrophs, and corticotrophs—secrete seven hormones (**Table 18.3**):

1. **Somatotrophs** (sō-MAT-ō-trōfs) secrete **growth hormone (GH)**, also known as *human growth hormone (hGH)* or *somatotropin* (sō′-ma-tō-TRŌ-pin; *somato-* = body; *-tropin* = change). Growth hormone stimulates general body growth and regulates aspects of metabolism.
2. **Thyrotrophs** (THĪ-rō-trōfs) secrete **thyroid-stimulating hormone (TSH)**, also known as *thyrotropin* (thī-rō-TRŌ-pin; *thyro-* = pertaining to the thyroid gland). TSH controls the secretions and other activities of the thyroid gland.
3. **Gonadotrophs** (gō-NAD-ō-trōfs; *gonado-* = seed) secrete two **gonadotropins**: **follicle-stimulating hormone (FSH)** and **luteinizing hormone (LH)** (LOO-tē-in′-īz-ing). FSH and LH both act on the gonads (testes and ovaries). In men, they stimulate the testes to produce sperm and to secrete testosterone. In women, they stimulate the ovaries to mature oocytes (eggs) and to secrete estrogens and progesterone.

FIGURE 18.5 **Hypothalamus and pituitary gland.**

Hypothalamic hormones are an important link between the nervous and endocrine systems.

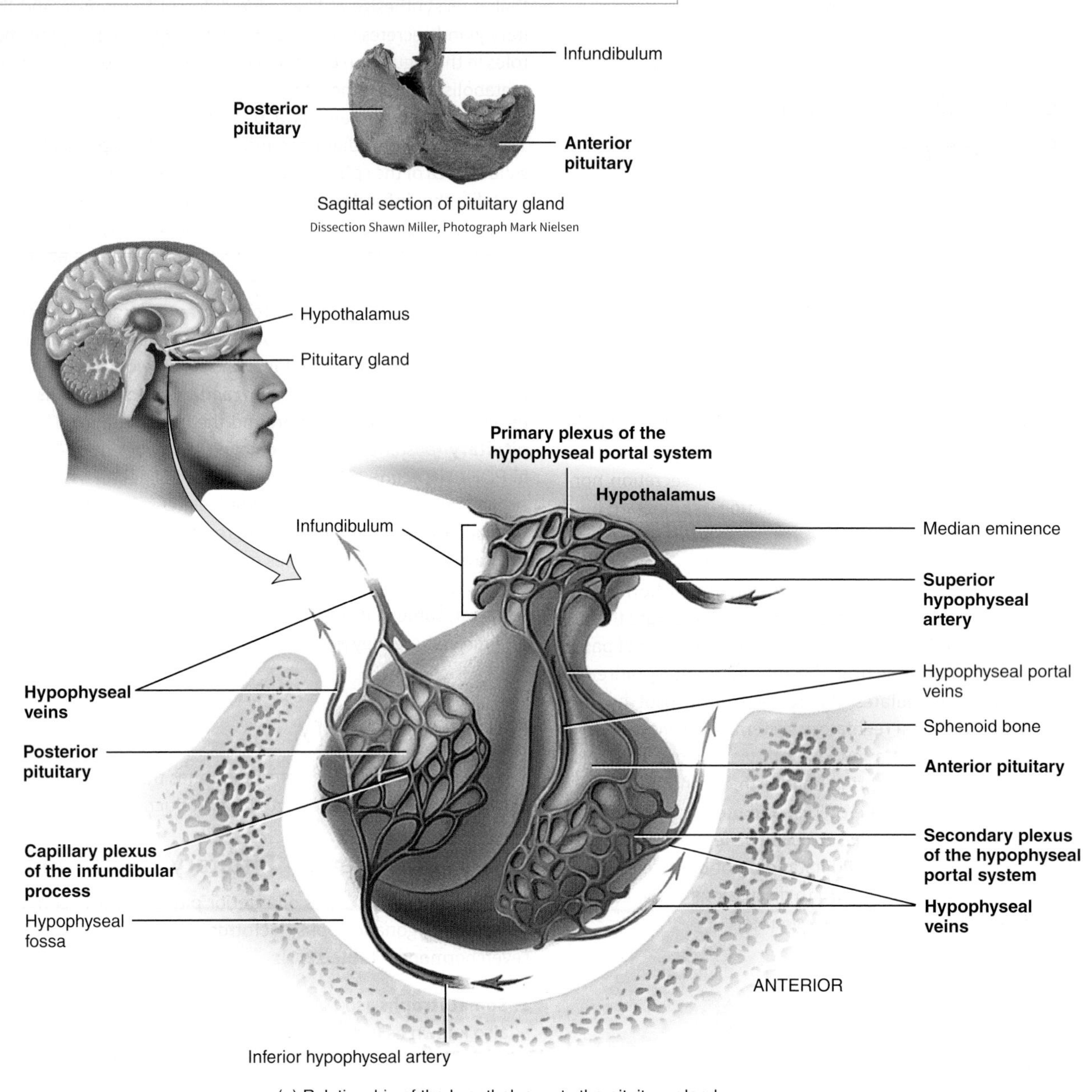

(a) Relationship of the hypothalamus to the pituitary gland

4. **Lactotrophs** (LAK-tō-trōfs; *lacto-* = milk) secrete **prolactin (PRL)**, which initiates milk production in the mammary glands.
5. **Corticotrophs** (KOR-ti-kō-trōfs) secrete **adrenocorticotropic hormone (ACTH)**, also known as *corticotropin* (kor′-ti-kō-TRŌ-pin; *cortico-* = rind or bark), which stimulates the adrenal cortex to secrete glucocorticoids such as cortisol. Some corticotrophs, remnants of the pars intermedia, also secrete **melanocyte-stimulating hormone (MSH)**.

Hypothalamic Control of the Anterior Pituitary

Release of anterior pituitary hormones is regulated in part by the hypothalamus. The hypothalamus secretes five **releasing hormones**, which stimulate secretion of anterior pituitary hormones (**Table 18.3**):

1. **Growth hormone-releasing hormone (GHRH)**, also known as *somatocrinin*, stimulates secretion of growth hormone.

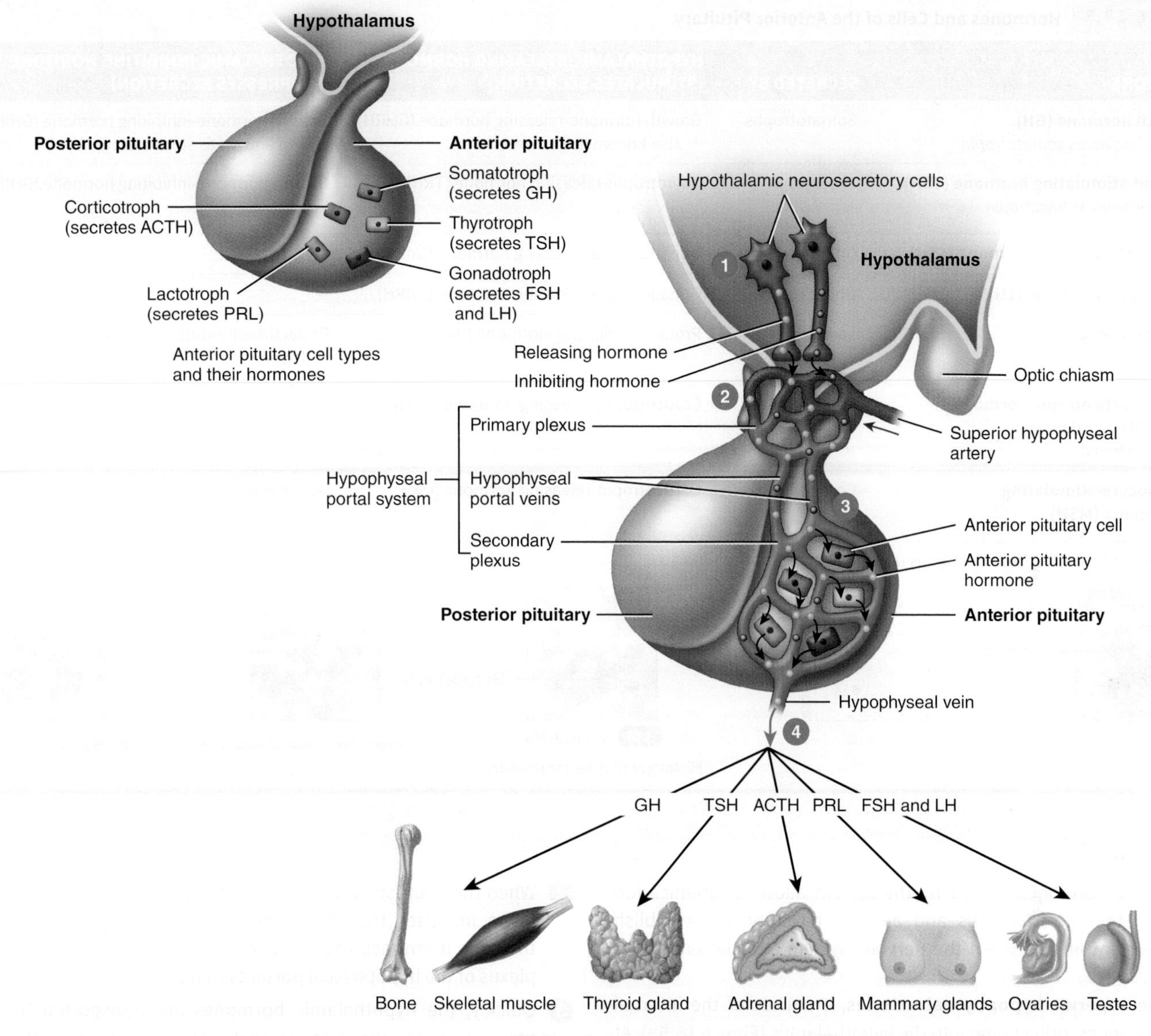

(b) Hypothalamic control of anterior pituitary hormone secretion

Q What is the functional importance of the hypophyseal portal veins?

2. **Thyrotropin-releasing hormone (TRH)** stimulates secretion of thyroid-stimulating hormone.
3. **Corticotropin-releasing hormone (CRH)** stimulates secretion of adrenocorticotropic hormone.
4. **Prolactin-releasing hormone (PRH)** stimulates secretion of prolactin.
5. **Gonadotropin-releasing hormone (GnRH)** stimulates secretion of FSH and LH.

The hypothalamus also produces two **inhibiting hormones**, which suppress secretion of anterior pituitary hormones:

1. **Growth hormone-inhibiting hormone (GHIH)**, also known as *somatostatin*, suppresses secretion of growth hormone.
2. **Prolactin-inhibiting hormone (PIH)**, which is dopamine, suppresses secretion of prolactin.

Hypophyseal Portal System Hypothalamic hormones that release or inhibit anterior pituitary hormones reach the anterior pituitary through a portal system. Usually, blood passes from the heart through an artery to a capillary to a vein and back to the heart. In a portal system, blood flows from one capillary network into a portal vein and then into a second capillary network before returning to the heart. The name of the portal system indicates the location of the second capillary network. In the **hypophyseal portal system** (hī′-pō-FIZ-ē-al), blood flows from capillaries in the hypothalamus into portal veins that carry blood to capillaries of the anterior pituitary. In other

TABLE 18.3 Hormones and Cells of the Anterior Pituitary

HORMONE	SECRETED BY	HYPOTHALAMIC RELEASING HORMONE (STIMULATES SECRETION)	HYPOTHALAMIC INHIBITING HORMONE (SUPPRESSES SECRETION)
Growth hormone (GH), also known as *somatotropin*	Somatotrophs	Growth hormone-releasing hormone (GHRH), also known as somatocrinin.	Growth hormone-inhibiting hormone (GHIH), also known as somatostatin.
Thyroid-stimulating hormone (TSH), also known as *thyrotropin*	Thyrotrophs	Thyrotropin-releasing hormone (TRH).	Growth hormone-inhibiting hormone (GHIH).
Follicle-stimulating hormone (FSH)	Gonadotrophs.	Gonadotropin-releasing hormone (GnRH).	—
Luteinizing hormone (LH)	Gonadotrophs.	Gonadotropin-releasing hormone (GnRH).	—
Prolactin (PRL)	Lactotrophs	Prolactin-releasing hormone (PRH).*	Prolactin-inhibiting hormone (PIH), which is dopamine.
Adrenocorticotropic hormone (ACTH), also known as *corticotropin*	Corticotrophs.	Corticotropin-releasing hormone (CRH).	—
Melanocyte-stimulating hormone (MSH)	Corticotrophs.	Corticotropin-releasing hormone (CRH).	Dopamine.

*Thought to exist, but exact nature is uncertain.

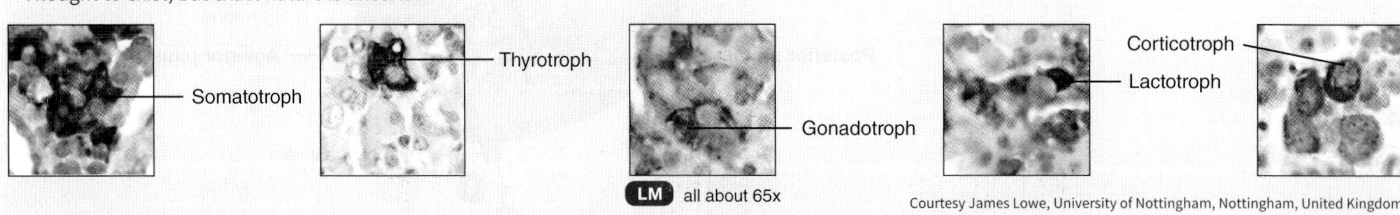

Histology of anterior pituitary

words, the hormones carried by the system allow communication between the hypothalamus and anterior pituitary and establish an important link between the nervous system and the endocrine system.

The **superior hypophyseal arteries**, branches of the internal carotid arteries, bring blood into the hypothalamus (**Figure 18.5a**). At the junction of the median eminence of the hypothalamus and the infundibulum, these arteries divide into a capillary network called the **primary plexus of the hypophyseal portal system**. From the primary plexus, blood drains into the **hypophyseal portal veins** that pass down the outside of the infundibulum. In the anterior pituitary, the hypophyseal portal veins divide again and form another capillary network called the **secondary plexus of the hypophyseal portal system**. **Hypophyseal veins** drain blood from the anterior pituitary.

Control of Anterior Pituitary Secretion

Regulation of anterior pituitary secretion by the hypothalamus occurs as follows (**Figure 18.5b**):

1. Above the optic chiasm are clusters of neurons called **neurosecretory cells.** They synthesize the hypothalamic releasing and inhibiting hormones in their cell bodies and package the hormones inside vesicles, which reach the axon terminals by fast axonal transport (see Section 12.2), where they are stored.
2. When the neurosecretory cells of the hypothalamus are excited, nerve impulses trigger exocytosis of the vesicles. The hypothalamic hormones then diffuse into the blood of the primary plexus of the hypophyseal portal system.
3. Quickly, the hypothalamic hormones are transported by the blood through the hypophyseal portal veins and into the secondary plexus. This direct route permits hypothalamic hormones to act immediately on anterior pituitary cells, before the hormones are diluted or destroyed in the general circulation. Within the secondary plexus the hypothalamic hormones diffuse out of the bloodstream and interact with anterior pituitary cells. When stimulated by the appropriate hypothalamic-releasing hormones, the anterior pituitary cells secrete hormones into the secondary plexus capillaries.
4. From the secondary plexus capillaries, the anterior pituitary hormones drain into the hypophyseal veins and out into the general circulation. Anterior pituitary hormones then travel to target tissues throughout the body. Those anterior pituitary hormones that act on other endocrine glands are called **tropic hormones** (TRŌ-pik) or *tropins*.

Release of anterior pituitary hormones is regulated not only by the hypothalamus (see **Table 18.3**) but also by negative feedback.

FIGURE 18.6 Negative feedback regulation of hypothalamic neurosecretory cells and anterior pituitary corticotrophs. Solid green arrows indicate stimulation of secretions; dashed red arrows indicate inhibition of secretion via negative feedback.

Cortisol secreted by the adrenal cortex suppresses secretion of CRH and ACTH.

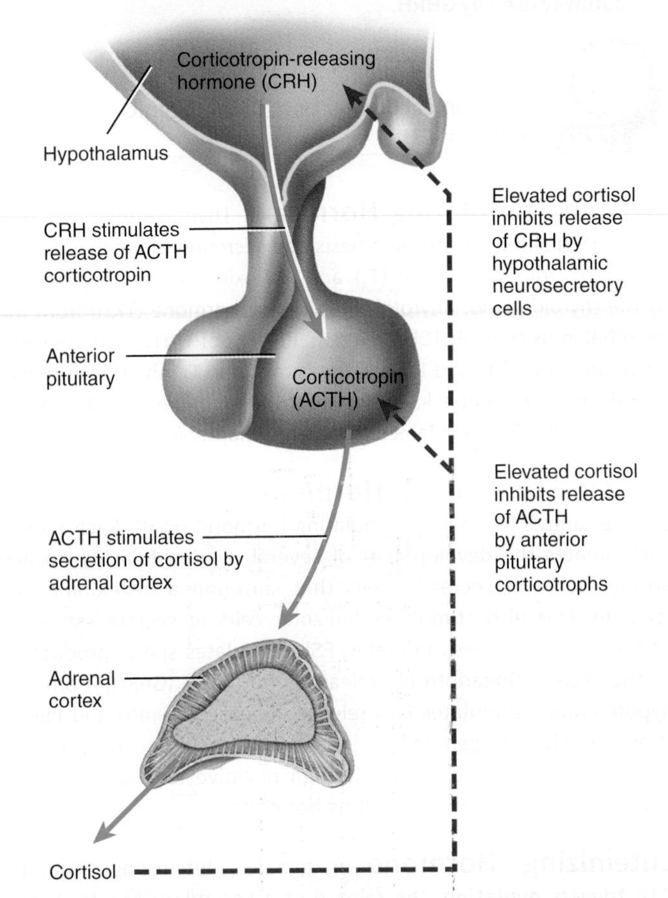

Q Which other target gland hormones suppress secretion of hypothalamic and anterior pituitary hormones by negative feedback?

The secretory activity of three types of anterior pituitary cells (thyrotrophs, corticotrophs, and gonadotrophs) decreases when blood levels of their target gland hormones rise. For example, adrenocorticotropic hormone (ACTH) stimulates the cortex of the adrenal gland to secrete glucocorticoids, mainly cortisol (**Figure 18.6**). In turn, an elevated blood level of cortisol decreases secretion of both ACTH (corticotropin) and corticotropin-releasing hormone (CRH) by suppressing the activity of the anterior pituitary corticotrophs and hypothalamic neurosecretory cells.

Growth Hormone Somatotrophs are the most numerous cells in the anterior pituitary, and growth hormone (GH) is the most plentiful anterior pituitary hormone. GH promotes growth of body tissues, including bones and skeletal muscles, and it regulates certain aspects of metabolism. GH exerts its growth-promoting effects indirectly through small protein hormones called **insulin-like growth factors (IGFs)** or *somatomedins* (sō′-ma-tō-MĒ-dins). In response to growth hormone, cells in the liver, skeletal muscle, cartilage, and bone secrete IGFs. IGFs synthesized in the liver enter the bloodstream as hormones that circulate to target cells throughout the body to cause growth. IGFs produced in skeletal muscle, cartilage, and bone act locally as autocrines or paracrines to cause growth of those tissues. Unlike the effects of GH on body growth, the effects of GH on metabolism are direct, meaning that GH interacts directly with target cells to cause specific metabolic reactions.

Using IGFs as mediators, GH causes growth of bones and other tissues of the body. Through direct effects, GH helps regulate certain metabolic reactions in body cells. The specific functions of IGFs and GH include the following:

1. ***Increase growth of bones and soft tissues.*** In bones, IGFs stimulate osteoblasts, promote cell division at the epiphyseal plate, and enhance synthesis of the proteins needed to build more bone matrix. In soft tissues such as skeletal muscle, the kidneys, and intestines, IGFs cause cells to grow by increasing uptake of amino acids into cells and accelerating protein synthesis. IGFs also decrease the breakdown of proteins and the use of amino acids for ATP production. Due to the effects of IGFs, GH increases growth of the skeleton and soft tissues during childhood and the teenage years. In adults, GH (acting via IGFs) helps maintain the mass of bones and soft tissues and promotes healing of injuries and tissue repair.
2. ***Enhance lipolysis.*** GH enhances lipolysis in adipose tissue, which results in increased use of the released fatty acids for ATP production by body cells.
3. ***Decrease glucose uptake.*** GH influences carbohydrate metabolism by decreasing glucose uptake, which decreases the use of glucose for ATP production by most body cells. This action spares glucose so that it is available to neurons for ATP production in times of glucose scarcity. GH also stimulates liver cells to release glucose into the blood.

Somatotrophs in the anterior pituitary release bursts of growth hormone every few hours, especially during sleep. Their secretory activity is controlled mainly by two hypothalamic hormones: (1) growth hormone-releasing hormone (GHRH) promotes secretion of growth hormone and (2) growth hormone-inhibiting hormone (GHIH) suppresses it. Regulation of growth hormone secretion by GHRH and GHIH occurs as follows (**Figure 18.7**).

1. GHRH is secreted from the hypothalamus. Factors that promote GHRH secretion include hypoglycemia (low blood glucose concentration); decreased blood levels of fatty acids; increased blood levels of amino acids; deep sleep (stages 3 and 4 of non-rapid eye movement sleep); increased activity of the sympathetic nervous system, such as might occur with stress or vigorous physical exercise; and other hormones, including testosterone, estrogens, thyroid hormones, and ghrelin.
2. Once secreted, GHRH enters the hypophyseal portal system and flows to the anterior pituitary, where it stimulates somatotrophs to secrete GH.

FIGURE 18.7 Regulation of growth hormone (GH) secretion. Each dashed arrow and negative sign indicates negative feedback.

Secretion of GH is stimulated by growth hormone–releasing hormone (GHRH) and inhibited by growth hormone–inhibiting hormone (GHIH).

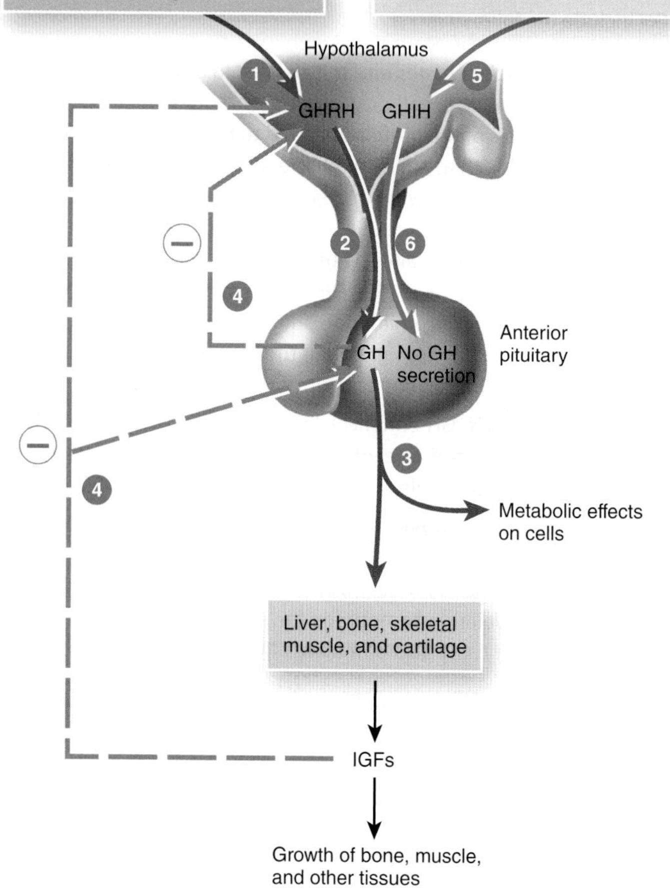

Q If a person has a pituitary tumor that secretes a large amount of GH and the tumor cells are not responsive to regulation by GHRH and GHIH, will hyperglycemia or hypoglycemia be more likely?

3. GH acts directly on various cells to promote certain metabolic reactions. In liver, bone, skeletal muscle, and cartilage, GH is converted to IGFs, which in turn promote growth of bones, skeletal muscle, and other tissues.

4. Elevated levels of GH and IGFs inhibit release of GHRH and GH (negative feedback inhibition).

5. GHIH is secreted from the hypothalamus. Factors that promote GHIH secretion include hyperglycemia (high blood glucose); increased blood levels of fatty acids; decreased blood levels of amino acids; obesity; aging; and high blood levels of GH and IGFs.

6. After being secreted, GHIH enters the hypophyseal portal system and flows to the anterior pituitary, where it prevents the somatotrophs from secreting GH by interfering with the signaling pathway used by GHRH.

See Clinical Connection: Diabetogenic Effect of GH

Thyroid-Stimulating Hormone

Thyroid-stimulating hormone (TSH) stimulates the synthesis and secretion of the two thyroid hormones, triiodothyronine (T_3) and thyroxine (T_4), both produced by the thyroid gland. Thyrotropin-releasing hormone (TRH) from the hypothalamus controls TSH secretion. Release of TRH in turn depends on blood levels of T_3 and T_4; high levels of T_3 and T_4 inhibit secretion of TRH via negative feedback. There is no thyrotropin-inhibiting hormone. The release of TRH is explained later in the chapter (see **Figure 18.12**).

Follicle-Stimulating Hormone

In females, the ovaries are the targets for follicle-stimulating hormone (FSH). Each month FSH initiates the development of several ovarian follicles, saclike arrangements of secretory cells that surround a developing egg (oocyte). FSH also stimulates follicular cells to secrete estrogens (female sex hormones). In males, FSH stimulates sperm production in the testes. Gonadotropin-releasing hormone (GnRH) from the hypothalamus stimulates FSH release. Release of GnRH and FSH is suppressed by estrogens in females and by testosterone (the principal male sex hormone) in males through negative feedback systems. There is no gonadotropin-inhibiting hormone.

Luteinizing Hormone

In females, luteinizing hormone (LH) triggers **ovulation**, the release of a secondary oocyte (future ovum) by an ovary. LH stimulates formation of the corpus luteum (structure formed after ovulation) in the ovary and the secretion of progesterone (another female sex hormone) by the corpus luteum. Together, FSH and LH also stimulate secretion of estrogens by ovarian cells. Estrogens and progesterone prepare the uterus for implantation of a fertilized ovum and help prepare the mammary glands for milk secretion. In males, LH stimulates cells in the testes to secrete testosterone. Secretion of LH, like that of FSH, is controlled by gonadotropin-releasing hormone (GnRH).

Prolactin

Prolactin (PRL), together with other hormones, initiates and maintains milk production by the mammary glands. By itself, prolactin has only a weak effect. Only after the mammary glands have been primed by estrogens, progesterone, glucocorticoids, growth hormone, thyroxine, and insulin, which exert permissive effects, does PRL bring about milk production. Ejection of milk from the mammary glands depends on the hormone oxytocin, which is released from the posterior pituitary. Together, milk production and ejection constitute *lactation*.

The hypothalamus secretes both inhibitory and excitatory hormones that regulate prolactin secretion. In females, prolactin-inhibiting hormone (PIH), which is dopamine, inhibits the release of prolactin from the anterior pituitary most of the time. Each month, just before menstruation begins, the secretion of PIH diminishes and the blood level of prolactin rises, but not enough to stimulate milk production. Breast tenderness just before menstruation may be caused by elevated prolactin. As the menstrual cycle begins anew, PIH is again secreted and the prolactin level drops. During pregnancy, the prolactin level rises, stimulated by prolactin-releasing hormone (PRH) from the hypothalamus. The sucking action of a nursing infant causes a reduction in hypothalamic secretion of PIH.

The function of prolactin is not known in males, but its hypersecretion causes erectile dysfunction (impotence, the inability to have an erection of the penis). In females, hypersecretion of prolactin causes galactorrhea (inappropriate lactation) and amenorrhea (absence of menstrual cycles).

Adrenocorticotropic Hormone Corticotrophs secrete mainly adrenocorticotropic hormone (ACTH). ACTH controls the production and secretion of cortisol and other glucocorticoids by the cortex (outer portion) of the adrenal glands. Corticotropin-releasing hormone (CRH) from the hypothalamus stimulates secretion of ACTH by corticotrophs. Stress-related stimuli, such as low blood glucose or physical trauma, and interleukin-1, a substance produced by macrophages, also stimulate release of ACTH. Glucocorticoids inhibit CRH and ACTH release via negative feedback.

Melanocyte-Stimulating Hormone Melanocyte-stimulating hormone (MSH) increases skin pigmentation in amphibians by stimulating the dispersion of melanin granules in melanocytes. Its exact role in humans is unknown, but the presence of MSH receptors in the brain suggests it may influence brain activity. There is little circulating MSH in humans. However, continued administration of MSH for several days does produce a darkening of the skin. Excessive levels of corticotropin-releasing hormone (CRH) can stimulate MSH release; dopamine inhibits MSH release.

Table 18.4 summarizes the principal actions of the anterior pituitary hormones.

TABLE 18.4 Summary of the Principal Actions of Anterior Pituitary Hormones

HORMONE	TARGET TISSUES	PRINCIPAL ACTIONS
Growth hormone (GH), also known as *somatotropin*	Liver (and other tissues)	Stimulates liver, muscle, cartilage, bone, and other tissues to synthesize and secrete insulin-like growth factors (IGFs), which in turn promote growth of body tissues. GH acts directly on target cells to enhance lipolysis and decrease glucose uptake.
Thyroid-stimulating hormone (TSH), also known as *thyrotropin*	Thyroid gland	Stimulates synthesis and secretion of thyroid hormones by thyroid gland.
Follicle-stimulating hormone (FSH)	Ovary, Testis	In females, initiates development of oocytes and induces ovarian secretion of estrogens. In males, stimulates testes to produce sperm.
Luteinizing hormone (LH)	Ovary, Testis	In females, stimulates secretion of estrogens and progesterone, ovulation, and formation of corpus luteum. In males, stimulates testes to produce testosterone.
Prolactin (PRL)	Mammary glands	Together with other hormones, promotes milk production by mammary glands.
Adrenocorticotropic hormone (ACTH), also known as *corticotropin*	Adrenal cortex	Stimulates secretion of glucocorticoids (mainly cortisol) by adrenal cortex.
Melanocyte-stimulating hormone (MSH)	Brain	Exact role in humans is unknown but may influence brain activity; when present in excess, can cause darkening of skin.

Posterior Pituitary

Although the **posterior pituitary** does not *synthesize* hormones, it does *store* and *release* two hormones. It consists of axons and axon terminals of more than 10,000 hypothalamic neurosecretory cells. The cell bodies of the neurosecretory cells are in the **paraventricular** and **supraoptic nuclei** of the hypothalamus; their axons form the **hypothalamic–hypophyseal tract** (hī′-pō-THAL-a-mik hī-pō-FIZ-ē-al). This tract begins in the hypothalamus and ends near blood capillaries in the posterior pituitary (**Figure 18.8a**). The neuronal cell bodies of both the paraventricular and the supraoptic nuclei synthesize the hormones **oxytocin (OT)** (ok′-sē-TŌ-sin; *okytoc* = quick birth) and **antidiuretic hormone (ADH)**, also called *vasopressin* (vā-sō-PRES-in; *vaso-* = blood; *-pressus* = to press). The axon terminals in the posterior pituitary are associated with specialized neuroglia called **pituicytes** (pi-TOO-i-sītz). These cells have a supporting role similar to that of astrocytes (see Chapter 12).

Blood is supplied to the posterior pituitary by the **inferior hypophyseal arteries**, which branch from the internal carotid arteries. In the posterior pituitary, the inferior hypophyseal arteries drain into the **capillary plexus of the infundibular process**, a capillary network that receives secreted oxytocin and antidiuretic hormone (see **Figure 18.5**). From this plexus, hormones pass into the **hypophyseal veins** for distribution to target cells in other tissues.

Control of Posterior Pituitary Secretion

Release of hormones from the posterior pituitary occurs as follows (**Figure 18.8b**):

1. Neurosecretory cells in the paraventricular and supraoptic nuclei of the hypothalamus synthesize oxytocin and antidiuretic hormone (ADH). The hormones are then packaged into vesicles.
2. The vesicles move by fast axonal transport along the hypothalamic–hypophyseal tract to the axon terminals in the posterior pituitary, where they are stored.
3. When the appropriate stimulus excites the hypothalamus, nerve impulses trigger exocytosis and release of oxytocin or ADH into the bloodstream (inferior hypophyseal artery, capillary plexus of the infundibular process, and hypophyseal vein).
4. The released oxytocin or ADH then travels to its target tissues in the body.

FIGURE 18.8 **The hypothalamic–hypophyseal tract and regulation of hormone release by the posterior pituitary.**

Oxytocin and antidiuretic hormone are synthesized in the hypothalamus and released into the capillary plexus of the infundibular process in the posterior pituitary.

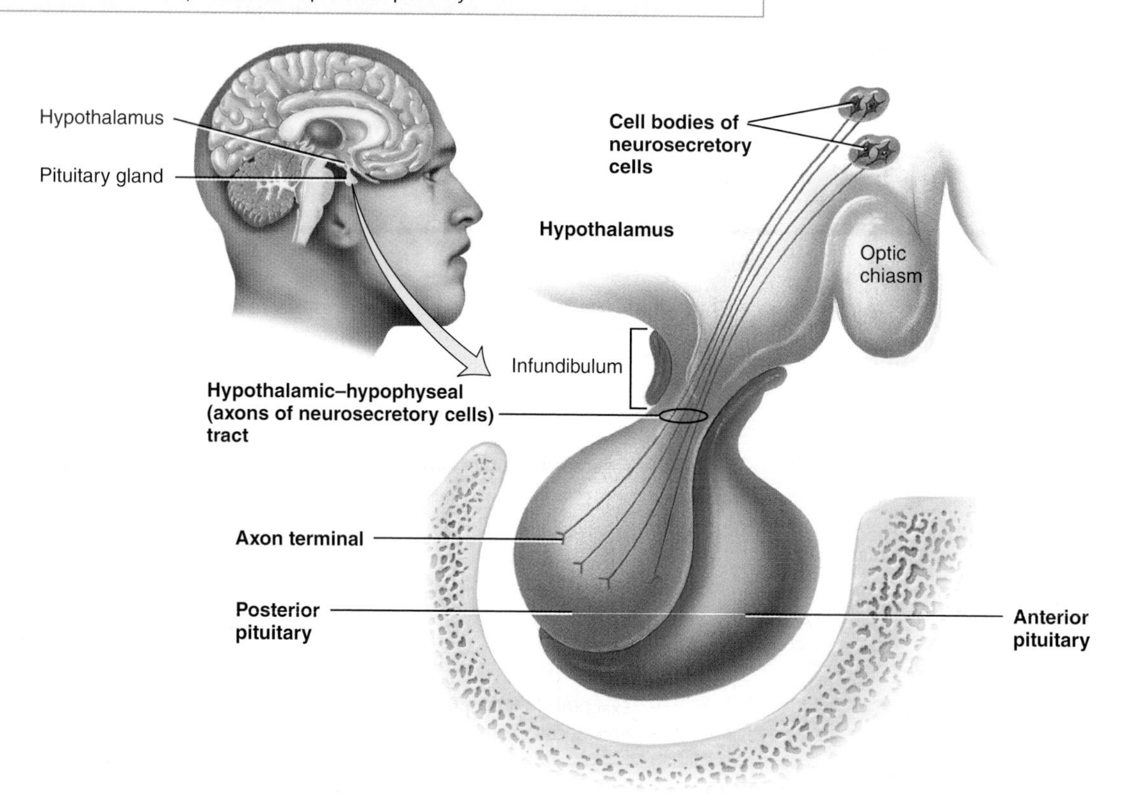

(a) Hypothalamic–hypophyseal tract

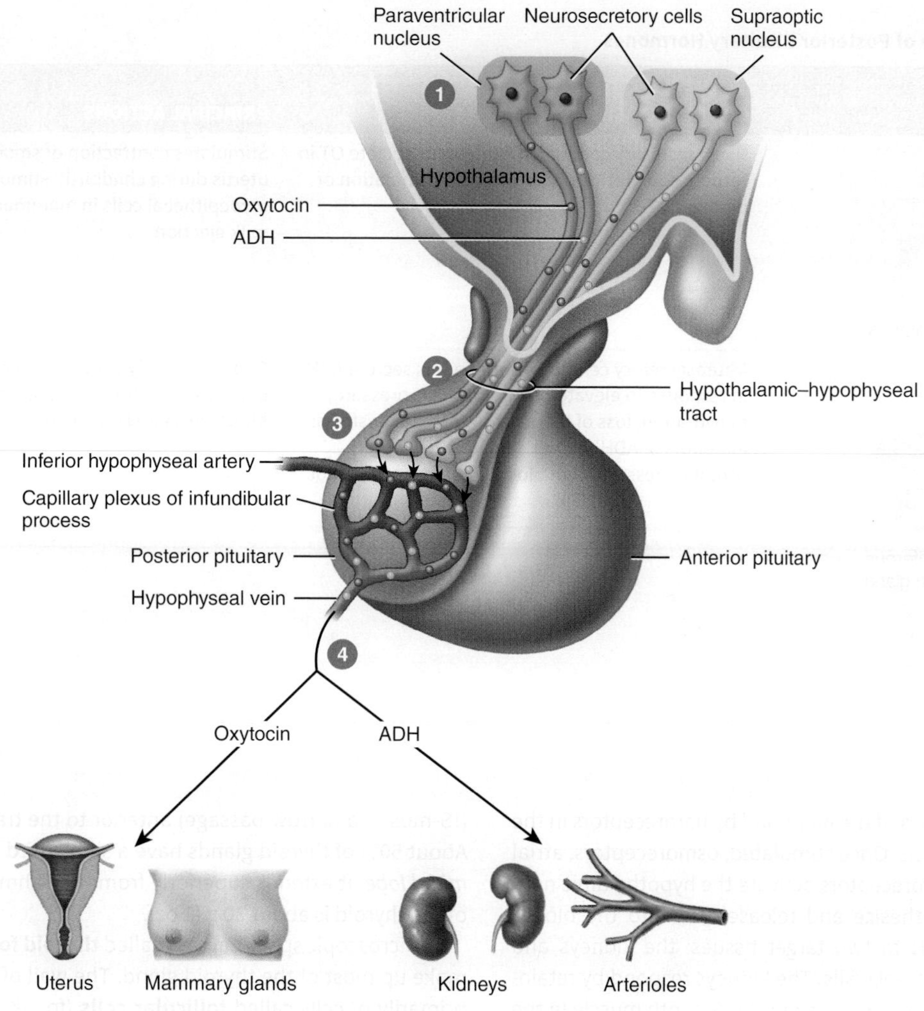

(b) Release of hormones from posterior pituitary

Q Functionally, how are the hypothalamic–hypophyseal tract and the hypophyseal portal veins similar? Structurally, how are they different?

Oxytocin During and after delivery of a baby, oxytocin affects two target tissues: the mother's uterus and breasts. During delivery, stretching of the cervix of the uterus stimulates the release of oxytocin which, in turn, enhances contraction of smooth muscle cells in the wall of the uterus (see **Figure 1.5**); after delivery, it stimulates milk ejection ("letdown") from the mammary glands in response to the mechanical stimulus provided by a suckling infant. The function of oxytocin in males and in nonpregnant females is not clear. Experiments with animals have suggested that it has actions within the brain that foster parental caretaking behavior toward young offspring. It may also be responsible, in part, for the feelings of sexual pleasure during and after intercourse.

See Clinical Connection: Oxytocin and Childbirth

Antidiuretic Hormone As its name implies, an **antidiuretic** (an-ti-dī-ū-RET-ik; *anti-* = against; *-dia-* = throughout; *-ouresis* = urination) is a substance that decreases urine production. ADH causes the kidneys to return more water to the blood, thus decreasing urine volume. In the absence of ADH, urine output increases more than tenfold, from the normal 1 to 2 liters to about 20 liters a day. Drinking alcohol often causes frequent and copious urination because alcohol inhibits secretion of ADH. (This dehydrating effect of alcohol may cause both the thirst and the headache typical of a hangover.) ADH also decreases the water lost through sweating and causes constriction of arterioles, which increases blood pressure. This hormone's other name, *vasopressin*, reflects this effect on blood pressure.

Two major stimuli promote ADH secretion: a rise in blood osmolarity and a decrease in blood volume. High blood osmolarity is detected by **osmoreceptors**, neurons in the hypothalamus that monitor changes in blood osmolarity. Decreased blood volume is detected by

TABLE 18.5 Summary of Posterior Pituitary Hormones

HORMONE AND TARGET TISSUES	CONTROL OF SECRETION	PRINCIPAL ACTIONS
Oxytocin (OT) Uterus, Mammary glands	Neurosecretory cells of hypothalamus secrete OT in response to uterine distension and stimulation of nipples.	Stimulates contraction of smooth muscle cells of uterus during childbirth; stimulates contraction of myoepithelial cells in mammary glands to cause milk ejection.
Antidiuretic hormone (ADH) or *vasopressin* Kidneys, Sudoriferous (sweat) glands, Arterioles	Neurosecretory cells of hypothalamus secrete ADH in response to elevated blood osmotic pressure, dehydration, loss of blood volume, pain, or stress; inhibitors of ADH secretion include low blood osmotic pressure, high blood volume, and alcohol.	Conserves body water by decreasing urine volume; decreases water loss through perspiration; raises blood pressure by constricting arterioles.

volume receptors in the atria of the heart and by baroreceptors in the walls of certain blood vessels. Once stimulated, osmoreceptors, atrial volume receptors, and baroreceptors activate the hypothalamic neurosecretory cells that synthesize and release ADH into the bloodstream. Blood carries ADH to two target tissues: the kidneys and smooth muscle in blood vessel walls. The kidneys respond by retaining more water, which decreases urine output. Smooth muscle in the walls of arterioles (small arteries) contracts in response to high levels of ADH, which constricts (narrows) the lumen of these blood vessels and increases blood pressure.

Secretion of ADH can be altered in other ways as well. Pain, stress, trauma, anxiety, acetylcholine, nicotine, and drugs such as morphine, tranquilizers, and some anesthetics stimulate ADH secretion.

Table 18.5 lists the posterior pituitary hormones, control of their secretion, and their principal actions.

18.7 Thyroid Gland

OBJECTIVE

- **Describe** the location, histology, hormones, and functions of the thyroid gland.

The butterfly-shaped **thyroid gland** is located just inferior to the larynx (voice box). It is composed of **right** and **left lateral lobes**, one on either side of the trachea, that are connected by an **isthmus** (IS-mus = a narrow passage) anterior to the trachea (Figure 18.9a). About 50% of thyroid glands have a small third lobe, called the *pyramidal lobe*. It extends superiorly from the isthmus. The normal mass of the thyroid is about 30 g (1 oz).

Microscopic spherical sacs called **thyroid follicles** (Figure 18.9b) make up most of the thyroid gland. The wall of each follicle consists primarily of cells called **follicular cells** (fo-LIK-ū-lar), most of which extend to the lumen (internal space) of the follicle. A **basement membrane** surrounds each follicle. When the follicular cells are inactive, their shape is low cuboidal to squamous, but under the influence of TSH they become active in secretion and range from cuboidal to low columnar in shape. The follicular cells produce two hormones: **thyroxine** (thī-ROK-sēn), which is also called *tetraiodothyronine (T_4)* (tet-ra-ī-ō-dō-THĪ-rō-nēn) because it contains four atoms of iodine, and **triiodothyronine (T_3)** (trī-ī′-ō-dō-THĪ-rō-nēn), which contains three atoms of iodine. T_3 and T_4 together are also known as **thyroid hormones**. A few cells called **parafollicular cells** (par′-a-fo-LIK-ū-lar) or *C cells* lie between follicles. They produce the hormone **calcitonin (CT)** (kal-si-TŌ-nin), which helps regulate calcium homeostasis.

Formation, Storage, and Release of Thyroid Hormones

The thyroid gland is the only endocrine gland that stores its secretory product in large quantities—normally about a 100-day supply. Synthesis and secretion of T_3 and T_4 occurs as follows (Figure 18.10):

1. ***Iodide trapping.*** Thyroid follicular cells trap iodide ions (I^-) by actively transporting them from the blood into the cytosol. As a result, the thyroid gland normally contains most of the iodide in the body.

FIGURE 18.9 **Location, blood supply, and histology of the thyroid gland.**

Thyroid hormones regulate (1) oxygen use and basal metabolic rate, (2) cellular metabolism, and (3) growth and development.

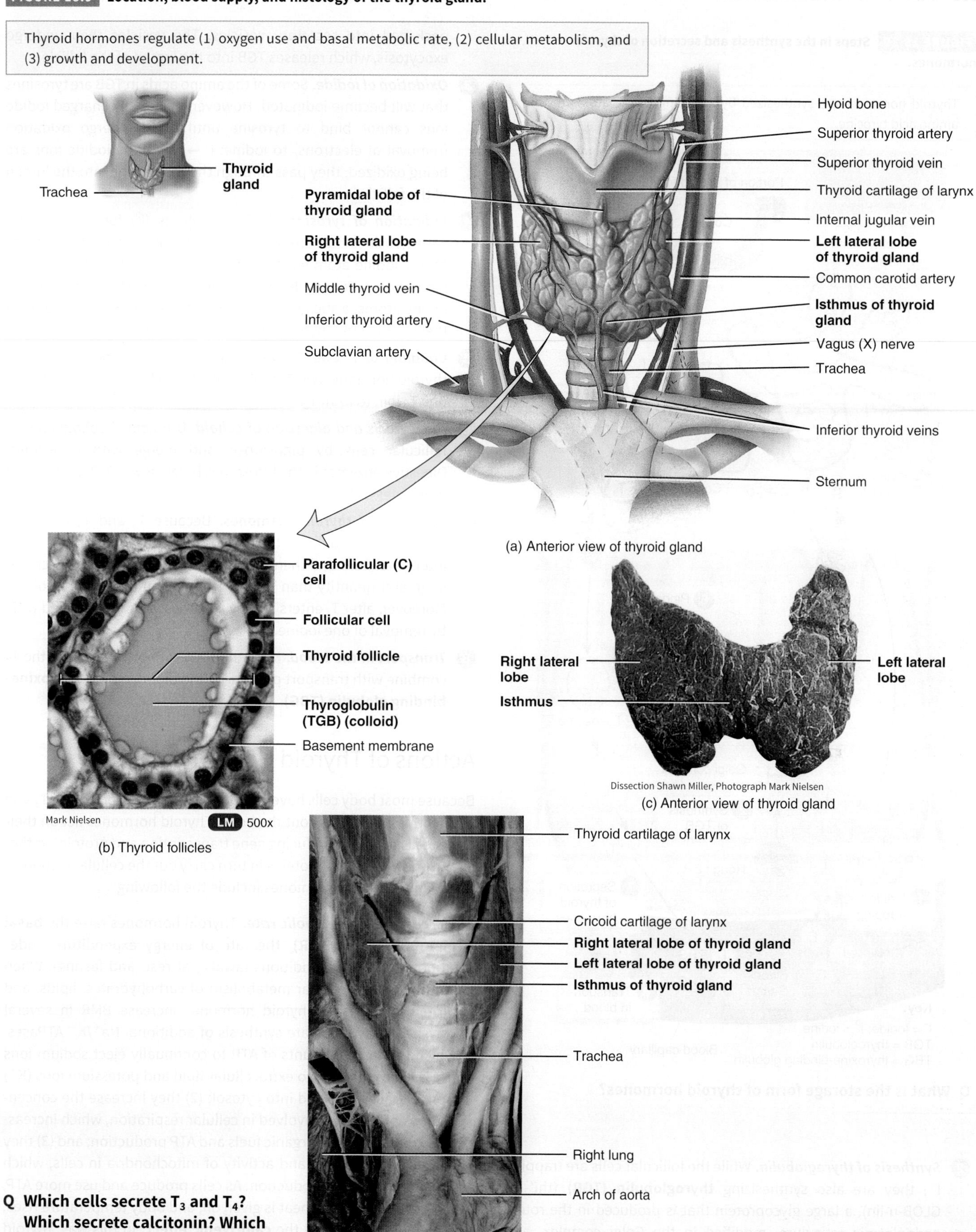

(a) Anterior view of thyroid gland

(b) Thyroid follicles

(c) Anterior view of thyroid gland

(d) Anterior view

Q Which cells secrete T_3 and T_4? Which secrete calcitonin? Which of these hormones are also called thyroid hormones?

FIGURE 18.10 Steps in the synthesis and secretion of thyroid hormones.

Thyroid hormones are synthesized by attaching iodine atoms to the amino acid tyrosine.

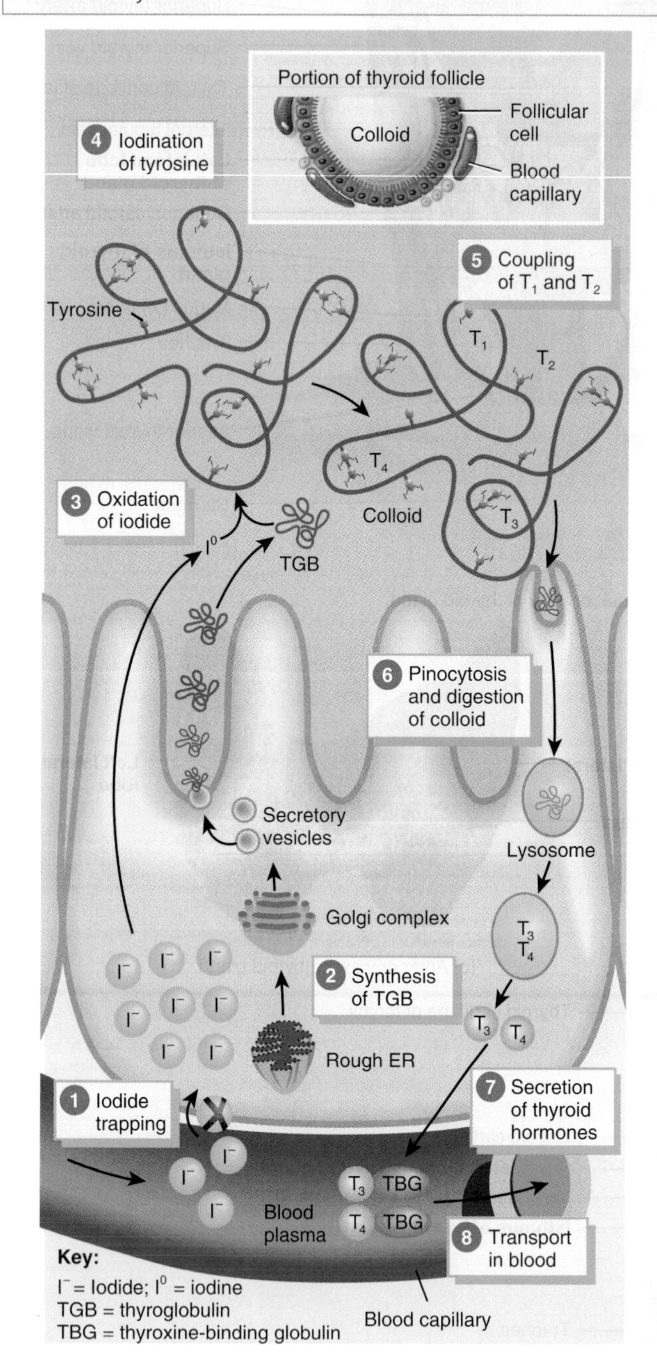

Q What is the storage form of thyroid hormones?

2 ***Synthesis of thyroglobulin.*** While the follicular cells are trapping I^-, they are also synthesizing **thyroglobulin (TGB)** (thī′-rō-GLOB-u-lin), a large glycoprotein that is produced in the rough endoplasmic reticulum, modified in the Golgi complex, and packaged into secretory vesicles. The vesicles then undergo exocytosis, which releases TGB into the lumen of the follicle.

3 ***Oxidation of iodide.*** Some of the amino acids in TGB are tyrosines that will become iodinated. However, negatively charged iodide ions cannot bind to tyrosine until they undergo oxidation (removal of electrons) to iodine: $I^- \rightarrow I^0$. As the iodide ions are being oxidized, they pass through the membrane into the lumen of the follicle.

4 ***Iodination of tyrosine.*** As iodine atoms (I^0) form, they react with tyrosines that are part of thyroglobulin molecules. Binding of one iodine atom yields monoiodotyrosine (T_1), and a second iodination produces diiodotyrosine (T_2). The TGB with attached iodine atoms, a sticky material that accumulates and is stored in the lumen of the thyroid follicle, is termed **colloid**.

5 ***Coupling of T_1 and T_2.*** During the last step in the synthesis of thyroid hormone, two T_2 molecules join to form T_4, or one T_1 and one T_2 join to form T_3.

6 ***Pinocytosis and digestion of colloid.*** Droplets of colloid reenter follicular cells by pinocytosis and merge with lysosomes. Digestive enzymes in the lysosomes break down TGB, cleaving off molecules of T_3 and T_4.

7 ***Secretion of thyroid hormones.*** Because T_3 and T_4 are lipid-soluble, they diffuse through the plasma membrane into interstitial fluid and then into the blood. T_4 normally is secreted in greater quantity than T_3, but T_3 is several times more potent. Moreover, after T_4 enters a body cell, most of it is converted to T_3 by removal of one iodine.

8 ***Transport in the blood.*** More than 99% of both the T_3 and the T_4 combine with transport proteins in the blood, mainly **thyroxine-binding globulin (TBG)**.

Actions of Thyroid Hormones

Because most body cells have receptors for thyroid hormones, T_3 and T_4 affect tissues throughout the body. Thyroid hormones act on their target cells mainly by inducing gene transcription and protein synthesis. The newly formed proteins in turn carry out the cellular response. Functions of thyroid hormones include the following:

1. ***Increase basal metabolic rate.*** Thyroid hormones raise the **basal metabolic rate (BMR)**, the rate of energy expenditure under standard or basal conditions (awake, at rest, and fasting). When BMR increases, cellular metabolism of carbohydrates, lipids, and proteins increases. Thyroid hormones increase BMR in several ways: (1) They stimulate synthesis of additional Na^+/K^+ ATPases, which use large amounts of ATP to continually eject sodium ions (Na^+) from cytosol into extracellular fluid and potassium ions (K^+) from extracellular fluid into cytosol; (2) they increase the concentrations of enzymes involved in cellular respiration, which increases the breakdown of organic fuels and ATP production; and (3) they increase the number and activity of mitochondria in cells, which also increases ATP production. As cells produce and use more ATP, BMR increases, more heat is given off, and body temperature rises, a phenomenon called the **calorigenic effect.** In this way, thyroid

hormones play an important role in the maintenance of normal body temperature. Normal mammals can survive in freezing temperatures, but those whose thyroid glands have been removed cannot.

2. ***Enhance actions of catechlolamines.*** Thyroid hormones have permissive effects on the catecholamines (epinephrine and norepinephrine) because they up-regulate β-adrenergic receptors. Recall that catecholamines bind to β-adrenergic receptors, promoting sympathetic responses. Therefore, symptoms of excess levels of thyroid hormone include increased heart rate, more forceful heartbeats, and increased blood pressure.
3. ***Regulate development and growth of nervous tissue and bones.*** Thyroid hormones are necessary for the development of the nervous system: They promote synapse formation, myelin production, and growth of dendrites. Thyroid hormones are also required for growth of the skeletal system: They promote formation of ossification centers in developing bones, synthesis of many bone proteins, and secretion of growth hormone (GH) and insulin-like growth factors (IGFs). Deficiency of thyroid hormones during fetal development, infancy, or childhood causes severe mental retardation and stunted bone growth.

Control of Thyroid Hormone Secretion

Thyrotropin-releasing hormone (TRH) from the hypothalamus and thyroid-stimulating hormone (TSH) from the anterior pituitary stimulate secretion of thyroid hormones, as shown in **Figure 18.11**:

1. Low blood levels of T_3 and T_4 or low metabolic rate stimulate the hypothalamus to secrete TRH.
2. TRH enters the hypothalamic–hypophyseal portal system and flows to the anterior pituitary, where it stimulates thyrotrophs to secrete TSH.
3. TSH stimulates virtually all aspects of thyroid follicular cell activity, including iodide trapping, hormone synthesis and secretion, and growth of the follicular cells (see **Figure 18.10**).
4. The thyroid follicular cells release T_3 and T_4 into the blood until the metabolic rate returns to normal.
5. An elevated level of T_3 inhibits release of TRH and TSH (negative feedback inhibition).

Conditions that increase ATP demand—a cold environment, hypoglycemia, high altitude, and pregnancy—increase the secretion of the thyroid hormones.

Calcitonin

The hormone produced by the **parafollicular cells** of the thyroid gland (see **Figure 18.9b**) is **calcitonin (CT)**. CT can decrease the level of calcium in the blood by inhibiting the action of osteoclasts, the cells that break down bone extracellular matrix. The secretion of CT is controlled by a negative feedback system (see **Figure 18.13**).

FIGURE 18.11 Regulation of secretion and actions of thyroid hormones. TRH = thyrotropin-releasing hormone, TSH = thyroid-stimulating hormone, T_3 = triiodothyronine, and T_4 = thyroxine (tetraiodothyronine).

TSH promotes release of thyroid hormones (T_3 and T_4) by the thyroid gland.

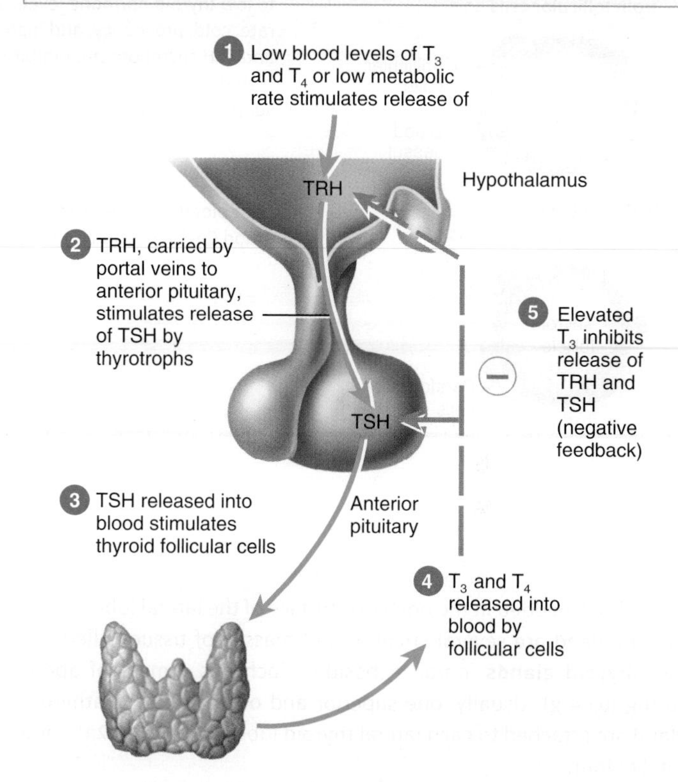

Q How could an iodine-deficient diet lead to goiter, which is an enlargement of the thyroid gland?

When its blood level is high, calcitonin lowers the amount of blood calcium and phosphates by inhibiting bone resorption (breakdown of bone extracellular matrix) by osteoclasts and by accelerating uptake of calcium and phosphates into bone extracellular matrix. Miacalcin, a calcitonin extract derived from salmon that is 10 times more potent than human calcitonin, is prescribed to treat osteoporosis.

Table 18.6 summarizes the hormones produced by the thyroid gland, control of their secretion, and their principal actions.

18.8 Parathyroid Glands

OBJECTIVE

- **Describe** the location, histology, hormone, and functions of the parathyroid glands.

TABLE 18.6 Summary of Thyroid Gland Hormones

HORMONE AND SOURCE	CONTROL OF SECRETION	PRINCIPAL ACTIONS
T_3 (triiodothyronine) and **T_4 (thyroxine)** or **thyroid hormones** from follicular cells	Secretion is increased by thyrotropin-releasing hormone (TRH), which stimulates release of thyroid-stimulating hormone (TSH) in response to low thyroid hormone levels, low metabolic rate, cold, pregnancy, and high altitudes; TRH and TSH secretions are inhibited in response to high thyroid hormone levels; high iodine level suppresses T_3/T_4 secretion.	Increase basal metabolic rate; stimulate synthesis of proteins; increase use of glucose and fatty acids for ATP production; increase lipolysis; enhance cholesterol excretion; accelerate body growth; contribute to development of nervous system.
Calcitonin (CT) from parafollicular cells	High blood Ca^{2+} levels stimulate secretion; low blood Ca^{2+} levels inhibit secretion.	Lowers blood levels of Ca^{2+} and HPO_4^{2-} by inhibiting bone resorption by osteoclasts and by accelerating uptake of calcium and phosphates into bone extracellular matrix.

Partially embedded in the posterior surface of the lateral lobes of the thyroid gland are several small, round masses of tissue called the **parathyroid glands** (*para-* = beside). Each has a mass of about 40 mg (0.04 g). Usually, one superior and one inferior parathyroid gland are attached to each lateral thyroid lobe (**Figure 18.12a**), for a total of four.

Microscopically, the parathyroid glands contain two kinds of epithelial cells (**Figure 18.12b, c**). The more numerous cells, called **chief cells** or principal cells, produce **parathyroid hormone (PTH)**, also called *parathormone*. The function of the other kind of cell, called an **oxyphil cell**, is not known in a normal parathyroid gland. However, its presence clearly helps to identify the parathyroid gland histologically due to its unique staining characteristics. Furthermore, in a cancer of the parathyroid glands, oxyphil cells secrete excess PTH.

Parathyroid Hormone

Parathyroid hormone is the major regulator of the levels of calcium (Ca^{2+}), magnesium (Mg^{2+}), and phosphate (HPO_4^{2-}) ions in the blood. The specific action of PTH is to increase the number and activity of osteoclasts. The result is elevated bone *resorption*, which releases ionic calcium (Ca^{2+}) and phosphates (HPO_4^{2-}) into the blood. PTH also acts on the kidneys. First, it slows the rate at which Ca^{2+} and Mg^{2+} are lost from blood into the urine. Second, it increases loss of HPO_4^{2-} from blood into the urine. Because more HPO_4^{2-} is lost in the urine than is gained from the bones, PTH decreases blood HPO_4^{2-} level and increases blood Ca^{2+} and Mg^{2+} levels. A third effect of PTH on the kidneys is to promote formation of the hormone **calcitriol** (kal′-si-TRĪ-ol), the active form of vitamin D. Calcitriol, also known as *1,25-dihydroxyvitamin D_3*, increases the rate of Ca^{2+}, HPO_4^{2-}, and Mg^{2+} *absorption* from the gastrointestinal tract into the blood.

Control of Secretion of Calcitonin and Parathyroid Hormone

The blood calcium level directly controls the secretion of both calcitonin and parathyroid hormone via negative feedback loops that do not involve the pituitary gland (**Figure 18.13**):

1. A higher-than-normal level of calcium ions (Ca^{2+}) in the blood stimulates parafollicular cells of the thyroid gland to release more calcitonin.
2. Calcitonin inhibits the activity of osteoclasts, thereby decreasing the blood Ca^{2+} level.
3. A lower-than-normal level of Ca^{2+} in the blood stimulates chief cells of the parathyroid gland to release more PTH.
4. PTH promotes resorption of bone extracellular matrix, which releases Ca^{2+} into the blood and slows loss of Ca^{2+} in the urine, raising the blood level of Ca^{2+}.
5. PTH also stimulates the kidneys to synthesize calcitriol, the active form of vitamin D.
6. Calcitriol stimulates increased absorption of Ca^{2+} from foods in the gastrointestinal tract, which helps increase the blood level of Ca^{2+}.

FIGURE 18.12 **Location, blood supply, and histology of the parathyroid gland.**

The parathyroid glands, normally four in number, are embedded in the posterior surface of the thyroid gland.

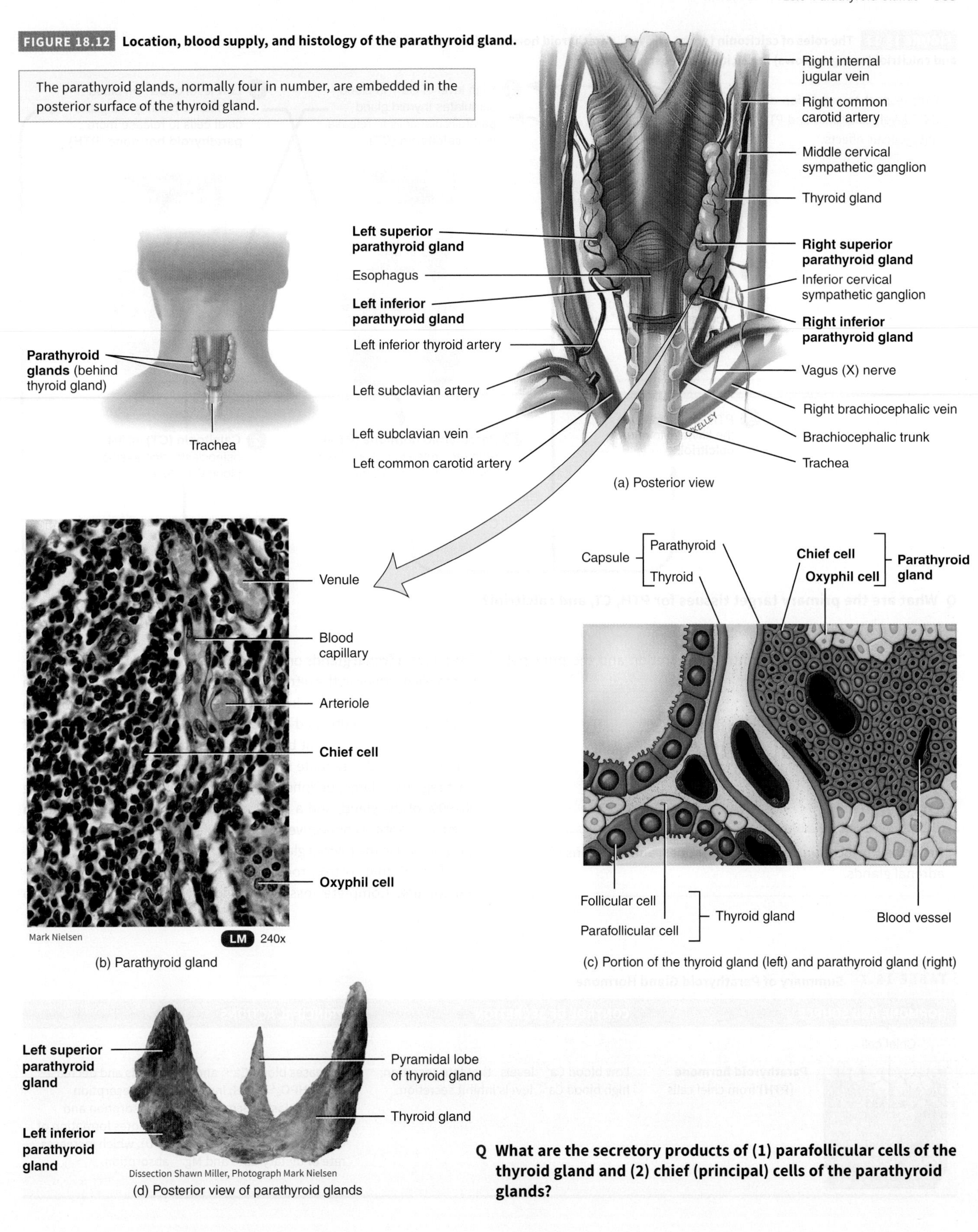

(a) Posterior view

(b) Parathyroid gland

(c) Portion of the thyroid gland (left) and parathyroid gland (right)

(d) Posterior view of parathyroid glands

Q What are the secretory products of (1) parafollicular cells of the thyroid gland and (2) chief (principal) cells of the parathyroid glands?

FIGURE 18.13 The roles of calcitonin (green arrows), parathyroid hormone (blue arrows), and calcitriol (orange arrows) in calcium homeostasis.

With respect to regulation of blood Ca^{2+} level, calcitonin and PTH have antagonistic effects.

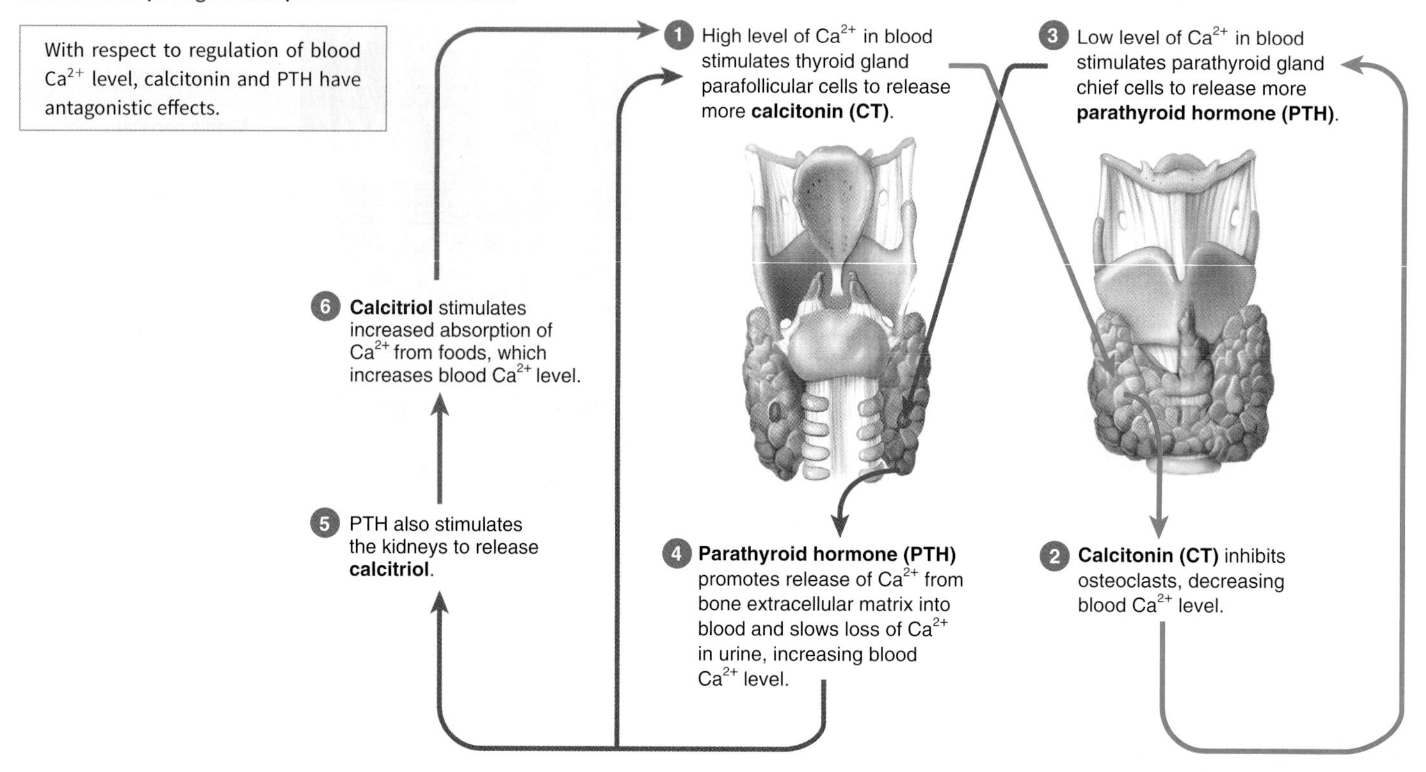

Q What are the primary target tissues for PTH, CT, and calcitriol?

Table 18.7 summarizes control of secretion and the principal actions of parathyroid hormone.

18.9 Adrenal Glands

OBJECTIVE

- **Describe** the location, histology, hormones, and functions of the adrenal glands.

The paired **adrenal glands** or *suprarenal glands*, one of which lies superior to each kidney in the retroperitoneal space (**Figure 18.14a**), have a flattened pyramidal shape. In an adult, each adrenal gland is 3–5 cm in height, 2–3 cm in width, and a little less than 1 cm thick, with a mass of 3.5–5 g, only half its size at birth. During embryonic development, the adrenal glands differentiate into two structurally and functionally distinct regions: a large, peripherally located **adrenal cortex**, comprising 80–90% of the gland, and a small, centrally located **adrenal medulla** (**Figure 18.14b**). A connective tissue capsule covers the gland. The adrenal glands, like the thyroid gland, are highly vascularized.

The adrenal cortex produces steroid hormones that are essential for life. Complete loss of adrenocortical hormones leads to

TABLE 18.7 Summary of Parathyroid Gland Hormone

HORMONE AND SOURCE	CONTROL OF SECRETION	PRINCIPAL ACTIONS
Chief cell **Parathyroid hormone (PTH)** from chief cells	Low blood Ca^{2+} levels stimulate secretion; high blood Ca^{2+} levels inhibit secretion.	Increases blood Ca^{2+} and Mg^{2+} levels and decreases blood HPO_4^{2-} level; increases bone resorption by osteoclasts; increases Ca^{2+} reabsorption and HPO_4^{2-} excretion by kidneys; promotes formation of calcitriol (active form of vitamin D), which increases rate of dietary Ca^{2+} and Mg^{2+} absorption.

death due to dehydration and electrolyte imbalances in a few days to a week, unless hormone replacement therapy begins promptly. The adrenal medulla produces three catecholamine hormones—norepinephrine, epinephrine, and a small amount of dopamine.

Adrenal Cortex

The adrenal cortex is subdivided into three zones, each of which secretes different hormones (Figure 18.14d). The outer zone, just deep to the connective tissue capsule, is the **zona glomerulosa** (glo-mer′-ū-LŌ-sa; *zona* = belt; *glomerul-* = little ball). Its cells, which are closely packed and arranged in spherical clusters and arched columns, secrete hormones called **mineralocorticoids** (min′-er-al-ō-KOR-ti-koyds) because they affect mineral homeostasis. The middle zone, or **zona fasciculata** (fa-sik′-ū-LA-ta; *fascicul-* = little bundle), is the widest of the three zones and consists of cells arranged in long, straight columns. The cells of the zona fasciculata secrete mainly **glucocorticoids** (gloo′-kō-KOR-ti-koyds), primarily cortisol, so named because they affect glucose homeostasis. The cells of the inner zone, the **zona reticularis** (re-tik′-ū-LAR-is; *reticul-* = network), are arranged in branching cords. They synthesize small amounts of weak **androgens** (*andro-* = a man), steroid hormones that have masculinizing effects.

Mineralocorticoids

Aldosterone (al-DOS-ter-ōn) is the major mineralocorticoid. It regulates homeostasis of two mineral ions—namely, sodium ions (Na^+) and potassium ions (K^+)—and helps adjust blood pressure and blood volume. Aldosterone also promotes excretion of H^+ in the urine; this removal of acids from the body can help prevent acidosis (blood pH below 7.35), which is discussed in Chapter 27.

Control of Aldosterone Secretion

The **renin–angiotensin–aldosterone (RAA) pathway** (RĒ-nin an′-jē-ō-TEN-sin) controls secretion of aldosterone (Figure 18.15):

1. Stimuli that initiate the renin–angiotensin–aldosterone pathway include dehydration, Na^+ deficiency, or hemorrhage.
2. These conditions cause a decrease in blood volume.
3. Decreased blood volume leads to decreased blood pressure.
4. Lowered blood pressure stimulates certain cells of the kidneys, called juxtaglomerular cells, to secrete the enzyme **renin**.
5. The level of renin in the blood increases.
6. Renin converts **angiotensinogen** (an′-jē-ō-ten-SIN-ō-jen), a plasma protein produced by the liver, into **angiotensin I**.

FIGURE 18.14 **Location, blood supply, and histology of the adrenal (suprarenal) glands.**

The adrenal cortex secretes steroid hormones that are essential for life; the adrenal medulla secretes norepinephrine and epinephrine.

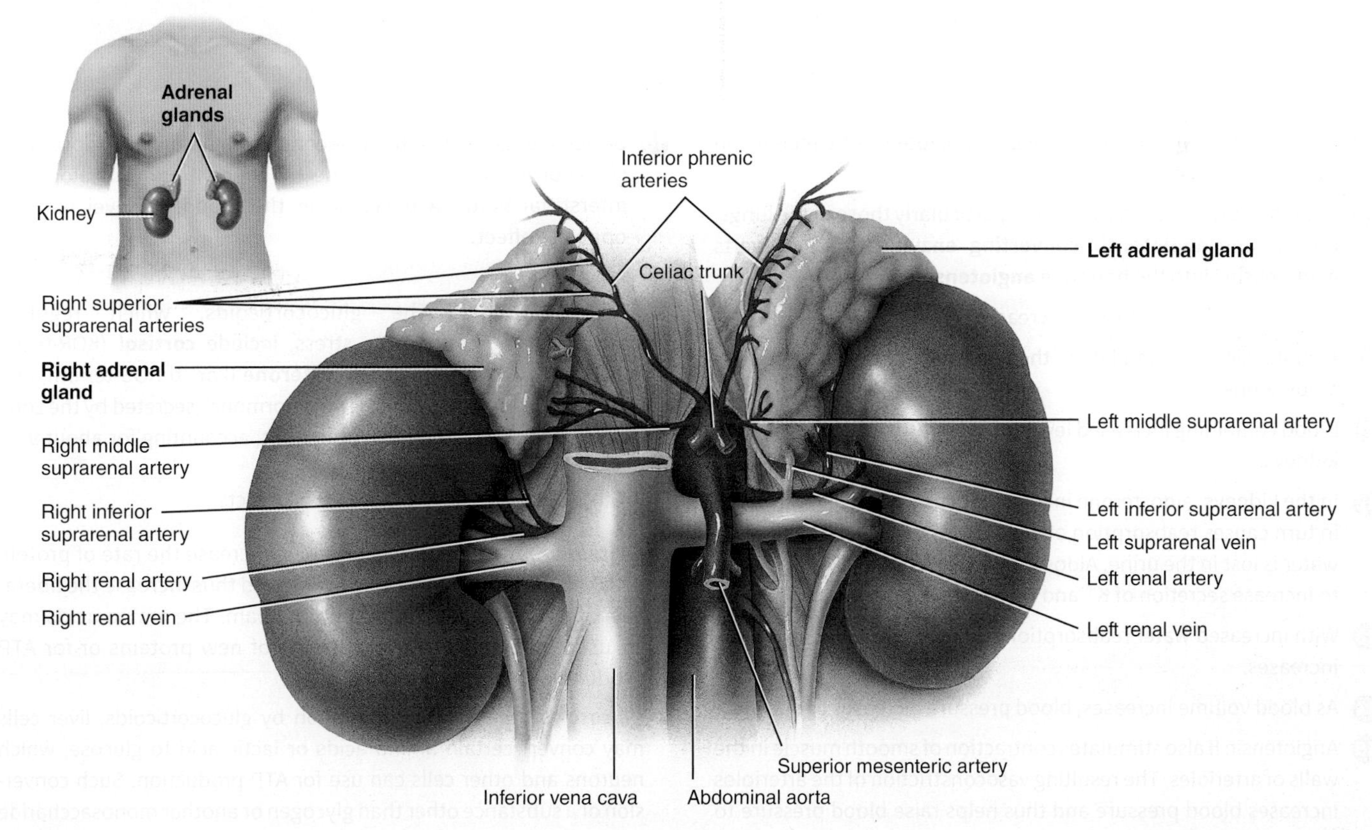

(a) Anterior view

Figure 18.14 *Continues*

FIGURE 18.14 Continued

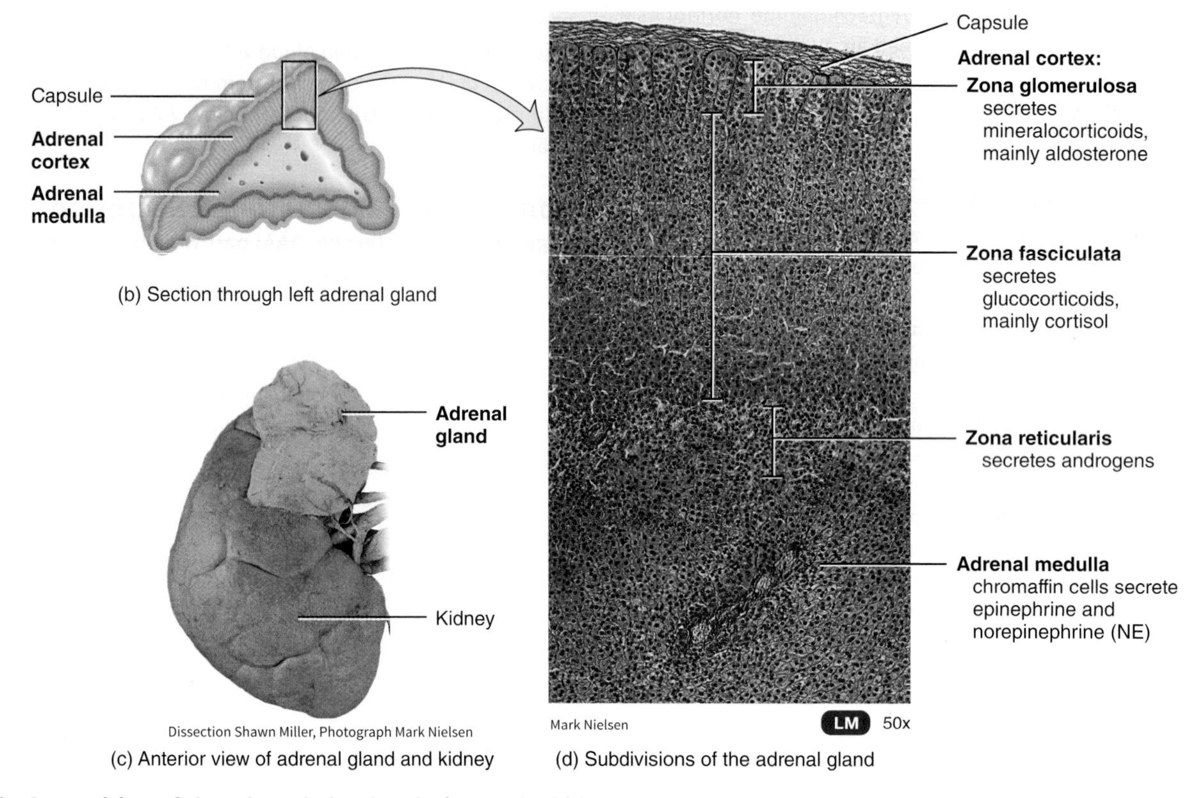

(b) Section through left adrenal gland

(c) Anterior view of adrenal gland and kidney

(d) Subdivisions of the adrenal gland

Q What is the position of the adrenal glands relative to the kidneys?

7. Blood containing increased levels of angiotensin I circulates in the body.
8. As blood flows through capillaries, particularly those of the lungs, the enzyme **angiotensin-converting enzyme (ACE)** converts angiotensin I into the hormone **angiotensin II**.
9. Blood level of angiotensin II increases.
10. Angiotensin II stimulates the adrenal cortex to secrete aldosterone.
11. Blood containing increased levels of aldosterone circulates to the kidneys.
12. In the kidneys, aldosterone increases reabsorption of Na^+, which in turn causes reabsorption of water by osmosis. As a result, less water is lost in the urine. Aldosterone also stimulates the kidneys to increase secretion of K^+ and H^+ into the urine.
13. With increased water reabsorption by the kidneys, blood volume increases.
14. As blood volume increases, blood pressure increases to normal.
15. Angiotensin II also stimulates contraction of smooth muscle in the walls of arterioles. The resulting vasoconstriction of the arterioles increases blood pressure and thus helps raise blood pressure to normal.
16. Besides angiotensin II, a second stimulator of aldosterone secretion is an increase in the K^+ concentration of blood (or interstitial fluid). A decrease in the blood K^+ level has the opposite effect.

Glucocorticoids The glucocorticoids, which regulate metabolism and resistance to stress, include **cortisol** (KOR-ti-sol; also called *hydrocortisone*), **corticosterone** (kor′-ti-KOS-ter-ōn), and **cortisone** (KOR-ti-sōn). Of these three hormones secreted by the zona fasciculata, cortisol is the most abundant, accounting for about 95% of glucocorticoid activity.

Glucocorticoids have the following effects:

1. ***Protein breakdown.*** Glucocorticoids increase the rate of protein breakdown, mainly in muscle fibers, and thus increase the liberation of amino acids into the bloodstream. The amino acids may be used by body cells for synthesis of new proteins or for ATP production.
2. ***Glucose formation.*** On stimulation by glucocorticoids, liver cells may convert certain amino acids or lactic acid to glucose, which neurons and other cells can use for ATP production. Such conversion of a substance other than glycogen or another monosaccharide into glucose is called **gluconeogenesis** (gloo′-ko-nē-ō-JEN-e-sis).

FIGURE 18.15 Regulation of aldosterone secretion by the renin–angiotensin–aldosterone (RAA) pathway.

Aldosterone helps regulate blood volume, blood pressure, and levels of Na^+, K^+, and H^+ in the blood.

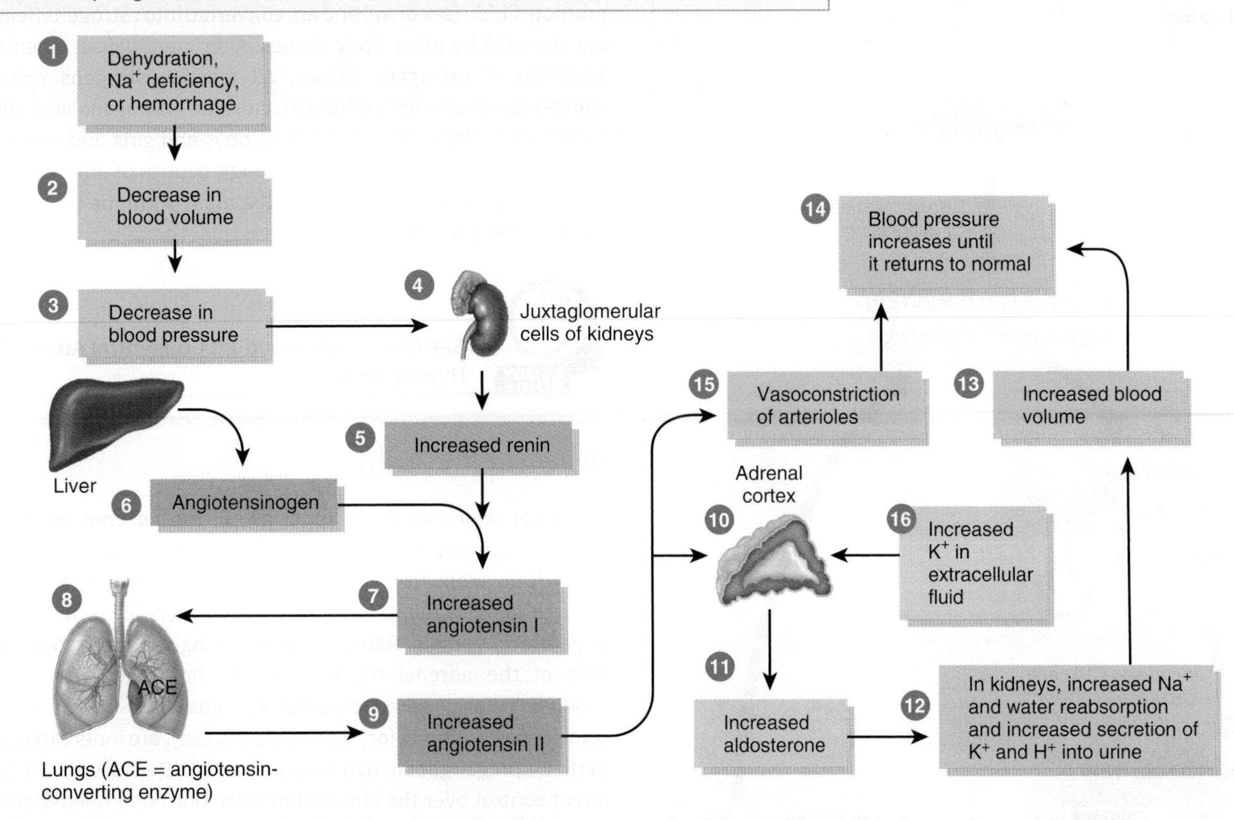

Q In what two ways can angiotensin II increase blood pressure, and what are its target tissues in each case?

3. ***Lipolysis.*** Glucocorticoids stimulate **lipolysis** (lī-POL-i-sis), the breakdown of triglycerides and release of fatty acids from adipose tissue into the blood.
4. ***Resistance to stress.*** Glucocorticoids work in many ways to provide resistance to stress. The additional glucose supplied by the liver cells provides tissues with a ready source of ATP to combat a range of stresses, including exercise, fasting, fright, temperature extremes, high altitude, bleeding, infection, surgery, trauma, and disease. Because glucocorticoids make blood vessels more sensitive to other hormones that cause vasoconstriction, they raise blood pressure. This effect would be an advantage in cases of severe blood loss, which causes blood pressure to drop.
5. ***Anti-inflammatory effects.*** Glucocorticoids inhibit white blood cells that participate in inflammatory responses. Unfortunately, glucocorticoids also retard tissue repair; as a result, they slow wound healing. Although high doses can cause severe mental disturbances, glucocorticoids are very useful in the treatment of chronic inflammatory disorders such as rheumatoid arthritis.
6. ***Depression of immune responses.*** High doses of glucocorticoids depress immune responses. For this reason, glucocorticoids are prescribed for organ transplant recipients to retard tissue rejection by the immune system.

Control of Glucocorticoid Secretion Control of glucocorticoid secretion occurs via a typical negative feedback system (**Figure 18.16**). Low blood levels of glucocorticoids, mainly cortisol, stimulate neurosecretory cells in the hypothalamus to secrete **corticotropin-releasing hormone (CRH)**. CRH (together with a low level of cortisol) promotes the release of ACTH from the anterior pituitary. ACTH flows in the blood to the adrenal cortex, where it stimulates glucocorticoid secretion. (To a much smaller extent, ACTH also stimulates secretion of aldosterone.) The discussion of stress at the end of the chapter describes how the hypothalamus also increases CRH release in response to a variety of physical and emotional stresses (see Section 18.14).

Androgens In both males and females, the adrenal cortex secretes small amounts of weak androgens. The major androgen secreted by the adrenal gland is **dehydroepiandrosterone (DHEA)** (dē-hī-drō-ep′-ē-an-DROS-ter-ōn). After puberty in males, the

FIGURE 18.16 Negative feedback regulation of glucocorticoid secretion.

A high level of CRH and a low level of glucocorticoids promote the release of ACTH, which stimulates glucocorticoid secretion by the adrenal cortex.

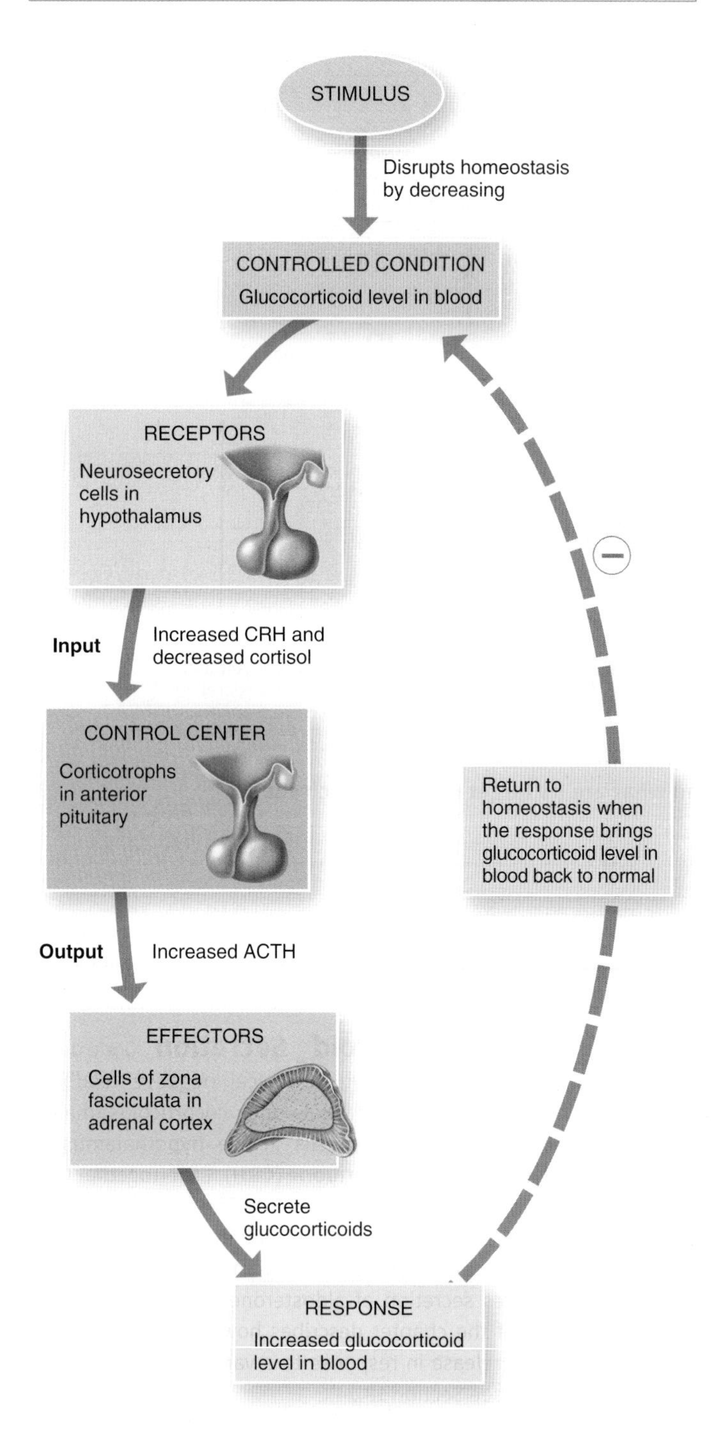

Q If a heart transplant patient receives prednisone (a glucocorticoid) to help prevent rejection of the transplanted tissue, will blood levels of ACTH and CRH be high or low? Explain.

androgen testosterone is also released in much greater quantity by the testes. Thus, the amount of androgens secreted by the adrenal gland in males is usually so low that their effects are insignificant. In females, however, adrenal androgens play important roles. They promote libido (sex drive) and are converted into estrogens (feminizing sex steroids) by other body tissues. After menopause, when ovarian secretion of estrogens ceases, all female estrogens come from conversion of adrenal androgens. Adrenal androgens also stimulate growth of axillary and pubic hair in boys and girls and contribute to the prepubertal growth spurt. Although control of adrenal androgen secretion is not fully understood, the main hormone that stimulates its secretion is ACTH.

See Clinical Connection: Congenital Adrenal Hyperplasia

Adrenal Medulla

The inner region of the adrenal gland, the **adrenal medulla**, is a modified sympathetic ganglion of the autonomic nervous system (ANS). It develops from the same embryonic tissue as all other sympathetic ganglia, but its cells, which lack axons, form clusters around large blood vessels. Rather than releasing a neurotransmitter, the cells of the adrenal medulla secrete hormones. The hormone-producing cells, called **chromaffin cells** (KRŌ-maf-in; *chrom-* = color; *-affin* = affinity for; see **Figure 18.14d**), are innervated by sympathetic preganglionic neurons of the ANS. Because the ANS exerts direct control over the chromaffin cells, hormone release can occur very quickly.

The two major hormones synthesized by the adrenal medulla are **epinephrine** (ep′-i-NEF-rin) and **norepinephrine (NE)**, also called *adrenaline* and *noradrenaline*, respectively. The chromaffin cells of the adrenal medulla secrete an unequal amount of these hormones—about 80% epinephrine and 20% norepinephrine. The hormones of the adrenal medulla intensify sympathetic responses that occur in other parts of the body.

Control of Secretion of Epinephrine and Norepinephrine In stressful situations and during exercise, impulses from the hypothalamus stimulate sympathetic preganglionic neurons, which in turn stimulate the chromaffin cells to secrete epinephrine and norepinephrine. These two hormones greatly augment the fight-or-flight response that you learned about in Chapter 15. By increasing heart rate and force of contraction, epinephrine and norepinephrine increase the output of the heart, which increases blood pressure. They also increase blood flow to the heart, liver, skeletal muscles, and adipose tissue; dilate airways to the lungs; and increase blood levels of glucose and fatty acids.

Table 18.8 summarizes the hormones produced by the adrenal glands, control of their secretion, and their principal actions.

TABLE 18.8 Summary of Adrenal Gland Hormones

HORMONE AND SOURCE	CONTROL OF SECRETION	PRINCIPAL ACTIONS
ADRENAL CORTEX HORMONES		
Mineralocorticoids (mainly aldosterone) from zona glomerulosa cells	Increased blood K^+ level and angiotensin II stimulate secretion.	Increase blood levels of Na^+ and water; decrease blood level of K^+.
Glucocorticoids (mainly cortisol) from zona fasciculata cells	ACTH stimulates release; corticotropin-releasing hormone (CRH) promotes ACTH secretion in response to stress and low blood levels of glucocorticoids.	Increase protein breakdown (except in liver), stimulate gluconeogenesis and lipolysis, provide resistance to stress, dampen inflammation, depress immune responses.
Androgens (mainly dehydroepiandrosterone, or DHEA) from zona reticularis cells	ACTH stimulates secretion.	Assist in early growth of axillary and pubic hair in both sexes; in females, contribute to libido and are source of estrogens after menopause.
Adrenal cortex		
ADRENAL MEDULLA HORMONES		
Epinephrine and norepinephrine from chromaffin cells	Sympathetic preganglionic neurons release acetylcholine, which stimulates secretion.	Enhance effects of sympathetic division of autonomic nervous system (ANS) during stress.
Adrenal medulla		

18.10 Pancreatic Islets

OBJECTIVE

- **Describe** the location, histology, hormones, and functions of the pancreatic islets.

The **pancreas** (*pan-* = all; *-creas* = flesh) is both an endocrine gland and an exocrine gland. We discuss its endocrine functions here and describe its exocrine functions in Chapter 24 in the coverage of the digestive system. A flattened organ that measures about 12.5–15 cm (5–6 in.) in length, the pancreas is located in the curve of the duodenum, the first part of the small intestine, and consists of a head, a body, and a tail (**Figure 18.17a**). Roughly 99% of the exocrine cells of the pancreas are arranged in clusters called **acini** (AS-i-nī). The acini produce digestive enzymes, which flow into the gastrointestinal tract through a network of ducts. Scattered among the exocrine acini are 1–2 million tiny clusters of endocrine tissue called **pancreatic islets** (Ī-lets) or *islets of Langerhans* (LAHNG-er-hanz; **Figure 18.17b, c**). Abundant capillaries serve both the exocrine and endocrine portions of the pancreas.

Cell Types in the Pancreatic Islets

Each pancreatic islet includes four types of hormone-secreting cells:

1. **Alpha** or *A* **cells** constitute about 17% of pancreatic islet cells and secrete **glucagon** (GLOO-ka-gon).
2. **Beta** or *B* **cells** constitute about 70% of pancreatic islet cells and secrete **insulin** (IN-soo-lin).
3. **Delta** or *D* **cells** constitute about 7% of pancreatic islet cells and secrete **somatostatin** (sō′-ma-tō-STAT-in).
4. **F cells** constitute the remainder of pancreatic islet cells and secrete **pancreatic polypeptide**.

The interactions of the four pancreatic hormones are complex and not completely understood. We do know that glucagon raises blood glucose level, and insulin lowers it. Somatostatin acts in a paracrine manner to inhibit both insulin and glucagon release from neighboring beta and alpha cells. It may also act as a circulating hormone to slow absorption of nutrients from the gastrointestinal tract. In addition, somatostatin inhibits the secretion of growth hormone. Pancreatic polypeptide inhibits somatostatin secretion, gallbladder contraction, and secretion of digestive enzymes by the pancreas.

Control of Secretion of Glucagon and Insulin

The principal action of glucagon is to increase blood glucose level when it falls below normal. Insulin, on the other hand, helps lower blood glucose level when it is too high. The level of blood glucose controls secretion of glucagon and insulin via negative feedback (**Figure 18.18**):

1. Low blood glucose level (hypoglycemia) stimulates secretion of glucagon from alpha cells of the pancreatic islets.

FIGURE 18.17 **Location, blood supply, and histology of the pancreas.**

Pancreatic hormones regulate blood glucose level.

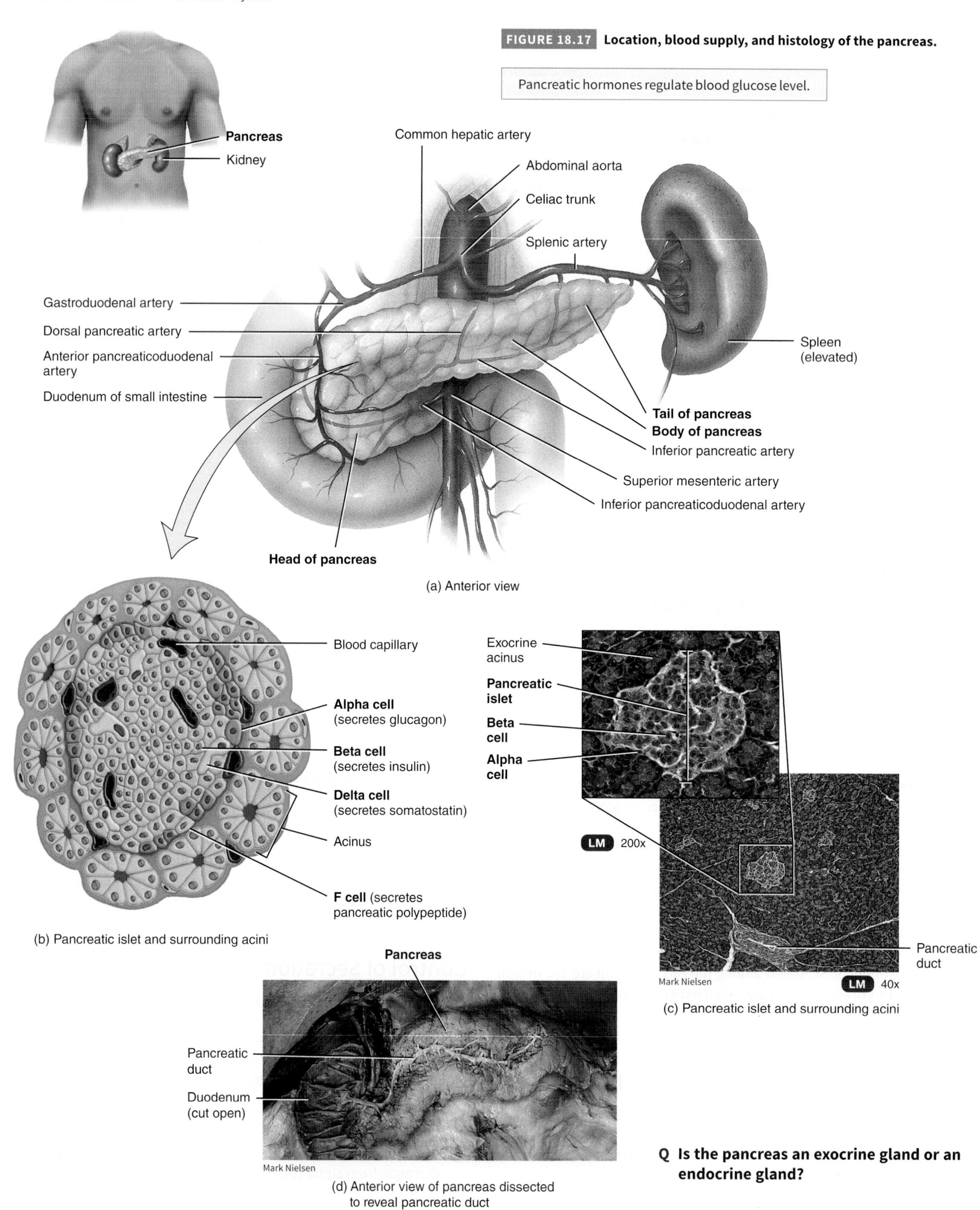

(a) Anterior view

(b) Pancreatic islet and surrounding acini

(c) Pancreatic islet and surrounding acini

(d) Anterior view of pancreas dissected to reveal pancreatic duct

Q Is the pancreas an exocrine gland or an endocrine gland?

FIGURE 18.18 Negative feedback regulation of the secretion of glucagon (blue arrows) and insulin (orange arrows).

Low blood glucose stimulates release of glucagon; high blood glucose stimulates secretion of insulin.

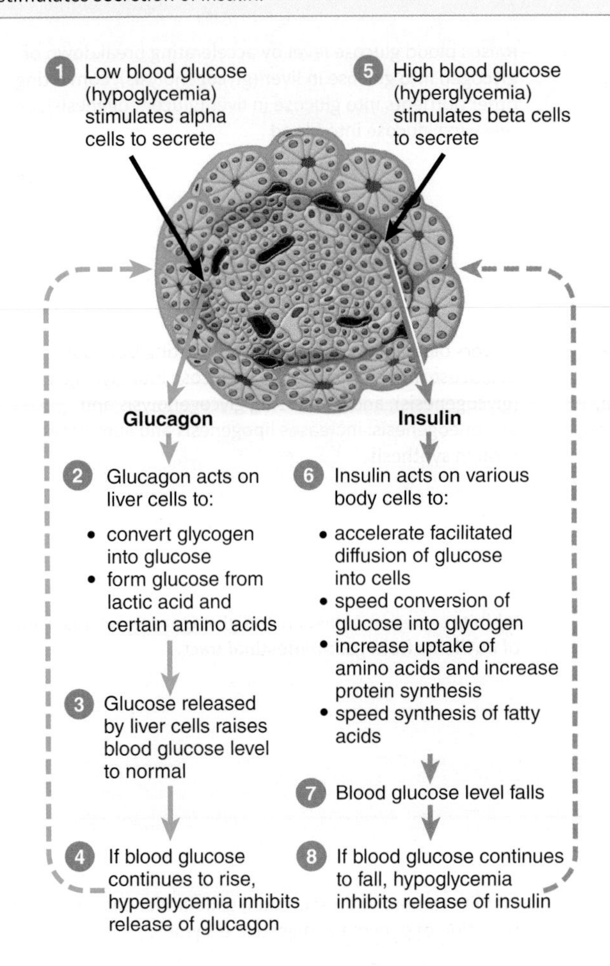

Q Does glycogenolysis increase or decrease blood glucose level?

2 Glucagon acts on hepatocytes (liver cells) to accelerate the conversion of glycogen into glucose (glycogenolysis) and to promote formation of glucose from lactic acid and certain amino acids (gluconeogenesis).

3 As a result, hepatocytes release glucose into the blood more rapidly, and blood glucose level rises.

4 If blood glucose continues to rise, high blood glucose level (hyperglycemia) inhibits release of glucagon (negative feedback).

5 High blood glucose (hyperglycemia) stimulates secretion of insulin by beta cells of the pancreatic islets.

6 Insulin acts on various cells in the body to accelerate facilitated diffusion of glucose into cells; to speed conversion of glucose into glycogen (glycogenesis); to increase uptake of amino acids by cells and to increase protein synthesis; to speed synthesis of fatty acids (lipogenesis); to slow the conversion of glycogen to glucose (glycogenolysis); and to slow the formation of glucose from lactic acid and amino acids (gluconeogenesis).

7 As a result, blood glucose level falls.

8 If blood glucose level drops below normal, low blood glucose inhibits release of insulin (negative feedback) and stimulates release of glucagon.

Although blood glucose level is the most important regulator of insulin and glucagon, several hormones and neurotransmitters also regulate the release of these two hormones. In addition to the responses to blood glucose level just described, glucagon stimulates insulin release directly; insulin has the opposite effect, suppressing glucagon secretion. As blood glucose level declines and less insulin is secreted, the alpha cells of the pancreas are released from the inhibitory effect of insulin so they can secrete more glucagon. Indirectly, growth hormone (GH) and adrenocorticotropic hormone (ACTH) stimulate secretion of insulin because they act to elevate blood glucose.

Insulin secretion is also stimulated by:

- Acetylcholine, the neurotransmitter liberated from axon terminals of parasympathetic vagus nerve fibers that innervate the pancreatic islets
- The amino acids arginine and leucine, which would be present in the blood at higher levels after a protein-containing meal
- Glucose-dependent insulinotropic peptide (GIP),* a hormone released by enteroendocrine cells of the small intestine in response to the presence of glucose in the gastrointestinal tract

Thus, digestion and absorption of food containing both carbohydrates and proteins provide strong stimulation for insulin release.

Glucagon secretion is stimulated by:

- Increased activity of the sympathetic division of the ANS, as occurs during exercise
- A rise in blood amino acids if blood glucose level is low, which could occur after a meal that contained mainly protein

Table 18.9 summarizes the hormones produced by the pancreas, control of their secretion, and their principal actions.

18.11 Ovaries and Testes

OBJECTIVE

- **Describe** the location, hormones, and functions of the male and female gonads.

Gonads are the organs that produce gametes—sperm in males and oocytes in females. In addition to their reproductive function, the

*GIP—previously called gastric inhibitory peptide—was renamed because at physiological concentration its inhibitory effect on stomach function is negligible.

TABLE 18.9 Summary of Pancreatic Islet Hormones

HORMONE AND SOURCE	CONTROL OF SECRETION	PRINCIPAL ACTIONS
Glucagon from alpha cells of pancreatic islets Alpha cell	Decreased blood level of glucose, exercise, and mainly protein meals stimulate secretion; somatostatin and insulin inhibit secretion.	Raises blood glucose level by accelerating breakdown of glycogen into glucose in liver (glycogenolysis), converting other nutrients into glucose in liver (gluconeogenesis), and releasing glucose into blood.
Insulin from beta cells of pancreatic islets Beta cell	Increased blood level of glucose, acetylcholine (released by parasympathetic vagus nerve fibers), arginine and leucine (two amino acids), glucagon, GIP, GH, and ACTH stimulate secretion; somatostatin inhibits secretion.	Lowers blood glucose level by accelerating transport of glucose into cells, converting glucose into glycogen (glycogenesis), and decreasing glycogenolysis and gluconeogenesis; increases lipogenesis and stimulates protein synthesis.
Somatostatin from delta cells of pancreatic islets Delta cell	Pancreatic polypeptide inhibits secretion.	Inhibits secretion of insulin and glucagon; slows absorption of nutrients from gastrointestinal tract.
Pancreatic polypeptide from F cells of pancreatic islets F cell	Meals containing protein, fasting, exercise, and acute hypoglycemia stimulate secretion; somatostatin and elevated blood glucose level inhibit secretion.	Inhibits somatostatin secretion, gallbladder contraction, and secretion of pancreatic digestive enzymes.

gonads secrete hormones. The **ovaries**, paired oval bodies located in the female pelvic cavity, produce several steroid hormones, including two **estrogens** (estradiol and estrone) and **progesterone**. These female sex hormones, along with follicle-stimulating hormone (FSH) and luteinizing hormone (LH) from the anterior pituitary, regulate the menstrual cycle, maintain pregnancy, and prepare the mammary glands for lactation. They also promote enlargement of the breasts and widening of the hips at puberty, and help maintain these female secondary sex characteristics. The ovaries also produce **inhibin**, a protein hormone that inhibits secretion of FSH. During pregnancy, the ovaries and placenta produce a peptide hormone called **relaxin (RLX)**, which increases the flexibility of the pubic symphysis during pregnancy and helps dilate the uterine cervix during labor and delivery. These actions help ease the baby's passage by enlarging the birth canal.

The male gonads, the **testes**, are oval glands that lie in the scrotum. The main hormone produced and secreted by the testes is **testosterone**, an **androgen** or male sex hormone. Testosterone stimulates descent of the testes before birth, regulates production of sperm, and stimulates the development and maintenance of male secondary sex characteristics, such as beard growth and deepening of the voice. The testes also produce inhibin, which inhibits secretion of FSH. The detailed structure of the ovaries and testes and the specific roles of sex hormones are discussed in Chapter 28.

Table 18.10 summarizes the hormones produced by the ovaries and testes and their principal actions.

TABLE 18.10 Summary of Hormones of the Ovaries and Testes

HORMONE	PRINCIPAL ACTIONS
OVARIAN HORMONES	
Estrogens and progesterone Ovary	Together with gonadotropic hormones of anterior pituitary, regulate female reproductive cycle, maintain pregnancy, prepare mammary glands for lactation, and promote development and maintenance of female secondary sex characteristics.
Relaxin (RLX)	Increases flexibility of pubic symphysis during pregnancy; helps dilate uterine cervix during labor and delivery.
Inhibin	Inhibits secretion of FSH from anterior pituitary.
TESTICULAR HORMONES	
Testosterone Testis	Stimulates descent of testes before birth; regulates sperm production; promotes development and maintenance of male secondary sex characteristics.
Inhibin	Inhibits secretion of FSH from anterior pituitary.

18.12 Pineal Gland and Thymus

OBJECTIVES

- **Describe** the location, histology, hormone, and functions of the pineal gland.
- **Describe** the role of the thymus in immunity.

The **pineal gland** (PĪN-ē-al = pinecone shape) is a small endocrine gland attached to the roof of the third ventricle of the brain at the midline (see **Figure 18.1**). Part of the epithalamus, it is positioned between the two superior colliculi, has a mass of 0.1–0.2 g, and is covered by a capsule formed by the pia mater. The gland consists of masses of neuroglia and secretory cells called **pinealocytes** (pin-ē-AL-ō-sīts).

The pineal gland secretes **melatonin**, an amine hormone derived from serotonin. Melatonin appears to contribute to the setting of the body's biological clock, which is controlled by the suprachiasmatic nucleus of the hypothalamus. As more melatonin is liberated during darkness than in light, this hormone is thought to promote sleepiness. In response to visual input from the eyes (retina), the suprachiasmatic nucleus stimulates sympathetic postganglionic neurons of the superior cervical ganglion, which in turn stimulate the pinealocytes of the pineal gland to secrete melatonin in a rhythmic pattern, with low levels of melatonin secreted during the day and significantly higher levels secreted at night. During sleep, plasma levels of melatonin increase tenfold and then decline to a low level again before awakening. Small doses of melatonin given orally can induce sleep and reset daily rhythms, which might benefit workers whose shifts alternate between daylight and nighttime hours. Melatonin also is a potent antioxidant that may provide some protection against damaging oxygen free radicals.

In animals that breed during specific seasons, melatonin inhibits reproductive functions, but it is unclear whether melatonin influences human reproductive function. Melatonin levels are higher in children and decline with age into adulthood, but there is no evidence that changes in melatonin secretion correlate with the onset of puberty and sexual maturation. Nevertheless, because melatonin causes atrophy of the gonads in several animal species, the possibility of adverse effects on human reproduction must be studied before its use to reset daily rhythms can be recommended.

See Clinical Connection: Seasonal Affective Disorder and Jet Lag

The **thymus** is located behind the sternum between the lungs. Because of the role of the thymus in immunity, the details of its structure and functions are discussed in Chapter 22. The hormones produced by the thymus—**thymosin**, **thymic humoral factor (THF)**, **thymic factor (TF)**, and **thymopoietin** (thī-mō-poy-Ē-tin)—promote the maturation of T cells (a type of white blood cell that destroys microbes and foreign substances) and may retard the aging process.

18.13 Other Endocrine Tissues and Organs, Eicosanoids, and Growth Factors

OBJECTIVES

- **Outline** the functions of each of the hormones secreted by cells in tissues and organs other than endocrine glands.
- **Describe** the actions of eicosanoids and growth factors.

Hormones from Other Endocrine Tissues and Organs

As you learned at the beginning of this chapter, cells in organs other than those usually classified as endocrine glands have an endocrine function and secrete hormones. You learned about several of these in this chapter: the hypothalamus, thymus, pancreas, ovaries, and testes. **Table 18.11** provides an overview of these organs and tissues and their hormones and actions.

TABLE 18.11 Summary of Hormones Produced by Other Organs and Tissues That Contain Endocrine Cells

HORMONE	PRINCIPAL ACTIONS
SKIN	
Cholecalciferol	Plays a role in the synthesis of calcitriol, the active form of vitamin D.
GASTROINTESTINAL TRACT	
Gastrin	Promotes secretion of gastric juice; increases movements of the stomach.
Glucose-dependent insulinotropic peptide (GIP)	Stimulates release of insulin by pancreatic beta cells.
Secretin	Stimulates secretion of pancreatic juice and bile.
Cholecystokinin (CCK)	Stimulates secretion of pancreatic juice; regulates release of bile from gallbladder; causes feeling of fullness after eating.
PLACENTA	
Human chorionic gonadotropin (hCG)	Stimulates corpus luteum in ovary to continue production of estrogens and progesterone to maintain pregnancy.
Estrogens and progesterone	Maintain pregnancy; help prepare mammary glands to secrete milk.
Human chorionic somatomammotropin (hCS)	Stimulates development of mammary glands for lactation.
KIDNEYS	
Renin	Part of reaction sequence that raises blood pressure by bringing about vasoconstriction and secretion of aldosterone.
Erythropoietin (EPO)	Increases rate of red blood cell formation.
Calcitriol* (active form of vitamin D)	Aids in absorption of dietary calcium and phosphorus
HEART	
Atrial natriuretic peptide (ANP)	Decreases blood pressure.
ADIPOSE TISSUE	
Leptin	Suppresses appetite; may increase FSH and LH activity.

*Synthesis begins in the skin, continues in the liver, and ends in the kidneys.

Eicosanoids

Two families of eicosanoid molecules—the **prostaglandins (PGs)** (pros′-ta-GLAN-dins) and the **leukotrienes (LTs)** (loo-kō-TRĪ-ēns)—are found in virtually all body cells except red blood cells, where they act as local hormones (paracrines or autocrines) in response to chemical or mechanical stimuli. They are synthesized by clipping a 20-carbon fatty acid called **arachidonic acid** (a-rak-i-DON-ik) from membrane phospholipid molecules. From arachidonic acid, different enzymatic reactions produce PGs or LTs. **Thromboxane (TX)** (throm-BOK-sān) is a modified PG that constricts blood vessels and promotes platelet activation. Appearing in the blood in minute quantities, eicosanoids are present only briefly due to rapid inactivation.

To exert their effects, eicosanoids bind to receptors on target-cell plasma membranes and stimulate or inhibit the synthesis of second messengers such as cyclic AMP. Leukotrienes stimulate chemotaxis (attraction to a chemical stimulus) of white blood cells and mediate inflammation. The prostaglandins alter smooth muscle contraction, glandular secretions, blood flow, reproductive processes, platelet function, respiration, nerve impulse transmission, lipid metabolism, and immune responses. They also have roles in promoting inflammation and fever, and in intensifying pain.

See Clinical Connection: Nonsteroidal Anti-inflammatory Drugs

Growth Factors

Several of the hormones we have described—insulinlike growth factor, thymosin, insulin, thyroid hormones, growth hormone, and prolactin—stimulate cell growth and division. In addition, several more recently discovered hormones called **growth factors** play important roles in tissue development, growth, and repair. Growth factors are *mitogenic* substances—they cause growth by stimulating cell division. Many growth factors act locally, as autocrines or paracrines.

A summary of sources and actions of six important growth factors is presented in Table 18.12.

18.14 The Stress Response

OBJECTIVE

- **Describe** how the body responds to stress.

It is impossible to remove all of the stress from our everyday lives. Some stress, called **eustress**, prepares us to meet certain challenges and thus is helpful. Other stress, called **distress**, is harmful. Any stimulus that produces a stress response is called a **stressor**. A stressor may be almost any disturbance of the human body—heat or cold, environmental poisons, toxins given off by bacteria, heavy bleeding from a wound or surgery, or a strong emotional reaction. The responses to stressors may be pleasant or unpleasant, and they vary among people and even within the same person at different times.

Your body's homeostatic mechanisms attempt to counteract stress. When they are successful, the internal environment remains within normal physiological limits. If stress is extreme, unusual, or long lasting, the normal mechanisms may not be enough. In 1936, Hans Selye, a pioneer in stress research, showed that a variety of stressful conditions or noxious agents elicit a similar sequence of

TABLE 18.12 Summary of Selected Growth Factors

GROWTH FACTOR	COMMENT
Epidermal growth factor (EGF)	Produced in submaxillary (salivary) glands; stimulates proliferation of epithelial cells, fibroblasts, neurons, and astrocytes; suppresses some cancer cells and secretion of gastric juice by stomach.
Platelet-derived growth factor (PDGF)	Produced in blood platelets; stimulates proliferation of neuroglia, smooth muscle fibers, and fibroblasts; appears to have role in wound healing; may contribute to atherosclerosis development.
Fibroblast growth factor (FGF)	Found in pituitary gland and brain; stimulates proliferation of many cells derived from embryonic mesoderm (fibroblasts, adrenocortical cells, smooth muscle fibers, chondrocytes, and endothelial cells); stimulates formation of new blood vessels (angiogenesis).
Nerve growth factor (NGF)	Produced in submandibular (salivary) glands and hippocampus of brain; stimulates growth of ganglia in embryo; maintains sympathetic nervous system; stimulates hypertrophy and differentiation of neurons.
Tumor angiogenesis factors (TAFs)	Produced by normal and tumor cells; stimulate growth of new capillaries, organ regeneration, and wound healing.
Transforming growth factors (TGFs)	Produced by various cells as separate molecules: TGF-alpha has activities similar to epidermal growth factor; TGF-beta inhibits proliferation of many cell types.

bodily changes. These changes, called the **stress response** or *general adaptation syndrome* (GAS), are controlled mainly by the hypothalamus. The stress response occurs in three stages: (1) an initial fight-or-flight response, (2) a slower resistance reaction, and eventually (3) exhaustion.

The Fight-or-Flight Response

The **fight-or-flight response**, initiated by nerve impulses from the hypothalamus to the sympathetic division of the autonomic nervous system (ANS), including the adrenal medulla, quickly mobilizes the body's resources for immediate physical activity (**Figure 18.19a**). It brings huge amounts of glucose and oxygen to the organs that are most active in warding off danger: the brain, which must become highly alert; the skeletal muscles, which may have to fight off an attacker or flee; and the heart, which must work vigorously to pump enough blood to the brain and muscles. During the fight-or-flight response, nonessential body functions such as digestive, urinary, and reproductive activities are inhibited. Reduction of blood flow to the kidneys promotes release of renin, which sets into motion the renin–angiotensin–aldosterone pathway (see **Figure 18.15**). Aldosterone causes the kidneys to retain Na^+, which leads to water retention and elevated blood pressure. Water retention also helps preserve body fluid volume in the case of severe bleeding.

The Resistance Reaction

The second stage in the stress response is the **resistance reaction** (**Figure 18.19b**). Unlike the short-lived fight-or-flight response, which is initiated by nerve impulses from the hypothalamus, the resistance reaction is initiated in large part by hypothalamic releasing hormones and is a longer-lasting response. The hormones involved are corticotropin-releasing hormone (CRH), growth hormone–releasing hormone (GHRH), and thyrotropin-releasing hormone (TRH).

CRH stimulates the anterior pituitary to secrete ACTH, which in turn stimulates the adrenal cortex to increase release of cortisol. Cortisol then stimulates gluconeogenesis by liver cells, breakdown of triglycerides into fatty acids (lipolysis), and catabolism of proteins into amino acids. Tissues throughout the body can use the resulting glucose, fatty acids, and amino acids to produce ATP or to repair damaged cells. Cortisol also reduces inflammation.

A second hypothalamic releasing hormone, GHRH, causes the anterior pituitary to secrete growth hormone (GH). Acting via insulin-like growth factors, GH stimulates lipolysis and glycogenolysis, the breakdown of glycogen to glucose, in the liver. A third hypothalamic releasing hormone, TRH, stimulates the anterior pituitary to secrete thyroid-stimulating hormone (TSH). TSH promotes secretion of thyroid hormones, which stimulate the increased use of glucose for ATP production. The combined actions of GH and TSH supply additional ATP for metabolically active cells throughout the body.

The resistance stage helps the body continue fighting a stressor long after the fight-or-flight response dissipates. This is why your heart continues to pound for several minutes even after the stressor is removed. Generally, it is successful in seeing us through a stressful episode, and our bodies then return to normal. Occasionally, however, the resistance stage fails to combat the stressor, and the body moves into the state of exhaustion.

Exhaustion

The resources of the body may eventually become so depleted that they cannot sustain the resistance stage, and **exhaustion** ensues. Prolonged exposure to high levels of cortisol and other hormones involved in the resistance reaction causes wasting of muscle, suppression of the immune system, ulceration of the gastrointestinal tract, and failure of pancreatic beta cells. In addition, pathological changes may occur because resistance reactions persist after the stressor has been removed.

Stress and Disease

Although the exact role of stress in human diseases is not known, it is clear that stress can lead to particular diseases by temporarily inhibiting certain components of the immune system. Stress-related disorders include gastritis, ulcerative colitis, irritable bowel syndrome, hypertension, asthma, rheumatoid arthritis (RA), migraine

FIGURE 18.19 Responses to stressors during the stress response. Red arrows (hormonal responses) and green arrows (neural responses) in (a) indicate immediate fight-or-flight reactions; black arrows in (b) indicate long-term resistance reactions.

Stressors stimulate the hypothalamus to initiate the stress response through the fight-or-flight response and the resistance reaction.

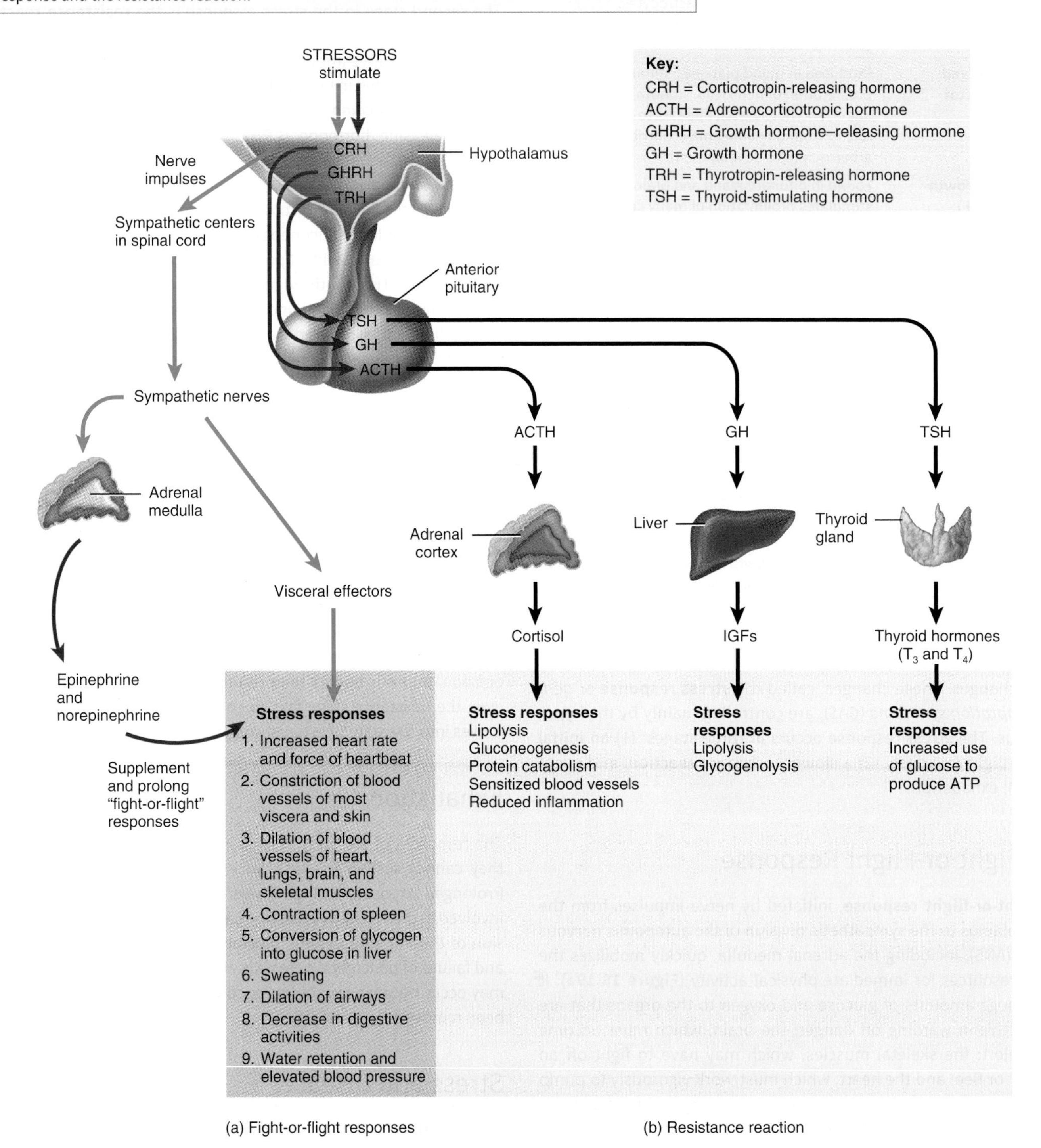

Q What is the basic difference between the stress response and homeostasis?

headaches, anxiety, and depression. People under stress are at a greater risk of developing chronic disease or dying prematurely.

Interleukin-1, a substance secreted by macrophages of the immune system (see the discussion of ACTH in Section 18.6), is an important link between stress and immunity. One action of interleukin-1 is to stimulate secretion of ACTH, which in turn stimulates the production of cortisol. Not only does cortisol provide resistance to stress and inflammation, but it also suppresses further production of interleukin-1. Thus, the immune system turns on the stress response, and the resulting cortisol then turns off one immune system mediator. This negative feedback system keeps the immune response in check once it has accomplished its goal. Because of this activity, cortisol and other glucocorticoids are used as immunosuppressive drugs for organ transplant recipients.

See Clinical Connection: Post-Traumatic Stress Disorder

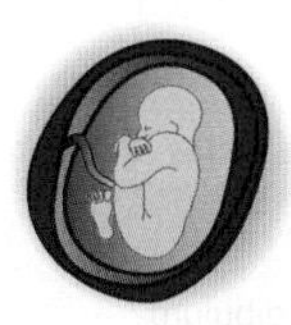

18.15 Development of the Endocrine System

OBJECTIVE

- **Describe** the development of endocrine glands.

The development of the endocrine system is not as localized as the development of other systems because, as you have already learned, endocrine organs are distributed throughout the body.

About 3 weeks after fertilization, the *pituitary gland (hypophysis)* begins to develop from two different regions of the **ectoderm**. The *posterior pituitary (neurohypophysis)* is derived from an outgrowth of ectoderm called the **neurohypophyseal bud** (noo′-rō-hī-pō-FIZ-ē-al), located on the floor of the hypothalamus (**Figure 18.20**). The *infundibulum*, also an outgrowth of the neurohypophyseal bud, connects the posterior pituitary to the hypothalamus. The *anterior pituitary (adenohypophysis)* is derived from an outgrowth of ectoderm from the roof of the mouth called the **hypophyseal pouch** or *Rathke's pouch*. The pouch grows toward the neurohypophyseal bud and eventually loses its connection with the roof of the mouth.

The *thyroid gland* develops during the fourth week as a midventral outgrowth of **endoderm**, called the **thyroid diverticulum** (dī′-ver-TIK-ū-lum), from the floor of the pharynx at the level of the second pair of pharyngeal pouches (**Figure 18.20a**). The outgrowth projects inferiorly and differentiates into the right and left lateral lobes and the isthmus of the gland.

The *parathyroid glands* develop during the fourth week from **endoderm** as outgrowths from the third and fourth **pharyngeal pouches** (fa-RIN-jē-al), which help to form structures of the head and neck.

The adrenal cortex and adrenal medulla develop during the fifth week and have completely different embryological origins. The *adrenal cortex* is derived from the same region of **mesoderm** that produces the gonads. Endocrine tissues that secrete steroid hormones all are derived from mesoderm. The *adrenal medulla* is derived from **ectoderm** from **neural crest** cells that migrate to the superior pole of the kidney. Recall that neural crest cells also give rise to sympathetic ganglia and other structures of the nervous system (see **Figure 14.27b**).

The *pancreas* develops during the fifth through seventh weeks from two outgrowths of **endoderm** from the part of the **foregut** that later becomes the duodenum (see **Figure 29.12c**). The two outgrowths eventually fuse to form the pancreas. The origin of the ovaries and testes is discussed in Section 28.5.

The *pineal gland* arises during the seventh week as an outgrowth between the thalamus and colliculi of the midbrain from **ectoderm** associated with the **diencephalon** (see **Figure 14.28d**).

The *thymus* arises during the fifth week from **endoderm** of the third pharyngeal pouches.

18.16 Aging and the Endocrine System

OBJECTIVE

- **Describe** the effects of aging on the endocrine system.

Although some endocrine glands shrink as we get older, their performance may or may not be compromised. Production of growth hormone by the anterior pituitary decreases, which is one cause of muscle atrophy as aging proceeds. The thyroid gland often decreases its output of thyroid hormones with age, causing a decrease in metabolic rate, an increase in body fat, and hypothyroidism, which is seen more often in older people. Because there is less negative feedback (lower levels of thyroid hormones), the level of thyroid-stimulating hormone increases with age (see **Figure 18.11**).

With aging, the blood level of PTH rises, perhaps due to inadequate dietary intake of calcium. In a study of older women who took 2400 mg/day of supplemental calcium, blood levels of PTH were as low as those of younger women. Both calcitriol and calcitonin levels are lower in older persons. Together, the rise in PTH and the fall in calcitonin level heighten the age-related decrease in bone mass that leads to osteoporosis and increased risk of fractures (see **Figure 18.13**).

The adrenal glands contain increasingly more fibrous tissue and produce less cortisol and aldosterone with advancing age. However, production of epinephrine and norepinephrine remains normal. The pancreas releases insulin more slowly with age, and receptor

FIGURE 18.20 Development of the endocrine system.

Glands of the endocrine system develop from all three primary germ layers: ectoderm, mesoderm, and endoderm.

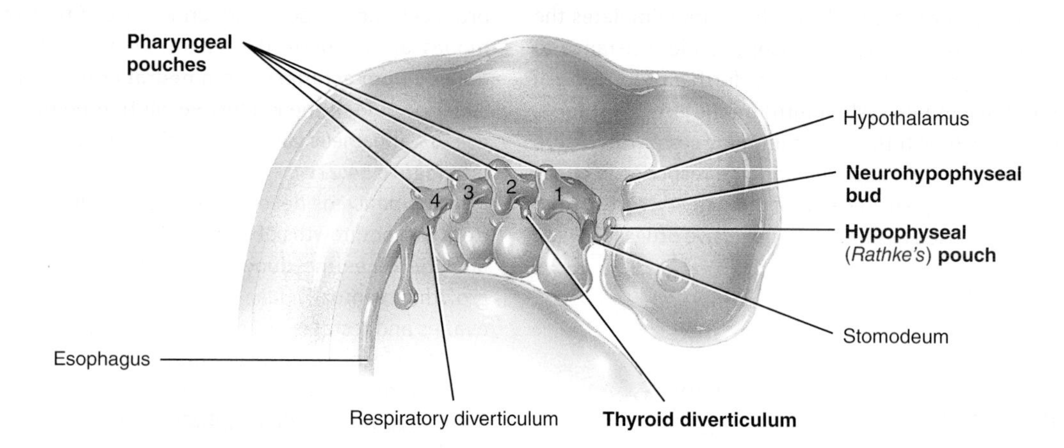

(a) Location of the neurohypophyseal bud, hypophyseal (Rathke's) pouch, thyroid diverticulum, and pharyngeal pouches in a 28-day embryo

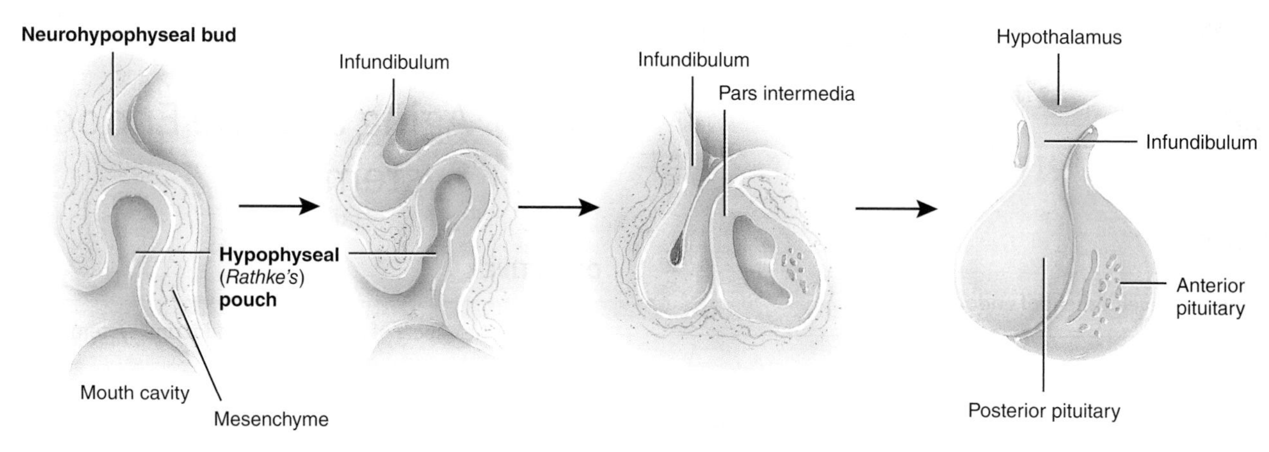

(b) Development of the pituitary gland between five and sixteen weeks

Q Which endocrine gland develops from tissues with two different embryological origins?

sensitivity to glucose declines. As a result, blood glucose levels in older people increase faster and return to normal more slowly than in younger individuals.

The thymus is largest in infancy. After puberty, its size begins to decrease, and thymic tissue is replaced by adipose and areolar connective tissue. In older adults, the thymus has atrophied significantly. However, it still produces new T cells for immune responses.

The ovaries decrease in size with age, and they no longer respond to gonadotropins. The resultant decreased output of estrogens leads to conditions such as osteoporosis, high blood cholesterol, and atherosclerosis. FSH and LH levels are high due to less negative feedback inhibition of estrogens. Although testosterone production by the testes decreases with age, the effects are not usually apparent until very old age; and many elderly males can still produce active sperm in normal numbers, even though there are higher numbers of morphologically abnormal sperm and decreased sperm motility.

• • •

To appreciate the many ways the endocrine system contributes to homeostasis of other body systems, examine *Focus on Homeostasis: Contributions of the Endocrine System*. Next, in Chapter 19, we will begin to explore the cardiovascular system, starting with a description of the composition and functions of blood.

FOCUS on HOMEOSTASIS

INTEGUMENTARY SYSTEM

- Androgens stimulate growth of axillary and pubic hair and activation of sebaceous glands
- Excess melanocyte-stimulating hormone (MSH) causes darkening of skin

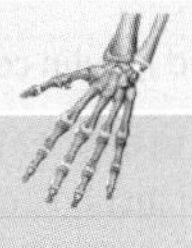

SKELETAL SYSTEM

- Growth hormone (GH) and insulin-like growth factors (IGFs) stimulate bone growth
- Estrogens cause closure of the epiphyseal plates at the end of puberty and help maintain bone mass in adults
- Parathyroid hormone (PTH) and calcitonin regulate levels of calcium and other minerals in bone matrix and blood
- Thyroid hormones are needed for normal development and growth of the skeleton

MUSCULAR SYSTEM

- Epinephrine and norepinephrine help increase blood flow to exercising muscle
- PTH maintains proper level of Ca^{2+}, needed for muscle contraction
- Glucagon, insulin, and other hormones regulate metabolism in muscle fibers
- GH, IGFs, and thyroid hormones help maintain muscle mass

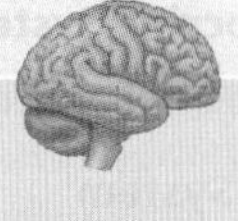

NERVOUS SYSTEM

- Several hormones, especially thyroid hormones, insulin, and growth hormone, influence growth and development of the nervous system
- PTH maintains proper level of Ca^{2+}, needed for generation and conduction of nerve impulses

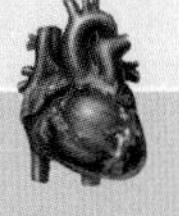

CARDIOVASCULAR SYSTEM

- Erythropoietin (EPO) promotes formation of red blood cells
- Aldosterone and antidiuretic hormone (ADH) increase blood volume
- Epinephrine and norepinephrine increase heart rate and force of contraction
- Several hormones elevate blood pressure during exercise and other stresses

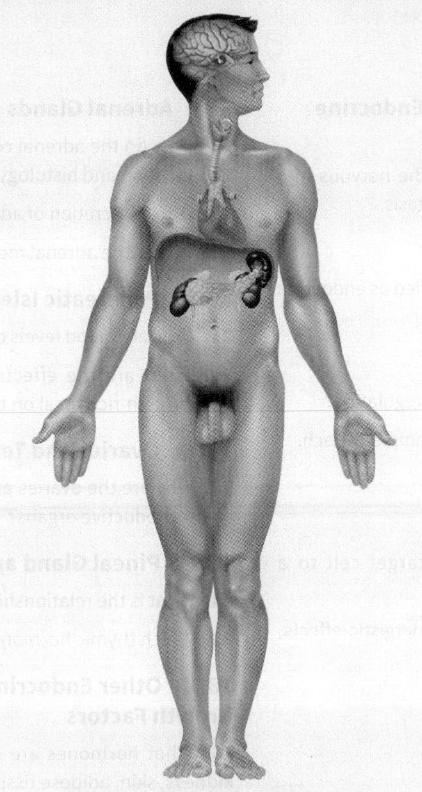

CONTRIBUTIONS OF THE ENDOCRINE SYSTEM FOR ALL BODY SYSTEMS

- Together with the nervous system, circulating and local hormones of the endocrine system regulate activity and growth of target cells throughout the body
- Several hormones regulate metabolism, uptake of glucose, and molecules used for ATP production by body cells

LYMPHATIC SYSTEM and IMMUNITY

- Glucocorticoids such as cortisol depress inflammation and immune responses
- Thymic hormones promote maturation of T cells (a type of white blood cell)

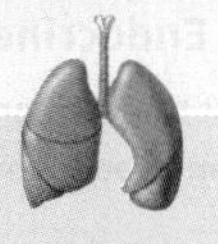

RESPIRATORY SYSTEM

- Epinephrine and norepinephrine dilate (widen) airways during exercise and other stresses
- Erythropoietin regulates amount of oxygen carried in blood by adjusting number of red blood cells

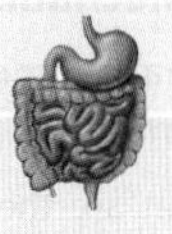

DIGESTIVE SYSTEM

- Epinephrine and norepinephrine depress activity of the digestive system
- Gastrin, cholecystokinin, secretin, and glucose-dependent insulinotropic peptide (GIP) help regulate digestion
- Calcitriol promotes absorption of dietary calcium
- Leptin suppresses appetite

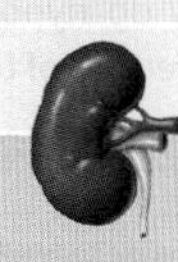

URINARY SYSTEM

- ADH, aldosterone, and atrial natriuretic peptide (ANP) adjust the rate of loss of water and ions in the urine, thereby regulating blood volume and ion content of the blood

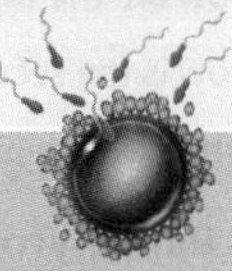

REPRODUCTIVE SYSTEMS

- Hypothalamic releasing and inhibiting hormones, follicle-stimulating hormone (FSH), and luteinizing hormone (LH) regulate development, growth, and secretions of the gonads (ovaries and testes)
- Estrogens and testosterone contribute to development of oocytes and sperm and stimulate development of secondary sex characteristics
- Prolactin promotes milk secretion in mammary glands
- Oxytocin causes contraction of the uterus and ejection of milk from the mammary glands

Review Questions

18.1 Comparison of Control by the Nervous and Endocrine Systems

1. List the similarities among and differences between the nervous and endocrine systems with regard to the control of homeostasis.

18.2 Endocrine Glands

2. List three organs or tissues that are not exclusively classified as endocrine glands but contain cells that secrete hormones.

18.3 Hormone Activity

3. What is the difference between down-regulation and up-regulation?
4. Identify the chemical classes of hormones, and give an example of each.
5. How are hormones transported in the blood?

18.4 Mechanisms of Hormone Action

6. What factors determine the responsiveness of a target cell to a hormone?
7. What are the differences among permissive effects, synergistic effects, and antagonistic effects of hormones?

18.5 Control of Hormone Secretion

8. What three types of signals control hormone secretion?

18.6 Hypothalamus and Pituitary Gland

9. In what respect is the pituitary gland actually two glands?
10. How do hypothalamic releasing and inhibiting hormones influence secretions of the anterior pituitary?
11. Describe the structure and importance of the hypothalamic–hypophyseal tract.

18.7 Thyroid Gland

12. Explain how blood levels of T_3/T_4, TSH, and TRH would change in a laboratory animal that has undergone a thyroidectomy (complete removal of its thyroid gland).
13. How are the thyroid hormones synthesized, stored, and secreted?
14. How is the secretion of T_3 and T_4 regulated?
15. What are the physiological effects of the thyroid hormones?

18.8 Parathyroid Glands

16. How is secretion of parathyroid hormone regulated?
17. In what ways are the actions of PTH and calcitriol similar? How are they different?

18.9 Adrenal Glands

18. How do the adrenal cortex and adrenal medulla compare with regard to location and histology?
19. How is secretion of adrenal cortex hormones regulated?
20. How is the adrenal medulla related to the autonomic nervous system?

18.10 Pancreatic Islets

21. How are blood levels of glucagon and insulin controlled?
22. What are the effects of exercise versus eating a carbohydrate-and protein-rich meal on the secretion of insulin and glucagon?

18.11 Ovaries and Testes

23. Why are the ovaries and testes classified as endocrine glands as well as reproductive organs?

18.12 Pineal Gland and Thymus

24. What is the relationship between melatonin and sleep?
25. Which thymic hormones play a role in immunity?

18.13 Other Endocrine Tissues and Organs, Eicosanoids, and Growth Factors

26. What hormones are secreted by the gastrointestinal tract, placenta, kidneys, skin, adipose tissue, and heart?
27. What are some functions of prostaglandins, leukotrienes, and growth factors?

18.14 The Stress Response

28. What is the central role of the hypothalamus during stress?
29. What body reactions occur during the fight-or-flight response, the resistance reaction, and exhaustion?
30. What is the relationship between stress and immunity?

18.15 Development of the Endocrine System

31. Compare the origins of the adrenal cortex and adrenal medulla.

18.16 Aging and the Endocrine System

32. Which hormone is related to the muscle atrophy that occurs with aging?

Critical Thinking Questions

1. Amanda hates her new student ID photo. Her hair looks dry, the extra weight that she's gained shows, and her neck looks fat. In fact, there's an odd butterfly-shaped swelling across the front of her neck, under her chin. Amanda's also been feeling very tired and mentally "dull" lately, but she figures all new A&P students feel that way. Should she visit the clinic or just wear turtlenecks?
2. Amanda (from question 1 above) goes to the clinic and blood is drawn. The results show that her T_4 levels are low and her TSH levels are low. Later she is given a TSH stimulation test in which TSH is injected and the T_4 levels are monitored. After TSH injection, her T_4 levels rise. Does Amanda have problems with her pituitary or with her thyroid gland? How did you come to your conclusion?
3. Mr. Hernandez visited his doctor complaining that he is constantly thirsty and is "in the bathroom day and night" relieving his bladder. The doctor ordered blood and urine tests to check for glucose and ketones, which were all negative. What is the doctor's diagnosis of Mr. Hernandez, and what gland(s) or organ(s) is(are) involved?

The Cardiovascular System: The Blood

CHAPTER **19**

Blood and Homeostasis

Blood contributes to homeostasis by transporting oxygen, carbon dioxide, nutrients, and hormones to and from your body's cells. It also helps regulate body pH and temperature, and provides protection against disease through phagocytosis and the production of antibodies.

The focus of this chapter is blood; the next two chapters will examine the heart and blood vessels, respectively. Blood transports various substances, helps regulate several life processes, and affords protection against disease. For all of its similarities in origin, composition, and functions, blood is as unique from one person to another as are skin, bone, and hair. Health-care professionals routinely examine and analyze its differences through various blood tests when trying to determine the cause of different diseases.

Q Did you ever wonder how analyzing blood can determine if we are healthy, detect a multitude of infections, and detect or rule out various diseases and injuries?

19.1 Functions and Properties of Blood

OBJECTIVES

- **Explain** the functions of blood.
- **Describe** the physical characteristics and principal components of blood.

The **cardiovascular system** (*cardio-* = heart; *vascular* = blood or blood vessels) consists of three interrelated components: blood, the heart, and blood vessels. The branch of science concerned with the study of blood, blood-forming tissues, and the disorders associated with them is **hematology** (hēm-a-TOL-ō-jē; *hema-* or *hemato-* = blood; *-logy* = study of).

Most cells of a multicellular organism cannot move around to obtain oxygen and nutrients or eliminate carbon dioxide and other wastes. Instead, these needs are met by two fluids: blood and interstitial fluid. **Blood** is a liquid connective tissue that consists of cells surrounded by a liquid extracellular matrix. The extracellular matrix is called blood plasma, and it suspends various cells and cell fragments. **Interstitial fluid** is the fluid that bathes body cells (see **Figure 27.1**) and is constantly renewed by the blood. Blood transports oxygen from the lungs and nutrients from the gastrointestinal tract, which diffuse from the blood into the interstitial fluid and then into body cells. Carbon dioxide and other wastes move in the reverse direction, from body cells to interstitial fluid to blood. Blood then transports the wastes to various organs—the lungs, kidneys, and skin—for elimination from the body.

Functions of Blood

Blood has three general functions:

1. ***Transportation.*** As you just learned, blood transports oxygen from the lungs to the cells of the body and carbon dioxide from the body cells to the lungs for exhalation. It carries nutrients from the gastrointestinal tract to body cells and hormones from endocrine glands to other body cells. Blood also transports heat and waste products to various organs for elimination from the body.
2. ***Regulation.*** Circulating blood helps maintain homeostasis of all body fluids. Blood helps regulate pH through the use of buffers (chemicals that convert strong acids or bases into weak ones). It also helps adjust body temperature through the heat-absorbing and coolant properties of the water (see Section 2.4) in blood plasma and its variable rate of flow through the skin, where excess heat can be lost from the blood to the environment. In addition, blood osmotic pressure influences the water content of cells, mainly through interactions of dissolved ions and proteins.
3. ***Protection.*** Blood can clot (become gel-like), which protects against its excessive loss from the cardiovascular system after an injury. In addition, its white blood cells protect against disease by carrying on phagocytosis. Several types of blood proteins, including antibodies, interferons, and complement, help protect against disease in a variety of ways.

Physical Characteristics of Blood

Blood is denser and more viscous (thicker) than water and feels slightly sticky. The temperature of blood is 38°C (100.4°F), about 1°C higher than oral or rectal body temperature, and it has a slightly alkaline pH ranging from 7.35 to 7.45 (average = 7.4). The color of blood varies with its oxygen content. When saturated with oxygen, it is bright red. When unsaturated with oxygen, it is dark red. Blood constitutes about 20% of extracellular fluid, amounting to 8% of the total body mass. The blood volume is 5 to 6 liters (1.5 gal) in an average-sized adult male and 4 to 5 liters (1.2 gal) in an average-sized adult female. The gender difference in volume is due to differences in body size. Several hormones, regulated by negative feedback, ensure that blood volume and osmotic pressure remain relatively constant. Especially important are the hormones aldosterone, antidiuretic hormone, and atrial natriuretic peptide, which regulate how much water is excreted in the urine (see Section 27.1).

See Clinical Connection: Withdrawing Blood

Components of Blood

Whole blood has two components: (1) blood plasma, a watery liquid extracellular matrix that contains dissolved substances, and (2) formed elements, which are cells and cell fragments. If a sample of blood is centrifuged (spun) in a small glass tube, the cells (which are more dense) sink to the bottom of the tube while the plasma (which is less dense) forms a layer on top (**Figure 19.1a**). Blood is about 45% formed

FIGURE 19.1 Components of blood in a normal adult.

Blood is a connective tissue that consists of blood plasma (liquid) plus formed elements (red blood cells, white blood cells, and platelets).

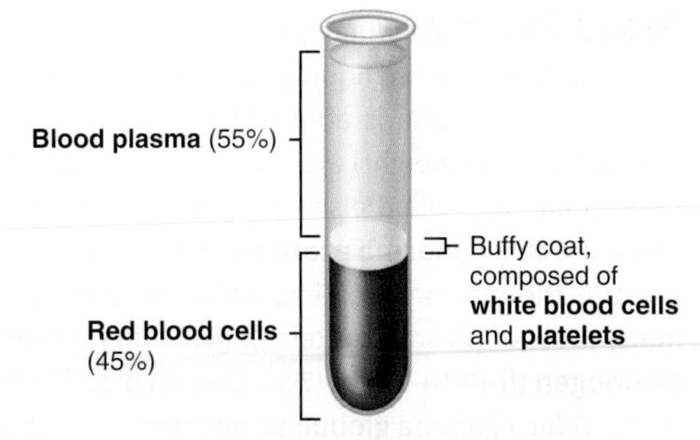

(a) Appearance of centrifuged blood

Functions of Blood

1. Transports oxygen, carbon dioxide, nutrients, hormones, heat, and wastes.
2. Regulates pH, body temperature, and water content of cells.
3. Protects against blood loss through clotting, and against disease through phagocytic white blood cells and proteins such as antibodies, interferons, and complement.

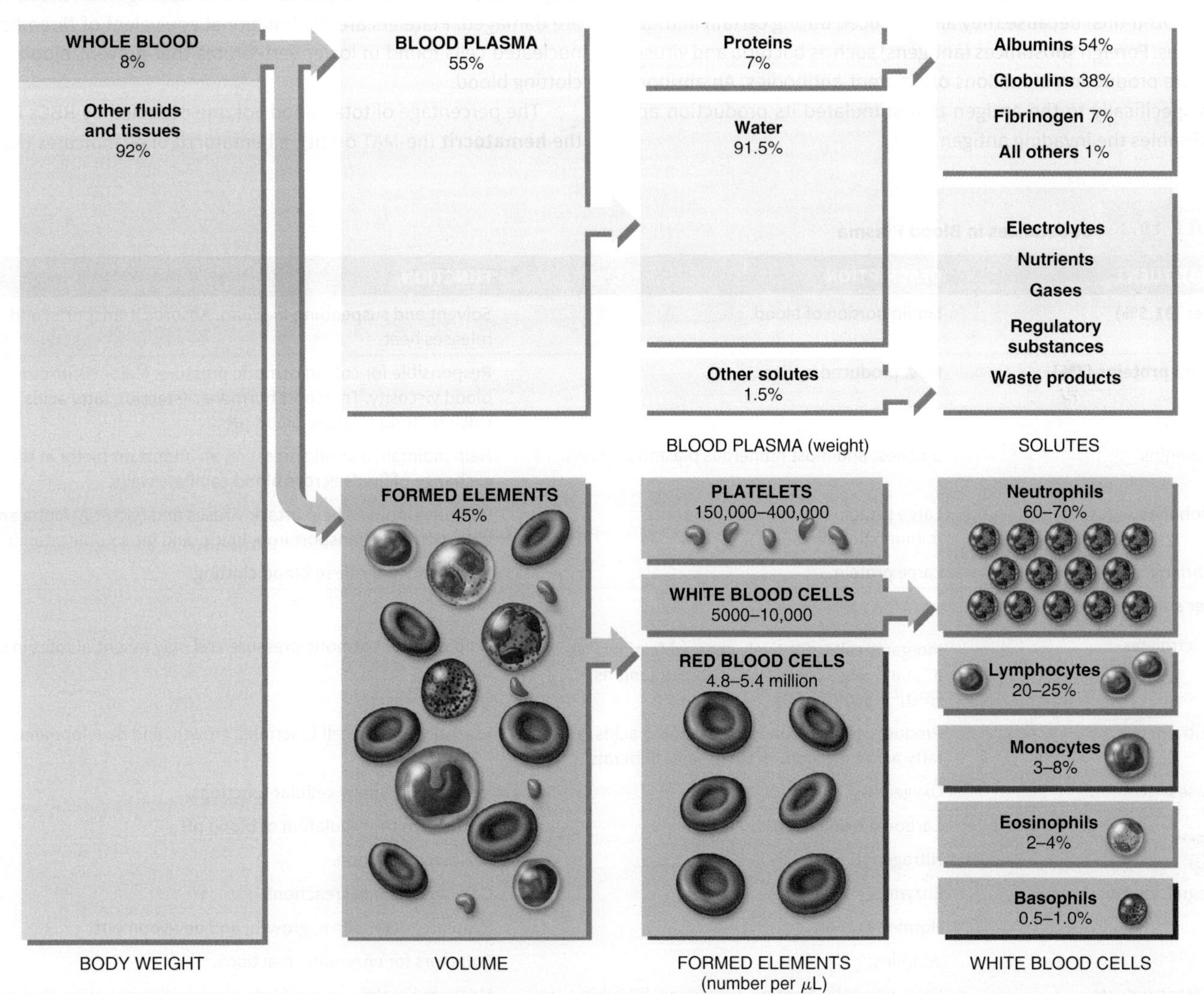

(b) Components of blood

Q What is the approximate volume of blood in your body?

elements and 55% blood plasma. Normally, more than 99% of the formed elements are cells named for their red color—red blood cells (RBCs). Pale, colorless white blood cells (WBCs) and platelets occupy less than 1% of the formed elements. Because they are less dense than red blood cells but more dense than blood plasma, they form a very thin **buffy coat** layer between the packed RBCs and plasma in centrifuged blood. Figure 19.1b shows the composition of blood plasma and the numbers of the various types of formed elements in blood.

Blood Plasma When the formed elements are removed from blood, a straw-colored liquid called **blood plasma** (or simply *plasma*) is left. Blood plasma is about 91.5% water and 8.5% solutes, most of which (7% by weight) are proteins. Some of the proteins in blood plasma are also found elsewhere in the body, but those confined to blood are called **plasma proteins**. Hepatocytes (liver cells) synthesize most of the plasma proteins, which include the **albumins** (al′-BŪ-mins) (54% of plasma proteins), **globulins** (GLOB-ū-lins) (38%), and **fibrinogen** (fī-BRIN-ō-jen) (7%). Certain blood cells develop into cells that produce gamma globulins, an important type of globulin. These plasma proteins are also called **antibodies** or *immunoglobulins* (im′-ū-nō-GLOB-ū-lins) because they are produced during certain immune responses. Foreign substances (antigens) such as bacteria and viruses stimulate production of millions of different antibodies. An antibody binds specifically to the antigen that stimulated its production and thus disables the invading antigen.

Besides proteins, other solutes in plasma include electrolytes, nutrients, regulatory substances such as enzymes and hormones, gases, and waste products such as urea, uric acid, creatinine, ammonia, and bilirubin.

Table 19.1 describes the chemical composition of blood plasma.

Formed Elements The **formed elements** of the blood include three principal components: red blood cells, white blood cells, and platelets (Figure 19.2). **Red blood cells (RBCs)** or *erythrocytes* transport oxygen from the lungs to body cells and deliver carbon dioxide from body cells to the lungs. **White blood cells (WBCs)** or *leukocytes* protect the body from invading pathogens and other foreign substances. There are several types of WBCs: *neutrophils*, *basophils*, *eosinophils*, *monocytes*, and *lymphocytes*. Lymphocytes are further subdivided into *B lymphocytes (B cells)*, *T lymphocytes (T cells)*, and *natural killer (NK) cells*. Each type of WBC contributes in its own way to the body's defense mechanisms. **Platelets**, the final type of formed element, are fragments of cells that do not have a nucleus. Among other actions, they release chemicals that promote blood clotting when blood vessels are damaged. Platelets are the functional equivalent of *thrombocytes*, nucleated cells found in lower vertebrates that prevent blood loss by clotting blood.

The percentage of total blood volume occupied by RBCs is called the **hematocrit** (hē-MAT-ō-krit); a hematocrit of 40 indicates that 40%

TABLE 19.1 Substances in Blood Plasma

CONSTITUENT	DESCRIPTION	FUNCTION
Water (91.5%)	Liquid portion of blood.	Solvent and suspending medium. Absorbs, transports, and releases heat.
Plasma proteins (7%)	Most produced by liver.	Responsible for colloid osmotic pressure. Major contributors to blood viscosity. Transport hormones (steroid), fatty acids, and calcium. Help regulate blood pH.
Albumins	Smallest and most numerous plasma proteins.	Help maintain osmotic pressure, an important factor in the exchange of fluids across blood capillary walls.
Globulins	Large proteins (plasma cells produce immunoglobulins).	Immunoglobulins help attack viruses and bacteria. Alpha and beta globulins transport iron, lipids, and fat-soluble vitamins.
Fibrinogen	Large protein.	Plays essential role in blood clotting.
Other solutes (1.5%)		
Electrolytes	Inorganic salts; positively charged (cations) Na^+, K^+, Ca^{2+}, Mg^{2+}; negatively charged (anions) Cl^-, HPO_4^{2-}, SO_4^{2-}, HCO_3^-.	Help maintain osmotic pressure and play essential roles in cell functions.
Nutrients	Products of digestion, such as amino acids, glucose, fatty acids, glycerol, vitamins, and minerals.	Essential roles in cell functions, growth, and development.
Gases	Oxygen (O_2).	Important in many cellular functions.
	Carbon dioxide (CO_2).	Involved in the regulation of blood pH.
	Nitrogen (N_2).	No known function.
Regulatory substances	Enzymes.	Catalyze chemical reactions.
	Hormones.	Regulate metabolism, growth, and development.
	Vitamins.	Cofactors for enzymatic reactions.
Waste products	Urea, uric acid, creatine, creatinine, bilirubin, ammonia.	Most are breakdown products of protein metabolism that are carried by the blood to organs of excretion.

FIGURE 19.2 Formed elements of blood.

The formed elements of blood are red blood cells (RBCs), white blood cells (WBCs), and platelets.

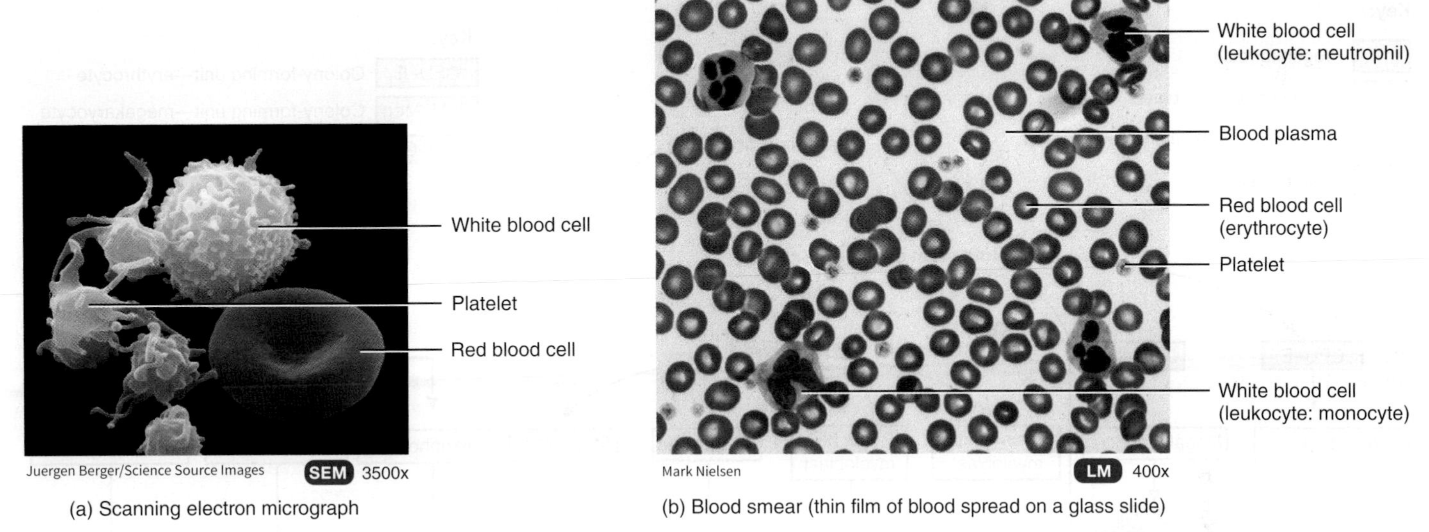

(a) Scanning electron micrograph

(b) Blood smear (thin film of blood spread on a glass slide)

Q Which formed elements of the blood are cell fragments?

of the volume of blood is composed of RBCs. The normal range of hematocrit for adult females is 38–46% (average = 42); for adult males, it is 40–54% (average = 47). The hormone testosterone, present in much higher concentration in males than in females, stimulates synthesis of erythropoietin (EPO), the hormone that in turn stimulates production of RBCs. Thus, testosterone contributes to higher hematocrits in males. Lower values in women during their reproductive years also may be due to excessive loss of blood during menstruation. A significant drop in hematocrit indicates *anemia*, a lower-than-normal number of RBCs. In **polycythemia** (pol′-ē-sī-THĒ-mē-a) the percentage of RBCs is abnormally high, and the hematocrit may be 65% or higher. This raises the viscosity of blood, which increases the resistance to flow and makes the blood more difficult for the heart to pump. Increased viscosity also contributes to high blood pressure and increased risk of stroke. Causes of polycythemia include abnormal increases in RBC production, tissue hypoxia, dehydration, blood doping, or the use of EPO by athletes.

19.2 Formation of Blood Cells

OBJECTIVE

- **Explain** the origin of blood cells.

Although some lymphocytes have a lifetime measured in years, most formed elements of the blood last only hours, days, or weeks, and must be replaced continually. Negative feedback systems regulate the total number of RBCs and platelets in circulation, and their numbers normally remain steady. The abundance of the different types of WBCs, however, varies in response to challenges by invading pathogens and other foreign antigens.

The process by which the formed elements of blood develop is called **hemopoiesis** (hēm-ō-poy-Ē-sis; *-poiesis* = making) or *hematopoiesis.* Before birth, hemopoiesis first occurs in the yolk sac of an embryo and later in the liver, spleen, thymus, and lymph nodes of a fetus. Red bone marrow becomes the primary site of hemopoiesis in the last 3 months before birth, and continues as the source of blood cells after birth and throughout life.

Red bone marrow is a highly vascularized connective tissue located in the microscopic spaces between trabeculae of spongy bone tissue. It is present chiefly in bones of the axial skeleton, pectoral and pelvic girdles, and the proximal epiphyses of the humerus and femur. About 0.05–0.1% of red bone marrow cells are called **pluripotent stem cells** (ploo-RI-pō-tent; *pluri-* = several) or *hemocytoblasts* and are derived from mesenchyme (tissue from which almost all connective tissues develop). These cells have the capacity to develop into many different types of cells (**Figure 19.3**). In newborns, all bone marrow is red and thus active in blood cell production. As an individual ages, the rate of blood cell formation decreases; red bone marrow in the medullary (marrow) cavity of long bones becomes inactive and is replaced by yellow bone marrow, which consists largely of fat cells. Under certain conditions, such as severe bleeding, yellow bone marrow can revert to red bone marrow; this occurs as blood-forming stem cells from red bone marrow move into yellow bone marrow, which is then repopulated by pluripotent stem cells.

Stem cells in red bone marrow reproduce themselves, proliferate, and differentiate into cells that give rise to blood cells,

FIGURE 19.3 Origin, development, and structure of blood cells. A few of the generations of some cell lines have been omitted.

Blood cell production, called hemopoiesis, occurs mainly in red bone marrow after birth.

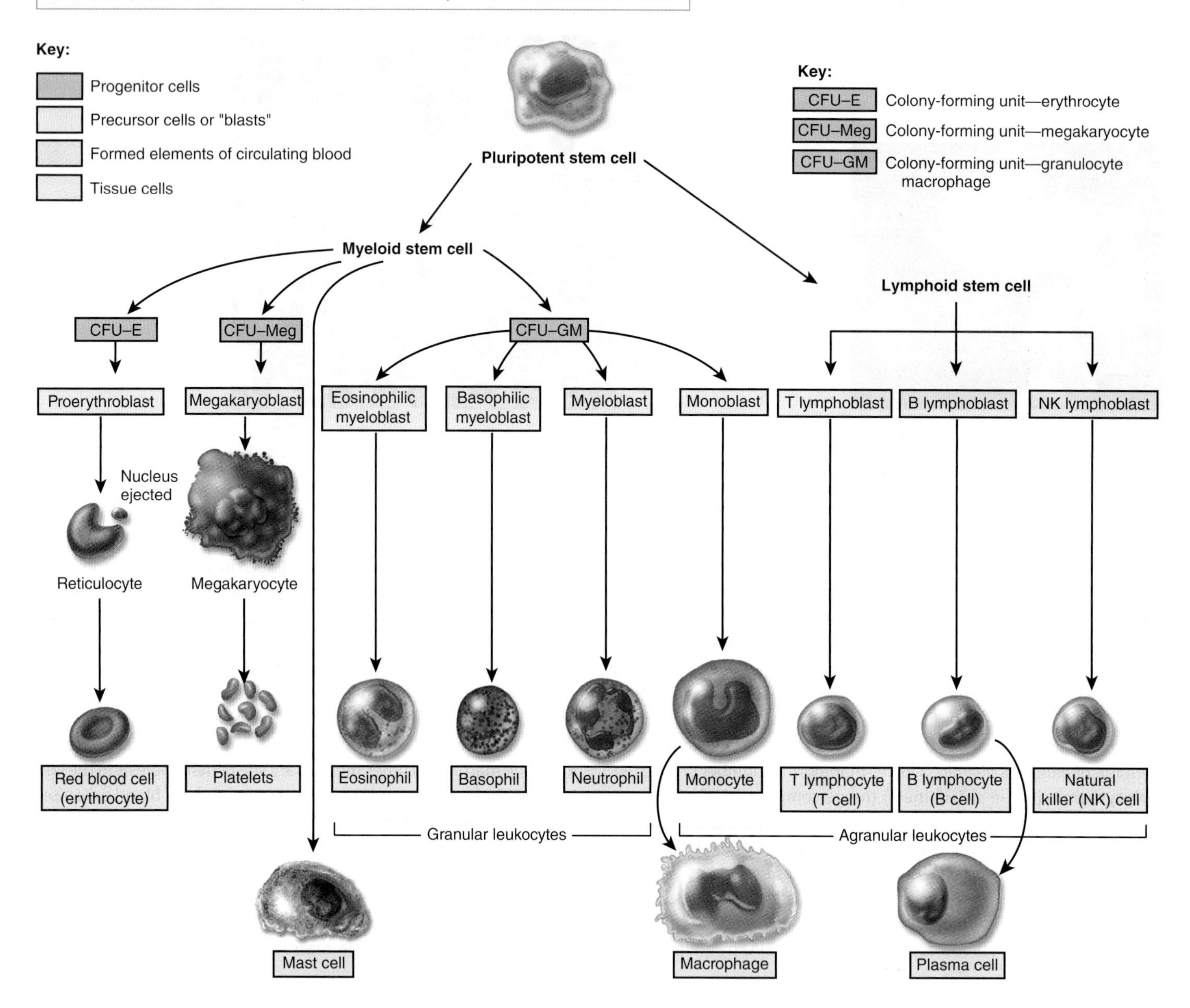

Q From which connective tissue cells do pluripotent stem cells develop?

macrophages, reticular cells, mast cells, and adipocytes. Some stem cells can also form osteoblasts, chondroblasts, and muscle cells, and may be destined for use as a source of bone, cartilage, and muscular tissue for tissue and organ replacement. The reticular cells produce reticular fibers, which form the stroma (framework) that supports red bone marrow cells. Blood from nutrient and metaphyseal arteries (see **Figure 6.4**) enters a bone and passes into the enlarged and leaky capillaries, called *sinuses*, that surround red bone marrow cells and fibers. After blood cells form, they enter the sinuses and other blood vessels and leave the bone through nutrient and periosteal veins (see **Figure 6.4**). With the exception of lymphocytes, formed elements do not divide once they leave red bone marrow.

See Clinical Connection: Bone Marrow Examination

In order to form blood cells, pluripotent stem cells in red bone marrow produce two further types of stem cells, which have the capacity to develop into several types of cells. These stem cells are called **myeloid stem cells** and **lymphoid stem cells**. Myeloid stem cells begin their development in red bone marrow and give rise to red blood cells, platelets, monocytes, neutrophils, eosinophils, basophils, and mast cells. Lymphoid stem cells, which give rise to lymphocytes, begin their development in red bone marrow but complete it in lymphatic tissues Lymphoid stem cells also give rise to natural killer (NK) cells. Although the various stem cells have distinctive cell identity markers in their plasma membranes, they cannot be distinguished histologically and resemble lymphocytes.

During hemopoiesis, some of the myeloid stem cells differentiate into **progenitor cells** (prō-JEN-i-tor). Other myeloid stem cells and the lymphoid stem cells develop directly into precursor cells (described shortly). Progenitor cells are no longer capable of reproducing themselves and are committed to giving rise to more specific elements of blood. Some progenitor cells are known as *colony-forming units (CFUs)*. Following the CFU designation is an abbreviation that indicates the mature elements in blood that they will produce: CFU–E ultimately produces erythrocytes (red blood cells); CFU–Meg produces megakaryocytes, the source of platelets; and CFU–GM ultimately produces granulocytes (specifically, neutrophils) and monocytes (see **Figure 19.3**). Progenitor cells, like stem cells, resemble lymphocytes and cannot be distinguished by their microscopic appearance alone.

In the next generation, the cells are called **precursor cells**, also known as **blasts**. Over several cell divisions they develop into the actual formed elements of blood. For example, monoblasts develop into monocytes, eosinophilic myeloblasts develop into eosinophils, and so on. Precursor cells have recognizable microscopic appearances.

Several hormones called **hemopoietic growth factors** (hē-mō-poy-ET-ik) regulate the differentiation and proliferation of particular progenitor cells. **Erythropoietin (EPO)** (e-rith′-rō-POY-ē-tin) increases the number of red blood cell precursors. EPO is produced primarily by cells in the kidneys that lie between the kidney tubules (peritubular interstitial cells). With renal failure, EPO release slows and RBC production is inadequate. This leads to a decreased hematocrit, which leads to a decreased ability to deliver oxygen to body tissues. **Thrombopoietin (TPO)** (throm′-bō-POY-ē-tin) is a hormone produced by the liver that stimulates the formation of platelets from megakaryocytes. Several different cytokines regulate development of different blood cell types. **Cytokines** (SĪ-tō-kīns) are small glycoproteins that are typically produced by cells such as red bone marrow cells, leukocytes, macrophages, fibroblasts, and endothelial cells. They generally act as local hormones (autocrines or paracrines; see Chapter 18). Cytokines stimulate proliferation of progenitor cells in red bone marrow and regulate the activities of cells involved in nonspecific defenses (such as phagocytes) and immune responses (such as B cells and T cells). Two important families of cytokines that stimulate white blood cell formation are **colony-stimulating factors (CSFs)** and **interleukins** (in′-ter-LOO-kins).

See Clinical Connection: Medical Uses of Hemopoietic Growth Factors

19.3 Red Blood Cells

OBJECTIVE

- **Describe** the structure, functions, life cycle, and production of red blood cells.

Red blood cells (RBCs) or *erythrocytes* (e-RITH-rō-sīts; *erythro-* = red; *-cyte* = cell) contain the oxygen-carrying protein **hemoglobin**, which is a pigment that gives whole blood its red color. A healthy adult male has about 5.4 million red blood cells per microliter (μL) of blood,* and a healthy adult female has about 4.8 million. (One drop of blood is about 50 μL.) To maintain normal numbers of RBCs, new mature cells must enter the circulation at the astonishing rate of at least 2 million per second, a pace that balances the equally high rate of RBC destruction.

RBC Anatomy

RBCs are biconcave discs with a diameter of 7–8 μm (**Figure 19.4a**). (Recall that 1 μm = 1/25,000 of an inch or 1/10,000 of a centimeter or 1/1000 of a millimeter.) Mature red blood cells have a simple structure. Their plasma membrane is both strong and flexible, which allows them to deform without rupturing as they squeeze through narrow blood capillaries. As you will see later, certain glycolipids in the plasma membrane of RBCs are antigens that account for the various blood groups such as the ABO and Rh groups. RBCs lack a nucleus and other organelles and can neither reproduce nor carry on extensive metabolic activities. The cytosol of RBCs contains hemoglobin molecules; these important molecules are synthesized before loss of the nucleus during RBC production and constitute about 33% of the cell's weight.

RBC Physiology

Red blood cells are highly specialized for their oxygen transport function. Because mature RBCs have no nucleus, all of their internal space is available for oxygen transport. Because RBCs lack mitochondria and generate ATP anaerobically (without oxygen), they do not use up any of the oxygen they transport. Even the shape of an RBC facilitates its function. A biconcave disc has a much greater surface area for the diffusion of gas molecules into and out of the RBC than would, say, a sphere or a cube.

Each RBC contains about 280 million hemoglobin molecules. A hemoglobin molecule consists of a protein called **globin**, composed of four polypeptide chains (two alpha and two beta chains); a ringlike nonprotein pigment called a **heme** (**Figure 19.4b**) is bound to each of the four chains. At the center of each heme ring is an iron ion (Fe^{2+})

*1 μL = 1 mm^3 = 10^{-6} liter.

FIGURE 19.4 **The shapes of a red blood cell (RBC) and a hemoglobin molecule.** In (b), note that each of the four polypeptide chains (blue) of a hemoglobin molecule has one heme group (gold), which contains an iron ion (Fe^{2+}), shown in red.

The iron portion of a heme group binds oxygen for transport by hemoglobin.

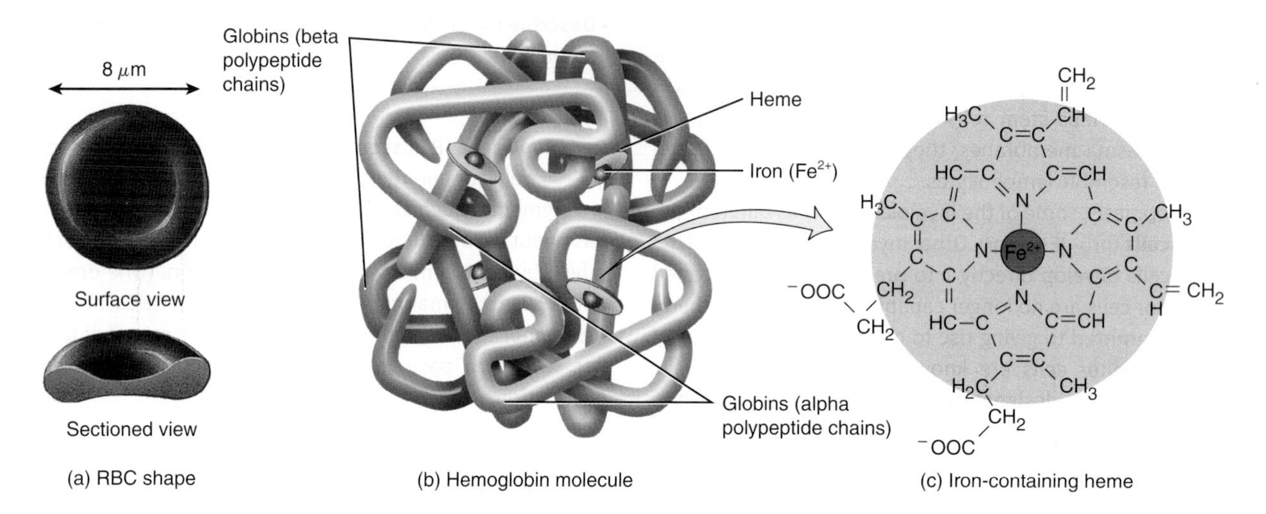

Q How many molecules of O_2 can one hemoglobin molecule transport?

that can combine reversibly with one oxygen molecule (**Figure 19.4c**), allowing each hemoglobin molecule to bind four oxygen molecules. Each oxygen molecule picked up from the lungs is bound to an iron ion. As blood flows through tissue capillaries, the iron–oxygen reaction reverses. Hemoglobin releases oxygen, which diffuses first into the interstitial fluid and then into cells.

Hemoglobin also transports about 23% of the total carbon dioxide, a waste product of metabolism. (The remaining carbon dioxide is dissolved in plasma or carried as bicarbonate ions.) Blood flowing through tissue capillaries picks up carbon dioxide, some of which combines with amino acids in the globin part of hemoglobin. As blood flows through the lungs, the carbon dioxide is released from hemoglobin and then exhaled.

In addition to its key role in transporting oxygen and carbon dioxide, hemoglobin also plays a role in the regulation of blood flow and blood pressure. The gaseous hormone **nitric oxide (NO)**, produced by the endothelial cells that line blood vessels, binds to hemoglobin. Under some circumstances, hemoglobin releases NO. The released NO causes *vasodilation*, an increase in blood vessel diameter that occurs when the smooth muscle in the vessel wall relaxes. Vasodilation improves blood flow and enhances oxygen delivery to cells near the site of NO release.

Red blood cells also contain the enzyme carbonic anhydrase (CA), which catalyzes the conversion of carbon dioxide and water to carbonic acid, which in turn dissociates into H^+ and HCO_3^-. The entire reaction is reversible and is summarized as follows:

$$\underset{\text{Carbon dioxide}}{CO_2} + \underset{\text{Water}}{H_2O} \overset{CA}{\rightleftharpoons} \underset{\text{Carbonic acid}}{H_2CO_3} \rightleftharpoons \underset{\text{Hydrogen ion}}{H^+} + \underset{\text{Bicarbonate ion}}{HCO_3^-}$$

This reaction is significant for two reasons: (1) It allows about 70% of CO_2 to be transported in blood plasma from tissue cells to the lungs in the form of HCO_3^- (see Chapter 23). (2) It also serves as an important buffer in extracellular fluid (see Chapter 27).

RBC Life Cycle

Red blood cells live only about 120 days because of the wear and tear their plasma membranes undergo as they squeeze through blood capillaries. Without a nucleus and other organelles, RBCs cannot synthesize new components to replace damaged ones. The plasma membrane becomes more fragile with age, and the cells are more likely to burst, especially as they squeeze through narrow channels in the spleen. Ruptured red blood cells are removed from circulation and destroyed by fixed phagocytic macrophages in the spleen and liver, and the breakdown products are recycled and used in numerous metabolic processes, including the formation of new red blood cells. The recycling occurs as follows (**Figure 19.5**):

1. Macrophages in the spleen, liver, or red bone marrow phagocytize ruptured and worn-out red blood cells.
2. The globin and heme portions of hemoglobin are split apart.
3. Globin is broken down into amino acids, which can be reused to synthesize other proteins.
4. Iron is removed from the heme portion in the form of Fe^{3+}, which associates with the plasma protein **transferrin** (trans-FER-in; *trans-* = across; *-ferr-* = iron), a transporter for Fe^{3+} in the bloodstream.

FIGURE 19.5 Formation and destruction of red blood cells, and the recycling of hemoglobin components. RBCs circulate for about 120 days after leaving red bone marrow before they are phagocytized by macrophages.

The rate of RBC formation by red bone marrow equals the rate of RBC destruction by macrophages.

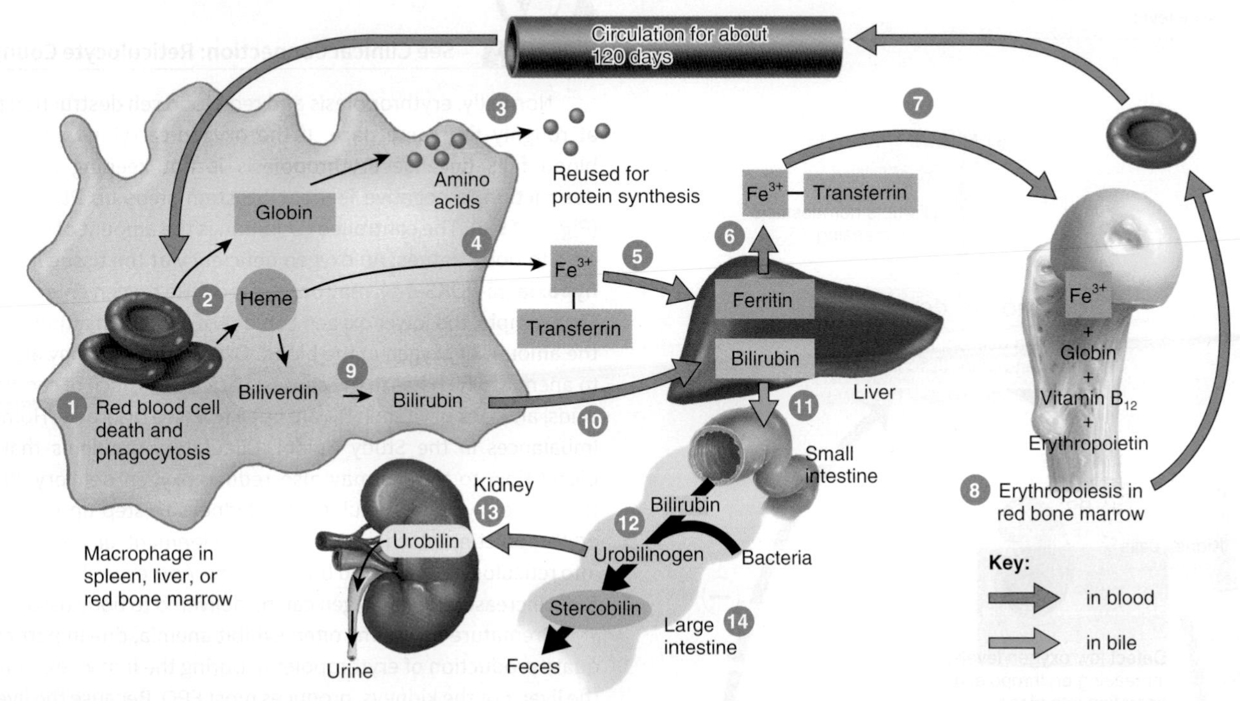

Q What is the function of transferrin?

5. In muscle fibers, liver cells, and macrophages of the spleen and liver, Fe^{3+} detaches from transferrin and attaches to an iron-storage protein called **ferritin** (FER-i-tin).
6. On release from a storage site or absorption from the gastrointestinal tract, Fe^{3+} reattaches to transferrin.
7. The Fe^{3+}–transferrin complex is then carried to red bone marrow, where RBC precursor cells take it up through receptor-mediated endocytosis (see **Figure 3.12**) for use in hemoglobin synthesis. Iron is needed for the heme portion of the hemoglobin molecule, and amino acids are needed for the globin portion. Vitamin B_{12} is also needed for the synthesis of hemoglobin.
8. Erythropoiesis in red bone marrow results in the production of red blood cells, which enter the circulation.
9. When iron is removed from heme, the non-iron portion of heme is converted to **biliverdin** (bil-ē-VER-din), a green pigment, and then into **bilirubin** (bil-ē-ROO-bin), a yellow-orange pigment.
10. Bilirubin enters the blood and is transported to the liver.
11. Within the liver, bilirubin is released by liver cells into bile, which passes into the small intestine and then into the large intestine.
12. In the large intestine, bacteria convert bilirubin into **urobilinogen** (ūr-ō-bī-LIN-ō-jen).
13. Some urobilinogen is absorbed back into the blood, converted to a yellow pigment called **urobilin** (ūr-ō-BĪ-lin), and excreted in urine.
14. Most urobilinogen is eliminated in feces in the form of a brown pigment called **stercobilin** (ster′-kō-BĪ-lin), which gives feces its characteristic color.

See Clinical Connection: Iron Overload and Tissue Damage

Erythropoiesis: Production of RBCs

Erythropoiesis (e-rith′-rō-poy-Ē-sis), the production of RBCs, starts in the red bone marrow with a precursor cell called a **proerythroblast** (prō-e-RITH-rō-blast) (see **Figure 19.3**). The proerythroblast divides several times, producing cells that begin to synthesize hemoglobin. Ultimately, a cell near the end of the development sequence ejects its nucleus and becomes a **reticulocyte** (re-TIK-ū-lō-sīt). Loss of the nucleus causes the center of the cell to indent, producing the red blood cell's distinctive biconcave shape. Reticulocytes retain some mitochondria, ribosomes, and endoplasmic reticulum. They pass

FIGURE 19.6 Negative feedback regulation of erythropoiesis (red blood cell formation). Lower oxygen content of air at high altitudes, anemia, and circulatory problems may reduce oxygen delivery to body tissues.

The main stimulus for erythropoiesis is hypoxia, an oxygen deficiency at the tissue level.

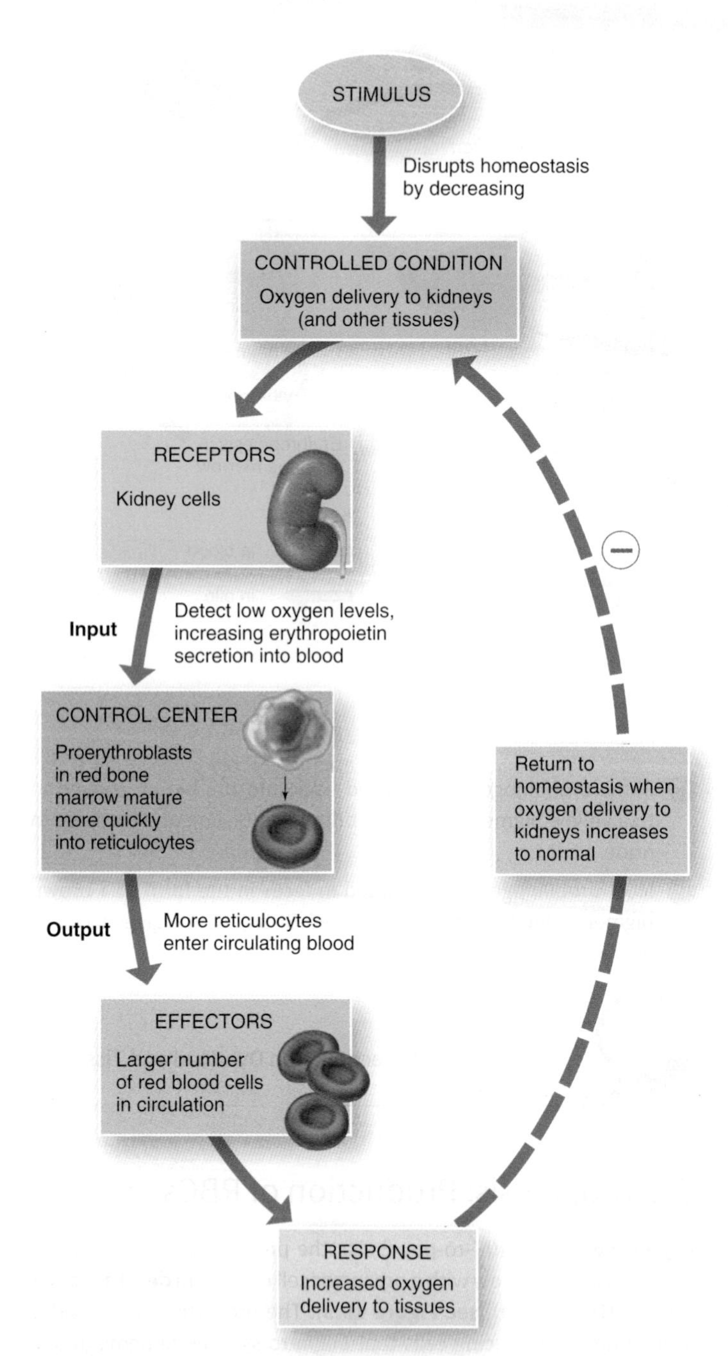

Q How might your hematocrit change if you moved from a town at sea level to a high mountain village?

from red bone marrow into the bloodstream by squeezing between plasma membranes of adjacent endothelial cells of blood capillaries. Reticulocytes develop into mature red blood cells within 1 to 2 days after their release from red bone marrow.

See Clinical Connection: Reticulocyte Count

Normally, erythropoiesis and red blood cell destruction proceed at roughly the same pace. If the oxygen-carrying capacity of the blood falls because erythropoiesis is not keeping up with RBC destruction, a negative feedback system steps up RBC production (**Figure 19.6**). The controlled condition is the amount of oxygen delivered to body tissues. An oxygen deficiency at the tissue level, called **hypoxia** (hī-POKS-ē-a), may occur if too little oxygen enters the blood. For example, the lower oxygen content of air at high altitudes reduces the amount of oxygen in the blood. Oxygen delivery may also fall due to anemia, which has many causes: Lack of iron, lack of certain amino acids, and lack of vitamin B_{12} are but a few (see Disorders: Homeostatic Imbalances in the Study Guide). Circulatory problems that reduce blood flow to tissues may also reduce oxygen delivery. Whatever the cause, hypoxia stimulates the kidneys to step up the release of erythropoietin, which speeds the development of proerythroblasts into reticulocytes in the red bone marrow. As the number of circulating RBCs increases, more oxygen can be delivered to body tissues.

Premature newborns often exhibit anemia, due in part to inadequate production of erythropoietin. During the first weeks after birth, the liver, not the kidneys, produces most EPO. Because the liver is less sensitive than the kidneys to hypoxia, newborns have a smaller EPO response to anemia than do adults. Because fetal hemoglobin (hemoglobin present at birth) carries up to 30% more oxygen, the loss of fetal hemoglobin, due to insufficient erythropoietin production, makes the anemia worse.

See Clinical Connection: Blood Doping

19.4 White Blood Cells

OBJECTIVE

- **Describe** the structure, functions, and production of white blood cells.

Types of White Blood Cells

Unlike red blood cells, **white blood cells (WBCs)** or *leukocytes* (LOO-kō-sīts; *leuko-* = white) have nuclei and a full complement of other organelles but they do not contain hemoglobin. WBCs are classified as either granular or agranular, depending on whether they contain conspicuous chemical-filled cytoplasmic granules (vesicles) that are made visible by staining when viewed through a light microscope. *Granular leukocytes* include neutrophils, eosinophils, and basophils; *agranular leukocytes* include lymphocytes and monocytes. As shown in **Figure 19.3**, monocytes and granular leukocytes develop from myeloid stem cells. In contrast, lymphocytes develop from lymphoid stem cells.

Granular Leukocytes

After staining, each of the three types of granular leukocytes displays conspicuous granules with distinctive coloration that can be recognized under a light microscope. Granular leukocytes can be distinguished as follows:

- **Neutrophil**. The granules of a **neutrophil** (NOO-trō-fil) are smaller than those of other granular leukocytes, evenly distributed, and pale lilac (**Figure 19.7a**). Because the granules do not strongly attract either the acidic (red) or basic (blue) stain, these WBCs are neutrophilic (= neutral loving). The nucleus has two to five lobes, connected by very thin strands of nuclear material. As the cells age, the number of nuclear lobes increases. Because older neutrophils thus have several differently shaped nuclear lobes, they are often called *polymorphonuclear leukocytes (PMNs)*, polymorphs, or "polys."
- **Eosinophil**. The large, uniform-sized granules within an **eosinophil** (ē-ō-SIN-ō-fil) are *eosinophilic* (= eosin-loving)—they stain red-orange with acidic dyes (**Figure 19.7b**). The granules usually do not cover or obscure the nucleus, which most often has two lobes connected by either a thin strand or a thick strand of nuclear material.
- **Basophil**. The round, variable-sized granules of a **basophil** (BĀ-sō-fil) are *basophilic* (= basic loving)—they stain blue-purple with basic dyes (**Figure 19.7c**). The granules commonly obscure the nucleus, which has two lobes.

Agranular Leukocytes

Even though so-called agranular leukocytes possess cytoplasmic granules, the granules are not visible under a light microscope because of their small size and poor staining qualities.

- **Lymphocyte**. The nucleus of a **lymphocyte** (LIM-fō-sīt) stains dark and is round or slightly indented (**Figure 19.7d**). The cytoplasm stains sky blue and forms a rim around the nucleus. The larger the cell, the more cytoplasm is visible. Lymphocytes are classified by cell diameter as large lymphocytes (10–14 μm) or small lymphocytes (6–9 μm). Although the functional significance of the size difference between small and large lymphocytes is unclear, the distinction is still clinically useful because an increase in the number of large lymphocytes has diagnostic significance in acute viral infections and in some immunodeficiency diseases.
- **Monocyte**. The nucleus of a **monocyte** (MON-ō-sīt′) is usually kidney-shaped or horseshoe-shaped, and the cytoplasm is blue-gray and has a foamy appearance (**Figure 19.7e**). The cytoplasm's color and appearance are due to very fine *azurophilic granules* (az′-ū-rō-FIL-ik; *azur-* = blue; *-philic* = loving), which are lysosomes. Blood is merely a conduit for monocytes, which migrate from the blood into the tissues, where they enlarge and differentiate into **macrophages** (MAK-rō-fā-jez = large eaters). Some become **fixed** (*tissue*) **macrophages**, which means they reside in a particular tissue; examples are alveolar macrophages in the lungs or macrophages in the spleen. Others become **wandering macrophages**, which roam the tissues and gather at sites of infection or inflammation.

White blood cells and all other nucleated cells in the body have proteins, called **major histocompatibility (MHC) antigens**, protruding from their plasma membrane into the extracellular fluid. These "cell identity markers" are unique for each person (except identical twins). Although RBCs possess blood group antigens, they lack the MHC antigens.

Functions of White Blood Cells

In a healthy body, some WBCs, especially lymphocytes, can live for several months or years, but most live only a few days. During a period

FIGURE 19.7 **Types of white blood cells.**

The shapes of their nuclei and the staining properties of their cytoplasmic granules distinguish white blood cells from one another.

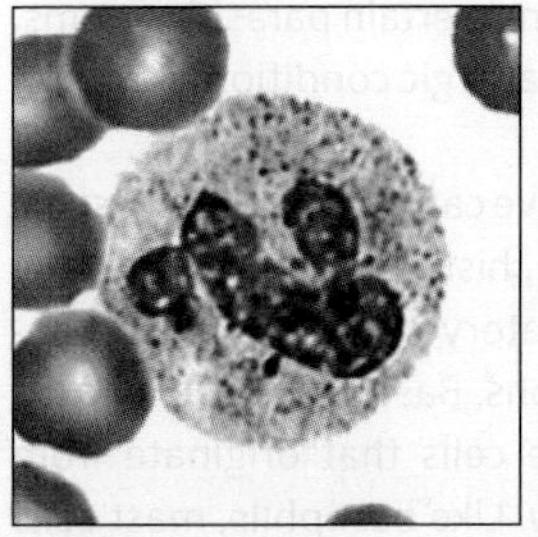

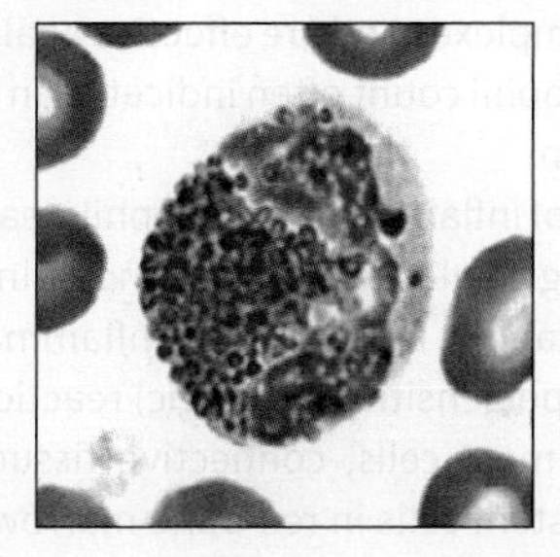

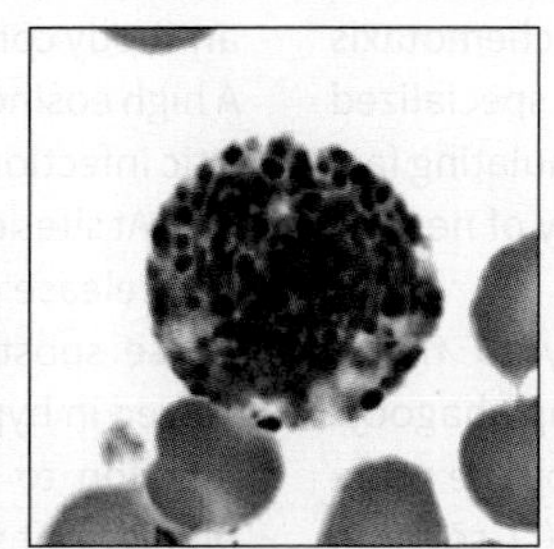

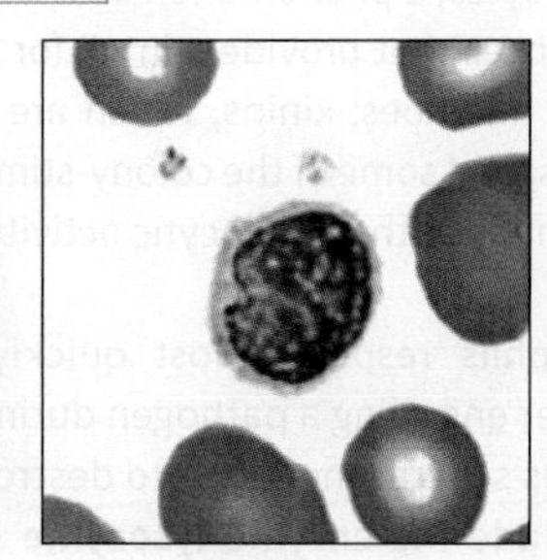

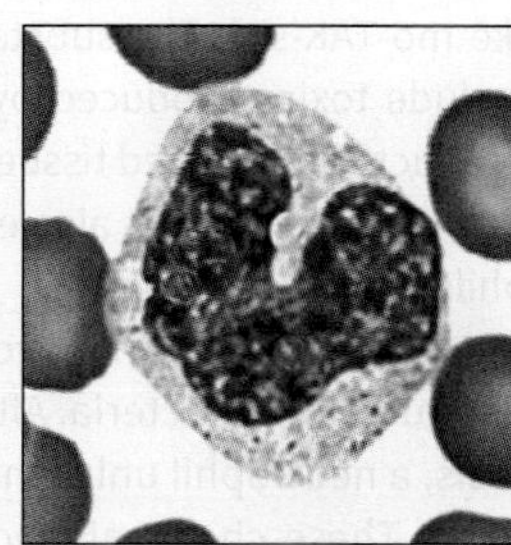

Courtesy Michael Ross, University of Florida

LM all 1600x

(a) Neutrophil (b) Eosinophil (c) Basophil (d) Lymphocyte (e) Monocyte

Q Which WBCs are called granular leukocytes? Why?

of infection, phagocytic WBCs may live only a few hours. WBCs are far less numerous than red blood cells; at about 5000–10,000 cells per microliter of blood, they are outnumbered by RBCs by about 700:1. **Leukocytosis** (loo′-kō-sī-TŌ-sis), an increase in the number of WBCs above 10,000/μL, is a normal, protective response to stresses such as invading microbes, strenuous exercise, anesthesia, and surgery. An abnormally low level of white blood cells (below 5000/μL) is termed **leukopenia** (loo′-kō-PĒ-nē-a). It is never beneficial and may be caused by radiation, shock, and certain chemotherapeutic agents.

The skin and mucous membranes of the body are continuously exposed to microbes and their toxins. Some of these microbes can invade deeper tissues to cause disease. Once pathogens enter the body, the general function of white blood cells is to combat them by phagocytosis or immune responses. To accomplish these tasks, many WBCs leave the bloodstream and collect at sites of pathogen invasion or inflammation. Once granular leukocytes and monocytes leave the bloodstream to fight injury or infection, they never return to it. Lymphocytes, on the other hand, continually recirculate—from blood to interstitial spaces of tissues to lymphatic fluid and back to blood. Only 2% of the total lymphocyte population is circulating in the blood at any given time; the rest is in lymphatic fluid and organs such as the skin, lungs, lymph nodes, and spleen.

RBCs are contained within the bloodstream, but WBCs leave the bloodstream by a process termed **emigration** (em′-i-GRĀ-shun; *e-* = out; *-migra-* = wander), also called *diapedesis* (dī-a-pe-DĒ-sis), in which they roll along the endothelium, stick to it, and then squeeze between endothelial cells (Figure 19.8). The precise signals that stimulate emigration through a particular blood vessel vary for the different types of WBCs. Molecules known as **adhesion molecules** help WBCs stick to the endothelium. For example, endothelial cells display adhesion molecules called *selectins* in response to nearby injury and inflammation. Selectins stick to carbohydrates on the surface of neutrophils, causing them to slow down and roll along the endothelial surface. On the neutrophil surface are other adhesion molecules called *integrins*, which tether neutrophils to the endothelium and assist their movement through the blood vessel wall and into the interstitial fluid of the injured tissue.

FIGURE 19.8 Emigration of white blood cells.

Adhesion molecules (selectins and integrins) assist the emigration of WBCs from the bloodstream into interstitial fluid.

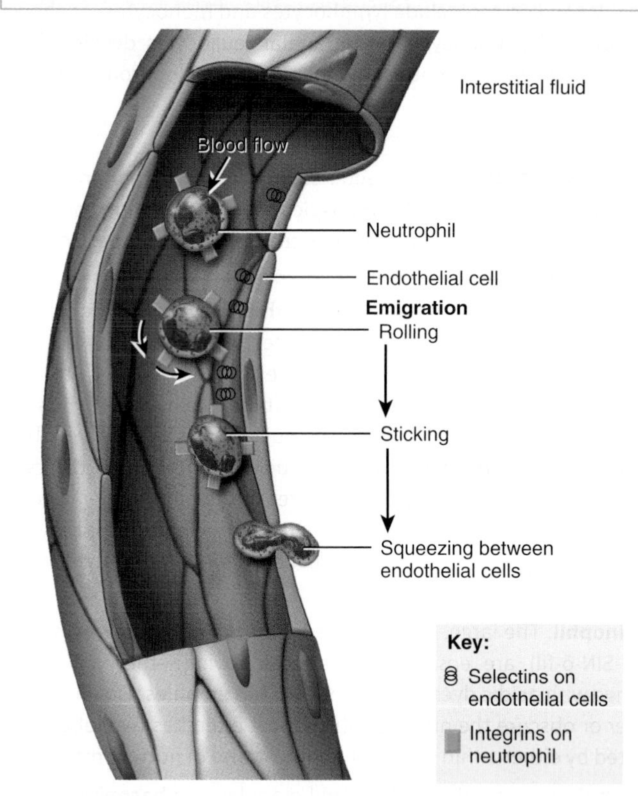

Q In what way is the "traffic pattern" of lymphocytes in the body different from that of other WBCs?

Neutrophils and macrophages are active in **phagocytosis** (fag′-ō-sī-TŌ-sis); they can ingest bacteria and dispose of dead matter (see Figure 3.13). Several different chemicals released by microbes and inflamed tissues attract phagocytes, a phenomenon called **chemotaxis** (kē-mō-TAK-sis). The substances that provide stimuli for chemotaxis include toxins produced by microbes; kinins, which are specialized products of damaged tissues; and some of the colony-stimulating factors (CSFs). The CSFs also enhance the phagocytic activity of neutrophils and macrophages.

Among WBCs, neutrophils respond most quickly to tissue destruction by bacteria. After engulfing a pathogen during phagocytosis, a neutrophil unleashes several chemicals to destroy the pathogen. These chemicals include the enzyme **lysozyme** (LĪ-sō-zīm), which destroys certain bacteria, and **strong oxidants**, such as the superoxide anion (O_2^-), hydrogen peroxide (H_2O_2), and the hypochlorite anion (OCl^-), which is similar to household bleach. Neutrophils also contain **defensins**, proteins that exhibit a broad range of antibiotic activity against bacteria and fungi. Within a neutrophil, vesicles containing defensins merge with phagosomes containing microbes. Defensins form peptide "spears" that poke holes in microbe membranes; the resulting loss of cellular contents kills the invader.

Eosinophils leave the capillaries and enter tissue fluid. They are believed to release enzymes, such as histaminase, that combat the effects of histamine and other substances involved in inflammation during allergic reactions. Eosinophils also phagocytize antigen–antibody complexes and are effective against certain parasitic worms. A high eosinophil count often indicates an allergic condition or a parasitic infection.

At sites of inflammation, basophils leave capillaries, enter tissues, and release granules that contain heparin, histamine, and serotonin. These substances intensify the inflammatory reaction and are involved in hypersensitivity (allergic) reactions. Basophils are similar in function to mast cells, connective tissue cells that originate from pluripotent stem cells in red bone marrow. Like basophils, mast cells release substances involved in inflammation, including heparin, histamine, and proteases. Mast cells are widely dispersed in the body, particularly in connective tissues of the skin and mucous membranes of the respiratory and gastrointestinal tracts.

Lymphocytes are the major soldiers in lymphatic system battles (described in detail in Chapter 22). Most lymphocytes continually move among lymphoid tissues, lymph, and blood, spending only a few hours at a time in blood. Thus, only a small proportion of the total lymphocytes are present in the blood at any given time. Three main types of lymphocytes are B cells, T cells, and natural killer (NK) cells. B cells are particularly effective in destroying bacteria and inactivating their toxins. T cells attack infected body cells and tumor cells, and are responsible for the rejection of transplanted organs. Immune responses carried out by both B cells and T cells help combat infection and provide protection against some diseases. Natural killer cells attack a wide variety of infected body cells and certain tumor cells.

Monocytes take longer to reach a site of infection than neutrophils, but they arrive in larger numbers and destroy more microbes. On their arrival, monocytes enlarge and differentiate into wandering macrophages, which clean up cellular debris and microbes by phagocytosis after an infection.

As you have already learned, an increase in the number of circulating WBCs usually indicates inflammation or infection. A physician may order a **differential white blood cell count**, or "**diff**", a count of each of the five types of white blood cells, to detect infection or inflammation, determine the effects of possible poisoning by chemicals or drugs, monitor blood disorders (for example, leukemia) and the effects of chemotherapy, or detect allergic reactions and parasitic infections. Because each type of white blood cell plays a different role, determining the *percentage* of each type in the blood assists in diagnosing the condition. Table 19.2 lists the significance of both high and low WBC counts.

TABLE 19.2 Significance of High and Low White Blood Cell Counts

WBC TYPE	HIGH COUNT MAY INDICATE	LOW COUNT MAY INDICATE
Neutrophils	Bacterial infection, burns, stress, inflammation.	Radiation exposure, drug toxicity, vitamin B_{12} deficiency, systemic lupus erythematosus (SLE).
Lymphocytes	Viral infections, some leukemias, infectious mononucleosis.	Prolonged illness, HIV infection, immunosuppression, treatment with cortisol.
Monocytes	Viral or fungal infections, tuberculosis, some leukemias, other chronic diseases.	Bone marrow suppression, treatment with cortisol.
Eosinophils	Allergic reactions, parasitic infections, autoimmune diseases.	Drug toxicity, stress, acute allergic reactions.
Basophils	Allergic reactions, leukemias, cancers, hypothyroidism.	Pregnancy, ovulation, stress, hypothyroidism.

Courtesy Michael Ross, University of Florida

19.5 Platelets

OBJECTIVE

- **Describe** the structure, function, and origin of platelets.

Besides the immature cell types that develop into erythrocytes and leukocytes, hemopoietic stem cells also differentiate into cells that produce platelets. Under the influence of the hormone thrombopoietin, myeloid stem cells develop into megakaryocyte colony-forming cells that in turn develop into precursor cells called *megakaryoblasts* (see Figure 19.3). Megakaryoblasts transform into megakaryocytes, huge cells that splinter into 2000 to 3000 fragments. Each fragment, enclosed by a piece of the plasma membrane, is a **platelet**. Platelets break off from the megakaryocytes in red bone marrow and then enter the blood circulation. Between 150,000 and 400,000 platelets are present in each microliter of blood. Each is irregularly disc-shaped, 2–4 μm in diameter, and has many vesicles but no nucleus.

Their granules contain chemicals that, once released, promote blood clotting. Platelets help stop blood loss from damaged blood vessels by forming a platelet plug. Platelets have a short life span, normally just 5 to 9 days. Aged and dead platelets are removed by fixed macrophages in the spleen and liver.

Table 19.3 summarizes the formed elements in blood.

See Clinical Connection: Complete Blood Count

19.6 Stem Cell Transplants from Bone Marrow and Cord Blood

OBJECTIVE

- **Explain** the importance of bone marrow transplants and stem cell transplants.

A **bone marrow transplant** is the replacement of cancerous or abnormal red bone marrow with healthy red bone marrow in order to establish normal blood cell counts. In patients with cancer or certain genetic diseases, the defective red bone marrow is destroyed by high doses of chemotherapy and whole body radiation just before the

TABLE 19.3 Summary of Formed Elements in Blood

NAME AND APPEARANCE	NUMBER	CHARACTERISTICS*	FUNCTIONS
RED BLOOD CELLS (RBCS) OR ERYTHROCYTES Juergen Berger/Science Source Images	4.8 million/μL in females; 5.4 million/μL in males.	7–8 μm diameter, biconcave discs, without nuclei; live for about 120 days.	Hemoglobin within RBCs transports most oxygen and part of carbon dioxide in blood.
WHITE BLOOD CELLS (WBCS) OR LEUKOCYTES	5000–10,000/μL.	Most live for a few hours to a few days.†	Combat pathogens and other foreign substances that enter body.
Granular leukocytes			
Neutrophils	60–70% of all WBCs.	10–12 μm diameter; nucleus has 2–5 lobes connected by thin strands of chromatin; cytoplasm has very fine, pale lilac granules.	Phagocytosis. Destruction of bacteria with lysozyme, defensins, and strong oxidants, such as superoxide anion, hydrogen peroxide, and hypochlorite anion.
Eosinophils	2–4% of all WBCs.	10–12 μm diameter; nucleus usually has 2 lobes connected by thick strand of chromatin; large, red-orange granules fill cytoplasm.	Combat effects of histamine in allergic reactions, phagocytize antigen–antibody complexes, and destroy certain parasitic worms.
Basophils Courtesy Michael Ross, University of Florida	0.5–1% of all WBCs.	8–10 μm diameter; nucleus has 2 lobes; large cytoplasmic granules appear deep blue-purple.	Liberate heparin, histamine, and serotonin in allergic reactions that intensify overall inflammatory response.
Agranular leukocytes			
Lymphocytes (T cells, B cells, and natural killer cells)	20–25% of all WBCs.	Small lymphocytes are 6–9 μm in diameter; large lymphocytes are 10–14 μm in diameter; nucleus is round or slightly indented; cytoplasm forms rim around nucleus that looks sky blue; the larger the cell, the more cytoplasm is visible.	Mediate immune responses, including antigen–antibody reactions. B cells develop into plasma cells, which secrete antibodies. T cells attack invading viruses, cancer cells, and transplanted tissue cells. Natural killer cells attack wide variety of infectious microbes and certain spontaneously arising tumor cells.
Monocytes Courtesy Michael Ross, University of Florida	3–8% of all WBCs.	12–20 μm diameter; nucleus is kidney-or horseshoe-shaped; cytoplasm is blue-gray and appears foamy.	Phagocytosis (after transforming into fixed or wandering macrophages).
PLATELETS Mark Nielsen	150,000–400,000/μL.	2–4 μm diameter cell fragments that live for 5–9 days; contain many vesicles but no nucleus.	Form platelet plug in hemostasis; release chemicals that promote vascular spasm and blood clotting.

*Colors are those seen when using Wright's stain.

†Some lymphocytes, called T and B memory cells, can live for many years once they are established.

transplant takes place. These treatments kill the cancer cells and destroy the patient's immune system in order to decrease the chance of transplant rejection.

Healthy red bone marrow for transplanting may be supplied by a donor or by the patient when the underlying disease is inactive, as when leukemia is in remission. The red bone marrow from a donor is usually removed from the iliac crest of the hip bone under general anesthesia with a syringe and is then injected into the recipient's vein, much like a blood transfusion. The injected marrow migrates to the recipient's red bone marrow cavities, where the donor's stem cells multiply. If all goes well, the recipient's red bone marrow is replaced entirely by healthy, noncancerous cells.

Bone marrow transplants have been used to treat aplastic anemia, certain types of leukemia, severe combined immunodeficiency disease (SCID), Hodgkin's disease, non-Hodgkin's lymphoma, multiple myeloma, thalassemia, sickle-cell disease, breast cancer, ovarian cancer, testicular cancer, and hemolytic anemia. However, there are some drawbacks. Since the recipient's white blood cells have been completely destroyed by chemotherapy and radiation, the patient is extremely vulnerable to infection. (It takes about 2–3 weeks for transplanted bone marrow to produce enough white blood cells to protect against infection.) In addition, transplanted red bone marrow may produce T cells that attack the recipient's tissues, a reaction called *graft-versus-host disease.* Similarly, any of the recipient's T cells that survived the chemotherapy and radiation can attack donor transplant cells. Another drawback is that patients must take immunosuppressive drugs for life. Because these drugs reduce the level of immune system activity, they increase the risk of infection. Immunosuppressive drugs also have side effects such as fever, muscle aches, headache, nausea, fatigue, depression, high blood pressure, and kidney and liver damage.

A more recent advance for obtaining stem cells involves a **cord-blood transplant**. The connection between the mother and embryo (and later the fetus) is the umbilical cord. Stem cells may be obtained from the umbilical cord shortly after birth. The stem cells are removed from the cord with a syringe and then frozen. Stem cells from the cord have several advantages over those obtained from red bone marrow:

1. They are easily collected following permission of the newborn's parents.
2. They are more abundant than stem cells in red bone marrow.
3. They are less likely to cause graft-versus-host disease, so the match between donor and recipient does not have to be as close as in a bone marrow transplant. This provides a larger number of potential donors.
4. They are less likely to transmit infections.
5. They can be stored indefinitely in cord-blood banks.

19.7 Hemostasis

OBJECTIVES

- **Describe** the three mechanisms that contribute to hemostasis.
- **Explain** the various factors that promote and inhibit blood clotting.

Hemostasis (hē-mō-STĀ-sis), not to be confused with the very similar term *homeostasis*, is a sequence of responses that stops bleeding. When blood vessels are damaged or ruptured, the hemostatic response must be quick, localized to the region of damage, and carefully controlled in order to be effective. Three mechanisms reduce blood loss: (1) vascular spasm, (2) platelet plug formation, and (3) blood clotting (coagulation). When successful, hemostasis prevents **hemorrhage** (HEM-o-rij; *-rhage* = burst forth), the loss of a large amount of blood from the vessels. Hemostatic mechanisms can prevent hemorrhage from smaller blood vessels, but extensive hemorrhage from larger vessels usually requires medical intervention.

Vascular Spasm

When arteries or arterioles are damaged, the circularly arranged smooth muscle in their walls contracts immediately, a reaction called **vascular spasm**. This reduces blood loss for several minutes to several hours, during which time the other hemostatic mechanisms go into operation. The spasm is probably caused by damage to the smooth muscle, by substances released from activated platelets, and by reflexes initiated by pain receptors.

Platelet Plug Formation

Considering their small size, platelets store an impressive array of chemicals. Within many vesicles are clotting factors, ADP, ATP, Ca^{2+}, and serotonin. Also present are enzymes that produce thromboxane A2, a prostaglandin; *fibrin-stabilizing factor*, which helps to strengthen a blood clot; lysosomes; some mitochondria; membrane systems that take up and store calcium and provide channels for release of the contents of granules; and glycogen. Also within platelets is **platelet-derived growth factor (PDGF)**, a hormone that can cause proliferation of vascular endothelial cells, vascular smooth muscle fibers, and fibroblasts to help repair damaged blood vessel walls.

Platelet plug formation occurs as follows (**Figure 19.9**):

1. Initially, platelets contact and stick to parts of a damaged blood vessel, such as collagen fibers of the connective tissue underlying the damaged endothelial cells. This process is called **platelet adhesion.**
2. Due to adhesion, the platelets become activated, and their characteristics change dramatically. They extend many projections that enable them to contact and interact with one another, and they begin to liberate the contents of their vesicles. This phase is called the **platelet release reaction**. Liberated ADP and thromboxane A2 play a major role by activating nearby platelets. Serotonin and thromboxane A2 function as vasoconstrictors, causing and sustaining contraction of vascular smooth muscle, which decreases blood flow through the injured vessel.
3. The release of ADP makes other platelets in the area sticky, and the stickiness of the newly recruited and activated platelets causes them to adhere to the originally activated platelets. This gathering of platelets is called **platelet aggregation**. Eventually, the accumulation and attachment of large numbers of platelets form a mass called a **platelet plug.**

FIGURE 19.9 Platelet plug formation.

A platelet plug can stop blood loss completely if the hole in a blood vessel is small enough.

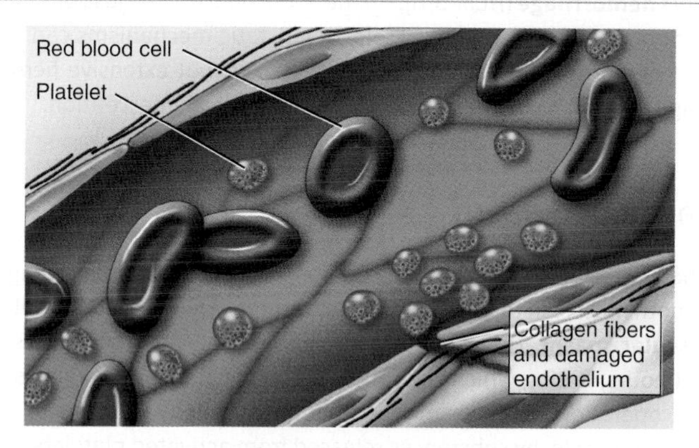

1 Platelet adhesion

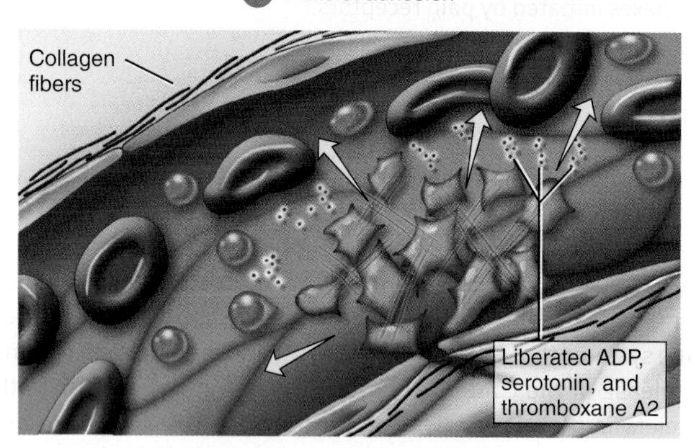

2 Platelet release reaction

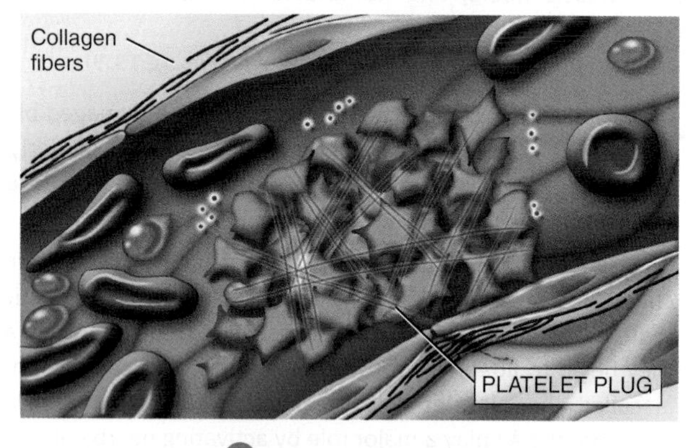

3 Platelet aggregation

Q Along with platelet plug formation, which two mechanisms contribute to hemostasis?

A platelet plug is very effective in preventing blood loss in a small vessel. Although initially the platelet plug is loose, it becomes quite tight when reinforced by fibrin threads formed during clotting (see Figure 19.10). A platelet plug can stop blood loss completely if the hole in a blood vessel is not too large.

FIGURE 19.10 Blood clot formation. Notice the platelet and red blood cells entrapped in fibrin threads.

A blood clot is a gel that contains formed elements of the blood entangled in fibrin threads.

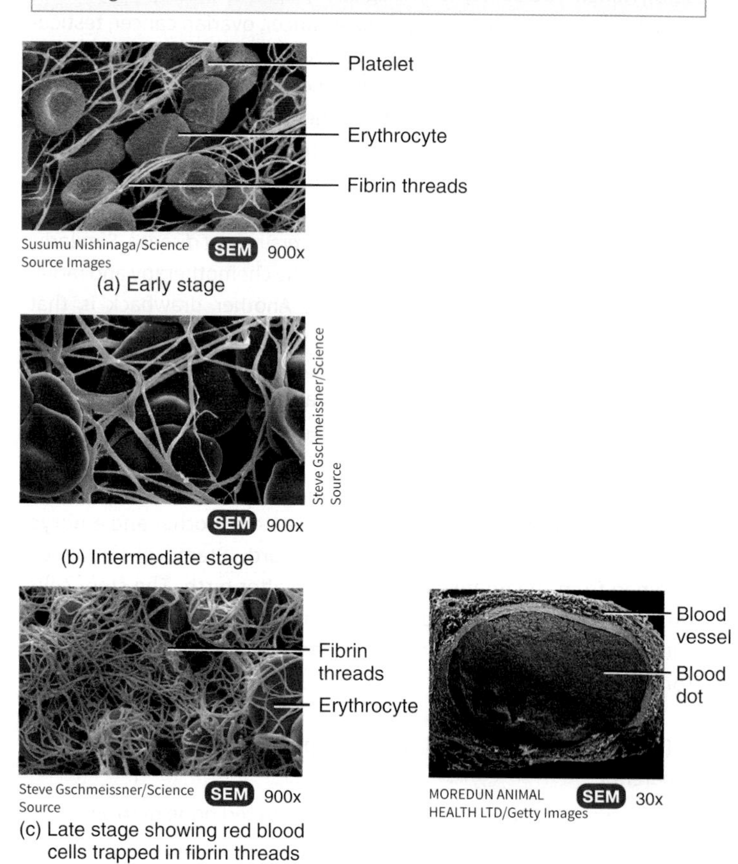

Q What is serum?

Blood Clotting

Normally, blood remains in its liquid form as long as it stays within its vessels. If it is drawn from the body, however, it thickens and forms a gel. Eventually, the gel separates from the liquid. The straw-colored liquid, called **serum**, is simply blood plasma minus the clotting proteins. The gel is called a **blood clot**. It consists of a network of insoluble protein fibers called fibrin in which the formed elements of blood are trapped (Figure 19.10).

The process of gel formation, called **clotting** or *coagulation* (kō-ag-u-LĀ-shun), is a series of chemical reactions that culminates in formation of fibrin threads. If blood clots too easily, the result can be **thrombosis** (throm-BŌ-sis; *thromb-* = clot; *-osis* = a condition of)—clotting in an undamaged blood vessel. If the blood takes too long to clot, hemorrhage can occur.

Clotting involves several substances known as **clotting** (*coagulation*) **factors**. These factors include calcium ions (Ca^{2+}), several inactive enzymes that are synthesized by hepatocytes (liver cells) and released into the bloodstream, and various molecules associated with platelets or released by damaged tissues. Most clotting factors are identified by

Roman numerals that indicate the order of their discovery (not necessarily the order of their participation in the clotting process).

Clotting is a complex cascade of enzymatic reactions in which each clotting factor activates many molecules of the next one in a fixed sequence. Finally, a large quantity of product (the insoluble protein fibrin) is formed. Clotting can be divided into three stages (**Figure 19.11**):

1. Two pathways, called the extrinsic pathway and the intrinsic pathway (**Figures 19.11a, b**), which will be described shortly, lead to the formation of prothrombinase. Once prothrombinase is formed, the steps involved in the next two stages of clotting are the same for both the extrinsic and intrinsic pathways, and together these two stages are referred to as the common pathway.
2. Prothrombinase converts prothrombin (a plasma protein formed by the liver) into the enzyme thrombin.
3. Thrombin converts soluble fibrinogen (another plasma protein formed by the liver) into insoluble fibrin. Fibrin forms the threads of the clot.

The Extrinsic Pathway The **extrinsic pathway** of blood clotting has fewer steps than the intrinsic pathway and occurs rapidly—within a matter of seconds if trauma is severe. It is so named because a tissue protein called **tissue factor (TF)**, also known as *thromboplastin* (throm′-bō-PLAS-tin), leaks into the blood from cells *outside (extrinsic to)* blood vessels and initiates the formation of prothrombinase. TF is a complex mixture of lipoproteins and phospholipids released from the surfaces of damaged cells. In the presence of Ca^{2+}, TF begins a sequence of reactions that ultimately activates clotting factor X (**Figure 19.11a**). Once factor X is activated, it combines with factor V in the presence of Ca^{2+} to form the active enzyme prothrombinase, completing the extrinsic pathway.

The Intrinsic Pathway The **intrinsic pathway** of blood clotting is more complex than the extrinsic pathway, and it occurs more slowly, usually requiring several minutes. The intrinsic pathway is so named because its activators are either in direct contact with blood or contained *within (intrinsic to)* the blood; outside tissue damage is not needed. If endothelial cells become roughened or damaged, blood can come in contact with collagen fibers in the connective tissue around the endothelium of the blood vessel. In addition, trauma to endothelial cells causes damage to platelets, resulting in the release of phospholipids by the platelets. Contact with collagen fibers (or with the glass sides of a blood collection tube) activates clotting factor XII (**Figure 19.11b**), which begins a sequence of reactions that eventually activates clotting factor X. Platelet phospholipids and Ca^{2+} can also participate in the activation of factor X. Once factor X is activated, it combines with factor V to form the active enzyme prothrombinase (just as occurs in the extrinsic pathway), completing the intrinsic pathway.

The Common Pathway The formation of prothrombinase marks the beginning of the **common pathway**. In the second stage of blood clotting (**Figure 19.11c**), prothrombinase and Ca^{2+} catalyze the conversion of prothrombin to thrombin. In the third stage, thrombin, in the presence of Ca^{2+}, converts fibrinogen, which is soluble, to loose fibrin threads, which are insoluble. Thrombin also activates factor XIII (fibrin stabilizing factor), which strengthens and stabilizes the fibrin threads into a sturdy clot. Plasma contains some factor XIII, which is also released by platelets trapped in the clot.

FIGURE 19.11 The blood-clotting cascade. Green arrows represent positive feedback cycles.

In blood clotting, coagulation factors are activated in sequence, resulting in a cascade of reactions that includes positive feedback cycles.

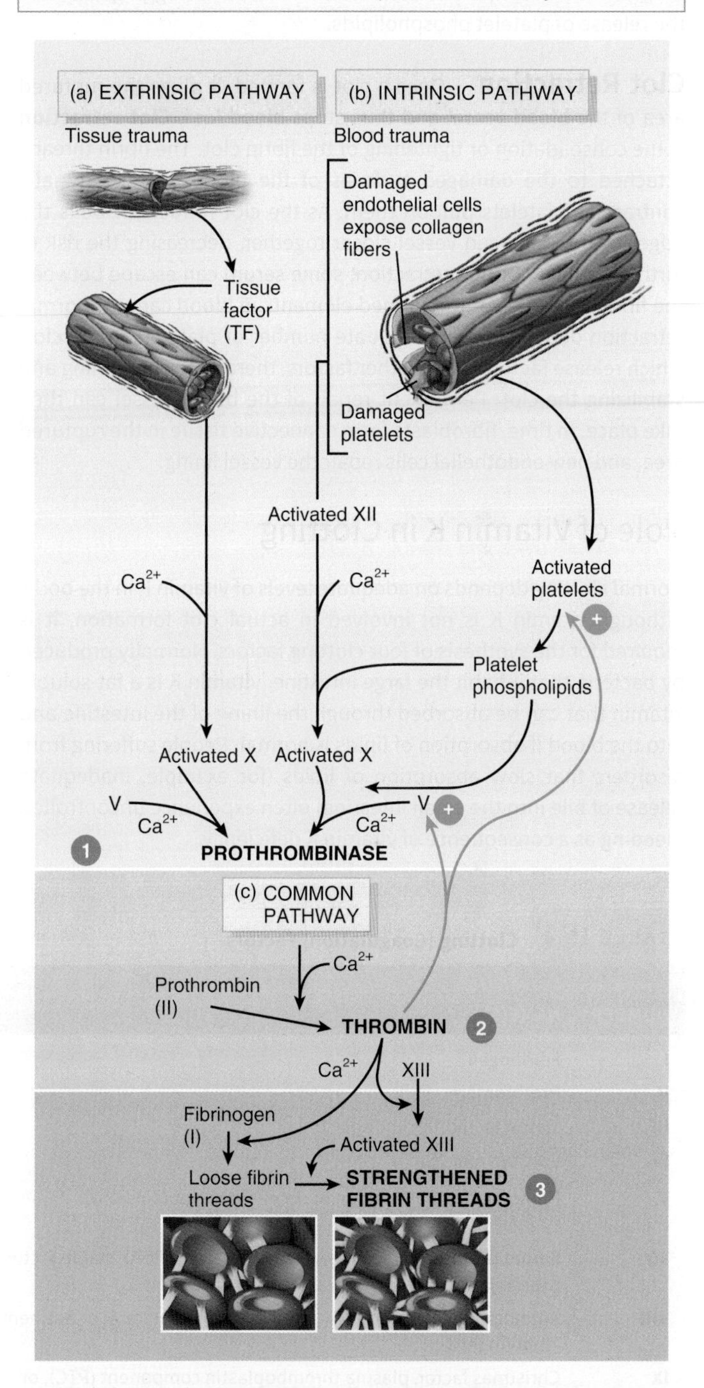

Q What is the outcome of the first stage of blood clotting?

Thrombin has two positive feedback effects. In the first positive feedback loop, which involves factor V, it accelerates the formation of prothrombinase. Prothrombinase in turn accelerates the production of more thrombin, and so on. In the second positive feedback loop, thrombin activates platelets, which reinforces their aggregation and the release of platelet phospholipids.

Clot Retraction Once a clot is formed, it plugs the ruptured area of the blood vessel and thus stops blood loss. **Clot retraction** is the consolidation or tightening of the fibrin clot. The fibrin threads attached to the damaged surfaces of the blood vessel gradually contract as platelets pull on them. As the clot retracts, it pulls the edges of the damaged vessel closer together, decreasing the risk of further damage. During retraction, some serum can escape between the fibrin threads, but the formed elements in blood cannot. Normal retraction depends on an adequate number of platelets in the clot, which release factor XIII and other factors, thereby strengthening and stabilizing the clot. Permanent repair of the blood vessel can then take place. In time, fibroblasts form connective tissue in the ruptured area, and new endothelial cells repair the vessel lining.

Role of Vitamin K in Clotting

Normal clotting depends on adequate levels of vitamin K in the body. Although vitamin K is not involved in actual clot formation, it is required for the synthesis of four clotting factors. Normally produced by bacteria that inhabit the large intestine, vitamin K is a fat-soluble vitamin that can be absorbed through the lining of the intestine and into the blood if absorption of lipids is normal. People suffering from disorders that slow absorption of lipids (for example, inadequate release of bile into the small intestine) often experience uncontrolled bleeding as a consequence of vitamin K deficiency.

The various clotting factors, their sources, and the pathways of activation are summarized in Table 19.4.

Homeostatic Control Mechanisms

Many times a day little clots start to form, often at a site of minor roughness or at a developing atherosclerotic plaque inside a blood vessel. Because blood clotting involves amplification and positive feedback cycles, a clot has a tendency to enlarge, creating the potential for impairment of blood flow through undamaged vessels. The **fibrinolytic system** (fī-bri-nō-LIT-ik) dissolves small, inappropriate clots; it also dissolves clots at a site of damage once the damage is repaired. Dissolution of a clot is called **fibrinolysis** (fī-bri-NOL-i-sis). When a clot is formed, an inactive plasma enzyme called **plasminogen** (plaz-MIN-o-jen) is incorporated into the clot. Both body tissues and blood contain substances that can activate plasminogen to **plasmin** or *fibrinolysin* (fī-brin-ō-LĪ-sin), an active plasma enzyme. Among these substances are thrombin, activated factor XII, and tissue plasminogen activator (t-PA), which is synthesized in endothelial cells of most tissues and liberated into the blood. Once plasmin is formed, it can dissolve the clot by digesting fibrin threads and inactivating substances such as fibrinogen, prothrombin, and factors V and XII.

Even though thrombin has a positive feedback effect on blood clotting, clot formation normally remains localized at the site of damage. A clot does not extend beyond a wound site into the general circulation, in part because fibrin absorbs thrombin into the clot. Another reason for localized clot formation is that because of the dispersal of some of the clotting factors by the blood, their concentrations are not high enough to bring about widespread clotting.

Several other mechanisms also control blood clotting. For example, endothelial cells and white blood cells produce a prostaglandin called **prostacyclin** (pros-ta-SĪ-klin) that opposes the actions of

TABLE 19.4 Clotting (Coagulation) Factors

NUMBER*	NAME(S)	SOURCE	PATHWAY(S) OF ACTIVATION
I	Fibrinogen.	Liver.	Common.
II	Prothrombin.	Liver.	Common.
III	Tissue factor (thromboplastin).	Damaged tissues and activated platelets.	Extrinsic.
IV	Calcium ions (Ca^{2+}).	Diet, bones, and platelets.	All.
V	Proaccelerin, labile factor, or accelerator globulin (AcG).	Liver and platelets.	Extrinsic and intrinsic.
VII	Serum prothrombin conversion accelerator (SPCA), stable factor, or proconvertin.	Liver.	Extrinsic.
VIII	Antihemophilic factor (AHF), antihemophilic factor A, or antihemophilic globulin (AHG).	Liver.	Intrinsic.
IX	Christmas factor, plasma thromboplastin component (PTC), or antihemophilic factor B.	Liver.	Intrinsic.
X	Stuart factor, Prower factor, or thrombokinase.	Liver.	Extrinsic and intrinsic.
XI	Plasma thromboplastin antecedent (PTA) or antihemophilic factor C.	Liver.	Intrinsic.
XII	Hageman factor, glass factor, contact factor, or antihemophilic factor D.	Liver.	Intrinsic.
XIII	Fibrin-stabilizing factor (FSF).	Liver and platelets.	Common.

*There is no factor VI. Prothrombinase (prothrombin activator) is a combination of activated factors V and X.

thromboxane A2. Prostacyclin is a powerful inhibitor of platelet adhesion and release.

In addition, substances that delay, suppress, or prevent blood clotting, called **anticoagulants** (an′-tī-kō-AG-ū-lants), are present in blood. These include **antithrombin**, which blocks the action of several factors, including XII, X, and II (prothrombin). **Heparin**, an anticoagulant that is produced by mast cells and basophils, combines with antithrombin and increases its effectiveness in blocking thrombin. Another anticoagulant, **activated protein C (APC)**, inactivates the two major clotting factors not blocked by antithrombin and enhances activity of plasminogen activators. Babies that lack the ability to produce APC due to a genetic mutation usually die of blood clots in infancy.

Intravascular Clotting

Despite the anticoagulating and fibrinolytic mechanisms, blood clots sometimes form within the cardiovascular system. Such clots may be initiated by roughened endothelial surfaces of a blood vessel resulting from atherosclerosis, trauma, or infection. These conditions induce adhesion of platelets. Intravascular clots may also form when blood flows too slowly (stasis), allowing clotting factors to accumulate locally in high enough concentrations to initiate coagulation. Clotting in an unbroken blood vessel (usually a vein) is called **thrombosis**. The clot itself, called a **thrombus** (THROM-bus), may dissolve spontaneously. If it remains intact, however, the thrombus may become dislodged and be swept away in the blood. A blood clot, bubble of air, fat from broken bones, or a piece of debris transported by the bloodstream is called an **embolus** (EM-bō-lus; *em-* = in; *-bolus* = a mass). An embolus that breaks away from an arterial wall may lodge in a smaller-diameter artery downstream and block blood flow to a vital organ. When an embolus lodges in the lungs, the condition is called **pulmonary embolism**.

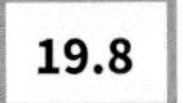

See Clinical Connection: Aspirin and Thrombolytic Agents

19.8 Blood Groups and Blood Types

OBJECTIVES

- **Distinguish** between the ABO and Rh blood groups.
- **Explain** why it is so important to match donor and recipient blood types before administering a transfusion.

The surfaces of erythrocytes contain a genetically determined assortment of **antigens** composed of glycoproteins and glycolipids. These antigens, called *agglutinogens* (a-gloo-TIN-ō-jens), occur in characteristic combinations. Based on the presence or absence of various antigens, blood is categorized into different **blood groups**. Within a given blood group, there may be two or more different **blood types**. There are at least 24 blood groups and more than 100 antigens that can be detected on the surface of red blood cells. Here we discuss two major blood groups—ABO and Rh. Other blood groups include the Lewis, Kell, Kidd, and Duffy systems. The incidence of ABO and Rh blood types varies among different population groups, as indicated in Table 19.5.

TABLE 19.5 Blood Types in the United States

	BLOOD TYPE (PERCENTAGE)				
POPULATION GROUP	O	A	B	AB	Rh^+
European-American	45	40	11	4	85
African-American	49	27	20	4	95
Korean-American	32	28	30	10	100
Japanese-American	31	38	21	10	100
Chinese-American	42	27	25	6	100
Native American	79	16	4	1	100

ABO Blood Group

The **ABO blood group** is based on two glycolipid antigens called A and B (Figure 19.12). People whose RBCs display *only antigen A* have **type A** blood. Those who have *only antigen B* are **type B**. Individuals who have *both A and B antigens* are **type AB**; those who have *neither antigen A nor B* are **type O**.

Blood plasma usually contains **antibodies** called *agglutinins* (a-GLOO-ti-nins) that react with the A or B antigens if the two are mixed. These are the **anti-A antibody**, which reacts with antigen A, and the **anti-B antibody**, which reacts with antigen B. The antibodies present in each of the four blood types are shown in Figure 19.12. You do not have antibodies that react with the antigens of your own RBCs, but you do have antibodies for any antigens that your RBCs lack. For example, if your blood type is B, you have B antigens on your red blood cells, and you have anti-A antibodies in your blood plasma. Although agglutinins start to appear in the blood within a few months after birth, the reason for their presence is not clear. Perhaps they are formed in response to bacteria that normally inhabit the gastrointestinal tract. Because the antibodies are large IgM-type antibodies (see Table 22.3) that do not cross the placenta, ABO incompatibility between a mother and her fetus rarely causes problems.

Transfusions

Despite the differences in RBC antigens reflected in the blood group systems, blood is the most easily shared of human tissues, saving many thousands of lives every year through transfusions. A **transfusion** (trans-FŪ-zhun) is the transfer of whole blood or blood components (red blood cells only or blood plasma only) into the bloodstream or directly into the red bone marrow. A transfusion is most often given to alleviate anemia, to increase blood volume (for example, after a severe hemorrhage), or to improve immunity. However, the normal components of one person's RBC plasma membrane can trigger damaging antigen–antibody responses in a transfusion recipient. In an incompatible blood transfusion, antibodies in the recipient's plasma bind to the antigens on the donated RBCs, which causes **agglutination** (a-gloo-ti-NĀ-shun), or *clumping*, of the RBCs. Agglutination is an

FIGURE 19.12 **Antigens and antibodies of the ABO blood types.**

The antibodies in your plasma do not react with the antigens on your red blood cells.

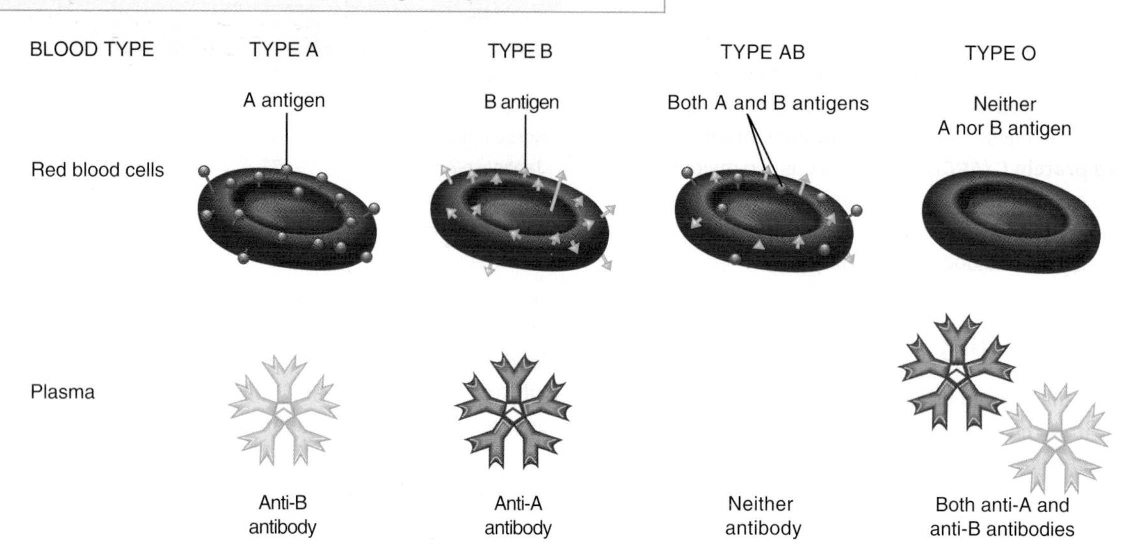

Q Which antibodies are usually present in type O blood?

antigen–antibody response in which RBCs become cross-linked to one another. (Note that agglutination is not the same as blood clotting.) When these antigen–antibody complexes form, they activate plasma proteins of the complement family (described in Section 22.6). In essence, complement molecules make the plasma membrane of the donated RBCs leaky, causing **hemolysis** (hē-MOL-i-sis) or rupture of the RBCs and the release of hemoglobin into the blood plasma. The liberated hemoglobin may cause kidney damage by clogging the filtration membranes. Although quite rare, it is possible for the viruses that cause AIDS and hepatitis B and C to be transmitted through transfusion of contaminated blood products.

Consider what happens if a person with type A blood receives a transfusion of type B blood. The recipient's blood (type A) contains A antigens on the red blood cells and anti-B antibodies in the plasma. The donor's blood (type B) contains B antigens and anti-A antibodies. In this situation, two things can happen. First, the anti-B antibodies in the recipient's plasma can bind to the B antigens on the donor's erythrocytes, causing agglutination and hemolysis of the red blood cells. Second, the anti-A antibodies in the donor's plasma can bind to the A antigens on the recipient's red blood cells, a less serious reaction because the donor's anti-A antibodies become so diluted in the recipient's plasma that they do not cause significant agglutination and hemolysis of the recipient's RBCs.

People with type AB blood do not have anti-A or anti-B antibodies in their blood plasma. They are sometimes called *universal recipients* because theoretically they can receive blood from donors of all four blood types. They have no antibodies to attack antigens on donated RBCs. People with type O blood have neither A nor B antigens on their RBCs and are sometimes called *universal donors* because theoretically they can donate blood to all four ABO blood types. Type O persons requiring blood may receive only type O blood (**Table 19.6**). In practice, use of the terms universal recipient and universal donor is misleading and dangerous. Blood contains antigens and antibodies other than those associated with the ABO system that can cause transfusion problems. Thus, blood should be carefully cross-matched or screened before transfusion. In about 80% of the population, soluble antigens of the ABO type appear in saliva and other body fluids, in which case blood type can be identified from a sample of saliva.

Rh Blood Group

The **Rh blood group** is so named because the Rh antigen, called **Rh factor**, was first found in the blood of the *Rhesus* monkey. The alleles of three genes may code for the Rh antigen. People whose RBCs have Rh antigens are designated Rh^+ (Rh positive); those who lack Rh antigens are designated Rh^- (Rh negative). **Table 19.5** shows the

TABLE 19.6 Summary of ABO Blood Group Interactions

	BLOOD TYPE			
CHARACTERISTIC	**A**	**B**	**AB**	**O**
Agglutinogen (antigen) on RBCs	A	B	Both A and B	Neither A nor B
Agglutinin (antibody) in plasma	Anti-B	Anti-A	Neither anti-A nor anti-B	Both anti-A and anti-B
Compatible donor blood types (no hemolysis)	A, O	B, O	A, B, AB, O	O
Incompatible donor blood types (hemolysis)	B, AB	A, AB	—	A, B, AB

incidence of Rh^+ and Rh^- in various populations. Normally, blood plasma does not contain anti-Rh antibodies. If an Rh^- person receives an Rh^+ blood transfusion, however, the immune system starts to make anti-Rh antibodies that will remain in the blood. If a second transfusion of Rh^+ blood is given later, the previously formed anti-Rh antibodies will cause agglutination and hemolysis of the RBCs in the donated blood, and a severe reaction may occur.

See Clinical Connection: Hemolytic Disease of the Newborn

Typing and Cross-Matching Blood for Transfusion

To avoid blood-type mismatches, laboratory technicians type the patient's blood and then either cross-match it to potential donor blood or screen it for the presence of antibodies. In the procedure for ABO blood typing, single drops of blood are mixed with different *antisera*, solutions that contain antibodies (**Figure 19.13**). One drop of blood is mixed with anti-A serum, which contains anti-A antibodies that will agglutinate red blood cells that possess A antigens. Another drop is mixed with anti-B serum, which contains anti-B antibodies that will agglutinate red blood cells that possess B antigens. If the red blood cells agglutinate only when mixed with anti-A serum, the blood is type A. If the red blood cells agglutinate only when mixed with anti-B serum, the blood is type B. The blood is type AB if both drops agglutinate; if neither drop agglutinates, the blood is type O.

In the procedure for determining Rh factor, a drop of blood is mixed with antiserum containing antibodies that will agglutinate RBCs displaying Rh antigens. If the blood agglutinates, it is Rh^+; no agglutination indicates Rh^-.

Once the patient's blood type is known, donor blood of the same ABO and Rh type is selected. In a **cross-match**, the possible donor RBCs are mixed with the recipient's serum. If agglutination does not occur, the recipient does not have antibodies that will attack the donor RBCs. Alternatively, the recipient's serum can be screened against a test panel of RBCs having antigens known to cause blood transfusion reactions to detect any antibodies that may be present.

See Clinical Connection: Anticoagulants

FIGURE 19.13 **ABO blood typing.** The boxed areas show agglutination (clumping) of red blood cells.

In the procedure for ABO blood typing, blood is mixed with anti-A serum and anti-B serum.

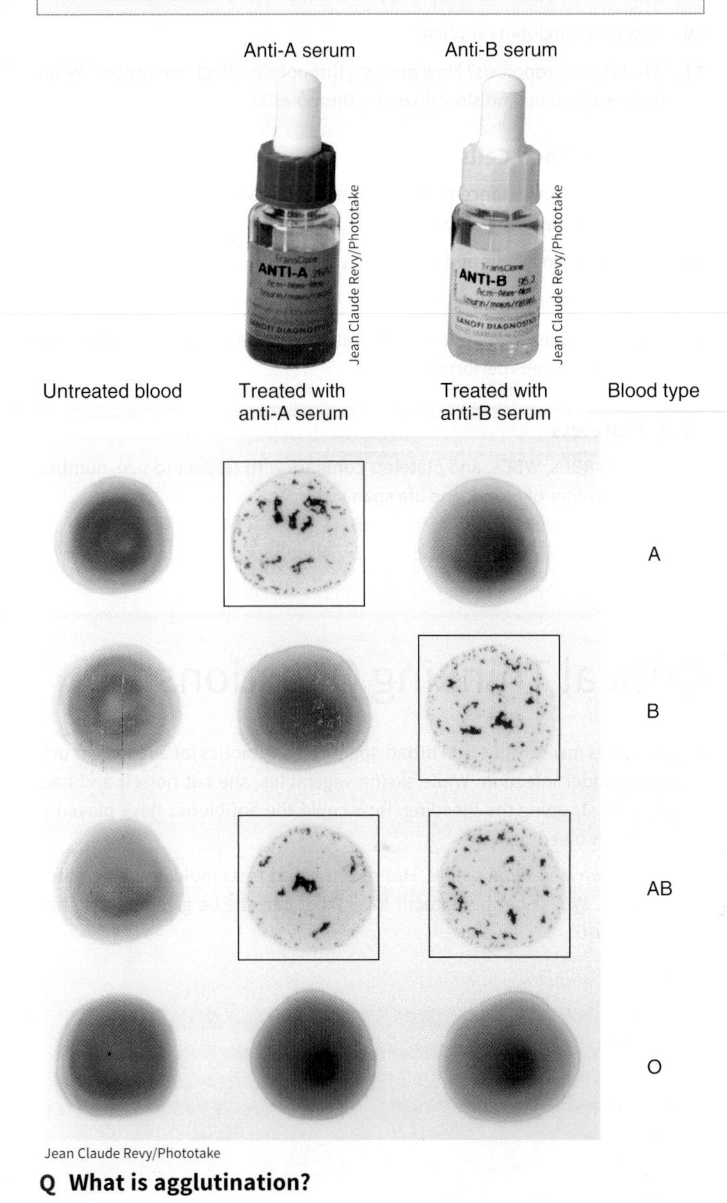

Jean Claude Revy/Phototake

Q What is agglutination?

Review Questions

19.1 Functions and Properties of Blood

1. In what ways is blood plasma similar to interstitial fluid? How does it differ?
2. What substances does blood transport?
3. How many kilograms or pounds of blood are there in your body?
4. How does the volume of blood plasma in your body compare to the volume of fluid in a 2-liter bottle of Coke?
5. List the formed elements in blood plasma and describe their functions.
6. What is the significance of lower-than-normal or higher-than-normal hematocrit?

19.2 Formation of Blood Cells

7. Which hemopoietic growth factor regulates differentiation and proliferation of red blood cell precursors?

8. Describe the formation of platelets from pluripotent stem cells, including the influence of hormones.

19.3 Red Blood Cells

9. Describe the size, microscopic appearance, and functions of RBCs.
10. How is hemoglobin recycled?
11. What is erythropoiesis? How does erythropoiesis affect hematocrit? What factors speed up and slow down erythropoiesis?

19.4 White Blood Cells

12. What is the importance of emigration, chemotaxis, and phagocytosis in fighting bacterial invaders?
13. How are leukocytosis and leukopenia different?
14. What is a differential white blood cell count?
15. What functions do granular leukocytes, macrophages, B cells, T cells, and natural killer cells perform?

19.5 Platelets

16. How do RBCs, WBCs, and platelets compare with respect to size, number per microliter of blood, and life span?

19.6 Stem Cell Transplants from Bone Marrow and Cord Blood

17. How are cord-blood transplants and bone marrow transplants similar? How do they differ?

19.7 Hemostasis

18. What is hemostasis?
19. How do vascular spasm and platelet plug formation occur?
20. What is fibrinolysis? Why does blood rarely remain clotted inside blood vessels?
21. How do the extrinsic and intrinsic pathways of blood clotting differ?
22. Define each of the following terms: anticoagulant, thrombus, embolus, and thrombolytic agent.

19.8 Blood Groups and Blood Types

23. What precautions must be taken before giving a blood transfusion?
24. What is hemolysis, and how can it occur after a mismatched blood transfusion?
25. Explain the conditions that may cause hemolytic disease of the newborn.

Critical Thinking Questions

1. Shilpa has recently been on broad-spectrum antibiotics for a recurrent urinary bladder infection. While slicing vegetables, she cut herself and had difficulty stopping the bleeding. How could the antibiotics have played a role in her bleeding?
2. Mrs. Brown is in renal failure. Her recent blood tests indicated a hematocrit of 22. Why is her hematocrit low? What can she be given to raise her hematocrit?
3. Thomas has hepatitis, which is disrupting his liver functions. What kinds of symptoms would he be experiencing based on the role(s) of the liver related to blood?

The Cardiovascular System: The Heart

CHAPTER **20**

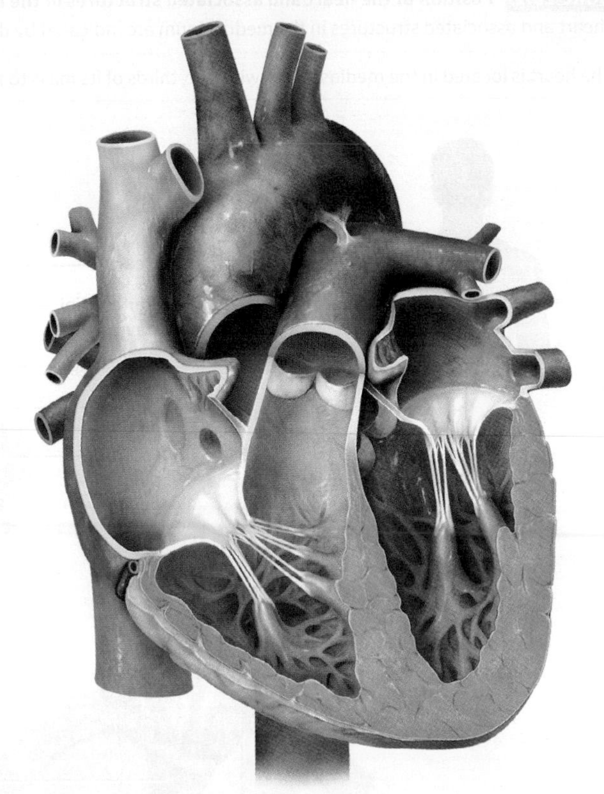

The Heart and Homeostasis

The heart contributes to homeostasis by pumping blood through blood vessels to the tissues of the body to deliver oxygen and nutrients and remove wastes.

As you learned in the previous chapter, the cardiovascular system consists of the blood, the heart, and blood vessels. We already examined the composition and functions of blood, and in this chapter you will learn about the pump that circulates it throughout the body—the heart. For blood to reach body cells and exchange materials with them, it must be pumped continuously by the heart through the body's blood vessels. The heart beats about 100,000 times every day, which adds up to about 35 million beats in a year, and approximately 2.5 billion times in an average lifetime. The left side of the heart pumps blood through an estimated 100,000 km (60,000 mi) of blood vessels, which is equivalent to traveling around the earth's equator about three times. The right side of the heart pumps blood through the lungs, enabling blood to pick up oxygen and unload carbon dioxide. Even while you are sleeping, your heart pumps 30 times its own weight each minute, which amounts to about 5 liters (5.3 qt) to the lungs and the same volume to the rest of the body. At this rate, your heart pumps more than about 14,000 liters (3600 gal) of blood in a day, or 5 million liters (1.3 million gal) in a year. You don't spend all of your time sleeping, however, and your heart pumps more vigorously when you are active. Thus, the actual blood volume your heart pumps in a single day is much larger. This chapter explores the structure of the heart and the unique properties that permit it to pump for a lifetime without rest.

Q Did you ever wonder about the difference between "good" and "bad" cholesterol?

20.1 Anatomy of the Heart

OBJECTIVES

- **Describe** the location of the heart.
- **Describe** the structure of the pericardium and the heart wall.
- **Discuss** the external and internal anatomy of the chambers of the heart.
- **Relate** the thickness of the chambers of the heart to their functions.

Location of the Heart

The scientific study of the normal heart and the diseases associated with it is known as **cardiology** (kar-dē-OL-ō-jē; *cardio-* = heart; *-logy* = study of).

For all its might, the **heart** is relatively small, roughly the same size (but not the same shape) as your closed fist. It is about 12 cm (5 in.) long, 9 cm (3.5 in.) wide at its broadest point, and 6 cm (2.5 in.) thick, with an average mass of 250 g (8 oz) in adult females and 300 g (10 oz) in adult males. The heart rests on the diaphragm, near the midline of the thoracic cavity. Recall that the midline is an imaginary vertical line that divides the body into unequal left and right sides. The heart lies in the **mediastinum** (mē′-dē-as-TĪ-num), an anatomical region that extends from the sternum to the vertebral column, from the first rib to the diaphragm, and between the lungs (**Figure 20.1a**). About two-thirds of the mass of the heart lies to the left of the body's midline (**Figure 20.1b**). You can visualize the heart as a cone lying on its side. The pointed **apex** is formed by the tip of the left ventricle (a lower chamber of the heart) and rests on the diaphragm. It is directed anteriorly, inferiorly, and to the left. The **base** of the heart is opposite the apex and is its posterior aspect. It is formed by the atria (upper chambers) of the heart, mostly the left atrium (see **Figure 20.3c**).

In addition to the apex and base, the heart has several distinct surfaces. The **anterior surface** is deep to the sternum and ribs. The

FIGURE 20.1 Position of the heart and associated structures in the mediastinum. The positions of the heart and associated structures in the mediastinum are indicated by dashed outlines.

The heart is located in the mediastinum, with two-thirds of its mass to the left of the midline.

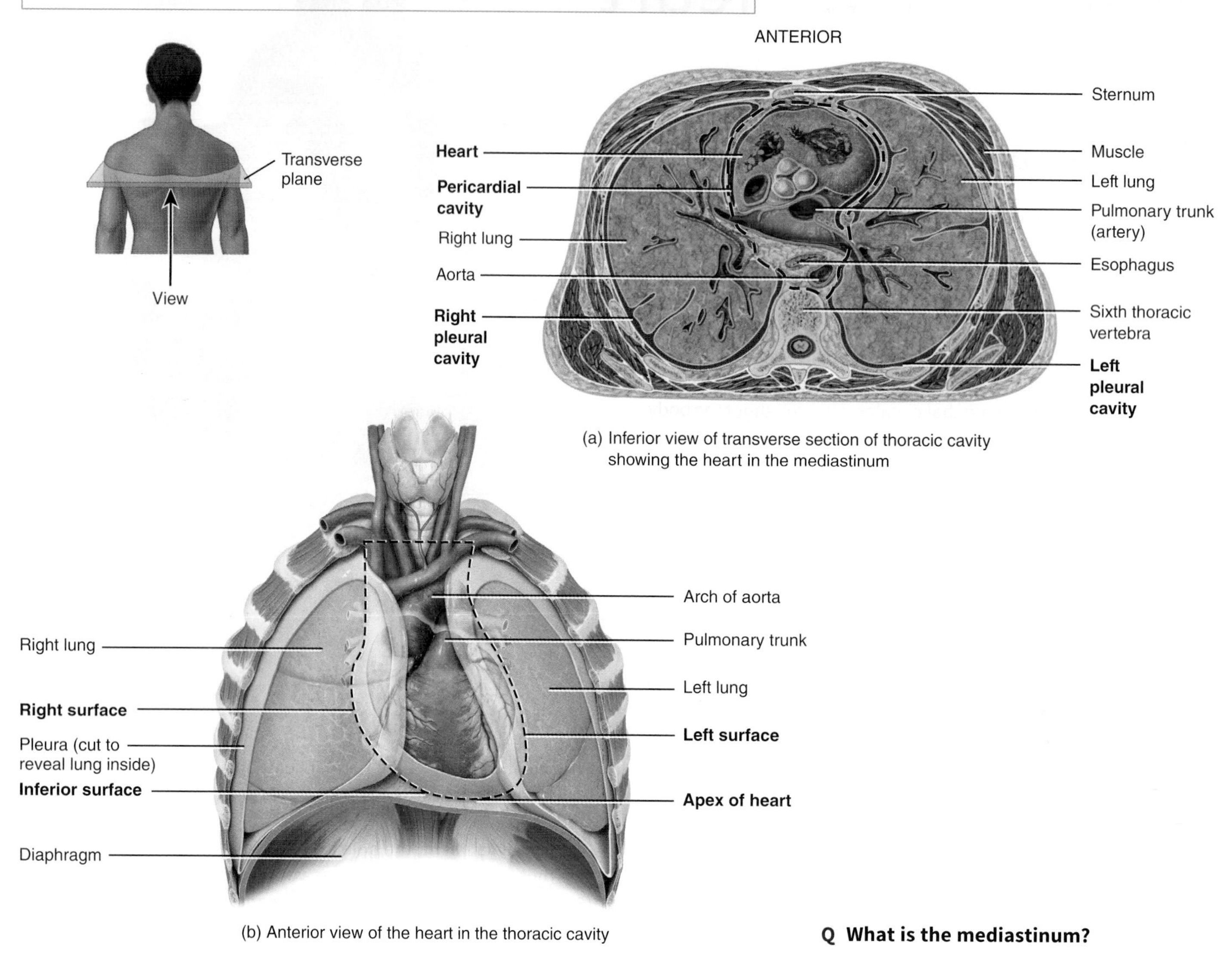

(a) Inferior view of transverse section of thoracic cavity showing the heart in the mediastinum

(b) Anterior view of the heart in the thoracic cavity

Q What is the mediastinum?

inferior surface is the part of the heart between the apex and right surface and rests mostly on the diaphragm (Figure 20.1b). The **right surface** faces the right lung and extends from the inferior surface to the base. The **left surface** faces the left lung and extends from the base to the apex.

Pericardium

The membrane that surrounds and protects the heart is the **pericardium** (per′-i-KAR-dē-um; *peri-* = around). It confines the heart to its position in the mediastinum, while allowing sufficient freedom of movement for vigorous and rapid contraction. The pericardium consists of two main parts: (1) the fibrous pericardium and (2) the serous pericardium (Figure 20.2a). The superficial **fibrous pericardium** is composed of tough, inelastic, dense irregular connective tissue. It resembles a bag that rests on and attaches to the diaphragm; its open end is fused to the connective tissues of the blood vessels entering and leaving the heart. The fibrous pericardium prevents overstretching of the heart, provides protection, and anchors the heart in the mediastinum. The fibrous pericardium near the apex of the heart is partially fused to the central tendon of the diaphragm and therefore movement of the diaphragm, as in deep breathing, facilitates the movement of blood by the heart.

See Clinical Connection: Cardiopulmonary Resuscitation

The deeper **serous pericardium** is a thinner, more delicate membrane that forms a double layer around the heart (Figure 20.2a). The outer **parietal layer of the serous pericardium** is fused to the fibrous pericardium. The inner **visceral layer of the serous pericardium**, which is also called the **epicardium** (ep′-i-KAR-dē-um;

epi- = on top of), is one of the layers of the heart wall and adheres tightly to the surface of the heart. Between the parietal and visceral layers of the serous pericardium is a thin film of lubricating serous fluid. This slippery secretion of the pericardial cells, known as **pericardial fluid**, reduces friction between the layers of the serous pericardium as the heart moves. The space that contains the few milliliters of pericardial fluid is called the **pericardial cavity**.

See Clinical Connection: Pericarditis

Layers of the Heart Wall

The wall of the heart consists of three layers (**Figure 20.2a**): the epicardium (external layer), the myocardium (middle layer), and the endocardium (inner layer). The **epicardium** is composed of two tissue layers. The outermost, as you just learned, is called the *visceral layer of the serous pericardium.* This thin, transparent outer layer of the heart wall is composed of mesothelium. Beneath the mesothelium is a variable layer of delicate fibroelastic tissue and adipose tissue. The adipose tissue predominates and becomes thickest over the ventricular surfaces, where it houses the major coronary and cardiac vessels of the heart. The amount of fat varies from person to person, corresponds to the general extent of body fat in an individual, and typically increases with age. The epicardium imparts a smooth, slippery texture to the outermost surface of the heart. The epicardium contains blood vessels, lymphatics, and vessels that supply the myocardium.

The middle **myocardium** (mī′-ō-KAR-dē-um; *myo-* = muscle) is responsible for the pumping action of the heart and is composed of cardiac muscle tissue. It makes up approximately 95% of the heart wall. The muscle fibers (cells), like those of striated skeletal muscle tissue, are wrapped and bundled with connective tissue sheaths composed of endomysium and perimysium. The cardiac muscle fibers are organized in bundles that

FIGURE 20.2 Pericardium and heart wall.

The pericardium is a triple-layered sac that surrounds and protects the heart.

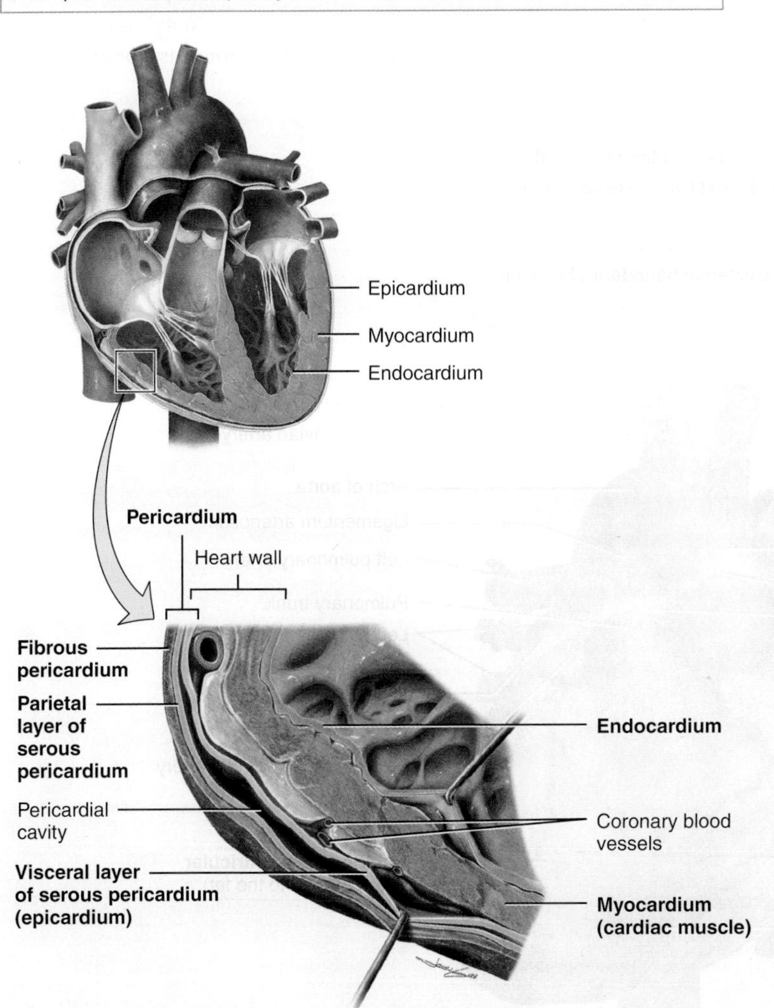

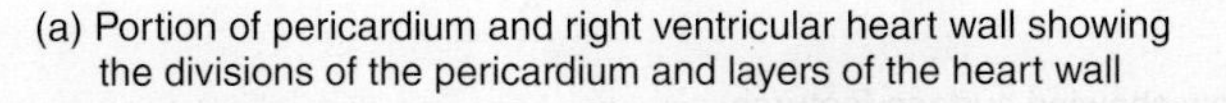
(a) Portion of pericardium and right ventricular heart wall showing the divisions of the pericardium and layers of the heart wall

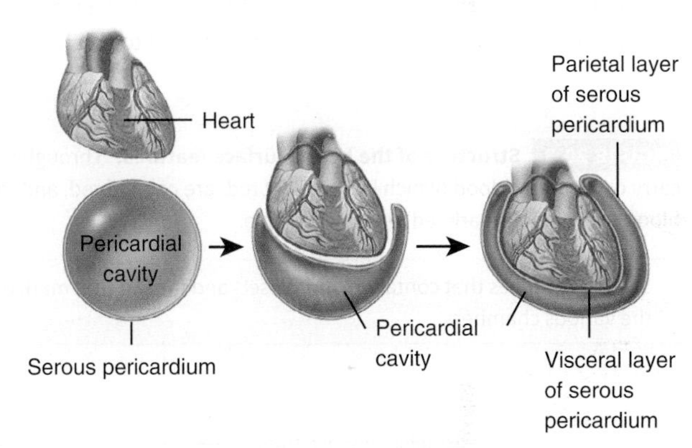

(b) Simplified relationship of the serous pericardium to the heart

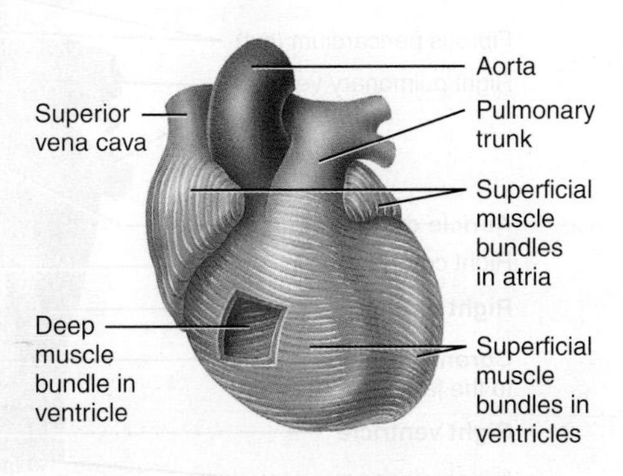

(c) Cardiac muscle bundles of the myocardium

Q Which layer is both a part of the pericardium and a part of the heart wall?

swirl diagonally around the heart and generate the strong pumping actions of the heart (Figure 20.2c). Although it is striated like skeletal muscle, recall that cardiac muscle is involuntary like smooth muscle.

The innermost **endocardium** (en′-dō-KAR-dē-um; *endo-* = within) is a thin layer of endothelium overlying a thin layer of connective tissue. It provides a smooth lining for the chambers of the heart and covers the valves of the heart. The smooth endothelial lining minimizes the surface friction as blood passes through the heart. The endocardium is continuous with the endothelial lining of the large blood vessels attached to the heart.

See Clinical Connection: Myocarditis and Endocarditis

Chambers of the Heart

The heart has four chambers. The two superior receiving chambers are the **atria** (= entry halls or chambers), and the two inferior pumping chambers are the **ventricles** (= little bellies). The paired atria receive blood from blood vessels returning blood to the heart, called veins, while the ventricles eject the blood from the heart into blood vessels called arteries. On the anterior surface of each atrium is a wrinkled pouchlike structure called an **auricle** (OR-i-kul; *auri-* = ear), so named because of its resemblance to a dog's ear (Figure 20.3). Each auricle slightly increases the capacity of an atrium so that it can hold a greater volume of blood. Also on the surface of the heart are a series of grooves, called **sulci** (SUL-sī), that contain coronary blood vessels and a variable amount of fat. Each *sulcus* (SUL-kus; singular) marks the external boundary between two chambers of the heart. The deep **coronary sulcus** (*coron-* = resembling a crown) encircles most of the heart and marks the external boundary between the superior atria and inferior ventricles. The **anterior interventricular sulcus** (in′-ter-ven-TRIK-ū-lar) is a shallow groove on the anterior surface of the heart that marks the external boundary between the right and left ventricles on the anterior aspect of the heart. This sulcus continues around to the posterior surface of the heart as the **posterior interventricular sulcus**, which marks the external boundary between the ventricles on the posterior aspect of the heart (Figure 20.3c).

Right Atrium

The **right atrium** forms the right surface of the heart and receives blood from three veins: the *superior vena cava*, *inferior vena cava*, and *coronary sinus* (Figure 20.4a). (Veins always carry blood toward the heart.) The right atrium is about 2–3 mm

FIGURE 20.3 Structure of the heart: surface features. Throughout this book, blood vessels that carry oxygenated blood (which looks bright red) are colored red, and those that carry deoxygenated blood (which looks dark red) are colored blue.

Sulci are grooves that contain blood vessels and fat and that mark the external boundaries between the various chambers.

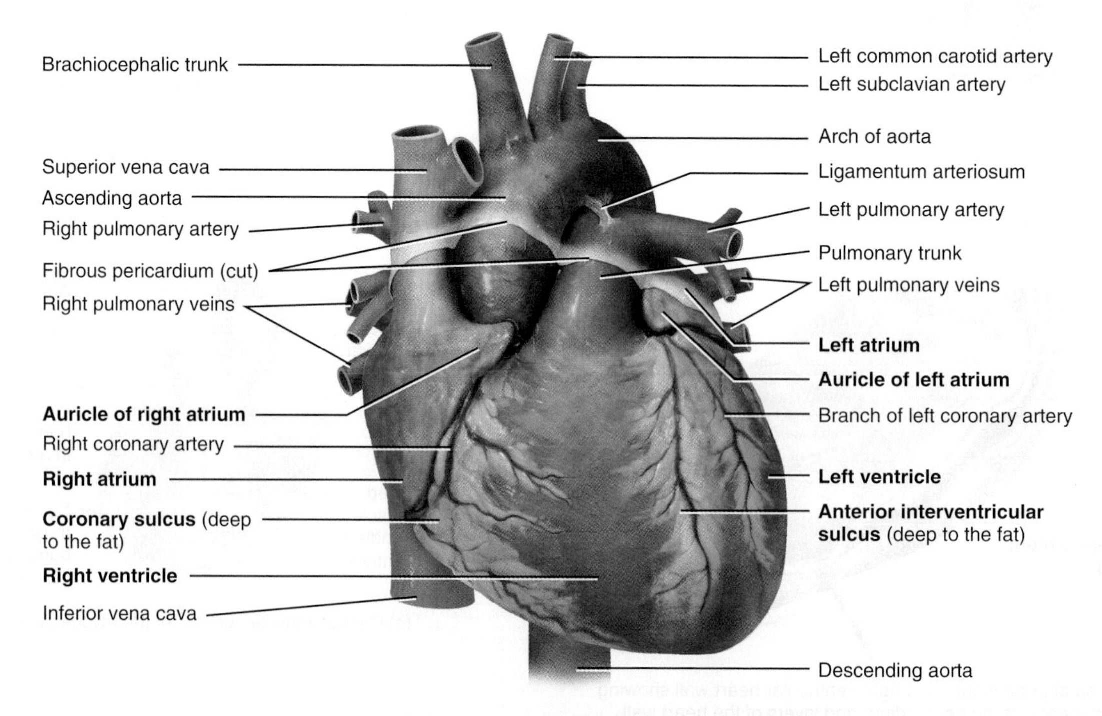

(a) Anterior external view showing surface features

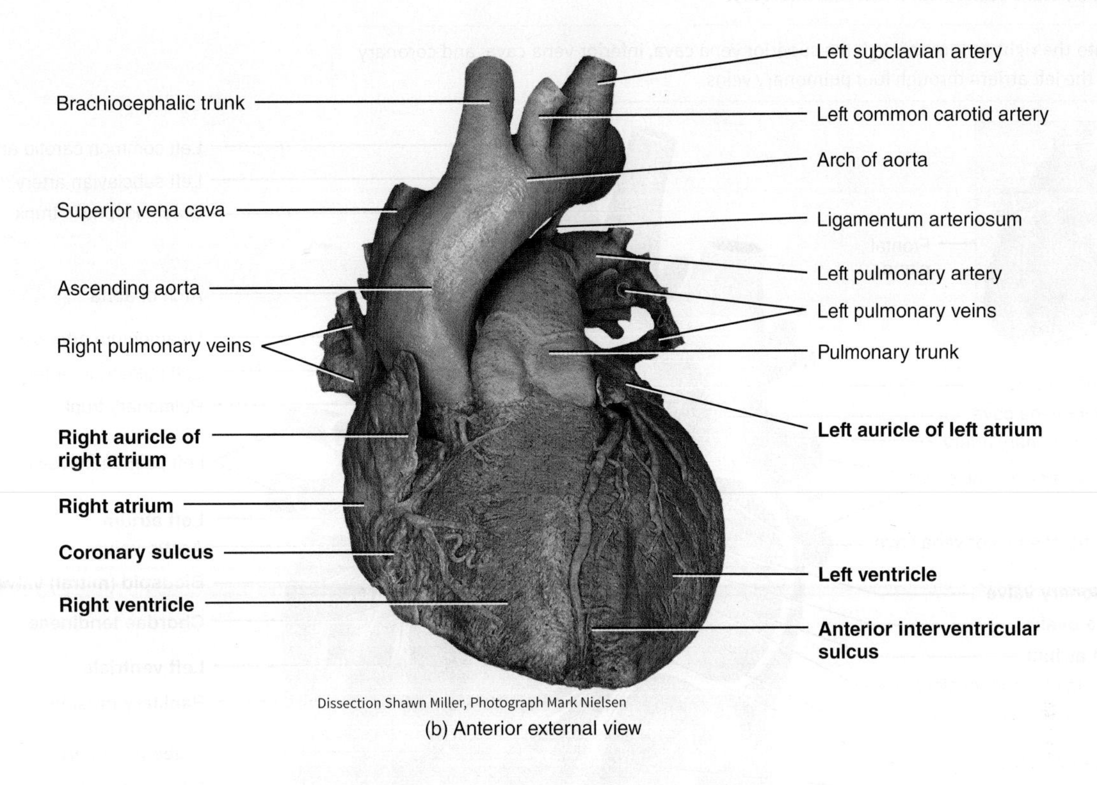

(b) Anterior external view

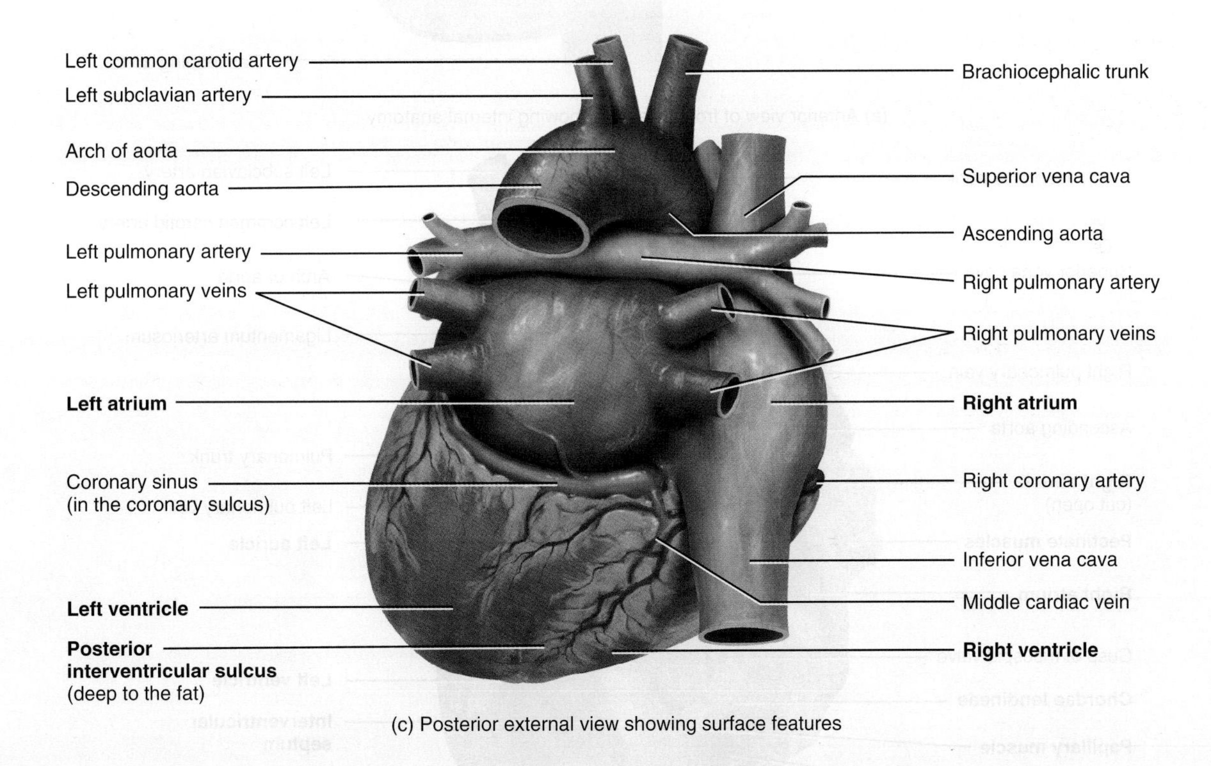

(c) Posterior external view showing surface features

Q The coronary sulcus forms an external boundary between which chambers of the heart?

FIGURE 20.4 **Structure of the heart: internal anatomy.**

Blood flows into the right atrium through the superior vena cava, inferior vena cava, and coronary sinus and into the left atrium through four pulmonary veins.

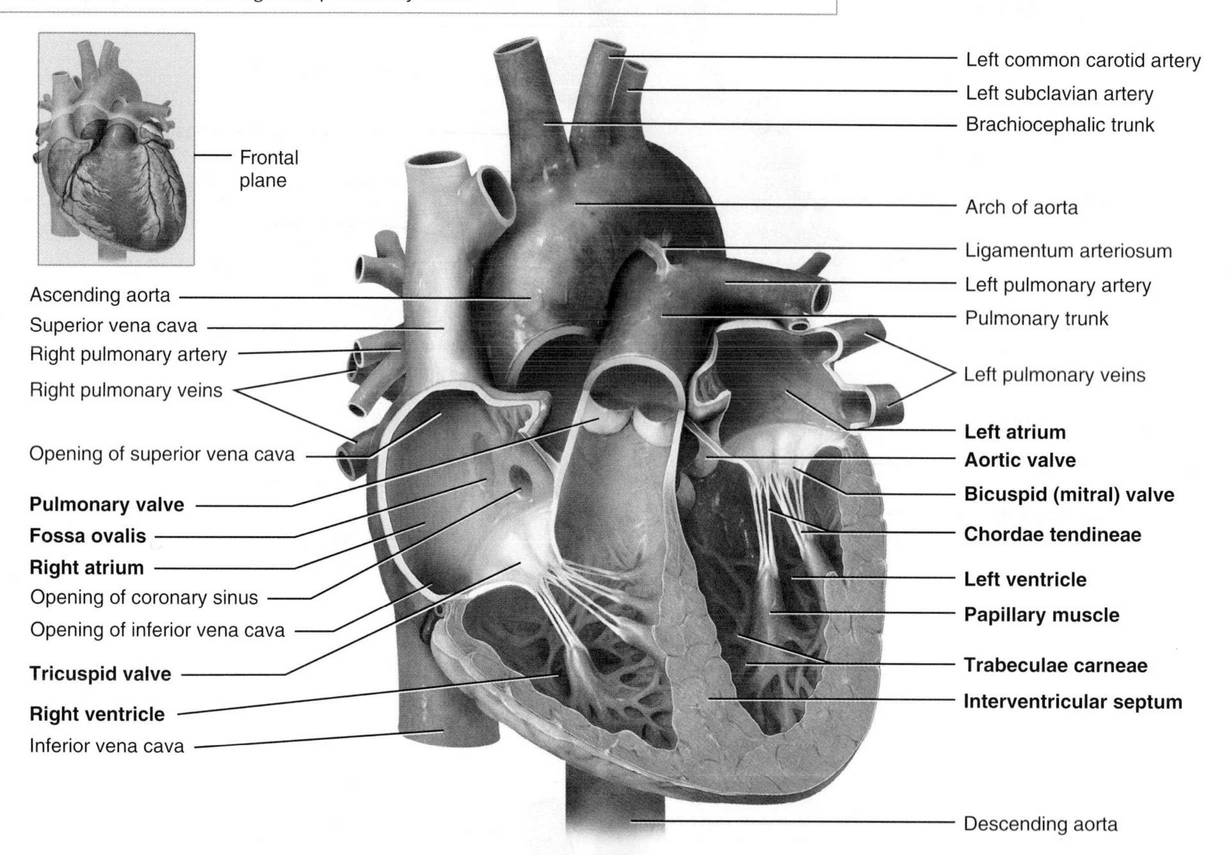

(a) Anterior view of frontal section showing internal anatomy

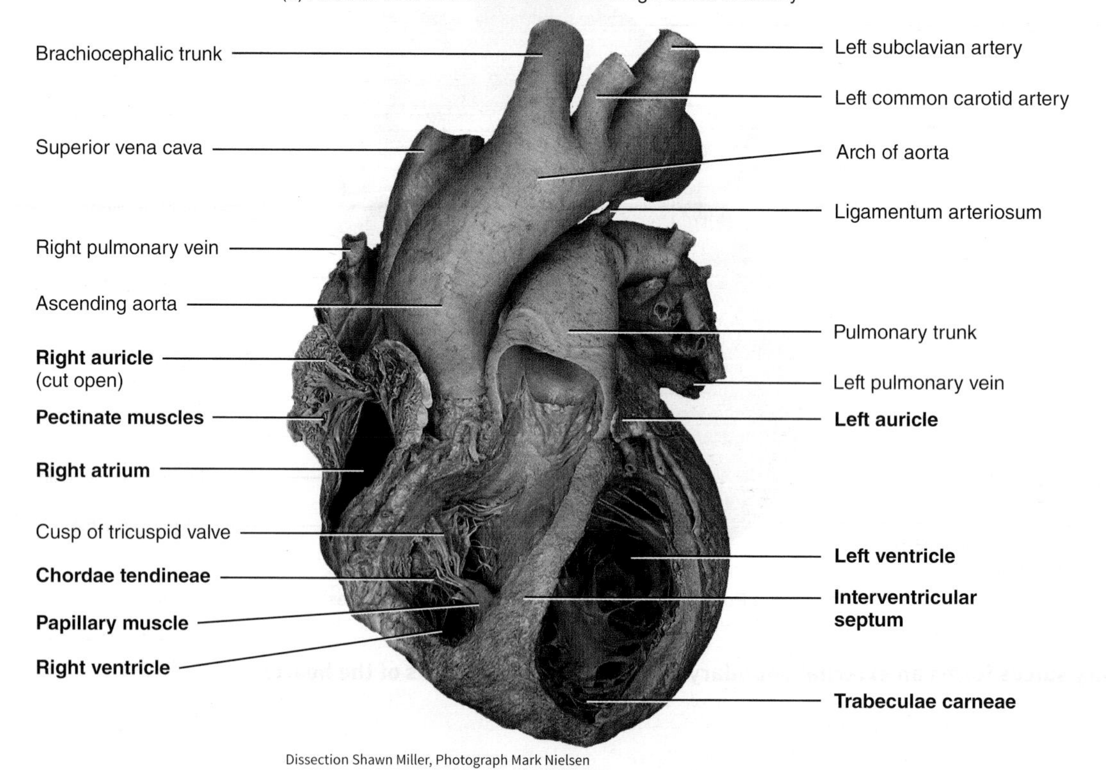

Dissection Shawn Miller, Photograph Mark Nielsen

(b) Anterior view of partially sectioned heart

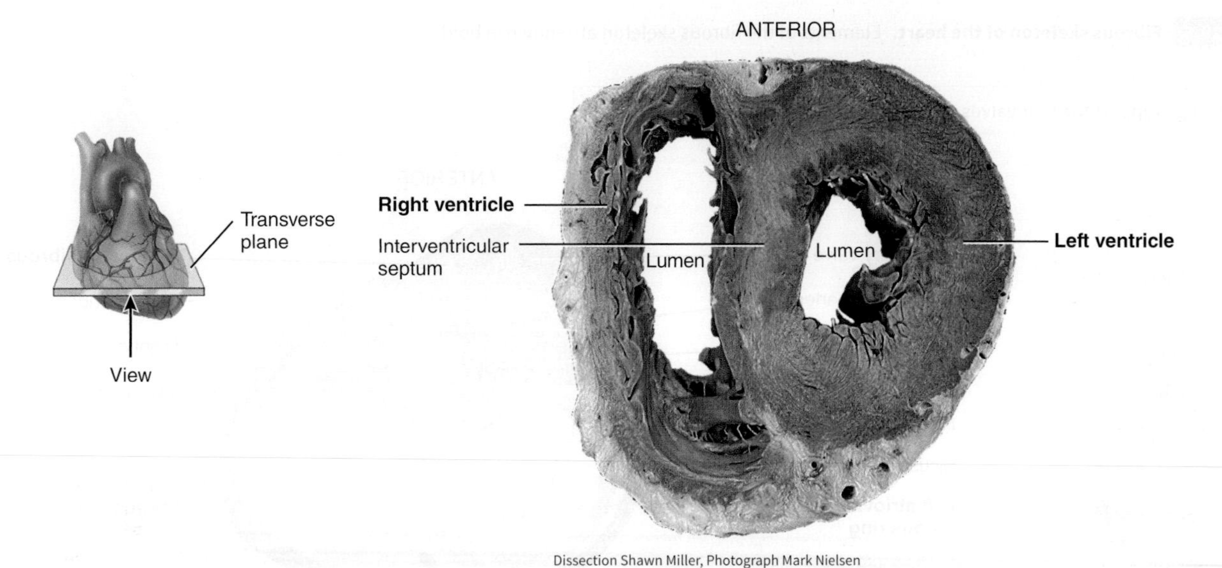

(c) Inferior view of transverse section showing differences in thickness of ventricular walls

Q How does thickness of the myocardium relate to the workload of a cardiac chamber?

(0.08–0.12 in.) in average thickness. The anterior and posterior walls of the right atrium are very different. The inside of the posterior wall is smooth; the inside of the anterior wall is rough due to the presence of muscular ridges called **pectinate muscles** (PEK-ti-nāt; *pectin* = comb), which also extend into the auricle (**Figure 20.4b**). Between the right atrium and left atrium is a thin partition called the **interatrial septum** (*inter-* = between; *septum* = a dividing wall or partition). A prominent feature of this septum is an oval depression called the **fossa ovalis**, the remnant of the *foramen ovale*, an opening in the interatrial septum of the fetal heart that normally closes soon after birth (see **Figure 21.31**). Blood passes from the right atrium into the right ventricle through a valve that is called the **tricuspid valve** (trē-KUS-pid; *tri-* = three; *-cuspid* = point) because it consists of three **cusps** or *leaflets* (**Figure 20.4a**). It is also called the *right atrioventricular valve* (ā′-trē-ō-ven-TRIK-ū-lar). The valves of the heart are composed of dense connective tissue covered by endocardium.

Right Ventricle The **right ventricle** is about 4–5 mm (0.16–0.2 in.) in average thickness and forms most of the anterior surface of the heart. The inside of the right ventricle contains a series of ridges formed by raised bundles of cardiac muscle fibers called **trabeculae carneae** (tra-BEK-ū-lē KAR-nē-ē; *trabeculae* = little beams; *carneae* = fleshy; see **Figure 20.2a**). Some of the trabeculae carneae convey part of the conduction system of the heart, which you will learn about later in this chapter (see Section 20.3). The cusps of the tricuspid valve are connected to tendonlike cords, the **chordae tendineae** (KOR-dē ten-DIN-ē-ē; *chord-* = cord; *tend-* = tendon), which in turn are connected to cone-shaped trabeculae carneae called **papillary muscles** (*papill-* = nipple). Internally, the right ventricle is separated from the left ventricle by a partition called the **interventricular septum**. Blood passes from the right ventricle through the **pulmonary valve** (*pulmonary semilunar valve*) into a large artery called the *pulmonary trunk*, which divides into right and left *pulmonary arteries* and carries blood to the lungs. Arteries always take blood away from the heart (a mnemonic to help you: artery = away).

Left Atrium The **left atrium** is about the same thickness as the right atrium and forms most of the base of the heart (**Figure 20.4a**). It receives blood from the lungs through four *pulmonary veins.* Like the right atrium, the inside of the left atrium has a smooth posterior wall. Because pectinate muscles are confined to the auricle of the left atrium, the anterior wall of the left atrium also is smooth. Blood passes from the left atrium into the left ventricle through the **bicuspid** (*mitral*) **valve** (*bi-* = two), which, as its name implies, has two cusps. The term *mitral* refers to the resemblance of the bicuspid valve to a bishop's miter (hat), which is two-sided. It is also called the *left atrioventricular valve*.

Left Ventricle The **left ventricle** is the thickest chamber of the heart, averaging 10–15 mm (0.4–0.6 in.), and forms the apex of the heart (see **Figure 20.1b**). Like the right ventricle, the left ventricle contains trabeculae carneae and has chordae tendineae that anchor the cusps of the bicuspid valve to papillary muscles. Blood passes from the left ventricle through the **aortic valve** *(aortic semilunar valve)* into the *ascending aorta* (*aorte* = to suspend, because the aorta once was believed to lift up the heart). Some of the blood in the aorta flows into the *coronary arteries*, which branch from the ascending aorta and carry blood to the heart wall. The remainder of the blood passes into the *arch of the aorta* and *descending aorta (thoracic aorta* and *abdominal aorta)*. Branches of the arch of the aorta and descending aorta carry blood throughout the body.

During fetal life, a temporary blood vessel, called the *ductus arteriosus*, shunts blood from the pulmonary trunk into the aorta. Hence, only a small amount of blood enters the nonfunctioning fetal lungs (see **Figure 21.31**). The ductus arteriosus normally closes shortly

FIGURE 20.5 **Fibrous skeleton of the heart.** Elements of the fibrous skeleton are shown in bold letters.

Fibrous rings support the four valves of the heart and are fused to one another.

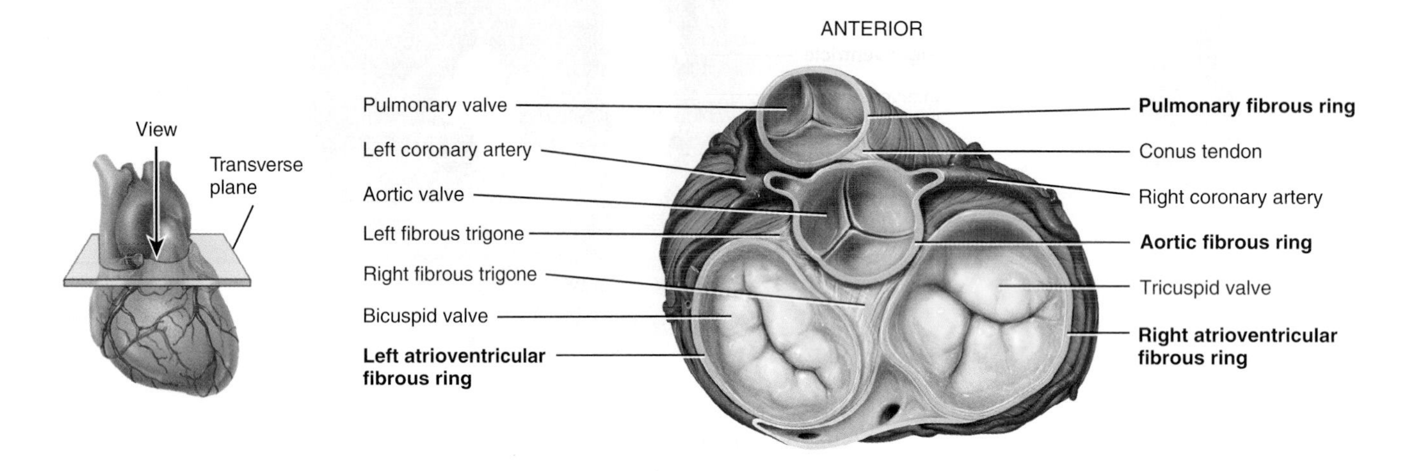

Superior view (the atria have been removed)

Q How does the fibrous skeleton contribute to the functioning of heart valves?

after birth, leaving a remnant known as the **ligamentum arteriosum** (lig′-a-MEN-tum ar-ter-ē-Ō-sum), which connects the arch of the aorta and pulmonary trunk (**Figure 20.4a**).

Myocardial Thickness and Function

The thickness of the myocardium of the four chambers varies according to each chamber's function. The thin-walled atria deliver blood under less pressure into the adjacent ventricles. Because the ventricles pump blood under higher pressure over greater distances, their walls are thicker (**Figure 20.4a**). Although the right and left ventricles act as two separate pumps that simultaneously eject equal volumes of blood, the right side has a much smaller workload. It pumps blood a short distance to the lungs at lower pressure, and the resistance to blood flow is small. The left ventricle pumps blood great distances to all other parts of the body at higher pressure, and the resistance to blood flow is larger. Therefore, the left ventricle works much harder than the right ventricle to maintain the same rate of blood flow. The anatomy of the two ventricles confirms this functional difference—the muscular wall of the left ventricle is considerably thicker than the wall of the right ventricle (**Figure 20.4c**). Note also that the perimeter of the lumen (space) of the left ventricle is roughly circular, in contrast to that of the right ventricle, which is somewhat crescent-shaped.

Fibrous Skeleton of the Heart

In addition to cardiac muscle tissue, the heart wall also contains dense connective tissue that forms the **fibrous skeleton of the heart** (**Figure 20.5**). Essentially, the fibrous skeleton consists of four dense connective tissue rings that surround the valves of the heart, fuse with one another, and merge with the interventricular septum. In addition to forming a structural foundation for the heart valves, the fibrous skeleton prevents overstretching of the valves as blood passes through them. It also serves as a point of insertion for bundles of cardiac muscle fibers and acts as an electrical insulator between the atria and ventricles.

20.2 Heart Valves and Circulation of Blood

OBJECTIVES

- **Describe** the structure and function of the valves of the heart.
- **Outline** the flow of blood through the chambers of the heart and through the systemic and pulmonary circulations.
- **Discuss** the coronary circulation.

As each chamber of the heart contracts, it pushes a volume of blood into a ventricle or out of the heart into an artery. Valves open and close in response to *pressure changes* as the heart contracts and relaxes. Each of the four valves helps ensure the one-way flow of blood by opening to let blood through and then closing to prevent its backflow.

Operation of the Atrioventricular Valves

Because they are located between an atrium and a ventricle, the tricuspid and bicuspid valves are termed **atrioventricular (AV) valves**. When an AV valve is open, the rounded ends of the cusps project into the ventricle. When the ventricles are relaxed, the papillary muscles are relaxed, the chordae tendineae are slack, and blood moves from a higher pressure in the atria to a lower pressure in the ventricles through open AV valves (**Figure 20.6a, d**). When the ventricles contract, the pressure of the blood drives the cusps upward until their edges meet and close the opening (**Figure 20.6b, e**). At the same time, the papillary muscles contract, which pulls on and tightens the chordae tendineae. This prevents the valve cusps from everting (opening into the atria) in response to the high ventricular pressure. If the AV valves or chordae tendineae are damaged, blood may regurgitate (flow back) into the atria when the ventricles contract.

Operation of the Semilunar Valves

The aortic and pulmonary valves are known as the **semilunar (SL) valves** (sem-ē-LOO-nar; *semi-* = half; *-lunar* = moon-shaped) because they are made up of three crescent moon–shaped cusps (**Figure 20.6d**). Each cusp attaches to the arterial wall by its convex outer margin. The SL valves allow ejection of blood from the heart into arteries but prevent backflow of blood into the ventricles. The free borders of the cusps project into the lumen of the artery. When the ventricles contract, pressure builds up within the chambers. The semilunar valves open when pressure in the ventricles exceeds the pressure in the arteries, permitting ejection of blood from the ventricles into the pulmonary trunk and aorta (**Figure 20.6e**). As the ventricles relax, blood starts to flow back toward the heart. This backflowing blood fills the valve cusps, which causes the free edges of the semilunar valves to contact each other tightly and close the opening between the ventricle and artery (**Figure 20.6d**).

FIGURE 20.6 Responses of the valves to the pumping of the heart.

Heart valves prevent the backflow of blood.

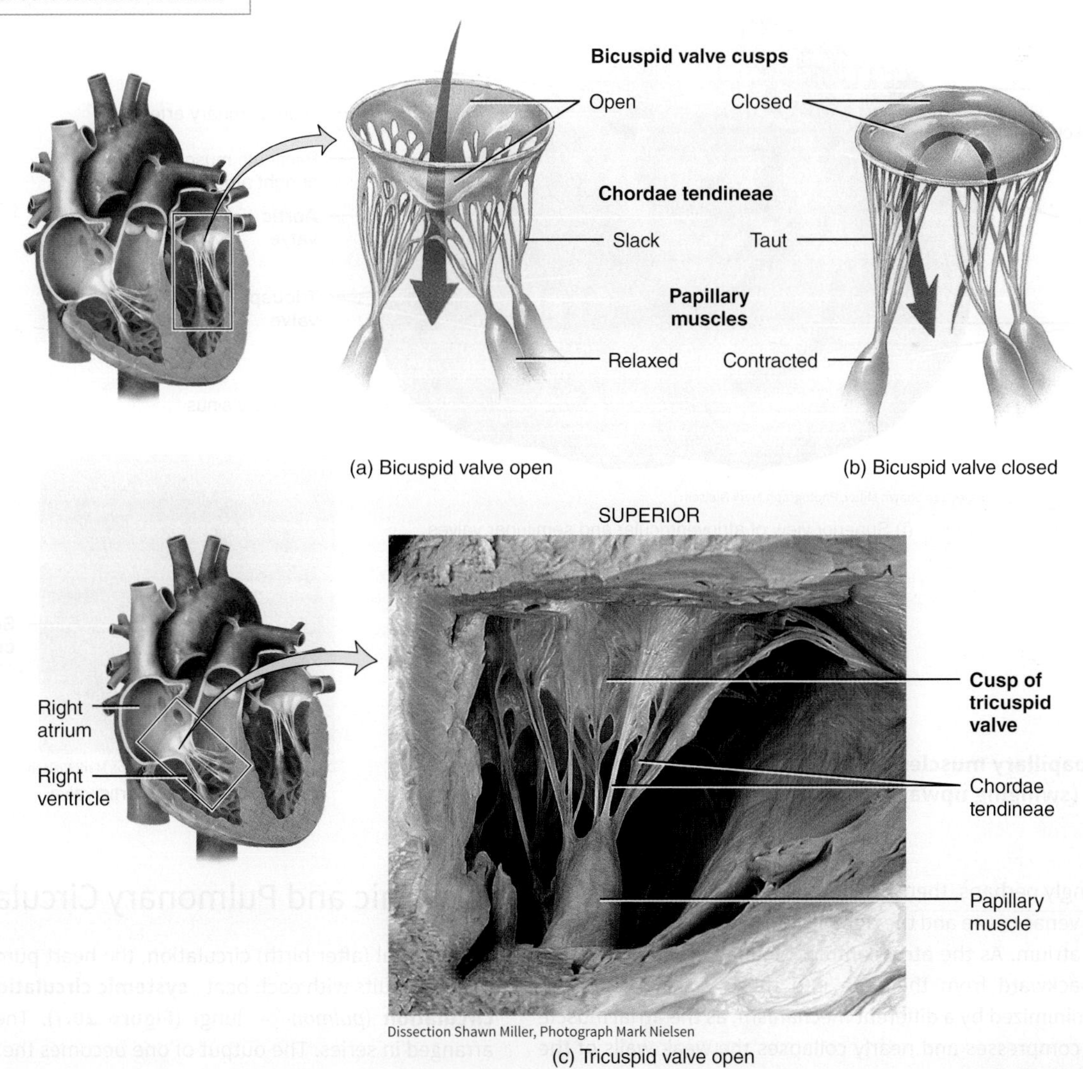

Figure 20.6 ***Continues***

FIGURE 20.6 Continued

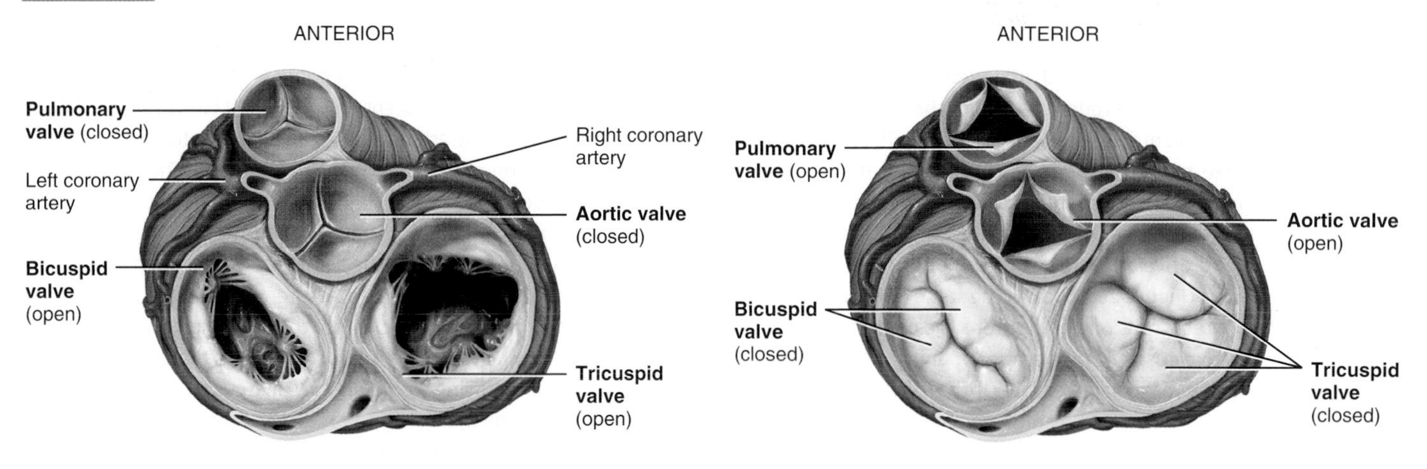

(d) Superior view with atria removed: pulmonary and aortic valves closed, bicuspid and tricuspid valves open

(e) Superior view with atria removed: pulmonary and aortic valves open, bicuspid and tricuspid valves closed

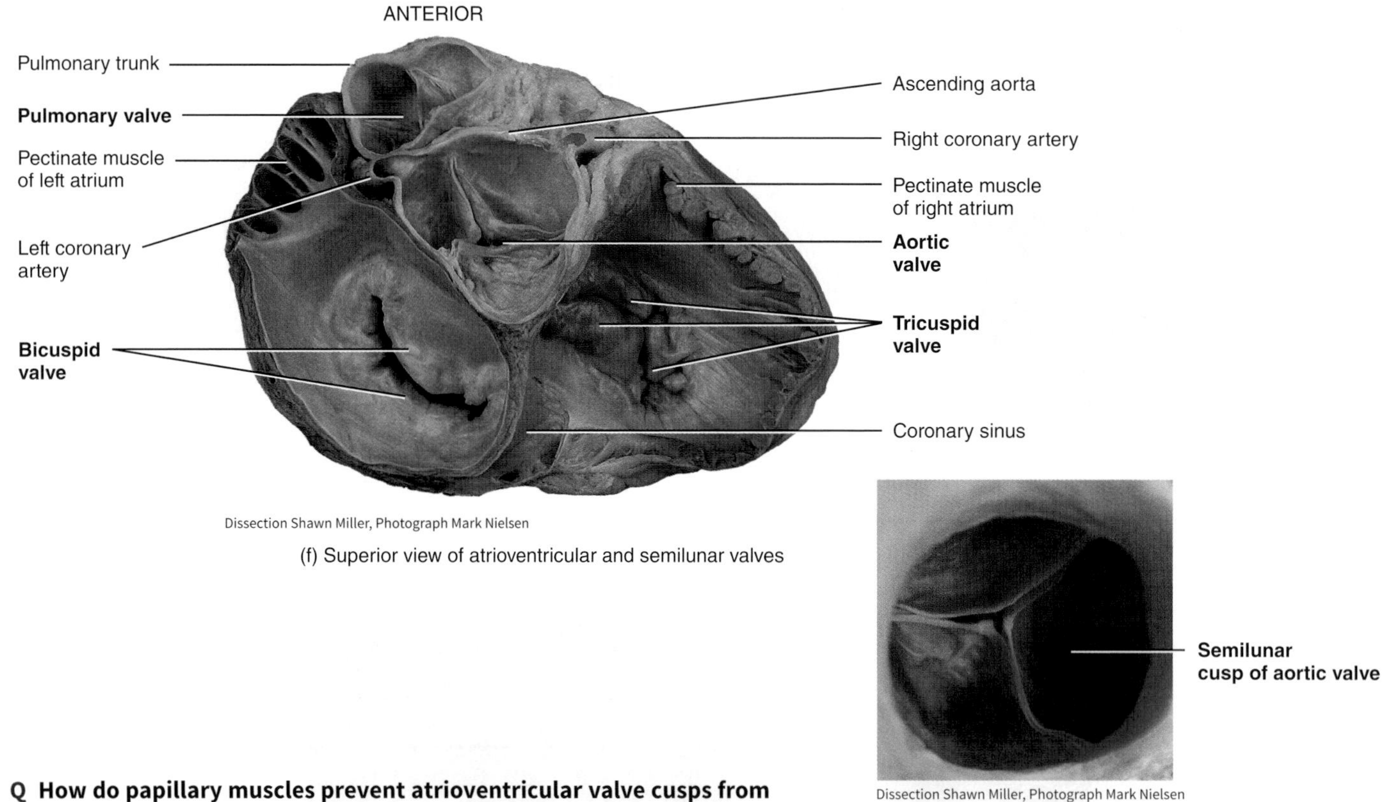

Dissection Shawn Miller, Photograph Mark Nielsen

(f) Superior view of atrioventricular and semilunar valves

Dissection Shawn Miller, Photograph Mark Nielsen

(g) Superior view of aortic valve

Q How do papillary muscles prevent atrioventricular valve cusps from everting (swinging upward) into the atria?

Surprisingly perhaps, there are no valves guarding the junctions between the venae cavae and the right atrium or the pulmonary veins and the left atrium. As the atria contract, a small amount of blood does flow backward from the atria into these vessels. However, backflow is minimized by a different mechanism; as the atrial muscle contracts, it compresses and nearly collapses the weak walls of the venous entry points.

Systemic and Pulmonary Circulations

In postnatal (after birth) circulation, the heart pumps blood into two closed circuits with each beat—**systemic circulation** and **pulmonary circulation** (*pulmon-* = lung) (**Figure 20.7**). The two circuits are arranged in series: The output of one becomes the input of the other, as would happen if you attached two garden hoses (see **Figure 21.17**).

FIGURE 20.7 Systemic and pulmonary circulations.

The left side of the heart pumps oxygenated blood into the systemic circulation to all tissues of the body except the air sacs (alveoli) of the lungs. The right side of the heart pumps deoxygenated blood into the pulmonary circulation to the air sacs.

(a) Path of blood flow through heart

(b) Path of blood flow through systemic and pulmonary circulation

Q Which numbers constitute the pulmonary circulation? Which constitute the systemic circulation?

The left side of the heart is the pump for systemic circulation; it receives bright red *oxygenated* (oxygen-rich) *blood* from the lungs. The left ventricle ejects blood into the *aorta* (**Figure 20.7**). From the aorta, the blood divides into separate streams, entering progressively smaller *systemic arteries* that carry it to all organs throughout the body—except for the air sacs (alveoli) of the lungs, which are supplied by the pulmonary circulation. In systemic tissues, arteries give rise to smaller-diameter *arterioles*, which finally lead into extensive beds of *systemic capillaries.* Exchange of nutrients and gases occurs across the thin capillary walls. Blood unloads O_2 (oxygen) and picks up CO_2 (carbon dioxide). In most cases, blood flows through only one capillary and then enters a *systemic venule.* Venules carry *deoxygenated* (oxygen-poor) *blood* away from tissues and merge to form larger *systemic veins.* Ultimately the blood flows back to the right atrium.

See Clinical Connection: Heart Valve Disorders

The right side of the heart is the pump for pulmonary circulation; it receives all of the dark-red deoxygenated blood returning from the systemic circulation. Blood ejected from the right ventricle flows into the *pulmonary trunk*, which branches into *pulmonary arteries* that carry blood to the right and left lungs. In pulmonary capillaries, blood unloads CO_2, which is exhaled, and picks up O_2 from inhaled air. The freshly oxygenated blood then flows into pulmonary veins and returns to the left atrium.

Coronary Circulation

Nutrients are not able to diffuse quickly enough from blood in the chambers of the heart to supply all layers of cells that make up the heart wall. For this reason, the myocardium has its own network of blood vessels, the **coronary circulation** or *cardiac circulation* (*coron-* = crown). The **coronary arteries** branch from the ascending aorta and encircle the heart like a crown encircles the head (Figure 20.8a). While the heart is contracting, little blood flows in the coronary arteries because they are squeezed shut. When the heart relaxes, however, the high pressure of blood in the aorta propels blood through the coronary arteries, into capillaries, and then into **coronary veins** (Figure 20.8b).

Coronary Arteries

Two coronary arteries, the left and right coronary arteries, branch from the ascending aorta and supply oxygenated blood to the myocardium (Figure 20.8a). The **left coronary artery** passes inferior to the left auricle and divides into the anterior interventricular and circumflex branches. The **anterior interventricular branch** or *left anterior descending (LAD) artery* is in the anterior interventricular sulcus and supplies oxygenated blood to the walls of both ventricles. The **circumflex branch** (SER-kum-fleks) lies in the coronary sulcus and distributes oxygenated blood to the walls of the left ventricle and left atrium.

The **right coronary artery** supplies small branches *(atrial branches)* to the right atrium. It continues inferior to the right auricle and ultimately divides into the posterior interventricular and marginal branches. The **posterior interventricular branch** follows the posterior interventricular sulcus and supplies the walls of the two ventricles with oxygenated blood. The **marginal branch** beyond the coronary sulcus runs along the right margin of the heart and transports oxygenated blood to the wall of the right ventricle.

Most parts of the body receive blood from branches of more than one artery, and where two or more arteries supply the same region, they usually connect. These connections, called **anastomoses** (a-nas′-tō-MŌ-sēs), provide alternate routes, called **collateral circulation**, for blood to reach a particular organ or tissue. The myocardium contains many anastomoses that connect branches of a given coronary artery or extend between branches of different coronary arteries. They provide detours for arterial blood if a main route becomes obstructed. This is important because the heart muscle may receive sufficient oxygen even if one of its coronary arteries is partially blocked.

Coronary Veins

After blood passes through the arteries of the coronary circulation, it flows into capillaries, where it delivers oxygen and nutrients to the heart muscle and collects carbon dioxide and waste, and then moves into coronary veins. Most of the deoxygenated blood from the myocardium drains into a large *vascular sinus* in the coronary sulcus on the posterior surface of the heart, called the **coronary sinus** (Figure 20.8b). (A *vascular sinus* is a thin-walled vein that has no smooth muscle to alter its diameter.) The deoxygenated blood in the coronary sinus empties into the right atrium. The principal tributaries carrying blood into the coronary sinus are the following:

- **Great cardiac vein** in the anterior interventricular sulcus, which drains the areas of the heart supplied by the left coronary artery (left and right ventricles and left atrium)
- **Middle cardiac vein** in the posterior interventricular sulcus, which drains the areas supplied by the posterior interventricular branch of the right coronary artery (left and right ventricles)
- **Small cardiac vein** in the coronary sulcus, which drains the right atrium and right ventricle
- **Anterior cardiac veins**, which drain the right ventricle and open directly into the right atrium

When blockage of a coronary artery deprives the heart muscle of oxygen, **reperfusion** (re′-per-FYŪ-zhun), the reestablishment of blood flow, may damage the tissue further. This surprising effect is due to the formation of oxygen **free radicals** from the reintroduced oxygen. As you learned in Chapter 2, free radicals are molecules that have an unpaired electron (see Figure 2.3b). These unstable, highly reactive molecules cause chain reactions that lead to cellular damage and death. To counter the effects of oxygen free radicals, body cells produce enzymes that convert free radicals to less reactive substances. Two such enzymes are *superoxide dismutase* (dis-MŪ-tās) and *catalase* (KAT-a-lās). In addition, nutrients such as vitamin E, vitamin C, beta-carotene, zinc, and selenium serve as antioxidants, which remove oxygen free radicals from circulation. Drugs that lessen reperfusion damage after a heart attack or stroke are currently under development.

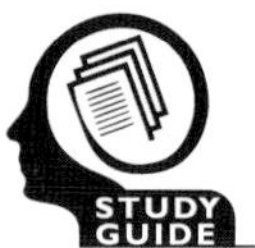

See Clinical Connection: Myocardial Ischemia and Infarction

20.3 Cardiac Muscle Tissue and the Cardiac Conduction System

OBJECTIVES

- **Describe** the structural and functional characteristics of cardiac muscle tissue and the cardiac conduction system.
- **Explain** how an action potential occurs in cardiac contractile fibers.
- **Describe** the electrical events of a normal electrocardiogram (ECG).

Histology of Cardiac Muscle Tissue

Compared with skeletal muscle fibers, cardiac muscle fibers are shorter in length and less circular in transverse section (Figure 20.9).

FIGURE 20.8 The coronary circulation. The views of the heart from the anterior aspect in (a) and (b) are drawn as if the heart were transparent to reveal blood vessels on the posterior aspect.

The left and right coronary arteries deliver blood to the heart; the coronary veins drain blood from the heart into the coronary sinus.

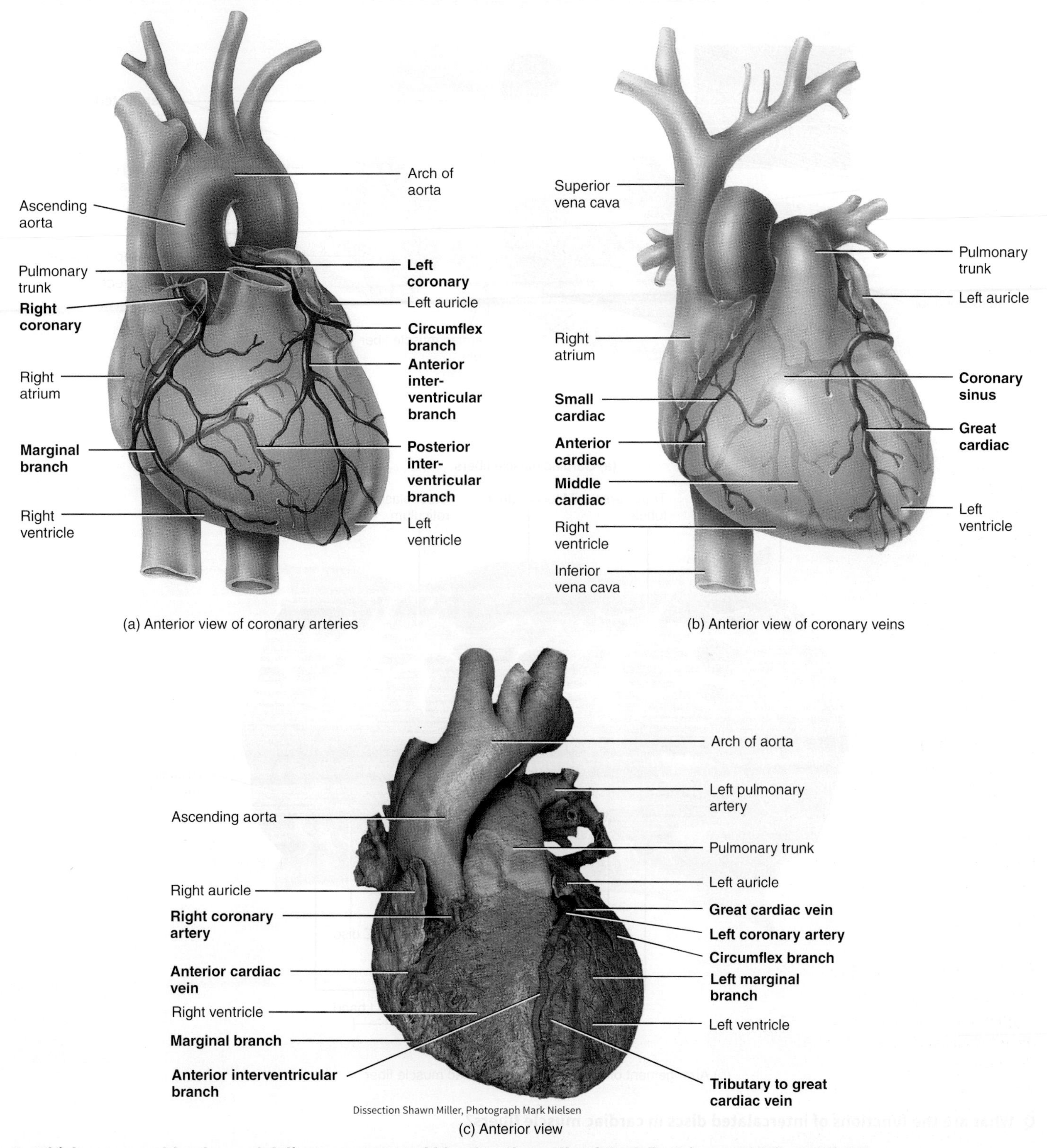

(a) Anterior view of coronary arteries

(b) Anterior view of coronary veins

(c) Anterior view

Q Which coronary blood vessel delivers oxygenated blood to the walls of the left atrium and left ventricle?

FIGURE 20.9 Histology of cardiac muscle tissue. (See Table 4.9 for a light micrograph of cardiac muscle.)

Cardiac muscle fibers connect to neighboring fibers by intercalated discs, which contain desmosomes and gap junctions.

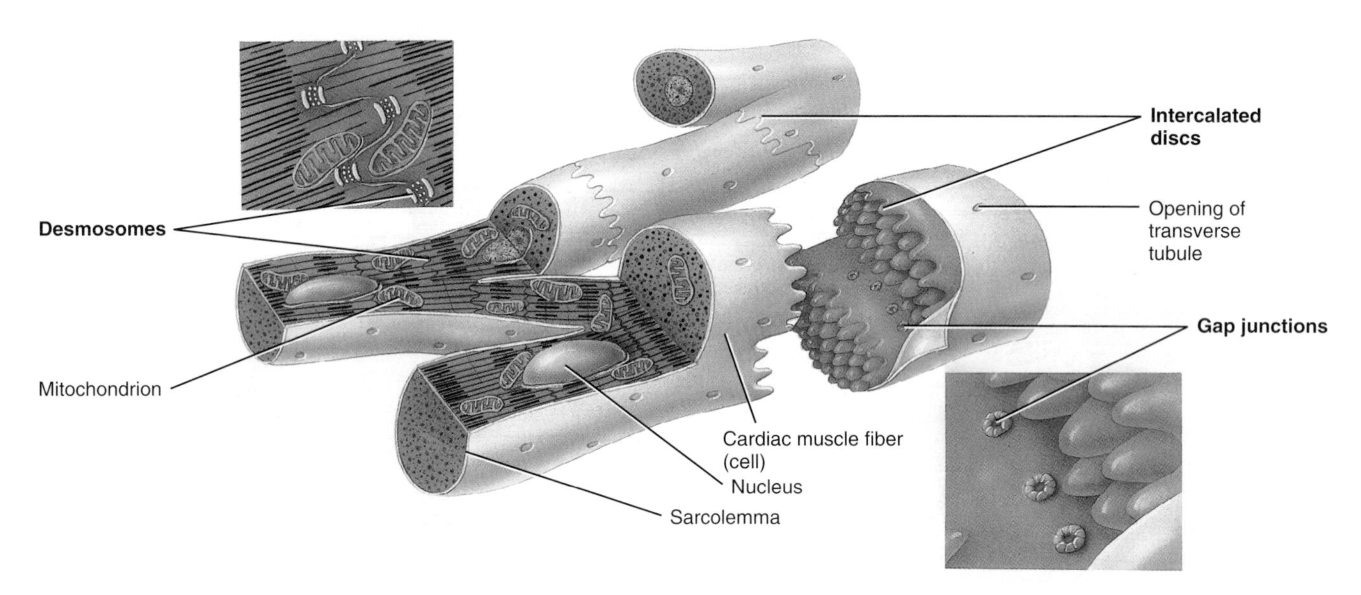

(a) Cardiac muscle fibers

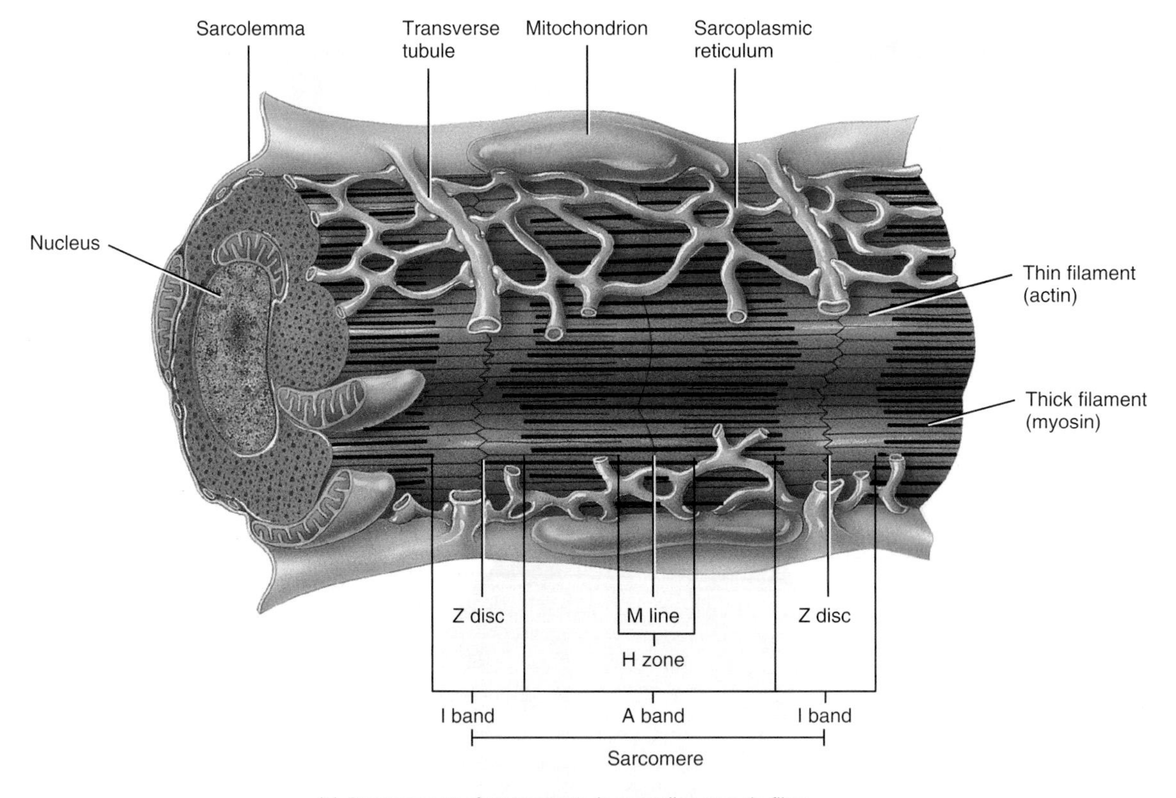

(b) Arrangement of components in a cardiac muscle fiber

Q What are the functions of intercalated discs in cardiac muscle fibers?

They also exhibit branching, which gives individual cardiac muscle fibers a "stair-step" appearance (see Table 4.9). A typical cardiac muscle fiber is 50–100 μm long and has a diameter of about 14 μm. Usually one centrally located nucleus is present, although an occasional cell may have two nuclei. The ends of cardiac muscle fibers connect to neighboring fibers by irregular transverse thickenings of the sarcolemma called **intercalated discs** (in-TER-ka-lāt-ed; *intercalat-* = to insert between). The discs contain **desmosomes**, which hold the fibers together, and **gap junctions**, which allow muscle action potentials to conduct from one muscle fiber to its neighbors. Gap junctions allow the entire myocardium of the atria or the ventricles to contract as a single, coordinated unit.

Mitochondria are larger and more numerous in cardiac muscle fibers than in skeletal muscle fibers. In a cardiac muscle fiber, they take up 25% of the cytosolic space; in a skeletal muscle fiber only 2% of the cytosolic space is occupied by mitochondria. Cardiac muscle fibers have the same arrangement of actin and myosin, and the same bands, zones, and Z discs, as skeletal muscle fibers. The transverse tubules of cardiac muscle are wider but less abundant than those of skeletal muscle; the one transverse tubule per sarcomere is located at the Z disc. The sarcoplasmic reticulum of cardiac muscle fibers is somewhat smaller than the SR of skeletal muscle fibers. As a result, cardiac muscle has a smaller intracellular reserve of Ca^{2+}.

See Clinical Connection: Regeneration of Heart Cells

Autorhythmic Fibers: The Conduction System

An inherent and rhythmical electrical activity is the reason for the heart's lifelong beat. The source of this electrical activity is a network of specialized cardiac muscle fibers called **autorhythmic fibers** (aw′-tō-RITH-mik; *auto-* = self) because they are self-excitable. Autorhythmic fibers repeatedly generate action potentials that trigger heart contractions. They continue to stimulate a heart to beat even after it is removed from the body—for example, to be transplanted into another person—and all of its nerves have been cut. (Note: Surgeons do not attempt to reattach heart nerves during heart transplant operations. For this reason, it has been said that heart surgeons are better "plumbers" than they are "electricians.")

During embryonic development, only about 1% of the cardiac muscle fibers become autorhythmic fibers; these relatively rare fibers have two important functions:

1. They act as a **pacemaker**, setting the rhythm of electrical excitation that causes contraction of the heart.
2. They form the **cardiac conduction system**, a network of specialized cardiac muscle fibers that provide a path for each cycle of cardiac excitation to progress through the heart. The conduction system ensures that cardiac chambers become stimulated to contract in a coordinated manner, which makes the heart an effective pump. As you will see later in the chapter, problems with autorhythmic fibers can result in arrhythmias (abnormal rhythms) in which the heart beats irregularly, too fast, or too slow.

Cardiac action potentials propagate through the conduction system in the following sequence (Figure 20.10a):

1. Cardiac excitation normally begins in the **sinoatrial (SA) node**, located in the right atrial wall just inferior and lateral to the opening of the superior vena cava. SA node cells do not have a stable resting potential. Rather, they repeatedly depolarize to threshold spontaneously. The spontaneous depolarization is a **pacemaker potential**. When the pacemaker potential reaches threshold, it triggers an action potential (Figure 20.10b). Each action potential from the SA node propagates throughout both atria via gap junctions in the intercalated discs of atrial muscle fibers. Following the action potential, the two atria contract at the same time.
2. By conducting along atrial muscle fibers, the action potential reaches the **atrioventricular (AV) node**, located in the interatrial septum, just anterior to the opening of the coronary sinus (Figure 20.10a). At the AV node, the action potential slows considerably as a result of various differences in cell structure in the AV node. This delay provides time for the atria to empty their blood into the ventricles.
3. From the AV node, the action potential enters the **atrioventricular (AV) bundle** (also known as the *bundle of His*, pronounced HIZ). This bundle is the only site where action potentials can conduct from the atria to the ventricles. (Elsewhere, the fibrous skeleton of the heart electrically insulates the atria from the ventricles.)
4. After propagating through the AV bundle, the action potential enters both the **right** and **left bundle branches**. The bundle branches extend through the interventricular septum toward the apex of the heart.
5. Finally, the large-diameter **Purkinje fibers** (pur-KIN-jē) rapidly conduct the action potential beginning at the apex of the heart upward to the remainder of the ventricular myocardium. Then the ventricles contract, pushing the blood upward toward the semilunar valves.

On their own, autorhythmic fibers in the SA node would initiate an action potential about every 0.6 second, or 100 times per minute. Thus, the SA node sets the rhythm for contraction of the heart—it is the **natural pacemaker**. This rate is faster than that of any other autorhythmic fibers. Because action potentials from the SA node spread through the conduction system and stimulate other areas before the other areas are able to generate an action potential at their own, slower rate, the SA node acts as the natural pacemaker of the heart. Nerve impulses from the autonomic nervous system (ANS) and blood-borne hormones (such as epinephrine) *modify the timing and strength* of each heartbeat, but they *do not establish the fundamental rhythm.* In a person at rest, for example, acetylcholine released by the parasympathetic division of the ANS slows SA node pacing to about every 0.8 second or 75 action potentials per minute (Figure 20.10b).

FIGURE 20.10 The conduction system of the heart. Autorhythmic fibers in the SA node, located in the right atrial wall (a), act as the heart's pacemaker, initiating cardiac action potentials (b) that cause contraction of the heart's chambers.

The conduction system ensures that the chambers of the heart contract in a coordinated manner.

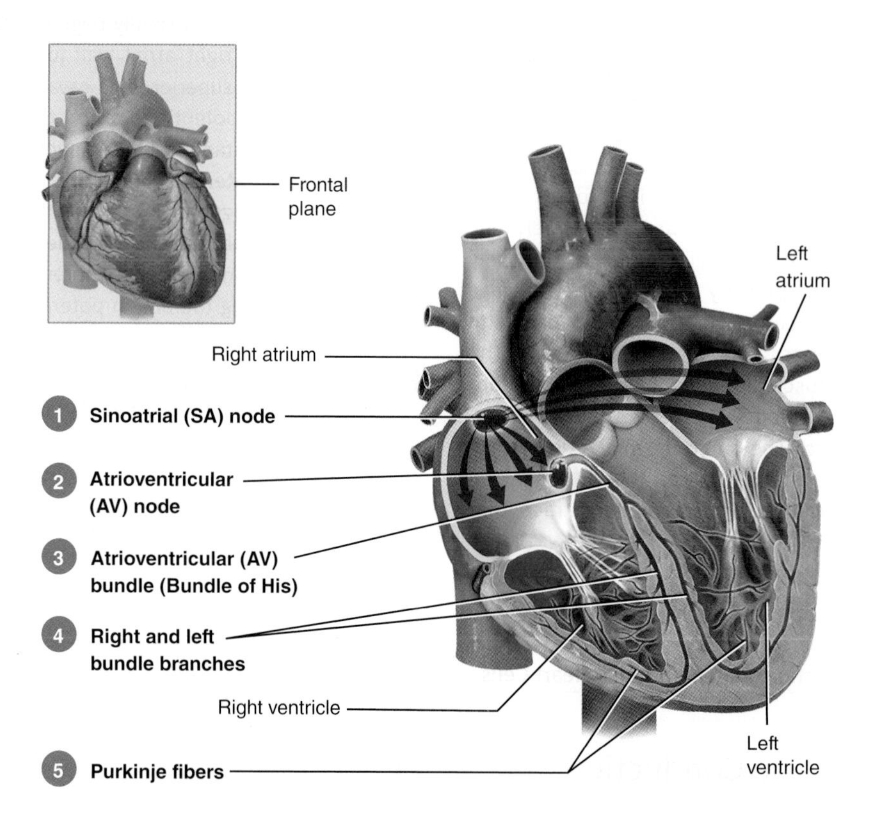

(a) Anterior view of frontal section

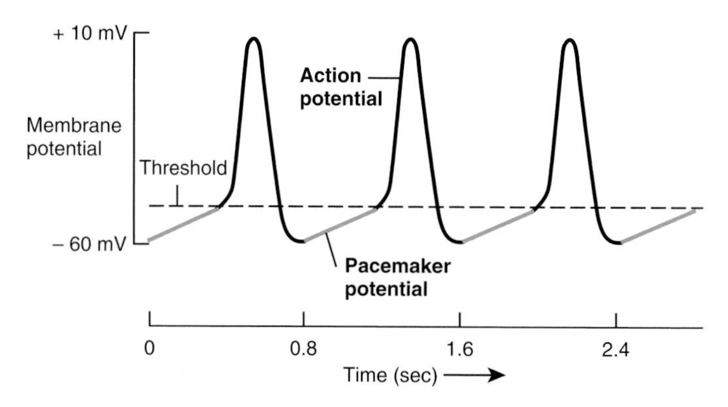

(b) Pacemaker potentials (green) and action potentials (black) in autorhythmic fibers of SA node

Q Which component of the conduction system provides the only electrical connection between the atria and the ventricles?

See Clinical Connection: Artificial Pacemakers

Action Potential and Contraction of Contractile Fibers

The action potential initiated by the SA node travels along the conduction system and spreads out to excite the "working" atrial and ventricular muscle fibers, called **contractile fibers**. An action potential occurs in a contractile fiber as follows (**Figure 20.11**):

1. ***Depolarization.*** Unlike autorhythmic fibers, contractile fibers have a stable resting membrane potential that is close to −90 mV. When a contractile fiber is brought to threshold by an action potential from neighboring fibers, its **voltage-gated fast Na^+ channels** open. These sodium ion channels are referred to as "fast" because they open very rapidly in response to a threshold-level depolarization. Opening of these channels allows Na^+ inflow because the cytosol of contractile fibers is electrically more negative than interstitial fluid and Na^+ concentration is higher in interstitial fluid. Inflow of Na^+ down the electrochemical gradient produces a **rapid depolarization** (dē′-pō-lar-i-ZĀ-shun). Within a few milliseconds, the fast Na^+ channels automatically inactivate and Na^+ inflow decreases.
2. ***Plateau.*** The next phase of an action potential in a contractile fiber is the **plateau**, a period of maintained depolarization. It is due in part to opening of **voltage-gated slow Ca^{2+} channels** in the sarcolemma. When these channels open, calcium ions move from the interstitial fluid (which has a higher Ca^{2+} concentration) into the cytosol. This inflow of Ca^{2+} causes even more Ca^{2+} to pour out of the sarcoplasmic reticulum into the cytosol through additional Ca^{2+} channels in the sarcoplasmic reticulum membrane. The increased Ca^{2+} concentration in the cytosol ultimately triggers contraction. Several different types of **voltage-gated K^+ channels** are also found in the sarcolemma of a contractile fiber. Just before the plateau phase begins, some of these K^+ channels open, allowing potassium ions to leave the contractile fiber. Therefore, depolarization is sustained during the plateau phase because Ca^{2+} inflow just balances K^+ outflow. The plateau phase lasts for about 0.2 sec, and the membrane potential of the contractile fiber is close to 0 mV. By comparison, depolarization in a neuron or skeletal muscle fiber is much briefer, about 1 msec (0.001 sec), because it lacks a plateau phase.
3. ***Repolarization.*** The recovery of the resting membrane potential during the **repolarization** (rē′-pō-lar-i-ZĀ-shun) phase of a cardiac

FIGURE 20.11 **Action potential in a ventricular contractile fiber.** The resting membrane potential is about −90 mV.

A long refractory period prevents tetanus in cardiac muscle fibers.

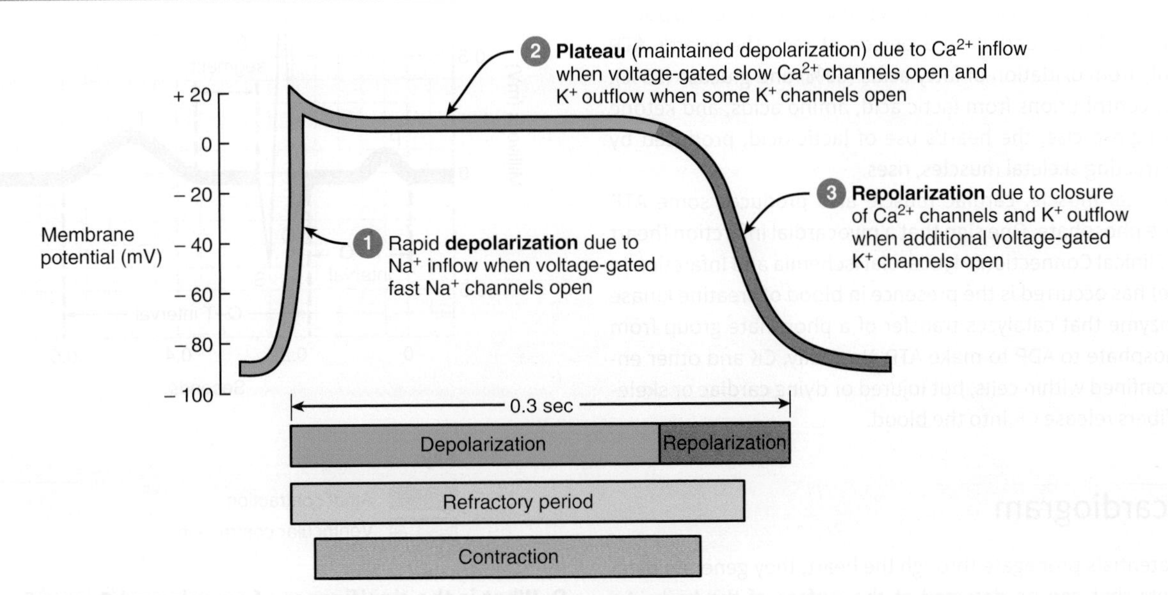

Q How does the duration of an action potential in a ventricular contractile fiber compare with that in a skeletal muscle fiber?

action potential resembles that in other excitable cells. After a delay (which is particularly prolonged in cardiac muscle), additional voltage-gated K^+ channels open. Outflow of K^+ restores the negative resting membrane potential (−90 mV). At the same time, the calcium channels in the sarcolemma and the sarcoplasmic reticulum are closing, which also contributes to repolarization.

The mechanism of contraction is similar in cardiac and skeletal muscle: The electrical activity (action potential) leads to the mechanical response (contraction) after a short delay. As Ca^{2+} concentration rises inside a contractile fiber, Ca^{2+} binds to the regulatory protein troponin, which allows the actin and myosin filaments to begin sliding past one another, and tension starts to develop. Substances that alter the movement of Ca^{2+} through slow Ca^{2+} channels influence the strength of heart contractions. Epinephrine, for example, increases contraction force by enhancing Ca^{2+} flow into the cytosol.

In muscle, the **refractory period** (re-FRAK-to-rē) is the time interval during which a second contraction cannot be triggered. The refractory period of a cardiac muscle fiber lasts longer than the contraction itself (**Figure 20.11**). As a result, another contraction cannot begin until relaxation is well under way. For this reason, tetanus (maintained contraction) cannot occur in cardiac muscle as it can in skeletal muscle. The advantage is apparent if you consider how the ventricles work. Their pumping function depends on alternating contraction (when they eject blood) and relaxation (when they refill). If heart muscle could undergo tetanus, blood flow would cease.

ATP Production in Cardiac Muscle

In contrast to skeletal muscle, cardiac muscle produces little of the ATP it needs by anaerobic cellular respiration (see **Figure 10.11**). Instead, it relies almost exclusively on aerobic cellular respiration in its numerous mitochondria. The needed oxygen diffuses from blood in the coronary circulation and is released from myoglobin inside cardiac muscle fibers. Cardiac muscle fibers use several fuels to power mitochondrial ATP production. In a person at rest, the heart's ATP comes mainly from oxidation of fatty acids (60%) and glucose (35%), with smaller contributions from lactic acid, amino acids, and ketone bodies. During exercise, the heart's use of lactic acid, produced by actively contracting skeletal muscles, rises.

Like skeletal muscle, cardiac muscle also produces some ATP from creatine phosphate. One sign that a myocardial infarction (heart attack; see Clinical Connection: Myocardial Ischemia and Infarction in Study Guide) has occurred is the presence in blood of creatine kinase (CK), the enzyme that catalyzes transfer of a phosphate group from creatine phosphate to ADP to make ATP. Normally, CK and other enzymes are confined within cells, but injured or dying cardiac or skeletal muscle fibers release CK into the blood.

Electrocardiogram

As action potentials propagate through the heart, they generate electrical currents that can be detected at the surface of the body. An **electrocardiogram** (e-lek′-trō-KAR-dē-ō-gram), abbreviated either ECG or EKG (from the German word *Elektrokardiogram*), is a recording of these electrical signals. The ECG is a composite record of action potentials produced by all of the heart muscle fibers during each heartbeat. The instrument used to record the changes is an **electrocardiograph**.

In clinical practice, electrodes are positioned on the arms and legs (limb leads) and at six positions on the chest (chest leads) to record the ECG. The electrocardiograph amplifies the heart's electrical signals and produces 12 different tracings from different combinations of limb and chest leads. Each limb and chest electrode records slightly different electrical activity because of the difference in its position relative to the heart. By comparing these records with one another and with normal records, it is possible to determine (1) if the conducting pathway is abnormal, (2) if the heart is enlarged, (3) if certain regions of the heart are damaged, and (4) the cause of chest pain.

In a typical record, three clearly recognizable waves appear with each heartbeat (**Figure 20.12**). The first, called the **P wave**, is a small upward deflection on the ECG. The P wave represents **atrial depolarization**, which spreads from the SA node through contractile fibers in both atria. The second wave, called the **QRS complex**,

FIGURE 20.12 Normal electrocardiogram (ECG). P wave = atrial depolarization; QRS complex = onset of ventricular depolarization; T wave = ventricular repolarization.

An ECG is a recording of the electrical activity that initiates each heartbeat.

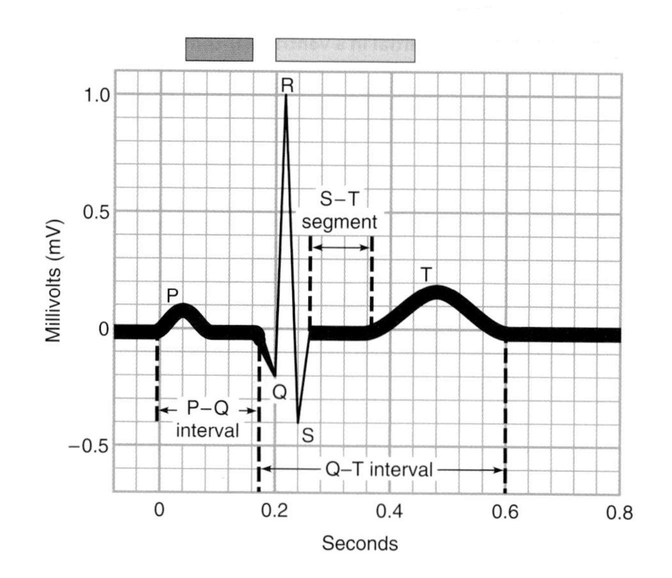

Q What is the significance of an enlarged Q wave?

begins as a downward deflection, continues as a large, upright, triangular wave, and ends as a downward wave. The QRS complex represents **rapid ventricular depolarization**, as the action potential spreads through ventricular contractile fibers. The third wave is a dome-shaped upward deflection called the **T wave**. It indicates **ventricular repolarization** and occurs just as the ventricles are starting to relax. The T wave is smaller and wider than the QRS complex because repolarization occurs more slowly than depolarization. During the plateau period of steady depolarization, the ECG tracing is flat.

In reading an ECG, the size of the waves can provide clues to abnormalities. Larger P waves indicate enlargement of an atrium; an enlarged Q wave may indicate a myocardial infarction; and an enlarged R wave generally indicates enlarged ventricles. The T wave is flatter than normal when the heart muscle is receiving insufficient oxygen—as, for example, in coronary artery disease. The T wave may be elevated in hyperkalemia (high blood K^+ level).

Analysis of an ECG also involves measuring the time spans between waves, which are called **intervals** or *segments*. For example, the **P–Q interval** is the time from the beginning of the P wave to the beginning of the QRS complex. It represents the conduction time from the beginning of atrial excitation to the beginning of ventricular excitation. Put another way, the P–Q interval is the time required for the action potential to travel through the atria, atrioventricular node, and the remaining fibers of the conduction system. As the action potential is forced to detour around scar tissue caused by disorders such as coronary artery disease and rheumatic fever, the P–Q interval lengthens.

The **S–T segment**, which begins at the end of the S wave and ends at the beginning of the T wave, represents the time when the ventricular contractile fibers are depolarized during the plateau phase of the action potential. The S–T segment is elevated (above the baseline) in acute myocardial infarction and depressed (below the baseline) when the heart muscle receives insufficient oxygen. The **Q–T interval** extends from the start of the QRS complex to the end of the T wave. It is the time from the beginning of ventricular depolarization to the end of ventricular repolarization. The Q–T interval may be lengthened by myocardial damage, myocardial ischemia (decreased blood flow), or conduction abnormalities.

Sometimes it is helpful to evaluate the heart's response to the stress of physical exercise (stress testing) (see Disorders: Homeostatic Imbalances in the Study Guide). Although narrowed coronary arteries may carry adequate oxygenated blood while a person is at rest, they will not be able to meet the heart's increased need for oxygen during strenuous exercise. This situation creates changes that can be seen on an electrocardiogram.

Abnormal heart rhythms and inadequate blood flow to the heart may occur only briefly or unpredictably. To detect these problems, **continuous ambulatory electrocardiography** is used. For this procedure, a person wears a battery-operated monitor (Holter monitor) that records an ECG continuously for 24 hours. Electrodes attached to the chest are connected to the monitor, and information on the heart's activity is stored in the monitor and retrieved later by medical personnel.

Correlation of ECG Waves with Atrial and Ventricular Systole

As you have learned, the atria and ventricles depolarize and then contract at different times because the conduction system routes cardiac action potentials along a specific pathway. The term **systole** (SIS-tō-lē = contraction) refers to the phase of contraction; the phase of relaxation is **diastole** (dī-AS-tō-lē = dilation or expansion). The ECG waves predict the timing of atrial and ventricular systole and diastole. At a heart rate of 75 beats per minute, the timing is as follows (**Figure 20.13**):

1. A cardiac action potential arises in the SA node. It propagates throughout the atrial muscle and down to the AV node in about 0.03 sec. As the atrial contractile fibers depolarize, the P wave appears in the ECG.
2. After the P wave begins, the atria contract (atrial systole). Conduction of the action potential slows at the AV node because the fibers there have much smaller diameters and fewer gap junctions. (Traffic slows in a similar way where a four-lane highway narrows to one lane in a construction zone!) The resulting 0.1-sec delay gives the atria time to contract, thus adding to the volume of blood in the ventricles, before ventricular systole begins.
3. The action potential propagates rapidly again after entering the AV bundle. About 0.2 sec after onset of the P wave, it has propagated through the bundle branches, Purkinje fibers, and the entire ventricular myocardium. Depolarization progresses down the septum, upward from the apex, and outward from the endocardial surface, producing the QRS complex. At the same time, atrial repolarization is occurring, but it is not usually evident in an ECG because the larger QRS complex masks it.
4. Contraction of ventricular contractile fibers (ventricular systole) begins shortly after the QRS complex appears and continues during the S–T segment. As contraction proceeds from the apex toward the base of the heart, blood is squeezed upward toward the semilunar valves.
5. Repolarization of ventricular contractile fibers begins at the apex and spreads throughout the ventricular myocardium. This produces the T wave in the ECG about 0.4 sec after the onset of the P wave.
6. Shortly after the T wave begins, the ventricles start to relax (ventricular diastole). By 0.6 sec, ventricular repolarization is complete and ventricular contractile fibers are relaxed.

During the next 0.2 sec, contractile fibers in both the atria and ventricles are relaxed. At 0.8 sec, the P wave appears again in the ECG, the atria begin to contract, and the cycle repeats.

As you have just learned, events in the heart occur in cycles that repeat for as long as you live. Next, we will see how the pressure changes associated with relaxation and contraction of the heart chambers allow the heart to alternately fill with blood and then eject blood into the aorta and pulmonary trunk.

FIGURE 20.13 Timing and route of action potential depolarization and repolarization through the conduction system and myocardium. Green indicates depolarization, and red indicates repolarization.

Depolarization causes contraction and repolarization causes relaxation of cardiac muscle fibers.

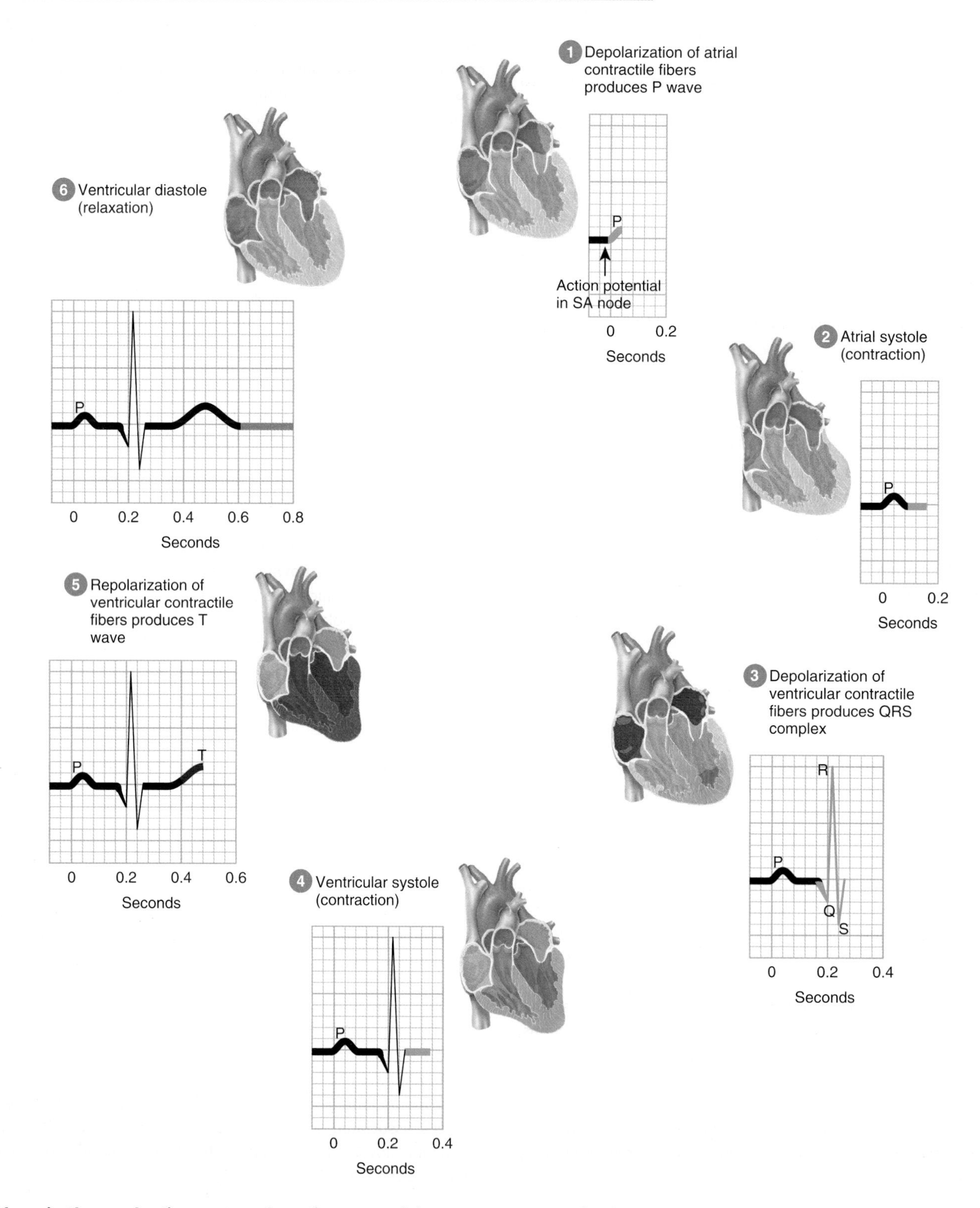

Q Where in the conduction system do action potentials propagate most slowly?

20.4 The Cardiac Cycle

OBJECTIVES

- **Describe** the pressure and volume changes that occur during a cardiac cycle.
- **Relate** the timing of heart sounds to the ECG waves and pressure changes during systole and diastole.

A single **cardiac cycle** includes all of the events associated with one heartbeat. Thus, a cardiac cycle consists of systole and diastole of the atria plus systole and diastole of the ventricles.

Pressure and Volume Changes during the Cardiac Cycle

In each cardiac cycle, the atria and ventricles alternately contract and relax, forcing blood from areas of higher pressure to areas of lower pressure. As a chamber of the heart contracts, blood pressure within it increases. Figure 20.14 shows the relationship between the heart's electrical signals (ECG) and changes in atrial pressure, ventricular pressure, aortic pressure, and ventricular volume during the cardiac cycle. The pressures given in the figure apply to the left side of the heart; pressures on the right side are considerably lower. Each ventricle, however, expels the same volume of blood per beat, and the same pattern exists for both pumping chambers. When heart rate is 75 beats/min, a cardiac cycle lasts 0.8 sec. To examine and correlate the events taking place during a cardiac cycle, we will begin with atrial systole.

Atrial Systole

During **atrial systole**, which lasts about 0.1 sec, the atria are contracting. At the same time, the ventricles are relaxed.

1. Depolarization of the SA node causes atrial depolarization, marked by the P wave in the ECG.
2. Atrial depolarization causes atrial systole. As the atria contract, they exert pressure on the blood within, which forces blood through the open AV valves into the ventricles.
3. Atrial systole contributes a final 25 mL of blood to the volume already in each ventricle (about 105 mL). The end of atrial systole is also the end of ventricular diastole (relaxation). Thus, each ventricle contains about 130 mL at the end of its relaxation period (diastole). This blood volume is called the **end-diastolic volume (EDV)**.
4. The QRS complex in the ECG marks the onset of ventricular depolarization.

Ventricular Systole

During **ventricular systole**, which lasts about 0.3 sec, the ventricles are contracting. At the same time, the atria are relaxed in **atrial diastole**.

5. Ventricular depolarization causes ventricular systole. As ventricular systole begins, pressure rises inside the ventricles and pushes blood up against the atrioventricular (AV) valves, forcing them shut. For about 0.05 seconds, both the SL (semilunar) and AV valves are closed. This is the period of **isovolumetric contraction** (ī-sō-VOL-ū-met′-rik; *iso-* = same). During this interval, cardiac muscle fibers are contracting and exerting force but are not yet shortening. Thus, the muscle contraction is isometric (same length). Moreover, because all four valves are closed, ventricular volume remains the same (isovolumic).
6. Continued contraction of the ventricles causes pressure inside the chambers to rise sharply. When left ventricular pressure surpasses aortic pressure at about 80 millimeters of mercury (mmHg) and right ventricular pressure rises above the pressure in the pulmonary trunk (about 20 mmHg), both SL valves open. At this point, ejection of blood from the heart begins. The period when the SL valves are open is **ventricular ejection** and lasts for about 0.25 sec. The pressure in the left ventricle continues to rise to about 120 mmHg, and the pressure in the right ventricle climbs to about 25–30 mmHg.
7. The left ventricle ejects about 70 mL of blood into the aorta and the right ventricle ejects the same volume of blood into the pulmonary trunk. The volume remaining in each ventricle at the end of systole, about 60 mL, is the **end-systolic volume (ESV)**. **Stroke volume**, the volume ejected per beat from each ventricle, equals end-diastolic volume minus end-systolic volume: SV = EDV − ESV. At rest, the stroke volume is about 130 mL − 60 mL = 70 mL (a little more than 2 oz).
8. The T wave in the ECG marks the onset of ventricular repolarization.

Relaxation Period

During the **relaxation period**, which lasts about 0.4 sec, the atria and the ventricles are both relaxed. As the heart beats faster and faster, the relaxation period becomes shorter and shorter, whereas the durations of atrial systole and ventricular systole shorten only slightly.

9. Ventricular repolarization causes **ventricular diastole**. As the ventricles relax, pressure within the chambers falls, and blood in the aorta and pulmonary trunk begins to flow backward toward the regions of lower pressure in the ventricles. Backflowing blood catches in the valve cusps and closes the SL valves. The aortic valve closes at a pressure of about 100 mmHg. Rebound of blood off the closed cusps of the aortic valve produces the **dicrotic wave** on the aortic pressure curve. After the SL valves close, there is a brief interval when ventricular blood volume does not change because all four valves are closed. This is the period of **isovolumetric relaxation**.
10. As the ventricles continue to relax, the pressure falls quickly. When ventricular pressure drops below atrial pressure, the AV valves open, and **ventricular filling** begins. The major part of ventricular filling occurs just after the AV valves open. Blood that has been flowing into and building up in the atria during ventricular systole then rushes rapidly into the ventricles. At the end of the relaxation period, the ventricles are about three-quarters full. The P wave appears in the ECG, signaling the start of another cardiac cycle.

FIGURE 20.14 Cardiac cycle. (a) ECG. (b) Changes in left atrial pressure (green line), left ventricular pressure (blue line), and aortic pressure (red line) as they relate to the opening and closing of heart valves. (c) Heart sounds. (d) Changes in left ventricular volume. (e) Phases of the cardiac cycle.

A cardiac cycle is composed of all of the events associated with one heartbeat.

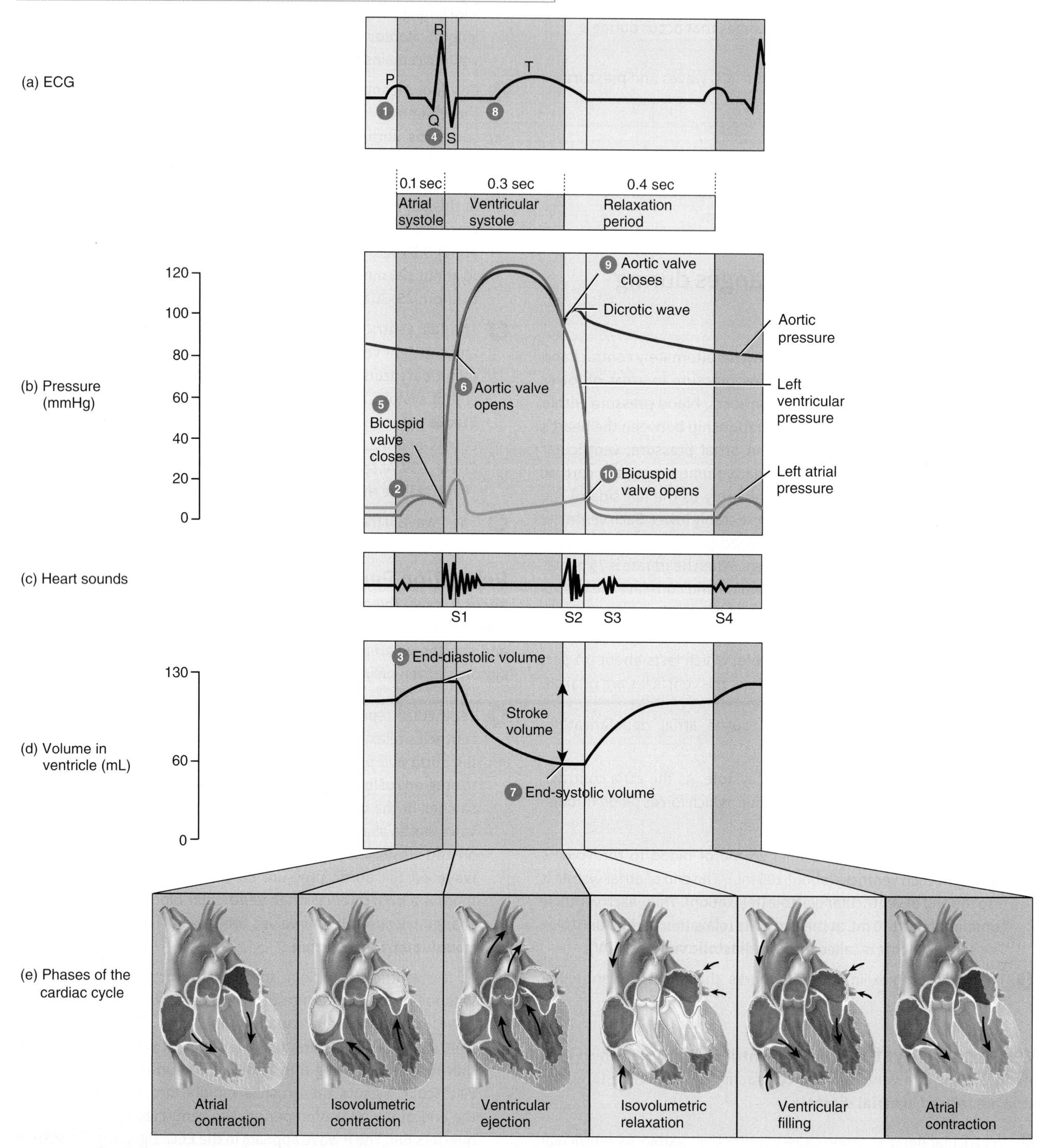

Q How much blood remains in each ventricle at the end of ventricular diastole in a resting person? What is this volume called?

Heart Sounds

Auscultation (aws-kul-TĀ-shun; *ausculta-* = listening), the act of listening to sounds within the body, is usually done with a stethoscope. The sound of the heartbeat comes primarily from blood turbulence caused by the closing of the heart valves. Smoothly flowing blood is silent. Compare the sounds made by white-water rapids or a waterfall with the silence of a smoothly flowing river. During each cardiac cycle, there are four **heart sounds**, but in a normal heart only the first and second heart sounds (S1 and S2) are loud enough to be heard through a stethoscope. Figure 20.14c shows the timing of heart sounds relative to other events in the cardiac cycle.

The first sound (S1), which can be described as a **lubb** sound, is louder and a bit longer than the second sound. S1 is caused by blood turbulence associated with closure of the AV valves soon after ventricular systole begins. The second sound (S2), which is shorter and not as loud as the first, can be described as a **dupp** sound. S2 is caused by blood turbulence associated with closure of the SL valves at the beginning of ventricular diastole. Although S1 and S2 are due to blood turbulence associated with the closure of valves, they are best heard at the surface of the chest in locations that are slightly different from the locations of the valves (Figure 20.15). This is because the sound is carried by the blood flow away from the valves. Normally not loud enough to be heard, S3 is due to blood turbulence during rapid ventricular filling, and S4 is due to blood turbulence during atrial systole.

FIGURE 20.15 Heart sounds. Location of valves (purple) and auscultation sites (red) for heart sounds.

Listening to sounds within the body is called auscultation; it is usually done with a stethoscope.

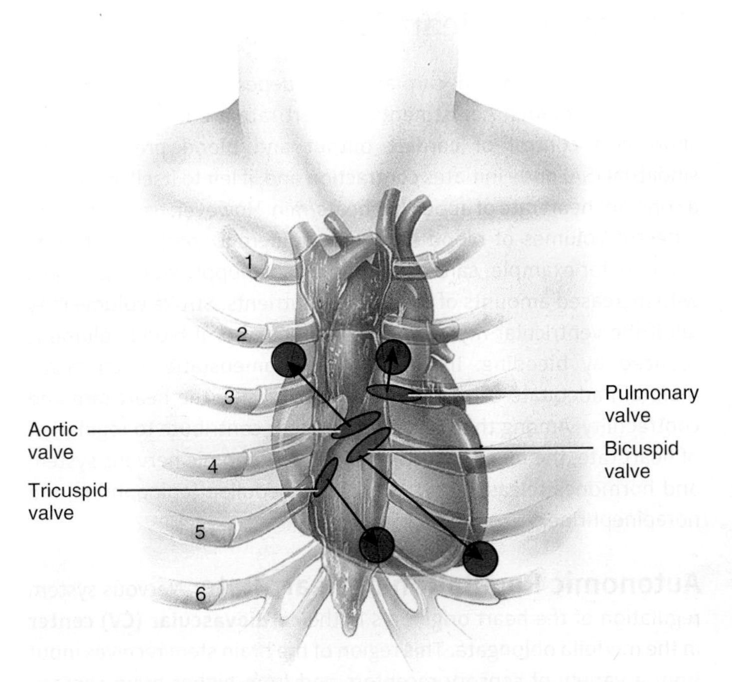

Q Which heart sound is related to blood turbulence associated with closure of the atrioventricular valves?

See Clinical Connection: Heart Murmurs

20.5 Cardiac Output

OBJECTIVES

- **Define** cardiac output.
- **Describe** the factors that affect regulation of stroke volume.
- **Outline** the factors that affect the regulation of heart rate.

Although the heart has autorhythmic fibers that enable it to beat independently, its operation is governed by events occurring throughout the body. Body cells must receive a certain amount of oxygen from blood each minute to maintain health and life. When cells are metabolically active, as during exercise, they take up even more oxygen from the blood. During rest periods, cellular metabolic need is reduced, and the workload of the heart decreases.

Cardiac output (CO) is the volume of blood ejected from the left ventricle (or the right ventricle) into the aorta (or pulmonary trunk) each minute. Cardiac output equals the **stroke volume (SV)**, the volume of blood ejected by the ventricle during each contraction, multiplied by the **heart rate (HR)**, the number of heartbeats per minute:

$$\underset{(\text{mL/min})}{\text{CO}} = \underset{(\text{mL/beat})}{\text{SV}} \times \underset{(\text{beats/min})}{\text{HR}}$$

In a typical resting adult male, stroke volume averages 70 mL/beat, and heart rate is about 75 beats/min. Thus, average cardiac output is

$$\begin{aligned}\text{CO} &= 70\ \text{mL/beat} \times 75\ \text{beats/min}\\ &= 5250\ \text{mL/min}\\ &= 5.25\ \text{L/min}\end{aligned}$$

This volume is close to the total blood volume, which is about 5 liters in a typical adult male. Thus, your entire blood volume flows through your pulmonary and systemic circulations each minute. Factors that increase stroke volume or heart rate normally increase CO. During mild exercise, for example, stroke volume may increase to 100 mL/beat, and heart rate to 100 beats/min. Cardiac output then would be 10 L/min. During intense (but still not maximal) exercise, the heart rate may accelerate to 150 beats/min, and stroke volume may rise to 130 mL/beat, resulting in a cardiac output of 19.5 L/min.

Cardiac reserve is the difference between a person's maximum cardiac output and cardiac output at rest. The average person has a cardiac reserve of four or five times the resting value. Top endurance athletes may have a cardiac reserve seven or eight times their resting CO. People with severe heart disease may have little or no cardiac reserve, which limits their ability to carry out even the simple tasks of daily living.

Regulation of Stroke Volume

A healthy heart will pump out the blood that entered its chambers during the previous diastole. In other words, if more blood returns to the heart during diastole, then more blood is ejected during the next systole. At rest, the stroke volume is 50–60% of the end-diastolic volume because 40–50% of the blood remains in the ventricles after each contraction (end-systolic volume). Three factors regulate stroke volume and ensure that the left and right ventricles pump equal volumes of blood: (1) **preload**, the degree of stretch on the heart before it contracts; (2) **contractility**, the forcefulness of contraction of individual ventricular muscle fibers; and (3) **afterload**, the pressure that must be exceeded before ejection of blood from the ventricles can occur.

Preload: Effect of Stretching

A greater preload (stretch) on cardiac muscle fibers prior to contraction increases their force of contraction. Preload can be compared to the stretching of a rubber band. The more the rubber band is stretched, the more forcefully it will snap back. Within limits, the more the heart fills with blood during diastole, the greater the force of contraction during systole. This relationship is known as the **Frank–Starling law of the heart**. The preload is proportional to the end-diastolic volume (EDV) (the volume of blood that fills the ventricles at the end of diastole). Normally, the greater the EDV, the more forceful the next contraction.

Two key factors determine EDV: (1) the duration of ventricular diastole and (2) **venous return**, the volume of blood returning to the right ventricle. When heart rate increases, the duration of diastole is shorter. Less filling time means a smaller EDV, and the ventricles may contract before they are adequately filled. By contrast, when venous return increases, a greater volume of blood flows into the ventricles, and the EDV is increased.

When heart rate exceeds about 160 beats/min, stroke volume usually declines due to the short filling time. At such rapid heart rates, EDV is less, and the preload is lower. People who have slow resting heart rates usually have large resting stroke volumes because filling time is prolonged and preload is larger.

The Frank–Starling law of the heart equalizes the output of the right and left ventricles and keeps the same volume of blood flowing to both the systemic and pulmonary circulations. If the left side of the heart pumps a little more blood than the right side, the volume of blood returning to the right ventricle (venous return) increases. The increased EDV causes the right ventricle to contract more forcefully on the next beat, bringing the two sides back into balance.

Contractility

The second factor that influences stroke volume is myocardial **contractility**, the strength of contraction at any given preload. Substances that increase contractility are **positive inotropic agents** (īn′-ō-TRŌ-pik); those that decrease contractility are **negative inotropic agents**. Thus, for a constant preload, the stroke volume increases when a positive inotropic substance is present. Positive inotropic agents often promote Ca^{2+} inflow during cardiac action potentials, which strengthens the force of the next contraction. Stimulation of the sympathetic division of the autonomic nervous system (ANS), hormones such as epinephrine and norepinephrine, increased Ca^{2+} level in the interstitial fluid, and the drug digitalis all have positive inotropic effects. In contrast, inhibition of the sympathetic division of the ANS, anoxia, acidosis, some anesthetics, and increased K^+ level in the interstitial fluid have negative inotropic effects. *Calcium channel blockers* are drugs that can have a negative inotropic effect by reducing Ca^{2+} inflow, thereby decreasing the strength of the heartbeat.

Afterload

Ejection of blood from the heart begins when pressure in the right ventricle exceeds the pressure in the pulmonary trunk (about 20 mmHg), and when the pressure in the left ventricle exceeds the pressure in the aorta (about 80 mmHg). At that point, the higher pressure in the ventricles causes blood to push the semilunar valves open. The pressure that must be overcome before a semilunar valve can open is termed the afterload. An increase in afterload causes stroke volume to decrease, so that more blood remains in the ventricles at the end of systole. Conditions that can increase afterload include hypertension (elevated blood pressure) and narrowing of arteries by atherosclerosis (see the entry on coronary artery disease in the Disorders: Homoeostatic Imbalances section in the Study Guide).

Regulation of Heart Rate

As you have just learned, cardiac output depends on both heart rate and stroke volume. Adjustments in heart rate are important in the short-term control of cardiac output and blood pressure. The sinoatrial (SA) node initiates contraction and, if left to itself, would set a constant heart rate of about 100 beats/min. However, tissues require different volumes of blood flow under different conditions. During exercise, for example, cardiac output rises to supply working tissues with increased amounts of oxygen and nutrients. Stroke volume may fall if the ventricular myocardium is damaged or if blood volume is reduced by bleeding. In these cases, homeostatic mechanisms maintain adequate cardiac output by increasing the heart rate and contractility. Among the several factors that contribute to regulation of heart rate, the most important are the autonomic nervous system and hormones released by the adrenal medullae (epinephrine and norepinephrine).

Autonomic Regulation of Heart Rate

Nervous system regulation of the heart originates in the **cardiovascular (CV) center** in the medulla oblongata. This region of the brain stem receives input from a variety of sensory receptors and from higher brain centers, such as the limbic system and cerebral cortex. The cardiovascular center then directs appropriate output by increasing or decreasing

FIGURE 20.16 **Nervous system control of the heart.**

The cardiovascular center in the medulla oblongata controls both sympathetic (blue) and parasympathetic nerves (red) that innervate the heart.

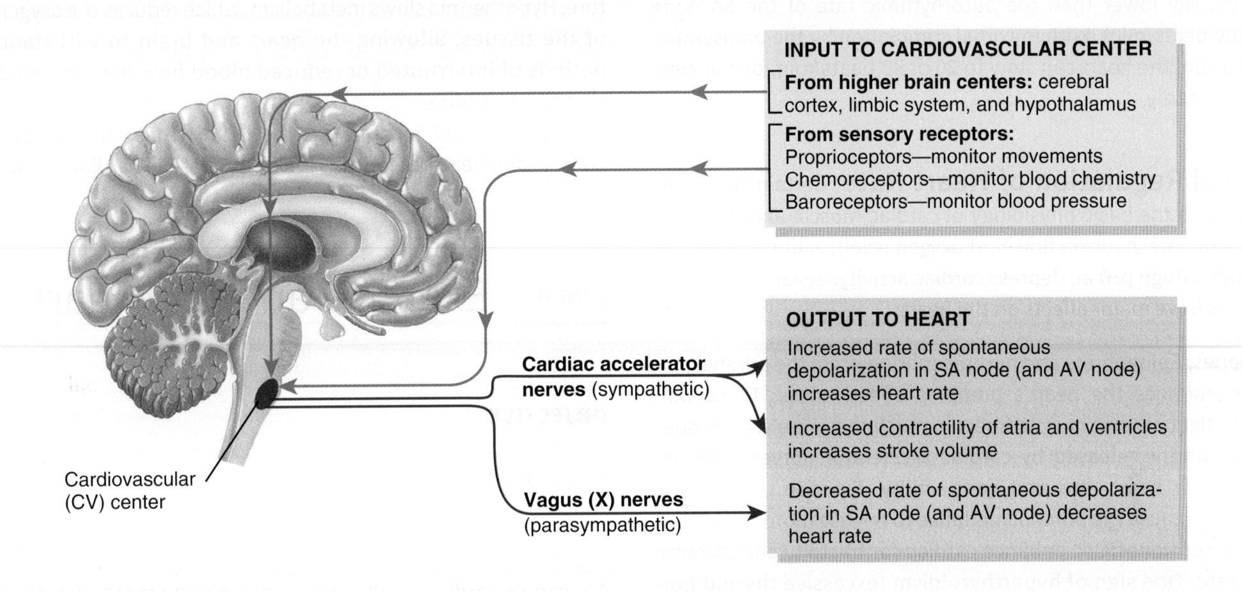

Q Which region of the heart is innervated by the sympathetic division but not by the parasympathetic division?

the frequency of nerve impulses in both the sympathetic and parasympathetic branches of the ANS (Figure 20.16).

Even before physical activity begins, especially in competitive situations, heart rate may climb. This anticipatory increase occurs because the limbic system sends nerve impulses to the cardiovascular center in the medulla. As physical activity begins, **proprioceptors** that are monitoring the position of limbs and muscles send nerve impulses at an increased frequency to the cardiovascular center. Proprioceptor input is a major stimulus for the quick rise in heart rate that occurs at the onset of physical activity. Other sensory receptors that provide input to the cardiovascular center include **chemoreceptors**, which monitor chemical changes in the blood, and **baroreceptors**, which monitor the stretching of major arteries and veins caused by the pressure of the blood flowing through them. Important baroreceptors located in the arch of the aorta and in the carotid arteries (see Figure 21.13) detect changes in blood pressure and provide input to the cardiovascular center when it changes. The role of baroreceptors in the regulation of blood pressure is discussed in detail in Chapter 21. Here we focus on the innervation of the heart by the sympathetic and parasympathetic branches of the ANS.

Sympathetic neurons extend from the medulla oblongata into the spinal cord. From the thoracic region of the spinal cord, sympathetic **cardiac accelerator nerves** extend out to the SA node, AV node, and most portions of the myocardium. Impulses in the cardiac accelerator nerves trigger the release of norepinephrine, which binds to beta-1 (β_1) receptors on cardiac muscle fibers. This interaction has two separate effects: (1) In SA (and AV) node fibers, norepinephrine speeds the rate of spontaneous depolarization so that these pacemakers fire impulses more rapidly and heart rate increases; (2) in contractile fibers throughout the atria and ventricles, norepinephrine enhances Ca^{2+} entry through the voltage-gated slow Ca^{2+} channels, thereby increasing contractility. As a result, a greater volume of blood is ejected during systole. With a moderate increase in heart rate, stroke volume does not decline because the increased contractility offsets the decreased preload. With maximal sympathetic stimulation, however, heart rate may reach 200 beats/min in a 20-year-old person. At such a high heart rate, stroke volume is lower than at rest due to the very short filling time. The maximal heart rate declines with age; as a rule, subtracting your age from 220 provides a good estimate of your maximal heart rate in beats per minute.

Parasympathetic nerve impulses reach the heart via the right and left **vagus (X) nerves**. Vagal axons terminate in the SA node, AV node, and atrial myocardium. They release acetylcholine, which decreases heart rate by slowing the rate of spontaneous depolarization in autorhythmic fibers. As only a few vagal fibers innervate ventricular muscle, changes in parasympathetic activity have little effect on contractility of the ventricles.

A continually shifting balance exists between sympathetic and parasympathetic stimulation of the heart. At rest, parasympathetic stimulation predominates. The resting heart rate—about 75 beats/min—is usually lower than the autorhythmic rate of the SA node (about 100 beats/min). With maximal stimulation by the parasympathetic division, the heart can slow to 20 or 30 beats/min, or can even stop momentarily.

Chemical Regulation of Heart Rate Certain chemicals influence both the basic physiology of cardiac muscle and the heart rate. For example, hypoxia (lowered oxygen level), acidosis (low pH), and alkalosis (high pH) all depress cardiac activity. Several hormones and cations have major effects on the heart:

1. ***Hormones.*** Epinephrine and norepinephrine (from the adrenal medullae) enhance the heart's pumping effectiveness. These hormones affect cardiac muscle fibers in much the same way as does norepinephrine released by cardiac accelerator nerves—they increase both heart rate and contractility. Exercise, stress, and excitement cause the adrenal medullae to release more hormones. Thyroid hormones also enhance cardiac contractility and increase heart rate. One sign of hyperthyroidism (excessive thyroid hormone) is **tachycardia** (tak′-i-KAR-dē-a), an elevated resting heart rate.
2. ***Cations.*** Given that differences between intracellular and extracellular concentrations of several cations (for example, Na^+ and K^+) are crucial for the production of action potentials in all nerve and muscle fibers, it is not surprising that ionic imbalances can quickly compromise the pumping effectiveness of the heart. In particular, the relative concentrations of three cations—K^+, Ca^{2+}, and Na^+—have a large effect on cardiac function. Elevated blood levels of K^+ or Na^+ decrease heart rate and contractility. Excess Na^+ blocks Ca^{2+} inflow during cardiac action potentials, thereby decreasing the force of contraction, whereas excess K^+ blocks generation of action potentials. A moderate increase in interstitial (and thus intracellular) Ca^{2+} level speeds heart rate and strengthens the heartbeat.

Other Factors in Heart Rate Regulation Age, gender, physical fitness, and body temperature also influence resting heart rate. A newborn baby is likely to have a resting heart rate over 120 beats/min; the rate then gradually declines throughout life. Adult females often have slightly higher resting heart rates than adult males, although regular exercise tends to bring resting heart rate down in both sexes. A physically fit person may even exhibit **bradycardia** (brād′-i-KAR-dē-a; *bradys-* = slow), a resting heart rate under 50 beats/min. This is a beneficial effect of endurance-type training because a slowly beating heart is more energy efficient than one that beats more rapidly.

Increased body temperature, as occurs during a fever or strenuous exercise, causes the SA node to discharge impulses more quickly, thereby increasing heart rate. Decreased body temperature decreases heart rate and strength of contraction.

During surgical repair of certain heart abnormalities, it is helpful to slow a patient's heart rate by **hypothermia** (hī′-pō-THER-mē-a), in which the person's body is deliberately cooled to a low core temperature. Hypothermia slows metabolism, which reduces the oxygen needs of the tissues, allowing the heart and brain to withstand short periods of interrupted or reduced blood flow during a medical or surgical procedure.

Figure 20.17 summarizes the factors that can increase stroke volume and heart rate to achieve an increase in cardiac output.

20.6 Exercise and the Heart

OBJECTIVE

- **Explain** how the heart is affected by exercise.

A person's cardiovascular fitness can be improved at any age with regular exercise. Some types of exercise are more effective than others for improving the health of the cardiovascular system. **Aerobics**, any activity that works large body muscles for at least 20 minutes, elevates cardiac output and accelerates metabolic rate. Three to five such sessions a week are usually recommended for improving the health of the cardiovascular system. Brisk walking, running, bicycling, cross-country skiing, and swimming are examples of aerobic activities.

Sustained exercise increases the oxygen demand of the muscles. Whether the demand is met depends mainly on the adequacy of cardiac output and proper functioning of the respiratory system. After several weeks of training, a healthy person increases maximal cardiac output (the amount of blood ejected from the ventricles into their respective arteries per minute), thereby increasing the maximal rate of oxygen delivery to the tissues. Oxygen delivery also rises because skeletal muscles develop more capillary networks in response to long-term training.

During strenuous activity, a well-trained athlete can achieve a cardiac output double that of a sedentary person, in part because training causes hypertrophy (enlargement) of the heart. This condition is referred to as **physiological cardiomegaly** (kar′-dē-ō-MEG-a-lē; *mega* = large). A **pathological cardiomegaly** is related to significant heart disease. Even though the heart of a well-trained athlete is larger, *resting* cardiac output is about the same as in a healthy untrained person, because *stroke volume* (volume of blood pumped by each beat of a ventricle) is increased while heart rate is decreased. The resting heart rate of a trained athlete often is only 40–60 beats per minute *(resting bradycardia).* Regular exercise also helps to reduce blood pressure, anxiety, and depression; control weight; and increase the body's ability to dissolve blood clots.

FIGURE 20.17 **Factors that increase cardiac output.**

Cardiac output equals stroke volume multiplied by heart rate.

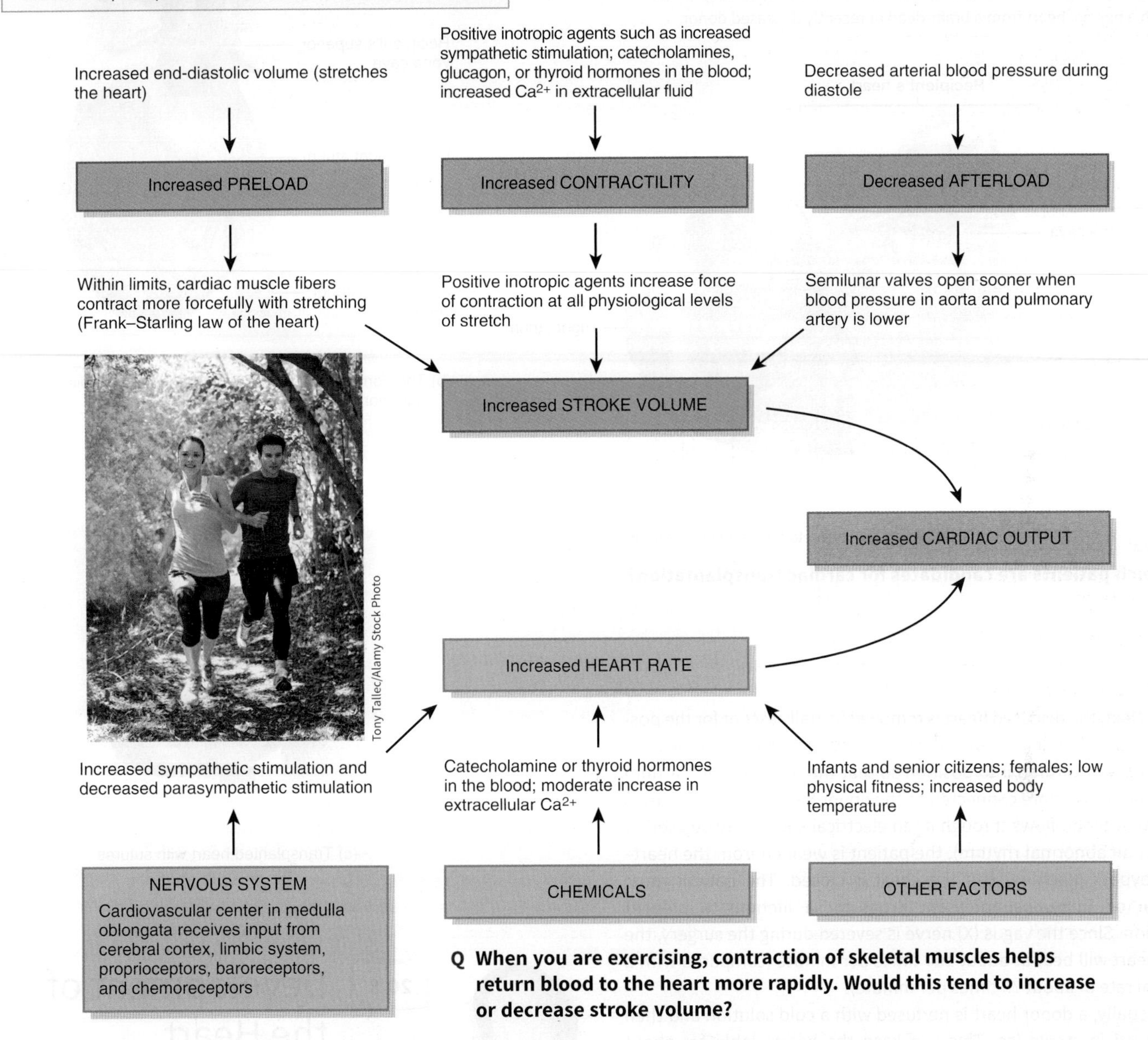

Q When you are exercising, contraction of skeletal muscles helps return blood to the heart more rapidly. Would this tend to increase or decrease stroke volume?

20.7 Help for Failing Hearts

OBJECTIVE

- **Describe** several techniques used for failing hearts.

As the heart fails, a person has decreasing ability to exercise or even to move around. A variety of surgical techniques and medical devices exist to aid a failing heart. For some patients, even a 10% increase in the volume of blood ejected from the ventricles can mean the difference between being bedridden and having limited mobility.

A **cardiac** (*heart*) **transplant** is the replacement of a severely damaged heart with a normal heart from a brain-dead or recently deceased donor. Cardiac transplants are performed on patients with end-stage heart failure or severe coronary artery disease. Once a suitable heart is located, the chest cavity is exposed through a midsternal cut. After the patient is placed on a heart–lung bypass machine, which oxygenates and circulates blood, the pericardium is cut to expose the

FIGURE 20.18 **Cardiac transplantation.**

Cardiac transplantation is the replacement of a severely damaged heart with a normal heart from a brain-dead or recently deceased donor.

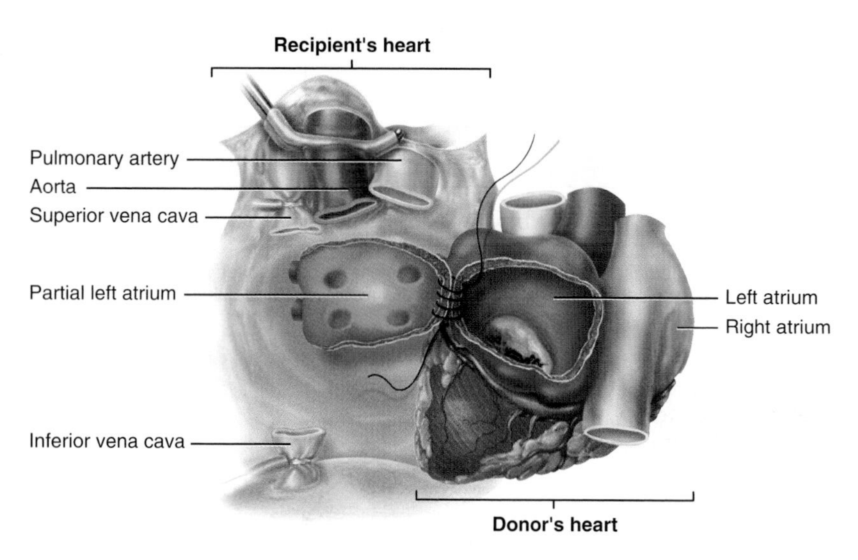

(a) The donor's left atrium is sutured to the recipient's left atrium

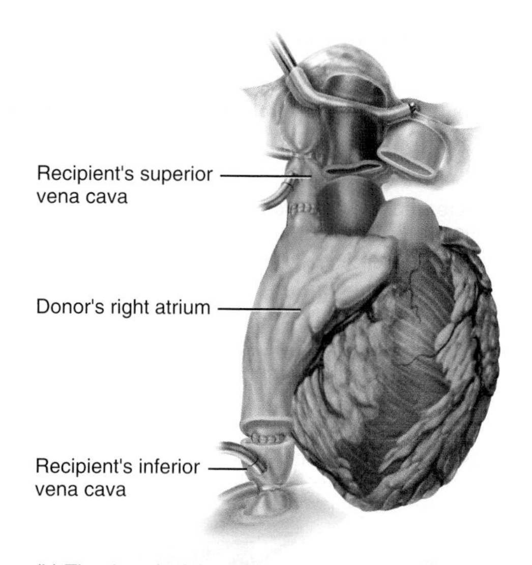

(b) The donor's right atrium is sutured to the recipient's superior and inferior venae cavae

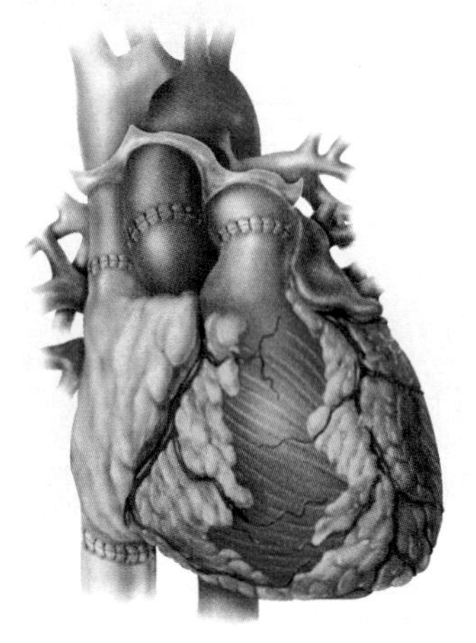

(c) Transplanted heart with sutures

Q Which patients are candidates for cardiac transplantation?

heart. Next, the diseased heart is removed (usually except for the posterior wall of the left atrium) (Figure 20.18) and the donor heart is trimmed and sutured into position so that the remaining left atrium and great vessels are connected to the donor heart. The new heart is started as blood flows through it (an electrical shock may be used to correct an abnormal rhythm), the patient is weaned from the heart-lung bypass machine, and the chest is closed. The patient must remain on immunosuppressant drugs for a lifetime to prevent rejection. Since the vagus (X) nerve is severed during the surgery, the new heart will beat at about 100 times per minute (compared with a normal rate of about 75 times per minute).

Usually, a donor heart is perfused with a cold solution and then preserved in sterile ice. This can keep the heart viable for about 4–5 hours. In May 2007, surgeons in the United States performed the first beating-heart transplant. The donor heart was maintained at normal body temperature and hooked up to an organ care system that allowed it to keep beating with warm, oxygenated blood flowing through it. This approach greatly prolongs the time between removal of the heart from the donor and transplantation into a recipient and decreases injury to the heart while being deprived of blood, which can lead to rejection.

Cardiac transplants are common today and produce good results, but the availability of donor hearts is very limited. Another approach is the use of cardiac assist devices and other surgical procedures that assist heart function without removing the heart. Table 20.1 describes several of these devices and procedures.

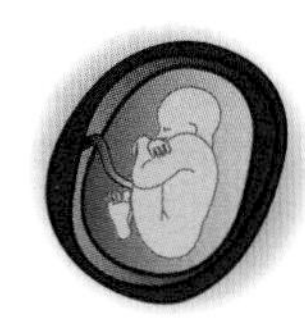

20.8 Development of the Heart

OBJECTIVE

- **Describe** the development of the heart.

Listening to a fetal heartbeat for the first time is an exciting moment for prospective parents, but it is also an important diagnostic tool. The cardiovascular system is one of the first systems to form in an embryo, and the heart is the first functional organ. This order of development is essential because of the need of the rapidly growing embryo to obtain oxygen and nutrients and get rid of wastes. As you will learn

TABLE 20.1 Cardiac Assist Devices and Procedures

DEVICE	DESCRIPTION
Intra-aortic balloon pump (IABP)	A 40-mL polyurethane balloon mounted on a catheter is inserted into an artery in the groin and threaded through the femoral artery into the thoracic aorta (see **Figure A**). An external pump inflates the balloon with helium gas at the beginning of ventricular diastole. As the balloon inflates, it pushes blood both backward toward the heart (improves coronary blood flow) and forward toward peripheral tissues. The balloon then is rapidly deflated just before the next ventricular systole, drawing blood out of the left ventricle (making it easier for the left ventricle to eject blood). Because the balloon is inflated between heartbeats, this technique is called *intra-aortic balloon counterpulsation*.
Ventricular assist device (VAD)	A mechanical pump helps a weakened ventricle pump blood throughout the body so the heart does not have to work as hard. A VAD may be used to help a patient survive until a heart transplant can be performed *(bridge to transplant)* or provide an alternative to heart transplantation *(destination therapy)*. VADs are classified according to the ventricle that requires support. A *left ventricular assist device (LVAD)*, the most common, helps the left ventricle pump blood into the aorta (see **Figure B**). A *right ventricular assist device (RVAD)* helps the right ventricle pump blood into the pulmonary trunk. A *biventricular assist device (BVAD)* helps both the left and right ventricles perform. The kind of VAD used depends on the patient's specific needs. To help you understand how a VAD works, see the LVAD (**Figure B**). An inflow tube attached to the apex of the left ventricle takes blood from the ventricle through a one-way valve into the pump unit. Once the pump fills with blood, an external control system triggers pumping, and blood flows through a one-way valve into an outflow tube that delivers blood into the aorta. The external control system is on a belt around the waist or on a shoulder strap. Some VADs pump at a constant rate; others are coordinated with the person's heartbeat.
Cardiomyoplasty	A large piece of the patient's own skeletal muscle (left latissimus dorsi) is partially freed from connective tissue attachments and wrapped around the heart, leaving the blood and nerve supply intact. An implanted pacemaker stimulates the skeletal muscle's motor neurons to cause contraction 10–20 times per minute, in synchrony with some of the heartbeats.
Skeletal muscle assist device	A piece of the patient's own skeletal muscle is used to fashion a pouch inserted between the heart and aorta, functioning as a booster heart. A pacemaker stimulates the muscle's motor neurons to elicit contraction.

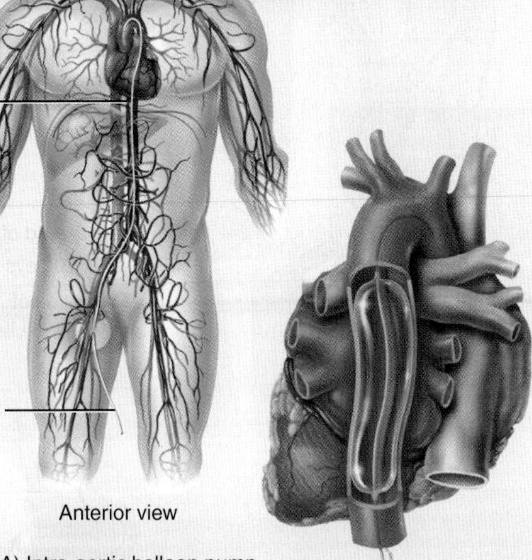

(A) Intra-aortic balloon pump

Outflow tube
Inflow tube
Outflow one-way valve
Inflow one-way valve
Pump unit
Parts of left ventricular assist device (LVAD)
Driveline
Aorta
Left ventricle
Implanted left ventricular assist device (LVAD)

(B) Left ventricular assist device (LVAD)

shortly, the development of the heart is a complex process, and any disruptions along the way can result in congenital (present at birth) disorders of the heart. Such disorders, described in Disorders: Homeostatic Imbalances in the Study Guide, are responsible for almost half of all deaths from birth defects.

The *heart* begins its development from **mesoderm** on day 18 or 19 following fertilization. In the head end of the embryo, the heart develops from a group of mesodermal cells called the **cardiogenic area** (kar-dē-ō-JEN-ik; *cardio-* = heart; *-genic* = producing) (**Figure 20.19a**). In response to signals from the underlying endoderm, the mesoderm in the cardiogenic area forms a pair of elongated strands called **cardiogenic cords**. Shortly after, these cords develop a hollow center and then become known as **endocardial tubes** (**Figure 20.19b**). With lateral folding of the embryo,

FIGURE 20.19 **Development of the heart.** Arrows within the structures indicate the direction of blood flow.

The heart begins its development from a group of mesodermal cells called the cardiogenic area during the third week after fertilization.

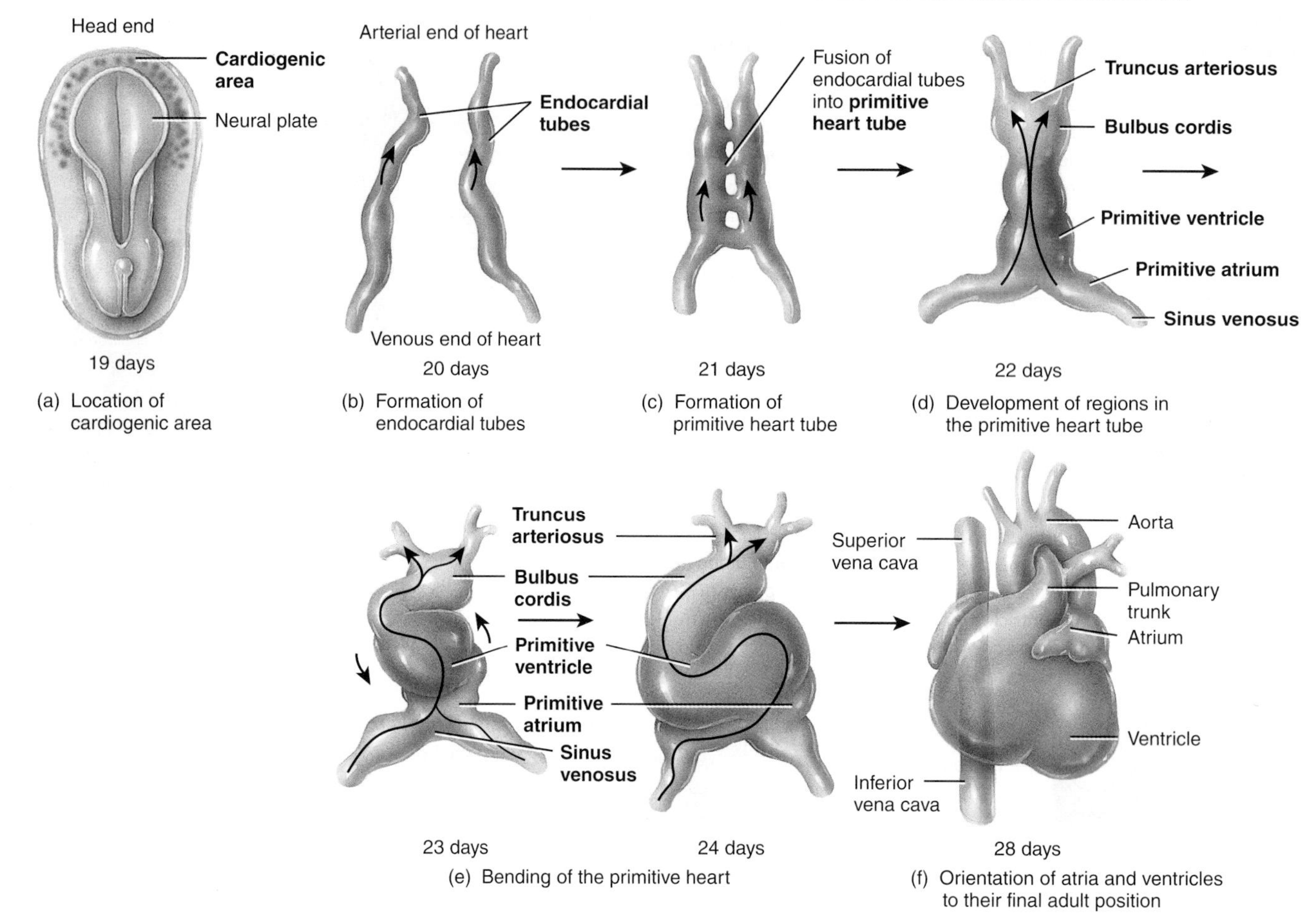

Q When during embryonic development does the primitive heart begin to contract?

the paired endocardial tubes approach each other and fuse into a single tube called the **primitive heart tube** on day 21 following fertilization (**Figure 20.19c**).

On the 22nd day, the primitive heart tube develops into five distinct regions and begins to pump blood. From tail end to head end (and in the same direction as blood flow) they are the (1) **sinus venosus**, (2) **primitive atrium**, (3) **primitive ventricle**, (4) **bulbus cordis**, and (5) **truncus arteriosus**. The sinus venosus initially receives blood from all veins in the embryo; contractions of the heart begin in this region and follow sequentially in the other regions. Thus, at this stage, the heart consists of a series of unpaired regions. The fates of the regions are as follows:

1. The sinus venosus develops into *part of the right atrium (posterior wall), coronary sinus,* and *sinoatrial (SA) node.*
2. The primitive atrium develops into *part of the right atrium (anterior wall), right auricle, part of the left atrium (anterior wall),* and the *left auricle.*
3. The primitive ventricle gives rise to the *left ventricle.*
4. The bulbus cordis develops into the *right ventricle.*
5. The truncus arteriosus gives rise to the *ascending aorta* and *pulmonary trunk.*

On day 23, the primitive heart tube elongates. Because the bulbus cordis and primitive ventricle grow more rapidly than other parts of the tube and because the atrial and venous ends of the tube are confined by the pericardium, the tube begins to loop and fold. At first, the primitive heart tube assumes a U-shape; later it becomes S-shaped (**Figure 20.19e**). As a result of these movements, which are

FIGURE 20.20 Partitioning of the heart into four chambers.

Partitioning of the heart begins on about the 28th day after fertilization.

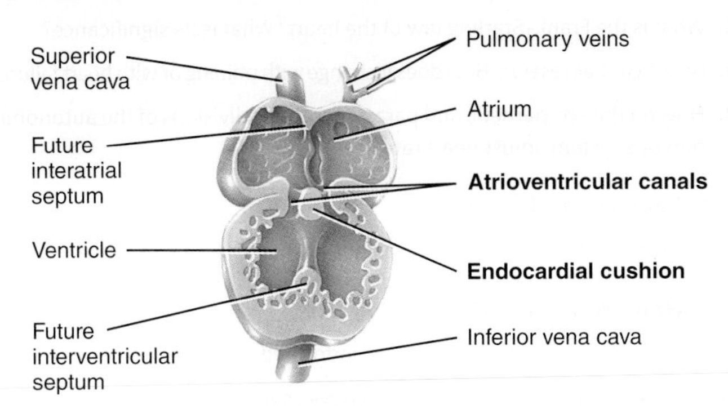

(a) Anterior view of frontal section at about 28 days

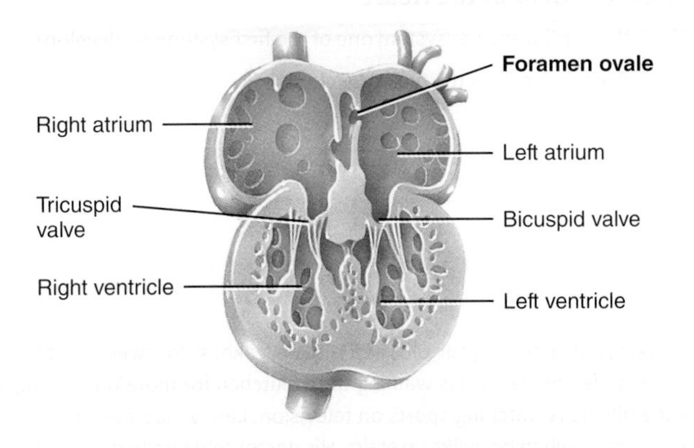

(b) Anterior view of frontal section at about 8 weeks

Q When is the partitioning of the heart complete?

completed by day 28, the primitive atria and ventricles of the future heart are reoriented to assume their final adult positions. The remainder of heart development consists of remodeling of the chambers and the formation of septa and valves to form a four-chambered heart.

On about day 28, thickenings of mesoderm of the inner lining of the heart wall, called **endocardial cushions**, appear (**Figure 20.20**). They grow toward each other, fuse, and divide the single **atrioventricular canal** (region between atria and ventricles) into smaller, separate left and right atrioventricular canals. Also, the *interatrial septum* begins its growth toward the fused endocardial cushions. Ultimately, the interatrial septum and endocardial cushions unite and an opening in the septum, the **foramen ovale** (ō-VAL-ē), develops. The interatrial septum divides the atrial region into a *right atrium* and a *left atrium.* Before birth, the foramen ovale allows most blood entering the right atrium to pass into the left atrium. After birth, it normally closes so that the interatrial septum is a complete partition. The remnant of the foramen ovale is the fossa ovalis (**Figure 20.4a**). Formation of the *interventricular septum* partitions the ventricular region into a *right ventricle* and a *left ventricle.* Partitioning of the atrioventricular canal, atrial region, and ventricular region is basically complete by the end of the fifth week. The *atrioventricular valves* form between the fifth and eighth weeks. The *semilunar valves* form between the fifth and ninth weeks.

Review Questions

20.1 Anatomy of the Heart

1. Define each of the following external features of the heart: auricle, coronary sulcus, anterior interventricular sulcus, and posterior interventricular sulcus.
2. Describe the structure of the pericardium and the layers of the wall of the heart.
3. What are the characteristic internal features of each chamber of the heart?
4. Which blood vessels deliver blood to the right and left atria?
5. What is the relationship between wall thickness and function among the various chambers of the heart?
6. What type of tissue composes the fibrous skeleton of the heart, and how is it organized?

20.2 Heart Valves and Circulation of Blood

7. What causes the heart valves to open and to close? What supporting structures ensure that the valves operate properly?
8. In correct sequence, which heart chambers, heart valves, and blood vessels would a drop of blood encounter as it flows from the right atrium to the aorta?
9. Which arteries deliver oxygenated blood to the myocardium of the left and right ventricles?

20.3 Cardiac Muscle Tissue and the Cardiac Conduction System

10. How do cardiac muscle fibers differ structurally and functionally from skeletal muscle fibers?

11. In what ways are autorhythmic fibers similar to and different from contractile fibers?
12. What happens during each of the three phases of an action potential in ventricular contractile fibers?
13. In what ways are ECGs helpful in diagnosing cardiac problems?
14. How does each ECG wave, interval, and segment relate to contraction (systole) and relaxation (diastole) of the atria and ventricles?

20.4 The Cardiac Cycle

15. Why must left ventricular pressure be greater than aortic pressure during ventricular ejection?
16. Does more blood flow through the coronary arteries during ventricular diastole or ventricular systole? Explain your answer.
17. During which two periods of the cardiac cycle do the heart muscle fibers exhibit isometric contractions?
18. What events produce the four normal heart sounds? Which ones can usually be heard through a stethoscope?

20.5 Cardiac Output

19. How is cardiac output calculated?
20. Define stroke volume (SV), and explain the factors that regulate it.
21. What is the Frank–Starling law of the heart? What is its significance?
22. Define cardiac reserve. How does it change with training or with heart failure?
23. How do the sympathetic and parasympathetic divisions of the autonomic nervous system adjust heart rate?

20.6 Exercise and the Heart

24. What are some of the cardiovascular benefits of regular exercise?

20.7 Help for Failing Hearts

25. Describe how a heart transplant is performed.
26. Explain four different cardiac assist devices and procedures.

20.8 Development of the Heart

27. Why is the cardiovascular system one of the first systems to develop?
28. From which tissue does the heart develop?

Critical Thinking Questions

1. Gerald recently visited the dentist. During the cleaning process, Gerald had some bleeding from his gums. A couple of days later, Gerald developed a fever, rapid heartbeat, sweating, and chills. He visited his family physician, who detected a slight heart murmur. Gerald was given antibiotics and continued to have his heart monitored. How was Gerald's dental visit related to his illness?
2. Unathletic Sylvia resolves to begin an exercise program. She tells you that she wants to make her heart "beat as fast as it can" during exercise. Explain why that may not be a good idea.
3. Mr. Perkins is a large, 62-year-old man with a weakness for sweets and fried foods. His idea of exercise is walking to the kitchen for more potato chips to eat while he is watching sports on television. Lately, he's been troubled by chest pains when he walks up stairs. His doctor told him to quit smoking and scheduled cardiac angiography for the next week. What is involved in performing this procedure? Why did the doctor order this test?

The Cardiovascular System: Blood Vessels and Hemodynamics

CHAPTER **21**

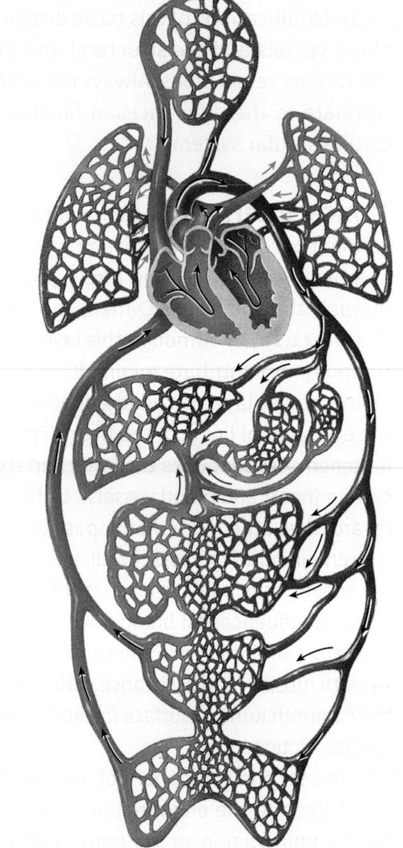

Blood Vessels, Hemodynamics, and Homeostasis

Blood vessels contribute to homeostasis by providing the structures for the flow of blood to and from the heart and the exchange of nutrients and wastes in tissues. They also play an important role in adjusting the velocity and volume of blood flow.

The cardiovascular system contributes to homeostasis of other body systems by transporting and distributing blood throughout the body to deliver materials (such as oxygen, nutrients, and hormones) and carry away wastes. The structures involved in these important tasks are the blood vessels, which form a closed system of tubes that carries blood away from the heart, transports it to the tissues of the body, and then returns it to the heart. The left side of the heart pumps blood through an estimated 100,000 km (60,000 mi) of blood vessels. The right side of the heart pumps blood through the lungs, enabling blood to pick up oxygen and unload carbon dioxide. Chapters 19 and 20 described the composition and functions of blood and the structure and function of the heart. In this chapter, we focus on the structure and functions of the various types of blood vessels; on the forces involved in circulating blood throughout the body; and on the blood vessels that constitute the major circulatory routes.

Q Did you ever wonder why untreated hypertension has so many damaging effects?

21.1 Structure and Function of Blood Vessels

OBJECTIVES

- **Contrast** the structure and function of arteries, arterioles, capillaries, venules, and veins.
- **Outline** the vessels through which the blood moves in its passage from the heart to the capillaries and back.
- **Distinguish** between pressure reservoirs and blood reservoirs.

The five main types of blood vessels are arteries, arterioles, capillaries, venules, and veins (see **Figure 21.17**). **Arteries** (AR-ter-ēz; *ar-* = air; *-ter-* = to carry) carry blood *away from the heart* to other organs. Large, elastic arteries leave the heart and divide into medium-sized, muscular arteries that branch out into the various regions of the body. Medium-sized arteries then divide into small arteries, which in turn divide into still smaller arteries called **arterioles** (ar-TĒR-ē-ōls). As the arterioles enter a tissue, they branch into numerous tiny vessels called **blood capillaries** (KAP-i-lar′-ēz = hairlike) or simply **capillaries.** The thin walls of capillaries allow the exchange of substances between the blood and body tissues. Groups of capillaries within a tissue reunite to form small veins called **venules** (VEN-ūls = little veins). These in turn merge to form progressively larger blood vessels called veins. **Veins** (VĀNZ) are the blood vessels that convey blood from the tissues *back to the heart.*

See Clinical Connection: Angiogenesis and Disease

Basic Structure of a Blood Vessel

The wall of a blood vessel consists of three layers, or tunics, of different tissues: an epithelial inner lining, a middle layer consisting of smooth muscle and elastic connective tissue, and a connective tissue outer covering. The three structural layers of a generalized blood vessel from innermost to outermost are the tunica interna (intima), tunica media, and tunica externa (adventitia) (**Figure 21.1**). Modifications of this basic design account for the five types of blood vessels and the structural and functional differences among the various vessel types. Always remember that structural variations correlate to the differences in function that occur throughout the cardiovascular system.

Tunica Interna

The **tunica interna** (*intima*) (TOO-ni-ka; *tunic* = garment or coat; *interna* or *intima* = innermost) forms the inner lining of a blood vessel and is in direct contact with the blood as it flows through the **lumen** (LOO-men), or interior opening, of the vessel (**Figure 21.1a, b**). Although this layer has multiple parts, these tissue components contribute minimally to the thickness of the vessel wall. Its innermost layer is called *endothelium*, which is continuous with the endocardial lining of the heart. The endothelium is a thin layer of flattened cells that lines the inner surface of the entire cardiovascular system (heart and blood vessels). Until recently, endothelial cells were regarded as little more than a passive barrier between the blood and the remainder of the vessel wall. It is now known that endothelial cells are active participants in a variety of vessel-related activities, including physical influences on blood flow, secretion of locally acting chemical mediators that influence the contractile state of the vessel's overlying smooth muscle, and assistance with capillary permeability. In addition, their smooth luminal surface facilitates efficient blood flow by reducing surface friction.

The second component of the tunica interna is a *basement membrane* deep to the endothelium. It provides a physical support base for the epithelial layer. Its framework of collagen fibers affords the basement membrane significant tensile strength, yet its properties also provide resilience for stretching and recoil. The basement membrane anchors the endothelium to the underlying connective tissue while also regulating molecular movement. It appears to play an important role in guiding cell movements during tissue repair of blood vessel walls. The outermost part of the tunica interna, which forms the boundary between the tunica interna and tunica media, is the *internal elastic lamina* (*lamina* = thin plate). The internal elastic lamina is a thin sheet of elastic fibers with a variable number of windowlike openings that give it the look of Swiss cheese. These openings facilitate diffusion of materials through the tunica interna to the thicker tunica media.

Tunica Media

The **tunica media** (*media* = middle) is a muscular and connective tissue layer that displays the greatest variation among the different vessel types (**Figure 21.1a, b**). In most vessels, it is a relatively thick layer comprising mainly smooth muscle cells and substantial amounts of elastic fibers. The primary role of the smooth muscle cells, which extend circularly around the lumen like a ring encircles your finger, is to regulate the diameter of the lumen. An increase in sympathetic stimulation typically stimulates the smooth muscle to contract, squeezing the vessel wall and narrowing the lumen. Such a decrease in the diameter of the lumen of a blood vessel is called **vasoconstriction** (vā-sō-kon-STRIK-shun). In contrast, when sympathetic stimulation decreases, or in the presence of certain chemicals (such as nitric oxide, H^+, and lactic acid) or in response to blood pressure, smooth muscle fibers relax. The resulting increase in lumen diameter is called **vasodilation** (vā-sō-dī-LĀ-shun). As you will learn in more detail shortly, the rate of blood flow through different parts of the body is regulated by the extent of smooth muscle contraction in the walls of particular vessels. Furthermore, the extent of smooth muscle contraction in particular vessel types is crucial in the regulation of blood pressure.

In addition to regulating blood flow and blood pressure, smooth muscle contracts when a small artery or arteriole is damaged (*vascular spasm*) to help limit loss of blood through the injured vessel. Smooth muscle cells also help produce the elastic fibers within the tunica media that allow the vessels to stretch and recoil under the applied pressure of the blood.

The tunica media is the most variable of the tunics. As you study the different types of blood vessels in the remainder of this chapter, you will see that the structural differences in this layer account for the many variations in function among the different vessel types. Separating the tunica media from the tunica externa is a network of elastic fibers, the *external elastic lamina*, which is part of the tunica media.

Tunica Externa

The outer covering of a blood vessel, the **tunica externa** (*externa* = outermost), consists of elastic and collagen fibers (**Figure 21.1a, b**). The tunica externa contains numerous nerves and, especially in larger vessels, tiny blood vessels that supply the tissue of the vessel wall. These small vessels that supply blood to the tissues of the vessel are called **vasa vasorum** (VĀ-sa va-SŌ-rum; *vas* = vessel), or vessels to the vessels. They are easily seen on large vessels such as the aorta. In addition to the important role of supplying the vessel wall with nerves and self-vessels, the tunica externa helps anchor the vessels to surrounding tissues.

Arteries

Because **arteries** were found empty at death, in ancient times they were thought to contain only air. The wall of an artery has the three layers of a typical blood vessel, but has a thick muscular-to-elastic tunica media (**Figure 21.1a**). Due to their plentiful elastic fibers, arteries normally have high *compliance*, which means that their walls stretch easily or expand without tearing in response to a small increase in pressure.

Elastic Arteries

Elastic arteries are the largest arteries in the body, ranging from the garden hose–sized aorta and pulmonary trunk to the finger-sized branches of the aorta. They have the largest diameter among arteries, but their vessel walls (approximately one-tenth of the vessel's total diameter) are relatively thin compared with the overall size of the vessel. These vessels are characterized by well-defined internal and external elastic laminae, along with a thick tunica media that is dominated by elastic fibers, called the **elastic lamellae** (la-MEL-ē = little plates). Elastic arteries include the two major trunks that exit the heart (the aorta and the pulmonary trunk), along with the aorta's major initial branches, such as the brachiocephalic,

FIGURE 21.1 Comparative structure of blood vessels. The capillary (c) is enlarged relative to the artery (a) and vein (b).

Arteries carry blood from the heart to tissues; veins carry blood from tissues to the heart.

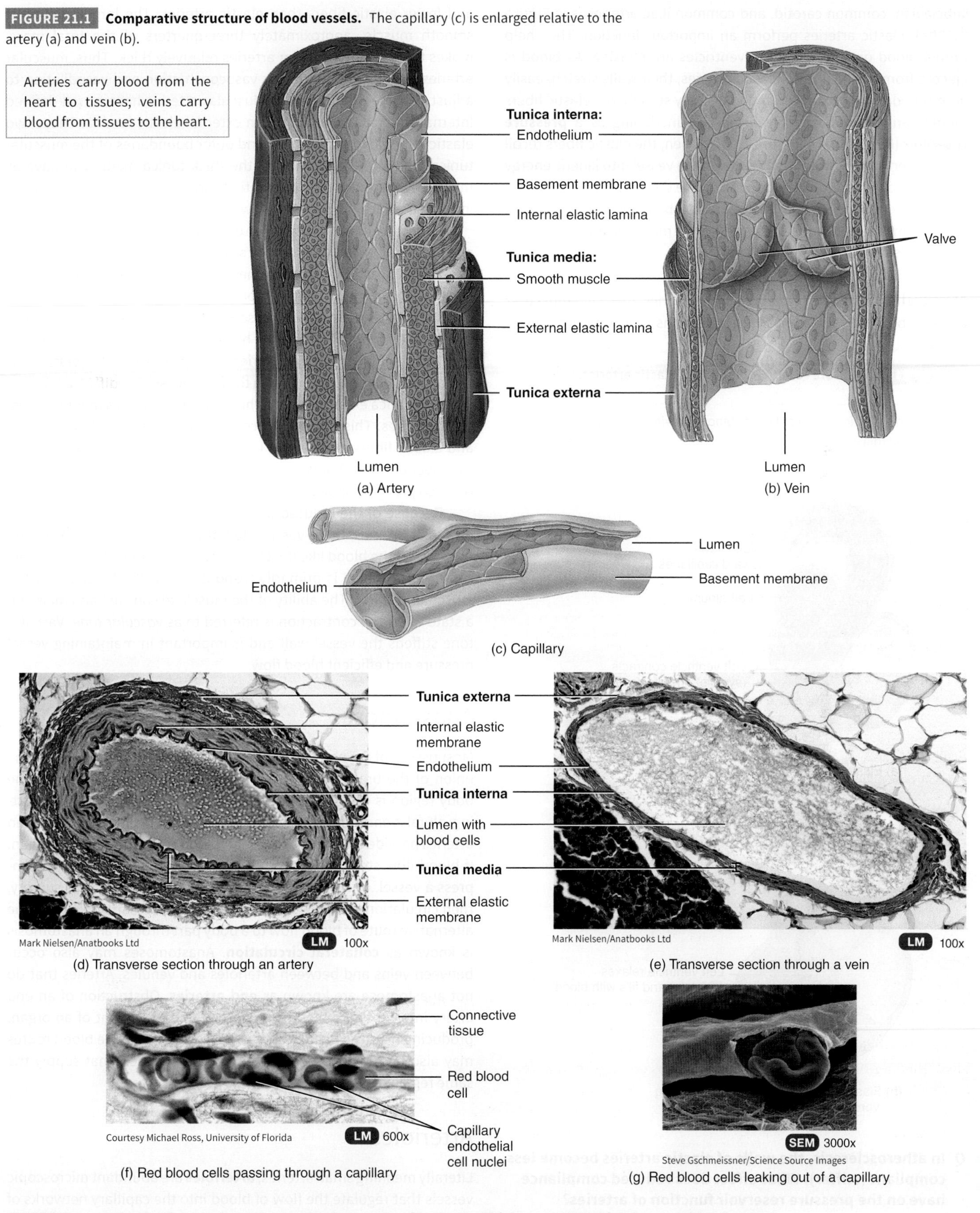

(a) Artery

(b) Vein

(c) Capillary

(d) Transverse section through an artery

(e) Transverse section through a vein

(f) Red blood cells passing through a capillary

(g) Red blood cells leaking out of a capillary

Q Which vessel—the femoral artery or the femoral vein—has a thicker wall? Which has a wider lumen?

subclavian, common carotid, and common iliac arteries (see **Figure 21.20a**). Elastic arteries perform an important function: They help propel blood onward while the ventricles are relaxing. As blood is ejected from the heart into elastic arteries, their walls stretch, easily accommodating the surge of blood. As they stretch, the elastic fibers momentarily store mechanical energy, functioning as a **pressure reservoir** (REZ-er-vwar) (**Figure 21.2a**). Then, the elastic fibers recoil and convert stored (potential) energy in the vessel into kinetic energy of the blood. Thus, blood continues to move through the arteries even while the ventricles are relaxed (**Figure 21.2b**). Because they conduct blood from the heart to medium-sized, more muscular arteries, elastic arteries also are called *conducting arteries*.

Muscular Arteries

Medium-sized arteries are called **muscular arteries** because their tunica media contains more smooth muscle and fewer elastic fibers than elastic arteries. The large amount of smooth muscle, approximately three-quarters of the total mass, makes the walls of muscular arteries relatively thick. Thus, muscular arteries are capable of greater vasoconstriction and vasodilation to adjust the rate of blood flow. Muscular arteries have a well-defined internal elastic lamina but a thin external elastic lamina. These two elastic laminae form the inner and outer boundaries of the muscular tunica media. In large arteries, the thick tunica media can have as many as 40 layers of circumferentially arranged smooth muscle cells; in smaller arteries there are as few as three layers.

Muscular arteries span a range of sizes from the pencil-sized femoral and axillary arteries to string-sized arteries that enter organs, measuring as little as 0.5 mm (1/64 inch) in diameter. Compared to elastic arteries, the vessel wall of muscular arteries comprises a larger percentage (25%) of the total vessel diameter. Because the muscular arteries continue to branch and ultimately distribute blood to each of the various organs, they are called **distributing arteries**. Examples include the brachial artery in the arm and radial artery in the forearm (see **Figure 21.20a**).

The tunica externa is often thicker than the tunica media in muscular arteries. This outer layer contains fibroblasts, collagen fibers, and elastic fibers all oriented longitudinally. The loose structure of this layer permits changes in the diameter of the vessel to take place but also prevents shortening or retraction of the vessel when it is cut.

Because of the reduced amount of elastic tissue in the walls of muscular arteries, these vessels do not have the ability to recoil and help propel the blood like the elastic arteries. Instead, the thick, muscular tunica media is primarily responsible for the functions of the muscular arteries. The ability of the muscle to contract and maintain a state of partial contraction is referred to as *vascular tone*. Vascular tone stiffens the vessel wall and is important in maintaining vessel pressure and efficient blood flow.

FIGURE 21.2 Pressure reservoir function of elastic arteries.

Recoil of elastic arteries keeps blood flowing during ventricular relaxation (diastole).

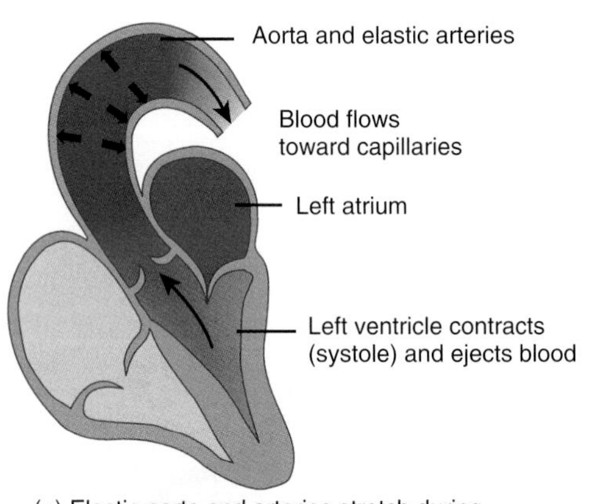

(a) Elastic aorta and arteries stretch during ventricular contraction

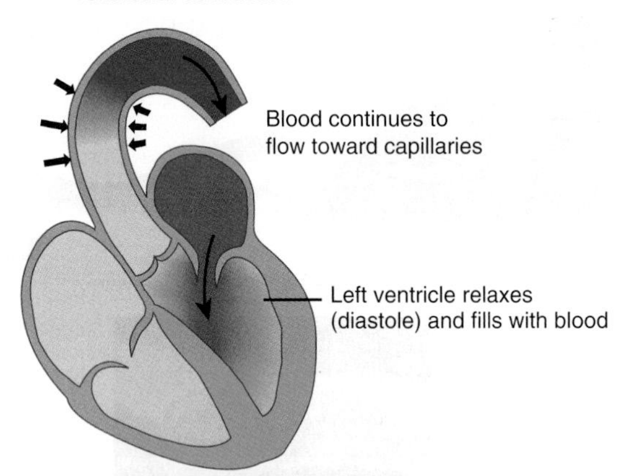

(b) Elastic aorta and arteries recoil during ventricular relaxation

Q In atherosclerosis, the walls of elastic arteries become less compliant (stiffer). What effect does reduced compliance have on the pressure reservoir function of arteries?

Anastomoses

Most tissues of the body receive blood from more than one artery. The union of the branches of two or more arteries supplying the same body region is called an **anastomosis** (a-nas′-tō-MŌ-sis = connecting; plural is *anastomoses*) (see **Figure 21.22c**). Anastomoses between arteries provide alternative routes for blood to reach a tissue or organ. If blood flow stops for a short time when normal movements compress a vessel, or if a vessel is blocked by disease, injury, or surgery, then circulation to a part of the body is not necessarily stopped. The alternative route of blood flow to a body part through an anastomosis is known as **collateral circulation**. Anastomoses may also occur between veins and between arterioles and venules. Arteries that do not anastomose are known as **end arteries**. Obstruction of an end artery interrupts the blood supply to a whole segment of an organ, producing necrosis (death) of that segment. Alternative blood routes may also be provided by nonanastomosing vessels that supply the same region of the body.

Arterioles

Literally meaning small arteries, **arterioles** are abundant microscopic vessels that regulate the flow of blood into the capillary networks of the body's tissues (see Figure 21.3). The approximately 400 million arterioles have diameters that range in size from 15 μm to 300 μm. The wall thickness of arterioles is one-half of the total vessel diameter.

Arterioles have a thin tunica interna with a thin, fenestrated (with small pores) internal elastic lamina that disappears at the terminal end. The tunica media consists of one to two layers of smooth muscle cells having a circular orientation in the vessel wall. The terminal end of the arteriole, the region called the **metarteriole** (met′-ar-TĒR-ē-ōl; *meta* = after), tapers toward the capillary junction. At the metarteriole–capillary junction, the distal-most muscle cell forms the **precapillary sphincter** (SFINGK-ter = to bind tight), which monitors the blood flow into the capillary; the other muscle cells in the arteriole regulate the resistance (opposition) to blood flow (see Figure 21.3).

The tunica externa of the arteriole consists of areolar connective tissue containing abundant unmyelinated sympathetic nerves. This sympathetic nerve supply, along with the actions of local chemical mediators, can alter the diameter of arterioles and thus vary the rate of blood flow and resistance through these vessels.

Arterioles play a key role in regulating blood flow from arteries into capillaries by regulating **resistance**, the opposition to blood flow due to friction between blood and the walls of blood vessels. Because of this they are known as *resistance vessels*. In a blood vessel, resistance is due mainly to friction between blood and the inner walls of blood vessels. When the blood vessel diameter is smaller, the friction is greater, so there is more resistance. Contraction of the smooth muscle of an arteriole causes vasoconstriction, which increases resistance even more and decreases blood flow into capillaries supplied by that arteriole. By contrast, relaxation of the smooth muscle of an arteriole causes vasodilation, which decreases resistance and increases blood flow into capillaries. A change in arteriole diameter can also affect blood pressure: Vasoconstriction of arterioles increases blood pressure, and vasodilation of arterioles decreases blood pressure.

Capillaries

Capillaries, the smallest of blood vessels, have diameters of 5–10 μm, and form the U-turns that connect the arterial outflow to the venous return (Figure 21.3). Since red blood cells have a diameter of 8 μm, they must often fold on themselves in order to pass single file through the lumens of these vessels. Capillaries form an extensive network, approximately 20 billion in number, of short (hundreds of micrometers in length), branched, interconnecting vessels that course among the individual cells of the body. This network forms an enormous surface area to make contact with the body's cells. The flow of blood from a

FIGURE 21.3 Arterioles, capillaries, and venules. Precapillary sphincters regulate the flow of blood through capillary beds.

In capillaries, nutrients, gases, and wastes are exchanged between the blood and interstitial fluid.

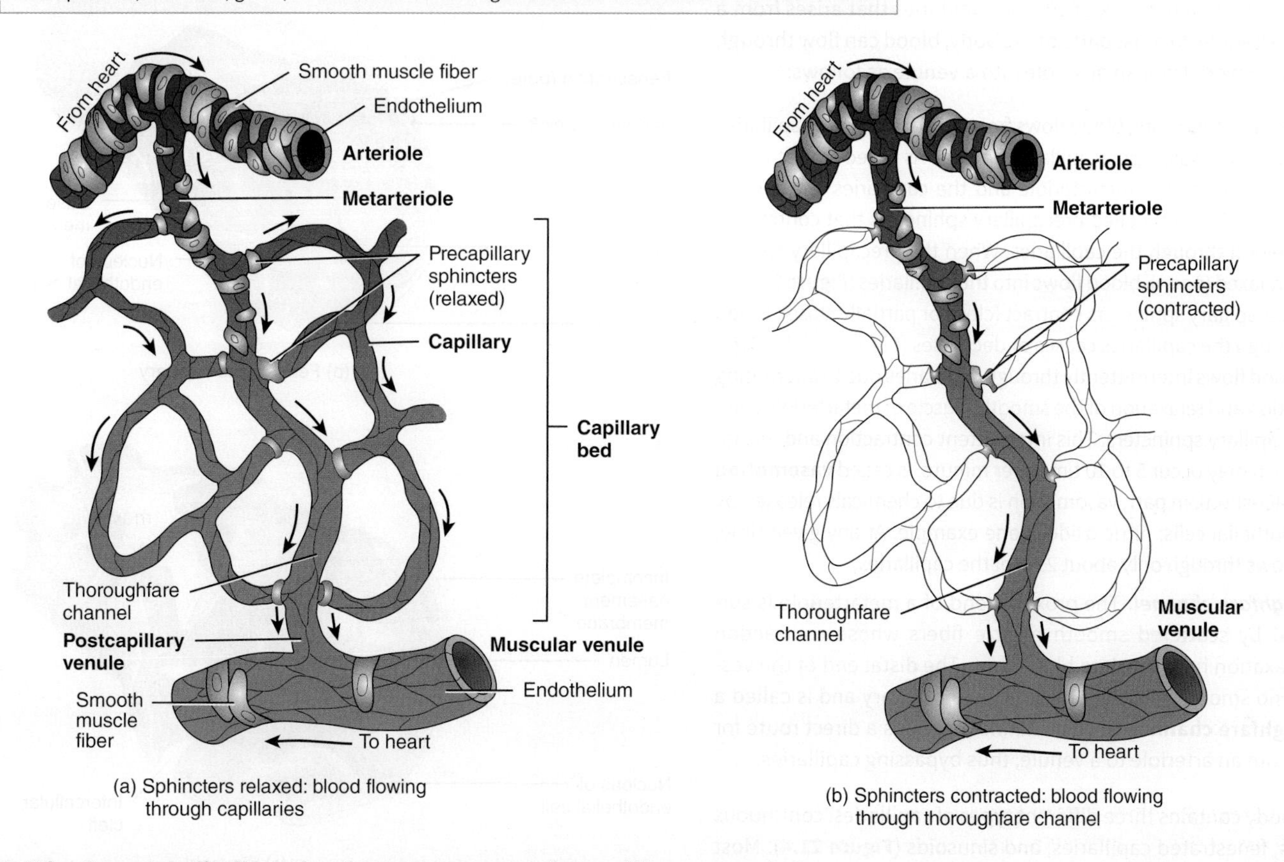

Q Why do metabolically active tissues have extensive capillary networks?

metarteriole through capillaries and into a **postcapillary venule** (venule that receives blood from a capillary) is called the **microcirculation** (*micro* = small) of the body. The primary function of capillaries is the exchange of substances between the blood and interstitial fluid. Because of this, these thin-walled vessels are referred to as *exchange vessels.*

Capillaries are found near almost every cell in the body, but their number varies with the metabolic activity of the tissue they serve. Body tissues with high metabolic requirements, such as muscles, the brain, the liver, the kidneys, and the nervous system, use more O_2 and nutrients and thus have extensive capillary networks. Tissues with lower metabolic requirements, such as tendons and ligaments, contain fewer capillaries. Capillaries are absent in a few tissues, such as all covering and lining epithelia, the cornea and lens of the eye, and cartilage.

The structure of capillaries is well suited to their function as exchange vessels because they lack both a tunica media and a tunica externa. Because capillary walls are composed of only a single layer of endothelial cells (see **Figure 21.1e**) and a basement membrane, a substance in the blood must pass through just one cell layer to reach the interstitial fluid and tissue cells. Exchange of materials occurs only through the walls of capillaries and the beginning of venules; the walls of arteries, arterioles, most venules, and veins present too thick a barrier. Capillaries form extensive branching networks that increase the surface area available for rapid exchange of materials. In most tissues, blood flows through only a small part of the capillary network when metabolic needs are low. However, when a tissue is active, such as contracting muscle, the entire capillary network fills with blood.

Throughout the body, capillaries function as part of a **capillary bed** (**Figure 21.3**), a network of 10–100 capillaries that arises from a single metarteriole. In most parts of the body, blood can flow through a capillary network from an arteriole into a venule as follows:

1. ***Capillaries.*** In this route, blood flows from an arteriole into capillaries and then into venules (postcapillary venules). As noted earlier, at the junctions between the metarteriole and the capillaries are rings of smooth muscle fibers called precapillary sphincters that control the flow of blood through the capillaries. When the precapillary sphincters are relaxed (open), blood flows into the capillaries (**Figure 21.3a**); when precapillary sphincters contract (close or partially close), blood flow through the capillaries ceases or decreases (**Figure 21.3b**). Typically, blood flows intermittently through capillaries due to alternating contraction and relaxation of the smooth muscle of metarterioles and the precapillary sphincters. This intermittent contraction and relaxation, which may occur 5 to 10 times per minute, is called **vasomotion** (vā-sō-MŌ-shun). In part, vasomotion is due to chemicals released by the endothelial cells; nitric oxide is one example. At any given time, blood flows through only about 25% of the capillaries.
2. ***Thoroughfare channel.*** The proximal end of a metarteriole is surrounded by scattered smooth muscle fibers whose contraction and relaxation help regulate blood flow. The distal end of the vessel has no smooth muscle; it resembles a capillary and is called a **thoroughfare channel**. Such a channel provides a direct route for blood from an arteriole to a venule, thus bypassing capillaries.

The body contains three different types of capillaries: continuous capillaries, fenestrated capillaries, and sinusoids (**Figure 21.4**). Most capillaries are **continuous capillaries**, in which the plasma membranes of endothelial cells form a continuous tube that is interrupted only by **intercellular clefts**, gaps between neighboring endothelial cells (**Figure 21.4a**). Continuous capillaries are found in the central nervous system, lungs, muscle tissue, and the skin.

Other capillaries of the body are **fenestrated capillaries** (fen′-es-TRĀ-ted; *fenestr-* = window). The plasma membranes of the endothelial cells in these capillaries have many **fenestrations** (fen′-es-TRĀ-shuns), small pores (holes) ranging from 70 to 100 nm in diameter

FIGURE 21.4 Types of capillaries.

Capillaries are microscopic blood vessels that connect arterioles and venules.

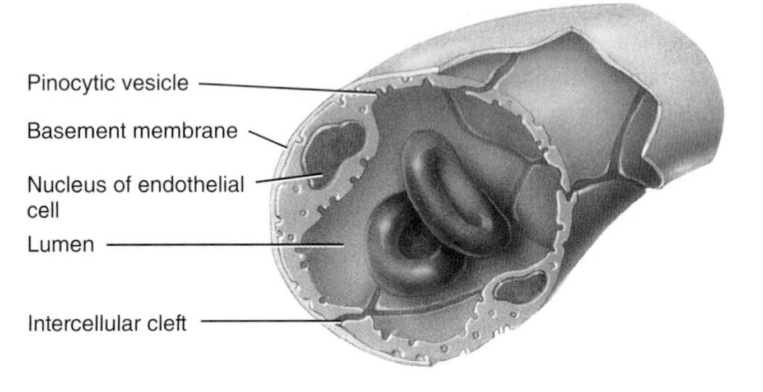

(a) Continuous capillary formed by endothelial cells

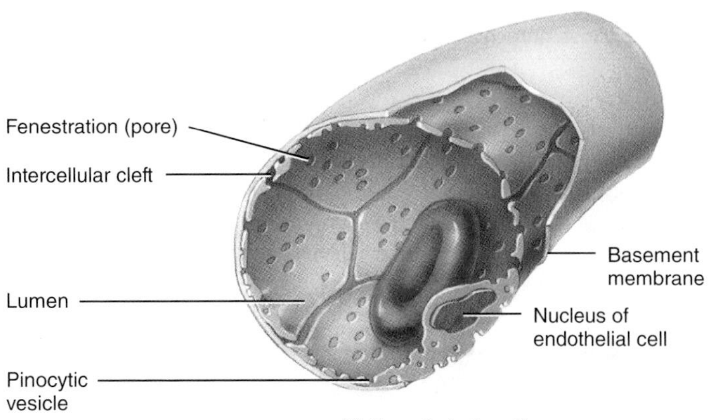

(b) Fenestrated capillary

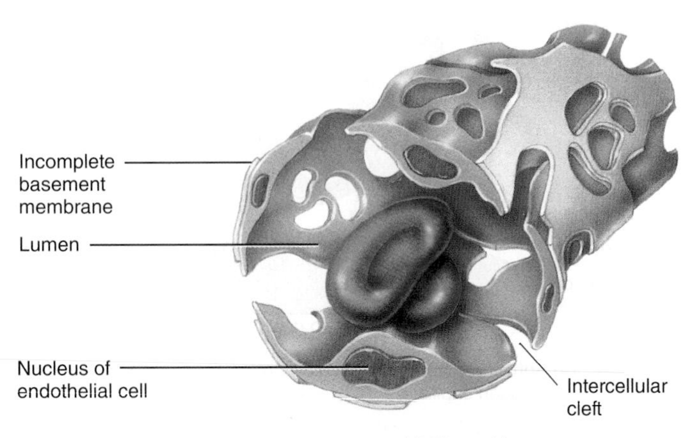

(c) Sinusoid

Q How do materials move through capillary walls?

(**Figure 21.4b**). Fenestrated capillaries are found in the kidneys, villi of the small intestine, choroid plexuses of the ventricles in the brain, ciliary processes of the eyes, and most endocrine glands.

Sinusoids (SĪ-nū-soyds; *sinus* = curve) are wider and more winding than other capillaries. Their endothelial cells may have unusually large fenestrations. In addition to having an incomplete or absent basement membrane (**Figure 21.4c**), sinusoids have very large intercellular clefts that allow proteins and in some cases even blood cells to pass from a tissue into the bloodstream. For example, newly formed blood cells enter the bloodstream through the sinusoids of red bone marrow. In addition, sinusoids contain specialized lining cells that are adapted to the function of the tissue. Sinusoids in the liver, for example, contain phagocytic cells that remove bacteria and other debris from the blood. The spleen, anterior pituitary, and parathyroid and adrenal glands also have sinusoids.

Usually blood passes from the heart and then in sequence through arteries, arterioles, capillaries, venules, and veins and then back to the heart. In some parts of the body, however, blood passes from one capillary network into another through a vein called a *portal vein*. Such a circulation of blood is called a **portal system**. The name of the portal system gives the name of the second capillary location. For example, there are portal systems associated with the liver (hepatic portal circulation; see **Figure 21.29**) and the pituitary gland (hypophyseal portal system; see **Figure 18.5**).

Venules

Unlike their thick-walled arterial counterparts, **venules** and veins have thin walls that do not readily maintain their shape. Venules drain the capillary blood and begin the return flow of blood back toward the heart (see **Figure 21.3**).

As noted earlier, venules that initially receive blood from capillaries are called **postcapillary venules**. They are the smallest venules, measuring 10 μm to 50 μm in diameter, and have loosely organized intercellular junctions (the weakest endothelial contacts encountered along the entire vascular tree) and thus are very porous. They function as significant sites of exchange of nutrients and wastes and white blood cell emigration, and for this reason form part of the microcirculatory exchange unit along with the capillaries.

As the postcapillary venules move away from capillaries, they acquire one or two layers of circularly arranged smooth muscle cells. These **muscular venules** (50 μm to 200 μm) have thicker walls across which exchanges with the interstitial fluid can no longer occur. The thin walls of the postcapillary and muscular venules are the most distensible elements of the vascular system; this allows them to expand and serve as excellent reservoirs for accumulating large volumes of blood. Blood volume increases of 360% have been measured in the postcapillary and muscular venules.

Veins

While **veins** do show structural changes as they increase in size from small to medium to large, the structural changes are not as distinct as they are in arteries. Veins, in general, have very thin walls relative to their total diameter (average thickness is less than one-tenth of the vessel diameter). They range in size from 0.5 mm in diameter for small veins to 3 cm in the large superior and interior venae cavae entering the heart.

Although veins are composed of essentially the same three layers as arteries, the relative thicknesses of the layers are different. The tunica interna of veins is thinner than that of arteries; the tunica media of veins is much thinner than in arteries, with relatively little smooth muscle and elastic fibers. The tunica externa of veins is the thickest layer and consists of collagen and elastic fibers. Veins lack the internal or external elastic laminae found in arteries (see **Figure 21.1b**). They are distensible enough to adapt to variations in the volume and pressure of blood passing through them, but are not designed to withstand high pressure. The lumen of a vein is larger than that of a comparable artery, and veins often appear collapsed (flattened) when sectioned.

The pumping action of the heart is a major factor in moving venous blood back to the heart. The contraction of skeletal muscles in the lower limbs also helps boost venous return to the heart (see **Figure 21.9**). The average blood pressure in veins is considerably lower than in arteries. The difference in pressure can be noticed when blood flows from a cut vessel. Blood leaves a cut vein in an even, slow flow but spurts rapidly from a cut artery. Most of the structural differences between arteries and veins reflect this pressure difference. For example, the walls of veins are not as strong as those of arteries.

Many veins, especially those in the limbs, also contain **valves**, thin folds of tunica interna that form flaplike cusps. The valve cusps project into the lumen, pointing toward the heart (**Figure 21.5**). The low blood pressure in veins allows blood returning to the heart to slow and even back up; the valves aid in venous return by preventing the backflow of blood.

FIGURE 21.5 Venous valves.

Valves in veins allow blood to flow in one direction only—toward the heart.

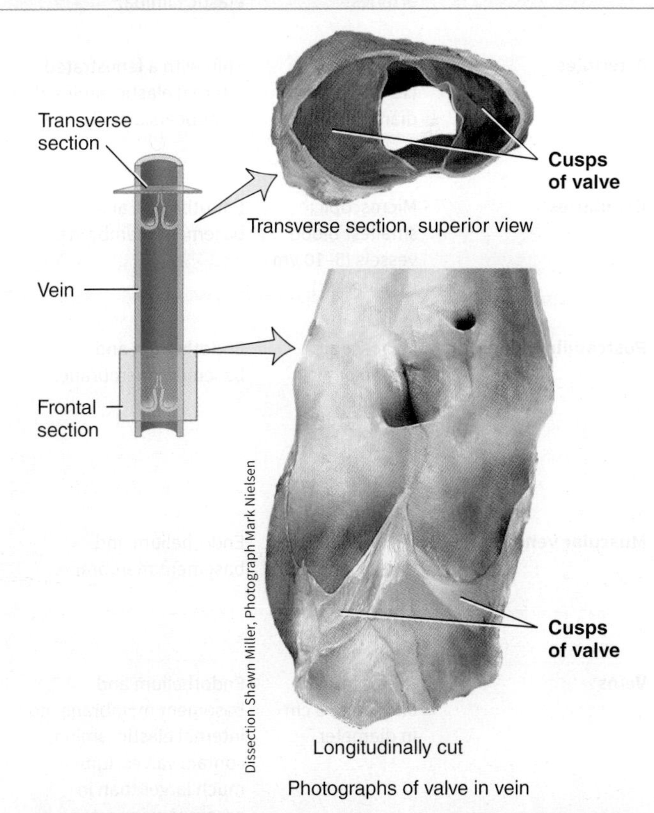

Q Why are valves more important in arm veins and leg veins than in neck veins?

A **vascular** (*venous*) **sinus** is a vein with a thin endothelial wall that has no smooth muscle to alter its diameter. In a vascular sinus, the surrounding dense connective tissue replaces the tunica media and tunica externa in providing support. For example, dural venous sinuses, which are supported by the dura mater, convey deoxygenated blood from the brain to the heart. Another example of a vascular sinus is the coronary sinus of the heart (see **Figure 20.3c**).

While veins follow paths similar to those of their arterial counterparts, they differ from arteries in a number of ways, aside from the structures of their walls. First, veins are more numerous than arteries for several reasons. Some veins are paired and accompany medium-to small-sized muscular arteries. These double sets of veins escort the arteries and connect with one another via venous channels called **anastomotic veins** (a-nas′-tō-MOT-ik). The anastomotic veins cross the accompanying artery to form ladderlike rungs between the paired veins (see **Figure 21.26c**). The greatest number of paired veins occurs within the limbs. The subcutaneous layer deep to the skin is another source of veins. These veins, called **superficial veins**, course through the subcutaneous layer unaccompanied by parallel arteries. Along their course, the superficial veins form small connections (anastomoses) with the **deep veins** that travel between the skeletal muscles. These connections allow communication between the deep and superficial flow of blood. The amount of blood flow through superficial veins varies from location to location within the body. In the upper limb, the superficial veins are much larger than the deep veins and serve as the major pathways from the capillaries of the upper limb back to the heart. In the lower limb, the opposite is true; the deep veins serve as the principal return pathways. In fact, one-way valves in small anastomosing vessels allow blood to pass from the superficial veins to the deep veins, but prevent the blood from passing in the reverse direction. This design has important implications in the development of varicose veins.

In some individuals the superficial veins can be seen as blue-colored tubes passing under the skin. While the venous blood is a deep dark red, the veins appear blue because their thin walls and the tissues of the skin absorb the red-light wavelengths, allowing the blue light to pass through the surface to our eyes where we see them as blue.

A summary of the distinguishing features of blood vessels is presented in **Table 21.1**.

TABLE 21.1 Distinguishing Features of Blood Vessels

BLOOD VESSEL	SIZE	TUNICA INTERNA	TUNICA MEDIA	TUNICA EXTERNA	FUNCTION
Elastic arteries	Largest arteries in the body.	Well-defined internal elastic lamina.	Thick and dominated by elastic fibers; well-defined external elastic lamina.	Thinner than tunica media.	Conduct blood from heart to muscular arteries.
Muscular arteries	Medium-sized arteries.	Well-defined internal elastic lamina.	Thick and dominated by smooth muscle; thin external elastic lamina.	Thicker than tunica media.	Distribute blood to arterioles.
Arterioles	Microscopic (15–300 μm in diameter).	Thin with a fenestrated internal elastic lamina that disappears distally.	One or two layers of circularly oriented smooth muscle; distalmost smooth muscle cell forms a precapillary sphincter.	Loose collagenous connective tissue and sympathetic nerves.	Deliver blood to capillaries and help regulate blood flow from arteries to capillaries.
Capillaries	Microscopic; smallest blood vessels (5–10 μm in diameter).	Endothelium and basement membrane.	None.	None.	Permit exchange of nutrients and wastes between blood and interstitial fluid; distribute blood to postcapillary venules.
Postcapillary venules	Microscopic (10–50 μm in diameter).	Endothelium and basement membrane.	None.	Sparse.	Pass blood into muscular venules; permit exchange of nutrients and wastes between blood and interstitial fluid and function in white blood cell emigration.
Muscular venules	Microscopic (50–200 μm in diameter).	Endothelium and basement membrane.	One or two layers of circularly oriented smooth muscle.	Sparse.	Pass blood into vein; act as reservoirs for accumulating large volumes of blood (along with postcapillary venules).
Veins	Range from 0.5 mm to 3 cm in diameter.	Endothelium and basement membrane; no internal elastic lamina; contain valves; lumen much larger than in accompanying artery.	Much thinner than in arteries; no external elastic lamina.	Thickest of the three layers.	Return blood to heart, facilitated by valves in limb veins.

Blood Distribution

The largest portion of your blood volume at rest—about 64%—is in systemic veins and venules (**Figure 21.6**). Systemic arteries and arterioles hold about 13% of the blood volume, systemic capillaries hold about 7%, pulmonary blood vessels hold about 9%, and the heart holds about 7%. Because systemic veins and venules contain a large percentage of the blood volume, they function as **blood reservoirs** from which blood can be diverted quickly if the need arises. For example, during increased muscular activity, the cardiovascular center in the brain stem sends a larger number of sympathetic impulses to veins. The result is *venoconstriction*, constriction of veins, which reduces the volume of blood in reservoirs and allows a greater blood volume to flow to skeletal muscles, where it is needed most. A similar mechanism operates in cases of hemorrhage, when blood volume and pressure decrease; in this case, venoconstriction helps counteract the drop in blood pressure. Among the principal blood reservoirs are the veins of the abdominal organs (especially the liver and spleen) and the veins of the skin.

See Clinical Connection: Varicose Veins

FIGURE 21.6 **Blood distribution in the cardiovascular system at rest.**

Because systemic veins and venules contain more than half of the total blood volume, they are called blood reservoirs.

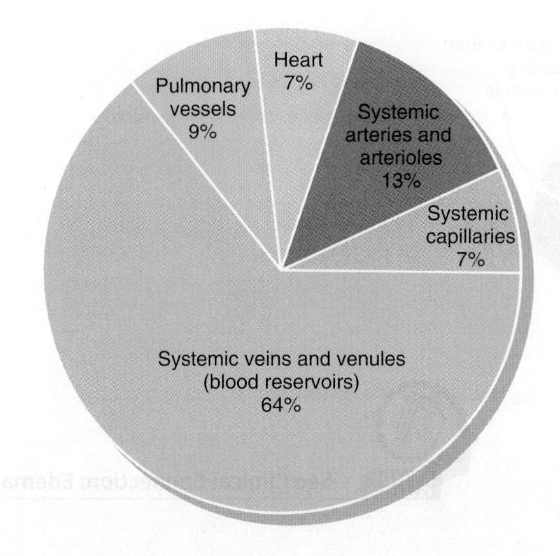

Q If your total blood volume is 5 liters, what volume is in your venules and veins right now? In your capillaries?

21.2 Capillary Exchange

OBJECTIVE

- **Discuss** the pressures that cause movement of fluids between capillaries and interstitial spaces.

The mission of the entire cardiovascular system is to keep blood flowing through capillaries to allow **capillary exchange**, the movement of substances between blood and interstitial fluid. The 7% of the blood in systemic capillaries at any given time is continually exchanging materials with interstitial fluid. Substances enter and leave capillaries by three basic mechanisms: diffusion, transcytosis, and bulk flow.

Diffusion

The most important method of capillary exchange is simple diffusion. Many substances, such as oxygen (O_2), carbon dioxide (CO_2), glucose, amino acids, and hormones, enter and leave capillaries by simple diffusion. Because O_2 and nutrients normally are present in higher concentrations in blood, they diffuse down their concentration gradients into interstitial fluid and then into body cells. CO_2 and other wastes released by body cells are present in higher concentrations in interstitial fluid, so they diffuse into blood.

Substances in blood or interstitial fluid can cross the walls of a capillary by diffusing through the intercellular clefts or fenestrations or by diffusing through the endothelial cells (see **Figure 21.4**). Water-soluble substances such as glucose and amino acids pass across capillary walls through intercellular clefts or fenestrations. Lipid-soluble materials, such as O_2, CO_2, and steroid hormones, may pass across capillary walls directly through the lipid bilayer of endothelial cell plasma membranes. Most plasma proteins and red blood cells cannot pass through capillary walls of continuous and fenestrated capillaries because they are too large to fit through the intercellular clefts and fenestrations.

In sinusoids, however, the intercellular clefts are so large that they allow even proteins and blood cells to pass through their walls. For example, hepatocytes (liver cells) synthesize and release many plasma proteins, such as fibrinogen (the main clotting protein) and albumin, which then diffuse into the bloodstream through sinusoids. In red bone marrow, blood cells are formed (hemopoiesis) and then enter the bloodstream through sinusoids.

In contrast to sinusoids, the capillaries of the brain allow only a few substances to move across their walls. Most areas of the brain contain continuous capillaries; however, these capillaries are very "tight." The endothelial cells of most brain capillaries are sealed together by tight junctions. The resulting blockade to movement of materials into and out of brain capillaries is known as the *blood–brain barrier* (see Section 14.1). In brain areas that lack the blood–brain barrier, for example, the hypothalamus, pineal gland, and pituitary gland, materials undergo capillary exchange more freely.

Transcytosis

A small quantity of material crosses capillary walls by **transcytosis** (tranz′-sī-TŌ-sis; *trans-* = across; *-cyt-* = cell; *-osis* = process). In this process, substances in blood plasma become enclosed within tiny pinocytic vesicles that first enter endothelial cells by endocytosis, then move across the cell and exit on the other side by exocytosis. This method of transport is important mainly for large, lipid-insoluble molecules that cannot cross capillary walls in any other way. For example, the hormone insulin (a small protein) enters the bloodstream by transcytosis, and certain antibodies (also proteins) pass from the maternal circulation into the fetal circulation by transcytosis.

Bulk Flow: Filtration and Reabsorption

Bulk flow is a passive process in which *large* numbers of ions, molecules, or particles in a fluid move together in the same direction. The substances move at rates far greater than can be accounted for by diffusion alone. Bulk flow occurs from an area of higher pressure to an area of lower pressure, and it continues as long as a pressure difference exists. Diffusion is more important for *solute exchange* between blood and interstitial fluid, but bulk flow is more important for regulation of the *relative volumes of blood and interstitial fluid.* Pressure-driven movement of fluid and solutes *from* blood capillaries *into* interstitial fluid is called **filtration**. Pressure-driven movement *from* interstitial fluid *into* blood capillaries is called **reabsorption**.

Two pressures promote filtration: blood hydrostatic pressure (BHP), the pressure generated by the pumping action of the heart, and interstitial fluid osmotic pressure (in′-ter-STISH-al). The main pressure promoting reabsorption of fluid is blood colloid osmotic pressure. The balance of these pressures, called **net filtration pressure (NFP)**, determines whether the volumes of blood and interstitial fluid remain steady or change. Overall, the volume of fluid and solutes reabsorbed normally is almost as large as the volume filtered. This near equilibrium is known as **Starling's law of the capillaries**. Let's see how these hydrostatic and osmotic pressures balance.

Within vessels, the hydrostatic pressure is due to the pressure that water in blood plasma exerts against blood vessel walls. The **blood hydrostatic pressure (BHP)** is about 35 millimeters of mercury (mmHg) at the arterial end of a capillary, and about 16 mmHg at the capillary's venous end (**Figure 21.7**). BHP "pushes" fluid out

FIGURE 21.7 Dynamics of capillary exchange (Starling's law of the capillaries). Excess filtered fluid drains into lymphatic capillaries.

Blood hydrostatic pressure pushes fluid out of capillaries (filtration), and blood colloid osmotic pressure pulls fluid into capillaries (reabsorption).

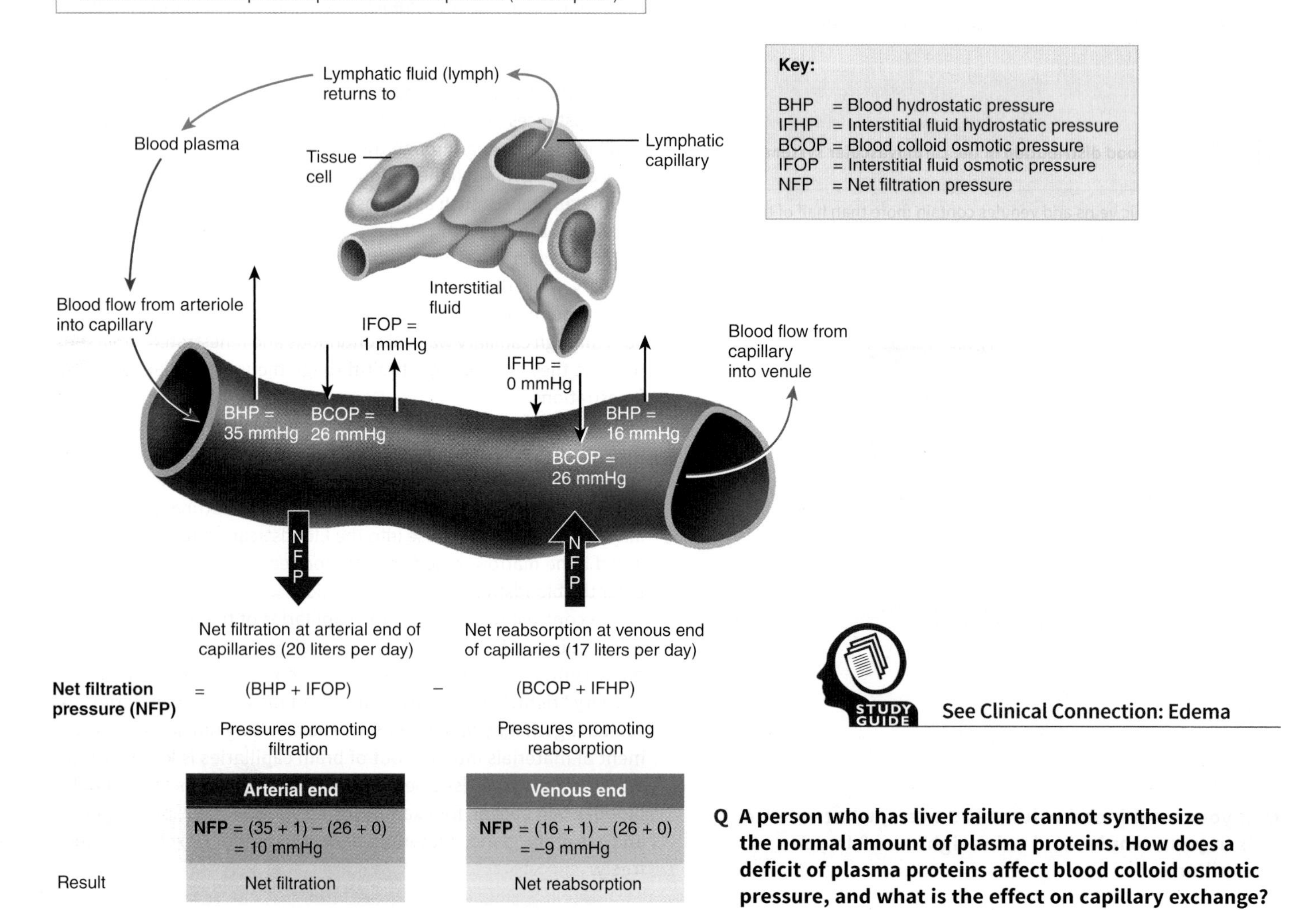

Q A person who has liver failure cannot synthesize the normal amount of plasma proteins. How does a deficit of plasma proteins affect blood colloid osmotic pressure, and what is the effect on capillary exchange?

of capillaries into interstitial fluid. The opposing pressure of the interstitial fluid, called **interstitial fluid hydrostatic pressure (IFHP)**, "pushes" fluid from interstitial spaces back into capillaries. However, IFHP is close to zero. (IFHP is difficult to measure, and its reported values vary from small positive values to small negative values.) For our discussion we assume that IFHP equals 0 mmHg all along the capillaries.

The difference in osmotic pressure across a capillary wall is due almost entirely to the presence in blood of plasma proteins, which are too large to pass through either fenestrations or gaps between endothelial cells. **Blood colloid osmotic pressure (BCOP)** is a force caused by the colloidal suspension of these large proteins in plasma that averages 26 mmHg in most capillaries. The effect of BCOP is to "pull" fluid from interstitial spaces into capillaries. Opposing BCOP is **interstitial fluid osmotic pressure (IFOP)**, which "pulls" fluid out of capillaries into interstitial fluid. Normally, IFOP is very small—0.1–5 mmHg—because only tiny amounts of protein are present in interstitial fluid. The small amount of protein that leaks from blood plasma into interstitial fluid does not accumulate there because it passes into lymph in lymphatic capillaries and is eventually returned to the blood. For discussion, we can use a value of 1 mmHg for IFOP.

Whether fluids leave or enter capillaries depends on the balance of pressures. If the pressures that push fluid out of capillaries exceed the pressures that pull fluid into capillaries, fluid will move from capillaries into interstitial spaces (filtration). If, however, the pressures that push fluid out of interstitial spaces into capillaries exceed the pressures that pull fluid out of capillaries, then fluid will move from interstitial spaces into capillaries (reabsorption).

The net filtration pressure (NFP), which indicates the direction of fluid movement, is calculated as follows:

$$\text{NFP} = \underbrace{(\text{BHP} + \text{IFOP})}_{\text{Pressures that promote filtration}} - \underbrace{(\text{BCOP} + \text{IFHP})}_{\text{Pressures that promote reabsorption}}$$

At the arterial end of a capillary,

$$\begin{aligned}\text{NFP} &= (35 + 1)\text{ mmHg} - (26 + 0)\text{ mmHg} \\ &= 36 - 26\text{ mmHg} = 10\text{ mmHg}\end{aligned}$$

Thus, at the arterial end of a capillary, there is a *net outward pressure* of 10 mmHg, and fluid moves out of the capillary into interstitial spaces (filtration).

At the venous end of a capillary,

$$\begin{aligned}\text{NFP} &= (16 + 1)\text{ mmHg} - (26 + 0)\text{ mmHg} \\ &= 17 - 26\text{ mmHg} = -9\text{ mmHg}\end{aligned}$$

At the venous end of a capillary, the negative value (−9 mmHg) represents a *net inward pressure*, and fluid moves into the capillary from tissue spaces (reabsorption).

On average, about 85% of the fluid filtered out of capillaries is reabsorbed. The excess filtered fluid and the few plasma proteins that do escape from blood into interstitial fluid enter lymphatic capillaries (see **Figure 22.2**). As lymph drains into the junction of the jugular and subclavian veins in the upper thorax (see **Figure 22.3**), these materials return to the blood. Every day about 20 liters of fluid filter out of capillaries in tissues throughout the body. Of this fluid, 17 liters are reabsorbed and 3 liters enter lymphatic capillaries (excluding filtration during urine formation).

21.3 Hemodynamics: Factors Affecting Blood Flow

OBJECTIVES

- **Explain** the factors that regulate the volume of blood flow.
- **Explain** how blood pressure changes throughout the cardiovascular system.
- **Describe** the factors that determine mean arterial pressure and systemic vascular resistance.
- **Describe** the relationship between cross-sectional area and velocity of blood flow.

Hemodynamics (hē-mō-dī-NAM-iks; *hemo-* = blood; *dynamics* = power) refers to the forces involved in circulating blood throughout the body. **Blood flow** is the volume of blood that flows through any tissue in a given time period (in mL/min). Total blood flow is cardiac output (CO), the volume of blood that circulates through systemic (or pulmonary) blood vessels each minute. In Chapter 20 we saw that cardiac output depends on heart rate and stroke volume: Cardiac output (CO) = heart rate (HR) × stroke volume (SV). How the cardiac output becomes distributed into circulatory routes that serve various body tissues depends on two more factors: (1) the *pressure difference* that drives the blood flow through a tissue and (2) the *resistance* to blood flow in specific blood vessels. Blood flows from regions of higher pressure to regions of lower pressure; the greater the pressure difference, the greater the blood flow. But the higher the resistance, the smaller the blood flow.

Blood Pressure

As you have just learned, blood flows from regions of higher pressure to regions of lower pressure; the greater the pressure difference, the greater the blood flow. Contraction of the ventricles generates **blood pressure (BP)**, the hydrostatic pressure exerted by blood on the walls of a blood vessel. BP is determined by cardiac output (see Section 20.5), blood volume, and vascular resistance (described shortly). BP is highest in the aorta and large systemic arteries; in a resting, young adult, BP rises to about 110 mmHg during systole (ventricular contraction) and drops to about 70 mmHg during diastole (ventricular relaxation). **Systolic blood pressure (SBP)** (sis-TOL-ik) is the highest pressure attained in arteries during systole, and **diastolic blood pressure (DBP)** (dī-a-STOL-ik) is the lowest arterial pressure during diastole (**Figure 21.8**). As blood leaves the aorta and flows through the systemic circulation, its pressure falls progressively as the distance from the left ventricle increases. Blood pressure decreases to about 35 mmHg as blood passes from systemic arteries through systemic arterioles and into capillaries, where the pressure fluctuations disappear. At the venous end of capillaries, blood pressure has dropped to about 16 mmHg. Blood pressure continues to drop as blood enters systemic venules and then veins because these vessels are farthest from the left ventricle. Finally, blood pressure reaches 0 mmHg as blood flows into the right ventricle.

FIGURE 21.8 Blood pressures in various parts of the cardiovascular system. The dashed line is the mean (average) blood pressure in the aorta, arteries, and arterioles.

Blood pressure rises and falls with each heartbeat in blood vessels leading to capillaries.

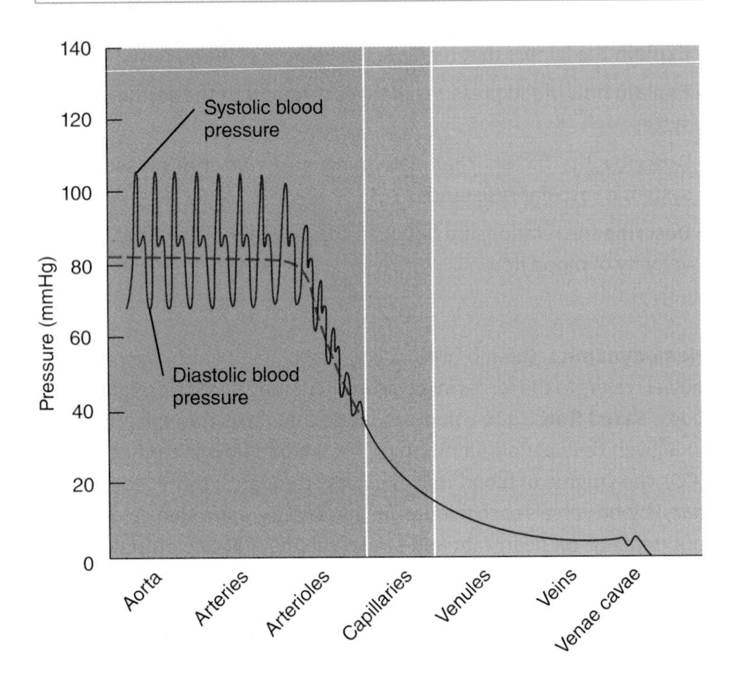

Q Is the mean blood pressure in the aorta closer to systolic or to diastolic pressure?

Mean arterial pressure (MAP), the average blood pressure in arteries, is roughly one-third of the way between the diastolic and systolic pressures. It can be estimated as follows:

$$\text{MAP} = \text{diastolic BP} + 1/3\,(\text{systolic BP} - \text{diastolic BP})$$

Thus, in a person whose BP is 110/70 mmHg, MAP is about 83 mmHg $[70 + 1/3(110 - 70)]$.

We have already seen that cardiac output equals heart rate multiplied by stroke volume. Another way to calculate cardiac output is to divide mean arterial pressure (MAP) by resistance (R): $\text{CO} = \text{MAP} \div \text{R}$. By rearranging the terms of this equation, you can see that $\text{MAP} = \text{CO} \times \text{R}$. If cardiac output rises due to an increase in stroke volume or heart rate, then the mean arterial pressure rises as long as resistance remains steady. Likewise, a decrease in cardiac output causes a decrease in mean arterial pressure if resistance does not change.

Blood pressure also depends on the total volume of blood in the cardiovascular system. The normal volume of blood in an adult is about 5 liters (5.3 qt). Any decrease in this volume, as from hemorrhage, decreases the amount of blood that is circulated through the arteries each minute. A modest decrease can be compensated for by homeostatic mechanisms that help maintain blood pressure (described in Section 21.4), but if the decrease in blood volume is greater than 10% of the total, blood pressure drops. Conversely, anything that increases blood volume, such as water retention in the body, tends to increase blood pressure.

Vascular Resistance

As noted earlier, **vascular resistance** is the opposition to blood flow due to friction between blood and the walls of blood vessels. Vascular resistance depends on (1) size of the blood vessel lumen, (2) blood viscosity, and (3) total blood vessel length.

1. ***Size of the lumen.*** The smaller the lumen of a blood vessel, the greater its resistance to blood flow. Resistance is inversely proportional to the fourth power of the diameter (d) of the blood vessel's lumen ($R \propto 1/d^4$). The smaller the diameter of the blood vessel, the greater the resistance it offers to blood flow. For example, if the diameter of a blood vessel decreases by one-half, its resistance to blood flow increases 16 times. Vasoconstriction narrows the lumen, and vasodilation widens it. Normally, moment-to-moment fluctuations in blood flow through a given tissue are due to vasoconstriction and vasodilation of the tissue's arterioles. As arterioles dilate, resistance decreases, and blood pressure falls. As arterioles constrict, resistance increases, and blood pressure rises.
2. ***Blood viscosity.*** The viscosity (vis-KOS-i-tē = thickness) of blood depends mostly on the ratio of red blood cells to plasma (fluid) volume, and to a smaller extent on the concentration of proteins in plasma. The higher the blood's viscosity, the higher the resistance. Any condition that increases the viscosity of blood, such as dehydration or polycythemia (an unusually high number of red blood cells), thus increases blood pressure. A depletion of plasma proteins or red blood cells, due to anemia or hemorrhage, decreases viscosity and thus decreases blood pressure.
3. ***Total blood vessel length.*** Resistance to blood flow through a vessel is directly proportional to the length of the blood vessel. The longer a blood vessel, the greater the resistance. Obese people often have hypertension (elevated blood pressure) because the additional blood vessels in their adipose tissue increase their total blood vessel length. An estimated 650 km (about 400 miles) of additional blood vessels develop for each extra kilogram (2.2 lb) of fat.

Systemic vascular resistance (SVR), also known as *total peripheral resistance (TPR)*, refers to all of the vascular resistances offered by systemic blood vessels. The diameters of arteries and veins are large, so their resistance is very small because most of the blood does not come into physical contact with the walls of the blood vessel. The smallest vessels—arterioles, capillaries, and venules—contribute the most resistance. A major function of arterioles is to control SVR—and therefore blood pressure and blood flow to particular tissues—by changing their diameters. Arterioles need to vasodilate or vasoconstrict only slightly to have a large effect on SVR. The main center for regulation of SVR is the vasomotor center in the brain stem (described shortly).

Venous Return

Venous return, the volume of blood flowing back to the heart through the systemic veins, occurs due to the pressure generated by contractions of the heart's left ventricle. The pressure difference from venules (averaging about 16 mmHg) to the right ventricle (0 mmHg), although small, normally is sufficient to cause venous return to the heart. If pressure increases in the right atrium or ventricle, venous return will decrease. One cause of increased pressure in the right atrium is an incompetent (leaky) tricuspid valve, which lets blood regurgitate (flow backward) as the ventricles contract. The result is decreased venous return and buildup of blood on the venous side of the systemic circulation.

When you stand up, for example, at the end of an anatomy and physiology lecture, the pressure pushing blood up the veins in your lower limbs is barely enough to overcome the force of gravity pushing it back down. Besides the heart, two other mechanisms "pump" blood from the lower body back to the heart: (1) the skeletal muscle pump and (2) the respiratory pump. Both pumps depend on the presence of valves in veins.

The **skeletal muscle pump** operates as follows (**Figure 21.9**):

1. While you are standing at rest, both the venous valve closer to the heart (proximal valve) and the one farther from the heart (distal valve) in this part of the leg are open, and blood flows upward toward the heart.
2. Contraction of leg muscles, such as when you stand on tiptoes or take a step, compresses the vein. The compression pushes blood through the proximal valve, an action called *milking*. At the same time, the distal valve in the uncompressed segment of the vein closes as some blood is pushed against it. People who are immobilized through injury or disease lack these contractions of leg muscles. As a result, their venous return is slower and they may develop circulation problems.
3. Just after muscle relaxation, pressure falls in the previously compressed section of vein, which causes the proximal valve to close. The distal valve now opens because blood pressure in the foot is higher than in the leg, and the vein fills with blood from the foot. The proximal valve then reopens.

FIGURE 21.9 Action of the skeletal muscle pump in returning blood to the heart.

Milking refers to skeletal muscle contractions that drive venous blood toward the heart.

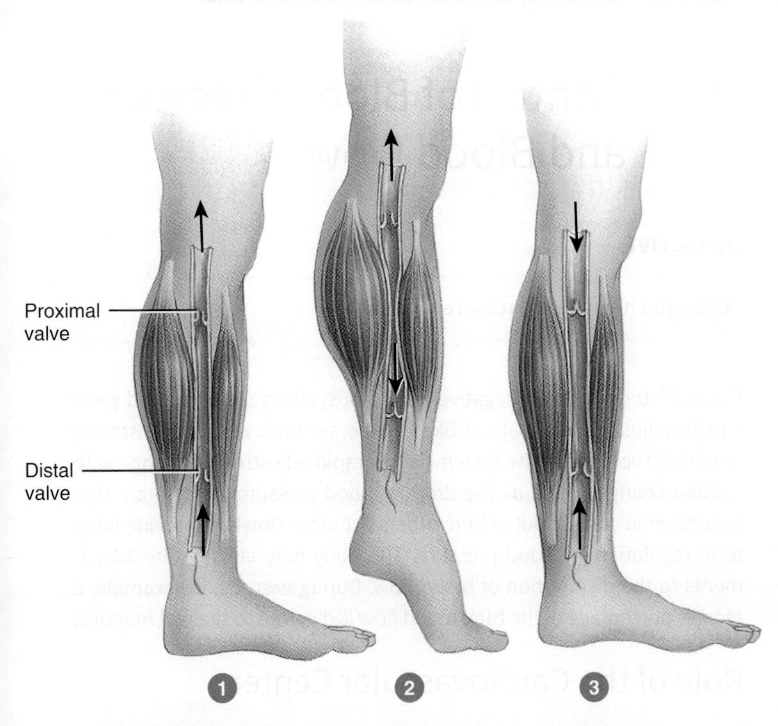

Q Aside from cardiac contractions, what mechanisms act as pumps to boost venous return?

The **respiratory pump** is also based on alternating compression and decompression of veins. During inhalation, the diaphragm moves downward, which causes a decrease in pressure in the thoracic cavity and an increase in pressure in the abdominal cavity. As a result, abdominal veins are compressed, and a greater volume of blood moves from the compressed abdominal veins into the decompressed thoracic veins and then into the right atrium. When the pressures reverse during exhalation, the valves in the veins prevent backflow of blood from the thoracic veins to the abdominal veins.

Figure 21.10 summarizes the factors that increase blood pressure through increasing cardiac output or systemic vascular resistance.

See Clinical Connection: Syncope

Velocity of Blood Flow

Earlier we saw that blood flow is the *volume* of blood that flows through any tissue in a given time period (in mL/min). The speed or *velocity* of blood flow (in cm/sec) is inversely related to the cross-sectional area. Velocity is slowest where the total cross-sectional area is greatest (**Figure 21.11**). Each time an artery branches, the total cross-sectional area of all of its branches is greater than the cross-sectional area of the original vessel, so blood flow becomes slower and slower as blood moves further away from the heart, and is slowest in the capillaries. Conversely, when venules unite to form veins, the total cross-sectional area becomes smaller and flow becomes faster. In an adult, the cross-sectional area of the aorta is only 3–5 cm^2, and the average velocity of the blood there is 40 cm/sec. In capillaries, the total cross-sectional area is 4500–6000 cm^2, and the velocity of blood flow is less than 0.1 cm/sec. In the two venae cavae combined, the cross-sectional area is about 14 cm^2, and the velocity is about 15 cm/sec. Thus, the velocity of blood flow decreases as blood flows from the aorta to arteries to arterioles to capillaries, and increases as it leaves capillaries and returns to the heart. The relatively slow rate of flow through capillaries aids the exchange of materials between blood and interstitial fluid.

Circulation time is the time required for a drop of blood to pass from the right atrium, through the pulmonary circulation, back to the left atrium, through the systemic circulation down to the foot, and back again to the right atrium. In a resting person, circulation time normally is about 1 minute.

FIGURE 21.10 Summary of factors that increase blood pressure. Changes noted within green boxes increase cardiac output; changes noted within blue boxes increase systemic vascular resistance.

Increases in cardiac output and increases in systemic vascular resistance will increase mean arterial pressure.

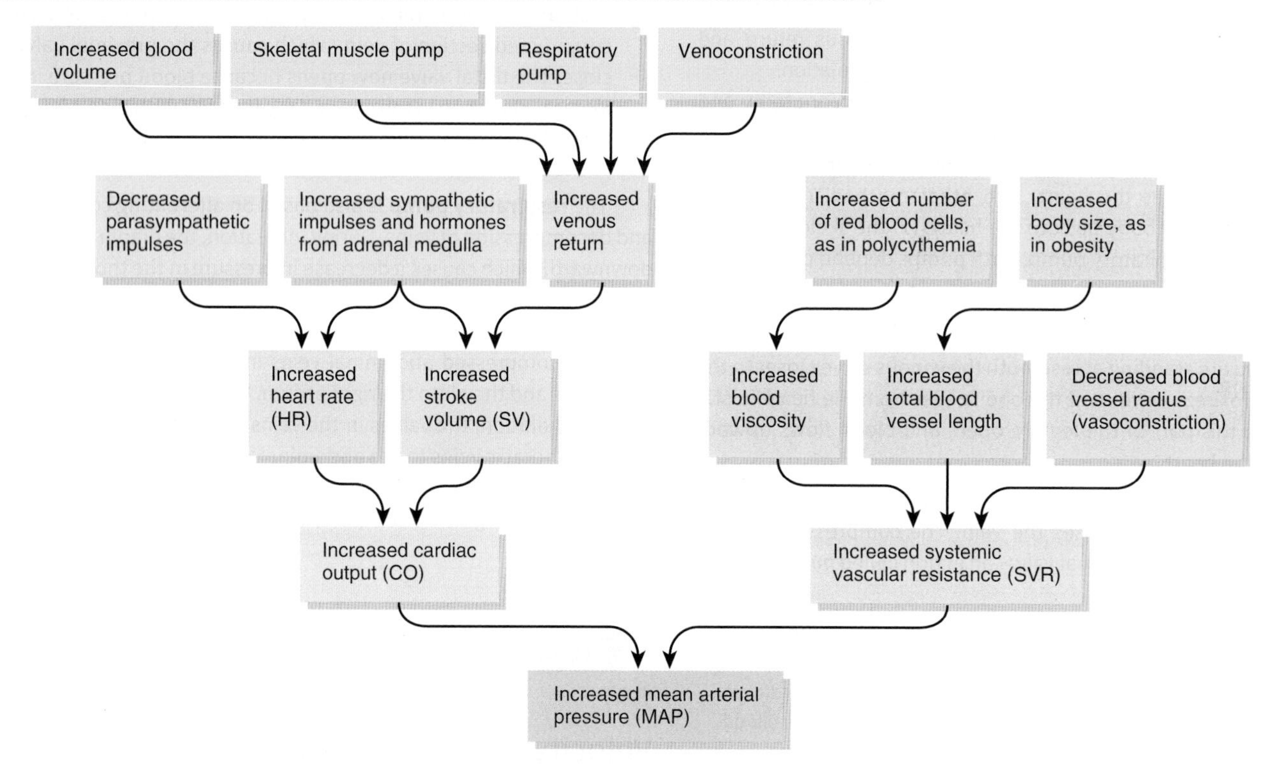

Q Which type of blood vessel exerts the major control of systemic vascular resistance, and how does it achieve this?

FIGURE 21.11 Relationship between velocity (speed) of blood flow and total cross-sectional area in different types of blood vessels.

Velocity of blood flow is slowest in the capillaries because they have the largest total cross-sectional area.

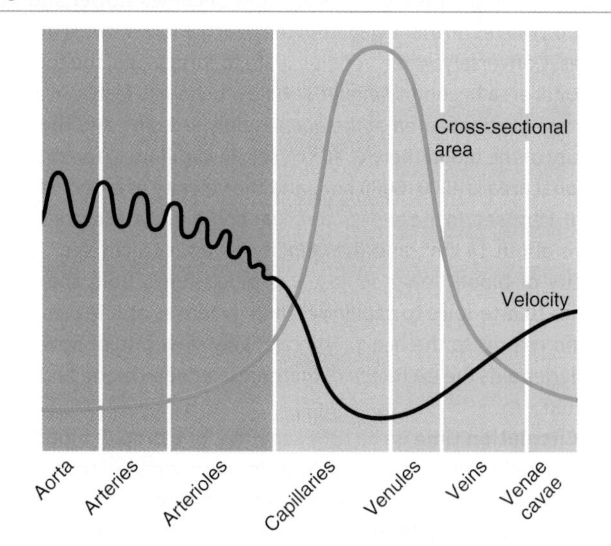

Q In which blood vessels is the velocity of flow fastest?

21.4 Control of Blood Pressure and Blood Flow

OBJECTIVE

- **Describe** how blood pressure is regulated.

Several interconnected negative feedback systems control blood pressure by adjusting heart rate, stroke volume, systemic vascular resistance, and blood volume. Some systems allow rapid adjustments to cope with sudden changes, such as the drop in blood pressure in the brain that occurs when you get out of bed; others act more slowly to provide long-term regulation of blood pressure. The body may also require adjustments to the distribution of blood flow. During exercise, for example, a greater percentage of the total blood flow is diverted to skeletal muscles.

Role of the Cardiovascular Center

In Chapter 20, we noted how the **cardiovascular (CV) center** in the medulla oblongata helps regulate heart rate and stroke volume. The

CV center also controls neural, hormonal, and local negative feedback systems that regulate blood pressure and blood flow to specific tissues. Groups of neurons scattered within the CV center regulate heart rate, contractility (force of contraction) of the ventricles, and blood vessel diameter. Some neurons stimulate the heart (cardiostimulatory center); others inhibit the heart (cardioinhibitory center). Still others control blood vessel diameter by causing constriction (vasoconstrictor center) or dilation (vasodilator center); these neurons are referred to collectively as the vasomotor center. Because the CV center neurons communicate with one another, function together, and are not clearly separated anatomically, we discuss them here as a group.

The cardiovascular center receives input both from higher brain regions and from sensory receptors (**Figure 21.12**). Nerve impulses descend from the cerebral cortex, limbic system, and hypothalamus to affect the cardiovascular center. For example, even before you start to run a race, your heart rate may increase due to nerve impulses conveyed from the limbic system to the CV center. If your body temperature rises during a race, the hypothalamus sends nerve impulses to the CV center. The resulting vasodilation of skin blood vessels allows heat to dissipate more rapidly from the surface of the skin. The three main types of sensory receptors that provide input to the cardiovascular center are proprioceptors, baroreceptors, and chemoreceptors. *Proprioceptors* (PRŌ-prē-ō-sep′-tors) monitor movements of joints and muscles and provide input to the cardiovascular center during physical activity. Their activity accounts for the rapid increase in heart rate at the beginning of exercise. *Baroreceptors* (bar′-ō-rē-SEP-tors) monitor changes in pressure and stretch in the walls of blood vessels, and *chemoreceptors* (kē′-mō-rē-SEP-tors) monitor the concentration of various chemicals in the blood.

Output from the cardiovascular center flows along sympathetic and parasympathetic neurons of the ANS (**Figure 21.12**). Sympathetic impulses reach the heart via the **cardiac accelerator nerves**. An increase in sympathetic stimulation increases heart rate and contractility; a decrease in sympathetic stimulation decreases heart rate and contractility. Parasympathetic stimulation, conveyed along the **vagus (X) nerves**, decreases heart rate. Thus, opposing sympathetic (stimulatory) and parasympathetic (inhibitory) influences control the heart.

The cardiovascular center also continually sends impulses to smooth muscle in blood vessel walls via **vasomotor nerves** (vā-sō-MŌ-tor). These sympathetic neurons exit the spinal cord through all thoracic and the first one or two lumbar spinal nerves and then pass into the sympathetic trunk ganglia (see **Figure 15.2**). From there, impulses propagate along sympathetic neurons that innervate blood vessels in viscera and peripheral areas. The vasomotor region of the cardiovascular center continually sends impulses over these routes to arterioles throughout the body, but especially to those in the skin and abdominal viscera. The result is a moderate state of tonic contraction or vasoconstriction, called **vasomotor tone**, that sets the resting level of systemic vascular resistance. Sympathetic stimulation of most veins causes constriction that moves blood out of venous blood reservoirs and increases blood pressure.

FIGURE 21.12 Location and function of the cardiovascular (CV) center in the medulla oblongata. The CV center receives input from higher brain centers, proprioceptors, baroreceptors, and chemoreceptors. Then, it provides output to the sympathetic and parasympathetic divisions of the autonomic nervous system (ANS).

The cardiovascular center is the main region for nervous system regulation of the heart and blood vessels.

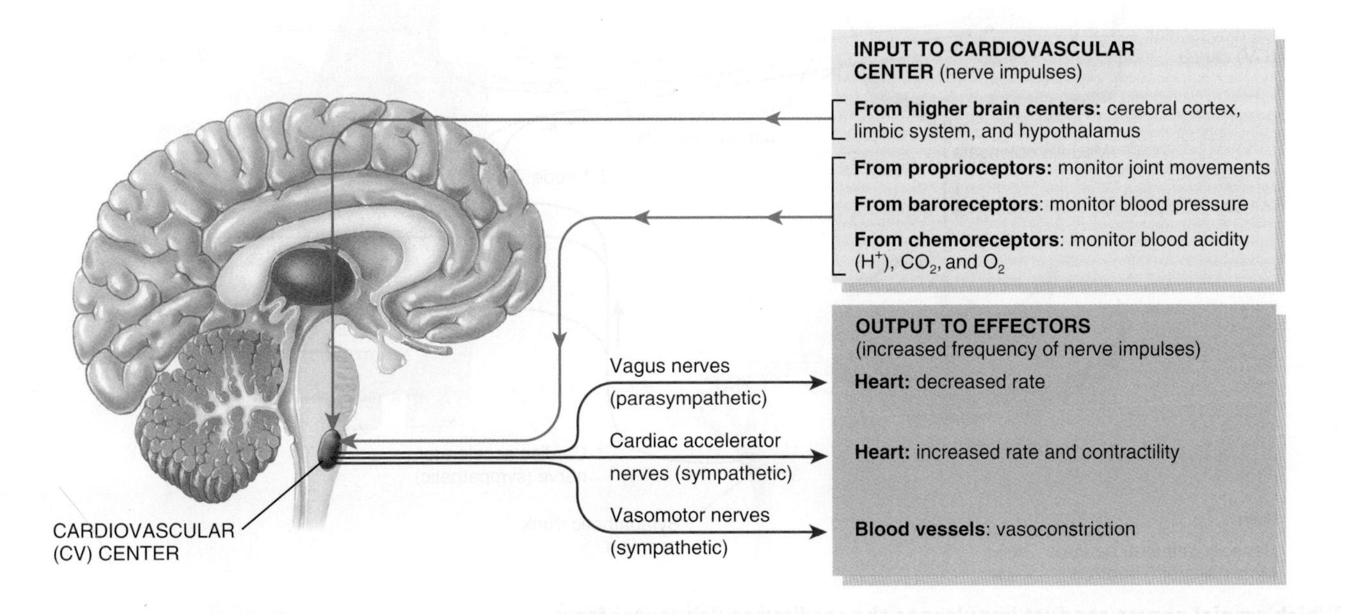

Q What types of effector tissues are regulated by the cardiovascular center?

Neural Regulation of Blood Pressure

The nervous system regulates blood pressure via negative feedback loops that occur as two types of reflexes: baroreceptor reflexes and chemoreceptor reflexes.

Baroreceptor Reflexes **Baroreceptors**, pressure-sensitive sensory receptors, are located in the aorta, internal carotid arteries (arteries in the neck that supply blood to the brain), and other large arteries in the neck and chest. They send impulses to the cardiovascular center to help regulate blood pressure. The two most important **baroreceptor reflexes** are the carotid sinus reflex and the aortic reflex.

Baroreceptors in the wall of the carotid sinuses initiate the **carotid sinus reflex** (ka-ROT-id), which helps regulate blood pressure in the brain. The **carotid sinuses** are small widenings of the right and left internal carotid arteries just above the point where they branch from the common carotid arteries (Figure 21.13). Blood pressure stretches the wall of the carotid sinus, which stimulates the baroreceptors. Nerve impulses propagate from the carotid sinus baroreceptors over sensory axons in the **glossopharyngeal (IX) nerves** (glos′-ō-fa-RIN-jē-al) to the cardiovascular center in the medulla oblongata. Baroreceptors in the wall of the ascending aorta and arch of the aorta initiate the **aortic reflex**, which regulates systemic blood pressure. Nerve impulses from aortic baroreceptors reach the cardiovascular center via sensory axons of the **vagus (X) nerves**.

When blood pressure falls, the baroreceptors are stretched less, and they send nerve impulses at a slower rate to the cardiovascular center (Figure 21.14). In response, the CV center decreases parasympathetic stimulation of the heart by way of motor axons of the vagus nerves and increases sympathetic stimulation of the heart via cardiac accelerator nerves. Another consequence of increased sympathetic stimulation is increased secretion of epinephrine and norepinephrine by the adrenal medulla. As the heart beats faster and more forcefully, and as systemic vascular resistance increases, cardiac output and systemic vascular resistance rise, and blood pressure increases to the normal level.

Conversely, when an increase in pressure is detected, the baroreceptors send impulses at a faster rate. The CV center responds by increasing parasympathetic stimulation and decreasing sympathetic

FIGURE 21.13 ANS innervation of the heart and the baroreceptor reflexes that help regulate blood pressure.

Baroreceptors are pressure-sensitive neurons that monitor stretching.

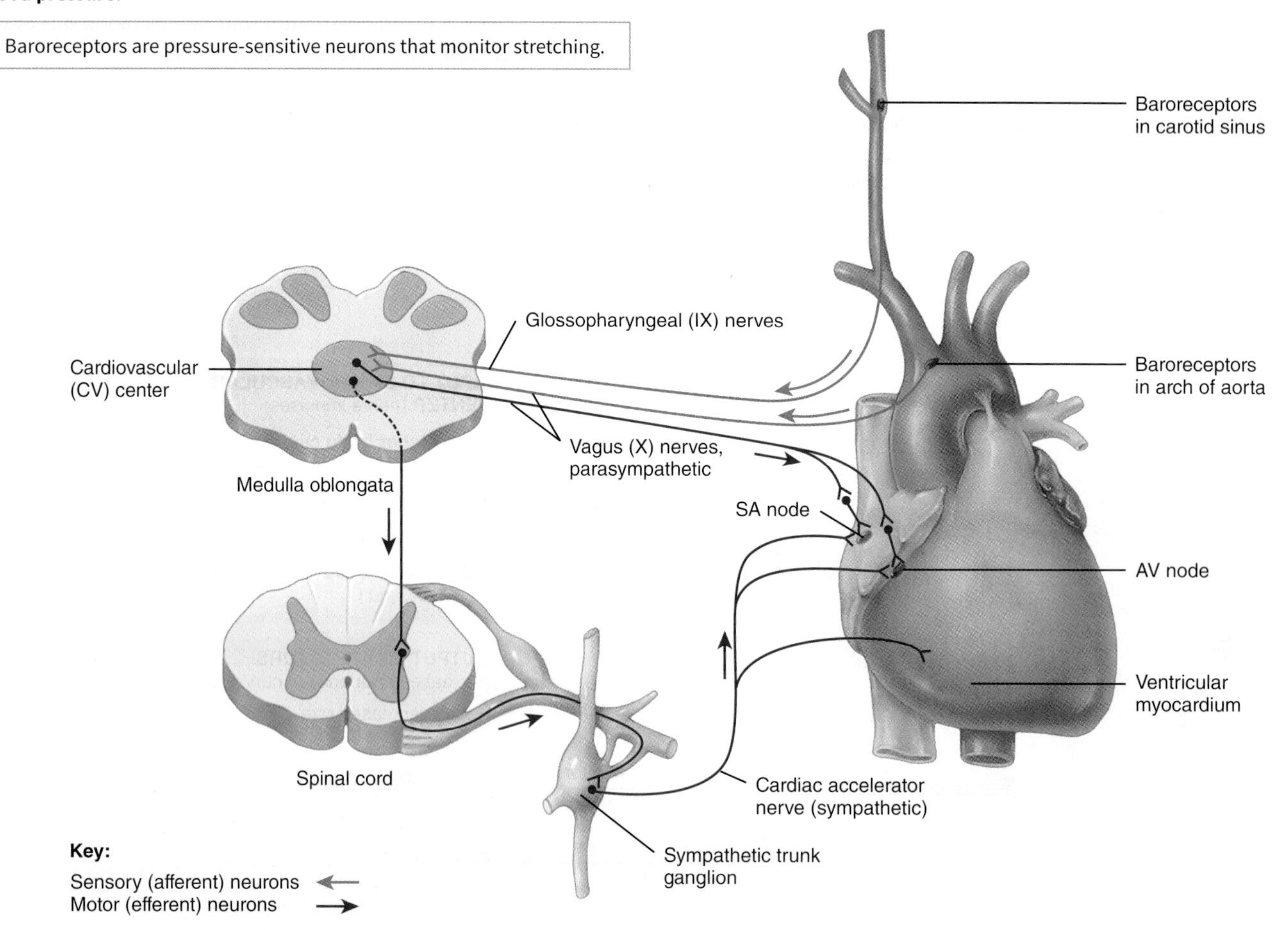

Q Which cranial nerves conduct impulses to the cardiovascular center from baroreceptors in the carotid sinuses and the arch of the aorta?

FIGURE 21.14 Negative feedback regulation of blood pressure via baroreceptor reflexes.

When blood pressure decreases, heart rate increases.

STIMULUS

Disrupts homeostasis by decreasing

CONTROLLED CONDITION
Blood pressure

RECEPTORS
Baroreceptors in carotid sinus and arch of aorta

Input: Stretch less, which decreases rate of nerve impulses

CONTROL CENTERS
CV center in medulla oblongata
Adrenal medulla

Output: Increased sympathetic, decreased para-sympathetic stimulation
Increased secretion of epinephrine and norepinephrine from adrenal medulla

EFFECTORS
Heart
Blood vessels

Increased stroke volume and heart rate lead to increased cardiac output (CO)

Constriction of blood vessels increases systemic vascular resistance (SVR)

RESPONSE
Increased blood pressure

Return to homeostasis when increased cardiac output and increased vascular resistance bring blood pressure back to normal

−

Q Does this negative feedback cycle represent the changes that occur when you lie down or when you stand up?

stimulation. The resulting decreases in heart rate and force of contraction reduce the cardiac output. The cardiovascular center also slows the rate at which it sends sympathetic impulses along vasomotor neurons that normally cause vasoconstriction. The resulting vasodilation lowers systemic vascular resistance. Decreased cardiac output and decreased systemic vascular resistance both lower systemic arterial blood pressure to the normal level.

Moving from a prone (lying down) to an erect position decreases blood pressure and blood flow in the head and upper part of the body. The baroreceptor reflexes, however, quickly counteract the drop in pressure. Sometimes these reflexes operate more slowly than normal, especially in the elderly, in which case a person can faint due to reduced brain blood flow after standing up too quickly.

See Clinical Connection: Carotid Sinus Massage and Carotid Sinus Syncope

Chemoreceptor Reflexes

Chemoreceptors, sensory receptors that monitor the chemical composition of blood, are located close to the baroreceptors of the carotid sinus and arch of the aorta in small structures called **carotid bodies** and **aortic bodies**, respectively. These chemoreceptors detect changes in blood level of O_2, CO_2, and H^+. *Hypoxia* (lowered O_2 availability), *acidosis* (an increase in H^+ concentration), or *hypercapnia* (excess CO_2) stimulates the chemoreceptors to send impulses to the cardiovascular center. In response, the CV center increases sympathetic stimulation to arterioles and veins, producing vasoconstriction and an increase in blood pressure. These chemoreceptors also provide input to the respiratory center in the brain stem to adjust the rate of breathing.

Hormonal Regulation of Blood Pressure

As you learned in Chapter 18, several hormones help regulate blood pressure and blood flow by altering cardiac output, changing systemic vascular resistance, or adjusting the total blood volume:

1. ***Renin–angiotensin–aldosterone (RAA) system.*** When blood volume falls or blood flow to the kidneys decreases, juxtaglomerular cells in the kidneys secrete **renin** into the bloodstream. In sequence, renin and angiotensin-converting enzyme (ACE) act on their substrates to produce the active hormone **angiotensin II** (an′-jē-ō-TEN-sin), which raises blood pressure in two ways. First, angiotensin II is a potent vasoconstrictor; it raises blood pressure by increasing systemic vascular resistance. Second, it stimulates secretion of **aldosterone**, which increases reabsorption of sodium ions (Na^+) and water by the kidneys. The water reabsorption increases total blood volume, which increases blood pressure. (See Section 21.6.)
2. ***Epinephrine and norepinephrine.*** In response to sympathetic stimulation, the adrenal medulla releases epinephrine and norepinephrine. These hormones increase cardiac output by increasing the rate and force of heart contractions. They also cause vasoconstriction of arterioles and veins in the skin and abdominal organs and vasodilation of arterioles in cardiac and skeletal muscle, which helps increase blood flow to muscle during exercise.

TABLE 21.2 Blood Pressure Regulation by Hormones

FACTOR INFLUENCING BLOOD PRESSURE	HORMONE	EFFECT ON BLOOD PRESSURE
CARDIAC OUTPUT		
Increased heart rate and contractility	Norepinephrine, epinephrine.	Increase.
SYSTEMIC VASCULAR RESISTANCE		
Vasoconstriction	Angiotensin II, antidiuretic hormone (ADH), norepinephrine,* epinephrine.†	Increase.
Vasodilation	Atrial natriuretic peptide (ANP), epinephrine,† nitric oxide.	Decrease.
BLOOD VOLUME		
Blood volume increase	Aldosterone, antidiuretic hormone.	Increase.
Blood volume decrease	Atrial natriuretic peptide.	Decrease.

*Acts at α_1 receptors in arterioles of abdomen and skin.
†Acts at β_2 receptors in arterioles of cardiac and skeletal muscle; norepinephrine has a much smaller vasodilating effect.

3. ***Antidiuretic hormone (ADH).*** **Antidiuretic hormone (ADH)** is produced by the hypothalamus and released from the posterior pituitary in response to dehydration or decreased blood volume. Among other actions, ADH causes vasoconstriction, which increases blood pressure. For this reason ADH is also called *vasopressin*. ADH also promotes movement of water from the lumen of kidney tubules into the bloodstream. This results in an increase in blood volume and a decrease in urine output.
4. ***Atrial natriuretic peptide (ANP).*** Released by cells in the atria of the heart, **atrial natriuretic peptide (ANP)** lowers blood pressure by causing vasodilation and by promoting the loss of salt and water in the urine, which reduces blood volume.

Table 21.2 summarizes the regulation of blood pressure by hormones.

Autoregulation of Blood Flow

In each capillary bed, local changes can regulate vasomotion. When vasodilators produce local dilation of arterioles and relaxation of precapillary sphincters, blood flow into capillary networks is increased, which increases O_2 level. Vasoconstrictors have the opposite effect. The ability of a tissue to automatically adjust its blood flow to match its metabolic demands is called **autoregulation** (aw′-tō-reg′-ū-LĀ-shun). In tissues such as the heart and skeletal muscle, where the demand for O_2 and nutrients and for the removal of wastes can increase as much as tenfold during physical activity, autoregulation is an important contributor to increased blood flow through the tissue. Autoregulation also controls regional blood flow in the brain; blood distribution to various parts of the brain changes dramatically for different mental and physical activities. During a conversation, for example, blood flow increases to your motor speech areas when you are talking and increases to the auditory areas when you are listening.

Two general types of stimuli cause autoregulatory changes in blood flow:

1. ***Physical changes.*** Warming promotes vasodilation, and cooling causes vasoconstriction. In addition, smooth muscle in arteriole walls exhibits a **myogenic response** (mī-ō-JEN-ik)—it contracts more forcefully when it is stretched and relaxes when stretching lessens. If, for example, blood flow through an arteriole decreases, stretching of the arteriole walls decreases. As a result, the smooth muscle relaxes and produces vasodilation, which increases blood flow.
2. ***Vasodilating and vasoconstricting chemicals.*** Several types of cells—including white blood cells, platelets, smooth muscle fibers, macrophages, and endothelial cells—release a wide variety of chemicals that alter blood-vessel diameter. Vasodilating chemicals released by metabolically active tissue cells include K^+, H^+, lactic acid (lactate), and adenosine (from ATP). Another important vasodilator released by endothelial cells is nitric oxide (NO). Tissue trauma or inflammation causes release of vasodilating kinins and histamine. Vasoconstrictors include thromboxane A2, superoxide radicals, serotonin (from platelets), and endothelins (from endothelial cells).

An important difference between the pulmonary and systemic circulations is their autoregulatory response to changes in O_2 level. The walls of blood vessels in the systemic circulation *dilate* in response to low O_2. With vasodilation, O_2 delivery increases, which restores the normal O_2 level. By contrast, the walls of blood vessels in the pulmonary circulation *constrict* in response to low levels of O_2. This response ensures that blood mostly bypasses those alveoli (air sacs) in the lungs that are poorly ventilated by fresh air. Thus, most blood flows to better-ventilated areas of the lung.

21.5 Checking Circulation

OBJECTIVE

- **Define** pulse, and systolic, diastolic, and pulse pressures.

Pulse

The alternate expansion and recoil of elastic arteries after each systole of the left ventricle creates a traveling pressure wave that is called the **pulse**. The pulse is strongest in the arteries closest to the heart, becomes weaker in the arterioles, and disappears altogether in the capillaries. The pulse may be felt in any artery that lies near the surface of the body that can be compressed against a bone or other firm structure. Table 21.3 depicts some common pulse points.

The pulse rate normally is the same as the heart rate, about 70 to 80 beats per minute at rest. **Tachycardia** (tak′-i-KAR-dē-a; *tachy-* = fast) is a rapid resting heart or pulse rate over 100 beats/min. **Bradycardia** (brād′-i-KAR-dē-a; *brady-* = slow) is a slow resting heart or pulse rate under 50 beats/min. Endurance-trained athletes normally exhibit bradycardia.

TABLE 21.3 Pulse Points

STRUCTURE	LOCATION	STRUCTURE	LOCATION
Superficial temporal artery	Medial to ear.	**Femoral artery**	Inferior to inguinal ligament.
Facial artery	Mandible (lower jawbone) on line with corners of mouth.	**Popliteal artery**	Posterior to knee.
Common carotid artery	Lateral to larynx (voice box).	**Radial artery**	Lateral aspect of wrist.
Brachial artery	Medial side of biceps brachii muscle.	**Dorsal artery of foot** (dorsalis pedis artery)	Superior to instep of foot.

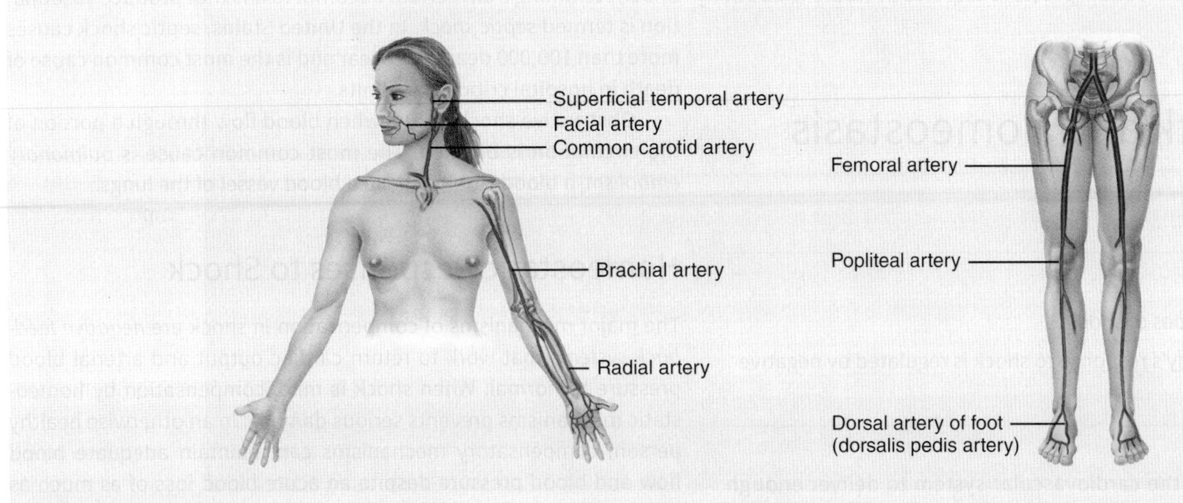

Measuring Blood Pressure

In clinical use, the term **blood pressure** usually refers to the pressure in arteries generated by the left ventricle during systole and the pressure remaining in the arteries when the ventricle is in diastole. Blood pressure is usually measured in the brachial artery in the left arm (**Table 21.3**). The device used to measure blood pressure is a **sphygmomanometer** (sfig′-mō-ma-NOM-e-ter; *sphygmo-* = pulse; *-manometer* = instrument used to measure pressure). It consists of a rubber cuff connected to a rubber bulb that is used to inflate the cuff and a meter that registers the pressure in the cuff. With the arm resting on a table so that it is about the same level as the heart, the cuff of the sphygmomanometer is wrapped around a bared arm. The cuff is inflated by squeezing the bulb until the brachial artery is compressed and blood flow stops, about 30 mmHg higher than the person's usual systolic pressure. The technician places a stethoscope below the cuff on the brachial artery, and slowly deflates the cuff. When the cuff is deflated enough to allow the artery to open, a spurt of blood passes through, resulting in the first sound heard through the stethoscope. This sound corresponds to **systolic blood pressure (SBP)**, the force of blood pressure on arterial walls just after ventricular contraction (**Figure 21.15**). As the cuff is deflated further, the sounds suddenly become too faint to be heard through the stethoscope. This level, called the **diastolic blood pressure (DBP),** represents the force exerted by the blood remaining in arteries during ventricular relaxation. At pressures below diastolic blood pressure, sounds disappear altogether. The various sounds that are heard while taking blood pressure are called **Korotkoff sounds** (kō-ROT-kof).

The normal blood pressure of an adult male is less than 120 mmHg systolic and less than 80 mmHg diastolic. For example,

FIGURE 21.15 Relationship of blood pressure changes to cuff pressure.

As the cuff is deflated, sounds first occur at the systolic blood pressure; the sounds suddenly become faint at the diastolic blood pressure.

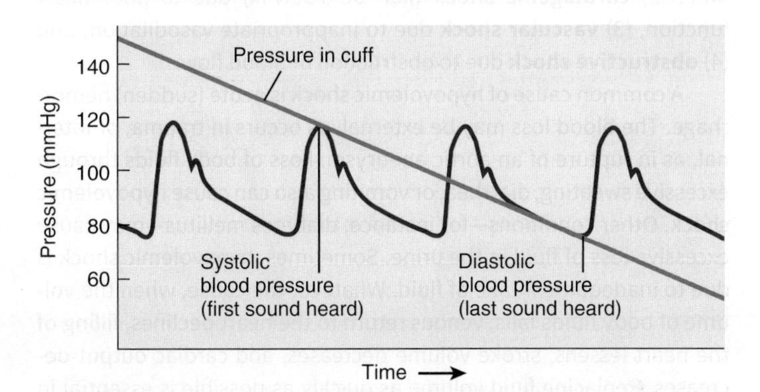

Q If a blood pressure is reported as "142 over 95," what are the diastolic, systolic, and pulse pressures? Does this person have hypertension as defined in Disorders: Homeostatic Imbalances at the end of the chapter?

"110 over 70" (written as 110/70) is a normal blood pressure. In young adult females, the pressures are 8 to 10 mmHg less. People who exercise regularly and are in good physical condition may have even lower blood pressures. Thus, blood pressure slightly lower than 120/80 may be a sign of good health and fitness.

The difference between systolic and diastolic pressure is called **pulse pressure**. This pressure, normally about 40 mmHg, provides information about the condition of the cardiovascular system. For example, conditions such as atherosclerosis and patent (open) ductus arteriosus greatly increase pulse pressure. The normal ratio of systolic pressure to diastolic pressure to pulse pressure is about 3:2:1.

21.6 Shock and Homeostasis

OBJECTIVES

- **Define** shock.
- **Describe** the four types of shock.
- **Explain** how the body's response to shock is regulated by negative feedback.

Shock is a failure of the cardiovascular system to deliver enough O_2 and nutrients to meet cellular metabolic needs. The causes of shock are many and varied, but all are characterized by inadequate blood flow to body tissues. With inadequate oxygen delivery, cells switch from aerobic to anaerobic production of ATP, and lactic acid accumulates in body fluids. If shock persists, cells and organs become damaged, and cells may die unless proper treatment begins quickly.

Types of Shock

Shock can be of four different types: (1) **hypovolemic shock** (hī-pō-vō-LĒ-mik; *hypo-* = low; *-volemic* = volume) due to decreased blood volume, (2) **cardiogenic shock** (kar′-dē-ō-JEN-ik) due to poor heart function, (3) **vascular shock** due to inappropriate vasodilation, and (4) **obstructive shock** due to obstruction of blood flow.

A common cause of hypovolemic shock is acute (sudden) hemorrhage. The blood loss may be external, as occurs in trauma, or internal, as in rupture of an aortic aneurysm. Loss of body fluids through excessive sweating, diarrhea, or vomiting also can cause hypovolemic shock. Other conditions—for instance, diabetes mellitus—may cause excessive loss of fluid in the urine. Sometimes, hypovolemic shock is due to inadequate intake of fluid. Whatever the cause, when the volume of body fluids falls, venous return to the heart declines, filling of the heart lessens, stroke volume decreases, and cardiac output decreases. Replacing fluid volume as quickly as possible is essential in managing hypovolemic shock.

In cardiogenic shock, the heart fails to pump adequately, most often because of a myocardial infarction (heart attack). Other causes of cardiogenic shock include poor perfusion of the heart (ischemia), heart valve problems, excessive preload or afterload, impaired contractility of heart muscle fibers, and arrhythmias.

Even with normal blood volume and cardiac output, shock may occur if blood pressure drops due to a decrease in systemic vascular resistance. A variety of conditions can cause inappropriate dilation of arterioles or venules. In *anaphylactic shock* (AN-a-fil-lak′-tik), a severe allergic reaction—for example, to a bee sting—releases histamine and other mediators that cause vasodilation. In *neurogenic shock*, vasodilation may occur following trauma to the head that causes malfunction of the cardiovascular center in the medulla. Shock stemming from certain bacterial toxins that produce vasodilation is termed *septic shock*. In the United States, septic shock causes more than 100,000 deaths per year and is the most common cause of death in hospital critical care units.

Obstructive shock occurs when blood flow through a portion of the circulation is blocked. The most common cause is *pulmonary embolism*, a blood clot lodged in a blood vessel of the lungs.

Homeostatic Responses to Shock

The major mechanisms of compensation in shock are *negative feedback systems* that work to return cardiac output and arterial blood pressure to normal. When shock is mild, compensation by homeostatic mechanisms prevents serious damage. In an otherwise healthy person, compensatory mechanisms can maintain adequate blood flow and blood pressure despite an acute blood loss of as much as 10% of total volume. Figure 21.16 shows several negative feedback systems that respond to hypovolemic shock.

1. ***Activation of the renin–angiotensin–aldosterone system.*** Decreased blood flow to the kidneys causes the kidneys to secrete renin and initiates the renin–angiotensin–aldosterone system (see Figure 18.15). Recall that angiotensin II causes vasoconstriction and stimulates the adrenal cortex to secrete aldosterone, a hormone that increases reabsorption of Na^+ and water by the kidneys. The increases in systemic vascular resistance and blood volume help raise blood pressure.
2. ***Secretion of antidiuretic hormone.*** In response to decreased blood pressure, the posterior pituitary releases more antidiuretic hormone (ADH). ADH enhances water reabsorption by the kidneys, which conserves remaining blood volume. It also causes vasoconstriction, which increases systemic vascular resistance.
3. ***Activation of the sympathetic division of the ANS.*** As blood pressure decreases, aortic and carotid baroreceptors initiate powerful sympathetic responses throughout the body. One result is marked vasoconstriction of arterioles and veins of the skin, kidneys, and other abdominal viscera. (Vasoconstriction does not occur in the brain or heart.) Constriction of arterioles increases systemic vascular resistance, and constriction of veins increases venous return. Both effects help maintain an adequate blood pressure. Sympathetic stimulation also increases heart rate and contractility and increases secretion of epinephrine and norepinephrine by the adrenal medulla. These hormones intensify vasoconstriction and increase heart rate and contractility, all of which help raise blood pressure.

FIGURE 21.16 Negative feedback systems that can restore normal blood pressure during hypovolemic shock.

Homeostatic mechanisms can compensate for an acute blood loss of as much as 10% of total blood volume.

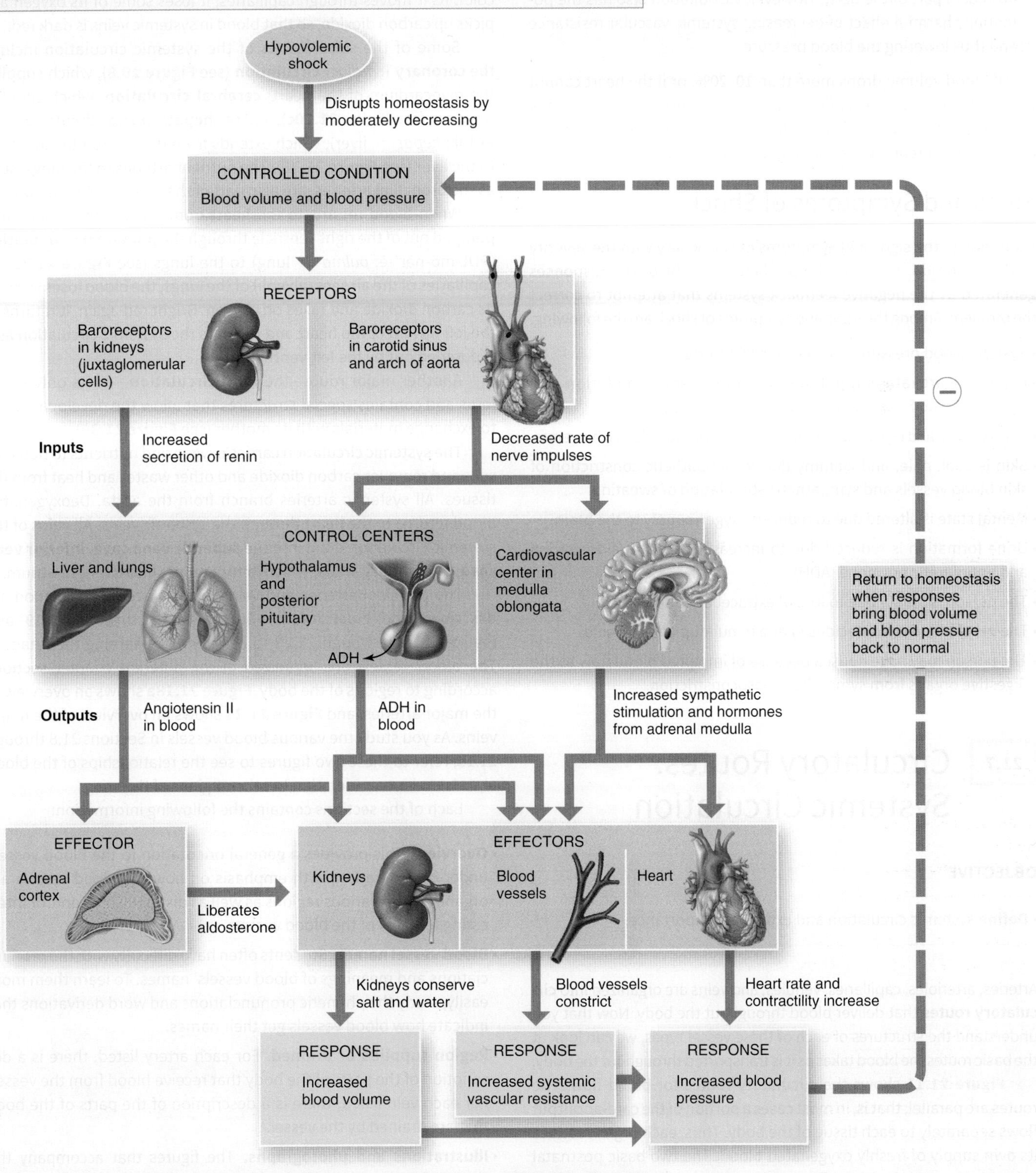

Q Does almost-normal blood pressure in a person who has lost blood indicate that the patient's tissues are receiving adequate perfusion (blood flow)?

4. ***Release of local vasodilators.*** In response to *hypoxia,* cells liberate vasodilators—including K^+, H^+, lactic acid, adenosine, and nitric oxide—that dilate arterioles and relax precapillary sphincters. Such vasodilation increases local blood flow and may restore O_2 level to normal in part of the body. However, vasodilation also has the potentially harmful effect of decreasing systemic vascular resistance and thus lowering the blood pressure.

If blood volume drops more than 10–20%, or if the heart cannot bring blood pressure up sufficiently, compensatory mechanisms may fail to maintain adequate blood flow to tissues. At this point, shock becomes life-threatening as damaged cells start to die.

Signs and Symptoms of Shock

Even though the signs and symptoms of shock vary with the severity of the condition, most can be predicted in light of the responses generated by the negative feedback systems that attempt to correct the problem. Among the signs and symptoms of shock are the following:

- Systolic blood pressure is lower than 90 mmHg.
- Resting heart rate is rapid due to sympathetic stimulation and increased blood levels of epinephrine and norepinephrine.
- Pulse is weak and rapid due to reduced cardiac output and fast heart rate.
- Skin is cool, pale, and clammy due to sympathetic constriction of skin blood vessels and sympathetic stimulation of sweating.
- Mental state is altered due to reduced oxygen supply to the brain.
- Urine formation is reduced due to increased levels of aldosterone and antidiuretic hormone (ADH).
- The person is thirsty due to loss of extracellular fluid.
- The pH of blood is low (acidosis) due to buildup of lactic acid.
- The person may have nausea because of impaired blood flow to the digestive organs from sympathetic vasoconstriction.

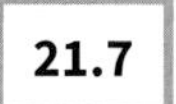

Circulatory Routes: Systemic Circulation

OBJECTIVE

- **Define** systemic circulation and explain its importance.

Arteries, arterioles, capillaries, venules, and veins are organized into **circulatory routes** that deliver blood throughout the body. Now that you understand the structures of each of these vessel types, we can look at the basic routes the blood takes as it is transported throughout the body.

Figure 21.17 shows the circulatory routes for blood flow. The routes are parallel; that is, in most cases a portion of the cardiac output flows separately to each tissue of the body. Thus, each organ receives its own supply of freshly oxygenated blood. The two basic postnatal (after birth) routes for blood flow are the systemic circulation and the pulmonary circulation. The **systemic circulation** includes all arteries and arterioles that carry oxygenated blood from the left ventricle to systemic capillaries, plus the veins and venules that return deoxygenated blood to the right atrium after flowing through body organs. Blood leaving the aorta and flowing through the systemic arteries is a bright red color. As it moves through capillaries, it loses some of its oxygen and picks up carbon dioxide, so that blood in systemic veins is dark red.

Some of the subdivisions of the systemic circulation include the **coronary** (*cardiac*) **circulation** (see Figure 20.8), which supplies the myocardium of the heart; **cerebral circulation**, which supplies the brain (see Figure 21.20c); and the **hepatic portal circulation** (he-PAT-ik; *hepat-* = liver), which extends from the gastrointestinal tract to the liver (see Figure 21.28). The nutrient arteries to the lungs, such as the bronchial arteries, are also part of the systemic circulation.

When blood returns to the heart from the systemic route, it is pumped out of the right ventricle through the **pulmonary circulation** (PUL-mō-nār′-ē; *pulmo-* = lung) to the lungs (see Figure 21.29). In capillaries of the air sacs (alveoli) of the lungs, the blood loses some of its carbon dioxide and takes on oxygen. Bright red again, it returns to the left atrium of the heart and reenters the systemic circulation as it is pumped out by the left ventricle.

Another major route—the **fetal circulation**—exists only in the fetus and contains special structures that allow the developing fetus to exchange materials with its mother (see Figure 21.30).

The systemic circulation carries oxygen and nutrients to body tissues and removes carbon dioxide and other wastes and heat from the tissues. All systemic arteries branch from the aorta. Deoxygenated blood returns to the heart through the systemic veins. All veins of the systemic circulation drain into the **superior vena cava**, **inferior vena cava**, or **coronary sinus**, which in turn empty into the right atrium.

The principal arteries and veins of the systemic circulation are described and illustrated in Sections 21.8 through 21.19 and Figures 21.18 through 21.28 to assist you in learning their names. The blood vessels are organized in the different chapter sections according to regions of the body. Figure 21.18a shows an overview of the major arteries, and Figure 21.23 shows an overview of the major veins. As you study the various blood vessels in Sections 21.8 through 21.19, refer to these two figures to see the relationships of the blood vessels under consideration to other regions of the body.

Each of the sections contains the following information:

- **Overview.** This provides a general orientation to the blood vessels under consideration, with emphasis on how the blood vessels are organized into various regions as well as distinguishing and/or interesting features of the blood vessels.
- **Blood vessel names.** Students often have difficulty with the pronunciations and meanings of blood vessels' names. To learn them more easily, study the phonetic pronunciations and word derivations that indicate how blood vessels get their names.
- **Region supplied or drained.** For each artery listed, there is a description of the parts of the body that receive blood from the vessel. For each vein listed, there is a description of the parts of the body that are drained by the vessel.
- **Illustrations and photographs.** The figures that accompany the Sections 21.8 through 21.19 contain several elements. Many include illustrations of the blood vessels under consideration and flow

FIGURE 21.17 Circulatory routes. Long black arrows indicate the systemic circulation, short blue arrows the pulmonary circulation (detailed in **Figure 21.29**), and red arrows the hepatic portal circulation (detailed in **Figure 21.28**). Refer to **Figure 20.8** for details of the coronary circulation, and to **Figure 21.30** for details of the fetal circulation.

Blood vessels are organized into various routes that deliver blood to tissues of the body.

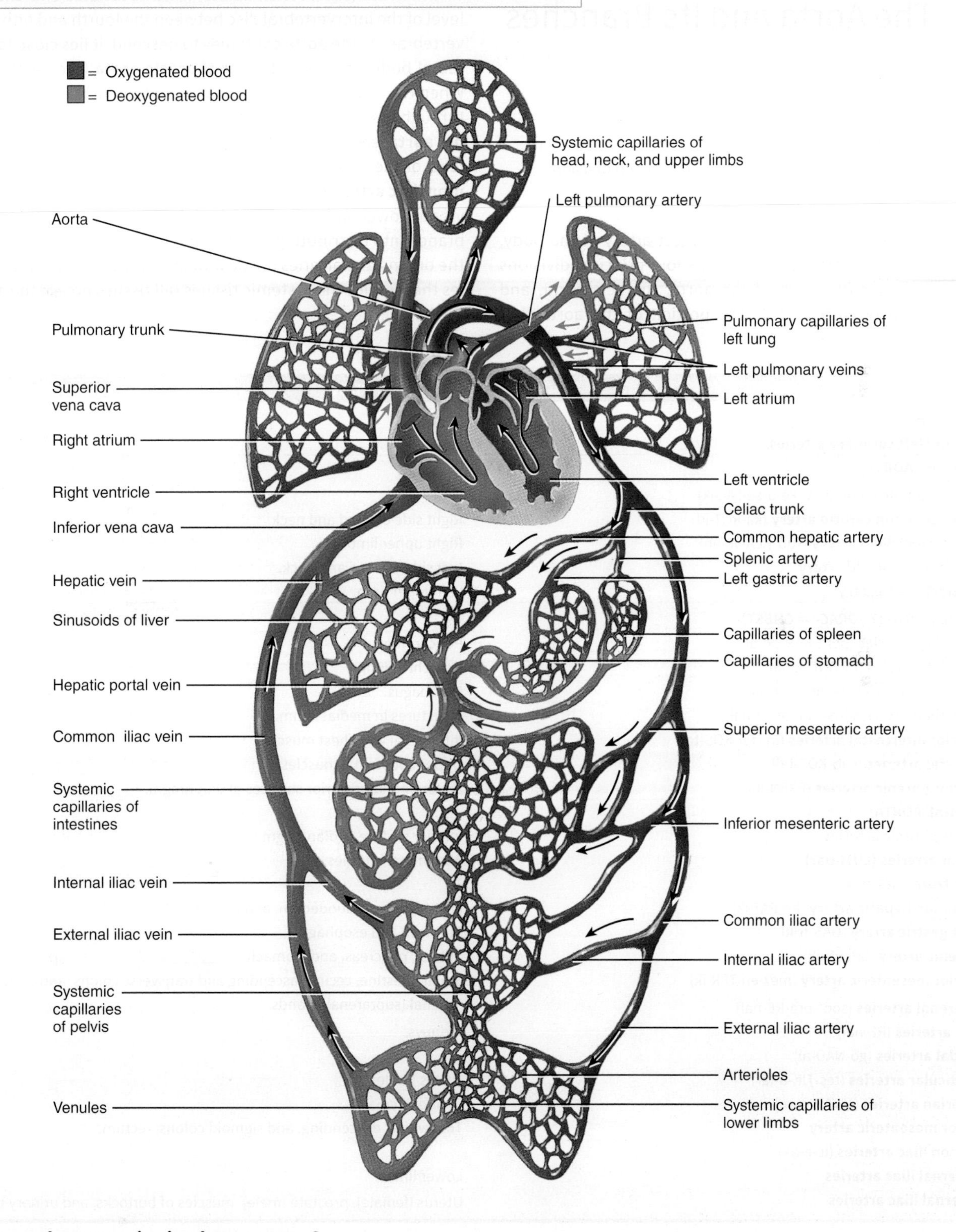

Q What are the two main circulatory routes?

diagrams to indicate the patterns of blood distribution or drainage. Cadaver photographs are also included in selected sections to provide more realistic views of the blood vessels.

21.8 The Aorta and Its Branches

OBJECTIVES

- **Identify** the four principal divisions of the aorta.
- **Locate** the major arterial branches arising from each division.

The **aorta** (ā-OR-ta = to lift up) is the largest artery of the body, with a diameter of 2–3 cm (about 1 in.). Its four principal divisions are the ascending aorta, arch of the aorta, thoracic aorta, and abdominal aorta (**Figure 21.18**). The portion of the aorta that emerges from the left ventricle posterior to the pulmonary trunk is the **ascending aorta** (see Section 21.9). The beginning of the aorta contains the aortic valve (see **Figure 20.4a**). The ascending aorta gives off two coronary arteries that supply the myocardium of the heart. Then the ascending aorta arches to the left, forming the **arch of the aorta** (see Section 21.10), which descends and ends at the level of the intervertebral disc between the fourth and fifth thoracic vertebrae. As the aorta continues to descend, it lies close to the vertebral bodies and is called the **thoracic aorta** (see Section 21.11). When the thoracic aorta reaches the bottom of the thorax it passes through the aortic hiatus of the diaphragm to become the **abdominal aorta** (see Section 21.12). The abdominal aorta descends to the level of the fourth lumbar vertebra where it divides into two **common iliac arteries** (see Section 21.13), which carry blood to the pelvis and lower limbs. Each division of the aorta gives off arteries that branch into distributing arteries that lead to various organs. Within the organs, the arteries divide into arterioles and then into capillaries that service the systemic tissues (all tissues except the alveoli of the lungs).

DIVISION AND BRANCHES	REGION SUPPLIED
ASCENDING AORTA	
Right and left coronary arteries	Heart.
ARCH OF THE AORTA	
Brachiocephalic trunk (brā′-kē-ō-se-FAL-ik)	
Right common carotid artery (ka-ROT-id)	Right side of head and neck.
Right subclavian artery (sub-KLĀ-vē-an)	Right upper limb.
Left common carotid artery	Left side of head and neck.
Left subclavian artery	Left upper limb.
THORACIC AORTA (*THORAC-* = CHEST)	
Pericardial arteries (per-i-KAR-dē-al)	Pericardium.
Bronchial arteries (BRONG-kē-al)	Bronchi of lungs.
Esophageal arteries (e-sof′-a-JĒ-al)	Esophagus.
Mediastinal arteries (mē′-dē-as-TĪ-nal)	Structures in mediastinum.
Posterior intercostal arteries (in′-ter-KOS-tal)	Intercostal and chest muscles.
Subcostal arteries (sub-KOS-tal)	Upper abdominal muscles.
Superior phrenic arteries (FREN-ik)	Superior and posterior surfaces of diaphragm.
ABDOMINAL AORTA	
Inferior phrenic arteries	Inferior surface of diaphragm.
Lumbar arteries (LUM-bar)	Abdominal muscles.
Celiac trunk (SĒ-lē-ak)	
Common hepatic artery (he-PAT-ik)	Liver, stomach, duodenum, and pancreas.
Left gastric artery (GAS-trik)	Stomach and esophagus.
Splenic artery (SPLĒN-ik)	Spleen, pancreas, and stomach.
Superior mesenteric artery (mez-en-TER-ik)	Small intestine, cecum, ascending and transverse colons, and pancreas.
Suprarenal arteries (soo′-pra-RĒ-nal)	Adrenal (suprarenal) glands.
Renal arteries (RĒ-nal)	Kidneys.
Gonadal arteries (gō-NAD-al)	
Testicular arteries (tes-TIK-ū-lar)	Testes (male).
Ovarian arteries (ō-VAR-ē-an)	Ovaries (female).
Inferior mesenteric artery	Transverse, descending, and sigmoid colons; rectum.
Common iliac arteries (IL-ē-ak)	
External iliac arteries	Lower limbs.
Internal iliac arteries	Uterus (female), prostate (male), muscles of buttocks, and urinary bladder.

FIGURE 21.18 Aorta and its principal branches.

All systemic arteries branch from the aorta.

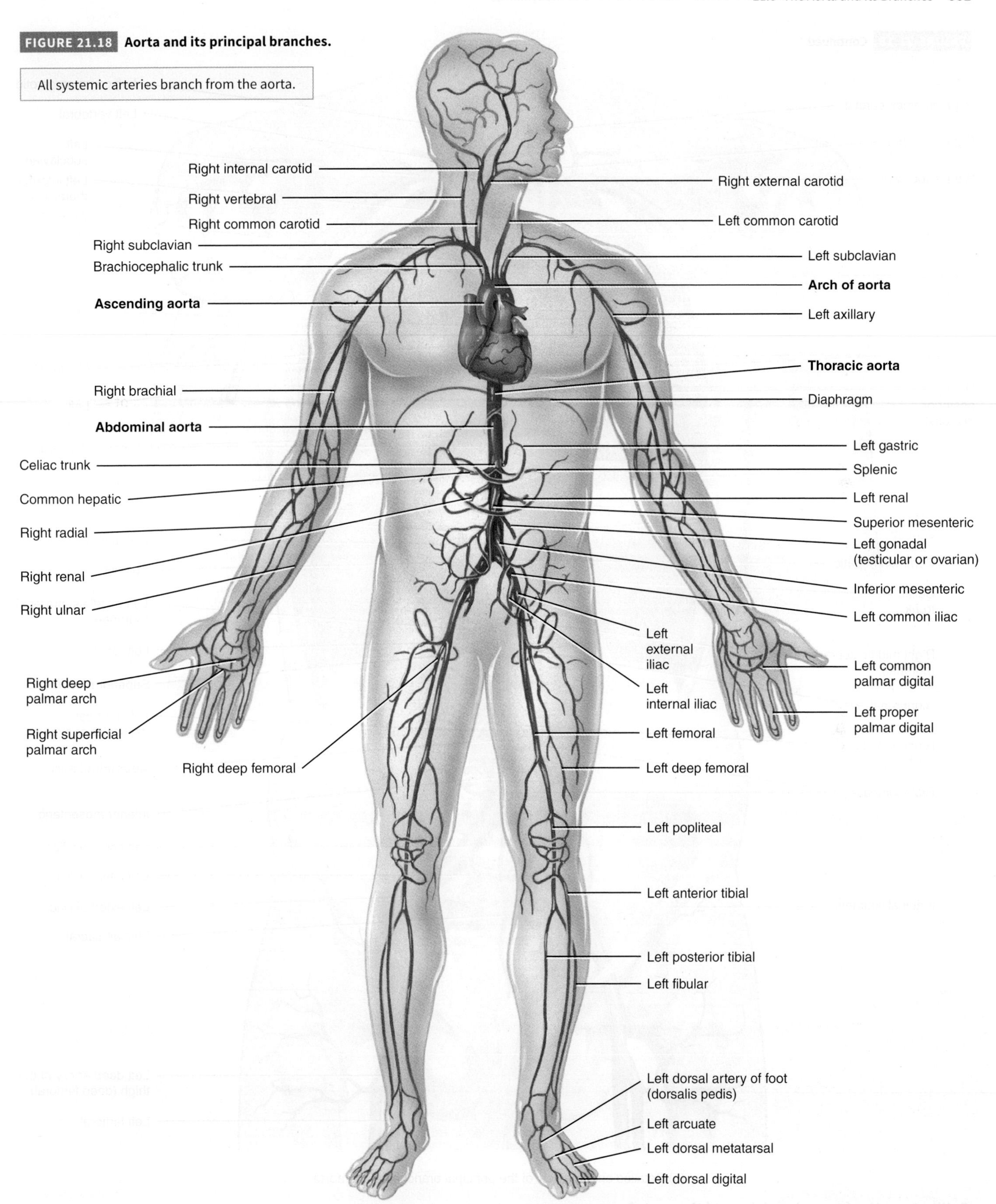

(a) Overall anterior view of the principal branches of the aorta

Figure 20.18 *Continues*

FIGURE 21.18 **Continued**

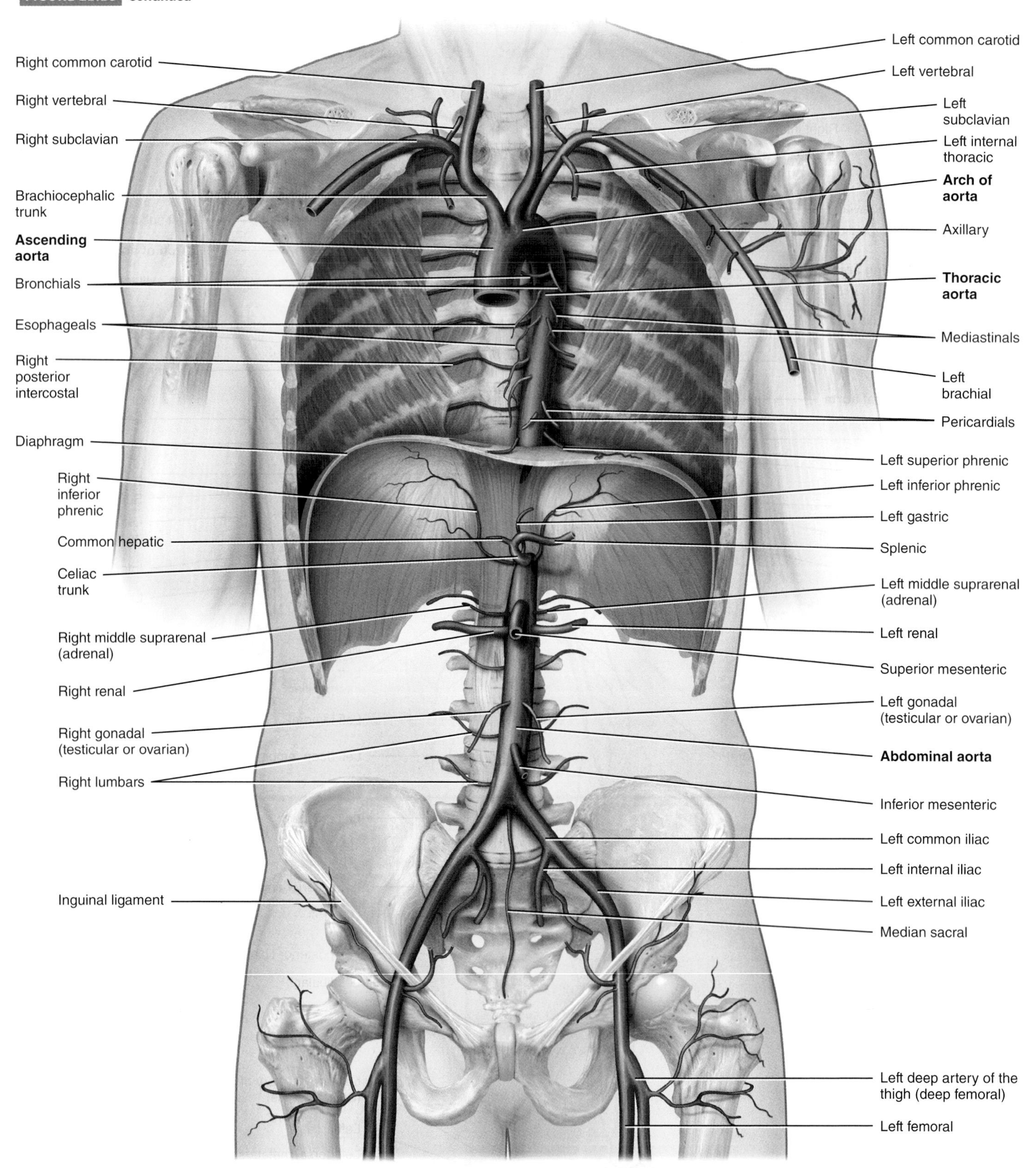

(b) Detailed anterior view of the principal branches of the aorta

Q What are the four subdivisions of the aorta?

21.9 Ascending Aorta

OBJECTIVE

- **Identify** the two primary arterial branches of the ascending aorta.

The **ascending aorta** is about 5 cm (2 in.) in length and begins at the aortic valve (see **Figure 20.8**). It is directed superiorly, slightly anteriorly, and to the right. It ends at the level of the sternal angle, where it becomes the arch of the aorta. The beginning of the ascending aorta is posterior to the pulmonary trunk and right auricle; the right pulmonary artery is posterior to it. At its origin, the ascending aorta contains three dilations called *aortic sinuses.* Two of these, the right and left sinuses, give rise to the right and left coronary arteries, respectively.

The right and left **coronary arteries** (*coron-* = crown) arise from the ascending aorta just superior to the aortic valve (see **Figure 21.19**). They form a crownlike ring around the heart, giving off branches to the atrial and ventricular myocardium. The **posterior interventricular branch** (in-ter-ven-TRIK-ū-lar; *inter-* = between) of the right coronary artery supplies both ventricles, and the **marginal branch** supplies the right ventricle. The **anterior interventricular branch**, also known as the **left anterior descending (LAD) branch**, of the left coronary artery supplies both ventricles, and the **circumflex branch** (SER-kum-flex; *circum-* = around; *-flex* = to bend) supplies the left atrium and left ventricle.

FIGURE 21.19 Arteries supplying the heart.

Coronary arteries are the first branches off the aorta.

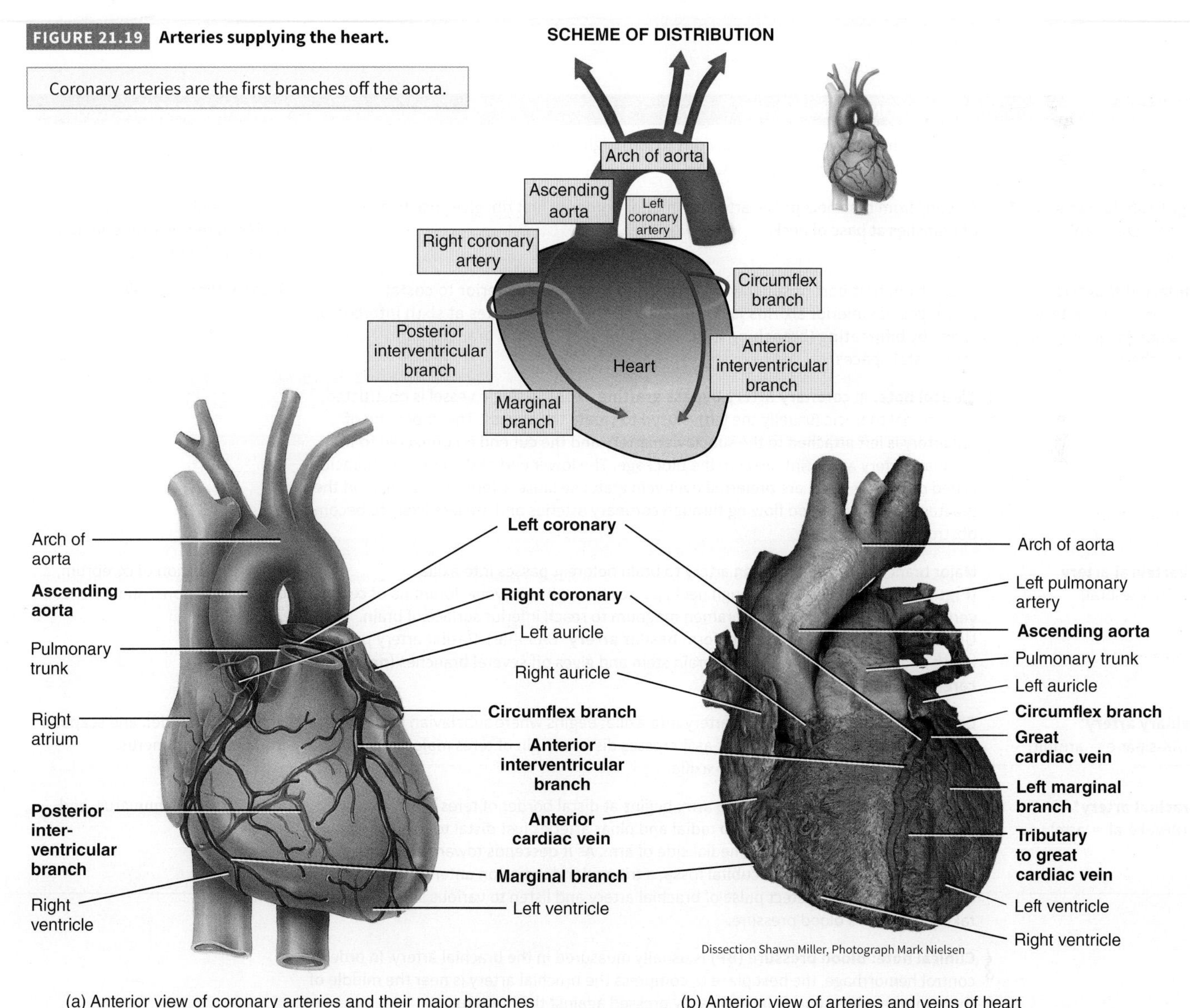

(a) Anterior view of coronary arteries and their major branches

(b) Anterior view of arteries and veins of heart

Q Why are these arteries called coronary arteries?

21.10 The Arch of the Aorta

OBJECTIVE

- **Identify** the three principal arteries that branch from the arch of the aorta.

The **arch of the aorta** is 4–5 cm (almost 2 in.) in length and is the continuation of the ascending aorta. It emerges from the pericardium posterior to the sternum at the level of the sternal angle (**Figure 21.20**). The arch of the aorta is directed superiorly and posteriorly to the left and then inferiorly; it ends at the intervertebral disc between the fourth and fifth thoracic vertebrae, where it becomes the thoracic aorta. Three major arteries branch from the superior aspect of the arch of the aorta: the brachiocephalic trunk, the left common carotid, and the left subclavian. The first and largest branch from the arch of the aorta is the **brachiocephalic trunk** (brā′-kē-ō-se-FAL-ik; *brachio-* = arm; *-cephalic* = head). It extends superiorly, bending slightly to the right, and divides at the right sternoclavicular joint to form the right subclavian artery and right common carotid artery. The second branch from the arch of the aorta is the **left common carotid artery** (ka-ROT-id), which divides into the same branches with the same names as the right common carotid artery. The third branch from the arch of the aorta is the **left subclavian artery** (sub-KLĀ-vē-an), which distributes blood to the left vertebral artery and vessels of the left upper limb. Arteries branching from the left subclavian artery are similar in distribution and name to those branching from the right subclavian artery.

BRANCH	DESCRIPTION AND BRANCHES	REGIONS SUPPLIED
Brachiocephalic	First branch of arch of the aorta; divides to form right subclavian artery and right common carotid artery (**Figure 21.20a**).	Head, neck, upper limb, and thoracic wall.
Right subclavian artery* (sub-KLĀ-vē-an)	Extends from brachiocephalic artery to inferior border of first rib; gives rise to a number of branches at base of neck.	Brain, spinal cord, neck, shoulder, thoracic muscle wall, and scapular muscles.
Internal thoracic (*mammary*) **artery** (thor-AS-ik; *thorac-* = chest)	Arises from first part of subclavian artery and descends posterior to costal cartilages of superior six ribs just lateral to sternum; terminates at sixth intercostal space by bifurcating (branching into two arteries) and sends branches into intercostal spaces. **Clinical note:** In **coronary artery bypass grafting**, if only a single vessel is obstructed, the internal thoracic (usually the left) is used to create the bypass. The upper end of the artery is left attached to the subclavian artery and the cut end is connected to the coronary artery at a point distal to the blockage. The lower end of the internal thoracic is tied off. Artery grafts are preferred over vein grafts because arteries can withstand the greater pressure of blood flowing through coronary arteries and are less likely to become obstructed over time.	Anterior thoracic wall.
Vertebral artery (VER-te-bral)	Major branch of right subclavian artery to brain before it passes into axilla (**Figure 21.20b**); ascends through neck, passes through transverse foramina of cervical vertebrae, and enters skull via foramen magnum to reach inferior surface of brain. Unites with left vertebral artery to form **basilar artery** (BĀS-i-lar). Basilar artery passes along midline of anterior aspect of brain stem and gives off several branches (**posterior cerebral** and **cerebellar arteries**).	Posterior portion of cerebrum, cerebellum, pons, and inner ear.
Axillary artery* (AK-sil-ār-ē = armpit)	Continuation of right subclavian artery into axilla; begins where subclavian artery passes inferior border of first rib and ends as it crosses distal margin of teres major muscle; gives rise to numerous branches in axilla.	Thoracic, shoulder, and scapular muscles and humerus.
Brachial artery* (BRĀ-kē-al = arm)	Continuation of axillary artery into arm; begins at distal border of teres major muscle and terminates by bifurcating into radial and ulnar arteries just distal to bend of elbow; superficial and palpable along medial side of arm. As it descends toward elbow it curves laterally and passes through cubital fossa, a triangular depression anterior to elbow, where you can easily detect pulse of brachial artery and listen to various sounds when taking a person's blood pressure. **Clinical note: Blood pressure (BP)** is usually measured in the brachial artery. In order to control hemorrhage, the best place to compress the brachial artery is near the middle of the arm where it is superficial and easily pressed against the humerus.	Muscles of arm, humerus, and elbow joint.

*This is an example of the practice of giving the same vessel different names as it passes through different regions. See the axillary and brachial arteries.

BRANCH	DESCRIPTION AND BRANCHES	REGIONS SUPPLIED
Radial artery (RĀ-dē-al = radius)	Smaller branch of brachial bifurcation; a direct continuation of brachial artery. Passes along lateral (radial) aspect of forearm and enters wrist where it bifurcates into superficial and deep branches that anastomose with corresponding branches of ulnar artery to form palmar arches of hand. Makes contact with distal end of radius at wrist, where it is covered only by fascia and skin. **Clinical note:** Because of its superficial location at this point, it is a common site for measuring **radial pulse**.	Major blood source to muscles of posterior compartment of forearm.
Ulnar artery (UL-nar = ulna)	Larger branch of brachial artery passes along medial (ulnar) aspect of forearm and then into wrist, where it branches into superficial and deep branches that enter hand. These branches anastomose with corresponding branches of radial artery to form palmar arches of hand.	Major blood source to muscles of anterior compartment of forearm.
Superficial palmar arch (*palma* = palm)	Formed mainly by superficial branch of ulnar artery, with contribution from superficial branch of radial artery; superficial to long flexor tendons of fingers and extends across palm at bases of metacarpals; gives rise to **common palmar digital arteries**, each of which divides into **proper palmar digital arteries**.	Muscles, bones, joints, and skin of palm and fingers.
Deep palmar arch	Arises mainly from deep branch of radial artery, but receives contribution from deep branch of ulnar artery; deep to long flexor tendons of fingers and extends across palm just distal to bases of metacarpals; gives rise to **palmar metacarpal arteries,** which anastomose with common palmar digital arteries from superficial arch.	Muscles, bones, and joints of palm and fingers.
Right common carotid	Begins at bifurcation of brachiocephalic trunk, posterior to right sternoclavicular joint; passes superiorly into neck to supply structures in head (Figure 21.20c); divides into right external and right internal carotid arteries at superior border of larynx (voice box). **Clinical note: Pulse** may be detected in the common carotid artery, just lateral to the larynx. It is convenient to detect a carotid pulse when exercising or when administering cardiopulmonary resuscitation.	Head and neck.
External carotid artery	Begins at superior border of larynx and terminates near temporomandibular joint of parotid gland, where it divides into two branches: **superficial temporal** and **maxillary arteries**. **Clinical note:** The **carotid pulse** can be detected in the external carotid artery just anterior to the sternocleidomastoid muscle at the superior border of the larynx.	Major blood source to all structures of head except brain. Supplies skin, connective tissues, muscles, bones, joints, dura and arachnoid mater in head, and much of neck anatomy.
Internal carotid artery	Arises from common carotid artery; enters cranial cavity through carotid foramen in temporal bone and emerges in cranial cavity near base of hypophyseal fossa of sphenoid bone; gives rise to numerous branches inside cranial cavity and terminates as anterior cerebral arteries. The **anterior cerebral artery** passes forward toward frontal lobe of cerebrum and **middle cerebral artery** passes laterally between temporal and parietal lobes of cerebrum. Inside cranium (Figure 21.20c), anastomoses of left and right internal carotid arteries via **anterior communicating artery** between two anterior cerebral arteries, along with internal carotid–basilar artery anastomoses, form an arrangement of blood vessels at base of brain called **cerebral arterial circle** (*circle of Willis*) (Figure 21.20c). Internal carotid–basilar anastomosis occurs where **posterior communicating arteries** arising from internal carotid artery anastomose with posterior cerebral arteries from basilar artery to link internal carotid blood supply with vertebral blood supply. Cerebral arterial circle equalizes blood pressure to brain and provides alternate routes for blood flow to brain, should arteries become damaged.	Eyeball and other orbital structures, ear, and parts of nose and nasal cavity. Frontal, temporal, parietal lobes of the cerebrum of brain, pituitary gland, and pia mater.
Left common carotid artery	Arises as second branch of arch of the aorta and ascends through mediastinum to enter neck deep to clavicle, then follows similar path to right common carotid artery.	Distribution similar to right common carotid artery.
Left subclavian artery	Arises as third and final branch of arch of the aorta; passes superior and lateral through mediastinum and deep to clavicle at base of neck as it courses toward upper limb; has similar course to right subclavian artery after leaving mediastinum.	Distribution similar to right subclavian artery.

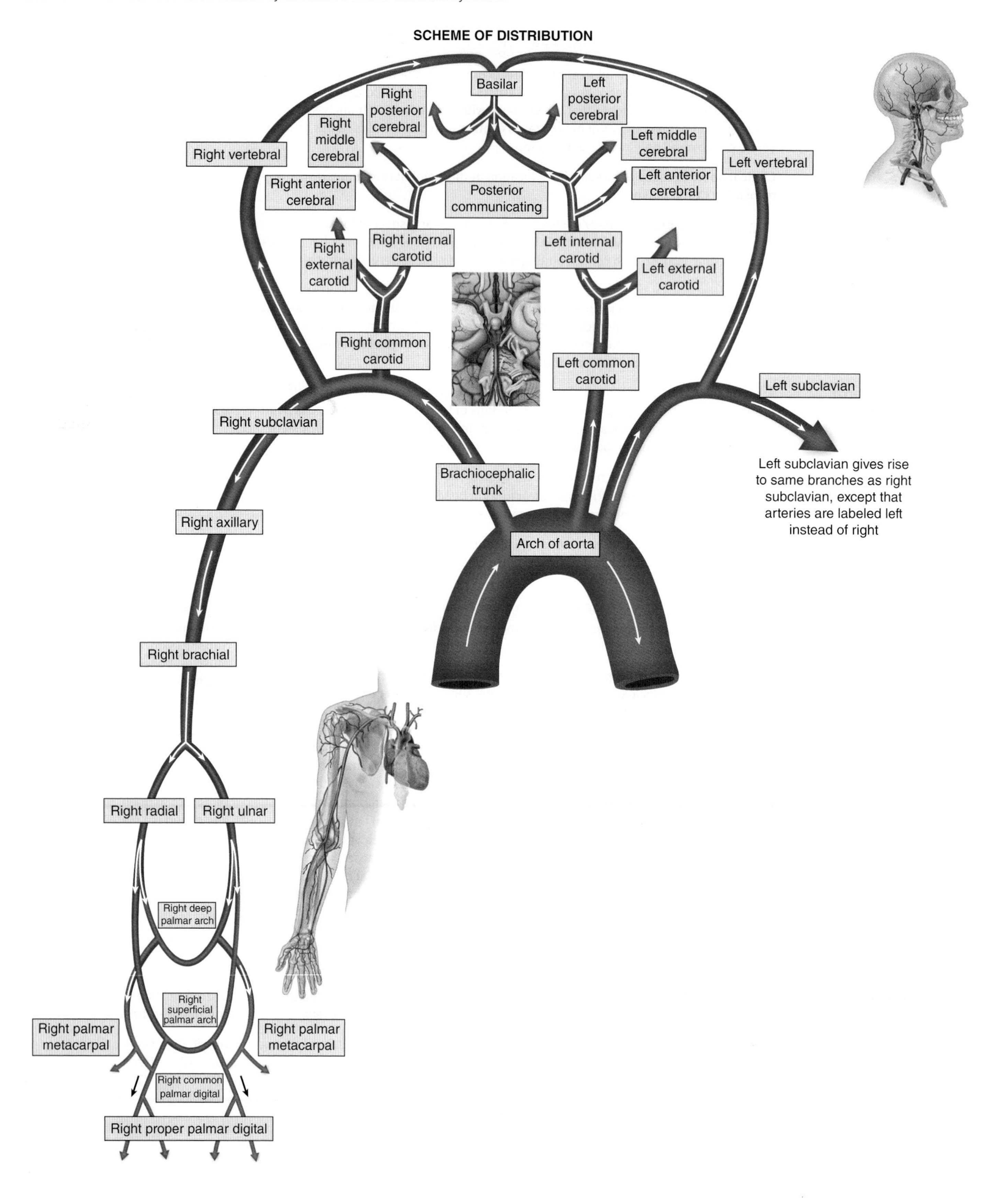
SCHEME OF DISTRIBUTION
Basilar
Right posterior cerebral
Left posterior cerebral
Right middle cerebral
Left middle cerebral
Right vertebral
Left vertebral
Right anterior cerebral
Left anterior cerebral
Posterior communicating
Right internal carotid
Left internal carotid
Right external carotid
Left external carotid
Right common carotid
Left common carotid
Left subclavian
Right subclavian
Left subclavian gives rise to same branches as right subclavian, except that arteries are labeled left instead of right
Brachiocephalic trunk
Right axillary
Arch of aorta
Right brachial
Right radial
Right ulnar
Right deep palmar arch
Right superficial palmar arch
Right palmar metacarpal
Right palmar metacarpal
Right common palmar digital
Right proper palmar digital

FIGURE 21.20 Arch of the aorta and its branches. Note in (c) the arteries that constitute the cerebral arterial circle (circle of Willis).

The arch of the aorta ends at the level of the intervertebral disc between the fourth and fifth thoracic vertebrae.

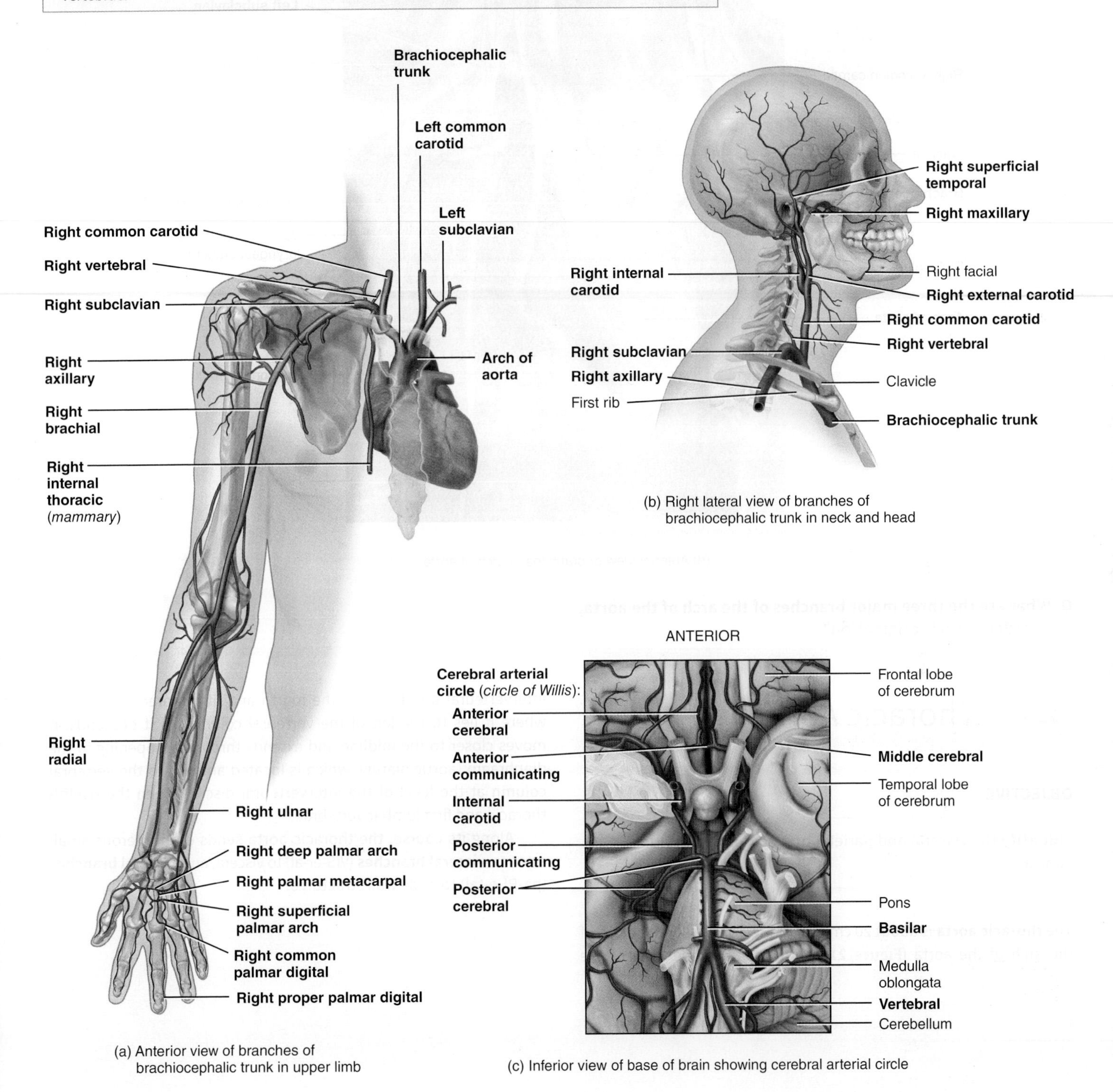

(a) Anterior view of branches of brachiocephalic trunk in upper limb

(b) Right lateral view of branches of brachiocephalic trunk in neck and head

(c) Inferior view of base of brain showing cerebral arterial circle

Figure 21.20 ***Continues***

Figure 21.20 ***Continued***

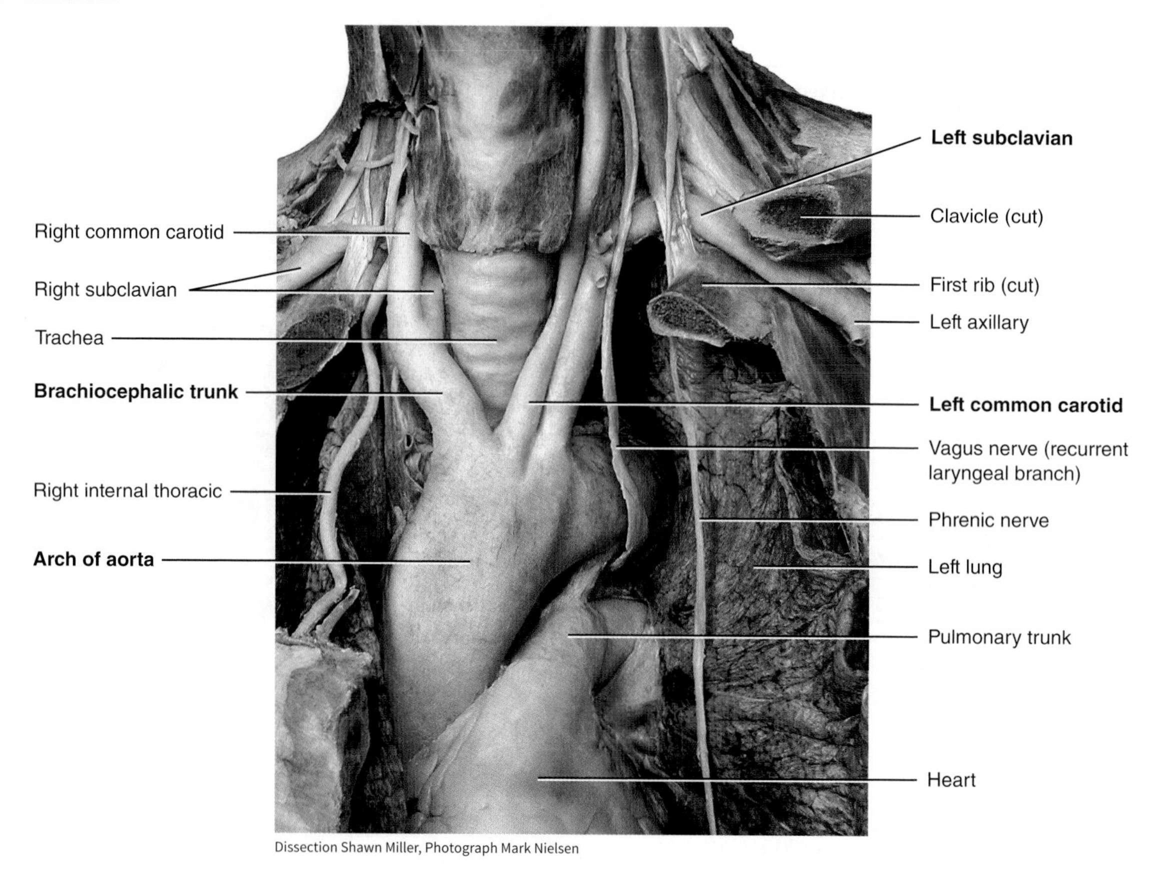

(d) Anterior view of branches of arch of aorta

Q What are the three major branches of the arch of the aorta, in order of their origination?

21.11 Thoracic Aorta

OBJECTIVE

- **Identify** the visceral and parietal branches of the thoracic aorta.

The **thoracic aorta** is about 20 cm (8 in.) long and is a continuation of the arch of the aorta (**Figure 21.21**). It begins at the level of the intervertebral disc between the fourth and fifth thoracic vertebrae, where it lies to the left of the vertebral column. As it descends, it moves closer to the midline and extends through an opening in the diaphragm (aortic hiatus), which is located anterior to the vertebral column at the level of the intervertebral disc between the twelfth thoracic and first lumbar vertebrae.

Along its course, the thoracic aorta sends off numerous small arteries, **visceral branches** (VIS-er-al) to viscera, and **parietal branches** (pa-RĪ-e-tal) to body wall structures.

BRANCH	DESCRIPTION AND BRANCHES	REGIONS SUPPLIED
VISCERAL BRANCHES		
Pericardial arteries (per-i-KAR-dē-al; *peri-* = around; *-cardia* = heart)	Two to three small arteries that arise from variable levels of thoracic aorta and pass forward to pericardial sac surrounding heart.	Tissues of pericardial sac.
Bronchial arteries (BRONG-kē-al = windpipe)	Arise from thoracic aorta or one of its branches. Right bronchial artery typically arises from third posterior intercostal artery; two left bronchial arteries arise from upper end of thoracic aorta. All follow bronchial tree into lungs.	Supply tissues of bronchial tree and surrounding lung tissue down to level of alveolar ducts.
Esophageal arteries (e-sof′-a-JĒ-al; *eso-* = to carry; *-phage* = food)	Four to five arteries that arise from anterior surface of thoracic aorta and pass forward to branch onto esophagus.	All tissues of esophagus.
Mediastinal arteries (mē′-dē-as-TĪ-nal)	Arise from various points on thoracic aorta.	Assorted tissues within mediastinum, primarily connective tissue and lymph nodes.
PARIETAL BRANCHES		
Posterior intercostal arteries (in′-ter-KOS-tal; *inter-* = between; *-costa* = rib)	Typically, nine pairs of arteries that arise from posterolateral aspect on each side of thoracic aorta. Each passes laterally and then anteriorly through intercostal space, where they will eventually anastomose with anterior branches from internal thoracic arteries.	Skin, muscles, and ribs of thoracic wall. Thoracic vertebrae, meninges, and spinal cord. Mammary glands.
Subcostal arteries (sub-KOS-tal; *sub-* = under)	The lowest segmental branches of thoracic aorta; one on each side passes into thoracic body wall inferior to 12th rib and courses forward into upper abdominal region of body wall.	Skin, muscles, and ribs. Twelfth thoracic vertebra, meninges, and spinal cord.
Superior phrenic arteries (FREN-ik = pertaining to diaphragm)	Arise from lower end of thoracic aorta and pass onto superior surface of diaphragm.	Diaphragm muscle and pleura covering diaphragm.

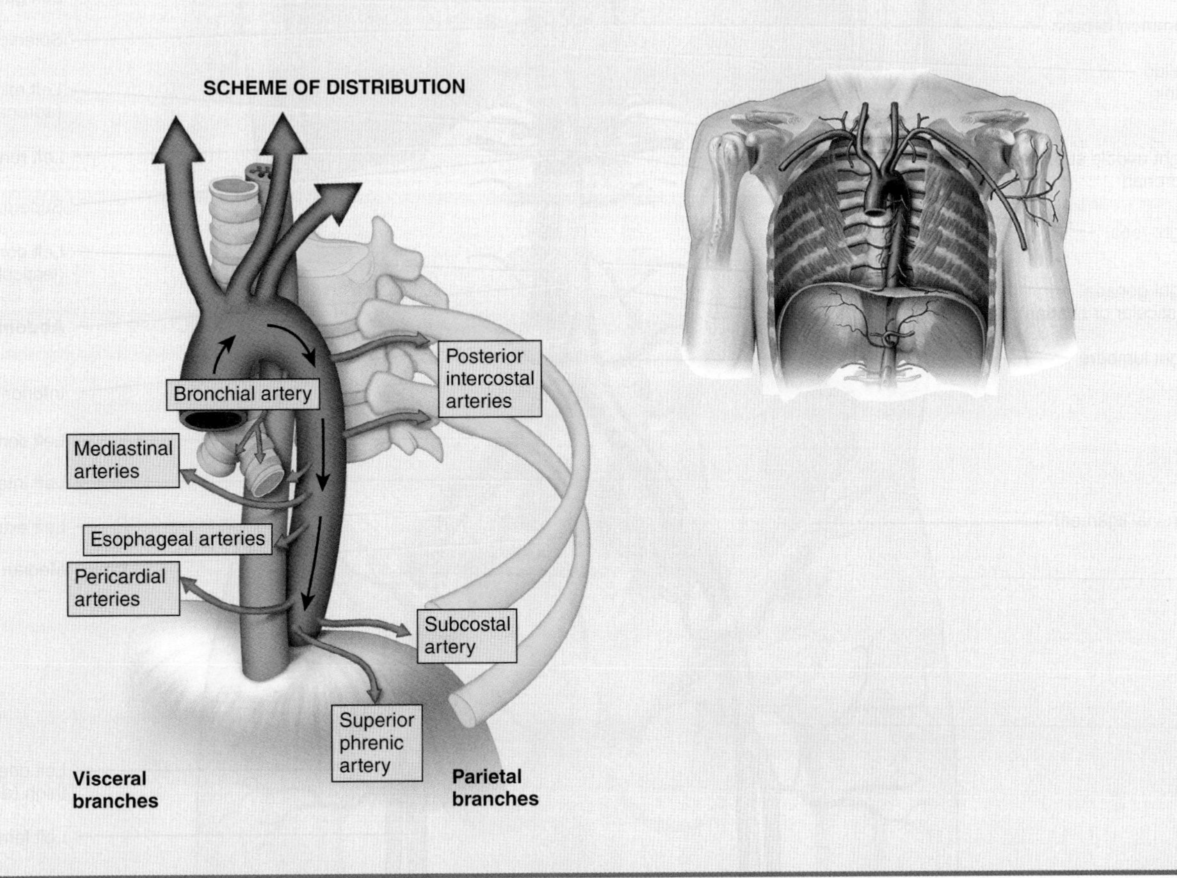

FIGURE 21.21 Thoracic aorta and abdominal aorta and their principal branches.

The thoracic aorta is the continuation of the ascending aorta.

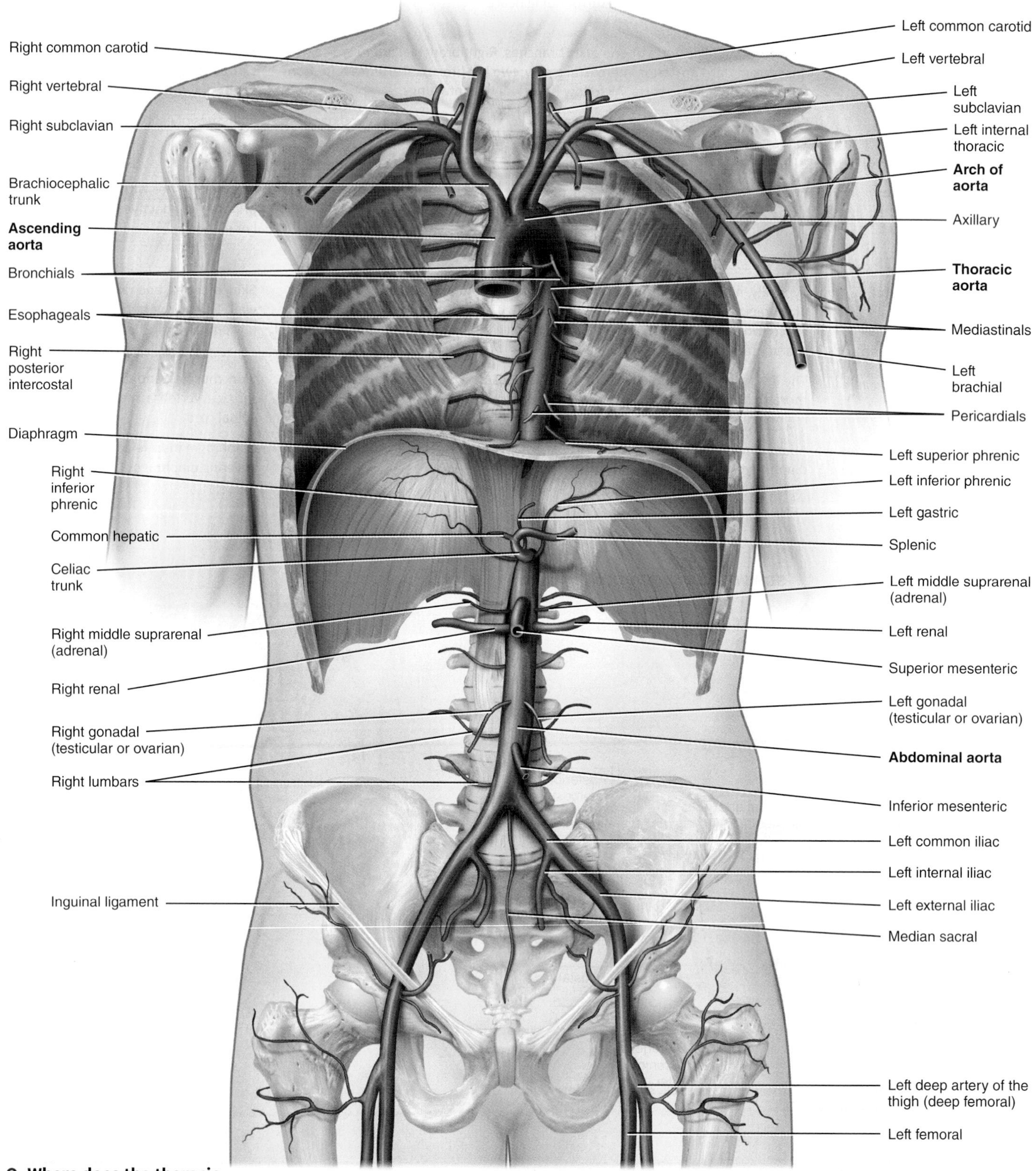

Detailed anterior view of the principal branches of the aorta

Q Where does the thoracic aorta begin?

21.12 Abdominal Aorta

OBJECTIVE

- **Identify** the visceral and parietal branches of the abdominal aorta.

The **abdominal aorta** is the continuation of the thoracic aorta after it passes through the diaphragm (see Figure 21.22). It begins at the aortic hiatus in the diaphragm and ends at about the level of the fourth lumbar vertebra, where it divides into the right and left common iliac arteries. The abdominal aorta lies anterior to the vertebral column.

As with the thoracic aorta, the abdominal aorta gives off **visceral** and **parietal branches**. The unpaired visceral branches arise from

BRANCH	DESCRIPTION AND BRANCHES	REGIONS SUPPLIED
UNPAIRED VISCERAL BRANCHES		
Celiac trunk (artery) (SĒ-lē-ak)	First visceral branch of aorta inferior to diaphragm; arises from abdominal aorta at level of twelfth thoracic vertebra as aorta passes through hiatus in diaphragm; divides into three branches: left gastric, splenic, and common hepatic arteries (Figure 21.22a).	Supplies all organs of gastrointestinal tract that arise from embryonic foregut, that is, from abdominal part of esophagus to duodenum, and also spleen.
	1. Left gastric artery (GAS-trik = stomach). Smallest of three celiac branches arises superiorly to left toward esophagus and then turns to follow lesser curvature of stomach. On lesser curvature of stomach it anastomoses with right gastric artery.	Abdominal part of esophagus lesser curvature of stomach, and lesser omentum.
	2. Splenic artery (SPLĒN-ik = spleen). Largest branch of celiac trunk arises from left side of celiac trunk distal to left gastric artery, and passes horizontally to left along pancreas. Before reaching spleen, it gives rise to three named arteries:	Spleen, pancreas, fundus and greater curvature of stomach, and greater omentum.
	• **Pancreatic arteries** (pan-krē-AT-ik), a series of small arteries that arise from splenic and descend into tissue of pancreas.	Pancreas.
	• **Left gastro-omental artery** (gas′-trō-ō-MEN-tal) or *gastro-epiploic artery* (gas′-trō-ep′-i-PLŪ-ik) arises from terminal end of splenic artery and passes from left to right along greater curvature of stomach.	Greater curvature of stomach and greater omentum.
	• **Short gastric arteries** arise from terminal end of splenic artery and pass onto fundus of stomach.	Fundus of stomach.
	3. Common hepatic artery (he-PAT-ik = liver). Intermediate in size between left gastric and splenic arteries; arises from right side of celiac trunk and gives rise to three arteries:	Liver, gallbladder, lesser omentum, stomach, pancreas, and duodenum.
	• **Proper hepatic artery** branches from common hepatic artery and ascends along bile ducts into liver and gallbladder.	Liver, gallbladder, and lesser omentum.
	• **Right gastric artery** arises from common hepatic artery and curves back to left along lesser curvature of stomach, where it anastomoses with left gastric artery.	Lesser curvature of stomach and lesser omentum.
	• **Gastroduodenal artery** (gas′-trō-doo′-ō-DĒ-nal) passes inferiorly toward stomach and duodenum and sends branches along greater curvature of stomach.	Lesser curvature of stomach, duodenum, and pancreas.
Superior mesenteric artery (mez-en-TER-ik; *meso-* = middle; *-enteric* = pertaining to intestines)	Arises from anterior surface of abdominal aorta about 1 cm inferior to celiac trunk at level of first lumbar vertebra (Figure 21.22b); extends inferiorly and anteriorly between layers of mesentery (portion of peritoneum that attaches small intestine to posterior abdominal wall). It anastomoses extensively and has five branches:	Supplies all organs of gastrointestinal tract that arise from embryonic midgut, that is, from duodenum to transverse colon.
	1. Inferior pancreaticoduodenal artery (pan-krē-at′-i-kō-doo′-ō-DĒ-nal) passes superiorly and to right toward head of pancreas and duodenum.	Pancreas and duodenum.
	2. Jejunal (je-JOO-nal) and **ileal arteries** (IL-ē-al) spread through mesentery to pass to loops of jejunum and ileum (small intestine).	Jejunum and ileum, which is majority of small intestine.
	3. Ileocolic artery (il′-ē-ō-KOL-ik) passes inferiorly and laterally toward right side toward terminal part of ileum, cecum, appendix, and first part of ascending colon.	Terminal part of ileum, cecum, appendix, and first part of ascending colon.
	4. Right colic artery (KOL-ik) passes laterally to right toward ascending colon.	Ascending colon and first part of transverse colon.
	5. Middle colic artery ascends slightly to right toward transverse colon.	Most of transverse colon.
Inferior mesenteric artery	Arises from anterior aspect of abdominal aorta at level of third lumbar vertebra and then passes inferiorly to left of aorta (Figure 21.22c). It anastomoses extensively and has three branches:	Supplies all organs of gastrointestinal tract that arise from embryonic hindgut from transverse colon to rectum.
	1. Left colic artery ascends laterally to left toward distal end of transverse colon and descending colon.	End of transverse colon and descending colon.
	2. Sigmoid arteries (SIG-moyd) descend laterally to left toward sigmoid colon.	Sigmoid colon.
	3. Superior rectal artery (REK-tal) passes inferiorly to superior part of rectum.	Upper part of rectum.

the anterior surface of the aorta and include the **celiac trunk** and the **superior mesenteric** and **inferior mesenteric arteries** (see Figure 21.21).

The paired visceral branches arise from the lateral surfaces of the aorta and include the **suprarenal**, **renal**, and **gonadal arteries**.

The lone unpaired parietal branch is the **median sacral artery**. The paired parietal branches arise from the posterolateral surfaces of the aorta and include the **inferior phrenic** and **lumbar arteries**.

BRANCH	DESCRIPTION AND BRANCHES	REGIONS SUPPLIED
PAIRED VISCERAL BRANCHES		
Suprarenal arteries (soo'-pra-RĒ-nal; *supra-* = above; *-ren-* = kidney)	There are typically three pairs (superior, middle, and inferior), but only middle pair originates directly from abdominal aorta (see Figure 21.21). **Middle suprarenal arteries** arise from abdominal aorta at level of first lumbar vertebra at or superior to renal arteries. **Superior suprarenal arteries** arise from inferior phrenic arteries, and **inferior suprarenal arteries** originate from renal arteries.	Suprarenal (adrenal) glands.
Renal arteries (RĒ-nal; *ren* = kidney)	Right and left renal arteries usually arise from lateral aspects of abdominal aorta at superior border of second lumbar vertebra, about 1 cm inferior to superior mesenteric artery (see Figure 21.21). Right renal artery, which is longer than left, arises slightly lower than left and passes posterior to right renal vein and inferior vena cava. Left renal artery is posterior to left renal vein and is crossed by inferior mesenteric vein.	All tissues of kidneys.
Gonadal (gō-NAD-al; *gon-* = seed) arteries [**testicular** (tes-TIK-ū-lar) or **ovarian** (ō-VAR-ē-an)]	Arise from anterior aspect of abdominal aorta at level of second lumbar vertebra just inferior to renal arteries (see Figure 21.21). In males, gonadal arteries are specifically referred to as **testicular arteries**. They descend along posterior abdominal wall to pass through inguinal canal and descend into scrotum. In females, gonadal arteries are called **ovarian arteries**. They are much shorter than testicular arteries and remain within abdominal cavity.	Males: testis, epididymis, ductus deferens, and ureters. Females: ovaries, uterine (fallopian) tubes, and ureters.
UNPAIRED PARIETAL BRANCH		
Median sacral artery (SĀ-kral = pertaining to sacrum)	Arises from posterior surface of abdominal aorta about 1 cm superior to *bifurcation* (division into two branches) of aorta into right and left common iliac arteries (see Figure 21.21).	Sacrum, coccyx, sacral spinal nerves, and piriformis muscle.
PAIRED PARIETAL BRANCH		
Inferior phrenic arteries (FREN-ik = pertaining to diaphragm)	First paired branches of abdominal aorta; arise immediately superior to origin of celiac trunk (see Figure 21.20). (They may also arise from renal arteries.)	Diaphragm and suprarenal (adrenal) glands.
Lumbar arteries (LUM-bar = pertaining to loin)	Four pairs arise from posterolateral surface of abdominal aorta to the pattern of the similar posterior intercostal arteries of thorax (see Figure 21.21); pass laterally into abdominal muscle wall and curve toward anterior aspect of wall.	Lumbar vertebrae, spinal cord and meninges, skin and muscles of posterior and lateral part of abdominal wall.

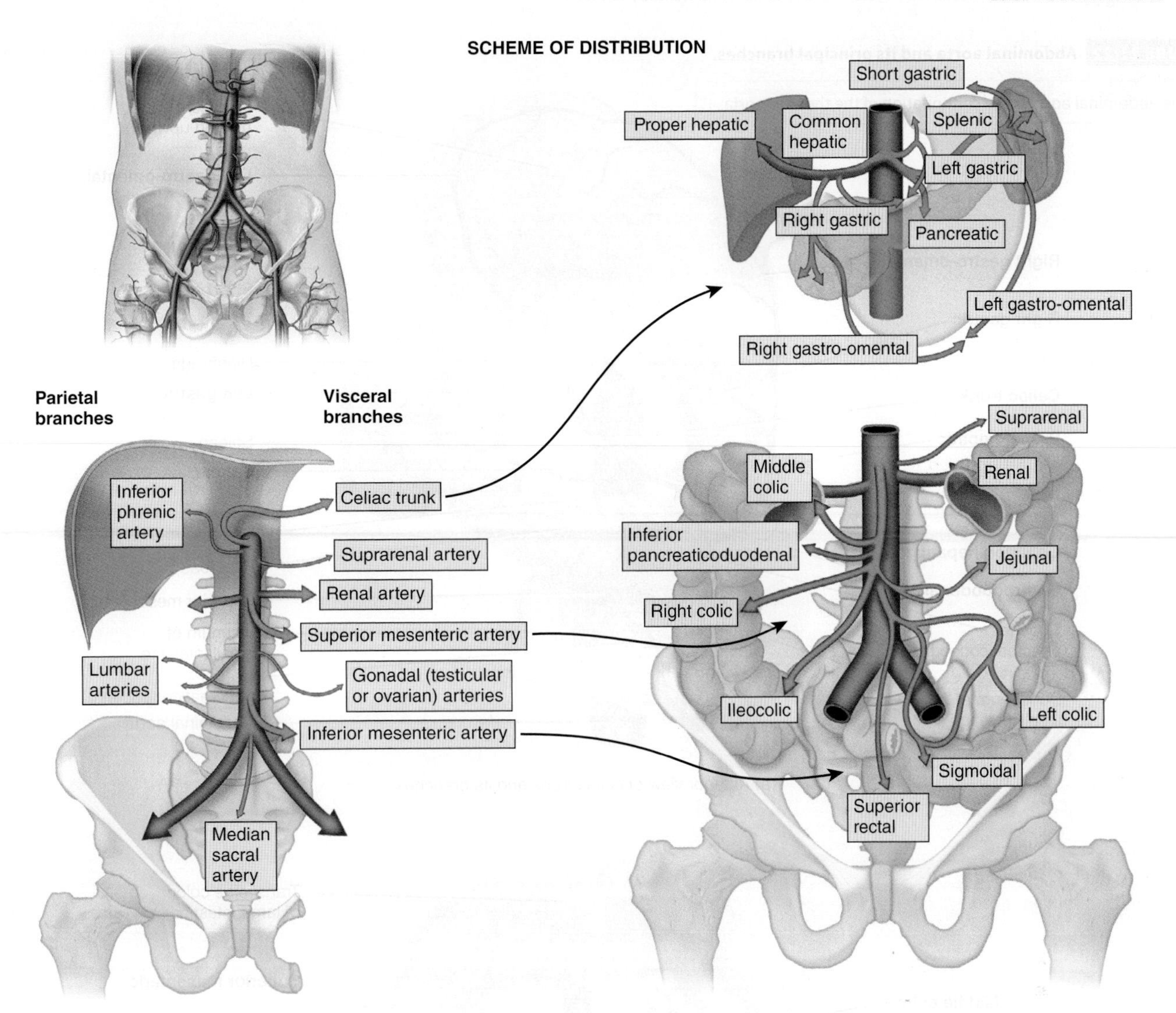
SCHEME OF DISTRIBUTION
Short gastric
Proper hepatic
Common hepatic
Splenic
Left gastric
Right gastric
Pancreatic
Left gastro-omental
Right gastro-omental
Parietal branches
Visceral branches
Suprarenal
Renal
Middle colic
Inferior phrenic artery
Celiac trunk
Inferior pancreaticoduodenal
Suprarenal artery
Jejunal
Renal artery
Right colic
Superior mesenteric artery
Lumbar arteries
Gonadal (testicular or ovarian) arteries
Ileocolic
Left colic
Inferior mesenteric artery
Sigmoidal
Superior rectal
Median sacral artery

FIGURE 21.22 Abdominal aorta and its principal branches.

The abdominal aorta is the continuation of the thoracic aorta.

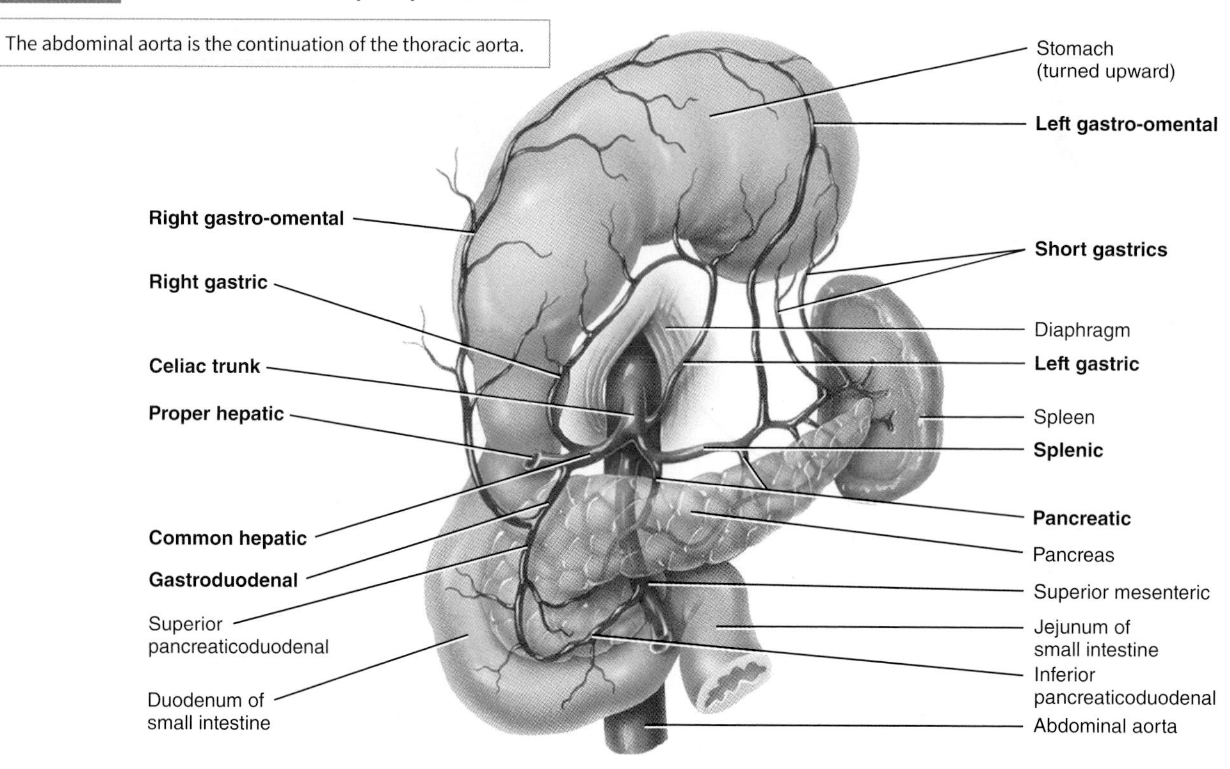

(a) Anterior view of celiac trunk and its branches

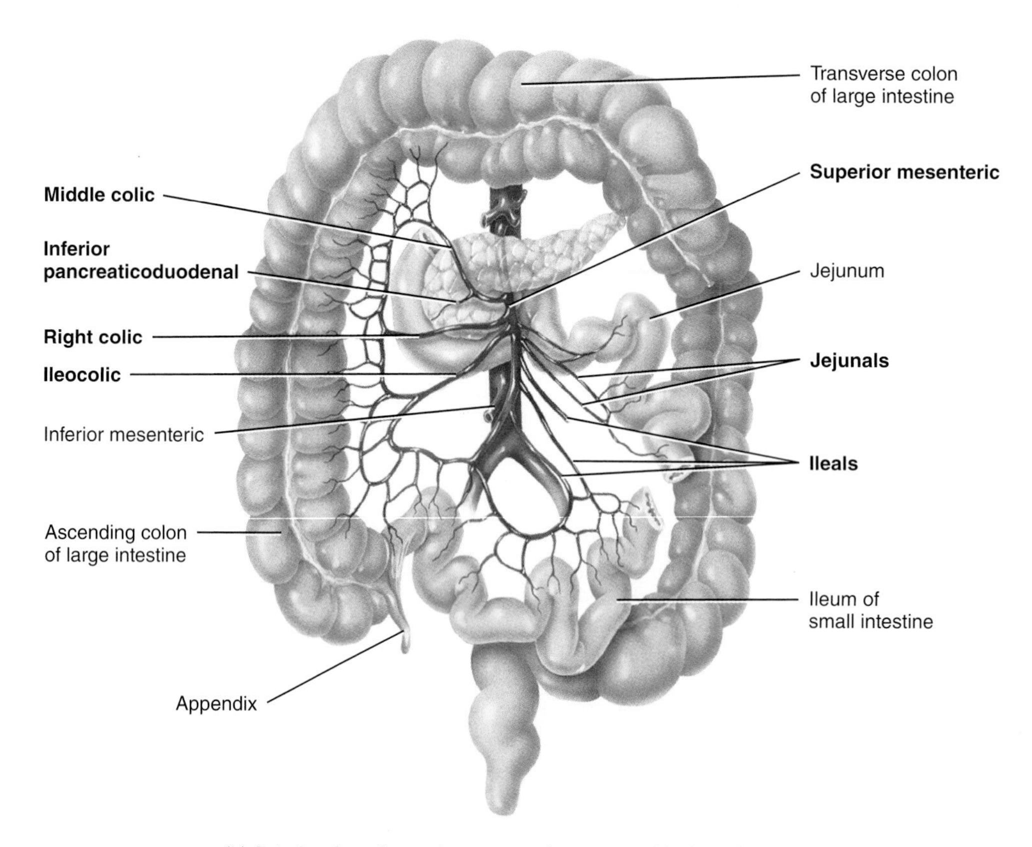

(b) Anterior view of superior mesenteric artery and its branches

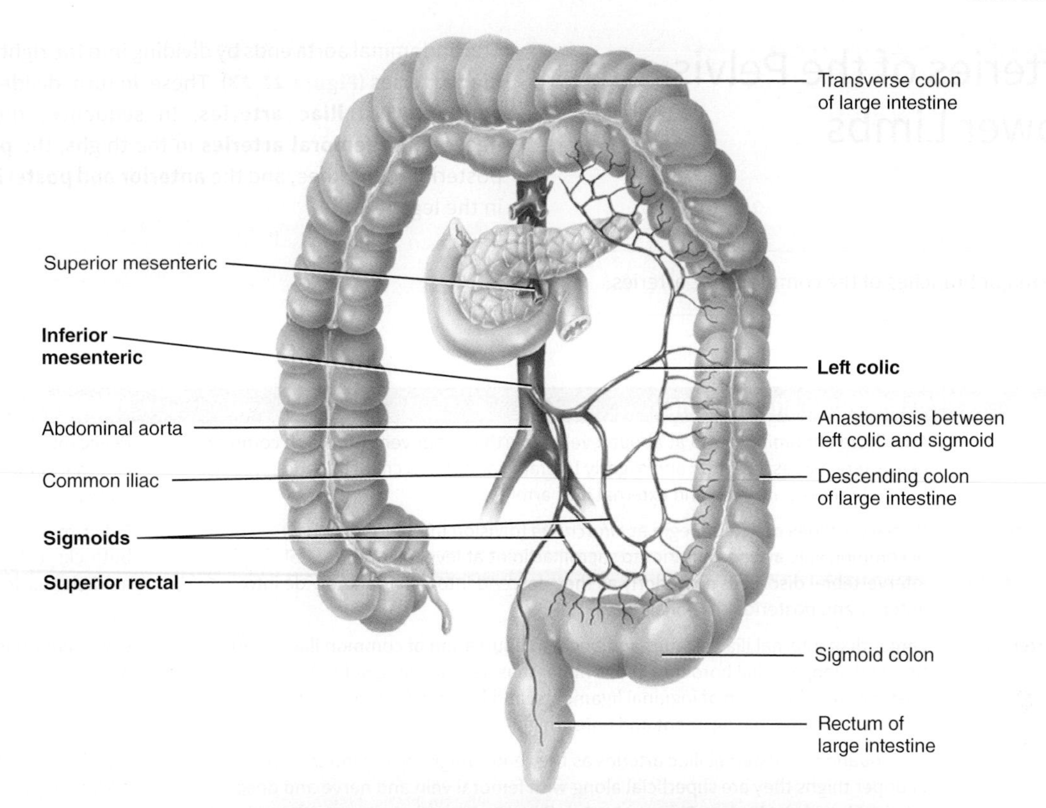

(c) Anterior view of inferior mesenteric artery and its branches

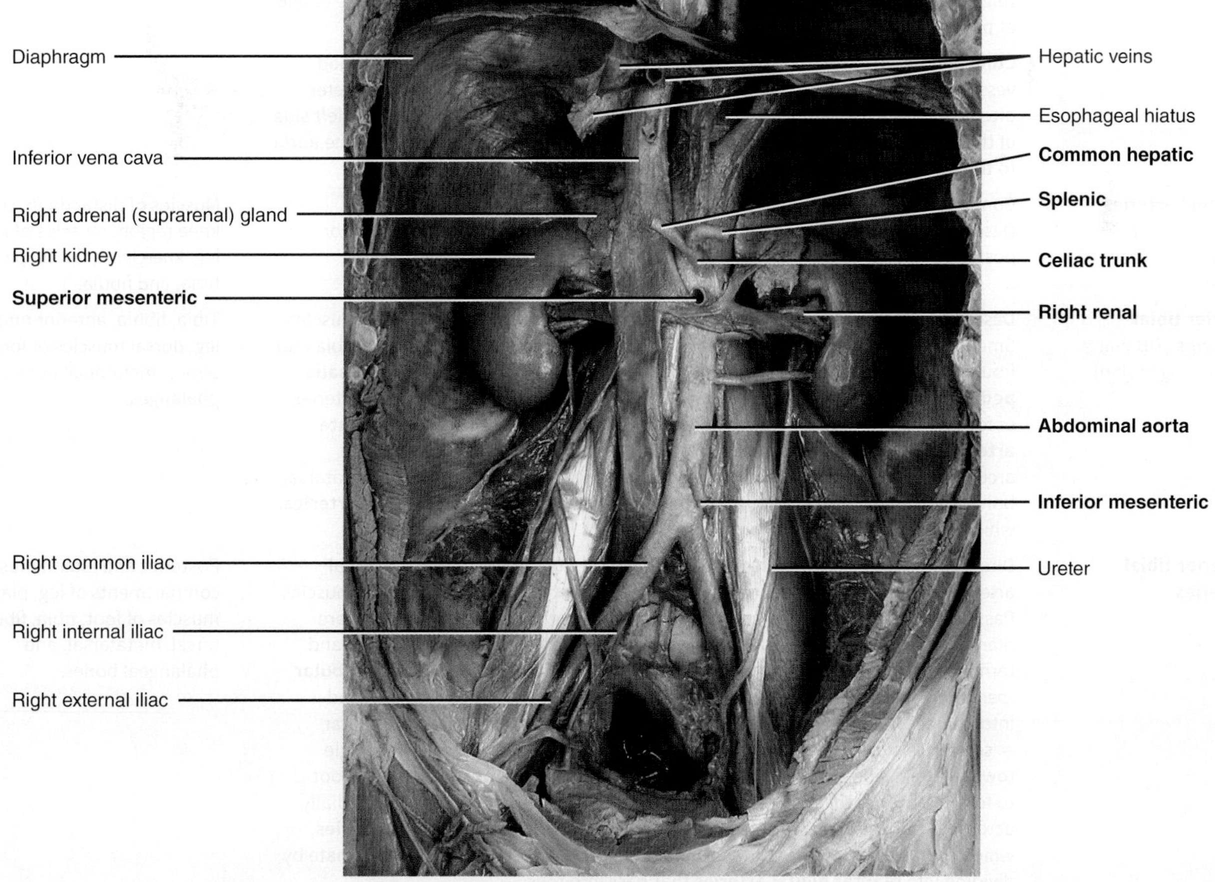

Dissection Shawn Miller, Photograph Mark Nielsen

(d) Anterior view of arteries of abdomen and pelvis

Q Where does the abdominal aorta begin?

21.13 Arteries of the Pelvis and Lower Limbs

OBJECTIVE

• **Identify** the two major branches of the common iliac arteries.

The abdominal aorta ends by dividing into the right and left **common iliac arteries** (Figure 21.23). These, in turn, divide into the **internal** and **external iliac arteries**. In sequence, the external iliacs become the **femoral arteries** in the thighs, the **popliteal arteries** posterior to the knee, and the **anterior** and **posterior tibial arteries** in the legs.

BRANCH	DESCRIPTION AND BRANCHES	REGIONS SUPPLIED
Common iliac arteries (IL-ē-ak = pertaining to ilium)	Arise from abdominal aorta at about level of fourth lumbar vertebra. Each common iliac artery passes inferiorly and slightly laterally for about 5 cm (2 in.) and gives rise to two branches: internal and external iliac arteries.	Pelvic muscle wall, pelvic organs, external genitals, and lower limbs.
Internal iliac arteries	Primary arteries of pelvis. Begin at *bifurcation* (division into two branches) of common iliac arteries anterior to sacroiliac joint at level of lumbosacral intervertebral disc. Pass posteriorly as they descend into pelvis and divide into anterior and posterior divisions.	Pelvic muscle wall, pelvic organs, buttocks, external genitals, and medial muscles of thigh.
External iliac arteries	Larger than internal iliac arteries and begin at bifurcation of common iliac arteries. Descend along medial border of psoas major muscles following pelvic brim, pass posterior to midportion of inguinal ligaments, and become femoral arteries as they pass beneath inguinal ligament and enter thigh.	Lower abdominal wall, cremaster muscle in males and round ligament of uterus in females, and lower limb.
Femoral arteries (FEM-o-ral = pertaining to thigh)	Continuations of external iliac arteries as they enter thigh. In *femoral triangle* of upper thighs they are superficial along with femoral vein and nerve and deep inguinal lymph nodes (see Figure 11.20a). Pass beneath sartorius muscle as they descend along anteromedial aspects of thighs and follow its course to distal end of thigh where they pass through opening in tendon of adductor magnus muscle to end at posterior aspect of knee, where they become popliteal arteries. **Clinical note:** In **cardiac catheterization**, a catheter is inserted through a blood vessel and advanced into the major vessels to access a heart chamber. A catheter often contains a measuring instrument or other device at its tip. To reach the left side of the heart, the catheter is inserted into the femoral artery and passed into the aorta to the coronary arteries or heart chamber.	Muscles of thigh (quadriceps, adductors, and hamstrings), femur, and ligaments and tendons around knee joint.
Popliteal arteries (pop′-li-TĒ-al = posterior surface of knee)	Continuation of femoral arteries through popliteal fossa (space behind knee). Descend to inferior border of popliteus muscles, where they divide into anterior and posterior tibial arteries.	Muscles of distal thigh, skin of knee region, muscles of proximal leg, knee joint, femur, patella, tibia, and fibula.
Anterior tibial arteries (TIB-ē-al = pertaining to shin)	Descend from bifurcation of popliteal arteries at distal border of popliteus muscles. Smaller than posterior tibial arteries; pass over interosseous membrane of tibia and fibula to descend through anterior muscle compartment of leg; become **dorsalis pedis arteries** (*dorsal arteries of foot*) at ankles. On dorsum of feet, dorsal arteries of foot give off transverse branch at first medial cuneiform bone called **arcuate arteries** (*arcuat-* = bowed) that run laterally over bases of metatarsals. From arcuate arteries branch **dorsal metatarsal arteries**, which course along metatarsal bones. Dorsal metatarsal arteries terminate by dividing into **dorsal digital arteries**, which pass into toes.	Tibia, fibula, anterior muscles of leg, dorsal muscles of foot, tarsal bones, metatarsal bones, and phalanges.
Posterior tibial arteries	Direct continuations of popliteal arteries, descend from bifurcation of popliteal arteries. Pass down posterior muscular compartment of legs deep to soleus muscles. Pass posterior to medial malleolus at distal end of leg and curve forward toward plantar surface of feet; pass deep to flexor retinaculum on medial side of feet and terminate by branching into medial and lateral plantar arteries. Give rise to **fibular** (*peroneal*) **arteries** in upper third of leg, which course laterally as they descend into lateral compartment of leg. The smaller **medial plantar arteries** (PLAN-tar = sole) pass along medial side of sole and larger **lateral plantar arteries** angle toward lateral side of sole and unite with branch of dorsalis pedis arteries of foot to form **plantar arch**. Arch begins at base of fifth metatarsal and extends medially across metatarsals. As arch crosses foot, it gives off **plantar metatarsal arteries**, which course along plantar surface of metatarsal bones. These arteries terminate by dividing into **plantar digital arteries** that pass into toes.	Posterior and lateral muscle compartments of leg, plantar muscles of foot, tibia, fibula, tarsal, metatarsal, and phalangeal bones.

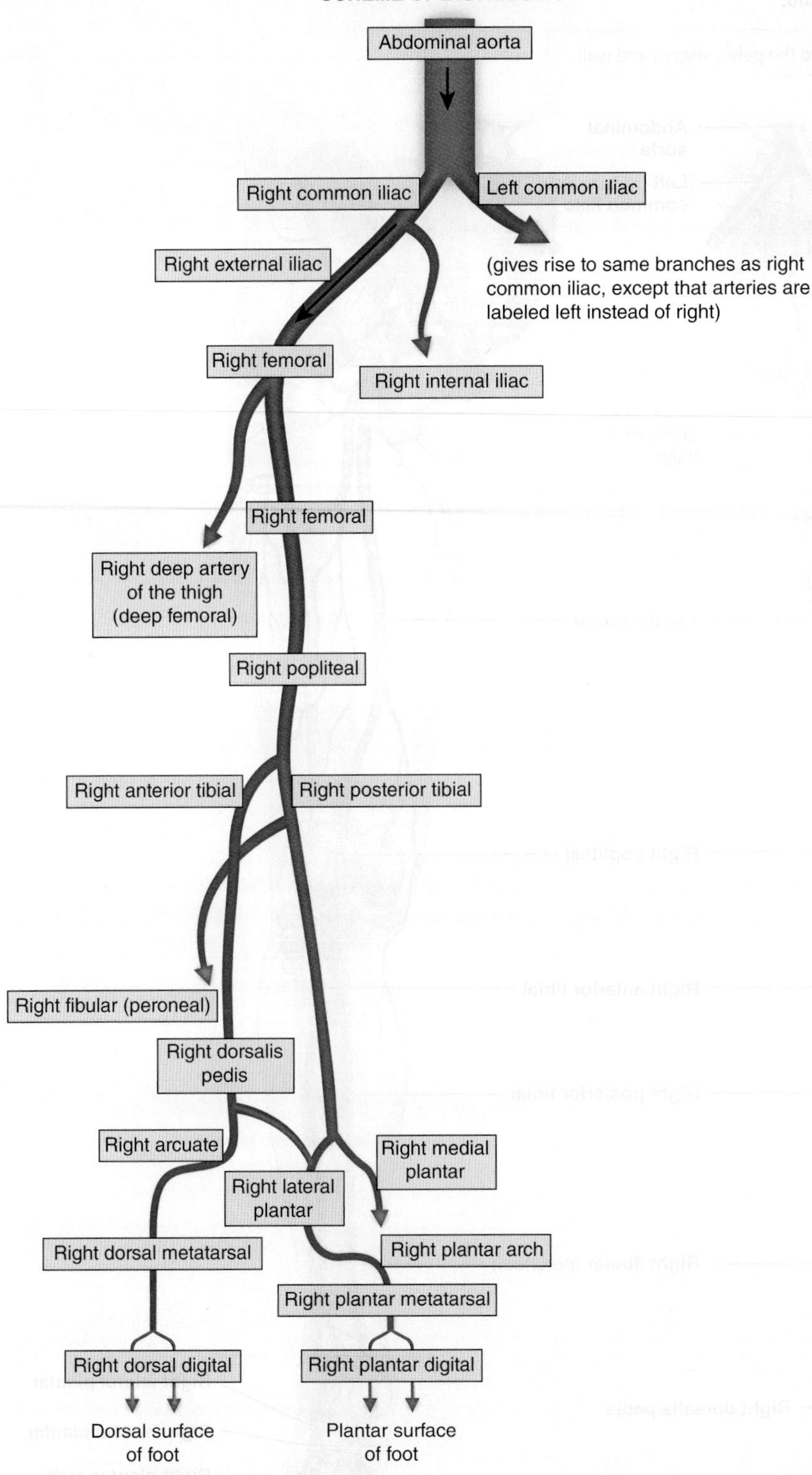
SCHEME OF DISTRIBUTION
Abdominal aorta
Right common iliac
Left common iliac
(gives rise to same branches as right common iliac, except that arteries are labeled left instead of right)
Right external iliac
Right femoral
Right internal iliac
Right femoral
Right deep artery of the thigh (deep femoral)
Right popliteal
Right anterior tibial
Right posterior tibial
Right fibular (peroneal)
Right dorsalis pedis
Right arcuate
Right medial plantar
Right lateral plantar
Right dorsal metatarsal
Right plantar arch
Right plantar metatarsal
Right dorsal digital
Right plantar digital
Dorsal surface of foot
Plantar surface of foot

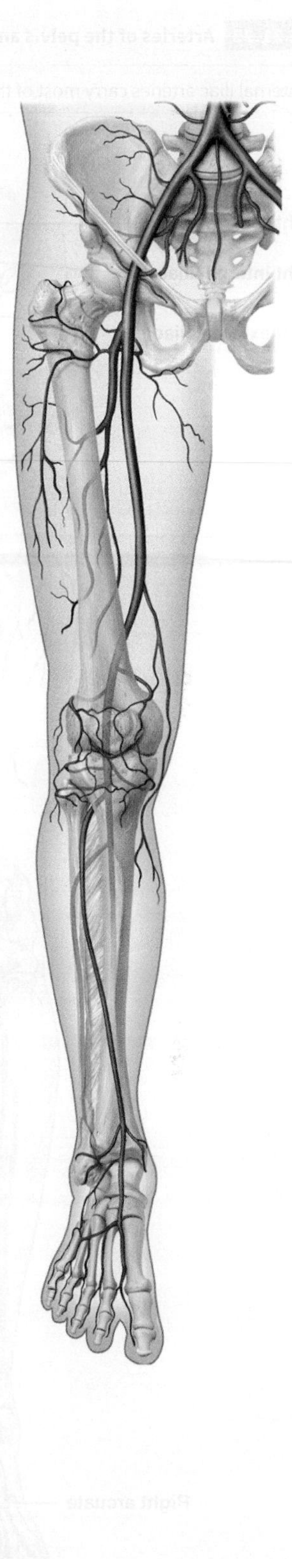

FIGURE 21.23 **Arteries of the pelvis and right lower limb.**

The internal iliac arteries carry most of the blood supply to the pelvic viscera and wall.

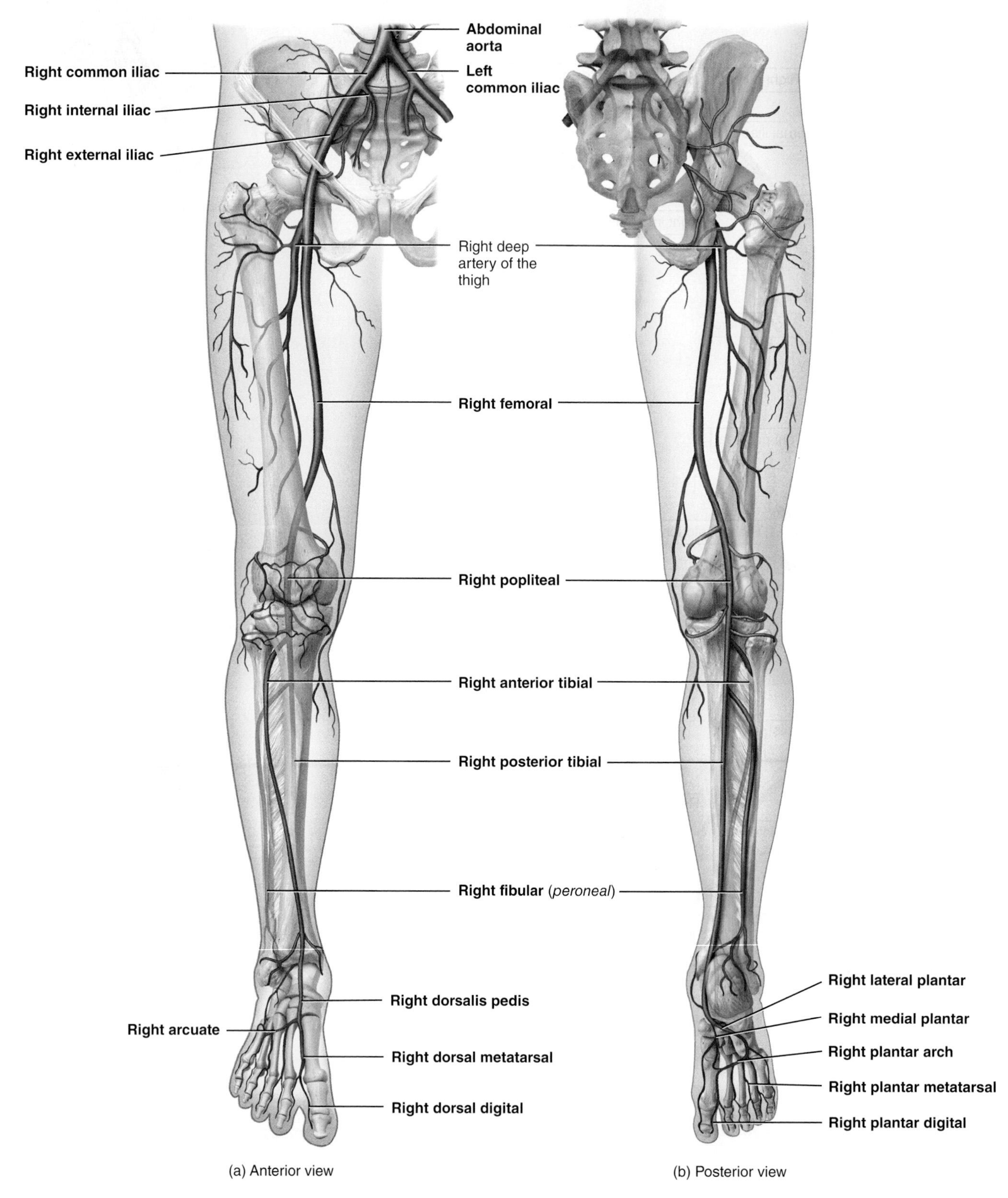

Q At what point does the abdominal aorta divide into the common iliac arteries?

21.14 Veins of the Systemic Circulation

OBJECTIVE

- **Identify** the three systemic veins that return deoxygenated blood to the heart.

As you have already learned, arteries distribute blood from the heart to various parts of the body, and veins drain blood away from the various parts and return the blood to the heart. In general, arteries are deep; veins may be superficial or deep. Superficial veins are located just beneath the skin and can be seen easily. Because there are no large superficial arteries, the names of superficial veins do not correspond to those of arteries. Superficial veins are clinically important as sites for withdrawing blood or giving injections. Deep veins generally travel alongside arteries and usually bear the same name. Arteries usually follow definite pathways; veins are more difficult to follow because they connect in irregular networks in which many tributaries merge to form a large vein. Although only one systemic artery, the aorta, takes oxygenated blood away from the heart (left ventricle), three systemic veins, the **coronary sinus**, **superior vena cava (SVC)** (VĒ-na KĀ-va), and **inferior vena cava (IVC)**, return deoxygenated blood to the heart (right atrium) (Figure 21.24). The coronary sinus receives blood from the cardiac veins that drain the heart; with few exceptions, the superior vena cava receives blood from other veins superior to the diaphragm, except the air sacs (alveoli) of the lungs; the inferior vena cava receives blood from veins inferior to the diaphragm.

VEINS	DESCRIPTION AND TRIBUTARIES	REGIONS DRAINED
Coronary sinus (KOR-ō-nar-ē; *corona* = crown)	Main vein of heart; receives almost all venous blood from myocardium; located in coronary sulcus (see Figure 20.3c) on posterior aspect of heart and opens into right atrium between orifice of inferior vena cava and tricuspid valve. Wide venous channel into which three veins drain. Receives **great cardiac vein** (from anterior interventricular sulcus) into its left end, and **middle cardiac vein** (from posterior interventricular sulcus) and **small cardiac vein** into its right end. Several **anterior cardiac veins** drain directly into right atrium.	All tissues of heart.
Superior vena cava (SVC) (VĒ-na KĀ-va; *vena* = vein; *cava* = cavelike)	About 7.5 cm (3 in.) long and 2 cm (1 in.) in diameter; empties its blood into superior part of right atrium. Begins posterior to right first costal cartilage by union of right and left brachiocephalic veins and ends at level of right third costal cartilage, where it enters right atrium.	Head, neck, upper limbs, and thorax.
Inferior vena cava (IVC)	Largest vein in body, about 3.5 cm (1.4 in.) in diameter. Begins anterior to fifth lumbar vertebra by union of common iliac veins, ascends behind peritoneum to right of midline, pierces caval opening of diaphragm at level of eighth thoracic vertebra, and enters inferior part of right atrium. **Clinical note:** The inferior vena cava is commonly **compressed during the later stages of pregnancy** by the enlarging uterus, producing edema of the ankles and feet and temporary varicose veins.	Abdomen, pelvis, and lower limbs.

FIGURE 21.24 Principal veins.

Deoxygenated blood returns to the heart via the superior vena cava, inferior vena cava, and coronary sinus.

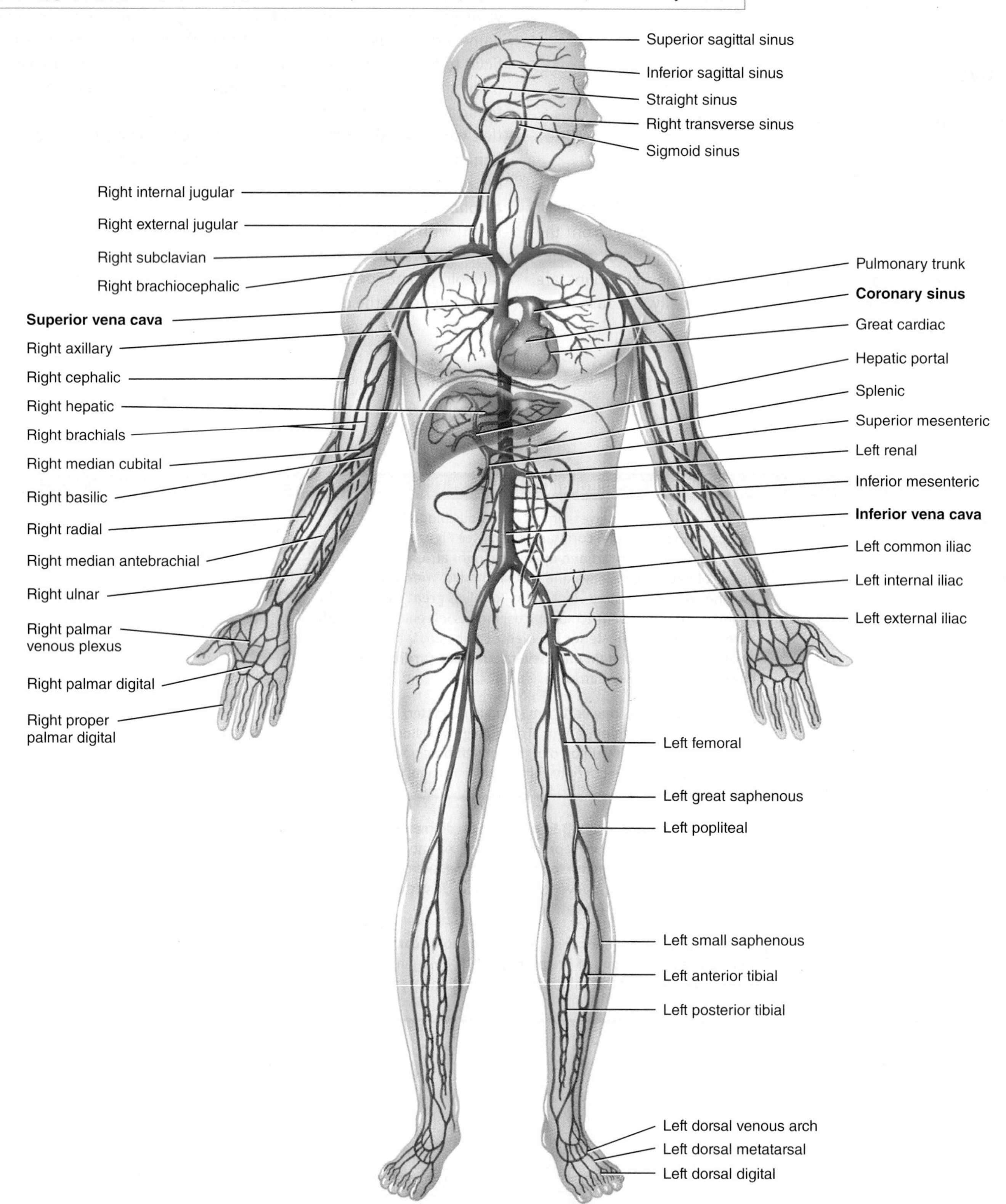

Overall anterior view of the principal veins

Q Which general regions of the body are drained by the superior vena cava and the inferior vena cava?

21.15 Veins of the Head and Neck

OBJECTIVE

- **Identify** the three major veins that drain blood from the head.

Most blood draining from the head passes into three pairs of veins: the **internal jugular** (JUG-ū-lar), **external jugular,** and **vertebral veins** (**Figure 21.25**). Within the cranial cavity, all veins drain into dural venous sinuses and then into the internal jugular veins. **Dural venous sinuses** are endothelial-lined venous channels between layers of the cranial dura mater.

VEINS	DESCRIPTION AND TRIBUTARIES	REGIONS DRAINED
Brachiocephalic veins	(See **Figure 21.27.**)	
Internal jugular veins (JUG-ū-lar = throat)	Begin at base of cranium as sigmoid sinus and inferior petrosal sinus; converge at opening of the jugular foramen. Descend within carotid sheath lateral to internal and common carotid arteries, deep to sternocleidomastoid muscles. Receive numerous tributaries from the face and neck. Internal jugular veins anastomose with subclavian veins to form brachiocephalic veins (brā′-kē-ō-se-FAL-ik; *brachi-* = arm; *-cephal-* = head) deep and slightly lateral to sternoclavicular joints. Major dural venous sinuses that contribute to internal jugular vein are as follows:	Brain, meninges, bones of cranium, muscles and tissues of face and neck.
	1. **Superior sagittal sinus** (SAJ-i-tal = arrow) begins at frontal bone, where it receives vein from nasal cavity, and passes posteriorly to occipital bone along midline of skull deep to sagittal suture. It usually angles to right and drains into right transverse sinus.	Nasal cavity; superior, lateral, and medial aspects of cerebrum; skull bones; meninges.
	2. **Inferior sagittal sinus** is much smaller than superior sagittal sinus. It begins posterior to attachment of falx cerebri and receives great cerebral vein to become straight sinus.	Medial aspects of cerebrum and diencephalon.
	3. **Straight sinus** runs in tentorium cerebelli and is formed by union of inferior sagittal sinus and great cerebral vein. It typically drains into left transverse sinus.	Medial and inferior aspects of cerebrum and the cerebellum.
	4. **Sigmoid sinuses** (SIG-moyd = S-shaped) are located along posterior aspect of petrous temporal bone. They begin where transverse sinuses and superior petrosal sinuses anastomose and terminate in internal jugular vein at jugular foramen.	Lateral and posterior aspect of cerebrum and the cerebellum.
	5. **Cavernous sinuses** (KAV-er-nus = cavelike) are located on both sides of the body of the sphenoid bone. Ophthalmic veins from orbits and cerebral veins from cerebral hemispheres, along with other small sinuses, empty into cavernous sinuses. They drain posteriorly to petrosal sinuses to eventually return to internal jugular veins. Cavernous sinuses are unique because they have major blood vessels and nerves passing through them on their way to orbit and face. Oculomotor (III) nerve, trochlear (IV) nerve, ophthalmic and maxillary branches of the trigeminal (V) nerve, abducens (VI) nerve, and internal carotid arteries pass through cavernous sinuses.	Orbits, nasal cavity, frontal regions of cerebrum, and superior aspect of brain stem.
Subclavian veins	(See **Figure 21.26.**)	
External jugular veins	Begin in parotid glands near angle of the mandible. Descend through neck across sternocleidomastoid muscles. Terminate at point opposite middle of clavicles, where they empty into subclavian veins. Become very prominent along side of neck when venous pressure rises, for example, during heavy coughing or straining or in cases of heart failure.	Scalp and skin of head and neck, muscles of face and neck, and oral cavity and pharynx.
Vertebral veins (VER-te-bral = of vertebrae)	Right and left vertebral veins originate inferior to occipital condyles. They descend through successive transverse foramina of first six cervical vertebrae and emerge from foramina of sixth cervical vertebra to enter brachiocephalic veins in root of neck.	Cervical vertebrae, cervical spinal cord and meninges, and some deep muscles in neck.

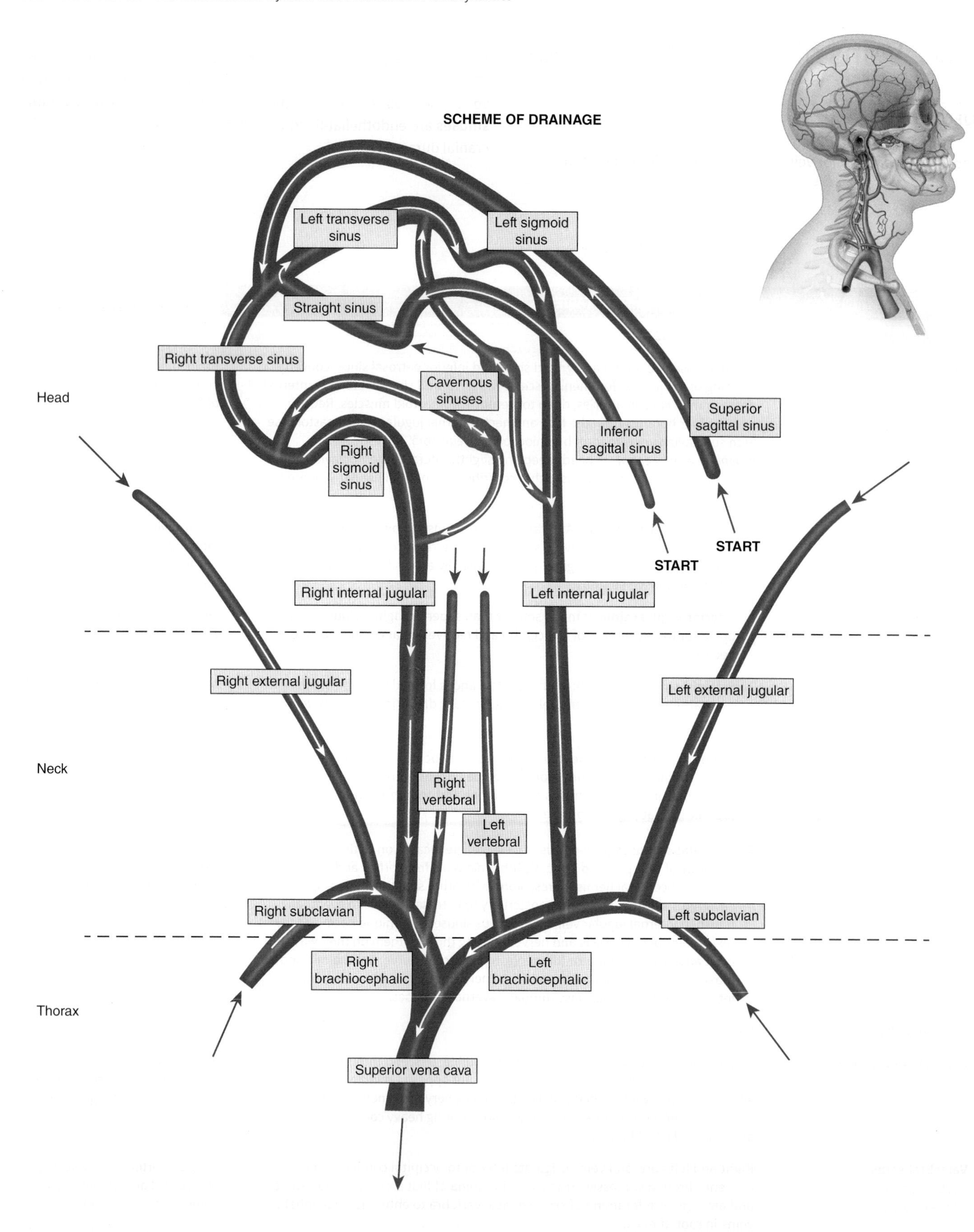
SCHEME OF DRAINAGE
Left transverse sinus
Left sigmoid sinus
Straight sinus
Right transverse sinus
Cavernous sinuses
Head
Superior sagittal sinus
Inferior sagittal sinus
Right sigmoid sinus
START
START
Right internal jugular
Left internal jugular
Right external jugular
Left external jugular
Neck
Right vertebral
Left vertebral
Right subclavian
Left subclavian
Right brachiocephalic
Left brachiocephalic
Thorax
Superior vena cava

FIGURE 21.25 Principal veins of the head and neck.

Blood draining from the head passes into the internal jugular, external jugular, and vertebral veins.

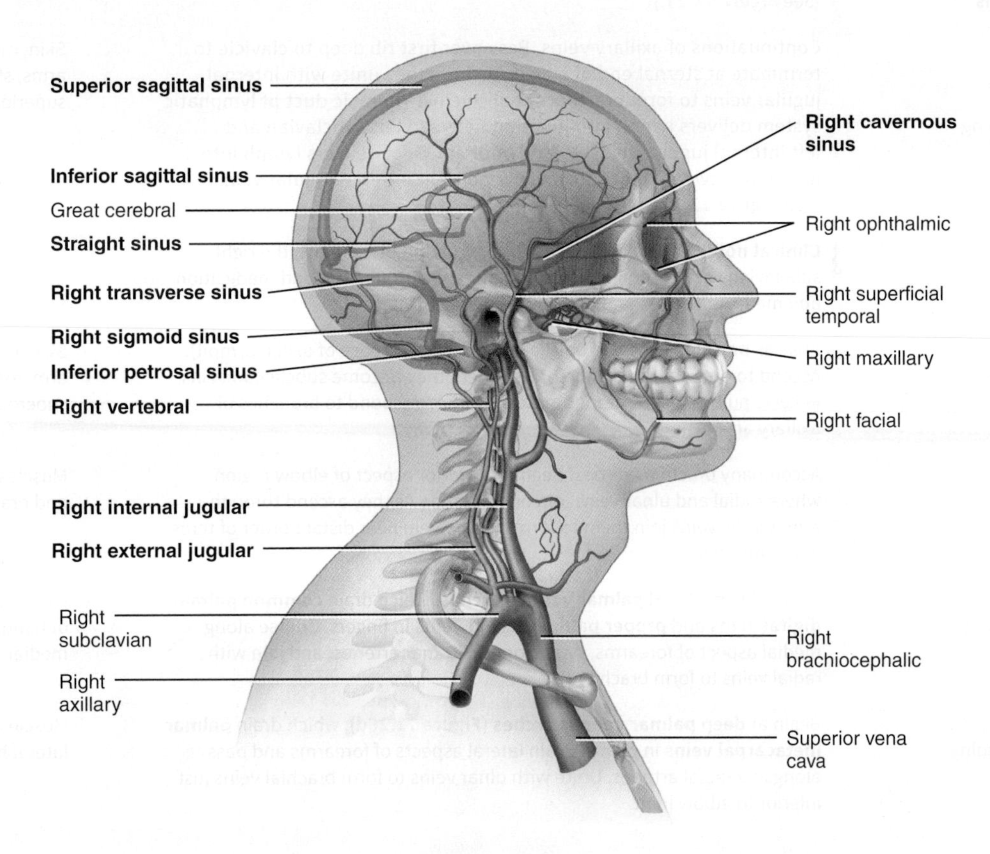

Right lateral view

Q Into which veins in the neck does all venous blood in the brain drain?

21.16 Veins of the Upper Limbs

OBJECTIVE

- **Identify** the principal veins that drain the upper limbs.

Both superficial and deep veins return blood from the upper limbs to the heart (**Figure 21.26**). Superficial veins are located just deep to the skin and are often visible. They anastomose extensively with one another and with deep veins, and they do not accompany arteries. Superficial veins are larger than deep veins and return most of the blood from the upper limbs. Deep veins are located deep in the body. They usually accompany arteries and have the same names as the corresponding arteries. Both superficial and deep veins have valves, but valves are more numerous in the deep veins.

VEINS	DESCRIPTION AND TRIBUTARIES	REGIONS DRAINED
DEEP VEINS		
Brachiocephalic veins	(See **Figure 21.27**.)	
Subclavian veins (sub-KLĀ-vē-an; *sub-* = under; *-clavian* = pertaining to clavicle)	Continuations of axillary veins. Pass over first rib deep to clavicle to terminate at sternal end of clavicle, where they unite with internal jugular veins to form brachiocephalic veins. Thoracic duct of lymphatic system delivers lymph into junction between left subclavian and left internal jugular veins. Right lymphatic duct delivers lymph into junction between right subclavian and right internal jugular veins (see **Figure 22.3**). **Clinical note:** In a procedure called **central line placement**, the right subclavian vein is frequently used to administer nutrients and medication and measure venous pressure.	Skin, muscles, bones of arms, shoulders, neck, and superior thoracic wall.
Axillary veins (AK-sil-ār-ē; *axilla* = armpit)	Arise as brachial veins and basilic veins unite near base of axilla (armpit). Ascend to outer borders of first ribs, where they become subclavian veins. Receive numerous tributaries in axilla that correspond to branches of axillary arteries.	Skin, muscles, bones of arm, axilla, shoulder, and superolateral chest wall.
Brachial veins (BRĀ-kē-al; *brachi-* = arm)	Accompany brachial arteries. Begin in anterior aspect of elbow region where radial and ulnar veins join one another. As they ascend through arm, basilic veins join them to form axillary vein near distal border of teres major muscle.	Muscles and bones of elbow and brachial regions.
Ulnar veins (UL-nar = pertaining to ulna)	Begin at **superficial palmar venous arches**, which drain **common palmar digital veins** and **proper palmar digital veins** in fingers. Course along medial aspect of forearms, pass alongside ulnar arteries, and join with radial veins to form brachial veins.	Muscles, bones, and skin of hand, and muscles of medial aspect of forearm.
Radial veins (RĀ-dē-al = pertaining to radius)	Begin at **deep palmar venous arches** (**Figure 21.26d**), which drain **palmar metacarpal veins** in palms. Drain lateral aspects of forearms and pass alongside radial arteries. Unite with ulnar veins to form brachial veins just inferior to elbow joint.	Muscles and bones of lateral hand and forearm.
SUPERFICIAL VEINS		
Cephalic veins (se-FAL-ik = pertaining to head)	Begin on lateral aspect of **dorsal venous networks of hands** (*dorsal venous arches*), networks of veins on dorsum of hands formed by dorsal metacarpal veins (**Figure 21.26b**). These veins in turn drain **dorsal digital veins**, which pass along sides of fingers. Arch around radial side of forearms to anterior surface and ascend through entire limbs along anterolateral surface. End where they join axillary veins, just inferior to clavicles. **Accessory cephalic veins** originate either from venous plexus on dorsum of forearms or from medial aspects of dorsal venous networks of hands, and unite with cephalic veins just inferior to elbow.	Integument and superficial muscles of lateral aspect of upper limb.
Basilic veins (ba-SIL-ik = royal, of prime importance)	Begin on medial aspects of dorsal venous networks of hands and ascend along posteromedial surface of forearm and anteromedial surface of arm (**Figure 21.26c**). Connected to cephalic veins anterior to elbow by **median cubital veins** (*cubital* = pertaining to elbow). After receiving median cubital veins, basilic veins continue ascending until they reach middle of arm. There they penetrate tissues deeply and run alongside brachial arteries until they join with brachial veins to form axillary veins. **Clinical note:** If veins must be **punctured** for an injection, transfusion, or removal of a blood sample, the median cubital veins are preferred.	Integument and superficial muscles of medial aspect of upper limb.
Median antebrachial veins (median veins of forearm) (an′-tē-BRĀ-kē-al; *ante-* = before, in front of)	Begin in **palmar venous plexuses**, networks of veins on palms. Drain **palmar digital veins** in fingers. Ascend anteriorly in forearms to join basilic or median cubital veins, sometimes both.	Integument and superficial muscles of palm and anterior aspect of upper limb.

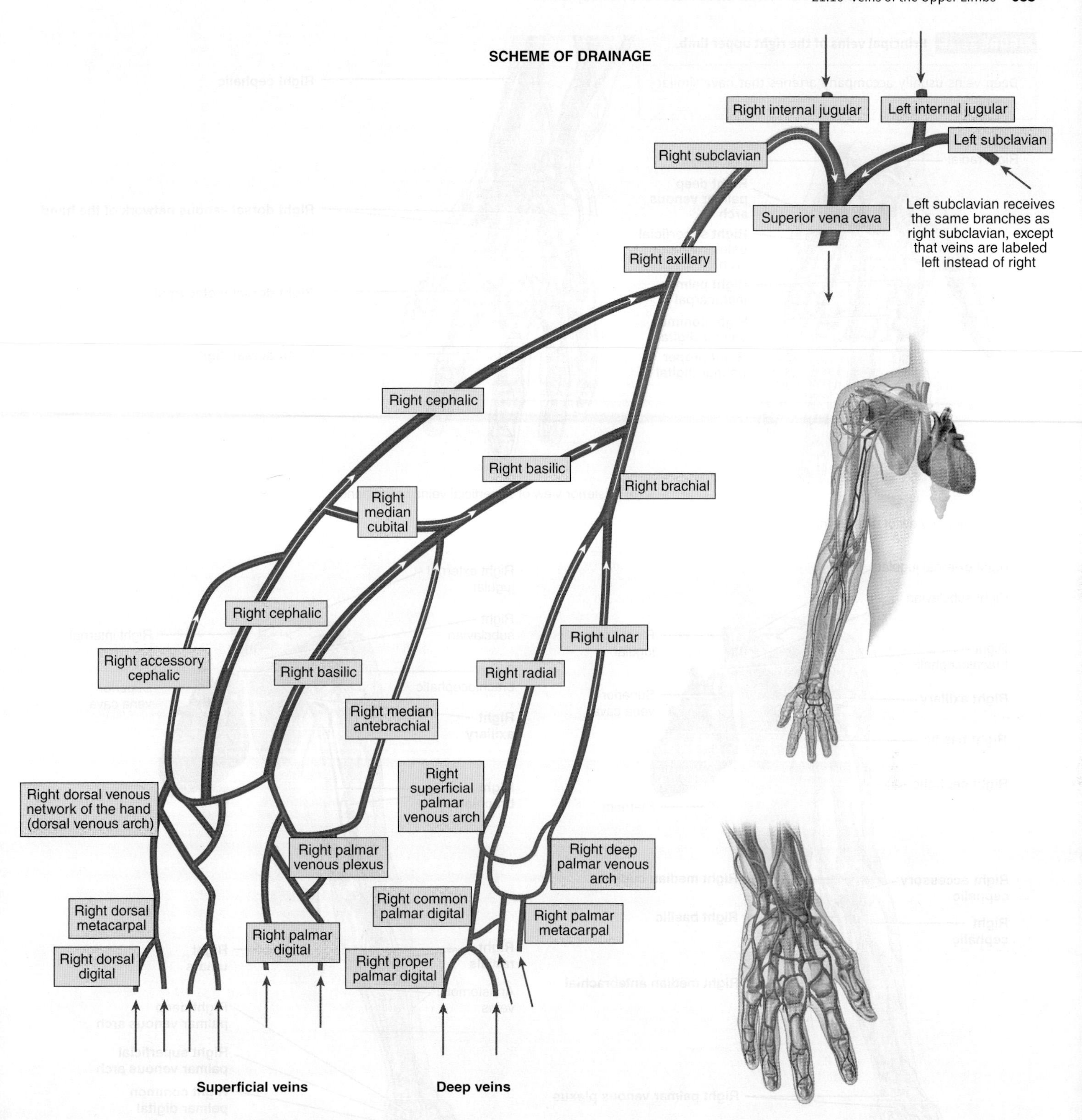
SCHEME OF DRAINAGE
Right internal jugular
Left internal jugular
Right subclavian
Left subclavian
Superior vena cava
Left subclavian receives the same branches as right subclavian, except that veins are labeled left instead of right
Right axillary
Right cephalic
Right basilic
Right brachial
Right median cubital
Right cephalic
Right accessory cephalic
Right basilic
Right ulnar
Right radial
Right median antebrachial
Right superficial palmar venous arch
Right dorsal venous network of the hand (dorsal venous arch)
Right palmar venous plexus
Right deep palmar venous arch
Right common palmar digital
Right dorsal metacarpal
Right palmar digital
Right palmar metacarpal
Right dorsal digital
Right proper palmar digital
Superficial veins
Deep veins

FIGURE 21.26 Principal veins of the right upper limb.

Deep veins usually accompany arteries that have similar names.

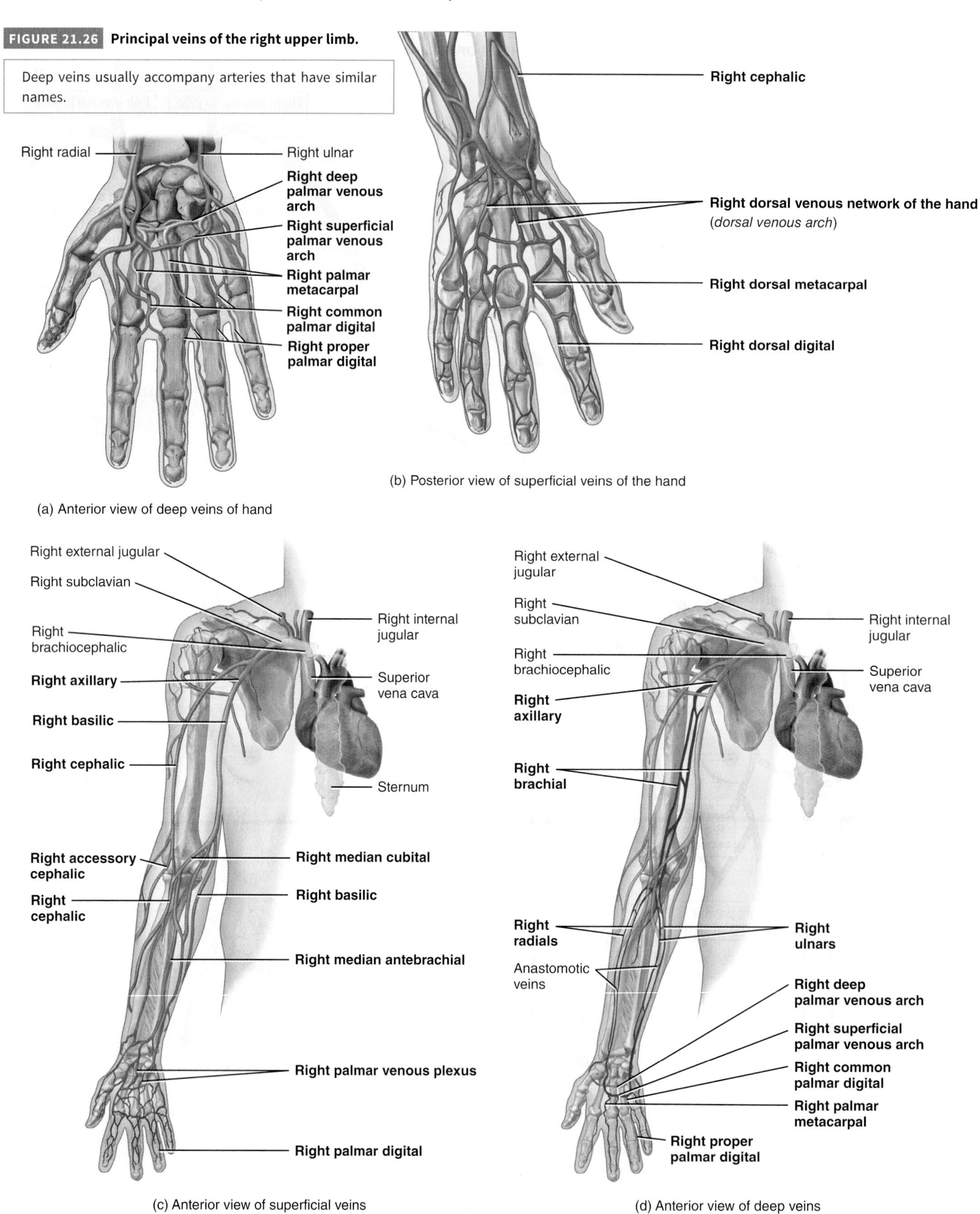

(a) Anterior view of deep veins of hand

(b) Posterior view of superficial veins of the hand

(c) Anterior view of superficial veins

(d) Anterior view of deep veins

Q From which vein in the upper limb is a blood sample often taken?

21.17 Veins of the Thorax

OBJECTIVE

- **Identify** the components of the azygos system of veins.

Although the brachiocephalic veins drain some portions of the thorax, most thoracic structures are drained by a network of veins, called the **azygos system** (az-Ī-gus or ā-ZĪ-gus), that runs on either side of the vertebral column (**Figure 21.27**). The system consists of three veins—the **azygos, hemiazygos**, and **accessory hemiazygos veins**—that show considerable variation in origin, course, tributaries, anastomoses, and termination. Ultimately they empty into the superior vena cava.

The azygos system, besides collecting blood from the thorax and abdominal wall, may serve as a bypass for the inferior vena cava, which drains blood from the lower body. Several small veins directly link the azygos system with the inferior vena cava. Larger

VEINS	DESCRIPTION AND TRIBUTARIES	REGIONS DRAINED
Brachiocephalic veins (brā′-kē-ō-se-FAL-ik; *brachi-* = arm; *-cephalic* = pertaining to head)	Form by union of subclavian and internal jugular veins. Two brachiocephalic veins unite to form superior vena cava. Because superior vena cava is to right of body's midline, left brachiocephalic vein is longer than right. Right brachiocephalic vein is anterior and to right of brachiocephalic trunk and follows more vertical course. Left brachiocephalic vein is anterior to brachiocephalic trunk, left common carotid and left subclavian arteries, trachea, left vagus (X) nerve, and phrenic nerve. It approaches more horizontal position as it passes from left to right.	Head, neck, upper limbs, mammary glands, and superior thorax.
Azygos vein (AZ-ī-gos = unpaired)	An unpaired vein that is anterior to vertebral column, slightly to right of midline. Usually begins at junction of **right ascending lumbar** and **right subcostal veins** near diaphragm. Arches over root of right lung at level of fourth thoracic vertebra to end in superior vena cava. Receives the following tributaries: **right posterior intercostal**, **hemiazygos**, **accessory hemiazygos**, **esophageal**, **mediastinal**, **pericardial**, and **bronchial veins**.	Right side of thoracic wall, thoracic viscera, and posterior abdominal wall.
Hemiazygos vein (hem′-ē-az-ī-gus; *hemi-* = half)	Anterior to vertebral column and slightly to left of midline. Often begins at junction of **left ascending lumbar and left subcostal veins**. Terminates by joining azygos vein at about level of ninth thoracic vertebra. Receives following tributaries: ninth through eleventh **left posterior intercostal**, **esophageal**, **mediastinal**, and sometimes **accessory hemiazygos veins**.	Left side of lower thoracic wall, thoracic viscera, and left posterior abdominal wall.
Accessory hemiazygos vein	Anterior to vertebral column and to left of midline. Begins at fourth or fifth intercostal space and descends from fifth to eighth thoracic vertebra or ends in hemiazygos vein. Terminates by joining azygos vein at about level of eighth thoracic vertebra. Receives the following tributaries: fourth through eighth **left posterior intercostal veins** (first through third posterior intercostal veins drain into left brachiocephalic vein), **left bronchial**, and **mediastinal veins**.	Left side of upper thoracic wall and thoracic viscera.

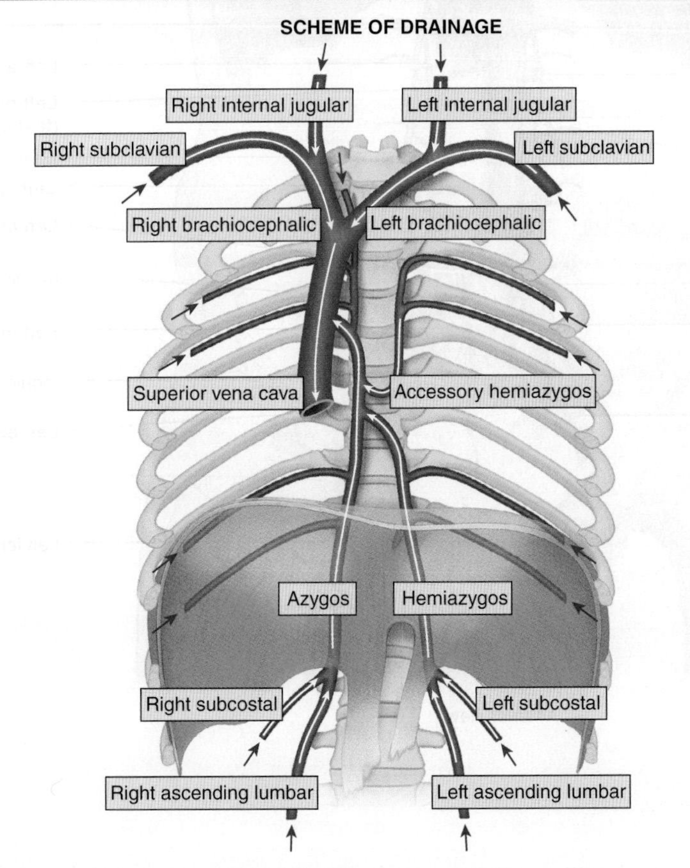

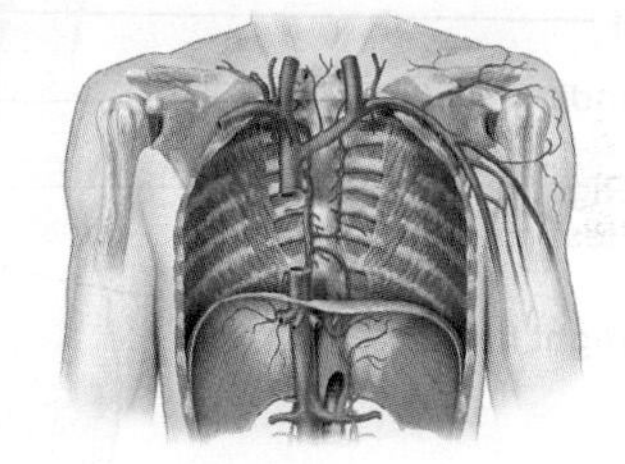

FIGURE 21.27 **Principal veins of the thorax, abdomen, and pelvis.**

Most thoracic structures are drained by the azygos system of veins.

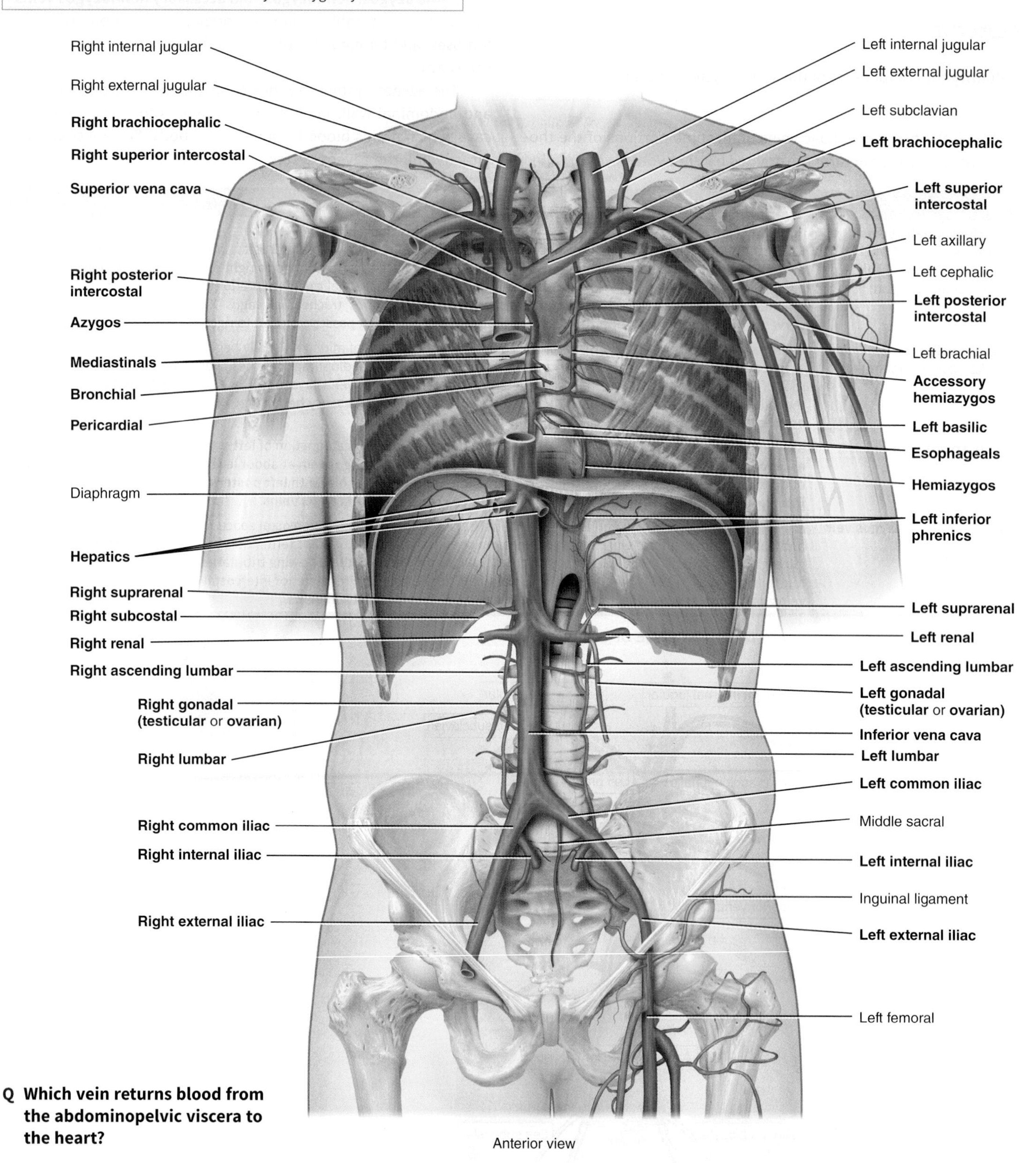

Q Which vein returns blood from the abdominopelvic viscera to the heart?

veins that drain the lower limbs and abdomen also connect into the azygos system. If the inferior vena cava or hepatic portal vein becomes obstructed, blood that typically passes through the inferior vena cava can detour into the azygos system to return blood from the lower body to the superior vena cava.

21.18 Veins of the Abdomen and Pelvis

OBJECTIVE

• **Identify** the principal veins that drain the abdomen and pelvis.

Blood from the abdominal and pelvic viscera and lower half of the abdominal wall returns to the heart via the inferior vena cava. Many small veins enter the inferior vena cava. Most carry return flow from parietal branches of the abdominal aorta, and their names correspond to the names of the arteries (see also Figure 21.27).

The inferior vena cava does not receive veins directly from the gastrointestinal tract, spleen, pancreas, and gallbladder. These organs pass their blood into a common vein, the **hepatic portal vein**, which delivers the blood to the liver. The superior mesenteric and splenic veins unite to form the hepatic portal vein (see Figure 21.29). This special flow of venous blood, called the *hepatic portal circulation*, is described shortly. After passing through the liver for processing, blood drains into the hepatic veins, which empty into the inferior vena cava.

VEINS	DESCRIPTION AND TRIBUTARIES	REGIONS DRAINED
Inferior vena cava	(See Figure 21.24.)	
Inferior phrenic veins (FREN-ik = pertaining to diaphragm)	Arise on inferior surface of diaphragm. Left inferior phrenic vein usually sends one tributary to left suprarenal vein, which empties into left renal vein, and another tributary into inferior vena cava. Right inferior phrenic vein empties into inferior vena cava.	Inferior surface of diaphragm and adjoining peritoneal tissues.
Hepatic veins (he-PAT-ik = pertaining to liver)	Typically two or three in number. Drain sinusoidal capillaries of liver. Capillaries of liver receive venous blood from capillaries of gastrointestinal organs via hepatic portal vein. **Hepatic portal vein** receives the following tributaries from gastrointestinal organs:	
	1. Left gastric vein arises from left side of lesser curvature of stomach and joins left side of hepatic portal vein in lesser omentum.	Terminal esophagus, stomach, liver, gallbladder, spleen, pancreas, small intestine, and large intestine.
	2. Right gastric vein arises from right aspect of lesser curvature of stomach and joins hepatic portal vein on its anterior surface within lesser omentum.	Lesser curvature of stomach, abdominal portion of esophagus, stomach, and duodenum.
	3. Splenic vein arises in spleen and crosses abdomen transversely posterior to stomach to anastomose with superior mesenteric vein to form hepatic portal vein. It receives near its junction with hepatic portal vein, it receives **inferior mesenteric vein,** which receives tributaries from second half of large intestine.	Spleen, fundus and greater curvature of stomach, pancreas, greater omentum, descending colon, sigmoid colon, and rectum.
	4. Superior mesenteric vein arises from numerous tributaries from most of small intestine and first half of large intestine and ascends to join splenic vein to form hepatic portal vein.	Duodenum, jejunum, ileum, cecum, appendix, ascending colon, and transverse colon.
Lumbar veins (LUM-bar = pertaining to loin)	Usually four on each side; course horizontally through posterior abdominal wall with lumbar arteries. Connect at right angles with right and left **ascending lumbar veins**, which form origin of corresponding azygos or hemiazygos vein. Join ascending lumbar veins and then connect from ascending lumbar veins to inferior vena cava.	Posterior and lateral abdominal muscle wall, lumbar vertebrae, spinal cord and spinal nerves (cauda equina) within vertebral canal, and meninges.
Suprarenal veins (soo′-pra-RĒ-nal; *supra-* = above)	Pass medially from adrenal (suprarenal) glands (**left suprarenal vein** joins left renal vein, and **right suprarenal vein** joins inferior vena cava).	Adrenal (suprarenal) glands.
Renal veins (RĒ-nal; *ren-* = kidney)	Pass anterior to renal arteries. Left renal vein is longer than right renal vein and passes anterior to abdominal aorta. It receives left testicular (or ovarian), left inferior phrenic, and usually left suprarenal veins. Right renal vein empties into inferior vena cava posterior to duodenum.	Kidneys.
Gonadal veins (gō-NAD-al; *gon-* = seed) [**testicular** (tes-TIK-ū-lar) or **ovarian** (ō-VAR-ē-an)]	Ascend with gonadal arteries along posterior abdominal wall. Called testicular veins in male. **Testicular veins** drain testes (left testicular vein joins left renal vein, and right testicular vein joins inferior vena cava). Called ovarian veins in female. **Ovarian veins** drain ovaries. Left ovarian vein joins left renal vein, and right ovarian vein joins inferior vena cava.	Testes, epididymis, ductus deferens, ovaries, and ureters.

VEINS	DESCRIPTION AND TRIBUTARIES	REGIONS DRAINED
Common iliac veins (IL-ē-ak = pertaining to ilium)	Formed by union of internal and external iliac veins anterior to sacroiliac joint and anastomose anterior to fifth lumbar vertebra to form inferior vena cava. Right common iliac is much shorter than left and is also more vertical, as inferior vena cava sits to right of midline.	Pelvis, external genitals, and lower limbs.
Internal iliac veins	Begin near superior portion of greater sciatic notch and run medial to their corresponding arteries.	Muscles of pelvic wall and gluteal region, pelvic viscera, and external genitals.
External iliac veins	Companions of internal iliac arteries. Begin at inguinal ligaments as continuations of femoral veins. End anterior to sacroiliac joints where they join with internal iliac veins to form common iliac veins.	Lower abdominal wall anteriorly, cremaster muscle in males, and external genitals and lower limb.

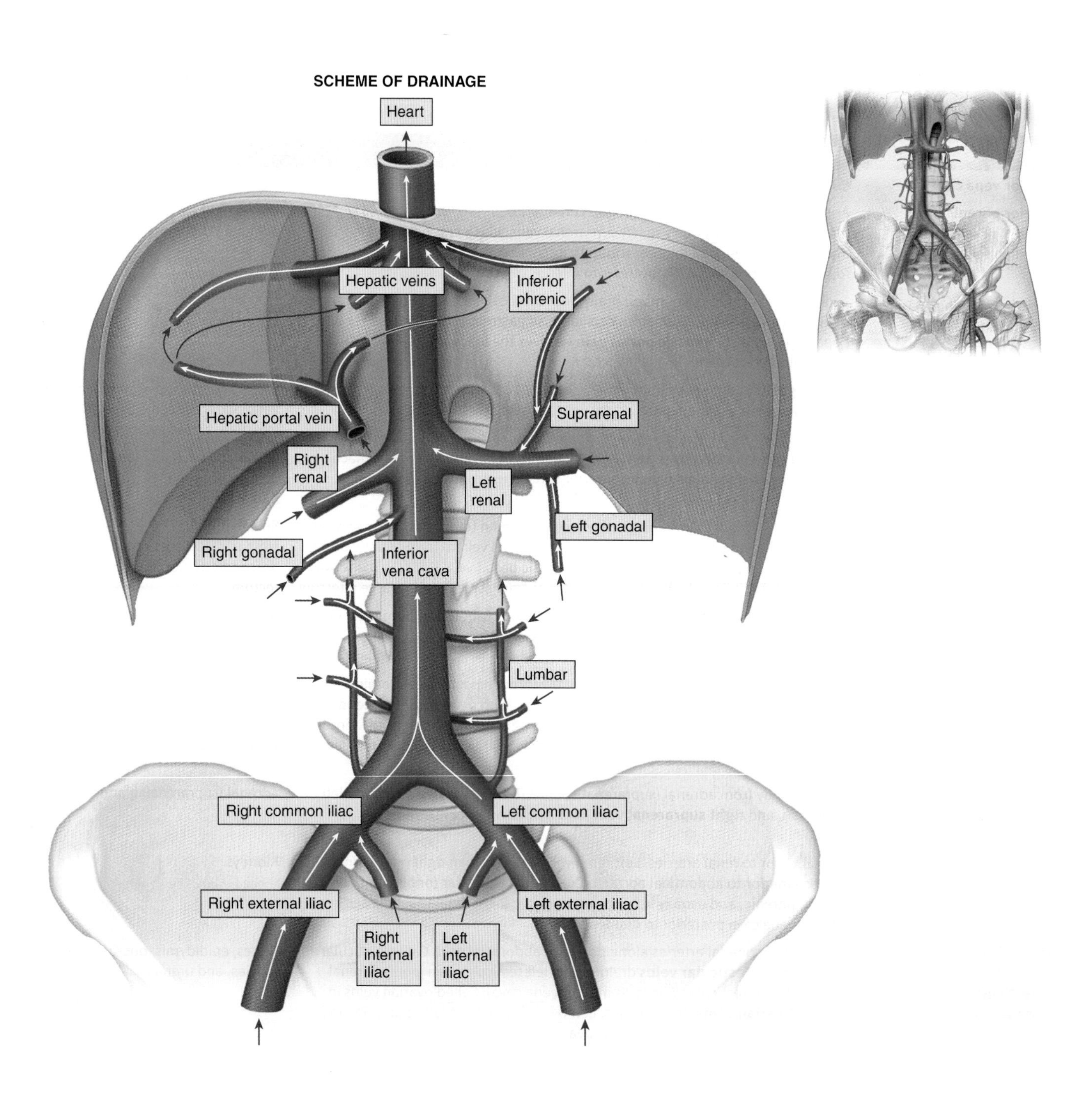

21.19 Veins of the Lower Limbs

OBJECTIVE

- **Identify** the principal superficial and deep veins that drain the lower limbs.

As with the upper limbs, blood from the lower limbs is drained by both superficial and deep veins. The superficial veins often anastomose with one another and with deep veins along their length. Deep veins, for the most part, have the same names as corresponding arteries (**Figure 21.28**). All veins of the lower limbs have valves, which are more numerous than in veins of the upper limbs.

VEINS	DESCRIPTION AND TRIBUTARIES	REGIONS DRAINED
DEEP VEINS		
Common iliac veins	(See **Figure 21.28**.)	
External iliac veins	(See **Figure 21.28**.)	
Femoral veins (FEM-o-ral)	Accompany femoral arteries and are continuations of popliteal veins just superior to knee where veins pass through opening in adductor magnus muscle. Ascend deep to sartorius muscle and emerge from beneath muscle in femoral triangle at proximal end of thigh. Receive **deep veins of thigh** (*deep femoral veins*) and great saphenous veins just before penetrating abdominal wall. Pass below inguinal ligament and enter abdominopelvic region to become external iliac veins.	Skin, lymph nodes, muscles, and bones of thigh, and external genitals.
	Clinical note: In order to take **blood samples** or **pressure recordings** from the right side of the heart, a catheter is inserted into the femoral vein as it passes through the femoral triangle. The catheter passes through the external and common iliac veins, then into the inferior vena cava, and finally into the right atrium.	
Popliteal veins (pop′-li-TĒ-al = pertaining to hollow behind knee)	Formed by union of anterior and posterior tibial veins at proximal end of leg; ascend through popliteal fossa with popliteal arteries and tibial nerve. Terminate where they pass through window in adductor magnus muscle and pass to front of knee to become femoral veins. Also receive blood from small saphenous veins and tributaries that correspond to branches of popliteal artery.	Knee joint and skin, muscles, and bones around knee joint.
Posterior tibial veins (TIB-ē-al)	Begin posterior to medial malleolus at union of **medial** and **lateral plantar veins** from plantar surface of foot. Ascend through leg with posterior tibial artery and tibial nerve deep to soleus muscle. Join posterior tibial veins about two-thirds of way up leg. Join anterior tibial veins near top of interosseous membrane to form popliteal veins. On plantar surface of foot, **plantar digital veins** unite to form **plantar metatarsal veins**, which parallel metatarsals. They in turn unite to form **deep plantar venous arches**. Medial and lateral plantar veins emerge from deep plantar venous arches.	Skin, muscles, and bones on plantar surface of foot, and skin, muscles, and bones from posterior and lateral aspects of leg.
Anterior tibial veins	Arise in dorsal venous arch and accompany anterior tibial artery. Ascend deep to tibialis anterior muscle on anterior surface of interosseous membrane. Pass through opening at superior end of interosseous membrane to join posterior tibial veins to form popliteal veins.	Dorsal surface of foot, ankle joint, anterior aspect of leg, knee joint, and tibiofibular joint.
SUPERFICIAL VEINS		
Great (*long*) **saphenous veins** (sa-FĒ-nus = clearly visible)	Longest veins in body; ascend from foot to groin in subcutaneous layer. Begin at medial end of dorsal venous arches of foot. **Dorsal venous arches** (VĒ-nus) are networks of veins on dorsum of foot formed by **dorsal digital veins**, which collect blood from toes, and then unite in pairs to form **dorsal metatarsal veins**, which parallel metatarsals. As dorsal metatarsal veins approach foot, they combine to form dorsal venous arches. Pass anterior to medial malleolus of tibia and then superiorly along medial aspect of leg and thigh just deep to skin. Receive tributaries from superficial tissues and connect with deep veins as well. Empty into femoral veins at groin. Have from 10 to 20 valves along their length, with more located in leg than thigh.	Integumentary tissues and superficial muscles of lower limbs, groin, and lower abdominal wall.
	Clinical note: These veins are more likely to be subject to **varicosities** than other veins in the lower limbs because they must support a long column of blood and are not well supported by skeletal muscles. The great saphenous veins are often used for prolonged administration of intravenous fluids. This is particularly important in very young children and in patients of any age who are in shock and whose veins are collapsed. In **coronary artery bypass grafting**, if multiple blood vessels need to be grafted, sections of the great saphenous vein are used along with at least one artery as a graft (see first **Clinical Note** in Section 21.10). After the great saphenous vein is removed and divided into sections, the sections are used to bypass the blockages. The vein grafts are reversed so that the valves do not obstruct the flow of blood.	
Small saphenous veins	Begin at lateral aspect of dorsal venous arches of foot. Pass posterior to lateral malleolus of fibula and ascend deep to skin along posterior aspect of leg. Empty into popliteal veins in popliteal fossa, posterior to knee. Have from 9 to 12 valves. May communicate with great saphenous veins in proximal leg.	Integumentary tissues and superficial muscles of foot and posterior aspect of leg.

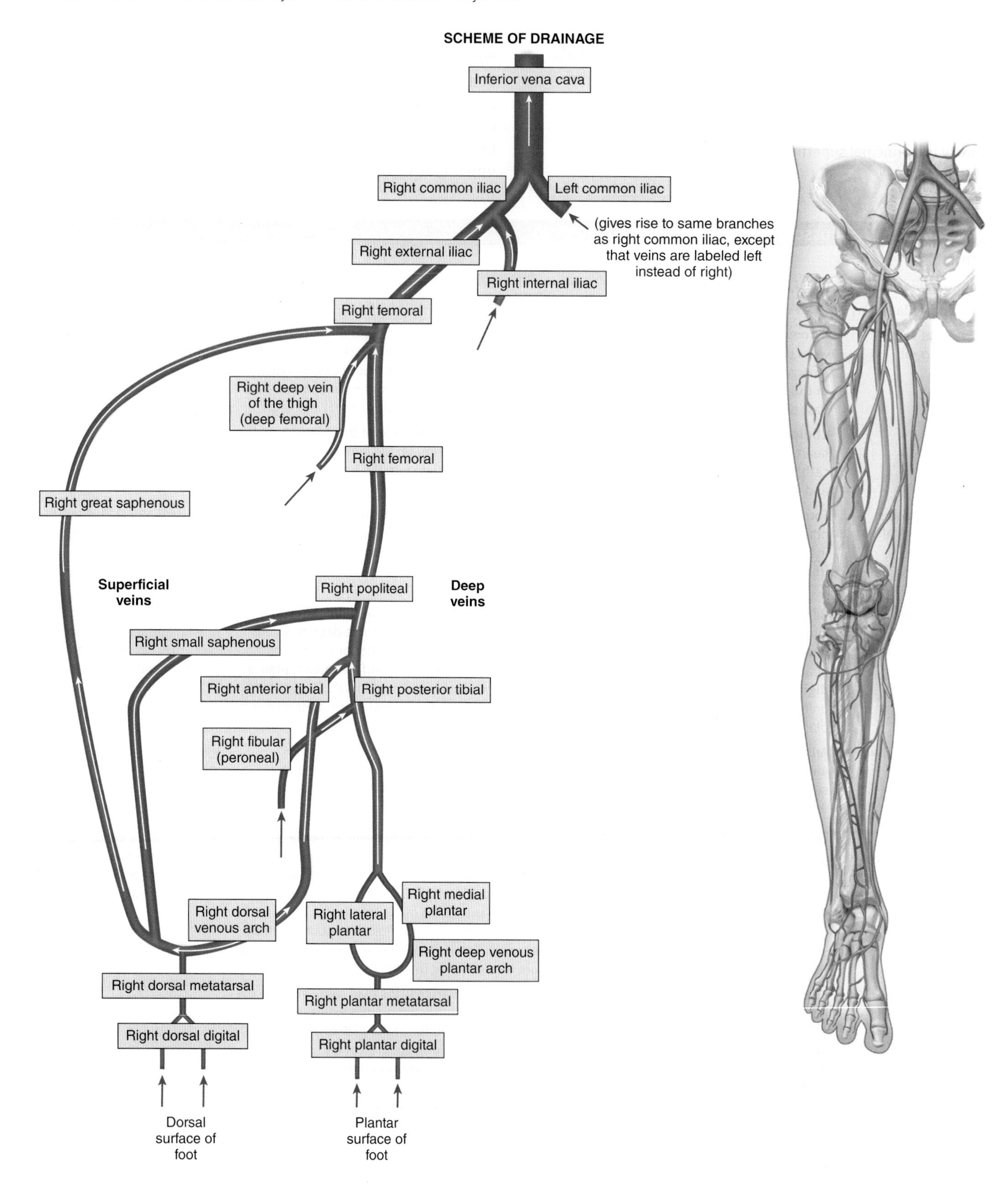
SCHEME OF DRAINAGE
Inferior vena cava
Right common iliac
Left common iliac
(gives rise to same branches as right common iliac, except that veins are labeled left instead of right)
Right external iliac
Right internal iliac
Right femoral
Right deep vein of the thigh (deep femoral)
Right femoral
Right great saphenous
Superficial veins
Right popliteal
Deep veins
Right small saphenous
Right anterior tibial
Right posterior tibial
Right fibular (peroneal)
Right medial plantar
Right dorsal venous arch
Right lateral plantar
Right deep venous plantar arch
Right dorsal metatarsal
Right plantar metatarsal
Right dorsal digital
Right plantar digital
Dorsal surface of foot
Plantar surface of foot

FIGURE 21.28 **Principal veins of the pelvis and lower limbs.**

Deep veins usually bear the names of their companion arteries.

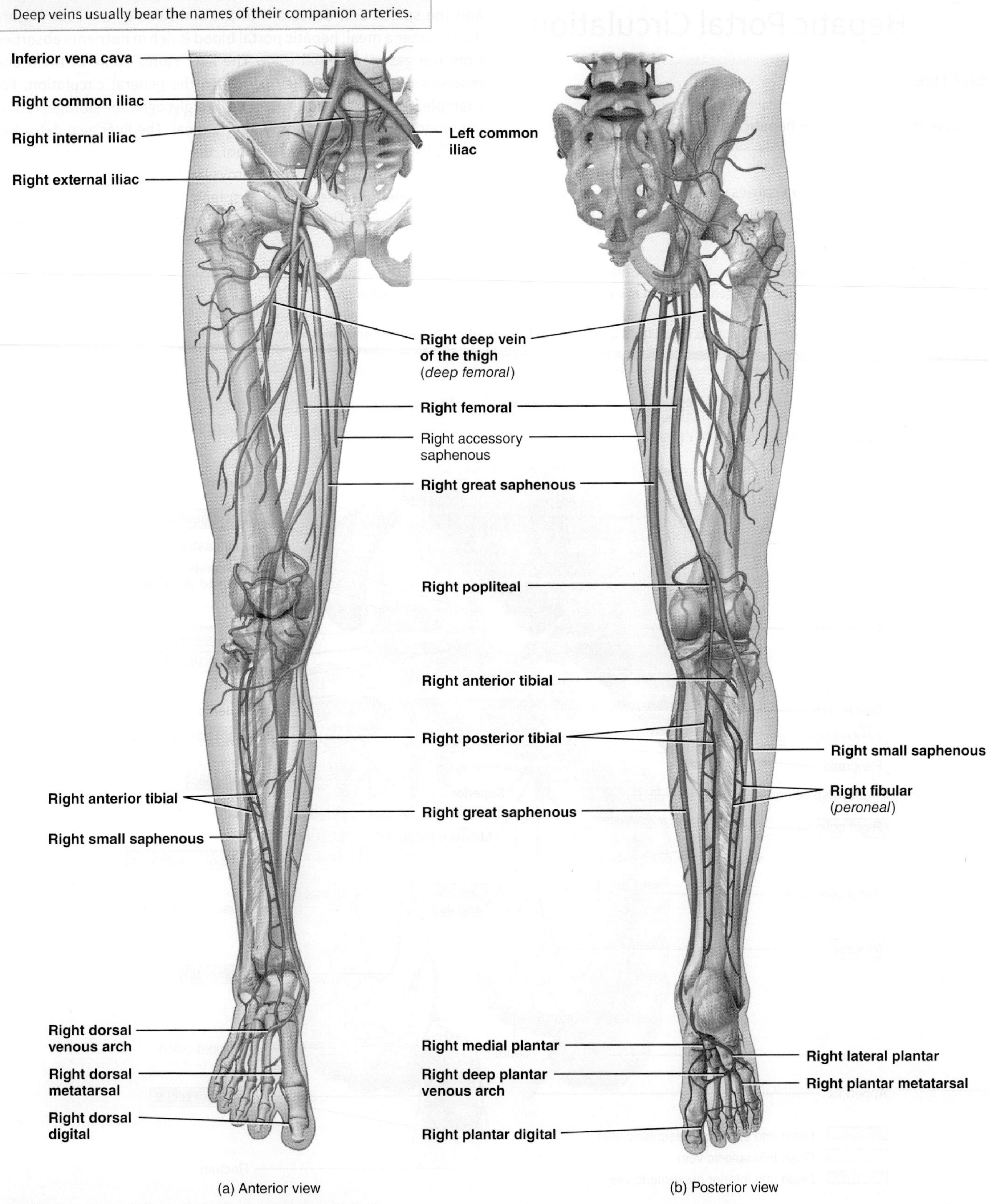

(a) Anterior view

(b) Posterior view

Q Which veins of the lower limb are superficial?

21.20 Circulatory Routes: The Hepatic Portal Circulation

OBJECTIVE

- **Describe** the importance of hepatic portal system.

The **hepatic portal circulation** carries venous blood from the gastrointestinal organs and spleen to the liver. A vein that carries blood from one capillary network to another is called a **portal vein**. The **hepatic portal vein** receives blood from capillaries of gastrointestinal organs and the spleen and delivers it to the sinusoids of the liver (**Figure 21.29**). After a meal, hepatic portal blood is rich in nutrients absorbed from the gastrointestinal tract. The liver stores some of them and modifies others before they pass into the general circulation. For example, the liver converts glucose into glycogen for storage, reducing blood glucose level shortly after a meal. The liver also detoxifies harmful substances, such as alcohol, that have been absorbed from the gastrointestinal tract and destroys bacteria by phagocytosis.

The superior mesenteric and splenic veins unite to form the hepatic portal vein. The **superior mesenteric vein** (mez-en-TER-ik)

FIGURE 21.29 Hepatic portal circulation. A schematic diagram of blood flow through the liver, including arterial circulation, is shown in (b). As usual, deoxygenated blood is indicated in blue, and oxygenated blood in red.

The hepatic portal circulation delivers venous blood from the organs of the gastrointestinal tract and spleen to the liver.

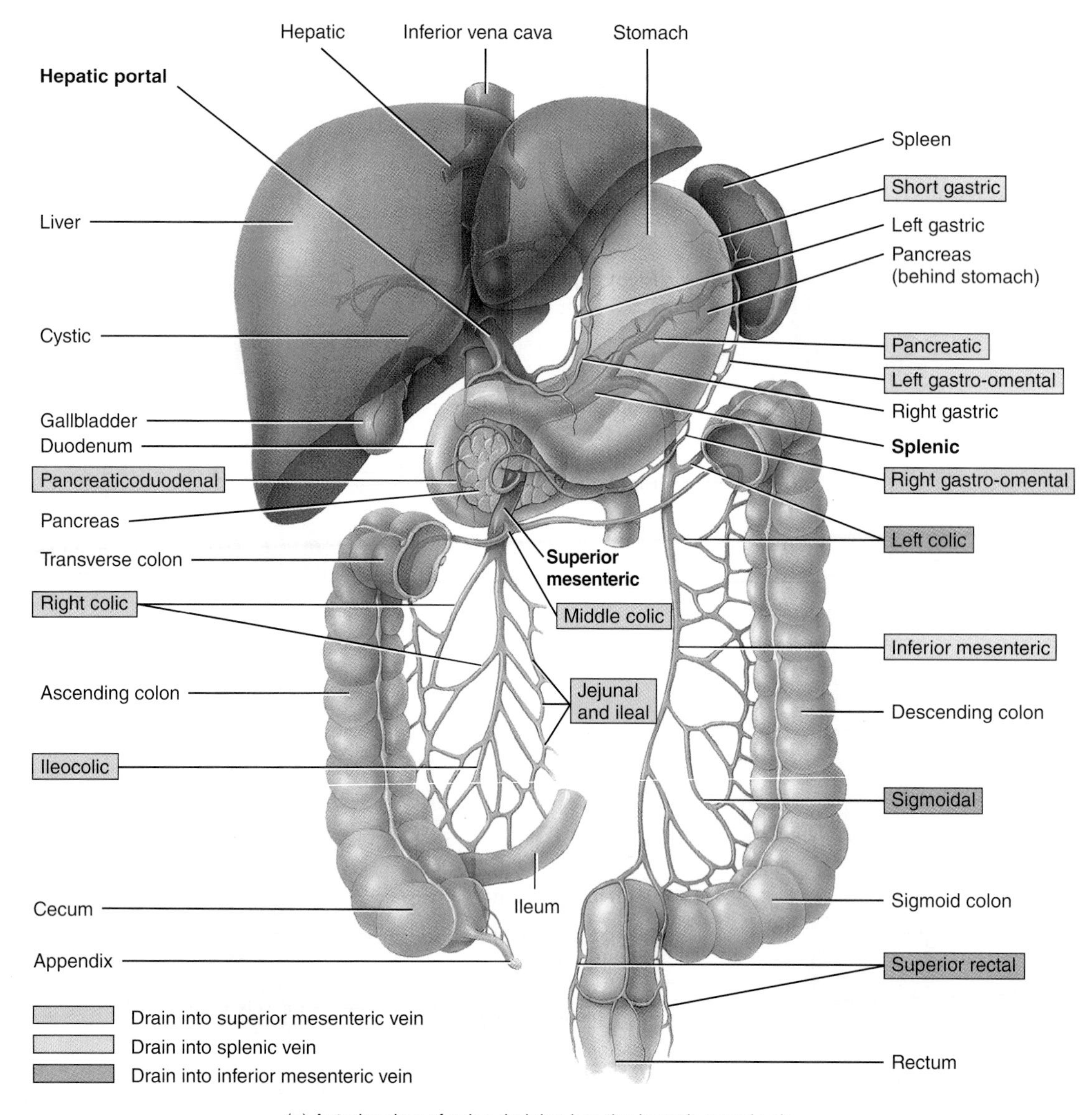

(a) Anterior view of veins draining into the hepatic portal vein

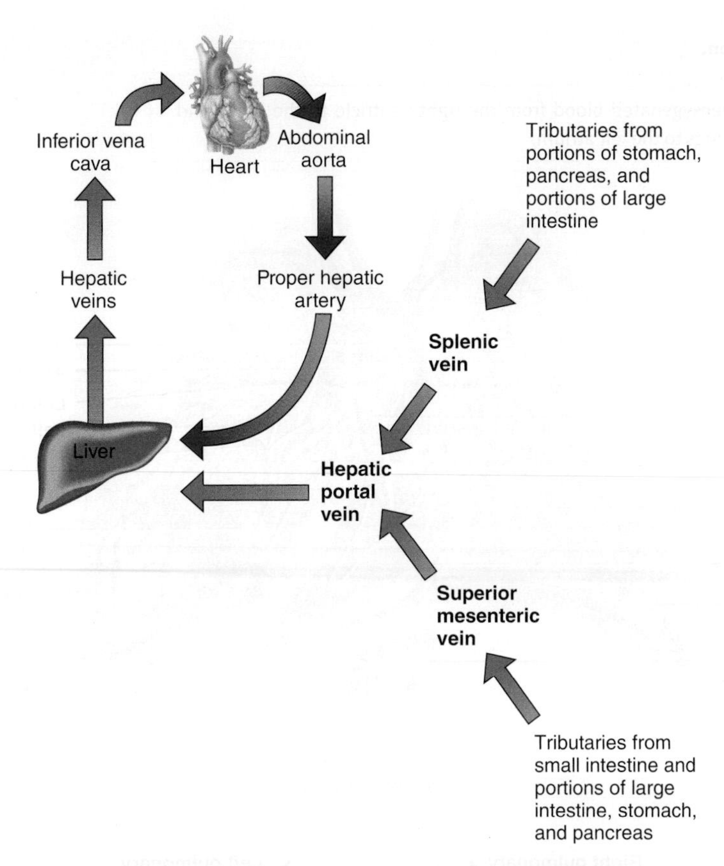

(b) Scheme of principal blood vessels of hepatic portal circulation and arterial supply and venous drainage of liver

Q Which veins carry blood away from the liver?

drains blood from the small intestine and portions of the large intestine, stomach, and pancreas through the *jejunal*, *ileal*, *ileocolic* (il′-ē-ō-KOL-ik), *right colic*, *middle colic*, *pancreaticoduodenal* (pan-krē-at′-i-kō-doo′-ō-DĒ-nal), and *right gastro-omental veins* (gas′-trō-ō-MEN-tal). The **splenic vein** drains blood from the stomach, pancreas, and portions of the large intestine through the *short gastric*, *left gastro-omental*, *pancreatic*, and *inferior mesenteric veins.* The inferior mesenteric vein, which passes into the splenic vein, drains portions of the large intestine through the *superior rectal*, *sigmoidal*, and *left colic veins.* The *right* and *left gastric veins*, which open directly into the hepatic portal vein, drain the stomach. The *cystic vein*, which also opens into the hepatic portal vein, drains the gallbladder.

At the same time the liver is receiving nutrient-rich but deoxygenated blood via the hepatic portal vein, it is also receiving oxygenated blood via the hepatic artery, a branch of the celiac trunk. The oxygenated blood mixes with the deoxygenated blood in sinusoids. Eventually, blood leaves the sinusoids of the liver through the **hepatic veins**, which drain into the inferior vena cava.

21.21 Circulatory Routes: The Pulmonary Circulation

OBJECTIVE

- **Explain** why pulmonary circulation is important

The **pulmonary circulation** carries deoxygenated blood from the right ventricle to the air sacs (alveoli) within the lungs and returns oxygenated blood from the air sacs to the left atrium (**Figure 21.30**). The **pulmonary trunk** emerges from the right ventricle and passes superiorly, posteriorly, and to the left. It then divides into two branches: the **right pulmonary artery** to the right lung and the **left pulmonary artery** to the left lung. After birth, the

FIGURE 21.30 **Pulmonary circulation.**

The pulmonary circulation brings deoxygenated blood from the right ventricle to the lungs and returns oxygenated blood from the lungs to the left atrium.

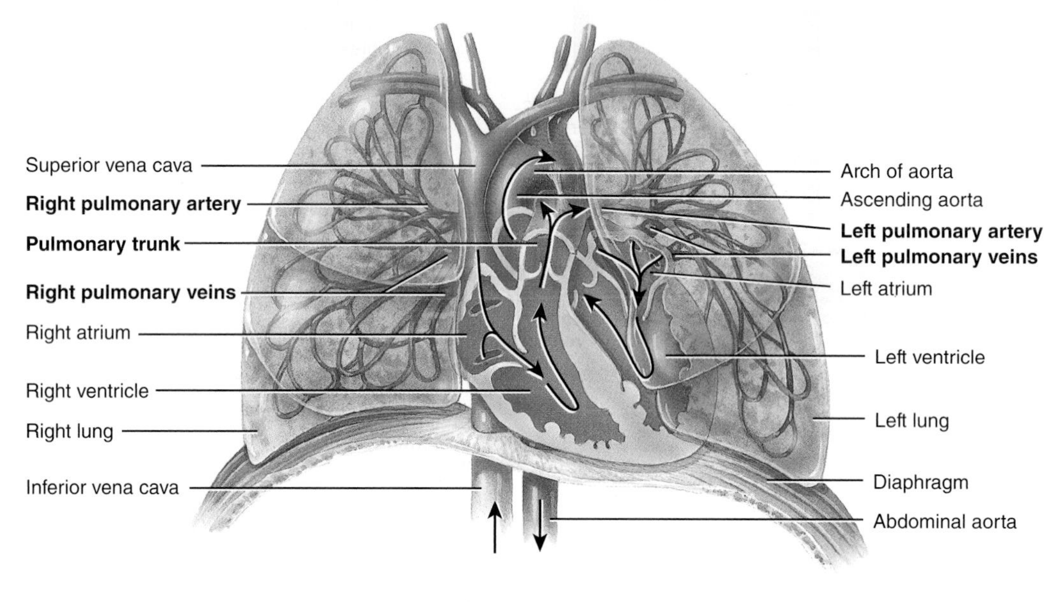

(a) Anterior view

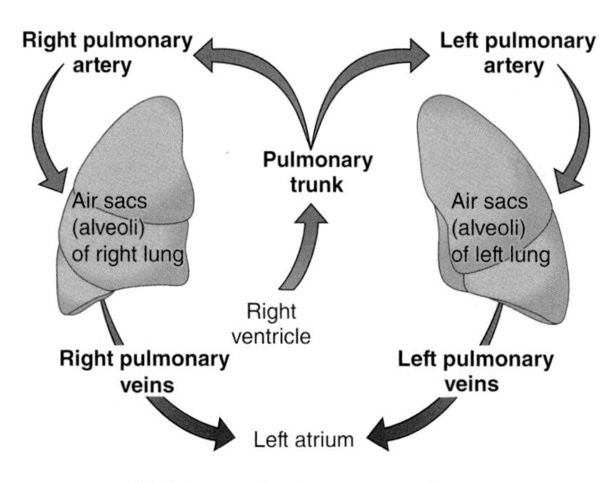

(b) Scheme of pulmonary circulation

Q After birth, which are the only arteries that carry deoxygenated blood?

pulmonary arteries are the only arteries that carry deoxygenated blood. On entering the lungs, the branches divide and subdivide until finally they form capillaries around the air sacs (alveoli) within the lungs. CO_2 passes from the blood into the air sacs and is exhaled. Inhaled O_2 passes from the air within the lungs into the blood. The pulmonary capillaries unite to form venules and eventually **pulmonary veins**, which exit the lungs and carry the oxygenated blood to the left atrium. Two left and two right pulmonary veins enter the left atrium. After birth, the pulmonary veins are the only veins that carry oxygenated blood. Contractions of the left ventricle then eject the oxygenated blood into the systemic circulation.

21.22 Circulatory Routes: The Fetal Circulation

OBJECTIVE

- **Describe** the fate of the fetal structures once postnatal circulation begins.

The circulatory system of a fetus, called the **fetal circulation**, exists only in the fetus and contains special structures that allow the developing fetus to exchange materials with its mother (**Figure 21.31**). It

FIGURE 21.31 Fetal circulation and changes at birth. The gold boxes between parts (a) and (b) describe the fate of certain fetal structures once postnatal circulation is established.

The lungs and gastrointestinal organs do not begin to function until birth.

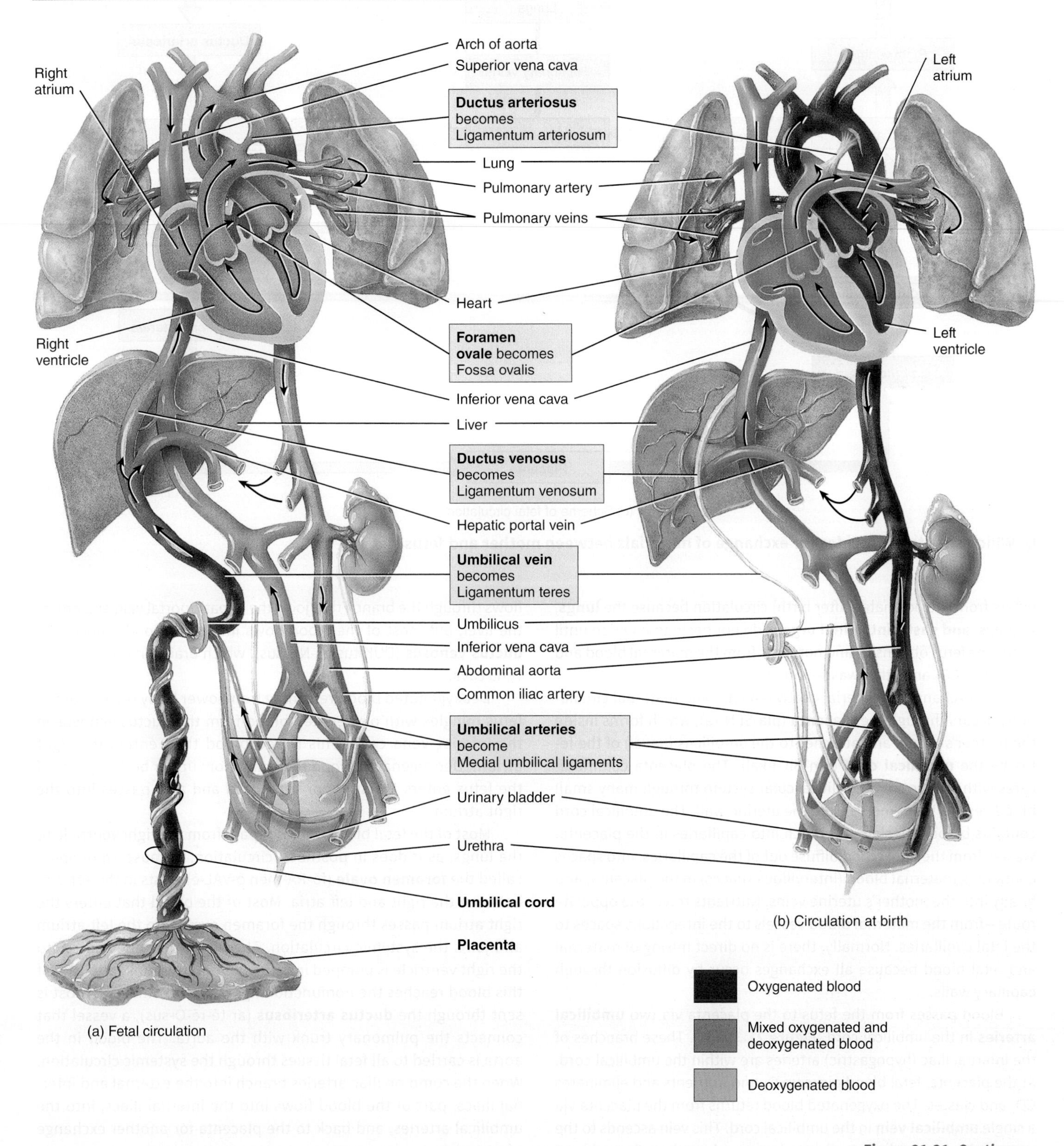

Figure 21.31 *Continues*

FIGURE 21.31 Continued

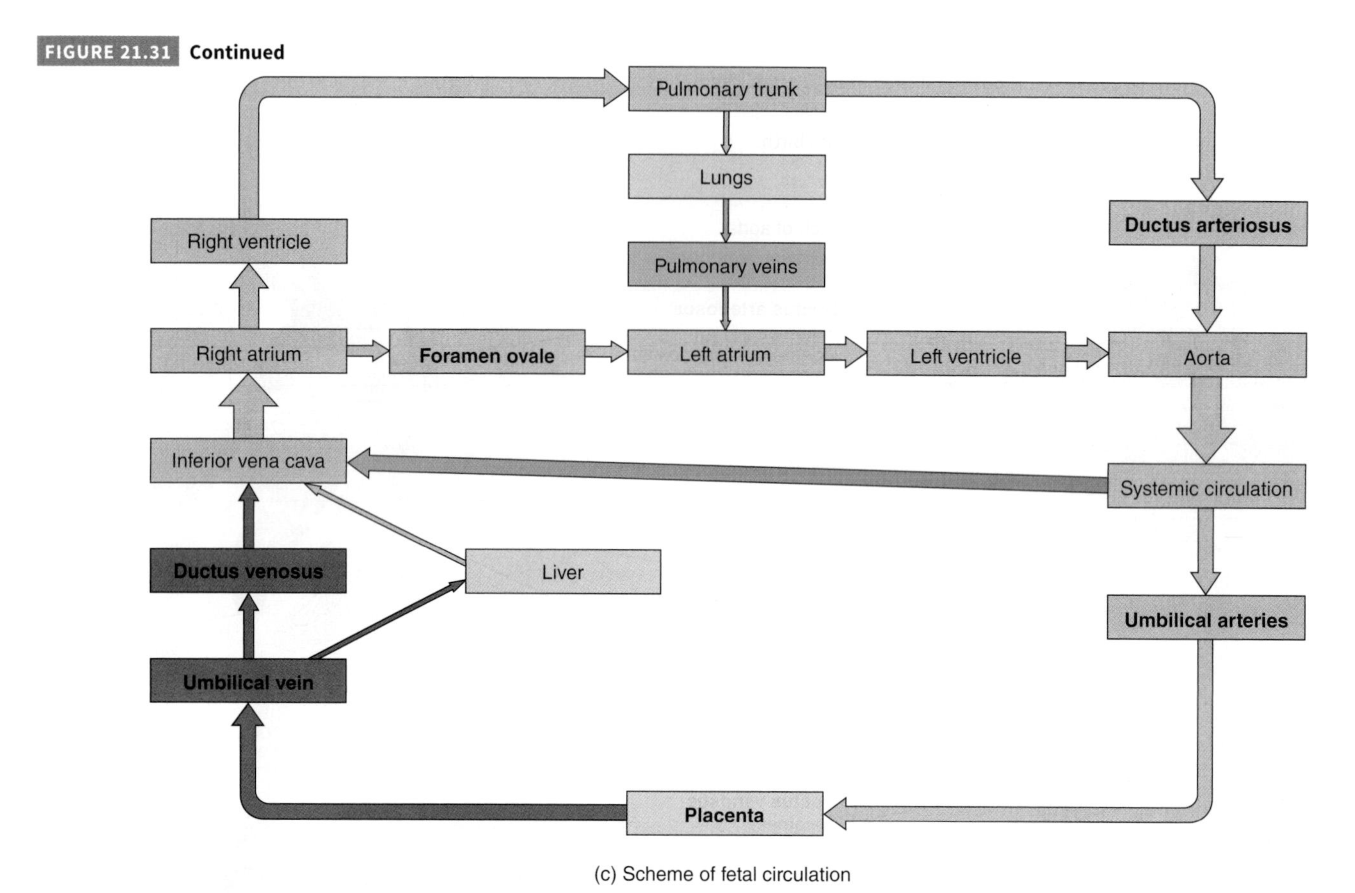

(c) Scheme of fetal circulation

Q Which structure provides for exchange of materials between mother and fetus?

differs from the postnatal (after birth) circulation because the **lungs, kidneys, and gastrointestinal organs** do not begin to function until birth. The fetus obtains O_2 and nutrients from the maternal blood and eliminates CO_2 and other wastes into it.

The exchange of materials between fetal and maternal circulations occurs through the **placenta** (pla-SEN-ta), which forms inside the mother's uterus and attaches to the umbilicus (navel) of the fetus by the **umbilical cord** (um-BIL-i-kal). The placenta communicates with the mother's cardiovascular system through many small blood vessels that emerge from the uterine wall. The umbilical cord contains blood vessels that branch into capillaries in the placenta. Wastes from the fetal blood diffuse out of the capillaries, into spaces containing maternal blood (intervillous spaces) in the placenta, and finally into the mother's uterine veins. Nutrients travel the opposite route—from the maternal blood vessels to the intervillous spaces to the fetal capillaries. Normally, there is no direct mixing of maternal and fetal blood because all exchanges occur by diffusion through capillary walls.

Blood passes from the fetus to the placenta via two **umbilical arteries** in the umbilical cord (Figure 21.31a, c). These branches of the internal iliac (hypogastric) arteries are within the umbilical cord. At the placenta, fetal blood picks up O_2 and nutrients and eliminates CO_2 and wastes. The oxygenated blood returns from the placenta via a single **umbilical vein** in the umbilical cord. This vein ascends to the liver of the fetus, where it divides into two branches. Some blood flows through the branch that joins the hepatic portal vein and enters the liver, but most of the blood flows into the second branch, the **ductus venosus** (DUK-tus ve-NŌ-sus), which drains into the inferior vena cava.

Deoxygenated blood returning from lower body regions of the fetus mingles with oxygenated blood from the ductus venosus in the inferior vena cava. This mixed blood then enters the right atrium. Deoxygenated blood returning from upper body regions of the fetus enters the superior vena cava and also passes into the right atrium.

Most of the fetal blood does not pass from the right ventricle to the lungs, as it does in postnatal circulation, because an opening called the **foramen ovale** (fō-RĀ-men ō-VAL-ē) exists in the septum between the right and left atria. Most of the blood that enters the right atrium passes through the foramen ovale into the left atrium and joins the systemic circulation. The blood that does pass into the right ventricle is pumped into the pulmonary trunk, but little of this blood reaches the nonfunctioning fetal lungs. Instead, most is sent through the **ductus arteriosus** (ar-tē-rē-Ō-sus), a vessel that connects the pulmonary trunk with the aorta. The blood in the aorta is carried to all fetal tissues through the systemic circulation. When the common iliac arteries branch into the external and internal iliacs, part of the blood flows into the internal iliacs, into the umbilical arteries, and back to the placenta for another exchange of materials.

After birth, when pulmonary (lung), renal (kidney), and digestive functions begin, the following vascular changes occur (Figure 21.31b):

1. When the umbilical cord is tied off, blood no longer flows through the umbilical arteries, they fill with connective tissue, and the distal portions of the umbilical arteries become fibrous cords called the **medial umbilical ligaments**. Although the arteries are closed functionally only a few minutes after birth, complete obliteration of the lumens may take 2 to 3 months.
2. The umbilical vein collapses but remains as the **ligamentum teres** (TE-rēz) (*round ligament*), a structure that attaches the umbilicus to the liver.
3. The ductus venosus collapses but remains as the **ligamentum venosum** (ve-NŌ-sum), a fibrous cord on the inferior surface of the liver.
4. The placenta is expelled as the **afterbirth**.
5. The foramen ovale normally closes shortly after birth to become the **fossa ovalis**, a depression in the interatrial septum. When an infant takes its first breath, the lungs expand and blood flow to the lungs increases. Blood returning from the lungs to the heart increases pressure in the left atrium. This closes the foramen ovale by pushing the valve that guards it against the interatrial septum. Permanent closure occurs in about a year.
6. The ductus arteriosus closes by vasoconstriction almost immediately after birth and becomes the **ligamentum arteriosum** (ar-tē′-rē-Ō-sum). Complete anatomical obliteration of the lumen takes 1 to 3 months.

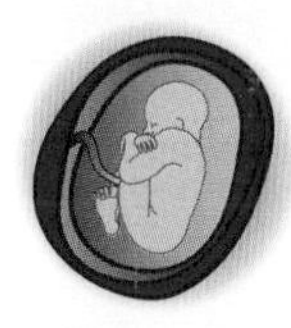

21.23 Development of Blood Vessels and Blood

OBJECTIVE

- **Describe** the development of blood vessels and blood.

The development of blood cells and the formation of blood vessels begins outside the embryo as early as 15 to 16 days in the **mesoderm** of the wall of the yolk sac, chorion, and connecting stalk. About 2 days later, blood vessels form within the embryo. The early formation of the cardiovascular system is linked to the small amount of yolk in the ovum and yolk sac. As the embryo develops rapidly during the third week, there is a greater need to develop a cardiovascular system to supply sufficient nutrients to the embryo and remove wastes from it.

Blood vessels and blood cells develop from the same precursor cell, called a **hemangioblast** (hē-MAN-jē-ō-blast; *hema-* = blood; *-blast* = immature stage). Once mesenchyme develops into hemangioblasts, they can give rise to cells that produce blood vessels (angioblasts) or cells that produce blood cells (pluripotent stem cells).

FIGURE 21.32 Development of blood vessels and blood cells from blood islands.

Blood vessel development begins in the embryo on about the 15th or 16th day.

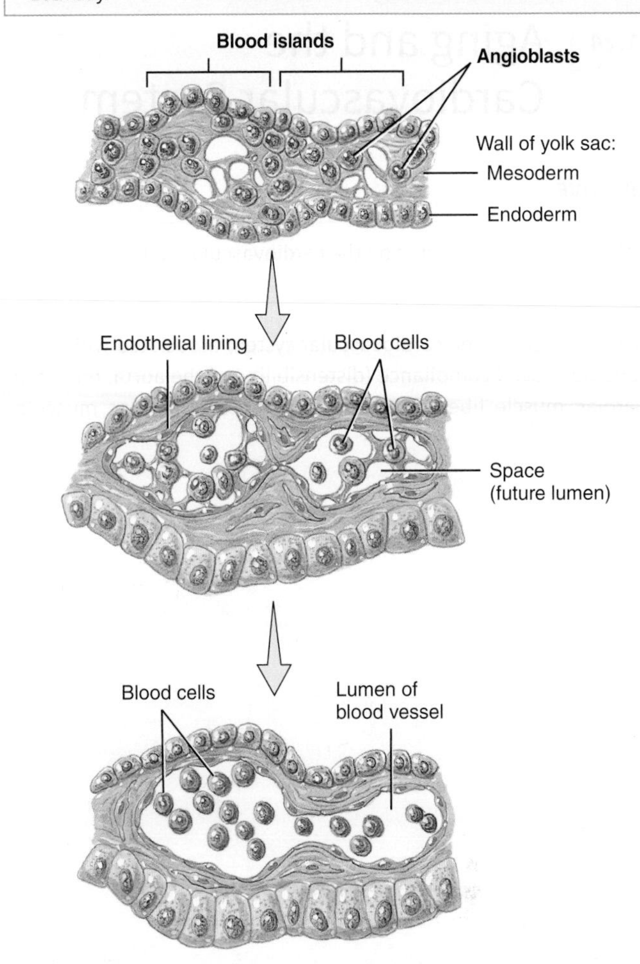

Q From which germ cell layer are blood vessels and blood derived?

Blood vessels develop from **angioblasts** (AN-jē-ō-blasts), which are derived from hemangioblasts. Angioblasts aggregate to form isolated masses and cords throughout the embryonic discs called **blood islands** (Figure 21.32). Spaces soon appear in the islands and become the lumens of the blood vessels. Some of the angioblasts immediately around the spaces give rise to the *endothelial lining of the blood vessels.* Angioblasts around the endothelium form the *tunics* (interna, media, and externa) of the larger blood vessels. Growth and fusion of blood islands form an extensive network of blood vessels throughout the embryo. By continuous branching, blood vessels outside the embryo connect with those inside the embryo, linking the embryo with the placenta.

Blood cells develop from **pluripotent stem cells** (ploo-RIP-ō-tent) derived from hemangioblasts. This development occurs in the walls of blood vessels in the yolk sac, chorion, and allantois at about 3 weeks after fertilization. Blood formation in the embryo itself begins at about the fifth week in the liver and the twelfth week in the spleen, red bone marrow, and thymus.

21.24 Aging and the Cardiovascular System

OBJECTIVE

- **Explain** the effects of aging on the cardiovascular system.

General changes in the cardiovascular system associated with aging include decreased compliance (distensibility) of the aorta, reduction in cardiac muscle fiber size, progressive loss of cardiac muscular strength, reduced cardiac output, a decline in maximum heart rate, and an increase in systolic blood pressure. Total blood cholesterol tends to increase with age, as does low-density lipoprotein (LDL); high-density lipoprotein (HDL) tends to decrease. There is an increase in the incidence of coronary artery disease (CAD), the major cause of heart disease and death in older Americans. Congestive heart failure (CHF), a set of symptoms associated with impaired pumping of the heart, is also prevalent in older individuals. Changes in blood vessels that serve brain tissue—for example, atherosclerosis—reduce nourishment to the brain and result in malfunction or death of brain cells. By age 80, cerebral blood flow is 20% less and renal blood flow is 50% less than in the same person at age 30 because of the effects of aging on blood vessels.

• • •

To appreciate the many ways the blood, heart, and blood vessels contribute to homeostasis of other body systems, examine *Focus on Homeostasis: Contributions of the Cardiovascular System.*

FOCUS on HOMEOSTASIS

INTEGUMENTARY SYSTEM

- Blood delivers clotting factors and white blood cells that aid in hemostasis when skin is damaged and contribute to repair of injured skin
- Changes in skin blood flow contribute to body temperature regulation by adjusting the amount of heat loss via the skin
- Blood flowing in skin may give skin a pink hue

SKELETAL SYSTEM

- Blood delivers calcium and phosphate ions that are needed for building bone extracellular matrix
- Blood transports hormones that govern building and breakdown of bone extracellular matrix, and erythropoietin that stimulates production of red blood cells by red bone marrow

MUSCULAR SYSTEM

- Blood circulating through exercising muscle removes heat and lactic acid

NERVOUS SYSTEM

- Endothelial cells lining choroid plexuses in brain ventricles help produce cerebrospinal fluid (CSF) and contribute to the blood–brain barrier

ENDOCRINE SYSTEM

- Circulating blood delivers most hormones to their target tissues
- Atrial cells secrete atrial natriuretic peptide

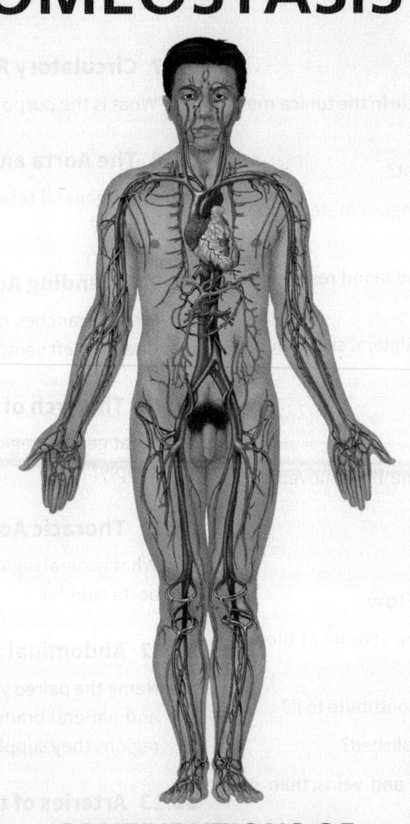

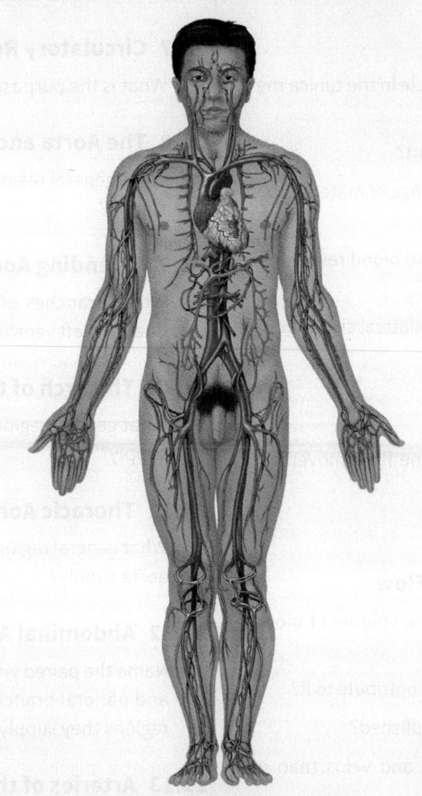

CONTRIBUTIONS OF THE CARDIOVASCULAR SYSTEM

FOR ALL BODY SYSTEMS

- The heart pumps blood through blood vessels to body tissues, delivering oxygen and nutrients and removing wastes by means of capillary exchange
- Circulating blood keeps body tissues at a proper temperature

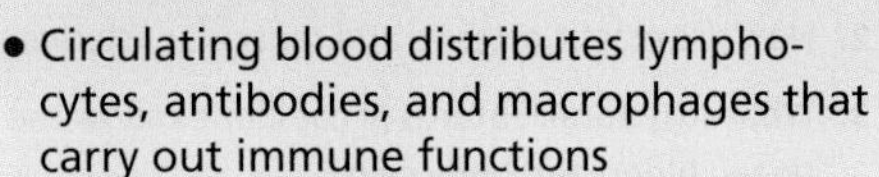

LYMPHATIC SYSTEM and IMMUNITY

- Circulating blood distributes lymphocytes, antibodies, and macrophages that carry out immune functions
- Lymph forms from excess interstitial fluid, which filters from blood plasma due to blood pressure generated by the heart

RESPIRATORY SYSTEM

- Circulating blood transports oxygen from the lungs to body tissues and carbon dioxide to the lungs for exhalation

DIGESTIVE SYSTEM

- Blood carries newly absorbed nutrients and water to the liver
- Blood distributes hormones that aid digestion

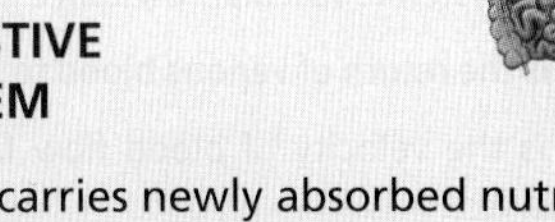

URINARY SYSTEM

- Heart and blood vessels deliver 20% of the resting cardiac output to the kidneys, where blood is filtered, needed substances are reabsorbed, and unneeded substances remain as part of urine, which is excreted

REPRODUCTIVE SYSTEMS

- Vasodilation of arterioles in penis and clitoris causes erection during sexual intercourse
- Blood distributes hormones that regulate reproductive functions

Review Questions

21.1 Structure and Function of Blood Vessels

1. What is the function of elastic fibers and smooth muscle in the tunica media of arteries?
2. How are elastic arteries and muscular arteries different?
3. What structural features of capillaries allow the exchange of materials between blood and body cells?
4. What is the difference between pressure reservoirs and blood reservoirs? Why is each important?
5. What is the relationship between anastomoses and collateral circulation?

21.2 Capillary Exchange

6. How can substances enter and leave blood plasma?
7. How do hydrostatic and osmotic pressures determine fluid movement across the walls of capillaries?
8. Define edema and describe how it develops.

21.3 Hemodynamics: Factors Affecting Blood Flow

9. Explain how blood pressure and resistance determine volume of blood flow.
10. What is systemic vascular resistance and what factors contribute to it?
11. How is the return of venous blood to the heart accomplished?
12. Why is the velocity of blood flow faster in arteries and veins than in capillaries?

21.4 Control of Blood Pressure and Blood Flow

13. What are the principal inputs to and outputs from the cardiovascular center?
14. Explain the operation of the carotid sinus reflex and the aortic reflex.
15. What is the role of chemoreceptors in the regulation of blood pressure?
16. How do hormones regulate blood pressure?
17. What is autoregulation, and how does it differ in the systemic and pulmonary circulations?

21.5 Checking Circulation

18. Where may the pulse be felt?
19. What do tachycardia and bradycardia mean?
20. How are systolic and diastolic blood pressures measured with a sphygmomanometer?

21.6 Shock and Homeostasis

21. Which symptoms of hypovolemic shock relate to actual body fluid loss, and which relate to the negative feedback systems that attempt to maintain blood pressure and blood flow?
22. Describe the types of shock and their causes and how a person in hypovolemic shock should be treated.

21.7 Circulatory Routes: Systemic Circulation

23. What is the purpose of systemic circulation?

21.8 The Aorta and Its Branches

24. What general regions do each of the four principal divisions of the aorta supply?

21.9 Ascending Aorta

25. Which branches of the coronary arteries supply the left ventricle? Why does the left ventricle have such an extensive arterial blood supply?

21.10 The Arch of the Aorta

26. What general regions do the arteries that arise from the arch of the aorta supply?

21.11 Thoracic Aorta

27. What general regions do the visceral and parietal branches of the thoracic aorta supply?

21.12 Abdominal Aorta

28. Name the paired visceral and parietal branches and the unpaired visceral and parietal branches of the abdominal aorta, and indicate the general regions they supply.

21.13 Arteries of the Pelvis and Lower Limbs

29. What general regions do the internal and external iliac arteries supply?

21.14 Veins of the Systemic Circulation

30. What are the three tributaries of the coronary sinus?

21.15 Veins of the Head and Neck

31. Which general areas are drained by the internal jugular, external jugular, and vertebral veins?

21.16 Veins of the Upper Limbs

32. Where do the cephalic, basilic, median antebrachial, radial, and ulnar veins originate?

21.17 Veins of the Thorax

33. What is the importance of the azygos system relative to the inferior vena cava?

21.18 Veins of the Abdomen and Pelvis

34. What structures do the lumbar, gonadal, renal, suprarenal, inferior phrenic, and hepatic veins drain?

21.19 Veins of the Lower Limbs

35. What is the clinical importance of the great saphenous veins?

21.20 Circulatory Routes: The Hepatic Portal Circulation

36. Diagram the hepatic portal circulation and describe its importance.

21.21 Circulatory Routes: The Pulmonary Circulation

37. Explain why pulmonary circulation is important.

21.22 Circulatory Routes: The Fetal Circulation

38. Discuss the anatomy and physiology of the fetal circulation. Indicate the function of the umbilical arteries, umbilical vein, ductus venosus, foramen ovale, and ductus arteriosus.

21.23 Development of Blood Vessels and Blood

39. What are the sites of blood cell production outside the embryo and within the embryo?

21.24 Aging and the Cardiovascular System

40. How does aging affect the heart?

Critical Thinking Questions

1. Kim Sung was told that her baby was born with a hole in the upper chambers of his heart. Is this something Kim Sung should worry about?

2. Michael was brought into the emergency room suffering from a gunshot wound. He is bleeding profusely and exhibits the following: systolic blood pressure is 40 mmHg; weak pulse of 200 beats per minute; cool, pale, and clammy skin. Michael is not producing urine but is asking for water. He is confused and disoriented. What is his diagnosis and what, specifically, is causing these symptoms?

3. Maureen's job entails standing on a concrete floor for 10-hour days on an assembly line. Lately she has noticed swelling in her ankles at the end of the day and some tenderness in her calves. What do you suspect is Maureen's problem and how could she help counteract the problem?

The Lymphatic System and Immunity

CHAPTER 22

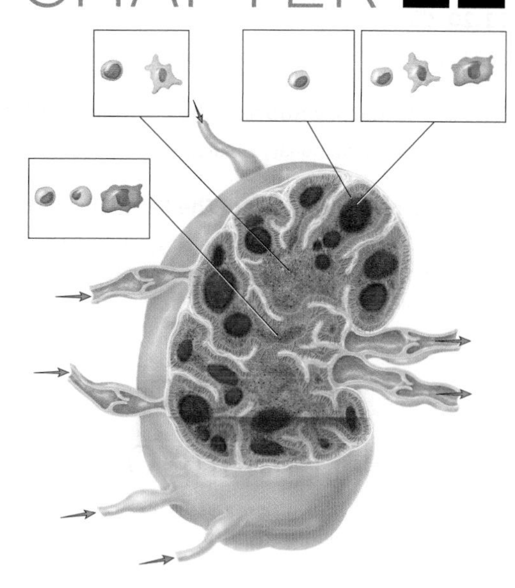

The Lymphatic System, Disease Resistance, and Homeostasis

The lymphatic system contributes to homeostasis by draining interstitial fluid as well as providing the mechanisms for defense against disease.

The environment in which we live is filled with microbes that have the ability to cause disease if given the right opportunity. If we did not resist these microbes, we would be ill constantly or even die. Fortunately, we have a number of defenses that keep microbes from either entering our bodies or combat them if they do gain entrance. The lymphatic system is one of the principal body systems that helps to defend us against disease-producing microbes. In this chapter you will learn about the organization and components of the lymphatic system and its role in keeping us healthy.

Q Did you ever wonder how cancer can spread from one part of the body to another?

22.1 The Concept of Immunity

OBJECTIVES

- **Define** immunity.
- **Compare** the two basic types of immunity.

Maintaining homeostasis in the body requires continual combat against harmful agents in our internal and external environments. Despite constant exposure to a variety of **pathogens** (PATH-ō-jens)—disease-producing microbes such as bacteria and viruses—most people remain healthy. The body surface also endures cuts and bumps, exposure to ultraviolet rays, chemical toxins, and minor burns with an array of defensive ploys.

Immunity (i-MŪ-ni-tē) or *resistance* is the ability to ward off damage or disease through our defenses. Vulnerability or lack of resistance is termed **susceptibility**. The two general types of immunity are (1) innate and (2) adaptive. **Innate** (*nonspecific*) **immunity** refers to defenses that are present at birth. Innate immunity does not involve specific recognition of a microbe and acts against all microbes in the same way. Among the components of innate immunity are the first line of defense (the physical and chemical barriers of the skin and mucous membranes) and the second line of defense (antimicrobial substances, natural killer cells, phagocytes, inflammation, and fever). Innate immune responses represent immunity's early warning system and are designed to prevent microbes from entering the body and to help eliminate those that do gain access.

Adaptive (*specific*) **immunity** refers to defenses that involve specific recognition of a microbe once it has breached the innate immunity defenses. Adaptive immunity is based on a specific response to a specific microbe; that is, it adapts or adjusts to handle a specific microbe. Adaptive immunity involves lymphocytes (a type of white blood cell) called T lymphocytes (T cells) and B lymphocytes (B cells).

The body system responsible for adaptive immunity (and some aspects of innate immunity) is the lymphatic system. This system is closely allied with the cardiovascular system, and it also functions with the digestive system in the absorption of fatty foods. In this chapter, we explore the mechanisms that provide defenses against intruders and promote the repair of damaged body tissues.

22.2 Overview of the Lymphatic System

OBJECTIVES

- **List** the components of the lymphatic system.
- **Describe** the functions of the lymphatic system.

Components of the Lymphatic System

The **lymphatic** or *lymphoid* **system** (lim-FAT-ik) consists of a fluid called lymph, vessels called lymphatic vessels that transport the lymph, a number of structures and organs containing lymphatic tissue (lymphocytes within a filtering tissue), and red bone marrow (Figure 22.1). The lymphatic system assists in circulating body fluids and helps defend the body against disease-causing agents. As you will see shortly, most components of blood plasma filter through blood capillary walls to form interstitial fluid. After interstitial fluid passes into lymphatic vessels, it is called **lymph** (LIMF = clear fluid). The major difference between interstitial fluid and lymph is location: Interstitial fluid is found between cells, and lymph is located within lymphatic vessels and lymphatic tissue.

Lymphatic tissue is a specialized form of reticular connective tissue (see Table 4.4) that contains large numbers of lymphocytes. Recall from Chapter 19 that lymphocytes are agranular white blood cells (see Section 19.4). Two types of lymphocytes participate in adaptive immune responses: B cells and T cells (described shortly).

Functions of the Lymphatic System

The lymphatic system has three primary functions:

1. ***Drains excess interstitial fluid.*** Lymphatic vessels drain excess interstitial fluid from tissue spaces and return it to the blood. This function closely links it with the cardiovascular system. In fact, without this function, the maintenance of circulating blood volume would not be possible.
2. ***Transports dietary lipids.*** Lymphatic vessels transport lipids and lipid-soluble vitamins (A, D, E, and K) absorbed by the gastrointestinal tract.
3. ***Carries out immune responses.*** Lymphatic tissue initiates highly specific responses directed against particular microbes or abnormal cells.

22.3 Lymphatic Vessels and Lymph Circulation

OBJECTIVES

- **Describe** the organization of lymphatic vessels.
- **Explain** the formation and flow of lymph.

Lymphatic vessels begin as **lymphatic capillaries**. These capillaries, which are located in the spaces between cells, are closed at one end (Figure 22.2). Just as blood capillaries converge to form venules and then veins, lymphatic capillaries unite to form larger **lymphatic vessels** (see Figure 22.1), which resemble small veins in structure but have thinner walls and more valves. At intervals along the lymphatic vessels, lymph flows through lymph nodes, encapsulated bean-shaped organs consisting of masses of B cells and T cells. In the skin, lymphatic vessels lie in the subcutaneous tissue and generally follow the same route as veins; lymphatic vessels of the viscera generally follow arteries, forming plexuses (networks) around them. Tissues that lack lymphatic capillaries include avascular tissues (such as cartilage, the epidermis, and the cornea of the eye), portions of the spleen, and red bone marrow.

Lymphatic Capillaries

Lymphatic capillaries have greater permeability than blood capillaries and thus can absorb large molecules such as proteins and lipids. Lymphatic capillaries are also slightly larger in diameter than blood capillaries and have a unique one-way structure that permits interstitial fluid to flow into them but not out. The ends of endothelial cells that make up the wall of a lymphatic capillary overlap (Figure 22.2b). When pressure is greater in the interstitial fluid than in lymph, the cells separate slightly, like the opening of a one-way swinging door, and interstitial fluid enters the lymphatic capillary. When pressure is greater inside the lymphatic capillary, the cells adhere more closely, and lymph cannot escape back into interstitial fluid. The pressure is relieved as lymph moves further down the lymphatic capillary. Attached to the lymphatic capillaries are *anchoring filaments*, which contain elastic fibers. They extend out from the lymphatic capillary, attaching lymphatic endothelial cells to surrounding tissues. When excess interstitial fluid accumulates and causes tissue swelling, the anchoring filaments are pulled, making the openings between cells even larger so that more fluid can flow into the lymphatic capillary.

In the small intestine, specialized lymphatic capillaries called **lacteals** (LAK-tē-als; *lact-* = milky) carry dietary lipids into lymphatic vessels and ultimately into the blood (see Figure 24.20). The presence of these lipids causes the lymph draining from the small intestine to appear creamy white; such lymph is referred to as **chyle** (KĪL = juice). Elsewhere, lymph is a clear, pale-yellow fluid.

Lymph Trunks and Ducts

As you have already learned, lymph passes from lymphatic capillaries into lymphatic vessels and then through lymph nodes. As lymphatic vessels exit lymph nodes in a particular region of the body, they unite to form **lymph trunks**. The principal trunks are the lumbar, intestinal, bronchomediastinal, subclavian, and jugular trunks (see Figure 22.3). The **lumbar trunks** drain lymph from the lower limbs, the wall and viscera of the pelvis, the kidneys, the adrenal glands, and the abdominal wall. The **intestinal trunk** drains lymph from the stomach, intestines, pancreas, spleen, and part of the liver. The **bronchomediastinal trunks** (brong-kō-mē′-dē-as-TĪ-nal) drain lymph from the thoracic wall, lung, and heart. The **subclavian trunks** drain the upper limbs. The **jugular trunks** drain the head and neck.

The lymph passage from the lymph trunks to the venous system differs on the right and left sides of the body. On the right side the

FIGURE 22.1 **Components of the lymphatic system.**

The lymphatic system consists of lymph, lymphatic vessels, lymphatic tissues, and red bone marrow.

Functions

1. Drains excess interstitial fluid.
2. Transports dietary lipids from the gastrointestinal tract to the blood.
3. Protects against invasion through immune responses.

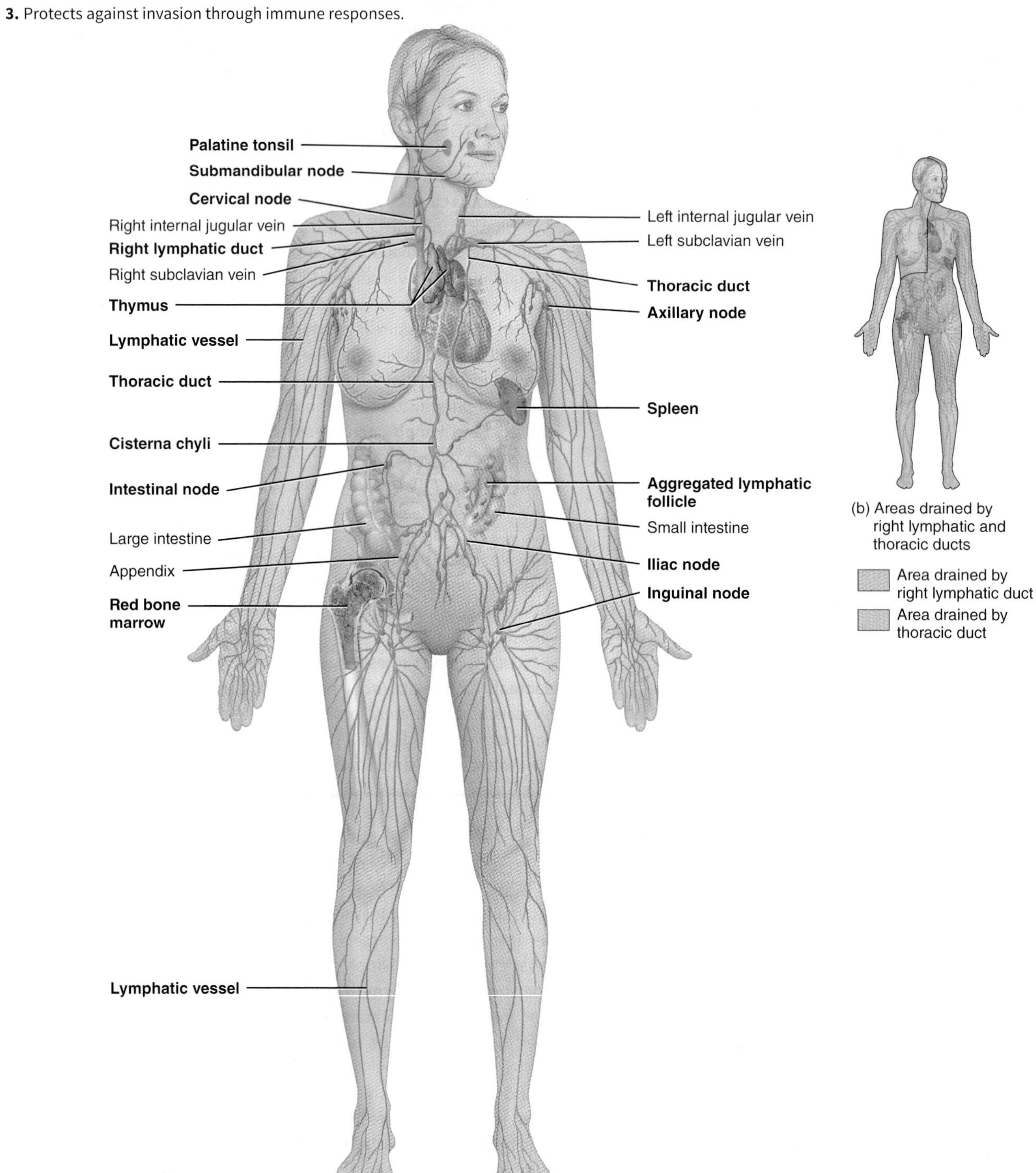

(a) Anterior view of principal components of lymphatic system

(b) Areas drained by right lymphatic and thoracic ducts

Q What tissue contains stem cells that develop into lymphocytes?

FIGURE 22.2 Lymphatic capillaries.

Lymphatic capillaries are found throughout the body except in avascular tissues, the central nervous system, portions of the spleen, and bone marrow.

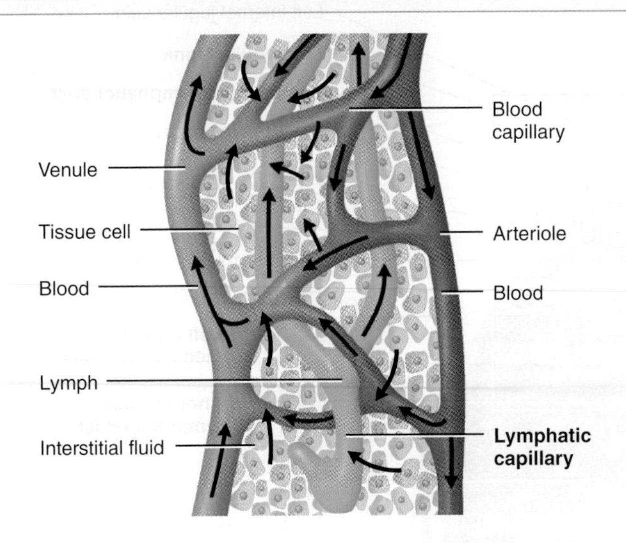

(a) Relationship of lymphatic capillaries to tissue cells and blood capillaries

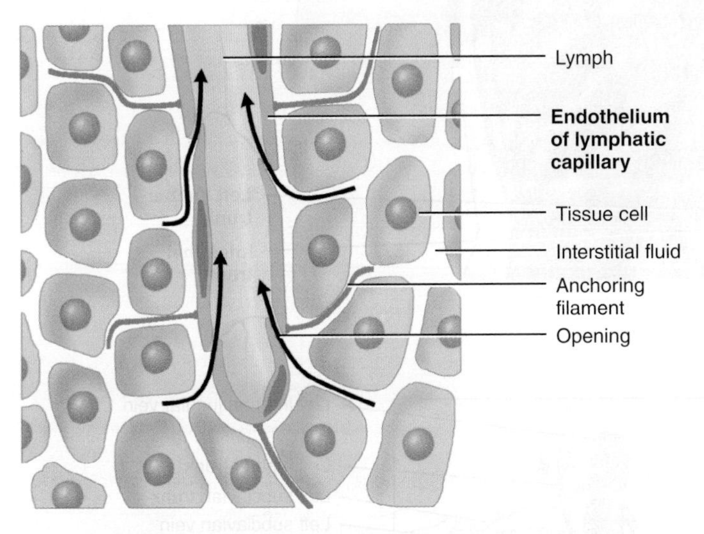

(b) Details of a lymphatic capillary

Q Is lymph more similar to blood plasma or to interstitial fluid? Why?

three lymph trunks (right jugular trunk, right subclavian trunk, and right bronchomediastinal trunk) usually open independently into the venous system on the anterior surface of the junction of the internal jugular and subclavian veins (Figure 22.3). Rarely, the three trunks will join to form a short **right lymphatic duct** that forms a single junction with the venous system. On the left side of the body, the largest lymph vessel, the **thoracic (left lymphatic) duct** forms the main duct for return of lymph to the blood. This long duct, approximately 38–45 cm (15–18 in.), begins as a dilation called the **cisterna chyli** (sis-TER-na KI-le; cisterna = cavity or reservoir) anterior to the second lumbar vertebra. The cisterna chyli receives lymph from the right and left lumbar trunks and from the intestinal trunk. In the neck, the thoracic duct also receives lymph from the left jugular and left subclavian trunks before opening into the anterior surface of the junction of the left internal jugular and subclavian veins. The left bronchomediastinal trunk joins the anterior surface of the subclavian vein independently and does not join the thoracic duct. As a result of these pathways, lymph from the upper right quadrant of the body returns to the superior vena cava from the right brachiocephalic vein, while all the lymph form the left upper side of the body and the entire body below the diaphragm returns to the superior vena cava via the left brachiocephalic vein.

Formation and Flow of Lymph

Most components of blood plasma, such as nutrients, gases, and hormones, filter freely through the capillary walls to form interstitial fluid, but more fluid filters out of blood capillaries than returns to them by reabsorption (see Figure 21.7). The excess filtered fluid—about 3 liters per day—drains into lymphatic vessels and becomes lymph. Because most plasma proteins are too large to leave blood vessels, interstitial fluid contains only a small amount of protein. Proteins that do leave blood plasma cannot return to the blood by diffusion because the concentration gradient (high level of proteins inside blood capillaries, low level outside) opposes such movement. The proteins can, however, move readily through the more permeable lymphatic capillaries into lymph. Thus, an important function of lymphatic vessels is to return the lost plasma proteins and plasma to the bloodstream.

Like some veins, lymphatic vessels contain valves, which ensure the one-way movement of lymph. As noted previously, lymph drains into venous blood through the right lymphatic duct and the thoracic duct at the junction of the internal jugular and subclavian veins (Figure 22.3). Thus, the sequence of fluid flow is blood capillaries (blood) → interstitial spaces (interstitial fluid) → lymphatic capillaries (lymph) → lymphatic vessels (lymph) → lymphatic trunks or ducts (lymph) → junction of the internal jugular and subclavian veins (blood). Figure 22.4 illustrates this sequence, along with the relationship of the lymphatic and cardiovascular systems. Both systems form a very efficient circulatory system.

The same two "pumps" that aid the return of venous blood to the heart maintain the flow of lymph.

1. ***Respiratory pump.*** Lymph flow is also maintained by pressure changes that occur during inhalation (breathing in). Lymph flows from the abdominal region, where the pressure is higher, toward the thoracic region, where it is lower. When the pressures reverse during exhalation (breathing out), the valves in lymphatic vessels prevent backflow of lymph. In addition, when a lymphatic vessel distends, the smooth muscle in its wall contracts, which helps move lymph from one segment of the vessel to the next.
2. ***Skeletal muscle pump.*** The "milking action" of skeletal muscle contractions (see Figure 21.9) compresses lymphatic vessels (as well as veins) and forces lymph toward the junction of the internal jugular and subclavian veins.

FIGURE 22.3 Routes for drainage of lymph from lymph trunks into the thoracic and right lymphatic ducts.

All lymph returns to the bloodstream through the thoracic (left) lymphatic duct and right lymphatic duct.

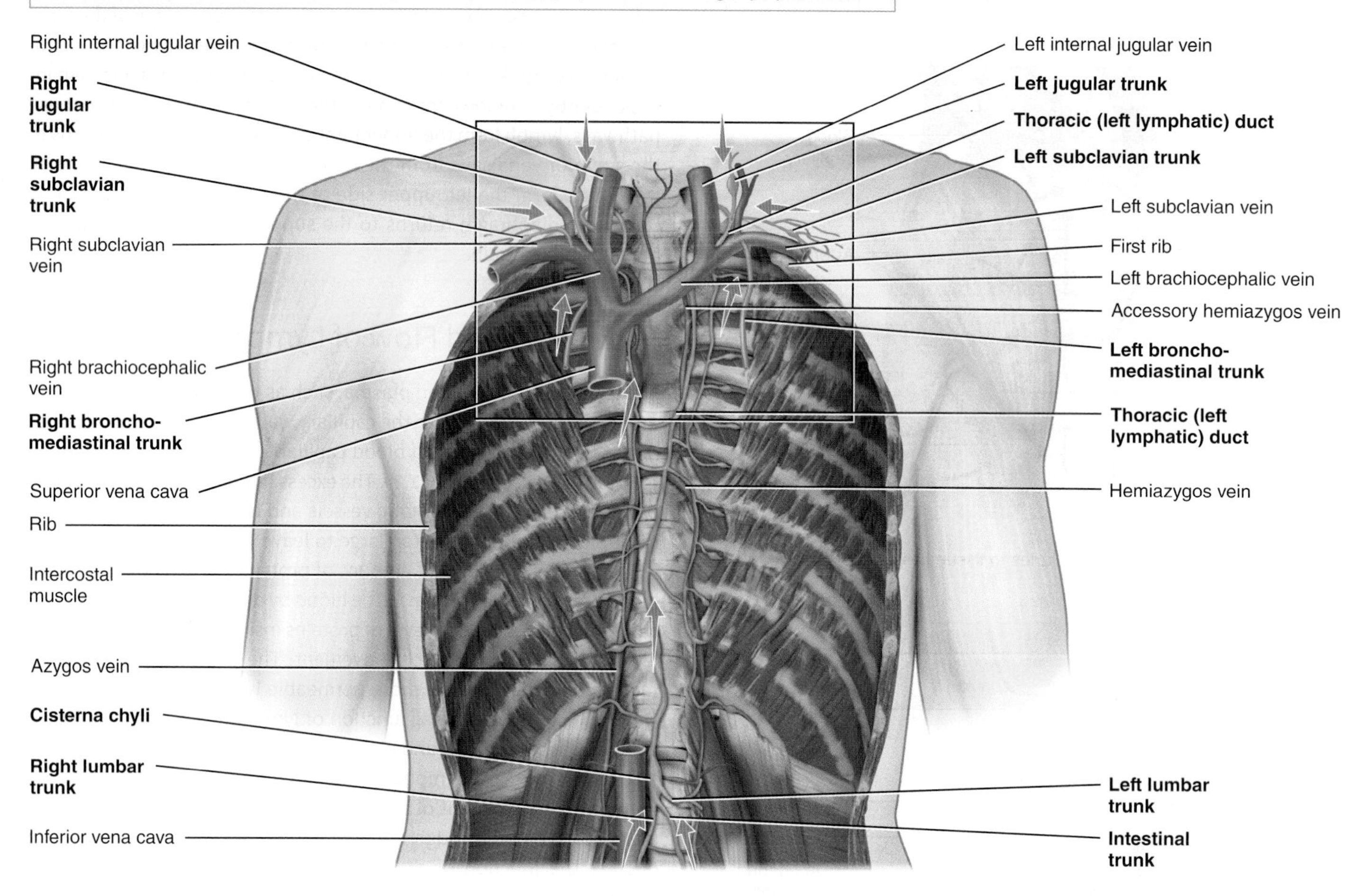

(a) Overall anterior view

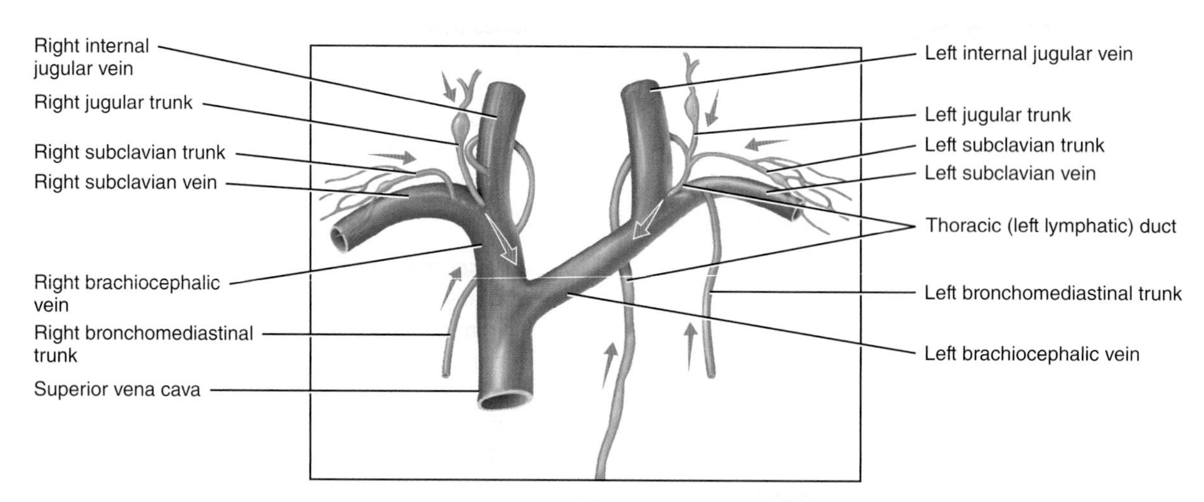

(b) Detailed anterior view of thoracic and right lymphatic duct

Q Which lymphatic vessels empty into the cisterna chyli, and which duct receives lymph from the cisterna chyli?

FIGURE 22.4 Schematic diagram showing the relationship of the lymphatic system to the cardiovascular system. Arrows indicate the direction of flow of lymph and blood.

> The sequence of fluid flow is blood capillaries (blood) → interstitial spaces (interstitial fluid) → lymphatic capillaries (lymph) → lymphatic vessels (lymph) → lymphatic trunks or ducts (lymph) → junction of the internal jugular and subclavian veins (blood).

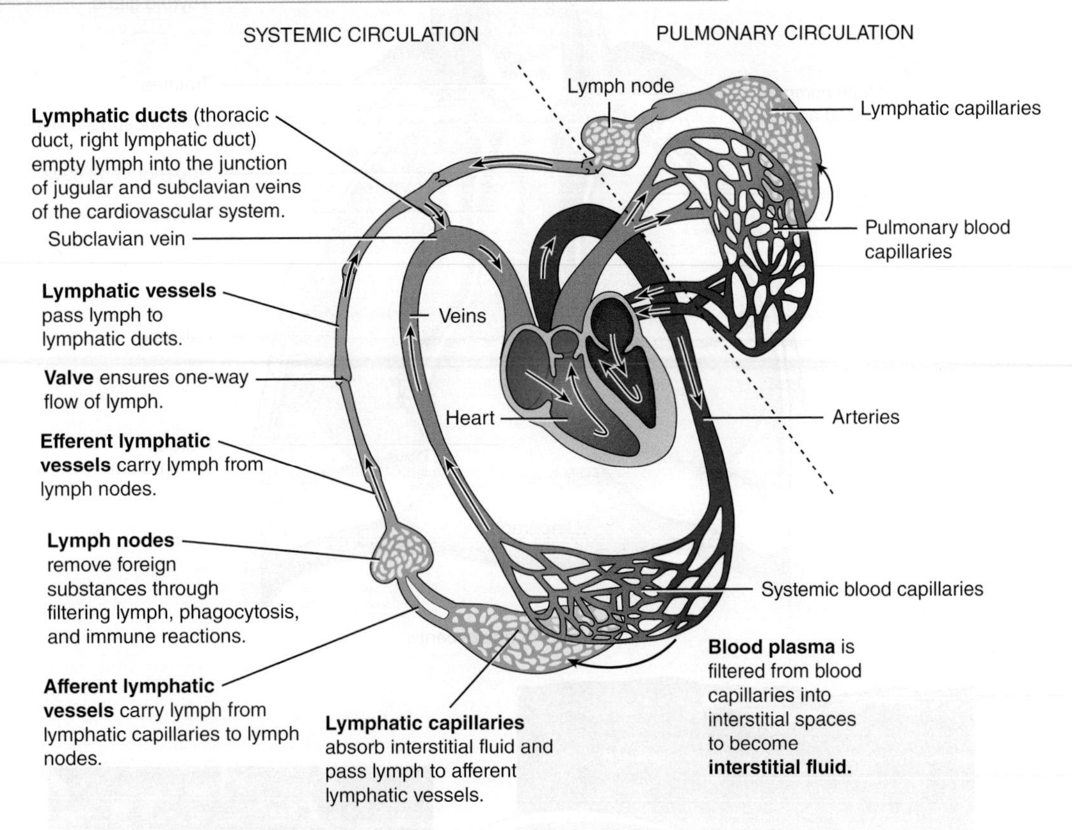

Q Does inhalation promote or hinder the flow of lymph?

22.4 Lymphatic Organs and Tissues

OBJECTIVE

- **Distinguish** between primary and secondary lymphatic organs.

The widely distributed lymphatic organs and tissues are classified into two groups based on their functions. **Primary lymphatic organs** are the sites where stem cells divide and become **immunocompetent** (im′-ū-nō-KOM-pe-tent), that is, capable of mounting an immune response. The primary lymphatic organs are the red bone marrow (in flat bones and the epiphyses of long bones of adults) and the thymus. Pluripotent stem cells in red bone marrow give rise to mature, immunocompetent B cells and to pre-T cells. The pre-T cells in turn migrate to the thymus, where they become immunocompetent T cells. The **secondary lymphatic organs** and **tissues** are the sites where most immune responses occur. They include lymph nodes, the spleen, and lymphatic nodules (follicles). The thymus, lymph nodes, and spleen are considered organs because each is surrounded by a connective tissue capsule; lymphatic nodules, in contrast, are not considered organs because they lack a capsule.

Thymus

The **thymus** is a bilobed organ located in the mediastinum between the sternum and the aorta. It extends from the top of the sternum or the inferior cervical region to the level of the fourth costal cartilages, anterior to the top of the heart and its great vessels (**Figure 23.5a**). An enveloping layer of connective tissue holds the two lobes closely together, but a connective tissue **capsule** encloses each lobe separately. Extensions of the capsule, called **trabeculae** (tra-BEK-ū-lē = little beams), penetrate inward and divide each lobe into **lobules** (**Figure 23.5b**).

Each thymic lobule consists of a deeply staining outer cortex and a lighter-staining central medulla (**Figure 22.5b**). The **cortex** is

FIGURE 22.5 **Thymus.**

The bilobed thymus is largest at puberty and then the functional portion atrophies with age.

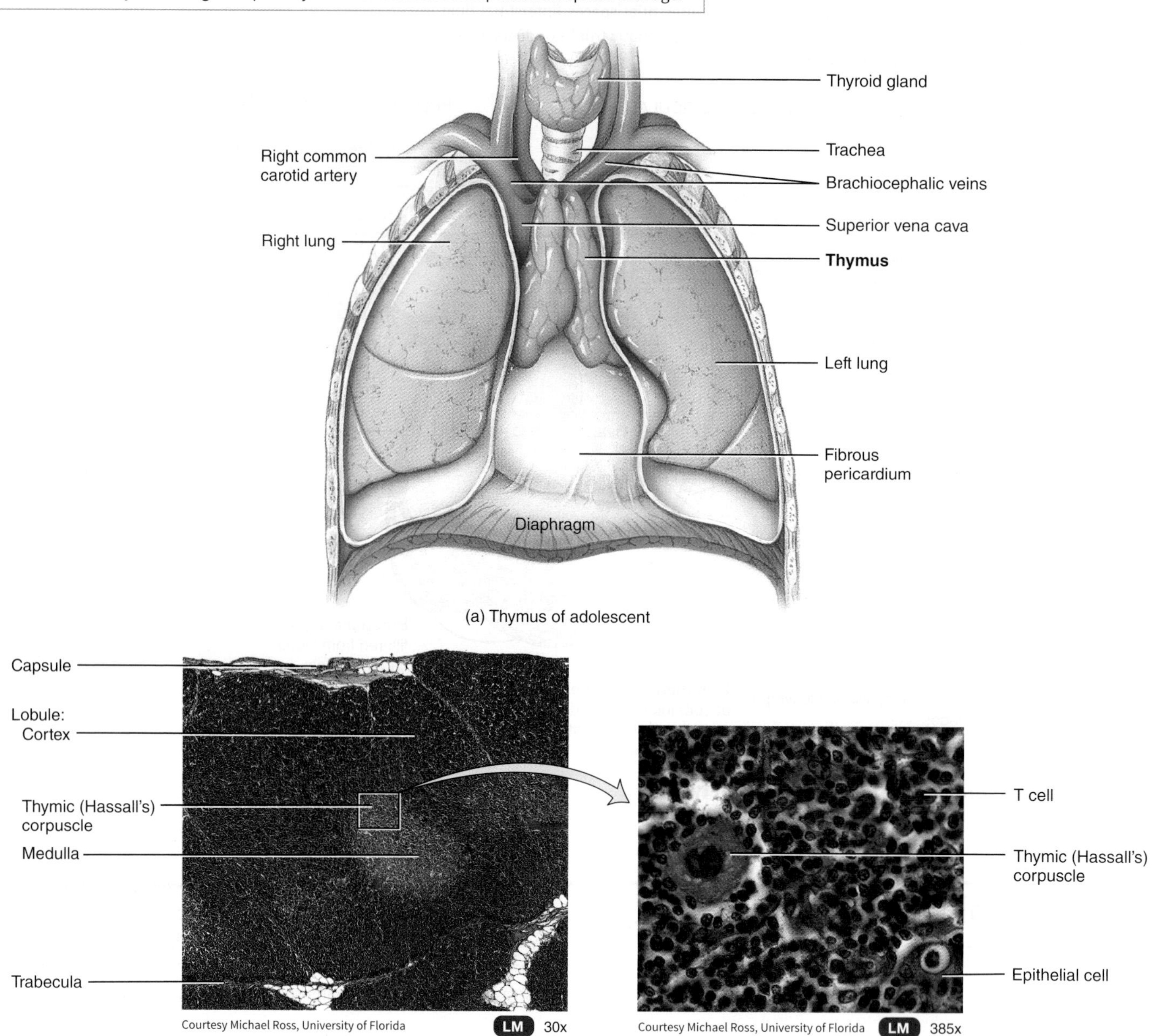

(a) Thymus of adolescent

Courtesy Michael Ross, University of Florida

(b) Thymic lobules

Courtesy Michael Ross, University of Florida

(c) Details of the thymic medulla

Q Which type of lymphocytes mature in the thymus?

composed of large numbers of T cells and scattered dendritic cells, epithelial cells, and macrophages. Immature T cells (pre-T cells) migrate from red bone marrow to the cortex of the thymus, where they proliferate and begin to mature. **Dendritic cells** (den-DRIT-ik; *dendr-* = a tree), which are derived from monocytes (and so named because they have long, branched projections that resemble the dendrites of a neuron), assist the maturation process. As you will see shortly, dendritic cells in other parts of the body, such as lymph nodes, play another key role in immune responses. Each of the specialized **epithelial cells** in the cortex has several long processes that surround and serve as a framework for as many as 50 T cells. These epithelial cells help "educate" the pre-T cells in a process known as positive selection (see Figure 22.22). Additionally, they produce thymic hormones that are thought to aid in the maturation of T cells. Only about 2% of developing T cells survive in the cortex. The remaining cells die via apoptosis (programmed cell death). Thymic **macrophages** (MAK-rō-fā-jez) help clear out the debris of dead and dying cells. The surviving T cells enter the medulla.

The **medulla** consists of widely scattered, more mature T cells, epithelial cells, dendritic cells, and macrophages (**Figure 22.5c**). Some of the epithelial cells become arranged into concentric layers of flat cells that degenerate and become filled with keratohyalin granules and keratin. These clusters are called **thymic** (*Hassall's*) **corpuscles**. Although their role is uncertain, they may serve as sites of T cell death in the medulla. T cells that leave the thymus via the blood migrate to lymph nodes, the spleen, and other lymphatic tissues, where they colonize parts of these organs and tissues.

Because of its high content of lymphoid tissue and a rich blood supply, the thymus has a reddish appearance in a living body. With age, however, fatty infiltrations replace the lymphoid tissue and the thymus takes on more of the yellowish color of the invading fat, giving the false impression of reduced size. However, the actual size of the thymus, defined by its connective tissue capsule, does not change. In infants, the thymus has a mass of about 70 g (2.3 oz). It is after puberty that adipose and areolar connective tissue begin to replace the thymic tissue. By the time a person reaches maturity, the functional portion of the gland is reduced considerably, and in old age the functional portion may weigh only 3 g (0.1 oz). Before the thymus atrophies, it populates the secondary lymphatic organs and tissues with T cells. However, some T cells continue to proliferate in the thymus throughout an individual's lifetime, but this number decreases with age.

Lymph Nodes

Located along lymphatic vessels are about 600 bean-shaped **lymph nodes**. They are scattered throughout the body, both superficially and deep, and usually occur in groups (see **Figure 22.1**). Large groups of lymph nodes are present near the mammary glands and in the axillae and groin.

Lymph nodes are 1–25 mm (0.04–1 in.) long and, like the thymus, are covered by a **capsule** of dense connective tissue that extends into the node (**Figure 22.6**). The capsular extensions, called **trabeculae**, divide the node into compartments, provide support, and provide a route for blood vessels into the interior of a node. Internal to the capsule is a supporting network of reticular fibers and fibroblasts. The capsule, trabeculae, reticular fibers, and fibroblasts constitute the *stroma* (supporting framework of connective tissue) of a lymph node.

The *parenchyma* (functioning part) of a lymph node is divided into a superficial cortex and a deep medulla. The cortex consists of an outer cortex and an inner cortex. Within the **outer cortex** are egg-shaped aggregates of B cells called **lymphatic nodules** (*follicles*). A lymphatic nodule consisting chiefly of B cells is called a *primary lymphatic nodule.* Most lymphatic nodules in the outer cortex are *secondary lymphatic nodules* (**Figure 22.6**), which form in response to an antigen (a foreign substance) and are sites of plasma cell and memory B cell formation. After B cells in a primary lymphatic nodule recognize an antigen, the primary lymphatic nodule develops into a secondary lymphatic nodule. The center of a secondary lymphatic nodule contains a region of light-staining cells called a *germinal center.* In the germinal center are B cells, follicular dendritic cells (a special type of dendritic cell), and macrophages. When follicular dendritic cells "present" an antigen (described later in the chapter), B cells proliferate and develop into antibody-producing plasma cells or develop into memory B cells. Memory B cells persist after an initial immune response and "remember" having encountered a specific antigen. B cells that do not develop properly undergo apoptosis (programmed cell death) and are destroyed by macrophages. The region of a secondary lymphatic nodule surrounding the germinal center is composed of dense accumulations of B cells that have migrated away from their site of origin within the nodule.

The **inner cortex** does not contain lymphatic nodules. It consists mainly of T cells and dendritic cells that enter a lymph node from other tissues. The dendritic cells present antigens to T cells, causing their proliferation. The newly formed T cells then migrate from the lymph node to areas of the body where there is antigenic activity.

The **medulla** of a lymph node contains B cells, antibody-producing plasma cells that have migrated out of the cortex into the medulla, and macrophages. The various cells are embedded in a network of reticular fibers and reticular cells.

As you have already learned, lymph flows through a node in one direction only (**Figure 22.6a**). It enters through several **afferent lymphatic vessels** (AF-er-ent; *afferent* = to carry toward), which penetrate the convex surface of the node at several points. The afferent vessels contain valves that open toward the center of the node, directing the lymph *inward.* Within the node, lymph enters **sinuses**, a series of irregular channels that contain branching reticular fibers, lymphocytes, and macrophages. From the afferent lymphatic vessels, lymph flows into the **subcapsular sinus** (sub-KAP-soo-lar), immediately beneath the capsule. From here the lymph flows through **trabecular sinuses** (tra-BEK-ū-lar), which extend through the cortex parallel to the trabeculae, and into **medullary sinuses**, which extend through the medulla. The medullary sinuses drain into one or two **efferent lymphatic vessels** (EF-er-ent; *efferent* = to carry away), which are wider and fewer in number than afferent vessels. They contain valves that open away from the center of the lymph node to convey lymph, antibodies secreted by plasma cells, and activated T cells *out* of the node. Efferent lymphatic vessels emerge from one side of the lymph node at a slight depression called a **hilum** (HĪ-lum). Blood vessels also enter and leave the node at the hilum.

Lymph nodes function as a type of filter. As lymph enters one end of a lymph node, foreign substances are trapped by the reticular fibers within the sinuses of the node. Then macrophages destroy some foreign substances by phagocytosis, while lymphocytes destroy others by immune responses. The filtered lymph then leaves the other end of the lymph node. Since there are many afferent lymphatic vessels that bring lymph into a lymph node and only one or two efferent lymphatic vessels that transport lymph out of a lymph node, the slow flow of lymph within the lymph nodes allows additional time for lymph to be filtered. Additionally, all lymph flows through multiple lymph nodes on its path through the lymph vessels. This exposes the lymph to multiple filtering events before returning to the blood.

See Clinical Connection: Metastasis through Lymphatic Vessels

Spleen The oval **spleen** is the largest single mass of lymphatic tissue in the body. It is a soft, encapsulated organ of variable size, but

FIGURE 22.6 **Structure of a lymph node.** Green arrows indicate the direction of lymph flow through a lymph node.

Lymph nodes are present throughout the body, usually clustered in groups.

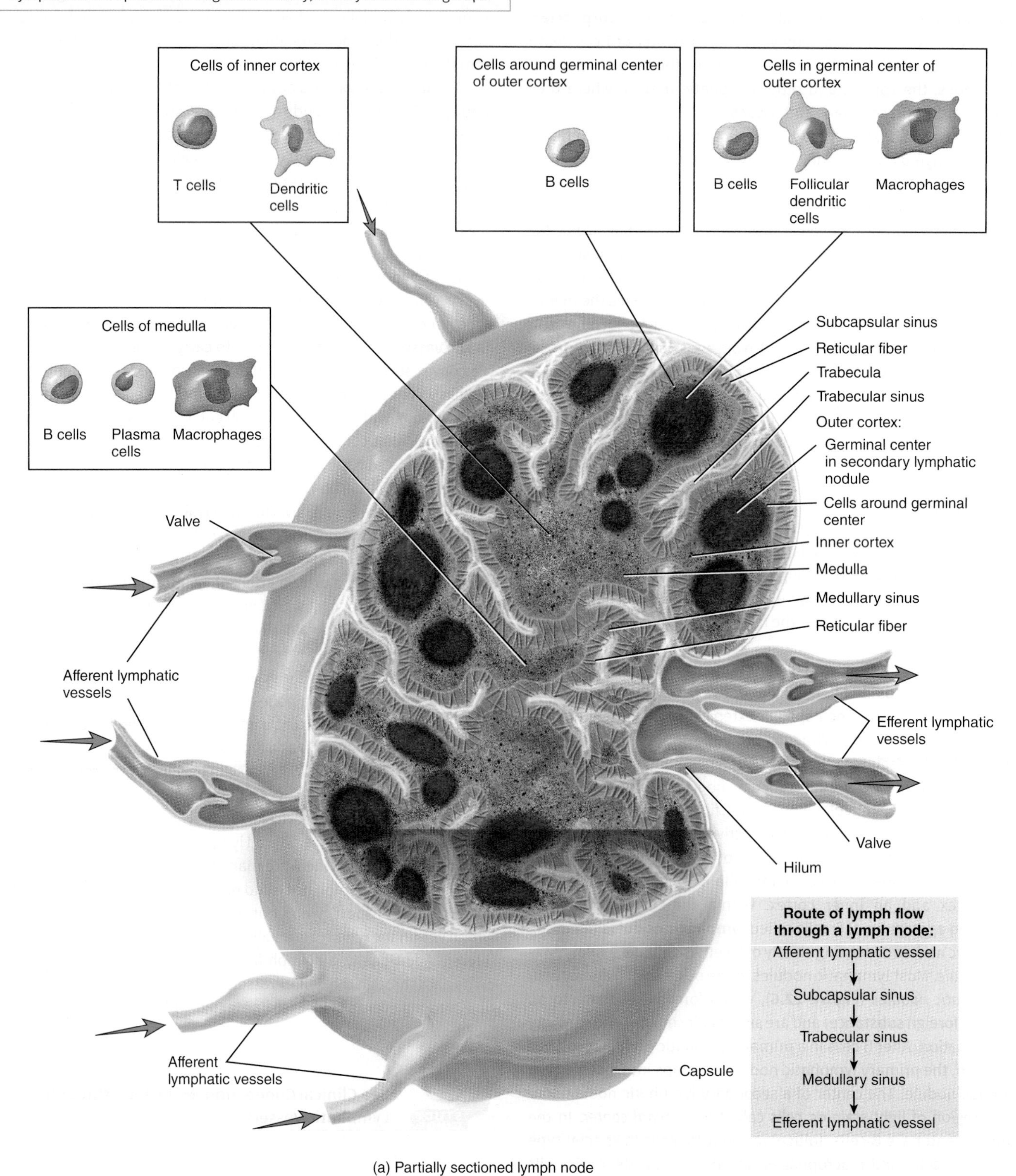

(a) Partially sectioned lymph node

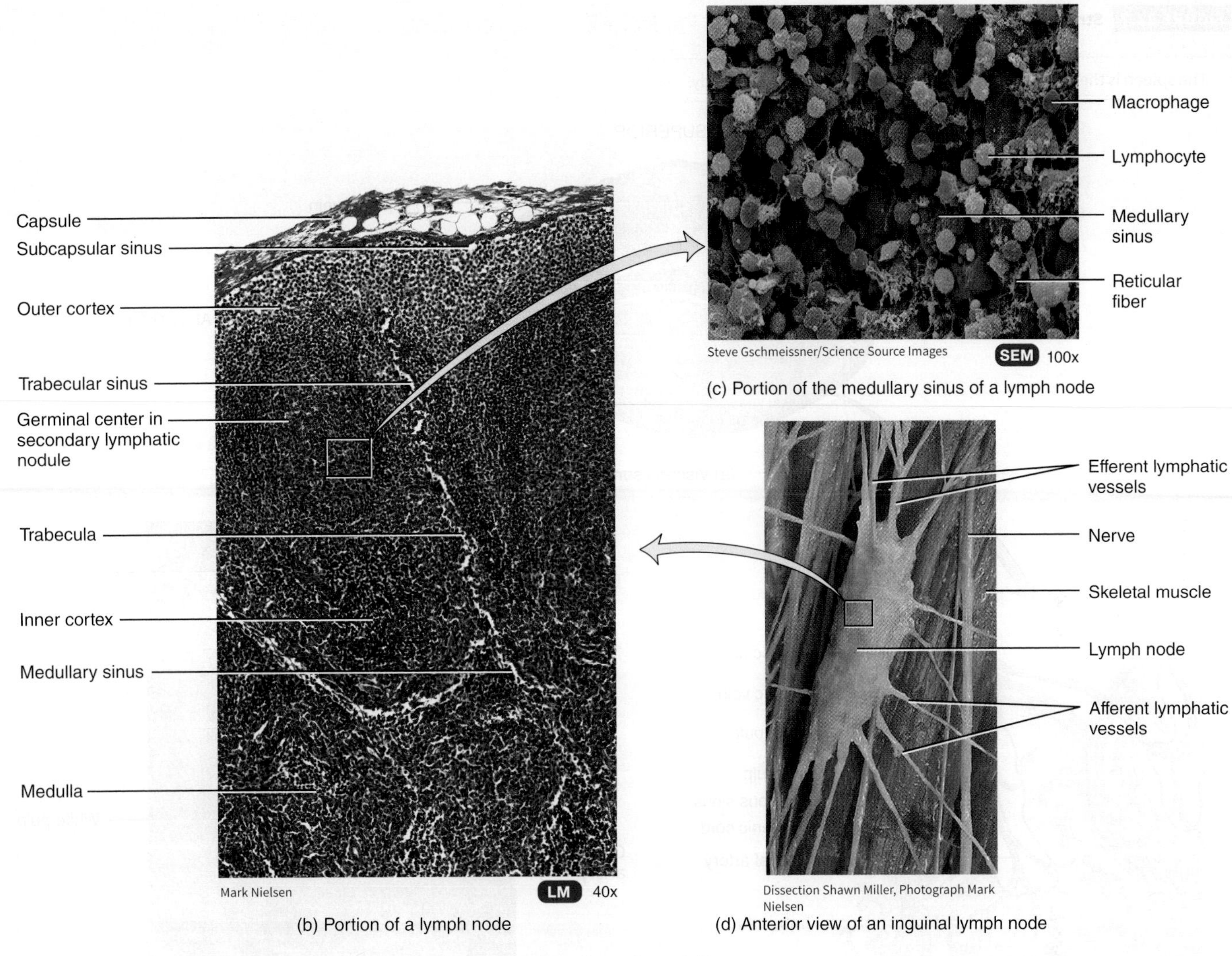

(b) Portion of a lymph node

(c) Portion of the medullary sinus of a lymph node

(d) Anterior view of an inguinal lymph node

Q What happens to foreign substances in lymph that enter a lymph node?

on average it fits in a person's open hand and measures about 12 cm (5 in.) in length (**Figure 22.7a**). It is located in the left hypochondriac region between the stomach and diaphragm. The superior surface of the spleen is smooth and convex and conforms to the concave surface of the diaphragm. Neighboring organs make indentations in the visceral surface of the spleen—the *gastric impression* (stomach), the *renal impression* (left kidney), and the *colic impression* (left colic flexure of large intestine). Like lymph nodes, the spleen has a hilum. Through it pass the splenic artery, splenic vein, and efferent lymphatic vessels.

A capsule of dense connective tissue surrounds the spleen and is covered in turn by a serous membrane, the visceral peritoneum. Trabeculae extend inward from the capsule. The capsule plus trabeculae, reticular fibers, and fibroblasts constitute the stroma of the spleen; the parenchyma of the spleen consists of two different kinds of tissue called white pulp and red pulp (**Figure 22.7b, c**). **White pulp** is lymphatic tissue, consisting mostly of lymphocytes and macrophages arranged around branches of the splenic artery called **central arteries**. The **red pulp** consists of blood-filled **venous sinuses** and cords of splenic tissue called **splenic cords** or *Billroth's cords*. Splenic cords consist of red blood cells, macrophages, lymphocytes, plasma cells, and granulocytes. Veins are closely associated with the red pulp.

Blood flowing into the spleen through the splenic artery enters the central arteries of the white pulp. Within the white pulp, B cells and T cells carry out immune functions, similar to lymph nodes, while spleen macrophages destroy blood-borne pathogens by phagocytosis. Within the red pulp, the spleen performs three functions related to blood cells: (1) removal by macrophages of ruptured, worn out, or defective blood cells and platelets; (2) storage of platelets, up to one-third of the body's supply; and (3) production of blood cells (hemopoiesis) during fetal life.

Lymphatic Nodules

Lymphatic nodules (*follicles*) are egg-shaped masses of lymphatic tissue that are not surrounded by a

FIGURE 22.7 Structure of the spleen.

The spleen is the largest single mass of lymphatic tissue in the body.

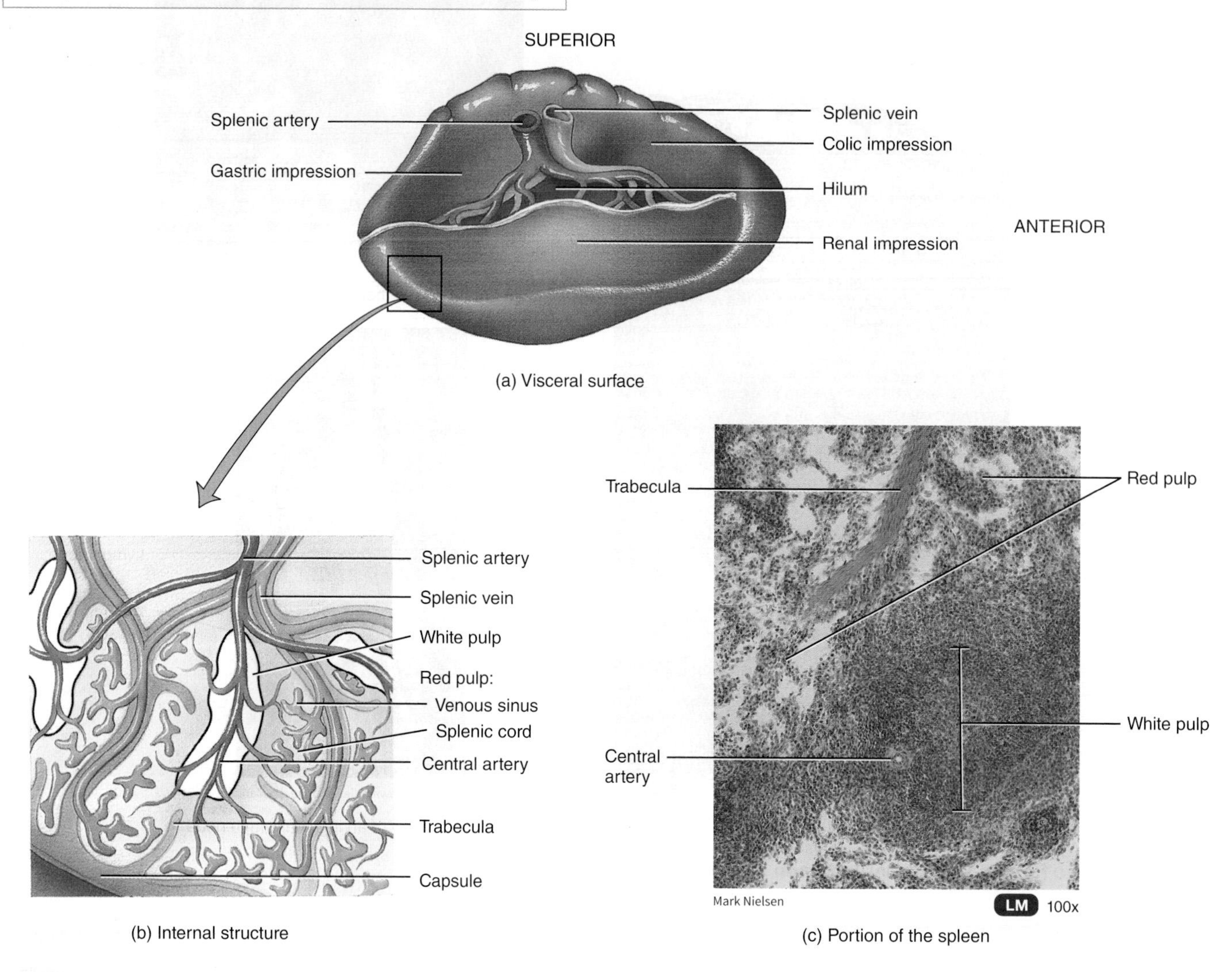

See Clinical Connection: Ruptured Spleen

Q After birth, what are the main functions of the spleen?

capsule. Because they are scattered throughout the lamina propria (connective tissue) of mucous membranes lining the gastrointestinal, urinary, and reproductive tracts and the respiratory airways, lymphatic nodules in these areas are also referred to as **mucosa-associated lymphatic tissue (MALT)**.

Although many lymphatic nodules are small and solitary, some occur in multiple large aggregations in specific parts of the body. Among these are the tonsils in the pharyngeal region and the aggregated lymphatic follicles (Peyer's patches) in the ileum of the small intestine. Aggregations of lymphatic nodules also occur in the appendix. Usually there are five **tonsils**, which form a ring at the junction of the oral cavity and oropharynx and at the junction of the nasal cavity and nasopharynx (see Figure 23.2b). The tonsils are strategically positioned to participate in immune responses against inhaled or ingested foreign substances. The single **pharyngeal tonsil** (fa-RIN-jē-al) or *adenoid* is embedded in the posterior wall of the nasopharynx. The two **palatine tonsils** (PAL-a-tīn) lie at the posterior region of the oral cavity, one on either side; these are the tonsils commonly removed in a tonsillectomy. The paired **lingual tonsils** (LIN-gwal), located at the base of the tongue, may also require removal during a tonsillectomy.

See Clinical Connection: Tonsillitis

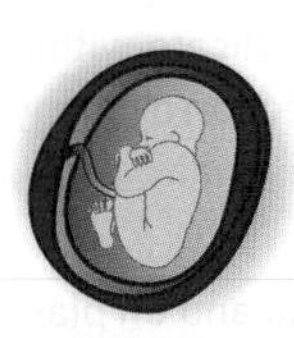

22.5 Development of Lymphatic Tissues

OBJECTIVE

• **Describe** the development of lymphatic tissues.

Lymphatic tissues begin to develop by the end of the fifth week of embryonic life. *Lymphatic vessels* develop from **lymph sacs** that arise from developing veins, which are derived from **mesoderm**.

The first lymph sacs to appear are the paired **jugular lymph sacs** at the junction of the internal jugular and subclavian veins (Figure 22.8). From the jugular lymph sacs, lymphatic capillary plexuses spread to the thorax, upper limbs, neck, and head. Some of the plexuses enlarge and form lymphatic vessels in their respective regions. Each jugular lymph sac retains at least one connection with its jugular vein, the left one developing into the superior portion of the thoracic duct (left lymphatic duct).

The next lymph sac to appear is the unpaired **retroperitoneal lymph sac** (re′-trō-per′-i-tō-NĒ-al) at the root of the mesentery of the intestine. It develops from the primitive vena cava and mesonephric (primitive kidney) veins. Capillary plexuses and lymphatic vessels spread from the retroperitoneal lymph sac to the abdominal viscera and diaphragm. The sac establishes connections with the cisterna chyli but loses its connections with neighboring veins.

At about the time the retroperitoneal lymph sac is developing, another lymph sac, the **cisterna chyli**, develops inferior to the diaphragm on the posterior abdominal wall. It gives rise to the inferior portion of the *thoracic duct* and the *cisterna chyli* of the thoracic duct. Like the retroperitoneal lymph sac, the cisterna chyli also loses its connections with surrounding veins.

The last of the lymph sacs, the paired **posterior lymph sacs**, develop from the iliac veins. The posterior lymph sacs produce capillary plexuses and lymphatic vessels of the abdominal wall, pelvic region, and lower limbs. The posterior lymph sacs join the cisterna chyli and lose their connections with adjacent veins.

With the exception of the anterior part of the sac from which the cisterna chyli develops, all lymph sacs become invaded by **mesenchymal cells** (me-SENG-kī-mal) and are converted into groups of *lymph nodes*.

The *spleen* develops from mesenchymal cells between layers of the dorsal mesentery of the stomach. The *thymus* arises as an outgrowth of the **third pharyngeal pouch** (see Figure 18.20a).

FIGURE 22.8 Development of lymphatic tissues.

Lymphatic tissues are derived from mesoderm.

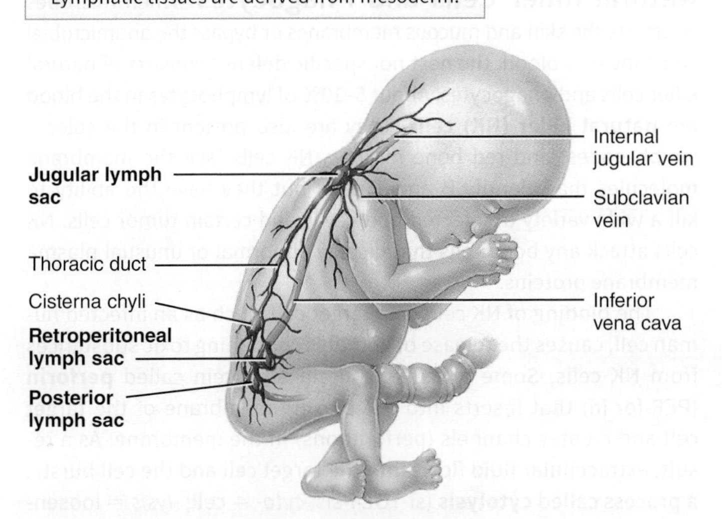

Q When do lymphatic tissues begin to develop?

22.6 Innate Immunity

OBJECTIVE

• **Describe** the components of innate immunity.

Innate (*nonspecific*) **immunity** includes the external physical and chemical barriers provided by the skin and mucous membranes. It also includes various internal defenses, such as antimicrobial substances, natural killer cells, phagocytes, inflammation, and fever.

First Line of Defense: Skin and Mucous Membranes

The skin and mucous membranes of the body are the first line of defense against pathogens. These structures provide both physical and chemical barriers that discourage pathogens and foreign substances from penetrating the body and causing disease.

With its many layers of closely packed, keratinized cells, the outer epithelial layer of the skin—the **epidermis**—provides a formidable physical barrier to the entrance of microbes (see Figure 5.1). In addition, periodic shedding of epidermal cells helps remove microbes at the skin surface. Bacteria rarely penetrate the intact surface of healthy epidermis. If this surface is broken by cuts, burns, or punctures, however, pathogens can penetrate the epidermis and invade adjacent tissues or circulate in the blood to other parts of the body.

The epithelial layer of **mucous membranes**, which line body cavities, secretes a fluid called **mucus** that lubricates and moistens the cavity surface. Because mucus is slightly viscous, it traps many microbes and foreign substances. The mucous membrane of the nose has mucus-coated **hairs** that trap and filter microbes, dust, and pollutants from inhaled air. The mucous membrane of the upper respiratory tract contains **cilia**, microscopic hairlike projections on the surface of the epithelial cells. The waving action of cilia propels inhaled dust and microbes that have become trapped in mucus toward the throat. Coughing and sneezing accelerate movement of mucus and its entrapped pathogens out of the body. Swallowing mucus sends pathogens to the stomach, where gastric juice destroys them.

Other fluids produced by various organs also help protect epithelial surfaces of the skin and mucous membranes. The **lacrimal apparatus** (LAK-ri-mal) of the eyes (see Figure 17.6) manufactures and drains away tears in response to irritants. Blinking spreads tears over the surface of the eyeball, and the continual washing action of tears helps to dilute microbes and keep them from settling on the surface of the eye. Tears also contain **lysozyme** (LĪ-sō-zīm), an enzyme capable of breaking down the cell walls of certain bacteria. Besides tears, lysozyme is present in saliva, perspiration, nasal secretions, and tissue fluids. **Saliva**, produced by the salivary glands, washes microbes from the surfaces of the teeth and from the mucous membrane of the mouth, much as tears wash the eyes. The flow of saliva reduces colonization of the mouth by microbes.

The cleansing of the urethra by the **flow of urine** retards microbial colonization of the urinary system. **Vaginal secretions** likewise move microbes out of the body in females. **Defecation** and **vomiting** also expel microbes. For example, in response to some microbial toxins, the smooth muscle of the lower gastrointestinal tract contracts vigorously; the resulting diarrhea rapidly expels many of the microbes.

Certain chemicals also contribute to the high degree of resistance of the skin and mucous membranes to microbial invasion. Sebaceous (oil) glands of the skin secrete an oily substance called **sebum** that forms a protective film over the surface of the skin. The unsaturated fatty acids in sebum inhibit the growth of certain pathogenic bacteria and fungi. The acidity of the skin (pH 3–5) is caused in part by the secretion of fatty acids and lactic acid. **Perspiration** helps flush microbes from the surface of the skin. **Gastric juice**, produced by the glands of the stomach, is a mixture of hydrochloric acid, enzymes, and mucus. The strong acidity of gastric juice (pH 1.2–3.0) destroys many bacteria and most bacterial toxins. Vaginal secretions also are slightly acidic, which discourages bacterial growth.

Second Line of Defense: Internal Defenses

When pathogens penetrate the physical and chemical barriers of the skin and mucous membranes, they encounter a second line of defense: internal antimicrobial substances, phagocytes, natural killer cells, inflammation, and fever.

Antimicrobial Substances

There are four main types of **antimicrobial substances** that discourage microbial growth: interferons, complement, iron-binding proteins, and antimicrobial proteins.

1. Lymphocytes, macrophages, and fibroblasts infected with viruses produce proteins called **interferons (IFNs)** (in′-ter-FĒR-ons). Once released by virus-infected cells, IFNs diffuse to uninfected neighboring cells, where they induce synthesis of antiviral proteins that interfere with viral replication. Although IFNs do not prevent viruses from attaching to and penetrating host cells, they do stop replication. Viruses can cause disease only if they can replicate within body cells. IFNs are an important defense against infection by many different viruses. The three types of interferons are alpha-, beta-, and gamma-IFN.
2. A group of normally inactive proteins in blood plasma and on plasma membranes makes up the **complement system**. When activated, these proteins "complement" or enhance certain immune reactions (see Section 22.9). The complement system causes cytolysis (bursting) of microbes, promotes phagocytosis, and contributes to inflammation.
3. **Iron-binding proteins** inhibit the growth of certain bacteria by reducing the amount of available iron. Examples include *transferrin* (found in blood and tissue fluids), *lactoferrin* (found in milk, saliva, and mucus), *ferritin* (found in the liver, spleen, and red bone marrow), and *hemoglobin* (found in red blood cells).
4. **Antimicrobial proteins (AMPs)** are short peptides that have a broad spectrum of antimicrobial activity. Examples of AMPs are *dermicidin* (der-ma-SĪ-din) (produced by sweat glands), *defensins* and *cathelicidins* (cath-el-i-SĪ-dins) (produced by neutrophils, macrophages, and epithelia), and *thrombocidin* (throm′-bō-SĪ-din) (produced by platelets). In addition to killing a wide range of microbes, AMPs can attract dendritic cells and mast cells, which participate in immune responses. Interestingly enough, microbes exposed to AMPs do not appear to develop resistance, as often happens with antibiotics.

Natural Killer Cells and Phagocytes

When microbes penetrate the skin and mucous membranes or bypass the antimicrobial substances in blood, the next nonspecific defense consists of natural killer cells and phagocytes. About 5–10% of lymphocytes in the blood are **natural killer (NK) cells**. They are also present in the spleen, lymph nodes, and red bone marrow. NK cells lack the membrane molecules that identify B and T cells, but they have the ability to kill a wide variety of infected body cells and certain tumor cells. NK cells attack any body cells that display abnormal or unusual plasma membrane proteins.

The binding of NK cells to a target cell, such as an infected human cell, causes the release of granules containing toxic substances from NK cells. Some granules contain a protein called **perforin** (PER-for-in) that inserts into the plasma membrane of the target cell and creates channels (perforations) in the membrane. As a result, extracellular fluid flows into the target cell and the cell bursts, a process called **cytolysis** (sī-TOL-i-sis; *cyto-* = cell; *-lysis* = loosening). Other granules of NK cells release **granzymes** (GRAN-zīms), which are protein-digesting enzymes that induce the target cell to undergo apoptosis, or self-destruction. This type of attack kills infected cells, but not the microbes inside the cells; the released microbes, which may or may not be intact, can be destroyed by phagocytes.

Phagocytes (FAG-ō-sīts; *phago-* = eat; *-cytes* = cells) are specialized cells that perform **phagocytosis** (fag-ō-sī-TŌ-sis; *-osis* = process), the ingestion of microbes or other particles such as cellular debris (see Figure 3.13). The two major types of phagocytes are **neutrophils** and **macrophages**. When an infection occurs, neutrophils and monocytes migrate to the infected area. During this migration, the monocytes enlarge and develop into actively phagocytic macrophages called **wandering macrophages**. Other macrophages, called **fixed macrophages**, stand guard in specific tissues. Among the fixed macrophages are *histiocytes* (HIS-tē-ō-sīts) (connective tissue macrophages), *stellate reticuloendothelial cells* (STEL-āt re-tik′-ū-lō-en-dō-THĒ-lē-al) or *Kupffer cells* (KOOP-fer) in the liver, *alveolar macrophages* in the lungs, *microglial cells* in the nervous system, and *tissue macrophages* in the spleen, lymph nodes, and red bone marrow. In addition to being an innate defense mechanism, phagocytosis plays a vital role in adaptive immunity, as discussed later in the chapter.

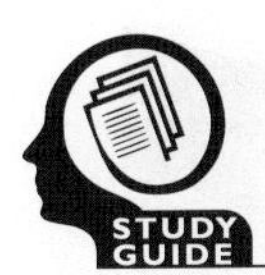

See Clinical Connection: Microbial Evasion of Phagocytosis

Phagocytosis occurs in five phases: chemotaxis, adherence, ingestion, digestion, and killing (Figure 22.9):

1. ***Chemotaxis.*** Phagocytosis begins with **chemotaxis** (kē-mō-TAK-sis), a chemically stimulated movement of phagocytes to a site of damage. Chemicals that attract phagocytes might come from invading microbes, white blood cells, damaged tissue cells, or activated complement proteins.
2. ***Adherence.*** Attachment of the phagocyte to the microbe or other foreign material is termed **adherence** (ad-HER-ents). The binding of complement proteins to the invading pathogen enhances adherence.
3. ***Ingestion.*** The plasma membrane of the phagocyte extends projections, called **pseudopods** (SOO-dō-pods), that engulf the microbe in a process called **ingestion**. When the pseudopods meet they fuse, surrounding the microorganism with a sac called a **phagosome** (FAG-ō-sōm).
4. ***Digestion.*** The phagosome enters the cytoplasm and merges with lysosomes to form a single, larger structure called a **phagolysosome** (fag-ō-LĪ-sō-sōm). The lysosome contributes lysozyme, which breaks down microbial cell walls, and other digestive enzymes that degrade carbohydrates, proteins, lipids, and nucleic acids. The phagocyte also forms lethal oxidants, such as superoxide anion (O_2^-), hypochlorite anion (OCl^-), and hydrogen peroxide (H_2O_2), in a process called an **oxidative burst**.

FIGURE 22.9 Phagocytosis of a microbe.

The major types of phagocytes are neutrophils and macrophages.

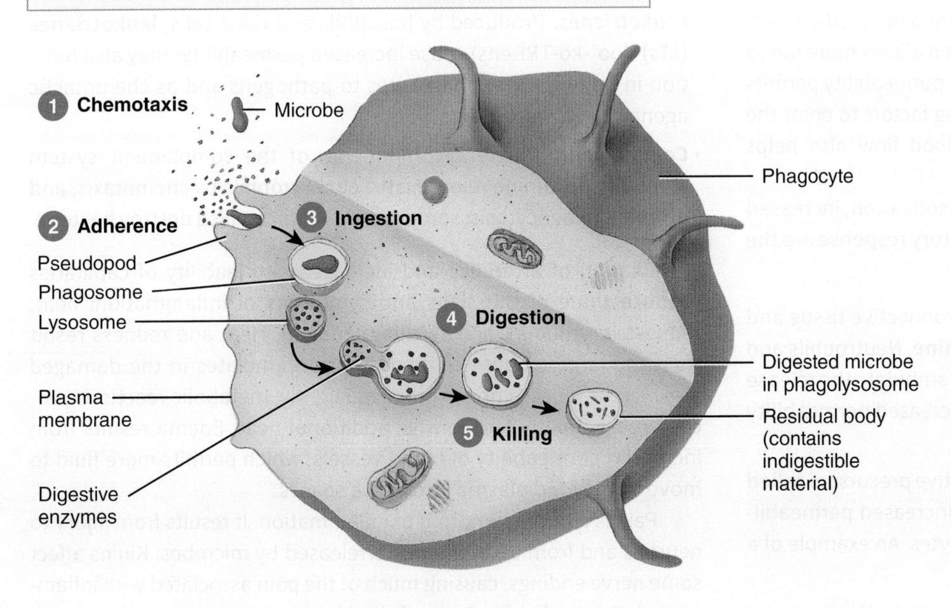

(a) Phases of phagocytosis

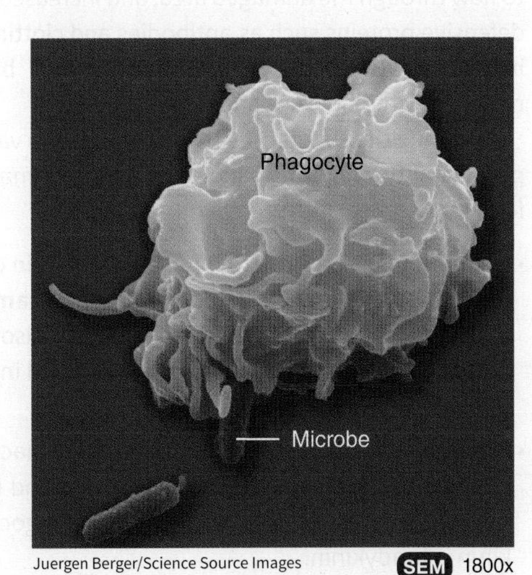

(b) Phagocyte (white blood cell) engulfing microbe.

Q What chemicals are responsible for killing ingested microbes?

5 ***Killing.*** The chemical onslaught provided by lysozyme, digestive enzymes, and oxidants within a phagolysosome quickly kills many types of microbes. Any materials that cannot be degraded further remain in structures called **residual bodies**.

Inflammation

Inflammation is a nonspecific, defensive response of the body to tissue damage. Among the conditions that may produce inflammation are pathogens, abrasions, chemical irritations, distortion or disturbances of cells, and extreme temperatures. Inflammation is an attempt to dispose of microbes, toxins, or foreign material at the site of injury, to prevent their spread to other tissues, and to prepare the site for tissue repair in an attempt to restore tissue homeostasis. There are certain signs-symptoms associated with inflammation and these can be recalled by using the following acronym: **PRISH.**

P is for pain due to the release of certain chemicals.
R is for redness because more blood is rushed to the affected area.
I is for immobility that results from some loss of function in severe inflammations.
S is for swelling caused by an accumulation of fluids.
H is for heat which is also due to more blood rushed to the affected area.

Because inflammation is one of the body's nonspecific defense mechanisms, the response of a tissue to a cut is similar to the response to damage caused by burns, radiation, or bacterial or viral invasion. In each case, the inflammatory response has three basic stages: (1) vasodilation and increased permeability of blood vessels, (2) emigration (movement) of phagocytes from the blood into interstitial fluid, and, ultimately, (3) tissue repair.

VASODILATION AND INCREASED BLOOD VESSEL PERMEABILITY Two immediate changes occur in the blood vessels in a region of tissue injury: **vasodilation** (increase in the diameter) of arterioles and increased permeability of capillaries (Figure 22.10). Increased permeability means that substances normally retained in blood are permitted to pass from the blood vessels. Vasodilation allows more blood to flow through the damaged area, and increased permeability permits defensive proteins such as antibodies and clotting factors to enter the injured area from the blood. The increased blood flow also helps remove microbial toxins and dead cells.

Among the substances that contribute to vasodilation, increased permeability, and other aspects of the inflammatory response are the following:

- ***Histamine.*** In response to injury, mast cells in connective tissue and basophils and platelets in blood release **histamine**. Neutrophils and macrophages attracted to the site of injury also stimulate the release of histamine, which causes vasodilation and increased permeability of blood vessels.
- ***Kinins.*** Polypeptides formed in blood from inactive precursors called kininogens, (**kinins**), induce vasodilation and increased permeability and serve as chemotactic agents for phagocytes. An example of a kinin is bradykinin.
- ***Prostaglandins***. **Prostaglandins (PGs)** (pros′-ta-GLAN-dins), especially those of the E series, are released by damaged cells and intensify the effects of histamine and kinins. PGs also may stimulate the emigration of phagocytes through capillary walls.

FIGURE 22.10 Inflammation.

The three stages of inflammation are as follows: (1) vasodilation and increased permeability of blood vessels, (2) phagocyte emigration, and (3) tissue repair.

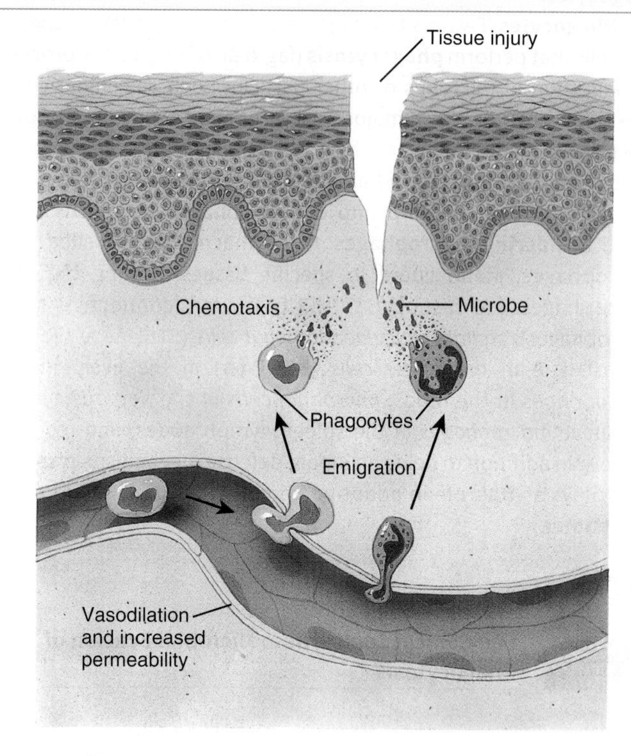

Phagocytes migrate from blood to site of tissue injury

Q What causes each of the following signs and symptoms of inflammation: redness, pain, heat, and swelling?

- ***Leukotrienes.*** Produced by basophils and mast cells, **leukotrienes (LTs)** (loo′-kō-TRĪ-ēns) cause increased permeability; they also function in adherence of phagocytes to pathogens and as chemotactic agents that attract phagocytes.
- ***Complement.*** Different components of the complement system stimulate histamine release, attract neutrophils by chemotaxis, and promote phagocytosis; some components can also destroy bacteria.

Dilation of arterioles and increased permeability of capillaries produce three of the signs and symptoms of inflammation: heat, redness (erythema), and swelling (edema). Heat and redness result from the large amount of blood that accumulates in the damaged area. As the local temperature rises slightly, metabolic reactions proceed more rapidly and release additional heat. Edema results from increased permeability of blood vessels, which permits more fluid to move from blood plasma into tissue spaces.

Pain is a prime symptom of inflammation. It results from injury to neurons and from toxic chemicals released by microbes. Kinins affect some nerve endings, causing much of the pain associated with inflammation. Prostaglandins intensify and prolong the pain associated with inflammation. Pain may also be due to increased pressure from edema.

The increased permeability of capillaries allows leakage of blood-clotting factors into tissues. The clotting sequence is set into motion, and fibrinogen is ultimately converted to an insoluble, thick mesh of fibrin threads that localizes and traps invading microbes and blocks their spread.

Emigration of Phagocytes Within an hour after the inflammatory process starts, phagocytes appear on the scene. As large amounts of blood accumulate, neutrophils begin to stick to the inner surface of the endothelium (lining) of blood vessels (**Figure 22.10**). Then the neutrophils begin to squeeze through the wall of the blood vessel to reach the damaged area. This process, called **emigration** (em′-i-GRĀ-shun), depends on chemotaxis. Neutrophils attempt to destroy the invading microbes by phagocytosis. A steady stream of neutrophils is ensured by the production and release of additional cells from red bone marrow. Such an increase in white blood cells in the blood is termed **leukocytosis** (loo-kō-sī-TŌ-sis).

Although neutrophils predominate in the early stages of infection, they die off rapidly. As the inflammatory response continues, monocytes follow the neutrophils into the infected area. Once in the tissue, monocytes transform into wandering macrophages that add to the phagocytic activity of the fixed macrophages already present. True to their name, macrophages are much more potent phagocytes than neutrophils. They are large enough to engulf damaged tissue, worn-out neutrophils, and invading microbes.

Eventually, macrophages also die. Within a few days, a pocket of dead phagocytes and damaged tissue forms; this collection of dead cells and fluid is called **pus**. Pus formation occurs in most inflammatory responses and usually continues until the infection subsides. At times, pus reaches the surface of the body or drains into an internal cavity and is dispersed; on other occasions the pus remains even after the infection is terminated. In this case, the pus is gradually destroyed over a period of days and is absorbed.

See Clinical Connection: Abscesses and Ulcers

Inflammation can be classified as acute or chronic depending on a number of factors. In **acute inflammation** the signs and symptoms develop rapidly and usually last for a few days or even a few weeks. It is usually mild and self-limiting and the principal defensive cells are neutrophils. Examples of acute inflammation are a sore throat, appendicitis, cold or flu, bacterial pneumonia, and a scratch on the skin. In **chronic inflammation** the signs and symptoms develop more slowly and can last for up to several months or years. It is often severe and progressive and the principal defensive cells are monocytes and macrophages. Examples of chronic inflammation are mononucleosis, peptic ulcers, tuberculosis, rheumatoid arthritis, and ulcerative colitis.

Fever

Fever is an abnormally high body temperature that occurs because the hypothalamic thermostat is reset. It commonly occurs during infection and inflammation. Many bacterial toxins elevate body temperature, sometimes by triggering release of fever-causing cytokines such as interleukin-1 from macrophages. Elevated body temperature intensifies the effects of interferons, inhibits the growth of some microbes, and speeds up body reactions that aid repair.

Table 22.1 summarizes the components of innate immunity.

TABLE 22.1 Summary of Innate Defenses

COMPONENT	FUNCTIONS
FIRST LINE OF DEFENSE: SKIN AND MUCOUS MEMBRANES	
Physical Factors	
Epidermis of skin	Forms physical barrier to entrance of microbes.
Mucous membranes	Inhibit entrance of many microbes, but not as effective as intact skin.
Mucus	Traps microbes in respiratory and gastrointestinal tracts.
Hairs	Filter out microbes and dust in nose.
Cilia	Together with mucus, trap and remove microbes and dust from upper respiratory tract.
Lacrimal apparatus	Tears dilute and wash away irritating substances and microbes.
Saliva	Washes microbes from surfaces of teeth and mucous membranes of mouth.
Urine	Washes microbes from urethra.
Defecation and vomiting	Expel microbes from body.
Chemical Factors	
Sebum	Forms protective acidic film over skin surface that inhibits growth of many microbes.
Lysozyme	Antimicrobial substance in perspiration, tears, saliva, nasal secretions, and tissue fluids.
Gastric juice	Destroys bacteria and most toxins in stomach.
Vaginal secretions	Slight acidity discourages bacterial growth; flush microbes out of vagina.
SECOND LINE OF DEFENSE: INTERNAL DEFENSES	
Antimicrobial Substances	
Interferons (IFNs)	Protect uninfected host cells from viral infection.
Complement system	Causes cytolysis of microbes; promotes phagocytosis; contributes to inflammation.
Iron-binding proteins	Inhibit growth of certain bacteria by reducing amount of available iron.
Antimicrobial proteins (AMPs)	Have broad-spectrum antimicrobial activities and attract dendritic cells and mast cells.
Natural killer (NK) cells	Kill infected target cells by releasing granules that contain perforin and granzymes; phagocytes then kill released microbes.
Phagocytes	Ingest foreign particulate matter.
Inflammation	Confines and destroys microbes; initiates tissue repair.
Fever	Intensifies effects of interferons; inhibits growth of some microbes; speeds up body reactions that aid repair.

22.7 Adaptive Immunity

OBJECTIVES

- **Describe** how T cells and B cells arise and function in adaptive immunity.
- **Explain** the relationship between an antigen and an antibody.
- **Compare** the functions of cell-mediated immunity and antibody-mediated immunity.

The ability of the body to defend itself against specific invading agents such as bacteria, toxins, viruses, and foreign tissues is called **adaptive** (*specific*) **immunity**. Substances that are recognized as foreign and provoke immune responses are called **antigens (Ags)** (AN-ti-jens), meaning *anti*body *gen*erators. Two properties distinguish adaptive immunity from innate immunity: (1) *specificity* for particular foreign molecules (antigens), which also involves distinguishing self from nonself molecules, and (2) *memory* for most previously encountered antigens so that a second encounter prompts an even more rapid and vigorous response. The branch of science that deals with the responses of the body when challenged by antigens is called **immunology** (im′-ū-NOL-ō-jē; *immuno-* = free from service or exempt; *-logy* = study of). The **immune system** includes the cells and tissues that carry out immune responses.

Maturation of T Cells and B Cells

Adaptive immunity involves lymphocytes called **B cells** and **T cells**. Both develop in primary lymphatic organs (red bone marrow and the thymus) from pluripotent stem cells that originate in red bone marrow (see **Figure 19.3**). B cells complete their development in red bone marrow, a process that continues throughout life. T cells develop from pre-T cells that migrate from red bone marrow into the thymus, where they mature (**Figure 22.11**). Most T cells arise before puberty, but they continue to mature and leave the thymus throughout life. B cells and T cells are named based on where they mature. In birds, B cells mature in an organ called the *bursa of Fabricius*. Although this organ is not present in humans, the term *B cell* is still used, but the letter *B* stands for *bursa equivalent*, which is the red bone marrow since that is the location in humans where B cells mature. T cells are so named because they mature in the *thymus* gland.

Before T cells leave the thymus or B cells leave red bone marrow, they develop **immunocompetence** (im′-ū-nō-KOM-pe-tens), the ability to carry out adaptive immune responses. This means that B cells and T cells begin to make several distinctive proteins that are inserted into their plasma membranes. Some of these proteins function as **antigen receptors**—molecules capable of recognizing specific antigens (**Figure 22.11**).

There are two major types of mature T cells that exit the thymus: **helper T cells** and **cytotoxic T cells** (sī-tō-TOK-sik) (**Figure 22.11**). Helper T cells are also known as **CD4 T cells**, which means that, in addition to antigen receptors, their plasma membranes include a protein called CD4. Cytotoxic T cells are also referred to as **CD8 T cells** because their plasma membranes contain not only antigen receptors but also a protein known as CD8. As we will see later in this chapter, these two types of T cells have very different functions.

Types of Adaptive Immunity

There are two types of adaptive immunity: cell-mediated immunity and antibody-mediated immunity. Both types of adaptive immunity are triggered by antigens. In **cell-mediated immunity**, cytotoxic T cells directly attack invading antigens. In **antibody-mediated immunity**, B cells transform into plasma cells, which synthesize and secrete specific proteins called **antibodies (Abs)** or *immunoglobulins* (*Igs*) (im′-ū-nō-GLOB-ū-lins). A given antibody can bind to and inactivate a specific antigen. Helper T cells aid the immune responses of both cell-mediated and antibody-mediated immunity.

Cell-mediated immunity is particularly effective against (1) intracellular pathogens, which include any viruses, bacteria, or fungi that are inside cells; (2) some cancer cells; and (3) foreign tissue transplants. Thus, cell-mediated immunity always involves cells attacking cells. Antibody-mediated immunity works mainly against extracellular pathogens, which include any viruses, bacteria, or fungi that are in body fluids outside cells. Since antibody-mediated immunity involves antibodies that bind to antigens in body *humors* or fluids (such as blood and lymph), it is also referred to as *humoral immunity*.

In most cases, when a particular antigen initially enters the body, there is only a small group of lymphocytes with the correct antigen receptors to respond to that antigen; this small group of cells includes a few helper T cells, cytotoxic T cells, and B cells. Depending on its location, a given antigen can provoke both types of adaptive immune responses. This is due to the fact that when a specific antigen invades the body, there are usually many copies of that antigen spread throughout the body's tissues and fluids. Some copies of the antigen may be present inside body cells (which provokes a cell-mediated immune response by cytotoxic T cells), while other copies of the antigen may be present in extracellular fluid (which provokes an antibody-mediated immune response by B cells). Thus, cell-mediated and antibody-mediated immune responses often work together to eliminate the large number of copies of a particular antigen from the body.

Clonal Selection: The Principle

As you just learned, when a specific antigen is present in the body, there are usually many copies of that antigen located throughout the body's tissues and fluids. The numerous copies of the antigen initially outnumber the small group of helper T cells, cytotoxic T cells, and B cells with the correct antigen receptors to respond to that antigen. Therefore, once each of these lymphocytes encounters a copy of the antigen and receives stimulatory cues, it subsequently undergoes clonal selection. **Clonal selection** is the process by which a lymphocyte *proliferates* (divides) and *differentiates* (forms more highly specialized cells) in response to a specific antigen. The result of clonal selection is

FIGURE 22.11 B cells and pre-T cells arise from pluripotent stem cells in red bone marrow. B cells and T cells develop in primary lymphatic tissues (red bone marrow and the thymus) and are activated in secondary lymphatic organs and tissues (lymph nodes, spleen, and lymphatic nodules). Once activated, each type of lymphocyte forms a clone of cells that can recognize a specific antigen. For simplicity, antigen receptors, CD4 proteins, and CD8 proteins are not shown in the plasma membranes of the cells of the lymphocyte clones.

> The two types of adaptive immunity are cell-mediated immunity and antibody-mediated immunity.

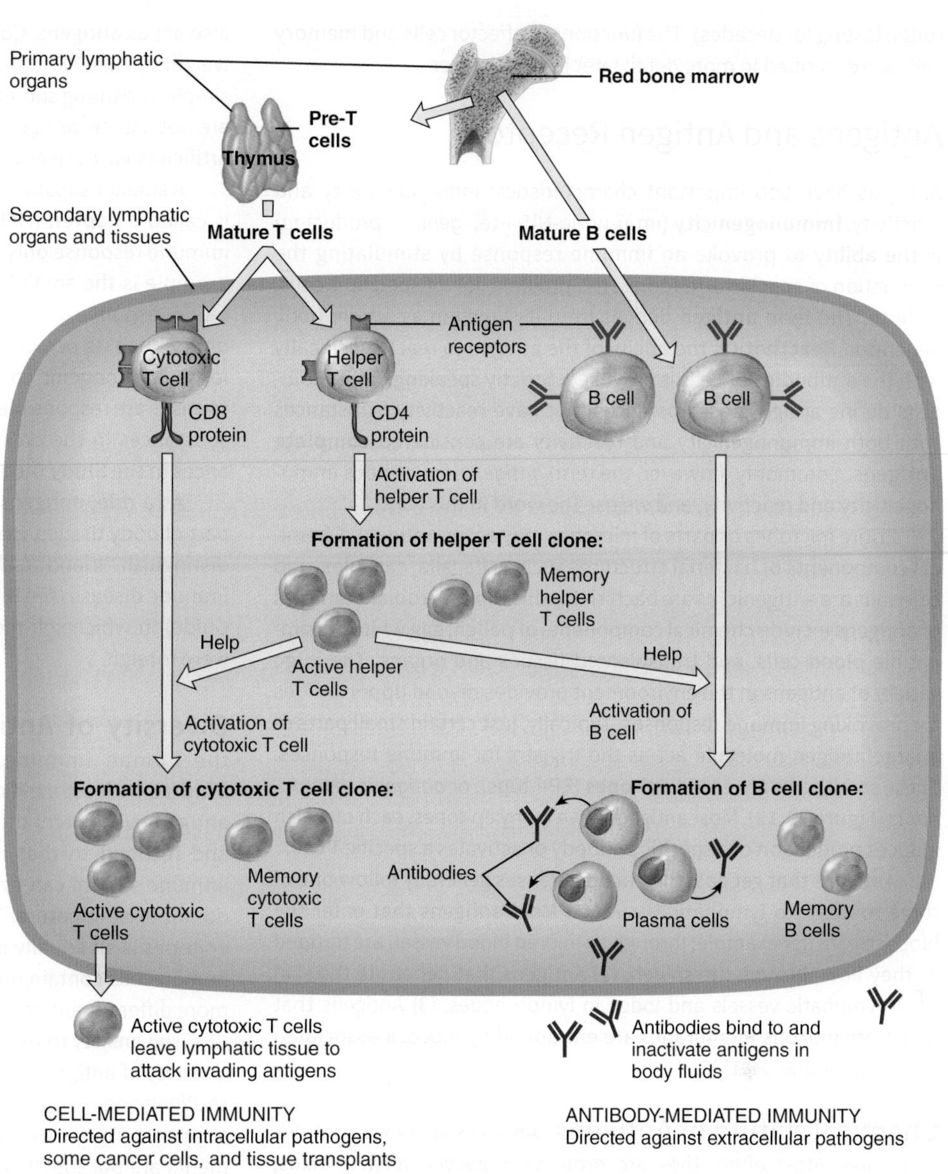

Q Which type of T cell participates in both cell-mediated and antibody-mediated immune responses?

the formation of a population of identical cells, called a **clone**, that can recognize the same specific antigen as the original lymphocyte (**Figure 22.11**). Before the first exposure to a given antigen, only a few lymphocytes are able to recognize it, but once clonal selection occurs, there are thousands of lymphocytes that can respond to that antigen. Clonal selection of lymphocytes occurs in the secondary lymphatic organs and tissues. The swollen tonsils or lymph nodes in your neck you experienced the last time you were sick were probably caused by clonal selection of lymphocytes participating in an immune response.

A lymphocyte that undergoes clonal selection gives rise to two major types of cells in the clone: effector cells and memory cells. The thousands of **effector cells** of a lymphocyte clone carry out immune responses that ultimately result in the destruction or inactivation of the antigen. Effector cells include **active helper T cells**, which are part of a helper T cell clone; **active cytotoxic T cells**, which are part of a cytotoxic T cell clone; and **plasma cells**, which are part of a B cell clone. Most effector cells eventually die after the immune response has been completed.

Memory cells do not actively participate in the initial immune response to the antigen. However, if the same antigen enters the body again in the future, the thousands of memory cells of a lymphocyte clone are available to initiate a far swifter reaction than occurred during the first invasion. The memory cells respond to the antigen by proliferating and differentiating into more effector cells and more memory cells. Consequently, the second response to the antigen is usually so fast and so vigorous that the antigen is destroyed before any signs or symptoms of disease can occur. Memory cells include **memory helper T cells**, which are part of a helper T cell clone; **memory cytotoxic T cells**, which are part of a cytotoxic T cell clone; and **memory B cells**, which are part of a B cell clone. Most memory cells do not die at the end of an immune response. Instead, they have long life spans (often lasting for decades). The functions of effector cells and memory cells are described in more detail later in this chapter.

Antigens and Antigen Receptors

Antigens have two important characteristics: immunogenicity and reactivity. **Immunogenicity** (im-ū-nō-je-NIS-i-tē; *-genic* = producing) is the ability to provoke an immune response by stimulating the production of specific antibodies, the proliferation of specific T cells, or both. The term **antigen** derives from its function as an *anti*body *gen*erator. **Reactivity** is the ability of the antigen to react specifically with the antibodies or cells it provoked. Strictly speaking, immunologists define antigens as substances that have reactivity; substances with both immunogenicity and reactivity are considered **complete antigens**. Commonly, however, the term *antigen* implies both immunogenicity and reactivity, and we use the word in this way.

Entire microbes or parts of microbes may act as antigens. Chemical components of bacterial structures such as flagella, capsules, and cell walls are antigenic, as are bacterial toxins. Nonmicrobial examples of antigens include chemical components of pollen, egg white, incompatible blood cells, and transplanted tissues and organs. The huge variety of antigens in the environment provides myriad opportunities for provoking immune responses. Typically, just certain small parts of a large antigen molecule act as the triggers for immune responses. These small parts are called **epitopes** (EP-i-tōps), or *antigenic determinants* (**Figure 22.12**). Most antigens have many epitopes, each of which induces production of a specific antibody or activates a specific T cell.

Antigens that get past the innate defenses generally follow one of three routes into lymphatic tissue: (1) Most antigens that enter the bloodstream (for example, through an injured blood vessel) are trapped as they flow through the spleen. (2) Antigens that penetrate the skin enter lymphatic vessels and lodge in lymph nodes. (3) Antigens that penetrate mucous membranes are entrapped by mucosa-associated lymphatic tissue (MALT).

Chemical Nature of Antigens Antigens are large, complex molecules. Most often, they are proteins. However, nucleic acids, lipoproteins, glycoproteins, and certain large polysaccharides may also act as antigens. Complete antigens usually have large molecular weights of 10,000 daltons or more, but large molecules that have simple, repeating subunits—for example, cellulose and most plastics—are not usually antigenic. This is why plastic materials can be used in artificial heart valves or joints.

A smaller substance that has reactivity but lacks immunogenicity is called a **hapten** (HAP-ten = to grasp). A hapten can stimulate an immune response only if it is attached to a larger carrier molecule. An example is the small lipid toxin in poison ivy, which triggers an immune response after combining with a body protein. Likewise, some drugs, such as penicillin, may combine with proteins in the body to form immunogenic complexes. Such hapten-stimulated immune responses are responsible for some allergic reactions to drugs and other substances in the environment (see Disorders: Homeostatic Imbalances in the Study Guide).

As a rule, antigens are foreign substances; they are not usually part of body tissues. However, sometimes the immune system fails to distinguish "friend" (self) from "foe" (nonself). The result is an autoimmune disease (see Disorders: Homeostatic Imbalances in the Study Guide) in which self-molecules or cells are attacked as though they were foreign.

FIGURE 22.12 Epitopes (antigenic determinants).

Most antigens have several epitopes that induce the production of different antibodies or activate different T cells.

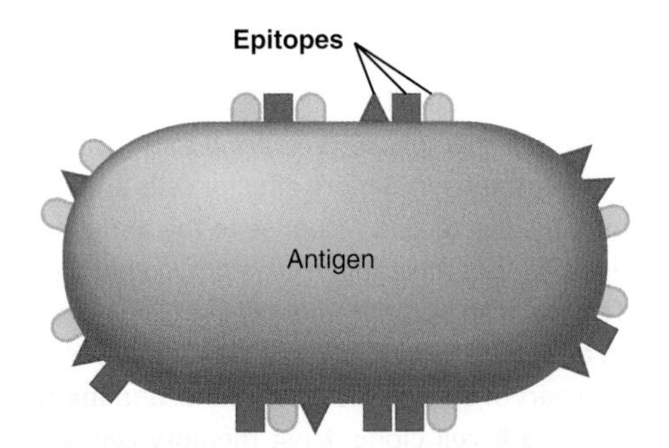

Q What is the difference between an epitope and a hapten?

Diversity of Antigen Receptors An amazing feature of the human immune system is its ability to recognize and bind to at least a billion (10^9) different epitopes. Before a particular antigen ever enters the body, T cells and B cells that can recognize and respond to that intruder are ready and waiting. Cells of the immune system can even recognize artificially made molecules that do not exist in nature. The basis for the ability to recognize so many epitopes is an equally large diversity of antigen receptors. Given that human cells contain only about 35,000 genes, how could a billion or more different antigen receptors possibly be generated?

The answer to this puzzle turned out to be simple in concept. The diversity of antigen receptors in both B cells and T cells is the result of shuffling and rearranging a few hundred versions of several small gene segments. This process is called **genetic recombination**. The gene segments are put together in different combinations as the lymphocytes are developing from stem cells in red bone marrow and the thymus. The situation is similar to shuffling a deck of 52 cards and then dealing out three cards. If you did this over and over, you could generate many more than 52 different sets of three cards. Because of genetic recombination, each B cell or T cell has a unique set of gene segments that codes for its unique antigen receptor. After transcription and translation, the receptor molecules are inserted into the plasma membrane.

Major Histocompatibility Complex Antigens

Located in the plasma membrane of body cells are "self-antigens," the **major histocompatibility complex (MHC) antigens** (his′-tō-kom-pat′-i-BIL-i-tē). These transmembrane glycoproteins are also called *human leukocyte antigens (HLA)* because they were first identified on white blood cells. Unless you have an identical twin, your MHC antigens are unique. Thousands to several hundred thousand MHC molecules mark the surface of each of your body cells except red blood cells. Although MHC antigens are the reason that tissues may be rejected when they are transplanted from one person to another, their normal function is to help T cells recognize that an antigen is foreign, not self. Such recognition is an important first step in any adaptive immune response.

The two types of major histocompatibility complex antigens are class I and class II. Class I MHC (MHC-I) molecules are built into the plasma membranes of all body cells except red blood cells. Class II MHC (MHC-II) molecules appear on the surface of antigen-presenting cells (described in the next section).

Pathways of Antigen Processing

For an immune response to occur, B cells and T cells must recognize that a foreign antigen is present. B cells can recognize and bind to antigens in lymph, interstitial fluid, or blood plasma. T cells only recognize fragments of antigenic proteins that are processed and presented in a certain way. In **antigen processing**, antigenic proteins are broken down into peptide fragments that then associate with MHC molecules. Next the antigen–MHC complex is inserted into the plasma membrane of a body cell. The insertion of the complex into the plasma membrane is called **antigen presentation**. When a peptide fragment comes from a *self-protein*, T cells ignore the antigen–MHC complex. However, if the peptide fragment comes from a *foreign protein*, T cells recognize the antigen–MHC complex as an intruder, and an immune response takes place. Antigen processing and presentation occur in two ways, depending on whether the antigen is located outside or inside body cells.

Processing of Exogenous Antigens Foreign antigens that are present in fluids *outside* body cells are termed **exogenous antigens** (ex-OG-e-nus). They include intruders such as bacteria and bacterial toxins, parasitic worms, inhaled pollen and dust, and viruses that have not yet infected a body cell. A special class of cells called **antigen-presenting cells (APCs)** process and present exogenous antigens. APCs include dendritic cells, macrophages, and B cells. They are strategically located in places where antigens are likely to penetrate the innate defenses and enter the body, such as the epidermis and dermis of the skin (intraepidermal macrophages are a type of dendritic cell); mucous membranes that line the respiratory, gastrointestinal, urinary, and reproductive tracts; and lymph nodes. After processing and presenting an antigen, APCs migrate from tissues via lymphatic vessels to lymph nodes.

The steps in the processing and presenting of an exogenous antigen by an antigen-presenting cell occur as follows (**Figure 22.13**):

1. ***Ingestion of the antigen.*** Antigen-presenting cells ingest exogenous antigens by phagocytosis or endocytosis. Ingestion could occur almost anywhere in the body that invaders, such as microbes, have penetrated the innate defenses.
2. ***Digestion of antigen into peptide fragments.*** Within the phagosome or endosome, protein-digesting enzymes split large antigens into short peptide fragments.

FIGURE 22.13 Processing and presenting of exogenous antigen by an antigen-presenting cell (APC).

Fragments of exogenous antigens are processed and then presented with MHC-II molecules on the surface of an antigen-presenting cell (APC).

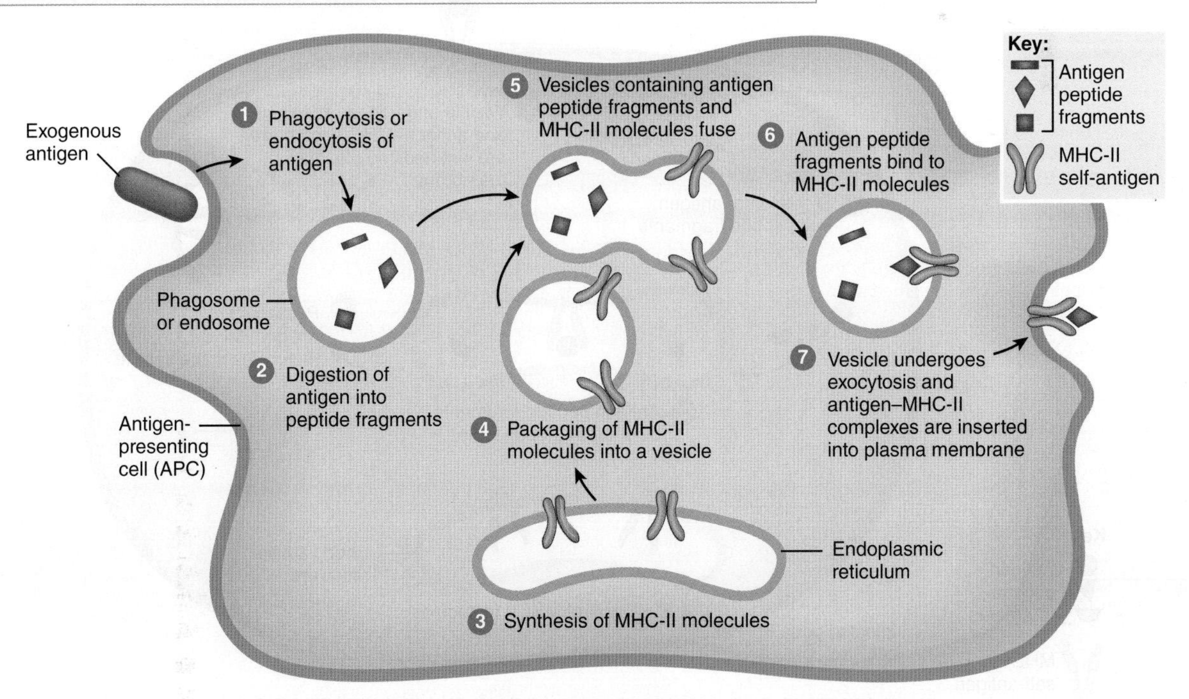

APCs present exogenous antigens in association with MHC-II molecules

Q What types of cells are APCs, and where in the body are they found?

3. ***Synthesis of MHC-II molecules.*** At the same time, the APC synthesizes MHC-II molecules at the endoplasmic reticulum (ER).
4. ***Packaging of MHC-II molecules.*** Once synthesized, the MHC-II molecules are packaged into vesicles.
5. ***Fusion of vesicles.*** The vesicles containing antigen peptide fragments and MHC-II molecules merge and fuse.
6. ***Binding of peptide fragments to MHC-II molecules.*** After fusion of the two types of vesicles, antigen peptide fragments bind to MHC-II molecules.
7. ***Insertion of antigen–MHC-II complexes into the plasma membrane.*** The combined vesicle that contains antigen–MHC-II complexes undergoes exocytosis. As a result, the antigen–MHC-II complexes are inserted into the plasma membrane.

After processing an antigen, the antigen-presenting cell migrates to lymphatic tissue to present the antigen to T cells. Within lymphatic tissue, a small number of T cells that have compatibly shaped receptors recognize and bind to the antigen fragment–MHC-II complex, triggering an adaptive immune response. The presentation of exogenous antigen together with MHC-II molecules by antigen-presenting cells informs T cells that intruders are present in the body and that combative action should begin.

Processing of Endogenous Antigens Foreign antigens that are present *inside* body cells are termed **endogenous antigens** (en-DOJ-e-nus). Such antigens may be viral proteins produced after a virus infects the cell and takes over the cell's metabolic machinery, toxins produced from intracellular bacteria, or abnormal proteins synthesized by a cancerous cell.

The steps in the processing and presenting of an endogenous antigen by an infected body cell occur as follows (**Figure 22.14**):

1. ***Digestion of antigen into peptide fragments.*** Within the infected cell, protein-digesting enzymes split the endogenous antigen into short peptide fragments.
2. ***Synthesis of MHC-I molecules.*** At the same time, the infected cell synthesizes MHC-I molecules at the endoplasmic reticulum (ER).
3. ***Binding of peptide fragments to MHC-I molecules.*** The antigen peptide fragments enter the ER and then bind to MHC-I molecules.
4. ***Packaging of antigen–MHC-I molecules.*** From the ER, antigen–MHC-I molecules are packaged into vesicles.
5. ***Insertion of antigen–MHC-I complexes into the plasma membrane.*** The vesicles that contain antigen–MHC-I complexes undergo exocytosis. As a result, the antigen–MHC-I complexes are inserted into the plasma membrane.

FIGURE 22.14 Processing and presenting of endogenous antigen by an infected body cell.

Fragments of endogenous antigens are processed and then presented with MHC-I proteins on the surface of an infected body cell.

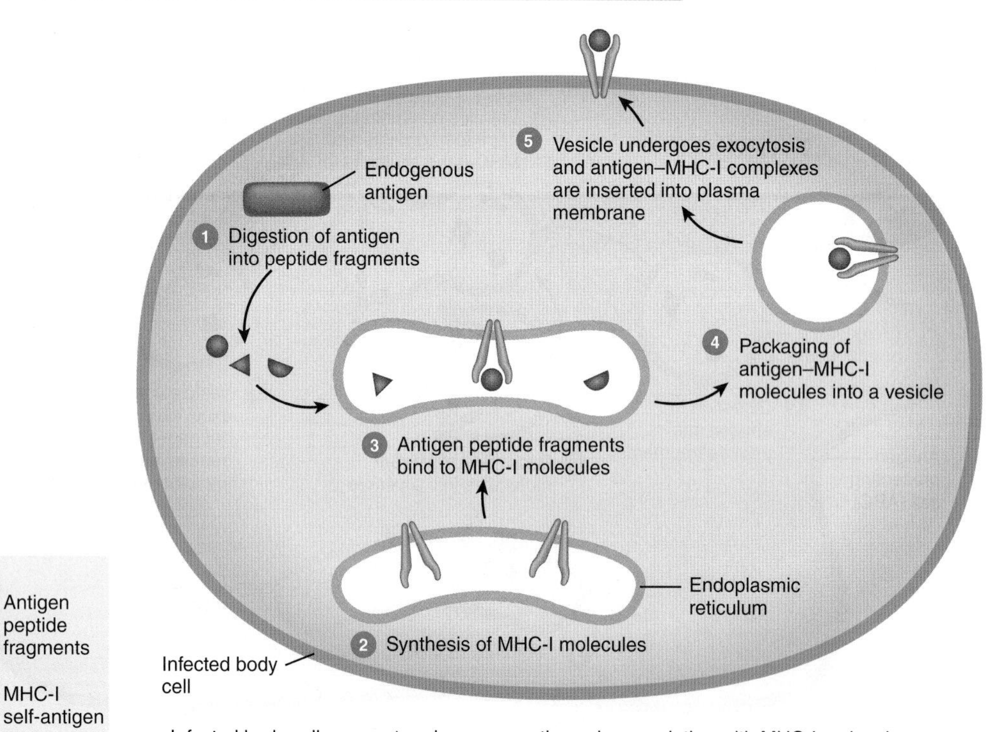

Q What are some examples of endogenous antigens?

TABLE 22.2 Summary of Cytokines Participating in Immune Responses

CYTOKINE	ORIGINS AND FUNCTIONS
Interleukin-1 (IL-1) (in′-ter-LOO-kin)	Produced by macrophages; promotes proliferation of helper T cells; acts on hypothalamus to cause fever.
Interleukin-2 (IL-2)	Secreted by helper T cells; costimulates proliferation of helper T cells, cytotoxic T cells, and B cells; activates NK cells.
Interleukin-4 (IL-4) (B cell stimulating factor)	Produced by helper T cells; costimulator for B cells; causes plasma cells to secrete IgE antibodies (see Table 22.3); promotes growth of T cells.
Interleukin-5 (IL-5)	Produced by some helper T cells and mast cells; costimulator for B cells; causes plasma cells to secrete IgA antibodies.
Interleukin-6 (IL-6)	Produced by helper T cells; enhances B cell proliferation, B cell differentiation into plasma cells, and secretion of antibodies by plasma cells.
Tumor necrosis factor (TNF) (ne-KRŌ-sis)	Produced mainly by macrophages; stimulates accumulation of neutrophils and macrophages at sites of inflammation and stimulates their killing of microbes.
Interferons (IFNs) (in′-ter-FĒR-ons)	Produced by virus-infected cells to inhibit viral replication in uninfected cells; activate cytotoxic T cells and natural killer cells, inhibit cell division, and suppress the formation of tumors.
Macrophage migration inhibiting factor	Produced by cytotoxic T cells; prevents macrophages from leaving site of infection.

Most cells of the body can process and present endogenous antigens. The display of an endogenous antigen bound to an MHC-I molecule signals that a cell has been infected and needs help.

Cytokines

Cytokines (SĪ-tō-kīns) are small protein hormones that stimulate or inhibit many normal cell functions, such as cell growth and differentiation. Lymphocytes and antigen-presenting cells secrete cytokines, as do fibroblasts, endothelial cells, monocytes, hepatocytes, and kidney cells. Some cytokines stimulate proliferation of progenitor blood cells in red bone marrow. Others regulate activities of cells involved in innate defenses or adaptive immune responses, as described in Table 22.2.

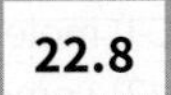

See Clinical Connection: Cytokine Therapy

22.8 Cell-Mediated Immunity

OBJECTIVES

- **Outline** the steps in a cell-mediated immune response.
- **Distinguish** between the action of natural killer cells and cytotoxic T cells.
- **Define** immunological surveillance.

A cell-mediated immune response begins with *activation* of a small number of T cells by a specific antigen. Once a T cell has been activated, it undergoes clonal selection. Recall that clonal selection is the process by which a lymphocyte proliferates (divides several times) and differentiates (forms more highly specialized cells) in response to a specific antigen. The result of clonal selection is the formation of a clone of cells that can recognize the same antigen as the original lymphocyte (see Figure 22.11). Some of the cells of a T cell clone become effector cells, while other cells of the clone become memory cells. The effector cells of a T cell clone carry out immune responses that ultimately result in *elimination* of the intruder.

Activation of T Cells

At any given time, most T cells are inactive. Antigen receptors on the surface of T cells, called **T-cell receptors (TCRs)**, recognize and bind to specific foreign antigen fragments that are presented in antigen–MHC complexes. There are millions of different T cells; each has its own unique TCRs that can recognize a specific antigen–MHC complex. When an antigen enters the body, only a few T cells have TCRs that can recognize and bind to the antigen. Antigen recognition also involves other surface proteins on T cells, the CD4 or CD8 proteins. These proteins interact with the MHC antigens and help maintain the TCR–MHC coupling. For this reason, they are referred to as *coreceptors.* Antigen recognition by a TCR with CD4 or CD8 proteins is the *first signal* in activation of a T cell.

A T cell becomes activated only if it binds to the foreign antigen and at the same time receives a *second signal*, a process known as **costimulation**. Of the more than 20 known costimulators, some are cytokines, such as **interleukin-2 (IL-2)**. Other costimulators include pairs of plasma membrane molecules, one on the surface of the T cell and a second on the surface of an antigen-presenting cell, that enable the two cells to adhere to one another for a period of time.

The need for two signals to activate a T cell is a little like starting and driving a car: When you insert the correct key (antigen) in the ignition (TCR) and turn it, the car starts (recognition of specific antigen), but it cannot move forward until you move the gear shift into drive (costimulation). The need for costimulation may prevent immune responses from occurring accidentally. Different costimulators affect the activated T cell in different ways, just as shifting a car into reverse has a different effect than shifting it into drive. Moreover, recognition (antigen binding to a receptor) without costimulation leads to a prolonged *state of inactivity* called **anergy** (AN-er-jē) in both T cells and B cells. Anergy is rather like leaving a car in neutral gear with its engine running until it's out of gas!

Once a T cell has received these two signals (antigen recognition and costimulation), it is activated. An activated T cell subsequently undergoes clonal selection.

Activation and Clonal Selection of Helper T Cells

Most T cells that display CD4 develop into **helper T cells**, also known as **CD4 T cells**. Inactive (resting) helper T cells recognize exogenous antigen fragments associated with major histocompatibility complex class II (MHC-II) molecules at the surface of an APC (**Figure 22.15**). With the aid of the CD4 protein, the helper T cell and APC interact with each other (antigenic recognition), costimulation occurs, and the helper T cell becomes activated.

Once activated, the helper T cell undergoes clonal selection (**Figure 22.15**). The result is the formation of a clone of helper T cells that consists of active helper T cells and memory helper T cells. Within hours after costimulation, **active helper T cells** start secreting a variety of cytokines (see **Table 22.2**). One very important cytokine produced by helper T cells is interleukin-2 (IL-2), which is needed for virtually all immune responses and is the prime trigger of T cell proliferation. IL-2 can act as a costimulator for resting helper T cells or cytotoxic T cells, and it enhances activation and proliferation of T cells, B cells, and natural killer cells. Some actions of interleukin-2 provide a good example of a beneficial positive feedback system. As noted earlier, activation of a helper T cell stimulates it to start secreting IL-2, which then acts in an autocrine manner by binding to IL-2 receptors on the plasma membrane of the cell that secreted it. One effect is stimulation of cell division. As the helper T cells proliferate, a positive feedback effect occurs because they secrete more IL-2, which causes further cell division. IL-2 may also act in a paracrine manner by binding to IL-2 receptors on neighboring helper T cells, cytotoxic T cells, or B cells. If any of these neighboring cells have already become bound to a copy of the same antigen, IL-2 serves as a costimulator.

The **memory helper T cells** of a helper T cell clone are not active cells. However, if the same antigen enters the body again in the future, memory helper T cells can quickly proliferate and differentiate into more active helper T cells and more memory helper T cells.

FIGURE 22.15 **Activation and clonal selection of a helper T cell.**

Once a helper T cell is activated, it forms a clone of active helper T cells and memory helper T cells.

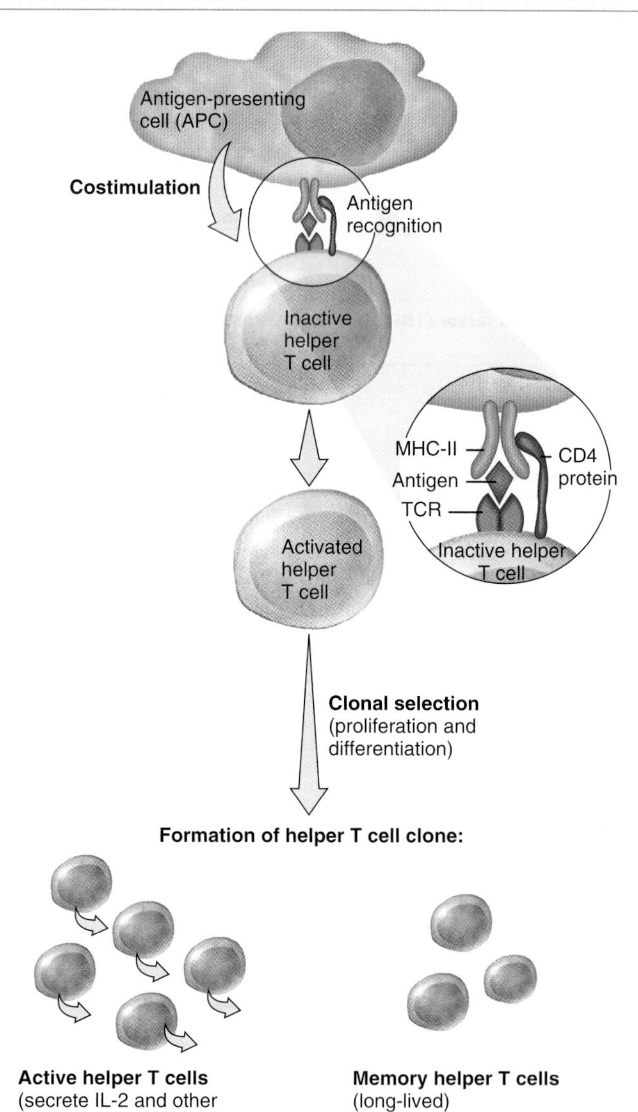

Q What are the first and second signals in activation of a T cell?

Activation and Clonal Selection of Cytotoxic T Cells

Most T cells that display CD8 develop into **cytotoxic T cells**, also termed **CD8 T cells**. Cytotoxic T cells recognize foreign antigens combined with major histocompatibility complex class I (MHC-I) molecules on the surface of (1) body cells infected by microbes, (2) some tumor cells, and (3) cells of a tissue transplant (**Figure 22.16**). Recognition requires the TCR and CD8 protein to maintain the coupling with MHC-I. Following antigenic recognition, costimulation occurs. In order to become activated, cytotoxic T cells require costimulation by interleukin-2 or other cytokines produced by active helper T cells that have already become bound to copies of the same antigen. (Recall that helper T cells are activated by antigen associated with MHC-II molecules.) Thus, *maximal activation* of cytotoxic T cells requires presentation of antigen associated with both MHC-I and MHC-II molecules.

FIGURE 22.16 **Activation and clonal selection of a cytotoxic T cell.**

Once a cytotoxic T cell is activated, it forms a clone of active cytotoxic T cells and memory cytotoxic T cells.

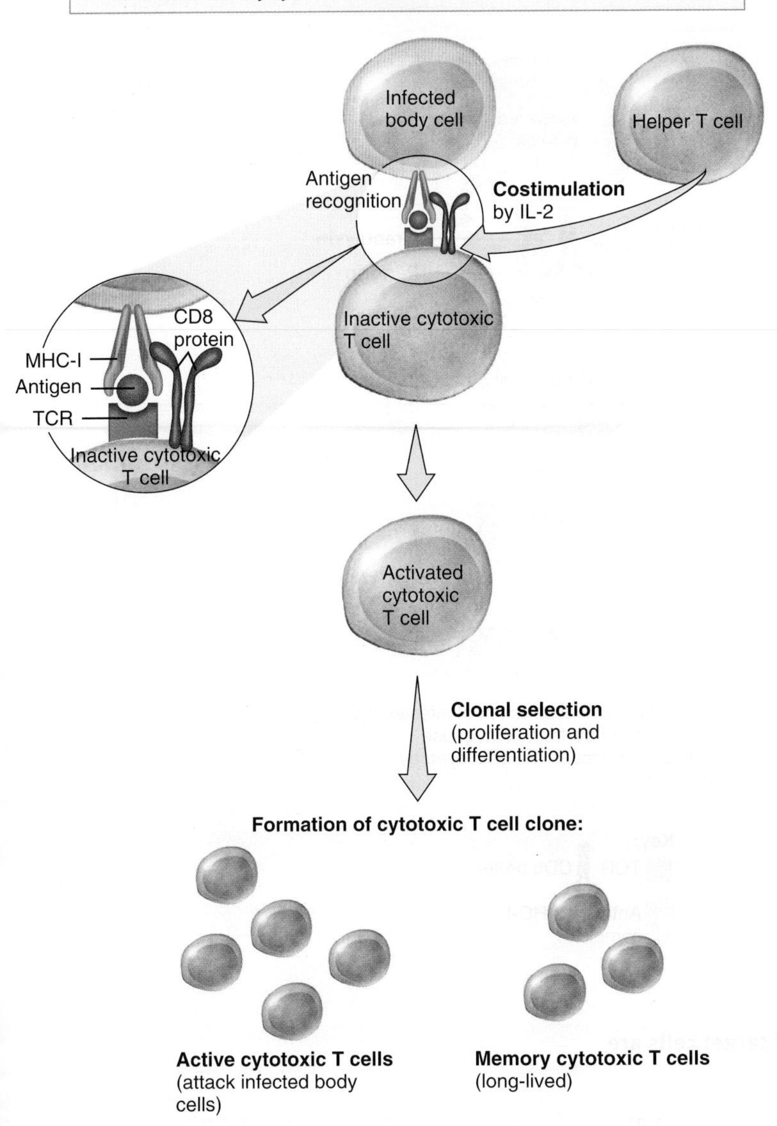

Q What is the function of the CD8 protein of a cytotoxic T cell?

Once activated, the cytotoxic T cell undergoes clonal selection. The result is the formation of a clone of cytotoxic T cells that consists of active cytotoxic T cells and memory cytotoxic T cells. **Active cytotoxic T cells** attack other body cells that have been infected with the antigen. **Memory cytotoxic T cells** do not attack infected body cells. Instead, they can quickly proliferate and differentiate into more active cytotoxic T cells and more memory cytotoxic T cells if the same antigen enters the body at a future time.

Elimination of Invaders

Cytotoxic T cells are the soldiers that march forth to do battle with foreign invaders in cell-mediated immune responses. They leave secondary lymphatic organs and tissues and migrate to seek out and destroy infected target cells, cancer cells, and transplanted cells (**Figure 22.17**). Cytotoxic T cells recognize and attach to target cells. Then, the cytotoxic T cells deliver a "lethal hit" that kills the target cells.

Cytotoxic T cells kill infected target body cells much like natural killer cells do. The major difference is that cytotoxic T cells have receptors specific for a particular microbe and thus kill only target body cells infected with *one* particular type of microbe; natural killer cells can destroy a wide variety of microbe-infected body cells. Cytotoxic T cells have two principal mechanisms for killing infected target cells.

1. Cytotoxic T cells, using receptors on their surfaces, recognize and bind to infected target cells that have microbial antigens displayed on their surface. The cytotoxic T cell then releases **granzymes**, protein-digesting enzymes that trigger apoptosis (**Figure 22.17a**). Once the infected cell is destroyed, the released microbes are killed by phagocytes.
2. Alternatively, cytotoxic T cells bind to infected body cells and release two proteins from their granules: perforin and granulysin. **Perforin** inserts into the plasma membrane of the target cell and creates channels in the membrane (**Figure 22.17b**). As a result, extracellular fluid flows into the target cell and cytolysis (cell bursting) occurs. Other granules in cytotoxic T cells release **granulysin** (gran′-ū-LĪ-sin), which enters through the channels and destroys the microbes by creating holes in their plasma membranes. Cytotoxic T cells may also destroy target cells by releasing a toxic molecule called **lymphotoxin** (lim′-fō-TOK-sin), which activates enzymes in the target cell. These enzymes cause the target cell's DNA to fragment, and the cell dies. In addition, cytotoxic T cells secrete gamma-interferon, which attracts and activates phagocytic cells, and macrophage migration inhibition factor, which prevents migration of phagocytes from the infection site. After detaching from a target cell, a cytotoxic T cell can seek out and destroy another target cell.

Immunological Surveillance

When a normal cell transforms into a cancerous cell, it often displays novel cell surface components called **tumor antigens**. These molecules are rarely, if ever, displayed on the surface of normal cells. If the immune system recognizes a tumor antigen as nonself, it can destroy any cancer cells carrying that antigen. Such immune responses, called **immunological surveillance** (im′-ū-nō-LOJ-i-kul sur-VĀ-lants), are carried out by cytotoxic T cells, macrophages, and natural killer cells. Immunological surveillance is most effective in eliminating tumor cells due to cancer-causing viruses. For this reason, transplant recipients who are taking immunosuppressive drugs to prevent transplant rejection have an increased incidence of virus-associated cancers. Their risk for other types of cancer is not increased.

See Clinical Connection: Graft Rejection and Tissue Typing

FIGURE 22.17 Activity of cytotoxic T cells. After delivering a "lethal hit," a cytotoxic T cell can detach and attack another infected target cell displaying the same antigen.

Cytotoxic T cells release granzymes that trigger apoptosis and perforin that triggers cytolysis of infected target cells.

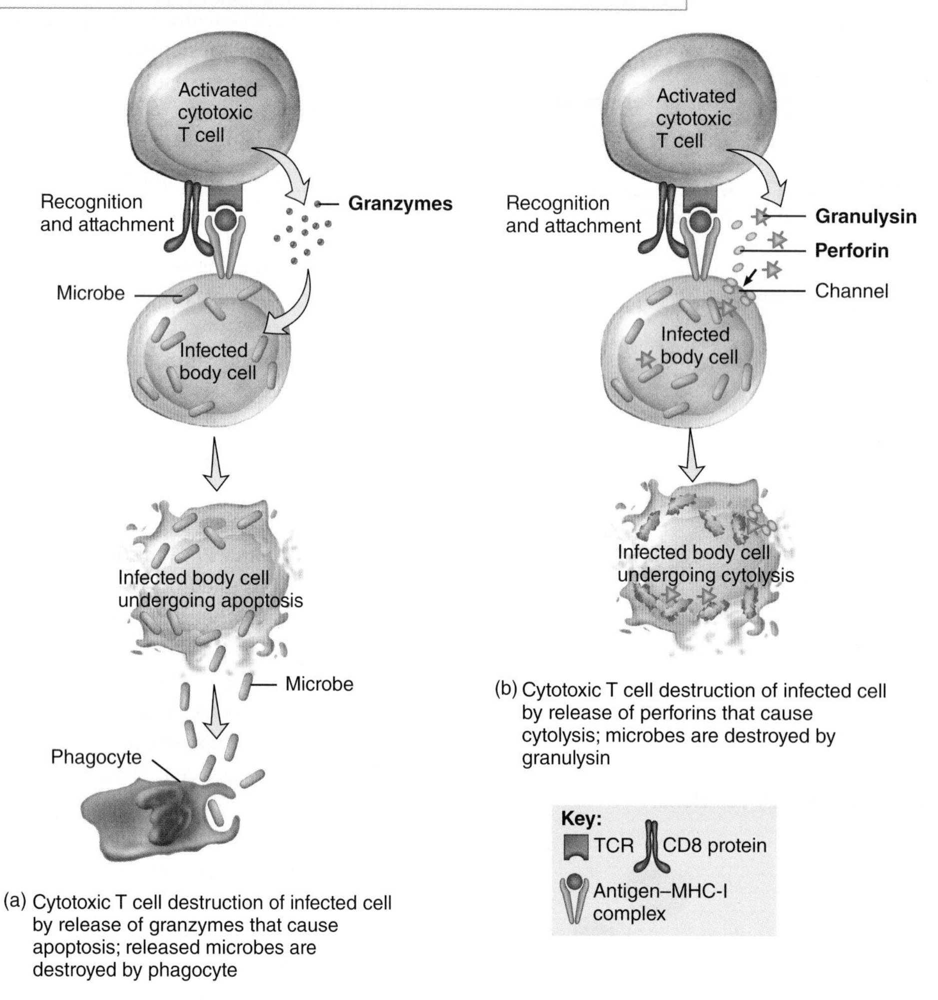

Q In addition to cells infected by microbes, what other types of target cells are attacked by cytotoxic T cells?

22.9 Antibody-Mediated Immunity

OBJECTIVES

- **Describe** the steps in an antibody-mediated immune response.
- **List** the chemical characteristics and actions of antibodies.
- **Explain** how the complement system operates.
- **Distinguish** between a primary response and a secondary response to infection.

The body contains not only millions of different T cells but also millions of different B cells, each capable of responding to a specific antigen. Cytotoxic T cells leave lymphatic tissues to seek out and destroy a foreign antigen, but B cells stay put. In the presence of a foreign antigen, a specific B cell in a lymph node, the spleen, or mucosa-associated lymphatic tissue becomes activated. Then it undergoes clonal selection, forming a clone of plasma cells and memory cells. Plasma cells are the effector cells of a B cell clone; they secrete specific antibodies, which in turn circulate in the lymph and blood to reach the site of invasion.

Activation and Clonal Selection of B Cells

During activation of a B cell, an antigen binds to **B-cell receptors (BCRs)** (**Figure 22.18**). These integral transmembrane proteins are chemically

FIGURE 22.18 Activation and clonal selection of B cells. Plasma cells are actually much larger than B cells.

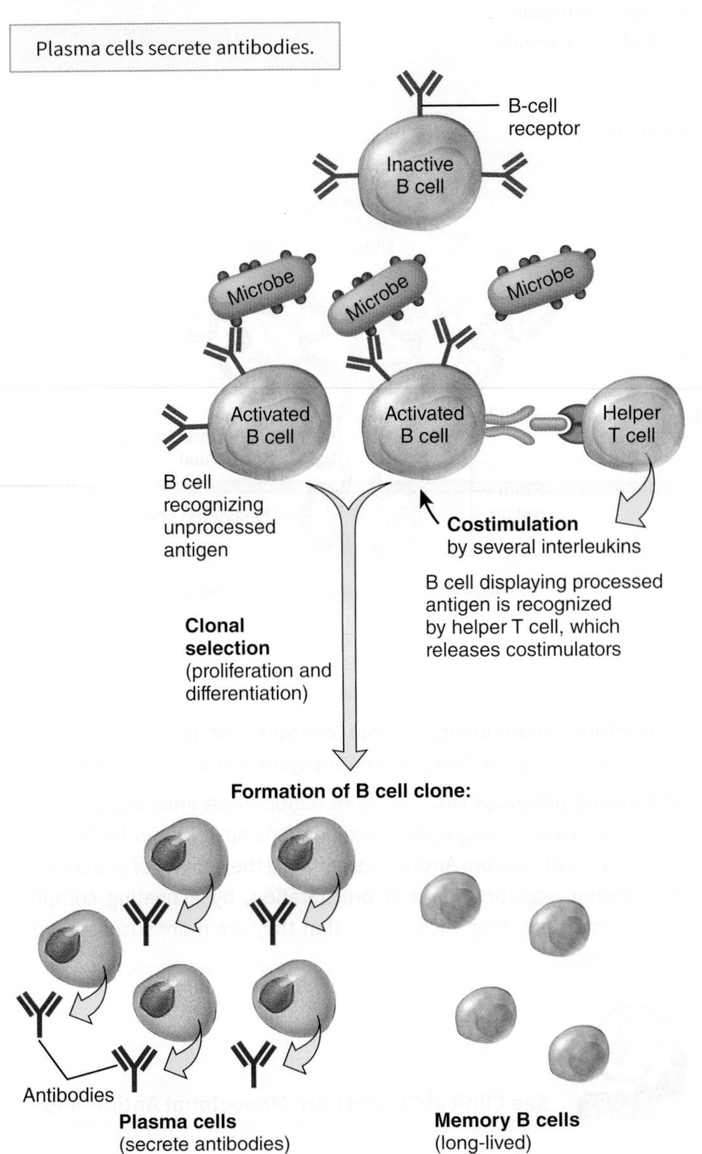

Q How many different kinds of antibodies will be secreted by the plasma cells in the clone shown here?

similar to the antibodies that eventually are secreted by plasma cells. Although B cells can respond to an unprocessed antigen present in lymph or interstitial fluid, their response is much more intense when they process the antigen. Antigen processing in a B cell occurs in the following way: The antigen is taken into the B cell, broken down into peptide fragments and combined with MHC-II self-antigens, and moved to the B cell plasma membrane. Helper T cells recognize the antigen–MHC-II complex and deliver the costimulation needed for B cell proliferation and differentiation. The helper T cell produces interleukin-2 and other cytokines that function as costimulators to activate B cells.

Once activated, a B cell undergoes clonal selection (Figure 22.18). The result is the formation of a clone of B cells that consists of plasma cells and memory B cells. **Plasma cells** secrete antibodies. A few days after exposure to an antigen, a plasma cell secretes hundreds of millions of antibodies each day for about 4 or 5 days, until the plasma cell dies. Most antibodies travel in lymph and blood to the invasion site. Interleukin-4 and interleukin-6, also produced by helper T cells, enhance B cell proliferation, B cell differentiation into plasma cells, and secretion of antibodies by plasma cells. **Memory B cells** do not secrete antibodies. Instead, they can quickly proliferate and differentiate into more plasma cells and more memory B cells should the same antigen reappear at a future time.

Different antigens stimulate different B cells to develop into plasma cells and their accompanying memory B cells. All of the B cells of a particular clone are capable of secreting only one type of antibody, which is identical to the antigen receptor displayed by the B cell that first responded. Each specific antigen activates only those B cells that are predestined (by the combination of gene segments they carry) to secrete antibody specific to that antigen. Antibodies produced by a clone of plasma cells enter the circulation and form antigen–antibody complexes with the antigen that initiated their production.

Antibodies

An **antibody (Ab)** can combine specifically with the epitope on the antigen that triggered its production. The antibody's structure matches its antigen much as a lock accepts a specific key. In theory, plasma cells could secrete as many different antibodies as there are different B-cell receptors because the same recombined gene segments code for both the BCR and the antibodies eventually secreted by plasma cells.

Antibody Structure Antibodies belong to a group of glycoproteins called globulins, and for this reason they are also known as **immunoglobulins (Igs)**. Most antibodies contain four polypeptide chains (Figure 22.19). Two of the chains are identical to each other and are called **heavy (H) chains**; each consists of about 450 amino acids. Short carbohydrate chains are attached to each heavy polypeptide chain. The two other polypeptide chains, also identical to each other, are called **light (L) chains**, and each consists of about 220 amino acids. A disulfide bond (S—S) holds each light chain to a heavy chain. Two disulfide bonds also link the midregion of the two heavy chains; this part of the antibody displays considerable flexibility and is called the **hinge region**. Because the antibody "arms" can move somewhat as the hinge region bends, an antibody can assume either a T shape or a Y shape (Figure 22.19a, b). Beyond the hinge region, parts of the two heavy chains form the **stem region**.

Within each H and L chain are two distinct regions. The tips of the H and L chains, called the **variable (V) regions**, constitute the **antigen-binding site**. The variable region, which is different for each kind of antibody, is the part of the antibody that recognizes and attaches specifically to a particular antigen. Because most antibodies have two antigen-binding sites, they are said to be *bivalent.* Flexibility at the hinge allows the antibody to simultaneously bind to two epitopes that are some distance apart—for example, on the surface of a microbe.

The remainder of each H and L chain, called the **constant (C) region**, is nearly the same in all antibodies of the same class and is responsible for the type of antigen–antibody reaction that occurs.

FIGURE 22.19 Chemical structure of the immunoglobulin G (IgG) class of antibody. Each molecule is composed of four polypeptide chains (two heavy and two light) plus a short carbohydrate chain attached to each heavy chain. In (a), each circle represents one amino acid. In (b), V_L = variable regions of light chain, C_L = constant region of light chain, V_H = variable region of heavy chain, and C_H = constant region of heavy chain.

An antibody combines only with the epitope on the antigen that triggered its production.

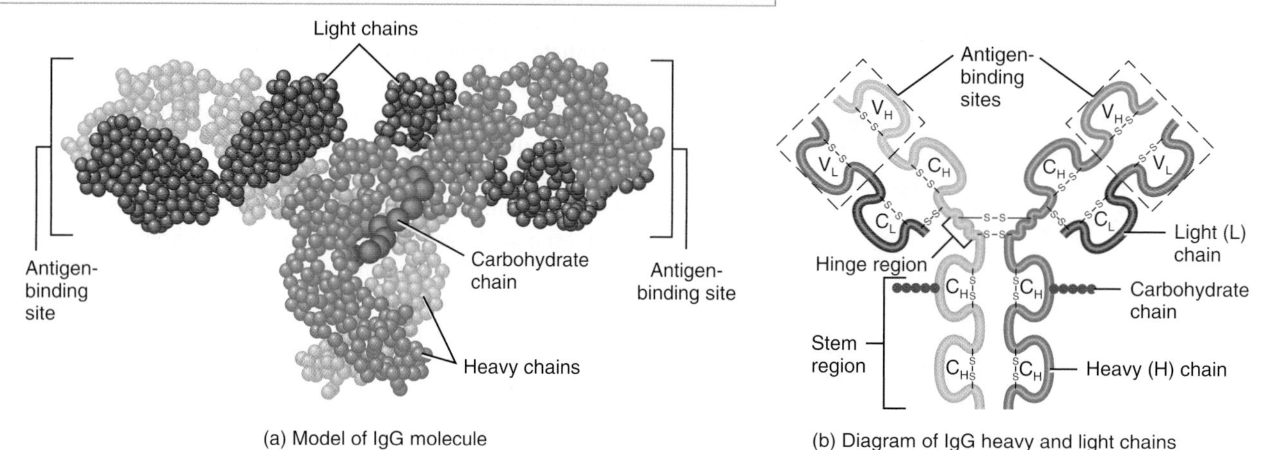

Q What is the function of the variable regions in an antibody molecule?

However, the constant region of the H chain differs from one class of antibody to another, and its structure serves as a basis for distinguishing five different classes, designated IgG, IgA, IgM, IgD, and IgE. Each class has a distinct chemical structure and a specific biological role. Because they appear first and are relatively short-lived, IgM antibodies indicate a recent invasion. In a sick patient, the responsible pathogen may be suggested by the presence of high levels of IgM specific to a particular organism. Resistance of the fetus and newborn baby to infection stems mainly from maternal IgG antibodies that cross the placenta before birth and IgA antibodies in breast milk after birth. Table 22.3 summarizes the structures and functions of the five classes of antibodies.

Antibody Actions

The actions of the five classes of immunoglobulins differ somewhat, but all of them act to disable antigens in some way. Actions of antibodies include the following:

- ***Neutralizing antigen.*** The reaction of antibody with antigen blocks or neutralizes some bacterial toxins and prevents attachment of some viruses to body cells.
- ***Immobilizing bacteria.*** If antibodies form against antigens on the cilia or flagella of motile bacteria, the antigen–antibody reaction may cause the bacteria to lose their motility, which limits their spread into nearby tissues.
- ***Agglutinating and precipitating antigen.*** Because antibodies have two or more sites for binding to antigen, the antigen–antibody reaction may cross-link pathogens to one another, causing agglutination (clumping together). Phagocytic cells ingest agglutinated microbes more readily. Likewise, soluble antigens may come out of solution and form a more easily phagocytized precipitate when cross-linked by antibodies.
- ***Activating complement.*** Antigen–antibody complexes initiate the classical pathway of the complement system (discussed shortly).
- ***Enhancing phagocytosis.*** The stem region of an antibody acts as a flag that attracts phagocytes once antigens have bound to the antibody's variable region. Antibodies enhance the activity of phagocytes by causing agglutination and precipitation, by activating complement, and by coating microbes so that they are more susceptible to phagocytosis.

See Clinical Connection: Monoclonal Antibodies

Role of the Complement System in Immunity

The **complement system** (KOM-ple-ment) is a defensive system made up of over 30 proteins produced by the liver and found circulating in blood plasma and within tissues throughout the body. Collectively, the complement proteins destroy microbes by causing phagocytosis, cytolysis, and inflammation; they also prevent excessive damage to body tissues.

Most complement proteins are designated by an uppercase letter C, numbered C1 through C9, named for the order in which they were discovered. The C1–C9 complement proteins are inactive and become activated only when split by enzymes into active fragments, which are indicated by lowercase letters *a* and *b*. For example, inactive complement protein C3 is split into the activated fragments, C3a and C3b. The active fragments carry out the destructive actions of the C1–C9 complement proteins. Other complement proteins are referred to as factors B, D, and P (properdin).

TABLE 22.3 Classes of Immunoglobulins (Igs)

NAME AND STRUCTURE	CHARACTERISTICS AND FUNCTIONS
IgG	Most abundant, about 80% of all antibodies in blood; found in blood, lymph, and intestines; monomer (one-unit) structure. Protects against bacteria and viruses by enhancing phagocytosis, neutralizing toxins, and triggering complement system. Is the only class of antibody to cross placenta from mother to fetus, conferring considerable immune protection in newborns.
IgA	Found mainly in sweat, tears, saliva, mucus, breast milk, and gastrointestinal secretions. Smaller quantities are present in blood and lymph. Makes up 10–15% of all antibodies in blood; occurs as monomers and dimers (two units). Levels decrease during stress, lowering resistance to infection. Provides localized protection of mucous membranes against bacteria and viruses.
IgM	About 5–10% of all antibodies in blood; also found in lymph. Occurs as pentamers (five units); first antibody class to be secreted by plasma cells after initial exposure to any antigen. Activates complement and causes agglutination and lysis of microbes. Also present as monomers on surfaces of B cells, where they serve as antigen receptors. In blood plasma, anti-A and anti-B antibodies of ABO blood group, which bind to A and B antigens during incompatible blood transfusions, are also IgM antibodies (see **Figure 19.12**).
IgD	Mainly found on surfaces of B cells as antigen receptors, where it occurs as monomers; involved in activation of B cells. About 0.2% of all antibodies in blood.
IgE	Less than 0.1% of all antibodies in blood; occurs as monomers; located on mast cells and basophils. Involved in allergic and hypersensitivity reactions; provides protection against parasitic worms.

Complement proteins act in a *cascade*—one reaction triggers another reaction, which in turn triggers another reaction, and so on. With each succeeding reaction, more and more product is formed so that the net effect is amplified many times.

Complement activation may begin by three different pathways (described shortly), all of which activate C3. Once activated, C3 begins a cascade of reactions that brings about phagocytosis, cytolysis, and inflammation as follows (**Figure 22.20**):

1. Inactivated C3 splits into activated C3a and C3b.
2. C3b binds to the surface of a microbe and receptors on phagocytes attach to the C3b. Thus C3b enhances **phagocytosis** by coating a microbe, a process called **opsonization** (op-sō-ni-ZĀ-shun). Opsonization promotes attachment of a phagocyte to a microbe.
3. C3b also initiates a series of reactions that bring about cytolysis. First, C3b splits C5. The C5b fragment then binds to C6 and C7, which attach to the plasma membrane of an invading microbe. Then C8 and several C9 molecules join the other complement proteins and together form a cylinder-shaped **membrane attack complex**, which inserts into the plasma membrane.
4. The membrane attack complex creates channels in the plasma membrane that result in **cytolysis**, the bursting of the microbial cells due to the inflow of extracellular fluid through the channels.
5. C3a and C5a bind to mast cells and cause them to release histamine that increases blood vessel permeability during **inflammation**. C5a also attracts phagocytes to the site of inflammation (chemotaxis).

C3 can be activated in three ways: (1) The **classical pathway** starts when antibodies bind to antigens (microbes). The antigen–antibody complex binds and activates C1. Eventually, C3 is activated and the C3 fragments initiate phagocytosis, cytolysis, and inflammation. (2) The **alternative pathway** does not involve antibodies. It is initiated by an interaction between lipid–carbohydrate complexes on the surface of microbes and complement protein factors B, D, and P. This interaction activates C3. (3) In the **lectin pathway**, macrophages that digest microbes release chemicals that cause the liver to produce proteins called **lectins**. Lectins bind to the carbohydrates on the surface of microbes, ultimately causing the activation of C3.

Once complement is activated, proteins in blood and on body cells such as blood cells break down activated C3. In this way, its destructive capabilities cease very quickly so that damage to body cells is minimized.

Immunological Memory

A hallmark of immune responses is memory for specific antigens that have triggered immune responses in the past. **Immunological**

FIGURE 22.20 Complement activation and results of activation. (Adapted from Tortora, Funke, and Case, *Microbiology: An Introduction, Eleventh Edition*, Figure 16.9, Pearson Benjamin-Cummings, 2013.)

When activated, complement proteins enhance phagocytosis, cytolysis, and inflammation.

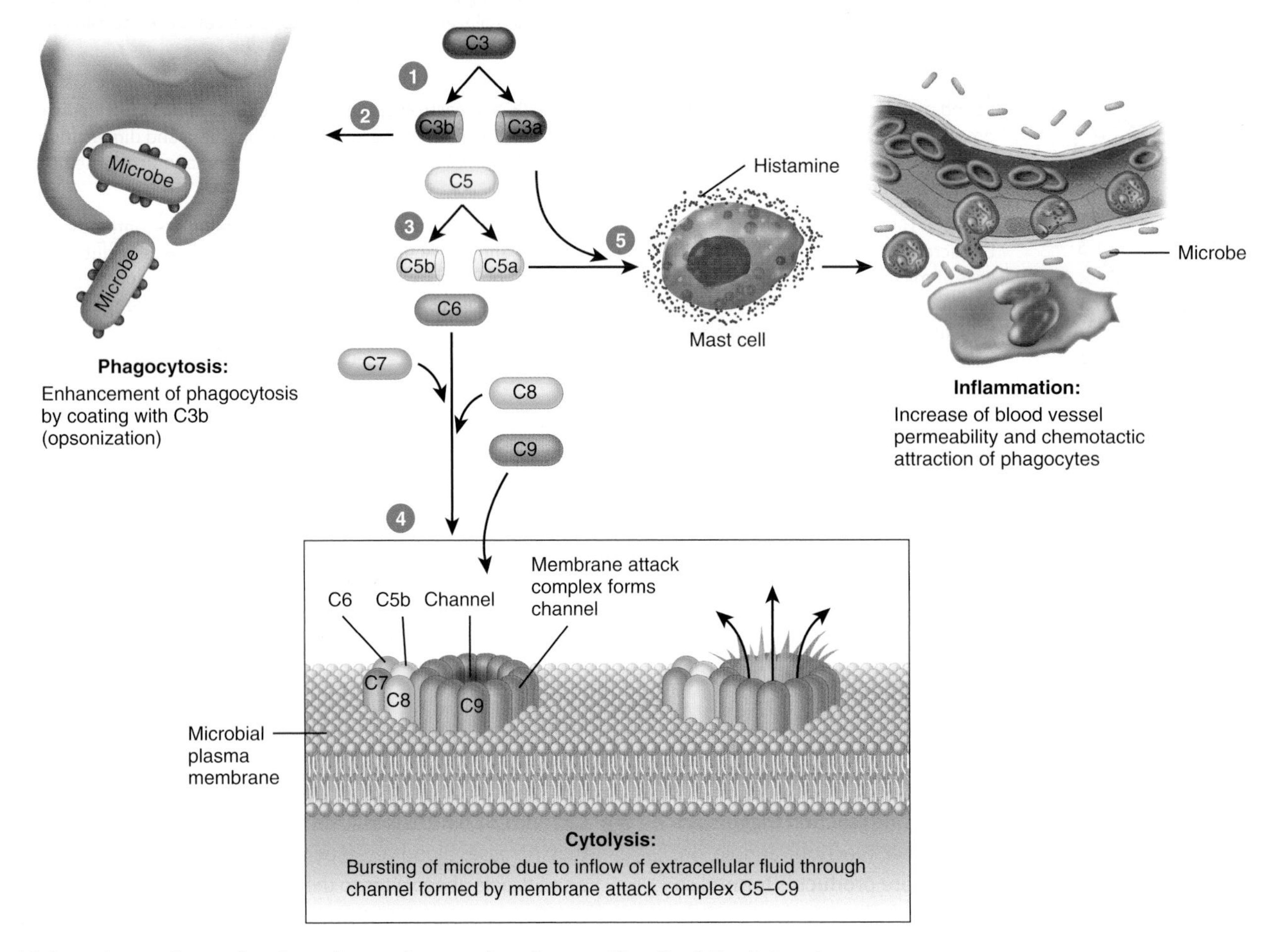

Q Which pathway for activation of complement involves antibodies? Explain why.

memory is due to the presence of long-lasting antibodies and very long-lived lymphocytes that arise during clonal selection of antigen-stimulated B cells and T cells.

Immune responses, whether cell-mediated or antibody-mediated, are much quicker and more intense after a second or subsequent exposure to an antigen than after the first exposure. Initially, only a few cells have the correct specificity to respond, and the immune response may take several days to build to maximum intensity. Because thousands of memory cells exist after an initial encounter with an antigen, the next time the same antigen appears they can proliferate and differentiate into helper T cells, cytotoxic T cells, or plasma cells within hours.

One measure of immunological memory is *antibody titer* (TĪ-ter), the amount of antibody in serum. After an initial contact with an antigen, no antibodies are present for a period of several days. Then, a slow rise in the antibody titer occurs, first IgM and then IgG, followed by a gradual decline in antibody titer (**Figure 22.21**). This is the **primary response**.

Memory cells may remain for decades. Every new encounter with the same antigen results in a rapid proliferation of memory cells. After subsequent encounters, the antibody titer is far greater than during a primary response and consists mainly of IgG antibodies. This accelerated, more intense response is called the **secondary response**. Antibodies produced during a secondary response have an even higher affinity for the antigen than those produced during a primary response, and thus they are more successful in disposing of it.

Primary and secondary responses occur during microbial infection. When you recover from an infection without taking antimicrobial drugs, it is usually because of the primary response. If the same microbe infects you later, the secondary response could be so swift that the microbes are destroyed before you exhibit any signs or symptoms of infection.

FIGURE 22.21 Production of antibodies in the primary and secondary responses to a given antigen.

Immunological memory is the basis for successful immunization by vaccination.

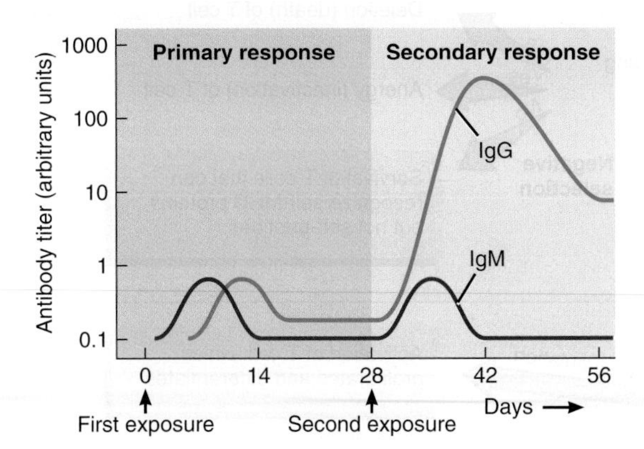

Q According to this graph, how much more IgG is circulating in the blood in the secondary response than in the primary response? (*Hint*: Notice that each mark on the antibody titer axis represents a 10-fold increase.)

Immunological memory provides the basis for immunization by vaccination against certain diseases (for example, polio). When you receive the vaccine, which may contain *attenuated* (weakened) or killed whole microbes or portions of microbes, your B cells and T cells are activated. Should you subsequently encounter the living pathogen as an infecting microbe, your body initiates a secondary response.

Table 22.4 summarizes the various ways to acquire adaptive immunity.

TABLE 22.4 Ways to Acquire Adaptive Immunity

METHOD	DESCRIPTION
Naturally acquired active immunity	Following exposure to a microbe, antigen recognition by B cells and T cells and costimulation lead to formation of antibody-secreting plasma cells, cytotoxic T cells, and B and T memory cells.
Naturally acquired passive immunity	IgG antibodies are transferred from mother to fetus across placenta, or IgA antibodies are transferred from mother to baby in milk during breast-feeding.
Artificially acquired active immunity	Antigens introduced during vaccination stimulate cell-mediated and antibody-mediated immune responses, leading to production of memory cells. Antigens are pretreated to be immunogenic but not pathogenic (they will trigger an immune response but not cause significant illness).
Artificially acquired passive immunity	Intravenous injection of immunoglobulins (antibodies).

22.10 Self-Recognition and Self-Tolerance

OBJECTIVE

- **Describe** how self-recognition and self-tolerance develop.

To function properly, your T cells must have two traits: (1) They must be able to *recognize* your own major histocompatibility complex (MHC) proteins, a process known as **self-recognition**, and (2) they must *lack reactivity* to peptide fragments from your own proteins, a condition known as **self-tolerance** (Figure 22.22). B cells also display self-tolerance. Loss of self-tolerance leads to the development of autoimmune diseases (see Disorders: Homeostatic Imbalances in the Study Guide).

Pre-T cells in the thymus develop the capability for self-recognition via **positive selection** (Figure 22.22a). In this process, some pre-T cells express T-cell receptors (TCRs) that interact with self-MHC proteins on epithelial cells in the thymic cortex. Because of this interaction, the T cells can recognize the MHC part of an antigen–MHC complex. These T cells survive. Other immature T cells that fail to interact with thymic epithelial cells are not able to recognize self-MHC proteins. These cells undergo apoptosis.

The development of self-tolerance occurs by a weeding-out process called **negative selection** in which the T cells interact with dendritic cells located at the junction of the cortex and medulla in the thymus. In this process, T cells with receptors that recognize self-peptide fragments or other self-antigens are eliminated or inactivated (Figure 22.22a). The T cells selected to survive do not respond to self-antigens, the fragments of molecules that are normally present in the body. Negative selection occurs via both deletion and anergy. In **deletion**, self-reactive T cells undergo apoptosis and die; in **anergy** they remain alive but are unresponsive to antigenic stimulation. Only 1–5% of the immature T cells in the thymus receive the proper signals to survive apoptosis during both positive and negative selection and emerge as mature, immunocompetent T cells.

Once T cells have emerged from the thymus, they may still encounter an unfamiliar self-protein; in such cases they may also become anergic if there is no costimulator (Figure 22.22b). Deletion of self-reactive T cells may also occur after they leave the thymus.

B cells also develop tolerance through deletion and anergy (Figure 22.22c). While B cells are developing in bone marrow, those cells exhibiting antigen receptors that recognize common self-antigens (such as MHC proteins or blood group antigens) are deleted. Once B cells are released into the blood, however, anergy appears to be the main mechanism for preventing responses to self-proteins. When B cells encounter an antigen not associated with an antigen-presenting cell, the necessary costimulation signal often is missing. In this case, the B cell is likely to become anergic (inactivated) rather than activated.

See Clinical Connection: Cancer Immunology

FIGURE 22.22 **Development of self-recognition and self-tolerance.** MHC = major histocompatibility complex; TCR = T-cell receptor.

Positive selection allows recognition of self-MHC proteins; negative selection provides self-tolerance of your own peptides and other self-antigens.

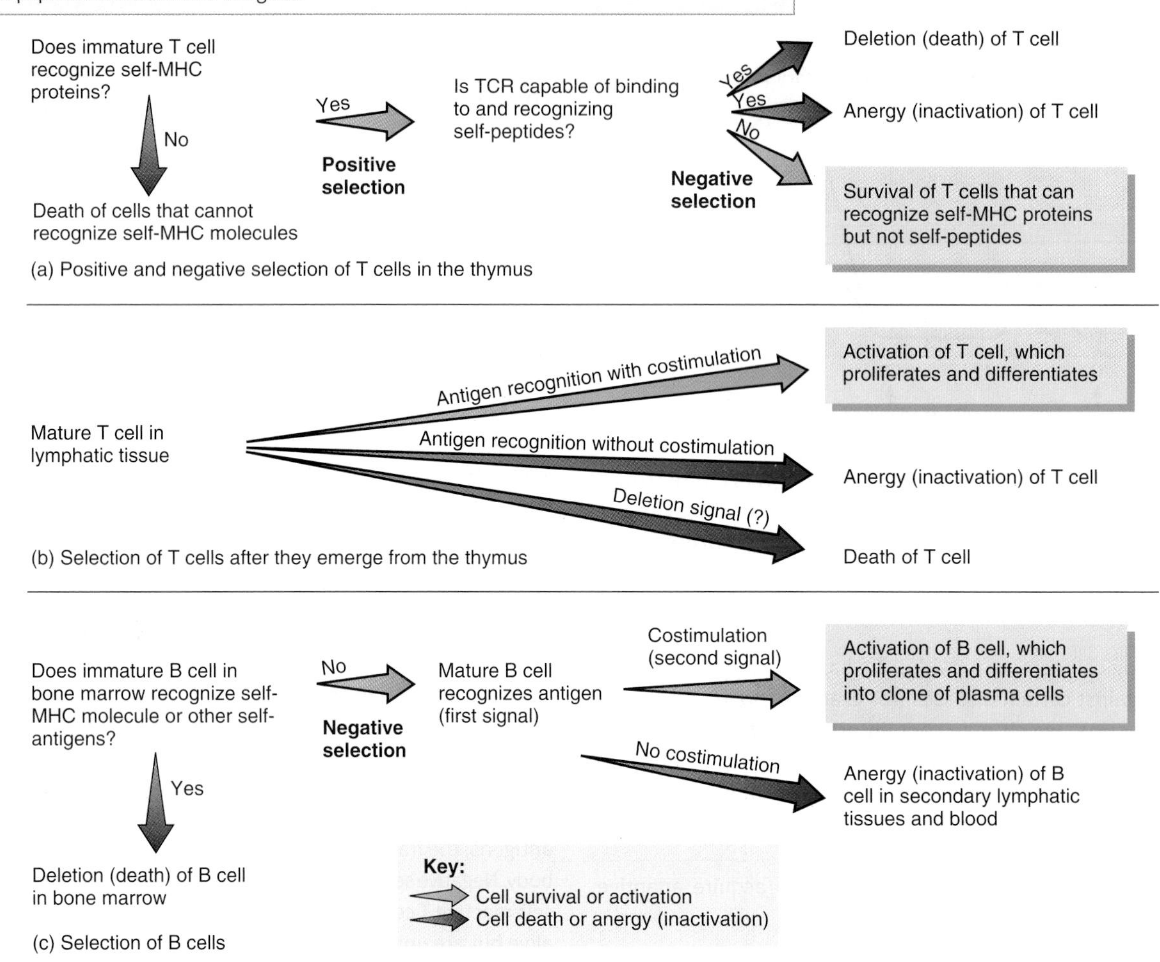

Q How does deletion differ from anergy?

Table 22.5 summarizes the activities of cells involved in adaptive immune responses.

22.11 Stress and Immunity

OBJECTIVE

- **Describe** the effects of stress on immunity.

The field of **psychoneuroimmunology (PNI)** deals with communication pathways that link the nervous, endocrine, and immune systems. PNI research appears to justify what people have long observed: Your thoughts, feelings, moods, and beliefs influence your level of health and the course of disease. For example, cortisol, a hormone secreted by the adrenal cortex in association with the stress response, inhibits immune system activity.

If you want to observe the relationship between lifestyle and immune function, visit a college campus. As the semester progresses and the workload accumulates, an increasing number of students can be found in the waiting rooms of student health services. When work and stress pile up, health habits can change. Many people smoke or consume more alcohol when stressed, two habits detrimental to optimal immune function. Under stress, people are less likely to eat well or exercise regularly, two habits that enhance immunity.

People resistant to the negative health effects of stress are more likely to experience a sense of control over the future, a commitment to their work, expectations of generally positive outcomes for themselves,

TABLE 22.5 Summary of Functions of Cells Participating in Adaptive Immune Responses

CELL	FUNCTIONS
ANTIGEN-PRESENTING CELLS (APCs)	
Macrophage	Processing and presentation of foreign antigens to T cells; secretion of interleukin-1, which stimulates secretion of interleukin-2 by helper T cells and induces proliferation of B cells; secretion of interferons that stimulate T cell growth.
Dendritic cell	Processes and presents antigen to T cells and B cells; found in mucous membranes, skin, lymph nodes.
B cell	Processes and presents antigen to helper T cells.
LYMPHOCYTES	
Cytotoxic T cell	Kills host target cells by releasing granzymes that induce apoptosis, perforin that forms channels to cause cytolysis, granulysin that destroys microbes, lymphotoxin that destroys target cell DNA, gamma-interferon that attracts macrophages and increases their phagocytic activity, and macrophage migration inhibition factor that prevents macrophage migration from site of infection.
Helper T cell	Cooperates with B cells to amplify antibody production by plasma cells and secretes interleukin-2, which stimulates proliferation of T cells and B cells. May secrete gamma-IFN and tumor necrosis factor (TNF), which stimulate inflammatory response.
Memory T cell	Remains in lymphatic tissue and recognizes original invading antigens, even years after first encounter.
B cell	Differentiates into antibody-producing plasma cell.
Plasma cell	Descendant of B cell that produces and secretes antibodies.
Memory B cell	Descendant of B cell that remains after immune response and is ready to respond rapidly and forcefully should the same antigen enter body in future.

and feelings of social support. To increase your stress resistance, cultivate an optimistic outlook, get involved in your work, and build good relationships with others.

Adequate sleep and relaxation are especially important for a healthy immune system. But when there aren't enough hours in the day, you may be tempted to steal some from the night. While skipping sleep may give you a few more hours of productive time in the short run, in the long run you end up even farther behind, especially if getting sick keeps you out of commission for several days, blurs your concentration, and blocks your creativity.

Even if you make time to get 8 hours of sleep, stress can cause insomnia. If you find yourself tossing and turning at night, it's time to improve your stress management and relaxation skills! Be sure to unwind from the day before going to bed.

22.12 Aging and the Immune System

OBJECTIVE

- **Describe** the effects of aging on the immune system.

With advancing age, most people become more susceptible to all types of infections and malignancies. Their response to vaccines is decreased, and they tend to produce more autoantibodies (antibodies against their body's own molecules). In addition, the immune system exhibits lowered levels of function. For example, T cells become less responsive to antigens, and fewer T cells respond to infections. This may result from age-related atrophy of the thymus or decreased production of thymic hormones. Because the T cell population decreases with age, B cells are also less responsive. Consequently, antibody levels do not increase as rapidly in response to a challenge by an antigen, resulting in increased susceptibility to various infections. It is for this key reason that elderly individuals are encouraged to get influenza (flu) vaccinations each year.

• • •

To appreciate the many ways that the lymphatic system contributes to homeostasis of other body systems, examine *Focus on Homeostasis: Contributions of the Lymphatic System and Immunity*.

Next, in Chapter 23, we will explore the structure and function of the respiratory system and see how its operation is regulated by the nervous system. Most importantly, the respiratory system provides for gas exchange—taking in oxygen and blowing off carbon dioxide. The cardiovascular system aids gas exchange by transporting blood containing these gases between the lungs and tissue cells.

FOCUS on HOMEOSTASIS

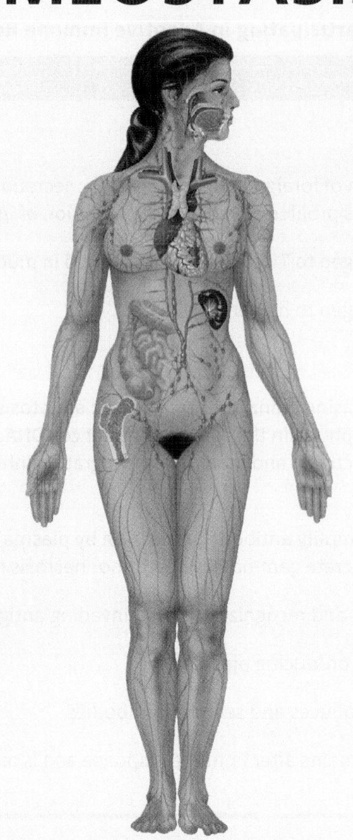

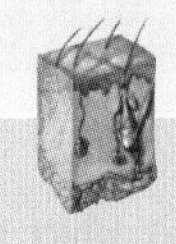

INTEGUMENTARY SYSTEM

- Lymphatic vessels drain excess interstitial fluid and leaked plasma proteins from dermis of skin
- Immune system cells (intraepidermal macrophages) in skin help protect skin
- Lymphatic tissue provides IgA antibodies in sweat

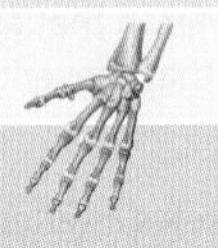

SKELETAL SYSTEM

- Lymphatic vessels drain excess interstitial fluid and leaked plasma proteins from connective tissue around bones

MUSCULAR SYSTEM

- Lymphatic vessels drain excess interstitial fluid and leaked plasma proteins from muscles

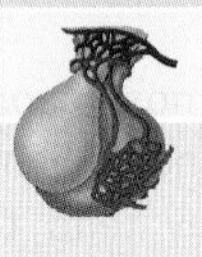

ENDOCRINE SYSTEM

- Flow of lymph helps distribute some hormones and cytokines
- Lymphatic vessels drain excess interstitial fluid and leaked plasma proteins from endocrine glands

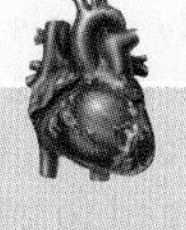

CARDIOVASCULAR SYSTEM

- Lymph returns excess fluid filtered from blood capillaries and leaked plasma proteins to venous blood
- Macrophages in spleen destroy aged red blood cells and remove debris in blood

CONTRIBUTIONS OF THE LYMPHATIC SYSTEM AND IMMUNITY

FOR ALL BODY SYSTEMS

- B cells, T cells, and antibodies protect all body systems from attack by harmful foreign invaders (pathogens), foreign cells, and cancer cells

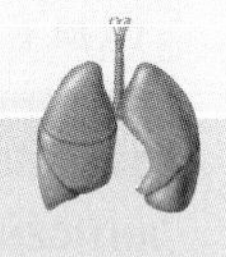

RESPIRATORY SYSTEM

- Tonsils, alveolar macrophages, and MALT (mucosa-associated lymphatic tissue) help protect lungs from pathogens
- Lymphatic vessels drain excess interstitial fluid from lungs

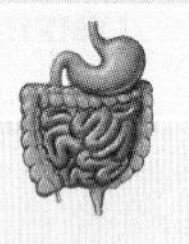

DIGESTIVE SYSTEM

- Tonsils and MALT help defend against toxins and pathogens that penetrate the body from the gastrointestinal tract
- Digestive system provides IgA antibodies in saliva and gastrointestinal secretions
- Lymphatic vessels pick up absorbed dietary lipids and fat-soluble vitamins from the small intestine and transport them to the blood
- Lymphatic vessels drain excess interstitial fluid and leaked plasma proteins from organs of the digestive system

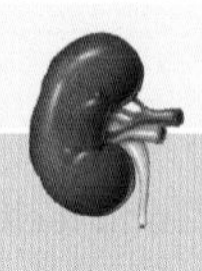

URINARY SYSTEM

- Lymphatic vessels drain excess interstitial fluid and leaked plasma proteins from organs of the urinary system
- MALT helps defend against toxins and pathogens that penetrate the body via the urethra

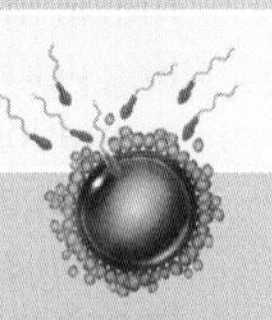

REPRODUCTIVE SYSTEMS

- Lymphatic vessels drain excess interstitial fluid and leaked plasma proteins from organs of the reproductive system
- MALT helps defend against toxins and pathogens that penetrate the body via the vagina and penis
- In females, sperm deposited in the vagina are not attacked as foreign invaders due to inhibition of immune responses
- IgG antibodies can cross the placenta to provide protection to a developing fetus
- Lymphatic tissue provides IgA antibodies in the milk of a nursing mother

Review Questions

22.1 The Concept of Immunity

1. What is a pathogen?
2. Now are innate and adaptive immunity different?

22.2 Overview of the Lymphatic System

3. What are the components and functions of the lymphatic system?

22.3 Lymphatic Vessels and Lymph Circulation

4. How do lymphatic vessels differ in structure form veins?
5. Diagram the route of lymph circulation.

22.4 Lymphatic Organs and Tissues

6. What is the role of the thymus in immunity?
7. What functions do lymph nodes, the spleen, and the tonsils serve?

22.5 Development of Lymphatic Tissues

8. What are the names of the four lymph sacs from which lymphatic vessels develop?

22.6 Innate Immunity

9. What physical and chemical factors provide protection from disease in the skin and mucous membranes?
10. What internal defenses provide protection against microbes that penetrate the skin and mucous membranes?
11. How are the activities of natural killer cells and phagocytes similar and different?
12. What are the main signs, symptoms, and stages of inflammation?

22.7 Adaptive Immunity

13. What is immunocompetence, and which body cells display it?
14. How do the major histocompatibility complex class I and class II self-antigens function?
15. How do antigens arrive at lymphatic tissues?
16. How do antigen-presenting cells process exogenous antigens?
17. What are cytokines, where do they arise, and how do they function?

22.8 Cell-Mediated Immunity

18. What are the functions of helper, cytotoxic, and memory T cells?
19. How do cytotoxic T cells kill infected target cells?
20. How is immunological surveillance useful?

22.9 Antibody-Mediated Immunity

21. How do the five classes of antibodies differ in structure and function?
22. How are cell-mediated and antibody-mediated immune responses similar and different?
23. In what ways does the complement system augment antibody-mediated immune responses?
24. How is the secondary response to an antigen different from the primary response?

22.10 Self-Recognition and Self-Tolerance

25. What do positive selection, negative selection, and anergy accomplish?

22.11 Stress and Immunity

26. Have you ever observed a connection between stress and illness in your own life?

22.12 Aging and the Immune System

27. How are T cells affected by aging?

Critical Thinking Questions

1. Esperanza watched as her mother got her flu shot. "Why do you need a shot if you're not sick?" she asked. "So I won't get sick," answered her mom. Explain how the influenza vaccination prevents illness.
2. Due to the presence of breast cancer, Mrs. Franco had a right radical mastectomy in which her right breast, underlying muscle, and right axillary lymph nodes and vessels were removed. Now she is experiencing severe swelling in her right arm. Why did the surgeon remove lymph tissue as well as the breast? Why is Mrs. Franco's right arm swollen?
3. Tariq's little sister has the mumps. Tariq can't remember if he has had mumps or not, but he is feeling slightly feverish. How could Tariq's doctor determine if he is getting sick with mumps or if he has previously had mumps?

The Respiratory System

CHAPTER **23**

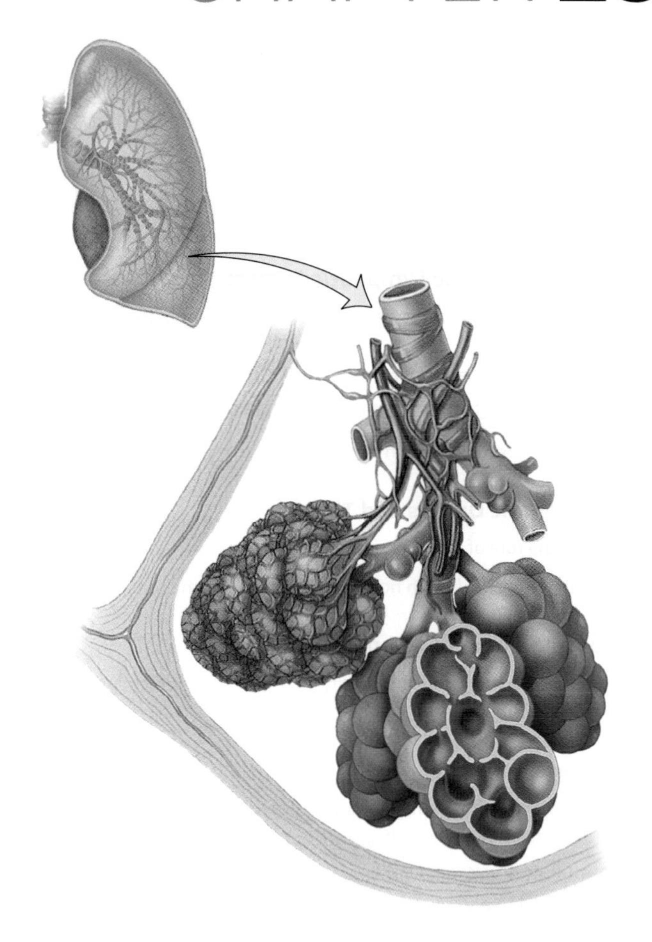

The Respiratory System and Homeostasis

The respiratory system contributes to homeostasis by providing for the exchange of gases—oxygen and carbon dioxide—between the atmospheric air, blood, and tissue cells. It also helps adjust the pH of body fluids.

Your body's cells continually use oxygen (O_2) for the metabolic reactions that generate ATP from the breakdown of nutrient molecules. At the same time, these reactions release carbon dioxide (CO_2) as a waste product. Because an excessive amount of CO_2 produces acidity that can be toxic to cells, excess CO_2 must be eliminated quickly and efficiently. The cardiovascular and respiratory systems cooperate to supply O_2 and eliminate CO_2. The respiratory system provides for gas exchange—intake of O_2 and elimination of CO_2—and the cardiovascular system transports blood containing the gases between the lungs and body cells. Failure of either system disrupts homeostasis by causing rapid death of cells from oxygen starvation and buildup of waste products. In addition to functioning in gas exchange, the respiratory system also participates in regulating blood pH, contains receptors for the sense of smell, filters inspired air, produces sounds, and rids the body of some water and heat in exhaled air. As in the digestive and urinary systems, which will be covered in subsequent chapters, in the respiratory system there is an extensive area of contact between the external environment and capillary blood vessels.

Q Did you ever wonder how smoking affects the respiratory system?

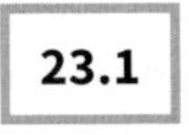

23.1 Overview of the Respiratory System

OBJECTIVES

- **Discuss** the steps that occur during respiration.
- **Define** the respiratory system.
- **Explain** how the respiratory organs are classified structurally and functionally.

The Steps Involved in Respiration

The process of supplying the body with O_2 and removing CO_2 is known as **respiration**, which has three basic steps (**Figure 23.1**):

1. **Pulmonary ventilation** (*pulmon-* = lung), or *breathing*, is the inhalation (inflow) and exhalation (outflow) of air and involves the exchange of air between the atmosphere and the alveoli of the lungs. Inhalation permits O_2 to enter the lungs and exhalation permits CO_2 to leave the lungs.
2. **External** (*pulmonary*) **respiration** is the exchange of gases between the alveoli of the lungs and the blood in pulmonary capillaries across the respiratory membrane. In this process, pulmonary capillary blood gains O_2 and loses CO_2.
3. **Internal** (*tissue*) **respiration** is the exchange of gases between blood in systemic capillaries and tissue cells. In this step the blood loses O_2 and gains CO_2. Within cells, the metabolic reactions that consume O_2 and give off CO_2 during the production of ATP are termed *cellular respiration* (discussed in Chapter 25).

FIGURE 23.1 **The three basic steps involved in respiration.**

During respiration, the body is supplied with O_2, and CO_2 is removed.

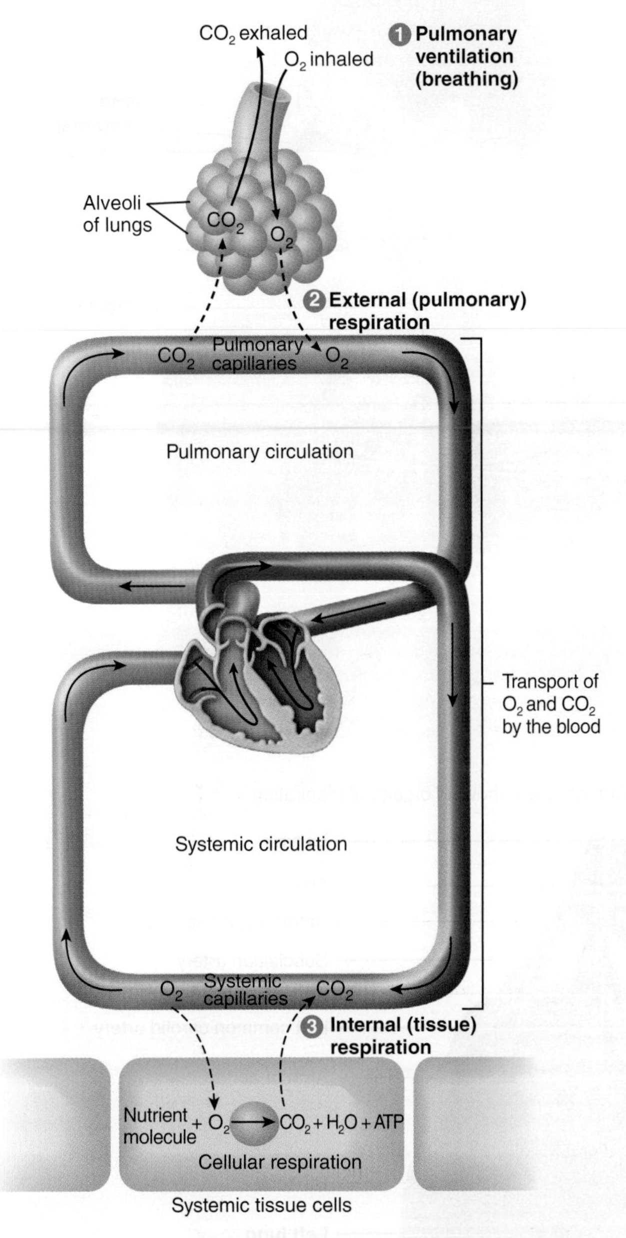

Q How does external respiration differ from internal respiration?

Components of the Respiratory System

The **respiratory system** (RES-pi-ra-tōr-ē) consists of the nose, pharynx (throat), larynx (voice box), trachea (windpipe), bronchi, and lungs (**Figure 23.2**). Its parts can be classified according to either structure or function. *Structurally*, the respiratory system consists of two parts: (1) The **upper respiratory system** includes the nose, nasal cavity, pharynx, and associated structures; (2) the **lower respiratory system** includes the larynx, trachea, bronchi, and lungs. *Functionally*, the respiratory system also consists of two parts. (1) The **conducting zone** consists of a series of interconnecting cavities and tubes both outside and within the lungs. These include the nose, nasal cavity, pharynx, larynx, trachea, bronchi, bronchioles, and terminal bronchioles; their function is to filter, warm, and moisten air and conduct it into the lungs. (2) The **respiratory zone** consists of tubes and tissues within the lungs where gas exchange occurs. These include the respiratory bronchioles, alveolar ducts, alveolar sacs, and alveoli and are the main sites of gas exchange between air and blood.

The branch of medicine that deals with the diagnosis and treatment of diseases of the ears, nose, and throat (ENT) is called **otorhinolaryngology** (ō′-tō-rī′-nō-lar-in-GOL-o-jē; *oto-* = ear; *-rhino-* = nose; *-laryngo-* = voice box; *-logy* = study of).

23.2 The Upper Respiratory System

OBJECTIVES

- **Describe** the anatomy and histology of the nose, pharynx and associated structures.
- **Identify** the functions of these respiratory structures.

Nose

The **nose** is a specialized organ at the entrance of the respiratory system that consists of a visible external portion (external nose) and an internal portion inside the skull called the nasal cavity (internal nose). The **external nose** is the portion of the nose visible on the face and consists of a supporting framework of bone and hyaline cartilage covered with muscle and skin and lined by a mucous membrane. The frontal bone, nasal bones, and maxillae form the *bony framework* of the external nose (**Figure 23.3a**). The *cartilaginous framework* of the external nose consists of several pieces of hyaline cartilage connected to each other and certain skull bones by fibrous connective tissue. The components of the cartilaginous framework are the **septal nasal cartilage**, which forms the anterior portion of the nasal septum; the **lateral nasal cartilages** inferior to the nasal bones; and the **alar cartilages** (Ā-lar), which form a portion of the walls of the nostrils. Because it consists of pliable hyaline cartilage, the cartilaginous framework of the external nose is somewhat flexible. On the undersurface of the external nose are two openings called the **external nares** (NĀ-rez; singular is **naris**) or *nostrils*, which lead into cavities called the **nasal vestibules**. **Figure 23.4** shows the surface anatomy of the nose.

The interior structures of the external nose have three functions: (1) warming, moistening, and filtering incoming air; (2) detecting olfactory stimuli; and (3) modifying speech vibrations as they pass through the large, hollow resonating chambers. *Resonance* refers to prolonging, amplifying, or modifying a sound by vibration.

FIGURE 23.2 **Structures of the respiratory system.**

The upper respiratory system includes the nose, nasal cavity, pharynx, and associated structures; the lower respiratory system includes the larynx, trachea, bronchi, and lungs.

Functions of the respiratory system

1. Provides for gas exchange: intake of O_2 for delivery to body cells and removal of CO_2 produced by body cells.
2. Helps regulate blood pH.
3. Contains receptors for sense of smell, filters inspired air, produces vocal sounds (phonation), and excretes small amounts of water and heat.

Nose:
External
Nasal cavity (internal)
Pharynx
Larynx
Trachea
Right main bronchus
Lungs

(a) Anterior view showing organs of respiration

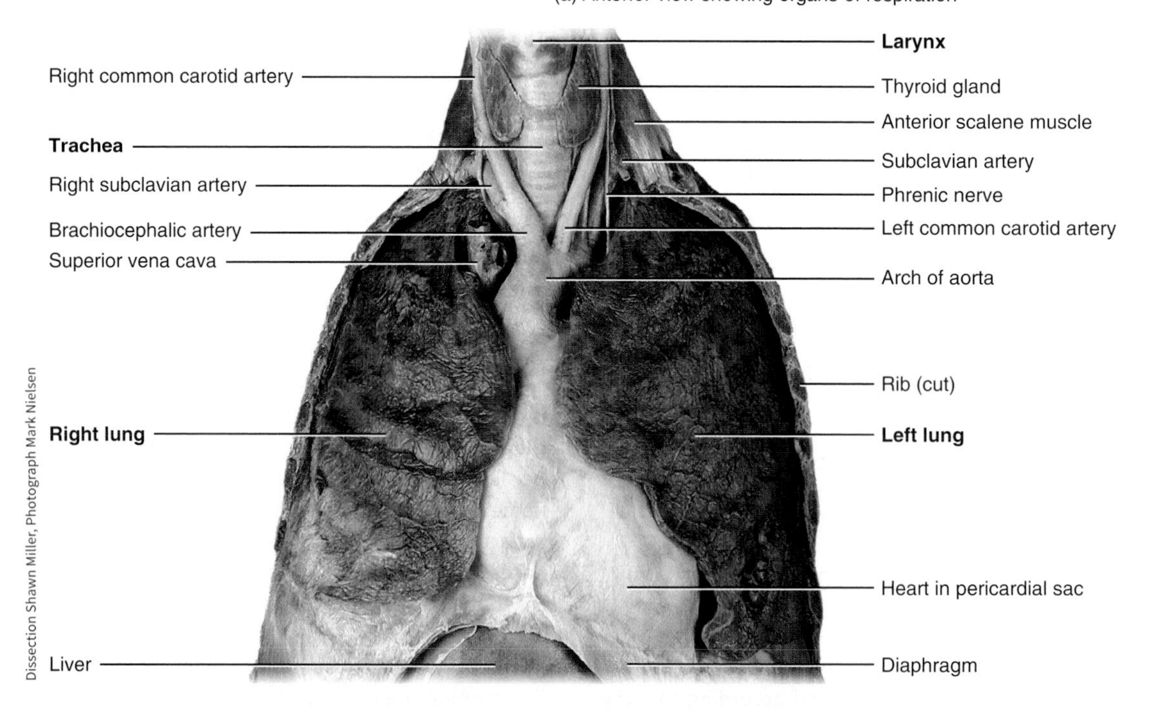

Dissection Shawn Miller, Photograph Mark Nielsen

(b) Anterior view of lungs and heart after removal of the anterolateral thoracic wall and pleura

Q Which structures are part of the conducting zone of the respiratory system?

FIGURE 23.3 Respiratory structures in the head and neck.

As air passes through the nose, it is warmed, filtered, and moistened; and olfaction occurs.

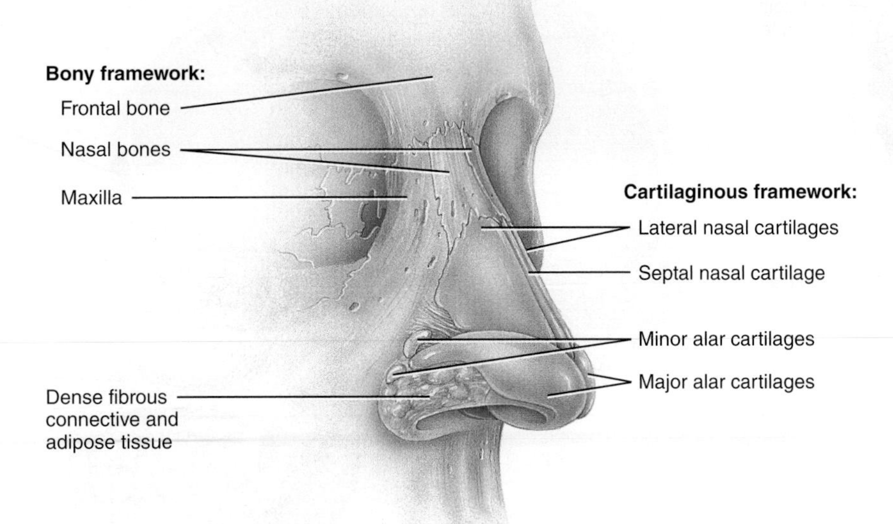

(a) Anterolateral view of nose showing cartilaginous and bony frameworks

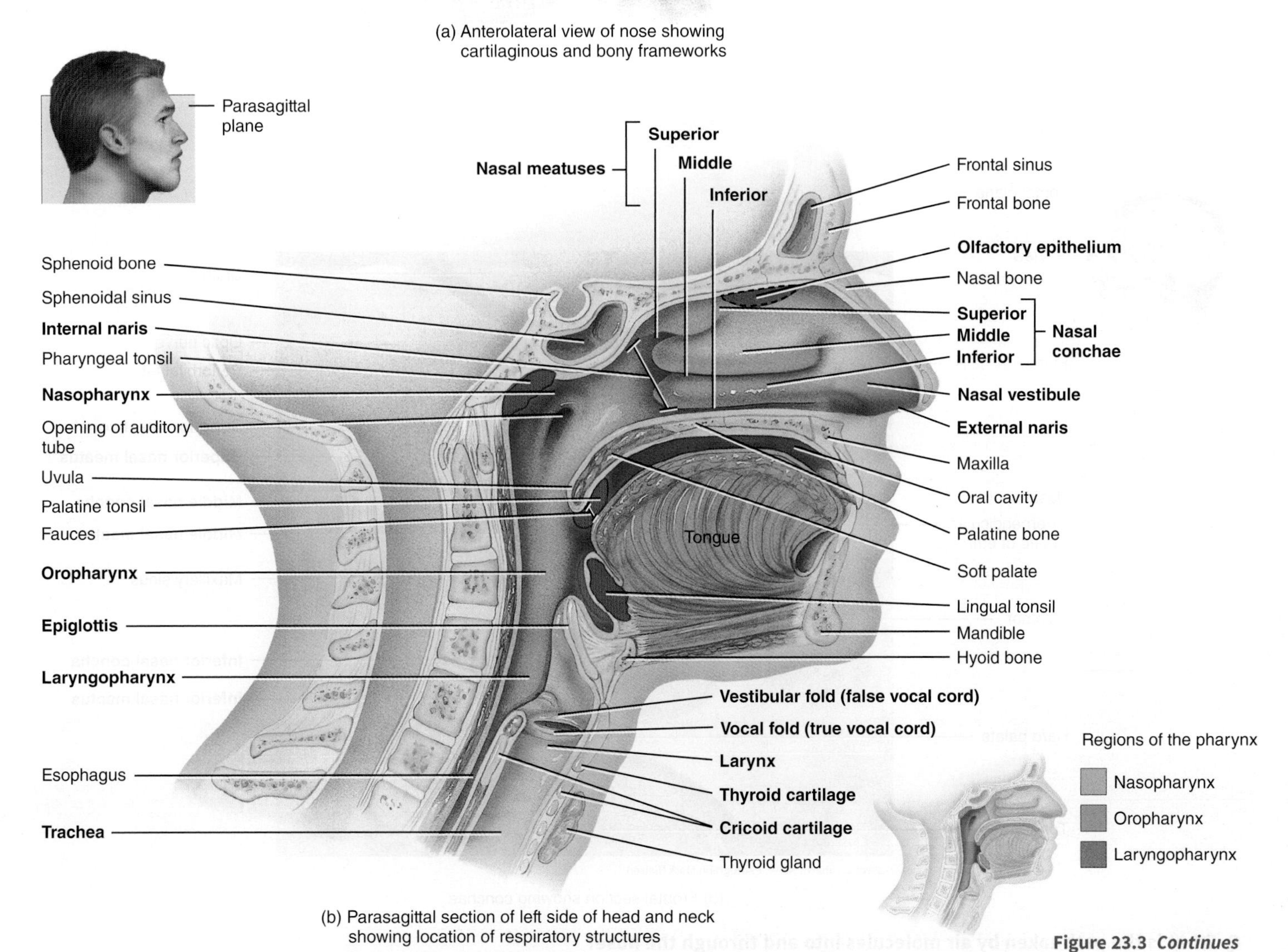

(b) Parasagittal section of left side of head and neck showing location of respiratory structures

Figure 23.3 *Continues*

FIGURE 23.3 Continued

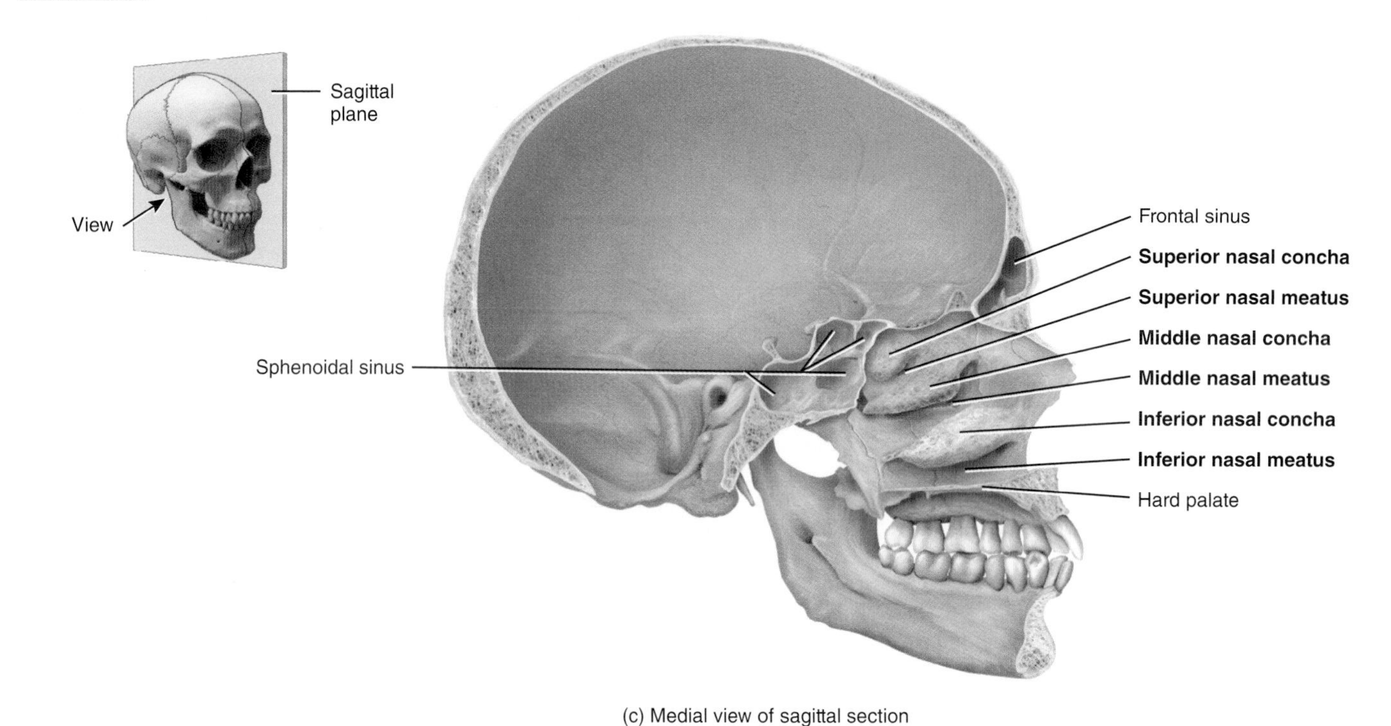

(c) Medial view of sagittal section

Frontal plane
View
Brain
Optic nerve
Periorbital fat
Ethmoidal cells
Superior nasal concha
Superior nasal meatus
Nasal septum:
Perpendicular plate of ethmoid
Middle nasal concha
Middle nasal meatus
Maxillary sinus
Vomer
Inferior nasal concha
Inferior nasal meatus
Hard palate
Tongue

Dissection Shawn Miller, Photograph Mark Nielsen

(d) Frontal section showing conchae

Q What is the path taken by air molecules into and through the nose?

FIGURE 23.4 **Surface anatomy of the nose.**

The external nose has a cartilaginous framework and a bony framework.

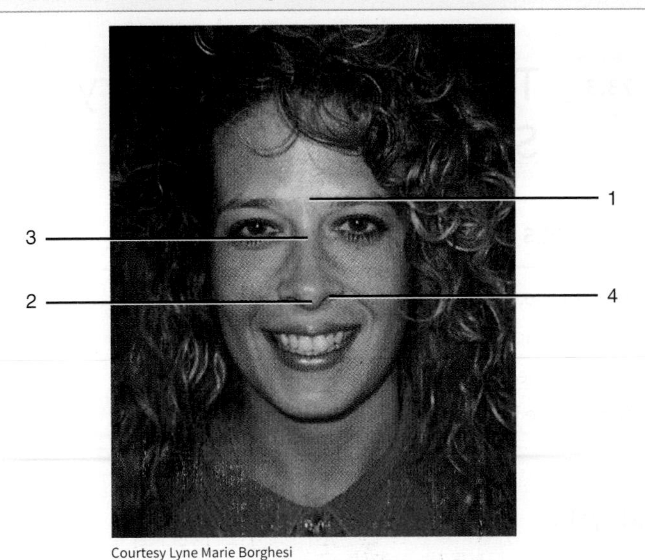

Courtesy Lyne Marie Borghesi

Anterior view

1. **Root**: Superior attachment of the nose to the frontal bone
2. **Apex**: Tip of nose
3. **Bridge**: Bony framework of nose formed by nasal bones
4. **External naris**: Nostril; external opening into nasal cavity

Q Which part of the nose is attached to the frontal bone?

See Clinical Connection: Rhinoplasty

The **nasal cavity** (*internal nose*) is a large space in the anterior aspect of the skull that lies inferior to the nasal bone and superior to the oral cavity; it is lined with muscle and mucous membrane. A vertical partition, the **nasal septum**, divides the nasal cavity into right and left sides. The anterior portion of the nasal septum consists primarily of hyaline cartilage; the remainder is formed by the vomer and the perpendicular plate of the ethmoid, maxillae, and palatine bones (see Figure 7.11).

Anteriorly, the nasal cavity merges with the external nose, and posteriorly it communicates with the pharynx through two openings called the **internal nares** or *choanae* (kō-Ā-nē) (see Figure 23.3b). Ducts from the *paranasal sinuses* (which drain mucus) and the *nasolacrimal ducts* (which drain tears) also open into the nasal cavity. Recall from Chapter 7 that the paranasal sinuses are cavities in certain cranial and facial bones lined with mucous membrane that are continuous with the lining of the nasal cavity. Skull bone, containing the paranasal sinuses are the frontal, sphenoid, ethmoid, and maxillae. Besides producing mucus, the paranasal sinuses serve as resonating chambers for sound as we speak or sing. The lateral walls of the internal nose are formed by the ethmoid, maxillae, lacrimal, palatine, and inferior nasal conchae bones (see Figure 7.9); the ethmoid bone also forms the roof. The palatine bones and palatine processes of the maxillae, which together constitute the hard palate, form the floor of the internal nose.

The bony and cartilaginous framework of the nose help to keep the vestibule and nasal cavity *patent*, that is, open or unobstructed. The nasal cavity is divided into a larger, inferior *respiratory region* and a smaller, superior *olfactory region*. The respiratory region is lined with ciliated pseudostratified columnar epithelium with numerous goblet cells, which is frequently called the **respiratory epithelium** (see Table 4.1). The anterior portion of the nasal cavity just inside the nostrils, called the **nasal vestibule**, is surrounded by cartilage; the superior part of the nasal cavity is surrounded by bone.

When air enters the nostrils, it passes first through the vestibule, which is lined by skin containing coarse hairs that filter out large dust particles. Three shelves formed by projections of the **superior**, **middle**, and **inferior nasal conchae** extend out of each lateral wall of the nasal cavity. The conchae, almost reaching the nasal septum, subdivide each side of the nasal cavity into a series of groovelike air passageways—the **superior**, **middle**, and **inferior nasal meatuses** (mē-Ā-tus-ēz = openings or passages; singular is **meatus**). Mucous membrane lines the nasal cavity and its shelves. The arrangement of conchae and meatuses increases surface area in the internal nose and prevents dehydration by trapping water droplets during exhalation.

See Clinical Connection: Tonsillectomy

As inhaled air whirls around the conchae and meatuses, it is warmed by blood in the capillaries. Mucus secreted by the goblet cells moistens the air and traps dust particles. Drainage from the nasolacrimal ducts also helps moisten the air, and is sometimes assisted by secretions from the paranasal sinuses. The cilia move the mucus and trapped dust particles toward the pharynx, at which point they can be swallowed or spit out, thus removing the particles from the respiratory tract.

The olfactory receptor cells, supporting cells, and basal cells lie in the respiratory region, which is near the superior nasal conchae and adjacent septum. These cells make up the **olfactory epithelium**. It contains cilia but no goblet cells.

Pharynx

The **pharynx** (FAR-inks), or throat, is a funnel-shaped tube about 13 cm (5 in.) long that starts at the internal nares and extends to the level of the cricoid cartilage, the most inferior cartilage of the larynx (voice box) (see Figure 23.3b). The pharynx lies just posterior to the nasal and oral cavities, superior to the larynx, and just anterior to the cervical vertebrae. Its wall is composed of skeletal muscles and is lined with a mucous membrane. Relaxed skeletal muscles help keep the pharynx patent. Contraction of the skeletal muscles assists in deglutition (swallowing). The pharynx functions as a passageway for air and food, provides a resonating chamber for speech sounds, and houses the tonsils, which participate in immunological reactions against foreign invaders.

The pharynx can be divided into three anatomical regions: (1) nasopharynx, (2) oropharynx, and (3) laryngopharynx. (See the lower orientation diagram in Figure 23.3b.) The muscles of the entire pharynx are arranged in two layers, an outer circular layer and an inner longitudinal layer.

The superior portion of the pharynx, called the **nasopharynx**, lies posterior to the nasal cavity and extends to the soft palate. The **soft palate**, which forms the posterior portion of the roof of the mouth, is an arch-shaped muscular partition between the nasopharynx and oropharynx that is lined by mucous membrane. There are five openings in its wall: two internal nares, two openings that lead into the *auditory* (*pharyngotympanic*) *tubes* (commonly known as the *eustachian tubes*), and the opening into the oropharynx. The posterior wall also contains the **pharyngeal tonsil** (fa-RIN-je-al), or *adenoid*. Through the internal nares, the nasopharynx receives air from the nasal cavity along with packages of dust-laden mucus. The nasopharynx is lined with ciliated pseudostratified columnar epithelium, and the cilia move the mucus down toward the most inferior part of the pharynx. The nasopharynx also exchanges small amounts of air with the auditory tubes to equalize air pressure between the middle ear and the atmosphere.

The intermediate portion of the pharynx, the **oropharynx**, lies posterior to the oral cavity and extends from the soft palate inferiorly to the level of the hyoid bone. It has only one opening into it, the **fauces** (FAW-sēz = throat), the opening from the mouth. This portion of the pharynx has both respiratory and digestive functions, serving as a common passageway for air, food, and drink. Because the oropharynx is subject to abrasion by food particles, it is lined with nonkeratinized stratified squamous epithelium. Two pairs of tonsils, the **palatine** and **lingual tonsils**, are found in the oropharynx.

The inferior portion of the pharynx, the **laryngopharynx** (la-RING-gō-far-ingks), or *hypopharynx*, begins at the level of the hyoid bone. At its inferior end it opens into the esophagus (food tube) posteriorly and the larynx (voice box) anteriorly. Like the oropharynx, the laryngopharynx is both a respiratory and a digestive pathway and is lined by nonkeratinized stratified squamous epithelium.

23.3 The Lower Respiratory System

OBJECTIVES

- **Identity** the features and purpose of the larynx.
- **List** the structures of voice production.
- **Describe** the anatomy and histology of the trachea.
- **Identify** the functions of each bronchial structure.

Larynx

The **larynx** (LAR-ingks), or voice box, is a short passageway that connects the laryngopharynx with the trachea. It lies in the midline of the neck anterior to the esophagus and the fourth through sixth cervical vertebrae (C4–C6).

The wall of the larynx is composed of nine pieces of cartilage (**Figure 23.5**). Three occur singly (thyroid cartilage, epiglottis, and cricoid cartilage), and three occur in pairs (arytenoid, cuneiform,

FIGURE 23.5 The larynx.

The larynx is composed of nine pieces of cartilage.

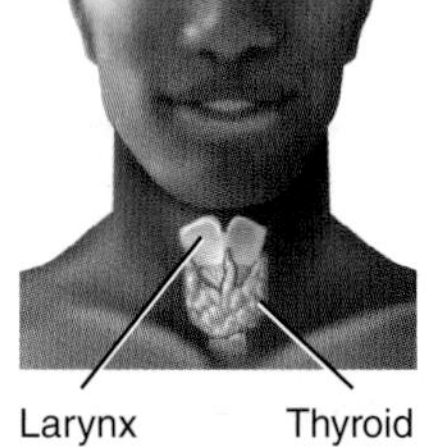

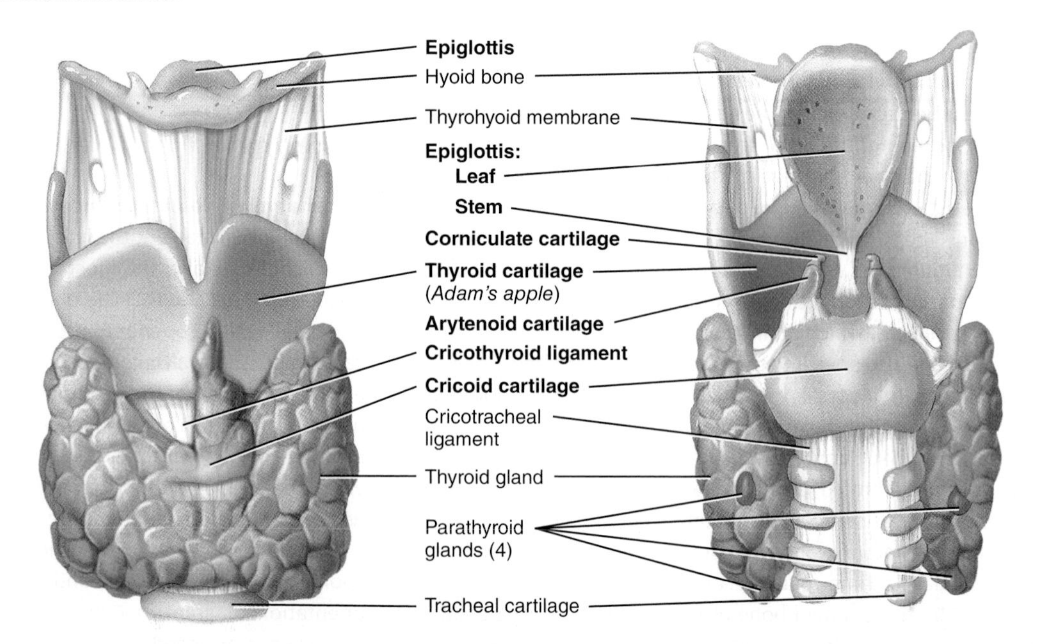

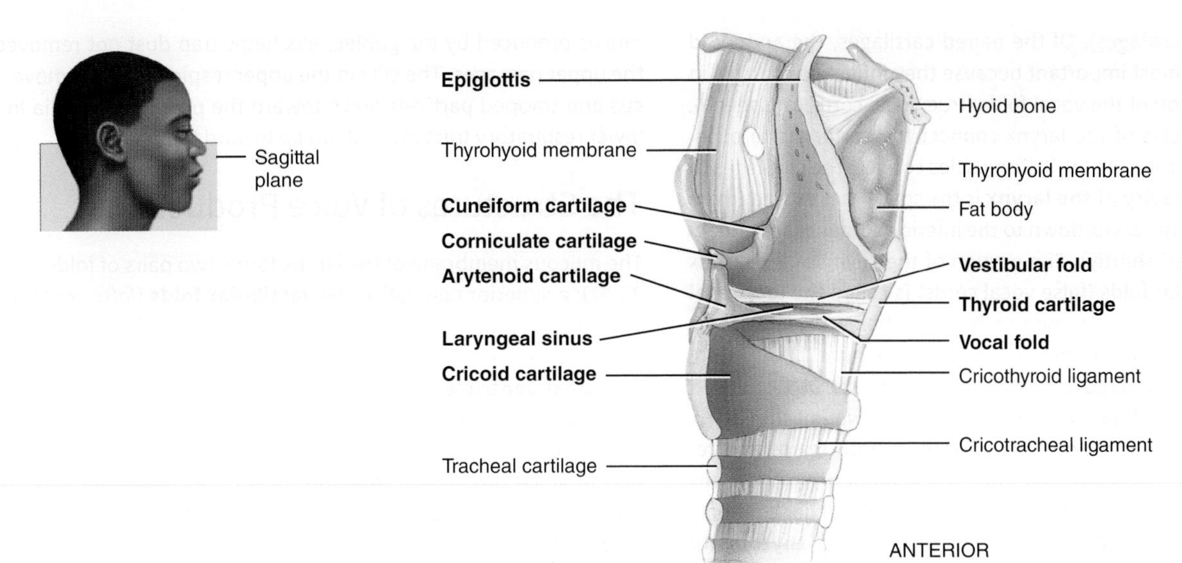

(c) Sagittal section

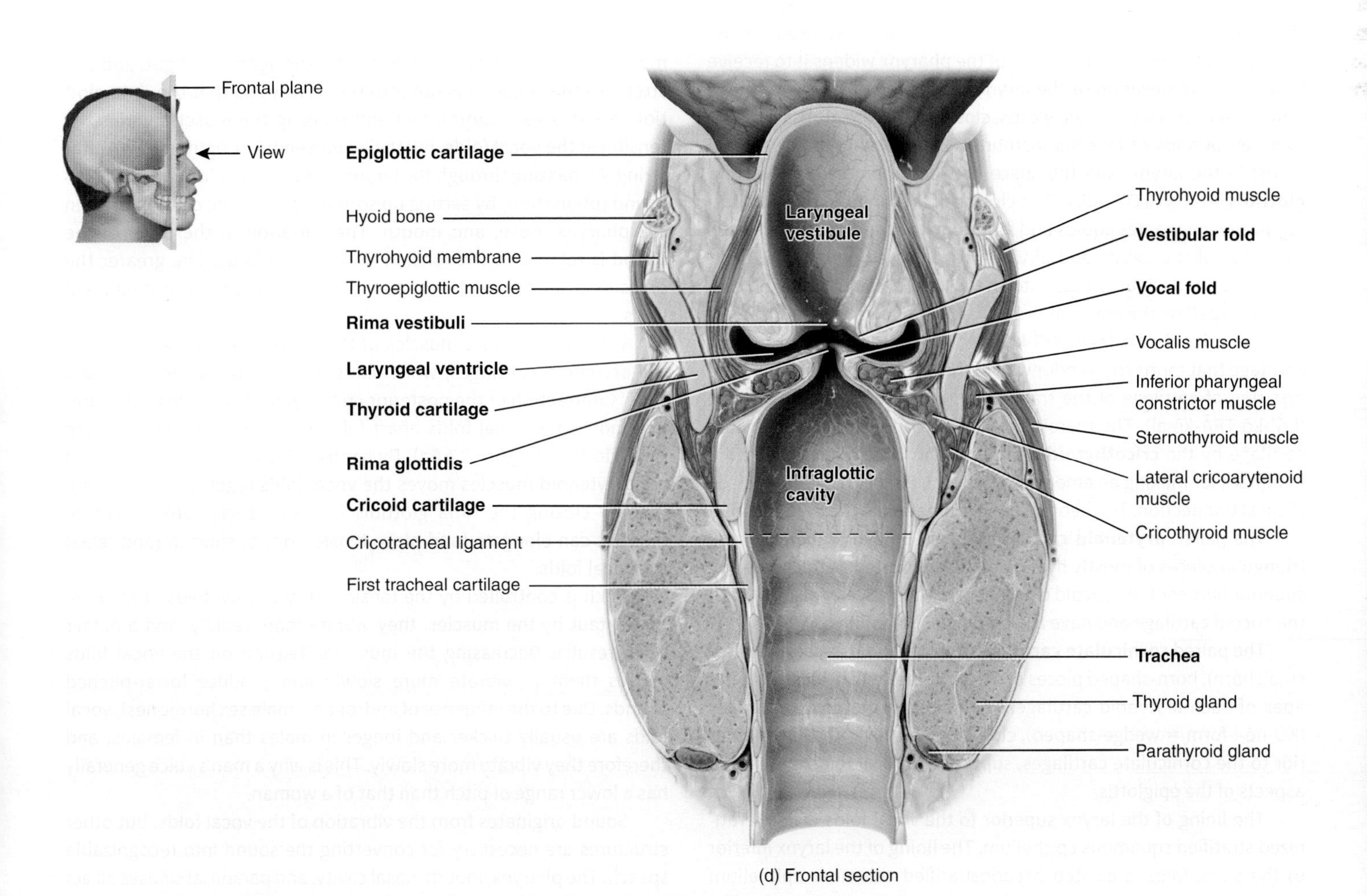

(d) Frontal section

Q How does the epiglottis prevent aspiration of foods and liquids?

and corniculate cartilages). Of the paired cartilages, the arytenoid cartilages are the most important because they influence changes in position and tension of the vocal folds (true vocal cords for speech). The extrinsic muscles of the larynx connect the cartilages to other structures in the throat; the intrinsic muscles connect the cartilages to one another. The **cavity of the larynx** is the space that extends from the entrance into the larynx down to the inferior border of the cricoid cartilage (described shortly). The portion of the cavity of the larynx above the vestibular folds (false vocal cords) is called the **laryngeal vestibule**. The portion of the cavity of the larynx below the vocal folds is called the **infraglottic cavity** (*infra-* = below) (Figure 23.5d).

The **thyroid cartilage** (*Adam's apple*) consists of two fused plates of hyaline cartilage that form the anterior wall of the larynx and give it a triangular shape. It is present in both males and females but is usually larger in males due to the influence of male sex hormones on its growth during puberty. The ligament that connects the thyroid cartilage to the hyoid bone is called the **thyrohyoid membrane**.

The **epiglottis** (*epi-* = over; *-glottis* = tongue) is a large, leaf-shaped piece of elastic cartilage that is covered with epithelium (see also Figure 23.3b). The "stem" of the epiglottis is the tapered inferior portion that is attached to the anterior rim of the thyroid cartilage. The broad superior "leaf" portion of the epiglottis is unattached and is free to move up and down like a trap door. During swallowing, the pharynx and larynx rise. Elevation of the pharynx widens it to receive food or drink; elevation of the larynx causes the epiglottis to move down and form a lid over the glottis, closing it off. The **glottis** consists of a pair of folds of mucous membrane, the vocal folds (true vocal cords) in the larynx, and the space between them called the **rima glottidis** (RĪ-ma GLOT-ti-dis). The closing of the larynx in this way during swallowing routes liquids and foods into the esophagus and keeps them out of the larynx and airways. When small particles of dust, smoke, food, or liquids pass into the larynx, a cough reflex occurs, usually expelling the material.

The **cricoid cartilage** (KRĪ-koyd = ringlike) is a ring of hyaline cartilage that forms the inferior wall of the larynx. It is attached to the first ring of cartilage of the trachea by the **cricotracheal ligament** (krī′-kō-TRĀ-kē-al). The thyroid cartilage is connected to the cricoid cartilage by the **cricothyroid ligament**. The cricoid cartilage is the landmark for making an emergency airway called a tracheotomy (see Clinical Connection: Tracheotomy and Intubation in the Study Guide).

The paired **arytenoid cartilages** (ar′-i-TĒ-noyd = ladlelike) are triangular pieces of mostly hyaline cartilage located at the posterior, superior border of the cricoid cartilage. They form synovial joints with the cricoid cartilage and have a wide range of mobility.

The paired **corniculate cartilages** (kor-NIK-ū-lāt = shaped like a small horn), horn-shaped pieces of elastic cartilage, are located at the apex of each arytenoid cartilage. The paired **cuneiform cartilages** (KŪ-nē-i-form = wedge-shaped), club-shaped elastic cartilages anterior to the corniculate cartilages, support the vocal folds and lateral aspects of the epiglottis.

The lining of the larynx superior to the vocal folds is nonkeratinized stratified squamous epithelium. The lining of the larynx inferior to the vocal folds is ciliated pseudostratified columnar epithelium consisting of ciliated columnar cells, goblet cells, and basal cells. The mucus produced by the goblet cells helps trap dust not removed in the upper passages. The cilia in the upper respiratory tract move mucus and trapped particles *down* toward the pharynx; the cilia in the lower respiratory tract move them *up* toward the pharynx.

The Structures of Voice Production

The mucous membrane of the larynx forms two pairs of folds (Figure 23.5c): a superior pair called the **vestibular folds** (*false vocal cords*) and an inferior pair called the **vocal folds** (*true vocal cords*). The space between the vestibular folds is known as the **rima vestibuli**. The **laryngeal ventricle** is a lateral expansion of the middle portion of the laryngeal cavity inferior to the vestibular folds and superior to the vocal folds (see Figure 23.3b). While the vestibular folds do not function in voice production, they do have other important functional roles. When the vestibular folds are brought together, they function in holding the breath against pressure in the thoracic cavity, such as might occur when a person strains to lift a heavy object.

The vocal folds are the principal structures of voice production. Deep to the mucous membrane of the vocal folds, which is nonkeratinized stratified squamous epithelium, are bands of elastic ligaments stretched between the rigid cartilages of the larynx like the strings on a guitar. Intrinsic laryngeal muscles attach to both the rigid cartilages and the vocal folds. When the muscles contract they move the cartilages, which pulls the elastic ligaments tight, and this stretches the vocal folds out into the airways so that the rima glottidis is narrowed. Contracting and relaxing the muscles varies the tension in the vocal folds, much like loosening or tightening a guitar string. Air passing through the larynx vibrates the folds and produces sound (phonation) by setting up sound waves in the column of air in the pharynx, nose, and mouth. The variation in the pitch of the sound is related to the tension in the vocal folds. The greater the pressure of air, the louder the sound produced by the vibrating vocal folds.

When the intrinsic muscles of the larynx contract, they pull on the arytenoid cartilages, which causes the cartilages to pivot and slide. Contraction of the posterior cricoarytenoid muscles, for example, moves the vocal folds apart (abduction), thereby opening the rima glottidis (Figure 23.6a). By contrast, contraction of the lateral cricoarytenoid muscles moves the vocal folds together (adduction), thereby closing the rima glottidis (Figure 23.6b). Other intrinsic muscles can elongate (and place tension on) or shorten (and relax) the vocal folds.

Pitch is controlled by the tension on the vocal folds. If they are pulled taut by the muscles, they vibrate more rapidly, and a higher pitch results. Decreasing the muscular tension on the vocal folds causes them to vibrate more slowly and produce lower-pitched sounds. Due to the influence of androgens (male sex hormones), vocal folds are usually thicker and longer in males than in females, and therefore they vibrate more slowly. This is why a man's voice generally has a lower range of pitch than that of a woman.

Sound originates from the vibration of the vocal folds, but other structures are necessary for converting the sound into recognizable speech. The pharynx, mouth, nasal cavity, and paranasal sinuses all act as resonating chambers that give the voice its human and individual

FIGURE 23.6 Movement of the vocal folds.

The glottis consists of a pair of folds of mucous membrane in the larynx (the vocal folds) and the space between them (the rima glottidis).

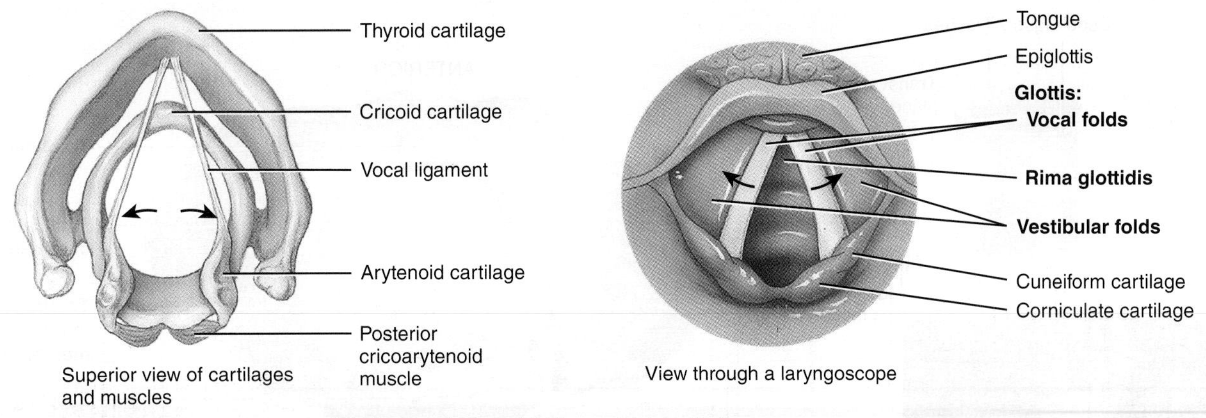

(a) Movement of vocal folds apart (abduction)

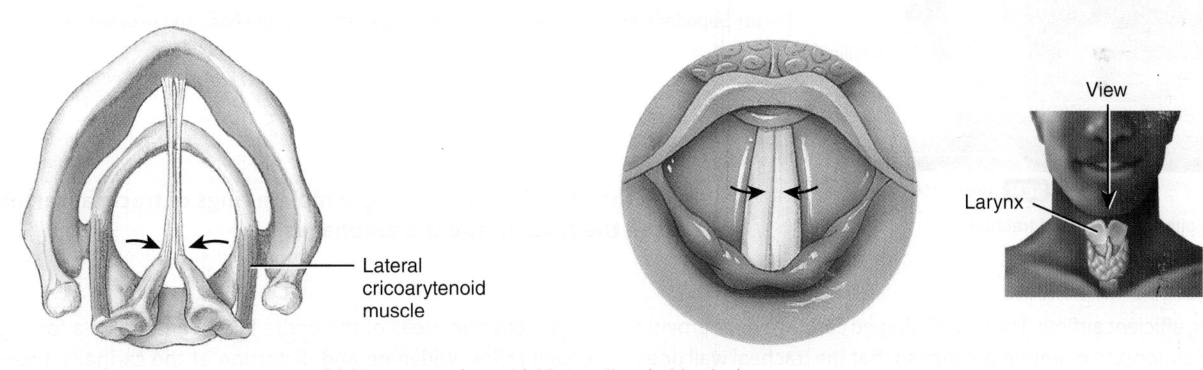

(b) Movement of vocal folds together (adduction)

Q What is the main function of the vocal folds?

quality. We produce the vowel sounds by constricting and relaxing the muscles in the wall of the pharynx. Muscles of the face, tongue, and lips help us enunciate words.

Whispering is accomplished by closing all but the posterior portion of the rima glottidis. Because the vocal folds do not vibrate during whispering, there is no pitch to this form of speech. However, we can still produce intelligible speech while whispering by changing the shape of the oral cavity as we enunciate. As the size of the oral cavity changes, its resonance qualities change, which imparts a vowel-like pitch to the air as it rushes toward the lips.

See Clinical Connection: Laryngitis and Cancer of the Larynx

Trachea

The **trachea** (TRĀ-kē-a = sturdy), or *windpipe*, is a tubular passageway for air that is about 12 cm (5 in.) long and 2.5 cm (1 in.) in diameter. It is located anterior to the esophagus (**Figure 23.7**) and extends from the larynx to the superior border of the fifth thoracic vertebra (T5), where it divides into right and left primary bronchi (see **Figure 23.8**).

The layers of the tracheal wall, from deep to superficial, are the (1) mucosa, (2) submucosa, (3) hyaline cartilage, and (4) adventitia (composed of areolar connective tissue). The mucosa of the trachea consists of an epithelial layer of ciliated pseudostratified columnar epithelium and an underlying layer of lamina propria that contains elastic and reticular fibers. It provides the same protection against dust as the membrane lining the nasal cavity and larynx. The submucosa consists of areolar connective tissue that contains seromucous glands and their ducts.

The 16–20 incomplete, horizontal rings of hyaline cartilage resemble the letter C, are stacked one above another, and are connected by dense connective tissue. They may be felt through the skin inferior to the larynx. The open part of each C-shaped cartilage ring faces posteriorly toward the esophagus (**Figure 23.7**) and is spanned by a *fibromuscular membrane*. Within this membrane are transverse smooth muscle fibers, called the *trachealis muscle* (trā-kē-Ā-lis), and elastic connective tissue that allow the diameter of the trachea to change subtly during inhalation and exhalation, which is important in

FIGURE 23.7 Location of the trachea in relation to the esophagus.

The trachea is anterior to the esophagus and extends from the larynx to the superior border of the fifth thoracic vertebra.

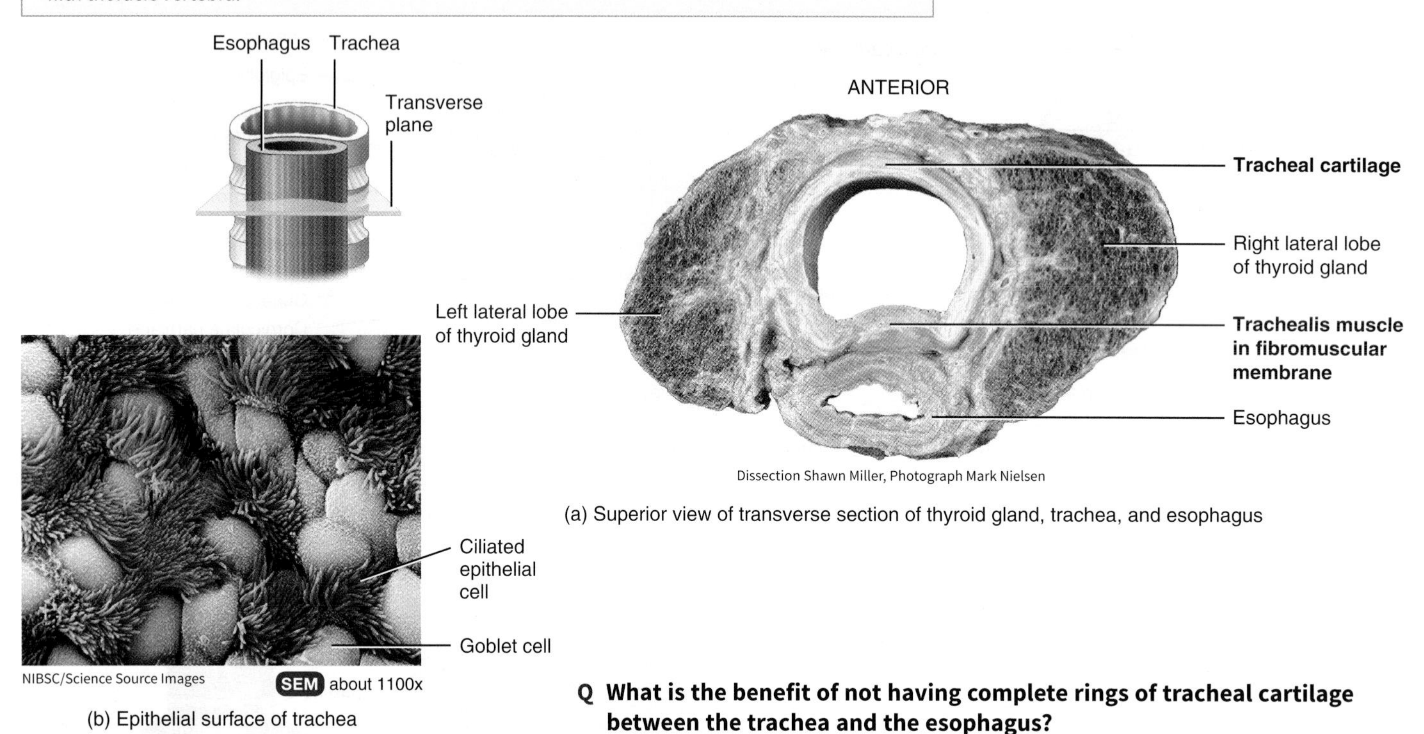

(a) Superior view of transverse section of thyroid gland, trachea, and esophagus

(b) Epithelial surface of trachea

Q What is the benefit of not having complete rings of tracheal cartilage between the trachea and the esophagus?

maintaining efficient airflow. The solid C-shaped cartilage rings provide a semirigid support to maintain patency so that the tracheal wall does not collapse inward (especially during inhalation) and obstruct the air passageway. The adventitia of the trachea consists of areolar connective tissue that joins the trachea to surrounding tissues.

See Clinical Connection: Tracheotomy and Intubation

Bronchi

At the superior border of the fifth thoracic vertebra, the trachea divides into a **right main** (*primary*) **bronchus** (BRONG-kus = windpipe), which goes into the right lung, and a **left main** (*primary*) **bronchus**, which goes into the left lung (Figure 23.8). The right main bronchus is more vertical, shorter, and wider than the left. As a result, an aspirated object is more likely to enter and lodge in the right main bronchus than the left. Like the trachea, the main bronchi (BRONG-kī) contain incomplete rings of cartilage and are lined by ciliated pseudostratified columnar epithelium.

At the point where the trachea divides into right and left main bronchi an internal ridge called the **carina** (ka-RĪ-na = keel of a boat) is formed by a posterior and somewhat inferior projection of the last tracheal cartilage. The mucous membrane of the carina is one of the most sensitive areas of the entire larynx and trachea for triggering a cough reflex. Widening and distortion of the carina is a serious sign because it usually indicates a carcinoma of the lymph nodes around the region where the trachea divides.

On entering the lungs, the main bronchi divide to form smaller bronchi—the **lobar** (*secondary*) **bronchi**, one for each lobe of the lung. (The right lung has three lobes; the left lung has two.) The lobar bronchi continue to branch, forming still smaller bronchi, called **segmental** (*tertiary*) **bronchi** (TER-shē-e-rē), that supply the specific bronchopulmonary segments within the lobes. The segmental bronchi then divide into **bronchioles**. Bronchioles in turn branch repeatedly, and the smallest ones branch into even smaller tubes called **terminal bronchioles**. These bronchioles contain *club* (*Clara*) *cells,* columnar, nonciliated cells interspersed among the epithelial cells. Club cells may protect against harmful effects of inhaled toxins and carcinogens, produce surfactant (discussed shortly), and function as stem cells (reserve cells), which give rise to various cells of the epithelium. The terminal bronchioles represent the end of the conducting zone of the respiratory system. This extensive branching from the trachea through the terminal bronchioles resembles an inverted tree and is commonly referred to as the **bronchial tree**. Beyond the terminal bronchioles of the bronchial tree, the branches become microscopic. These branches are called the respiratory bronchioles and alveolar ducts, which will be described shortly (see Figure 23.11).

The respiratory passages from the trachea to the alveolar ducts contain about 23 generations of branching; branching from the

FIGURE 23.8 Branching of airways from the trachea.

The bronchial tree consists of macroscopic airways that begin at the trachea and continue through the terminal bronchioles.

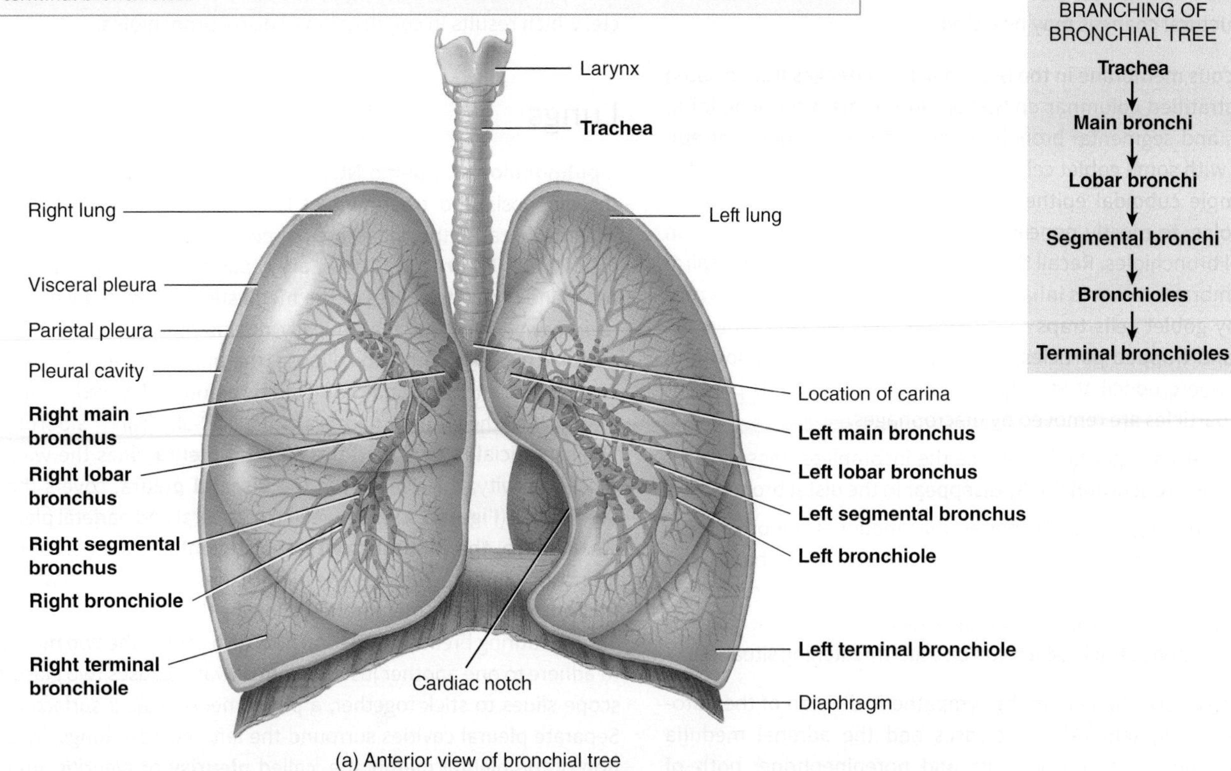

(a) Anterior view of bronchial tree

Airway branching			
		Names of branches	Generation #
Conducting zone		Trachea	0
		Main bronchi	1
		Lobar and segmental bronchi	2–10
		Bronchioles and terminal bronchioles	11–16
Respiratory zone		Respiratory bronchioles	17–19
		Alveolar ducts	20–22
		Alveolar sacs	23

(b) Airway branching

Q How many lobes and secondary bronchi are present in each lung?

trachea into main bronchi is called first-generation branching, that from main bronchi into lobar bronchi is called second-generation branching, and so on down to the alveolar ducts (**Figure 23.8b**).

As the branching becomes more extensive in the bronchial tree, several structural changes may be noted.

1. The mucous membrane in the bronchial tree changes from ciliated pseudostratified columnar epithelium in the main bronchi, lobar bronchi, and segmental bronchi to ciliated simple columnar epithelium with some goblet cells in larger bronchioles, to mostly ciliated simple cuboidal epithelium with no goblet cells in smaller bronchioles, to mostly nonciliated simple cuboidal epithelium in terminal bronchioles. Recall that ciliated epithelium of the respiratory membrane removes inhaled particles in two ways; mucus produced by goblet cells traps the particles, and the cilia move the mucus and trapped particles toward the pharynx for removal. In regions where nonciliated simple cuboidal epithelium is present, inhaled particles are removed by macrophages.
2. Plates of cartilage gradually replace the incomplete rings of cartilage in main bronchi and finally disappear in the distal bronchioles.
3. As the amount of cartilage decreases, the amount of smooth muscle increases. Smooth muscle encircles the lumen in spiral bands and helps maintain patency. However, because there is no supporting cartilage, muscle spasms can close off the airways. This is what happens during an asthma attack, which can be a life-threatening situation.

During exercise, activity in the sympathetic division of the autonomic nervous system (ANS) increases and the adrenal medulla releases the hormones epinephrine and norepinephrine; both of these events cause relaxation of smooth muscle in the bronchioles, which dilates the airways. Because air reaches the alveoli more quickly, lung ventilation improves. The parasympathetic division of the ANS and mediators of allergic reactions such as histamine have the opposite effect, causing contraction of bronchiolar smooth muscle, which results in constriction of distal bronchioles.

Lungs

A **pulmonologist** (pul-mō-NOL-ō-gist; *pulmo-* = lung) is a specialist in the diagnosis and treatment of lung diseases. The **lungs** (= lightweights, because they float) are paired cone-shaped organs in the thoracic cavity (**Figure 23.9**). They are separated from each other by the heart and other structures of the mediastinum, which divides the thoracic cavity into two anatomically distinct chambers. As a result, if trauma causes one lung to collapse, the other may remain expanded. Each lung is enclosed and protected by a double-layered serous membrane called the **pleural membrane** (PLOOR-al; *pleur-* = side) or *pleura*. The superficial layer, called the **parietal pleura**, lines the wall of the thoracic cavity; the deep layer, the **visceral pleura**, covers the lungs themselves (**Figure 23.9**). Between the visceral and parietal pleurae is a small space, the **pleural cavity**, which contains a small amount of lubricating fluid secreted by the membranes. This pleural fluid reduces friction between the membranes, allowing them to slide easily over one another during breathing. Pleural fluid also causes the two membranes to adhere to one another just as a film of water causes two glass microscope slides to stick together, a phenomenon called surface tension. Separate pleural cavities surround the left and right lungs. Inflammation of the pleural membrane, called **pleurisy** or *pleuritis*, may in its early stages cause pain due to friction between the parietal and visceral

FIGURE 23.9 **Relationship of the pleural membranes to the lungs.**

The parietal pleura lines the thoracic cavity, and the visceral pleura covers the lungs.

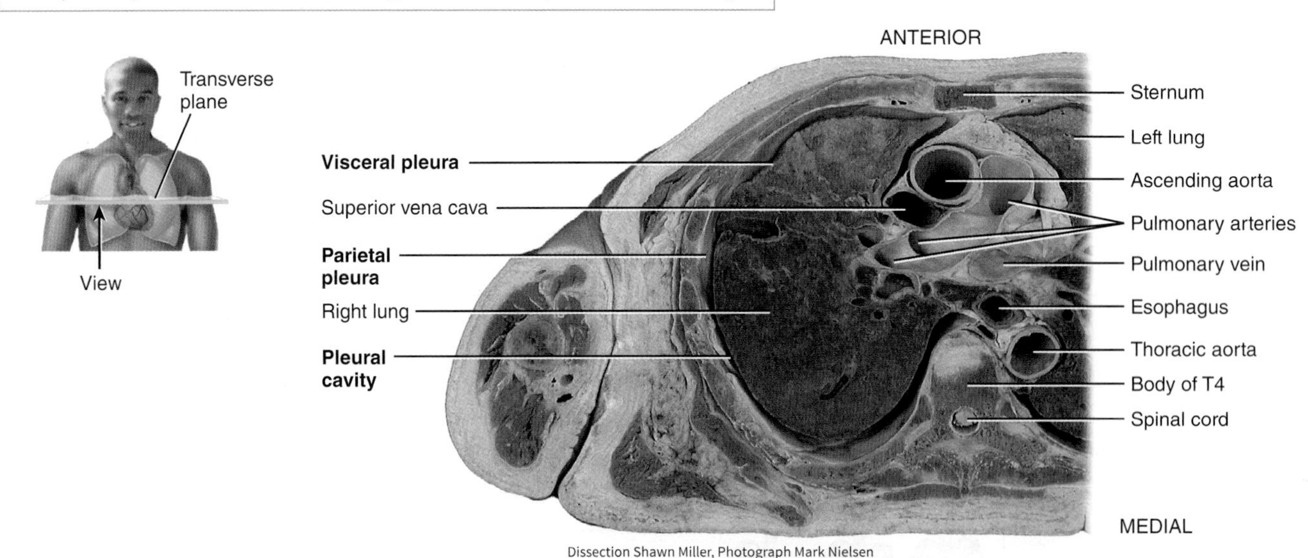

Inferior view of a transverse section through the thoracic cavity showing the pleural cavity and pleural membranes

Q What type of membrane is the pleural membrane?

layers of the pleura. If the inflammation persists, excess fluid accumulates in the pleural space, a condition known as **pleural effusion**.

See Clinical Connection: Pneumothorax and Hemothorax

The lungs extend from the diaphragm to just slightly superior to the clavicles and lie against the ribs anteriorly and posteriorly (**Figure 23.10a**). The broad inferior portion of the lung, the **base**, is concave and fits over the convex area of the diaphragm. The narrow superior portion of the lung is the **apex**. The surface of the lung lying against the ribs, the **costal surface**, matches the rounded curvature of the ribs.

FIGURE 23.10 **Surface anatomy of the lungs.**

The oblique fissure divides the left lung into two lobes. The oblique and horizontal fissures divide the right lung into three lobes.

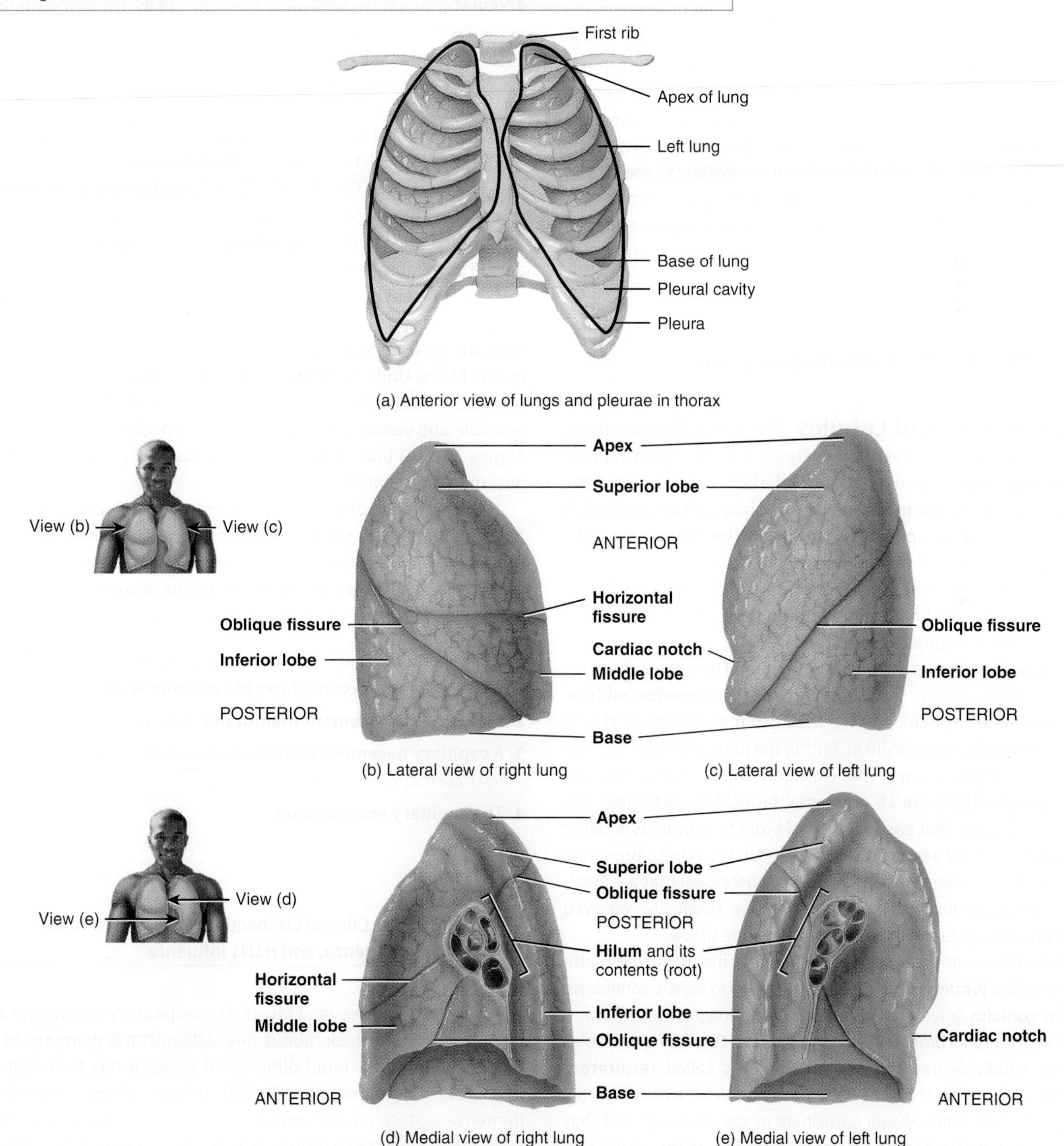

Q Why are the right and left lungs slightly different in size and shape?

The **mediastinal** (*medial*) **surface** of each lung contains a region, the **hilum**, through which bronchi, pulmonary blood vessels, lymphatic vessels, and nerves enter and exit (Figure 23.10e). These structures are held together by the pleura and connective tissue and constitute the **root** of the lung. Medially, the left lung also contains a concavity, the **cardiac notch**, in which the apex of the heart lies. Due to the space occupied by the heart, the left lung is about 10% smaller than the right lung. Although the right lung is thicker and broader, it is also somewhat shorter than the left lung because the diaphragm is higher on the right side, accommodating the liver that lies inferior to it.

The lungs almost fill the thorax (Figure 23.10a). The apex of the lungs lies superior to the medial third of the clavicles, and this is the only area that can be palpated. The anterior, lateral, and posterior surfaces of the lungs lie against the ribs. The base of the lungs extends from the sixth costal cartilage anteriorly to the spinous process of the tenth thoracic vertebra posteriorly. The pleura extends about 5 cm (2 in.) below the base from the sixth costal cartilage anteriorly to the twelfth rib posteriorly. Thus, the lungs do not completely fill the pleural cavity in this area. Removal of excessive fluid in the pleural cavity can be accomplished without injuring lung tissue by inserting a needle anteriorly through the seventh intercostal space, a procedure called **thoracentesis** (thor′-a-sen-TĒ-sis; *-centesis* = puncture). The needle is passed along the superior border of the lower rib to avoid damage to the intercostal nerves and blood vessels. Inferior to the seventh intercostal space there is danger of penetrating the diaphragm.

Lobes, Fissures, and Lobules

One or two **fissures** divide each lung into sections called **lobes** (Figure 23.10b–e). Both lungs have an **oblique fissure,** which extends inferiorly and anteriorly; the right lung also has a **horizontal fissure**. The oblique fissure in the left lung separates the **superior lobe** from the **inferior lobe**. In the right lung, the superior part of the oblique fissure separates the superior lobe from the inferior lobe; the inferior part of the oblique fissure separates the inferior lobe from the **middle lobe,** which is bordered superiorly by the horizontal fissure.

Each lobe receives its own lobar bronchus. Thus, the right main bronchus gives rise to three lobar bronchi called the **superior**, **middle**, and **inferior lobar bronchi**, and the left main bronchus gives rise to superior and inferior lobar bronchi. Within the lung, the lobar bronchi give rise to the segmental bronchi, which are constant in both origin and distribution—there are 10 segmental bronchi in each lung. The portion of lung tissue that each segmental bronchus supplies is called a **bronchopulmonary segment** (brong-kō-PUL-mō-nār-ē). Bronchial and pulmonary disorders (such as tumors or abscesses) that are localized in a bronchopulmonary segment may be surgically removed without seriously disrupting the surrounding lung tissue.

Each bronchopulmonary segment of the lungs has many small compartments called **lobules**; each lobule is wrapped in elastic connective tissue and contains a lymphatic vessel, an arteriole, a venule, and a branch from a terminal bronchiole (Figure 23.11a). Terminal bronchioles and lobule subdivide into microscopic branches called **respiratory bronchioles** (Figure 23.11b). They also have alveoli (described shortly) budding from their walls. Alveoli participate in gas exchange, and thus respiratory bronchioles begin the respiratory zone of the respiratory system. As the respiratory bronchioles penetrate more deeply into the lungs, the epithelial lining changes from simple cuboidal to simple squamous. Respiratory bronchioles in turn subdivide into several (2–11) **alveolar ducts** (al-VĒ-ō-lar), which consist of simple squamous epithelium.

Alveolar Sacs and Alveoli

The terminal dilation of an alveolar duct is called an **alveolar sac** (al-vē-Ō-lar) and is analogous to a cluster of grapes. Each alveolar sac is composed of outpouchings called **alveoli** (al-vē-Ō-lī), analogous to individual grapes (Figure 23.11). The wall of each alveolus (singular) consists of two types of alveolar epithelial cells (Figure 23.12). The more numerous **type I alveolar** (*squamous pulmonary epithelial*) **cells** are simple squamous epithelial cells that form a nearly continuous lining of the alveolar wall. **Type II alveolar cells**, also called *septal cells*, are fewer in number and are found between type I alveolar cells. The thin type I alveolar cells are the main sites of gas exchange. Type II alveolar cells, rounded or cuboidal epithelial cells with free surfaces containing microvilli, secrete **alveolar fluid**, which keeps the surface between the cells and the air moist. Included in the alveolar fluid is **surfactant** (sur-FAK-tant), a complex mixture of phospholipids and lipoproteins. Surfactant lowers the surface tension of alveolar fluid, which reduces the tendency of alveoli to collapse and thus maintains their patency (described later).

Also present in the alveolar wall are **alveolar macrophages** (*dust cells*), phagocytes that remove fine dust particles and other debris from the alveolar spaces, and fibroblasts that produce reticular and elastic fibers. Underlying the layer of type I alveolar cells is an elastic basement membrane. On the outer surface of the alveoli, the lobule's arteriole and venule disperse into a network of blood capillaries (see Figure 23.11a) that consist of a single layer of endothelial cells and basement membrane.

The exchange of O_2 and CO_2 between the air spaces in the lungs and the blood takes place by diffusion across the alveolar and capillary walls, which together form the **respiratory membrane**. Extending from the alveolar air space to blood plasma, the respiratory membrane consists of four layers (Figure 23.12b):

1. A layer of type I and type II alveolar cells and associated alveolar macrophages that constitutes the **alveolar wall**
2. An **epithelial basement membrane** underlying the alveolar wall
3. A **capillary basement membrane** that is often fused to the epithelial basement membrane
4. The **capillary endothelium**

See Clinical Connection: Coryza, Seasonal Influenza, and H1N1 Influenza

Despite having several layers, the respiratory membrane is very thin—only 0.5 μm thick, about one-sixteenth the diameter of a red blood cell—to allow rapid diffusion of gases. It has been estimated that both lungs contain 300–500 million alveoli, providing an immense surface area of about 75 m^2 (807 ft^2)—about the size of a racquetball court or slightly larger—for gas exchange. The hundreds of millions of alveoli account for the spongy texture of the lungs.

FIGURE 23.11 Microscopic anatomy of a lobule of the lungs.

An alveolar sac is the terminal dilation of an alveolar duct and is composed of alveoli.

MICROSCOPIC AIRWAYS
Respiratory bronchioles
↓
Alveolar ducts
↓
Alveolar sacs
↓
Alveoli

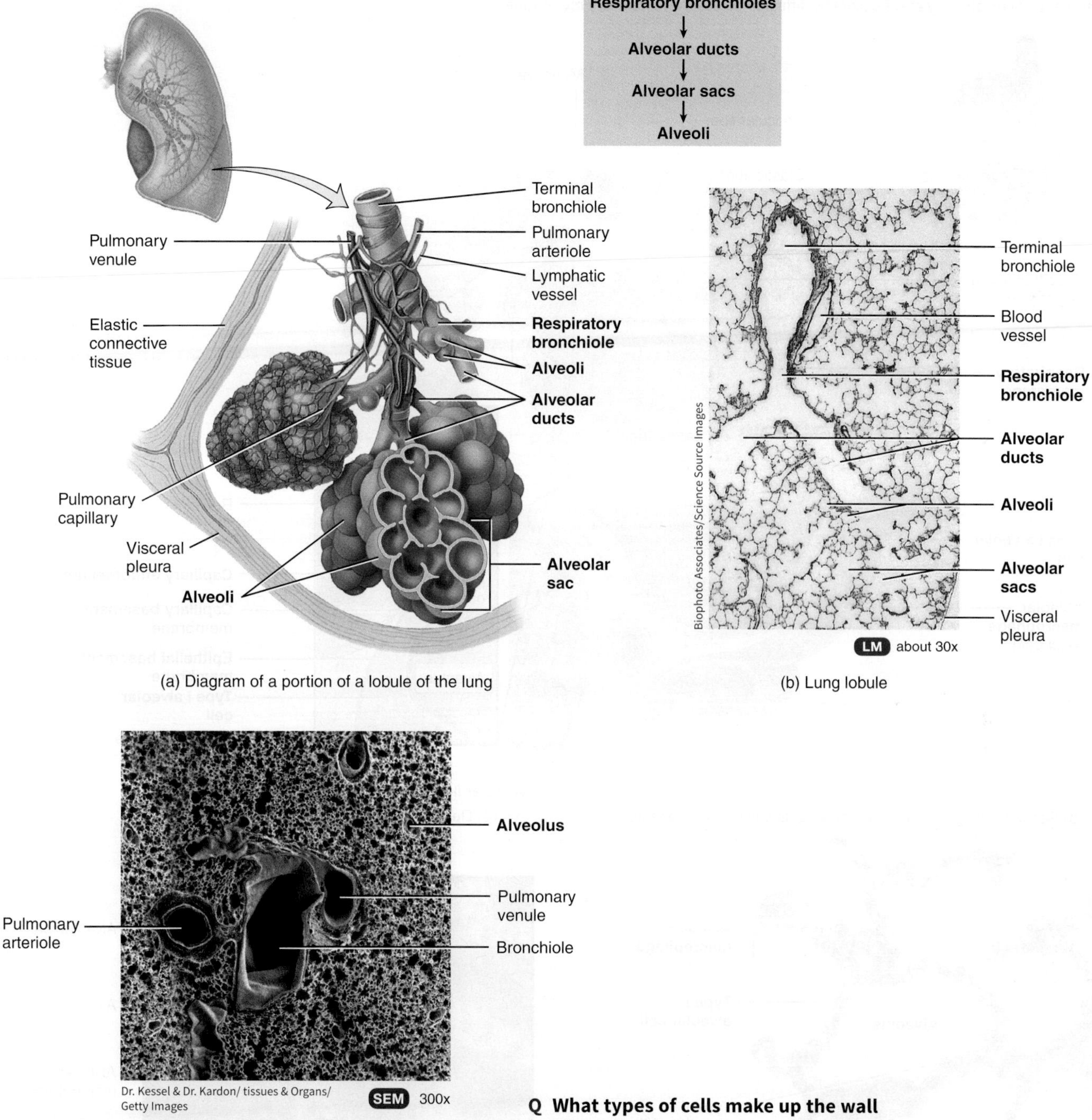

(a) Diagram of a portion of a lobule of the lung

Biophoto Associates/Science Source Images

LM about 30x

(b) Lung lobule

Dr. Kessel & Dr. Kardon/ tissues & Organs/ Getty Images

SEM 300x

(c) Section of lung lobule

Q What types of cells make up the wall of an alveolus?

Blood Supply to the Lungs The lungs receive blood via two sets of arteries: pulmonary arteries and bronchial arteries. Deoxygenated blood passes through the pulmonary trunk, which divides into a left pulmonary artery that enters the left lung and a right pulmonary artery that enters the right lung. (The pulmonary arteries are the only arteries in the body that carry deoxygenated blood.) Return of the oxygenated blood to the heart occurs by way of the four pulmonary veins, which drain into the left atrium (see **Figure 21.30**). A unique feature of pulmonary blood vessels is their constriction in response to localized hypoxia (low O_2 level). In all other body tissues,

FIGURE 23.12 Structural components of an alveolus. The respiratory membrane consists of a layer of type I and type II alveolar cells, an epithelial basement membrane, a capillary basement membrane, and the capillary endothelium.

> The exchange of respiratory gases occurs by diffusion across the respiratory membrane.

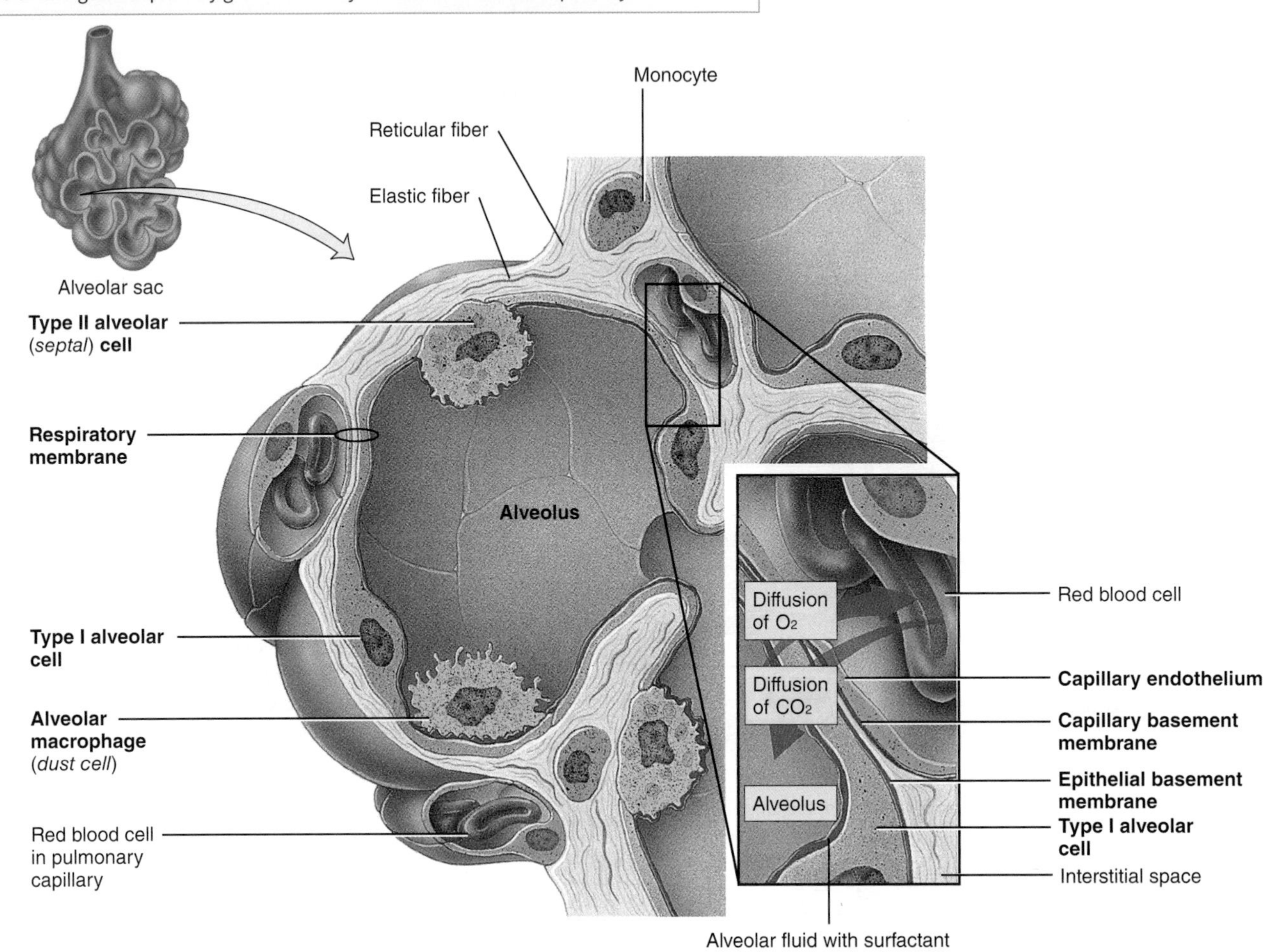

(a) Section through an alveolus showing its cellular components

(b) Details of respiratory membrane

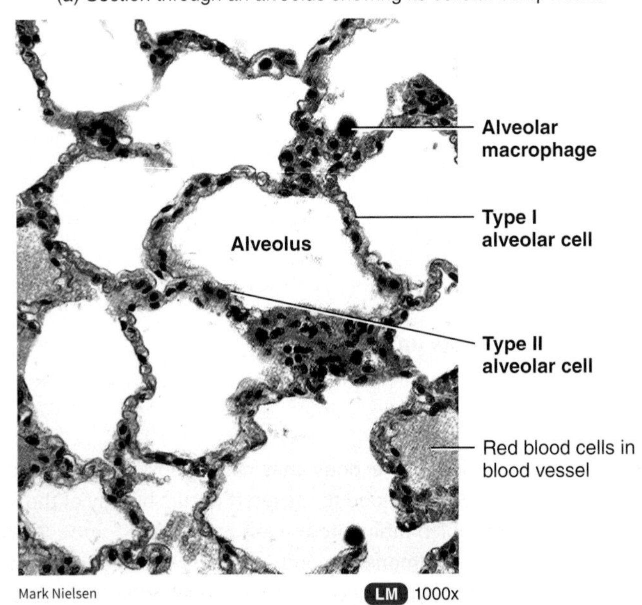

Mark Nielsen

(c) Details of several alveoli

Alveolar wall

Alveoli

Alveolar macrophage

D. Phillips/Getty Images

SEM 2200x

Section of alveoli

Q How thick is the respiratory membrane?

TABLE 23.1 Summary of the Structures of the Respiratory System

STRUCTURE	EPITHELIUM	CILIA	GOBLET CELLS	SPECIAL FEATURES
NOSE				
Vestibule	Nonkeratinized stratified squamous.	No.	No.	Contains numerous hairs.
Respiratory region	Pseudostratified ciliated columnar.	Yes.	Yes.	Contains conchae and meatuses.
Olfactory region	Olfactory epithelium (olfactory receptors).	Yes.	No.	Functions in olfaction.
PHARYNX				
Nasopharynx	Pseudostratified ciliated columnar.	Yes.	Yes.	Passageway for air; contains internal nares, openings for auditory tubes, and pharyngeal tonsil.
Oropharynx	Nonkeratinized stratified squamous.	No.	No.	Passageway for both air and food and drink; contains opening from mouth (fauces).
Laryngopharynx	Nonkeratinized stratified squamous.	No.	No.	Passageway for both air and food and drink.
LARYNX	Nonkeratinized stratified squamous above the vocal folds; pseudostratified ciliated columnar below the vocal folds.	No above folds; yes below folds.	No above folds; yes below folds.	Passageway for air; contains vocal folds for voice production.
TRACHEA	Pseudostratified ciliated columnar.	Yes.	Yes.	Passageway for air; contains C-shaped rings of cartilage to keep trachea open.
BRONCHI				
Main bronchi	Pseudostratified ciliated columnar.	Yes.	Yes.	Passageway for air; contain C-shaped rings of cartilage to maintain patency.
Lobar bronchi	Pseudostratified ciliated columnar.	Yes.	Yes.	Passageway for air; contain plates of cartilage to maintain patency.
Segmental bronchi	Pseudostratified ciliated columnar.	Yes.	Yes.	Passageway for air; contain plates of cartilage to maintain patency.
Larger bronchioles	Ciliated simple columnar.	Yes.	Yes.	Passageway for air; contain more smooth muscle than in the bronchi.
Smaller bronchioles	Ciliated simple columnar.	Yes.	No.	Passageway for air; contain more smooth muscle than in the larger bronchioles.
Terminal bronchioles	Nonciliated simple columnar.	No.	No.	Passageway for air; contain more smooth muscle than in the smaller bronchioles.
LUNGS				
Respiratory bronchioles	Simple cuboidal to simple squamous.	No.	No.	Passageway for air; gas exchange.
Alveolar ducts	Simple squamous.	No.	No.	Passageway for air; gas exchange; produce surfactant.
Alveoli	Simple squamous.	No.	No.	Passageway for air; gas exchange; produce surfactant to maintain patency.

☐ Conducting structures ☐ Gas exchange structures

hypoxia causes dilation of blood vessels to increase blood flow. In the lungs, however, vasoconstriction in response to hypoxia diverts pulmonary blood from poorly ventilated areas of the lungs to well-ventilated regions for more efficient gas exchange. This phenomenon is known as **ventilation–perfusion coupling** (per-FYU-zhun) because the perfusion (blood flow) to each area of the lungs matches the extent of ventilation (airflow) to alveoli in that area.

Bronchial arteries, which branch from the aorta, deliver oxygenated blood to the lungs. This blood mainly perfuses the muscular walls of the bronchi and bronchioles. Connections do exist between branches of the bronchial arteries and branches of the pulmonary arteries, however; most blood returns to the heart via pulmonary veins. Some blood drains into bronchial veins, branches of the azygos system, and returns to the heart via the superior vena cava.

Patency of the Respiratory System

Throughout the discussion of the respiratory organs, several examples were given of structures or secretions that help to maintain patency of the system so that air passageways are kept free of obstruction. These included the bony and cartilaginous frameworks of the nose, skeletal muscles of the pharynx, cartilages of the larynx, C-shaped rings of cartilage in the trachea and bronchi, smooth muscle in the bronchioles, and surfactant in the alveoli.

Unfortunately, there are also factors that can compromise patency. These include crushing injuries to bone and cartilage, a deviated nasal septum, nasal polyps, inflammation of mucous membranes, spasms of smooth muscle, and a deficiency of surfactant.

A summary of the epithelial linings and special features of the organs of the respiratory system is presented in **Table 23.1**.

23.4 Pulmonary Ventilation

OBJECTIVE

- **Describe** the events that cause inhalation and exhalation.

Pulmonary ventilation, or *breathing*, is the flow of air into and out of the lungs. In pulmonary ventilation, air flows between the atmosphere and the alveoli of the lungs because of alternating pressure differences created by contraction and relaxation of respiratory muscles. The rate of airflow and the amount of effort needed for breathing are also influenced by alveolar surface tension, compliance of the lungs, and airway resistance.

Pressure Changes during Pulmonary Ventilation

Air moves into the lungs when the air pressure inside the lungs is less than the air pressure in the atmosphere. Air moves out of the lungs when the air pressure inside the lungs is greater than the air pressure in the atmosphere.

Inhalation Breathing in is called **inhalation** (*inspiration*). Just before each inhalation, the air pressure inside the lungs is equal to the air pressure of the atmosphere, which at sea level is about 760 millimeters of mercury (mmHg), or 1 atmosphere (atm). For air to flow into the lungs, the pressure inside the alveoli must become lower than the atmospheric pressure. This condition is achieved by increasing the size of the lungs.

The pressure of a gas in a closed container is inversely proportional to the volume of the container. This means that if the size of a closed container is increased, the pressure of the gas inside the container decreases, and that if the size of the container is decreased, then the pressure inside it increases. This inverse relationship between volume and pressure, called **Boyle's law**, may be demonstrated

FIGURE 23.13 **Boyle's law.**

The volume of a gas varies inversely with its pressure.

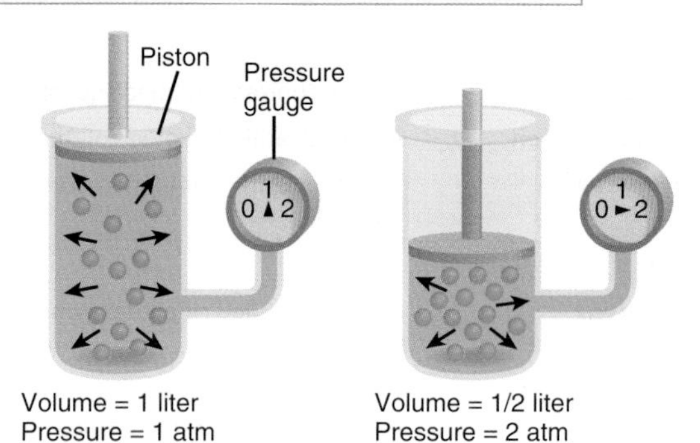

Q If the volume is decreased from 1 liter to 1/4 liter, how would the pressure change?

as follows (**Figure 23.13**): Suppose we place a gas in a cylinder that has a movable piston and a pressure gauge, and that the initial pressure created by the gas molecules striking the wall of the container is 1 atm. If the piston is pushed down, the gas is compressed into a smaller volume, so that the same number of gas molecules strikes less wall area. The gauge shows that the pressure doubles as the gas is compressed to half its original volume. In other words, the same number of molecules in half the volume produces twice the pressure. Conversely, if the piston is raised to increase the volume, the pressure decreases. Thus, the pressure of a gas varies inversely with volume.

Differences in pressure caused by changes in lung volume force air into our lungs when we inhale and out when we exhale. For inhalation to occur, the lungs must expand, which increases lung volume and thus decreases the pressure in the lungs to below atmospheric pressure. The first step in expanding the lungs during normal quiet inhalation involves contraction of the main muscle of inhalation, the diaphragm, with resistance from external intercostals (**Figure 23.14**).

The most important muscle of inhalation is the diaphragm, the dome-shaped skeletal muscle that forms the floor of the thoracic cavity. It is innervated by fibers of the phrenic nerves, which emerge from the spinal cord at cervical levels 3, 4, and 5. Contraction of the diaphragm causes it to flatten, lowering its dome. This increases the vertical diameter of the thoracic cavity. During normal quiet inhalation, the diaphragm descends about 1 cm (0.4 in.), producing a pressure difference of 1–3 mmHg and the inhalation of about 500 mL of air. In strenuous breathing, the diaphragm may descend 10 cm (4 in.), which produces a pressure difference of 100 mmHg and the inhalation of 2–3 liters of air. Contraction of the diaphragm is responsible for about 75% of the air that enters the lungs during quiet breathing. Advanced pregnancy, excessive obesity, or confining abdominal clothing can prevent complete descent of the diaphragm.

The next most important muscles of inhalation are the external intercostals. When these muscles contract, they elevate the ribs. As a result, there is an increase in the anteroposterior and lateral diameters of the chest cavity. Contraction of the external intercostals is responsible for about 25% of the air that enters the lungs during normal quiet breathing.

FIGURE 23.14 Muscles of inhalation and exhalation. The pectoralis minor muscle (not shown here) is illustrated in **Figure 11.14a**.

During normal, quiet inhalation, the diaphragm and external intercostals contract, the lungs expand, and air moves into the lungs; during normal, quiet exhalation, the diaphragm and external intercostals relax and the lungs recoil, forcing air out of the lungs.

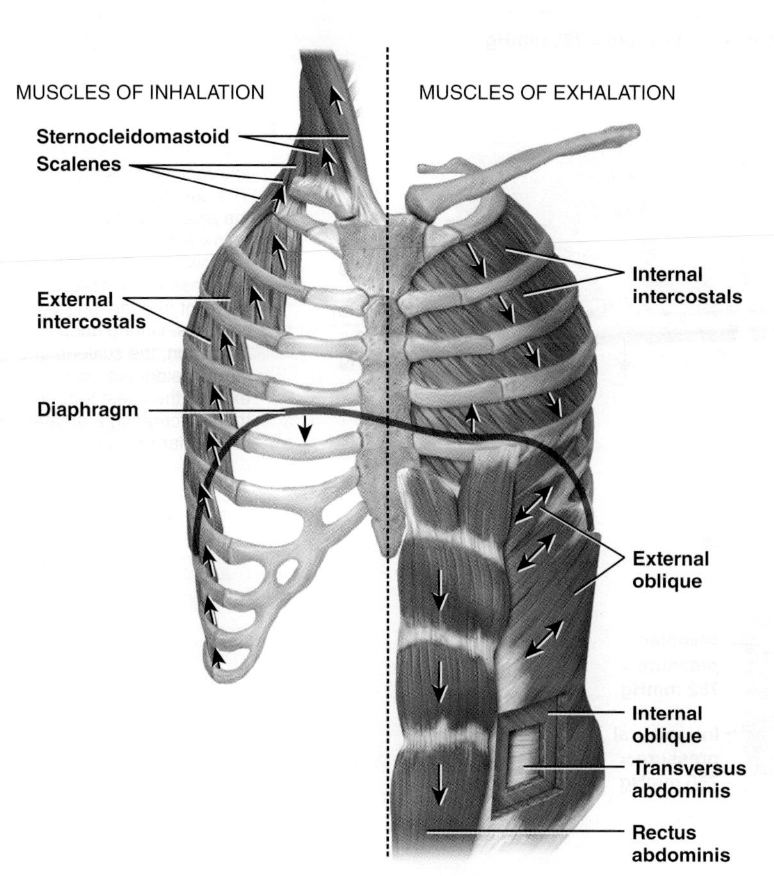

(a) Muscles of inhalation (left); muscles of exhalation (right); arrows indicate the direction of muscle contraction

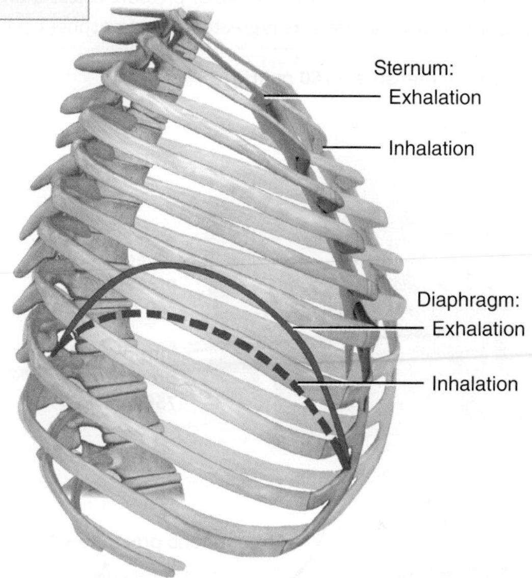

(b) Changes in size of thoracic cavity during inhalation and exhalation

(c) During inhalation, the lower ribs (7–10) move upward and outward like the handle on a bucket

Q Right now, what is the main muscle that is powering your breathing?

Intrapleural pressure is the pressure within the pleural cavity. Recall that the pleural cavity is the space between the parietal pleura and visceral pleura (see **Figure 23.15**). A small amount of lubricating fluid is present in this space. Intrapleural pressure is always a negative pressure (lower than atmospheric pressure), ranging from 754–756 mmHg during normal quiet breathing. Because the pleural cavity has a negative pressure, it essentially functions as a vacuum. The suction of this vacuum attaches the visceral pleura to the chest wall. Thus, if the thoracic cavity increases in size, the lungs also expand; if the thoracic cavity decreases in size, the lungs recoil (become smaller). Just before inhalation, intrapleural pressure is about 4 mmHg less than atmospheric pressure, or about 756 mmHg at an atmospheric pressure of 760 mmHg (**Figure 23.15**). As the diaphragm and external intercostals contract and the overall size of the thoracic cavity increases, the volume of the pleural cavity also increases, which causes intrapleural pressure to decrease to about 754 mmHg. As the thoracic cavity expands, the parietal pleura lining the cavity is pulled outward in all directions, and the visceral pleura and lungs and pulled along with it.

As the volume of the lungs increases in this way, the pressure of air within the alveoli of the lungs, called the **alveolar** (*intrapulmonic*) **pressure**, drops from 760 to 758 mmHg. A pressure difference is thus established between the atmosphere and the alveoli. Because air always flows from a region of higher pressure to a region of lower pressure, inhalation takes place. Air continues to flow into the lungs as long as a pressure difference exists. Although the lungs enlarge in all directions during inhalation, most of the increase in volume appears to be due to the lengthening and expansion of the alveolar ducts and the increase in size of the openings into the alveoli. During deep, forceful inhalations, accessory muscles of inspiration also participate in increasing the size of the thoracic cavity (see **Figure 23.14a**). The muscles are so named because they make little, if any, contribution during normal quiet inhalation, but during exercise or

FIGURE 23.15 Pressure changes in pulmonary ventilation. During inhalation, the diaphragm contracts, the chest expands, the lungs are pulled outward, and alveolar pressure decreases. During exhalation, the diaphragm relaxes, the lungs recoil inward, and alveolar pressure increases, forcing air out of the lungs.

Air moves into the lungs when alveolar pressure is less than atmospheric pressure, and out of the lungs when alveolar pressure is greater than atmospheric pressure.

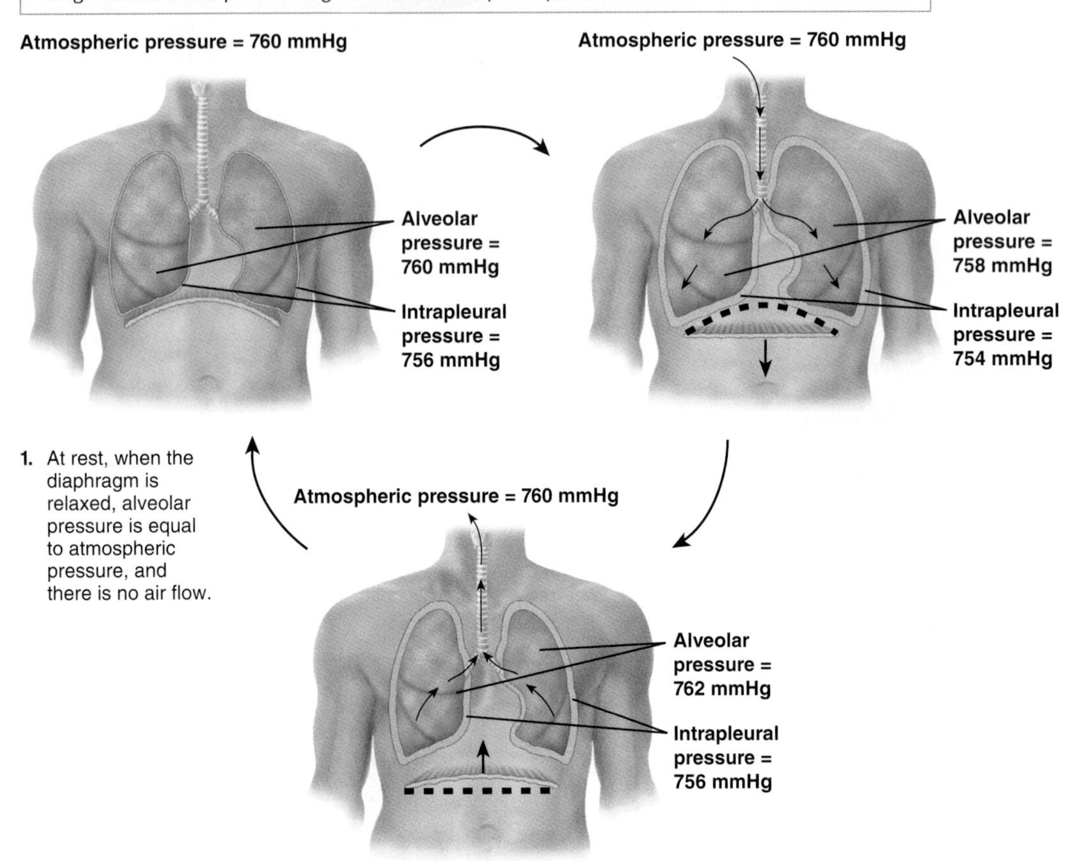

2. During inhalation, the diaphragm contracts and the external intercostals contract. The chest cavity expands, and the alveolar pressure drops below atmospheric pressure. Air flows into the lungs in response to the pressure gradient and the lung volume expands. During deep inhalation, the scalene and sternocleidomastoid muscles expand the chest further, thereby creating a greater drop in alveolar pressure.

3. During exhalation, the diaphragm relaxes and the external intercostals relax. The chest and lungs recoil, the chest cavity contracts, and the alveolar pressure increases above atmospheric pressure. Air flows out of the lungs in response to the pressure gradient, and the lung volume decreases. During forced exhalations, the internal intercostals and abdominal muscles contract, thereby reducing the size of the chest cavity further and creating a greater increase in alveolar pressure.

Q How does the intrapleural pressure change during a normal, quiet breath?

forced breathing they may contract vigorously. The accessory muscles of inhalation include the sternocleidomastoid muscles, which elevate the sternum; the scalene muscles, which elevate the first two ribs; and the pectoralis minor muscles, which elevate the third through fifth ribs. Because both normal quiet inhalation and inhalation during exercise or forced breathing involve muscular contraction, the process of inhalation is said to be *active.*

Exhalation Breathing out, called **exhalation** (*expiration*), is also due to a pressure gradient, but in this case the gradient is in the opposite direction: The pressure in the lungs is greater than the pressure of the atmosphere. Normal exhalation during quiet breathing, unlike inhalation, is a *passive process* because no muscular contractions are involved. Instead, exhalation results from **elastic recoil** of the chest wall and lungs, both of which have a natural tendency to spring back after they have been stretched. Two inwardly directed forces contribute to elastic recoil: (1) the recoil of elastic fibers that were stretched during inhalation and (2) the inward pull of surface tension due to the film of intrapleural fluid between the visceral and parietal pleurae.

Exhalation starts when the inspiratory muscles relax. As the diaphragm relaxes, its dome moves superiorly owing to its elasticity. As the external intercostals relax, the ribs are depressed. These movements decrease the vertical, lateral, and anteroposterior diameters of the thoracic cavity, which decreases lung volume. In turn, the alveolar pressure increases to about 762 mmHg. Air then flows from the area of higher pressure in the alveoli to the area of lower pressure in the atmosphere (see **Figure 23.15**).

Exhalation becomes active only during forceful breathing, as occurs while playing a wind instrument or during exercise. During these times, muscles of exhalation—the abdominal and internal intercostals (see **Figure 23.14a**)—contract, which increases pressure in the abdominal region and thorax. Contraction of the abdominal muscles moves the inferior ribs downward and compresses the abdominal viscera, thereby forcing the diaphragm superiorly. Contraction of the internal intercostals, which extend inferiorly and posteriorly between adjacent ribs, pulls the ribs inferiorly. Although intrapleural pressure is always less than alveolar pressure, it may briefly exceed atmospheric pressure during a forceful exhalation, such as during a cough.

Other Factors Affecting Pulmonary Ventilation

As you have just learned, air pressure differences drive airflow during inhalation and exhalation. However, three other factors affect the rate of airflow and the ease of pulmonary ventilation: surface tension of the alveolar fluid, compliance of the lungs, and airway resistance.

Surface Tension of Alveolar Fluid As noted earlier, a thin layer of alveolar fluid coats the luminal surface of alveoli and exerts a force known as **surface tension**. Surface tension arises at all air–water interfaces because the polar water molecules are more strongly attracted to each other than they are to gas molecules in the air. When liquid surrounds a sphere of air, as in an alveolus or a soap bubble, surface tension produces an inwardly directed force. Soap bubbles "burst" because they collapse inward due to surface tension. In the lungs, surface tension causes the alveoli to assume the smallest possible diameter. During breathing, surface tension must be overcome to expand the lungs during each inhalation. Surface tension also accounts for two-thirds of lung elastic recoil, which decreases the size of alveoli during exhalation.

The **surfactant** (a mixture of phospholipids and lipoproteins) present in alveolar fluid reduces its surface tension below the surface tension of pure water. A deficiency of surfactant in premature infants causes *respiratory distress syndrome*, in which the surface tension of alveolar fluid is greatly increased, so that many alveoli collapse at the end of each exhalation. Great effort is then needed at the next inhalation to reopen the collapsed alveoli.

See Clinical Connection: Respiratory Distress Syndrome

Compliance of the Lungs **Compliance** refers to how much effort is required to stretch the lungs and chest wall. High compliance means that the lungs and chest wall expand easily; low compliance means that they resist expansion. By analogy, a thin balloon that is easy to inflate has high compliance, and a heavy and stiff balloon that takes a lot of effort to inflate has low compliance. In the lungs, compliance is related to two principal factors: elasticity and surface tension. The lungs normally have high compliance and expand easily because elastic fibers in lung tissue are easily stretched and surfactant in alveolar fluid reduces surface tension. Decreased compliance is a common feature in pulmonary conditions that (1) scar lung tissue (for example, tuberculosis), (2) cause lung tissue to become filled with fluid (pulmonary edema), (3) produce a deficiency in surfactant, or (4) impede lung expansion in any way (for example, paralysis of the intercostal muscles). Increased lung compliance occurs in emphysema (see Disorders: Homeostatic Imbalances in the Study Guide) due to destruction of elastic fibers in alveolar walls.

Airway Resistance Like the flow of blood through blood vessels, the rate of airflow through the airways depends on both the pressure difference and the resistance: Airflow equals the pressure difference between the alveoli and the atmosphere divided by the resistance. The walls of the airways, especially the bronchioles, offer some resistance to the normal flow of air into and out of the lungs. As the lungs expand during inhalation, the bronchioles enlarge because their walls are pulled outward in all directions. Larger-diameter airways have decreased resistance. Airway resistance then increases during exhalation as the diameter of bronchioles decreases. Airway diameter is also regulated by the degree of contraction or relaxation of smooth muscle in the walls of the airways. Signals from the sympathetic division of the autonomic nervous system (ANS) cause relaxation of bronchiolar smooth muscle (bronchodilation) which results in decreased resistance. Signals from the parasympathetic division of the ANS cause contraction of bronchiolar smooth muscle (bronchoconstriction) resulting in increased resistance.

Any condition that narrows or obstructs the airways increases resistance, so that more pressure is required to maintain the same airflow. The hallmark of asthma or chronic obstructive pulmonary disease (COPD)—emphysema or chronic bronchitis—is increased airway resistance due to obstruction or collapse of airways.

Breathing Patterns and Modified Breathing Movements

The term for the normal pattern of quiet breathing is **eupnea** (ūp-NĒ-a; *eu-* = good, easy, or normal; *-pnea* = breath). Eupnea can consist of shallow, deep, or combined shallow and deep breathing. A pattern of shallow (chest) breathing, called **costal breathing**, consists of an upward and outward movement of the chest due to contraction of the external intercostal muscles. A pattern of deep (abdominal) breathing, called **diaphragmatic breathing** (dī′-a-frag-MAT-ik), consists of the outward movement of the abdomen due to the contraction and descent of the diaphragm.

Breathing also provides humans with methods for expressing emotions such as laughing, sighing, and sobbing and can be used to expel foreign matter from the lower air passages through actions such as sneezing and coughing. Breathing movements are also modified and controlled during talking and singing. Some of the modified breathing movements that express emotion or clear the airways are listed in **Table 23.2**. All of these movements are reflexes, but some of them also can be initiated voluntarily.

TABLE 23.2 Modified Breathing Movements

MOVEMENT	DESCRIPTION
Coughing	A long-drawn and deep inhalation followed by a complete closure of the rima glottidis, which results in a strong exhalation that suddenly pushes the rima glottidis open and sends a blast of air through the upper respiratory passages. Stimulus for this reflex act may be a foreign body lodged in the larynx, trachea, or epiglottis.
Sneezing	Spasmodic contraction of muscles of exhalation that forcefully expels air through the nose and mouth. Stimulus may be an irritation of the nasal mucosa.
Sighing	A long-drawn and deep inhalation immediately followed by a shorter but forceful exhalation.
Yawning	A deep inhalation through the widely opened mouth producing an exaggerated depression of the mandible. It may be stimulated by drowsiness, or someone else's yawning, but the precise cause is unknown.
Sobbing	A series of convulsive inhalations followed by a single prolonged exhalation. The rima glottidis closes earlier than normal after each inhalation so only a little air enters the lungs with each inhalation.
Crying	An inhalation followed by many short convulsive exhalations, during which the rima glottidis remains open and the vocal folds vibrate; accompanied by characteristic facial expressions and tears.
Laughing	The same basic movements as crying, but the rhythm of the movements and the facial expressions usually differ from those of crying. Laughing and crying are sometimes indistinguishable.
Hiccupping	Spasmodic contraction of the diaphragm followed by a spasmodic closure of the rima glottidis, which produces a sharp sound on inhalation. Stimulus is usually irritation of the sensory nerve endings of the gastrointestinal tract.
Valsalva (val-SAL-va) **maneuver**	Forced exhalation against a closed rima glottidis as may occur during periods of straining while defecating.
Pressurizing the middle ear	The nose and mouth are held closed and air from the lungs is forced through the auditory tube into the middle ear. Employed by those snorkeling or scuba diving during descent to equalize the pressure of the middle ear with that of the external environment.

23.5 Lung Volumes and Capacities

OBJECTIVES

- **Explain** the differences among tidal volume, inspiratory reserve volume, expiratory reserve volume, and residual volume.
- **Differentiate** among inspiratory capacity, functional residual capacity, vital capacity, and total lung capacity.

During inhalation and exhalation, varying amounts of air move into and out of the lungs. These amounts depend on many factors related to various characteristics of healthy individuals and pulmonary disorders. The different amounts of air can be classified into two types: (1) **lung volumes**, which can be measured directly by use of a spirometer (described shortly) and (2) **lungs capacities**, which are combinations of different lung volumes. The apparatus used to measure volumes and capacities is called a **spirometer** (spī-ROM-e-ter; *spiro-* = breathe; *-meter* = measuring device) or *respirometer* (res′-pi-ROM-e-ter). The record is called a **spirogram**. Inhalation is recorded as an upward deflection, and exhalation is recorded as a downward deflection (Figure 23.16). In general, lung volumes and capacities are larger in males, taller individuals, younger adults, people who live at higher altitudes, and those who are not obese. Various disorders may be diagnosed by comparison of actual and predicted normal values for a person's gender, height, and age.

Lung Volumes

While at rest, a healthy adult averages 12 breaths a minute, with each inhalation and exhalation moving about 500 mL of air into and out of the lungs. The volume of one breath is called the **tidal volume (V_T)**.

Tidal volume varies considerably from one person to another and in the same person at different times. In a typical adult, about 70% of the tidal volume (350 mL) actually reaches the respiratory zone of the respiratory system—the respiratory bronchioles, alveolar ducts, alveolar sacs, and alveoli—and participates in external respiration. The other 30% (150 mL) remains in the conducting airways of the nose, pharynx, larynx, trachea, bronchi, bronchioles, and terminal bronchioles. Collectively, the conducting airways with air that does not undergo respiratory exchange are known as the **anatomic** (*respiratory*) **dead space**. (An easy rule of thumb for determining the volume of your anatomic dead space is that it is about the same in milliliters as your ideal weight in pounds.) Not all of the inhaled air can be used in gas exchange because some of it remains in the anatomic dead space.

By taking a very deep breath, you can inhale a good deal more than 500 mL. This additional inhaled air, called the **inspiratory reserve volume (IRV)**, is about 3100 mL in an average adult male and 1900 mL in an average adult female (Figure 23.16). Even more air can be inhaled if inhalation follows forced exhalation. If you inhale normally and then exhale as forcibly as possible, you should be able to push out considerably more air in addition to the 500 mL

FIGURE 23.16 Spirogram of lung volumes and capacities. The average values for a healthy adult male and female are indicated, with the values for a female in parentheses. Note that the spirogram is read from right (start of record) to left (end of record).

Lung capacities are combinations of various lung volumes.

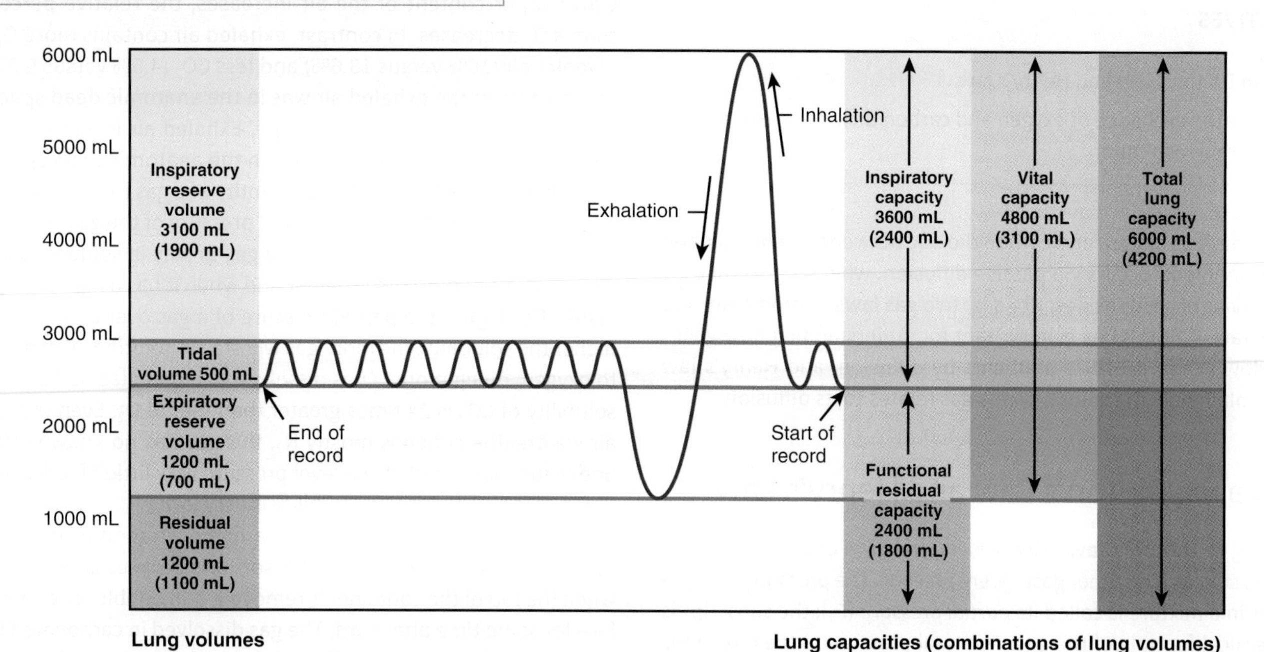

Q If you breathe in as deeply as possible and then exhale as much air as you can, which lung capacity have you demonstrated?

of tidal volume. The extra 1200 mL in males and 700 mL in females is called the **expiratory reserve volume (ERV)**. The **forced expiratory volume in 1 second, (FEV_1)** is the volume of air that can be exhaled from the lungs in 1 second with maximal effort following a maximal inhalation. Typically, chronic obstructive pulmonary disease (COPD) greatly reduces FEV_1 because COPD increases airway resistance.

Even after the expiratory reserve volume is exhaled, considerable air remains in the lungs because the subatmospheric intrapleural pressure keeps the alveoli slightly inflated, and some air remains in the noncollapsible airways. This volume, which cannot be measured by spirometry, is called the **residual volume** (re-ZID-u-al) **(RV)** and amounts to about 1200 mL in males and 1100 mL in females.

If the thoracic cavity is opened, the intrapleural pressure rises to equal the atmospheric pressure and forces out some of the residual volume. The air remaining is called the **minimal volume**. Minimal volume provides a medical and legal tool for determining whether a baby is born dead (stillborn) or died after birth. The presence of minimal volume can be demonstrated by placing a piece of lung in water and observing if it floats. Fetal lungs contain no air, so the lung of a stillborn baby will not float in water.

Lung Capacities

Lung capacities are combinations of specific lung volumes (**Figure 23.16**). **Inspiratory capacity (IC)** is the sum of tidal volume and inspiratory reserve volume (500 mL + 3100 mL = 3600 mL in males and 500 mL + 1900 mL = 2400 mL in females). **Functional residual capacity (FRC)** is the sum of residual volume and expiratory reserve volume (1200 mL + 1200 mL = 2400 mL in males and 1100 mL + 700 mL = 1800 mL in females). **Vital capacity (VC)** is the sum of inspiratory reserve volume, tidal volume, and expiratory reserve volume (4800 mL in males and 3100 mL in females). Finally, **total lung capacity (TLC)** is the sum of vital capacity and residual volume (4800 mL + 1200 mL = 6000 mL in males and 3100 mL + 1100 mL = 4200 mL in females).

Another way to assess pulmonary function is to determine the amount of air that flows into and out of the lungs each minute. The **minute ventilation ($\dot{V}$)**—the total volume of air inspired and expired each minute—is tidal volume multiplied by respiratory rate. In a typical adult at rest, minute ventilation is about 6000 mL/min ($\dot{V}$ = 12 breaths per minute × 500 mL = 6000 mL/min). A lower-than-normal minute ventilation usually is a sign of pulmonary malfunction.

As noted earlier, not all of inhaled air (500 mL) actually reaches the respiratory zone of the respiration system. The 150 mL in the conducting zone is the anatomic dead space. Hence, not all of the minute ventilation can be used in gas exchange because some of it remains in the anatomic dead space. The **alveolar ventilation ($\dot{V}_A$)** is the volume of air per minute that actually reaches the respiratory zone (350 mL). Alveolar ventilation is typically about 4200 mL/min ($\dot{V}_A$ = 12 breaths per minute × 350 mL = 4200 mL/min).

23.6 Exchange of Oxygen and Carbon Dioxide

OBJECTIVES

- **Explain** Dalton's law and Henry's law.
- **Describe** the exchange of oxygen and carbon dioxide in external and internal respiration.

The exchange of oxygen and carbon dioxide between alveolar air and pulmonary blood occurs via passive diffusion, which is governed by the behavior of gases as described by two gas laws, Dalton's law and Henry's law. Dalton's law is important for understanding how gases move down their pressure gradients by diffusion, and Henry's law helps explain how the solubility of a gas relates to its diffusion.

Gas Laws: Dalton's Law and Henry's Law

According to **Dalton's law**, each gas in a mixture of gases exerts its own pressure as if no other gases were present. The pressure of a specific gas in a mixture is called its *partial pressure* (P_x); the subscript is the chemical formula of the gas. The total pressure of the mixture is calculated simply by adding all of the partial pressures. Atmospheric air is a mixture of gases—nitrogen (N_2), oxygen (O_2), argon (Ar), carbon dioxide (CO_2), variable amounts of water vapor (H_2O), plus other gases present in small quantities. Atmospheric pressure is the sum of the pressures of all of these gases:

$$\text{Atmospheric pressure (760 mmHg)} = P_{N_2} + P_{O_2} + P_{Ar} + P_{H_2O} + P_{CO_2} + P_{\text{other gases}}$$

We can determine the partial pressure exerted by each component in the mixture by multiplying the percentage of the gas in the mixture by the total pressure of the mixture. Atmospheric air is 78.6% nitrogen, 20.9% oxygen, 0.093% argon, 0.04% carbon dioxide, and 0.06% other gases; a variable amount of water vapor is also present. The amount of water varies from practically 0% over a desert to 4% over the ocean, to about 0.3% on a cool, dry day. Thus, the partial pressures of the gases in inhaled air are as follows:

$$\begin{aligned}
P_{N_2} &= 0.786 \times 760\text{ mmHg} = 597.4\text{ mmHg}\\
P_{O_2} &= 0.209 \times 760\text{ mmHg} = 158.8\text{ mmHg}\\
P_{Ar} &= 0.0009 \times 760\text{ mmHg} = 0.7\text{ mmHg}\\
P_{H_2O} &= 0.003 \times 760\text{ mmHg} = 2.3\text{ mmHg}\\
P_{CO_2} &= 0.0004 \times 760\text{ mmHg} = 0.3\text{ mmHg}\\
P_{\text{other gases}} &= 0.0006 \times 760\text{ mmHg} = 0.5\text{ mmHg}\\
\text{Total} &= 760.0\text{ mmHg}
\end{aligned}$$

These partial pressures determine the movement of O_2 and CO_2 between the atmosphere and lungs, between the lungs and blood, and between the blood and body cells. Each gas diffuses across a permeable membrane from the area where its partial pressure is greater to the area where its partial pressure is less. The greater the difference in partial pressure, the faster the rate of diffusion.

Compared with inhaled air, alveolar air has less O_2 (13.6% versus 20.9%) and more CO_2 (5.2% versus 0.04%) for two reasons. First, gas exchange in the alveoli increases the CO_2 content and decreases the O_2 content of alveolar air. Second, when air is inhaled it becomes humidified as it passes along the moist mucosal linings. As water vapor content of the air increases, the relative percentage that is O_2 decreases. In contrast, exhaled air contains more O_2 than alveolar air (16% versus 13.6%) and less CO_2 (4.5% versus 5.2%) because some of the exhaled air was in the anatomic dead space and did not participate in gas exchange. Exhaled air is a mixture of alveolar air and inhaled air that was in the anatomic dead space.

Henry's law states that the quantity of a gas that will dissolve in a liquid is proportional to the partial pressure of the gas and its solubility. In body fluids, the ability of a gas to stay in solution is greater when its partial pressure is higher and when it has a high solubility in water. The higher the partial pressure of a gas over a liquid and the higher the solubility, the more gas will stay in solution. In comparison to oxygen, much more CO_2 is dissolved in blood plasma because the solubility of CO_2 is 24 times greater than that of O_2. Even though the air we breathe contains mostly N_2, this gas has no known effect on bodily functions, and at sea level pressure very little of it dissolves in blood plasma because its solubility is very low.

An everyday experience gives a demonstration of Henry's law. You have probably noticed that a soft drink makes a hissing sound when the top of the container is removed, and bubbles rise to the surface for some time afterward. The gas dissolved in carbonated beverages is CO_2. Because the soft drink is bottled or canned under high pressure and capped, the CO_2 remains dissolved as long as the container is unopened. Once you remove the cap, the pressure decreases and the gas begins to bubble out of solution.

See Clinical Connection: Hyperbaric Oxygenation

Henry's law explains two conditions resulting from changes in the solubility of nitrogen in body fluids. As the total air pressure increases, the partial pressures of all of its gases increase. When a scuba diver breathes air under high pressure, the nitrogen in the mixture can have serious negative effects. Because the partial pressure of nitrogen is higher in a mixture of compressed air than in air at sea level pressure, a considerable amount of nitrogen dissolves in plasma and interstitial fluid. Excessive amounts of dissolved nitrogen may produce giddiness and other symptoms similar to alcohol intoxication. The condition is called **nitrogen narcosis** or "rapture of the deep."

If a diver comes to the surface slowly, the dissolved nitrogen can be eliminated by exhaling it. However, if the ascent is too rapid, nitrogen comes out of solution too quickly and forms gas bubbles in the tissues, resulting in **decompression sickness** (*the bends*). The effects of decompression sickness typically result from bubbles in nervous tissue and can be mild or severe, depending on the number of bubbles formed. Symptoms include joint pain, especially in the arms and legs, dizziness, shortness of breath, extreme fatigue, paralysis, and unconsciousness.

External Respiration

External respiration or *pulmonary gas exchange* is the diffusion of O_2 from air in the alveoli of the lungs to blood in pulmonary capillaries and the diffusion of CO_2 in the opposite direction (**Figure 23.17a**). External respiration in the lungs converts **deoxygenated blood** (depleted of some O_2) coming from the right side of the heart into **oxygenated blood** (saturated with O_2) that returns to the left side of the heart (see **Figure 21.30**). As blood flows through the pulmonary capillaries, it picks up O_2 from alveolar air and unloads CO_2 into alveolar air. Although this process is commonly called an "exchange" of gases, each gas diffuses independently from the area where its partial pressure is higher to the area where its partial pressure is lower.

As **Figure 23.17a** shows, O_2 diffuses from alveolar air, where its partial pressure is 105 mmHg, into the blood in pulmonary capillaries, where P_{O_2} is only 40 mmHg in a resting person. If you have been exercising, the P_{O_2} will be even lower because contracting muscle fibers are using more O_2. Diffusion continues until the P_{O_2} of pulmonary capillary blood increases to match the P_{O_2} of alveolar air, 105 mmHg. Because blood leaving pulmonary capillaries near alveolar air spaces mixes with a small volume of blood that has flowed through conducting portions of the respiratory system, where gas exchange does not occur, the P_{O_2} of blood in the pulmonary veins is slightly less than the P_{O_2} in pulmonary capillaries, about 100 mmHg.

While O_2 is diffusing from alveolar air into deoxygenated blood, CO_2 is diffusing in the opposite direction. The P_{CO_2} of deoxygenated blood is 45 mmHg in a resting person, and the P_{CO_2} of alveolar air is 40 mmHg. Because of this difference in P_{CO_2}, carbon dioxide diffuses from deoxygenated blood into the alveoli until the P_{CO_2} of the blood decreases to 40 mmHg. Exhalation keeps alveolar P_{CO_2} at 40 mmHg. Oxygenated blood returning to the left side of the heart in the pulmonary veins thus has a P_{CO_2} of 40 mmHg.

The number of capillaries near alveoli in the lungs is very large, and blood flows slowly enough through these capillaries that it picks up a maximal amount of O_2. During vigorous exercise, when cardiac output is increased, blood flows more rapidly through both the systemic and pulmonary circulations. As a result, blood's transit time in the pulmonary capillaries is shorter. Still, the P_{O_2} of blood in the pulmonary veins normally reaches 100 mmHg. In diseases that decrease the rate of gas diffusion, however, the blood may not come into full equilibrium with alveolar air, especially during exercise. When this happens, the P_{O_2} declines and P_{CO_2} rises in systemic arterial blood.

Internal Respiration

The left ventricle pumps oxygenated blood into the aorta and through the systemic arteries to systemic capillaries. The exchange of O_2 and CO_2 between systemic capillaries and tissue cells is called **internal respiration** or *systemic gas exchange* (**Figure 23.17b**). As O_2 leaves the bloodstream, oxygenated blood is converted into deoxygenated blood. Unlike external respiration, which occurs only in the lungs, internal respiration occurs in tissues throughout the body.

The P_{O_2} of blood pumped into systemic capillaries is higher (100 mmHg) than the P_{O_2} in tissue cells (40 mmHg at rest) because the cells constantly use O_2 to produce ATP. Due to this pressure difference, oxygen diffuses out of the capillaries into tissue cells and blood P_{O_2} drops to 40 mmHg by the time the blood exits systemic capillaries.

While O_2 diffuses from the systemic capillaries into tissue cells, CO_2 diffuses in the opposite direction. Because tissue cells are constantly producing CO_2, the P_{CO_2} of cells (45 mmHg at rest) is higher than that of systemic capillary blood (40 mmHg). As a result, CO_2 diffuses from tissue cells through interstitial fluid into systemic capillaries until the P_{CO_2} in the blood increases to 45 mmHg. The deoxygenated blood then returns to the heart and is pumped to the lungs for another cycle of external respiration.

In a person at rest, tissue cells, on average, need only 25% of the available O_2 in oxygenated blood; despite its name, deoxygenated blood retains 75% of its O_2 content. During exercise, more O_2 diffuses from the blood into metabolically active cells, such as contracting skeletal muscle fibers. Active cells use more O_2 for ATP production, causing the O_2 content of deoxygenated blood to drop below 75%.

The *rate* of pulmonary and systemic gas exchange depends on several factors.

- ***Partial pressure difference of the gases.*** Alveolar P_{O_2} must be higher than blood P_{O_2} for oxygen to diffuse from alveolar air into the blood. The rate of diffusion is faster when the difference between P_{O_2} in alveolar air and pulmonary capillary blood is larger; diffusion is slower when the difference is smaller. The differences between P_{O_2} and P_{CO_2} in alveolar air versus pulmonary blood increase during exercise. The larger partial pressure differences accelerate the rates of gas diffusion. The partial pressures of O_2 and CO_2 in alveolar air also depend on the rate of airflow into and out of the lungs. Certain drugs (such as morphine) slow ventilation, thereby decreasing the amount of O_2 and CO_2 that can be exchanged between alveolar air and blood. With increasing altitude, the total atmospheric pressure decreases, as does the partial pressure of O_2—from 159 mmHg at sea level, to 110 mmHg at 10,000 ft, to 73 mmHg at 20,000 ft. Although O_2 still is 20.9% of the total, the P_{O_2} of inhaled air decreases with increasing altitude. Alveolar P_{O_2} decreases correspondingly, and O_2 diffuses into the blood more slowly. The common signs and symptoms of **high altitude sickness**—shortness of breath, headache, fatigue, insomnia, nausea, and dizziness—are due to a lower level of oxygen in the blood.
- ***Surface area available for gas exchange.*** As you learned earlier in the chapter, the surface area of the alveoli is huge (about 75 m^2 or 807 ft^2). In addition, many capillaries surround each alveolus, so many that as much as 900 mL of blood is able to participate in gas exchange at any instant. Any pulmonary disorder that decreases the functional surface area of the respiratory membranes decreases the rate of external respiration. In emphysema (see Disorders: Homeostatic Imbalances in the Study Guide), for example, alveolar walls disintegrate, so surface area is smaller than normal and pulmonary gas exchange is slowed.
- ***Diffusion distance.*** The respiratory membrane is very thin, so diffusion occurs quickly. Also, the capillaries are so narrow that the red blood cells must pass through them in single file, which minimizes the diffusion distance from an alveolar air space to hemoglobin inside red blood cells. Buildup of interstitial fluid between alveoli, as occurs in pulmonary edema (see Disorders: Homeostatic imbalances in the Study Guide), slows the rate of gas exchange because it increases diffusion distance.

FIGURE 23.17 Changes in partial pressures of oxygen and carbon dioxide (in mmHg) during external and internal respiration.

Gases diffuse from areas of higher partial pressure to areas of lower partial pressure.

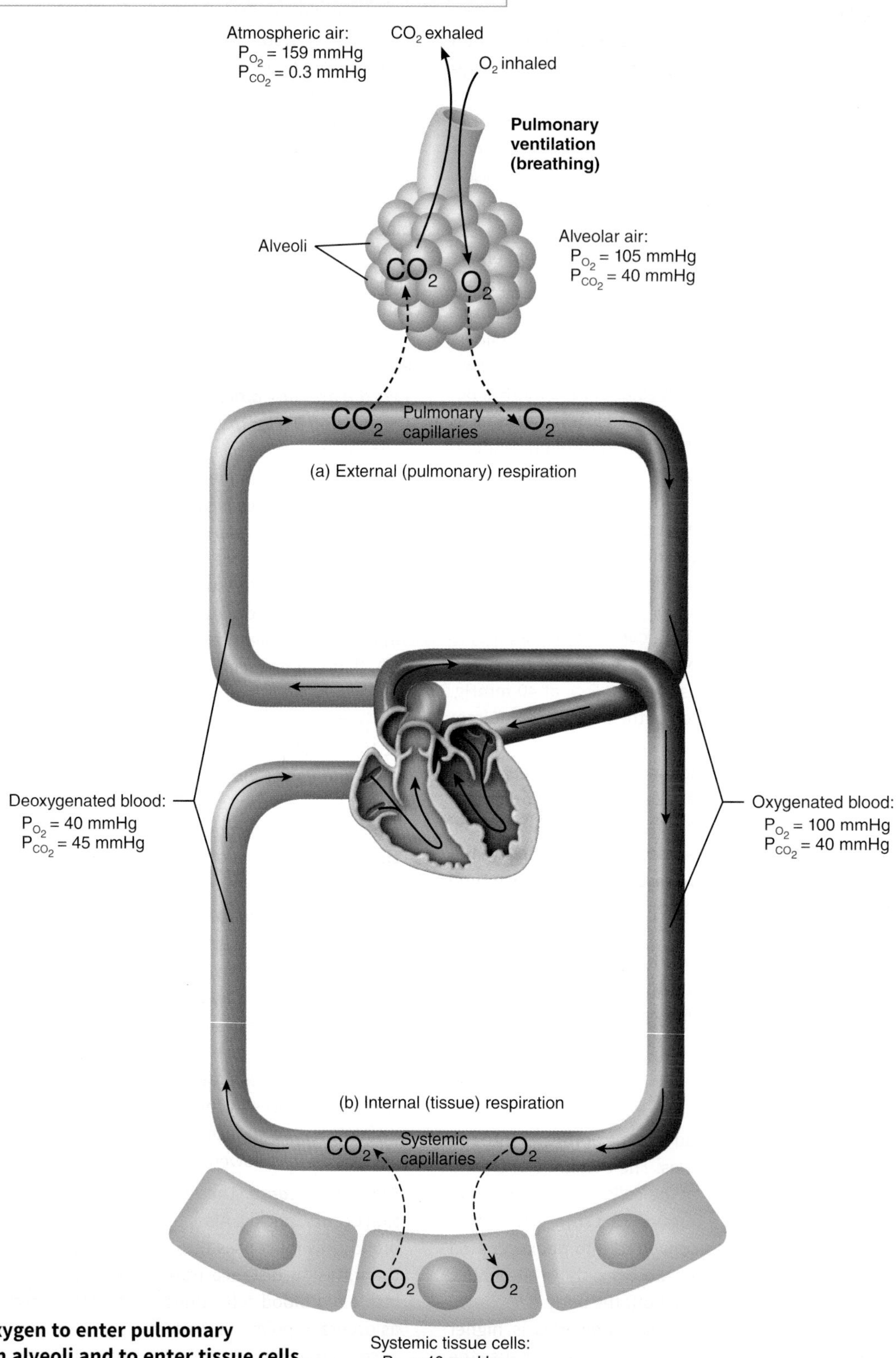

Q What causes oxygen to enter pulmonary capillaries from alveoli and to enter tissue cells from systemic capillaries?

- ***Molecular weight and solubility of the gases.*** Because O_2 has a lower molecular weight than CO_2, it could be expected to diffuse across the respiratory membrane about 1.2 times faster. However, the solubility of CO_2 in the fluid portions of the respiratory membrane is about 24 times greater than that of O_2. Taking both of these factors into account, net outward CO_2 diffusion occurs 20 times more rapidly than net inward O_2 diffusion. Consequently, when diffusion is slower than normal—for example, in emphysema or pulmonary edema—O_2 insufficiency (hypoxia) typically occurs before there is significant retention of CO_2 (hypercapnia).

23.7 Transport of Oxygen and Carbon Dioxide

OBJECTIVE

- **Describe** how the blood transports oxygen and carbon dioxide.

As you have already learned, the blood transports gases between the lungs and body tissues. When O_2 and CO_2 enter the blood, certain chemical reactions occur that aid in gas transport and gas exchange.

Oxygen Transport

Oxygen does not dissolve easily in water, so only about 1.5% of inhaled O_2 is dissolved in blood plasma, which is mostly water. About 98.5% of blood O_2 is bound to hemoglobin in red blood cells (**Figure 23.18**). Each 100 mL of oxygenated blood contains the equivalent of 20 mL of gaseous O_2. Using the percentages just given, the amount dissolved in the plasma is 0.3 mL and the amount bound to hemoglobin is 19.7 mL.

The heme portion of hemoglobin contains four atoms of iron, each capable of binding to a molecule of O_2 (see **Figure 19.4b, c**). Oxygen and hemoglobin bind in an easily reversible reaction to form **oxyhemoglobin**:

$$\underset{\text{Reduced hemoglobin (deoxyhemoglobin)}}{Hb} + \underset{\text{Oxygen}}{O_2} \underset{\text{Dissociation of } O_2}{\overset{\text{Binding of } O_2}{\rightleftharpoons}} \underset{\text{Oxyhemoglobin}}{Hb\text{–}O_2}$$

The 98.5% of the O_2 that is bound to hemoglobin is trapped inside RBCs, so only the dissolved O_2 (1.5%) can diffuse out of tissue capillaries into tissue cells. Thus, it is important to understand the factors that promote O_2 binding to and dissociation (separation) from hemoglobin.

The Relationship between Hemoglobin and Oxygen Partial Pressure

The most important factor that determines how much O_2 binds to hemoglobin is the P_{O_2}; the higher the P_{O_2}, the more O_2 combines with Hb. When reduced hemoglobin (Hb) is completely converted to oxyhemoglobin (Hb–O_2), the hemoglobin is said to be **fully saturated**; when hemoglobin consists of a mixture of Hb and Hb–O_2, it is **partially saturated**. The **percent saturation of hemoglobin** expresses the average saturation of hemoglobin with oxygen. For instance, if each hemoglobin molecule has bound two O_2 molecules, then the hemoglobin is 50% saturated because each Hb can bind a maximum of four O_2.

The relationship between the percent saturation of hemoglobin and P_{O_2} is illustrated in the oxygen–hemoglobin dissociation curve in **Figure 23.19**. Note that when the P_{O_2} is high, hemoglobin binds with large amounts of O_2 and is almost 100% saturated. When P_{O_2} is low, hemoglobin is only partially saturated. In other words, the greater the P_{O_2}, the more O_2 will bind to hemoglobin, until all the available hemoglobin molecules are saturated. Therefore, in pulmonary capillaries, where P_{O_2} is high, a lot of O_2 binds to hemoglobin. In tissue capillaries, where the P_{O_2} is lower, hemoglobin does not hold as much O_2, and the dissolved O_2 is unloaded via diffusion into tissue cells (see **Figure 23.18b**). Note that hemoglobin is still 75% saturated with O_2 at a P_{O_2} of 40 mmHg, the average P_{O_2} of tissue cells in a person at rest. This is the basis for the earlier statement that only 25% of the available O_2 unloads from hemoglobin and is used by tissue cells under resting conditions.

When the P_{O_2} is between 60 and 100 mmHg, hemoglobin is 90% or more saturated with O_2 (**Figure 23.19**). Thus, blood picks up a nearly full load of O_2 from the lungs even when the P_{O_2} of alveolar air is as low as 60 mmHg. The Hb–P_{O_2} curve explains why people can still perform well at high altitudes or when they have certain cardiac and pulmonary diseases, even though P_{O_2} may drop as low as 60 mmHg. Note also in the curve that at a considerably lower P_{O_2} of 40 mmHg, hemoglobin is still 75% saturated with O_2. However, oxygen saturation of Hb drops to 35% at 20 mmHg. Between 40 and 20 mmHg, large amounts of O_2 are released from hemoglobin in response to only small decreases in P_{O_2}. In active tissues such as contracting muscles, P_{O_2} may drop well below 40 mmHg. Then, a large percentage of the O_2 is released from hemoglobin, providing more O_2 to metabolically active tissues.

Other Factors Affecting the Affinity of Hemoglobin for Oxygen

Although P_{O_2} is the most important factor that determines the percent O_2 saturation of hemoglobin, several other factors influence the tightness or **affinity** with which hemoglobin binds O_2. In effect, these factors shift the entire curve either to the left (higher affinity) or to the right (lower affinity). The changing affinity of hemoglobin for O_2 is another example of how homeostatic mechanisms adjust body activities to cellular needs. Each one makes sense if you keep in mind that metabolically active tissue cells need O_2 and produce acids, CO_2, and heat as wastes. The following four factors affect the affinity of hemoglobin for O_2:

1. ***Acidity (pH).*** As acidity increases (pH decreases), the affinity of hemoglobin for O_2 decreases, and O_2 dissociates more readily from hemoglobin (**Figure 23.20a**). In other words, increasing acidity enhances the unloading of oxygen from hemoglobin. The main acids produced by metabolically active tissues are lactic acid and carbonic acid. When pH decreases, the entire oxygen–hemoglobin dissociation curve shifts to the right; at any given P_{O_2}, Hb is less saturated with O_2, a change termed the **Bohr effect** (BŌR). The Bohr effect works both ways: An increase in H^+ in blood causes O_2 to unload

FIGURE 23.18 Transport of oxygen (O_2) and carbon dioxide (CO_2) in the blood.

Most O_2 is transported by hemoglobin as oxyhemoglobin (Hb–O_2) within red blood cells; most CO_2 is transported in blood plasma as bicarbonate ions (HCO_3^-).

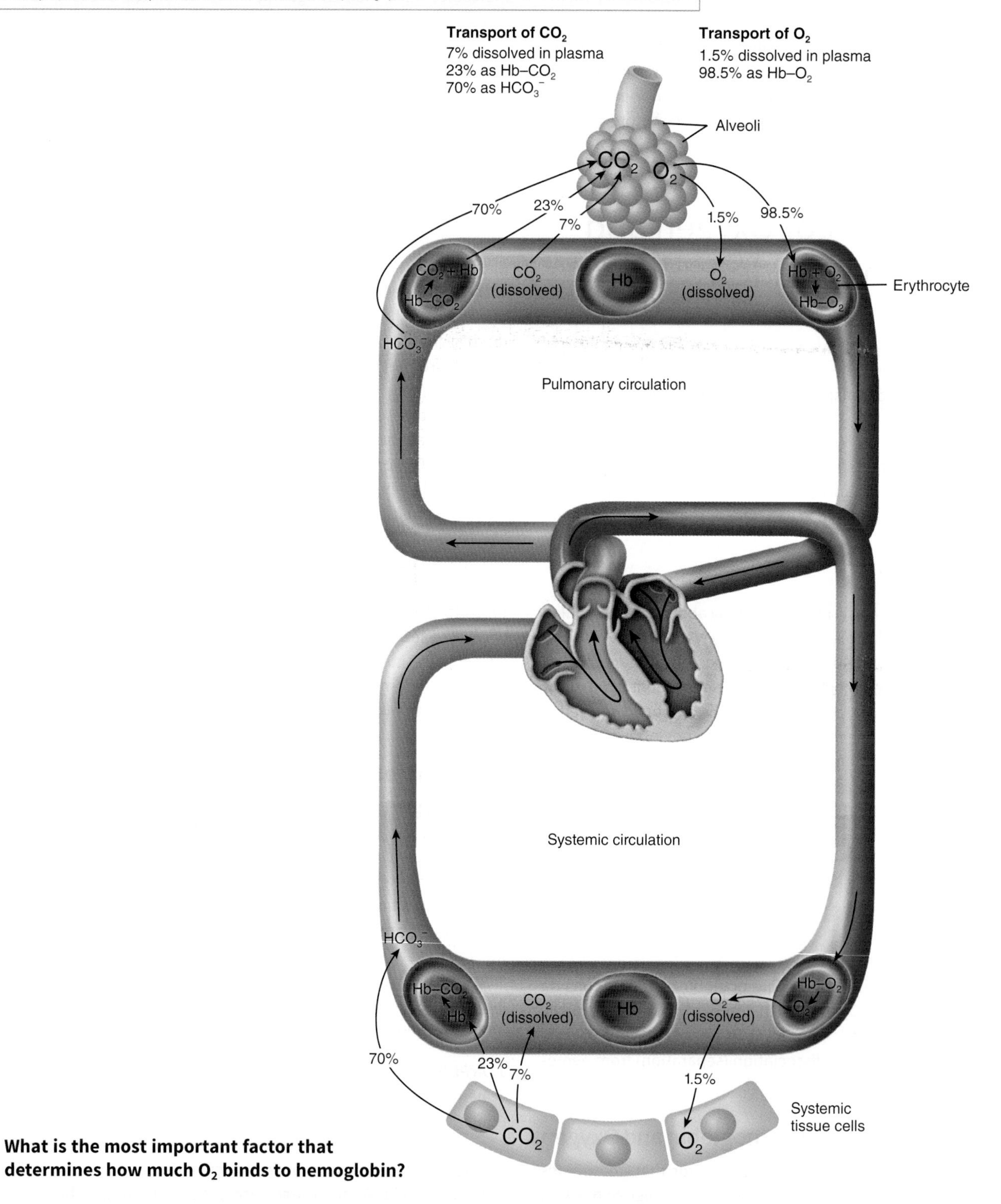

Q What is the most important factor that determines how much O_2 binds to hemoglobin?

from hemoglobin, and the binding of O_2 to hemoglobin causes unloading of H^+ from hemoglobin. The explanation for the Bohr effect is that hemoglobin can act as a buffer for hydrogen ions (H^+). But when H^+ ions bind to amino acids in hemoglobin, they alter its structure slightly, decreasing its oxygen-carrying capacity. Thus, lowered pH drives O_2 off hemoglobin, making more O_2 available for tissue cells. By contrast, elevated pH increases the affinity of hemoglobin for O_2 and shifts the oxygen–hemoglobin dissociation curve to the left.

FIGURE 23.19 Oxygen–hemoglobin dissociation curve showing the relationship between hemoglobin saturation and PO_2 at normal body temperature.

As P_{O_2} increases, more O_2 combines with hemoglobin.

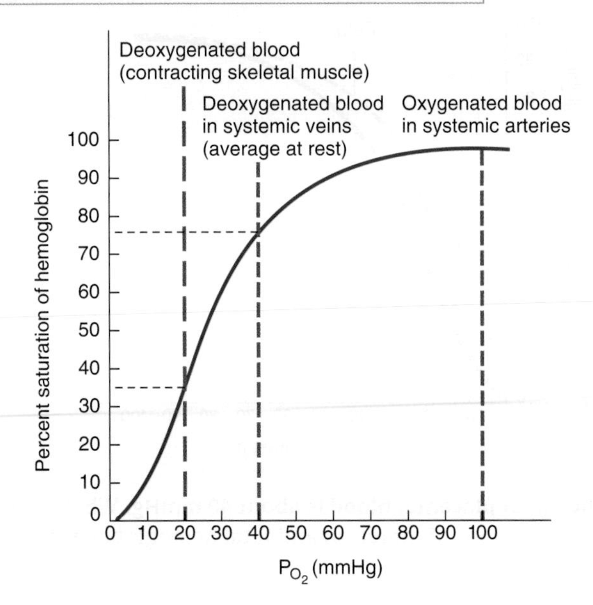

Q What point on the curve represents blood in your pulmonary veins right now? In your pulmonary veins if you were jogging?

2. ***Partial pressure of carbon dioxide.*** CO_2 also can bind to hemoglobin, and the effect is similar to that of H^+ (shifting the curve to the right). As P_{CO_2} rises, hemoglobin releases O_2 more readily (**Figure 23.20b**). P_{CO_2} and pH are related factors because low blood pH (acidity) results from high P_{CO_2}. As CO_2 enters the blood, much of it is temporarily converted to carbonic acid (H_2CO_3), a reaction catalyzed by an enzyme in red blood cells called *carbonic anhydrase (CA):*

$$\underset{\text{Carbon dioxide}}{CO_2} + \underset{\text{Water}}{H_2O} \overset{CA}{\rightleftharpoons} \underset{\text{Carbonic acid}}{H_2CO_3} \rightleftharpoons \underset{\text{Hydrogen ion}}{H^+} + \underset{\text{Bicarbonate ion}}{HCO_3^-}$$

The carbonic acid thus formed in red blood cells dissociates into hydrogen ions and bicarbonate ions. As the H^+ concentration increases, pH decreases. Thus, an increased P_{CO_2} produces a more acidic environment, which helps release O_2 from hemoglobin. During exercise, lactic acid—a by-product of anaerobic metabolism within muscles—also decreases blood pH. Decreased P_{CO_2} (and elevated pH) shifts the saturation curve to the left.

3. ***Temperature.*** Within limits, as temperature increases, so does the amount of O_2 released from hemoglobin (**Figure 23.21**). Heat is a by-product of the metabolic reactions of all cells, and the heat released by contracting muscle fibers tends to raise body temperature. Metabolically active cells require more O_2 and liberate more acids and heat. The acids and heat in turn promote release of O_2 from oxyhemoglobin. Fever produces a similar result. In contrast, during hypothermia (lowered body temperature) cellular metabolism slows, the need for O_2 is reduced, and more O_2 remains bound to hemoglobin (a shift to the left in the saturation curve).

FIGURE 23.20 Oxygen–hemoglobin dissociation curves showing the relationship of (a) pH and (b) P_{CO_2} to hemoglobin saturation at normal body temperature. As pH increases or P_{CO_2} decreases, O_2 combines more tightly with hemoglobin, so that less is available to tissues. The broken lines emphasize these relationships.

As pH decreases or P_{CO_2} increases, the affinity of hemoglobin for O_2 declines, so less O_2 combines with hemoglobin and more is available to tissues.

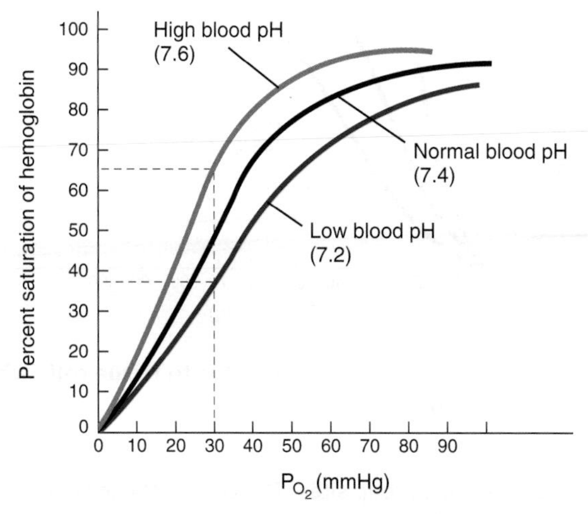

(a) Effect of pH on affinity of hemoglobin for oxygen

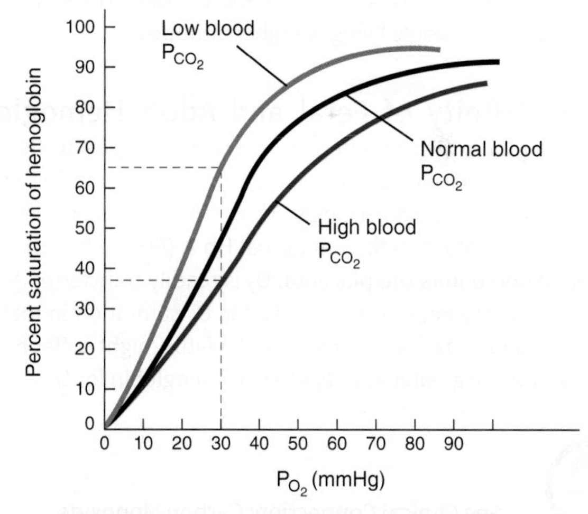

(b) Effect of P_{CO_2} on affinity of hemoglobin for oxygen

Q In comparison to the value when you are sitting, is the affinity of your hemoglobin for O_2 higher or lower when you are exercising? How does this benefit you?

4. ***BPG.*** A substance found in red blood cells called **2,3-bisphosphoglycerate (BPG)** (bis′-fos-fō-GLIS-e-rāt), formerly called *diphosphoglycerate (DPG)*, decreases the affinity of hemoglobin for O_2 and thus helps unload O_2 from hemoglobin. BPG is formed in red blood cells when they break down glucose to produce ATP in a process called glycolysis (described in Section 25.3). When BPG combines with hemoglobin by binding to the terminal amino groups of the two beta globin chains, the hemoglobin binds O_2 less

FIGURE 23.21 Oxygen–hemoglobin dissociation curves showing the effect of temperature changes.

As temperature increases, the affinity of hemoglobin for O_2 decreases.

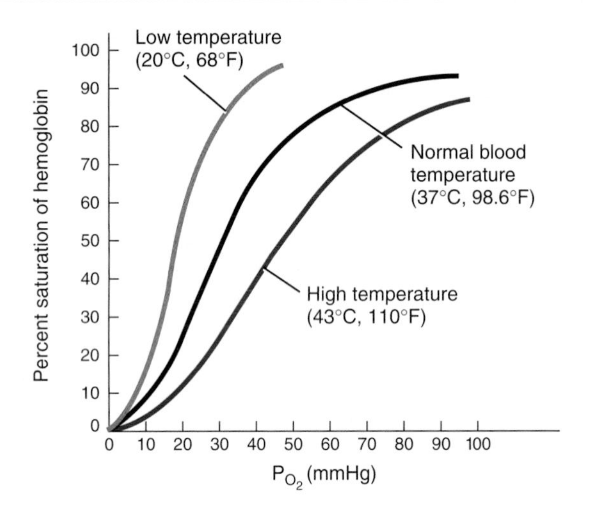

Q Is O_2 more available or less available to tissue cells when you have a fever? Why?

FIGURE 23.22 Oxygen–hemoglobin dissociation curves comparing fetal and maternal hemoglobin.

Fetal hemoglobin has a higher affinity for O_2 than does adult hemoglobin.

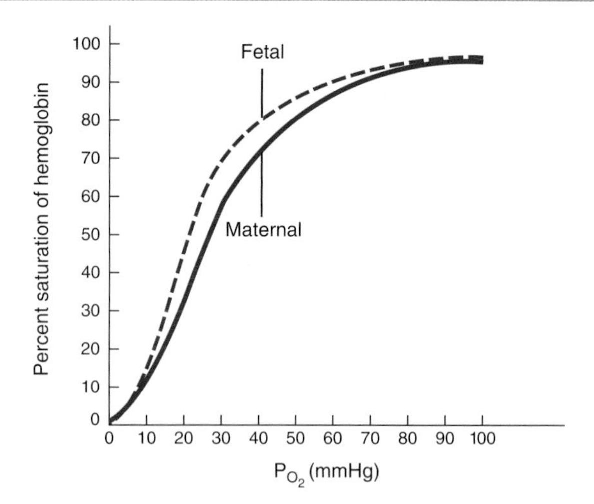

Q The P_{O_2} of placental blood is about 40 mmHg. What are the O_2 saturations of maternal and fetal hemoglobin at this P_{O_2}?

tightly at the heme group sites. The greater the level of BPG, the more O_2 is unloaded from hemoglobin. Certain hormones, such as thyroxine, human growth hormone, epinephrine, norepinephrine, and testosterone, increase the formation of BPG. The level of BPG also is higher in people living at higher altitudes.

Oxygen Affinity of Fetal and Adult Hemoglobin

Fetal hemoglobin (Hb-F) differs from **adult hemoglobin (Hb-A)** in structure and in its affinity for O_2. Hb-F has a higher affinity for O_2 because it binds BPG less strongly. Thus, when P_{O_2} is low, Hb-F can carry up to 30% more O_2 than maternal Hb-A (**Figure 23.22**). As the maternal blood enters the placenta, O_2 is readily transferred to fetal blood. This is very important because the O_2 saturation in maternal blood in the placenta is quite low, and the fetus might suffer hypoxia were it not for the greater affinity of fetal hemoglobin for O_2.

See Clinical Connection: Carbon Monoxide Poisoning

Carbon Dioxide Transport

Under normal resting conditions, each 100 mL of deoxygenated blood contains the equivalent of 53 mL of gaseous CO_2, which is transported in the blood in three main forms (see **Figure 23.18**):

1. ***Dissolved CO_2.*** The smallest percentage—about 7%—is dissolved in blood plasma. On reaching the lungs, it diffuses into alveolar air and is exhaled.

2. ***Carbamino compounds.*** A somewhat higher percentage, about 23%, combines with the amino groups of amino acids and proteins in blood to form **carbamino compounds** (kar-BAM-i-nō). Because the most prevalent protein in blood is hemoglobin (inside red blood cells), most of the CO_2 transported in this manner is bound to hemoglobin. The main CO_2 binding sites are the terminal amino acids in the two alpha and two beta globin chains. Hemoglobin that has bound CO_2 is termed **carbaminohemoglobin (Hb–CO_2)**:

$$\underset{\text{Hemoglobin}}{\text{Hb}} + \underset{\text{Carbon dioxide}}{CO_2} \rightleftharpoons \underset{\text{Carbaminohemoglobin}}{\text{Hb–}CO_2}$$

The formation of carbaminohemoglobin is greatly influenced by P_{CO_2}. For example, in tissue capillaries P_{CO_2} is relatively high, which promotes formation of carbaminohemoglobin. But in pulmonary capillaries, P_{CO_2} is relatively low, and the CO_2 readily splits apart from globin and enters the alveoli by diffusion.

3. ***Bicarbonate ions.*** The greatest percentage of CO_2—about 70%—is transported in blood plasma as **bicarbonate ions** (HCO_3^-) (bī′-KAR-bo-nāt). As CO_2 diffuses into systemic capillaries and enters red blood cells, it reacts with water in the presence of the enzyme carbonic anhydrase (CA) to form carbonic acid, which dissociates into H^+ and HCO_3^-:

$$\underset{\substack{\text{Carbon}\\\text{dioxide}}}{CO_2} + \underset{\text{Water}}{H_2O} \overset{CA}{\rightleftharpoons} \underset{\substack{\text{Carbonic}\\\text{acid}}}{H_2CO_3} \rightleftharpoons \underset{\substack{\text{Hydrogen}\\\text{ion}}}{H^+} + \underset{\substack{\text{Bicarbonate}\\\text{ion}}}{HCO_3^-}$$

Thus, as blood picks up CO_2, HCO_3^- accumulates inside RBCs. Some HCO_3^- moves out into the blood plasma, down its concentration gradient. In exchange, chloride ions (Cl^-) move from plasma into the RBCs. This exchange of negative ions, which maintains the electrical balance between blood plasma and RBC cytosol, is known as the **chloride shift** (see Figure 23.23b). The net effect of these reactions is that CO_2 is removed from tissue cells and transported in blood plasma as HCO_3^-. As blood passes through pulmonary capillaries in the lungs, all of these reactions reverse and CO_2 is exhaled.

The amount of CO_2 that can be transported in the blood is influenced by the percent saturation of hemoglobin with oxygen. The lower the amount of oxyhemoglobin (Hb–O_2), the higher the CO_2-carrying capacity of the blood, a relationship known as the **Haldane effect**. Two characteristics of deoxyhemoglobin give rise to the Haldane effect: (1) Deoxyhemoglobin binds to and thus transports more CO_2 than does Hb–O_2. (2) Deoxyhemoglobin also buffers more H^+ than does Hb–O_2, thereby removing H^+ from solution and promoting conversion of CO_2 to HCO_3^- via the reaction catalyzed by carbonic anhydrase.

Summary of Gas Exchange and Transport in Lungs and Tissues

Deoxygenated blood returning to the pulmonary capillaries in the lungs (Figure 23.23a) contains CO_2 dissolved in blood plasma, CO_2 combined

FIGURE 23.23 Summary of chemical reactions that occur during gas exchange. (a) As carbon dioxide (CO_2) is exhaled, hemoglobin (Hb) inside red blood cells in pulmonary capillaries unloads CO_2 and picks up O_2 from alveolar air. Binding of O_2 to Hb–H releases hydrogen ions (H^+). Bicarbonate ions (HCO_3^-) pass into the RBC and bind to released H^+, forming carbonic acid (H_2CO_3). The H_2CO_3 dissociates into water (H_2O) and CO_2, and the CO_2 diffuses from blood into alveolar air. To maintain electrical balance, a chloride ion (Cl^-) exits the RBC for each HCO_3^- that enters (reverse chloride shift). (b) CO_2 diffuses out of tissue cells that produce it and enters red blood cells, where some of it binds to hemoglobin, forming carbaminohemoglobin (Hb–CO_2). This reaction causes O_2 to dissociate from oxyhemoglobin (Hb–O_2). Other molecules of CO_2 combine with water to produce bicarbonate ions (HCO_3^-) and hydrogen ions (H^+). As Hb buffers H^+, the Hb releases O_2 (Bohr effect). To maintain electrical balance, a chloride ion (Cl^-) enters the RBC for each HCO_3^- that exits (chloride shift).

Hemoglobin inside red blood cells transports O_2, CO_2, and H^+.

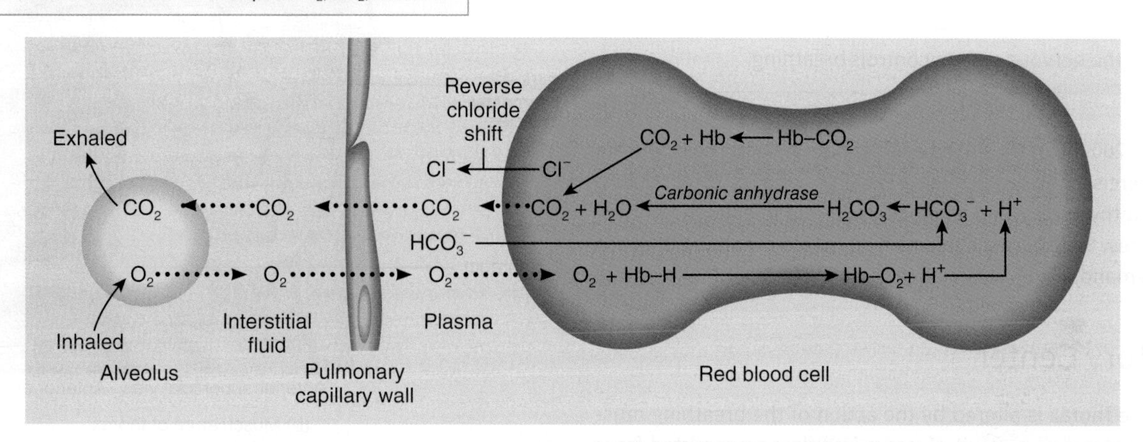

(a) Exchange of O_2 and CO_2 in pulmonary capillaries (external respiration)

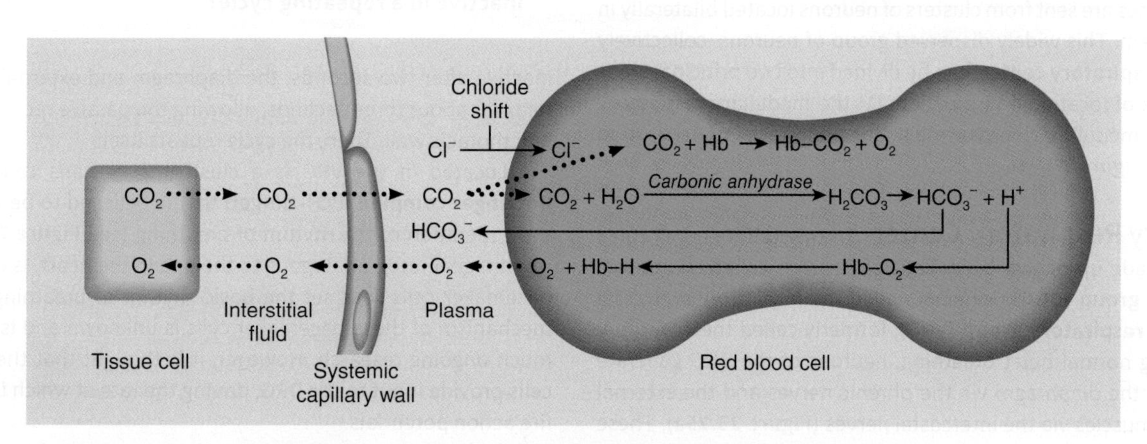

(b) Exchange of O_2 and CO_2 in systemic capillaries (internal respiration)

Q Would you expect the concentration of HCO_3^- to be higher in blood plasma taken from a systemic artery or a systemic vein?

with globin as carbaminohemoglobin ($Hb–CO_2$), and CO_2 incorporated into HCO_3^- within RBCs. The RBCs have also picked up H^+, some of which binds to and therefore is buffered by hemoglobin ($Hb–H$). As blood passes through the pulmonary capillaries, molecules of CO_2 dissolved in blood plasma and CO_2 that dissociates from the globin portion of hemoglobin diffuse into alveolar air and are exhaled. At the same time, inhaled O_2 is diffusing from alveolar air into RBCs and is binding to hemoglobin to form oxyhemoglobin ($Hb–O_2$). Carbon dioxide also is released from HCO_3^- when H^+ combines with HCO_3^- inside RBCs. The H_2CO_3 formed from this reaction then splits into CO_2, which is exhaled, and H_2O. As the concentration of HCO_3^- declines inside RBCs in pulmonary capillaries, HCO_3^- diffuses in from the blood plasma, in exchange for Cl^-. In sum, oxygenated blood leaving the lungs has increased O_2 content and decreased amounts of CO_2 and H^+. In systemic capillaries, as cells use O_2 and produce CO_2, the chemical reactions reverse (Figure 23.23b).

23.8 Control of Breathing

OBJECTIVE

- **Explain** how the nervous system controls breathing.

At rest, about 200 mL of O_2 is used each minute by body cells. During strenuous exercise, however, O_2 use typically increases 15- to 20-fold in normal healthy adults, and as much as 30-fold in elite endurance-trained athletes. Several mechanisms help match breathing effort to metabolic demand.

Respiratory Center

The size of the thorax is altered by the action of the breathing muscles, which contract as a result of nerve impulses transmitted from centers in the brain and relax in the absence of nerve impulses. These nerve impulses are sent from clusters of neurons located bilaterally in the brain stem. This widely dispersed group of neurons, collectively called the **respiratory center**, can be divided into two principal areas on the basis of location and function: (1) the medullary respiratory center in the medulla oblongata and (2) the pontine respiratory group in the pons (Figure 23.24).

FIGURE 23.24 Locations of areas of the respiratory center.

The respiratory center is composed of neurons in the medullary respiratory center in the medulla plus the pontine respiratory group in the pons.

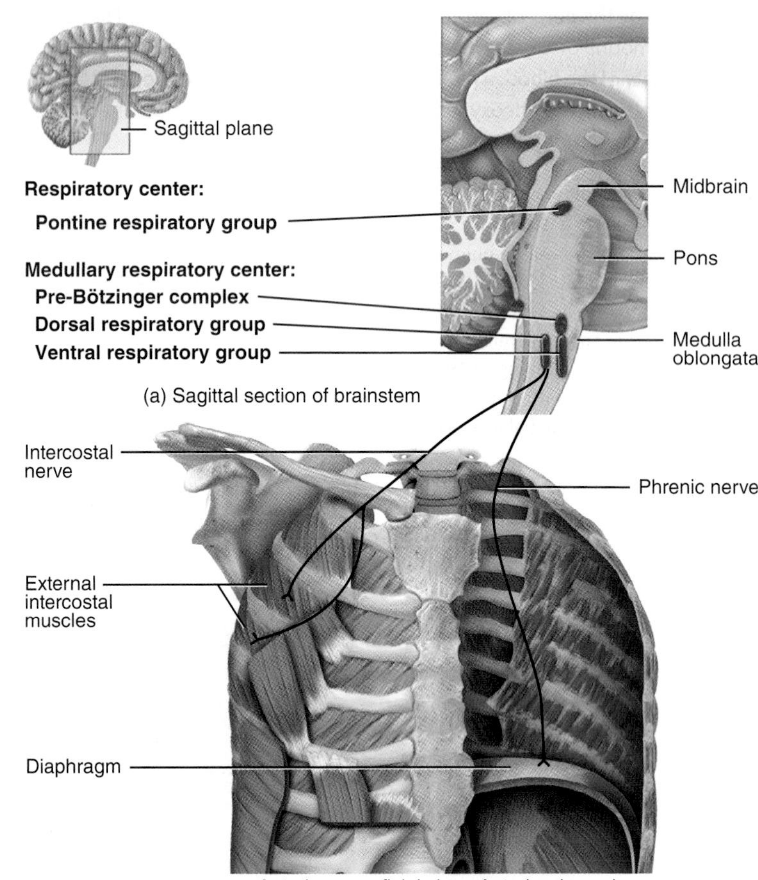

(b) Musculature of thorax

Q Which area contains neurons that are active and then inactive in a repeating cycle?

Medullary Respiratory Center

The **medullary respiratory center** is made up of two collections of neurons called the **dorsal respiratory group (DRG)**, formerly called the *inspiratory area*, and the **ventral respiratory group (VRG)**, formerly called the *expiratory area*. During normal quiet breathing, neurons of the DRG generate impulses to the diaphragm via the phrenic nerves and the external intercostal muscles via the intercostal nerves (Figure 23.25a). These impulses are released in bursts, which begin weakly, increase in strength for about two seconds, and then stop altogether. When the nerve impulses reach the diaphragm and external intercostals, the muscles contract and inhalation occurs. When the DRG becomes inactive after two seconds, the diaphragm and external intercostals relax for about three seconds, allowing the passive recoil of the lungs and thoracic wall. Then, the cycle repeats itself.

Located in the VRG is a cluster of neurons called the **pre-Bötzinger complex** (BOT-zin-ger) that is believed to be important in the generation of the rhythm of breathing (see Figure 23.24a). This rhythm generator, analogous to the one in the heart, is composed of pacemaker cells that set the basic rhythm of breathing. The exact mechanism of these pacemaker cells is unknown and is the topic of much ongoing research. However, it is thought that the pacemaker cells provide input to the DRG, driving the rate at which DRG neurons fire action potentials.

The remaining neurons of the VRG do not participate in normal quiet breathing. The VRG becomes activated when forceful breathing is required, such as during exercise, when playing a wind instrument, or at high altitudes. During forceful inhalation (Figure 23.25b), nerve

FIGURE 23.25 Roles of the medullary respiratory center in controlling (a) normal quiet breathing and (b) forceful breathing.

During normal quiet breathing, the ventral respiratory group is inactive; during forceful breathing, the dorsal respiratory group activates the ventral respiratory group.

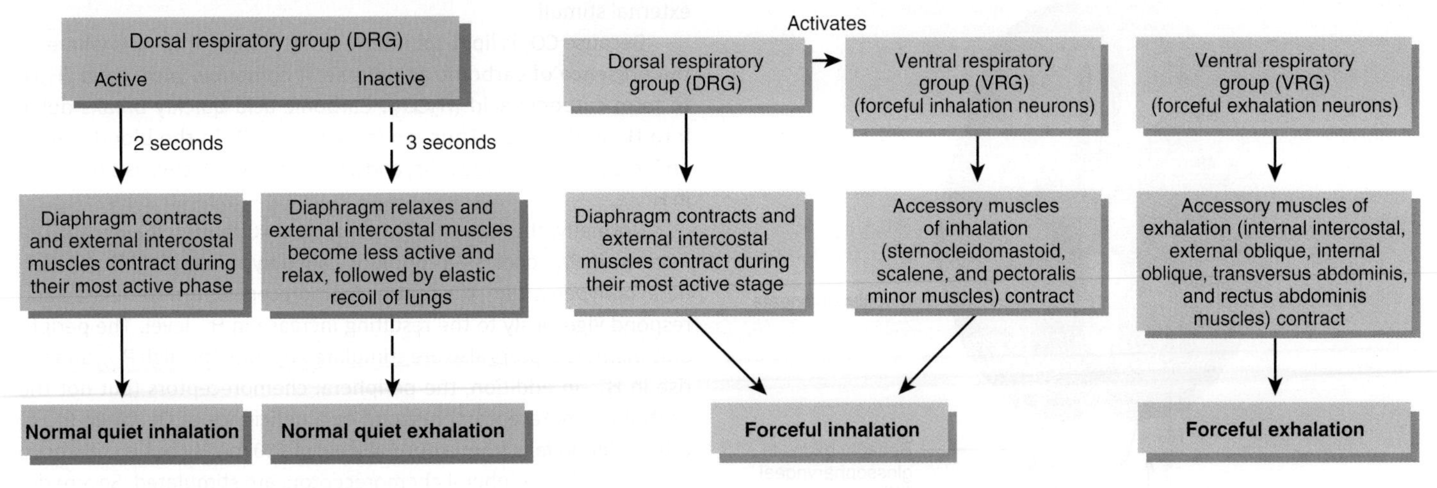

(a) During normal quiet breathing (b) During forceful breathing

Q Which nerves convey impulses from the respiratory center to the diaphragm?

impulses from the DRG not only stimulate the diaphragm and external intercostal muscles to contract, they also activate neurons of the VRG involved in forceful inhalation to send impulses to the accessory muscles of inhalation (sternocleidomastoid, scalenes, and pectoralis minor). Contraction of these muscles results in forceful inhalation.

During forceful exhalation (**Figure 23.25b**), the DRG is inactive along with the neurons of the VRG that result in forceful inhalation. However, those neurons of the VRG involved in forceful exhalation send nerve impulses to the accessory muscles of exhalation (internal intercostals, external oblique, internal oblique, transversus abdominis, and rectus abdominis). Contraction of these muscles results in forceful exhalation.

Pontine Respiratory Group The **pontine respiratory group** (**PRG**) (PON-tēn), formerly called the *pneumotaxic area*, is a collection of neurons in the pons (see **Figure 23.24a**). The neurons in the PRG are active during inhalation and exhalation. The PRG transmits nerve impulses to the DRG in the medulla. The PRG may play a role in both inhalation and exhalation by modifying the basic rhythm of breathing generated by the VRG, as when exercising, speaking, or sleeping.

Regulation of the Respiratory Center

Activity of the respiratory center can be modified in response to inputs from other brain regions, receptors in the peripheral nervous system, and other factors in order to maintain the homeostasis of breathing.

Cortical Influences on Breathing Because the cerebral cortex has connections with the respiratory center, we can voluntarily alter our pattern of breathing. We can even refuse to breathe at all for a short time. Voluntary control is protective because it enables us to prevent water or irritating gases from entering the lungs. The ability to not breathe, however, is limited by the buildup of CO_2 and H^+ in the body. When P_{CO_2} and H^+ concentrations increase to a certain level, the DRG neurons of the medullary respiratory center are strongly stimulated, nerve impulses are sent along the phrenic and intercostal nerves to inspiratory muscles, and breathing resumes, whether the person wants it to or not. It is impossible for small children to kill themselves by voluntarily holding their breath, even though many have tried in order to get their way. If breath is held long enough to cause fainting, breathing resumes when consciousness is lost. Nerve impulses from the hypothalamus and limbic system also stimulate the respiratory center, allowing emotional stimuli to alter breathing as, for example, in laughing and crying.

Chemoreceptor Regulation of Breathing Certain chemical stimuli modulate how quickly and how deeply we breathe. The respiratory system functions to maintain proper levels of CO_2 and O_2 and is very responsive to changes in the levels of these gases in body fluids. We introduced sensory neurons that are responsive to chemicals, called **chemoreceptors** (kē′-mō-rē-SEP-tors), in Chapter 21. Chemoreceptors in two locations of the respiratory system monitor levels of CO_2, H^+, and O_2 and provide input to the respiratory center (**Figure 23.26**). **Central chemoreceptors** are located in or near the medulla oblongata in the *central* nervous system. They respond to changes in H^+ concentration or P_{CO_2}, or both, in cerebrospinal fluid. **Peripheral chemoreceptors** are located in the **aortic bodies**, clusters of chemoreceptors located in the wall of the arch of the aorta, and in the **carotid bodies**, which are oval nodules in the wall of the left and

FIGURE 23.26 Locations of peripheral chemoreceptors.

Chemoreceptors are sensory neurons that respond to changes in the levels of certain chemicals in the body.

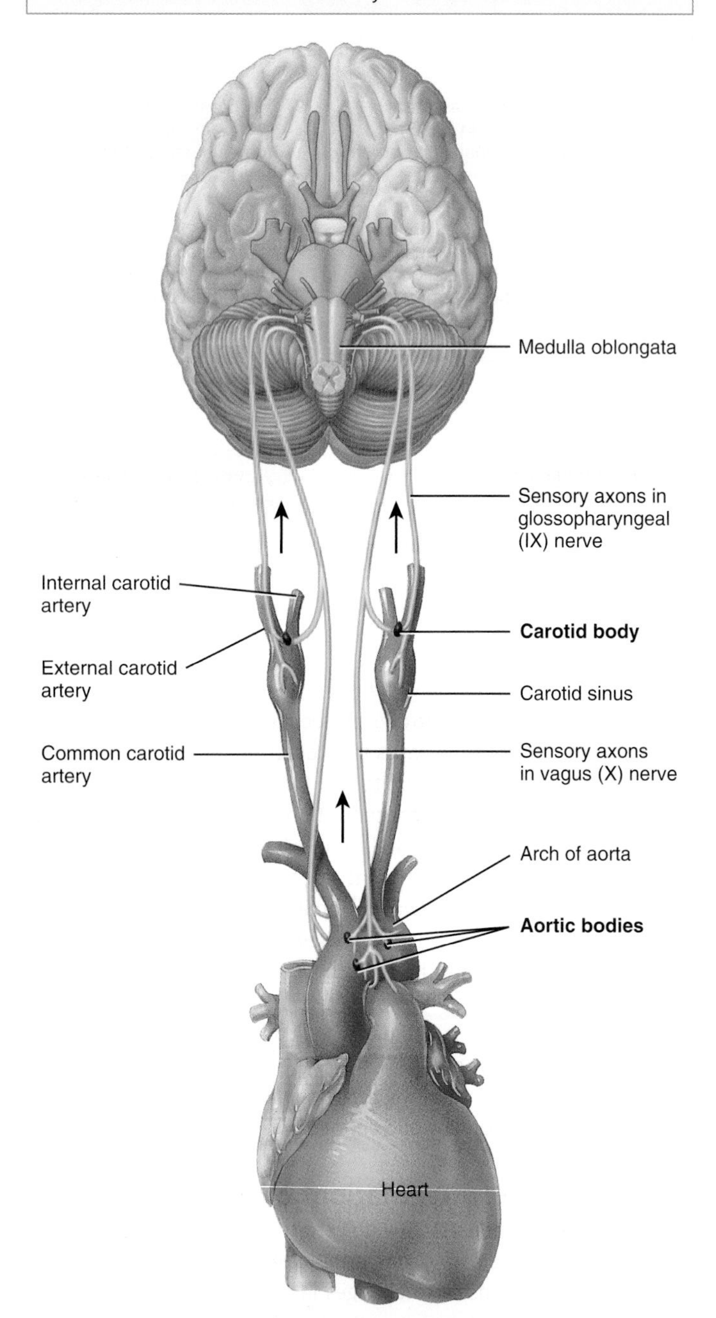

Q Which chemicals stimulate peripheral chemoreceptors?

right common carotid arteries where they divide into the internal and external carotid arteries. (The chemoreceptors of the aortic bodies are located close to the aortic baroreceptors, and the carotid bodies are located close to the carotid sinus baroreceptors. Recall from Chapter 21 that baroreceptors are sensory receptors that monitor blood pressure.) These chemoreceptors are part of the *peripheral* nervous system and are sensitive to changes in P_{O_2}, H^+, and P_{CO_2} in the blood. Axons of sensory neurons from the aortic bodies are part of the vagus (X) nerves, and those from the carotid bodies are part of the right and left glossopharyngeal (IX) nerves. Recall from Chapter 17 that olfactory receptors for the sense of smell and gustatory receptor cells for the sense of taste are also chemoreceptors. Both respond to external stimuli.

Because CO_2 is lipid-soluble, it easily diffuses into cells where, in the presence of carbonic anhydrase, it combines with water (H_2O) to form carbonic acid (H_2CO_3). Carbonic acid quickly breaks down into H^+ and HCO_3^-. Thus, an increase in CO_2 in the blood causes an increase in H^+ inside cells, and a decrease in CO_2 causes a decrease in H^+.

Normally, the P_{CO_2} in arterial blood is 40 mmHg. If even a slight increase in P_{CO_2} occurs—a condition called **hypercapnia** (hī′-per-KAP-nē-a) or *hypercarbia*—the central chemoreceptors are stimulated and respond vigorously to the resulting increase in H^+ level. The peripheral chemoreceptors also are stimulated by both the high P_{CO_2} and the rise in H^+. In addition, the peripheral chemoreceptors (but not the central chemoreceptors) respond to a deficiency of O_2. When P_{O_2} in arterial blood falls from a normal level of 100 mmHg but is still above 50 mmHg, the peripheral chemoreceptors are stimulated. Severe deficiency of O_2 depresses activity of the central chemoreceptors and DRG, which then do not respond well to any inputs and send fewer impulses to the muscles of inhalation. As the breathing rate decreases or breathing ceases altogether, P_{O_2} falls lower and lower, establishing a positive feedback cycle with a possibly fatal result.

The chemoreceptors participate in a negative feedback system that regulates the levels of CO_2, O_2, and H^+ in the blood (**Figure 23.27**). As a result of increased P_{CO_2}, decreased pH (increased H^+), or decreased P_{O_2}, input from the central and peripheral chemoreceptors causes the DRG to become highly active, and the rate and depth of breathing increase. Rapid and deep breathing, called **hyperventilation**, allows the inhalation of more O_2 and exhalation of more CO_2 until P_{CO_2} and H^+ are lowered to normal.

If arterial P_{CO_2} is lower than 40 mmHg—a condition called **hypocapnia** or *hypocarbia*—the central and peripheral chemoreceptors are not stimulated, and stimulatory impulses are not sent to the DRG. As a result, DRG neurons set their own moderate pace until CO_2 accumulates and the P_{CO_2} rises to 40 mmHg. DRG neurons are more strongly stimulated when P_{CO_2} is rising above normal than when P_{O_2} is falling below normal. As a result, people who hyperventilate voluntarily and cause hypocapnia can hold their breath for an unusually long period. Swimmers were once encouraged to hyperventilate just before diving in to compete. However, this practice is risky because the O_2 level may fall dangerously low and cause fainting before the P_{CO_2} rises high enough to stimulate inhalation. If you faint on land you may suffer bumps and bruises, but if you faint in the water you could drown.

Proprioceptor Stimulation of Breathing

As soon as you start exercising, your rate and depth of breathing increase, even before changes in P_{O_2}, P_{CO_2}, or H^+ level occur. The main stimulus for these quick changes in respiratory effort is input from proprioceptors, which monitor movement of joints and muscles. Nerve impulses from the proprioceptors stimulate the DRG of the medulla. At the same time, axon collaterals (branches) of upper motor neurons that

FIGURE 23.27 **Regulation of breathing in response to changes in blood P_{CO_2}, PO_2, and pH (H^+) via negative feedback control.**

An increase in arterial blood P_{CO_2} stimulates the dorsal respiratory group (DRG).

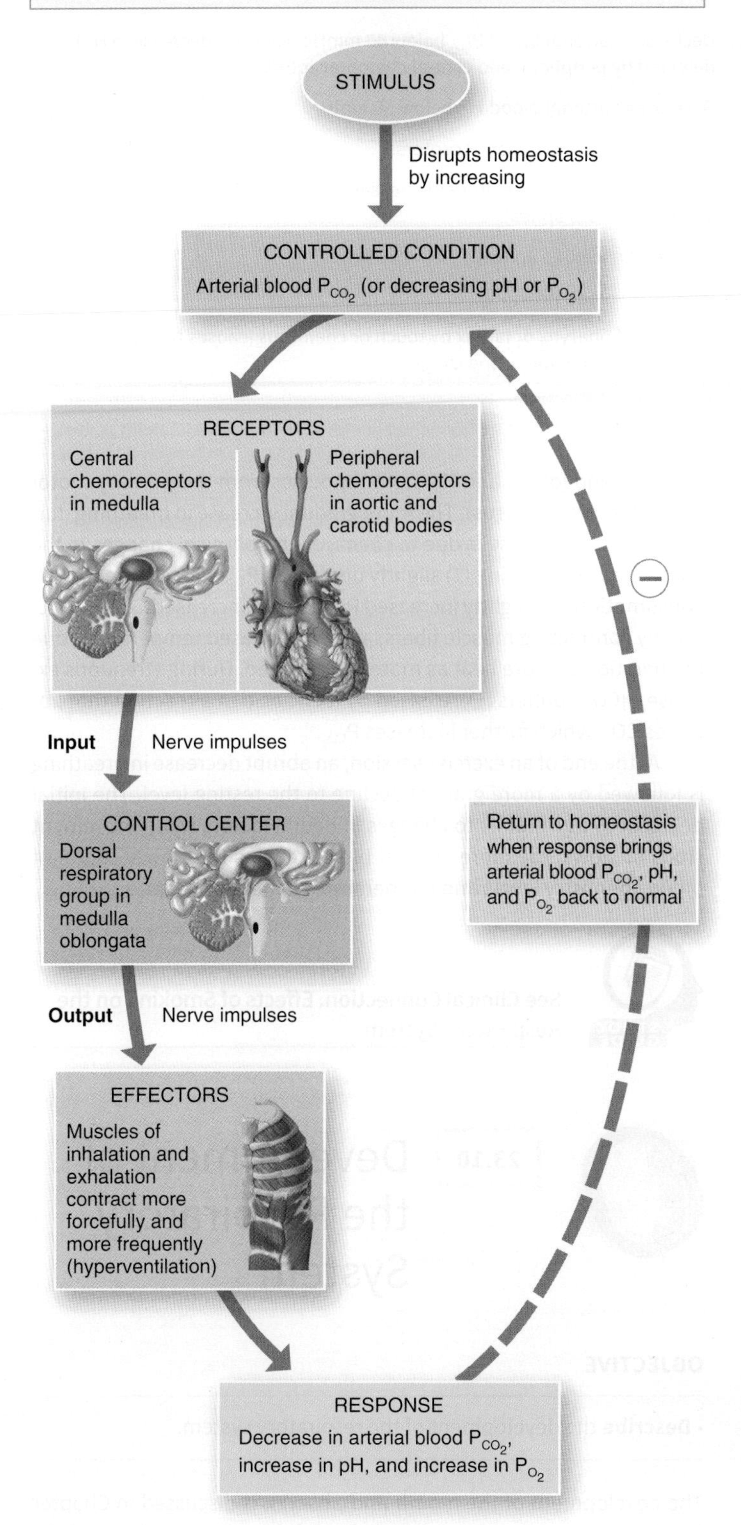

Q What is the normal arterial blood P_{CO_2}?

originate in the primary motor cortex (precentral gyrus) also feed excitatory impulses into the DRG.

See Clinical Connection: Hypoxia

The Inflation Reflex

Similar to those in the blood vessels, stretch-sensitive receptors called **baroreceptors** or *stretch receptors* are located in the walls of bronchi and bronchioles. When these receptors become stretched during overinflation of the lungs, nerve impulses are sent along the vagus (X) nerves to the dorsal respiratory group (DRG) in the medullary respiratory center. In response, the DRG is inhibited and the diaphragm and external intercostals relax. As a result, further inhalation is stopped and exhalation begins. As air leaves the lungs during exhalation, the lungs deflate and the stretch receptors are no longer stimulated. Thus, the DRG is no longer inhibited, and a new inhalation begins. This reflex is referred to as the **inflation reflex** or *Hering–Breuer reflex* (HER-ing BROY-er). In infants, the reflex appears to function in normal breathing. In adults, however, the reflex is not activated until tidal volume (normally 500 mL) reaches more than 1500 mL. Therefore, the reflex in adults is a protective mechanism that prevents excessive inflation of the lungs, for example, during severe exercise, rather than a key component in the normal control of breathing.

Other Influences on Breathing

Other factors that contribute to regulation of breathing include the following:

- ***Limbic system stimulation.*** Anticipation of activity or emotional anxiety may stimulate the limbic system, which then sends excitatory input to the DRG, increasing the rate and depth of breathing.
- ***Temperature.*** An increase in body temperature, as occurs during a fever or vigorous muscular exercise, increases the rate of breathing. A decrease in body temperature decreases breathing rate. A sudden cold stimulus (such as plunging into cold water) causes temporary **apnea** (AP-nē-a; *a-* = without; *-pnea* = breath), an absence of breathing.
- ***Pain.*** A sudden, severe pain brings about brief apnea, but a prolonged somatic pain increases breathing rate. Visceral pain may slow the rate of breathing.
- ***Stretching the anal sphincter muscle.*** This action increases the breathing rate and is sometimes used to stimulate respiration in a newborn baby or a person who has stopped breathing.
- ***Irritation of airways.*** Physical or chemical irritation of the pharynx or larynx brings about an immediate cessation of breathing followed by coughing or sneezing.
- ***Blood pressure.*** The carotid and aortic baroreceptors that detect changes in blood pressure have a small effect on breathing. A sudden rise in blood pressure decreases the rate of breathing, and a drop in blood pressure increases the breathing rate.

Table 23.3 summarizes the stimuli that affect the rate and depth of breathing.

TABLE 23.3 Summary of Stimuli That Affect Breathing Rate and Depth

STIMULI THAT INCREASE BREATHING RATE AND DEPTH	STIMULI THAT DECREASE BREATHING RATE AND DEPTH
Voluntary hyperventilation controlled by cerebral cortex and anticipation of activity by stimulation of limbic system.	Voluntary hypoventilation controlled by cerebral cortex.
Increase in arterial blood P_{CO_2} above 40 mmHg (causes an increase in H^+) detected by peripheral and central chemoreceptors.	Decrease in arterial blood P_{CO_2} below 40 mmHg (causes a decrease in H^+) detected by peripheral and central chemoreceptors.
Decrease in arterial blood P_{O_2} from 105 mmHg to 50 mmHg.	Decrease in arterial blood P_{O_2} below 50 mmHg.
Increased activity of proprioceptors.	Decreased activity of proprioceptors.
Increase in body temperature	Decrease in body temperature (decreases respiration rate), sudden cold stimulus (causes apnea).
Prolonged pain.	Severe pain (causes apnea).
Decrease in blood pressure.	Increase in blood pressure.
Stretching of anal sphincter.	Irritation of pharynx or larynx by touch or chemicals (causes brief apnea followed by coughing or sneezing).

23.9 Exercise and the Respiratory System

OBJECTIVE

- **Describe** the effects of exercise on the respiratory system.

The respiratory and cardiovascular systems make adjustments in response to both the intensity and duration of exercise. The effects of exercise on the heart are discussed in Chapter 20. Here we focus on how exercise affects the respiratory system.

Recall that the heart pumps the same amount of blood to the lungs as to all the rest of the body. Thus, as cardiac output rises, the blood flow to the lungs, termed **pulmonary perfusion**, increases as well. In addition, the **O_2 diffusing capacity**, a measure of the rate at which O_2 can diffuse from alveolar air into the blood, may increase threefold during maximal exercise because more pulmonary capillaries become maximally perfused. As a result, there is a greater surface area available for diffusion of O_2 into pulmonary blood capillaries.

When muscles contract during exercise, they consume large amounts of O_2 and produce large amounts of CO_2. During vigorous exercise, O_2 consumption and breathing both increase dramatically. At the onset of exercise, an abrupt increase in breathing is followed by a more gradual increase. With moderate exercise, the increase is due mostly to an increase in the depth of breathing rather than to increased breathing rate. When exercise is more strenuous, the frequency of breathing also increases.

The abrupt increase in breathing at the start of exercise is due to *neural* changes that send excitatory impulses to the dorsal respiratory group (DRG) of the medullary respiratory center in the medulla. These changes include (1) anticipation of the activity, which stimulates the limbic system; (2) sensory impulses from proprioceptors in muscles, tendons, and joints; and (3) motor impulses from the primary motor cortex (precentral gyrus). The more gradual increase in breathing during moderate exercise is due to *chemical* and *physical* changes in the bloodstream, including (1) slightly decreased P_{O_2}, due to increased O_2 consumption; (2) slightly increased P_{CO_2}, due to increased CO_2 production by contracting muscle fibers; and (3) increased temperature, due to liberation of more heat as more O_2 is utilized. During strenuous exercise, HCO_3^- buffers H^+ released by lactic acid in a reaction that liberates CO_2, which further increases P_{CO_2}.

At the end of an exercise session, an abrupt decrease in breathing is followed by a more gradual decline to the resting level. The initial decrease is due mainly to changes in neural factors when movement stops or slows; the more gradual phase reflects the slower return of blood chemistry levels and temperature to the resting state.

See Clinical Connection: Effects of Smoking on the Respiratory System

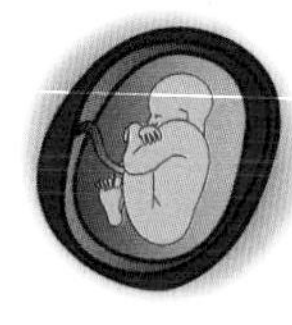

23.10 Development of the Respiratory System

OBJECTIVE

- **Describe** the development of the respiratory system.

The development of the mouth and pharynx is discussed in Chapter 24. Here we consider the development of the other structures of the respiratory system that you learned about in this chapter.

At about 4 weeks of development, the respiratory system begins as an outgrowth of the foregut (precursor of some digestive organs) just inferior to the pharynx. This outgrowth is called the **respiratory diverticulum** (dī-ver-TIK-ū-lum) or *lung bud* (**Figure 23.28**). The **endoderm** lining the respiratory diverticulum gives rise to the epithelium and glands of the trachea, bronchi, and alveoli. **Mesoderm** surrounding the respiratory diverticulum gives rise to the connective tissue, cartilage, and smooth muscle of these structures.

FIGURE 23.28 **Development of the bronchial tubes and lungs.**

The respiratory system develops from endoderm and mesoderm.

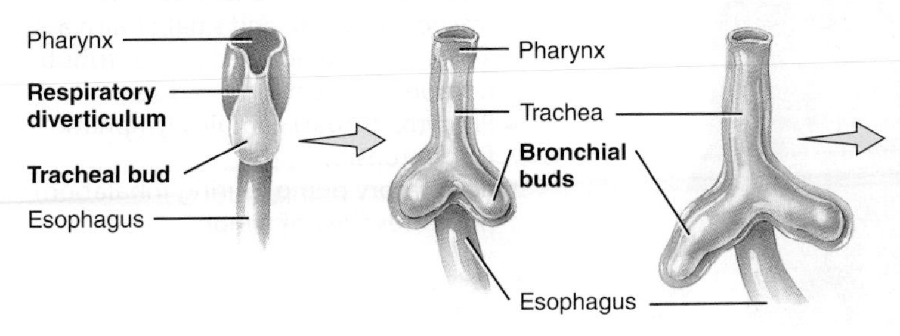

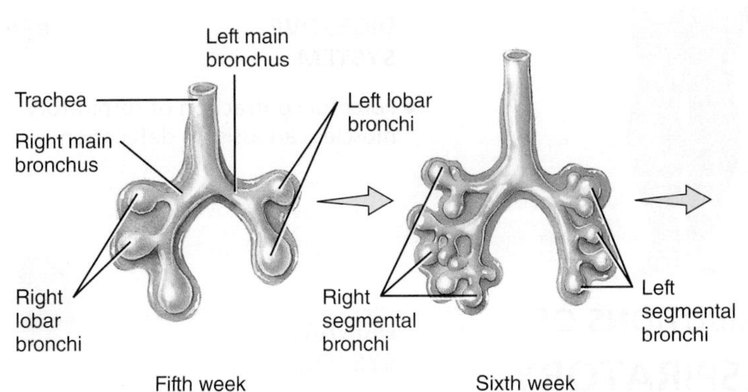

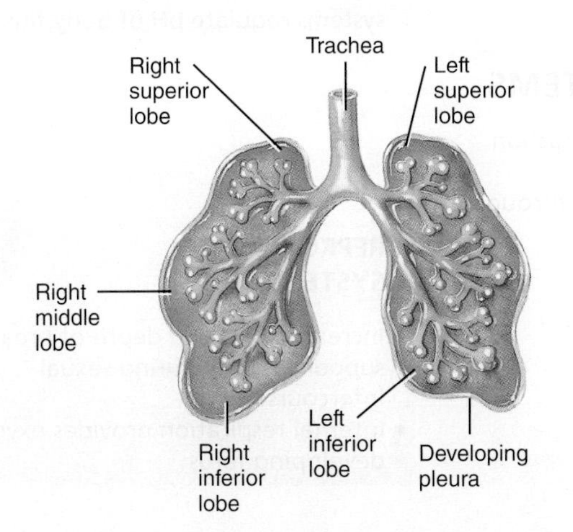

Q When does the respiratory system begin to develop in an embryo?

The epithelial lining of the *larynx* develops from the endoderm of the respiratory diverticulum; the cartilages and muscles originate from the **fourth** and **sixth pharyngeal arches**, swellings on the surface of the embryo (see **Figure 29.13**).

As the respiratory diverticulum elongates, its distal end enlarges to form a globular **tracheal bud**, which gives rise to the *trachea*. Soon after, the tracheal bud divides into **bronchial buds**, which branch repeatedly and develop with the *bronchi*. By 24 weeks, 17 orders of branches have formed and *respiratory bronchioles* have developed.

During weeks 6 to 16, all major elements of the *lungs* have formed, except for those involved in gaseous exchange (respiratory bronchioles, alveolar ducts, and alveoli). Since breathing is not possible at this stage, fetuses born during this time cannot survive.

During weeks 16 to 26, lung tissue becomes highly vascular and respiratory bronchioles, alveolar ducts, and some primitive alveoli develop. Although it is possible for a fetus born near the end of this period to survive if given intensive care, death frequently occurs due to the immaturity of the respiratory and other systems.

From 26 weeks to birth, many more primitive alveoli develop; they consist of type I alveolar cells (main sites of gaseous exchange) and type II surfactant-producing cells. Blood capillaries also establish close contact with the primitive alveoli. Recall that surfactant is necessary to lower surface tension of alveolar fluid and thus reduce the tendency of alveoli to collapse on exhalation. Although surfactant production begins by 20 weeks, it is present in only small quantities. Amounts sufficient to permit survival of a premature (preterm) infant are not produced until 26 to 28 weeks gestation. Infants born before 26 to 28 weeks are at high risk of respiratory distress syndrome (RDS), in which the alveoli collapse during exhalation and must be reinflated during inhalation (refer Clinical Connection: Respiratory Distress Syndrome in Section 23.4).

At about 30 weeks, mature alveoli develop. However, it is estimated that only about one-sixth of the full complement of alveoli develop before birth; the remainder develop after birth during the first 8 years.

As the lungs develop, they acquire their *pleural sacs*. The *visceral pleura* and the *parietal pleura* develop from mesoderm. The space between the pleural layers is the *pleural cavity*.

During development, breathing movements of the fetus cause the aspiration of fluid into the lungs. This fluid is a mixture of amniotic fluid, mucus from bronchial glands, and surfactant. At birth, the lungs are about half-filled with fluid. When breathing begins at birth, most of the fluid is rapidly reabsorbed by blood and lymph capillaries and a small amount is expelled through the nose and mouth during delivery.

23.11 Aging and the Respiratory System

OBJECTIVE

- **Describe** the effects of aging on the respiratory system.

FOCUS on HOMEOSTASIS

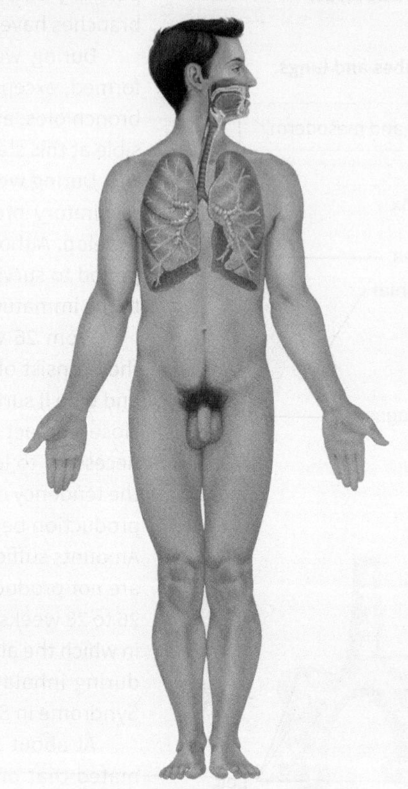

MUSCULAR SYSTEM

- Increased rate and depth of breathing support increased activity of skeletal muscles during exercise

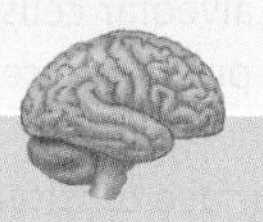

NERVOUS SYSTEM

- Nose contains receptors for sense of smell (olfaction)
- Vibrations of air flowing across vocal folds produce sounds for speech

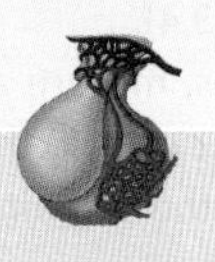

ENDOCRINE SYSTEM

- Angiotensin-converting enzyme (ACE) in lungs catalyzes formation of the hormone angiotensin II from angiotensin I

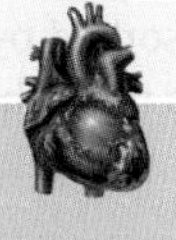

CARDIOVASCULAR SYSTEM

- During inhalations, respiratory pump aids return of venous blood to the heart

CONTRIBUTIONS OF THE RESPIRATORY SYSTEM

FOR ALL BODY SYSTEMS

- Provides oxygen and removes carbon dioxide
- Helps adjust pH of body fluids through exhalation of carbon dioxide

LYMPHATIC SYSTEM and IMMUNITY

- Hairs in nose, cilia and mucus in trachea, bronchi, and smaller airways, and alveolar macrophages contribute to nonspecific resistance to disease
- Pharynx (throat) contains lymphatic tissue (tonsils)
- Respiratory pump (during inhalation) promotes flow of lymph

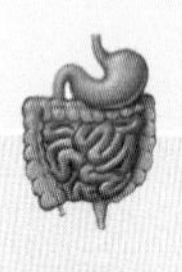

DIGESTIVE SYSTEM

- Forceful contraction of respiratory muscles can assist in defecation

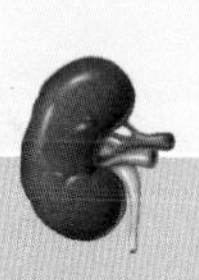

URINARY SYSTEM

- Together, respiratory and urinary systems regulate pH of body fluids

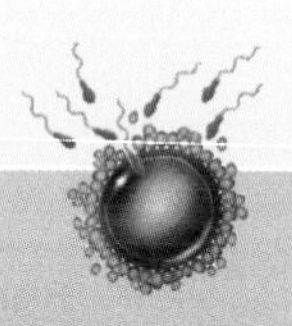

REPRODUCTIVE SYSTEMS

- Increased rate and depth of breathing support activity during sexual intercourse
- Internal respiration provides oxygen to developing fetus

With advancing age, the airways and tissues of the respiratory tract, including the alveoli, become less elastic and more rigid; the chest wall becomes more rigid as well. The result is a decrease in lung capacity. In fact, vital capacity (the maximum amount of air that can be exhaled after maximal inhalation) can decrease as much as 35% by age 70. A decrease in blood level of O_2, decreased activity of alveolar macrophages, and diminished ciliary action of the epithelium lining the respiratory tract occur. Because of these age-related factors, elderly people are more susceptible to pneumonia, bronchitis, emphysema, and other pulmonary disorders. Age-related changes in the structure and functions of the lung can also contribute to an older person's reduced ability to perform vigorous exercises, such as running.

• • •

To appreciate the many ways that the respiratory system contributes to homeostasis of other body systems, examine *Focus on Homeostasis: Contributions of the Respiratory System.* Next, in Chapter 24, we will see how the digestive system makes nutrients available to body cells so that oxygen provided by the respiratory system can be used for ATP production.

Review Questions

23.1 Overview of the Respiratory System

1. What are the three basic steps involved in respiration?
2. What are the components of the respiratory system?
3. Why is the respiratory zone important?

23.2 The Upper Respiratory System

4. Compare the structure and functions of the external nose and the internal nose.
5. What are the functions of the three subdivisions of the pharynx.

23.3 The Lower Respiratory System

6. How does the larynx function in respiration and voice production?
7. Describe the location, structure, and function of the trachea.
8. Describe the structure of the bronchial tree.
9. Where are the lungs located? Distinguish the parietal pleura from the visceral pleura.
10. Define each of the following parts of a lung: base, apex, costal surface, medial surface, hilum, root, cardiac notch, lobe, and lobule.
11. What is a bronchopulmonary segment?
12. Describe the histology and function of the respiratory membrane.

23.4 Pulmonary Ventilation

13. What are the basic differences among pulmonary ventilation, external respiration, and internal respiration?
14. Compare what happens during quiet versus forceful breathing.
15. Describe how alveolar surface tension, compliance, and airway resistance affect breathing.
16. Demonstrate the various types of modified breathing movements.

23.5 Lung Volumes and Capacities

17. What is a spirometer?
18. What is the difference between a lung volume and a lung capacity?
19. How is minute ventilation calculated?
20. Define alveolar ventilation rate and FEV_1.

23.6 Exchange of Oxygen and Carbon Dioxide

21. Distinguish between Dalton's law and Henry's law and give a practical application of each.
22. How does the partial pressure of oxygen change as altitude changes?
23. What are the diffusion paths of oxygen and carbon dioxide during external and internal respiration?
24. What factors affect the rate of diffusion of oxygen and carbon dioxide?

23.7 Transport of Oxygen and Carbon Dioxide

25. In a resting person, how many O_2 molecules are attached to each hemoglobin molecule, on average, in blood in the pulmonary arteries? In blood in the pulmonary veins?
26. What is the relationship between hemoglobin and P_{O_2}? How do temperature, H^+, P_{CO_2}, and BPG influence the affinity of Hb for O_2?
27. Why can hemoglobin unload more oxygen as blood flows through capillaries of metabolically active tissues, such as skeletal muscle during exercise, than is unloaded at rest?

23.8 Control of Breathing

28. How does the medullary respiratory center regulate breathing?
29. How is the pontine respiratory group related to the control of breathing?
30. How do the cerebral cortex, levels of CO_2 and O_2, proprioceptors, inflation reflex, temperature changes, pain, and irritation of the airways modify breathing?

23.9 Exercise and the Respiratory System

31. How does exercise affect the DRG?

23.10 Development of the Respiratory System

32. What structures develop from the laryngotracheal bud?

23.11 Aging and the Respiratory System

33. What accounts for the decrease in lung capacity with aging?

Critical Thinking Questions

1. Aretha loves to sing. Right now she has a cold, a severely runny nose, and a "sore throat" that is affecting her ability to sing and talk. What structures are involved and how are they affected by her cold?
2. Ms. Brown has smoked cigarettes for years and is having breathing difficulties. She has been diagnosed with emphysema. Describe specific kinds of structural changes you would expect to observe in Ms. Brown's respiratory system. How are air flow and gas exchange affected by these structural changes?
3. The Robinson family went to bed one frigid winter night and were found deceased the next day. A squirrel's nest was found in their chimney. What happened to the Robinsons?

The Digestive System CHAPTER 24

The Digestive System and Homeostasis

The digestive system contributes to homeostasis by breaking down food into forms that can be absorbed and used by body cells. It also absorbs water, vitamins, and minerals, and it eliminates wastes from the body.

The food we eat contains a variety of nutrients, which are used for building new body tissues and repairing damaged tissues. Food is also vital to life because it is our only source of chemical energy. However, most of the food we eat consists of molecules that are too large to be used by body cells. Therefore, foods must be broken down into molecules that are small enough to enter body cells for their use. This is accomplished by the digestive system, which forms an extensive surface area in contact with the external environment, and is closely associated with the cardiovascular system. The combination of extensive environmental exposure and close association with blood vessels is essential for processing the food that we eat.

Q Did you ever wonder why some people are sensitive to dairy products?

24.1 Overview of the Digestive System

OBJECTIVES

- **Identify** the organs of the digestive system.
- **Describe** the basic processes performed by the digestive system.

The **digestive system** (*dis* = apart; *gerere* = to carry) consists of a group of organs that break down the food we eat into smaller molecules that can be used by body cells. Two groups of organs compose the digestive system (**Figure 24.1**): the gastrointestinal (GI) tract and the accessory digestive organs. The **gastrointestinal (GI) tract**, or *alimentary canal* (*alimentary* = nourishment), is a continuous tube that extends from the mouth to the anus through the thoracic and abdominopelvic cavities. Organs of the gastrointestinal tract include the mouth, most of the pharynx, esophagus, stomach, small intestine, and large intestine. The length of the GI tract is about 5–7 meters (16.5–23 ft) in a living person when the muscles along the wall of the GI tract organs are in a state of *tonus* (sustained contraction). It is longer in a cadaver (about 7–9 meters or 23–29.5 ft) because of the loss of muscle tone after death. The **accessory digestive organs** include the teeth, tongue, salivary glands, liver, gallbladder, and pancreas. Teeth aid in the physical breakdown of food, and the tongue assists in chewing and swallowing. The other accessory digestive organs, however, never come into direct contact with food. They produce or store secretions that flow into the GI tract through ducts; the secretions aid in the chemical breakdown of food.

The GI tract contains food from the time it is eaten until it is digested and absorbed or eliminated. Muscular contractions in the wall of the GI tract physically break down the food by churning it and propel the food along the tract, from the esophagus to the anus. The contractions also help to dissolve foods by mixing them with fluids secreted into the tract. Enzymes secreted by accessory digestive organs and cells that line the tract break down the food chemically.

Overall, the digestive system performs six basic processes (**Figure 24.2**):

1. ***Ingestion.*** This process involves taking foods and liquids into the mouth (eating).
2. ***Secretion.*** Each day, cells within the walls of the GI tract and accessory digestive organs secrete a total of about 7 liters of water, acid, buffers, and enzymes into the lumen (interior space) of the tract.
3. ***Motility.*** Alternating contractions and relaxations of smooth muscle in the walls of the GI tract mix food and secretions and move them toward the anus. This capability of the GI tract to mix and move material along its length is called **motility** (mō-TIL-i-tē).
4. ***Digestion.*** **Digestion** is the process of breaking down ingested food into small molecules that can be used by body cells. In **mechanical digestion** the teeth cut and grind food before it is swallowed, and then smooth muscles of the stomach and small intestine churn the

FIGURE 24.1 **Organs of the digestive system.**

Organs of the gastrointestinal (GI) tract are the mouth, pharynx, esophagus, stomach, small intestine, and large intestine. Accessory digestive organs include the teeth, tongue, salivary glands, liver, gallbladder, and pancreas and are indicated in red.

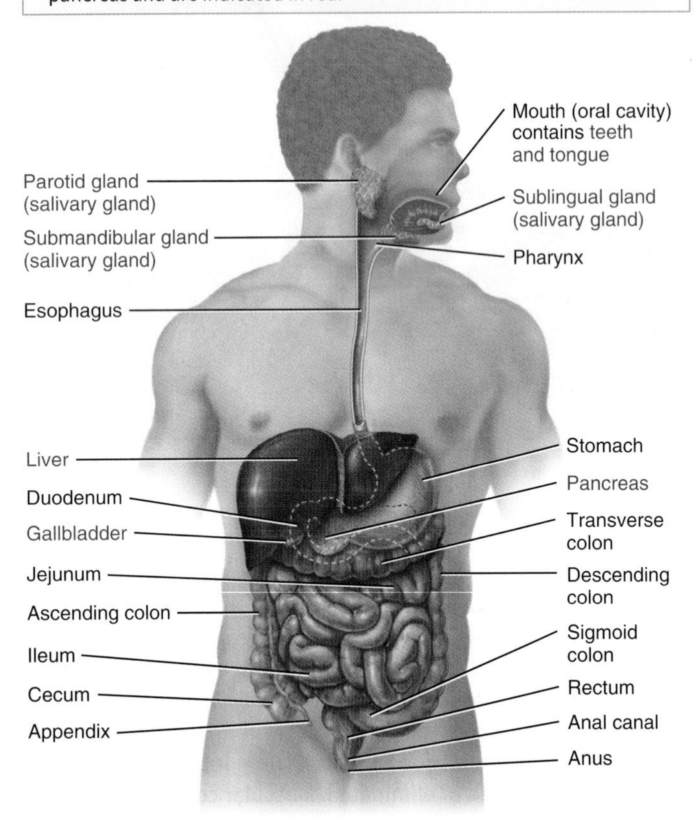

(a) Right lateral view of head and neck and anterior view of trunk

Functions of the Digestive System

1. Ingestion: taking food into mouth.
2. Secretion: release of water, acid, buffers, and enzymes into lumen of GI tract.
3. Mixing and propulsion: churning and movement of food through GI tract.
4. Digestion: mechanical and chemical breakdown of food.
5. Absorption: passage of digested products from GI tract into blood and lymph.
6. Defecation: elimination of feces from GI tract.

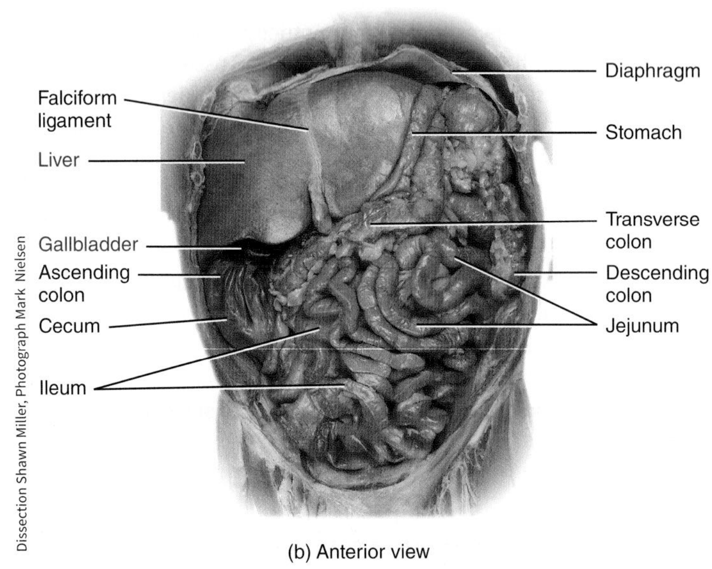

(b) Anterior view

Q Which structures of the digestive system secrete digestive enzymes?

FIGURE 24.2 **Digestive processes.**

The digestive system performs six basic processes: ingestion, secretion, motility, digestion, absorption, and defecation.

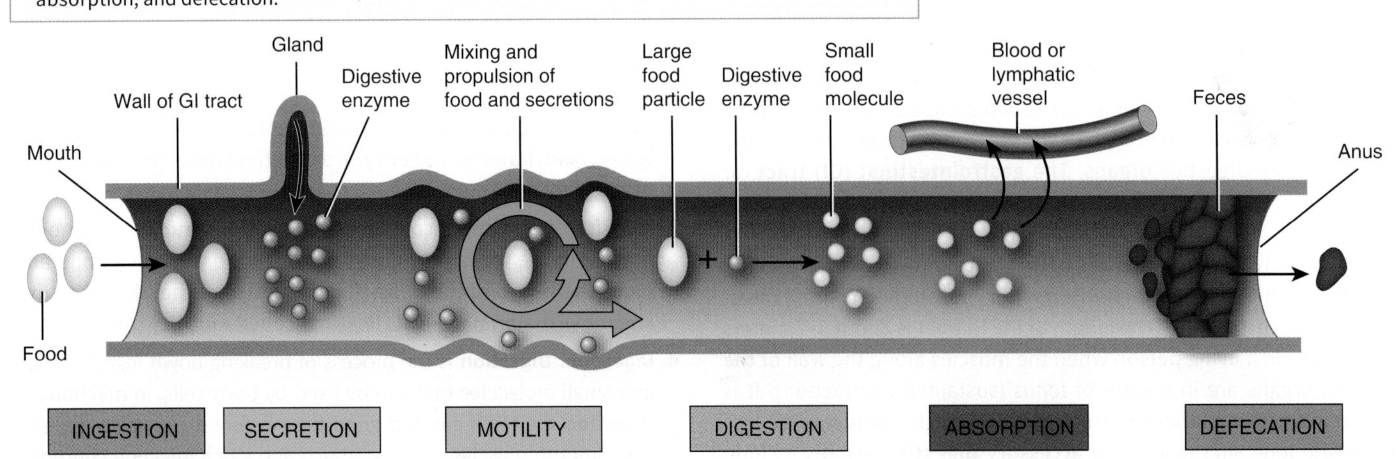

Q What is absorption?

food to further assist the process. As a result, food molecules become dissolved and thoroughly mixed with digestive enzymes. In **chemical digestion** the large carbohydrate, lipid, protein, and nucleic acid molecules in food are split into smaller molecules by hydrolysis (see **Figure 2.15**). Digestive enzymes produced by the salivary glands, tongue, stomach, pancreas, and small intestine catalyze these catabolic reactions.

5. ***Absorption.*** The movement of the products of digestion from the lumen of the GI tract into blood or lymph is called **absorption** (ab-SŌRP-shun). Once absorbed, these substances circulate to cells throughout the body. A few substances in food can be absorbed without undergoing digestion. These include vitamins, ions, cholesterol, and water.
6. ***Defecation.*** Wastes, indigestible substances, bacteria, cells sloughed from the lining of the GI tract, and digested materials that were not absorbed in their journey through the digestive tract leave the body through the anus in a process called **defecation** (def-e-KĀ-shun). The eliminated material is termed **feces** (FĒ-sēz) or *stool*.

24.2 Layers of the GI Tract

OBJECTIVE

- **Describe** the structure and function of the layers that form the wall of the gastrointestinal tract.

The wall of the GI tract from the lower esophagus to the anal canal has the same basic, four-layered arrangement of tissues. The four layers of the tract, from deep to superficial, are the mucosa, submucosa, muscularis, and serosa/adventitia (**Figure 24.3**).

Mucosa

The **mucosa**, or inner lining of the GI tract, is a mucous membrane. It is composed of (1) a layer of epithelium in direct contact with the

FIGURE 24.3 Layers of the gastrointestinal tract. Variations in this basic plan may be seen in the esophagus (**Figure 24.10**), stomach (**Figure 24.13**), small intestine (**Figure 24.20**), and large intestine (**Figure 24.25**).

The four layers of the GI tract, from deep to superficial, are the mucosa, submucosa, muscularis, and serosa.

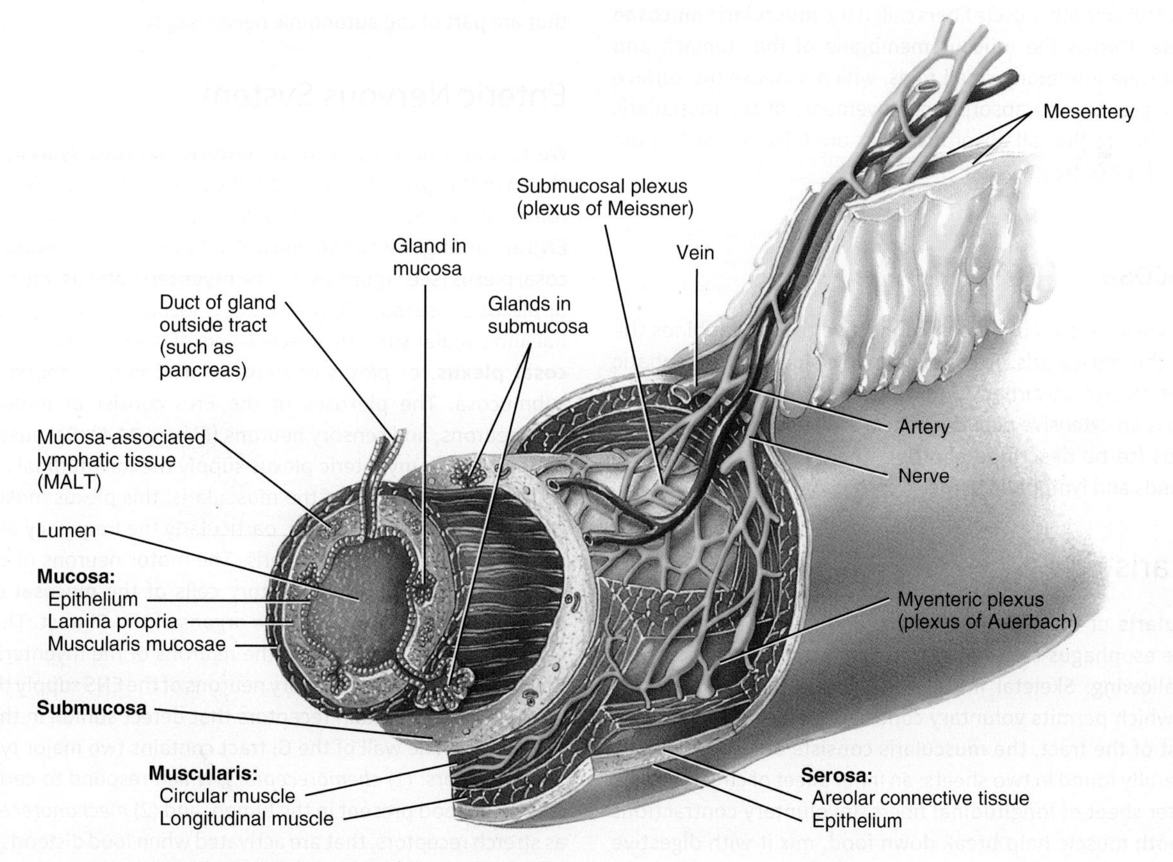

Q What are the functions of the lamina propria?

contents of the GI tract, (2) a layer of connective tissue called the lamina propria, and (3) a thin layer of smooth muscle (muscularis mucosae).

1. The **epithelium** in the mouth, pharynx, esophagus, and anal canal is mainly nonkeratinized stratified squamous epithelium that serves a protective function. Simple columnar epithelium, which functions in secretion and absorption, lines the stomach and intestines. The tight junctions that firmly seal neighboring simple columnar epithelial cells to one another restrict leakage between the cells. The rate of renewal of GI tract epithelial cells is rapid: Every 5 to 7 days they slough off and are replaced by new cells. Located among the epithelial cells are exocrine cells that secrete mucus and fluid into the lumen of the tract, and several types of endocrine cells, collectively called **enteroendocrine cells** (en′-ter-ō-EN-dō-krin), which secrete hormones.
2. The **lamina propria** (*lamina* = thin, flat plate; *propria* = one's own) is areolar connective tissue containing many blood and lymphatic vessels, which are the routes by which nutrients absorbed into the GI tract reach the other tissues of the body. This layer supports the epithelium and binds it to the muscularis mucosae (discussed next). The lamina propria also contains the majority of the cells of the **mucosa-associated lymphatic tissue (MALT)**. These prominent lymphatic nodules contain immune system cells that protect against disease (see Chapter 22). MALT is present all along the GI tract, especially in the tonsils, small intestine, appendix, and large intestine.
3. A thin layer of smooth muscle fibers called the **muscularis mucosae** (mū-KŌ-sē) throws the mucous membrane of the stomach and small intestine into many small folds, which increase the surface area for digestion and absorption. Movements of the muscularis mucosae ensure that all absorptive cells are fully exposed to the contents of the GI tract.

Submucosa

The **submucosa** consists of areolar connective tissue that binds the mucosa to the muscularis. It contains many blood and lymphatic vessels that receive absorbed food molecules. Also located in the submucosa is an extensive network of neurons known as the submucosal plexus (to be described shortly). The submucosa may also contain glands and lymphatic tissue.

Muscularis

The **muscularis** of the mouth, pharynx, and superior and middle parts of the esophagus contains *skeletal muscle* that produces voluntary swallowing. Skeletal muscle also forms the external anal sphincter, which permits voluntary control of defecation. Throughout the rest of the tract, the muscularis consists of *smooth muscle* that is generally found in two sheets: an inner sheet of circular fibers and an outer sheet of longitudinal fibers. Involuntary contractions of the smooth muscle help break down food, mix it with digestive secretions, and propel it along the tract. Between the layers of the muscularis is a second plexus of neurons—the myenteric plexus (to be described shortly).

Serosa

Those portions of the GI tract that are suspended in the abdominal cavity have a superficial layer called the **serosa**. As its name implies, the serosa is a serous membrane composed of areolar connective tissue and simple squamous epithelium (mesothelium). The serosa is also called the *visceral peritoneum* because it forms a portion of the peritoneum, which we examine in detail shortly. The esophagus lacks a serosa; instead, only a single layer of areolar connective tissue called the *adventitia* forms the superficial layer of this organ.

24.3 Neural Innervation of the GI Tract

OBJECTIVE

- **Describe** the nerve supply of the GI tract.

The gastrointestinal tract is regulated by an intrinsic set of nerves known as the enteric nervous system and by an extrinsic set of nerves that are part of the autonomic nervous system.

Enteric Nervous System

We first introduced you to the **enteric nervous system (ENS)**, the "brain of the gut," in Chapter 12. It consists of about 100 million neurons that extend from the esophagus to the anus. The neurons of the ENS are arranged into two plexuses: the myenteric plexus and submucosal plexus (see Figure 24.3). The **myenteric plexus** (*myo-* = muscle), or *plexus of Auerbach* (OW-er-bak), is located between the longitudinal and circular smooth muscle layers of the muscularis. The **submucosal plexus**, or *plexus of Meissner* (MĪZ-ner), is found within the submucosa. The plexuses of the ENS consist of motor neurons, interneurons, and sensory neurons (Figure 24.4). Because the motor neurons of the myenteric plexus supply the longitudinal and circular smooth muscle layers of the muscularis, this plexus mostly controls GI tract motility (movement), particularly the frequency and strength of contraction of the muscularis. The motor neurons of the submucosal plexus supply the secretory cells of the mucosal epithelium, controlling the secretions of the organs of the GI tract. The interneurons of the ENS interconnect the neurons of the myenteric and submucosal plexuses. The sensory neurons of the ENS supply the mucosal epithelium and contain receptors that detect stimuli in the lumen of the GI tract. The wall of the GI tract contains two major types of sensory receptors: (1) *chemoreceptors*, which respond to certain chemicals in the food present in the lumen, and (2) *mechanoreceptors*, such as stretch receptors, that are activated when food distends (stretches) the wall of a GI organ.

FIGURE 24.4 **Organization of the enteric nervous system.**

The enteric nervous system consists of neurons arranged into the myenteric and submucosal plexuses.

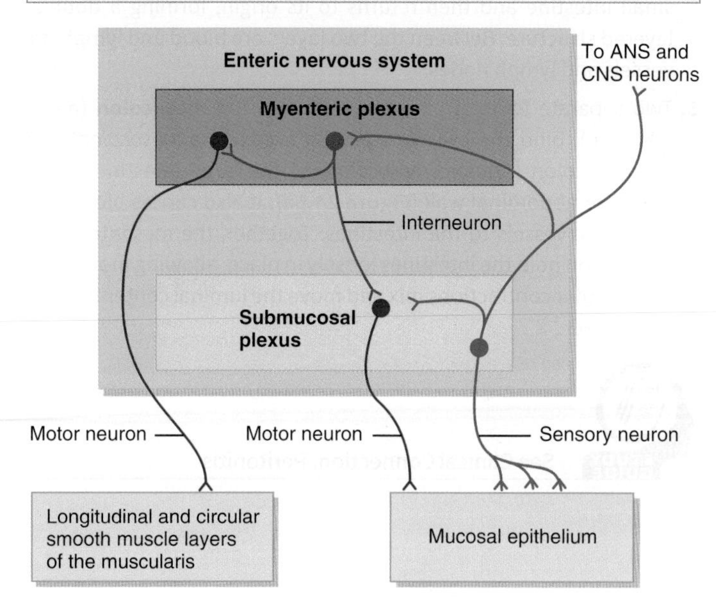

Q What are the functions of the myenteric and submucosal plexuses of the enteric nervous system?

Autonomic Nervous System

Although the neurons of the ENS can function independently, they are subject to regulation by the neurons of the autonomic nervous system. The vagus (X) nerves supply parasympathetic fibers to most parts of the GI tract, with the exception of the last half of the large intestine, which is supplied with parasympathetic fibers from the sacral spinal cord. The parasympathetic nerves that supply the GI tract form neural connections with the ENS. Parasympathetic preganglionic neurons of the vagus or pelvic splanchnic nerves synapse with parasympathetic postganglionic neurons located in the myenteric and submucosal plexuses. Some of the parasympathetic postganglionic neurons in turn synapse with neurons in the ENS; others directly innervate smooth muscle and glands within the wall of the GI tract. In general, stimulation of the parasympathetic nerves that innervate the GI tract causes an increase in GI secretion and motility by increasing the activity of ENS neurons.

Sympathetic nerves that supply the GI tract arise from the thoracic and upper lumbar regions of the spinal cord. Like the parasympathetic nerves, these sympathetic nerves form neural connections with the ENS. Sympathetic postganglionic neurons synapse with neurons located in the myenteric plexus and the submucosal plexus. In general, the sympathetic nerves that supply the GI tract cause a decrease in GI secretion and motility by inhibiting the neurons of the ENS. Emotions such as anger, fear, and anxiety may slow digestion because they stimulate the sympathetic nerves that supply the GI tract.

Gastrointestinal Reflex Pathways

Many neurons of the ENS are components of *GI (gastrointestinal) reflex pathways* that regulate GI secretion and motility in response to stimuli present in the lumen of the GI tract. The initial components of a typical GI reflex pathway are sensory receptors (such as chemoreceptors and stretch receptors) that are associated with the sensory neurons of the ENS. The axons of these sensory neurons can synapse with other neurons located in the ENS, CNS, or ANS, informing these regions about the nature of the contents and the degree of distension (stretching) of the GI tract. The neurons of the ENS, CNS, or ANS subsequently activate or inhibit GI glands and smooth muscle, altering GI secretion and motility.

24.4 Peritoneum

OBJECTIVE

- **Describe** the peritoneum and its folds.

The **peritoneum** (per′-i-tō-NĒ-um; *peri-* = around) is the largest serous membrane of the body; it consists of a layer of simple squamous epithelium (mesothelium) with an underlying supporting layer of areolar connective tissue. The peritoneum is divided into the **parietal peritoneum**, which lines the wall of the abdominal cavity, and the **visceral peritoneum**, which covers some of the organs in the cavity and is their serosa (**Figure 24.5a**). The slim space containing lubricating serous fluid that is between the parietal and visceral portions of the peritoneum is called the **peritoneal cavity**. In certain diseases, the peritoneal cavity may become distended by the accumulation of several liters of fluid, a condition called **ascites** (a-SĪ-tēz).

As you will see shortly, some organs lie on the posterior abdominal wall and are covered by peritoneum only on their anterior surfaces; they are not in the peritoneal cavity. Such organs, including the kidneys, ascending and descending colons of the large intestine, duodenum of the small intestine, and pancreas, are said to be **retroperitoneal** (*retro-* = behind).

Unlike the pericardium and pleurae, which smoothly cover the heart and lungs, the peritoneum contains large folds that weave between the viscera. The folds bind the organs to one another and to the walls of the abdominal cavity. They also contain blood vessels, lymphatic vessels, and nerves that supply the abdominal organs. There are five major peritoneal folds: the greater omentum, falciform ligament, lesser omentum, mesentery, and mesocolon:

1. The **greater omentum** (ō-MEN-tum = fat skin), the longest peritoneal fold, drapes over the transverse colon and coils of the small intestine like a "fatty apron" (**Figure 24.5a, d**). The greater omentum is a double sheet that folds back on itself, giving it a total of four layers. From attachments along the stomach and duodenum, the greater omentum extends downward anterior to the small

intestine, then turns and extends upward and attaches to the transverse colon. The greater omentum normally contains a considerable amount of adipose tissue. Its adipose tissue content can greatly expand with weight gain, contributing to the characteristic "beer belly" seen in some overweight individuals. The many lymph nodes of the greater omentum contribute macrophages and antibody-producing plasma cells that help combat and contain infections of the GI tract.

2. The **falciform ligament** (FAL-si-form; *falc-* = sickle-shaped) attaches the liver to the anterior abdominal wall and diaphragm (**Figure 24.5b**). The liver is the only digestive organ that is attached to the anterior abdominal wall.
3. The **lesser omentum** arises as an anterior fold in the serosa of the stomach and duodenum, and it connects the stomach and duodenum to the liver (**Figure 24.5a, c**). It is the pathway for blood vessels entering the liver and contains the hepatic portal vein, common hepatic artery, and common bile duct, along with some lymph nodes.
4. A fan-shaped fold of the peritoneum, called the **mesentery** (MEZ-en-ter′-ē; *mes-* = middle), binds the jejunum and ileum of the small intestine to the posterior abdominal wall (**Figure 24.5a, d**). This is the most massive peritoneal fold, is typically laden with fat, and contributes extensively to the large abdomen in obese individuals. It extends from the posterior abdominal wall to wrap around the small intestine and then returns to its origin, forming a double-layered structure. Between the two layers are blood and lymphatic vessels and lymph nodes.
5. Two separate folds of peritoneum, called the **mesocolon** (mez′-ō-KŌ-lon), bind the transverse colon (*transverse mesocolon*) and sigmoid colon (*sigmoid mesocolon*) of the large intestine to the posterior abdominal wall (**Figure 24.5a**). It also carries blood and lymphatic vessels to the intestines. Together, the mesentery and mesocolon hold the intestines loosely in place, allowing movement as muscular contractions mix and move the luminal contents along the GI tract.

See Clinical Connection: Peritonitis

FIGURE 24.5 Relationship of the peritoneal folds to one another and to organs of the gastrointestinal tract. The size of the peritoneal cavity in (a) is exaggerated for emphasis.

The peritoneum is the largest serous membrane in the body.

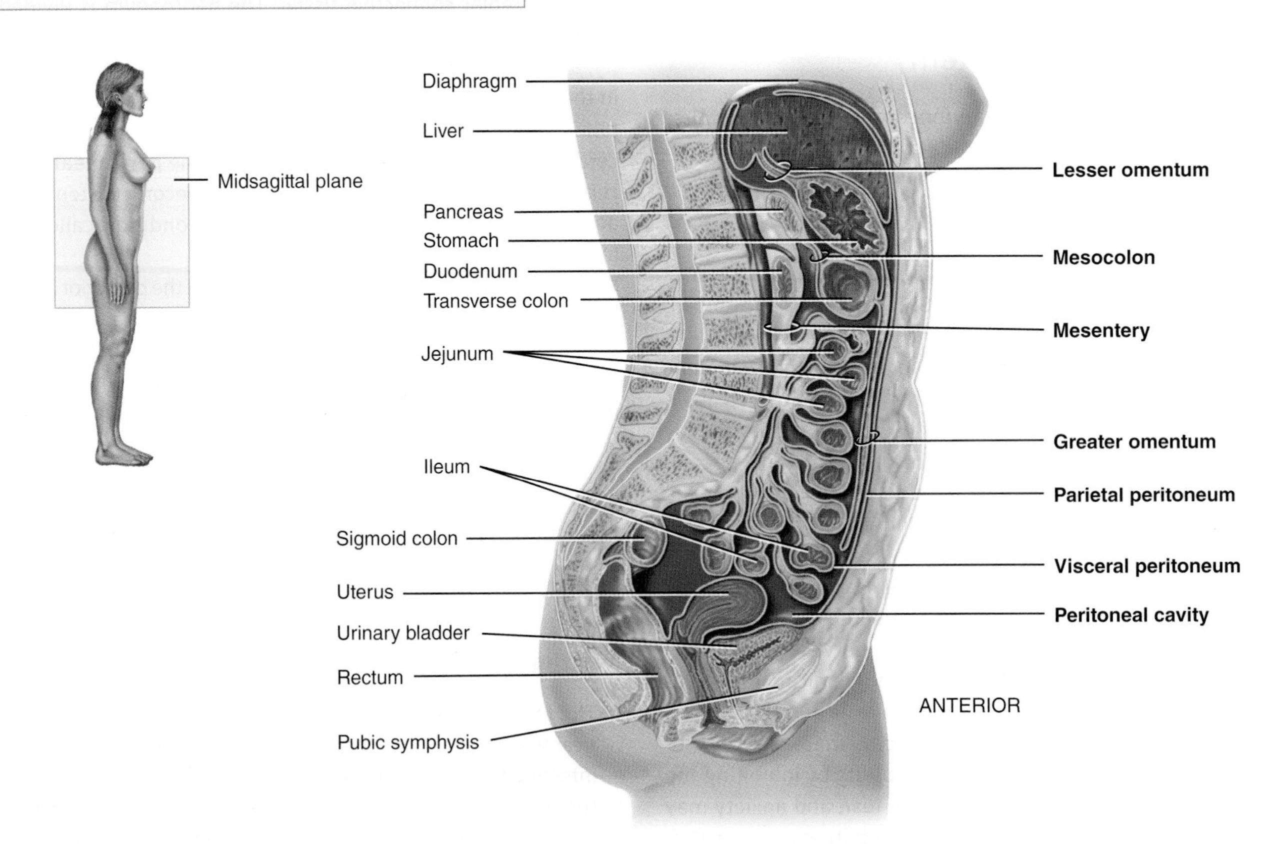

(a) Midsagittal section showing the peritoneal folds

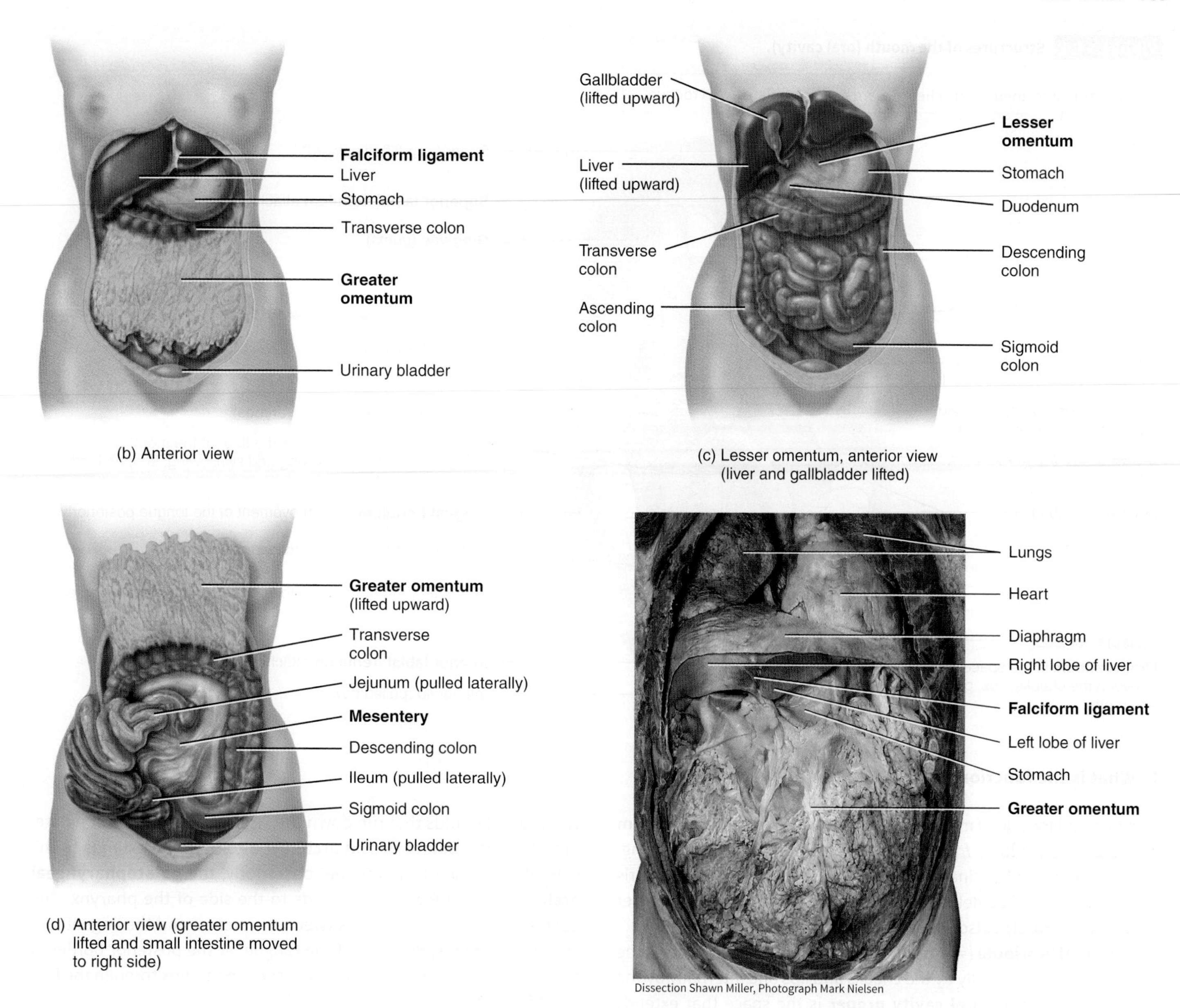

(b) Anterior view

(c) Lesser omentum, anterior view (liver and gallbladder lifted)

(d) Anterior view (greater omentum lifted and small intestine moved to right side)

(e) Anterior view

Q Which peritoneal fold binds the small intestine to the posterior abdominal wall?

24.5 Mouth

OBJECTIVES

- **Identify** the locations of the salivary glands, and describe the functions of their secretions.
- **Describe** the structure and functions of the tongue.
- **Identify** the parts of a typical tooth, and compare deciduous and permanent dentitions.

The **mouth**, also referred to as the *oral* or *buccal cavity* (BUK-al; *bucca* = cheeks), is formed by the cheeks, hard and soft palates, and tongue (**Figure 24.6**). The **cheeks** form the lateral walls of the oral cavity. They are covered externally by skin and internally by a mucous membrane, which consists of nonkeratinized stratified squamous epithelium. Buccinator muscles and connective tissue lie between the skin and mucous membranes of the cheeks. The anterior portions of the cheeks end at the lips.

The **lips** or *labia* (= fleshy borders) are fleshy folds surrounding the opening of the mouth. They contain the orbicularis oris muscle and are covered externally by skin and internally by a mucous membrane. The inner surface of each lip is attached to its corresponding

FIGURE 24.6 **Structures of the mouth (oral cavity).**

The mouth is formed by the cheeks, hard and soft palates, and tongue.

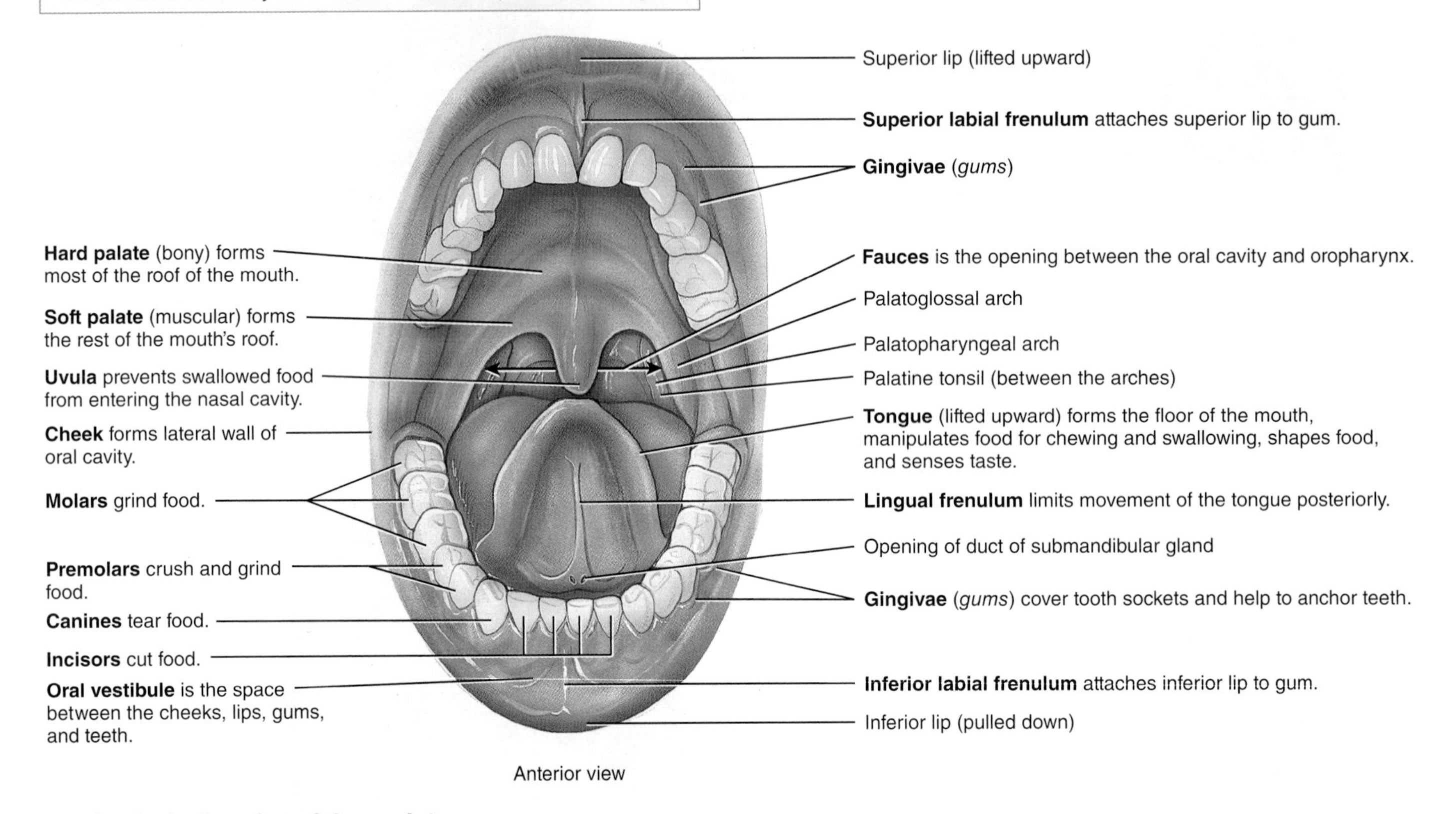

Q What is the function of the uvula?

gum by a midline fold of mucous membrane called the **labial frenulum** (LĀ-bē-al FREN-ū-lum; *frenulum* = small bridle). During chewing, contraction of the buccinator muscles in the cheeks and orbicularis oris muscle in the lips helps keep food between the upper and lower teeth. These muscles also assist in speech.

The **oral vestibule** (= entrance to a canal) of the oral cavity is the space bounded externally by the cheeks and lips and internally by the gums and teeth. The **oral cavity proper** is the space that extends from the gums and teeth to the **fauces** (FAW-sēz = passages), the opening between the oral cavity and the oropharynx (throat).

The **palate** is a wall or septum that separates the oral cavity from the nasal cavity, and forms the roof of the mouth. This important structure makes it possible to chew and breathe at the same time. The **hard palate**—the anterior portion of the roof of the mouth—is formed by the maxillae and palatine bones and is covered by a mucous membrane; it forms a bony partition between the oral and nasal cavities. The **soft palate**, which forms the posterior portion of the roof of the mouth, is an arch-shaped muscular partition between the oropharynx and nasopharynx that is lined with mucous membrane.

Hanging from the free border of the soft palate is a fingerlike muscular structure called the **uvula** (Ū-vū-la = little grape). During swallowing, the soft palate and uvula are drawn superiorly, closing off the nasopharynx and preventing swallowed foods and liquids from entering the nasal cavity. Lateral to the base of the uvula are two muscular folds that run down the lateral sides of the soft palate: Anteriorly, the **palatoglossal arch** (pal-a-tō-GLOS-al) extends to the side of the base of the tongue; posteriorly, the **palatopharyngeal arch** (pal-a-tō-fa-RIN-jē-al) extends to the side of the pharynx. The palatine tonsils are situated between the arches, and the lingual tonsils are situated at the base of the tongue. At the posterior border of the soft palate, the mouth opens into the oropharynx through the fauces (**Figure 24.6**).

Salivary Glands

A **salivary gland** (SAL-i-vār-ē) is a gland that releases a secretion called saliva into the oral cavity. Ordinarily, just enough saliva is secreted to keep the mucous membranes of the mouth and pharynx moist and to cleanse the mouth and teeth. When food enters the mouth, however, secretion of saliva increases, and it lubricates, dissolves, and begins the chemical breakdown of the food.

The mucous membrane of the mouth and tongue contains many small salivary glands that open directly, or indirectly via short ducts, to the oral cavity. These glands include *labial*, *buccal*, and *palatal glands* in the lips, cheeks, and palate, respectively, and *lingual glands* in the tongue, all of which make a small contribution to saliva.

However, most saliva is secreted by the **major salivary glands**, which lie beyond the oral mucosa, into ducts that lead to the oral cavity. There are three pairs of major salivary glands: the parotid,

FIGURE 24.7 The three major salivary glands—parotid, sublingual, and submandibular. The submandibular glands, shown in the light micrograph (b), consist mostly of serous acini (serous fluid–secreting portions of gland) and a few mucous acini (mucus-secreting portions of gland); the parotid glands consist of serous acini only; and the sublingual glands consist of mostly mucous acini and a few serous acini.

Saliva lubricates and dissolves foods and begins the chemical breakdown of carbohydrates and lipids.

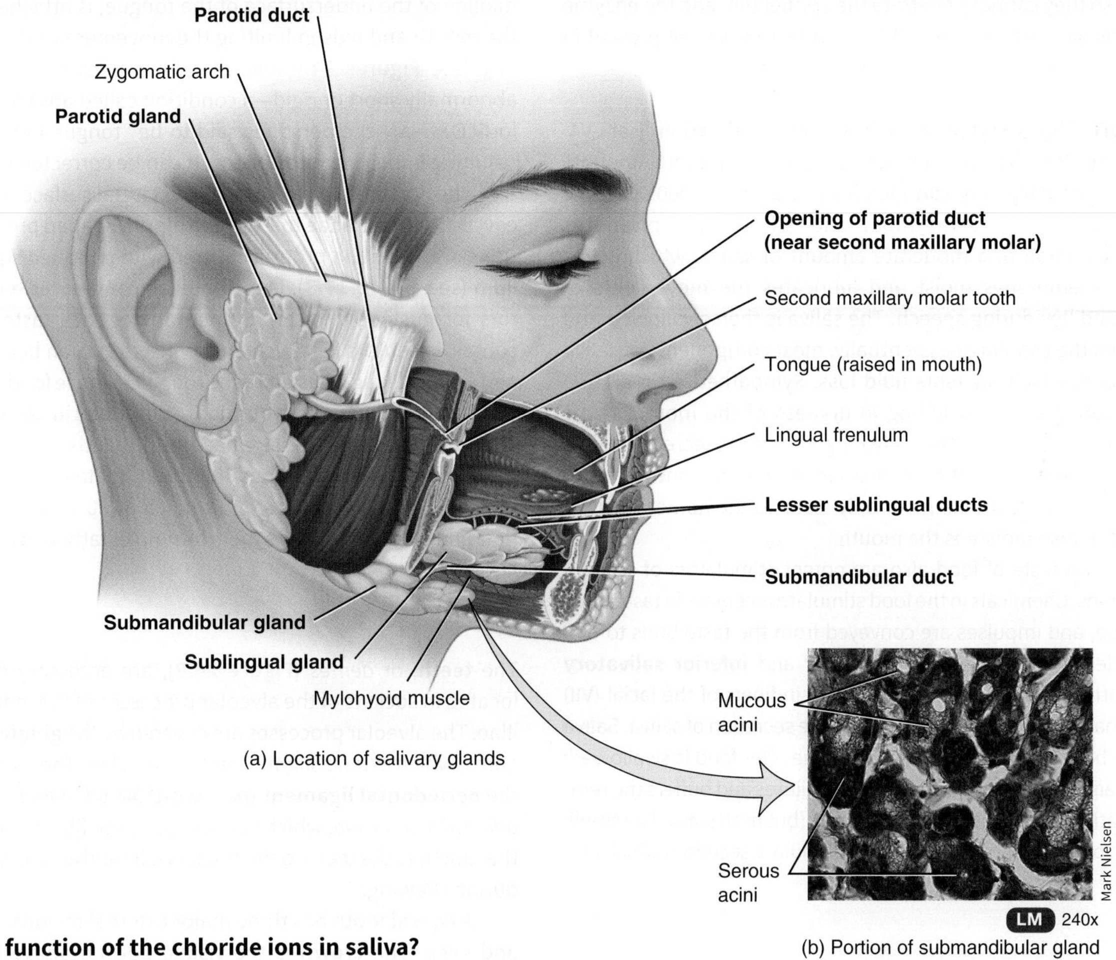

(a) Location of salivary glands

(b) Portion of submandibular gland

Q What is the function of the chloride ions in saliva?

submandibular, and sublingual glands (**Figure 24.7a**). The **parotid glands** (pa-ROT-id; *par-* = near; *oto-* = ear) are located inferior and anterior to the ears, between the skin and the masseter muscle. Each secretes saliva into the oral cavity via a **parotid duct** that pierces the buccinator muscle to open into the vestibule opposite the second maxillary (upper) molar tooth. The **submandibular glands** (sub′-man-DIB-ū-lar) are found in the floor of the mouth; they are medial and partly inferior to the body of the mandible. Their ducts, the **submandibular ducts**, run under the mucosa on either side of the midline of the floor of the mouth and enter the oral cavity proper lateral to the lingual frenulum. The **sublingual glands** (sub-LING-gwal) are beneath the tongue and superior to the submandibular glands. Their ducts, the **lesser sublingual ducts**, open into the floor of the mouth in the oral cavity proper.

Composition and Functions of Saliva

Chemically, **saliva** is 99.5% water and 0.5% solutes. Among the solutes are ions, including sodium, potassium, chloride, bicarbonate, and phosphate. Also present are some dissolved gases and various organic substances, including urea and uric acid, mucus, immunoglobulin A, the bacteriolytic enzyme lysozyme, and salivary amylase, a digestive enzyme that acts on starch.

Not all salivary glands supply the same ingredients. The parotid glands secrete a watery (serous) liquid containing salivary amylase. Because the submandibular glands contain cells similar to those found in the parotid glands, plus some mucous cells, they secrete a fluid that contains amylase but is thickened with mucus. The sublingual glands contain mostly mucous cells, so they secrete a much thicker fluid that contributes only a small amount of salivary amylase.

The water in saliva provides a medium for dissolving foods so that they can be tasted by gustatory receptors and so that digestive reactions can begin. Chloride ions in the saliva activate **salivary amylase** (AM-i-lās), an enzyme that starts the breakdown of starch in the mouth into maltose, maltotriose, and α-dextrin. Bicarbonate and phosphate ions buffer acidic foods that enter the mouth, so saliva is only slightly acidic (pH 6.35–6.85). Salivary glands (like the sweat glands of the skin) help remove waste molecules from the body, which accounts for the presence of urea and uric acid in saliva. Mucus lubricates food so it can be moved around easily in the mouth, formed into a ball, and swallowed. Immunoglobulin A (IgA) prevents attachment of microbes so they cannot penetrate the epithelium, and the enzyme lysozyme kills bacteria; however, these substances are not present in large enough quantities to eliminate all oral bacteria.

Salivation The secretion of saliva, called **salivation** (sal-i-VĀ-shun), is controlled by the autonomic nervous system. Amounts of saliva secreted daily vary considerably but average 1000–1500 mL (1–1.6 qt). Normally, parasympathetic stimulation promotes continuous secretion of a moderate amount of saliva, which keeps the mucous membranes moist and lubricates the movements of the tongue and lips during speech. The saliva is then swallowed and helps moisten the esophagus. Eventually, most components of saliva are reabsorbed, which prevents fluid loss. Sympathetic stimulation dominates during stress, resulting in dryness of the mouth. If the body becomes dehydrated, the salivary glands stop secreting saliva to conserve water; the resulting dryness in the mouth contributes to the sensation of thirst. Drinking not only restores the homeostasis of body water but also moistens the mouth.

The feel and taste of food also are potent stimulators of salivary gland secretions. Chemicals in the food stimulate receptors in taste buds on the tongue, and impulses are conveyed from the taste buds to two salivary nuclei in the brain stem (**superior** and **inferior salivatory nuclei**). Returning parasympathetic impulses in fibers of the facial (VII) and glossopharyngeal (IX) nerves stimulate the secretion of saliva. Saliva continues to be secreted heavily for some time after food is swallowed; this flow of saliva washes out the mouth and dilutes and buffers the remnants of irritating chemicals such as that tasty (but hot!) salsa. The smell, sight, sound, or thought of food may also stimulate secretion of saliva.

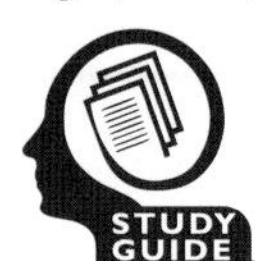

See Clinical Connection: Mumps

Tongue

The **tongue** is an accessory digestive organ composed of skeletal muscle covered with mucous membrane. Together with its associated muscles, it forms the floor of the oral cavity. The tongue is divided into symmetrical lateral halves by a median septum that extends its entire length, and it is attached inferiorly to the hyoid bone, styloid process of the temporal bone, and mandible. Each half of the tongue consists of an identical complement of extrinsic and intrinsic muscles.

The **extrinsic muscles of the tongue**, which originate outside the tongue (attach to bones in the area) and insert into connective tissues in the tongue, include the *hyoglossus*, *genioglossus*, and *styloglossus muscles* (see **Figure 11.7**). The extrinsic muscles move the tongue from side to side and in and out to maneuver food for chewing, shape the food into a rounded mass, and force the food to the back of the mouth for swallowing. They also form the floor of the mouth and hold the tongue in position. The **intrinsic muscles of the tongue** originate in and insert into connective tissue within the tongue. They alter the shape and size of the tongue for speech and swallowing. The intrinsic muscles include the *longitudinalis superior*, *longitudinalis inferior*, *transversus linguae*, and *verticalis linguae muscles.* The **lingual frenulum** (*lingua* = the tongue), a fold of mucous membrane in the midline of the undersurface of the tongue, is attached to the floor of the mouth and aids in limiting the movement of the tongue posteriorly (see **Figures 24.6** and **24.7**). If a person's lingual frenulum is abnormally short or rigid—a condition called **ankyloglossia** (ang′-ki-lō-GLOS-ē-a)—the person is said to be "tongue-tied" because of the resulting impairment to speech. It can be corrected surgically.

The dorsum (upper surface) and lateral surfaces of the tongue are covered with **papillae** (pa-PIL-ē = nipple-shaped projections), projections of the lamina propria covered with stratified squamous epithelium (see **Figure 17.3**). Many papillae contain taste buds, the receptors for gustation (taste). Some papillae lack taste buds, but they contain receptors for touch and increase friction between the tongue and food, making it easier for the tongue to move food in the oral cavity. The different types of taste buds are described in detail in Section 17.2. **Lingual glands** in the lamina propria of the tongue secrete both mucus and a watery serous fluid that contains the enzyme **lingual lipase** (LĪ-pās), which acts on as much as 30% of dietary triglycerides (fats and oils) and converts them to simpler fatty acids and diglycerides.

Teeth

The **teeth**, or *dentes* (**Figure 24.8**), are accessory digestive organs located in sockets of the alveolar processes of the mandible and maxillae. The alveolar processes are covered by the **gingivae** (JIN-ji-vē), or gums, which extend slightly into each socket. The sockets are lined by the **periodontal ligament** (per′-ē-ō-DON-tal; *odont-* = tooth) or *periodontal membrane*, which consists of dense fibrous connective tissue that anchors the teeth to the socket walls and acts as a shock absorber during chewing.

A typical tooth has three major external regions: the crown, root, and neck. The **crown** is the visible portion above the level of the gums. Embedded in the socket are one to three **roots**. The **neck** is the constricted junction of the crown and root near the gum line.

See Clinical Connection: Root Canal Therapy

Internally, **dentin** forms the majority of the tooth. Dentin consists of a calcified connective tissue that gives the tooth its basic shape and rigidity. It is harder than bone because of its higher content of hydroxyapatite (70% versus 55% of dry weight).

The dentin of the crown is covered by **enamel**, which consists primarily of calcium phosphate and calcium carbonate. Enamel is also harder than bone because of its even higher content of calcium salts (about 95% of dry weight). In fact, enamel is the hardest substance in the body. It serves to protect the tooth from the wear and tear of chewing. It also protects against acids that can easily dissolve dentin.

FIGURE 24.8 **A typical tooth and surrounding structures.**

Teeth are anchored in sockets of the alveolar processes of the mandible and maxillae.

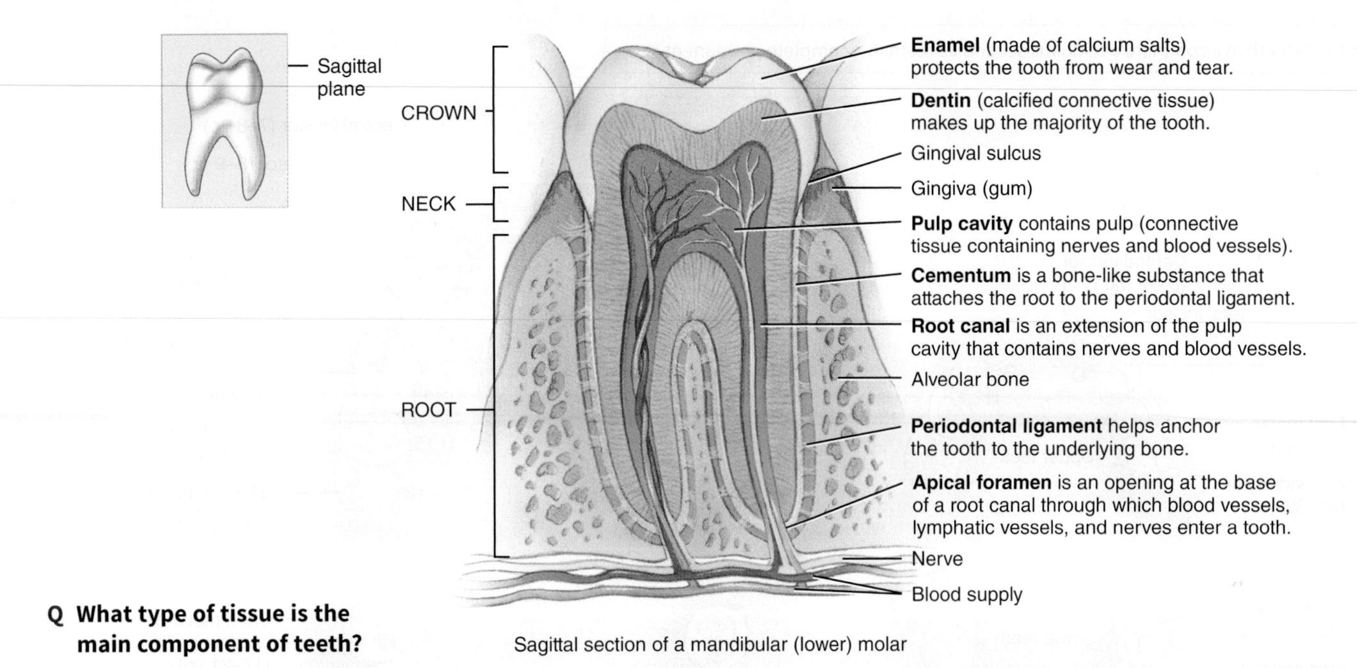

Sagittal section of a mandibular (lower) molar

Q What type of tissue is the main component of teeth?

The dentin of the root is covered by **cementum**, another bonelike substance, which attaches the root to the periodontal ligament.

The dentin of a tooth encloses a space. The enlarged part of the space, the **pulp cavity**, lies within the crown and is filled with **pulp**, a connective tissue containing blood vessels, nerves, and lymphatic vessels. Narrow extensions of the pulp cavity, called **root canals**, run through the root of the tooth. Each root canal has an opening at its base, the **apical foramen**, through which blood vessels, lymphatic vessels, and nerves enter a tooth. The blood vessels bring nourishment, the lymphatic vessels offer protection, and the nerves provide sensation.

The branch of dentistry that is concerned with the prevention, diagnosis, and treatment of diseases that affect the pulp, root, periodontal ligament, and alveolar bone is known as **endodontics** (en′-dō-DON-tiks; *endo-* = within). **Orthodontics** (or′-thō-DON-tiks; *ortho-* = straight) is a branch of dentistry that is concerned with the prevention and correction of abnormally aligned teeth; **periodontics** (per′-ē-ō-DON-tiks) is a branch of dentistry concerned with the treatment of abnormal conditions of the tissues immediately surrounding the teeth, such as gingivitis (gum disease).

Humans have two **dentitions**, or sets of teeth: deciduous and permanent. The first of these—the **deciduous teeth** (*decidu-* = falling out), also called *primary teeth, milk teeth*, or *baby teeth*—begin to erupt at about 6 months of age, and approximately two teeth appear each month thereafter, until all 20 are present (**Figure 24.9a**). The **incisors**, which are closest to the midline, are chisel-shaped and adapted for cutting into food. They are referred to as either **central** or **lateral incisors** based on their position. Next to the incisors, moving posteriorly, are the **canines**, which have a pointed surface called a *cusp*. Canines are used to tear and shred food. Incisors and canines have only one root apiece. Posterior to the canines lie the **first** and **second deciduous molars**, which have four cusps. Maxillary (upper) molars have three roots; mandibular (lower) molars have two roots. The molars crush and grind food to prepare it for swallowing.

All of the deciduous teeth are lost—generally between ages 6 and 12 years—and are replaced by the **permanent** (*secondary*) **teeth** (**Figure 24.9b**). The permanent dentition contains 32 teeth that erupt between age 6 and adulthood. The pattern resembles the deciduous dentition, with the following exceptions. The deciduous molars are replaced by the **first** and **second premolars** (*bicuspids*), which have two cusps and one root and are used for crushing and grinding. The permanent molars, which erupt into the mouth posterior to the premolars, do not replace any deciduous teeth and erupt as the jaw grows to accommodate them—the **first permanent molars** at age 6 (six-year molars), the **second permanent molars** at age 12 (twelve-year molars), and the **third permanent molars** (*wisdom teeth*) after age 17 or not at all.

Often the human jaw does not have enough room posterior to the second molars to accommodate the eruption of the third molars. In this case, the third molars remain embedded in the alveolar bone and are said to be *impacted*. They often cause pressure and pain and must be removed surgically. In some people, third molars may be dwarfed in size or may not develop at all.

Mechanical and Chemical Digestion in the Mouth

Mechanical digestion in the mouth results from chewing, or **mastication** (mas′-ti-KĀ-shun = to chew), in which food is manipulated by

FIGURE 24.9 Dentitions and times of eruption. A designated letter (deciduous teeth) or number (permanent teeth) uniquely identifies each tooth. Deciduous teeth begin to erupt at 6 months of age, and approximately two teeth appear each month thereafter, until all 20 are present. Times of eruption are indicated in parentheses.

There are 20 teeth in a complete deciduous set and 32 teeth in a complete permanent set.

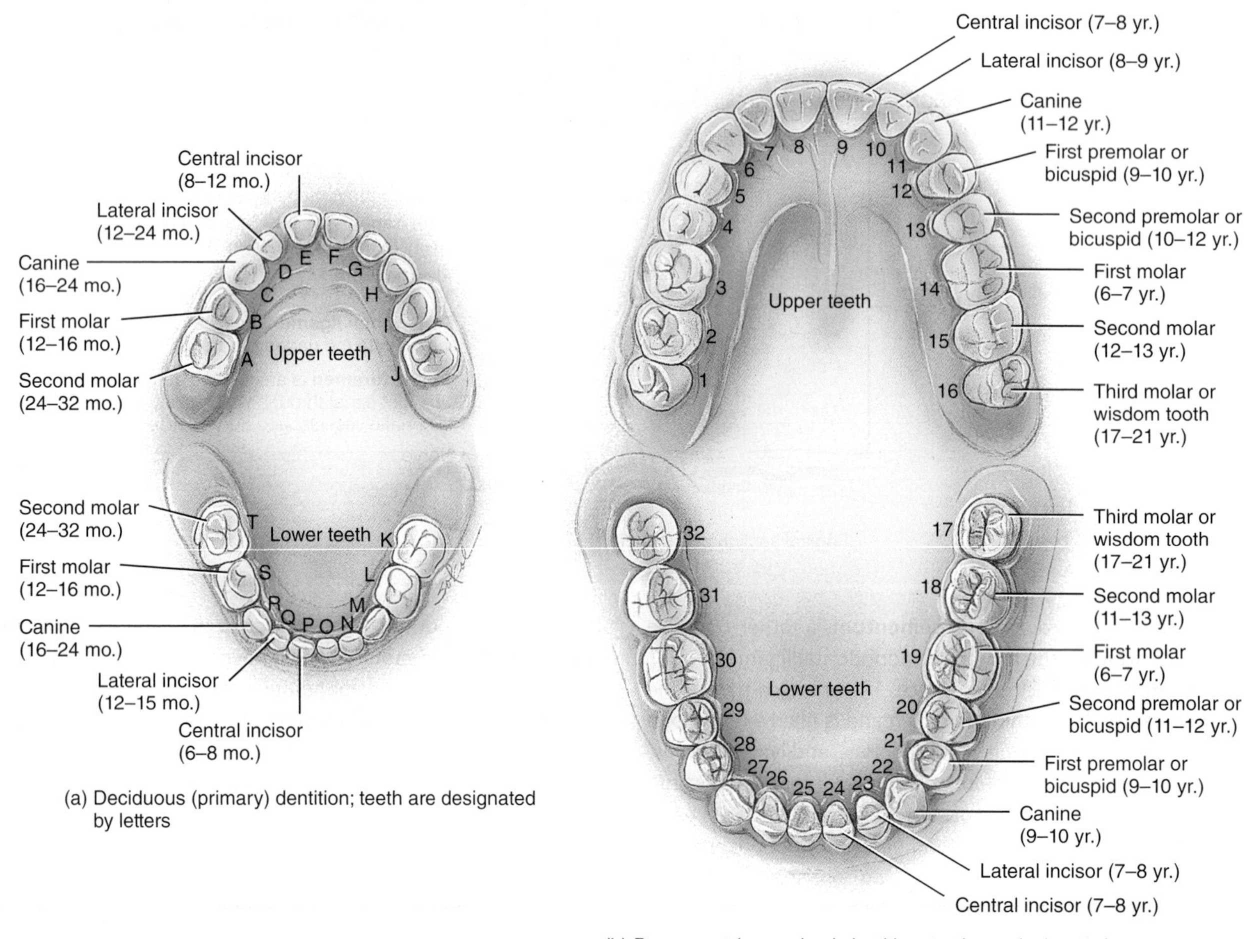

(a) Deciduous (primary) dentition; teeth are designated by letters

(b) Permanent (secondary) dentition; teeth are designated by numbers

Q Which permanent teeth do not replace any deciduous teeth?

the tongue, ground by the teeth, and mixed with saliva. As a result, the food is reduced to a soft, flexible, easily swallowed mass called a **bolus** (= lump). Food molecules begin to dissolve in the water in saliva, an important activity because enzymes can react with food molecules in a liquid medium only.

Two enzymes, salivary amylase and lingual lipase, contribute to chemical digestion in the mouth. Salivary amylase, which is secreted by the salivary glands, initiates the breakdown of starch. Dietary carbohydrates are either monosaccharide and disaccharide sugars or complex polysaccharides such as starches. Most of the carbohydrates we eat are starches, but only monosaccharides can be absorbed into the bloodstream. Thus, ingested disaccharides and starches must be broken down into monosaccharides. The function of salivary amylase is to begin starch digestion by breaking down starch into smaller molecules such as the disaccharide maltose, the trisaccharide maltotriose, and short-chain glucose polymers called α-dextrins. Even though food is usually swallowed too quickly for all starches to be broken down in the mouth, salivary amylase in the swallowed food continues to act on the starches for about another hour, at which time stomach acids inactivate it. Saliva also contains *lingual lipase*, which is secreted by lingual glands in the tongue. This enzyme becomes activated in the acidic environment of the stomach and thus starts to work after food is swallowed. It breaks down dietary triglycerides (fats and oils) into fatty acids and diglycerides. A diglyceride consists of a glycerol molecule that is attached to two fatty acids.

Table 24.1 summarizes the digestive activities in the mouth.

TABLE 24.1 Summary of Digestive Activities in the Mouth

STRUCTURE	ACTIVITY	RESULT
Cheeks and lips	Keep food between teeth.	Foods uniformly chewed during mastication.
Salivary glands	Secrete saliva.	Lining of mouth and pharynx moistened and lubricated. Saliva softens, moistens, and dissolves food and cleanses mouth and teeth. Salivary amylase splits starch into smaller fragments (maltose, maltotriose, and α-dextrins).
Tongue		
Extrinsic tongue muscles	Move tongue from side to side and in and out.	Food maneuvered for mastication, shaped into bolus, and maneuvered for swallowing.
Intrinsic tongue muscles	Alter shape of tongue.	Swallowing and speech.
Taste buds	Serve as receptors for gustation (taste) and presence of food in mouth.	Secretion of saliva stimulated by nerve impulses from taste buds to salivatory nuclei in brain stem to salivary glands.
Lingual glands	Secrete lingual lipase.	Triglycerides broken down into fatty acids and diglycerides.
Teeth	Cut, tear, and pulverize food.	Solid foods reduced to smaller particles for swallowing.

24.6 Pharynx

OBJECTIVE

- **Describe** the location and function of the pharynx.

When food is first swallowed, it passes from the mouth into the **pharynx** (= throat) or *throat*, a funnel-shaped tube that extends from the internal nares to the esophagus posteriorly and to the larynx anteriorly (see **Figure 23.2**). The pharynx is composed of skeletal muscle and lined by mucous membrane, and is divided into three parts: the nasopharynx, the oropharynx, and the laryngopharynx. The nasopharynx functions only in respiration, but both the oropharynx and laryngopharynx have digestive as well as respiratory functions. Swallowed food passes from the mouth into the oropharynx and laryngopharynx; the muscular contractions of these areas help propel food into the esophagus and then into the stomach.

24.7 Esophagus

OBJECTIVE

- **Describe** the location, anatomy, histology, and functions of the esophagus.

The **esophagus** (e-SOF-a-gus = eating gullet) is a collapsible muscular tube, about 25 cm (10 in.) long, that lies posterior to the trachea. The esophagus begins at the inferior end of the laryngopharynx, passes through the inferior aspect of the neck, and enters the mediastinum anterior to the vertebral column. Then it pierces the diaphragm through an opening called the **esophageal hiatus** (e-sof-a-JĒ-al hī-Ā-tus), and ends in the superior portion of the stomach (see **Figure 24.1**). Sometimes, part of the stomach protrudes above the diaphragm through the esophageal hiatus. This condition, termed a **hiatus hernia** (HER-nē-a), is described in the Medical Terminology section at the end of the chapter.

Histology of the Esophagus

The mucosa of the esophagus consists of nonkeratinized stratified squamous epithelium, lamina propria (areolar connective tissue), and a muscularis mucosae (smooth muscle) (**Figure 24.10**). Near the stomach, the mucosa of the esophagus also contains mucous glands. The stratified squamous epithelium associated with the lips, mouth, tongue, oropharynx, laryngopharynx, and esophagus affords considerable protection against abrasion and wear and tear from food particles that are chewed, mixed with secretions, and swallowed. The submucosa contains areolar connective tissue, blood vessels, and mucous glands. The muscularis of the superior third of the esophagus is skeletal muscle, the intermediate third is skeletal and smooth muscle, and the inferior third is smooth muscle. At each end of the esophagus, the muscularis becomes slightly more prominent and forms two sphincters—the **upper esophageal sphincter (UES)** (e-sof′-a-JĒ-al), which consists of skeletal muscle, and the **lower esophageal** (*cardiac*) **sphincter (LES)**, which consists of smooth muscle and is near the heart. The upper esophageal sphincter regulates the movement of food from the pharynx into the esophagus; the lower esophageal sphincter regulates the movement of food from the esophagus into the stomach. The superficial layer of the esophagus is known as the **adventitia** (ad-ven-TISH-a), rather than the serosa as in the stomach and intestines, because the areolar connective tissue of this layer is not covered by mesothelium and because the connective tissue merges with the connective tissue of surrounding structures of the mediastinum through which it passes. The adventitia attaches the esophagus to surrounding structures.

Physiology of the Esophagus

The esophagus secretes mucus and transports food into the stomach. It does not produce digestive enzymes, and it does not carry on absorption.

FIGURE 24.10 Histology of the esophagus. A higher-magnification view of nonkeratinized stratified squamous epithelium is shown in **Table 4.1G**.

The esophagus secretes mucus and transports food to the stomach.

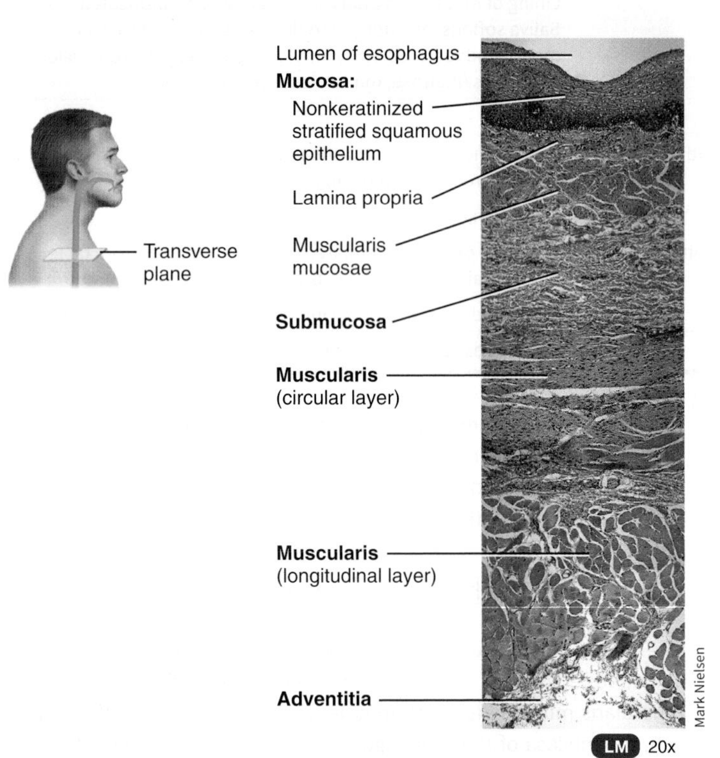

(a) Wall of the esophagus

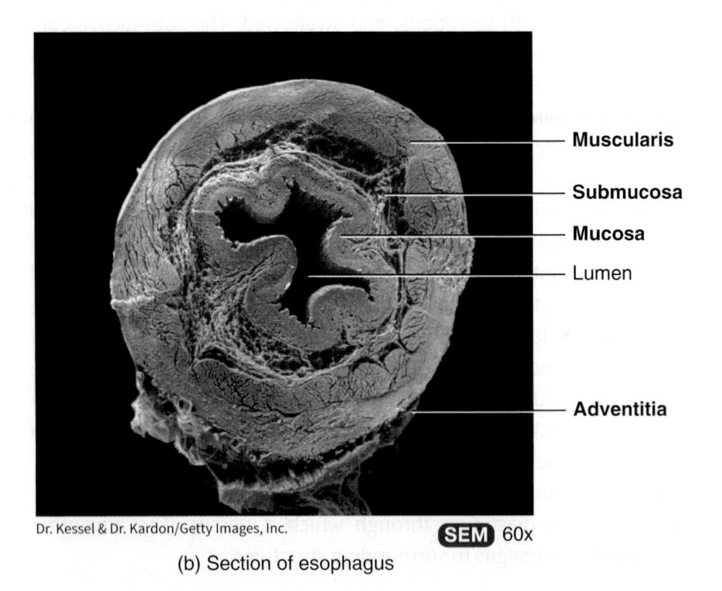

(b) Section of esophagus

Q In which layers of the esophagus are the glands that secrete lubricating mucus located?

24.8 Deglutition

OBJECTIVE

- **Describe** the three phases of deglutition.

The movement of food from the mouth into the stomach is achieved by the act of **deglutition** (dē-gloo-TISH-un) or *swallowing* (**Figure 24.11**). Deglutition is facilitated by the secretion of saliva and mucus and involves the mouth, pharynx, and esophagus. Swallowing occurs in three stages: (1) the voluntary stage, in which the bolus is passed into the oropharynx; (2) the pharyngeal stage, the involuntary passage of the bolus through the pharynx into the esophagus; and (3) the esophageal stage, the involuntary passage of the bolus through the esophagus into the stomach.

Swallowing starts when the bolus is forced to the back of the oral cavity and into the oropharynx by the movement of the tongue upward and backward against the palate; these actions constitute the **voluntary stage** of swallowing. With the passage of the bolus into the oropharynx, the involuntary **pharyngeal stage** of swallowing begins (**Figure 24.11b**). The bolus stimulates receptors in the oropharynx, which send impulses to the **deglutition center** in the medulla oblongata and lower pons of the brain stem. The returning impulses cause the soft palate and uvula to move upward to close off the nasopharynx, which prevents swallowed foods and liquids from entering the nasal cavity. In addition, the epiglottis closes off the opening to the larynx, which prevents the bolus from entering the rest of the respiratory tract. The bolus moves through the oropharynx and the laryngopharynx. Once the upper esophageal sphincter relaxes, the bolus moves into the esophagus.

The **esophageal stage** of swallowing begins once the bolus enters the esophagus. During this phase, **peristalsis** (per′-i-STAL-sis; *stalsis* = constriction), a progression of coordinated contractions and relaxations of the circular and longitudinal layers of the muscularis, pushes the bolus onward (**Figure 24.11c**). (Peristalsis occurs in other tubular structures, including other parts of the GI tract to the anus and the ureters, bile ducts, and uterine tubes; in the esophagus it is controlled by the medulla oblongata.)

1. In the section of the esophagus just superior to the bolus, the circular muscle fibers contract, constricting the esophageal wall and squeezing the bolus toward the stomach.
2. Longitudinal fibers inferior to the bolus also contract, which shortens this inferior section and pushes its walls outward so it can receive the bolus. The contractions are repeated in waves that push the food toward the stomach. Steps 1 and 2 repeat until the bolus reaches the lower esophageal sphincter muscles.
3. The lower esophageal sphincter relaxes and the bolus moves into the stomach.

Mucus secreted by esophageal glands lubricates the bolus and reduces friction. The passage of solid or semisolid food from the mouth to the stomach takes 4 to 8 seconds; very soft foods and liquids pass through in about 1 second.

Table 24.2 summarizes the digestive activities of the pharynx and esophagus.

FIGURE 24.11 Deglutition (swallowing). During the pharyngeal stage (b) the tongue rises against the palate, the nasopharynx is closed off, the larynx rises, the epiglottis seals off the larynx, and the bolus is passed into the esophagus. During the esophageal stage (c), food moves through the esophagus into the stomach via peristalsis.

Deglutition is a mechanism that moves food from the mouth into the stomach.

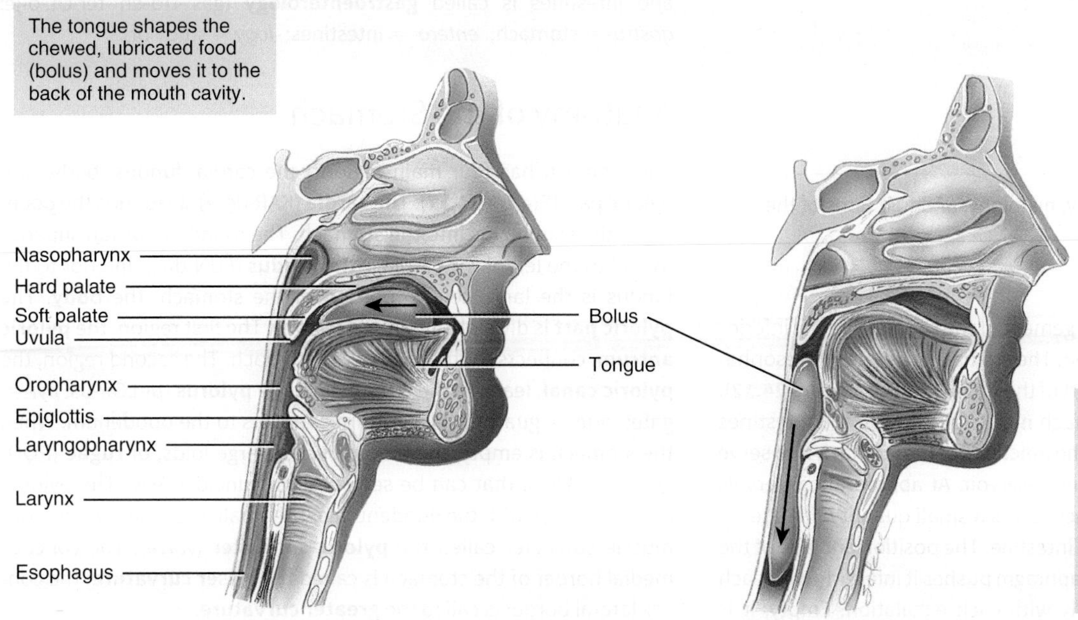

(a) Position of structures during voluntary stage

(b) Pharyngeal stage of swallowing

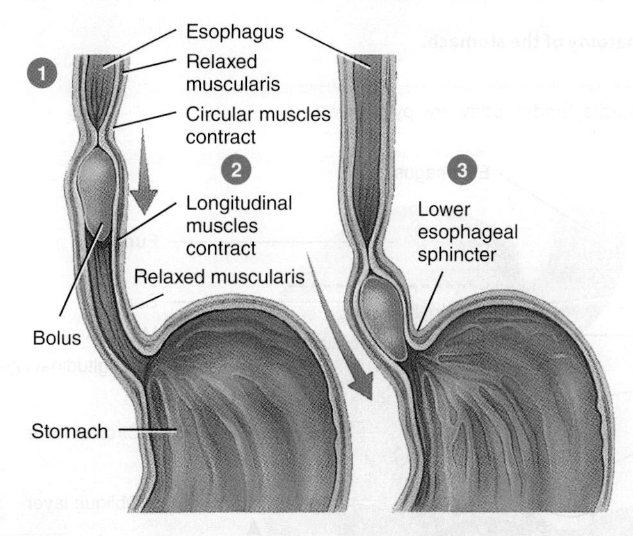

(c) Esophageal stage of swallowing

Q Is swallowing a voluntary action or an involuntary action?

TABLE 24.2 Summary of Digestive Activities in the Pharynx and Esophagus

STRUCTURE	ACTIVITY	RESULT
Pharynx	Pharyngeal stage of deglutition.	Moves bolus from oropharynx to laryngopharynx and into esophagus; closes air passageways.
Esophagus	Relaxation of upper esophageal sphincter.	Permits entry of bolus from laryngopharynx into esophagus.
	Esophageal stage of deglutition (peristalsis).	Pushes bolus down esophagus.
	Relaxation of lower esophageal sphincter.	Permits entry of bolus into stomach.
	Secretion of mucus.	Lubricates esophagus for smooth passage of bolus.

See Clinical Connection: Gastroesophageal Reflux Disease

24.9 Stomach

OBJECTIVE

- **Describe** the location, anatomy, histology, and functions of the stomach.

The **stomach** is a J-shaped enlargement of the GI tract directly inferior to the diaphragm in the abdomen. The stomach connects the esophagus to the duodenum, the first part of the small intestine (Figure 24.12). Because a meal can be eaten much more quickly than the intestines can digest and absorb it, one of the functions of the stomach is to serve as a mixing chamber and holding reservoir. At appropriate intervals after food is ingested, the stomach forces a small quantity of material into the first portion of the small intestine. The position and size of the stomach vary continually; the diaphragm pushes it inferiorly with each inhalation and pulls it superiorly with each exhalation. Empty, it is about the size of a large sausage, but it is the most distensible part of the GI tract and can accommodate a large quantity of food. In the stomach, digestion of starch and triglycerides continues, digestion of proteins begins, the semisolid bolus is converted to a liquid, and certain substances are absorbed. The medical specialty that deals with the structure, function, diagnosis, and treatment of diseases of the stomach and intestines is called **gastroenterology** (gas′-trō-en′-ter-OL-ō-jē; *gastro-* = stomach; *-entero-* = intestines; *-logy* = study of).

Anatomy of the Stomach

The stomach has four main regions: the cardia, fundus, body, and pyloric part (Figure 24.12). The **cardia** (KAR-dē-a) surrounds the opening of the esophagus into the stomach. The rounded portion superior to and to the left of the cardia is the **fundus** (FUN-dus). Inferior to the fundus is the large central portion of the stomach, the **body**. The **pyloric part** is divisible into three regions. The first region, the **pyloric antrum**, connects to the body of the stomach. The second region, the **pyloric canal**, leads to the third region, the **pylorus** (pī-LOR-us; *pyl-* = gate; *-orus* = guard), which in turn connects to the duodenum. When the stomach is empty, the mucosa lies in large folds, or **rugae** (ROO-gē = wrinkles), that can be seen with the unaided eye. The pylorus communicates with the duodenum of the small intestine via a smooth muscle sphincter called the **pyloric sphincter** (*valve*). The concave medial border of the stomach is called the **lesser curvature**; the convex lateral border is called the **greater curvature**.

FIGURE 24.12 External and internal anatomy of the stomach.

The four regions of the stomach are the cardia, fundus, body, and pyloric part.

(a) Anterior view of regions of stomach

Functions of the Stomach

1. Mixes saliva, food, and gastric juice to form chyme.
2. Serves as reservoir for food before release into small intestine.
3. Secretes gastric juice, which contains HCl (kills bacteria and denatures proteins), pepsin (begins the digestion of proteins), intrinsic factor (aids absorption of vitamin B_{12}), and gastric lipase (aids digestion of triglycerides).
4. Secretes gastrin into blood.

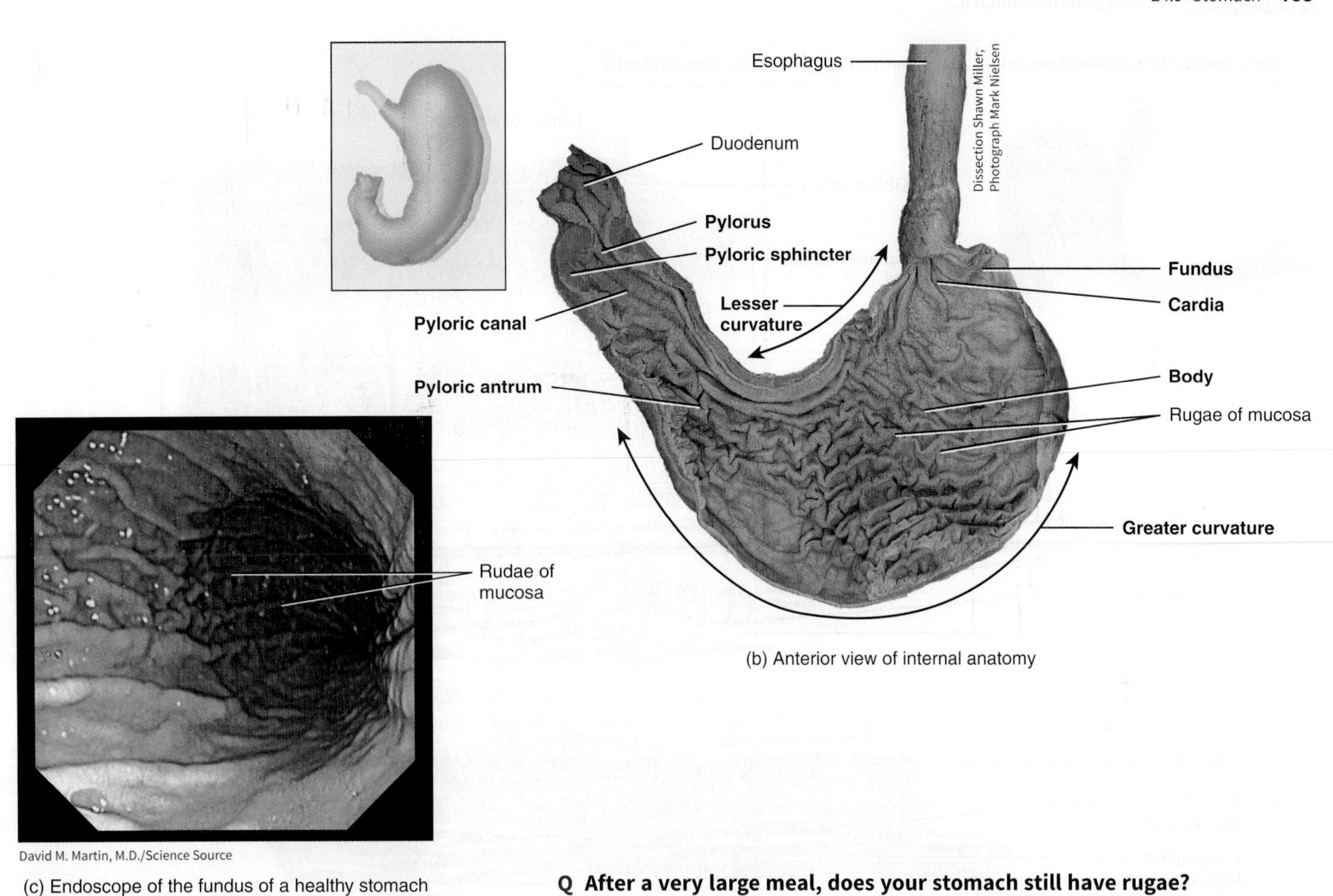

(b) Anterior view of internal anatomy

(c) Endoscope of the fundus of a healthy stomach

Q After a very large meal, does your stomach still have rugae?

See Clinical Connection: Pylorospasm and Pyloric Stenosis

Histology of the Stomach

The stomach wall is composed of the same basic layers as the rest of the GI tract, with certain modifications. The surface of the mucosa is a layer of simple columnar epithelial cells called **surface mucous cells** (Figure 24.13). The mucosa contains a lamina propria (areolar connective tissue) and a muscularis mucosae (smooth muscle) (Figure 24.13). Epithelial cells extend down into the lamina propria, where they form columns of secretory cells called **gastric glands**. Several gastric glands open into the bottom of narrow channels called **gastric pits**. Secretions from several gastric glands flow into each gastric pit and then into the lumen of the stomach.

The gastric glands contain three types of *exocrine gland cells* that secrete their products into the stomach lumen: mucous neck cells, chief cells, and parietal cells. Both surface mucous cells and **mucous neck cells** secrete mucus (Figure 24.13b). **Parietal cells** produce intrinsic factor (needed for absorption of vitamin B_{12}) and hydrochloric acid. The **chief** (*zymogenic*) **cells** secrete pepsinogen and gastric lipase. The secretions of the mucous, parietal, and chief cells form **gastric juice**, which totals 2000–3000 mL (roughly 2–3 qt) per day. In addition, gastric glands include a type of enteroendocrine cell, the **G cell**, which is located mainly in the pyloric antrum and secretes the hormone gastrin into the bloodstream. As we will see shortly, this hormone stimulates several aspects of gastric activity.

Three additional layers lie deep to the mucosa. The submucosa of the stomach is composed of areolar connective tissue. The muscularis has three layers of smooth muscle (rather than the two found in the esophagus and small and large intestines): an outer longitudinal layer, a middle circular layer, and an inner oblique layer. The oblique layer is limited primarily to the body of the stomach. The serosa is composed of simple squamous epithelium (mesothelium) and areolar connective tissue; the portion of the serosa covering the stomach is part of the visceral peritoneum. At the lesser curvature of the stomach, the visceral peritoneum extends upward to the liver as the lesser omentum. At the greater curvature of the stomach, the visceral peritoneum continues downward as the greater omentum and drapes over the intestines.

Mechanical and Chemical Digestion in the Stomach

Several minutes after food enters the stomach, waves of peristalsis pass over the stomach every 15 to 25 seconds. Few peristaltic waves are observed in the fundus, which primarily has a storage function. Instead, most waves begin at the body of the stomach and intensify as

FIGURE 24.13 **Histology of the stomach.**

Gastric juice is the combined secretions of mucous cells, parietal cells, and chief cells.

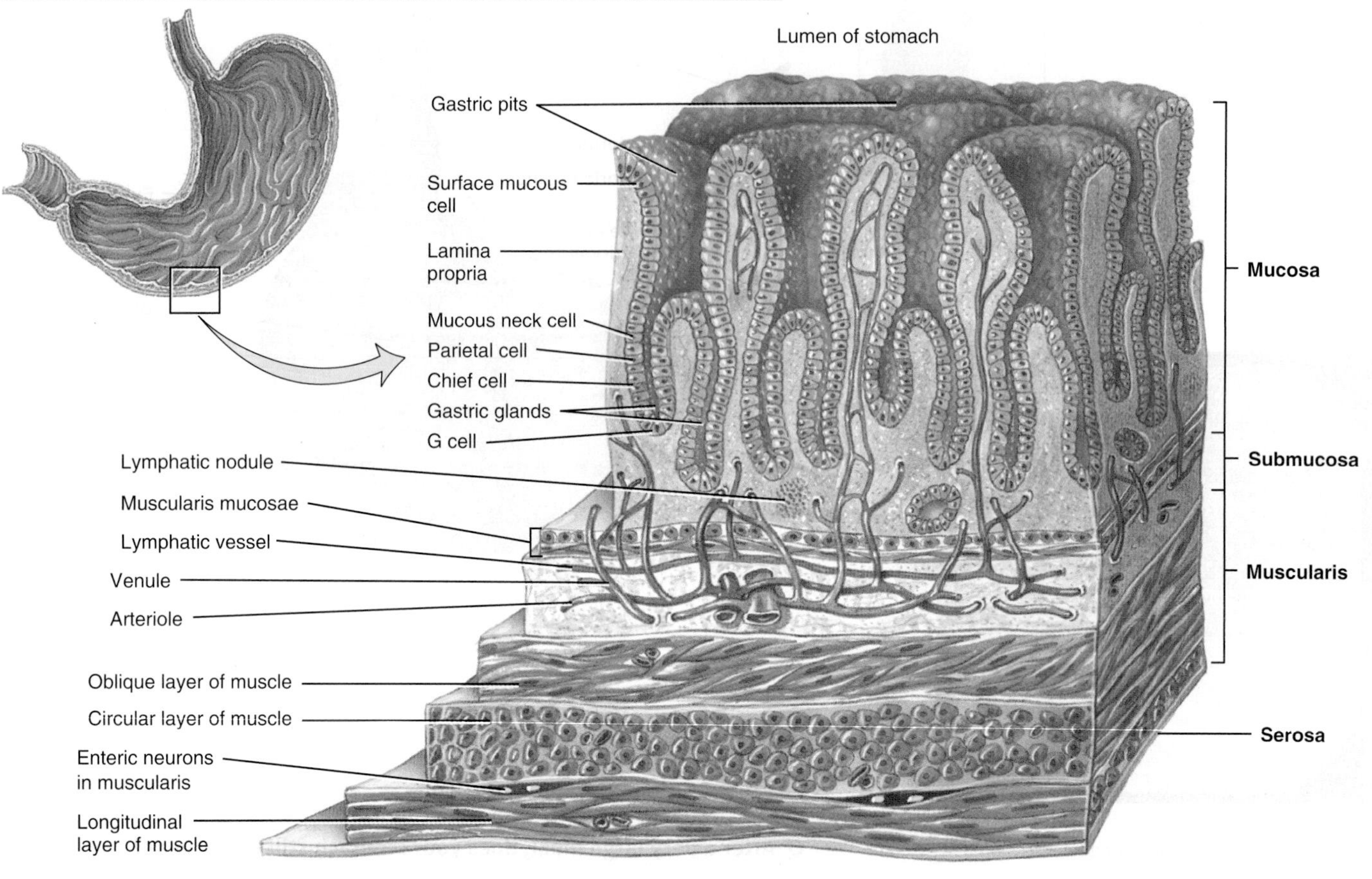

(a) Three-dimensional view of layers of stomach

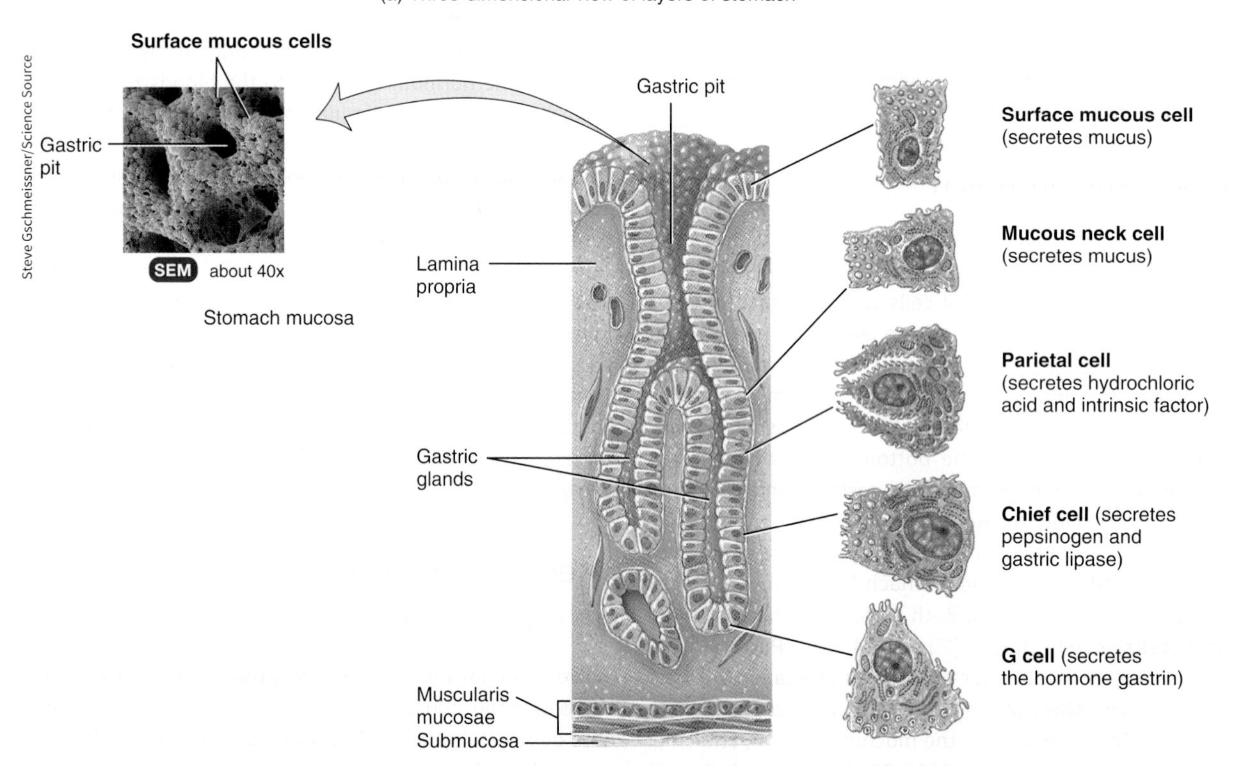

(b) Sectional view of the stomach mucosa showing gastric glands and cell types

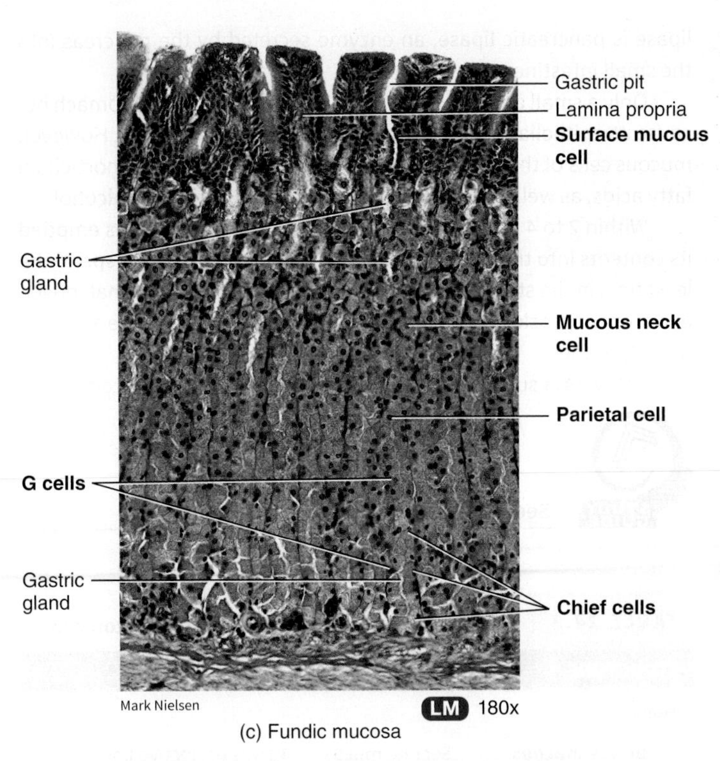

(c) Fundic mucosa

Q Where is HCl secreted, and what are its functions?

FIGURE 24.14 Secretion of HCl (hydrochloric acid) by parietal cells in the stomach.

Proton pumps, powered by ATP, secrete the H^+; Cl^- diffuses into the stomach lumen through Cl^- channels.

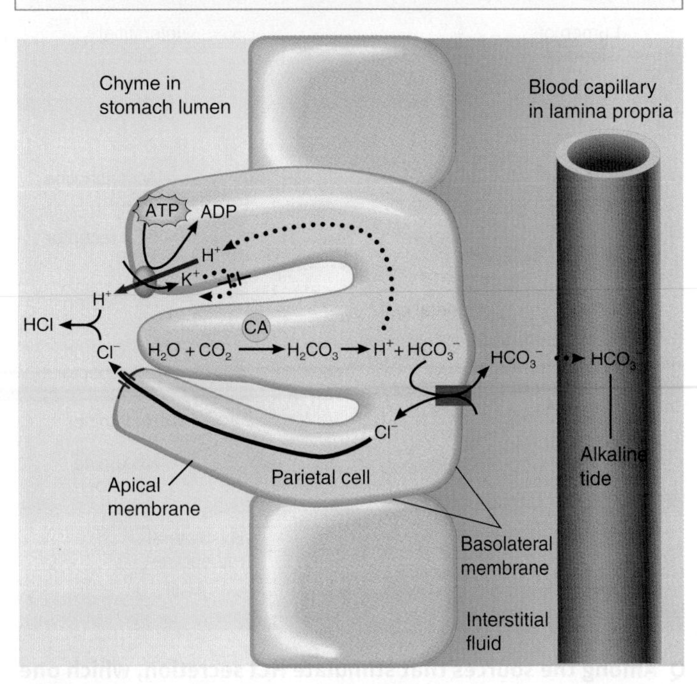

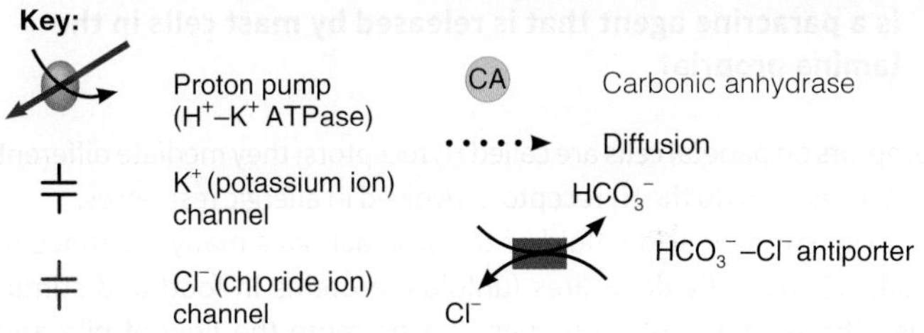

Q What molecule is the source of the hydrogen ions that are secreted into gastric juice?

they reach the antrum. Each peristaltic wave moves gastric contents from the body of the stomach down into the antrum, a process known as **propulsion**. The pyloric sphincter normally remains almost, but not completely, closed. Because most food particles in the stomach initially are too large to fit through the narrow pyloric sphincter, they are forced back into the body of the stomach, a process referred to as **retropulsion**. Another round of propulsion then occurs, moving the food particles back down into the antrum. If the food particles are still too large to pass through the pyloric sphincter, retropulsion occurs again as the particles are squeezed back into the body of the stomach. Then yet another round of propulsion occurs and the cycle continues to repeat. The net result of these movements is that gastric contents are mixed with gastric juice, eventually becoming reduced to a soupy liquid called **chyme** (KĪM = juice). Once the food particles in chyme are small enough, they can pass through the pyloric sphincter, a phenomenon known as **gastric emptying**. Gastric emptying is a slow process: only about 3 mL of chyme moves through the pyloric sphincter at a time.

Foods may remain in the fundus for about an hour without becoming mixed with gastric juice. During this time, digestion by salivary amylase from the salivary glands continues. Soon, however, the churning action mixes chyme with acidic gastric juice, inactivating salivary amylase and activating lingual lipase produced by the tongue, which starts to digest triglycerides into fatty acids and diglycerides.

Although parietal cells secrete hydrogen ions (H^+) and chloride ions (Cl^-) separately into the stomach lumen, the net effect is secretion of hydrochloric acid (HCl). **Proton pumps** powered by H^+–K^+ ATPases actively transport H^+ into the lumen while bringing potassium ions (K^+) into the cell (**Figure 24.14**). At the same time, Cl^- and K^+ diffuse out into the lumen through Cl^- and K^+ channels in the apical membrane. The enzyme *carbonic anhydrase*, which is especially plentiful in parietal cells, catalyzes the formation of carbonic acid (H_2CO_3) from water (H_2O) and carbon dioxide (CO_2). As carbonic acid dissociates, it provides a ready source of H^+ for the proton pumps but also generates bicarbonate ions (HCO_3^-). As HCO_3^- builds up in the cytosol, it exits the parietal cell in exchange for Cl^- via Cl^-–HCO_3^- antiporters in the basolateral membrane (next to the lamina propria). HCO_3^- diffuses into nearby blood capillaries. This "alkaline tide" of bicarbonate ions entering the bloodstream after a meal may be large enough to elevate blood pH slightly and make urine more alkaline.

HCl secretion by parietal cells can be stimulated by several sources: acetylcholine (ACh) released by parasympathetic neurons, gastrin secreted by G cells, and histamine, which is a paracrine substance released by mast cells in the nearby lamina propria (**Figure 24.15**). Acetylcholine and gastrin stimulate parietal cells to secrete more HCl in the presence of histamine. In other words, histamine acts synergistically, enhancing the effects of acetylcholine and gastrin. Receptors for all three substances are present in the plasma membrane of parietal cells. The histamine

FIGURE 24.15 Regulation of HCl secretion.

HCl secretion by parietal cells can be stimulated by several sources: acetylcholine (ACh), gastrin, and histamine.

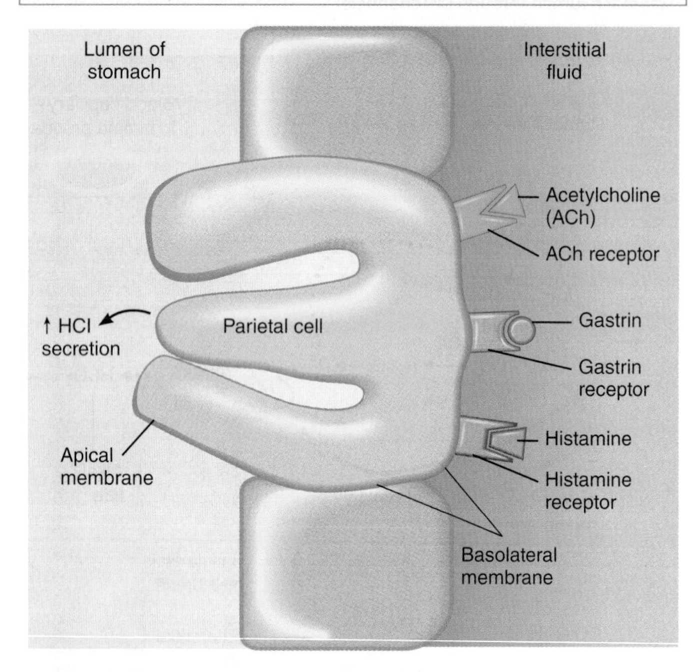

Q Among the sources that stimulate HCl secretion, which one is a paracrine agent that is released by mast cells in the lamina propria?

receptors on parietal cells are called H_2 receptors; they mediate different responses than do the H_1 receptors involved in allergic responses.

The strongly acidic fluid of the stomach kills many microbes in food. HCl partially denatures (unfolds) proteins in food and stimulates the secretion of hormones that promote the flow of bile and pancreatic juice. Enzymatic digestion of proteins also begins in the stomach. The only proteolytic (protein-digesting) enzyme in the stomach is **pepsin**, which is secreted by chief cells. Pepsin severs certain peptide bonds between amino acids, breaking down a protein chain of many amino acids into smaller peptide fragments. Pepsin is most effective in the very acidic environment of the stomach (pH 2); it becomes inactive at a higher pH.

What keeps pepsin from digesting the protein in stomach cells along with the food? First, pepsin is secreted in an inactive form called *pepsinogen*; in this form, it cannot digest the proteins in the chief cells that produce it. Pepsinogen is not converted into active pepsin until it comes in contact with hydrochloric acid secreted by parietal cells or active pepsin molecules. Second, the stomach epithelial cells are protected from gastric juices by a layer 1–3 mm thick of alkaline mucus secreted by surface mucous cells and mucous neck cells.

Another enzyme of the stomach is **gastric lipase**, which splits triglycerides (fats and oils) in fat molecules (such as those found in milk) into fatty acids and monoglycerides. A monoglyceride consists of a glycerol molecule that is attached to one fatty acid molecule. This enzyme, which has a limited role in the adult stomach, operates best at a pH of 5–6. More important than either lingual lipase or gastric lipase is pancreatic lipase, an enzyme secreted by the pancreas into the small intestine.

Only a small amount of nutrients are absorbed in the stomach because its epithelial cells are impermeable to most materials. However, mucous cells of the stomach absorb some water, ions, and short-chain fatty acids, as well as certain drugs (especially aspirin) and alcohol.

Within 2 to 4 hours after eating a meal, the stomach has emptied its contents into the duodenum. Foods rich in carbohydrate spend the least time in the stomach; high-protein foods remain somewhat longer, and emptying is slowest after a fat-laden meal containing large amounts of triglycerides.

Table 24.3 summarizes the digestive activities of the stomach.

See Clinical Connection: Vomiting

TABLE 24.3 Summary of Digestive Activities in the Stomach

STRUCTURE	ACTIVITY	RESULT
Mucosa		
Surface mucous cells and mucous neck cells	Secrete mucus.	Forms protective barrier that prevents digestion of stomach wall.
	Absorption.	Small quantity of water, ions, short-chain fatty acids, and some drugs enter bloodstream.
Parietal cells	Secrete intrinsic factor.	Needed for absorption of vitamin B_{12} (used in red blood cell formation, or erythropoiesis).
	Secrete hydrochloric acid.	Kills microbes in food; denatures proteins; converts pepsinogen into pepsin.
Chief cells	Secrete pepsinogen.	Pepsin (activated form) breaks down proteins into peptides.
	Secrete gastric lipase.	Splits triglycerides into fatty acids and monoglycerides.
G cells	Secrete gastrin.	Stimulates parietal cells to secrete HCl and chief cells to secrete pepsinogen; contracts lower esophageal sphincter, increases motility of stomach, and relaxes pyloric sphincter.
Muscularis	Mixing waves (gentle peristaltic movements).	Churns and physically breaks down food and mixes it with gastric juice, forming chyme. Forces chyme through pyloric sphincter.
Pyloric sphincter	Opens to permit passage of chyme into duodenum.	Regulates passage of chyme from stomach to duodenum; prevents backflow of chyme from duodenum to stomach.

24.10 Pancreas

OBJECTIVE

- **Describe** the location, anatomy, histology, and function of the pancreas.

From the stomach, chyme passes into the small intestine. Because chemical digestion in the small intestine depends on activities of the pancreas, liver, and gallbladder, we first consider the activities of these accessory digestive organs and their contributions to digestion in the small intestine.

Anatomy of the Pancreas

The **pancreas** (*pan-* = all; *-creas* = flesh), a retroperitoneal gland that is about 12–15 cm (5–6 in.) long and 2.5 cm (1 in.) thick, lies posterior to the greater curvature of the stomach. The pancreas consists of a head, a body, and a tail and is usually connected to the duodenum of the small intestine by two ducts (Figure 24.16a). The **head** is the expanded portion of the organ near the curve of the duodenum; superior to and to the left of the head are the central **body** and the tapering **tail**.

Pancreatic juices are secreted by exocrine cells into small ducts that ultimately unite to form two larger ducts, the pancreatic duct and the accessory duct. These in turn convey the secretions into the small intestine. The **pancreatic duct**, or *duct of Wirsung* (VĒR-sung), is the larger of the two ducts. In most people, the pancreatic duct joins the common bile duct from the liver and gallbladder and enters the duodenum as a dilated common duct called the **hepatopancreatic ampulla** (hep′-a-tō-pan-krē-A-tik), or *ampulla of Vater* (FAH-ter). The ampulla opens on an elevation of the duodenal mucosa known as the **major duodenal papilla**, which lies about 10 cm (4 in.) inferior to the pyloric sphincter of the stomach. The passage of pancreatic juice and bile through the hepatopancreatic ampulla into the duodenum of the small intestine is regulated by a mass of smooth muscle surrounding the ampulla known as the **sphincter of the hepatopancreatic ampulla**, or *sphincter of Oddi* (OD-ē). The other major duct of the pancreas, the **accessory duct** (*duct of Santorini*), leads from the pancreas and empties into the duodenum about 2.5 cm (1 in.) superior to the hepatopancreatic ampulla.

Histology of the Pancreas

The pancreas is made up of small clusters of glandular epithelial cells. About 99% of the clusters, called **acini** (AS-i-nī), constitute the *exocrine*

FIGURE 24.16 Relationship of the pancreas to the liver, gallbladder, and duodenum. The inset (b) shows details of the common bile duct and pancreatic duct forming the hepatopancreatic ampulla and emptying into the duodenum.

Pancreatic enzymes digest starches (polysaccharides), proteins, triglycerides, and nucleic acids.

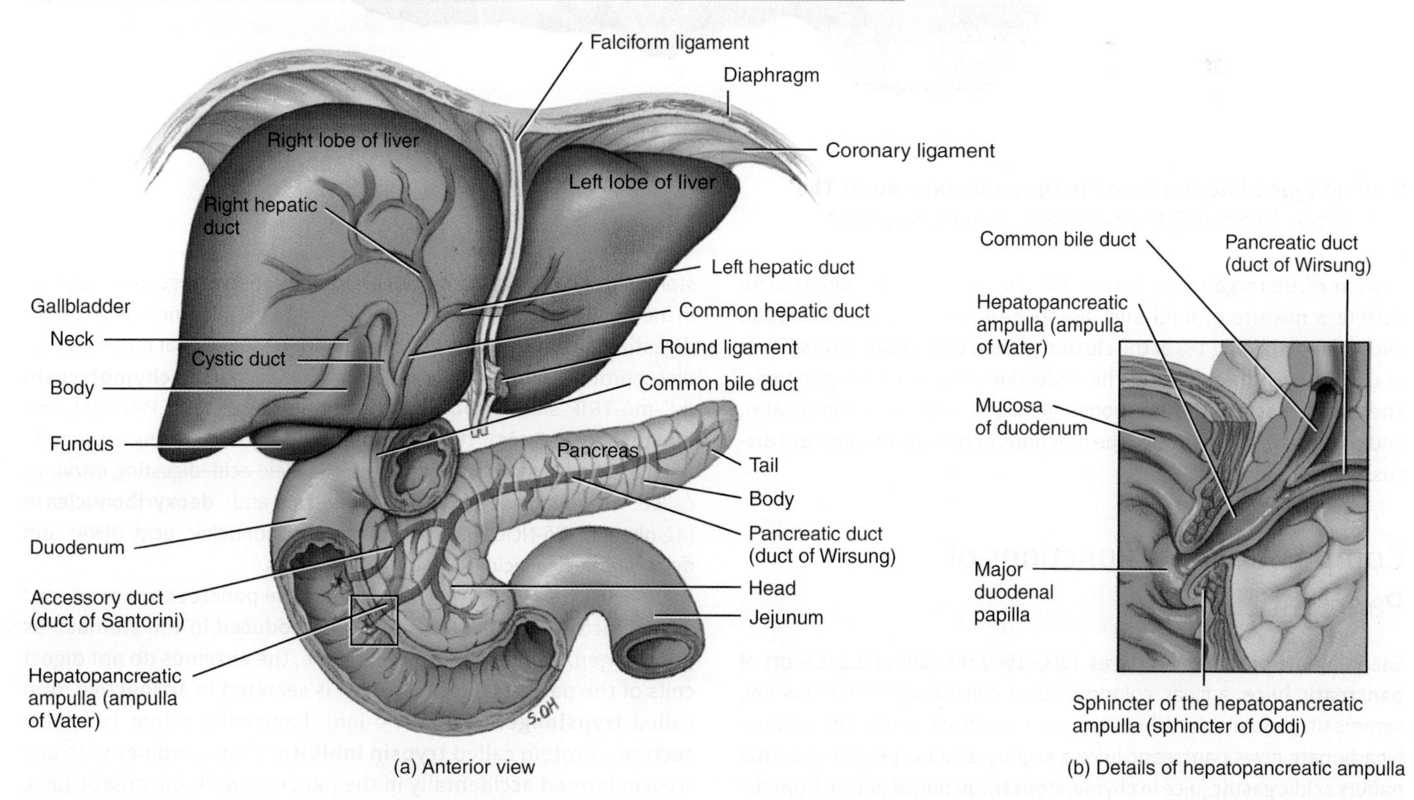

Figure 24.16 *Continues*

FIGURE 24.16 Continued

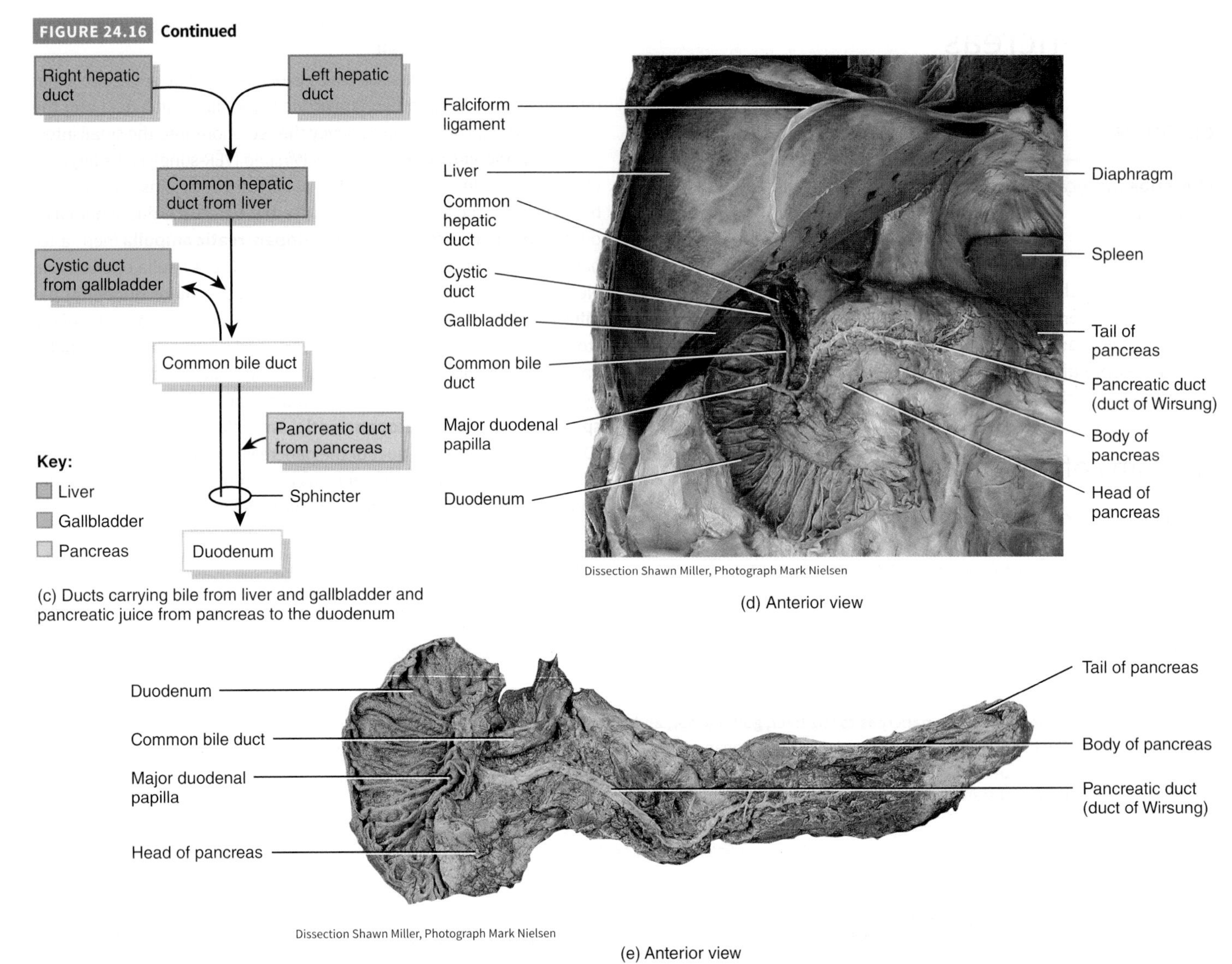

(c) Ducts carrying bile from liver and gallbladder and pancreatic juice from pancreas to the duodenum

(d) Anterior view

(e) Anterior view

Q What type of fluid is found in the pancreatic duct? The common bile duct? The hepatopancreatic ampulla?

portion of the organ (see **Figure 18.17b, c**). The cells within acini secrete a mixture of fluid and digestive enzymes called pancreatic juice. The remaining 1% of the clusters, called **pancreatic islets** (*islets of Langerhans*) (ī-lets), form the *endocrine* portion of the pancreas. These cells secrete the hormones glucagon, insulin, somatostatin, and pancreatic polypeptide. The functions of these hormones are discussed in Chapter 18.

Composition and Functions of Pancreatic Juice

Each day the pancreas produces 1200–1500 mL (about 1.2–1.5 qt) of **pancreatic juice**, a clear, colorless liquid consisting mostly of water, some salts, sodium bicarbonate, and several enzymes. The sodium bicarbonate gives pancreatic juice a slightly alkaline pH (7.1–8.2) that buffers acidic gastric juice in chyme, stops the action of pepsin from the stomach, and creates the proper pH for the action of digestive enzymes in the small intestine. The enzymes in pancreatic juice include a starch-digesting enzyme called **pancreatic amylase**; several enzymes that digest proteins into peptides called **trypsin** (TRIP-sin), **chymotrypsin** (kī′-mō-TRIP-sin), **carboxypeptidase** (kar-bok′-sē-PEP-ti-dās), and **elastase** (ē-LAS-tās); the principal triglyceride–digesting enzyme in adults, called **pancreatic lipase**; and nucleic acid–digesting enzymes called **ribonuclease** (rī′-bō-NOO-klē-ās) and **deoxyribonuclease** (dē-oks-ē-rī′-bō-NOO-klē-ās) that digest ribonucleic acid (RNA) and deoxyribonucleic acid (DNA) into nucleotides.

The protein-digesting enzymes of the pancreas are produced in an inactive form just as pepsin is produced in the stomach as pepsinogen. Because they are inactive, the enzymes do not digest cells of the pancreas itself. Trypsin is secreted in an inactive form called **trypsinogen** (trip-SIN-ō-jen). Pancreatic acinar cells also secrete a protein called **trypsin inhibitor** that combines with any trypsin formed accidentally in the pancreas or in pancreatic juice and blocks its enzymatic activity. When trypsinogen reaches the lumen of the small intestine, it encounters an activating brush-border enzyme called **enterokinase** (en′-ter-ō-KĪ-nās), which splits off part of the trypsinogen molecule to form trypsin. In turn, trypsin acts on the inactive precursors (called **chymotrypsinogen**, **procarboxypeptidase**, and **proelastase**) to produce chymotrypsin, carboxypeptidase, and elastase, respectively.

See Clinical Connection: Pancreatitis and Pancreatic Cancer

24.11 Liver and Gallbladder

OBJECTIVE

- **Describe** the location, anatomy, histology, and functions of the liver and gallbladder.

The **liver** is the heaviest gland of the body, weighing about 1.4 kg (about 3 lb) in an average adult. Of all of the organs of the body, it is second only to the skin in size. The liver is inferior to the diaphragm and occupies most of the right hypochondriac and part of the epigastric regions of the abdominopelvic cavity (see Figure 1.13b).

The **gallbladder** (*gall-* = bile) is a pear-shaped sac that is located in a depression of the posterior surface of the liver. It is 7–10 cm (3–4 in.) long and typically hangs from the anterior inferior margin of the liver (Figure 24.16a).

Anatomy of the Liver and Gallbladder

The liver is almost completely covered by visceral peritoneum and is completely covered by a dense irregular connective tissue layer that lies deep to the peritoneum. The liver is divided into two principal lobes—a large **right lobe** and a smaller **left lobe**—by the falciform ligament, a fold of the mesentery (Figure 24.16a). Although the right lobe is considered by many anatomists to include an inferior **quadrate lobe** (kwa-DRĀT) and a posterior **caudate lobe** (KAW-dāt), based on internal morphology (primarily the distribution of blood vessels), the quadrate and caudate lobes more appropriately belong to the left lobe. The falciform ligament extends from the undersurface of the diaphragm between the two principal lobes of the liver to the superior surface of the liver, helping to suspend the liver in the abdominal cavity. In the free border of the falciform ligament is the **ligamentum teres** (*round ligament*), a remnant of the umbilical vein of the fetus (see 21.31a, b); this fibrous cord extends from the liver to the umbilicus. The right and left **coronary ligaments** are narrow extensions of the parietal peritoneum that suspend the liver from the diaphragm.

The parts of the gallbladder include the broad **fundus**, which projects inferiorly beyond the inferior border of the liver; the **body**, the central portion; and the **neck**, the tapered portion. The body and neck project superiorly.

Histology of the Liver and Gallbladder

Histologically, the liver is composed of several components (Figure 24.17a–c):

1. **Hepatocytes. Hepatocytes** (*hepat-* = liver; *-cytes* = cells) are the major functional cells of the liver and perform a wide array of metabolic, secretory, and endocrine functions. These are specialized epithelial cells with 5 to 12 sides that make up about 80% of the volume of the liver. Hepatocytes form complex three-dimensional arrangements called **hepatic laminae** (LAM-i-nē). The hepatic laminae are plates of hepatocytes one cell thick bordered on either side by the endothelial-lined vascular spaces called hepatic sinusoids. The hepatic laminae are highly branched, irregular structures. Grooves in the cell membranes between neighboring hepatocytes provide spaces for canaliculi (described next) into which the hepatocytes secrete bile. Bile, a yellow, brownish, or olive-green liquid secreted by hepatocytes, serves as both an excretory product and a digestive secretion.
2. **Bile canaliculi**. **Bile canaliculi** (kan-a-LIK-ū-li = small canals) are small ducts between hepatocytes that collect bile produced by the hepatocytes. From bile canaliculi, bile passes into **bile ductules** and then **bile ducts**. The bile ducts merge and eventually form the larger **right** and **left hepatic ducts**, which unite and exit the liver as the **common hepatic duct** (see Figure 24.16). The common hepatic duct joins the **cystic duct** (*cystic* = bladder) from the gallbladder to form the **common bile duct**. From here, bile enters the duodenum of the small intestine to participate in digestion.
3. **Hepatic sinusoids**. **Hepatic sinusoids** are highly permeable blood capillaries between rows of hepatocytes that receive oxygenated blood from branches of the hepatic artery and nutrient-rich deoxygenated blood from branches of the hepatic portal vein. Recall that the hepatic portal vein brings venous blood from the gastrointestinal organs and spleen into the liver. Hepatic sinusoids converge and deliver blood into a **central vein**. From central veins the blood flows into the **hepatic veins**, which drain into the inferior vena cava (see Figure 21.29). In contrast to blood, which flows toward a central vein, bile flows in the opposite direction. Also present in the hepatic sinusoids are fixed phagocytes called **stellate reticuloendothelial cells** (STEL-āt re-tik′-ū-lō-en′-dō-THĒ-lē-al) or *hepatic macrophages*, which destroy worn-out white and red blood cells, bacteria, and other foreign matter in the venous blood draining from the gastrointestinal tract.

Together, a bile duct, branch of the hepatic artery, and branch of the hepatic vein are referred to as a **portal triad** (*tri-* = three).

The hepatocytes, bile duct system, and hepatic sinusoids can be organized into anatomical and functional units in three different ways:

1. **Hepatic lobule.** For years, anatomists described the **hepatic lobule** as the functional unit of the liver. According to this model, each hepatic lobule is shaped like a hexagon (six-sided structure) (Figure 24.17d, left). At its center is the central vein, and radiating out from it are rows of hepatocytes and hepatic sinusoids. Located at three corners of the hexagon is a portal triad. This model is based on a description of the liver of adult pigs. In the human liver it is difficult to find such well-defined hepatic lobules surrounded by thick layers of connective tissue.

FIGURE 24.17 **Histology of the liver.**

Histologically, the liver is composed of hepatocytes, bile canaliculi, and hepatic sinusoids.

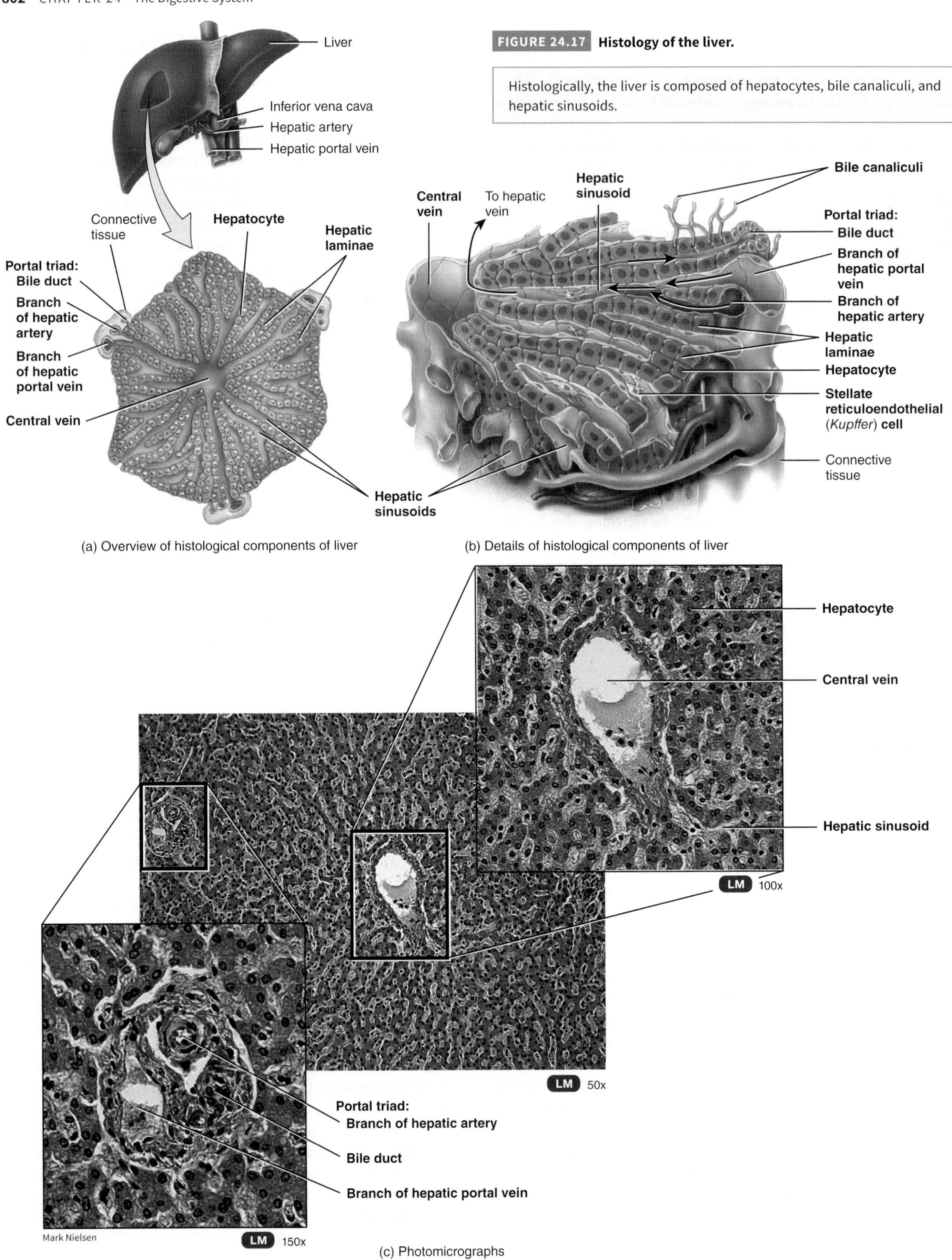

(a) Overview of histological components of liver

(b) Details of histological components of liver

(c) Photomicrographs

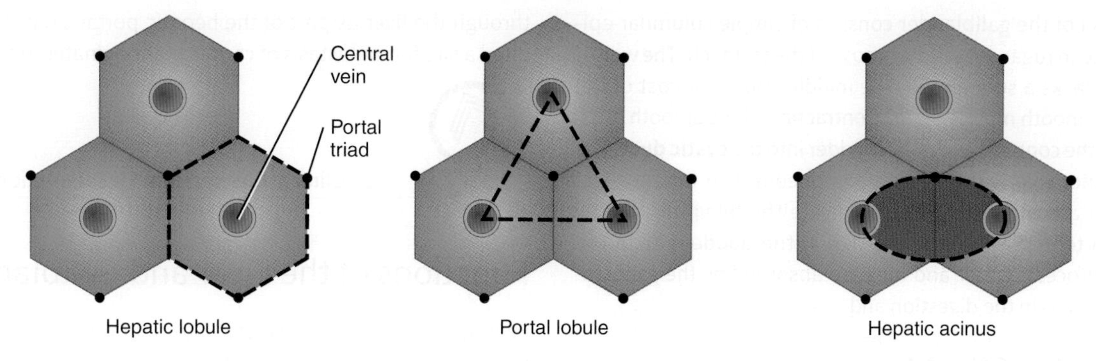

(d) Comparison of three units of liver structure and function

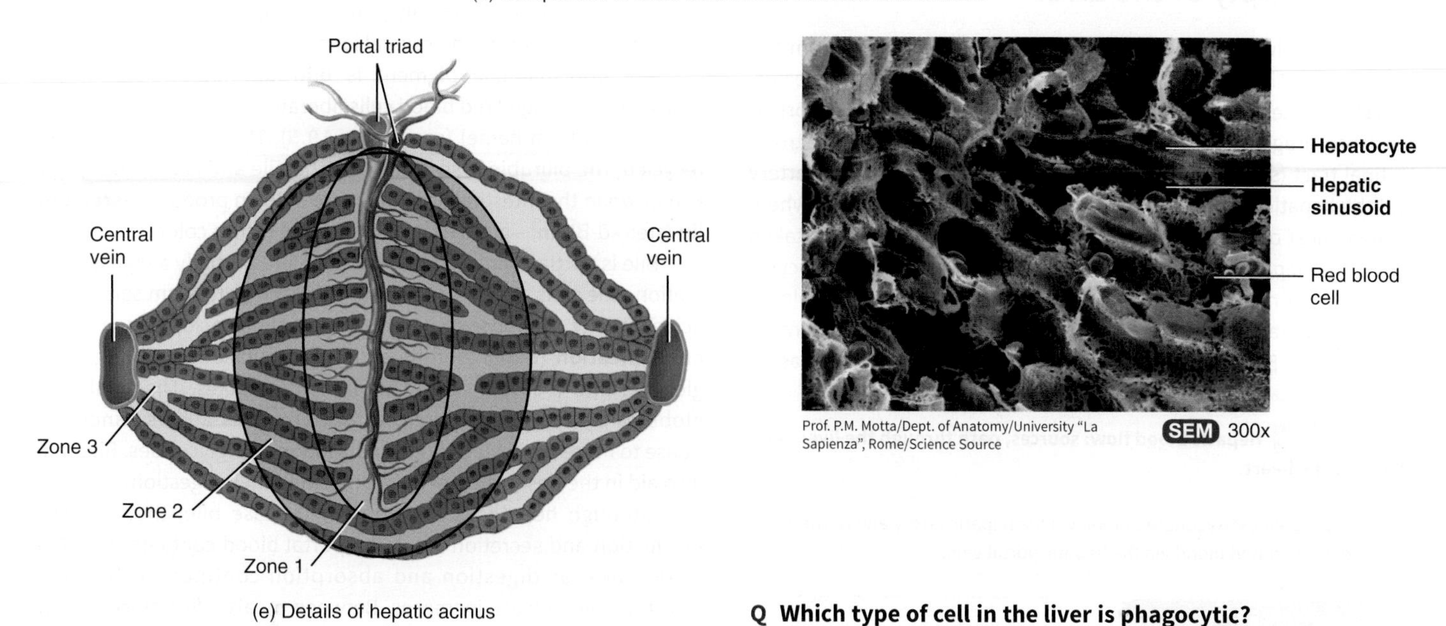

(e) Details of hepatic acinus

Q Which type of cell in the liver is phagocytic?

2. **Portal lobule.** This model emphasizes the exocrine function of the liver, that is, bile secretion. Accordingly, the bile duct of a portal triad is taken as the center of the **portal lobule**. The portal lobule is triangular in shape and is defined by three imaginary straight lines that connect three central veins that are closest to the portal triad (Figure 24.17d, center). This model has not gained widespread acceptance.
3. **Hepatic acinus.** In recent years, the preferred structural and functional unit of the liver is the **hepatic acinus** (AS-i-nus). Each hepatic acinus is an approximately oval mass that includes portions of two neighboring hepatic lobules. The short axis of the hepatic acinus is defined by branches of the portal triad—branches of the hepatic artery, vein, and bile ducts—that run along the border of the hepatic lobules. The long axis of the acinus is defined by two imaginary curved lines, which connect the two central veins closest to the short axis (Figure 24.17d, bottom). Hepatocytes in the hepatic acinus are arranged in three zones around the short axis, with no sharp boundaries between them (Figure 24.17e). Cells in zone 1 are closest to the branches of the portal triad and the first to receive incoming oxygen, nutrients, and toxins from incoming blood. These cells are the first ones to take up glucose and store it as glycogen after a meal and break down glycogen to glucose during fasting. They are also the first to show morphological changes following bile duct obstruction or exposure to toxic substances. Zone 1 cells are the last ones to die if circulation is impaired and the first ones to regenerate. Cells in zone 3 are farthest from branches of the portal triad and are the last to show the effects of bile obstruction or exposure to toxins, the first ones to show the effects of impaired circulation, and the last ones to regenerate. Zone 3 cells also are the first to show evidence of fat accumulation. Cells in zone 2 have structural and functional characteristics intermediate between the cells in zones 1 and 3.

The hepatic acinus is the smallest structural and functional unit of the liver. Its popularity and appeal are based on the fact that it provides a logical description and interpretation of (1) patterns of glycogen storage and release and (2) toxic effects, degeneration, and regeneration relative to the proximity of the acinar zones to branches of the portal triad.

See Clinical Connection: Jaundice

The mucosa of the gallbladder consists of simple columnar epithelium arranged in rugae resembling those of the stomach. The wall of the gallbladder lacks a submucosa. The middle, muscular coat of the wall consists of smooth muscle fibers. Contraction of the smooth muscle fibers ejects the contents of the gallbladder into the **cystic duct**. The gallbladder's outer coat is the visceral peritoneum. The functions of the gallbladder are to store and concentrate the bile produced by the liver (up to tenfold) until it is needed in the duodenum. In the concentration process, water and ions are absorbed by the gallbladder mucosa. Bile aids in the digestion and absorption of fats.

Blood Supply of the Liver

The liver receives blood from two sources (Figure 24.18). From the hepatic artery it obtains oxygenated blood, and from the hepatic portal vein it receives deoxygenated blood containing newly absorbed nutrients, drugs, and possibly microbes and toxins from the gastrointestinal tract (see Figure 21.29). Branches of both the hepatic artery and the hepatic portal vein carry blood into hepatic sinusoids, where oxygen, most of the nutrients, and certain toxic substances are taken up by the hepatocytes. Products manufactured by the hepatocytes and nutrients needed by other cells are secreted back into the blood, which then drains into the central vein and eventually passes into a hepatic vein. Because blood from the gastrointestinal tract passes through the liver as part of the hepatic portal circulation, the liver is often a site for metastasis of cancer that originates in the GI tract.

FIGURE 24.18 Hepatic blood flow: sources, path through the liver, and return to the heart.

The liver receives oxygenated blood via the hepatic artery and nutrient-rich deoxygenated blood via the hepatic portal vein.

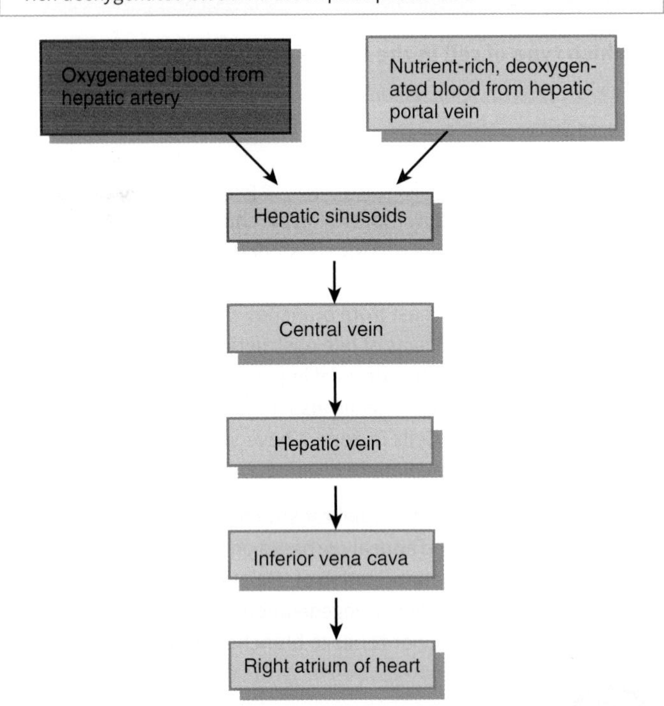

Q During the first few hours after a meal, how does the chemical composition of blood change as it flows through the liver sinusoids?

See Clinical Connection: Liver Function Tests

Functions of the Liver and Gallbladder

Each day, hepatocytes secrete 800–1000 mL (about 1 qt) of **bile**, a yellow, brownish, or olive-green liquid. It has a pH of 7.6–8.6 and consists mostly of water, bile salts, cholesterol, a phospholipid called lecithin, bile pigments, and several ions.

The principal bile pigment is **bilirubin** (bil-i-ROO-bin). The phagocytosis of aged red blood cells liberates iron, globin, and bilirubin (derived from heme) (see Figure 19.5). The iron and globin are recycled; the bilirubin is secreted into the bile and is eventually broken down in the intestine. One of its breakdown products—**stercobilin** (ster-kō-BĪ-lin)—gives feces their normal brown color.

Bile is partially an excretory product and partially a digestive secretion. Bile salts, which are sodium salts and potassium salts of bile acids (mostly chenodeoxycholic acid and cholic acid), play a role in **emulsification** (e-mul′-si-fi-KĀ-shun), the breakdown of large lipid globules into a suspension of small lipid globules. The small lipid globules present a very large surface area that allows pancreatic lipase to more rapidly accomplish digestion of triglycerides. Bile salts also aid in the absorption of lipids following their digestion.

Although hepatocytes continually release bile, they increase production and secretion when the portal blood contains more bile acids; thus, as digestion and absorption continue in the small intestine, bile release increases. Between meals, after most absorption has occurred, bile flows into the gallbladder for storage because the sphincter of the hepatopancreatic ampulla (sphincter of Oddi; see Figure 24.16) closes off the entrance to the duodenum. The sphincter surrounds the hepatopancreatic ampulla.

In addition to secreting bile, which is needed for absorption of dietary fats, the liver performs many other vital functions:

- ***Carbohydrate metabolism.*** The liver is especially important in maintaining a normal blood glucose level. When blood glucose is low, the liver can break down glycogen to glucose and release the glucose into the bloodstream. The liver can also convert certain amino acids and lactic acid to glucose, and it can convert other sugars, such as fructose and galactose, into glucose. When blood glucose is high, as occurs just after eating a meal, the liver converts glucose to glycogen and triglycerides for storage.

See Clinical Connection: Gallstones

- ***Lipid metabolism.*** Hepatocytes store some triglycerides; break down fatty acids to generate ATP; synthesize lipoproteins, which transport fatty acids, triglycerides, and cholesterol to and from body cells; synthesize cholesterol; and use cholesterol to make bile salts.

- ***Protein metabolism.*** Hepatocytes *deaminate* (remove the amino group, NH_2, from) amino acids so that the amino acids can be used for ATP production or converted to carbohydrates or fats. The resulting toxic ammonia (NH_3) is then converted into the much less toxic urea, which is excreted in urine. Hepatocytes also synthesize most plasma proteins, such as alpha and beta globulins, albumin, prothrombin, and fibrinogen.
- ***Processing of drugs and hormones.*** The liver can detoxify substances such as alcohol and excrete drugs such as penicillin, erythromycin, and sulfonamides into bile. It can also chemically alter or excrete thyroid hormones and steroid hormones such as estrogens and aldosterone.
- ***Excretion of bilirubin.*** As previously noted, bilirubin, derived from the heme of aged red blood cells, is absorbed by the liver from the blood and secreted into bile. Most of the bilirubin in bile is metabolized in the small intestine by bacteria and eliminated in feces.
- ***Synthesis of bile salts.*** Bile salts are used in the small intestine for the emulsification and absorption of lipids.
- ***Storage.*** In addition to glycogen, the liver is a prime storage site for certain vitamins (A, B_{12}, D, E, and K) and minerals (iron and copper), which are released from the liver when needed elsewhere in the body.
- ***Phagocytosis.*** The stellate reticuloendothelial (Kupffer) cells of the liver phagocytize aged red blood cells, white blood cells, and some bacteria.
- ***Activation of vitamin D.*** The skin, liver, and kidneys participate in synthesizing the active form of vitamin D.

The liver functions related to metabolism are discussed more fully in Chapter 25.

24.12 Small Intestine

OBJECTIVES

- **Describe** the location and structure of the small intestine.
- **Identify** the functions of the small intestine.

Most digestion and absorption of nutrients occur in a long tube called the **small intestine**. Because of this, its structure is specially adapted for these functions. Its length alone provides a large surface area for digestion and absorption, and that area is further increased by circular folds, villi, and microvilli. The small intestine begins at the pyloric sphincter of the stomach, coils through the central and inferior part of the abdominal cavity, and eventually opens into the large intestine. It averages 2.5 cm (1 in.) in diameter; its length is about 3 m (10 ft) in a living person and about 6.5 m (21 ft) in a cadaver due to the loss of smooth muscle tone after death.

Anatomy of the Small Intestine

The small intestine is divided into three regions (**Figure 24.19**). The first part of the small intestine is the **duodenum** (doo′-ō-DĒ-num or

FIGURE 24.19 Anatomy of the small intestine. (a) Regions of the small intestine are the duodenum, jejunum, and ileum. (b) Circular folds increase the surface area for digestion and absorption in the small intestine.

Most digestion and absorption occur in the small intestine.

Functions of the Small Intestine

1. Segmentations mix chyme with digestive juices and bring food into contact with mucosa for absorption; peristalsis propels chyme through small intestine.
2. Completes digestion of carbohydrates, proteins, and lipids; begins and completes digestion of nucleic acids.
3. Absorbs about 90% of nutrients and water that pass through digestive system.

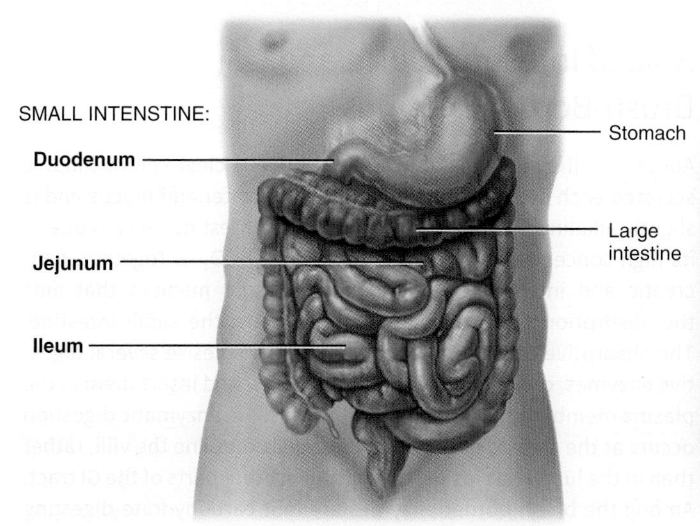

(a) Anterior view of external anatomy

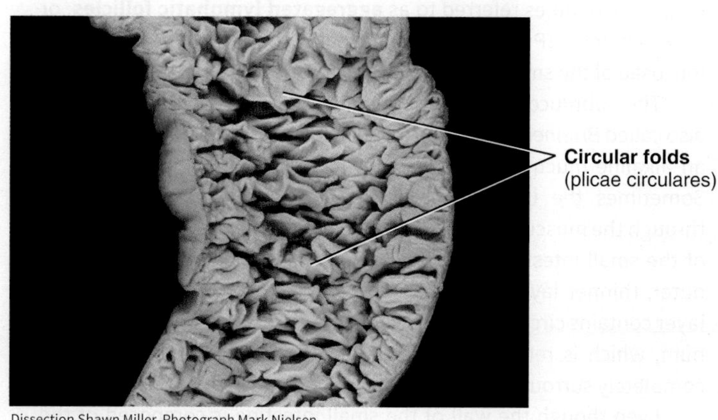

Dissection Shawn Miller, Photograph Mark Nielsen

(b) Internal anatomy of jejunum

Q Which portion of the small intestine is the longest?

doo-OD-e-num), the shortest region, and is retroperitoneal. It starts at the pyloric sphincter of the stomach and is in the form of a C-shaped tube that extends about 25 cm (10 in.) until it merges with the jejunum. *Duodenum* means "12"; it is so named because it is about as long as the width of 12 fingers. The **jejunum** (je-JOO-num) is the next portion and is about 1 m (3 ft) long and extends to the ileum. *Jejunum* means "empty," which is how it is found at death. The final and longest region of the small intestine, the **ileum** (IL-ē-um = twisted), measures about 2 m (6 ft) and joins the large intestine at a smooth muscle sphincter called the **ileocecal sphincter** (*valve*) (il′-ē-ō-SĒ-kal).

Histology of the Small Intestine

The wall of the small intestine is composed of the same four layers that make up most of the GI tract: mucosa, submucosa, muscularis, and serosa (**Figure 24.20b**). The mucosa is composed of a layer of epithelium, lamina propria, and muscularis mucosae. The epithelial layer of the small intestinal mucosa consists of simple columnar epithelium that contains many types of cells (**Figure 24.20c**). **Absorptive cells** of the epithelium contain enzymes that digest food and possess microvilli that absorb nutrients in small intestinal chyme. Also present in the epithelium are **goblet cells**, which secrete mucus. The small intestinal mucosa contains many deep crevices lined with glandular epithelium. Cells lining the crevices form the **intestinal glands**, or *crypts of Lieberkühn* (LĒ-ber-kūn), and secrete intestinal juice (to be discussed shortly). Besides absorptive cells and goblet cells, the intestinal glands also contain paneth cells and enteroendocrine cells. **Paneth cells** secrete lysozyme, a bactericidal enzyme, and are capable of phagocytosis. Paneth cells may have a role in regulating the microbial population in the small intestine. Three types of enteroendocrine cells are found in the intestinal glands of the small intestine: **S cells**, **CCK cells**, and **K cells**, which secrete the hormones **secretin** (se-KRĒ-tin), **cholecystokinin (CCK)** (kō-lē-sis′-tō-KĪN-in), and **glucose-dependent insulinotropic peptide (GIP)** (in-soo-lin′-ō-TRŌ-pik), respectively.

The lamina propria of the small intestinal mucosa contains areolar connective tissue and has an abundance of mucosa-associated lymphoid tissue (MALT). **Solitary lymphatic nodules** are most numerous in the distal part of the ileum (see **Figure 24.21c**). Groups of lymphatic nodules referred to as **aggregated lymphatic follicles**, or *Peyer's patches* (PĪ-erz), are also present in the ileum. The muscularis mucosae of the small intestinal mucosa consists of smooth muscle.

The submucosa of the duodenum contains **duodenal glands**, also called *Brunner's glands* (BRUN-erz) (**Figure 24.21a**), which secrete an alkaline mucus that helps neutralize gastric acid in the chyme. Sometimes the lymphatic tissue of the lamina propria extends through the muscularis mucosae into the submucosa. The muscularis of the small intestine consists of two layers of smooth muscle. The outer, thinner layer contains longitudinal fibers; the inner, thicker layer contains circular fibers. Except for a major portion of the duodenum, which is retroperitoneal, the serosa (or visceral peritoneum) completely surrounds the small intestine.

Even though the wall of the small intestine is composed of the same four basic layers as the rest of the GI tract, special structural features of the small intestine facilitate the process of digestion and absorption. These structural features include circular folds, villi, and microvilli. **Circular folds** or *plicae circulares* are folds of the mucosa and submucosa (see **Figures 24.19b** and **24.20a**). These permanent ridges, which are about 10 mm (0.4 in.) long, begin near the proximal portion of the duodenum and end at about the midportion of the ileum. Some extend all the way around the circumference of the intestine; others extend only part of the way around. Circular folds enhance absorption by increasing surface area and causing the chyme to spiral, rather than move in a straight line, as it passes through the small intestine.

Also present in the small intestine are **villi** (= tufts of hair), which are fingerlike projections of the mucosa that are 0.5–1 mm long (see **Figure 24.20b, c**). The large number of villi (20–40 per square millimeter) vastly increases the surface area of the epithelium available for absorption and digestion and gives the intestinal mucosa a velvety appearance. Each villus (singular form) is covered by epithelium and has a core of lamina propria; embedded in the connective tissue of the lamina propria are an arteriole, a venule, a blood capillary network, and a **lacteal** (LAK-tē-al = milky), which is a lymphatic capillary (see **Figure 24.20c**). Nutrients absorbed by the epithelial cells covering the villus pass through the wall of a capillary or a lacteal to enter blood or lymph, respectively.

Besides circular folds and villi, the small intestine also has **microvilli** (mī-krō-VIL-ī; *micro-* = small), which are projections of the apical (free) membrane of the absorptive cells. Each microvillus is a 1-μm-long cylindrical, membrane-covered projection that contains a bundle of 20–30 actin filaments. When viewed through a light microscope, the microvilli are too small to be seen individually; instead they form a fuzzy line, called the **brush border**, extending into the lumen of the small intestine (**Figure 24.21d**). There are an estimated 200 million microvilli per square millimeter of small intestine. Because the microvilli greatly increase the surface area of the plasma membrane, larger amounts of digested nutrients can diffuse into absorptive cells in a given period. The brush border also contains several brush-border enzymes that have digestive functions (discussed shortly).

Role of Intestinal Juice and Brush-Border Enzymes

About 1–2 liters (1–2 qt) of **intestinal juice**, a clear yellow fluid, is secreted each day. Intestinal juice contains water and mucus and is slightly alkaline (pH 7.6). The alkaline pH of intestinal juice is due to its high concentration of bicarbonate ions (HCO_3^-). Together, pancreatic and intestinal juices provide a liquid medium that aids the absorption of substances from chyme in the small intestine. The absorptive cells of the small intestine synthesize several digestive enzymes, called **brush-border enzymes**, and insert them in the plasma membrane of the microvilli. Thus, some enzymatic digestion occurs at the surface of the absorptive cells that line the villi, rather than in the lumen exclusively, as occurs in other parts of the GI tract. Among the brush-border enzymes are four carbohydrate-digesting enzymes called α-dextrinase, maltase, sucrase, and lactase; protein-digesting enzymes called peptidases (aminopeptidase and dipeptidase); and two types of nucleotide-digesting enzymes, nucleosidases

FIGURE 24.20 **Histology of the small intestine.**

Circular folds, villi, and microvilli increase the surface area of the small intestine for digestion and absorption.

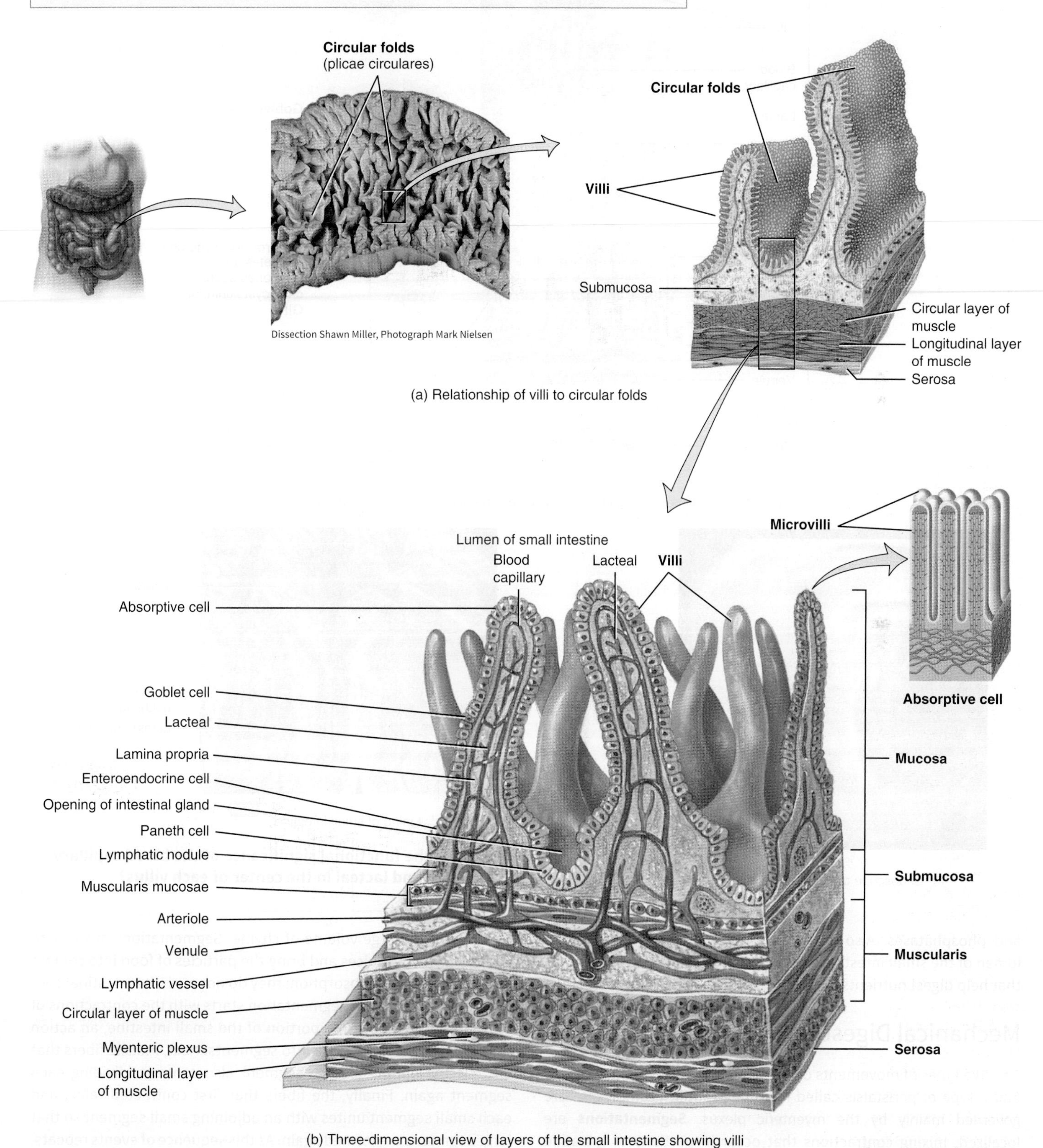

(a) Relationship of villi to circular folds

(b) Three-dimensional view of layers of the small intestine showing villi

Figure 24.20 *Continues*

FIGURE 24.20 Continued

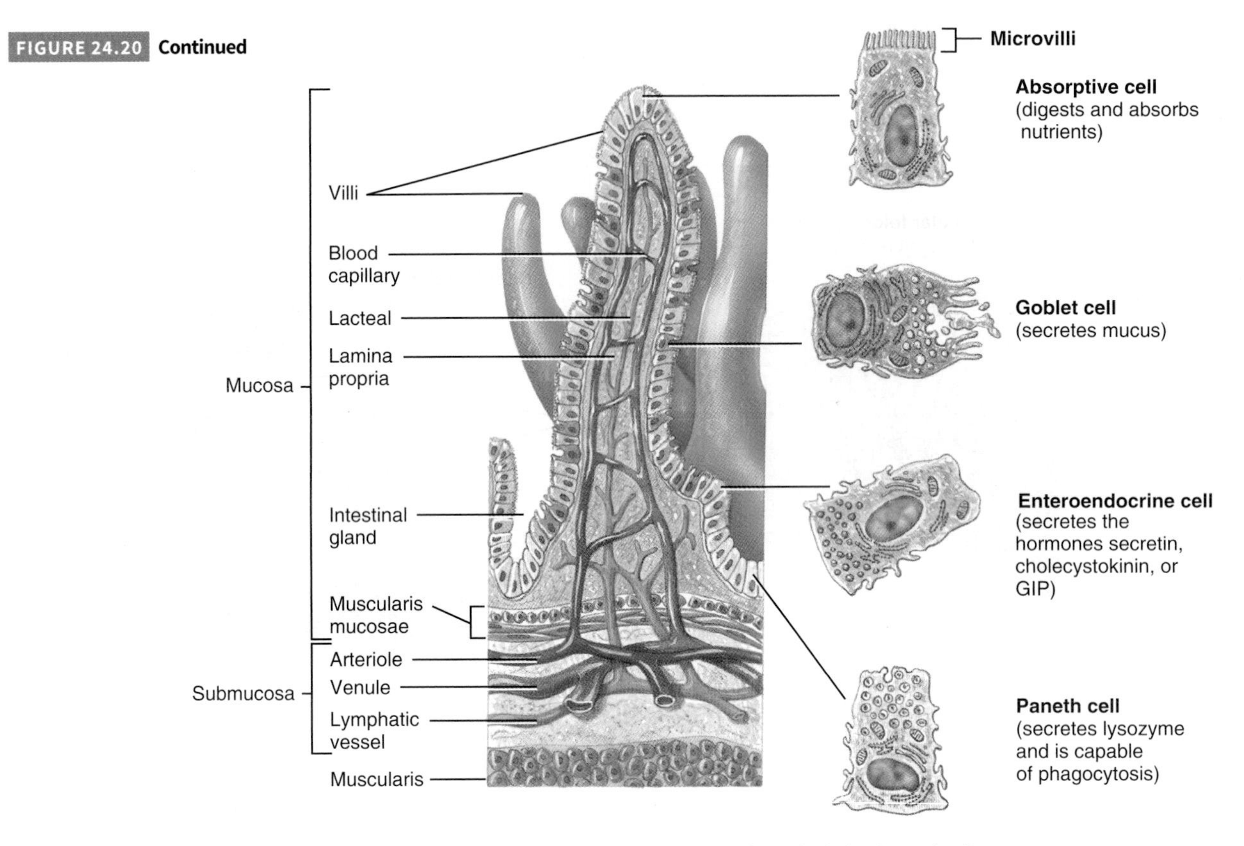

(c) Enlarged villus showing lacteal, capillaries, intestinal glands, and cell types

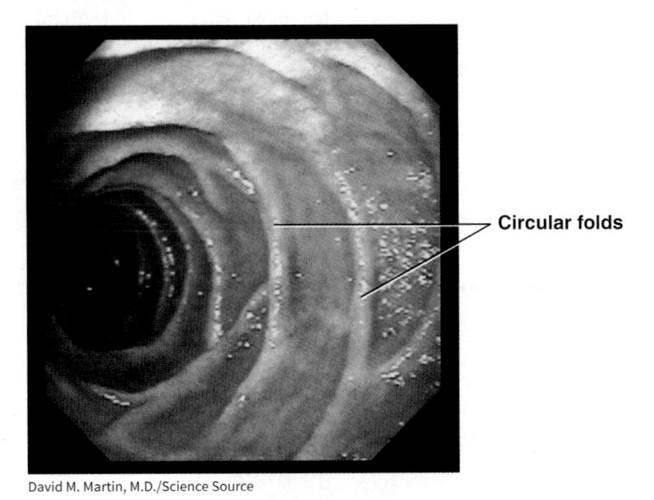

David M. Martin, M.D./Science Source

(d) Endoscope of healthy duodenum

Food
Villus
Nonciliated simple columnar epithelium
Lamina propria

SEM 80x
Lining of small intestine

Q What is the functional significance of the blood capillary network and lacteal in the center of each villus?

and phosphatases. Also, as absorptive cells slough off into the lumen of the small intestine, they break apart and release enzymes that help digest nutrients in the chyme.

Mechanical Digestion in the Small Intestine

The two types of movements of the small intestine—segmentations and a type of peristalsis called migrating motility complexes—are governed mainly by the myenteric plexus. **Segmentations** are localized, mixing contractions that occur in portions of intestine distended by a large volume of chyme. Segmentations mix chyme with the digestive juices and bring the particles of food into contact with the mucosa for absorption; they do not push the intestinal contents along the tract. A segmentation starts with the contractions of circular muscle fibers in a portion of the small intestine, an action that constricts the intestine into segments. Next, muscle fibers that encircle the middle of each segment also contract, dividing each segment again. Finally, the fibers that first contracted relax, and each small segment unites with an adjoining small segment so that large segments are formed again. As this sequence of events repeats,

FIGURE 24.21 Histology of the duodenum and ileum.

Microvilli in the small intestine contain several brush-border enzymes that help digest nutrients.

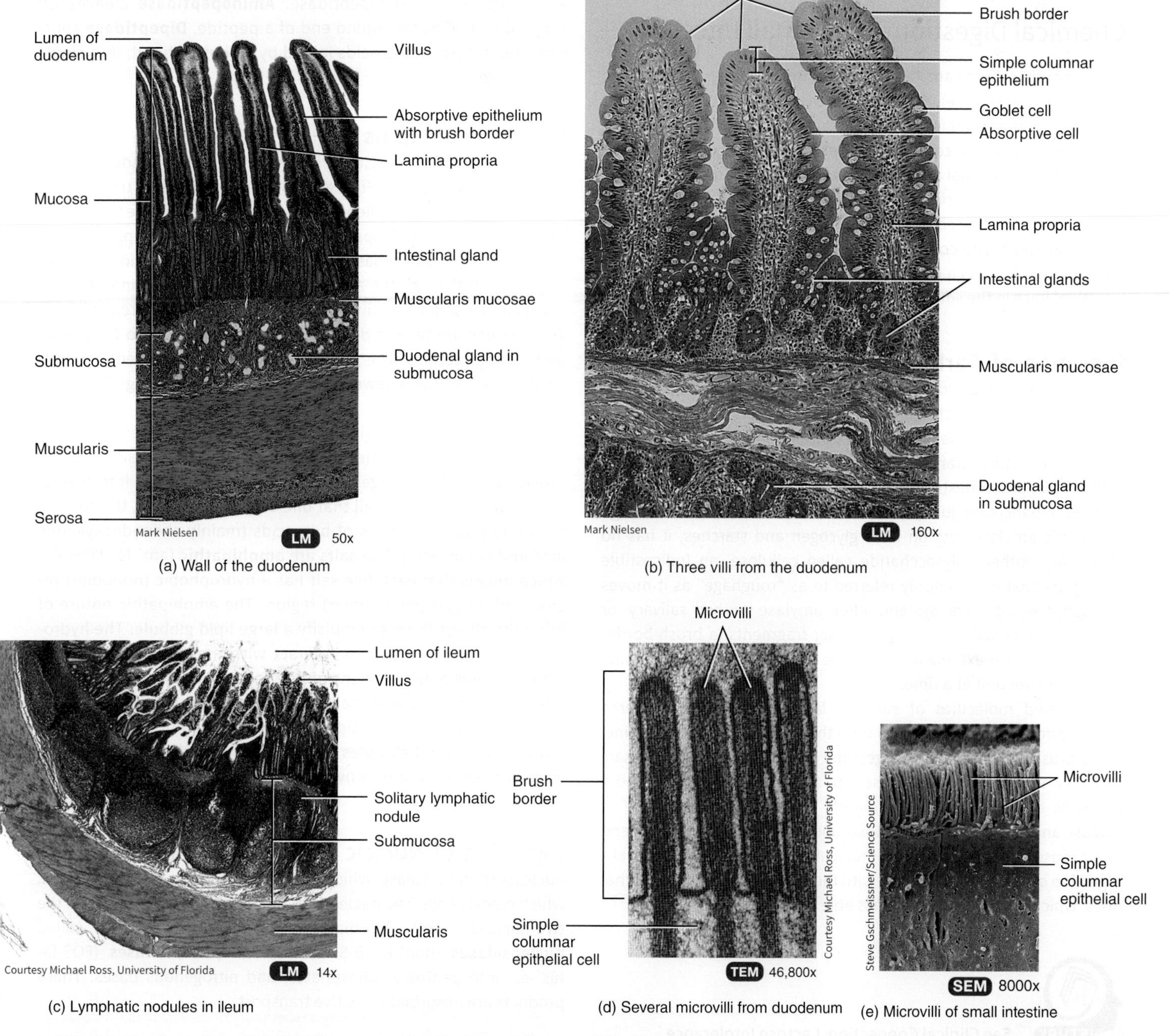

(a) Wall of the duodenum

(b) Three villi from the duodenum

(c) Lymphatic nodules in ileum

(d) Several microvilli from duodenum

(e) Microvilli of small intestine

Q What is the function of the fluid secreted by duodenal (Brunner's) glands?

the chyme sloshes back and forth. Segmentations occur most rapidly in the duodenum, about 12 times per minute, and progressively slow to about 8 times per minute in the ileum. This movement is similar to alternately squeezing the middle and then the ends of a capped tube of toothpaste.

After most of a meal has been absorbed, which lessens distension of the wall of the small intestine, segmentation stops and peristalsis begins. The type of peristalsis that occurs in the small intestine, termed a **migrating motility complex (MMC)**, begins in the lower portion of the stomach and pushes chyme forward along a short stretch of small intestine before dying out. The MMC slowly migrates down the small intestine, reaching the end of the ileum in 90–120 minutes. Then another MMC begins in the stomach. Altogether, chyme remains in the small intestine for 3–5 hours.

Chemical Digestion in the Small Intestine

In the mouth, salivary amylase converts starch (a polysaccharide) to maltose (a disaccharide), maltotriose (a trisaccharide), and α-dextrins (short-chain, branched fragments of starch with 5–10 glucose units). In the stomach, pepsin converts proteins to peptides (small fragments of proteins), and lingual and gastric lipases convert some triglycerides into fatty acids, diglycerides, and monoglycerides. Thus, chyme entering the small intestine contains partially digested carbohydrates, proteins, and lipids. The completion of the digestion of carbohydrates, proteins, and lipids is a collective effort of pancreatic juice, bile, and intestinal juice in the small intestine.

Digestion of Carbohydrates Even though the action of salivary amylase may continue in the stomach for a while, the acidic pH of the stomach destroys salivary amylase and ends its activity. Thus, only a few starches are broken down by the time chyme leaves the stomach. Those starches not already broken down into maltose, maltotriose, and α-dextrins are cleaved by **pancreatic amylase**, an enzyme in pancreatic juice that acts in the small intestine. Although pancreatic amylase acts on both glycogen and starches, it has no effect on another polysaccharide called cellulose, an indigestible plant fiber that is commonly referred to as "roughage" as it moves through the digestive system. After amylase (either salivary or pancreatic) has split starch into smaller fragments, a brush-border enzyme called **α-dextrinase** acts on the resulting α-dextrins, clipping off one glucose unit at a time.

Ingested molecules of sucrose, lactose, and maltose—three disaccharides—are not acted on until they reach the small intestine. Three brush-border enzymes digest the disaccharides into monosaccharides. **Sucrase** breaks sucrose into a molecule of glucose and a molecule of fructose; **lactase** digests lactose into a molecule of glucose and a molecule of galactose; and **maltase** splits maltose and maltotriose into two or three molecules of glucose, respectively. Digestion of carbohydrates ends with the production of monosaccharides, which the digestive system is able to absorb.

See Clinical Connection: Lactose Intolerance

Digestion of Proteins Recall that protein digestion starts in the stomach, where proteins are fragmented into peptides by the action of pepsin. Enzymes in pancreatic juice—trypsin, chymotrypsin, carboxypeptidase, and elastase—continue to break down proteins into peptides. Although all of these enzymes convert whole proteins into peptides, their actions differ somewhat because each splits peptide bonds between different amino acids. Trypsin, chymotrypsin, and elastase all cleave the peptide bond between a specific amino acid and its neighbor; carboxypeptidase splits off the amino acid at the carboxyl end of a peptide. Protein digestion is completed by two **peptidases** in the brush border: aminopeptidase and dipeptidase. **Aminopeptidase** cleaves off the amino acid at the amino end of a peptide. **Dipeptidase** splits dipeptides (two amino acids joined by a peptide bond) into single amino acids.

Digestion of Lipids The most abundant lipids in the diet are triglycerides, which consist of a molecule of glycerol bonded to three fatty acid molecules (see Figure 2.17). Enzymes that split triglycerides and phospholipids are called **lipases**. Recall that there are three types of lipases that can participate in lipid digestion: lingual lipase, gastric lipase, and pancreatic lipase. Although some lipid digestion occurs in the stomach through the action of lingual and gastric lipases, most occurs in the small intestine through the action of pancreatic lipase. Triglycerides are broken down by pancreatic lipase into fatty acids and monoglycerides. The liberated fatty acids can be either short-chain fatty acids (with fewer than 10–12 carbons) or long-chain fatty acids.

Before a large lipid globule containing triglycerides can be digested in the small intestine, it must first undergo emulsification—a process in which the large lipid globule is broken down into several small lipid globules. Recall that bile contains bile salts, the sodium salts and potassium salts of bile acids (mainly chenodeoxycholic acid and cholic acid). Bile salts are **amphipathic** (am′-fē-PATH-ik), which means that each bile salt has a hydrophobic (nonpolar) region and a hydrophilic (polar) region. The amphipathic nature of bile salts allows them to emulsify a large lipid globule: The hydrophobic regions of bile salts interact with the large lipid globule, while the hydrophilic regions of bile salts interact with the watery intestinal chyme. Consequently, the large lipid globule is broken apart into several small lipid globules, each about 1 μm in diameter. The small lipid globules formed from emulsification provide a large surface area that allows pancreatic lipase to function more effectively.

Digestion of Nucleic Acids Pancreatic juice contains two nucleases: ribonuclease, which digests RNA, and deoxyribonuclease, which digests DNA. The nucleotides that result from the action of the two nucleases are further digested by brush-border enzymes called **nucleosidases** (noo′-klē-ō-SĪ-dās-ez) and **phosphatases** (FOS-fa-tās′-ez) into pentoses, phosphates, and nitrogenous bases. These products are absorbed via active transport.

Absorption in the Small Intestine

All of the chemical and mechanical phases of digestion from the mouth through the small intestine are directed toward changing food into forms that can pass through the absorptive epithelial cells lining the mucosa and into the underlying blood and lymphatic vessels. These forms are monosaccharides (glucose, fructose, and galactose) from carbohydrates; single amino acids, dipeptides, and tripeptides from proteins; and fatty acids, glycerol, and monoglycerides from triglycerides. Passage of these digested nutrients from the gastrointestinal tract into the blood or lymph is called absorption.

Absorption of materials occurs via diffusion, facilitated diffusion, osmosis, and active transport. About 90% of all absorption of nutrients occurs in the small intestine; the other 10% occurs in the stomach and large intestine. Any undigested or unabsorbed material left in the small intestine passes on to the large intestine.

Absorption of Monosaccharides All carbohydrates are absorbed as monosaccharides. The capacity of the small intestine to absorb monosaccharides is huge—an estimated 120 grams per hour. As a result, all dietary carbohydrates that are digested normally are absorbed, leaving only indigestible cellulose and fibers in the feces. Monosaccharides pass from the lumen through the apical membrane via *facilitated diffusion* or *active transport*. Fructose, a monosaccharide found in fruits, is transported via *facilitated diffusion*; glucose and galactose are transported into absorptive cells of the villi via *secondary active transport* that is coupled to the active transport of Na^+ (**Figure 24.22a**). The transporter has binding sites for one glucose molecule and two sodium ions; unless all three sites are filled, neither substance is transported. Galactose competes with glucose to ride the same transporter. (Because both Na^+ and glucose or galactose move in the same direction, this is a *symporter*.) Monosaccharides then move out of the absorptive cells through their basolateral surfaces via *facilitated diffusion* and enter the capillaries of the villi (**Figure 24.22b**).

Absorption of Amino Acids, Dipeptides, and Tripeptides Most proteins are absorbed as amino acids via *active transport* processes that occur mainly in the duodenum and jejunum. About half of the absorbed amino acids are present in food; the other half come from the body itself as proteins in digestive juices and dead cells that slough off the mucosal surface! Normally, 95–98% of the protein present in the small intestine is digested and absorbed. Different transporters carry different types of amino acids. Some amino acids enter absorptive cells of the villi via Na^+-dependent secondary active transport processes that are similar to the glucose transporter; other amino acids are actively transported by themselves. At least one symporter brings in dipeptides and tripeptides together with H^+; the peptides then are hydrolyzed to single amino acids inside the absorptive cells. Amino acids move out of the absorptive cells via diffusion and enter capillaries of the villus (**Figure 24.22**). Both monosaccharides and amino acids are transported in the blood to the liver by way of the hepatic portal system. If not removed by hepatocytes, they enter the general circulation.

Absorption of Lipids and Bile Salts All dietary lipids are absorbed via *simple diffusion*. Adults absorb about 95% of the lipids present in the small intestine; due to their lower production of bile, newborn infants absorb only about 85% of lipids. As a result of their emulsification and digestion, triglycerides are mainly broken down into monoglycerides and fatty acids, which can be either short-chain fatty acids or long-chain fatty acids. Small short-chain fatty acids are hydrophobic, contain less than 10–12 carbon atoms, and are more water-soluble. Thus, they can dissolve in the watery intestinal chyme, pass through the absorptive cells via simple diffusion, and follow the same route taken by monosaccharides and amino acids into a blood capillary of a villus (**Figure 24.22a**).

Large short-chain fatty acids (with more than 10–12 carbon atoms), long-chain fatty acids, and monoglycerides are larger and hydrophobic, and since they are not water-soluble, they have difficulty being suspended in the watery environment of the intestinal chyme. Besides their role in emulsification, bile salts also help to make these large short-chain fatty acids, long-chain fatty acids, and monoglycerides more soluble. The bile salts in intestinal chyme surround them, forming tiny spheres called **micelles** (mī-SELZ = small morsels), each of which is 2–10 nm in diameter and includes 20–50 bile salt molecules (**Figure 24.22a**). Micelles are formed due to the amphipathic nature of bile salts: The hydrophobic regions of bile salts interact with the large short-chain fatty acids, long-chain fatty acids, and monoglycerides, and the hydrophilic regions of bile salts interact with the watery intestinal chyme. Once formed, the micelles move from the interior of the small intestinal lumen to the brush border of the absorptive cells. At that point, the large short-chain fatty acids, long-chain fatty acids, and monoglycerides diffuse out of the micelles into the absorptive cells, leaving the micelles behind in the chyme. The micelles continually repeat this ferrying function as they move from the brush border back through the chyme to the interior of the small intestinal lumen to pick up more of the large short-chain fatty acids, long-chain fatty acids, and monoglycerides. Micelles also solubilize other large hydrophobic molecules such as fat-soluble vitamins (A, D, E, and K) and cholesterol that may be present in intestinal chyme, and aid in their absorption. These fat-soluble vitamins and cholesterol molecules are packed in the micelles along with the long-chain fatty acids and monoglycerides.

Once inside the absorptive cells, long-chain fatty acids and monoglycerides are recombined to form triglycerides, which aggregate into globules along with phospholipids and cholesterol and become coated with proteins. These large spherical masses, about 80 nm in diameter, are called **chylomicrons** (kī-lō-MĪ-kronz). Chylomicrons leave the absorptive cell via exocytosis. Because they are so large and bulky, chylomicrons cannot enter blood capillaries—the pores in the walls of blood capillaries are too small. Instead, chylomicrons enter lacteals, which have much larger pores than blood capillaries. From lacteals, chylomicrons are transported by way of lymphatic vessels to the thoracic duct and enter the blood at the junction of the left internal jugular and left subclavian veins (**Figure 24.22b**). The hydrophilic protein coat that surrounds each chylomicron keeps the chylomicrons suspended in blood and prevents them from sticking to each other.

Within 10 minutes after absorption, about half of the chylomicrons have already been removed from the blood as they pass through blood capillaries in the liver and adipose tissue. This removal is accomplished by an enzyme attached to the apical surface of capillary endothelial cells, called **lipoprotein lipase**, that breaks down triglycerides in chylomicrons and other lipoproteins

FIGURE 24.22 Absorption of digested nutrients in the small intestine. For simplicity, all digested foods are shown in the lumen of the small intestine, even though some nutrients are digested by brush-border enzymes.

Long-chain fatty acids and monoglycerides are absorbed into lacteals; other products of digestion enter blood capillaries.

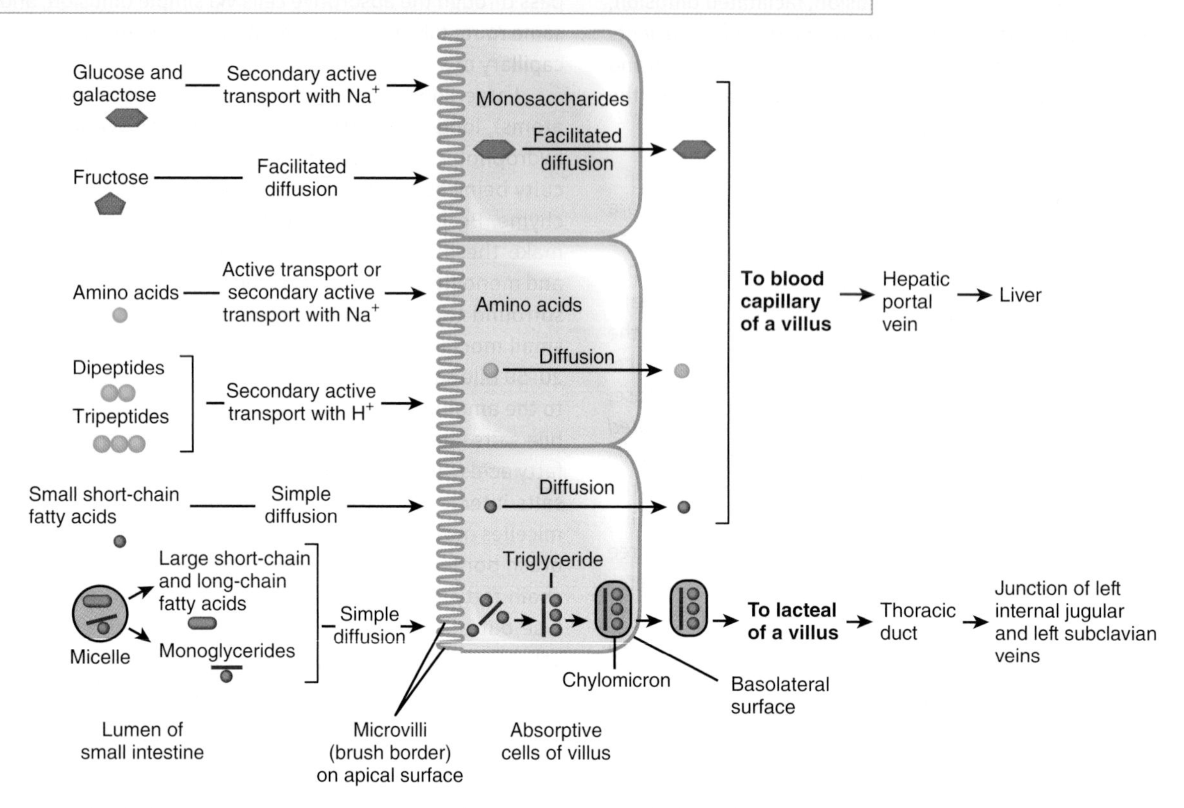

(a) Mechanisms for movement of nutrients through absorptive epithelial cells of villi

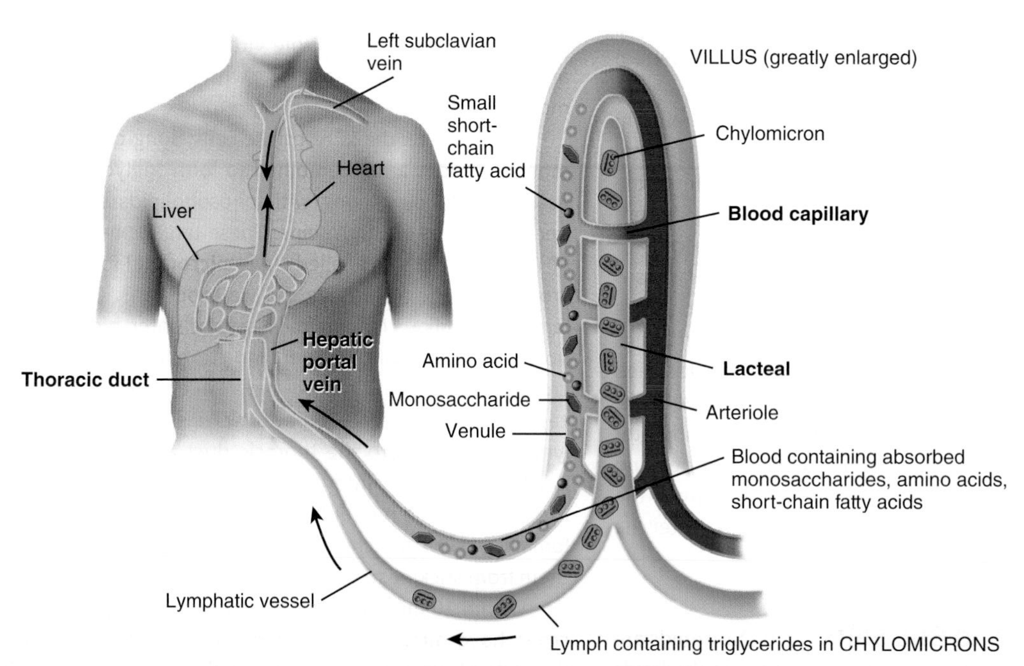

(b) Movement of absorbed nutrients into blood and lymph

Q A monoglyceride may be larger than an amino acid. Why can monoglycerides be absorbed by simple diffusion, but amino acids cannot?

into fatty acids and glycerol. The fatty acids diffuse into hepatocytes and adipose cells and combine with glycerol during resynthesis of triglycerides. Two or three hours after a meal, few chylomicrons remain in the blood.

After participating in the emulsification and absorption of lipids, most of the bile salts are reabsorbed by active transport in the final segment of the small intestine (ileum) and returned by the blood to the liver through the hepatic portal system for recycling. This cycle of bile salt secretion by hepatocytes into bile, reabsorption by the ileum, and resecretion into bile is called the **enterohepatic circulation** (en′-ter-ō-he-PAT-ik). Insufficient bile salts, due either to obstruction of the bile ducts or removal of the gallbladder, can result in the loss of up to 40% of dietary lipids in feces due to diminished lipid absorption. There are several benefits to including some healthy fats in the diet. For example, fats delay gastric emptying, which helps a person feel full. Fats also enhance the feeling of fullness by triggering the release of a hormone called cholecystokinin. Finally, fats are necessary for the absorption of fat-soluble vitamins.

Absorption of Electrolytes Many of the electrolytes absorbed by the small intestine come from gastrointestinal secretions, and some are part of ingested foods and liquids. Recall that electrolytes are compounds that separate into ions in water and conduct electricity. Sodium ions are actively transported out of absorptive cells by basolateral sodium–potassium pumps (Na^+–K^+ ATPases) after they have moved into absorptive cells via diffusion and secondary active transport. Thus, most of the sodium ions (Na^+) in gastrointestinal secretions are reclaimed and not lost in the feces. Negatively charged bicarbonate, chloride, iodide, and nitrate ions can passively follow Na^+ or be actively transported. Calcium ions also are absorbed actively in a process stimulated by calcitriol. Other electrolytes such as iron, potassium, magnesium, and phosphate ions also are absorbed via active transport mechanisms.

Absorption of Vitamins As you have just learned, the fat-soluble vitamins A, D, E, and K are included with ingested dietary lipids in micelles and are absorbed via simple diffusion. Most water-soluble vitamins, such as most B vitamins and vitamin C, also are absorbed via simple diffusion. Vitamin B_{12}, however, combines with intrinsic factor produced by the stomach, and the combination is absorbed in the ileum via an active transport mechanism.

Absorption of Water The total volume of fluid that enters the small intestine each day—about 9.3 liters (9.8 qt)—comes from ingestion of liquids (about 2.3 liters) and from various gastrointestinal secretions (about 7.0 liters). Figure 24.23 depicts the amounts of fluid ingested, secreted, absorbed, and excreted by the GI tract. The small intestine absorbs about 8.3 liters of the fluid; the remainder passes into the large intestine, where most of the rest of it—about 0.9 liter—is also absorbed. Only 0.1 liter (100 mL) of water is excreted in the feces each day.

FIGURE 24.23 Daily volumes of fluid ingested, secreted, absorbed, and excreted from the GI tract.

All water absorption in the GI tract occurs via osmosis.

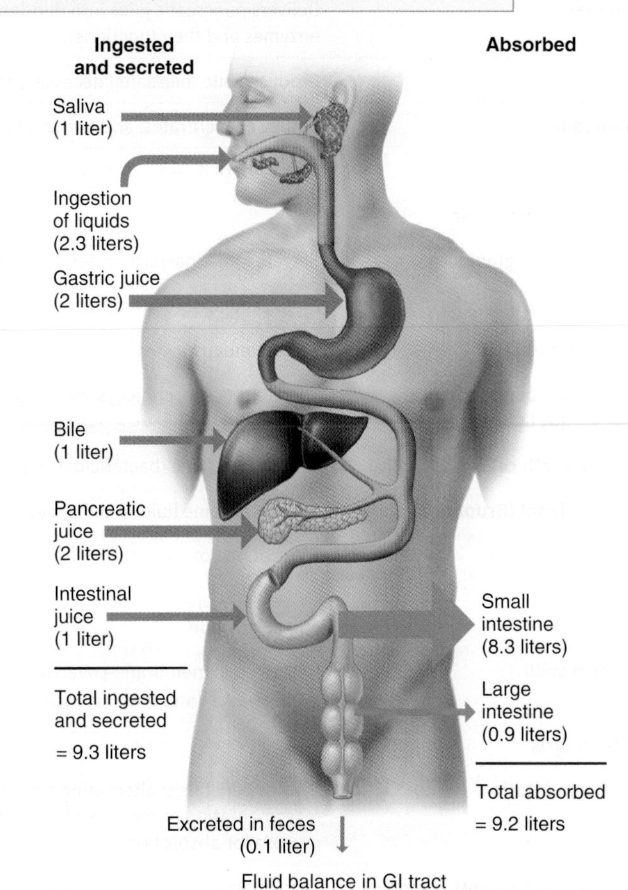

Q Which two organs of the digestive system secrete the most fluid?

See Clinical Connection: Absorption of Alcohol

All water absorption in the GI tract occurs via *osmosis* from the lumen of the intestines through absorptive cells and into blood capillaries. Because water can move across the intestinal mucosa in both directions, the absorption of water from the small intestine depends on the absorption of electrolytes and nutrients to maintain an osmotic balance with the blood. The absorbed electrolytes, monosaccharides, and amino acids establish a concentration gradient for water that promotes water absorption via osmosis.

Table 24.4 summarizes the digestive activities of the pancreas, liver, gallbladder, and small intestine and Table 24.5 summarizes the digestive enzymes and their functions in the digestive system.

TABLE 24.4 Summary of Digestive Activities in the Pancreas, Liver, Gallbladder, and Small Intestine

STRUCTURE	ACTIVITY
Pancreas	Delivers pancreatic juice into duodenum via pancreatic duct to assist absorption (see Table 24.5 for pancreatic enzymes and their functions).
Liver	Produces bile (bile salts) necessary for emulsification and absorption of lipids.
Gallbladder	Stores, concentrates, and delivers bile into duodenum via common bile duct.
Small intestine	Major site of digestion and absorption of nutrients and water in gastrointestinal tract.
Mucosa/submucosa	
Intestinal glands	Secrete intestinal juice to assist absorption.
Absorptive cells	Digest and absorb nutrients.
Goblet cells	Secrete mucus.
Enteroendocrine cells (S, CCK, K)	Secrete secretin, cholecystokinin, and glucose-dependent insulinotropic peptide.
Paneth cells	Secrete lysozyme (bactericidal enzyme), and phagocytosis.
Duodenal (Brunner's) glands	Secrete alkaline fluid to buffer stomach acids and mucus for protection and lubrication.
Circular folds	Folds of mucosa and submucosa that increase surface area for digestion and absorption.
Villi	Fingerlike projections of mucosa that are sites of absorption of digested food and increase surface area for digestion and absorption.
Microvilli	Microscopic, membrane-covered projections of absorptive epithelial cells that contain brush-border enzymes (listed in Table 24.5) and that increase surface area for digestion and absorption.
Muscularis	
Segmentation	Type of peristalsis: alternating contractions of circular smooth muscle fibers that produce segmentation and resegmentation of sections of small intestine; mixes chyme with digestive juices and brings food into contact with mucosa for absorption.
Migrating motility complex (MMC)	Type of peristalsis: waves of contraction and relaxation of circular and longitudinal smooth muscle fibers passing the length of the small intestine; moves chyme toward ileocecal sphincter.

24.13 Large Intestine

OBJECTIVE

- **Describe** the anatomy, histology, and functions of the large intestine.

The large intestine is the terminal portion of the GI tract. The overall functions of the large intestine are the completion of absorption, the production of certain vitamins, the formation of feces, and the expulsion of feces from the body. The medical specialty that deals with the diagnosis and treatment of disorders of the rectum and anus is called **proctology** (prok-TOL-ō-jē; *proct-* = rectum).

Anatomy of the Large Intestine

The **large intestine** (Figure 24.24), which is about 1.5 m (5 ft) long and 6.5 cm (2.5 in.) in diameter in living humans and cadavers, extends from the ileum to the anus. It is attached to the posterior abdominal wall by its mesocolon, which is a double layer of peritoneum (see Figure 24.5a). Structurally, the four major regions of the large intestine are the cecum, colon, rectum, and anal canal (Figure 24.24a).

The opening from the ileum into the large intestine is guarded by a fold of mucous membrane called the ileocecal sphincter (valve), which allows materials from the small intestine to pass into the large intestine. Hanging inferior to the ileocecal valve is the **cecum**, a small pouch about 6 cm (2.4 in.) long. Attached to the cecum is a twisted, coiled tube, measuring about 8 cm (3 in.) in length, called the **appendix** or *vermiform appendix* (VER-mi-form; *vermiform* = worm-shaped; *appendix* = appendage). The mesentery of the appendix, called the **mesoappendix** (mez-ō-a-PEN-diks), attaches the appendix to the inferior part of the mesentery of the ileum.

See Clinical Connection: Appendicitis

The open end of the cecum merges with a long tube called the **colon** (= food passage), which is divided into ascending, transverse, descending, and sigmoid portions. Both the ascending and descending

TABLE 24.5 **Summary of Digestive Enzymes**

ENZYME	SOURCE	SUBSTRATES	PRODUCTS
SALIVA			
Salivary amylase	Salivary glands.	Starches (polysaccharides).	Maltose (disaccharide), maltotriose (trisaccharide), and α-dextrins.
Lingual lipase	Lingual glands in tongue.	Triglycerides (fats and oils) and other lipids.	Fatty acids and diglycerides.
GASTRIC JUICE			
Pepsin (activated from pepsinogen by pepsin and hydrochloric acid)	Stomach chief cells.	Proteins.	Peptides.
Gastric lipase	Stomach chief cells.	Triglycerides (fats and oils).	Fatty acids and monoglycerides.
PANCREATIC JUICE			
Pancreatic amylase	Pancreatic acinar cells.	Starches (polysaccharides).	Maltose (disaccharide), maltotriose (trisaccharide), and α-dextrins.
Trypsin (activated from trypsinogen by enterokinase)	Pancreatic acinar cells.	Proteins.	Peptides.
Chymotrypsin (activated from chymotrypsinogen by trypsin)	Pancreatic acinar cells.	Proteins.	Peptides.
Elastase (activated from proelastase by trypsin)	Pancreatic acinar cells.	Proteins.	Peptides.
Carboxypeptidase (activated from procarboxypeptidase by trypsin)	Pancreatic acinar cells.	Amino acid at carboxyl end of peptides.	Amino acids and peptides.
Pancreatic lipase	Pancreatic acinar cells.	Triglycerides (fats and oils) that have been emulsified by bile salts.	Fatty acids and monoglycerides.
Nucleases			
Ribonuclease	Pancreatic acinar cells.	Ribonucleic acid.	Nucleotides.
Deoxyribonuclease	Pancreatic acinar cells.	Deoxyribonucleic acid.	Nucleotides.
BRUSH-BORDER ENZYMES IN MICROVILLI PLASMA MEMBRANE			
α-Dextrinase	Small intestine.	α-Dextrins.	Glucose.
Maltase	Small intestine.	Maltose.	Glucose.
Sucrase	Small intestine.	Sucrose.	Glucose and fructose.
Lactase	Small intestine.	Lactose.	Glucose and galactose.
Enterokinase	Small intestine.	Trypsinogen.	Trypsin.
Peptidases			
Aminopeptidase	Small intestine.	Amino acid at amino end of peptides.	Amino acids and peptides.
Dipeptidase	Small intestine.	Dipeptides.	Amino acids.
Nucleosidases and phosphatases	Small intestine.	Nucleotides.	Nitrogenous bases, pentoses, and phosphates.

colon are retroperitoneal; the transverse and sigmoid colon are not. True to its name, the **ascending colon** ascends on the right side of the abdomen, reaches the inferior surface of the liver, and turns abruptly to the left to form the **right colic** (*hepatic*) **flexure**. The colon continues across the abdomen to the left side as the **transverse colon**. It curves beneath the inferior end of the spleen on the left side as the **left colic** (*splenic*) **flexure** and passes inferiorly to the level of the iliac crest as the **descending colon**. The **sigmoid colon** (*sigm-* = S-shaped) begins near the left iliac crest, projects medially to the midline, and terminates as the rectum at about the level of the third sacral vertebra.

The **rectum** is about 15 cm (6 in.) in length and lies anterior to the sacrum and coccyx. The terminal 2–3 cm (1 in.) of the large intestine is called the **anal canal** (**Figure 24.24b**). The mucous membrane of the anal canal is arranged in longitudinal folds called **anal columns** that contain a network of arteries and veins. The opening of the anal canal to the exterior, called the **anus**, is guarded by an **internal anal sphincter** of smooth muscle (involuntary) and an **external anal sphincter** of skeletal muscle (voluntary). Normally these sphincters keep the anus closed except during the elimination of feces.

FIGURE 24.24 **Anatomy of the large intestine.**

The regions of the large intestine are the cecum, colon, rectum, and anal canal.

Functions of the Large Intestine

1. Haustral churning, peristalsis, and mass peristalsis drive contents of colon into rectum.
2. Bacteria in large intestine convert proteins to amino acids, break down amino acids, and produce some B vitamins and vitamin K.
3. Absorption of some water, ions, and vitamins.
4. Formation of feces.
5. Defecation (emptying rectum).

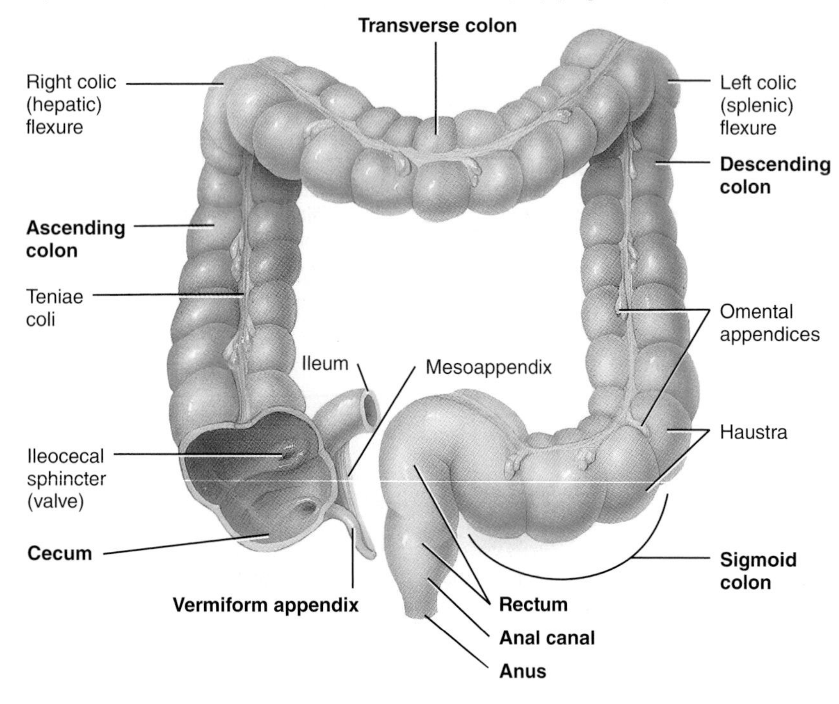

(a) Anterior view of large intestine showing major regions

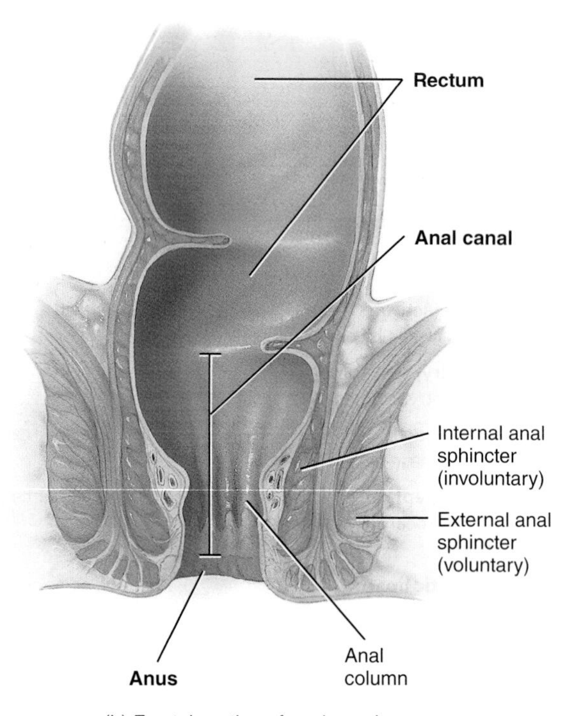

(b) Frontal section of anal canal

Q Which portions of the colon are retroperitoneal?

Histology of the Large Intestine

The wall of the large intestine contains the typical four layers found in the rest of the GI tract: mucosa, submucosa, muscularis, and serosa. The mucosa consists of simple columnar epithelium, lamina propria (areolar connective tissue), and muscularis mucosae (smooth muscle) (Figure 24.25a). The epithelium contains mostly absorptive and goblet cells (Figure 24.25b, d). The absorptive cells function primarily in water absorption; the goblet cells secrete mucus that lubricates the passage of the colonic contents. Both absorptive and goblet cells are located in long, straight, tubular intestinal glands (crypts of Lieberkühn) that extend the full thickness of the mucosa. Solitary lymphatic nodules are also found in the lamina propria of the mucosa and may extend through the muscularis mucosae into the submucosa. Compared to the small intestine, the mucosa of the large intestine does not have as many structural adaptations that increase surface area. There are no circular folds or villi; however, microvilli are present on the absorptive cells. Consequently, much more absorption occurs in the small intestine than in the large intestine.

The submucosa of the large intestine consists of areolar connective tissue. The muscularis consists of an external layer of longitudinal smooth muscle and an internal layer of circular smooth muscle. Unlike other parts of the GI tract, portions of the longitudinal muscles are thickened, forming three conspicuous bands called the **teniae coli** (TĒ-nē-ē KŌ-lī; *teniae* = flat bands) that run most of the length of the large intestine (see Figure 24.24a). The teniae coli are separated by portions of the wall with less or no longitudinal muscle. Tonic contractions of the bands gather the colon into a series of pouches called **haustra** (HAWS-tra = shaped like pouches; singular is *haustrum*), which give the colon a puckered appearance. A single layer of circular smooth muscle lies between teniae coli. The serosa of the large intestine is part of the visceral peritoneum. Small pouches of visceral peritoneum filled with fat are attached to teniae coli and are called **omental** (*fatty*) **appendices**.

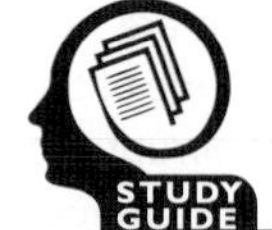

See Clinical Connection: Polyps in the Colon

Mechanical Digestion in the Large Intestine

The passage of chyme from the ileum into the cecum is regulated by the action of the ileocecal sphincter. Normally, the valve remains partially closed so that the passage of chyme into the cecum usually occurs slowly.

FIGURE 24.25 Histology of the large intestine.

Intestinal glands formed by simple columnar epithelial cells and goblet cells extend the full thickness of the mucosa.

David M. Martin/Science Source

Malignant tumor in colon

Lumen of large intestine

Openings of intestinal glands

Absorptive cell

Goblet cell

Intestinal gland

Lamina propria

Lymphatic nodule

Muscularis mucosae

Lymphatic vessel

Arteriole

Venule

Circular layer of muscle

Myenteric plexus

Longitudinal layer of muscle

Mucosa

Submucosa

Muscularis

Serosa

(a) Three-dimensional view of layers of the large intestine

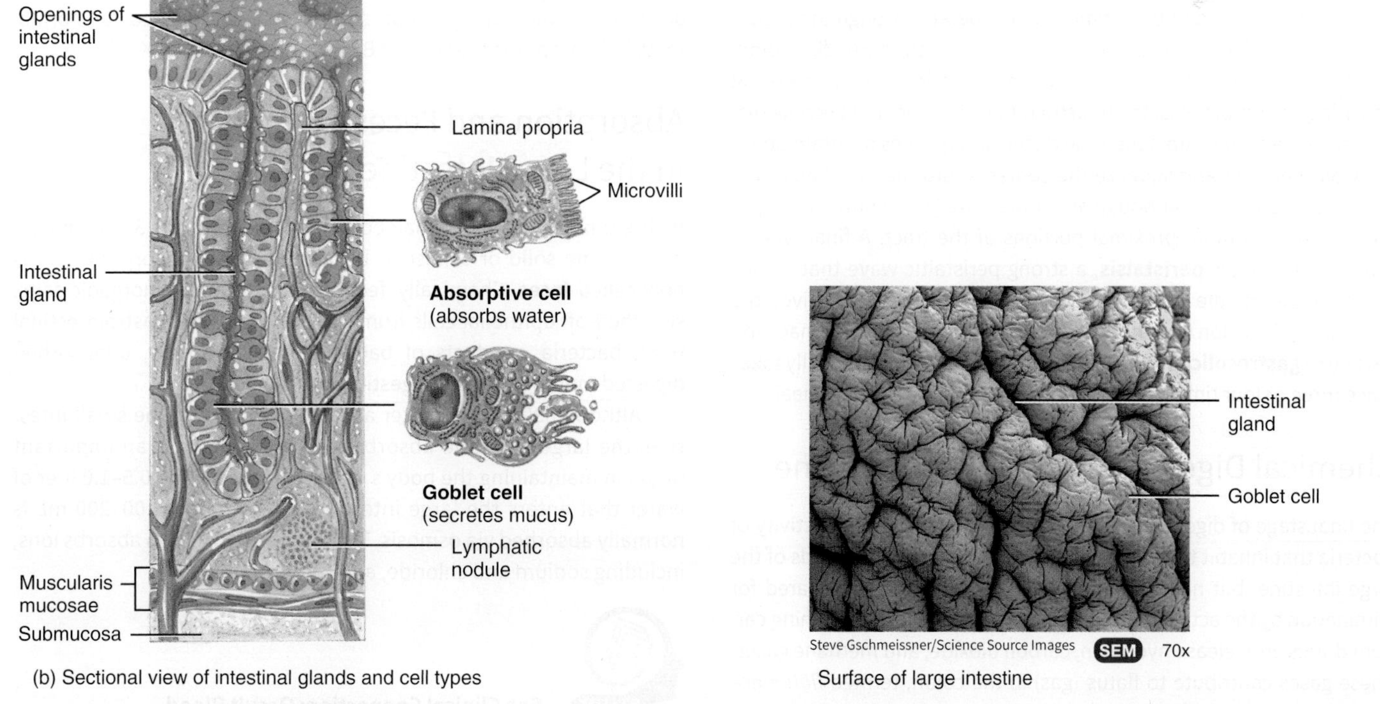

(b) Sectional view of intestinal glands and cell types

Surface of large intestine

Figure 24.25 *Continues*

FIGURE 24.25 **Continued**

Mucosa
Submucosa
Muscularis
Serosa
Lumen of large intestine
Lamina propria
Intestinal gland
Lymphatic nodule
Muscularis mucosae

Courtesy Michael Ross, University of Florida

LM 30x

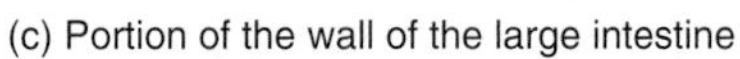
(c) Portion of the wall of the large intestine

Opening of intestinal gland
Lumen of large intestine
Absorptive cell
Goblet cell
Lamina propria
Intestinal gland

Courtesy Michael Ross, University of Florida

LM 300x

(d) Details of mucosa of large intestine

Q What is the function of the goblet cells in the large intestine?

Immediately after a meal, a **gastroileal reflex** (gas′-trō-IL-ē-al) intensifies peristalsis in the ileum and forces any chyme into the cecum. The hormone gastrin also relaxes the sphincter. Whenever the cecum is distended, the degree of contraction of the ileocecal sphincter intensifies.

Movements of the colon begin when substances pass the ileocecal sphincter. Because chyme moves through the small intestine at a fairly constant rate, the time required for a meal to pass into the colon is determined by gastric emptying time. As food passes through the ileocecal sphincter, it fills the cecum and accumulates in the ascending colon.

One movement characteristic of the large intestine is **haustral churning**. In this process, the haustra remain relaxed and become distended while they fill up. When the distension reaches a certain point, the walls contract and squeeze the contents into the next haustrum. Peristalsis also occurs, although at a slower rate (3–12 contractions per minute) than in more proximal portions of the tract. A final type of movement is **mass peristalsis**, a strong peristaltic wave that begins at about the middle of the transverse colon and quickly drives the contents of the colon into the rectum. Because food in the stomach initiates this **gastrocolic reflex** in the colon, mass peristalsis usually takes place three or four times a day, during or immediately after a meal.

Chemical Digestion in the Large Intestine

The final stage of digestion occurs in the colon through the activity of bacteria that inhabit the lumen. Mucus is secreted by the glands of the large intestine, but no enzymes are secreted. Chyme is prepared for elimination by the action of bacteria, which ferment any remaining carbohydrates and release hydrogen, carbon dioxide, and methane gases. These gases contribute to flatus (gas) in the colon, termed *flatulence* when it is excessive. Bacteria also convert any remaining proteins to amino acids and break down the amino acids into simpler substances: indole, skatole, hydrogen sulfide, and fatty acids. Some of the indole and skatole is eliminated in the feces and contributes to their odor; the rest is absorbed and transported to the liver, where these compounds are converted to less toxic compounds and excreted in the urine. Bacteria also decompose bilirubin to simpler pigments, including stercobilin, which gives feces their brown color. Bacterial products that are absorbed in the colon include several vitamins needed for normal metabolism, among them some B vitamins and vitamin K.

Absorption and Feces Formation in the Large Intestine

By the time chyme has remained in the large intestine 3–10 hours, it has become solid or semisolid because of water absorption and is now called **feces**. Chemically, feces consist of water, inorganic salts, sloughed-off epithelial cells from the mucosa of the gastrointestinal tract, bacteria, products of bacterial decomposition, unabsorbed digested materials, and indigestible parts of food.

Although 90% of all water absorption occurs in the small intestine, the large intestine absorbs enough to make it an important organ in maintaining the body's water balance. Of the 0.5–1.0 liter of water that enters the large intestine, all but about 100–200 mL is normally absorbed via osmosis. The large intestine also absorbs ions, including sodium and chloride, and some vitamins.

See Clinical Connection: Occult Blood

The Defecation Reflex

Mass peristaltic movements push fecal material from the sigmoid colon into the rectum. The resulting distension of the rectal wall stimulates stretch receptors, which initiates a **defecation reflex** that results in **defecation**, the elimination of feces from the rectum through the anus. The defecation reflex occurs as follows: In response to distension of the rectal wall, the receptors send sensory nerve impulses to the sacral spinal cord. Motor impulses from the cord travel along parasympathetic nerves back to the descending colon, sigmoid colon, rectum, and anus. The resulting contraction of the longitudinal rectal muscles shortens the rectum, thereby increasing the pressure within it. This pressure, along with voluntary contractions of the diaphragm and abdominal muscles, plus parasympathetic stimulation, opens the internal anal sphincter.

See Clinical Connection: Dietary Fiber

The external anal sphincter is voluntarily controlled. If it is voluntarily relaxed, defecation occurs and the feces are expelled through the anus; if it is voluntarily constricted, defecation can be postponed. Voluntary contractions of the diaphragm and abdominal muscles aid defecation by increasing the pressure within the abdomen, which pushes the walls of the sigmoid colon and rectum inward. If defecation does not occur, the feces back up into the sigmoid colon until the next wave of mass peristalsis stimulates the stretch receptors, again creating the urge to defecate. In infants, the defecation reflex causes automatic emptying of the rectum because voluntary control of the external anal sphincter has not yet developed.

The amount of bowel movements that a person has over a given period of time depends on various factors such as diet, health, and stress. The normal range of bowel activity varies from two or three bowel movements per day to three or four bowel movements per week.

Diarrhea (dī-a-RĒ-a; *dia-* = through; *-rrhea* = flow) is an increase in the frequency, volume, and fluid content of the feces caused by increased motility of and decreased absorption by the intestines. When chyme passes too quickly through the small intestine and feces pass too quickly through the large intestine, there is not enough time for absorption. Frequent diarrhea can result in dehydration and electrolyte imbalances. Excessive motility may be caused by lactose intolerance, stress, and microbes that irritate the gastrointestinal mucosa.

Constipation (kon-sti-PĀ-shun; *con-* = together; *-stip-* = to press) refers to infrequent or difficult defecation caused by decreased motility of the intestines. Because the feces remain in the colon for prolonged periods, excessive water absorption occurs, and the feces become dry and hard. Constipation may be caused by poor habits (delaying defecation), spasms of the colon, insufficient fiber in the diet, inadequate fluid intake, lack of exercise, emotional stress, and certain drugs. A common treatment is a mild laxative, such as milk of magnesia, which induces defecation. However, many physicians maintain that laxatives are habit-forming, and that adding fiber to the diet, increasing the amount of exercise, and increasing fluid intake are safer ways of controlling this common problem.

Table 24.6 summarizes the digestive activities in the large intestine, and **Table 24.7** summarizes the functions of all digestive system organs.

TABLE 24.6 Summary of Digestive Activities in the Large Intestine

STRUCTURE	ACTIVITY	FUNCTION(S)
Lumen	Bacterial activity.	Breaks down undigested carbohydrates, proteins, and amino acids into products that can be expelled in feces or absorbed and detoxified by liver; synthesizes certain B vitamins and vitamin K.
Mucosa	Secretes mucus.	Lubricates colon; protects mucosa.
	Absorption.	Water absorption solidifies feces and contributes to body's water balance; solutes absorbed include ions and some vitamins.
Muscularis	Haustral churning.	Moves contents from haustrum to haustrum by muscular contractions.
	Peristalsis.	Moves contents along length of colon by contractions of circular and longitudinal muscles.
	Mass peristalsis.	Forces contents into sigmoid colon and rectum.
	Defecation reflex.	Eliminates feces by contractions in sigmoid colon and rectum.

24.14 Phases of Digestion

OBJECTIVE

- **Explain** the three phases of digestion.
- **Describe** the major hormones regulating digestive activities.

Digestive activities occur in three overlapping phases: the cephalic phase, the gastric phase, and the intestinal phase.

Cephalic Phase

During the **cephalic phase** of digestion, the smell, sight, thought, or initial taste of food activates neural centers in the cerebral cortex, hypothalamus, and brain stem. The brain stem then activates the facial (VII), glossopharyngeal (IX), and vagus (X) nerves. The facial and glossopharyngeal nerves stimulate the salivary glands to secrete saliva, while the vagus nerves stimulate the gastric glands to secrete gastric juice. The purpose of the cephalic phase of digestion is to prepare the mouth and stomach for food that is about to be eaten.

Gastric Phase

Once food reaches the stomach, the **gastric phase** of digestion begins. Neural and hormonal mechanisms regulate the gastric phase of digestion to promote gastric secretion and gastric motility.

- ***Neural regulation.*** Food of any kind distends the stomach and stimulates stretch receptors in its walls. Chemoreceptors in the stomach

TABLE 24.7 Summary of Organs of the Digestive System and Their Functions

ORGAN	FUNCTION(S)
Tongue	Maneuvers food for mastication, shapes food into a bolus, maneuvers food for deglutition, detects sensations for taste, and initiates digestion of triglycerides.
Salivary glands	Saliva produced by these glands softens, moistens, and dissolves foods; cleanses mouth and teeth; initiates the digestion of starch.
Teeth	Cut, tear, and pulverize food to reduce solids to smaller particles for swallowing.
Pancreas	Pancreatic juice buffers acidic gastric juice in chyme, stops the action of pepsin from the stomach, creates the proper pH for digestion in the small intestine, and participates in the digestion of carbohydrates, proteins, triglycerides, and nucleic acids.
Liver	Produces bile, which is required for the emulsification and absorption of lipids in the small intestine.
Gallbladder	Stores and concentrates bile and releases it into the small intestine.
Mouth	See the functions of the tongue, salivary glands, and teeth, all of which are in the mouth. Additionally, the lips and cheeks keep food between the teeth during mastication, and buccal glands lining the mouth produce saliva.
Pharynx	Receives a bolus from the oral cavity and passes it into the esophagus.
Esophagus	Receives a bolus from the pharynx and moves it into the stomach; this requires relaxation of the upper esophageal sphincter and secretion of mucus.
Stomach	Mixing waves combine saliva, food, and gastric juice, which activates pepsin, initiates protein digestion, kills microbes in food, helps absorb vitamin B_{12}, contracts the lower esophageal sphincter, increases stomach motility, relaxes the pyloric sphincter, and moves chyme into the small intestine.
Small intestine	Segmentation mixes chyme with digestive juices; peristalsis propels chyme toward the ileocecal sphincter; digestive secretions from the small intestine, pancreas, and liver complete the digestion of carbohydrates, proteins, lipids, and nucleic acids; circular folds, villi, and microvilli help absorb about 90% of digested nutrients.
Large intestine	Haustral churning, peristalsis, and mass peristalsis drive the colonic contents into the rectum; bacteria produce some B vitamins and vitamin K; absorption of some water, ions, and vitamins occurs; defecation.

monitor the pH of the stomach chyme. When the stomach walls are distended or pH increases because proteins have entered the stomach and buffered some of the stomach acid, the stretch receptors and chemoreceptors are activated, and a neural negative feedback loop is set in motion (Figure 24.26). From the stretch receptors and chemoreceptors, nerve impulses propagate to the submucosal plexus, where they activate parasympathetic and enteric neurons. The resulting nerve impulses cause waves of peristalsis and continue to stimulate the flow of gastric juice from gastric glands. The peristaltic waves mix the food with gastric juice; when the waves become strong enough, a small quantity of chyme undergoes gastric emptying into the duodenum. The pH of the stomach chyme decreases (becomes more acidic) and the distension of the stomach walls lessens because chyme has passed into the small intestine, suppressing secretion of gastric juice.

- ***Hormonal regulation.*** Gastric secretion during the gastric phase is also regulated by the hormone **gastrin**. Gastrin is released from the G cells of the gastric glands in response to several stimuli: distension of the stomach by chyme, partially digested proteins in chyme, the high pH of chyme due to the presence of food in the stomach, caffeine in gastric chyme, and acetylcholine released from parasympathetic neurons. Once it is released, gastrin enters the bloodstream, makes a round-trip through the body, and finally reaches its target organs in the digestive system. Gastrin stimulates gastric glands to secrete large amounts of gastric juice. It also strengthens the contraction of the lower esophageal sphincter to prevent reflux of acid chyme into the esophagus, increases motility of the stomach, and relaxes the pyloric sphincter, which promotes gastric emptying. Gastrin secretion is inhibited when the pH of gastric juice drops below 2.0 and is stimulated when the pH rises. This negative feedback mechanism helps provide an optimal low pH for the functioning of pepsin, the killing of microbes, and the denaturing of proteins in the stomach.

Intestinal Phase

The **intestinal phase** of digestion begins once food enters the small intestine. In contrast to reflexes initiated during the cephalic and gastric phases, which stimulate stomach secretory activity and motility, those occurring during the intestinal phase have inhibitory effects that slow the exit of chyme from the stomach. This prevents the duodenum from being overloaded with more chyme than it can handle. In addition, responses occurring during the intestinal phase promote the continued digestion of foods that have reached the small intestine. These activities of the intestinal phase of digestion are regulated by neural and hormonal mechanisms.

- ***Neural regulation.*** Distension of the duodenum by the presence of chyme causes the **enterogastric reflex** (en′-ter-ō-GAS-trik). Stretch receptors in the duodenal wall send nerve impulses to the medulla oblongata, where they inhibit parasympathetic stimulation and stimulate the sympathetic nerves to the stomach. As a result, gastric motility is inhibited and there is an increase in the contraction of the pyloric sphincter, which decreases gastric emptying.
- ***Hormonal regulation.*** The intestinal phase of digestion is mediated by two major hormones secreted by the small intestine: cholecystokinin and secretin. Cholecystokinin (CCK) is secreted by the CCK cells of intestinal glands in the small intestine in response to chyme containing amino acids from partially digested proteins and fatty acids from

FIGURE 24.26 Neural negative feedback regulation of the pH of gastric juice and gastric motility during the gastric phase of digestion.

Food entering the stomach stimulates secretion of gastric juice and causes vigorous waves of peristalsis.

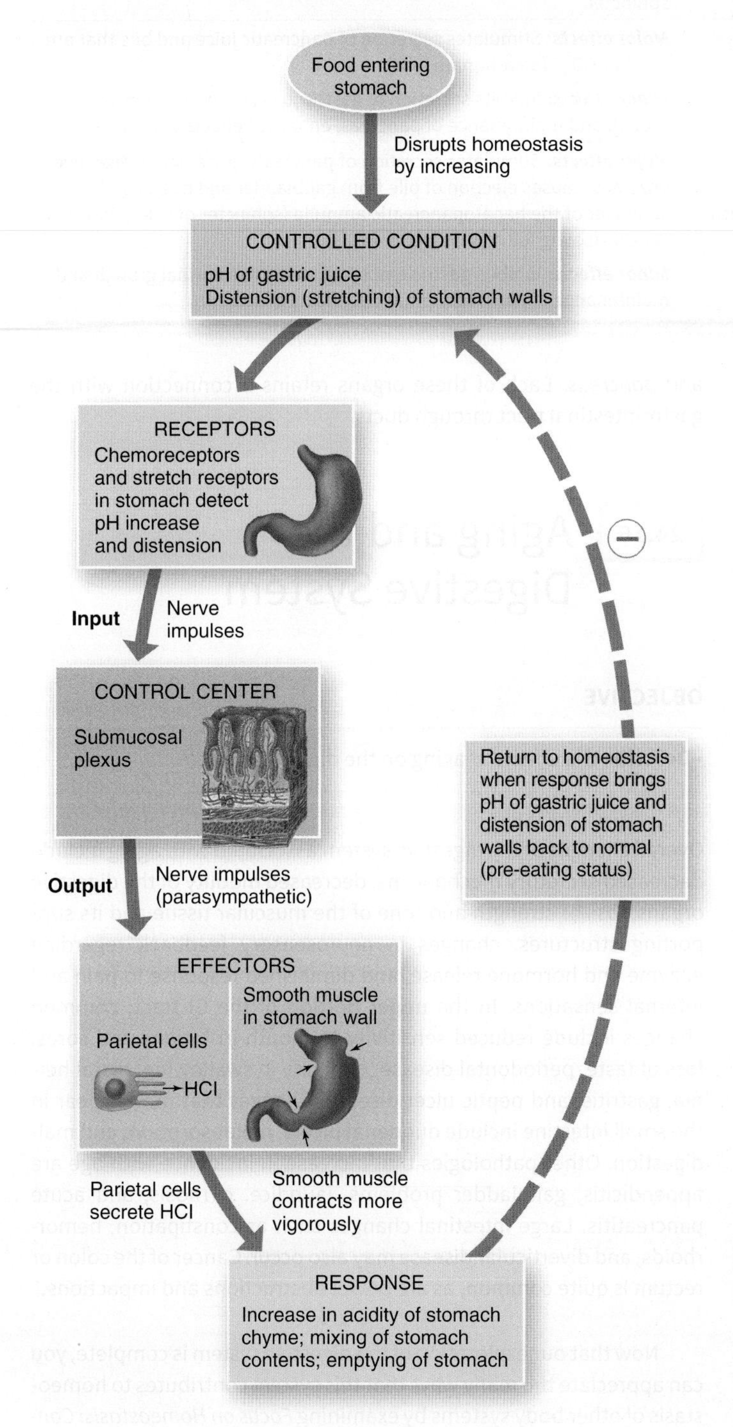

Q Why does food initially cause the pH of the gastric juice to rise?

partially digested triglycerides. CCK stimulates secretion of pancreatic juice that is rich in digestive enzymes. It also causes contraction of the wall of the gallbladder, which squeezes stored bile out of the gallbladder into the cystic duct and through the common bile duct. In addition, CCK causes relaxation of the sphincter of the hepatopancreatic ampulla (sphincter of Oddi), which allows pancreatic juice and bile to flow into the duodenum. CCK also slows gastric emptying by promoting contraction of the pyloric sphincter, produces satiety (a feeling of fullness) by acting on the hypothalamus in the brain, promotes normal growth and maintenance of the pancreas, and enhances the effects of secretin. Acidic chyme entering the duodenum stimulates the release of **secretin** from the S cells of the intestinal glands in the small intestine. In turn, secretin stimulates the flow of pancreatic juice that is rich in bicarbonate (HCO_3^-) ions to buffer the acidic chyme that enters the duodenum from the stomach. In addition to this major effect, secretin inhibits secretion of gastric juice, promotes normal growth and maintenance of the pancreas, and enhances the effects of CCK. Overall, secretin causes buffering of acid in chyme that reaches the duodenum and slows production of acid in the stomach.

Table 24.8 summarizes the major hormones that control digestion.

Other Hormones of the Digestive System

Besides gastrin, CCK, and secretin, there are many other hormones of the digestive system. For example, **ghrelin**, which is secreted by the stomach, plays a role in increasing appetite. **Glucose-dependent insulinotropic peptide (GIP)** and **glucagon-like peptide (GLP)**, which are secreted by the small intestine in response to the presence of food, stimulate the release of insulin from the pancreas, thereby increasing the blood glucose concentration. GIP and GLP are collectively referred to as **incretins**; they provide a type of feedforward control that anticipates the increase in blood glucose occurring after a typical meal. At least 10 other so-called gut hormones are secreted by and have effects on the GI tract. They include **motilin**, **substance P**, and **bombesin**, which stimulate motility of the intestines; **vasoactive intestinal polypeptide (VIP)**, which stimulates secretion of ions and water by the intestines and inhibits gastric acid secretion; **gastrin-releasing peptide**, which stimulates release of gastrin; and **somatostatin**, which inhibits gastrin release. Some of these hormones are thought to act as local hormones (paracrines), whereas others are secreted into the blood or even into the lumen of the GI tract. The physiological roles of these and other gut hormones are still under investigation.

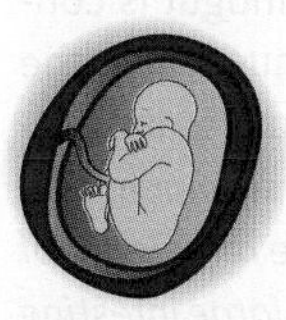

24.15 Development of the Digestive System

OBJECTIVE

- **Describe** the development of the digestive system.

TABLE 24.8 Major Hormones That Control Digestion

HORMONE	STIMULUS AND SITE OF SECRETION	ACTIONS
Gastrin	Distension of stomach, partially digested proteins and caffeine in stomach, and high pH of stomach chyme stimulate gastrin secretion by enteroendocrine G cells, located mainly in mucosa of pyloric antrum of stomach.	***Major effects:*** Promotes secretion of gastric juice, increases gastric motility, promotes growth of gastric mucosa. ***Minor effects:*** Constricts lower esophageal sphincter, relaxes pyloric sphincter.
Secretin	Acidic (high H^+ level) chyme that enters small intestine stimulates secretion of secretin by enteroendocrine S cells in the mucosa of duodenum.	***Major effects:*** Stimulates secretion of pancreatic juice and bile that are rich in HCO_3^- (bicarbonate ions). ***Minor effects:*** Inhibits secretion of gastric juice, promotes normal growth and maintenance of pancreas, enhances effects of CCK.
Cholecystokinin (CCK)	Partially digested proteins (amino acids), triglycerides, and fatty acids that enter small intestine stimulate secretion of CCK by enteroendocrine CCK cells in mucosa of small intestine; CCK is also released in brain.	***Major effects:*** Stimulates secretion of pancreatic juice rich in digestive enzymes, causes ejection of bile from gallbladder and opening of sphincter of the hepatopancreatic ampulla (sphincter of Oddi), induces satiety (feeling full to satisfaction). ***Minor effects:*** Inhibits gastric emptying, promotes normal growth and maintenance of pancreas, enhances effects of secretin.

During the fourth week of development, the cells of the **endoderm** form a cavity called the **primitive gut**, the forerunner of the gastrointestinal tract (see Figure 29.12b). Soon afterward the mesoderm forms and splits into two layers (somatic and splanchnic), as shown in Figure 29.9d. The splanchnic mesoderm associates with the endoderm of the primitive gut; as a result, the primitive gut has a double-layered wall. The **endodermal layer** gives rise to the *epithelial lining* and *glands* of most of the gastrointestinal tract; the **mesodermal layer** produces the *smooth muscle* and *connective tissue* of the tract.

The primitive gut elongates and differentiates into an anterior **foregut**, an intermediate **midgut**, and a posterior **hindgut** (see Figure 29.12c). Until the fifth week of development, the midgut opens into the yolk sac; after that time, the yolk sac constricts and detaches from the midgut, and the midgut seals. In the region of the foregut, a depression consisting of ectoderm, the **stomodeum** (stō-mō-DĒ-um), appears (see Figure 29.12d), which develops into the *oral cavity.* The **oropharyngeal membrane** (or′-ō-fa-RIN-jē-al) is a depression of fused ectoderm and endoderm on the surface of the embryo that separates the foregut from the stomodeum. The membrane ruptures during the fourth week of development, so that the foregut is continuous with the outside of the embryo through the oral cavity. Another depression consisting of ectoderm, the **proctodeum** (prok-tō-DĒ-um), forms in the hindgut and goes on to develop into the *anus* (see Figure 29.12d). The **cloacal membrane** (klō-Ā-kul) is a fused membrane of ectoderm and endoderm that separates the hindgut from the proctodeum. After it ruptures during the seventh week, the hindgut is continuous with the outside of the embryo through the anus. Thus, the gastrointestinal tract forms a continuous tube from mouth to anus.

The foregut develops into the *pharynx*, *esophagus*, *stomach*, and *part of the duodenum*. The midgut is transformed into the *remainder of the duodenum*, the *jejunum*, the *ileum*, and *portions of the large intestine* (cecum, appendix, ascending colon, and most of the transverse colon). The hindgut develops into the *remainder of the large intestine*, except for a portion of the anal canal that is derived from the proctodeum.

As development progresses, the endoderm at various places along the foregut develops into hollow buds that grow into the mesoderm. These buds will develop into the *salivary glands*, *liver*, *gallbladder*, and *pancreas.* Each of these organs retains a connection with the gastrointestinal tract through ducts.

24.16 Aging and the Digestive System

OBJECTIVE

- **Describe** the effects of aging on the digestive system.

Overall changes of the digestive system associated with aging include decreased secretory mechanisms, decreased motility of the digestive organs, loss of strength and tone of the muscular tissue and its supporting structures, changes in neurosensory feedback regarding enzyme and hormone release, and diminished response to pain and internal sensations. In the upper portion of the GI tract, common changes include reduced sensitivity to mouth irritations and sores, loss of taste, periodontal disease, difficulty in swallowing, hiatal hernia, gastritis, and peptic ulcer disease. Changes that may appear in the small intestine include duodenal ulcers, malabsorption, and maldigestion. Other pathologies that increase in incidence with age are appendicitis, gallbladder problems, jaundice, cirrhosis, and acute pancreatitis. Large intestinal changes such as constipation, hemorrhoids, and diverticular disease may also occur. Cancer of the colon or rectum is quite common, as are bowel obstructions and impactions.

• • •

Now that our exploration of the digestive system is complete, you can appreciate the many ways that this system contributes to homeostasis of other body systems by examining *Focus on Homeostasis: Contributions of the Digestive System*. Next, in Chapter 25, you will discover how the nutrients absorbed by the GI tract enter into metabolic reactions in the body tissues.

FOCUS on HOMEOSTASIS

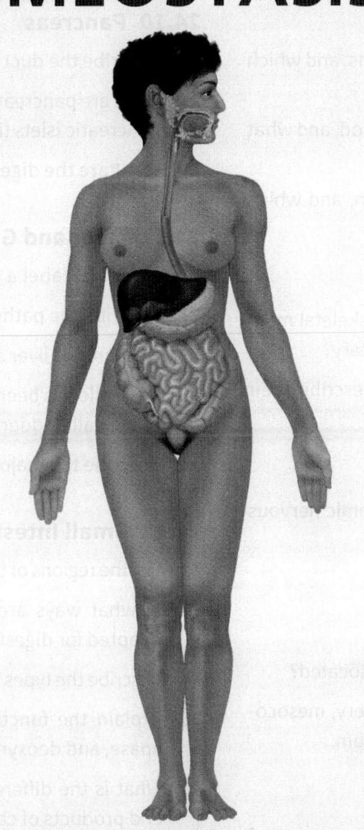

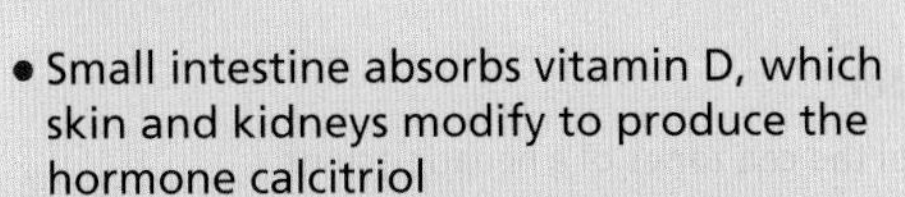

INTEGUMENTARY SYSTEM

- Small intestine absorbs vitamin D, which skin and kidneys modify to produce the hormone calcitriol
- Excess dietary calories are stored as triglycerides in adipose cells in dermis and subcutaneous layer

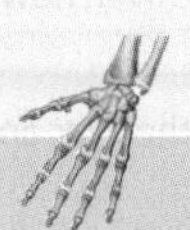

SKELETAL SYSTEM

- Small intestine absorbs dietary calcium and phosphorus salts needed to build bone extracellular matrix

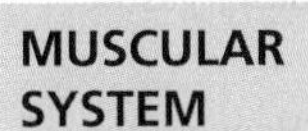

MUSCULAR SYSTEM

- Liver can convert lactic acid (produced by muscles during exercise) to glucose

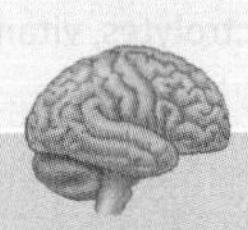

NERVOUS SYSTEM

- Gluconeogenesis (synthesis of new glucose molecules) in liver plus digestion and absorption of dietary carbohydrates provide glucose, needed for ATP production by neurons

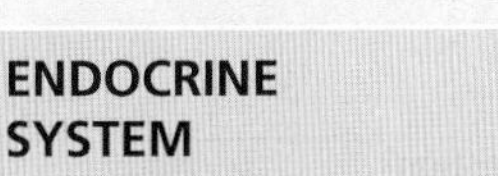

ENDOCRINE SYSTEM

- Liver inactivates some hormones, ending their activity
- Pancreatic islets release insulin and glucagon
- Cells in mucosa of stomach and small intestine release hormones that regulate digestive activities
- Liver produces angiotensinogen

CONTRIBUTIONS OF THE DIGESTIVE SYSTEM FOR ALL BODY SYSTEMS

- The digestive system breaks down dietary nutrients into forms that can be absorbed and used by body cells for producing ATP and building body tissues
- Absorbs water, minerals, and vitamins needed for growth and function of body tissues
- Eliminates wastes from body tissues in feces

CARDIOVASCULAR SYSTEM

- GI tract absorbs water that helps maintain blood volume and iron that is needed for synthesis of hemoglobin in red blood cells
- Bilirubin from hemoglobin breakdown is partially excreted in feces
- Liver synthesizes most plasma proteins

LYMPHATIC SYSTEM and IMMUNITY

- Acidity of gastric juice destroys bacteria and most toxins in stomach
- Lymphatic nodules in lamina propria of mucosa of gastrointestinal tract (MALT) destroy microbes

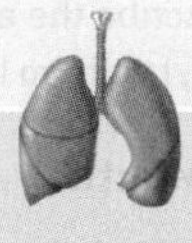

RESPIRATORY SYSTEM

- Pressure of abdominal organs against diaphragm helps expel air quickly during forced exhalation

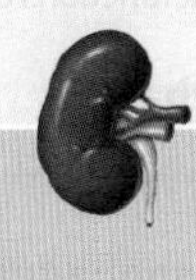

URINARY SYSTEM

- Absorption of water by GI tract provides water needed to excrete waste products in urine

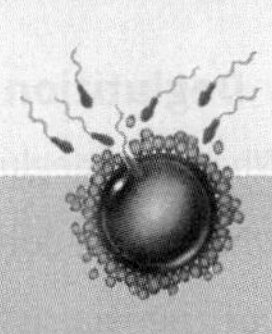

REPRODUCTIVE SYSTEMS

- Digestion and absorption provide adequate nutrients, including fats, for normal development of reproductive structures, for production of gametes (oocytes and sperm), and for fetal growth and development during pregnancy

Review Questions

24.1 Overview of the Digestive System

1. Which components of the digestive system are GI tract organs, and which are accessory digestive organs?

2. Which organs of the digestive system come in contact with food, and what are some of their digestive functions?

3. Which kinds of food molecules undergo chemical digestion, and which do not?

24.2 Layers of the GI Tract

4. Where along the GI tract is the muscularis composed of skeletal muscle? Is control of this skeletal muscle voluntary or involuntary?

5. Name the four layers of the gastrointestinal tract, and describe their functions.

24.3 Neural Innervation of the GI Tract

6. How is the enteric nervous system regulated by the autonomic nervous system?

7. What is a gastrointestinal reflex pathway?

24.4 Peritoneum

8. Where are the visceral peritoneum and parietal peritoneum located?

9. Describe the attachment sites and functions of the mesentery, mesocolon, falciform ligament, lesser omentum, and greater omentum.

24.5 Mouth

10. What structures form the mouth?

11. How are the major salivary glands distinguished on the basis of location?

12. How is the secretion of saliva regulated?

13. What functions do incisors, cuspids, premolars, and molars perform?

24.6 Pharynx

14. To which two organ systems does the pharynx belong?

24.7 Esophagus

15. Describe the location and histology of the esophagus. What is its role in digestion?

16. What are the functions of the upper and lower esophageal sphincters?

24.8 Deglutition

17. What does deglutition mean?

18. What occurs during the voluntary and pharyngeal phases of swallowing?

19. Does peristalsis "push" or "pull" food along the gastrointestinal tract?

24.9 Stomach

20. Compare the epithelium of the esophagus with that of the stomach. How is each adapted to the function of the organ?

21. What is the importance of rugae, surface mucous cells, mucous neck cells, chief cells, parietal cells, and G cells in the stomach?

22. What is the role of pepsin? Why is it secreted in an inactive form?

23. What are the functions of gastric lipase and lingual lipase in the stomach?

24.10 Pancreas

24. Describe the duct system connecting the pancreas to the duodenum.

25. What are pancreatic acini? How do their functions differ from those of the pancreatic islets (islets of Langerhans)?

26. What are the digestive functions of the components of pancreatic juice?

24.11 Liver and Gallbladder

27. Draw and label a diagram of the cell zones of a hepatic acinus.

28. Describe the pathways of blood flow into, through, and out of the liver.

29. How are the liver and gallbladder connected to the duodenum?

30. Once bile has been formed by the liver, how is it collected and transported to the gallbladder for storage?

31. Describe the major functions of the liver and gallbladder.

24.12 Small Intestine

32. List the regions of the small intestine and describe their functions.

33. In what ways are the mucosa and submucosa of the small intestine adapted for digestion and absorption?

34. Describe the types of movement that occur in the small intestine.

35. Explain the functions of pancreatic amylase, aminopeptidase, gastric lipase, and deoxyribonuclease.

36. What is the difference between digestion and absorption? How are the end products of carbohydrate, protein, and lipid digestion absorbed?

37. By what routes do absorbed nutrients reach the liver?

38. Describe the absorption of electrolytes, vitamins, and water by the small intestine.

24.13 Large Intestine

39. What are the major regions of the large intestine?

40. How does the muscularis of the large intestine differ from that of the rest of the gastrointestinal tract? What are haustra?

41. Describe the mechanical movements that occur in the large intestine.

42. What is defecation, and how does it occur?

43. What activities occur in the large intestine to change its contents into feces?

24.14 Phases of Digestion

44. What is the purpose of the cephalic phase of digestion?

45. Describe the role of gastrin in the gastric phase of digestion.

46. Outline the steps of the enterogastric reflex.

47. Explain the roles of CCK and secretin in the intestinal phase of digestion.

24.15 Development of the Digestive System

48. What structures develop from the foregut, midgut, and hindgut?

24.16 Aging and the Digestive System

49. What are the general effects of aging on the digestive system?

Critical Thinking Questions

1. Why would you *not* want to completely suppress HCl secretion in the stomach?
2. Trey has cystic fibrosis, a genetic disorder that is characterized by the production of excessive mucus, affecting several body systems (e.g., respiratory, digestive, reproductive). In the digestive system, the excess mucus blocks bile ducts in the liver and pancreatic ducts. How would this affect Trey's digestive processes?
3. Antonio had dinner at his favorite Italian restaurant. His menu consisted of a salad, a large plate of spaghetti, garlic bread, and wine. For dessert, he consumed "death by chocolate" cake and a cup of coffee. He topped off his evening with a cigarette and brandy. He returned home and, while lying on his couch watching television, he experienced a pain in his chest. He called 911 because he was certain he was having a heart attack. Antonio was told his heart was fine, but he needed to watch his diet. What happened to Antonio?

Metabolism and Nutrition

CHAPTER **25**

Metabolism, Nutrition, and Homeostasis

Metabolic reactions contribute to homeostasis by harvesting chemical energy from consumed nutrients for use in the body's growth, repair, and normal functioning.

The food we eat is our only source of energy for running, walking, and even breathing. Many molecules needed to maintain cells and tissues can be made from simpler precursors by the body's metabolic reactions; others—the essential amino acids, essential fatty acids, vitamins, and minerals—must be obtained from our food. As you learned in Chapter 24, carbohydrates, lipids, and proteins in food are digested by enzymes and absorbed in the gastrointestinal tract. The products of digestion that reach body cells are monosaccharides, fatty acids, glycerol, monoglycerides, and amino acids. Some minerals and many vitamins are part of enzyme systems that catalyze the breakdown and synthesis of carbohydrates, lipids, and proteins. Food molecules absorbed by the gastrointestinal (GI) tract have three main fates:

1. Most food molecules are used to *supply energy* for sustaining life processes, such as active transport, DNA replication, protein synthesis, muscle contraction, maintenance of body temperature, and mitosis.
2. Some food molecules *serve as building blocks* for the synthesis of more complex structural or functional molecules, such as muscle proteins, hormones, and enzymes.
3. Other food molecules are *stored for future use.* For example, glycogen is stored in liver cells, and triglycerides are stored in adipose cells.

In this chapter we discuss how metabolic reactions harvest the chemical energy stored in foods; how each group of food molecules contributes to the body's growth, repair, and energy needs; how energy balance is maintained in the body; and how body temperature is regulated. Finally, we explore some aspects of nutrition to discover why you should opt for fish instead of a burger the next time you eat out.

Q Did you ever wonder how fasting and starvation affect the body?

25.1 Metabolic Reactions

OBJECTIVES

- **Define** metabolism.
- **Explain** the role of ATP in anabolism and catabolism.

Metabolism (me-TAB-ō-lizm; *metabol-* = change) refers to all of the chemical reactions that occur in the body. There are two types of metabolism: catabolism and anabolism. Those chemical reactions that break down complex organic molecules into simpler ones are collectively known as **catabolism** (ka-TAB-ō-lizm; *cata-* = downward). Overall, catabolic (decomposition) reactions are *exergonic*; they produce more energy than they consume, releasing the chemical energy stored in organic molecules. Important sets of catabolic reactions occur in glycolysis, the Krebs cycle, and the electron transport chain, each of which will be discussed later in the chapter.

Chemical reactions that combine simple molecules and monomers to form the body's complex structural and functional components are collectively known as **anabolism** (a-NAB-ō-lizm; *ana-* = upward). Examples of anabolic reactions are the formation of peptide bonds between amino acids during protein synthesis, the building of fatty acids into phospholipids that form the plasma membrane bilayer, and the linkage of glucose monomers to form glycogen. Anabolic reactions are *endergonic*; they consume more energy than they produce.

Metabolism is an energy-balancing act between catabolic (decomposition) reactions and anabolic (synthesis) reactions. The molecule that participates most often in energy exchanges in living cells is **ATP (adenosine triphosphate)**, which couples energy-releasing catabolic reactions to energy-requiring anabolic reactions.

The metabolic reactions that occur depend on which enzymes are active in a particular cell at a particular time, or even in a particular part of the cell. Catabolic reactions can be occurring in the mitochondria of a cell at the same time as anabolic reactions are taking place in the endoplasmic reticulum.

A molecule synthesized in an anabolic reaction has a limited lifetime. With few exceptions, it will eventually be broken down and its component atoms recycled into other molecules or excreted from the body. Recycling of biological molecules occurs continuously in living tissues, more rapidly in some than in others. Individual cells may be refurbished molecule by molecule, or a whole tissue may be rebuilt cell by cell.

Coupling of Catabolism and Anabolism by ATP

The chemical reactions of living systems depend on the efficient transfer of manageable amounts of energy from one molecule to another. The molecule that most often performs this task is ATP, the "energy currency" of a living cell. Like money, it is readily available to "buy" cellular activities; it is spent and earned over and over. A typical cell has about a billion molecules of ATP, each of which typically lasts for less than a minute before being used. Thus, ATP is not a long-term storage form of currency, like gold in a vault, but rather convenient cash for moment-to-moment transactions.

Recall from Chapter 2 that a molecule of ATP consists of an adenine molecule, a ribose molecule, and three phosphate groups bonded to one another (see **Figure 2.26**). **Figure 25.1** shows how ATP links anabolic and catabolic reactions. When the terminal phosphate group is split off ATP, adenosine diphosphate (ADP) and a phosphate group (symbolized as Ⓟ) are formed. Some of the energy released is used to drive anabolic reactions such as the formation of glycogen from glucose. In addition, energy from complex molecules is used in catabolic reactions to combine ADP and a phosphate group to resynthesize ATP:

$$ADP + Ⓟ + energy \longrightarrow ATP$$

About 40% of the energy released in catabolism is used for cellular functions; the rest is converted to heat, some of which helps maintain normal body temperature. Excess heat is lost to the environment. Compared with machines, which typically convert only 10–20% of energy into work, the 40% efficiency of the body's metabolism is impressive. Still, the body has a continuous need to take in and process external sources of energy so that cells can synthesize enough ATP to sustain life.

FIGURE 25.1 Role of ATP in linking anabolic and catabolic reactions. When complex molecules and polymers are split apart (catabolism, at left), some of the energy is transferred to form ATP and the rest is given off as heat. When simple molecules and monomers are combined to form complex molecules (anabolism, at right), ATP provides the energy for synthesis, and again some energy is given off as heat.

The coupling of energy-releasing and energy-requiring reactions is achieved through ATP.

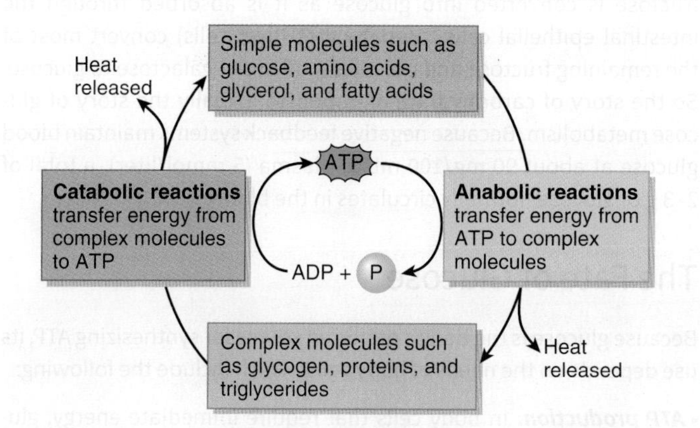

Q In a pancreatic cell that produces digestive enzymes, does anabolism or catabolism predominate?

25.2 Energy Transfer

OBJECTIVES

- **Describe** oxidation–reduction reactions.
- **Explain** the role of ATP in metabolism.

Various catabolic reactions transfer energy into the "high-energy" phosphate bonds of ATP. Although the amount of energy in these bonds is not exceptionally large, it can be released quickly and easily. Before discussing metabolic pathways, it is important to understand how this transfer of energy occurs. Two important aspects of energy transfer are oxidation–reduction reactions and mechanisms of ATP generation.

Oxidation–Reduction Reactions

Oxidation (ok′-si-DĀ-shun) is the *removal of electrons* from an atom or molecule; the result is a *decrease* in the potential energy of the atom or molecule. Because most biological oxidation reactions involve the loss of hydrogen atoms, they are called *dehydrogenation reactions.* An example of an oxidation reaction is the conversion of lactic acid into pyruvic acid:

$$\text{H–C(COOH)(CH}_3\text{)–OH (Lactic acid)} \xrightarrow[\text{Remove 2 H (H}^+ + \text{H}^-)]{\text{Oxidation}} \text{C(COOH)(CH}_3\text{)=O (Pyruvic acid)}$$

In the preceding reaction, 2H ($H^+ + H^-$) means that two neutral hydrogen atoms (2H) are removed as one hydrogen ion (H^+) plus one hydride ion (H^-).

Reduction (rē-DUK-shun) is the opposite of oxidation; it is the *addition of electrons* to a molecule. Reduction results in an *increase* in the potential energy of the molecule. An example of a reduction reaction is the conversion of pyruvic acid into lactic acid:

$$\underset{\text{Pyruvic acid}}{\begin{matrix} COOH \\ | \\ C{=}O \\ | \\ CH_3 \end{matrix}} \xrightarrow[\text{Add 2 H (H}^+ + \text{H}^-)]{\text{Reduction}} \underset{\text{Lactic acid}}{\begin{matrix} COOH \\ | \\ H-C-OH \\ | \\ CH_3 \end{matrix}}$$

When a substance is oxidized, the liberated hydrogen atoms do not remain free in the cell but are transferred immediately by coenzymes to another compound. Two coenzymes are commonly used by animal cells to carry hydrogen atoms: **nicotinamide adenine dinucleotide (NAD)**, a derivative of the B vitamin niacin, and **flavin adenine dinucleotide (FAD)**, a derivative of vitamin B_2 (riboflavin). The oxidation and reduction states of NAD^+ and FAD can be represented as follows:

$$\underset{\text{Oxidized}}{NAD^+} \underset{-2\,H\,(H^+ + H^-)}{\overset{+2\,H\,(H^+ + H^-)}{\rightleftharpoons}} \underset{\text{Reduced}}{NADH + H^+}$$

$$\underset{\text{Oxidized}}{FAD} \underset{-2\,H\,(H^+ + H^-)}{\overset{+2\,H\,(H^+ + H^-)}{\rightleftharpoons}} \underset{\text{Reduced}}{FADH_2}$$

When NAD^+ is reduced to $NADH + H^+$, the NAD^+ gains a hydride ion (H^-), neutralizing its charge, and the H^+ is released into the surrounding solution. When NADH is oxidized to NAD^+, the loss of the hydride ion results in one less hydrogen atom and an additional positive charge. FAD is reduced to $FADH_2$ when it gains a hydrogen ion and a hydride ion, and $FADH_2$ is oxidized to FAD when it loses the same two ions.

Oxidation and reduction reactions are always coupled; each time one substance is oxidized, another is simultaneously reduced. Such paired reactions are called **oxidation–reduction** or *redox reactions*. For example, when lactic acid is *oxidized* to form pyruvic acid, the two hydrogen atoms removed in the reaction are used to *reduce* NAD^+. This coupled redox reaction may be written as follows:

Lactic acid (Reduced) → Pyruvic acid (Oxidized)

NAD^+ (Oxidized) → $NADH + H^+$ (Reduced)

An important point to remember about oxidation–reduction reactions is that oxidation is usually an exergonic (energy-releasing) reaction. Cells use multistep biochemical reactions to release energy from energy-rich, highly reduced compounds (with many hydrogen atoms) to lower-energy, highly oxidized compounds (with many oxygen atoms or multiple bonds). For example, when a cell oxidizes a molecule of glucose ($C_6H_{12}O_6$), the energy in the glucose molecule is removed in a stepwise manner. Ultimately, some of the energy is captured by transferring it to ATP, which then serves as an energy source for energy-requiring reactions within the cell. Compounds with many hydrogen atoms such as glucose contain more chemical potential energy than oxidized compounds. For this reason, glucose is a valuable nutrient.

Mechanisms of ATP Generation

Some of the energy released during oxidation reactions is captured within a cell when ATP is formed. Briefly, a phosphate group Ⓟ is added to ADP, with an input of energy, to form ATP. The two high-energy phosphate bonds that can be used to transfer energy are indicated by "squiggles" (~):

$$\underset{\text{ADP}}{\text{Adenosine} - Ⓟ \sim Ⓟ} + Ⓟ + \text{energy} \longrightarrow \underset{\text{ATP}}{\text{Adenosine} - Ⓟ \sim Ⓟ \sim Ⓟ}$$

The high-energy phosphate bond that attaches the third phosphate group contains the energy stored in this reaction. The addition of a phosphate group to a molecule, called **phosphorylation** (fos′-for-i-LĀ-shun), increases its potential energy. Organisms use three mechanisms of phosphorylation to generate ATP:

1. **Substrate-level phosphorylation** generates ATP by transferring a high-energy phosphate group from an intermediate phosphorylated metabolic compound—a substrate—directly to ADP. In human cells, this process occurs in the cytosol.
2. **Oxidative phosphorylation** removes electrons from organic compounds and passes them through a series of electron acceptors, called the **electron transport chain**, to molecules of oxygen (O_2). This process occurs in the inner mitochondrial membrane of cells.
3. **Photophosphorylation** occurs only in chlorophyll-containing plant cells or in certain bacteria that contain other light-absorbing pigments.

25.3 Carbohydrate Metabolism

OBJECTIVE

- **Describe** the fate, metabolism, and functions of carbohydrates.

As you learned in Chapter 24, both polysaccharides and disaccharides are hydrolyzed into the monosaccharides **glucose** (about 80%), fructose, and galactose during the digestion of **carbohydrates**. (Some fructose is converted into glucose as it is absorbed through the intestinal epithelial cells.) Hepatocytes (liver cells) convert most of the remaining fructose and practically all of the galactose to glucose. So the story of carbohydrate metabolism is really the story of glucose metabolism. Because negative feedback systems maintain blood glucose at about 90 mg/100 mL of plasma (5 mmol/liter), a total of 2–3 g of glucose normally circulates in the blood.

The Fate of Glucose

Because glucose is the body's preferred source for synthesizing ATP, its use depends on the needs of body cells, which include the following:

- ***ATP production.*** In body cells that require immediate energy, glucose is oxidized to produce ATP. Glucose not needed for immediate ATP production can enter one of several other metabolic pathways.

- ***Amino acid synthesis.*** Cells throughout the body can use glucose to form several amino acids, which then can be incorporated into proteins.
- ***Glycogen synthesis.*** Hepatocytes and muscle fibers can perform **glycogenesis** (glī′-kō-JEN-e-sis; *glyco-* = sugar or sweet; *-genesis* = to generate), in which hundreds of glucose monomers are combined to form the polysaccharide glycogen. Total storage capacity of glycogen is about 125 g in the liver and 375 g in skeletal muscles.
- ***Triglyceride synthesis.*** When the glycogen storage areas are filled up, hepatocytes can transform the glucose to glycerol and fatty acids that can be used for **lipogenesis** (lip-ō-JEN-e-sis), the synthesis of triglycerides. Triglycerides then are deposited in adipose tissue, which has virtually unlimited storage capacity.

Glucose Movement into Cells

Before glucose can be used by body cells, it must first pass through the plasma membrane and enter the cytosol. Glucose absorption in the gastrointestinal tract (and kidney tubules) is accomplished via secondary active transport (Na^+–glucose symporters). Glucose entry into most other body cells occurs via GluT molecules, a family of transporters that bring glucose into cells via facilitated diffusion (see Section 3.3). A high level of insulin increases the insertion of one type of GluT, called GluT4, into the plasma membranes of most body cells, thereby increasing the rate of facilitated diffusion of glucose into cells. In neurons and hepatocytes, however, another type of GluT is always present in the plasma membrane, so glucose entry is always "turned on." On entering a cell, glucose becomes phosphorylated. Because GluT cannot transport phosphorylated glucose, this reaction traps glucose within the cell.

Glucose Catabolism

The oxidation of glucose to produce ATP is also known as **cellular respiration**, and it involves four sets of reactions: glycolysis, the formation of acetyl coenzyme A, the Krebs cycle, and the electron transport chain (**Figure 25.2**).

1. ***Glycolysis.*** A set of reactions in which one glucose molecule is oxidized and two molecules of pyruvic acid are produced. The reactions also produce two molecules of ATP and two energy-containing NADH + H^+.
2. ***Formation of acetyl coenzyme A.*** A transition step that prepares pyruvic acid for entrance into the Krebs cycle. This step also produces energy-containing NADH + H^+ plus carbon dioxide (CO_2).
3. ***Krebs cycle reactions.*** These reactions oxidize acetyl coenzyme A and produce CO_2, ATP, NADH + H^+, and $FADH_2$.
4. ***Electron transport chain reactions.*** These reactions oxidize NADH + H^+ and $FADH_2$ and transfer their electrons through a series of electron carriers.

Because glycolysis does not require oxygen, it can occur under **aerobic** (with oxygen) or **anaerobic** (without oxygen) conditions. By contrast, the reactions of the Krebs cycle and electron transport chain require oxygen and are collectively referred to as **aerobic respiration**.

FIGURE 25.2 **Overview of cellular respiration (oxidation of glucose).** A modified version of this figure appears in several places in this chapter to indicate the relationships of particular reactions to the overall process of cellular respiration.

The oxidation of glucose involves glycolysis, the formation of acetyl coenzyme A, the Krebs cycle, and the electron transport chain.

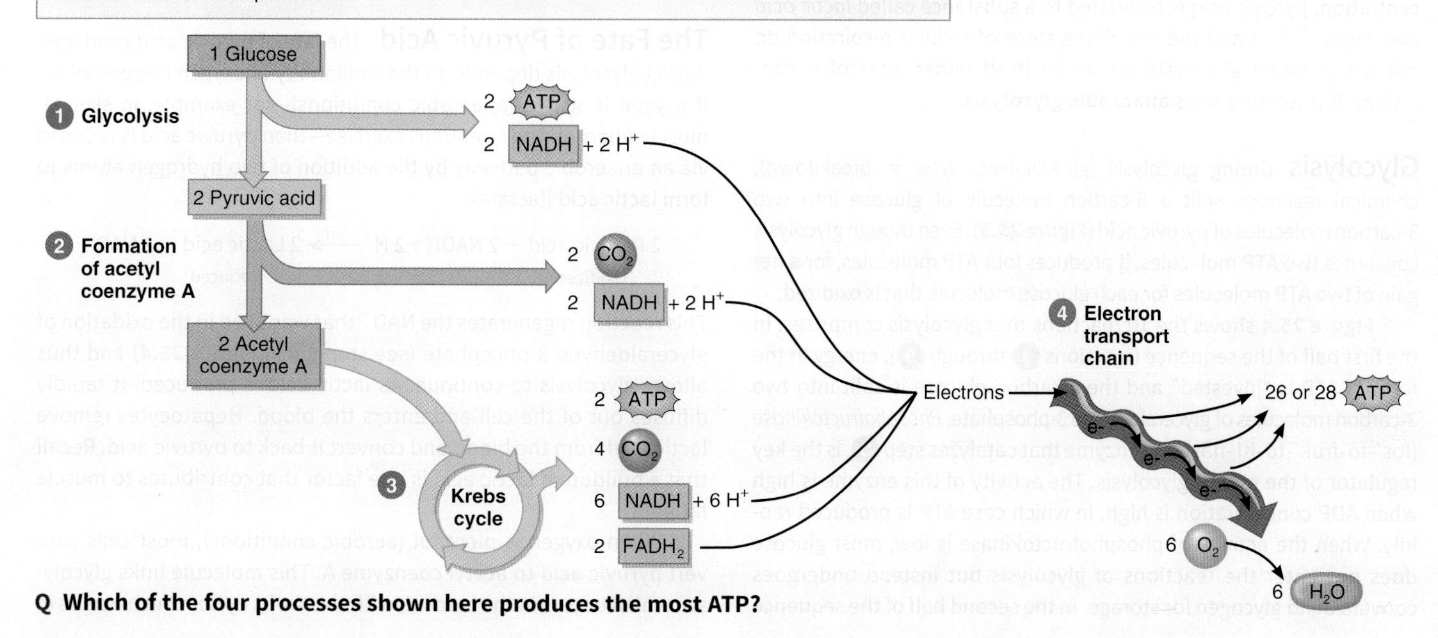

Q Which of the four processes shown here produces the most ATP?

FIGURE 25.3 **The role of glycolysis in cellular respiration.**

During glycolysis, each molecule of glucose is converted to two molecules of pyruvic acid.

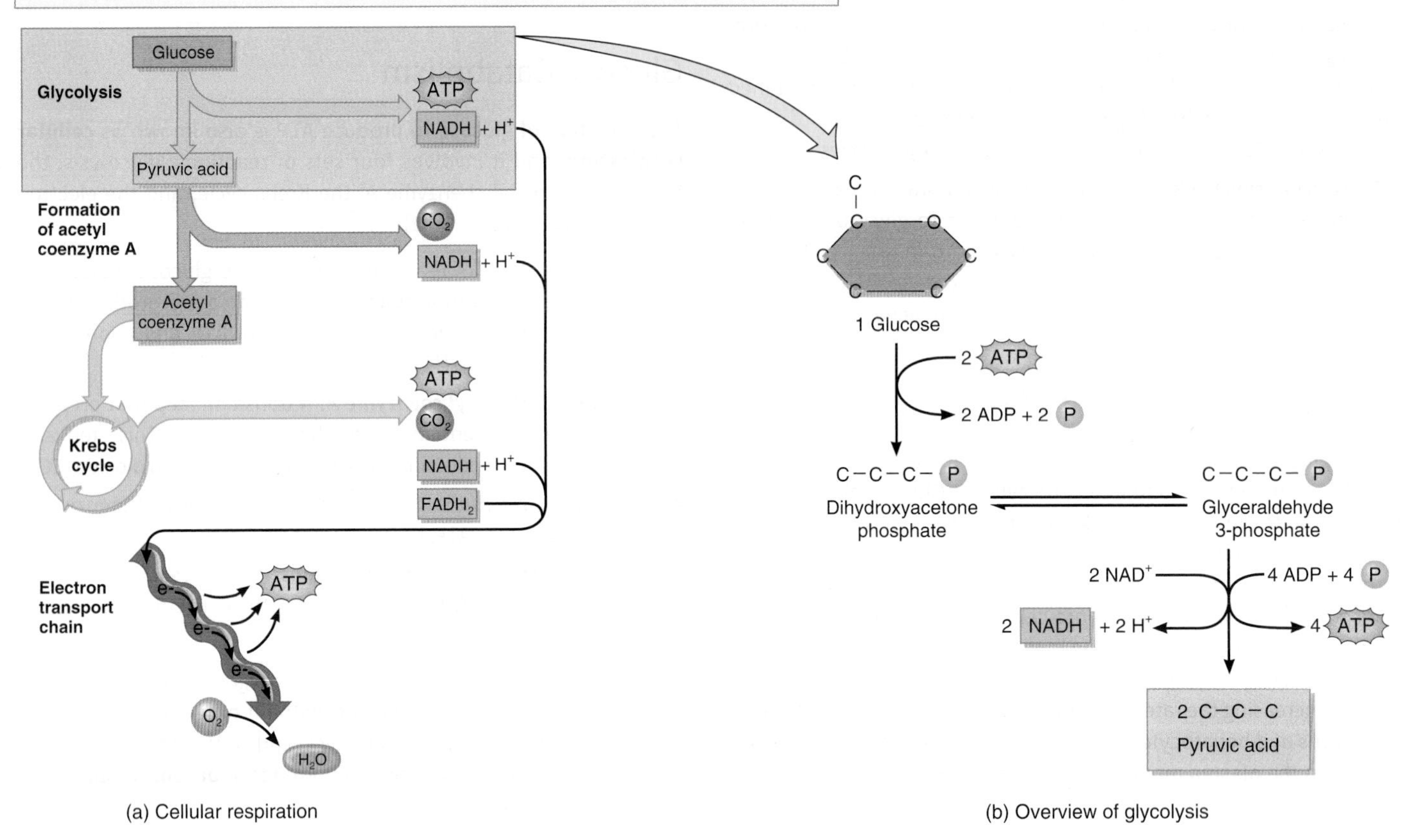

(a) Cellular respiration (b) Overview of glycolysis

Q For each glucose molecule that undergoes glycolysis, how many ATP molecules are generated?

Thus, when oxygen is present, all four phases occur: glycolysis, formation of acetyl coenzyme A, the Krebs cycle, and the electron transport chain. However, if oxygen is not available or at a low concentration, pyruvic acid is converted to a substance called *lactic acid* (see **Figure 25.5**) and the remaining steps of cellular respiration do not occur. When glycolysis occurs by itself under anaerobic conditions, it is referred to as **anaerobic glycolysis**.

Glycolysis

During glycolysis (glī-KOL-i-sis; *-lysis* = breakdown), chemical reactions split a 6-carbon molecule of glucose into two 3-carbon molecules of pyruvic acid (**Figure 25.3**). Even though glycolysis consumes two ATP molecules, it produces four ATP molecules, for a net gain of two ATP molecules for each glucose molecule that is oxidized.

Figure 25.4 shows the 10 reactions that glycolysis comprises. In the first half of the sequence (reactions 1 through 5), energy in the form of ATP is "invested" and the 6-carbon glucose is split into two 3-carbon molecules of glyceraldehyde 3-phosphate. *Phosphofructokinase* (fos′-fō-fruk′-tō-KĪ-nās), the enzyme that catalyzes step 3, is the key regulator of the rate of glycolysis. The activity of this enzyme is high when ADP concentration is high, in which case ATP is produced rapidly. When the activity of phosphofructokinase is low, most glucose does not enter the reactions of glycolysis but instead undergoes conversion to glycogen for storage. In the second half of the sequence (reactions 6 through 10), the two glyceraldehyde 3-phosphate molecules are converted to two pyruvic acid molecules and ATP is generated.

The Fate of Pyruvic Acid

The fate of pyruvic acid produced during glycolysis depends on the availability of oxygen (**Figure 25.5**). If oxygen is scarce (anaerobic conditions)—for example, in skeletal muscle fibers during strenuous exercise—then pyruvic acid is reduced via an anaerobic pathway by the addition of two hydrogen atoms to form lactic acid (lactate):

$$\underset{\text{Oxidized}}{\text{2 Pyruvic acid}} + \text{2 NADH} + 2\,H^+ \longrightarrow \text{2 Lactic acid} + \underset{\text{Reduced}}{2\,NAD^+}$$

This reaction regenerates the NAD^+ that was used in the oxidation of glyceraldehyde 3-phosphate (see step 6 in **Figure 25.4**) and thus allows glycolysis to continue. As lactic acid is produced, it rapidly diffuses out of the cell and enters the blood. Hepatocytes remove lactic acid from the blood and convert it back to pyruvic acid. Recall that a buildup of lactic acid is one factor that contributes to muscle fatigue.

When oxygen is plentiful (aerobic conditions), most cells convert pyruvic acid to acetyl coenzyme A. This molecule links glycolysis, which occurs in the cytosol, with the Krebs cycle, which occurs in

FIGURE 25.4 The 10 reactions of glycolysis. (1) Glucose is phosphorylated, using a phosphate group from an ATP molecule to form glucose 6-phosphate. (2) Glucose 6-phosphate is converted to fructose 6-phosphate. (3) A second ATP is used to add a second phosphate group to fructose 6-phosphate to form fructose 1,6-bisphosphate. (4) and (5) Fructose splits into two 3-carbon molecules, glyceraldehyde 3-phosphate (G 3-P) and dihydroxyacetone phosphate, each having one phosphate group. (6) Oxidation occurs as two molecules of NAD^+ accept two pairs of electrons and hydrogen ions from two molecules of G 3-P to form two molecules of NADH. Body cells use the two NADH produced in this step to generate ATP in the electron transport chain. A second phosphate group attaches to G 3-P, forming 1,3-bisphosphoglyceric acid (BPG). (7) through (10) These reactions generate four molecules of ATP and produce two molecules of pyruvic acid (pyruvate*).

Glycolysis results in a net gain of two ATP, two NADH, and two H^+.

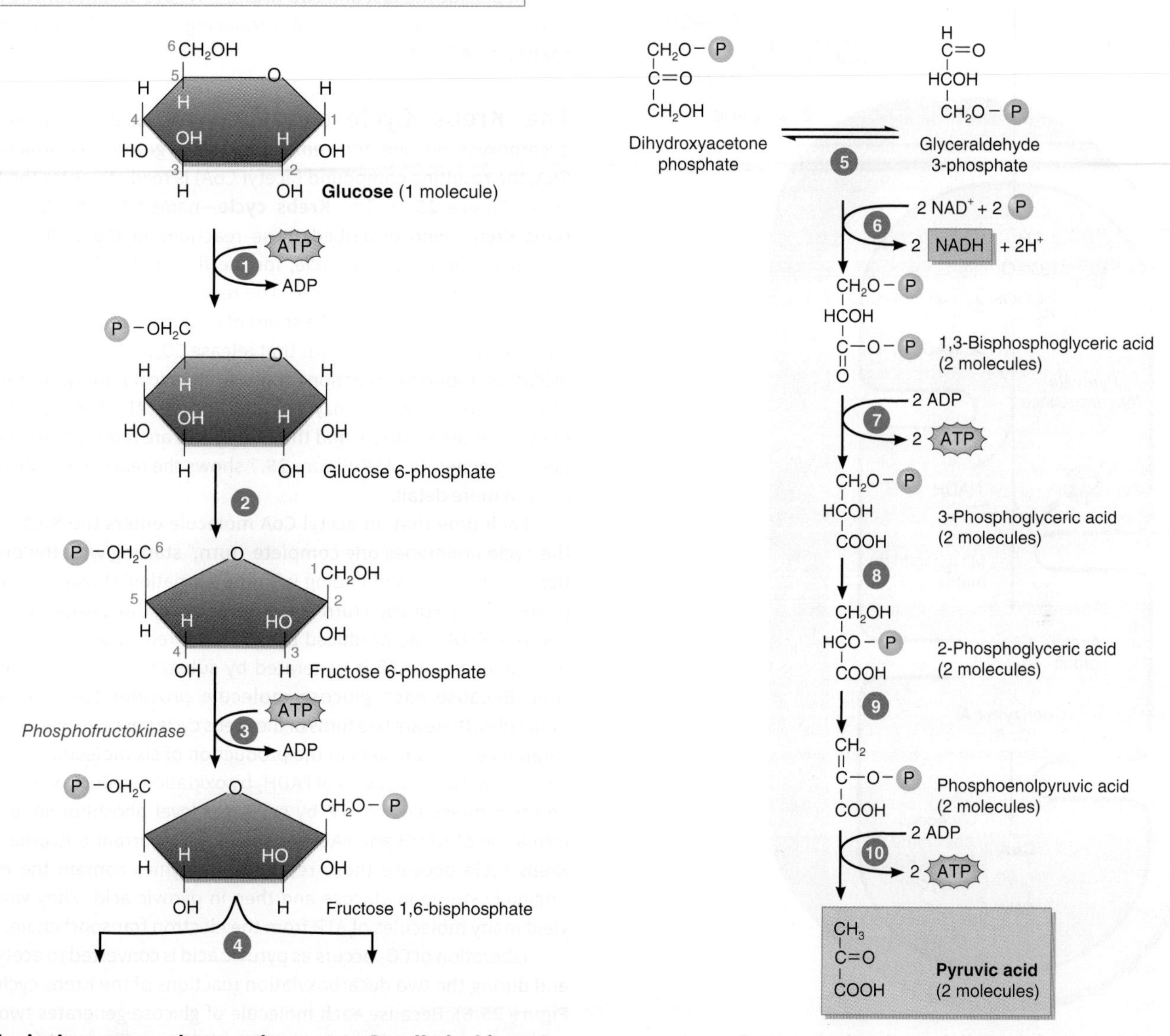

Q Why is the enzyme that catalyzes step (3) called a kinase?

*The carboxyl groups (–COOH) of intermediates in glycolysis and in the citric acid cycle are mostly ionized at the pH of body fluids to $-COO^-$. The suffix "-ic acid" indicates the non-ionized form, whereas the ending "-ate" indicates the ionized form. Although the "-ate" names are more correct, we will use the "acid" names because these terms are more familiar.

the matrix of mitochondria. Pyruvic acid enters the mitochondrial matrix with the help of a special transporter protein. Because they lack mitochondria, red blood cells can only produce ATP through glycolysis.

Formation of Acetyl Coenzyme A Each step in the oxidation of glucose requires a different enzyme, and often a coenzyme as well. The coenzyme used at this point in cellular respiration is **coenzyme A (CoA),** which is derived from pantothenic acid,

FIGURE 25.5 Fate of pyruvic acid.

When oxygen is plentiful, pyruvic acid enters mitochondria, is converted to acetyl coenzyme A, and enters the Krebs cycle (aerobic pathway). When oxygen is scarce, most pyruvic acid is converted to lactic acid via an anaerobic pathway.

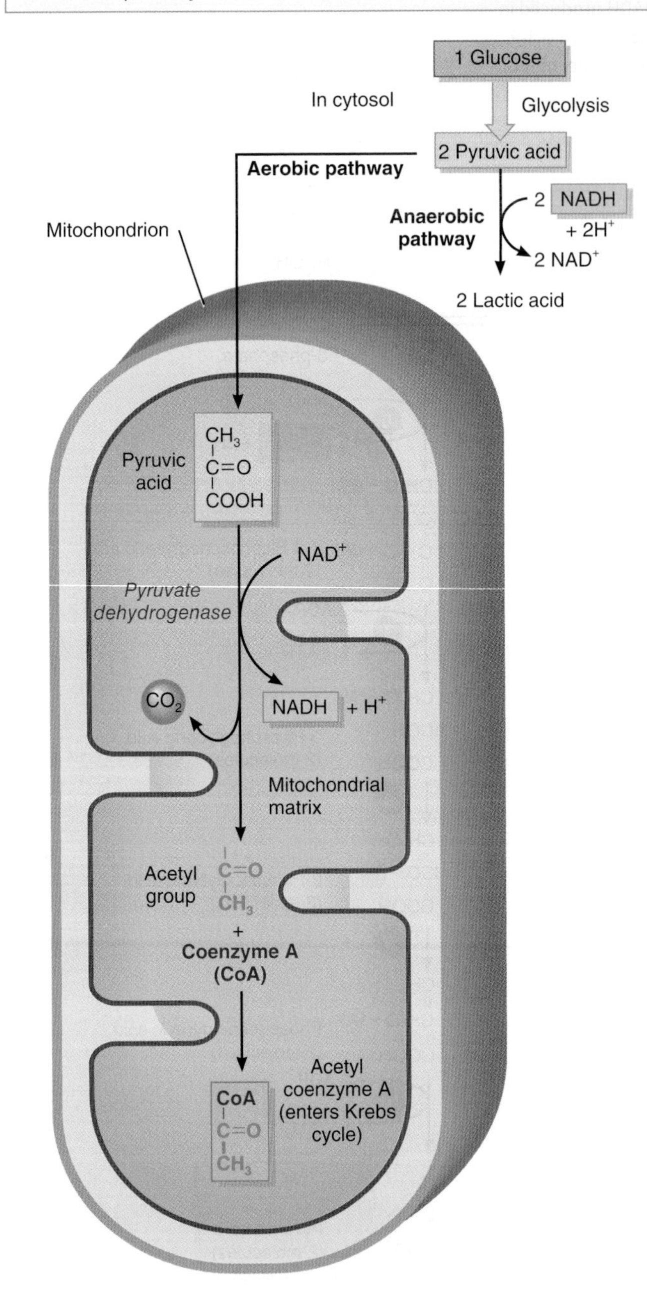

Q In which part of the cell does glycolysis occur?

a B vitamin. During the transitional step between glycolysis and the Krebs cycle, pyruvic acid is prepared for entrance into the cycle. The enzyme *pyruvate dehydrogenase* (pī-ROO-vāt dē-HĪ-drō-jen-ās), which is located exclusively in the mitochondrial matrix, converts pyruvic acid to a 2-carbon fragment called an **acetyl group** (AS-e-til) by removing a molecule of carbon dioxide (Figure 25.5). The loss of a molecule of CO_2 by a substance is called **decarboxylation** (dē-kar-bok′-si-LĀ-shun). This is the first reaction in cellular respiration that releases CO_2. During this reaction, pyruvic acid is also oxidized. Each pyruvic acid loses two hydrogen atoms in the form of one hydride ion (H^-) plus one hydrogen ion (H^+). The coenzyme NAD^+ is reduced as it picks up the H^- from pyruvic acid; the H^+ is released into the mitochondrial matrix. The reduction of NAD^+ to NADH + H^+ is indicated in Figure 25.5 by the curved arrow entering and then leaving the reaction. Recall that the oxidation of one glucose molecule produces two molecules of pyruvic acid, so for each molecule of glucose, two molecules of carbon dioxide are lost and two NADH + H^+ are produced. The acetyl group attaches to coenzyme A, producing a molecule called **acetyl coenzyme A** (*acetyl CoA*).

The Krebs Cycle Once the pyruvic acid has undergone decarboxylation and the remaining acetyl group has attached to CoA, the resulting compound (acetyl CoA) is ready to enter the Krebs cycle (Figure 25.6). The **Krebs cycle**—named for the biochemist Hans Krebs, who described these reactions in the 1930s—is also known as the *citric acid cycle*, for the first molecule formed when an acetyl group joins the cycle. The reactions occur in the matrix of mitochondria and consist of a series of oxidation–reduction reactions and decarboxylation reactions that release CO_2. In the Krebs cycle, the oxidation–reduction reactions transfer chemical energy, in the form of electrons, to two coenzymes—NAD^+ and FAD. The pyruvic acid derivatives are oxidized, and the coenzymes are reduced. In addition, one step generates ATP. Figure 25.7 shows the reactions of the Krebs cycle in more detail.

Each time that an acetyl CoA molecule enters the Krebs cycle, the cycle undergoes one complete "turn," starting with the production of citric acid and ending with the formation of oxaloacetic acid (Figure 25.7). For each turn of the Krebs cycle, three NADH, three H^+, and one $FADH_2$ are produced by oxidation–reduction reactions, and one molecule of ATP is generated by substrate-level phosphorylation. Because each glucose molecule provides two acetyl CoA molecules, there are two turns of the Krebs cycle per molecule of glucose catabolized. This results in the production of six molecules of NADH, six H^+, and two molecules of $FADH_2$ by oxidation–reduction reactions, and two molecules of ATP by substrate-level phosphorylation. The formation of NADH and $FADH_2$ is the most important outcome of the Krebs cycle because these reduced coenzymes contain the energy originally stored in glucose and then in pyruvic acid. They will later yield many molecules of ATP from the electron transport chain.

Liberation of CO_2 occurs as pyruvic acid is converted to acetyl CoA and during the two decarboxylation reactions of the Krebs cycle (see Figure 25.6). Because each molecule of glucose generates two molecules of pyruvic acid, six molecules of CO_2 are liberated from each original glucose molecule catabolized along this pathway. The molecules of CO_2 diffuse out of the mitochondria, through the cytosol and plasma membrane, and then into the blood. Blood transports the CO_2 to the lungs, where it eventually is exhaled.

The Electron Transport Chain The **electron transport chain** is a series of **electron carriers**, integral membrane proteins

FIGURE 25.6 After formation of acetyl coenzyme A, the next stage of cellular respiration is the Krebs cycle.

Reactions of the Krebs cycle occur in the matrix of mitochondria.

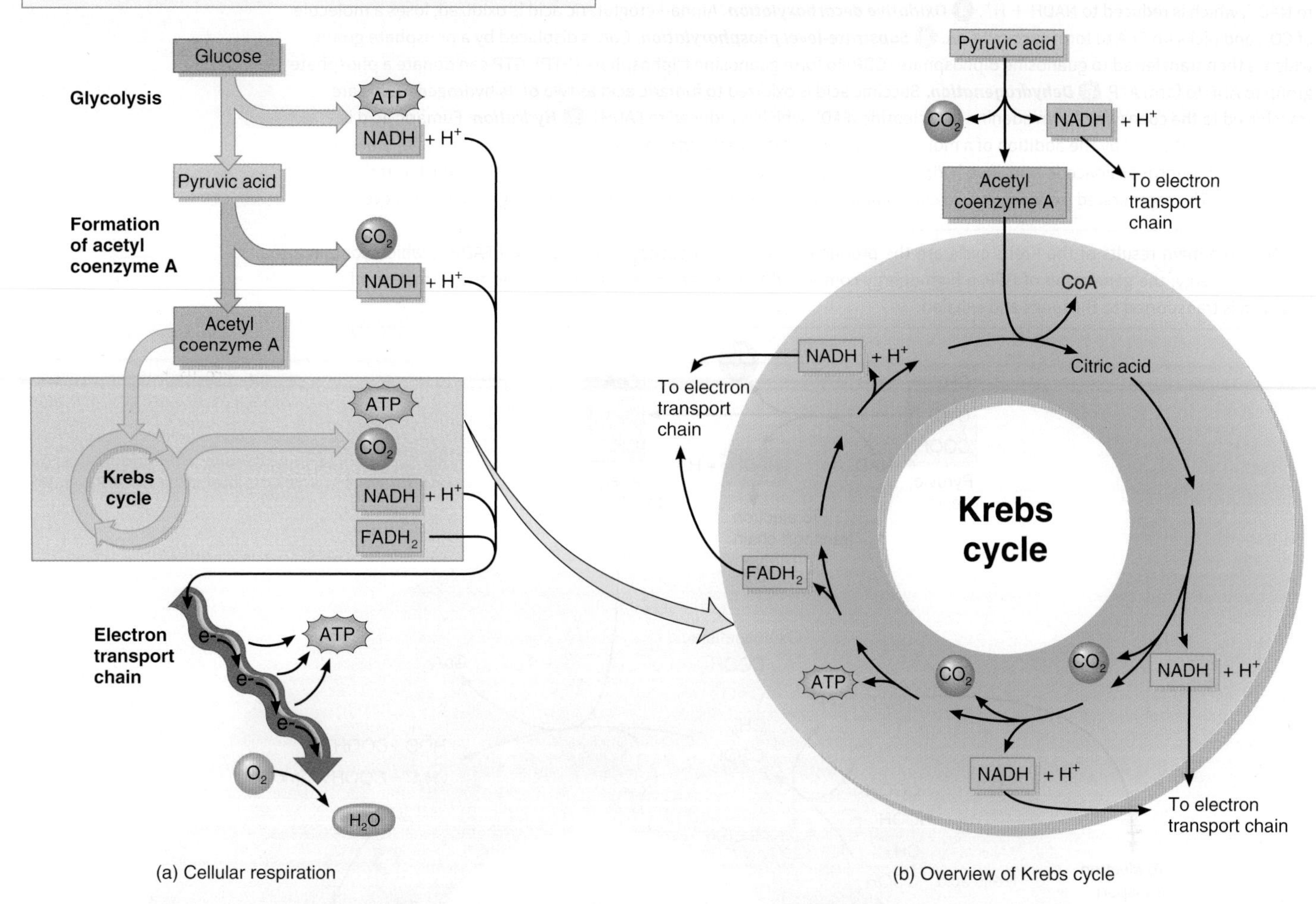

(a) Cellular respiration

(b) Overview of Krebs cycle

Q When in cellular respiration is carbon dioxide given off? What happens to this gas?

in the inner mitochondrial membrane. This membrane is folded into cristae that increase its surface area, accommodating thousands of copies of the transport chain in each mitochondrion. Each carrier in the chain is reduced as it picks up electrons and oxidized as it gives up electrons. As electrons pass through the chain, a series of exergonic reactions release small amounts of energy; this energy is used to form ATP. In cellular respiration, the final electron acceptor of the chain is oxygen. Because this mechanism of ATP generation links chemical reactions (the passage of electrons along the transport chain) with the pumping of hydrogen ions, it is called **chemiosmosis** (kem′-ē-oz-MŌ-sis; *chemi-* = chemical; *-osmosis* = pushing). Together, chemiosmosis and the electron transport chain constitute oxidative phosphorylation.

Briefly, chemiosmosis works as follows (**Figure 25.8**):

1. Energy from NADH + H^+ passes along the electron transport chain and is used to pump H^+ from the matrix of the mitochondrion into the space between the inner and outer mitochondrial membranes. This mechanism is called a **proton pump** because H^+ ions consist of a single proton.
2. A high concentration of H^+ accumulates between the inner and outer mitochondrial membranes.
3. ATP synthesis then occurs as hydrogen ions flow back into the mitochondrial matrix through a special type of H^+ channel in the inner membrane.

Electron Carriers Several types of molecules and atoms serve as electron carriers:

- **Flavin mononucleotide (FMN)** (FLĀ-vin mon′-ō-NOO-klē-ō-tīd) is a flavoprotein derived from riboflavin (vitamin B_2).
- **Cytochromes** (SĪ-tō-krōmz) are proteins with an iron-containing group (heme) capable of existing alternately in a reduced form (Fe^{2+}) and an oxidized form (Fe^{3+}). The cytochromes involved in the electron transport chain include cytochrome *b* (cyt *b*), cytochrome c_1 (cyt c_1), cytochrome *c* (cyt *c*), cytochrome *a* (cyt *a*), and cytochrome a_3 (cyt a_3).

FIGURE 25.7 **The eight reactions of the Krebs cycle.** ① ***Entry of the acetyl group.*** The chemical bond that attaches the acetyl group to coenzyme A (CoA) breaks, and the 2-carbon acetyl group attaches to a 4-carbon molecule of oxaloacetic acid to form a 6-carbon molecule called citric acid. CoA is free to combine with another acetyl group from pyruvic acid and repeat the process. ② ***Isomerization.*** Citric acid undergoes isomerization to isocitric acid, which has the same molecular formula as citrate. Notice, however, that the hydroxyl group (–OH) is attached to a different carbon. ③ ***Oxidative decarboxylation.*** Isocitric acid is oxidized and loses a molecule of CO_2, forming alpha-ketoglutaric acid. The H^+ from the oxidation is passed on to NAD^+, which is reduced to NADH + H^+. ④ ***Oxidative decarboxylation.*** Alpha-ketoglutaric acid is oxidized, loses a molecule of CO_2, and picks up CoA to form succinyl-CoA. ⑤ ***Substrate-level phosphorylation.*** CoA is displaced by a phosphate group, which is then transferred to guanosine diphosphate (GDP) to form guanosine triphosphate (GTP). GTP can donate a phosphate group to ADP to form ATP. ⑥ ***Dehydrogenation.*** Succinic acid is oxidized to fumaric acid as two of its hydrogen atoms are transferred to the coenzyme flavin adenine dinucleotide (FAD), which is reduced to $FADH_2$. ⑦ ***Hydration.*** Fumaric acid is converted to malic acid by the addition of a molecule of water. ⑧ ***Dehydrogenation.*** In the final step in the cycle, malic acid is oxidized to re-form oxaloacetic acid. Two hydrogen atoms are removed and one is transferred to NAD^+, which is reduced to NADH + H^+. The regenerated oxaloacetic acid can combine with another molecule of acetyl CoA, beginning a new cycle.

The three main results of the Krebs cycle are the production of reduced coenzymes (NADH and $FADH_2$), which contain stored energy; the generation of GTP, a high-energy compound that is used to produce ATP; and the formation of CO_2, which is transported to the lungs and exhaled.

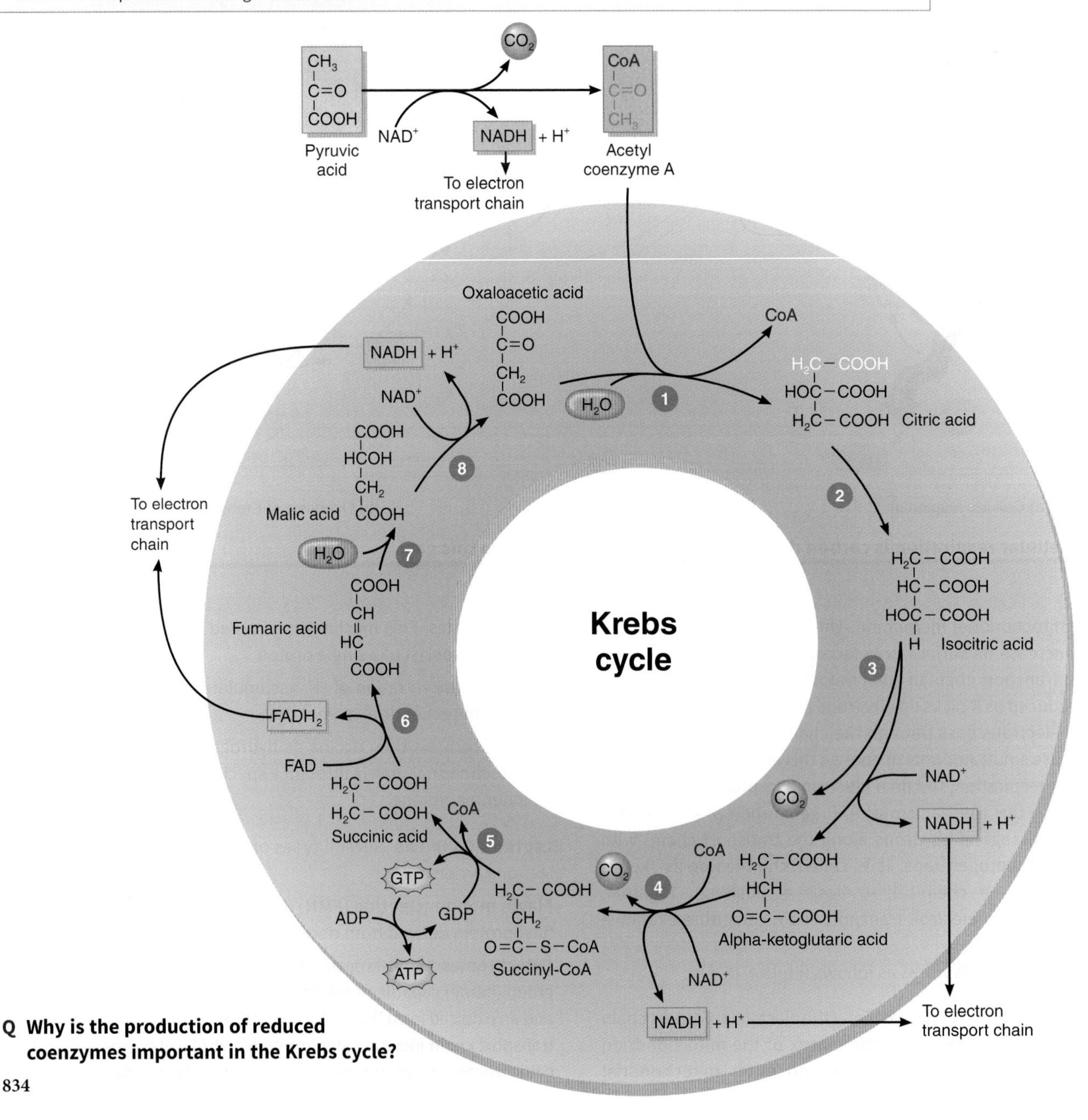

Q Why is the production of reduced coenzymes important in the Krebs cycle?

FIGURE 25.8 **Chemiosmosis.**

In chemiosmosis, ATP is produced when hydrogen ions diffuse back into the mitochondrial matrix.

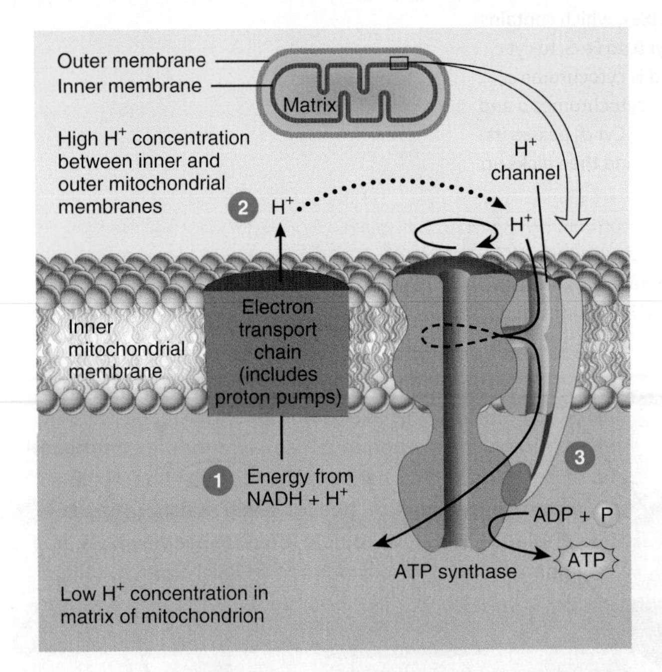

Q What is the energy source that powers the proton pumps?

- **Iron–sulfur (Fe-S) centers** contain either two or four iron atoms bound to sulfur atoms that form an electron transfer center within a protein.
- **Copper (Cu) atoms** bound to two proteins in the chain also participate in electron transfer.
- **Coenzyme Q (Q)**, is a nonprotein, low-molecular-weight carrier that is mobile in the lipid bilayer of the inner membrane.

STEPS IN ELECTRON TRANSPORT AND CHEMIOSMOTIC ATP GENERATION Within the inner mitochondrial membrane, the carriers of the electron transport chain are clustered into three complexes, each of which acts as a proton pump that expels H^+ from the mitochondrial matrix and helps create an electrochemical gradient of H^+. Each of the three proton pumps transports electrons and pumps H^+, as shown in Figure 25.9. Notice that oxygen is used to help form water in step ③. This is the only point in aerobic cellular respiration where O_2 is consumed. **Cyanide** is a deadly poison because it binds to the cytochrome oxidase complex and blocks this last step in electron transport.

The pumping of H^+ produces both a concentration gradient of protons and an electrical gradient. The buildup of H^+ makes one side of the inner mitochondrial membrane positively charged compared with the other side. The resulting electrochemical gradient has potential energy, called the *proton motive force.* Proton channels in the inner mitochondrial membrane allow H^+ to flow back across the membrane, driven by the proton motive force. As H^+ flow back, they generate ATP because the H^+ channels also include an enzyme called **ATP synthase** (SIN-thās). The enzyme uses the proton motive force to synthesize ATP from ADP and Ⓟ. The process of chemiosmosis is responsible for most of the ATP produced during cellular respiration.

For every molecule of NADH + H^+ that drops off hydrogen atoms to the electron transport chain, two or three molecules of ATP (average = 2.5) are produced via oxidative phosphorylation. For every molecule of $FADH_2$ that drops off hydrogen atoms to the electron transport chain, only one or two molecules of ATP (average = 1.5) are produced via oxidative phosphorylation. This is due to the fact that $FADH_2$ drops off its hydrogen atoms at a lower step along the electron transport chain than NADH + H^+.

Summary of Cellular Respiration

The various electron transfers in the electron transport chain generate either 26 or 28 ATP molecules from each molecule of glucose that is catabolized: either 23 or 25 from the 10 molecules of NADH + H^+ and three from the two molecules of $FADH_2$. The discrepancy in the number of ATP formed from NADH + H^+ via oxidative phosphorylation is due to the fact that the two NADH + H^+ molecules produced in the cytosol during glycolysis cannot enter mitochondria. Instead they donate their electrons to one of two transfer shuttles, known as the *malate shuttle* and the *glycerol phosphate shuttle*. In cells of the liver, kidneys, and heart, use of the malate shuttle results in an average of 2.5 molecules of ATP synthesized for each molecule of NADH + H^+. In other body cells, such as skeletal muscle fibers and neurons, use of the glycerol phosphate shuttle results in an average of 1.5 molecules of ATP synthesized for each molecule of NADH + H^+.

Recall that four ATP molecules are produced via substrate-level phosphorylation (two from glycolysis and two from the Krebs cycle). If the four ATP produced via substrate-level phosphorylation are added to the 26 or 28 ATP produced via oxidative phosphorylation, a total of either 30 or 32 ATP is generated for each molecule of glucose catabolized during cellular respiration. The overall reaction is

$$\underset{\text{Glucose}}{C_6H_{12}O_6} + \underset{\text{Oxygen}}{6\,O_2} + 30 \text{ or } 32 \text{ ADPs} + 30 \text{ or } 32 \text{ Ⓟ} \longrightarrow \underset{\text{Carbon dioxide}}{6\,CO_2} + \underset{\text{Water}}{6\,H_2O} + 30 \text{ or } 32 \text{ ATPs}$$

Table 25.1 summarizes the ATP yield during cellular respiration. A schematic depiction of the principal reactions of cellular respiration is presented in Figure 25.10.

Glycolysis, the Krebs cycle, and especially the electron transport chain provide all of the ATP for cellular activities. Because the Krebs cycle and electron transport chain are aerobic processes, cells cannot carry on their activities for long if oxygen is lacking.

Glucose Anabolism

Even though most of the glucose in the body is catabolized to generate ATP, glucose may take part in or be formed via several anabolic reactions. One is the synthesis of glycogen; another is the synthesis of new glucose molecules from some of the products of protein and lipid breakdown.

Glucose Storage: Glycogenesis

If glucose is not needed immediately for ATP production, it combines with many other

FIGURE 25.9 **The actions of the three proton pumps and ATP synthase in the inner membrane of mitochondria.** Each pump is a complex of three or more electron carriers. (1) The first proton pump is the *NADH dehydrogenase complex*, which contains flavin mononucleotide (FMN) and five or more Fe-S centers. NADH + H^+ is oxidized to NAD^+, and FMN is reduced to $FMNH_2$, which in turn is oxidized as it passes electrons to the iron–sulfur centers. Q, which is mobile in the membrane, shuttles electrons to the second pump complex. (2) The second proton pump is the *cytochrome* b–c_1 *complex*, which contains cytochromes and an iron–sulfur center. Electrons are passed successively from Q to cyt *b*, to Fe-S, to cyt c_1. The mobile shuttle that passes electrons from the second pump complex to the third is cytochrome *c* (cyt *c*). (3) The third proton pump is the *cytochrome oxidase complex*, which contains cytochromes *a* and a_3 and two copper atoms. Electrons pass from cyt *c*, to Cu, to cyt *a*, and finally to cyt a_3. Cyt a_3 passes its electrons to one-half of a molecule of oxygen (O_2), which becomes negatively charged and then picks up two H^+ from the surrounding medium to form H_2O.

As the three proton pumps pass electrons from one carrier to the next, they also move protons (H^+) from the matrix into the space between the inner and outer mitochondrial membranes. As protons flow back into the mitochondrial matrix through the H^+ channel in ATP synthase, ATP is synthesized.

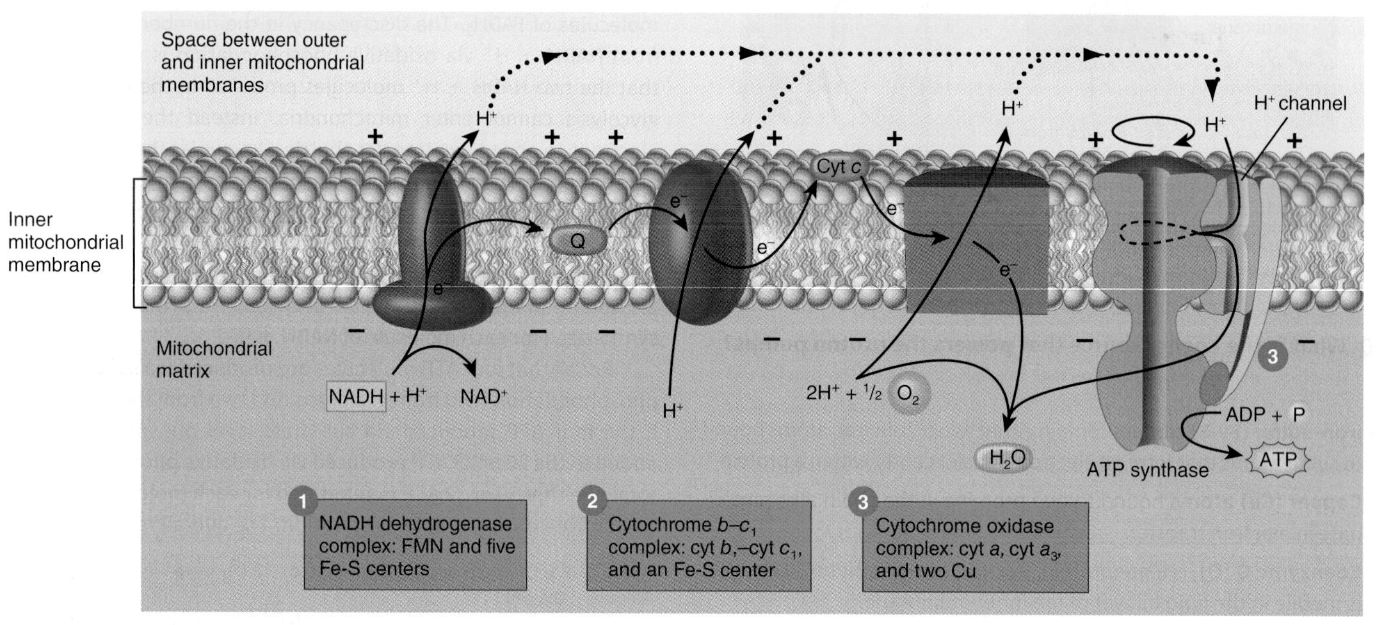

Q Where is the concentration of H_1 highest?

molecules of glucose to form **glycogen**, a polysaccharide that is the only stored form of carbohydrate in the body. The hormone insulin, from pancreatic beta cells, stimulates hepatocytes and skeletal muscle cells to carry out **glycogenesis** (glī′-kō-JEN-e-sis), the synthesis of glycogen (**Figure 25.11**). The body can store about 500 g (about 1.1 lb) of glycogen, roughly 75% in skeletal muscle fibers and the rest in liver cells. During glycogenesis, glucose is first phosphorylated to glucose 6-phosphate by hexokinase. Glucose 6-phosphate is converted to glucose 1-phosphate, then to uridine diphosphate glucose, and finally to glycogen.

Glucose Release: Glycogenolysis

When body activities require ATP, glycogen stored in hepatocytes is broken down into glucose and released into the blood to be transported to cells, where it will be catabolized by the processes of cellular respiration already described. The process of splitting glycogen into its glucose subunits is called **glycogenolysis** (glī′-kō-je-NOL-e-sis). (Note: Do not confuse *glycogenolysis*, the breakdown of glycogen to glucose, with *glycolysis*, the 10 reactions that convert glucose to pyruvic acid.)

Glycogenolysis is not a simple reversal of the steps of glycogenesis (**Figure 25.11**). It begins by splitting off glucose molecules from the branched glycogen molecule via phosphorylation to form glucose 1-phosphate. Phosphorylase, the enzyme that catalyzes this reaction, is activated by glucagon from pancreatic alpha cells and epinephrine from the adrenal medullae. Glucose 1-phosphate is then converted to glucose 6-phosphate and finally to glucose, which leaves hepatocytes via glucose transporters (GluT) in the plasma membrane. Phosphorylated glucose molecules cannot ride aboard the GluT transporters, however, and *phosphatase*, the enzyme that converts glucose 6-phosphate into glucose, is absent in skeletal muscle cells. Thus, hepatocytes, which have phosphatase, can release glucose derived from glycogen to the bloodstream, but skeletal muscle cells cannot. In skeletal muscle cells, glycogen is broken down into

TABLE 25.1 Summary of ATP Produced in Cellular Respiration

SOURCE	ATP YIELD PER GLUCOSE MOLECULE (PROCESS)
GLYCOLYSIS	
Oxidation of one glucose molecule to two pyruvic acid molecules	2 ATPs (substrate-level phosphorylation).
Production of 2 NADH + H^+	3 or 5 ATPs (oxidative phosphorylation).
FORMATION OF TWO MOLECULES OF ACETYL COENZYME A	
2 NADH + 2 H^+	5 ATPs (oxidative phosphorylation).
KREBS CYCLE AND ELECTRON TRANSPORT CHAIN	
Oxidation of succinyl-CoA to succinic acid	2 GTPs that are converted to 2 ATPs (substrate-level phosphorylation).
Production of 6 NADH + 6 H^+	15 ATPs (oxidative phosphorylation).
Production of 2 $FADH_2$	3 ATPs (oxidative phosphorylation).
Total	30 or 32 ATPs per glucose molecule.

glucose 1-phosphate, which is then catabolized for ATP production via glycolysis and the Krebs cycle. However, the lactic acid produced by glycolysis in muscle cells can be converted to glucose in the liver. In this way, muscle glycogen can be an indirect source of blood glucose.

See Clinical Connection: Carbohydrate Loading

Formation of Glucose from Proteins and Fats: Gluconeogenesis

When your liver runs low on glycogen, it is time to eat. If you don't, your body starts catabolizing triglycerides (fats) and proteins. Actually, the body normally catabolizes some of its triglycerides and proteins, but large-scale triglyceride and protein catabolism does not happen unless you are starving, eating very few carbohydrates, or suffering from an endocrine disorder.

The glycerol part of triglycerides, lactic acid, and certain amino acids can be converted in the liver to glucose (**Figure 25.12**). The process by which glucose is formed from these noncarbohydrate sources is called **gluconeogenesis** (gloo′-kō-nē′-ō-JEN-e-sis; *neo-* = new). An easy way to distinguish this term from glycogenesis or glycogenolysis is to remember that in this case glucose is not converted back from glycogen, but is instead *newly formed.* About 60% of the amino acids in the body can be used for gluconeogenesis. Lactic acid and amino acids such as alanine, cysteine, glycine, serine, and threonine are converted to pyruvic acid, which then may be synthesized into glucose or enter the Krebs cycle. Glycerol may be converted into glyceraldehyde 3-phosphate, which may form pyruvic acid or be used to synthesize glucose.

Gluconeogenesis is stimulated by cortisol, the main glucocorticoid hormone of the adrenal cortex, and by glucagon from the pancreas. In addition, cortisol stimulates the breakdown of proteins into amino acids, thus expanding the pool of amino acids available for gluconeogenesis. Thyroid hormones (thyroxine and triiodothyronine) also mobilize proteins and may mobilize triglycerides from adipose tissue, thereby making glycerol available for gluconeogenesis.

FIGURE 25.10 Summary of the principal reactions of cellular respiration. ETC = electron transport chain and chemiosmosis.

Except for glycolysis, which occurs in the cytosol, all other reactions of cellular respiration occur within mitochondria.

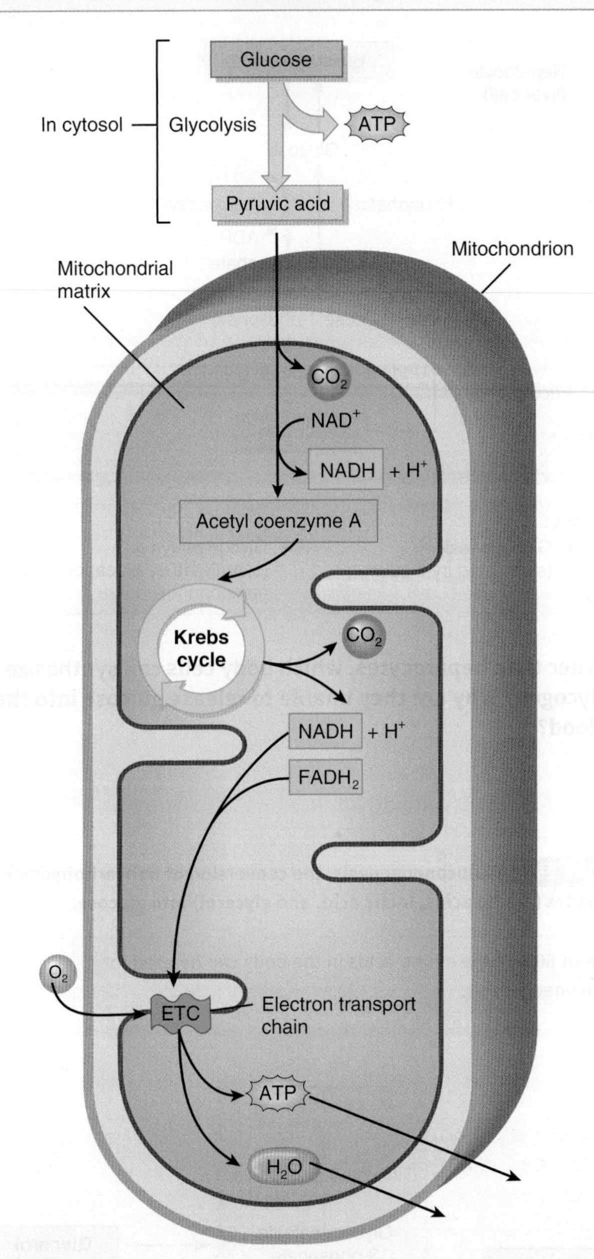

Q How many molecules of O_2 are used, and how many molecules of CO_2 are produced during the complete oxidation of one glucose molecule?

FIGURE 25.11 **Glycogenesis and glycogenolysis.**

The glycogenesis pathway converts glucose into glycogen; the glycogenolysis pathway breaks down glycogen into glucose.

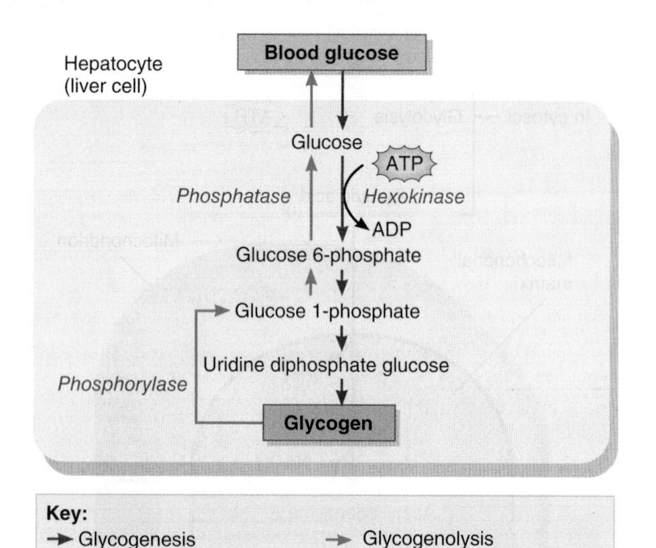

Q Other than hepatocytes, which body cells can synthesize glycogen? Why are they unable to release glucose into the blood?

FIGURE 25.12 **Gluconeogenesis, the conversion of noncarbohydrate molecules (amino acids, lactic acid, and glycerol) into glucose.**

About 60% of the amino acids in the body can be used for gluconeogenesis.

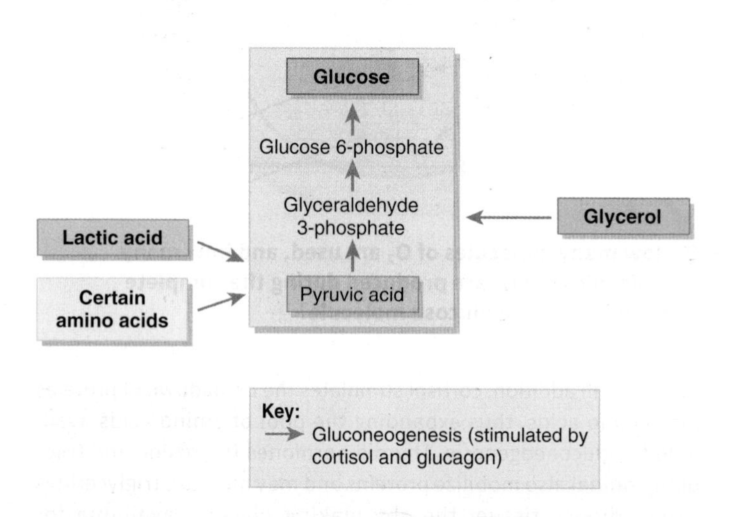

Q What cells can carry out gluconeogenesis and glycogenesis?

25.4 Lipid Metabolism

OBJECTIVES

- **Describe** the lipoproteins that transport lipids in the blood.
- **Discuss** the fate, metabolism, and functions of lipids.

Transport of Lipids by Lipoproteins

Most **lipids**, such as triglycerides, are nonpolar and therefore very hydrophobic molecules. They do not dissolve in water. To be transported in watery blood, such molecules first must be made more water-soluble by combining them with proteins produced by the liver and intestine. The lipid and protein combinations thus formed are **lipoproteins** (lip′-ō-PRŌ-tēns), spherical particles with an outer shell of proteins, phospholipids, and cholesterol molecules surrounding an inner core of triglycerides and other lipids (**Figure 25.13**). The proteins in the outer shell are called **apoproteins (apo)** (ap-ō-PRŌ-tēns) and are designated by the letters A, B, C, D, and E plus a number. In addition to helping solubilize the lipoprotein in body fluids, each apoprotein has specific functions.

Each of the several types of lipoproteins has different functions, but all are essentially transport vehicles. They provide delivery and pickup services so that lipids can be available when cells need them or removed from circulation when they are not needed. Lipoproteins are categorized and named mainly according to their density, which varies with the ratio of lipids (which have a low density) to proteins

FIGURE 25.13 **A lipoprotein.** Shown here is a VLDL.

A single layer of amphipathic phospholipids, cholesterol, and proteins surrounds a core of nonpolar lipids.

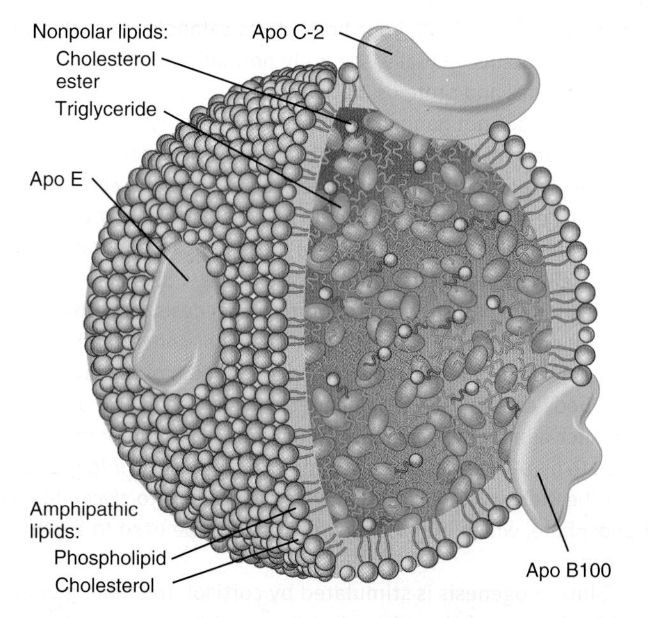

Q Which type of lipoprotein delivers cholesterol to body cells?

(which have a high density). From largest and lightest to smallest and heaviest, the four major classes of lipoproteins are chylomicrons, very-low-density lipoproteins (VLDLs), low-density lipoproteins (LDLs), and high-density lipoproteins (HDLs).

Chylomicrons (kī-lō-MI-krons), which form in mucosal epithelial cells of the small intestine, transport *dietary* (ingested) lipids to adipose tissue for storage. They contain about 1–2% proteins, 85% triglycerides, 7% phospholipids, and 6–7% cholesterol, plus a small amount of fat-soluble vitamins. Chylomicrons enter lacteals of intestinal villi and are carried by lymph into venous blood and then into the systemic circulation. Their presence gives blood plasma a milky appearance, but they remain in the blood for only a few minutes. As chylomicrons circulate through the capillaries of adipose tissue, one of their apoproteins, **apo C-2**, activates *endothelial lipoprotein lipase*, an enzyme that removes fatty acids from chylomicron triglycerides. The free fatty acids are then taken up by adipocytes for synthesis and storage as triglycerides and by muscle cells for ATP production. Hepatocytes remove chylomicron remnants from the blood via receptor-mediated endocytosis, in which another chylomicron apoprotein, **apo E**, is the docking protein.

Very-low-density lipoproteins (VLDLs), which form in hepatocytes, contain mainly *endogenous* (made in the body) lipids. VLDLs contain about 10% proteins, 50% triglycerides, 20% phospholipids, and 20% cholesterol. VLDLs transport triglycerides synthesized in hepatocytes to adipocytes for storage. Like chylomicrons, they lose triglycerides as their apo C-2 activates endothelial lipoprotein lipase, and the resulting fatty acids are taken up by adipocytes for storage and by muscle cells for ATP production. As they deposit some of their triglycerides in adipose cells, VLDLs are converted to LDLs.

Low-density lipoproteins (LDLs) contain 25% proteins, 5% triglycerides, 20% phospholipids, and 50% cholesterol. They carry about 75% of the total cholesterol in blood and deliver it to cells throughout the body for use in repair of cell membranes and synthesis of steroid hormones and bile salts. LDLs contain a single apoprotein, **apo B100**, which is the docking protein that binds to LDL receptors on the plasma membrane of body cells so that LDL can enter the cell via receptor-mediated endocytosis. Within the cell, the LDL is broken down, and the cholesterol is released to serve the cell's needs. Once a cell has sufficient cholesterol for its activities, a negative feedback system inhibits the cell's synthesis of new LDL receptors.

When present in excessive numbers, LDLs also deposit cholesterol in and around smooth muscle fibers in arteries, forming fatty plaques that increase the risk of coronary artery disease (see Disorders: Homeostatic Imbalances in Chapter 20 of the Study Guide). For this reason, the cholesterol in LDLs, called LDL-cholesterol, is known as "bad" cholesterol. Because some people have too few LDL receptors, their body cells remove LDL from the blood less efficiently; as a result, their plasma LDL level is abnormally high, and they are more likely to develop fatty plaques. Eating a high-fat diet increases the production of VLDLs, which elevates the LDL level and increases the formation of fatty plaques.

High-density lipoproteins (HDLs), which contain 40–45% proteins, 5–10% triglycerides, 30% phospholipids, and 20% cholesterol, remove excess cholesterol from body cells and the blood and transport it to the liver for elimination. Because HDLs prevent accumulation of cholesterol in the blood, a high HDL level is associated with decreased risk of coronary artery disease. For this reason, HDL-cholesterol is known as "good" cholesterol.

Sources and Significance of Blood Cholesterol

There are two sources of cholesterol in the body. Some is present in foods (eggs, dairy products, organ meats, beef, pork, and processed luncheon meats), but most is synthesized by hepatocytes. Fatty foods that don't contain any cholesterol at all can still dramatically increase blood cholesterol level in two ways. First, a high intake of dietary fats stimulates reabsorption of cholesterol-containing bile back into the blood, so less cholesterol is lost in the feces. Second, when saturated fats are broken down in the body, hepatocytes use some of the breakdown products to make cholesterol.

A lipid profile test usually measures total cholesterol (TC), HDL-cholesterol, and triglycerides (VLDLs). LDL-cholesterol then is calculated by using the following formula: LDL-cholesterol = TC − HDL-cholesterol − (triglycerides/5). In the United States, blood cholesterol is usually measured in milligrams per deciliter (mg/dL); a deciliter is 0.1 liter or 100 mL. For adults, desirable levels of blood cholesterol are total cholesterol under 200 mg/dL, LDL-cholesterol under 130 mg/dL, and HDL-cholesterol over 40 mg/dL. Normally, triglycerides are in the range of 10–190 mg/dL.

As total cholesterol level increases, the risk of coronary artery disease begins to rise. When total cholesterol is above 200 mg/dL (5.2 mmol/liter), the risk of a heart attack doubles with every 50 mg/dL (1.3 mmol/liter) increase in total cholesterol. Total cholesterol of 200–239 mg/dL and LDL of 130–159 mg/dL are borderline-high; total cholesterol above 239 mg/dL and LDL above 159 mg/dL are classified as high blood cholesterol. The ratio of total cholesterol to HDL-cholesterol predicts the risk of developing coronary artery disease. For example, a person with a total cholesterol of 180 mg/dL and HDL of 60 mg/dL has a risk ratio of 3. Ratios above 4 are considered undesirable; the higher the ratio, the greater the risk of developing coronary artery disease.

Among the therapies used to reduce blood cholesterol level are exercise, diet, and drugs. Regular physical activity at aerobic and nearly aerobic levels raises HDL level. Dietary changes are aimed at reducing the intake of total fat, saturated fats, and cholesterol. Drugs used to treat high blood cholesterol levels include cholestyramine (Questran) and colestipol (Colestid), which promote excretion of bile in the feces; nicotinic acid (Liponicin); and the "statin" drugs—atorvastatin (Lipitor), lovastatin (Mevacor), and simvastatin (Zocor), which block the key enzyme (HMG-CoA reductase) needed for cholesterol synthesis.

The Fate of Lipids

Lipids, like carbohydrates, may be oxidized to produce ATP. If the body has no immediate need to use lipids in this way, they are stored in adipose tissue (fat depots) throughout the body and in the liver. A few lipids are used as structural molecules or to synthesize other essential substances. Some examples include phospholipids, which are constituents of plasma membranes; lipoproteins, which are used to transport cholesterol throughout the body; thromboplastin, which is needed for blood clotting; and myelin sheaths, which speed up nerve impulse conduction. Two **essential fatty acids** that the body cannot synthesize are linoleic acid and linolenic acid. Dietary sources include vegetable oils and leafy vegetables. The various functions of lipids in the body may be reviewed in Table 2.7.

Triglyceride Storage

A major function of adipose tissue is to remove triglycerides from chylomicrons and VLDLs and store them until they are needed for ATP production in other parts of the body. Triglycerides stored in adipose tissue constitute 98% of all body energy reserves. They are stored more readily than glycogen, in part because triglycerides are hydrophobic and do not exert osmotic pressure on cell membranes. Adipose tissue also insulates and protects various parts of the body. Adipocytes in the subcutaneous layer contain about 50% of the stored triglycerides. Other adipose tissues account for the other half: about 12% around the kidneys, 10–15% in the omenta, 15% in genital areas, 5–8% between muscles, and 5% behind the eyes, in the sulci of the heart, and attached to the outside of the large intestine. Triglycerides in adipose tissue are continually broken down and resynthesized. Thus, the triglycerides stored in adipose tissue today are not the same molecules that were present last month because they are continually released from storage, transported in the blood, and redeposited in other adipose tissue cells.

Lipid Catabolism: Lipolysis

In order for muscle, liver, and adipose tissue to oxidize the fatty acids derived from triglycerides to produce ATP, the triglycerides must first be split into glycerol and fatty acids, a process called **lipolysis** (li-POL-i-sis). Lipolysis is catalyzed by enzymes called **lipases**. Epinephrine and norepinephrine enhance triglyceride breakdown into fatty acids and glycerol. These hormones are released when sympathetic tone increases, as occurs, for example, during exercise. Other lipolytic hormones include cortisol, thyroid hormones, and insulinlike growth factors. By contrast, insulin inhibits lipolysis.

The glycerol and fatty acids that result from lipolysis are catabolized via different pathways (**Figure 25.14**). Glycerol is converted by many cells of the body to glyceraldehyde 3-phosphate, one of the compounds also formed during the catabolism of glucose. If ATP supply in a cell is high, glyceraldehyde 3-phosphate is converted into glucose, an example of gluconeogenesis. If ATP supply in a cell is low, glyceraldehyde 3-phosphate enters the catabolic pathway to pyruvic acid.

Fatty acids are catabolized differently than glycerol and yield more ATP. The first stage in fatty acid catabolism is a series of reactions, collectively called **beta oxidation** (BĀ-ta), that occurs in the matrix of mitochondria. Enzymes remove two carbon atoms at a time from the long chain of carbon atoms composing a fatty acid and attach the resulting two-carbon fragment to coenzyme A, forming acetyl CoA. Then, acetyl CoA enters the Krebs cycle (**Figure 25.14**). A 16-carbon fatty acid such as palmitic acid can yield as many as 129 ATPs on its complete oxidation via beta oxidation, the Krebs cycle, and the electron transport chain.

As part of normal fatty acid catabolism, hepatocytes can take two acetyl CoA molecules at a time and condense them to form **acetoacetic acid** (as′-ē-tō-a-SĒ-tik). This reaction liberates the bulky CoA portion,

FIGURE 25.14 Pathways of lipid metabolism. Glycerol may be converted to glyceraldehyde 3-phosphate, which can then be converted to glucose or enter the Krebs cycle for oxidation. Fatty acids undergo beta oxidation and enter the Krebs cycle via acetyl coenzyme A. The synthesis of lipids from glucose or amino acids is called lipogenesis.

Glycerol and fatty acids are catabolized in separate pathways.

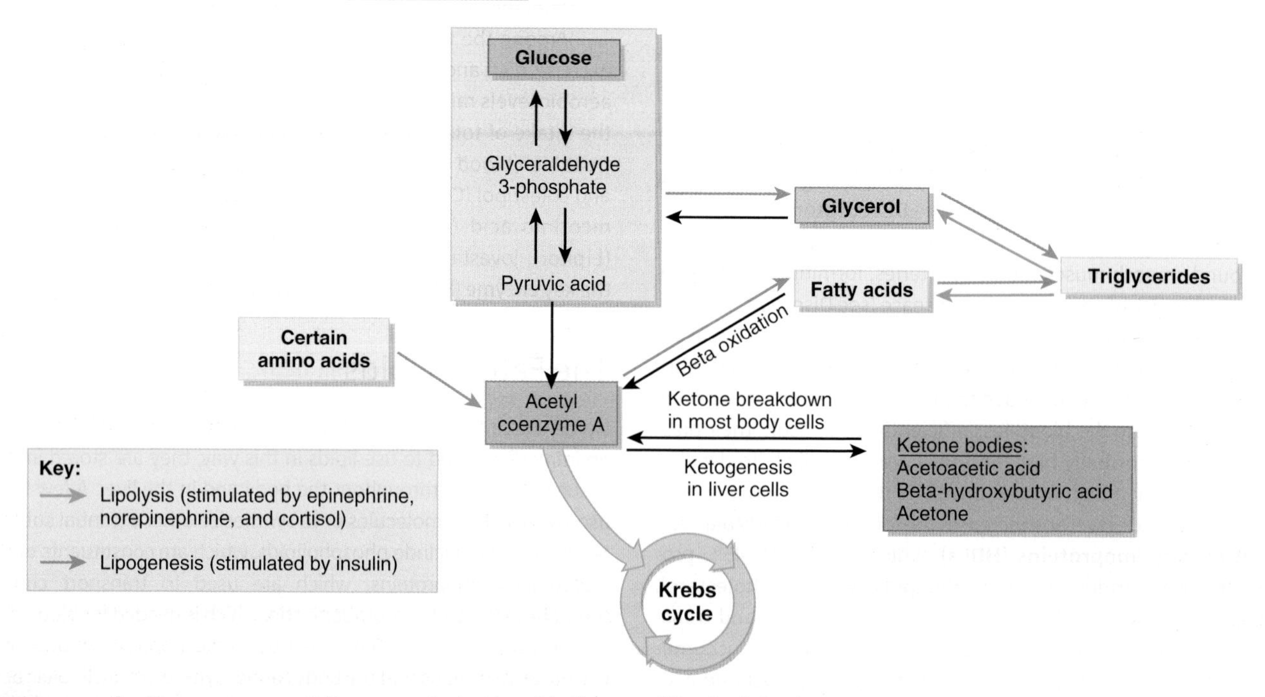

Q What types of cells can carry out lipogenesis, beta oxidation, and lipolysis? What type of cell can carry out ketogenesis?

which cannot diffuse out of cells. Some acetoacetic acid is converted into **beta-hydroxybutyric acid** (hī-drok-sē-bū-TIR-ik) and **acetone** (AS-e-tōn). The formation of these three substances, collectively known as **ketone bodies** (KĒ-tōn), is called **ketogenesis** (kē-tō-JEN-e-sis) (**Figure 25.14**). Because ketone bodies freely diffuse through plasma membranes, they leave hepatocytes and enter the bloodstream.

Other cells take up acetoacetic acid and attach its four carbons to two coenzyme A molecules to form two acetyl CoA molecules, which can then enter the Krebs cycle for oxidation. Heart muscle and the cortex (outer part) of the kidneys use acetoacetic acid in preference to glucose for generating ATP. Hepatocytes, which make acetoacetic acid, cannot use it for ATP production because they lack the enzyme that transfers acetoacetic acid back to coenzyme A.

Lipid Anabolism: Lipogenesis

Liver cells and adipose cells can synthesize lipids from glucose or amino acids through **lipogenesis** (**Figure 25.14**), which is stimulated by insulin. Lipogenesis occurs when individuals consume more calories than are needed to satisfy their ATP needs. Excess dietary carbohydrates, proteins, and fats all have the same fate—they are converted into triglycerides. Certain amino acids can undergo the following reactions: amino acids → acetyl CoA → fatty acids → triglycerides. The use of glucose to form lipids takes place via two pathways: (1) glucose → glyceraldehyde 3-phosphate → glycerol and (2) glucose → glyceraldehyde 3-phosphate → acetyl CoA → fatty acids. The resulting glycerol and fatty acids can undergo anabolic reactions to become stored triglycerides, or they can go through a series of anabolic reactions to produce other lipids such as lipoproteins, phospholipids, and cholesterol.

See Clinical Connection: Ketosis

25.5 Protein Metabolism

OBJECTIVE

- **Describe** the fate, metabolism, and functions of proteins.

During digestion, **proteins** are broken down into amino acids. Unlike carbohydrates and triglycerides, which are stored, proteins are not warehoused for future use. Instead, amino acids are either oxidized to produce ATP or used to synthesize new proteins for body growth and repair. Excess dietary amino acids are not excreted in the urine or feces but instead are converted into glucose (gluconeogenesis) or triglycerides (lipogenesis).

The Fate of Proteins

The active transport of amino acids into body cells is stimulated by insulinlike growth factors (IGFs) and insulin. Almost immediately after digestion, amino acids are reassembled into proteins. Many proteins function as enzymes; others are involved in transportation (hemoglobin) or serve as antibodies, clotting chemicals (fibrinogen), hormones (insulin), or contractile elements in muscle fibers (actin and myosin). Several proteins serve as structural components of the body (collagen, elastin, and keratin). The various functions of proteins in the body may be reviewed in **Table 2.8**.

Protein Catabolism

A certain amount of protein catabolism occurs in the body each day, stimulated mainly by cortisol from the adrenal cortex. Proteins from worn-out cells (such as red blood cells) are broken down into amino acids. Some amino acids are converted into other amino acids, peptide bonds are re-formed, and new proteins are synthesized as part of the recycling process. Hepatocytes convert some amino acids to fatty acids, ketone bodies, or glucose. Cells throughout the body oxidize a small amount of amino acids to generate ATP via the Krebs cycle and the electron transport chain. However, before amino acids can be oxidized, they must first be converted to molecules that are part of the Krebs cycle or can enter the Krebs cycle, such as acetyl CoA (**Figure 25.15**). Before amino acids can enter the Krebs cycle, their amino group (NH_2) must first be removed—a process called **deamination** (dē-am′-i-NĀ-shun). Deamination occurs in hepatocytes and produces ammonia (NH_3). The liver cells then convert the highly toxic ammonia to urea, a relatively harmless substance that is excreted in the urine. The conversion of amino acids into glucose (gluconeogenesis) may be reviewed in **Figure 25.12**; the conversion of amino acids into fatty acids (lipogenesis) or ketone bodies (ketogenesis) is shown in **Figure 25.14**.

Protein Anabolism

Protein anabolism, the formation of peptide bonds between amino acids to produce new proteins, is carried out on the ribosomes of almost every cell in the body, directed by the cells' DNA and RNA (see **Figure 3.29**). Insulinlike growth factors, thyroid hormones (T_3 and T_4), insulin, estrogen, and testosterone all stimulate protein synthesis. Because proteins are a main component of most cell structures, adequate dietary protein is especially essential during the growth years, during pregnancy, and when tissue has been damaged by disease or injury. Once dietary intake of protein is adequate, eating more protein will not increase bone or muscle mass; only a regular program of forceful, weight-bearing muscular activity accomplishes that goal.

Of the 20 amino acids in the human body, 10 are **essential amino acids**: They must be present in the diet because they cannot be synthesized in the body in adequate amounts. It is *essential* to include them in your diet. Humans are unable to synthesize eight amino acids (isoleucine, leucine, lysine, methionine, phenylalanine, threonine, tryptophan, and valine) and synthesize two others (arginine and histidine) in inadequate amounts, especially in childhood. A **complete protein** contains sufficient amounts of all essential amino acids. Beef, fish, poultry, eggs, and milk are examples of foods that contain complete proteins. An **incomplete protein** does not contain all essential amino acids. Examples of incomplete proteins are leafy green vegetables, legumes (beans and peas), and grains. **Nonessential amino**

FIGURE 25.15 **Points at which amino acids (yellow boxes) enter the Krebs cycle for oxidation.**

Before amino acids can be catabolized, they must first be converted to various substances that can enter the Krebs cycle.

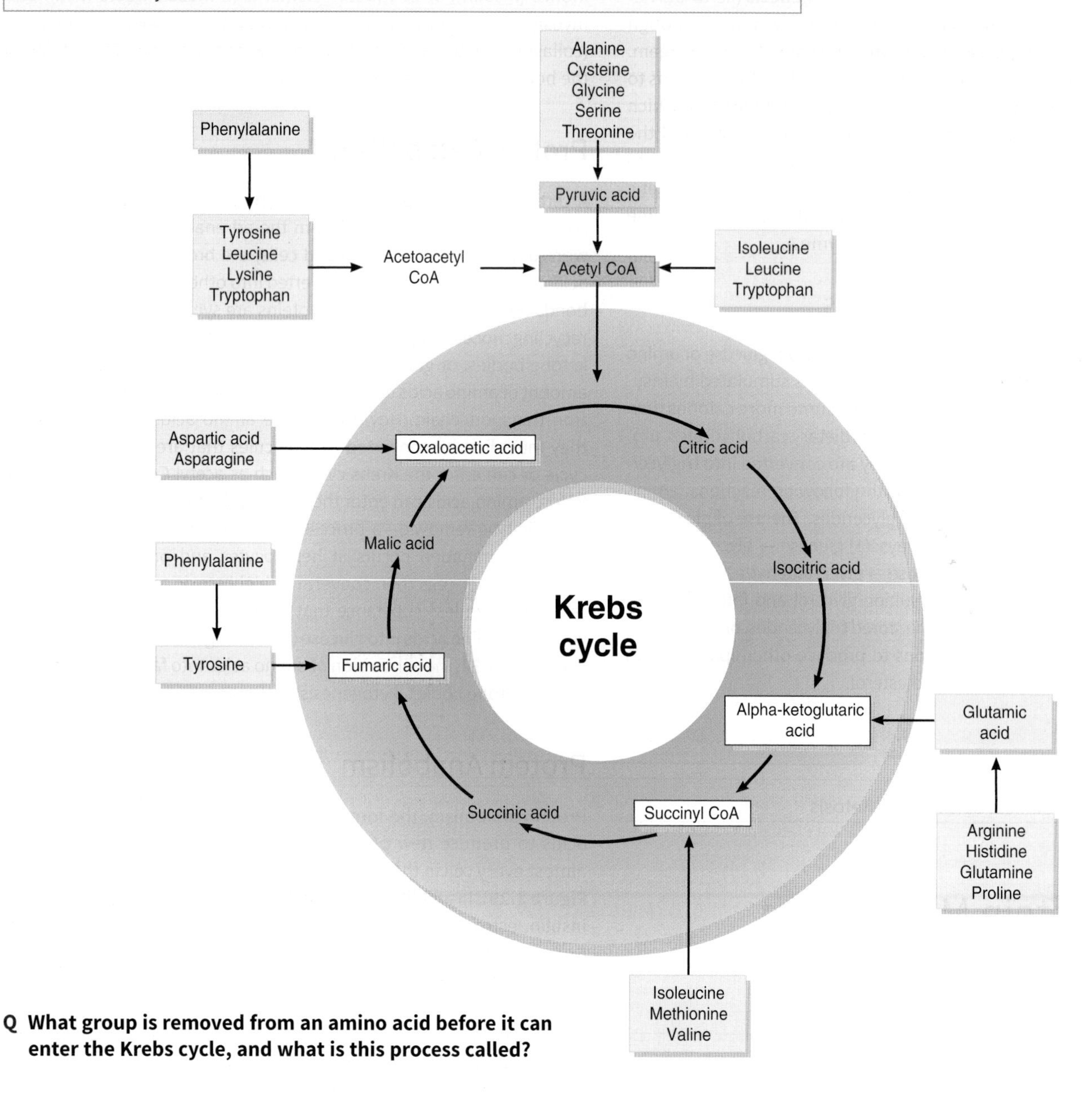

Q What group is removed from an amino acid before it can enter the Krebs cycle, and what is this process called?

acids can be synthesized by body cells. They are formed by **transamination** (trans′-am-i-NĀ-shun), the transfer of an amino group from an amino acid to pyruvic acid or to an acid in the Krebs cycle. Once the appropriate essential and nonessential amino acids are present in cells, protein synthesis occurs rapidly.

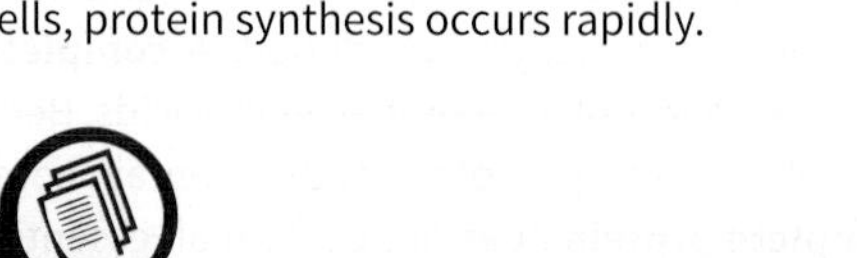

See Clinical Connection: Phenylketonuria

25.6 Key Molecules at Metabolic Crossroads

OBJECTIVE

- **Describe** the reactions of key molecules and the products formed during metabolism.

FIGURE 25.16 **Summary of the roles of the key molecules in metabolic pathways.** Double-headed arrows indicate that reactions between two molecules may proceed in either direction, if the appropriate enzymes are present and the conditions are favorable; single-headed arrows signify the presence of an irreversible step.

Three molecules—glucose 6-phosphate, pyruvic acid, and acetyl coenzyme A—stand at "metabolic crossroads." They can undergo different reactions depending on your nutritional or activity status.

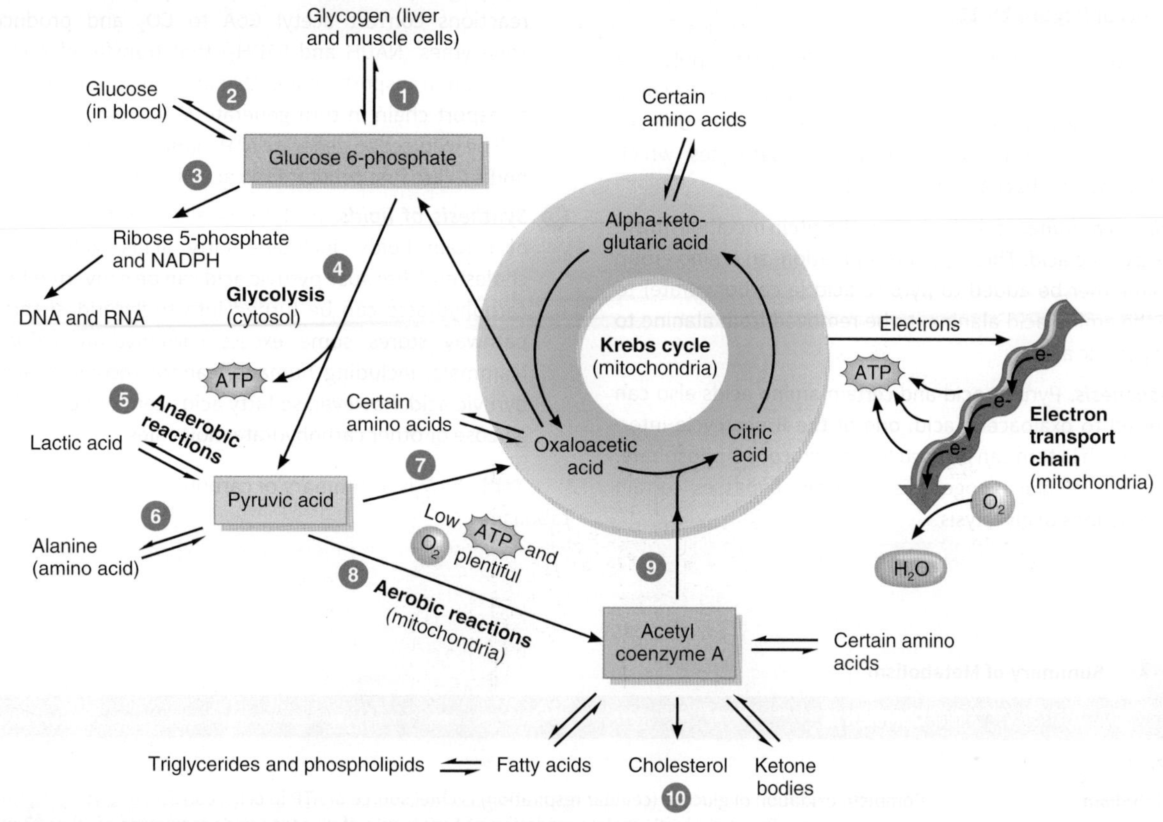

Q Which substance is the gateway into the Krebs cycle for molecules that are being oxidized to generate ATP?

Although there are thousands of different chemicals in cells, three molecules—glucose 6-phosphate, pyruvic acid, and acetyl coenzyme A—play pivotal roles in metabolism (**Figure 25.16**). These molecules stand at "metabolic crossroads"; as you will learn shortly, the reactions that occur (or do not occur) depend on the nutritional or activity status of the individual. Reactions 1 through 7 in **Figure 25.16** occur in the cytosol, reactions 8 and 9 occur inside mitochondria, and reactions indicated by 10 occur on smooth endoplasmic reticulum.

The Role of Glucose 6-Phosphate

Shortly after glucose enters a body cell, a kinase converts it to **glucose 6-phosphate**. Four possible fates await glucose 6-phosphate (see **Figure 25.16**):

1. ***Synthesis of glycogen.*** When glucose is abundant in the bloodstream, as it is just after a meal, a large amount of glucose 6-phosphate is used to synthesize glycogen, the storage form of carbohydrate in animals. Subsequent breakdown of glycogen into glucose 6-phosphate occurs through a slightly different series of reactions. Synthesis and breakdown of glycogen occur mainly in skeletal muscle fibers and hepatocytes.
2. ***Release of glucose into the bloodstream.*** If the enzyme glucose 6-phosphatase is present and active, glucose 6-phosphate can be dephosphorylated to glucose. Once glucose is released from the phosphate group, it can leave the cell and enter the bloodstream. Hepatocytes are the main cells that can provide glucose to the bloodstream in this way.
3. ***Synthesis of nucleic acids.*** Glucose 6-phosphate is the precursor used by cells throughout the body to make ribose 5-phosphate, a 5-carbon sugar that is needed for synthesis of RNA (ribonucleic acid) and DNA (deoxyribonucleic acid). The same sequence of reactions also produces NADPH. This molecule is a hydrogen and electron donor in certain reduction reactions, such as synthesis of fatty acids and steroid hormones.
4. ***Glycolysis.*** Some ATP is produced anaerobically via glycolysis, in which glucose 6-phosphate is converted to pyruvic acid, another key molecule in metabolism. Most body cells carry out glycolysis.

The Role of Pyruvic Acid

Each 6-carbon molecule of glucose that undergoes glycolysis yields two 3-carbon molecules of **pyruvic acid** (pī-ROO-vik). This molecule, like glucose 6-phosphate, stands at a metabolic crossroads: Given enough oxygen, the aerobic (oxygen-consuming) reactions of cellular respiration can proceed; if oxygen is in short supply, anaerobic reactions can occur (Figure 25.16):

5. ***Production of lactic acid.*** When oxygen is in short supply in a tissue, as in actively contracting skeletal or cardiac muscle, some pyruvic acid is changed to lactic acid. The lactic acid then diffuses into the bloodstream and is taken up by hepatocytes, which eventually convert it back to pyruvic acid.
6. ***Production of alanine.*** Carbohydrate and protein metabolism are linked by pyruvic acid. Through transamination, an amino group ($-NH_2$) can either be added to pyruvic acid (a carbohydrate) to produce the amino acid alanine, or be removed from alanine to generate pyruvic acid.
7. ***Gluconeogenesis.*** Pyruvic acid and certain amino acids also can be converted to oxaloacetic acid, one of the Krebs cycle intermediates, which in turn can be used to form glucose 6-phosphate. This sequence of gluconeogenesis reactions bypasses certain one-way reactions of glycolysis.

The Role of Acetyl Coenzyme A

8. When the ATP level in a cell is low but oxygen is plentiful, most pyruvic acid streams toward ATP-producing reactions—the Krebs cycle and electron transport chain—via conversion to **acetyl coenzyme A**.
9. ***Entry into the Krebs cycle.*** Acetyl CoA is the vehicle for 2-carbon acetyl groups to enter the Krebs cycle. Oxidative Krebs cycle reactions convert acetyl CoA to CO_2 and produce reduced coenzymes (NADH and $FADH_2$) that transfer electrons into the electron transport chain. Oxidative reactions in the electron transport chain in turn generate ATP. Most fuel molecules that will be oxidized to generate ATP—glucose, fatty acids, and ketone bodies—are first converted to acetyl CoA.
10. ***Synthesis of lipids.*** Acetyl CoA also can be used for synthesis of certain lipids, including fatty acids, ketone bodies, and cholesterol. Because pyruvic acid can be converted to acetyl CoA, carbohydrates can be turned into triglycerides; this metabolic pathway stores some excess carbohydrate calories as fat. Mammals, including humans, cannot reconvert acetyl CoA to pyruvic acid, however, so fatty acids cannot be used to generate glucose or other carbohydrate molecules.

Table 25.2 is a summary of carbohydrate, lipid, and protein metabolism.

TABLE 25.2 Summary of Metabolism

PROCESS	COMMENTS
CARBOHYDRATES	
Glucose catabolism	Complete oxidation of glucose (cellular respiration) is chief source of ATP in cells; consists of glycolysis, Krebs cycle, and electron transport chain. Complete oxidation of 1 molecule of glucose yields maximum of 30 or 32 molecules of ATP.
Glycolysis	Conversion of glucose into pyruvic acid results in production of some ATP. Reactions do not require oxygen.
Krebs cycle	Cycle includes series of oxidation–reduction reactions in which coenzymes (NAD^+ and FAD) pick up hydrogen ions and hydride ions from oxidized organic acids; some ATP produced. CO_2 and H_2O are by-products. Reactions are aerobic.
Electron transport chain	Third set of reactions in glucose catabolism: another series of oxidation–reduction reactions, in which electrons are passed from one carrier to next; most ATP produced. Reactions require oxygen (aerobic cellular respiration).
Glucose anabolism	Some glucose is converted into glycogen (glycogenesis) for storage if not needed immediately for ATP production. Glycogen can be reconverted to glucose (glycogenolysis). Conversion of amino acids, glycerol, and lactic acid into glucose is called gluconeogenesis.
LIPIDS	
Triglyceride catabolism	Triglycerides are broken down into glycerol and fatty acids. Glycerol may be converted into glucose (gluconeogenesis) or catabolized via glycolysis. Fatty acids are catabolized via beta oxidation into acetyl coenzyme A that can enter Krebs cycle for ATP production or be converted into ketone bodies (ketogenesis).
Triglyceride anabolism	Synthesis of triglycerides from glucose and fatty acids is called lipogenesis. Triglycerides are stored in adipose tissue.
PROTEINS	
Protein catabolism	Amino acids are oxidized via Krebs cycle after deamination. Ammonia resulting from deamination is converted into urea in liver, passed into blood, and excreted in urine. Amino acids may be converted into glucose (gluconeogenesis), fatty acids, or ketone bodies.
Protein anabolism	Protein synthesis is directed by DNA and utilizes cells' RNA and ribosomes.

25.7 Metabolic Adaptations

OBJECTIVE

- **Compare** metabolism during the absorptive and postabsorptive states.

Regulation of metabolic reactions depends both on the chemical environment within body cells, such as the levels of ATP and oxygen, and on signals from the nervous and endocrine systems. Some aspects of metabolism depend on how much time has passed since the last meal. During the **absorptive state**, ingested nutrients are entering the bloodstream, and glucose is readily available for ATP production. During the **postabsorptive state**, absorption of nutrients from the GI tract is complete, and energy needs must be met by fuels already in the body. A typical meal requires about 4 hours for complete absorption; given three meals a day, the absorptive state exists for about 12 hours each day. Assuming no between-meal snacks, the other 12 hours—typically late morning, late afternoon, and most of the night—are spent in the postabsorptive state.

Because the nervous system and red blood cells continue to depend on glucose for ATP production during the postabsorptive state, maintaining a steady blood glucose level is critical during this period. Hormones are the major regulators of metabolism in each state. The effects of insulin dominate in the absorptive state; several other hormones regulate metabolism in the postabsorptive state. During fasting and starvation, many body cells turn to ketone bodies for ATP production, as noted in the Clinical Connection on Ketosis in (see Section 25.4 for reference).

Metabolism during the Absorptive State

Soon after a meal, nutrients start to enter the blood. Recall that ingested food reaches the bloodstream mainly as glucose, amino acids, and triglycerides (in chylomicrons).

Absorptive State Reactions During the absorptive state, some of the absorbed nutrients are catabolized for the body's energy needs or are used to synthesize proteins. The following reactions of the absorptive state reflect this function (Figure 25.17):

1. ***Catabolism of glucose***. Most cells of the body produce the majority of their ATP by catabolizing glucose via cellular respiration. Hence glucose is the body's main energy source during the absorptive state. About 50% of the glucose absorbed from a typical meal is catabolized by cells throughout the body to produce ATP.
2. ***Catabolism of amino acids***. Some amino acids enter hepatocytes (liver cells), where they are deaminated to keto acids. The keto acids in turn can either enter the Krebs cycle for ATP production or be used to synthesize glucose or fatty acids.
3. ***Protein synthesis***. Many amino acids enter body cells, such as muscle cells and hepatocytes, for synthesis of proteins.
4. ***Catabolism of few dietary lipids.*** During the absorptive state, only a small portion of dietary lipids are catabolized for energy; most dietary lipids are stored in adipose tissue.

Another key event of the absorptive state is that absorbed nutrients in excess of the body's energy needs are converted into **nutrient stores**—namely glycogen and fat. This function is reflected by the following absorptive state reactions (Figure 25.17):

5. ***Glycogenesis.*** Some of the glucose that may be in excess of the body's needs in taken up by the liver and skeletal muscle and then converted into glycogen (glycogenesis).
6. ***Lipogenesis***. The liver can also convert excess glucose or amino acids to fatty acids for use in the synthesis of triglycerides (lipogenesis). Adipocytes also take up glucose not picked up by the liver and convert it into triglycerides for storage. Overall, about 40% of the glucose absorbed from a meal is converted to triglycerides, and about 10% is stored as glycogen in skeletal muscles and the liver.
7. ***Transport of triglycerides from liver to adipose tissue.*** Some fatty acids and triglycerides synthesized in the liver remain there, but hepatocytes package most into very low-density lipoproteins (VLDLs), which carry lipids to adipose tissue for storage.

Regulation of Metabolism during the Absorptive State Soon after a meal, glucose-dependent insulinotropic peptide (GIP), plus the rising blood levels of glucose and certain amino acids, stimulates pancreatic beta cells to release the hormone insulin. In general, insulin increases the activity of enzymes needed for anabolism and the synthesis of storage molecules; at the same time, it decreases the activity of enzymes needed for catabolic or breakdown reactions. Insulin promotes the entry of glucose and amino acids into cells of many tissues, and it stimulates the conversion of glucose to glycogen (glycogenesis) in both liver and muscle cells. In liver and adipose tissue, insulin enhances the synthesis of triglycerides (lipogenesis), and in cells throughout the body, insulin stimulates protein synthesis. (See Section 18.10 to review the effects of insulin.) Insulin-like growth factors and the thyroid hormones (T_3 and T_4) also stimulate protein synthesis.

Before glucose can be used by body cells, it must first pass through the plasma membrane and enter the cytosol. Glucose entry into most body cells occurs via **glucose transporter (GLUT)** molecules, a family of transporters that bring glucose into cells via facilitated diffusion. A high level of insulin increases the insertion of one type of GLUT, called **GLUT4**, into the plasma membranes of most body cells (especially muscle fiber and adipocytes), increasing the rate of facilitated diffusion of glucose into cells. In neurons and hepatocytes, however, other types of GLUTs are always present in the plasma membrane, so glucose entry is always "turned on." Upon entering a cell, glucose becomes phosphorylated. Because GLUT cannot transport phosphorylated glucose, this reaction traps glucose within the cell. Table 25.3 summarizes the hormonal regulation of metabolism in the absorptive state.

FIGURE 25.17 Principal metabolic pathways during the absorptive state.

During the absorptive state, most body cells produce ATP by catabolizing glucose to CO_2 and H_2O.

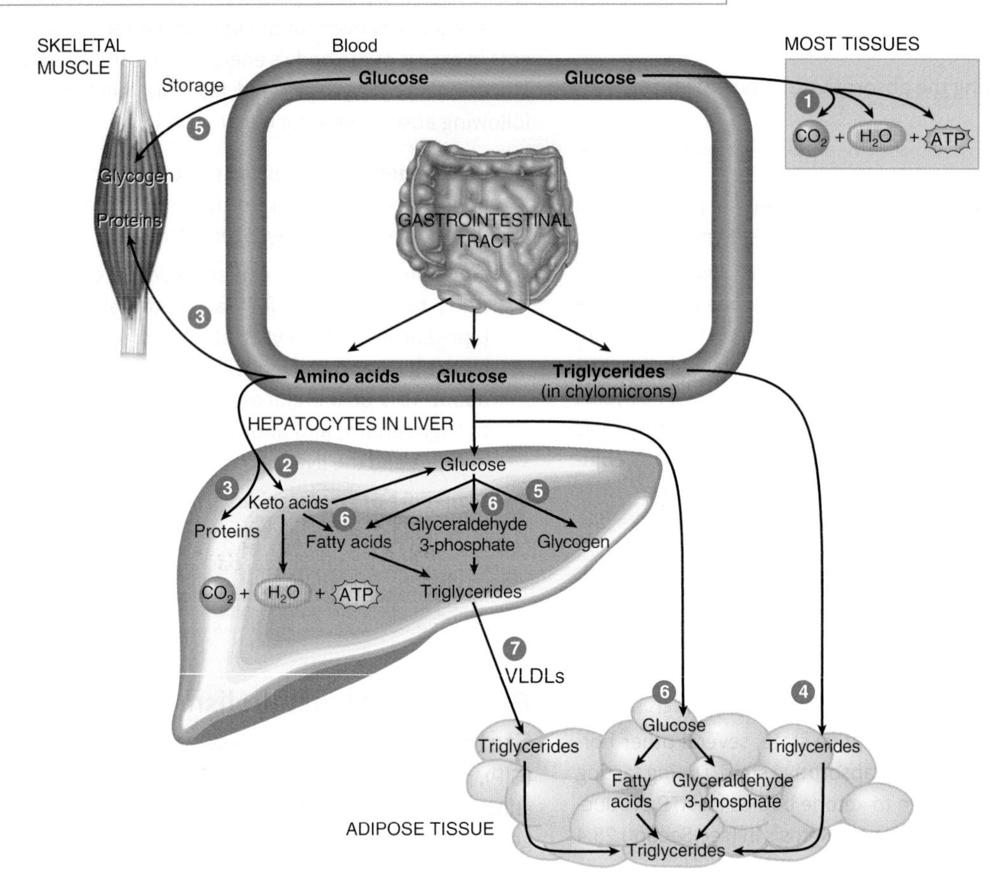

Q Are the reactions shown in this figure mainly anabolic or catabolic?

TABLE 25.3 Hormonal Regulation of Metabolism in the Absorptive State

PROCESS	LOCATION(S)	MAIN STIMULATING HORMONE(S)
Facilitated diffusion of glucose into cells	Most cells.	Insulin.*
Active transport of amino acids into cells	Most cells.	Insulin.
Glycogenesis (glycogen synthesis)	Hepatocytes and muscle fibers.	Insulin.
Protein synthesis	All body cells.	Insulin, thyroid hormones, and insulinlike growth factors.
Lipogenesis (triglyceride synthesis)	Adipose cells and hepatocytes.	Insulin.

*Facilitated diffusion of glucose into hepatocytes (liver cells) and neurons is always "turned on" and does not require insulin.

Metabolism during the Postabsorptive State

About 4 hours after the last meal, absorption of nutrients from the small intestine is complete, and the blood glucose level starts to fall because glucose continues to leave the bloodstream and enter body cells while none is being absorbed from the GI tract. Thus, the main metabolic challenge during the postabsorptive state is to maintain the normal blood glucose level of 70–110 mg/100 mL (3.9–6.1 mmol/liter). Homeostasis of blood glucose concentration is especially important for the nervous system and for red blood cells for the following reasons:

- The dominant fuel molecule for ATP production in the nervous system is glucose because fatty acids are unable to pass the blood–brain barrier.
- Red blood cells derive all of their ATP from glycolysis of glucose because they have no mitochondria, so the Krebs cycle and the electron transport chain are not available to them.

Postabsorptive State Reactions A key feature of the postabsorptive state is that the blood glucose concentration is maintained at a normal level due to the breakdown of the body's nutrient stores (glycogen and fat) and the formation of new glucose from noncarbohydrate sources (gluconeogenesis). The reactions of the postabsorptive state that produce glucose are as follows (**Figure 25.18**):

1. ***Glycogenolysis in the liver.*** During the postabsorptive state, a major source of blood glucose is liver glycogenolysis, which can provide about 4-hour supply of glucose. Once glycogenolysis occurs in the liver, the glucose is released into the blood.
2. ***Glycogenolysis in muscle.*** Glycogenolysis can also occur in skeletal muscle. However, in skeletal muscle, the glucose that is formed from glycogenolysis is catabolized to provide ATP for muscle contraction: Glycogen is broken down to glucose 6-phosphate, which undergoes glycolysis. If anaerobic conditions exist in the skeletal muscle, the pyruvic acid is converted to lactic acid, which is released into the blood. The liver takes up the lactic acid, converts it back to glucose, and then releases glucose into the blood.
3. ***Lipolysis.*** In adipose tissue, triglycerides are broken down into fatty acids and glycerol, which are released into the blood. The glycerol is taken up by the liver and then converted into glucose, which in turn is released into the bloodstream.
4. ***Protein catabolism.*** Modest breakdown of proteins in skeletal muscle and other tissues releases amino acids, which then can be converted to glucose by the liver. The glucose in turn is released into the bloodstream.
5. ***Gluconeogenesis.*** During the postabsorptive state, new glucose is formed from noncarbohydrate sources. Examples of gluconeogenesis include the formation of glucose from lactic acid, glycerol, or an amino acid.

Another hallmark feature of the postabsorptive state is that glucose sparing occurs. **Glucose sparing** means that most body cells switch to other fuels besides glucose as their main source of energy, leaving more glucose in the blood for the brain and red blood cells. The following reactions produce ATP without using glucose (**Figure 25.18**):

6. ***Catabolism of fatty acids.*** The fatty acids released by lipolysis of triglycerides cannot be used for glucose production because acetyl CoA cannot be readily converted to pyruvic acid. But most

FIGURE 25.18 Principal metabolic pathways during the postabsorptive state.

The principal function of postabsorptive state reactions is to maintain a normal blood glucose level.

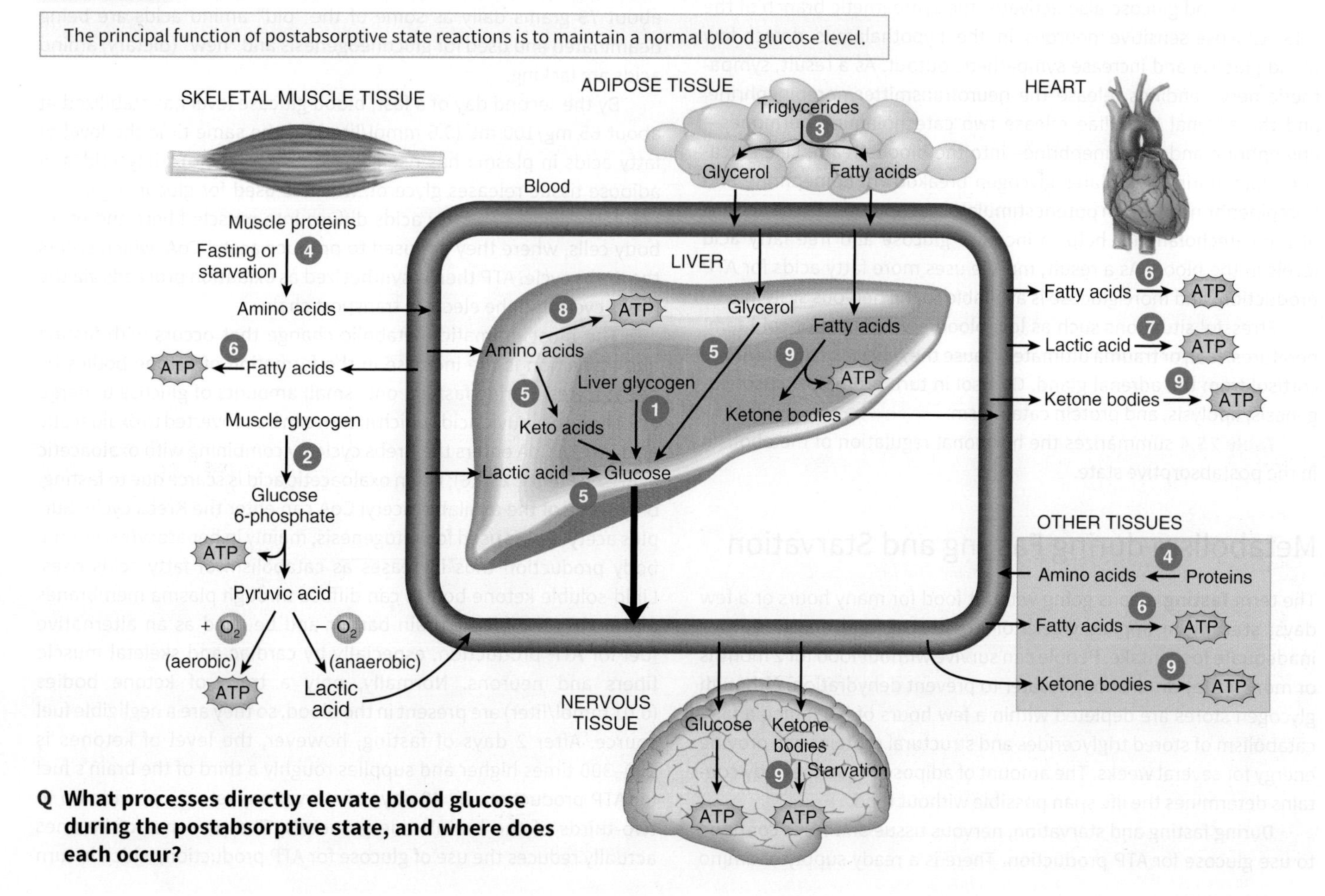

Q What processes directly elevate blood glucose during the postabsorptive state, and where does each occur?

cells can catabolize the fatty acids directly, feed them into the Krebs cycle as acetyl CoA, and produce ATP through the electron transport chain.

7. ***Catabolism of lactic acid.*** Cardiac muscle can produce ATP aerobically from lactic acid.
8. ***Catabolism of amino acids.*** In hepatocytes, amino acids may be catabolized directly to produce ATP.
9. ***Catabolism of ketone bodies.*** Hepatocytes also convert fatty acids to ketone bodies (acetoacetic acid, beta-hydroxybutyric acid, and acetone), which can be used by the heart, kidneys, and other tissues for ATP production.

Regulation of Metabolism during the Postabsorptive State

Both hormones and the sympathetic division of the autonomic nervous system (ANS) regulate metabolism during the postabsorptive state. The hormones that regulate postabsorptive state metabolism sometimes are called anti-insulin hormones because they counter the effects of insulin during the absorptive state. As blood glucose level declines, the secretion of insulin falls and the release of anti-insulin hormones rises.

When blood glucose concentration starts to drop, the pancreatic alpha cells release the hormone glucagon. The primary target tissue of glucagon is the liver; the major effect is increased release of glucose into the bloodstream due to gluconeogenesis and glycogenolysis.

Low blood glucose also activates the sympathetic branch of the ANS. Glucose-sensitive neurons in the hypothalamus detect low blood glucose and increase sympathetic output. As a result, sympathetic nerve endings release the neurotransmitter norepinephrine, and the adrenal medullae release two catecholamine hormones—epinephrine and norepinephrine—into the bloodstream. Like glucagon, epinephrine stimulates glycogen breakdown. Epinephrine and norepinephrine are both potent stimulators of lipolysis. These actions of the catecholamines help to increase glucose and free fatty acid levels in the blood. As a result, muscle uses more fatty acids for ATP production, and more glucose is available to the nervous system.

Stressful situations such as low blood glucose, hot or cold temperatures, fear, or trauma ultimately cause the release of the hormone cortisol from the adrenal gland. Cortisol in turn promotes gluconeogenesis, lipolysis, and protein catabolism.

Table 25.4 summarizes the hormonal regulation of metabolism in the postabsorptive state.

TABLE 25.4 Hormonal Regulation of Metabolism in the Postabsorptive State

PROCESS	LOCATION(S)	MAIN STIMULATING HORMONE(S)
Glycogenolysis (glycogen breakdown)	Hepatocytes and skeletal muscle fibers.	Glucagon and epinephrine.
Lipolysis (triglyceride breakdown)	Adipocytes.	Epinephrine, norepinephrine, cortisol, insulinlike growth factors, thyroid hormones, and others.
Protein breakdown	Most body cells, but especially skeletal muscle fibers.	Cortisol.
Gluconeogenesis (synthesis of glucose from noncarbohydrates)	Hepatocytes and kidney cortex cells.	Glucagon and cortisol.

Metabolism during Fasting and Starvation

The term **fasting** means going without food for many hours or a few days; **starvation** implies weeks or months of food deprivation or inadequate food intake. People can survive without food for 2 months or more if they drink enough water to prevent dehydration. Although glycogen stores are depleted within a few hours of beginning a fast, catabolism of stored triglycerides and structural proteins can provide energy for several weeks. The amount of adipose tissue the body contains determines the life span possible without food.

During fasting and starvation, nervous tissue and RBCs continue to use glucose for ATP production. There is a ready supply of amino acids for gluconeogenesis because lowered insulin and increased cortisol levels slow the pace of protein synthesis and promote protein catabolism. Most cells in the body, especially skeletal muscle cells (because of their high protein content), can spare a fair amount of protein before their performance is adversely affected. During the first few days of fasting, protein catabolism outpaces protein synthesis by about 75 grams daily as some of the "old" amino acids are being deaminated and used for gluconeogenesis and "new" (dietary) amino acids are lacking.

By the second day of a fast, blood glucose level has stabilized at about 65 mg/100 mL (3.6 mmol/liter); at the same time the level of fatty acids in plasma has risen fourfold. Lipolysis of triglycerides in adipose tissue releases glycerol, which is used for gluconeogenesis, and fatty acids. The fatty acids diffuse into muscle fibers and other body cells, where they are used to produce acetyl CoA, which enters the Krebs cycle. ATP then is synthesized as oxidation proceeds via the Krebs cycle and the electron transport chain.

The most dramatic metabolic change that occurs with fasting and starvation is the increase in the formation of ketone bodies by hepatocytes. During fasting, only small amounts of glucose undergo glycolysis to pyruvic acid, which in turn can be converted to oxaloacetic acid. Acetyl CoA enters the Krebs cycle by combining with oxaloacetic acid (see Figure 25.16); when oxaloacetic acid is scarce due to fasting, only some of the available acetyl CoA can enter the Krebs cycle. Surplus acetyl CoA is used for ketogenesis, mainly in hepatocytes. Ketone body production thus increases as catabolism of fatty acids rises. Lipid-soluble ketone bodies can diffuse through plasma membranes and across the blood–brain barrier and be used as an alternative fuel for ATP production, especially by cardiac and skeletal muscle fibers and neurons. Normally, only a trace of ketone bodies (0.01 mmol/liter) are present in the blood, so they are a negligible fuel source. After 2 days of fasting, however, the level of ketones is 100–300 times higher and supplies roughly a third of the brain's fuel for ATP production. By 40 days of starvation, ketones provide up to two-thirds of the brain's energy needs. The presence of ketones actually reduces the use of glucose for ATP production, which in turn decreases the demand for gluconeogenesis and slows the catabolism of muscle proteins later in starvation to about 20 grams daily.

25.8 Energy Balance

OBJECTIVES

- **Explain** what is meant by the term *energy balance.*
- **Discuss** the various factors that affect metabolic rate.
- **Describe** the role of the hypothalamus in the regulation of food intake.

Energy balance refers to the precise matching of energy intake (in food) to energy expenditure over time. When the energy content of food balances the energy used by all cells of the body, body weight remains constant (unless there is a gain or loss of water). In many people, weight stability persists despite large day-to-day variations in activity and food intake. In the more affluent nations, however, a large fraction of the population is overweight. Easy access to tasty, high-calorie foods and a "couch-potato" lifestyle both promote weight gain. Being overweight increases the risk of dying from a variety of cardiovascular and metabolic disorders, including hypertension, varicose veins, diabetes mellitus, arthritis, and certain cancers.

Food Calories

As you learned in Chapter 4, when catabolic reactions occur, energy is released. About 40% of this energy is used to perform biological work, such as active transport and muscle contraction. The remaining 60% is converted to heat, some of which helps maintain normal body temperature. Excess heat is lost to the environment. When the body catabolizes the organic compounds in food, the heat energy released can be measured in units called calories. A **calorie (cal)** is defined as the amount of energy in the form of heat required to raise the temperature of 1 gram of water 1°C. Because the calorie is a relatively small unit, the **kilocalorie (kcal)** or *Calorie (Cal)* (always spelled with an uppercase C) is often used to express the energy content of foods. A kilocalorie equals 1000 calories. Thus, when we say that a particular food item contains 500 Calories, we are actually referring to kilocalories.

Essentially all of the kilocalories in our food come from the catabolism of carbohydrates, proteins, and fats. The catabolism of carbohydrates or proteins yields about the same amount of energy—about 4 kcal/g. The catabolism of fat yields much more energy—about 9 kcal/g. Some foods or beverages may contains alcohol, and the catabolism of alcohol also yields energy—about 7 kcal/g. The energy content of carbohydrates, proteins, fats, and alcohol is summarized in **Table 25.5**.

The number of kilocalories from a component in a particular food can be calculated by multiplying the number of grams of that component by its energy content. For example, suppose that one slice of pizza contains 27 g of carbohydrate, 14 g of fat, and 12 g of protein. To calculate the number of kcal from carbohydrate in this slice of pizza, multiply the number of grams of carbohydrate in the pizza by the energy content of carbohydrates: 27 g carbohydrate × 4 kcal/g = 108 kcal. To calculate the number of kcal from fat in the slice of pizza, multiple the number of grams of fat in the pizza by the energy content of fat: 14 g fat × 9 kcal/g = 126 kcal. To calculate the number of kcal from protein in the slice of pizza, multiply the number of grams of protein in the pizza by the energy content of protein: 12 g protein × 4 kcal/g = 48 kcal. Finally, to calculate the total kcal in the slice of pizza, add together all of the kcal from carbohydrate, fat, and protein: 108 kcal + 126 kcal + 48 kcal = 282 kcal.

TABLE 25.5 Energy Content of Various Nutrients and Alcohol

NUTRIENT	ENERGY CONTENT
Carbohydrate	4 kcal/g
Protein	4 kcal/g
Fat	9 kcal/g
Alcohol	7 kcal/g

Table 25.6 lists the caloric content of several familiar foods. The higher the caloric content of a particular food, the greater the amount of energy released as it is catabolized. For example, the energy content of one medium apple is 80 kcal; this means that 80 kcal is the amount of energy released as the apple is catabolized. The energy content of a slice of chocolate cake is 247 kcal; this means that 247 kcal is the amount of energy released as the chocolate cake is catabolized. Suppose that you eat the apple or the chocolate cake. Based on the caloric content of these foods, your body will have to work harder (via exercise, for example) to release more energy in order to catabolize the chocolate cake compared with the apple.

Beverages can also be a source of calories. For example, a cola soft drink (12 ounces) contains 40 g of carbohydrate, 0 g of protein, and 0 g of fat, so the energy content of this soda is 160 kcal (40 g carbohydrate × 4 kcal/g). A typical serving of vodka (1.5 ounces) contains 0 g of carbohydrate, 0 g of protein, 0 g of fat, and 14 g of alcohol, so the energy content of this drink is 98 calories (14 g × 7 kcal/g). If juice, soda, or cocktail mix is added to the vodka, these solutions usually contain carbohydrates that contribute additional calories. **Table 25.7** lists the caloric content of several beverages.

Metabolic Rate

The overall rate at which metabolic reactions use energy is termed the **metabolic rate.** As you have already learned, some of the energy is used to produce ATP, and some is released as heat. Thus, the higher the metabolic rate, the higher the rate of heat production.

Several factors affect the metabolic rate:

- ***Hormones.*** Thyroid hormones (thyroxine and triiodothyronine) are the main regulators of basal metabolic rate (BMR), the metabolic rate under basal conditions (described shortly). BMR increases as the blood levels of thyroid hormones rise. The response to changing levels of thyroid hormones is slow, however, taking several days

TABLE 25.6 Caloric Content of Various Foods

FOOD	SERVING SIZE	ENERGY (kcal)	CARBOHYDRATE (g)	FAT (g)	PROTEIN (g)
Apple	1	80	19	0	1
Broccoli (raw)	1/2 cup	16	3	0	1
Baked potato (plain)	1	160	35	0	5
Wheat bread	1 slice	65	12	1	2
Vegetable soup	1 cup	100	20	0	5
Baked chicken	3 ounces	158	0	6	26
Lean ground beef (10% fat)	3 ounces	178	0	10	22
Baked trout	3 ounces	101	0	1	23
McDonald's® Big Mac	1	541	45	29	25
Wendy's® Biggie Fry	1	530	68	25	6
Chick-fil-A® chicken sandwich (fried)	1	408	38	16	28
Burger King® Whopper	1	710	52	42	31
Pizza Hut® super supreme pizza	1 slice	282	27	14	12
Cinnabon® roll	1	808	115	32	15
Chocolate cake	1 slice	247	35	11	2
Butter	1 tablespoon	108	0	12	0
Sour cream	2 tablespoons	62	1	6	1
Mayonnaise	1 tablespoon	99	0	11	0

to appear. Thyroid hormones increase BMR in part by stimulating cellular respiration. As cells use more oxygen to produce ATP, more heat is given off, and body temperature rises. This effect of thyroid hormones on BMR is called the **calorigenic effect**. Other hormones have minor effects on BMR. Testosterone, insulin, and growth hormone can increase the metabolic rate by 5–15%.

- ***Exercise.*** During strenuous exercise, the metabolic rate may increase to as much as 15 times the basal rate. In well-trained athletes, the rate may increase up to 20 times.
- ***Nervous system.*** During exercise or in a stressful situation, the sympathetic division of the autonomic nervous system is stimulated. Its postganglionic neurons release norepinephrine (NE), and it also stimulates release of the hormones epinephrine and norepinephrine by the adrenal medulla. Both epinephrine and norepinephrine increase the metabolic rate of body cells.
- ***Body temperature.*** The higher the body temperature, the higher the metabolic rate. Each 1°C rise in core temperature increases the rate of biochemical reactions by about 10%. As a result, metabolic rate may be increased substantially during a fever.
- ***Ingestion of food.*** The ingestion of food raises the metabolic rate 10–20% due to the energy "costs" of digesting, absorbing, and storing nutrients. This effect, **food-induced thermogenesis**, is greatest after eating a high-protein meal and is less after eating carbohydrates and lipids.

TABLE 25.7 Caloric Content of Various Beverages

BEVERAGE	SERVING SIZE	ENERGY (kcal)	CARBOHYDRATE (g)	(FAT) (g)	PROTEIN (g)	ALCOHOL (g)
Cola soft drink	12 ounces	160	40	0	0	0
Whole milk	1 cup	148	11	8	8	0
Orange juice	1 cup	108	25	0	2	0
White wine	5 ounces	102	1	0	0	14
Red wine	5 ounces	110	3	0	0	14
Beer	12 ounces	143	13	0	0	13
Vodka	1.5 ounces	98	0	0	0	14
Whiskey	1.5 ounces	98	0	0	0	14
Bourbon	1.5 ounces	98	0	0	0	14

• ***Age.*** The metabolic rate of a child, in relation to its size, is about double that of an elderly person due to the high rates of reactions related to growth.

• ***Other factors.*** Other factors that affect metabolic rate include gender (lower in females, except during pregnancy and lactation), climate (lower in tropical regions), sleep (lower), and malnutrition (lower).

Basal Metabolic Rate

Because many factors affect metabolic rate, it is measured under standard conditions, with the body in a quiet, resting, and fasting condition called the **basal state**. The measurement obtained under these conditions is the **basal metabolic rate (BMR)**. The most common way to determine BMR is by measuring the amount of oxygen used per kilocalorie of food metabolized. When the body uses 1 liter of oxygen to catabolize a typical dietary mixture of triglycerides, carbohydrates, and proteins, about 4.8 kcal of energy is released. BMR is 1200–1800 kcal/day in adults, or about 24 kcal/kg of body mass in adult males and 22 kcal/kg in adult females. The added calories needed to support daily activities, such as digestion and walking, range from 500 kcal for a small, relatively sedentary person to over 3000 kcal for a person in training for Olympic-level competitions or mountain climbing.

Total Metabolic Rate

The **total metabolic rate (TMR)** is the total energy expenditure by the body per unit of time. Three components contribute to the TMR:

1. ***Basal metabolic rate.*** The basal metabolic rate accounts for about 60% of the TMR.
2. ***Physical activity.*** Physical activity typically adds 30–35% but can be lower in sedentary people. The energy expenditure is partly from voluntary exercise, such as walking, and partly from **non-exercise activity thermogenesis (NEAT)**, the energy costs for maintaining muscle tone, posture while sitting or standing, and involuntary fidgeting movements. **Table 25.8** lists various activities and the calories that they burn per hour.
3. ***Food-induced thermogenesis.*** Food-induced thermogenesis—the heat produced while food is being digested, absorbed, and stored—represents 5–10% of the TMR.

Adipose Tissue and Stored Chemical Energy

The major site of stored chemical energy in the body is adipose tissue. When energy use exceeds energy input, triglycerides in adipose tissue are catabolized to provide the extra energy, and when energy input exceeds energy expenditure, triglycerides are stored. Over time, the amount of stored triglycerides indicates the excess of energy intake over energy expenditure. Even small differences add up over time. A gain of 20 lb (9 kg) between ages 25 and 55 represents only a tiny imbalance, about 0.3% more energy intake in food than energy expenditure.

TABLE 25.8 Various Activities and the Calories Released

ACTIVITY	ENERGY EXPENDITURE (kcal/hr)
Aerobics	419
Canoeing	248
Dancing	332
House cleaning	202
Office work	105
Playing the piano	170
Reading	86
Walking (3 mph)	250
Running (5 mph)	570
Sitting	102
Standing	132
Studying at desk	128
Swimming	572
Talking on phone	71
Weightlifting	224
Writing	122
Texting	40

Regulation of Food Intake

Negative feedback mechanisms regulate both our energy intake and our energy expenditure. But no sensory receptors exist to monitor our weight or size. How then is food intake regulated? The answer to this question is incomplete, but important advances in understanding regulation of food intake have occurred in the past decade. It depends on many factors, including neural and endocrine signals, levels of certain nutrients in the blood, psychological elements such as stress or depression, signals from the GI track and the special senses, and neural connections between the hypothalamus and other parts of the brain.

Within the hypothalamus are clusters of neurons that play key roles in regulating food intake. **Satiety** is a feeling of fullness accompanied by lack of desire to eat. Two hypothalamic areas involved in regulation of food intake are the *arcuate nucleus* and the *paraventricular nucleus* (see **Figure 14.10**). In 1994, the first experiments were reported on a mouse gene, named *obese*, that causes overeating and severe obesity in its mutated form. The product of this gene is the hormone **leptin**. In both mice and humans, leptin helps decrease **adiposity**, total body-fat mass. Leptin is synthesized and secreted by adipocytes in proportion to adiposity; as more triglycerides are stored, more leptin is secreted into the bloodstream. Leptin acts on the hypothalamus to inhibit circuits that stimulate eating while also activating circuits that increase energy expenditure. The hormone insulin has a similar but smaller effect. Both leptin and insulin are able to pass through the blood–brain barrier.

When leptin and insulin levels are *low*, neurons that extend from the arcuate nucleus to the paraventricular nucleus release a neurotransmitter called **neuropeptide Y** that stimulates food intake. Other neurons that extend between the arcuate and paraventricular nuclei release a neurotransmitter called **melanocortin**, which is similar to melanocyte-stimulating hormone (MSH). Leptin stimulates release of melanocortin, which acts to inhibit food intake. Another hormone involved in the regulation of food intake is **ghrelin**, which is produced by endocrine cells of the stomach. Ghrelin plays a role in increasing appetite. It is thought that ghrelin performs this function by stimulating the release of neuropeptide Y from hypothalamic neurons. Although leptin, neuropeptide Y, melanocortin, and ghrelin are key signaling molecules for maintaining energy balance, several other hormones and neurotransmitters also contribute. Other areas of the hypothalamus plus nuclei in the brainstem, limbic system, and cerebral cortex take part. An understanding of the brain circuits involved is still far from complete.

Achieving energy balance requires regulation of energy intake. Most increases and decreases in food intake are due to changes in meal size rather than changes in number of meals. Many experiments have demonstrated the presence of satiety signals, chemical or neural changes that help terminate eating when "fullness" is attained. For example, an increase in blood glucose level, as occurs after a meal, decreases appetite. Several hormones, such as glucagon, cholecystokinin, estrogens, and epinephrine (acting via beta receptors) act to signal satiety and to increase energy expenditure. Distension of the GI tract, particularly the stomach and duodenum, also contributes to termination of food intake. Other hormones increase appetite and decrease energy expenditure. These include growth hormone–releasing hormone (GHRH), androgens, glucocoticoids, epinephrine (acting via alpha receptors), and progesterone.

See Clinical Connection: Emotion Eating

25.9 Regulation of Body Temperature

OBJECTIVES

- **Describe** the various mechanisms of heat transfer.
- **Explain** how normal body temperature is maintained by negative feedback loops involving the hypothalamic thermostat.

Your body produces more or less heat depending on the rates of its metabolic reactions. Because homeostasis of body temperature can be maintained only if the rate of heat loss from the body equals the rate of heat production by metabolism, it is important to understand the ways in which heat can be lost, gained, or conserved. **Heat** is a form of energy that can be measured as **temperature**. Despite wide fluctuations in environmental temperature, homeostatic mechanisms can maintain a normal range for internal body temperature. If the rate of body heat production equals the rate of heat loss, the body maintains a constant core temperature near 37°C (98.6°F). **Core temperature** is the temperature in body structures deep to the skin and subcutaneous layer. **Shell temperature** is the temperature near the body surface—in the skin and subcutaneous layer. Depending on the environmental temperature, shell temperature is 1–6°C lower than core temperature. A core temperature that is too high kills by denaturing body proteins; a core temperature that is too low causes cardiac arrhythmias that result in death.

Mechanisms of Heat Transfer

Maintaining normal body temperature depends on the ability to lose heat to the environment at the same rate as it is produced by metabolic reactions. Heat can be transferred between the body and its surroundings in four ways: via conduction, convection, radiation, and evaporation.

1. **Conduction** is the heat exchange that occurs between molecules of two materials that are in direct contact with each other. At rest, about 3% of body heat is lost via conduction to cooler, solid materials in contact with the body, such as a chair, clothing, and jewelry. Heat can also be gained via conduction—for example, while soaking in a hot tub. Because water conducts heat 20 times more effectively than air, heat loss or heat gain via conduction is much greater when the body is submerged in cold or hot water.
2. **Convection** is the transfer of heat by the movement of air or water between areas of different temperatures. The contact of air or water with your body results in heat transfer by both conduction and convection. When cool air makes contact with the body, the air becomes warmed and therefore less dense and is carried away by convection currents created as the less dense air rises. The faster the air moves—for example, by a breeze or a fan—the faster the rate of convection. At rest, about 15% of body heat is lost to the air via conduction and convection.
3. **Radiation** is the transfer of heat in the form of infrared rays between a warmer object and a cooler one without physical contact. Your body loses heat by radiating more infrared waves than it absorbs from cooler objects. If surrounding objects are warmer than you are, you absorb more heat than you lose by radiation. In a room at 21°C (70°F), about 60% of heat loss occurs via radiation in a resting person.
4. **Evaporation** is the conversion of a liquid to a vapor. Every milliliter of evaporating water takes with it a great deal of heat—about 0.58 kcal/mL. Under typical resting conditions, about 22% of heat loss occurs through evaporation of about 700 mL of water per day—300 mL in exhaled air and 400 mL from the skin surface. Because we are not normally aware of this water loss through the skin and mucous membranes of the mouth and respiratory system, it is termed **insensible water loss**. The rate of evaporation is inversely related to relative humidity, the ratio of the actual amount of moisture in the air to the maximum amount it can hold at a given temperature. The higher the relative humidity, the lower the rate of evaporation. At 100% humidity, heat is gained via condensation of water on the skin surface as fast as heat is lost via evaporation.

Evaporation provides the main defense against overheating during exercise. Under extreme conditions, a maximum of about 3 liters of sweat can be produced each hour, removing more than 1700 kcal of heat if all of it evaporates. (Note: Sweat that drips off the body rather than evaporating removes very little heat.)

Hypothalamic Thermostat

The control center that functions as the body's thermostat is a group of neurons in the anterior part of the hypothalamus, the **preoptic area**. This area receives input from thermoreceptors in the skin (*peripheral thermoreceptors*) and in the hypothalamus itself (*central thermoreceptors*). Neurons of the preoptic area generate action potentials at a higher frequency when blood temperature increases and at a lower frequency when blood temperature decreases.

Action potentials from the preoptic area propagate to two other parts of the hypothalamus known as the **heat-losing center** and the **heat-promoting center**, which, when stimulated by the preoptic area, set into operation a series of responses that lower body temperature and raise body temperature, respectively.

Thermoregulation

If core temperature declines, mechanisms that help conserve heat and increase heat production act via negative feedback to raise the body temperature to normal (**Figure 25.19**). Peripheral thermoreceptors and central thermoreceptors send input to the preoptic area of the hypothalamus, which in turn activates the heat-promoting center. In response, the hypothalamus discharges action potentials and secretes thyrotropin-releasing hormone (TRH), which in turn stimulates thyrotrophs in the anterior pituitary gland to release thyroid-stimulating hormone (TSH). Action potentials from the hypothalamus and TSH then activate several effectors, which respond in the following ways to increase the core temperature to the normal value:

- ***Vasoconstriction.*** Action potentials from the heat-promoting center stimulate sympathetic nerves that cause blood vessels of the skin to constrict. Vasoconstriction decreases the flow of warm blood, and thus the transfer of heat, from the internal organs to the skin. Slowing the rate of heat loss allows the internal body temperature to increase as metabolic reactions continue to produce heat.
- ***Release of epinephrine and norepinephrine.*** Action potentials in sympathetic nerves leading to the adrenal medulla stimulate the release of epinephrine and norepinephrine into the blood. The hormones in turn bring about an increase in cellular metabolism, which increases heat production.
- ***Shivering.*** The heat-promoting center stimulates parts of the brain that increase muscle tone and hence heat production. As muscle tone increases in one muscle (the agonist), the small contractions stretch muscle spindles in its antagonist, initiating a stretch reflex. The resulting contraction in the antagonist stretches muscle spindles in the agonist, and it too develops a stretch reflex. This repetitive cycle—called **shivering**—greatly increases the rate of heat production. During maximal shivering, body heat production can rise to about four times the basal rate in just a few minutes.
- ***Release of thyroid hormones.*** The thyroid gland responds to TSH by releasing more thyroid hormones into the blood. As increased levels of thyroid hormones slowly increase the metabolic rate, body temperature rises.

If core body temperature rises above normal, a negative feedback loop opposite to the one depicted in **Figure 25.19** goes into action. The higher temperature of the blood stimulates peripheral and central thermoreceptors that send input to the preoptic area, which in turn stimulates the heat-losing center and inhibits the heat-promoting center. Action potentials from the heat-losing center cause dilation of blood vessels in the skin. The skin becomes warm, and the excess heat is lost to the environment via radiation and conduction as an increased volume of blood flows from the warmer core of the body into the cooler skin. At the same time, metabolic rate decreases, and shivering does not occur. The high temperature of the blood stimulates sweat glands of the skin via hypothalamic activation of sympathetic nerves. As the water in perspiration evaporates from the surface of the skin, the skin is cooled. All of these responses counteract heat-promoting effects and help return body temperature to normal.

See Clinical Connection: Hypothermia

25.10 Nutrition

OBJECTIVES

- **Describe** how to select foods to maintain a healthy diet.
- **Compare** the sources, functions, and importance of minerals and vitamins in metabolism.

Nutrients are chemical substances in food that body cells use for growth, maintenance, and repair. The six main types of nutrients are water, carbohydrates, lipids, proteins, minerals, and vitamins. Water is the nutrient needed in the largest amount—about 2–3 liters per day. As the most abundant compound in the body, water provides the medium in which most metabolic reactions occur, and it also participates in some reactions (for example, hydrolysis reactions). The important roles of water in the body can be reviewed in Section 2.4. Three organic nutrients—carbohydrates, lipids, and proteins—provide the energy needed for metabolic reactions and serve as building blocks to make body structures. Some minerals and many vitamins are components of the enzyme systems that catalyze metabolic reactions. *Essential nutrients* are specific nutrient molecules that the body cannot make in sufficient quantity to meet its needs and thus must be obtained from the diet. Some amino acids, fatty acids, vitamins, and minerals are essential nutrients.

Next, we discuss some guidelines for healthy eating and the roles of minerals and vitamins in metabolism.

FIGURE 25.19 Negative feedback mechanisms that conserve heat and increase heat production.

Core temperature is the temperature in body structures deep to the skin and subcutaneous layer; shell temperature is the temperature near the body surface.

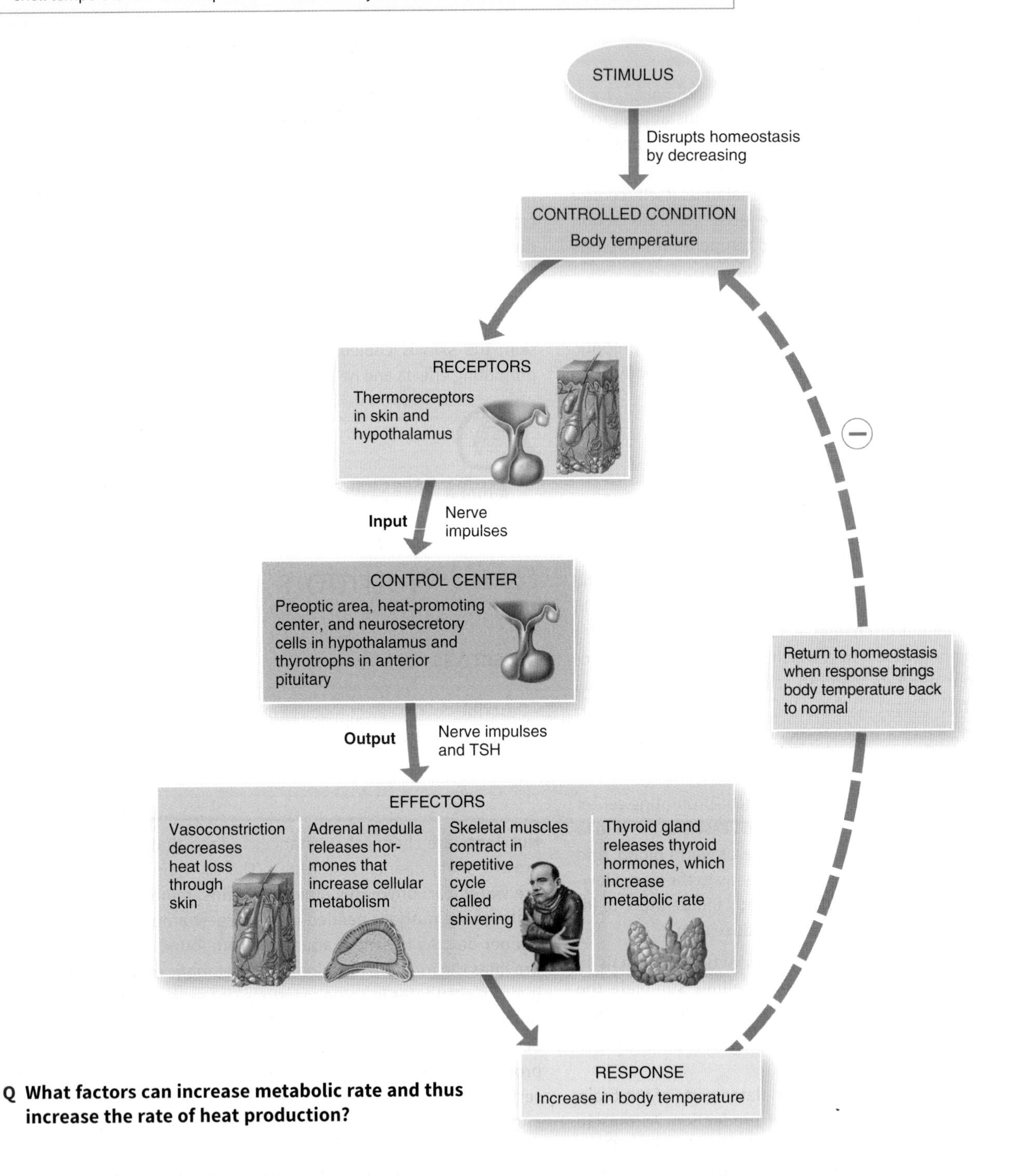

Q What factors can increase metabolic rate and thus increase the rate of heat production?

Guidelines for Healthy Eating

Each gram of protein or carbohydrate in food provides about 4 Calories; 1 gram of fat (lipids) provides about 9 Calories. We do not know with certainty what levels and types of carbohydrate, fat, and protein are optimal in the diet. Different populations around the world eat radically different diets that are adapted to their particular lifestyles. However, many experts recommend the following distribution of calories: 50–60% from carbohydrates, with less than 15% from simple sugars; less than 30% from fats (triglycerides are the main type of dietary fat), with no more than 10% as saturated fats; and about 12–15% from proteins.

On June 2, 2011, the United States Department of Agriculture (USDA) introduced a revised icon called **MyPlate** based on revised guidelines for healthy eating. It replaces the USDA MyPyramid, which first appeared in 2005. As shown in Figure 25.20, the plate is divided into four different-sized colored sections:

- Green (vegetables)
- Red (fruits)
- Orange (grains)
- Purple (protein)

The blue cup (dairy) adjacent to the plate icon is a reminder to include three daily servings of dairy.

The Dietary Guidelines for Americans released in January 2011 are the basis of MyPlate. Among the guidelines are the following:

- Enjoy food but balance calories by eating less.
- Avoid oversized portions, and make half of your plate vegetables and fruits.
- Switch to fat-free or low-fat milk.
- Make at least half of your grains whole grains.
- Choose foods that have a lower sodium content.
- Drink water instead of sugary drinks.

MyPlate places of a lot of emphasis on proportionality, variety, moderation, and nutrient density in a healthy diet. Proportionality simply means eating more of some types of foods than others. The MyPlate icon shows how much of your plate should be filled with foods from various food groups. Note that the vegetables and fruits take up one half of the plate, while protein and grains take up the other half. Note also that vegetables and grains represent the largest portions.

FIGURE 25.20 **MyPlate.**

The different colored sections are meant to be visual cues to help make healthier eating choices.

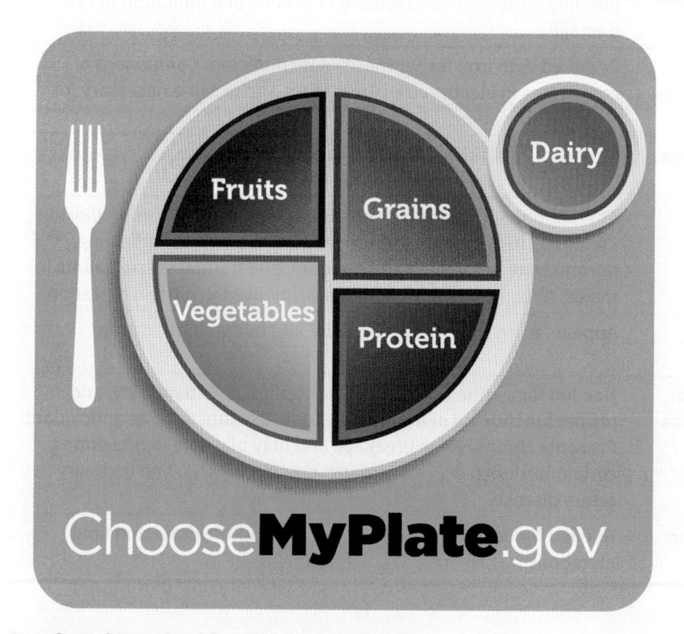

Q What does the blue cup represent?

Variety is important for a healthy diet because no one food or food group provides all of the nutrients and food types that the body needs. Accordingly, a variety of foods should be selected from within each food group. Vegetable choices should be varied to include dark green vegetables such as broccoli, collard greens, and kale; red and orange vegetables such as carrots, sweet potatoes, and red peppers; starchy vegetables such as corn, green peas, and potatoes; other vegetables such as cabbage, asparagus, and artichokes; and beans and peas such as lentils, chickpeas, and black beans. Beans and peas are good sources of the nutrients found in both vegetables and protein foods so they can be counted in either food group. Protein food choices are extremely varied, and include meat, poultry, seafood, beans and peas, eggs, processed soy products, nuts, and seeds. Grains include whole grains such as whole-wheat bread, oatmeal, and brown rice as well as refined grains such as white bread, white rice, and white pasta. Fruits include fresh, canned, or dried fruit and 100% fruit juice. Dairy includes all fluid milk products and many foods made from milk such as cheese, yogurt, and pudding, as well as calcium-fortified soy products.

Choosing nutrient-dense foods helps individuals practice moderation to balance calories consumed with calories expended. Tips include making half of your grains whole grains, choosing whole or cut-up fruits more often than juice, selecting fat-free or low-fat dairy products, and keeping meat and poultry portions small and lean.

Minerals

Minerals are inorganic elements that occur naturally in the earth's crust. In the body they appear in combination with one another, in combination with organic compounds, or as ions in solution. Minerals constitute about 4% of total body mass and are concentrated most heavily in the skeleton. Minerals with known functions in the body include calcium, phosphorus, potassium, sulfur, sodium, chloride, magnesium, iron, iodide, manganese, copper, cobalt, zinc, fluoride, selenium, and chromium. Table 25.9 describes the vital functions of these minerals. Note that the body generally uses the ions of the minerals rather than the non-ionized form. Some minerals, such as chlorine, are toxic or even fatal if ingested in the non-ionized form. Other minerals—aluminum, boron, silicon, and molybdenum—are present but their functions are unclear. Typical diets supply adequate amounts of potassium, sodium, chloride, and magnesium. Some attention must be paid to eating foods that provide enough calcium, phosphorus, iron, and iodide. Excess amounts of most minerals are excreted in the urine and feces.

Calcium and phosphorus form part of the matrix of bone. Because minerals do not form long-chain compounds, they are otherwise poor building materials. A major role of minerals is to help regulate enzymatic reactions. Calcium, iron, magnesium, and manganese are constituents of some coenzymes. Magnesium also serves as a catalyst for the conversion of ADP to ATP. Minerals such as sodium and phosphorus work in buffer systems, which help control the pH of body fluids. Sodium also helps regulate the osmosis of water and, along with other ions, is involved in the generation of nerve impulses.

TABLE 25.9 Minerals Vital to the Body

MINERAL	COMMENTS	IMPORTANCE
Calcium	Most abundant mineral in body. Appears in combination with phosphates. About 99% stored in bone and teeth. Blood Ca^{2+} level controlled by parathyroid hormone (PTH). Calcitriol promotes absorption of dietary calcium. Excess excreted in feces and urine. Sources: milk, egg yolk, shellfish, leafy green vegetables.	Formation of bones and teeth, blood clotting, normal muscle and nerve activity, endocytosis and exocytosis, cellular motility, chromosome movement during cell division, glycogen metabolism, release of neurotransmitters and hormones.
Phosphorus	About 80% found in bones and teeth as phosphate salts. Blood phosphate level controlled by parathyroid hormone (PTH). Excess excreted in urine; small amount eliminated in feces. Sources: dairy products, meat, fish, poultry, nuts.	Formation of bones and teeth. Phosphates ($H_2PO_4^-$, HPO_4^{2-}, and PO_4^{3-}) constitute a major buffer system of blood. Role in muscle contraction and nerve activity. Component of many enzymes. Involved in energy transfer (ATP). Component of DNA and RNA.
Potassium	Major cation (K^+) in intracellular fluid. Excess excreted in urine. Present in most foods (meats, fish, poultry, fruits, nuts).	Needed for generation and conduction of action potentials in neurons and muscle fibers.
Sulfur	Component of many proteins (such as insulin and chondroitin sulfate), electron carriers in electron transport chain, and some vitamins (thiamine and biotin). Excreted in urine. Sources: beef, liver, lamb, fish, poultry, eggs, cheese, beans.	As component of hormones and vitamins, regulates various body activities. Needed for ATP production by electron transport chain.
Sodium	Most abundant cation (Na^+) in extracellular fluids; some found in bones. Excreted in urine and perspiration. Normal intake of NaCl (table salt) supplies more than required amounts.	Strongly affects distribution of water through osmosis. Part of bicarbonate buffer system. Functions in nerve and muscle action potential conduction.
Chloride	Major anion (Cl^-) in extracellular fluid. Excess excreted in urine. Sources: table salt (NaCl), soy sauce, processed foods.	Role in acid–base balance of blood, water balance, and formation of HCl in stomach.
Magnesium	Important cation (Mg^{2+}) in intracellular fluid. Excreted in urine and feces. Widespread in various foods, such as green leafy vegetables, seafood, and whole-grain cereals.	Required for normal functioning of muscle and nervous tissue. Participates in bone formation. Constituent of many coenzymes.
Iron	About 66% found in hemoglobin of blood. Normal losses of iron occur by shedding of hair, epithelial cells, and mucosal cells, and in sweat, urine, feces, bile, and blood lost during menstruation. Sources: meat, liver, shellfish, egg yolk, beans, legumes, dried fruits, nuts, cereals.	As component of hemoglobin, reversibly binds O_2. Component of cytochromes involved in electron transport chain.
Iodide	Essential component of thyroid hormones. Excreted in urine. Sources: seafood, iodized salt, vegetables grown in iodine-rich soils.	Required by thyroid gland to synthesize thyroid hormones, which regulate metabolic rate.
Manganese	Some stored in liver and spleen. Most excreted in feces. Sources: spinach, romaine lettuce, pineapple.	Activates several enzymes. Needed for hemoglobin synthesis, urea formation, growth, reproduction, lactation, bone formation, and possibly production and release of insulin, and inhibition of cell damage.
Copper	Some stored in liver and spleen. Most excreted in feces. Sources: eggs, whole-wheat flour, beans, beets, liver, fish, spinach, asparagus.	Required with iron for synthesis of hemoglobin. Component of coenzymes in electron transport chain and enzyme necessary for melanin formation.
Cobalt	Constituent of vitamin B_{12}. Sources: liver, kidney, milk, eggs, cheese, meat.	As part of vitamin B_{12}, required for erythropoiesis.
Zinc	Important component of certain enzymes. Widespread in many foods, especially meats.	As component of carbonic anhydrase, important in carbon dioxide metabolism. Necessary for normal growth and wound healing, normal taste sensations and appetite, and normal sperm counts in males. As component of peptidases, involved in protein digestion.
Fluoride	Component of bones, teeth, other tissues. Sources: seafood, tea, gelatin.	Appears to improve tooth structure and inhibit tooth decay.
Selenium	Important component of certain enzymes. Sources: seafood, meat, chicken, tomatoes, egg yolk, milk, mushrooms, garlic, cereal grains grown in selenium-rich soil.	Needed for synthesis of thyroid hormones, sperm motility, and proper functioning of immune system. Also functions as antioxidant. Prevents chromosome breakage and may play role in preventing certain birth defects, miscarriage, prostate cancer, and coronary artery disease.
Chromium	Found in high concentrations in brewer's yeast. Also found in wine and some brands of beer.	Needed for normal activity of insulin in carbohydrate and lipid metabolism.

Vitamins

Organic nutrients required in small amounts to maintain growth and normal metabolism are called **vitamins**. Unlike carbohydrates, lipids, or proteins, vitamins do not provide energy or serve as the body's building materials. Most vitamins with known functions are coenzymes.

Most vitamins cannot be synthesized by the body and must be ingested in food. Other vitamins, such as vitamin K, are produced by bacteria in the GI tract and then absorbed. The body can assemble some vitamins if the raw materials, called **provitamins,** are provided. For example, vitamin A is produced by the body from the provitamin beta-carotene, a chemical present in yellow vegetables such as carrots and in dark green vegetables such as spinach. No single food contains all of the required vitamins—one of the best reasons to eat a varied diet.

Vitamins are divided into two main groups: fat-soluble and water-soluble. The **fat-soluble vitamins**, vitamins A, D, E, and K, are absorbed along with other dietary lipids in the small intestine and packaged into chylomicrons. They cannot be absorbed in adequate quantity unless they are ingested with other lipids. Fat-soluble vitamins may be stored in cells, particularly hepatocytes. The **water-soluble vitamins**, including several B vitamins and vitamin C, are dissolved in body fluids. Excess quantities of these vitamins are not stored but instead are excreted in the urine.

In addition to their other functions, three vitamins—C, E, and beta-carotene (a provitamin)—are termed **antioxidant vitamins** because they inactivate oxygen free radicals. Recall that free radicals are highly reactive ions or molecules that carry an unpaired electron in their outermost electron shell (see Figure 2.3). Free radicals damage cell membranes, DNA, and other cellular structures and contribute to the formation of artery-narrowing atherosclerotic plaques. Some free radicals arise naturally in the body, and others come from environmental hazards such as tobacco smoke and radiation. Antioxidant vitamins are thought to play a role in protecting against some kinds of cancer, reducing the buildup of atherosclerotic plaque, delaying some effects of aging, and decreasing the chance of cataract formation in the lens of the eyes. Table 25.10 lists the major vitamins, their sources, their functions, and related deficiency disorders.

See Clinical Connection: Vitamin and Mineral Supplements

TABLE 25.10 The Principal Vitamins

VITAMIN	COMMENT AND SOURCE	FUNCTIONS	DEFICIENCY SYMPTOMS AND DISORDERS
Fat-soluble	All require bile salts and some dietary lipids for adequate absorption.		
A	Formed from provitamin beta-carotene (and other provitamins) in GI tract. Stored in liver. Sources of carotene and other provitamins: orange, yellow, and green vegetables. Sources of vitamin A: liver, milk.	Maintains general health and vigor of epithelial cells. Beta-carotene acts as antioxidant to inactivate free radicals. Essential for formation of light-sensitive pigments in photoreceptors of retina. Aids in growth of bones and teeth by helping to regulate activity of osteoblasts and osteoclasts.	Deficiency results in atrophy and keratinization of epithelium, leading to dry skin and hair; increased incidence of ear, sinus, respiratory, urinary, and digestive system infections; inability to gain weight; drying of cornea; and skin sores. **Night blindness** (decreased ability for dark adaptation). Slow and faulty development of bones and teeth.
D	Sunlight converts 7-dehydrocholesterol in skin to cholecalciferol (vitamin D_3). A liver enzyme then converts cholecalciferol to 25-hydroxycholecalciferol. A second enzyme in kidneys converts 25-hydroxycholecalciferol to calcitriol (1,25-dihydroxycalciferol), the active form of vitamin D. Most excreted in bile. Dietary sources: fish-liver oils, egg yolk, fortified milk.	Essential for absorption of calcium and phosphorus from GI tract. Works with parathyroid hormone (PTH) to maintain Ca^{2+} homeostasis.	Defective utilization of calcium by bones leads to **rickets** in children and **osteomalacia** in adults. Possible loss of muscle tone.
E (tocopherols)	Stored in liver, adipose tissue, and muscles. Sources: fresh nuts and wheat germ, seed oils, green leafy vegetables.	Inhibits catabolism of certain fatty acids that help form cell structures, especially membranes. Involved in formation of DNA, RNA, and red blood cells. May promote wound healing, contribute to normal structure and functioning of nervous system, and prevent scarring. May help protect liver from toxic chemicals such as carbon tetrachloride. Acts as antioxidant to inactivate free radicals.	May cause oxidation of monounsaturated fats, resulting in abnormal structure and function of mitochondria, lysosomes, and plasma membranes. Possible consequence is hemolytic anemia.

Table 25.10 *Continues*

TABLE 25.10 The Principal Vitamins (*Continued*)

VITAMIN	COMMENT AND SOURCE	FUNCTIONS	DEFICIENCY SYMPTOMS AND DISORDERS
K	Produced by intestinal bacteria. Stored in liver and spleen. Dietary sources: spinach, cauliflower, cabbage, liver.	Coenzyme essential for synthesis of several clotting factors by liver, including prothrombin.	Delayed clotting time results in excessive bleeding.
Water-soluble	Dissolved in body fluids. Most not stored in body. Excess intake eliminated in urine.		
B_1 (thiamine)	Rapidly destroyed by heat. Sources: whole-grain products, eggs, pork, nuts, liver, yeast.	Acts as coenzyme for many different enzymes that break carbon-to-carbon bonds and are involved in carbohydrate metabolism of pyruvic acid to CO_2 and H_2O. Essential for synthesis of neurotransmitter acetylcholine.	Improper carbohydrate metabolism leads to buildup of pyruvic and lactic acids and insufficient production of ATP for muscle and nerve cells. Deficiency leads to (1) **beriberi**, partial paralysis of smooth muscle of GI tract, causing digestive disturbances; skeletal muscle paralysis; and atrophy of limbs; (2) **polyneuritis**, due to degeneration of myelin sheaths; impaired reflexes, impaired sense of touch, stunted growth in children, and poor appetite.
B_2 (riboflavin)	Small amounts supplied by bacteria of GI tract. Dietary sources: yeast, liver, beef, veal, lamb, eggs, whole-grain products, asparagus, peas, beets, peanuts.	Component of certain coenzymes (for example, FAD and FMN) in carbohydrate and protein metabolism, especially in cells of eye, integument, mucosa of intestine, and blood.	Deficiency may lead to improper utilization of oxygen, resulting in blurred vision, cataracts, and corneal ulcerations. Also dermatitis and cracking of skin, lesions of intestinal mucosa, and one type of anemia.
Niacin (nicotinamide)	Derived from amino acid tryptophan. Sources: yeast, meats, liver, fish, whole-grain products, peas, beans, nuts.	Essential component of NAD and NADP, coenzymes in oxidation–reduction reactions. In lipid metabolism, inhibits production of cholesterol and assists in triglyceride breakdown.	Principal deficiency is **pellagra**, characterized by dermatitis, diarrhea, and psychological disturbances.
B_6 (pyridoxine)	Synthesized by bacteria of GI tract. Stored in liver, muscle, and brain. Other sources: salmon, yeast, tomatoes, yellow corn, spinach, whole grain products, liver, yogurt.	Essential coenzyme for normal amino acid metabolism. Assists production of circulating antibodies. May function as coenzyme in triglyceride metabolism.	Most common deficiency symptom is dermatitis of eyes, nose, and mouth. Other symptoms are retarded growth and nausea.
B_{12} (cyanocobalamin)	Only B vitamin not found in vegetables; only vitamin containing cobalt. Absorption from GI tract depends on intrinsic factor secreted by gastric mucosa. Sources: liver, kidney, milk, eggs, cheese, meat.	Coenzyme necessary for red blood cell formation, formation of amino acid methionine, entrance of some amino acids into Krebs cycle, and manufacture of choline (used to synthesize acetylcholine).	Pernicious anemia, neuropsychiatric abnormalities (ataxia, memory loss, weakness, personality and mood changes, and abnormal sensations), and impaired activity of osteoblasts.
Pantothenic acid	Some produced by bacteria of GI tract. Stored primarily in liver and kidneys. Other sources: kidneys, liver, yeast, green vegetables, cereal.	Constituent of coenzyme A, which is essential for transfer of acetyl group from pyruvic acid into Krebs cycle, conversion of lipids and amino acids into glucose, and synthesis of cholesterol and steroid hormones.	Fatigue, muscle spasms, insufficient production of adrenal steroid hormones, vomiting, and insomnia.
Folic acid (folate, folacin)	Synthesized by bacteria of GI tract. Dietary sources: green leafy vegetables, broccoli, asparagus, breads, dried beans, citrus fruits.	Component of enzyme systems synthesizing nitrogenous bases of DNA and RNA. Essential for normal production of red and white blood cells.	Production of abnormally large red blood cells (macrocytic anemia). Higher risk of neural tube defects in babies born to folate-deficient mothers.
Biotin	Synthesized by bacteria of GI tract. Dietary sources include yeast, liver, egg yolk, kidneys.	Essential coenzyme for conversion of pyruvic acid to oxaloacetic acid and synthesis of fatty acids and purines.	Mental depression, muscular pain, dermatitis, fatigue, and nausea.
C (ascorbic acid)	Rapidly destroyed by heat. Some stored in glandular tissue and plasma. Sources: citrus fruits, tomatoes, green vegetables.	Promotes protein synthesis, including laying down of collagen in formation of connective tissue. As coenzyme, may combine with poisons, rendering them harmless until excreted. Works with antibodies, promotes wound healing, and functions as an antioxidant.	Scurvy; anemia; many symptoms related to poor collagen formation, including tender swollen gums, loosening of teeth (alveolar processes also deteriorate), poor wound healing, bleeding (vessel walls are fragile because of connective tissue degeneration), and retardation of growth.

Review Questions

25.1 Metabolic Reactions

1. What is metabolism? Distinguish between anabolism and catabolism, and give examples of each.
2. How does ATP link anabolism and catabolism?

25.2 Energy Transfer

3. How is a hydride ion different from a hydrogen ion? What is the involvement of both ions in redox reactions?
4. What are three ways that ATP can be generated?

25.3 Carbohydrate Metabolism

5. How does glucose move into or out of body cells?
6. What happens during glycolysis?
7. How is acetyl coenzyme A formed?
8. Outline the principal events and outcomes of the Krebs cycle.
9. What happens in the electron transport chain and why is this process called chemiosmosis?
10. Which reactions produce ATP during the complete oxidation of a molecule of glucose?
11. Under what circumstances do glycogenesis and glycogenolysis occur?
12. What is gluconeogenesis, and why is it important?

25.4 Lipid Metabolism

13. What are the functions of the apoproteins in lipoproteins?
14. Which lipoprotein particles contain "good" and "bad" cholesterol, and why are these terms used?
15. Where are triglycerides stored in the body?
16. Explain the principal events of the catabolism of glycerol and fatty acids.
17. What are ketone bodies? What is ketosis?
18. Define lipogenesis and explain its importance.

25.5 Protein Metabolism

19. What is deamination and why does it occur?
20. What are the possible fates of the amino acids from protein catabolism?
21. How are essential and nonessential amino acids different?

25.6 Key Molecules at Metabolic Crossroads

22. What are the possible fates of glucose 6-phosphate, pyruvic acid, and acetyl coenzyme A in a cell?

25.7 Metabolic Adaptations

23. What are the roles of insulin, glucagon, epinephrine, insulinlike growth factors, thyroxine, cortisol, estrogen, and testosterone in regulation of metabolism?
24. Why is ketogenesis more significant during fasting or starvation than during normal absorptive and postabsorptive states?

25.8 Energy Balance

25. What is a calorie? Why is the kilocalorie often used more than the calorie to express the energy content of food?
26. What are the three components that contribute to the total metabolic rate?
27. What are the functions of leptin, neuropeptide Y, melanocortin, and ghrelin?

25.9 Regulation of Body Temperature

28. Distinguish between core temperature and shell temperature.
29. In what ways can a person lose heat to or gain heat from the surroundings? How is it possible for a person to lose heat on a sunny beach when the temperature is 40°C (104°F) and the humidity is 85%?
30. Describe how each of the following parts of the hypothalamus plays a role in thermoregulation: preoptic area, heat-promoting center, and heat-losing center.

25.10 Nutrition

31. What is a nutrient?
32. Briefly describe the USDA's MyPlate and give examples of foods from each food group.
33. What is a mineral? Briefly describe the functions of the following minerals: calcium, phosphorus, potassium, sulfur, sodium, chloride, magnesium, iron, iodine, copper, zinc, fluoride, manganese, cobalt, chromium, and selenium.
34. Define vitamin. Explain how we obtain vitamins. Distinguish between a fat-soluble vitamin and a water-soluble vitamin.
35. For each of the following vitamins, indicate its principal function and the effect(s) of deficiency: A, D, E, K, B_1, B_2, niacin, B_6, B_{12}, pantothenic acid, folic acid, biotin, and C.

Critical Thinking Questions

1. Jane Doe's deceased body was found at her dining room table. Her death was considered suspicious. Lab results from the medical investigation revealed cyanide in her blood. How did the cyanide cause her death?
2. During a recent physical, 55-year-old Glenn's blood serum lab results showed the following: total cholesterol = 300 mg/dL; LDL = 175 mg/dL; HDL = 20 mg/dL. Interpret these results for Glenn and indicate to him what changes, if any, he needs to make in his lifestyle. Why are these changes important?
3. Marissa has joined a weight loss program. As part of the program, she regularly submits a urine sample which is tested for ketones. She went to the clinic today, had her urine checked, and was confronted by the nurse who accused Marissa of "cheating" on her diet. How did the nurse know Marissa was not following her diet?

The Urinary System

CHAPTER **26**

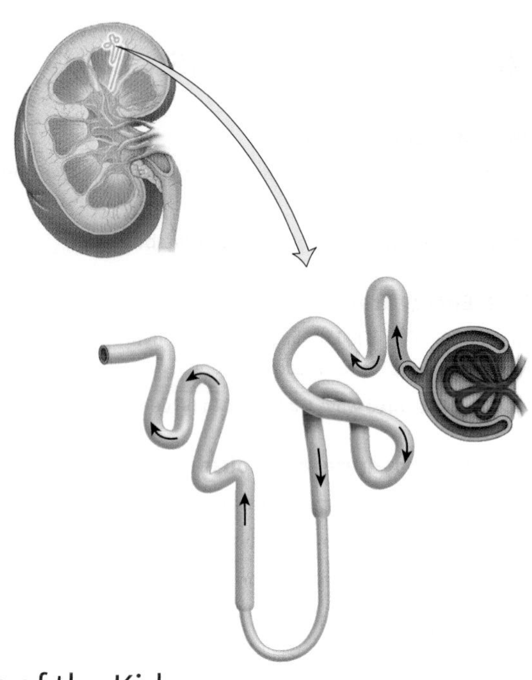

The Urinary System and Homeostasis

The urinary system contributes to homeostasis by excreting wastes; altering blood composition, pH, volume, and pressure; maintaining blood osmolarity; and producing hormones.

As body cells carry out metabolic activities, they consume oxygen and nutrients and produce waste products such as carbon dioxide, urea, and uric acid. Wastes must be eliminated from the body because they can be toxic to cells if they accumulate. While the respiratory system rids the body of carbon dioxide, the urinary system disposes of most other wastes. The urinary system performs this function by removing wastes from the blood and excreting them into urine. Disposal of wastes through the release of urine is not the only purpose of the urinary system. The urinary system also helps regulate blood composition, pH, volume, and pressure; maintains blood osmolarity; and produces hormones.

Q Did you ever wonder how diuretics work and why they are used?

26.1 Overview of the Urinary System

OBJECTIVE

- **Describe** the major structures of the urinary system and the functions they perform.

Components of the Urinary System

The **urinary system** consists of two kidneys, two ureters, one urinary bladder, and one urethra (**Figure 26.1**). The kidneys filter blood of wastes and excrete them into a fluid called **urine**. Once formed, urine passes through the ureters and is stored in the urinary bladder until it is excreted from the body through the urethra. **Nephrology** (nef-ROL-ō-jē; *nephr-* = kidney; *-ology* = study of) is the scientific study of the anatomy, physiology, and pathology of the kidneys. The branch of medicine that deals with the male and female urinary systems and the male reproductive system is called **urology** (ū-ROL-ō-jē; *uro-* = urine). A physician who specializes in this branch of medicine is called a **urologist** (ū-ROL-ō-jist).

Functions of the Kidneys

The kidneys do the major work of the urinary system. The other parts of the system are mainly passageways and storage areas. Functions of the kidneys include the following:

- ***Excretion of wastes.*** By forming urine, the kidneys help excrete wastes from the body. Some wastes excreted in urine result from metabolic reactions. These include urea and ammonia from the deamination of amino acids; creatinine from the breakdown of creatine phosphate; uric acid from the catabolism of nucleic acids; and urobilin from the breakdown of hemoglobin. Urea, ammonia, creatinine, uric acid, and urobilin are collectively known as **nitrogenous wastes** because they are waste products that contain nitrogen. Other wastes excreted in the urine are foreign substances that have entered the body, such as drugs and environmental toxins.
- ***Regulation of blood ionic composition.*** The kidneys help regulate the blood levels of several ions, most importantly sodium ions (Na^+), potassium ions (K^+), calcium ions (Ca^{2+}), chloride ions (Cl^-), and phosphate ions (HPO_4^{2-}). The kidneys accomplish this task by adjusting the amounts of these ions that are excreted into the urine.
- ***Regulation of blood pH.*** The kidneys excrete a variable amount of hydrogen ions (H^+) into the urine and conserve bicarbonate ions (HCO_3^-), which are an important buffer of H^+ in the blood. Both of these activities help regulate blood pH.
- ***Regulation of blood volume.*** The kidneys adjust blood volume by conserving or eliminating water in the urine. An increase in blood volume increases blood pressure; a decrease in blood volume decreases blood pressure.
- ***Regulation of blood pressure.*** The kidneys also help regulate blood pressure by secreting the enzyme renin, which activates the renin–angiotensin–aldosterone pathway (see **Figure 18.15**). Increased renin causes an increase in blood pressure.

FIGURE 26.1 Organs of the urinary system in a female.

Urine formed by the kidneys passes first into the ureters, then to the urinary bladder for storage, and finally through the urethra for elimination from the body.

Functions of the Urinary System

1. Kidneys regulate blood volume and composition; help regulate blood pressure, pH, and glucose levels; produce two hormones (calcitriol and erythropoietin); and excrete wastes in urine.
2. Ureters transport urine from kidneys to urinary bladder.
3. Urinary bladder stores urine and expels it into urethra.
4. Urethra discharges urine from body.

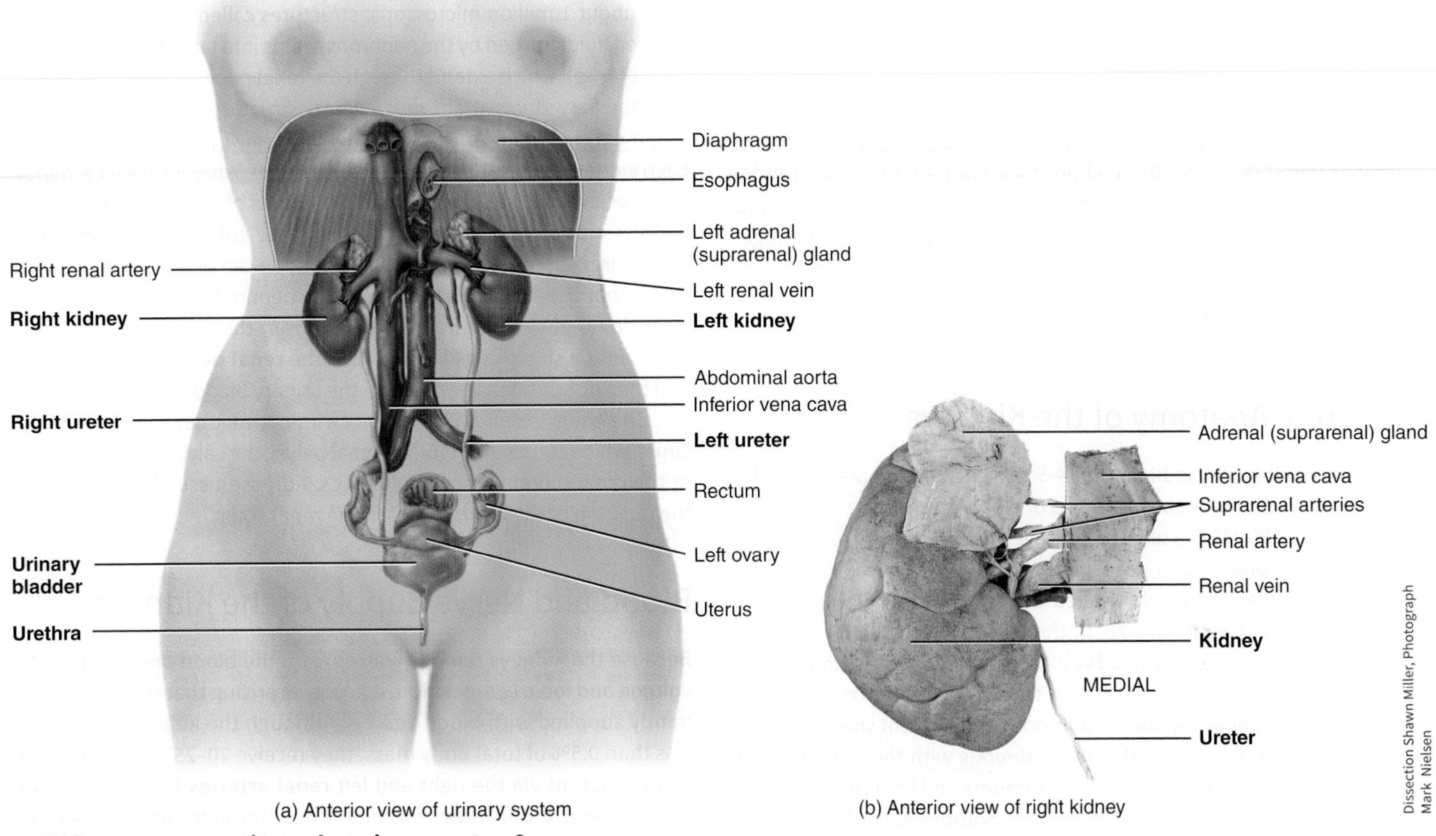

(a) Anterior view of urinary system

(b) Anterior view of right kidney

Q Which organs constitute the urinary system?

- ***Maintenance of blood osmolarity.*** By separately regulating loss of water and loss of solutes in the urine, the kidneys maintain a relatively constant blood osmolarity close to 300 milliosmoles per liter (mOsm/liter).*
- ***Production of hormones.*** The kidneys produce two hormones. *Calcitriol*, the active form of vitamin D, helps regulate calcium homeostasis (see **Figure 18.13**), and *erythropoietin* stimulates the production of red blood cells (see **Figure 19.5**).
- ***Regulation of blood glucose level.*** Like the liver, the kidneys can use the amino acid glutamine in *gluconeogenesis*, the synthesis of new glucose molecules. They can then release glucose into the blood to help maintain a normal blood glucose level.

As is evident from the functions listed, urine contains more than just waste products. It also contains water and other substances, such ions, that have important roles in the body, but are in excess of the body's needs. You will learn more about the composition of urine in Section 26.8.

*The **osmolarity** of a solution is a measure of the total number of dissolved particles per liter of solution. The particles may be molecules, ions, or a mixture of both. To calculate osmolarity, multiply molarity (see Section 2.4) by the number of particles per molecule, once the molecule dissolves. A similar term, *osmolality*, is the number of particles of solute per *kilogram* of water. Because it is easier to measure volumes of solutions than to determine the mass of water they contain, osmolarity is used more commonly than osmolality. Most body fluids and solutions used clinically are dilute, in which case there is less than a 1% difference between the two measures.

26.2 Anatomy of the Kidneys

OBJECTIVES

- **Describe** the external and internal gross anatomical features of the kidneys.
- **Trace** the path of blood flow through the kidneys.

The paired **kidneys** are reddish, kidney bean–shaped organs located just above the waist between the peritoneum and the posterior wall of the abdomen. Because their position is posterior to the peritoneum of the abdominal cavity, the organs are said to be **retroperitoneal** (re′-trō-per-i-tō-NĒ-al; *retro-* = behind) (Figure 26.2). The kidneys are located between the levels of the last thoracic and third lumbar vertebrae, a position where they are partially protected by ribs 11 and 12. If these lower ribs are fractured, they can puncture the kidneys and cause significant, even life-threatening damage. The right kidney is slightly lower than the left (see Figure 26.1) because the liver occupies considerable space on the right side superior to the kidney.

External Anatomy of the Kidneys

A typical adult kidney is 10–12 cm (4–5 in.) long, 5–7 cm (2–3 in.) wide, and 3 cm (1 in.) thick—about the size of a bar of bath soap—and has a mass of 135–150 g (4.5–5 oz). The concave medial border of each kidney faces the vertebral column (see Figure 26.1). Near the center of the concave border is an indentation called the **renal hilum** (RĒ-nal HĪ-lum; *renal* = kidney) (see Figure 26.3), through which the ureter emerges from the kidney along with blood vessels, lymphatic vessels, and nerves.

Three layers of tissue surround each kidney (Figure 26.2). The deep layer, the **renal capsule**, is a smooth, transparent sheet of dense irregular connective tissue that is continuous with the outer coat of the ureter. It serves as a barrier against trauma and helps maintain the shape of the kidney. The middle layer, the **adipose capsule**, is a mass of fatty tissue surrounding the renal capsule. It also protects the kidney from trauma and holds it firmly in place within the abdominal cavity. The superficial layer, the **renal fascia** (FASH-ē-a), is another thin layer of dense irregular connective tissue that anchors the kidney to the surrounding structures and to the abdominal wall. On the anterior surface of the kidneys, the renal fascia is deep to the peritoneum.

See Clinical Connection: Nephroptosis (Floating Kidney)

Internal Anatomy of the Kidneys

A frontal section through the kidney reveals two distinct regions: a superficial, light red region called the **renal cortex** (*cortex* = rind or bark) and a deep, darker reddish-brown inner region called the **renal medulla** (*medulla* = inner portion) (Figure 26.3). The renal medulla consists of several cone-shaped **renal pyramids**. The base (wider end) of each pyramid faces the renal cortex, and its apex (narrower end), called a **renal papilla**, points toward the renal hilum. The renal cortex is the smooth-textured area extending from the renal capsule to the bases of the renal pyramids and into the spaces between them. It is divided into an outer *cortical zone* and an inner *juxtamedullary zone* (juks′-ta-MED-ū-la-rē). Those portions of the renal cortex that extend between renal pyramids are called **renal columns**.

Together, the renal cortex and renal pyramids of the renal medulla constitute the **parenchyma** (pa-RENG-kī-ma) or functional portion of the kidney. Within the parenchyma are the functional units of the kidney—about 1 million microscopic structures called **nephrons**. Filtrate (filtered fluid) formed by the nephrons drains into large **papillary ducts** (PAP-i-lar′-ē), which extend through the renal papillae of the pyramids. The papillary ducts drain into cuplike structures called **minor** and **major calyces** (KĀ-li-sēz = cups; singular is *calyx*, pronounced KĀ-liks). Each kidney has 8 to 18 minor calyces and 2 or 3 major calyces. A minor calyx receives filtrate from the papillary ducts of one renal papilla and delivers it to a major calyx. Once the filtrate enters the calyces it becomes urine because no further reabsorption can occur. The reason for this is that the simple epithelium of the nephron and ducts becomes transitional epithelium in the calyces. From the major calyces, urine drains into a single large cavity called the **renal pelvis** (*pelv-* = basin) and then out through the ureter to the urinary bladder.

The hilum expands into a cavity within the kidney called the **renal sinus**, which contains part of the renal pelvis, the calyces, and branches of the renal blood vessels and nerves. Adipose tissue helps stabilize the position of these structures in the renal sinus.

Blood and Nerve Supply of the Kidneys

Because the kidneys remove wastes from the blood and regulate its volume and ionic composition, it is not surprising that they are abundantly supplied with blood vessels. Although the kidneys constitute less than 0.5% of total body mass, they receive 20–25% of the resting cardiac output via the right and left **renal arteries** (Figure 26.4). In adults, **renal blood flow**, the blood flow through both kidneys, is about 1200 mL per minute.

Within the kidney, the renal artery divides into several **segmental arteries** (seg-MEN-tal), which supply different segments (areas) of the kidney. Each segmental artery gives off several branches that enter the parenchyma and pass through the renal columns between the renal lobes as the **interlobar arteries** (in′-ter-LŌ-bar). A **renal lobe** consists of a renal pyramid, some of the renal column on either side of the renal pyramid, and the renal cortex at the base of the renal pyramid (see Figure 26.3a). At the bases of the renal pyramids, the interlobar arteries arch between the renal medulla and cortex; here they are known as the **arcuate arteries** (AR-kū-āt = shaped like a bow). Divisions of the arcuate arteries produce a series of **cortical radiate** (*interlobular*) **arteries** (KOR-ti-kal RĀ-dē-at). These arteries radiate outward and enter the renal cortex. Here, they give off branches called **afferent arterioles** (AF-er-ent; *af-* = toward; *-ferrent* = to carry).

Each nephron receives one afferent arteriole, which divides into a tangled, ball-shaped capillary network called the **glomerulus**

FIGURE 26.2 Position and coverings of the kidneys.

The kidneys are surrounded by a renal capsule, adipose capsule, and renal fascia.

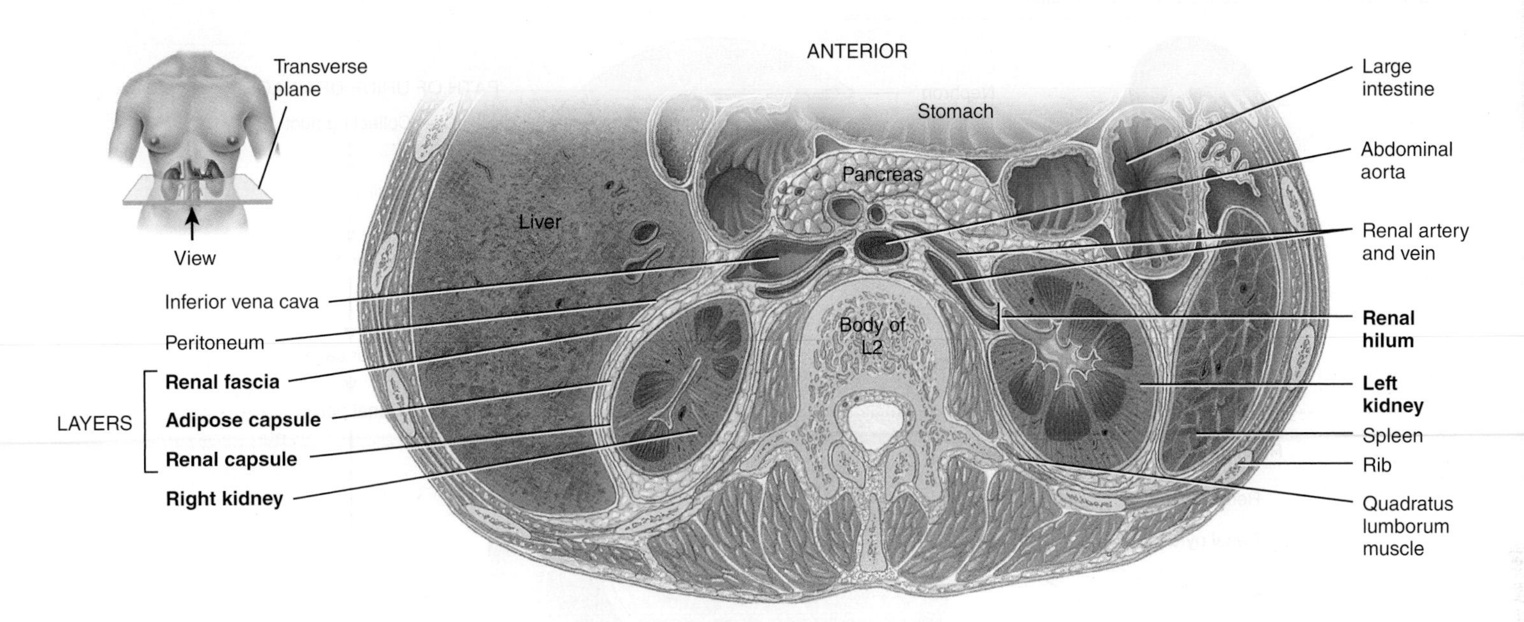

(a) Inferior view of transverse section of abdomen (L2)

(b) Sagittal section through the right kidney

Q Why are the kidneys said to be retroperitoneal?

FIGURE 26.3 **Internal anatomy of the kidneys.**

The two main regions of the kidney are the superficial, light red region called the renal cortex and the deep, dark red region called the renal medulla.

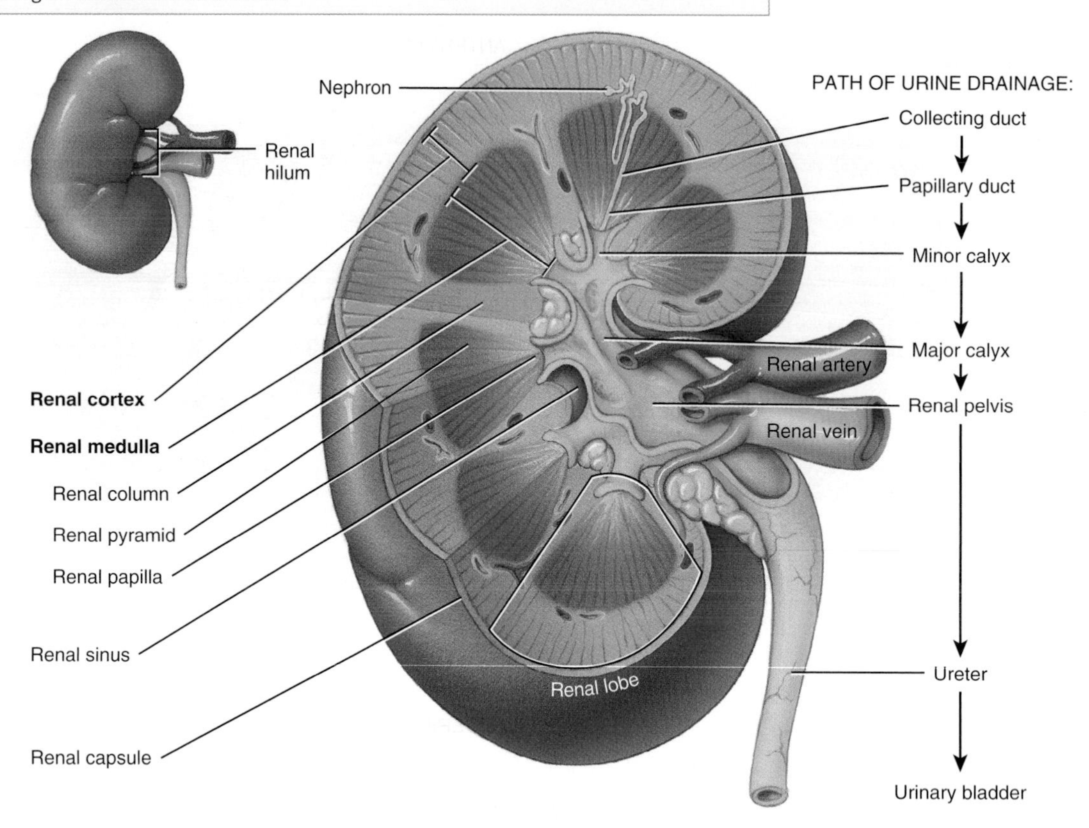

(a) Anterior view of dissection of right kidney

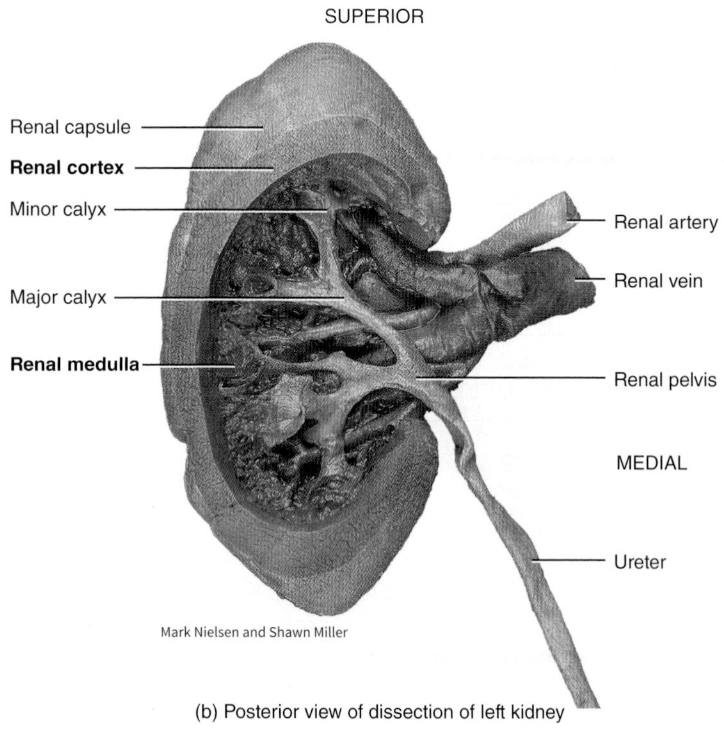

(b) Posterior view of dissection of left kidney

Q What structures pass through the renal hilum?

(glō-MER-ū-lus = little ball; plural is *glomeruli*). The glomerular capillaries then reunite to form an **efferent arteriole** (EF-er-ent; *ef-* = out) that carries blood out of the glomerulus. Glomerular capillaries are unique among capillaries in the body because they are positioned between two arterioles, rather than between an arteriole and a venule. Because they are capillary networks and they also play an important role in urine formation, the glomeruli are considered part of both the cardiovascular and the urinary systems.

The efferent arterioles divide to form the **peritubular capillaries** (per-i-TOOB-ū-lar; *peri-* = around), which surround tubular parts of the nephron in the renal cortex. Extending from some efferent arterioles are long, loop-shaped capillaries called **vasa recta** (VĀ-sa REK-ta; *vasa* = vessels; *recta* = straight) that supply tubular portions of the nephron in the renal medulla (see Figure 26.4a).

The peritubular capillaries eventually reunite to form **cortical radiate** (*interlobular*) **veins**, which also receive blood from the vasa recta. Then the blood drains through the **arcuate veins** to the **interlobar veins** running between the renal pyramids. Blood leaves the kidney through a single **renal vein** that exits at the renal hilum and carries venous blood to the inferior vena cava.

Many renal nerves originate in the *renal ganglion* and pass through the *renal plexus* into the kidneys along with the renal arteries.

FIGURE 26.4 **Blood supply of the kidneys.**

The renal arteries deliver 20–25% of the resting cardiac output to the kidneys.

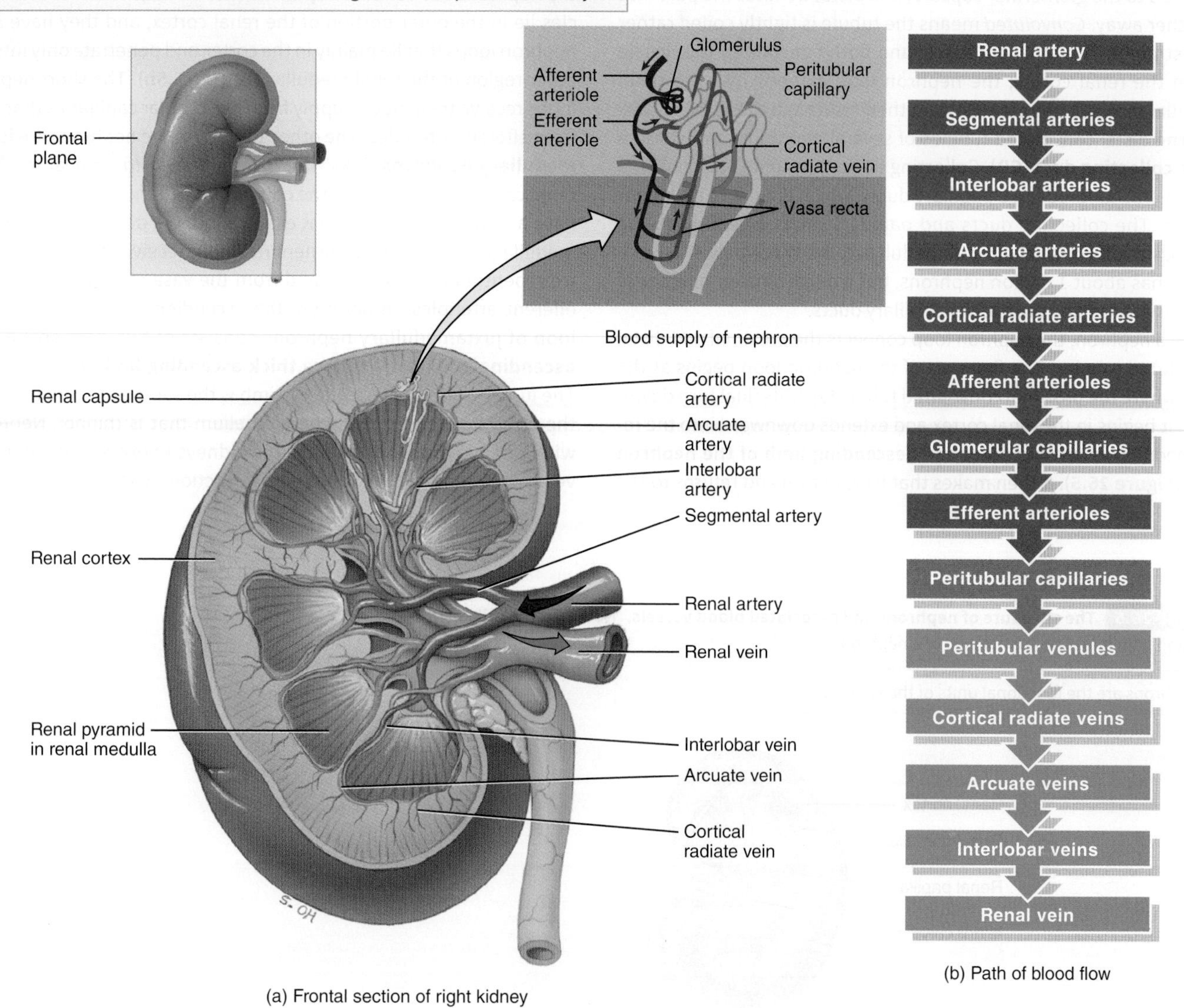

(a) Frontal section of right kidney

(b) Path of blood flow

Q What volume of blood enters the renal arteries per minute?

Renal nerves are part of the sympathetic division of the autonomic nervous system. Most are vasomotor nerves that regulate the flow of blood through the kidney by causing vasodilation or vasoconstriction of renal arterioles.

26.3 The Nephron

OBJECTIVES

- **Describe** the parts of a nephron.
- **Explain** the histology of a nephron and collecting duct.

Parts of a Nephron

Nephrons (NEF-rons) are the functional units of the kidneys. Each nephron consists of two parts: a **renal corpuscle** (KOR-pus-el = tiny body), where blood plasma is filtered, and a renal tubule into which the filtered fluid (glomerular filtrate) passes (**Figure 26.5**). Closely associated with a nephron is its blood supply, which was just described. The two components of a renal corpuscle are the **glomerulus** (capillary network) and the **glomerular capsule** or *Bowman's capsule*, a double-walled epithelial cup that surrounds the glomerular capillaries. Blood plasma is filtered in the glomerular capsule, and then the filtered fluid passes into the renal tubule, which has three main sections. In the order that fluid passes through them, the renal tubule consists of a (1) **proximal convoluted tubule** (**PCT**) (kon′-vō-LOOT-ed), (2) **nephron loop** (*loop of Henle*), and (3) **distal convoluted tubule** (**DCT**). *Proximal* denotes the part of the tubule attached to the glomerular capsule, and *distal* denotes the part that is further away. *Convoluted* means the tubule is tightly coiled rather than straight. The renal corpuscle and both convoluted tubules lie within the renal cortex; the nephron loop extends into the renal medulla, makes a hairpin turn, and then returns to the renal cortex.

The distal convoluted tubules of several nephrons empty into a single **collecting duct (CD)**. Collecting ducts then unite and converge into several hundred large papillary ducts, which drain into the minor calyces. The collecting ducts and papillary ducts extend from the renal cortex through the renal medulla to the renal pelvis. So one kidney has about 1 million nephrons, but a much smaller number of collecting ducts and even fewer papillary ducts.

In a nephron, the nephron loop connects the proximal and distal convoluted tubules. The first part of the nephron loop begins at the point where the proximal convoluted tubule takes its final turn downward. It begins in the renal cortex and extends downward into the renal medulla, where it is called the **descending limb of the nephron loop** (Figure 26.5). It then makes that hairpin turn and returns to the renal cortex where it terminates at the distal convoluted tubule and is known as the **ascending limb of the nephron loop**. About 80–85% of the nephrons are **cortical nephrons** (KOR-ti-kul). Their renal corpuscles lie in the outer portion of the renal cortex, and they have *short* nephron loops that lie mainly in the cortex and penetrate only into the outer region of the renal medulla (Figure 26.5b). The short nephron loops receive their blood supply from peritubular capillaries that arise from efferent arterioles. The other 15–20% of the nephrons are **juxtamedullary nephrons** (juks′-ta-MED-ū-lar′-ē; *juxta-* = near to). Their renal corpuscles lie deep in the cortex, close to the medulla, and they have a *long* nephron loop that extends into the deepest region of the medulla (Figure 26.5c). Long nephron loops receive their blood supply from peritubular capillaries and from the vasa recta that arise from efferent arterioles. In addition, the ascending limb of the nephron loop of juxtamedullary nephrons consists of two portions: a **thin ascending limb** followed by a **thick ascending limb** (Figure 26.5c). The lumen of the thin ascending limb is the same as in other areas of the renal tubule; it is only the epithelium that is thinner. Nephrons with long nephron loops enable the kidneys to excrete very dilute or very concentrated urine (described in Section 26.7).

FIGURE 26.5 The structure of nephrons and associated blood vessels. Note that the collecting duct and papillary duct are not part of a nephron.

Nephrons are the functional units of the kidneys.

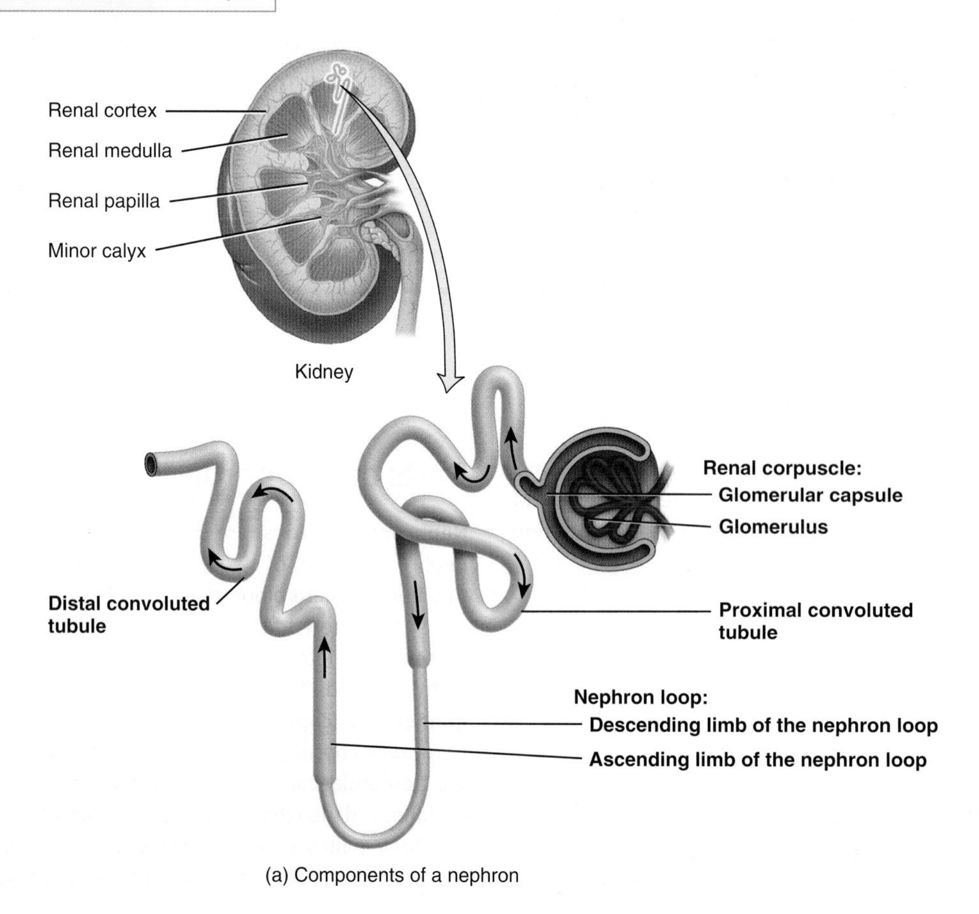

(a) Components of a nephron

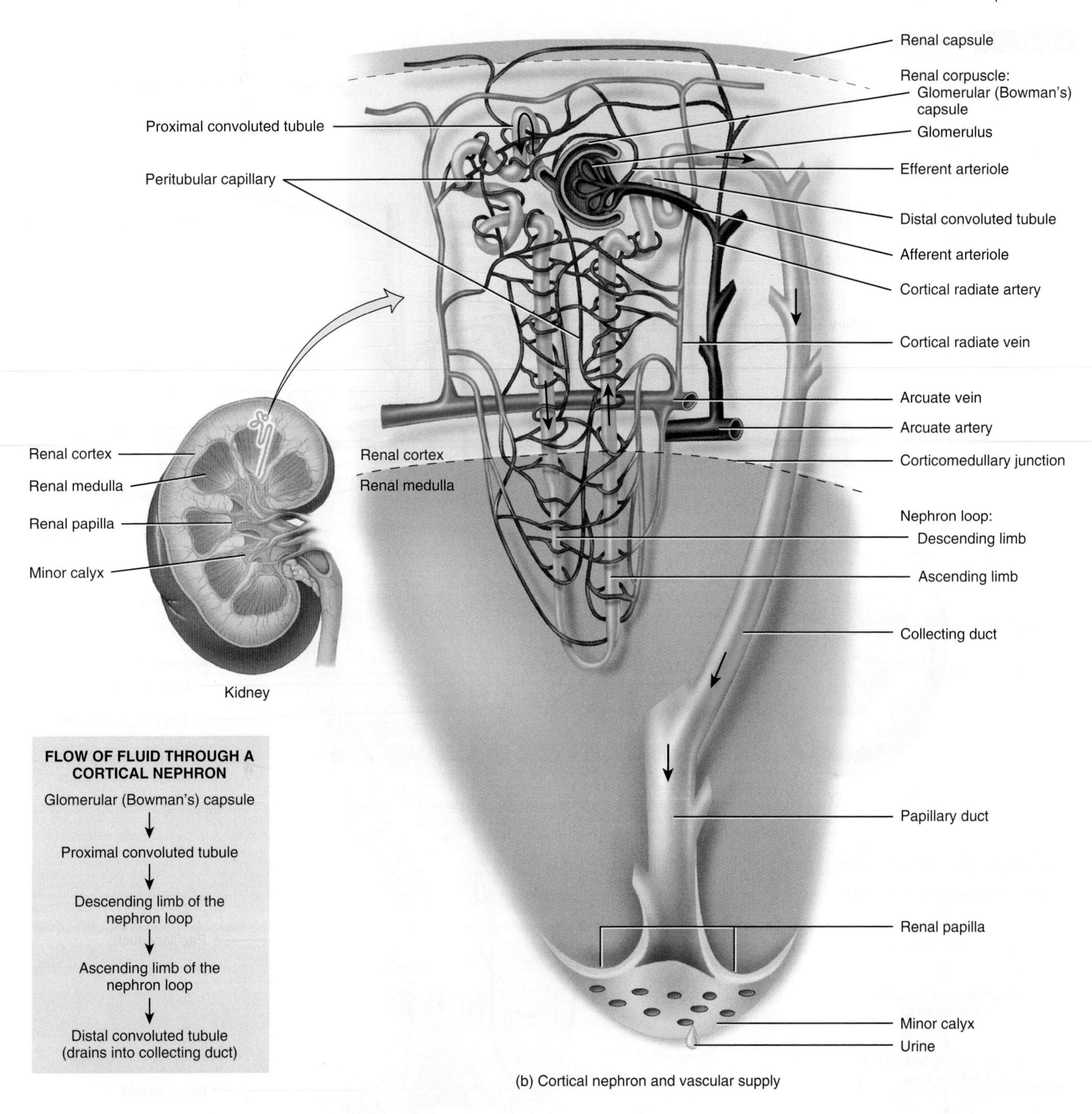

(b) Cortical nephron and vascular supply

Figure 26.5 ***Continues***

FIGURE 26.5 Continued

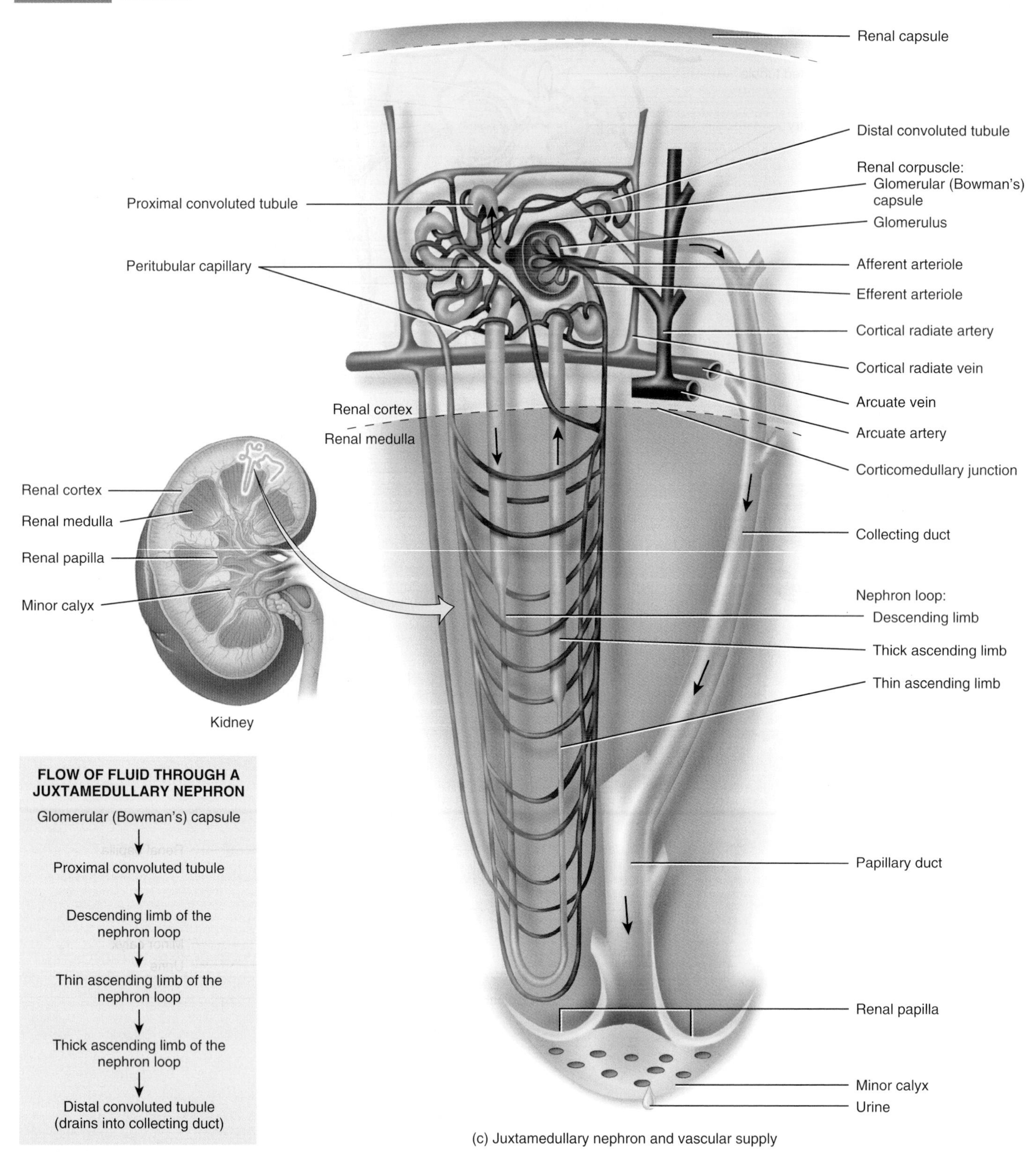

(c) Juxtamedullary nephron and vascular supply

Q What are the basic differences between cortical and juxtamedullary nephrons?

Histology of the Nephron and Collecting Duct

A single layer of epithelial cells forms the entire wall of the glomerular capsule, renal tubule, and ducts (**Figure 26.6**). However, each part has distinctive histological features that reflect its particular functions. We will discuss them in the order that fluid flows through them: glomerular capsule, renal tubule, and collecting duct.

Glomerular Capsule The glomerular (Bowman's) capsule consists of visceral and parietal layers (**Figure 26.6a**). The visceral layer consists of modified simple squamous epithelial cells called **podocytes** (PŌD-ō-sīts; *podo-* = foot; *-cytes* = cells). The many footlike projections of these cells (pedicels) wrap around the single layer of endothelial cells of the glomerular capillaries and form the inner wall of the capsule. The parietal layer of the glomerular capsule consists of simple squamous epithelium and forms the outer wall of the capsule. Fluid filtered from the glomerular capillaries enters the **capsular space**, the space between the two layers of the glomerular capsule, which is continuous with the lumen of the renal tubule. Think of the relationship between the glomerulus and glomerular capsule in the following way; the glomerulus is a fist punched into a limp balloon (the glomerular capsule) until the fist is covered by two layers of balloon (the layer of the balloon touching the fist is the visceral layer and the layer not against the fist is the parietal layer) with a space in between (the inside of the balloon), the capsular space.

Renal Tubule and Collecting Duct **Table 26.1** illustrates the histology of the cells that form the renal tubule and collecting duct. In the proximal convoluted tubule, the cells are simple cuboidal epithelial cells with a prominent brush border of microvilli on their apical surface (surface facing the lumen). These microvilli, like those of the small intestine, increase the surface area for reabsorption and secretion. The descending limb of the nephron loop and the first part of the ascending limb of the nephron loop (the thin ascending limb) are composed of simple squamous epithelium. (Recall that cortical or short-loop nephrons lack the thin ascending limb.) The thick ascending limb of the nephron loop is composed of simple cuboidal to low columnar epithelium.

In each nephron, the final part of the ascending limb of the nephron loop makes contact with the afferent arteriole serving that renal corpuscle (**Figure 26.6b**). Because the columnar tubule cells in this region are crowded together, they are known as the **macula densa** (MAK-ū-la DEN-sa; *macula* = spot; *densa* = dense). Alongside the macula densa, the wall of the afferent arteriole (and sometimes the efferent arteriole) contains modified smooth muscle fibers called **juxtaglomerular cells (JG)** (juks′-ta-glō-MER-ū-lar). Together with the macula densa, they constitute the **juxtaglomerular apparatus (JGA)**. As you will see later, the JGA helps regulate blood pressure within the kidneys. The distal convoluted tubule (DCT) begins a short distance past the macula densa. In the last part of the DCT and continuing into the collecting ducts, two different types of cells are present. Most are **principal cells**, which have receptors for both antidiuretic hormone (ADH) and aldosterone, two hormones that regulate their functions. A smaller number are **intercalated cells** (in-TER-ka-lā-ted), which play a role in the homeostasis of blood pH. The collecting ducts drain into large papillary ducts, which are lined by simple columnar epithelium.

The number of nephrons is constant from birth. Any increase in kidney size is due solely to the growth of individual nephrons. If nephrons are injured or become diseased, new ones do not form. Signs of kidney dysfunction usually do not become apparent until function declines to less than 25% of normal because the remaining functional nephrons adapt to handle a larger-than-normal load. Surgical removal of one kidney, for example, stimulates hypertrophy (enlargement) of the remaining kidney, which eventually is able to filter blood at 80% of the rate of two normal kidneys.

26.4 Overview of Renal Physiology

OBJECTIVE

- **Identify** the three basic functions performed by nephrons and collecting ducts, and indicate where each occurs.

To produce urine, nephrons and collecting ducts perform three basic processes—glomerular filtration, tubular reabsorption, and tubular secretion (**Figure 26.7**):

1. ***Glomerular filtration.*** In the first step of urine production, water and most solutes in blood plasma move across the wall of glomerular capillaries, where they are filtered and move into the glomerular capsule and then into the renal tubule.
2. ***Tubular reabsorption.*** As filtered fluid flows through the renal tubules and through the collecting ducts, tubule cells reabsorb about 99% of the filtered water and many useful solutes. The water and solutes return to the blood as it flows through the peritubular capillaries and vasa recta. Note that the term *reabsorption* refers to the return of substances to the bloodstream. The term *absorption*, by contrast, means entry of new substances into the body, as occurs in the gastrointestinal tract.
3. ***Tubular secretion.*** As filtered fluid flows through the renal tubules and collecting ducts, the renal tubule and duct cells secrete other materials, such as wastes, drugs, and excess ions, into the fluid. Notice that tubular secretion *removes a substance from the blood.*

Solutes and the fluid that drain into the minor and major calyces and renal pelvis constitute urine and are excreted. The rate of urinary excretion of any solute is equal to its rate of glomerular filtration, plus its rate of secretion, minus its rate of reabsorption.

By filtering, reabsorbing, and secreting, nephrons help maintain homeostasis of the blood's volume and composition. The situation is somewhat analogous to a recycling center: Garbage trucks dump garbage into an input hopper, where the smaller garbage passes onto a conveyor belt (glomerular filtration of plasma). As the conveyor belt carries the garbage along, workers remove useful items, such as

FIGURE 26.6 **Histology of a renal corpuscle.**

A renal corpuscle consists of a glomerulus and a glomerular (Bowman's) capsule.

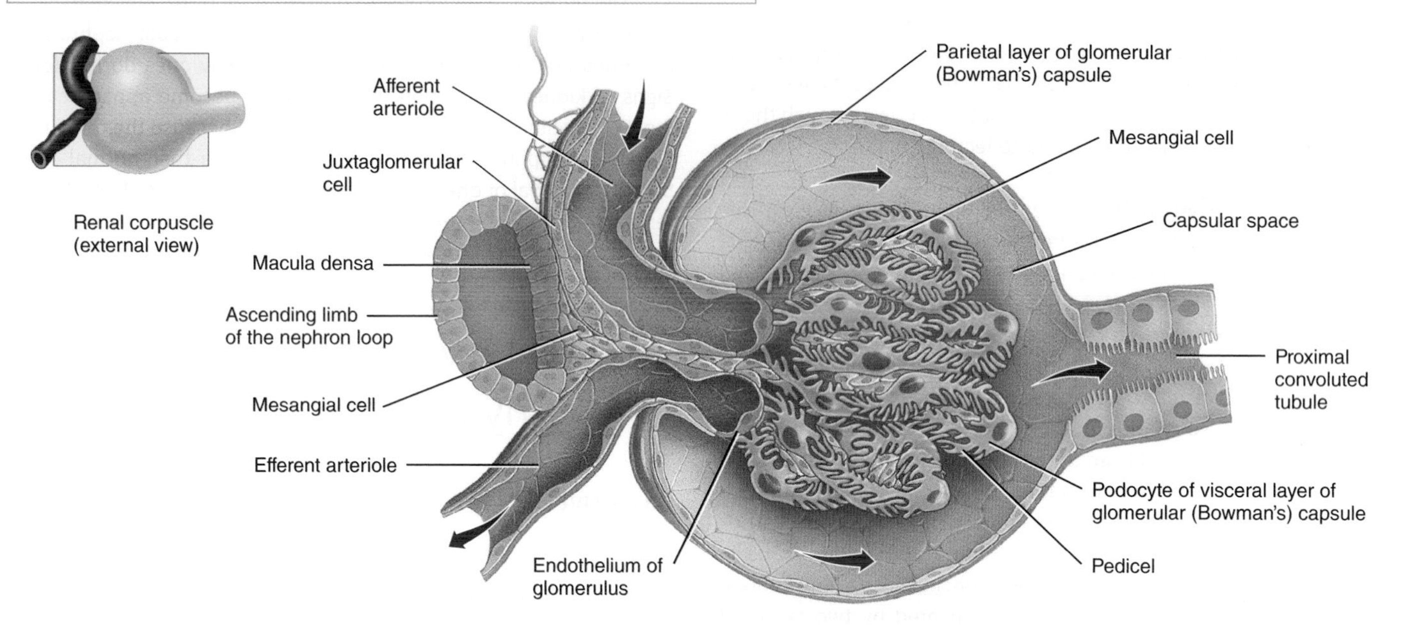

(a) Renal corpuscle (internal view)

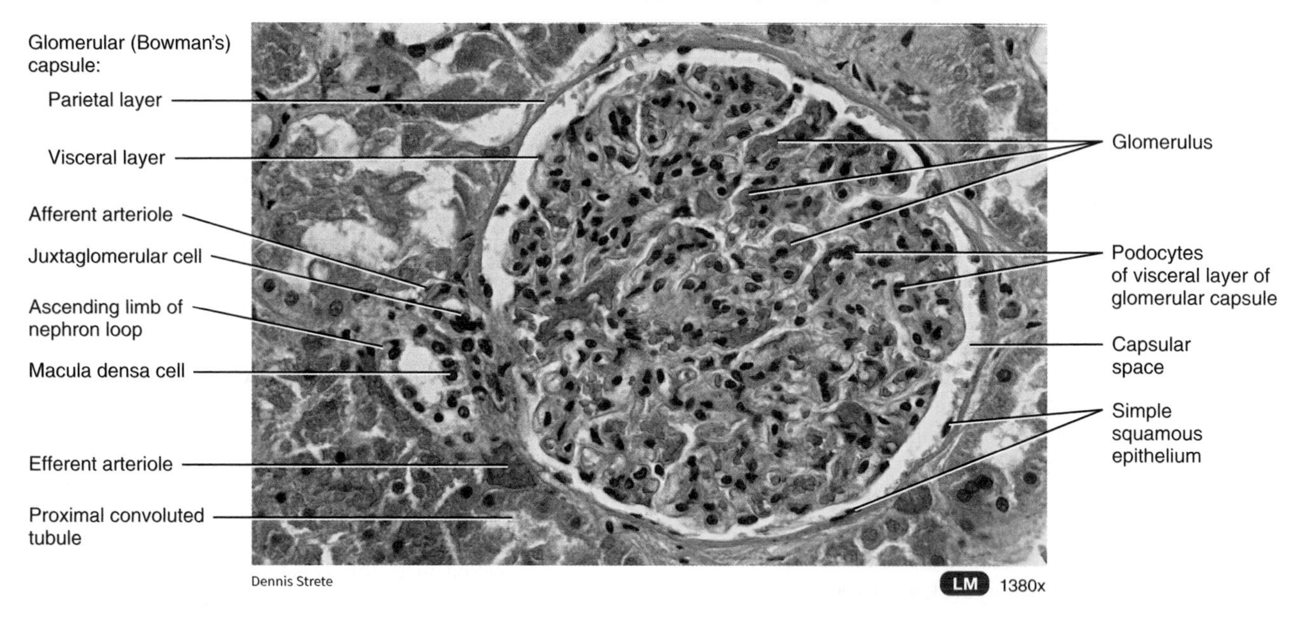

Dennis Strete

(b) Renal corpuscle

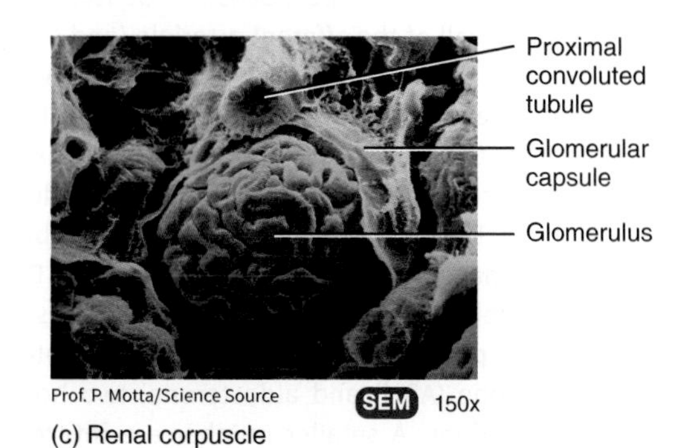

Prof. P. Motta/Science Source

(c) Renal corpuscle

Q Is the photomicrograph in (b) from a section through the renal cortex or renal medulla? How can you tell?

TABLE 26.1 **Histological Features of the Renal Tubule and Collecting Duct**

REGION AND HISTOLOGY		DESCRIPTION
Proximal convoluted tubule (PCT)	Microvilli; Mitochondrion; Apical surface	Simple cuboidal epithelial cells with prominent brush borders of microvilli.
Nephron loop: descending limb and thin ascending limb		Simple squamous epithelial cells.
Nephron loop: thick ascending limb		Simple cuboidal to low columnar epithelial cells.
Most of distal convoluted tubule (DCT)		Simple cuboidal epithelial cells.
Last part of DCT and all of collecting duct (CD)	Intercalated cell; Principal cell	Simple cuboidal epithelium consisting of principal cells and intercalated cells.

aluminum cans, plastics, and glass containers (reabsorption). Other workers place additional garbage left at the center and larger items onto the conveyor belt (secretion). At the end of the belt, all remaining garbage falls into a truck for transport to the landfill (excretion of wastes in urine).

26.5 Glomerular Filtration

OBJECTIVES

- **Describe** the filtration membrane.
- **Discuss** the pressures that promote and oppose glomerular filtration.

The fluid that enters the capsular space is called the **glomerular filtrate**. The fraction of blood plasma in the afferent arterioles of the kidneys that becomes glomerular filtrate is the **filtration fraction**. Although a filtration fraction of 0.16–0.20 (16–20%) is typical, the value varies considerably in both health and disease. On average, the daily volume of glomerular filtrate in adults is 150 liters in females and 180 liters in males. More than 99% of the glomerular filtrate returns to the bloodstream via tubular reabsorption, so only 1–2 liters (about 1–2 qt) is excreted as urine.

The Filtration Membrane

Together, the glomerular capillaries and the podocytes, which completely encircle the capillaries, form a leaky barrier known as the **filtration** (*endothelial-capsular*) **membrane**. This sandwichlike assembly permits filtration of water and small solutes but prevents

FIGURE 26.7 Relationship of a nephron's structure to its three basic functions: glomerular filtration, tubular reabsorption, and tubular secretion. Excreted substances remain in the urine and subsequently leave the body. For any substance S, excretion rate of S = filtration rate of S − reabsorption rate of S + secretion rate of S.

Glomerular filtration occurs in the renal corpuscle. Tubular reabsorption and tubular secretion occur all along the renal tubule and collecting duct.

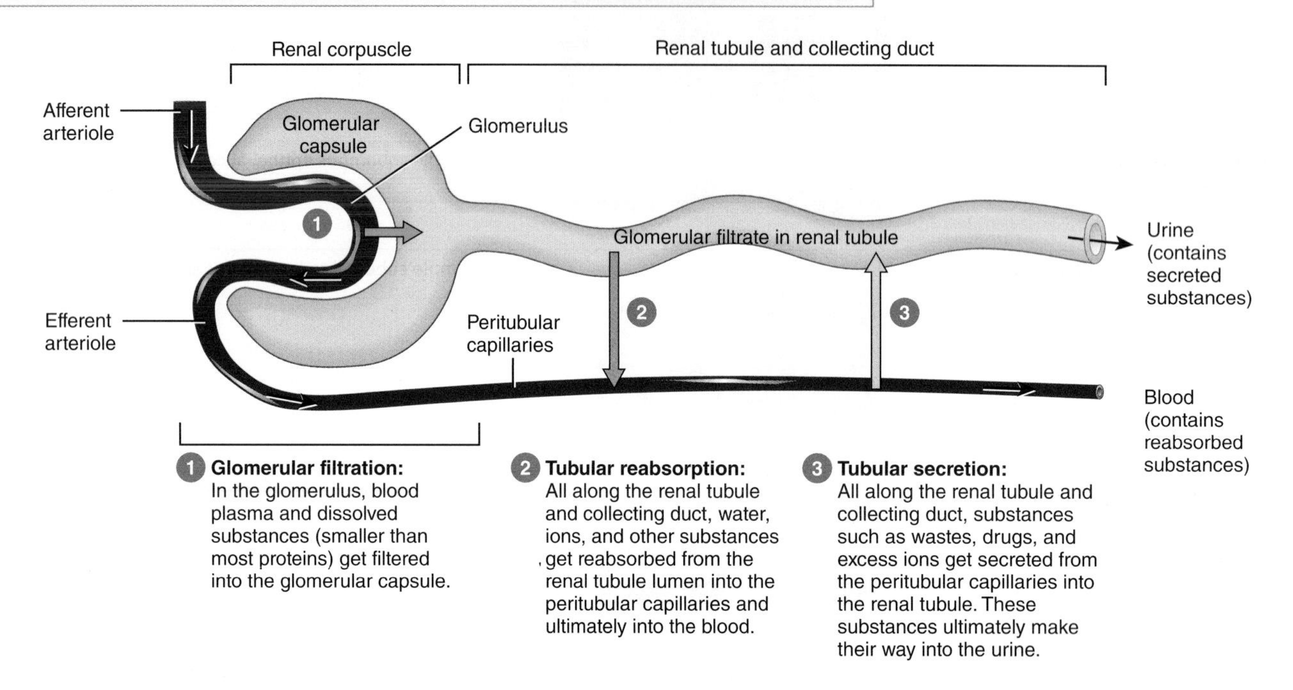

Q When cells of the renal tubules secrete the drug penicillin, is the drug being added to or removed from the bloodstream?

filtration of most plasma proteins and blood cells. Substances filtered from the blood cross three filtration barriers—a glomerular endothelial cell, the basement membrane, and a filtration slit formed by a podocyte (**Figure 26.8**):

1. Glomerular endothelial cells are quite leaky because they have large **fenestrations** (fen′-es-TRĀ-shuns) (pores) that measure 0.07–0.1 μm in diameter. This size permits all solutes in blood plasma to exit glomerular capillaries but prevents filtration of blood cells. Located among the glomerular capillaries and in the cleft between afferent and efferent arterioles are **mesangial cells** (mes-AN-jē-al; *mes-* = in the middle; *-angi-* = blood vessel) (see **Figure 26.6a**). These contractile cells help regulate glomerular filtration.
2. The **basement membrane**, a layer of acellular material between the endothelium and the podocytes, consists of minute collagen fibers and negatively charged glycoproteins. The pores within the basement membrane allow water and most small solutes to pass through. However, the negative charges of the glycoproteins repel plasma proteins, most of which are anionic; the repulsion hinders filtration of these proteins.
3. Extending from each podocyte are thousands of footlike processes termed **pedicels** (PED-i-sels = little feet) that wrap around glomerular capillaries. The spaces between pedicels are the **filtration slits**. A thin membrane, the **slit membrane**, extends across each filtration slit; it permits the passage of molecules having a diameter smaller than 0.006–0.007 μm, including water, glucose, vitamins, amino acids, very small plasma proteins, ammonia, urea, and ions. Less than 1% of albumin, the most plentiful plasma protein, passes the slit membrane because, with a diameter of 0.007 μm, it is slightly too big to get through.

The principle of *filtration*—the use of pressure to force fluids and solutes through a membrane—is the same in glomerular capillaries as in blood capillaries elsewhere in the body (see Starling's law of the capillaries, Section 21.2). However, the volume of fluid filtered by the renal corpuscle is much larger than in other blood capillaries of the body for three reasons:

1. Glomerular capillaries present a large surface area for filtration because they are long and extensive. Mesangial cells regulate how much surface area is available. When mesangial cells are relaxed, surface area is maximal, and glomerular filtration is very high. Contraction of mesangial cells reduces the available surface area, and glomerular filtration decreases.

FIGURE 26.8 The filtration membrane. The size of the endothelial fenestrations and filtration slits have been exaggerated for emphasis.

During glomerular filtration, water and solutes pass from blood plasma into the capsular space.

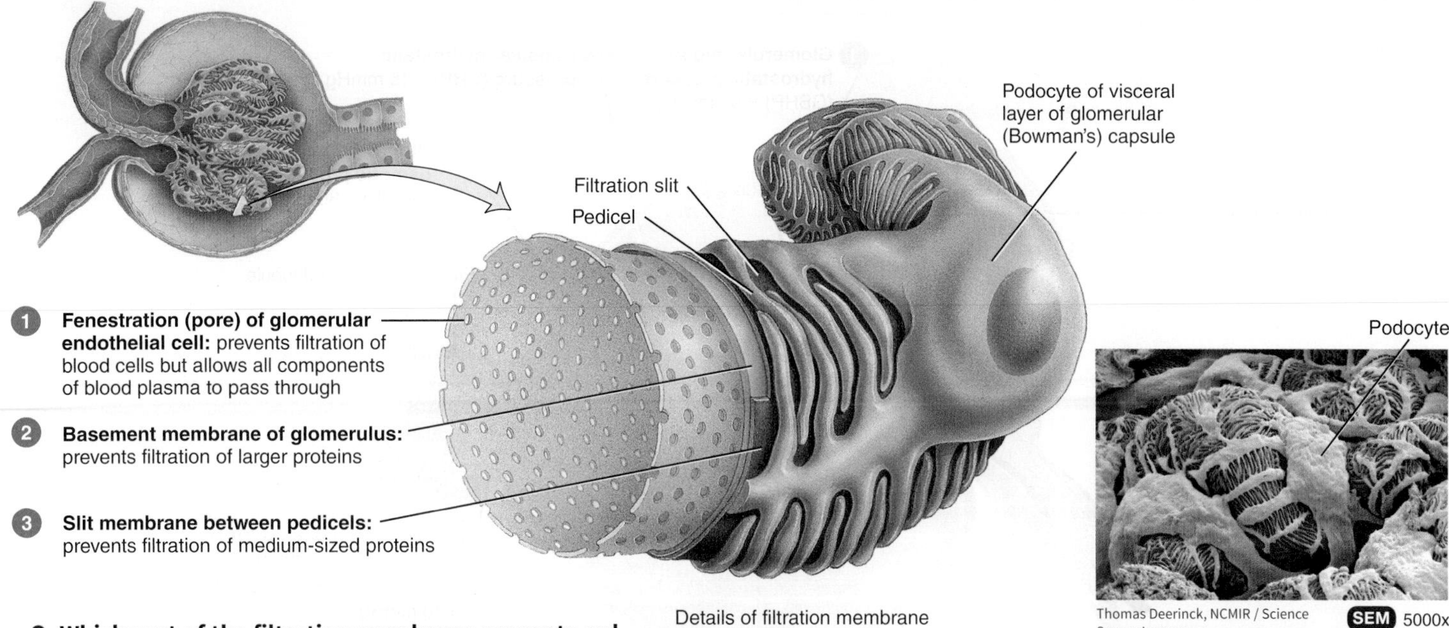

Q Which part of the filtration membrane prevents red blood cells from entering the capsular space?

2. The filtration membrane is thin and porous. Despite having several layers, the thickness of the filtration membrane is only 0.1 mm. Glomerular capillaries also are about 50 times leakier than blood capillaries in most other tissues, mainly because of their large fenestrations.
3. Glomerular capillary blood pressure is high. Because the efferent arteriole is smaller in diameter than the afferent arteriole, resistance to the outflow of blood from the glomerulus is high. As a result, blood pressure in glomerular capillaries is considerably higher than in blood capillaries elsewhere in the body.

Net Filtration Pressure

Glomerular filtration depends on three main pressures. One pressure *promotes* filtration and two pressures *oppose* filtration (**Figure 26.9**):

1. **Glomerular blood hydrostatic pressure (GBHP)** is the blood pressure in glomerular capillaries. Generally, GBHP is about 55 mmHg. It promotes filtration by forcing water and solutes in blood plasma through the filtration membrane.
2. **Capsular hydrostatic pressure (CHP)** is the hydrostatic pressure exerted against the filtration membrane by fluid already in the capsular space and renal tubule. CHP opposes filtration and represents a "back pressure" of about 15 mmHg.
3. **Blood colloid osmotic pressure (BCOP)**, which is due to the presence of proteins such as albumin, globulins, and fibrinogen in blood plasma, also opposes filtration. The average BCOP in glomerular capillaries is 30 mmHg.

Net filtration pressure (NFP), the total pressure that promotes filtration, is determined as follows:

$$\text{Net filtration pressure (NFP)} = \text{GBHP} - \text{CHP} - \text{BCOP}$$

By substituting the values just given, normal NFP may be calculated:

$$\begin{aligned}\text{NFP} &= 55 \text{ mmHg} - 15 \text{ mmHg} - 30 \text{ mmHg}\\ &= 10 \text{ mmHg}\end{aligned}$$

Thus, a pressure of only 10 mmHg causes a normal amount of blood plasma (minus plasma proteins) to filter from the glomerulus into the capsular space.

See Clinical Connection: Loss of Plasma Proteins in Urine Causes Edema

Glomerular Filtration Rate

The amount of filtrate formed in all renal corpuscles of both kidneys each minute is the **glomerular filtration rate (GFR)**. In adults, the GFR averages 125 mL/min in males and 105 mL/min in females. Homeostasis of body fluids requires that the kidneys maintain a relatively constant GFR. If the GFR is too high, needed substances may

FIGURE 26.9 **The pressures that drive glomerular filtration.** Taken together, these pressures determine net filtration pressure (NFP).

Glomerular blood hydrostatic pressure promotes filtration, whereas capsular hydrostatic pressure and blood colloid osmotic pressure oppose filtration.

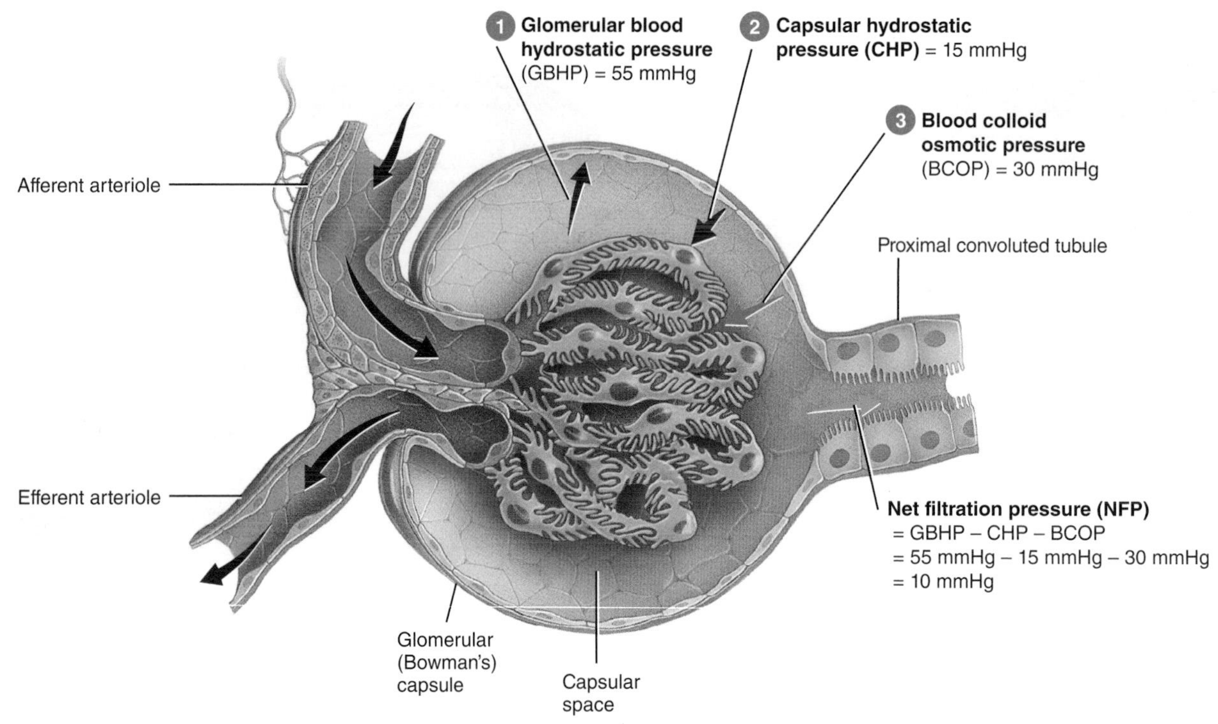

Q Suppose a tumor is pressing on and obstructing the right ureter. What effect might this have on CHP and thus on NFP in the right kidney? Would the left kidney also be affected?

pass so quickly through the renal tubules that some are not reabsorbed and are lost in the urine. If the GFR is too low, nearly all the filtrate may be reabsorbed and certain waste products may not be adequately excreted.

GFR is directly related to the pressures that determine net filtration pressure; any change in net filtration pressure will affect GFR. Severe blood loss, for example, reduces mean arterial blood pressure and decreases the glomerular blood hydrostatic pressure. Filtration ceases if glomerular blood hydrostatic pressure drops to 45 mmHg because the opposing pressures add up to 45 mmHg. Amazingly, when systemic blood pressure rises above normal, net filtration pressure and GFR increase very little. GFR is nearly constant when the mean arterial blood pressure is anywhere between 80 and 180 mmHg.

The mechanisms that regulate glomerular filtration rate operate in two main ways: (1) by adjusting blood flow into and out of the glomerulus and (2) by altering the glomerular capillary surface area available for filtration. GFR increases when blood flow into the glomerular capillaries increases. Coordinated control of the diameter of both afferent and efferent arterioles regulates glomerular blood flow. Constriction of the afferent arteriole decreases blood flow into the glomerulus; dilation of the afferent arteriole increases it. Three mechanisms control GFR: renal autoregulation, neural regulation, and hormonal regulation.

Renal Autoregulation of GFR

The kidneys themselves help maintain a constant renal blood flow and GFR despite normal, everyday changes in blood pressure, like those that occur during exercise. This capability is called renal autoregulation (aw′-tō-reg′-ū-LĀ-shun) and consists of two mechanisms—the myogenic mechanism and tubuloglomerular feedback. Working together, they can maintain nearly constant GFR over a wide range of systemic blood pressures.

The **myogenic mechanism** (mī-ō-JEN-ik; *myo-* = muscle; *-genic* = producing) occurs when stretching triggers contraction of smooth muscle cells in the walls of afferent arterioles. As blood pressure rises, GFR also rises because renal blood flow increases. However, the elevated blood pressure stretches the walls of the afferent arterioles. In response, smooth muscle fibers in the wall of the afferent arteriole contract, which narrows the arteriole's lumen. As a result, renal blood flow decreases, thus reducing GFR to its previous level. Conversely, when arterial blood pressure drops, the smooth muscle cells are stretched less and thus relax. The afferent arterioles dilate, renal blood flow increases, and GFR increases. The myogenic mechanism normalizes renal blood flow and GFR within seconds after a change in blood pressure.

The second contributor to renal autoregulation, **tubuloglomerular feedback** (too′-bū-lō-glō-MER-ū-lar), is so named because part of the renal tubules—the macula densa—provides feedback to the glomerulus (**Figure 26.10**). When GFR is above normal due to elevated systemic blood pressure, filtered fluid flows more rapidly along the renal tubules. As a result, the proximal convoluted tubule and nephron loop have less time to reabsorb Na^+, Cl^-, and water. Macula densa cells are thought to detect the increased delivery of Na^+, Cl^-, and water and to inhibit release of nitric oxide (NO) from cells in the juxtaglomerular apparatus (JGA). Because NO causes vasodilation, afferent arterioles constrict when the level of NO declines. As a result, less blood flows into the glomerular capillaries, and GFR decreases. When blood pressure falls, causing GFR to be lower than normal, the opposite sequence of events occurs, although to a lesser degree. Tubuloglomerular feedback operates more slowly than the myogenic mechanism.

Neural Regulation of GFR

Like most blood vessels of the body, those of the kidneys are supplied by sympathetic ANS fibers that release norepinephrine. Norepinephrine causes vasoconstriction through the activation of α_1 receptors, which are particularly plentiful in the smooth muscle fibers of afferent arterioles. At rest, sympathetic stimulation is moderately low, the afferent and efferent arterioles are dilated, and renal autoregulation of GFR prevails. With moderate sympathetic stimulation, both afferent and efferent arterioles constrict to the same degree. Blood flow into and out of the glomerulus is restricted to the same extent, which decreases GFR only slightly. With greater sympathetic stimulation, however, as occurs during exercise or hemorrhage, vasoconstriction of the afferent arterioles predominates. As a result, blood flow into glomerular capillaries is greatly decreased, and GFR drops. This lowering of renal blood flow has two consequences: (1) It reduces urine output, which helps conserve blood volume. (2) It permits greater blood flow to other body tissues.

Hormonal Regulation of GFR

Two hormones contribute to regulation of GFR. Angiotensin II reduces GFR; atrial natriuretic peptide (ANP) increases GFR. **Angiotensin** II (an′-jē-ō-TEN-sin) is a very potent vasoconstrictor that narrows both afferent and efferent arterioles and reduces renal blood flow, thereby decreasing GFR. Cells in the atria of the heart secrete **atrial natriuretic peptide (ANP)** (nā′-trē-ū-RET-ik). Stretching of the atria, as occurs when blood volume increases, stimulates secretion of ANP. By causing relaxation of the glomerular mesangial cells, ANP increases the capillary surface area available for filtration. Glomerular filtration rate rises as the surface area increases.

Table 26.2 summarizes the regulation of glomerular filtration rate.

FIGURE 26.10 Tubuloglomerular feedback.

Macula densa cells of the juxtaglomerular apparatus (JGA) provide negative feedback regulation of the glomerular filtration rate.

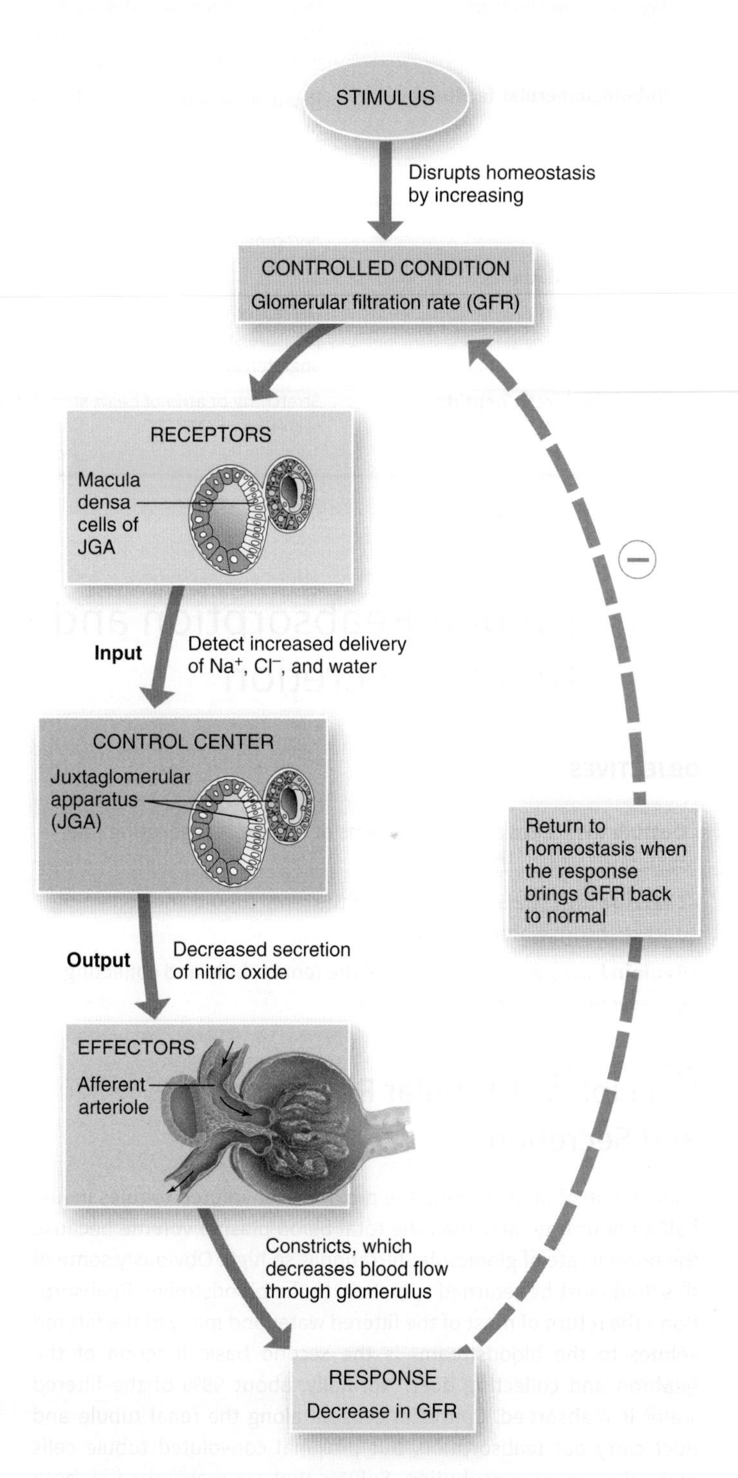

Q Why is this process termed autoregulation?

TABLE 26.2 Regulation of Glomerular Filtration Rate (GFR)

TYPE OF REGULATION	MAJOR STIMULUS	MECHANISM AND SITE OF ACTION	EFFECT ON GFR
Renal autoregulation			
Myogenic mechanism	Increased stretching of smooth muscle fibers in afferent arteriole walls due to increased blood pressure.	Stretched smooth muscle fibers contract, thereby narrowing lumen of afferent arterioles.	Decrease.
Tubuloglomerular feedback	Rapid delivery of Na^+ and Cl^- to the macula densa due to high systemic blood pressure.	Decreased release of nitric oxide (NO) by juxtaglomerular apparatus causes constriction of afferent arterioles.	Decrease.
Neural regulation	Increase in activity level of renal sympathetic nerves releases norepinephrine.	Constriction of afferent arterioles through activation of α_1 receptors and increased release of renin.	Decrease.
Hormone regulation			
Angiotensin II	Decreased blood volume or blood pressure stimulates production of angiotensin II.	Constriction of afferent and efferent arterioles.	Decrease.
Atrial natriuretic peptide (ANP)	Stretching of atria of heart stimulates secretion of ANP.	Relaxation of mesangial cells in glomerulus increases capillary surface area available for filtration.	Increase.

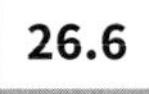

26.6 Tubular Reabsorption and Tubular Secretion

OBJECTIVES

- **Outline** the routes and mechanisms of tubular reabsorption and secretion.
- **Describe** how specific segments of the renal tubule and collecting duct reabsorb water and solutes.
- **Explain** how specific segments of the renal tubule and collecting duct secrete solutes into the urine.

Principles of Tubular Reabsorption and Secretion

The volume of fluid entering the proximal convoluted tubules in just half an hour is greater than the total blood plasma volume because the normal rate of glomerular filtration is so high. Obviously some of this fluid must be returned somehow to the bloodstream. Reabsorption—the return of most of the filtered water and many of the filtered solutes to the bloodstream—is the second basic function of the nephron and collecting duct. Normally, about 99% of the filtered water is reabsorbed. Epithelial cells all along the renal tubule and duct carry out reabsorption, but proximal convoluted tubule cells make the largest contribution. Solutes that are reabsorbed by both active and passive processes include glucose, amino acids, urea, and ions such as Na^+ (sodium), K^+ (potassium), Ca^{2+} (calcium), Cl^- (chloride), HCO_3^- (bicarbonate), and HPO_4^{2-} (phosphate). Once fluid passes through the proximal convoluted tubule, cells located more distally fine-tune the reabsorption processes to maintain homeostatic balances of water and selected ions. Most small proteins and peptides that pass through the filter also are reabsorbed, usually via pinocytosis. To appreciate the magnitude of tubular reabsorption, look at Table 26.3 and compare the amounts of substances that are filtered, reabsorbed, and secreted in urine.

The third function of nephrons and collecting ducts is tubular secretion, the transfer of materials from the blood and tubule cells into glomerular filtrate. Secreted substances include hydrogen ions (H^+), K^+, ammonium ions (NH_4^+), creatinine, and certain drugs such as penicillin. Tubular secretion has two important outcomes: (1) The secretion of H^+ helps control blood pH. (2) The secretion of other substances helps eliminate them from the body in urine.

As a result of tubular secretion, certain substances pass from blood into urine and may be detected by a urinalysis (see Section 26.8). It is especially important to test athletes for the presence of performance-enhancing drugs such as anabolic steroids, plasma expanders, erythropoietin, hCG, hGH, and amphetamines. Urine tests can also be used to detect the presence of alcohol or illegal drugs such as marijuana, cocaine, and heroin.

Reabsorption Routes A substance being reabsorbed from the fluid in the tubule lumen can take one of two routes before entering a peritubular capillary: It can move *between* adjacent tubule cells or *through* an individual tubule cell (Figure 26.11). Along the renal tubule, tight junctions surround and join neighboring cells to one another, much like the plastic rings that hold a six-pack of soda cans together. The **apical membrane** (the tops of the soda cans) contacts the tubular fluid, and the **basolateral membrane** (the bottoms and sides of the soda cans) contacts interstitial fluid at the base and sides of the cell.

TABLE 26.3 Substances Filtered, Reabsorbed, and Secreted per Day

SUBSTANCE	FILTERED* (ENTERS GLOMERULAR CAPSULE)	REABSORBED (RETURNED TO BLOOD)	SECRETED (TO BECOME URINE)
Water	180 liters	178–178.5 liters	1.5–2 liters
Proteins	2.0 g	1.9 g	0.1 g
Sodium ions (Na^+)	579 g	575 g	4 g
Chloride ions (Cl^-)	640 g	633.7 g	6.3 g
Bicarbonate ions (HCO_3^-)	275 g	274.97 g	0.03 g
Glucose	162 g	162 g	0 g
Urea	54 g	24 g	30 g†
Potassium ions (K^+)	29.6 g	29.6 g	2.0 g‡
Uric acid	8.5 g	7.7 g	0.8 g
Creatinine	1.6 g	0 g	1.6 g

*Assuming GFR is 180 liters per day.
†In addition to being filtered and reabsorbed, urea is secreted.
‡After virtually all filtered K^+ is reabsorbed in the convoluted tubules and nephron loop, a variable amount of K^+ is secreted by principal cells in the collecting duct.

FIGURE 26.11 Reabsorption routes: paracellular reabsorption and transcellular reabsorption.

In paracellular reabsorption, water and solutes in tubular fluid return to the bloodstream by moving between tubule cells; in transcellular reabsorption, solutes and water in tubular fluid return to the bloodstream by passing through a tubule cell.

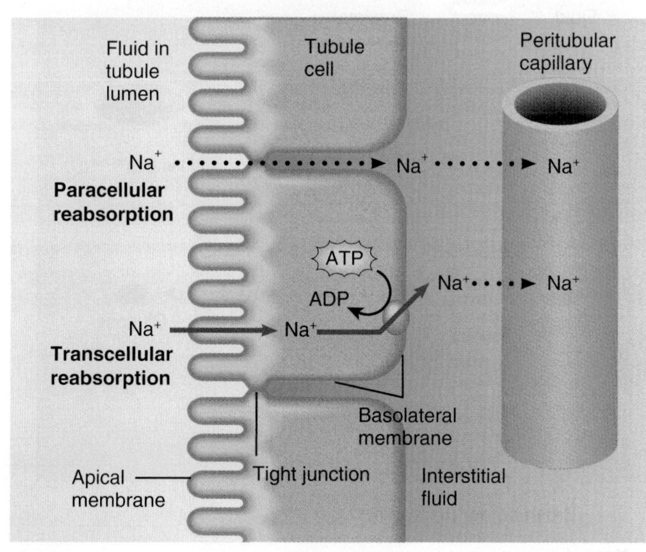

Q What is the main function of the tight junctions between tubule cells?

Fluid can leak *between* the cells in a passive process known as **paracellular reabsorption** (par′-a-SEL-ū-lar; *para-* = beside). Even though the epithelial cells are connected by tight junctions, the tight junctions between cells in the proximal convoluted tubules are "leaky" and permit some reabsorbed substances to pass between cells into peritubular capillaries. In some parts of the renal tubule, the paracellular route is thought to account for up to 50% of the reabsorption of certain ions and the water that accompanies them via osmosis. In **transcellular reabsorption** (trans′-SEL-ū-lar; *trans-* = across), a substance passes from the fluid in the tubular lumen *through* the apical membrane of a tubule cell, across the cytosol, and out into interstitial fluid through the basolateral membrane.

Transport Mechanisms

When renal cells transport solutes out of or into tubular fluid, they move specific substances in one direction only. Not surprisingly, different types of transport proteins are present in the apical and basolateral membranes. The tight junctions form a barrier that prevents mixing of proteins in the apical and basolateral membrane compartments. Reabsorption of Na^+ by the renal tubules is especially important because of the large number of sodium ions that pass through the glomerular filters.

Cells lining the renal tubules, like other cells throughout the body, have a low concentration of Na^+ in their cytosol due to the activity of sodium–potassium pumps (Na^+–K^+ ATPases). These pumps are located in the basolateral membranes and eject Na^+ from the renal tubule cells (**Figure 26.11**). The absence of sodium–potassium pumps in the apical membrane ensures that reabsorption of Na^+ is a one-way process. Most sodium ions that cross the apical membrane will be pumped into interstitial fluid at the base and sides of the cell. The amount of ATP used by sodium–potassium pumps in the renal tubules is about 6% of the total ATP consumption of the body at rest. This may not sound like much, but it is about the same amount of energy used by the diaphragm as it contracts during quiet breathing.

As we noted in Chapter 3, transport of materials across membranes may be either active or passive. Recall that in **primary active transport** the energy derived from hydrolysis of ATP is used to "pump" a substance across a membrane; the sodium–potassium pump is one such pump. In **secondary active transport** the energy stored in an ion's electrochemical gradient, rather than hydrolysis of ATP, drives another substance across a membrane. Secondary active transport couples movement of an ion down its electrochemical gradient to the "uphill" movement of a second substance against its electrochemical gradient. *Symporters* are membrane proteins that move two or more substances in the same direction across a membrane. *Antiporters* move two or more substances in opposite directions across a membrane. Each type of transporter has an upper limit on how fast it can work, just as an escalator has a limit on how many people it can carry from one level to another in a given period. This limit, called the **transport maximum (T_m)**, is measured in mg/min.

Solute reabsorption drives water reabsorption because all water reabsorption occurs via osmosis. About 90% of the reabsorption of water filtered by the kidneys occurs along with the reabsorption of solutes such as Na^+, Cl^-, and glucose. Water reabsorbed with solutes in tubular fluid is termed **obligatory water reabsorption** (ob-LIG-a-tor′-ē) because the water is "obliged" to follow the solutes when they are reabsorbed. This type of water reabsorption occurs in the proximal convoluted tubule and the descending limb of the nephron loop because these segments of the nephron are always permeable to water. Reabsorption of the final 10% of the water, a total of 10–20 liters per day, is termed **facultative water reabsorption** (FAK-ul-tā′-tiv). The word *facultative* means "capable of adapting to a need." Facultative water reabsorption is regulated by antidiuretic hormone and occurs mainly in the collecting ducts.

See Clinical Connection: Glucosuria

Now that we have discussed the principles of renal transport, we will follow the filtered fluid from the proximal convoluted tubule, into the nephron loop, on to the distal convoluted tubule, and through the collecting ducts. In each segment, we will examine where and how specific substances are reabsorbed and secreted. The filtered fluid becomes *tubular fluid* once it enters the proximal convoluted tubule. The composition of tubular fluid changes as it flows along the nephron tubule and through the collecting duct due to reabsorption and secretion. The fluid that drains from papillary ducts into the renal pelvis is *urine*.

Reabsorption and Secretion in the Proximal Convoluted Tubule

The largest amount of solute and water reabsorption from filtered fluid occurs in the proximal convoluted tubules, which reabsorb 65% of the filtered water, Na^+, and K^+; 100% of most filtered organic solutes such as glucose and amino acids; 50% of the filtered Cl^-; 80–90% of the filtered HCO_3^-; 50% of the filtered urea; and a variable amount of the filtered Ca^{2+}, Mg^{2+}, and HPO_4^{2-} (phosphate). In addition, proximal convoluted tubules secrete a variable amount of H^+, ammonium ions (NH_4^+), and urea.

Most solute reabsorption in the proximal convoluted tubule (PCT) involves Na^+. Na^+ transport occurs via symport and antiport mechanisms in the proximal convoluted tubule. Normally, filtered glucose, amino acids, lactic acid, water-soluble vitamins, and other nutrients are not lost in the urine. Rather, they are completely reabsorbed in the first half of the proximal convoluted tubule by several types of **Na^+ symporters** located in the apical membrane. Figure 26.12 depicts the operation of one such symporter, the **Na^+–glucose symporter** in the apical membrane of a cell in the PCT. Two Na^+ and a molecule of glucose attach to the symporter protein, which carries them from the tubular fluid into the tubule cell. The glucose molecules then exit the basolateral membrane via facilitated diffusion and

FIGURE 26.12 Reabsorption of glucose by Na^+–glucose symporters in cells of the proximal convoluted tubule (PCT).

Normally, all filtered glucose is reabsorbed in the PCT.

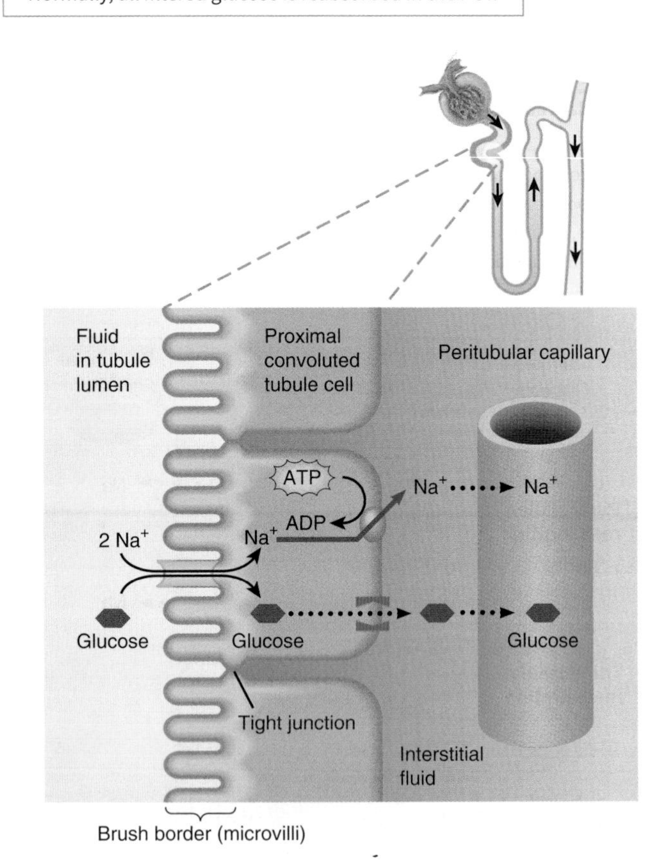

Q How does filtered glucose enter and leave a PCT cell?

they diffuse into peritubular capillaries. Other Na^+ symporters in the PCT reclaim filtered HPO_4^{2-} (phosphate) and SO_4^{2-} (sulfate) ions, all amino acids, and lactic acid in a similar way.

In another secondary active transport process, the **Na^+–H^+ antiporters** carry filtered Na^+ down its concentration gradient into a PCT cell as H^+ is moved from the cytosol into the lumen (**Figure 26.13a**), causing Na^+ to be reabsorbed into blood and H^+ to be secreted into tubular fluid. PCT cells produce the H^+ needed to keep the antiporters running in the following: way: Carbon dioxide (CO_2) diffuses from peritubular blood or tubular fluid or is produced by metabolic reactions within the cells. As also occurs in red blood cells (see **Figure 23.24**), the enzyme *carbonic anhydrase (CA)* (an-HĪ-drās) catalyzes the reaction of CO_2 with water (H_2O) to form carbonic acid (H_2CO_3), which then dissociates into H^+ and HCO_3^-:

$$CO_2 + H_2O \xrightarrow{\text{Carbonic anhydrase}} H_2CO_3 \longrightarrow H^+ + HCO_3^-$$

Most of the HCO_3^- in filtered fluid is reabsorbed in proximal convoluted tubules, thereby safeguarding the body's supply of an important buffer (**Figure 26.13b**). After H^+ is secreted into the fluid within the lumen of the proximal convoluted tubule, it reacts with filtered HCO_3^- to form H_2CO_3, which readily dissociates into CO_2 and H_2O. Carbon dioxide then diffuses into the tubule cells and joins with H_2O to form H_2CO_3, which dissociates into H^+ and HCO_3^-. As the level of HCO_3^- rises in the cytosol, it exits via facilitated diffusion transporters in the basolateral membrane and diffuses into the blood with Na^+. Thus, for every H^+ secreted into the tubular fluid of the proximal convoluted tubule, one HCO_3^- and one Na^+ are reabsorbed.

Solute reabsorption in proximal convoluted tubules promotes osmosis of water. Each reabsorbed solute increases the osmolarity, first inside the tubule cell, then in interstitial fluid, and finally in the blood. Water thus moves rapidly from the tubular fluid, via both the paracellular and transcellular routes, into the peritubular capillaries and restores osmotic balance (**Figure 26.14**). In other words, reabsorption of the solutes creates an osmotic gradient that promotes the reabsorption of water via osmosis. Cells lining the proximal convoluted tubule and the descending limb of the nephron loop are especially permeable to water because they have many molecules of **aquaporin-1** (ak-kwa-PŌR-in). This integral protein in the plasma membrane is a water channel that greatly increases the rate of water movement across the apical and basolateral membranes.

As water leaves the tubular fluid, the concentrations of the remaining filtered solutes increase. In the second half of the PCT, electrochemical gradients for Cl^-, K^+, Ca^{2+}, Mg^{2+}, and urea promote their passive diffusion into peritubular capillaries via both paracellular and transcellular routes. Among these ions, Cl^- is present in the highest concentration. Diffusion of negatively charged Cl^- into interstitial fluid via the paracellular route makes the interstitial fluid electrically more negative than the tubular fluid. This negativity promotes passive paracellular reabsorption of cations, such as K^+, Ca^{2+}, and Mg^{2+}.

Ammonia (NH_3) is a poisonous waste product derived from the deamination (removal of an amino group) of various amino acids, a reaction that occurs mainly in hepatocytes (liver cells). Hepatocytes convert most of this ammonia to urea, a less-toxic compound.

FIGURE 26.13 Actions of Na^+–H^+ antiporters in proximal convoluted tubule cells. (a) Reabsorption of sodium ions (Na^+) and secretion of hydrogen ions (H^+) via secondary active transport through the apical membrane. (b) Reabsorption of bicarbonate ions (HCO_3^-) via facilitated diffusion through the basolateral membrane. CO_2 = carbon dioxide; H_2CO_3 = carbonic acid; CA = carbonic anhydrase.

Na^+–H^+ antiporters promote transcellular reabsorption of Na^+ and secretion of H^+.

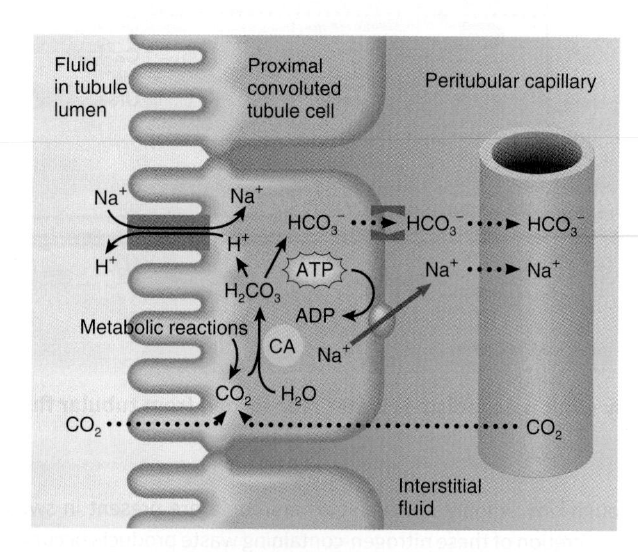

(a) Na^+ reabsorption and H^+ secretion

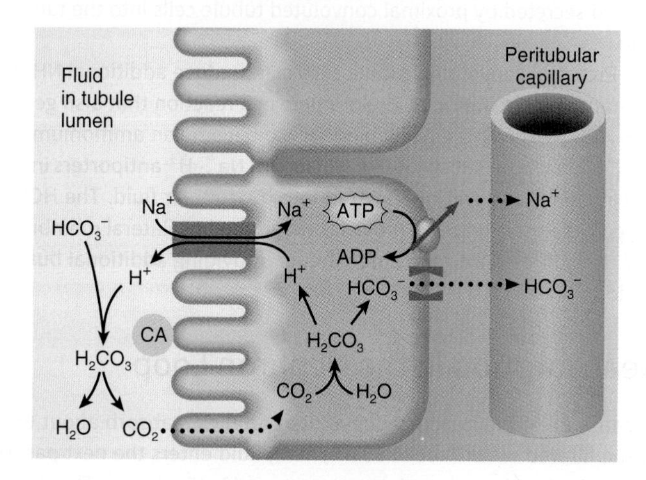

(b) HCO_3^- reabsorption

Key:

- Na^+–H^+ antiporter
- HCO_3^- facilitated diffusion transporter
- Diffusion
- Sodium–potassium pump

Q Which step in Na^+ movement in part (a) is promoted by the electrochemical gradient?

FIGURE 26.14 Passive reabsorption of Cl^-, K^+, Ca^{2+}, Mg^{2+}, urea, and water in the second half of the proximal convoluted tubule.

Electrochemical gradients promote passive reabsorption of solutes via both paracellular and transcellular routes.

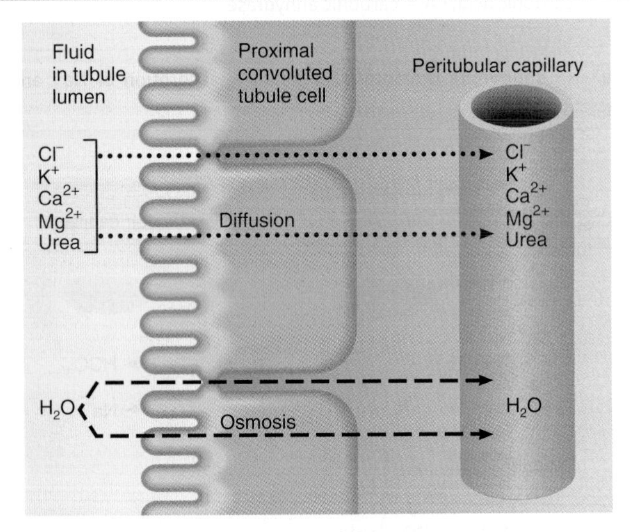

Q By what mechanism is water reabsorbed from tubular fluid?

Although tiny amounts of urea and ammonia are present in sweat, most excretion of these nitrogen-containing waste products occurs via the urine. Urea and ammonia in blood are both filtered at the glomerulus and secreted by proximal convoluted tubule cells into the tubular fluid.

Proximal convoluted tubule cells can produce additional NH_3 by deaminating the amino acid glutamine in a reaction that also generates HCO_3^-. The NH_3 quickly binds H^+ to become an ammonium ion (NH_4^+), which can substitute for H^+ aboard Na^+–H^+ antiporters in the apical membrane and be secreted into the tubular fluid. The HCO_3^- generated in this reaction moves through the basolateral membrane and then diffuses into the bloodstream, providing additional buffers in blood plasma.

Reabsorption in the Nephron Loop

Because all of the proximal convoluted tubules reabsorb about 65% of the filtered water (about 80 mL/min), fluid enters the next part of the nephron, the nephron loop, at a rate of 40–45 mL/min. The chemical composition of the tubular fluid now is quite different from that of glomerular filtrate because glucose, amino acids, and other nutrients are no longer present. The osmolarity of the tubular fluid is still close to the osmolarity of blood, however, because reabsorption of water by osmosis keeps pace with reabsorption of solutes all along the proximal convoluted tubule.

The nephron loop reabsorbs about 15% of the filtered water, 20–30% of the filtered Na^+ and K^+, 35% of the filtered Cl^-, 10–20% of the filtered HCO_3^-, and a variable amount of the filtered Ca^{2+} and Mg^{2+}. Here, for the first time, reabsorption of water via osmosis is *not* automatically coupled to reabsorption of filtered solutes because part of the nephron loop is relatively impermeable to water. The nephron loop thus sets the stage for *independent* regulation of both the *volume* and *osmolarity* of body fluids.

The apical membranes of cells in the thick ascending limb of the nephron loop have **Na^+–K^+–$2Cl^-$ symporters** that simultaneously reclaim one Na^+, one K^+, and two Cl^- from the fluid in the tubular lumen (**Figure 26.15**). Na^+ that is actively transported into interstitial fluid at the base and sides of the cell diffuses into the vasa recta. Cl^- moves through leakage channels in the basolateral membrane into interstitial fluid and then into the vasa recta. Because many K^+ leakage channels are present in the apical membrane, most K^+ brought in by the symporters moves down its concentration gradient back

FIGURE 26.15 Na^+–K^+–$2Cl^-$ symporter in the thick ascending limb of the nephron loop.

Cells in the thick ascending limb have symporters that simultaneously reabsorb one Na^+, one K^+, and two Cl^-.

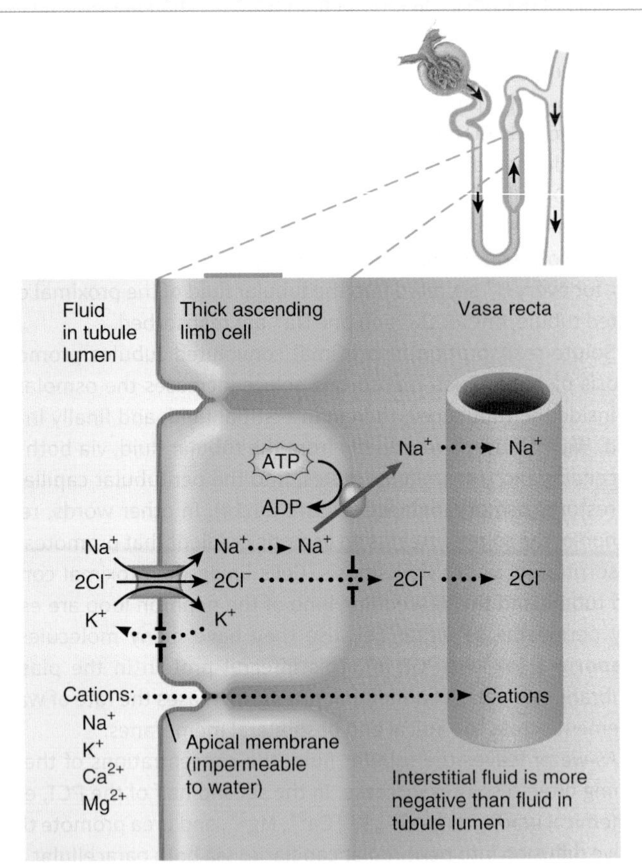

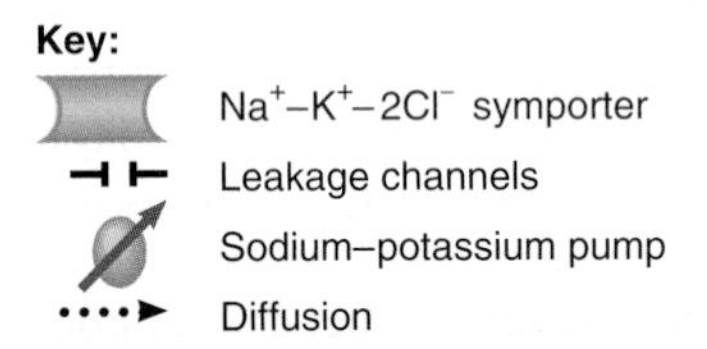

Q Why is this process considered secondary active transport? Does water reabsorption accompany ion reabsorption in this region of the nephron?

into the tubular fluid. Thus, the main effect of the $Na^+-K^+-2Cl^-$ symporters is reabsorption of Na^+ and Cl^-.

The movement of positively charged K^+ into the tubular fluid through the apical membrane channels leaves the interstitial fluid and blood with more negative charges relative to fluid in the ascending limb of the nephron loop. This relative negativity promotes reabsorption of cations—Na^+, K^+, Ca^{2+}, and Mg^{2+}—via the paracellular route.

Although about 15% of the filtered water is reabsorbed in the *descending* limb of the nephron loop, little or no water is reabsorbed in the *ascending* limb. In this segment of the tubule, the apical membranes are virtually impermeable to water. Because ions but not water molecules are reabsorbed, the osmolarity of the tubular fluid decreases progressively as fluid flows toward the end of the ascending limb.

Reabsorption in the Early Distal Convoluted Tubule

Fluid enters the distal convoluted tubules at a rate of about 25 mL/min because 80% of the filtered water has now been reabsorbed. The early or initial part of the distal convoluted tubule (DCT) reabsorbs about 10–15% of the filtered water, 5% of the filtered Na^+, and 5% of the filtered Cl^-. Reabsorption of Na^+ and Cl^- occurs by means of **Na^+–Cl^- symporters** in the apical membranes. Sodium–potassium pumps and Cl^- leakage channels in the basolateral membranes then permit reabsorption of Na^+ and Cl^- into the peritubular capillaries. The early DCT also is a major site where parathyroid hormone (PTH) stimulates reabsorption of Ca^{2+}. The amount of Ca^{2+} reabsorption in the early DCT varies depending on the body's needs.

Reabsorption and Secretion in the Late Distal Convoluted Tubule and Collecting Duct

By the time fluid reaches the end of the distal convoluted tubule, 90–95% of the filtered solutes and water have returned to the bloodstream. Recall that two different types of cells—principal cells and intercalated cells—are present at the late or terminal part of the distal convoluted tubule and throughout the collecting duct. The principal cells reabsorb Na^+ and secrete K^+. These cells also have receptors for aldosterone and antidiuretic hormone (ADH). The intercalated cells reabsorb HCO_3^- and secrete H^+, thereby playing a role in blood pH regulation. In addition, the intercalated cells reabsorb K^+. In the late distal convoluted tubules and collecting ducts, the amount of water and solute reabsorption and the amount of solute secretion vary depending on the body's needs.

In contrast to earlier segments of the nephron, Na^+ passes through the apical membrane of principal cells via Na^+ leakage channels rather than by means of symporters or antiporters (**Figure 26.16**). The concentration of Na^+ in the cytosol remains low, as usual, because the sodium–potassium pumps actively transport Na^+ across the basolateral membranes. Then Na^+ passively diffuses into the peritubular capillaries from the interstitial spaces around the tubule cells.

Normally, transcellular and paracellular reabsorption in the proximal convoluted tubule and nephron loop return most filtered K^+ to the bloodstream. To adjust for varying dietary intake of potassium and to maintain a stable level of K^+ in body fluids, principal cells

FIGURE 26.16 Reabsorption of Na^+ and secretion of K^+ by principal cells in the last part of the distal convoluted tubule and in the collecting duct.

In the apical membrane of principal cells, Na^+ leakage channels allow entry of Na^+ while K^+ leakage channels allow exit of K^+ into the tubular fluid.

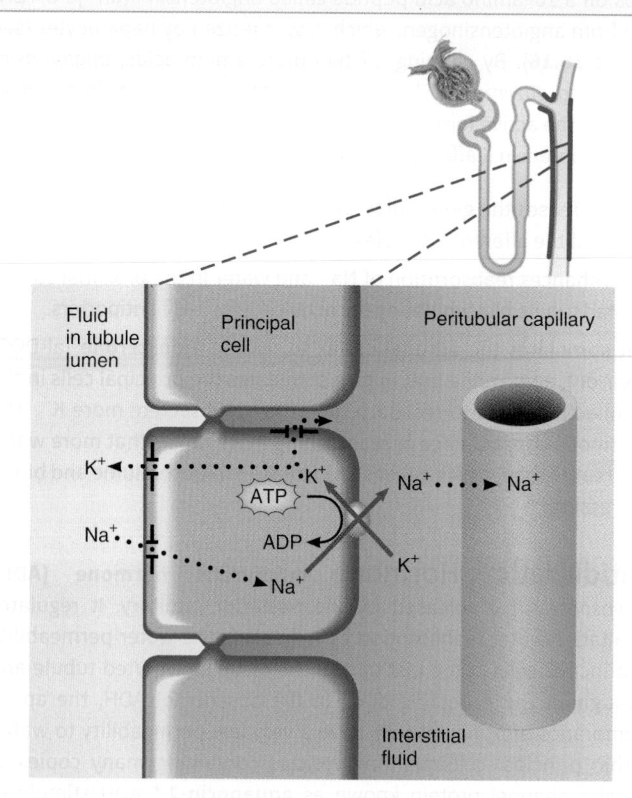

Key:
Diffusion
Leakage channels
Sodium–potassium pump

Q Which hormone stimulates reabsorption and secretion by principal cells, and how does this hormone exert its effect?

secrete a variable amount of K^+ (**Figure 26.16**). Because the basolateral sodium–potassium pumps continually bring K^+ into principal cells, the intracellular concentration of K^+ remains high. K^+ leakage channels are present in both the apical and basolateral membranes. Thus, some K^+ diffuses down its concentration gradient into the tubular fluid, where the K^+ concentration is very low. This secretion mechanism is the main source of K^+ excreted in the urine.

Homeostatic Regulation of Tubular Reabsorption and Tubular Secretion

Five hormones affect the extent of Na^+, Ca^{2+}, and water reabsorption as well as K^+ secretion by the renal tubules. These hormones include angiotensin II, aldosterone, antidiuretic hormone, atrial natriuretic peptide, and parathyroid hormone.

Renin–Angiotensin–Aldosterone System When blood volume and blood pressure decrease, the walls of the afferent arterioles are stretched less, and the juxtaglomerular cells secrete the enzyme **renin** (RĒ-nin) into the blood. Sympathetic stimulation also directly stimulates release of renin from juxtaglomerular cells. Renin clips off a 10–amino acid peptide called angiotensin I (an′-jē-ō-TEN-sin) from angiotensinogen, which is synthesized by hepatocytes (see Figure 18.16). By clipping off two more amino acids, *angiotensin-converting enzyme (ACE)* converts angiotensin I to **angiotensin II**, which is the active form of the hormone.

Angiotensin II affects renal physiology in three main ways:

1. It decreases the glomerular filtration rate by causing vasoconstriction of the afferent arterioles.
2. It enhances reabsorption of Na^+ and water in the proximal convoluted tubule by stimulating the activity of Na^+–H^+ antiporters.
3. It stimulates the adrenal cortex to release **aldosterone** (al-DOS-ter-ōn), a hormone that in turn stimulates the principal cells in the collecting ducts to reabsorb more Na^+ and secrete more K^+. The osmotic consequence of reabsorbing more Na^+ is that more water is reabsorbed, which causes an increase in blood volume and blood pressure.

Antidiuretic Hormone **Antidiuretic hormone (ADH)** or *vasopressin* is released by the posterior pituitary. It regulates facultative water reabsorption by increasing the water permeability of principal cells in the last part of the distal convoluted tubule and throughout the collecting duct. In the absence of ADH, the apical membranes of principal cells have a very low permeability to water. Within principal cells are tiny vesicles containing many copies of a water channel protein known as **aquaporin-2**.* ADH stimulates insertion of the aquaporin-2–containing vesicles into the apical membranes via exocytosis. As a result, the water permeability of the principal cell's apical membrane increases, and water molecules move more rapidly from the tubular fluid into the cells. Because the basolateral membranes are always relatively permeable to water, water molecules then move rapidly into the blood. This results in an increase in blood volume and blood pressure. When the ADH level declines, the aquaporin-2 channels are removed from the apical membrane via endocytosis, and water permeability of the principal cells decreases.

A negative feedback system involving ADH regulates facultative water reabsorption (Figure 26.17). When the osmolarity or osmotic pressure of plasma and interstitial fluid increases—that is, when water concentration decreases—by as little as 1%, osmoreceptors in the hypothalamus detect the change. Their nerve impulses stimulate secretion of more ADH into the blood, and the principal cells become more permeable to water. As facultative water reabsorption increases, plasma osmolarity decreases to normal. A second powerful stimulus for ADH secretion is a decrease in blood volume, as occurs in hemorrhaging or severe dehydration. In the pathological absence of ADH activity, a condition known as *diabetes insipidus*, a person may excrete up to 20 liters of very dilute urine daily.

*ADH does not govern the previously mentioned water channel (aquaporin-1).

FIGURE 26.17 Negative feedback regulation of facultative water reabsorption by ADH.

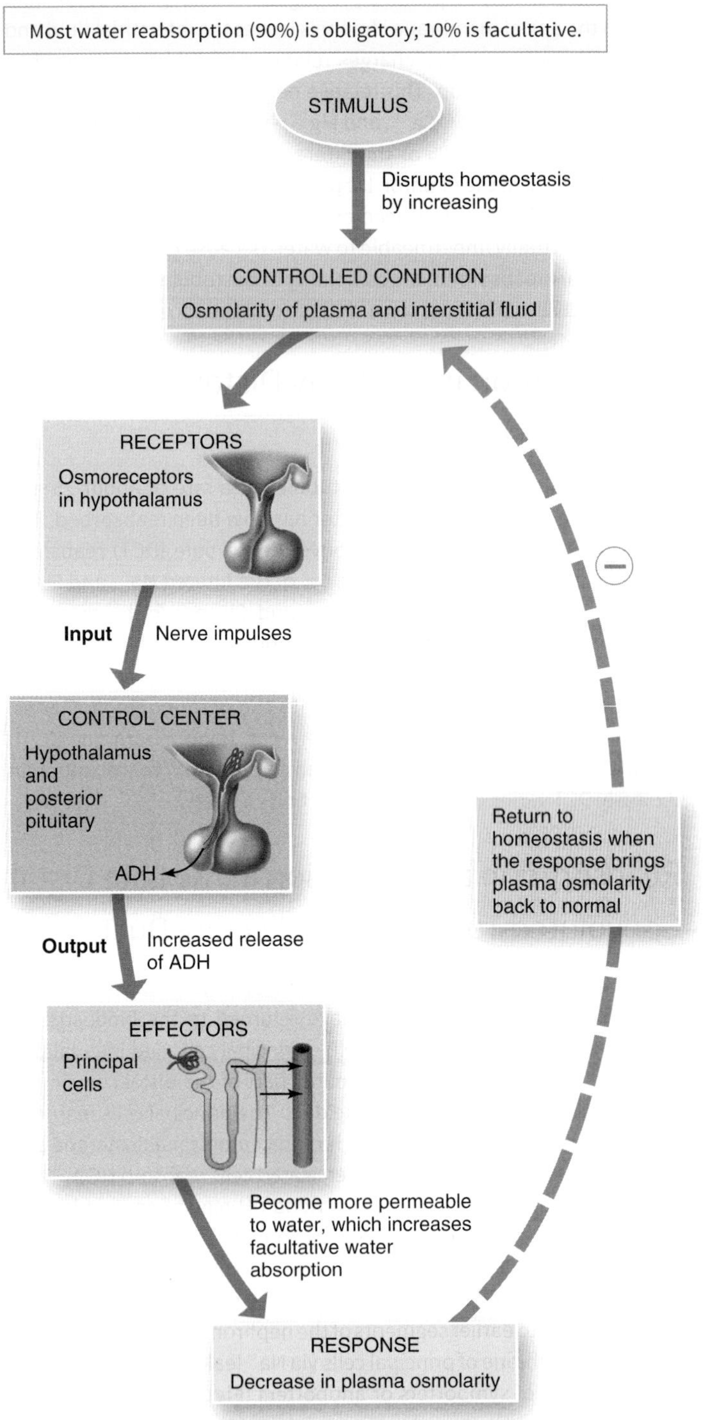

Most water reabsorption (90%) is obligatory; 10% is facultative.

Q In addition to ADH, which other hormones contribute to the regulation of water reabsorption?

The degree of facultative water reabsorption caused by ADH in the late distal tubule and collecting duct depends on whether the body is normally hydrated, dehydrated, or overhydrated.

- ***Normal hydration.*** Under conditions of normal body hydration (adequate water intake), enough ADH is present in the blood to cause reabsorption of 19% of the filtered water in the late distal tubule and the collecting duct. This means that the total amount of filtered water reabsorbed in the renal tubule and collecting duct is 99%: 65% in the proximal tubule + 15% in the nephron loop + 19% in the late distal tubule and collecting duct. The remaining 1% of the filtered water (about 1.5–2 L/day) is excreted in urine. Therefore, when the body is normally hydrated, the kidneys produce about 1.5–2 L of urine on a daily basis and the urine is slightly hyperosmotic (slightly concentrated) compared to blood.

- ***Dehydration.*** When the body is dehydrated, the concentration of ADH in the blood increases. This in turn causes an increase in the amount of filtered water that is reabsorbed in the late distal tubule and collecting duct. Depending on how much the blood ADH level increases, the amount of filtered water that is reabsorbed in the late distal tubule and collecting duct can increase from just above 19% to as high as 19.8%. As a result, less than 1% of filtered water remains unreabsorbed in the late distal tubule and collecting duct, which corresponds to a urine output *below* the normal 1.5–2 L/day. The urine produced under these circumstances is very hyperosmotic (highly concentrated) compared to blood because it contains less water than normal. In the case of severe dehydration, the amount of filtered water that is reabsorbed in the late distal tubule and collecting duct reaches a maximum limit of 19.8%. This means that the total amount of filtered water reabsorbed in the renal tubule and collecting duct is 99.8%: 65% in the proximal tubule + 15% in the nephron loop + 19.8% in the late distal tubule and collecting duct. The remaining 0.2% of the filtered water (about 400 mL/day) is excreted in urine. Thus, the kidneys produce a small volume of highly concentrated urine when the body is dehydrated.
- ***Overhydration.*** When the body is overhydrated (too much water intake), the concentration of ADH in the blood decreases. This in turn causes a decrease in the amount of filtered water that is reabsorbed in the late distal tubule and collecting duct. Depending on how much the blood ADH level decreases, the amount of filtered water that is reabsorbed in the late distal tubule and collecting duct can decrease from just below 19% to as low as 0%. As a result, more than 1% of filtered water remains unreabsorbed in the late distal tubule and collecting duct, which corresponds to a urine output *above* the normal 1.5–2 L/day. The urine produced under these conditions is hypoosmotic (dilute) compared to blood because it contains more water than normal. In the case of severe overhydration, no ADH is present in the blood, and the amount of water reabsorbed in the late distal tubule and collecting duct is 0%. This means that the total amount of filtered water that is reabsorbed in the renal tubule and collecting duct is 80%: 65% in the proximal tubule + 15% in the nephron loop + 0% in the late distal tubule and collecting duct. The remaining 20% of filtered water (about 36 L/day) is excreted in urine. Hence, the kidneys produce a large volume of dilute urine when the body is overhydrated.

Atrial Natriuretic Peptide

A large increase in blood volume promotes release of atrial natriuretic peptide (ANP) from the heart. Although the importance of ANP in normal regulation of tubular function is unclear, it can inhibit reabsorption of Na^+ and water in the proximal convoluted tubule and collecting duct. ANP also suppresses the secretion of aldosterone and ADH. These effects increase the excretion of Na^+ in urine (natriuresis) and increase urine output (diuresis), which decreases blood volume and blood pressure.

Parathyroid Hormone

Although the hormones mentioned thus far involve regulation of water loss as urine, the kidney tubules also respond to a hormone that regulates ionic composition. For example, a lower than normal level of Ca^{2+} in the blood stimulates the parathyroid glands to release **parathyroid hormone (PTH)**. PTH in turn stimulates cells in the early distal convoluted tubules to reabsorb more Ca^{2+} into the blood. PTH also inhibits HPO_4^{2-} (phosphate) reabsorption in proximal convoluted tubules, thereby promoting phosphate excretion.

Table 26.4 summarizes hormonal regulation of tubular reabsorption and tubular secretion.

TABLE 26.4 Hormonal Regulation of Tubular Reabsorption and Tubular Secretion

HORMONE	MAJOR STIMULI THAT TRIGGER RELEASE	MECHANISM AND SITE OF ACTION	EFFECTS
Angiotensin II	Low blood volume or low blood pressure stimulates renin-induced production of angiotensin II.	Stimulates activity of Na^+–H^+ antiporters in proximal tubule cells.	Increases reabsorption of Na^+ and water, which increases blood volume and blood pressure.
Aldosterone	Increased angiotensin II level and increased level of plasma K^+ promote release of aldosterone by adrenal cortex.	Enhances activity of sodium–potassium pumps in basolateral membrane and Na^+ channels in apical membrane of principal cells in collecting duct.	Increases secretion of K^+ and reabsorption of Na^+; increases reabsorption of water, which increases blood volume and blood pressure.
Antidiuretic hormone (ADH)	Increased osmolarity of extracellular fluid or decreased blood volume promotes release of ADH from posterior pituitary gland.	Stimulates insertion of water channel proteins (aquaporin-2) into apical membranes of principal cells.	Increases facultative reabsorption of water, which decreases osmolarity of body fluids.
Atrial natriuretic peptide (ANP)	Stretching of atria of heart stimulates ANP secretion.	Suppresses reabsorption of Na^+ and water in proximal tubule and collecting duct; inhibits secretion of aldosterone and ADH.	Increases excretion of Na^+ in urine (natriuresis); increases urine output (diuresis) and thus decreases blood volume and blood pressure.
Parathyroid hormone (PTH)	Decreased level of plasma Ca^{2+} promotes release of PTH from parathyroid glands.	Stimulates opening of Ca^{2+} channels in apical membranes of early distal tubule cells.	Increases reabsorption of Ca^{2+}.

26.7 Production of Dilute and Concentrated Urine

OBJECTIVE

- **Describe** how the renal tubule and collecting ducts produce dilute and concentrated urine.

Even though your fluid intake can be highly variable, the total volume of fluid in your body normally remains stable. Homeostasis of body fluid volume depends in large part on the ability of the kidneys to regulate the rate of water loss in urine. Normally functioning kidneys produce a large volume of dilute urine when fluid intake is high, and a small volume of concentrated urine when fluid intake is low or fluid loss is large. ADH controls whether dilute urine or concentrated urine is formed. In the absence of ADH, urine is very dilute. However, a high level of ADH stimulates reabsorption of more water into blood, producing a concentrated urine.

Formation of Dilute Urine

Glomerular filtrate has the same ratio of water and solute particles as blood; its osmolarity is about 300 mOsm/liter. As previously noted, fluid leaving the proximal convoluted tubule is still isotonic to plasma. When *dilute* urine is being formed (**Figure 26.18**), the osmolarity of the fluid in the tubular lumen *increases* as it flows down the descending limb of the nephron loop, *decreases* as it flows up the ascending limb, and *decreases* still more as it flows through the rest of the nephron and collecting duct. These changes in osmolarity result from the following conditions along the path of tubular fluid:

FIGURE 26.18 Formation of dilute urine. Numbers indicate osmolarity in milliosmoles per liter (mOsm/liter). Heavy brown lines in the ascending limb of the nephron loop and in the distal convoluted tubule indicate impermeability to water; heavy blue lines indicate the last part of the distal convoluted tubule and the collecting duct, which are impermeable to water in the absence of ADH; light blue areas around the nephron represent interstitial fluid.

When the ADH level is low, urine is dilute and has an osmolarity less than the osmolarity of blood.

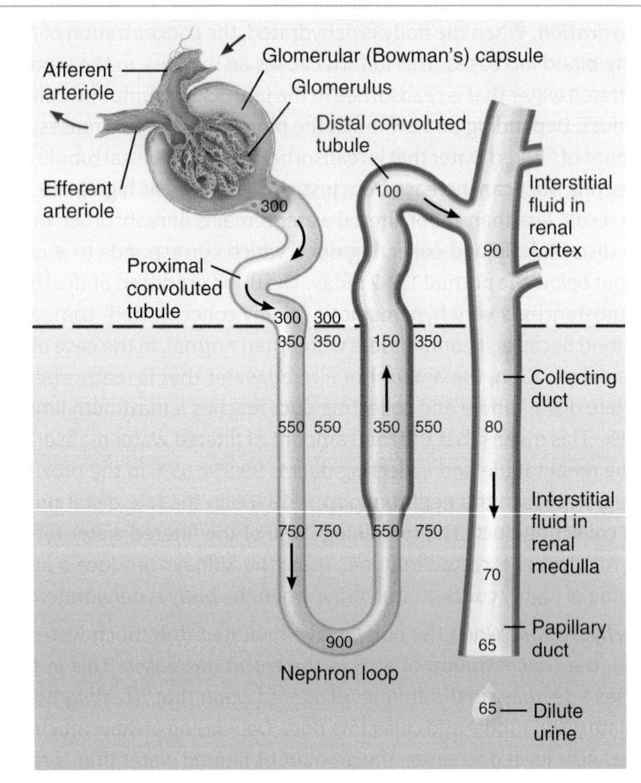

Q Which portions of the renal tubule and collecting duct reabsorb more solutes than water to produce dilute urine?

1. Because the osmolarity of the interstitial fluid of the renal medulla becomes progressively greater, more and more water is reabsorbed by osmosis as tubular fluid flows along the descending limb toward the tip of the nephron loop. (The source of this medullary osmotic gradient is explained shortly.) As a result, the fluid remaining in the lumen becomes progressively more concentrated.
2. Cells lining the thick ascending limb of the loop have symporters that actively reabsorb Na^+, K^+, and Cl^- from the tubular fluid (see **Figure 26.15**). The ions pass from the tubular fluid into thick ascending limb cells, then into interstitial fluid, and finally some diffuse into the blood inside the vasa recta.
3. Although solutes are being reabsorbed in the thick ascending limb, the water permeability of this portion of the nephron is always quite low, so water cannot follow by osmosis. As solutes—but not water molecules—are leaving the tubular fluid, its osmolarity drops to about 150 mOsm/liter. The fluid entering the distal convoluted tubule is thus more dilute than plasma.
4. While the fluid continues flowing along the distal convoluted tubule, additional solutes but only a few water molecules are reabsorbed. The early distal convoluted tubule cells are not very permeable to water and are not regulated by ADH.
5. Finally, the principal cells of the late distal convoluted tubules and collecting ducts are impermeable to water when the ADH level is very low. Thus, tubular fluid becomes progressively more dilute as it flows onward. By the time the tubular fluid drains into the renal pelvis, its concentration can be as low as 65–70 mOsm/liter. This is four times more dilute than blood plasma or glomerular filtrate.

Formation of Concentrated Urine

When water intake is low or water loss is high (such as during heavy sweating), the kidneys must conserve water while still eliminating wastes and excess ions. Under the influence of ADH, the kidneys produce a small volume of highly concentrated urine. Urine can be four times more concentrated (up to 1200 mOsm/liter) than blood plasma or glomerular filtrate (300 mOsm/liter).

The ability of ADH to cause excretion of concentrated urine depends on the presence of an **osmotic gradient** of solutes in the interstitial fluid of the renal medulla. Notice in **Figure 26.19** that the solute concentration of the interstitial fluid in the kidney increases from about 300 mOsm/liter in the renal cortex to about 1200 mOsm/liter deep in the renal medulla. The three major solutes that contribute to this high osmolarity are Na^+, Cl^-, and urea. Two main factors contribute to building and maintaining this osmotic gradient: (1) differences in solute and water permeability and reabsorption in different sections of the long nephron loops and the collecting ducts, and (2) the countercurrent flow of fluid through tube-shaped structures in the renal medulla. *Countercurrent flow* refers to the flow of fluid in opposite directions. This occurs when fluid flowing in one tube runs counter (opposite) to fluid flowing in a nearby parallel tube. Examples

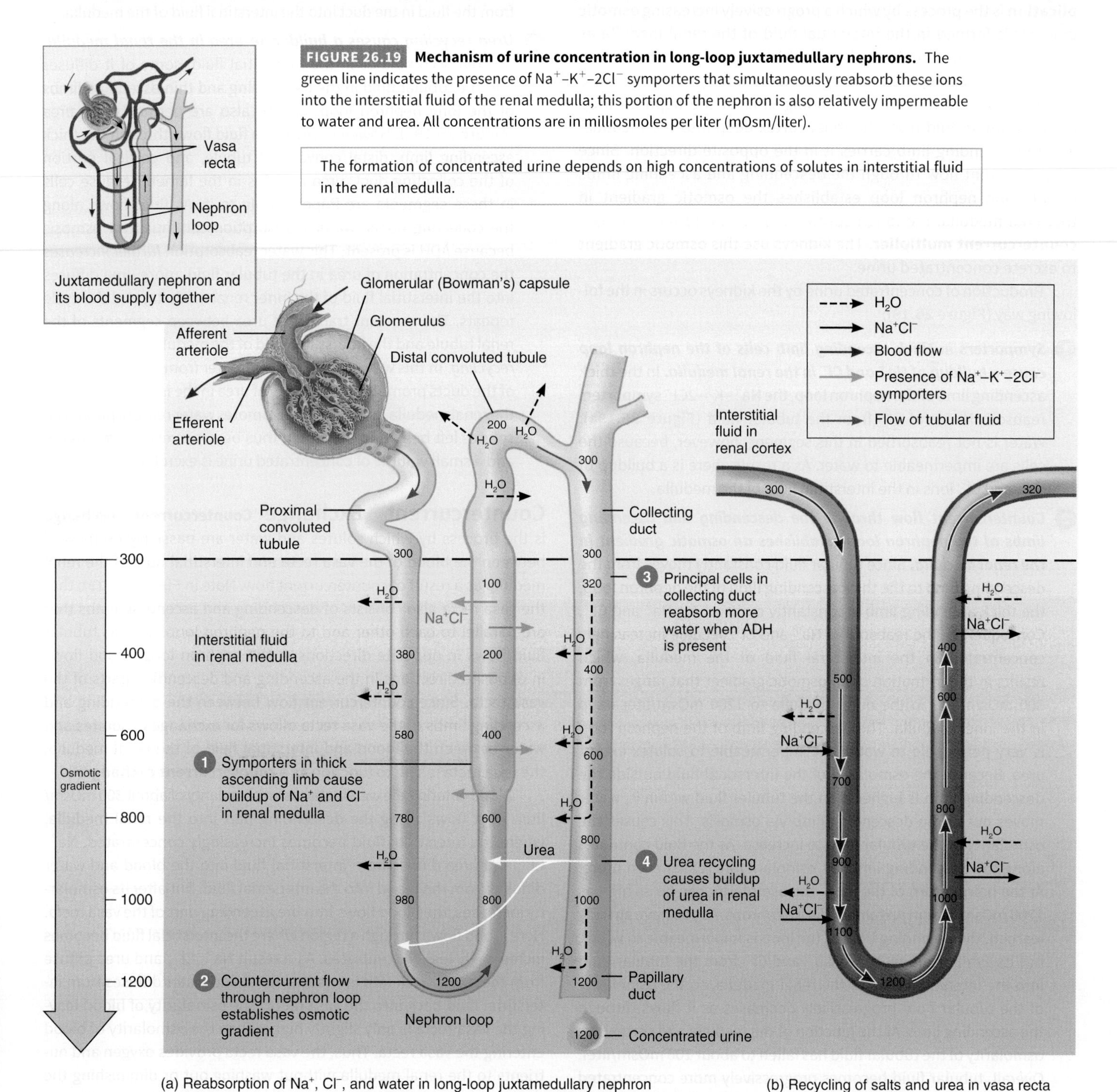

FIGURE 26.19 Mechanism of urine concentration in long-loop juxtamedullary nephrons. The green line indicates the presence of Na^+–K^+–$2Cl^-$ symporters that simultaneously reabsorb these ions into the interstitial fluid of the renal medulla; this portion of the nephron is also relatively impermeable to water and urea. All concentrations are in milliosmoles per liter (mOsm/liter).

The formation of concentrated urine depends on high concentrations of solutes in interstitial fluid in the renal medulla.

Q Which solutes are the main contributors to the high osmolarity of interstitial fluid in the renal medulla?

of countercurrent flow include the flow of tubular fluid through the descending and ascending limbs of the nephron loop and the flow of blood through the ascending and descending parts of the vasa recta. Two types of **countercurrent mechanisms** exist in the kidneys: countercurrent multiplication and countercurrent exchange.

Countercurrent Multiplication

Countercurrent multiplication is the process by which a progressively increasing osmotic gradient is formed in the interstitial fluid of the renal medulla as a result of countercurrent flow. Countercurrent multiplication involves the long nephron loops of juxtamedullary nephrons. Note in Figure 26.19a that the descending limb of the nephron loop carries tubular fluid from the renal cortex deep into the medulla, and the ascending limb carries it in the opposite direction. Since countercurrent flow through the descending and ascending limbs of the long nephron loop establishes the osmotic gradient in the renal medulla, the long nephron loop is said to function as a **countercurrent multiplier**. The kidneys use this osmotic gradient to excrete concentrated urine.

Production of concentrated urine by the kidneys occurs in the following way (Figure 26.19):

1. ***Symporters in thick ascending limb cells of the nephron loop cause a buildup of Na^+ and Cl^- in the renal medulla.*** In the thick ascending limb of the nephron loop, the Na^+–K^+–$2Cl^-$ symporters reabsorb Na^+ and Cl^- from the tubular fluid (Figure 26.19a). Water is not reabsorbed in this segment, however, because the cells are impermeable to water. As a result, there is a buildup of Na^+ and Cl^- ions in the interstitial fluid of the medulla.
2. ***Countercurrent flow through the descending and ascending limbs of the nephron loop establishes an osmotic gradient in the renal medulla.*** Since tubular fluid constantly moves from the descending limb to the thick ascending limb of the nephron loop, the thick ascending limb is constantly reabsorbing Na^+ and Cl^-. Consequently, the reabsorbed Na^+ and Cl^- become increasingly concentrated in the interstitial fluid of the medulla, which results in the formation of an osmotic gradient that ranges from 300 mOsm/liter in the outer medulla to 1200 mOsm/liter deep in the inner medulla. The descending limb of the nephron loop is very permeable to water but impermeable to solutes except urea. Because the osmolarity of the interstitial fluid outside the descending limb is higher than the tubular fluid within it, water moves out of the descending limb via osmosis. This causes the osmolarity of the tubular fluid to increase. As the fluid continues along the descending limb, its osmolarity increases even more: At the hairpin turn of the loop, the osmolarity can be as high as 1200 mOsm/liter in juxtamedullary nephrons. As you have already learned, the ascending limb of the loop is impermeable to water, but its symporters reabsorb Na^+ and Cl^- from the tubular fluid into the interstitial fluid of the renal medulla, so the osmolarity of the tubular fluid progressively decreases as it flows through the ascending limb. At the junction of the medulla and cortex, the osmolarity of the tubular fluid has fallen to about 100 mOsm/liter. Overall, tubular fluid becomes progressively more concentrated as it flows along the descending limb and progressively more dilute as it moves along the ascending limb.
3. ***Cells in the collecting ducts reabsorb more water and urea.*** When ADH increases the water permeability of the principal cells, water quickly moves via osmosis out of the collecting duct tubular fluid, into the interstitial fluid of the inner medulla, and then into the vasa recta. With loss of water, the urea left behind in the tubular fluid of the collecting duct becomes increasingly concentrated. Because duct cells deep in the medulla are permeable to it, urea diffuses from the fluid in the duct into the interstitial fluid of the medulla.
4. ***Urea recycling causes a buildup of urea in the renal medulla.*** As urea accumulates in the interstitial fluid, some of it diffuses into the tubular fluid in the descending and thin ascending limbs of the long nephron loops, which also are permeable to urea (Figure 26.19a). However, while the fluid flows through the thick ascending limb, distal convoluted tubule, and cortical portion of the collecting duct, urea remains in the lumen because cells in these segments are impermeable to it. As fluid flows along the collecting ducts, water reabsorption continues via osmosis because ADH is present. This water reabsorption *further increases* the concentration of urea in the tubular fluid, more urea diffuses into the interstitial fluid of the inner renal medulla, and the cycle repeats. The constant transfer of urea between segments of the renal tubule and the interstitial fluid of the medulla is termed *urea recycling.* In this way, reabsorption of water from the tubular fluid of the ducts promotes the buildup of urea in the interstitial fluid of the renal medulla, which in turn promotes water reabsorption. The solutes left behind in the lumen thus become very concentrated, and a small volume of concentrated urine is excreted.

Countercurrent Exchange

Countercurrent exchange is the process by which solutes and water are passively exchanged between the blood of the vasa recta and interstitial fluid of the renal medulla as a result of countercurrent flow. Note in Figure 26.19b that the vasa recta also consists of descending and ascending limbs that are parallel to each other and to the nephron loop. Just as tubular fluid flows in opposite directions in the nephron loop, blood flows in opposite directions in the ascending and descending parts of the vasa recta. Since countercurrent flow between the descending and ascending limbs of the vasa recta allows for exchange of solutes and water between the blood and interstitial fluid of the renal medulla, the vasa recta is said to function as a **countercurrent exchanger**.

Blood entering the vasa recta has an osmolarity of about 300 mOsm/liter. As it flows along the descending part into the renal medulla, where the interstitial fluid becomes increasingly concentrated, Na^+, Cl^-, and urea diffuse from interstitial fluid into the blood and water diffuses from the blood into the interstitial fluid. But after its osmolarity increases, the blood flows into the ascending part of the vasa recta. Here blood flows through a region where the interstitial fluid becomes increasingly less concentrated. As a result Na^+, Cl^-, and urea diffuse from the blood back into interstitial fluid, and water diffuses from interstitial fluid back into the vasa recta. The osmolarity of blood leaving the vasa recta is only slightly higher than the osmolarity of blood entering the vasa recta. Thus, the vasa recta provides oxygen and nutrients to the renal medulla without washing out or diminishing the osmotic gradient. The long nephron loop *establishes* the osmotic gradient in the renal medulla by countercurrent multiplication, but the vasa recta *maintains* the osmotic gradient in the renal medulla by countercurrent exchange.

Figure 26.20 summarizes the processes of filtration, reabsorption, and secretion in each segment of the nephron and collecting duct.

See Clinical Connection: Diuretics

FIGURE 26.20 Summary of filtration, reabsorption, and secretion in the nephron and collecting duct.

Filtration occurs in the renal corpuscle; reabsorption occurs all along the renal tubule and collecting ducts.

RENAL CORPUSCLE

Glomerular filtration rate: 105–125 mL/min of fluid that is isotonic to blood

Filtered substances: water and all solutes present in blood (except proteins) including ions, glucose, amino acids, creatinine, uric acid

PROXIMAL CONVOLUTED TUBULE

Reabsorption (into blood) of filtered:

Water	65% (osmosis)
Na^+	65% (sodium–potassium pumps, symporters, antiporters)
K^+	65% (diffusion)
Glucose	100% (symporters and facilitated diffusion)
Amino acids	100% (symporters and facilitated diffusion)
Cl^-	50% (diffusion)
HCO_3^-	80–90% (facilitated diffusion)
Urea	50% (diffusion)
Ca^{2+}, Mg^{2+}	variable (diffusion)

Secretion (into urine) of:

H^+	variable (antiporters)
NH_4^+	variable, increases in acidosis (antiporters)
Urea	variable (diffusion)
Creatinine	small amount

At end of PCT, tubular fluid is still isotonic to blood (300 mOsm/liter).

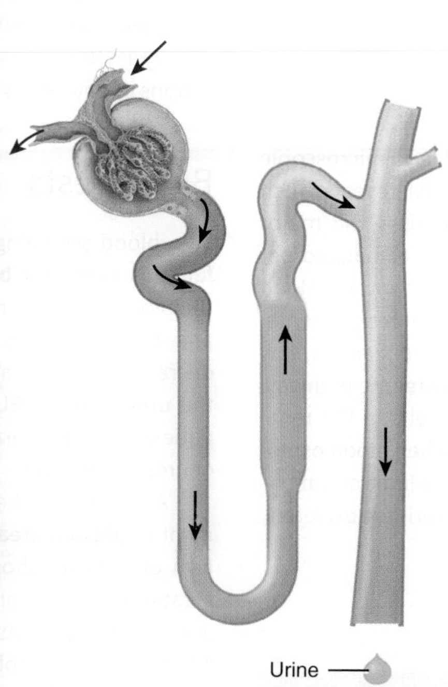

NEPHRON LOOP

Reabsorption (into blood) of:

Water	15% (osmosis in descending limb)
Na^+	20–30% (symporters in ascending limb)
K^+	20–30% (symporters in ascending limb)
Cl^-	35% (symporters in ascending limb)
HCO_3^-	10–20% (facilitated diffusion)
Ca^{2+}, Mg^{2+}	variable (diffusion)

Secretion (into urine) of:

Urea	variable (recycling from collecting duct)

At end of nephron loop, tubular fluid is hypotonic (100–150 mOsm/liter).

EARLY DISTAL CONVOLUTED TUBULE

Reabsorption (into blood) of:

Water	10–15% (osmosis)
Na^+	5% (symporters)
Cl^-	5% (symporters)
Ca^{2+}	variable (stimulated by parathyroid hormone)

LATE DISTAL CONVOLUTED TUBULE AND COLLECTING DUCT

Reabsorption (into blood) of:

Water	5–9% (insertion of water channels stimulated by ADH)
Na^+	1–4% (sodium–potassium pumps and sodium channels stimulated by aldosterone)
HCO_3^-	variable amount, depends on H^+ secretion (antiporters)
Urea	variable (recycling to nephron loop)

Secretion (into urine) of:

K^+	variable amount to adjust for dietary intake (leakage channels)
H^+	variable amounts to maintain acid–base homeostasis (H^+ pumps)

Tubular fluid leaving the collecting duct is dilute when ADH level is low and concentrated when ADH level is high.

Q In which segments of the nephron and collecting duct does secretion occur?

26.8 Evaluation of Kidney Function

OBJECTIVES

- **Define** urinalysis and describe its importance.
- **Define** renal plasma clearance and describe its importance.

Routine assessment of kidney function involves evaluating both the quantity and quality of urine and the levels of wastes in the blood.

Urinalysis

An analysis of the volume and physical, chemical, and microscopic properties of urine, called a **urinalysis** (ū-ri-NAL-i-sis), reveals much about the state of the body. Table 26.5 summarizes the major characteristics of normal urine. The volume of urine eliminated per day in a normal adult is 1–2 liters (about 1–2 qt). Fluid intake, blood pressure, blood osmolarity, diet, body temperature, diuretics, mental state, and general health influence urine volume. For example, low blood pressure triggers the renin–angiotensin–aldosterone pathway. Aldosterone increases reabsorption of water and salts in the renal tubules and decreases urine volume. By contrast, when blood osmolarity decreases—for example, after drinking a large volume of water—secretion of ADH is inhibited and a larger volume of urine is excreted.

TABLE 26.5 Characteristics of Normal Urine

CHARACTERISTIC	DESCRIPTION
Volume	One to two liters in 24 hours; varies considerably.
Color	Yellow or amber; varies with urine concentration and diet. Color due to urochrome (pigment produced from breakdown of bile) and urobilin (from breakdown of hemoglobin). Concentrated urine is darker in color. Color affected by diet (reddish from beets), medications, and certain diseases. Kidney stones may produce blood in urine.
Turbidity	Transparent when freshly voided; becomes turbid (cloudy) on standing.
Odor	Mildly aromatic; becomes ammonia-like on standing. Some people inherit ability to form methylmercaptan from digested asparagus, which gives characteristic odor. Urine of diabetics has fruity odor due to presence of ketone bodies.
pH	Ranges between 4.6 and 8.0; average 6.0; varies considerably with diet. High-protein diets increase acidity; vegetarian diets increase alkalinity.
Specific gravity (density)	Specific gravity (density) is ratio of weight of volume of substance to weight of equal volume of distilled water. In urine, 1.001–1.035. The higher the concentration of solutes, the higher the specific gravity.

Water accounts for about 95% of the total volume of urine. The remaining 5% consists of electrolytes, solutes derived from cellular metabolism, and exogenous substances such as drugs. Normal urine is virtually protein-free. Typical solutes normally present in urine include filtered and secreted electrolytes that are not reabsorbed, urea (from breakdown of proteins), creatinine (from breakdown of creatine phosphate in muscle fibers), uric acid (from breakdown of nucleic acids), urobilinogen (from breakdown of hemoglobin), and small quantities of other substances, such as fatty acids, pigments, enzymes, and hormones.

If disease alters body metabolism or kidney function, traces of substances not normally present may appear in the urine, or normal constituents may appear in abnormal amounts. Table 26.6 lists several abnormal constituents in urine that may be detected as part of a urinalysis. Normal values of urine components and the clinical implications of deviations from normal are listed in Appendix D.

Blood Tests

Two blood-screening tests can provide information about kidney function. One is the **blood urea nitrogen (BUN)** test, which measures the blood nitrogen that is part of the urea resulting from catabolism and deamination of amino acids. When glomerular filtration rate decreases severely, as may occur with renal disease or obstruction of the urinary tract, BUN rises steeply. One strategy in treating such patients is to minimize their protein intake, thereby reducing the rate of urea production.

Another test often used to evaluate kidney function is measurement of **plasma creatinine** (krē-AT-i-nin), which results from catabolism of creatine phosphate in skeletal muscle. Normally, the blood creatinine level remains steady because the rate of creatinine excretion in the urine equals its discharge from muscle. A creatinine level above 1.5 mg/dL (135 mmol/liter) usually is an indication of poor renal function. Normal values for selected blood tests are listed in Appendix C along with situations that may cause the values to increase or decrease.

Renal Plasma Clearance

Even more useful than BUN and blood creatinine values in the diagnosis of kidney problems is an evaluation of how effectively the kidneys are removing a given substance from blood plasma. **Renal plasma clearance** is the volume of blood that is "cleaned" or cleared of a substance per unit of time, usually expressed in units of *milliliters per minute.* High renal plasma clearance indicates efficient excretion of a substance in the urine; low clearance indicates inefficient excretion. For example, the clearance of glucose normally is zero because it is completely reabsorbed (see Table 26.3); therefore, glucose is not excreted at all. Knowing a drug's clearance is essential for determining the correct dosage. If clearance is high (one example is penicillin), then the dosage must also be high, and the drug must be given several times a day to maintain an adequate therapeutic level in the blood.

The following equation is used to calculate clearance:

$$\text{Renal plasma clearance of substance S} = \left(\frac{U \times V}{P}\right)$$

TABLE 26.6 Summary of Abnormal Constituents in Urine

ABNORMAL CONSTITUENT	COMMENTS
Albumin	Normal constituent of plasma; usually appears in only very small amounts in urine because it is too large to pass through capillary fenestrations. Presence of excessive albumin in urine—**albuminuria** (al′-bū-mi-NOO-rē-a)—indicates increase in permeability of filtration membranes due to injury or disease, increased blood pressure, or irritation of kidney cells by substances such as bacterial toxins, ether, or heavy metals.
Glucose	Presence of glucose in urine—**glucosuria** (gloo-kō-SOO-rē-a)—usually indicates diabetes mellitus. Occasionally caused by stress, which can cause excessive epinephrine secretion. Epinephrine stimulates breakdown of glycogen and liberation of glucose from liver.
Red blood cells (erythrocytes)	Presence of red blood cells in urine—**hematuria** (hēm-a-TOO-rē-a)—generally indicates pathological condition. One cause is acute inflammation of urinary organs due to disease or irritation from kidney stones. Other causes: tumors, trauma, kidney disease, contamination of sample by menstrual blood.
Ketone bodies	High levels of ketone bodies in urine—**ketonuria** (kē-tō-NOO-rē-a)—may indicate diabetes mellitus, anorexia, starvation, or too little carbohydrate in diet.
Bilirubin	When red blood cells are destroyed by macrophages, the globin portion of hemoglobin is split off and heme is converted to biliverdin. Most biliverdin is converted to bilirubin, which gives bile its major pigmentation. Above-normal level of bilirubin in urine is called **bilirubinuria** (bil′-ē-roo-bi-NOO-rē-a).
Urobilinogen	Presence of urobilinogen (breakdown product of hemoglobin) in urine is called **urobilinogenuria** (ū′-rō-bi-lin′-ō-je-NOO-rē-a). Trace amounts are normal, but elevated urobilinogen may be due to hemolytic or pernicious anemia, infectious hepatitis, biliary obstruction, jaundice, cirrhosis, congestive heart failure, or infectious mononucleosis.
Casts	Casts are tiny masses of material that have hardened and assumed shape of lumen of tubule in which they formed, from which they are flushed when filtrate builds up behind them. Casts are named after cells or substances that compose them or based on appearance (for example, white blood cell casts, red blood cell casts, and epithelial cell casts that contain cells from walls of tubules).
Microbes	Number and type of bacteria vary with specific urinary tract infections. One of the most common is *E. coli*. Most common fungus is yeast *Candida albicans*, cause of vaginitis. Most frequent protozoan is *Trichomonas vaginalis*, cause of vaginitis in females and urethritis in males.

where U and P are the concentrations of the substance in urine and plasma, respectively (both expressed in the same units, such as mg/mL), and V is the urine flow rate in mL/min.

The clearance of a solute depends on the three basic processes of a nephron: glomerular filtration, tubular reabsorption, and tubular secretion. Consider a substance that is filtered but neither reabsorbed nor secreted. Its clearance equals the glomerular filtration rate because all molecules that pass the filtration membrane appear in the urine. This is the situation for the plant polysaccharide **inulin** (IN-ū-lin); it easily passes the filter, it is not reabsorbed, and it is not secreted. (Do not confuse inulin with the hormone insulin, which is produced by the pancreas.) Typically, the clearance of inulin is about 125 mL/min, which equals the GFR. Clinically, the clearance of inulin can be used to determine the GFR. The clearance of inulin is obtained in the following way: Inulin is administered intravenously and then the concentrations of inulin in plasma and urine are measured along with the urine flow rate. Although using the clearance of inulin is an accurate method for determining the GFR, it has its drawbacks: Inulin is not produced by the body and it must be infused continuously while clearance measurements are being determined. Measuring the creatinine clearance is an easier way to assess the GFR because creatinine is a substance that is naturally produced by the body as an end product of muscle metabolism. Once creatinine is filtered, it is not reabsorbed, and is secreted only to a very small extent. Because there is a small amount of creatinine secretion, the creatinine clearance is only a close estimate of the GFR and is not as accurate as using the inulin clearance. The creatinine clearance is normally about 120–140 mL/min.

The clearance of the organic anion **para-aminohippuric acid (PAH)** (par′-a-a-mē′-nō-hi-PYOOR-ik) is also of clinical importance. After PAH is administered intravenously, it is filtered and secreted in a single pass through the kidneys. Thus, the clearance of PAH is used to measure **renal plasma flow**, the amount of plasma that passes through the kidneys in one minute. Typically, the renal plasma flow is 650 mL per minute, which is about 55% of the renal blood flow (1200 mL per minute).

See Clinical Connection: Dialysis

26.9 Urine Transportation, Storage, and Elimination

OBJECTIVE

- **Describe** the anatomy, histology, and physiology of the ureters, urinary bladder, and urethra.

From collecting ducts, urine drains into the minor calyces, which join to become major calyces that unite to form the renal pelvis (see **Figure 26.3**). From the renal pelvis, urine first drains into the ureters and then into the urinary bladder. Urine is then discharged from the body through the single urethra (see **Figure 26.1**).

Ureters

Each of the two **ureters** (Ū-rē-ters) transports urine from the renal pelvis of one kidney to the urinary bladder. Peristaltic contractions of the muscular walls of the ureters push urine toward the urinary bladder, but hydrostatic pressure and gravity also contribute. Peristaltic waves that pass from the renal pelvis to the urinary bladder vary in frequency from one to five per minute, depending on how fast urine is being formed.

The ureters are 25–30 cm (10–12 in.) long and are thick-walled, narrow tubes that vary in diameter from 1 mm to 10 mm along their course between the renal pelvis and the urinary bladder. Like the kidneys, the ureters are retroperitoneal. At the base of the urinary bladder, the ureters curve medially and pass obliquely through the wall of the posterior aspect of the urinary bladder (**Figure 26.21**).

Even though there is no anatomical valve at the opening of each ureter into the urinary bladder, a physiological one is quite effective. As the urinary bladder fills with urine, pressure within it compresses the oblique openings into the ureters and prevents the backflow of urine. When this physiological valve is not operating properly, it is possible for microbes to travel up the ureters from the urinary bladder to infect one or both kidneys.

Three layers of tissue form the wall of the ureters. The deepest coat, the **mucosa**, is a mucous membrane with **transitional epithelium** (see **Table 4.1**) and an underlying **lamina propria** of areolar connective tissue with considerable collagen, elastic fibers, and lymphatic tissue. Transitional epithelium is able to stretch—a marked advantage for any organ that must accommodate a variable volume of fluid. Mucus secreted by the goblet cells of the mucosa prevents the cells from coming in contact with urine, the solute concentration and pH of which may differ drastically from the cytosol of cells that form the wall of the ureters.

Throughout most of the length of the ureters, the intermediate coat, the **muscularis**, is composed of inner longitudinal and outer circular layers of smooth muscle fibers. This arrangement is opposite to that of the gastrointestinal tract, which contains inner circular and outer longitudinal layers. The muscularis of the distal third of the ureters also contains an outer layer of longitudinal muscle fibers. Thus, the muscularis in the distal third of the ureter is inner longitudinal, middle circular, and outer longitudinal. Peristalsis is the major function of the muscularis.

The superficial coat of the ureters is the **adventitia**, a layer of areolar connective tissue containing blood vessels, lymphatic vessels, and nerves that serve the muscularis and mucosa. The adventitia blends in with surrounding connective tissue and anchors the ureters in place.

Urinary Bladder

The **urinary bladder** is a hollow, distensible muscular organ situated in the pelvic cavity posterior to the pubic symphysis. In males, it is directly anterior to the rectum; in females, it is anterior to the vagina and inferior to the uterus (see **Figure 26.22**). Folds of the peritoneum hold the urinary bladder in position. When slightly distended due to the accumulation of urine, the urinary bladder is spherical. When it is empty, it collapses. As urine volume increases, it becomes pear-shaped and rises into the abdominal cavity. Urinary bladder capacity

FIGURE 26.21 **Ureters, urinary bladder, and urethra in a female.**

Urine is stored in the urinary bladder before being expelled by micturition.

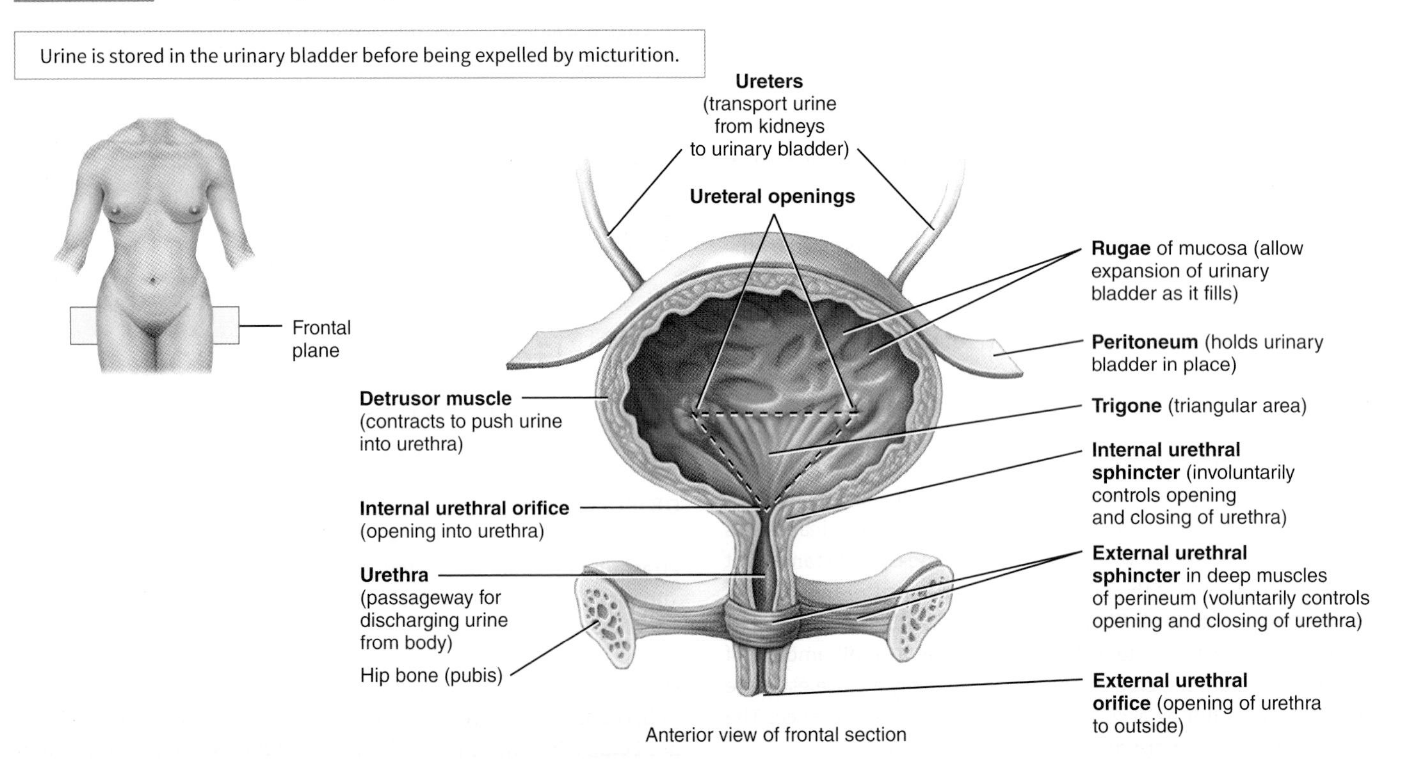

Q What is a lack of voluntary control over micturition called?

FIGURE 26.22 **Comparison between male and female urethras.**

The male urethra is about 20 cm (8 in.) in length, while the female urethra is about 4 cm (1.5 in.) in length.

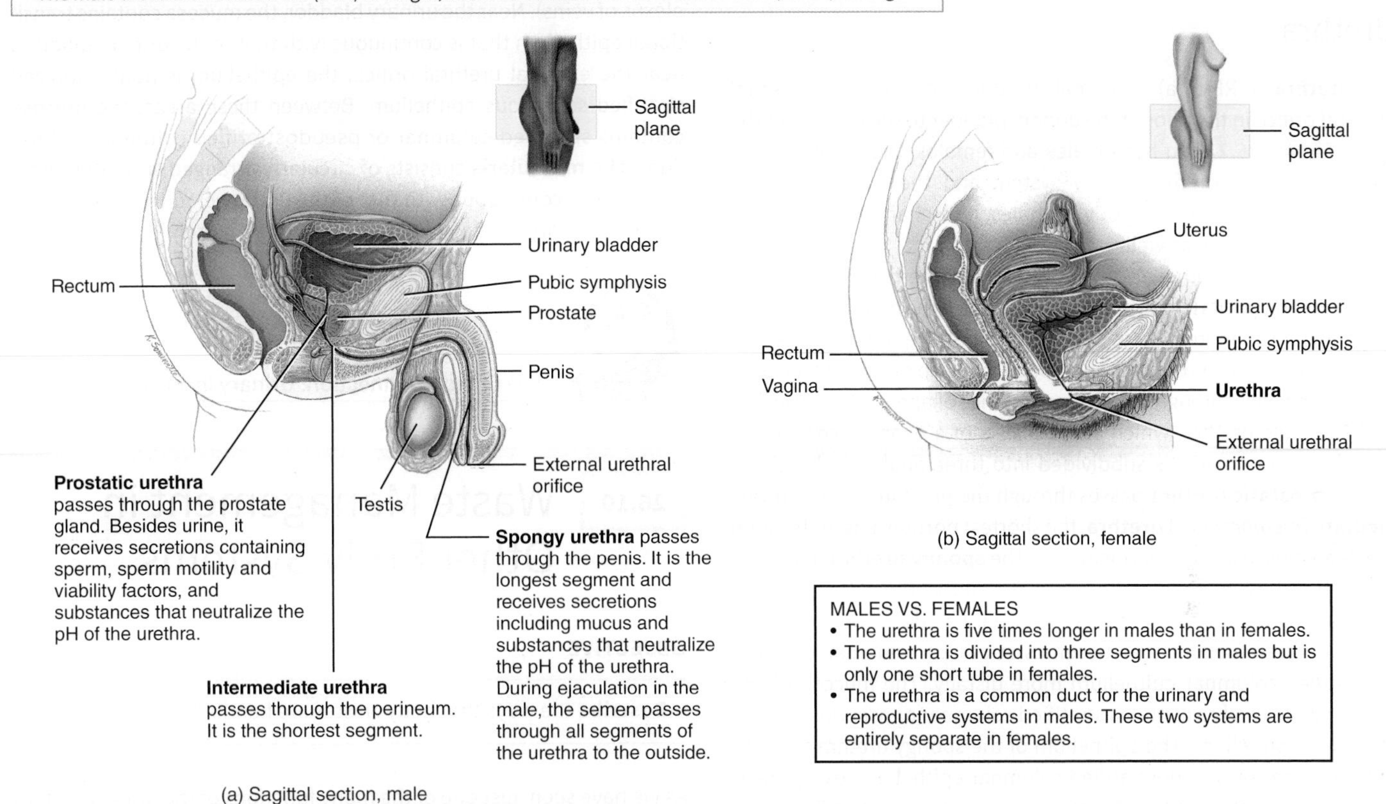

Q What are the three subdivisions of the male urethra?

averages 700–800 mL. It is smaller in females because the uterus occupies the space just superior to the urinary bladder.

Anatomy and Histology of the Urinary Bladder

In the floor of the urinary bladder is a small triangular area called the **trigone** (TRĪ-gōn = triangle). The two posterior corners of the trigone contain the two ureteral openings; the opening into the urethra, the **internal urethral orifice** (OR-i-fis), lies in the anterior corner (see Figure 26.21). Because its mucosa is firmly bound to the muscularis, the trigone has a smooth appearance.

Three coats make up the wall of the urinary bladder. The deepest is the **mucosa**, a mucous membrane composed of **transitional epithelium** and an underlying **lamina propria** similar to that of the ureters. The transitional epithelium permits stretching. Rugae (the folds in the mucosa) are also present to permit expansion of the urinary bladder. Surrounding the mucosa is the intermediate **muscularis**, also called the **detrusor muscle** (de-TROO-ser = to push down), which consists of three layers of smooth muscle fibers: the inner longitudinal, middle circular, and outer longitudinal layers. Around the opening to the urethra the circular fibers form an **internal urethral sphincter**; inferior to it is the **external urethral sphincter**, which is composed of skeletal muscle and is a modification of the deep muscles of the perineum (see Figure 11.12). The most superficial coat of the urinary bladder on the posterior and inferior surfaces is the **adventitia**, a layer of areolar connective tissue that is continuous with that of the ureters. Over the superior surface of the urinary bladder is the **serosa**, a layer of visceral peritoneum.

The Micturition Reflex

Discharge of urine from the urinary bladder, called **micturition** (mik′-choo-RISH-un; *mictur-* = urinate), is also known as *urination* or *voiding.* Micturition occurs via a combination of involuntary and voluntary muscle contractions. When the volume of urine in the urinary bladder exceeds 200–400 mL, pressure within the bladder increases considerably, and stretch receptors in its wall transmit nerve impulses into the spinal cord. These impulses propagate to the **micturition center** in sacral spinal cord segments S2 and S3 and trigger a spinal reflex called the **micturition reflex**. In this reflex arc, parasympathetic impulses from the micturition center propagate to the urinary bladder wall and internal urethral sphincter. The nerve impulses cause *contraction* of the detrusor muscle and *relaxation* of the internal urethral sphincter muscle. Simultaneously, the micturition center inhibits somatic motor neurons that innervate skeletal muscle in the external urethral sphincter. On contraction of the urinary bladder wall and relaxation of the sphincters, urination takes place. Urinary bladder filling causes a sensation of fullness that initiates a conscious desire to urinate before the micturition reflex actually occurs. Although emptying of the urinary bladder is a reflex, in early childhood we learn to initiate it and stop it voluntarily. Through learned control of the external urethral sphincter muscle and certain muscles of the pelvic floor, the cerebral cortex can initiate micturition or delay its occurrence for a limited period.

Urethra

The **urethra** (ū-RĒ-thra) is a small tube leading from the internal urethral orifice in the floor of the urinary bladder to the exterior of the body (Figure 26.22). In both males and females, the urethra is the terminal portion of the urinary system and the passageway for discharging urine from the body. In males, it discharges semen (fluid that contains sperm) as well.

In males, the urethra also extends from the internal urethral orifice to the exterior, but its length and passage through the body are considerably different than in females (Figure 26.22a). The male urethra first passes through the prostate, then through the deep muscles of the perineum, and finally through the penis, a distance of about 20 cm (8 in.).

The male urethra, which also consists of a deep mucosa and a superficial muscularis, is subdivided into three anatomical regions: (1) The **prostatic urethra** passes through the prostate. (2) The **intermediate (*membranous*) urethra**, the shortest portion, passes through the deep muscles of the perineum. (3) The **spongy urethra**, the longest portion, passes through the penis. The epithelium of the prostatic urethra is continuous with that of the urinary bladder and consists of transitional epithelium that becomes stratified columnar or pseudostratified columnar epithelium more distally. The mucosa of the intermediate urethra contains stratified columnar or pseudostratified columnar epithelium. The epithelium of the spongy urethra is stratified columnar or pseudostratified columnar epithelium, except near the external urethral orifice. There it is nonkeratinized stratified squamous epithelium. The lamina propria of the male urethra is areolar connective tissue with elastic fibers and a plexus of veins.

The muscularis of the prostatic urethra is composed of mostly circular smooth muscle fibers superficial to the lamina propria; these circular fibers help form the internal urethral sphincter of the urinary bladder. The muscularis of the intermediate (membranous) urethra consists of circularly arranged skeletal muscle fibers of the deep muscles of the perineum that help form the external urethral sphincter of the urinary bladder.

Several glands and other structures associated with reproduction deliver their contents into the male urethra (see Figure 28.9). The prostatic urethra contains the openings of (1) ducts that transport secretions from the **prostate** and (2) the **seminal vesicles** and **ductus** (*vas*) **deferens**, which deliver sperm into the urethra and provide secretions that both neutralize the acidity of the female reproductive tract and contribute to sperm motility and viability. The openings of the ducts of the **bulbourethral glands** (bul′-bō-ū-RĒ-thral) or *Cowper's glands* empty into the spongy urethra. They deliver an alkaline substance prior to ejaculation that neutralizes the acidity of the urethra. The glands also secrete mucus, which lubricates the end of the penis during sexual arousal. Throughout the urethra, but especially in the spongy urethra, the openings of the ducts of **urethral glands** or *Littré glands* (LĒ-trē) discharge mucus during sexual arousal and ejaculation.

In females, the urethra lies directly posterior to the pubic symphysis; is directed obliquely, inferiorly, and anteriorly; and has a length of 4 cm (1.5 in.) (Figure 26.22b). The opening of the urethra to the exterior, the **external urethral orifice**, is located between the clitoris and the vaginal opening (see Figure 28.11a). The wall of the female urethra consists of a deep mucosa and a superficial muscularis. The **mucosa** is a mucous membrane composed of **epithelium** and **lamina propria** (areolar connective tissue with elastic fibers and a plexus of veins). Near the urinary bladder, the mucosa contains transitional epithelium that is continuous with that of the urinary bladder; near the external urethral orifice, the epithelium is nonkeratinized stratified squamous epithelium. Between these areas, the mucosa contains stratified columnar or pseudostratified columnar epithelium. The **muscularis** consists of circularly arranged smooth muscle fibers and is continuous with that of the urinary bladder.

A summary of the organs of the urinary system is presented in Table 26.7.

See Clinical Connection: Urinary Incontinence

26.10 Waste Management in Other Body Systems

OBJECTIVE

- **Describe** the ways that body wastes are handled.

As we have seen, just one of the many functions of the urinary system is to help rid the body of some kinds of waste materials. Besides the kidneys, several other tissues, organs, and processes contribute to the temporary confinement of wastes, the transport of waste materials for disposal, the recycling of materials, and the excretion of excess or toxic substances in the body. These waste management systems include the following:

- ***Body buffers.*** Buffers in body fluids bind excess hydrogen ions (H^+), thereby preventing an increase in the acidity of body fluids. Buffers, like wastebaskets, have a limited capacity; eventually the H^+, like the paper in a wastebasket, must be eliminated from the body by excretion.
- ***Blood.*** The bloodstream provides pickup and delivery services for the transport of wastes, in much the same way that garbage trucks and sewer lines serve a community.
- ***Liver.*** The liver is the primary site for metabolic recycling, as occurs, for example, in the conversion of amino acids into glucose or of glucose into fatty acids. The liver also converts toxic substances into less toxic ones, such as ammonia into urea. These functions of the liver are described in Chapters 24 and 25.
- ***Lungs.*** With each exhalation, the lungs excrete CO_2, and expel heat and a little water vapor.
- ***Sweat (sudoriferous) glands.*** Especially during exercise, sweat glands in the skin help eliminate excess heat, water, and CO_2, plus small quantities of salts and urea as well.
- ***Gastrointestinal tract.*** Through defecation, the gastrointestinal tract excretes solid, undigested foods; wastes; some CO_2; water; salts; and heat.

TABLE 26.7 Summary of Urinary System Organs

STRUCTURE	LOCATION	DESCRIPTION	FUNCTION
Kidneys	Posterior abdomen between last thoracic and third lumbar vertebrae posterior to peritoneum (retroperitoneal). Lie against ribs 11 and 12.	Solid, reddish, bean-shaped organs. Internal structure: three tubular systems (arteries, veins, urinary tubes).	Regulate blood volume and composition, help regulate blood pressure, synthesize glucose, release erythropoietin, participate in vitamin D synthesis, excrete wastes in urine.
Ureters	Posterior to peritoneum (retroperitoneal); descend from kidney to urinary bladder along anterior surface of psoas major muscle and cross back of pelvis to reach inferoposterior surface of urinary bladder anterior to sacrum.	Thick, muscular walled tubes with three structural layers: mucosa of transitional epithelium, muscularis with circular and longitudinal layers of smooth muscle, adventitia of areolar connective tissue.	Transport tubes that move urine from kidneys to urinary bladder.
Urinary bladder	In pelvic cavity anterior to sacrum and rectum in males and sacrum, rectum, and vagina in females and posterior to pubis in both sexes. In males, superior surface covered with parietal peritoneum; in females, uterus covers superior aspect.	Hollow, distensible, muscular organ with variable shape depending on how much urine it contains. Three basic layers: inner mucosa of transitional epithelium, middle smooth muscle coat (detrusor muscle), outer adventitia or serosa over superior aspect in males.	Storage organ that temporarily stores urine until convenient to discharge from body.
Urethra	Exits urinary bladder in both sexes. In females, runs through perineal floor of pelvis to exit between labia minora. In males, passes through prostate, then perineal floor of pelvis, and then penis to exit at its tip.	Thin-walled tubes with three structural layers: inner mucosa that consists of transitional, stratified columnar, and stratified squamous epithelium; thin middle layer of circular smooth muscle; thin connective tissue exterior.	Drainage tube that transports stored urine from body.

26.11 Development of the Urinary System

OBJECTIVE

- **Describe** the development of the urinary system.

Starting in the third week of fetal development, a portion of the mesoderm along the posterior aspect of the embryo, the **intermediate mesoderm**, differentiates into the kidneys. The intermediate mesoderm is located in paired elevations called **urogenital ridges** (ū-rō-JEN-i-tal). Three pairs of kidneys form within the intermediate mesoderm in succession: the pronephros, the mesonephros, and the metanephros (**Figure 26.23**). Only the last pair remains as the functional kidneys of the newborn.

The first kidney to form, the **pronephros** (prō-NEF-rōs; *pro-* = before; *-nephros* = kidney), is the most superior of the three and has an associated **pronephric duct**. This duct empties into the **cloaca** (klō-Ā-ka), the expanded terminal part of the hindgut, which functions as a common outlet for the urinary, digestive, and reproductive ducts. The pronephros begins to degenerate during the fourth week and is completely gone by the sixth week.

The second kidney, the **mesonephros** (mez′-ō-NEF-rōs; *meso-* = middle), replaces the pronephros. The retained portion of the pronephric duct, which connects to the mesonephros, develops into the **mesonephric duct**. The mesonephros begins to degenerate by the sixth week and is almost gone by the eighth week.

At about the fifth week, a mesodermal outgrowth, called a **ureteric bud** (ū-rē-TER-ik), develops from the distal portion of the mesonephric duct near the cloaca. The **metanephros** (met-a-NEF-rōs; *meta-* = after), or ultimate kidney, develops from the ureteric bud and metanephric mesoderm. The ureteric bud forms the *collecting ducts, calyces, renal pelvis,* and *ureter*. The **metanephric mesoderm** (met′-a-NEF-rik) forms the *nephrons* of the kidneys. By the third month, the fetal kidneys begin excreting urine into the surrounding amniotic fluid; indeed, fetal urine makes up most of the amniotic fluid.

During development, the cloaca divides into a **urogenital sinus**, into which urinary and genital ducts empty, and a *rectum* that discharges into the anal canal. The *urinary bladder* develops from the urogenital sinus. In females, the *urethra* develops as a result of lengthening of the short duct that extends from the urinary bladder to the urogenital sinus. In males, the urethra is considerably longer and more complicated, but it is also derived from the urogenital sinus.

Although the metanephric kidneys form in the pelvis, they ascend to their ultimate destination in the abdomen. As they do so, they receive renal blood vessels. Although the inferior blood vessels usually degenerate as superior ones appear, sometimes the inferior vessels do not degenerate. Consequently, some individuals (about 30%) develop multiple renal vessels.

In a condition called **unilateral renal agenesis** (ā-JEN-e-sis; *a-* = without; *-genesis* = production; *unilateral* = one side) only one kidney develops (usually the right) due to the absence of a ureteric bud. The condition occurs once in every 1000 newborn infants and usually affects males more than females. Other kidney abnormalities that occur during development are **malrotated kidneys** (the hilum faces anteriorly, posteriorly, or laterally instead of medially); **ectopic kidney** (one or both kidneys may be in an abnormal position, usually

FIGURE 26.23 Development of the urinary system.

Three pairs of kidneys form within intermediate mesoderm in succession: pronephros, mesonephros, and metanephros.

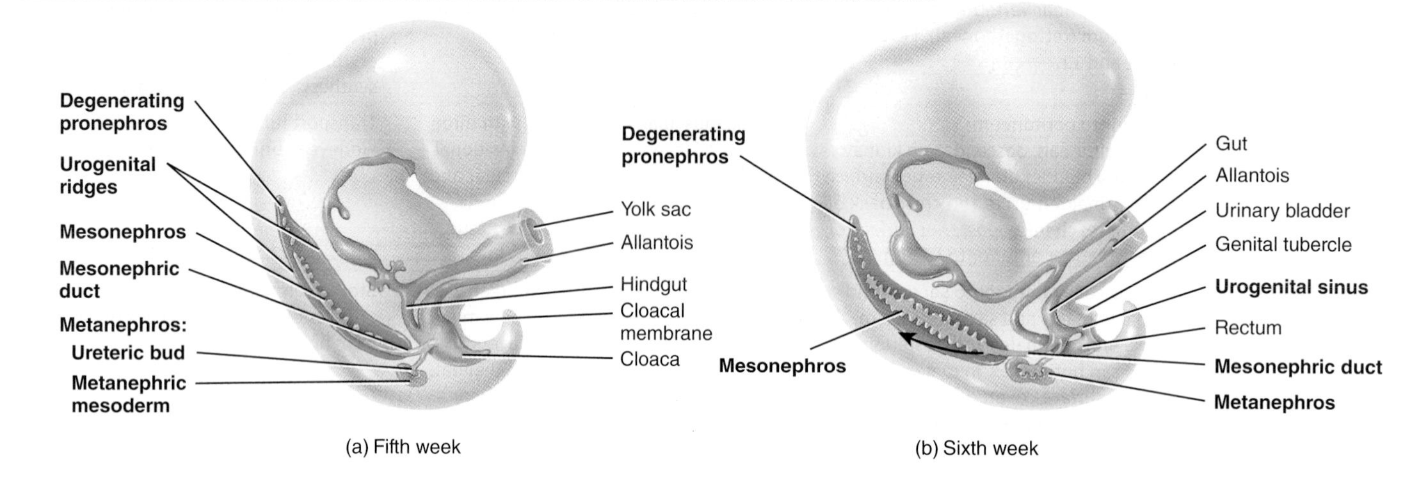

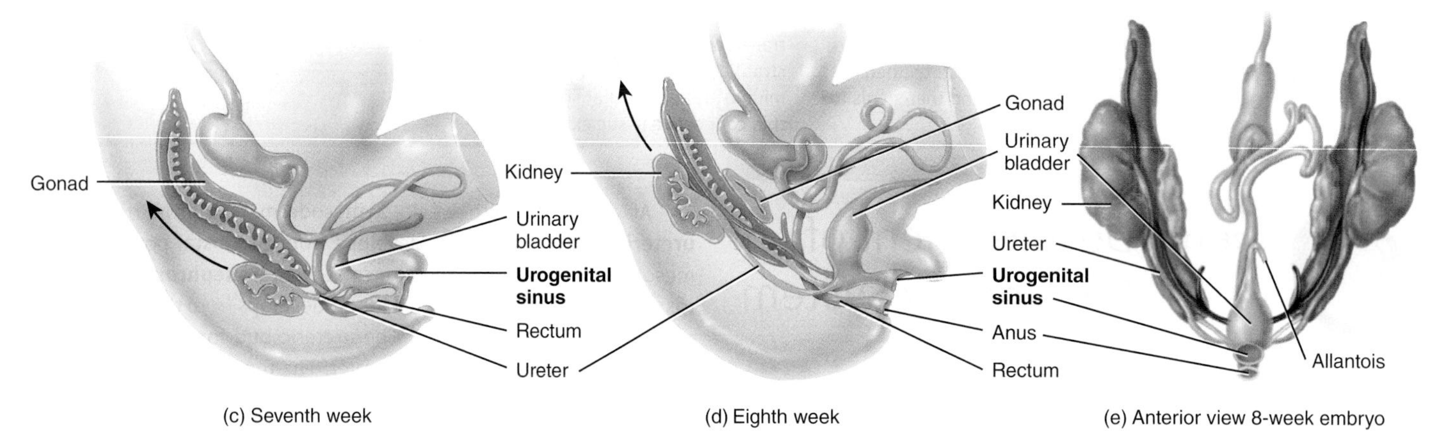

Q When do the kidneys begin to develop?

inferior); and **horseshoe kidney** (the fusion of the two kidneys, usually inferiorly, into a single U-shaped kidney).

26.12 Aging and the Urinary System

OBJECTIVE

- **Describe** the effects of aging on the urinary system.

With aging, the kidneys shrink in size, have a decreased blood flow, and filter less blood. These age-related changes in kidney size and function seem to be linked to a progressive reduction in blood supply to the kidneys as an individual gets older; for example, blood vessels such as the glomeruli become damaged or decrease in number. The mass of the two kidneys decreases from an average of nearly 300 g in 20-year-olds to less than 200 g by age 80, a decrease of about one-third. Likewise, renal blood flow and filtration rate decline by 50% between ages 40 and 70. By age 80, about 40% of glomeruli are not functioning and thus filtration, reabsorption, and secretion decrease. Kidney diseases that become more common with age include acute and chronic kidney inflammations and renal calculi (kidney stones). Because the sensation of thirst diminishes with age, older individuals also are susceptible to dehydration. Urinary bladder changes that occur with aging include a reduction in size and capacity and weakening of the muscles. Urinary tract infections are more common among the elderly, as are polyuria (excessive urine production), nocturia (excessive urination at night), increased frequency of urination, dysuria (painful urination), urinary retention or incontinence, and hematuria (blood in the urine).

To appreciate the many ways that the urinary system contributes to homeostasis of other body systems, examine *Focus on Homeostasis: Contributions of the Urinary System*. Next, in Chapter 27, we will see how the kidneys and lungs contribute to maintenance of homeostasis of body fluid volume, electrolyte levels in body fluids, and acid–base balance.

FOCUS on HOMEOSTASIS

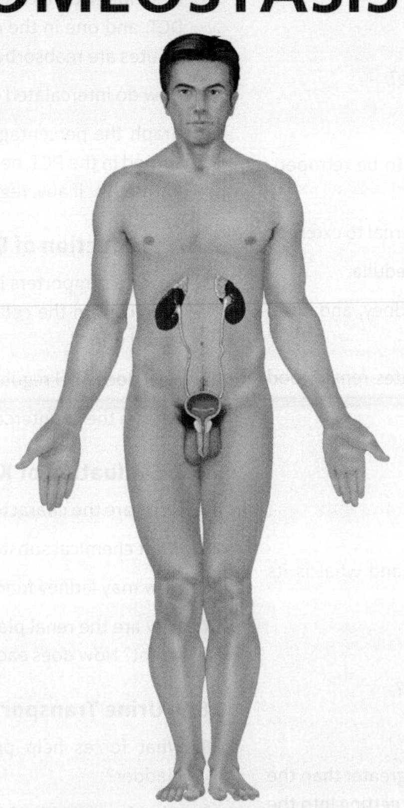

INTEGUMENTARY SYSTEM

- Kidneys and skin both contribute to synthesis of calcitriol, the active form of vitamin D

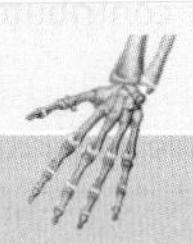

SKELETAL SYSTEM

- Kidneys help adjust levels of blood calcium and phosphates, needed for building extracellular bone matrix

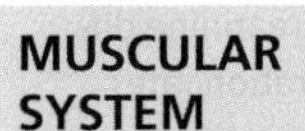

MUSCULAR SYSTEM

- Kidneys help adjust level of blood calcium, needed for contraction of muscle

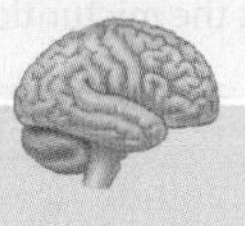

NERVOUS SYSTEM

- Kidneys perform gluconeogenesis, which provides glucose for ATP production in neurons, especially during fasting or starvation

ENDOCRINE SYSTEM

- Kidneys participate in synthesis of calcitriol, the active form of vitamin D
- Kidneys release erythropoietin, the hormone that stimulates production of red blood cells

CONTRIBUTIONS OF THE URINARY SYSTEM FOR ALL BODY SYSTEMS

- Kidneys regulate volume, composition, and pH of body fluids by removing wastes and excess substances from blood and excreting them in urine
- Ureters transport urine from kidneys to urinary bladder, which stores urine until it is eliminated through urethra

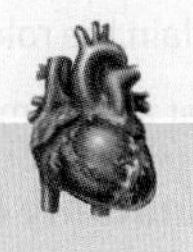

CARDIOVASCULAR SYSTEM

- By increasing or decreasing their reabsorption of water filtered from blood, kidneys help adjust blood volume and blood pressure
- Renin released by juxtaglomerular cells in kidneys raises blood pressure
- Some bilirubin from hemoglobin breakdown is converted to a yellow pigment (urobilin), which is excreted in urine

LYMPHATIC SYSTEM and IMMUNITY

- By increasing or decreasing their reabsorption of water filtered from blood, kidneys help adjust volume of interstitial fluid and lymph; urine flushes microbes out of urethra

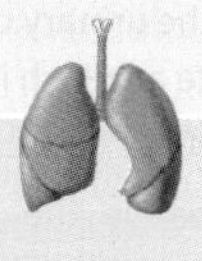

RESPIRATORY SYSTEM

- Kidneys and lungs cooperate in adjusting pH of body fluids

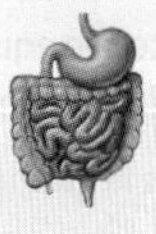

DIGESTIVE SYSTEM

- Kidneys help synthesize calcitriol, the active form of vitamin D, which is needed for absorption of dietary calcium

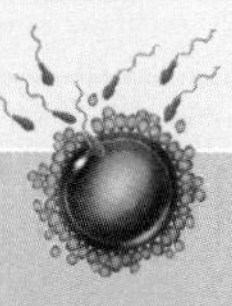

REPRODUCTIVE SYSTEMS

- In males, portion of urethra that extends through prostate and penis is passageway for semen as well as urine

Review Questions

26.1 Overview of the Urinary System

1. Explain the role of each organ of the urinary system.
2. What are examples of wastes that may be present in urine?

26.2 Anatomy of the Kidneys

3. Describe the location of the kidneys. Why are they said to be retroperitoneal?
4. Identify the three layers that surround the kidney from internal to external.
5. Describe the components of the renal cortex and renal medulla.
6. Trace a drop of blood into a renal artery, through the kidney, and out a renal vein.
7. Which branch of the autonomic nervous system innervates renal blood vessels?

26.3 The Nephron

8. What are the two main parts of a nephron?
9. What are the components of the renal tubule?
10. Where is the juxtaglomerular apparatus (JGA) located, and what is its structure?

26.4 Overview of Renal Physiology

11. How do tubular reabsorption and tubular secretion differ?

26.5 Glomerular Filtration

12. If the urinary excretion rate of a drug such as penicillin is greater than the rate at which it is filtered at the glomerulus, how else is it getting into the urine?
13. What is the major chemical difference between blood plasma and glomerular filtrate?
14. Why is there much greater filtration through glomerular capillaries than through capillaries elsewhere in the body?
15. Write the equation for the calculation of net filtration pressure (NFP), and explain the meaning of each term.
16. How is glomerular filtration rate regulated?

26.6 Tubular Reabsorption and Tubular Secretion

17. Diagram the reabsorption of substances via the transcellular and paracellular routes. Label the apical membrane and the basolateral membrane. Where are the sodium–potassium pumps located?
18. Describe two mechanisms in the PCT, one in the nephron loop, one in the DCT, and one in the collecting duct for reabsorption of Na^+. What other solutes are reabsorbed or secreted with Na^+ in each mechanism?
19. How do intercalated cells secrete hydrogen ions?
20. Graph the percentages of filtered water and filtered Na^+ that are reabsorbed in the PCT, nephron loop, DCT, and collecting duct. Indicate which hormones, if any, regulate reabsorption in each segment.

26.7 Production of Dilute and Concentrated Urine

21. How do symporters in the ascending limb of the nephron loop and principal cells in the collecting duct contribute to the formation of concentrated urine?
22. How does ADH regulate facultative water reabsorption?
23. What is the countercurrent mechanism? Why is it important?

26.8 Evaluation of Kidney Function

24. What are the characteristics of normal urine?
25. What chemical substances normally are present in urine?
26. How may kidney function be evaluated?
27. Why are the renal plasma clearances of glucose, urea, and creatinine different? How does each clearance compare to glomerular filtration rate?

26.9 Urine Transportation, Storage, and Elimination

28. What forces help propel urine from the renal pelvis to the urinary bladder?
29. What is micturition? How does the micturition reflex occur?
30. How do the location, length, and histology of the urethra compare in males and females?

26.10 Waste Management in Other Body Systems

31. What roles do the liver and lungs play in the elimination of wastes?

26.11 Development of the Urinary System

32. Which type of embryonic tissue develops into nephrons?
33. Which tissue gives rise to collecting ducts, calyces, renal pelves, and ureters?

26.12 Aging and the Urinary System

34. To what extent do kidney mass and filtration rate decrease with age?

Critical Thinking Questions

1. Imagine the discovery of a new toxin that blocks renal tubule reabsorption but does not affect filtration. Predict the short-term effects of this toxin.
2. For each of the following urinalysis results, indicate whether you should be concerned or not and why: (a) dark yellow urine that is turbid; (b) ammonia-like odor of the urine; (c) presence of excessive albumin; (d) presence of epithelial cell casts; (e) pH of 5.5; (f) hematuria.
3. Bruce is experiencing sudden, rhythmic waves of pain in his groin area. He has noticed that, although he is consuming fluids, his urine output has decreased. From what condition is Bruce suffering? How is it treated? How can he prevent future episodes?

Fluid, Electrolyte, and Acid–Base Homeostasis

CHAPTER 27

Fluid, Electrolyte, and Acid–Base Homeostasis

Regulating the volume and composition of body fluids, controlling their distribution throughout the body, and balancing the pH of body fluids are crucial to maintaining overall homeostasis and health.

In Chapter 26 you learned how the kidneys form urine. One important function of the kidneys is to help maintain fluid balance in the body. Regulatory mechanisms involving the kidneys and other organs normally maintain homeostasis of the body fluids. Malfunction in any or all of them may seriously endanger the functioning of organs throughout the body. In this chapter, we will explore the mechanisms that regulate the volume and distribution of body fluids and examine the factors that determine the concentrations of solutes and the pH of body fluids.

Q Did you ever wonder how breathing can affect your body's pH?

27.1 Fluid Compartments and Fluid Homeostasis

OBJECTIVES

- **Compare** the locations of intracellular fluid (ICF) and extracellular fluid (ECF).
- **Describe** the various fluid compartments of the body.
- **Discuss** the sources and regulation of water and solute gain and loss.
- **Explain** how fluids move between compartments.

A **body fluid** is a substance, usually a liquid, that is produced by the body and consists of water and dissolved solutes. In lean adults, body fluids constitute between 55% and 60% of total body mass in females and males, respectively (**Figure 27.1**). Body fluids are present in two main "compartments"—inside cells and outside cells. About two-thirds of body fluid is **intracellular fluid (ICF)** (*intra-* = within) or *cytosol*, the fluid within cells. The other third, called **extracellular fluid (ECF)** (*extra-* = outside), is outside cells and includes all other body fluids. About 80% of the ECF is **interstitial fluid** (*inter-* = between), which occupies the microscopic spaces between tissue cells, and 20% of the ECF is **blood plasma**, the liquid portion of the blood. Other extracellular fluids that are grouped with interstitial fluid include lymph in lymphatic vessels; cerebrospinal fluid in the nervous system; synovial fluid in joints; aqueous humor and vitreous body in the eyes; endolymph and perilymph in the ears; and pleural, pericardial, and peritoneal fluids between serous membranes.

Two general "barriers" separate intracellular fluid, interstitial fluid, and blood plasma.

1. The *plasma membrane* of individual cells separates intracellular fluid from the surrounding interstitial fluid. You learned in Chapter 3 that the plasma membrane is a selectively permeable barrier: It allows some substances to cross but blocks the movement of other substances. In addition, active transport pumps work continuously to maintain different concentrations of certain ions in the cytosol and interstitial fluid.
2. *Blood vessel walls* divide the interstitial fluid from blood plasma. Only in capillaries, the smallest blood vessels, are the walls thin enough and leaky enough to permit the exchange of water and solutes between blood plasma and interstitial fluid.

The body is in **fluid balance** when the required amounts of water and solutes are present and are correctly proportioned among the various compartments. **Water** is by far the largest single component of the body, making up 45–75% of total body mass, depending on age, gender, and the amount of adipose tissue (fat) present in the body. Obese people have proportionally less water than leaner people because water comprises less than 20% of the

FIGURE 27.1 **Body fluid compartments.**

The term body fluid refers to body water and its dissolved substances.

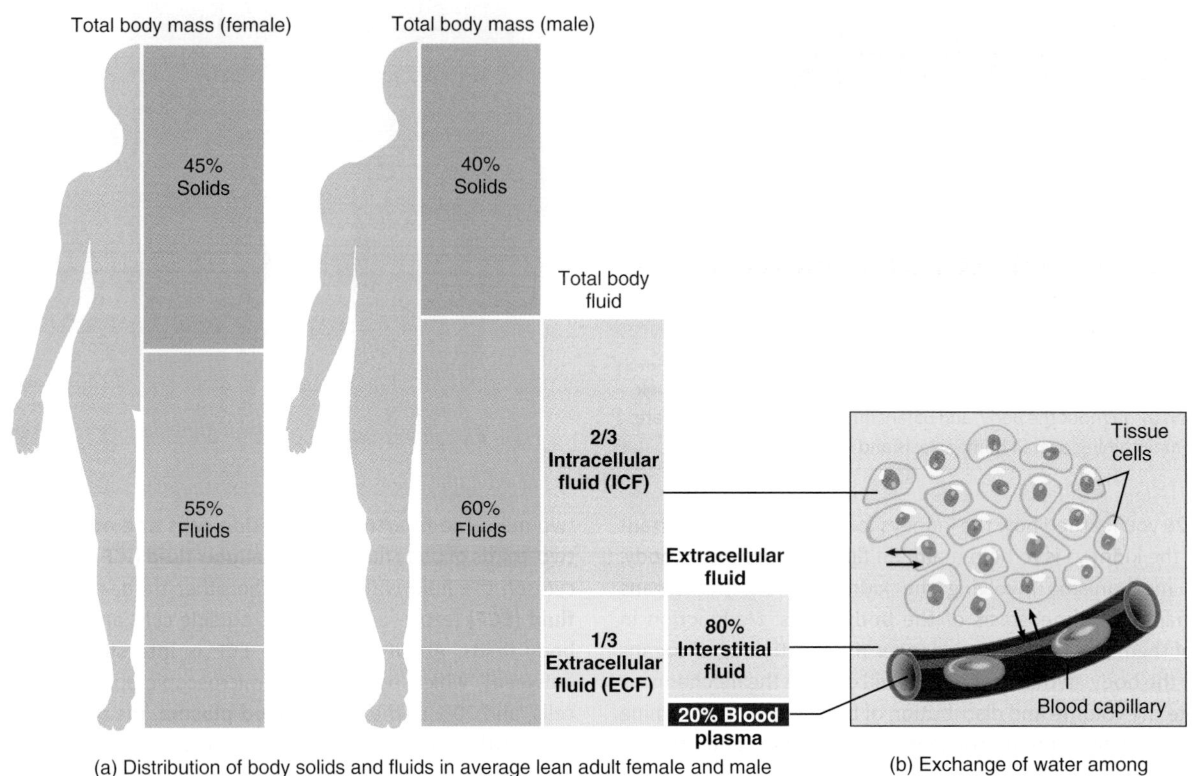

(a) Distribution of body solids and fluids in average lean adult female and male

(b) Exchange of water among body fluid compartments

Q What is the approximate volume of blood plasma in a lean 60-kg male? In a lean 60-kg female? (Note: One liter of body fluid has a mass of 1 kilogram.)

mass of adipose tissue. Skeletal muscle tissue, by contrast, is about 65% water. Infants have the highest percentage of water, up to 75% of body mass. The percentage of body mass that is water decreases until about 2 years of age. Until puberty, water accounts for about 60% of body mass in boys and girls. In lean adult males, water still accounts for about 60% of body mass. However, lean adult females have more subcutaneous fat than do lean adult males. Thus, their percentage of total body water is lower, accounting for about 55% of body mass.

The processes of filtration, reabsorption, diffusion, and osmosis allow continual exchange of water and solutes among body fluid compartments (Figure 27.1b). Yet the volume of fluid in each compartment remains remarkably stable. The pressures that promote filtration of fluid from blood capillaries and reabsorption of fluid back into capillaries can be reviewed in Figure 21.7. Because osmosis is the primary means of water movement between intracellular fluid and interstitial fluid, the concentration of solutes in these fluids determines the *direction* of water movement. Because most solutes in body fluids are **electrolytes**, inorganic compounds that dissociate into ions, fluid balance is closely related to electrolyte balance. Because intake of water and electrolytes rarely occurs in exactly the same proportions as their presence in body fluids, the ability of the kidneys to excrete excess water by producing dilute urine, or to excrete excess electrolytes by producing concentrated urine, is of utmost importance in the maintenance of homeostasis.

Sources of Body Water Gain and Loss

The body can gain water by ingestion and by metabolic synthesis (Figure 27.2). The main sources of body water are ingested liquids (about 1600 mL) and moist foods (about 700 mL) absorbed from the gastrointestinal (GI) tract, which total about 2300 mL/day. The other source of water is **metabolic water** that is produced in the body mainly when electrons are accepted by oxygen during aerobic respiration (see Figure 25.2) and to a smaller extent during dehydration synthesis reactions (see Figure 2.15). Metabolic water gain accounts for only 200 mL/day. Daily water gain from these two sources totals about 2500 mL.

FIGURE 27.2 Sources of daily water gain and loss under normal conditions. Numbers are average volumes for adults.

Normally, daily water loss equals daily water gain.

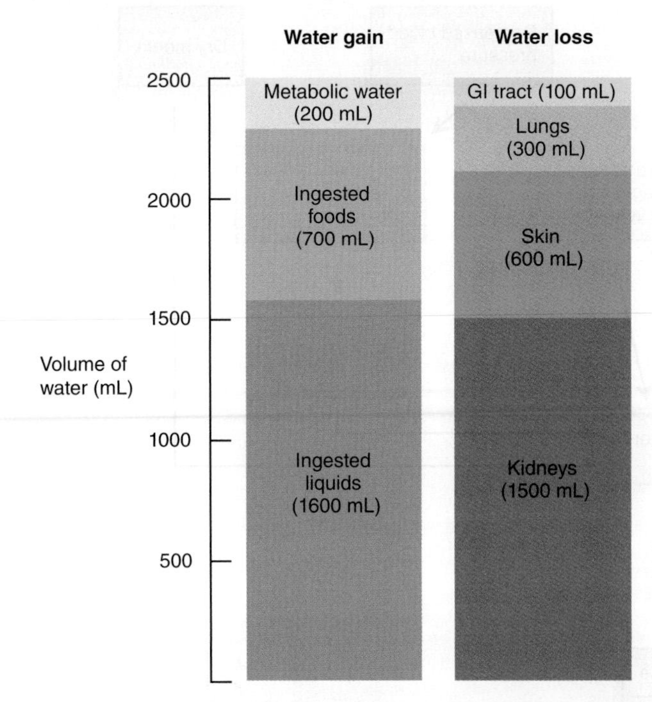

Q How does each of the following affect fluid balance: Hyperventilation? Vomiting? Fever? Diuretics?

Normally, body fluid volume remains constant because water loss equals water gain. Water loss occurs in four ways (Figure 27.2). Each day the kidneys excrete about 1500 mL in urine, the skin evaporates about 600 mL (400 mL through insensible perspiration—sweat that evaporates before it is perceived as moisture—and 200 mL as sweat), the lungs exhale about 300 mL as water vapor, and the gastrointestinal tract eliminates about 100 mL in feces. In women of reproductive age, additional water is lost in menstrual flow. On average, daily water loss totals about 2500 mL. The amount of water lost by a given route can vary considerably over time. For example, water may literally pour from the skin in the form of sweat during strenuous exertion. In other cases, water may be lost in diarrhea during a GI tract infection.

Regulation of Body Water Gain

The volume of metabolic water formed in the body depends entirely on the level of aerobic respiration, which reflects the demand for ATP in body cells. When more ATP is produced, more water is formed. Body water gain is regulated mainly by the volume of water intake, or how much fluid you drink. An area in the hypothalamus known as the **thirst center** governs the urge to drink.

When water loss is greater than water gain, **dehydration**—a decrease in volume and an increase in osmolarity of body fluids—occurs. A decrease in blood volume causes blood pressure to fall. Increased activity from osmoreceptors in the hypothalamus, triggered by increased blood osmolarity, stimulates the thirst center in the hypothalamus (Figure 27.3). Other signals that stimulate the thirst center come from (1) volume receptors in the atria that detect the decrease in blood volume, (2) baroreceptors in blood vessels that detect the decrease in blood pressure, (3) angiotensin II that is formed due to activation of the renin-angiotensin-aldosterone pathway by the decrease in blood pressure, and (4) neurons in the mouth that detect dryness due to a decreased flow of saliva. As a result of these stimuli, the sensation of thirst increases, which usually leads to increased fluid intake (as long as fluids are available) and restoration of normal fluid volume. Overall, fluid gain balances fluid loss. Sometimes, however, the sensation of thirst does not occur quickly enough or access to fluids is restricted, and significant dehydration ensues. This happens most often in elderly people, in infants, and in those who are in a confused mental state. When heavy sweating or fluid loss from diarrhea or vomiting occurs, it is wise to start replacing body fluids by drinking fluids even before the sensation of thirst occurs.

Regulation of Water and Solute Loss

Even though the loss of water and solutes through sweating and exhalation increases during exercise, elimination of *excess* body water or solutes occurs mainly by control of their loss in urine. The extent of *urinary salt (NaCl) loss* is the main factor that determines body fluid *volume.* The reason for this is that "water follows solutes" in osmosis, and the two main solutes in extracellular fluid (and in urine) are sodium ions (Na^+) and chloride ions (Cl^-). In a similar way, the main factor that determines body fluid *osmolarity* is the extent of *urinary water loss.*

The major hormone that regulates water loss is antidiuretic hormone (ADH). This hormone, also known as *vasopressin*, is produced by neurosecretory cells in the hypothalamus and stored in the posterior pituitary gland. When the osmolarity of body fluids increases, osmoreceptors in the hypothalamus not only stimulate thirst; they also increase the synthesis and release of ADH (Figure 27.4). ADH promotes the insertion of water-channel proteins (aquaporin-2) into the apical membranes of principal cells in the late distal tubules and collecting ducts of the kidneys. As a result, the permeability of these cells to water increases. Water molecules move by osmosis from the renal tubular fluid into the cells and then from the cells into the bloodstream. This results in a decrease in blood osmolarity, an increase in blood volume and blood pressure, and the production of a small volume of concentrated urine. Once the body has adequate water, the ADH level in the bloodstream decreases. As the amount of ADH in the blood declines, some of the aquaporin-2 channels are removed from the apical membrane via endocytosis. Consequently, the water permeability of the principal cells decreases and more water is lost in the urine.

Factors other than blood osmolarity influence ADH secretion (Figure 27.4). A decrease in blood volume or blood pressure also

FIGURE 27.3 **Pathways involved in the thirst response.**

A major stimulus that promotes the sensation of thirst is an increase in the osmolarity of body fluids.

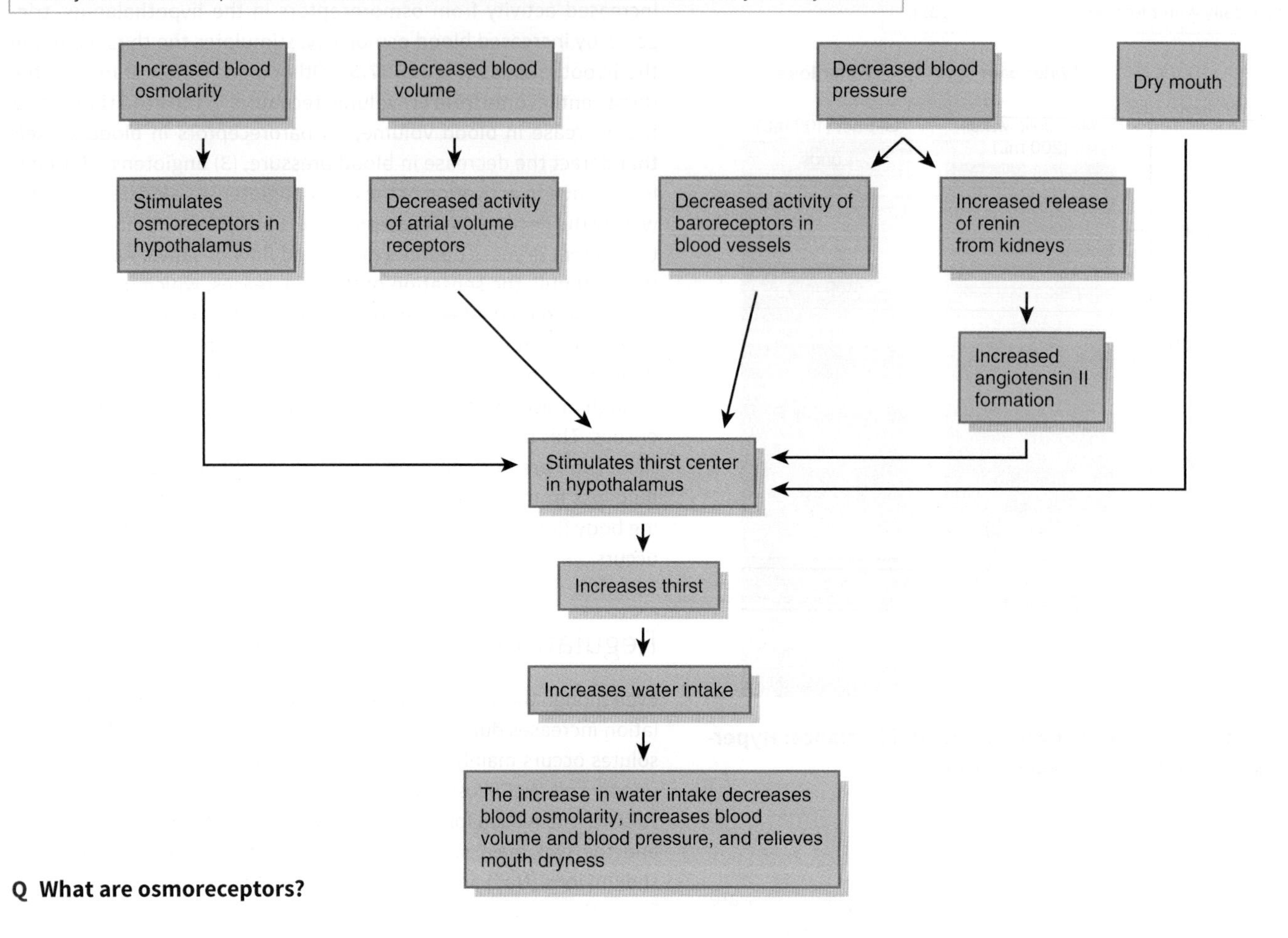

Q What are osmoreceptors?

stimulates ADH release. Atrial volume receptors detect the decrease in blood volume, and baroreceptors in blood vessels detect the decrease in blood pressure. ADH release is also stimulated by factors that are unrelated to water balance, such as pain, nausea, and stress. Secretion of ADH is inhibited by alcohol, which is why consumption of alcoholic beverages promotes diuresis (voiding large amounts of urine).

Because our daily diet contains a highly variable amount of NaCl, urinary excretion of Na^+ and Cl^- must also vary to maintain homeostasis. Hormones regulate the urinary loss of Na^+ ions, Cl^- ions usually follow Na^+ ions because of electrical attraction or because they are transported along with Na^+ ions via symporters. The two most important hormones that regulate the extent of renal Na^+ reabsorption (and thus how much is lost in the urine) are aldosterone and atrial natriuretic peptide.

1. ***Aldosterone.*** When there is a decrease in blood pressure, which occurs in response to a decrease in blood volume, or when there is a deficiency of Na^+ in the plasma, the kidneys release renin, which activates the renin-angiotensin-aldosterone pathway (Figure 27.5). Once aldosterone is formed, it increases Na^+ reabsorption in the late distal tubules and collecting ducts of the kidneys, which relieves the Na^+ deficiency in the plasma. Because antidiuretic hormone (ADH) is also released when blood pressure is low, water reabsorption accompanies Na^+ reabsorption via osmosis. This conserves the volume of body fluids by reducing urinary loss of water.
2. ***Atrial natriuretic peptide.*** An increase in blood volume, as might occur after you finish one or more supersized drinks, stretches the atria of the heart and promotes release of **atrial natriuretic peptide (ANP)** (Figure 27.6). ANP promotes **natriuresis**, elevated excretion of Na^+ into the urine. The osmotic consequence of excreting more Na^+ is loss of more water in urine, which decreases blood volume and blood pressure. In addition to stimulating the release of ANP, an increase in blood volume also slows the release of renin from the kidneys. When the renin level declines, less aldosterone is formed, which causes reabsorption of filtered Na^+ to slow in the late distal tubules and collecting ducts of the kidneys. More

FIGURE 27.4 **Role of antidiuretic hormone (ADH) in water balance.**

ADH increases the amount of water reabsorption in the kidneys.

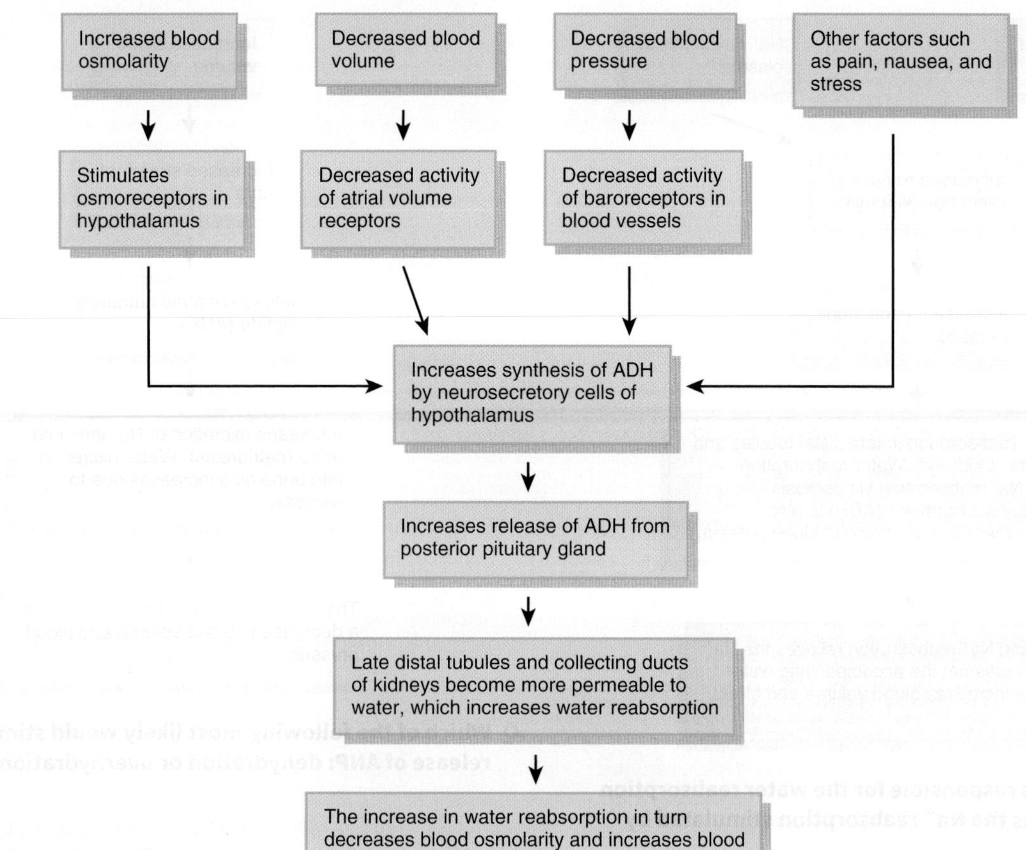

Q What effect does alcohol have on ADH secretion?

filtered Na^+ and water (due to osmosis) thus remain in the tubular fluid to be excreted in the urine.

Table 27.1 summarizes the factors that maintain body water balance.

Movement of Water between Body Fluid Compartments

Normally, the cells of the body neither shrink nor swell because the extracellular fluid that surrounds them is isotonic. This means that intracellular fluid and extracellular fluid have the same osmolarity (concentration of solutes). Changes in the osmolarity of extracellular fluid, however, cause fluid imbalances. If extracellular fluid becomes hypertonic (i.e., it has a greater concentration of solutes than intracellular fluid because its osmolarity has increased), water moves from cells into extracellular fluid by osmosis, causing the cells to shrink. If extracellular fluid becomes hypotonic (i.e., it has a lower concentration of solutes than intracellular fluid because its osmolarity has decreased) water moves from

TABLE 27.1 **Summary of Factors That Maintain Body Water Balance**

FACTOR	MECHANISM	EFFECT
Thirst center in hypothalamus	Stimulates desire to drink fluids.	Water gained if thirst is quenched.
Antidiuretic hormone (ADH), also known as *vasopressin*	Promotes insertion of water-channel proteins (aquaporin-2) into apical membranes of principal cells in collecting ducts of kidneys. As a result, water permeability of these cells increases and more water is reabsorbed.	Reduces loss of water in urine.
Aldosterone	By promoting urinary reabsorption of Na^+, increases water reabsorption via osmosis.	Reduces loss of water in urine.
Atrial natriuretic peptide (ANP)	Promotes natriuresis, elevated urinary excretion of Na^+, accompanied by water.	Increases loss of water in urine.

FIGURE 27.5 Role of aldosterone in sodium balance.

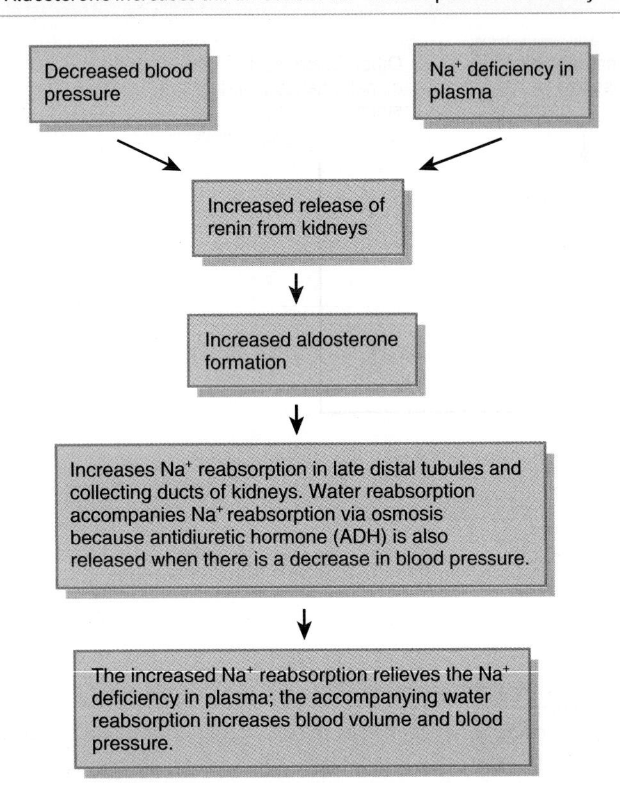

Q What hormone is responsible for the water reabsorption that accompanies the Na^+ reabsorption stimulated by aldosterone?

FIGURE 27.6 Role of atrial natriuretic peptide (ANP) in sodium balance.

ANP increases the excretion of Na^+ ions into urine (natriuresis)

Increased blood volume

↓

Increased stretch of atria

↓

Release of atrial natriuretic peptide (ANP)

↓

Increases excretion of Na^+ ions into urine (natriuresis). Water excretion into urine also increases due to osmosis.

↓

The increase in water excretion causes a decrease in blood volume and blood pressure.

Q Which of the following most likely would stimulate the release of ANP: dehydration or overhydration?

extracellular fluid into cells by osmosis, causing the cells to swell. Changes in osmolarity most often result from changes in the concentrations of Na^+ and Cl^- (the major contributors to osmolarity of extracellular fluid).

An *increase* in the osmolarity of extracellular fluid can occur, for example, after you eat a salty meal. The increased intake of NaCl produces an increase in the levels of Na^+ and Cl^- in extracellular fluid. As a result, the osmolarity of extracellular fluid increases, which causes net movement of water from cells into extracellular fluid. Such water movement shrinks the cells of the body. If neurons of the brain remain in this state for a significant period of time, mental confusion, convulsions, coma, and even death can occur. Body cells usually shrink only slightly and only for a short duration in response to an increase in the osmolarity of extracellular fluid because corrective measures such as the thirst mechanism and secretion of antidiuretic hormone increase the amount of body water, thereby reducing the concentration of solutes in extracellular fluid back to normal levels.

A *decrease* in the osmolarity of extracellular fluid can occur, for example, after drinking a large volume of water. This dilution causes the levels of Na^+ and Cl^- in extracellular fluid to fall below the normal range. When the extracellular concentrations of Na^+ and Cl^- decrease, the osmolarity of extracellular fluid also decreases. The net result is movement of water from extracellular fluid into cells, which causes the cells to swell. Usually when the osmolarity of extracellular fluid decreases, secretion of ADH is inhibited and the kidneys excrete a large volume of dilute urine, which restores the osmolarity of body fluids back to normal. As a result, body cells swell only slightly and only for a brief period. But when a person steadily consumes water faster than the kidneys can excrete it (the maximum urine flow rate is about 15 mL/min) or when renal function is poor, the result may be **water intoxication**, a state in which excessive body water causes cells to swell dangerously (Figure 27.7). As is the case when neurons of the brain shrink, swelling of the brain's neurons can result in mental confusion, seizures, coma, and possibly death. To prevent this dire sequence of events in cases of severe electrolyte and water loss, solutions given for intravenous or oral rehydration therapy (ORT) include a small amount of table salt (NaCl).

See Clinical Connection: Enemas and Fluid Balance

FIGURE 27.7 **Series of events in water intoxication.**

Water intoxication is a state in which excessive body water causes cells to swell.

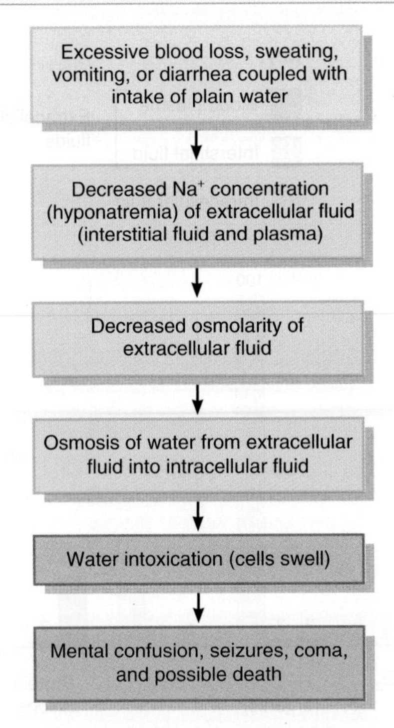

Q Why do solutions used for oral rehydration therapy contain a small amount of table salt (NaCl)?

27.2 Electrolytes in Body Fluids

OBJECTIVES

- **Compare** the electrolyte composition of the three major fluid compartments: plasma, interstitial fluid, and intracellular fluid.
- **Discuss** the functions and regulation of sodium, chloride, potassium, bicarbonate, calcium, phosphate, and magnesium ions.

The ions formed when electrolytes dissolve and dissociate serve four general functions in the body. (1) Because they are largely confined to particular fluid compartments and are more numerous than nonelectrolytes, certain ions *control the osmosis of water between fluid compartments.* (2) Ions *help maintain the acid–base balance* required for normal cellular activities. (3) Ions *carry electrical current,* which allows production of action potentials and graded potentials. (4) Several ions *serve as cofactors* needed for optimal activity of enzymes.

Concentrations of Electrolytes in Body Fluids

To compare the charge carried by ions in different solutions, the concentration of ions is typically expressed in units of **milliequivalents per liter (mEq/liter)** (mil′-i-ē-KWIV-a-lents). These units give the concentration of cations or anions in a given volume of solution. One equivalent is the positive or negative charge equal to the amount of charge in one mole of H^+; a milliequivalent is one one-thousandth of an equivalent. Recall that a mole of a substance is its molecular weight expressed in grams. For ions such as sodium (Na^+), potassium (K^+), and bicarbonate (HCO_3^-), which have a single positive or negative charge, the number of mEq/liter is equal to the number of mmol/liter. For ions such as calcium (Ca^{2+}) or phosphate (HPO_4^{2-}), which have two positive or negative charges, the number of mEq/liter is twice the number of mmol/liter.

Figure 27.8 compares the concentrations of the main electrolytes and protein anions in blood plasma, interstitial fluid, and intracellular fluid. The chief difference between the two extracellular fluids—blood plasma and interstitial fluid—is that blood plasma contains many protein anions, in contrast to interstitial fluid, which has very few. Because normal capillary membranes are virtually impermeable to proteins, only a few plasma proteins leak out of blood vessels into the interstitial fluid. This difference in protein concentration is largely responsible for the blood colloid osmotic pressure exerted by blood plasma. In other respects, the two fluids are similar.

The electrolyte content of intracellular fluid differs considerably from that of extracellular fluid. In extracellular fluid, the most abundant cation is Na^+, and the most abundant anion is Cl^-. In intracellular fluid, the most abundant cation is K^+, and the most abundant anions are proteins and phosphates (HPO_4^{2-}). By actively transporting Na^+ out of cells and K^+ into cells, sodium–potassium pumps (Na^+–K^+ ATPases) play a major role in maintaining the high intracellular concentration of K^+ and high extracellular concentration of Na^+.

Sodium

Sodium ions (Na^+) are the most abundant ions in extracellular fluid, accounting for 90% of the extracellular cations. The normal blood plasma Na^+ concentration is 136–148 mEq/liter. As we have already learned, Na^+ plays a pivotal role in fluid and electrolyte balance because it accounts for almost half of the osmolarity of extracellular fluid (142 of about 300 mOsm/liter). The flow of Na^+ through voltage-gated channels in the plasma membrane also is necessary for the generation and conduction of action potentials in neurons and muscle fibers. The typical daily intake of Na^+ in North America often far exceeds the body's normal daily requirements, due largely to excess dietary salt. The kidneys excrete excess Na^+, but they also can conserve it during periods of shortage.

The Na^+ level in the blood is controlled by aldosterone, antidiuretic hormone (ADH), and atrial natriuretic peptide (ANP). Aldosterone increases renal reabsorption of Na^+. When the blood plasma concentration of Na^+ drops below 135 mEq/liter, a condition called *hyponatremia*, ADH release ceases. The lack of ADH in turn permits

FIGURE 27.8 Electrolyte and protein anion concentrations in plasma, interstitial fluid, and intracellular fluid. The height of each column represents milliequivalents per liter (mEq/liter).

The electrolytes present in extracellular fluids are different from those present in intracellular fluid.

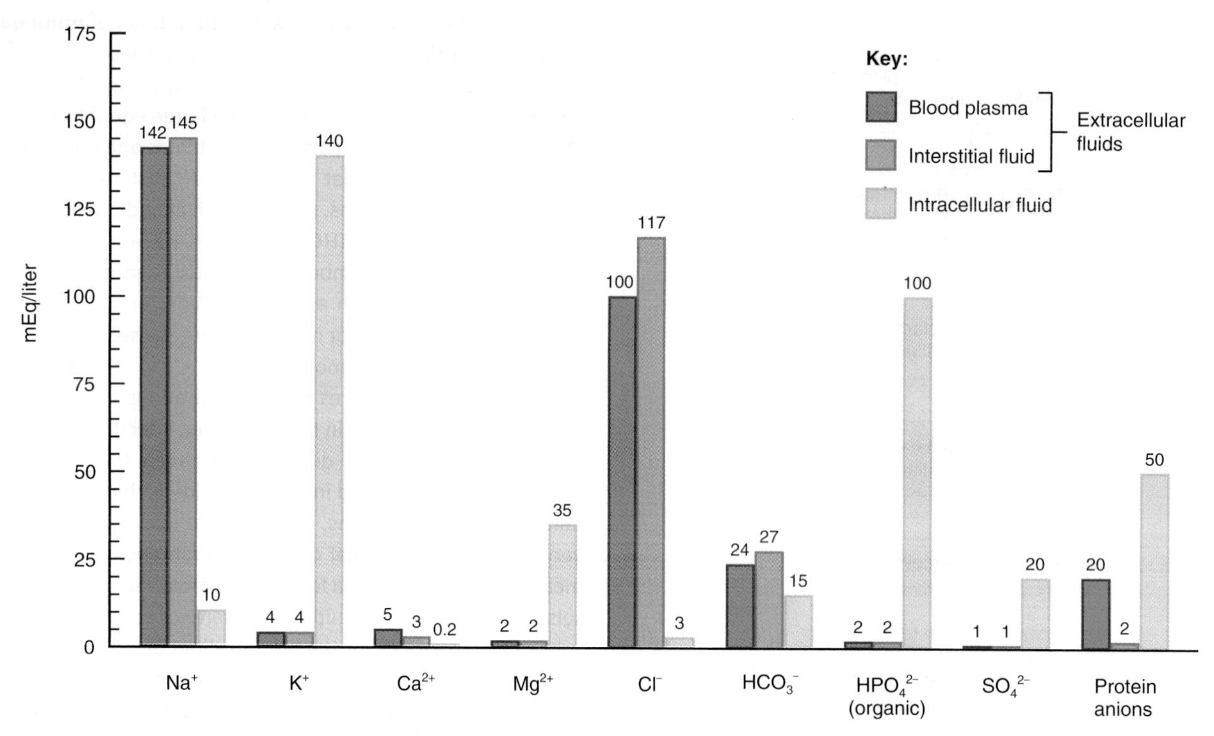

Q What cation and two anions are present in the highest concentrations in ECF and ICF?

greater excretion of water in urine and restoration of the normal Na^+ level in ECF. Atrial natriuretic peptide increases Na^+ excretion by the kidneys when the Na^+ level is above normal, a condition called *hypernatremia.*

See Clinical Connection: Indicators of Na^+ Imbalance

Chloride

Chloride ions (Cl^-) are the most prevalent anions in extracellular fluid. The normal blood plasma Cl^- concentration is 95–105 mEq/ liter. Cl^- moves relatively easily between the extracellular and intracellular compartments because most plasma membranes contain many Cl^- leakage channels and antiporters. For this reason, Cl^- can help balance the level of anions in different fluid compartments. One example is the chloride shift that occurs between red blood cells and blood plasma as the blood level of carbon dioxide either increases or decreases (see Figure 23.23b). In this case, the antiporter exchange of Cl^- for HCO_3^- maintains the correct balance of anions between ECF and ICF. Chloride ions also are part of the hydrochloric acid secreted into gastric juice. ADH helps regulate Cl^- balance in body fluids because it governs the extent of water loss in urine. Processes that increase or decrease renal reabsorption of sodium ions also affect reabsorption of chloride ions. (Recall that reabsorption of Na^+ and Cl^- occurs by means of Na^+–Cl^- symporters.)

Potassium

Potassium ions (K^+) are the most abundant cations in intracellular fluid (140 mEq/liter). K^+ plays a key role in establishing the resting membrane potential and in the repolarization phase of action potentials in neurons and muscle fibers; K^+ also helps maintain normal intracellular fluid volume. When K^+ moves into or out of cells, it often is exchanged for H^+ and thereby helps regulate the pH of body fluids.

The normal blood plasma K^+ concentration is 3.5–5.0 mEq/liter and is controlled mainly by aldosterone. When blood plasma K^+ concentration is high, more aldosterone is secreted into the blood. Aldosterone then stimulates principal cells of the renal collecting ducts to secrete more K^+ so excess K^+ is lost in the urine. Conversely, when blood plasma K^+ concentration is low, aldosterone secretion decreases and less K^+ is excreted in urine. Because K^+ is needed during the repolarization phase of action potentials, abnormal K^+ levels can be lethal. For instance, *hyperkalemia* (above-normal concentration of K^+ in blood) can cause death due to ventricular fibrillation.

Bicarbonate

Bicarbonate ions (HCO_3^-) are the second most prevalent extracellular anions. Normal blood plasma HCO_3^- concentration is 22–26 mEq/liter in systemic arterial blood and 23–27 mEq/liter in systemic venous blood. HCO_3^- concentration increases as blood flows through systemic capillaries because the carbon dioxide released by metabolically active cells combines with water to form carbonic acid; the carbonic acid then dissociates into H^+ and HCO_3^-. As blood flows through pulmonary capillaries, however, the concentration of HCO_3^- decreases again as carbon dioxide is exhaled. (**Figure 23.23** shows these reactions.) Intracellular fluid also contains a small amount of HCO_3^-. As previously noted, the exchange of Cl^- for HCO_3^- helps maintain the correct balance of anions in extracellular fluid and intracellular fluid.

The kidneys are the main regulators of blood HCO_3^- concentration. The intercalated cells of the renal tubule can either form HCO_3^- and release it into the blood when the blood level is low (see **Figure 27.10**) or excrete excess HCO_3^- in the urine when the level in blood is too high. Changes in the blood level of HCO_3^- are considered later in this chapter in the section on acid–base balance.

Calcium

Because such a large amount of calcium is stored in bone, it is the most abundant mineral in the body. About 98% of the calcium in adults is located in the skeleton and teeth, where it is combined with phosphates to form a crystal lattice of mineral salts. In body fluids, calcium is mainly an extracellular cation (Ca^{2+}). The normal concentration of free or unattached Ca^{2+} in blood plasma is 4.5–5.5 mEq/liter. About the same amount of Ca^{2+} is attached to various plasma proteins. Besides contributing to the hardness of bones and teeth, Ca^{2+} plays important roles in blood clotting, neurotransmitter release, maintenance of muscle tone, and excitability of nervous and muscle tissue.

The most important regulator of Ca^{2+} concentration in blood plasma is parathyroid hormone (PTH) (see **Figure 18.13**). A low level of Ca^{2+} in blood plasma promotes release of more PTH, which stimulates osteoclasts in bone tissue to release calcium (and phosphate) from bone extracellular matrix. Thus, PTH increases bone *resorption*. Parathyroid hormone also enhances *reabsorption* of Ca^{2+} from glomerular filtrate through renal tubule cells and back into blood, and increases production of calcitriol (the form of vitamin D that acts as a hormone), which in turn increases Ca^{2+} *absorption* from food in the gastrointestinal tract. Recall that calcitonin (CT) produced by the thyroid gland inhibits the activity of osteoclasts, accelerates Ca^{2+} deposition into bones, and thus lowers blood Ca^{2+} levels.

Phosphate

About 85% of the phosphate in adults is present as calcium phosphate salts, which are structural components of bone and teeth. The remaining 15% is ionized. Three phosphate ions ($H_2PO_4^-$, HPO_4^{2-}, and PO_4^{3-}) are important intracellular anions. At the normal pH of body fluids, HPO_4^{2-} is the most prevalent form. Phosphates contribute about 100 mEq/liter of anions to intracellular fluid. HPO_4^{2-} is an important buffer of H^+, both in body fluids and in the urine. Although some are "free," most phosphate ions are covalently bound to organic molecules such as lipids (phospholipids), proteins, carbohydrates, nucleic acids (DNA and RNA), and adenosine triphosphate (ATP).

The normal blood plasma concentration of ionized phosphate is only 1.7–2.6 mEq/liter. The same two hormones that govern calcium homeostasis—parathyroid hormone (PTH) and calcitriol—also regulate the level of HPO_4^{2-} in blood plasma. PTH stimulates resorption of bone extracellular matrix by osteoclasts, which releases both phosphate and calcium ions into the bloodstream. In the kidneys, however, PTH inhibits reabsorption of phosphate ions while stimulating reabsorption of calcium ions by renal tubular cells. Thus, PTH increases urinary excretion of phosphate and lowers blood phosphate level. Calcitriol promotes absorption of both phosphates and calcium from the gastrointestinal tract. Fibroblast growth factor 23 (FGF 23) is a polypeptide paracrine (local hormone) that also helps regulate blood plasma levels of HPO_4^{2-}. This hormone decreases HPO_4^{2-} blood levels by increasing HPO_4^{2-} excretion by the kidneys and decreasing absorption of HPO_4^{2-} by the gastrointestinal tract.

Magnesium

In adults, about 54% of the total body magnesium is part of bone matrix as magnesium salts. The remaining 46% occurs as magnesium ions (Mg^{2+}) in intracellular fluid (45%) and extracellular fluid (1%). Mg^{2+} is the second most common intracellular cation (35 mEq/liter). Functionally, Mg^{2+} is a cofactor for certain enzymes needed for the metabolism of carbohydrates and proteins and for the sodium–potassium pump. Mg^{2+} is essential for normal neuromuscular activity, synaptic transmission, and myocardial functioning. In addition, secretion of parathyroid hormone (PTH) depends on Mg^{2+}.

Normal blood plasma Mg^{2+} concentration is low, only 1.3–2.1 mEq/liter. Several factors regulate the blood plasma level of Mg^{2+} by varying the rate at which it is excreted in the urine. The kidneys increase urinary excretion of Mg^{2+} in response to hypercalcemia, hypermagnesemia, increases in extracellular fluid volume, decreases in parathyroid hormone, and acidosis. The opposite conditions decrease renal excretion of Mg^{2+}.

Table 27.2 describes the imbalances that result from the deficiency or excess of several electrolytes.

People at risk for fluid and electrolyte imbalances include those who depend on others for fluid and food, such as infants, the elderly, and the hospitalized; individuals undergoing medical treatment that involves intravenous infusions, drainages or suctions, and urinary catheters; and people who receive diuretics, experience excessive fluid losses and require increased fluid intake, or experience fluid retention and have fluid restrictions. Finally, athletes and military personnel in extremely hot environments, postoperative individuals, severe burn or trauma cases, individuals with chronic diseases (congestive heart failure, diabetes, chronic obstructive lung disease, and cancer), people in confinement, and individuals with altered levels of consciousness who may be unable to communicate needs or respond to thirst are also subject to fluid and electrolyte imbalances.

TABLE 27.2 Blood Electrolyte Imbalances

	DEFICIENCY		EXCESS	
ELECTROLYTE*	**NAME AND CAUSES**	**SIGNS AND SYMPTOMS**	**NAME AND CAUSES**	**SIGNS AND SYMPTOMS**
Sodium (Na^+) **136–148 mEq/liter**	**Hyponatremia** (hī′-po-na-TRĒ-mē-a) may be due to decreased sodium intake; increased sodium loss through vomiting, diarrhea, aldosterone deficiency, or taking certain diuretics; and excessive water intake.	Muscular weakness; dizziness, headache, and hypotension; tachycardia and shock; mental confusion, stupor, and coma.	**Hypernatremia** may occur with dehydration, water deprivation, or excessive sodium in diet or intravenous fluids; causes hypertonicity of ECF, which pulls water out of body cells into ECF, causing cellular dehydration.	Intense thirst, hypertension, edema, agitation, and convulsions.
Chloride (Cl^-) **95–105 mEq/liter**	**Hypochloremia** (hī′-pō-klō-RĒ-mē-a) may be due to excessive vomiting, overhydration, aldosterone deficiency, congestive heart failure, and therapy with certain diuretics such as furosemide (Lasix®).	Muscle spasms, metabolic alkalosis, shallow respirations, hypotension, and tetany.	**Hyperchloremia** may result from dehydration due to water loss or water deprivation; excessive chloride intake; or severe renal failure, hyperaldosteronism, certain types of acidosis, and some drugs.	Lethargy, weakness, metabolic acidosis, and rapid, deep breathing.
Potassium (K^+) **3.5–5.0 mEq/liter**	**Hypokalemia** (hī′-pō-ka-LĒ-mē-a) may result from excessive loss due to vomiting or diarrhea, decreased potassium intake, hyperaldosteronism, kidney disease, and therapy with some diuretics.	Muscle fatigue, flaccid paralysis, mental confusion, increased urine output, shallow respirations, and changes in electrocardiogram, including flattening of T wave.	**Hyperkalemia** may be due to excessive potassium intake, renal failure, aldosterone deficiency, crushing injuries to body tissues, or transfusion of hemolyzed blood.	Irritability, nausea, vomiting, diarrhea, muscular weakness; can cause death by inducing ventricular fibrillation.
Calcium (Ca^{2+}) **Total = 9.0–10.5 mg/dL; ionized = 4.5–5.5 mEq/liter**	**Hypocalcemia** (hī′-po-kal-SĒ-mē-a) may be due to increased calcium loss, reduced calcium intake, elevated phosphate levels, or hypoparathyroidism.	Numbness and tingling of fingers; hyperactive reflexes, muscle cramps, tetany, and convulsions; bone fractures; spasms of laryngeal muscles that can cause death by asphyxiation.	**Hypercalcemia** may result from hyperparathyroidism, some cancers, excessive intake of vitamin D, and Paget's disease of bone.	Lethargy, weakness, anorexia, nausea, vomiting, polyuria, itching, bone pain, depression, confusion, paresthesia, stupor, and coma.
Phosphate (HPO_4^{2-}) **1.7–2.6 mEq/liter**	**Hypophosphatemia** (hī′-po-fos-fa-TĒ-mē-a) may occur through increased urinary losses, decreased intestinal absorption, or increased utilization.	Confusion, seizures, coma, chest and muscle pain, numbness and tingling of fingers, decreased coordination, memory loss, and lethargy.	**Hyperphosphatemia** occurs when kidneys fail to excrete excess phosphate, as in renal failure; can also result from increased intake of phosphates or destruction of body cells, which releases phosphates into blood.	Anorexia, nausea, vomiting, muscular weakness, hyperactive reflexes, tetany, and tachycardia.
Magnesium (Mg^{2+}) **1.3–2.1 mEq/liter**	**Hypomagnesemia** (hī′-po-mag-ne-SĒ-mē-a) may be due to inadequate intake or excessive loss in urine or feces; also occurs in alcoholism, malnutrition, diabetes mellitus, and diuretic therapy.	Weakness, irritability, tetany, delirium, convulsions, confusion, anorexia, nausea, vomiting, paresthesia, and cardiac arrhythmias.	**Hypermagnesemia** occurs in renal failure or due to increased intake of Mg^{2+}, such as Mg^{2+}-containing antacids; also occurs in aldosterone deficiency and hypothyroidism.	Hypotension, muscular weakness or paralysis, nausea, vomiting, and altered mental functioning.

*Values are normal ranges of blood plasma levels in adults.

27.3 Acid–Base Balance

OBJECTIVES

- **Compare** the roles of buffers, exhalation of carbon dioxide, and kidney excretion of H^+ in maintaining pH of body fluids.
- **Describe** the different types of acid–base imbalances.

From our discussion thus far, it should be clear that various ions play different roles that help maintain homeostasis. A major homeostatic challenge is keeping the H^+ concentration (pH) of body fluids at an appropriate level. This task—the maintenance of acid–base balance—is of critical importance to normal cellular function. For example, the three-dimensional shape of all body proteins, which enables them to perform specific functions, is very sensitive to pH changes. When the diet contains a large amount of protein, as is typical in North America, cellular metabolism produces more acids than bases, which tends to acidify the blood. Before proceeding with this section of the chapter, you may wish to review the discussion of acids, bases, and pH in Section 2.4.

In a healthy person, several mechanisms help maintain the pH of systemic arterial blood between 7.35 and 7.45. (A pH of 7.4 corresponds to a H^+ concentration of 0.00004 mEq/liter = 40 nEq/liter.) Because metabolic reactions often produce a huge excess of H^+, the lack of any mechanism for the disposal of H^+ would cause H^+ in body fluids to rise quickly to a lethal level. Homeostasis of H^+ concentration within a narrow range is thus essential to survival. The removal of H^+ from body fluids and its subsequent elimination from the body depend on the following three major mechanisms:

1. ***Buffer systems.*** Buffers act quickly to temporarily bind H^+, removing the highly reactive, excess H^+ from solution. Buffers thus raise pH of body fluids but do not remove H^+ from the body.
2. ***Exhalation of carbon dioxide.*** By increasing the rate and depth of breathing, more carbon dioxide can be exhaled. Within minutes this reduces the level of carbonic acid in blood, which raises the blood pH (reduces blood H^+ level).
3. ***Kidney excretion of H^+.*** The slowest mechanism, but the only way to eliminate acids other than carbonic acid, is through their excretion in urine.

We will examine each of these mechanisms in more detail in the following sections.

The Actions of Buffer Systems

Most **buffer systems** in the body consist of a weak acid and the salt of that acid, which functions as a weak base. Buffers prevent rapid, drastic changes in the pH of body fluids by converting strong acids and bases into weak acids and weak bases within fractions of a second. Strong acids lower pH more than weak acids because strong acids release H^+ more readily and thus contribute more free hydrogen ions. Similarly, strong bases raise pH more than weak ones. The principal buffer systems of the body fluids are the protein buffer system, the carbonic acid–bicarbonate buffer system, and the phosphate buffer system.

Protein Buffer System The **protein buffer system** is the most abundant buffer in intracellular fluid and blood plasma. For example, the protein hemoglobin is an especially good buffer within red blood cells, and albumin is the main protein buffer in blood plasma. Proteins are composed of amino acids, organic molecules that contain at least one carboxyl group (—COOH) and at least one amino group (—NH_2); these groups are the functional components of the protein buffer system. The free carboxyl group at one end of a protein acts like an acid by releasing H^+ when pH rises; it dissociates as follows:

$$NH_2-\underset{\underset{H}{|}}{\overset{\overset{R}{|}}{C}}-COOH \longrightarrow NH_2-\underset{\underset{H}{|}}{\overset{\overset{R}{|}}{C}}-COO^- + H^+$$

The H^+ is then able to react with any excess OH^- in the solution to form water. The free amino group at the other end of a protein can act as a base by combining with H^+ when pH falls, as follows:

$$NH_2-\underset{\underset{H}{|}}{\overset{\overset{R}{|}}{C}}-COOH + H^+ \longrightarrow {}^+NH_3-\underset{\underset{H}{|}}{\overset{\overset{R}{|}}{C}}-COOH$$

So proteins can buffer both acids and bases. In addition to the terminal carboxyl and amino groups, side chains that can buffer H^+ are present on 7 of the 20 amino acids.

As we have already noted, the protein hemoglobin is an important buffer of H^+ in red blood cells (see **Figure 23.23**). As blood flows through the systemic capillaries, carbon dioxide (CO_2) passes from tissue cells into red blood cells, where it combines with water (H_2O) to form carbonic acid (H_2CO_3). Once formed, H_2CO_3 dissociates into H^+ and HCO_3^-. At the same time that CO_2 is entering red blood cells, oxyhemoglobin (Hb–O_2) is giving up its oxygen to tissue cells. Reduced hemoglobin (deoxyhemoglobin) picks up most of the H^+. For this reason, reduced hemoglobin usually is written as Hb–H. The following reactions summarize these relationships:

$$\underset{\text{Water}}{H_2O} + \underset{\substack{\text{Carbon dioxide}\\\text{(entering RBCs)}}}{CO_2} \longrightarrow \underset{\text{Carbonic acid}}{H_2CO_3}$$

$$\underset{\text{Carbonic acid}}{H_2CO_3} \longrightarrow \underset{\text{Hydrogen ion}}{H^+} + \underset{\text{Bicarbonate ion}}{HCO_3^-}$$

$$\underset{\substack{\text{Oxyhemoglobin}\\\text{(in RBCs)}}}{Hb{-}O_2} + \underset{\substack{\text{Hydrogen ion}\\\text{(from carbonic}\\\text{acid)}}}{H^+} \longrightarrow \underset{\substack{\text{Reduced}\\\text{hemoglobin}}}{Hb{-}H} + \underset{\substack{\text{Oxygen}\\\text{(released to}\\\text{tissue cells)}}}{O_2}$$

Carbonic Acid–Bicarbonate Buffer System The **carbonic acid–bicarbonate buffer system** is based on the *bicarbonate ion* (HCO_3^-), which can act as a weak base, and *carbonic acid* (H_2CO_3), which can act as a weak acid. As you have already learned, HCO_3^- is a significant anion in both intracellular and extracellular fluids (see Figure 27.8). Because the kidneys also synthesize new HCO_3^- and reabsorb filtered HCO_3^-, this important buffer is not lost in the urine. If there is an excess of H^+, the HCO_3^- can function as a weak base and remove the excess H^+ as follows:

$$\underset{\text{Hydrogen ion}}{H^+} + \underset{\substack{\text{Bicarbonate ion}\\\text{(weak base)}}}{HCO_3^-} \longrightarrow \underset{\text{Carbonic acid}}{H_2CO_3}$$

Then, H_2CO_3 dissociates into water and carbon dioxide, and the CO_2 is exhaled from the lungs.

Conversely, if there is a shortage of H^+, the H_2CO_3 can function as a weak acid and provide H^+ as follows:

$$\underset{\substack{\text{Carbonic acid}\\\text{(weak acid)}}}{H_2CO_3} \longrightarrow \underset{\text{Hydrogen ion}}{H^+} + \underset{\text{Bicarbonate ion}}{HCO_3^-}$$

At a pH of 7.4, HCO_3^- concentration is about 24 mEq/liter and H_2CO_3 concentration is about 1.2 mmol/liter, so bicarbonate ions outnumber carbonic acid molecules by 20 to 1. Because CO_2 and H_2O combine to form H_2CO_3, this buffer system cannot protect against pH changes due to respiratory problems in which there is an excess or shortage of CO_2.

Phosphate Buffer System

The **phosphate buffer system** acts via a mechanism similar to the one for the carbonic acid–bicarbonate buffer system. The components of the phosphate buffer system are the ions *dihydrogen phosphate* ($H_2PO_4^-$) and *monohydrogen phosphate* (HPO_4^{2-}). Recall that phosphates are major anions in intracellular fluid and minor ones in extracellular fluids (see Figure 27.8). The dihydrogen phosphate ion acts as a weak acid and is capable of buffering strong bases such as OH^-, as follows:

$$\underset{\substack{\text{Hydroxide ion}\\\text{(strong base)}}}{OH^+} + \underset{\substack{\text{Dihydrogen}\\\text{phosphate}\\\text{(weak acid)}}}{H_2PO_4^-} \longrightarrow \underset{\text{Water}}{H_2O} + \underset{\substack{\text{Monohydrogen}\\\text{phosphate}\\\text{(weak base)}}}{HPO_4^{2-}}$$

The monohydrogen phosphate ion is capable of buffering the H^+ released by a strong acid such as hydrochloric acid (HCl) by acting as a weak base:

$$\underset{\substack{\text{Hydrogen ion}\\\text{(strong acid)}}}{H^+} + \underset{\substack{\text{Monohydrogen}\\\text{phosphate}\\\text{(weak base)}}}{HPO_4^{2-}} \longrightarrow \underset{\substack{\text{Dihydrogen}\\\text{phosphate}\\\text{(weak acid)}}}{H_2PO_4^-}$$

Because the concentration of phosphates is highest in intracellular fluid, the phosphate buffer system is an important regulator of pH in the cytosol. It also acts to a smaller degree in extracellular fluids and buffers acids in urine. $H_2PO_4^-$ is formed when excess H^+ in the kidney tubule fluid combines with HPO_4^{2-} (see Figure 27.10). The H^+ that becomes part of the $H_2PO_4^-$ passes into the urine. This reaction is one way the kidneys help maintain blood pH by excreting H^+ in the urine.

Exhalation of Carbon Dioxide

The simple act of breathing also plays an important role in maintaining the pH of body fluids. An increase in the carbon dioxide (CO_2) concentration in body fluids increases H^+ concentration and thus lowers the pH (makes body fluids more acidic). Because H_2CO_3 can be eliminated by exhaling CO_2, it is called a **volatile acid**. Conversely, a decrease in the CO_2 concentration of body fluids raises the pH (makes body fluids more alkaline). This chemical interaction is illustrated by the following reversible reactions:

$$\underset{\substack{\text{Carbon}\\\text{dioxide}}}{CO_2} + \underset{\text{Water}}{H_2O} \rightleftharpoons \underset{\substack{\text{Carbonic}\\\text{acid}}}{H_2CO_3} \rightleftharpoons \underset{\substack{\text{Hydrogen}\\\text{ion}}}{H^+} + \underset{\substack{\text{Bicarbonate}\\\text{ion}}}{HCO_3^-}$$

Changes in the rate and depth of breathing can alter the pH of body fluids within a couple of minutes. With increased ventilation, more CO_2 is exhaled. When CO_2 levels decrease, the reaction is driven to the left, H^+ concentration falls, and blood pH increases. Doubling the breathing increases pH by about 0.23 units, from 7.4 to 7.63. If ventilation is slower than normal, less carbon dioxide is exhaled. When CO_2 levels increase, the reaction is driven to the right, the H^+ concentration increases, and blood pH decreases. Reducing ventilation to one-quarter of normal lowers the pH by 0.4 units, from 7.4 to 7.0. These examples show the powerful effect of alterations in breathing on the pH of body fluids.

The pH of body fluids and the rate and depth of breathing interact via a negative feedback loop (Figure 27.9). When the blood acidity increases, the decrease in pH (increase in concentration of H^+) is detected by central chemoreceptors in the medulla oblongata and peripheral chemoreceptors in the aortic and carotid bodies, both of which stimulate the dorsal respiratory group in the medulla oblongata. As a result, the diaphragm and other respiratory muscles contract more forcefully and frequently, so more CO_2 is exhaled. As less H_2CO_3 forms and fewer H^+ are present, blood pH increases. When the response brings blood pH (H^+ concentration) back to normal, there is a return to acid–base homeostasis. The same negative feedback loop operates if the blood level of CO_2 increases. Ventilation increases, which removes more CO_2, reducing the H^+ concentration and increasing the blood's pH.

By contrast, if the pH of the blood increases, the respiratory center is inhibited and the rate and depth of breathing decrease. A decrease in the CO_2 concentration of the blood has the same effect. When breathing decreases, CO_2 accumulates in the blood so its H^+ concentration increases.

Kidney Excretion of H^+

Metabolic reactions produce **nonvolatile acids** such as sulfuric acid at a rate of about 1 mEq of H^+ per day for every kilogram of body mass. The only way to eliminate this huge acid load is to excrete H^+ in the urine. Given the magnitude of these contributions to acid–base balance, it's not surprising that renal failure can quickly cause death.

As you learned in Chapter 26, cells in both the proximal convoluted tubules (PCT) and the collecting ducts of the kidneys secrete hydrogen ions into the tubular fluid. In the PCT, Na^+–H^+ antiporters secrete H^+ as they reabsorb Na^+ (see Figure 26.13). Even more important for regulation of pH of body fluids, however, are the intercalated cells of the collecting duct. The *apical* membranes of some intercalated cells include **proton pumps (H^+ ATPases)** that secrete H^+ into the tubular fluid (Figure 27.10). Intercalated cells can secrete H^+ against a concentration gradient so effectively that urine can be up to 1000 times (3 pH units) more acidic than blood. HCO_3- produced by

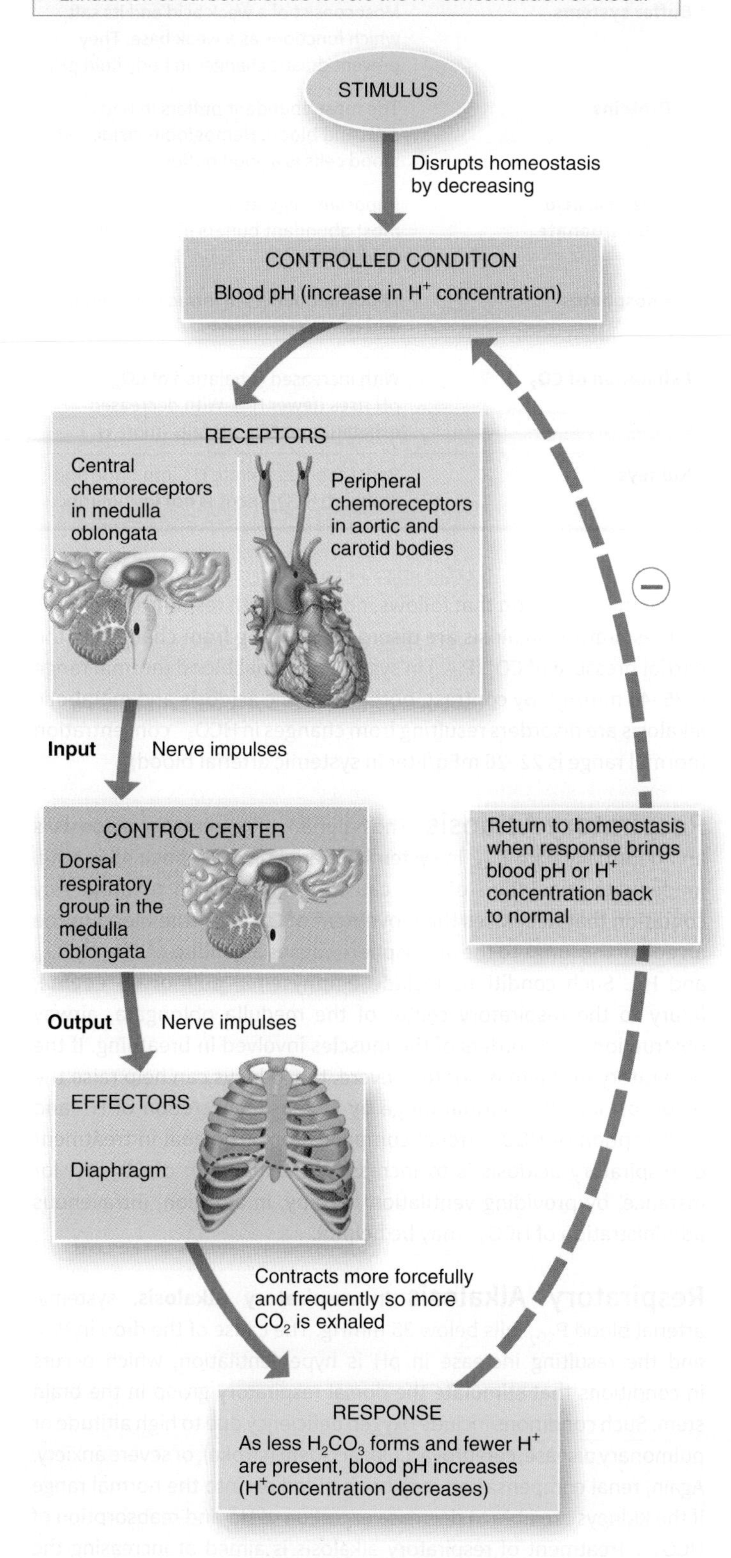

FIGURE 27.9 **Negative feedback regulation of blood pH by the respiratory system.**

Q If you hold your breath for 30 seconds, what is likely to happen to your blood pH?

dissociation of H_2CO_3 inside intercalated cells crosses the basolateral membrane by means of **Cl^-–HCO_3^- antiporters** and then diffuses into peritubular capillaries (**Figure 27.10a**). The HCO_3^- that enters the blood in this way is *new* (not filtered). For this reason, blood leaving the kidney in the renal vein may have a higher HCO_3^- concentration than blood entering the kidney in the renal artery.

Interestingly, a second type of intercalated cell has proton pumps in its *basolateral* membrane and Cl^-–HCO_3^- antiporters in its apical membrane. These intercalated cells secrete HCO_3^- and reabsorb H^+. Thus, the two types of intercalated cells help maintain the pH of body fluids in two ways—by excreting excess H^+ when pH of body fluids is too low and by excreting excess HCO_3^- when pH is too high.

Some H^+ secreted into the tubular fluid of the collecting duct is buffered, but not by HCO_3^-, most of which has been filtered and reabsorbed. Two other buffers combine with H^+ in the collecting duct (**Figure 27.10b**). The most plentiful buffer in the tubular fluid of the collecting duct is HPO_4^{2-} (monohydrogen phosphate ion). In addition, a small amount of NH_3 (ammonia) also is present. H^+ combines with HPO_4^{2-} to form $H_2PO_4^-$ (dihydrogen phosphate ion) and with NH_3 to form NH_4^+ (ammonium ion). Because these ions cannot diffuse back into tubule cells, they are excreted in the urine.

Table 27.3 summarizes the mechanisms that maintain the pH of body fluids.

Acid–Base Imbalances

The normal pH range of systemic arterial blood is between 7.35 (= 45 nEq of H^+/liter) and 7.45 (= 35 nEq of H^+/liter). **Acidosis** (or *acidemia*) is a condition in which blood pH is below 7.35; **alkalosis** (or *alkalemia*) is a condition in which blood pH is higher than 7.45.

The major physiological effect of acidosis is depression of the central nervous system through depression of synaptic transmission. If the systemic arterial blood pH falls below 7, depression of the nervous system is so severe that the individual becomes disoriented, then comatose, and may die. Patients with severe acidosis usually die while in a coma. A major physiological effect of alkalosis, by contrast, is overexcitability in both the central nervous system and peripheral nerves. Neurons conduct impulses repetitively, even when not stimulated by normal stimuli; the results are nervousness, muscle spasms, and even convulsions and death.

A change in blood pH that leads to acidosis or alkalosis may be countered by **compensation**, the physiological response to an acid–base imbalance that acts to normalize arterial blood pH. Compensation may be either *complete*, if pH indeed is brought within the normal range, or *partial*, if systemic arterial blood pH is still lower than 7.35 or higher than 7.45. If a person has altered blood pH due to metabolic causes, hyperventilation or hypoventilation can help bring blood pH back toward the normal range; this form of compensation, termed **respiratory compensation**, occurs within minutes and reaches its maximum within hours. If, however, a person has altered blood pH due to respiratory causes, then **renal compensation**—changes in secretion of H^+ and reabsorption of HCO_3^- by the kidney tubules—can help reverse the change. Renal compensation may begin in minutes, but it takes days to reach maximum effectiveness.

FIGURE 27.10 Secretion of H^+ by intercalated cells in the collecting duct. HCO_3^- = bicarbonate ion; CO_2 = carbon dioxide; H_2O = water; H_2CO_3 = carbonic acid; Cl^- = chloride ion; NH_3 = ammonia; NH_4^+ = ammonium ion; HPO_4^{2-} = monohydrogen phosphate ion; $H_2PO_4^-$ = dihydrogen phosphate ion.

Urine can be up to 1000 times more acidic than blood due to the operation of the proton pumps in the collecting ducts of the kidneys.

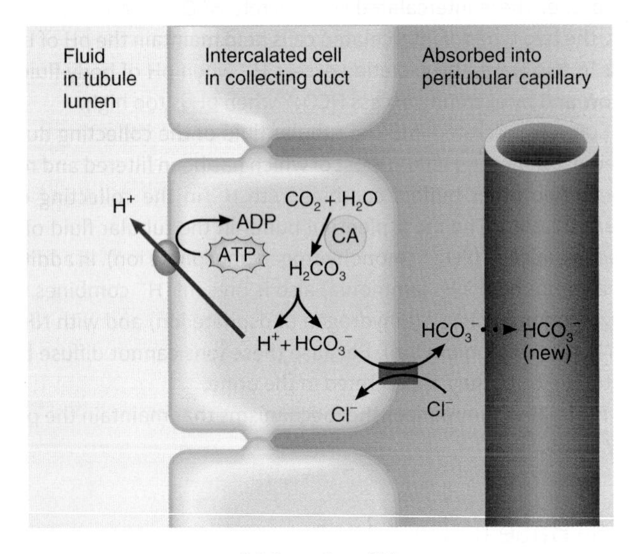

(a) Secretion of H^+

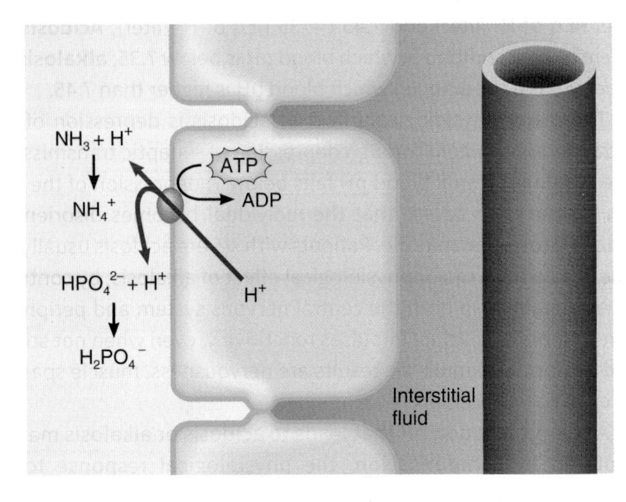

(b) Buffering of H^+ in urine

Key:

 Proton pump (H^+ ATPase) in apical membrane

 HCO_3^-–Cl^- antiporter in basolateral membrane

Diffusion

Q What would be the effects of a drug that blocks the activity of carbonic anhydrase?

TABLE 27.3 Mechanisms That Maintain pH of Body Fluids

MECHANISM	COMMENTS
Buffer systems	Most consist of a weak acid and its salt, which functions as a weak base. They prevent drastic changes in body fluid pH.
Proteins	The most abundant buffers in body cells and blood. Hemoglobin inside red blood cells is a good buffer.
Carbonic acid–bicarbonate	Important regulator of blood pH. The most abundant buffers in extracellular fluid (ECF).
Phosphates	Important buffers in intracellular fluid and urine.
Exhalation of CO_2	With increased exhalation of CO_2, pH rises (fewer H^+). With decreased exhalation of CO_2, pH falls (more H^+).
Kidneys	Renal tubules secrete H^+ into urine and reabsorb HCO_3^- so it is not lost in urine.

In the discussion that follows, note that both respiratory acidosis and respiratory alkalosis are disorders resulting from changes in the partial pressure of CO_2 (P_{CO_2}) in systemic arterial blood (normal range is 35–45 mmHg). By contrast, both metabolic acidosis and metabolic alkalosis are disorders resulting from changes in HCO_3^- concentration (normal range is 22–26 mEq/liter in systemic arterial blood).

Respiratory Acidosis

The hallmark of **respiratory acidosis** is an abnormally high P_{CO_2} in systemic arterial blood—above 45 mmHg. Inadequate exhalation of CO_2 causes the blood pH to drop. Any condition that decreases the movement of CO_2 from the blood to the alveoli of the lungs to the atmosphere causes a buildup of CO_2, H_2CO_3, and H^+. Such conditions include emphysema, pulmonary edema, injury to the respiratory center of the medulla oblongata, airway obstruction, or disorders of the muscles involved in breathing. If the respiratory problem is not too severe, the kidneys can help raise the blood pH into the normal range by increasing excretion of H^+ and reabsorption of HCO_3^- (renal compensation). The goal in treatment of respiratory acidosis is to increase the exhalation of CO_2, as, for instance, by providing ventilation therapy. In addition, intravenous administration of HCO_3^- may be helpful.

Respiratory Alkalosis

In **respiratory alkalosis**, systemic arterial blood P_{CO_2} falls below 35 mmHg. The cause of the drop in P_{CO_2} and the resulting increase in pH is hyperventilation, which occurs in conditions that stimulate the dorsal respiratory group in the brain stem. Such conditions include oxygen deficiency due to high altitude or pulmonary disease, cerebrovascular accident (stroke), or severe anxiety. Again, renal compensation may bring blood pH into the normal range if the kidneys are able to decrease excretion of H^+ and reabsorption of HCO_3^-. Treatment of respiratory alkalosis is aimed at increasing the level of CO_2 in the body. In cases where respiratory alkalosis is caused by severe anxiety, a simple treatment is to have the person inhale and

TABLE 27.4 Summary of Acidosis and Alkalosis

CONDITION	DEFINITION	COMMON CAUSES	COMPENSATORY MECHANISM
Respiratory acidosis	Increased P_{CO_2} (above 45 mmHg) and decreased pH (below 7.35) if no compensation.	Hypoventilation due to emphysema, pulmonary edema, trauma to respiratory center, airway obstructions, or dysfunction of muscles of respiration.	Renal: increased excretion of H^+; increased reabsorption of HCO_3^-. If compensation is complete, pH will be within normal range but P_{CO2} will be high.
Respiratory alkalosis	Decreased P_{CO_2} (below 35 mmHg) and increased pH (above 7.45) if no compensation.	Hyperventilation due to oxygen deficiency, pulmonary disease, cerebrovascular accident (CVA), or severe anxiety.	Renal: decreased excretion of H^+; decreased reabsorption of HCO_3^-. If compensation is complete, pH will be within normal range but P_{CO2} will be low.
Metabolic acidosis	Decreased HCO_3^- (below 22 mEq/liter) and decreased pH (below 7.35) if no compensation.	Loss of bicarbonate ions due to diarrhea, accumulation of acid (ketosis), renal dysfunction.	Respiratory: hyperventilation, which increases loss of CO_2. If compensation is complete, pH will be within normal range but HCO_3^- will be low.
Metabolic alkalosis	Increased HCO_3^- (above 26 mEq/liter) and increased pH (above 7.45) if no compensation.	Loss of acid due to vomiting, gastric suctioning, or use of certain diuretics; excessive intake of alkaline drugs.	Respiratory: hypoventilation, which slows loss of CO_2. If compensation is complete, pH will be within normal range but HCO_3^- will be high.

exhale into a paper bag for a short period; as a result, the person inhales air containing a higher-than-normal concentration of CO_2.

Metabolic Acidosis In **metabolic acidosis**, the systemic arterial blood HCO_3^- level drops below 22 mEq/liter. Such a decline in this important buffer causes the blood pH to decrease. Three situations may lower the blood level of HCO_3^-: (1) actual loss of HCO_3^-, such as may occur with severe diarrhea or renal dysfunction; (2) accumulation of an acid other than carbonic acid, as may occur in ketosis (described in Clinical Connection: Ketosis in chapter 25 in the Study Guide); or (3) failure of the kidneys to excrete H^+ from metabolism of dietary proteins. If the problem is not too severe, hyperventilation can help bring blood pH into the normal range (respiratory compensation). Treatment of metabolic acidosis consists of administering intravenous solutions of sodium bicarbonate and correcting the cause of the acidosis.

Metabolic Alkalosis In **metabolic alkalosis**, the systemic arterial blood HCO_3^- concentration is above 26 mEq/liter. A nonrespiratory loss of acid or excessive intake of alkaline drugs causes the blood pH to increase above 7.45. Excessive vomiting of gastric contents, which results in a substantial loss of hydrochloric acid, is probably the most frequent cause of metabolic alkalosis. Other causes include gastric suctioning, use of certain diuretics, endocrine disorders, excessive intake of alkaline drugs (antacids), and severe dehydration. Respiratory compensation through hypoventilation may bring blood pH into the normal range. Treatment of metabolic alkalosis consists of giving fluid solutions to correct Cl^-, K^+, and other electrolyte deficiencies plus correcting the cause of alkalosis.

Table 27.4 summarizes respiratory and metabolic acidosis and alkalosis.

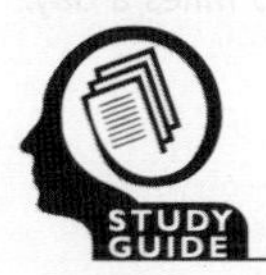

See Clinical Connection: Diagnosis of Acid–Base Imbalances

27.4 Aging and Fluid, Electrolyte, and Acid–Base Homeostasis

OBJECTIVE

- **Describe** the changes in fluid, electrolyte, and acid–base balance that may occur with aging.

There are significant differences between adults and infants, especially premature infants, with respect to fluid distribution, regulation of fluid and electrolyte balance, and acid–base homeostasis. Accordingly, infants experience more problems than adults in these areas. The differences are related to the following conditions:

- ***Proportion and distribution of water.*** A newborn's total body mass is about 75% water (and can be as high as 90% in a premature infant); an adult's total body mass is about 55–60% water. (The "adult" percentage is achieved at about 2 years of age.) Adults have twice as much water in ICF as ECF, but the opposite is true in premature infants. Because ECF is subject to more changes than ICF, rapid losses or gains of body water are much more critical in infants. Given that the rate of fluid intake and output is about seven times higher in infants than in adults, the slightest changes in fluid balance can result in severe abnormalities.
- ***Metabolic rate.*** The metabolic rate of infants is about double that of adults. This results in the production of more metabolic wastes and acids, which can lead to the development of acidosis in infants.
- ***Functional development of the kidneys.*** Infant kidneys are only about half as efficient in concentrating urine as those of adults.

(Functional development is not complete until close to the end of the first month after birth.) As a result, the kidneys of newborns can neither concentrate urine nor rid the body of excess acids as effectively as those of adults.

- ***Body surface area.*** The ratio of body surface area to body volume of infants is about three times greater than that of adults. Water loss through the skin is significantly higher in infants than in adults.
- ***Breathing rate.*** The higher breathing rate of infants (about 30 to 80 times a minute) causes greater water loss from the lungs. Respiratory alkalosis may occur because greater ventilation eliminates more CO_2 and lowers the P_{CO_2}.
- ***Ion concentrations.*** Newborns have higher K^+ and Cl^- concentrations than adults. This creates a tendency toward metabolic acidosis.

By comparison with children and younger adults, older adults often have an impaired ability to maintain fluid, electrolyte, and acid–base balance. With increasing age, many people have a decreased volume of intracellular fluid and decreased total body K^+ due to declining skeletal muscle mass and increasing mass of adipose tissue (which contains very little water). Age-related decreases in respiratory and renal functioning may compromise acid–base balance by slowing the exhalation of CO_2 and the excretion of excess acids in urine. Other kidney changes, such as decreased blood flow, decreased glomerular filtration rate, and reduced sensitivity to antidiuretic hormone, have an adverse effect on the ability to maintain fluid and electrolyte balance. Due to a decrease in the number and efficiency of sweat glands, water loss from the skin declines with age. Because of these age-related changes, older adults are susceptible to several fluid and electrolyte disorders:

- *Dehydration* and *hypernatremia* often occur due to inadequate fluid intake or loss of more water than Na^+ in vomit, feces, or urine.
- *Hyponatremia* may occur due to inadequate intake of Na^+; elevated loss of Na^+ in urine, vomit, or diarrhea; or impaired ability of the kidneys to produce dilute urine.
- *Hypokalemia* often occurs in older adults who chronically use laxatives to relieve constipation or who take K^+-depleting diuretic drugs for treatment of hypertension or heart disease.
- *Acidosis* may occur due to impaired ability of the lungs and kidneys to compensate for acid–base imbalances. One cause of acidosis is decreased production of ammonia (NH_3) by renal tubule cells, which then is not available to combine with H^+ and be excreted in urine as NH_4^+; another cause is reduced exhalation of CO_2.

Review Questions

27.1 Fluid Compartments and Fluid Homeostasis

1. What is the approximate volume of each of your body fluid compartments?
2. How are the routes of water gain and loss from the body regulated?
3. By what mechanism does thirst help regulate water intake?
4. How do aldosterone, atrial natriuretic peptide, and antidiuretic hormone regulate the volume and osmolarity of body fluids?
5. What factors control the movement of water between interstitial fluid and intracellular fluid?

27.2 Electrolytes in Body Fluids

6. What are the functions of electrolytes in the body?
7. Name three important extracellular electrolytes, and three important intracellular electrolytes, and indicate how each is regulated.

27.3 Acid–Base Balance

8. Explain how each of the following buffer systems helps to maintain the pH of body fluids: proteins, carbonic acid–bicarbonate buffers, and phosphates.
9. Define acidosis and alkalosis. Distinguish among respiratory and metabolic acidosis and alkalosis.
10. What are the principal physiological effects of acidosis and alkalosis?

27.4 Aging and Fluid, Electrolyte, and Acid–Base Homeostasis

11. Why do infants experience greater problems with fluid, electrolyte, and acid–base balance than adults?

Critical Thinking Questions

1. Robin is in the early stages of pregnancy and has been vomiting excessively for several days. She became weak, was confused, and was taken to the emergency room. What do you suspect has happened to Robin's acid–base balance? How would her body attempt to compensate? What electrolytes would be affected by her vomiting, and how do her symptoms reflect those imbalances?
2. Henry is in the intensive care unit because he suffered a severe myocardial infarction three days ago. The lab reports the following values from an arterial blood sample: pH 7.30, HCO_3^- = 20 mEq/liter, P_{CO_2} = 32 mmHg. Diagnose Henry's acid-base status and decide whether compensation is occurring.
3. This summer, Sam is training for a marathon by running 10 miles a day. Describe changes in his fluid balance as he trains.

The Reproductive Systems

CHAPTER 28

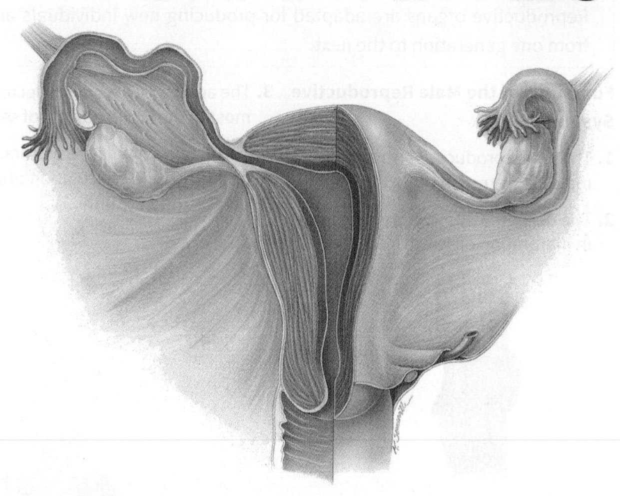

The Reproductive Systems and Homeostasis

> The male and female reproductive organs work together to produce offspring. In addition, the female reproductive organs contribute to sustaining the growth of embryos and fetuses.

Humans produce offspring by a process called sexual reproduction in which haploid sperm cells produced by the testes of males fertilize the haploid secondary oocytes produced by the ovaries of females. As a result of fertilization, the resulting diploid cell is called a zygote and contains one set of chromosomes from each parent. Males and females have anatomically distinct reproductive organs that are designed to produce, nourish, and transport the haploid cells, facilitate fertilization and, in females, sustain the growth of the embryo and fetus.

Q Did you ever wonder how breast augmentation and breast reduction are performed?

28.1 Male Reproductive System

OBJECTIVES

- **Describe** the location, structure, and functions of the organs of the male reproductive system.
- **Discuss** the process of spermatogenesis in the testes.

The male and female reproductive organs can be grouped by function. The **gonads**—testes in males and ovaries in females—produce gametes and secrete sex hormones. Various **ducts** then store and transport the gametes, and **accessory sex glands** produce substances that protect the gametes and facilitate their movement. Finally, **supporting structures**, such as the penis in males and the uterus in females, assist the delivery of gametes, and the uterus is also the site for the growth of the embryo and fetus during pregnancy.

The organs of the **male reproductive system** include the testes, a system of ducts (epididymis, ductus deferens, ejaculatory ducts, and urethra), accessory sex glands (seminal vesicles, prostate, and bulbourethral glands), and several supporting structures, including the scrotum and the penis (Figure 28.1). The testes (male gonads) produce sperm and secrete hormones. The duct system transports and stores sperm, assists in their maturation, and conveys them to the exterior. Semen contains sperm plus the secretions provided by the accessory sex glands. The supporting structures have various functions. The penis delivers sperm into the female reproductive tract and the scrotum supports the testes.

As noted in Chapter 26, **urology** (ū-ROL-ō-jē) is the study of the urinary system. Urologists also diagnose and treat diseases and disorders of the male reproductive system. The branch of medicine that deals with male disorders, especially infertility and sexual dysfunction, is called **andrology** (an-DROL-ō-jē; *andro-* = masculine).

Scrotum

The **scrotum** (SKRŌ-tum = bag), the supporting structure for the testes, consists of loose skin and underlying subcutaneous layer that hangs from the root (attached portion) of the penis (Figure 28.1a). Externally, the scrotum looks like a single pouch of skin separated into lateral portions by a median ridge called the **raphe** (RĀ-fē = seam). Internally, the **scrotal septum** divides the scrotum into two compartments, each containing a single testis (Figure 28.2). The septum is made up of a subcutaneous layer and muscle tissue called the **dartos muscle** (DAR-tōs = skinned), which is composed of bundles of smooth muscle fibers. The dartos muscle is also found in the subcutaneous layer of the scrotum. Associated with each testis in the scrotum is the **cremaster muscle** (krē-MAS-ter = suspender), a series of small bands of skeletal muscle that descend as an extension of the internal oblique muscle through the spermatic cord to surround the testes.

The location of the scrotum and the contraction of its muscle fibers regulate the temperature of the testes. Normal sperm production requires a temperature about 2–3°C below core body temperature. This lowered temperature is maintained within the scrotum because it is outside the pelvic cavity. In response to cold temperatures, the cremaster and dartos muscles contract. Contraction of the cremaster muscles moves the testes closer to the body, where they can absorb body heat. Contraction of the dartos muscle causes the scrotum to become tight (wrinkled in appearance), which reduces heat loss. Exposure to warmth reverses these actions.

FIGURE 28.1 **Male organs of reproduction and surrounding structures.**

Reproductive organs are adapted for producing new individuals and passing on genetic material from one generation to the next.

Functions of the Male Reproductive System

1. The testes produce sperm and the male sex hormone testosterone.
2. The ducts transport, store, and assist in maturation of sperm.
3. The accessory sex glands secrete most of the liquid portion of semen.
4. The penis contains the urethra, a passageway for ejaculation of semen and excretion of urine.

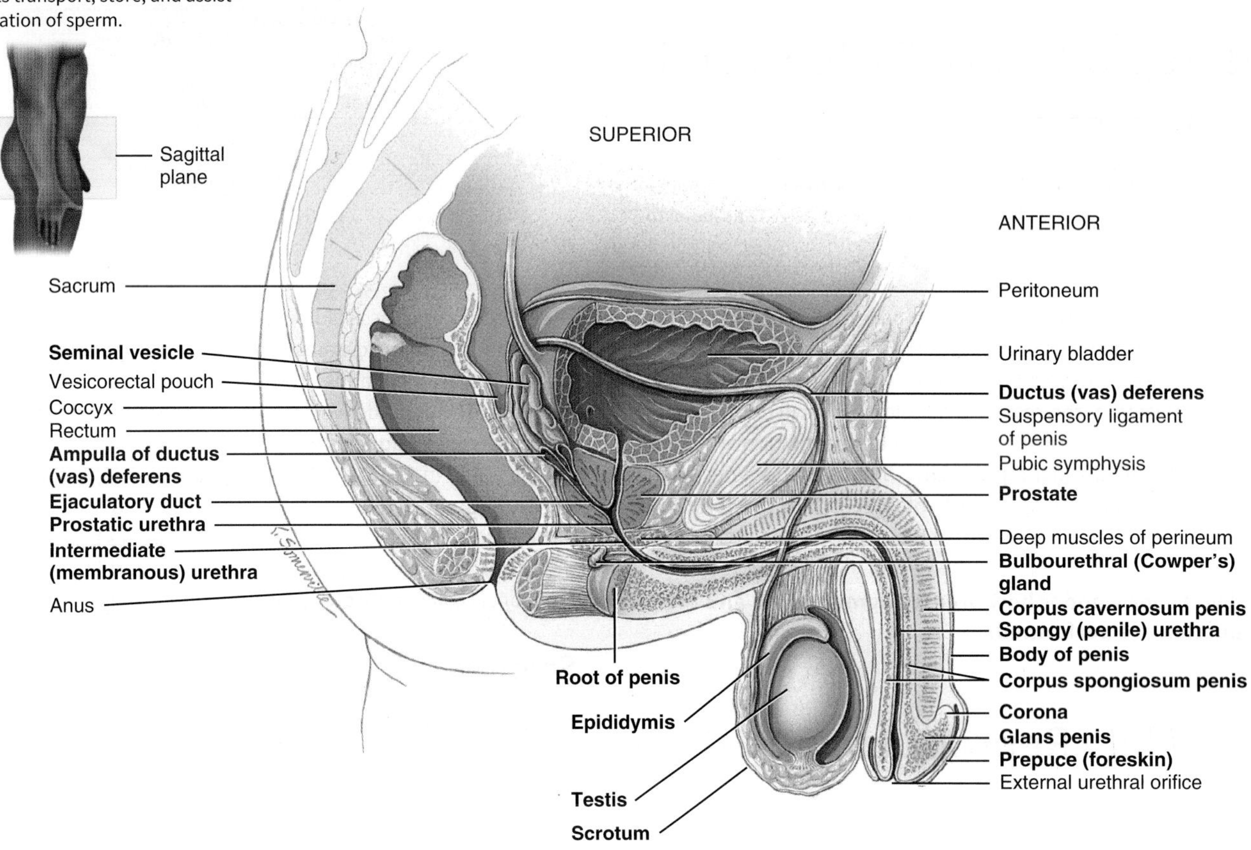

(a) Sagittal section

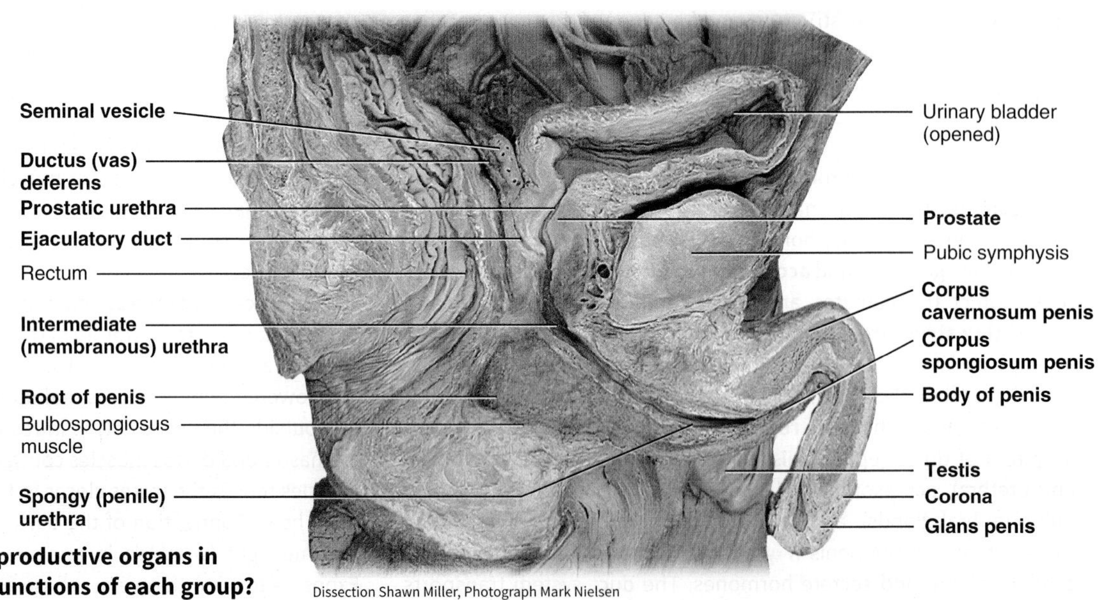

Dissection Shawn Miller, Photograph Mark Nielsen

(b) Sagittal section

Q What are the groups of reproductive organs in males, and what are the functions of each group?

FIGURE 28.2 **The scrotum, the supporting structure for the testes.**

The scrotum consists of loose skin and an underlying subcutaneous layer and supports the testes.

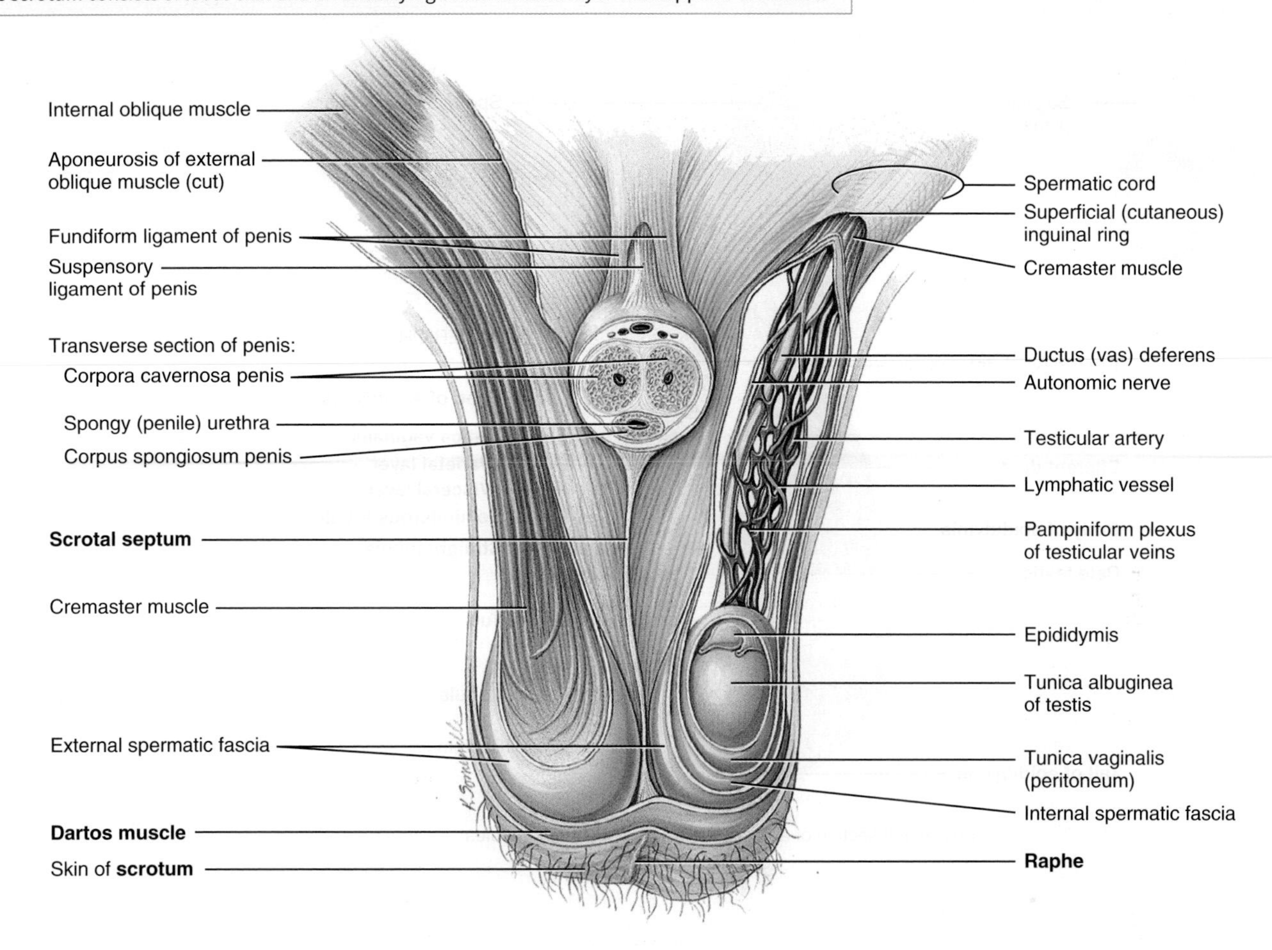

Anterior view of scrotum and testes and transverse section of penis

Q Which muscles help regulate the temperature of the testes?

Testes

The **testes** (TES-tēz = witness), or *testicles*, are paired oval glands in the scrotum measuring about 5 cm (2 in.) long and 2.5 cm (1 in.) in diameter (Figure 28.3). Each testis (singular) has a mass of 10–15 grams. The testes develop near the kidneys, in the posterior portion of the abdomen, and they usually begin their descent into the scrotum through the inguinal canals (passageways in the lower anterior abdominal wall) during the latter half of the seventh month of fetal development.

A serous membrane called the **tunica vaginalis** (TOO-ni-ka vaj-i-NAL-is; *tunica* = sheath), which is derived from the peritoneum and forms during the descent of the testes, partially covers the testes. A collection of serous fluid in the tunica vaginalis is called a **hydrocele** (HĪ-drō-sēl; *hydro-* = water; *-kele* = hernia). It may be caused by injury to the testes or inflammation of the epididymis. Usually, no treatment is required. Internal to the tunica vaginalis the testis is surrounded by a white fibrous capsule composed of dense irregular connective tissue, the **tunica albuginea** (al′-bū-JIN-ē-a; *albu-* = white); it extends inward, forming septa that divide the testis into a series of internal compartments called **lobules**. Each of the 200–300 lobules contains one to three tightly coiled tubules, the **seminiferous tubules** (sem′-i-NIF-er-us; *semin-* = seed; *-fer-* = to carry), where sperm are produced. The process by which the seminiferous tubules of the testes produce sperm is called **spermatogenesis** (sper′-ma-tō-JEN-e-sis; *genesis* = to be born).

The seminiferous tubules contain two types of cells: **spermatogenic cells** (sper′-ma-tō-JEN-ik), the sperm-forming cells, and **sustentacular cells** (sus′-ten-TAK-ū-lar) or *Sertoli cells* (ser-TŌ-lē), which have several functions in supporting spermatogenesis (Figure 28.4). Stem cells called **spermatogonia** (sper′-ma-tō-GŌ-nē-a; *-gonia* = offspring; singular is *spermatogonium*) develop from **primordial germ cells** (prī-MŌR-dē-al = primitive or early form) that arise from the yolk sac and enter the testes during the fifth week of development. In the embryonic testes, the primordial germ cells differentiate into spermatogonia,

FIGURE 28.3 **Internal and external anatomy of a testis.**

The testes are the male gonads, which produce haploid sperm.

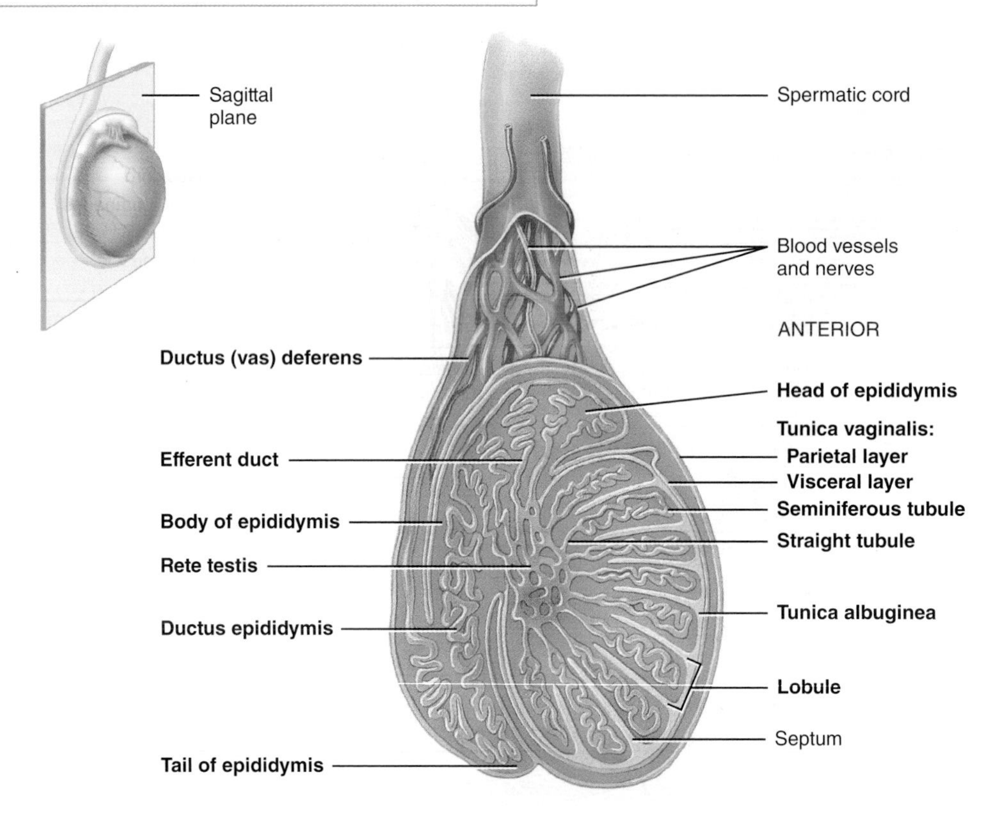

(a) Sagittal section of a testis showing seminiferous tubules

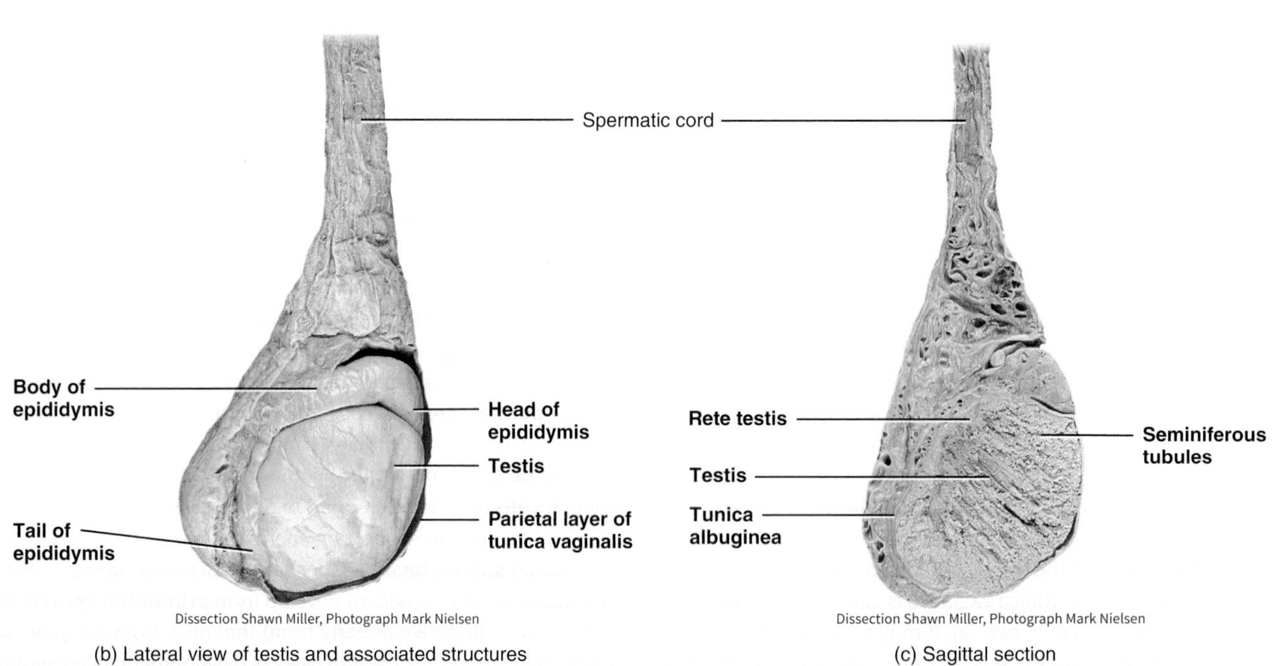

(b) Lateral view of testis and associated structures

(c) Sagittal section

Q What tissue layers cover and protect the testes?

FIGURE 28.4 Microscopic anatomy of the seminiferous tubules and stages of sperm production (spermatogenesis). Arrows indicate the progression of spermatogenic cells from least mature to most mature. The (*n*) and (2*n*) refer to haploid and diploid numbers of chromosomes, respectively.

Spermatogenesis occurs in the seminiferous tubules of the testes.

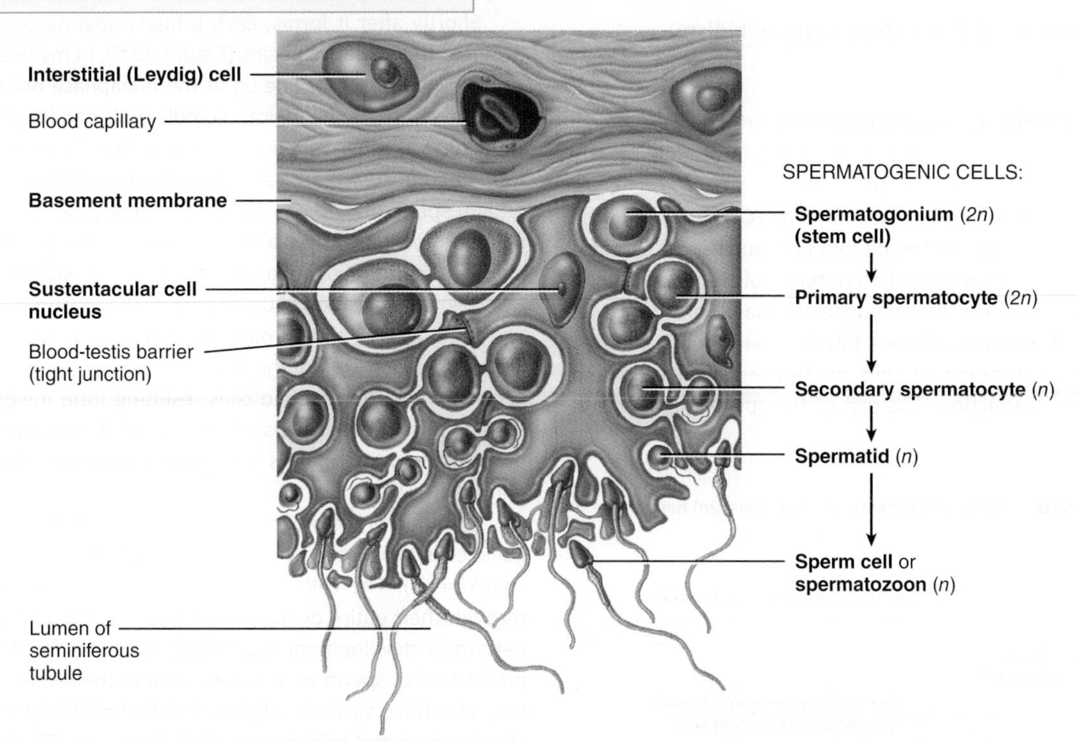

Transverse section of a part of a seminiferous tubule

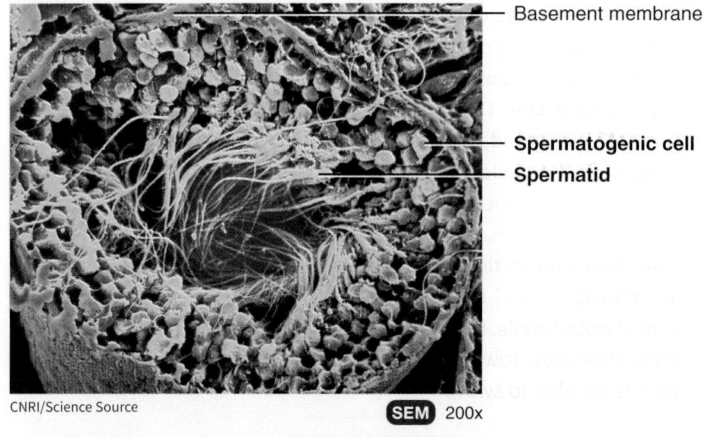

CNRI/Science Source

Transverse section of seminiferous tubule

Q Which cells secrete testosterone?

which remain dormant during childhood and actively begin producing sperm at puberty. Toward the lumen of the seminiferous tubule are layers of progressively more mature cells. In order of advancing maturity, these are primary spermatocytes, secondary spermatocytes, spermatids, and sperm cells. After a **sperm cell**, or *spermatozoon* (sper′-ma-tō-ZŌ-on; *zoon* = life), has formed, it is released into the lumen of the seminiferous tubule. (The plural terms are *sperm* and *spermatozoa*.)

Embedded among the spermatogenic cells in the seminiferous tubules are large **sustentacular cells** or *Sertoli cells*, which extend from the basement membrane to the lumen of the tubule. Internal to the basement membrane and spermatogonia, tight junctions join neighboring sustentacular cells to one another. These junctions form an obstruction known as the **blood–testis barrier** because substances must first pass through the sustentacular cells before they can reach the developing sperm. By isolating the developing gametes from the blood, the blood–testis barrier prevents an immune response against the spermatogenic cell's surface antigens, which are recognized as "foreign" by the immune system. The blood–testis barrier does not include spermatogonia.

Sustentacular cells support and protect developing spermatogenic cells in several ways. They nourish spermatocytes, spermatids, and sperm; phagocytize excess spermatid cytoplasm as development proceeds; and control movements of spermatogenic cells and the release of sperm into the lumen of the seminiferous tubule. They also produce fluid for sperm transport, secrete the hormone inhibin, and regulate the effects of testosterone and FSH (follicle-stimulating hormone).

In the spaces between adjacent seminiferous tubules are clusters of cells called **interstitial cells** or *Leydig cells* (LĪ-dig) (**Figure 28.4**). These cells secrete testosterone, the most prevalent androgen. An **androgen** is a hormone that promotes the development of masculine characteristics. Testosterone also promotes a man's *libido* (sexual drive).

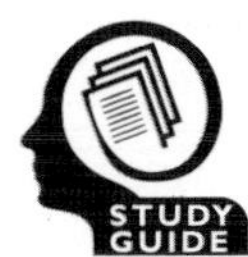

See Clinical Connection: Cryptorchidism

Spermatogenesis

Before you read this section, please review the topic of reproductive cell division in Chapter 3 in Section 3.7. Pay particular attention to Figures 3.33 and 3.34.

In humans, spermatogenesis takes 65–75 days. It begins with the spermatogonia, which contain the diploid (2*n*) number of chromosomes (Figure 28.5). Spermatogonia are types of *stem cells*; when they undergo mitosis, some spermatogonia remain near the basement membrane of the seminiferous tubule in an undifferentiated state to serve as a reservoir of cells for future cell division and subsequent sperm production. The rest of the spermatogonia lose contact with the basement membrane, squeeze through the tight junctions of the blood–testis barrier, undergo developmental changes, and differentiate into **primary spermatocytes** (SPER-ma-tō-sītz′). Primary spermatocytes, like spermatogonia, are diploid (2*n*); that is, they have 46 chromosomes.

Shortly after it forms, each primary spermatocyte replicates its DNA and then meiosis begins (Figure 28.5). In meiosis I, homologous pairs of chromosomes line up at the metaphase plate, and crossing-over occurs. Then, the meiotic spindle pulls one (duplicated) chromosome of each pair to an opposite pole of the dividing cell. The two cells formed by meiosis I are called **secondary spermatocytes**. Each secondary spermatocyte has 23 chromosomes, the haploid number (*n*). Each chromosome within a secondary spermatocyte, however, is made up of two chromatids (two copies of the DNA) still attached by a centromere. No replication of DNA occurs in the secondary spermatocytes.

In meiosis II, the chromosomes line up in single file along the metaphase plate, and the two chromatids of each chromosome separate. The four haploid cells resulting from meiosis II are called **spermatids** (SPER-ma-tids). A single primary spermatocyte therefore produces four spermatids via two rounds of cell division (meiosis I and meiosis II).

A unique process occurs during spermatogenesis. As spermatogenic cells proliferate, they fail to complete cytoplasmic separation (cytokinesis). The cells remain in contact via cytoplasmic bridges through their entire development (see Figures 28.4 and 28.5). This pattern of development most likely accounts for the synchronized production of sperm in any given area of the seminiferous tubule. It may also have survival value in that half of the sperm contain an X chromosome and half contain a Y chromosome. The larger X chromosome may carry genes needed for spermatogenesis that are lacking on the smaller Y chromosome.

The final stage of spermatogenesis, **spermiogenesis** (sper′-mē-ō-JEN-e-sis), is the development of haploid spermatids into sperm. No cell division occurs in spermiogenesis; each spermatid becomes a single **sperm cell**. During this process, spherical spermatids transform into elongated, slender sperm. An acrosome (described shortly) forms atop the nucleus, which condenses and elongates, a flagellum develops, and mitochondria multiply. Sustentacular cells dispose of the excess cytoplasm that sloughs off. Finally, sperm are released from their connections to sustentacular cells, an event known as **spermiation** (sper′-mē-Ā-shun). Sperm then enter the lumen of the seminiferous tubule. Fluid secreted by sustentacular cells pushes sperm along their way, toward the ducts of the testes. At this point, sperm are not yet able to swim.

FIGURE 28.5 Events in spermatogenesis. Diploid cells (2*n*) have 46 chromosomes; haploid cells (*n*) have 23 chromosomes.

Spermiogenesis involves the maturation of spermatids into sperm.

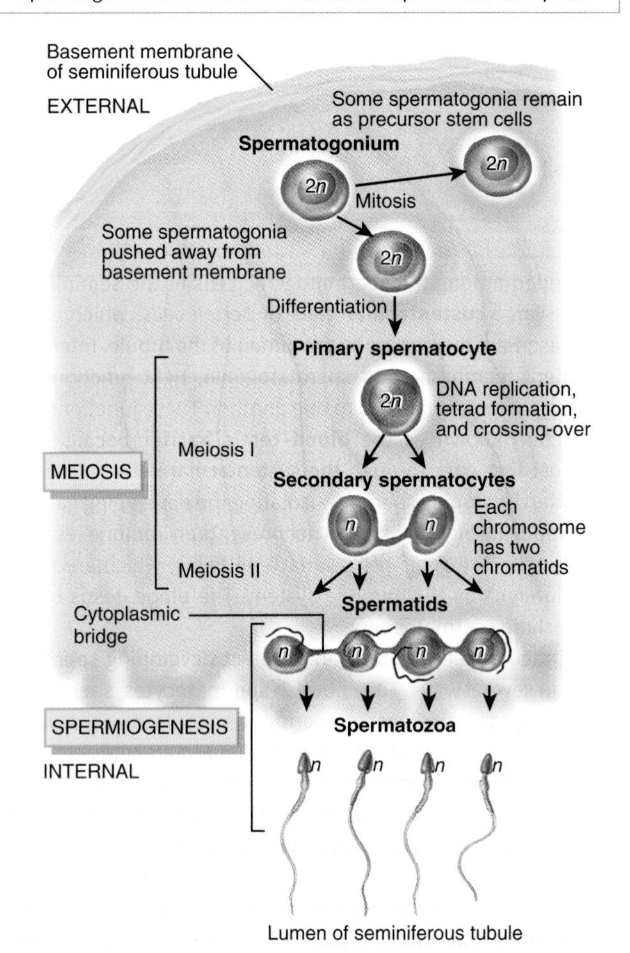

Q What is the outcome of meiosis I?

Sperm

Each day about 300 million sperm complete the process of spermatogenesis. A sperm is about 60 μm long and contains several structures that are highly adapted for reaching and penetrating a secondary oocyte (Figure 28.6). The major parts of a sperm are the head and the tail. The flattened, pointed **head** of the sperm is about 4–5 μm long. It contains a **nucleus** with 23 highly condensed chromosomes. Covering the anterior two-thirds of the nucleus is the **acrosome** (AK-rō-sōm; *acro-* = atop; *-some* = body), a caplike vesicle filled with enzymes that help a sperm to penetrate a secondary oocyte to bring about fertilization. Among the enzymes are hyaluronidase and proteases.

FIGURE 28.6 **Parts of a sperm cell.**

About 300 million sperm mature each day.

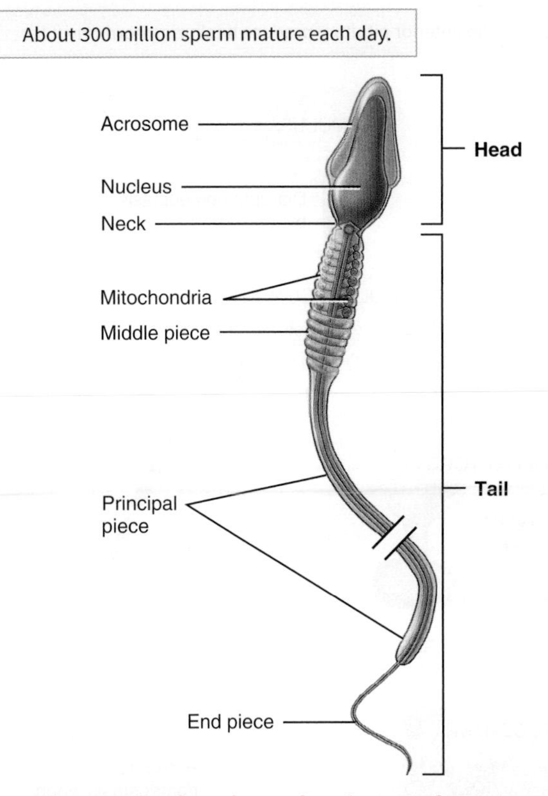

Q What are the functions of each part of a sperm cell?

The **tail** of a sperm is subdivided into four parts: neck, middle piece, principal piece, and end piece. The **neck** is the constricted region just behind the head that contains centrioles. The centrioles form the microtubules that comprise the remainder of the tail. The **middle piece** contains mitochondria arranged in a spiral, which provide the energy (ATP) for locomotion of sperm to the site of fertilization and for sperm metabolism. The **principal piece** is the longest portion of the tail, and the **end piece** is the terminal, tapering portion of the tail. Once ejaculated, most sperm do not survive more than 48 hours within the female reproductive tract.

Hormonal Control of Testicular Function

Although the initiating factors are unknown, at puberty certain hypothalamic neurosecretory cells increase their secretion of **gonadotropin-releasing hormone (GnRH)** (gō′-nad-ō-TRŌ-pin). This hormone in turn stimulates gonadotrophs in the anterior pituitary to increase their secretion of the two gonadotropins, **luteinizing hormone (LH)** (LOO-tē-in′-īz-ing) and **follicle-stimulating hormone (FSH)**. Figure 28.7 shows the hormones and negative feedback loops that control secretion of testosterone and spermatogenesis.

LH stimulates interstitial cells, which are located between seminiferous tubules, to secrete the hormone **testosterone** (tes-TOS-te-rōn). This steroid hormone is synthesized from cholesterol in the testes and is the principal androgen. It is lipid-soluble and readily diffuses out of interstitial cells into the interstitial fluid and then into blood. Via negative feedback, testosterone suppresses secretion of LH

FIGURE 28.7 **Hormonal control of spermatogenesis and actions of testosterone and dihydrotestosterone (DHT).** In response to stimulation by FSH and testosterone, sustentacular cells secrete androgen-binding protein (ABP). Dashed red lines indicate negative feedback inhibition.

Release of FSH is stimulated by GnRH and inhibited by inhibin; release of LH is stimulated by GnRH and inhibited by testosterone.

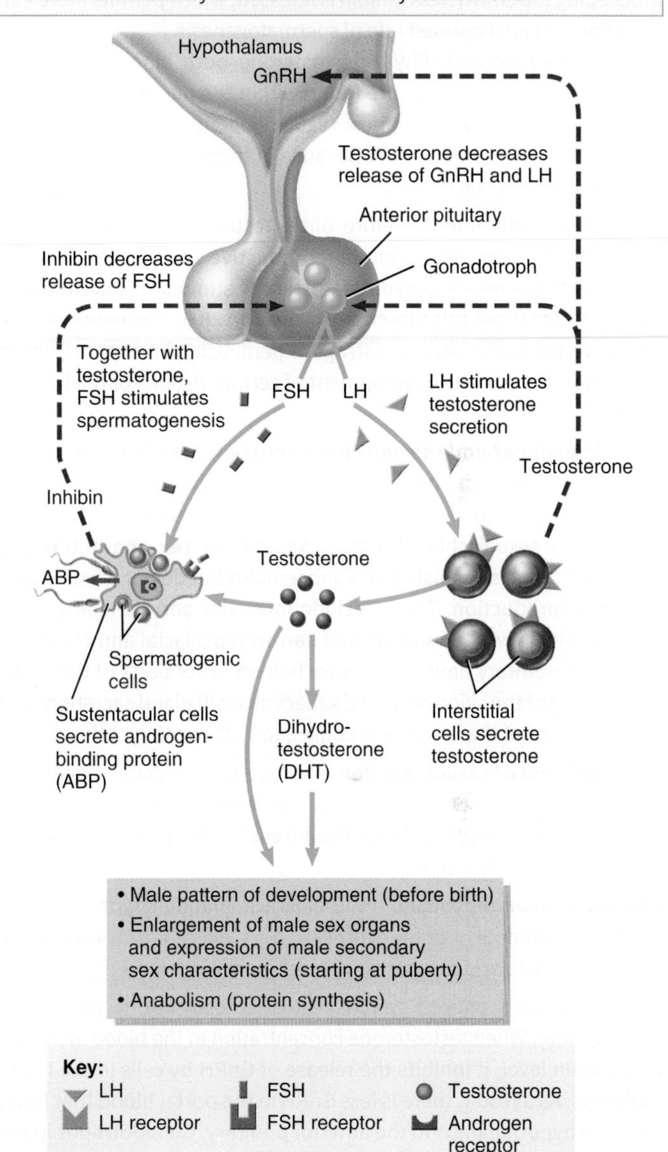

Q Which cells secrete inhibin?

by anterior pituitary gonadotrophs and suppresses secretion of GnRH by hypothalamic neurosecretory cells. In some target cells, such as those in the external genitals and prostate, the enzyme 5 alpha-reductase converts testosterone to another androgen called **dihydrotestosterone (DHT)** (dī-hī′-drō-tes-TOS-ter-ōn).

FSH acts indirectly to stimulate spermatogenesis (Figure 28.7). FSH and testosterone act synergistically on the sustentacular cells to stimulate secretion of **androgen-binding protein (ABP)** into the lumen of the seminiferous tubules and into the interstitial fluid around the spermatogenic cells. ABP binds to testosterone, keeping its concentration high. Testosterone stimulates the final steps of spermatogenesis in the seminiferous tubules. Once the degree of spermatogenesis required for male reproductive functions has been achieved, sustentacular cells release **inhibin**, a protein hormone named for its role in inhibiting FSH secretion by the anterior pituitary (Figure 28.7). If spermatogenesis is proceeding too slowly, less inhibin is released, which permits more FSH secretion and an increased rate of spermatogenesis.

Testosterone and dihydrotestosterone both bind to the same androgen receptors, which are found within the nuclei of target cells. The hormone–receptor complex regulates gene expression, turning some genes on and others off. Because of these changes, the androgens produce several effects:

- ***Prenatal development.*** Before birth, testosterone stimulates the male pattern of development of reproductive system ducts and the descent of the testes. Dihydrotestosterone stimulates development of the external genitals (described in Section 28.6). Testosterone also is converted in the brain to estrogens (feminizing hormones), which may play a role in the development of certain regions of the brain in males.
- ***Development of male sexual characteristics.*** At puberty, testosterone and dihydrotestosterone bring about development and enlargement of the male sex organs and the development of masculine secondary sexual characteristics. **Secondary sex characteristics** are traits that distinguish males and females but do not have a direct role in reproduction. These include muscular and skeletal growth that results in wide shoulders and narrow hips; facial and chest hair (within hereditary limits) and more hair on other parts of the body; thickening of the skin; increased sebaceous (oil) gland secretion; and enlargement of the larynx and consequent deepening of the voice.
- ***Development of sexual function.*** Androgens contribute to male sexual behavior and spermatogenesis and to sex drive (libido) in both males and females. Recall that the adrenal cortex is the main source of androgens in females.
- ***Stimulation of anabolism.*** Androgens are anabolic hormones; that is, they stimulate protein synthesis. This effect is obvious in the heavier muscle and bone mass of most men as compared to women.

A negative feedback system regulates testosterone production (Figure 28.8). When testosterone concentration in the blood increases to a certain level, it inhibits the release of GnRH by cells in the hypothalamus. As a result, there is less GnRH in the portal blood that flows from the hypothalamus to the anterior pituitary. Gonadotrophs in the anterior pituitary then release less LH, so the concentration of LH in systemic blood falls. With less stimulation by LH, the interstitial cells in the testes secrete less testosterone, and there is a return to homeostasis. If the testosterone concentration in the blood falls too low, however, GnRH is again released by the hypothalamus and stimulates secretion of LH by the anterior pituitary. LH in turn stimulates testosterone production by the testes.

FIGURE 28.8 Negative feedback control of blood level of testosterone.

Gonadotrophs of the anterior pituitary produce luteinizing hormone (LH).

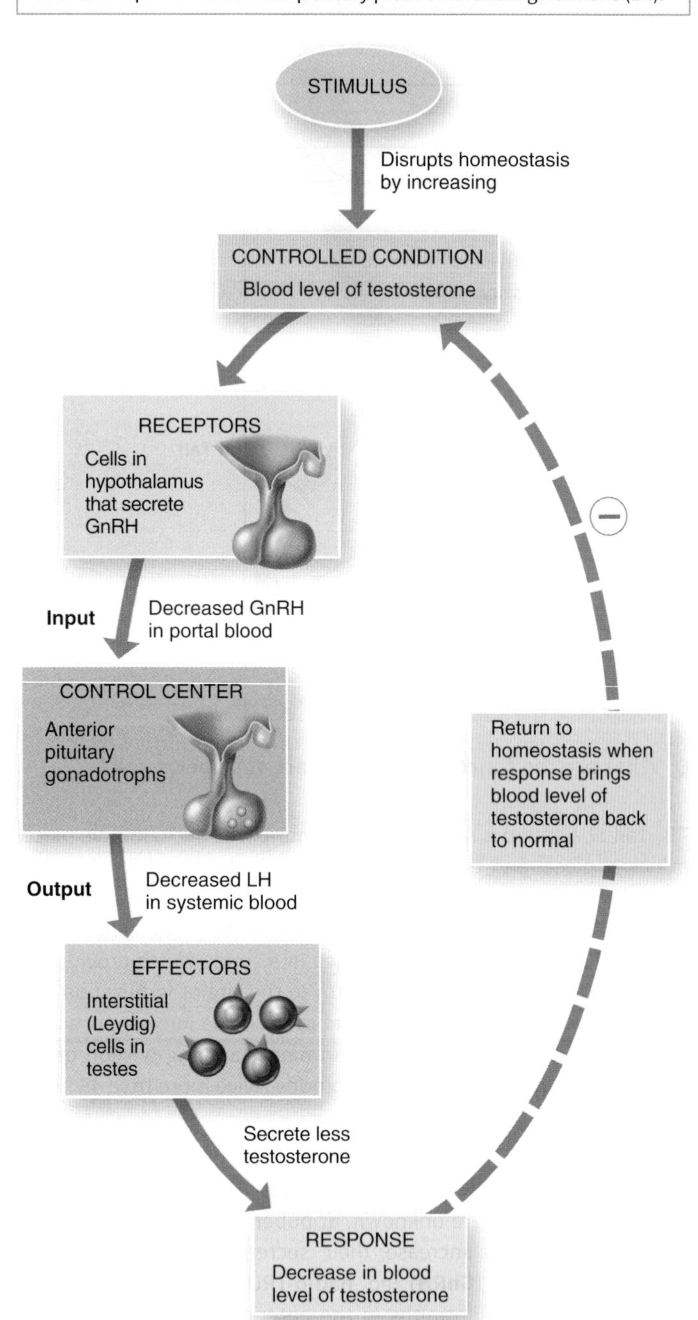

Q Which hormones inhibit secretion of FSH and LH by the anterior pituitary?

Reproductive System Ducts in Males

Ducts of the Testis Pressure generated by the fluid secreted by sustentacular cells pushes sperm and fluid along the lumen of seminiferous tubules and then into a series of very short ducts called **straight tubules** (see Figure 28.3a). The straight tubules lead to a network of ducts in the testis called the **rete testis** (RĒ-tē = network). From the rete testis, sperm move into a series of coiled **efferent ducts** (EF-er-ent) in the epididymis that empty into a single tube called the **ductus epididymis**.

Epididymis The **epididymis** (ep′-i-DID-i-mis; *epi-* = above or over; *-didymis* = testis) is an organ about 4 cm (1.5 in.) long that curves along the superior and posterior border of each testis having a comma shape in profile (see Figure 28.3a). The plural is *epididymides* (ep′-i-di-DIM-i-dēz). Each epididymis consists mostly of the tightly coiled **ductus epididymis**. The efferent ducts from the testis join the ductus epididymis at the larger, superior portion of the epididymis called the **head**. The **body** is the narrow midportion of the epididymis, and the tail is the smaller, inferior portion. At its distal end, the **tail** of the epididymis continues as the ductus (vas) deferens (discussed shortly).

The ductus epididymis would measure about 6 m (20 ft) in length if it were uncoiled. It is lined with pseudostratified columnar epithelium and encircled by layers of smooth muscle. The free surfaces of the columnar cells contain **stereocilia** (ster′-ē-ō-SIL-ē-a), which despite their name are long, branching microvilli (not cilia) that increase the surface area for the reabsorption of degenerated sperm. Connective tissue around the muscle layer attaches the loops of the ductus epididymis and carries blood vessels and nerves.

Functionally, the epididymis is the site of **sperm maturation**, the process by which sperm acquire motility and the ability to fertilize an ovum. This occurs over a period of about 14 days. The epididymis also helps propel sperm into the ductus (vas) deferens during sexual arousal by peristaltic contraction of its smooth muscle. In addition, the epididymis stores sperm, which remain viable here for up to several months. Any stored sperm that are not ejaculated by that time are eventually reabsorbed.

Ductus Deferens Within the tail of the epididymis, the ductus epididymis becomes less convoluted, and its diameter increases. Beyond this point, the duct is known as the **ductus deferens** or *vas deferens* (DEF-er-enz) (see Figure 28.3a). The ductus deferens, which is about 45 cm (18 in.) long, ascends along the posterior border of the epididymis through the spermatic cord and then enters the pelvic cavity. There it loops over the ureter and passes over the side and down the posterior surface of the urinary bladder (see Figure 28.1a). The dilated terminal portion of the ductus deferens is the **ampulla** (am-PUL-la = little jar; see Figure 28.9). The mucosa of the ductus deferens consists of pseudostratified columnar epithelium and lamina propria (areolar connective tissue). The muscularis is composed of three layers of smooth muscle; the inner and outer layers are longitudinal, and the middle layer is circular.

Functionally, the ductus deferens conveys sperm during sexual arousal from the epididymis toward the urethra by peristaltic contractions of its muscular coat. Like the epididymis, the ductus deferens also can store sperm for several months. Any stored sperm that are not ejaculated by that time are eventually reabsorbed.

Spermatic Cord The **spermatic cord** is a supporting structure of the male reproductive system that ascends out of the scrotum (see Figure 28.2). Each spermatic cord consists of a ductus (vas) deferens as it ascends through the scrotum, the testicular artery, veins that drain the testis and carry testosterone into circulation (the pampiniform plexus), autonomic nerves, lymphatic vessels, and the cremaster muscle. The spermatic cord and ilioinguinal nerve pass through the **inguinal canal** (ING-gwi-nal = groin), an oblique passageway in the anterior abdominal wall just superior and parallel to the medial half of the inguinal ligament. The canal, which is about 4–5 cm (about 2 in.) long, originates at the **deep** (*abdominal*) **inguinal ring**, a slitlike opening in the aponeurosis of the transversus abdominis muscle; the canal ends at the **superficial** (*subcutaneous*) **inguinal ring** (see Figure 28.2), a somewhat triangular opening in the aponeurosis of the external oblique muscle. In females, the round ligament of the uterus and ilioinguinal nerve pass through the inguinal canal.

The term **varicocele** (VAR-i-kō-sēl; *varico-* = varicose; *-kele* = hernia) refers to a swelling in the scrotum due to a dilation of the veins that drain the testes. It is usually more apparent when the person is standing and typically does not require treatment.

Ejaculatory Ducts Each **ejaculatory duct** (ē-JAK-ū-la-tōr-ē; *ejacul-* = to expel) is about 2 cm (1 in.) long and is formed by the union of the duct from the seminal vesicle and the ampulla of the ductus (vas) deferens (Figure 28.9). The short ejaculatory ducts form just superior to the base (superior portion) of the prostate and pass inferiorly and anteriorly through the prostate. They terminate in the prostatic urethra, where they eject sperm and seminal vesicle secretions just before the release of semen from the urethra to the exterior.

Urethra In males, the **urethra** (ū-RĒ-thra) is the shared terminal duct of the reproductive and urinary systems; it serves as a passageway for both semen and urine. About 20 cm (8 in.) long, it passes through the prostate, the deep muscles of the perineum, and the penis, and is subdivided into three parts (see Figures 28.1 and 26.22). The **prostatic urethra** (pros-TAT-ik) is 2–3 cm (1 in.) long and passes through the prostate. As this duct continues inferiorly, it passes through the deep muscles of the perineum, where it is known as the **intermediate** (*membranous*) **urethra** (MEM-bra-nus). The intermediate urethra is about 1 cm (0.5 in.) in length. As this duct passes through the corpus spongiosum of the penis, it is known as the **spongy urethra**, which is about 15–20 cm (6–8 in.) long. The spongy urethra ends at the **external urethral orifice**. The histology of the male urethra may be reviewed in Section 26.8.

Accessory Sex Glands

The ducts of the male reproductive system store and transport sperm cells, but the accessory sex glands secrete most of the liquid portion of semen. The accessory sex glands include the seminal vesicles, the prostate, and the bulbourethral glands.

Seminal Vesicles The paired **seminal vesicles** (VES-i-kuls) or *seminal glands* are convoluted pouchlike structures, about 5 cm (2 in.) in length, lying posterior to the base of the urinary bladder and anterior to the rectum (Figure 28.9). Through the seminal vesicle ducts they secrete an alkaline, viscous fluid that contains fructose (a monosaccharide sugar), prostaglandins, and clotting proteins that are different from those in blood. The alkaline nature of the seminal fluid helps to neutralize the acidic environment of the male urethra and female reproductive tract that otherwise would inactivate and kill sperm. The fructose is used for ATP production by sperm. Prostaglandins contribute to sperm motility and viability and may stimulate smooth muscle contractions within the female reproductive tract. The clotting proteins help semen coagulate after ejaculation. It is thought that coagulation occurs in order to keep sperm cells from leaking from the vagina. Fluid secreted by the seminal vesicles normally constitutes about 60% of the volume of semen.

Prostate The **prostate** (PROS-tāt; *prostata* = one who stands before) is a single, doughnut-shaped gland about the size of a golf ball. It measures about 4 cm (1.6 in.) from side to side, about 3 cm (1.2 in.) from top to bottom, and about 2 cm (0.8 in.) from front to back. It is inferior to the urinary bladder and surrounds the prostatic urethra

FIGURE 28.9 Locations of several accessory reproductive organs in males. The prostate, urethra, and penis have been sectioned to show internal details.

The male urethra has three subdivisions: the prostatic, membranous, and spongy (penile) urethra.

Functions of Accessory Sex Gland Secretions

1. The seminal vesicles secrete an alkaline, viscous fluid that helps neutralize acid in the female reproductive tract, provides fructose for ATP production by sperm, contributes to sperm motility and viability, and helps semen coagulate after ejaculation.
2. The prostate secretes a milky, slightly acidic fluid that contains enzymes that break down clotting proteins from the seminal vesicles.
3. The bulbourethral glands secrete an alkaline fluid that neutralizes the acidic environment of the urethra and mucus that lubricates the lining of the urethra and the tip of the penis during sexual intercourse.

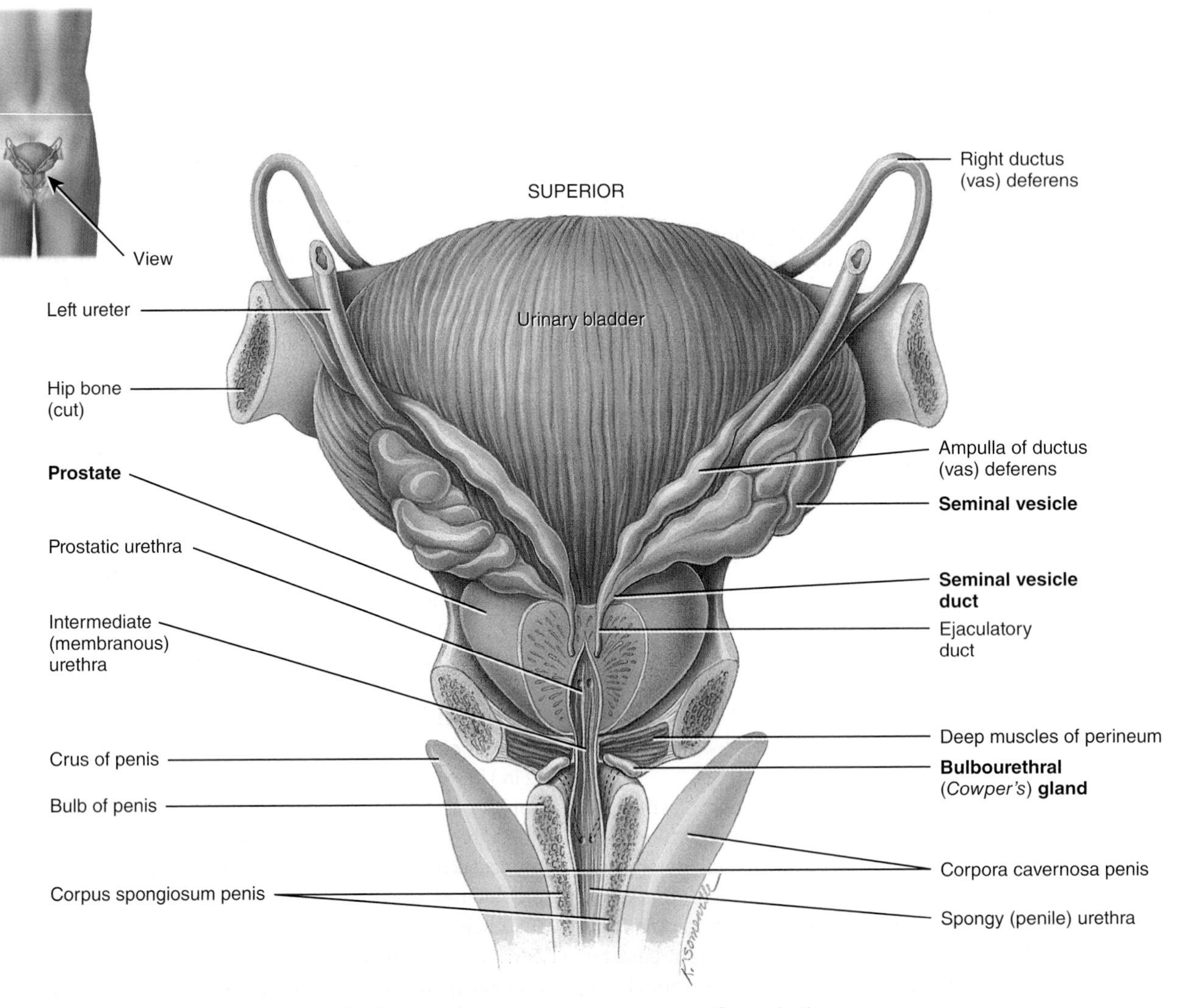

(a) Posterior view of male accessory organs of reproduction

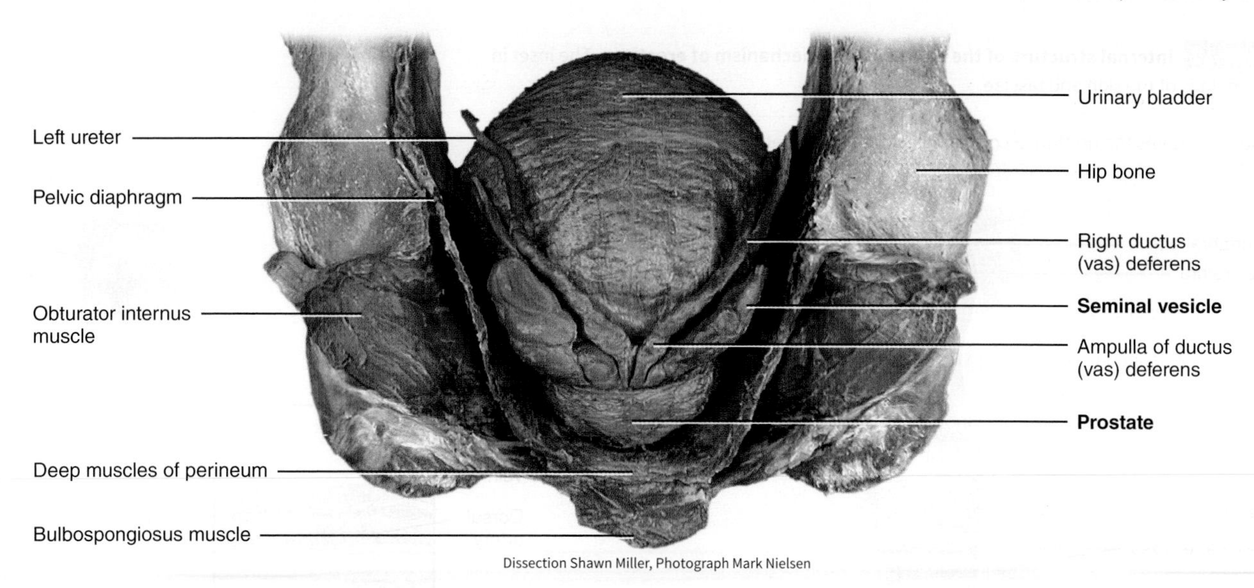

(b) Posterior view of male accessory organs of reproduction

Q What accessory sex gland contributes the majority of the seminal fluid?

(Figure 28.9). The prostate slowly increases in size from birth to puberty. It then expands rapidly until about age 30, after which time its size typically remains stable until about age 45, when further enlargement may occur, constricting the urethra and interfering with urine flow.

The prostate secretes a milky, slightly acidic fluid (pH about 6.5) that contains several substances. (1) *Citric acid* in prostatic fluid is used by sperm for ATP production via the Krebs cycle. (2) Several *proteolytic enzymes*, such as *prostate-specific antigen (PSA)*, pepsinogen, lysozyme, amylase, and hyaluronidase, eventually break down the clotting proteins from the seminal vesicles. (3) The function of the *acid phosphatase* secreted by the prostate is unknown. (4) *Seminalplasmin* in prostatic fluid is an antibiotic that can destroy bacteria. Seminalplasmin may help decrease the number of naturally occurring bacteria in semen and in the lower female reproductive tract. Secretions of the prostate enter the prostatic urethra through many prostatic ducts. Prostatic secretions make up about 25% of the volume of semen and contribute to sperm motility and viability.

Bulbourethral Glands The paired **bulbourethral glands** (bul′-bō-ū-RĒ-thral), or *Cowper's glands* (KOW-pers), are about the size of peas. They are located inferior to the prostate on either side of the membranous urethra within the deep muscles of the perineum, and their ducts open into the spongy urethra (Figure 28.9). During sexual arousal, the bulbourethral glands secrete an alkaline fluid into the urethra that protects the passing sperm by neutralizing acids from urine in the urethra. They also secrete mucus that lubricates the end of the penis and the lining of the urethra, decreasing the number of sperm damaged during ejaculation. Some males release a drop or two of this mucus upon sexual arousal and erection. The fluid does not contain sperm cells.

Semen

Semen (= seed) is a mixture of sperm and **seminal fluid**, a liquid that consists of the secretions of the seminiferous tubules, seminal vesicles, prostate, and bulbourethral glands. The volume of semen in a typical ejaculation is 2.5–5 milliliters (mL), with 50–150 million sperm per mL. When the number falls below 20 million/mL, the male is likely to be infertile. A very large number of sperm is required for successful fertilization because only a tiny fraction ever reaches the secondary oocyte, whereas too many sperm without sufficient dilution from seminal fluid results in infertility because the sperm tails tangle and lose mobility.

Despite the slight acidity of prostatic fluid, semen has a slightly alkaline pH of 7.2–7.7 due to the higher pH and larger volume of fluid from the seminal vesicles. The prostatic secretion gives semen a milky appearance, and fluids from the seminal vesicles and bulbourethral glands give it a sticky consistency. Seminal fluid provides sperm with a transportation medium, nutrients, and protection from the hostile acidic environment of the male's urethra and the female's vagina.

Once ejaculated, liquid semen coagulates within 5 minutes due to the presence of clotting proteins from the seminal vesicles. The functional role of semen coagulation is not known, but the proteins involved are different from those that cause blood coagulation. After about 10 to 20 minutes, semen reliquefies because prostate-specific antigen (PSA) and other proteolytic enzymes produced by the prostate break down the clot. Abnormal or delayed liquefaction of clotted semen may cause complete or partial immobilization of sperm, thereby inhibiting their movement through the cervix of the uterus. After passing through the uterus and uterine tube, the sperm are affected by secretions of the uterine tube in a process called **capacitation** (see Section 28.2). The presence of blood in semen is called **hemospermia** (hē-mō-SPER-mē-a; *hemo-* = blood; *-sperma* = seed). In most cases, it is caused by inflammation of the blood vessels lining the seminal vesicles; it is usually treated with antibiotics.

Penis

The **penis** (= tail) contains the urethra and is a passageway for the ejaculation of semen and the excretion of urine (Figure 28.10). It is

FIGURE 28.10 Internal structure of the penis and the mechanism of erection. The inset in (b) shows details of the skin and fasciae.

The penis contains the urethra, a common pathway for semen and urine.

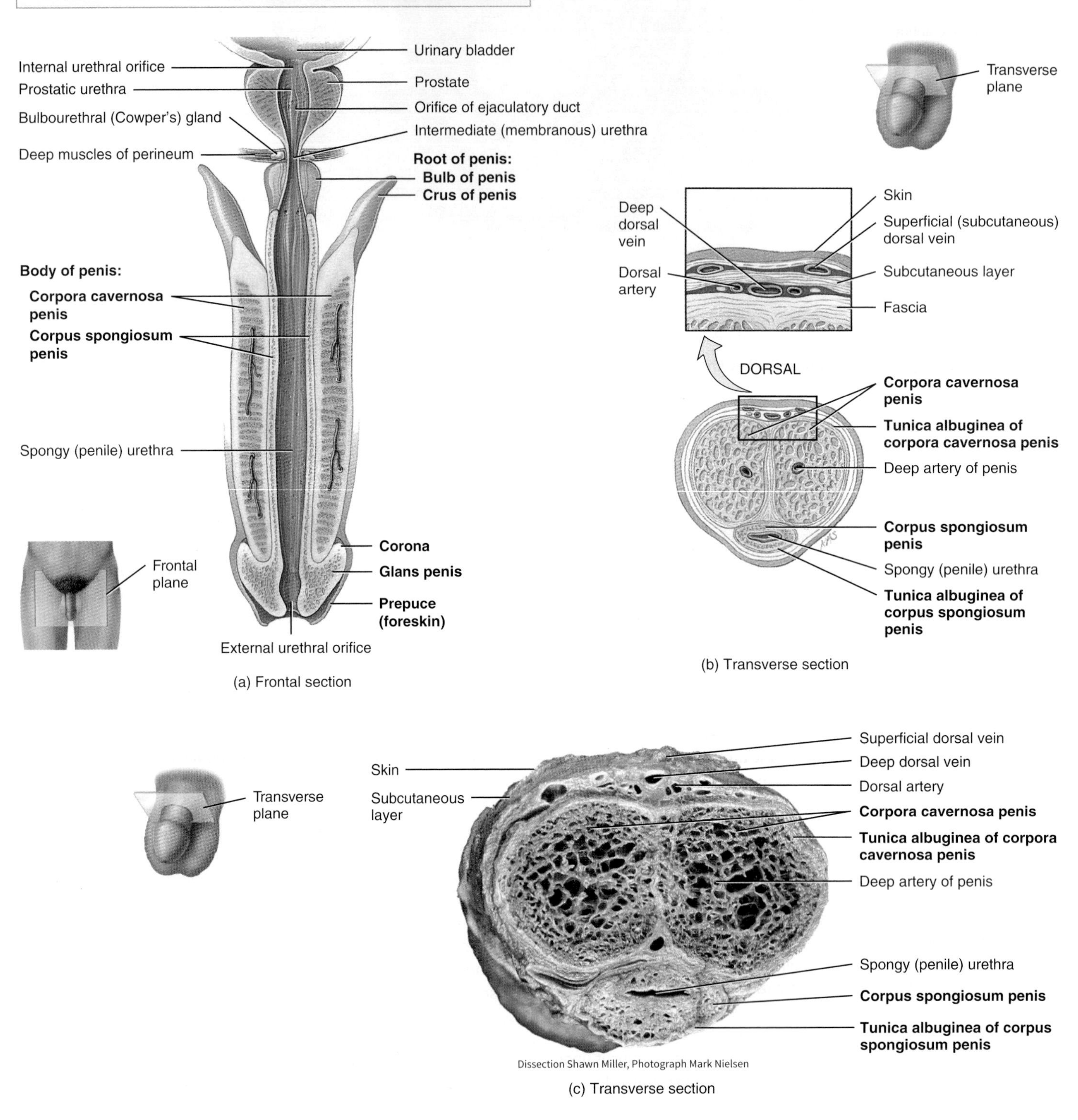

See Clinical Connection: Circumcision

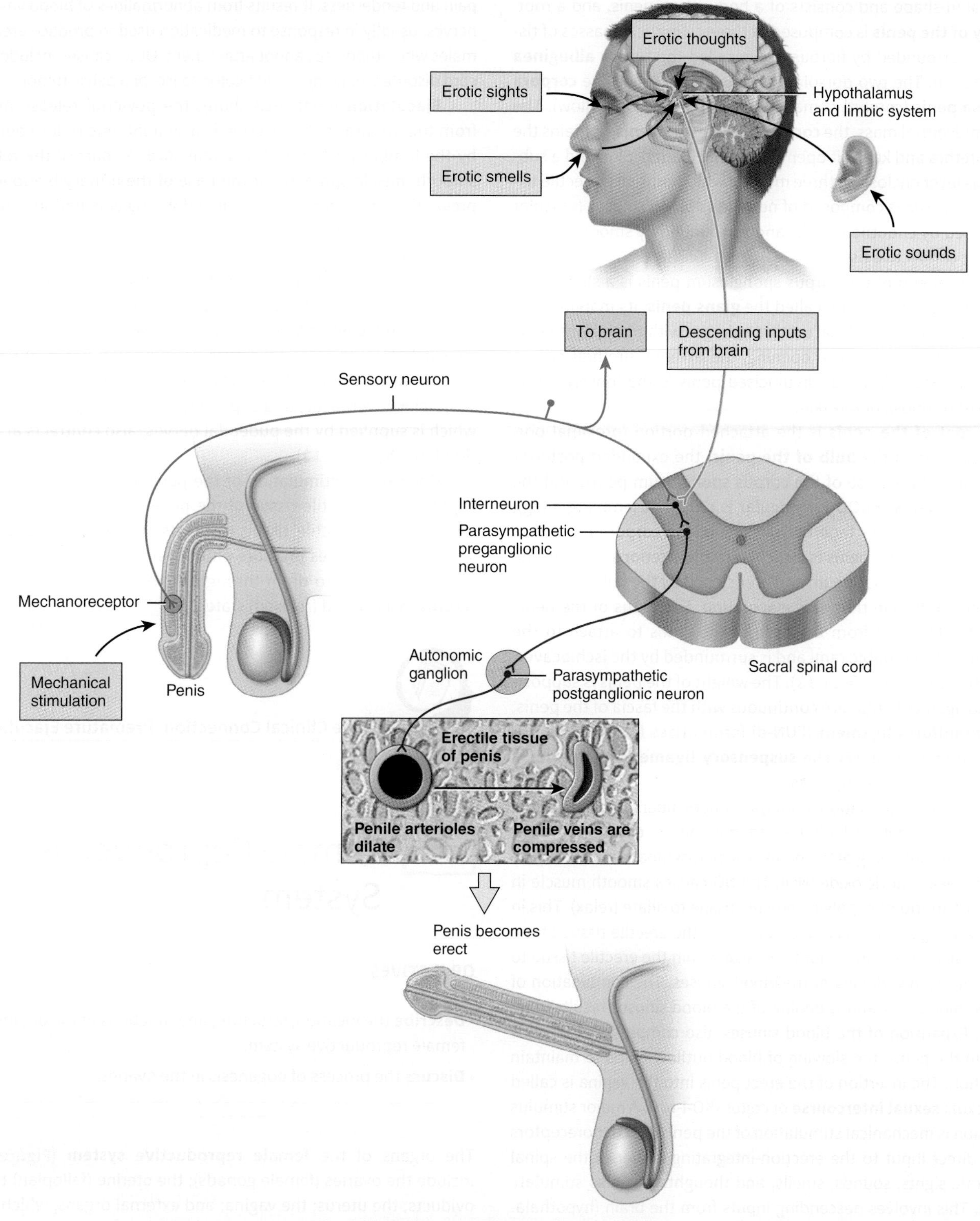

(d) Neural circuits involved in erection

Q Which tissue masses form the erectile tissue in the penis, and why do they become rigid during sexual arousal?

cylindrical in shape and consists of a body, glans penis, and a root. The **body of the penis** is composed of three cylindrical masses of tissue, each surrounded by fibrous tissue called the **tunica albuginea** (Figure 28.10). The two dorsolateral masses are called the **corpora cavernosa penis** (*corpora* = main bodies; *cavernosa* = hollow). The smaller midventral mass, the **corpus spongiosum penis**, contains the spongy urethra and keeps it open during ejaculation. Skin and a subcutaneous layer enclose all three masses, which consist of erectile tissue. *Erectile tissue* is composed of numerous blood sinuses (vascular spaces) lined by endothelial cells and surrounded by smooth muscle and elastic connective tissue.

The distal end of the corpus spongiosum penis is a slightly enlarged, acorn-shaped region called the **glans penis**; its margin is the **corona** (kō-RŌ-na). The distal urethra enlarges within the glans penis and forms a terminal slitlike opening, the external urethral orifice. Covering the glans in an uncircumcised penis is the loosely fitting **prepuce** (PRĒ-poos), or *foreskin*.

The **root of the penis** is the attached portion (proximal portion). It consists of the **bulb of the penis**, the expanded posterior continuation of the base of the corpus spongiosum penis, and the **crura of the penis** (KROO-ra; singular is *crus* = resembling a leg), the two separated and tapered portions of the corpora cavernosa penis. The bulb of the penis is attached to the inferior surface of the deep muscles of the perineum and is enclosed by the bulbospongiosus muscle, a muscle that aids ejaculation. Each crus of the penis bends laterally away from the bulb of the penis to attach to the ischial and inferior pubic rami and is surrounded by the ischiocavernosus muscle (see Figure 11.13). The weight of the penis is supported by two ligaments that are continuous with the fascia of the penis. (1) The **fundiform ligament** (FUN-di-form) arises from the inferior part of the linea alba. (2) The **suspensory ligament of the penis** arises from the pubic symphysis.

Upon sexual stimulation, parasympathetic fibers from the sacral portion of the spinal cord initiate and maintain an **erection**, the enlargement and stiffening of the penis. The parasympathetic fibers produce and release nitric oxide (NO). The NO causes smooth muscle in the walls of arterioles supplying erectile tissue to dilate (relax). This in turn causes large amounts of blood to enter the erectile tissue of the penis. NO also causes the smooth muscle within the erectile tissue to relax, resulting in widening of the blood sinuses. The combination of increased blood flow and widening of the blood sinuses results in an erection. Expansion of the blood sinuses also compresses the veins that drain the penis; the slowing of blood outflow helps to maintain the erection. The insertion of the erect penis into the vagina is called heterosexual **sexual intercourse** or *coitus* (KŌ-i-tus). A major stimulus for erection is mechanical stimulation of the penis. Mechanoreceptors provide direct input to the erection-integrating center in the spinal cord. Erotic sights, sounds, smells, and thoughts can also stimulate erection. This involves descending inputs from the brain (hypothalamus and limbic system) to the spinal cord. Negative stimuli (a bad mood, depression, anxiety, etc.) can also inhibit erection through these descending pathways.

The term **priapism** (PRĪ-a-pizm) refers to a persistent and usually painful erection of the penis that does not involve sexual desire or excitement. The condition may last up to several hours and is accompanied by pain and tenderness. It results from abnormalities of blood vessels and nerves, usually in response to medication used to produce erections in males who otherwise cannot attain them. Other causes include a spinal cord disorder, leukemia, sickle-cell disease, or a pelvic tumor.

Ejaculation (ē-jak-ū-LĀ-shun), the powerful release of semen from the urethra to the exterior, is a sympathetic reflex coordinated by the lumbar portion of the spinal cord. As part of the reflex, the smooth muscle sphincter at the base of the urinary bladder closes, preventing urine from being expelled during ejaculation, and semen from entering the urinary bladder. Even before ejaculation occurs, peristaltic contractions in the epididymis, ductus (vas) deferens, seminal vesicles, ejaculatory ducts, and prostate propel semen into the penile portion of the urethra (spongy urethra). Typically, this leads to **emission** (ē-MISH-un), the discharge of a small volume of semen before ejaculation. Emission may also occur during sleep (nocturnal emission). The musculature of the penis (bulbospongiosus, ischiocavernosus, and superficial transverse perineal muscles), which is supplied by the pudendal nerves, also contracts at ejaculation (see Figure 11.13).

Once sexual stimulation of the penis has ended, the arterioles supplying the erectile tissue of the penis constrict and the smooth muscle within erectile tissue contracts, making the blood sinuses smaller. This relieves pressure on the veins supplying the penis and allows the blood to drain through them. Consequently, the penis returns to its flaccid (relaxed) state.

See Clinical Connection: Premature Ejaculation

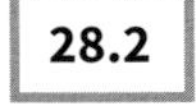

28.2 Female Reproductive System

OBJECTIVES

- **Describe** the location, structure, and functions of the organs of the female reproductive system.
- **Discuss** the process of oogenesis in the ovaries.

The organs of the **female reproductive system** (Figure 28.11) include the ovaries (female gonads); the uterine (fallopian) tubes, or oviducts; the uterus; the vagina; and external organs, which are collectively called the vulva, or pudendum. The mammary glands are considered part of both the integumentary system and the female reproductive system. **Gynecology** (gī-ne-KOL-ō-jē; *gyneco-* = woman; *-logy* = study of) is the specialized branch of medicine concerned with the diagnosis and treatment of diseases of the female reproductive system.

Ovaries

The **ovaries** (= egg receptacles), which are the female gonads, are paired glands that resemble unshelled almonds in size and shape; they are homologous to the testes. (Here *homologous* means that two organs have the same embryonic origin.) The ovaries produce (1) gametes, secondary oocytes that develop into mature ova (eggs) after fertilization, and (2) hormones, including progesterone and estrogens (the female sex hormones), inhibin, and relaxin.

The ovaries, one on either side of the uterus, descend to the brim of the superior portion of the pelvic cavity during the third month of development. A series of ligaments holds them in position (**Figure 28.12**). The **broad ligament** of the uterus, which is a fold of the parietal peritoneum, attaches to the ovaries by a double-layered fold

FIGURE 28.11 Female organs of reproduction and surrounding structures.

The organs of reproduction in females include the ovaries, uterine (fallopian) tubes, uterus, vagina, vulva, and mammary glands.

Functions of the Female Reproductive System

1. The ovaries produce secondary oocytes and hormones, including progesterone and estrogens (female sex hormones), inhibin, and relaxin.
2. The uterine tubes transport a secondary oocyte to the uterus and normally are the sites where fertilization occurs.
3. The uterus is the site of implantation of a fertilized ovum, development of the fetus during pregnancy, and labor.
4. The vagina receives the penis during sexual intercourse and is a passageway for childbirth.
5. The mammary glands synthesize, secrete, and eject milk for nourishment of the newborn.

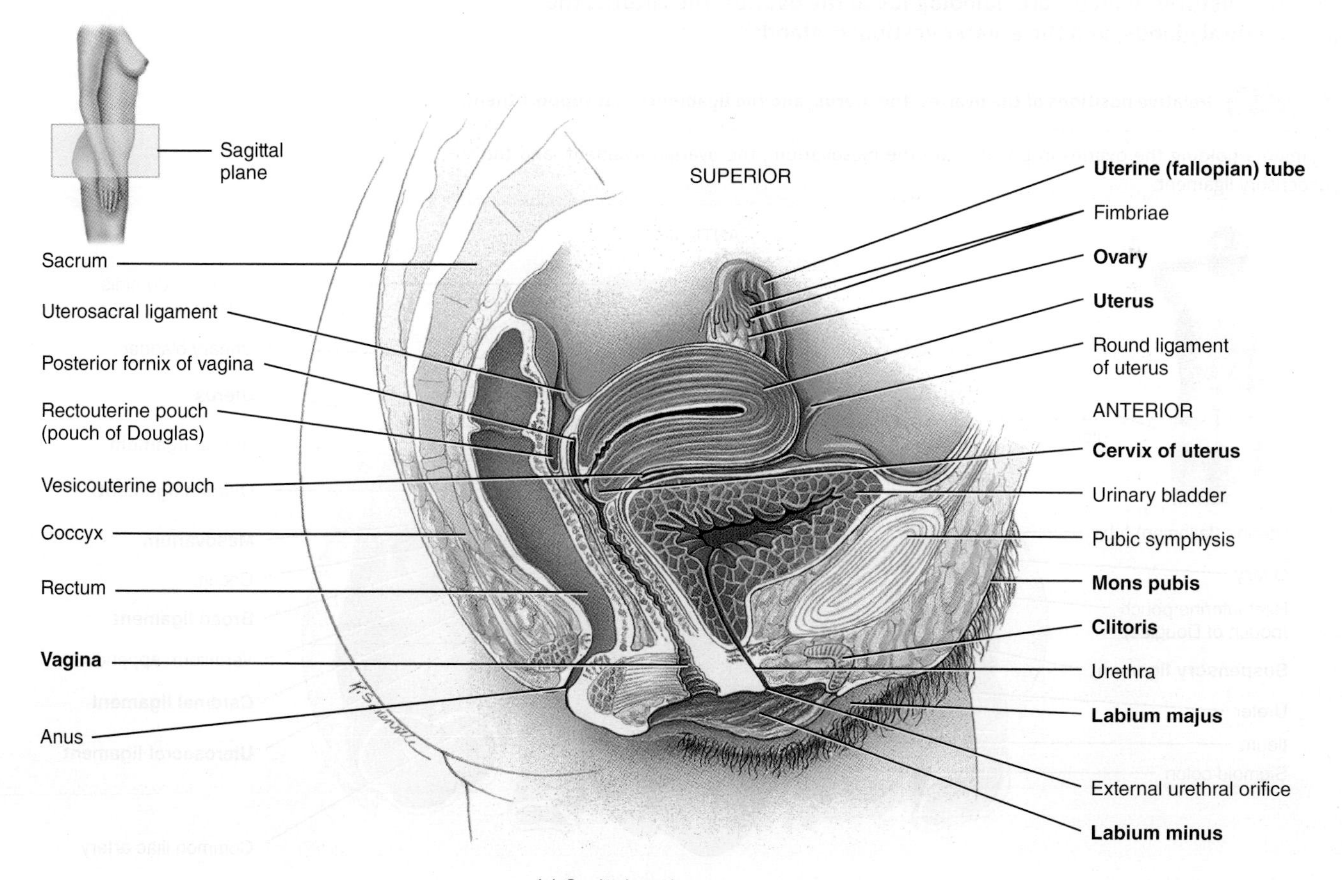

(a) Sagittal section

Figure 28.11 *Continues*

FIGURE 28.11 Continued

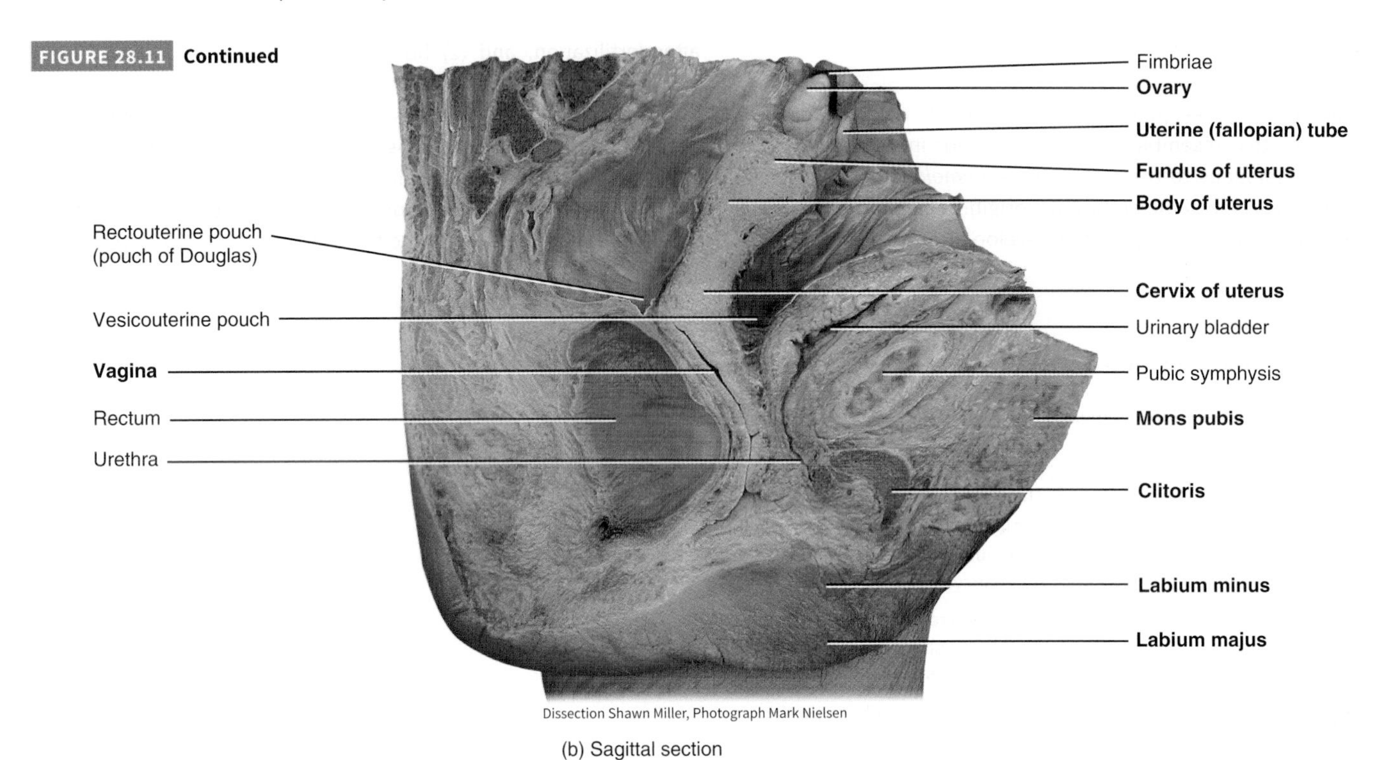

(b) Sagittal section

Q Which structures in males are homologous to the ovaries, the clitoris, the paraurethral glands, and the greater vestibular glands?

FIGURE 28.12 Relative positions of the ovaries, the uterus, and the ligaments that support them.

Ligaments holding the ovaries in position are the mesovarium, the ovarian ligament, and the suspensory ligament.

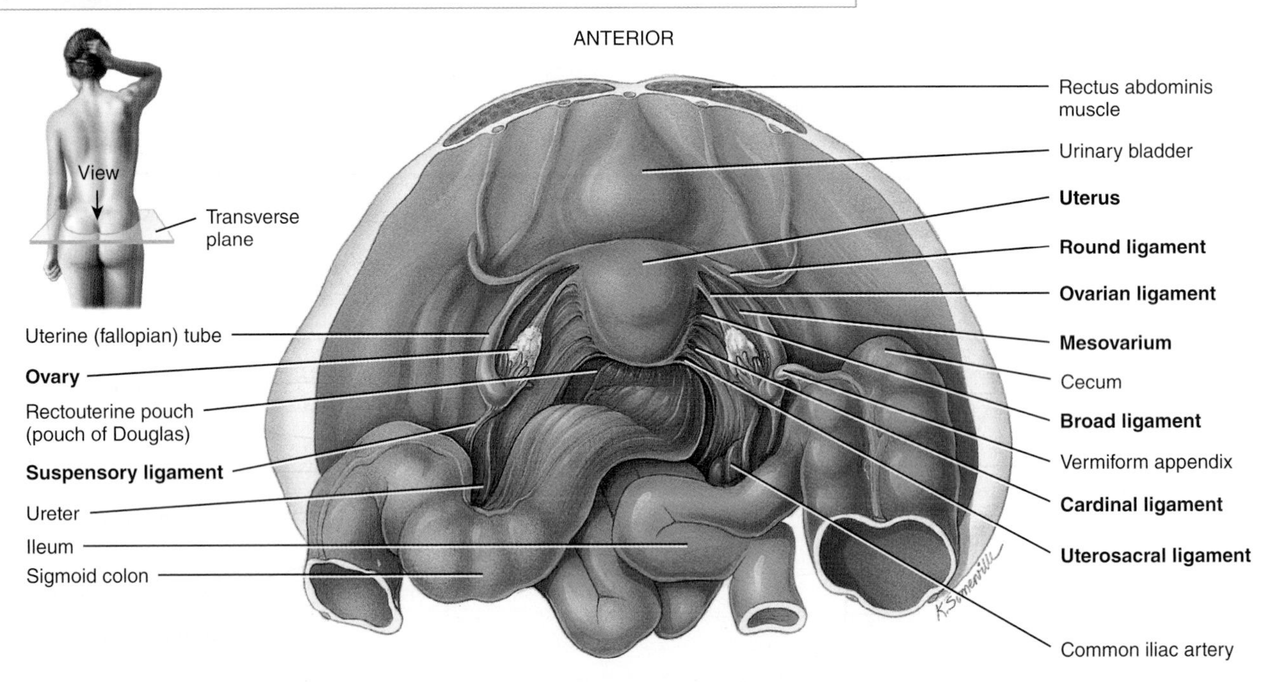

Superior view of transverse section

Q To which structures do the mesovarium, ovarian ligament, and suspensory ligament anchor the ovary?

of peritoneum called the **mesovarium** (mez′-ō-VĀ-rē-um). The **ovarian ligament** anchors the ovaries to the uterus, and the **suspensory ligament** attaches them to the pelvic wall. Each ovary contains a **hilum** (HĪ-lum), the point of entrance and exit for blood vessels and nerves along which the mesovarium is attached.

Histology of the Ovary

Each ovary consists of the following parts (Figure 28.13):

- The **ovarian mesothelium** or *surface epithelium* is a layer of simple epithelium (low cuboidal or squamous) that covers the surface of the ovary.
- The **tunica albuginea** is a whitish capsule of dense irregular connective tissue located immediately deep to the ovarian mesothelium.
- The **ovarian cortex** is a region just deep to the tunica albuginea. It consists of ovarian follicles (described shortly) surrounded by dense irregular connective tissue that contains collagen fibers and fibroblast-like cells called *stromal cells.*
- The **ovarian medulla** is deep to the ovarian cortex. The border between the cortex and medulla is indistinct, but the medulla consists of more loosely arranged connective tissue and contains blood vessels, lymphatic vessels, and nerves.
- **Ovarian follicles** (*folliculus* = little bag) are in the cortex and consist of **oocytes** (Ō-ō-sīts) in various stages of development, plus the cells surrounding them. When the surrounding cells form a single layer, they are called **follicular cells** (fo-LIK-ū-lar); later in development, when they form several layers, they are referred to as **granulosa cells** (gran′-u-LŌ-sa). The surrounding cells nourish the developing oocyte and begin to secrete estrogens as the follicle grows larger.
- A **mature** (*graafian*) **follicle** (GRĀ-fē-an) is a large, fluid-filled follicle that is ready to rupture and expel its secondary oocyte, a process known as **ovulation** (ov′-ū-LĀ-shun).
- A **corpus luteum** (= yellow body) contains the remnants of a mature follicle after ovulation. The corpus luteum produces progesterone, estrogens, relaxin, and inhibin until it degenerates into fibrous scar tissue called the **corpus albicans** (AL-bi-kanz = white body).

Oogenesis and Follicular Development

The formation of gametes in the ovaries is termed **oogenesis** (ō-ō-JEN-e-sis; *oo-* = egg). In contrast to spermatogenesis, which begins in males at puberty, oogenesis begins in females before they are even born. Oogenesis occurs in essentially the same manner as spermatogenesis; meiosis (see Chapter 3) takes place and the resulting germ cells undergo maturation.

During early fetal development, primordial (primitive) germ cells migrate from the yolk sac to the ovaries. There, germ cells differentiate within the ovaries into **oogonia** (ō-ō-GŌ-nē-a; singular is *oogonium*). Oogonia are diploid (2*n*) stem cells that divide mitotically to produce millions of germ cells. Even before birth, most of these germ cells degenerate in a process known as **atresia** (a-TRĒ-zē-a). A few, however, develop into larger cells called **primary oocytes** that enter prophase of meiosis I during fetal development but do not complete that phase until after puberty. During this arrested stage of development, each primary oocyte is surrounded by a single layer of flat follicular cells, and the entire structure is called a **primordial follicle** (Figure 28.14a). The ovarian cortex surrounding the primordial follicles consists of collagen fibers and fibroblast-like **stromal cells**. At birth, approximately 200,000 to 2,000,000 primary oocytes remain in each ovary. Of these, about 40,000 are still present at puberty, and around 400 will mature and ovulate during a woman's reproductive lifetime. The remainder of the primary oocytes undergo atresia.

Each month after puberty until menopause, gonadotropins (FSH and LH) secreted by the anterior pituitary further stimulate the development of several primordial follicles, although only one will typically reach the maturity needed for ovulation. A few primordial follicles start to grow, developing into **primary follicles** (Figure 28.14b). Each primary follicle consists of a primary oocyte that is surrounded in a later stage of development by several layers of cuboidal and low-columnar cells called granulosa cells. The outermost granulosa cells rest on a basement membrane. As the primary follicle grows, it forms a clear glycoprotein layer called the **zona pellucida** (pe-LOO-si-da) between the primary oocyte and the granulosa cells. In addition, stromal cells surrounding the basement membrane begin to form an organized layer called the **theca folliculi** (THĒ-ka fo-LIK-ū-lī).

With continuing maturation, a primary follicle develops into a secondary follicle (Figure 28.14c). In a **secondary follicle**, the theca differentiates into two layers: (1) the **theca interna**, a highly vascularized internal layer of cuboidal secretory cells that secrete estrogens, and (2) the **theca externa**, an outer layer of stromal cells and collagen fibers. In addition, the granulosa cells begin to secrete follicular fluid, which builds up in a cavity called the **antrum** in the center of the secondary follicle. The innermost layer of granulosa cells becomes firmly attached to the zona pellucida and is now called the **corona radiata** (*corona* = crown; *radiata* = radiation) (Figure 28.14c).

The secondary follicle eventually becomes larger, turning into a mature (graafian) follicle (Figure 28.14d). While in this follicle, and just before ovulation, the diploid primary oocyte completes meiosis I, producing two haploid (*n*) cells of unequal size—each with 23 chromosomes (Figure 28.15). The smaller cell produced by meiosis I, called the **first polar body**, is essentially a packet of discarded nuclear material. The larger cell, known as the **secondary oocyte**, receives most of the cytoplasm. Once a secondary oocyte is formed, it begins meiosis II but then stops in metaphase. The mature (graafian) follicle soon ruptures and releases its secondary oocyte, a process known as ovulation.

At ovulation, the secondary oocyte is expelled into the pelvic cavity together with the first polar body and corona radiata. Normally these cells are swept into the uterine tube. If fertilization does not occur, the cells degenerate. If sperm are present in the uterine tube and one penetrates the secondary oocyte, however, meiosis II resumes. The secondary oocyte splits into two haploid cells, again of unequal size. The larger cell is the **ovum**, or mature egg; the smaller one is the **second polar body**. The nuclei of the sperm cell and the ovum then unite, forming a diploid **zygote**. If the first polar body undergoes another division to produce two polar bodies, then the primary oocyte ultimately gives rise to three haploid polar bodies,

FIGURE 28.13 Histology of the ovary. The arrows indicate the sequence of developmental stages that occur as part of the maturation of an ovum during the ovarian cycle.

The ovaries are the female gonads; they produce haploid oocytes.

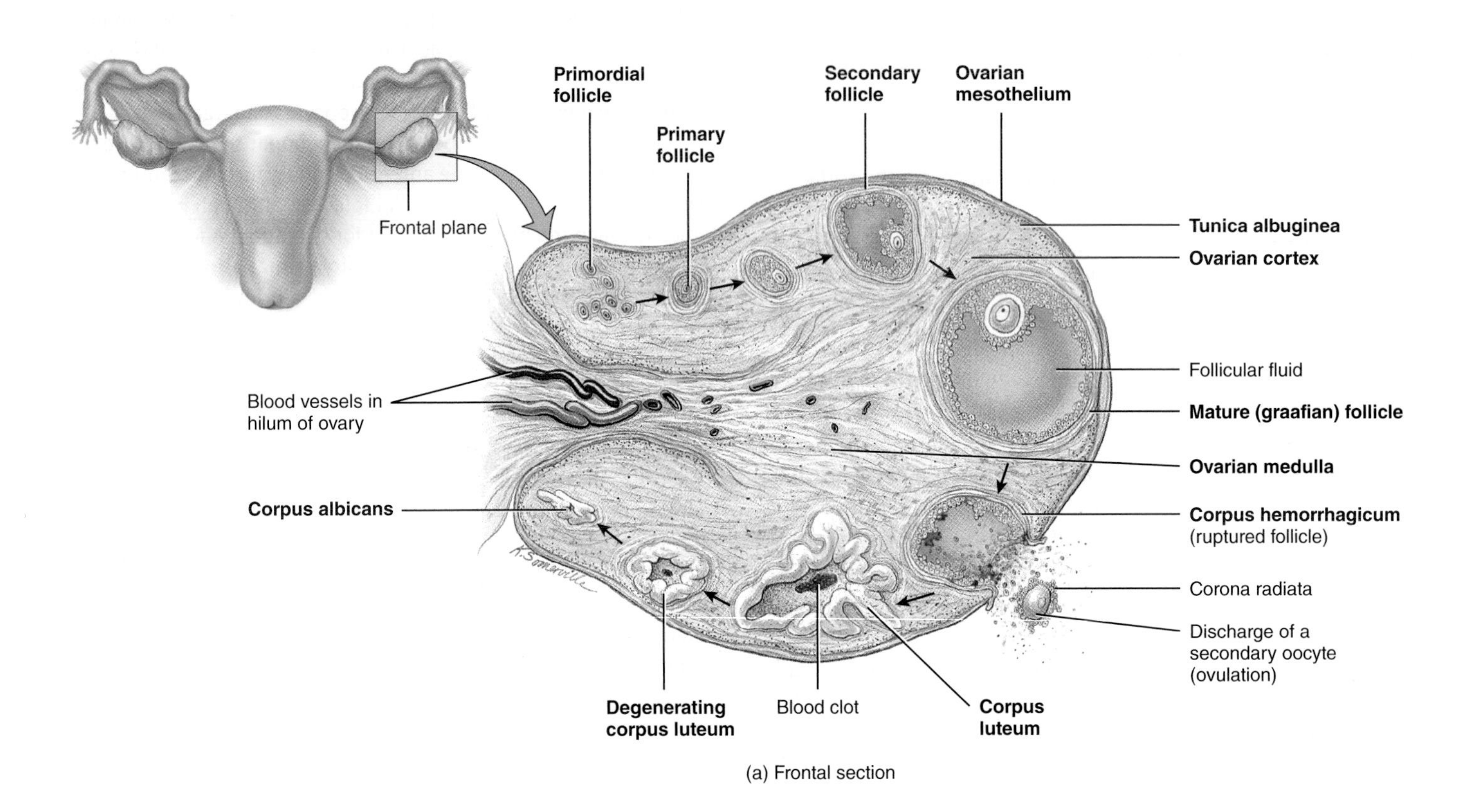

(a) Frontal section

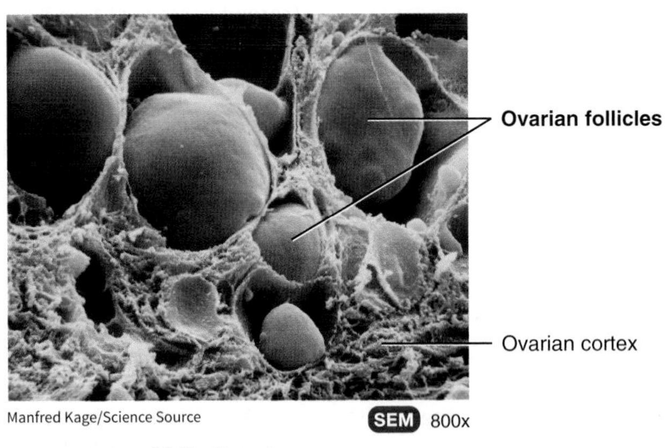

Manfred Kage/Science Source

(b) Section of ovary

Q What structures in the ovary contain endocrine tissue, and what hormones do they secrete?

FIGURE 28.14 Ovarian follicles.

As an ovarian follicle enlarges, follicular fluid accumulates in a cavity called the antrum.

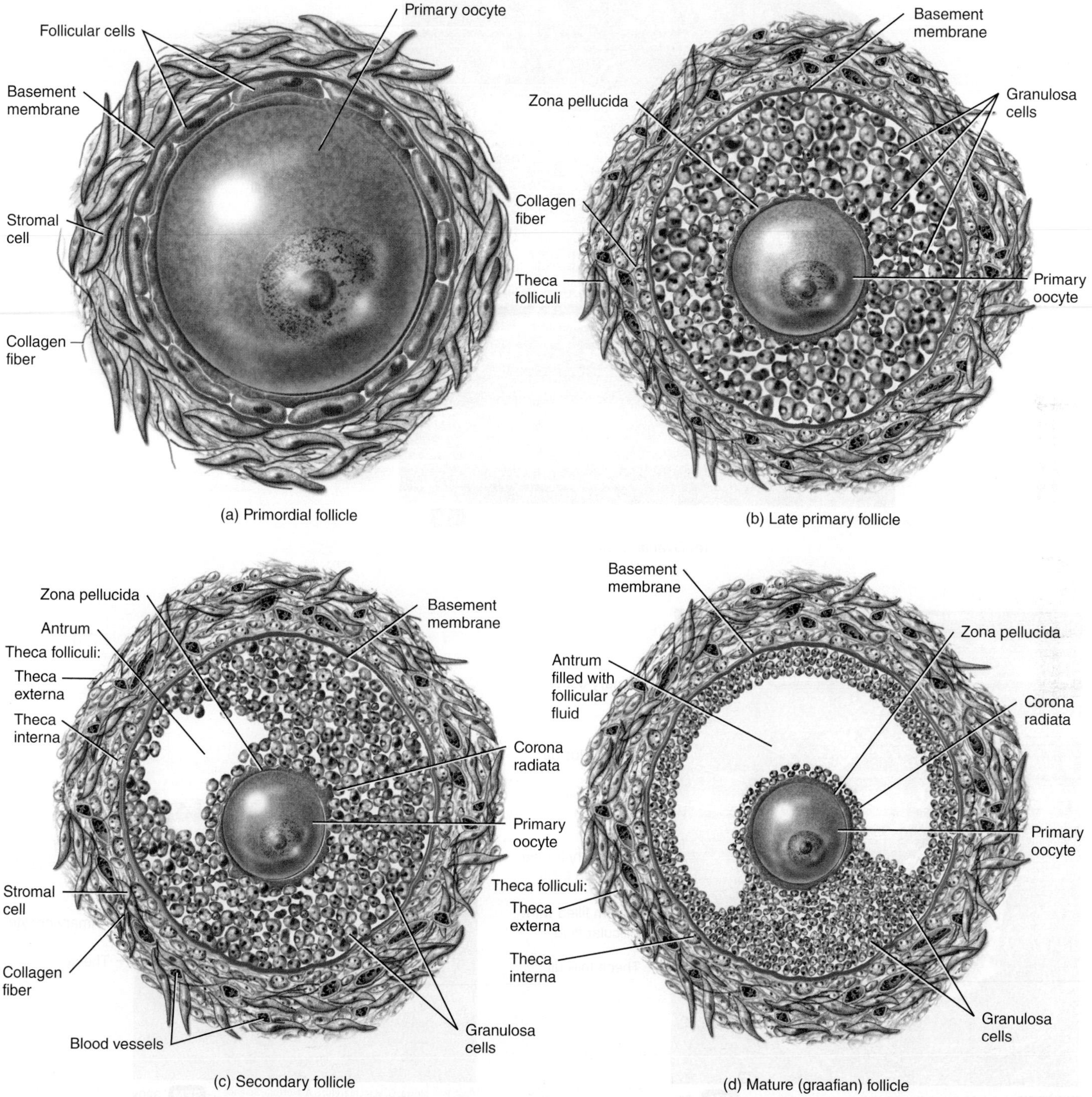

(a) Primordial follicle

(b) Late primary follicle

(c) Secondary follicle

(d) Mature (graafian) follicle

Figure 28.14 ***Continues***

FIGURE 28.14 Continued

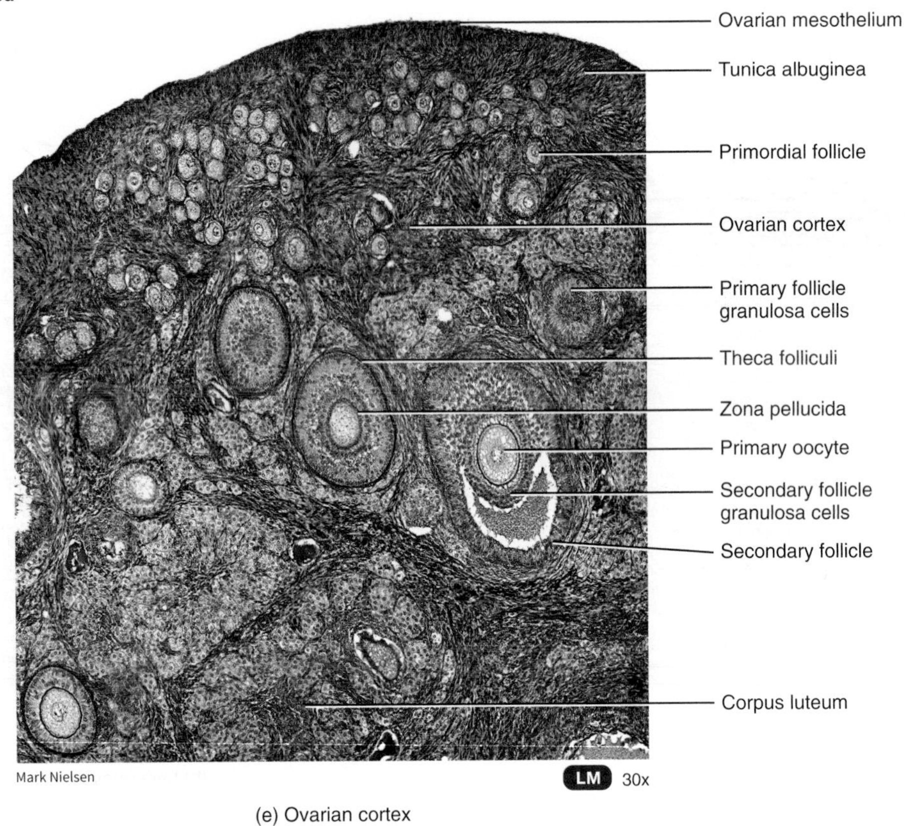

(e) Ovarian cortex

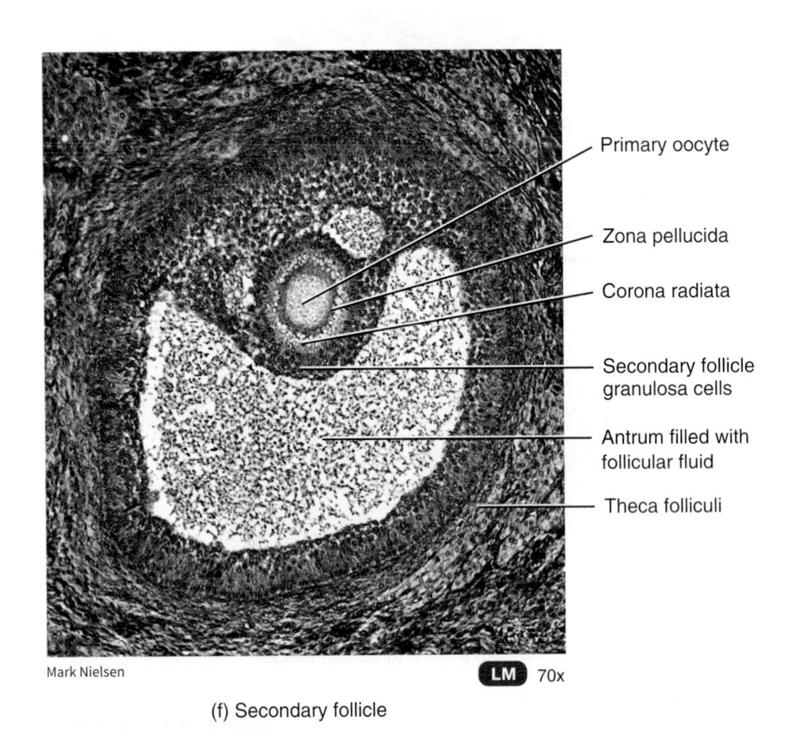

(f) Secondary follicle

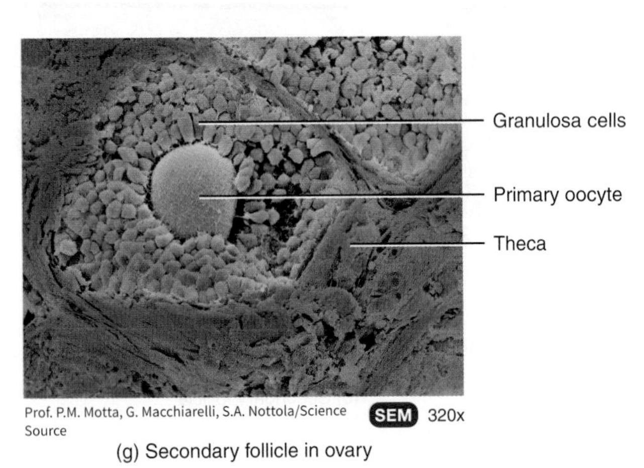

(g) Secondary follicle in ovary

Q What happens to most ovarian follicles?

FIGURE 28.15 Oogenesis. Diploid cells (2*n*) have 46 chromosomes; haploid cells (*n*) have 23 chromosomes.

In a secondary oocyte, meiosis II is completed only if fertilization occurs.

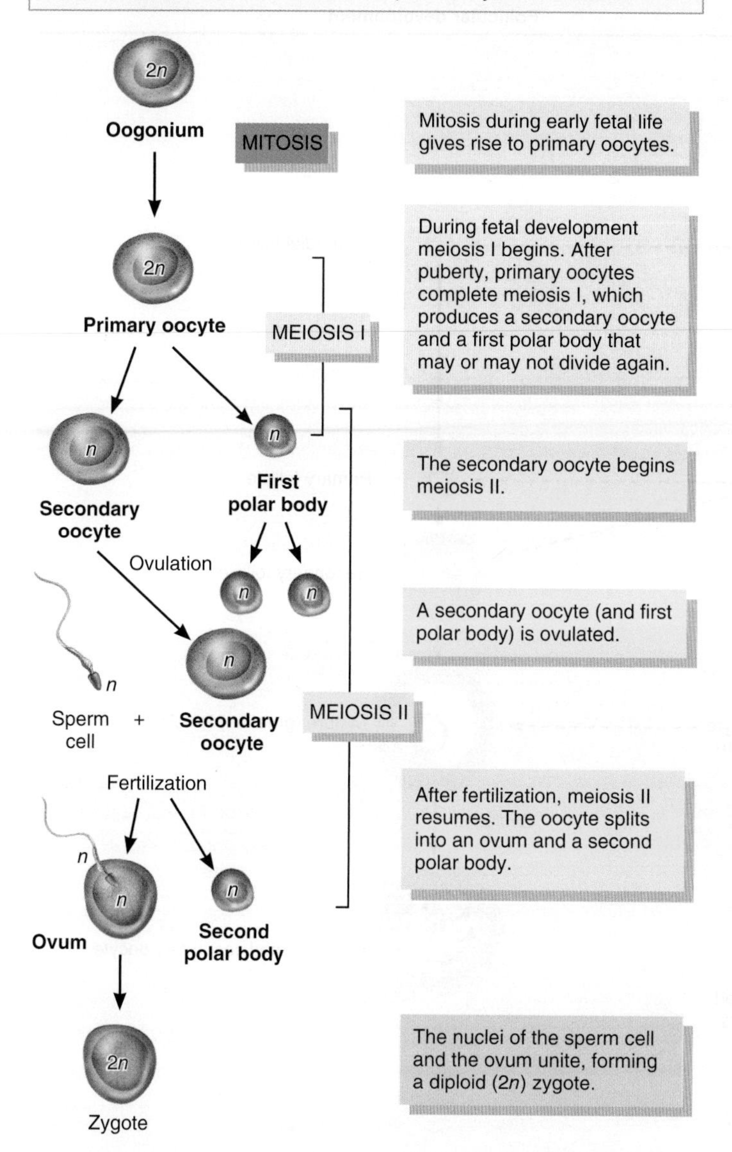

Q How does the age of a primary oocyte in a female compare with the age of a primary spermatocyte in a male?

which all degenerate, and a single haploid ovum. Thus, one primary oocyte gives rise to a single gamete (an ovum). By contrast, recall that in males one primary spermatocyte produces four gametes (sperm).

Table 28.1 summarizes the events of oogenesis and follicular development.

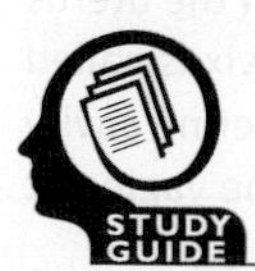

See Clinical Connection: Ovarian Cysts

Uterine Tubes

Females have two **uterine tubes**, also called *fallopian tubes* or *oviducts*, that extend laterally from the uterus (Figure 28.16). The tubes, which measure about 10 cm (4 in.) long, lie within the folds of the broad ligaments of the uterus. They provide a route for sperm to reach an ovum and transport secondary oocytes and fertilized ova from the ovaries to the uterus. The funnel-shaped portion of each tube, called the **infundibulum** (in-fun-DIB-ū-lum), is close to the ovary but is open to the pelvic cavity. It ends in a fringe of fingerlike projections called **fimbriae** (FIM-brē-ē = fringe), one of which is attached to the lateral end of the ovary. From the infundibulum, the uterine tube extends medially and eventually inferiorly and attaches to the superior lateral angle of the uterus. The **ampulla** (am-PUL-la) of the uterine tube is the widest, longest portion, making up about the lateral two-thirds of its length. The **isthmus** (IS-mus) of the uterine tube is the more medial, short, narrow, thick-walled portion that joins the uterus.

Histologically, the uterine tubes are composed of three layers: mucosa, muscularis, and serosa. The mucosa consists of epithelium and lamina propria (areolar connective tissue). The epithelium contains ciliated simple columnar cells, which function as a "ciliary conveyor belt" to help move a fertilized ovum (or secondary oocyte) within the uterine tube toward the uterus, and nonciliated cells called **peg cells**, which have microvilli and secrete a fluid that provides nutrition for the ovum (Figure 28.17). The middle layer, the muscularis, is composed of an inner, thick, circular ring of smooth muscle and an outer, thin region of longitudinal smooth muscle. Peristaltic contractions of the muscularis and the ciliary action of the mucosa help move the oocyte or fertilized ovum toward the uterus. The outer layer of the uterine tubes is a serous membrane, the serosa.

After ovulation, local currents are produced by movements of the fimbriae, which surround the surface of the mature follicle just before ovulation occurs. These currents sweep the ovulated secondary oocyte from the peritoneal cavity into the uterine tube. A sperm cell usually encounters and fertilizes a secondary oocyte in the ampulla of the uterine tube, although fertilization in the peritoneal cavity is not uncommon. Fertilization can occur up to about 24 hours after ovulation. Some hours after fertilization, the nuclear materials of the haploid ovum and sperm unite. The diploid fertilized ovum is now called a **zygote** and begins to undergo cell divisions while moving toward the uterus. It arrives in the uterus 6 to 7 days after ovulation. Unfertilized secondary oocytes disintegrate.

Uterus

The **uterus** (womb) serves as part of the pathway for sperm deposited in the vagina to reach the uterine tubes. It is also the site of implantation of a fertilized ovum, development of the fetus during pregnancy, and labor. During reproductive cycles when implantation does not occur, the uterus is the source of menstrual flow.

Anatomy of the Uterus Situated between the urinary bladder and the rectum, the uterus is the size and shape of an inverted pear (see Figure 28.16). In females who have never been pregnant, it

TABLE 28.1 Summary of Oogenesis and Follicular Developments

AGE	OOGENESIS	FOLLICULAR DEVELOPMENT

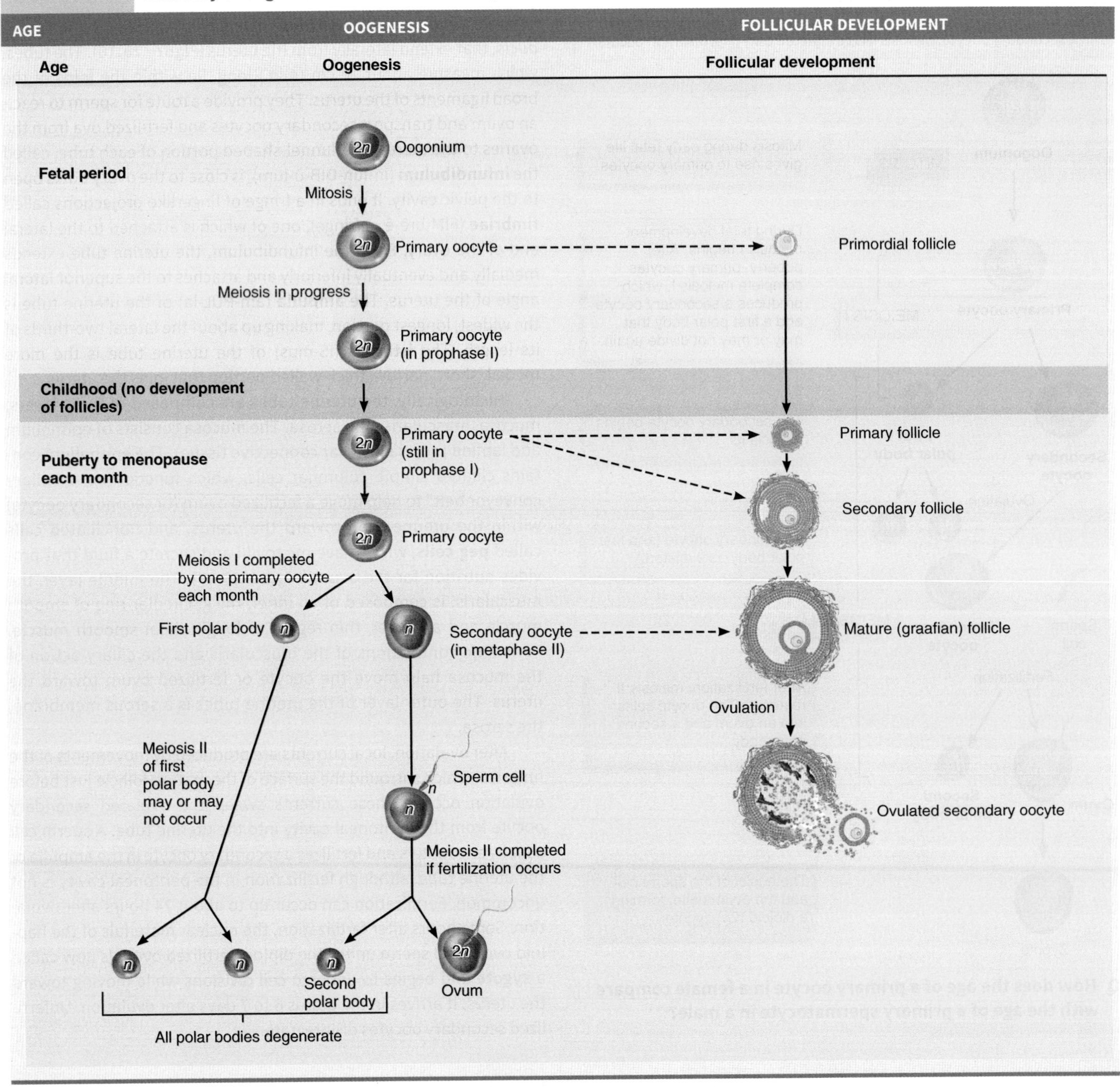

is about 7.5 cm (3 in.) long, 5 cm (2 in.) wide, and 2.5 cm (1 in.) thick. The uterus is larger in females who have recently been pregnant, and smaller (atrophied) when sex hormone levels are low, as occurs after menopause.

Anatomical subdivisions of the uterus include (1) a dome-shaped portion superior to the uterine tubes called the **fundus**, (2) a tapering central portion called the **body**, and (3) an inferior narrow portion called the **cervix** that opens into the vagina. Between the body of the uterus and the cervix is the **isthmus**, a constricted region about 1 cm (0.5 in.) long. The interior of the body of the uterus is called the **uterine cavity**, and the interior of the cervix is called the **cervical canal**. The cervical canal opens into the uterine cavity at the **internal os** (*os* = mouthlike opening) and into the vagina at the **external os**.

FIGURE 28.16 Relationship of the uterine tubes to the ovaries, uterus, and associated structures. In the left side of the drawing, the uterine tube and uterus have been sectioned to show internal structures.

After ovulation, a secondary oocyte and its corona radiata move from the pelvic cavity into the infundibulum of the uterine tube. The uterus is the site of menstruation, implantation of a fertilized ovum, development of the fetus, and labor.

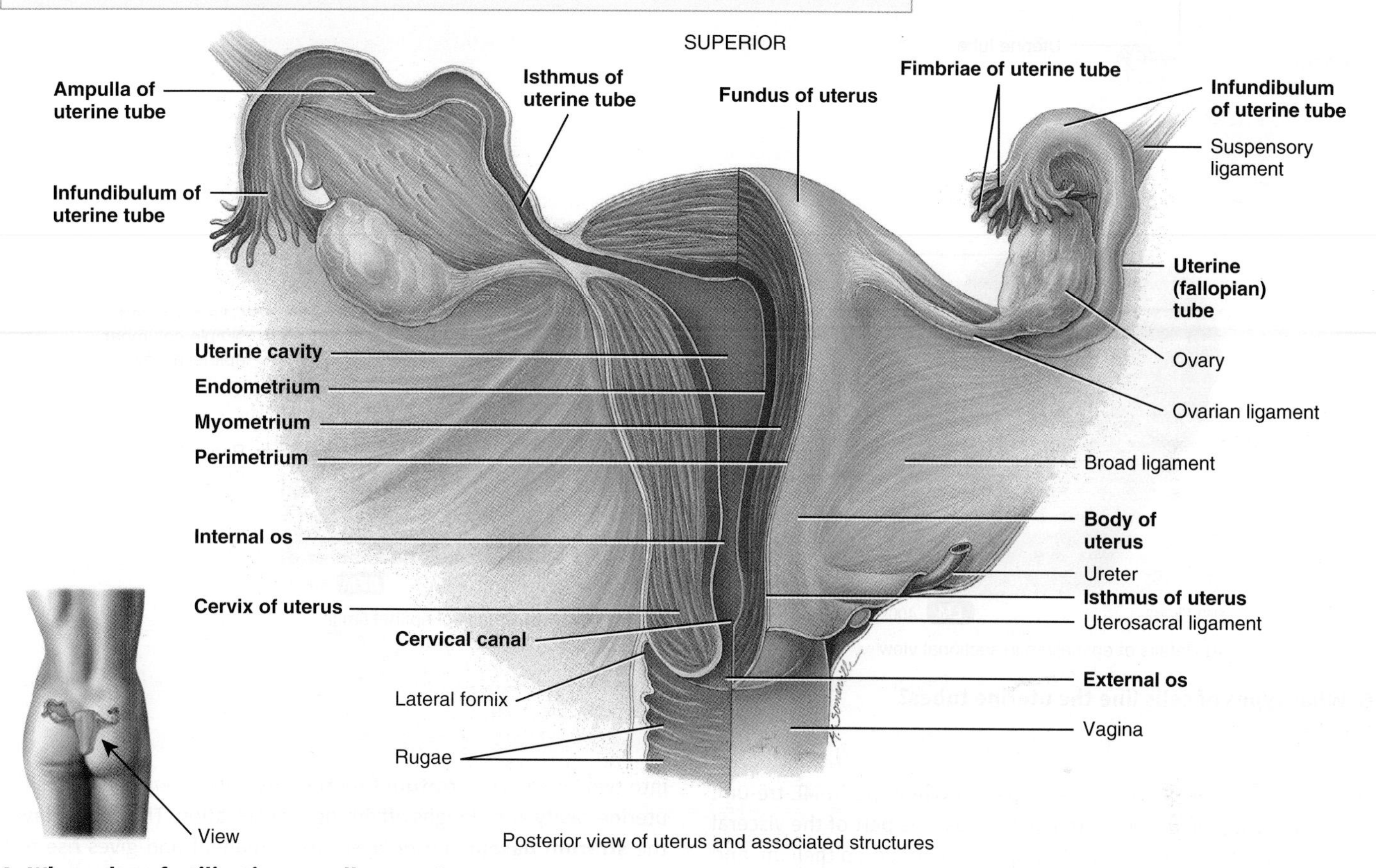

Q Where does fertilization usually occur?

Normally, the body of the uterus projects anteriorly and superiorly over the urinary bladder in a position called **anteflexion** (an′-te-FLEK-shun; *ante-* = before). The cervix projects inferiorly and posteriorly and enters the anterior wall of the vagina at nearly a right angle (see Figure 28.11). Several ligaments that are either extensions of the parietal peritoneum or fibromuscular cords maintain the position of the uterus (see Figure 28.12). The paired **broad ligaments** are double folds of peritoneum attaching the uterus to either side of the pelvic cavity. The paired **uterosacral ligaments** (ū′-ter-ō-SĀ-kral), also peritoneal extensions, lie on either side of the rectum and connect the uterus to the sacrum. The **cardinal** (*lateral cervical*) **ligaments** are located inferior to the bases of the broad ligaments and extend from the pelvic wall to the cervix and vagina. The **round ligaments** are bands of fibrous connective tissue between the layers of the broad ligament; they extend from a point on the uterus just inferior to the uterine tubes to a portion of the labia majora of the external genitalia. Although the ligaments normally maintain the anteflexed position of the uterus, they also allow the uterine body enough movement such that the uterus may become malpositioned. A posterior tilting of the uterus, called **retroflexion** (RET-rō-flek-shun; *retro-* = backward or behind), is a harmless variation of the normal position of the uterus. There is often no cause for the condition, but it may occur after childbirth.

See Clinical Connection: Uterine Prolapse

Histology of the Uterus Histologically, the uterus consists of three layers of tissue: perimetrium, myometrium, and endometrium

FIGURE 28.17 **Histology of the uterine tube.**

Peristaltic contractions of the muscularis and ciliary action of the mucosa of the uterine tube help move the oocyte or fertilized ovum toward the uterus.

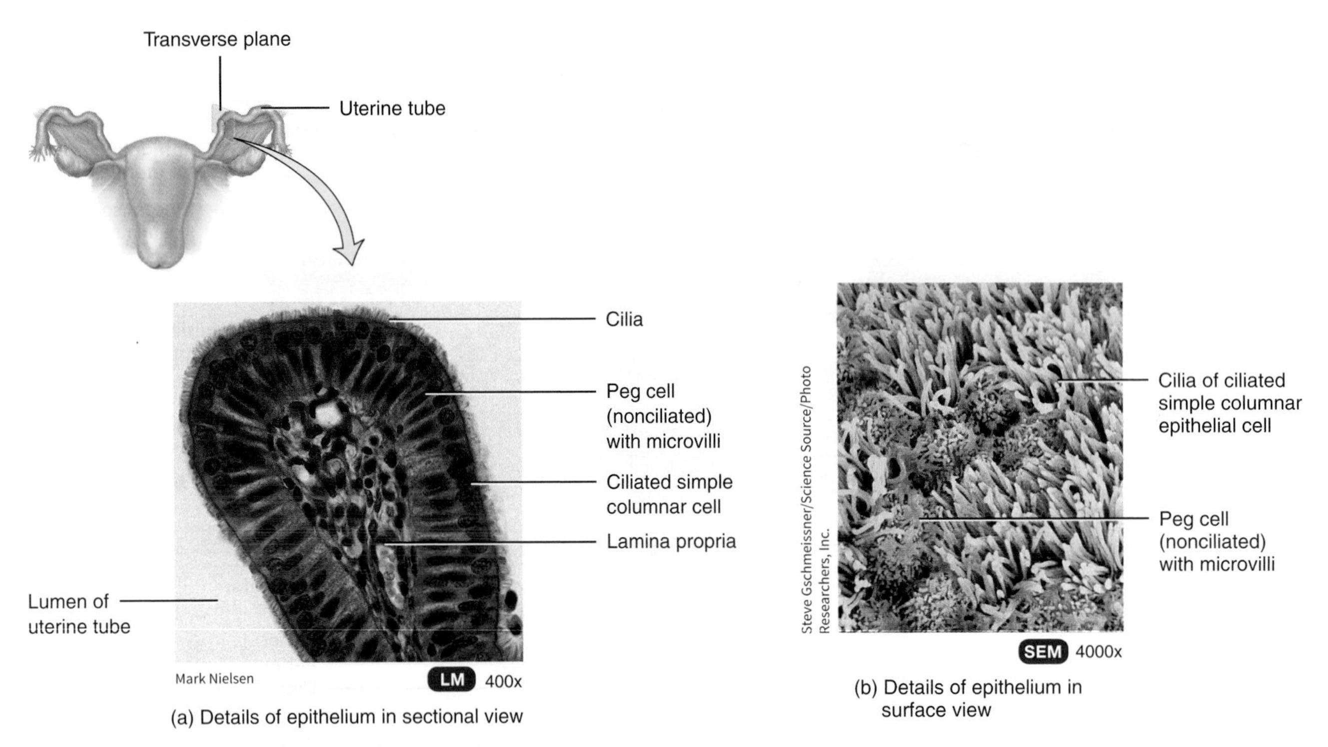

(a) Details of epithelium in sectional view

(b) Details of epithelium in surface view

Q What types of cells line the uterine tubes?

(Figure 28.18). The outer layer—the **perimetrium** (per′-i-MĒ-trē-um; *peri-* = around; *-metrium* = uterus) or serosa—is part of the visceral peritoneum; it is composed of simple squamous epithelium and areolar connective tissue. Laterally, it becomes the broad ligament. Anteriorly, it covers the urinary bladder and forms a shallow pouch, the **vesicouterine pouch** (ves′-i-kō-Ū -ter-in; *vesico-* = bladder; see Figure 28.11). Posteriorly, it covers the rectum and forms a deep pouch between the uterus and rectum, the **rectouterine pouch** (rek-tō-Ū-ter-in; *recto-* = rletum) or *pouch of Douglas*—the most inferior point in the pelvic cavity.

The middle layer of the uterus, the **myometrium** (*myo-* = muscle), consists of three layers of smooth muscle fibers that are thickest in the fundus and thinnest in the cervix. The thicker middle layer is circular; the inner and outer layers are longitudinal or oblique. During labor and childbirth, coordinated contractions of the myometrium in response to oxytocin from the posterior pituitary help expel the fetus from the uterus.

The inner layer of the uterus, the **endometrium** (*endo-* = within), is highly vascularized and has three components: (1) An innermost layer composed of simple columnar epithelium (ciliated and secretory cells) lines the lumen. (2) An underlying endometrial stroma is a very thick region of lamina propria (areolar connective tissue). (3) Endometrial (uterine) glands develop as invaginations of the luminal epithelium and extend almost to the myometrium. The endometrium is divided into two layers. The **stratum functionalis** (*functional layer*) lines the uterine cavity and sloughs off during menstruation. The deeper layer, the **stratum basalis** (*basal layer*), is permanent and gives rise to a new stratum functionalis after each menstruation.

Branches of the internal iliac artery called **uterine arteries** (Figure 28.19) supply blood to the uterus. Uterine arteries give off branches called **arcuate arteries** (AR-kū-āt = shaped like a bow) that are arranged in a circular fashion in the myometrium. These arteries branch into **radial arteries** that penetrate deeply into the myometrium. Just before the branches enter the endometrium, they divide into two kinds of arterioles: **Straight arterioles** supply the stratum basalis with the materials needed to regenerate the stratum functionalis; **spiral arterioles** supply the stratum functionalis and change markedly during the menstrual cycle. Blood leaving the uterus is drained by the **uterine veins** into the internal iliac veins. The extensive blood supply of the uterus is essential to support regrowth of a new stratum functionalis after menstruation, implantation of a fertilized ovum, and development of the placenta.

Cervical Mucus The secretory cells of the mucosa of the cervix produce a secretion called **cervical mucus**, a mixture of water, glycoproteins, lipids, enzymes, and inorganic salts. During their reproductive years, females secrete 20–60 mL of cervical mucus per day. Cervical mucus is more hospitable to sperm at or near the

FIGURE 28.18 **Histology of the uterus.**

The three layers of the uterus from superficial to deep are the perimetrium (serosa), the myometrium, and the endometrium.

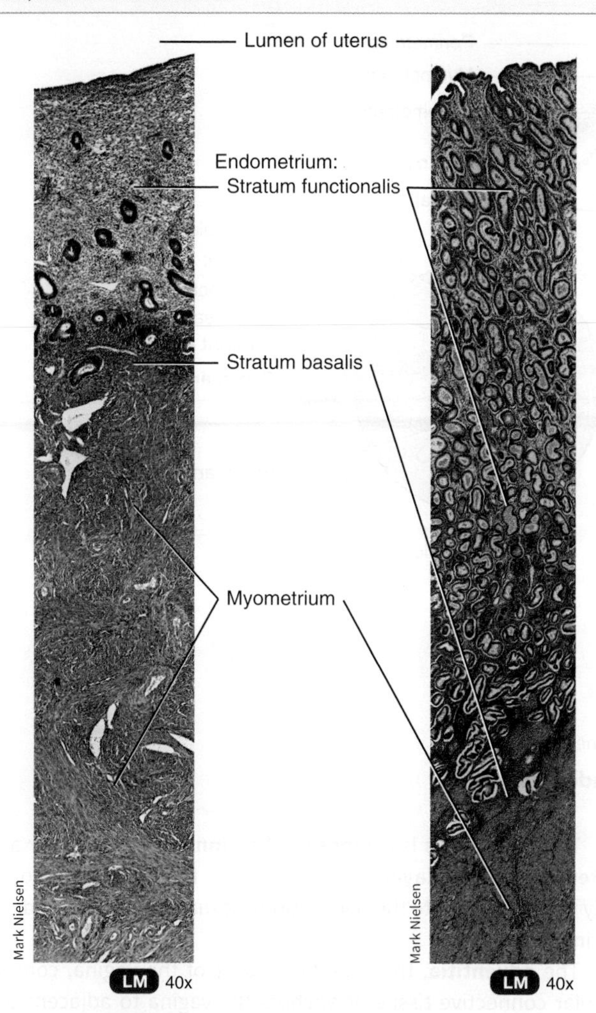

(a) Transverse section through the uterine wall: second week of menstrual cycle (left) and third week of menstrual cycle (right)

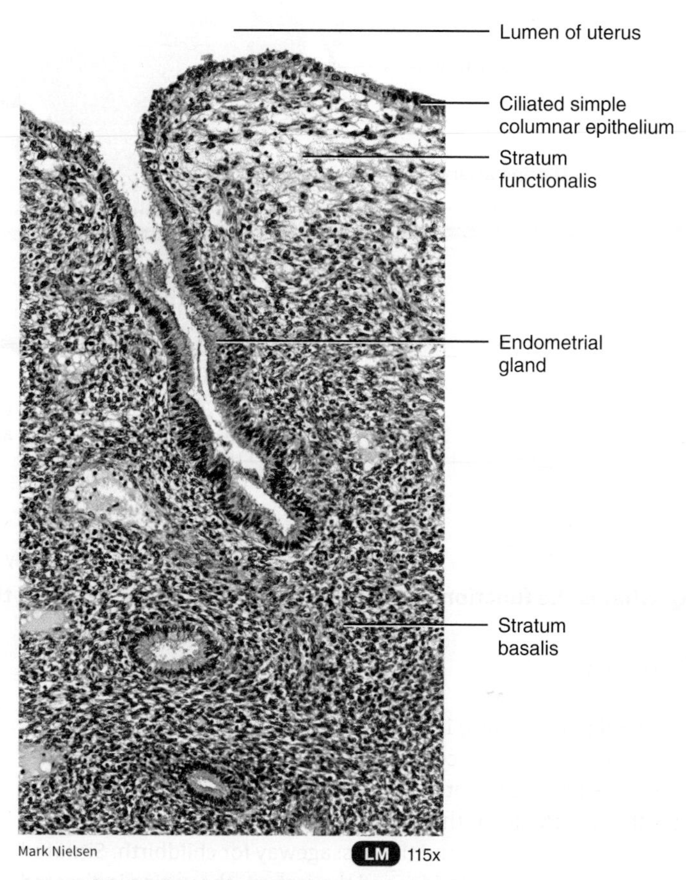

(b) Details of endometrium

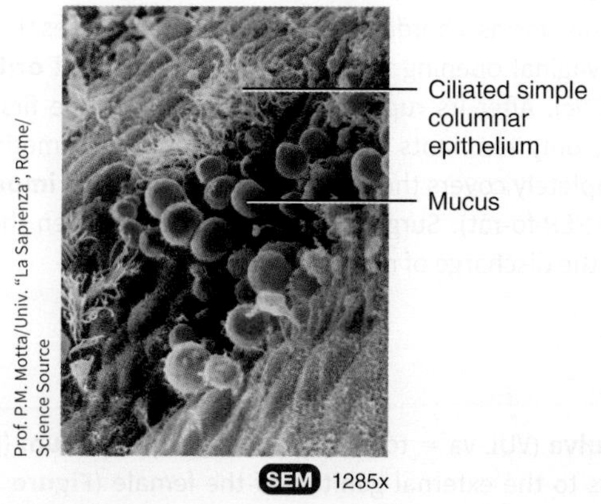

(c) Endometrium during secretory phase

Q What structural features of the endometrium and myometrium contribute to their functions?

time of ovulation because it is then less viscous and more alkaline (pH 8.5). At other times, a more viscous mucus forms a cervical plug that physically impedes sperm penetration. Cervical mucus supplements the energy needs of sperm, and both the cervix and cervical mucus protect sperm from phagocytes and the hostile environment of the vagina and uterus. Cervical mucus may also play a role in **capacitation** (ka-pas′-i-TĀ-shun)—a series of functional changes that sperm undergo in the female reproductive tract before they are able to fertilize a secondary oocyte. Capacitation causes a sperm cell's tail to beat even more vigorously, and it prepares the sperm cell's plasma membrane to fuse with the oocyte's plasma membrane.

See Clinical Connection: Hysterectomy

FIGURE 28.19 **Blood supply of the uterus.** The inset shows histological details of the blood vessels of the endometrium.

Straight arterioles supply the materials needed for regeneration of the stratum functionalis.

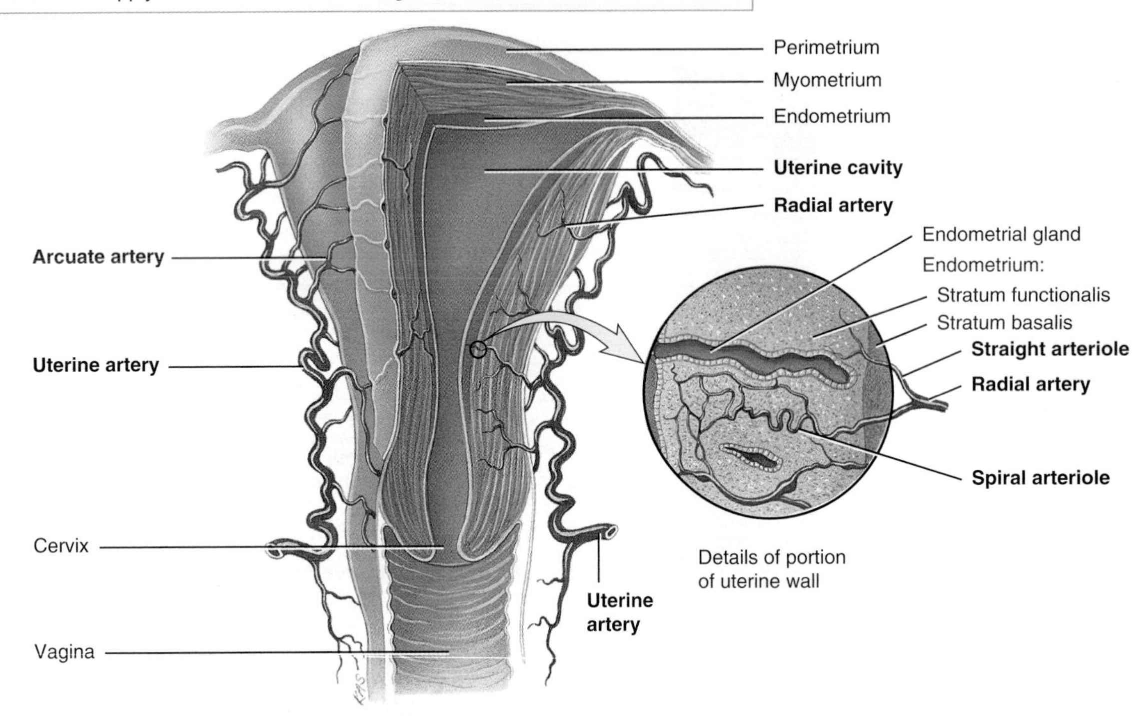

Anterior view with left side of uterus partially sectioned

Q What is the functional significance of the stratum basalis of the endometrium?

Vagina

The **vagina** (= sheath) is a tubular, 10-cm (4-in.) long fibromuscular canal lined with mucous membrane that extends from the exterior of the body to the uterine cervix (see Figures 28.11 and 28.16). It is the receptacle for the penis during sexual intercourse, the outlet for menstrual flow, and the passageway for childbirth. Situated between the urinary bladder and the rectum, the vagina is directed superiorly and posteriorly, where it attaches to the uterus. A recess called the **fornix** (= arch or vault) surrounds the vaginal attachment to the cervix. When properly inserted, a contraceptive diaphragm rests in the fornix, where it is held in place as it covers the cervix.

The **mucosa** of the vagina is continuous with that of the uterus (Figure 28.20a, b). Histologically, it consists of nonkeratinized stratified squamous epithelium and areolar connective tissue that lies in a series of transverse folds called **rugae** (ROO-gē). Dendritic cells in the mucosa are antigen-presenting cells (described in Section 22.4). Unfortunately, they also participate in the transmission of viruses—for example, HIV (the virus that causes AIDS)—to a female during intercourse with an infected male. The mucosa of the vagina contains large stores of glycogen, the decomposition of which produces organic acids. The resulting acidic environment retards microbial growth, but it also is harmful to sperm. Alkaline components of semen, mainly from the seminal vesicles, raise the pH of fluid in the vagina and increase viability of the sperm.

The **muscularis** is composed of an **inner circular layer** and an **outer longitudinal layer** of smooth muscle that can stretch considerably to accommodate the penis during sexual intercourse and a child during birth.

The **adventitia**, the superficial layer of the vagina, consists of areolar connective tissue. It anchors the vagina to adjacent organs such as the urethra and urinary bladder anteriorly and the rectum and anal canal posteriorly.

A thin fold of vascularized mucous membrane, called the **hymen** (= membrane), forms a border around and partially closes the inferior end of the vaginal opening to the exterior, the **vaginal orifice** (see Figure 28.20c). After its rupture, usually following the first sexual intercourse, only remnants of the hymen remain. Sometimes the hymen completely covers the orifice, a condition called **imperforate hymen** (im-PER-fō-rāt). Surgery may be needed to open the orifice and permit the discharge of menstrual flow.

Vulva

The term **vulva** (VUL-va = to wrap around), or **pudendum** (pū-DEN-dum), refers to the external genitals of the female (Figure 28.20a). The following components make up the vulva:

- Anterior to the vaginal and urethral openings is the **mons pubis** (MONZ PŪ-bis; *mons* = mountain), an elevation of adipose tissue covered by skin and coarse pubic hair that cushions the pubic symphysis.

FIGURE 28.20 **The vagina and components of the vulva (pudendum).**

The vulva refers to the external genitals of the female.

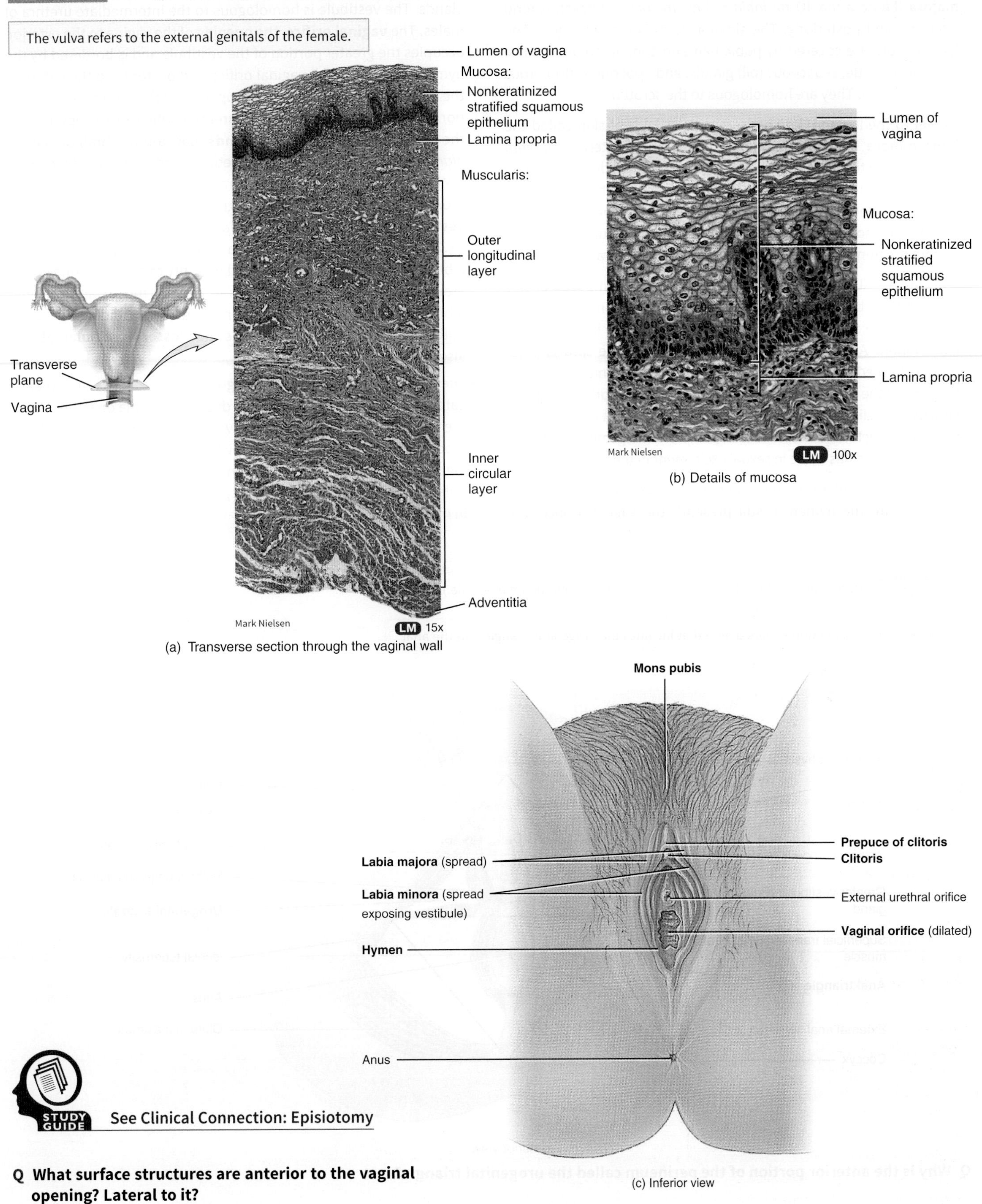

(a) Transverse section through the vaginal wall

(b) Details of mucosa

(c) Inferior view

See Clinical Connection: Episiotomy

Q What surface structures are anterior to the vaginal opening? Lateral to it?

- From the mons pubis, two longitudinal folds of skin, the **labia majora** (LĀ-bē-a ma-JŌ-ra; *labia* = lips; *majora* = larger), extend inferiorly and posteriorly. The singular term is *labium majus.* The labia majora are covered by pubic hair and contain an abundance of adipose tissue, sebaceous (oil) glands, and apocrine sudoriferous (sweat) glands. They are homologous to the scrotum.
- Medial to the labia majora are two smaller folds of skin called the **labia minora** (min-OR-a = smaller). The singular term is *labium minus.* Unlike the labia majora, the labia minora are devoid of pubic hair and fat and have few sudoriferous glands, but they do contain many sebaceous glands which produce antimicrobial substances and provide some lubrication during sexual intercourse. The labia minora are homologous to the spongy (penile) urethra.
- The **clitoris** (KLI-to-ris) is a small cylindrical mass composed of two small erectile bodies, the *corpora cavernosa*, and numerous nerves and blood vessels. The clitoris is located at the anterior junction of the labia minora. A layer of skin called the **prepuce of the clitoris** (PRĒ-poos) is formed at the point where the labia minora unite and covers the body of the clitoris. The exposed portion of the clitoris is the **glans clitoris**. The clitoris is homologous to the glans penis in males. Like the male structure, the clitoris is capable of enlargement on tactile stimulation and has a role in sexual excitement in the female.
- The region between the labia minora is the **vestibule**. Within the vestibule are the hymen (if still present), the vaginal orifice, the external urethral orifice, and the openings of the ducts of several glands. The vestibule is homologous to the intermediate urethra of males. The **vaginal orifice**, the opening of the vagina to the exterior, occupies the greater portion of the vestibule and is bordered by the hymen. Anterior to the vaginal orifice and posterior to the clitoris is the **external urethral orifice**, the opening of the urethra to the exterior. On either side of the external urethral orifice are the openings of the ducts of the **paraurethral glands** (par′-a-ū-RĒ-thral) or *Skene's glands* (SKĒNZ). These mucus-secreting glands are embedded in the wall of the urethra. The paraurethral glands are homologous to the prostate. On either side of the vaginal orifice itself are the **greater vestibular glands** or *Bartholin's glands* (BAR-to-linz) (see Figure 28.21), which open by ducts into a groove between the hymen and labia minora. They produce a small quantity of mucus during sexual arousal and intercourse that adds to cervical mucus and provides lubrication. The greater vestibular glands are homologous to the bulbourethral glands in males. Several **lesser vestibular glands** also open into the vestibule.
- The **bulb of the vestibule** (see Figure 28.21) consists of two elongated masses of erectile tissue just deep to the labia on either side of the vaginal orifice. The bulb of the vestibule becomes engorged with blood during sexual arousal, narrowing the vaginal orifice and placing pressure on the penis during intercourse. The bulb of the vestibule is homologous to the corpus spongiosum and bulb of the penis in males.

FIGURE 28.21 Perineum of a female. (Figure 11.13 shows the perineum of a male.)

The perineum is a diamond-shaped area that includes the urogenital triangle and the anal triangle.

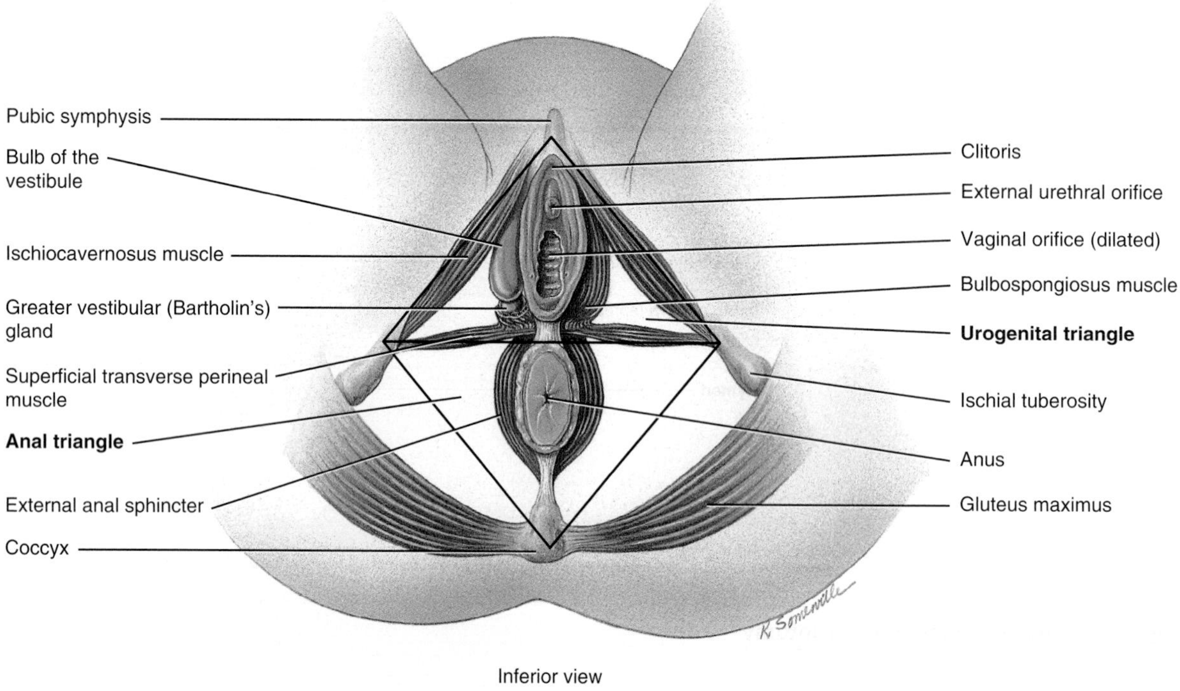

Q Why is the anterior portion of the perineum called the urogenital triangle?

TABLE 28.2 Summary of Homologous Structures of the Female and Male Reproductive Systems

FEMALE STRUCTURES	MALE STRUCTURES
Ovaries	Testes
Ovum	Sperm cell
Labia majora	Scrotum
Labia minora	Spongy urethra
Vestibule	Intermediate urethra
Bulb of vestibule	Corpus spongiosum penis and bulb of penis
Clitoris	Glans penis and corpora cavernosa
Paraurethral glands	Prostate
Greater vestibular glands	Bulbourethral glands

Table 28.2 summarizes the homologous structures of the female and male reproductive systems.

Perineum

The **perineum** (per′-i-NĒ-um) is the diamond-shaped area medial to the thighs and buttocks of both males and females (Figure 28.21). It contains the external genitals and anus. The perineum is bounded anteriorly by the pubic symphysis, laterally by the ischial tuberosities, and posteriorly by the coccyx. A transverse line drawn between the ischial tuberosities divides the perineum into an anterior **urogenital triangle** (ū′-rō-JEN-i-tal) that contains the external genitals and a posterior **anal triangle** that contains the anus.

Mammary Glands

Each **breast** is a hemispheric projection of variable size anterior to the pectoralis major and serratus anterior muscles and attached to them by a layer of fascia composed of dense irregular connective tissue.

Each breast has one pigmented projection, the **nipple**, that has a series of closely spaced openings of ducts called **lactiferous ducts** (lak-TIF-e-rus), where milk emerges. The circular pigmented area of skin surrounding the nipple is called the **areola** (a-RĒ-ō-la = small space); it appears rough because it contains modified sebaceous (oil) glands. Strands of connective tissue called the **suspensory ligaments of the breast** (*Cooper's ligaments*) run between the skin and fascia and support the breast. These ligaments become looser with age or with the excessive strain that can occur in long-term jogging or high-impact aerobics. Wearing a supportive bra can slow this process and help maintain the strength of the suspensory ligaments.

Within each breast is a **mammary gland**, a modified sudoriferous (sweat) gland that produces milk (Figure 28.22). A mammary

FIGURE 28.22 Mammary glands within the breasts.

The mammary glands function in the synthesis, secretion, and ejection of milk (lactation).

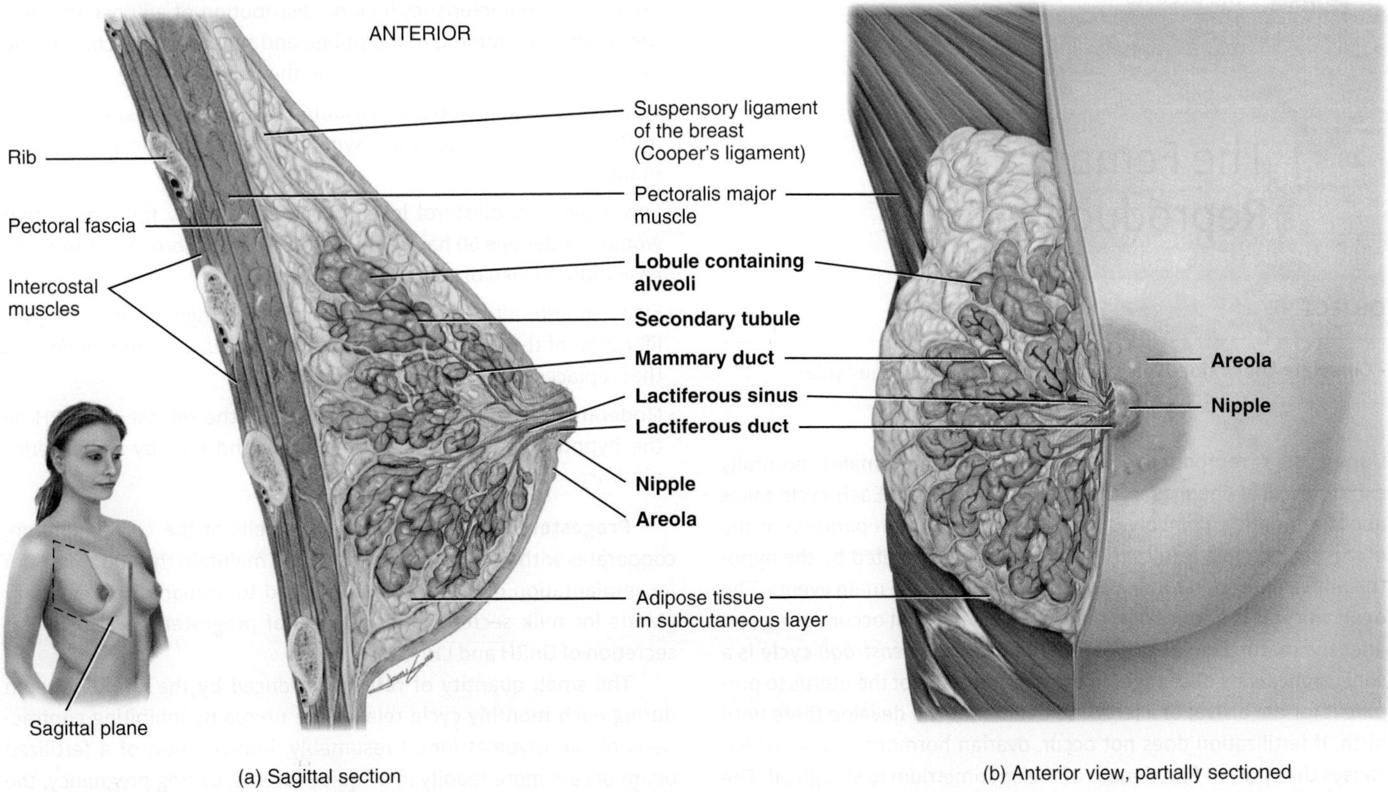

(a) Sagittal section

(b) Anterior view, partially sectioned

Q What hormones regulate the synthesis and ejection of milk?

gland consists of 15 to 20 **lobes**, or compartments, separated by a variable amount of adipose tissue. In each lobe are several smaller compartments called **lobules**, composed of grapelike clusters of milk-secreting glands termed **alveoli** (al-VĒ-o-lī = small cavities) embedded in connective tissue. Contraction of **myoepithelial cells** (mī′-ō-ep′-i-THĒ-lē-al) surrounding the alveoli helps propel milk toward the nipples. When milk is being produced, it passes from the alveoli into a series of **secondary tubules** and then into the **mammary ducts**. Near the nipple, the mammary ducts expand slightly to form sinuses called **lactiferous sinuses** (*lact-* = milk), where some milk may be stored before draining into a **lactiferous duct**. Each lactiferous duct typically carries milk from one of the lobes to the exterior.

The functions of the mammary glands are the synthesis, secretion, and ejection of milk; these functions, called **lactation** (lak-TĀ-shun), are associated with pregnancy and childbirth. Milk production is stimulated largely by the hormone prolactin from the anterior pituitary, with contributions from progesterone and estrogens. The ejection of milk is stimulated by oxytocin, which is released from the posterior pituitary in response to the sucking of an infant on the mother's nipple (suckling).

See Clinical Connection: Breast Augmentation and Reduction

See Clinical Connection: Fibrocystic Disease of the Breasts

28.3 The Female Reproductive Cycle

OBJECTIVE

- **Compare** the major events of the ovarian and uterine cycles.

During their reproductive years, nonpregnant females normally exhibit cyclical changes in the ovaries and uterus. Each cycle takes about a month and involves both oogenesis and preparation of the uterus to receive a fertilized ovum. Hormones secreted by the hypothalamus, anterior pituitary, and ovaries control the main events. The **ovarian cycle** is a series of events in the ovaries that occur during and after the maturation of an oocyte. The **uterine** (*menstrual*) **cycle** is a concurrent series of changes in the endometrium of the uterus to prepare it for the arrival of a fertilized ovum that will develop there until birth. If fertilization does not occur, ovarian hormones wane, which causes the stratum functionalis of the endometrium to slough off. The general term **female reproductive cycle** encompasses the ovarian and uterine cycles, the hormonal changes that regulate them, and the related cyclical changes in the breasts and cervix.

Hormonal Regulation of the Female Reproductive Cycle

Gonadotropin-releasing hormone (GnRH) secreted by the hypothalamus controls the ovarian and uterine cycles (Figure 28.23). GnRH stimulates the release of follicle-stimulating hormone (FSH) and luteinizing hormone (LH) from the anterior pituitary. FSH initiates follicular growth, while LH stimulates further development of the ovarian follicles. In addition, both FSH and LH stimulate the ovarian follicles to secrete estrogens. LH stimulates the theca cells of a developing follicle to produce androgens. Under the influence of FSH, the androgens are taken up by the granulosa cells of the follicle and then converted into estrogens. At midcycle, LH triggers ovulation and then promotes formation of the corpus luteum, the reason for the name luteinizing hormone. Stimulated by LH, the corpus luteum produces and secretes estrogens, progesterone, relaxin, and inhibin.

At least six different estrogens have been isolated from the plasma of human females, but only three are present in significant quantities: *beta* (*β*)-*estradiol* (es-tra-DĪ-ol), *estrone*, and *estriol* (ES-trē-ol). In a nonpregnant woman, the most abundant estrogen is *β*-estradiol, which is synthesized from cholesterol in the ovaries.

Estrogens secreted by ovarian follicles have several important functions (Figure 28.23): They:

- Promote the development and maintenance of female reproductive structures, secondary sex characteristics, and the breasts. The secondary sex characteristics include distribution of adipose tissue in the breasts, abdomen, mons pubis, and hips; voice pitch; a broad pelvis; and pattern of hair growth on the head and body.
- Increase protein anabolism, including the building of strong bones. In this regard, estrogens are synergistic with human growth hormone (hGH).
- Lower blood cholesterol level, which is probably the reason that women under age 50 have a much lower risk of coronary artery disease than do men of comparable age.
- Every month, after menstruation occurs, estrogens stimulate proliferation of the stratum basalis to form a new stratum functionalis that replaces the one that has sloughed off.
- Moderate levels in the blood inhibit both the release of GnRH by the hypothalamus and secretion of LH and FSH by the anterior pituitary.

Progesterone, secreted mainly by cells of the corpus luteum, cooperates with estrogens to prepare and maintain the endometrium for implantation of a fertilized ovum and to prepare the mammary glands for milk secretion. High levels of progesterone also inhibit secretion of GnRH and LH.

The small quantity of **relaxin** produced by the corpus luteum during each monthly cycle relaxes the uterus by inhibiting contractions of the myometrium. Presumably, implantation of a fertilized ovum occurs more readily in a "quiet" uterus. During pregnancy, the placenta produces much more relaxin, and it continues to relax uterine

FIGURE 28.23 Secretion and physiological effects of estrogens, progesterone, relaxin, and inhibin in the female reproductive cycle. Dashed red lines indicate negative feedback inhibition.

The uterine and ovarian cycles are controlled by gonadotropin-releasing hormone (GnRH) and ovarian hormones (estrogens and progesterone).

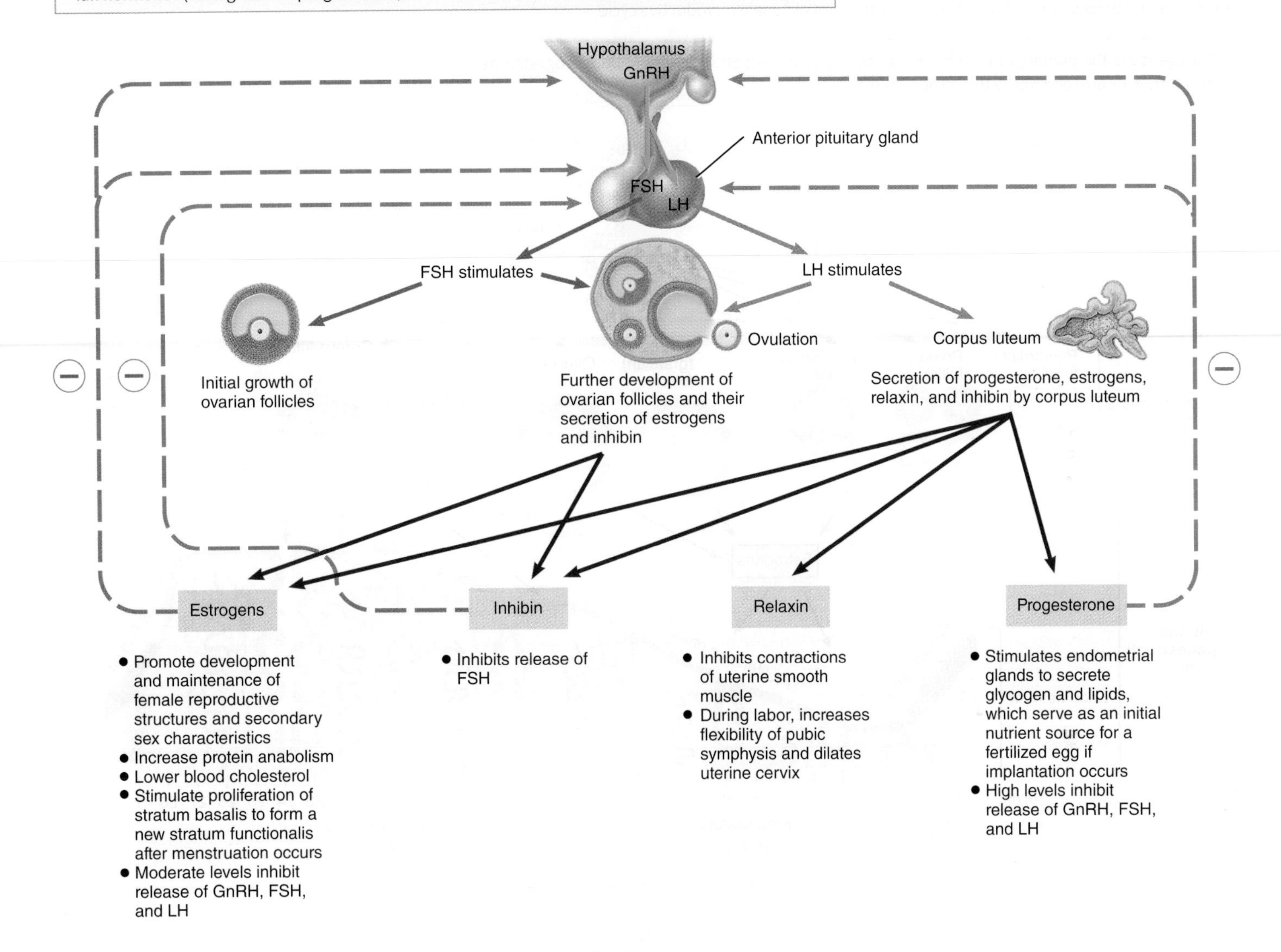

Q Of the several estrogens, which one exerts the major effect?

smooth muscle. At the end of pregnancy, relaxin also increases the flexibility of the pubic symphysis and may help dilate the uterine cervix, both of which ease delivery of the baby.

Inhibin is secreted by granulosa cells of growing follicles and by the corpus luteum after ovulation. It inhibits secretion of FSH and, to a lesser extent, LH.

Phases of the Female Reproductive Cycle

The duration of the female reproductive cycle typically ranges from 24 to 36 days. For this discussion, we assume a duration of 28 days and divide it into four phases: the menstrual phase, the preovulatory phase, ovulation, and the postovulatory phase (Figure 28.24).

Menstrual Phase

The **menstrual phase** (MEN-stroo-al), also called **menstruation** (men′-stroo-Ā-shun) or *menses* (MEN-sēz = month), lasts for roughly the first 5 days of the cycle. (By convention, the first day of menstruation is day 1 of a new cycle.)

Events in the Ovaries Under the influence of FSH, several primordial follicles develop into primary follicles and then into secondary follicles. This developmental process may take several months to occur. Therefore, a follicle that begins to develop at the beginning of a

FIGURE 28.24 The female reproductive cycle. The length of the female reproductive cycle typically is 24 to 36 days; the preovulatory phase is more variable in length than the other phases. (a) Events in the ovarian and uterine cycles and the release of anterior pituitary hormones are correlated with the sequence of the cycle's four phases. In the cycle shown, fertilization and implantation have not occurred. (b) Relative concentrations of anterior pituitary hormones (FSH and LH) and ovarian hormones (estrogens and progesterone) during the phases of a normal female reproductive cycle.

Estrogens are the primary ovarian hormones before ovulation; after ovulation, both progesterone and estrogens are secreted by the corpus luteum.

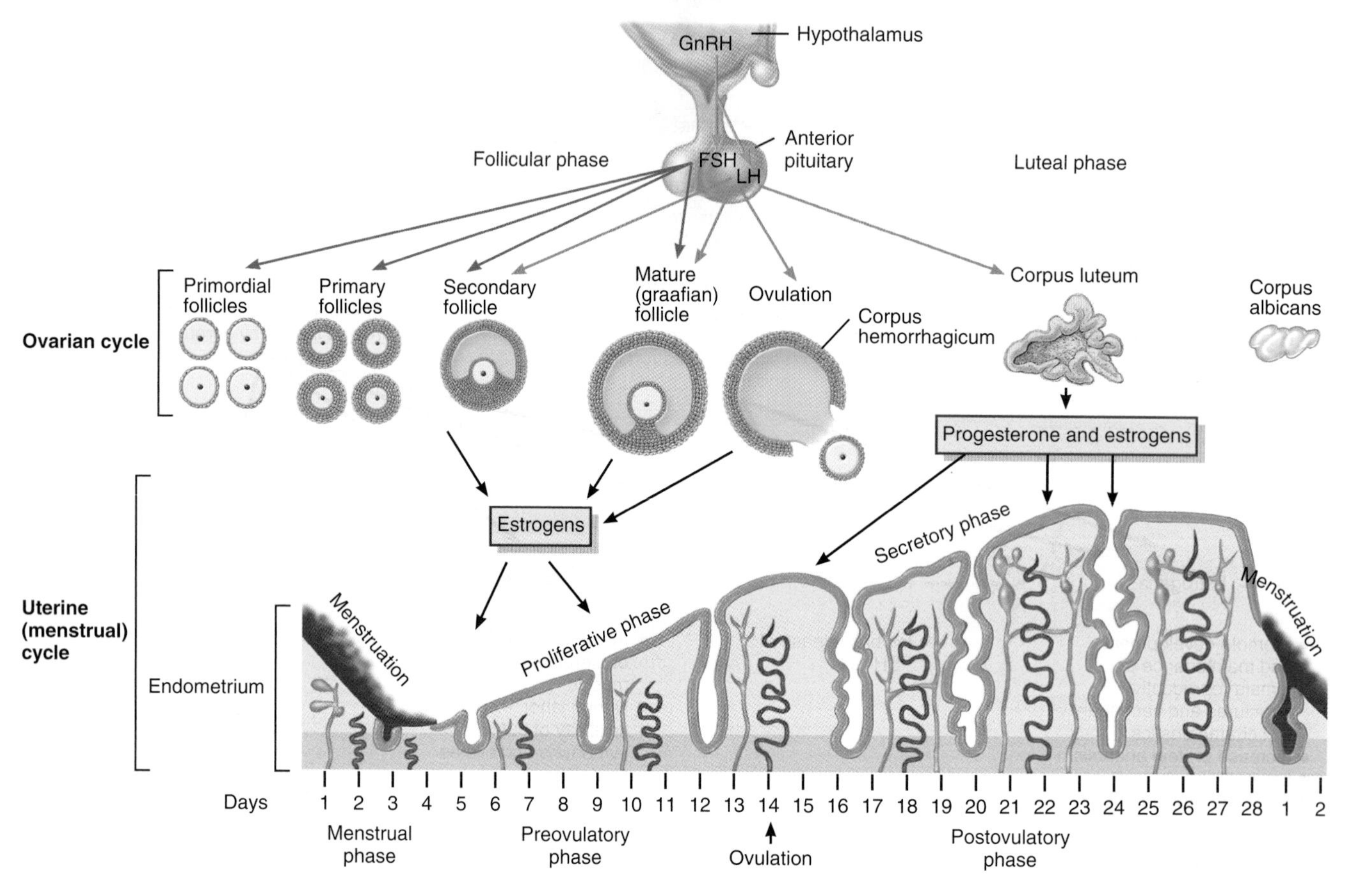

(a) Hormonal regulation of changes in the ovary and uterus

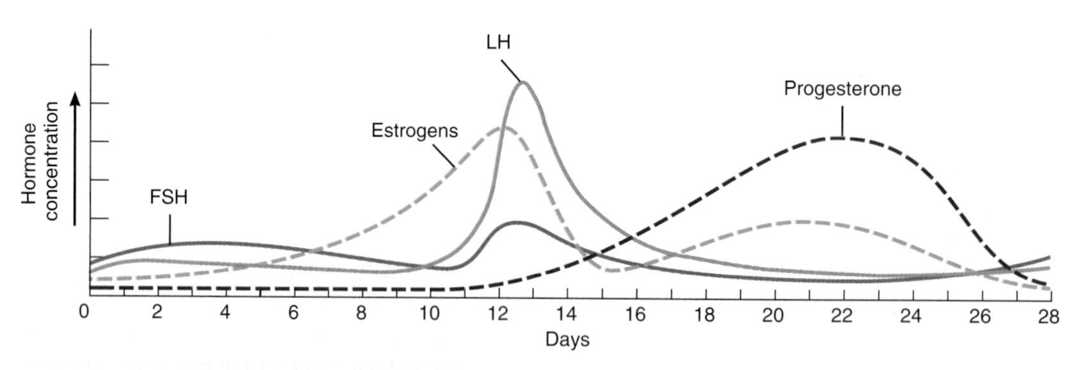

(b) Changes in concentration of anterior pituitary and ovarian hormones

Q Which hormones are responsible for the proliferative phase of endometrial growth, for ovulation, for growth of the corpus luteum, and for the surge of LH at midcycle?

particular menstrual cycle may not reach maturity and ovulate until several menstrual cycles later.

Events in the Uterus Menstrual flow from the uterus consists of 50–150 mL of blood, tissue fluid, mucus, and epithelial cells shed from the endometrium. This discharge occurs because the declining levels of progesterone and estrogens stimulate release of prostaglandins that cause the uterine spiral arterioles to constrict. As a result, the cells they supply become oxygen-deprived and start to die. Eventually, the entire stratum functionalis sloughs off. At this time the endometrium is very thin, about 2–5 mm, because only the stratum basalis remains. The menstrual flow passes from the uterine cavity through the cervix and vagina to the exterior.

Preovulatory Phase

The **preovulatory phase** (prē-OV-ū-la-tō-rē) is the time between the end of menstruation and ovulation. The preovulatory phase of the cycle is more variable in length than the other phases and accounts for most of the differences in length of the cycle. It lasts from days 6 to 13 in a 28-day cycle.

Events in the Ovaries Some of the secondary follicles in the ovaries begin to secrete estrogens and inhibin. By about day 6, a single secondary follicle in one of the two ovaries has outgrown all of the others to become the **dominant follicle**. Estrogens and inhibin secreted by the dominant follicle decrease the secretion of FSH, which causes other, less well-developed follicles to stop growing and degenerate. Fraternal (nonidentical) twins or triplets result when two or three secondary follicles become codominant and later are ovulated and fertilized at about the same time.

Normally, the one dominant secondary follicle becomes the **mature** *(graafian)* **follicle**, which continues to enlarge until it is more than 20 mm in diameter and ready for ovulation (see **Figure 28.13**). This follicle forms a blisterlike bulge due to the swelling antrum on the surface of the ovary. During the final maturation process, the mature follicle continues to increase its production of estrogens (**Figure 28.24**).

With reference to the ovarian cycle, the menstrual and preovulatory phases together are termed the **follicular phase** (fo-LIK-ū-lar) because ovarian follicles are growing and developing.

Events in the Uterus Estrogens liberated into the blood by growing ovarian follicles stimulate the repair of the endometrium; cells of the stratum basale undergo mitosis and produce a new stratum functionalis. As the endometrium thickens, the short, straight endometrial glands develop, and the arterioles coil and lengthen as they penetrate the stratum functionalis. The thickness of the endometrium approximately doubles, to about 4–10 mm. With reference to the uterine cycle, the preovulatory phase is also termed the **proliferative phase** (prō-LIF-er-a-tiv) because the endometrium is proliferating.

Ovulation

Ovulation, the rupture of the mature (graafian) follicle and the release of the secondary oocyte into the pelvic cavity, usually occurs on day 14 in a 28-day cycle. During ovulation, the secondary oocyte remains surrounded by its zona pellucida and corona radiata.

FIGURE 28.25 **High levels of estrogens exert a positive feedback effect (red arrows) on the hypothalamus and anterior pituitary, thereby increasing secretion of GnRH and LH.**

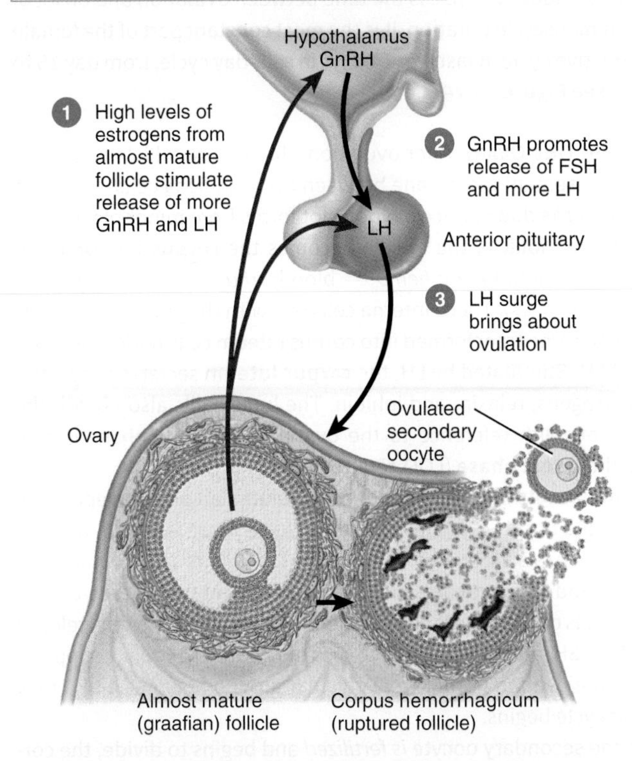

Q What is the effect of rising but still moderate levels of estrogens on the secretion of GnRH, LH, and FSH?

The *high levels of estrogens* during the last part of the preovulatory phase exert a *positive* feedback effect on the cells that secrete LH and gonadotropin-releasing hormone (GnRH) and cause ovulation, as follows (**Figure 28.25**):

1. A high concentration of estrogens stimulates more frequent release of GnRH from the hypothalamus. It also directly stimulates gonadotrophs in the anterior pituitary to secrete LH.
2. GnRH promotes the release of FSH and additional LH by the anterior pituitary.
3. LH causes rupture of the mature (graafian) follicle and expulsion of a secondary oocyte about 9 hours after the peak of the LH surge. The ovulated oocyte and its corona radiata cells are usually swept into the uterine tube.

From time to time, an oocyte is lost into the pelvic cavity, where it later disintegrates. The small amount of blood that sometimes leaks into the pelvic cavity from the ruptured follicle can cause pain, known as **mittelschmerz** (MIT-el-shmārts = pain in the middle), at the time of ovulation.

An over-the-counter home test that detects a rising level of LH can be used to predict ovulation a day in advance.

Postovulatory Phase

The **postovulatory phase** of the female reproductive cycle is the time between ovulation and onset of the next menses. In duration, it is the most constant part of the female reproductive cycle. It lasts for 14 days in a 28-day cycle, from day 15 to day 28 (see Figure 28.24).

EVENTS IN ONE OVARY After ovulation, the mature follicle collapses, and the basement membrane between the granulosa cells and theca interna breaks down. Once a blood clot forms from minor bleeding of the ruptured follicle, the follicle becomes the **corpus hemorrhagicum** (hem′-o-RAJ-i-kum; *hemo-* = blood; *rrhagic-* = bursting forth) (see Figure 28.13). Theca interna cells mix with the granulosa cells as they all become transformed into corpus luteum cells under the influence of LH. Stimulated by LH, the **corpus luteum** secretes progesterone, estrogens, relaxin, and inhibin. The luteal cells also absorb the blood clot. With reference to the ovarian cycle, this phase is also called the **luteal phase** (LOO-tē-al).

Later events in an ovary that has ovulated an oocyte depend on whether the oocyte is fertilized. If the oocyte *is not fertilized*, the corpus luteum has a life span of only 2 weeks. Then, its secretory activity declines, and it degenerates into a corpus albicans (see Figure 28.13). As the levels of progesterone, estrogens, and inhibin decrease, release of GnRH, FSH, and LH rises due to loss of negative feedback suppression by the ovarian hormones. Follicular growth resumes and a new ovarian cycle begins.

If the secondary oocyte *is fertilized* and begins to divide, the corpus luteum persists past its normal 2-week life span. It is "rescued" from degeneration by **human chorionic gonadotropin (hCG)** (kō-rē-ON-ik). This hormone is produced by the chorion of the embryo beginning about 8 days after fertilization. Like LH, hCG stimulates the secretory activity of the corpus luteum. The presence of hCG in maternal blood or urine is an indicator of pregnancy and is the hormone detected by home pregnancy tests.

EVENTS IN THE UTERUS Progesterone and estrogens produced by the corpus luteum promote growth and coiling of the endometrial glands, vascularization of the superficial endometrium, and thickening of the endometrium to 12–18 mm (0.48–0.72 in.). Because of the secretory activity of the endometrial glands, which begin to secrete glycogen, this period is called the **secretory phase** of the uterine cycle. These preparatory changes peak about 1 week after ovulation, at the time a fertilized ovum might arrive in the uterus. If fertilization does not occur, the levels of progesterone and estrogens decline due to degeneration of the corpus luteum. Withdrawal of progesterone and estrogens causes menstruation.

Figure 28.26 summarizes the hormonal interactions and cyclical changes in the ovaries and uterus during the ovarian and uterine cycles.

See Clinical Connection: Female Athlete Triad: Disordered Eating, Amenorrhea, and Premature Osteoporosis

28.4 The Human Sexual Response

OBJECTIVE

- **Compare** the sexual responses of males and females.

During heterosexual **sexual intercourse**, also called *copulation* or *coitus* (KŌ-i-tus), the erect penis is inserted into the vagina. The similar sequence of physiological and emotional changes experienced by both males and females before, during, and after intercourse is termed the **human sexual response**. William Masters and Virginia Johnson, who began their pioneering research on human sexuality in the late 1950s, divided the human sexual response into four phases: excitement, plateau, orgasm, and resolution.

During the **excitement** phase, there is **vasocongestion**—engorgement with blood—of genital tissues, resulting in erection of the penis in men and erection of the clitoris and swelling of the labia and vagina in women. In addition, vasocongestion causes the breasts to swell and the nipples to become erect. The excitement phase is also associated with an increase in the secretion of fluid that lubricates the walls of the vagina. When the connective tissue of the vagina becomes engorged with blood, lubricating fluid oozes from the capillaries and seeps through the epithelial lining via a process called **transudation**. Glands within the cervical mucosa and the greater vestibular (Bartholin's) glands contribute a small quantity of lubricating mucus. Without satisfactory lubrication, sexual intercourse is difficult and painful for both partners and inhibits orgasm. Other changes that occur during the excitement phase include increased heart rate and blood pressure, increased skeletal muscle tone throughout the body, and hyperventilation. Direct physical contact (as in kissing or touching), especially of the penis, clitoris, nipples of the breasts, and earlobes is a potent initiator of excitement. However, anticipation or fear; memories; visual, olfactory, and auditory sensations; and fantasies can enhance or diminish the likelihood that excitement will occur.

The changes that begin during excitement are sustained at an intense level in the **plateau** phase, which may last for only a few seconds or for many minutes. During this phase, many females and some males display a **sex flush**, a rashlike redness of the face and chest due to vasodilation of blood vessels in those parts of the body. The head of the penis increases in diameter and the testes swell. Late in the plateau phase, pronounced vasocongestion of the lower third of the vagina swells the tissue and narrows the opening. Because of this response, the vagina grips the penis more firmly.

Generally, the briefest phase is **orgasm** (*climax*), during which both sexes experience several rhythmic muscular contractions about 0.8 sec apart, accompanied by intense, pleasurable sensations and a further increase in blood pressure, heart rate, and respiratory rate. The sex flush is also most prominent at this time. In males, contraction of smooth muscle in the walls of the epididymis, vas deferens, and ejaculatory ducts as well as secretion of fluid by the accessory sex glands cause semen to move into the urethra (emission). Then, rhythmic contractions of skeletal muscles at the base of the penis propel semen out of the penis (ejaculation). In males, orgasm usually

FIGURE 28.26 Summary of hormonal interactions in the ovarian and uterine cycles.

Hormones from the anterior pituitary regulate ovarian function, and hormones from the ovaries regulate the changes in the endometrial lining of the uterus.

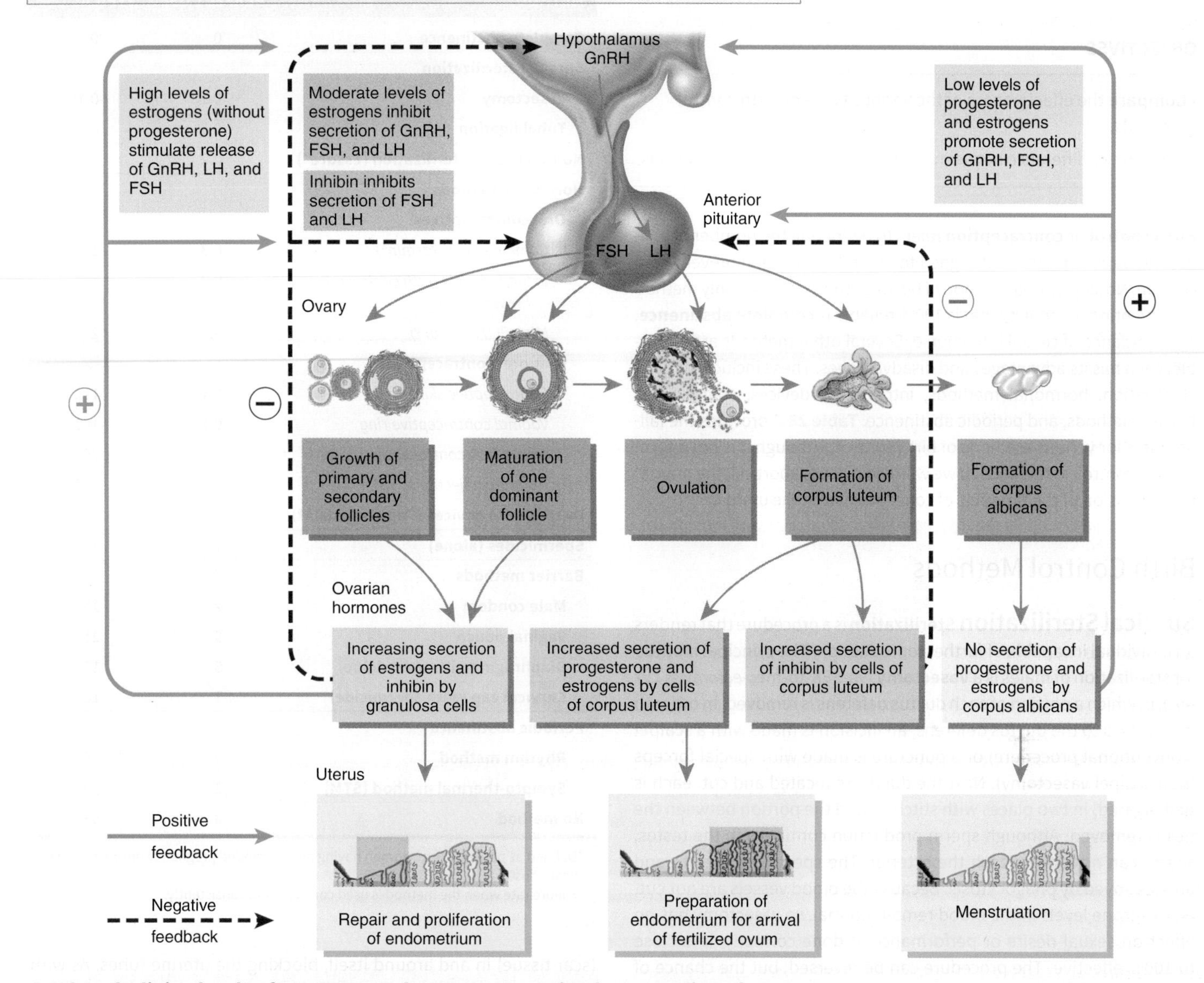

Q When declining levels of estrogens and progesterone stimulate secretion of GnRH, is this a positive or a negative feedback effect? Why?

accompanies ejaculation. In women, if effective sexual stimulation continues, orgasm may occur, associated with 3–12 rhythmic contractions of the skeletal muscles that underlie the vulva. Reception of the ejaculate provides little stimulus for a female, especially if she is not already at the plateau phase; this is why a female partner does not automatically experience orgasm simultaneously with her partner. In both males and females, orgasm is a total body response that may produce milder sensations on some occasions and more intense, explosive sensations at other times. Whereas females may experience two or more orgasms in rapid succession, males enter a **refractory period**, a recovery time during which a second ejaculation and orgasm is physiologically impossible. In some males, the refractory period lasts only a few minutes; in others it lasts for several hours. A female does not have to experience an orgasm for fertilization to occur.

In the final phase—**resolution**, which begins with a sense of profound relaxation—genital tissues heart rate, blood pressure, breathing, and muscle tone return to the unaroused state. If sexual excitement has been intense but orgasm has not occurred, resolution takes place more slowly.

The four phases of the human sexual response are not always clearly separated from one another and may vary considerably among different people, and even in the same person at different times.

28.5 Birth Control Methods and Abortion

OBJECTIVES

- **Compare** the effectiveness of the various types of birth control methods.
- **Explain** the difference between induced and spontaneous abortions.

Birth control or **contraception** refers to restricting the number of children by various methods designed to control fertility and prevent conception. No single, ideal method of birth control exists. The only method of preventing pregnancy that is 100% reliable is **complete abstinence**, the avoidance of sexual intercourse. Several other methods are available; each has its advantages and disadvantages. These include surgical sterilization, hormonal methods, intrauterine devices, spermicides, barrier methods, and periodic abstinence. Table 28.3 provides the failure rates for various methods of birth control. Although it is not a form of birth control, in this section we will also discuss abortion, the premature expulsion of the products of conception from the uterus.

TABLE 28.3 Failure Rates for Several Birth Control Methods

METHOD	FAILURE RATES* (%) PERFECT USE†	TYPICAL USE
Complete abstinence	0	0
Surgical sterilization		
Vasectomy	0.10	0.15
Tubal ligation	0.5	0.5
Non-incisional sterilization (Essure®)	0.2	0.2
Hormonal methods		
Oral contraceptives		
Combined pill (Yasmin®)	0.3	1–2
Extended cycle birth control pill (Seasonale®)	0.3	1–2
Minipill (Micronar®)	0.5	2
Non-oral contraceptives		
Contraceptive skin patch	0.1	1–2
Vaginal contraceptive ring	0.1	1–2
Emergency contraception	25	25
Hormone injections	0.3	1–2
Intrauterine devices (Copper T 380A®)	0.6	0.8
Spermicides (alone)	15	29
Barrier methods		
Male condom	2	15
Vaginal pouch	5	21
Diaphragm (with spermicide)	6	16
Cervical cap (with spermicide)	9	16
Periodic abstinence		
Rhythm method	9	25
Sympto-thermal method (STM)	2	20
No method	85	85

*Defined as percentage of women having an unintended pregnancy during the first year of use.

†Failure rate when the method is used correctly and consistently.

Birth Control Methods

Surgical Sterilization **Sterilization** is a procedure that renders an individual incapable of further reproduction. The principal method for sterilization of males is a **vasectomy** (va-SEK-tō-mē; *-ectomy* = cut out), in which a portion of each ductus deferens is removed. In order to gain access to the ductus deferens, an incision is made with a scalpel (conventional procedure) or a puncture is made with special forceps (non-scalpel vasectomy). Next the ducts are located and cut, each is tied (ligated) in two places with stitches, and the portion between the ties is removed. Although sperm production continues in the testes, sperm can no longer reach the exterior. The sperm degenerate and are destroyed by phagocytosis. Because the blood vessels are not cut, testosterone levels in the blood remain normal, so vasectomy has no effect on sexual desire or performance. If done correctly, it is close to 100% effective. The procedure can be reversed, but the chance of regaining fertility is only 30–40%. Sterilization in females most often is achieved by performing a **tubal ligation** (lī-GĀ-shun), in which both uterine tubes are tied closed and then cut. This can be achieved in a few different ways. "Clips" or "clamps" can be placed on the uterine tubes, the tubes can be tied and/or cut, and sometimes they are cauterized. In any case, the result is that the secondary oocyte cannot pass through the uterine tubes, and sperm cannot reach the oocyte.

Non-Incisional Sterilization Essure® is one means of **non-incisional sterilization** that is an alternative to tubal ligation. In the Essure® procedure, a soft micro-insert coil made of polyester fibers and metals (nickel–titanium and stainless steel) is inserted with a catheter into the vagina, through the uterus, and into each uterine tube. Over a three-month period, the insert stimulates tissue growth (scar tissue) in and around itself, blocking the uterine tubes. As with tubal ligation, the secondary oocyte cannot pass through the uterine tubes, and sperm cannot reach the oocyte. Unlike tubal ligation, non-incisional sterilization does not require general anesthesia.

Hormonal Methods Aside from complete abstinence or surgical sterilization, hormonal methods are the most effective means of birth control. Oral contraceptives (the pill) contain hormones designed to prevent pregnancy. Some, called *combined oral contraceptives (COCs)*, contain both progestin (hormone with actions similar to progesterone) and estrogens. The primary action of COCs is to inhibit ovulation by suppressing the gonadotropins FSH and LH. The low levels of FSH and LH usually prevent the development of a dominant follicle in the ovary. As a result, levels of estrogens do not rise, the midcycle LH surge does not occur, and ovulation does not take place. Even if ovulation does occur, as it does in some cases, COCs may also block implantation in the uterus and inhibit the transport of ova and sperm in the uterine tubes.

Progestins thicken cervical mucus and make it more difficult for sperm to enter the uterus. *Progestin-only pills* thicken cervical mucus and may block implantation in the uterus, but they do not consistently inhibit ovulation.

Among the noncontraceptive benefits of oral contraceptives are regulation of the length of menstrual cycle and decreased menstrual flow (and therefore decreased risk of anemia). The pill also provides protection against endometrial and ovarian cancers and reduces the risk of endometriosis. However, oral contraceptives may not be advised for women with a history of blood clotting disorders, cerebral blood vessel damage, migraine headaches, hypertension, liver malfunction, or heart disease. Women who take the pill and smoke face far higher odds of having a heart attack or stroke than do nonsmoking pill users. Smokers should quit smoking or use an alternative method of birth control.

Following are several variations of *oral* hormonal methods of contraception:

- **Combined pill.** The **combined pill** contains both progestin and estrogens and is typically taken once a day for 3 weeks to prevent pregnancy and regulate the menstrual cycle. The pills taken during the fourth week are inactive (do not contain hormones) and permit menstruation to occur. An example is Yasmin®.
- **Extended cycle birth control pill.** Containing both progestin and estrogens, the **extended cycle birth control pill** is taken once a day in 3-month cycles of 12 weeks of hormone-containing pills followed by 1 week of inactive pills. Menstruation occurs during the thirteenth week. An example is Seasonale®.
- **Minipill.** The **minipill** contains low dose progestin only and is taken every day of the month. An example is Micronar®.

Non-oral hormonal methods of contraception are also available. Among these are the following:

- **Contraceptive skin patch.** The **contraceptive skin patch** (Ortho Evra®) contains both progestin and estrogens delivered in a skin patch placed on the upper outer arm, back, lower abdomen, or buttocks once a week for 3 weeks. After 1 week, the patch is removed from one location and then a new one is placed elsewhere. During the fourth week no patch is used.
- **Vaginal contraceptive ring.** A flexible doughnut-shaped ring about 5 cm (2 in.) in diameter, the **vaginal contraceptive ring** (NuvaRing®) contains estrogens and progesterone and is inserted by the female herself into the vagina. It is left in the vagina for 3 weeks to prevent conception and then removed for one week to permit menstruation.
- **Emergency contraception (EC). Emergency contraception (EC)**, also known as the *morning-after pill*, consists of progestin and estrogens or progestin alone to prevent pregnancy following unprotected sexual intercourse. The relatively high levels of progestin and estrogens in EC pills provide inhibition of FSH and LH secretion. Loss of the stimulating effects of these gonadotropic hormones causes the ovaries to cease secretion of their own estrogens and progesterone. In turn, declining levels of estrogens and progesterone induce shedding of the uterine lining, thereby blocking implantation. One pill is taken as soon as possible but within 72 hours of unprotected sexual intercourse. The second pill must be taken 12 hours after the first. The pills work in the same way as regular birth control pills.
- **Hormone injections. Hormone injections** are injectable progestins such as Depo-provera® given intramuscularly by a health-care practitioner once every 3 months.

Intrauterine Devices An **intrauterine device (IUD)** is a small object made of plastic, copper, or stainless steel that is inserted by a health-care professional into the cavity of the uterus. IUDs prevent fertilization from taking place by blocking sperm from entering the uterine tubes. The IUD most commonly used in the United States today is the Copper T 380A®, which is approved for up to 10 years of use and has long-term effectiveness comparable to that of tubal ligation. Some women cannot use IUDs because of expulsion, bleeding, or discomfort.

Spermicides Various foams, creams, jellies, suppositories, and douches that contain sperm-killing agents, or **spermicides** (SPER-mi-sīds), make the vagina and cervix unfavorable for sperm survival and are available without prescription. They are placed in the vagina before sexual intercourse. The most widely used spermicide is *nonoxynol-9*, which kills sperm by disrupting their plasma membranes. A spermicide is more effective when used with a barrier method such as a male condom, vaginal pouch, diaphragm, or cervical cap.

Barrier Methods **Barrier methods** use a physical barrier and are designed to prevent sperm from gaining access to the uterine cavity and uterine tubes. In addition to preventing pregnancy, certain barrier methods (male condom and vaginal pouch) may also provide some protection against sexually transmitted diseases (STDs) such as AIDS. In contrast, oral contraceptives and IUDs confer no such protection. Among the barrier methods are the male condom, vaginal pouch, diaphragm, and cervical cap.

A **male condom** is a nonporous, latex covering placed over the penis that prevents deposition of sperm in the female reproductive tract. A **vaginal pouch**, sometimes called a **female condom**, is designed to prevent sperm from entering the uterus. It is made of two flexible rings connected by a polyurethane sheath. One ring lies inside the sheath and is inserted to fit over the cervix; the other ring remains outside the vagina and covers the female external genitals. A **diaphragm** is a rubber, dome-shaped structure that fits over the cervix and is used in conjunction with a spermicide. It can be inserted by the female up to 6 hours before intercourse. The diaphragm stops most sperm from passing into the cervix and the spermicide kills most sperm that do get by. Although diaphragm use does decrease the risk of some STDs, it does not fully protect against HIV infection because the vagina is still exposed. A **cervical cap** resembles a diaphragm but is smaller and more rigid. It fits snugly over the cervix and must be fitted by a health-care professional. Spermicides should be used with the cervical cap.

Periodic Abstinence A couple can use their knowledge of the physiological changes that occur during the female reproductive cycle to decide either to abstain from intercourse on those days when pregnancy is a likely result, or to plan intercourse on those days if they wish to conceive a child. In females with normal and regular menstrual cycles, these physiological events help to predict the day on which ovulation is likely to occur.

The first physiologically based method, developed in the 1930s, is known as the **rhythm method**. It involves abstaining from sexual activity on the days that ovulation is likely to occur in each reproductive cycle. During this time (3 days before ovulation, the day of ovulation, and 3 days after ovulation) the couple abstains from intercourse. The effectiveness of the rhythm method for birth control is poor in many women due to the irregularity of the female reproductive cycle.

Another system is the **sympto-thermal method (STM)**, a natural, fertility-awareness-based method of family planning that is used to either avoid or achieve pregnancy. STM utilizes normally fluctuating physiological markers to determine ovulation such as increased basal body temperature and the production of abundant, clear, stretchy cervical mucus that resembles uncooked egg white. These indicators, reflecting the hormonal changes that govern female fertility, provide a double-check system by which a female knows when she is or is not fertile. Sexual intercourse is avoided during the fertile time to avoid pregnancy. STM users observe and chart these changes and interpret them according to precise rules.

Abortion

Abortion refers to the premature expulsion of the products of conception from the uterus, usually before the 20th week of pregnancy. An abortion may be *spontaneous* (naturally occurring; also called a *miscarriage*) or *induced* (intentionally performed).

There are several types of induced abortions. One involves **mifepristone** (MIF-pris-tōn), also known as **RU 486**. It is a hormone approved only for pregnancies 9 weeks or less when taken with misoprostol (a prostaglandin). Mifepristone is an antiprogestin; it blocks the action of progesterone by binding to and blocking progesterone receptors. Progesterone prepares the uterine endometrium for implantation and then maintains the uterine lining after implantation. If the level of progesterone falls during pregnancy or if the action of the hormone is blocked, menstruation occurs, and the embryo sloughs off along with the uterine lining. Within 12 hours after taking mifepristone, the endometrium starts to degenerate, and within 72 hours it begins to slough off. Misoprostol stimulates uterine contractions and is given after mifepristone to aid in expulsion of the endometrium.

Another type of induced abortion is called **vacuum aspiration** (suction) and can be performed up to the 16th week of pregnancy. A small, flexible tube attached to a vacuum source is inserted into the uterus through the vagina. The embryo or fetus, placenta, and lining of the uterus are then removed by suction. For pregnancies between 13 and 16 weeks, a technique called **dilation and evacuation** is commonly used. After the cervix is dilated, suction and forceps are used to remove the fetus, placenta, and uterine lining. From the 16th to 24th week, a **late-stage abortion** may be employed using surgical methods similar to dilation and evacuation or through nonsurgical methods using a saline solution or medications to induce abortion. Labor may be induced by using vaginal suppositories, intravenous infusion, or injections into the amniotic fluid through the uterus.

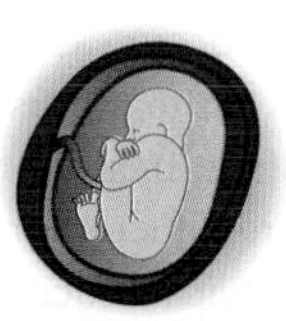

28.6 Development of the Reproductive Systems

OBJECTIVES

- **Explain** how genetic sex is determined.
- **Describe** the development of the male and female reproductive systems.

Recall from Chapter 3 that somatic cells are diploid ($2n$): They contain 23 pairs of homologous chromosomes, for a total of 46 chromosomes. Of these chromosomes, there are 22 pairs of autosomes and one pair of sex chromosomes. Autosomes code for the overall form of the human body and for specific traits such as eye color and height. The two sex chromosomes—a large **X chromosome** and a smaller **Y chromosome**—determine the genetic sex of an individual. In a genetic female, somatic cells contain two X chromosomes. In a genetic male, somatic cells contain one X and one Y chromosome. Determination of genetic sex by the sex chromosomes is known as **sex determination**.

In gametes (sperm or eggs), which are haploid (n), there are only 23 total chromosomes. Of these chromosomes, there are 22 autosomes and 1 sex chromosome. In sperm cells, the sex chromosome is either X or Y—approximately half of the sperm cells produced by meiosis contain an X and the other half a Y. In an egg, the sex chromosome is always an X. Genetic sex is established at the moment of conception by the type of sperm cell (X-bearing or Y-bearing) that fertilizes the egg. If an X-bearing sperm fertilizes the egg, the embryo formed will be a genetic female (XX). If a Y-bearing sperm fertilizes the egg, the embryo formed will be a genetic male (XY).

The early embryo is **bipotential**, which means that it has the ability to form either male or female reproductive organs. The first step in the development of the reproductive organs occurs in response to the genetic sex of the embryo. If the embryo is genetically male, testes develop; if the embryo is genetically female, ovaries develop. Once testes form in a male embryo, they begin to secrete androgens (masculinizing hormones), which cause a male reproductive tract and male external genitalia to develop. Female embryos, which contain ovaries instead of testes, do not produce testicular androgens. The lack of testicular androgens in a female embryo causes a female reproductive tract and female external genitalia to develop by default. Such a default pathway is ideal because both male and female embryos are exposed to high levels of estrogens and progesterone from the mother's placenta and ovaries during pregnancy. If female sex hormones played a role in sex differentiation, then all embryos (whether genetically male or female) would develop female reproductive organs. **Sex differentiation** is the process by which reproductive organs develop along male or female lines. To understand the steps involved in sex differentiation, you will first examine how the internal reproductive organs are formed and then you will discover how the external genitalia are developed.

The *gonads* develop from **gonadal ridges** that arise from growth of **intermediate mesoderm**. During the fifth week of development, the gonadal ridges appear as bulges just medial to the mesonephros (intermediate kidney) (**Figure 28.27**). Adjacent to the gonadal ridges

FIGURE 28.27 Development of the internal reproductive systems.

The gonads develop from intermediate mesoderm.

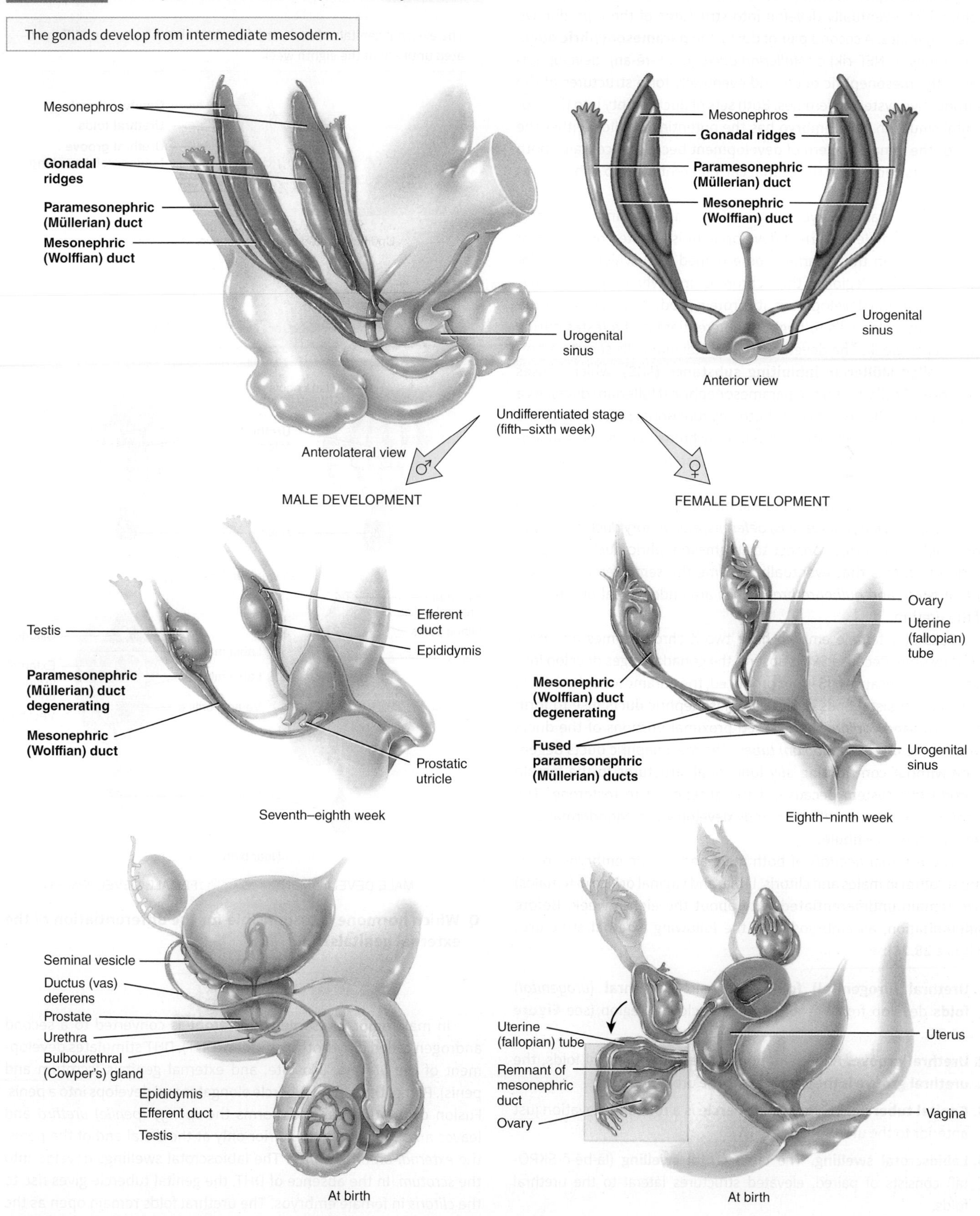

Q Which gene is responsible for the development of the gonads into testes?

are the **mesonephric ducts** (mez′-o-NEF-rik) or *Wolffian ducts* (WULF-ē-an), which eventually develop into structures of the reproductive system in males. A second pair of ducts, the **paramesonephric ducts** (par′-a-mes′-o-NEF-rik) or *Müllerian ducts* (mil-E-rē-an), develop lateral to the mesonephric ducts and eventually form structures of the reproductive system in females. Both sets of ducts empty into the urogenital sinus. An early embryo has the potential to follow either the male or the female pattern of development because it contains both sets of ducts and genital ridges that can differentiate into either testes or ovaries.

Cells of a male embryo have one X chromosome and one Y chromosome. The male pattern of development is initiated by a "master switch" gene on the Y chromosome named ***SRY***, which stands for *S*ex-determining *R*egion of the *Y* chromosome. When the *SRY* gene is expressed during development, its protein product causes the primitive sustentacular cells to begin to differentiate in the *testes* during the seventh week. The developing sustentacular cells secrete a hormone called **Müllerian-inhibiting substance (MIS)**, which causes apoptosis of cells within the paramesonephric (Müllerian) ducts. As a result, those cells do not contribute any functional structures to the male reproductive system. Stimulated by human chorionic gonadotropin (hCG), primitive interstitial cells in the testes begin to secrete the androgen **testosterone** during the eighth week. Testosterone then stimulates development of the mesonephric duct on each side into the *epididymis*, *ductus (vas) deferens*, *ejaculatory duct*, and *seminal vesicle*. The *testes* connect to the mesonephric duct through a series of tubules that eventually become the *seminiferous tubules.* The *prostate* and *bulbourethral glands* are endodermal outgrowths of the urethra.

Cells of a female embryo have two X chromosomes and no Y chromosome. Because *SRY* is absent, the gonadal ridges develop into *ovaries*, and because MIS is not produced, the paramesonephric ducts flourish. The distal ends of the paramesonephric ducts fuse to form the *uterus* and *vagina*; the unfused proximal portions of the ducts become the *uterine (fallopian) tubes.* The mesonephric ducts degenerate without contributing any functional structures to the female reproductive system because of the absence of testosterone. The *greater* and *lesser vestibular glands* develop from endodermal outgrowths of the vestibule.

The *external genitals* of both male and female embryos (penis and scrotum in males and clitoris, labia, and vaginal orifice in females) also remain undifferentiated until about the eighth week. Before differentiation, all embryos have the following external structures (**Figure 28.28**):

1. **Urethral (urogenital) folds.** The paired **urethral** *(urogenital)* **folds** develop from mesoderm in the cloacal region (see **Figure 26.23**).
2. **Urethral groove.** An indentation between the urethral folds, the **urethral groove** is the opening into the urogenital sinus.
3. **Genital tubercle.** The **genital tubercle** is a rounded elevation just anterior to the urethral folds.
4. **Labioscrotal swelling.** The **labioscrotal swelling** (lā-bē-ō-SKRŌ-tal) consists of paired, elevated structures lateral to the urethral folds.

FIGURE 28.28 Development of the external genitals.

The external genitals of male and female embryos remain undifferentiated until about the eighth week.

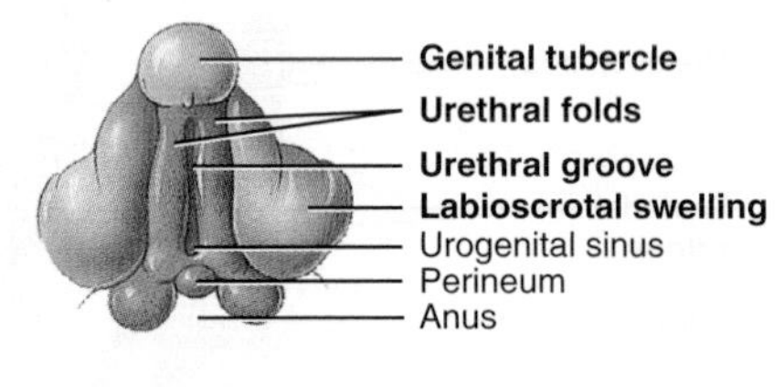

Undifferentiated stage (about five-week embryo)

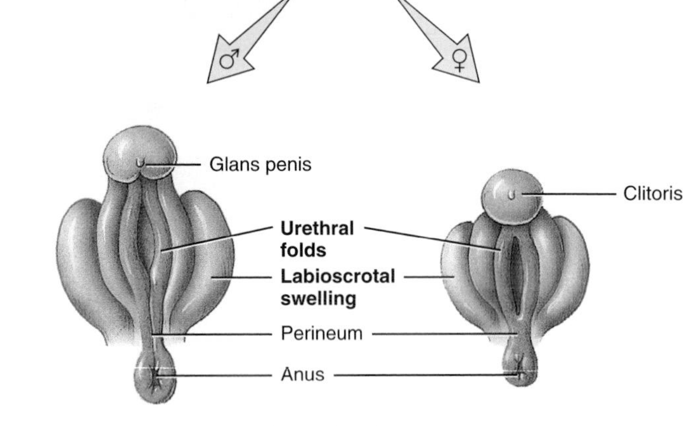

Ten-week embryo

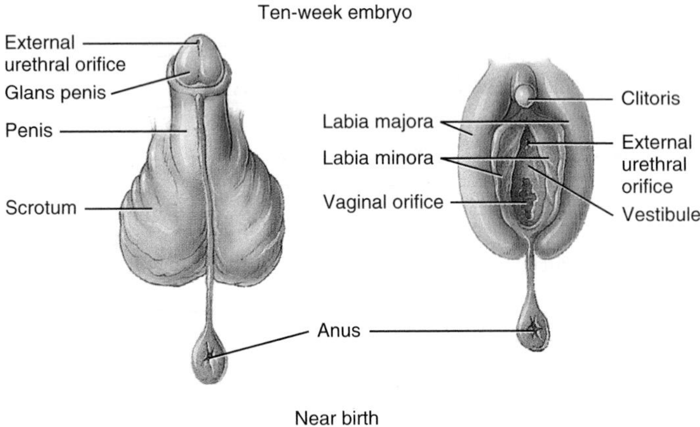

Near birth

MALE DEVELOPMENT FEMALE DEVELOPMENT

Q Which hormone is responsible for the differentiation of the external genitals?

In male embryos, some testosterone is converted to a second androgen called dihydrotestosterone (DHT). DHT stimulates development of the urethra, prostate, and external genitals (scrotum and penis). Part of the genital tubercle elongates and develops into a penis. Fusion of the urethral folds forms the *spongy (penile) urethra* and leaves an opening to the exterior only at the distal end of the penis, the *external urethral orifice.* The labioscrotal swellings develop into the *scrotum.* In the absence of DHT, the genital tubercle gives rise to the *clitoris* in female embryos. The urethral folds remain open as the *labia minora*, and the labioscrotal swellings become the *labia majora.* The urethral groove becomes the *vestibule.* After birth, androgen levels decline because hCG is no longer present to stimulate secretion of testosterone.

28.7 Aging and the Reproductive Systems

OBJECTIVE

- **Describe** the effects of aging on the reproductive systems.

During the first decade of life, the reproductive system is in a juvenile state. At about age 10, hormone-directed changes start to occur in both sexes. **Puberty** (PŪ-ber-tē = a ripe age) is the period when secondary sexual characteristics begin to develop and the potential for sexual reproduction is reached. The onset of puberty is marked by pulses or bursts of LH and FSH secretion, each triggered by a pulse of GnRH. Most pulses occur during sleep. As puberty advances, the hormone pulses occur during the day as well as at night. The pulses increase in frequency during a 3- to 4-year period until the adult pattern is established. The stimuli that cause the GnRH pulses are still unclear, but a role for the hormone leptin is starting to unfold. Just before puberty, leptin levels rise in proportion to adipose tissue mass. Interestingly, leptin receptors are present in both the hypothalamus and anterior pituitary. Mice that lack a functional leptin gene from birth are sterile and remain in a prepubertal state. Giving leptin to such mice elicits secretion of gonadotropins, and they become fertile. Leptin may signal the hypothalamus that long-term energy stores (triglycerides in adipose tissue) are adequate for reproductive functions to begin.

In females, the reproductive cycle normally occurs once each month from **menarche** (me-NAR-kē), the first menses, to **menopause**, the permanent cessation of menses. Thus, the female reproductive system has a time-limited span of fertility between menarche and menopause. For the first 1 to 2 years after menarche, ovulation only occurs in about 10% of the cycles and the luteal phase is short. Gradually, the percentage of ovulatory cycles increases, and the luteal phase reaches its normal duration of 14 days. With age, fertility declines. Between the ages of 40 and 50 the pool of remaining ovarian follicles becomes exhausted. As a result, the ovaries become less responsive to hormonal stimulation. The production of estrogens declines, despite copious secretion of FSH and LH by the anterior pituitary. Many women experience hot flashes and heavy sweating, which coincide with bursts of GnRH release. Other symptoms of menopause are headache, hair loss, muscular pains, vaginal dryness, insomnia, depression, weight gain, and mood swings. Some atrophy of the ovaries, uterine tubes, uterus, vagina, external genitalia, and breasts occurs in postmenopausal women. Due to loss of estrogens, most women experience a decline in bone mineral density after menopause. Sexual desire (libido) does not show a parallel decline; it may be maintained by adrenal sex steroids. The risk of having uterine cancer peaks at about 65 years of age, but cervical cancer is more common in younger women.

In males, declining reproductive function is much more subtle than in females. Healthy men often retain reproductive capacity into their eighties or nineties. At about age 55 a decline in testosterone synthesis leads to reduced muscle strength, fewer viable sperm, and decreased sexual desire. Although sperm production decreases 50–70% between ages 60 and 80, abundant sperm may still be present even in old age.

Enlargement of the prostate to two to four times its normal size occurs in most males over age 60. This condition, called **benign prostatic hyperplasia (BPH)** (hī-per-PLĀ-zē-a), decreases the size of the prostatic urethra and is characterized by frequent urination, nocturia (having to urinate at night), hesitancy in urination, decreased force of urinary stream, postvoiding dribbling, and a sensation of incomplete emptying.

• • •

To appreciate the many ways that the reproductive systems contribute to homeostasis of other body systems, examine *Focus on Homeostasis: Contributions of the Reproductive Systems.* Next, in Chapter 29, you will explore the major events that occur during pregnancy and you will discover how genetics (inheritance) plays a role in the development of a child.

FOCUS on HOMEOSTASIS

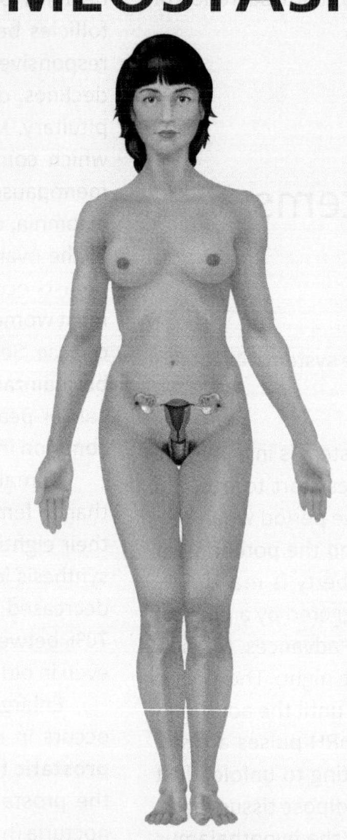

CONTRIBUTIONS OF THE REPRODUCTIVE SYSTEMS

FOR ALL BODY SYSTEMS

- The male and female reproductive systems produce gametes (oocytes and sperm) that unite to form embryos and fetuses, which contain cells that divide and differentiate to form all of the organ systems of the body

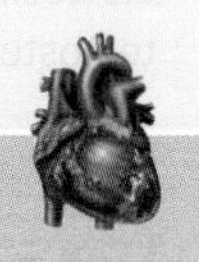

INTEGUMENTARY SYSTEMS

- Androgens promote the growth of body hair
- Estrogens stimulate the deposition of fat in the breasts, abdomen, and hips
- Mammary glands produce milk
- Skin stretches during pregnancy as the fetus enlarges

SKELETAL SYSTEM

- Androgens and estrogens stimulate the growth and maintenance of bones of the skeletal system

MUSCULAR SYSTEM

- Androgens stimulate the growth of skeletal muscles

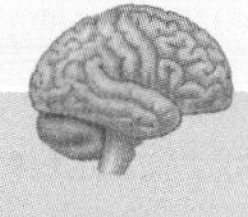

NERVOUS SYSTEM

- Androgens influence libido (sex drive)
- Estrogens may play a role in the development of certain regions of the brain in males

ENDOCRINE SYSTEM

- Testosterone and estrogens exert feedback effects on the hypothalamus and anterior pituitary gland

CARDIOVASCULAR SYSTEM

- Estrogens lower blood cholesterol level and may reduce the risk of coronary artery disease in women under age 50

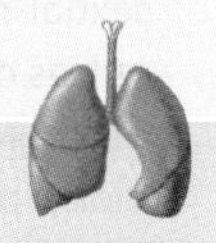

LYMPHATIC SYSTEMS and IMMUNITY

- The presence of an antibiotic-like chemical in semen and the acidic pH of vaginal fluid provide innate immunity against microbes in the reproductive tract

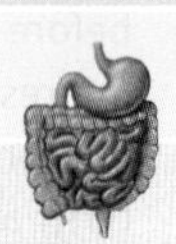

RESPIRATORY SYSTEM

- Sexual arousal increases the rate and depth of breathing

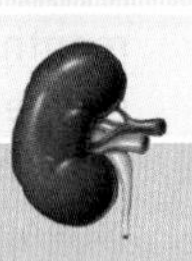

DIGESTIVE SYSTEM

- The presence of the fetus during pregnancy crowds the digestive organs, which leads to heartburn and constipation

URINARY SYSTEM

- In males, the portion of the urethra that extends through the prostate and penis is a passageway for urine as well as semen

Review Questions

28.1 Male Reproductive System

1. Describe the function of the scrotum in protecting the testes from temperature fluctuations.
2. Describe the internal structure of a testis. Where are sperm cells produced? What are the functions of sustentacular cells and interstitial (Leydig) cells?
3. Describe the principal events of spermatogenesis.
4. Which part of a sperm cell contains enzymes that help the sperm cell fertilize a secondary oocyte?
5. What are the roles of FSH, LH, testosterone, and inhibin in the male reproductive system? How is secretion of these hormones controlled?
6. Which ducts transport sperm within the testes?
7. Describe the location, structure, and functions of the ductus epididymis, ductus (vas) deferens, and ejaculatory duct.
8. Give the locations of the three subdivisions of the male urethra.
9. Trace the course of sperm through the system of ducts from the seminiferous tubules to the urethra.
10. List the structures within the spermatic cord.
11. Briefly explain the locations and functions of the seminal vesicles, the prostate, and the bulbourethral (Cowper's) glands.
12. What is semen? What is its function?
13. Explain the physiological processes involved in erection and ejaculation.

28.2 Female Reproductive System

14. How are the ovaries held in position in the pelvic cavity?
15. Describe the microscopic structure and functions of an ovary.
16. Describe the principal events of oogenesis.
17. Where are the uterine tubes located, and what is their function?
18. What are the principal parts of the uterus? Where are they located in relation to one another?
19. Describe the arrangement of ligaments that hold the uterus in its normal position.
20. Describe the histology of the uterus.
21. Why is an abundant blood supply important to the uterus?
22. How does the histology of the vagina contribute to its function?
23. What are the structures and functions of each part of the vulva?
24. Describe the components of the mammary glands and the structures that support them.
25. Outline the route milk takes from the alveoli of the mammary gland to the nipple.

28.3 The Female Reproductive Cycle

26. Describe the function of each of the following hormones in the uterine and ovarian cycles: GnRH, FSH, LH, estrogens, progesterone, and inhibin.
27. Briefly outline the major events of each phase of the uterine cycle, and correlate them with the events of the ovarian cycle.
28. Prepare a labeled diagram of the major hormonal changes that occur during the uterine and ovarian cycles.

28.4 The Human Sexual Response

29. What happens during each of the four phases of the human sexual response?

28.5 Birth Control Methods and Abortion

30. How do oral contraceptives reduce the likelihood of pregnancy?
31. How do some methods of birth control protect against sexually transmitted diseases?
32. What is the problem with developing an oral contraceptive pill for males?

28.6 Development of the Reproductive Systems

33. How does the type of sperm cell (X-bearing or Y-bearing) determine the genetic sex of the embryo?
34. Describe the role of hormones in the differentiation of the Wolffian ducts, the Müllerian ducts, and the external genitalia.

28.7 Aging and the Reproductive Systems

35. What changes occur in males and females at puberty?
36. What do the terms menarche and menopause mean?

Critical Thinking Questions

1. Twenty-three-year-old Monica and her husband Bill are ready to start a family. They are both avid bicyclists and weight-lifters who carefully watch what they eat and pride themselves on their "buff" bodies. However, Monica is having difficulty becoming pregnant. Monica hasn't had a menstrual period for some time but informs the doctor that is normal for her. After consulting with her physician, the doctor tells Monica that she needs to cut back on her exercise routine and "put on some weight" in order to get pregnant. Monica is outraged because she figures she will gain enough weight when she is pregnant! Explain to Monica what has happened to her and why weight gain could help her achieve her goal of pregnancy.
2. The term "progesterone" means "for gestation (or pregnancy)." Describe how progesterone helps prepare the female body for pregnancy and helps maintain pregnancy.
3. After having borne five children, Mark's wife, Isabella, insists that he have a vasectomy. Mark is afraid that he will "dry up" and won't be able to perform sexually. How can you reassure him that his reproductive organs will function fine?

Development and Inheritance

CHAPTER **29**

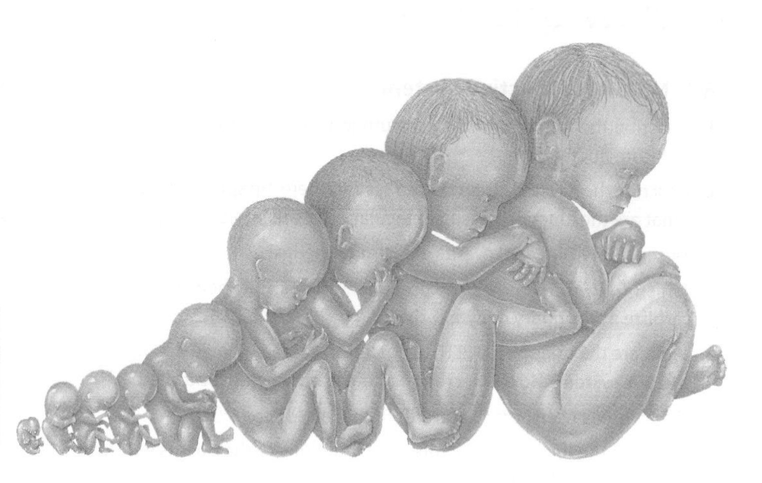

Development, Inheritance, and Homeostasis

Both the genetic material inherited from parents (heredity) and normal development in the uterus (environment) play important roles in determining the homeostasis of a developing embryo and fetus and the subsequent birth of a healthy child.

In this chapter we will study the sequence of events from the fertilization of a secondary oocyte by a sperm cell to the formation of an adult organism. In particular, we focus on the developmental sequence from fertilization through implantation, embryonic and fetal development, labor, and birth. We will also examine the principles of inheritance (the passage of hereditary traits from one generation to another).

Q Did you ever wonder why the heart, blood vessels, and blood begin to form so early in the developmental process?

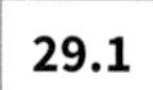

29.1 Overview of Development

OBJECTIVES

- **Describe** the sequence of events involved in development.
- **Describe** the trimesters of prenatal development.

As you learned in Chapter 28, sexual reproduction is the process by which organisms produce offspring by making sex cells called **gametes** (GAM-ēts = spouses). Male gametes are called **sperm** (spermatozoa) and female gametes are called **secondary oocytes**. The organs that produce gametes are called **gonads**; these are the testes in the male and the ovaries in the female. Once sperm have been deposited in the female reproductive tract and a secondary oocyte has been released from the ovary, fertilization can occur. This process initiates a cascade of developmental events that, when completed properly, produces a healthy newborn baby.

Pregnancy is a sequence of events that begins with fertilization, proceeds to implantation, embryonic development, and fetal development, and ideally ends with birth about 38 weeks later, or 40 weeks after the mother's last menstrual period.

Development biology is the study of the sequence of events from the fertilization of a secondary oocyte by a sperm cell to the formation of an adult organism. From fertilization through the eighth week of development, the **embryonic period**, the developing human is called an **embryo** (*em-* = into; *-bryo* = grow). **Embryology** (em-brē-OL-ō-jē) is the study of development from the fertilized egg through the eighth week. The **fetal period** begins at week nine and continues until birth. During this time, the developing human is called a **fetus** (FĒ-tus = offspring).

Prenatal development (prē-NĀ-tal; *pre-* = before; *natal* = birth) is the time from fertilization to birth and includes both the embryonic and fetal periods. Prenatal development is divided into periods of three calendar months each, called **trimesters**.

1. During the **first trimester**, the most critical stage of development, all of the major organ-systems begin to form. Because of the extensive, widespread activity, it is also the period when the developing organism is most vulnerable to the effects of drugs, radiation, and microbes.
2. The **second trimester** is characterized by the nearly complete development of organ systems. By the end of this stage, the fetus assumes distinctively human features.
3. The **third trimester** represents a period of rapid fetal growth in which the weight of the fetus doubles. During the early stages of this period, most of the organ systems become fully functional.

29.2 The First Two Weeks of the Embryonic Period

OBJECTIVE

- **Explain** the major events that occur during the first and second weeks of development.

First Week of Development

The **embryonic period** extends from fertilization through the eighth week. The first week of development is characterized by several significant events including fertilization, cleavage of the zygote, blastocyst formation, and implantation.

Fertilization During **fertilization** (fer′-ti-li-ZĀ-shun; *fertil-* = fruitful), the genetic material from a haploid sperm cell (spermatozoon) and a haploid secondary oocyte merges into a single diploid nucleus. Of the 200 million sperm introduced into the vagina, fewer than 2 million (1%) reach the cervix of the uterus and only about 200 reach the secondary oocyte. Fertilization normally occurs in the uterine (fallopian) tube within 12 to 24 hours after ovulation. Sperm can remain viable for about 48 hours after deposition in the vagina, although a secondary oocyte is viable for only about 24 hours after ovulation. Thus, pregnancy is *most likely* to occur if intercourse takes place during a 3-day window—from 2 days before ovulation to 1 day after ovulation.

Sperm swim from the vagina into the cervical canal by the whiplike movements of their tails (flagella). The passage of sperm through the rest of the uterus and then into the uterine tube results mainly from contractions of the walls of these organs. Prostaglandins in semen are believed to stimulate uterine motility at the time of intercourse and to aid in the movement of sperm through the uterus and into the uterine tube. Sperm that reach the vicinity of the oocyte within minutes after ejaculation *are not capable* of fertilizing it until about 7 hours later. During this time in the female reproductive tract, mostly in the uterine tube, sperm undergo **capacitation** (ka-pas-i-TĀ-shun; *capacit-* = capable of), a series of functional changes that cause the sperm's tail to beat even more vigorously and prepare its plasma membrane to fuse with the oocyte's plasma membrane. During capacitation, sperm are acted on by secretions in the female reproductive tract that result in the removal of cholesterol, glycoproteins, and proteins from the plasma membrane around the head of the sperm cell. Only capacitated sperm are capable of being attracted by and responding to chemical factors produced by the surrounding cells of the ovulated oocyte.

For fertilization to occur, a sperm cell first must penetrate two layers: the **corona radiata** (kō-RŌ-na = crown; rā-dē-A-ta = to shine), the granulosa cells that surround the secondary oocyte, and the **zona pellucida** (ZŌ-na = zone; pe-LOO-si-da = allowing passage of light), the clear glycoprotein layer between the corona radiata and the oocyte's plasma membrane (**Figure 29.1a**). The **acrosome** (AK-rō-sōm), a helmetlike structure that covers the head of a sperm (see **Figure 28.6**), contains several enzymes. Acrosomal enzymes and strong tail movements by the sperm help it penetrate the cells of the corona radiata and come in contact with the zona pellucida. One of the glycoproteins in the zona pellucida, called ZP3, acts as a sperm receptor. Its binding to specific membrane proteins in the sperm head triggers the **acrosomal reaction**, the release of the contents of the acrosome. The acrosomal enzymes digest a path through the zona pellucida as the lashing sperm tail pushes the sperm cell onward. Although many sperm bind to ZP3 molecules and undergo acrosomal reactions, only the first sperm cell to penetrate the entire zona pellucida and reach the oocyte's plasma membrane fuses with the oocyte.

The fusion of a sperm cell with a secondary oocyte sets in motion events that block **polyspermy** (POL-ē-sper′-mē), fertilization by more than one sperm cell. Within a few seconds, the cell membrane of the oocyte depolarizes, which acts as a *fast block to polyspermy*—the inability of a depolarized oocyte to fuse with another sperm. Depolarization also triggers the intracellular release of calcium ions, which stimulate exocytosis of secretory vesicles from the oocyte. The molecules released by exocytosis inactivate ZP3 and harden the entire zona pellucida, events called the *slow block to polyspermy.*

Once a sperm cell enters a secondary oocyte, the oocyte first must complete meiosis II. It divides into a larger ovum (mature egg) and a smaller second polar body that fragments and disintegrates (see **Figure 28.15**). The nucleus in the head of the sperm develops into the **male pronucleus**, and the nucleus of the fertilized ovum

FIGURE 29.1 Selected structures and events in fertilization.

During fertilization, genetic material from a sperm cell and a secondary oocyte merge to form a single diploid nucleus.

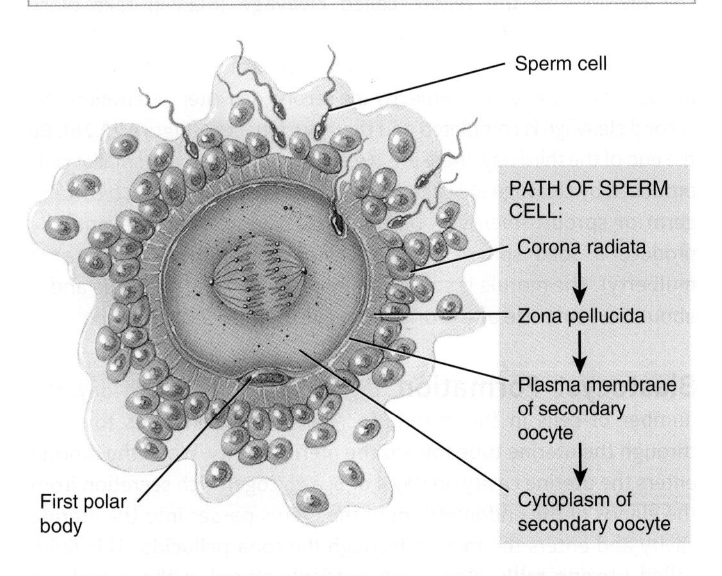

(a) Sperm cell penetrating a secondary oocyte

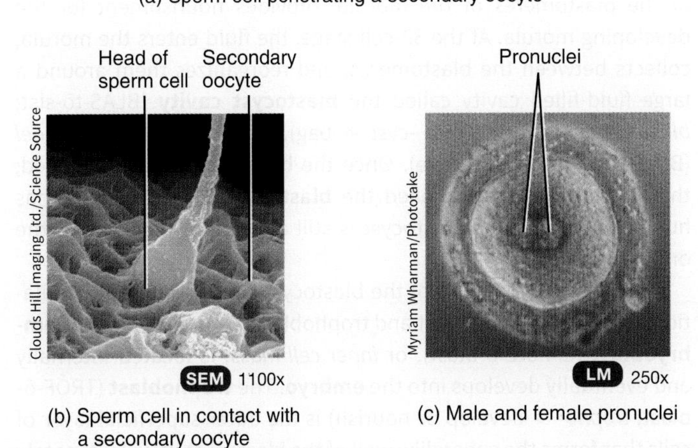

(b) Sperm cell in contact with a secondary oocyte

(c) Male and female pronuclei

Q What is capacitation?

develops into the **female pronucleus** (Figure 29.1c). After the male and female pronuclei form, they fuse, producing a single diploid nucleus, a process known as **syngamy** (SIN-ga-mē). Thus, the fusion of the haploid (n) pronuclei restores the diploid number ($2n$) of 46 chromosomes. The fertilized ovum now is called a **zygote** (*zygon* = yolk).

Dizygotic (*fraternal*) **twins** are produced from the independent release of two secondary oocytes and the subsequent fertilization of each by different sperm. They are the same age and in the uterus at the same time, but genetically they are as dissimilar as any other siblings. Dizygotic twins may or may not be the same sex. Because **monozygotic** (*identical*) **twins** develop from a single fertilized ovum, they contain exactly the same genetic material and are always the same sex. Monozygotic twins arise from separation of the developing cells into two embryos, which in 99% of the cases occurs before 8 days have passed. Separations that occur later than 8 days are likely to produce **conjoined twins**, a situation in which the twins are joined together and share some body structures.

Cleavage of the Zygote

After fertilization, rapid mitotic cell divisions of the zygote called **cleavage** (KLĒV-ij) take place (Figure 29.2). The first division of the zygote begins about 24 hours after fertilization and is completed about 6 hours later. Each succeeding division takes slightly less time. By the second day after fertilization, the second cleavage is completed and there are four cells (Figure 29.2b). By the end of the third day, there are 16 cells. The progressively smaller cells produced by cleavage are called **blastomeres** (BLAS-tō-mērz; *blasto-* = germ or sprout; *-meres* = parts). Successive cleavages eventually produce a solid sphere of cells called the **morula** (MOR-ū-la = mulberry). The morula is still surrounded by the zona pellucida and is about the same size as the original zygote (Figure 29.2c).

Blastocyst Formation

By the end of the fourth day, the number of cells in the morula increases as it continues to move through the uterine tube toward the uterine cavity. When the morula enters the uterine cavity on day 4 or 5, a glycogen-rich secretion from the glands of the endometrium of the uterus passes into the uterine cavity and enters the morula through the zona pellucida. This fluid, called **uterine milk**, along with nutrients stored in the cytoplasm of the blastomeres of the morula, provides nourishment for the developing morula. At the 32-cell stage, the fluid enters the morula, collects between the blastomeres, and reorganizes them around a large fluid-filled cavity called the **blastocyst cavity** (BLAS-tō-sist; *blasto-* = germ or sprout; *-cyst* = bag), also called the *blastocoel* (BLAS-tō-sēl) (Figure 29.2e). Once the blastocyst cavity is formed, the developing mass is called the **blastocyst**. Though it now has hundreds of cells, the blastocyst is still about the same size as the original zygote.

During the formation of the blastocyst two distinct cell populations arise: the embryoblast and trophoblast (Figure 29.2e). The **embryoblast** (EM-brē-ō-blast), or *inner cell mass*, is located internally and eventually develops into the **embryo**. The **trophoblast** (TRŌF-ō-blast; *tropho-* = develop or nourish) is the outer superficial layer of cells that forms the spherelike wall of the blastocyst. It will ultimately develop into the outer chorionic sac that surrounds the fetus and the

FIGURE 29.2 Cleavage and the formation of the morula and blastocyst.

Cleavage refers to the early, rapid mitotic divisions of a zygote.

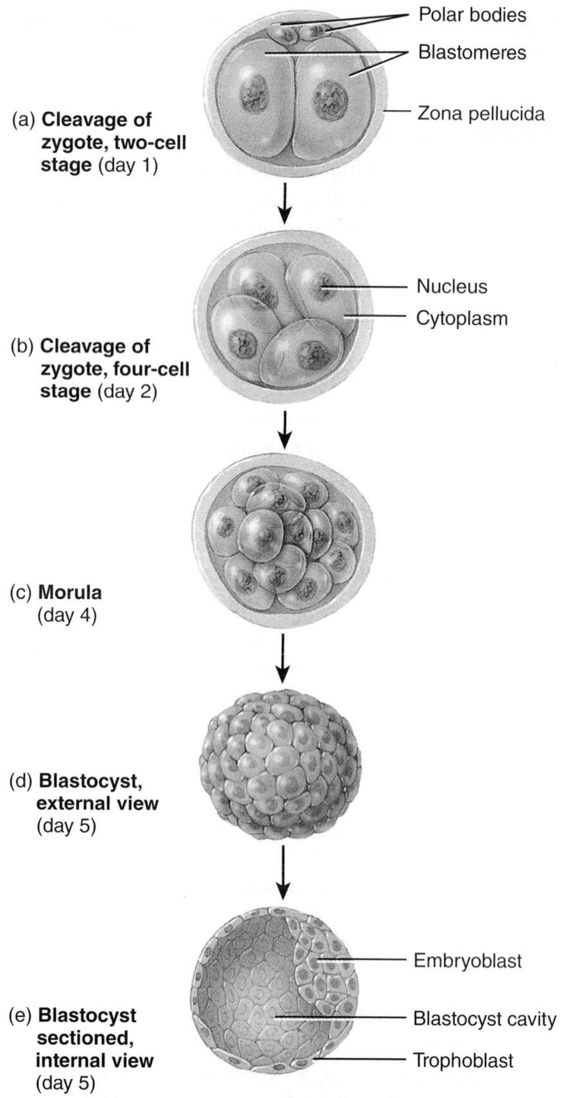

(a) **Cleavage of zygote, two-cell stage** (day 1)

Polar bodies
Blastomeres
Zona pellucida

(b) **Cleavage of zygote, four-cell stage** (day 2)

Nucleus
Cytoplasm

(c) **Morula** (day 4)

(d) **Blastocyst, external view** (day 5)

(e) **Blastocyst sectioned, internal view** (day 5)

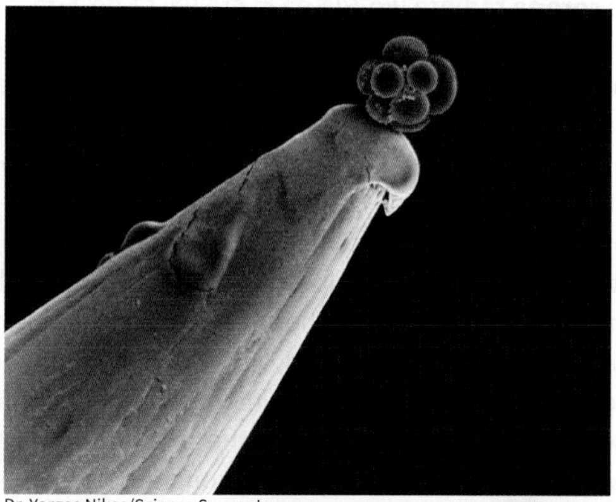

Embryoblast
Blastocyst cavity
Trophoblast

Dr. Yorgos Nikas/Science Source Images

SEM 130x

16-cell human embryo on the top of a pin

Q What is the histological difference between a morula and a blastocyst?

fetal portion of the placenta, the site of exchange of nutrients and wastes between the mother and fetus. Around the fifth day after fertilization, the blastocyst "hatches" from the zona pellucida by digesting a hole in it with an enzyme, and then squeezing through the hole. This shedding of the zona pellucida is necessary in order to permit the next step, implantation (attachment) into the vascular, glandular endometrial lining of the uterus.

Implantation The blastocyst remains free within the uterine cavity for about 2 days before it attaches to the uterine wall. At this time the endometrium is in its secretory phase. About 6 days after fertilization, the blastocyst loosely attaches to the endometrium in a process called **implantation** (im-plan-TĀ-shun) (**Figure 29.3**). As the blastocyst implants, usually in either the posterior portion of the fundus or the body of the uterus, it orients with the inner cell mass toward the endometrium (**Figure 29.3b**). About 7 days after fertilization, the blastocyst attaches to the endometrium more firmly, endometrial glands in the vicinity enlarge, and the endometrium becomes more vascularized (forms new blood vessels). The blastocyst eventually secretes enzymes and burrows into the endometrium, and becomes surrounded by it.

See Clinical Connection: Stem Cell Research and Therapeutic Cloning

FIGURE 29.3 Relationship of a blastocyst to the endometrium of the uterus at the time of implantation.

Implantation, the attachment of a blastocyst to the endometrium, occurs about 6 days after fertilization.

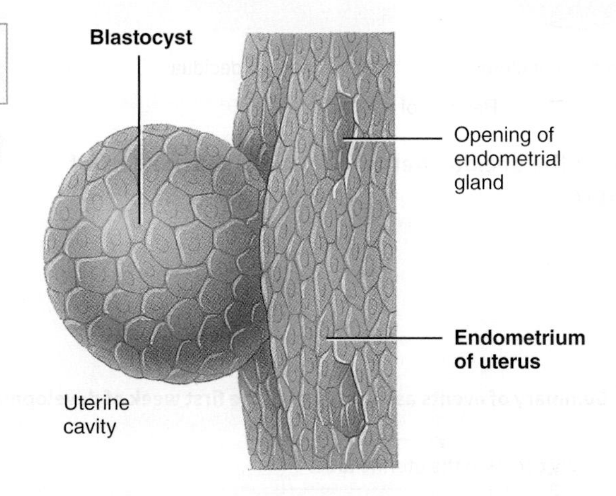

(a) External view of blastocyst, about 6 days after fertilization

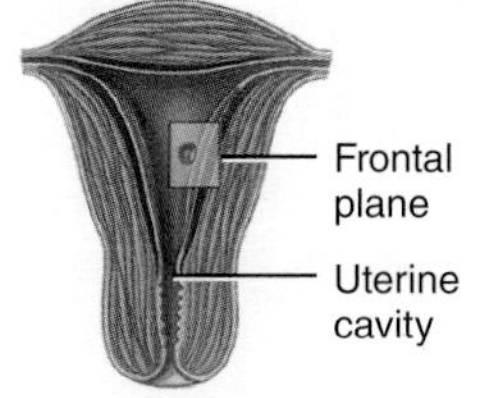

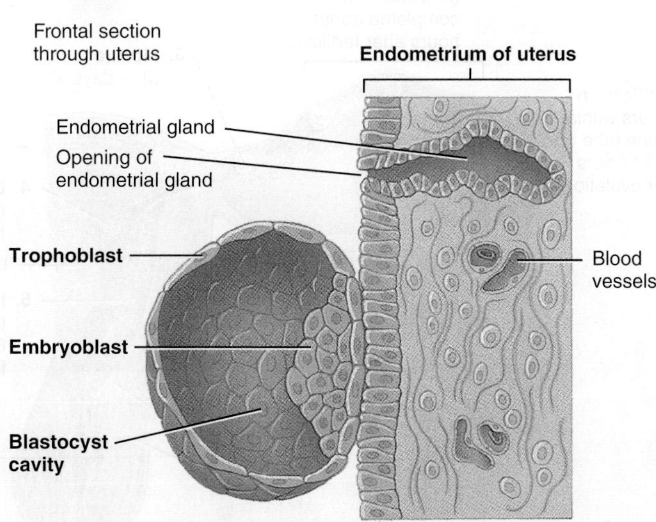

(b) Frontal section through endometrium of uterus and blastocyst, about 6 days after fertilization

Q How does the blastocyst merge with and burrow into the endometrium?

FIGURE 29.4 **Regions of the decidua.**

The decidua is a modified portion of the endometrium that develops after implantation.

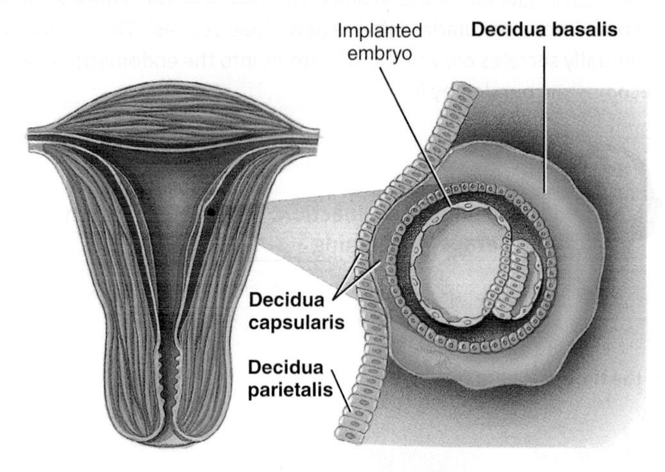

Regions of the decidua

Q Which part of the decidua helps form the maternal part of the placenta?

Following implantation, the endometrium is known as the **decidua** (dē-SID-ū-a = falling off). The decidua separates from the endometrium after the fetus is delivered, much as it does in normal menstruation. Different regions of the decidua are named based on their positions relative to the site of the implanted blastocyst (**Figure 29.4**). The **decidua basalis** is the portion of the endometrium between the embryo and the stratum basale of the uterus; it provides large amounts of glycogen and lipids for the developing embryo and fetus and later becomes the maternal part of the placenta. The **decidua capsularis** is the portion of the endometrium located between the embryo and the uterine cavity. The **decidua parietalis** (par-ri-e-TAL-is) is the remaining modified endometrium that lines the noninvolved areas of the rest of the uterus. As the embryo and later the fetus enlarges, the decidua capsularis bulges into the uterine cavity and fuses with the decidua parietalis, thereby obliterating the uterine cavity. By about 27 weeks, the decidua capsularis degenerates and disappears.

The major events associated with the first week of development are summarized in **Figure 29.5**.

See Clinical Connection: Ectopic Pregnancy

FIGURE 29.5 **Summary of events associated with the first week of development.**

Fertilization usually occurs in the uterine tube.

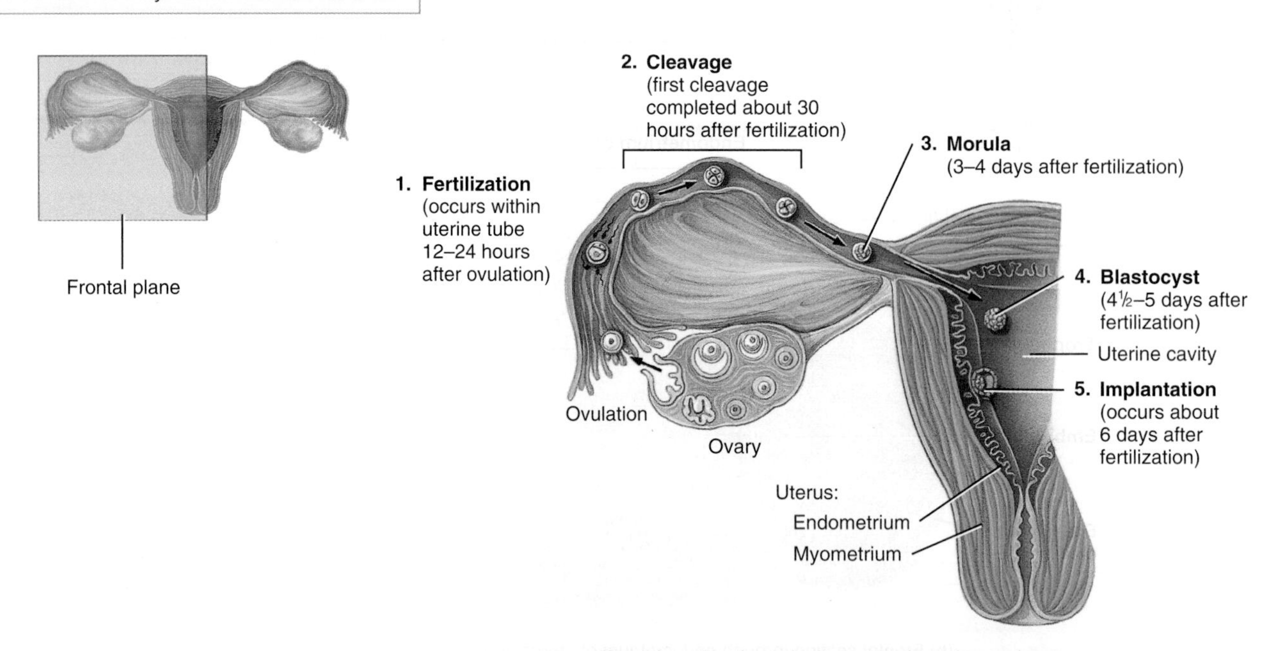

Frontal section through uterus, uterine tube, and ovary

Q In which phase of the uterine cycle does implantation occur?

Second Week of Development

Development of the Trophoblast About 8 days after fertilization, the trophoblast develops into two layers in the region of contact between the blastocyst and endometrium. These are a **syncytiotrophoblast** (sin-sīt′-ē-ō-TRŌF-ō-blast) that contains no distinct cell boundaries and a **cytotrophoblast** (sī-tō-TRŌF-ō-blast) between the embryoblast and syncytiotrophoblast that is composed of distinct cells (**Figure 29.6a**). The two layers of trophoblast become part of the chorion (one of the fetal membranes) as they undergo further growth (see **Figure 29.11a** inset). During implantation, the syncytiotrophoblast secretes enzymes that enable the blastocyst to penetrate the uterine lining by digesting and liquefying the endometrial cells. Eventually, the blastocyst becomes buried in the endometrium and inner one-third of the myometrium. Another secretion of the trophoblast is human chorionic gonadotropin (hCG), which has actions similar to LH. Human chorionic gonadotropin rescues the corpus luteum from degeneration and sustains its secretion of progesterone and estrogens. These hormones maintain the uterine lining in a secretory state, preventing menstruation. Peak secretion of hCG occurs about the ninth week of pregnancy, at which time the placenta is fully developed and produces the progesterone and estrogens that continue to sustain the pregnancy. The presence of hCG in maternal blood or urine is an indicator of pregnancy and is detected by home pregnancy tests.

Development of the Bilaminar Embryonic Disc Like those of the trophoblast, cells of the embryoblast also differentiate into two layers around 8 days after fertilization: a **hypoblast** (*primitive endoderm*) and **epiblast** (*primitive ectoderm*) (**Figure 29.6a**). Cells of the hypoblast and epiblast together form a flat disc referred to as the **bilaminar embryonic disc** (bī-LAM-in-ar = two-layered). Soon, a small cavity appears within the epiblast and eventually enlarges to form the **amniotic cavity** (am-nē-OT-ik; *amnio-* = lamb).

Development of the Amnion As the amniotic cavity enlarges, a single layer of squamous cells forms a domelike roof above the epiblast cells called the **amnion** (AM-nē-on) (**Figure 29.6a**). Thus, the amnion forms the roof of the amniotic cavity, and the epiblast forms the floor. Initially, the amnion overlies only the bilaminar embryonic disc. However, as the embryonic disc increases in size and begins to fold, the amnion eventually surrounds the entire embryo (see **Figure 29.11a** inset), creating the amniotic cavity that becomes filled with **amniotic fluid** (am′-nē-OT-ik). Most amniotic fluid is initially derived from maternal blood. Later, the fetus contributes to the fluid by excreting urine into the amniotic cavity. Amniotic fluid serves as a shock absorber for the fetus, helps regulate fetal body temperature, helps prevent the fetus from drying out, and prevents adhesions between the skin of the fetus and surrounding tissues. The amnion usually ruptures just before birth; it and its fluid constitute the "bag of waters." Embryonic cells are normally sloughed off into amniotic fluid. They can be examined in a procedure called *amniocentesis,* which involves withdrawing some of the amniotic fluid that bathes the developing fetus and analyzing the fetal cells and dissolved substances (see Section 29.6).

FIGURE 29.6 **Principal events of the second week of development.**

About 8 days after fertilization, the trophoblast develops into a syncytiotrophoblast and a cytotrophoblast; the embryoblast develops into a hypoblast and epiblast (bilaminar embryonic disc).

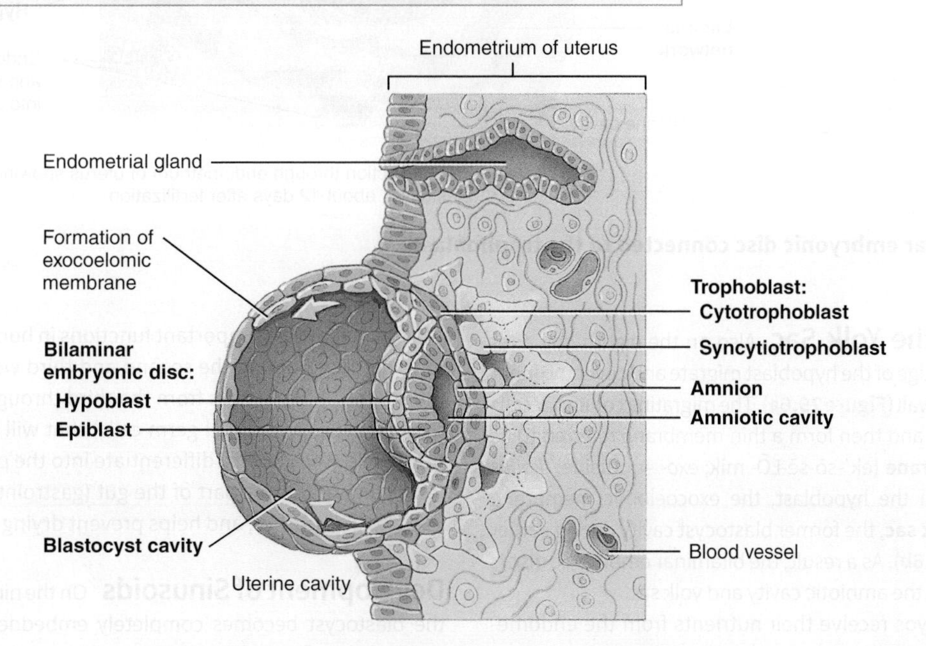

(a) Frontal section through endometrium of uterus showing blastocyst, about 8 days after fertilization

Figure 29.6 ***Continues***

FIGURE 29.6 Continued

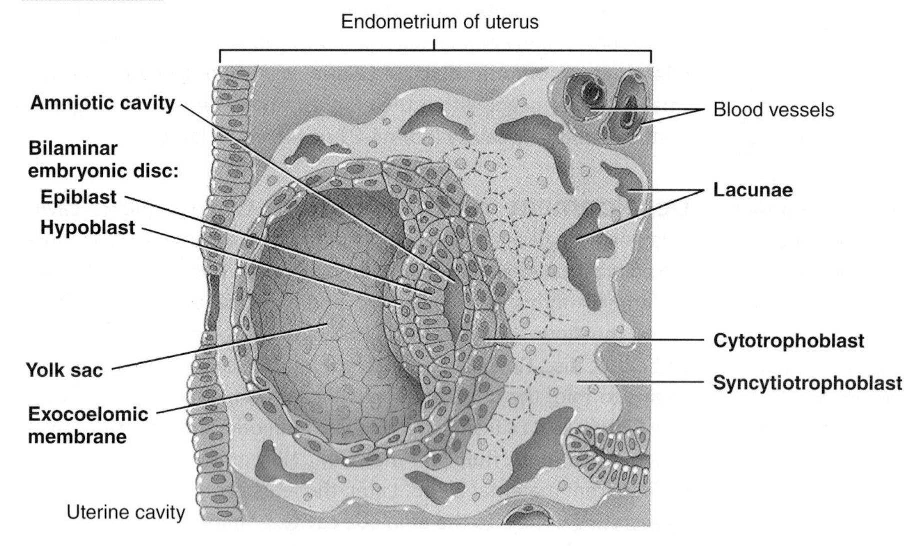

(b) Frontal section through endometrium of uterus showing blastocyst, about 9 days after fertilization

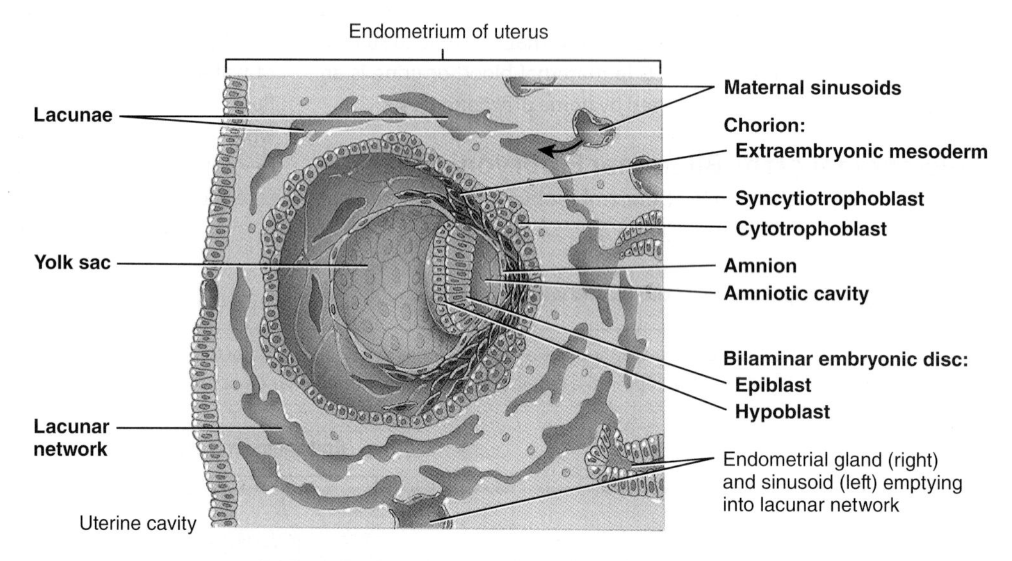

(c) Frontal section through endometrium of uterus showing blastocyst, about 12 days after fertilization

Q How is the bilaminar embryonic disc connected to the trophoblast?

Development of the Yolk Sac Also on the eighth day after fertilization, cells at the edge of the hypoblast migrate and cover the inner surface of the blastocyst wall (**Figure 29.6a**). The migrating columnar cells become squamous (flat) and then form a thin membrane referred to as the **exocoelomic membrane** (ek′-sō-sē-LŌ- mik; *exo-* = outside; *-koilos* = space). Together with the hypoblast, the exocoelomic membrane forms the wall of the **yolk sac**, the former blastocyst cavity during earlier development (**Figure 29.6b**). As a result, the bilaminar embryonic disc is now positioned between the amniotic cavity and yolk sac.

Since human embryos receive their nutrients from the endometrium, the yolk sac is relatively empty and small, and decreases in size as development progresses (see **Figure 29.11a**). Nevertheless, the yolk sac has several important functions in humans: supplies nutrients to the embryo during the second and third weeks of development; is the source of blood cells from the third through sixth weeks; contains the first cells (primordial germ cells) that will eventually migrate into the developing gonads, differentiate into the primitive germ cells, and form gametes; forms part of the gut (gastrointestinal tract); functions as a shock absorber; and helps prevent drying out of the embryo.

Development of Sinusoids On the ninth day after fertilization, the blastocyst becomes completely embedded in the endometrium. As the syncytiotrophoblast expands, small spaces called **lacunae** (la-KOO-nē = little lakes) develop within it (**Figure 29.6b**).

By the twelfth day of development, the lacunae fuse to form larger, interconnecting spaces called **lacunar networks** (Figure 29.6c). Endometrial capillaries around the developing embryo become dilated and are referred to as **maternal sinusoids** (SĪ-nū-soyds). As the syncytiotrophoblast erodes some of the maternal sinusoids and endometrial glands, maternal blood and secretions from the glands enter the lacunar networks and flow through them. Maternal blood is both a rich source of materials for embryonic nutrition and a disposal site for the embryo's wastes.

Development of the Extraembryonic Coelom About the twelfth day after fertilization, the **extraembryonic mesoderm** develops. These mesodermal cells are derived from the yolk sac and form a connective tissue layer (mesenchyme) around the amnion and yolk sac (Figure 29.6c). Soon a number of large cavities develop in the extraembryonic mesoderm, which then fuse to form a single, larger cavity called the **extraembryonic coelom** (SĒ-lom).

Development of the Chorion The extraembryonic mesoderm, together with the two layers of the trophoblast (the cytotrophoblast and syncytiotrophoblast), forms the **chorion** (KŌ-rē-on = membrane) (Figure 29.6c). The chorion surrounds the embryo and, later, the fetus (see Figure 29.11a). Eventually it becomes the principal embryonic part of the placenta, the structure for exchange of materials between mother and fetus. The chorion also protects the embryo and fetus from the immune responses of the mother in two ways: (1) It secretes proteins that block antibody production by the mother. (2) It promotes the production of T lymphocytes that suppress the normal immune response in the uterus. Finally, the chorion produces human chorionic gonadotropin (hCG), an important hormone of pregnancy (see Figure 29.16).

The inner layer of the chorion eventually fuses with the amnion. With the development of the chorion, the extraembryonic coelom is now referred to as the **chorionic cavity**. By the end of the second week of development, the bilaminar embryonic disc becomes connected to the trophoblast by a band of extraembryonic mesoderm called the **connecting** (*body*) **stalk** (see Figure 29.7). The connecting stalk is the future umbilical cord.

29.3 The Remaining Weeks of the Embryonic Period

OBJECTIVE

- **Describe** the major events that occur during the third through the eighth weeks of development.

Third Week of Development

The third embryonic week begins a 6-week period of very rapid development and differentiation. During the third week, the three primary germ layers are established and lay the groundwork for organ development in weeks 4 through 8.

Gastrulation The first major event of the third week of development, **gastrulation** (gas-troo-LĀ-shun), occurs about 15 days after fertilization. In this process, the bilaminar (two-layered) embryonic disc, consisting of epiblast and hypoblast, transforms into a **trilaminar** (three-layered) **embryonic disc** consisting of three layers: the ectoderm, mesoderm, and endoderm. These **primary germ layers** are the major embryonic tissues from which the various tissues and organs of the body develop.

Gastrulation involves the rearrangement and migration of cells from the epiblast. The first evidence of gastrulation is the formation of the **primitive streak**, a faint groove on the dorsal surface of the epiblast that elongates from the posterior to the anterior part of the embryo (Figure 29.7a). The primitive streak clearly establishes the head

FIGURE 29.7 Gastrulation.

Gastrulation involves the rearrangement and migration of cells from the epiblast.

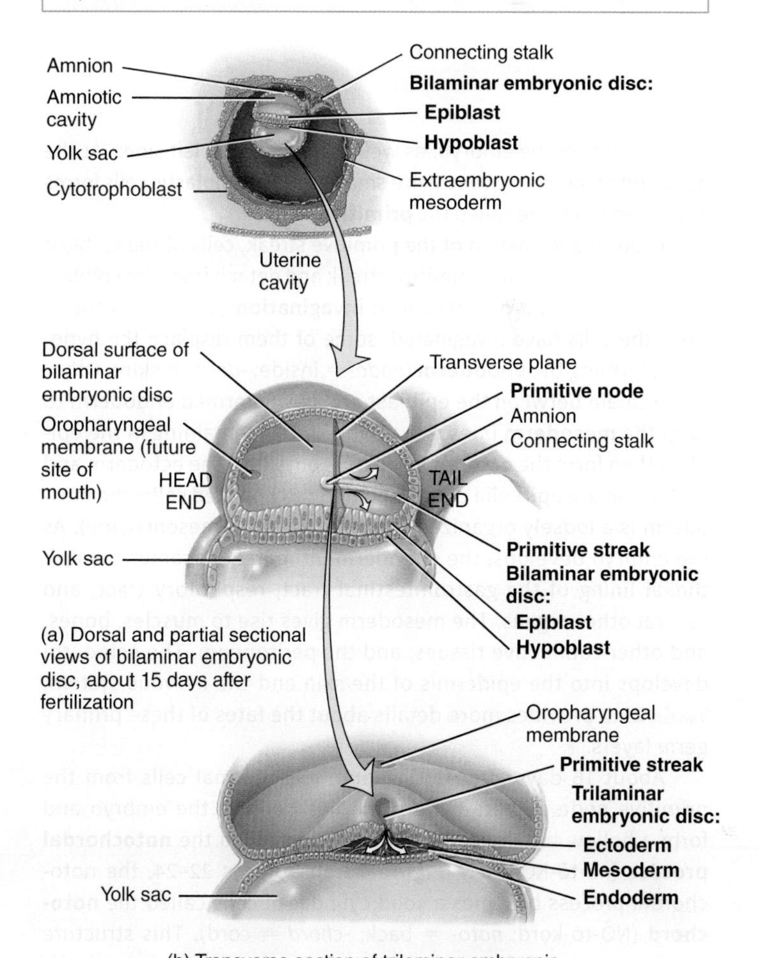

(a) Dorsal and partial sectional views of bilaminar embryonic disc, about 15 days after fertilization

(b) Transverse section of trilaminar embryonic disc, about 16 days after fertilization

Q What is the significance of gastrulation?

TABLE 29.1 Structures Produced by the Three Primary Germ Layers

ENDODERM	MESODERM	ECTODERM
Epithelial lining of gastrointestinal tract (except oral cavity and anal canal) and epithelium of its glands. Epithelial lining of urinary bladder, gallbladder, and liver. Epithelial lining of pharynx, auditory (eustachian) tubes, tonsils, tympanic (middle ear) cavity, larynx, trachea, bronchi, and lungs. Epithelium of thyroid gland, parathyroid glands, pancreas, and thymus. Epithelial lining of prostate and bulbourethral (Cowper's) glands, vagina, vestibule, urethra, and associated glands such as greater (Bartholin's) vestibular glands and lesser vestibular glands. Gametes (sperm and oocytes).	All skeletal and cardiac muscle tissue and most smooth muscle tissue. Cartilage, bone, and other connective tissues. Blood, red bone marrow, and lymphatic tissue. Blood vessels and lymphatic vessels. Dermis of skin. Fibrous tunic and vascular tunic of eye. Mesothelium of thoracic, abdominal, and pelvic cavities. Kidneys and ureters. Adrenal cortex. Gonads and genital ducts (except germ cells). Dura mater.	All nervous tissue. Epidermis of skin. Hair follicles, arrector pili muscles, nails, epithelium of skin glands (sebaceous and sudoriferous), and mammary glands. Lens, cornea, and internal eye muscles. Internal and external ear. Neuroepithelium of sense organs. Epithelium of oral cavity, nasal cavity, paranasal sinuses, salivary glands, and anal canal. Epithelium of pineal gland, pituitary gland, and adrenal medullae. Melanocytes (pigment cells). Almost all skeletal and connective tissue components of head. Arachnoid mater and pia mater.

and tail ends of the embryo, as well as its right and left sides. At the head end of the primitive streak a small group of epiblastic cells forms a rounded structure called the **primitive node**.

Following formation of the primitive streak, cells of the epiblast move inward below the primitive streak and detach from the epiblast (Figure 29.7b) in a process called **invagination** (in-vaj′-i-NĀ-shun). Once the cells have invaginated, some of them displace the hypoblast, forming the **endoderm** (*endo-* = inside; *-derm* = skin). Other cells remain between the epiblast and newly formed endoderm to form the **mesoderm** (*meso-* = middle). Cells remaining in the epiblast then form the **ectoderm** (*ecto-* = outside). The ectoderm and endoderm are epithelia composed of tightly packed cells; the mesoderm is a loosely organized connective tissue (mesenchyme). As the embryo develops, the endoderm ultimately becomes the epithelial lining of the gastrointestinal tract, respiratory tract, and several other organs. The mesoderm gives rise to muscles, bones, and other connective tissues, and the peritoneum. The ectoderm develops into the epidermis of the skin and the nervous system. Table 29.1 provides more details about the fates of these primary germ layers.

About 16 days after fertilization, mesodermal cells from the primitive node migrate toward the head end of the embryo and form a hollow tube of cells in the midline called the **notochordal process** (nō-tō-KOR-dal) (Figure 29.8). By days 22–24, the notochordal process becomes a solid cylinder of cells called the **notochord** (NŌ-tō-kord; *noto-* = back; *-chord* = cord). This structure plays an extremely important role in **induction** (in-DUK-shun), the process by which one tissue (*inducing tissue*) stimulates the development of an adjacent unspecialized tissue (*responding tissue*) into a specialized one. An inducing tissue usually produces a chemical substance that influences the responding tissue. The notochord induces certain mesodermal cells to develop into the vertebral bodies. It also forms the nucleus pulposus of the intervertebral discs (see Figure SG7.1).

Also during the third week of development, two faint depressions appear on the dorsal surface of the embryo where the ectoderm and endoderm make contact but lack mesoderm between them. The structure closer to the head end is called the **oropharyngeal membrane** (or-ō-fa-RIN-jē-al; *oro-* = mouth; *-pharyngeal* = pertaining to the pharynx) (Figure 29.8a, b). It breaks down during the fourth week to connect the mouth cavity to the pharynx and the remainder of the gastrointestinal tract. The structure closer to the tail end is called the **cloacal membrane** (klō-Ā-kul = sewer), which degenerates in the seventh week to form the openings of the anus and urinary and reproductive tracts.

When the cloacal membrane appears, the wall of the yolk sac forms a small vascularized outpouching called the **allantois** (a-LAN-tō-is; *allant-* = sausage) that extends into the connecting stalk (Figure 29.8b). In nonmammalian organisms enclosed in an amnion, the allantois is used for gas exchange and waste removal. Because of the role of the human placenta in these activities, the allantois is not a prominent structure in humans (see Figure 29.11a). Nevertheless, it does function in the early formation of blood and blood vessels, and it is associated with the development of the urinary bladder.

Neurulation

In addition to inducing mesodermal cells to develop into vertebral bodies, the notochord also induces

FIGURE 29.8 **Development of the notochordal process.**

The notochordal process develops from the primitive node and later becomes the notochord.

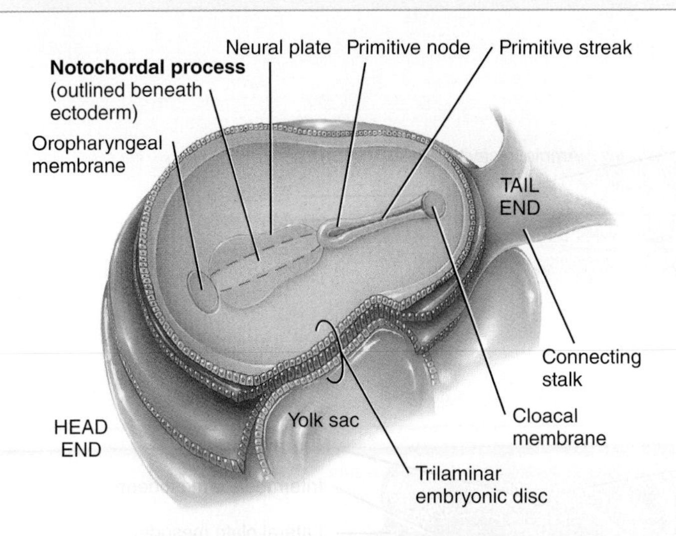

(a) Dorsal and partial sectional views of trilaminar embryonic disc, about 16 days after fertilization

(b) Sagittal section of trilaminar embryonic disc, about 16 days after fertilization

Q What is the significance of the notochord?

ectodermal cells over it to form the **neural plate** (**Figure 29.9a**). (Also see **Figure 14.27**.) By the end of the third week, the lateral edges of the neural plate become more elevated and form the **neural fold** (**Figure 29.9b**). The depressed midregion is called the **neural groove** (**Figure 29.9c**). Generally, the neural folds approach each other and fuse, thus converting the neural plate into a **neural tube** (**Figure 29.9d**). This occurs first near the middle of the embryo and then progresses toward the head and tail ends. Neural tube cells then develop into the brain and spinal cord. The process by which the neural plate, neural folds, and neural tube form is called **neurulation** (noor-oo-LĀ-shun).

As the neural tube forms, some of the ectodermal cells from the tube migrate to form several layers of cells called the **neural crest** (see **Figure 14.27b**). Neural crest cells give rise to all sensory neurons and postganglionic neurons of the peripheral nerves, the adrenal medullae, melanocytes (pigment cells) of the skin, arachnoid mater, and pia mater of the brain and spinal cord, and almost all of the skeletal and connective tissue components of the head.

At about 4 weeks after fertilization, the head end of the neural tube develops into three enlarged areas called **primary brain vesicles** (see **Figure 14.28**): the **prosencephalon** (pros′-en-SEF-a-lon) or **forebrain**, **mesencephalon** (mes′-en-SEF-a-lon) or **midbrain**, and **rhombencephalon** (rom′-ben-SEF-a-lon) or **hindbrain**. At about 5 weeks, the prosencephalon develops into **secondary brain vesicles** called the **telencephalon** (tel′-en-SEF-a-lon) and **diencephalon** (dī-en-SEF-a-lon), and the rhombencephalon develops into secondary brain vesicles called the **metencephalon** (met′-en-SEF-a-lon) and **myelencephalon** (mi-el-en-SEF-a-lon). The areas of the neural tube adjacent to the myelencephalon develop into the spinal cord. The parts of the brain that develop from the various brain vesicles are described in Section 14.1.

See Clinical Connection: Anencephaly

Development of Somites By about the 17th day after fertilization, the mesoderm adjacent to the notochord and neural tube forms paired longitudinal columns of **paraxial mesoderm** (par-AK-sē-al; para- = near) (**Figure 29.9b**). The mesoderm lateral to the paraxial mesoderm forms paired cylindrical masses called **intermediate mesoderm**. The mesoderm lateral to the intermediate mesoderm consists of a pair of flattened sheets called **lateral plate mesoderm**. The paraxial mesoderm soon segments into a series of paired, cube-shaped structures called **somites** (SŌ-mīts = little bodies). By the end of the fifth week, 42–44 pairs of somites are present. The number of somites that develop over a given period can be correlated to the approximate age of the embryo.

Each somite differentiates into three regions: a **myotome** (MĪ-ō-tōm), a **dermatome**, and a **sclerotome** (SKLER-ō-tōm) (see **Figure 10.17b**). The myotomes develop into the skeletal muscles of the neck, trunk, and limbs; the dermatomes form connective tissue, including the dermis of the skin; and the sclerotomes give rise to the vertebrae and ribs.

Development of the Intraembryonic Coelom In the third week of development, small spaces appear in the lateral

FIGURE 29.9 **Neurulation and the development of somites.**

Neurulation is the process by which the neural plate, neural folds, and neural tube form.

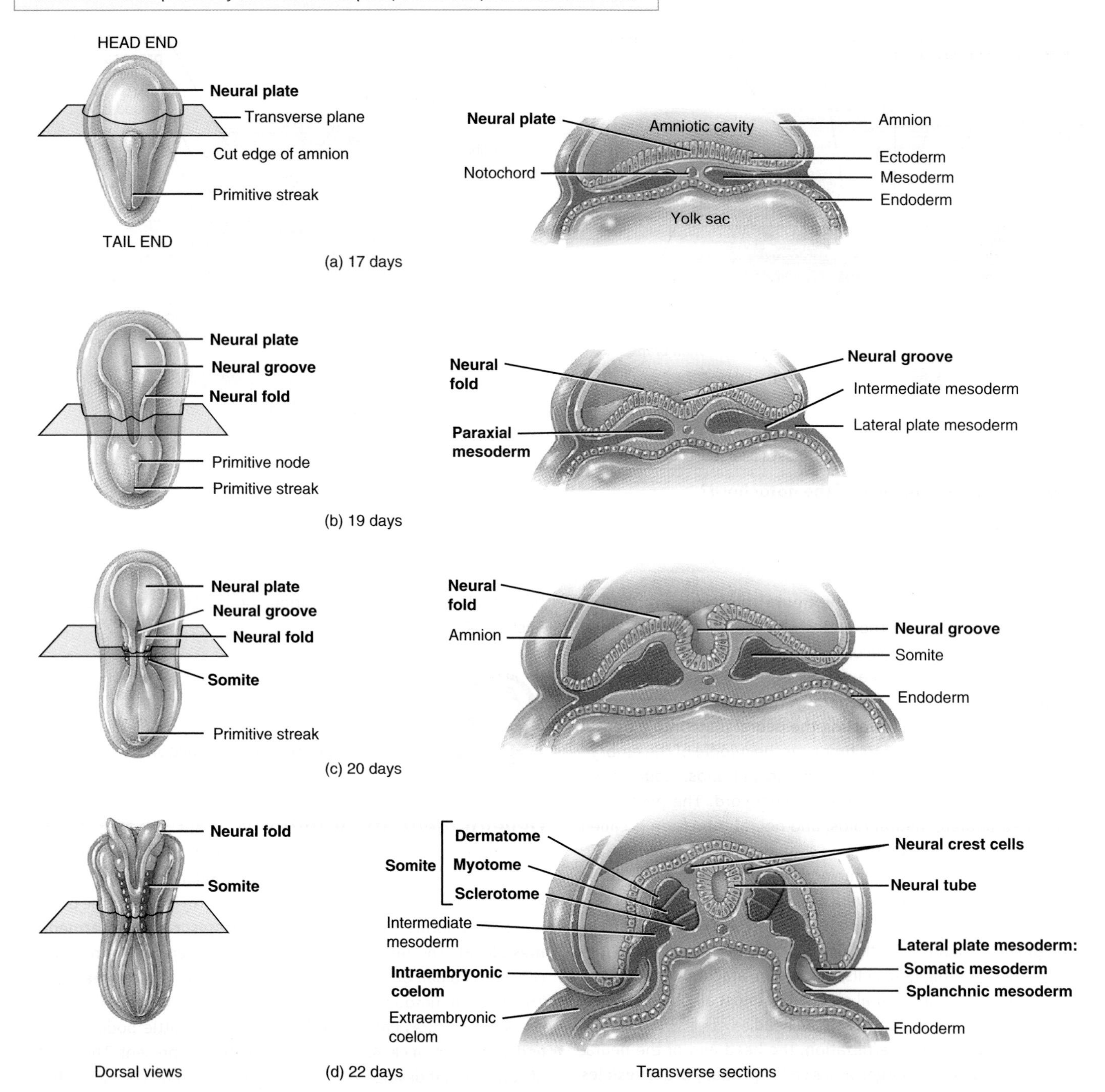

Q Which structures develop from the neural tube and somites?

plate mesoderm. These spaces soon merge to form a larger cavity called the **intraembryonic coelom** (SĒ-lom = cavity). This cavity splits the lateral plate mesoderm into two parts called the splanchnic mesoderm and somatic mesoderm (**Figure 29.9d**). **Splanchnic mesoderm** (SPLANK-nik = visceral) forms the heart and the visceral layer of the serous pericardium, blood vessels, the smooth muscle and connective tissues of the respiratory and digestive organs, and the visceral layer of the serous membrane of the pleurae and peritoneum. **Somatic mesoderm** (sō-MAT-ik; *soma-* = body) gives rise to the bones, ligaments, blood vessels, and connective tissue of the limbs and the parietal layer of the serous membrane of the pericardium, pleurae, and peritoneum.

Development of the Cardiovascular System At the beginning of the third week, **angiogenesis** (an′-jē-ō-JEN-e-sis; *angio-* = vessel; *-genesis* = production), the formation of blood vessels, begins in the extraembryonic mesoderm in the yolk sac, connecting stalk, and chorion. This early development is necessary because there is insufficient yolk in the yolk sac and ovum to provide adequate nutrition for the rapidly developing embryo. Angiogenesis is initiated when mesodermal cells differentiate into **hemangioblasts** (hē-MAN-jē-ō-blasts). These then develop into cells called **angioblasts**, which aggregate to form isolated masses of cells referred to as **blood islands** (see Figure 21.32). Spaces soon develop in the blood islands and form the lumens of blood vessels. Some angioblasts arrange themselves around each space to form the endothelium and the tunics (layers) of the developing blood vessels. As the blood islands grow and fuse, they soon form an extensive system of blood vessels throughout the embryo.

About 3 weeks after fertilization, blood cells and blood plasma begin to develop *outside* the embryo from hemangioblasts in the blood vessels in the walls of the yolk sac, allantois, and chorion. These then develop into pluripotent stem cells that form blood cells. Blood formation begins *within* the embryo at about the fifth week in the liver and the twelfth week in the spleen, red bone marrow, and thymus.

The heart forms from splanchnic mesoderm in the head end of the embryo on days 18 and 19. This region of mesodermal cells is called the **cardiogenic area** (kar-dē-ō-JEN-ik; *cardio-* = heart; *-genic* = producing). In response to induction signals from the underlying endoderm, these mesodermal cells form a pair of **endocardial tubes** (see Figure 20.19). The tubes then fuse to form a single **primitive heart tube**. By the end of the third week, the primitive heart tube bends on itself, becomes S-shaped, and begins to beat. It then joins blood vessels in other parts of the embryo, connecting stalk, chorion, and yolk sac to form a primitive cardiovascular system.

Development of the Chorionic Villi and Placenta

As the embryonic tissue invades the uterine wall, maternal uterine vessels are eroded and maternal blood fills spaces, called **lacunae** (Figure 29.10) within the invading tissue. By the end of the second week of development, **chorionic villi** (kō-rē-ON-ik VIL-ī) begin to develop. These fingerlike projections consist of chorion (syncytiotrophoblast surrounded by cytotrophoblast) that projects into the endometrial wall of the uterus (Figure 29.10a). By the end of the third week, blood capillaries develop in the chorionic villi (Figure 29.10b). Blood vessels in the chorionic villi connect to the embryonic heart by way of the umbilical arteries and umbilical vein through the connecting (body) stalk, which will eventually become the umbilical cord (Figure 29.10c). The fetal blood capillaries within the chorionic villi project into the lacunae, which unite to form the **intervillous spaces** (in′-ter-VIL-us) that bathe the chorionic villi with maternal blood. As a result, maternal blood bathes the chorion-covered fetal blood vessels. Note, however, that maternal and fetal blood vessels do not join, and the blood they carry does not normally mix. Instead, oxygen and nutrients in the blood of the mother's *intervillous spaces*, the spaces between chorionic villi, diffuse across the cell membranes into the capillaries of the villi. Waste products such as carbon dioxide diffuse in the opposite direction.

FIGURE 29.10 Development of chorionic villi.

Blood vessels in chorionic villi connect to the embryonic heart via the umbilical arteries and umbilical vein.

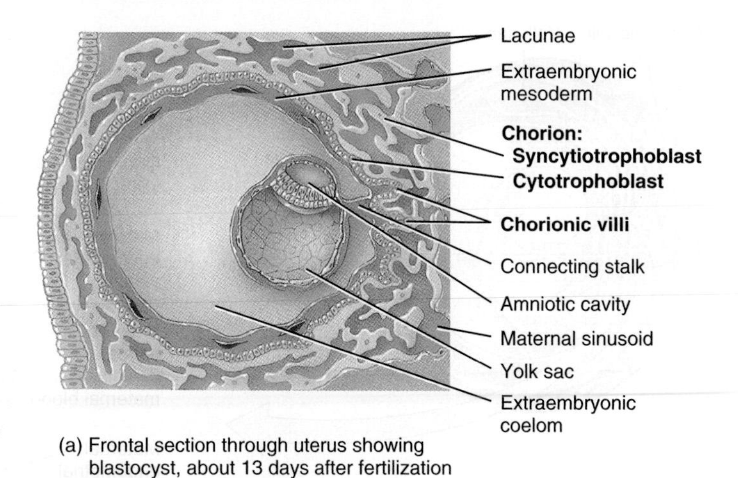

(a) Frontal section through uterus showing blastocyst, about 13 days after fertilization

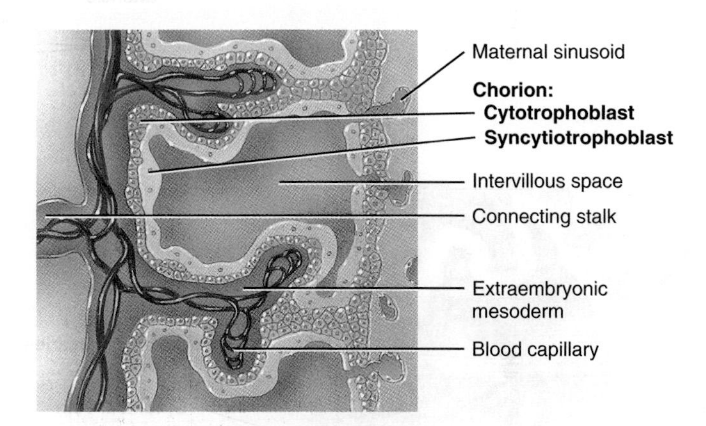

(b) Details of two chorionic villi, about 21 days after fertilization

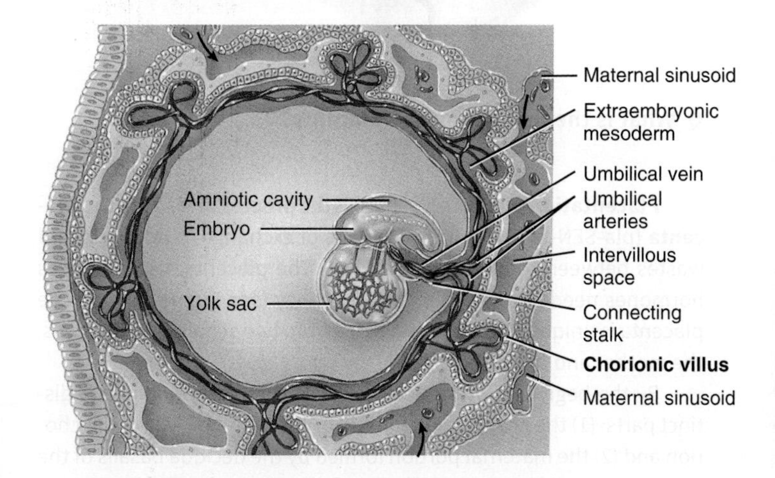

(c) Frontal section through uterus showing an embryo and its vascular supply, about 21 days after fertilization

Q Why is development of chorionic villi important?

FIGURE 29.11 Placenta and umbilical cord.

The placenta is formed by the chorionic villi of the embryo and the decidua basalis of the endometrium of the mother.

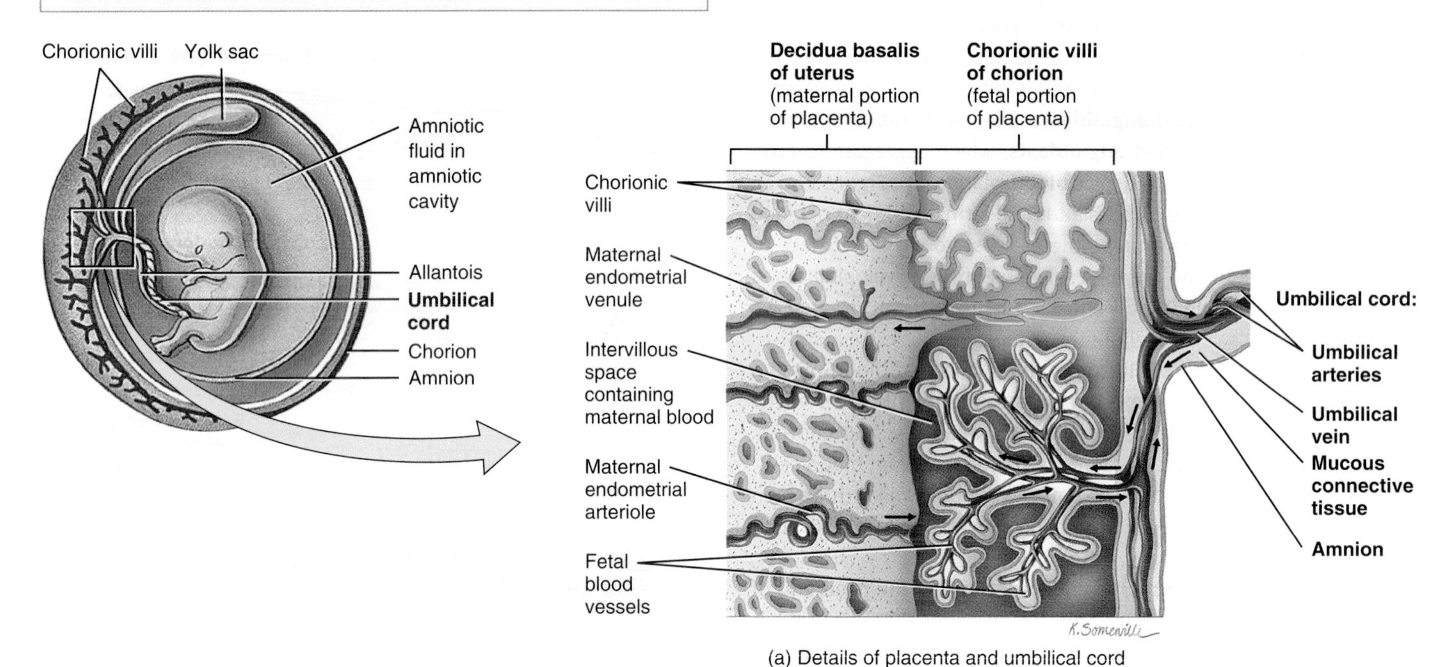

(a) Details of placenta and umbilical cord

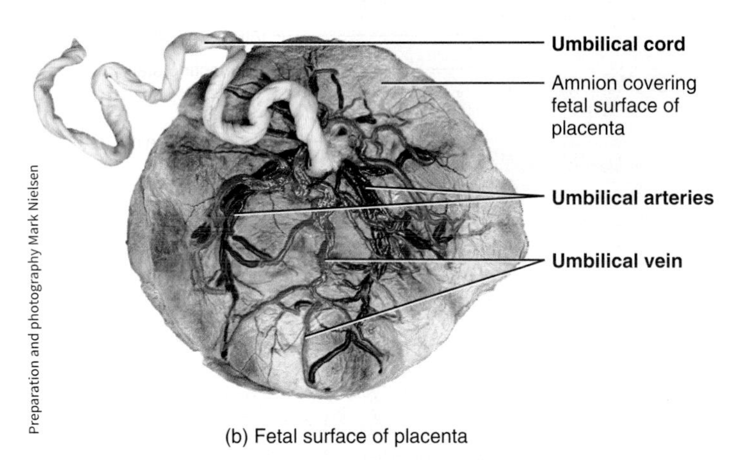

(b) Fetal surface of placenta

Q What is the function of the placenta?

Placentation (plas-en-TĀ-shun) is the process of forming the **placenta** (pla-SEN-ta = flat cake), the site of exchange of nutrients and wastes between the mother and fetus. The placenta also produces hormones needed to sustain the pregnancy (see Figure 29.16). The placenta is unique because it develops from two separate individuals, the mother and the fetus.

By the beginning of the twelfth week, the placenta has two distinct parts: (1) the fetal portion formed by the chorionic villi of the chorion and (2) the maternal portion formed by the decidua basalis of the endometrium (Figure 29.11a). When fully developed, the placenta is shaped like a pancake (Figure 29.11b). Functionally, the placenta allows oxygen and nutrients to diffuse from maternal blood into fetal blood while carbon dioxide and wastes diffuse from fetal blood into maternal blood. The placenta also is a protective barrier because most microorganisms cannot pass through it. However, certain viruses, such as those that cause AIDS, German measles, chickenpox, measles, encephalitis, and poliomyelitis, can cross the placenta. Many drugs, alcohol, and some substances that can cause birth defects also pass freely. The placenta stores nutrients such as carbohydrates, proteins, calcium, and iron, which are released into fetal circulation as required.

The actual connection between the placenta and embryo, and later the fetus, is through the **umbilical cord** (um-BIL-i-kal = navel), which develops from the connecting stalk and is usually about 2 cm (1 in.) wide and about 50–60 cm (20–24 in.) in length. The umbilical cord consists of two umbilical arteries that carry deoxygenated fetal blood to the placenta, one umbilical vein that carries oxygen and nutrients acquired from the mother's intervillous spaces into the fetus, and supporting mucous connective tissue called **Wharton's jelly** (WOR-tons) derived from the allantois. A layer of amnion surrounds the entire umbilical cord and gives it a shiny appearance (Figure 29.11). In some cases, the umbilical vein is used to transfuse blood into a fetus or to introduce drugs for various medical treatments.

In about 1 in 200 newborns, only one of the two umbilical arteries is present in the umbilical cord. It may be due to failure of the artery to develop or degeneration of the vessel early in development. Nearly 20% of infants with this condition develop cardiovascular defects.

After the birth of the baby, the placenta detaches from the uterus and is therefore termed the **afterbirth**. At this time, the umbilical cord is tied off and then severed. The small portion (about an inch) of the cord that remains attached to the infant begins to wither and falls off, usually within 12 to 15 days after birth. The area where the cord was attached becomes covered by a thin layer of skin, and scar tissue forms. The scar is the **umbilicus** (um-BIL-i-kus) or navel.

Pharmaceutical companies use human placentas as a source of hormones, drugs, and blood; portions of placentas are also used for burn coverage. The placental and umbilical cord veins can also be used in blood vessel grafts, and cord blood can be frozen to provide a future source of pluripotent stem cells, for example, to repopulate red bone marrow following radiotherapy for cancer.

See Clinical Connection: Placenta Previa

Fourth Week of Development

The fourth through eighth weeks of development are very significant in embryonic development because all major organs appear during this time. The term **organogenesis** (or′-ga-nō-JEN-e-sis) refers to the formation of body organs and systems. By the end of the eighth week, all of the major body systems have begun to develop, although their functions for the most part are minimal. Organogenesis requires the presence of blood vessels to supply developing organs with oxygen and other nutrients. However, recent studies suggest that blood vessels play a significant role in organogenesis even before blood begins to flow within them. The endothelial cells of blood vessels apparently provide some type of developmental signal, either a secreted substance or a direct cell-to-cell interaction, that is necessary for organogenesis.

During the fourth week after fertilization, the embryo undergoes very dramatic changes in shape and size, nearly tripling its size. It is essentially converted from a flat, two-dimensional trilaminar embryonic disc to a three-dimensional cylinder, a process called **embryonic folding** (Figure 29.12a–d). The cylinder consists of endoderm in the center (gut), ectoderm on the outside (epidermis), and mesoderm in between. The main force responsible for embryonic folding is the different rates of growth of various parts of the embryo, especially the rapid longitudinal growth of the nervous system (neural tube). Folding in the median plane produces a **head fold** and a **tail fold**; folding in the horizontal plane results in the two **lateral folds**. Overall, due to the foldings, the embryo curves into a C-shape.

The head fold brings the developing heart and mouth into their eventual adult positions. The tail fold brings the developing anus into its eventual adult position. The lateral folds form as the lateral margins of the trilaminar embryonic disc bend ventrally. As they move toward the midline, the lateral folds incorporate the dorsal part of the yolk sac into the embryo as the **primitive gut**, the forerunner of the gastrointestinal tract (Figure 29.12b). The primitive gut differentiates into an anterior **foregut**, an intermediate **midgut**, and a posterior **hindgut** (Figure 29.12c). The fates of the foregut, midgut, and hindgut are described in Section 24.16. Recall that the oropharyngeal membrane is located in the head end of the embryo (see Figure 29.8). It separates the future pharyngeal (throat) region of the foregut from the **stomodeum** (stō-mō-DĒ-um; *stomo-* = mouth), the future oral cavity. Because of head folding, the oropharyngeal membrane moves downward and the foregut and stomodeum move closer to their final positions. When the oropharyngeal membrane ruptures during the fourth week, the pharyngeal region of the pharynx is brought into contact with the stomodeum.

In a developing embryo, the last part of the hindgut expands into a cavity called the **cloaca** (klo-Ā-ka) (see Figure 26.23). On the outside of the embryo is a small cavity in the tail region called the **proctodeum** (prok-tō-DĒ-um; *procto-* = anus) (Figure 29.12c). Separating the cloaca from the proctodeum is the **cloacal membrane** (see Figure 29.8). During embryonic development, the cloaca divides into a ventral urogenital sinus and a dorsal anorectal canal. As a result of tail folding, the cloacal membrane moves downward and the urogenital sinus, anorectal canal, and proctodeum move closer to their final positions. When the cloacal membrane ruptures during the seventh week of development, the urogenital and anal openings are created.

In addition to embryonic folding, development of somites, and development of the neural tube, four pairs of **pharyngeal arches** (fa-RIN-jē-al) or *branchial arches* (BRANG-kē-al; *branch* = gill) begin to develop on each side of the future head and neck regions (Figure 29.13) during the fourth week. These four paired structures begin their development on the 22nd day after fertilization and form swellings on the surface of the embryo. Each pharyngeal arch consists of an outer covering of ectoderm and an inner covering of endoderm, with mesoderm in between. Within each pharyngeal arch there is an artery, a cranial nerve, cartilaginous skeletal rods that support the arch, and skeletal muscle tissue that attaches to and moves the cartilage rods. On the ectodermal surface of the pharyngeal region, each pharyngeal arch is separated by a groove called a **pharyngeal cleft** (Figure 29.13a). The pharyngeal clefts meet corresponding balloonlike outgrowths of the endodermal pharyngeal lining called **pharyngeal** (*branchial*) **pouches**. Where the pharyngeal cleft and pouch meet to separate the arches, the outer ectoderm of the cleft contacts the inner endoderm of the pouch and there is no mesoderm between (Figure 29.13b).

Just as the somite gives rise to specified structures in the body wall, each pharyngeal arch, cleft, and pouch gives rise to specified structures in the head and neck. Each pharyngeal arch is a developmental unit and includes a skeletal component, muscle, nerve, and blood vessels. In the human embryo, there are four obvious pharyngeal arches. Each of these arches develops into a specific and unique component of the head and neck region. For example, the first pharyngeal arch is often called the *mandibular arch* because it forms the jaws (the *mandible* is the lower jawbone).

The first sign of a developing ear is a thickened area of ectoderm, the **otic placode** (PLAK-ōd), or future internal ear, which can be distinguished about 22 days after fertilization. A thickened area of ectoderm called the **lens placode**, which will become the eye, also appears at this time (see Figure 29.13a).

By the middle of the fourth week, the upper limbs begin their development as outgrowths of mesoderm covered by ectoderm called **upper limb buds** (see Figure 8.16b). By the end of the fourth week, the **lower limb buds** develop. The heart also forms a distinct projection on the ventral surface of the embryo called the **heart prominence** (see Figure 8.16b). At the end of the fourth week the embryo has a distinctive **tail** (see Figure 8.16b).

FIGURE 29.12 **Embryonic folding.**

Embryonic folding converts the two-dimensional trilaminar embryonic disc into a three-dimensional cylinder.

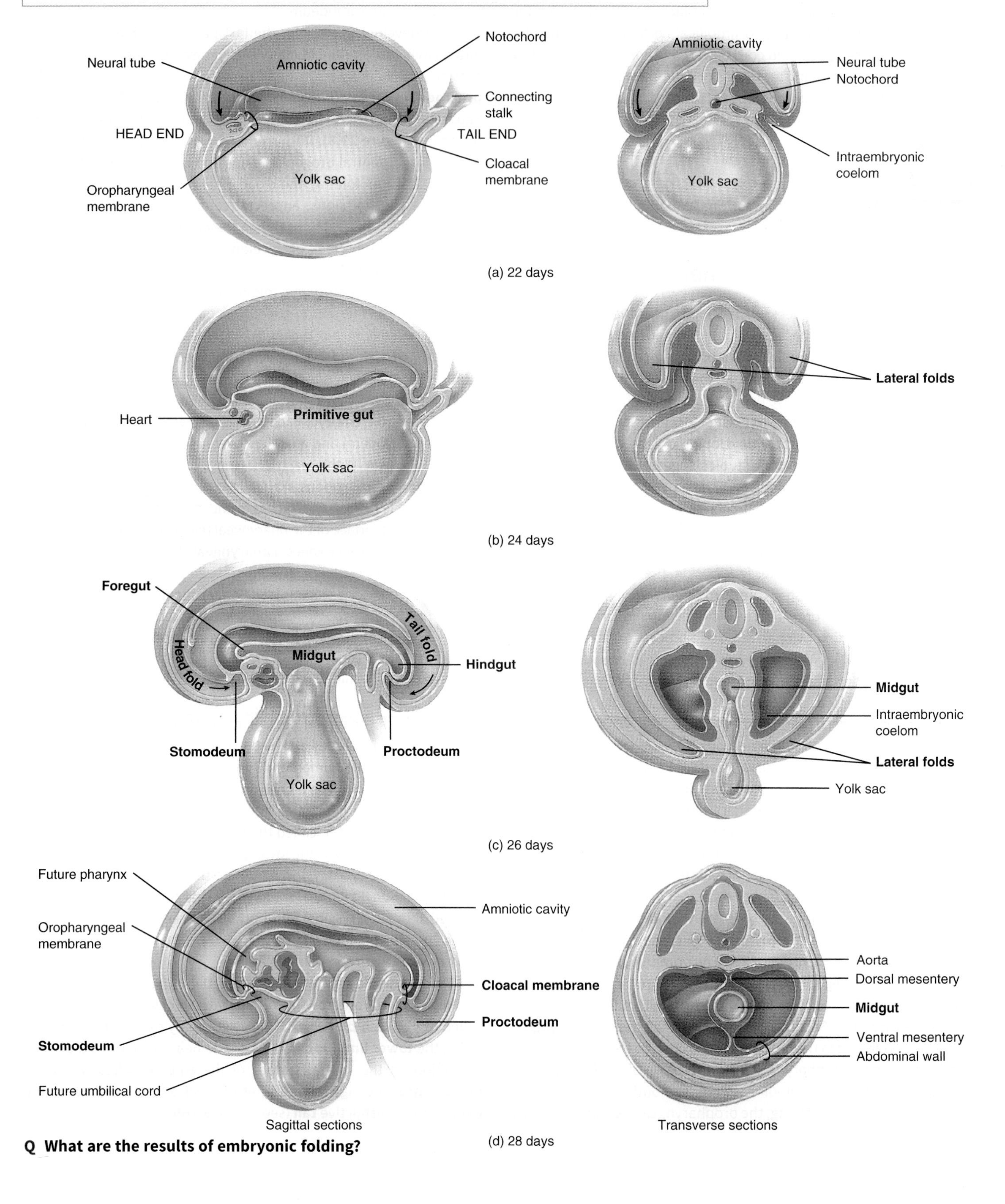

Q What are the results of embryonic folding?

FIGURE 29.13 Development of pharyngeal arches, pharyngeal clefts, and pharyngeal pouches.

The four pairs of pharyngeal pouches consist of ectoderm, mesoderm, and endoderm and contain blood vessels, cranial nerves, cartilage, and muscular tissue.

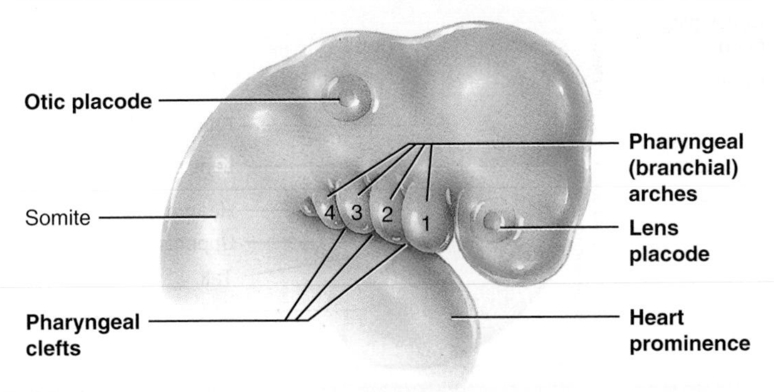

(a) External view, about 28-day embryo

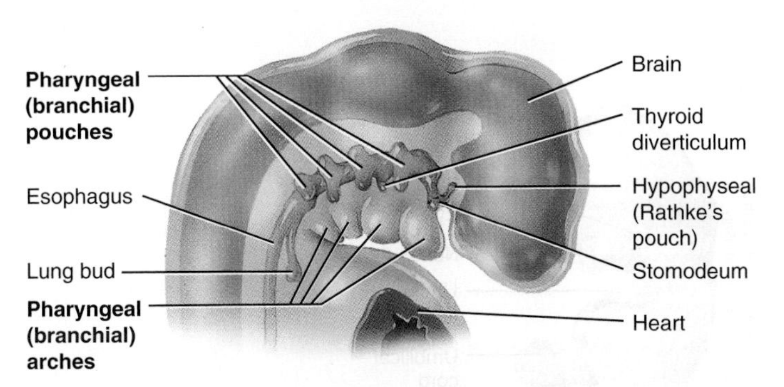

(b) Sagittal section, about 28-day embryo

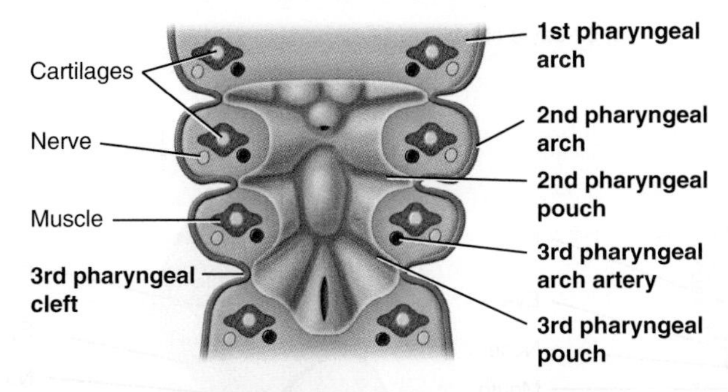

(c) Transverse section of the pharynx, about 28-day embryo

Q Why are pharyngeal arches, clefts, and pouches important?

Fifth through Eighth Weeks of Development

During the fifth week of development, there is a very rapid development of the brain, so growth of the head is considerable. By the end of the sixth week, the head grows even larger relative to the trunk, and the limbs show substantial development (see **Figure 8.16c**). In addition, the neck and trunk begin to straighten, and the heart is now four-chambered. By the seventh week, the various regions of the limbs become distinct and the beginnings of digits appear (see **Figure 8.16d**). At the start of the eighth week (the final week of the embryonic period), the digits of the hands are short and webbed, the tail is shorter but still visible, the eyes are open, and the auricles of the ears are visible (see **Figure 8.16c**). By the end of the eighth week, all regions of limbs are apparent; the digits are distinct and no longer webbed due to removal of cells via apoptosis. Also, the eyelids come together and may fuse, the tail disappears, and the external genitals begin to differentiate. The embryo now has clearly human characteristics.

29.4 Fetal Period

OBJECTIVE

- **Describe** the major events of the fetal period.

During the **fetal period** (from the ninth week until birth), tissues and organs that developed during the embryonic period grow and differentiate. Very few new structures appear during the fetal period, but the rate of body growth is remarkable, especially during the second half of intrauterine life. For example, during the last 2.5 months of intrauterine life, half of the full-term weight is added. At the beginning of the fetal period, the head is half the length of the body. By the end of the fetal period, the head size is only one-quarter the length of the body. During the same period, the limbs also increase in size from one-eighth to one-half the fetal length. The **fetus** is also less vulnerable to the damaging effects of drugs, radiation, and microbes than it was as an embryo.

A summary of the major developmental events of the embryonic and fetal periods is illustrated in **Figure 29.14** and presented in **Table 29.2**.

Throughout the text we have discussed the developmental anatomy of the various body systems in their respective chapters. The following list of these sections is presented here for your review.

- Integumentary System (Section 5.6)
- Skeletal System (Section 8.7)
- Muscular System (Section 10.11)
- Nervous System (Section 14.19)
- Endocrine System (Section 18.15)
- Heart (Section 20.8)
- Blood and Blood Vessels (Section 21.23)
- Lymphatic System and Immunity (Section 22.5)
- Respiratory System (Section 23.10)
- Digestive System (Section 24.15)
- Urinary System (Section 26.11)
- Reproductive Systems (Section 28.6)

FIGURE 29.14 Summary of representative developmental events of the embryonic and fetal periods. The embryos and fetuses are not shown at their actual sizes.

Development during the fetal period is mostly concerned with the growth and differentiation of tissues and organs formed during the embryonic period.

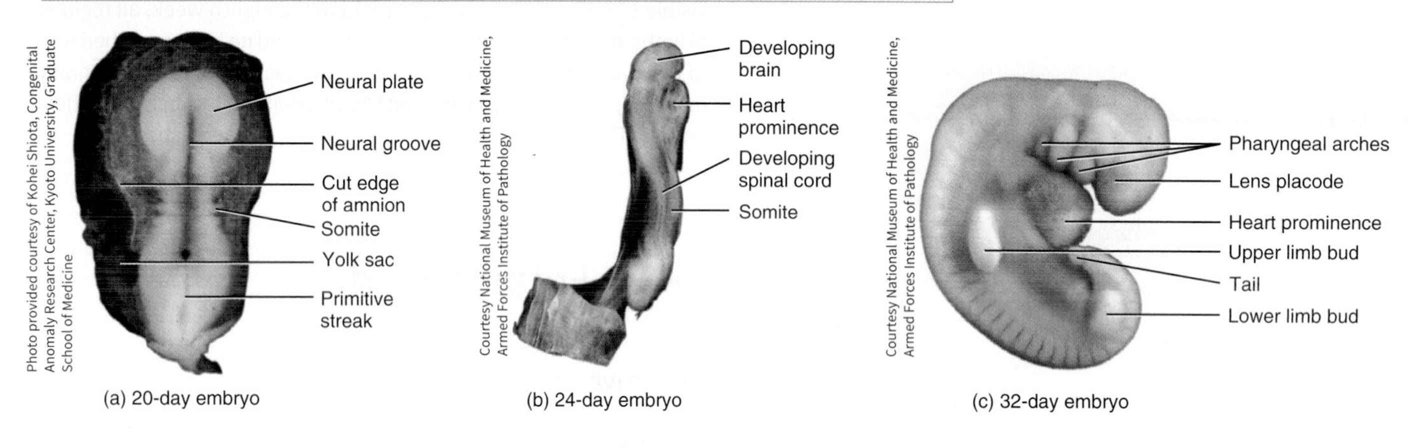

(a) 20-day embryo (b) 24-day embryo (c) 32-day embryo

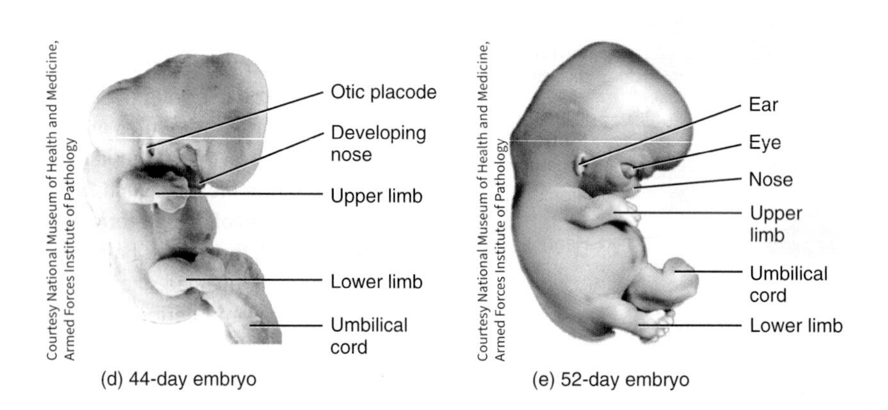

(d) 44-day embryo (e) 52-day embryo

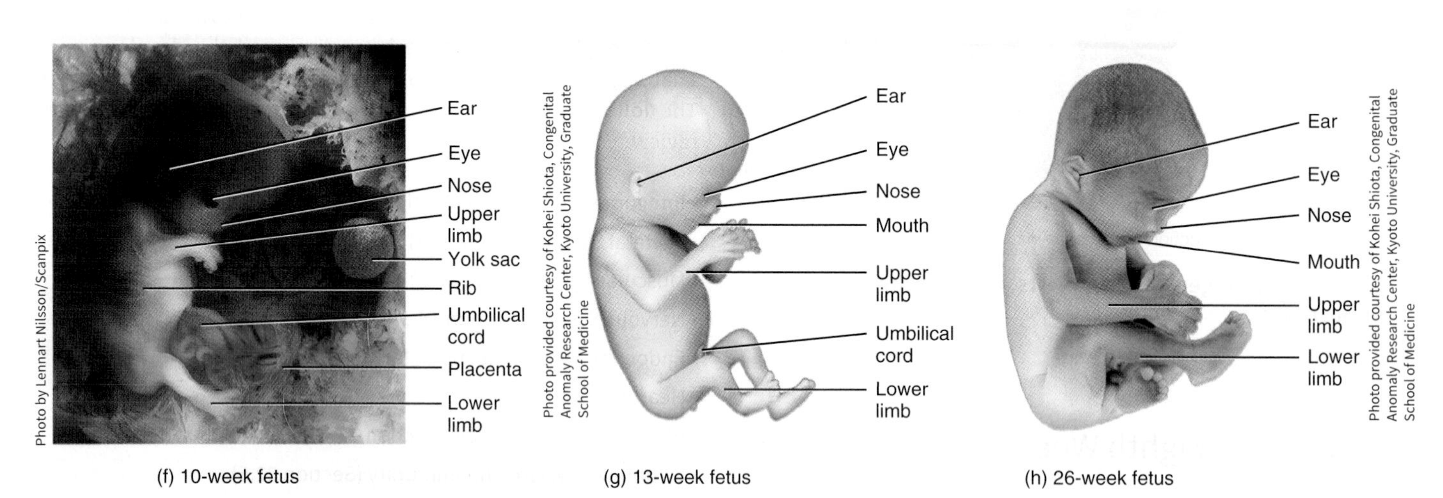

(f) 10-week fetus (g) 13-week fetus (h) 26-week fetus

Q How does mid-fetal weight compare to end-fetal weight?

TABLE 29.2 Summary of Changes during Embryonic and Fetal Development

TIME	APPROXIMATE SIZE AND WEIGHT	REPRESENTATIVE CHANGES
EMBRYONIC PERIOD		
1–4 weeks	0.6 cm (3/16 in.)	Primary germ layers and notochord develop. Neurulation occurs. Primary brain vesicles, somites, and intraembryonic coelom develop. Blood vessel formation begins and blood forms in yolk sac, allantois, and chorion. Heart forms and begins to beat. Chorionic villi develop and placental formation begins. The embryo folds. The primitive gut, pharyngeal arches, and limb buds develop. Eyes and ears begin to develop, tail forms, and body systems begin to form.
5–8 weeks	3 cm (1.25 in.) 1 g (1/30 oz)	Limbs become distinct and digits appear. Heart becomes four-chambered. Eyes are far apart and eyelids are fused. Nose develops and is flat. Face is more humanlike. Bone formation begins. Blood cells start to form in liver. External genitals begin to differentiate. Tail disappears. Major blood vessels form. Many internal organs continue to develop.
FETAL PERIOD		
9–12 weeks	7.5 cm (3 in.) 30 g (1 oz)	Head constitutes about half the length of fetal body, and fetal length nearly doubles. Brain continues to enlarge. Face is broad, with eyes fully developed, closed, and widely separated. Nose develops a bridge. External ears develop and are low set. Bone formation continues. Upper limbs almost reach final relative length but lower limbs are not quite as well developed. Heartbeat can be detected. Gender is distinguishable from external genitals. Urine secreted by fetus is added to amniotic fluid. Red bone marrow, thymus, and spleen participate in blood cell formation. Fetus begins to move, but its movements cannot be felt yet by the mother. Body systems continue to develop.
13–16 weeks	18 cm (6.5–7 in.) 100 g (4 oz)	Head is relatively smaller than rest of body. Eyes move medially to final positions, and ears move to final positions on sides of head. Lower limbs lengthen. Fetus appears even more humanlike. Rapid development of body systems occurs.
17–20 weeks	25–30 cm (10–12 in.) 200–450 g (0.5–1 lb)	Head is more proportionate to rest of body. Eyebrows and head hair are visible. Growth slows but lower limbs continue to lengthen. Vernix caseosa (fatty secretions of oil glands and dead epithelial cells) and lanugo (delicate fetal hair) cover fetus. Brown fat forms and is the site of heat production. Fetal movements are commonly felt by mother (quickening).
21–25 weeks	27–35 cm (11–14 in.) 550–800 g (1.25–1.5 lb)	Head becomes even more proportionate to rest of body. Weight gain is substantial, and skin is pink and wrinkled. Fetuses 24 weeks and older usually survive if born prematurely.
26–29 weeks	32–42 cm (13–17 in.) 1100–1350 g (2.5–3 lb)	Head and body are more proportionate and eyes are open. Toenails are visible. Body fat is 3.5% of total body mass and additional subcutaneous fat smoothes out some wrinkles. Testes begin to descend toward scrotum at 28 to 32 weeks. Red bone marrow is major site of blood cell production. Many fetuses born prematurely during this period survive if given intensive care because lungs can provide adequate ventilation and central nervous system is developed enough to control breathing and body temperature.
30–34 weeks	41–45 cm (16.5–18 in.) 2000–2300 g (4.5–5 lb)	Skin is pink and smooth. Fetus assumes upside-down position. Body fat is 8% of total body mass.
35–38 weeks	50 cm (20 in.) 3200–3400 g (7–7.5 lb)	By 38 weeks, circumference of fetal abdomen is greater than that of head. Skin is usually bluish-pink, and growth slows as birth approaches. Body fat is 16% of total body mass. Testes are usually in scrotum in full-term male infants. Even after birth, an infant is not completely developed; an additional year is required, especially for complete development of nervous system.

Table 29.2 *Continues*

TABLE 29.2 Summary of Changes during Embryonic and Fetal Development (*Continued*)

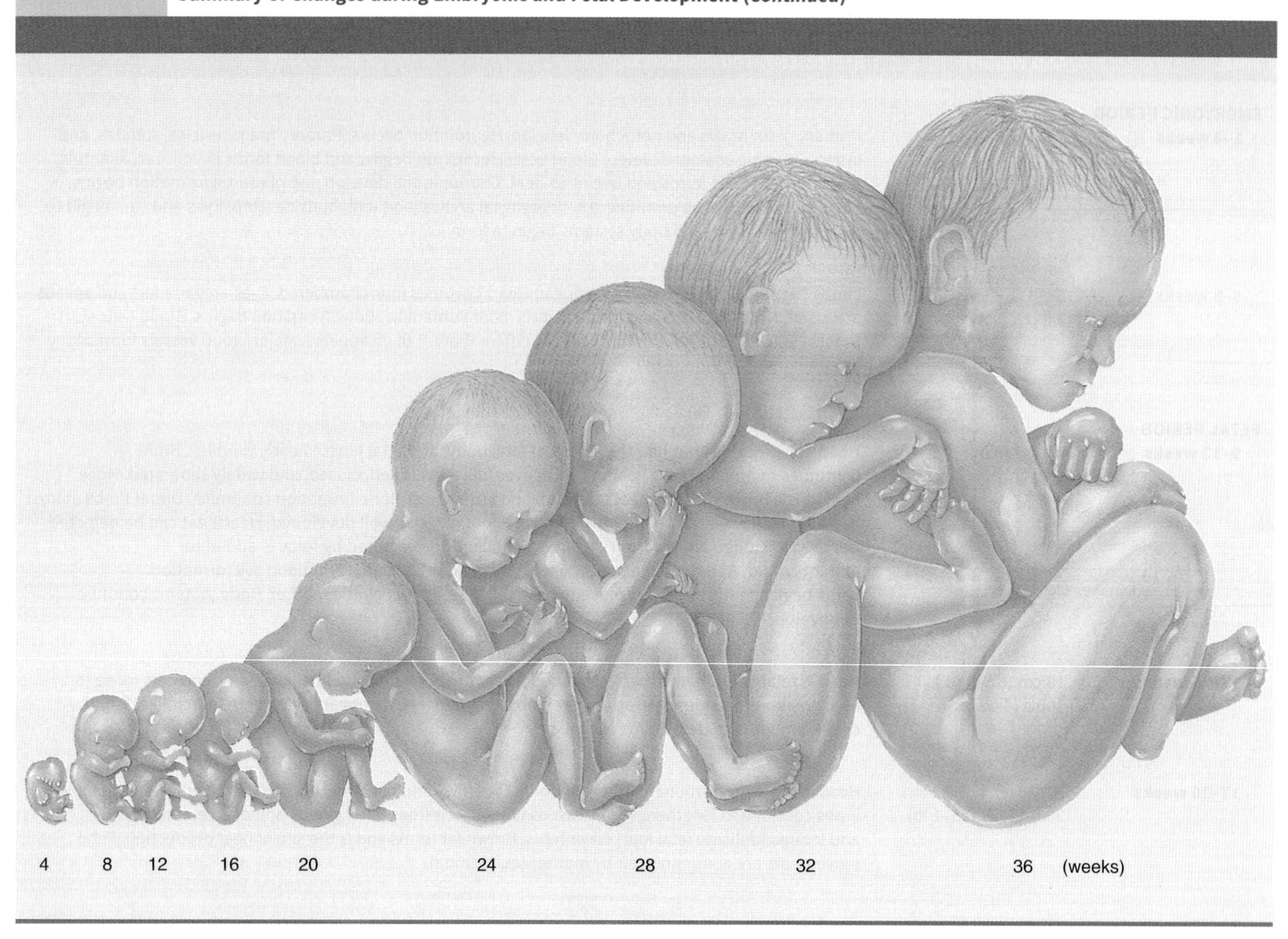

29.5 Teratogens

OBJECTIVE

- **Define** a teratogen and **provide** several examples of teratogens.

Exposure of a developing embryo or fetus to certain environmental factors can damage the developing organism or even cause death. A **teratogen** (TER-a-tō-jen; *terato-* = monster; *-gen* = creating) is any agent or influence that causes developmental defects in the embryo. In the following sections we briefly discuss several examples.

Chemicals and Drugs

Because the placenta is not an absolute barrier between the maternal and fetal circulations, any drug or chemical that is dangerous to an infant should be considered potentially dangerous to the fetus when given to the mother. Alcohol is by far the number-one fetal teratogen. Intrauterine exposure to even a small amount of alcohol may result in **fetal alcohol syndrome (FAS)**, one of the most common causes of mental retardation and the most common preventable cause of birth defects in the United States. The symptoms of FAS may include slow growth before and after birth, characteristic facial features (short palpebral fissures, a thin upper lip, and sunken nasal bridge), defective heart and other organs, malformed limbs, genital abnormalities, and central nervous system damage. Behavioral problems, such as hyperactivity, extreme nervousness, reduced ability to concentrate, and an inability to appreciate cause-and-effect relationships, are common.

Other teratogens include certain viruses (hepatitis B and C and certain papilloma viruses that cause sexually transmitted diseases); pesticides; defoliants (chemicals that cause plants to shed their leaves prematurely); industrial chemicals; some hormones; antibiotics; oral anticoagulants, anticonvulsants, antitumor agents, thyroid drugs, thalidomide, diethylstilbestrol (DES), and numerous other prescription drugs; LSD; and cocaine. A pregnant woman who uses cocaine, for example, subjects the fetus to higher risk of retarded growth, attention and orientation problems, hyperirritability, a tendency to stop breathing, malformed or missing organs, strokes, and seizures. The risks of spontaneous abortion, premature birth, and stillbirth also increase with fetal exposure to cocaine.

Cigarette Smoking

Strong evidence implicates cigarette smoking during pregnancy as a cause of low infant birth weight; there is also a strong association between smoking and a higher fetal and infant mortality rate. Women who smoke have a much higher risk of an ectopic pregnancy. Cigarette smoke may be teratogenic and may cause cardiac abnormalities as well as anencephaly (for reference see Clinical Connection: Anencephaly in Section 29.1). Maternal smoking also is a significant factor in the development of cleft lip and palate and has been linked with sudden infant death syndrome (SIDS). Infants nursing from smoking mothers have also been found to have an increased incidence of gastrointestinal disturbances. Even a mother's exposure to secondhand cigarette smoke (breathing air containing tobacco smoke) during pregnancy or while nursing predisposes her baby to increased incidence of respiratory problems, including bronchitis and pneumonia, during the first year of life.

Irradiation

Ionizing radiation of various kinds is a potent teratogen. Exposure of pregnant mothers to x-rays or radioactive isotopes during the embryo's susceptible period of development may cause microcephaly (small head size relative to the rest of the body), mental retardation, and skeletal malformations. Caution is advised, especially during the first trimester of pregnancy.

29.6 Prenatal Diagnostic Tests

OBJECTIVE

- **Describe** the procedures for fetal ultrasonography, amniocentesis, and chorionic villi sampling.

Several tests are available to detect genetic disorders and assess fetal well-being. Here we describe fetal ultrasonography, amniocentesis, and chorionic villi sampling (CVS).

Fetal Ultrasonography

If there is a question about the normal progress of a pregnancy, **fetal ultrasonography** (ul-tra-son-OG-ra-fē) may be performed. By far the most common use of diagnostic ultrasound is to determine a more accurate fetal age when the date of conception is unclear. It is also used to confirm pregnancy, evaluate fetal viability and growth, determine fetal position, identify multiple pregnancies, identify fetal–maternal abnormalities, and serve as an adjunct to special procedures such as amniocentesis. During fetal ultrasonography, a transducer, an instrument that emits high-frequency sound waves, is passed back and forth over the abdomen. The reflected sound waves from the developing fetus are picked up by the transducer and converted to an on-screen image called a **sonogram** (see Table 1.3). Because the urinary bladder serves as a landmark during the procedure, the patient needs to drink liquids before the procedure and not void urine to maintain a full bladder.

Amniocentesis

Amniocentesis (am′-nē-ō-sen-TĒ-sis; *amnio-* = amnion; *-centesis* = puncture to remove fluid) involves withdrawing some of the amniotic fluid that bathes the developing fetus and analyzing the fetal cells and dissolved substances. It is used to test for the presence of certain genetic disorders, such as Down syndrome (DS), hemophilia, Tay-Sachs disease, sickle cell disease, and certain muscular dystrophies. It is also used to help determine survivability of the fetus. The test is usually done at 14–18 weeks of gestation. All gross chromosomal abnormalities and over 50 biochemical defects can be detected through amniocentesis. It can also reveal the baby's gender; this is important information for the diagnosis of sex-linked disorders, in which an abnormal gene carried by the mother affects her male offspring only (described in Section 29.12).

During amniocentesis, the position of the fetus and placenta is first identified using ultrasound and palpation. After the skin is prepared with an antiseptic and a local anesthetic is given, a hypodermic needle is inserted through the mother's abdominal wall and into the amniotic cavity within the uterus. Then, 10 to 30 mL of fluid and suspended cells are aspirated (Figure 29.15a) for microscopic examination and biochemical testing. Elevated levels of alpha-fetoprotein (AFP) and acetylcholinesterase may indicate failure of the nervous system to develop properly, as occurs in spina bifida or anencephaly (absence of the cerebrum), or may be due to other developmental or chromosomal problems. Chromosome studies, which require growing the cells for 2–4 weeks in a culture medium, may reveal rearranged, missing, or extra chromosomes. Amniocentesis is performed only when a risk for genetic defects is suspected, because there is about a 0.5% chance of spontaneous abortion after the procedure.

Chorionic Villi Sampling

In **chorionic villi sampling (CVS)**, a catheter is guided through the vagina and cervix of the uterus and then advanced to the chorionic villi under ultrasound guidance (Figure 29.15b). About 30 milligrams of tissue is suctioned out and prepared for chromosomal analysis. Alternatively, the chorionic villi can be sampled by inserting a needle through the abdominal cavity, as performed in amniocentesis.

CVS can identify the same defects as amniocentesis because chorion cells and fetal cells contain the same genome. CVS offers several advantages over amniocentesis: It can be performed as early as 8 weeks of gestation, and test results are available in only a few days, permitting an earlier decision on whether to continue the pregnancy. However, CVS is slightly riskier than amniocentesis; after the procedure there is a 1–2% chance of spontaneous abortion.

Noninvasive Prenatal Tests

Currently, chorionic villi testing and amniocentesis are the only useful ways to obtain fetal tissue for prenatal testing of gene defects. While these invasive procedures pose relatively little risk when performed by experts, much work has been done to develop **noninvasive prenatal tests**, which do not require the penetration of any embryonic structure. The goal is to develop accurate, safe, more efficient, and less expensive tests for screening a large population.

FIGURE 29.15 **Amniocentesis and chorionic villi sampling.**

To detect genetic abnormalities, amniocentesis is performed at 14–16 weeks of gestation; chorionic villi sampling may be performed as early as 8 weeks of gestation.

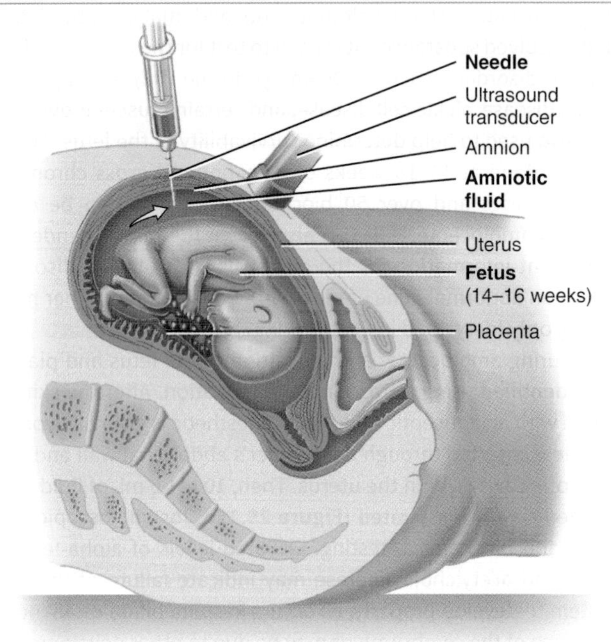

(a) Amniocentesis

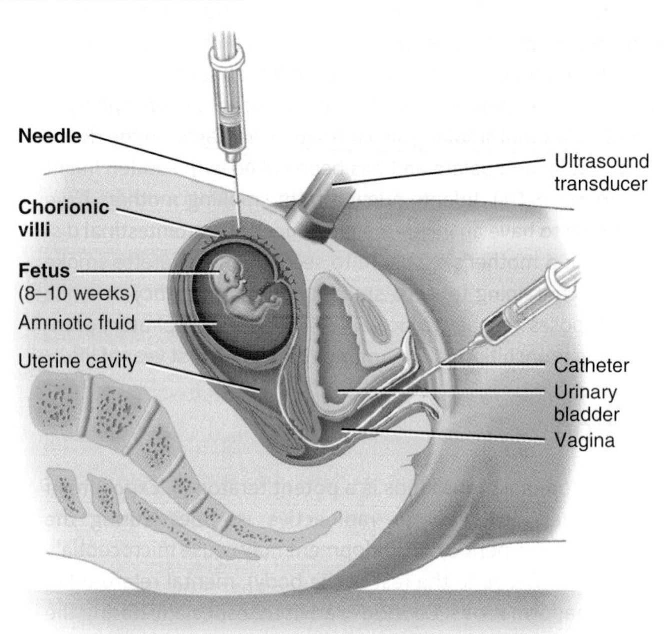

(b) Chorionic villi sampling (CVS)

Q What information can be provided by amniocentesis?

The first such test developed was the **maternal alpha-fetoprotein (AFP) test** (AL-fa fē′-tō-PRŌ-tēn). In this test, the mother's blood is analyzed for the presence of AFP, a protein synthesized in the fetus that passes into the maternal circulation. The highest levels of AFP normally occur during weeks 12 through 15 of pregnancy. Later, AFP is not produced, and its concentration decreases to a very low level both in the fetus and in maternal blood. A high level of AFP after week 16 usually indicates that the fetus has a neural tube defect, such as spina bifida or anencephaly. Because the test is 95% accurate, it is now recommended that all pregnant women be tested for AFP. A newer test (Quad AFP Plus) probes maternal blood for AFP and three other molecules. The test permits prenatal screening for Down syndrome, trisomy 18, and neural tube defects; it also helps predict the delivery date and may reveal the presence of twins.

29.7 Maternal Changes during Pregnancy

OBJECTIVES

- **Describe** the sources and functions of the hormones secreted during pregnancy.
- **Discuss** the hormonal, anatomical, and physiological changes in the mother during pregnancy.

Hormones of Pregnancy

During the first 3 to 4 months of pregnancy, the corpus luteum in the ovary continues to secrete **progesterone** and **estrogens**, which maintain the lining of the uterus during pregnancy and prepare the mammary glands to secrete milk. The amounts secreted by the corpus luteum, however, are only slightly more than those produced after ovulation in a normal menstrual cycle. From the third month through the remainder of the pregnancy, the placenta itself provides the high levels of progesterone and estrogens required. As noted previously, the chorion of the placenta secretes **human chorionic gonadotropin (hCG)** (kō-rē-ON-ik gō′-nad-ō-TRŌ-pin) into the blood. In turn, hCG stimulates the corpus luteum to continue production of progesterone and estrogens—an activity required to prevent menstruation and for the continued attachment of the embryo and fetus to the lining of the uterus (Figure 29.16a). By the eighth day after fertilization, hCG can be detected in the blood and urine of a pregnant woman. Peak secretion of hCG occurs at about the ninth week of pregnancy (Figure 29.16b). During the fourth and fifth months the hCG level decreases sharply and then levels off until childbirth.

The chorion begins to secrete estrogens after the first 3 or 4 weeks of pregnancy and progesterone by the sixth week. These hormones are secreted in increasing quantities until the time of birth (Figure 29.16b). By the fourth month, when the placenta is fully established, the secretion of hCG is greatly reduced, and the secretions of the corpus luteum are no longer essential. A high level of progesterone ensures that the uterine myometrium is relaxed and that the cervix is tightly closed. After delivery, estrogens and progesterone in the blood decrease to normal levels.

FIGURE 29.16 Hormones during pregnancy.

The corpus luteum produces progesterone and estrogens during the first 3–4 months of pregnancy, after which time the placenta assumes this function.

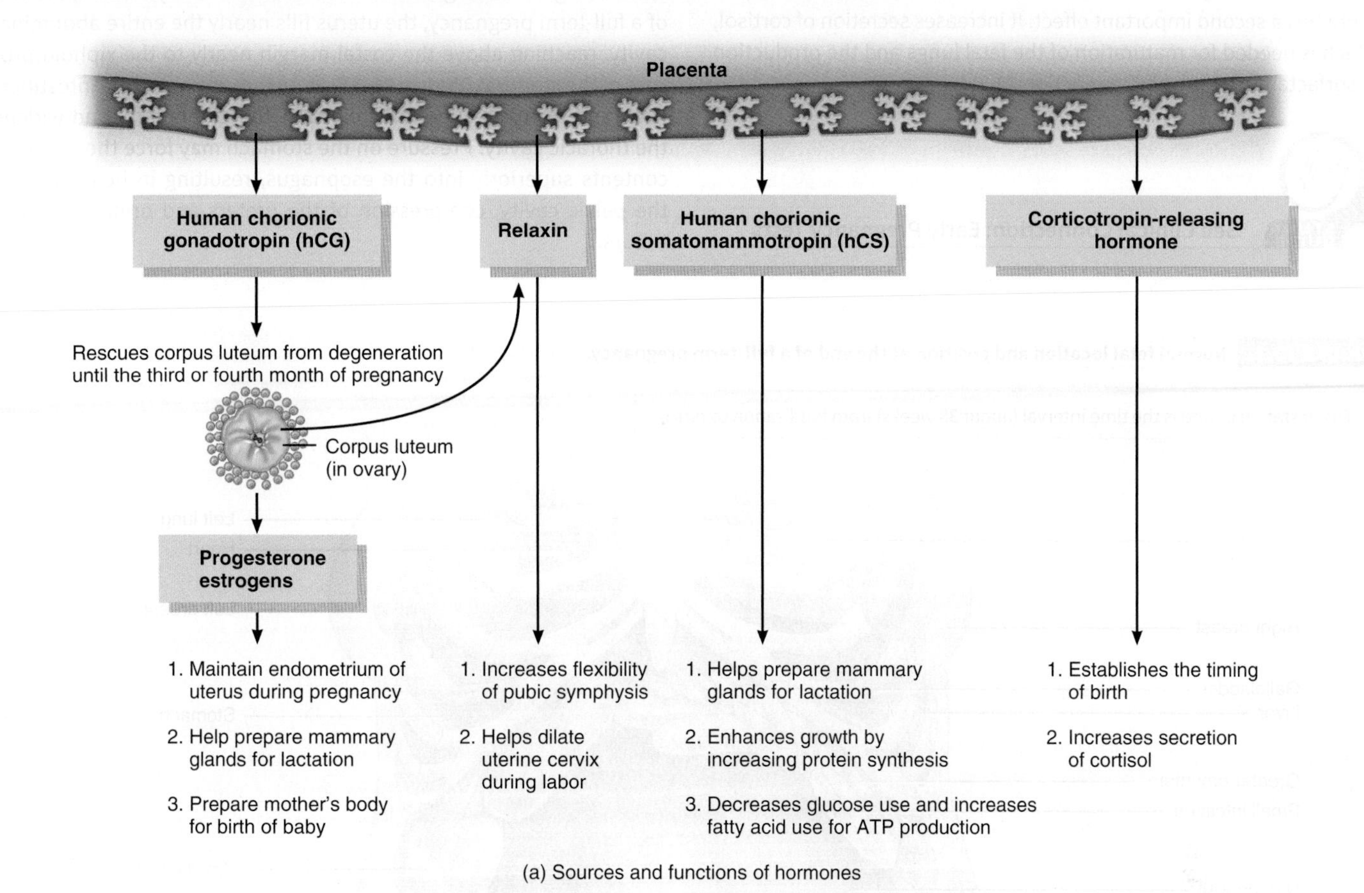

(a) Sources and functions of hormones

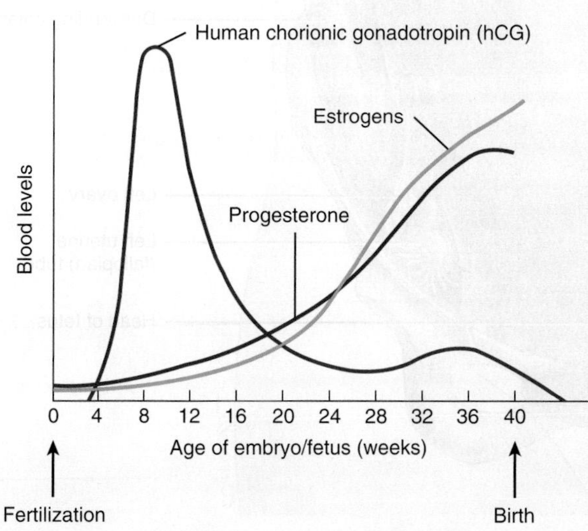

(b) Blood levels of hormones during pregnancy

Q Which hormone is detected by early pregnancy tests?

Relaxin, a hormone produced first by the corpus luteum of the ovary and later by the placenta, increases the flexibility of the pubic symphysis and ligaments of the sacroiliac and sacrococcygeal joints and helps dilate the uterine cervix during labor. Both of these actions ease delivery of the baby.

A third hormone produced by the chorion of the placenta is **human chorionic somatomammotropin (hCS)** (sō′-ma-tō-MAM-ō-trō-pin), also known as *human placental lactogen (hPL)*. The rate of secretion of hCS increases in proportion to placental mass, reaching maximum levels after 32 weeks and remaining relatively constant after that. It is thought to help prepare the mammary glands for lactation, enhance maternal growth by increasing protein synthesis, and regulate certain aspects of metabolism in both mother and fetus. For example, hCS decreases the use of glucose by the mother and promotes the release of fatty acids from her adipose tissue, making more glucose available to the fetus.

The hormone most recently found to be produced by the placenta is **corticotropin-releasing hormone (CRH)** (kor′-ti-kō-TRŌ-pin), which in nonpregnant people is secreted only by neurosecretory cells in the hypothalamus. CRH is now thought to be part of the "clock" that establishes the timing of birth. Secretion of CRH by the placenta begins at about 12 weeks and increases enormously toward the end of pregnancy. Women who have higher levels of CRH earlier in pregnancy are more likely to deliver prematurely; those who have low levels are more likely to deliver after their due date. CRH from the placenta has a second important effect: It increases secretion of cortisol, which is needed for maturation of the fetal lungs and the production of surfactant (see "Alveoli" in Section 23.3).

See Clinical Connection: Early Pregnancy Tests

Changes during Pregnancy

Near the end of the third month of pregnancy, the uterus occupies most of the pelvic cavity. As the fetus continues to grow, the uterus extends higher and higher into the abdominal cavity. Toward the end of a full-term pregnancy, the uterus fills nearly the entire abdominal cavity, reaching above the costal margin nearly to the xiphoid process of the sternum (**Figure 29.17**). It pushes the maternal intestines, liver, and stomach superiorly, elevates the diaphragm, and widens the thoracic cavity. Pressure on the stomach may force the stomach contents superiorly into the esophagus, resulting in heartburn. In the pelvic cavity, compression of the ureters and urinary bladder occurs.

FIGURE 29.17 Normal fetal location and position at the end of a full-term pregnancy.

The gestation period is the time interval (about 38 weeks) from fertilization to birth.

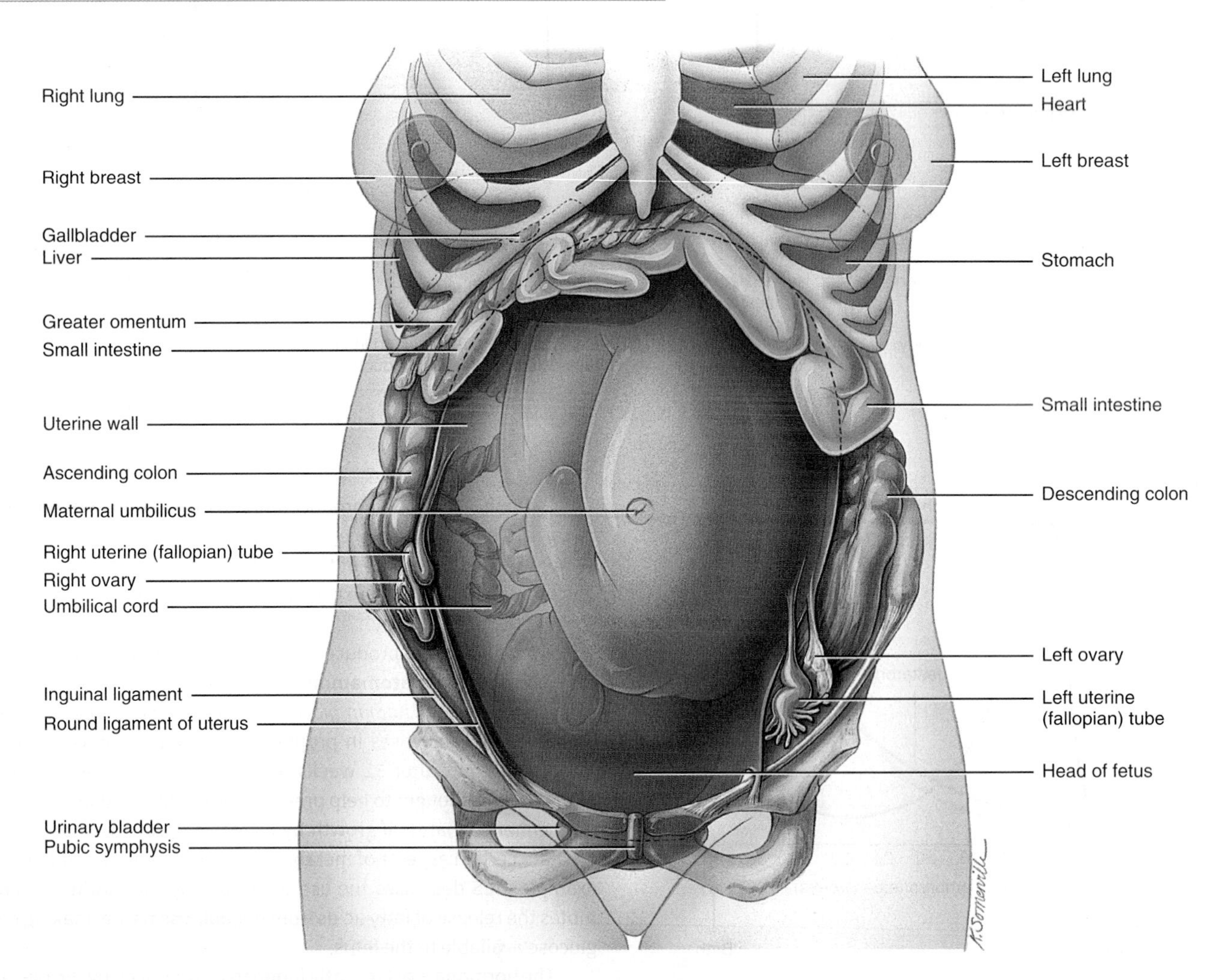

Anterior view of position of organs at end of full-term pregnancy

Q What hormone increases the flexibility of the pubic symphysis and helps dilate the cervix of the uterus to ease delivery of the baby?

Pregnancy-induced physiological changes also occur, including weight gain due to the fetus, amniotic fluid, the placenta, uterine enlargement, and increased total body water; increased storage of proteins, triglycerides, and minerals; marked breast enlargement in preparation for lactation; and lower back pain due to lordosis (hollow back).

Several changes occur in the maternal cardiovascular system. Stroke volume increases by about 30% and cardiac output rises by 20–30% due to increased maternal blood flow to the placenta and increased metabolism. Heart rate increases 10–15% and blood volume increases 30–50%, mostly during the second half of pregnancy. These increases are necessary to meet the additional demands of the fetus for nutrients and oxygen. When a pregnant woman is lying on her back, the enlarged uterus may compress the aorta, resulting in diminished blood flow to the uterus. Compression of the inferior vena cava also decreases venous return, which leads to edema in the lower limbs and may produce varicose veins. Compression of the renal artery can lead to renal hypertension.

Respiratory function is also altered during pregnancy to meet the added oxygen demands of the fetus. Tidal volume can increase by 30–40%, expiratory reserve volume can be reduced by up to 40%, functional residual capacity can decline by up to 25%, minute ventilation (the total volume of air inhaled and exhaled each minute) can increase by up to 40%, airway resistance in the bronchial tree can decline by 30–40%, and total body oxygen consumption can increase by about 10–20%. Dyspnea (difficult breathing) also occurs.

The digestive system also undergoes changes. Pregnant women experience an increase in appetite due to the added nutritional demands of the fetus. A general decrease in GI tract motility can cause constipation, delay gastric emptying time, and produce nausea, vomiting, and heartburn.

Pressure on the urinary bladder by the enlarging uterus can produce urinary symptoms, such as increased frequency and urgency of urination, and stress incontinence. An increase in renal plasma flow up to 35% and an increase in glomerular filtration rate up to 40% increase the renal filtering capacity, which allows faster elimination of the extra wastes produced by the fetus.

Changes in the skin during pregnancy are more apparent in some women than in others. Some women experience increased pigmentation around the eyes and cheekbones in a masklike pattern (*chloasma*), in the areolae of the breasts, and in the linea alba of the lower abdomen (*linea nigra*). *Striae* (stretch marks) over the abdomen can occur as the uterus enlarges, and hair loss increases.

Changes in the reproductive system include edema and increased vascularity of the vulva and increased pliability and vascularity of the vagina. The uterus increases from its nonpregnant mass of 60–80 g to 900–1200 g at term because of hyperplasia of muscle fibers in the myometrium in early pregnancy and hypertrophy of muscle fibers during the second and third trimesters.

See Clinical Connection: Pregnancy-Induced Hypertension

29.8 Exercise and Pregnancy

OBJECTIVE

- **Explain** the effects of pregnancy on exercise and of exercise on pregnancy.

Only a few changes in early pregnancy affect the ability to exercise. A pregnant woman may tire more easily than usual, or morning sickness may interfere with regular exercise. As the pregnancy progresses, weight is gained and posture changes, so more energy is needed to perform activities, and certain maneuvers (sudden stopping, changes in direction, rapid movements) are more difficult to execute. In addition, certain joints, especially the pubic symphysis, become less stable in response to the increased level of the hormone relaxin. As compensation, many mothers-to-be walk with widely spread legs and a shuffling motion.

Although blood shifts from viscera (including the uterus) to the muscles and skin during exercise, there is no evidence of inadequate blood flow to the placenta. The heat generated during exercise may cause dehydration and further increase body temperature. Especially during early pregnancy, excessive exercise and heat buildup should be avoided because elevated body temperature has been implicated in neural tube defects. Exercise has no known effect on lactation, provided a woman remains hydrated and wears a bra that provides good support. Overall, moderate physical activity does not endanger the fetus of a healthy woman who has a normal pregnancy. However, any physical activity that might endanger the fetus should be avoided.

Among the benefits of exercise to the mother during pregnancy are a greater sense of well-being and fewer physical complaints.

29.9 Labor

OBJECTIVE

- **Explain** the events associated with the three stages of labor.

Labor is the process by which the fetus is expelled from the uterus through the vagina, also referred to as giving birth. A synonym for labor is *parturition* (par-toor-ISH-un; *parturit-* = childbirth).

The onset of labor is determined by complex interactions of several placental and fetal hormones. Because progesterone inhibits uterine contractions, labor cannot take place until the effects of progesterone are diminished. Toward the end of gestation, the levels of estrogens in the mother's blood rise sharply, producing changes that overcome the inhibiting effects of progesterone. The rise in estrogens results from increasing secretion by the placenta of corticotropin-releasing hormone, which stimulates the anterior pituitary gland of the fetus to secrete ACTH (adrenocorticotropic hormone). In turn, ACTH stimulates the fetal adrenal gland to secrete cortisol and dehydroepiandrosterone (DHEA) (dē-hī-drō-ep-ē-an-DROS-ter-ōn), the major adrenal androgen. The placenta then converts DHEA into an estrogen. High levels of estrogens cause the number of receptors for oxytocin on uterine muscle fibers to increase, and cause uterine muscle fibers to form gap junctions with one another. Oxytocin released by the posterior pituitary stimulates uterine contractions, and relaxin from the placenta assists by increasing the flexibility of the pubic symphysis and helping dilate the uterine cervix. Estrogen also stimulates the placenta to release prostaglandins, which induce production of enzymes that digest collagen fibers in the cervix, causing it to soften.

Control of labor contractions during parturition occurs via a positive feedback cycle (see **Figure 1.5**). Contractions of the uterine myometrium force the baby's head or body into the cervix, distending (stretching) the cervix. Stretch receptors in the cervix send nerve impulses to neurosecretory cells in the hypothalamus, causing them to release oxytocin into blood capillaries of the posterior pituitary gland. Oxytocin then is carried by the blood to the uterus, where it stimulates the myometrium to contract more forcefully. As the contractions intensify, the baby's body stretches the cervix still more, and the resulting nerve impulses stimulate the secretion of yet more oxytocin. With birth of the infant, the positive feedback cycle is broken because cervical distension suddenly lessens.

Uterine contractions occur in waves (quite similar to the peristaltic waves of the gastrointestinal tract) that start at the top of the uterus and move downward, eventually expelling the fetus. **True labor** begins when uterine contractions occur at regular intervals, usually producing pain. As the interval between contractions shortens, the contractions intensify. Another symptom of true labor in some women is localization of pain in the back that is intensified by walking. The most reliable indicator of true labor is dilation of the cervix and the "show," a discharge of a blood-containing mucus into the cervical canal. In **false labor**, pain is felt in the abdomen at irregular intervals, but it does not intensify and walking does not alter it significantly. There is no "show" and no cervical dilation.

True labor can be divided into three stages (**Figure 29.18**):

1. ***Stage of dilation***. The time from the onset of labor to the complete dilation of the cervix is the **stage of dilation**. This stage, which typically lasts 6–12 hours, features regular contractions of the uterus, usually a rupturing of the amniotic sac, and complete dilation (to 10 cm) of the cervix. If the amniotic sac does not rupture spontaneously, it is ruptured intentionally.
2. ***Stage of expulsion***. The time (10 minutes to several hours) from complete cervical dilation to delivery of the baby is the **stage of expulsion**.
3. ***Placental stage***. The time (5–30 minutes or more) after delivery until the placenta or "afterbirth" is expelled by powerful uterine contractions is the **placental stage**. These contractions also constrict blood vessels that were torn during delivery, reducing the likelihood of hemorrhage.

As a rule, labor lasts longer with first babies, typically about 14 hours. For women who have previously given birth, the average duration of labor is about 8 hours—although the time varies enormously among births. Because the fetus may be squeezed through the birth canal (cervix and vagina) for up to several hours, the fetus is stressed during childbirth: The fetal head is compressed, and the fetus undergoes some degree of intermittent hypoxia due to compression of the umbilical cord and the placenta during uterine contractions. In response to this stress, the fetal adrenal medullae secrete very high levels of epinephrine and norepinephrine, the "fight-or-flight" hormones. Much of the protection against the stresses of parturition, as well as preparation of the infant for surviving extrauterine life, is

FIGURE 29.18 Stages of true labor.

The term *parturition* refers to birth.

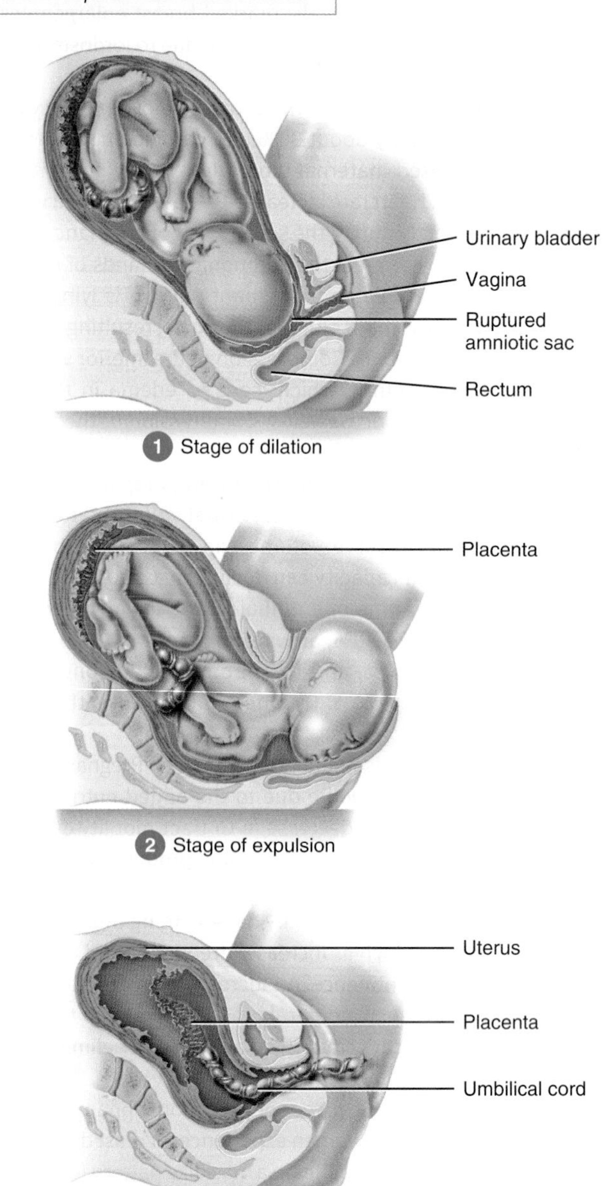

Q What event marks the beginning of the stage of expulsion?

provided by these hormones. Among other functions, epinephrine and norepinephrine clear the lungs and alter their physiology in readiness for breathing air, mobilize readily usable nutrients for cellular metabolism, and promote an increased flow of blood to the brain and heart.

About 7% of pregnant women do not deliver by 2 weeks after their due date. Such cases carry an increased risk of brain damage to the fetus, and even fetal death, due to inadequate supplies of oxygen and nutrients from an aging placenta. Post-term deliveries may be facilitated by inducing labor, initiated by administration of oxytocin (Pitocin®), or by surgical delivery (cesarean section).

Following the delivery of the baby and placenta is a 6-week period during which the maternal reproductive organs and physiology return to the prepregnancy state. This period is called the **puerperium** (pū-er-PER-ē-um). Through a process of tissue catabolism, the uterus undergoes a remarkable reduction in size, called **involution** (in-vō-LOO-shun), especially in lactating women. The cervix loses its elasticity and regains its prepregnancy firmness. For 2–4 weeks after delivery, women have a uterine discharge called **lochia** (LŌ-kē-a), which consists initially of blood and later of serous fluid derived from the former site of the placenta.

See Clinical Connection: Dystocia and Cesarean Section

29.10 Adjustments of the Infant at Birth

OBJECTIVE

- **Explain** the respiratory and cardiovascular adjustments that occur in an infant at birth.

During pregnancy, the embryo (and later the fetus) is totally dependent on the mother for its existence. The mother supplies the fetus with oxygen and nutrients, eliminates its carbon dioxide and other wastes, protects it against shocks and temperature changes, and provides antibodies that confer protection against certain harmful microbes. At birth, a physiologically mature baby becomes much more self-supporting, and the newborn's body systems must make various adjustments. The most dramatic changes occur in the respiratory and cardiovascular systems.

Respiratory Adjustments

The reason that the fetus depends entirely on the mother for obtaining oxygen and eliminating carbon dioxide is that the fetal lungs are either collapsed or partially filled with amniotic fluid. The production of surfactant begins by the end of the sixth month of development. Because the respiratory system is fairly well developed at least 2 months before birth, premature babies delivered at 7 months are able to breathe and cry. After delivery, the baby's supply of oxygen from the mother ceases, and any amniotic fluid in the fetal lungs is absorbed. Because carbon dioxide is no longer being removed, it builds up in the blood. A rising CO_2 level stimulates the respiratory center in the medulla oblongata, causing the respiratory muscles to contract, and the baby to draw his or her first breath. Because the first inspiration is unusually deep, as the lungs contain no air, the baby also exhales vigorously and naturally cries. A full-term baby may breathe 45 times a minute for the first 2 weeks after birth. Breathing rate gradually declines until it approaches a normal rate of 12 breaths per minute.

Cardiovascular Adjustments

After the baby's first inspiration, the cardiovascular system must make several adjustments (see **Figure 21.31**). Closure of the foramen ovale between the atria of the fetal heart, which occurs at the moment of birth, diverts deoxygenated blood to the lungs for the first time. The foramen ovale is closed by two flaps of septal heart tissue that fold together and permanently fuse. The remnant of the foramen ovale is the fossa ovalis.

Once the lungs begin to function, the ductus arteriosus shuts off due to contractions of smooth muscle in its wall, and it becomes the ligamentum arteriosum. The muscle contraction is probably mediated by the polypeptide bradykinin, released from the lungs during their initial inflation. The ductus arteriosus generally does not close completely until about 3 months after birth. Prolonged incomplete closure results in a condition called **patent ductus arteriosus** (see **Figure SG20.3b**).

After the umbilical cord is tied off and severed and blood no longer flows through the umbilical arteries, they fill with connective tissue, and their distal portions become the medial umbilical ligaments. The umbilical vein then becomes the ligamentum teres (round ligament) of the liver.

In the fetus, the ductus venosus connects the umbilical vein directly with the inferior vena cava, allowing blood from the placenta to bypass the fetal liver. When the umbilical cord is severed, the ductus venosus collapses, and venous blood from the viscera of the fetus flows into the hepatic portal vein to the liver and then via the hepatic vein to the inferior vena cava. The remnant of the ductus venosus becomes the ligamentum venosum.

At birth, an infant's pulse may range from 120 to 160 beats per minute and may go as high as 180 on excitation. After birth, oxygen use increases, which stimulates an increase in the rate of red blood cell and hemoglobin production. The white blood cell count at birth is very high—sometimes as much as 45,000 cells per microliter—but the count decreases rapidly by the seventh day. Recall that the white blood cell count of an adult is 5000–10,000 cells per microliter.

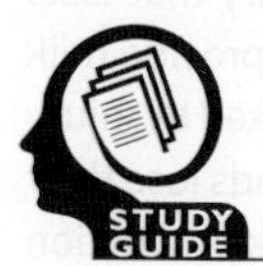

See Clinical Connection: Premature Infants

29.11 The Physiology of Lactation

OBJECTIVE

- **Discuss** the physiology and hormonal control of lactation.

Lactation (lak-TĀ-shun) is the production and ejection of milk from the mammary glands. A principal hormone in promoting milk production is **prolactin (PRL)**, which is secreted from the anterior pituitary gland. Even though prolactin levels increase as the pregnancy progresses, no milk production occurs because progesterone inhibits the effects of prolactin. After delivery, the levels of estrogens and progesterone in the mother's blood decrease, and the inhibition is removed. The principal stimulus in maintaining prolactin secretion during lactation is the sucking action of the infant. Suckling initiates nerve impulses from stretch receptors in the nipples to the hypothalamus; the impulses decrease hypothalamic release of prolactin-inhibiting hormone (PIH) and increase release of prolactin-releasing hormone (PRH), so more prolactin is released by the anterior pituitary.

Oxytocin causes release of milk into the mammary ducts via the **milk ejection reflex** (Figure 29.19). Milk formed by the glandular cells of the breasts is stored until the baby begins active suckling. Stimulation of touch receptors in the nipple initiates sensory nerve impulses that are relayed to the hypothalamus. In response, secretion of oxytocin from the posterior pituitary increases. Carried by the bloodstream to the mammary glands, oxytocin stimulates contraction of myoepithelial (smooth muscle–like) cells surrounding the glandular cells and ducts. The resulting compression moves the milk from the alveoli of the mammary glands into the mammary ducts, where it can be suckled. This process is termed **milk ejection** (*let-down*). Even though the actual ejection of milk does not occur until 30–60 seconds after nursing begins (the latent period), some milk stored in lactiferous sinuses near the nipple is available during the latent period. Stimuli other than suckling, such as hearing a baby's cry or touching the mother's genitals, also can trigger oxytocin release and milk ejection. The suckling stimulation that produces the release of oxytocin also inhibits the release of PIH; this results in increased secretion of prolactin, which maintains lactation.

During late pregnancy and the first few days after birth, the mammary glands secrete a cloudy fluid called **colostrum**. Although it is not as nutritious as milk—it contains less lactose and virtually no fat—colostrum serves adequately until the appearance of true milk on about the fourth day. Colostrum and maternal milk contain important antibodies that protect the infant during the first few months of life.

Following birth of the infant, the prolactin level starts to return to the nonpregnant level. However, each time the mother nurses the infant, nerve impulses from the nipples to the hypothalamus increase the release of PRH (and decrease the release of PIH), resulting in a tenfold increase in prolactin secretion by the anterior pituitary that lasts about an hour. Prolactin acts on the mammary glands to provide milk for the next nursing period. If this surge of prolactin is blocked by injury or disease, or if nursing is discontinued, the mammary glands lose their ability to produce milk in only a few days. Even though milk production

FIGURE 29.19 The milk ejection reflex, a positive feedback cycle.

Oxytocin stimulates contraction of myoepithelial cells in the breasts, which squeezes the glandular and duct cells and causes milk ejection.

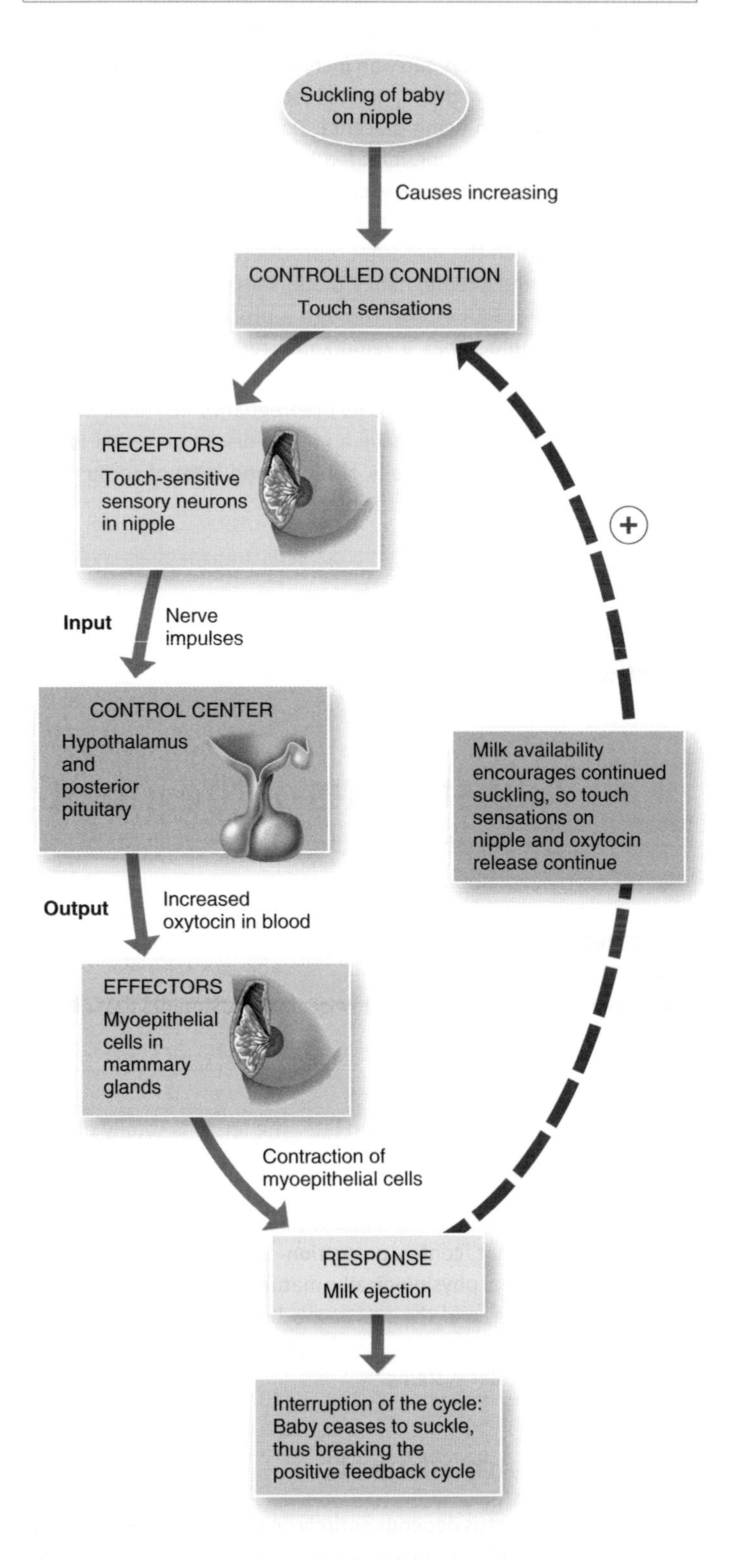

Q What is another function of oxytocin?

normally decreases considerably within 7–9 months after birth, it can continue for several years if nursing or **breastfeeding** continues.

Lactation often blocks ovarian cycles for the first few months following delivery, if the frequency of sucking is about 8–10 times a day. This effect is inconsistent, however, and ovulation commonly precedes the first menstrual period after delivery of a baby. As a result, the mother can never be certain she is not fertile. Breastfeeding is therefore an unreliable birth control measure. The suppression of ovulation during lactation is believed to occur as follows: During breastfeeding, neural input from the nipple reaches the hypothalamus and causes it to produce neurotransmitters that suppress the release of gonadotropin-releasing hormone (GnRH). As a result, production of LH and FSH decreases, and ovulation is inhibited.

A primary benefit of breastfeeding is nutritional: Human milk is a sterile solution that contains amounts of fatty acids, lactose, amino acids, minerals, vitamins, and water that are ideal for the baby's digestion, brain development, and growth. Breastfeeding also benefits infants by providing the following:

- ***Beneficial cells.*** Several types of white blood cells are present in breast milk. Neutrophils and macrophages serve as phagocytes, ingesting microbes in the baby's gastrointestinal tract. Macrophages also produce lysozyme and other immune system components. Plasma cells, which develop from B lymphocytes, produce antibodies against specific microbes, and T lymphocytes kill microbes directly or help mobilize other defenses.
- ***Beneficial molecules.*** Breast milk also contains an abundance of beneficial molecules. Maternal IgA antibodies in breast milk bind to microbes in the baby's gastrointestinal tract and prevent their migration into other body tissues. Because a mother produces antibodies to whatever disease-causing microbes are present in her environment, her breast milk affords protection against the specific infectious agents to which her baby is also exposed. Additionally, two milk proteins bind to nutrients that many bacteria need to grow and survive: B_{12}-binding protein ties up vitamin B_{12}, and lactoferrin ties up iron. Some fatty acids can kill certain viruses by disrupting their membranes, and lysozyme kills bacteria by disrupting their cell walls. Finally, interferons enhance the antimicrobial activity of immune cells.
- ***Decreased incidence of diseases later in life.*** Breastfeeding provides children with a slight reduction in risk of lymphoma, heart disease, allergies, respiratory and gastrointestinal infections, ear infections, diarrhea, diabetes mellitus, and meningitis.
- ***Miscellaneous benefits.*** Breastfeeding supports optimal infant growth, enhances intellectual and neurological development, and fosters mother–infant relations by establishing early and prolonged contact between them. Compared to cow's milk, the fats and iron in breast milk are more easily absorbed, the proteins in breast milk are more readily metabolized, and the lower sodium content of breast milk is more suited to an infant's needs. Premature infants benefit even more from breast-feeding because the milk produced by mothers of premature infants seems to be specially adapted to the infant's needs; it has a higher protein content than the milk of mothers of full-term infants. Finally, a baby is less likely to have an allergic reaction to its mother's milk than to milk from another source.

Years before oxytocin was discovered, it was common practice in midwifery to let a first-born twin nurse at the mother's breast to speed the birth of the second child. Now we know why this practice is helpful—it stimulates the release of oxytocin. Even after a single birth, nursing promotes expulsion of the placenta (afterbirth) and helps the uterus return to its normal size. Synthetic oxytocin (Pitocin) is often given to induce labor or to increase uterine tone and control hemorrhage just after parturition.

29.12 Inheritance

OBJECTIVE

- **Explain** the inheritance of dominant, recessive, complex, and sex-linked traits.

As previously indicated, the genetic material of a father and a mother unite when a sperm cell fuses with a secondary oocyte to form a zygote. Children resemble their parents because they inherit traits passed down from both parents. We now examine some of the principles involved in that process, called inheritance.

Inheritance is the passage of hereditary traits from one generation to the next. It is the process by which you acquired your characteristics from your parents and may transmit some of your traits to your children. The branch of biology that deals with inheritance is called **genetics** (je-NET-iks). The area of health care that offers advice on genetic problems (or potential problems) is called **genetic counseling**.

Genotype and Phenotype

As you have already learned, the nuclei of all human cells except gametes contain 23 pairs of chromosomes—the diploid number ($2n$). One chromosome in each pair came from the mother, and the other came from the father. Each of these two homologues contains genes that control the same traits. If one chromosome of the pair contains a gene for body hair, for example, its homologue will contain a gene for body hair in the same position. Alternative forms of a gene that code for the same trait and are at the same location on homologous chromosomes are called **alleles** (a-LĒLZ). One allele of the previously mentioned body hair gene might code for coarse hair, and another might code for fine hair. A **mutation** (mū-TĀ-shun; *muta-* = change) is a permanent heritable change in an allele that produces a different variant of the same trait.

The relationship of genes to heredity is illustrated by examining the alleles involved in a disorder called **phenylketonuria (PKU)** (fen′-il-kē′-tō-NOO-rē-a). People with PKU (for reference see Clinical Connection: Phenylketonuria in Section 25.5) are unable to manufacture the enzyme phenylalanine hydroxylase. The allele that codes for phenylalanine hydroxylase is symbolized as *P*; the mutated allele that fails to produce a functional enzyme is represented by *p*. The chart in **Figure 29.20**, which shows the possible combinations of gametes from two

FIGURE 29.20 Inheritance of phenylketonuria (PKU).

Genotype refers to genetic makeup; phenotype refers to the physical or outward expression of a gene.

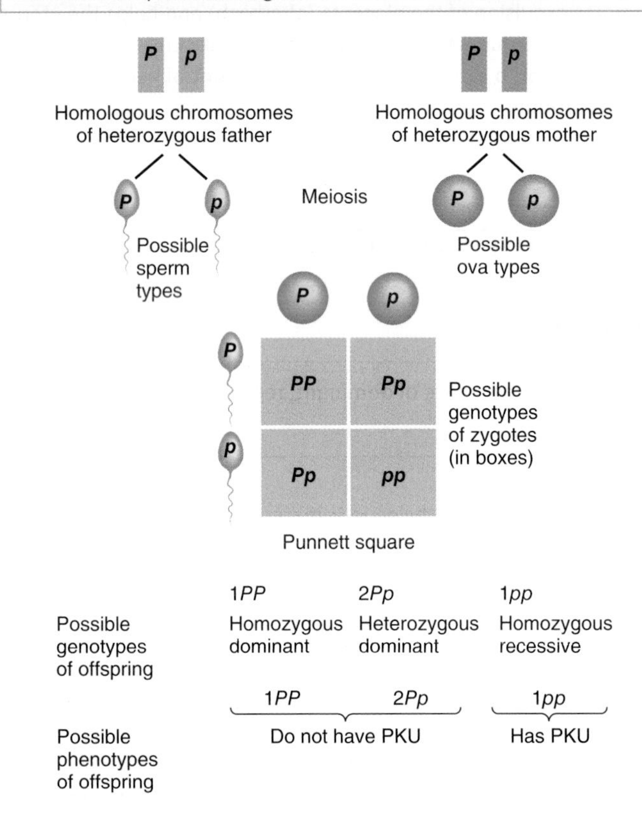

Q If parents have the genotypes shown here, what is the chance that their first child will have PKU? What is the chance of PKU occurring in their second child?

parents who each have one *P* and one *p* allele, is called a **Punnett square**. In constructing a Punnett square, the possible paternal alleles in sperm are written at the left side and the possible maternal alleles in ova (or secondary oocytes) are written at the top. The four spaces on the chart show how the alleles can combine in zygotes formed by the union of these sperm and ova to produce the three different combinations of genes, or **genotypes** (JĒ-nō-tīps): *PP*, *Pp*, or *pp*. Notice from the Punnett square that 25% of the offspring will have the *PP* genotype, 50% will have the *Pp* genotype, and 25% will have the *pp* genotype. (These percentages are probabilities only; parents who have four children won't necessarily end up with one with PKU.) People who inherit *PP* or *Pp* genotypes do not have PKU; those with a *pp* genotype suffer from the disorder. Although people with a *Pp* genotype have one PKU allele (*p*), the allele that codes for the normal trait (*P*) masks the presence of the PKU allele. An allele that dominates or masks the presence of another allele and is fully expressed (*P* in this example) is said to be a **dominant allele**, and the trait expressed is called a dominant trait. The allele whose presence is completely masked (*p* in this example) is said to be a **recessive allele**, and the trait it controls is called a recessive trait.

By tradition, the symbols for genes are written in italics, with dominant alleles written in capital letters and recessive alleles in lowercase letters. A person with the same alleles on homologous chromosomes (for example, *PP* or *pp*) is said to be **homozygous** (hō-mō-ZĪ-gus) for the trait. *PP* is homozygous dominant, and *pp* is homozygous recessive. An individual with different alleles on homologous chromosomes (for example, *Pp*) is said to be **heterozygous** (het′-er-ō-ZĪ-gus) for the trait.

Phenotype (FĒ-nō-tīp; *pheno-* = showing) refers to how the genetic makeup is expressed in the body; it is the physical or outward expression of a gene. A person with *Pp* (a heterozygote) has a different *genotype* from a person with *PP* (a homozygote), but both have the same *phenotype*—normal production of phenylalanine hydroxylase. Heterozygous individuals who carry a recessive gene but do not express it (*Pp*) can pass the gene on to their offspring. Such individuals are called **carriers** of the recessive gene.

Most genes give rise to the same phenotype whether they are inherited from the mother or the father. In a few cases, however, the phenotype is dramatically different, depending on the parental origin. This surprising phenomenon, first appreciated in the 1980s, is called **genomic imprinting**. In humans, the abnormalities most clearly associated with mutation of an imprinted gene are *Angelman syndrome* (mental retardation, ataxia, seizures, and minimal speech), which results when the gene for a particular abnormal trait is inherited from the mother, and *Prader-Willi syndrome* (short stature, mental retardation, obesity, poor responsiveness to external stimuli, and sexual immaturity), which results when it is inherited from the father.

Alleles that code for normal traits do not always dominate over those that code for abnormal ones, but dominant alleles for severe disorders usually are lethal and cause death of the embryo or fetus. One exception is Huntington disease (HD) (for reference see Clinical Connection: Disorders of the Basal Nuclei in Section 16.4), which is caused by a dominant allele with effects that are not manifested until adulthood. Both homozygous dominant and heterozygous people exhibit the disease; homozygous recessive people are normal. HD causes progressive degeneration of the nervous system and eventual death, but because symptoms typically do not appear until after age 30 or 40, many afflicted individuals will already have passed on the allele for the condition to their children by the time they discover they have the disease.

Occasionally an error in cell division, called **nondisjunction** (non′-dis-JUNK-shun), results in an abnormal number of chromosomes. In this situation, homologous chromosomes (during meiosis I) or sister chromatids (during anaphase of mitosis or meiosis II) fail to separate properly. See **Figure 3.34**. A cell from which one or more chromosomes has been added or deleted is called an **aneuploid** (AN-ū-ployd). A monosomic cell ($2n - 1$) is missing a chromosome; a trisomic cell ($2n + 1$) has an extra chromosome. Most cases of Down syndrome (see Disorders: Homeostatic Imbalances in Study Guide) are aneuploid disorders in which there is trisomy of chromosome 21. Nondisjunction usually occurs during gametogenesis (meiosis), but about 2% of Down syndrome cases result from nondisjunction during mitotic divisions in early embryonic development.

Another error in meiosis is a **translocation**. In this case, two chromosomes that are *not* homologous break and interchange portions.

TABLE 29.3 Selected Hereditary Traits in Humans

DOMINANT	RECESSIVE
Normal skin pigmentation	Albinism
Near- or farsightedness	Normal vision
PTC taster*	PTC nontaster
Polydactyly (extra digits)	Normal digits
Brachydactyly (short digits)	Normal digits
Syndactylism (webbed digits)	Normal digits
Diabetes insipidus	Normal urine excretion
Huntington disease	Normal nervous system
Widow's peak	Straight hairline
Curved (hyperextended) thumb	Straight thumb
Normal Cl^- transport	Cystic fibrosis
Hypercholesterolemia (familial)	Normal cholesterol level

*Ability to taste a chemical compound called phenylthiocarbamide (PTC).

The individual who has a translocation may be perfectly normal if no loss of genetic material took place when the rearrangement occurred. However, some of the person's gametes may not contain the correct amount and type of genetic material. About 3% of Down syndrome cases result from a translocation of part of chromosome 21 to another chromosome, usually chromosome 14 or 15. The individual who has this translocation is normal and does not even know that he or she is a "carrier." When such a carrier produces gametes, however, some gametes end up with a whole chromosome 21 plus another chromosome with the translocated fragment of chromosome 21. On fertilization, the zygote then has three, rather than two, copies of that part of chromosome 21.

Table 29.3 lists some dominant and recessive inherited structural and functional traits in humans.

Variations on Dominant–Recessive Inheritance

Most patterns of inheritance do not conform to the simple **dominant–recessive inheritance** we have just described, in which only dominant and recessive alleles interact. The phenotypic expression of a particular gene may be influenced not only by which alleles are present, but also by other genes and by the environment. Most inherited traits are influenced by more than one gene, and, to complicate matters, most genes can influence more than one trait. Variations on dominant–recessive inheritance include incomplete dominance, multiple-allele inheritance, and complex inheritance.

Incomplete Dominance In **incomplete dominance,** neither member of a pair of alleles is dominant over the other, and the heterozygote has a phenotype intermediate between the homozygous dominant and the homozygous recessive phenotypes. An example of incomplete dominance in humans is the inheritance of **sickle cell disease (SCD)** (Figure 29.21). People with the homozygous dominant genotype Hb^AHb^A form normal hemoglobin; those with the homozygous recessive genotype Hb^SHb^S have sickle cell disease and severe anemia. Although they are usually healthy, those with the heterozygous genotype Hb^AHb^S have minor problems with anemia because half of their hemoglobin is normal and half is not. Heterozygotes are carriers, and they are said to have *sickle cell trait.*

FIGURE 29.21 Inheritance of sickle cell disease.

Sickle cell disease is an example of incomplete dominance.

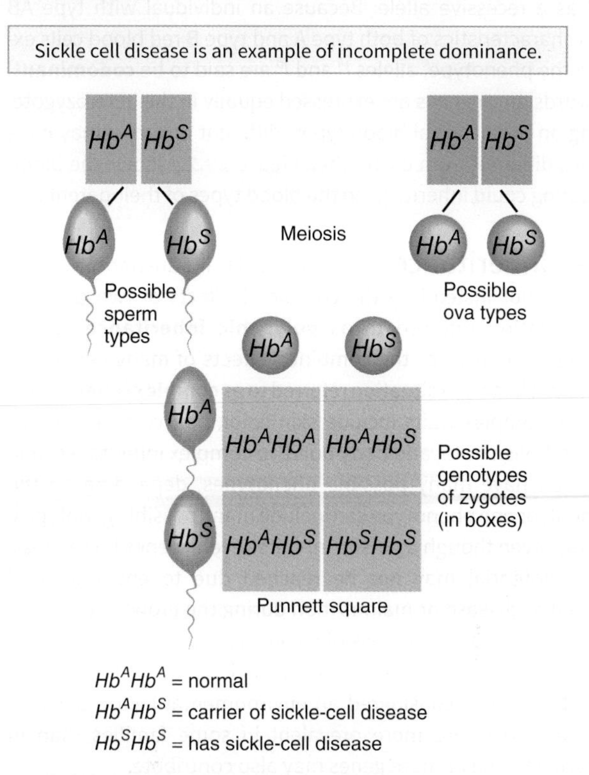

Q What are the distinguishing features of incomplete dominance?

Multiple-Allele Inheritance Although a single individual inherits only two alleles for each gene, some genes may have more than two alternative forms; this is the basis for **multiple-allele inheritance.** One example of multiple-allele inheritance is the inheritance of the ABO blood group. The four blood types (phenotypes) of the ABO group—A, B, AB, and O—result from the inheritance of six combinations of three different alleles of a single gene called the *I* gene: (1) allele I^A produces the A antigen, (2) allele I^B produces the B antigen, and (3) allele *i* produces neither A nor B antigen. Each person inherits two *I*-gene alleles, one from each parent, that give rise to the various phenotypes. The six possible genotypes produce four blood types, as follows:

Genotype	*Blood type (phenotype)*
I^AI^A or I^Ai	A
I^BI^B or I^Bi	B
I^AI^B	AB
ii	O

Notice that both I^A and I^B are inherited as dominant alleles, and *i* is inherited as a recessive allele. Because an individual with type AB blood has characteristics of both type A and type B red blood cells expressed in the phenotype, alleles I^A and I^B are said to be **codominant**. In other words, both genes are expressed equally in the heterozygote. Depending on the parental blood types, different offspring may have blood types different from each other. Figure 29.22 shows the blood types offspring could inherit, given the blood types of their parents.

Complex Inheritance Most inherited traits are not controlled by one gene, but instead by the combined effects of two or more genes, a situation referred to as **polygenic inheritance** (pol-ē-JĒN-ik; poly- = many), or the combined effects of many genes and environmental factors, a situation referred to as **complex inheritance**. Examples of complex traits include skin color, hair color, eye color, height, metabolism rate, and body build. In complex inheritance, one genotype can have many possible phenotypes, depending on the environment, or one phenotype can include many possible genotypes. For example, even though a person inherits several genes for tallness, full height potential may not be reached due to environmental factors, such as disease or malnutrition during the growth years. You have already learned that the risk of having a child with a neural tube defect is greater in pregnant women who lack adequate folic acid in their diet; this is also considered an environmental factor. Because neural tube defects are more prevalent in some families than in others, however, one or more genes may also contribute.

Often, a complex trait shows a continuous gradation of small differences between extremes among individuals. It is relatively easy to predict the risk of passing on an undesirable trait that is due to a single dominant or recessive gene, but it is very difficult to make this prediction when the trait is complex. Such traits are difficult to follow in a family because the range of variation is large, the number of different genes involved usually is not known, and the impact of environmental factors may be incompletely understood.

Skin color is a good example of a complex trait. It depends on environmental factors such as sun exposure and nutrition, as well as on several genes. Suppose that skin color is controlled by three separate genes, each having two alleles: *A*, *a*; *B*, *b*; and *C*, *c* (Figure 29.23). A person with the genotype *AABBCC* is very dark skinned, an individual with the genotype *aabbcc* is very light skinned, and a person with the genotype *AaBbCc* has an intermediate skin color. Parents having an intermediate skin color may have children with very light, very dark, or intermediate skin color. Note that the **P generation** (parental generation) is the starting generation, the **F_1 generation** (first filial generation) is produced from the P generation, and the **F_2 generation** (second filial generation) is produced from the F_1 generation.

Autosomes, Sex Chromosomes, and Sex Determination

When viewed under a microscope, the 46 human chromosomes in a normal somatic cell can be identified by their size, shape, and staining

FIGURE 29.22 The 10 possible combinations of parental ABO blood types and the blood types their offspring could inherit. For each possible set of parents, the blue letters represent the blood types their offspring could inherit.

Inheritance of ABO blood types is an example of multiple-allele inheritance.

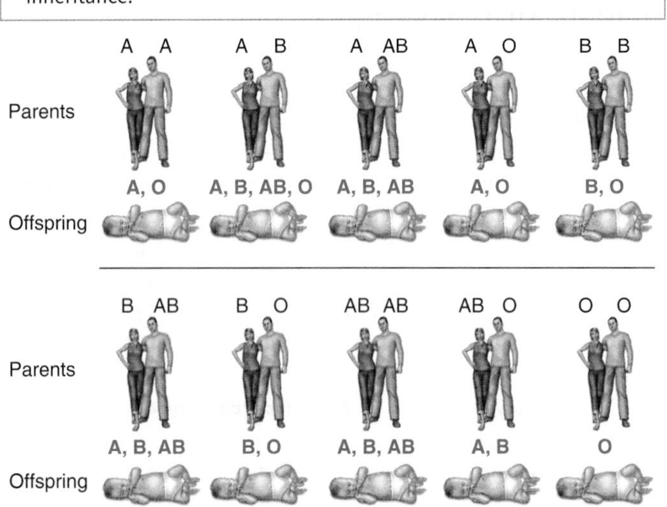

Q How is it possible for a baby to have type O blood if neither parent is type O?

FIGURE 29.23 Complex inheritance of skin color.

In complex inheritance, a trait is controlled by the combined effects of many genes and environmental factors.

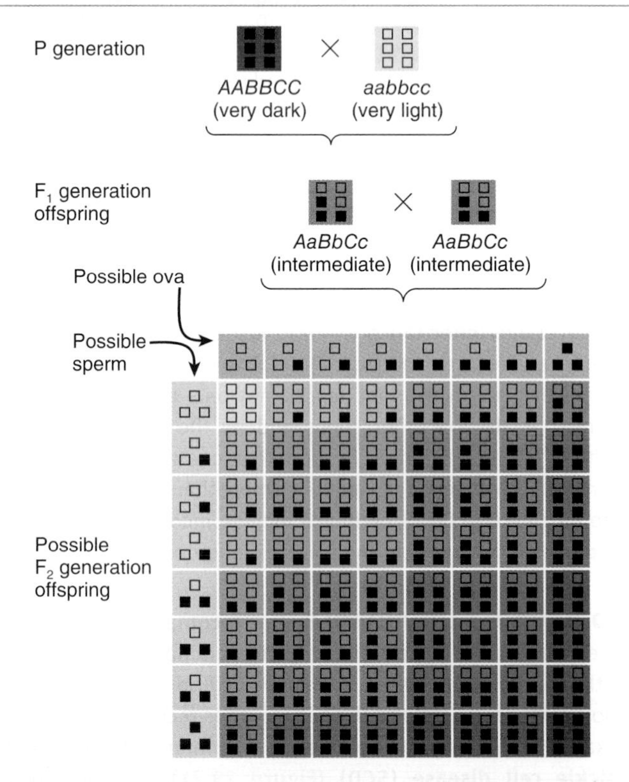

Q What other traits are transmitted by complex inheritance?

FIGURE 29.24 Human karyotype showing autosomes and sex chromosomes. The white circles are the centromeres.

Human somatic cells contain 23 different pairs of chromosomes.

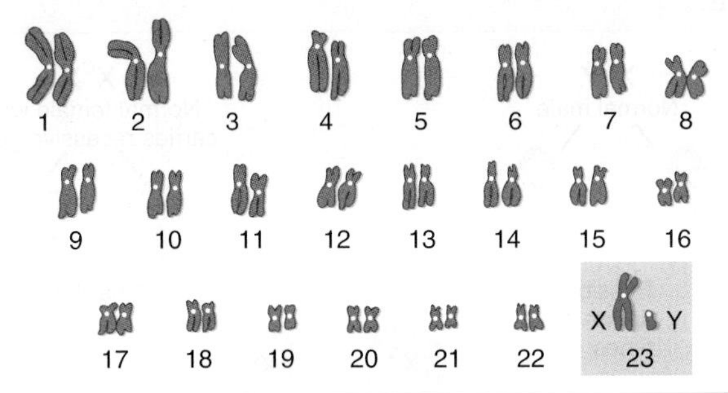

Q What are the two sex chromosomes in females and males?

FIGURE 29.25 Sex determination.

Sex is determined at the time of fertilization by the presence or absence of a Y chromosome in the sperm.

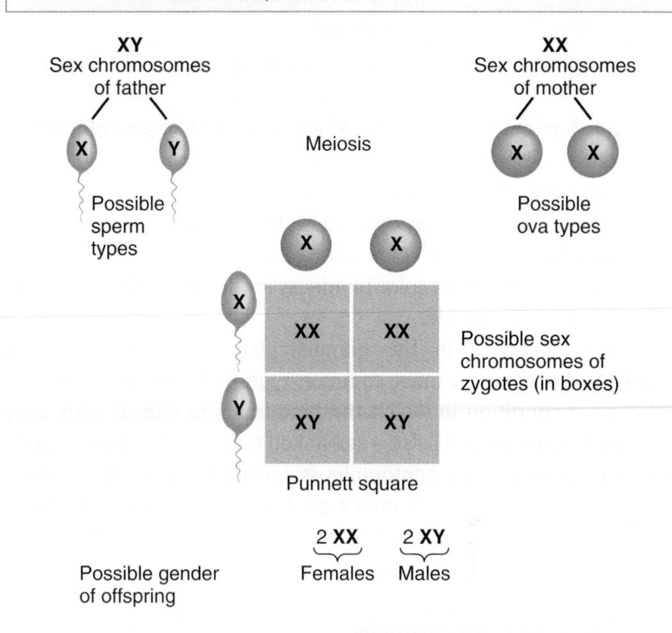

Q What are all chromosomes other than the sex chromosomes called?

pattern to be members of 23 different pairs. An entire set of chromosomes arranged in decreasing order of size and according to the position of the centromere is called a **karyotype** (KAR-ē-ō-tīp; *karyo-* = nucleus; *-typos* = model) (**Figure 29.24**). In 22 of the pairs, the homologous chromosomes look alike and have the same appearance in both males and females; these 22 pairs are called **autosomes** (AW-tō-sōms). The two members of the 23rd pair are termed the **sex chromosomes**; they look different in males and females. In females, the pair consists of two chromosomes called X chromosomes. One X chromosome is also present in males, but its mate is a much smaller chromosome called a Y chromosome. The Y chromosome has only 231 genes, less than 10% of the 2968 genes present on chromosome 1, the largest autosome.

When a spermatocyte undergoes meiosis to reduce its chromosome number, it gives rise to two sperm that contain an X chromosome and two sperm that contain a Y chromosome. Oocytes have no Y chromosomes and produce only X-containing gametes. If the secondary oocyte is fertilized by an X-bearing sperm, the offspring normally is female (XX). Fertilization by a Y-bearing sperm produces a male (XY). Thus, an individual's sex is determined by the father's chromosomes (**Figure 29.25**).

Both female and male embryos develop identically until about 7 weeks after fertilization. At that point, one or more genes set into motion a cascade of events that leads to the development of a male; in the absence of normal expression of the gene or genes, the female pattern of development occurs. It has been known since 1959 that the Y chromosome is needed to initiate male development. Experiments published in 1991 established that the prime male-determining gene is one called ***SRY*** (sex-determining region of the Y chromosome). When a small DNA fragment containing this gene was inserted into 11 female mouse embryos, three of them developed as males. (The researchers suspected that the gene failed to be integrated into the genetic material in the other eight.) *SRY* acts as a molecular switch to turn on the male pattern of development. Only if the *SRY* gene is present and functional in a fertilized ovum will the fetus develop testes and differentiate into a male; in the absence of *SRY*, the fetus will develop ovaries and differentiate into a female.

Case studies have confirmed the key role of *SRY* in directing the male pattern of development in humans. In some cases, phenotypic females with an XY genotype were found to have mutated *SRY* genes. These individuals failed to develop normally as males because their *SRY* gene was defective. In other cases, phenotypic males with an XX genotype were found to have a small piece of the Y chromosome, including the *SRY* gene, inserted into one of their X chromosomes.

Sex-Linked Inheritance

In addition to determining the sex of the offspring, the sex chromosomes are responsible for the transmission of several nonsexual traits. Many of the genes for these traits are present on X chromosomes but are absent from Y chromosomes. This feature produces a pattern of heredity called **sex-linked inheritance** that is different from the patterns already described.

Red–Green Color Blindness One example of sex-linked inheritance is **red–green color blindness**, the most common type of color blindness. This condition is characterized by a deficiency in either red- or green-sensitive cones, so red and green are seen as the same color (either red or green, depending on which cone is present). The gene for red–green color blindness is a recessive one designated *c*. Normal color vision, designated *C*, dominates. The *C/c* genes are located only on the X chromosome, so the ability to see colors depends entirely on the X chromosomes. The possible combinations are as follows:

Genotype	*Phenotype*
X^CX^C	Normal female
X^CX^c	Normal female (but carrier of recessive gene)
X^cX^c	Red–green color-blind female
X^CY	Normal male
X^cY	Red–green color-blind male

Only females who have two X^c genes are red–green color blind. This rare situation can result only from the mating of a color-blind male and a color-blind or carrier female. Because males do not have a second X chromosome that could mask the trait, all males with an X^c gene will be red–green color blind. Figure 29.26 illustrates the inheritance of red–green color blindness in the offspring of a normal male and a carrier female.

Traits inherited in the manner just described are called **sex-linked traits**. The most common type of **hemophilia** (hē-mō-FIL-ē-a)—a condition in which the blood fails to clot or clots very slowly after an injury—is also a sex-linked trait. Like the trait for red–green color blindness, hemophilia is caused by a recessive gene. Other sex-linked traits in humans are fragile X syndrome, nonfunctional sweat glands, certain forms of diabetes, some types of deafness, uncontrollable rolling of the eyeballs, absence of central incisors, night blindness, one form of cataract, juvenile glaucoma, and juvenile muscular dystrophy.

FIGURE 29.26 An example of the inheritance of red–green color blindness.

Red–green color blindness and hemophilia are examples of sex-linked traits.

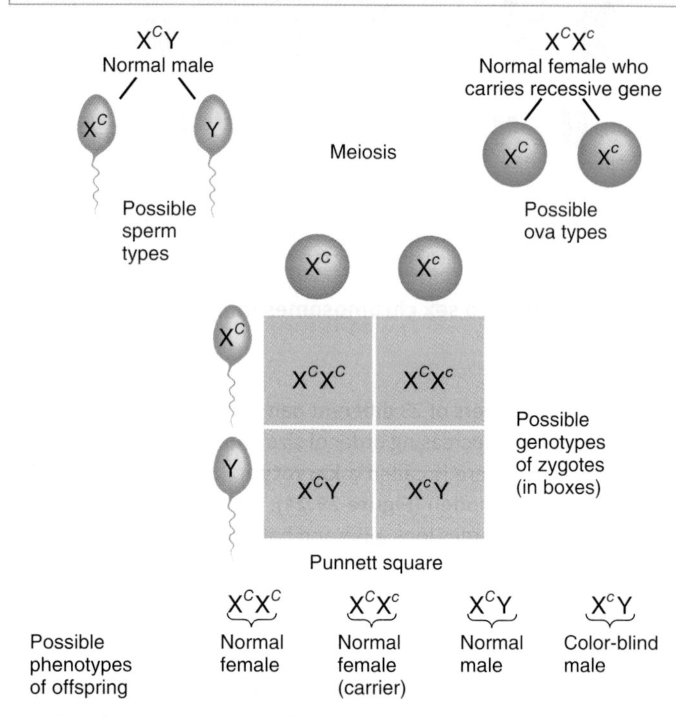

Q What is the genotype of a red–green color-blind female?

X-Chromosome Inactivation

Because they have two X chromosomes in every cell (except developing oocytes), females have a double set of all genes on the X chromosome. A mechanism termed **X-chromosome inactivation** (*lyonization*) in effect reduces the X-chromosome genes to a single set in females. In each cell of a female's body, one X chromosome is randomly and permanently inactivated early in development, and most of the genes of the inactivated X chromosome are not expressed (transcribed and translated). The nuclei of cells in female mammals contain a dark-staining body, called a **Barr body**, that is not present in the nuclei of cells in males. Geneticist Mary Lyon correctly predicted in 1961 that the Barr body is the inactivated X chromosome. During inactivation, chemical groups that prevent transcription into RNA are added to the X chromosome's DNA. As a result, an inactivated X chromosome reacts differently to histological stains and has a different appearance than the rest of the DNA. In nondividing (interphase) cells, it remains tightly coiled and can be seen as a dark-staining body within the nucleus. In a blood smear, the Barr body of neutrophils is termed a "drumstick" because it looks like a tiny drumstick-shaped projection of the nucleus.

Review Questions

29.1 Overview of Development

1. What is pregnancy?
2. What are the major events of each trimester?

29.2 The First Two Weeks of the Embryonic Period

3. Where does fertilization normally occur?
4. How is polyspermy prevented?
5. What is a morula, and how is it formed?
6. Describe the layers of a blastocyst and their eventual fates.
7. When, where, and how does implantation occur?
8. What are the functions of the trophoblast?
9. How is the bilaminar embryonic disc formed?
10. Describe the formation of the amnion, yolk sac, and chorion, and explain their functions.
11. Why are sinusoids important during embryonic development?

29.3 The Remaining Weeks of the Embryonic Period

12. When does gastrulation occur?
13. How do the three primary germ layers form? Why are they important?
14. What is meant by the term *induction*?

15. Describe how neurulation occurs. Why is it important?
16. What are the functions of somites?
17. How does the cardiovascular system develop?
18. How does the placenta form?
19. How does embryonic folding occur?
20. How does the primitive gut form, and what is its significance?
21. What is the origin of the structures of the head and neck?
22. What are limb buds?
23. What changes occur in the limbs during the second half of the embryonic period?

29.4 Fetal Period

24. What are the general developmental trends during the fetal period?
25. Using **Table 29.2** as a guide, select any one body structure in weeks 9 through 12 and trace its development through the remainder of the fetal period.

29.5 Teratogens

26. What are some of the symptoms of fetal alcohol syndrome?
27. How does cigarette smoking affect embryonic and fetal development?

29.6 Prenatal Diagnostic Tests

28. What conditions can be detected using fetal ultrasonography, amniocentesis, and chorionic villi sampling? What are the advantages of noninvasive prenatal tests?

29.7 Maternal Changes during Pregnancy

29. List the hormones involved in pregnancy, and describe the functions of each.
30. What structural and functional changes occur in the mother during pregnancy?

29.8 Exercise and Pregnancy

31. Which changes in pregnancy have an effect on the ability to exercise?

29.9 Labor

32. What hormonal changes induce labor?
33. What is the difference between false labor and true labor?
34. What happens during the stage of dilation, the stage of expulsion, and the placental stage of true labor?

29.10 Adjustments of the Infant at Birth

35. Why are respiratory and cardiovascular adjustments so important at birth?

29.11 The Physiology of Lactation

36. Which hormones contribute to lactation? What is the function of each?
37. What are the benefits of breast-feeding over bottle-feeding?

29.12 Inheritance

38. What do the terms genotype, phenotype, dominant, recessive, homozygous, and heterozygous mean?
39. What are genomic imprinting and nondisjunction?
40. Give an example of incomplete dominance.
41. What is multiple-allele inheritance? Give an example.
42. Define complex inheritance and give an example.
43. Why does X-chromosome inactivation occur?

Critical Thinking Questions

1. Kathy is breastfeeding her infant and is experiencing what feels like early labor pains. What is causing these painful feelings? Is there a benefit to them?
2. Jack has hemophilia, which is a sex-linked blood-clotting disorder. He blames his father for passing on the gene for hemophilia. Explain to Jack why his reasoning is wrong. How can Jack have hemophilia if his parents do not?
3. Alisa has asked her obstetrician to save and freeze her baby's cord blood after delivery in case the child needs a future bone marrow transplant. What is in the baby's cord blood that could be used to treat future disorders in the child?

Appendix A

Measurements

U.S. Customary System

PARAMETER	UNIT	RELATION TO OTHER U.S. UNITS	SI (METRIC) EQUIVALENT
Length	inch	1/12 foot	2.54 centimeters
	foot	12 inches	0.305 meter
	yard	36 inches	0.914 meter
	mile	5,280 feet	1.609 kilometers
Mass	grain	1/1,000 pound	64.799 milligrams
	dram	1/16 ounce	1.772 grams
	ounce	16 drams	28.350 grams
	pound	16 ounces	453.6 grams
	ton	2,000 pounds	907.18 kilograms
Volume (Liquid)	ounce	1/16 pint	29.574 milliliters
	pint	16 ounces	0.473 liter
	quart	2 pints	0.946 liter
	gallon	4 quarts	3.785 liters
Volume (Dry)	pint	1/2 quart	0.551 liter
	quart	2 pints	1.101 liters
	peck	8 quarts	8.810 liters
	bushel	4 pecks	35.239 liters

International System (SI)

BASE UNITS			PREFIXES		
UNIT	**QUANTITY**	**SYMBOL**	**PREFIX**	**MULTIPLIER**	**SYMBOL**
meter	length	m	tera-	$10^{12} = 1,000,000,000,000$	T
kilogram	mass	kg	giga-	$10^{9} = 1,000,000,000$	G
second	time	s	mega-	$10^{6} = 1,000,000$	M
liter	volume	L	kilo-	$10^{3} = 1,000$	k
mole	amount of matter	mol	hecto-	$10^{2} = 100$	h
			deca-	$10^{1} = 10$	da
			deci-	$10^{-1} = 0.1$	d
			centi-	$10^{-2} = 0.01$	c
			milli-	$10^{-3} = 0.001$	m
			micro-	$10^{-6} = 0.000,001$	μ
			nano-	$10^{-9} = 0.000,000,001$	n
			pico-	$10^{-12} = 0.000,000,000,001$	p

Temperature Conversion

FAHRENHEIT (F) TO CELSIUS (C)
°C = (°F − 32) ÷ 1.8

CELSIUS (C) TO FAHRENHEIT (F)
°F = (°C × 1.8) + 32

U.S. to SI (Metric) Conversion

WHEN YOU KNOW	MULTIPLY BY	TO FIND
inches	2.54	centimeters
feet	30.48	centimeters
yards	0.91	meters
miles	1.61	kilometers
ounces	28.35	grams
pounds	0.45	kilograms
tons	0.91	metric tons
fluid ounces	29.57	milliliters
pints	0.47	liters
quarts	0.95	liters
gallons	3.79	liters

SI (Metric) to U.S. Conversion

WHEN YOU KNOW	MULTIPLY BY	TO FIND
millimeters	0.04	inches
centimeters	0.39	inches
meters	3.28	feet
kilometers	0.62	miles
liters	1.06	quarts
cubic meters	35.31	cubic feet
grams	0.035	ounces
kilograms	2.21	pounds

Periodic Table

The periodic table lists the known **chemical elements**, the basic units of matter. The elements in the table are arranged left to right in rows in order of their **atomic number**, the number of protons in the nucleus. Each horizontal row, numbered from 1 to 7, is a **period**. All elements in a given period have the same number of electron shells as their period number. For example, an atom of hydrogen or helium each has one electron shell, while an atom of potassium or calcium each has four electron shells. The elements in each column, or **group**, share chemical properties. For example, the elements in column IA are very chemically reactive, whereas the elements in column VIIIA have full electron shells and thus are chemically inert.
Scientists now recognize 117 different elements; 92 occur naturally on Earth, and the rest are produced from the natural elements using particle accelerators or nuclear reactors. Elements are designated by **chemical symbols**, which are the first one or two letters of the element's name in English, Latin, or another language.

Twenty-six of the 92 naturally occurring elements normally are present in your body. Of these, just four elements—oxygen (O), carbon (C), hydrogen (H), and nitrogen (N) (coded blue)—constitute about 96% of the body's mass. Eight others—calcium (Ca), phosphorus (P), potassium (K), sulfur (S), sodium (Na), chlorine (Cl), magnesium (Mg), and iron (Fe) (coded pink)—contribute 3.8% of the body's mass. An additional 14 elements, called **trace elements** because they are present in tiny amounts, account for the remaining 0.2% of the body's mass. The trace elements are aluminum, boron, chromium, cobalt, copper, fluorine, iodine, manganese, molybdenum, selenium, silicon, tin, vanadium, and zinc (coded yellow). Table 2.1 on page 27 provides information about the main chemical elements in the body.

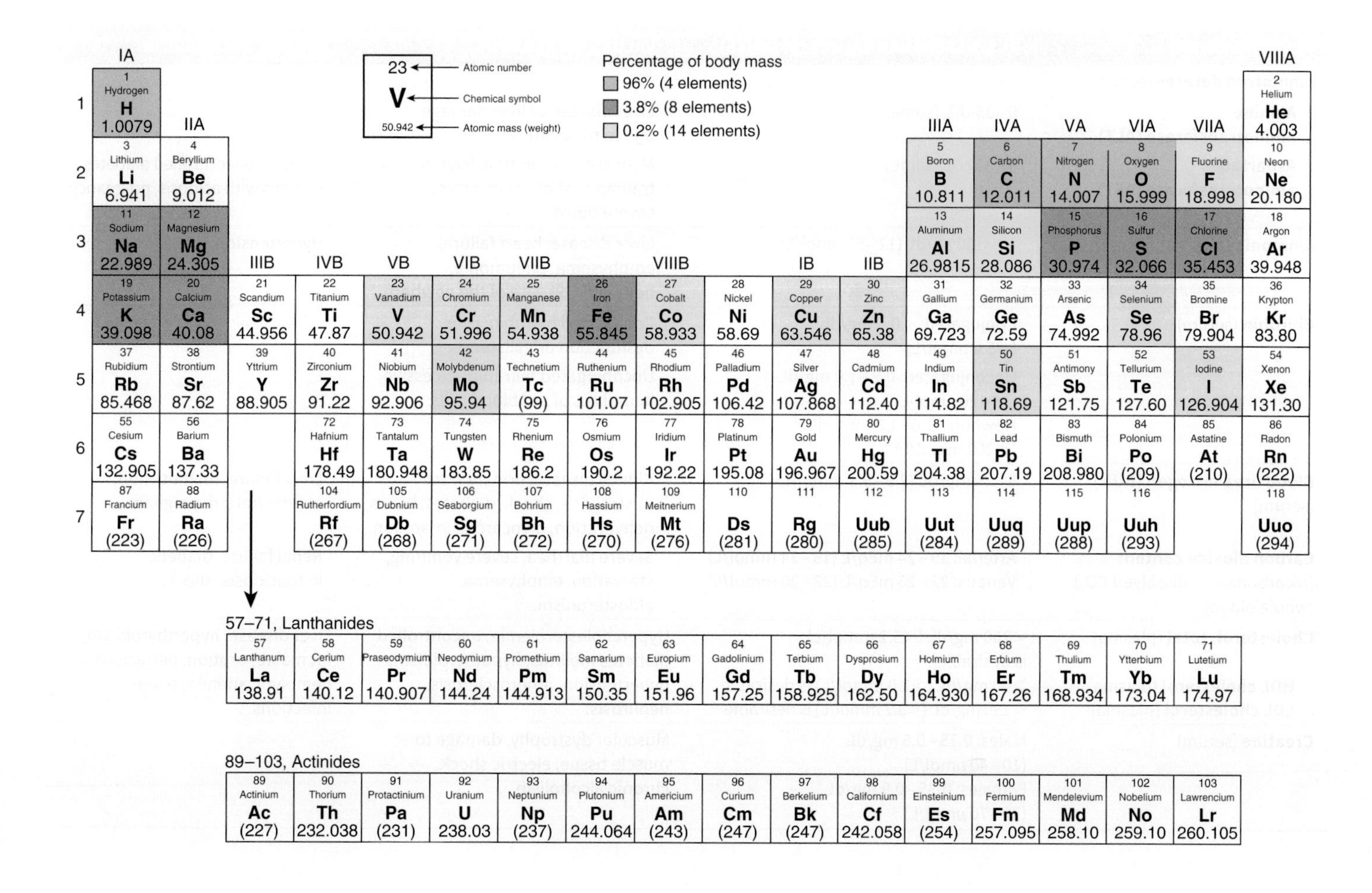

Appendix C

Normal Values for Selected Blood Tests

The system of international (SI) units (Système Internationale d'Unités) is used in most countries and in many medical and scientific journals. Clinical laboratories in the United States, by contrast, usually report values for blood and urine tests in conventional units. The laboratory values in this Appendix give conventional units first, followed by SI equivalents in parentheses. Values listed for various blood tests should be viewed as reference values rather than absolute "normal" values for all well people. Values may vary due to age, gender, diet, and environment of the subject or the equipment, methods, and standards of the lab performing the measurement.

Key To Symbols

g = gram	mL = milliliter
mg = milligram = 10^{-3} gram	μL = microliter
μg = microgram = 10^{-6} gram	mEq/L = milliequivalents per liter
U = units	mmol/L = millimoles per liter
L = liter	μmol/L = micromoles per liter
dL = deciliter	> = greater than; < = less than

Blood Tests

TEST (SPECIMEN)	U.S. REFERENCE VALUES (SI UNITS)	VALUES INCREASE IN	VALUES DECREASE IN
Aminotransferases (serum)			
Alanine aminotransferase (ALT)	0–35 U/L (same)	Liver disease or liver damage due to toxic drugs.	
Aspartate aminotransferase (AST)	0–35 U/L (same)	Myocardial infarction, liver disease, trauma to skeletal muscles, severe burns.	Beriberi, uncontrolled diabetes mellitus with acidosis, pregnancy.
Ammonia (plasma)	20–120 μg/dL (12–55 μmol/L)	Liver disease, heart failure, emphysema, pneumonia, hemolytic disease of the newborn.	Hypertension.
Bilirubin (serum)	Conjugated: <0.5 mg/dL (<5.0 μmol/L) Unconjugated: 0.2–1.0 mg/dL (18–20 μmol/L) Newborn: 1.0–12.0 mg/dL (<200 mmol/L)	Conjugated bilirubin: liver dysfunction or gallstones. Unconjugated bilirubin: excessive hemolysis of red blood cells.	
Blood urea nitrogen (BUN) (serum)	8–26 mg/dL (2.9–9.3 mmol/L)	Kidney disease, urinary tract obstruction, shock, diabetes, burns, dehydration, myocardial infarction.	Liver failure, malnutrition, overhydration, pregnancy.
Carbon dioxide content (bicarbonate + dissolved CO_2) (whole blood)	Arterial: 19–24 mEq/L (19–24 mmol/L) Venous: 22–26 mEq/L (22–26 mmol/L)	Severe diarrhea, severe vomiting, starvation, emphysema, aldosteronism.	Renal failure, diabetic ketoacidosis, shock.
Cholesterol, total (plasma)	<200 mg/dL (<5.2 mmol/L) is desirable	Hypercholesterolemia, uncontrolled diabetes mellitus, hypothyroidism, hypertension, atherosclerosis, nephrosis.	Liver disease, hyperthyroidism, fat malabsorption, pernicious or hemolytic anemia, severe infections.
HDL cholesterol (plasma)	>40 mg/dL (>1.0 mmol/L) is desirable		
LDL cholesterol (plasma)	<130 mg/dL (<3.2 mmol/L) is desirable		
Creatine (serum)	Males: 0.15–0.5 mg/dL (10–40 μmol/L) Females: 0.35–0.9 mg/dL (30–70 μmol/L)	Muscular dystrophy, damage to muscle tissue, electric shock, chronic alcoholism.	

Blood Tests Continued

TEST (SPECIMEN)	U.S. REFERENCE VALUES (SI UNITS)	VALUES INCREASE IN	VALUES DECREASE IN
Creatine kinase (CK), also known as **creatine phosphokinase (CPK)** (serum)	0–130 U/L (same)	Myocardial infarction, progressive muscular dystrophy, hypothyroidism, pulmonary edema.	
Creatinine (serum)	0.5–1.2 mg/dL (45–105 μmol/L)	Impaired renal function, urinary tract obstruction, giantism, acromegaly.	Decreased muscle mass, as occurs in muscular dystrophy or myasthenia gravis.
Electrolytes (plasma)	See Table 27.2 on page 1043		
Gamma-glutamyl transferase (GGT) (serum)	0–30 U/L (same)	Bile duct obstruction, cirrhosis, alcoholism, metastatic liver cancer, congestive heart failure.	
Glucose (plasma)	70–110 mg/dL (3.9–6.1 mmol/L)	Diabetes mellitus, acute stress, hyperthyroidism, chronic liver disease, Cushing's syndrome.	Addison's disease, hypothyroidism, hyperinsulinism.
Hemoglobin (whole blood)	Males: 14–18 g/100 mL (140–180 g/L) Females: 12–16 g/100 mL (120–160 g/L) Newborns: 14–20 g/100 mL (140–200 g/L)	Polycythemia, congestive heart failure, chronic obstructive pulmonary disease, living at high altitude.	Anemia, severe hemorrhage, cancer, hemolysis, Hodgkin disease, nutritional deficiency of vitamin B_{12}, systemic lupus erythematosus, kidney disease.
Iron, total (serum)	Males: 80–180 μg/dL (14–32 μmol/L) Females: 60–160 μg/dL (11–29 μmol/L)	Liver disease, hemolytic anemia, iron poisoning.	Iron-deficiency anemia, chronic blood loss, pregnancy (late), chronic heavy menstruation.
Lactic dehydrogenase (LDH) (serum)	71–207 U/L (same)	Myocardial infarction, liver disease, skeletal muscle necrosis, extensive cancer.	
Lipids (serum) **Total** **Triglycerides**	 400–850 mg/dL (4.0–8.5 g/L) 10–190 mg/dL (0.1–1.9 g/L)	Hyperlipidemia, diabetes mellitus.	Fat malabsorption, hypothyroidism.
Platelet (thrombocyte) count (whole blood)	150,000–400,000/μL	Cancer, trauma, leukemia, cirrhosis.	Anemias, allergic conditions, hemorrhage.
Protein (serum) **Total** **Albumin** **Globulin**	 6–8 g/dL (60–80 g/L) 4–6 g/dL (40–60 g/L) 2.3–3.5 g/dL (23–35 g/L)	Dehydration, shock, chronic infections.	Liver disease, poor protein intake, hemorrhage, diarrhea, malabsorption, chronic renal failure, severe burns.
Red blood cell (erythrocyte) count (whole blood)	Males: 4.5–6.5 million/μL Females: 3.9–5.6 million/μL	Polycythemia, dehydration, living at high altitude.	Hemorrhage, hemolysis, anemias, cancer, overhydration.
Uric acid (urate) (serum)	2.0–7.0 mg/dL (120–420 μmol/L)	Impaired renal function, gout, metastatic cancer, shock, starvation.	
White blood cell (leukocyte) count, total (whole blood)	5,000–10,000/μL (See Table 19.3 on page 682 for relative percentages of different types of WBCs.)	Acute infections, trauma, malignant diseases, cardiovascular diseases. (See also Table 19.2 on page 595.)	Diabetes mellitus, anemia. (See also Table 19.2 on page 595.)

Appendix D

Normal Values for Selected Urine Tests

Urine Tests

TEST (SPECIMEN)	U.S. REFERENCE VALUES (SI UNITS)	CLINICAL IMPLICATIONS
Amylase (2 hour)	35–260 Somogyi units/hr (6.5–48.1 units/hr)	Values increase in inflammation of the pancreas (pancreatitis) or salivary glands, obstruction of the pancreatic duct, and perforated peptic ulcer.
Bilirubin* (random)	Negative	Values increase in liver disease and obstructive biliary disease.
Blood* (random)	Negative	Values increase in renal disease, extensive burns, transfusion reactions, and hemolytic anemia.
Calcium (Ca^{2+}) (random)	10 mg/dL (2.5 mmol/L); up to 300 mg/24 hr (7.5 mmol/24 hr)	Amount depends on dietary intake; values increase in hyperparathyroidism, metastatic malignancies, and primary cancer of breasts and lungs; values decrease in hypoparathyroidism and vitamin D deficiency.
Casts (24 hour)		
Epithelial	Occasional	Values increase in nephrosis and heavy metal poisoning.
Granular	Occasional	Values increase in nephritis and pyelonephritis.
Hyaline	Occasional	Values increase in kidney infections.
Red blood cell	Occasional	Values increase in glomerular membrane damage and fever.
White blood cell	Occasional	Values increase in pyelonephritis, kidney stones, and cystitis.
Chloride (Cl^-) (24 hour)	140–250 mEq/24 hr (140–250 mmol/24 hr)	Amount depends on dietary salt intake; values increase in Addison's disease, dehydration, and starvation; values decrease in pyloric obstruction, diarrhea, and emphysema.
Color (random)	Yellow, straw, amber	Varies with many disease states, hydration, and diet.
Creatinine (24 hour)	Males: 1.0–2.0 g/24 hr (9–18 mmol/24 hr) Females: 0.8–1.8 g/24 hr (7–16 mmol/24 hr)	Values increase in infections; values decrease in muscular atrophy, anemia, and kidney diseases.
Glucose*	Negative	Values increase in diabetes mellitus, brain injury, and myocardial infarction.
Hydroxycorticosteroids (17-hydroxysteroids) (24 hour)	Males: 5–15 mg/24 hr (13–41 μmol/24 hr) Females: 2–13 mg/24 hr (5–36 μmol/24 hr)	Values increase in Cushing's syndrome, burns, and infections; values decrease in Addison's disease.
Ketone bodies* (random)	Negative	Values increase in diabetic acidosis, fever, anorexia, fasting, and starvation.
17-Ketosteroids (24 hour)	Males: 8–25 mg/24 hr (28–87 μmol/24 hr) Females: 5–15 mg/24 hr (17–53 μmol/24 hr)	Values decrease in surgery, burns, infections, adrenogenital syndrome, and Cushing's syndrome.

Urine Tests Continued

TEST (SPECIMEN)	U.S. REFERENCE VALUES (SI UNITS)	CLINICAL IMPLICATIONS
Odor (random)	Aromatic	Becomes acetonelike in diabetic ketosis.
Osmolality (24 hour)	500 – 800 mOsm/kg water (500 – 800 mmol/kg water)	Values increase in cirrhosis, congestive heart failure (CHF), and high-protein diets; values decrease in aldosteronism, diabetes insipidus, and hypokalemia.
pH* (random)	4.6 – 8.0	Values increase in urinary tract infections and severe alkalosis; values decrease in acidosis, emphysema, starvation, and dehydration.
Phenylpyruvic acid (random)	Negative	Values increase in phenylketonuria (PKU).
Potassium (K^+) (24 hour)	40 – 80 mEq/24 hr (40 – 80 mmol/24 hr)	Values increase in chronic renal failure, dehydration, starvation, and Cushing's syndrome; values decrease in diarrhea, malabsorption syndrome, and adrenal cortical insufficiency.
Protein* (albumin) (random)	Negative	Values increase in nephritis, fever, severe anemias, trauma, and hyperthyroidism.
Sodium (Na^+) (24 hour)	75–200 mEq/24 hr (75–200 mmol/24 hr)	Amount depends on dietary salt intake; values increase in dehydration, starvation, and diabetic acidosis; values decrease in diarrhea, acute renal failure, emphysema, and Cushing's syndrome.
Specific gravity* (random)	1.001–1.035 (same)	Values increase in diabetes mellitus and excessive water loss; values decrease in absence of antidiuretic hormone (ADH) and severe renal damage.
Urea (24 hour)	25–35 g/24 hr (420–580 mmol/24 hr)	Values increase in response to increased protein intake; values decrease in impaired renal function.
Uric acid (24 hour)	0.4–1.0 g/24 hr (1.5–4.0 mmol/24 hr)	Values increase in gout, leukemia, and liver disease; values decrease in kidney disease.
Urobilinogen* (24 hour)	1.7–6.0 μmol/24 hr	Values increase in anemias, hepatitis A (infectious), biliary disease, and cirrhosis; values decrease in cholelithiasis and renal insufficiency.
Volume, total (24 hour)	1000–2000 mL/24 hr (1.0–2.0 L/24 hr)	Varies with many factors.

* Test often performed using a **dipstick**, a plastic strip impregnated with chemicals that is dipped into a urine specimen to detect particular substances. Certain colors indicate the presence or absence of a substance and sometimes give a rough estimate of the amount(s) present.

Appendix E: Answers

Answers to Critical Thinking Questions

Chapter 1

1. No. Computed tomography is used to look at differences in tissue density. To assess activity in an organ such as the brain, a positron emission tomography (PET) scan or a single-photo–emission computerized tomography (SPECT) scan would provide a colorized visual assessment of brain activity.

2. Stem cells are undifferentiated cells. Research using stem cells has shown that these undifferentiated cells may be prompted to differentiate into the specific cells needed to replace those which are damaged or malfunctioning.

3. Homeostasis is the relative constancy (or dynamic equilibrium) of the body's internal environment. Homeostasis is maintained as the body changes in response to shifting external and internal conditions, including those of temperature, pressure, fluid, electrolytes, and other chemicals.

Chapter 2

1. Neither butter nor margarine is a particularly good choice for frying eggs. Butter contains saturated fats that are associated with heart disease. However, many margarines contain hydrogenated or partially hydrogenated *trans*-fatty acids that also increase the risk of heart disease. An alternative would be frying the eggs in any of the mono- or polyunsaturated fats such as olive oil, peanut oil, or corn oil. Boiling or poaching eggs instead of frying them would reduce the fat content of his breakfast, as would eating only the egg whites (not the high-fat yolks).

2. High body temperatures can be life-threatening, especially in infants. The increased temperature can cause denaturing of structural proteins and vital enzymes. When this happens, the proteins become nonfunctional. If the denatured enzymes are required for reactions that are necessary for life, then the infant could die.

3. Simply adding water to the table sugar does not cause it to break apart into monosaccharides. The water acts as a solvent, dissolving the sucrose and forming a sugar–water solution. To complete the breakdown of table sugar to glucose and fructose would require the presence of the enzyme sucrase.

Chapter 3

1. Synthesis of mucin by ribosomes on rough endoplasmic reticulum, to transport vesicle, to entry face of Golgi complex, to transfer vesicle, to medial cisternae where protein is modified, to transfer vesicle, to exit face, to secretory vesicle, to plasma membrane where it undergoes exocytosis.

2. Since smooth ER inactivates or detoxifies drugs, and peroxisomes also destroy harmful substances such as alcohol, we would expect to see increased numbers of these organelles in Sebastian's liver cells.

3. In order to restore water balance to the cells, the runners need to consume hypotonic solutions. The water in the hypotonic solution will move from the blood, into the interstitial fluid, and then into the cells. Plain water works well; sports drinks contain water and some electrolytes (which may have been lost due to sweating) but will still be hypotonic in relation to the body cells.

Chapter 4

1. There are many possible adaptations, including: more adipose tissue for insulation; thicker bones for support; more red blood cells for oxygen transport; increased thickness of skin to prevent water loss; etc.

2. Infants tend to have a high proportion of brown fat, which contains many mitochondria and is highly vascularized. When broken down, brown fat produces heat that helps to maintain infants' body temperatures. This heat can also warm the blood, which then distributes the heat throughout the body.

3. Your bread-and-water diet is not providing you with the necessary nutrients to encourage tissue repair. You need proper amounts of many essential vitamins, especially vitamin C, which is required for repair of the matrix and blood vessels. Vitamin A is needed to help properly maintain epithelial tissue. Adequate protein is also needed in order to synthesize the structural proteins of the damaged tissue.

Chapter 5

1. The dust particles are primarily keratinocytes that are shed from the stratum corneum of the skin.

2. Tattoos are created by depositing ink into the dermis, which does not undergo shedding as the epidermis does. Although the tattoo will fade due to exposure to sunlight and the flushing away of ink particles by the lymphatic system, the tattoo is indeed permanent.

3. Chef Eduardo has damaged the nail matrix—the part of the nail that produces growth. Because the damaged area has not regrown properly, the nail matrix may be permanently damaged.

Chapter 6

1. Due to the strenuous, repetitive activity, Taryn has probably developed a stress fracture of her right tibia (lower leg bone). Stress fractures are due to repeated stress on a bone that causes microscopic breaks in the bone without any evidence of injury to other tissue. An x-ray would not reveal the stress fracture, but a bone scan would. Thus the bone scan would either confirm or negate the physician's diagnosis.

2. When Marcus broke his arm as a child, he injured his epiphyseal (growth) plate. Damage to the cartilage in the epiphyseal plate resulted in premature closure of the plate, which interfered with the lengthwise growth of the arm bone.

3. Exercise causes mechanical stress on bones, but because there is effectively zero gravity in space, the pull of gravity on bones is missing. The lack of stress from gravity results in bone demineralization and weakness.

Chapter 7

1. Inability to open mouth—damage to the mandible, probably at temporomandibular joint; black eye—trauma to the ridge over the supraorbital margin; broken nose—probably damage to the nasal septum (includes the vomer, septal cartilage, and perpendicular plate of the ethmoid) and possibly the nasal bones; broken cheek—fracture of zygomatic bone; broken upper jaw—fracture of maxilla; damaged eye socket—fracture of parts of the sphenoid, frontal, ethmoid, palatine, zygomatic, lacrimal, and maxilla (all compose the eye socket); punctured lung—damage to the thoracic vertebrae, which have punctured the lung.

2. Due to the repeated and extensive tension on his bone surfaces, Bubba would experience deposition of new bone tissue. His arm bones would be thicker and with increased raised areas (projections) where the tendons attach his muscles to bone.

3. The "soft area" being referred to is the anterior fontanel, located between the parietal and frontal bones. This is one of several areas of fibrous connective tissue in the skull that has

not ossified; it should complete its ossification at 18–24 months after birth. Fontanels allow flexibility of the skull for childbirth and for brain growth after birth. The connective tissue will not allow passage of water; thus no brain damage will occur through simply washing the baby's hair.

Chapter 8

1. There are several characteristics of the bony pelves that can be used to differentiate male from female: (1) The pelvis in the female is wider and more shallow than the male's; (2) the pelvic brim of the female is larger and more oval; (3) the pubic arch has an angle greater than 90º; (4) the pelvic outlet is wider than in a male; (5) the female's iliac crest is less curved and the ilium less vertical. Table 8.1 provides additional differences between female and male pelves. Age of the skeleton can be determined by the size of the bones, the presence or absence of epiphyseal plates, the degree of demineralization of the bones, and the general appearance of the "bumps" and ridges of bones.

2. Infants do have "flat feet" because their arches have not yet developed. As they begin to stand and walk, the arches should begin to develop in order to accommodate and support their body weight. The arches are usually fully developed by age 12 or 13, so Dad doesn't need to worry yet!

3. There are 14 phalanges in each hand: two bones in the thumb and three in each of the other fingers. Farmer White has lost five phalanges on his left hand (two in his thumb and three in his index finger), so he has nine remaining on his left and 14 remaining on his right for a total of 23.

Chapter 9

1. Katie's vertebral column, head, thighs, lower legs, lower arms, and fingers are flexed. Her forearms and shoulders are medially rotated. Her thighs and arms are adducted.

2. The knee joint is commonly injured, especially among athletes. The twisting of Jeremiah's leg could have resulted in a multitude of internal injuries to the knee joint but often football players suffer tearing of the anterior cruciate ligament and medial meniscus. The immediate swelling is due to blood from damaged blood vessels, damaged synovial membranes, and the torn meniscus. Continued swelling is a result of a buildup of synovial fluid, which can result in pain and decreased mobility. Jeremiah's doctor may aspirate some of the fluid ("draining the water off his knee") and might want to perform arthroscopy to check for the extent of the knee damage.

3. The condylar processes of the mandible passed anteriorly to the articular tubercles of the temporal bones, and this dislocated Antonio's mandible. It could be corrected by pressing the thumbs downward on the lower molar teeth and pushing the mandible backward.

Chapter 10

1. Muscle cells lose their ability to undergo cell division after birth. Therefore, the increase in size is not due to an increase in the number of muscle cells but rather is due to enlargement of the existing muscle fibers (hypertrophy). This enlargement can occur from forceful, repetitive muscular activity. It will cause the muscle fibers to increase their production of internal structures such as mitochondria and myofibrils and produce an increase in the muscle fiber diameter.

2. The "dark meat" of both chickens and ducks is composed primarily of slow oxidative (SO) muscle fibers. These fibers contain large amounts of myoglobin and capillaries, which accounts for their dark color. In addition, these fibers contain large numbers of mitochondria and generate ATP by aerobic respiration. SO fibers are resistant to fatigue and can produce sustained contractions for many hours. The legs of chickens and ducks are used for support, walking, and swimming (in ducks), all activities in which endurance is needed. In addition, migrating ducks require SO fibers in their breasts to enable them to have enough energy to fly for extremely long distances while migrating. There may be some fast oxidative–glycolytic (FOG) fibers in the dark meat. FOG fibers also contain large amounts of myoglobin and capillaries, contributing to the dark color. They can use aerobic or anaerobic cellular respiration to generate ATP and have high-to-moderate resistance to fatigue. These fibers would be good for the occasional "sprint" that ducks and chickens undergo to escape dangerous situations. In contrast, the white meat of a chicken breast is composed primarily of fast glycolytic (FG) fibers. FG fibers have lower amounts of myoglobin and capillaries that give the meat its white color. There are also few mitochondria in FG fibers, so these fibers generate ATP mainly by glycolysis. These fibers contract strongly and quickly and are adapted for intense anaerobic movements of short duration. Chickens occasionally use their breasts for flying extremely short distances, usually to escape prey or perceived danger, so FG fibers are appropriate for their breast muscle.

3. Destruction of the somatic motor neurons to skeletal muscle fibers will result in a loss of stimulation to the skeletal muscles. When not stimulated on a regular basis, a muscle begins to lose its muscle tone. Through lack of use, the muscle fibers will weaken, begin to decrease in size, and can be replaced by fibrous connective tissue, resulting in a type of denervation atrophy. A lack of stimulation of the breathing muscles (especially the diaphragm) from motor neurons can result in inability of the breathing muscles to contract, thus causing respiratory paralysis and possibly death of the individual from respiratory failure.

Chapter 11

1. All of the following could occur on the affected (right) side of the face: (1) drooping of eyelid—levator palpebrae superioris; (2) drooping of the mouth, drooling, keeping food in mouth—orbicularis oris, buccinator; (3) uneven smile—zygomaticus major, levator labii superioris, risorius; (4) inability to wrinkle forehead—occipitofrontalis; (5) trouble sucking through a straw—buccinator.

2. Bulbospongiosus, external urethral sphincter, and deep transverse perineal.

3. The rotator cuff is formed by a combination of the tendons of four deep muscles of the shoulder—subscapularis, supraspinatus, infraspinatus, and teres minor. These muscles add strength and stability to the shoulder joint. Although any of the muscles' tendons can be injured, the subscapularis is most often damaged. Dependent upon the injured muscle, Jose may have trouble medially rotating his arm (subscapularis), abducting his arm (supraspinatus), laterally rotating his arm (infraspinatus, teres minor), or extending his arm (teres minor).

Chapter 12

1. Smelling the coffee and hearing the alarm are somatic sensory, stretching and yawning are somatic motor, salivating is autonomic (parasympathetic) motor, stomach rumble is enteric motor.

2. Demyelinaton or destruction of the myelin sheath can lead to multiple problems, especially in infants and children whose myelin sheaths are still in the process of developing. The affected axons deteriorate, which will interfere with function in both the CNS and PNS. There will be lack of sensation and loss of motor control with less rapid and less coordinated body responses. Damage to the axons in the CNS can be permanent and Ming's brain development may be irreversibly affected.

3. Dr. Moro could develop a drug that: (1) is an agonist of substance P; (2) blocks the breakdown of substance P; (3) blocks the reuptake of substance P; (4) promotes the release of substance P; (5) suppresses the release of enkephalins.

Chapter 13

1. The needles will pierce the epidermis, dermis, and subcutaneous layer and then go between the vertebrae through the epidural space, the dura mater, the subdural space, the arachnoid mater, and into the CSF in the subarachnoid space. CSF is produced in the brain, and the spinal meninges are continuous with the cranial meninges.

2. The anterior gray horns contain cell bodies of somatic motor neurons and motor nuclei that are responsible for the nerve impulses for skeletal muscle contraction. Because the lower cervical region is affected (brachial plexus, C5–C8), you would expect that Sunil may have trouble with movement in his shoulder, arm, and hand on the affected side.

3. Allyson has damaged her posterior columns in the lower (lumbar) region of the spinal cord. The posterior columns are responsible for transmitting nerve impulses responsible for awareness of muscle position (proprioception) and touch—which are affected in Allyson—as well as other functions such as light pressure

sensations and vibration sensations. Relating Allyson's symptoms to the distribution of dermatomes, it is likely that regions L4, L5, and S1 of her spinal cord were compressed.

Chapter 14

1. Movement of the right arm is controlled by the left hemisphere's primary motor area, located in the precentral gyrus. Speech is controlled by the motor speech area in the left hemisphere's frontal lobe just superior to the lateral cerebral sulcus.

2. Nicky's right facial (vii) nerve has been affected; she is suffering from Bell's palsy due to the viral infection. The facial nerve controls contraction of skeletal muscles of the face, tear gland and salivary gland secretion, as well as conveying sensory impulses from many of the taste buds on the tongue.

3. You will need to design a drug that can get through the brain's blood–brain barrier (BBB). The drug should be lipid- or water-soluble. If the drug can open a gap between the tight junctions of the endothelial cells of the brain capillaries, it would be more likely to pass through the BBB. Targeting the drug to enter the brain in certain areas near the third ventricle (the circumventricular organs) might be an option as the BBB is entirely absent in those areas and the capillary endothelium is more permeable, allowing the blood-borne drug to more readily enter the brain tissue.

Chapter 15

1. Digestion and relaxation are controlled by increased stimulation of the parasympathetic division of the ANS. The salivary glands, pancreas, and liver will show increased secretion; the stomach and intestines will have increased activity; the gallbladder will have increased contractions; heart contractions will have decreased force and rate. Following is the nerve supply to each listed organ: salivary glands—facial (VII) nerves and glossopharyngeal (IX) nerves; pancreas, liver, stomach, gallbladder, intestines, and heart—vagus (X) nerves.

2. Ciara experienced one of the "E situations" (emergency in her case), which has activated the fight-or-flight response. Some noticeable effects of increased sympathetic activity include an increase in heart rate, sweating on the palms, and contraction of the arrector pili muscles, which causes the goose flesh. Secretion of epinephrine and norepinephrine from the adrenal medullae will intensify and prolong the responses.

3. Mrs. Young needs to slow down the activity of her digestive system, which seems to be experiencing increased parasympathetic response. A parasympathetic blocking agent is needed. Because the stomach and intestines have muscarinic receptors, she needs to be provided with a muscarinic blocking agent (such as atropine), which will result in decreased motility in the stomach and intestines.

Chapter 16

1. Chemoreceptors in the nose detect odors. Proprioceptors detect body position and are involved in equilibrium. The chemoreceptors in the nose are rapidly adapting, whereas proprioceptors are slowly adapting. Thus the smell faded while the sensation of motion remained.

2. Thermal (heat) receptors in her left hand detect the stimulus. A nerve impulse is transmitted to the spinal cord through first-order neurons with cell bodies in posterior root ganglia. The impulses travel into the spinal cord, where the first-order neurons synapse with second-order neurons, whose cell bodies are located in the posterior gray horn of the spinal cord. The axons of the second-order neurons decussate to the right side in the spinal cord and then the impulses ascend through the lateral spinothalamic tract. The axons of the second-order neurons end in the ventral posterior nucleus of the right side of the thalamus, where they synapse with the third-order neurons. Axons of the third-order neurons transmit impulses to the specific primary somatosensory areas in the postcentral gyrus of the right parietal lobe.

3. When Marvin settled down for the night, he passed through stages 1–3 of NREM sleep. Sleepwalking occurred when he was in stage 4 (slow-wave sleep). Because this is the deepest stage of sleep, his mother was able to return him to his bed without awakening him. Marvin then cycled through REM and NREM sleep. His dreaming occurred during the REM phases of sleep. The noise of the alarm clock provided the sensory stimulus that stimulated the reticular activating system. Activation of this system sends numerous nerve impulses to widespread areas of the cerebral cortex, both directly and via the thalamus. The result is the state of wakefulness.

Chapter 17

1. Damage to the facial (vii) nerve would affect smell, taste, and hearing. Within the nasal epithelium and connective tissue, both the supporting cells and olfactory glands are innervated by branches of the facial nerve. Without input from the facial nerve, there will be a lack of mucus production required to dissolve odorants. The facial nerve also serves taste buds in the anterior two-thirds of the tongue, so damage can affect taste sensations. Hearing will be affected by a damaged facial nerve because the stapedius muscle, which is attached to the stapes, is innervated by the facial nerve. Contraction of the stapedius muscle helps to protect the inner ear from prolonged loud noises. Damage to the facial nerve will result in sounds that are excessively loud, resulting in more susceptibility to damage by prolonged loud noises.

2. With age, Gertrude has lost much of her sense of smell and taste due to a decline in olfactory and gustatory receptors. Since smell and taste are intimately linked, food no longer smells nor tastes as good to Gertrude. Gertrude has presbyopia, a loss of lens elasticity, which makes it difficult to read. She may also be experiencing age-related loss of sharpness of vision and depth perception. Gertrude's hearing difficulties could be a result of damage to hair cells in the spiral organ or degeneration of the nerve pathway for hearing. The "buzzing" Gertrude hears may be tinnitus, which also occurs more frequently in the elderly.

3. Some of the eye drops placed in the eye may pass through the nasolacrimal duct into the nasal cavity where olfactory receptors are stimulated. Because most "tastes" are actually smells, the child will "taste" the medicine from her eye.

Chapter 18

1. Yes, Amanda should visit the clinic, as these are serious signs and symptoms. She has an enlarged thyroid gland, or goiter. The goiter is probably due to hypothyroidism, which is causing the weight gain, fatigue, mental dullness, and other symptoms.

2. Amanda's problem is her pituitary gland, which is not secreting normal levels of TSH. Rising thyroxine (T_4) levels after the TSH injection indicates that her thyroid is functioning normally and able to respond to the increased TSH levels. If the thyroxine levels had not risen, then the problem would have been her thyroid gland.

3. Mr. Hernandez has diabetes insipidus caused by either insufficient production or release of ADH due to hypothalamus or posterior pituitary damage. He also could have defective ADH receptors in the kidneys. Diabetes insipidus is characterized by production of large volumes of urine, dehydration, and increased thirst, but with no glucose or ketones present in the urine (which would be indicative of diabetes mellitus rather than diabetes insipidus).

Chapter 19

1. The broad-spectrum antibiotics may have destroyed the bacteria that caused Shilpa's bladder infection but also destroyed the naturally occurring large intestine bacteria that produce vitamin K. Vitamin K is required for the synthesis of four clotting factors (II, VII, IX, and X). Without these clotting factors present in normal amounts, Shilpa will experience clotting problems until the intestinal bacteria reach normal levels and produce additional vitamin K.

2. Mrs. Brown's kidney failure is interfering with her ability to produce erythropoietin (EPO). Her physician can prescribe Epoetin alfa, a recombinant EPO, which is very effective in treating the decline in RBC production with renal failure.

3. A primary problem Thomas may experience is with clotting. Clotting time becomes longer because the liver is responsible for producing many of the clotting factors and clotting proteins such as fibrinogen. Thrombopoietin, which stimulates the formation of platelets, is also produced in the liver. In addition, the liver is responsible for eliminating bilirubin, produced from the breakdown of RBCs. With a malfunctioning liver, the bilirubin will accumulate, resulting

in jaundice. In addition, there can be decreased concentrations of the plasma protein albumin, which can affect blood pressure.

Chapter 20

1. The dental procedures introduced bacteria into Gerald's blood. The bacteria colonized his endocardium and heart valves, resulting in bacterial endocarditis. Gerald may have had a previously undetected heart murmur, or the heart murmur may have resulted from his endocarditis. His physician will want to monitor his heart to watch for further damage to the valve.

2. Extremely rapid heart rates can result in a decreased stroke volume due to insufficient ventricular filling. As a result, the cardiac output will decline to the point where there may not be enough blood reaching the central nervous system. She initially may experience light-headedness but could lose consciousness if the cardiac output declines dramatically.

3. Mr. Perkins is suffering from angina pectoris and has several risk factors for coronary artery disease such as smoking, obesity, sedentary lifestyle, and male gender. Cardiac angiography involves the use of a cardiac catheter to inject a radiopaque medium into the heart and its vessels. The angiogram may reveal blockages such as atherosclerotic plaques in his coronary arteries.

Chapter 21

1. The hole in the heart was the foramen ovale, which is an opening between the right and left atria. In fetal circulation it allows blood to bypass the right ventricle, enter the left atrium, and join systemic circulation. The "hole" should close shortly after birth to become the fossa ovalis. Closure of the foramen ovale after birth will allow the deoxgenated blood from the right atrium to enter pulmonary circulation so that the blood can become oxygenated prior to entering systemic circulation. If closure doesn't occur, surgery may be required.

2. Michael is suffering from hypovolemic shock due to the loss of blood. The low blood pressure is a result of low blood volume and a subsequent decrease in cardiac output. His rapid, weak pulse is an attempt of the heart to compensate for the decrease in cardiac output through sympathetic stimulation of the heart and increased blood levels of epinephrine and norepinephrine. His pale, cool, and clammy skin is a result of sympathetic constriction of the blood vessels of the skin and sympathetic stimulation of sweat glands. The lack of urine production is due to increased secretion of aldosterone and ADH, both of which are produced to increase blood volume in order to compensate for Michael's hypotension. The fluid loss from his bleeding results in activation of the thirst center in the hypothalamus. His confusion and disorientation are caused by a reduced oxygen supply to the brain from the decreased cardiac output.

3. Maureen has varicose veins, a condition in which the venous valves become leaky. The leaking valves allow the backflow of blood and an increased pressure that distends the veins and allows fluid to leak into the surrounding tissue. Standing on hard surfaces for long periods of time can cause varicosities to develop. Maureen needs to elevate her legs when possible to counteract the effects of gravity on the blood flow in the lower legs. She could also utilize support hose, which add external support for the superficial veins, much like skeletal muscle does for deeper veins. If the varices become severe, Maureen may require more extensive treatment such as sclerotherapy, radiofrequency endovenous occlusion, laser occlusion, or stripping.

Chapter 22

1. The influenza vaccination introduces a weakened or killed virus (which will not cause the disease) to the body. The immune system recognizes the antigen and mounts a primary immune response. Upon exposure to the same flu virus that was in the vaccine, the body will produce a secondary response, which will usually prevent a case of the flu. This is artificially acquired active immunity.

2. Mrs. Franco's lymph nodes were removed because metastasis of cancerous cells can occur through the lymph nodes and lymphatic vessels. Mrs. Franco's swelling is a lymphedema that is occurring due to the buildup of interstitial fluid from interference with drainage in the lymph vessels.

3. Tariq's physician would need to perform an antibody titer, which is a measure of the amount of antibody in the serum. If Tariq has previously been exposed to mumps (or been vaccinated for mumps), his blood should have elevated levels of IgG antibodies after this exposure from his sister. His immune system would be experiencing a secondary response. If he has not previously had mumps and has contracted mumps from his sister, his immune system would initiate a primary response. In that case, his blood would show an elevated titer of IgM antibodies, which are secreted by plasma cells after an initial exposure to the mumps antigen.

Chapter 23

1. Aretha's excess mucus production is causing blockage of the paranasal sinuses, which are used as hollow resonating chambers for singing and speech. In addition, her sore throat could be due to inflammation of the pharynx and larynx, which will affect their normal functions. Normally, the pharynx also acts as a resonating chamber and the vocal folds, located in the larynx, vibrate for speech and singing. Inflammation of the vocal folds (laryngitis) interferes with their ability to freely vibrate, which will affect both singing and speech.

2. In emphysema, there is destruction of the alveolar walls, producing abnormally large air spaces that remain filled with air during exhalation. The destruction of alveoli decreases the surface area for gas exchange across the respiratory membrane, resulting in a decreased blood O_2 level. Damage to the alveolar walls also causes a loss of elasticity, making exhalation more difficult. This can result in a buildup of CO_2. Cigarette smoke contains nicotine, carbon monoxide, and a variety of irritants, all of which affect the lungs. Nicotine constricts terminal bronchioles, decreasing the air flow into and out of the lungs; carbon monoxide binds to hemoglobin, reducing its ability to carry oxygen; irritants such as tar and fine particulate matter destroy cilia and increase mucus secretion, interfering with the ability of the respiratory passages to cleanse themselves.

3. The squirrel's nest blocked the passage of exhaust gas from the furnace, causing an accumulation of carbon monoxide (CO), a colorless, odorless gas, in the home. As they were sleeping, their blood was saturated with CO, which has a stronger affinity for hemoglobin than oxygen. As a result, the Robinsons became oxygen deficient. Without adequate oxygenation of the brain, the Robinsons died during their sleep.

Chapter 24

1. HCl has several important roles in digestion. HCl stimulates the secretion of hormones that promote the flow of bile and pancreatic juice. The presence of HCl destroys certain microbes that may have been ingested with food. HCl begins denaturing proteins in food, and provides the proper chemical environment for activating pepsinogen into pepsin, which breaks apart certain peptide bonds in proteins. It also helps in the action of gastric lipase, which splits triglycerides in fat molecules found in milk into fatty acids and monoglycerides.

2. Blockage of the pancreatic and bile ducts prevents pancreatic digestive enzymes and bile from reaching the duodenum. As a consequence, there will be problems digesting carbohydrates, proteins, nucleic acids, and lipids. Of particular concern is lipid digestion since the pancreatic juices contain the primary lipid-digesting enzyme. Fats will not be adequately digested, and Trey's feces will contain larger than normal amounts of lipids. In addition, the lack of bile salts will affect the body's ability to emulsify lipids and to form micelles required for absorption of fatty acids and monoglycerides (from lipid breakdown). When lipids are not absorbed properly, then there will be malabsorption of the lipid-soluble vitamins (A, D, E, and K).

3. Antonio experienced gastroesophageal reflux. The stomach contents backed up (refluxed) into Antonio's esophagus due to a failure of the lower esophageal sphincter to fully close. The HCl from the stomach irritated the esophageal wall, which resulted in the burning sensation he felt; this is commonly known as "heartburn," even though it is not related to the heart. Antonio's recent meal worsened the problem. Alcohol and smoking both can cause the sphincter to relax, while certain foods such as tomatoes, chocolate, and coffee can stimulate stomach acid

secretion. In addition, lying down immediately after a meal can exacerbate the problem.

Chapter 25

1. Ingestion of cyanide affects cellular respiration. The cyanide binds to the cytochrome oxidase complex in the inner membrane of mitochondria. Blocking this complex interferes with the last step in electron transport in aerobic ATP production. Jane Doe's body would quickly run out of energy to perform vital functions, resulting in her death.

2. Glenn's total cholesterol and LDL levels are very high, while his HDL levels are low. Total cholesterol above 239 mg/dL and LDL above 159 mg/dL are considered high. The ratio of total cholesterol (TC) to HDL-cholesterol is a predictor of the risk of developing coronary artery disease. Glenn's TC to HDL is 15; a ratio above 4 is undesirable. His ratio places him at high risk of developing coronary artery disease. In addition, for every 50 mg/dL TC over 200 mg/dL, the risk of a heart attack doubles. Glenn needs to reduce his TC and LDL-cholesterol while raising his HDL-cholesterol levels. LDLs contribute to fatty plaque formation on coronary artery walls. On the other hand, HDLs help remove excess cholesterol from the blood, which helps decrease the risk of coronary artery disease. Glenn will need to reduce his dietary intake of total fat, saturated fats, and cholesterol, all of which contribute to raising LDL levels. Exercise will raise HDL levels. If those changes are not successful, drug therapy may be required.

3. The goal of weight loss programs is to reduce caloric intake so that the body utilizes stored lipids as an energy source. As part of that desired lipid metabolism, ketone bodies are produced. Some of these ketone bodies will be excreted in the urine. If no ketones are present, then Marissa's body is not breaking down lipids. Only through using fewer calories than needed will her body break down the stored fat and release ketones. Thus, she must be eating more calories than needed to support her daily activities—she is "cheating."

Chapter 26

1. Without reabsorption, initially 105–125 mL of filtrate would be lost per minute, assuming normal glomerular filtration rate. Fluid loss from the blood would cause a decrease in blood pressure, and therefore a decrease in GBHP. When GBHP dropped below 45 mmHg, filtration would stop (assuming normal CHP and BCOP) because NFP would be zero.

2. a. Although normally pale yellow, urine color can vary based upon concentration, diet, drugs, and disease. A dark yellow color would not necessarily indicate a problem, but further investigation may be needed. Turbidity or cloudiness can be caused by urine that has been standing for a period of time, from certain foods, or from bacterial infections. Further investigation is needed. b. Ammonia-like odor occurs when the urine sample is allowed to stand. c. Albumin should not be present in urine (or be present only in very small amounts) because it is too large to pass through the filtration membranes. The presence of high levels of albumin is cause for concern as it indicates damage to the filtration membranes. d. Casts are hardened masses of material that are flushed out in the urine. The presence of casts is not normal and indicates a pathology. e. The pH of normal urine ranges from 4.8 to 8.0. A pH of 5.5 is in normal range. f. Hematuria is the presence of red blood cells in the urine. It can occur with certain pathological conditions or from kidney trauma. Hematuria may occur if the urine sample was contaminated with menstrual blood.

3. Bruce has developed renal calculi (kidney stones), which are blocking his ureters and interfering with the flow of urine from the kidneys to the bladder. The rhythmic pains are a result of the peristaltic contractions of the ureters as they attempt to move the stones toward the bladder. Bruce can wait for the stones to pass, can have them surgically removed, or can use shock-wave lithotripsy to break apart the stones into smaller fragments that can be eliminated with urine. To prevent future episodes, Bruce needs to watch his diet (limit calcium) and drink fluids, and may need drug intervention.

Chapter 27

1. The loss of stomach acids can result in metabolic alkalosis. Robin's HCO_3^- levels would be higher than normal. She would be hypoventilating in order to decrease her pH by slowing the loss of CO_2. Excessive vomiting can result in hyponatremia, hypokalemia, and hypochloremia. Both hyponatremia and hypokalemia can cause mental confusion.

2. (Step 1) pH = 7.30 indicates slight acidosis, which could be caused by elevated P_{CO_2} or lowered HCO_3^-. (Step 2) The HCO_3^- is lower than normal (20 mEq/liter), so (Step 3) the cause is metabolic. (Step 4) The P_{CO_2} is lower than normal (32 mmHg), so hyperventilation is providing some compensation. Diagnosis: Henry has partially compensated metabolic acidosis. A possible cause is kidney damage that resulted from interruption of blood flow during the heart attack.

3. Sam will experience increased fluid loss through increased evaporation from the skin and water vapor from the respiratory system through his increased respiratory rate. His insensible water loss will also increase (loss of water from mucous membranes of the mouth and respiratory system). Sam will have a decrease in urine formation.

Chapter 28

1. Monica's excessive training has resulted in an abnormally low amount of body fat. A certain amount of body fat is needed in order to produce the hormones required for the ovarian cycle. Several hormones are involved. Her amenorrhea is due to a lack of gonadotropin-releasing hormone, which in turn reduces the release of LH and FSH. Her follicles with their enclosed ova fail to develop and ovulation will not occur. In addition, synthesis of estrogens and progesterone declines from the lack of hormonal feedback. Usually a gain of weight will allow normal hormonal feedback mechanisms to return.

2. Along with estrogens, progesterone helps to prepare the endometrium for possible implantation of a zygote by promoting growth of the endometrium. The endometrial glands secrete glycogen, which will help sustain an embryo if one should implant. If implantation occurs, progesterone helps maintain the endometrium for the developing fetus. In addition, it helps prepare mammary glands to secrete milk. It inhibits the release of GnRH and LH, which stops a new ovarian cycle from occurring.

3. The ductus deferens is cut and tied in a vasectomy. This stops the release of sperm into the ejaculatory duct and urethra. Mark will still produce the secretions from his accessory glands (prostate, seminal vesicles, bulbourethral glands) in his ejaculate. In addition, a vasectomy does not affect sexual performance; he will be able to achieve erection and ejaculation, as those events are nervous system responses.

Chapter 29

1. As part of the feedback mechanism for lactation, oxytocin is released from the posterior pituitary. It is carried to the mammary glands where it causes milk to be released into the mammary ducts (milk ejection). The oxytocin is also transported in the blood to the uterus, which contains oxytocin receptors on the myometrium. The oxytocin causes contraction of the myometrium, resulting in the painful sensations that Kathy is experiencing. The uterine contractions can help return the uterus back to its prepregnancy size.

2. Sex-linked genetic traits, such as hemophilia, are present on the X chromosomes but not on the Y chromosomes. In males, the X chromosome is always inherited from the mother, and the Y chromosome from the father. Thus, Jack's hemophilia gene was inherited from his mother on his X chromosome. The gene for hemophilia is a recessive gene. His mother would need two recessive genes, one on each of her X chromosomes, to be hemophiliac. His father must carry the dominant (nonhemophiliac) gene on his X chromosome, so he also would not have hemophilia.

3. The cord blood is a source of pluripotent stem cells, which are unspecialized cells that have the potential to specialize into cells with specific functions. The hope is that stem cells can be used to generate cells and tissues to treat a variety of disorders. It is assumed that the tissues would not be rejected since they would contain the same genetic material as the patient—in this case Alisa's baby.

Answers to Figure Questions

CHAPTER 1

1.1 Organs are composed of two or more different types of tissues that work together to perform a specific function.

1.2 A nutrient moves from the external environment into plasma via the digestive system, then into the interstitial fluid, and then to a body cell.

1.3 The difference between negative and positive feedback systems is that in negative feedback systems the response reverses the original stimulus, but in positive feedback systems the response enhances the original stimulus.

1.4 When something causes blood pressure to decrease, then heart rate increases due to operation of this negative feedback system.

1.5 Because positive feedback systems continually intensify or reinforce the original stimulus, some mechanism is needed to end the response.

1.6 Having one standard anatomical position allows directional terms to be clearly defined so that any body part can be described in relation to any other part.

1.7 No, the radius is *distal* to the humerus. No, the esophagus is *posterior* to the trachea. Yes, the ribs are superficial to the lungs. Yes, the urinary bladder is medial to the ascending colon. No, the sternum is medial to the descending colon.

1.8 The frontal plane divides the heart into anterior and posterior portions.

1.9 The parasagittal plane (not shown in the figure) divides the brain into unequal right and left portions.

1.10 Urinary bladder = P, stomach = A, heart = T, small intestine = A, lungs = T, internal female reproductive organs = P, thymus = T, spleen = A, liver = A.

1.11 The pericardial cavity surrounds the heart, and the pleural cavities surround the lungs.

1.12 The illustrated abdominal cavity organs all belong to the digestive system (liver, gallbladder, stomach, small intestine, and most of the large intestine). Illustrated pelvic cavity organs belong to the urinary system (the urinary bladder) and the digestive system (part of the large intestine).

1.13 The liver is mostly in the epigastric region; the ascending colon is in the right lumbar region; the urinary bladder is in the hypogastric region; most of the small intestine is in the umbilical region. The pain associated with appendicitis would be felt in the right lower quadrant (RLQ).

CHAPTER 2

2.1 In carbon, the first shell contains two electrons and the second shell contains four electrons.

2.2 The four most plentiful elements in living organisms are oxygen, carbon, hydrogen, and nitrogen.

2.3 Antioxidants such as selenium, zinc, beta-carotene, vitamin C, and vitamin E can inactivate free radicals derived from oxygen.

2.4 A cation is a positively charged ion; an anion is a negatively charged ion.

2.5 An ionic bond involves the *loss* and *gain* of electrons; a covalent bond involves the *sharing* of pairs of electrons.

2.6 The N atom in ammonia is electronegative. Because it attracts electrons more strongly than do the H atoms, the nitrogen end of ammonia acquires a slight negative charge, allowing H atoms in water molecules (or in other ammonia molecules) to form hydrogen bonds with it. Likewise, O atoms in water molecules can form hydrogen bonds with H atoms in ammonia molecules.

2.7 The number of hydrogen atoms in the reactants must equal the number in the products—in this case, four hydrogen atoms total. Put another way, two molecules of H_2 are needed to react with each molecule of O_2 so that the number of H atoms and O atoms in the reactants is the same as the number of H atoms and O atoms in the products.

2.8 This reaction is exergonic because the reactants have more potential energy than the products.

2.9 No. A catalyst does not change the potential energies of the products and reactants; it only lowers the activation energy needed to get the reaction going.

2.10 No. Because sugar easily dissolves in a polar solvent (water), you can correctly predict that it has several polar covalent bonds.

2.11 $CaCO_3$ is a salt, and H_2SO_4 is an acid.

2.12 At pH = 6, $[H^+] = 10^{-6}$ mol/L and $[OH^-] = 10^{-8}$ mol/L. A pH of 6.82 is more acidic than a pH of 6.91. Both pH = 8.41 and pH = 5.59 are 1.41 pH units from neutral (pH = 7).

2.13 Glucose has five —OH groups and six carbon atoms.

2.14 Hexoses are six-carbon sugars; examples include glucose, fructose, and galactose.

2.15 There are 6 carbons in fructose and 12 in sucrose.

2.16 Cells in the liver and in skeletal muscle store glycogen.

2.17 The oxygen in the water molecule comes from a fatty acid.

2.18 The polar head is hydrophilic, and the nonpolar tails are hydrophobic.

2.19 The only differences between estradiol and testosterone are the number of double bonds and the types of functional groups attached to ring A.

2.20 An amino acid has a minimum of two carbon atoms and one nitrogen atom.

2.21 Hydrolysis occurs during catabolism of proteins.

2.22 No. Proteins consisting of a single polypeptide chain do not have a quaternary structure.

2.23 Sucrase has specificity for the sucrose molecule and thus would not "recognize" glucose and fructose.

2.24 Cytosine, thymine, adenine, and guanine are the nitrogenous bases present in DNA; cytosine, uracil, adenine, and guanine are the nitrogenous bases present in RNA.

2.25 In DNA, thymine always pairs with adenine, and cytosine always pairs with guanine.

2.26 Cellular activities that depend on energy supplied by ATP include muscular contractions, movement of chromosomes, transport of substances across cell membranes, and synthesis (anabolic) reactions.

CHAPTER 3

3.1 The three main parts of a cell are the plasma membrane, cytoplasm, and nucleus.

3.2 The glycocalyx is the sugary coat on the extracellular surface of the plasma membrane. It is composed of the carbohydrate portions of membrane glycolipids and glycoproteins.

3.3 The membrane protein that binds to insulin acts as a receptor.

3.4 Because fever involves an increase in body temperature, the rates of all diffusion processes would increase.

3.5 Nonpolar, hydrophobic molecules (oxygen, carbon dioxide, and nitrogen gases; fatty acids; steroids; and fat-soluble vitamins) plus small, uncharged polar molecules (water, urea, and small alcohols) move across the lipid bilayer of the plasma membrane through the process of simple diffusion.

3.6 The concentration of K^+ is higher in the cytosol of body cells than in extracellular fluids.

3.7 Yes. Insulin promotes insertion of glucose transporter (GluT) in the plasma membrane, which increases cellular glucose uptake by carrier-mediated facilitated diffusion.

3.8 No. The water concentrations can never be the same in the two arms because the left arm contains pure water and the right arm contains a solution that is less than 100% water.

3.9 A 2% solution of NaCl will cause crenation of RBCs because it is hypertonic.

3.10 ATP adds a phosphate group to the pump protein, which changes the pump's three-dimensional shape. ATP transfers energy to power the pump.

3.11 In secondary active transport, hydrolysis of ATP is used indirectly to drive the activity of symporter or antiporter proteins; this reaction directly powers the pump protein in primary active transport.

3.12 Transferrin, vitamins, and hormones are other examples of ligands that can undergo receptor-mediated endocytosis.

3.13 The binding of particles to a plasma membrane receptor triggers pseudopod formation.

3.14 Receptor-mediated endocytosis and phagocytosis involve receptor proteins; bulk-phase endocytosis does not.

3.15 Microtubules help to form centrioles, cilia, and flagella.

3.16 A cell without a centrosome probably would not be able to undergo cell division.

3.17 Cilia move fluids across cell surfaces; flagella move an entire cell.

3.18 Large and small ribosomal subunits are synthesized separately in the nucleolus within the nucleus, and are then assembled in the cytoplasm.

3.19 Rough ER has attached ribosomes; smooth ER does not. Rough ER synthesizes proteins that will be exported from the cell; smooth ER is associated with lipid synthesis and other metabolic reactions.

3.20 The entry face receives and modifies proteins from rough ER; the exit face modifies, sorts, and packages molecules for transport to other destinations.

3.21 Some proteins are secreted from the cell by exocytosis, some are incorporated into the plasma membrane, and some occupy storage vesicles that become lysosomes.

3.22 Digestion of worn-out organelles by lysosomes is called autophagy.

3.23 Mitochondrial cristae increase the surface area available for chemical reactions and contain some of the enzymes needed for ATP production.

3.24 Chromatin is a complex of DNA, proteins, and some RNA.

3.25 A nucleosome is a double-stranded molecule of DNA wrapped twice around a core of eight histones (proteins).

3.26 Proteins determine the physical and chemical characteristics of cells.

3.27 The DNA base sequence AGCT would be transcribed into the mRNA base sequence UCGA by RNA polymerase.

3.28 The P site holds the tRNA attached to the growing polypeptide. The A site holds the tRNA carrying the next amino acid to be added to the growing polypeptide.

3.29 When a ribosome encounters a stop codon at the A site, it releases the completed protein from the final tRNA.

3.30 DNA replicates during the S phase of interphase of the cell cycle.

3.31 DNA replication must occur before cytokinesis so that each of the new cells will have a complete genome.

3.32 Cytokinesis usually starts in late anaphase.

3.33 The result of crossing-over is that four haploid gametes are genetically unlike each other and genetically unlike the starting cell that produced them.

3.34 During anaphase I of meiosis, the paired chromatids are held together by a centromere and do not separate. During anaphase of mitosis, the paired chromatids separate and the centromeres split.

3.35 Sperm, which use the flagella for locomotion, are the only body cells required to move considerable distances.

CHAPTER 4

4.1 Epithelial tissue covers the body, lines various structures, and forms glands. Connective tissue protects, supports, binds organs together, stores energy, and helps provide immunity. Muscular tissue contracts and generates force and heat. Nervous tissue detects changes in the environment and generates nerve impulses that activate muscular contraction and glandular secretion.

4.2 Gap junctions allow cellular communication via passage of electrical and chemical signals between adjacent cells.

4.3 Since epithelial tissue is avascular, it depends on blood vessels in connective tissue for oxygen, nutrients, and waste disposal.

4.4 The basement membrane provides physical support for the epithelial tissue and plays a part in growth and wound healing, restriction of molecular movement between tissues, and blood filtration in the kidneys.

4.5 Because the cells are so thin, substances move most rapidly through squamous cells.

4.6 Simple multicellular exocrine glands have a nonbranched duct; compound multicellular exocrine glands have a branched duct.

4.7 Sebaceous (oil) glands are holocrine glands, and salivary glands are merocrine glands.

4.8 Fibroblasts secrete fibers and ground substance of extracellular matrix.

4.9 An epithelial membrane is a membrane that consists of an epithelial layer and an underlying layer of connective tissue.

CHAPTER 5

5.1 The epidermis is composed of epithelial tissue; the dermis is made up of connective tissue.

5.2 Melanin protects DNA of keratinocytes from the damaging effects of UV light.

5.3 The stratum basale is the layer of the epidermis with stem cells that continually undergo cell division.

5.4 Plucking a hair stimulates hair root plexuses in the dermis, some of which are sensitive to pain. Because the cells of a hair shaft are already dead and the hair shaft lacks nerves, cutting hair is not painful.

5.5 Nails are hard because they are composed of tightly packed, hard, keratinized epidermal cells.

5.6 Since the epidermis is avascular, an epidermal wound would not produce any bleeding.

5.7 Vernix caseosa consists of secretions from sebaceous glands, sloughed off peridermal cells, and hairs.

CHAPTER 6

6.1 The periosteum is essential for growth in bone thickness, bone repair, and bone nutrition. It also serves as a point of attachment for ligaments and tendons.

6.2 Bone resorption is necessary for the development, maintenance, and repair of bone.

6.3 The osteonic (haversian) canals are the main blood supply to the osteocytes of an osteon (haversian system), so their blockage would lead to death of the osteocytes.

6.4 Periosteal arteries enter bone tissue through perforating interosteonic (perforating or Volkmann's) canals.

6.5 Flat bones of the skull, most facial bones, the mandible (lower jawbone), and the medial part of the clavicle develop by intramembranous ossification.

6.6 Secondary ossification centers develop in the regions of the cartilage model that will give rise to the epiphyses.

6.7 The lengthwise growth of the diaphysis is caused by cell divisions in the zone of proliferating cartilage and replacement of the zone of calcified cartilage with bone (new diaphysis).

6.8 The medullary cavity enlarges by activity of the osteoclasts in the endosteum.

6.9 Healing of bone fractures can take months because calcium and phosphorus deposition is a slow process, and bone cells generally grow and reproduce slowly.

6.10 Heartbeat, respiration, nerve cell functioning, enzyme functioning, and blood clotting all depend on proper levels of calcium.

CHAPTER 7

7.1 The skull and vertebral column are part of the axial skeleton. The clavicle, shoulder girdle, humerus, pelvic girdle, and femur are part of the appendicular skeleton.

7.2 Flat bones protect underlying organs and provide a large surface area for muscle attachment.

7.3 The frontal, parietal, sphenoid, ethmoid, and temporal bones are all cranial bones (the occipital bone is not shown).

7.4 The parietal and temporal bones are joined by the squamous suture, the parietal and occipital bones are joined by the lambdoid suture, and the parietal and frontal bones are joined by the coronal suture.

7.5 The temporal bone articulates with the mandible and the parietal, sphenoid, zygomatic, and occipital bones.

7.6 The parietal bones form the posterior, lateral portion of the cranium.

7.7 The medulla oblongata of the brain connects with the spinal cord in the foramen magnum.

7.8 From the crista galli of the ethmoid bone, the sphenoid articulates with the frontal, parietal, temporal, occipital, temporal, parietal, and frontal bones, ending again at the crista galli of the ethmoid bone.

7.9 The perpendicular plate of the ethmoid bone forms the superior part of the nasal septum, and the lateral masses compose most of the medial walls of the orbits.

7.10 The mandible is the only movable skull bone, other than the auditory ossicles.

7.11 The nasal septum divides the nasal cavity into right and left sides.

7.12 Bones forming the orbit are the frontal, sphenoid, zygomatic, maxilla, lacrimal, ethmoid, and palatine.

7.13 The paranasal sinuses produce mucus and serve as resonating chambers for vocalizations.

7.14 The paired anterolateral fontanels are bordered by four different skull bones: the frontal, parietal, temporal, and sphenoid bones.

7.15 The hyoid bone is the only bone of the body that does not articulate with any other bone.

7.16 The thoracic and sacral curves of the vertebral column are concave relative to the anterior of the body.

7.17 The vertebral foramina enclose the spinal cord; the intervertebral foramina provide spaces through which spinal nerves exit the vertebral column.

7.18 The atlas moving on the axis at the atlanto-axial joint permits movement of the head to signify "no."

7.19 The facets and demifacets on the vertebral bodies of the thoracic vertebrae articulate with the heads of the ribs, and the facets on the transverse processes of these vertebrae articulate with the tubercles of the ribs.

7.20 The lumbar vertebrae are the largest and strongest in the body because the amount of weight supported by vertebrae increases toward the inferior end of the vertebral column.

7.21 There are four pairs of sacral foramina, for a total of eight. Each anterior sacral foramen joins a posterior sacral foramen at the intervertebral foramen. Nerves and blood vessels pass through these tunnels in the bone.

7.22 The body of the sternum articulates directly or indirectly with ribs 2–10.

7.23 The facet on the head of a rib fits into a facet or demifacet on the body of a vertebra, and the articular part of the tubercle of a rib articulates with the facet of the transverse process of a vertebra.

CHAPTER 8

8.1 The pectoral girdles attach the upper limbs to the axial skeleton.

8.2 The weakest part of the clavicle is its midregion at the junction of the two curves.

8.3 The acromion of the scapula forms the high point of the shoulder.

8.4 The radius articulates at the elbow with the capitulum and radial fossa of the humerus. The ulna articulates at the elbow with the trochlea, coronoid fossa, and olecranon fossa of the humerus.

8.5 The olecranon is the "elbow" part of the ulna.

8.6 The radius and ulna form the proximal and distal radioulnar joints. Their shafts are also connected by the interosseous membrane.

8.7 The scaphoid is the most frequently fractured wrist bone.

8.8 The bony pelvis attaches the lower limbs to the axial skeleton and supports the backbone and pelvic viscera.

8.9 The femur articulates with the acetabulum of the hip bone; the sacrum articulates with the auricular surface of the hip bone.

8.10 The pelvic axis is the course taken by a baby's head as it descends through the pelvis during childbirth.

8.11 The angle of convergence of the femurs is greater in females than males because the female pelvis is broader.

8.12 The patella is classified as a sesamoid bone because it develops in a tendon (the tendon of the quadriceps femoris muscle of the thigh).

8.13 The tibia is the weight-bearing bone of the leg.

8.14 The talus is the only tarsal bone that articulates with the tibia and the fibula.

8.15 Because the arches are not rigid, they yield when weight is applied and spring back when weight is lifted, allowing them to absorb the shock of walking.

8.16 Most of the skeletal system arises from embryonic mesoderm.

CHAPTER 9

9.1 Functionally, sutures are classified as synarthroses because they are immovable; syndesmoses are classified as amphiarthroses because they are slightly movable.

9.2 A synchondrosis is held together by hyaline cartilage, a symphysis is held together by fibrocartilage, and epiphyseal cartilage is a hyaline cartilage growth center during endochondral bone formation.

9.3 Functionally, synovial joints are diarthroses, freely movable joints.

9.4 Gliding movements occur at intercarpal joints and at intertarsal joints.

9.5 Two examples of flexion that do not occur along the sagittal plane are flexion of the thumb and lateral flexion of the trunk.

9.6 When you adduct your arm or leg, you bring it closer to the midline of the body, thus "adding" it to the trunk.

9.7 Circumduction involves flexion, abduction, extension, adduction, and rotation in a continuous sequence (or in the opposite order).

9.8 The anterior surface of a bone or limb rotates toward the midline in medial rotation, and away from the midline in lateral rotation.

9.9 Bringing your arms forward until the elbows touch is an example of protraction.

9.10 Other examples of pivot joints include the atlanto-axial joints.

9.11 The lateral ligament prevents displacement of the mandible.

9.12 The shoulder joint is the most freely movable joint in the body because of the looseness of its articular capsule and the shallowness of the glenoid cavity in relation to the size of the head of the humerus.

9.13 A hinge joint permits flexion and extension.

9.14 Tension in three ligaments—iliofemoral, pubofemoral, and ischiofemoral—limits the degree of extension at the hip joint.

9.15 Contraction of the quadriceps femoris muscle causes extension at the knee joint.

9.16 The purpose of arthroplasty is to relieve joint pain and permit greater range of motion.

CHAPTER 10

10.1 Perimysium bundles groups of muscle fibers into fascicles.

10.2 The sarcoplasmic reticulum releases calcium ions to trigger muscle contraction.

10.3 The following are arranged from smallest to largest: thick filament, myofibril, muscle fiber.

10.4 Actin and titin anchor into the Z disc. A bands contain myosin, actin, troponin, tropomyosin, and titin; I bands contain actin, troponin, tropomyosin, and titin.

10.5 The I bands and H zones disappear during muscle contraction; the lengths of the thin and thick filaments do not change.

10.6 If ATP were not available, the cross-bridges would not be able to detach from actin. The muscles would remain in a state of rigidity, as occurs in rigor mortis.

10.7 Three functions of ATP in muscle contraction are the following: (1) Its hydrolysis by an ATPase activates the myosin head so it can bind to actin and rotate; (2) its binding to myosin causes detachment from actin after the power stroke; and (3) it powers the pumps that transport Ca^{2+} from the sarcoplasm back into the sarcoplasmic reticulum.

10.8 A sarcomere length of 2.2 μm gives a generous zone of overlap between the parts of the thick filaments that have myosin heads and the thin filaments without the overlap being so extensive that sarcomere shortening is limited.

10.9 The part of the sarcolemma that contains acetylcholine receptors is the motor end plate.

10.10 Steps 4 through 6 are part of excitation–contraction coupling (muscle action potential through binding of myosin heads to actin).

10.11 Glycolysis, exchange of phosphate between creatine phosphate and ADP, and glycogen breakdown occur in the cytosol. Oxidation of pyruvic acid, amino acids, and fatty acids (aerobic respiration) occurs in mitochondria.

10.12 Motor units having many muscle fibers are capable of more forceful contractions than those having only a few fibers.

10.13 During the latent period, the muscle action potential sweeps over the sarcolemma and calcium ions are released from the sarcoplasmic reticulum.

10.14 If the second stimulus were applied a little later, the second contraction would be smaller than the one illustrated in part (b).

10.15 Holding your head upright without movement involves mainly isometric contractions.

10.16 Visceral smooth muscle is more like cardiac muscle; both contain gap junctions, which allow action potentials to spread from each cell to its neighbors.

10.17 The myotome of a somite differentiates into skeletal muscle.

CHAPTER 11

11.1 The belly of the muscle that extends the forearm, the triceps brachii, is located posterior to the humerus.

11.2 Second-class levers produce the most force.

11.3 For muscles named after their various characteristics, here are possible correct responses (for others, see **Table 11.2**): direction of fibers: external oblique; shape: deltoid; action: extensor digitorum; size: gluteus maximus; origin and insertion: sternocleidomastoid; location: tibialis anterior; number of tendons of origin: biceps brachii.

11.4 The corrugator supercilii muscle is involved in frowning; the zygomaticus major muscle contracts when you smile; the mentalis muscle contributes to pouting; the orbicularis oculi muscle contributes to squinting.

11.5 The inferior oblique muscle moves the eyeball superiorly and laterally because it originates at the anteromedial aspect of the floor of the orbit and inserts on the posterolateral aspect of the eyeball.

11.6 The masseter is the strongest muscle of mastication.

11.7 Functions of the tongue include chewing, detection of taste, swallowing, and speech.

11.8 The suprahyoid and infrahyoid muscles stabilize the hyoid bone to assist in tongue movements.

11.9 The triangles in the neck formed by the sternocleidomastoid muscles are important anatomically and surgically because of the structures that lie within their boundaries.

11.10 The rectus abdominis muscle aids in urination.

11.11 The diaphragm is innervated by the phrenic nerve.

11.12 The borders of the pelvic diaphragm are the pubic symphysis anteriorly, the coccyx posteriorly, and the walls of the pelvis laterally.

11.13 The borders of the perineum are the pubic symphysis anteriorly, the coccyx posteriorly, and the ischial tuberosities laterally.

11.14 The main action of the muscles that move the pectoral girdle is to stabilize the scapula to assist in movements of the humerus.

11.15 The rotator cuff consists of the flat tendons of the subscapularis, supraspinatus, infraspinatus, and teres minor muscles that form a nearly complete circle around the shoulder joint.

11.16 The brachialis is the most powerful forearm flexor; the triceps brachii is the most powerful forearm extensor.

11.17 Flexor tendons of the digits and wrist and the median nerve pass deep to the flexor retinaculum.

11.18 Muscles of the thenar eminence act on the thumb (pollex).

11.19 The splenius muscles arise from the midline and extend laterally and superiorly to their insertions.

11.20 Upper limb muscles exhibit diversity of movement; lower limb muscles function in stability, locomotion, and maintenance of posture. In addition, lower limb muscles usually cross two joints and act equally on both.

11.21 The quadriceps femoris consists of the rectus femoris, vastus lateralis, vastus medialis, and vastus intermedius; the hamstrings consist of the biceps femoris, semitendinosus, and semimembranosus.

11.22 The superior and inferior extensor retinacula firmly hold the tendons of the anterior compartment muscles to the ankle.

11.23 The plantar aponeurosis (fascia) supports the longitudinal arch and encloses the flexor tendons of the foot.

CHAPTER 12

12.1 The CNS processes many different kinds of sensory information; is the source of thoughts, emotions, and memories; and gives rise to signals that stimulate muscles to contract and glands to secrete.

12.2 Dendrites and the cell body receive input; the axon conducts nerve impulses (action potentials) and transmits the message to another neuron or effector cell by releasing a neurotransmitter at its synaptic end bulbs.

12.3 Most neurons in the CNS are multipolar neurons.

12.4 The cell body of a pyramidal cell is shaped like a pyramid.

12.5 Interneurons are responsible for integration.

12.6 Microglia function as phagocytes in the central nervous system.

12.7 One Schwann cell myelinates a single axon; one oligodendrocyte myelinates several axons.

12.8 Myelination increases the speed of nerve impulse conduction.

12.9 Myelin makes white matter look shiny and white.

12.10 Perception primarily occurs in the cerebral cortex.

12.11 A touch on the arm activates mechanically-gated channels.

12.12 The −70 mV resting membrane potential of a neuron means that the inside of the neuron is 70 mV more negative than the outside when the neuron is at rest (not excited by a stimulus).

12.13 More Na^+ ions would leak into the cell and fewer K^+ ions would leak out of the cell, which would make the resting membrane potential more positive.

12.14 A change in membrane potential from −70 to −60 mV is a depolarizing graded potential since the membrane potential is inside less negative than at rest. A change in membrane potential from −70 to −80 mV is a hyperpolarizing graded potential since the membrane potential is inside more negative than at rest.

12.15 Ligand-gated channels and mechanically-gated channels can be present in the dendrites of sensory neurons, and ligand-gated channels are numerous in the dendrites and cell bodies of interneurons and motor neurons.

12.16 A stronger stimulus opens more mechanically-gated channels or ligand-gated channels than a weaker stimulus.

12.17 Since individual graded potentials undergo decremental conduction, they would die out as they spread through the dendrites and cell body if summation did not occur and an action potential would not be generated at the trigger zone of the axon.

12.18 Voltage-gated Na^+ channels are open during the depolarizing phase, and voltage-gated K^+ channels are open during the repolarizing phase.

12.19 An action potential will not occur in response to a hyperpolarizing graded potential because a hyperpolarizing graded potential causes the membrane potential to become inside more negative and, therefore, farther away from threshold (−55 mV).

12.20 Yes, because the leak channels would still allow K^+ to exit more rapidly than Na^+ could enter the axon. Some mammalian myelinated axons have only a few voltage-gated K^+ channels.

12.21 The diameter of an axon, presence or absence of a myelin sheath, and temperature determine the speed of propagation of an action potential.

12.22 A synapse is a region of contact between two neurons or between a neuron and an effector.

12.23 In some electrical synapses (gap junctions), ions may flow equally well in either direction, so either neuron may be the presynaptic one. At a chemical synapse, one neuron releases neurotransmitter and the other neuron has receptors that bind this chemical. Thus, the signal can proceed in only one direction.

12.24 At some excitatory synapses, ACh binds to ionotropic receptors with cation channels that open and subsequently generate EPSPs in the postsynaptic cell. At some inhibitory synapses, ACh binds to metabotropic receptors coupled to G proteins that open K^+ channels, resulting in the formation of IPSPs in the postsynaptic cell.

12.25 This is an example of spatial summation since the summation results from the buildup of neurotransmitter released simultaneously by several presynaptic end bulbs.

12.26 Since −60 mV is below threshold, an action potential will not occur in the postsynaptic neuron.

12.27 Norepinephrine, epinephrine, dopamine, and serotonin are classified as biogenic amines because they are derived from amino acids that have been chemically modified.

12.28 A motor neuron receiving input from several other neurons is an example of convergence.

12.29 The neurolemma provides a regeneration tube that guides regrowth of a severed axon.

CHAPTER 13

13.1 The superior boundary of the spinal dura mater is the foramen magnum of the occipital bone. The inferior boundary is the second sacral vertebra.

13.2 The cervical enlargement connects with sensory and motor nerves of the upper limbs.

13.3 A horn is an area of gray matter, and a column is a region of white matter in the spinal cord.

13.4 Lateral gray horns are found in the thoracic and upper lumbar segments of the spinal cord.

13.5 All spinal nerves are classified as mixed because their posterior roots contain sensory axons and their anterior roots contain motor axons.

13.6 The anterior rami serve the upper and lower limbs.

13.7 The only spinal nerve without a corresponding dermatome is C1.

13.8 Severing the spinal cord at level C2 causes respiratory arrest because it prevents descending nerve impulses from reaching the phrenic nerve, which stimulates contraction of the diaphragm, the main muscle needed for breathing.

13.9 The axillary, musculocutaneous, radial, median, and ulnar nerves are five important nerves that arise from the brachial plexus.

13.10 Signs of femoral nerve injury include inability to extend the leg and loss of sensation in the skin over the anterolateral aspect of the thigh.

13.11 The origin of the sacral plexus is the anterior rami of spinal nerves L4–L5 and S1–S4.

13.12 The spinothalamic tract originates in the spinal cord and ends in the thalamus (a region of the brain). Because "spinal" comes first in the name, you know it contains ascending axons and thus is a sensory tract.

13.13 A sensory receptor produces a generator potential, which triggers a nerve impulse if the generator potential reaches threshold. Reflex integrating centers are in the CNS.

13.14 In an ipsilateral reflex, the sensory and motor neurons are on the same side of the spinal cord.

13.15 Reciprocal innervation is a type of arrangement of a neural circuit involving simultaneous contraction of one muscle and relaxation of its antagonist.

13.16 The flexor reflex is intersegmental because impulses go out over motor neurons located in several spinal nerves, each arising from a different segment of the spinal cord.

13.17 The crossed extensor reflex is a contralateral reflex arc because the motor impulses leave the spinal cord on the side opposite the entry of sensory impulses.

CHAPTER 14

14.1 The largest part of the brain is the cerebrum.

14.2 From superficial to deep, the three cranial meninges are the dura mater, arachnoid mater, and pia mater.

14.3 The brainstem is anterior to the fourth ventricle, and the cerebellum is posterior to it.

14.4 Cerebrospinal fluid is reabsorbed by the arachnoid villi that project into the dural venous sinuses.

14.5 The medulla oblongata contains the pyramids; the midbrain contains the cerebral peduncles; *pons* means "bridge."

14.6 Decussation means crossing to the opposite side. The functional consequence of decussation of the pyramids is that each side of the cerebrum controls muscles on the opposite side of the body.

14.7 The cerebral peduncles are the main sites through which tracts extend and nerve impulses are conducted between the superior parts of the brain and the inferior parts of the brain and the spinal cord.

14.8 The cerebellar peduncles carry information into and out of the cerebellum.

14.9 In about 70% of human brains, the intermediate mass connects the right and left halves of the thalamus.

14.10 From posterior to anterior, the four major regions of the hypothalamus are the mammillary, tuberal, supraoptic, and preoptic regions.

14.11 The gray matter enlarges more rapidly during development, in the process producing convolutions or gyri (folds), sulci (shallow grooves), and fissures (deep grooves).

14.12 Association tracts connect gyri of the same hemisphere; commissural tracts connect gyri in opposite hemispheres; projection tracts connect the cerebrum with the thalamus, brainstem, and spinal cord.

14.13 The basal nuclei are lateral, superior, and inferior to the thalamus.

14.14 The hippocampus is the part of the limbic system that functions with the cerebrum in memory.

14.15 The common integrative area integrates interpretation of visual, auditory, and somatic sensations; Broca's speech area translates thoughts into speech; the premotor area controls skilled muscular movements; the primary gustatory area interprets sensations related to taste; the primary auditory area allows you to interpret pitch and rhythm; the primary visual area allows you to interpret shape, color, and movement of objects; and the frontal eye field area controls voluntary scanning movements of the eyes.

14.16 In an EEG, theta waves indicate emotional stress.

14.17 Axons in the olfactory tracts terminate in the primary olfactory area in the temporal lobe of the cerebral cortex.

14.18 Most axons in the optic tracts terminate in the lateral geniculate nucleus of the thalamus.

14.19 The superior branch of the oculomotor nerve is distributed to the superior rectus muscle; the trochlear nerve is the smallest cranial nerve.

14.20 The trigeminal nerve is the largest cranial nerve.

14.21 Motor axons of the facial nerve originate in the pons.

14.22 The vestibular ganglion contains cell bodies from sensory axons that arise in the semicircular canals, saccule, and utricle; the spiral ganglion contains cell bodies from axons that arise in the spiral organ of the cochlea.

14.23 The glossopharyngeal nerve exits the skull through the jugular foramen.

14.24 The vagus nerve is located medial and posterior to the internal jugular vein and common carotid artery in the neck.

14.25 The accessory nerve is the only cranial nerve that originates from both the brain and spinal cord.

14.26 Two important motor functions of the hypoglossal nerve are speech and swallowing.

14.27 The gray matter of the nervous system derives from the mantle layer cells of the neural tube.

14.28 The mesencephalon does not develop into a secondary brain vesicle.

CHAPTER 15

15.1 Dual innervation means that a body organ receives neural innervation from both sympathetic and parasympathetic neurons of the ANS.

15.2 Most parasympathetic preganglionic axons are longer than most sympathetic preganglionic axons because most parasympathetic ganglia are in the walls of visceral organs, but most sympathetic ganglia are close to the spinal cord in the sympathetic trunk.

15.3 Terminal ganglia are associated with the parasympathetic division; sympathetic trunk and prevertebral ganglia are associated with the sympathetic division.

15.4 Sympathetic trunk ganglia contain sympathetic postganglionic neurons that lie in a vertical row on either side of the vertebral column.

15.5 The largest autonomic plexus is the celiac (solar) plexus.

15.6 Pelvic splanchnic nerves branch from the second through fourth sacral spinal nerves.

15.7 Most (but not all) sympathetic postganglionic neurons are adrenergic. Muscarinic receptors are present in the plasma membranes of all effectors (smooth muscle, cardiac muscle, and glands) innervated by parasympathetic postganglionic neurons and in sweat glands innervated by cholinergic sympathetic postganglionic neurons.

CHAPTER 16

16.1 The special senses of vision, taste, hearing, and equilibrium are served by separate sensory cells.

16.2 Pain, thermal sensations, tickle, and itch arise with activation of different free nerve endings.

16.3 The kidneys have the broadest area for referred pain.

16.4 Muscle spindles are activated when the central areas of the intrafusal fibers are stretched.

16.5 The posterior columns consist of the cuneate fasciculus and the gracile fasciculus.

16.6 Damage to the right spinothalamic tract could result in loss of pain, thermal, itch, and tickle sensations on the left side of the body.

16.7 The left trigeminal (V) nerve conveys nerve impulses for most somatic sensations from the left side of the face into the pons.

16.8 The hand has a larger representation in the motor area than in the somatosensory area, which implies greater precision in the hand's movement control than fine ability in its sensation.

16.9 Cerebral cortex UMNs are essential for the execution of voluntary movements of the body. Brainstem UMNs regulate muscle tone, control postural muscles, and help maintain balance and orientation of the head and body.

16.10 The lateral corticospinal tract conducts impulses that result in contractions of the muscles in the distal parts of the limbs.

16.11 The axons of the corticobulbar tract terminate in the motor nuclei of the following cranial nerves: oculomotor (III), trochlear (IV), trigeminal (V), abducens (VI), facial (VII), glossopharyngeal (IX), vagus (X), accessory (XI), and hypoglossal (XII).

16.12 The rubrospinal tract helps promote voluntary contractions of the upper limbs, whereas the rest of the indirect motor pathways cause involuntary contractions of muscles in the body.

16.13 The anterior and posterior spinocerebellar tracts carry information from proprioceptors in joints and muscles to the cerebellum.

CHAPTER 17

17.1 An olfactory receptor cell has a life span of about one month.

17.2 Olfactory transduction occurs in the olfactory cilia of an olfactory receptor cell.

17.3 Basal cells develop into gustatory receptor cells.

17.4 Visible light that has a wavelength of 700 nm is red.

17.5 The conjunctiva is continuous with the inner lining of the eyelids.

17.6 Lacrimal fluid, or tears, is a watery solution containing salts, some mucus, and lysozyme that protects, cleans, lubricates, and moistens the eyeball.

17.7 The fibrous tunic consists of the cornea and sclera; the vascular tunic consists of the choroid, ciliary body, and iris.

17.8 The parasympathetic division of the ANS causes pupillary constriction; the sympathetic division causes pupillary dilation.

17.9 An ophthalmoscopic examination of the blood vessels of the eye can reveal evidence of hypertension, diabetes mellitus, cataracts, and age-related macular disease.

17.10 The two types of photoreceptors are rods and cones. Rods provide black-and-white vision in dim light; cones provide high visual acuity and color vision in bright light.

17.11 After its secretion by the ciliary process, aqueous humor flows into the posterior chamber, around the iris, into the anterior chamber, and out of the eyeball through the scleral venous sinus.

17.12 During accommodation the ciliary muscle contracts, causing the zonular fibers to slacken. The lens then becomes more convex, increasing its focusing power.

17.13 Presbyopia is the loss of lens elasticity that occurs with aging.

17.14 Both rods and cones transduce light into receptor potentials, use a photopigment embedded in outer segment discs or folds, and release neurotransmitter at synapses with bipolar cells and horizontal cells.

17.15 The conversion of *cis*-retinal to *trans*-retinal is called isomerization.

17.16 Cyclic GMP is the ligand that opens Na^+ channels in photoreceptors, causing the dark current to flow.

17.17 Light rays from an object in the temporal half of the visual field fall on the nasal half of the retina.

17.18 The malleus of the middle ear is attached to the eardrum, which is part of the external ear.

17.19 The oval and round windows separate the middle ear from the internal ear.

17.20 The two sacs in the membranous labyrinth of the vestibule are the utricle and saccule.

17.21 The three subdivisions of the bony labyrinth are the semicircular canals, vestibule, and cochlea.

17.22 The region of the basilar membrane close to the oval and round windows vibrates most vigorously in response to high-frequency sounds.

17.23 A tip link protein connects a cation channel in a stereocilium to the tip of its taller stereocilium neighbor. When the stereocilia bend toward the tallest stereocilium, the tip link is stretched and tugs on the cation channel, causing it to completely open. This allows large amount of KI to enter the hair cell, resulting in the formation of a strong depolarizing receptor potential.

17.24 The superior olivary nucleus of the pons is the part of the auditory pathway that allows a person to locate the source of a sound.

17.25 The utricle detects linear acceleration or deceleration that occurs in a horizontal direction and also head tilt; the saccule detects linear acceleration or deceleration that occurs in a vertical direction.

17.26 The semicircular ducts detect rotational acceleration or deceleration.

17.27 The vestibular nuclei are located in the medulla and the pons.

17.28 The optic cup forms the neural and pigmented layers of the retina.

17.29 The internal ear develops from surface ectoderm, the middle ear develops from pharyngeal pouches, and the external ear develops from a pharyngeal cleft.

CHAPTER 18

18.1 Secretions of endocrine glands diffuse into interstitial fluid and then into the blood; exocrine secretions flow into ducts that lead into body cavities or to the body surface.

18.2 In the stomach, histamine is a paracrine because it acts on nearby parietal cells without entering the blood.

18.3 The receptor–hormone complex alters gene expression by turning specific genes of nuclear DNA on or off.

18.4 Cyclic AMP is termed a second messenger because it translates the presence of the first messenger, the water-soluble hormone, into a response inside the cell.

18.5 The hypophyseal portal veins carry blood from the median eminence of the hypothalamus, where hypothalamic releasing and inhibiting hormones are secreted, to the anterior pituitary, where these hormones act.

18.6 Thyroid hormones suppress secretion of TSH by thyrotrophs and of TRH by hypothalamic neurosecretory cells; gonadal hormones suppress secretion of FSH and LH by gonadotrophs and of GnRH by hypothalamic neurosecretory cells.

18.7 Excess levels of GH would cause hyperglycemia.

18.8 Functionally, both the hypothalamic–hypophyseal tract and the hypophyseal portal veins carry hypothalamic hormones to the pituitary gland. Structurally, the tract is composed of axons of neurons that extend from the hypothalamus to the posterior pituitary; the portal veins are blood vessels that extend from the hypothalamus to the anterior pituitary.

18.9 Follicular cells secrete T_3 and T_4, also known as thyroid hormones. Parafollicular cells secrete calcitonin.

18.10 The storage form of thyroid hormones is thyroglobulin.

18.11 Lack of iodine in the diet → diminished production of T_3 and T_4 → increased release of TSH → growth (enlargement) of the thyroid gland → goiter.

18.12 Parafollicular cells of the thyroid gland secrete calcitonin; chief (principal) cells of the parathyroid gland secrete PTH.

18.13 Target tissues for PTH are bones and the kidneys; target tissue for CT is bone; target tissue for calcitriol is the GI tract.

18.14 The adrenal glands are superior to the kidneys in the retroperitoneal space.

18.15 Angiotensin II acts to constrict blood vessels by causing contraction of vascular smooth muscle, and it stimulates secretion of aldosterone (by zona glomerulosa cells of the adrenal cortex), which in turn causes the kidneys to conserve water and thereby increase blood volume.

18.16 A transplant recipient who takes prednisone will have low blood levels of ACTH and CRH due to negative feedback suppression of the anterior pituitary and hypothalamus by the prednisone.

18.17 The pancreas is both an endocrine and an exocrine gland.

18.18 Glycogenolysis is the conversion of glycogen into glucose and therefore it increases blood glucose level.

18.19 Homeostasis maintains controlled conditions typical of a normal internal environment; the stress response resets controlled conditions at a different level to cope with various stressors.

18.20 The adrenal cortex of the adrenal gland is derived from mesoderm, while the adrenal medulla of the adrenal gland is derived from ectoderm.

CHAPTER 19

19.1 Blood volume is about 8% of your body mass, roughly 5–6 liters in males and 4–5 liters in females. For instance, a 70-kg (150-lb) person has a blood volume of 5.6 liters (70 kg × 8% × 1 liter/kg).

19.2 Platelets are cell fragments.

19.3 Pluripotent stem cells develop from mesenchyme.

19.4 One hemoglobin molecule can transport a maximum of four O_2 molecules, one O_2 bound to each heme group.

19.5 Transferrin is a plasma protein that transports iron in the blood.

19.6 Once you moved to high altitude, your hematocrit would increase due to increased secretion of erythropoietin.

19.7 Neutrophils, eosinophils, and basophils are called granular leukocytes because all have cytoplasmic granules that are visible through a light microscope when stained.

19.8 Lymphocytes recirculate from blood to tissues and back to blood. After leaving the blood, other WBCs remain in the tissues until they die.

19.9 Along with platelet plug formation, vascular spasm and blood clotting contribute to hemostasis.

19.10 Serum is blood plasma minus the clotting proteins.

19.11 The outcome of the first stage of clotting is the formation of prothrombinase.

19.12 Type O blood usually contains both anti-A and anti-B antibodies.

19.13 Because the mother is most likely to start making anti-Rh antibodies after the first baby is already born, that baby suffers no damage.

CHAPTER 20

20.1 The mediastinum is the anatomical region that extends from the sternum to the vertebral column, from the first rib to the diaphragm, and between the lungs.

20.2 The visceral layer of the serous pericardium (epicardium) is both a part of the pericardium and a part of the heart wall.

20.3 The coronary sulcus forms a boundary between the atria and ventricles.

20.4 The greater the workload of a heart chamber, the thicker its myocardium.

20.5 The fibrous skeleton attaches to the heart valves and prevents overstretching of the valves as blood passes through them.

20.6 The papillary muscles contract, which pulls on the chordae tendineae and prevents cusps of the atrioventricular valves from everting and letting blood flow back into the atria.

20.7 Numbers 2 through 6 depict the pulmonary circulation; numbers 7 through 1 depict the systemic circulation.

20.8 The circumflex artery delivers oxygenated blood to the left atrium and left ventricle.

20.9 The intercalated discs hold the cardiac muscle fibers together and enable action potentials to propagate from one muscle fiber to another.

20.10 The only electrical connection between the atria and the ventricles is the atrioventricular bundle.

20.11 The duration of an action potential is much longer in a ventricular contractile fiber (0.3 sec = 300 msec) than in a skeletal muscle fiber (1–2 msec).

20.12 An enlarged Q wave may indicate a myocardial infarction.

20.13 Action potentials propagate most slowly through the AV node.

20.14 The amount of blood in each ventricle at the end of ventricular diastole—called the end-diastolic volume—is about 130 mL in a resting person.

20.15 The first heart sound (S1), or lubb, is associated with closure of the atrioventricular valves.

20.16 The ventricular myocardium receives innervation from the sympathetic division only.

20.17 The skeletal muscle contraction increases stroke volume by increasing preload (end-diastolic volume).

20.18 Individuals with end-stage heart failure or severe coronary artery disease are candidates for cardiac transplantation.

20.19 The heart begins to contract by the 22nd day of gestation.

20.20 Partitioning of the heart is complete by the end of the fifth week.

CHAPTER 21

21.1 The femoral artery has the thicker wall; the femoral vein has the wider lumen.

21.2 Due to atherosclerosis, less energy is stored in the less-compliant elastic arteries during systole; thus, the heart must pump harder to maintain the same rate of blood flow.

21.3 Metabolically active tissues use O_2 and produce wastes more rapidly than inactive tissues, so they require more extensive capillary networks.

21.4 Materials cross capillary walls through intercellular clefts and fenestrations, via transcytosis in pinocytic vesicles, and through the plasma membranes of endothelial cells.

21.5 Valves are more important in arm veins and leg veins than in neck veins because, when you are standing, gravity causes pooling of blood in the veins of the free limbs but aids the flow of blood in neck veins back toward the heart.

21.6 Blood volume in venules and veins is about 64% of 5 liters, or 3.2 liters; blood volume in capillaries is about 7% of 5 liters, or 350 mL.

21.7 Blood colloid osmotic pressure is lower than normal in a person with a low level of plasma proteins, and therefore capillary reabsorption is low. The result is edema.

21.8 Mean blood pressure in the aorta is closer to diastolic than to systolic pressure.

21.9 The skeletal muscle pump and respiratory pump also aid venous return.

21.10 Vasodilation and vasoconstriction of arterioles are the main regulators of systemic vascular resistance.

21.11 Velocity of blood flow is fastest in the aorta and arteries.

21.12 The effector tissues regulated by the cardiovascular center are cardiac muscle in the heart and smooth muscle in blood vessel walls.

21.13 Impulses to the cardiovascular center pass from baroreceptors in the carotid sinuses via the glossopharyngeal (IX) nerves and from baroreceptors in the arch of the aorta via the vagus (X) nerves.

21.14 It represents a change that occurs when you stand up because gravity causes pooling of blood in leg veins once you are upright, decreasing the blood pressure in your upper body.

21.15 Diastolic blood pressure = 95 mmHg; systolic blood pressure = 142 mmHg; pulse pressure = 47 mmHg. This person has stage I hypertension because the systolic blood pressure is greater than 140 mmHg and the diastolic blood pressure is greater than 90 mmHg.

21.16 Almost-normal blood pressure in a person who has lost blood does not necessarily indicate that the patient's tissues are receiving adequate blood flow; if systemic vascular resistance has increased greatly, tissue perfusion may be inadequate.

21.17 The two main circulatory routes are the systemic circulation and the pulmonary circulation.

21.18 The four subdivisions of the aorta are the ascending aorta, arch of the aorta, thoracic aorta, and abdominal aorta.

21.19 The arteries supplying the heart are called coronary arteries because they form a crown above the ventricles of the heart.

21.20 Branches of the arch of the aorta (in order of origination) are the brachiocephalic trunk, left common carotid artery, and left subclavian artery.

21.21 The thoracic aorta begins at the level of the intervertebral disc between T4 and T5.

21.22 The abdominal aorta begins at the aortic hiatus in the diaphragm.

21.23 The abdominal aorta divides into the common iliac arteries at about the level of L4.

21.24 The superior vena cava drains regions above the diaphragm, and the inferior vena cava drains regions below the diaphragm.

21.25 All venous blood in the brain drains into the internal jugular veins.

21.26 The median cubital vein of the upper limb is often used for withdrawing blood.

21.27 The inferior vena cava returns blood from abdominopelvic viscera to the heart.

21.28 Superficial veins of the lower limbs are the dorsal venous arches and the great saphenous and small saphenous veins.

21.29 The hepatic veins carry blood away from the liver.

21.30 After birth, the pulmonary arteries are the only arteries that carry deoxygenated blood.

21.31 Exchange of materials between mother and fetus occurs across the placenta.

21.32 Blood vessels and blood are derived from mesoderm.

CHAPTER 22

22.1 Red bone marrow contains stem cells that develop into lymphocytes.

22.2 Lymph is more similar to interstitial fluid than to blood plasma because the protein content of lymph is low.

22.3 The left and right lumbar trunks and the intestinal trunk empty into the cisterna chyli, which then drains into the thoracic duct.

22.4 Inhalation promotes the movement of lymph from abdominal lymphatic vessels toward the thoracic region because the pressure in the vessels of the thoracic region is lower than the pressure in the abdominal region when a person inhales.

22.5 T cells mature in the thymus.

22.6 Foreign substances in lymph that enter a lymph node may be phagocytized by macrophages or attacked by lymphocytes that mount immune responses.

22.7 White pulp of the spleen functions in immunity; red pulp of the spleen performs functions related to blood cells.

22.8 Lymphatic tissues begin to develop by the end of the fifth week of gestation.

22.9 Lysozyme, digestive enzymes, and oxidants can kill microbes ingested during phagocytosis.

22.10 Redness results from increased blood flow due to vasodilation; pain, from injury of nerve fibers, irritation by microbial toxins, kinins, and prostaglandins, and pressure due to edema; heat, from increased blood flow and heat released by locally increased metabolic reactions; swelling, from leakage of fluid from capillaries due to increased permeability.

22.11 Helper T cells participate in both cell-mediated and antibody-mediated immune responses.

22.12 Epitopes are small immunogenic parts of a larger antigen; haptens are small molecules that become immunogenic only when they attach to a body protein.

22.13 APCs include macrophages in tissues throughout the body, B cells in blood and lymphatic tissue, and dendritic cells in mucous membranes and the skin.

22.14 Endogenous antigens include viral proteins, toxins from intracellular bacteria, and abnormal proteins synthesized by a cancerous cell.

22.15 The first signal in T cell activation is antigen binding to a TCR; the second signal is a costimulator, such as a cytokine or another pair of plasma membrane molecules.

22.16 The CD8 protein of a cytotoxic T cell binds to the MHC-I molecule of an infected body cell to help anchor the T-cell receptor (TCR)–antigen interaction so that antigen recognition can occur.

22.17 Cytotoxic T cells attack some tumor cells and transplanted tissue cells, as well as cells infected by microbes.

22.18 Since all of the plasma cells in this figure are part of the same clone, they secrete just one kind of antibody.

22.19 The variable regions recognize and bind to a specific antigen.

22.20 The classical pathway for the activation of complement is linked to antibody-mediated immunity because Ag–Ab complexes activate C1.

22.21 At peak secretion, approximately 1000 times more IgG is produced in the secondary response than in the primary response.

22.22 In deletion, self-reactive T cells or B cells die; in anergy, T cells or B cells are alive but are unresponsive to antigenic stimulation.

CHAPTER 23

23.1 External respiration involves the exchange of O_2 and CO_2 between the alveoli of the lungs and the blood in pulmonary capillaries; internal respiration involves the exchange of O_2 and CO_2 between the blood in systemic capillaries and tissue cells of the body.

23.2 The conducting zone of the respiratory system includes the nose, pharynx, larynx, trachea, bronchi, and bronchioles (except the respiratory bronchioles).

23.3 The path of air is external nares → vestibule → nasal cavity → internal nares.

23.4 The root of the nose attaches it to the frontal bone.

23.5 During swallowing, the epiglottis closes over the rima glottidis, the entrance to the trachea, to prevent aspiration of food and liquids into the lungs.

23.6 The main function of the vocal folds is voice production.

23.7 Because the tissues between the esophagus and trachea are soft, the esophagus can bulge and press against the trachea during swallowing.

23.8 The left lung has two lobes and two lobar bronchi; the right lung has three of each.

23.9 The pleural membrane is a serous membrane.

23.10 Because two-thirds of the heart lies to the left of the midline, the left lung contains a cardiac notch to accommodate the presence of the heart. The right lung is shorter than the left because the diaphragm is higher on the right side to accommodate the liver.

23.11 The wall of an alveolus is made up of type I alveolar cells, type II alveolar cells, and associated alveolar macrophages.

23.12 The respiratory membrane averages 0.5 μm in thickness.

23.13 The pressure would increase fourfold, to 4 atm.

23.14 If you are at rest while reading, your diaphragm is responsible for about 75% of each inhalation.

23.15 At the start of inhalation, intrapleural pressure is about 756 mmHg. With contraction of the diaphragm, it decreases to about 754 mmHg as the volume of the space between the two pleural layers expands. With relaxation of the diaphragm, it increases back to 756 mmHg.

23.16 Breathing in and then exhaling as much air as possible demonstrates vital capacity.

23.17 A difference in P_{O_2} promotes oxygen diffusion into pulmonary capillaries from alveoli and into tissue cells from systemic capillaries.

23.18 The most important factor that determines how much O_2 binds to hemoglobin is the P_{O_2}.

23.19 Both during exercise and at rest, hemoglobin in your pulmonary veins would be fully saturated with O_2, a point that is at the upper right of the curve.

23.20 Because lactic acid (lactate) and CO_2 are produced by active skeletal muscles, blood pH decreases slightly and P_{CO_2} increases when you are actively exercising. The result is lowered affinity of hemoglobin for O_2, so more O_2 is available to the working muscles.

23.21 O_2 is more available to your tissue cells when you have a fever because the affinity of hemoglobin for O_2 decreases with increasing temperature.

23.22 At a P_{O_2} of 40 mmHg, fetal Hb is 80% saturated with O_2 and maternal Hb is about 75% saturated.

23.23 Blood in a systemic vein would have a higher concentration of HCO_3^-.

23.24 The medullary respiratory center in the medulla contains neurons that are active and then inactive in a repeating cycle.

23.25 The phrenic nerves innervate the diaphragm.

23.26 Peripheral chemoreceptors are responsive to changes in blood levels of oxygen, carbon dioxide, and H^+.

23.27 Normal arterial blood P_{CO_2} is 40 mmHg.

23.28 The respiratory system begins to develop about 4 weeks after fertilization.

CHAPTER 24

24.1 Digestive enzymes are produced by the salivary glands, tongue, stomach, pancreas, and small intestine.

24.2 In the context of the digestive system, absorption is the movement of the products of digestion from the lumen of the GI tract into blood or lymph.

24.3 The lamina propria has the following functions: (1) It contains blood vessels and lymphatic vessels, which are the routes by which nutrients are absorbed from the GI tract; (2) it supports the mucosal epithelium and binds it to the muscularis mucosae; and (3) it contains mucosa-associated lymphatic tissue (MALT), which helps protect against disease.

24.4 The neurons of the myenteric plexus regulate GI tract motility, and the neurons of the submucosal plexus regulate GI secretion.

24.5 Mesentery binds the small intestine to the posterior abdominal wall.

24.6 The uvula helps prevent foods and liquids from entering the nasal cavity during swallowing.

24.7 Chloride ions in saliva activate salivary amylase.

24.8 The main component of teeth is connective tissue, specifically dentin.

24.9 The first, second, and third molars do not replace any deciduous teeth.

24.10 The esophageal mucosa and submucosa contain mucus-secreting glands.

24.11 Both. Initiation of swallowing is voluntary and the action is carried out by skeletal muscles. Completion of swallowing—moving a bolus along the esophagus and into the stomach—is involuntary and involves peristalsis by smooth muscle.

24.12 After a large meal, the rugae stretch and disappear as the stomach fills.

24.13 Parietal cells in gastric glands secrete HCl, which is a component of gastric juice. HCl kills microbes in food, denatures proteins, and converts pepsinogen into pepsin.

24.14 Hydrogen ions secreted into gastric juice are derived from carbonic acid (H_2CO_3).

24.15 Histamine is a paracrine agent released by mast cells in the lamina propria.

24.16 The pancreatic duct contains pancreatic juice (fluid and digestive enzymes); the common bile duct contains bile; the hepatopancreatic ampulla contains pancreatic juice and bile.

24.17 The phagocytic cell in the liver is the stellate reticuloendothelial (Kupffer) cell.

24.18 While a meal is being absorbed, nutrients, O_2, and certain toxic substances are removed by hepatocytes from blood flowing through liver sinusoids.

24.19 The ileum is the longest part of the small intestine.

24.20 Nutrients being absorbed in the small intestine enter the blood via capillaries or the lymph via lacteals.

24.21 The fluid secreted by duodenal (Brunner's) glands—alkaline mucus—neutralizes gastric acid and protects the mucosal lining of the duodenum.

24.22 Because monoglycerides are hydrophobic (nonpolar) molecules, they can dissolve in and diffuse through the lipid bilayer of the plasma membrane.

24.23 The stomach and pancreas are the two digestive system organs that secrete the largest volumes of fluid.

24.24 The ascending and descending portions of the colon are retroperitoneal.

24.25 Goblet cells in the large intestine secrete mucus to lubricate colonic contents.

24.26 The pH of gastric juice rises due to the buffering action of some amino acids in food proteins.

CHAPTER 25

25.1 In pancreatic acinar cells, anabolism predominates because the primary activity is synthesis of complex molecules (digestive enzymes).

25.2 The electron transport chain produces the most ATP.

25.3 The reactions of glycolysis consume two molecules of ATP but generate four molecules of ATP, for a net gain of two.

25.4 Kinases are enzymes that phosphorylate (add phosphate to) their substrate.

25.5 Glycolysis occurs in the cytosol.

25.6 CO_2 is given off during the production of acetyl coenzyme A and during the Krebs cycle. It diffuses into the blood, is transported by the blood to the lungs, and is exhaled.

25.7 The production of reduced coenzymes is important in the Krebs cycle because they will subsequently yield ATP in the electron transport chain.

25.8 The energy source that powers the proton pumps is electrons provided by NADH + H^+.

25.9 The concentration of H^+ is highest in the space between the inner and outer mitochondrial membranes.

25.10 During the complete oxidation of one glucose molecule, six molecules of O_2 are used and six molecules of CO_2 are produced.

25.11 Skeletal muscle fibers can synthesize glycogen, but they cannot release glucose into the blood because they lack the enzyme phosphatase required to remove the phosphate group from glucose.

25.12 Hepatocytes can carry out gluconeogenesis and glycogenesis.

25.13 LDLs deliver cholesterol to body cells.

25.14 Hepatocytes and adipose cells carry out lipogenesis, beta oxidation, and lipolysis; hepatocytes carry out ketogenesis.

25.15 Before an amino acid can enter the Krebs cycle, an amino group must be removed via deamination.

25.16 Acetyl coenzyme A is the gateway into the Krebs cycle for molecules being oxidized to generate ATP.

25.17 Reactions of the absorptive state are mainly anabolic.

25.18 Processes that directly elevate blood glucose during the postabsorptive state include lipolysis (in adipocytes and hepatocytes), gluconeogenesis (in hepatocytes), and glycogenolysis (in hepatocytes).

25.19 Exercise, the sympathetic nervous system, hormones (epinephrine, norepinephrine, thyroxine, testosterone, growth hormone), elevated body temperature, and ingestion of food increase metabolic rate, which results in an increase in body temperature.

25.20 The blue cup is a reminder to include three daily servings of dairy such as milk, yogurt, and cheese.

CHAPTER 26

26.1 The kidneys, ureters, urinary bladder, and urethra are the components of the urinary system.

26.2 The kidneys are retroperitoneal because they are posterior to the peritoneum.

26.3 Blood vessels, lymphatic vessels, nerves, and a ureter pass through the renal hilum.

26.4 About 1200 mL of blood enters the renal arteries each minute.

26.5 Cortical nephrons have glomeruli in the superficial renal cortex, and their short nephron loops penetrate only into the superficial renal medulla. Juxtamedullary nephrons have glomeruli deep in the renal cortex, and their long nephron loops extend through the renal medulla nearly to the renal papilla.

26.6 This section must pass through the renal cortex because there are no renal corpuscles in the renal medulla.

26.7 Secreted penicillin is being removed from the bloodstream.

26.8 Endothelial fenestrations (pores) in glomerular capillaries are too small for red blood cells to pass through.

26.9 Obstruction of the right ureter would increase CHP and thus decrease NFP in the right kidney; the obstruction would have no effect on the left kidney.

26.10 *Auto-* means self; tubuloglomerular feedback is an example of autoregulation because it takes place entirely within the kidneys.

26.11 The tight junctions between tubule cells form a barrier that prevents diffusion of transporter, channel, and pump proteins between the apical and basolateral membranes.

26.12 Glucose enters a PCT cell via a Na^+–glucose symporter in the apical membrane and leaves via facilitated diffusion through the basolateral membrane.

26.13 The electrochemical gradient promotes movement of Na^+ into the tubule cell through the apical membrane antiporters.

26.14 Reabsorption of the solutes creates an osmotic gradient that promotes the reabsorption of water via osmosis.

26.15 This is considered secondary active transport because the symporter uses the energy stored in the concentration gradient of Na^+ between extracellular fluid and the cytosol. No water is reabsorbed here because the thick ascending limb of the nephron loop is virtually impermeable to water.

26.16 In principal cells, aldosterone stimulates secretion of K^+ and reabsorption of Na^+ by increasing the activity of sodium–potassium pumps and number of leakage channels for Na^+ and K^+.

26.17 Aldosterone and atrial natriuretic peptide influence renal water reabsorption along with ADH.

26.18 Dilute urine is produced when the thick ascending limb of the nephron loop, the distal convoluted tubule, and the collecting duct reabsorb more solutes than water.

26.19 The high osmolarity of interstitial fluid in the renal medulla is due mainly to Na^+, Cl^-, and urea.

26.20 Secretion occurs in the proximal convoluted tubule, the nephron loop, and the collecting duct.

26.21 Lack of voluntary control over micturition is termed urinary incontinence.

26.22 The three subdivisions of the male urethra are the prostatic urethra, membranous urethra, and spongy urethra.

26.23 The kidneys start to form during the third week of development.

CHAPTER 27

27.1 Plasma volume equals body mass × percent of body mass that is body fluid × proportion of body fluid that is ECF × proportion of ECF that is plasma × a conversion factor (1 liter/kg). For males, blood plasma volume = 60 kg × 0.60 × 1/3 × 0.20 × 1 liter/kg = 2.4 liters. Using similar calculations, female blood plasma volume is 2.2 liters.

27.2 Hyperventilation, vomiting, fever, and diuretics all increase fluid loss.

27.3 Osmoreceptors are receptors that detect changes in the osmolarity (concentration of dissolved solutes) of body fluids.

27.4 Alcohol inhibits secretion of ADH.

27.5 ADH is responsible for the water reabsorption that accompanies aldosterone-mediated Na^+ reabsorption.

27.6 Overhydration would most likely stimulate the release of ANP.

27.7 If a solution used for oral rehydration therapy contains a small amount of salt, both the salt and water are absorbed in the gastrointestinal tract, blood volume increases without a decrease in osmolarity, and water intoxication does not occur.

27.8 In ECF, the major cation is Na^+, and the major anions are Cl^- and HCO_3^-. In ICF, the major cation is K^+, and the major anions are proteins and organic phosphates (for example, ATP).

27.9 Holding your breath causes blood pH to decrease slightly as CO_2 and H^+ accumulate in the blood.

27.10 A carbonic anhydrase inhibitor reduces secretion of H^+ into the urine and reduces reabsorption of Na^+ and HCO_3^- into the blood. It has a diuretic effect and can cause acidosis (lowered pH of the blood) due to loss of HCO_3^- in the urine.

CHAPTER 28

28.1 The gonads (testes) produce gametes (sperm) and hormones; the ducts transport, store, and receive gametes; the accessory sex glands secrete materials that support gametes; and the penis assists in the delivery and joining of gametes.

28.2 The cremaster and dartos muscles help regulate the temperature of the testes.

28.3 The tunica vaginalis and tunica albuginea are tissue layers that cover and protect the testes.

28.4 The interstitial (Leydig) cells of the testes secrete testosterone.

28.5 As a result of meiosis I, the number of chromosomes in each cell is reduced by half.

28.6 The sperm head contains the nucleus with 23 highly condensed chromosomes and an acrosome that contains enzymes for penetration of a secondary oocyte; the neck contains centrioles that produce microtubules for the rest of the tail; the midpiece contains mitochondria for ATP production for locomotion and metabolism; the principal and end pieces of the tail provide motility.

28.7 Sustentacular cells secrete inhibin.

28.8 Testosterone inhibits secretion of LH, and inhibin inhibits secretion of FSH.

28.9 The seminal vesicles are the accessory sex glands that contribute the largest volume to seminal fluid.

28.10 Two tissue masses called the corpora cavernosa penis and one corpus spongiosum penis contain blood sinuses that fill with blood that cannot flow out of the penis as quickly as it flows in. The trapped blood engorges and stiffens the tissue, producing an erection. The corpus spongiosum penis keeps the spongy urethra open so that ejaculation can occur.

28.11 The testes are homologous to the ovaries; the glans penis is homologous to the clitoris; the prostate is homologous to the paraurethral glands; and the bulbourethral glands are homologous to the greater vestibular glands (see Table 28.2).

28.12 The mesovarium anchors the ovary to the broad ligament of the uterus and the uterine tube; the ovarian ligament anchors it to the uterus; the suspensory ligament anchors it to the pelvic wall.

28.13 Ovarian follicles secrete estrogens; the corpus luteum secretes progesterone, estrogens, relaxin, and inhibin.

28.14 Most ovarian follicles undergo atresia (degeneration).

28.15 Primary oocytes are present in the ovary at birth, so they are as old as the woman. In males, primary spermatocytes are continually being formed from stem cells (spermatogonia) and thus are only a few days old.

28.16 Fertilization most often occurs in the ampulla of the uterine tube.

28.17 Ciliated columnar epithelial cells and nonciliated (peg) cells with microvilli line the uterine tubes.

28.18 The endometrium is a highly vascularized, secretory epithelium that provides the oxygen and nutrients needed to sustain a fertilized egg; the myometrium is a thick smooth muscle layer that supports the uterine wall during pregnancy and contracts to expel the fetus at birth.

28.19 The stratum basalis of the endometrium provides cells to replace those that are shed (the stratum functionalis) during each menstruation.

28.20 Anterior to the vaginal opening are the mons pubis, clitoris, prepuce, and external urethral orifice. Lateral to the vaginal opening are the labia minora and labia majora.

28.21 The anterior portion of the perineum is called the urogenital triangle because its borders form a triangle that encloses the urethral (uro-) and vaginal (-genital) orifices.

28.22 Prolactin, estrogens, and progesterone regulate the synthesis of milk. Oxytocin regulates the ejection of milk.

28.23 The principal estrogen is β-estradiol.

28.24 The hormones responsible for the proliferative phase of endometrial growth are estrogens; for ovulation, LH; for growth of the corpus luteum, LH; and for the midcycle surge of LH, estrogens.

28.25 The effect of rising but moderate levels of estrogens is negative feedback inhibition of the secretion of GnRH, LH, and FSH.

28.26 This is negative feedback, because the response is opposite to the stimulus. A reduced amount of negative feedback due to declining levels of estrogens and progesterone stimulates release of GnRH, which in turn increases the production and release of FSH and LH, ultimately stimulating the secretion of estrogens.

28.27 The *SRY* gene on the Y chromosome is responsible for the development of the gonads into testes.

28.28 The presence of dihydrotestosterone (DHT) stimulates differentiation of the external genitals in males; its absence allows differentiation of the external genitals in females.

CHAPTER 29

29.1 Capacitation is the group of functional changes in sperm that enable them to fertilize a secondary oocyte; the changes occur after the sperm have been deposited in the female reproductive tract.

29.2 A morula is a solid ball of cells; a blastocyst consists of a rim of cells (trophoblast) surrounding a cavity (blastocyst cavity) and an inner cell mass.

29.3 The blastocyst secretes digestive enzymes that eat away the endometrial lining at the site of implantation.

29.4 The decidua basalis helps form the maternal part of the placenta.

29.5 Implantation occurs during the secretory phase of the uterine cycle.

29.6 The bilaminar embryonic disc is attached to the trophoblast by the connecting stalk.

29.7 Gastrulation converts a bilaminar embryonic disc into a trilaminar embryonic disc.

29.8 The notochord induces mesodermal cells to develop into vertebral bodies and forms the nucleus pulposus of intervertebral discs.

29.9 The neural tube forms the brain and spinal cord; somites develop into skeletal muscles, connective tissue, and the vertebrae.

29.10 Chorionic villi help to bring the fetal and maternal blood vessels into close proximity.

29.11 The placenta participates in the exchange of materials between fetus and mother, serves as a protective barrier against many microbes, and stores nutrients.

29.12 As a result of embryonic folding, the embryo curves into a C-shape, various organs are brought into their eventual adult positions, and the primitive gut is formed.

29.13 Pharyngeal arches, clefts, and pouches give rise to structures of the head and neck.

29.14 Fetal weight doubles between the midfetal period and birth.

29.15 Amniocentesis is used primarily to detect genetic disorders, but it also provides information concerning the maturity (and survivability) of the fetus.

29.16 Early pregnancy tests detect elevated levels of human chorionic gonadotropin (hCG).

29.17 Relaxin increases the flexibility of the pubic symphysis and helps dilate the cervix of the uterus to ease delivery.

29.18 Complete dilation of the cervix marks the onset of the stage of expulsion.

29.19 Oxytocin also stimulates contraction of the uterus during delivery of a baby.

29.20 The odds that a child will have PKU are the same for each child—25%.

29.21 In incomplete dominance, neither member of an allelic pair is dominant; the heterozygote has a phenotype intermediate between the homozygous dominant and the homozygous recessive phenotypes.

29.22 A baby can have blood type O if each parent is heterozygous and has one *i* allele.

29.23 Hair color, height, and body build are some of the traits passed on by complex inheritance.

29.24 The female sex chromosomes are XX, and the male sex chromosomes are XY.

29.25 The chromosomes that are not sex chromosomes are called autosomes.

29.26 A red–green color-blind female has an X^cX^c genotype.

Glossary

Pronunciation Key

1. The most strongly accented syllable appears in capital letters, for example, bilateral (bī-LAT-er-al) and diagnosis (dī-ag-NŌ-sis).

2. If there is a secondary accent, it is noted by a prime ('), for example, constitution (kon'-sti-TOO-shun) and physiology (fiz'-ē-OL-ō-jē). Any additional secondary accents are also noted by a prime, for example, decarboxylation (dē'-kar-bok'-si-LĀ-shun).

3. Vowels marked by a line above the letter are pronounced with the long sound, as in the following common words:

ā as in *māke* | *ō* as in *pōle*
ē as in *bē* | *ū* as in *cūte*
ī as in *īvy*

4. Vowels not marked by a line above the letter are pronounced with the short sound, as in the following words:

a as in *above* or *at* | *o* as in *not*
e as in *bet* | *u* as in *bud*
i as in *sip*

5. Other vowel sounds are indicated as follows:

oy as in *oil*
oo as in *root*

6. Consonant sounds are pronounced as in the following words:

b as in *bat* | *m* as in *mother*
ch as in *chair* | *n* as in *no*
d as in *dog* | *p* as in *pick*
f as in *father* | *r* as in *rib*
g as in *get* | *s* as in *so*
h as in *hat* | *t* as in *tea*
j as in *jump* | *v* as in *very*
k as in *can* | *w* as in *welcome*
ks as in *tax* | *z* as in *zero*
kw as in *quit* | *zh* as in *lesion*
l as in *let*

Abdominal cavity Superior portion of the abdominopelvic cavity that contains the stomach, spleen, liver, gallbladder, most of the small intestine, and part of the large intestine.

Abdominopelvic cavity A cavity inferior to the diaphragm that is subdivided into a superior abdominal cavity and an inferior pelvic cavity.

Abduction (ab-DUK-shun) Movement away from the midline of the body.

Abortion (a-BOR-shun) The premature loss (spontaneous) or removal (induced) of the embryo or nonviable fetus; miscarriage due to a failure in the normal process of developing or maturing.

Abscess (AB-ses) A localized collection of pus and liquefied tissue in a cavity.

Absorption (ab-SORP-shun) Intake of fluids or other substances by cells of the skin or mucous membranes; the passage of digested foods from the gastrointestinal tract into blood or lymph.

Accessory duct A duct of the pancreas that empties into the duodenum about 2.5 cm (1 in.) superior to the ampulla of Vater (hepatopancreatic ampulla). Also called the duct of Santorini (san'-tō-RĒ-nē).

Acetabulum (as'-e-TAB-ū-lum) The rounded cavity on the external surface of the hip bone that receives the head of the femur.

Acetylcholine (as'-ē-til-KŌ-lēn) **(ACh)** A neurotransmitter liberated by many peripheral nervous system neurons and some central nervous system neurons. It is excitatory at neuromuscular junctions but inhibitory at some other synapses.

Achalasia (ak'-a-LĀ-zē-a) A condition, caused by malfunction of the myenteric plexus, in which the lower esophageal sphincter fails to relax normally as food approaches. A whole meal may become lodged in the esophagus and enter the stomach very slowly. Distension of the esophagus results in chest pain that is often confused with pain originating from the heart.

Acini (AS-i-nī) Groups of cells in the pancreas that secrete digestive enzymes.

Acquired immunodeficiency syndrome (AIDS) A fatal disease caused by the human immunodeficiency virus (HIV). Characterized by a positive HIV-antibody test, low helper T cell count, and certain indicator diseases (for example Kaposi's sarcoma, pneumocystis carinii pneumonia, tuberculosis, fungal diseases). Other symptoms include fever or night sweats, coughing, sore throat, fatigue, body aches, weight loss, and enlarged lymph nodes.

Acrosome (AK-rō-sōm) A lysosomelike organelle in the head of a sperm cell containing enzymes that facilitate the penetration of a sperm cell into a secondary oocyte.

Actin (AK-tin) A contractile protein that is part of thin filaments in muscle fibers.

Action potential (AP) An electrical signal that propagates along the membrane of a neuron or muscle fiber (cell); a rapid change in membrane potential that involves a depolarization followed by a repolarization. Also called a nerve action potential or nerve impulse as it relates to a neuron, and a muscle action potential as it relates to a muscle fiber.

Activation (ak'-ti-VĀ-shun) **energy** The minimum amount of energy required for a chemical reaction to occur.

Active transport The movement of substances across cell membranes against a concentration gradient, requiring the expenditure of cellular energy (ATP).

Adaptation (ad'-ap-TĀ-shun) The adjustment of the pupil of the eye to changes in light intensity. The property by which a sensory neuron relays a decreased frequency of action potentials from a receptor, even though the strength of the stimulus remains constant; the decrease in perception of a sensation over time while the stimulus is still present.

Adduction (ad-DUK-shun) Movement toward the midline of the body.

Adenosine (a-DEN-ō-sēn trī-FOS-fāt) **triphosphate (ATP)** The main energy currency in living cells; used to transfer the chemical energy needed for metabolic reactions. ATP consists of the purine base adenine and the five-carbon sugar ribose, to which are added, in linear array, three phosphate groups.

Adhesion (ad-HĒ-zhun) Abnormal joining of parts to each other.

Adipocyte (AD-i-pō-sīt) Fat cell, derived from a fibroblast.

Adipose (AD-i-pōz) **tissue** Tissue composed of adipocytes specialized for triglyceride storage and present in the form of soft pads between various organs for support, protection, and insulation.

Adrenal cortex (a-DRĒ-nal KOR-teks) The outer portion of an adrenal gland, divided into three zones; the zona glomerulosa secretes mineralocorticoids, the zona fasciculata secretes glucocorticoids, and the zona reticularis secretes androgens.

Adrenal glands Two glands located superior to each kidney. Also called the suprarenal (soo'-pra-RĒ-nal) glands.

Adrenal medulla (me-DUL-la) The inner part of an adrenal gland, consisting of cells that secrete epinephrine, norepinephrine, and a small amount of dopamine in response to stimulation by sympathetic preganglionic neurons.

Adrenergic (ad′-ren-ER-jik) **neuron** A neuron that releases epinephrine (adrenaline) or norepinephrine (noradrenaline) as its neurotransmitter.

Adrenocorticotropic hormone (ad-rē-nō-kor-ti-kō-TRŌP-ik) **(ACTH)** A hormone produced by the anterior pituitary that influences the production and secretion of certain hormones of the adrenal cortex.

Adventitia (ad-ven-TISH-a) The outermost covering of a structure or organ; the superficial coat of the ureters and the posterior and inferior surfaces of the urinary bladder.

Aerobic (ār-Ō-bik) Requiring molecular oxygen.

Aerobic (ār-Ō-bik) **respiration** The production of ATP (30 or 32 molecules) from the complete oxidation of pyruvic acid in mitochondria. Carbon dioxide, water, and heat are also produced.

Afferent arteriole A blood vessel of a kidney that divides into the capillary network called a glomerulus; there is one afferent arteriole for each glomerulus.

Agglutination (a-gloo-ti-NĀ-shun) Clumping of microorganisms or blood cells, typically due to an antigen–antibody reaction.

Aggregated lymphatic follicles Clusters of lymph nodules that are most numerous in the ileum. Also called Peyer's (PĪ-erz) patches.

Albinism (AL-bin-izm) Abnormal, nonpathological, partial, or total absence of pigment in skin, hair, and eyes.

Aldosterone (al-DOS-ter-ōn) A mineralocorticoid produced by the adrenal cortex that promotes sodium and water reabsorption by the kidneys and potassium excretion in urine.

Allantois (a-LAN-tō-is) A small, vascularized outpouching of the yolk sac that serves as an early site for blood formation and development of the urinary bladder.

Alleles (a-LĒLZ) Alternate forms of a single gene that control the same inherited trait (such as type A blood) and are located at the same position on homologous chromosomes.

Allergen (AL-er-jen) An antigen that evokes a hypersensitivity reaction.

All-or-none-principle If a stimulus depolarizes a neuron to threshold, the neuron fires at its maximum voltage (all); if threshold is not reached, the neuron does not fire at all (none). Given above threshold, stronger stimuli do not produce stronger action potentials.

Alopecia (al-ō-PĒ-shē-a) The partial or complete lack of hair as a result of factors such as genetics, aging, endocrine disorders, chemotherapy, and skin diseases.

Alpha (AL-fa) **cell** A type of cell in the pancreatic islets (islets of Langerhans) in the pancreas that secretes the hormone glucagon. Also termed an A cell.

Alpha (α) receptor A type of receptor for norepinephrine and epinephrine; present on visceral effectors innervated by sympathetic postganglionic neurons.

Alveolar duct Branch of a respiratory bronchiole around which alveoli and alveolar sacs are arranged.

Alveolar macrophage Highly phagocytic cell found in the alveolar walls of the lungs. Also called a dust cell.

Alveolar sac A cluster of alveoli that share a common opening.

Alveolus (al-VĒ-ō-lus) A small hollow or cavity; an air sac in the lungs; milk-secreting portion of a mammary gland. Plural is alveoli (al-VĒ-ō-lī).

Alzheimer's (ALTZ-hī-mers) **disease (AD)** Disabling neurological disorder characterized by dysfunction and death of specific cerebral neurons, resulting in widespread intellectual impairment, personality changes, and fluctuations in alertness.

Amenorrhea (ā-men-ō-RĒ-a) Absence of menstruation.

Amnesia (am-NĒ-zē-a) A lack or loss of memory.

Amnion (AM-nē-on) A thin, protective fetal membrane that develops from the epiblast; holds the fetus suspended in amniotic fluid. Also called the "bag of waters."

Amniotic (am′-nē-OT-ic) **fluid** Fluid within the amniotic cavity derived from maternal blood and wastes from the fetus.

Amphiarthrosis (am′-fē-ar-THRŌ-sis) A slightly movable joint, in which the articulating bony surfaces are separated by fibrous connective tissue or fibrocartilage to which both are attached; types are syndesmosis and symphysis.

Ampulla (am-PUL-la) A saclike dilation of a canal or duct. Dilated terminal portion of the ductus deferens. Widest, longest portion of the uterine tube.

Anabolism (a-NAB-ō-lizm) Synthetic, energy-requiring reactions whereby small molecules are built up into larger ones.

Anaerobic (an-ar-Ō-bik) Not requiring oxygen.

Anal canal The last 2 or 3 cm (1 in.) of the rectum; opens to the exterior through the anus.

Anal column A longitudinal fold in the mucous membrane of the anal canal that contains a network of arteries and veins.

Anal triangle The subdivision of the female or male perineum that contains the anus.

Analgesia (an-an-JĒ-zē-a) Pain relief; absence of the sensation of pain.

Anaphase (AN-a-fāz) The third stage of mitosis in which the chromatids that have separated at the centromeres move to opposite poles of the cell.

Anastomosis (a-nas′-tō-MŌ-sis) An end-to-end union or joining of blood vessels, lymphatic vessels, or nerves. The plural is anastomoses.

Anatomic (respiratory) dead space Spaces of the nose, pharynx, larynx, trachea, bronchi, and bronchioles totaling about 150 mL of the 500 mL in a quiet breath (tidal volume); air in the anatomic dead space does not reach the alveoli to participate in gas exchange.

Anatomical position A position of the body universally used in anatomical descriptions in which the body is erect, the head is level, the eyes face forward, the upper limbs are at the sides, the palms face forward, and the feet are flat on the floor.

Anatomy The structure or study of the structure of the body and the relationship of its parts to each other.

Androgens (AN-drō-jenz) Masculinizing sex hormones produced by the testes in males and the adrenal cortex in both sexes; responsible for libido (sexual desire); the two main androgens are testosterone and dihydrotestosterone.

Anemia (a-NĒ-mē-a) Condition of the blood in which the number of functional red blood cells or their hemoglobin content is below normal.

Angina pectoris A pain in the chest related to reduced coronary circulation due to coronary artery disease (CAD) or spasms of vascular smooth muscle in coronary arteries.

Angiogenesis (an′-jē-ō-JEN-e-sis) The formation of blood vessels in the extraembryonic mesoderm of the yolk sac, connecting stalk, and chorion at the beginning of the third week of development.

Antagonist (an-TAG-ō-nist) A muscle that has an action opposite that of the prime mover (agonist) and yields to the movement of the prime mover.

Antagonistic (an-tag-ō-NIST-ik) **effect** A hormonal interaction in which the effect of one hormone on a target cell is opposed by another hormone.

Anterior pituitary Anterior lobe of the pituitary gland. Also called the adenohypophysis (ad′-e-nō-hī-POF-i-sis).

Anterior root The structure composed of axons of motor (efferent) neurons that emerges from the anterior aspect of the spinal cord and extends laterally to join a posterior root, forming a spinal nerve. Also called a ventral root.

Anterolateral (an′-ter-ō-LAT-er-al) **pathway** Sensory pathway that conveys information related to pain, temperature, tickle, and itch. Also called the spinothalamic pathway.

Antibody (AN-ti-bod′-ē) **(Ab)** A protein produced by plasma cells in response to a specific antigen; the antibody combines with that antigen to neutralize, inhibit, or destroy it. Also called an immunoglobulin (im-ū-nō-GLOB-ū-lin) or Ig.

Anticoagulant (an-tī-cō-AG-ū-lant) A substance that can delay, suppress, or prevent the clotting of blood.

Antidiuretic (an′-ti-dī-ū-RET-ik) Substance that inhibits urine formation.

Antidiuretic hormone (ADH) Hormone produced by neurosecretory cells in the paraventricular and supraoptic nuclei of the hypothalamus that stimulates water reabsorption from kidney tubule cells into the blood and vasoconstriction of arterioles. Also called vasopressin (vāz-ō-PRES-in).

Antigen (AN-ti-jen) **(Ag)** A substance that has immunogenicity (the ability to provoke an immune response) and reactivity (the ability to react with the antibodies or cells that result from the immune response); derived from the term antibody generator. Also termed a complete antigen.

Antigen-presenting cell (APC) Special class of migratory cell that processes and presents antigens to T cells during an immune response; APCs include macrophages, B cells, and dendritic cells, which are present in the skin, mucous membranes, and lymph nodes.

Antioxidant A substance that inactivates oxygen derived free radicals. Examples are selenium, zinc, beta carotene, and vitamins C and E.

Antioxidant vitamins Vitamins that inactivate oxygen free radicals; vitamins C and E and the provitamin beta-carotene.

Antrum (AN-trum) Any nearly closed cavity or chamber, especially one within a bone, such as a sinus. Cavity in the center of a secondary follicle.

Anus (Ā-nus) The distal end and outlet of the rectum.

Aortic (ā-OR-tik) **body** Cluster of chemoreceptors on or near the arch of the aorta that respond to changes in blood levels of oxygen, carbon dioxide, and hydrogen ions (H^+).

Aortic reflex A reflex that helps maintain normal systemic blood pressure; initiated by baroreceptors in the wall of the ascending aorta and arch of the aorta. Nerve impulses from aortic baroreceptors reach the cardiovascular center via sensory axons of the vagus (X) nerves.

Apex (Ā-peks) The pointed end of a conical structure, such as the apex of the heart.

Aphasia (a-FĀ-zē-a) Loss of ability to express oneself properly through speech or loss of verbal comprehension.

Apnea (AP-nē-a) Temporary cessation of breathing.

Apocrine (AP-ō-krin) **sweat gland** A type of gland in which the secretory products gather at the free end of the secreting cell and are pinched off, along with some of the cytoplasm, to become the secretion, as in mammary glands.

Aponeurosis (ap-ō-noo-RŌ-sis) A sheetlike tendon joining one muscle with another or with bone.

Apoptosis (ap′-ōp-TŌ-sis *or* ap-ō-TŌ-sis) Programmed cell death; a normal type of cell death that removes unneeded cells during embryological development, regulates the number of cells in tissues, and eliminates many potentially dangerous cells such as cancer cells.

Appositional (a-pō-ZISH-o-nal) **growth** Growth due to surface deposition of material, as in the growth in diameter of cartilage and bone. Also called exogenous (eks-OJ-e-nus) growth.

Aqueduct (AK-we-dukt) **of the midbrain** A channel through the midbrain connecting the third and fourth ventricles and containing cerebrospinal fluid. Also called the cerebral aqueduct.

Aqueous humor (ĀK-wē-us HŪ-mer) The watery fluid, similar in composition to cerebrospinal fluid, that fills the anterior cavity of the eye.

Arachnoid (a-RAK-noyd) **mater** The middle of the three meninges (coverings) of the brain and spinal cord. Also termed the arachnoid.

Arachnoid villus (VIL-us) Berrylike tuft of the arachnoid mater that protrudes into the superior sagittal sinus and through which cerebrospinal fluid is reabsorbed into the bloodstream.

Arbor vitae (AR-bor VĪ-tē) The white matter tracts of the cerebellum, which have a treelike appearance when seen in midsagittal section.

Arch of the aorta The most superior portion of the aorta, lying between the ascending and descending segments of the aorta.

Areola (a-RĒ-ō-la) The pigmented ring around the nipple of the breast. Any tiny space in a tissue.

Arousal (a-ROW-zal) Awakening from sleep, a response due to stimulation of the reticular activating system (RAS).

Arrector pili (a-REK-tor PĪ-lē) Smooth muscles attached to hairs; contraction pulls the hairs into a vertical position, resulting in "goose bumps."

Arrhythmia An irregular heart rhythm. Also called a dysrhythmia.

Arteriole (ar-TĒ-rē-ōl) A small, almost microscopic, artery that delivers blood to a capillary.

Arteriosclerosis (ar-tē-rē-ō-skle-RŌ-sis) Group of diseases characterized by thickening of the walls of arteries and loss of elasticity.

Artery (AR-ter-ē) A blood vessel that carries blood away from the heart.

Arthritis (ar-THRĪ-tis) Inflammation of a joint.

Arthrology (ar-THROL-ō-jē) The study or description of joints.

Arthroplasty (AR-thrō-plas′-tē) Surgical replacement of joints, for example, the hip and knee joints.

Arthroscopy (ar-THROS-kō-pē) A procedure for examining the interior of a joint, usually the knee, by inserting an arthroscope into a small incision; used to determine extent of damage, remove torn cartilage, repair cruciate ligaments, and obtain samples for analysis.

Arthrosis (ar-THRŌ-sis) A joint or articulation.

Articular (ar-TIK-ū-lar) **capsule** Sleevelike structure around a synovial joint composed of a fibrous capsule and a synovial membrane. Also called a joint capsule.

Articular cartilage (KAR-ti-lij) Hyaline cartilage attached to articular bone surfaces.

Articular disc Fibrocartilage pad between articular surfaces of bones of some synovial joints. Also called a meniscus (men-IS-kus).

Articulation (ar-tik-ū-LĀ-shun) A joint; a point of contact between bones, cartilage and bones, or teeth and bones.

Arytenoid (ar′-i-TĒ-noyd) **cartilages** A pair of small, pyramidal cartilages of the larynx that attach to the vocal folds and intrinsic pharyngeal muscles and can move the vocal folds.

Ascending colon (KŌ-lon) The part of the large intestine that passes superiorly from the cecum to the inferior border of the liver, where it bends at the right colic (hepatic) flexure to become the transverse colon.

Ascites (a-SĪ-tēz) Abnormal accumulation of serous fluid in the peritoneal cavity.

Association areas Large cortical regions on the lateral surfaces of the occipital, parietal, and temporal lobes and on the frontal lobes anterior to the motor areas connected by many motor and sensory axons to other parts of the cortex; concerned with motor patterns, memory, concepts of word-hearing and word-seeing, reasoning, will, judgment, and personality traits.

Asthma (AZ-ma) Usually allergic reaction characterized by smooth muscle spasms in bronchi resulting in wheezing and difficult breathing. Also called bronchial asthma.

Astigmatism (a-STIG-ma-tizm) An irregularity of the lens or cornea of the eye causing the image to be out of focus and producing faulty vision.

Astrocyte (AS-trō-sīt) A neuroglial cell having a star shape that participates in brain development and the metabolism of neurotransmitters, helps form the blood–brain barrier, helps maintain the proper balance of K^+ for generation of nerve impulses, and provides a link between neurons and blood vessels.

Ataxia (a-TAK-sē-a) A lack of muscular coordination, lack of precision.

Atherosclerosis (ath-er-ō-skle-RŌ-sis) A progressive disease characterized by the formation in the walls of large and medium-sized arteries of lesions called atherosclerotic plaques.

Atherosclerotic (ath-er-ō-skle-RO-tik) **plaque** A lesion that results from accumulated cholesterol and smooth muscle fibers (cells) of the tunica media of an artery; may become obstructive.

Atom Unit of matter that makes up a chemical element; consists of a nucleus (containing positively charged protons and uncharged neutrons) and negatively charged electrons that orbit the nucleus.

Atresia (a-TRĒ-zē-a) Degeneration and reabsorption of an ovarian follicle before it fully matures and ruptures; abnormal closure of a passage, or absence of a normal body opening.

Atrial fibrillation (Ā-trē-al fib-ri-LĀ-shun) **(AF)** Asynchronous contraction of cardiac muscle fibers in the atria that results in the cessation of atrial pumping.

Atrial natriuretic peptide (ANP) Peptide hormone, produced by the atria of the heart in response to stretching, that inhibits aldosterone production and thus lowers blood pressure; causes natriuresis, increased urinary excretion of sodium.

Atrioventricular (ā′-trē-ō-ven-TRIK-ū-lar) **(AV) bundle** The part of the conduction system of the heart that begins at the atrioventricular (AV) node, passes through the cardiac skeleton separating the atria and the ventricles, then extends a short distance down the interventricular septum before splitting into right and left bundle branches. Also called the bundle of His (HIZ).

Atrioventricular (AV) node The part of the conduction system of the heart made up of a compact mass of conducting cells located in the septum between the two atria.

Atrioventricular (AV) valve A heart valve made up of membranous flaps or cusps that allows blood to flow in one direction only, from an atrium into a ventricle.

Atrium (Ā-trē-um) A superior chamber of the heart. Plural is atria.

Auditory ossicle (Aw-di-tō-rē OS-si-kul) One of the three small bones of the middle ear called the malleus, incus, and stapes.

Auditory tube The tube that connects the middle ear with the nose and nasopharynx region of the throat. Also called the eustachian (ū-STĀ-shun or ū-STĀ-kē-an) tube or pharyngotympanic tube.

Auscultation (aws-kul-TĀ-shun) Examination by listening to sounds in the body.

Autoimmunity An immunological response against a person's own tissues.

Autolysis (aw-TOL-i-sis) Self-destruction of cells by their own lysosomal digestive enzymes after death or in a pathological process.

Autonomic ganglion (aw′-tō-NOM-ik GANG-lē-on) A cluster of cell bodies of sympathetic or parasympathetic neurons located outside the central nervous system.

Autonomic nervous system (ANS) The part of the peripheral nervous system that conveys output to smooth muscle, cardiac muscle, and glands. Consists of two main divisions (sympathetic nervous system and parasympathetic nervous system) and an enteric nervous system. So named because this part of the nervous system was thought to be self-governing or spontaneous.

Autonomic plexus (PLEK-sus) A network of sympathetic and parasympathetic axons; examples are the cardiac, celiac, and pelvic plexuses, which are located in the thorax, abdomen, and pelvis, respectively.

Autophagy (aw-TOF-a-jē) Process by which worn-out organelles are digested within lysosomes.

Autopsy The examination of the body after death.

Autorhythmic (aw′-tō-RITH-mik) **fibers** Cells that repeatedly and rhythmically generate action potentials.

Autorhythmicity (aw′-tō-rith-MISS-i-tē) The ability to repeatedly generate spontaneous action potentials.

Autosome (AW-tō-sōm) Any chromosome other than the X and Y chromosomes (sex chromosomes).

Axon (AK-son) The usually single, long process of a nerve cell that propagates a nerve impulse toward the axon terminals.

Axon terminal Terminal branch of an axon where synaptic vesicles undergo exocytosis to release neurotransmitter molecules. Also called telodendria (tel′-o-DEN-drea).

Axoplasm Cytoplasm of an axon.

Axosomatic From axon to cell body.

B cell Lymphocyte that begins development in primary lymphatic organs and completes it in red bone marrow, a process that occurs throughout life.

Babinski sign Extension of the great toe, with or without fanning of the other toes, in response to stimulation of the outer margin of the sole; normal up to 18 months of age and indicative of damage to descending motor pathways such as the corticospinal tracts after that age.

Ball-and-socket joint A synovial joint in which the rounded surface of one bone moves within a cup-shaped depression or socket of another bone, as in the shoulder or hip joint. Also called a spheroid (SFĒ-royd) joint.

Baroreceptor (bar′-ō-rē-SEP-tor) Neuron capable of responding to changes in blood, air, or fluid pressure. Also called a stretch receptor.

Basal nuclei Paired clusters of gray matter deep in each cerebral hemisphere including the globus pallidus, putamen, and caudate nucleus.

Base Posterior aspect of the heart opposite the apex and formed by the atria.

Basement membrane Thin, extracellular layer between epithelium and connective tissue consisting of a basal lamina and a reticular lamina.

Basilar (BĀS-i-lar) **membrane** A membrane in the cochlea of the internal ear that separates the cochlear duct from the scala tympani and on which the spiral organ (organ of Corti) rests.

Basophil (BĀ-sō-fil) A type of white blood cell characterized by a pale nucleus and large granules that stain blue-purple with basic dyes.

Beta (BĀ-ta) **cell** A type of cell in the pancreatic islets (islets of Langerhans) in the pancreas that secretes the hormone insulin. Also called a B cell.

Beta (β) receptor A type of adrenergic receptor for epinephrine and norepinephrine; found on visceral effectors innervated by sympathetic postganglionic neurons.

Bicuspid (mitral) valve Atrioventricular (AV) valve on the left side of the heart. Also called the mitral valve or left atrioventricular valve.

Bile (BĪL) A secretion of the liver consisting of water, bile salts, bile pigments, cholesterol, lecithin, and several ions that emulsifies lipids prior to their digestion.

Bilirubin (bil-ē-ROO-bin) An orange pigment that is one of the end products of hemoglobin breakdown in the hepatocytes and is excreted as a waste material in bile.

Blastocyst (BLAS-tō-sist) In the development of an embryo, a hollow ball of cells that consists of a blastocele (the internal cavity), trophoblast (outer cells), and inner cell mass.

Blastocyst (BLAS-tō-sist) **cavity** The fluid-filled cavity within the blastocyst.

Blastomere (BLAS-tō-mērz) One of the cells resulting from the cleavage of a fertilized ovum.

Blind spot Area in the retina at the end of the optic (II) nerve in which there are no photoreceptors. Also called the optic disc.

Blood The fluid that circulates through the heart, arteries, capillaries, and veins and that constitutes the chief means of transport within the body.

Blood clot A gel that consists of the formed elements of blood trapped in a network of insoluble protein fibers.

Blood island Isolated mass of mesoderm derived from angioblasts and from which blood vessels develop.

Blood plasma The extracellular fluid found in blood vessels; blood minus the formed elements.

Blood pressure (BP) Force exerted by blood against the walls of blood vessels due to contraction of the heart and influenced by the elasticity of the vessel walls; clinically, a measure of the pressure in arteries during ventricular systole and ventricular diastole.

Blood reservoir (REZ-er-vwar) Systemic veins and venules that contain large amounts of blood that can be moved quickly to parts of the body requiring the blood.

Blood–brain barrier (BBB) A barrier consisting of specialized brain capillaries and astrocytes that prevents the passage of materials from the blood to the cerebrospinal fluid and brain.

Blood–testis barrier A barrier formed by Sertoli cells that prevents an immune response against antigens produced by spermatogenic cells by isolating the cells from the blood.

Body cavity A space within the body that contains various internal organs.

Bolus (BŌ-lus) A soft, rounded mass, usually food, that is swallowed.

Bone remodeling Replacement of old bone by new bone tissue.

Bony labyrinth (LAB-i-rinth) A series of cavities within the petrous portion of the temporal bone forming the vestibule, cochlea, and semicircular canals of the inner ear.

Brachial plexus (BRĀ-kē-al PLEK-sus) A network of nerve axons of the anterior rami of spinal nerves C5, C6, C7, C8, and T1. The nerves that emerge from the brachial plexus supply the upper limb.

Bradycardia (brād′-i-KAR-dē-a) A slow resting heart or pulse rate (under 50 beats per minute).

Brain The part of the central nervous system contained within the cranial cavity.

Brainstem The portion of the brain immediately superior to the spinal cord, made up of the medulla oblongata, pons, and midbrain.

Brain waves Electrical signals that can be recorded from the skin of the head due to electrical activity of brain neurons.

Broad ligament A double fold of parietal peritoneum attaching the uterus to the side of the pelvic cavity.

Bronchi (BRON-kī) Division of the trachea at the superior border of the fifth thoracic vertebra that extends to the right lung.

Bronchial (BRON-kē-al) **tree** The trachea, bronchi, and their branching structures up to and including the terminal bronchioles.

Bronchiole (BRONG-kē-ōl) Branch of a tertiary bronchus further dividing into terminal bronchioles (distributed to lobules of the lung), which divide into respiratory bronchioles (distributed to alveolar sacs).

Bronchitis (brong-KĪ-tis) Inflammation of the mucous membrane of the bronchial tree; characterized by hypertrophy and hyperplasia of seromucous glands and goblet cells that line the bronchi, which results in a productive cough.

Bronchopulmonary (brong′-kō-PUL-mō-ner-ē) **segment** One of the smaller divisions of a lobe of a lung supplied by its own segmental bronchus.

Buccal Pertaining to the cheek or mouth.

Buffer system A weak acid and the salt of that acid (which functions as a weak base). Buffers prevent drastic changes in pH by converting strong acids and bases to weak acids and bases.

Bulb of the penis Expanded portion of the base of the corpus spongiosum penis.

Bulbourethral (bul′-bō-ū-RĒ-thral) **gland** One of a pair of glands located inferior to the prostate on either side of the urethra that secretes an alkaline fluid into the cavernous urethra. Also called a Cowper's (KOW-perz) gland.

Bulimia (boo-LĒ-mē-a) A disorder characterized by overeating at least twice a week followed by purging by self-induced vomiting, strict dieting or fasting, vigorous exercise, or use of laxatives or diuretics. Also called binge–purge syndrome.

Bulk-phase endocytosis A process by which most body cells can ingest membrane-surrounded droplets of interstitial fluid.

Burn Tissue damage caused by excessive heat, electricity, radioactivity, or corrosive chemicals that denature (break down) proteins in the skin.

Bursa (BUR-sa) A sac or pouch of synovial fluid located at friction points, especially around joints.

Bursitis (bur-SĪ-tis) Inflammation of a bursa.

Calcaneal tendon (kal-KĀ-nē-al) The tendon of the soleus, gastrocnemius, and plantaris muscles at the back of the heel. Also called the Achilles (a-KIL-ēz) tendon.

Calcification (kal′-si-fi-KĀ-shun) Deposition of mineral salts, primarily hydroxyapatite, in a framework formed by collagen fibers in which the tissue hardens. Also called mineralization (min′-e-ral-i-ZĀ-shun).

Calcitonin (kal-si-TŌ-nin) **(CT)** A hormone produced by the parafollicular cells of the thyroid gland that can lower the amount of blood calcium and phosphates by inhibiting bone resorption (breakdown of bone extracellular matrix) and by accelerating uptake of calcium and phosphates into bone matrix.

Callus (KAL-lus) An abnormal thickening of the stratum corneum.

Canaliculi (kan′-a-LIK-ū-lī). Small channels or canals, as in bones, where they connect lacunae. Singular is called canaliculus (kan′-a-LIK-ū -lus).

Cancer A group of diseases characterized by uncontrolled or abnormal cell division.

Capacitation (ka-pas′-i-TĀ-shun) The functional changes that sperm undergo in the female reproductive tract that allow them to fertilize a secondary oocyte.

Capillary (KAP-i-lar′-ē) A microscopic blood vessel located between an arteriole and venule through which materials are exchanged between blood and interstitial fluid.

Carbohydrate Organic compound consisting of carbon, hydrogen, and oxygen; the ratio of hydrogen to oxygen atoms is usually 2:1. Examples include sugars, glycogen, starches, and glucose.

Cardiac circulation The pathway followed by the blood from the ascending aorta through the blood vessels supplying the heart and returning to the right atrium. Also called coronary circulation.

Cardiac conduction system A group of autorhythmic cardiac muscle fibers that generates and distributes electrical impulses to stimulate coordinated contraction of the heart chambers; includes the sinoatrial (SA) node, the atrioventricular (AV) node, the atrioventricular (AV) bundle, the right and left bundle branches, and the Purkinje fibers.

Cardiac cycle A complete heartbeat consisting of systole (contraction) and diastole (relaxation) of both atria plus systole and diastole of both ventricles.

Cardiac muscle tissue Striated muscle fibers (cells) that form the wall of the heart; stimulated by an intrinsic conduction system and autonomic motor neurons.

Cardiac notch An angular notch in the anterior border of the left lung into which part of the heart fits.

Cardiac output (CO) Volume of blood ejected from the left ventricle (or the right ventricle) into the aorta (or pulmonary trunk) each minute.

Cardinal ligament A ligament of the uterus, extending laterally from the cervix and vagina as a continuation of the broad ligament. Also called the lateral cervical ligament.

Cardiogenic (kar-dē-ō-JEN-ik) **area** A group of mesodermal cells in the head end of an embryo that gives rise to the heart.

Cardiology (kar-dē-OL-ō-jē) The study of the heart and diseases associated with it.

Cardiovascular (kar-dē-ō-VAS-kū-lar) **(CV) center** Groups of neurons scattered within the medulla oblongata that regulate heart rate, force of contraction, and blood vessel diameter.

Cardiovascular (kar-dē-ō-VAS-kū-lar) **system** Body system that consists of blood, the heart, and blood vessels.

Carotene (KAR-ō-tēn) Antioxidant precursor of vitamin A, which is needed for synthesis of photopigments; yellow-orange pigment present in the stratum corneum of the epidermis. Accounts for the yellowish coloration of skin. Also termed beta-carotene.

Carotid (ka-ROT-id) **body** Cluster of chemoreceptors on or near the carotid sinus that respond to changes in blood levels of oxygen, carbon dioxide, and hydrogen ions.

Carotid sinus (SĪ-nus) A dilated region of the internal carotid artery just superior to where it branches from the common carotid artery; it contains baroreceptors that monitor blood pressure.

Carpals The eight bones of the wrist. Also called carpal bones.

Carpus (KAR-pus) A collective term for the eight bones of the wrist.

Cartilage (KAR-ti-lij) A type of connective tissue consisting of chondrocytes in lacunae embedded in a dense network of collagen and elastic fibers and an extracellular matrix of chondroitin sulfate.

Cartilaginous (kar′-til-LAJ-in-us) **joint** A joint without a synovial (joint) cavity where the articulating bones are held tightly together by cartilage, allowing little or no movement.

Catabolism (ka-TAB-ō-lizm) Chemical reactions that break down complex organic compounds into simple ones, with the net release of energy.

Catalyst Chemical compounds that speed up chemical reactions by lowering the activation energy needed for a reaction to occur.

Cataract (KAT-a-rakt) Loss of transparency of the lens of the eye or its capsule or both.

Cauda equina (KAW-da ē-KWĪ-na) A tail-like array of roots of spinal nerves at the inferior end of the spinal cord.

Cecum (SĒ-kum) A blind pouch at the proximal end of the large intestine that attaches to the ileum.

Celiac plexus (SĒ-lē-ak PLEK-sus) A large mass of autonomic ganglia and axons located at the level of the superior part of the first lumbar vertebra. Also called the solar plexus.

Cell The basic structural and functional unit of all organisms; the smallest structure capable of performing all activities vital to life.

Cell biology The study of cellular structure and function. Also called cytology.

Cell cycle Growth and division of a single cell into two identical cells; consists of interphase and cell division.

Cell division Process by which a cell reproduces itself that consists of a nuclear division (mitosis) and a cytoplasmic division (cytokinesis); types include somatic and reproductive cell division.

Cell junction Point of contact between plasma membranes of tissue cells.

Cellular respiration Oxidation of glucose to produce ATP; consists of glycolysis, formation of acetyl coenzyme A, the Krebs cycle, and the electron transport chain.

Cementum (se-MEN-tum) Calcified tissue covering the root of a tooth.

Central canal A microscopic tube running the length of the spinal cord in the gray commissure. A circular channel running longitudinally in the center of an osteon (haversian system) of mature compact bone, containing blood and lymphatic vessels and nerves. Also called a haversian (ha-VER-shun) canal.

Central nervous system (CNS) That portion of the nervous system that consists of the brain and spinal cord.

Centrioles (SEN-trē-ōlz) Paired, cylindrical structures of a centrosome, each consisting of a ring of microtubules and arranged at right angles to each other.

Centromere (SEN-trō-mēr) The constricted portion of a chromosome where the two chromatids are joined; serves as the point of attachment for the microtubules that pull chromatids during anaphase of cell division.

Centrosome (SEN-trō-sōm) A dense network of small protein fibers near the nucleus of a cell, containing a pair of centrioles and pericentriolar material.

Cerebellar peduncle (ser-e-BEL-ar ped-DUNG-kul) A bundle of nerve axons connecting the cerebellum with the brainstem.

Cerebellum (ser′-e-BEL-um) The part of the brain lying posterior to the medulla oblongata and pons; governs balance and coordinates skilled movements.

Cerebral cortex The surface of the cerebral hemispheres, 2–4 mm thick, consisting of gray matter; arranged in six layers of neuronal cell bodies in most areas.

Cerebral peduncle One of a pair of nerve axon bundles located on the anterior surface of the midbrain, conducting nerve impulses between the pons and the cerebral hemispheres.

Cerebrospinal (se-rē′-brō-SPĪ-nal) **fluid (CSF)** A fluid produced by ependymal cells that cover choroid plexuses in the ventricles of the brain; the fluid circulates in the ventricles, the central canal, and the subarachnoid space around the brain and spinal cord.

Cerebrovascular (se-rē-brō-VAS-kū-lar) **accident (CVA)** Destruction of brain tissue (infarction) resulting from obstruction or rupture of blood vessels that supply the brain. Also called a stroke or brain attack.

Cerebrum (se-RĒ-brum) The two hemispheres of the forebrain (derived from the telencephalon), making up the largest part of the brain.

Cerumen (se-ROO-men) Waxlike secretion produced by ceruminous glands in the external auditory meatus (ear canal). Also termed ear wax.

Ceruminous (se-RŪ-min-us) **gland** A modified sudoriferous (sweat) gland in the external auditory meatus that secretes cerumen (ear wax).

Cervical plexus A network formed by nerve axons from the anterior rami of the first four cervical nerves and receiving gray rami communicantes from the superior cervical ganglion.

Cervix (SER-viks) Neck; any constricted portion of an organ, such as the inferior cylindrical part of the uterus.

Chemical reaction The formation of new chemical bonds or the breaking of old chemical bonds between atoms.

Chemistry (KEM-is-trē) The science of the structure and interactions of matter.

Chemoreceptor (kē′-mō-rē-SEP-tor) Sensory receptor that detects the presence of a specific chemical.

Chief cell The secreting cell of a gastric gland that produces pepsinogen, the precursor of the enzyme pepsin, and the enzyme gastric lipase. Also called a zymogenic (zī′-mō-JEN-ik) cell. Cell in the parathyroid glands that secretes parathyroid hormone (PTH). Also called a principal cell.

Cholecystectomy (kō′-lē-sis-TEK-tō-mē) Surgical removal of the gallbladder.

Cholesterol (kō-LES-te-rol) Classified as a lipid, the most abundant steroid in animal tissues; located in cell membranes and used for the synthesis of steroid hormones and bile salts.

Cholinergic (kō-lin-ER-jik) **neuron** A neuron that liberates acetylcholine as its neurotransmitter.

Chondrocyte (KON-drō-sīt) Cell of mature cartilage.

Chondroitin (kon-DROY-tin) **sulfate** An amorphous extracellular matrix material found outside connective tissue cells.

Chordae tendineae (KOR-dē TEN-din-ē-ē) Tendonlike, fibrous cords that connect atrioventricular valves of the heart with papillary muscles.

Chorion (KŌ-rē-on) The most superficial fetal membrane that becomes the principal embryonic portion of the placenta; serves a protective and nutritive function.

Chorionic villi (kō-rē-ON-ik VIL-lī) Fingerlike projections of the chorion that grow into the decidua basalis of the endometrium and contain fetal blood vessels.

Chorionic villi sampling (CVS) The removal of a sample of chorionic villus tissue by means of a catheter to analyze the tissue for prenatal genetic defects.

Choroid (KŌ-royd) One of the vascular coats of the eyeball.

Choroid plexus (PLEK-sus) A network of capillaries located in the roof of each of the four ventricles of the brain; ependymal cells around choroid plexuses produce cerebrospinal fluid.

Chromatid (KRŌ-ma-tid) One of a pair of identical connected nucleoprotein strands that are joined at the centromere and separate during cell division, each becoming a chromosome of one of the two daughter cells.

Chromaffin (KRŌ-maf-in) **cell** Cell that has an affinity for chrome salts, due in part to the presence of the precursors of the neurotransmitter epinephrine; found, among other places, in the adrenal medulla.

Chromatin (KRŌ-ma-tin) The threadlike mass of genetic material, consisting of DNA and histone proteins, that is present in the nucleus of a nondividing or interphase cell.

Chromatolysis (kro′-ma-TOL-i-sis) The breakdown of Nissl bodies into finely granular masses in the cell body of a neuron whose axon has been damaged.

Chromosome (KRŌ-mō-sōm) One of the small, threadlike structures in the nucleus of a cell, normally 46 in a human diploid cell, that bears the genetic material; composed of DNA and proteins (histones) that form a delicate chromatin thread during interphase; becomes packaged into compact rodlike structures that are visible under the light microscope during cell division.

Chronic obstructive pulmonary disease (COPD) A disease, such as bronchitis or emphysema, in which there is some degree of obstruction of airways and consequent increase in airway resistance.

Chyle (KĪL) The milky-appearing fluid found in the lacteals of the small intestine after absorption of lipids in food.

Chyme (KĪM) The semifluid mixture of partly digested food and digestive secretions found in the stomach and small intestine during digestion of a meal.

Cilia (SIL-ē-a) A hair or hairlike process projecting from a cell that may be used to move the entire cell or to move substances along the surface of the cell. Singular is cilium.

Ciliary (SIL-ē-ar′-ē) **body** One of the three parts of the vascular tunic of the eyeball, the others being the choroid and the iris; includes the ciliary muscle and the ciliary processes.

Ciliary (SIL-ē-ar′-ē) **ganglion** A very small parasympathetic ganglion with preganglionic axons from the oculomotor (III) nerve and postganglionic axons that carry nerve impulses to the ciliary muscle and the sphincter muscle of the iris.

Cilium (SIL-ē-um) A hair or hairlike process projecting from a cell that may be used to move the entire cell or to move substances along the surface of the cell. Plural is cilia.

Circadian (ser-KĀ-dē-an) **rhythm** The pattern of biological activity on a 24-hour cycle, such as the sleep–wake cycle.

Circular folds Permanent, deep, transverse folds in the mucosa and submucosa of the small intestine that increase the surface area for absorption. Also called plicae circulares (PLĪ-kē SER-kū-lar-ēs).

Circulation time Time required for a drop of blood to pass through the pulmonary and systemic circulations; normally about 1 minute.

Circumduction (ser-kum-DUK-shun) A movement at a synovial joint in which the distal end of a bone moves in a circle while the proximal end remains relatively stable.

Cirrhosis (si-RŌ-sis) A liver disorder in which the parenchymal cells are destroyed and replaced by connective tissue.

Cisterna chyli (sis-TER-na KĪ-lē) The origin of the thoracic duct.

Cleavage (KLĒV-ij) The rapid mitotic divisions following the fertilization of a secondary oocyte, resulting in an increased number of progressively smaller cells, called blastomeres.

Clitoris (KLI-to-ris) An erectile organ of the female, located at the anterior junction of the labia minora, that is homologous to the male penis.

Clone (KLŌN) A population of identical cells.

Coarctation (kō′-ark-TĀ-shun) **of the aorta** A congenital heart defect in which a segment of the aorta is too narrow. As a result, the flow of oxygenated blood to the body is reduced, the left ventricle is forced to pump harder, and high blood pressure develops.

Coccyx (KOK-siks) The fused bones at the inferior end of the vertebral column.

Cochlea (KOK-lē-a) A winding, cone-shaped tube forming a portion of the inner ear and containing the spiral organ (organ of Corti).

Cochlear duct The membranous cochlea consisting of a spirally arranged tube enclosed in the bony cochlea and lying along its outer wall. Also called the scala media (SCA-la MĒ-dē-a).

Collateral circulation The alternate route taken by blood through an anastomosis.

Colon The portion of the large intestine consisting of ascending, transverse, descending, and sigmoid portions.

Colony-stimulating factor (CSF) One of a group of molecules that stimulates development of white blood cells.

Colostrum (kō-LOS-trum) A thin, cloudy fluid secreted by the mammary glands a few days prior to or after delivery, before true milk is produced.

Column (KOL-um) Group of white matter tracts in the spinal cord.

Common bile duct A tube formed by the union of the common hepatic duct and the cystic duct that empties bile into the duodenum at the hepatopancreatic ampulla (ampulla of Vater).

Compact bone tissue Bone tissue that contains few spaces between osteons (haversian systems); forms the external portion of all bones and the bulk of the diaphysis (shaft) of long bones; is found immediately deep to the periosteum and external to spongy bone.

Compartment A group of skeletal muscles, their associated blood vessels, and associated nerves with a common function.

Concussion (kon-KUSH-un) Traumatic injury to the brain that produces no visible bruising but may result in abrupt, temporary loss of consciousness.

Condyloid (KON-di-loyd) **joint** A synovial joint structured so that an oval-shaped condyle of one bone fits into an elliptical cavity of another bone, permitting side-to-side and back-and-forth movements, such as the joint at the wrist between the radius and carpals. Also called an ellipsoidal (ē-lip-SOYD-al) joint.

Cone The type of photoreceptor in the retina that is specialized for highly acute color vision in bright light.

Conjunctiva (kon′-junk-TĪ-va) The delicate membrane covering the eyeball and lining the eyes.

Connective tissue One of the most abundant of the four basic tissue types in the body, performing the functions of binding and supporting; consists of relatively few cells in a generous matrix (the ground substance and fibers between the cells).

Consciousness (KON-shus-nes) A state of wakefulness in which an individual is fully alert, aware, and oriented, partly as a result of feedback between the cerebral cortex and reticular activating system.

Continuous conduction Propagation of an action potential (nerve impulse) in a step-by-step depolarization of each adjacent area of an axon membrane.

Contractility (kon′-trak-TIL-i-tē) The ability of cells or parts of cells to actively generate force to undergo shortening for movements. Muscle fibers (cells) exhibit a high degree of contractility.

Control center Part of a feedback system that sets the range of values within which a controlled condition should be maintained, evaluates input from receptors, and generates output commands.

Conus medullaris (KŌ-nus med-ū-LAR-is) The tapered portion of the spinal cord inferior to the lumbar enlargement.

Convergence (con-VER-jens) A synaptic arrangement in which the synaptic end bulbs of several presynaptic neurons terminate on one postsynaptic neuron. The medial movement of the two eyeballs so that both are directed toward a near object being viewed in order to produce a single image.

Cornea (KOR-nē-a) The nonvascular, transparent fibrous coat through which the iris of the eye can be seen.

Corona (kō-RŌ-na) Margin of the glans penis.

Corona radiata (kō-RŌ-na rā-dē-A-ta) The innermost layer of granulosa cells that is firmly attached to the zona pellucida around a secondary oocyte.

Coronary artery disease (CAD) A condition such as atherosclerosis that causes narrowing of coronary arteries so that blood flow to the heart is reduced. The result is coronary heart disease (CHD), in which the heart muscle receives inadequate blood flow due to an interruption of its blood supply.

Coronary circulation The pathway followed by the blood from the ascending aorta through the blood vessels supplying the heart and returning to the right atrium. Also called cardiac circulation.

Coronary sinus (SĪ-nus) A wide venous channel on the posterior surface of the heart that collects the blood from the myocardium.

Corpus albicans (KOR-pus AL-bi-kanz) A white fibrous patch in the ovary that forms after the corpus luteum regresses.

Corpus callosum (kal-LŌ-sum) The great commissure of the brain between the cerebral hemispheres.

Corpus luteum (LOO-tē-um) A yellowish body in the ovary formed when a follicle has discharged its secondary oocyte; secretes estrogens, progesterone, relaxin, and inhibin.

Corpus striatum (strī-Ā-tum) An area in the interior of each cerebral hemisphere composed of the lentiform and caudate nuclei.

Corpuscle of touch A sensory receptor for touch; found in dermal papillae, especially in the palms and soles. Also called a Meissner corpuscle.

Cortex (KOR-teks) An outer layer of an organ. The convoluted layer of gray matter covering each cerebral hemisphere.

Cramp A spasmodic, usually painful contraction of a muscle.

Cranial cavity A subdivision of the dorsal body cavity formed by the cranial bones and containing the brain.

Cranial nerve One of 12 pairs of nerves that leave the brain; pass through foramina in the skull; and supply sensory and motor neurons to the head, neck, part of the trunk, and viscera of the thorax and abdomen. Each is designated by a Roman numeral and a name.

Craniosacral (krā-nē-ō-SĀK-ral) **outflow** The axons of parasympathetic preganglionic neurons, which have their cell bodies located in nuclei in the brainstem and in the lateral gray matter of the sacral portion of the spinal cord.

Crista (KRIS-ta) A crest or ridged structure. A small elevation in the ampulla of each semicircular duct that contains receptors for dynamic equilibrium. Plural is cristae.

Crossing-over The exchange of a portion of one chromatid with another during meiosis. It permits an exchange of genes among chromatids and is one factor that results in genetic variation of progeny.

Crura (KROO-ra) **of the penis** Separated, tapered portion of the corpora cavernosa penis. Singular is crus (KROOS).

Cryptorchidism (krip-TOR-ki-dizm) The condition of undescended testes.

Cuneate (KŪ-nē-āt) **nucleus** A group of neurons in the inferior part of the medulla oblongata in which axons of the cuneate fasciculus terminate.

Cupula (KU-pū-la) A mass of gelatinous material covering the hair cells of a crista; a sensory receptor in the ampulla of a semicircular canal stimulated when the head moves.

Cushing's syndrome Condition caused by a hypersecretion of glucocorticoids characterized by spindly legs, "moon face," "buffalo hump," pendulous abdomen, flushed facial skin, poor wound healing, hyperglycemia, osteoporosis, hypertension, and increased susceptibility to disease.

Cutaneous (kū-TĀ-nē-us) Pertaining to the skin.

Cystic (SIS-tik) **duct** The duct that carries bile from the gallbladder to the common bile duct.

Cytokinesis (sī′-tō-ki-NĒ-sis) Distribution of the cytoplasm into two separate cells during cell division; coordinated with nuclear division (mitosis).

Cytolysis (sī-TOL-i-sis) The rupture of living cells in which the contents leak out.

Cytoplasm (SĪ-tō-plasm) Cytosol plus all organelles except the nucleus.

Cytoskeleton Complex internal structure of cytoplasm consisting of microfilaments, microtubules, and intermediate filaments.

Cytosol (SĪ-tō-sol) Semifluid portion of cytoplasm in which organelles and inclusions are suspended and solutes are dissolved. Also called intracellular fluid.

Dartos (DAR-tōs) **muscle** Muscle tissue composed of bundles of smooth muscle fibers that makes up the scrotal septum.

Decidua (dē-SID-ū-a) That portion of the endometrium of the uterus (all but the deepest layer) that is modified during pregnancy and shed after childbirth.

Deciduous (dē-SID-ū-us) **teeth** First set of teeth. Also called primary teeth, milk teeth, or baby teeth.

Decussation (dē′-ku-SĀ-shun) **of pyramids** Crossing of 90% of the axons in the large motor tracts to opposite sides in the medullary pyramids.

Deep Away from the surface of the body or an organ.

Deep (abdominal) inguinal ring A slitlike opening in the aponeurosis of the transversus abdominis muscle that represents the origin of the inguinal canal.

Deep vein thrombosis (DVT) The presence of a thrombus in a vein, usually a deep vein of the lower limbs.

Defecation (def-e-KĀ-shun) The discharge of feces from the rectum.

Deglutition (dē-gloo-TISH-un) The act of swallowing.

Dehydration (dē-hī-DRĀ-shun) Excessive loss of water from the body or its parts.

Delta cell A cell in the pancreatic islets (islets of Langerhans) in the pancreas that secretes somatostatin. Also termed a D cell.

Demineralization (dē-min′-er-al-i-ZĀ-shun) Loss of calcium and phosphorus from bones.

Dendrite (DEN-drīt) A neuronal process that carries electrical signals, usually graded potentials, toward the cell body.

Dendritic (den-DRIT-ik) **cell** One type of antigen-presenting cell with long branchlike projections that commonly is present in mucosal linings such as the vagina, in the skin (intraepidermal macrophages in the epidermis), and in lymph nodes (follicular dendritic cells).

Dental caries (KĀR-ēz) Gradual demineralization of the enamel and dentin of a tooth that may invade the pulp and alveolar bone. Also called tooth decay.

Dentin (DEN-tin) The bony tissues of a tooth enclosing the pulp cavity.

Dentition (den-TI-shun) The number, shape, and arrangement of teeth. The eruption of teeth.

Deoxyribonucleic (dē-ok′-sē-rī-bō-nū-KLĒ-ik) **acid (DNA)** A nucleic acid constructed of nucleotides consisting of one of four bases (adenine, cytosine, guanine, or thymine), deoxyribose, and a phosphate group; encoded in the nucleotides is genetic information.

Depression (de-PRESH-un) Movement in which a part of the body moves inferiorly.

Dermal papillae (pa-PIL-ē) Fingerlike projection of the papillary region of the dermis that may contain blood capillaries or corpuscles of touch (Meissner corpuscles); singular is dermal papilla.

Dermatology (der′-ma-TOL-ō-jē) The medical specialty dealing with diseases of the skin.

Dermatome (DER-ma-tōm) The cutaneous area developed from one embryonic spinal cord segment and receiving most of its sensory innervation from one spinal nerve. An instrument for incising the skin or cutting thin transplants of skin.

Dermis (DER-mis) A layer of dense irregular connective tissue lying deep to the epidermis.

Descending colon (KŌ-lon) The part of the large intestine descending from the left colic (splenic) flexure to the level of the left iliac crest.

Detrusor muscle Smooth muscle that forms the wall of the urinary bladder.

Developmental biology The study of development from the fertilized egg to the adult form.

Deviated nasal septum A nasal septum that does not run along the midline of the nasal cavity. It deviates (bends) to one side.

Diabetes mellitus (dī-a-BĒ-tēz MEL-i-tus) An endocrine disorder caused by an inability to produce or use insulin. It is characterized by the three "polys": polyuria (excessive urine production), polydipsia (excessive thirst), and polyphagia (excess eating).

Diagnosis Distinguishing one disease from another or determining the nature of a disease from signs and symptoms by inspection, palpation, laboratory tests, and other means.

Dialysis The removal of waste products from blood by diffusion through a selectively permeable membrane.

Diaphragm (DĪ-a-fram) Any partition that separates one area from another, especially the dome-shaped skeletal muscle between the thoracic and abdominal cavities. A dome-shaped device that is placed over the cervix, usually with a spermicide, to prevent conception.

Diaphysis (dī-AF-i-sis) The shaft of a long bone.

Diarrhea (dī-a-RĒ-a) Frequent defecation of liquid caused by increased motility of the intestines.

Diarthrosis (dī-ar-THRŌ-sis) A freely movable joint; types are plane, hinge, pivot, condyloid, saddle, and ball-and-socket.

Diastole (dī-AS-tō-lē) In the cardiac cycle, the phase of relaxation or dilation of the heart muscle, especially of the ventricles.

Diastolic (dī-as-TOL-ik) **blood pressure** The force exerted by blood on arterial walls during ventricular relaxation; the lowest blood pressure measured in the large arteries, normally less than 80 mmHg in a young adult.

Diencephalon (dī-en-SEF-a-lon) A part of the brain consisting of the thalamus, hypothalamus, and epithalamus.

Differentiation The development of a cell from an unspecialized state to a specialized state.

Diffusion (di-FŪ-zhun) A passive process in which there is a net or greater movement of molecules or ions from a region of high concentration to a region of low concentration until equilibrium is reached.

Digestion (dī-JES-chun) The mechanical and chemical breakdown of food to simple molecules that can be absorbed and used by body cells.

Digestive system Body system that ingests food, breaks it down, processes it, and eliminates wastes from the body.

Diploid cell (2n) (DIP-loyd) Having two sets of chromosomes.

Direct motor pathways Collections of upper motor neurons with cell bodies in the motor cortex that project axons into the spinal cord, where they synapse with lower motor neurons or interneurons in the anterior horns. Also called the pyramidal pathways.

Disease An illness characterized by a recognizable set of signs and symptoms.

Dislocation (dis′-lō-KĀ-shun) Displacement of a bone from a joint with tearing of ligaments, tendons, and articular capsules. Also called luxation (luks-Ā-shun).

Divergence (dī-VER-jens) A synaptic arrangement in which the synaptic end bulbs of one presynaptic neuron terminate on several postsynaptic neurons.

Diverticulum (dī′-ver-TIK-ū-lum) A sac or pouch in the wall of a canal or organ, especially in the colon.

Dorsiflexion (dor-si-FLEK-shun) Bending the foot in the direction of the dorsum (upper surface).

Down-regulation Phenomenon in which there is a decrease in the number of receptors in response to an excess of a hormone or neurotransmitter.

Dual innervation The concept by which most organs of the body receive impulses from sympathetic and parasympathetic neurons.

Ductus (vas) deferens The duct that carries sperm from the epididymis to the ejaculatory duct. Also called the seminal duct.

Ductus arteriosus (DUK-tus ar-tē-rē-Ō-sus) A small vessel connecting the pulmonary trunk with the aorta; found only in the fetus.

Ductus deferens (DEF-er-ens) The duct that carries sperm from the epididymis to the ejaculatory duct. Also called the vas deferens.

Ductus epididymis (ep′-i-DID-i-mis) A tightly coiled tube inside the epididymis, distinguished into a head, body, and tail, in which sperm undergo maturation.

Ductus venosus (ve-NŌ-sus) A small vessel in the fetus that helps the circulation bypass the liver.

Duodenal (doo-ō-DĒ-nal) **gland** Gland in the submucosa of the duodenum that secretes an alkaline mucus to protect the lining of the small intestine from the action of enzymes and to help neutralize the acid in chyme. Also called Brunner's (BRUN-erz) gland.

Duodenum (doo′-ō-DĒ-num or doo-OD-e-num) The first 25 cm (10 in.) of the small intestine, which connects the stomach and the ileum.

Dura mater (DOO-ra MĀ-ter) The outermost of the three meninges (coverings) of the brain and spinal cord.

Dysmenorrhea (dis-men-ō-RĒ-a) Painful menstruation.

Dyspnea (DISP-nē-a) Shortness of breath; painful or labored breathing.

Ectoderm The primary germ layer that gives rise to the nervous system and the epidermis of skin and its derivatives.

Ectopic (ek-TOP-ik) **pregnancy** The development of an embryo or fetus outside the uterine cavity.

Edema (e-DĒ-ma) An abnormal accumulation of interstitial fluid.

Effector (e-FEK-tor) An organ of the body, either a muscle or a gland, that is innervated by somatic or autonomic motor neurons.

Efferent arteriole A vessel of the renal vascular system that carries blood from a glomerulus to a peritubular capillary.

Efferent (EF-er-ent) **ducts** A series of coiled tubes that transport sperm from the rete testis to the epididymis.

Ejaculation (ē-jak-ū-LĀ-shun) The reflex ejection or expulsion of semen from the penis.

Ejaculatory (ē-JAK-ū-la-tō-rē) **duct** A tube that transports sperm from the ductus (vas) deferens to the prostatic urethra.

Elasticity (e-las-TIS-i-tē) The ability of tissue to return to its original shape after contraction or extension.

Electrical excitability (ek-sīt′-a-BIL-i-tē) Ability to respond to certain stimuli by producing electrical signals.

Electrocardiogram (e-lek′-trō-KAR-dē-ō-gram) A recording of the electrical changes that accompany the cardiac cycle that can be detected at the surface of the body; may be resting, stress, or ambulatory.

Elevation (el-e-VĀ-shun) Movement in which a part of the body moves superiorly.

Embolus (EM-bō-lus) A blood clot, bubble of air or fat from broken bones, mass of bacteria, or other debris or foreign material transported by the blood.

Embryo (EM-brē-ō) The young of any organism in an early stage of development; in humans, the developing organism from fertilization to the end of the eighth week of development.

Embryoblast (EM-brē-ō-blast) A region of cells of a blastocyst that differentiates into the three primary germ layers—ectoderm, mesoderm, and endoderm—from which all tissues and organs develop; also called an inner cell mass.

Embryology The study of development from the fertilized egg to the end of the eighth week of development.

Emigration (em′-i-GRĀ-shun) Process whereby white blood cells (WBCs) leave the bloodstream by rolling along the endothelium, sticking to it, and squeezing between the endothelial cells. Also known as migration or extravasation.

Emission (ē-MISH-un) Propulsion of sperm into the urethra due to peristaltic contractions of the ducts of the testes, epididymides, and ductus (vas) deferens as a result of sympathetic stimulation.

Emphysema (em-fi-SĒ-ma) A lung disorder in which alveolar walls disintegrate, producing abnormally large air spaces and loss of elasticity in the lungs; typically caused by exposure to cigarette smoke.

Emulsification (e-mul-si-fi-KĀ-shun) The dispersion of large lipid globules into smaller, uniformly distributed particles in the presence of bile.

Enamel (e-NAM-el) The hard, white substance covering the crown of a tooth.

Endocardium (en-dō-KAR-dē-um) The layer of the heart wall, composed of endothelium and smooth muscle, that lines the inside of the heart and covers the valves and tendons that hold the valves open.

Endochondral (en′-dō-KON-dral) **ossification** The replacement of cartilage by bone. Also called intracartilaginous (in′-tra-kar′-ti-LAJ-i-nus) ossification.

Endocrine (EN-dō-krin) **gland** A gland that secretes hormones into interstitial fluid and then the blood; a ductless gland.

Endocrine (EN-dō-krin) **system** All endocrine glands and hormone-secreting cells.

Endocrinology (en′-dī-kri-NOL-ō-jē) The science concerned with the structure and functions of endocrine glands and the diagnosis and treatment of disorders of the endocrine system.

Endocytosis (en′-dō-sī-TŌ-sis) The uptake into a cell of large molecules and particles by vesicles formed from the plasma membrane.

Endoderm A primary germ layer of the developing embryo; gives rise to the gastrointestinal tract, urinary bladder, urethra, and respiratory tract.

Endodontics (en′-dō-DON-tiks) The branch of dentistry concerned with the prevention, diagnosis, and treatment of diseases that affect the pulp, root, periodontal ligament, and alveolar bone.

Endolymph (EN-dō-limf′) The fluid within the membranous labyrinth of the internal ear.

Endometriosis (en′-dō-MĒ-trē-ō-sis) The growth of endometrial tissue outside the uterus.

Endometrium The mucous membrane lining the uterus.

Endomysium (en′-dō-MIZ-ē-um) Invagination of the perimysium separating each individual muscle fiber (cell).

Endoneurium (en′-dō-NOO-rē-um) Connective tissue wrapping around individual nerve axons.

Endoplasmic reticulum (en′-dō PLAS-mik re-TIK-ū-lum) **(ER)** A network of channels running through the cytoplasm of a cell that serves in intracellular transportation, support, storage, synthesis, and packaging of molecules. Portions of ER where ribosomes are attached to the outer surface are called rough ER; portions that have no ribosomes are called smooth ER.

Endosteum (end-OS-tē-um) The membrane that lines the medullary (marrow) cavity of bones, consisting of osteogenic cells and scattered osteoclasts.

Endothelium (en′-dō-THĒ-lē-um) The layer of simple squamous epithelium that lines the cavities of the heart, blood vessels, and lymphatic vessels.

Energy The capacity to do work.

Enteric (en-TER-ik) **nervous system (ENS)** The part of the nervous system that is embedded in the submucosa and muscularis of the gastrointestinal (GI) tract; governs motility and secretions of the GI tract.

Enteroendocrine (en-ter-ō-EN-dō-krin) **cell** A cell of the mucosa of the gastrointestinal tract that secretes a hormone that governs function of the GI tract.

Enzyme (EN-zīm) A substance that accelerates chemical reactions; an organic catalyst, usually a protein.

Eosinophil (ē-ō-SIN-ō-fil) A type of white blood cell characterized by granules that stain red or pink with acid dyes.

Ependymal (ep-EN-de-mal) **cells** Neuroglial cells that cover choroid plexuses and produce cerebrospinal fluid (CSF); they also line the ventricles of the brain and probably assist in the circulation of CSF.

Epicardium (ep′-i-KAR-dē-um) The thin outer layer of the heart wall, composed of serous tissue and mesothelium. Also called the visceral pericardium.

Epidemiology (ep′-i-dē-mē-OL-ō-jē) Study of the occurrence and transmission of diseases and disorders in human populations.

Epidermis (ep′-i-DERM-is) The superficial, thinner layer of skin, composed of keratinized stratified squamous epithelium.

Epididymis (ep′-i-DID-i-mis) A comma-shaped organ that lies along the posterior border of the testis and contains the ductus epididymis, in which sperm undergo maturation. Plural is epididymides (ep′-i-di-DIM-i-dēz).

Epidural (eo′-i-DOO-ral) **space** A space between the spinal dura mater and the vertebral canal, containing areolar connective tissue and a plexus of veins.

Epiglottis (ep′-i-GLOT-is) A large, leaf-shaped piece of cartilage lying on top of the larynx, attached to the thyroid cartilage; its unattached portion is free to move up and down to cover the glottis (vocal folds and rima glottidis) during swallowing.

Epimysium (ep-i-MĪZ-ē-um) Fibrous connective tissue around muscles.

Epinephrine (ep-ē-NEF-rin) Hormone secreted by the adrenal medulla that produces actions similar to those that result from sympathetic stimulation. Also called adrenaline (a-DREN-a-lin).

Epineurium (ep′-i-NOO-rē-um) The superficial connective tissue covering around an entire nerve.

Epiphyseal (ep′-i-FIZ-ē-al) **line** The remnant of the epiphyseal plate in the metaphysis of a long bone.

Epiphyseal cartilage ep′-i-FIZ-ē-al) Hyaline cartilage growth center formed during endochondral ossification; not a joint associated with movement.

Epiphyseal plate The hyaline cartilage plate in the metaphysis of a long bone; site of lengthwise growth of long bones. Also called the growth plate.

Epiphysis (e-PIF-i-sis) The end of a long bone, usually larger in diameter than the shaft (diaphysis).

Episiotomy (e-piz-ē-OT-ō-mē) A cut made with surgical scissors to avoid tearing of the perineum at the end of the second stage of labor.

Epithalamus (ep′-i-THAL-a-mus) Part of the diencephalon superior and posterior to the thalamus, comprising the pineal gland and associated structures.

Epithelial (ep-i-THĒ-lē-al) **tissue** The tissue that forms the innermost and outermost surfaces of body structures and forms glands. Also called epithelium.

Eponychium (ep′-ō-NIK-ē-um) Narrow band of stratum corneum at the proximal border of a nail that extends from the margin of the nail wall. Also called the cuticle.

Equilibrium (ē-kwi-LIB-rē-um) The state of being balanced.

Erectile dysfunction Failure to maintain an erection long enough for sexual intercourse. Previously known as impotence (IM-pō-tens).

Erection (ē-REK-shun) The enlarged and stiff state of the penis or clitoris resulting from the engorgement of the spongy erectile tissue with blood.

Eructation (e-ruk′-TĀ-shun) The forceful expulsion of gas from the stomach. Also called belching.

Erythema (er-e-THĒ-ma) Skin redness usually caused by dilation of the capillaries.

Erythropoietin (e-rith′-rō-POY-ē-tin) **(EPO)** A hormone released by the juxtaglomerular cells of the kidneys that stimulates red blood cell production.

Esophagus (e-SOF-a-gus) The hollow muscular tube that connects the pharynx and the stomach.

Estrogens (ES-trō-jenz) Feminizing sex hormones produced by the ovaries; govern development of oocytes, maintenance of female reproductive structures, and appearance of secondary sex characteristics; also affect fluid and electrolyte balance, and protein anabolism.

Eupnea (ŪP-nē-a) Normal quiet breathing.

Eversion (ē-VER-zhun) The movement of the sole laterally at the ankle joint or of an atrioventricular valve into an atrium during ventricular contraction.

Excretion (eks-KRĒ-shun) The process of eliminating waste products from the body; also the products excreted.

Exhalation (eks-ha-LĀ-shun) Breathing out; expelling air from the lungs into the atmosphere. Also called expiration.

Exocrine (EK-sō-krin) **gland** A gland that secretes its products into ducts that carry the secretions into body cavities, into the lumen of an organ, or to the outer surface of the body.

Exocytosis (ek-sō-sī-TŌ-sis) A process in which membrane-enclosed secretory vesicles form inside the cell, fuse with the plasma membrane, and release their contents into the interstitial fluid; achieves secretion of materials from a cell.

Extensibility (ek-sten′-si-BIL-i-tē) The ability of muscle tissue to stretch when it is pulled.

Extension (eks-TEN-shun) An increase in the angle between two bones; restoring a body part to its anatomical position after flexion.

External auditory canal A curved tube in the temporal bone that leads to the middle ear. Also called a meatus.

External ear The outer ear, consisting of the pinna, external auditory canal, and tympanic membrane (eardrum).

External nares (NĀ-rez) The openings into the nasal cavity on the exterior of the body. Also called the nostrils.

External respiration The exchange of respiratory gases between the lungs and blood. Also called pulmonary respiration.

Exteroceptor (EKS-ter-ō-sep′-tor) A sensory receptor adapted for the reception of stimuli from outside the body.

Extracellular fluid (ECF) Fluid outside body cells, such as interstitial fluid and plasma.

Extracellular matrix (MĀ-triks) The ground substance and fibers between cells in a connective tissue.

Eyebrow The hairy ridge superior to the eye.

F cell A cell in the pancreatic islets (islets of Langerhans) that secretes pancreatic polypeptide.

Falciform ligament (FAL-si-form LIG-a-ment) A sheet of parietal peritoneum between the two principal lobes of the liver. The ligamentum teres, or remnant of the umbilical vein, lies within its fold.

Falx cerebelli (FALKS′ ser-e-BEL-lī) A small triangular process of the dura mater attached to the occipital bone in the posterior cranial fossa and projecting inward between the two cerebellar hemispheres.

Falx cerebri (FALKS SER-e-brē) A fold of the dura mater extending deep into the longitudinal fissure between the two cerebral hemispheres.

Fascia (FASH-ē-a) Large connective tissue sheets that wrap around groups of muscles.

Fascicle (FAS-i-kul) A small bundle or cluster, especially of nerve or muscle fibers (cells).

Fasciculation (fa-sik-ū-LĀ-shun) Abnormal, spontaneous twitch of all skeletal muscle fibers in one motor unit that is visible at the skin surface; not associated with movement of the affected muscle; present in progressive diseases of motor neurons, for example, poliomyelitis.

Fat A triglyceride that is a solid at room temperature.

Fatty acid A simple lipid that consists of a carboxyl group and a hydrocarbon chain; used to synthesize triglycerides and phospholipids.

Fauces (FAW-sēs) The opening from the mouth into the pharynx.

Feces (FĒ-sēz) Material discharged from the rectum and made up of bacteria, excretions, and food residue. Also called stool.

Feedback system Cycle of events in which the status of a body condition is monitored, evaluated, changed, remonitored, and reevaluated.

Female reproductive cycle General term for the ovarian and uterine cycles, the hormonal changes that accompany them, and cyclic changes in the breasts and cervix; includes changes in the endometrium of a nonpregnant female that prepares the lining of the uterus to receive a fertilized ovum. Less correctly termed the menstrual cycle.

Female reproductive system Reproductive system in the female, including the ovaries, uterine tubes, uterus, vulva, and mammary glands.

Fertilization (fer-til-i-ZĀ-shun) Penetration of a secondary oocyte by a sperm cell, meiotic division of secondary oocyte to form an ovum, and subsequent union of the nuclei of the gametes.

Fetal circulation The cardiovascular system of the fetus, including the placenta and special blood vessels involved in the exchange of materials between fetus and mother.

Fetus (FĒ-tus) In humans, the developing organism in utero from the beginning of the third month to birth.

Fever An elevation in body temperature above the normal temperature of (37°C, 98.6°F) due to a resetting of the hypothalamic thermostat.

Fibroblast (FĪ-brō-blast) A large, flat cell that secretes most of the extracellular matrix of areolar and dense connective tissues.

Fibrosis The process by which fibroblasts synthesize collagen fibers and other extracellular matrix materials that aggregate to form scar tissue.

Fibrous (FĪ-brus) **joint** A joint that allows little or no movement, such as a suture, syndesmosis, or interosseous membrane.

Fibrous tunic (TOO-nik) The superficial coat of the eyeball, made up of the posterior sclera and the anterior cornea.

Fight-or-flight response The effects produced upon stimulation of the sympathetic division of the autonomic nervous system. First of three stages of the stress response.

Filiform papilla (FIL-i-form pa-PIL-a) One of the conical projections that are distributed in parallel rows over the anterior two-thirds of the tongue and lack taste buds.

Filtration (fil-TRĀ-shun) The flow of a liquid through a filter (or membrane that acts like a filter) due to a hydrostatic pressure; occurs in capillaries due to blood pressure.

Filum terminale (FĪ-lum ter-mi-NAL-ē) Non-nervous fibrous tissue of the spinal cord that extends inferiorly from the conus medullaris to the coccyx.

Fimbriae (FIM-brē-ē) Fingerlike structures, especially the lateral ends of the uterine (fallopian) tubes.

Fissure (FISH-ur) A groove, fold, or slit that may be normal or abnormal.

Fixator A muscle that stabilizes the origin of the prime mover so that the prime mover can act more efficiently.

Fixed macrophage (MAK-rō-fāj) Stationary phagocytic cell found in the liver, lungs, brain, spleen, lymph nodes, subcutaneous tissue, and red bone marrow. Also called a histiocyte (HIS-tē-ō-sīt).

Flaccid (FLAK-sid) Relaxed, flabby, or soft; lacking muscle tone.

Flagella (fla-JEL-a) Hairlike, motile processes on the extremity of a bacterium, protozoan, or sperm cell. Singular is flagellum.

Flatus (FLĀ-tus) Gas in the stomach or intestines; commonly used to denote expulsion of gas through the anus.

Flexion (FLEK-shun) Movement in which there is a decrease in the angle between two bones.

Follicle-stimulating hormone (FSH) Hormone secreted by the anterior pituitary; it initiates development of ova and stimulates the ovaries to secrete estrogens in females, and initiates sperm production in males.

Fontanel (font-ta-NEL) A mesenchyme-filled space where bone formation is not yet complete, especially between the cranial bones of an infant's skull.

Foramen ovale (fō-RĀ-men ō-VAL-ē) An opening in the fetal heart in the septum between the right and left atria. A hole in the greater wing of the sphenoid bone that transmits the mandibular branch of the trigeminal (V) nerve.

Foramina (fō-RAM-i-na) Passages or openings; means of communication between two cavities of an organ, or holes in bones for passage of vessels or nerves. Singular is foramen (fō-RAM-in).

Fornix (FOR-niks) An arch or fold; a tract in the brain made up of association fibers, connecting the hippocampus with the mammillary bodies; a recess around the cervix of the uterus where it protrudes into the vagina.

Fourth ventricle (VEN-tri-kul) A cavity filled with cerebrospinal fluid within the brain lying between the cerebellum and the medulla oblongata and pons.

Fovea (FŌ-vē-a) **centralis** A depression in the center of the macula lutea of the retina, containing cones only and lacking blood vessels; the area of highest visual acuity (sharpness of vision).

Fracture (FRAK-choor) Any break in a bone.

Free radical An atom or group of atoms with an unpaired electron in the outermost shell. It is unstable, highly reactive, and destroys nearby molecules.

Frontal plane A plane at a right angle to a midsagittal plane that divides the body or organs into anterior and posterior portions. Also called a coronal (kō-RŌ-nal) plane.

Fundus (FUN-dus) The part of a hollow organ farthest from the opening; the rounded portion of the stomach superior and to the left of the cardia; the broad portion of the gallbladder that projects downward beyond the inferior border of the liver.

Fungiform papilla (FUN-ji-form pa-PIL-a) A mushroomlike elevation on the upper surface of the tongue appearing as a red dot; most contain taste buds.

Gallbladder A small pouch, located inferior to the liver, that stores bile and empties by means of the cystic duct.

Gallstone A solid mass, usually containing cholesterol, in the gallbladder or a bile-containing duct; formed anywhere between bile canaliculi in the liver and the hepatopancreatic ampulla (ampulla of Vater), where bile enters the duodenum. Also called a biliary calculus.

Gamete (GAM-ēt) A male or female reproductive cell; a sperm cell or secondary oocyte.

Ganglion (GANG-glē-on) A group of neuronal cell bodies lying outside the central nervous system (CNS). Plural is ganglia (GANG-glē-a).

Gastric (GAS-trik) **glands** Glands in the mucosa of the stomach composed of cells that empty their secretions into narrow channels called gastric pits.

Gastroenterology (gas′-trō-en-ter-OL-ō-jē) The medical specialty that deals with the structure, function, diagnosis, and treatment of diseases of the stomach and intestines.

Gastrointestinal (gas-trō-in-TES-tin-al) **(GI) tract** A continuous tube running through the ventral body cavity extending from the mouth to the anus. Also called the alimentary (al′-i-MEN-tar-ē) canal.

Gastrulation (gas-trū-LĀ-shun) The migration of groups of cells from the epiblast that transform a bilaminar embryonic disc into a trilaminar embryonic disc with three primary germ layers; transformation of the blastula into the gastrula.

Gene (JĒN) Biological unit of heredity; a segment of DNA located in a definite position on a particular chromosome; a sequence of DNA that codes for a particular mRNA, rRNA, or tRNA.

Genetics The study of genes and heredity.

Genome (JĒ-nōm) The complete set of genes of an organism.

Genotype (JĒ-nō-tīp) The genetic makeup of an individual; the combination of alleles present at one or more chromosomal locations, as distinguished from the appearance, or phenotype, that results from those alleles.

Geriatrics (jer′-ē-AT-riks) The branch of medicine devoted to the medical problems and care of elderly persons.

Germ cell A gamete (sperm or oocyte) or any precursor cell destined to become a gamete.

Gingivae (JIN-ji-vē) Tissue covering the alveolar processes of the mandible and maxilla and extending slightly into each socket. Also called gums.

Gland Specialized epithelial cell or cells that secrete substances; may be exocrine or endocrine.

Glans penis (glanz PĒ-nis) The slightly enlarged region at the distal end of the penis.

Glaucoma (glaw-KŌ-ma) An eye disorder in which there is increased intraocular pressure due to an excess of aqueous humor.

Glomerular (Bowman's) capsule A double-walled epithelial cup at the proximal end of a nephron that encloses the glomerular capillaries. Also called Bowman's (BŌ-manz) capsule.

Glomerular filtrate The fluid produced when blood is filtered by the filtration membrane in the glomeruli of the kidneys.

Glomerulus A rounded mass of nerves or blood vessels, especially the microscopic tuft of capillaries that is surrounded by the glomerular (Bowman's) capsule of each kidney tubule. Plural is glomeruli.

Glottis (GLOT-is) The vocal folds (true vocal cords) in the larynx plus the space between them (rima glottidis).

Glucagon (GLOO-ka-gon) A hormone produced by the alpha cells of the pancreatic islets (islets of Langerhans) that increases blood glucose level.

Glucocorticoids (gloo′-kō-KOR-ti-koyds) Hormones secreted by the cortex of the adrenal gland, especially cortisol, that influence glucose metabolism.

Glucose (GLOO-kōs) A hexose (six-carbon sugar), $C_6H_{12}O_6$, that is a major energy source for the production of ATP by body cells.

Glucosuria The presence of glucose in the urine; may be temporary or pathological. Also called glycosuria.

Glycogen (GLĪ-kō-jen) A highly branched polymer of glucose containing thousands of subunits; functions as a compact store of glucose molecules in liver and muscle fibers (cells).

Goblet cell A goblet-shaped unicellular gland that secretes mucus; present in epithelium of the airways and intestines.

Goiter (GOY-ter) An enlarged thyroid gland.

Golgi (GOL-jē) **complex** An organelle in the cytoplasm of cells consisting of four to six flattened sacs (cisternae), stacked on one another, with expanded areas at their ends; functions in processing, sorting, packaging, and delivering proteins and lipids to the plasma membrane, lysosomes, and secretory vesicles.

Gomphosis (gom-FŌ-sis) A fibrous joint in which a cone-shaped peg fits into a socket.

Gonad (GŌ-nad) A gland that produces gametes and hormones; the ovary in the female and the testis in the male.

Gout (GOWT) Hereditary condition associated with excessive uric acid in the blood; the acid crystallizes and deposits in joints, kidneys, and soft tissue.

Gracile (GRAS-īl) **nucleus** A group of nerve cells in the inferior part of the medulla oblongata in which axons of the gracile fasciculus terminate.

Gray commissure (KOM-mi-shur) A narrow strip of gray matter connecting the two lateral gray masses within the spinal cord.

Gray matter Areas in the central nervous system and ganglia containing neuronal cell bodies, dendrites, unmyelinated axons, axon terminals, and neuroglia; Nissl bodies impart a gray color and there is little or no myelin in gray matter.

Gray ramus communicans (RĀ-mus kō-MŪ-ni-kans) A short nerve containing axons of sympathetic postganglionic neurons; the cell bodies of the neurons are in a sympathetic chain ganglion, and the unmyelinated axons extend via the gray ramus to a spinal nerve and then to the periphery to supply smooth muscle in blood vessels, arrector pili muscles, and sweat glands. Plural is rami communicantes (RĀ-mē–kō-mū-ni-KAN-tēz).

Greater omentum (ō-MEN-tum) A large fold in the serosa of the stomach that hangs down like an apron anterior to the intestines.

Greater vestibular (ves-TIB-ū-lar) **glands** A pair of glands on either side of the vaginal orifice that open by a duct into the space between the hymen and the labia minora. Also called Bartholin's (BAR-to-linz) glands.

Growth An increase in size due to an increase in (1) the number of cells, (2) the size of existing cells as internal components increase in size, or (3) the size of intercellular substances.

Growth hormone (GH) Hormone secreted by the anterior pituitary that stimulates growth of body tissues, especially skeletal and muscular tissues. Also known as somatotropin.

Gustation (gus-TĀ-shun) The sense of taste.

Gynecology (gī′-ne-KOL-ō-jē) The branch of medicine dealing with the study and treatment of disorders of the female reproductive system.

Gyrus (JĪ-rus) One of the folds of the cerebral cortex of the brain. Plural is gyri (JĪ-rī). Also called a convolution.

Hair A threadlike structure produced by hair follicles that develops in the dermis. Also called a pilus (PĪ-lus); plural is pili (PĪ -lī).

Hair follicle Structure composed of epithelium and surrounding the root of a hair from which hair develops.

Hair root plexus (PLEK-sus) A network of dendrites arranged around the root of a hair as free or naked nerve endings that are stimulated when a hair shaft is moved.

Haploid (HAP-loyd) ***(n)*** **cell** Having half the number of chromosomes characteristically found in the somatic cells of an organism; characteristic of mature gametes. Symbolized *n*.

Hard palate (PAL-at) The anterior portion of the roof of the mouth, formed by the maxillae and palatine bones and lined by mucous membrane.

Haustra (HAWS-tra) A series of pouches that characterize the colon; caused by tonic contractions of the teniae coli. Singular is haustrum.

Haversian canal See Central canal.

Haversian system See Osteon.

Head The superior part of a human, cephalic to the neck. The superior or proximal part of a structure.

Hearing The ability to perceive sound.

Heart Organ of the cardiovascular system responsible for pumping blood throughout the body; located in the thoracic cavity superior to the diaphragm.

Heart block An arrhythmia (dysrhythmia) of the heart in which the atria and ventricles contract independently because of a blocking of electrical impulses through the heart at some point in the conduction system.

Heart murmur An abnormal sound that consists of a flow noise that is heard before, between, or after the normal heart sounds, or that may mask normal heart sounds.

Hemangioblast (hē-MAN-jē-ō-blast) A precursor mesodermal cell that develops into blood and blood vessels.

Hematocrit (he-MAT-ō-krit) **(Hct)** The percentage of blood made up of red blood cells. Usually measured by centrifuging a blood sample in a graduated tube and then reading the volume of red blood cells and dividing it by the total volume of blood in the sample.

Hematology (hēm-a-TOL-ō-jē) The study of blood.

Hemiplegia Paralysis of the upper limb, trunk, and lower limb on one side of the body.

Hemodynamics (hē-mō-dī-NAM-iks) The forces involved in circulating blood throughout the body.

Hemoglobin (hē-mō-GLŌ-bin) A substance in red blood cells consisting of the protein globin and the iron-containing red pigment heme that transports most of the oxygen and some carbon dioxide in blood.

Hemolysis (hē-MOL-i-sis) The escape of hemoglobin from the interior of a red blood cell into the surrounding medium; results from disruption of the cell membrane by toxins or drugs, freezing or thawing, or hypotonic solutions.

Hemolytic disease of the newborn (HDN) A hemolytic anemia of a newborn child that results from the destruction of the infant's erythrocytes (red blood cells) by antibodies produced by the mother; usually the antibodies are due to an Rh blood type incompatibility. Also called erythroblastosis fetalis (e-rith′-rō-blas-TŌ-sis fe-TAL-is).

Hemophilia (hē′-mō-FIL-ē-a) A hereditary blood disorder where there is a deficient production of certain factors involved in blood clotting, resulting in excessive bleeding into joints, deep tissues, and elsewhere.

Hemopoiesis (hēm-ō-poy-Ē-sis) Blood cell production, which occurs in red bone marrow after birth. Also called hematopoiesis (hem′-a-tō-poy-E-sis).

Hemorrhage (HEM-o-rij) Bleeding; the escape of blood from blood vessels, especially when the loss is profuse.

Hemorrhoids (HEM-ō-royds) Dilated or varicosed blood vessels (usually veins) in the anal region. Also called piles.

Hepatic portal circulation The flow of blood from the gastrointestinal organs to the liver before returning to the heart.

Hepatocyte (he-PAT-ō-cīt) A liver cell.

Hepatopancreatic (hep′-a-tō-pan′-krē-A-tik) **ampulla** A small, raised area in the duodenum where the combined common bile duct and main pancreatic duct empty into the duodenum. Also called the ampulla of Vater (VA-ter).

Hernia (HER-nē-a) The protrusion or projection of an organ or part of an organ through a membrane or cavity wall, usually the abdominal cavity.

Herniated (HER-nē-ā-ted) **disc** A rupture of an intervertebral disc so that the nucleus pulposus protrudes into the vertebral cavity. Also called a slipped disc.

Hilum (HĪ-lum) An area, depression, or pit where blood vessels and nerves enter or leave an organ. Also called a hilus.

Hinge joint A synovial joint in which a convex surface of one bone fits into a concave surface of another bone, such as the elbow, knee, ankle, and interphalangeal joints. Also called a ginglymus (JIN-gli-mus) joint.

Hirsutism (HER-soo-tizm) An excessive growth of hair in females and children, with a distribution similar to that in adult males, due to the conversion of vellus hairs into large terminal hairs in response to higher-than-normal levels of androgens.

Histology (his′-TOL-ō-jē) Microscopic study of the structure of tissues.

Holocrine (HŌ-lō-krin) **gland** A type of gland in which entire secretory cells, along with their accumulated secretions, make up the secretory product of the gland, as in the sebaceous (oil) glands.

Homeostasis The condition in which the body's internal environment remains relatively constant within physiological limits.

Homologous (hō-MOL-ō-gus) **chromosomes** Two chromosomes that belong to a pair. Also called homologs.

Hormone (HOR-mōn) A secretion of endocrine cells that alters the physiological activity of target cells of the body.

Horn An area of gray matter (anterior, lateral, or posterior) in the spinal cord.

Human chorionic gonadotropin (kō-rē-ON-ik gō-nad-ō-TRŌ-pin) **(hCG)** A hormone produced by the developing placenta that maintains the corpus luteum.

Human chorionic somatomammotropin (sō-mat-ō-mam-ō-TRŌ-pin) **(hCS)** Hormone produced by the chorion of the placenta that stimulates breast tissue for lactation, enhances body growth, and regulates metabolism. Also called human placental lactogen (hPL).

Hyaluronic acid (hī′-a-loo-RON-ik) A viscous, amorphous extracellular material that binds cells together, lubricates joints, and maintains the shape of the eyeballs.

Hymen A thin fold of vascularized mucous membrane at the vaginal orifice.

Hyperextension (hī′-per-ek-STEN-shun) Continuation of extension beyond the anatomical position, as in bending the head backward.

Hyperplasia (hī-per-PLĀ-zē-a) An abnormal increase in the number of normal cells in a tissue or organ, increasing its size.

Hypersecretion (hī′-per-se-KRĒ-shun) Overactivity of glands resulting in excessive secretion.

Hypersensitivity (hī′-per-sen-si-TI-vi-tē) Overreaction to an allergen that results in pathological changes in tissues. Also called allergy.

Hypertension (hī′-per-TEN-shun) High blood pressure.

Hypertonia (hī′-per-TŌ-nē-a) Increased muscle tone that is expressed as spasticity or rigidity.

Hypertonic (hī′-per-TON-ik) **solution** Solution that causes cells to shrink due to loss of water by osmosis.

Hypertrophy (hī-per-TRŌ-fē) An excessive enlargement or overgrowth of tissue without cell division.

Hyperventilation (hī′-per-ven-til-LĀ-shun) A rate of inhalation and exhalation higher than that required to maintain a normal partial pressure of carbon dioxide in the blood.

Hyponychium (hi′-pō-NIK-ē-um) Portion of the nail beneath the free edge composed of a thickened region of stratum corneum.

Hypophyseal (hī′pō-FIZ-ē-al) **pouch** An outgrowth of ectoderm from the roof of the mouth from which the anterior pituitary develops. Also called Rathke's pouch.

Hyposecretion (hī′-pō-se-KRĒ-shun) Underactivity of glands resulting in diminished secretion.

Hypothalamohypophyseal (hī′-pō-thal′-a-mō-hī-pō-FIZ-ē-al) **tract** A bundle of axons containing secretory vesicles filled with oxytocin or antidiuretic hormone that extends from the hypothalamus to the posterior pituitary.

Hypothalamus (hī′-pō-THAL-a-mus) A portion of the diencephalon, lying beneath the thalamus and forming the floor and part of the wall of the third ventricle.

Hypothermia (hī′-pō-THER-mē-a) Lowering of body temperature below 35°C (95°F); in surgical procedures, it refers to deliberate cooling of the body to slow down metabolism and reduce oxygen needs of tissues.

Hypotonia (hī′-pō-TŌ-nē-a) Decreased or lost muscle tone in which muscles appear flaccid.

Hypotonic (hī′-pō-TON-ik) **solution** Solution that causes cells to swell and perhaps rupture due to gain of water by osmosis.

Hypoventilation (hī-pō-ven-ti-LĀ-shun) A rate of inhalation and exhalation lower than that required to maintain a normal partial pressure of carbon dioxide in plasma.

Hypoxia (hī-POKS-ē-a) Lack of adequate oxygen at the tissue level.

Hysterectomy (hiss-te-REK-tō-mē) The surgical removal of the uterus.

Ileocecal sphincter (valve) A fold of mucous membrane that guards the opening from the ileum into the large intestine. Also called the ileocecal valve.

Ileum (IL-ē-um) The terminal part of the small intestine.

Immunity (i-MŪ-ni-tē) The state of being resistant to injury, particularly by poisons, foreign proteins, and invading pathogens. Also called resistance.

Immunoglobulin (im-ū-nō-GLOB-ū-lin) **(Ig)** A protein synthesized by plasma cells derived from B lymphocytes in response to a specific antigen. Also called an antibody.

Immunology (im′-ū-NOL-ō-jē) The study of the responses of the body when challenged by antigens.

Implantation (im′-plan-TĀ-shun) The insertion of a tissue or a part into the body. The attachment of the blastocyst to the stratum basalis of the endometrium about 6 days after fertilization.

Indirect motor pathway Motor tracts that convey information from the brain down the spinal cord for automatic movements, coordination of body movements with visual stimuli, skeletal muscle tone and posture, and balance. Also known as extrapyramidal pathways.

Induction (in-DUK-shun) The process by which one tissue (inducting tissue) stimulates the development of an adjacent unspecialized tissue (responding tissue) into a specialized one.

Inferior Away from the head or toward the lower part of a structure. Also called caudal (KAW-dal).

Inferior vena cava (VĒ-na KĀ-va) **(IVC)** Large vein that collects blood from parts of the body inferior to the heart and returns it to the right atrium.

Infertility Inability to conceive or to cause conception. Also called sterility.

Inflammation (in′-fla-MĀ-shun) Localized, protective response to tissue injury designed to destroy, dilute, or wall off the infecting agent or injured tissue; characterized by redness, pain, heat, swelling, and sometimes loss of function.

Infundibulum (in-fun-DIB-ū-lum) The stalklike structure that attaches the pituitary gland to the hypothalamus of the brain. The funnel-shaped, open, distal end of the uterine (fallopian) tube.

Ingestion (in-JES-chun) The taking in of food, liquids, or drugs, by mouth. Process by which phagocytes engulf microbes.

Inguinal canal An oblique passageway in the anterior abdominal wall just superior and parallel to the medial half of the inguinal ligament that transmits the spermatic cord and ilioinguinal nerve in the male and round ligament of the uterus and ilioinguinal nerve in the female.

Inhalation (in-ha-LĀ-shun) The act of drawing air into the lungs. Also called inspiration.

Inheritance The acquisition of body traits by transmission of genetic information from parents to offspring.

Inhibin A hormone secreted by the gonads that inhibits release of follicle-stimulating hormone (FSH) by the anterior pituitary.

Inhibiting hormone Hormone secreted by the hypothalamus that can suppress secretion of hormones by the anterior pituitary.

Insertion (in-SER-shun) The attachment of a muscle tendon to a movable bone or the end opposite the origin.

Insula (IN-soo-la) A triangular area of the cerebral cortex that lies deep within the lateral cerebral fissure, under the parietal, frontal, and temporal lobes.

Insulin (IN-soo-lin) A hormone produced by the beta cells of a pancreatic islet (islet of Langerhans) that decreases the blood glucose level.

Integrins (IN-te-grinz) A family of transmembrane glycoproteins in plasma membranes that function in cell adhesion; they are present in hemidesmosomes, which anchor cells to a basement membrane, and they mediate adhesion of neutrophils to endothelial cells during emigration.

Integumentary (in-teg-ū-MEN-tar-ē) **system** Body system composed of the skin, hair, oil and sweat glands, nails, and sensory receptors that helps maintain body temperature, protects the body, and provides sensory information.

Intercalated (in-TER-ka-lāt-ed) **disc** An irregular transverse thickening of sarcolemma that contains desmosomes, which hold cardiac muscle fibers (cells) together, and gap junctions, which aid in conduction of muscle action potentials from one fiber to the next.

Intercostal (in′-ter-KOS-tal) **nerve** A nerve supplying a muscle located between the ribs. Also called thoracic nerve.

Intermediate filament Protein filament, ranging from 8 to 12 nm in diameter, that may provide structural reinforcement, hold organelles in place, and give shape to a cell.

Internal capsule A large tract of projection fibers lateral to the thalamus that is the major connection between the cerebral cortex and the brainstem and spinal cord; contains axons of sensory neurons carrying auditory, visual, and somatic sensory signals to the cerebral cortex plus axons of motor neurons descending from the cerebral cortex to the thalamus, subthalamus, brainstem, and spinal cord.

Internal ear The inner ear or labyrinth, lying inside the temporal bone, containing the organs of hearing and balance.

Internal nares (NĀ-rez) The two openings posterior to the nasal cavities opening into the nasopharynx. Also called the choanae (kō-Ā-nē).

Internal respiration The exchange of respiratory gases between blood and body cells. Also called tissue respiration or systemic gas exchange.

Interneurons (in′-ter-NOO-ronz) Neurons whose axons extend only for a short distance and contact nearby neurons in the brain, spinal cord, or a ganglion; they comprise the vast majority of neurons in the body. Also called association neurons.

Interoceptor (IN-ter-ō-sep′-tor) Sensory receptor located in blood vessels and viscera that provides information about the body's internal environment. Also called a visceroceptor.

Interphase (IN-ter-fāz) The period of the cell cycle between cell divisions, consisting of the G_1 (gap or growth) phase, when the cell is engaged in growth, metabolism, and production of substances required for division; S (synthesis) phase, during which chromosomes are replicated; and G_2 phase.

Interstitial (in′-ter-STISH-al) **cell** A type of cell that secretes testosterone; located in the connective tissue between seminiferous tubules in a mature testis. Also known as a Leydig cell.

Interstitial fluid The portion of extracellular fluid that fills the microscopic spaces between the cells of tissues; the internal environment of the body. Also called intercellular or tissue fluid.

Interstitial growth Growth from within, as in the growth of cartilage.

Interventricular (in′-ter-ven-TRIK-ū-lar) **foramen** A narrow, oval opening through which the lateral ventricles of the brain communicate with the third ventricle.

Intervertebral (in′-ter-VER-te-bral) **disc** A pad of fibrocartilage located between the bodies of two vertebrae.

Intestinal gland A gland that opens onto the surface of the intestinal mucosa and secretes digestive enzymes. Also called a crypt of Lieberkühn (LĒ-ber-kūn).

Intracellular (in′-tra-SEL-ū-lar) **fluid (ICF)** Fluid located within cells. Also called cytosol.

Intraepidermal macrophage Epidermal dendritic cell that functions as an antigen-presenting cell (APC) during an immune response. Also called a Langerhans (LANG-er-hans) cell.

Intrafusal (in′-tra-FŪ-sal) **fibers** Three to ten specialized muscle fibers (cells), partially enclosed in a spindle-shaped connective tissue capsule, that make up a muscle spindle.

Intramembranous (in′-tra-MEM-bra-nus) **ossification** The method of bone formation in which the bone is formed directly in mesenchyme arranged in sheetlike layers that resemble membranes.

Intramuscular (IM) injection An injection that penetrates the skin and subcutaneous layer to enter a skeletal muscle. Common sites are the deltoid, gluteus medius, and vastus lateralis muscles.

Intraocular (in′-tra-OK-ū-lar) **pressure** Pressure in the eyeball, produced mainly by aqueous humor.

Invagination (in-vaj′-i-NĀ-shun) The pushing of the wall of a cavity into the cavity itself.

Inversion (in-VER-zhun) The movement of the sole medially at the ankle joint.

Iris The colored portion of the vascular tunic of the eyeball seen through the cornea that contains circular and radial smooth muscle; the hole in the center of the iris is the pupil.

Irritable bowel syndrome (IBS) Disease of the entire gastrointestinal tract in which a person reacts to stress by developing symptoms (such as cramping and abdominal pain) associated with alternating patterns of diarrhea and constipation. Excessive amounts of mucus may appear in feces, and other symptoms include flatulence, nausea, and loss of appetite. Also known as irritable colon or spastic colitis.

Ischemia (is-KĒ-mē-a) A lack of sufficient blood to a body part due to obstruction or constriction of a blood vessel.

Isotonic (ī′-sō-TON-ik) **solution** A solution having the same concentration of impermeable solutes as cytosol.

Isthmus (IS-mus) A narrow strip of tissue or narrow passage connecting two larger parts. The medial, short, narrow, thick-walled portion of the uterine tube that joins the uterus. Constricted region of the uterus between the body and cervix.

Jaundice (JON-dis) A condition characterized by yellowness of the skin, the white of the eyes, mucous membranes, and body fluids because of a buildup of bilirubin.

Jejunum (je-JOO-num) The middle part of the small intestine.

Joint A point of contact between two bones, between bone and cartilage, or between bone and teeth. Also called an articulation or arthrosis.

Joint kinesthetic (kin′-es-THET-ik) **receptor** A proprioceptive receptor located in a joint, stimulated by joint movement.

Juxtaglomerular apparatus (JGA) Consists of the macula densa (cells of the distal convoluted tubule adjacent to the afferent and efferent arterioles) and juxtaglomerular cells (modified cells of the afferent and sometimes efferent arterioles); secretes renin when blood pressure starts to fall.

Keratin (KER-a-tin) An insoluble protein found in the hair, nails, and other keratinized tissues of the epidermis.

Keratinocyte (ker-a-TIN-ō-sīt) The most numerous of the epidermal cells; produces keratin.

Kidney One of the paired reddish organs located in the lumbar region that regulates the composition, volume, and pressure of blood and produces urine.

Kinesiology (ki-nē-sē-OL-ō-jē) The study of the movement of body parts.

Kinesthesia (kin′-es-THĒ-zē-a) The perception of the extent and direction of movement of body parts; this sense is possible due to nerve impulses generated by proprioceptors.

Kinetochore (ki-NET-ō-kor) Protein complex attached to the outside of a centromere to which kinetochore microtubules attach.

Kyphosis (kī-FŌ-sis) An exaggeration of the thoracic curve of the vertebral column, resulting in a "round-shouldered" appearance. Also called hunchback.

Labia majora (LĀ-bē-a ma-JŌ-ra) Two longitudinal folds of skin extending downward and backward from the mons pubis of the female.

Labia minora (min-OR-a) Two small folds of mucous membrane lying medial to the labia majora of the female.

Labial frenulum (LĀ-bē-al FREN-ū-lum) A medial fold of mucous membrane between the inner surface of the lip and the gums.

Labor The process of giving birth in which a fetus is expelled from the uterus through the vagina. Also called parturition.

Labyrinth Intricate communicating passageway, especially in the internal ear. Another name for the internal (inner) ear.

Lacrimal canaliculus A duct, one on each eyelid, beginning at the punctum at the medial margin of an eyelid and conveying tears medially into the nasolacrimal sac. Plural is canaliculi.

Lacrimal gland Secretory cells, located at the superior anterolateral portion of each orbit, that secrete tears into excretory ducts that open onto the surface of the conjunctiva.

Lacrimal sac The superior expanded portion of the nasolacrimal duct that receives the tears from a lacrimal canal.

Lactation (lak-TĀ-shun) The secretion and ejection of milk by the mammary glands.

Lacteal (LAK-tē-al) One of many lymphatic vessels in villi of the intestines that absorb triglycerides and other lipids from digested food.

Lacuna (la-KOO-na) A small, hollow space, such as that found within the syncytiotrophoblast. Plural is lacunae (la-KOO-nē).

Lambdoid (LAM-doyd) **suture** The joint in the skull between the parietal bones and the occipital bone; sometimes contains sutural bones.

Lamellae (la-MEL-ē) Concentric rings of hard, calcified extracellular matrix found in compact bone.

Lamellated corpuscle Oval-shaped vibration receptor located in the dermis or subcutaneous tissue and consisting of concentric layers of a connective tissue wrapped around the dendrites of a sensory neuron. Also called a pacinian corpuscle (pa-SIN-ē-an).

Lamina propria (PRŌ-prē-a) Areolar connective tissue with elastic fibers and a plexus of veins; part of the mucosa of the organs such as the ureters, urinary bladder, and urethra.

Lanugo (la-NOO-gō) Fine downy hairs that cover the fetus.

Large intestine The portion of the gastrointestinal tract extending from the ileum of the small intestine to the anus, divided structurally into the cecum, colon, rectum, and anal canal.

Laryngopharynx (la-RING-gō-far-ingks) The inferior portion of the pharynx, extending downward from the level of the hyoid bone that divides posteriorly into the esophagus and anteriorly into the larynx. Also called the hypopharynx.

Larynx (LAR-ingks) The voice box, a short passageway that connects the pharynx with the trachea.

Lateral ventricle (VEN-tri-kul) A cavity within a cerebral hemisphere that communicates with the lateral ventricle in the other cerebral hemisphere and with the third ventricle by way of the interventricular foramen.

Lens A transparent organ constructed of proteins (crystallins) lying posterior to the pupil and iris of the eyeball and anterior to the vitreous body.

Lesser omentum (ō-MEN-tum) A fold of the peritoneum that extends from the liver to the lesser curvature of the stomach and the first part of the duodenum.

Lesser vestibular (ves-TIB-ū-lar) **gland** One of the paired mucus-secreting glands with ducts that open on either side of the urethral orifice in the vestibule of the female.

Leukemia (loo-KĒ-mē-a) A malignant disease of the blood-forming tissues characterized by either uncontrolled production and accumulation of immature leukocytes in which many cells fail to reach maturity (acute) or an accumulation of mature leukocytes in the blood because they do not die at the end of their normal life span (chronic).

Leukocyte (LOO-kō-sīt) A white blood cell.

Ligament (LIG-a-ment) Dense regular connective tissue that attaches bone to bone.

Ligamentum teres (TE-rēz) A band of fibrous connective tissue enclosed between the folds of the broad ligament of the uterus, emerging from the uterus just inferior to the uterine tube, extending laterally along the pelvic wall and through the deep inguinal ring to end in the labia majora. Also called the round ligament.

Ligand (LĪ-gand) A chemical substance that binds to a specific receptor.

Limbic system A part of the forebrain, sometimes termed the visceral brain, concerned with various aspects of emotion and behavior; includes the limbic lobe, dentate gyrus, amygdala, septal nuclei, mammillary bodies, anterior thalamic nucleus, olfactory bulbs, and bundles of myelinated axons.

Lingual (LIN-gwal FREN-ū-lum) **frenulum** A fold of mucous membrane that connects the tongue to the floor of the mouth.

Lipases Enzymes that split triglycerides and phospholipids.

Lipid (LIP-id) An organic compound composed of carbon, hydrogen, and oxygen that is usually insoluble in water, but soluble in alcohol, ether, and chloroform; examples include triglycerides (fats and oils), phospholipids, steroids, and eicosanoids.

Lipid bilayer Arrangement of phospholipid, glycolipid, and cholesterol molecules in two parallel sheets in which the hydrophilic "heads" face outward and the hydrophobic "tails" face inward; found in cellular membranes.

Lipoprotein (lip′-ō-PRŌ-tēn) One of several types of particles containing lipids (cholesterol and triglycerides) and proteins that make it water soluble for transport in the blood; high levels of low-density lipoproteins (LDLs) are associated with increased risk of atherosclerosis, whereas high levels of high-density lipoproteins (HDLs) are associated with decreased risk of atherosclerosis.

Liver Large organ under the diaphragm that occupies most of the right hypochondriac region and part of the epigastric region. Functionally, it produces bile and synthesizes most plasma proteins; interconverts nutrients; detoxifies substances; stores glycogen, iron, and vitamins; carries on phagocytosis of worn-out blood cells and bacteria; and helps synthesize the active form of vitamin D.

Long-term potentiation (pō-ten′-shē-Ā-shun) **(LTP)** Prolonged, enhanced synaptic transmission that occurs at certain synapses within the hippocampus of the brain; believed to underlie some aspects of memory.

Lordosis (lor-DŌ-sis) An exaggeration of the lumbar curve of the vertebral column. Also called hollow back.

Lower limb The appendage attached at the pelvic (hip) girdle, consisting of the thigh, knee, leg, ankle, foot, and toes. Also called the lower extremity.

Lumbar plexus A network formed by the anterior branches of spinal nerves L1 through L4.

Lumen (LOO-men) The space within an artery, vein, intestine, renal tubule, or other tubular structure.

Lungs Main organs of respiration that lie on either side of the heart in the thoracic cavity.

Lunula (LOO-noo-la) The moon-shaped white area at the base of a nail.

Luteinizing (LOO-tē-in′-īz-ing) **hormone (LH)** A hormone secreted by the anterior pituitary that stimulates ovulation, stimulates progesterone secretion by the corpus luteum, and readies the mammary glands for milk secretion in females; stimulates testosterone secretion by the testes in males.

Lymph (LIMF) Fluid confined in lymphatic vessels and flowing through the lymphatic system until it is returned to the blood.

Lymph node An oval or bean-shaped structure located along lymphatic vessels.

Lymphatic (lim-FAT-ik) **capillary** Closed-ended microscopic lymphatic vessel that begins in spaces between cells and converges with other lymphatic capillaries to form lymphatic vessels.

Lymphatic system A system consisting of a fluid called lymph, vessels called lymphatics that transport lymph, a number of organs containing lymphatic tissue (lymphocytes within a filtering tissue), and red bone marrow.

Lymphatic tissue A specialized form of reticular tissue that contains large numbers of lymphocytes.

Lymphatic vessel A large vessel that collects lymph from lymphatic capillaries and converges with other lymphatic vessels to form the thoracic and right lymphatic ducts.

Lymphocyte (LIM-fō-sīt) A type of white blood cell that helps carry out cell-mediated and antibody-mediated immune responses; found in blood and in lymphatic tissues.

Lysosome (LĪ-sō-sōm) An organelle in the cytoplasm of a cell, enclosed by a single membrane and containing powerful digestive enzymes.

Lysozyme (LĪ-sō-zīm) A bactericidal enzyme found in tears, saliva, and perspiration.

Macrophage (MAK-rō-fāj) Phagocytic cell derived from a monocyte; may be fixed or wandering.

Macula (MAK-ū-la) A discolored spot or a colored area. A small, thickened region on the wall of the utricle and saccule that contains receptors for linear acceleration or deceleration and head tilt.

Macula lutea (LOO-tē-a) The yellow spot in the center of the retina.

Major histocompatibility complex (MHC) antigens Surface proteins on white blood cells and other nucleated cells that are unique for each person (except for identical siblings); used to type tissues and help prevent rejection of transplanted tissues. Also known as human leukocyte antigens (HLA).

Malleus One of the three small bones of the middle ear called the auditory ossicles.

Mammary (MAM-ar-ē) **gland** Modified sudoriferous (sweat) gland of the female that produces milk for the nourishment of the young.

Mammillary (MAM-i-ler-ē) **bodies** Two small rounded bodies on the inferior aspect of the hypothalamus that are involved in reflexes related to the sense of smell.

Mast cell A cell found in areolar connective tissue that releases histamine, a dilator of small blood vessels, during inflammation.

Mastication (mas′-ti-KĀ-shun) Chewing.

Mature follicle A large, fluid-filled follicle containing a secondary oocyte and surrounding granulosa cells that secrete estrogens. Also called a Graafian (GRĀF-ē-an) follicle.

Meatus (mē-Ā-tus) A passage or opening, especially the external portion of a canal.

Mechanoreceptor (me-KAN-ō-rē-sep-tor) Sensory receptor that detects mechanical deformation of the receptor itself or adjacent cells; stimuli so detected include those related to touch, pressure, vibration, proprioception, hearing, equilibrium, and blood pressure.

Medial lemniscus (lem-NIS-kus) A white matter tract that originates in the gracile and cuneate nuclei of the medulla oblongata and extends to the thalamus on the same side; sensory axons in this tract conduct nerve impulses for the sensations of proprioception, touch, vibration, hearing, and equilibrium.

Median aperture (AP-er-choor) One of the three openings in the roof of the fourth ventricle through which cerebrospinal fluid enters the subarachnoid space of the brain and spinal cord.

Median plane A vertical plane dividing the body into right and left halves. Situated in the middle.

Mediastinum (mē′-dē-as-TĪ-num) The broad, median partition between the pleurae of the lungs that extends from the sternum to the vertebral column in the thoracic cavity.

Medulla (me-DOOL-la) An inner layer of an organ, such as the medulla of the kidneys; alternate term for the medulla oblongata.

Medulla oblongata (me-DOOL-la ob′-long-GA-ta) The most inferior part of the brainstem. Also termed the medulla.

Medullary (MED-ū-lar′-ē) **cavity** The space within the diaphysis of a bone that contains yellow bone marrow. Also called the marrow cavity.

Medullary respiratory center The neurons of the respiratory center in the medulla oblongata that consist of the dorsal respiratory group that is active during normal quiet breathing and the ventral respiratory group that is active during forceful breathing.

Meiosis (mī-Ō-sis) A type of cell division that occurs during production of gametes, involving two successive nuclear divisions that result in cells with the haploid (*n*) number of chromosomes.

Meissner corpuscle See Corpuscle of touch.

Melanin (MEL-a-nin) A dark black, brown, or yellow pigment found in some parts of the body such as the skin, hair, and pigmented layer of the retina.

Melanocyte (MEL-a-nō-sīt′) A pigmented cell, located between or beneath cells of the deepest layer of the epidermis, that synthesizes melanin.

Melanocyte-stimulating hormone (MSH) A hormone secreted by the anterior pituitary that stimulates the dispersion of melanin granules in melanocytes in amphibians; continued administration produces darkening of skin in humans.

Melatonin (me-a-TŌN-in) A hormone secreted by the pineal gland that helps set the timing of the body's biological clock.

Membrane A thin, flexible sheet of tissue composed of an epithelial layer and an underlying connective tissue layer, as in an epithelial membrane, or of areolar connective tissue only, as in a synovial membrane.

Membranous labyrinth (MEM-bra-nus LAB-i-rinth) The part of the labyrinth of the internal ear that is located inside the bony labyrinth and separated from it by the perilymph; made up of the semicircular ducts, the saccule and utricle, and the cochlear duct.

Membranous (MEM-bra-nus) **urethra** Subdivision of the male urethra that passes through the deep muscles of the perineum.

Memory The ability to recall thoughts; commonly classifed as short-term (activated) and long-term.

Menarche (me-NAR-kē) The first menses (menstrual flow) and beginning of ovarian and uterine cycles.

Meninges (me-NIN-jēz) Three membranes covering the brain and spinal cord, called the dura mater, arachnoid mater, and pia mater. Singular is meninx (MEN-inks).

Menopause (MEN-ō-pawz) The termination of the menstrual cycles.

Menstruation (men′-stroo-Ā-shun) Periodic discharge of blood, tissue fluid, mucus, and epithelial cells that usually lasts for 5 days; caused by a sudden reduction in estrogens and progesterone. Also called the menstrual phase or menses.

Merocrine (MER-ō-krin) **gland** Gland made up of secretory cells that remain intact throughout the process of formation and discharge of the secretory product, as in the salivary and pancreatic glands.

Mesenchyme (MEZ-en-kīm) An embryonic connective tissue from which all other connective tissues arise.

Mesentery (MEZ-en-ter′-ē) A fold of peritoneum attaching the small intestine to the posterior abdominal wall.

Mesocolon (mez′-ō-KŌ-lon) A fold of peritoneum attaching the colon to the posterior abdominal wall.

Mesoderm The middle primary germ layer that gives rise to connective tissues, blood and blood vessels, and muscles.

Mesothelium (mez′-ō-THĒ-lē-um) The layer of simple squamous epithelium that lines serous membranes.

Mesovarium (mez′-ō-VĀ-rē-um) A short fold of peritoneum that attaches an ovary to the broad ligament of the uterus.

Metabolism (me-TAB-ō-lizm) All the biochemical reactions that occur within an organism, including the synthetic (anabolic) reactions and decomposition (catabolic) reactions.

Metacarpus A collective term for the five bones that make up the palm.

Metaphase (MET-a-fāz) The second stage of mitosis, in which chromatid pairs line up on the metaphase plate of the cell.

Metaphysis (me-TAF-i-sis) Region of a long bone between the diaphysis and epiphysis that contains the epiphyseal plate in a growing bone.

Metarteriole (met′-ar-Ē-rē-ōl) A blood vessel that emerges from an arteriole, traverses a capillary network, and empties into a venule.

Metastasis (me-TAS-ta-sis) The spread of cancer to surrounding tissues (local) or to other body sites (distant).

Metatarsus (met′-a-TAR-sus) A collective term for the five bones located in the foot between the tarsals and the phalanges.

Microglial cells Neuroglial cells that carry on phagocytosis. Also called microglia (mī-KROG-lē-a).

Microtubule (mī-krō-TOO-būl) Cylindrical protein filament, from 18 to 30 nm in diameter, consisting of the protein tubulin; provides support, structure, and transportation.

Microvilli (mī-krō-VIL-ī) Microscopic, fingerlike projections of the plasma membranes of cells that increase surface area for absorption, especially in the small intestine and proximal convoluted tubules of the kidneys.

Micturition The act of expelling urine from the urinary bladder. Also called urination (ū-ri-NĀ-shun).

Midbrain The part of the brain between the pons and the diencephalon. Also called the mesencephalon (mes′-en-SEF-a-lon).

Middle ear A small, epithelial-lined cavity hollowed out of the temporal bone, separated from the external ear by the eardrum and from the internal ear by a thin bony partition containing the oval and round windows; extending across the middle ear are the three auditory ossicles. Also called the tympanic (tim-PAN-ik) cavity.

Midline An imaginary vertical line that divides the body into equal left and right sides.

Midsagittal plane A vertical plane through the midline of the body that divides the body or organs into equal right and left sides. Also called a median plane.

Mineralocorticoids (min′-er-al-ō-KORT-ti-koyds) A group of hormones of the adrenal cortex that help regulate sodium and potassium balance.

Mitochondrion (mī-tō-KON-drē-on) A double-membraned organelle that plays a central role in the production of ATP; known as the "powerhouse" of the cell. *Plural* is mitochondria.

Mitosis (mī-TŌ-sis) The orderly division of the nucleus of a cell that ensures that each new nucleus has the same number and kind of chromosomes as the original nucleus. The process includes the replication of chromosomes and the distribution of the two sets of chromosomes into two separate and equal nuclei.

Mitotic spindle Collective term for a football-shaped assembly of microtubules (nonkinetochore, kinetochore, and aster) that is responsible for the movement of chromosomes during cell division.

Modiolus (mō-DĪ-ō′-lus) The central pillar or column of the cochlea.

Molecule (mol′-e-KŪL) A combination of two or more atoms that share electrons.

Monocyte (MON-ō-sīt′) The largest type of white blood cell, characterized by agranular cytoplasm.

Monounsaturated fat A fatty acid that contains one double covalent bond between its carbon atoms; it is not completely saturated with hydrogen atoms. Plentiful in triglycerides of olive and peanut oils.

Mons pubis (MONZ PŪ-bis) The rounded, fatty prominence over the pubic symphysis, covered by coarse pubic hair.

Morula (MOR-ū-la) A solid sphere of cells produced by successive cleavages of a fertilized ovum about four days after fertilization.

Motor area The region of the cerebral cortex that governs muscular movement, particularly the precentral gyrus of the frontal lobe.

Motor end plate Region of the sarcolemma of a muscle fiber (cell) that includes acetylcholine (ACh) receptors, which bind ACh released by synaptic end bulbs of somatic motor neurons.

Motor neurons (NOO-ronz) Neurons that conduct impulses from the brain toward the spinal cord or out of the brain and spinal cord into cranial or spinal nerves to effectors that may be either muscles or glands. Also called efferent neurons.

Motor speech area Motor area of the brain in the frontal lobe that translates thoughts into speech. Also called Broca's speech area.

Motor unit A motor neuron together with the muscle fibers (cells) it stimulates.

Mucosa (mū-KŌ-sa) A membrane that lines a body cavity that opens to the exterior. Also called the mucous membrane.

Mucosa-associated lymphatic tissue (MALT) Lymphatic nodules scattered throughout the lamina propria (connective tissue) of mucous membranes lining the gastrointestinal tract, respiratory airways, urinary tract, and reproductive tract.

Mucous membrane A membrane that lines a body cavity that opens to the exterior. Also called the mucosa (mū-KŌ-sa).

Mucus The thick fluid secretion of goblet cells, mucous cells, mucous glands, and mucous membranes.

Muscarinic (mus′-ka-RIN-ik) **receptor** Receptor for the neurotransmitter acetylcholine found on all effectors innervated by parasympathetic postganglionic axons and on sweat glands innervated by cholinergic sympathetic postganglionic axons; so named because muscarine activates these receptors but does not activate nicotinic receptors for acetylcholine.

Muscle action potential A stimulating impulse that propagates along the sarcolemma and transverse tubules; in skeletal muscle, it is generated by acetylcholine, which increases the permeability of the sarcolemma to cations, especially sodium ions (Na^+).

Muscle fatigue (fa-TĒG) Inability of a muscle to maintain its strength of contraction or tension; may be related to insufficient oxygen, depletion of glycogen, and/or lactic acid buildup.

Muscle spindle An encapsulated proprioceptor in a skeletal muscle, consisting of specialized intrafusal muscle fibers and nerve endings; stimulated by changes in length or tension of muscle fibers.

Muscle tone A sustained, partial contraction of portions of a skeletal or smooth muscle in response to activation of stretch receptors or a baseline level of action potentials in the innervating motor neurons.

Muscular dystrophy (DIS-trō-fē) Inherited muscle destroying diseases, characterized by degeneration of muscle fibers (cells), which causes progressive atrophy of the skeletal muscle.

Muscular system Usually refers to the approximately 100 voluntary muscles of the body that are composed of skeletal muscle tissue.

Muscular tissue A tissue specialized to produce motion in response to muscle action potentials by its qualities of contractility, extensibility, elasticity, and excitability; types include skeletal, cardiac, and smooth.

Muscularis (MUS-kū-lar′-is) A muscular layer (coat or tunic) of an organ, such as the muscularis of the vagina.

Muscularis mucosae (mū-KŌ-sē) A thin layer of smooth muscle fibers that underlie the lamina propria of the mucosa of the gastrointestinal tract.

Musculoskeletal (mus′-kyū-lō-SKEL-e-tal) **system** An integrated body system consisting of bones, joints, and muscles.

Mutation (mū-TĀ-shun) Any change in the sequence of bases in a DNA molecule resulting in a permanent alteration in some inheritable trait.

Myasthenia (mī-as-THĒ-nē-a) **gravis** Weakness and fatigue of skeletal muscles caused by antibodies directed against acetylcholine receptors.

Myelin (MĪ-e-lin) **sheath** Multilayered lipid and protein covering, formed by Schwann cells and oligodendrocytes, around axons of many peripheral and central nervous system neurons.

Myenteric (mī-en-TER-ik) **plexus** A network of autonomic axons and postganglionic cell bodies located in the muscularis of the gastrointestinal tract. Also called the plexus of Auerbach (OW-er-bak).

Myocardial infarction (mī′-ō-KAR-dē-al in-FARK-shun) Gross necrosis of myocardial tissue due to interrupted blood supply. Also called a heart attack.

Myocardium (mī′-ō-KAR-dē-um) The middle layer of the heart wall, made up of cardiac muscle tissue, lying between the epicardium and the endocardium and constituting the bulk of the heart.

Myofibrils (mī-ō-FĪ-brils) Threadlike structures extending longitudinally through a muscle fiber (cell) consisting mainly of thick filaments (myosin) and thin filaments (actin, troponin, and tropomyosin).

Myoglobin (mī-ō-GLŌB-in) The oxygen-binding, iron-containing protein present in the sarcoplasm of muscle fibers (cells); contributes the red color to muscle.

Myogram (MĪ-ō-gram) The record or tracing produced by a myograph, an apparatus that measures and records the force of muscular contractions.

Myology (mī-OL-ō-jē) The study of muscles.

Myometrium (mī′-ō-MĒ-trē-um) The smooth muscle layer of the uterus.

Myopathy (mī-OP-a-thē) Any abnormal condition or disease of muscle tissue.

Myopia (mī-Ō-pē-a) Defect in vision in which objects can be seen distinctly only when very close to the eyes; nearsightedness.

Myosin (MĪ-ō-sin) The contractile protein that makes up the thick filaments of muscle fibers.

Myotome (MĪ-ō-tōm) A group of muscles innervated by the motor neurons of a single spinal segment. In an embryo, the portion of a somite that develops into some skeletal muscles.

Nail A hard plate, composed largely of keratin, that develops from the epidermis of the skin to form a protective covering on the dorsal surface of the distal phalanges of the fingers and toes.

Nail matrix The portion of the epithelium proximal to the nail root.

Nasal (NĀ-zal) **cavity** A mucosa-lined cavity on either side of the nasal septum that opens onto the face at the external nares and into the nasopharynx at the internal nares.

Nasal septum (SEP-tum) A vertical partition composed of bone (perpendicular plate of ethmoid and vomer) and cartilage, covered with a mucous membrane, separating the nasal cavity into left and right sides.

Nasolacrimal duct A canal that transports the lacrimal secretion (tears) from the nasolacrimal sac into the nose.

Nasopharynx (nā′-zō-FAR-inks) The superior portion of the pharynx, lying posterior to the nose and extending inferiorly to the soft palate.

Neck The part of the body connecting the head and the trunk. A constricted portion of an organ, such as the neck of the femur or uterus.

Necrosis (ne-KRŌ-sis) A pathological type of cell death that results from disease, injury, or lack of blood supply in which many adjacent cells swell, burst, and spill their contents into the interstitial fluid, triggering an inflammatory response.

Negative feedback system A feedback system that reverses a change in a controlled condition.

Neonatal (nē-ō-NĀ-tal) **period** The first four weeks after birth.

Neoplasm (NĒ-ō-plazm) A new growth that may be benign or malignant.

Nephron The functional unit of the kidney.

Nephron loop The part of the renal tubule that receives fluid from the proximal convoluted tubule and transmits it to the distal convoluted tubule. Also called the loop of Henle.

Nerve A cordlike bundle of neuronal axons and/or dendrites and associated connective tissue coursing together outside the central nervous system.

Nerve action potential A wave of depolarization and repolarization that self-propagates along the plasma membrane of a neuron. Also called a nerve impulse.

Nerve fiber General term for any process (axon or dendrite) projecting from the cell body of a neuron.

Nervous system A network of billions of neurons and even more neuroglia that is organized into two main divisions: central nervous system (brain and spinal cord) and peripheral nervous system (nerves, ganglia, enteric plexuses, and sensory receptors outside the central nervous system).

Nervous tissue Tissue containing neurons that initiate and conduct nerve impulses to coordinate homeostasis, and neuroglia that provide support and nourishment to neurons.

Neural plate A thickening of ectoderm, induced by the notochord, that forms early in the third week of development and represents the beginning of the development of the nervous system.

Neural tube defect (NTD) A developmental abnormality in which the neural tube does not close properly. Examples are spina bifida and anencephaly.

Neuroglia (noo-RŌG-lē-a) Cells of the nervous system that perform various supportive functions. The neuroglia of the central nervous system are the astrocytes, oligodendrocytes, microglia, and ependymal cells; neuroglia of the peripheral nervous system include Schwann (SCHWON) cells and satellite cells. Also called glia (GLĒ-a).

Neurohypophyseal (noo′-rō-hī′pō-FIZ-ē-al) **bud** An outgrowth of ectoderm located on the floor of the hypothalamus that gives rise to the posterior pituitary.

Neurolemma (noo-rō-LEM-a) The peripheral, nucleated cytoplasmic layer of the Schwann cell. Also called sheath of Schwann.

Neurology (noo-ROL-ō-jē) The study of the normal functioning and disorders of the nervous system.

Neuromuscular (noo-rō-MUS-kū-lar) **junction (NMJ)** A synapse between the axon terminals of a motor neuron and the sarcolemma of a muscle fiber (cell).

Neuron (NOO-ron) A nerve cell, consisting of a cell body, dendrites, and an axon.

Neurosecretory (noo-rō-SĒK-re-tō-rē) **cell** A neuron that secretes a hypothalamic releasing hormone or inhibiting hormone into blood capillaries of the hypothalamus; a neuron that secretes oxytocin or antidiuretic hormone into blood capillaries of the posterior pituitary.

Neurotransmitter (noo-rō-trans′-MIT-er) One of a variety of molecules within axon terminals that are released into the synaptic cleft in response to a nerve impulse and that change the membrane potential of the postsynaptic neuron.

Neurulation (noor-oo-LĀ-shun) The process by which the neural plate, neural folds, and neural tube develop.

Neutrophil (NOO-trō-fil) A type of white blood cell characterized by granules that stain pale lilac with a combination of acidic and basic dyes.

Nicotinic (nik′-ō-TIN-ik) **receptor** Receptor for the neurotransmitter acetylcholine found on both sympathetic and parasympathetic postganglionic neurons and on skeletal muscle in the motor end plate; so named because nicotine activates these receptors but does not activate muscarinic receptors for acetylcholine.

Nipple A pigmented, wrinkled projection on the surface of the breast that in the female is the location of the openings of the lactiferous ducts for milk release.

Nociceptor (nō′-sē-SEP-tor) A free (naked) nerve ending that detects painful stimuli.

Node of Ranvier (RON-vē-a) A space along a myelinated axon between the individual Schwann cells that form the myelin sheath and the neurolemma.

Norepinephrine (nor′-ep-ē-NEF-rin) **(NE)** A hormone secreted by the adrenal medulla that produces actions similar to those that result from sympathetic stimulation. Also called noradrenaline (nor-a-DREN-a-lin).

Notochord (NŌ-tō-kord) A flexible rod of mesodermal tissue that lies where the future vertebral column will develop and plays a role in induction.

Nucleic (noo-KLĒ-ik) **acid** An organic compound that is a long polymer of nucleotides, with each nucleotide containing a pentose sugar, a phosphate group, and one of four possible nitrogenous bases (adenine, cytosine, guanine, and thymine or uracil).

Nucleoli Spherical bodies within a cell nucleus composed of protein, DNA, and RNA that are the sites of the assembly of small and large ribosomal subunits. Singular is nucleolus.

Nucleosome (NOO-klē-ō-sōm) Structural subunit of a chromosome consisting of histones and DNA.

Nucleus (NOO-klē-us) A spherical or oval organelle of a cell that contains the hereditary factors of the cell, called genes. A cluster of unmyelinated nerve cell bodies in the central nervous system. The central part of an atom made up of protons and neutrons.

Nutrient A chemical substance in food that provides energy, forms new body components, or assists in various body functions.

Obesity (ō-BĒS-i-tē) Body weight more than 20% above a desirable standard due to excessive accumulation of fat.

Oblique plane A plane that passes through the body or an organ at an angle between the transverse plane and the midsagittal, parasagittal, or frontal plane.

Obstetrics (ob-STET-riks) The specialized branch of medicine that deals with pregnancy, labor, and the period of time immediately after delivery (about 6 weeks).

Olfaction (ōl-FAK-shun) The sense of smell.

Olfactory bulb A mass of gray matter containing cell bodies of neurons that form synapses with neurons of the olfactory (I) nerve, lying inferior to the frontal lobe of the cerebrum on either side of the crista galli of the ethmoid bone.

Olfactory receptor cell A bipolar neuron with its cell body lying between supporting cells located in the mucous membrane lining the superior portion of each nasal cavity; transduces odors into neural signals.

Olfactory tract A bundle of axons that extends posteriorly from the olfactory bulb to olfactory regions of the cerebral cortex.

Oligodendrocyte (OL-i-gō-den′-drō-sīt) A neuroglial cell that supports neurons and produces a myelin sheath around axons of neurons of the central nervous system.

Olive A prominent oval mass on each lateral surface of the superior part of the medulla oblongata.

Oncogene (ON-kō-jēn) Cancer-causing gene; it derives from a normal gene, termed a protooncogene, that encodes proteins involved in cell growth or cell regulation but has the ability to transform a normal cell into a cancerous cell when it is mutated or inappropriately activated.

Oncology (on-KOL-ō-jē) The study of tumors.

Oogenesis (ō-ō-JEN-e-sis) Formation and development of female gametes (oocytes).

Oophorectomy (ō′-of-ō-REK-tō-mē) Surgical removal of the ovaries.

Ophthalmology (of-thal-MOL-ō-jē) The study of the structure, function, and diseases of the eye.

Opposition Movement of the thumb at the carpometacarpal joint in which the thumb moves across the palm to touch the tips of the fingers on the same hand.

Optic chiasm (kī-AZM) A crossing point of the two branches of the optic (II) nerve, anterior to the pituitary gland.

Optic disc A small area of the retina containing openings through which the axons of the ganglion cells emerge as the optic (II) nerve. Also called the blind spot.

Optic tract A bundle of axons that carry nerve impulses from the retina of the eye between the optic chiasm and the thalamus.

Ora serrata (Ō-ra ser-RĀ-ta) The irregular margin of the retina lying internal and slightly posterior to the junction of the choroid and ciliary body.

Orbit (OR-bit) The bony, pyramidal-shaped cavity of the skull that holds the eyeball.

Organ A structure composed of two or more different kinds of tissues with a specific function and usually a recognizable shape.

Organelle (or-ga-NEL) A permanent structure within a cell with characteristic morphology that is specialized to serve a specific function in cellular activities.

Organism A total living form; one individual.

Organogenesis (or′-ga-nō-JEN-e-sis) The formation of body organs and systems. By the end of the eighth week of development, all major body systems have begun to develop.

Origin (OR-i-jin) The attachment of a muscle tendon to a stationary bone or the end opposite the insertion.

Oropharynx (or′-ō-FAR-inks) The intermediate portion of the pharynx, lying posterior to the mouth and extending from the soft palate to the hyoid bone.

Orthopedics (or′-thō-PĒ-diks) The branch of medicine that deals with the preservation and restoration of the skeletal system, articulations, and associated structures.

Osmoreceptor (oz′-mō-rē-SEP-tor) Receptor in the hypothalamus that is sensitive to changes in blood osmolarity and, in response to high osmolarity (low water concentration), stimulates synthesis and release of antidiuretic hormone (ADH).

Osmosis (oz-MŌ-sis) The net movement of water molecules through a selectively permeable membrane from an area of higher water concentration to an area of lower water concentration until equilibrium is reached.

Ossification (os′-i-fi-KĀ-shun) Formation of bone. Also called osteogenesis.

Ossification center An area in the cartilage model of a future bone where the cartilage cells hypertrophy, secrete enzymes that calcify their extracellular matrix, and die, and the area they occupied is invaded by osteoblasts that then lay down bone.

Osteoblast (OS-tē-ō-blast′) Cell formed from an osteogenic cell that participates in bone formation by secreting some organic components and inorganic salts.

Osteoclast (OS-tē-ō-klast′) A large, multinuclear cell that resorbs (destroys) bone matrix.

Osteocyte (OS-tē-ō-sīt′) A mature bone cell that maintains the daily activities of bone tissue.

Osteoprogenitor (os-tē-ō-prō-JEN-i-tor) **cell** Stem cell derived from mesenchyme that has mitotic potential and the ability to differentiate into an osteoblast.

Osteology (os-tē-OL-ō-jē) The study of bones.

Osteon (OS-tē-on) The basic unit of structure in adult compact bone, consisting of a central (haversian) canal with its concentrically arranged lamellae, lacunae, osteocytes, and canaliculi. Also called a haversian (ha-VER-shan) system.

Osteoporosis (os′-tē-ō-pō-RŌ-sis) Age-related disorder characterized by decreased bone mass and increased susceptibility to fractures, often as a result of decreased levels of estrogens.

Otolith (Ō-tō-lith) A particle of calcium carbonate embedded in the otolithic membrane that functions in maintaining static equilibrium.

Otolithic (ō-tō-LITH-ik) **membrane** Thick, gelatinous, glycoprotein layer located directly over hair cells of the macula in the saccule and utricle of the internal ear.

Otorhinolaryngology (ō-tō-rī′-nō-lar-in-GOL-ō-jē) The branch of medicine that deals with the diagnosis and treatment of diseases of the ears, nose, and throat.

Oval window A small, membrane-covered opening between the middle ear and inner ear into which the footplate of the stapes fits.

Ovarian (ō-VAR-ē-an) **cycle** A monthly series of events in the ovary associated with the maturation of a secondary oocyte.

Ovarian follicle (FOL-i-kul) A general name for oocytes (immature ova) in any stage of development, along with their surrounding epithelial cells.

Ovarian ligament (LIG-a-ment) A rounded cord of connective tissue that attaches the ovary to the uterus.

Ovary (Ō-var-ē) Female gonad that produces oocytes and the hormones estrogens, progesterone, inhibin, and relaxin.

Ovulation (ov′-ū-LĀ-shun) The rupture of a mature ovarian (Graafian) follicle with discharge of a secondary oocyte into the pelvic cavity.

Ovum (Ō-vum) The female reproductive or germ cell; an egg cell; arises through completion of meiosis in a secondary oocyte after penetration by a sperm.

Oxyhemoglobin (ok′-sē-HĒ-mō-glō-bin) Hemoglobin combined with oxygen.

Oxytocin (ok-sē-TŌ-sin) **(OT)** A hormone secreted by neurosecretory cells in the paraventricular and supraoptic nuclei of the hypothalamus that stimulates contraction of smooth muscle in the pregnant uterus and myoepithelial cells around the ducts of mammary glands.

P wave The deflection wave of an electrocardiogram that signifies atrial depolarization.

Pacinian corpuscle See Lamellated corpuscle.

Palate (PAL-at) The horizontal structure separating the oral and the nasal cavities; the roof of the mouth.

Pancreas (PAN-krē-as) A soft, oblong organ lying along the greater curvature of the stomach and connected by a duct to the duodenum. It is both an exocrine gland (secreting pancreatic juice) and an endocrine gland (secreting insulin, glucagon, somatostatin, and pancreatic polypeptide).

Pancreatic (pan′-krē-AT-ik) **duct** A single large tube that unites with the common bile duct from the liver and gallbladder and drains pancreatic juice into the duodenum at the hepatopancreatic ampulla (ampulla of Vater). Also called the duct of Wirsung.

Pancreatic islet (Ī-let) A cluster of endocrine gland cells in the pancreas that secretes insulin, glucagon, somatostatin, and pancreatic polypeptide. Also called an islet of Langerhans (LANG-er-hanz).

Papanicolaou (pa-pa-NI-kō-lō) **test** A cytological staining test for the detection and diagnosis of premalignant and malignant conditions of the female genital tract. Cells scraped from the epithelium of the cervix of the uterus are examined microscopically. Also called a Pap test or Pap smear.

Papillae (pa-PIL-ē) Projections of the lamina propria covered with stratified squamous epithelium that cover the dorsal and lateral surfaces of the tongue.

Paranasal sinus (par′-a-NĀ-zal SĪ-nus) A mucus-lined air cavity in a skull bone that communicates with the nasal cavity. Paranasal sinuses are located in the frontal, maxillary, ethmoid, and sphenoid bones.

Paraplegia (par-a-PLĒ-jē-a) Paralysis of both lower limbs.

Parasagittal plane A vertical plane that does not pass through the midline and that divides the body or organs into unequal left and right portions.

Parasympathetic (par′-a-sim-pa-THET-ik) **nervous system** One of the two main subdivisions of the autonomic nervous system, having cell bodies of preganglionic neurons in nuclei in the brainstem and in the lateral gray horn of the sacral portion of the spinal cord; primarily concerned with activities that conserve and restore body energy. Also known as the craniosacral division.

Parathyroid (par′-a-THĪ-royd) **gland** One of usually four small endocrine glands embedded in the posterior surfaces of the lateral lobes of the thyroid gland.

Parathyroid hormone (PTH) A hormone secreted by the chief (principal) cells of the parathyroid glands that increases blood calcium level and decreases blood phosphate level. Also called parathormone.

Paraurethral (par′-a-ū-RĒ-thral) **gland** Gland embedded in the wall of the urethra with a duct that opens on either side of the urethral orifice and secretes mucus. Also called Skene's (SKE–NZ) gland.

Parenchyma (pa-RENG-ki-ma) The functional parts of any organ, as opposed to tissue that forms its stroma or framework.

Parietal cell A type of secretory cell in gastric glands that produces hydrochloric acid and intrinsic factor.

Parietal pleura (PLOO-ra) The outer layer of the serous pleural membrane that encloses and protects the lungs; the layer that is attached to the wall of the pleural cavity.

Parkinson's disease (PD) Progressive degeneration of the basal nuclei and substantia nigra of the cerebrum resulting in decreased production of dopamine (DA) that leads to tremor, slowing of voluntary movements, and muscle weakness.

Parotid (pa-ROT-id) **gland** One of the paired salivary glands located inferior and anterior to the ears and connected to the oral cavity via a duct (parotid) that opens into the inside of the cheek opposite the maxillary (upper) second molar tooth.

Pars intermedia A small avascular zone between the anterior and posterior pituitary glands.

Patent (PĀ-tent) **ductus arteriosus (PDA)** A congenital heart defect in which the ductus arteriosus remains open. As a result, aortic blood flows into the lower-pressure pulmonary trunk, increasing pulmonary trunk pressure and overworking both ventricles.

Pathogen (PATH-ō-jen) A disease-producing microbe.

Pectinate (PEK-ti-nāt) **muscles** Projecting muscle bundles of the anterior atrial walls and the lining of the auricles.

Pectoral (PEK-tō-ral) Pertaining to the chest or breast.

Pedicel Footlike structure, as on podocytes of a glomerulus.

Pelvic cavity Inferior portion of the abdominopelvic cavity that contains the urinary bladder, sigmoid colon, rectum, and internal female and male reproductive structures.

Pelvic splanchnic (PEL-vic SPLANGK-nik) **nerves** Consist of preganglionic parasympathetic axons from the levels of S2, S3, and S4 that supply the urinary bladder, reproductive organs, and the descending and sigmoid colon and rectum.

Penis (PĒ-nis) The organ of urination and copulation in males; used to deposit semen into the female vagina.

Pepsin Protein-digesting enzyme secreted by chief cells of the stomach in the inactive form pepsinogen, which is converted to active pepsin by hydrochloric acid.

Peptic ulcer An ulcer that develops in areas of the gastrointestinal tract exposed to hydrochloric acid; classified as a gastric ulcer if in the lesser curvature of the stomach and as a duodenal ulcer if in the first part of the duodenum.

Percussion The act of striking (percussing) an underlying part of the body with short, sharp taps as an aid in diagnosing the part by the quality of the sound produced.

Perforating canal A minute passageway by means of which blood vessels and nerves from the periosteum penetrate into compact bone. Also called Volkmann's (FŌLK-mans) canal.

Perforating fibers Thick bundles of collagen that extend from the periosteum into the bone extracellular matrix to attach the periosteum to the underlying bone. Also called Sharpey's fibers.

Pericardial (per′-i-KAR-dē-al) **cavity** Small potential space between the visceral and parietal layers of the serous pericardium that contains pericardial fluid.

Pericardium (per′-i-KAR-dē-um) A loose-fitting membrane that encloses the heart, consisting of a superficial fibrous layer and a deep serous layer.

Perichondrium (per′-i-KON-drē-um) The membrane that covers cartilage.

Perilymph (PER-i-limf) The fluid contained between the bony and membranous labyrinths of the inner ear.

Perimetrium (per′-i-MĒ-trē-um) The serosa of the uterus.

Perimysium (per-i-MĪZ-ē-um) Invagination of the epimysium that divides muscles into bundles.

Perineum (per′-i-NĒ-um) The pelvic floor; the space between the anus and the scrotum in the male and between the anus and the vulva in the female.

Perineurium Connective tissue wrapping around fascicles in a nerve.

Periodontal (per′-ē-ō-DON-tal) **disease** A collective term for conditions characterized by degeneration of gingivae, alveolar bone, periodontal ligament, and cementum.

Periodontal ligament The periosteum lining the alveoli (sockets) for the teeth in the alveolar processes of the mandible and maxillae. Also called the periodontal membrane.

Periosteum (per-ē-OS-tē-um) The membrane that covers bone and consists of connective tissue, osteogenic cells, and osteoblasts; is essential for bone growth, repair, and nutrition.

Peripheral nervous system (PNS) The part of the nervous system that lies outside the central nervous system, consisting of nerves and ganglia.

Peristalsis (per′-i-STAL-sis) Successive muscular contractions along the wall of a hollow muscular structure.

Peritoneum (per′-i-tō-NĒ-um) The largest serous membrane of the body that lines the abdominal cavity and covers the viscera within it.

Peritonitis (per′-i-tō-NĪ-tis) Inflammation of the peritoneum.

Peroxisome (pe-ROKS-i-sōm) Organelle similar in structure to a lysosome that contains enzymes that use molecular oxygen to oxidize various organic compounds; such reactions produce hydrogen peroxide; abundant in liver cells.

Perspiration Sweat; produced by sudoriferous (sweat) glands and containing water, salts, urea, uric acid, amino acids, ammonia, sugar, lactic acid, and ascorbic acid.

pH A measure of the concentration of hydrogen ions (H^+) in a solution. The pH scale extends from 0 to 14, with a value of 7 expressing neutrality, values lower than 7 expressing increasing acidity, and values higher than 7 expressing increasing alkalinity.

Phagocytes Body cells that engulf large solid particles.

Phagocytosis (fag′-ō-sī-TŌ-sis) The process by which phagocytes ingest and destroy microbes, cell debris, and other foreign matter.

Phalanges (fa-LAN-jēz) Bones of fingers or toes. Singular is phalanx (FĀ-lanks).

Pharmacology (far′-ma-KOL-ō-jē) The science of the effects and uses of drugs in the treatment of disease.

Pharyngeal tonsil Single tonsil embedded in the posterior wall of the nasopharynx. Also called the adenoid.

Pharynx (FAR-inks) The throat; a tube that starts at the internal nares and runs partway down the neck, where it opens into the esophagus posteriorly and the larynx anteriorly.

Phenotype (FĒ-nō-tīp) The observable expression of genotype; physical characteristics of an organism determined by genetic makeup and influenced by interaction between genes and internal and external environmental factors.

Phlebitis (fle-BĪ-tis) Inflammation of a vein, usually in a lower limb.

Photopigment A substance that can absorb light and undergo structural changes that can lead to the development of a receptor potential. In the eye, also called visual pigment.

Photoreceptor Receptor that detects light shining on the retina of the eye.

Physiology Science that deals with the functions of an organism or its parts.

Pia mater (PĒ-a MĀ-ter) The innermost of the three meninges (coverings) of the brain and spinal cord.

Pineal (PĪN-ē-al) **gland** A cone-shaped gland located in the roof of the third ventricle that secretes melatonin.

Pinealocyte (pin-ē-AL-ō-sīt) Secretory cell of the pineal gland that releases melatonin.

Pinna (PIN-na) The projecting part of the external ear composed of elastic cartilage and covered by skin and shaped like the flared end of a trumpet. Also called the auricle (OR-i-kul).

Pituicyte (pi-TOO-i-sīt) Supporting cell of the posterior pituitary.

Pituitary (pi-TOO-i-tār-ē) **gland** A small endocrine gland occupying the hypophyseal fossa of the sphenoid bone and attached to the hypothalamus by the infundibulum. Also called the hypophysis (hī-POF-i-sis).

Pivot joint A synovial joint in which a rounded, pointed, or conical surface of one bone articulates with a ring formed partly by another bone and partly by a ligament, as in the joint between the atlas and axis and between the proximal ends of the radius and ulna. Also called a trochoid (TRŌ-koyd) joint.

Placenta (pla-SEN-ta) The special structure through which the exchange of materials between fetal and maternal circulations occurs. Called the afterbirth following birth.

Plane joint Joint in which the articulating surfaces are flat or slightly curved that permits back-and-forth and side-to-side movements and rotation between the flat surfaces.

Plantar flexion (PLAN-tar FLEK-shun) Bending the foot in the direction of the plantar surface (sole).

Plasma (PLAZ-ma) The extracellular fluid found in blood vessels; blood minus the formed elements.

Plasma cell Cell that develops from a B cell (lymphocyte) and produces antibodies.

Plasma membrane Outer, limiting membrane that separates the cell's internal parts from extracellular fluid or the external environment.

Platelet (PLĀT-let) A fragment of cytoplasm enclosed in a cell membrane and lacking a nucleus; found in the circulating blood; plays a role in hemostasis.

Platelet plug Aggregation of platelets (thrombocytes) at a site where a blood vessel is damaged that helps stop or slow blood loss.

Pleura (PLOO-ra) The serous membrane that covers the lungs and lines the walls of the chest and the diaphragm.

Pleural cavity Small potential space between the visceral and parietal pleurae.

Plexus (PLEK-sus) A network of nerves, veins, or lymphatic vessels.

Pluripotent (ploo-RI-pō-tent) **stem cell** Immature stem cell in red bone marrow that gives rise to precursors of all the different mature blood cells.

Polycythemia (pol′ -ē-sī-THĒ-mē-a) Disorder characterized by an above-normal hematocrit (above 55%) in which hypertension, thrombosis, and hemorrhage can occur.

Polyunsaturated fat A fatty acid that contains more than one double covalent bond between its carbon atoms; abundant in triglycerides of corn oil, safflower oil, and cottonseed oil.

Pons (PONZ) The part of the brainstem that forms a "bridge" between the medulla oblongata and the midbrain, anterior to the cerebellum.

Pontine respiratory group A collection of neurons in the pons that transmits nerve impulses to the dorsal respiratory group, and may modify the basic rhythm of breathing. Formerly called the pneumotaxic area.

Portal system The circulation of blood from one capillary network into another through a vein.

Positive feedback system Feedback system that strengthens a change in one of the body's controlled conditions.

Postcentral gyrus Gyrus of cerebral cortex located immediately posterior to the central sulcus; contains the primary somatosensory area.

Posterior column–medial lemniscus pathways Sensory pathways that carry information related to proprioception, touch, pressure, and vibration. First-order neurons project from the spinal cord to the ipsilateral medulla in the posterior columns (gracile fasciculus and cuneate fasciculus). Second-order neurons project from the medulla to the contralateral thalamus in the medial lemniscus. Third-order neurons project from the thalamus to the somatosensory cortex (postcentral gyrus) on the same side.

Posterior pituitary Posterior lobe of the pituitary gland. Also called the neurohypophysis (noo-rō-hī-POF-i-sis).

Posterior root The structure composed of sensory axons lying between a spinal nerve and the dorsolateral aspect of the spinal cord. Also called the dorsal root.

Posterior root ganglion (GANG-glē-on) A group of cell bodies of sensory neurons and their supporting cells located along the posterior root of a spinal nerve. Also called a dorsal root ganglion.

Postganglionic neuron (post′-gang-lē-ON-ik NOO-ron) The second autonomic motor neuron in an autonomic pathway, having its cell body and dendrites located in an autonomic ganglion and its unmyelinated axon ending at cardiac muscle, smooth muscle, or a gland.

Postsynaptic (post-sin-AP-tik) **neuron** The nerve cell that is activated by the release of a neurotransmitter from another neuron and carries nerve impulses away from the synapse.

Precapillary sphincter (SFINGK-ter) A ring of smooth muscle fibers (cells) at the site of origin of true capillaries that regulate blood flow into true capillaries.

Precentral gyrus Gyrus of cerebral cortex located immediately anterior to the central sulcus; contains the primary motor area.

Preganglionic (prē-gang-lē-ON-ik) **neuron** The first autonomic motor neuron in an autonomic pathway, with its cell body and dendrites in the brain or spinal cord and its myelinated axon ending at an autonomic ganglion, where it synapses with a postganglionic neuron.

Pregnancy Sequence of events that normally includes fertilization, implantation, embryonic growth, and fetal growth and terminates in birth.

Premenstrual syndrome Severe physical and emotional stress occurring late in the postovulatory phase of the menstrual cycle and sometimes overlapping with menstruation.

Prepuce (PRĒ-poos) The loose-fitting skin covering the glans of the penis and clitoris. Also called the foreskin.

Presbyopia (prez-bē-Ō-pē-a) A loss of elasticity of the lens of the eye due to advancing age, with resulting inability to focus clearly on near objects.

Presynaptic (prē-sin-AP-tik) **neuron** A neuron that propagates nerve impulses toward a synapse.

Prevertebral ganglion (prē-VER-te-bral GANG-glē-on) A cluster of cell bodies of postganglionic sympathetic neurons anterior to the spinal column and close to large abdominal arteries. Also called a collateral ganglion.

Primary germ layers The major embryonic tissues from which the various tissues and organs of the body develop: ectoderm, mesoderm, and endoderm.

Primary motor area A region of the cerebral cortex in the precentral gyrus of the frontal lobe of the cerebrum that controls specific muscles or groups of muscles.

Primary somatosensory area A region of the cerebral cortex posterior to the central sulcus in the postcentral gyrus of the parietal lobe of the cerebrum that localizes exactly the points of the body where somatic sensations originate.

Prime mover The muscle directly responsible for producing a desired motion. Also called an agonist (AG-ō-nist).

Primitive gut Embryonic structure formed from the dorsal part of the yolk sac that gives rise to most of the gastrointestinal tract.

Primordial (prī-MŌR-dē-al) **germ cells** Cells that arise from the yolk sac endoderm and enter the testes during the fifth week of development.

Proctology (prok-TOL-ō-jē) The branch of medicine concerned with the rectum and its disorders.

Progesterone (prō-JES-te-rōn) A female sex hormone produced by the ovaries that helps prepare the endometrium of the uterus for implantation of a fertilized ovum and the mammary glands for milk secretion.

Prolactin (prō-LAK-tin) **(PRL)** A hormone secreted by the anterior pituitary that initiates and maintains milk secretion by the mammary glands.

Pronation (prō-NĀ-shun) A movement of the forearm in which the palm is turned posteriorly.

Prophase The first stage of mitosis during which chromatid pairs are formed and aggregate around the metaphase plate of the cell.

Proprioceptor (PRO-prē-ōsep′-tor) A receptor located in muscles, tendons, joints, or the internal ear (muscle spindles, tendon organs, joint kinesthetic receptors, and hair cells of the vestibular apparatus) that provides information about body position and movements. Also called a visceroceptor.

Prostaglandins (pros′-ta-GLAN-dins) **(PG)** Lipids released by damaged cells that intensify the effects of histamine and kinins.

Prostate (PROS-tāt) A doughnut-shaped gland inferior to the urinary bladder that surrounds the superior portion of the male urethra and secretes a slightly acidic solution that contributes to sperm motility and viability.

Proteasome (PRŌ-tē-a-sōm) Tiny cellular organelle in cytosol and nucleus containing proteases that destroy unneeded, damaged, or faulty proteins.

Protein An organic compound consisting of carbon, hydrogen, oxygen, nitrogen, and sometimes sulfur and phosphorus; synthesized on ribosomes and made up of amino acids linked by peptide bonds.

Proto-oncogene (prō′-tō-ON-kō-jēn) Gene responsible for some aspect of normal growth and development; it may transform into an oncogene, a gene capable of causing cancer.

Protraction (prō-TRAK-shun) The movement of the mandible or shoulder girdle forward on a plane parallel with the ground.

Pseudopods (SOO-dō-pods) Temporary protrusions of the leading edge of a migrating cell; cellular projections that surround a particle undergoing phagocytosis.

Pterygopalatine (ter-i-gō-PAL-a-tīn) **ganglion** A cluster of cell bodies of parasympathetic postganglionic neurons ending at the lacrimal and nasal glands.

Ptosis (TŌ-sis) Drooping, as of the eyelid or the kidney.

Puberty (PŪ-ber-tē) The time of life during which the secondary sex characteristics begin to appear and the capability for sexual reproduction is possible; usually occurs between the ages of 10 and 17.

Pubic symphysis A slightly movable cartilaginous joint between the anterior surfaces of the hip bones.

Puerperium (pū-er-PER-ē-um) The period immediately after childbirth, usually 4–6 weeks.

Pulmonary circulation The flow of deoxygenated blood from the right ventricle to the lungs and the return of oxygenated blood from the lungs to the left atrium.

Pulmonary edema (e-DĒ-ma) An abnormal accumulation of interstitial fluid in the tissue spaces and alveoli of the lungs due to increased pulmonary capillary permeability or increased pulmonary capillary pressure.

Pulmonary embolism (EM-bō-lizm) The presence of a blood clot or a foreign substance in a pulmonary arterial blood vessel that obstructs circulation to lung tissue.

Pulmonary ventilation The inflow (inhalation) and outflow (exhalation) of air between the atmosphere and the lungs. Also called breathing.

Pulp cavity A cavity within the crown and neck of a tooth, which is filled with pulp, a connective tissue containing blood vessels, nerves, and lymphatic vessels.

Pulse (Puls) The rhythmic expansion and elastic recoil of a systemic artery after each contraction of the left ventricle.

Pupil The hole in the center of the iris, the area through which light enters the posterior cavity of the eyeball.

Purkinje (pur-KIN-jē) **cell** Neuron in the cerebellum named for the histologist who first described it them.

Purkinje (pur-KIN-jē) **fiber** Muscle fiber (cell) in the ventricular tissue of the heart specialized for conducting an action potential to the myocardium; part of the conduction system of the heart.

Pus The liquid product of inflammation containing leukocytes or their remains and debris of dead cells.

Pyloric (pī-LOR-ik) **sphincter** A thickened ring of smooth muscle through which the pylorus of the stomach communicates with the duodenum.

Pylorus (pī-LOR-us) Region of the pyloric part of the stomach that connects to the duodenum.

Pyorrhea (pī-ō-RĒ-a) A discharge or flow of pus, especially in the alveoli (sockets) and the tissues of the gums.

Pyramid (PIR-a-mid) A pointed or cone-shaped structure. One of two roughly triangular structures on the anterior aspect of the medulla oblongata composed of the largest motor tracts that run from the cerebral cortex to the spinal cord. A triangular structure in the renal medulla.

QRS complex The deflection waves of an electrocardiogram that represent onset of ventricular depolarization.

Quadrant One of four parts.

Quadriplegia (kwod′-ri-PLĒ-jē-a) Paralysis of four limbs: two upper and two lower.

Rami communicantes (RĀ-mē kō-mū-ni-KAN-tēz) Branches of a spinal nerve that are components of the autonomic nervous system. Singular is ramus communicans (RĀ-mus kō-MŪ-ni-kans).

Receptor A specialized cell or a distal portion of a neuron that responds to a specific sensory modality, such as touch, pressure, cold, light, or sound, and converts it to an electrical signal (generator or receptor potential). A specific molecule or cluster of molecules that recognizes and binds a particular ligand.

Receptor-mediated endocytosis A highly selective process whereby cells take up specific ligands, which usually are large molecules or particles, by enveloping them within a sac of plasma membrane.

Rectouterine (rek-tō-Ū-ter-in) **pouch** A pocket formed by the parietal peritoneum as it moves posteriorly from the surface of the uterus and is reflected onto the rectum; the most inferior point in the pelvic cavity. Also called the pouch of Douglas.

Rectum (REK-tum) The last 20 cm (8 in.) of the gastrointestinal tract, from the sigmoid colon to the anus.

Red blood cells (RBCs) Blood cells without nuclei that contain the oxygen-carrying protein hemoglobin; responsible for oxygen transport throughout the body.

Red bone marrow A highly vascularized connective tissue located in microscopic spaces between trabeculae of spongy bone tissue.

Red nucleus A cluster of cell bodies in the midbrain, occupying a large part of the tectum from which axons extend into the rubroreticular and rubrospinal tracts.

Red pulp That portion of the spleen that consists of venous sinuses filled with blood and thin plates of splenic tissue called splenic (Billroth's) cords.

Referred pain Pain that is felt at a site remote from the place of origin.

Reflex Fast response to a change (stimulus) in the internal or external environment that attempts to restore homeostasis.

Reflex arc The most basic conduction pathway through the nervous system, connecting a receptor and an effector and consisting of a receptor, a sensory neuron, an integrating center in the central nervous system, a motor neuron, and an effector. Also called a reflex circuit.

Relaxin (RLX) A female hormone produced by the ovaries and placenta that increases flexibility of the pubic symphysis and helps dilate the uterine cervix to ease delivery of a baby.

Releasing hormone Hormone secreted by the hypothalamus that can stimulate secretion of hormones of the anterior pituitary.

Renal corpuscle A glomerular (Bowman's) capsule and its enclosed glomerulus.

Renal pelvis A cavity in the center of the kidney formed by the expanded, proximal portion of the ureter, lying within the kidney, and into which the major calyces open.

Renal pyramid A triangular structure in the renal medulla containing the straight segments of renal tubules and the vasa recta.

Reproduction The formation of new cells for growth, repair, or replacement; the production of a new individual.

Reproductive cell divisio Type of cell division in which gametes (sperm and oocytes) are produced; consists of meiosis and cytokinesis.

Respiration (res-pi-RĀ-shun) Overall exchange of gases between the atmosphere, blood, and body cells consisting of pulmonary ventilation, external respiration, and internal respiration.

Respiratory center Neurons in the pons and medulla oblongata of the brainstem that regulate breathing. It is divided into the medullary respiratory center and the pontine respiratory center.

Respiratory (RES-pi-ra-tō-rē) **system** Body system consisting of the nose, nasal cavity, pharynx, larynx, trachea, bronchi, and lungs.

Rete (RĒ-tē) **testis** The network of ducts in the testes.

Reticular (re-TIK-ū-lar) **activating system (RAS)** A portion of the reticular formation that has many ascending connections with the cerebral cortex; when this area of the brainstem is active, nerve impulses pass to the thalamus and widespread areas of the cerebral cortex, resulting in generalized alertness or arousal from sleep.

Reticular formation A network of small groups of neuronal cell bodies scattered among bundles of axons (mixed gray and white matter) beginning in the medulla oblongata and extending superiorly through the central part of the brainstem.

Retina (RET-i-na) The deep coat of the posterior portion of the eyeball consisting of nervous tissue (where the process of vision begins) and a pigmented layer of epithelial cells that contact the choroid.

Retinaculum (ret-i-NAK-ū-lum) A thickening of fascia that holds structures in place, for example, the superior and inferior retinacula of the ankle.

Retraction (rē-TRAK-shun) The movement of a protracted part of the body posteriorly on a plane parallel to the ground, as in pulling the lower jaw back in line with the upper jaw.

Retroperitoneal (re′-trō-per-i-tō-NĒ-al) External to the peritoneal lining of the abdominal cavity.

Rh factor Rh antigen.

Ribonucleic (rī-bō-noo-KLĒ-ik) **acid (RNA)** A single-stranded nucleic acid made up of nucleotides, each consisting of a nitrogenous base (adenine, cytosine, guanine, or uracil), ribose, and a phosphate group; three types are messenger RNA (mRNA), transfer RNA (tRNA), and ribosomal RNA (rRNA), each of which has a specific role during protein synthesis.

Ribosome (RĪ-bō-sōm) A cellular structure in the cytoplasm of cells, composed of a small subunit and a large subunit that contain ribosomal RNA and ribosomal proteins; the site of protein synthesis.

Right lymphatic duct A vessel of the lymphatic system that drains lymph from the upper right side of the body and empties it into the right subclavian vein.

Rigidity (ri-JID-i-tē) Hypertonia characterized by increased muscle tone, but reflexes are not affected.

Rigor mortis State of partial contraction of muscles after death due to lack of ATP; myosin heads (cross-bridges) remain attached to actin, thus preventing relaxation.

Rod One of two types of photoreceptor in the retina of the eye; specialized for vision in dim light.

Root canal A narrow extension of the pulp cavity lying within the root of a tooth.

Root of the penis Attached portion of penis that consists of the bulb and crura.

Rotation (rō-TĀ-shun) Moving a bone around its own axis, with no other movement.

Rotator cuff Refers to the tendons of four deep shoulder muscles (subscapularis, supraspinatus, infraspinatus, and teres minor) that form a complete circle around the shoulder; they strengthen and stabilize the shoulder joint.

Round ligament A band of fibrous connective tissue enclosed between the folds of the broad ligament of the uterus, emerging from the uterus just inferior to the uterine tube, extending laterally along the pelvic wall and through the deep inguinal ring to end in the labia majora.

Round window A small opening between the middle and internal ear, directly inferior to the oval window, covered by the secondary tympanic membrane.

Rugae (ROO-gē) Large folds in the mucosa of an empty hollow organ, such as the stomach and vagina.

Saccule (SAK-ūl) The inferior and smaller of the two chambers in the membranous labyrinth inside the vestibule of the internal ear containing a receptor organ for linear acceleration or deceleration that occurs in a vertical direction.

Sacral plexus (SĀ-kral PLEK-sus) A network formed by the anterior rami branches of spinal nerves L4 through S3.

Saddle joint A synovial joint in which the articular surface of one bone is saddle-shaped and the articular surface of the other bone is shaped like the legs of the rider sitting in the saddle, as in the joint between the trapezium and the metacarpal of the thumb.

Sagittal plane A plane that divides the body or organs into left and right portions. Such a plane may be midsagittal (median), in which the divisions are equal, or parasagittal, in which the divisions are unequal.

Saliva (sa-LĪ-va) A clear, alkaline, somewhat viscous secretion produced mostly by the three pairs of salivary glands; contains various salts, mucin, lysozyme, salivary amylase, and lingual lipase (produced by glands in the tongue).

Salivary amylase (SAL-i-ver-ē AM-i-lās) An enzyme in saliva that initiates the chemical breakdown of starch.

Salivary gland One of three pairs of glands that lie external to the mouth and pour their secretory product (saliva) into ducts that empty into the oral cavity; the parotid, submandibular, and sublingual glands.

Sarcolemma (sar′-kō-LEM-ma) The cell membrane of a muscle fiber (cell), especially of a skeletal muscle fiber.

Sarcomere (SAR-kō-mēr) A contractile unit in a striated muscle fiber (cell) extending from one Z disc to the next Z disc.

Sarcoplasm (SAR-kō-plazm) The cytoplasm of a muscle fiber (cell).

Sarcoplasmic reticulum (sar′-kō-PLAZ-mik re-TIK-ū-lum) **(SR)** A network of saccules and tubes surrounding myofibrils of a muscle fiber (cell), comparable to endoplasmic reticulum; functions to reabsorb calcium ions during relaxation and to release them to cause contraction.

Satellite (SAT-i-līt) **cells** Flat neuroglial cells that surround cell bodies of peripheral nervous system ganglia to provide structural support and regulate the exchange of material between a neuronal cell body and interstitial fluid.

Saturated fat A fatty acid that contains only single bonds (no double bonds) between its carbon atoms; all carbon atoms are bonded to the maximum number of hydrogen atoms; prevalent in triglycerides of animal products such as meat, milk, milk products, and eggs.

Scala tympani (SKA-la TIM-pan-ē) The inferior spiral-shaped channel of the bony cochlea, filled with perilymph.

Scala vestibuli (ves-TIB-ū-lē) The superior spiral-shaped channel of the bony cochlea, filled with perilymph.

Schwann cell (SCHVON or SCHWON) A neuroglial cell of the peripheral nervous system that forms the myelin sheath and neurolemma around a nerve axon by wrapping around the axon in a jelly-roll fashion.

Sciatica Inflammation and pain along the sciatic nerve; felt along the posterior aspect of the thigh extending down the inside of the leg.

Sclera (SKLE-ra) The white coat of fibrous tissue that forms the superficial protective covering over the eyeball except in the most anterior portion; the posterior portion of the fibrous tunic.

Scleral venous sinus A circular venous sinus located at the junction of the sclera and the cornea through which aqueous humor drains from the anterior chamber of the eyeball into the blood. Also called the canal of Schlemm (SHLEM).

Scoliosis (skō-lē-Ō-sis) An abnormal lateral curvature from the normal vertical line of the backbone.

Scrotum (SKRŌ-tum) A skin-covered pouch that contains the testes and their accessory structures.

Sebaceous ciliary glands Glands at the base of the hair follicles of the eyelashes that release a lubricating fluid into the follicles.

Sebaceous gland (se-BĀ-shus) An exocrine gland in the dermis of the skin, almost always associated with a hair follicle, that secretes sebum. Also called an oil gland.

Sebum (SĒ-bum) Secretion of sebaceous (oil) glands.

Secondary sex characteristics Traits that distinguish males and females but do not have a direct role in reproduction.

Secretion (se-KRĒ-shun) Production and release from a cell or a gland of a physiologically active substance.

Selective permeability (per′-mē-a-BIL-i-tē) The property of a membrane by which it permits the passage of certain substances but restricts the passage of others.

Semen (SĒ-men) A fluid discharged at ejaculation by a male that consists of a mixture of sperm and the secretions of the seminiferous tubules, seminal vesicles, prostate, and bulbourethral (Cowper's) glands.

Semicircular canals Three bony channels (anterior, posterior, lateral), filled with perilymph, in which lie the membranous semicircular canals filled with endolymph. They contain receptors for equilibrium.

Semicircular ducts The membranous semicircular canals filled with endolymph and floating in the perilymph of the bony semicircular canals; they contain cristae that are concerned with rotational acceleration or deceleration.

Semilunar (sem′-ē-LOO-nar) **(SL) valve** A valve between the aorta or the pulmonary trunk and a ventricle of the heart.

Seminal vesicle (SEM-i-nal VES-i-kul) One of a pair of convoluted, pouchlike structures, lying posterior and inferior to the urinary bladder and anterior to the rectum, that secrete a component of semen into the ejaculatory ducts. Also termed seminal gland.

Seminiferous tubule (sem′-i-NI-fer us TOO-būl) A tightly coiled duct, located in the testis, where sperm are produced.

Sensation A state of awareness of external or internal conditions of the body.

Sensory area A region of the cerebral cortex concerned with the interpretation of sensory impulses.

Sensory modality (mō-DAL-i-tē) Any of the specific sensory entities, such as vision, smell, taste, or touch.

Sensory neuron (NOO-ron) Neuron that carries sensory information from cranial and spinal nerves into the brain and spinal cord or from a lower to a higher level in the spinal cord and brain. Also called an afferent neuron.

Septal defect An opening in the atrial septum (atrial septal defect) because the foramen ovale fails to close, or the ventricular septum (ventricular septal defect) due to incomplete development of the ventricular septum.

Serosa (se-RŌ-sa) Superficial layer of the portions of the GI tract that are suspended in the abdominal cavity. Also called the serous membrane.

Serous (SĒR-us) **membrane** A membrane that lines a body cavity that does not open to the exterior. The external layer of an organ formed by a serous membrane. The membrane that lines the pleural, pericardial, and peritoneal cavities. Also called a serosa (se-RŌ-sa).

Serum Blood plasma minus its clotting proteins.

Sesamoid (SES-a-moyd) **bones** Small bones usually found in tendons that develop where there is considerable friction, tension, and physical stress; numbers vary from person to person.

Sex chromosomes The twenty-third pair of chromosomes, designated X and Y, which determine the genetic sex of an individual; in males, the pair is XY; in females, XX.

Sexual reproduction The process by which organisms produce offspring by making sex cells called gametes.

Shock Failure of the cardiovascular system to deliver adequate amounts of oxygen and nutrients to meet the metabolic needs of the body due to inadequate cardiac output.

Shoulder joint A synovial joint where the humerus articulates with the scapula.

Sigmoid colon (SIG-moyd KŌ-lon) The S-shaped part of the large intestine that begins at the level of the left iliac crest, projects medially, and terminates at the rectum at about the level of the third sacral vertebra.

Sign Any objective evidence of disease that can be observed or measured, such as a lesion, swelling, or fever.

Sinoatrial (si-nō-Ā-trē-al) **(SA) node** A small mass of cardiac muscle fibers (cells) located in the right atrium inferior to the opening of the superior vena cava that spontaneously depolarize and generate a cardiac action potential about 100 times per minute. Also called the natural pacemaker.

Sinus (SĪ-nus) A hollow in a bone (paranasal sinus) or other tissue; a channel for blood (vascular sinus); any cavity having a narrow opening.

Sinusoid (SĪ-nū-soyd) A large, thin-walled, and leaky type of capillary, having large intercellular clefts that may allow proteins and blood cells to pass from a tissue into the bloodstream; present in the liver, spleen, anterior pituitary, parathyroid glands, and red bone marrow.

Skeletal muscle tissue Tissue of the skeletal muscle, composed of striated muscle fibers (cells), supported by connective tissue, attached to a bone by a tendon or an aponeurosis, and stimulated by somatic motor neurons.

Skeletal system Framework of bones and their associated cartilages, ligaments, and tendons.

Skin The external covering of the body that consists of a superficial, thinner epidermis (epithelial tissue) and a deep, thicker dermis (connective tissue) that is anchored to the subcutaneous layer. Also called cutaneous membrane.

Skull The skeleton of the head consisting of the cranial and facial bones.

Sleep A state of partial unconsciousness from which a person can be aroused; associated with a low level of activity in the reticular activating system.

Small intestine A long tube of the gastrointestinal tract that begins at the pyloric sphincter of the stomach, coils through the central and inferior part of the abdominal cavity, and ends at the large intestine; divided into three segments: duodenum, jejunum, and ileum.

Smooth muscle tissue A tissue specialized for contraction, composed of smooth muscle fibers (cells), located in the walls of hollow internal organs, and innervated by autonomic motor neurons.

Sodium–potassium pump An active transport pump located in the plasma membrane that transports sodium ions out of the cell and potassium ions into the cell at the expense of cellular ATP. It functions to keep the ionic concentrations of these ions at physiological levels. Also called the Na^+–K^+ ATPase.

Soft palate (PAL-at) The posterior portion of the roof of the mouth, extending from the palatine bones to the uvula. It is a muscular partition lined with mucous membrane.

Somatic (sō-MAT-ik) **cell division** Type of cell division in which a single starting cell duplicates itself to produce two identical cells; consists of mitosis and cytokinesis.

Somatic motor pathway Pathway that carries information from the cerebral cortex, basal nuclei, and cerebellum that stimulates contraction of skeletal muscles.

Somatic nervous system (SNS) The portion of the peripheral nervous system that conveys output to skeletal muscles.

Somatic sensory pathway Pathway that carries information from somatic sensory receptor to the primary somatosensory area in the cerebral cortex and cerebellum.

Somite (SO-mīt) Block of mesodermal cells in a developing embryo that is distinguished into a myotome (which forms most of the skeletal muscles), dermatome (which forms connective tissues), and sclerotome (which forms the vertebrae).

Spasm (SPAZM) A sudden, involuntary contraction of large groups of muscles.

Spasticity (spas-TIS-i-tē) Hypertonia characterized by increased muscle tone, increased tendon reflexes, and pathological reflexes (Babinski sign).

Sperm cell A mature male gamete. Also called a spermatozoon (sper′-ma-tō-ZŌ-on).

Spermatic (sper-MAT-ik) **cord** A supporting structure of the male reproductive system, extending from a testis to the deep inguinal ring, that includes the ductus (vas) deferens, arteries, veins, lymphatic vessels, nerves, cremaster muscle, and connective tissue.

Spermatogenesis (sper′-ma-tō-JEN-e-sis) The formation and development of sperm in the seminiferous tubules of the testes.

Spermiogenesis (sper′-mē-ō-JEN-e-sis) The maturation of spermatids into sperm.

Sphincter of the hepatopancreatic ampulla A circular muscle at the opening of the common bile and main pancreatic ducts in the duodenum. Also called the sphincter of Oddi (OD-ē).

Spinal (SPĪ-nal) **cord** A mass of nerve tissue located in the vertebral canal from which 31 pairs of spinal nerves originate.

Spinal nerve One of the 31 pairs of nerves that originate on the spinal cord from posterior and anterior roots.

Spinal shock A period from several days to several weeks following transection of the spinal cord that is characterized by the abolition of all reflex activity.

Spinothalamic (spī-nō-tha-LAM-ik) **tract** Sensory (ascending) tract that conveys information up the spinal cord to the thalamus for sensations of pain, temperature, itch, and tickle.

Spiral organ The organ of hearing, consisting of supporting cells and hair cells that rest on the basilar membrane and extend into the endolymph of the cochlear duct. Also called the organ of Corti (KOR-tē).

Spleen (SPLĒN) Large mass of lymphatic tissue between the fundus of the stomach and the diaphragm that functions in formation of blood cells during early fetal development, phagocytosis of ruptured blood cells, and proliferation of B cells during immune responses.

Spongy bone tissue Bone tissue that consists of an irregular latticework of thin plates of bone called trabeculae; spaces between trabeculae of some bones are filled with red bone marrow; found inside short, flat, and irregular bones and in the epiphyses (ends) of long bones.

Sprain Forcible wrenching or twisting of a joint with partial rupture or other injury to its attachments without dislocation.

Starvation (star-VĀ-shun) The loss of energy stores in the form of glycogen, triglycerides, and proteins due to inadequate intake of nutrients or inability to digest, absorb, or metabolize ingested nutrients.

Stellate reticuloendothelial (STEL-āt re-tik′-ū-lō-en-dō-THĒ-lē-al) **cell** Phagocytic cell bordering a sinusoid of the liver. Also called a hepatic macrophage or Kupffer (KOOP-fer) cell.

Stem cell An unspecialized cell that has the ability to divide for indefinite periods and give rise to a specialized cell.

Stenosis (sten-Ō-sis) An abnormal narrowing or constriction of a duct or opening.

Stereocilia (ste′-rē-ō-SIL-ē-a) Groups of extremely long, slender, nonmotile microvilli projecting from epithelial cells lining the epididymis.

Sterilization (ster′-i-li-ZĀ-shun) Elimination of all living microorganisms. Any procedure that renders an individual incapable of reproduction (for example, castration, vasectomy, hysterectomy, or oophorectomy).

Stimulus Any stress that changes a controlled condition; any change in the internal or external environment that excites a sensory receptor, a neuron, or a muscle fiber.

Stomach The J-shaped enlargement of the gastrointestinal tract directly inferior to the diaphragm in the epigastric, umbilical, and left hypochondriac regions of the abdomen, between the esophagus and small intestine.

Straight tubule (TOO-būl) A duct in a testis leading from a convoluted seminiferous tubule to the rete testis.

Stratum (STRĀ-tum) A layer.

Stratum basale (ba-SA-lē) The deepest layer of the epidermis; also called the stratum germinativum.

Stratum basalis The layer of the endometrium next to the myometrium that is maintained during menstruation and gestation and produces a new stratum functionalis following menstruation or parturition.

Stratum functionalis (funk′-shun-AL-is) The layer of the endometrium next to the uterine cavity that is shed during menstruation and that forms the maternal portion of the placenta during gestation.

Stretch receptor Receptor in the walls of blood vessels, airways, or organs that monitors the amount of stretching. Also termed baroreceptor.

Striae (STRĪ-ē) Internal scarring due to overstretching of the skin in which collagen fibers and blood vessels in the dermis are damaged. Also called stretch marks.

Stroma (STRŌ-ma) The tissue that forms the ground substance, foundation, or framework of an organ, as opposed to its functional parts (parenchyma).

Subarachnoid (sub′-a-RAK-noyd) **space** A space between the arachnoid mater and the pia mater that surrounds the brain and spinal cord and through which cerebrospinal fluid circulates.

Subatomic particles Components of an atom.

Subcutaneous (sub′-kū-TĀ-nē-us) **(subQ) layer** A continuous sheet of areolar connective tissue and adipose tissue between the dermis of the skin and the deep fascia of the muscles. Also called the hypodermis.

Subdural (sub-DOO-ral) **space** A space between the dura mater and the arachnoid mater of the brain and spinal cord that contains a small amount of fluid.

Sublingual (sub-LING-gwal) **gland** One of a pair of salivary glands situated in the floor of the mouth deep to the mucous membrane and to the side of the lingual frenulum, with a duct (Rivinus') that opens into the floor of the mouth.

Submandibular (sub′-man-DIB-ū-lar) **gland** One of a pair of salivary glands found inferior to the base of the tongue deep to the mucous membrane in the posterior part of the floor of the mouth, posterior to the sublingual glands, with a duct (submandibular) situated to the side of the lingual frenulum.

Submucosa (sub-mū-KŌ-sa) A layer of connective tissue located deep to a mucous membrane, as in the gastrointestinal tract or the urinary bladder; the submucosa connects the mucosa to the muscularis layer.

Submucosal plexus A network of autonomic nerve fibers located in the superficial part of the submucous layer of the small intestine. Also called the plexus of Meissner (MĪZ-ner).

Substrate A reactant molecule upon which an enzyme acts.

Sudoriferous (soo′-dor-IF-er-us) **gland** An apocrine or eccrine exocrine gland in the dermis or subcutaneous layer that produces perspiration. Also called a sweat gland.

Sulcus (SUL-kus) A groove or depression between parts, especially between the convolutions of the brain. Plural is sulci (SUL-sī).

Superficial (soo′-per-FISH-al) Located on or near the surface of the body or an organ. Also called external.

Superficial (subcutaneous) inguinal ring A triangular opening in the aponeurosis of the external oblique muscle that represents the termination of the inguinal canal.

Superior Toward the head or upper part of a structure. Also called cephalic or cranial.

Superior vena cava (VĒ-na KĀ-va) **(SVC)** Large vein that collects blood from parts of the body superior to the heart and returns it to the right atrium.

Supination (soo-pi-NĀ-shun) A movement of the forearm in which the palm is turned anteriorly.

Surfactant (sur-FAK-tant) Complex mixture of phospholipids and lipoproteins, produced by type II alveolar (septal) cells in the lungs, that decreases surface tension.

Suspensory ligament (sus-PEN-sō-rē LIG-a-ment) A fold of peritoneum extending laterally from the surface of the ovary to the pelvic wall.

Sustentacular (sus′-ten-TAK-ū-lar) **cell** A supporting cell in the seminiferous tubules that secretes fluid for supplying nutrients to sperm and the hormone inhibin, removes excess cytoplasm from spermatogenic cells, and mediates the effects of FSH and testosterone on spermatogenesis. Also called a Sertoli cell.

Sutural (SOO-chur-al) **bone** A small bone located within a suture between certain cranial bones.

Suture (SOO-chur) An immovable fibrous joint that joins skull bones.

Sympathetic (sim′-pa-THET-ik) **nervous system** One of the two main subdivisions of the autonomic nervous system, having cell bodies of preganglionic neurons in the lateral gray columns of the thoracic segment and the first two or three lumbar segments of the spinal cord; primarily concerned with processes involving the expenditure of energy. Also called the thoracolumbar division.

Sympathetic trunk ganglion (GANG-glē-on) A cluster of cell bodies of sympathetic postganglionic neurons lateral to the vertebral column, close to the body of a vertebra. These ganglia extend inferiorly through the neck, thorax, and abdomen to the coccyx on both sides of the vertebral column and are connected to one another to form a chain on each side of the vertebral column. Also called vertebral chain ganglia or paravertebral ganglia.

Symphysis (SIM-fi-sis) A line of union. A slightly movable cartilaginous joint such as the pubic symphysis.

Symptoms Subjective changes in body functions that are not apparent to an observer.

Synapse (SIN-aps) The functional junction between two neurons or between a neuron and an effector, such as a muscle or gland; may be electrical or chemical.

Synapsis (sin-AP-sis) The pairing of homologous chromosomes during prophase I of meiosis.

Synaptic (sin-AP-tik) **cleft** The narrow gap at a chemical synapse that separates the axon terminal of one neuron from another neuron or muscle fiber (cell) and across which a neurotransmitter diffuses to affect the postsynaptic cell.

Synaptic end bulb Expanded distal end of an axon terminal that contains synaptic vesicles. Also called a synaptic knob.

Synaptic vesicle Membrane-enclosed sac in a synaptic end bulb that stores neurotransmitters.

Synarthrosis (sin′-ar-THRŌ-sis) An immovable joint such as a suture, gomphosis, or synchondrosis.

Synchondrosis (sin′-kon-DRŌ-sis) A cartilaginous joint in which the connecting material is hyaline cartilage.

Syndesmosis (sin′-dez-MŌ-sis) A slightly movable joint in which articulating bones are united by fibrous connective tissue.

Synergist (SIN-er-jist) A muscle that assists the prime mover by reducing undesired action or unnecessary movement.

Synergistic (sin-er-JIS-tik) **effect** A hormonal interaction in which the effects of two or more hormones acting together is greater or more extensive than the effect of each hormone acting alone.

Synostosis (sin′-os-TŌ-sis) A joint in which the dense fibrous connective tissue that unites bones at a suture has been replaced by bone, resulting in a complete fusion across the suture line.

Synovial (si-NŌ-vē-al) **cavity** The space between the articulating bones of a synovial joint, filled with synovial fluid. Also called a joint cavity.

Synovial fluid Secretion of synovial membranes that lubricates joints and nourishes articular cartilage.

Synovial joint A fully movable or diarthrotic joint in which a synovial (joint) cavity is present between the two articulating bones.

Synovial membrane The deeper of the two layers of the articular capsule of a synovial joint, composed of areolar connective tissue that secretes synovial fluid into the synovial (joint) cavity.

System An association of organs that have a common function.

Systemic circulation The routes through which oxygenated blood flows from the left ventricle through the aorta to all the organs of the body and deoxygenated blood returns to the right atrium.

Systole (SIS-tō-lē) In the cardiac cycle, the phase of contraction of the heart muscle, especially of the ventricles.

Systolic (sis-TOL-ik) **blood pressure (SBP)** The force exerted by blood on arterial walls during ventricular contraction; the highest pressure measured in the large arteries, less than 120 mmHg under normal conditions for a young adult.

T cell Lymphocyte that begins development in primary lymphatic organs and completes it in the thymus.

T wave The deflection wave of an electrocardiogram that represents ventricular repolarization.

Tachycardia (tak′-i-KAR-dē-a) An abnormally rapid resting heartbeat or pulse rate (over 100 beats per minute).

Tactile (TAK-tīl) Pertaining to the sense of touch.

Tactile disc Saucer-shaped free nerve endings that make contact with tactile epithelial cells in the epidermis and function as touch receptors. Also called a Merkel disc.

Tactile epithelial cell Type of cell in the epidermis of hairless skin that makes contact with a tactile disc, which functions in touch. Also called a Merkel cell.

Tarsal bones The seven bones of the ankle. Also called tarsals.

Tarsal gland Sebaceous (oil) gland that opens on the edge of each eyelid. Also called a Meibomian (mī-BŌ-mē-an) gland.

Tarsal plate A thin, elongated sheet of connective tissue, one in each eyelid, giving the eyelid form and support. The aponeurosis of the levator palpebrae superioris is attached to the tarsal plate of the superior eyelid.

Tarsus (TAR-sus) A collective term for the seven bones of the ankle.

Tectorial (tek-TŌ-rē-al) **membrane** A gelatinous membrane projecting over and in contact with the hair cells of the spiral organ (organ of Corti) in the cochlear duct.

Teeth (TĒTH) Accessory structures of digestion, composed of calcified connective tissue and embedded in bony sockets of the mandible and maxilla, that cut, shred, crush, and grind food. Also called dentes (DEN-tēz).

Telophase (TEL-ō-fāz) The final stage of mitosis.

Tendon (TEN-don) A white fibrous cord of dense regular connective tissue that attaches muscle to bone.

Tendon organ A proprioceptive receptor, sensitive to changes in muscle tension and force of contraction, found chiefly near the junctions of tendons and muscles. Also called a Golgi (GOL-jē) tendon organ.

Tendon reflex A polysynaptic, ipsilateral reflex that protects tendons and their associated muscles from damage that might be brought about by excessive tension. The receptors involved are called tendon organs.

Teniae coli (TĒ-nē-ē KŌ-lī) The three flat bands of thickened, longitudinal smooth muscle running the length of the large intestine, except in the rectum. Singular is tenia coli.

Tentorium cerebelli A transverse shelf of dura mater that forms a partition between the occipital lobe of the cerebral hemispheres and the cerebellum and that covers the cerebellum.

Teratogen (TER-a-tō-jen) Any agent or factor that causes physical defects in a developing embryo.

Terminal ganglion (TER-min-al GANG-lē-on) A cluster of cell bodies of parasympathetic postganglionic neurons either lying very close to the visceral effectors or located within the walls of the visceral effectors supplied by the postganglionic neurons. Also called intramural ganglion.

Testis (TES-tis) Male gonad that produces sperm and the hormones testosterone and inhibin. Plural is testes. Also called a testicle.

Testosterone (tes-TOS-te-rōn) A male sex hormone (androgen) secreted by interstitial endocrinocytes (Leydig cells) of a mature testis; needed for development of sperm; together with a second androgen termed dihydrotestosterone (DHT), controls the growth and development of male reproductive organs, secondary sex characteristics, and body growth.

Tetralogy of Fallot (tet-RAL-ō-jē of fal-Ō) A combination of four congenital heart defects: (1) constricted pulmonary semilunar valve, (2) interventricular septal opening, (3) emergence of the aorta from both ventricles instead of from the left only, and (4) enlarged right ventricle.

Thalamus (THAL-a-mus) A large, oval structure located bilaterally on either side of the third ventricle, consisting of two masses of gray matter organized into nuclei; main relay center for sensory impulses ascending to the cerebral cortex.

Thermoreceptor (THER-mō-rē-sep-tor) Sensory receptor that detects changes in temperature.

Thermoregulation Homeostatic regulation of body temperature through sweating and adjustment of blood flow in the dermis.

Third ventricle (VEN-tri-kul) A slitlike cavity between the right and left halves of the thalamus and between the lateral ventricles of the brain.

Thoracic cavity Cavity superior to the diaphragm that contains two pleural cavities, the mediastinum, and the pericardial cavity.

Thoracic duct A lymphatic vessel that begins as a dilation called the cisterna chyli, receives lymph from the left side of the head, neck, and chest, left arm, and the entire body below the ribs, and empties into the junction between the internal jugular and left subclavian veins. Also called the left lymphatic (lim-FAT-ik) duct.

Thoracolumbar (thōr′-a-kō-LUM-bar) **outflow** The axons of sympathetic preganglionic neurons, which have their cell bodies in the lateral gray columns of the thoracic segments and first two or three lumbar segments of the spinal cord.

Thorax (THŌ-raks) The chest region.

Thrombopoietin (TPO) Hormone produced by the liver that stimulates formation of platelets (thrombocytes) from megakaryocytes.

Thrombosis (throm-BŌ-sis) The formation of a clot in an unbroken blood vessel, usually a vein.

Thrombus (Throm-bus) A stationary clot formed in an unbroken blood vessel, usually a vein.

Thymus (THĪ-mus) A bilobed organ, located in the superior mediastinum posterior to the sternum and between the lungs, in which T cells develop immunocompetence.

Thyroid cartilage (THĪ-royd KAR-ti-lij) The largest single cartilage of the larynx, consisting of two fused plates that form the anterior wall of the larynx. Also called the Adam's apple.

Thyroid follicle (FOL-i-kul) Spherical sac that forms the parenchyma of the thyroid gland and consists of follicular cells that produce thyroxine (T_4) and triiodothyronine (T_3).

Thyroid gland An endocrine gland with right and left lateral lobes on either side of the trachea connected by an isthmus; located anterior to the trachea just inferior to the cricoid cartilage; secretes thyroxine (T_4), triiodothyronine (T_3), and calcitonin.

Thyroid-stimulating hormone (TSH) A hormone secreted by the anterior pituitary that stimulates the synthesis and secretion of thyroxine (T_4) and triiodothyronine (T_3). Also known as thyrotropin.

Thyroxine (thī-ROK-sēn) **(T_4)** A hormone secreted by the thyroid gland that regulates metabolism, growth and development, and the activity of the nervous system. Also called tetraiodothyronine.

Tissue A group of similar cells and their intercellular substance joined together to perform a specific function.

Tongue A large skeletal muscle covered by a mucous membrane located on the floor of the oral cavity.

Tonsil (TON-sil) An aggregation of large lymphatic nodules embedded in the mucous membrane of the throat.

Torn cartilage A tearing of an articular disc (meniscus) in the knee.

Trabecula (tra-BEK-ū-la) Irregular latticework of thin plates of spongy bone tissue. Fibrous cord of connective tissue serving as supporting fiber by forming a septum extending into an organ from its wall or capsule. Plural is trabeculae (tra-BEK-ū-lē).

Trabeculae carneae (tra-BEK-ū-lē KAR-nē-ē) Ridges and folds of the myocardium in the ventricles.

Trachea (TRĀ-kē-a) Tubular air passageway extending from the larynx to the fifth thoracic vertebra. Also called the windpipe.

Tract A bundle of nerve axons in the central nervous system.

Transcription The process of copying the information represented by the sequence of base triplets in DNA into a complementary sequence of codons.

Translation Process in which the nucleotide sequence in an mRNA molecule specifies the amino acid sequence of a protein.

Transverse colon (trans-VERS KŌ-lon) The portion of the large intestine extending across the abdomen from the right colic (hepatic) flexure to the left colic (splenic) flexure.

Transverse fissure (FISH-er) The deep cleft that separates the cerebrum from the cerebellum.

Transverse plane A plane that divides the body or organs into superior and inferior portions. Also called a cross-sectional or horizontal plane.

Transverse (T) tubules (TOO-būls) Small, cylindrical invaginations of the sarcolemma of striated muscle fibers (cells) that conduct muscle action potentials toward the center of the muscle fiber.

Tremor (TREM-or) Rhythmic, involuntary, purposeless contraction of opposing muscle groups.

Triad (TRĪ-ad) A complex of three units in a muscle fiber composed of a transverse tubule and the sarcoplasmic reticulum terminal cisterns on both sides of it.

Tricuspid (trī-KUS-pid) **valve** Atrioventricular (AV) valve on the right side of the heart.

Triglyceride (trī-GLI-ser-īd) A lipid formed from one molecule of glycerol and three molecules of fatty acids that may be either solid (fats) or liquid (oils) at room temperature; the body's most highly concentrated source of chemical potential energy. Found mainly within adipocytes. Also called a neutral fat or a triacylglycerol.

Trigone A triangular region at the base of the urinary bladder.

Triiodothyronine (trī-ī-ō-dō-THĪ-rō-nēn) **(T_3)** A hormone produced by the thyroid gland that regulates metabolism, growth and development, and the activity of the nervous system.

Trophoblast (TRŌF-ō-blast) The superficial covering of cells of the blastocyst.

Tropic (TRŌ-pik) **hormone** A hormone whose target is another endocrine gland.

Trunk The part of the body to which the upper and lower limbs are attached.

Tubal ligation (lī-GĀ-shun) A sterilization procedure in which the uterine (fallopian) tubes are tied and cut.

Tunica albuginea (TOO-ni-kaal′-bū-JIN-ē-a) A dense white fibrous capsule covering a testis, the penis, or deep to the surface of an ovary.

Tunica externa (eks-TER-na) The superficial coat of an artery or vein, composed mostly of elastic and collagen fibers. Also called the adventitia.

Tunica interna (in-TER-na) The deep coat of an artery or vein, consisting of a lining of endothelium, basement membrane, and internal elastic lamina. Also called the tunica intima (IN-ti-ma).

Tunica media (MĒ-dē-a) The intermediate coat of an artery or vein, composed of smooth muscle and elastic fibers.

Tympanic (tim-PAN-ik) **membrane** A thin, semi-transparent partition of fibrous connective tissue between the external auditory meatus and the middle ear. Also called the eardrum.

Type I cutaneous mechanoreceptor Slowly adapting touch receptor for continuous touch; also called a tactile disc or Merkel disc.

Type II cutaneous mechanoreceptor A sensory receptor embedded deeply in the dermis and deeper tissues that detects stretching of skin. Also called a Ruffini corpuscle.

Umbilical (um-BIL-i-kul) **cord** The long, ropelike structure containing the umbilical arteries and vein that connect the fetus to the placenta.

Umbilicus (um-BIL-i-kus) A small scar on the abdomen that marks the former attachment of the umbilical cord to the fetus. Also called the navel.

Upper limb The appendage attached at the shoulder girdle, consisting of the arm, forearm, wrist, hand, and fingers. Also called upper extremity.

Ureter One of two tubes that connect the kidney with the urinary bladder.

Urethra (ū-RĒ-thra) The duct from the urinary bladder to the exterior of the body that conveys urine in females and urine and semen in males.

Urinalysis An analysis of the volume and physical, chemical, and microscopic properties of urine.

Urinary bladder A hollow, muscular organ situated in the pelvic cavity posterior to the pubic symphysis; receives urine via two ureters and stores urine until it is excreted through the urethra.

Urinary system The body system consisting of the kidneys, ureters, urinary bladder, and urethra.

Urine The fluid produced by the kidneys that contains wastes and excess materials; excreted from the body through the urethra.

Urogenital (ū′-rō-JEN-i-tal) **triangle** The region of the pelvic floor inferior to the pubic symphysis, bounded by the pubic symphysis and the ischial tuberosities, and containing the external genitalia.

Urology (ū-ROL-ō-jē) The specialized branch of medicine that deals with the structure, function, and diseases of the male and female urinary systems and the male reproductive system.

Uterine cycle A series of changes in the endometrium of the uterus to prepare it for the arrival and development of a fertilized ovum. Also called the menstrual cycle.

Uterine prolapse (PRŌ-laps) A dropping or falling down of the uterus.

Uterine (Ū-ter-in) **tube** Duct that transports ova from the ovary to the uterus. Also called the fallopian (fal-LŌ-pē-an) tube or oviduct.

Uterosacral ligament A fibrous band of tissue extending from the cervix of the uterus laterally to the sacrum.

Uterus (Ū-te-rus) The hollow, muscular organ in females that is the site of menstruation, implantation, development of the fetus, and labor. Also called the womb.

Utricle (Ū-tri-kul) The larger of the two divisions of the membranous labyrinth located inside the vestibule of the inner ear, containing a receptor organ for linear acceleration or deceleration that occurs in a horizontal direction and also head tilt.

Uvula (Ū-vū-la) A soft, fleshy mass, especially the V-shaped pendant part, descending from the soft palate.

Vagina (va-JĪ-na) A muscular, tubular organ that leads from the uterus to the vestibule, situated between the urinary bladder and the rectum of the female.

Vallate papilla (VAL-āt pa-PIL-a) One of the circular projections that is arranged in an inverted V-shaped row at the back of the tongue; the largest of the elevations on the upper surface of the tongue containing taste buds. Also called circumvallate papilla.

Varicocele (VAR-i-kō-sēl) A twisted vein; especially, the accumulation of blood in the veins of the spermatic cord.

Varicose (VAR-i-kōs) Pertaining to an unnatural swelling, as in the case of a varicose vein.

Vas deferens (DEF-er-ens) The duct that carries sperm from the epididymis to the ejaculatory duct. Also called the ductus deferens.

Vasa recta Extensions of the efferent arteriole of a juxtamedullary nephron that run alongside the nephron loop in the medullary region of the kidney.

Vasa vasorum (va-SŌ-rum) Blood vessels that supply nutrients to the larger arteries and veins.

Vascular (venous) sinus A vein with a thin endothelial wall that lacks a tunica media and externa and is supported by surrounding tissue.

Vascular spasm Contraction of the smooth muscle in the wall of a damaged blood vessel to prevent blood loss.

Vascular tunic (TOO-nik) The middle layer of the eyeball, composed of the choroid, ciliary body, and iris. Also called the uvea (Ū-vē-a).

Vasectomy (va-SEK-tō-mē) A means of sterilization of males in which a portion of each ductus (vas) deferens is removed.

Vasoconstriction (vāz-ō-kon-STRIK-shun) A decrease in the size of the lumen of a blood vessel caused by contraction of the smooth muscle in the wall of the vessel.

Vasodilation (vāz-ō-dī-LĀ-shun) An increase in the size of the lumen of a blood vessel caused by relaxation of the smooth muscle in the wall of the vessel.

Vein A blood vessel that conveys blood from tissues back to the heart.

Ventricle (VEN-tri-kul) A cavity in the brain filled with cerebrospinal fluid. An inferior chamber of the heart.

Ventricular fibrillation (ven-TRIK-ū-lar fib-ri-LĀ-shun) (**VF** or **V-fib**) Asynchronous ventricular contractions; unless reversed by defibrillation, results in heart failure.

Venule (VEN-ūl) A small vein that collects blood from capillaries and delivers it to a vein.

Vermiform appendix (VER-mi-form a-PEN-diks) A twisted, coiled tube attached to the cecum. Also called the appendix.

Vermis (VER-mis) The central constricted area of the cerebellum that separates the two cerebellar hemispheres.

Vertebrae (VER-te-brē) Bones that make up the vertebral column.

Vertebral (spinal) canal A cavity within the vertebral column formed by the vertebral foramina of all vertebrae and containing the spinal cord. Also called the spinal canal.

Vertebral column The 26 vertebrae of an adult and 33 vertebrae of a child; encloses and protects the spinal cord and serves as a point of attachment for the ribs and back muscles. Also called the backbone, spine, or spinal column.

Vesicle (VES-i-kul) A small bladder or sac containing liquid.

Vesicouterine (ves′-ik-ō-Ū-ter-in) **pouch** A shallow pouch formed by the reflection of the peritoneum from the anterior surface of the uterus, at the junction of the cervix and the body, to the posterior surface of the urinary bladder.

Vestibular (ves-TIB-ū-lar) **apparatus** Collective term for the organs of equilibrium, which includes the saccule, utricle, and semicircular ducts.

Vestibular membrane The membrane that separates the cochlear duct from the scala vestibuli.

Vestibule (VES-ti-būl) A small space or cavity at the beginning of a canal, especially the inner ear, larynx, mouth, nose, and vagina.

Villus (VIL-lus) A projection of the intestinal mucosal cells containing connective tissue, blood vessels, and a lymphatic vessel; functions in the absorption of the end products of digestion. Plural is villi (VIL-ī).

Viscera The organs inside the ventral body cavity.

Vision The act of seeing.

Vitamin An organic molecule necessary in trace amounts that acts as a catalyst in normal metabolic processes in the body.

Vitreous (VIT-rē-us) **body** A soft, jellylike substance that fills the vitreous chamber of the eyeball, lying between the lens and the retina.

Vocal folds Pair of mucous membrane folds below the ventricular folds that function in voice production. Also called true vocal cords.

Vulva (VUL-va) Collective designation for the external genitalia of the female. Also called the pudendum (poo-DEN-dum).

Wallerian (wal-LE-rē-an) **degeneration** Degeneration of the portion of the axon and myelin sheath of a neuron distal to the site of injury.

Wandering macrophage (MAK-rō-fāj) Phagocytic cell that develops from a monocyte, leaves the blood, and migrates to infected tissues.

White blood cells (WBCs) Nucleated blood cells that are responsible for protecting the body from foreign substances via phagocytosis or immune reactions.

White matter Aggregations or bundles of myelinated and unmyelinated axons located in the brain and spinal cord.

White pulp The regions of the spleen composed of lymphatic tissue, mostly B lymphocytes.

White ramus communicans (RĀ-mus kō-MŪ-ni-kanzs) The portion of a preganglionic sympathetic axon that branches from the anterior ramus of a spinal nerve to enter the nearest sympathetic trunk ganglion. Plural is white rami communicantes.

Xiphoid (ZĪ-foyd) **process** The inferior portion of the sternum (breastbone).

Yolk sac An extraembryonic membrane composed of the exocoelomic membrane and hypoblast. It transfers nutrients to the embryo, is a source of blood cells, contains primordial germ cells that migrate into the gonads to form primitive germ cells, forms part of the gut, and helps prevent desiccation of the embryo.

Zona fasciculata (ZŌ-na fa-sik′-ū-LA-ta) The middle zone of the adrenal cortex consisting of cells arranged in long, straight cords that secrete glucocorticoid hormones, mainly cortisol.

Zona glomerulosa (glo-mer′-ū-LŌ-sa) The outer zone of the adrenal cortex, directly under the connective tissue covering, consisting of cells arranged in arched loops or round balls that secrete mineralocorticoid hormones, mainly aldosterone.

Zona pellucida (ZŌ-na pe-LOO-si-da) Clear glycoprotein layer between a secondary oocyte and the surrounding granulosa cells of the corona radiata.

Zona reticularis (ret-ik′-ū-LAR-is) The inner zone of the adrenal cortex, consisting of cords of branching cells that secrete sex hormones, chiefly androgens.

Zygote (ZĪ-gōt) The single cell resulting from the union of male and female gametes; the fertilized ovum.

Index

Note: Exhibits, figures, tables, and Study Guide are indicated by italic *e*, *f*, *t*, and SG respectively, following the page reference.

C

F

G

H

M

O

Q

R

S

T

EPONYMS USED IN THIS TEXT

EPONYM	CURRENT TERMINOLOGY
Achilles tendon	calcaneal tendon
Adam's apple	thyroid cartilage
ampulla of Vater (VA-ter)	hepatopancreatic ampulla
Bartholin's (BAR-tō-linz) gland	greater vestibular gland
Billroth's (BIL-rōtz) cord	splenic cord
Bowman's (BŌ-manz) capsule	glomerular capsule
Bowman's (BŌ-manz) gland	olfactory gland
Broca's (BRŌ-kaz) area	motor speech area
Brunner's (BRUN-erz) gland	duodenal gland
bundle of His (HISS)	atrioventricular (AV) bundle
canal of Schlemm (SHLEM)	scleral venous sinus
circle of Willis (WIL-is)	cerebral arterial circle
Cooper's (KOO-perz) ligament	suspensory ligament of the breast
Cowper's (KOW-perz) gland	bulbourethral gland
crypt of Lieberkühn (LĒ-ber-kūn)	intestinal gland
duct of Rivinus (re-VĒ-nus)	lesser sublingual duct
duct of Santorini (san′-tō-RĒ-nē)	accessory duct
duct of Wirsung (VĒR-sung)	pancreatic duct
Eustachian (ū-STĀ-kē-an) tube	auditory tube
fallopian (fal-LŌ-pē-an) tube	uterine tube
gland of Littré (LĒ-trā)	urethral gland
Golgi (GOL-jē) tendon organ	tendon organ
graafian (GRAF-ē-an) follicle	mature ovarian follicle
Hassall's (HAS-alz) corpuscle	thymic corpuscle
haversian (ha-VĒR-shun) canal	central canal
haversian (ha-VĒR-shun) system	osteon
Heimlich (HĪM-lik) maneuver	abdominal thrust maneuver
interstitial cell of Leydig (LĪ-dig)	interstitial cell
islet of Langerhans (LANG-er-hanz)	pancreatic islet
Kupffer (KOOP-fer) cell	stellate reticuloendothelial cell

EPONYM	CURRENT TERMINOLOGY
loop of Henle (HEN-lē)	nephron loop
Meibomian (mī-BŌ-mē-an) gland	tarsal gland
Meissner (MĪS-ner) corpuscle	corpuscle of touch
Merkel (MER-kel) disc	type I cutaneous mechanoreceptor
Müllerian (mil-E-rē-an) duct	paramesonephric duct
Nissl (NIS-l) bodies	chromatophilic substances
organ of Corti (KOR-tē)	spiral organ
Pacinian (pa-SIN-ē-an) corpuscle	lamellated corpuscle
Peyer's (PĪ-erz) patch	aggregated lymphatic follicles
plexus of Auerbach (OW-er-bak)	myenteric plexus
plexus of Meissner (MĪS-ner)	submucosal plexus
pouch of Douglas	rectouterine pouch
Rathke's (rath-KĒZ) pouch	hypophyseal pouch
Ruffini corpuscle (roo-FĒ-nē)	type II cutaneous mechanoreceptor
Sertoli (ser-TŌ-lē) cell	sustentacular cell
Sharpey's (SHAR-pēz) fiber	perforating fiber
Sheath of Schwann (SCHVON)	neurolemma
Skene's (SKĒNZ) gland	paraurethral gland
sphincter of Oddi (OD-dē)	sphincter of the hepatopancreatic ampulla
Stensen's (STEN-senz) duct	parotid duct
Volkmann's (FŌLK-manz) canal	perforating canal
Wernicke's (VER-ni-kēz) area	auditory association area
Wharton's (HWAR-tunz) duct	submandibular duct
Wharton's (HWAR-tunz) jelly	mucous connective tissue
Wormian (WER-mē-an) bone	sutural bone

COMBINING FORMS, WORD ROOTS, PREFIXES, AND SUFFIXES

Many of the terms used in anatomy and physiology are compound words; that is, they are made up of word roots and one or more prefixes or suffixes. For example, *leukocyte* is formed from the word roots *leuk-* meaning "white," a connecting vowel-*o*-, and *-cyte* meaning "cell." Thus, a leukocyte is a white blood cell. The following list includes some of the most commonly used combining forms, word roots, prefixes, and suffixes used in the study of anatomy and physiology. Each entry includes a usage example. Learning the meanings of these fundamental word parts will help you remember terms that, at first glance may seem long or complicated.

COMBINING FORMS AND WORD ROOTS

Acous-, Acu- hearing Acoustics.
Acr- extremity Acromegaly.
Aden- gland Adenoma.
Alg-, Algia- pain Neuralgia.
Angi- vessel Angiocardiography.
Arthr- joint Arthropathy.
Audit- hearing Auditory canal.
Aut-, Auto- self Autolysis.

Bio- life, living Biopsy.
Blast- germ, bud Blastula.
Blephar- eyelid Blepharitis.
Brachi- arm Brachial plexus.
Bronch- trachea, windpipe Bronchoscopy.
Bucc- cheek Buccal.

Capit- head Decapitate.
Carcin- cancer Carcinogenic.
Cardi-, Cardia-, Cardio- heart Cardiogram.
Cephal- head Hydrocephalus.
Cerebro- brain Cerebrospinal fluid.
Chole- bile, gall Cholecystogram.
Chondr-, cartilage Chondrocyte.
Cor-, Coron- heart Coronary.
Cost- rib Costal.
Crani- skull Craniotomy.
Cut- skin Subcutaneous.
Cyst- sac, bladder Cystoscope.

Derma-, Dermato- skin Dermatosis.
Dura- hard Dura mater.

Enter- intestine Enteritis.
Erythr- red Erythrocyte.

Gastr- stomach Gastrointestinal.
Gloss- tongue Hypoglossal.
Glyco- sugar Glycogen.
Gyn-, Gynec- female, woman Gynecology.

Hem-, Hemat- blood Hematoma.
Hepar-, Hepat- liver Hepatitis.
Hist-, Histio- tissue Histology.
Hydr- water Dehydration.
Hyster- uterus Hysterectomy.

Ischi- hip, hip joint Ischium.

Kines- motion Kinesiology.

Labi- lip Labial.
Lacri- tears Lacrimal glands.
Laparo- loin, flank, abdomen Laparoscopy.
Leuko- white Leukocyte.
Lingu- tongue Sublingual glands.
Lip- fat Lipid.
Lumb- lower back, loin Lumbar.

Macul- spot, blotch Macula.
Malign- bad, harmful Malignant.
Mamm-, Mast- breast Mammography, Mastitis.
Meningo- membrane Meningitis.
My-, Myo- muscle Myocardium.
Myel- marrow, spinal cord Myeloblast.

Necro- corpse, dead Necrosis.
Nephro- kidney Nephron.
Neuro- nerve Neurotransmitter.

Ocul- eye Binocular.
Odont- tooth Orthodontic.
Onco- mass, tumor Oncology.
Oo- egg Oocyte.
Ophthalm- eye Ophthalmology.
Or- mouth Oral.
Os-, Osseo-, Osteo- bone Osteocyte.
Osm- odor, sense of small Anosmia.
Ot- ear Otitis media.

Palpebr- eyelid Palpebra.
Patho- disease Pathogen.
Pelv- basin Renal pelvis.
Phag- to eat Phagocytosis.
Phleb- vein Phlebitis.
Phren- diaphragm Phrenic.
Pilo- hair Depilatory.
Pneumo- lung, air Pneumothorax.
Pod- foot Podocyte.
Procto- anus, rectum Proctology.
Pulmon- lung Pulmonary.

Ren- kidneys Renal artery.
Rhin- nose Rhinitis.

Scler-, Sclero- hard Atherosclerosis.
Sep-, Septic- toxic condition due to microorganisms Septicemia.
Soma-, Somato- body Somatotropin.
Stasis-, Stat- standstill Homeostasis.
Sten- narrow Stenosis.

Tegument- skin, covering Integumentary.
Therm- heat Thermogenesis.
Thromb- clot, lump Thrombus.

Vas- vessel, duct Vasoconstriction.

Zyg- joined Zygote.

PREFIXES

A-, An- without, lack of, deficient Anesthesia.
Ab- away from, from Abnormal.
Ad-, Af- to, toward Adduction, Afferent neuron.
Alb- white Albino.
Alveol- cavity, socket Alveolus.
Andro- male, masculine Androgen.
Ante- before Antebrachial vein.
Anti- against Anticoagulant.

Bas- base, foundation Basal nuclei.
Bi- two, double Biceps.
Brady- slow Bradycardia.

Cata- down, lower, under Catabolism.
Circum- around Circumduction.
Cirrh- yellow Cirrhosis of the liver.
Co-, Con-, Com with, together Congenital.
Contra- against, opposite Contraception.
Crypt- hidden, concealed Cryptorchidism.
Cyano- blue Cyanosis.

De- down, from Deciduous.
Demi-, hemi- half Hemiplegia.
Di-, Diplo- two Diploid.
Dis- separation, apart, away from Dissection.
Dys- painful, difficult Dyspnea.

E-, Ec-, Ef- out from, out of Efferent neuron.
Ecto-, Exo- outside Ectopic pregnancy.
Em-, En- in, on Emmetropia.
End-, Endo- within, inside Endocardium.
Epi- upon, on, above Epidermis.
Eu- good, easy, normal Eupnea.
Ex-, Exo- outside, beyond Exocrine gland.
Extra- outside, beyond, in addition to Extracellular fluid.

Fore- before, in front of Forehead.

Gen- originate, produce, form Genitalia.
Gingiv- gum Gingivitis.

Hemi- half Hemiplegia.
Heter-, Hetero- other, different Heterozygous.
Homeo-, Homo- unchanging, the same, steady Homeostasis.
Hyper- over, above, excessive Hyperglycemia.
Hypo- under, beneath, deficient Hypothalamus.

Im-, In- in, inside, not Incontinent.
Infra- beneath Infraorbital.
Inter- among, between Intercostal.
Intra- within, inside Intracellular fluid.
Ipsi- same Ipsilateral.
Iso- equal, like Isotonic.

Juxta- near to Juxtaglomerular apparatus.

Later- side Lateral.

Macro- large, great Macrophage.
Mal- bad, abnormal Malnutrition.
Medi-, Meso- middle Medial.
Mega-, Megalo- great, large Megakaryocyte.
Melan- black Melanin.
Meta- after, beyond Metacarpus.
Micro- small Microfilament.
Mono- one Monounsaturated fat.

Neo- new Neonatal.